# Microelectronic Circuit Analysis and Design

Microelectronics:
Circuit Analysis
and Design

# Microelectronics: Circuit Analysis and Design

## Fourth Edition

## Donald A. Neamen
University of New Mexico

Mc Graw Hill

*Connect*
*Learn*
*Succeed*™

The McGraw·Hill Companies

MICROELECTRONICS: CIRCUIT ANALYSIS AND DESIGN, FOURTH EDITION
International Edition 2010

**When ordering this title, use ISBN 978-007-128947-4 or MHID 007-128947-X**

Printed in Singapore

www.mhhe.com

# Dedication

To the many students I've had the privilege of teaching over the years who have contributed in many ways to the broad field of electrical engineering, and to future students who will contribute in ways we cannot now imagine.

# About the Author

Donald A. Neamen is a professor emeritus in the Department of Electrical and Computer Engineering at the University of New Mexico where he taught for more than 25 years. He received his Ph.D. degree from the University of New Mexico and then became an electronics engineer at the Solid State Sciences Laboratory at Hanscom Air Force Base. In 1976, he joined the faculty in the ECE department at the University of New Mexico, where he specialized in teaching semiconductor physics and devices courses and electronic circuits courses. He is still a part-time instructor in the department. He also just recently taught for a semester at the University of Michigan–Shanghai Jiao Tong University (UM-SJTU) Joint Institute in Shanghai.

In 1980, Professor Neamen received the Outstanding Teacher Award for the University of New Mexico. In 1990, and each year from 1994 through 2001, he received the Faculty Recognition Award, presented by graduating ECE students. He was also honored with the Teaching Excellence Award in the College of Engineering in 1994.

In addition to his teaching, Professor Neamen served as Associate Chair of the ECE department for several years and has also worked in industry with Martin Marietta, Sandia National Laboratories, and Raytheon Company. He has published many papers and is the author of *Semiconductor Physics and Devices: Basic Principles,* third edition and *An Introduction to Semiconductor Devices.*

# Brief Table of Contents

# Contents

## PROLOGUE II

## PROLOGUE TO ELECTRONIC DESIGN　615

# Preface

## PHILOSOPHY AND GOALS

*Microelectronics: Circuit Analysis and Design* is intended as a core text in electronics for undergraduate electrical and computer engineering students. The purpose of the fourth edition of the book is to continue to provide a foundation for analyzing and designing both analog and digital electronic circuits. A goal is to make this book very readable and student-friendly.

Most electronic circuit design today involves integrated circuits (ICs), in which the entire circuit is fabricated on a single piece of semiconductor material. The IC can contain millions of semiconductor devices and other elements and can perform complex functions. The microprocessor is a classic example of such a circuit. The ultimate goal of this text is to clearly present the operation, characteristics, and limitations of the basic circuits that form these complex integrated circuits. Although most engineers will use existing ICs in specialized design applications, they must be aware of the fundamental circuit's characteristics in order to understand the operation and limitations of the IC.

Initially, discrete transistor circuits are analyzed and designed. The complexity of circuits being studied increases throughout the text so that, eventually, the reader should be able to analyze and design the basic elements of integrated circuits, such as linear amplifiers and digital logic gates.

This text is an introduction to the complex subject of electronic circuits. Therefore, more advanced material is not included. Specific technologies, such as gallium arsenide, which is used in special applications, are also not included, although reference may be made to a few specialized applications. Finally, the layout and fabrication of ICs are not covered, since these topics alone can warrant entire texts.

## DESIGN EMPHASIS

Design is the heart of engineering. Good design evolves out of considerable experience with analysis. In this text, we point out various characteristics and properties of circuits as we go through the analysis. The objective is to develop an intuition that can be applied to the design process.

Many design examples, design exercise problems, and end-of-chapter design problems are included in this text. The end-of-chapter design problems are designated with a "D". Many of these examples and problems have a set of specifications that lead to a unique solution. Although engineering design in its truest sense does not lead to a unique solution, these initial design examples and problems are a first step, the author believes, in learning the design process. A separate section, Design Problems, found in the end-of-chapter problems, contains open-ended design problems.

## COMPUTER-AIDED ANALYSIS AND DESIGN

Computer analysis and computer-aided-design (CAD) are significant factors in electronics. One of the most prevalent electronic circuit simulation programs is Simulation Program with Integrated Circuit Emphasis (SPICE), developed at the University of California. A version of SPICE tailored for personal computers is PSpice, which is used in this text.

The text emphasizes hand analysis and design in order to concentrate on basic circuit concepts. However, in several places in the text, PSpice results are included and are correlated with the hand analysis results. Obviously, at the instructor's discretion, computer simulation may be incorporated at any point in the text. A separate section, Computer Simulation Problems, is found in the end-of-chapter problems.

In some chapters, particularly the chapters on frequency response and feedback, computer analysis is used more heavily. Even in these situations, however, computer analysis is considered only after the fundamental properties of the circuit have been covered. The computer is a tool that can aid in the analysis and design of electronic circuits, but is not a substitute for a thorough understanding of the basic concepts of circuit analysis.

## PREREQUISITES

This book is intended for junior undergraduates in electrical and computer engineering. The prerequisites for understanding the material include dc analysis and steady-state sinusoidal analysis of electric circuits and the transient analysis of RC circuits. Various network concepts, such as Thevenin's and Norton's theorems, are used extensively. Some background in Laplace transform techniques may also be useful. Prior knowledge of semiconductor device physics is not required.

## ORGANIZATION

The book is divided into three parts. Part 1, consisting of the first eight chapters, covers semiconductor materials, the basic diode operation and diode circuits, and basic transistor operations and transistor circuits. Part 2 addresses more advanced analog electronics, such as operational amplifier circuits, biasing techniques used in integrated circuits, and other analog circuits applications. Part 3 covers digital electronics including CMOS integrated circuits. Five appendices are included at the end of the text.

### Content
**Part 1.** Chapter 1 introduces the semiconductor material and pn junction, which leads to diode circuits and applications given in Chapter 2. Chapter 3 covers the field-effect transistor, with strong emphasis on the metal-oxide-semiconductor FET (MOSFET), and Chapter 4 presents basic FET linear amplifiers. Chapter 5 discusses the bipolar junction transistor, with basic bipolar linear amplifier applications given in Chapter 6.

The frequency response of transistors and transistor circuits is covered in a separate Chapter 7. The emphasis in Chapters 3 through 6 was on the analysis and design techniques, so mixing the two transistor types within a given chapter would introduce unnecessary confusion. However, starting with Chapter 7, both MOSFET circuits and bipolar circuits are discussed within the same chapter. Finally, Chapter 8, covering output stages and power amplifiers, completes Part 1 of the text.

**Part 2.** Chapters 9 through 15 are included in Part 2, which addresses more advanced analog electronics. In this portion of the text, the emphasis is placed on the operational amplifier and on circuits that form the basic building blocks of integrated circuits (ICs). The ideal operational amplifier and ideal op-amp circuits are covered in Chapter 9. Chapter 10 presents constant-current source biasing circuits and introduces the active load, both of which are used extensively in ICs. The differential amplifier, the heart of the op-amp, is discussed in Chapter 11, and feedback is considered in Chapter 12. Chapter 13 presents the analysis and design of various circuits that form operational amplifiers. Nonideal effects in analog ICs are addressed in Chapter 14, and applications, such as active filters and oscillators, are covered in Chapter 15.

**Part 3.** Chapters 16 and 17 form Part 3 of the text, and cover the basics of digital electronics. The analysis and design of MOS digital electronics is discussed in Chapter 16. The emphasis in this chapter is on CMOS circuits, which form the basis of most present-day digital circuits. Basic digital logic gate circuits are initially covered, then shift registers, flip-flops, and then basic A/D and D/A converters are presented. Chapter 17 introduces bipolar digital electronics, including emitter-coupled logic and classical transistor-transistor logic circuits.

**Appendices.** Five appendices are included at the end of the text. Appendix A contains physical constants and conversion factors. Manufacturers' data sheets for several devices and circuits are included in Appendix B. Standard resistor and capacitor values are given in Appendix C, and references and other reading sources are listed in Appendix D. Finally, answers to selected end-of chapter problems are given in Appendix E.

## Order of Presentation

The book is written with a certain degree of flexibility so that instructors can design their own order of presentation of topics.

1. *Op-Amp Circuits:* For those instructors who wish to present ideal op-amp circuits as a first topic in electronics, Chapter 9 is written such that sections 9.1 through 9.5.5 can be studied as a first chapter in electronics.

> Chapter Presentation
> Ideal Op-Amp Circuits:
>     1. Chapter 9, Sections 9.1–9.5.5.
>     2. Chapters 1, 2, etc.

2. *MOSFETs versus Bipolars:* The chapters covering MOSFETs (3 and 4) and the chapters covering bipolars (5 and 6) are written independently of each other. Instructors, therefore, have the option of discussing MOSFETs before bipolars, as

given in the text, or discussing bipolars before MOSFETs in the more traditional manner.

<table>
<tr><th colspan="2">Chapter Presentation</th><th colspan="2"></th></tr>
<tr><th colspan="2">Text</th><th colspan="2">Traditional</th></tr>
<tr><th>Chapter</th><th>Topic</th><th>Chapter</th><th>Topic</th></tr>
<tr><td>1</td><td>pn Junctions</td><td>1</td><td>pn Junctions</td></tr>
<tr><td>2</td><td>Diode Circuits</td><td>2</td><td>Diode Circuits</td></tr>
<tr><td>3</td><td>MOS Transistors</td><td>5</td><td>Bipolar Transistors</td></tr>
<tr><td>4</td><td>MOSFET Circuits</td><td>6</td><td>Bipolar Circuits</td></tr>
<tr><td>5</td><td>Bipolar Transistors</td><td>3</td><td>MOS Transistors</td></tr>
<tr><td>6</td><td>Bipolar Circuits</td><td>4</td><td>MOSFET Circuits</td></tr>
<tr><td>etc.</td><td></td><td>etc.</td><td></td></tr>
</table>

3. *Digital versus Analog:* For those instructors who wish to present digital electronics before analog electronics, Part 3 is written to be independent of Part 2. Therefore, instructors may cover Chapters 1, 2, 3, and then jump to Chapter 16.

<table>
<tr><th colspan="2">Chapter Presentation:</th></tr>
<tr><th>Chapter</th><th>Topic</th></tr>
<tr><td>1</td><td>pn Junctions</td></tr>
<tr><td>2</td><td>Diode Circuits</td></tr>
<tr><td>3</td><td>MOS Transistors</td></tr>
<tr><td>16</td><td>MOSFET Digital Circuits</td></tr>
<tr><td>5</td><td>Bipolar Transistors</td></tr>
<tr><td>17</td><td>Bipolar Digital Circuits</td></tr>
<tr><td>etc.</td><td>Analog Circuits</td></tr>
</table>

## NEW TO THE FOURTH EDITION

- Addition of over 250 new Exercise and Test Your Understanding Problems.
- Addition of over 580 new end-of-chapter problems.
- Addition of over 50 new open-ended Design Problems in the end-of-chapter problems sections.
- Addition of over 65 new Computer Simulation Problems in the end-of-chapter problems sections.
- Voltage levels in circuits were updated to more closely match modern day electronics.
- MOSFET device parameters were updated to more closely match modern day electronics.
- Chapter 9 was rewritten such that ideal op-amp circuits can be studied as a first topic in electronics.
- Maintained the mathematical rigor necessary to more clearly understand basic circuit operation and characteristics.

# RETAINED FEATURES OF THE TEXT

- A short introduction at the beginning of each chapter links the new chapter to the material presented in previous chapters. The objectives of the Chapter, i.e., what the reader should gain from the chapter, are presented in the Preview section and are listed in bullet form for easy reference.
- Each major section of a chapter begins with a restatement of the objective for this portion of the chapter.
- An extensive number of worked examples are used throughout the text to reinforce the theoretical concepts being developed. These examples contain all the details of the analysis or design, so the reader does not have to fill in missing steps.
- An Exercise Problem follows each example. The exercise problem is very similar to the worked example so that readers can immediately test their understanding of the material just covered. Answers are given for each exercise problem so readers do not have to search for an answer at the end of the book. These exercise problems will reinforce readers' grasp of the material before they move on to the next section.
- Test Your Understanding exercise problems are included at the end of most major sections of the chapter. These exercise problems are, in general, more comprehensive that those presented at the end of an example. These problems will also reinforce readers' grasp of the material before they move on to the next section. Answers to these exercise problems are also given.
- Problem Solving Techniques are given throughout each chapter to assist the reader in analyzing circuits. Although there can be more than one method of solving a problem, these Problem Solving Techniques are intended to help the reader get started in the analysis of a circuit.
- A Design Application is included as the last section of each chapter. A specific electronic design related to that chapter is presented. Over the course of the book, students will learn to build circuits for an electronic thermometer. Though not every Design Application deals with the thermometer, each application illustrates how students will use design in the real world.
- A Summary section follows the text of each chapter. This section summarizes the overall results derived in the chapter and reviews the basic concepts developed. The summary section is written in bullet form for easy reference.
- A Checkpoint section follows the Summary section. This section states the goals that should have been met and states the abilities the reader should have gained. The Checkpoints will help assess progress before moving to the next chapter.
- A list of review questions is included at the end of each chapter. These questions serve as a self-test to help the reader determine how well the concepts developed in the chapter have been mastered.
- A large number of problems are given at the end of each chapter, organized according to the subject of each section. Many new problems have been incorporated into the fourth edition. Design oriented problems are included as well as problems with varying degrees of difficulty. A "D" indicates design-type problems, and an asterisk (*) indicates more difficult problems. Separate computer simulation problems and open-ended design problems are also included.
- Answers to selected problems are given in Appendix E. Knowing the answer to a problem can aid and reinforce the problem solving ability.
- Manufacturers' data sheets for selected devices and circuits are given in Appendix B. These data sheets should allow the reader to relate the basic concepts and circuit characteristics studied to real circuit characteristics and limitations.

## SUPPLEMENTS

The website for Microeletronics features tools for students and teachers. Professors can benefit from McGraw-Hill's COSMOS electronic solutions manual. COSMOS enables instructors to generate a limitless supply of problem material for assignment, as well as transfer and integrate their own problems into the software. For students, there are profiles of electrical engineers that give students insight into the real world of electrical engineering by presenting interviews with engineers working at a number of businesses, from Fairchild Semiconductor to Apple. In addition, the website boasts PowerPoint slides, an image library, the complete Instructor's Solution Manual (password protected), data sheets, laboratory manual, and links to other important websites. You can find the site at **www.mhhe.com/neamen.**

## ELECTRONIC TEXTBOOK OPTIONS

This text is offered through CourseSmart for both instructors and students. Course-Smart is an online resource where students can purchase the complete text online at almost half the cost of a traditional text. Purchasing the eTextbook allows students to take advantage of CourseSmart's Web tools for learning, which include full text search, notes and highlighting, and email tools for sharing notes between classmates. To learn more about CourseSmart options, contact your sales representative or visit **www.CourseSmart.com.**

## ACKNOWLEDGMENTS

I am indebted to the many students I have taught over the years who have helped in the evolution of this text. Their enthusiasm and constructive criticism have been invaluable, and their delight when they think they have found an error their professor may have made is priceless. I also want to acknowledge Professor Hawkins, Professor Fleddermann, and Dr. Ed Graham of the University of New Mexico who have taught from the third edition and who have made excellent suggestions for improvement.

I want to thank the many people at McGraw-Hill for their tremendous support. To Raghu Srinivasan, publisher, and Lora Neyens, development editor, I am grateful for their encouragement and support. I also want to thank Mr. John Griffith for his many constructive suggestions. I also appreciate the efforts of project managers who guided the work through its final phase toward publication. This effort included gently, but firmly, pushing me through proofreading.

Let me express my continued appreciation to those reviewers who read the original manuscript in its various phases, a focus group who spent an entire precious weekend discussing and evaluating the original project, and the accuracy checkers who worked through the original examples, exercises, and problems to minimize any errors I may have introduced. My thanks also go out to those individuals who have continued to review the book prior to new editions being published. Their contributions and suggestions for continued improvement are incredibly valuable.

# REVIEWERS FOR THE FOURTH EDITION

Doran Baker
*Utah State University*
Marc Cahay
*University of Cincinatti*
Richard H. Cockrum
*California State University, Pomona*
Norman R. Cox
*Missouri University of Science and
    Technology Engineering*
Stephen M. Goodnick
*Arizona State University*
Rongqing Hui
*University of Kansas*
Syed K Islam
*University of Tennessee*

Richard Kwor
*University of Colorado, Colorado
    Springs*
Juin J. Liou
*University of Central Florida*
Sannasi Ramanan
*Rochester Institute of Technology*
Ron Roscoe
*Massachusetts Institute of Technology*
John Scalzo
*Louisiana State University*
Mark J. Wharton
*Pennsylvania State University*
Weizhong Wang
*University of Wisconsin, Milwaukee*

Donald A. Neamen

# Prologue to Electronics

When most of us hear the word electronics, we think of televisions, laptop computers, cell phones, or iPods. Actually, these items are electronic *systems* composed of subsystems or electronic circuits, which include amplifiers, signal sources, power supplies, and digital logic circuits.

**Electronics** is defined as the science of the motion of charges in a gas, vacuum, or semiconductor. (Note that the charge motion in a metal is excluded from this definition.) This definition was used early in the 20th century to separate the field of electrical engineering, which dealt with motors, generators, and wire communications, from the new field of electronic engineering, which at that time dealt with vacuum tubes. Today, electronics generally involves transistors and transistor circuits. **Microelectronics** refers to integrated circuit (IC) technology, which can produce a circuit with multimillions of components on a single piece of semiconductor material.

A typical electrical engineer will perform many diverse functions, and is likely to use, design, or build systems incorporating some form of electronics. Consequently, the division between electrical and electronic engineering is no longer as clear as originally defined.

## BRIEF HISTORY

The development of the transistor and the integrated circuit has led to remarkable electronic capabilities. The IC permeates almost every facet of our daily lives, from instant communications by cellular phone to the automobile. One dramatic example of IC technology is the small laptop computer, which today has more capability than the equipment that just a few years ago would have filled an entire room. The cell phone has shown dramatic changes. It not only provides for instant messaging, but also includes a camera so that pictures can be instantly sent to virtually every point on earth.

A fundamental breakthrough in electronics came in December 1947, when the first transistor was demonstrated at Bell Telephone Laboratories by William Shockley, John Bardeen, and Walter Brattain. From then until approximately 1959, the transistor was available only as a discrete device, so the fabrication of circuits required that the transistor terminals be soldered directly to the terminals of other components.

In September 1958, Jack Kilby of Texas Instruments demonstrated the first integrated circuit fabricated in germanium. At about the same time, Robert Noyce of Fairchild Semiconductor introduced the integrated circuit in silicon. The development of the IC continued at a rapid rate through the 1960s, using primarily bipolar transistor technology. Since then, the metal-oxide-semiconductor field-effect transistor (MOSFET) and MOS integrated circuit technology have emerged as a dominant force, especially in digital integrated circuits.

Since the first IC, circuit design has become more sophisticated and the integrated circuit more complex. Device size continues to shrink and the number of devices fabricated on a single chip continues to increase at a rapid rate. Today, an IC can contain arithmatic, logic, and memory functions on a single semiconductor chip. The primary example of this type of integrated circuit is the microprocessor.

## PASSIVE AND ACTIVE DEVICES

In a passive electrical device, the time average power delivered to the device over an infinite time period is always greater than or equal to zero. Resistors, capacitors, and inductors, are examples of **passive devices.** Inductors and capacitors can store energy, but they cannot deliver an average power greater than zero over an infinite time interval.

Active devices, such as dc power supplies, batteries, and ac signal generators, are capable of supplying particular types of power. Transistors are also considered to be **active devices** in that they are capable of supplying more signal power to a load than they receive. This phenomenon is called amplification. The additional power in the output signal is a result of a redistribution of ac and dc power within the device.

## ELECTRONIC CIRCUITS

In most electronic circuits, there are two inputs (Figure PRl.1).One input is from a power supply that provides dc voltages and currents to establish the proper biasing for transistors. The second input is a signal. Time-varying signals from a particular source very often need to be **amplified** before the signal is capable of being "useful." For example, Figure PR1.l shows a signal source that is the output of a compact disc system. The output music signal from the compact disc system consists of a small time-varying voltage and current, which means that the signal power is relatively small. The power required to drive the speakers is larger than the output signal from the compact disc, so the compact disc signal must be amplified before it is capable of driving the speakers in order that sound can be heard.

Other examples of signals that must be amplified before they are capable of driving loads include the output of a microphone, voice signals received on earth from an orbiting manned shuttle, video signals from an orbiting weather satellite, and the output of an electrocardiograph (EKG). Although the output signal can be larger than the input signal, the output power can never exceed the dc input power. Therefore, the magnitude of the dc power supply is one limitation to the output signal response.

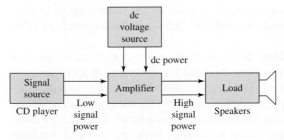

**Figure PR1.1** Schematic of an electronic circuit with two input signals: the dc power supply input, and the signal input

The analysis of electronic circuits, then, is divided into two parts: one deals with the dc input and its circuit response, and the other deals with the signal input and the resulting circuit response. Dependent voltage and current sources are used to model the active devices and to represent the amplification or signal gain. In general, different equivalent circuit models must be used for the dc and ac analyses.

## DISCRETE AND INTEGRATED CIRCUITS

In this text, we will deal principally with discrete electronic circuits, that is, circuits that contain discrete components, such as resistors, capacitors, and transistors. We will focus on the types of circuits that are the building blocks of the IC. For example, we will look at the various circuits that make up the operational amplifier, an important IC in analog electronics. We will also discuss various logic circuits used in digital ICs.

## ANALOG AND DIGITAL SIGNALS

### Analog signals

The voltage signal shown graphically in Figure PR1.2(a) is called an analog signal. The magnitude of an analog signal can take on any value within limits and may vary continuously with time. Electronic circuits that process analog signals are called **analog circuits.** One example of an analog circuit is a linear amplifier. A **linear amplifier** magnifies an input signal and produces an output signal whose amplitude is larger and directly proportional to the input signal.

The vast majority of signals in the "real world" are analog. Voice communications and music are just two examples. The amplification of such signals is a large part of electronics, and doing so with little or no distortion is a major consideration. Therefore, in signal amplifiers, the output should be a linear function of the input. An example is the power amplifier circuit in a stereo system. This circuit provides sufficient power to "drive" the speaker system. Yet, it must remain linear in order to reproduce the sound without distortion.

### Digital signals

An alternative signal is at one of two distinct levels and is called a digital signal (Figure PR1.2(b)). Since the digital signal has discrete values, it is said to be quantized. Electronic circuits that process digital signals are called **digital circuits.**

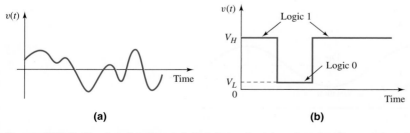

**(a)**                **(b)**

**Figure PR1.2** Graphs of analog and digital signals: (a) analog signal versus time and (b) digital signal versus time

| Table PR1.1 | Summary of notation |
|---|---|
| **Variable** | **Meaning** |
| $i_B$, $v_{BE}$ | Total instantaneous values |
| $I_B$, $V_{BE}$ | dc values |
| $i_b$, $v_{be}$ | Total instantaneous ac values |
| $I_b$, $V_{be}$ | Phasor values |

In many electronic systems, signals are processed, transmitted, and received in digital form. Digital systems and signal processing are now a large part of electronics because of the tremendous advances made in the design and fabrication of digital circuits. Digital processing allows a wide variety of functions to be performed that would be impractical using analog means. In many cases, digital signals must be converted from and to analog signals. These signals need to be processed through analog-to-digital (A/D) converters and digital-to-analog (D/A) converters. A significant part of electronics deals with these conversions.

## NOTATION

The following notation, summarized in Table PR1.1, is used throughout this text. A lowercase letter with an uppercase subscript, such as $i_B$ and $v_{BE}$, indicates a *total instantaneous value*. An uppercase letter with an uppercase subscript, such as $I_B$ and $V_{BE}$, indicates a *dc quantity*. A lowercase letter with a lowercase subscript, such as $i_b$ and $v_{be}$, indicates an *instantaneous value* of a time-varying signal. Finally, an uppercase letter with a lowercase subscript, such as $I_b$ and $V_{be}$, indicates a *phasor quantity*.

As an example, Figure PR1.3 shows a sinusoidal voltage superimposed on a dc voltage. Using our notation, we would write

$$v_{BE} = V_{BE} + v_{be} = V_{BE} + V_M \cos(\omega t + \phi_m)$$

The phasor concept is rooted in Euler's identity and relates the exponential function to the trigonometric function. We can write the sinusoidal voltage as

$$v_{be} = V_M \cos(\omega t + \phi_m) = V_M \, \mathrm{Re}\{e^{j(\omega t + \phi_m)}\} = \mathrm{Re}\{V_M e^{j\phi_m} e^{j\omega t}\}$$

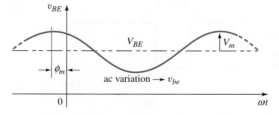

**Figure PR1.3** Sinusoidal voltage superimposed on dc voltage, showing notation used throughout this text

where Re stands for "the real part of." The coefficient of $e^{j\omega t}$ is a complex number that represents the amplitude and phase angle of the sinusoidal voltage. This complex number, then, is the phasor of that voltage, or

$$V_{be} = V_M e^{j\phi_m}$$

In some cases throughout the text, the input and output signals will be quasistatic quantities. For these situations, we may use either the total instantaneous notation, such as $i_B$ and $v_{BE}$, or the dc notation, $I_B$ and $V_{BE}$.

## SUMMARY

Semiconductor devices are the basic components in electronic circuits. The electrical characteristics of these devices provide the controlled switching required for signal processing, for example. Most electrical engineers are users of electronics rather than designers of electronic circuits and ICs. As with any discipline, however, the basics must be mastered before the overall system characteristics and limitations can be understood. In electronics, the discrete circuit must be thoroughly studied and analyzed before the operation, properties, and limitations of an IC can be fully appreciated.

# Semiconductor Devices and Basic Applications

In the first part of the text, we introduce the physical and electrical characteristics of the major semiconductor devices. Various basic circuits in which these devices are used are analyzed. This introduction will illustrate how the device characteristics are utilized in switching, digital, and amplification applications.

Chapter 1 briefly discusses semiconductor material characteristics and then introduces the semiconductor diode. Chapter 2 looks at various diode circuits that demonstrate how the nonlinear characteristics of the diode itself are used in switching and waveshaping applications. Chapter 3 introduces the metal-oxide-semiconductor field-effect transistor (MOSFET), presents the dc analysis of MOS transistor circuits, and discusses basic applications of this transistor. In Chapter 4, we analyze and design fundamental MOS transistor circuits, including amplifiers.

Chapter 5 introduces the bipolar transistor, presents the dc analysis of bipolar transistor circuits, and discusses basic applications of this transistor. Various bipolar transistor circuits, including amplifiers, are analyzed and designed in Chapter 6. Chapter 7 considers the frequency response of both MOS and bipolar transistor circuits. Finally, Chapter 8 discusses the designs and applications of these basic electronic circuits, including power amplifiers and various output stages.

# Semiconductor Materials and Diodes

This text deals with the analysis and design of circuits containing electronic devices, such as diodes and transistors. These electronic devices are fabricated using semiconductor materials, so we begin Chapter 1 with a brief discussion of the properties and characteristics of semiconductors. The intent of this brief discussion is to become familiar with some of the semiconductor material terminology, and to gain an understanding of the mechanisms that generate currents in a semiconductor.

A basic electronic device is the pn junction diode. The diode is a two-terminal device, but the $i$–$v$ relationship is nonlinear. Since the diode is a nonlinear element, the analysis of circuits containing diodes is not as straightforward as the analysis of simple linear resistor circuits. A goal of the chapter is to become familiar with the analysis of diode circuits.

## PREVIEW

In this chapter, we will:

- Gain a basic understanding of a few semiconductor material properties including the two types of charged carriers that exist in a semiconductor and the two mechanisms that generate currents in a semiconductor.
- Determine the properties of a pn junction including the ideal current–voltage characteristics of the pn junction diode.
- Examine dc analysis techniques for diode circuits using various models to describe the nonlinear diode characteristics.
- Develop an equivalent circuit for a diode that is used when a small, time-varying signal is applied to a diode circuit.
- Gain an understanding of the properties and characteristics of a few specialized diodes.
- As an application, design a simple electronic thermometer using the temperature characteristics of a diode.

## 1.1   SEMICONDUCTOR MATERIALS AND PROPERTIES

**Objective:** • Gain a basic understanding of a few semiconductor material properties including the two types of charged carriers that exist in a semiconductor and the two mechanisms that generate currents in a semiconductor.

Most electronic devices are fabricated by using semiconductor materials along with conductors and insulators. To gain a better understanding of the behavior of the electronic devices in circuits, we must first understand a few of the characteristics of the semiconductor material. Silicon is by far the most common semiconductor material used for semiconductor devices and integrated circuits. Other semiconductor materials are used for specialized applications. For example, gallium arsenide and related compounds are used for very high speed devices and optical devices. A list of some semiconductor materials is given in Table 1.1.

### 1.1.1   Intrinsic Semiconductors

An atom is composed of a nucleus, which contains positively charged protons and neutral neutrons, and negatively charged electrons that, in the classical sense, orbit the nucleus. The electrons are distributed in various "shells" at different distances from the nucleus, and electron energy increases as shell radius increases. Electrons in the outermost shell are called **valence electrons,** and the chemical activity of a material is determined primarily by the number of such electrons.

Elements in the periodic table can be grouped according to the number of valence electrons. Table 1.2 shows a portion of the periodic table in which the more common semiconductors are found. Silicon (Si) and germanium (Ge) are in group IV and are **elemental semiconductors.** In contrast, gallium arsenide is a group III–V **compound semiconductor.** We will show that the elements in group III and group V are also important in semiconductors.

Figure 1.1(a) shows five noninteracting silicon atoms, with the four valence electrons of each atom shown as dashed lines emanating from the atom. As silicon

| Table 1.1 | A list of some semiconductor materials | |
|---|---|---|
| **Elemental semiconductors** | | **Compound semiconductors** |
| Si | Silicon | GaAs | Gallium arsenide |
| Ge | Germanium | GaP | Gallium phosphide |
| | | AlP | Aluminum phosphide |
| | | AlAs | Aluminum arsenide |
| | | InP | Indium phosphide |

| Table 1.2 | A portion of the periodic table | |
|---|---|---|
| **III** | **IV** | **V** |
| 5 **B** Boron | 6 **C** Carbon | |
| 13 **Al** Aluminum | 14 **Si** Silicon | 15 **P** Phosphorus |
| 31 **Ga** Gallium | 32 **Ge** Germanium | 33 **As** Arsenic |
| 49 **In** Indium | | 51 **Sb** Antimony |

atoms come into close proximity to each other, the valence electrons interact to form a crystal. The final crystal structure is a tetrahedral configuration in which each silicon atom has four nearest neighbors, as shown in Figure 1.1(b). The valence electrons are shared between atoms, forming what are called **covalent bonds.** Germanium, gallium arsenide, and many other semiconductor materials have the same tetrahedral configuration.

Figure 1.1(c) is a two-dimensional representation of the lattice formed by the five silicon atoms in Figure 1.1(a). An important property of such a lattice is that valence electrons are always available on the outer edge of the silicon crystal so that additional atoms can be added to form very large single-crystal structures.

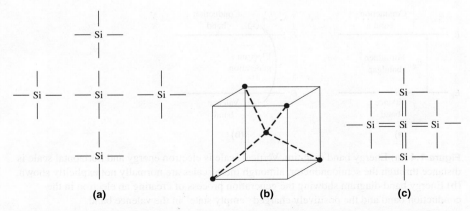

(a)    (b)    (c)

**Figure 1.1** Silicon atoms in a crystal matrix: (a) five noninteracting silicon atoms, each with four valence electrons, (b) the tetrahedral configuration, (c) a two-dimensional representation showing the covalent bonding

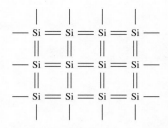

**Figure 1.2** Two-dimensional representation of single crystal silicon at $T = 0$ K; all valence electrons are bound to the silicon atoms by covalent bonding

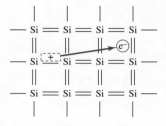

**Figure 1.3** The breaking of a covalent bond for $T > 0$ K creating an electron in the conduction band and a positively charged "empty state"

A two-dimensional representation of a silicon single crystal is shown in Figure 1.2, for $T = 0$ K, where $T =$ temperature. Each line between atoms represents a valence electron. At $T = 0$ K, each electron is in its lowest possible energy state, so each covalent bonding position is filled. If a small electric field is applied to this material, the electrons will not move, because they will still be bound to their individual atoms. Therefore, at $T = 0$ K, silicon is an **insulator;** that is, no charge flows through it.

When silicon atoms come together to form a crystal, the electrons occupy particular allowed energy bands. At $T = 0$ K, all valence electrons occupy the valence energy band. If the temperature increases, the valence electrons may gain thermal energy. Any such electron may gain enough thermal energy to break the covalent bond and move away from its original position as schematically shown in Figure 1.3. In order to break the covalent bond, the valence electron must gain a minimum energy, $E_g$, called the **bandgap energy.** The electrons that gain this minimum energy now exist in the conduction band and are said to be free electrons. These free electrons in the conduction band can move throughout the crystal. The net flow of electrons in the conduction band generates a current.

An energy band diagram is shown in Figure 1.4(a). The energy $E_v$ is the maximum energy of the valence energy band and the energy $E_c$ is the minimum energy of the conduction energy band. The bandgap energy $E_g$ is the difference between $E_c$ and $E_v$, and the region between these two energies is called the **forbidden bandgap.** Electrons cannot exist within the forbidden bandgap. Figure 1.4(b) qualitatively shows an electron from the valence band gaining enough energy and moving into the conduction band. This process is called generation.

Materials that have large bandgap energies, in the range of 3 to 6 electron–volts[1] (eV), are insulators because, at room temperature, essentially no free electrons exist in the conduction band. In contrast, materials that contain very large numbers of free electrons at room temperature are conductors. In a *semiconductor,* the bandgap energy is on the order of 1 eV.

The net charge in a semiconductor is zero; that is, the semiconductor is neutral. If a negatively charged electron breaks its covalent bond and moves away from its original position, a positively charged "empty" state is created at that position

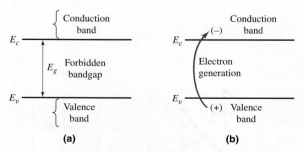

**Figure 1.4** (a) Energy band diagram. Vertical scale is electron energy and horizontal scale is distance through the semiconductor, although these scales are normally not explicitly shown. (b) Energy band diagram showing the generation process of creating an electron in the conduction band and the positively charged "empty state" in the valence band.

---

[1]An electron–volt is the energy of an electron that has been accelerated through a potential difference of 1 volt, and 1 eV $= 1.6 \times 10^{-19}$ joules.

| Table 1.3 | Semiconductor constants | |
| --- | --- | --- |
| **Material** | **$E_g$ (eV)** | **$B$ (cm$^{-3}$ K$^{-3/2}$)** |
| Silicon (Si) | 1.1 | $5.23 \times 10^{15}$ |
| Gallium arsenide (GaAs) | 1.4 | $2.10 \times 10^{14}$ |
| Germanium (Ge) | 0.66 | $1.66 \times 10^{15}$ |

(Figure 1.3). As the temperature increases, more covalent bonds are broken, and more free electrons and positive empty states are created.

A valence electron that has a certain thermal energy and is adjacent to an empty state may move into that position, as shown in Figure 1.5, making it appear as if a positive charge is moving through the semiconductor. This positively charged "particle" is called a **hole.** In semiconductors, then, two types of charged particles contribute to the current: the negatively charged free electron, and the positively charged hole. (This description of a hole is greatly oversimplified, and is meant only to convey the concept of the moving positive charge.) We may note that the charge of a hole has the same magnitude as the charge of an electron.

The concentrations (#/cm$^3$) of electrons and holes are important parameters in the characteristics of a semiconductor material, because they directly influence the magnitude of the current. An **intrinsic semiconductor** is a single-crystal semiconductor material with no other types of atoms within the crystal. In an intrinsic semiconductor, the densities of electrons and holes are equal, since the thermally generated electrons and holes are the only source of such particles. Therefore, we use the notation $n_i$ as the **intrinsic carrier concentration** for the concentration of the free electrons, as well as that of the holes. The equation for $n_i$ is as follows:

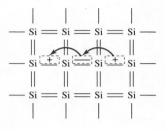

**Figure 1.5** A two-dimensional representation of the silicon crystal showing the movement of the positively charged "empty state"

$$n_i = BT^{3/2}e^{\left(\frac{-E_g}{2kT}\right)} \tag{1.1}$$

where $B$ is a coefficient related to the specific semiconductor material, $E_g$ is the bandgap energy (eV), $T$ is the temperature (K), $k$ is Boltzmann's constant ($86 \times 10^{-6}$ eV/K), and $e$, in this context, represents the exponential function. The values for $B$ and $E_g$ for several semiconductor materials are given in Table 1.3. The bandgap energy $E_g$ and coefficient $B$ are not strong functions of temperature. The intrinsic concentration $n_i$ is a parameter that appears often in the current–voltage equations for semiconductor devices.

## EXAMPLE 1.1

**Objective:** Calculate the intrinsic carrier concentration in silicon at $T = 300$ K.

**Solution:** For silicon at $T = 300$ K, we can write

$$n_i = BT^{3/2}e^{\left(\frac{-E_g}{2kT}\right)}$$

$$= (5.23 \times 10^{15})(300)^{3/2}e^{\left(\frac{-1.1}{2(86 \times 10^{-6})(300)}\right)}$$

or

$$n_i = 1.5 \times 10^{10}\text{cm}^{-3}$$

**Comment:** An intrinsic electron concentration of $1.5 \times 10^{10}$ cm$^{-3}$ may appear to be large, but it is relatively small compared to the concentration of silicon atoms, which is $5 \times 10^{22}$ cm$^{-3}$.

---

EXERCISE PROBLEM

**Ex 1.1:** Calculate the intrinsic carrier concentration in gallium arsenide and germanium at $T = 300$ K. (Ans. GaAs, $n_i = 1.80 \times 10^6$ cm$^{-3}$; Ge, $n_i = 2.40 \times 10^{13}$ cm$^{-3}$)

---

### 1.1.2    Extrinsic Semiconductors

Since the electron and hole concentrations in an intrinsic semiconductor are relatively small, only very small currents are possible. However, these concentrations can be greatly increased by adding controlled amounts of certain impurities. A desirable impurity is one that enters the crystal lattice and replaces (i.e., substitutes for) one of the semiconductor atoms, even though the impurity atom does not have the same valence electron structure. For silicon, the desirable substitutional impurities are from the group III and V elements (see Table 1.2).

The most common group V elements used for this purpose are phosphorus and arsenic. For example, when a phosphorus atom substitutes for a silicon atom, as shown in Figure 1.6(a), four of its valence electrons are used to satisfy the covalent bond requirements. The fifth valence electron is more loosely bound to the phosphorus atom. At room temperature, this electron has enough thermal energy to break the bond, thus being free to move through the crystal and contribute to the electron current in the semiconductor. When the fifth phosphorus valence electron moves into the conduction band, a positively charged phosphorus ion is created as shown in Figure 1.6(b).

The phosphorus atom is called a **donor impurity,** since it donates an electron that is free to move. Although the remaining phosphorus atom has a net positive charge, the atom is immobile in the crystal and cannot contribute to the current. Therefore, when a donor impurity is added to a semiconductor, free electrons are created without generating holes. This process is called **doping,** and it allows us to control the concentration of free electrons in a semiconductor.

A semiconductor that contains donor impurity atoms is called an **n-type semiconductor** (for the negatively charged electrons) and has a preponderance of electrons compared to holes.

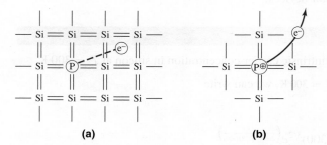

(a)                                      (b)

**Figure 1.6** (a) Two-dimensional representation of a silicon lattice doped with a phosphorus atom showing the fifth phosphorus valence electron, (b) the resulting positively charged phosphorus ion after the fifth valence electron has moved into the conduction band

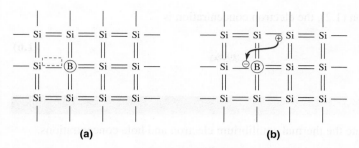

**Figure 1.7** (a) Two-dimensional representation of a silicon lattice doped with a boron atom showing the vacant covalent bond position, (b) the resulting negatively charged boron ion after it has accepted an electron from the valence band. A positively charged hole is created.

The most common group III element used for silicon doping is boron. When a boron atom replaces a silicon atom, its three valence electrons are used to satisfy the covalent bond requirements for three of the four nearest silicon atoms (Figure 1.7(a)). This leaves one bond position open. At room temperature, adjacent silicon valence electrons have sufficient thermal energy to move into this position, thereby creating a hole. This effect is shown in Figure 1.7(b). The boron atom then has a net negative charge, but cannot move, and a hole is created that can contribute to a hole current.

Because the boron atom has accepted a valence electron, the boron is therefore called an **acceptor impurity.** Acceptor atoms lead to the creation of holes without electrons being generated. This process, also called doping, can be used to control the concentration of holes in a semiconductor.

A semiconductor that contains acceptor impurity atoms is called a **p-type semiconductor** (for the positively charged holes created) and has a preponderance of holes compared to electrons.

The materials containing impurity atoms are called **extrinsic semiconductors, or doped semiconductors.** The doping process, which allows us to control the concentrations of free electrons and holes, determines the conductivity and currents in the material.

A fundamental relationship between the electron and hole concentrations in a semiconductor *in thermal equilibrium* is given by

$$n_o p_o = n_i^2 \tag{1.2}$$

where $n_o$ is the thermal equilibrium concentration of free electrons, $p_o$ is the thermal equilibrium concentration of holes, and $n_i$ is the intrinsic carrier concentration.

At room temperature ($T = 300$ K), each donor atom donates a free electron to the semiconductor. If the donor concentration $N_d$ is much larger than the intrinsic concentration, we can approximate

$$n_o \cong N_d \tag{1.3}$$

Then, from Equation (1.2), the hole concentration is

$$p_o = \frac{n_i^2}{N_d} \tag{1.4}$$

Similarly, at room temperature, each acceptor atom accepts a valence electron, creating a hole. If the acceptor concentration $N_a$ is much larger than the intrinsic concentration, we can approximate

$$p_o \cong N_a \tag{1.5}$$

Then, from Equation (1.2), the electron concentration is

$$n_o = \frac{n_i^2}{N_a}$$ (1.6)

## EXAMPLE 1.2

**Objective:** Calculate the thermal equilibrium electron and hole concentrations.

**(a)** Consider silicon at $T = 300$ K doped with phosphorus at a concentration of $N_d = 10^{16}$ cm$^{-3}$. Recall from Example 1.1 that $n_i = 1.5 \times 10^{10}$ cm$^{-3}$.

**Solution:** Since $N_d \gg n_i$, the electron concentration is

$$n_o \cong N_d = 10^{16} \text{ cm}^{-3}$$

and the hole concentration is

$$p_o = \frac{n_i^2}{N_d} = \frac{(1.5 \times 10^{10})^2}{10^{16}} = 2.25 \times 10^4 \text{ cm}^{-3}$$

**(b)** Consider silicon at $T = 300$ K doped with boron at a concentration of $N_a = 5 \times 10^{16}$ cm$^{-3}$.

**Solution:** Since $N_a \gg n_i$, the hole concentration is

$$p_o \cong N_a = 5 \times 10^{16} \text{ cm}^{-3}$$

and the electron concentration is

$$n_o = \frac{n_i^2}{N_a} = \frac{(1.5 \times 10^{10})^2}{5 \times 10^{16}} = 4.5 \times 10^3 \text{ cm}^{-3}$$

**Comment:** We see that in a semiconductor doped with donors, the concentration of electrons is far greater than that of the holes. Conversely, in a semiconductor doped with acceptors, the concentration of holes is far greater than that of the electrons. It is also important to note that the difference in the concentrations between electrons and holes in a particular semiconductor is many orders of magnitude.

## EXERCISE PROBLEM

**Ex 1.2:** (a) Calculate the majority and minority carrier concentrations in silicon at $T = 300$ K for (i) $N_d = 2 \times 10^{16}$ cm$^{-3}$ and (ii) $N_a = 10^{15}$ cm$^{-3}$. (b) Repeat part (a) for GaAs. (Ans. (a) (i) $n_o = 2 \times 10^{16}$ cm$^{-3}$, $p_o = 1.125 \times 10^4$ cm$^{-3}$; (ii) $p_o = 10^{15}$ cm$^{-3}$, $n_o = 2.25 \times 10^5$ cm$^{-3}$; (b) (i) $n_o = 2 \times 10^{16}$ cm$^{-3}$, $p_o = 1.62 \times 10^{-4}$ cm$^{-3}$; (ii) $p_o = 10^{15}$ cm$^{-3}$, $n_o = 3.24 \times 10^{-3}$ cm$^{-3}$).

In an n-type semiconductor, electrons are called the **majority carrier** because they far outnumber the holes, which are termed the **minority carrier.** The results obtained in Example 1.2 clarify this definition. In contrast, in a p-type semiconductor, the holes are the majority carrier and the electrons are the minority carrier.

### 1.1.3 Drift and Diffusion Currents

We've described the creation of negatively charged electrons and positively charged holes in the semiconductor. If these charged particles move, a current is generated. These charged electrons and holes are simply referred to as **carriers.**

The two basic processes which cause electrons and holes to move in a semiconductor are: (a) **drift,** which is the movement caused by electric fields, and (b) **diffusion,** which is the flow caused by variations in the concentration, that is, concentration gradients. Such gradients can be caused by a nonhomogeneous doping distribution, or by the injection of a quantity of electrons or holes into a region, using methods to be discussed later in this chapter.

#### Drift Current Density

To understand drift, assume an electric field is applied to a semiconductor. The field produces a force that acts on free electrons and holes, which then experience a net drift velocity and net movement. Consider an n-type semiconductor with a large number of free electrons (Figure 1.8(a)). An electric field $E$ applied in one direction produces a force on the electrons in the *opposite* direction, because of the electrons' negative charge. The electrons acquire a drift velocity $v_{dn}$ (in cm/s) which can be written as

$$v_{dn} = -\mu_n E \qquad (1.7)$$

where $\mu_n$ is a constant called the **electron mobility** and has units of $cm^2/V-s$. For low-doped silicon, the value of $\mu_n$ is typically 1350 $cm^2/V-s$. The mobility can be thought of as a parameter indicating how well an electron can move in a semiconductor. The negative sign in Equation (1.7) indicates that the electron drift velocity is opposite to that of the applied electric field as shown in Figure 1.8(a). The electron drift produces a drift current density $J_n$ ($A/cm^2$) given by

$$J_n = -env_{dn} = -en(-\mu_n E) = +en\mu_n E \qquad (1.8)$$

where $n$ is the electron concentration ($\#/cm^3$) and $e$, in this context, is the magnitude of the electronic charge. The conventional drift current is in the opposite direction from the flow of negative charge, which means that the drift current in an n-type semiconductor is in the same direction as the applied electric field.

Next consider a p-type semiconductor with a large number of holes (Figure 1.8(b)). An electric field $E$ applied in one direction produces a force on the holes in the *same* direction, because of the positive charge on the holes. The holes acquire a drift velocity $v_{dp}$ (in cm/s), which can be written as

$$v_{dp} = +\mu_p E \qquad (1.9)$$

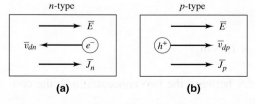

|        | (a)    |        | (b)    |
|--------|--------|--------|--------|

**Figure 1.8** Directions of applied electric field and resulting carrier drift velocity and drift current density in (a) an n-type semiconductor and (b) a p-type semiconductor

where $\mu_p$ is a constant called the **hole mobility,** and again has units of $cm^2/V-s$. For low-doped silicon, the value of $\mu_p$ is typically 480 $cm^2/V-s$, which is less than half the value of the electron mobility. The positive sign in Equation (1.9) indicates that the hole drift velocity is in the same direction as the applied electric field as shown in Figure 1.8(b). The hole drift produces a drift current density $J_p$ $(A/cm^2)$ given by

$$J_p = +epv_{dp} = +ep(+\mu_p E) = +ep\mu_p E \qquad (1.10)$$

where $p$ is the hole concentration $(\#/cm^3)$ and $e$ is again the magnitude of the electronic charge. The conventional drift current is in the same direction as the flow of positive charge, which means that the drift current in a p-type material is also in the same direction as the applied electric field.

Since a semiconductor contains both electrons and holes, the total drift current density is the sum of the electron and hole components. The total drift current density is then written as

$$J = en\mu_n E + ep\mu_p E = \sigma E = \frac{1}{\rho} E \qquad (1.11(a))$$

where

$$\sigma = en\mu_n + ep\mu_p \qquad (1.11(b))$$

and where $\sigma$ is the **conductivity** of the semiconductor in $(\Omega-cm)^{-1}$ and where $\rho = 1/\sigma$ is the **resistivity** of the semiconductor in $(\Omega-cm)$. The conductivity is related to the concentration of electrons and holes. If the electric field is the result of applying a voltage to the semiconductor, then Equation (1.11(a)) becomes a linear relationship between current and voltage and is one form of Ohm's law.

From Equation (1.11(b)), we see that the conductivity can be changed from strongly n-type, $n \gg p$, by donor impurity doping to strongly p-type, $p \gg n$, by acceptor impurity doping. **Being able to control the conductivity of a semiconductor by selective doping is what enables us to fabricate the variety of electronic devices that are available.**

## EXAMPLE 1.3

**Objective:** Calculate the drift current density for a given semiconductor.

Consider silicon at $T = 300$ K doped with arsenic atoms at a concentration of $N_d = 8 \times 10^{15}$ $cm^{-3}$. Assume mobility values of $\mu_n = 1350$ $cm^2/V-s$ and $\mu_p = 480$ $cm^2/V-s$. Assume the applied electric field is 100 V/cm.

**Solution:** The electron and hole concentrations are

$$n \cong N_d = 8 \times 10^{15} \ cm^{-3}$$

and

$$p = \frac{n_i^2}{N_d} = \frac{(1.5 \times 10^{10})^2}{8 \times 10^{15}} = 2.81 \times 10^4 \ cm^{-3}$$

Because of the difference in magnitudes between the two concentrations, the conductivity is given by

$$\sigma = e\mu_n n + e\mu_p p \cong e\mu_n n$$

or

$$\sigma = (1.6 \times 10^{-19})(1350)(8 \times 10^{15}) = 1.73(\Omega\text{–cm})^{-1}$$

The drift current density is then

$$J = \sigma E = (1.73)(100) = 173 \text{ A/cm}^2$$

**Comment:** Since $n \gg p$, the conductivity is essentially a function of the electron concentration and mobility only. We may note that a current density of a few hundred amperes per square centimeter can be generated in a semiconductor.

## EXERCISE PROBLEM

**Ex 1.3:** Consider n-type GaAs at $T = 300$ K doped to a concentration of $N_d = 2 \times 10^{16}$ cm$^{-3}$. Assume mobility values of $\mu_n = 6800$ cm$^2$/V–s and $\mu_p = 300$ cm$^2$/V–s. (a) Determine the resistivity of the material. (b) Determine the applied electric field that will induce a drift current density of 175 A/cm$^2$. (Ans. (a) 0.0460 $\Omega$–cm, (b) 8.04 V/cm).

**Note:** Two factors need to be mentioned concerning drift velocity and mobility. Equations (1.7) and (1.9) imply that the carrier drift velocities are linear functions of the applied electric field. This is true for relatively small electric fields. As the electric field increases, the carrier drift velocities will reach a maximum value of approximately $10^7$ cm/s. Any further increase in electric field will not produce an increase in drift velocity. This phenomenon is called drift velocity saturation.

Electron and hole mobility values were given in Example 1.3. The mobility values are actually functions of donor and/or acceptor impurity concentrations. As the impurity concentration increases, the mobility values will decrease. This effect then means that the conductivity, Equation (1.11(b)), is not a linear function of impurity doping.

These two factors are important in the design of semiconductor devices, but will not be considered in detail in this text.

### Diffusion Current Density

In the diffusion process, particles flow from a region of high concentration to a region of lower concentration. This is a statistical phenomenon related to kinetic theory. To explain, the electrons and holes in a semiconductor are in continuous motion, with an average speed determined by the temperature, and with the directions randomized by interactions with the lattice atoms. Statistically, we can assume that, at any particular instant, approximately half of the particles in the high-concentration region are moving *away* from that region toward the lower-concentration region. We can also assume that, at the same time, approximately half of the particles in the lower-concentration region are moving *toward* the high-concentration region. However, by definition, there are fewer particles in the lower-concentration region than there are in the high-concentration region. Therefore, the net result is a flow of particles away from the high-concentration region and toward the lower-concentration region. This is the basic diffusion process.

For example, consider an electron concentration that varies as a function of distance $x$, as shown in Figure 1.9(a). The diffusion of electrons from a high-concentration region

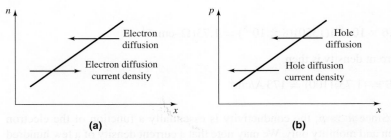

**Figure 1.9** (a) Assumed electron concentration versus distance in a semiconductor, and the resulting electron diffusion and electron diffusion current density, (b) assumed hole concentration versus distance in a semiconductor, and the resulting hole diffusion and hole diffusion current density

to a low-concentration region produces a flow of electrons in the negative $x$ direction. Since electrons are negatively charged, the conventional current direction is in the positive $x$ direction.

The diffusion current density due to the diffusion of electrons can be written as (for one dimension)

$$J_n = eD_n \frac{dn}{dx} \tag{1.12}$$

where $e$, in this context, is the magnitude of the electronic charge, $dn/dx$ is the gradient of the electron concentration, and $D_n$ is the **electron diffusion coefficient.**

In Figure 1.9(b), the hole concentration is a function of distance. The diffusion of holes from a high-concentration region to a low-concentration region produces a flow of holes in the negative $x$ direction. (Conventional current is in the direction of the flow of positive charge.)

The diffusion current density due to the diffusion of holes can be written as (for one dimension)

$$J_p = -eD_p \frac{dp}{dx} \tag{1.13}$$

where $e$ is still the magnitude of the electronic charge, $dp/dx$ is the gradient of the hole concentration, and $D_p$ is the **hole diffusion coefficient.** Note the change in sign between the two diffusion current equations. This change in sign is due to the difference in sign of the electronic charge between the negatively charged electron and the positively charged hole.

### EXAMPLE 1.4

**Objective:** Calculate the diffusion current density for a given semiconductor.

Consider silicon at $T = 300$ K. Assume the electron concentration varies linearly from $n = 10^{12}$ cm$^{-3}$ to $n = 10^{16}$ cm$^{-3}$ over the distance from $x = 0$ to $x = 3$ $\mu$m. Assume $D_n = 35$ cm$^2$/s.

**Solution:** We have

$$J_n = eD_n \frac{dn}{dx} = eD_n \frac{\Delta n}{\Delta x} = (1.6 \times 10^{-19})(35)\left(\frac{10^{12} - 10^{16}}{0 - 3 \times 10^{-4}}\right)$$

or

$$J_n = 187 \text{ A/cm}^2$$

**Comment:** Diffusion current densities on the order of a few hundred amperes per square centimeter can also be generated in a semiconductor.

**EXERCISE PROBLEM**

**Ex 1.4:** Consider silicon at $T = 300$ K. Assume the hole concentration is given by $p = 10^{16} e^{-x/L_p}$ (cm$^{-3}$), where $L_p = 10^{-3}$ cm. Calculate the hole diffusion current density at (a) $x = 0$ and (b) $x = 10^{-3}$ cm. Assume $D_p = 10$ cm$^2$/s. (Ans. (a) 16 A/cm$^2$, (b) 5.89 A/cm$^2$)

The mobility values in the drift current equations and the diffusion coefficient values in the diffusion current equations are not independent quantities. They are related by the **Einstein relation,** which is

$$\frac{D_n}{\mu_n} = \frac{D_p}{\mu_p} = \frac{kT}{e} \cong 0.026 \text{ V} \tag{1.14}$$

at room temperature.

The *total* current density is the sum of the drift and diffusion components. Fortunately, in most cases only one component dominates the current at any one time in a given region of a semiconductor.

**DESIGN POINTER**

In the previous two examples, current densities on the order of 200 A/cm$^2$ have been calculated. This implies that if a current of 1 mA, for example, is required in a semiconductor device, the size of the device is small. The total current is given by $I = JA$, where $A$ is the cross-sectional area. For $I = 1$ mA $= 1 \times 10^{-3}$ A and $J = 200$ A/cm$^2$, the cross-sectional area is $A = 5 \times 10^{-6}$ cm$^2$. This simple calculation again shows why semiconductor devices are small in size.

### 1.1.4    Excess Carriers

Up to this point, we have assumed that the semiconductor is in thermal equilibrium. In the discussion of drift and diffusion currents, we implicitly assumed that equilibrium was not significantly disturbed. Yet, when a voltage is applied to, or a current exists in, a semiconductor device, the semiconductor is really not in equilibrium. In this section, we will discuss the behavior of nonequilibrium electron and hole concentrations.

Valence electrons may acquire sufficient energy to break the covalent bond and become free electrons if they interact with high-energy photons incident on the semiconductor. When this occurs, both an electron and a hole are produced, thus generating an electron–hole pair. These additional electrons and holes are called **excess electrons** and **excess holes.**

When these excess electrons and holes are created, the concentrations of free electrons and holes increase above their thermal equilibrium values. This may be represented by

$$n = n_o + \delta n \qquad\qquad (1.15(a))$$

and

$$p = p_o + \delta p \qquad\qquad (1.15(b))$$

where $n_o$ and $p_o$ are the thermal equilibrium concentrations of electrons and holes, and $\delta n$ and $\delta p$ are the excess electron and hole concentrations.

If the semiconductor is in a steady-state condition, the creation of excess electrons and holes will not cause the carrier concentration to increase indefinitely, because a free electron may recombine with a hole, in a process called **electron–hole recombination.** Both the free electron and the hole disappear causing the excess concentration to reach a steady-state value. The mean time over which an excess electron and hole exist before recombination is called the **excess carrier lifetime.**

Excess carriers are involved in the current mechanisms of, for example, solar cells and photodiodes. These devices are discussed in Section 1.5.

## Test Your Understanding

**TYU 1.1** Determine the intrinsic carrier concentration in silicon, germanium, and GaAs at (a) $T = 400$ K and (b) $T = 250$ K. (Ans. (a) Si: $n_i = 4.76 \times 10^{12}$ cm$^{-3}$, Ge: $n_i = 9.06 \times 10^{14}$ cm$^{-3}$, GaAs: $n_i = 2.44 \times 10^9$ cm$^{-3}$; (b) Si: $n_i = 1.61 \times 10^8$ cm$^{-3}$, Ge: $n_i = 1.42 \times 10^{12}$ cm$^{-3}$, GaAs: $n_i = 6.02 \times 10^3$ cm$^{-3}$)

**TYU 1.2** (a) Consider silicon at $T = 300$ K. Assume that $\mu_n = 1350$ cm$^2$/V–s and $\mu_p = 480$ cm$^2$/V–s. Determine the conductivity and resistivity if (a) $N_a = 2 \times 10^{15}$ cm$^{-3}$ and (b) $N_d = 2 \times 10^{17}$ cm$^{-3}$. (Ans. (a) $\sigma = 0.154$ ($\Omega$–cm)$^{-1}$, $\rho = 6.51$ $\Omega$–cm; (b) $\sigma = 43.2$ ($\Omega$–cm)$^{-1}$, $\rho = 0.0231$ $\Omega$–cm).

**TYU 1.3** Using the results of TYU1.2, determine the drift current density if an electric field of 4 V/cm is applied to the semiconductor. (Ans. (a) 0.616 A/cm$^2$, (b) 172.8 A/cm$^2$).

**TYU 1.4** The electron and hole diffusion coefficients in silicon are $D_n = 35$ cm$^2$/s and $D_p = 12.5$ cm$^2$/s, respectively. Calculate the electron and hole diffusion current densities (a) if an electron concentration varies linearly from $n = 10^{15}$ cm$^{-3}$ to $n = 10^{16}$ cm$^{-3}$ over the distance from $x = 0$ to $x = 2.5$ $\mu$m and (b) if a hole concentration varies linearly from $p = 10^{14}$ cm$^{-3}$ to $p = 5 \times 10^{15}$ cm$^{-3}$ over the distance from $x = 0$ to $x = 4.0$ $\mu$m. (Ans. (a) $J_n = 202$ A/cm$^2$, (b) $J_p = -24.5$ A/cm$^2$)

**TYU 1.5** A sample of silicon at $T = 300$ K is doped to $N_d = 8 \times 10^{15}$ cm$^{-3}$. (a) Calculate $n_o$ and $p_o$. (b) If excess holes and electrons are generated such that their respective concentrations are $\delta n = \delta p = 10^{14}$ cm$^{-3}$, determine the total concentrations of holes and electrons. (Ans. (a) $n_o = 8 \times 10^{15}$ cm$^{-3}$, $p_o = 2.81 \times 10^4$ cm$^{-3}$; (b) $n = 8.1 \times 10^{15}$ cm$^{-3}$, $p \approx 10^{14}$ cm$^{-3}$)

## 1.2     THE pn JUNCTION

**Objective:** • Determine the properties of a pn junction including the ideal current-voltage characteristics of the pn junction diode.

In the preceding sections, we looked at characteristics of semiconductor materials. The real power of semiconductor electronics occurs when p- and n-regions are directly adjacent to each other, forming a **pn junction.** One important concept to remember is that in most integrated circuit applications, the entire semiconductor material is a single crystal, with one region doped to be p-type and the adjacent region doped to be n-type.

### 1.2.1     The Equilibrium pn Junction

Figure 1.10(a) is a simplified block diagram of a pn junction. Figure 1.10(b) shows the respective p-type and n-type doping concentrations, assuming uniform doping in each region, as well as the minority carrier concentrations in each region, assuming thermal equilibrium. Figure 1.10(c) is a three-dimensional diagram of the pn junction showing the cross-sectional area of the device.

The interface at $x = 0$ is called the **metallurgical junction.** A large density gradient in both the hole and electron concentrations occurs across this junction. Initially, then, there is a diffusion of holes from the p-region into the n-region, and a diffusion of electrons from the n-region into the p-region (Figure 1.11). The flow of holes from the p-region uncovers negatively charged acceptor ions, and the flow of

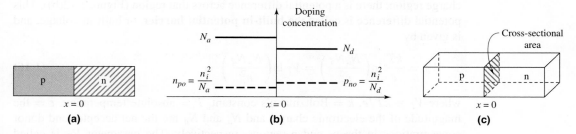

**Figure 1.10** (a) The pn junction: (a) simplified one-dimensional geometry, (b) doping profile of an ideal uniformly doped pn junction, and (c) three-dimensional representation showing the cross-sectional area

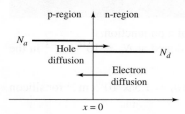

**Figure 1.11** Initial diffusion of electrons and holes across the metallurgical junction at the "instant in time" that the p- and n-regions are joined together

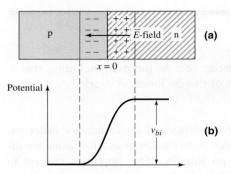

**Figure 1.12** The pn junction in thermal equilibrium. (a) The space charge region with negatively charged acceptor ions in the p-region and positively charged donor ions in the n-region; the resulting electric field from the n- to the p-region. (b) The potential through the junction and the built-in potential barrier $V_{bi}$ across the junction.

electrons from the n-region uncovers positively charged donor ions. This action creates a charge separation (Figure 1.12(a)), which sets up an electric field oriented in the direction from the positive charge to the negative charge.

If no voltage is applied to the pn junction, the diffusion of holes and electrons must eventually cease. The direction of the induced electric field will cause the resulting force to repel the diffusion of holes from the p-region and the diffusion of electrons from the n-region. Thermal equilibrium occurs when the force produced by the electric field and the "force" produced by the density gradient exactly balance.

The positively charged region and the negatively charged region comprise the **space-charge** region, or **depletion region,** of the pn junction, in which there are essentially no mobile electrons or holes. Because of the electric field in the space-charge region, there is a potential difference across that region (Figure 1.12(b)). This potential difference is called the **built-in potential barrier,** or built-in voltage, and is given by

$$V_{bi} = \frac{kT}{e} \ln\left(\frac{N_a N_d}{n_i^2}\right) = V_T \ln\left(\frac{N_a N_d}{n_i^2}\right) \tag{1.16}$$

where $V_T \equiv kT/e$, $k$ = Boltzmann's constant, $T$ = absolute temperature, $e$ = the magnitude of the electronic charge, and $N_a$ and $N_d$ are the net acceptor and donor concentrations in the p- and n-regions, respectively. The parameter $V_T$ is called the **thermal voltage** and is approximately $V_T = 0.026$ V at room temperature, $T = 300$ K.

## EXAMPLE 1.5

**Objective:** Calculate the built-in potential barrier of a pn junction.

Consider a silicon pn junction at $T = 300$ K, doped at $N_a = 10^{16}$ cm$^{-3}$ in the p-region and $N_d = 10^{17}$cm$^{-3}$ in the n-region.

**Solution:** From the results of Example 1.1, we have $n_i = 1.5 \times 10^{10}$cm$^{-3}$ for silicon at room temperature. We then find

$$V_{bi} = V_T \ln\left(\frac{N_a N_d}{n_i^2}\right) = (0.026) \ln\left[\frac{(10^{16})(10^{17})}{(1.5 \times 10^{10})^2}\right] = 0.757 \text{ V}$$

**Comment:** Because of the log function, the magnitude of $V_{bi}$ is not a strong function of the doping concentrations. Therefore, the value of $V_{bi}$ for silicon pn junctions is usually within 0.1 to 0.2 V of this calculated value.

The potential difference, or built-in potential barrier, across the space-charge region cannot be measured by a voltmeter because new potential barriers form between the probes of the voltmeter and the semiconductor, canceling the effects of $V_{bi}$. In essence, $V_{bi}$ maintains equilibrium, so no current is produced by this voltage. However, the magnitude of $V_{bi}$ becomes important when we apply a forward-bias voltage, as discussed later in this chapter.

### 1.2.2 Reverse-Biased pn Junction

Assume a positive voltage is applied to the n-region of a pn junction, as shown in Figure 1.13. The applied voltage $V_R$ induces an applied electric field, $E_A$, in the semiconductor. The direction of this applied field is the same as that of the $E$-field in the space-charge region. The magnitude of the electric field *in* the space-charge region increases above the thermal equilibrium value. This increased electric field holds back the holes in the p-region and the electrons in the n-region, so there is essentially no current across the pn junction. By definition, this applied voltage polarity is called **reverse bias.**

When the electric field in the space-charge region increases, the number of positive and negative charges must increase. If the doping concentrations are not changed, the increase in the fixed charge can only occur if the width $W$ of the space-charge region increases. Therefore, with an increasing reverse-bias voltage $V_R$, space-charge width $W$ also increases. This effect is shown in Figure 1.14.

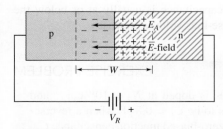

**Figure 1.13** A pn junction with an applied reverse-bias voltage, showing the direction of the electric field induced by $V_R$ and the direction of the original space-charge electric field. Both electric fields are in the same direction, resulting in a larger net electric field and a larger barrier between the p- and n-regions.

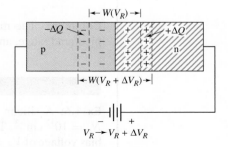

**Figure 1.14** Increase in space-charge width with an increase in reverse bias voltage from $V_R$ to $V_R + \Delta V_R$. Creation of additional charges $+\Delta Q$ and $-\Delta Q$ leads to a junction capacitance.

Because of the additional positive and negative charges induced in the space-charge region with an increase in reverse-bias voltage, a capacitance is associated with the pn junction when a reverse-bias voltage is applied. This **junction capacitance,** or depletion layer capacitance, can be written in the form

$$C_j = C_{jo}\left(1 + \frac{V_R}{V_{bi}}\right)^{-1/2} \tag{1.17}$$

where $C_{jo}$ is the junction capacitance at zero applied voltage.

The junction capacitance will affect the switching characteristics of the pn junction, as we will see later in the chapter. The voltage across a capacitance cannot change instantaneously, so changes in voltages in circuits containing pn junctions will not occur instantaneously.

The capacitance–voltage characteristics can make the pn junction useful for electrically tunable resonant circuits. Junctions fabricated specifically for this purpose are called **varactor diodes.** Varactor diodes can be used in electrically tunable oscillators, such as a Hartley oscillator, discussed in Chapter 15, or in tuned amplifiers, considered in Chapter 8.

## EXAMPLE 1.6

**Objective:** Calculate the junction capacitance of a pn junction.

Consider a silicon pn junction at $T = 300$ K, with doping concentrations of $N_a = 10^{16}$ cm$^{-3}$ and $N_d = 10^{15}$ cm$^{-3}$. Assume that $n_i = 1.5 \times 10^{10}$ cm$^{-3}$ and let $C_{jo} = 0.5$ pF. Calculate the junction capacitance at $V_R = 1$ V and $V_R = 5$ V.

**Solution:** The built-in potential is determined by

$$V_{bi} = V_T \ln\left(\frac{N_a N_d}{n_i^2}\right) = (0.026)\ln\left[\frac{(10^{16})(10^{15})}{(1.5 \times 10^{10})^2}\right] = 0.637 \text{ V}$$

The junction capacitance for $V_R = 1$ V is then found to be

$$C_j = C_{jo}\left(1 + \frac{V_R}{V_{bi}}\right)^{-1/2} = (0.5)\left(1 + \frac{1}{0.637}\right)^{-1/2} = 0.312 \text{ pF}$$

For $V_R = 5$ V

$$C_j = (0.5)\left(1 + \frac{5}{0.637}\right)^{-1/2} = 0.168 \text{ pF}$$

**Comment:** The magnitude of the junction capacitance is usually at or below the picofarad range, and it decreases as the reverse-bias voltage increases.

## EXERCISE PROBLEM

**Ex 1.6:** A silicon pn junction at $T = 300$ K is doped at $N_d = 10^{16}$ cm$^{-3}$ and $N_a = 10^{17}$ cm$^{-3}$. The junction capacitance is to be $C_j = 0.8$ pF when a reverse-bias voltage of $V_R = 5$ V is applied. Find the zero-biased junction capacitance $C_{jo}$. (Ans. $C_{jo} = 2.21$ pF)

As implied in the previous section, the magnitude of the electric field in the space-charge region increases as the reverse-bias voltage increases, and the maximum electric field occurs at the metallurgical junction. However, neither the electric field in the space-charge region nor the applied reverse-bias voltage can increase indefinitely because at some point, breakdown will occur and a large reverse bias current will be generated. This concept will be described in detail later in this chapter.

### 1.2.3 Forward-Biased pn Junction

We have seen that the n-region contains many more free electrons than the p-region; similarly, the p-region contains many more holes than the n-region. With zero applied voltage, the built-in potential barrier prevents these majority carriers from diffusing across the space-charge region; thus, the barrier maintains equilibrium between the carrier distributions on either side of the pn junction.

If a positive voltage $v_D$ is applied to the p-region, the potential barrier decreases (Figure 1.15). The electric fields in the space-charge region are very large compared to those in the remainder of the p- and n-regions, so essentially all of the applied voltage exists across the pn junction region. The applied electric field, $E_A$, induced by the applied voltage is in the opposite direction from that of the thermal equilibrium space-charge $E$-field. However, the net electric field is *always* from the n- to the p-region. The net result is that the electric field in the space-charge region is lower than the equilibrium value. This upsets the delicate balance between diffusion and the $E$-field force. Majority carrier electrons from the n-region diffuse into the p-region, and majority carrier holes from the p-region diffuse into the n-region. The process continues as long as the voltage $v_D$ is applied, thus creating a current in the pn junction. This process would be analogous to lowering a dam wall slightly. A slight drop in the wall height can send a large amount of water (current) over the barrier.

This applied voltage polarity (i.e., bias) is known as **forward bias.** The forward-bias voltage $v_D$ must always be less than the built-in potential barrier $V_{bi}$.

As the majority carriers cross into the opposite regions, they become minority carriers in those regions, causing the minority carrier concentrations to increase. Figure 1.16 shows the resulting excess minority carrier concentrations at the space-charge region edges. These excess minority carriers diffuse into the neutral n- and p-regions, where they recombine with majority carriers, thus establishing a steady-state condition, as shown in Figure 1.16.

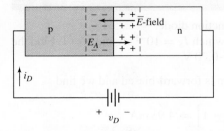

**Figure 1.15** A pn junction with an applied forward-bias voltage showing the direction of the electric field induced by $V_D$ and the direction of the original space-charge electric field. The two electric fields are in opposite directions resulting in a smaller net electric field and a smaller barrier between the p- and n-regions. The net electric field is always from the n- to the p-region.

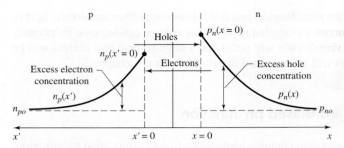

**Figure 1.16** Steady-state minority carrier concentrations in a pn junction under forward bias. The gradients in the minority carrier concentrations generate diffusion currents in the device.

### 1.2.4    Ideal Current–Voltage Relationship

As shown in Figure 1.16, an applied voltage results in a gradient in the minority carrier concentrations, which in turn causes diffusion currents. The theoretical relationship between the voltage and the current in the pn junction is given by

$$i_D = I_S \left[ e^{\left(\frac{v_D}{nV_T}\right)} - 1 \right] \tag{1.18}$$

The parameter $I_S$ is the **reverse-bias saturation current.** For silicon pn junctions, typical values of $I_S$ are in the range of $10^{-18}$ to $10^{-12}$ A. The actual value depends on the doping concentrations and is also proportional to the cross-sectional area of the junction. The parameter $V_T$ is the thermal voltage, as defined in Equation (1.16), and is approximately $V_T = 0.026$ V at room temperature. The parameter $n$ is usually called the emission coefficient or ideality factor, and its value is in the range $1 \leq n \leq 2$.

The emission coefficient $n$ takes into account any recombination of electrons and holes in the space-charge region. At very low current levels, recombination may be a significant factor and the value of $n$ may be close to 2. At higher current levels, recombination is less a factor, and the value of $n$ will be 1. Unless otherwise stated, we will assume the emission coefficient is $n = 1$.

This pn junction, with nonlinear rectifying current characteristics, is called a **pn junction diode.**

### EXAMPLE 1.7

**Objective:** Determine the current in a pn junction diode.

Consider a pn junction at $T = 300$ K in which $I_S = 10^{-14}$ A and $n = 1$. Find the diode current for $v_D = +0.70$ V and $v_D = -0.70$ V.

**Solution:** For $v_D = +0.70$ V, the pn junction is forward-biased and we find

$$i_D = I_S \left[ e^{\left(\frac{v_D}{V_T}\right)} - 1 \right] = (10^{-14}) \left[ e^{\left(\frac{+0.70}{0.026}\right)} - 1 \right] \Rightarrow 4.93 \, \text{mA}$$

For $v_D = -0.70$ V, the pn junction is reverse-biased and we find

$$i_D = I_S \left[ e^{\left(\frac{v_D}{V_T}\right)} - 1 \right] = (10^{-14}) \left[ e^{\left(\frac{-0.70}{0.026}\right)} - 1 \right] \cong -10^{-14} \, \text{A}$$

**Comment:** Although $I_S$ is quite small, even a relatively small value of forward-bias voltage can induce a moderate junction current. With a reverse-bias voltage applied, the junction current is virtually zero.

EXERCISE PROBLEM

**Ex 1.7:** (a) A silicon pn junction at $T = 300$ K has a reverse-saturation current of $I_S = 2 \times 10^{-14}$ A. Determine the required forward-bias voltage to produce a current of (i) $I_D = 50\ \mu$A and (ii) $I_D = 1$ mA. (b) Repeat part (a) for $I_S = 2 \times 10^{-12}$ A. (Ans. (a) (i) 0.563 V, (ii) 0.641 V; (b) (i) 0.443 V, (ii) 0.521 V).

## 1.2.5    pn Junction Diode

Figure 1.17 is a plot of the derived current–voltage characteristics of a pn junction. For a forward-bias voltage, the current is an exponential function of voltage. Figure 1.18 depicts the forward-bias current plotted on a log scale. With only a small change in the forward-bias voltage, the corresponding forward-bias current increases by orders of magnitude. For a forward-bias voltage $v_D > +0.1$ V, the $(-1)$ term in Equation (1.18) can be neglected. In the reverse-bias direction, the current is almost zero.

Figure 1.19 shows the diode circuit symbol and the conventional current direction and voltage polarity. The diode can be thought of and used as a voltage controlled switch that is "off" for a reverse-bias voltage and "on" for a forward-bias voltage. In the forward-bias or "on" state, a relatively large current is produced by a fairly small applied voltage; in the reverse-bias, or "off" state, only a very small current is created.

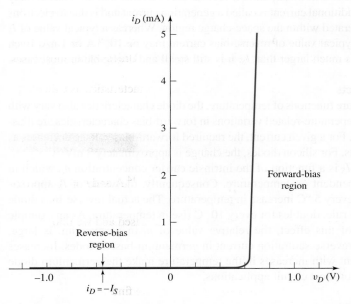

**Figure 1.17** Ideal $I$–$V$ characteristics of a pn junction diode for $I_S = 10^{-14}$ A. The diode current is an exponential function of diode voltage in the forward-bias region and is very nearly zero in the reverse-bias region. The pn junction diode is a nonlinear electronic device.

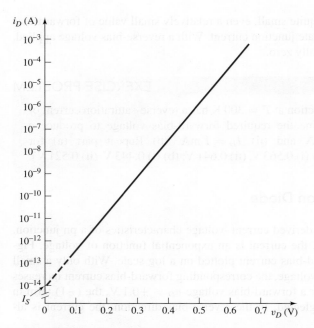

**Figure 1.18** Ideal forward-biased *I–V* characteristics of a pn junction diode, with the current plotted an a log scale for $I_S = 10^{-14}$ A and $n = 1$. The diode current increases approximately one order of magnitude for every 60-mV increase in diode voltage.

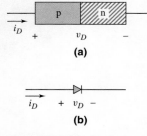

**Figure 1.19** The basic pn junction diode: (a) simplified geometry and (b) circuit symbol, and conventional current direction and voltage polarity

When a diode is reverse-biased by at least 0.1 V, the diode current is $i_D = -I_S$. The current is in the reverse direction and is a constant, hence the name reverse-bias saturation current. Real diodes, however, exhibit reverse-bias currents that are considerably larger than $I_S$. This additional current is called a generation current and is due to electrons and holes being generated within the space-charge region. Whereas a typical value of $I_S$ may be $10^{-14}$ A, a typical value of reverse-bias current may be $10^{-9}$ A or 1 nA. Even though this current is much larger than $I_S$, it is still small and negligible in most cases.

### Temperature Effects

Since both $I_S$ and $V_T$ are functions of temperature, the diode characteristics also vary with temperature. The temperature-related variations in forward-bias characteristics are illustrated in Figure 1.20. For a given current, the required forward-bias voltage decreases as temperature increases. For silicon diodes, the change is approximately 2 mV/°C.

The parameter $I_S$ is a function of the intrinsic carrier concentration $n_i$, which in turn is strongly dependent on temperature. Consequently, the value of $I_S$ approximately doubles for every 5 °C increase in temperature. The actual reverse-bias diode current, as a general rule, doubles for every 10 °C rise in temperature. As an example of the importance of this effect, the relative value of $n_i$ in germanium, is large, resulting in a large reverse-saturation current in germanium-based diodes. Increases in this reverse current with increases in the temperature make the germanium diode highly impractical for most circuit applications.

### Breakdown Voltage

When a reverse-bias voltage is applied to a pn junction, the electric field in the space-charge region increases. The electric field may become large enough that covalent

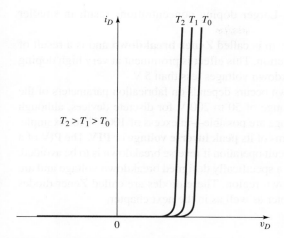

**Figure 1.20** Forward-biased pn junction characteristics versus temperature. The required diode voltage to produce a given current decreases with an increase in temperature.

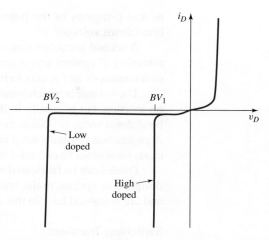

**Figure 1.21** Reverse-biased diode characteristics showing breakdown for a low-doped pn junction and a high-doped pn junction. The reverse-bias current increases rapidly once breakdown has occurred.

bonds are broken and electron–hole pairs are created. Electrons are swept into the n-region and holes are swept into the p-region by the electric field, generating a large reverse bias current. This phenomenon is called **breakdown.** The reverse-bias current created by the breakdown mechanism is limited only by the external circuit. If the current is not sufficiently limited, a large power can be dissipated in the junction that may damage the device and cause burnout. The current–voltage characteristic of a diode in breakdown is shown in Figure 1.21.

The most common breakdown mechanism is called **avalanche breakdown,** which occurs when carriers crossing the space charge region gain sufficient kinetic energy from the high electric field to be able to break covalent bonds during a collision process. The basic avalanche multiplication process is demonstrated in Figure 1.22. The generated electron–hole pairs can themselves be involved in a collision process generating additional electron–hole pairs, thus the avalanche process. The breakdown voltage is a function of the doping concentrations in the

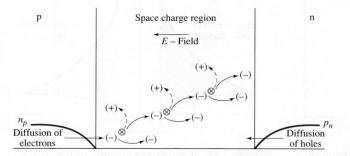

**Figure 1.22** The avalanche multiplication process in the space charge region. Shown are the collisions of electrons creating additional electron–hole pairs. Holes can also be involved in collisions creating additional electron–hole pairs.

n- and p-regions of the junction. Larger doping concentrations result in smaller breakdown voltages.

A second breakdown mechanism is called **Zener breakdown** and is a result of tunneling of carriers across the junction. This effect is prominent at very high doping concentrations and results in breakdown voltages less than 5 V.

The voltage at which breakdown occurs depends on fabrication parameters of the pn junction, but is usually in the range of 50 to 200 V for discrete devices, although breakdown voltages outside this range are possible—in excess of 1000 V, for example. A pn junction is usually rated in terms of its **peak inverse voltage** or **PIV.** The PIV of a diode must never be exceeded in circuit operation if reverse breakdown is to be avoided.

Diodes can be fabricated with a specifically designed breakdown voltage and are designed to operate in the breakdown region. These diodes are called Zener diodes and are discussed later in this chapter as well as in the next chapter.

### Switching Transient

Since the pn junction diode can be used as an electrical switch, an important parameter is its transient response, that is, its speed and characteristics, as it is switched from one state to the other. Assume, for example, that the diode is switched from the forward-bias "on" state to the reverse-bias "off" state. Figure 1.23 shows a simple circuit that will switch the applied voltage at time $t = 0$. For $t < 0$, the forward-bias current $i_D$ is

$$i_D = I_F = \frac{V_F - v_D}{R_F} \tag{1.19}$$

The minority carrier concentrations for an applied forward-bias voltage and an applied reverse-bias voltage are shown in Figure 1.24. Here, we neglect the change in the space charge region width. When a forward-bias voltage is applied, excess minority carrier charge is stored in both the p- and n-regions. The excess charge is the difference between the minority carrier concentrations for a forward-bias voltage and those for a reverse-bias voltage as indicated in the figure. This charge must be removed when the diode is switched from the forward to the reverse bias.

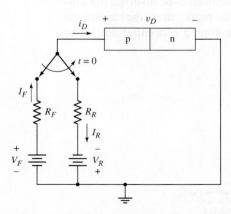

**Figure 1.23** Simple circuit for switching a diode from forward to reverse bias

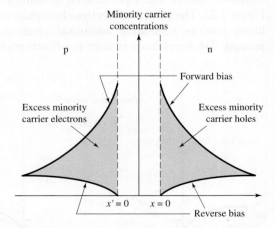

**Figure 1.24** Stored excess minority carrier charge under forward bias compared to reverse bias. This charge must be removed as the diode is switched from forward to reverse bias.

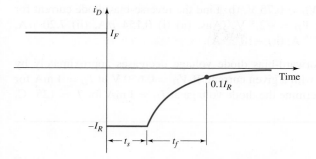

**Figure 1.25** Current characteristics versus time during diode switching

As the forward-bias voltage is removed, relatively large diffusion currents are created in the reverse-bias direction. This happens because the excess minority carrier electrons flow back across the junction into the n-region, and the excess minority carrier holes flow back across the junction into the p-region.

The large reverse-bias current is initially limited by resistor $R_R$ to approximately

$$i_D = -I_R \cong \frac{-V_R}{R_R} \qquad (1.20)$$

The junction capacitances do not allow the junction voltage to change instantaneously. The reverse current $I_R$ is approximately constant for $0^+ < t < t_s$, where $t_s$ is the **storage time,** which is the length of time required for the minority carrier concentrations at the space-charge region edges to reach the thermal equilibrium values. After this time, the voltage across the junction begins to change. The fall time $t_f$ is typically defined as the time required for the current to fall to 10 percent of its initial value. The total **turn-off time** is the sum of the storage time and the fall time. Figure 1.25 shows the current characteristics as this entire process takes place.

In order to switch a diode quickly, the diode must have a small excess minority carrier lifetime, and we must be able to produce a large reverse current pulse. Therefore, in the design of diode circuits, we must provide a path for the transient reverse-bias current pulse. These same transient effects impact the switching of transistors. For example, the switching speed of transistors in digital circuits will affect the speed of computers.

The turn-on transient occurs when the diode is switched from the "off" state to the forward-bias "on" state, which can be initiated by applying a forward-bias current pulse. The transient **turn-on time** is the time required to establish the forward-bias minority carrier distributions. During this time, the voltage across the junction gradually increases toward its steady-state value. Although the turn-on time for the pn junction diode is not zero, it is usually less than the transient turn-off time.

## Test Your Understanding

**TYU 1.6** (a) Determine $V_{bi}$ for a silicon pn junction at $T = 300$ K for $N_a = 10^{15}$ cm$^{-3}$ and $N_d = 5 \times 10^{16}$ cm$^{-3}$. (b) Repeat part (a) for a GaAs pn junction. (c) Repeat part (a) for a Ge pn junction. (Ans. (a) 0.679 V, (b) 1.15 V, (c) 0.296 V).

**TYU 1.7** A silicon pn junction diode at $T = 300$ K has a reverse-saturation current of $I_S = 10^{-16}$ A. (a) Determine the forward-bias diode current for (i) $V_D = 0.55$ V,

(ii) $V_D = 0.65$ V, and (iii) $V_D = 0.75$ V, (b) Find the reverse-bias diode current for (i) $V_D = -0.55$ V and (ii) $V_D = -2.5$ V. (Ans. (a) (i) 0.154 $\mu$A, (ii) 7.20 $\mu$A, (iii) 0.337 mA; (b) (i) $-10^{-16}$ A, (ii) $-10^{-16}$ A).

**TYU 1.8** Recall that the forward-bias diode voltage decreases approximately by 2 mV/°C for silicon diodes with a given current. If $V_D = 0.650$ V at $I_D = 1$ mA for a temperature of 25 °C, determine the diode voltage at $I_D = 1$ mA for $T = 125$ °C. (Ans. $V_D = 0.450$ V)

## 1.3    DIODE CIRCUITS: DC ANALYSIS AND MODELS

**Objective:** • Examine dc analysis techniques for diode circuits using various models to describe the diode characteristics.

In this section, we begin to study the diode in various circuit configurations. As we have seen, the diode is a two-terminal device with nonlinear $i$–$v$ characteristics, as opposed to a two-terminal resistor, which has a linear relationship between current and voltage. The analysis of nonlinear electronic circuits is not as straightforward as the analysis of linear electric circuits. However, there are electronic functions that can be implemented only by nonlinear circuits. Examples include the generation of dc voltages from sinusoidal voltages and the implementation of logic functions.

Mathematical relationships, or **models,** that describe the current–voltage characteristics of electrical elements allow us to analyze and design circuits without having to fabricate and test them in the laboratory. An example is Ohm's law, which describes the properties of a resistor. In this section, we will develop the dc analysis and modeling techniques of diode circuits.

This section considers the current–voltage characteristics of the pn junction diode in order to construct various circuit models. Large-signal models are initially developed that describe the behavior of the device with relatively large changes in voltages and currents. These models simplify the analysis of diode circuits and make the analysis of relatively complex circuits much easier. In the next section, we will consider a small-signal model of the diode that will describe the behavior of the pn junction with small changes in voltages and currents. It is important to understand the difference between large-signal and small-signal models and the conditions when they are used.

To begin to understand diode circuits, consider a simple diode application. The current–voltage characteristics of the pn junction diode were given in Figure 1.17. An **ideal diode** (as opposed to a diode with ideal I–V characteristics) has the characteristics shown in Figure 1.26(a). When a reverse-bias voltage is applied, the current through the diode is zero (Figure 1.26(b)); when current through the diode is greater than zero, the voltage across the diode is zero (Figure 1.26(c)). An external circuit connected to the diode must be designed to control the forward current through the diode.

One diode circuit is the **rectifier** circuit shown in Figure 1.27(a). Assume that the input voltage $v_I$ is a sinusoidal signal, as shown in Figure 1.27(b), and the diode is an ideal diode (see Figure 1.26(a)). During the positive half-cycle of the sinusoidal

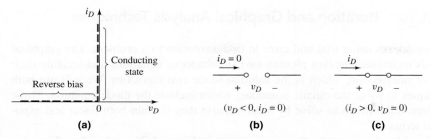

**(a)**          **(b)**          **(c)**

**Figure 1.26** The ideal diode: (a) the *I–V* characteristics of the ideal diode, (b) equivalent circuit under reverse bias (an open circuit), and (c) equivalent circuit in the conducting state (a short circuit)

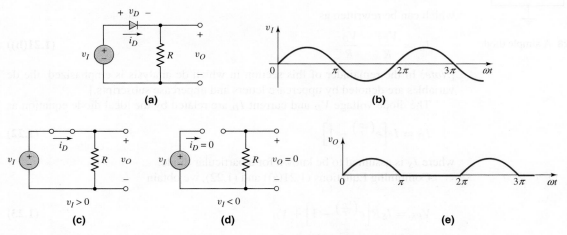

**Figure 1.27** The diode rectifier: (a) circuit, (b) sinusoidal input signal, (c) equivalent circuit for $v_I > 0$, (d) equivalent circuit for $v_I < 0$, and (e) rectified output signal

input, a forward-bias current exists in the diode and the voltage across the diode is zero. The equivalent circuit for this condition is shown in Figure 1.27(c). The output voltage $v_O$ is then equal to the input voltage. During the negative half-cycle of the sinusoidal input, the diode is reverse biased. The equivalent circuit for this condition is shown in Figure 1.27(d). In this part of the cycle, the diode acts as an open circuit, the current is zero, and the output voltage is zero. The output voltage of the circuit is shown in Figure 1.27(e).

Over the entire cycle, the input signal is sinusoidal and has a zero average value; however, the output signal contains only positive values and therefore has a positive average value. Consequently, this circuit is said to **rectify** the input signal, which is the first step in generating a dc voltage from a sinusoidal (ac) voltage. A dc voltage is required in virtually all electronic circuits.

As mentioned, the analysis of nonlinear circuits is not as straightforward as that of linear circuits. In this section, we will look at four approaches to the dc analysis of diode circuits: (a) iteration; (b) graphical techniques; (c) a piecewise linear modeling method; and (d) a computer analysis. Methods (a) and (b) are closely related and are therefore presented together.

### 1.3.1    Iteration and Graphical Analysis Techniques

Iteration means using trial and error to find a solution to a problem. The graphical analysis technique involves plotting two simultaneous equations and locating their point of intersection, which is the solution to the two equations. We will use both techniques to solve the circuit equations, which include the diode equation. These equations are difficult to solve by hand because they contain both linear and exponential terms.

Consider, for example, the circuit shown in Figure 1.28, with a dc voltage $V_{PS}$ applied across a resistor and a diode. **Kirchhoff's voltage law** applies both to nonlinear and linear circuits, so we can write

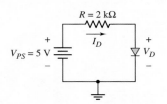

$R = 2$ k$\Omega$

$V_{PS} = 5$ V

$I_D$

$V_D$

**Figure 1.28** A simple diode circuit

$$V_{PS} = I_D R + V_D \qquad \textbf{(1.21(a))}$$

which can be rewritten as

$$I_D = \frac{V_{PS}}{R} - \frac{V_D}{R} \qquad \textbf{(1.21(b))}$$

[*Note:* In the remainder of this section in which dc analysis is emphasized, the dc variables are denoted by uppercase letters and uppercase subscripts.]

The diode voltage $V_D$ and current $I_D$ are related by the ideal diode equation as

$$I_D = I_S\left[e^{\left(\frac{V_D}{V_T}\right)} - 1\right] \qquad \textbf{(1.22)}$$

where $I_S$ is assumed to be known for a particular diode.

Combining Equations (1.21(a)) and (1.22), we obtain

$$V_{PS} = I_S R\left[e^{\left(\frac{V_D}{V_T}\right)} - 1\right] + V_D \qquad \textbf{(1.23)}$$

which contains only one unknown, $V_D$. However, Equation (1.23) is a transcendental equation and cannot be solved directly. The use of iteration to find a solution to this equation is demonstrated in the following example.

### EXAMPLE 1.8

**Objective:** Determine the diode voltage and current for the circuit shown in Figure 1.28.

Consider a diode with a given reverse-saturation current of $I_S = 10^{-13}$ A.

**Solution:** We can write Equation (1.23) as

$$5 = (10^{-13})(2 \times 10^3)\left[e^{\left(\frac{V_D}{0.026}\right)} - 1\right] + V_D \qquad \textbf{(1.24)}$$

If we first try $V_D = 0.60$ V, the right side of Equation (1.24) is 2.7 V, so the equation is not balanced and we must try again. If we next try $V_D = 0.65$ V, the right side of Equation (1.24) is 15.1 V. Again, the equation is not balanced, but we can see that the solution for $V_D$ is between 0.6 and 0.65 V. If we continue refining our guesses, we will be able to show that, when $V_D = 0.619$ V, the right side of Equation (1.29) is 4.99 V, which is essentially equal to the value of the left side of the equation.

The current in the circuit can then be determined by dividing the voltage difference across the resistor by the resistance, or

$$I_D = \frac{V_{PS} - V_D}{R} = \frac{5 - 0.619}{2} = 2.19 \text{ mA}$$

**Comment:** Once the diode voltage is known, the current can also be determined from the ideal diode equation. However, dividing the voltage difference across a resistor by the resistance is usually easier, and this approach is used extensively in the analysis of diode and transistor circuits.

## EXERCISE PROBLEM

**Ex 1.8:** Consider the circuit in Figure 1.28. Let $V_{PS} = 4$ V, $R = 4$ k$\Omega$, and $I_S = 10^{-12}$A. Determine $V_D$ and $I_D$, using the ideal diode equation and the iteration method. (Ans. $V_D = 0.535$ V, $I_D = 0.866$ mA)

To use a graphical approach to analyze the circuit, we go back to Kirchhoff's voltage law, as expressed by Equation (1.21(a)), which was $V_{PS} = I_D R + V_D$. Solving for the current $I_D$, we have

$$I_D = \frac{V_{PS}}{R} - \frac{V_D}{R}$$

which was also given by Equation (1.21(b)). This equation gives a linear relation between the diode current $I_D$ and the diode voltage $V_D$ for a given power supply voltage $V_{PS}$ and resistance $R$. This equation is referred to as the circuit **load line,** and is usually plotted on a graph with the current $I_D$ as the vertical axis and the voltage $V_D$ as the horizontal axis.

From Equation (1.21(b)), we see that if $I_D = 0$, then $V_D = V_{PS}$ which is the horizontal axis intercept. Also from this equation, if $V_D = 0$, then $I_D = V_{PS}/R$ which is the vertical axis intercept. The load line can be drawn between these two points. From Equation (1.21(b)), we see that the slope of the load line is $-1/R$.

Using the values given in Example (1.8), we can plot the straight line shown in Figure 1.29. The second plot in the figure is that of Equation (1.22), which is the ideal diode equation relating the diode current and voltage. The intersection of the load line and the device characteristics curve provides the dc current $I_D \approx 2.2$ mA through the diode and the dc voltage $V_D \approx 0.62$ V across the diode. This point is referred to as the **quiescent point,** or the **Q-point.**

The graphical analysis method can yield accurate results, but it is somewhat cumbersome. However, the concept of the load line and the graphical approach are useful for "visualizing" the response of a circuit, and the load line is used extensively in the evaluation of electronic circuits.

## 1.3.2    Piecewise Linear Model

Another, simpler way to analyze diode circuits is to *approximate* the diode's current–voltage characteristics, using linear relationships or straight lines. Figure 1.30, for example, shows the ideal current–voltage characteristics and two linear approximations.

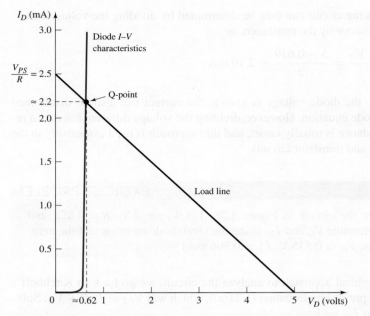

**Figure 1.29** The diode and load line characteristics for the circuit shown in Figure 1.28

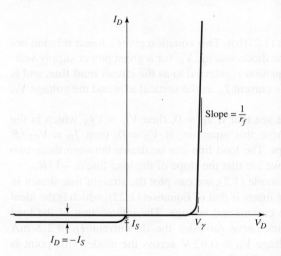

**Figure 1.30** The diode $I$–$V$ characteristics and two linear approximations. The linear approximations form the piecewise linear model of the diode.

For $V_D \geq V_\gamma$, we assume a straight-line approximation whose slope is $1/r_f$, where $V_\gamma$ is the **turn-on, or cut-in, voltage** of the diode, and $r_f$ is the **forward diode resistance.** The equivalent circuit for this linear approximation is a constant-voltage source in series with a resistor (Figure 1.31(a)).[2] For $V_D < V_\gamma$, we assume a straight-line

---

[2]It is important to keep in mind that the voltage source in Figure 1.31(a) only represents a voltage drop for $V_D \geq V_\gamma$. When $V_D < V_\gamma$, the $V_\gamma$ source does *not* produce a negative diode current. For $V_D < V_\gamma$, the equivalent circuit in Figure 1.31(b) must be used.

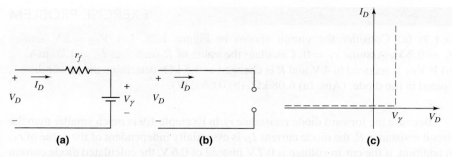

**Figure 1.31** The diode piecewise equivalent circuit (a) in the "on" condition when $V_D \geq V_\gamma$, (b) in the "off" condition when $V_D < V_\gamma$, and (c) piecewise linear approximation when $r_f = 0$. When $r_f = 0$, the voltage across the diode is a constant at $V_D = V_\gamma$ when the diode is conducting.

approximation parallel with the $V_D$ axis at the zero current level. In this case, the equivalent circuit is an open circuit (Figure 1.31(b)).

This method models the diode with segments of straight lines; thus the name **piecewise linear model.** If we assume $r_f = 0$, the piecewise linear diode characteristics are shown in Figure 1.31(c).

## EXAMPLE 1.9

**Objective:** Determine the diode voltage and current in the circuit shown in Figure 1.28, using a piecewise linear model. Also determine the power dissipated in the diode.

Assume piecewise linear diode parameters of $V_\gamma = 0.6$ V and $r_f = 10\,\Omega$.

**Solution:** With the given input voltage polarity, the diode is forward biased or "turned on," so $I_D > 0$. The equivalent circuit is shown in Figure 1.31(a). The diode current is determined by

$$I_D = \frac{V_{PS} - V_\gamma}{R + r_f} = \frac{5 - 0.6}{2 \times 10^3 + 10} \Rightarrow 2.19\,\text{mA}$$

and the diode voltage is

$$V_D = V_\gamma + I_D r_f = 0.6 + (2.19 \times 10^{-3})(10) = 0.622\,\text{V}$$

The power dissipated in the diode is given by

$$P_D = I_D V_D$$

We then find

$$P_D = (2.19)(0.622) = 1.36\,\text{mW}$$

**Comment:** This solution, obtained using the piecewise linear model, is nearly equal to the solution obtained in Example 1.8, in which the ideal diode equation was used. However, the analysis using the piecewise-linear model in this example is by far easier than using the actual diode $I$–$V$ characteristics as was done in Example 1.8. In general, we are willing to accept some slight analysis inaccuracy for ease of analysis.

Because the forward diode resistance $r_f$ in Example 1.9 is much smaller than the circuit resistance $R$, the diode current $I_D$ is essentially independent of the value of $r_f$. In addition, if the cut-in voltage is 0.7 V instead of 0.6 V, the calculated diode current will be 2.15 mA, which is not significantly different from the previous results. Therefore, the calculated diode current is not a strong function of the cut-in voltage. Consequently, we will often assume a cut-in voltage of 0.7 V for silicon pn junction diodes.

The concept of the load line and the piecewise linear model can be combined in diode circuit analyses. From Kirchhoff's voltage law, the load line for the circuit shown in Figure 1.28 and for the piecewise linear model of the diode can be written as

$$V_{PS} = I_D R + V_\gamma$$

where $V_\gamma$ is the diode cut-in voltage. We can assume $V_\gamma = 0.7$ V. Various load lines can be determined and plotted for the following circuit conditions:

A: $V_{PS} = 5$ V,    $R = 2$ kΩ
B: $V_{PS} = 5$ V,    $R = 4$ kΩ
C: $V_{PS} = 2.5$ V, $R = 2$ kΩ
D: $V_{PS} = 2.5$ V, $R = 4$ kΩ

The load line for condition A is plotted in Figure 1.32(a). Also plotted in the figure are the piecewise linear characteristics of the diode. The intersection of the two curves corresponds to the Q-point. For this case, the quiescent diode current is $I_{DQ} \cong 2.15$ mA.

Figure 1.32(b) shows the same piecewise linear characteristics of the diode. In addition, all four load lines, defined by the conditions listed above in A, B, C, and D are plotted on the figure. We see that the Q-point of the diode is a function of the load line. The Q-point changes for each load line.

The load line concept is also useful when the diode is reverse biased. Figure 1.33 (a) shows the same diode circuit as before, but with the direction of the diode reversed.

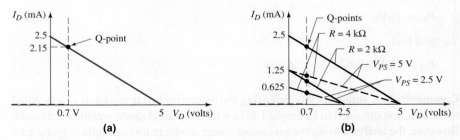

**Figure 1.32** Piecewise linear diode approximation superimposed on (a) load line for $V_{PS} = 5$ V, $R = 2$ kΩ and (b) several load lines. The Q-point of the diode changes when the load line changes.

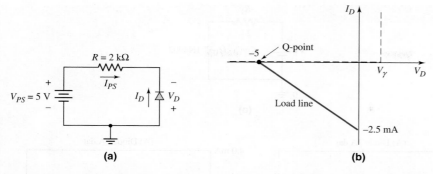

**Figure 1.33** Reverse-biased diode (a) circuit and (b) piecewise linear approximation and load line

The diode current $I_D$ and voltage $V_D$ shown are the usual forward-biased parameters. Applying Kirchhoff's voltage law, we can write

$$V_{PS} = I_{PS}R - V_D = -I_D R - V_D \qquad \textbf{(1.25(a))}$$

or

$$I_D = -\frac{V_{PS}}{R} - \frac{V_D}{R} \qquad \textbf{(1.25(b))}$$

where $I_D = -I_{PS}$. Equation (1.25(b)) is the load line equation. The two end points are found by setting $I_D = 0$, which yields $V_D = -V_{PS} = -5\,\text{V}$, and by setting $V_D = 0$, which yields $I_D = -V_{PS}/R = -5/2 = -2.5\,\text{mA}$. The diode characteristics and the load line are plotted in Figure 1.33(b). We see that the load line is now in the third quadrant, where it intersects the diode characteristics curve at $V_D = -5\,\text{V}$ and $I_D = 0$, demonstrating that the diode is reverse biased.

Although the piecewise linear model may yield solutions that are less accurate than those obtained with the ideal diode equation, the analysis is much easier.

### 1.3.3     Computer Simulation and Analysis

Today's computers are capable of using detailed simulation models of various components and performing complex circuit analyses quickly and relatively easily. Such models can factor in many diverse conditions, such as the temperature dependence of various parameters. One of the earliest, and now the most widely used, circuit analysis programs is the *simulation program with integrated circuit emphasis* (SPICE). This program, developed at the University of California at Berkeley, was first released about 1973, and has been continuously refined since that time. One outgrowth of SPICE is PSpice, which is designed for use on personal computers.

### EXAMPLE **1.10**

**Objective:** Determine the diode current and voltage characteristics of the circuit shown in Figure 1.28 using a PSpice analysis.

**Solution:** Figure 1.34(a) is the PSpice circuit schematic diagram. A standard 1N4002 diode from the PSpice library was used in the analysis. The input voltage $V1$ was varied (dc sweep) from 0 to 5 V. Figure 1.34(b) and (c) shows the diode voltage and diode current characteristics versus the input voltage.

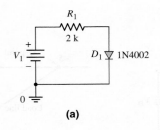

**(a)**

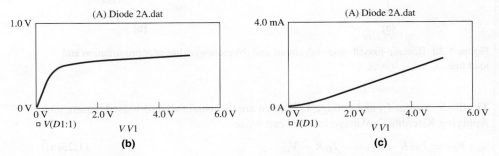

**(b)**          **(c)**

**Figure 1.34**  (a) PSpice circuit schematic, (b) diode voltage, and (c) diode current for Example 1.10

**Discussion:** Several observations may be made from the results. The diode voltage increases at almost a linear rate up to approximately 400 mV without any discernible (mA) current being measured. For an input voltage greater than approximately 500 mV, the diode voltage increases gradually to a value of about 610 mV at the maximum input voltage. The current also increases to a maximum value of approximately 2.2 mA at the maximum input voltage. The piecewise linear model predicts quite accurate results at the maximum input voltage. However, these results show that there is definitely a non-linear relation between the diode current and diode voltage. We must keep in mind that the piecewise linear model is an approximation technique that works very well in many applications.

## EXERCISE PROBLEM

**Ex 1.10:** The resistor parameter in the circuit shown in Figure 1.28 is changed to $R = 20 \, k\Omega$. Using a PSpice analysis, plot the diode current $I_D$ and diode voltage $V_D$ versus the power supply voltage $V_{PS}$ over the range $0 \leq V_{PS} \leq 10 \, V$.

### 1.3.4   Summary of Diode Models

The two dc diode models used in the hand analysis of diode circuits are: the ideal diode equation and the piecewise linear approximation. For the ideal diode equation, the reverse-saturation current $I_S$ must be specified. For the piecewise linear model, the cut-in voltage $V_\gamma$ and forward diode resistance $r_f$ must be specified. In most cases, however, $r_f$ is assumed to be zero unless otherwise given.

The particular model that should be used in a specific application or situation is a compromise between accuracy and ease of calculation. This decision comes with experience. In general, a simple model can be used in an initial design for ease of calculation. In a final design, we may want to use a computer simulation for better accuracy. However, it is very important to understand that the diode model or diode parameters used in the computer simulation must correspond to the actual diode parameters used in the circuit to ensure that the results are meaningful.

## Test Your Understanding

**TYU 1.9** Consider the diode and circuit in Exercise EX 1.8. Determine $V_D$ and $I_D$, using the graphical technique. (Ans. $V_D \cong 0.54$ V, $I_D \cong 0.87$ mA)

**TYU 1.10** Consider the circuit in Figure 1.28. Let $R = 4$ k$\Omega$ and $V_\gamma = 0.7$ V. Determine $I_D$ for (a) $V_{PS} = 0.5$ V, (b) $V_{PS} = 2$ V, (c) $V_{PS} = 5$ V, (d) $V_{PS} = -1$ V, and (e) $V_{PS} = -5$ V. (Ans. (a) 0, (b) 0.325 mA, (c) 1.075 mA, (d) 0, (e) 0).

**TYU 1.11** The power supply (input) voltage in the circuit of Figure 1.28 is $V_{PS} = 10$ V and the diode cut-in voltage is $V_\gamma = 0.7$ V (assume $r_f = 0$). The power dissipated in the diode is to be no more than 1.05 mW. Determine the maximum diode current and the minimum value of $R$ to meet the power specification. (Ans. $I_D = 1.5$ mA, $R = 6.2$ k$\Omega$)

---

 ## 1.4    DIODE CIRCUITS: AC EQUIVALENT CIRCUIT

> **Objective:** • Develop an equivalent circuit for a diode that is used when a small, time-varying signal is applied to a diode circuit.

Up to this point, we have only looked at the dc characteristics of the pn junction diode. When semiconductor devices with pn junctions are used in linear amplifier circuits, the time-varying, or ac, characteristics of the pn junction become important, because sinusoidal signals may be superimposed on the dc currents and voltages. The following sections examine these ac characteristics.

### 1.4.1    Sinusoidal Analysis

In the circuit shown in Figure 1.35(a), the voltage source $v_i$ is assumed to be a sinusoidal, or time-varying, signal. The total input voltage $v_I$ is composed of a dc component $V_{PS}$ and an ac component $v_i$ superimposed on the dc value. To investigate this circuit, we will look at two types of analyses: a dc analysis involving only the dc voltages and currents, and an ac analysis involving only the ac voltages and currents.

#### Current–Voltage Relationships
Since the input voltage contains a dc component with an ac signal superimposed, the diode current will also contain a dc component with an ac signal superimposed, as shown in Figure 1.35(b). Here, $I_{DQ}$ is the dc quiescent diode current. In addition, the diode voltage will contain a dc value with an ac signal superimposed, as shown in Figure 1.35(c). For this analysis, assume that the ac signal is small compared to the dc component, so that a linear ac model can be developed from the nonlinear diode.

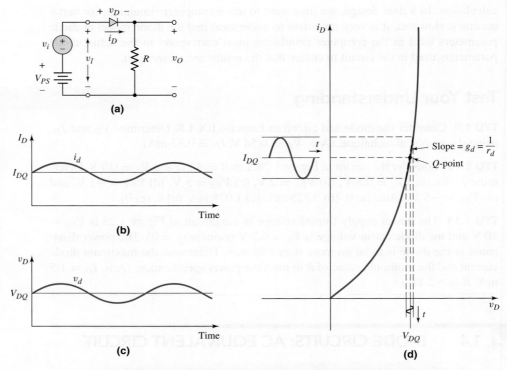

**Figure 1.35** AC circuit analysis: (a) circuit with combined dc and sinusoidal input voltages, (b) sinusoidal diode current superimposed on the quiescent current, (c) sinusoidal diode voltage superimposed on the quiescent value, and (d) forward-biased diode $I–V$ characteristics with a sinusoidal current and voltage superimposed on the quiescent values

The relationship between the diode current and voltage can be written as

$$i_D \cong I_S e^{\left(\frac{v_D}{V_T}\right)} = I_S e^{\left(\frac{V_{DQ}+v_d}{V_T}\right)} \tag{1.26}$$

where $V_{DQ}$ is the dc quiescent voltage and $v_d$ is the ac component. We are neglecting the $-1$ term in the diode equation given by Equation (1.22). Equation (1.26) can be rewritten as

$$i_D = I_S \left[ e^{\left(\frac{V_{DQ}}{V_T}\right)} \right] \cdot \left[ e^{\left(\frac{v_d}{V_T}\right)} \right] \tag{1.27}$$

If the ac signal is "small," then $v_d \ll V_T$, and we can expand the exponential function into a linear series, as follows:

$$e^{\left(\frac{v_d}{V_T}\right)} \cong 1 + \frac{v_d}{V_T} \tag{1.28}$$

We may also write the quiescent diode current as

$$I_{DQ} = I_S e^{\left(\frac{V_{DQ}}{V_T}\right)} \tag{1.29}$$

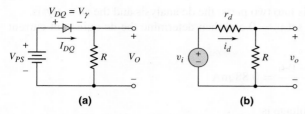

**Figure 1.36**  Equivalent circuits: (a) dc and (b) ac

The diode current–voltage relationship from Equation (1.27) can then be written as

$$i_D = I_{DQ}\left(1 + \frac{v_d}{V_T}\right) = I_{DQ} + \frac{I_{DQ}}{V_T} \cdot v_d = I_{DQ} + i_d \qquad (1.30)$$

where $i_d$ is the ac component of the diode current. The relationship between the ac components of the diode voltage and current is then

$$i_d = \left(\frac{I_{DQ}}{V_T}\right) \cdot v_d = g_d \cdot v_d \qquad (1.31(a))$$

or

$$v_d = \left(\frac{V_T}{I_{DQ}}\right) \cdot i_d = r_d \cdot i_d \qquad (1.31(b))$$

The parameters $g_d$ and $r_d$, respectively, are the diode **small-signal incremental conductance** and **resistance,** also called the **diffusion conductance** and **diffusion resistance.** We see from these two equations that

$$r_d = \frac{1}{g_d} = \frac{V_T}{I_{DQ}} \qquad (1.32)$$

This equation tells us that the incremental resistance is a function of the dc bias current $I_{DQ}$ and is inversely proportional to the slope of the $I$–$V$ characteristics curve, as shown in Figure 1.35(d).

### Circuit Analysis

To analyze the circuit shown in Figure 1.35(a), we first perform a dc analysis and then an ac analysis. These two types of analyses will use two equivalent circuits. Figure 1.36(a) is the dc equivalent circuit that we have seen previously. If the diode is forward biased, then the voltage across the diode is the piecewise linear turn-on voltage.

Figure 1.36(b) is the ac equivalent circuit. The diode has been replaced by its equivalent resistance $r_d$. All parameters in this circuit are the small-signal time-varying parameters.

## EXAMPLE **1.11**

**Objective:**  Analyze the circuit shown in Figure 1.35(a).

Assume circuit and diode parameters of $V_{PS} = 5$ V, $R = 5$ k$\Omega$, $V_\gamma = 0.6$ V, and $v_i = 0.1 \sin \omega t$ (V).

**Solution:** Divide the analysis into two parts: the dc analysis and the ac analysis.

For *the dc analysis,* we set $v_i = 0$ and then determine the dc quiescent current from Figure 1.36(a) as

$$I_{DQ} = \frac{V_{PS} - V_\gamma}{R} = \frac{5 - 0.6}{5} = 0.88 \text{ mA}$$

The dc value of the output voltage is

$$V_o = I_{DQ}R = (0.88)(5) = 4.4 \text{ V}$$

For *the ac analysis,* we consider only the ac signals and parameters in the circuit in Figure 1.36(b). In other words, we effectively set $V_{PS} = 0$. The ac Kirchhoff voltage law (KVL) equation becomes

$$v_i = i_d r_d + i_d R = i_d(r_d + R)$$

where $r_d$ is again the small-signal diode diffusion resistance. From Equation (1.32), we have

$$r_d = \frac{V_T}{I_{DQ}} = \frac{0.026}{0.88} = 0.0295 \text{ k}\Omega$$

The ac diode current is

$$i_d = \frac{v_i}{r_d + R} = \frac{0.1 \sin \omega t}{0.0295 + 5} \Rightarrow 19.9 \sin \omega t \ (\mu A)$$

The ac component of the output voltage is

$$v_o = i_d R = 0.0995 \sin \omega t \ (V)$$

**Comment:** Throughout the text, we will divide the circuit analysis into a dc analysis and an ac analysis. To do so, we will use separate equivalent circuit models for each analysis.

---

### EXERCISE PROBLEM

**Ex 1.11:** (a) The circuit and diode parameters for the circuit shown in Figure 1.35(a) are $V_{PS} = 8$ V, $R = 20$ k$\Omega$, $V_\gamma = 0.7$ V, and $v_i = 0.25 \sin \omega t$ (V). Determine the quiescent diode current and the time-varying diode current. (b) Repeat part (a) if the resistor is changed to $R = 10$ k$\Omega$. (Ans. (a) $I_{DQ} = 0.365$ mA, $i_d = 12.5 \sin \omega t \, (\mu A)$; (b) $I_{DQ} = 0.730$ mA, $i_d = 24.9 \sin \omega t \, (\mu A)$).

### Frequency Response

In the previous analysis, we implicitly assumed that the frequency of the ac signal was small enough that capacitance effects in the circuit would be negligible. If the frequency of the ac input signal increases, the **diffusion capacitance** associated with a forward-biased pn junction becomes important. The source of the diffusion capacitance is shown in Figure 1.37.

Consider the minority carrier hole concentration on the right side of the figure. At the quiescent diode voltage, $V_{DQ}$, the minority carrier hole concentration is shown as the solid line and indicated by $p_{n|V_{DQ}}$.

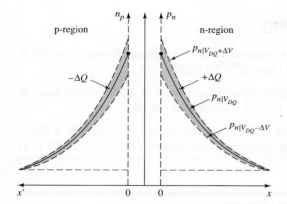

**Figure 1.37** Change in minority carrier stored charge with a time-varying voltage superimposed on a dc quiescent diode voltage. The change in stored charge leads to a diode diffusion capacitance.

If the total diode voltage increases by $\Delta V$ during the positive half cycle of a sinusoidal signal superimposed on the quiescent value, the hole concentration will increase to that shown by the dotted line indicated by $p_{n|V_{DQ}+\Delta V}$. Now, if the total diode voltage decreases by $\Delta V$ during the negative half cycle of a sinusoidal signal superimposed on the quiescent value, the hole concentration will decrease to that shown by the dotted line indicated by $p_{n|V_{DQ}-\Delta V}$. The $+\Delta Q$ charge is alternately being charged and discharged through the pn junction as the voltage across the junction changes.

The same process is occurring with the minority carrier electrons in the p-region.

The diffusion capacitance is the change in the stored minority carrier charge that is caused by a change in the voltage, or

$$C_d = \frac{dQ}{dV_D} \tag{1.33}$$

The diffusion capacitance $C_d$ is normally much larger than the junction capacitance $C_j$, because of the magnitude of the charges involved.

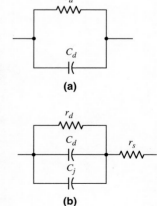

**Figure 1.38** Small-signal equivalent circuit of the diode: (a) simplified version and (b) complete circuit

## 1.4.2   Small-Signal Equivalent Circuit

The small-signal equivalent circuit of the forward-biased pn junction is shown in Figure 1.38 and is developed partially from the equation for the **admittance,** which is given by

$$Y = g_d + j\omega C_d \tag{1.34}$$

where $g_d$ and $C_d$ are the diffusion conductance and capacitance, respectively. We must also add the junction capacitance, which is in parallel with the diffusion resistance and capacitance, and a series resistance, which is required because of the finite resistances in the neutral n- and p-regions.

The small-signal equivalent circuit of the pn junction is used to obtain the ac response of a diode circuit subjected to ac signals superimposed on the Q-point values. Small-signal equivalent circuits of pn junctions are also used to develop small-signal models of transistors, and these models are used in the analysis and design of transistor amplifiers.

## Test Your Understanding

**TYU 1.12** Determine the diffusion conductance of a pn junction diode at $T = 300$ K and biased at a current of 0.8 mA. (Ans. $g_d = 30.8$ mS)

**TYU 1.13** Determine the small-signal diffusion resistance of a pn junction diode at $I_D = 10\ \mu\text{A}$, $100\ \mu\text{A}$, and 1 mA. (Ans. 2.6 k$\Omega$, 260 $\Omega$, 26 $\Omega$).

**TYU 1.14** The diffusion resistance of a pn junction diode at $T = 300$ K is determined to be $r_d = 50\ \Omega$. What is the quiescent diode current? (Ans. $I_{DQ} = 0.52$ mA)

## 1.5    OTHER DIODE TYPES

> **Objective:** • Gain an understanding of the properties and characteristics of a few specialized diodes.

There are many other types of diodes with specialized characteristics that are useful in particular applications. We will briefly consider only a few of these diodes. We will consider the solar cell, photodiode, light-emitting diode, Schottky diode, and Zener diode.

### 1.5.1    Solar Cell

A **solar cell** is a pn junction device with no voltage directly applied across the junction. The pn junction, which converts solar energy into electrical energy, is connected to a load as indicated in Figure 1.39. When light hits the space-charge region, electrons and holes are generated. They are quickly separated and swept out of the space-charge region by the electric field, thus creating a **photocurrent.** The generated photocurrent will produce a voltage across the load, which means that the solar cell has supplied power. Solar cells are usually fabricated from silicon, but may be made from GaAs or other III–V compound semiconductors.

Solar cells have long been used to power the electronics in satellites and space vehicles, and also as the power supply to some calculators. Solar cells are also used to power race cars in a Sunrayce event. Collegiate teams in the United States design,

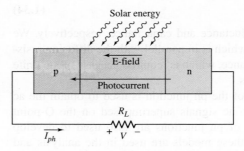

**Figure 1.39** A pn junction solar cell connected to load

build and drive the race cars. Typically, a Sunrayce car has 8 m$^2$ of solar cell arrays that can produce 800 W of power on a sunny day at noon. The power from the solar array can be used either to power an electric motor or to charge a battery pack.

### 1.5.2  Photodiode

**Photodetectors** are devices that convert optical signals into electrical signals. An example is the **photodiode,** which is similar to a solar cell except that the pn junction is operated with a reverse-bias voltage. Incident photons or light waves create excess electrons and holes in the space-charge region. These excess carriers are quickly separated and swept out of the space-charge region by the electric field, thus creating a "photocurrent." This generated photocurrent is directly proportional to the incident photon flux.

### 1.5.3  Light-Emitting Diode

The **light-emitting diode (LED)** converts current to light. As previously explained, when a forward-bias voltage is applied across a pn junction, electrons and holes flow across the space-charge region and become excess minority carriers. These excess minority carriers diffuse into the neutral semiconductor regions, where they recombine with majority carriers. If the semiconductor is a **direct bandgap material,** such as GaAs, the electron and hole can recombine with no change in momentum, and a photon or light wave can be emitted. Conversely, in an **indirect bandgap material,** such as silicon, when an electron and hole recombine, both energy and momentum must be conserved, so the emission of a photon is very unlikely. Therefore, LEDs are fabricated from GaAs or other compound semiconductor materials. In an LED, the diode current is directly proportional to the recombination rate, which means that the output light intensity is also proportional to the diode current.

Monolithic arrays of LEDs are fabricated for numeric and alphanumeric displays, such as the readout of a digital voltmeter.

An LED may be integrated into an optical cavity to produce a coherent photon output with a very narrow bandwidth. Such a device is a laser diode, which is used in optical communications applications.

The LED can be used in conjunction with a photodiode to create an optical system such as that shown in Figure 1.40. The light signal created may travel over

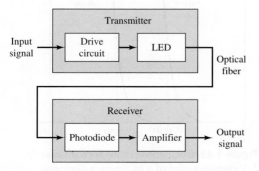

**Figure 1.40** Basic elements in an optical transmission system

relatively long distances through the optical fiber, because of the low optical absorption in high-quality optical fibers.

### 1.5.4    Schottky Barrier Diode

A **Schottky barrier diode,** or simply a Schottky diode, is formed when a metal, such as aluminum, is brought into contact with a *moderately* doped n-type semiconductor to form a rectifying junction. Figure 1.41(a) shows the metal-semiconductor contact, and Figure 1.41(b) shows the circuit symbol with the current direction and voltage polarity.

The current–voltage characteristics of a Schottky diode are very similar to those of a pn junction diode. The same ideal diode equation can be used for both devices. However, there are two important differences between the two diodes that directly affect the response of the Schottky diode.

First, the current mechanism in the two devices is different. The current in a pn junction diode is controlled by the diffusion of minority carriers. The current in a Schottky diode results from the flow of majority carriers over the potential barrier at the metallurgical junction. This means that there is no minority carrier storage in the Schottky diode, so the switching time from a forward bias to a reverse bias is very short compared to that of a pn junction diode. The storage time, $t_s$, for a Schottky diode is essentially zero.

Second, the reverse-saturation current $I_S$ for a Schottky diode is larger than that of a pn junction diode for comparable device areas. This property means that it takes less forward bias voltage to induce a particular current compared to a pn junction diode. We will see an application of this in Chapter 17.

Figure 1.42 compares the characteristics of the two diodes. Applying the piecewise linear model, we can determine that the Schottky diode has a smaller turn-on

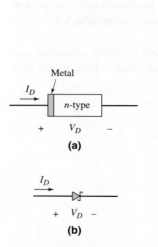

**Figure 1.41** Schottky barrier diode: (a) simplified geometry and (b) circuit symbol

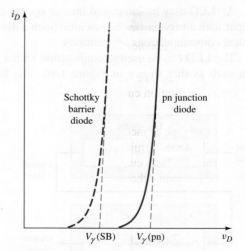

**Figure 1.42** Comparison of the forward-bias *I–V* characteristics of a pn junction diode and a Schottky barrier diode

voltage than the pn junction diode. In later chapters, we will see how this lower turn-on voltage and the shorter switching time make the Schottky diode useful in integrated-circuit applications.

## EXAMPLE 1.12

**Objective:** Determine diode voltages.

The reverse saturation currents of a pn junction diode and a Schottky diode are $I_S = 10^{-12}$ A and $10^{-8}$ A, respectively. Determine the forward-bias voltages required to produce 1 mA in each diode.

**Solution:** The diode current-voltage relationship is given by

$$I_D = I_S e^{V_D/V_T}$$

Solving for the diode voltage, we obtain

$$V_D = V_T \ln\left(\frac{I_D}{I_S}\right)$$

We then find, for the pn junction diode

$$V_D = (0.026) \ln\left(\frac{1 \times 10^{-3}}{10^{-12}}\right) = 0.539 \text{ V}$$

and, for the Schottky diode

$$V_D = (0.026) \ln\left(\frac{1 \times 10^{-3}}{10^{-8}}\right) = 0.299 \text{ V}$$

**Comment:** Since the reverse-saturation current for the Schottky diode is relatively large, less voltage across this diode is required to produce a given current compared to the pn junction diode.

## EXERCISE PROBLEM

**Ex 1.12:** A pn junction diode and a Schottky diode both have forward-bias currents of 1.2 mA. The reverse-saturation current of the pn junction diode is $I_S = 4 \times 10^{-15}$ A. The difference in forward-bias voltages is 0.265 V. Determine the reverse-saturation current of the Schottky diode. (Ans. $I_S = 1.07 \times 10^{-10}$ A)

Another type of metal–semiconductor junction is also possible. A metal applied to a heavily doped semiconductor forms, in most cases, an *ohmic contact:* that is, a contact that conducts current equally in both directions, with very little voltage drop across the junction. Ohmic contacts are used to connect one semiconductor device to another on an IC, or to connect an IC to its external terminals.

### 1.5.5   Zener Diode

As mentioned earlier in this chapter, the applied reverse-bias voltage cannot increase without limit. At some point, breakdown occurs and the current in the reverse-bias

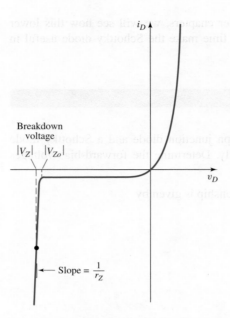

**Figure 1.43** Diode *I–V* characteristics showing breakdown effects

direction increases rapidly. The voltage at this point is called the breakdown voltage. The diode *I–V* characteristics, including breakdown, are shown in Figure 1.43.

Diodes, called **Zener diodes,** can be designed and fabricated to provide a specified breakdown voltage $V_{Zo}$. (Although the breakdown voltage is on the negative voltage axis (reverse-bias), its value is given as a positive quantity.) The large current that may exist at breakdown can cause heating effects and catastrophic failure of the diode due to the large power dissipation in the device. However, diodes can be operated in the breakdown region by limiting the current to a value within the capabilities of the device. Such a diode can be used as a constant-voltage reference in a circuit. The diode breakdown voltage is essentially constant over a wide range of currents and temperatures. The slope of the *I–V* characteristics curve in breakdown is quite large, so the incremental resistance $r_z$ is small. Typically, $r_z$ is in the range of a few ohms or tens of ohms.

The circuit symbol of the Zener diode is shown in Figure 1.44. (Note the subtle difference between this symbol and the Schottky diode symbol.) The voltage $V_Z$ is the Zener breakdown voltage, and the current $I_Z$ is the reverse-bias current when the diode is operating in the breakdown region. We will see applications of the Zener diode in the next chapter.

**Figure 1.44** Circuit symbol of the Zener diode

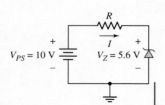

**Figure 1.45** Simple circuit containing a Zener diode in which the Zener diode is biased in the breakdown region

## DESIGN EXAMPLE 1.13

**Objective:** Consider a simple constant-voltage reference circuit and design the value of resistance required to limit the current in this circuit.

Consider the circuit shown in Figure 1.45. Assume that the Zener diode breakdown voltage is $V_Z = 5.6$ V and the Zener resistance is $r_z = 0$. The current in the diode is to be limited to 3 mA.

**Solution:** As before, we can determine the current from the voltage difference across $R$ divided by the resistance. That is,

$$I = \frac{V_{PS} - V_Z}{R}$$

The resistance is then

$$R = \frac{V_{PS} - V_Z}{I} = \frac{10 - 5.6}{3} = 1.47\,\text{k}\Omega$$

The power dissipated in the Zener diode is

$$P_Z = I_Z V_Z = (3)(5.6) = 16.8\,\text{mW}$$

The Zener diode must be able to dissipate 16.8 mW of power without being damaged.

**Comment:** The resistance external to the Zener diode limits the current when the diode is operating in the breakdown region. In the circuit shown in the figure, the output voltage will remain constant at 5.6 V, even though the power supply voltage and the resistance may change over a limited range. Hence, this circuit provides a constant output voltage. We will see further applications of the Zener diode in the next chapter.

---

EXERCISE PROBLEM

**Ex 1.13:** Consider the circuit shown in Figure 1.45. Determine the value of resistance $R$ required to limit the power dissipated in the Zener diode to 10 mW. (Ans. $R = 2.46\,\text{k}\Omega$)

## Test Your Understanding

**TYU 1.15**  Consider the circuit shown in Figure 1.46. The diode can be either a pn junction diode or a Schottky diode. Assume the cut-in voltages are $V_\gamma = 0.7$ V and $V_\gamma = 0.3$ V for the pn junction diode and Schottky diode, respectively. Let $r_f = 0$ for both diodes. Calculate the current $I_D$ when each diode is inserted in the circuit. (Ans. pn diode, 0.825 mA; Schottky diode, 0.925 mA).

**TYU 1.16**  A Zener diode has an equivalent series resistance of 20 Ω. If the voltage across the Zener diode is 5.20 V at $I_Z = 1$ mA, determine the voltage across the diode at $I_Z = 10$ mA. (Ans. $V_Z = 5.38$ V)

**TYU 1.17**  The resistor in the circuit shown in Figure 1.45 has a value of $R = 4$ kΩ, the Zener diode breakdown voltage is $V_Z = 3.6$ V, and the power rating of the Zener diode is $P = 6.5$ mW. Determine the maximum diode current and the maximum power supply voltage that can be applied without damaging the diode. (Ans. 1.81 mA, 10.8 V).

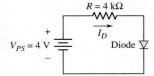

**Figure 1.46**  Circuit for exercise problem TYU 1.15. The diode can be either a pn junction diode or a Schottky diode.

## 1.6    DESIGN APPLICATION: DIODE THERMOMETER

**Objective:** • Design a simple electronic thermometer using the temperature characteristics of a diode.

**Specifications:** The temperature range is to be 0 to 100 °F.

**Design Approach:** We will use the forward-bias diode temperature characteristics as shown in Figure 1.20. If the diode current is held constant, the variation in diode voltage is a function of temperature.

**Choices:** Assume that a silicon pn junction diode with a reverse-saturation current of $I_S = 10^{-13}$ A at $T = 300$ K is available.

**Solution:** Neglecting the $(-1)$ term in the diode I–V relation, we have

$$I_D = I_S e^{V_D/V_T} \propto n_i^2 e^{V_D/V_T} \propto e^{-E_g/kT} \cdot e^{V_D/V_T}$$

The reverse-saturation current $I_S$ is proportional to $n_i^2$ and in turn $n_i^2$ is proportional to the exponential function involving the bandgap energy $E_g$ and temperature.

Taking the ratio of the diode current at two temperature values and using the definition of thermal voltage, we have[3]

$$\frac{I_{D1}}{I_{D2}} = \frac{e^{-E_g/kT_1} \cdot e^{eV_{D1}/kT_1}}{e^{-E_g/kT_2} \cdot e^{eV_{D2}/kT_2}} \tag{1.35}$$

where $V_{D1}$ and $V_{D2}$ are the diode voltages at temperatures $T_1$ and $T_2$, respectively. If the diode current is held constant at the different temperatures, Equation (1.35) can be written as

$$e^{eV_{D2}/kT_2} = e^{-E_g/kT_1} e^{+E_g/kT_2} e^{eV_{D1}/kT_1} \tag{1.36}$$

Taking the natural logarithm of both sides, we obtain

$$\frac{eV_{D2}}{kT_2} = \frac{-E_g}{kT_1} + \frac{E_g}{kT_2} + \frac{eV_{D1}}{kT_1} \tag{1.37}$$

or

$$V_{D2} = \frac{-E_g}{e}\left(\frac{T_2}{T_1}\right) + \frac{E_g}{e} + V_{D1}\left(\frac{T_2}{T_1}\right) \tag{1.38}$$

For silicon, the bandgap energy is $E_g/e = 1.12$ V. Then, assuming the bandgap energy does not vary over the temperature range, we have

$$V_{D2} = 1.12\left(1 - \frac{T_2}{T_1}\right) + V_{D1}\left(\frac{T_2}{T_1}\right) \tag{1.39}$$

---

[3]Note that $e$ in, for example, $e^{-E_g/kT}$ represents the exponential function whereas $e$ in the exponent, for example, $eV_{D1}/kT_1$ is the magnitude of the electronic charge. The context in which $e$ is used should make the meaning clear.

Consider the circuit shown in Figure 1.47. Assume that the diode has a reverse-saturation current of $I_S = 10^{-13}$ A at $T = 300$ K. From the circuit, we can write

$$I_D = \frac{15 - V_D}{R} = I_S e^{V_D/V_T}$$

or

$$\frac{15 - V_D}{15 \times 10^3} = 10^{-13} e^{V_D/0.026}$$

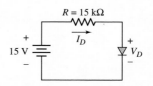

**Figure 1.47** Circuit of diode thermometer

By trial and error, we find

$$V_D = 0.5976 \text{ V}$$

and

$$I_D = \frac{15 - 0.5976}{15 \times 10^3} \Rightarrow 0.960 \text{ mA}$$

In Equation (1.39), we can set $T_1 = 300$ K and let $T_2 \equiv T$ be a variable temperature. We find

$$V_D = 1.12 - 0.522 \left( \frac{T}{300} \right) \qquad \textbf{(1.40)}$$

so the diode voltage is a linear function of temperature. If the temperature range is to be from 0 to 100 °F, for example, the corresponding change in kelvins is from 255.2 to 310.8. The diode voltage versus temperature is plotted in Figure 1.48.

A simple circuit that can be used was shown in Figure 1.47. With a power supply voltage of 15 V, a change in diode voltage of approximately 0.1 V over the temperature range produces only an approximately 0.67 percent change in diode current. Thus the preceding analysis is valid.

**Comment:** This design example shows that a diode connected in a simple circuit can be used as a sensing element in an electronic thermometer. We assumed a diode reverse-saturation current of $I_S = 10^{-13}$ A at $T = 300$ K(80 °F). The actual reverse-saturation current of a particular diode may be different. This difference simply means that the diode voltage versus temperature curve shown in Figure 1.48 would slide up or down to match the actual diode voltage at room temperature.

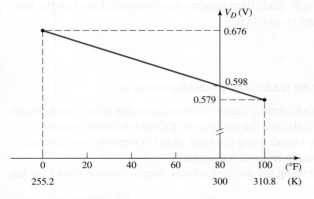

**Figure 1.48** Diode voltage versus temperature

**Design Pointer:** In order to complete this design, two additional components or electronic systems must be added to the circuit shown in Figure 1.47. First, we must add a circuit to measure the diode voltage. Adding this circuit must not alter the diode characteristics and there must be no loading effects. An op-amp circuit that will be described in Chapter 9 can be used for this purpose. A second electronic system required is to convert the diode voltage to a temperature reading. An analog-to-digital converter that will be described in Chapter 16 can be used to provide a digital temperature reading.

##  1.7  SUMMARY

- We initially considered some of the characteristics and properties of semiconductor materials, such as the concept of electrons (negative charge) and holes (positive charge) as two distinct charge carriers in a semiconductor. The doping process produces either n-type or p-type semiconductor materials. The concepts of n-type and p-type materials are used throughout the text.
- A pn junction diode is formed when an n-doped region and a p-doped region are directly adjacent to each other. The diode current is an exponential function of voltage in the forward-bias condition, and is essentially zero in the reverse-bias condition.
- A piecewise-linear model of the diode was developed so that approximate hand calculation results can be easily obtained. The $i - v$ characteristics of the diode are broken into linear segments, which are valid over particular regions of operation. The concept of a diode turn-on voltage was introduced.
- Time-varying, or ac signals, may be superimposed on a dc diode current and voltage. A small-signal linear equivalent circuit was developed and is used to determine the relationship between the ac current and ac voltage. This same equivalent circuit will be applied extensively when the frequency response of transistors is discussed.
- Specialized pn junction devices were discussed. In particular, pn junction solar cells are used to convert solar energy to electrical energy. Schottky barrier diodes are metal–semiconductor rectifying junctions that, in general, have smaller turn-on voltages than pn junctions. Zener diodes operate in the reverse breakdown region and are used in constant-voltage circuits. Photodiodes and LEDs were also briefly discussed.
- As an application, a simple diode thermometer was designed, based on the temperature properties of the pn junction.

## CHECKPOINT

After studying this chapter, the reader should have the ability to:

✓ Understand the concept of intrinsic carrier concentration, the difference between n-type and p-type materials, and the concept of drift and diffusion currents.
✓ Analyze a simple diode circuit using the ideal diode current–voltage characteristics and using the iteration analysis technique.
✓ Analyze a diode circuit using the piecewise linear approximation model for the diode.

✓ Determine the small-signal characteristics of a diode using the small-signal equivalent circuit.

✓ Understand the general characteristics of a solar cell, light-emitting diode, Schottky barrier diode, and Zener diode.

##  REVIEW QUESTIONS

1. Describe an intrinsic semiconductor material. What is meant by the intrinsic carrier concentration?
2. Describe the concept of an electron and a hole as charge carriers in the semiconductor material.
3. Describe an extrinsic semiconductor material. What is the electron concentration in terms of the donor impurity concentration? What is the hole concentration in terms of the acceptor impurity concentration?
4. Describe the concepts of drift current and diffusion current in a semiconductor material.
5. How is a pn junction formed? What is meant by a built-in potential barrier, and how is it formed?
6. How is a junction capacitance created in a reverse-biased pn junction diode?
7. Write the ideal diode current–voltage relationship. Describe the meaning of $I_S$ and $V_T$.
8. Describe the iteration method of analysis and when it must be used to analyze a diode circuit.
9. Describe the piecewise linear model of a diode and why it is useful. What is the diode turn-on voltage?
10. Define a load line in a simple diode circuit.
11. Under what conditions is the small-signal model of a diode used in the analysis of a diode circuit?
12. Describe the operation of a simple solar cell circuit.
13. How do the $i$–$v$ characteristics of a Schottky barrier diode differ from those of a pn junction diode?
14. What characteristic of a Zener diode is used in the design of a Zener diode circuit?
15. Describe the characteristics of a photodiode and a photodiode circuit.

##  PROBLEMS

[Note: Unless otherwise specified, assume that $T = 300$ K in the following problems. Also, assume the emission coefficient is $n = 1$ unless otherwise stated.]

### Section 1.1  Semiconductor Materials and Properties

1.1   (a) Calculate the intrinsic carrier concentration in silicon at (i) $T = 250$ K and (ii) $T = 350$ K. (b) Repeat part (a) for gallium arsenide.

1.2   (a) The intrinsic carrier concentration in silicon is to be no larger than $n_i = 10^{12}$ cm$^{-3}$. Determine the maximum allowable temperature. (b) Repeat part (a) for $n_i = 10^9$ cm$^{-3}$.

1.3   Calculate the intrinsic carrier concentration in silicon and germanium at (a) $T = 100$ K, (b) $T = 300$ K, and (c) $T = 500$ K.

1.4    (a) Find the concentration of electrons and holes in a sample of germanium that has a concentration of donor atoms equal to $10^{15}$ cm$^{-3}$. Is the semiconductor n-type or p-type? (b) Repeat part (a) for silicon.

1.5    Gallium arsenide is doped with acceptor impurity atoms at a concentration of $10^{16}$ cm$^{-3}$. (a) Find the concentration of electrons and holes. Is the semiconductor n-type or p-type? (b) Repeat part (a) for germanium.

1.6    Silicon is doped with $5 \times 10^{16}$ arsenic atoms/cm$^3$. (a) Is the material n- or p-type? (b) Calculate the electron and hole concentrations at $T = 300$ K. (c) Repeat part (b) for $T = 350$ K.

1.7    (a) Calculate the concentration of electrons and holes in silicon that has a concentration of acceptor atoms equal to $5 \times 10^{16}$ cm$^{-3}$. Is the semiconductor n-type or p-type? (b) Repeat part (a) for GaAs.

1.8    A silicon sample is fabricated such that the hole concentration is $p_o = 2 \times 10^{17}$ cm$^{-3}$. (a) Should boron or arsenic atoms be added to the intrinsic silicon? (b) What concentration of impurity atoms must be added? (c) What is the concentration of electrons?

1.9    The electron concentration in silicon at $T = 300$ K is $n_o = 5 \times 10^{15}$ cm$^{-3}$. (a) Determine the hole concentration. (b) Is the material n-type or p-type? (c) What is the impurity doping concentration?

1.10   (a) A silicon semiconductor material is to be designed such that the majority carrier electron concentration is $n_o = 7 \times 10^{15}$ cm$^{-3}$. Should donor or acceptor impurity atoms be added to intrinsic silicon to achieve this electron concentration? What concentration of dopant impurity atoms is required? (b) In this silicon material, the minority carrier hole concentration is to be no larger than $p_o = 10^6$ cm$^{-3}$. Determine the maximum allowable temperature.

1.11   (a) The applied electric field in p-type silicon is $E = 10$ V/cm. The semiconductor conductivity is $\sigma = 1.5$ ($\Omega$–cm)$^{-1}$ and the cross-sectional area is $A = 10^{-5}$ cm$^2$. Determine the drift current. (b) The cross-sectional area of a semiconductor is $A = 2 \times 10^{-4}$ cm$^2$ and the resistivity is $\rho = 0.4$ ($\Omega$–cm). If the drift current is $I = 1.2$ mA, what applied electric field must be applied?

1.12   A drift current density of 120 A/cm$^2$ is established in n-type silicon with an applied electric field of 18 V/cm. If the electron and hole mobilities are $\mu_n = 1250$ cm$^2$/V–s and $\mu_p = 450$ cm$^2$/V–s, respectively, determine the required doping concentration.

1.13   An n-type silicon material has a resistivity of $\rho = 0.65$ $\Omega$–cm. (a) If the electron mobility is $\mu_n = 1250$ cm$^2$/V–s, what is the concentration of donor atoms? (b) Determine the required electric field to establish a drift current density of $J = 160$ A/cm$^2$.

1.14   (a) The required conductivity of a silicon material must be $\sigma = 1.5$ ($\Omega$–cm)$^{-1}$. If $\mu_n = 1000$ cm$^2$/V–s and $\mu_p = 375$ cm$^2$/V–s, what concentration of donor atoms must be added? (b) The required conductivity of a silicon material must be $\sigma = 0.8$ ($\Omega$–cm)$^{-1}$. If $\mu_n = 1200$ cm$^2$/V–s and $\mu_p = 400$ cm$^2$/V–s, what concentration of acceptor atoms must be added?

1.15   In GaAs, the mobilities are $\mu_n = 8500$ cm$^2$/V–s and $\mu_p = 400$ cm$^2$/V–s. (a) Determine the range in conductivity for a range in donor concentration of $10^{15} \le N_d \le 10^{19}$ cm$^{-3}$. (b) Using the results of part (a), determine the range in drift current density if the applied electric field is $E = 0.10$ V/cm.

1.16   The electron and hole concentrations in a sample of silicon are shown in Figure P1.16. Assume the electron and hole mobilities are the same as in Problem 1.12. Determine the total diffusion current density versus distance $x$ for $0 \leq x \leq 0.001$ cm.

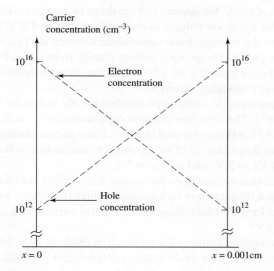

Carrier concentration (cm$^{-3}$)

Electron concentration

Hole concentration

$x = 0$              $x = 0.001$ cm

Figure P1.16

1.17   The hole concentration in silicon is given by

$$p(x) = 10^4 + 10^{15} \exp(-x/L_p) \qquad x \geq 0$$

The value of $L_p$ is $10\,\mu$m. The hole diffusion coefficient is $D_p = 15\ \text{cm}^2/\text{s}$. Determine the hole diffusion current density at (a) $x = 0$, (b) $x = 10\,\mu$m, and (c) $x = 30\,\mu$m.

1.18   GaAs is doped to $N_a = 10^{17}\ \text{cm}^{-3}$. (a) Calculate $n_o$ and $p_o$. (b) Excess electrons and holes are generated such that $\delta n = \delta p = 10^{15}\ \text{cm}^{-3}$. Determine the total concentration of electrons and holes.

## Section 1.2  The pn Junction

1.19   (a) Determine the built-in potential barrier $V_{bi}$ in a silicon pn junction for (i) $N_d = N_a = 5 \times 10^{15}\ \text{cm}^{-3}$; (ii) $N_d = 5 \times 10^{17}\ \text{cm}^{-3}$ and $N_a = 10^{15}\ \text{cm}^{-3}$; (iii) $N_a = N_d = 10^{18}\ \text{cm}^{-3}$. (b) Repeat part (a) for GaAs.

1.20   Consider a silicon pn junction. The n-region is doped to a value of $N_d = 10^{16}\ \text{cm}^{-3}$. The built-in potential barrier is to be $V_{bi} = 0.712$ V. Determine the required p-type doping concentration.

1.21   The donor concentration in the n-region of a silicon pn junction is $N_d = 10^{16}\ \text{cm}^{-3}$. Plot $V_{bi}$ versus $N_a$ over the range $10^{15} \leq N_a \leq 10^{18}\ \text{cm}^{-3}$ where $N_a$ is the acceptor concentration in the p-region.

1.22   Consider a uniformly doped GaAs pn junction with doping concentrations of $N_a = 5 \times 10^{18}\ \text{cm}^{-3}$ and $N_d = 5 \times 10^{16}\ \text{cm}^{-3}$. Plot the built-in potential barrier $V_{bi}$ versus temperature for $200\,\text{K} \leq T \leq 500\,\text{K}$.

1.23   The zero-biased junction capacitance of a silicon pn junction is $C_{jo} = 0.4$ pF. The doping concentrations are $N_a = 1.5 \times 10^{16}\ \text{cm}^{-3}$ and

$N_d = 4 \times 10^{15} \ \mathrm{cm}^{-3}$. Determine the junction capacitance at (a) $V_R = 1 \ \mathrm{V}$, (b) $V_R = 3 \ \mathrm{V}$, and (c) $V_R = 5 \ \mathrm{V}$.

*1.24 The zero-bias capacitance of a silicon pn junction diode is $C_{jo} = 0.02 \ \mathrm{pF}$ and the built-in potential is $V_{bi} = 0.80 \ \mathrm{V}$. The diode is reverse biased through a 47-k$\Omega$ resistor and a voltage source. (a) For $t < 0$, the applied voltage is 5 V and, at $t = 0$, the applied voltage drops to zero volts. Estimate the time it takes for the diode voltage to change from 5 V to 1.5 V. (As an approximation, use the average diode capacitance between the two voltage levels.) (b) Repeat part (a) for an input voltage change from 0 V to 5 V and a diode voltage change from 0 V to 3.5 V. (Use the average diode capacitance between these two voltage levels.)

1.25 The doping concentrations in a silicon pn junction are $N_d = 5 \times 10^{15} \ \mathrm{cm}^{-3}$ and $N_a = 10^{17} \ \mathrm{cm}^{-3}$. The zero-bias junction capacitance is $C_{jo} = 0.60 \ \mathrm{pF}$. An inductance of 1.50 mH is connected in parallel with the pn junction. Calculate the resonant frequency $f_o$ of the circuit for reverse-bias voltages of (a) $V_R = 1 \ \mathrm{V}$, (b) $V_R = 3 \ \mathrm{V}$, and (c) $V_R = 5 \ \mathrm{V}$.

1.26 (a) At what reverse-bias voltage does the reverse-bias current in a silicon pn junction diode reach 90 percent of its saturation value? (b) What is the ratio of the current for a forward-bias voltage of 0.2 V to the current for a reverse-bias voltage of 0.2 V?

1.27 (a) The reverse-saturation current of a pn junction diode is $I_S = 10^{-11} \ \mathrm{A}$. Determine the diode current for diode voltages of $0.3, 0.5, 0.7, -0.02, -0.2$, and $-2$ V. (b) Repeat part (a) for $I_S = 10^{-13} \ \mathrm{A}$.

1.28 (a) The reverse-saturation current of a pn junction diode is $I_S = 10^{-11} \ \mathrm{A}$. Determine the diode voltage to produce currents of (i) 10 $\mu$A, 100 $\mu$A, 1 mA, and (ii) $-5 \times 10^{-12} \ \mathrm{A}$. (b) Repeat part (a) for $I_S = 10^{-13} \ \mathrm{A}$ and part (a) (ii) for $-10^{-14} \ \mathrm{A}$.

1.29 A silicon pn junction diode has an emission coefficient of $n = 1$. The diode current is $I_D = 1 \ \mathrm{mA}$ when $V_D = 0.7 \ \mathrm{V}$. (a) What is the reverse-bias saturation current? (b) Plot, on the same graph, $\log_{10} I_D$ versus $V_D$ over the range $0.1 \leq V_D \leq 0.7 \ \mathrm{V}$ when the emission coefficient is (i) $n = 1$ and (ii) $n = 2$.

1.30 Plot $\log_{10} I_D$ versus $V_D$ over the range $0.1 \leq V_D \leq 0.7 \ \mathrm{V}$ for (a) $I_S = 10^{-12}$ and (b) $I_S = 10^{-14} \ \mathrm{A}$.

1.31 (a) Consider a silicon pn junction diode operating in the forward-bias region. Determine the increase in forward-bias voltage that will cause a factor of 10 increase in current. (b) Repeat part (a) for a factor of 100 increase in current.

1.32 A pn junction diode has $I_S = 2 \ \mathrm{nA}$. (a) Determine the diode voltage if (i) $I_D = 2 \ \mathrm{A}$ and (ii) $I_D = 20 \ \mathrm{A}$. (b) Determine the diode current if (i) $V_D = 0.4 \ \mathrm{V}$ and (ii) $V_D = 0.65 \ \mathrm{V}$.

1.33 The reverse-bias saturation current for a set of diodes varies between $5 \times 10^{-14} \leq I_S \leq 5 \times 10^{-12} \ \mathrm{A}$. The diodes are all to be biased at $I_D = 2 \ \mathrm{mA}$. What is the range of forward-bias voltages that must be applied?

1.34 (a) A germanium pn junction has a diode current of $I_D = 1.5 \ \mathrm{mA}$ when biased at $V_D = 0.30 \ \mathrm{V}$. What is the reverse-bias saturation current? (b) Using the results of part (a), determine the diode current when the diode is biased at (i) $V_D = 0.35 \ \mathrm{V}$ and (ii) $V_D = 0.25 \ \mathrm{V}$.

1.35 (a) The reverse-saturation current of a gallium arsenide pn junction diode is $I_S = 10^{-22} \ \mathrm{A}$. Determine the diode current for diode voltages of 0.8, 1.0, 1.2, $-0.02$, $-0.2$, and $-2$ V. (b) Repeat part (a) for $I_S = 5 \times 10^{-24} \ \mathrm{A}$.

*1.36   The reverse-saturation current of a silicon pn junction diode at $T = 300\,\text{K}$ is $I_S = 10^{-12}\,\text{A}$. Determine the temperature range over which $I_S$ varies from $0.5 \times 10^{-12}\,\text{A}$ to $50 \times 10^{-12}\,\text{A}$.

*1.37   A silicon pn junction diode has an applied forward-bias voltage of 0.6 V. Determine the ratio of current at $100\,^\circ\text{C}$ to that at $-55\,^\circ\text{C}$.

## Section 1.3 DC Diode Analysis

1.38   A pn junction diode is in series with a 1 MΩ resistor and a 2.8 V power supply. The reverse-saturation current of the diode is $I_S = 5 \times 10^{-11}\,\text{A}$. (a) Determine the diode current and voltage if the diode is forward biased. (b) Repeat part (a) if the diode is reverse biased.

1.39   Consider the diode circuit shown in Figure P1.39. The diode reverse-saturation current is $I_S = 10^{-12}\,\text{A}$. Determine the diode current $I_D$ and diode voltage $V_D$.

*1.40   The diode in the circuit shown in Figure P1.40 has a reverse-saturation current of $I_S = 5 \times 10^{-13}\,\text{A}$. Determine the diode voltage and current.

1.41   (a) For the circuit shown in Figure P1.41(a), determine $I_{D1}, I_{D2}, V_{D1},$ and $V_{D2}$ for (i) $I_{S1} = I_{S2} = 10^{-13}\,\text{A}$ and (ii) $I_{S1} = 5 \times 10^{-14}\,\text{A}$, $I_{S2} = 5 \times 10^{-13}\,\text{A}$. (b) Repeat part (a) for the circuit shown in Figure P1.41(b).

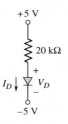

**Figure P1.39**

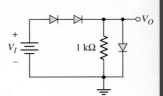

**Figure P1.40**

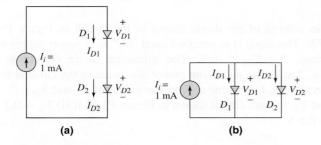

(a)                     (b)

**Figure P1.41**

1.42   (a) The reverse-saturation current of each diode in the circuit shown in Figure P1.42 is $I_S = 6 \times 10^{-14}\,\text{A}$. Determine the input voltage $V_I$ required to produce an output voltage of $V_O = 0.635\,\text{V}$. (b) Repeat part (a) if the 1 kΩ resistor is changed to $R = 500\,\Omega$.

1.43   (a) Consider the circuit shown in Figure P1.40. The value of $R_1$ is reduced to $R_1 = 10\,\text{k}\Omega$ and the cut-in voltage of the diode is $V_\gamma = 0.7\,\text{V}$. Determine $I_D$ and $V_D$. (b) Repeat part (a) if $R_1 = 50\,\text{k}\Omega$.

1.44   Consider the circuit shown in Figure P1.44. Determine the diode current $I_D$ and diode voltage $V_D$ for (a) $V_\gamma = 0.6\,\text{V}$ and (b) $V_\gamma = 0.7\,\text{V}$.

1.45   The diode cut-in voltage is $V_\gamma = 0.7\,\text{V}$ for the circuits shown in Figure P1.45. Plot $V_O$ and $I_D$ versus $I_I$ over the range $0 \le I_I \le 2\,\text{mA}$ for the circuit shown in (a) Figure P1.45(a), (b) Figure P1.45(b), and (c) Figure P1.45(c).

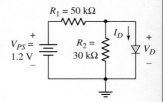

**Figure P1.42**

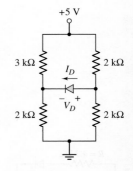

**Figure P1.44**

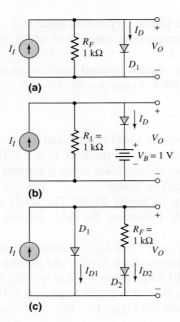

(a)

(b)

(c)

Figure P1.45

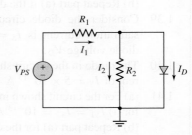

Figure P1.46

*1.46   The cut-in voltage of the diode shown in the circuit in Figure P1.46 is $V_\gamma = 0.7\,\text{V}$. The diode is to remain biased "on" for a power supply voltage in the range $5 \leq V_{PS} \leq 10\,\text{V}$. The minimum diode current is to be $I_D(\text{min}) = 2\,\text{mA}$. The maximum power dissipated in the diode is to be no more than 10 mW. Determine appropriate values of $R_1$ and $R_2$.

1.47   Find $I$ and $V_O$ in each circuit shown in Figure P1.47 if (i) $V_\gamma = 0.7\,\text{V}$ and (ii) $V_\gamma = 0.6\,\text{V}$.

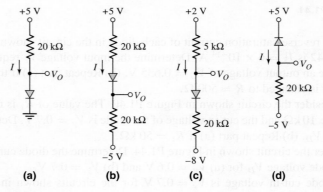

(a)

(b)

(c)

(d)

Figure P1.47

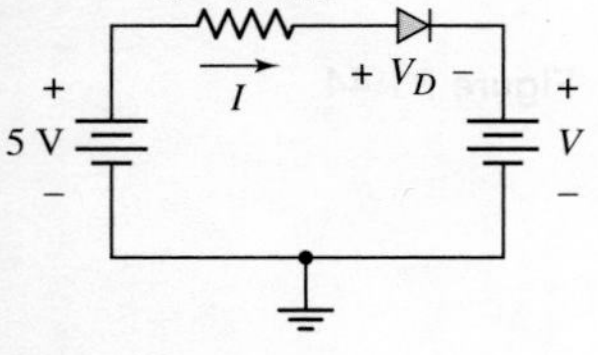

*1.48   Repeat Problem 1.47 if the reverse-saturation current for each diode is $I_S = 5 \times 10^{-14}\,\text{A}$. What is the voltage across each diode?

1.49   (a) In the circuit shown in Figure P1.49, find the diode voltage $V_D$ and the supply voltage $V$ such that the current is $I_D = 0.4\,\text{mA}$. Assume the diode cut-in voltage is $V_\gamma = 0.7\,\text{V}$. (b) Using the results of part (a), determine the power dissipated in the diode.

$R = 4.7\,\text{k}\Omega$

$+ V_D -$

5 V

$I$

$V$

Figure P1.49

1.50 Assume each diode in the circuit shown in Figure P1.50 has a cut-in voltage of $V_\gamma = 0.65$ V. (a) The input voltage is $V_I = 5$ V. Determine the value of $R_1$ required such that $I_{D1}$ is one-half the value of $I_{D2}$. What are the values of $I_{D1}$ and $I_{D2}$? (b) If $V_I = 8$ V and $R_1 = 2$ k$\Omega$, determine $I_{D1}$ and $I_{D2}$.

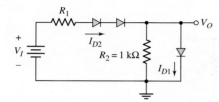

**Figure P1.50**

## Section 1.4 Small-Signal Diode Analysis

1.51 (a) Consider a pn junction diode biased at $I_{DQ} = 1$ mA. A sinusoidal voltage is superimposed on $V_{DQ}$ such that the peak-to-peak sinusoidal current is $0.05 I_{DQ}$. Find the value of the applied peak-to-peak sinusoidal voltage. (b) Repeat part (a) if $I_{DQ} = 0.1$ mA.

1.52 Determine the small-signal diffusion resistance $r_d$ for a diode biased at (a) $I_D = 26$ $\mu$A, (b) $I_D = 260$ $\mu$A, and (c) $I_D = 2.6$ mA.

*1.53 The diode in the circuit shown in Figure P1.53 is biased with a constant current source $I$. A sinusoidal signal $v_s$ is coupled through $R_S$ and $C$. Assume that $C$ is large so that it acts as a short circuit to the signal. (a) Show that the sinusoidal component of the diode voltage is given by

$$v_o = v_s \left( \frac{V_T}{V_T + I R_S} \right)$$

(b) If $R_S = 260$ $\Omega$, find $v_o/v_s$, for $I = 1$ mA, $I = 0.1$ mA, and $I = 0.01$ mA.

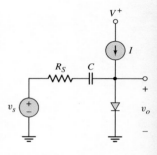

**Figure P1.53**

## Section 1.5 Other Types of Diodes

1.54 The forward-bias currents in a pn junction diode and a Schottky diode are 0.72 mA. The reverse-saturation currents are $I_S = 5 \times 10^{-13}$ A and $I_S = 5 \times 10^{-8}$ A, respectively. Determine the forward-bias voltage across each diode.

1.55 A pn junction diode and a Schottky diode have equal cross-sectional areas and have forward-bias currents of 0.5 mA. The reverse-saturation current of the Schottky diode is $I_S = 5 \times 10^{-7}$ A. The difference in forward-bias voltages between the two diodes is 0.30 V. Determine the reverse-saturation current of the pn junction diode.

1.56 The reverse-saturation currents of a Schottky diode and a pn junction diode are $I_S = 5 \times 10^{-8}$ A and $10^{-12}$ A, respectively. (a) The diodes are connected in parallel and the parallel combination is driven by a constant current of 0.5 mA. (i) Determine the current in each diode. (ii) Determine the voltage across each diode. (b) Repeat part (a) for the diodes connected in series, with a voltage of 0.90 V connected across the series combination.

*1.57   Consider the Zener diode circuit shown in Figure P1.57. The Zener break-down voltage is $V_Z = 5.6\,\text{V}$ at $I_Z = 0.1\,\text{mA}$, and the incremental Zener resistance is $r_z = 10\,\Omega$. (a) Determine $V_O$ with no load ($R_L = \infty$). (b) Find the change in the output voltage if $V_{PS}$ changes by $\pm 1$ V. (c) Find $V_O$ if $V_{PS} = 10$ V and $R_L = 2\,\text{k}\Omega$.

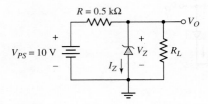

**Figure P1.57**

1.58   (a) The Zener diode in Figure P1.57 is ideal with $V_Z = 6.8$ V. Determine the maximum current and maximum power dissipated in the diode ($R_L = \infty$). (b) Determine the value of $R_L$ such that $I_Z$ is reduced to 0.1 of its maximum value.

*1.59   Consider the Zener diode circuit shown in Figure P1.57. The Zener diode voltage is $V_Z = 6.8$ V at $I_Z = 0.1\,\text{mA}$ and the incremental Zener resistance is $r_z = 20\,\Omega$. (a) Calculate $V_O$ with no load ($R_L = \infty$). (b) Find the change in the output voltage when a load resistance of $R_L = 1\,\text{k}\Omega$ is connected.

1.60   The output current of a pn junction diode used as a solar cell can be given by

$$I_D = 0.2 - 5 \times 10^{-14}\left[\exp\left(\frac{V_D}{V_T}\right) - 1\right] \quad \text{A}$$

The short-circuit current is defined as $I_{SC} = I_D$ when $V_D = 0$ and the open-circuit voltage is defined as $V_{OC} = V_D$ when $I_D = 0$. Find the values of $I_{SC}$ and $V_{OC}$.

1.61   Using the current–voltage characteristics of the solar cell described in Problem 1.60, plot $I_D$ versus $V_D$.

1.62   (a) Using the current–voltage characteristics of the solar cell described in problem 1.60, determine $V_D$ when $I_D = 0.8 I_{SC}$. (b) Using the results of part (a), determine the power supplied by the solar cell.

## COMPUTER SIMULATION PROBLEMS

1.63   Use a computer simulation to generate the ideal current–voltage characteristics of a diode from a reverse-bias voltage of 5 V to a forward-bias current of 1 mA, for an $I_S$ parameter value of (a) $10^{-15}$ A and (b) $10^{-13}$ A. Use the default values for all other parameters.

1.64   Use a computer simulation to find the diode current and voltage for the circuit described in Problem 1.38.

1.65   The reverse-saturation current for each diode in Figure P1.42 is $I_S = 10^{-14}$ A. Use a computer simulation to plot the output voltage $V_O$ versus the input voltage $V_I$ over the range $0 \leq V_I \leq 2.0$ V.

1.66   Use a computer simulation to find the diode current, diode voltage, and output voltage for each circuit shown in Figure P1.47. Assume $I_S = 10^{-13}$ A for each diode.

## DESIGN PROBLEMS

[Note: Each design should be verified by a computer simulation.]

*D1.67   Design a diode circuit to produce the load line and $Q$-point shown in Figure P1.67. Assume diode piecewise linear parameters of $V_\gamma = 0.7$ V and $r_f = 0$.

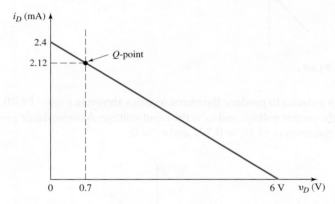

**Figure P1.67**

*D1.68   Design a circuit to produce the characteristics shown in Figure P1.68, where $i_D$ is the diode current and $v_I$ is the input voltage. Assume diode piecewise linear parameters of $V_\gamma = 0.7$ V and $r_f = 0$.

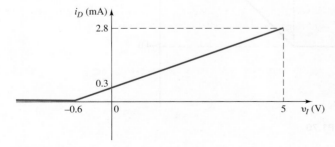

**Figure P1.68**

*D1.69   Design a circuit to produce the characteristics shown in Figure P1.69, where $v_O$ is the output voltage and $v_I$ is the input voltage. Assume diode piecewise linear parameters of $V_\gamma = 0.7$ V and $r_f = 0$.

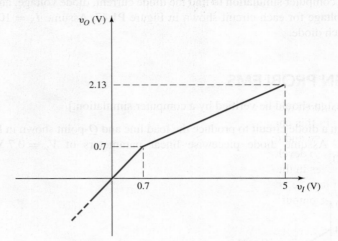

**Figure P1.69**

*D1.70   Design a circuit to produce the characteristics shown in Figure P1.70, where $v_O$ is the output voltage and $v_I$ is the input voltage. Assume diode piecewise linear parameters of $V_\gamma = 0.7$ V and $r_f = 0$.

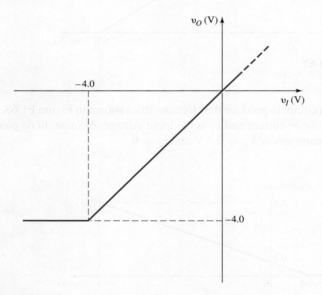

**Figure P1.70**

# Diode Circuits

In the last chapter, we discussed some of the properties of semiconductor materials and introduced the diode. We presented the ideal current–voltage relationship of the diode and considered the piecewise linear model, which simplifies the dc analysis of diode circuits. In this chapter, the techniques and concepts developed in Chapter 1 are used to analyze and design electronic circuits containing diodes. A general goal of this chapter is to develop the ability to use the piecewise linear model and approximation techniques in the hand analysis and design of various diode circuits.

Each circuit to be considered accepts an input signal at a set of input terminals and produces an output signal at a set of output terminals. This process is called **signal processing.** The circuit "processes" the input signal and produces an output signal that is a different shape or a different function compared to the input signal. We will see in this chapter how diodes are used to perform these various signal processing functions.

Although diodes are useful electronic devices, we will begin to see the limitations of these devices and the desirability of having some type of "amplifying" device.

## PREVIEW

In this chapter, we will:

- Determine the operation and characteristics of diode rectifier circuits, which, in general, form the first stage of the process of converting an ac signal into a dc signal in the electronic power supply.
- Apply the characteristics of the Zener diode to a Zener diode voltage regulator circuit.
- Apply the nonlinear characteristics of diodes to create waveshaping circuits known as clippers and clampers.
- Examine the techniques used to analyze circuits that contain more than one diode.
- Understand the operation and characteristics of specialized photodiode and light-emitting diode circuits.
- Design a basic dc power supply incorporating a filtered rectifier circuit and a Zener diode.

## 2.1    RECTIFIER CIRCUITS

**Objective:** • Determine the operation and characteristics of diode rectifier circuits, which form the first stage in the process of converting an ac signal into a dc signal in the electronic power supply.

One application of diodes is in the design of rectifier circuits. A diode rectifier forms the first stage of a dc power supply. A dc voltage is required to power essentially every electronic device, including personal computers, televisions, and stereo systems. An electrical cord that is plugged into a wall socket and attached to a television, for example, is connected to a rectifier circuit inside the TV. In addition, battery chargers for portable electronic devices such as cell phones and laptop computers contain rectifier circuits.

Figure 2.1 is a diagram of a dc power supply. The output voltage[1] $v_O$ is usually in the range of 3 to 24 V depending on the particular electronics application. Throughout the first part of this chapter, we will analyze and design various stages in the power supply circuit.

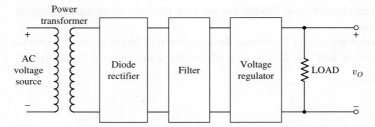

**Figure 2.1** Diagram of an electronic power supply. The circuits that characterize each block diagram are considered in this chapter.

Rectification is the process of converting an alternating (ac) voltage into one that is limited to one polarity. The diode is useful for this function because of its nonlinear characteristics, that is, current exists for one voltage polarity, but is essentially zero for the opposite polarity. Rectification is classified as **half-wave** or **full-wave,** with half-wave being the simpler and full-wave being more efficient.

### 2.1.1    Half-Wave Rectification

Figure 2.2(a) shows a power transformer with a diode and resistor connected to the secondary of the transformer. We will use the piecewise linear approach in analyzing this circuit, assuming the diode forward resistance is $r_f = 0$ when the diode is "on."

The input signal, $v_I$, is, in general, a 120 V (rms), 60 Hz ac signal. Recall that the secondary voltage, $v_S$, and primary voltage, $v_I$, of an ideal transformer are related by

$$\frac{v_I}{v_S} = \frac{N_1}{N_2} \tag{2.1}$$

---

[1]Ideally, the output voltage of a rectifier circuit is a dc voltage. However, as we will see, there may be an ac ripple voltage superimposed on a dc value. For this reason, we will use the notation $v_O$ as the instantaneous value of output voltage.

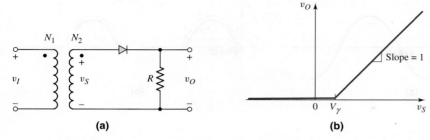

**Figure 2.2** Half-wave rectifier (a) circuit and (b) voltage transfer characteristics

where $N_1$ and $N_2$ are the number of primary and secondary turns, respectively. The ratio $N_1/N_2$ is called the **transformer turns ratio.** The transformer turns ratio will be designed to provide a particular secondary voltage, $v_S$, which in turn will produce a particular output voltage $v_O$.

### Problem-Solving Technique: Diode Circuits

In using the piecewise linear model of the diode, the first objective is to determine the linear region (conducting or not conducting) in which the diode is operating. To do this, we can:

1.  Determine the input voltage condition such that a diode is conducting (on). Then find the output signal for this condition.
2.  Determine the input voltage condition such that a diode is not conducting (off). Then find the output signal for this condition.

[Note: Item 2 can be performed before item 1 if desired.]

Figure 2.2(b) shows the voltage transfer characteristics, $v_O$ versus $v_S$, for the circuit. For $v_S < 0$, the diode is reverse biased, which means that the current is zero and the output voltage is zero. As long as $v_S < V_\gamma$, the diode will be nonconducting, so the output voltage will remain zero. When $v_S > V_\gamma$, the diode becomes forward biased and a current is induced in the circuit. In this case, we can write

$$i_D = \frac{v_S - V_\gamma}{R} \qquad\qquad (2.2(\text{a}))$$

and

$$v_O = i_D R = v_S - V_\gamma \qquad\qquad (2.2(\text{b}))$$

For $v_S > V_\gamma$, the slope of the transfer curve is 1.

If $v_S$ is a sinusoidal signal, as shown in Figure 2.3(a), the output voltage can be found using the voltage transfer curve in Figure 2.2(b). For $v_S \leq V_\gamma$ the output voltage is zero; for $v_S > V_\gamma$, the output is given by Equation (2.2(b)), or

$$v_O = v_S - V_\gamma$$

and is shown in Figure 2.3(b). We can see that while the input signal $v_S$ alternates polarity and has a time-average value of zero, the output voltage $v_O$ is unidirectional and has an average value that is not zero. The input signal is therefore rectified. Also, since the output voltage appears only during the positive cycle of the input signal, the circuit is called a **half-wave rectifier.**

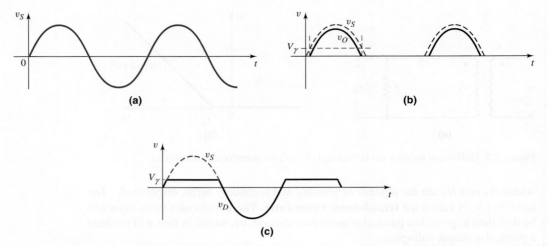

(a)

(b)

(c)

**Figure 2.3** Signals of the half-wave rectifier circuit: (a) sinusoidal input voltage, (b) rectified output voltage, and (c) diode voltage

When the diode is cut off and nonconducting, no voltage drop occurs across the resistor $R$; therefore, the entire input signal voltage appears across the diode (Figure 2.3(c)). Consequently, the diode must be capable of handling the peak current in the forward direction and sustaining the largest peak inverse voltage (PIV) without breakdown. For the circuit shown in Figure 2.2(a), the value of PIV is equal to the peak value of $v_S$.

We can use a half-wave rectifier circuit to charge a battery as shown in Figure 2.4(a). Charging current exists whenever the instantaneous ac source voltage is greater than the battery voltage plus the diode cut-in voltage as shown in Figure 2.4(b). The resistance $R$ in the circuit is to limit the current. When the ac source voltage is less than $V_B$, the current is zero. Thus current flows only in the direction to charge the battery. One disadvantage of the half-wave rectifier is that we "waste" the negative half-cycles. The current is zero during the negative half-cycles, so there is no energy dissipated, but at the same time, we are not making use of any possible available energy.

## EXAMPLE 2.1

**Objective:** Determine the currents and voltages in a half-wave rectifier circuit.

Consider the circuit shown in Figure 2.4. Assume $V_B = 12$ V, $R = 100\ \Omega$, and $V_\gamma = 0.6$ V. Also assume $v_S(t) = 24 \sin \omega t$. Determine the peak diode current,

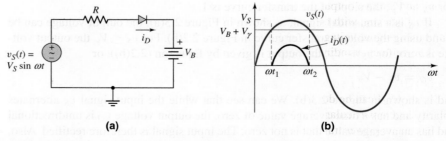

(a)

(b)

**Figure 2.4** (a) Half-wave rectifier used as a battery charger; (b) input voltage and diode current waveforms

maximum reverse-bias diode voltage, and the fraction of the cycle over which the diode is conducting.

**Solution:** Peak diode current:

$$i_D(peak) = \frac{V_S - V_B - V_\gamma}{R} = \frac{24 - 12 - 0.6}{0.10} = 114 \text{ mA}$$

Maximum reverse-bias diode voltage:

$$v_R(\max) = V_S + V_B = 24 + 12 = 36 \text{ V}$$

Diode conduction cycle:

$$v_I = 24 \sin \omega t_1 = 12.6$$

or

$$\omega t_1 = \sin^{-1}\left(\frac{12.6}{24}\right) \Rightarrow 31.7°$$

By symmetry,

$$\omega t_2 = 180 - 31.7 = 148.3°$$

Then

$$\text{Percent time} = \frac{148.3 - 31.7}{360} \times 100\% = 32.4\%$$

**Comment:** This example shows that the diode conducts only approximately one-third of the time, which means that the efficiency of this battery charger is quite low.

### EXERCISE PROBLEM

**Ex 2.1:** Repeat Example 2.1 if the input voltage is $v_s(t) = 12 \sin \omega t$ (V), $V_B = 4.5$ V, and $R = 250 \, \Omega$. (Ans. $i_D(peak) = 27.6$ mA, $v_R(\max) = 16.5$ V, 36.0 %)

### 2.1.2 Full-Wave Rectification

The full-wave rectifier inverts the negative portions of the sine wave so that a unipolar output signal is generated during both halves of the input sinusoid. One example of a full-wave rectifier circuit appears in Figure 2.5(a). The input to the rectifier consists of a power transformer, in which the input is normally a 120 V (rms), 60 Hz ac signal, and the two outputs are from a center-tapped secondary winding that provides equal voltages $v_S$, with the polarities shown. When the input line voltage is positive, both output signal voltages $v_S$ are also positive.

The primary winding connected to the 120 V ac source has $N_1$ windings, and each half of the secondary winding has $N_2$ windings. The value of the $v_S$ output voltage is 120 $(N_2/N_1)$ volts (rms). The **turns ratio** of the transformer, usually designated $(N_1/N_2)$ can be designed to "step down" the input line voltage to a value that will produce a particular dc output voltage from the rectifier.

The input power transformer also provides electrical isolation between the powerline circuit and the electronic circuits to be biased by the rectifier circuit. This isolation reduces the risk of electrical shock.

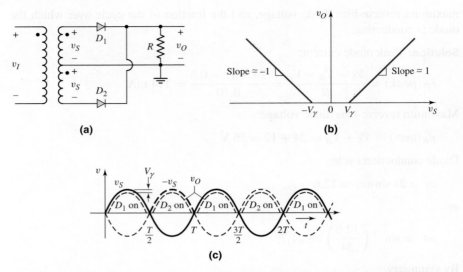

**Figure 2.5** Full-wave rectifier: (a) circuit with center-tapped transformer, (b) voltage transfer characteristics, and (c) input and output waveforms

During the positive half of the input voltage cycle, both output voltages $v_S$ are positive; therefore, diode $D_1$ is forward biased and conducting and $D_2$ is reverse biased and cut off. The current through $D_1$ and the output resistance produce a positive output voltage. During the negative half cycle, $D_1$ is cut off and $D_2$ is forward biased, or "on," and the current through the output resistance again produces a positive output voltage. If we assume that the forward diode resistance $r_f$ of each diode is small and negligible, we obtain the voltage transfer characteristics, $v_O$ versus $v_S$, shown in Figure 2.5(b).

For a sinusoidal input voltage, we can determine the output voltage versus time by using the voltage transfer curve shown in Figure 2.5(b). When $v_S > V_\gamma$, $D_1$ is on and the output voltage is $v_O = v_S - V_\gamma$. When $v_S$ is negative, then for $v_S < -V_\gamma$ or $-v_S > V_\gamma$, $D_2$ is on and the output voltage is $v_O = -v_S - V_\gamma$. The corresponding input and output voltage signals are shown in Figure 2.5(c). Since a rectified output voltage occurs during both the positive and negative cycles of the input signal, this circuit is called a **full-wave rectifier.**

Another example of a full-wave rectifier circuit appears in Figure 2.6(a). This circuit is a **bridge rectifier,** which still provides electrical isolation between the input ac powerline and the rectifier output, but does not require a center-tapped secondary winding. However, it does use four diodes, compared to only two in the previous circuit.

During the positive half of the input voltage cycle, $v_S$ is positive, $D_1$ and $D_2$ are forward biased, $D_3$ and $D_4$ are reverse biased, and the direction of the current is as shown in Figure 2.6(a). During the negative half-cycle of the input voltage, $v_S$ is negative, and $D_3$ and $D_4$ are forward biased. The direction of the current, shown in Figure 2.6(b), produces the same output voltage polarity as before.

Figure 2.6(c) shows the sinusoidal voltage $v_S$ and the rectified output voltage $v_O$. Because two diodes are in series in the conduction path, the magnitude of $v_O$ is two diode drops less than the magnitude of $v_S$.

One difference to be noted in the bridge rectifier circuit in Figure 2.6(a) and the rectifier in Figure 2.5(a) is the ground connection. Whereas the center tap of the secondary winding of the circuit in Figure 2.5(a) is at ground potential, the secondary

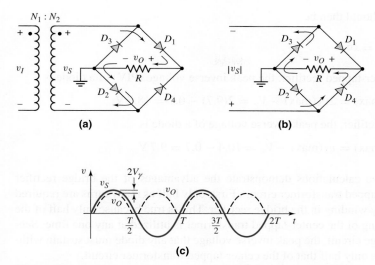

**(c)**

**Figure 2.6** A full-wave bridge rectifier: (a) circuit showing the current direction for a positive input cycle, (b) current direction for a negative input cycle, and (c) input and output voltage waveforms

winding of the bridge circuit (Figure 2.6(a)) is not directly grounded. One side of the load $R$ is grounded, but the secondary of the transformer is not.

## EXAMPLE 2.2

**Objective:** Compare voltages and the transformer turns ratio in two full-wave rectifier circuits.

Consider the rectifier circuits shown in Figures 2.5(a) and 2.6(a). Assume the input voltage is from a 120 V (rms), 60 Hz ac source. The desired peak output voltage $v_O$ is 9 V, and the diode cut-in voltage is assumed to be $V_\gamma = 0.7$ V.

**Solution:** For the center-tapped transformer circuit shown in Figure 2.5(a), a peak voltage of $v_O(\text{max}) = 9$ V means that the peak value of $v_S$ is

$$v_S(\text{max}) = v_O(\text{max}) + V_\gamma = 9 + 0.7 = 9.7 \text{ V}$$

For a sinusoidal signal, this produces an rms value of

$$v_{S,\text{rms}} = \frac{9.7}{\sqrt{2}} = 6.86 \text{ V}$$

The turns ratio of the primary to each secondary winding must then be

$$\frac{N_1}{N_2} = \frac{120}{6.86} \cong 17.5$$

For the bridge circuit shown in Figure 2.6(a), a peak voltage of $v_O(\text{max}) = 9$ V means that the peak value of $v_S$ is

$$v_S(\text{max}) = v_O(\text{max}) + 2V_\gamma = 9 + 2(0.7) = 10.4 \text{ V}$$

For a sinusoidal signal, this produces an rms value of

$$v_{S,\text{rms}} = \frac{10.4}{\sqrt{2}} = 7.35 \text{ V}$$

The turns ratio should then be

$$\frac{N_1}{N_2} = \frac{120}{7.35} \cong 16.3$$

For the center-tapped rectifier, the peak inverse voltage (PIV) of a diode is

$$\text{PIV} = v_R(\max) = 2v_S(\max) - V_\gamma = 2(9.7) - 0.7 = 18.7\,\text{V}$$

For the bridge rectifier, the peak inverse voltage of a diode is

$$\text{PIV} = v_R(\max) = v_S(\max) - V_\gamma = 10.4 - 0.7 = 9.7\,\text{V}$$

**Comment:** These calculations demonstrate the advantages of the bridge rectifier over the center-tapped transformer circuit. First, only half as many turns are required for the secondary winding in the bridge rectifier. This is true because only half of the secondary winding of the center-tapped transformer is utilized at any one time. Second, for the bridge circuit, the peak inverse voltage that any diode must sustain without breakdown is only half that of the center-tapped transformer circuit.

**EXERCISE PROBLEM**

**Ex 2.2:** Consider the bridge circuit shown in Figure 2.6(a) with an input voltage $v_S = V_M \sin \omega t$. Assume a diode cut-in voltage of $V_\gamma = 0.7\,\text{V}$. Determine the fraction (percent) of time that the diode $D_1$ is conducting for peak sinusoidal voltages of (a) $V_M = 12\,\text{V}$ and (b) $V_M = 4\,\text{V}$. (Ans. (a) 46.3% (b) 38.6%)

Because of the advantages demonstrated in Example 2.2 the bridge rectifier circuit is used more often than the center-tapped transformer circuit.

Both full-wave rectifier circuits discussed (Figures 2.5 and 2.6) produce a positive output voltage. As we will see in the next chapter discussing transistor circuits, there are times when a negative dc voltage is also required. We can produce negative rectification by reversing the direction of the diodes in either circuit. Figure 2.7(a) shows the bridge circuit with the diodes reversed compared to those in Figure 2.6. The direction of current is shown during the positive half cycle of $v_S$. The output voltage is now negative with respect to ground potential. During the negative half cycle of $v_S$, the complementary diodes turn on and the direction of current through the load is the same, producing a negative output voltage. The input and output voltages are shown in Figure 2.7(b).

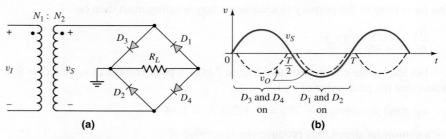

**Figure 2.7** (a) Full-wave bridge rectifier circuit to produce negative output voltages. (b) Input and output waveforms.

## 2.1.3    Filters, Ripple Voltage, and Diode Current

If a capacitor is added in parallel with the load resistor of a half-wave rectifier to form a simple filter circuit (Figure 2.8(a)), we can begin to transform the half-wave sinusoidal output into a dc voltage. Figure 2.8(b) shows the positive half of the output sine wave, and the beginning portion of the voltage across the capacitor, assuming the capacitor is initially uncharged. If we assume that the diode forward resistance is $r_f = 0$, which means that the $r_f C$ time constant is zero, the voltage across the capacitor follows this initial portion of the signal voltage. When the signal voltage reaches its peak and begins to decrease, the voltage across the capacitor also starts to decrease, which means the capacitor starts to discharge. The only discharge current path is through the resistor. If the $RC$ time constant is large, the voltage across the capacitor discharges exponentially with time (Figure 2.8(c)). During this time period, the diode is cut off.

A more detailed analysis of the circuit response when the input voltage is near its peak value indicates a subtle difference between actual circuit operation and the qualitative description. If we assume that the diode turns off immediately when the input voltage starts to decrease from its peak value, then the output voltage will decrease exponentially with time, as previously indicated. An exaggerated sketch of these two voltages is shown in Figure 2.8(d). The output voltage decreases at a faster rate than the input voltage, which means that at time $t_1$, the difference between $v_I$ and

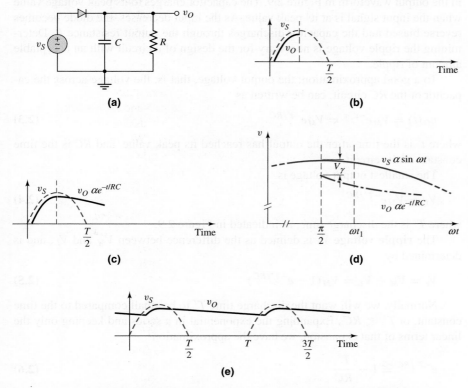

**Figure 2.8** Simple filter circuit: (a) half-wave rectifier with an RC filter, (b) positive input voltage and initial portion of output voltage, (c) output voltage resulting from capacitor discharge, (d) expanded view of input and output voltages assuming capacitor discharge begins at $\omega t = \pi/2$, and (e) steady-state input and output voltages

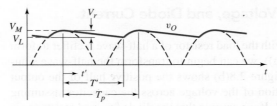

**Figure 2.9** Output voltage of a full-wave rectifier with an $RC$ filter showing the ripple voltage

$v_O$, that is, the voltage across the diode, is greater than $V_\gamma$. However, this condition cannot exist. Therefore, the diode does not turn off immediately. If the $RC$ time constant is large, there is only a small difference between the time of the peak input voltage and the time the diode turns off. In this situation, a computer analysis may provide more accurate results than an approximate hand analysis.

During the next positive cycle of the input voltage, there is a point at which the input voltage is greater than the capacitor voltage, and the diode turns back on. The diode remains on until the input reaches its peak value and the capacitor voltage is completely recharged.

Since the capacitor filters out a large portion of the sinusoidal signal, it is called a **filter capacitor.** The steady-state output voltage of the $RC$ filter is shown in Figure 2.8(e).

The ripple effect in the output from a full-wave filtered rectifier circuit can be seen in the output waveform in Figure 2.9. The capacitor charges to its peak voltage value when the input signal is at its peak value. As the input decreases, the diode becomes reverse biased and the capacitor discharges through the output resistance $R$. Determining the ripple voltage is necessary for the design of a circuit with an acceptable amount of ripple.

To a good approximation, the output voltage, that is, the voltage across the capacitor or the $RC$ circuit, can be written as

$$v_O(t) = V_M e^{-t'/\tau} = V_M e^{-t'/RC} \tag{2.3}$$

where $t'$ is the time after the output has reached its peak value, and $RC$ is the time constant of the circuit.

The smallest output voltage is

$$V_L = V_M e^{-T'/RC} \tag{2.4}$$

where $T'$ is the discharge time, as indicated in Figure 2.9.

The **ripple voltage** $V_r$ is defined as the difference between $V_M$ and $V_L$, and is determined by

$$V_r = V_M - V_L = V_M(1 - e^{-T'/RC}) \tag{2.5}$$

Normally, we will want the discharge time $T'$ to be small compared to the time constant, or $T' \ll RC$. Expanding the exponential in a series and keeping only the linear terms of that expansion, we have the approximation[2]

$$e^{-T'/RC} \cong 1 - \frac{T'}{RC} \tag{2.6}$$

---

[2]We can show that the difference between the exponential function and the linear approximation given by Equation (2.6) is less than 0.5 percent for $RC = 10T'$. We need a relatively large $RC$ time constant for this application.

The ripple voltage can now be written as

$$V_r \cong V_M \left( \frac{T'}{RC} \right) \tag{2.7}$$

Since the discharge time $T'$ depends on the $RC$ time constant, Equation (2.7) is difficult to solve. However, if the ripple effect is small, then as an approximation, we can let $T' = T_p$, so that

$$V_r \cong V_M \left( \frac{T_p}{RC} \right) \tag{2.8}$$

where $T_p$ is the time between peak values of the output voltage. For a full-wave rectifier, $T_p$ is one-half the signal period. Therefore, we can relate $T_p$ to the signal frequency,

$$f = \frac{1}{2T_p}$$

The ripple voltage is then

$$V_r = \frac{V_M}{2fRC} \tag{2.9}$$

For a half-wave rectifier, the time $T_p$ corresponds to a full period (not a half period) of the signal, so the factor 2 does not appear in Equation (2.9). The factor of 2 shows that the full-wave rectifier has half the ripple voltage of the half-wave rectifier.

Equation (2.9) can be used to determine the capacitor value required for a particular ripple voltage.

## EXAMPLE 2.3

**Objective:** Determine the capacitance required to yield a particular ripple voltage.

Consider a full-wave rectifier circuit with a 60 Hz input signal and a peak output voltage of $V_M = 10$ V. Assume the output load resistance is $R = 10$ k$\Omega$ and the ripple voltage is to be limited to $V_r = 0.2$ V.

**Solution:** From Equation (2.9), we can write

$$C = \frac{V_M}{2fRV_r} = \frac{10}{2(60)(10 \times 10^3)(0.2)} \Rightarrow 41.7\,\mu\text{F}$$

**Comment:** If the ripple voltage is to be limited to a smaller value, a larger filter capacitor must be used. Note that the size of the ripple voltage and the size of filter capacitor are related to the load resistance $R$.

## EXERCISE PROBLEM

**Ex 2.3:** Assume the input signal to a rectifier circuit has a peak value of $V_M = 12$ V and is at a frequency of 60 Hz. Assume the output load resistance is $R = 2$ k$\Omega$ and the ripple voltage is to be limited to $V_r = 0.4$ V. Determine the capacitance required to yield this specification for a (a) full-wave rectifier and (b) half-wave rectifier. (Ans. (a) $C = 125\,\mu$F, (b) $C = 250\,\mu$F).

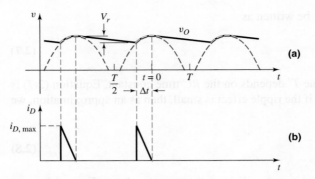

**Figure 2.10** Output of a full-wave rectifier with an *RC* filter: (a) diode conduction time and (b) diode current

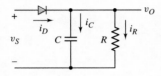

**Figure 2.11** Equivalent circuit of a full-wave rectifier during capacitor charging cycle

The diode in a filtered rectifier circuit conducts for a brief interval $\Delta t$ near the peak of the sinusoidal input signal. The diode current supplies the charge lost by the capacitor during the discharge time. Figure 2.10 shows the rectified output of a full-wave rectifier and the filtered output assuming ideal diodes ($V_\gamma = 0$) in the rectifier circuit. We will use this approximate model to estimate the diode current during the diode conduction time. Figure 2.11 shows the equivalent circuit of the full-wave rectifier during the charging time. We see that

$$i_D = i_C + i_R = C\frac{dv_O}{dt} + \frac{v_O}{R} \tag{2.10}$$

During the diode conduction time near $t = 0$ (Figure 2.10), we can write

$$v_O = V_M \cos \omega t \tag{2.11}$$

For small ripple voltages, the diode conduction time is small, so we can approximate the output voltage as

$$v_O = V_M \cos \omega t \cong V_M\left[1 - \frac{1}{2}(\omega t)^2\right] \tag{2.12}$$

The charging current through the capacitor is

$$i_C = C\frac{dv_O}{dt} = CV_M\left[-\frac{1}{2}(2)(\omega t)(\omega)\right] = -\omega CV_M\omega t \tag{2.13}$$

From Figure 2.10, the diode conduction occurs during the time $-\Delta t < t < 0$, so that the capacitor current is positive and is a linear function of time. We note that at $t = 0$, the capacitor current is $i_C = 0$. At $t = -\Delta t$, the capacitor charging current is at a peak value and is given by

$$i_{C,\text{peak}} = -\omega CV_M[\omega(-\Delta t)] = +\omega CV_M\omega\Delta t \tag{2.14}$$

The capacitor current during the diode charging time is approximately triangular and is shown in Figure 2.10(b).

From Equation (2.11), we can write that the voltage $V_L$ is given by

$$V_L = V_M \cos[\omega(-\Delta t)] \cong V_M\left[1 - \frac{1}{2}(\omega\Delta t)^2\right] \tag{2.15}$$

Solving for $\omega \Delta t$, we find

$$\omega \Delta t = \sqrt{\frac{2V_r}{V_M}} \tag{2.16}$$

where $V_r = V_M - V_L$.

From Equation (2.9), we can write

$$fC = \frac{V_M}{2RV_r} \tag{2.17(a)}$$

or

$$2\pi f C = \omega C = \frac{\pi V_M}{RV_r} \tag{2.17(b)}$$

Substituting Equations (2.17(b)) and (2.16) into Equation (2.14), we have

$$i_{C,\text{peak}} = \left(\frac{\pi V_M}{RV_r}\right) V_M \left(\sqrt{\frac{2V_r}{V_M}}\right) \tag{2.18(a)}$$

or

$$i_{C,\text{peak}} = \pi \frac{V_M}{R} \sqrt{\frac{2V_M}{V_r}} \tag{2.18(b)}$$

Since the charging current through the capacitor is triangular, we have that the average capacitor current during the diode charging time is

$$i_{C,\text{avg}} = \frac{\pi}{2} \frac{V_M}{R} \sqrt{\frac{2V_M}{V_r}} \tag{2.19}$$

During the capacitor charging time, there is still a current through the load. This current is also being supplied through the diode. Neglecting the ripple voltage, the load current is approximately

$$i_L \cong \frac{V_M}{R} \tag{2.20}$$

Therefore, the peak diode current during the diode conduction time for a full-wave rectifier is approximately

$$i_{D,\text{peak}} \cong \frac{V_M}{R}\left(1 + \pi \sqrt{\frac{2V_M}{V_r}}\right) \tag{2.21}$$

and the average diode current during the diode conduction time is

$$i_{D,\text{avg}} \cong \frac{V_M}{R}\left(1 + \frac{\pi}{2}\sqrt{\frac{2V_M}{V_r}}\right) \tag{2.22}$$

The average diode current over the entire input signal period is

$$i_D(\text{avg}) = \frac{V_M}{R}\left(1 + \frac{\pi}{2}\sqrt{\frac{2V_M}{V}}\right)\frac{\Delta t}{T} \tag{2.23}$$

For the full-wave rectifier, we have $1/2T = f$, so

$$\Delta t = \frac{1}{\omega}\sqrt{\frac{2V_r}{V_M}} = \frac{1}{2\pi f}\sqrt{\frac{2V_r}{V_M}} \tag{2.24(a)}$$

Then

$$\frac{\Delta t}{T} = \frac{1}{2\pi f}\sqrt{\frac{2V_r}{V_M}}2f = \frac{1}{\pi}\sqrt{\frac{2V_r}{V_M}} \tag{2.24(b)}$$

Then the average current through the diode during the entire cycle for a full-wave rectifier is

$$i_D(\text{avg}) = \frac{1}{\pi}\sqrt{\frac{2V_r}{V_M}}\frac{V_M}{R}\left(1 + \frac{\pi}{2}\sqrt{\frac{2V_M}{V_r}}\right) \tag{2.25}$$

## DESIGN EXAMPLE 2.4

**Objective:** Design a full-wave rectifier to meet particular specifications.

A full-wave rectifier is to be designed to produce a peak output voltage of 12 V, deliver 120 mA to the load, and produce an output with a ripple of not more than 5 percent. An input line voltage of 120 V (rms), 60 Hz is available.

**Solution:** A full-wave bridge rectifier will be used, because of the advantages previously discussed. The effective load resistance is

$$R = \frac{V_O}{I_L} = \frac{12}{0.12} = 100\,\Omega$$

Assuming a diode cut-in voltage of 0.7 V, the peak value of $v_S$ is

$$v_S(\text{max}) = v_O(\text{max}) + 2V_\gamma = 12 + 2(0.7) = 13.4\,\text{V}$$

For a sinusoidal signal, this produces an rms voltage value of

$$v_{S,\text{rms}} = \frac{13.4}{\sqrt{2}} = 9.48\,\text{V}$$

The transformer turns ratio is then

$$\frac{N_1}{N_2} = \frac{120}{9.48} = 12.7$$

For a 5 percent ripple, the ripple voltage is

$$V_r = (0.05)V_M = (0.05)(12) = 0.6\,\text{V}$$

The required filter capacitor is found to be

$$C = \frac{V_M}{2fRV_r} = \frac{12}{2(60)(100)(0.6)} \Rightarrow 1667\,\mu\text{F}$$

The peak diode current, from Equation (2.21), is

$$i_{D,\text{peak}} = \frac{12}{100}\left[1 + \pi\sqrt{\frac{2(12)}{0.6}}\right] = 2.50\,\text{A}$$

and the average diode current over the entire signal period, from Equation (2.25), is

$$i_D(\text{avg}) = \frac{1}{\pi}\sqrt{\frac{2(0.6)}{12}}\left(\frac{12}{100}\right)\left(1 + \frac{\pi}{2}\sqrt{\frac{2(12)}{0.6}}\right) \Rightarrow 132\,\text{mA}$$

Finally, the peak inverse voltage that each diode must sustain is

$$\text{PIV} = v_R(\text{max}) = v_S(\text{max}) - V_\gamma = 13.4 - 0.7 = 12.7\,\text{V}$$

**Comment:** The minimum specifications for the diodes in this full-wave rectifier circuit are: a peak current of 2.50 A, an average current of 132 mA, and a peak inverse voltage of 12.7 V. In order to meet the desired ripple specification, the required filter capacitance must be large, since the effective load resistance is small.

**Design Pointer:** (1) A particular turns ratio was determined for the transformer. However, this particular transformer design is probably not commercially available. This means an expensive custom transformer design would be required, or if a standard transformer is used, then additional circuit design is required to meet the output voltage specification. (2) A constant 120 V (rms) input voltage is assumed to be available. However, this voltage can fluctuate, so the output voltage will also fluctuate.

We will see later how more sophisticated designs will solve these two problems.

**Computer Verification:** Since we simply used an assumed cut-in voltage for the diode and used approximations in the development of the ripple voltage equations, we can use PSpice to give us a more accurate evaluation of the circuit. The PSpice circuit schematic and the steady-state output voltage are shown in Figure 2.12. We see that the peak output voltage is 11.6 V, which is close to the desired 12 V. One reason for the slight discrepancy is that the diode voltage drop for the maximum input voltage is slightly greater than 0.8 V rather than the assumed 0.7 V. The ripple voltage is approximately 0.5 V, which is within the 0.6 V specification.

**Discussion:** In the PSpice simulation, a standard diode, 1N4002, was used. In order for the computer simulation to be valid, the diode used in the simulation and in the actual circuit must match. In this example, to reduce the diode voltage and increase the peak output voltage, a diode with a larger cross-sectional area should be used.

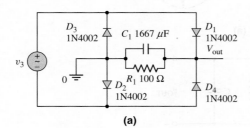

**(a)**

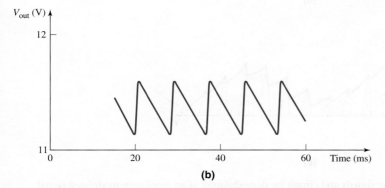

**(b)**

**Figure 2.12** (a) PSpice circuit schematic of diode bridge circuit with an RC filter; (b) Steady-state output voltage of PSpice analysis of diode bridge circuit for a 60 Hz input sine wave with a peak value of 13.4 V

**Ex 2.4:** The input voltage to the half-wave rectifier in Figure 2.8(a) is $v_S = 75 \sin[2\pi(60)t]$ V. Assume a diode cut-in voltage of $V_\gamma = 0$. The ripple voltage is to be no more than $V_r = 4$ V. If the filter capacitor is 50 $\mu$F, determine the minimum load resistance that can be connected to the output. (Ans. $R = 6.25$ k$\Omega$)

### 2.1.4    Detectors

One of the first applications of semiconductor diodes was as a detector for amplitude-modulated (AM) radio signals. An amplitude-modulated signal consists of a radio-frequency carrier wave whose amplitude varies with an audio frequency as shown in Figure 2.13(a). The detector circuit is shown in Figure 2.13(b) and is a half-wave rec-

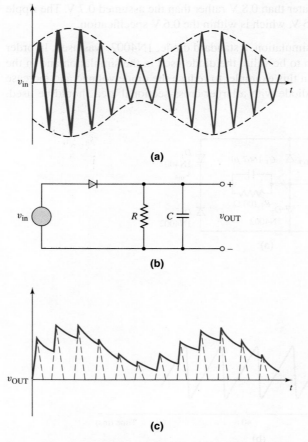

**(a)**

**(b)**

**(c)**

**Figure 2.13** The signals and circuit for demodulation of an amplitude-modulated signal. (a) The amplitude-modulated input signal. (b) The detector circuit. (c) The demodulated output signal.

tifier circuit with an *RC* filter on the output. For this application, the *RC* time constant should be approximately equal to the period of the carrier signal, so that the output voltage can follow each peak value of the carrier signal. If the time constant is too large, the output will not be able to change fast enough and the output will not represent the audio output. The output of the detector is shown in Figure 2.13(c).

The output of the detector circuit is then coupled to an amplifier through a capacitor to remove the dc component of the signal, and the output of the amplifier is then fed to a speaker.

### 2.1.5 Voltage Doubler Circuit

A **voltage doubler circuit** is very similar to the full-wave rectifier, except that two diodes are replaced by capacitors, and it can produce a voltage equal to approximately twice the peak output of a transformer (Figure 2.14).

Figure 2.15(a) shows the equivalent circuit when the voltage polarity at the "top" of the transformer is negative; Figure 2.15(b) shows the equivalent circuit for the opposite polarity. In the circuit in Figure 2.15(a), the forward diode resistance of $D_2$ is small; therefore, the capacitor $C_1$ will charge to almost the peak value of $v_S$. Terminal 2 on $C_1$ is positive with respect to terminal 1. As the magnitude of $v_S$ decreases from its peak value, $C_1$ discharges through $R_L$ and $C_2$. We assume that the time constant $R_L C_2$ is very long compared to the period of the input signal.

As the polarity of $v_S$ changes to that shown in Figure 2.15(b), the voltage across $C_1$ is essentially constant at $V_M$, with terminal 2 remaining positive. As $v_S$ reaches its maximum value, the voltage across $C_2$ essentially becomes $V_M$. By Kirchhoff's voltage law,

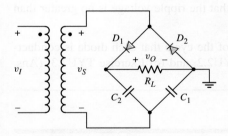

**Figure 2.14** A voltage doubler circuit

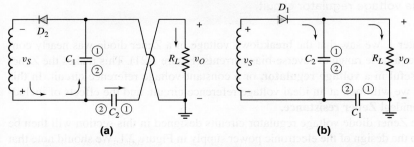

**(a)**      **(b)**

**Figure 2.15** Equivalent circuit of the voltage doubler circuit: (a) negative input cycle and (b) positive input cycle

the peak voltage across $R_L$ is now essentially equal to $2V_M$, or twice the peak output of the transformer. The same ripple effect occurs as in the output voltage of the rectifier circuits, but if $C_1$ and $C_2$ are relatively large, then the ripple voltage $V_r$, is quite small.

There are also voltage tripler and voltage quadrupler circuits. These circuits provide a means by which multiple dc voltages can be generated from a single ac source and power transformer.

## Test Your Understanding

**TYU 2.1** Consider the circuit in Figure 2.4. The input voltage is $v_s(t) = 15 \sin \omega t$ (V) and the diode cut-in voltage is $V_\gamma = 0.7$ V. The voltage $V_B$ varies between $4 \le V_B \le 8$ V. The peak current is to be limited to $i_D(\text{peak}) = 18$ mA. (a) Determine the minimum value of $R$. (b) Using the results of part (a), determine the range in peak current and the range in duty cycle. (Ans. (a) $R = 572\,\Omega$; (b) $11 \le i_D(\text{peak}) \le 18$ mA, $30.3 \le$ duty cycle $\le 39.9\,\%$).

**TYU 2.2** The circuit in Figure 2.5(a) is used to rectify a sinusoidal input signal with a peak voltage of 120 V and a frequency of 60 Hz. A filter capacitor is connected in parallel with $R$. If the output voltage cannot drop below 100 V, determine the required value of the capacitance $C$. The transformer has a turns ratio of $N_1 : N_2 = 1 : 1$, where $N_2$ is the number of turns on each of the secondary windings. Assume the diode cut-in voltage is 0.7 V and the output resistance is 2.5 k$\Omega$. (Ans. $C = 20.6\ \mu$F)

**TYU 2.3** The secondary transformer voltage of the rectifier circuit shown in Figure 2.6(a) is $v_S = 50 \sin[2\pi(60)t]$ V. Each diode has a cut-in voltage of $V_\gamma = 0.7$ V, and the load resistance is $R = 10$ k$\Omega$. Determine the value of the filter capacitor that must be connected in parallel with $R$ such that the ripple voltage is no greater than $V_r = 2$ V. (Ans. $C = 20.3\ \mu$F)

**TYU 2.4** Determine the fraction (percent) of the cycle that each diode is conducting in (a) Exercise EX2.4, (b) Exercise TYU2.2, and (c) Exercise TYU2.3. (Ans. (a) 5.2%, (b) 18.1%, (c) 9.14%)

## 2.2   ZENER DIODE CIRCUITS

**Objective:** • Apply the characteristics of the Zener diode to a Zener diode voltage regulator circuit.

In Chapter 1, we saw that the breakdown voltage of a Zener diode was nearly constant over a wide range of reverse-bias currents (Figure 1.21). This makes the Zener diode useful in a **voltage regulator,** or a constant-voltage reference circuit. In this chapter, we will look at an ideal voltage reference circuit, and the effects of including a nonideal **Zener resistance.**

The Zener diode voltage regulator circuits designed in this section will then be added to the design of the electronic power supply in Figure 2.1. We should note that in actual power supply designs, the voltage regulator will be a more sophisticated integrated circuit rather than the simpler Zener diode design that will be developed here. One reason is that a standard Zener diode with a particular desired breakdown voltage may not be available. However, this section will provide the basic concept of a voltage regulator.

**Ideal Voltage Reference Circuit**

Figure 2.16 shows a Zener voltage regulator circuit. For this circuit, the output voltage should remain constant, even when the output load resistance varies over a fairly wide range, and when the input voltage varies over a specific range. The variation in $V_{PS}$ may be the ripple voltage from a rectifier circuit.

We determine, initially, the proper input resistance $R_i$. The resistance $R_i$ limits the current through the Zener diode and drops the "excess" voltage between $V_{PS}$ and $V_Z$. We can write

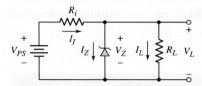

**Figure 2.16**  A Zener diode voltage regulator circuit

$$R_i = \frac{V_{PS} - V_Z}{I_I} = \frac{V_{PS} - V_Z}{I_Z + I_L} \tag{2.26}$$

which assumes that the Zener resistance is zero for the ideal diode. Solving this equation for the diode current, $I_Z$, we get

$$I_Z = \frac{V_{PS} - V_Z}{R_i} - I_L \tag{2.27}$$

where $I_L = V_Z / R_L$, and the variables are the input voltage source $V_{PS}$ and the load current $I_L$.

For proper operation of this circuit, the diode must remain in the breakdown region and the power dissipation in the diode must not exceed its rated value. In other words:

1. The current in the diode is a minimum, $I_Z(\text{min})$, when the load current is a maximum, $I_L(\text{max})$, and the source voltage is a minimum, $V_{PS}(\text{min})$.
2. The current in the diode is a maximum, $I_Z(\text{max})$, when the load current is a minimum, $I_L(\text{min})$, and the source voltage is a maximum, $V_{PS}(\text{max})$.

Inserting these two specifications into Equation (2.26), we obtain

$$R_i = \frac{V_{PS}(\text{min}) - V_Z}{I_Z(\text{min}) + I_L(\text{max})} \tag{2.28(a)}$$

and

$$R_i = \frac{V_{PS}(\text{max}) - V_Z}{I_Z(\text{max}) + I_L(\text{min})} \tag{2.28(b)}$$

Equating these two expressions, we then obtain

$$[V_{PS}(\text{min}) - V_Z] \cdot [I_Z(\text{max}) + I_L(\text{min})]$$
$$= [V_{PS}(\text{max}) - V_Z] \cdot [I_Z(\text{min}) + I_L(\text{max})] \tag{2.29}$$

Reasonably, we can assume that we know the range of input voltage, the range of output load current, and the Zener voltage. Equation (2.29) then contains two unknowns, $I_Z(\text{min})$ and $I_Z(\text{max})$. Further, as a minimum requirement, we can set the minimum Zener current to be one-tenth the maximum Zener current, or $I_Z(\text{min}) = 0.1 I_Z(\text{max})$. (More stringent design requirements may require the minimum Zener current to be 20 to 30 percent of the maximum value.) We can then solve for $I_Z(\text{max})$, using Equation (2.29), as follows:

$$I_Z(\text{max}) = \frac{I_L(\text{max}) \cdot [V_{PS}(\text{max}) - V_Z] - I_L(\text{min}) \cdot [V_{PS}(\text{min}) - V_Z]}{V_{PS}(\text{min}) - 0.9 V_Z - 0.1 V_{PS}(\text{max})} \tag{2.30}$$

Using the maximum current thus obtained from Equation (2.30), we can determine the maximum required power rating of the Zener diode. Then, combining Equation (2.30) with either Equation (2.28(a)) or (2.28(b)), we can determine the required value of the input resistance $R_i$.

## DESIGN EXAMPLE 2.5

**Objective:** Design a voltage regulator using the circuit in Figure 2.16.

The voltage regulator is to power a car radio at $V_L = 9$ V from an automobile battery whose voltage may vary between 11 and 13.6 V. The current in the radio will vary between 0 (off) to 100 mA (full volume).

The equivalent circuit is shown in Figure 2.17.

**Solution:** The maximum Zener diode current can be determined from Equation (2.30) as

$$I_Z(\text{max}) = \frac{(100)[13.6 - 9] - 0}{11 - (0.9)(9) - (0.1)(13.6)} \cong 300 \, \text{mA}$$

The maximum power dissipated in the Zener diode is then

$$P_Z(\text{max}) = I_Z(\text{max}) \cdot V_Z = (300)(9) \Rightarrow 2.7 \, \text{W}$$

The value of the current-limiting resistor $R_i$, from Equation (2.28(b)), is

$$R_i = \frac{13.6 - 9}{0.3 + 0} = 15.3 \, \Omega$$

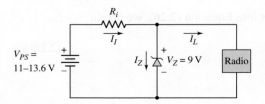

**Figure 2.17** Circuit for Design Example 2.5

The maximum power dissipated in this resistor is

$$P_{Ri}(\max) = \frac{(V_{PS}(\max) - V_Z)^2}{R_i} = \frac{(13.6 - 9)^2}{15.3} \cong 1.4\,\text{W}$$

We find

$$I_Z(\min) = \frac{11 - 9}{15.3} - 0.10 \Rightarrow 30.7\,\text{mA}$$

**Comment:** From this design, we see that the minimum power ratings of the Zener diode and input resistor are 2.7 W and 1.4 W, respectively. The minimum Zener diode current occurs for $V_{PS}(\min)$ and $I_L(\max)$. We find $I_Z(\min) = 30.7$ mA, which is approximately 10 percent of $I_Z(\max)$ as specified by the design equations.

**Design Pointer:** (1) The variable input in this example was due to a variable battery voltage. However, referring back to Example 2.4, the variable input could also be a function of using a standard transformer with a given turns ratio as opposed to a custom design with a particular turns ratio and/or having a 120 V (rms) input voltage that is not exactly constant.

(2) The 9 V output is a result of using a 9 V Zener diode. However, a Zener diode with exactly a 9 V breakdown voltage may also not be available. We will again see later how more sophisticated designs can solve this problem.

EXERCISE PROBLEM

**Ex 2.5:** The Zener diode regulator circuit shown in Figure 2.16 has an input voltage that varies between 10 and 14 V, and a load resistance that varies between $R_L = 20$ and $100\,\Omega$. Assume a 5.6 Zener diode is used, and assume $I_Z(\min) = 0.1 I_Z(\max)$. Find the value of $R_i$ required and the minimum power rating of the diode. (Ans. $P_Z = 3.31$ W, $R_i \cong 13\,\Omega$)

The operation of the Zener diode circuit shown in Figure 2.17 can be visualized by using load lines. Summing currents at the Zener diode, we have

$$\frac{v_{PS} - V_Z}{R_i} = I_Z + \frac{V_Z}{R_L} \tag{2.31}$$

Solving for $V_Z$, we obtain

$$V_Z = v_{PS}\left(\frac{R_L}{R_i + R_L}\right) - I_Z\left(\frac{R_i R_L}{R_i + R_L}\right) \tag{2.32}$$

which is the load line equation. Using the parameters of Example 2.5, the load resistance varies from $R_L = \infty(I_L = 0)$ to $R_L = 9/0.1 = 90\,\Omega(I_L = 100$ mA). The current limiting resistor is $R_i = 15\,\Omega$ and the input voltage varies over the range $11 \le v_{PS} \le 13.6$ V.

We may write load line equations for the various circuit conditions.

A: $v_{PS} = 11$ V,  $R_L = \infty$;  $V_Z = 11 - I_Z(15)$
B: $v_{PS} = 11$ V,  $R_L = 90\,\Omega$; $V_Z = 9.43 - I_Z(12.9)$
C: $v_{PS} = 13.6$ V, $R_L = \infty$;  $V_Z = 13.6 - I_Z(15)$
D: $v_{PS} = 13.6$ V, $R_L = 90\,\Omega$; $V_Z = 11.7 - I_Z(12.9)$

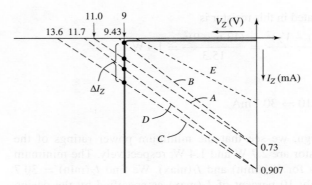

**Figure 2.18** Zener diode $I$–$V$ characteristics with various load lines superimposed

Figure 2.18 shows the Zener diode $I$–$V$ characteristics. Superimposed on the figure are the four load lines designated as A, B, C, and D. Each load line intersects the diode characteristics in the breakdown region, which is the required condition for proper diode operation. The variation in Zener diode current $\Delta I_Z$ for the various combinations of input voltage and load resistance is shown on the figure.

If we were to choose the input resistance to be $R_i = 25\ \Omega$ and let $v_{PS} = 11$ V and $R_L = 90\ \Omega$, the load line equation (Equation (2.32)) becomes

$$V_Z = 8.61 - I_Z(19.6) \tag{2.33}$$

This load line is plotted as curve E on Figure 2.18. We see that this load line does not intersect the diode characteristics in the breakdown region. For this condition, the output voltage will not equal the breakdown voltage of $V_Z = 9$ V; the circuit does not operate "properly."

### 2.2.2    Zener Resistance and Percent Regulation

In the ideal Zener diode, the Zener resistance is zero. In actual Zener diodes, however, this is not the case. The result is that the output voltage will fluctuate slightly with a fluctuation in the input voltage, and will fluctuate with changes in the output load resistance.

Figure 2.19 shows the equivalent circuit of the voltage regulator including the Zener resistance. Because of the Zener resistance, the output voltage will change with a change in the Zener diode current.

Two figures of merit can be defined for a voltage regulator. The first is the **source regulation** and is a measure of the change in output voltage with a change in source

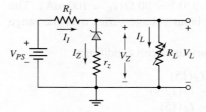

**Figure 2.19**  A Zener diode voltage regulator circuit with a nonzero Zener resistance

voltage. The second is the **load regulation** and is a measure of the change in output voltage with a change in load current.

The source regulation is defined as

$$\text{Source regulation} = \frac{\Delta v_L}{\Delta v_{PS}} \times 100\% \tag{2.34}$$

where $\Delta v_L$ is the change in output voltage with a change of $\Delta v_{PS}$ in the input voltage.

The load regulation is defined as

$$\text{Load regulation} = \frac{v_{L,\text{no load}} - v_{L,\text{full load}}}{v_{L,\text{full load}}} \times 100\% \tag{2.35}$$

where $v_{L,\text{no load}}$ is the output voltage for zero load current and $v_{L,\text{full load}}$ is the output voltage for the maximum rated output current.

The circuit approaches that of an ideal voltage regulator as the source and load regulation factors approach zero.

## EXAMPLE 2.6

**Objective:** Determine the source regulation and load regulation of a voltage regulator circuit.

Consider the circuit described in Example 2.5 and assume a Zener resistance of $r_z = 2 \, \Omega$.

**Solution:** Consider the effect of a change in input voltage for a no-load condition ($R_L = \infty$). For $v_{PS} = 13.6$ V, we find

$$I_Z = \frac{13.6 - 9}{15.3 + 2} = 0.2659 \, \text{A}$$

Then

$$v_{L,\text{max}} = 9 + (2)(0.2659) = 9.532 \, \text{V}$$

For $v_{PS} = 11$ V, we find

$$I_Z = \frac{11 - 9}{15.3 + 2} = 0.1156 \, \text{A}$$

Then

$$v_{L,\text{min}} = 9 + (2)(0.1156) = 9.231 \, \text{V}$$

We obtain

$$\text{Source regulation} = \frac{\Delta v_L}{\Delta v_{PS}} \times 100\% = \frac{9.532 - 9.231}{13.6 - 11} \times 100\% = 11.6\%$$

Now consider the effect of a change in load current for $v_{PS} = 13.6$ V. For $I_L = 0$, we find

$$I_Z = \frac{13.6 - 9}{15.3 + 2} = 0.2659 \, \text{A}$$

and

$$v_{L,\text{no load}} = 9 + (2)(0.2659) = 9.532 \, \text{V}$$

For a load current of $I_L = 100$ mA, we find

$$I_Z = \frac{13.6 - [9 + I_Z(2)]}{15.3} - 0.10$$

which yields

$$I_Z = 0.1775 \text{ A}$$

Then

$$v_{L,\text{full load}} = 9 + (2)(0.1775) = 9.355 \text{ V}$$

We now obtain

$$\text{Load regulation} = \frac{v_{L,\text{no load}} - v_{L,\text{full load}}}{v_{L,\text{full load}}} \times 100\%$$

$$= \frac{9.532 - 9.355}{9.355} \times 100\% = 1.89\%$$

**Comment:** The ripple voltage on the input of 2.6 V is reduced by approximately a factor of 10. The change in output load results in a small percentage change in the output voltage.

## EXERCISE PROBLEM

**Ex 2.6:** Repeat Example 2.6 for $r_z = 4 \, \Omega$. Assume all other parameters are the same as listed in the example. (Ans. Source regulation = 20.7%, load regulation = 3.29%)

## Test Your Understanding

**TYU 2.5** Consider the circuit shown in Figure 2.19. Let $V_{PS} = 12$ V, $V_{ZO} = 6.2$ V, and $r_z = 3 \, \Omega$. The power rating of the diode is $P = 1$ W. (a) Determine $I_Z(\text{max})$ and $R_i$. (b) If $I_Z(\text{min}) = 0.1 I_Z(\text{max})$, determine $R_L(\text{min})$ and the load regulation. (Ans. (a) $I_Z(\text{max}) = 150$ mA, $R_i = 35.7 \, \Omega$; (b) $R_L(\text{min}) = 42.7 \, \Omega$, 6.09 %).

**TYU 2.6** Suppose the current-limiting resistor in Example 2.5 is replaced by one whose value is $R_i = 20 \, \Omega$. Determine the minimum and maximum Zener diode current. Does the circuit operate "properly"? (Ans. $I_Z(\text{min}) = 0$, $I_Z(\text{max}) = 230$ mA).

**TYU 2.7** Suppose the power supply voltage in the circuit shown in Figure 2.17 drops to $V_{PS} = 10$ V. Let $R_i = 15.3 \, \Omega$. What is the maximum load current in the radio if the minimum Zener diode current is to be maintained at $I_Z(\text{min}) = 30$ mA? (Ans. $I_L(\text{max}) = 35.4$ mA).

## 2.3   CLIPPER AND CLAMPER CIRCUITS

**Objective:** • Apply the nonlinear characteristics of diodes to create waveshaping circuits known as clippers and clampers.

In this section, we continue our discussion of nonlinear circuit applications of diodes. Diodes can be used in waveshaping circuits that either limit or "clip" portions of a signal, or shift the dc voltage level. The circuits are called **clippers** and **clampers,** respectively.

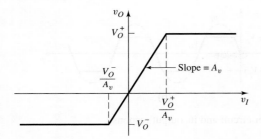

**Figure 2.20** General voltage transfer characteristics of a limiter circuit

### 2.3.1    Clippers

Clipper circuits, also called **limiter circuits,** are used to eliminate portions of a signal that are above or below a specified level. For example, the half-wave rectifier is a clipper circuit, since all voltages below zero are eliminated. A simple application of a clipper is to limit the voltage at the input to an electronic circuit so as to prevent breakdown of the transistors in the circuit. The circuit may be used to measure the frequency of the signal, if the amplitude is not an important part of the signal.

Figure 2.20 shows the general voltage transfer characteristics of a limiter circuit. The limiter is a linear circuit if the input signal is in the range $V_O^-/A_v \leq v_I \leq V_O^+/A_v$, where $A_v$ is the slope of the transfer curve. If $A_v \leq 1$, as in diode circuits, the circuit is a **passive limiter.** If $v_I > V_O^+/A_v$, the output is limited to a maximum value of $V_O^+$. Similarly, if $v_I < V_O^-/A_v$, the output is limited to a minimum value of $V_O^-$. Figure 2.20 shows the general transfer curve of a double limiter, in which both the positive and negative peak values of the input signal are clipped.

Various combinations of $V_O^+$ and $V_O^-$ are possible. Both parameters may be positive, both negative, or one may be positive while the other negative, as indicated in the figure. If either $V_O^-$ approaches minus infinity or $V_O^+$ approaches plus infinity, then the circuit reverts to a single limiter.

Figure 2.21(a) is a single-diode clipper circuit. The diode $D_1$ is off as long as $v_I < V_B + V_\gamma$. With $D_1$ off, the current is approximately zero, the voltage drop across $R$ is essentially zero, and the output voltage follows the input voltage. When $v_I > V_B + V_\gamma$, the diode turns on, the output voltage is clipped, and $v_O$ equals $V_B + V_\gamma$. The output signal is shown in Figure 2.21(b). In this circuit, the output is clipped above $V_B + V_\gamma$.

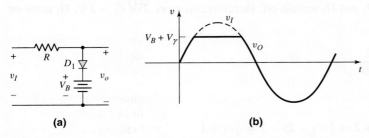

**Figure 2.21** Single-diode clipper: (a) circuit and (b) output response

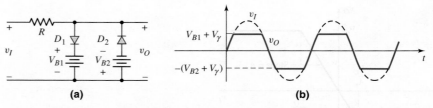

**Figure 2.22** A parallel-based diode clipper circuit and its output response

The resistor $R$ in Figure 2.21 is selected to be large enough so that the forward diode current is limited to be within reasonable values (usually in the milliampere range), but small enough so that the reverse diode current produces a negligible voltage drop. Normally, a wide range of resistor values will result in satisfactory performance of a given circuit.

Other clipping circuits can be constructed by reversing the diode, the polarity of the voltage source, or both.

Positive and negative clipping can be performed simultaneously by using a double limiter or a **parallel-based clipper,** such as the circuit shown in Figure 2.22. The input and output signals are also shown in the figure. The parallel-based clipper is designed with two diodes and two voltage sources oriented in opposite directions.

## EXAMPLE 2.7

**Objective:** Find the output of the parallel-based clipper in Figure 2.23(a).

For simplicity, assume that $V_\gamma = 0$ and $r_f = 0$ for both diodes.

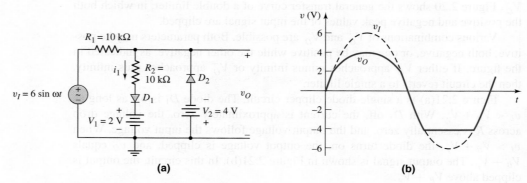

**Figure 2.23** Figure for Example 2.7

**Solution:** For $t = 0$, we see that $v_I = 0$ and both $D_1$ and $D_2$ are reverse biased. For $0 < v_I \leq 2\,\text{V}$, $D_1$ and $D_2$ remain off; therefore, $v_O = v_I$. For $v_I > 2\,\text{V}$, $D_1$ turns on and

$$i_1 = \frac{v_I - 2}{10 + 10}$$

Also,

$$v_O = i_1 R_2 + 2 = \tfrac{1}{2}(v_I - 2) + 2 = \tfrac{1}{2}v_I + 1$$

If $v_I = 6$ V, then $v_O = 4$ V.

For $-4 < v_I < 0$ V, both $D_1$ and $D_2$ are off and $v_O = v_I$. For $v_I \leq -4$ V, $D_2$ turns on and the output is constant at $v_O = -4$ V. The input and output waveforms are plotted in Figure 2.23(b).

**Comment:** If we assume that $V_\gamma \neq 0$, the output will be very similar to the results calculated here. The only difference will be the points at which the diodes turn on.

<hr>

**EXERCISE PROBLEM**

**Ex 2.7:** Design a parallel-based clipper that will yield the voltage transfer function shown in Figure 2.24. Assume diode cut-in voltages of $V_\gamma = 0.7$ V. (Ans. For Figure 2.23(b), $V_2 = 4.3$, $V_1 = 1.8$ V, and $R_1 = 2R_2$)

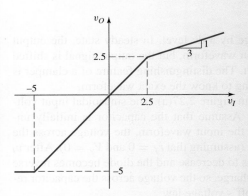

**Figure 2.24** Figure for Exercise Ex 2.7

Diode clipper circuits can also be designed such that the dc power supply is in series with the input signal. Figure 2.25 shows one example. The battery in series with the input signal causes the input signal to be superimposed on the $V_B$ dc voltage. The resulting conditioned input signal and corresponding output signal is also shown in Figure 2.25.

In all of the clipper circuits considered, we have included batteries that basically set the limits of the output voltage. However, batteries need periodic replacement, so that these circuits are not practical. Zener diodes, operated in the reverse breakdown region, provide essentially a constant voltage drop. We can replace the batteries by Zener diodes.

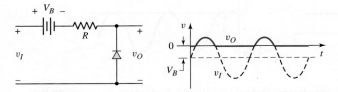

**Figure 2.25** Series-based diode clipper circuit and resulting output response

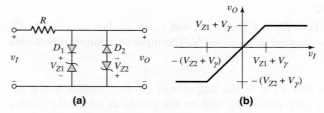

**Figure 2.26** (a) Parallel-based clipper circuit using Zener diodes; (b) voltage transfer characteristics

Figure 2.26(a) shows a parallel based clipper circuit using Zener diodes. The voltage transfer characteristics are shown in Figure 2.26(b). The performance of the circuit in Figure 2.26(a) is essentially the same as that shown in Figure 2.22.

### 2.3.2    Clampers

Clamping shifts the entire signal voltage by a dc level. In steady state, the output waveform is an exact replica of the input waveform, but the output signal is shifted by a dc value that depends on the circuit. The distinguishing feature of a clamper is that it adjusts the dc level without needing to know the exact waveform.

An example of clamping is shown in Figure 2.27(a). The sinusoidal input voltage signal is shown in Figure 2.27(b). Assume that the capacitor is initially uncharged. During the first 90 degrees of the input waveform, the voltage across the capacitor follows the input, and $v_C = v_I$ (assuming that $r_f = 0$ and $V_\gamma = 0$). After $v_I$ and $v_C$ reach their peak values, $v_I$ begins to decrease and the diode becomes reverse biased. Ideally, the capacitor cannot discharge, so the voltage across the capacitor remains constant at $v_C = V_M$. By Kirchhoff's voltage law

$$v_O = -v_C + v_I = -V_M + V_M \sin \omega t \tag{2.36(a)}$$

or

$$v_O = V_M(\sin \omega t - 1) \tag{2.36(b)}$$

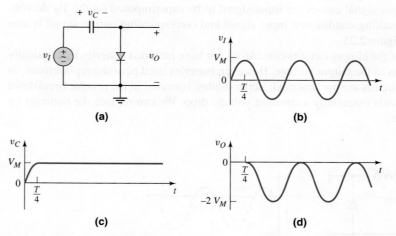

**Figure 2.27** Action of a diode clamper circuit: (a) a typical diode clamper circuit, (b) the sinusoidal input signal, (c) the capacitor voltage, and (d) the output voltage

The capacitor and output voltages are shown in Figures 2.27(c) and (d). The output voltage is "clamped" at zero volts, that is, $v_O \leq 0$. In steady state, the waveshapes of the input and output signals are the same, and the output signal is shifted by a certain dc level compared to the input signal.

A clamping circuit that includes an independent voltage source $V_B$ is shown in Figure 2.28(a). In this circuit, the $R_L C$ time constant is assumed to be large, where $R_L$ is the load resistance connected to the output. If we assume, for simplicity, that $r_f = 0$ and $V_\gamma = 0$, then the output is clamped at $V_B$. Figure 2.28(b) shows an example of a sinusoidal input signal and the resulting output voltage signal. When the polarity of $V_B$ is as shown, the output is shifted in a negative voltage direction. Similarly, Figure 2.28(c) shows a square-wave input signal and the resulting output voltage signal. For the square-wave signal, we have neglected the diode capacitance effects and assume the voltage can change instantaneously.

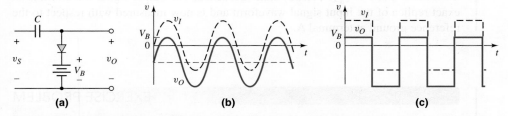

**(a)**                    **(b)**                    **(c)**

**Figure 2.28**  Action of a diode clamper circuit with a voltage source assuming an ideal diode ($V_r = 0$): (a) the circuit, (b) steady-state sinusoidal input and output signals, and (c) steady-state square-wave input and output signals

Electronic signals tend to lose their dc levels during signal transmission. For example, the dc level of a TV signal may be lost during transmission, so that the dc level must be restored at the TV receiver. The following example illustrates this effect.

## EXAMPLE 2.8

**Objective:** Find the steady-state output of the diode-clamper circuit shown in Figure 2.29(a).

The input $v_I$ is assumed to be a sinusoidal signal whose dc level has been shifted with respect to a receiver ground by a value $V_B$ during transmission. Assume $V_\gamma = 0$ and $r_f = 0$ for the diode.

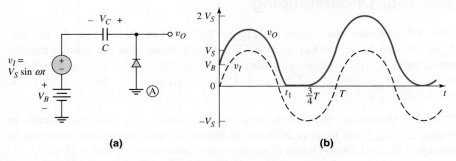

**(a)**                         **(b)**

**Figure 2.29**  (a) Circuit for Example 2.8; (b) input and output waveforms

**Solution:** Figure 2.29(b) shows the sinusoidal input signal. If the capacitor is initially uncharged, then the output voltage is $v_O = V_B$ at $t = 0$ (diode reverse-biased). For $0 \leq t \leq t_1$, the effective $RC$ time constant is infinite, the voltage across the capacitor does not change, and $v_O = v_I + V_B$.

At $t = t_1$, the diode becomes forward biased; the output cannot go negative, so the voltage across the capacitor changes (the $r_f C$ time constant is zero).

At $t = (\frac{3}{4})T$, the input signal begins increasing and the diode becomes reverse biased, so the voltage across the capacitor now remains constant at $V_C = V_S - V_B$ with the polarity shown. The output voltage is now given by

$$v_O = (V_S - V_B) + v_I + V_B = (V_S - V_B) + V_S \sin \omega t + V_B$$

or

$$v_O = V_S(1 + \sin \omega t)$$

**Comment:** For $t > (\frac{3}{4})T$, steady state is reached. The output signal waveform is an exact replica of the input signal waveform and is now measured with respect to the reference ground at terminal A.

EXERCISE PROBLEM

**Ex 2.8:** Sketch the steady-state output voltage for the input signal given for the circuit shown in Figure 2.30. Assume $V_\gamma = r_f = 0$. (Ans. Square wave between $+2\,\text{V}$ and $-8\,\text{V}$)

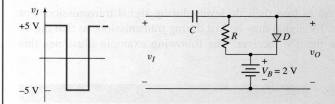

**Figure 2.30**  Figure for Exercise Ex 2.8

## Test Your Understanding

**TYU 2.8** Consider the circuit in Figure 2.23(a). Let $R_1 = 5\,\text{k}\Omega$, $R_2 = 2\,\text{k}\Omega$, $V_1 = 1\,\text{V}$, and $V_2 = 3\,\text{V}$. Let $V_\gamma = 0.7\,\text{V}$ for each diode. Plot the voltage transfer characteristics ($v_O$ versus $v_I$) for $-5 \leq v_I \leq 5\,\text{V}$. (Ans. For $v_I \leq -3.7\,\text{V}$, $v_O = -3.7\,\text{V}$; for $-3.7 \leq v_I \leq 1.7\,\text{V}$, $v_O = v_I$; for $v_I \geq 1.7\,\text{V}$, $v_O = 0.286\,v_I + 1.21$)

**TYU 2.9** Determine the steady-state output voltage $v_O$ for the circuit in Figure 2.31(a), if the input is as shown in Figure 2.31(b). Assume the diode cut-in voltage is $V_\gamma = 0$. (Ans. Output is a square wave between $+5\,\text{V}$ and $+35\,\text{V}$)

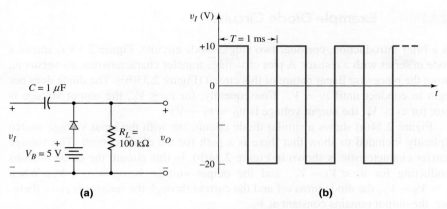

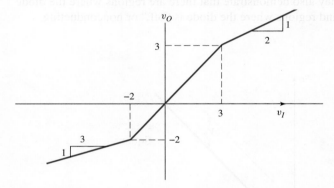

**Figure 2.31** Figure for Exercise TYU 2.9: (a) the circuit and (b) input signal

**Figure 2.32** Figure for Exercise TYU 2.10

**TYU 2.10** Design a parallel-based clipper circuit that will yield the voltage transfer characteristics shown in Figure 2.32. Assume a diode cut-in voltage of $V_\gamma = 0.7$ V. (Ans. From Figure 2.23(a), $V_1 = 2.3$ V, $V_2 = 1.3$ V, $R_1 = R_2$, include $R_3$ in series with $D_2$, where $R_3 = 0.5R_1$)

## 2.4    MULTIPLE-DIODE CIRCUITS

**Objective: •** Examine the techniques used to analyze circuits that contain more than one diode.

Since a diode is a nonlinear device, part of the analysis of a diode circuit involves determining whether the diode is on or off. If a circuit contains more than one diode, the analysis is complicated by the various possible combinations of on and off.

In this section, we will look at several multiple-diode circuits. We will see, for example, how diode circuits can be used to perform logic functions. This section serves as an introduction to digital logic circuits that will be considered in detail in Chapters 16 and 17.

2.4.1    **Example Diode Circuits**

As a brief introduction, consider two single-diode circuits. Figure 2.33(a) shows a diode in series with a resistor. A plot of voltage transfer characteristics, $v_O$ versus $v_I$, shows the piecewise linear nature of this circuit (Figure 2.33(b)). The diode does not begin to conduct until $v_I = V_\gamma$. Consequently, for $v_I \leq V_\gamma$, the output voltage is zero; for $v_I > V_r$, the output voltage is $v_O = v_I - V\gamma$.

Figure 2.34(a) shows a similar diode circuit, but with the input voltage source explicitly included to show that there is a path for the diode current. The voltage transfer characteristic is shown in Figure 2.34(b). In this circuit, the diode remains conducting for $v_I < V_S - V_\gamma$, and the output voltage is $v_O = v_I + V_\gamma$. When $v_I > V_S - V_\gamma$, the diode turns off and the current through the resistor is zero; therefore, the output remains constant at $V_S$.

These two examples demonstrate the piecewise linear nature of the diode and the diode circuit. They also demonstrate that there are regions where the diode is "on," or conducting, and regions where the diode is "off," or nonconducting.

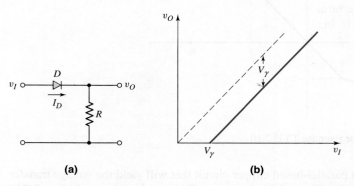

**Figure 2.33**  Diode and resistor in series: (a) circuit and (b) voltage transfer characteristics

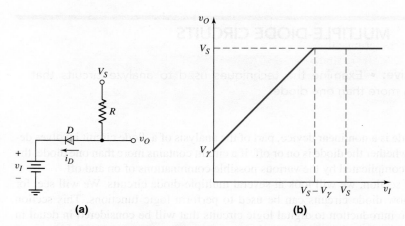

**Figure 2.34**  Diode with input voltage source: (a) circuit and (b) voltage transfer characteristics

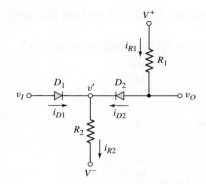

Figure 2.35  A two-diode circuit

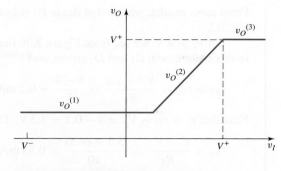

Figure 2.36  Voltage transfer characteristics for the two-diode circuit in Figure 2.35

In multidiode circuits, each diode may be either on or off. Consider the two-diode circuit in Figure 2.35. Since each diode may be either on or off, the circuit has four possible states. However, some of these states may not be feasible because of diode directions and voltage polarities.

If we assume that $V^+ > V^-$ and that $V^+ - V^- > V_\gamma$, there is at least a possibility that $D_2$ can be turned on. First, $v'$ cannot be less than $V^-$. Then, for $v_I = V^-$, diode $D_1$ must be off. In this case, $D_2$ is on, $i_{R1} = i_{D2} = i_{R2}$, and

$$v_O = V^+ - i_{R1}R_1 \tag{2.37}$$

where

$$i_{R1} = \frac{V^+ - V_\gamma - V^-}{R_1 + R_2} \tag{2.38}$$

Voltage $v'$ is one diode drop below $v_O$, and $D_1$ remains off as long as $v_I$ is less than the output voltage. As $v_I$ increases and becomes equal to $v_O$, both $D_1$ and $D_2$ turn on. This condition or state is valid as long as $v_I < V^+$. When $v_I = V^+$, $i_{R1} = i_{D2} = 0$, at which point $D_2$ turns off and $v_O$ cannot increase any further.

Figure 2.36 shows the resulting plot of $v_O$ versus $v_I$. Three distinct regions, $v_O^{(1)}$, $v_O^{(2)}$, and $v_O^{(3)}$, correspond to the various conducting states of $D_1$ and $D_2$. The fourth possible state, corresponding to both $D_1$ and $D_2$ being off, is not feasible in this circuit.

## EXAMPLE 2.9

**Objective:** Determine the output voltage and diode currents for the circuit shown in Figure 2.35, for two values of input voltage.

Assume the circuit parameters are $R_1 = 5\,\text{k}\Omega$, $R_2 = 10\,\text{k}\Omega$, $V_\gamma = 0.7\,\text{V}$, $V^+ = +5\,\text{V}$, and $V^- = -5\,\text{V}$. Determine $v_O$, $i_{D1}$, and $i_{D2}$ for $v_I = 0$ and $v_I = 4\,\text{V}$.

**Solution:** For $v_I = 0$, assume initially that $D_1$ is off. The currents are then

$$i_{R1} = i_{D2} = i_{R2} = \frac{V^+ - V_\gamma - V^-}{R_1 + R_2} = \frac{5 - 0.7 - (-5)}{5 + 10} = 0.62\,\text{mA}$$

The output voltage is

$$v_O = V^+ - i_{R1}R_1 = 5 - (0.62)(5) = 1.9\,\text{V}$$

and $v'$ is

$$v' = v_O - V_\gamma = 1.9 - 0.7 = 1.2\,\text{V}$$

From these results, we see that diode $D_1$ is indeed cut off, $i_{D1} = 0$, and our analysis is valid.

For $v_I = 4$ V, we see from Figure 2.36 that $v_O = v_I$; therefore, $v_O = v_I = 4$ V. In this region, both $D_1$ and $D_2$ are on, and

$$i_{R1} = i_{D2} = \frac{V^+ - v_O}{R_1} = \frac{5 - 4}{5} = 0.2 \,\text{mA}$$

Note that $v' = v_O - V_\gamma = 4 - 0.7 = 3.3$ V. Thus,

$$i_{R2} = \frac{v' - V^-}{R_2} = \frac{3.3 - (-5)}{10} = 0.83 \,\text{mA}$$

The current through $D_1$ is found from $i_{D1} + i_{D2} = i_{R2}$ or

$$i_{D1} = i_{R2} - i_{D2} = 0.83 - 0.2 = 0.63 \,\text{mA}$$

**Comment:** For $v_I = 0$, we see that $v_O = 1.9$ V and $v' = 1.2$ V. This means that $D_1$ is reverse biased, or off, as we initially assumed. For $v_I = 4$ V, we have $i_{D1} > 0$ and $i_{D2} > 0$, indicating that both $D_1$ and $D_2$ are forward biased, as we assumed.

**Computer analysis:** For multidiode circuits, a PSpice analysis may be useful in determining the conditions under which the various diodes are conducting or not conducting. This avoids guessing the conducting state of each diode in a hand analysis. Figure 2.37 is the PSpice circuit schematic of the diode circuit in Figure 2.35.

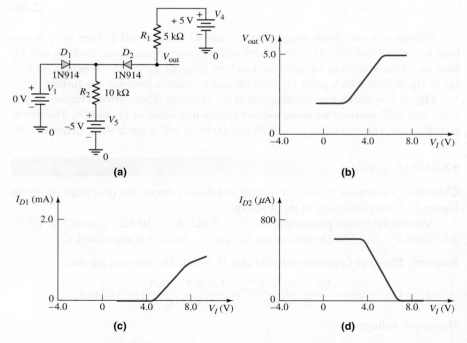

**Figure 2.37** (a) PSpice circuit schematic; (b) output voltage; (c) current in diode 1, and (d) current in diode 2 for the diode circuit in Example 2.9

Figure 2.37 also shows the output voltage and the two diode currents as the input is varied between $-1$ V and $+7$ V. From these curves, we can determine when the diodes turn on and off.

**Comment:** The hand analysis results, based on the piecewise linear model for the diode, agree very well with the computer simulation results. This gives us confidence in the piecewise linear model when quick hand calculations are made.

---

### EXERCISE PROBLEM

**Ex 2.9:** Consider the circuit shown in Figure 2.38, in which the diode cut-in voltages are $V_\gamma = 0.6$ V. Plot $v_O$ versus $v_I$ for $0 \leq v_I \leq 10$ V. (Ans. For $0 \leq v_I \leq 3.5$ V, $v_O = 4.4$ V; for $v_I > 3.5$ V, $D_2$ turns off; and for $v_I \geq 9.4$ V, $v_O = 10$ V)

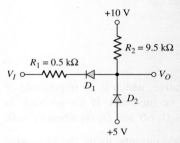

**Figure 2.38** Figure for Exercise Ex 2.9

---

### Problem-Solving Technique: Multiple Diode Circuits

Analyzing multidiode circuits requires determining if the individual devices are "on" or "off." In many cases, the choice is not obvious, so we must initially guess the state of each device, then analyze the circuit to determine if we have a solution consistent with our initial guess. To do this, we can:

1.  Assume the state of a diode. If a diode is assumed on, the voltage across the diode is assumed to be $V_\gamma$. If a diode is assumed to be off, the current through the diode is assumed to be zero.
2.  Analyze the "linear" circuit with the assumed diode states.
3.  Evaluate the resulting state of each diode. If the initial assumption were that a diode is "off" and the analysis shows that $I_D = 0$ and $V_D \leq V_\gamma$, then the assumption is correct. If, however, the analysis actually shows that $I_D > 0$ and/or $V_D > V_\gamma$, then the initial assumption is incorrect. Similarly, if the initial assumption were that a diode is "on" and the analysis shows that $I_D \geq 0$ and $V_D = V_\gamma$, then the initial assumption is correct. If, however, the analysis shows that $I_D < 0$ and/or $V_D < V_\gamma$, then the initial assumption is incorrect.
4.  If any initial assumption is proven incorrect, then a new assumption must be made and the new "linear" circuit must be analyzed. Step 3 must then be repeated.

## EXAMPLE 2.10

**Objective:** Demonstrate how inconsistencies develop in a solution with incorrect assumptions.

For the circuit shown in Figure 2.35, assume that parameters are the same as those given in Example 2.9. Determine $v_O$, $i_{D1}$, $i_{D2}$, and $i_{R2}$ for $v_I = 0$.

**Solution:** Assume initially that both $D_1$ and $D_2$ are conducting (i.e., on). Then, $v' = -0.7$ V and $v_O = 0$. The two currents are

$$i_{R1} = i_{D2} = \frac{V^+ - v_O}{R_1} = \frac{5 - 0}{5} = 1.0 \text{ mA}$$

and

$$i_{R2} = \frac{v' - V^-}{R_2} = \frac{-0.7 - (-5)}{10} = 0.43 \text{ mA}$$

Summing the currents at the $v'$ node, we find that

$$i_{D1} = i_{R2} - i_{D2} = 0.43 - 1.0 = -0.57 \text{ mA}$$

Since this analysis shows the $D_1$ current to be negative, which is an impossible or inconsistent solution, our initial assumption must be incorrect. If we go back to Example 2.9, we will see that the correct solution is $D_1$ off and $D_2$ on when $v_I = 0$.

**Comment:** We can perform linear analyses on diode circuits, using the piecewise linear model. However, we must first determine if each diode in the circuit is operating in the "on" linear region or the "off" linear region.

---

### EXERCISE PROBLEM

**Ex 2.10:** Consider the circuit shown in Figure 2.39. The cut-in voltage of each diode is $V_\gamma = 0.7$ V. (a) Let $v_I = 5$ V. Assume both diodes are conducting. Is this a correct assumption? Why or why not? Determine $I_{R1}$, $I_{D1}$, $I_{D2}$, and $v_O$. (b) Repeat part (a) for $v_I = 10$ V. (Ans. (a) $D_1$ is off, $I_{D1} = 0$, $I_{R1} = I_{D2} = 0.754$ mA, $v_O = 3.72$ V; (b) $I_{D1} = 0.9$ mA, $I_{D2} = 1.9$ mA, $I_{R1} = 1.0$ mA, $v_O = 8.3$ V)

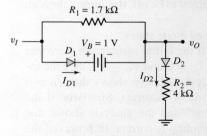

**Figure 2.39** Figure for Exercise Ex 2.10

---

## EXAMPLE 2.11

**Objective:** Determine the current in each diode and the voltages $V_A$ and $V_B$ in the multidiode circuit shown in Figure 2.40. Let $V_\gamma = 0.7$ V for each diode.

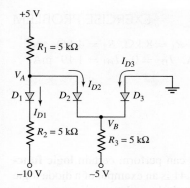

**Figure 2.40** Diode circuit for Example 2.11

**Solution:** Initially assume each diode is in its conducting state. Starting with $D_3$ and considering the voltages, we see that

$$V_B = -0.7\text{ V} \qquad \text{and} \qquad V_A = 0$$

Summing currents at the $V_A$ node, we find

$$\frac{5 - V_A}{5} = I_{D2} + \frac{(V_A - 0.7) - (-10)}{5}$$

Since $V_A = 0$, we obtain

$$\frac{5}{5} = I_{D2} + \frac{9.3}{5} \Rightarrow I_{D2} = -0.86\text{ mA}$$

which is inconsistent with the assumption that all diodes are "on" (an "on" diode would have a positive diode current).

Now assume that $D_1$ and $D_3$ are on and $D_2$ is off. We see that

$$I_{D1} = \frac{5 - 0.7 - (-10)}{5 + 5} = 1.43\text{ mA}$$

and

$$I_{D3} = \frac{(0 - 0.7) - (-5)}{5} = 0.86\text{ mA}$$

We find the voltages as

$$V_B = -0.7\text{ V}$$

and

$$V_A = 5 - (1.43)(5) = -2.15\text{ V}$$

From the values of $V_A$ and $V_B$, the diode $D_2$ is indeed reverse biased and off, so $I_{D2} = 0$.

**Comment:** With more diodes in a circuit, the number of combinations of diodes being either on or off increases, which may increase the number of times a circuit must be analyzed before a correct solution is obtained. In the case of multiple diode circuits, a computer simulation might save time.

**Ex 2.11:** Repeat Example 2.11 for the case when $R_1 = 8$ k$\Omega$, $R_2 = 4$ k$\Omega$, and $R_3 = 2$k$\Omega$. (Ans. $V_B = -0.7$ V, $I_{D3} = 2.15$ mA, $I_{D2} = 0$, $I_{D1} = 1.19$ mA, $V_A = -4.53$ V)

### 2.4.2    Diode Logic Circuits

Diodes in conjunction with other circuit elements can perform certain **logic functions,** such as AND and OR. The circuit in Figure 2.41 is an example of a diode logic circuit. The four conditions of operation of this circuit depend on various combinations of input voltages. If $V_1 = V_2 = 0$, there is no excitation to the circuit so both diodes are off and $V_O = 0$. If at least one input goes to 5 V, for example, at least one diode turns on and $V_O = 4.3$ V, assuming $V_\gamma = 0.7$ V.

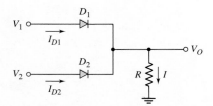

**Figure 2.41**  A two-input diode OR logic circuit

These results are shown in Table 2.1. By definition, in a positive logic system, a voltage near zero corresponds to a logic 0 and a voltage close to the supply voltage of 5 V corresponds to a logic 1. The results shown in Table 2.1 indicate that this circuit performs the OR logic function. The circuit of Figure 2.41, then, is a two-input diode OR logic circuit.

Next, consider the circuit in Figure 2.42. Assume a diode cut-in voltage of $V_\gamma = 0.7$ V. Again, there are four possible states, depending on the combination of input voltages. If at least one input is at zero volts, then at least one diode is conducting and $V_O = 0.7$ V. If both $V_1 = V_2 = 5$ V, there is no potential difference between the supply voltage and the input voltage. All currents are zero and $V_O = 5$ V.

These results are shown in Table 2.2. This circuit performs the AND logic function. The circuit of Figure 2.42 is a two-input diode AND logic circuit.

| Table 2.1 | Two-diode OR logic circuit response | |
| --- | --- | --- |
| $V_1$ (V) | $V_2$ (V) | $V_o$ (V) |
| 0 | 0 | 0 |
| 5 | 0 | 4.3 |
| 0 | 5 | 4.3 |
| 5 | 5 | 4.3 |

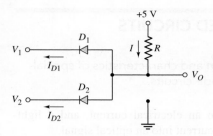

**Figure 2.42** A two-input diode AND logic circuit

If we examine Tables 2.1 and 2.2, we see that the input "low" and "high" voltages may not be the same as the output "low" and "high" voltages. As an example, for the AND circuit (Table 2.2), the input "low" is 0 V, but the output "low" is 0.7 V. This can create a problem because the output of one logic gate is often the input to another logic gate. Another problem occurs when diode logic circuits are connected in cascade; that is, the output of one OR gate is connected to the input of a second OR gate. The logic 1 levels of the two OR gates are not the same (see Problems 2.61 and 2.62). The logic 1 level degrades or decreases as additional logic gates are connected. However, these problems may be overcome with the use of amplifying devices (transistors) in digital logic systems.

| Table 2.2 | Two-diode AND logic circuit response | |
|---|---|---|
| $V_1$ (V) | $V_2$ (V) | $V_o$ (V) |
| 0 | 0 | 0.7 |
| 5 | 0 | 0.7 |
| 0 | 5 | 0.7 |
| 5 | 5 | 5 |

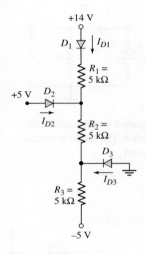

**Figure 2.43** Figure for Exercise TYU 2.11

## Test Your Understanding

**TYU 2.11** The cut-in voltage of each diode in the circuit shown in Figure 2.43 is $V_\gamma = 0.7$ V. Determine $I_{D1}$, $I_{D2}$, $I_{D3}$, $V_A$, and $V_B$. (Ans. $I_{D1} = 1.22$ mA, $I_{D2} = I_{D3} = 0$, $V_A = 7.2$ V, $V_B = 1.1$ V)

**TYU 2.12** Repeat Exercise TYU 2.11 for $R_1 = 8$ k$\Omega$, $R_2 = 12$ k$\Omega$, and $R_3 = 2.5$ k$\Omega$. (Ans. $I_{D1} = 0.7$ mA, $I_{D2} = 0$, $I_{D3} = 1.02$ mA, $V_A = 7.7$ V, $V_B = -0.7$ V)

**TYU 2.13** Consider the OR logic circuit shown in Figure 2.41. Assume a diode cut-in voltage of $V_\gamma = 0.6$ V. (a) Plot $V_O$ versus $V_1$ for $0 \le V_1 \le 5$ V, if $V_2 = 0$. (b) Repeat part (a) if $V_2 = 3$ V. (Ans. (a) $V_O = 0$ for $V_1 \le 0.6$ V, $V_O = V_1 - 0.6$ for $0.6 \le V_1 \le 5$ V; (b) $V_O = 2.4$ V for $0 \le V_1 \le 3$ V, $V_O = V_1 - 0.6$ for $3 \le V_1 \le 5$ V)

**TYU 2.14** Consider the AND logic circuit shown in Figure 2.42. Assume a diode cut-in voltage of $V_\gamma = 0.6$ V. (a) Plot $V_O$ versus $V_1$ for $0 \le V_1 \le 5$ V, if $V_2 = 0$. (b) Repeat part (a) if $V_2 = 3$ V. (Ans. (a) $V_O = 0.6$ V for all $V_1$, (b) $V_O = V_1 + 0.6$ for $0 \le V_1 \le 3$ V, $V_O = 3.6$ V for $V_1 \ge 3$ V)

**Objective:** • Understand the operation and characteristics of specialized photodiode and light-emitting diode circuits.

A photodiode converts an optical signal into an electrical current, and a light-emitting diode (LED) transforms an electrical current into an optical signal.

### 2.5.1    Photodiode Circuit

Figure 2.44 shows a typical photodiode circuit in which a reverse-bias voltage is applied to the photodiode. If the photon intensity is zero, the only current through the diode is the reverse-saturation current, which is normally very small. Photons striking the diode create excess electrons and holes in the space-charge region. The electric field quickly separates these excess carriers and sweeps them out of the space-charge region, thus creating a **photocurrent** in the reverse-bias direction. The photocurrent is

$$I_{ph} = \eta e \Phi A \qquad\qquad (2.39)$$

where $\eta$ is the quantum efficiency, $e$ is the electronic charge, $\Phi$ is the photon flux density (#/cm$^2$−s), and $A$ is the junction area. This linear relationship between photocurrent and photon flux is based on the assumption that the reverse-bias voltage across the diode is constant. This in turn means that the voltage drop across $R$ induced by the photocurrent must be small, or that the resistance $R$ is small.

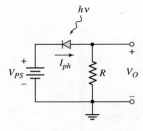

**Figure 2.44** A photodiode circuit. The diode is reverse biased

### EXAMPLE 2.12

**Objective:** Calculate the photocurrent generated in a photodiode.

For the photodiode shown in Figure 2.44 assume the quantum efficiency is 1, the junction area is $10^{-2}$ cm$^2$, and the incident photon flux is $5 \times 10^{17}$ cm$^{-2} - $s$^{-1}$.

**Solution:** From Equation (2.39), the photocurrent is

$$I_{ph} = \eta e \Phi A = (1)(1.6 \times 10^{-19})(5 \times 10^{17})(10^{-2}) \Rightarrow 0.8 \, \text{mA}$$

**Comment:** The incident photon flux is normally given in terms of light intensity, in lumens, foot-candles, or W/cm$^2$. The light intensity includes the energy of the photons, as well as the photon flux.

### EXERCISE PROBLEM

**Ex 2.12:** (a) Photons with an energy of $h\nu = 2$ eV are incident on the photodiode shown in Figure 2.44. The junction area is $A = 0.5$ cm$^2$, the quantum efficiency is $\eta = 0.8$, and the light intensity is $6.4 \times 10^{-2}$ W/cm$^2$. Determine the photocurrent $I_{ph}$. (b) If $R = 1$ k$\Omega$, determine the minimum power supply voltage $V_{PS}$ needed to ensure that the diode is reverse biased. (Ans. (a) $I_{ph} = 12.8$ mA, (b) $V_{PS}(\text{min}) = 12.8$ V)

2.5.2 ## LED Circuit

A light-emitting diode (LED) is the inverse of a photodiode; that is, a current is converted into an optical signal. If the diode is forward biased, electrons and holes are injected across the space-charge region, where they become excess minority carriers. These excess minority carriers diffuse into the neutral n- and p-regions, where they recombine with majority carriers, and the recombination can result in the emission of a photon.

LEDs are fabricated from compound semiconductor materials, such as gallium arsenide or gallium arsenide phosphide. These materials are direct-bandgap semiconductors. Because these materials have higher bandgap energies than silicon, the forward-bias junction voltage is larger than that in silicon-based diodes.

It is common practice to use a seven-segment LED for the numeric readout of digital instruments, such as a digital voltmeter. The **seven-segment display** is sketched in Figure 2.45. Each segment is an LED normally controlled by IC logic gates.

Figure 2.46 shows one possible circuit connection, known as a common-anode display. In this circuit, the anodes of all LEDs are connected to a 5 V source and the inputs are controlled by logic gates. If $V_{I1}$ is "high," for example, $D_1$ is off and there is no light output. When $V_{I1}$ goes "low," $D_1$ becomes forward biased and produces a light output.

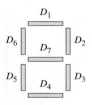

**Figure 2.45** Seven-segment LED display

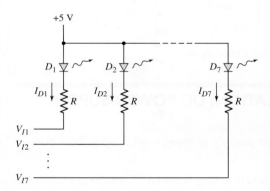

**Figure 2.46**  Control circuit for the seven-segment LED display

---

## EXAMPLE 2.13

**Objective:** Determine the value of $R$ required to limit the current in the circuit in Figure 2.46 when the input is in the low state.

Assume that a diode current of 10 mA produces the desired light output, and that the corresponding forward-bias voltage drop is 1.7 V.

**Solution:** If $V_I = 0.2$ V in the "low" state, then the diode current is

$$I = \frac{5 - V_\gamma - V_I}{R}$$

The resistance $R$ is then determined as

$$R = \frac{5 - V_\gamma - V_I}{I} = \frac{5 - 1.7 - 0.2}{10} \Rightarrow 310 \ \Omega$$

**Comment:** Typical LED current-limiting resistor values are in the range of 300 to 350 $\Omega$.

**Ex 2.13:** Determine the value of resistance $R$ required to limit the current in the circuit shown in Figure 2.46 to $I = 15$ mA. Assume $V_\gamma = 1.7$ V, $r_f = 15 \ \Omega$, and $V_I = 0.2$ V in the "low" state. (Ans. $R = 192 \ \Omega$)

One application of LEDs and photodiodes is in **optoisolators,** in which the input signal is electrically decoupled from the output (Figure 2.47). An input signal applied to the LED generates light, which is subsequently detected by the photodiode. The photodiode then converts the light back to an electrical signal. There is no electrical feedback or interaction between the output and input portions of the circuit.

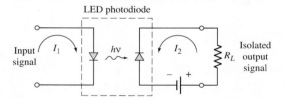

**Figure 2.47** Optoisolator using an LED and a photodiode

## 2.6    DESIGN APPLICATION: DC POWER SUPPLY

**Objective:** • Design a dc power supply to meet a set of specifications.

**Specifications:** The output load current is to vary between 25 and 50 mA while the output voltage is to remain in the range $12 \le v_O \le 12.2$ V.

**Design Approach:** The circuit configuration to be designed is shown in Figure 2.48. A diode bridge circuit with an $RC$ filter will be used and a Zener diode will be in parallel with the output load.

**Choices:** An ac input voltage with an rms value in the range $110 \le v_I \le 120$ V and at 60 Hz is available. A Zener diode with a Zener voltage of $V_{ZO} = 12$ V and a Zener resistance of 2 $\Omega$ that can operate over a current range of $10 \le I_Z \le 100$ mA is available. Also, a transformer with an 8:1 turns ratio is available.

**Solution:** With an 8:1 transformer turns ratio, the peak value of $v_S$ is in the range $19.4 \le v_S \le 21.2$ V. Assuming diode turn-on voltages of $V_\gamma = 0.7$ V, the peak value of $v_{O1}$ is in the range $18.0 \le v_{O1} \le 19.8$ V.

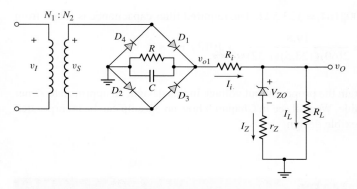

**Figure 2.48** DC power supply circuit for design application

For $v_{O1}(\max)$ and minimum load current, let $I_Z = 90\,\text{mA}$. Then

$$v_O = V_{ZO} + I_Z r_z = 12 + (0.090)(2) = 12.18\,\text{V}$$

The input current is

$$I_i = I_Z + I_L = 90 + 25 = 115\,\text{mA}$$

The input resistance $R_i$ must then be

$$R_i = \frac{v_{O1} - v_O}{I_i} = \frac{19.8 - 12.18}{0.115} = 66.3\,\Omega$$

The minimum Zener current occurs for $I_L(\max)$ and $v_{O1}(\min)$. The voltage $v_{O1}(\min)$ occurs for $v_S(\min)$ and must also take into account the ripple voltage. Let $I_Z(\min) = 20\,\text{mA}$. Then the output voltage is

$$v_O = V_{ZO} + I_Z r_Z = 12 + (0.020)(2) = 12.04\,\text{V}$$

The output voltage is within the specified range of output voltage.

We now find

$$I_i = I_Z + I_L = 20 + 50 = 70\,\text{mA}$$

and

$$v_{O1}(\min) = I_i R_i + v_O = (0.070)(66.3) + 12.04$$

or

$$v_{O1}(\min) = 16.68\,\text{V}$$

The minimum ripple voltage of the filter is then

$$V_r = v_S(\min) - 1.4 - v_{O1}(\min) = 19.4 - 1.4 - 16.68$$

or

$$V_r = 1.32\,\text{V}$$

Now, let $R_1 = 500\,\Omega$. The effective resistance to ground from $v_{O1}$ is $R_1 \| R_{i,\text{eff}}$ where $R_{i,\text{eff}}$ is the effective resistance to ground through $R_i$ and the other circuit elements. We can approximate

$$R_{i,\text{eff}} \approx \frac{v_S(\text{avg}) - 1.4}{I_i(\max)} = \frac{20.3 - 1.4}{0.115} = 164\,\Omega$$

Then $R_1 \| R_{i,\text{eff}} = 500 \| 164 = 123.5 \ \Omega$. The required filter capacitance is found from

$$C = \frac{V_M}{2fRV_r} = \frac{19.8}{2(60)(123.5)(1.32)} \Rightarrow 1012 \ \mu\text{F}$$

**Comments:** To obtain the proper output voltage in this design, an appropriate Zener diode must be available. We will see in Chapter 9 how an op-amp can be incorporated to provide a more flexible design.

 ## 2.7    SUMMARY

- Diode circuits, taking advantage of the nonlinear $i$–$v$ characteristics of the pn junction, were analyzed and designed in this chapter.
- Half-wave and full-wave rectifier circuits convert a sinusoidal (i.e., ac) signal to an approximate dc signal. A dc power supply, which is used to bias electronic circuits and systems, utilizes these types of circuits. An $RC$ filter can be connected to the output of the rectifier circuit to reduce the ripple effect.
- Zener diodes operate in the reverse-breakdown region and are used in voltage reference or regulator circuits. The percent regulation, a figure of merit for regulator circuits, was defined and determined for various regulator circuits.
- Techniques used to analyze multidiode circuits were developed. The technique requires making assumptions as to whether a diode is conducting (on) or not conducting (off). After analyzing the circuit using these assumptions, we must go back and verify that the assumptions made were valid.
- Diode circuits can be designed to perform basic digital logic functions, such as the AND and OR function. However, there are some inconsistencies between input and output logic values as well as some loading effects, which will severely limit the use of diode logic gates as stand-alone circuits.
- The LED converts electrical current to light and is used extensively in such applications as the seven-segment alphanumeric display. Conversely, the photodiode detects an incident light signal and transforms it into an electrical current.
- As an application, a simple dc power supply was designed using a rectifier circuit in conjunction with a Zener diode.

 ## CHECKPOINT

After studying this chapter, the reader should have the ability to:

- ✓ In general, apply the diode piecewise linear model in the analysis of diode circuits.
- ✓ Analyze diode rectifier circuits, including the calculation of ripple voltage.
- ✓ Analyze Zener diode circuits, including the effect of a Zener resistance.
- ✓ Determine the output signal for a given input signal of diode clipper and clamper circuits.
- ✓ Analyze circuits with multiple diodes by making initial assumptions and then verifying these initial assumptions.

##  REVIEW QUESTIONS

1. What characteristic of a diode is used in the design of diode signal processing circuits?
2. Describe a simple half-wave diode rectifier circuit and sketch the output voltage versus time.
3. Describe a simple full-wave diode rectifier circuit and sketch the output voltage versus time.
4. What is the advantage of connecting an $RC$ filter to the output of a diode rectifier circuit?
5. Define ripple voltage. How can the magnitude of the ripple voltage be reduced?
6. Describe a simple Zener diode voltage reference circuit.
7. What effect does the Zener diode resistance have on the voltage reference circuit operation? Define load regulation.
8. What are the general characteristics of diode clipper circuits?
9. Describe a simple diode clipper circuit that limits the negative portion of a sinusoidal input voltage to a specified value.
10. What are the general characteristics of diode clamper circuits?
11. What one circuit element, besides a diode, is present in all diode clamper circuits?
12. Describe the procedure used in the analysis of a circuit containing two diodes. How many initial assumptions concerning the state of the circuit are possible?
13. Describe a diode OR logic circuit. Compare a logic 1 value at the output compared to a logic 1 value at the input. Are they the same value?
14. Describe a diode AND logic circuit. Compare a logic 0 value at the output compared to a logic 0 value at the input. Are they the same value?
15. Describe a simple circuit that can be used to turn an LED on or off with a high or low input voltage.

##  PROBLEMS

[Note: In the following problems, assume $r_f = 0$ unless otherwise specified.]

### Section 2.1  Rectifier Circuits

2.1 Consider the circuit shown in Figure P2.1. Let $R = 1\,k\Omega$, $V_\gamma = 0.6$ V, and $r_f = 20\,\Omega$. (a) Plot the voltage transfer characteristics $v_O$ versus $v_I$ over the range $-10 \le v_I \le 10$ V. (b) Assume $v_I = 10\sin\omega t$ (V). (i) Sketch $v_O$ versus time for the sinusoidal input. (ii) Find the average value of $v_O$. (iii) Determine the peak diode current. (iv) What is the PIV of the diode?

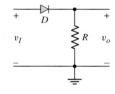

**Figure P2.1**

2.2 For the circuit shown in Figure P2.1, show that for $v_I \ge 0$, the output voltage is approximately given by

$$v_O = v_I - V_T \ln\left(\frac{v_O}{I_S R}\right)$$

2.3 A half-wave rectifier such as shown in Figure 2.2(a) has a 2 k$\Omega$ load. The input is a 120 V (rms), 60 Hz signal and the transformer is a 10:1 step-down transformer. The diode has a cut-in voltage of $V_\gamma = 0.7$ V ($r_f = 0$). (a) What is the peak output voltage? (b) Determine the peak diode current.

(c) What is the fraction (percent) of a cycle that $v_O > 0$. (d) Determine the average output voltage. (e) Find the average current in the load.

2.4   Consider the battery charging circuit shown in Figure 2.4(a). Assume that $V_B = 9$ V, $V_S = 15$ V, and $\omega = 2\pi(60)$. (a) Determine the value of $R$ such that the average battery charging current is $i_D = 0.8$ A. (b) Find the fraction of time that the diode is conducting.

2.5   Figure P2.5 shows a simple full-wave battery charging circuit. Assume $V_B = 9$ V, $V_\gamma = 0.7$ V, and $v_S = 15 \sin[2\pi(60)t]$ (V). (a) Determine $R$ such that the peak battery charging current is 1.2 A. (b) Determine the average battery charging current. (c) Determine the fraction of time that each diode is conducting.

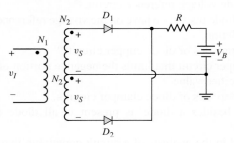

**Figure P2.5**

2.6   The full-wave rectifier circuit shown in Figure 2.5(a) in the text is to deliver 0.2 A and 12 V (peak values) to a load. The ripple voltage is to be limited to 0.25 V. The input signal is 120 V (rms) at 60 Hz. Assume diode cut-in voltages of 0.7 V. (a) Determine the required turns ratio of the transformer. (b) Find the required value of the capacitor. (c) What is the PIV rating of the diode?

2.7   The input signal voltage to the full-wave rectifier circuit in Figure 2.6(a) in the text is $v_I = 160 \sin[2\pi(60)t]$ V. Assume $V_\gamma = 0.7$ V for each diode. Determine the required turns ratio of the transformer to produce a peak output voltage of (a) 25 V, and (b) 100 V. (c) What must be the diode PIV rating for each case?

2.8   The output resistance of the full-wave rectifier in Figure 2.6(a) in the text is $R = 150\ \Omega$. A filter capacitor is connected in parallel with $R$. Assume $V_\gamma = 0.7$ V. The peak output voltage is to be 12 V and the ripple voltage is to be no more than 0.3 V. The input frequency is 60 Hz. (a) Determine the required rms value of $v_S$. (b) Determine the required filter capacitance value. (c) Determine the peak current through each diode.

2.9   Repeat Problem 2.8 for the half-wave rectifier in Figure 2.2(a).

2.10  Consider the half-wave rectifier circuit shown in Figure 2.8(a) in the text. Assume $v_S = 10 \sin[2\pi(60)t]$ (V), $V_\gamma = 0.7$ V, and $R = 500\ \Omega$. (a) What is the peak output voltage? (b) Determine the value of capacitance required such that the ripple voltage is no more that $V_r = 0.5$ V. (c) What is the PIV rating of the diode?

2.11  The parameters of the half-wave rectifier circuit in Figure 2.8(a) in the text are $R = 1\ \text{k}\Omega$, $C = 350\ \mu\text{F}$, and $V_\gamma = 0.7$ V. Assume $v_S(t) = V_S \sin[2\pi(60)t]$ (V) where $V_S$ is in the range of $11 \leq V_S \leq 13$ V. (a) What is the range in output voltage? (b) Determine the range in ripple voltage. (c) If the ripple voltage is to be limited to $V_r = 0.4$ V, determine the minimum value of capacitance required.

2.12  The full-wave rectifier circuit shown in Figure P2.12 has an input signal whose frequency is 60 Hz. The rms value of $v_S = 8.5$ V. Assume each diode cut-in voltage is $V_\gamma = 0.7$ V. (a) What is the maximum value of $V_O$? (b) If $R = 10\,\Omega$, determine the value of $C$ such that the ripple voltage is no larger than 0.25 V. (c) What must be the PIV rating of each diode?

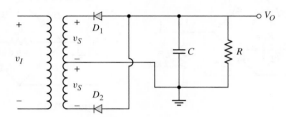

**Figure P2.12**

2.13  Consider the full-wave rectifier circuit in Figure 2.7 of the text. The output resistance is $R_L = 125\,\Omega$, each diode cut-in voltage is $V_\gamma = 0.7$ V, and the frequency of the input signal is 60 Hz. A filter capacitor is connected in parallel with $R_L$. The magnitude of the peak output voltage is to be 15 V and the ripple voltage is to be no more than 0.35 V. (a) Determine the rms value of $v_S$ and (b) the required value of the capacitor.

2.14  The circuit in Figure P2.14 is a complementary output rectifier. If $v_s = 26 \sin [2\pi (60)t]$ V, sketch the output waveforms $v_o^+$ and $v_o^-$ versus time, assuming $V_\gamma = 0.6$ V for each diode.

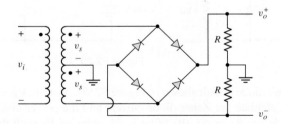

**Figure P2.14**

2.15  A full-wave rectifier is to be designed using the center-tapped transformer configuration. The peak output voltage is to be 12 V, the nominal load current is to be 0.5 A, and the ripple voltage is to be limited to 3 percent. Assume $V_\gamma = 0.8$ V and let $v_I = 120\sqrt{2}\sin[2\pi (60)t]$ V. (a) What is the transformer turns ratio? (b) What is the minimum value of $C$ required? (c) What is the peak diode current? (d) Determine the average diode current. (e) What is the PIV rating of the diodes.

2.16  A full-wave rectifier is to be designed using the bridge circuit configuration. The peak output voltage is to be 9 V, the nominal load current is to be 100 mA, and the ripple voltage is to be limited to $V_r = 0.2$ V. Assume $V_\gamma = 0.8$ V and let $v_I = 120\sqrt{2}\sin[2\pi (60)t]$ (V). (a) What is the transformer turns ratio? (b) What is the minimum value of $C$ required? (c) What is the peak diode current? (d) Determine the average diode current. (e) What is the PIV rating of the diodes.

*2.17    Sketch $v_o$ versus time for the circuit in Figure P2.17 with the input shown. Assume $V_\gamma = 0$.

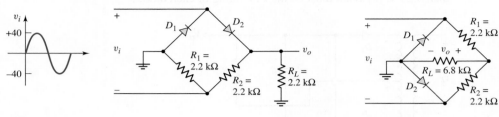

Figure P2.17                                              Figure P2.18

*2.18    (a) Sketch $v_o$ versus time for the circuit in Figure P2.18. The input is a sine wave given by $v_i = 10 \sin \omega t$ V. Assume $V_\gamma = 0$. (b) Determine the rms value of the output voltage.

## Section 2.2 Zener Diode Circuits

2.19    Consider the circuit shown in Figure P2.19. The Zener diode voltage is $V_Z = 3.9$ V and the Zener diode incremental resistance is $r_z = 0$. (a) Determine $I_Z$, $I_L$, and the power dissipated in the diode. (b) Repeat part (a) if the 4 kΩ load resistor is increased to 10 kΩ.

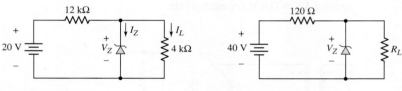

Figure P2.19                                              Figure P2.20

2.20    Consider the Zener diode circuit shown in Figure P2.20. Assume $V_Z = 12$ V and $r_z = 0$. (a) Calculate the Zener diode current and the power dissipated in the Zener diode for $R_L = \infty$. (b) What is the value of $R_L$ such that the current in the Zener diode is one-tenth of the current supplied by the 40 V source? (c) Determine the power dissipated in the Zener diode for the conditions of part (b).

2.21    Consider the Zener diode circuit shown in Figure P2.21. Let $V_I = 60$ V, $R_i = 150 \,\Omega$, and $V_{ZO} = 15.4$ V. Assume $r_z = 0$. The power rating of the diode is 4 W and the minimum diode current is to be 15 mA. (a) Determine the range of diode currents. (b) Determine the range of load resistance.

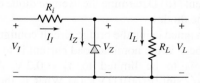

Figure P2.21

*2.22   In the voltage regulator circuit in Figure P2.21, $V_I = 20$ V, $V_Z = 10$ V, $R_i = 222\,\Omega$, and $P_Z(\text{max}) = 400$ mW. (a) Determine $I_L$, $I_Z$, and $I_I$, if $R_L = 380\,\Omega$. (b) Determine the value of $R_L$ that will establish $P_Z(\text{max})$ in the diode. (c) Repeat part (b) if $R_i = 175\,\Omega$.

2.23   A Zener diode is connected in a voltage regulator circuit as shown in Figure P2.21. The Zener voltage is $V_Z = 10$ V and the Zener resistance is assumed to be $r_z = 0$. (a) Determine the value of $R_i$ such that the Zener diode remains in breakdown if the load current varies from $I_L = 50$ to 500 mA and if the input voltage varies from $V_I = 15$ to 20 V. Assume $I_Z(\text{min}) = 0.1 I_Z(\text{max})$. (b) Determine the power rating required for the Zener diode and the load resistor.

2.24   Consider the Zener diode circuit in Figure 2.19 in the text. Assume parameter values of $V_{ZO} = 5.6$ V (diode voltage when $I_Z \cong 0$), $r_z = 3\,\Omega$, and $R_i = 50\,\Omega$. Determine $V_L$, $I_Z$, $I_L$, and the power dissipated in the diode for (a) $V_{PS} = 10$ V, $R_L = \infty$; (b) $V_{PS} = 10$ V, $R_L = 200\,\Omega$; (c) $V_{PS} = 12$ V, $R_L = \infty$; and (d) $V_{PS} = 12$ V, $R_L = 200\,\Omega$.

D2.25   Design a voltage regulator circuit such as shown in Figure P2.21 so that $V_L = 7.5$ V. The Zener diode voltage is $V_Z = 7.5$ V at $I_Z = 10$ mA. The incremental diode resistance is $r_z = 12\,\Omega$. The nominal supply voltage is $V_I = 12$ V and the nominal load resistance is $R_L = 1$ k$\Omega$. (a) Determine $R_i$. (b) If $V_I$ varies by $\pm 10$ percent, calculate the source regulation. What is the variation in output voltage? (c) If $R_L$ varies over the range of $1\,\text{k}\Omega \le R_L \le \infty$, what is the variation in output voltage? Determine the load regulation.

2.26   The percent regulation of the Zener diode regulator shown in Figure 2.16 is 5 percent. The Zener voltage is $V_{ZO} = 6$ V and the Zener resistance is $r_z = 3\,\Omega$. Also, the load resistance varies between 500 and 1000 $\Omega$, the input resistance is $R_i = 280\,\Omega$, and the minimum power supply voltage is $V_{PS}(\text{min}) = 15$ V. Determine the maximum power supply voltage allowed.

*2.27   A voltage regulator is to have a nominal output voltage of 10 V. The specified Zener diode has a rating of 1 W, has a 10 V drop at $I_Z = 25$ mA, and has a Zener resistance of $r_z = 5\,\Omega$. The input power supply has a nominal value of $V_{PS} = 20$ V and can vary by $\pm 25$ percent. The output load current is to vary between $I_L = 0$ and 20 mA. (a) If the minimum Zener current is to be $I_Z = 5$ mA, determine the required $R_i$. (b) Determine the maximum variation in output voltage. (c) Determine the percent regulation.

*2.28   Consider the circuit in Figure P2.28. Let $V_\gamma = 0$. The secondary voltage is given by $v_s = V_s \sin \omega t$, where $V_s = 24$ V. The Zener diode has parameters $V_Z = 16$ V at $I_Z = 40$ mA and $r_z = 2\,\Omega$. Determine $R_i$ such that the load current can vary over the range $40 \le I_L \le 400$ mA with $I_Z(\text{min}) = 40$ mA and find $C$ such that the ripple voltage is no larger than 1 V.

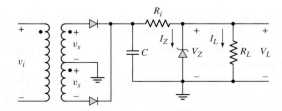

**Figure P2.28**

*2.29   The secondary voltage in the circuit in Figure P2.28 is $v_s = 12 \sin \omega t$ V. The Zener diode has parameters $V_Z = 8$ V at $I_Z = 100$ mA and $r_z = 0.5\ \Omega$. Let $V_\gamma = 0$ and $R_i = 3\ \Omega$. Determine the percent regulation for load currents between $I_L = 0.2$ and 1 A. Find $C$ such that the ripple voltage is no larger than 0.8 V.

## Section 2.3 Clipper and Clamper Circuits

2.30   The parameters in the circuit shown in Figure P2.30 are $V_\gamma = 0.7$ V, $V_{Z1} = 2.3$ V, and $V_{Z2} = 5.6$ V. Plot $v_O$ versus $v_I$ over the range of $-10 \le v_I \le +10$ V.

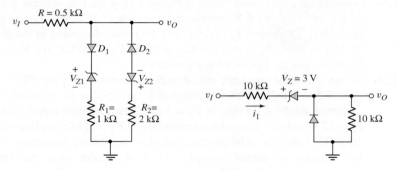

Figure P2.30                    Figure P2.31

2.31   Consider the circuit in Figure P2.31. Let $V_\gamma = 0$. (a) Plot $v_O$ versus $v_I$ over the range $-10 \le v_I \le +10$ V. (b) Plot $i_1$ over the same input voltage range as part (a).

2.32   For the circuit in Figure P2.32, (a) plot $v_O$ versus $v_I$ for $0 \le v_I \le 15$ V. Assume $V_\gamma = 0.7$ V. Indicate all breakpoints. (b) Plot $i_D$ over the same range of input voltage. (c) Compare the results of parts (a) and (b) with a computer simulation.

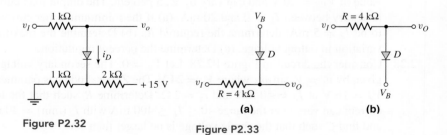

Figure P2.32                    Figure P2.33

2.33   Each diode cut-in voltage is 0.7 V for the circuits shown in Figure P2.33. (a) Plot $v_O$ versus $v_I$ over the range $-5 \le v_I \le +5$ V for the circuit in Figure P2.33(a) for (i) $V_B = 1.8$ V and (ii) $V_B = -1.8$ V. (b) Repeat part (a) for the circuit shown in Figure P2.33(b).

*2.34   The diode in the circuit of Figure P2.34(a) has piecewise linear parameters $V_\gamma = 0.7$ V and $r_f = 10\ \Omega$. (a) Plot $v_O$ versus $v_I$ for $-30 \le v_I \le 30$ V. (b) If the triangular wave, shown in Figure P2.34(b), is applied, plot the output versus time.

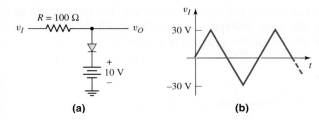

**(a)**                                                        **(b)**

**Figure P2.34**

2.35   Consider the circuits shown in Figure P2.35. Each diode cut-in voltage is $V_\gamma = 0.7$ V. (a) Plot $v_O$ versus $v_I$ over the range $-10 \le v_I \le +10$ V for the circuit in Figure P2.35(a) for (i) $V_B = 5$ V and (ii) $V_B = -5$ V. (b) Repeat part (a) for the circuit in Figure P2.35(b).

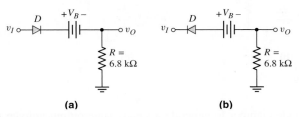

**(a)**                                 **(b)**

**Figure P2.35**

2.36   Plot $v_O$ for each circuit in Figure P2.36 for the input shown. Assume (a) $V_\gamma = 0$ and (b) $V_\gamma = 0.6$ V.

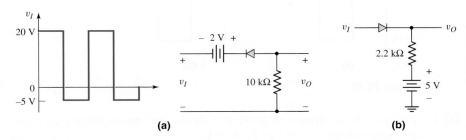

**(a)**                                                        **(b)**

**Figure P2.36**

2.37   Consider the parallel clipper circuit in Figure 2.26 in the text. Assume $V_{Z1} = 6$ V, $V_{Z2} = 4$ V, and $V_\gamma = 0.7$ V for all diodes. For $v_I = 10 \sin \omega t$, sketch $v_O$ versus time over two periods of the input signal.

*2.38   A car's radio may be subjected to voltage spikes induced by coupling from the ignition system. Pulses on the order of $\pm 250$ V and lasting for 120 $\mu$s may exist. Design a clipper circuit using resistors, diodes, and Zener diodes to limit the input voltage between $+14$ V and $-0.7$ V. Specify power ratings of the components.

2.39    Sketch the steady-state output voltage $v_O$ versus time for each circuit in Figure P2.39 with the input voltage given in Figure P2.39(a). Assume $V_\gamma = 0$ and assume the $RC$ time constant is large.

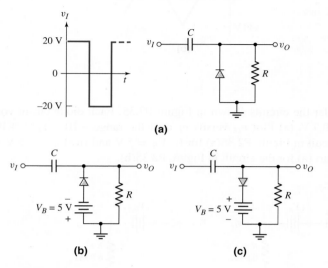

**(a)**

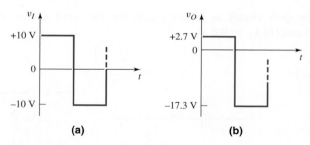

**(b)**                                **(c)**

**Figure P2.39**

D2.40    Design a diode clamper to generate a steady-state output voltage $v_O$ from the input voltage $v_I$ shown in Figure P2.40 if (a) $V_\gamma = 0$ and (b) $V_\gamma = 0.7$ V.

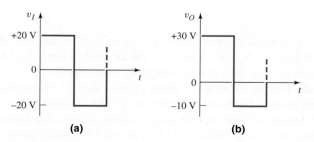

**(a)**                                **(b)**

**Figure P2.40**

D2.41    Design a diode clamper to generate a steady-state output voltage $v_O$ from the input voltage $v_I$ in Figure P2.41 if $V_\gamma = 0$.

**Figure P2.41**

2.42 For the circuit in Figure P2.39(b), let $V_\gamma = 0$ and $v_I = 10 \sin \omega t$ (V). Plot $v_O$ versus time over three cycles of input voltage. Assume the initial voltage across the capacitor is zero and assume the $RC$ time constant is very large.

2.43 Repeat Problem 2.42 for the circuit in Figure P2.39(c) for (i) $V_B = 5$ V and (ii) $V_B = -5$ V.

## Section 2.4 Multiple Diode Circuits

2.44 The diodes in the circuit in Figure P2.44 have piecewise linear parameters of $V_\gamma = 0.6$ V and $r_f = 0$. Determine the output voltage $V_O$ and the diode currents $I_{D1}$ and $I_{D2}$ for the following input conditions: (a) $V_1 = 10$ V, $V_2 = 0$; (b) $V_1 = 5$ V, $V_2 = 0$; (c) $V_1 = 10$ V, $V_2 = 5$ V; and (d) $V_1 = V_2 = 10$ V. (e) Compare the results of parts (a) through (d) with a computer simulation analysis.

2.45 In the circuit in Figure P2.45 the diodes have the same piecewise linear parameters as described in Problem 2.44. Calculate the output voltage $V_O$ and the currents $I_{D1}$, $I_{D2}$, and $I$ for the following input conditions: (a) $V_1 = V_2 = 10$ V; (b) $V_1 = 10$ V, $V_2 = 0$; (c) $V_1 = 10$ V, $V_2 = 5$ V; and (d) $V_1 = V_2 = 0$.

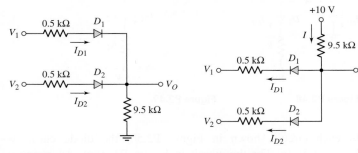

**Figure P2.44**          **Figure P2.45**

2.46 The diodes in the circuit in Figure P2.46 have the same piecewise linear parameters as described in Problem 2.44. Determine the output voltage $V_O$ and the currents $I_{D1}$, $I_{D2}$, $I_{D3}$, and $I$ for the following input conditions: (a) $V_1 = V_2 = 0$; (b) $V_1 = V_2 = 5$ V; (c) $V_1 = 5$ V, $V_2 = 0$; and (d) $V_1 = 5$ V, $V_2 = 2$ V.

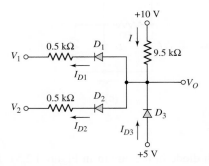

**Figure P2.46**

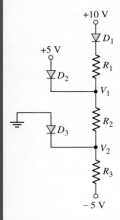

**Figure P2.47**

2.47   Consider the circuit shown in Figure P2.47. Assume each diode cut-in voltage is $V_\gamma = 0.6$ V. (a) Determine $R_1$, $R_2$, and $R_3$ such that $I_{D1} = 0.2$ mA, $I_{D2} = 0.3$ mA, and $I_{D3} = 0.5$ mA. (b) Find $V_1$, $V_2$, and each diode current for $R_1 = 10$ k$\Omega$, $R_2 = 4$ k$\Omega$, and $R_3 = 2.2$ k$\Omega$. (c) Repeat part (b) for $R_1 = 3$ k$\Omega$, $R_2 = 6$ k$\Omega$, and $R_3 = 2.5$ k$\Omega$. (d) Repeat part (b) for $R_1 = 6$ k$\Omega$, $R_2 = 3$ k$\Omega$, and $R_3 = 6$ k$\Omega$.

2.48   The diode cut-in voltage for each diode in the circuit shown in Figure P2.48 is 0.7 V. Determine the value of $R$ such that (a) $I_{D1} = I_{D2}$, (b) $I_{D1} = 0.2I_{D2}$, and (c) $I_{D1} = 5I_{D2}$.

2.49   Consider the circuit in Figure P2.49. Each diode cut-in voltage is $V_\gamma = 0.7$ V. (a) For $R_2 = 1.1$ k$\Omega$, determine $I_{D1}$, $I_{D2}$, and $V_A$. (b) Repeat part (a) for $R_2 = 2.5$ k$\Omega$. (c) Find $R_2$ such that $V_A = 0$. What are the values of $I_{D1}$ and $I_{D2}$?

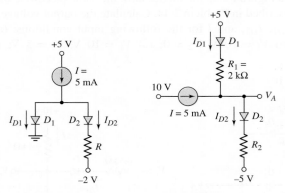

**Figure P2.48**          **Figure P2.49**

2.50   In each circuit shown in Figure P2.50, the diode cut-in voltage is $V_\gamma = 0.6$ V. (a) For the circuit in Figure P2.50(a), determine $v_O$ for (i) $v_I = +5$ V and (ii) $v_I = -5$ V. (b) Repeat part (a) for the circuit in Figure P2.50(b). (c) Plot the voltage transfer characteristics, $v_O$ versus $v_I$, of each circuit over the range $-5 \le v_I \le +5$ V.

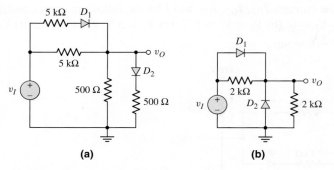

**(a)**                          **(b)**

**Figure P2.50**

*2.51   Assume $V_\gamma = 0.7$ V for each diode in the circuit in Figure P2.51. Plot $v_O$ versus $v_I$ for $-10 \le v_I \le +10$ V.

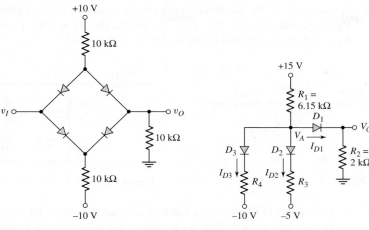

Figure P2.51

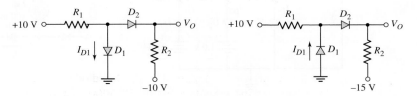

Figure P2.52

2.52 The cut-in voltage of each diode in the circuit shown in Figure P2.52 is $V_\gamma = 0.7$ V. Determine $I_{D1}$, $I_{D2}$, $I_{D3}$, and $V_A$ for (a) $R_3 = 14\,k\Omega$, $R_4 = 24\,k\Omega$; (b) $R_3 = 3.3\,k\Omega$, $R_4 = 5.2\,k\Omega$; and (c) $R_3 = 3.3\,k\Omega$, $R_4 = 1.32\,k\Omega$.

2.53 Let $V_\gamma = 0.7$ V for each diode in the circuit in Figure P2.53. (a) Find $I_{D1}$ and $V_O$ for $R_1 = 5\,k\Omega$ and $R_2 = 10\,k\Omega$. (b) Repeat part (a) for $R_1 = 10\,k\Omega$ and $R_2 = 5\,k\Omega$.

Figure P2.53                    Figure P2.54

2.54 For the circuit shown in Figure P2.54, let $V_\gamma = 0.7$ V for each diode. Calculate $I_{D1}$ and $V_O$ for (a) $R_1 = 10\,k\Omega$, $R_2 = 5\,k\Omega$ and for (b) $R_1 = 5\,k\Omega$, $R_2 = 10\,k\Omega$.

2.55 Assume each diode cut-in voltage is $V_\gamma = 0.7$ V for the circuit in Figure P2.55. Determine $I_{D1}$ and $V_O$ for (a) $R_1 = 10\,k\Omega$, $R_2 = 5\,k\Omega$ and (b) $R_1 = 5\,k\Omega$, $R_2 = 10\,k\Omega$.

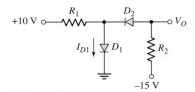

Figure P2.55

2.56 If $V_\gamma = 0.7$ V for the diode in the circuit in Figure P2.56 determine $I_D$ and $V_O$.

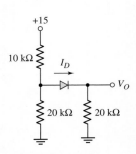

Figure P2.56

2.57 Let $V_\gamma = 0.7$ V for the diode in the circuit in Figure P2.57. Determine $I_D$, $V_D$, $V_A$, and $V_B$ for (a) $V_1 = V_2 = 6$ V; (b) $V_1 = 2$ V, $V_2 = 5$ V; (c) $V_1 = 5$ V, $V_2 = 4$ V; and (d) $V_1 = 2$ V, $V_2 = 8$ V.

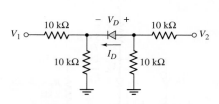

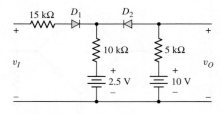

**Figure P2.57**                              **Figure P2.58**

2.58 (a) Each diode in the circuit in Figure P2.58 has piecewise linear parameters of $V_\gamma = 0$ and $r_f = 0$. Plot $v_O$ versus $v_I$ for $0 \leq v_I \leq 30$ V. Indicate the breakpoints and give the state of each diode in the various regions of the plot. (b) Compare the results of part (a) with a computer simulation analysis.

2.59 Each diode cut-in voltage in the circuit in Figure P2.59 is 0.7 V. Determine $I_{D1}$, $I_{D2}$, $I_{D3}$, and $v_O$ for (a) $v_I = 0.5$ V, (b) $v_I = 1.5$ V, (c) $v_I = 3.0$ V, and (d) $v_I = 5.0$ V.

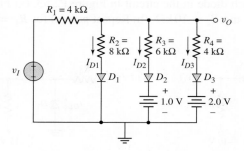

**Figure P2.59**                              **Figure P2.60**

2.60 Let $V_\gamma = 0.7$ V for each diode in the circuit shown in Figure P2.60. Plot $I_{D2}$ versus $v_I$ over the range $0 \leq v_I \leq 12$ V for (a) $V_B = 4.5$ V and (b) $V_B = 9$ V.

2.61 Consider the circuit in Figure P2.61. The output of a diode OR logic gate is connected to the input of a second diode OR logic gate. Assume $V_\gamma = 0.6$ V for each diode. Determine the outputs $V_{O1}$ and $V_{O2}$ for: (a) $V_1 = V_2 = 0$; (b) $V_1 = 5$ V, $V_2 = 0$; and (c) $V_1 = V_2 = 5$ V. What can be said about the relative values of $V_{O1}$ and $V_{O2}$ in their "high" state?

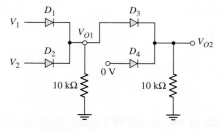

**Figure P2.61**

2.62    Consider the circuit in Figure P2.62. The output of a diode AND logic gate is connected to the input of a second diode AND logic gate. Assume $V_\gamma = 0.6$ V for each diode. Determine the outputs $V_{O1}$ and $V_{O2}$ for: (a) $V_1 = V_2 = 5$ V; (b) $V_1 = 0$, $V_2 = 5$ V; and (c) $V_1 = V_2 = 0$. What can be said about the relative values of $V_{O1}$ and $V_{O2}$ in their "low" state?

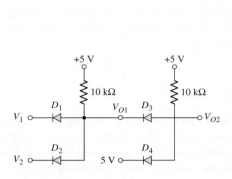

**Figure P2.62**

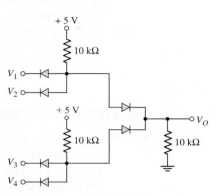

**Figure P2.63**

2.63    Determine the Boolean expression for $V_O$ in terms of the four input voltages for the circuit in Figure P2.63 (Hint: A truth table might be helpful.)

## Section 2.5 LED and Photodiode Circuits

2.64    Consider the circuit shown in Figure P2.64. The forward-bias cut-in voltage of the diode is 1.5 V and the forward-bias resistance is $r_f = 10\,\Omega$. Determine the value of $R$ required to limit the current to $I = 12$ mA when $V_I = 0.2$ V.

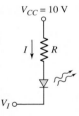

**Figure P2.64**

2.65    The light-emitting diode in the circuit shown in Figure P2.64 has parameters $V_\gamma = 1.7$ V and $r_f = 0$. Light will first be detected when the current is $I = 8$ mA. If $R = 750\,\Omega$, determine the value of $V_I$ at which light will first be detected.

2.66    The parameters of $D_1$ and $D_2$ in the circuit shown in Figure P2.66 are $V_\gamma = 1.7$ V and $r_f = 20\,\Omega$. The current in each diode is to be limited to $I_D = 15$ mA for $V_I = \pm 5$ V. Determine the required value of $R$.

2.67    If the resistor in Example 2.12 is $R = 2\,k\Omega$ and the diode is to be reverse biased by at least 1 V, determine the minimum power supply voltage required.

2.68    Consider the photodiode circuit shown in Figure 2.44. Assume the quantum efficiency is 1. A photocurrent of 0.6 mA is required for an incident photon flux of $\Phi = 10^{17}$ cm$^{-2}$–s$^{-1}$. Determine the required cross-sectional area of the diode.

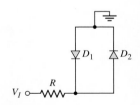

**Figure P2.66**

## COMPUTER SIMULATION PROBLEMS

2.69    Consider the voltage doubler circuit in Figure 2.14. Assume a 60 Hz, 120 V (rms) signal is applied at the input of the transformer with a 20:1 turns ratio. Let $R = 10\,k\Omega$ and $C_1 = C_2 = 200\,\mu F$. Using a computer simulation, plot the output voltage over four cycles of input voltage.

2.70    Consider the parameters and results of Example 2.2. Use a computer simulation to plot the output voltage of each rectifier over four cycles of input voltage. Also determine the PIV of each diode. How do the computer results compare with the results of the example?

2.71    (a) Using a computer simulation, verify the results of Exercise TYU2.3. (b) Determine the ripple voltage if a filter capacitance of $C = 50\,\mu F$ is connected in parallel with the load resistance.

2.72    (a) Using a computer simulation, determine each diode current and voltage in the circuit shown in Figure 2.40. (b) Repeat part (a) using the circuit parameters given in Exercise 2.11.

## DESIGN PROBLEMS

*D2.73    Consider the full-wave bridge rectifier circuit. The input signal is 120 V (rms) at 60 Hz. The load resistance is $R_L = 250\,\Omega$. The peak output voltage is to be 9 V and the ripple voltage is to be no more than 5 percent. Determine the required turns ratio and the required value of filter capacitance.

*D2.74    Design a simple dc voltage source using a 120 V (rms), 60 Hz input signal to a nominal 10 V output signal. A Zener diode with parameters $V_{ZO} = 10\,V$ and $r_z = 3\,\Omega$ is available. The rated power of the Zener diode is 5 W. The source regulation is to be limited to 2 percent.

*D2.75    A clipper is to be designed such that $v_O = 2.5\,V$ for $v_I \geq 2.5\,V$ and $v_O = -1.25\,V$ for $v_I \leq -1.25\,V$.

*D2.76    Design a circuit to provide the voltage transfer characteristics shown in Figure P2.76. Use diodes and Zener diodes with appropriate breakdown voltages in the design. The maximum current in the circuit is to be limited to 1mA.

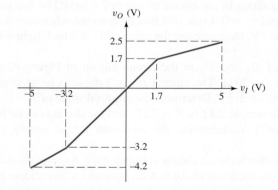

**Figure P2.76**

# The Field-Effect Transistor

In this chapter, we introduce a major type of transistor, the metal-oxide-semiconductor field-effect transistor (MOSFET). The MOSFET led to the electronics revolution of the 1970s and 1980s, in which the microprocessor made possible powerful desktop computers, laptop computers, sophisticated handheld calculators, iPods, and a plethora of other electronic systems. The MOSFET can be made very small, so high-density very large scale integration (VLSI) circuits and high-density memories are possible.

Two complementary devices, the n-channel MOSFET (NMOS) and the p-channel MOSFET (PMOS), exist. Each device is equally important and allows a high degree of flexibility in electronic circuit design. The $i-v$ characteristics of these devices are introduced, and the dc analysis and design techniques of MOSFET circuits are developed.

Another type of field-effect transistor is the junction FET. There are two general categories of junction field-effect transistors (JFETs)—the pn junction FET (pn JFET) and the metal-semiconductor field-effect transistor (MESFET), which is fabricated with a Schottky barrier junction. JFETs were developed before MOSFETs, but the applications and uses of MOSFETs have far surpassed those of the JFET. However, we will consider a few JFET circuits in this chapter.

## PREVIEW

In this chapter, we will:

- Study and understand the structure, operation, and characteristics of the various types of MOSFETs.
- Understand and become familiar with the dc analysis and design techniques of MOSFET circuits.
- Examine three applications of MOSFET circuits.
- Investigate current source biasing of MOSFET circuits, such as those used in integrated circuits.
- Analyze the dc biasing of multistage or multitransistor circuits.
- Understand the operation and characteristics of the junction field-effect transistor, and analyze the dc response of JFET circuits.
- As an application, incorporate a MOS transistor in a circuit design that enhances the simple diode electronic thermometer discussed in Chapter 1.

# 3.1    MOS FIELD-EFFECT TRANSISTOR

**Objective:** • Understand the operation and characteristics of the various types of metal-oxide semiconductor field-effect transistors (MOSFETs).

The **metal-oxide-semiconductor field-effect transistor (MOSFET)** became a practical reality in the 1970s. The MOSFET, compared to BJTs, can be made very small (that is, it occupies a very small area on an IC chip). Since digital circuits can be designed using only MOSFETs, with essentially no resistors or diodes required, high-density VLSI circuits, including microprocessors and memories, can be fabricated. The MOSFET has made possible the handheld calculator, the powerful personal computer, and the laptop computer. MOSFETs can also be used in analog circuits, as we will see in the next chapter.

In the MOSFET, the current is controlled by an electric field applied perpendicular to both the semiconductor surface and to the direction of current. The phenomenon used to modulate the conductance of a semiconductor, or control the current in a semiconductor, by applying an electric field perpendicular to the surface is called the **field effect.** The basic transistor principle is that the voltage between two terminals controls the current through the third terminal.

In the following two sections, we will discuss the various types of MOSFETs, develop the $i$–$v$ characteristics, and then consider the dc biasing of various MOSFET circuit configurations. After studying these sections, you should be familiar and comfortable with the MOSFET and MOSFET circuits.

## 3.1.1    Two-Terminal MOS Structure

The heart of the MOSFET is the metal-oxide-semiconductor capacitor shown in Figure 3.l. The metal may be aluminum or some other type of metal. In most cases, the metal is replaced by a high-conductivity polycrystalline silicon layer deposited on the oxide. However, the term metal is usually still used in referring to MOSFETs. In the figure, the parameter $t_{ox}$ is the thickness of the oxide and $\epsilon_{ox}$ is the oxide permittivity.

The physics of the MOS structure can be explained with the aid of a simple parallel-plate capacitor.[1] Figure 3.2(a) shows a parallel-plate capacitor with the top plate at a negative voltage with respect to the bottom plate. An insulator material separates the two plates. With this bias, a negative charge exists on the top plate, a positive charge exists on the bottom plate, and an electric field is induced between the two plates, as shown.

A MOS capacitor with a p-type semiconductor substrate is shown in Figure 3.2(b). The top metal terminal, also called the **gate,** is at a negative voltage with respect to the semiconductor substrate. From the example of the parallel-plate capacitor, we can see that a negative charge will exist on the top metal plate and an electric field will be induced in the direction shown in the figure. If the electric field penetrates the

---

[1]The capacitance of a parallel plate capacitor, neglecting fringing fields, is $C = \epsilon A/d$, where $A$ is the area of one plate, $d$ is the distance between plates, and $\epsilon$ is the permittivity of the medium between the plates.

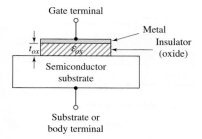

Figure 3.1 The basic MOS capacitor structure

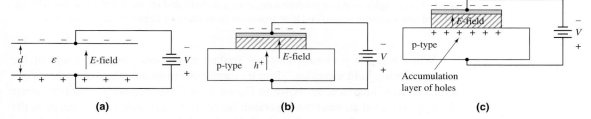

**Figure 3.2** (a) A parallel-plate capacitor, showing the electric field and conductor charges, (b) a corresponding MOS capacitor with a negative gate bias, showing the electric field and charge flow, and (c) the MOS capacitor with an accumulation layer of holes

semiconductor, the holes in the p-type semiconductor will experience a force toward the oxide-semiconductor interface. The equilibrium distribution of charge in the MOS capacitor with this particular applied voltage is shown in Figure 3.2(c). An accumulation layer of positively charged holes at the oxide-semiconductor interface corresponds to the positive charge on the bottom "plate" of the MOS capacitor.

Figure 3.3(a) shows the same MOS capacitor, but with the polarity of the applied voltage reversed. A positive charge now exists on the top metal plate and the induced electric field is in the opposite direction, as shown. In this case, if the electric field penetrates the semiconductor, holes in the p-type material will experience a force away from the oxide-semiconductor interface. As the holes are pushed away from the interface, a negative space-charge region is created, because of the fixed acceptor impurity atoms. The negative charge in the induced depletion region corresponds to the negative charge on the bottom "plate" of the MOS capacitor. Figure 3.3(b) shows the equilibrium distribution of charge in the MOS capacitor with this applied voltage.

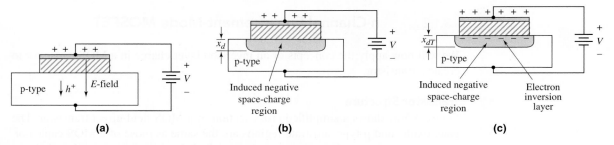

**Figure 3.3** The MOS capacitor with p-type substrate: (a) effect of positive gate bias, showing the electric field and charge flow, (b) the MOS capacitor with an induced space-charge region due to a moderate positive gate bias, and (c) the MOS capacitor with an induced space-charge region and electron inversion layer due to a larger positive gate bias

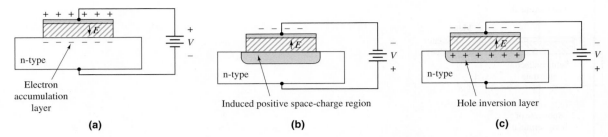

**Figure 3.4** The MOS capacitor with n-type substrate: (a) effect of a positive gate bias and the formation of an electron accumulation layer, (b) the MOS capacitor with an induced space-charge region due to a moderate negative gate bias, and (c) the MOS capacitor with an induced space-charge region and hole inversion layer due to a larger negative gate bias

When a larger positive voltage is applied to the gate, the magnitude of the induced electric field increases. Minority carrier electrons are attracted to the oxide-semiconductor interface, as shown in Figure 3.3(c). This region of minority carrier electrons is called an **electron inversion layer.** The magnitude of the charge in the inversion layer is a function of the applied gate voltage.

The same basic charge distributions can be obtained in a MOS capacitor with an n-type semiconductor substrate. Figure 3.4(a) shows this MOS capacitor structure, with a positive voltage applied to the top gate terminal. A positive charge is created on the top gate and an electric field is induced in the direction shown. In this situation, an accumulation layer of electrons is induced in the n-type semiconductor.

Figure 3.4(b) shows the case when a negative voltage is applied to the gate terminal. A positive space-charge region is induced in the n-type substrate by the induced electric field. When a larger negative voltage is applied, a region of positive charge is created at the oxide-semiconductor interface, as shown in Figure 3.4(c). This region of minority carrier holes is called a **hole inversion layer.** The magnitude of the positive charge in the inversion layer is a function of the applied gate voltage.

The term **enhancement mode** means that a voltage must be applied to the gate to create an inversion layer. For the MOS capacitor with a p-type substrate, a positive gate voltage must be applied to create the electron inversion layer; for the MOS capacitor with an n-type substrate, a negative gate voltage must be applied to create the hole inversion layer.

### 3.1.2    n-Channel Enhancement-Mode MOSFET

We will now apply the concepts of an inversion layer charge in a MOS capacitor to create a transistor.

**Transistor Structure**

Figure 3.5(a) shows a simplified cross section of a MOS field-effect transistor. The gate, oxide, and p-type substrate regions are the same as those of a MOS capacitor. In addition, we now have two n-regions, called the **source terminal** and **drain terminal.** The current in a MOSFET is the result of the flow of charge in the inversion layer, also called the **channel region,** adjacent to the oxide–semiconductor interface.

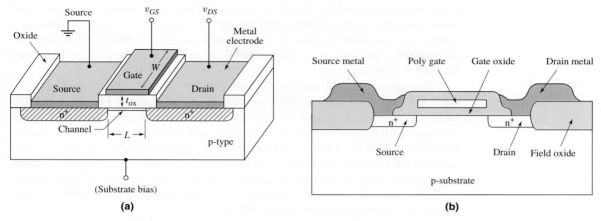

**Figure 3.5** (a) Schematic diagram of an n-channel enhancement-mode MOSFET and (b) an n-channel MOSFET, showing the field oxide and polysilicon gate

The channel length $L$ and channel width $W$ are defined on the figure. The channel length of a typical integrated circuit MOSFET is less than 1 $\mu$m ($10^{-6}$ m), which means that MOSFETs are small devices. The oxide thickness $t_{ox}$ is typically on the order of 400 angstroms, or less.

The diagram in Figure 3.5(a) is a simplified sketch of the basic structure of the transistor. Figure 3.5(b) shows a more detailed cross section of a MOSFET fabricated into an integrated circuit configuration. A thick oxide, called the **field oxide,** is deposited outside the area in which the metal interconnect lines are formed. The gate material is usually heavily doped polysilicon. Even though the actual structure of a MOSFET may be fairly complex, the simplified diagram may be used to develop the basic transistor characteristics.

### Basic Transistor Operation

With zero bias applied to the gate, the source and drain terminals are separated by the p-region, as shown in Figure 3.6(a). This is equivalent to two back-to-back diodes, as shown in Figure 3.6(b). The current in this case is essentially zero. If a large enough positive gate voltage is applied, an electron inversion layer is created at the oxide–semiconductor interface and this layer "connects" the n-source to the n-drain,

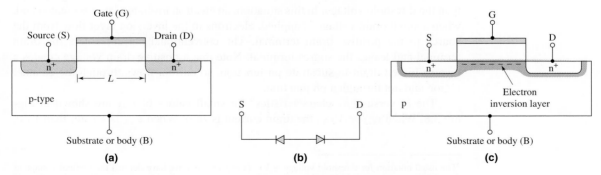

**Figure 3.6** (a) Cross section of the n-channel MOSFET prior to the formation of an electron inversion layer, (b) equivalent back-to-back diodes between source and drain when the transistor is in cutoff, and (c) cross section after the formation of an electron inversion layer

as shown in Figure 3.6(c). A current can then be generated between the source and drain terminals. Since a voltage must be applied to the gate to create the inversion charge, this transistor is called an **enhancement-mode MOSFET.** Also, since the carriers in the inversion layer are electrons, this device is also called an **n-channel MOSFET (NMOS).**

The source terminal supplies carriers that flow through the channel, and the drain terminal allows the carriers to drain from the channel. For the n-channel MOSFET, electrons flow from the source to the drain with an applied drain-to-source voltage, which means the conventional current enters the drain and leaves the source. The magnitude of the current is a function of the amount of charge in the inversion layer, which in turn is a function of the applied gate voltage. Since the gate terminal is separated from the channel by an oxide or insulator, there is no gate current. Similarly, since the channel and substrate are separated by a space-charge region, there is essentially no current through the substrate.

### 3.1.3    Ideal MOSFET Current–Voltage Characteristics—NMOS Device

The **threshold voltage** of the n-channel MOSFET, denoted as $V_{TN}$, is defined[2] as the applied gate voltage needed to create an inversion charge in which the density is equal to the concentration of majority carriers in the semiconductor substrate. In simple terms, we can think of the threshold voltage as the gate voltage required to "turn on" the transistor.

For the n-channel enhancement-mode MOSFET, the threshold voltage is positive because a positive gate voltage is required to create the inversion charge. If the gate voltage is less than the threshold voltage, the current in the device is essentially zero. If the gate voltage is greater than the threshold voltage, a drain-to-source current is generated as the drain-to-source voltage is applied. The gate and drain voltages are measured with respect to the source.

Figure 3.7(a) shows an n-channel enhancement-mode MOSFET with the source and substrate terminals connected to ground. The gate-to-source voltage is less than the threshold voltage, and there is a small drain-to-source voltage. With this bias configuration, there is no electron inversion layer, the drain-to-substrate pn junction is reverse biased, and the drain current is zero (neglecting pn junction leakage currents).

Figure 3.7(b) shows the same MOSFET with an applied gate voltage greater than the threshold voltage. In this situation, an electron inversion layer is created and, when a small drain voltage is applied, electrons in the inversion layer flow from the source to the positive drain terminal. The conventional current enters the drain terminal and leaves the source terminal. Note that a positive drain voltage creates a reverse-biased drain-to-substrate pn junction, so current flows through the channel region and not through a pn junction.

The $i_D$ versus $v_{DS}$ characteristics[3] for small values of $v_{DS}$ are shown in Figure 3.8. When $v_{GS} < V_{TN}$, the drain current is zero. When $v_{GS}$ is greater than $V_{TN}$,

---

[2]The usual notation for threshold voltage is $V_T$. However, since we have defined the thermal voltage as $V_T = kT/q$, we will use $V_{TN}$ for the threshold voltage of the n-channel device.

[3]The voltage notation $v_{DS}$ and $v_{GS}$, with the dual subscript, denotes the voltage between the drain (D) and source (S) and between the gate (G) and source (S), respectively. Implicit in the notation is that the first subscript is positive with respect to the second subscript.

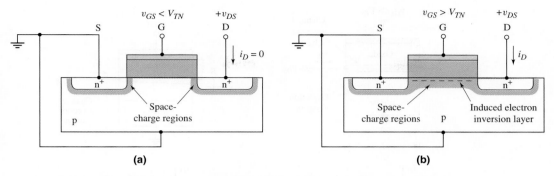

**Figure 3.7** The n-channel enhancement-mode MOSFET (a) with an applied gate voltage $v_{GS} < V_{TN}$, and (b) with an applied gate voltage $v_{GS} > V_{TN}$

the channel inversion charge is formed and the drain current increases with $v_{DS}$. Then, with a larger gate voltage, a larger inversion charge density is created, and the drain current is greater for a given value of $v_{DS}$.

Figure 3.9(a) shows the basic MOS structure for $v_{GS} > V_{TN}$ and a small applied $v_{DS}$. In the figure, the thickness of the inversion channel layer qualitatively indicates the relative charge density, which for this case is essentially constant along the entire channel length. The corresponding $i_D$ versus $v_{DS}$ curve is also shown in the figure.

Figure 3.9(b) shows the situation when $v_{DS}$ increases. As the drain voltage increases, the voltage drop across the oxide near the drain terminal decreases, which means that the induced inversion charge density near the drain also decreases. The incremental conductance of the channel at the drain then decreases, which causes the slope of the $i_D$ versus $v_{DS}$ curve to decrease. This effect is shown in the $i_D$ versus $v_{DS}$ curve in the figure.

As $v_{DS}$ increases to the point where the potential difference, $v_{GS} - v_{DS}$, across the oxide at the drain terminal is equal to $V_{TN}$, the induced inversion charge density at the drain terminal is zero. This effect is shown schematically in Figure 3.9(c). For this condition, the incremental channel conductance at the drain is zero, which means that the slope of the $i_D$ versus $v_{DS}$ curve is zero. We can write

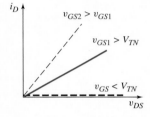

**Figure 3.8** Plot of $i_D$ versus $v_{DS}$ characteristic for small values of $v_{DS}$ at three $v_{GS}$ voltages

$$v_{GS} - v_{DS}(\text{sat}) = V_{TN} \qquad \textbf{(3.1(a))}$$

or

$$v_{DS}(\text{sat}) = v_{GS} - V_{TN} \qquad \textbf{(3.1(b))}$$

where $v_{DS}(\text{sat})$ is the drain-to-source voltage that produces zero inversion charge density at the drain terminal.

When $v_{DS}$ becomes larger than $v_{DS}(\text{sat})$, the point in the channel at which the inversion charge is just zero moves toward the source terminal. In this case, electrons enter the channel at the source, travel through the channel toward the drain, and then, at the point where the charge goes to zero, are injected into the space-charge region, where they are swept by the $E$-field to the drain contact. In the ideal MOSFET, the drain current is constant for $v_{DS} > v_{DS}(\text{sat})$. This region of the $i_D$ versus $v_{DS}$ characteristic is referred to as the **saturation region,** which is shown in Figure 3.9(d).

As the applied gate-to-source voltage changes, the $i_D$ versus $v_{DS}$ curve changes. In Figure 3.8, we saw that the initial slope of $i_D$ versus $v_{DS}$ increases as $v_{GS}$ increases. Also, Equation (3.1(b)) shows that $v_{DS}(\text{sat})$ is a function of $v_{GS}$. Therefore, we can

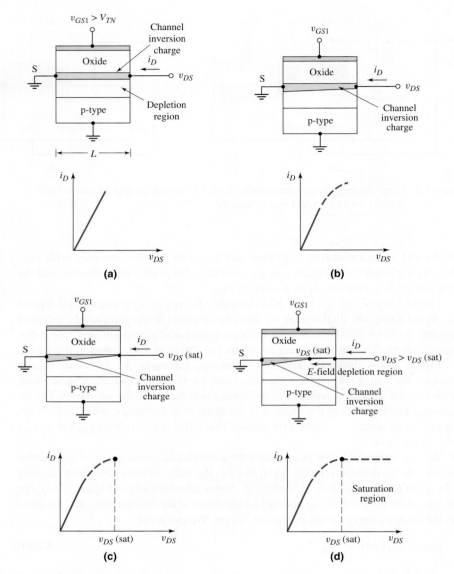

**Figure 3.9** Cross section and $i_D$ versus $v_{DS}$ curve for an n-channel enhancement-mode MOSFET when $v_{GS} > V_{TN}$ for (a) a small $v_{DS}$ value, (b) a larger $v_{DS}$ value but for $v_{DS} < v_{DS}(\text{sat})$, (c) $v_{DS} = v_{DS}(\text{sat})$, and (d) $v_{DS} > v_{DS}(\text{sat})$

generate the family of curves for this n-channel enhancement mode MOSFET as shown in Figure 3.10.

Although the derivation of the current–voltage characteristics of the MOSFET is beyond the scope of this text, we can define the relationships. The region for which $v_{DS} < v_{DS}(\text{sat})$ is known as the **nonsaturation** or **triode region.** The ideal current–voltage characteristics in this region are described by the equation

$$i_D = K_n \left[ 2(v_{GS} - V_{TN})v_{DS} - v_{DS}^2 \right] \tag{3.2(a)}$$

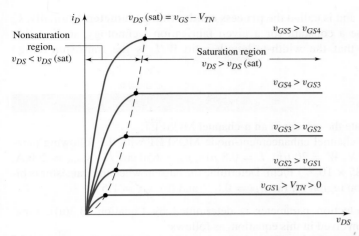

**Figure 3.10** Family of $i_D$ versus $v_{DS}$ curves for an n-channel enhancement-mode MOSFET. Note that the $v_{DS}(\text{sat})$ voltage is a single point on each of the curves. This point denotes the transition between the nonsaturation region and the saturation region

In the saturation region, the ideal current–voltage characteristics for $v_{GS} > V_{TN}$ are described by the equation

$$i_D = K_n(v_{GS} - V_{TN})^2 \qquad \text{(3.2(b))}$$

In the saturation region, since the ideal drain current is independent of the drain-to-source voltage, the incremental or small-signal resistance is infinite. We see that

$$r_0 = \Delta v_{DS}/\Delta i_D|_{v_{GS}=\text{const.}} = \infty$$

The parameter $K_n$ is sometimes called the transconduction parameter for the n-channel device. However, this term is not to be confused with the small-signal transconductance parameter introduced in the next chapter. For simplicity, we will refer to this parameter as the **conduction parameter,** which for an n-channel device is given by

$$K_n = \frac{W\mu_n C_{\text{ox}}}{2L} \qquad \text{(3.3(a))}$$

where $C_{\text{ox}}$ is the oxide capacitance per unit area. The capacitance is given by

$$C_{\text{ox}} = \epsilon_{\text{ox}}/t_{\text{ox}}$$

where $t_{\text{ox}}$ is the oxide thickness and $\epsilon_{\text{ox}}$ is the oxide permittivity. For silicon devices, $\epsilon_{\text{ox}} = (3.9)(8.85 \times 10^{-14})$ F/cm. The parameter $\mu_n$ is the mobility of the electrons in the inversion layer. The channel width $W$ and channel length $L$ were shown in Figure 3.5(a).

As Equation (3.3(a)) indicates, the conduction parameter is a function of both electrical and geometric parameters. The oxide capacitance and carrier mobility are essentially constants for a given fabrication technology. However, the geometry, or width-to-length ratio $W/L$, is a variable in the design of MOSFETs that is used to produce specific current–voltage characteristics in MOSFET circuits.

We can rewrite the conduction parameter in the form

$$K_n = \frac{k_n'}{2} \cdot \frac{W}{L} \qquad \text{(3.3(b))}$$

where $k_n' = \mu_n C_{ox}$ and is called the **process conduction parameter.** Normally, $k_n'$ is considered to be a constant for a given fabrication technology, so Equation (3.3(b)) indicates that the width-to-length ratio $W/L$ is the transistor design variable.

## EXAMPLE 3.1

**Objective:** Calculate the current in an n-channel MOSFET.

Consider an n-channel enhancement-mode MOSFET with the following parameters: $V_{TN} = 0.4$ V, $W = 20$ $\mu$m, $L = 0.8$ $\mu$m, $\mu_n = 650$ cm$^2$/V–s, $t_{ox} = 200$ Å, and $\epsilon_{ox} = (3.9)(8.85 \times 10^{-14})$ F/cm. Determine the current when the transistor is biased in the saturation region for (a) $v_{GS} = 0.8$ V and (b) $v_{GS} = 1.6$ V.

**Solution:** The conduction parameter is determined by Equation (3.3(a)). First, consider the units involved in this equation, as follows:

$$K_n = \frac{W(\text{cm}) \cdot \mu_n \left(\dfrac{\text{cm}^2}{\text{V–s}}\right) \epsilon_{ox} \left(\dfrac{\text{F}}{\text{cm}}\right)}{2L(\text{cm}) \cdot t_{ox}(\text{cm})} = \frac{\text{F}}{\text{V–s}} = \frac{(\text{C/V})}{\text{V–s}} = \frac{\text{A}}{\text{V}^2}$$

The value of the conduction parameter is therefore

$$K_n = \frac{W \mu_n \epsilon_{ox}}{2L t_{ox}} = \frac{(20 \times 10^{-4})(650)(3.9)(8.85 \times 10^{-14})}{2(0.8 \times 10^{-4})(200 \times 10^{-8})}$$

or

$$K_n = 1.40 \text{ mA/V}^2$$

From Equation (3.2(b)), we find:

(a) For $v_{GS} = 0.8$ V,

$$i_D = K_n(v_{GS} - V_{TN})^2 = (1.40)(0.8 - 0.4)^2 = 0.224 \text{ mA}$$

(b) For $v_{GS} = 1.6$ V,

$$i_D = (1.40)(1.6 - 0.4)^2 = 2.02 \text{ mA}$$

**Comment:** The current capability of a transistor can be increased by increasing the conduction parameter. For a given fabrication technology, $K_n$ is adjusted by varying the transistor width $W$.

## EXERCISE PROBLEM

**Ex 3.1:** An NMOS transistor with $V_{TN} = 1$ V has a drain current $i_D = 0.8$ mA when $v_{GS} = 3$ V and $v_{DS} = 4.5$ V. Calculate the drain current when: (a) $v_{GS} = 2$ V, $v_{DS} = 4.5$ V; and (b) $v_{GS} = 3$ V, $v_{DS} = 1$ V. (Ans. (a) 0.2 mA (b) 0.6 mA)

### 3.1.4  p-Channel Enhancement-Mode MOSFET

The complementary device of the n-channel enhancement-mode MOSFET is the p-channel enhancement-mode MOSFET.

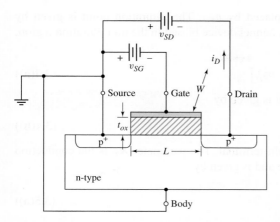

**Figure 3.11**  Cross section of p-channel enhancement-mode MOSFET. The device is cut off for $v_{SG} = 0$. The dimension $W$ extends into the plane of the page.

## Transistor Structure

Figure 3.11 shows a simplified cross section of the p-channel enhancement-mode transistor. The substrate is now n-type and the source and drain areas are p-type. The channel length, channel width, and oxide thickness parameter definitions are the same as those for the NMOS device shown in Figure 3.5(a).

## Basic Transistor Operation

The operation of the p-channel device is the same as that of the n-channel device, except the hole is the charge carrier rather than the electron. A negative gate bias is required to induce an inversion layer of holes in the channel region directly under the oxide. The threshold voltage for the p-channel device is denoted as $V_{TP}$.[4] Since the threshold voltage is defined as the gate voltage required to induce the inversion layer, then $V_{TP} < 0$ for the p-channel enhancement-mode device.

Once the inversion layer has been created, the p-type source region is the source of the charge carrier so that holes flow from the source to the drain. A negative drain voltage is therefore required to induce an electric field in the channel forcing the holes to move from the source to the drain. The conventional current direction, then, for the PMOS transistor is into the source and out of the drain. The conventional current direction and voltage polarity for the PMOS device are reversed compared to the NMOS device.

Note in Figure 3.11 the reversal of the voltage subscripts. For $v_{SG} > 0$, the gate voltage is negative with respect to that at the source. Similarly, for $v_{SD} > 0$, the drain voltage is negative with respect to that at the source.

### 3.1.5    Ideal MOSFET Current–Voltage Characteristics—PMOS Device

The ideal current–voltage characteristics of the p-channel enhancement-mode device are essentially the same as those shown in Figure 3.10, noting that the drain current

---

[4]Using a different threshold voltage parameter for a PMOS device compared to the NMOS device is for clarity only.

is out of the drain and $v_{DS}$ is replaced by $v_{SD}$. The saturation point is given by $v_{SD}(\text{sat}) = v_{SG} + V_{TP}$. For the p-channel device biased in the nonsaturation region, the current is given by

$$i_D = K_p\left[2(v_{SG} + V_{TP})v_{SD} - v_{SD}^2\right] \tag{3.4(a)}$$

In the saturation region, the current is given by

$$i_D = K_p(v_{SG} + V_{TP})^2 \tag{3.4(b)}$$

and the drain current exits the drain terminal. The parameter $K_p$ is the conduction parameter for the p-channel device and is given by

$$K_p = \frac{W\mu_p C_{\text{ox}}}{2L} \tag{3.5(a)}$$

where $W$, $L$, and $C_{\text{ox}}$ are the channel width, length, and oxide capacitance per unit area, as previously defined. The parameter $\mu_p$ is the mobility of holes in the hole inversion layer. In general, the hole inversion layer mobility is less than the electron inversion layer mobility.

We can also rewrite Equation (3.5(a)) in the form

$$K_p = \frac{k_p'}{2} \cdot \frac{W}{L} \tag{3.5(b)}$$

where $k_p' = \mu_p C_{\text{ox}}$.

For a p-channel MOSFET biased in the saturation region, we have

$$v_{SD} > v_{SD}(\text{sat}) = v_{SG} + V_{TP} \tag{3.6}$$

## EXAMPLE 3.2

**Objective:** Determine the source-to-drain voltage required to bias a p-channel enhancement-mode MOSFET in the saturation region.

Consider an enhancement-mode p-channel MOSFET for which $K_p = 0.2\,\text{mA/V}^2$, $V_{TP} = -0.50$ V, and $i_D = 0.50$ mA.

**Solution:** In the saturation region, the drain current is given by

$$i_D = K_p(v_{SG} + V_{TP})^2$$

or

$$0.50 = 0.2(v_{SG} - 0.50)^2$$

which yields

$$v_{SG} = 2.08 \text{ V}$$

To bias this p-channel MOSFET in the saturation region, the following must apply:

$$v_{SD} > v_{SD}(\text{sat}) = v_{SG} + V_{TP} = 2.08 - 0.5 = 1.58 \text{ V}$$

**Comment:** Biasing a transistor in either the saturation or the nonsaturation region depends on both the gate-to-source voltage and the drain-to-source voltage.

**Ex 3.2:** A PMOS device with $V_{TP} = -1.2$ V has a drain current $i_D = 0.5$ mA when $v_{SG} = 3$ V and $v_{SD} = 5$ V. Calculate the drain current when (a) $v_{SG} = 2$ V, $v_{SD} = 3$ V; and (b) $v_{SG} = 5$ V, $v_{SD} = 2$ V. (Ans. (a) 0.0986 mA, (b) 1.72 mA)

### 3.1.6    Circuit Symbols and Conventions

The conventional circuit symbol for the n-channel enhancement-mode MOSFET is shown in Figure 3.12(a). The vertical solid line denotes the gate electrode, the vertical broken line denotes the channel (the broken line indicates the device is enhancement mode), and the separation between the gate line and channel line denotes the oxide that insulates the gate from the channel. The polarity of the pn junction between the substrate and the channel is indicated by the arrowhead on the body or substrate terminal. The direction of the arrowhead indicates the type of transistor, which in this case is an n-channel device. This symbol shows the four-terminal structure of the MOSFET device.

In most applications in this text, we will implicitly assume that the source and substrate terminals are connected together. Explicitly drawing the substrate terminal for each transistor in a circuit becomes redundant and makes the circuits appear more complex. Instead, we will use the circuit symbol for the n-channel MOSFET shown in Figure 3.12(b). In this symbol, the arrowhead is on the source terminal and it indicates the direction of current, which for the n-channel device is out of the source. By including the arrowhead in the symbol, we do not need to explicitly indicate the source and drain terminals. We will use this circuit symbol throughout the text except in specific applications.

In more advanced texts and journal articles, the circuit symbol of the n-channel MOSFET shown in Figure 3.12(c) is generally used. The gate terminal is obvious and it is implicitly understood that the "top" terminal is the drain and the "bottom" terminal is the source. The top terminal, in this case the drain, is usually at a more positive voltage than the bottom terminal. In this introductory text, we will use the symbol shown in Figure 3.12(b) for clarity.

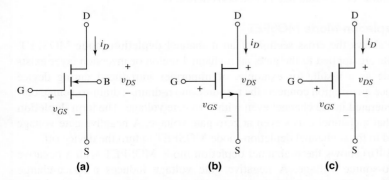

**Figure 3.12** The n-channel enhancement-mode MOSFET: (a) conventional circuit symbol, (b) circuit symbol that will be used in this text, and (c) a simplified circuit symbol used in more advanced texts

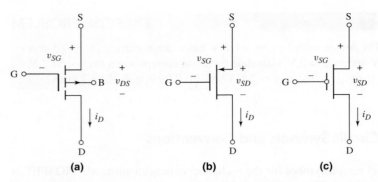

**Figure 3.13** The p-channel enhancement-mode MOSFET: (a) conventional circuit symbol, (b) circuit symbol that will be used in this text, and (c) a simplified circuit symbol used in more advanced texts

The conventional circuit symbol for the p-channel enhancement-mode MOSFET appears in Figure 3.13(a). Note that the arrowhead direction on the substrate terminal is reversed from that in the n-channel enhancement-mode device. This circuit symbol again shows the four terminal structure of the MOSFET device.

The circuit symbol for the p-channel enhancement-mode device shown in Figure 3.13(b) will be used in this text. The arrowhead is on the source terminal indicating the direction of the current, which for the p-channel device is into the source terminal.

In more advanced texts and journal articles, the circuit symbol of the p-channel MOSFET shown in Figure 3.13(c) is generally used. Again, the gate terminal is obvious but includes the O symbol to indicate that this is a PMOS device. It is implicitly understood that the "top" terminal is the source and the "bottom" terminal is the drain. The top terminal, in this case the source, is normally at a higher potential than the bottom terminal. Again, in this text, we will use the symbol shown in Figure 3.13(b) for clarity.

### 3.1.7    Additional MOSFET Structures and Circuit Symbols

Before we start analyzing MOSFET circuits, there are two other MOSFET structures in addition to the n-channel enhancement-mode device and the p-channel enhancement-mode device that need to be considered.

#### n-Channel Depletion-Mode MOSFET

Figure 3.14(a) shows the cross section of an n-channel **depletion-mode** MOSFET. When zero volts are applied to the gate, an n-channel region or inversion layer exists under the oxide as a result, for example, of impurities introduced during device fabrication. Since an n-region connects the n-source and n-drain, a drain-to-source current may be generated in the channel even with zero gate voltage. The term **depletion mode** means that a channel exists even at zero gate voltage. A negative gate voltage must be applied to the n-channel depletion-mode MOSFET to turn the device off.

Figure 3.14(b) shows the n-channel depletion mode MOSFET with a negative applied gate-to-source voltage. A negative gate voltage induces a space-charge region under the oxide, thereby reducing the thickness of the n-channel region. The reduced thickness decreases the channel conductance, which in turn reduces the drain current. When the gate voltage is equal to the threshold voltage, which is

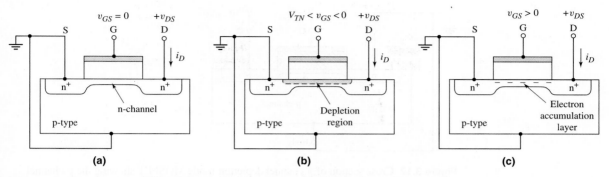

**Figure 3.14** Cross section of an n-channel depletion mode MOSFET for: (a) $v_{GS} = 0$, (b) $v_{GS} < 0$, and (c) $v_{GS} > 0$

negative for this device, the induced space-charge region extends completely through the n-channel region, and the current goes to zero. A positive gate voltage creates an electron accumulation layer, as shown in Figure 3.14(c) which increases the drain current. The general $i_D$ versus $v_{DS}$ family of curves for the n-channel depletion-mode MOSFET is shown in Figure 3.15.

The current–voltage characteristics defined by Equations (3.2(a)) and (3.2(b)) apply to both enhancement- and depletion-mode n-channel devices. The only difference is that the threshold voltage $V_{TN}$ is positive for the enhancement-mode MOSFET and negative for the depletion-mode MOSFET. Even though the current–voltage characteristics of enhancement- and depletion-mode devices are described by the same equations, different circuit symbols are used, simply for purposes of clarity.

The conventional circuit symbol for the n-channel depletion-mode MOSFET is shown in Figure 3.16(a). The vertical solid line denoting the channel indicates the device is depletion mode. A comparison of Figures 3.12(a) and 3.16(a) shows that the only difference between the enhancement- and depletion-mode symbols is the broken versus the solid line representing the channel.

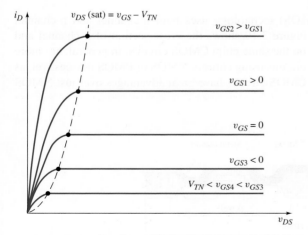

**Figure 3.15** Family of $i_D$ versus $v_{DS}$ curves for an n-channel depletion-mode MOSFET. Note again that the $v_{DS}$(sat) voltage is a single point on each curve.

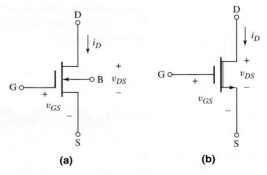

**Figure 3.16** The n-channel depletion-mode MOSFET: (a) conventional circuit symbol and (b) simplified circuit symbol

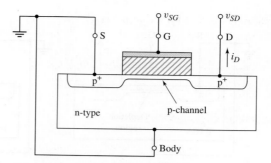

**Figure 3.17** Cross section of p-channel depletion-mode MOSFET showing the p-channel under the oxide at zero gate voltage

A simplified symbol for the n-channel depletion-mode MOSFET is shown in Figure 3.16(b). The arrowhead is again on the source terminal and indicates the direction of current, which for the n-channel device is out of the source. The heavy solid line represents the depletion-mode channel region. Again, using a different circuit symbol for the depletion-mode device compared to the enhancement-mode device is simply for clarity in a circuit diagram.

### p-Channel Depletion-Mode MOSFET

Figure 3.17 shows the cross section of a p-channel depletion-mode MOSFET, as well as the biasing configuration and current direction. In the depletion-mode device, a channel region of holes already exists under the oxide, even with zero gate voltage. A positive gate voltage is required to turn the device off. Hence the threshold voltage of a p-channel depletion-mode MOSFET is positive ($V_{TP} > 0$).

The conventional and simplified circuit symbols for the p-channel depletion-mode device are shown in Figure 3.18. The heavy solid line in the simplified symbol represents the channel region and denotes the depletion-mode device. The arrowhead is again on the source terminal and it indicates the current direction.

### Complementary MOSFETs

**Complementary MOS (CMOS)** technology uses both n-channel and p-channel devices in the same circuit. Figure 3.19 shows the cross section of n-channel and p-channel devices fabricated on the same chip. CMOS circuits, in general, are more complicated to fabricate than circuits using entirely NMOS or PMOS devices. Yet, as we will see in later chapters, CMOS circuits have great advantages over just NMOS or PMOS circuits.

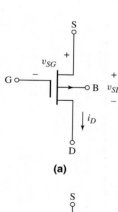

**(a)**

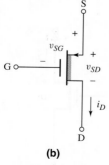

**(b)**

**Figure 3.18** The p-channel depletion mode MOSFET: (a) conventional circuit symbol and (b) simplified circuit symbol

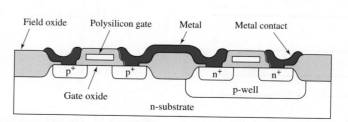

**Figure 3.19** Cross sections of n-channel and p-channel transistors fabricated with a p-well CMOS technology

In order to fabricate n-channel and p-channel devices that are electrically equivalent, the magnitude of the threshold voltages must be equal, and the n-channel and p-channel conduction parameters must be equal. Since, in general, $\mu_n$, and $\mu_p$ are not equal, the design of equivalent transistors involves adjusting the width-to-length ratios of the transistors.

### 3.1.8 Summary of Transistor Operation

We have presented a first-order model of the operation of the MOS transistor. For an n-channel enhancement-mode MOSFET, a positive gate-to-source voltage, greater than the threshold voltage $V_{TN}$, must be applied to induce an electron inversion layer. For $v_{GS} > V_{TN}$, the device is turned on. For an n-channel depletion-mode device, a channel between the source and drain exists even for $v_{GS} = 0$. The threshold voltage is negative, so that a negative value of $v_{GS}$ is required to turn the device off.

For a p-channel device, all voltage polarities and current directions are reversed compared to the NMOS device. For the p-channel enhancement-mode transistor, $V_{TP} < 0$ and for the depletion-mode PMOS transistor, $V_{TP} > 0$.

Table 3.1 lists the first-order equations that describe the $i$–$v$ relationships in MOS devices. We note that $K_n$ and $K_p$ are positive values and that the drain current $i_D$ is positive into the drain for the NMOS device and positive out of the drain for the PMOS device.

| **Table 3.1** | Summary of the MOSFET current–voltage relationships |
|---|---|
| **NMOS** | **PMOS** |
| Nonsaturation region ($v_{DS} < v_{DS}(\text{sat})$) | Nonsaturation region ($v_{SD} < v_{SD}(\text{sat})$) |
| $i_D = K_n[2(v_{GS} - V_{TN})v_{DS} - v_{DS}^2]$ | $i_D = K_p[2(v_{SG} + V_{TP})v_{SD} - v_{SD}^2]$ |
| Saturation region ($v_{DS} > v_{DS}(\text{sat})$) | Saturation region ($v_{SD} > v_{SD}(\text{sat})$) |
| $i_D = K_n(v_{GS} - V_{TN})^2$ | $i_D = K_p(v_{SG} + V_{TP})^2$ |
| Transition point | Transition point |
| $v_{DS}(\text{sat}) = v_{GS} - V_{TN}$ | $v_{SD}(\text{sat}) = v_{SG} + V_{TP}$ |
| Enhancement mode | Enhancement mode |
| $V_{TN} > 0$ | $V_{TP} < 0$ |
| Depletion mode | Depletion mode |
| $V_{TN} < 0$ | $V_{TP} > 0$ |

### 3.1.9 Short-Channel Effects

The current–voltage relations given by Equations (3.2(a)) and (3.2(b)) for the n-channel device and Equations (3.4(a)) and (3.4(b)) for the p-channel device are the ideal relations for long-channel devices. A long-channel device is generally one whose channel length is greater than 2 $\mu$m. In this device, the horizontal electric field in the channel induced by the drain voltage and the vertical electric field induced by the gate voltage can be treated independently. However, the channel length of present-day devices is on the order of 0.2 $\mu$m or less.

There are several effects in these short-channel devices that influence and change the long-channel current–voltage characteristics. One such effect is a variation in threshold voltage. The value of threshold voltage is a function of the channel length. This variation must be considered in the design and fabrication of these devices. The threshold voltage also becomes a function of the drain voltage. As the drain voltage increases, the effective threshold voltage decreases. This effect also influences the current–voltage characteristics.

The process conduction parameters, $k'_n$ and $k'_p$, are directly related to the carrier mobility. We have assumed that the carrier mobilities and corresponding process conduction parameters are constant. However, the carrier mobility values are functions of the vertical electric field in the inversion layer. As the gate voltage and vertical electric field increase, the carrier mobility decreases. This result, again, directly influences the current–voltage characteristics of the device.

Another effect that occurs in short-channel devices is velocity saturation. As the horizontal electric field increases, the velocity of the carriers reaches a constant value and will no longer increase with an increase in drain voltage. Velocity saturation will lower the $V_{DS}(\text{sat})$ voltage value. The drain current will reach its saturation value at a smaller $V_{DS}$ voltage. The drain current also becomes approximately a linear function of the gate voltage in the saturation region rather than the quadratic function of gate voltage in the long-channel characteristics.

Although the analysis of modern MOSFET circuits must take into account these short-channel effects, we will use the long-channel current–voltage relations in this introductory text. We will still be able to obtain a good basic understanding of the operation and characteristics of these devices, and we can still obtain a good basic understanding of the operation and characteristics of MOSFET circuits using the ideal long-channel current–voltage relations.

### 3.1.10    Additional Nonideal Current–Voltage Characteristics

The five nonideal effects in the current–voltage characteristics of MOS transistors are: the finite output resistance in the saturation region, the body effect, subthreshold conduction, breakdown effects, and temperature effects. This section will examine each of these effects.

### Finite Output Resistance

In the ideal case, when a MOSFET is biased in the saturation region, the drain current $i_D$ is independent of drain-to-source voltage $v_{DS}$. However, in actual MOSFET $i_D$ versus $v_{DS}$ characteristics, a nonzero slope does exist beyond the saturation point. For $v_{DS} > v_{DS}(\text{sat})$, the actual point in the channel at which the inversion charge goes to zero moves away from the drain terminal (see Figure 3.9(d)). The effective channel length decreases, producing the phenomenon called **channel length modulation.**

An exaggerated view of the current–voltage characteristics is shown in Figure 3.20. The curves can be extrapolated so that they intercept the voltage axis at a point $v_{DS} = -V_A$. The voltage $V_A$ is usually defined as a positive quantity. The slope of the curve in the saturation region can be described by expressing the $i_D$ versus $v_{DS}$ characteristic in the form, for an n-channel device,

$$i_D = K_n[(v_{GS} - V_{TN})^2(1 + \lambda v_{DS})] \tag{3.7}$$

where $\lambda$ is a positive quantity called the channel-length modulation parameter.

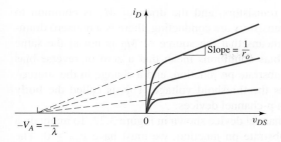

**Figure 3.20** Family of $i_D$ versus $v_{DS}$ curves showing the effect of channel length modulation producing a finite output resistance

The parameters $\lambda$ and $V_A$ are related. From Equation (3.7), we have $(1 + \lambda v_{DS}) = 0$ at the extrapolated point where $i_D = 0$. At this point, $v_{DS} = -V_A$, which means that $V_A = 1/\lambda$.

The output resistance due to the channel length modulation is defined as

$$r_o = \left( \frac{\partial i_D}{\partial v_{DS}} \right)^{-1} \Bigg|_{v_{GS}=\text{const.}} \tag{3.8}$$

From Equation (3.7), the output resistance, evaluated at the $Q$-point, is

$$r_o = [\lambda K_n (V_{GSQ} - V_{TN})^2]^{-1} \tag{3.9(a)}$$

or

$$r_o \cong [\lambda I_{DQ}]^{-1} = \frac{1}{\lambda I_{DQ}} = \frac{V_A}{I_{DQ}} \tag{3.9(b)}$$

The output resistance $r_o$ is also a factor in the small-signal equivalent circuit of the MOSFET, which is discussed in the next chapter.

## Body Effect

Up to this point, we have assumed that the substrate, or body, is connected to the source. For this bias condition, the threshold voltage is a constant.

In integrated circuits, however, the substrates of all n-channel MOSFETs are usually common and are tied to the most negative potential in the circuit. An example of two n-channel MOSFETs in series is shown in Figure 3.21. The p-type

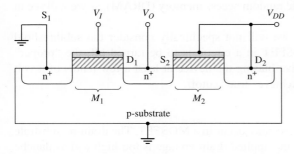

**Figure 3.21** Two n-channel MOSFETs fabricated in series in the same substrate. The source terminal, $S_2$, of the transistor $M_2$ is more than likely not at ground potential.

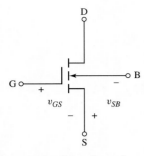

**Figure 3.22** An n-channel enhancement-mode MOSFET with a substrate voltage

substrate is common to the two transistors, and the drain of $M_1$ is common to the source of $M_2$. When the two transistors are conducting, there is a nonzero drain-to-source voltage on $M_1$, which means that the source of $M_2$ is not at the same potential as the substrate. These bias conditions mean that a zero or reverse-bias voltage exists across the source–substrate pn junction, and a change in the source–substrate junction voltage changes the threshold voltage. This is called the **body effect.** The same situation exists in p-channel devices.

For example, consider the n-channel device shown in Figure 3.22. To maintain a zero- or reverse-biased source–substrate pn junction, we must have $v_{SB} \geq 0$. The threshold voltage for this condition is given by

$$V_{TN} = V_{TNO} + \gamma \left[ \sqrt{2\phi_f + v_{SB}} - \sqrt{2\phi_f} \right] \tag{3.10}$$

where $V_{TNO}$ is the threshold voltage for $v_{SB} = 0$; $\gamma$, called the bulk threshold or **body-effect parameter,** is related to device properties, and is typically on the order of 0.5 $V^{1/2}$; and $\phi_f$ is a semiconductor parameter, typically on the order of 0.35 V, and is a function of the semiconductor doping. We see from Equation (3.10) that the threshold voltage in n-channel devices increases due to this body effect.

The body effect can cause a degradation in circuit performance because of the changing threshold voltage. However, we will generally neglect the body effect in our circuit analyses, for simplicity.

### Subthreshold Conduction

If we consider the ideal current-voltage relationship for the n-channel MOSFET biased in the saturation region, we have, from Equation (3.2(b)),

$$i_D = K_n(v_{GS} - V_{TN})^2$$

Taking the square root of both sides of the equation, we obtain

$$\sqrt{i_D} = \sqrt{K_n}(v_{GS} - V_{TN}) \tag{3.11}$$

From Equation (3.11), we see that $\sqrt{i_d}$ is a linear function of $v_{GS}$. Figure 3.23 shows a plot of this ideal relationship.

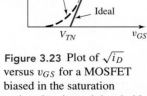

**Figure 3.23** Plot of $\sqrt{i_D}$ versus $v_{GS}$ for a MOSFET biased in the saturation region showing subthreshold conduction. Experimentally, a subthreshold current exists even for $v_{GS} < V_{TN}$.

Also plotted in Figure 3.23 are experimental results, which show that when $v_{GS}$ is slightly less than $V_{TN}$, the drain current is not zero, as previously assumed. This current is called the **subthreshold current.** The effect may not be significant for a single device, but if thousands or millions of devices on an integrated circuit are biased just slightly below the threshold voltage, the power supply current will not be zero but may contribute to significant power dissipation in the integrated circuit. One example of this is a dynamic random access memory (DRAM), as we will see in Chapter 16.

In this text, for simplicity we will not specifically consider the subthreshold current. However, when a MOSFET in a circuit is to be turned off, the "proper" design of the circuit must involve biasing the device at least a few tenths of a volt below the threshold voltage to achieve "true" cutoff.

### Breakdown Effects

Several possible breakdown effects may occur in a MOSFET. The drain-to-substrate pn junction may break down if the applied drain voltage is too high and avalanche multiplication occurs. This breakdown is the same reverse-biased pn junction breakdown discussed in Chapter 1 in Section 1.2.5.

As the size of the device becomes smaller, another breakdown mechanism, called *punch-through,* may become significant. **Punch-through** occurs when the drain voltage is large enough for the depletion region around the drain to extend completely through the channel to the source terminal. This effect also causes the drain current to increase rapidly with only a small increase in drain voltage.

A third breakdown mechanism is called **near-avalanche** or **snapback breakdown.** This breakdown process is due to second-order effects within the MOSFET. The source-substrate-drain structure is equivalent to that of a bipolar transistor. As the device size shrinks, we may begin to see a parasitic bipolar transistor action with increases in the drain voltage. This parasitic action enhances the breakdown effect.

If the electric field in the oxide becomes large enough, breakdown can also occur in the oxide, which can lead to catastrophic failure. In silicon dioxide, the electric field at breakdown is on the order of $6 \times 10^6$ V/cm, which, to a first approximation, is given by $E_{ox} = V_G / t_{ox}$. A gate voltage of approximately 30 V would produce breakdown in an oxide with a thickness of $t_{ox} = 500$ Å. However, a safety margin of a factor of 3 is common, which means that the maximum safe gate voltage for $t_{ox} = 500$ Å would be 10 V. A safety margin is necessary since there may be defects in the oxide that lower the breakdown field. We must also keep in mind that the input impedance at the gate is very high, and a small amount of static charge accumulating on the gate can cause the breakdown voltage to be exceeded. To prevent the accumulation of static charge on the gate capacitance of a MOSFET, a gate protection device, such as a reverse-biased diode, is usually included at the input of a MOS integrated circuit.

### Temperature Effects

Both the threshold voltage $V_{TN}$ and conduction parameter $K_n$ are functions of temperature. The magnitude of the threshold voltage decreases with temperature, which means that the drain current increases with temperature at a given $V_{GS}$. However, the conduction parameter is a direct function of the inversion carrier mobility, which decreases as the temperature increases. Since the temperature dependence of mobility is larger than that of the threshold voltage, the net effect of increasing temperature is a decrease in drain current at a given $V_{GS}$. This particular result provides a negative feedback condition in power MOSFETs. A decreasing value of $K_n$ inherently limits the channel current and provides stability for a power MOSFET.

## Test Your Understanding

**TYU 3.1** (a) An n-channel enhancement-mode MOSFET has a threshold voltage of $V_{TN} = 1.2$ V and an applied gate-to-source voltage of $v_{GS} = 2$ V. Determine the region of operation when: (i) $v_{DS} = 0.4$ V; (ii) $v_{DS} = 1$ V; and (iii) $v_{DS} = 5$ V. (b) Repeat part (a) for an n-channel depletion-mode MOSFET with a threshold voltage of $V_{TN} = -1.2$ V. (Ans. (a) (i) nonsaturation, (ii) saturation, (iii) saturation; (b) (i) nonsaturation, (ii) nonsaturation, (iii) saturation)

**TYU 3.2** The NMOS devices described in Exercise TYU 3.1 have parameters $W = 20\,\mu$m, $L = 0.8\,\mu$m, $t_{ox} = 200$ Å, $\mu_n = 500$ cm$^2$/V–s, and $\lambda = 0$. (a) Calculate the conduction parameter $K_n$ for each device. (b) Calculate the drain current for each bias condition listed in Exercise TYU 3.1. (Ans. (a) $K_n = 1.08$ mA/V$^2$; (b) $i_D = 0.518$ mA, 0.691 mA, and 0.691 mA; $i_D = 2.59$ mA, 5.83 mA, and 11.1 mA)

**TYU 3.3** (a) A p-channel enhancement-mode MOSFET has a threshold voltage of $V_{TP} = -1.2$ V and an applied source-to-gate voltage of $v_{SG} = 2$ V. Determine the region of operation when (i) $v_{SD} = 0.4$ V, (ii) $v_{SD} = 1$ V, and (iii) $v_{SD} = 5$ V. (b) Repeat part (a) for a p-channel depletion-mode MOSFET with a threshold voltage of $V_{TP} = +1.2$ V. (Ans. (a) (i) nonsaturation, (ii) saturation, (iii) saturation; (b) (i) nonsaturation, (ii) nonsaturation, (iii) saturation)

**TYU 3.4** The PMOS devices described in Exercise TYU 3.3 have parameters $W = 10 \, \mu\text{m}$, $L = 0.8 \, \mu\text{m}$, $t_{\text{ox}} = 200 \, \text{Å}$, $\mu_p = 300 \, \text{cm}^2/\text{V–s}$, and $\lambda = 0$. (a) Calculate the conduction parameter $K_p$ for each device. (b) Calculate the drain current for each bias condition listed in Exercise TYU 3.3. (Ans. (a) $K_p = 0.324$ mA/V$^2$; (b) $i_D = 0.156$ mA, 0.207 mA; and 0.207 mA; $i_D = 0.778$ mA, 1.75 mA, and 3.32 mA)

**TYU 3.5** The parameters of an NMOS enhancement-mode device are $V_{TN} = 0.25$ V and $K_n = 10 \, \mu\text{A/V}^2$. The device is biased at $v_{GS} = 0.5$ V. Calculate the drain current when (i) $v_{DS} = 0.5$ V and (ii) $v_{DS} = 1.2$ V for (a) $\lambda = 0$ and (b) $\lambda = 0.03 \, \text{V}^{-1}$. (c) Calculate the output resistance $r_o$ for parts (a) and (b). (Ans. (a) (i) and (ii) $i_D = 0.625 \, \mu\text{A}$; (b) (i) $i_D = 0.6344 \, \mu\text{A}$, (ii) $i_D = 0.6475 \, \mu\text{A}$; (c) (i) $r_o = \infty$, (ii) $r_o = 53.3$ MΩ).

**TYU 3.6** An NMOS transistor has parameters $V_{TNO} = 0.4$ V, $\gamma = 0.15 \, \text{V}^{1/2}$, and $\phi_f = 0.35$ V. Calculate the threshold voltage when (a) $v_{SB} = 0$, (b) $v_{SB} = 0.5$ V, and (c) $v_{SB} = 1.5$ V. (Ans. (a) 0.4 V, (b) 0.439 V, (c) 0.497 V)

## 3.2 MOSFET DC CIRCUIT ANALYSIS

**Objective:** • Understand and become familiar with the dc analysis and design techniques of MOSFET circuits.

In the last section, we considered the basic MOSFET characteristics and properties. We now start analyzing and designing the dc biasing of MOS transistor circuits. A primary purpose of the rest of the chapter is to continue to become familiar and comfortable with the MOS transistor and MOSFET circuits. The dc biasing of MOSFETs, the focus of this chapter, is an important part of the design of amplifiers. MOSFET amplifier design is the focus of the next chapter.

In most of the circuits presented in this chapter, resistors are used in conjunction with the MOS transistors. In a real MOSFET integrated circuit, however, the resistors are generally replaced by other MOSFETs, so the circuit is composed entirely of MOS devices. In general, a MOSFET device requires a smaller area than a resistor. As we go through the chapter, we will begin to see how this is accomplished and as we finish the text, we will indeed analyze and design circuits containing only MOSFETs.

In the dc analysis of MOSFET circuits, we can use the ideal current–voltage equations listed in Table 3.1 in Section 3.1.

### 3.2.1 Common-Source Circuit

One of the basic MOSFET circuit configurations is called the **common-source circuit.** Figure 3.24 shows one example of this type of circuit using an n-channel

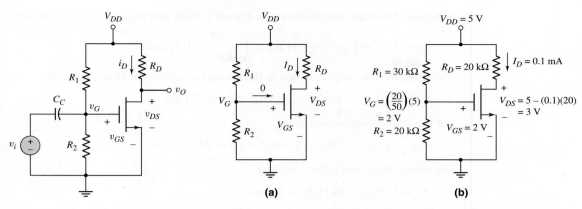

**Figure 3.24** An NMOS common-source circuit

**Figure 3.25** (a) The dc equivalent circuit of the NMOS common-source circuit and (b) the NMOS circuit for Example 3.3, showing current and voltage values

enhancement-mode MOSFET. The source terminal is at ground potential and is common to both the input and output portions of the circuit. The coupling capacitor $C_C$ acts as an open circuit to dc but it allows the signal voltage to be coupled to the gate of the MOSFET.

The dc equivalent circuit is shown in Figure 3.25(a). In the following dc analyses, we again use the notation for dc currents and voltages. Since the gate current into the transistor is zero, the voltage at the gate is given by a voltage divider, which can be written as

$$V_G = V_{GS} = \left( \frac{R_2}{R_1 + R_2} \right) V_{DD} \qquad (3.12)$$

Assuming that the gate-to-source voltage given by Equation (3.12) is greater than $V_{TN}$, and that the transistor is biased in the saturation region, the drain current is

$$I_D = K_n (V_{GS} - V_{TN})^2 \qquad (3.13)$$

The drain-to-source voltage is

$$V_{DS} = V_{DD} - I_D R_D \qquad (3.14)$$

If $V_{DS} > V_{DS}(\text{sat}) = V_{GS} - V_{TN}$, then the transistor is biased in the saturation region, as we initially assumed, and our analysis is correct. If $V_{DS} < V_{DS}(\text{sat})$, then the transistor is biased in the nonsaturation region, and the drain current is given by the more complicated characteristic Equation (3.2(a)).

The power dissipated in the transistor, since there is no gate current, is simply given by

$$P_T = I_D V_{DS} \qquad (3.15)$$

## EXAMPLE 3.3

**Objective:** Calculate the drain current and drain-to-source voltage of a common-source circuit with an n-channel enhancement-mode MOSFET. Find the power dissipated in the transistor.

For the circuit shown in Figure 3.25(a), assume that $R_1 = 30\,\text{k}\Omega$, $R_2 = 20\,\text{k}\Omega$, $R_D = 20\,\text{k}\Omega$, $V_{DD} = 5\,\text{V}$, $V_{TN} = 1\,\text{V}$, and $K_n = 0.1\,\text{mA/V}^2$.

**Solution:** From the circuit shown in Figure 3.25(b) and Equation (3.12), we have

$$V_G = V_{GS} = \left(\frac{R_2}{R_1 + R_2}\right) V_{DD} = \left(\frac{20}{20 + 30}\right)(5) = 2\,\text{V}$$

Assuming the transistor is biased in the saturation region, the drain current is

$$I_D = K_n(V_{GS} - V_{TN})^2 = (0.1)(2 - 1)^2 = 0.1\,\text{mA}$$

and the drain-to-source voltage is

$$V_{DS} = V_{DD} - I_D R_D = 5 - (0.1)(20) = 3\,\text{V}$$

The power dissipated in the transistor is

$$P_T = I_D V_{DS} = (0.1)(3) = 0.3\,\text{mW}$$

**Comment:** Because $V_{DS} = 3\,\text{V} > V_{DS}(\text{sat}) = V_{GS} - V_{TN} = 2 - 1 = 1\,\text{V}$, the transistor is indeed biased in the saturation region and our analysis is valid.

The dc analysis produces the quiescent values ($Q$-points) of drain current and drain-to-source voltage, usually indicated by $I_{DQ}$ and $V_{DSQ}$.

### EXERCISE PROBLEM

**Ex 3.3:** The transistor in Figure 3.25(a) has parameters $V_{TN} = 0.35$ V and $K_n = 25\,\mu\text{A/V}^2$. The circuit parameters are $V_{DD} = 2.2$ V, $R_1 = 355\,\text{k}\Omega$, $R_2 = 245\,\text{k}\Omega$, and $R_D = 100\,\text{k}\Omega$. Find $I_D$, $V_{GS}$, and $V_{DS}$. (Ans. $I_D = 7.52\,\mu\text{A}$, $V_{GS} = 0.898$ V, $V_{DS} = 1.45$ V)

Figure 3.26 (a) shows a common-source circuit with a p-channel enhancement-mode MOSFET. The source terminal is tied to $+V_{DD}$, which becomes signal ground in the ac equivalent circuit. Thus the terminology common-source applies to this circuit.

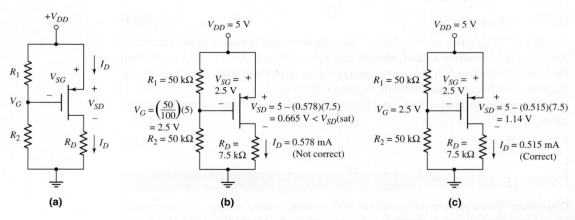

**Figure 3.26** (a) A PMOS common-source circuit, (b) the PMOS common-source circuit for Example 3.4 showing current and voltage values when the saturation-region bias assumption is incorrect, and (c) the circuit for Example 3.4 showing current and voltage values when the nonsaturation-region bias assumption is correct

The dc analysis is essentially the same as for the n-channel MOSFET circuit. The gate voltage is

$$V_G = \left( \frac{R_2}{R_1 + R_2} \right)(V_{DD}) \tag{3.16}$$

and the source-to-gate voltage is

$$V_{SG} = V_{DD} - V_G \tag{3.17}$$

Assuming that $V_{GS} < V_{TP}$, or $V_{SG} > |V_{TP}|$, and that the device is biased in the saturation region, the drain current is given by

$$I_D = K_p(V_{SG} + V_{TP})^2 \tag{3.18}$$

and the source-to-drain voltage is

$$V_{SD} = V_{DD} - I_D R_D \tag{3.19}$$

If $V_{SD} > V_{SD}(\text{sat}) = V_{SG} + V_{TP}$, then the transistor is indeed biased in the saturation region, as we have assumed. However, if $V_{SD} < V_{SD}(\text{sat})$, the transistor is biased in the nonsaturation region.

## EXAMPLE 3.4

**Objective:** Calculate the drain current and source-to-drain voltage of a common-source circuit with a p-channel enhancement-mode MOSFET.

Consider the circuit shown in Figure 3.26(a). Assume that $R_1 = R_2 = 50\,\text{k}\Omega$, $V_{DD} = 5\,\text{V}$, $R_D = 7.5\,\text{k}\Omega$, $V_{TP} = -0.8\,\text{V}$, and $K_p = 0.2\,\text{mA/V}^2$.

**Solution:** From the circuit shown in Figure 3.26(b) and Equation (3.16), we have

$$V_G = \left( \frac{R_2}{R_1 + R_2} \right)(V_{DD}) = \left( \frac{50}{50 + 50} \right)(5) = 2.5\,\text{V}$$

The source-to-gate voltage is therefore

$$V_{SG} = V_{DD} - V_G = 5 - 2.5 = 2.5\,\text{V}$$

Assuming the transistor is biased in the saturation region, the drain current is

$$I_D = K_p(V_{SG} + V_{TP})^2 = (0.2)(2.5 - 0.8)^2 = 0.578\,\text{mA}$$

and the source-to-drain voltage is

$$V_{SD} = V_{DD} - I_D R_D = 5 - (0.578)(7.5) = 0.665\,\text{V}$$

Since $V_{SD} = 0.665$ V is not greater than $V_{SD}(\text{sat}) = V_{SG} + V_{TP} = 2.5 - 0.8 = 1.7$ V, the p-channel MOSFET is **not** biased in the saturation region, as we initially assumed.

In the nonsaturation region, the drain current is given by

$$I_D = K_p\big[2(V_{SG} + V_{TP})V_{SD} - V_{SD}^2\big]$$

and the source-to-drain voltage is

$$V_{SD} = V_{DD} - I_D R_D$$

Combining these two equations, we obtain

$$I_D = K_p[2(V_{SG} + V_{TP})(V_{DD} - I_D R_D) - (V_{DD} - I_D R_D)^2]$$

or

$$I_D = (0.2)[2(2.5 - 0.8)(5 - I_D(7.5)) - (5 - I_D(7.5))^2]$$

Solving this quadratic equation for $I_D$, we find

$$I_D = 0.515\,\text{mA}$$

We also find that

$$V_{SD} = 1.14\,\text{V}$$

Therefore, $V_{SD} < V_{SD}(\text{sat})$, which verifies that the transistor is biased in the nonsaturation region.

**Comment:** In solving the quadratic equation for $I_D$, we find a second solution that yields $V_{SD} = 2.93$ V. However, this value of $V_{SD}$ is greater than $V_{SD}(\text{sat})$, so it is not a valid solution since we assumed the transistor to be biased in the nonsaturation region.

### EXERCISE PROBLEM

**Ex 3.4:** The transistor in Figure 3.26(a) has parameters $V_{TP} = -0.6$ V and $K_p = 0.2$ mA/V$^2$. The circuit is biased at $V_{DD} = 3.3$ V. Assume $R_1 \| R_2 = 300$ k$\Omega$. Design the circuit such that $I_{DQ} = 0.5$ mA and $V_{SDQ} = 2.0$ V. (Ans. $R_1 = 885$ k$\Omega$, $R_2 = 454$ k$\Omega$, $R_D = 2.6$ k$\Omega$)

### COMPUTER ANALYSIS EXERCISE

**PS 3.1:** Verify the results of Example 3.4 with a PSpice analysis.

As Example 3.4 illustrated, we may not know initially whether a transistor is biased in the saturation or nonsaturation region. The approach involves making an educated guess and then verifying that assumption. If the assumption proves incorrect, we must then change it and reanalyze the circuit.

In linear amplifiers containing MOSFETs, the transistors are biased in the saturation region.

### DESIGN EXAMPLE 3.5

**Objective:** Design a MOSFET circuit biased with both positive and negative voltages to meet a set of specifications.

**Specifications:** The circuit configuration to be designed is shown in Figure 3.27. Design the circuit such that $I_{DQ} = 0.5$ mA and $V_{DSQ} = 4$ V.

**Choices:** Standard resistors are to be used in the final design. A transistor with nominal parameters of $k'_n = 80\ \mu\text{A/V}^2$, $(W/L) = 6.25$, and $V_{TN} = 1.2$ V is available.

**Solution:** Assuming the transistor is biased in the saturation region, we have $I_{DQ} = K_n(V_{GS} - V_{TN})^2$. The conduction parameter is

$$K_n = \frac{k'_n}{2} \cdot \frac{W}{L} = \left(\frac{0.080}{2}\right)(6.25) = 0.25\,\text{mA/V}^2$$

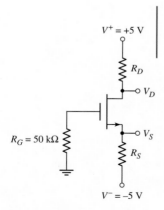

**Figure 3.27** Circuit configuration for Example 3.5

Solving for the gate-to-source voltage, we find the required gate-to-source voltage to induce the specified drain current.

$$V_{GS} = \sqrt{\frac{I_{DQ}}{K_n}} + V_{TN} = \sqrt{\frac{0.5}{0.25}} + 1.2$$

or

$$V_{GS} = 2.614 \text{ V}$$

Since the gate current is zero, the gate is at ground potential. The voltage at the source terminal is then $V_S = -V_{GS} = -2.614$ V. The value of the source resistor is found from

$$R_S = \frac{V_S - V^-}{I_{DQ}} = \frac{-2.614 - (-5)}{0.5}$$

or

$$R_S = 4.77 \text{ k}\Omega$$

The voltage at the drain terminal is determined to be

$$V_D = V_S + V_{DS} = -2.614 + 4 = 1.386 \text{ V}$$

The value of the drain resistor is

$$R_D = \frac{V^+ - V_D}{I_{DQ}} = \frac{5 - 1.386}{0.5}$$

or

$$R_D = 7.23 \text{ k}\Omega$$

We may note that

$$V_{DS} = 4 \text{ V} > V_{DS}(\text{sat}) = V_{GS} - V_{TN} = 2.61 - 1.2 = 1.41 \text{ V}$$

which means that the transistor is indeed biased in the saturation region.

**Trade-offs:** The closest standard resistor values are $R_S = 4.7 \text{ k}\Omega$ and $R_D = 7.5 \text{ k}\Omega$. We may find the gate-to-source voltage from

$$V_{GS} + I_D R_S - 5 = 0$$

where

$$I_D = K_n(V_{GS} - V_{TN})^2$$

Using the standard resistor values, we find $V_{GS} = 2.622$ V, $I_{DQ} = 0.506$ mA, and $V_{DSQ} = 3.83$ V. In this case, the drain current is within 1.2 percent of the design specification and the drain-to-source voltage is within 4.25 percent of the design specification.

**Comment:** It is important to keep in mind that the current into the gate terminal is zero. In this case, then, there is zero voltage drop across the $R_G$ resistor.

**Design Pointer:** In an actual circuit design using discrete elements, we need to choose standard resistor values that are closest to the design values. In addition, the discrete resistors have tolerances that need to be taken into account. In the final design, then, the actual drain current and drain-to-source voltage are somewhat different from the specified values. In many applications, this slight deviation from the specified values will not cause a problem.

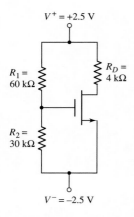

$V^+ = +2.5$ V

$R_1 = 60$ kΩ

$R_D = 4$ kΩ

$R_2 = 30$ kΩ

$V^- = -2.5$ V

**Figure 3.28** Circuit for Exercise Ex 3.5

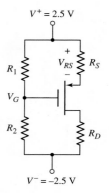

$V^+ = 2.5$ V

$R_1$

$+$
$V_{RS}$
$-$

$R_S$

$V_G$

$R_2$

$R_D$

$V^- = -2.5$ V

**Figure 3.29** Circuit configuration for Example 3.6

**EXERCISE PROBLEM**

**Ex 3.5:** For the transistor in the circuit in Figure 3.28, the nominal parameter values are $V_{TN} = 0.6$ V and $K_n = 0.5$ mA/V$^2$. (a) Determine the quiescent values $V_{GSQ}$, $I_{DQ}$, and $V_{DSQ}$. (b) Determine the range in $I_D$ and $V_{DS}$ values for a $\pm 5$ percent variation in $V_{TN}$ and $K_n$. (Ans. (a) $V_{GSQ} = 1.667$ V, $I_{DQ} = 0.5689$ mA, $V_{DSQ} = 2.724$ V; (b) $0.5105 \leq I_D \leq 0.6314$ mA, $2.474 \leq V_{DS} \leq 2.958$ V)

Now consider an example of a p-channel device biased with both positive and negative voltages.

## DESIGN EXAMPLE **3.6**

**Objective:** Design a circuit with a p-channel MOSFET that is biased with both negative and positive supply voltages and that contains a source resistor $R_S$ to meet a set of specifications.

**Specifications:** The circuit to be designed is shown in Figure 3.29. Design the circuit such that $I_{DQ} = 100$ μA, $V_{SDQ} = 3$ V, and $V_{RS} = 0.8$ V. Note that $V_{RS}$ is the voltage across the source resistor $R_S$. The value of the larger bias resistor, either $R_1$ or $R_2$, is to be 200 kΩ.

**Choices:** A transistor with nominal parameter values of $K_p = 100$ μA/V$^2$ and $V_{TP} = -0.4$ V is available. The conduction parameter may vary by $\pm 5$ percent.

**Solution:** Assuming that the transistor is biased in the saturation region, we have $I_{DQ} = K_p(V_{SG} + V_{TP})^2$. Solving for the source-to-gate voltage, we find the required value of source-to-gate voltage to be

$$V_{SG} = \sqrt{\frac{I_{DQ}}{K_p}} - V_{TP} = \sqrt{\frac{100}{100}} - (-0.4)$$

or

$$V_{SG} = 1.4 \text{ V}$$

We may note that the design value of

$$V_{SDQ} = 3 \text{ V} > V_{SDQ} \text{ (sat)} = V_{SGQ} + V_{TP} = 1.4 - 0.4 = 1 \text{ V}$$

so that the transistor will be biased in the saturation region.
The voltage at the gate with respect to ground potential is found to be

$$V_G = V^+ - V_{RS} - V_{SG} = 2.5 - 0.8 - 1.4 = 0.3 \text{ V}$$

With $V_G > 0$, the resistor $R_2$ will be the larger of the two bias resistors, so set $R_2 = 200$ kΩ. The current through $R_2$ is then

$$I_{\text{Bias}} = \frac{V_G - V^-}{R_2} = \frac{0.3 - (-2.5)}{200} = 0.014 \text{ mA}$$

Since the current through $R_1$ is the same, we can find the value of $R_1$ to be

$$R_1 = \frac{V^+ - V_G}{I_{\text{Bias}}} = \frac{2.5 - 0.3}{0.014}$$

which yields

$$R_1 = 157 \text{ k}\Omega$$

The source resistor value is found from

$$R_S = \frac{V_{RS}}{I_{DQ}} = \frac{0.8}{0.1}$$

or

$$R_S = 8 \, k\Omega$$

The voltage at the drain terminal is

$$V_D = V^+ - V_{RS} - V_{SD} = 2.5 - 0.8 - 3 = -1.3 \, V$$

Then the drain resistor value is found as

$$R_D = \frac{V_D - V^-}{I_{DQ}} = \frac{-1.3 - (-2.5)}{0.1}$$

or

$$R_D = 12 \, k\Omega$$

**Trade-offs:** If the conduction parameter $K_p$ varies by $\pm 5\%$, the quiescent drain current $I_{DQ}$ and the source-to-drain voltage $V_{SDQ}$ will change. Using the resistor values found in the previous design, we find the following:

| $K_p$ | $V_{SGQ}$ | $I_{DQ}$ | $V_{SDQ}$ |
|---|---|---|---|
| 95 $\mu A/V^2$ | 1.416 V | 98.0 $\mu A$ | 3.04 V |
| 105 $\mu A/V^2$ | 1.385 V | 101.9 $\mu A$ | 2.962 V |
| $\pm 5\%$ | $\pm 1.14\%$ | $\pm 2\%$ | $\pm 1.33\%$ |

**Comment:** We may note that the variation in the $Q$-point values is smaller that the variation in $K_p$. Including the source resistor $R_S$ tends to stabilize the $Q$-point.

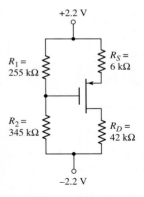

**Figure 3.30** Figure for Exercise Ex 3.6

**EXERCISE PROBLEM**

**Ex 3.6:** Consider the circuit shown in Figure 3.30. The nominal transistor parameters are $V_{TP} = -0.30 \, V$ and $K_p = 120 \, \mu A/V^2$. (a) Calculate $V_{SG}$, $I_D$, and $V_{SD}$. (b) Determine the variation in $I_D$ if the threshold voltage varies by $\pm 5$ percent. (Ans. (a) $V_{SG} = 1.631 \, V$, $I_D = 0.2126 \, mA$, $V_{SD} = 3.295 \, V$; (b) $0.2091 \leq I_D \leq 0.2160 \, mA$)

**COMPUTER ANALYSIS EXERCISE**

**PS 3.2** Verify the circuit design in Example 3.6 with a PSpice simulation. Also investigate the change in $Q$-point values with $\pm 10$ percent variations in resistor values.

### 3.2.2    Load Line and Modes of Operation

The load line is helpful in visualizing the region in which the MOSFET is biased. Consider again the common-source circuit shown in Figure 3.25(b). Writing a Kirchhoff's voltage law equation around the drain-source loop results in Equation (3.14), which is the load line equation, showing a linear relationship between the drain current and drain-to-source voltage.

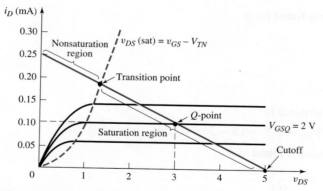

**Figure 3.31** Transistor characteristics, $v_{DS}$(sat) curve, load line, and $Q$-point for the NMOS common-source circuit in Figure 3.25(b)

Figure 3.31 shows the $v_{DS}$(sat) characteristic for the transistor described in Example 3.3. The load line is given by

$$V_{DS} = V_{DD} - I_D R_D = 5 - I_D(20) \qquad (3.20(a))$$

or

$$I_D = \frac{5}{20} - \frac{V_{DS}}{20} \text{(mA)} \qquad (3.20(b))$$

and is also plotted in the figure. The two end points of the load line are determined in the usual manner. If $I_D = 0$, then $V_{DS} = 5\,\text{V}$; if $V_{DS} = 0$, then $I_D = 5/20 = 0.25\,\text{mA}$. The $Q$-point of the transistor is given by the dc drain current and drain-to-source voltage, and it is always on the load line, as shown in the figure. A few transistor characteristics are also shown on the figure.

If the gate-to-source voltage is less than $V_{TN}$, the drain current is zero and the transistor is in cutoff. As the gate-to-source voltage becomes just greater than $V_{TN}$, the transistor turns on and is biased in the saturation region. As $V_{GS}$ increases, the $Q$-point moves up the load line. The **transition point** is the boundary between the saturation and nonsaturation regions and is defined as the point where $V_{DS} = V_{DS}$(sat) $= V_{GS} - V_{TN}$. As $V_{GS}$ increases above the transition point value, the transistor becomes biased in the nonsaturation region.

## EXAMPLE 3.7

**Objective:** Determine the transition point parameters for a common-source circuit.

Consider the circuit shown in Figure 3.25(b). Assume transistor parameters of $V_{TN} = 1\,\text{V}$ and $K_n = 0.1\,\text{mA/V}^2$.

**Solution:** At the transition point,

$$V_{DS} = V_{DS}(\text{sat}) = V_{GS} - V_{TN} = V_{DD} - I_D R_D$$

The drain current is still

$$I_D = K_n(V_{GS} - V_{TN})^2$$

Combining the last two equations, we obtain

$$V_{GS} - V_{TN} = V_{DD} - K_n R_D(V_{GS} - V_{TN})^2$$

Rearranging this equation produces

$$K_n R_D (V_{GS} - V_{TN})^2 + (V_{GS} - V_{TN}) - V_{DD} = 0$$

or

$$(0.1)(20)(V_{GS} - V_{TN})^2 + (V_{GS} - V_{TN}) - 5 = 0$$

Solving the quadratic equation, we find that

$$V_{GS} - V_{TN} = 1.35 \, \text{V} = V_{DS}$$

Therefore,

$$V_{GS} = 2.35 \, \text{V}$$

and

$$I_D = (0.1)(2.35 - 1)^2 = 0.182 \, \text{mA}$$

**Comment:** For $V_{GS} < 2.35 \, \text{V}$, the transistor is biased in the saturation region; for $V_{GS} > 2.35 \, \text{V}$, the transistor is biased in the nonsaturation region.

## EXERCISE PROBLEM

**Ex 3.7:** Consider the circuit in Figure 3.30. Using the nominal transistor parameters described in Exercise Ex 3.6, draw the load line and determine the transition point parameters. (Ans. $V_{SG} = 2.272 \, \text{V}$, $I_D = 0.4668 \, \text{mA}$, $V_{SD} = 1.972 \, \text{V}$)

### Problem-Solving Technique: MOSFET DC Analysis

Analyzing the dc response of a MOSFET circuit requires knowing the bias condition (saturation or nonsaturation) of the transistor. In some cases, the bias condition may not be obvious, which means that we have to guess the bias condition, then analyze the circuit to determine if we have a solution consistent with our initial guess. To do this, we can:

1.  Assume that the transistor is biased in the saturation region, in which case $V_{GS} > V_{TN}$, $I_D > 0$, and $V_{DS} \geq V_{DS}(\text{sat})$.
2.  Analyze the circuit using the saturation current-voltage relations.
3.  Evaluate the resulting bias condition of the transistor. If the assumed parameter values in step 1 are valid, then the initial assumption is correct. If $V_{GS} < V_{TN}$, then the transistor is probably cutoff, and if $V_{DS} < V_{DS}(\text{sat})$, the transistor is likely biased in the nonsaturation region.
4.  If the initial assumption is proved incorrect, then a new assumption must be made and the circuit reanalyzed. Step 3 must then be repeated.

### 3.2.3    Additional MOSFET Configurations: DC Analysis

There are other MOSFET circuits, in addition to the basic common-source circuits just considered, that are biased with the basic four-resistor configuration.

However, MOSFET integrated circuit amplifiers are generally biased with constant current sources. Example 3.8 demonstrates this technique using an ideal current source.

## DESIGN EXAMPLE 3.8

**Objective:**  Design a MOSFET circuit biased with a constant-current source to meet a set of specifications.

**Specifications:**  The circuit configuration to be designed is shown in Figure 3.32(a). Design the circuit such that the quiescent values are $I_{DQ} = 250\ \mu\text{A}$ and $V_D = 2.5$ V.

**Choices:**  A transistor with nominal values of $V_{TN} = 0.8$ V, $k_n' = 80\ \mu\text{A/V}^2$, and $W/L = 3$ is available. Assume $k_n'$ varies by $\pm 5$ percent.

**Solution:**  The dc equivalent circuit is shown in Figure 3.32(b). Since $v_i = 0$, the gate is at ground potential and there is no current through $R_G$. We have that $I_Q = I_{DQ} = 250\ \mu\text{A}$.

Assuming the transistor is biased in the saturation region, we have

$$I_D = \frac{k_n'}{2} \cdot \frac{W}{L}(V_{GS} - V_{TN})^2$$

or

$$250 = \left(\frac{80}{2}\right) \cdot (3)(V_{GS} - 0.8)^2$$

which yields

$$V_{GS} = 2.24\ \text{V}$$

The voltage at the source terminal is $V_S = -V_{GS} = -2.24$ V.

The drain current can also be written as

$$I_D = \frac{5 - V_D}{R_D}$$

For $V_D = 2.5$ V, we have

$$R_D = \frac{5 - 2.5}{0.25} = 10\ \text{k}\Omega$$

The drain-to-source voltage is

$$V_{DS} = V_D - V_S = 2.5 - (-2.24) = 4.74\ \text{V}$$

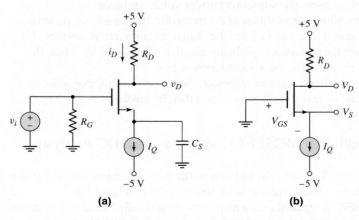

**(a)**                                   **(b)**

**Figure 3.32**  (a) NMOS common-source circuit biased with a constant-current source and (b) equivalent dc circuit

Since $V_{DS} = 4.74\,\text{V} > V_{DS}(\text{sat}) = V_{GS} - V_{TN} = 2.24 - 0.8 = 1.44\,\text{V}$, the transistor is biased in the saturation region, as initially assumed.

**Trade-offs:** Note that even if $k'_n$ changes, the drain current remains constant. For $76 \leq k'_n \leq 84\,\mu\text{A/V}^2$, the variation in $V_{GSQ}$ is $2.209 \leq V_{GSQ} \leq 2.281$ V and the variation in $V_{DSQ}$ is $4.709 \leq V_{DSQ} \leq 4.781$ V. The variation in $V_{DSQ}$ is $\pm 0.87$ percent even with a $\pm 5$ percent variation in $k'_n$. This stability effect is one advantage of using constant current biasing.

**Comment:** MOSFET circuits can be biased by using constant-current sources, which in turn are designed by using other MOS transistors, as we will see. Biasing with current sources tends to stabilize circuits against variations in device or circuit parameters.

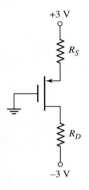

**Figure 3.33** Circuit for Exercise Ex 3.8

### EXERCISE PROBLEM

**Ex 3.8:** (a) Consider the circuit shown in Figure 3.33. The transistor parameters are $V_{TP} = -0.40$ V and $K_p = 30\,\mu\text{A/V}^2$. Design the circuit such that $I_{DQ} = 60\,\mu\text{A}$ and $V_{SDQ} = 2.5$ V. (b) Determine the variation in $Q$-point values if the $V_{TP}$ and $K_p$ parameters vary by $\pm 5$ percent. (Ans. (a) $R_S = 19.77\,\text{k}\Omega$, $R_D = 38.57\,\text{k}\Omega$; (b) $58.2 \leq I_{DQ} \leq 61.08\,\mu\text{A}$, $2.437 \leq V_{SDQ} \leq 2.605$ V)

### n-Channel Enhancement-Load Device

An enhancement-mode MOSFET connected in a configuration such as that shown in Figure 3.34 can be used as a nonlinear resistor. A transistor with this connection is called an **enhancement-load device.** Since the transistor is an enhancement mode device, $V_{TN} > 0$. Also, for this circuit, $v_{DS} = v_{GS} > v_{DS}(\text{sat}) = v_{GS} - V_{TN}$, which means that the transistor is always biased in the saturation region. The general $i_D$ versus $v_{DS}$ characteristics can then be written as

$$i_D = K_n(v_{GS} - V_{TN})^2 = K_n(v_{DS} - V_{TN})^2 \tag{3.21}$$

Figure 3.35 shows a plot of Equation (3.21) for the case when $K_n = 1\,\text{mA/V}^2$ and $V_{TN} = 1$ V.

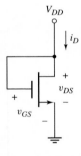

**Figure 3.34** Enhancement-mode NMOS device with the gate connected to the drain

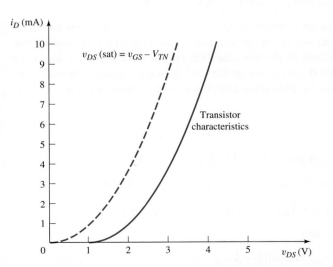

**Figure 3.35** Current–voltage characteristic of an enhancement load device

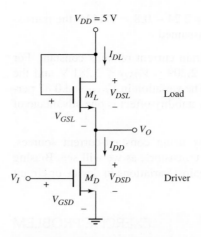

**Figure 3.36** Circuit with enhancement-load device and NMOS driver

If an enhancement-load device is connected in a circuit with another MOSFET in the configuration shown in Figure 3.36, the circuit can be used as an amplifier or as an inverter in a digital logic circuit. The load device, $M_L$, is always biased in the saturation region, and the transistor $M_D$, called the **driver transistor,** can be biased in either the saturation or nonsaturation region, depending on the value of the input voltage. The next example addresses the dc analysis of this circuit for dc input voltages to the gate of $M_D$.

## EXAMPLE 3.9

**Objective:** Determine the dc transistor currents and voltages in a circuit containing an enhancement load device.

The transistors in the circuit shown in Figure 3.36 have parameters $V_{TND} = V_{TNL} = 1 \text{ V}$, $K_{nD} = 50 \ \mu\text{A/V}^2$, and $K_{nL} = 10 \ \mu\text{A/V}^2$. Also assume $\lambda_{nD} = \lambda_{nL} = 0$. (The subscript $D$ applies to the driver transistor and the subscript $L$ applies to the load transistor.) Determine $V_O$ for $V_I = 5 \text{ V}$ and $V_I = 1.5 \text{ V}$.

**Solution:** $(V_I = 5 \text{ V})$ For an inverter circuit with a resistive load, when the input voltage is large, the output voltage drops to a low value. Therefore, we assume that the driver transistor is biased in the nonsaturation region since the drain-to-source voltage will be small. The drain current in the load device is equal to the drain current in the driver transistor. Writing these currents in generic form, we have

$$I_{DD} = I_{DL}$$

or

$$K_{nD}\left[2(V_{GSD} - V_{TND})V_{DSD} - V_{DSD}^2\right] = K_{nL}[V_{GSL} - V_{TNL}]^2$$

Since $V_{GSD} = V_I$, $V_{DSD} = V_O$, and $V_{GSL} = V_{DSL} = V_{DD} - V_O$, then

$$K_{nD}\left[2(V_I - V_{TND})V_O - V_O^2\right] = K_{nL}[V_{DD} - V_O - V_{TNL}]^2$$

Substituting numbers, we find

$$(50)\left[2(5 - 1)V_O - V_O^2\right] = (10)[5 - V_O - 1]^2$$

Rearranging the terms provides

$$3V_O^2 - 24V_O + 8 = 0$$

Using the quadratic formula, we obtain two possible solutions:

$$V_O = 7.65 \text{ V} \quad \text{or} \quad V_O = 0.349 \text{ V}$$

Since the output voltage cannot be greater than the supply voltage $V_{DD} = 5$ V, the valid solution is $V_O = 0.349$ V.

Also, since $V_{DSD} = V_O = 0.349 \text{ V} < V_{GSD} - V_{TND} = 5 - 1 = 4 \text{ V}$, the driver $M_D$ is biased in the nonsaturation region, as initially assumed.

The current can be determined from

$$I_D = K_{nL}(V_{GSL} - V_{TNL})^2 = K_{nL}(V_{DD} - V_O - V_{TNL})^2$$

or

$$I_D = (10)(5 - 0.349 - 1)^2 = 133 \ \mu\text{A}$$

**Solution:** $(V_I = 1.5 \text{ V})$ Since the threshold voltage of the driver transistor is $V_{TN} = 1$ V, an input voltage of 1.5 V means the transistor current is going to be relatively small so the output voltage should be relatively large. For this reason, we will assume that the driver transistor $M_D$ is biased in the saturation region. Equating the currents in the two transistors and writing the current equations in generic form, we have

$$I_{DD} = I_{DL}$$

or

$$K_{nD}[V_{GSD} - V_{TND}]^2 = K_{nL}[V_{GSL} - V_{TNL}]^2$$

Again, since $V_{GSD} = V_I$ and $V_{GSL} = V_{DSL} = V_{DD} - V_O$, then

$$K_{nD}[V_I - V_{TND}]^2 = K_{nL}[V_{DD} - V_O - V_{TNL}]^2$$

Substituting numbers and taking the square root, we find

$$\sqrt{50}[1.5 - 1] = \sqrt{10}[5 - V_O - 1]$$

which yields $V_O = 2.88$ V.

Since $V_{DSD} = V_O = 2.88 \text{ V} > V_{GSD} - V_{TND} = 1.5 - 1 = 0.5 \text{ V}$, the driver transistor $M_D$ is biased in the saturation region, as initially assumed.

The current is

$$I_D = K_{nD}(V_{GSD} - V_{TND})^2 = (50)(1.5 - 1)^2 = 12.5 \ \mu\text{A}$$

**Comment:** For this example, we made an initial guess as to whether the driver transistor was biased in the saturation or nonsaturation region. A more analytical approach is shown following this example.

**Computer Simulation:** The voltage transfer characteristics of the NMOS inverter with enhancement load shown in Figure 3.36 were obtained by a PSpice analysis. These results are shown in Figure 3.37. As the input voltage decreases from its high state, the output voltage increases, charging and discharging capacitances in the transistors. The current in the circuit goes to zero when the driver transistor is cutoff. This occurs when $V_I = V_{GSD} = V_{TN} = 1$ V. At this point, the output voltage is $V_O = 4$ V. Since there is no current, the capacitances cease charging and discharging so the output voltage cannot get to the full $V_{DD} = 5$ V value. The maximum output voltage is $V_O(\text{max}) = V_{DD} - V_{TNL} = 5 - 1 = 4$ V.

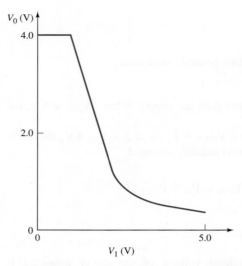

**Figure 3.37** Voltage transfer characteristics of NMOS inverter with enhancement load device

When the input voltage is just greater than 1 V, both transistors are biased in the saturation region as the previous analysis for $V_I = 1.5$ V showed. The output voltage is a linear function of input voltage as we will see in Equation (3.24).

For an input voltage greater than approximately 2.25 V, the driver transistor is biased in the nonsaturation region and the output voltage is a nonlinear function of input voltage.

EXERCISE PROBLEM

**Ex 3.9:** Consider the NMOS inverter shown in Figure 3.36 with transistor parameters described in Example 3.9. Determine the output voltage $V_O$ for input voltages (a) $V_I = 4$ V and (b) $V_I = 2$ V. (Ans. (a) 0.454 V, (b) 1.76 V)

COMPUTER ANALYSIS EXERCISE

**PS 3.3:** Consider the NMOS circuit shown in Figure 3.36. Plot the voltage transfer characteristics, using a PSpice simulation. Use transistor parameters similar to those in Example 3.9. What are the values of $V_O$ for $V_I = 1.5$ V and $V_I = 5$ V?

In the circuit shown in Figure 3.36, we can determine the transition point for the driver transistor that separates the saturation and nonsaturation regions. The transition point is determined by the equation

$$V_{DSD}(\text{sat}) = V_{GSD} - V_{TND} \tag{3.22}$$

Again, the drain currents in the two transistors are equal. Using the saturation drain current relationship for the driver transistor, we have

$$I_{DD} = I_{DL} \tag{3.23(a)}$$

or

$$K_{nD}[V_{GSD} - V_{TND}]^2 = K_{nL}[V_{GSL} - V_{TNL}]^2 \tag{3.23(b)}$$

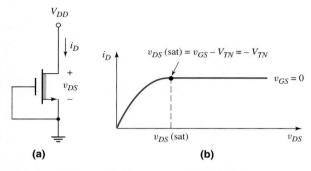

**Figure 3.38** (a) Depletion-mode NMOS device with the gate connected to the source and (b) current–voltage characteristics

Again, noting that $V_{GSD} = V_I$ and $V_{GSL} = V_{DSL} = V_{DD} - V_O$, and taking the square root, we have

$$\sqrt{\frac{K_{nD}}{K_{nL}}}(V_I - V_{TND}) = (V_{DD} - V_O - V_{TNL}) \tag{3.24}$$

At the transition point, we can define the input voltage as $V_I = V_{It}$ and the output voltage as $V_{Ot} = V_{DSD}(\text{sat}) = V_{It} - V_{TND}$. Then, from Equation (3.24), the input voltage at the transition point is

$$V_{It} = \frac{V_{DD} - V_{TNL} + V_{TND}(1 + \sqrt{K_{nD}/K_{nL}})}{1 + \sqrt{K_{nD}/K_{nL}}} \tag{3.25}$$

If we apply Equation (3.25) to the previous example, we can show that our initial assumptions were correct.

### n-Channel Depletion-Load Device

An n-channel depletion-mode MOSFET can also be used as a load device. Consider the depletion-mode MOSFET with the gate and source connected together shown in Figure 3.38(a). The current–voltage characteristics are shown in Figure 3.38(b). The transistor may be biased in either the saturation or nonsaturation regions. The transition point is also shown on the plot. The threshold voltage of the n-channel depletion-mode MOSFET is negative so that $v_{DS}(\text{sat})$ is positive.

A depletion-load device can be used in conjunction with another MOSFET, as shown in Figure 3.39, to create a circuit that can be used as an amplifier or as an inverter in a digital logic circuit. Both the load device $M_L$ and driver transistor $M_D$ may be biased in either the saturation or nonsaturation region, depending on the value of the input voltage. We will perform the dc analysis of this circuit for a particular dc input voltage to the gate of the driver transistor.

## EXAMPLE 3.10

**Objective:** Determine the dc transistor currents and voltages in a circuit containing a depletion load device.

Consider the circuit shown in Figure 3.39. Let $V_{DD} = 5$ V and assume transistor parameters of $V_{TND} = 1$ V, $V_{TNL} = -2$ V, $K_{nD} = 50\,\mu\text{A/V}^2$, and $K_{nL} = 10\,\mu\text{A/V}^2$. Determine $V_O$ for $V_I = 5$ V.

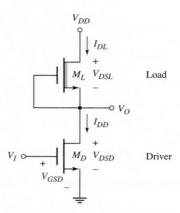

**Figure 3.39**  Circuit with depletion-load device and NMOS driver

**Solution:** Assume the driver transistor $M_D$ is biased in the nonsaturation region and the load transistor $M_L$ is biased in the saturation region. The drain currents in the two transistors are equal. In generic form, these currents are

$$I_{DD} = I_{DL}$$

or

$$K_{nD}\left[2(V_{GSD} - V_{TND})V_{DSD} - V_{DSD}^2\right] = K_{nL}[V_{GSL} - V_{TNL}]^2$$

Since $V_{GSD} = V_I$, $V_{DSD} = V_O$, and $V_{GSL} = 0$, then

$$K_{nD}\left[2(V_I - V_{TND})V_O - V_O^2\right] = K_{nL}[-V_{TNL}]^2$$

Substituting numbers, we find

$$(50)\left[2(5 - 1)V_O - V_O^2\right] = (10)[-(-2)]^2$$

Rearranging the terms produces

$$5V_O^2 - 40V_O + 4 = 0$$

Using the quadratic formula, we obtain two possible solutions:

$$V_O = 7.90\,\text{V} \quad \text{or} \quad V_O = 0.10\,\text{V}$$

Since the output voltage cannot be greater than the supply voltage $V_{DD} = 5$ V, the valid solution is $V_O = 0.10$ V.

The current is

$$I_D = K_{nL}(-V_{TNL})^2 = (10)[-(-2)]^2 = 40\,\mu\text{A}$$

**Comment:** Since $V_{DSD} = V_O = 0.10\,\text{V} < V_{GSD} - V_{TND} = 5 - 1 = 4\,\text{V}$, $M_D$ is biased in the nonsaturation region, as assumed. Similarly, since $V_{DSL} = V_{DD} - V_O = 4.9\,\text{V} > V_{GSL} - V_{TNL} = 0 - (-2) = 2\,\text{V}$, $M_L$ is biased in the saturation region, as originally assumed.

**Computer Simulation:** The voltage transfer characteristics of the NMOS inverter circuit with depletion load in Figure 3.39 were obtained using a PSpice analysis. These results are shown in Figure 3.40. For an input voltage less than 1 V, the driver is cut off and the output voltage is $V_O = V_{DD} = 5$ V.

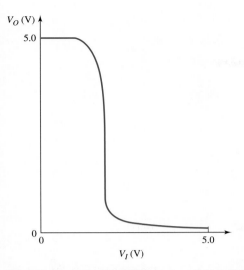

**Figure 3.40** Voltage transfer characteristics of NMOS inverter with depletion load device

When the input voltage is just greater than 1 V, the driver transistor is biased in the saturation region and the load device in the nonsaturation region. When the input voltage is approximately 1.9 V, both transistors are biased in the saturation region. If the channel length modulation parameter $\lambda$ is assumed to be zero as in this example, there is no change in the input voltage during this transition region. As the input voltage becomes larger than 1.9 V, the driver is biased in the nonsaturation region and the load in the saturation region.

## EXERCISE PROBLEM

**Ex 3.10:** Consider the circuit shown in Figure 3.39 with transistor parameters $V_{TND} = 1$ V and $V_{TNL} = -2$ V. (a) Design the ratio $K_{nD}/K_{nL}$ that will produce an output voltage of $V_O = 0.25$ V at $V_I = 5$ V. (b) Find $K_{nD}$ and $K_{nL}$ if the transistor currents are 0.2 mA when $V_I = 5$ V. (Ans. (a) $K_{nD}/K_{nL} = 2.06$ (b) $K_{nL} = 50\ \mu\text{A/V}^2$, $K_{nD} = 103\ \mu\text{A/V}^2$)

## COMPUTER ANALYSIS EXERCISE

**PS 3.4:** Consider the NMOS circuit shown in Figure 3.39. Plot the voltage transfer characteristics using a PSpice simulation. Use transistor parameters similar to those in Example 3.10. What are the values of $V_O$ for $V_I = 1.5$ V and $V_I = 5$ V?

### p-Channel Enhancement-Load Device
A p-channel enhancement-mode transistor can also be used as a load device to form a **complementary MOS (CMOS)** inverter. The term complementary implies that both n-channel and p-channel transistors are used in the same circuit. The CMOS technology is used extensively in both analog and digital electronic circuits.

Figure 3.41 shows one example of a CMOS inverter. The NMOS transistor is used as the amplifying device, or the driver, and the PMOS device is the load, which is referred to as an active load. This configuration is typically used in analog applications.

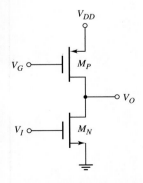

**Figure 3.41** Example of CMOS inverter

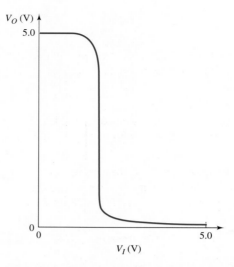

**Figure 3.42** Voltage transfer characteristics of CMOS inverter in Figure 3.41

In another configuration, the two gates are tied together and form the input. This configuration will be discussed in detail in Chapter 16.

As with the previous two NMOS inverters, the two transistors shown in Figure 3.41 may be biased in either the saturation or nonsaturation region, depending on the value of the input voltage. The voltage transfer characteristic is most easily determined from a PSpice analysis.

## EXAMPLE 3.11

**Objective:** Determine the voltage transfer characteristic of the CMOS inverter using a PSpice analysis.

For the circuit shown in Figure 3.41, assume transistor parameters of $V_{TN} = 1$ V, $V_{TP} = -1$ V, and $K_n = K_p$. Also assume $V_{DD} = 5$ V and $V_G = 3.25$ V.

**Solution:** The voltage transfer characteristics are shown in Figure 3.42. In this case, there is a region, as was the case for an NMOS inverter with depletion load, in which both transistors are biased in the saturation region, and the input voltage is a constant over this transition region for the assumption that the channel length modulation parameter $\lambda$ is zero.

**Comment:** In this example, the source-to-gate voltage of the PMOS device is only $V_{SG} = 1.75$ V. The effective resistance looking into the drain of the PMOS device is then relatively large. This is a desirable characteristic for an amplifier, as we will see in the next chapter.

## EXERCISE PROBLEM

**Ex 3.11:** Consider the circuit in Figure 3.41. Assume the same transistor parameters and circuit parameters as given in Example 3.11. Determine the transition point parameters for the transistors $M_N$ and $M_P$. (Ans. $M_P$: $V_{Ot} = 4.25$ V, $V_{It} = 1.75$ V; $M_N$: $V_{Ot} = 0.75$ V, $V_{It} = 1.75$ V)

## Test Your Understanding

**TYU 3.7** The transistor in the circuit shown in Figure 3.25(a) has parameters $V_{TN} = 0.25$ V and $K_n = 30\,\mu\text{A/V}^2$. The circuit is biased at $V_{DD} = 2.2$ V. Let $R_1 + R_2 = 500\,\text{k}\Omega$. Redesign the circuit such that $I_{DQ} = 70\,\mu\text{A}$ and $V_{DSQ} = 1.2$ V. (Ans. $R_1 = 96\,\text{k}\Omega$, $R_2 = 404\,\text{k}\Omega$, $R_D = 14.3\,\text{k}\Omega$)

**TYU 3.8** Consider the circuit in Figure 3.43. The transistor parameters are $V_{TN} = 0.4$ V and $k'_n = 100\,\mu\text{A/V}^2$. Design the transistor width-to-length ratio such that $V_{DS} = 1.6$ V. (Ans. 2.36)

**TYU 3.9** For the circuit shown in Figure 3.36, use the transistor parameters given in Example 3.9. (a) Determine $V_I$ and $V_O$ at the transition point for the driver transistor. (b) Calculate the transistor currents at the transition point. (Ans. (a) $V_{It} = 2.236$ V, $V_{Ot} = 1.236$ V; (b) $I_D = 76.4\,\mu\text{A}$)

**TYU 3.10** Consider the circuit shown in Figure 3.44. The transistor parameters are $V_{TN} = -1.2$ V and $k'_n = 80\,\mu\text{A/V}^2$. (a) Design the transistor width-to-length ratio such that $V_{DS} = 1.8$ V. Is the transistor biased in the saturation or nonsaturation region? (b) Repeat part (a) for $V_{DS} = 0.8$ V. (Ans. (a) 3.26, (b) 6.10)

**TYU 3.11** For the circuit shown in Figure 3.39, use the transistor parameters given in Example 3.10. (a) Determine $V_I$ and $V_O$ at the transition point for the load transistor. (b) Determine $V_I$ and $V_O$ at the transition point for the driver transistor. (Ans. (a) $V_{It} = 1.89$ V, $V_{Ot} = 3$ V; (b) $V_{It} = 1.89$ V, $V_{Ot} = 0.89$ V)

**Figure 3.43** Circuit for Exercise TYU 3.8

**Figure 3.44** Circuit for Exercise TYU 3.10

## 3.3   BASIC MOSFET APPLICATIONS: SWITCH, DIGITAL LOGIC GATE, AND AMPLIFIER

**Objective:** • Examine three applications of MOSFET circuits: a switch circuit, digital logic circuit, and an amplifier circuit.

MOSFETs may be used to: switch currents, voltages, and power; perform digital logic functions; and amplify small time-varying signals. In this section, we will examine the switching properties of an NMOS transistor, analyze a simple NMOS transistor digital logic circuit, and discuss how the MOSFET can be used to amplify small signals.

### 3.3.1   NMOS Inverter

The MOSFET can be used as a switch in a wide variety of electronic applications. The transistor switch provides an advantage over mechanical switches in both speed and reliability. The transistor switch considered in this section is also called an inverter. Two other switch configurations, the NMOS transmission gate and the CMOS transmission gate, are discussed in Chapter 16.

Figure 3.45 shows the n-channel enhancement-mode MOSFET inverter circuit. If $v_I < V_{TN}$, the transistor is in cutoff and $i_D = 0$. There is no voltage drop across $R_D$, and the output voltage is $v_O = V_{DD}$. Also, since $i_D = 0$, no power is dissipated in the transistor.

If $v_I > V_{TN}$, the transistor is on and initially is biased in the saturation region, since $v_{DS} > v_{GS} - V_{TN}$. As the input voltage increases, the drain-to-source voltage

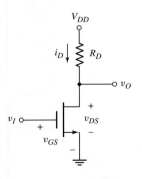

**Figure 3.45** NMOS inverter circuit

decreases, and the transistor eventually becomes biased in the nonsaturation region. When $v_I = V_{DD}$, the transistor is biased in the nonsaturation region, $v_O$ reaches a minimum value, and the drain current reaches a maximum value. The current and voltage are given by

$$i_D = K_n \left[ 2(v_I - V_{TN})v_O - v_O^2 \right] \tag{3.26}$$

and

$$v_O = v_{DD} - i_D R_D \tag{3.27}$$

where $v_O = v_{DS}$ and $v_I = v_{GS}$.

## DESIGN EXAMPLE 3.12

**Objective:** Design the size of a power MOSFET to meet the specification of a particular switch application.

The load in the inverter circuit in Figure 3.45 is a coil of an electromagnet that requires a current of 0.5 A when turned on. The effective load resistance varies between 8 and 10 $\Omega$, depending on temperature and other variables. A 10 V power supply is available. The transistor parameters are $k_n' = 80 \ \mu A/V^2$ and $V_{TN} = 1$ V.

**Solution:** One solution is to bias the transistor in the saturation region so that the current is constant, independent of the load resistance.

The minimum $V_{DS}$ value is 5 V. We need $V_{DS} > V_{DS}(\text{sat}) = V_{GS} - V_{TN}$. If we bias the transistor at $V_{GS} = 5$ V, then the transistor will always be biased in the saturation region. We can then write

$$I_D = \frac{k_n'}{2} \cdot \frac{W}{L}(V_{GS} - V_{TN})^2$$

or

$$0.5 = \frac{80 \times 10^{-6}}{2} \left( \frac{W}{L} \right) \cdot (5 - 1)^2$$

which yields $W/L = 781$.

The maximum power dissipation in the transistor occurs when the load resistance is 8 $\Omega$ and $V_{DS} = 6$ V. Then

$$P(\text{max}) = V_{DS}(\text{max}) \cdot I_D = (6) \cdot (0.5) = 3 \text{ W}$$

**Comment:** We see that we can switch a relatively large drain current with essentially no input current to the transistor. The size of the transistor required is fairly large, which implies a power transistor is necessary. If a transistor with a slightly different width-to-length ratio is available, the applied $V_{GS}$ can be changed to meet the specification.

## EXERCISE PROBLEM

**Ex 3.12:** For the MOS inverter circuit shown in Figure 3.45, assume the circuit values are $V_{DD} = 5$ V and $R_D = 500 \ \Omega$. The threshold voltage of the transistor is $V_{TN} = 1$ V. (a) Determine the value of the conduction parameter $K_n$ such that $v_O = 0.2$ V when $v_I = 5$ V. (b) What is the power dissipated in the transistor? (Ans. (a) $K_n = 6.15 \text{ mA/V}^2$, (b) $P = 1.92 \text{ mW}$)

### 3.3.2 Digital Logic Gate

For the transistor inverter circuit in Figure 3.45, when the input is low and approximately zero volts, the transistor is cut off, and the output is high and equal to $V_{DD}$. When the input is high and equal to $V_{DD}$, the transistor is biased in the nonsaturation region and the output reaches a low value. Since the input voltages will be either high or low, we can analyze the circuit in terms of dc parameters.

Now consider the case when a second transistor is connected in parallel, as shown in Figure 3.46. If the two inputs are zero, both $M_1$ and $M_2$ are cut off, and $V_O = 5$ V. When $V_1 = 5$ V and $V_2 = 0$, the transistor $M_1$ turns on and $M_2$ is still cut off. Transistor $M_1$ is biased in the nonsaturation region, and $V_O$ reaches a low value. If we reverse the input voltages such that $V_1 = 0$ and $V_2 = 5$ V, then $M_1$ is cut off and $M_2$ is biased in the nonsaturation region. Again, $V_O$ is at a low value. If both inputs are high, at $V_1 = V_2 = 5$ V, then both transistors are biased in the nonsaturation region and $V_O$ is low.

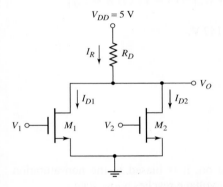

**Figure 3.46** A two-input NMOS NOR logic gate

Table 3.2 shows these various conditions for the circuit in Figure 3.46. In a positive logic system, these results indicate that this circuit performs the NOR logic function, and, it is therefore called a two-input NOR logic circuit. In actual NMOS logic circuits, the resistor $R_D$ is replaced by another NMOS transistor.

| Table 3.2 | NMOS NOR logic circuit response | |
|---|---|---|
| $V_1$(V) | $V_2$(V) | $V_O$(V) |
| 0 | 0 | High |
| 5 | 0 | Low |
| 0 | 5 | Low |
| 5 | 5 | Low |

### EXAMPLE 3.13

**Objective:** Determine the currents and voltages in a digital logic gate, for various input conditions.

Consider the circuit shown in Figure 3.46 with circuit and transistor parameters $R_D = 20\,\text{k}\Omega$, $K_n = 0.1\,\text{mA/V}^2$, $V_{TN} = 0.8$ V, and $\lambda = 0$.

**Solution:** For $V_1 = V_2 = 0$, both $M_1$ and $M_2$ are cut off and $V_O = V_{DD} = 5\,\text{V}$. For $V_1 = 5\,\text{V}$ and $V_2 = 0$, the transistor $M_1$ is biased in the nonsaturation region, and we can write

$$I_R = I_{D1} = \frac{5 - V_O}{R_D} = K_n \big[ 2(V_1 - V_{TN})V_O - V_O^2 \big]$$

Solving for the output voltage $V_O$, we obtain $V_O = 0.29\,\text{V}$.

The currents are

$$I_R = I_{D1} = \frac{5 - 0.29}{20} = 0.236\,\text{mA}$$

For $V_1 = 0$ and $V_2 = 5\,\text{V}$, we have $V_O = 0.29\,\text{V}$ and $I_R = I_{D2} = 0.236\,\text{mA}$. When both inputs go high to $V_1 = V_2 = 5\,\text{V}$, we have $I_R = I_{D1} + I_{D2}$, or

$$\frac{5 - V_O}{R_D} = K_n \big[ 2(V_1 - V_{TN})V_O - V_O^2 \big] + K_n \big[ 2(V_2 - V_{TN})V_O - V_O^2 \big]$$

which can be solved for $V_O$ to yield $V_O = 0.147\,\text{V}$.

The currents are

$$I_R = \frac{5 - 0.147}{20} = 0.243\,\text{mA}$$

and

$$I_{D1} = I_{D2} = \frac{I_R}{2} = 0.121\,\text{mA}$$

**Comment:** When either transistor is biased on, it is biased in the nonsaturation region, since $V_{DS} < V_{DS}(\text{sat})$, and the output voltage reaches a low state.

### EXERCISE PROBLEM

**Ex 3.13:** For the circuit in Figure 3.46, assume the circuit and transistor parameters are: $R_D = 30\,\text{k}\Omega$, $V_{TN} = 1\,\text{V}$, and $K_n = 50\,\mu\text{A/V}^2$. Determine $V_O$, $I_R$, $I_{D1}$, and $I_{D2}$ for: (a) $V_1 = 5\,\text{V}$, $V_2 = 0$; and (b) $V_1 = V_2 = 5\,\text{V}$. (Ans. (a) $V_O = 0.40\,\text{V}$, $I_R = I_{D1} = 0.153\,\text{mA}$, $I_{D2} = 0$ (b) $V_O = 0.205\,\text{V}$, $I_R = 0.16\,\text{mA}$, $I_{D1} = I_{D2} = 0.080\,\text{mA}$)

This example and discussion illustrates that MOS transistors can be configured in a circuit to perform logic functions. A more detailed analysis and design of MOSFET logic gates and circuits is presented in Chapter 16. As we will see in that chapter, most MOS logic gate circuits are fabricated by using CMOS, which means designing circuits with both n-channel and p-channel transistors and no resistors.

### 3.3.3    MOSFET Small-Signal Amplifier

The MOSFET, in conjunction with other circuit elements, can amplify small time-varying signals. Figure 3.47(a) shows the MOSFET small-signal amplifier, which is a common-source circuit in which a time-varying signal is coupled to the gate through a coupling capacitor. Figure 3.47(b) shows the transistor characteristics and the load line. The load line is determined for $v_i = 0$.

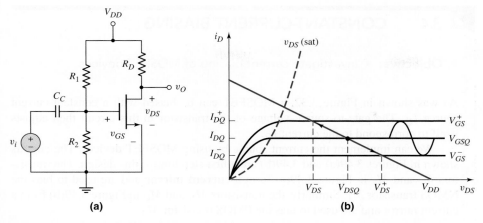

**Figure 3.47** (a) An NMOS common-source circuit with a time-varying signal coupled to the gate and (b) transistor characteristics, load line, and superimposed sinusoidal signals

We can establish a particular $Q$-point on the load line by designing the ratio of the bias resistors $R_1$ and $R_2$. If we assume that $v_i = V_i \sin \omega t$, the gate-to-source voltage will have a sinusoidal signal superimposed on the dc quiescent value. As the gate-to-source voltage changes over time, the $Q$-point will move up and down the line, as indicated in the figure.

Moving up and down the load line translates into a sinusoidal variation in the drain current and in the drain-to-source voltage. The variation in output voltage can be larger than the input signal voltage, which means the input signal is amplified. The actual signal gain depends on both the transistor parameters and the circuit element values.

In the next chapter, we will develop an equivalent circuit for the transistor used to determine the time-varying small-signal gain and other characteristics of the circuit.

## Test Your Understanding

**TYU 3.12** The circuit shown in Figure 3.45 is biased at $V_{DD} = 10$ V, and the transistor parameters are $V_{TN} = 0.7$ V and $K_n = 4$ mA/V$^2$. Design the value of $R_D$ such that the output voltage will be $v_O = 0.20$ V when $v_I = 10$ V. (Ans. 0.666 k$\Omega$)

**TYU 3.13** The transistor in the circuit shown in Figure 3.48 has parameters $K_n = 4$ mA/V$^2$ and $V_{TN} = 0.8$ V, and is used to switch the LED on and off. The LED cutin voltage is $V_\gamma = 1.5$ V. The LED is turned on by applying an input voltage of $v_I = 5$ V. (a) Determine the value of $R$ such that the diode current is 12 mA. (b) From the results of part (a), what is the value of $v_{DS}$? (Ans. (a) $R = 261$ $\Omega$, (b) $v_{DS} = 0.374$ V)

**TYU 3.14** In the circuit in Figure 3.46, let $R_D = 25$ k$\Omega$ and $V_{TN} = 1$ V. (a) Determine the value of the conduction parameter $K_n$ required such that $V_O = 0.10$ V when $V_1 = 0$ and $V_2 = 5$ V. (b) Using the results of part (a), find the value of $V_O$ when $V_1 = V_2 = 5$ V. (Ans. (a) $K_n = 0.248$ mA/V$^2$, (b) $V_O = 0.0502$ V)

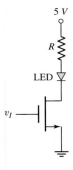

**Figure 3.48**

## 3.4    CONSTANT-CURRENT BIASING

**Objective:** • Investigate current biasing of MOSFET devices.

As was shown in Figure 3.32, a MOSFET can be biased with a constant-current source $I_Q$. The gate-to-source voltage of the transistor in this circuit then adjusts itself to correspond to the current $I_Q$.

We can implement the current source by using MOSFET devices. The circuits shown in Figures 3.49(a) and 3.49(b) are a first step toward this design. The transistors $M_2$ and $M_3$ in Figure 3.49(a) form a **current mirror** and are used to bias the NMOS transistor $M_1$. Similarly, the transistors $M_B$ and $M_C$ in Figure 3.49(b) form a current mirror and are used to bias the PMOS transistor $M_A$.

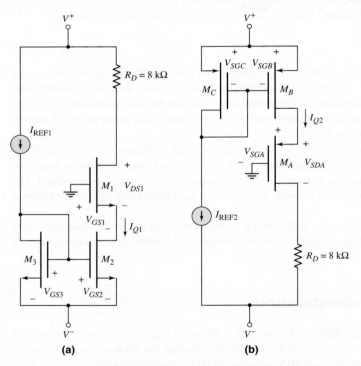

**Figure 3.49**  (a) NMOS current mirror and (b) PMOS current mirror

The operation and characteristics of these circuits are demonstrated in the following two examples.

### EXAMPLE 3.14

**Objective:** Analyze the circuit shown in Figure 3.49(a). Determine the bias current $I_{Q1}$, the gate-to-source voltages of the transistors, and the drain-to-source voltage of $M_1$.

Assume circuit parameters of $I_{REF1} = 200 \ \mu\text{A}$, $V^+ = 2.5$ V, and $V^- = -2.5$ V. Assume transistor parameters of $V_{TN} = 0.4$ V (all transistors), $\lambda = 0$ (all transistors), $K_{n1} = 0.25 \ \text{mA/V}^2$, and $K_{n2} = K_{n3} = 0.15 \ \text{mA/V}^2$.

**Solution:** The drain current in $M_3$ is $I_{D3} = I_{REF1} = 200 \, \mu A$ and is given by the relation $I_{D3} = K_{n3}(V_{GS3} - V_{TN})^2$ (the transistor is biased in the saturation region). Solving for the gate-to-source voltage, we find

$$V_{GS3} = \sqrt{\frac{I_{D3}}{K_{n3}}} + V_{TN} = \sqrt{\frac{0.2}{0.15}} + 0.4$$

or

$$V_{GS3} = 1.555 \text{ V}$$

We note that $V_{GS3} = V_{GS2} = 1.555$ V. We can write

$$I_{D2} = I_{Q1} = K_{n2}(V_{GS2} - V_{TN})^2 = 0.15(1.555 - 0.4)^2$$

or

$$I_{Q1} = 200 \, \mu A$$

The gate-to-source voltage $V_{GS1}$ (assuming $M_1$ is biased in the saturation region) can be written as

$$V_{GS1} = \sqrt{\frac{I_{Q1}}{K_{n1}}} + V_{TN} = \sqrt{\frac{0.2}{0.25}} + 0.4$$

or

$$V_{GS1} = 1.29 \text{ V}$$

The drain-to-source voltage is found from

$$V_{DS1} = V^+ - I_{Q1}R_D - (-V_{GS1})$$
$$= 2.5 - (0.2)(8) - (-1.29)$$

or

$$V_{DS1} = 2.19 \text{ V}$$

We may note that $M_1$ is indeed biased in the saturation region.

**Comment:** Since the current mirror transistors $M_2$ and $M_3$ are matched (identical parameters) and since the gate-to-source voltages are the same in the two transistors, the bias current, $I_{Q1}$, is equal to (i.e., mirrors) the reference current, $I_{REF1}$.

## EXERCISE PROBLEM

**Ex 3.14:** For the circuit shown in Figure 3.49(a), assume circuit parameters of $I_{REF1} = 120 \, \mu A$, $V^+ = 3$ V, and $V^- = -3$ V; and assume transistor parameters of $V_{TN} = 0.4$ V, $\lambda = 0$, $K_{n1} = 50 \, \mu A/V^2$, $K_{n2} = 30 \, \mu A/V^2$, and $K_{n3} = 60$ $\mu A/V^2$. Determine $I_{Q1}$ and all gate-to source voltages. (Ans. $I_{Q1} = 60 \, \mu A$, $V_{GS1} = 1.495$ V, $V_{GS2} = V_{GS3} = 1.814$ V)

We will now consider a current mirror in which the bias current and reference current are not equal.

## EXAMPLE **3.15**

**Objective:** Design the circuit shown in Figure 3.49(b) to provide a bias current of $I_{Q2} = 150 \, \mu A$.

Assume circuit parameters of $I_{REF2} = 250 \ \mu A$, $V^+ = 3$ V, and $V^- = -3$ V. Assume transistor parameters of $V_{TP} = -0.6$ V (all transistors), $\lambda = 0$ (all transistors), $k_p' = 40 \ \mu A/V^2$ (all transistors), $W/L_C = 15$, and $W/L_A = 25$.

**Solution:** Since the bias current $I_{Q2}$ and reference current $I_{REF2}$ are not equal, the $W/L$ ratios of the current mirror transistors, $M_C$ and $M_B$, will not be the same.

For $M_C$, since the transistor is biased in the saturation region, we have

$$I_{DC} = I_{REF2} = \frac{k_p'}{2} \cdot \left( \frac{W}{L} \right)_C (V_{SGC} + V_{TP})^2$$

or

$$250 = \frac{40}{2}(15)[V_{SGC} + (-0.6)]^2 = 300(V_{SGC} - 0.6)^2$$

Then

$$V_{SGC} = \sqrt{\frac{250}{300}} + 0.6$$

or

$$V_{SGC} = 1.513 \text{ V}$$

Since $V_{SGC} = V_{SGB} = 1.513$ V, we obtain

$$I_B = I_{Q2} = \frac{k_p'}{2} \cdot \left( \frac{W}{L} \right)_B (V_{SGB} + V_{TP})^2$$

or

$$150 = \frac{40}{2} \cdot \left( \frac{W}{L} \right)_B [1.513 + (-0.6)]^2$$

We find

$$\left( \frac{W}{L} \right)_B = 9$$

For $M_A$, we have

$$I_{DA} = I_{Q2} = \frac{k_p'}{2} \cdot \left( \frac{W}{L} \right)_A (V_{SGA} + V_{TP})^2$$

or

$$150 = \frac{40}{2}(25)(V_{SGA} + (-0.6))^2 = 500(V_{SGA} - 0.6)^2$$

Now

$$V_{SGA} = \sqrt{\frac{150}{500}} + 0.6$$

or

$$V_{SGA} = 1.148 \text{ V}$$

The source-to-drain voltage of $M_A$ is found from

$$V_{SDA} = V_{SGA} - I_{Q2}R_D - V^- = 1.148 - (0.15)(8) - (-3)$$

or

$$V_{SDA} = 2.95 \text{ V}$$

We may note that the transistor $M_A$ is biased in the saturation region.

**Comment:** By designing the $W/L$ ratios of the current mirror transistors, we can obtain different reference current and bias current values.

### EXERCISE PROBLEM

**Ex 3.15:** Consider the circuit shown in Figure 3.49(b). Assume circuit parameters of $I_{\text{REF2}} = 0.1$ mA, $V^+ = 5$ V, and $V^- = -5$ V. The transistor parameters are the same as given in Example 3.15. Design the circuit such that $I_{Q2} = 0.2$ mA. Also determine all source-to-gate voltages. (Ans. $V_{SGC} = V_{SGB} = 1.18$ V, $(W/L)_B = 30$, $V_{SGA} = 1.23$ V)

The constant-current source can be implemented by using MOSFETs as shown in Figure 3.50. The transistors $M_2$, $M_3$, and $M_4$ form the current source. Transistors $M_3$ and $M_4$ are each connected in a diode-type configuration, and they establish a reference current. We noted in the last section that this diode-type connection implies the transistor is always biased in the saturation region. Transistors $M_3$ and $M_4$ are therefore biased in the saturation region, and $M_2$ is assumed to be biased in the saturation region. The resulting gate-to-source voltage on $M_3$ is applied to $M_2$, and this establishes the bias current $I_Q$.

Since the reference current is the same in transistors $M_3$ and $M_4$, we can write

$$K_{n3}(V_{GS3} - V_{TN3})^2 = K_{n4}(V_{GS4} - V_{TN4})^2 \tag{3.28}$$

We also know that

$$V_{GS4} + V_{GS3} = (-V^-) \tag{3.29}$$

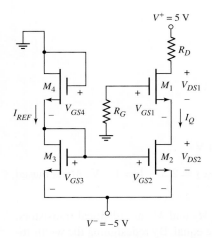

Figure 3.50  Implementation of a MOSFET constant-current source

Solving Equation (3.29) for $V_{GS4}$ and substituting the result into Equation (3.28) yields

$$V_{GS3} = \frac{\sqrt{\dfrac{K_{n4}}{K_{n3}}}[(-V^-) - V_{TN4}] + V_{TN3}}{1 + \sqrt{\dfrac{K_{n4}}{K_{n3}}}} \tag{3.30}$$

Since $V_{GS3} = V_{GS2}$, the bias current is

$$I_Q = K_{n2}(V_{GS3} - V_{TN2})^2 \tag{3.31}$$

## EXAMPLE 3.16

**Objective:** Determine the currents and voltages in a MOSFET constant-current source.

For the circuit shown in Figure 3.50, the transistor parameters are: $K_{n1} = 0.2$ mA/V$^2$, $K_{n2} = K_{n3} = K_{n4} = 0.1$ mA/V$^2$, and $V_{TN1} = V_{TN2} = V_{TN3} = V_{TN4} = 1$V.

**Solution:** From Equation (3.30), we can determine $V_{GS3}$, as follows:

$$V_{GS3} = \frac{\sqrt{\dfrac{0.1}{0.1}}[5 - 1] + 1}{1 + \sqrt{\dfrac{0.1}{0.1}}} = 2.5 \text{ V}$$

Since $M_3$ and $M_4$ are identical transistors, $V_{GS3}$ should be one-half of the bias voltage. The bias current $I_Q$ is then

$$I_Q = (0.1) \cdot (2.5 - 1)^2 = 0.225 \text{ mA}$$

The gate-to-source voltage on $M_1$ is found from

$$I_Q = K_{n1}(V_{GS1} - V_{TN1})^2$$

or

$$0.225 = (0.2) \cdot (V_{GS1} - 1)^2$$

which yields

$$V_{GS1} = 2.06 \text{ V}$$

The drain-to-source voltage on $M_2$ is

$$V_{DS2} = (-V^-) - V_{GS1} = 5 - 2.06 = 2.94 \text{ V}$$

Since $V_{DS2} = 2.94$ V $> V_{DS}(\text{sat}) = V_{GS2} - V_{TN2} = 2.5 - 1 = 1.5$ V, $M_2$ is biased in the saturation region.

**Design Consideration:** Since in this example $M_2$ and $M_3$ are identical transistors, the reference current $I_{REF}$ and bias current $I_Q$ are equal. By redesigning the width-to-length ratios of $M_2$, $M_3$, and $M_4$, we can obtain a specific bias current $I_Q$. If $M_2$ and

$M_3$ are not identical, then $I_Q$ and $I_{REF}$ will not be equal. A variety of design options are possible with such a circuit configuration.

## Test Your Understanding

**TYU 3.15** Consider the circuit in Figure 3.49(b). Assume circuit parameters of $I_{REF2} = 40$ $\mu$A, $V^+ = 2.5$ V, $V^- = -2.5$ V, and $R_D = 20$ k$\Omega$. The transistor parameters are $V_{TP} = -0.30$ V, $K_{pC} = 40$ $\mu$A/V$^2$, $K_{pB} = 60$ $\mu$A/V$^2$, and $K_{pA} = 75$ $\mu$A/V$^2$. Determine $I_{Q2}$ and all source-to-gate voltages. (Ans. $I_{Q2} = 60$ $\mu$A, $V_{SGC} = V_{SGB} = 1.30$ V, $V_{SGA} = 1.19$ V)

**TYU 3.16** Consider the circuit shown in Figure 3.50. Assume all transistor threshold voltages are 0.7 V. Determine the values of $K_{n1}$, $K_{n2}$, $K_{n3}$, and $K_{n4}$ such that $I_{REF} = 80$ $\mu$A, $I_Q = 120$ $\mu$A, $V_{GS3} = 2$ V, and $V_{GS1} = 1.5$ V. (Ans. $K_{n1} = 187.5$ $\mu$A/V$^2$, $K_{n2} = 71.0$ $\mu$A/V$^2$, $K_{n3} = 47.3$ $\mu$A/V$^2$, $K_{n4} = 15.12$ $\mu$A/V$^2$)

## 3.5 MULTISTAGE MOSFET CIRCUITS

**Objective:** • Consider the dc biasing of multistage or multitransistor circuits.

In most applications, a single-transistor amplifier will not be able to meet the combined specifications of a given amplification factor, input resistance, and output resistance. For example, the required voltage gain may exceed that which can be obtained in a single-transistor circuit.

Transistor amplifier circuits can be connected in series, or **cascaded,** as shown in Figure 3.51. This may be done either to increase the overall small-signal voltage gain, or provide an overall voltage gain greater than 1, with a very low output resistance.

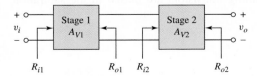

**Figure 3.51** Generalized two-stage amplifier

The overall voltage gain may not simply be the product of the individual amplification factors. Loading effects, in general, need to be taken into account.

There are many possible multistage configurations; we will examine a few here, in order to understand the type of analysis required.

### 3.5.1    Multitransistor Circuit: Cascade Configuration

The circuit shown in Figure 3.52 is a cascade of a common-source amplifier followed by a source-follower amplifier. We will show in the next chapter that the common-source amplifier provides a small-signal voltage gain and the source-follower has a low output impedance.

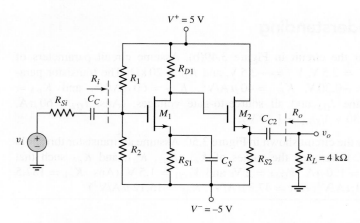

**Figure 3.52**  Common-source amplifier in cascade with source follower

### DESIGN EXAMPLE 3.17

**Objective:** Design the biasing of a multistage MOSFET circuit to meet specific requirements.

Consider the circuit shown in Figure 3.52 with transistor parameters $K_{n1} = 500 \ \mu\text{A/V}^2$, $K_{n2} = 200 \ \mu\text{A/V}^2$, $V_{TN1} = V_{TN2} = 1.2$ V, and $\lambda_1 = \lambda_2 = 0$. Design the circuit such that $I_{DQ1} = 0.2$ mA, $I_{DQ2} = 0.5$ mA, $V_{DSQ1} = V_{DSQ2} = 6$ V, and $R_i = 100 \ \text{k}\Omega$. Let $R_{Si} = 4 \ \text{k}\Omega$.

**Solution:** For output transistor $M_2$, we have

$$V_{DSQ2} = 5 - (-5) - I_{DQ2}R_{S2}$$

or

$$6 = 10 - (0.5)R_{S2}$$

which yields $R_{S2} = 8 \ \text{k}\Omega$. Also, assuming transistors are biased in the saturation region,

$$I_{DQ2} = K_{n2}(K_{GS2} - V_{TN2})^2$$

or

$$0.5 = 0.2(V_{GS2} - 1.2)^2$$

which yields

$$V_{GS2} = 2.78 \text{ V}$$

Since $V_{DSQ2} = 6$ V, the source voltage of $M_2$ is $V_{S2} = -1$ V. With $V_{GS2} = 2.78$ V, the gate voltage on $M_2$ must be

$$V_{G2} = -1 + 2.78 = 1.78 \text{ V}$$

The resistor $R_{D1}$ is then

$$R_{D1} = \frac{5 - 1.78}{0.2} = 16.1 \text{ k}\Omega$$

For $V_{DSQ1} = 6$ V, the source voltage of $M_1$ is

$$V_{S1} = 1.78 - 6 = -4.22 \text{ V}$$

The resistor $R_{S1}$ is then

$$R_{S1} = \frac{-4.22 - (-5)}{0.2} = 3.9 \text{ k}\Omega$$

For transistor $M_1$, we have

$$I_{DQ1} = K_{n1}(V_{GS1} - V_{TN1})^2$$

or

$$0.2 = 0.50(V_{GS1} - 1.2)^2$$

which yields

$$V_{GS1} = 1.83 \text{ V}$$

To find $R_1$ and $R_2$, we can write

$$V_{GS1} = \left(\frac{R_2}{R_1 + R_2}\right)(10) - I_{DQ1}R_{S1}$$

Since

$$\frac{R_2}{R_1 + R_2} = \frac{1}{R_1} \cdot \left(\frac{R_1 R_2}{R_1 + R_2}\right) = \frac{1}{R_1} \cdot R_i$$

then, since the input resistance is specified to be 100 kΩ, we have

$$1.83 = \frac{1}{R_1}(100)(10) - (0.2)(3.9)$$

which yields $R_1 = 383$ kΩ. From $R_i = 100$ kΩ, we find that $R_2 = 135$ kΩ.

**Comment:** Both transistors are biased in the saturation region, as assumed, which is desired for linear amplifiers as we will see in the next chapter.

## EXERCISE PROBLEM

**Ex 3.17:** The transistor parameters for the circuit shown in Figure 3.52 are the same as described in Example 3.17. Design the circuit such that $I_{DQ1} = 0.1$ mA, $I_{DQ2} = 0.3$ mA, $V_{DSQ1} = V_{DSQ2} = 5$ V, and $R_i = 200$ kΩ. (Ans. $R_{S2} = 16.7$ kΩ, $R_{D1} = 25.8$ kΩ, $R_{S1} = 24.3$ kΩ, $R_1 = 491$ kΩ, and $R_2 = 337$ kΩ)

| 3.5.2 | **Multitransistor Circuit: Cascode Configuration** |
| --- | --- |

Figure 3.53 shows a **cascode** circuit with n-channel MOSFETs. Transistor $M_1$ is connected in a common-source configuration and $M_2$ is connected in a common-gate configuration. The advantage of this type of circuit is a higher frequency response, which is discussed in a later chapter.

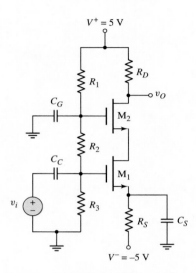

**Figure 3.53** NMOS cascode circuit

---

## DESIGN EXAMPLE 3.18

**Objective:** Design the biasing of the cascode circuit to meet specific requirements.

For the circuit shown in Figure 3.53, the transistor parameters are: $V_{TN1} = V_{TN2} = 1.2$ V, $K_{n1} = K_{n2} = 0.8$ mA/V$^2$, and $\lambda_1 = \lambda_2 = 0$. Let $R_1 + R_2 + R_3 = 300$ k$\Omega$ and $R_S = 10$ k$\Omega$. Design the circuit such that $I_{DQ} = 0.4$ mA and $V_{DSQ1} = V_{DSQ2} = 2.5$ V.

**Solution:** The dc voltage at the source of $M_1$ is

$$V_{S1} = I_{DQ}R_S - 5 = (0.4)(10) - 5 = -1 \text{ V}$$

Since $M_1$ and $M_2$ are identical transistors, and since the same current exists in the two transistors, the gate-to-source voltage is the same for both devices. We have

$$I_D = K_n(V_{GS} - V_{TN})^2$$

or

$$0.4 = 0.8(V_{GS} - 1.2)^2$$

which yields

$$V_{GS} = 1.907 \text{ V}$$

Then,

$$V_{G1} = \left( \frac{R_3}{R_1 + R_2 + R_3} \right)(5) = V_{GS} + V_{S1}$$

or

$$\left(\frac{R_3}{300}\right)(5) = 1.907 - 1 = 0.907$$

which yields

$R_3 = 54.4 \text{ k}\Omega$

The voltage at the source of $M_2$ is

$V_{S2} = V_{DSQ1} + V_{S1} = 2.5 - 1 = 1.5 \text{ V}$

Then,

$$V_{G2} = \left(\frac{R_2 + R_3}{R_1 + R_2 + R_3}\right)(5) = V_{GS} + V_{S2}$$

or

$$\left(\frac{R_2 + R_3}{300}\right)(5) = 1.907 + 1.5 = 3.407 \text{ V}$$

which yields

$R_2 + R_3 = 204.4 \text{ k}\Omega$

and

$R_2 = 150 \text{ k}\Omega$

Therefore

$R_1 = 95.6 \text{ k}\Omega$

The voltage at the drain of $M_2$ is

$V_{D2} = V_{DSQ2} + V_{S2} = 2.5 + 1.5 = 4 \text{ V}$

The drain resistor is therefore

$$R_D = \frac{5 - V_{D2}}{I_{DQ}} = \frac{5 - 4}{0.4} = 2.5 \text{ k}\Omega$$

Comment: Since $V_{DS} = 2.5 \text{ V} > V_{GS} - V_{TN} = 1.91 - 1.2 = 0.71 \text{ V}$, each transistor is biased in the saturation region.

EXERCISE PROBLEM

Ex 3.18: The transistor parameters for the circuit shown in Figure 3.53 are $V_{TN1} = V_{TN2} = 0.8 \text{ V}$, $K_{n1} = K_{n2} = 0.5 \text{ mA/V}^2$, and $\lambda_1 = \lambda_2 = 0$. Let $R_1 + R_2 + R_3 = 500 \text{ k}\Omega$ and $R_S = 16 \text{ k}\Omega$. Design the circuit such that $I_{DQ} = 0.25 \text{ mA}$ and $V_{DSQ1} = V_{DSQ2} = 2.5 \text{ V}$. (Ans. $R_3 = 50.7 \text{ k}\Omega$, $R_2 = 250 \text{ k}\Omega$, $R_1 = 199.3 \text{ k}\Omega$, $R_D = 4 \text{ k}\Omega$)

We will encounter many more examples of multitransistor and multistage amplifiers in later chapters of this text. Specifically in Chapter 11, we will consider the differential amplifier and in Chapter 13, we will analyze circuits that form the operational amplifier.

# 3.6   JUNCTION FIELD-EFFECT TRANSISTOR

**Objective:** • Understand the operation and characteristics of the pn junction FET (JFET) and the Schottky barrier junction FET (MESFET), and understand the dc analysis techniques of JFET and MESFET circuits.

The two general categories of **junction field-effect transistor (JFET)** are the pn junction FET, or **pn JFET,** and the **metal-semiconductor field-effect transistor (MESFET),** which is fabricated with a Schottky barrier junction.

The current in a JFET is through a semiconductor region known as the channel, with ohmic contacts at each end. The basic transistor action is the modulation of the channel conductance by an electric field perpendicular to the channel. Since the modulating electric field is induced in the space-charge region of a reverse-biased pn junction or Schottky barrier junction, the field is a function of the gate voltage. Modulation of the channel conductance by the gate voltage modulates the channel current.

JFETs were developed before MOSFETs, but the applications and uses of the MOSFET have far surpassed those of the JFET. One reason is that the voltages applied to the gate and drain of a MOSFET are the same polarity (both positive or both negative), whereas the voltages applied to the gate and drain of most JFETs must have opposite polarities. Since the JFET is used only in specialized applications, our discussion will be brief.

## 3.6.1   pn JFET and MESFET Operation

### pn JFET

A simplified cross section of a symmetrical pn JFET is shown in Figure 3.54. In the n-region channel between the two p-regions, majority carrier electrons flow from the source to the drain terminal; thus, the JFET is called a majority-carrier device. The two gate terminals shown in Figure 3.54 are connected to form a single gate.

In a p-channel JFET, the p- and n-regions are reversed from those of the n-channel device, and holes flow in the channel from the source to the drain. The current direction and voltage polarities in the p-channel JFET are reversed from

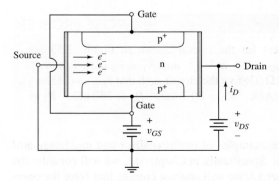

**Figure 3.54** Cross section of a symmetrical n-channel pn junction field-effect transistor

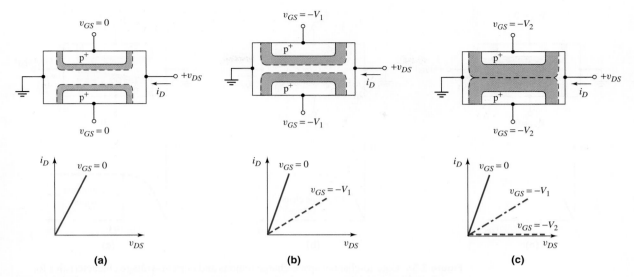

**Figure 3.55** Gate-to-channel space-charge regions and current–voltage characteristics for small drain-to-source voltages and for: (a) zero gate voltage, (b) small reverse-biased gate voltage, and (c) a gate voltage that achieves pinchoff

those in the n-channel device. Also, the p-channel JFET is generally a lower-frequency device than the n-channel JFET, because hole mobility is lower than electron mobility.

Figure 3.55(a) shows an n-channel JFET with zero volts applied to the gate. If the source is at ground potential, and if a small positive drain voltage is applied, a drain current $i_D$ is produced between the source and drain terminals. Since the n-channel acts essentially as a resistance, the $i_D$ versus $v_{DS}$ characteristic for small $v_{DS}$ values is approximately linear, as shown in the figure.

If a voltage is applied to the gate of a pn JFET, the channel conductance changes. If a negative gate voltage is applied to the n-channel pn JFET in Figure 3.55, the gate-to-channel pn junction becomes reverse biased. The space-charge region widens, the channel region narrows, the resistance of the n-channel increases, and the slope of the $i_D$ versus $v_{DS}$ curve, for small $v_{DS}$, decreases. These effects are shown in Figure 3.55(b). If a larger negative gate voltage is applied, the condition shown in Figure 3.55(c) can be achieved. The reverse-biased gate-to-channel space-charge region completely fills the channel region. This condition is known as **pinchoff.** Since the depletion region isolates the source and drain terminals, the drain current at pinchoff is essentially zero. The $i_D$ versus $v_{DS}$ curves are shown in Figure 3.55(c). The current in the channel is controlled by the gate voltage. The control of the current in one part of the device by a voltage in another part of the device is the basic transistor action. The pn JFET is a "normally on," or depletion mode, device; that is, a voltage must be applied to the gate terminal to turn the device off.

Consider the situation in which the gate voltage is zero, $v_{GS} = 0$, and the drain voltage changes, as shown in Figure 3.56(a). As the drain voltage increases (positive), the gate-to-channel pn junction becomes reverse biased near the drain terminal, and the space-charge region widens, extending farther into the channel. The channel acts essentially as a resistor, and the effective channel resistance increases as the space-charge region widens; therefore, the slope of the $i_D$ versus $v_{DS}$ characteristic

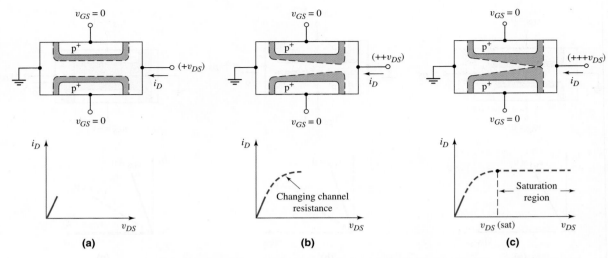

**Figure 3.56** Gate-to-channel space-charge regions and current–voltage characteristics for zero gate voltage and for: (a) a small drain voltage, (b) a larger drain voltage, and (c) a drain voltage that achieves pinchoff at the drain terminal

decreases, as shown in Figure 3.56(b). The effective channel resistance now varies along the channel length, and, since the channel current must be constant, the voltage drop through the channel becomes dependent on position.

If the drain voltage increases further, the condition shown in Figure 3.56(c) can result. The channel is pinched off at the drain terminal. Any further increase in drain voltage will not increase the drain current. The $i_D$–$v_{DS}$ characteristic for this condition is also shown in the figure. The drain voltage at pinchoff is $v_{DS}$(sat). Therefore, for $v_{DS} > v_{DS}$(sat), the transistor is biased in the saturation region, and the drain current for this ideal case is independent of $v_{DS}$.

### MESFET

In the MESFET, the gate junction is a Schottky barrier junction, instead of a pn junction. Although MESFETs can be fabricated in silicon, they are usually associated with gallium arsenide or other compound-semiconductor materials.

A simplified cross section of a GaAs MESFET is shown in Figure 3.57. A thin, epitaxial layer of GaAs is used for the active region; the substrate is a very high resistivity GaAs material, referred to as a semi-insulating substrate. The advantages of

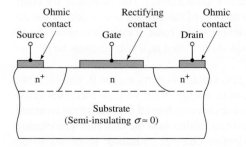

**Figure 3.57** Cross section of an n-channel MESFET with a semi-insulating substrate

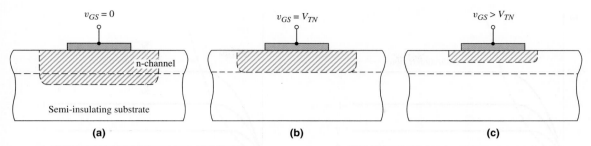

$v_{GS} = 0$    $v_{GS} = V_{TN}$    $v_{GS} > V_{TN}$

n-channel

Semi-insulating substrate

(a)    (b)    (c)

**Figure 3.58** Channel space-charge region of an enhancement-mode MESFET for: (a) $v_{GS} = 0$, (b) $v_{GS} = V_{TN}$, and (c) $v_{GS} > V_{TN}$

these devices include: higher electron mobility in GaAs, hence smaller transit time and faster response; and decreased parasitic capacitance and a simplified fabrication process, resulting from the semi-insulating GaAs substrate.

In the MESFET in Figure 3.57, a reverse-bias gate-to-source voltage induces a space-charge region under the metal gate, which modulates the channel conductance, as in the case of the pn JFET. If a negative applied gate voltage is sufficiently large, the space-charge region will eventually reach the substrate. Again, pinchoff will occur. Also, the device shown in the figure is a depletion mode device, since a gate voltage must be applied to pinch off the channel, that is, to turn the device off.

In another type of MESFET, the channel is pinched off even at $v_{GS} = 0$, as shown in Figure 3.58(a). For this MESFET, the channel thickness is smaller than the zero-biased space-charge width. To open a channel, the depletion region must be reduced; that is, a forward-biased voltage must be applied to the gate–semiconductor junction. When a slightly forward-bias voltage is applied, the depletion region extends just to the width of the channel as shown in Figure 3.58(b). The threshold voltage is the gate-to-source voltage required to create the pinchoff condition. The threshold voltage for this n-channel MESFET is positive, in contrast to the negative threshold voltage of the n-channel depletion-mode device. If a larger forward-bias voltage is applied, the channel region opens, as shown in Figure 3.58(c). The applied forward-bias gate voltage is limited to a few tenths of a volt before a significant gate current occurs.

This device is an **n-channel enhancement-mode MESFET.** Enhancement-mode p-channel MESFETs and enhancement-mode pn JFETs have also been fabricated. The advantage of enhancement-mode MESFETs is that circuits can be designed in which the voltage polarities on the gate and drain are the same. However, the output voltage swing of these devices is quite small.

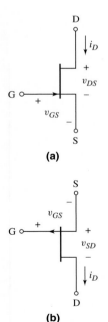

(a)

(b)

**Figure 3.59** Circuit symbols for: (a) n-channel JFET and (b) p-channel JFET

### 3.6.2    Current–Voltage Characteristics

The circuit symbols for the n-channel and p-channel JFETs are shown in Figure 3.59, along with the gate-to-source voltages and current directions. The ideal current–voltage characteristics, when the transistor is biased in the saturation region, are described by

$$i_D = I_{DSS} \left( 1 - \frac{v_{GS}}{V_P} \right)^2 \tag{3.32}$$

where $I_{DSS}$ is the saturation current when $v_{GS} = 0$, and $V_P$ is the **pinchoff voltage.**

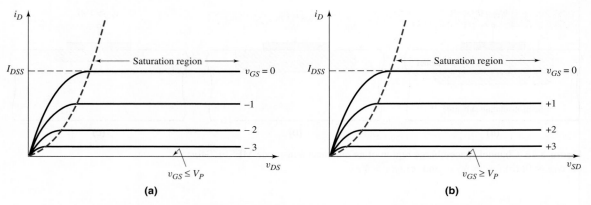

**Figure 3.60** Current–voltage characteristics for: (a) n-channel JFET and (b) p-channel JFET

The current–voltage characteristics for n-channel and p-channel JFETs are shown in Figures 3.60(a) and 3.60(b), respectively. Note that the pinchoff voltage $V_P$ for the n-channel JFET is negative and the gate-to-source voltage $v_{GS}$ is usually negative; therefore, the ratio $v_{GS}/V_P$ is positive. Similarly, the pinchoff voltage $V_P$ for the p-channel JFET is positive and the gate-to-source voltage $v_{GS}$ must be positive, and therefore the ratio $v_{GS}/V_P$ is positive.

For the n-channel device, the saturation region occurs when $v_{DS} \geq v_{DS}(\text{sat})$ where

$$v_{DS}(\text{sat}) = v_{GS} - V_P \qquad (3.33)$$

For the p-channel device, the saturation region occurs when $v_{SD} \geq v_{SD}(\text{sat})$ where

$$v_{SD}(\text{sat}) = V_P - v_{GS} \qquad (3.34)$$

## EXAMPLE 3.19

**Objective:** Calculate $i_D$ and $v_{DS}(\text{sat})$ in an n-channel pn JFET.

Assume the saturation current is $I_{DSS} = 2\,\text{mA}$ and the pinchoff voltage is $V_P = -3.5\,\text{V}$. Calculate $i_D$ and $v_{DS}(\text{sat})$ for $v_{GS} = 0$, $V_P/4$, and $V_P/2$.

**Solution:** From Equation (3.32), we have

$$i_D = I_{DSS}\left(1 - \frac{v_{GS}}{V_P}\right)^2 = (2)\left(1 - \frac{v_{GS}}{(-3.5)}\right)^2$$

Therefore, for $v_{GS} = 0$, $V_P/4$, and $V_P/2$, we obtain

$$i_D = 2,\ 1.13,\ \text{and}\ 0.5\,\text{mA}$$

From Equation (3.33), we have

$$v_{DS}(\text{sat}) = v_{GS} - V_P = v_{GS} - (-3.5)$$

Therefore, for $v_{GS} = 0$, $V_P/4$, and $V_P/2$, we obtain

$$v_{DS}(\text{sat}) = 3.5,\ 2.63,\ \text{and}\ 1.75\,\text{V}$$

**Comment:** The current capability of a JFET can be increased by increasing the value of $I_{DSS}$, which is a function of the transistor width.

**Ex 3.19:** The parameters of an n-channel JFET are $I_{DSS} = 12$ mA, $V_P = -4.5$ V, and $\lambda = 0$. Determine $V_{DS}(\text{sat})$ for $V_{GS} = -1.2$ V, and calculate $I_D$ for $V_{DS} > V_{DS}(\text{sat})$. (Ans. $V_{DS}(\text{sat}) = 3.3$ V, $I_D = 6.45$ mA)

As in the case of the MOSFET, the $i_D$ versus $v_{DS}$ characteristic for the JFET may have a nonzero slope beyond the saturation point. This nonzero slope can be described through the following equation:

$$i_D = I_{DSS}\left(1 - \frac{v_{GS}}{V_P}\right)^2 (1 + \lambda v_{DS}) \tag{3.35}$$

The output resistance $r_o$ is defined as

$$r_o = \left(\frac{\partial i_D}{\partial v_{DS}}\right)^{-1}\Bigg|_{v_{GS}=\text{const.}} \tag{3.36}$$

Using Equation (3.35), we find that

$$r_o = \left[\lambda I_{DSS}\left(1 - \frac{V_{GSQ}}{V_P}\right)^2\right]^{-1} \tag{3.37(a)}$$

or

$$r_o \cong [\lambda I_{DQ}]^{-1} = \frac{1}{\lambda I_{DQ}} \tag{3.37(b)}$$

The output resistance will be considered again when we discuss the small-signal equivalent circuit of the JFET in the next chapter.

Enhancement-mode GaAs MESFETs can be fabricated with current–voltage characteristics much like those of the enhancement-mode MOSFET. Therefore, for the ideal enhancement-mode MESFET biased in the saturation region, we can write

$$i_D = K_n(v_{GS} - V_{TN})^2 \tag{3.38(a)}$$

For the ideal enhancement-mode MESFET biased in the nonsaturation region,

$$i_D = K_n\left[2(v_{GS} - V_{TN})v_{DS} - v_{DS}^2\right] \tag{3.38(b)}$$

where $K_n$ is the conduction parameter and $V_{TN}$ is the threshold voltage, which in this case is equivalent to the pinchoff voltage. For an n-channel enhancement-mode MESFET, the threshold voltage is positive.

### 3.6.3 Common JFET Configurations: dc Analysis

There are several common JFET circuit configurations. We will look at a few of these, using examples, and will illustrate the dc analysis and design of such circuits.

**Objective:** Design a JFET circuit with a voltage divider biasing circuit.

Consider the circuit shown in Figure 3.61(a) with transistor parameters $I_{DSS} = 12$ mA, $V_P = -3.5$ V, and $\lambda = 0$. Let $R_1 + R_2 = 100$ k$\Omega$. Design the circuit such that the dc drain current is $I_D = 5$ mA and the dc drain-to-source voltage is $V_{DS} = 5$ V.

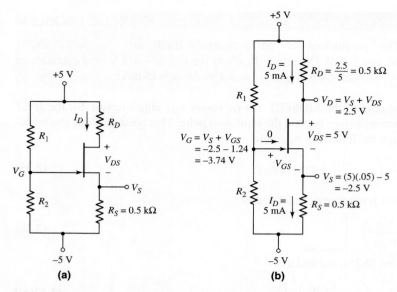

**Figure 3.61** (a) An n-channel JFET circuit with voltage divider biasing and (b) the n-channel JFET circuit for Example 3.20

**Solution:** Assume the transistor is biased in the saturation region. The dc drain current is then given by

$$I_D = I_{DSS}\left(1 - \frac{V_{GS}}{V_P}\right)^2$$

Therefore,

$$5 = 12\left(1 - \frac{V_{GS}}{(-3.5)}\right)^2$$

which yields

$$V_{GS} = -1.24 \, \text{V}$$

From Figure 3.61(b), the voltage at the source terminal is

$$V_S = I_D R_S - 5 = (5)(0.5) - 5 = -2.5 \, \text{V}$$

which means that the gate voltage is

$$V_G = V_{GS} + V_S = -1.24 - 2.5 = -3.74 \, \text{V}$$

We can also write the gate voltage as

$$V_G = \left(\frac{R_2}{R_1 + R_2}\right)(10) - 5$$

or

$$-3.74 = \frac{R_2}{100}(10) - 5.$$

Therefore,

$$R_2 = 12.6\,k\Omega$$

and

$$R_1 = 87.4\,k\Omega$$

The drain-to-source voltage is

$$V_{DS} = 5 - I_D R_D - I_D R_S - (-5)$$

Therefore,

$$R_D = \frac{10 - V_{DS} - I_D R_S}{I_D} = \frac{10 - 5 - (5)(0.5)}{5} = 0.5\,k\Omega$$

We also see that

$$V_{DS} = 5\,V > V_{GS} - V_P = -1.24 - (-3.5) = 2.26\,V$$

which shows that the JFET is indeed biased in the saturation region, as initially assumed.

**Comment:** The dc analysis of the JFET circuit is essentially the same as that of the MOSFET circuit, since the gate current is assumed to be zero.

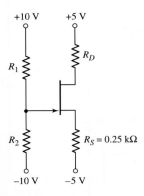

Figure 3.62  Circuit for Exercise Ex 3.20

EXERCISE PROBLEM

**Ex 3.20:** The transistor in the circuit in Figure 3.62 has parameters $I_{DSS} = 6$ mA, $V_P = -4$ V, and $\lambda = 0$. Design the circuit such that $I_{DQ} = 2.5$ mA and $V_{DS} = 6$ V, and the total power dissipated in $R_1$ and $R_2$ is 2 mW. (Ans. $R_D = 1.35$ kΩ, $R_1 = 158$ kΩ, $R_2 = 42$ kΩ)

## EXAMPLE 3.21

**Objective:** Calculate the quiescent current and voltage values in a p-channel JFET circuit.

The parameters of the transistor in the circuit shown in Figure 3.63 are: $I_{DSS} = 2.5$ mA, $V_P = +2.5$ V, and $\lambda = 0$. The transistor is biased with a constant-current source.

**Solution:** From Figure 3.63, we can write the dc drain current as

$$I_D = I_Q = 0.8\,mA = \frac{V_D - (-9)}{R_D}$$

which yields

$$V_D = (0.8)(4) - 9 = -5.8\,V$$

Now, assume the transistor is biased in the saturation region. We then have

$$I_D = I_{DSS}\left(1 - \frac{V_{GS}}{V_P}\right)^2$$

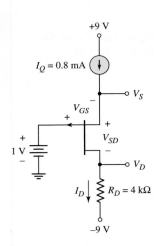

Figure 3.63  A p-channel JFET circuit biased with a constant-current source

or

$$0.8 = 2.5\left(1 - \frac{V_{GS}}{2.5}\right)^2$$

which yields

$$V_{GS} = 1.086 \text{ V}$$

Then

$$V_S = 1 - V_{GS} = 1 - 1.086 = -0.086 \text{ V}$$

and

$$V_{SD} = V_S - V_D = -0.086 - (-5.8) = 5.71 \text{ V}$$

Again, we see that

$$V_{SD} = 5.71 \text{ V} > V_P - V_{GS} = 2.5 - 1.086 = 1.41 \text{ V}$$

which verifies that the transistor is biased in the saturation region, as assumed.

**Comment:** In the same way as bipolar or MOS transistors, junction field-effect transistors can be biased using constant-current sources.

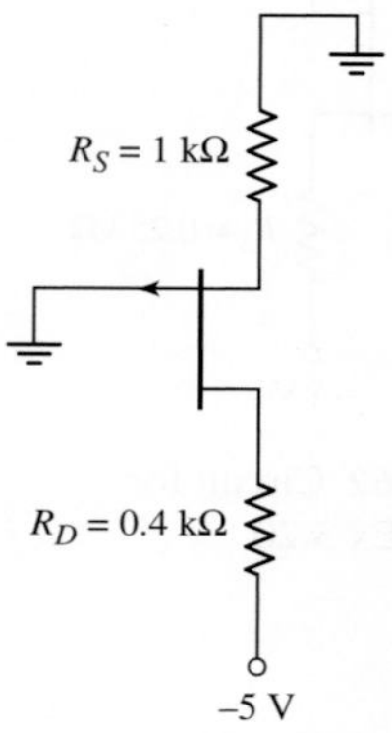

$R_S = 1 \text{ k}\Omega$

$R_D = 0.4 \text{ k}\Omega$

$-5 \text{ V}$

**Figure 3.64** Circuit for Exercise Ex 3.21

## EXERCISE PROBLEM

**Ex 3.21:** For the p-channel transistor in the circuit in Figure 3.64, the parameters are: $I_{DSS} = 6 \text{ mA}$, $V_P = 4 \text{ V}$, and $\lambda = 0$. Calculate the quiescent values of $I_D$, $V_{GS}$, and $V_{SD}$. Is the transistor biased in the saturation or nonsaturation region? (Ans. $V_{GS} = 1.81 \text{ V}$, $I_D = 1.81 \text{ mA}$, $V_{SD} = 2.47 \text{ V}$, saturation region)

## DESIGN EXAMPLE 3.22

**Objective:** Design a circuit with an enhancement-mode MESFET.

Consider the circuit shown in Figure 3.65(a). The transistor parameters are: $V_{TN} = 0.24 \text{ V}$, $K_n = 1.1 \text{ mA/V}^2$, and $\lambda = 0$. Let $R_1 + R_2 = 50 \text{ k}\Omega$. Design the circuit such that $V_{GS} = 0.50 \text{ V}$ and $V_{DS} = 2.5 \text{ V}$.

**Solution:** From Equation (3.38(a)) the drain current is

$$I_D = K_n(V_{GS} - V_{TN})^2 = (1.1)(0.5 - 0.24)^2 = 74.4 \ \mu\text{A}$$

From Figure 3.65(b), the voltage at the drain is

$$V_D = V_{DD} - I_D R_D = 4 - (0.0744)(6.7) = 3.5 \text{ V}$$

Therefore, the voltage at the source is

$$V_S = V_D - V_{DS} = 3.5 - 2.5 = 1 \text{ V}$$

The source resistance is then

$$R_S = \frac{V_S}{I_D} = \frac{1}{0.0744} = 13.4 \text{ k}\Omega$$

The voltage at the gate is

$$V_G = V_{GS} + V_S = 0.5 + 1 = 1.5 \text{ V}$$

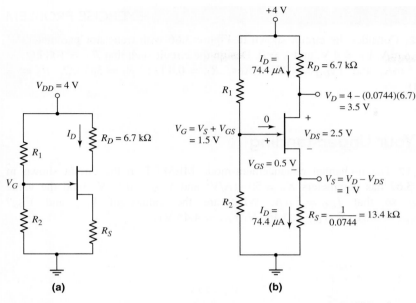

**Figure 3.65**  (a) An n-channel enhancement-mode MESFET circuit and (b) the n-channel MESFET circuit for Example 3.22

Since the gate current is zero, the gate voltage is also given by a voltage divider equation, as follows:

$$V_G = \left( \frac{R_2}{R_1 + R_2} \right)(V_{DD})$$

or

$$1.5 = \left( \frac{R_2}{50} \right)(4)$$

which yields

$$R_2 = 18.75\,\text{k}\Omega$$

and

$$R_1 = 31.25\,\text{k}\Omega$$

Again, we see that

$$V_{DS} = 2.5\,\text{V} > V_{GS} - V_{TN} = 0.5 - 0.24 = 0.26\,\text{V}$$

which confirms that the transistor is biased in the saturation region, as initially assumed.

**Comment:**  The dc analysis and design of an enhancement-mode MESFET circuit is similar to that of MOSFET circuits, except that the gate-to-source voltage of the MESFET must be held to no more than a few tenths of a volt.

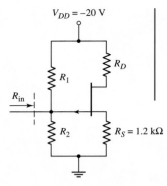

**Figure 3.66** Circuit for Exercise Ex3.22

EXERCISE PROBLEM

**Ex 3.22:** Consider the circuit shown in Figure 3.66 with transistor parameters $I_{DSS} = 8$ mA, $V_P = 4$ V, and $\lambda = 0$. Design the circuit such that $R_{in} = 100$ kΩ, $I_{DQ} = 5$ mA, and $V_{SDQ} = 12$ V. (Ans. $R_D = 0.4$ kΩ, $R_1 = 387$ kΩ, $R_2 = 135$ kΩ)

## Test Your Understanding

**TYU 3.17** The n-channel enhancement-mode MESFET in the circuit shown in Figure 3.67 has parameters $K_n = 50$ μA/V$^2$ and $V_{TN} = 0.15$ V. Find the value of $V_{GG}$ so that $I_{DQ} = 5$ μA. What are the values of $V_{GS}$ and $V_{DS}$? (Ans. $V_{GG} = 0.516$ V, $V_{GS} = 0.466$ V, $V_{DS} = 4.45$ V)

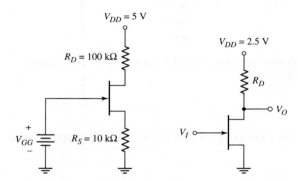

**Figure 3.67** Circuit for Exercise TYU 3.17

**Figure 3.68** Circuit for Exercise TYU 3.18

**TYU 3.18** For the inverter circuit shown in Figure 3.68, the n-channel enhancement-mode MESFET parameters are $K_n = 100$ μA/V$^2$ and $V_{TN} = 0.2$ V. Determine the value of $R_D$ required to produce $V_O = 0.10$ V when $V_I = 0.7$ V. (Ans. $R_D = 267$ kΩ)

## 3.7    DESIGN APPLICATION: DIODE THERMOMETER WITH AN MOS TRANSISTOR

**Objective:** • Incorporate an MOS transistor in a design application that enhances the simple diode thermometer design discussed in Chapter 1.

**Specifications:** The electronic thermometer is to operate over a temperature range of 0 to 100 °F.

**Design Approach:** The output diode voltage developed in the diode thermometer in Figure 1.47 is to be applied between the gate–source terminals of an NMOS transistor to enhance the voltage over the temperature range. The NMOS transistor is to be held at a constant temperature.

**Choices:** Assume an n-channel, depletion-mode MOSFET is available with the parameters $k'_n = 80 \ \mu A/V^2$, $W/L = 10$, and $V_{TN} = -1$ V.

**Solution:** From the design in Chapter 1, the diode voltage is given by

$$V_D = 1.12 - 0.522 \left( \frac{T}{300} \right)$$

where $T$ is in kelvins.

Consider the circuit shown in Figure 3.69. We assume that the diode is in a variable temperature environment while the rest of the circuit is held at room temperature.

From the circuit, we see that $V_{GS} = V_D$, where $V_D$ is the diode voltage and not the drain voltage. We want the MOSFET biased in the saturation region, so

$$I_D = K_n (V_{GS} - V_{TN})^2 = \frac{k'_n}{2} \cdot \frac{W}{L} (V_D - V_{TN})^2$$

We find the output voltage as

$$V_O = 15 - I_D R_D$$

$$= 15 - \frac{k'_n}{2} \cdot \frac{W}{L} \cdot R_D (V_D - V_{TN})^2$$

The diode current and output voltage can be written as

$$I_D = \frac{0.080}{2} \cdot \frac{10}{1} (V_D + 1)^2 = 0.4(V_D + 1)^2 \ (mA)$$

and

$$V_O = 15 - [0.4(V_D + 1)^2](10) = 15 - 4(V_D + 1)^2 \ (V)$$

From Chapter 1, we have the following:

| $T$ (°F) | $V_D$ (V) |
|---|---|
| 0 | 0.6760 |
| 40 | 0.6372 |
| 80 | 0.5976 |
| 100 | 0.5790 |

We find the circuit response as:

| $T$ (°F) | $I_D$ (mA) | $V_O$ (V) |
|---|---|---|
| 0 | 1.124 | 3.764 |
| 40 | 1.072 | 4.278 |
| 80 | 1.021 | 4.791 |
| 100 | 0.9973 | 5.027 |

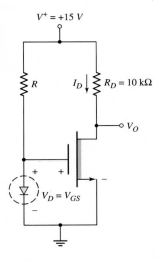

$V^+ = +15$ V

$R_D = 10 \ k\Omega$

$V_O$

$V_D = V_{GS}$

**Figure 3.69** Design application circuit to measure output voltage of diode versus temperature

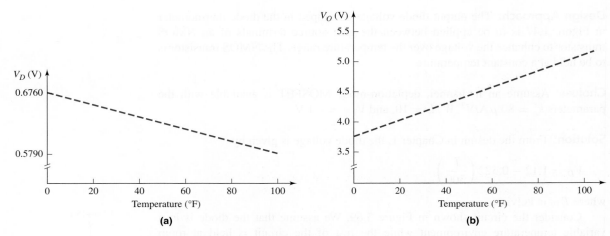

Figure 3.70   (a) Diode voltage versus temperature and (b) circuit output voltage versus temperature

**Comment:** Figure 3.70(a) shows the diode voltage versus temperature and Figure 3.70(b) now shows the output voltage versus temperature from the MOSFET circuit. We can see that the transistor circuit provides a voltage gain. This voltage gain is the desired characteristic of the transistor circuit.

**Discussion:** We can see from the equations that the diode current and output voltage are not linear functions of the diode voltage. This effect implies that the transistor output voltage is also not a linear function of temperature. We will see a better circuit design using operational amplifiers in Chapter 9.

We can note from the results that $V_O = V_{DS} > V_{DS}(\text{sat})$ in all cases, so the transistor is biased in the saturation region as desired.

## 3.8   SUMMARY

- In this chapter, we have considered the physical structure and dc electrical characteristics of the MOSFET.
- The current in the MOSFET is controlled by the gate voltage. In the nonsaturation bias region of operation, the drain current is also a function of drain voltage, whereas in the saturation bias region of operation the drain current is essentially independent of the drain voltage. The drain current is directly proportional to the width-to-length ratio of the transistor, so this parameter becomes the primary design variable in MOSFET circuit design.
- The dc analysis and design techniques of MOSFET circuits were emphasized in this chapter. The use of MOSFETs in place of resistors was investigated. This leads to the design of all-MOSFET circuits.
- Basic applications of the MOSFET include switching currents and voltages, performing digital logic functions, and amplifying time-varying signals. The amplifying characteristics will be considered in the next chapter and the digital applications will be considered in Chapter 16.
- MOSFET circuits that provide constant-current biasing to other MOSFET circuits were analyzed and designed. This current biasing technique is used in integrated circuits.

- The dc analysis and design of multistage MOSFET circuits were considered.
- The physical structure and dc electrical characteristics of JFET and MESFET devices as well as the analysis and design of JFET and MESFET circuits were considered.
- As an application, a MOSFET transistor was incorporated in a circuit design that enhances the simple diode electronic thermometer discussed in Chapter 1.

##  CHECKPOINT

After studying this chapter, the reader should have the ability to:

✓ Understand and describe the structure and general operation of n-channel and p-channel enhancement-mode and depletion-mode MOSFETs.

✓ Apply the ideal current–voltage relations in the dc analysis and design of various MOSFET circuits using any of the four basic MOSFETs.

✓ Understand how MOSFETs can be used in place of resistor load devices to create all-MOSFET circuits.

✓ Qualitatively understand how MOSFETs can be used to switch currents and voltages, to perform digital logic functions, and to amplify time-varying signals.

✓ Understand the basic operation of a MOSFET constant-current circuit.

✓ Understand the dc analysis and design of a multistage MOSFET circuit.

✓ Understand the general operation and characteristics of junction FETs.

##  REVIEW QUESTIONS

1. Describe the basic structure and operation of a MOSFET. Define enhancement mode and depletion mode.
2. Sketch the general current-voltage characteristics for both enhancement-mode and depletion-mode MOSFETs. Define the saturation and nonsaturation bias regions.
3. Describe what is meant by threshold voltage, width-to-length ratio, and drain-to-source saturation voltage.
4. Describe the channel length modulation effect and define the parameter $\lambda$. Describe the body effect and define the parameter $\gamma$.
5. Describe a simple common-source MOSFET circuit with an n-channel enhancement-mode device and discuss the relation between the drain-to-source voltage and gate-to-source voltage.
6. How do you prove that a MOSFET is biased in the saturation region?
7. In the dc analysis of some MOSFET circuits, quadratic equations in gate-to-source voltage are developed. How do you determine which of the two possible solutions is the correct one?
8. How can the $Q$-point be stabilized against variations in transistor parameters?
9. Describe the current–voltage relation of an n-channel enhancement-mode MOSFET with the gate connected to the drain.
10. Describe the current–voltage relation of an n-channel depletion-mode MOSFET with the gate connected to the source.
11. Describe a MOSFET NOR logic circuit.
12. Describe how a MOSFET can be used to amplify a time-varying voltage.
13. Describe the basic operation of a junction FET.
14. What is the difference between a MESFET and a pn junction FET?

**PROBLEMS**

[Note: In all problems, assume the transistor parameter $\lambda = 0$, unless otherwise stated.]

## Section 3.1 MOS Field-Effect Transistor

3.1    (a) Calculate the drain current in an NMOS transistor with parameters $V_{TN} = 0.4$ V, $k'_n = 120\,\mu\text{A/V}^2$, $W = 10\,\mu$m, $L = 0.8\,\mu$m, and with applied voltages of $V_{DS} = 0.1$ V and (i) $V_{GS} = 0$, (ii) $V_{GS} = 1$ V, (iii) $V_{GS} = 2$ V, and (iv) $V_{GS} = 3$ V. (b) Repeat part (a) for $V_{DS} = 4$ V.

3.2    The current in an NMOS transistor is 0.5 mA when $V_{GS} - V_{TN} = 0.6$ V and 1.0 mA when $V_{GS} - V_{TN} = 1.0$ V. The device is operating in the nonsaturation region. Determine $V_{DS}$ and $K_n$.

3.3    The transistor characteristics $i_D$ versus $v_{DS}$ for an NMOS device are shown in Figure P3.3. (a) Is this an enhancement-mode or depletion-mode device? (b) Determine the values for $K_n$ and $V_{TN}$. (c) Determine $i_D$(sat) for $v_{GS} = 3.5$ V and $v_{GS} = 4.5$ V.

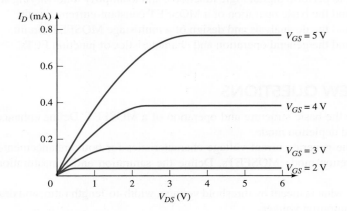

**Figure P3.3**

3.4    For an n-channel depletion-mode MOSFET, the parameters are $V_{TN} = -2.5$ V and $K_n = 1.1$ mA/V$^2$. (a) Determine $I_D$ for $V_{GS} = 0$; and: (i) $V_{DS} = 0.5$ V, (ii) $V_{DS} = 2.5$ V, and (iii) $V_{DS} = 5$ V. (b) Repeat part (a) for $V_{GS} = 2$ V.

3.5    The threshold voltage of each transistor in Figure P3.5 is $V_{TN} = 0.4$ V. Determine the region of operation of the transistor in each circuit.

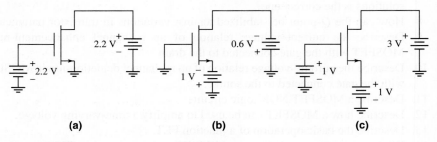

**(a)**                      **(b)**                      **(c)**

**Figure P3.5**

3.6    The threshold voltage of each transistor in Figure P3.6 is $V_{TP} = -0.4$ V. Determine the region of operation of the transistor in each circuit.

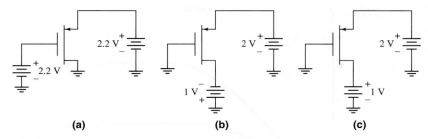

(a)                    (b)                    (c)

**Figure P3.6**

3.7    Consider an n-channel depletion-mode MOSFET with parameters $V_{TN} = -1.2$ V and $k'_n = 120\ \mu\text{A/V}^2$. The drain current is $I_D = 0.5$ mA at $V_{GS} = 0$ and $V_{DS} = 2$ V. Determine the $W/L$ ratio.

3.8    Determine the value of the process conduction parameter $k'_n$ for an NMOS transistor with $\mu_n = 600$ cm$^2$/V–s and for an oxide thickness $t_{\text{ox}}$ of (a) 500 Å, (b) 250 Å, (c)100 Å, (d) 50 Å, and (e) 25 Å.

3.9    An n-channel enhancement-mode MOSFET has parameters $V_{TN} = 0.4$ V, $W = 20\ \mu$m, $L = 0.8\ \mu$m, $t_{\text{ox}} = 200$ Å, and $\mu_n = 650$ cm$^2$/V–s. (a) Calculate the conduction parameter $K_n$.(b) Determine the drain current when $V_{GS} = V_{DS} = 2$ V. (c) With $V_{GS} = 2$ V, what value of $V_{DS}$ puts the device at the edge of saturation?

3.10   An NMOS device has parameters $V_{TN} = 0.8$ V, $L = 0.8\ \mu$m, and $k'_n = 120\ \mu\text{A/V}^2$. When the transistor is biased in the saturation region with $V_{GS} = 1.4$ V, the drain current is $I_D = 0.6$ mA. (a) What is the channel width $W$? (b) Determine the drain current when $V_{DS} = 0.4$ V. (c) What value of $V_{DS}$ puts the device at the edge of saturation?

3.11   A particular NMOS device has parameters $V_{TN} = 0.6$ V, $L = 0.8\ \mu$m, $t_{\text{ox}} = 200$ Å, and $\mu_n = 600$ cm$^2$/V–s. A drain current of $I_D = 1.2$ mA is required when the device is biased in the saturation region at $V_{GS} = 3$ V. Determine the required channel width of the device.

3.12   MOS transistors with very short channels do not exhibit the square law voltage relation in saturation. The drain current is instead given by

$$I_D = WC_{\text{ox}}(V_{GS} - V_{TN})v_{\text{sat}}$$

where $v_{\text{sat}}$ is a saturation velocity. Assuming $v_{\text{sat}} = 2 \times 10^7$ cm/s and using the parameters in Problem 3.11, determine the current.

3.13   For a p-channel enhancement-mode MOSFET, $k'_p = 50\ \mu\text{A/V}^2$. The device has drain currents of $I_D = 0.225$ mA at $V_{SG} = V_{SD} = 2$ V and $I_D = 0.65$ mA at $V_{SG} = V_{SD} = 3$ V. Determine the $W/L$ ratio and the value of $V_{TP}$.

3.14   For a p-channel enhancement-mode MOSFET, the parameters are $K_P = 2$ mA/V$^2$ and $V_{TP} = -0.5$ V. The gate is at ground potential, and the source and substrate terminals are at +5 V. Determine $I_D$ when the drain terminal voltage is: (a) $V_D = 0$ V, (b) $V_D = 2$ V, (c) $V_D = 4$ V, and (d) $V_D = 5$ V.

3.15   The transistor characteristics $i_D$ versus $v_{SD}$ for a PMOS device are shown in Figure P3.15. (a) Is this an enhancement-mode or depletion-mode device?

(b) Determine the values for $K_p$ and $V_{TP}$. (c) Determine $i_D$(sat) for $v_{SG} = 3.5$ V and $v_{SG} = 4.5$ V.

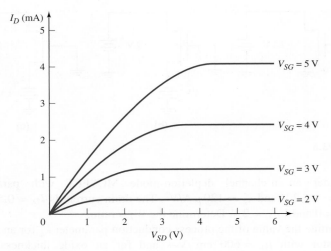

**Figure P3.15**

3.16 A p-channel depletion-mode MOSFET has parameters $V_{TP} = +2$ V, $k_p' = 40 \ \mu A/V^2$, and $W/L = 6$. Determine $V_{SD}$(sat) for: (a) $V_{SG} = -1$ V, (b) $V_{SG} = 0$, and (c) $V_{SG} = +1$ V. If the transistor is biased in the saturation region, calculate the drain current for each value of $V_{SG}$.

3.17 Calculate the drain current in a PMOS transistor with parameters $V_{TP} = -0.5$ V, $k_p' = 50 \ \mu A/V^2$, $W = 12 \ \mu m$, $L = 0.8 \ \mu m$, and with applied voltages of $V_{SG} = 2$ V and (a) $V_{SD} = 0.2$ V, (b) $V_{SD} = 0.8$ V, (c) $V_{SD} = 1.2$ V, (d) $V_{SD} = 2.2$ V, and (e) $V_{SD} = 3.2$ V.

3.18 Determine the value of the process conduction parameter $k_p'$ for a PMOS transistor with $\mu_p = 250 \ cm^2/V-s$ and for an oxide thickness $t_{ox}$ of (a) 500 Å, (b) 250 Å, (c) 100 Å, (d) 50 Å, and (e) 25 Å.

3.19 Enhancement-mode NMOS and PMOS devices both have parameters $L = 4 \ \mu m$ and $t_{ox} = 500$ Å. For the NMOS transistor, $V_{TN} = +0.6$ V, $\mu_n = 675 \ cm^2/V-s$, and the channel width is $W_n$; for the PMOS transistor, $V_{TP} = -0.6$ V, $\mu_p = 375 \ cm^2/V-s$, and the channel width is $W_p$. Design the widths of the two transistors such that they are electrically equivalent and the drain current in the PMOS transistor is $I_D = 0.8$ mA when it is biased in the saturation region at $V_{SG} = 5$ V. What are the values of $K_n$, $K_p$, $W_n$, and $W_p$?

3.20 For an NMOS enhancement-mode transistor, the parameters are: $V_{TN} = 1.2$ V, $K_n = 0.20 \ mA/V^2$, and $\lambda = 0.01 \ V^{-1}$. Calculate the output resistance $r_o$ for $V_{GS} = 2.0$ V and for $V_{GS} = 4.0$ V. What is the value of $V_A$?

3.21 The parameters of an n-channel enhancement-mode MOSFET are $V_{TN} = 0.5$ V, $k_n' = 120 \ \mu A/V^2$, and $W/L = 4$. What is the maximum value of $\lambda$ and the minimum value of $V_A$ such that for $V_{GS} = 2$ V, $r_o \geq 200 \ k\Omega$?

3.22 An enhancement-mode NMOS transistor has parameters $V_{TNO} = 0.8$ V, $\gamma = 0.8 \ V^{1/2}$, and $\phi_f = 0.35$ V. At what value of $V_{SB}$ will the threshold voltage change by 2V due to the body effect?

3.23 An NMOS transistor has parameters $V_{TO} = 0.75$ V, $k_n' = 80 \ \mu A/V^2$, $W/L = 15$, $\phi_f = 0.37$ V, and $\gamma = 0.6 \ V^{1/2}$. (a) The transistor is biased at

$V_{GS} = 2.5$ V, $V_{SB} = 3$ V, and $V_{DS} = 3$ V. Determine the drain current $I_D$.
(b) Repeat part (a) for $V_{DS} = 0.25$ V.

3.24 (a) A silicon dioxide gate insulator of an MOS transistor has a thickness of $t_{ox} = 120$ Å. (i) Calculate the ideal oxide breakdown voltage. (ii) If a safety factor of three is required, determine the maximum safe gate voltage that may be applied. (b) Repeat part (a) for an oxide thickness of $t_{ox} = 200$ Å.

3.25 In a power MOS transistor, the maximum applied gate voltage is 24 V. If a safety factor of three is specified, determine the minimum thickness necessary for the silicon dioxide gate insulator.

## Section 3.2 Transistor dc Analysis

3.26 In the circuit in Figure P3.26, the transistor parameters are $V_{TN} = 0.8$ V and $K_n = 0.5$ mA/V$^2$. Calculate $V_{GS}$, $I_D$, and $V_{DS}$.

3.27 The transistor in the circuit in Figure P3.27 has parameters $V_{TN} = 0.8$ V and $K_n = 0.25$ mA/V$^2$. Sketch the load line and plot the $Q$-point for (a) $V_{DD} = 4$ V, $R_D = 1$ kΩ and (b) $V_{DD} = 5$ V, $R_D = 3$ kΩ. What is the operating bias region for each condition?

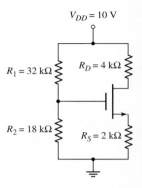

Figure P3.26

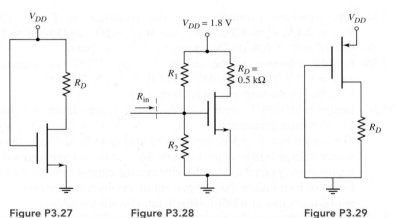

Figure P3.27    Figure P3.28    Figure P3.29    Figure P3.30

D3.28 The transistor in Figure P3.28 has parameters $V_{TN} = 0.4$ V, $k'_n = 120$ $\mu$A/V$^2$, and $W/L = 80$. Design the circuit such that $I_Q = 0.8$ mA and $R_{in} = 200$ kΩ.

3.29 The transistor in the circuit in Figure P3.29 has parameters $V_{TP} = -0.8$ V and $K_p = 0.20$ mA/V$^2$. Sketch the load line and plot the $Q$-point for (a) $V_{DD} = 3.5$ V, $R_D = 1.2$ kΩ and (b) $V_{DD} = 5$ V, $R_D = 4$ kΩ. What is the operating bias region for each condition?

3.30 Consider the circuit in Figure P3.30. The transistor parameters are $V_{TP} = -0.8$ V and $K_p = 0.5$ mA/V$^2$. Determine $I_D$, $V_{SG}$, and $V_{SD}$.

3.31 For the circuit in Figure P3.31, the transistor parameters are $V_{TP} = -0.8$ V and $K_p = 200$ $\mu$A/V$^2$. Determine $V_S$ and $V_{SD}$.

D3.32 Design a MOSFET circuit in the configuration shown in Figure P3.26. The transistor parameters are $V_{TN} = 0.4$ V and $k'_n = 120 \mu$A/V$^2$, and $\lambda = 0$. The circuit parameters are $V_{DD} = 3.3$ V and $R_D = 5$ kΩ. Design the circuit so that $V_{DSQ} \cong 1.6$ V and the voltage across $R_S$ is approximately 0.8 V. Set $V_{GS} = 0.8$ V. The current through the bias resistors is to be approximately 5 percent of the drain current.

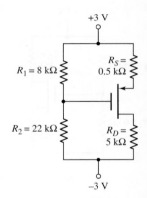

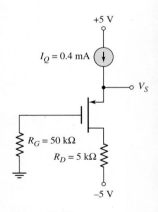

Figure P3.31

3.33   Consider the circuit shown in Figure P3.33. The transistor parameters are $V_{TN} = 0.4\,\text{V}$ and $k'_n = 120\,\mu\text{A/V}^2$. The voltage drop across $R_S$ is to be 0.20 V. Design the transistor $W/L$ ratio such that $V_{DS} = V_{DS}\,(\text{sat}) + 0.4\,\text{V}$, and determine $R_1$ and $R_2$ such that $R_{\text{in}} = 200\,\text{k}\Omega$.

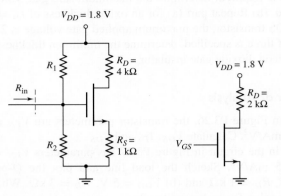

Figure P3.33                           Figure P3.34

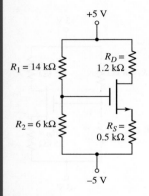

Figure P3.35

3.34   The transistor parameters for the transistor in Figure P3.34 are $V_{TN} = 0.4\,\text{V}$, $k'_n = 120\,\mu\text{A/V}^2$, and $W/L = 50$. (a) Determine $V_{GS}$ such that $I_D = 0.35\,\text{mA}$. (b) Determine $V_{DS}$ and $V_{DS}\,(\text{sat})$.

3.35   For the transistor in the circuit in Figure P3.35, the parameters are $V_{TN} = 0.4\,\text{V}$, $k'_n = 120\,\mu\text{A/V}^2$, and $W/L = 25$. Determine $V_{GS}$, $I_D$, and $V_{DS}$. Sketch the load line and plot the $Q$-point.

D3.36  Design a MOSFET circuit with the configuration shown in Figure P3.30. The transistor parameters are $V_{TP} = -0.6\,\text{V}$, $k'_p = 50\,\mu\text{A/V}^2$, and $\lambda = 0$. The circuit bias is $\pm 3\,\text{V}$, the drain current is to be 0.2 mA, the drain-to-source voltage is to be approximately 3 V, and the voltage across $R_S$ is to be approximately equal to $V_{SG}$. In addition, the current through the bias resistors is to be no more than 10 percent of the drain current. (*Hint*: choose a reasonable value of width-to-length ratio for the transistor.)

3.37   The parameters of the transistors in Figures P3.37 (a) and (b) are $K_n = 0.5\,\text{mA/V}^2$, $V_{TN} = 1.2\,\text{V}$, and $\lambda = 0$. Determine $v_{GS}$ and $v_{DS}$ for each transistor when (i) $I_Q = 50\,\mu\text{A}$ and (ii) $I_Q = 1\,\text{mA}$.

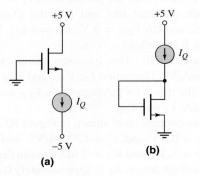

Figure P3.37

3.38  For the circuit in Figure P3.38, the transistor parameters are $V_{TN} = 0.6$ V and $K_n = 200 \ \mu A/V^2$. Determine $V_S$ and $V_D$.

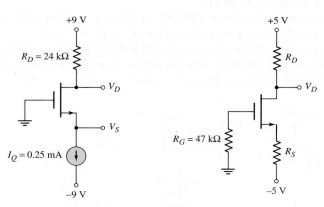

**Figure P3.38**

**Figure P3.39**

*3.39  (a) Design the circuit in Figure P3.39 such that $I_{DQ} = 0.50$ mA and $V_D = 1$ V. The transistor parameters are $K_n = 0.25 \ mA/V^2$ and $V_{TN} = 1.4$ V. Sketch the load line and plot the $Q$-point. (b) Choose standard resistor values that are closest to the ideal designed values. What are the resulting $Q$-point values? (c) If the resistors in part (b) have tolerances of $\pm 10$ percent, determine the maximum and minimum values of $I_{DQ}$.

3.40  The PMOS transistor in Figure P3.40 has parameters $V_{TP} = -0.7$ V, $k'_p = 50 \ \mu A/V^2$, $L = 0.8 \ \mu m$, and $\lambda = 0$. Determine the values of $W$ and $R$ such that $I_D = 0.1$ mA and $V_{SD} = 2.5$ V.

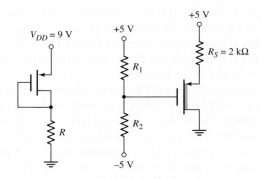

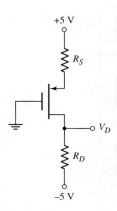

**Figure P3.40**     **Figure P3.41**

3.41  Design the circuit in Figure P3.41 so that $V_{SD} = 2.5$ V. The current in the bias resistors should be no more than 10 percent of the drain current. The transistor parameters are $V_{TP} = +1.5$ V and $K_p = 0.5 \ mA/V^2$.

*3.42  (a) Design the circuit in Figure P3.42 such that $I_{DQ} = 0.25$ mA and $V_D = -2$ V. The nominal transistor parameters are $V_{TP} = -1.2$ V, $k'_p = 35 \ \mu A/V^2$, and $W/L = 15$. Sketch the load line and plot the $Q$-point. (b) Determine the maximum and minimum $Q$-point values if the tolerance of the $k'_p$ parameter is $\pm 5$ percent.

**Figure P3.42**

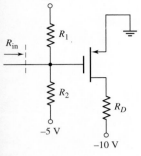

**Figure P3.43**

3.43  The parameters of the transistor in the circuit in Figure P3.43 are $V_{TP} = -1.75$ V and $K_p = 3$ mA/V$^2$. Design the circuit such that $I_D = 5$ mA, $V_{SD} = 6$ V, and $R_{in} = 80$ k$\Omega$.

3.44  For each transistor in the circuit in Figure P3.44, $k'_n = 120\,\mu$A/V$^2$. Also for $M_1$, $W/L = 4$ and $V_{TN} = 0.4$ V, and for $M_2$, $W/L = 1$ and $V_{TN} = -0.6$ V. (a) Determine the input voltage such that both $M_1$ and $M_2$ are biased in the saturation region. (b) Determine the region of operation of each transistor and the output voltage $v_O$ for: (i) $v_I = 0.6$ V and (ii) $v_I = 1.5$ V.

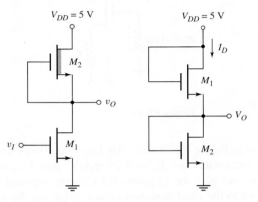

**Figure P3.44**                      **Figure P3.46**

3.45  Consider the circuit in Figure P3.44. The transistor parameters for $M_1$ are $V_{TN} = 0.4$ V and $k'_n = 120\,\mu$A/V$^2$, and for $M_2$ are $V_{TN} = -0.6$ V, $k'_n = 120\,\mu$A/V$^2$, and $W/L = 1$. Determine the $W/L$ ratio of $M_1$ such that $v_O = 0.025$ V when $v_I = 3$ V.

3.46  The transistors in the circuit in Figure P3.46 both have parameters $V_{TN} = 0.4$ V and $k'_n = 120\,\mu$A/V$^2$. (a) If the width-to-length ratios of $M_1$ and $M_2$ are $(W/L)_1 = (W/L)_2 = 30$, determine $V_{GS1}$, $V_{GS2}$, $V_O$, and $I_D$. (b) Repeat part (a) if the width-to-length ratios are changed to $(W/L)_1 = 30$ and $(W/L)_2 = 15$. (c) Repeat part (a) if the width-to-length ratios are changed to $(W/L)_1 = 15$ and $(W/L)_2 = 30$.

3.47  Consider the circuit in Figure P3.47. (a) The nominal transistor parameters are $V_{TN} = 0.6$ V and $k'_n = 120\,\mu$A/V$^2$. Design the width-to-length ratio required in each transistor such that $I_{DQ} = 0.8$ mA, $V_1 = 2.5$ V, and $V_2 = 6$ V. (b) Determine the change in the values of $V_1$ and $V_2$ if the $k'_n$ parameter in each transistor changes by (i) $+5$ percent and (ii) $-5$ percent. (c) Determine the values of $V_1$ and $V_2$ if the $k'_n$ parameter of $M_1$ decreases by 5 percent and the $k'_n$ parameter of $M_2$ and $M_3$ increases by 5 percent.

3.48  The transistors in the circuit in Figure 3.36 in the text have parameters $V_{TN} = 0.6$ V, $k'_n = 120\,\mu$A/V$^2$, and $\lambda = 0$. The width-to-length ratio of $M_L$ is $(W/L)_L = 2$. Design the width-to-length ratio of the driver transistor such that $V_O = 0.15$ V when $V_I = 5$ V.

3.49  For the circuit in Figure 3.39 in the text, the transistor parameters are: $V_{TND} = 0.6$ V, $V_{TNL} = -1.2$ V, $\lambda = 0$, and $k'_n = 120\,\mu$A/V$^2$. Let $V_{DD} = 5$ V. The width-to-length ratio of $M_L$ is $(W/L)_L = 2$. Design the width-to-length ratio of the driver transistor such that $V_O = 0.10$ V when $V_I = 5$ V.

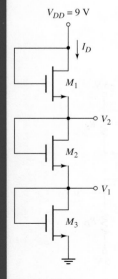

**Figure P3.47**

## Section 3.3 MOSFET Switch and Amplifier

3.50   Consider the circuit in Figure P3.50. The circuit parameters are $V_{DD} = 3$ V and $R_D = 30 \, k\Omega$. The transistor parameters are $V_{TN} = 0.4$ V and $k'_n = 120 \, \mu A/V^2$. (a) Determine the transistor width-to-length ratio such that $V_O = 0.08$ V when $V_I = 2.6$ V. (b) Repeat part (a) for $V_I = 3$ V.

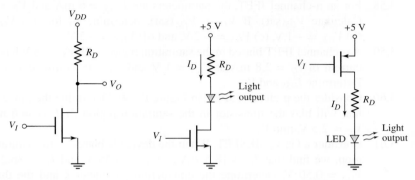

Figure P3.50          Figure P3.51          Figure P3.52

3.51   The transistor in the circuit in Figure P3.51 is used to turn the LED on and off. The transistor parameters are $V_{TN} = 0.6$ V, $k'_n = 80 \, \mu A/V^2$, and $\lambda = 0$. The diode cut-in voltage is $V_\gamma = 1.6$ V. Design $R_D$ and the transistor width-to-length ratio such that $I_D = 12$ mA for $V_I = 5$ V and $V_{DS} = 0.15$ V.

3.52   The circuit in Figure P3.52 is another configuration used to switch an LED on and off. The transistor parameters are $V_{TP} = -0.6$ V, $k'_p = 40 \, \mu A/V^2$, and $\lambda = 0$. The diode cut-in voltage is $V_\gamma = 1.6$ V. Design $R_D$ and the transistor width-to-length ratio such that $I_D = 15$ mA for $V_I = 0$ V and $V_{SD} = 0.20$ V.

3.53   For the two-input NMOS NOR logic gate in Figure 3.46 in the text, the transistor parameters are $V_{TN1} = V_{TN2} = 0.6$ V, $\lambda_1 = \lambda_2 = 0$, and $k'_{n1} = k'_{n2} = 120 \, \mu A/V^2$. The drain resistor is $R_D = 50 \, k\Omega$. (a) Determine the width-to-length ratios of the transistors so that $V_O = 0.15$ V when $V_1 = V_2 = 5$ V. Assume that $(W/L)_1 = (W/L)_2$. (b) Using the results of part (a), find $V_O$ when $V_1 = 5$ V and $V_2 = 0.2$ V.

3.54   All transistors in the current-source circuit shown in Figure 3.49(a) in the text have parameters $V_{TN} = 0.4$ V, $k'_n = 120 \, \mu A/V^2$, and $\lambda = 0$. Transistors $M_1$ and $M_2$ are matched. The bias sources are $V^+ = 2.5$ V and $V^- = -2.5$ V. The currents are to be $I_{Q1} = 125 \, \mu A$ and $I_{REF1} = 225 \, \mu A$. For $M_2$, we require $V_{DS2}$ (sat) $= 0.5$ V, and for $M_1$ we require $V_{DS1} = 2$ V. (a) Find the $W/L$ ratios of the transistors. (b) Find $R_D$.

3.55   All transistors in the current-source circuit shown in Figure 3.49(b) in the text have parameters $V_{TP} = -0.4$ V, $k'_p = 50 \, \mu A/V^2$, and $\lambda = 0$. The bias sources are $V^+ = 5$ V and $V^- = -5$ V. The currents are to be $I_{Q2} = 200 \, \mu A$ and $I_{REF2} = 125 \, \mu A$. For $M_B$, we require $V_{SDB}$ (sat) $= 0.8$ V, and for $M_A$, we require $V_{SDA} = 4$ V. Transistors $M_A$ and $M_B$ are matched. (a) Find the $W/L$ ratios of the transistors. (b) Find the value of $R_D$.

3.56   Consider the circuit shown in Figure 3.50 in the text. The threshold voltage and process conduction parameter for each transistor is $V_{TN} = 0.6$ V and $k'_n = 120 \, \mu A/V^2$. Let $\lambda = 0$ for all transistors. Assume that $M_1$ and $M_2$ are matched. Design width-to-length ratios such that $I_Q = 0.35$ mA, $I_{REF} = 0.15$ mA, and $V_{DS2}$ (sat) $= 0.5$ V. Find $R_D$ such that $V_{DS1} = 3.5$ V.

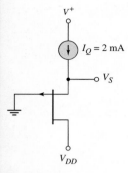

**Figure P3.60**

## Section 3.6 Junction Field-Effect Transistor

3.57    The gate and source of an n-channel depletion-mode JFET are connected together. What value of $V_{DS}$ will ensure that this two-terminal device is biased in the saturation region. What is the drain current for this bias condition?

3.58    For an n-channel JFET, the parameters are $I_{DSS} = 6$ mA and $V_P = -3$ V. Calculate $V_{DS}(\text{sat})$. If $V_{DS} > V_{DS}(\text{sat})$, determine $I_D$ for: (a) $V_{GS} = 0$, (b) $V_{GS} = -1$ V, (c) $V_{GS} = -2$ V, and (d) $V_{GS} = -3$ V.

3.59    A p-channel JFET biased in the saturation region with $V_{SD} = 5$ V has a drain current of $I_D = 2.8$ mA at $V_{GS} = 1$ V and $I_D = 0.30$ mA at $V_{GS} = 3$ V. Determine $I_{DSS}$ and $V_P$.

3.60    Consider the p-channel JFET in Figure P3.60. Determine the range of $V_{DD}$ that will bias the transistor in the saturation region. If $I_{DSS} = 6$ mA and $V_P = 2.5$ V, find $V_S$.

3.61    Consider a GaAs MESFET. When the device is biased in the saturation region, we find that $I_D = 18.5$ $\mu$A at $V_{GS} = 0.35$ V and $I_D = 86.2$ $\mu$A at $V_{GS} = 0.50$ V. Determine the conduction parameter $k$ and the threshold voltage $V_{TN}$.

3.62    The threshold voltage of a GaAs MESFET is $V_{TN} = 0.24$ V. The maximum allowable gate-to-source voltage is $V_{GS} = 0.75$ V. When the transistor is biased in the saturation region, the maximum drain current is $I_D = 250$ $\mu$A. What is the value of the conduction parameter $k$?

*3.63    For the transistor in the circuit in Figure P3.63, the parameters are: $I_{DSS} = 10$ mA and $V_P = -5$ V. Determine $I_{DQ}$, $V_{GSQ}$, and $V_{DSQ}$.

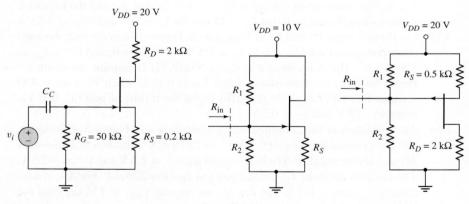

**Figure P3.63**          **Figure P3.64**          **Figure P3.65**

3.64    Consider the source follower with the n-channel JFET in Figure P3.64. The input resistance is to be $R_{in} = 500$ k$\Omega$. We wish to have $I_{DQ} = 5$ mA, $V_{DSQ} = 8$ V, and $V_{GSQ} = -1$ V. Determine $R_S$, $R_1$, and $R_2$, and the required transistor values of $I_{DSS}$ and $V_P$.

3.65    The transistor in the circuit in Figure P3.65 has parameters $I_{DSS} = 8$ mA and $V_P = 4$ V. Design the circuit such that $I_D = 5$ mA. Assume $R_{in} = 100$ k$\Omega$. Determine $V_{GS}$ and $V_{SD}$.

3.66    For the circuit in Figure P3.66, the transistor parameters are $I_{DSS} = 7$ mA and $V_P = 3$ V. Let $R_1 + R_2 = 100$ k$\Omega$. Design the circuit such that $I_{DQ} = 5.0$ mA and $V_{SDQ} = 6$ V.

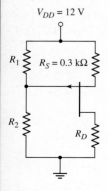

**Figure P3.66**

3.67   The transistor in the circuit in Figure P3.67 has parameters $I_{DSS} = 8$ mA and $V_P = -4$ V. Determine $V_G$, $I_{DQ}$, $V_{GSQ}$, and $V_{DSQ}$.

3.68   Consider the circuit in Figure P3.68. The quiescent value of $V_{DS}$ is found to be $V_{DSQ} = 5$ V. If $I_{DSS} = 10$ mA, determine $I_{DQ}$, $V_{GSQ}$, and $V_P$.

3.69   For the circuit in Figure P3.69, the transistor parameters are $I_{DSS} = 4$ mA and $V_P = -3$ V. Design $R_D$ such that $V_{DS} = |V_P|$. What is the value of $I_D$?

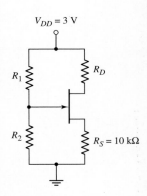

Figure P3.67

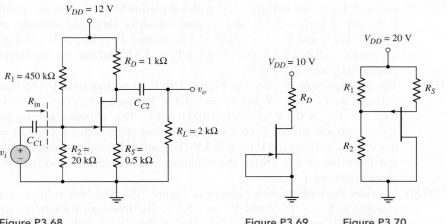

Figure P3.68          Figure P3.69     Figure P3.70

3.70   Consider the source-follower circuit in Figure P3.70. The transistor parameters are $I_{DSS} = 2$ mA and $V_P = 2$ V. Design the circuit such that $I_{DQ} = 1$ mA, $V_{SDQ} = 10$ V, and the current through $R_1$ and $R_2$ is 0.1 mA.

3.71   The GaAs MESFET in the circuit in Figure P3.71 has parameters $k = 250$ $\mu$A/V$^2$ and $V_{TN} = 0.20$ V. Let $R_1 + R_2 = 150$ k$\Omega$. Design the circuit such that $I_D = 40$ $\mu$A and $V_{DS} = 2$ V.

3.72   For the circuit in Figure P3.72, the GaAs MESFET threshold voltage is $V_{TN} = 0.15$ V. Let $R_D = 50$ k$\Omega$. Determine the value of the conduction parameter required so that $V_O = 0.70$ V when $V_I = 0.75$ V.

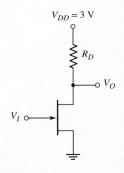

Figure P3.71

## COMPUTER SIMULATION PROBLEMS

3.73   Using a computer simulation, verify the results of Exercise Ex 3.5.

3.74   (a) Using a computer simulation, plot the voltage transfer characteristics of the CMOS circuit shown in Figure 3.41. Use the parameters given in Example 3.11. (b) Repeat part (a) for the case when the width-to-length ratio of $M_N$ is doubled.

3.75   (a) Using a computer simulation, plot the voltage transfer characteristics of the NMOS circuit shown in Figure 3.46 for $V_2 = 0$ and $0 \leq V_1 \leq 5$ V. Use the circuit and transistor parameters given in Example 3.13. (b) Repeat part (a) for $0 \leq V_1 = V_2 \leq 5$ V.

3.76   Using a computer simulation, verify the results of Example 3.17 for the multitransistor circuit shown in Figure 3.52.

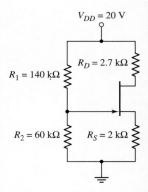

Figure P3.72

**DESIGN PROBLEMS**

[Note: All design should be correlated with a computer simulation.]

*D3.77  Consider the PMOS circuit shown in Figure 3.30. The circuit is to be re-designed such that $I_{DQ} = 100\,\mu\text{A}$ and the Q-point is in the center of the saturation region of the load line. Assume $R_1 + R_2 = 500\,\text{k}\Omega$ and assume the same transistor parameters as given in Exercise Ex 3.6.

*D3.78  Consider the circuit in Figure 3.39 with a depletion load. Assume the circuit is biased at $V_{DD} = 3.3\,\text{V}$, and assume transistor threshold voltages of $V_{TND} = 0.4\,\text{V}$ and $V_{TNL} = -0.75\,\text{V}$. Also assume $k_n' = 80\,\mu\text{A/V}^2$. Design the circuit such that $V_O = 0.05\,\text{V}$ when $V_I = 3.3\,\text{V}$ and that the maximum power dissipation is $150\,\mu\text{W}$.

*D3.79  The constant-current source in Figure 3.50 is to be redesigned. The bias voltages are $V^+ = 3.3\,\text{V}$ and $V^- = -3.3\,\text{V}$. The parameters of all transistors are $V_{TN} = 0.4\,\text{V}$ and $k_n' = 100\,\mu\text{A/V}^2$. The currents are to be $I_{REF} = 100\,\mu\text{A}$ and $I_Q = 60\,\mu\text{A}$. We will also specify that $V_{DS2}$ (sat) $= 0.6\,\text{V}$, $V_{GS1} = V_{GS2}$, and $V_{DS1} = 2.5\,\text{V}$. Determine all width-to-length ratios and the value of $R_D$. (Note: the minimum width-to-length ratio is to be greater than one.)

*D3.80  Consider the multitransistor circuit in Figure 3.52. The bias voltages are changed to $V^+ = 3.3\,\text{V}$ and $V^- = -3.3\,\text{V}$. The transistor parameters are $V_{TN} = 0.4\,\text{V}$ and $k_n' = 100\,\mu\text{A/V}^2$. Design the circuit such that $I_{DQ1} = 100\,\mu\text{A}$, $I_{DQ2} = 250\,\mu\text{A}$, $V_{DSQ1} = V_{DSQ2} = 3.3\,\text{V}$, and $R_i = 200\,\text{k}\Omega$.

# Basic FET Amplifiers

In Chapter 3, we described the structure and operation of the FET, in particular the MOSFET, and analyzed and designed the dc response of circuits containing these devices. In this chapter, we emphasize the use of the FETs in linear amplifier applications. Linear amplifiers imply that, for the most part, we are dealing with analog signals. The magnitude of an analog signal may have any value, within limits, and may vary continuously with respect to time. Although a major use of MOSFETs is in digital applications, they are also used in linear amplifier circuits.

We will begin to see how all-transistor circuits, that is, circuits with no resistors, can be designed. Since MOS transistors are small devices, high-density all-transistor circuits can be fabricated as integrated circuits.

## PREVIEW

In this chapter, we will:

- Investigate the process by which a single-MOS transistor circuit can amplify a small, time-varying input signal.
- Develop the small-signal models of the transistor that are used in the analysis of linear amplifiers.
- Discuss the three basic transistor amplifier configurations.
- Analyze the common-source, source-follower, and common-gate amplifiers, and become familiar with the general characteristics of these circuits.
- Compare the general characteristics of the three basic amplifier configurations.
- Analyze all-MOS transistor circuits that become the foundation of integrated circuits.
- Analyze multitransistor or multistage amplifiers and understand the advantages of these circuits over single-transistor amplifiers.
- Develop the small-signal model of JFET devices and analyze basic JFET amplifiers.
- As an application, incorporate MOS transistors in a design of a two-stage amplifier.

## 4.1    THE MOSFET AMPLIFIER

**Objective:** • Investigate the process by which a single-transistor circuit can amplify a small, time-varying input signal and develop the small-signal models of the transistor that are used in the analysis of linear amplifiers.

In this chapter, we will be considering **signals, analog circuits,** and **amplifiers.** A signal contains some type of information. For example, sound waves produced by a speaking human contain the information the person is conveying to another person. A sound wave is an analog signal. The magnitude of an **analog signal** can take on any value, within limits, and may vary continuously with time. Electronic circuits that process analog signals are called analog circuits. One example of an analog circuit is a linear amplifier. A **linear amplifier** magnifies an input signal and produces an output signal whose magnitude is larger and directly proportional to the input signal.

In this chapter, we analyze and design linear amplifiers that use field-effect transistors as the amplifying device. The term **small signal** means that we can linearize the ac equivalent circuit. We will define what is meant by small signal in the case of MOSFET circuits. The term linear amplifiers means that we can use superposition so that the dc analysis and ac analysis of the circuits can be performed separately and the total response is the sum of the two individual responses.

The mechanism with which MOSFET circuits amplify small time-varying signals was introduced in the last chapter. In this section, we will expand that discussion using the graphical technique, dc load line, and ac load line. In the process, we will develop the various small-signal parameters of linear circuits and the corresponding equivalent circuits.

### 4.1.1    Graphical Analysis, Load Lines, and Small-Signal Parameters

Figure 4.1 shows an NMOS common-source circuit with a time-varying voltage source in series with the dc source. We assume the time-varying input signal is sinusoidal. Figure 4.2 shows the transistor characteristics, dc load line, and $Q$-point, where the dc load line and $Q$-point are functions of $v_{GS}$, $V_{DD}$, $R_D$, and the transistor parameters. For the output voltage to be a linear function of the input voltage, the transistor must be biased in the saturation region. (Note that, although we primarily use n-channel, enhancement-mode MOSFETs in our discussions, the same results apply to the other MOSFETs.)

Also shown in Figure 4.2 are the sinusoidal variations in the gate-to-source voltage, drain current, and drain-to-source voltage, as a result of the sinusoidal source $v_i$. The total gate-to-source voltage is the sum of $V_{GSQ}$ and $v_i$. As $v_i$ increases, the instantaneous value of $v_{GS}$ increases, and the bias point moves up the load line. A larger value of $v_{GS}$ means a larger drain current and a smaller value of $v_{DS}$. For a negative $v_i$ (the negative portion of the sine wave), the instantaneous value of $v_{GS}$ decreases below the quiescent value, and the bias point moves down the load line. A smaller $v_{GS}$ value means a smaller drain current and increased value of $v_{DS}$. Once the

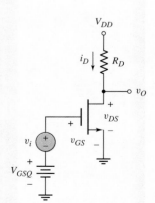

**Figure 4.1** NMOS common-source circuit with time-varying signal source in series with gate dc source

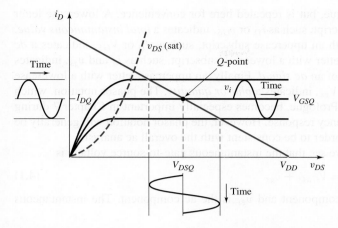

**Figure 4.2** Common-source transistor characteristics, dc load line, and sinusoidal variation in gate-to-source voltage, drain current, and drain-to-source voltage

$Q$-point is established, we can develop a mathematical model for the sinusoidal, or small-signal, variations in gate-to-source voltage, drain-to-source voltage, and drain current.

The time-varying signal source $v_i$ in Figure 4.1 generates a time-varying component of the gate-to-source voltage. In this case, $v_{gs} = v_i$, where $v_{gs}$ is the time-varying component of the gate-to-source voltage. For the FET to operate as a linear amplifier, the transistor must be biased in the saturation region, and the instantaneous drain current and drain-to-source voltage must also be confined to the saturation region.

When symmetrical sinusoidal signals are applied to the input of an amplifier, symmetrical sinusoidal signals are generated at the output, as long as the amplifier operation remains linear. We can use the load line to determine the maximum output symmetrical swing. If the output exceeds this limit, a portion of the output signal will be clipped and signal distortion will occur.

In the case of FET amplifiers, the output signal must avoid cutoff ($i_D = 0$) and must stay in the saturation region ($v_{DS} > v_{DS}(\text{sat})$). This maximum range of output signal can be determined from the load line in Figure 4.2.

## Transistor Parameters

We will be dealing with time-varying as well as dc currents and voltages in this chapter. Table 4.1 gives a summary of notation that will be used. This notation was

| Table 4.1 | Summary of notation |
| --- | --- |
| **Variable** | **Meaning** |
| $i_D$, $v_{GS}$ | Total instantaneous values |
| $I_D$, $V_{GS}$ | DC values |
| $i_d$, $v_{gs}$ | Instantaneous ac values |
| $I_d$, $V_{gs}$ | Phasor values |

discussed in the Prologue, but is repeated here for convenience. A lowercase letter with an upper case subscript, such as $i_D$ or $v_{GS}$, indicates a *total instantaneous value*. An uppercase letter with an uppercase subscript, such as $I_D$ or $V_{GS}$, indicates a *dc quantity*. A lowercase letter with a lowercase subscript, such as $i_d$ and $v_{gs}$, indicates an instantaneous value of an *ac signal*. Finally, an uppercase letter with a lowercase subscript, such as $I_d$ or $V_{gs}$, indicates a *phasor quantity*. The phasor notation, which is also reviewed in the Prologue, becomes especially important in Chapter 7 during the discussion of frequency response. However, the phasor notation will generally be used in this chapter in order to be consistent with the overall ac analysis.

From Figure 4.1, we see that the instantaneous gate-to-source voltage is

$$v_{GS} = V_{GSQ} + v_i = V_{GSQ} + v_{gs} \tag{4.1}$$

where $V_{GSQ}$ is the dc component and $v_{gs}$ is the ac component. The instantaneous drain current is

$$i_D = K_n(v_{GS} - V_{TN})^2 \tag{4.2}$$

Substituting Equation (4.1) into (4.2) produces

$$i_D = K_n[V_{GSQ} + v_{gs} - V_{TN}]^2 = K_n[(V_{GSQ} - V_{TN}) + v_{gs}]^2 \tag{4.3(a)}$$

or

$$i_D = K_n(V_{GSQ} - V_{TN})^2 + 2K_n(V_{GSQ} - V_{TN})v_{gs} + K_n v_{gs}^2 \tag{4.3(b)}$$

The first term in Equation (4.3(b)) is the dc or quiescent drain current $I_{DQ}$, the second term is the time-varying drain current component that is linearly related to the signal $v_{gs}$, and the third term is proportional to the square of the signal voltage. For a sinusoidal input signal, the squared term produces undesirable harmonics, or non-linear distortion, in the output voltage. To minimize these harmonics, we require

$$v_{gs} \ll 2(V_{GSQ} - V_{TN}) \tag{4.4}$$

which means that the third term in Equation (4.3(b)) will be much smaller than the second term. *Equation (4.4) represents the small-signal condition that must be satisfied for linear amplifiers.*

Neglecting the $v_{gs}^2$ term, we can write Equation (4.3(b)) as

$$i_D = I_{DQ} + i_d \tag{4.5}$$

Again, small-signal implies linearity so that the total current can be separated into a dc component and an ac component. The ac component of the drain current is given by

$$i_d = 2K_n(V_{GSQ} - V_{TN})v_{gs} \tag{4.6}$$

The small-signal drain current is related to the small-signal gate-to-source voltage by the transconductance $g_m$. The relationship is

$$g_m = \frac{i_d}{v_{gs}} = 2K_n(V_{GSQ} - V_{TN}) \tag{4.7}$$

The transconductance is a transfer coefficient relating output current to input voltage and can be thought of as representing the gain of the transistor.

The transconductance can also be obtained from the derivative

$$g_m = \frac{\partial i_D}{\partial v_{GS}}\bigg|_{v_{GS}=V_{GSQ}=\text{const.}} = 2K_n(V_{GSQ} - V_{TN}) \tag{4.8(a)}$$

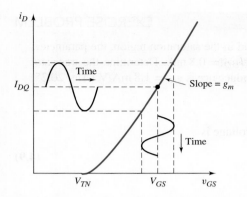

**Figure 4.3** Drain current versus gate-to-source voltage characteristics, with superimposed sinusoidal signals

which can be written as

$$g_m = 2\sqrt{K_n I_{DQ}} \qquad\qquad\qquad \textbf{(4.8(b))}$$

The drain current versus gate-to-source voltage for the transistor biased in the saturation region is given in Equation (4.2) and is shown in Figure 4.3. The transconductance $g_m$ is the slope of the curve. If the time-varying signal $v_{gs}$ is sufficiently small, the transconductance $g_m$ is a constant. With the $Q$-point in the saturation region, the transistor operates as a current source that is linearly controlled by $v_{gs}$. If the $Q$-point moves into the nonsaturation region, the transistor no longer operates as a linearly controlled current source.

As shown in Equation (4.8(a)), the transconductance is directly proportional to the conduction parameter $K_n$, which in turn is a function of the width-to-length ratio. Therefore, increasing the width of the transistor increases the transconductance, or gain, of the transistor.

## EXAMPLE 4.1

**Objective:** Calculate the transconductance of an n-channel MOSFET.

Consider an n-channel MOSFET with parameters $V_{TN} = 0.4$ V, $k'_n = 100\,\mu\text{A/V}^2$, and $W/L = 25$. Assume the drain current is $I_D = 0.40$ mA.

**Solution:** The conduction parameter is

$$K_n = \frac{k'_n}{2} \cdot \frac{W}{L} = (\frac{0.1}{2})(25) = 1.25\,\text{mA/V}^2$$

Assuming the transistor is biased in the saturation region, the transconductance is determined from Equation (4.8(b)) as

$$g_m = 2\sqrt{K_n I_{DQ}} = 2\sqrt{(1.25)(0.4)} = 1.41\,\text{mA/V}$$

**Comment:** The value of the transconductance can be increased by increasing the transistor $W/L$ ratio and also by increasing the quiescent drain current.

**Ex 4.1:** For an n-channel MOSFET biased in the saturation region, the parameters are $k'_n = 100\,\mu\text{A/V}^2$, $V_{TN} = 0.6\,\text{V}$, and $I_{DQ} = 0.8\,\text{mA}$. Determine the transistor width-to-length ratio such that the transconductance is $g_m = 1.8\,\text{mA/V}$. (Ans. 20.25)

## AC Equivalent Circuit

From Figure 4.1, we see that the output voltage is

$$v_{DS} = v_O = V_{DD} - i_D R_D \tag{4.9}$$

Using Equation (4.5), we obtain

$$v_O = V_{DD} - (I_{DQ} + i_d)R_D = (V_{DD} - I_{DQ}R_D) - i_d R_D \tag{4.10}$$

The output voltage is also a combination of dc and ac values. The time-varying output signal is the time-varying drain-to-source voltage, or

$$v_o = v_{ds} = -i_d R_D \tag{4.11}$$

Also, from Equations (4.6) and (4.7), we have

$$i_d = g_m v_{gs} \tag{4.12}$$

In summary, the following relationships exist between the time-varying signals for the circuit in Figure 4.1. The equations are given in terms of the instantaneous ac values, as well as the phasors. We have

$$v_{gs} = v_i \qquad \text{or} \qquad V_{gs} = V_i \tag{4.13}$$

and

$$i_d = g_m v_{gs} \qquad \text{or} \qquad I_d = g_m V_{gs} \tag{4.14}$$

and

$$v_{ds} = -i_d R_D \qquad \text{or} \qquad V_{ds} = -I_d R_D \tag{4.15}$$

The ac equivalent circuit in Figure 4.4 is developed by setting the dc sources in Figure 4.1 equal to zero. The small-signal relationships are given in Equations (4.13), (4.14), and (4.15). As shown in Figure 4.1, the drain current, which is composed of ac signals superimposed on the quiescent value, flows through the voltage source $V_{DD}$. Since the voltage across this source is assumed to be constant, the sinusoidal current produces no sinusoidal voltage component across this element. The equivalent ac impedance is therefore zero, or a short circuit. Consequently, in the ac equivalent circuit, the dc voltage sources are equal to zero. We say that the node connecting $R_D$ and $V_{DD}$ is at signal ground.

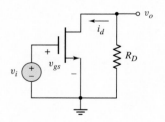

**Figure 4.4** AC equivalent circuit of common-source amplifier with NMOS transistor

## 4.1.2    Small-Signal Equivalent Circuit

Now that we have the ac equivalent circuit for the NMOS amplifier circuit, (Figure 4.4), we must develop a small-signal equivalent circuit for the transistor.

Initially, we assume that the signal frequency is sufficiently low so that any capacitance at the gate terminal can be neglected. The input to the gate thus appears as an open circuit, or an infinite resistance. Equation (4.14) relates the small-signal

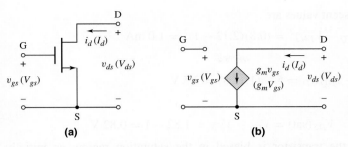

**Figure 4.5** (a) Common-source NMOS transistor with small-signal parameters and (b) simplified small-signal equivalent circuit for NMOS transistor

drain current to the small-signal input voltage, and Equation (4.7) shows that the transconductance $g_m$ is a function of the Q-point. The resulting simplified small-signal equivalent circuit for the NMOS device is shown in Figure 4.5. (The phasor components are in parentheses.)

This small-signal equivalent circuit can also be expanded to take into account the finite output resistance of a MOSFET biased in the saturation region. This effect, discussed in the last chapter, is a result of the nonzero slope in the $i_D$ versus $v_{DS}$ curve.

We know that

$$i_D = K_n[(v_{GS} - V_{TN})^2(1 + \lambda v_{DS})] \tag{4.16}$$

where $\lambda$ is the channel-length modulation parameter and is a positive quantity. The small-signal output resistance, as previously defined, is

$$r_o = \left(\frac{\partial i_D}{\partial v_{DS}}\right)^{-1}\Bigg|_{v_{GS}=V_{GSQ}=\text{const.}} \tag{4.17}$$

or

$$r_o = [\lambda K_n(V_{GSQ} - V_{TN})^2]^{-1} \cong [\lambda I_{DQ}]^{-1} \tag{4.18}$$

This small-signal output resistance is also a function of the Q-point parameters.

The expanded small-signal equivalent circuit of the n-channel MOSFET is shown in Figure 4.6 in phasor notation. Note that this equivalent circuit is a transconductance amplifier in that the input signal is a voltage and the output signal is a current. This equivalent circuit can now be inserted into the amplifier ac equivalent circuit in Figure 4.4 to produce the circuit in Figure 4.7.

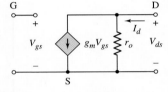

**Figure 4.6** Expanded small-signal equivalent circuit, including output resistance, for NMOS transistor

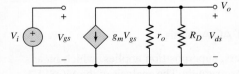

**Figure 4.7** Small-signal equivalent circuit of common-source circuit with NMOS transistor model

## EXAMPLE 4.2

**Objective:** Determine the small-signal voltage gain of a MOSFET circuit.

For the circuit in Figure 4.1, assume parameters are: $V_{GSQ} = 2.12$ V, $V_{DD} = 5$ V, and $R_D = 2.5$ kΩ. Assume transistor parameters are: $V_{TN} = 1$ V. $K_n = 0.80$ mA/V$^2$, and $\lambda = 0.02$ V$^{-1}$. Assume the transistor is biased in the saturation region.

**Solution:** The quiescent values are

$$I_{DQ} \cong K_n(V_{GSQ} - V_{TN})^2 = (0.8)(2.12 - 1)^2 = 1.0 \text{ mA}$$

and

$$V_{DSQ} = V_{DD} - I_{DQ}R_D = 5 - (1)(2.5) = 2.5 \text{ V}$$

Therefore,

$$V_{DSQ} = 2.5 \text{ V} > V_{DS}(\text{sat}) = V_{GS} - V_{TN} = 1.82 - 1 = 0.82 \text{ V}$$

which means that the transistor is biased in the saturation region, as initially assumed, and as required for a linear amplifier. The transconductance is

$$g_m = 2K_n(V_{GSQ} - V_{TN}) = 2(0.8)(2.12 - 1) = 1.79 \text{ mA/V}$$

and the output resistance is

$$r_o = [\lambda I_{DQ}]^{-1} = [(0.02)(1)]^{-1} = 50 \text{ k}\Omega$$

From Figure 4.7, the output voltage is

$$V_o = -g_m V_{gs}(r_o \| R_D)$$

Since $V_{gs} = V_i$, the small-signal voltage gain is

$$A_v = \frac{V_o}{V_i} = -g_m(r_o \| R_d) = -(1.79)(50\|2.5) = -4.26$$

**Comment:** The magnitude of the ac output voltage is 4.26 times larger than the magnitude of the input voltage. Hence, we have an amplifier. Note that the small-signal voltage gain contains a minus sign, which means that the sinusoidal output voltage is 180 degrees out of phase with respect to the input sinusoidal signal.

## EXERCISE PROBLEM

**Ex 4.2:** For the circuit shown in Figure 4.1, $V_{DD} = 3.3$ V and $R_D = 10$ k$\Omega$. The transistor parameters are $V_{TN} = 0.4$ V, $k'_n = 100 \,\mu\text{A/V}^2$, $W/L = 50$, and $\lambda = 0.025$ V$^{-1}$. Assume the transistor is biased such that $I_{DQ} = 0.25$ mA. (a) Verify that the transistor is biased in the saturation region. (b) Determine the small-signal parameters $g_m$ and $r_o$. (c) Determine the small-signal voltage gain. (Ans. (a) $V_{GSQ} = 0.716$ V and $V_{DSQ} = 0.8$ V so that $V_{DS} > V_{DS}$ (sat); (b) $g_m = 1.58$ mA/V, $r_o = 160$ k$\Omega$; (c) $-14.9$)

### Problem-Solving Technique: MOSFET AC Analysis

Since we are dealing with linear amplifiers, superposition applies, which means that we can perform the dc and ac analyses separately. The analysis of the MOSFET amplifier proceeds as follows:

1. Analyze the circuit with only the dc sources present. This solution is the dc or quiescent solution. The transistor must be biased in the saturation region in order to produce a linear amplifier.
2. Replace each element in the circuit with its small-signal model, which means replacing the transistor by its small-signal equivalent circuit.
3. Analyze the small-signal equivalent circuit, setting the dc source components equal to zero, to produce the response of the circuit to the time-varying input signals only.

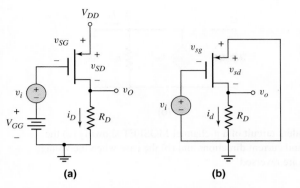

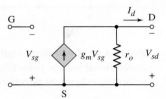

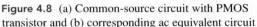

**(a)**          **(b)**

**Figure 4.8** (a) Common-source circuit with PMOS transistor and (b) corresponding ac equivalent circuit

**Figure 4.9** Small-signal equivalent circuit of PMOS transistor

The previous discussion was for an n-channel MOSFET amplifier. The same basic analysis and equivalent circuit also applies to the p-channel transistor. Figure 4.8(a) shows a circuit containing a p-channel MOSFET. Note that the power supply voltage $V_{DD}$ is connected to the source. (The subscript $DD$ can be used to indicate that the supply is connected to the drain terminal. Here, however, $V_{DD}$ is simply the usual notation for the power supply voltage in MOSFET circuits.) Also note the change in current directions and voltage polarities compared to the circuit containing the NMOS transistor. Figure 4.8(b) shows the ac equivalent circuit, with the dc voltage sources replaced by ac short circuits, and all currents and voltages shown are the time-varying components.

In the circuit of Figure 4.8(b), the transistor can be replaced by the equivalent circuit in Figure 4.9. The equivalent circuit of the p-channel MOSFET is the same as that of the n-channel device, except that all current directions and voltage polarities are reversed.

The final small-signal equivalent circuit of the p-channel MOSFET amplifier is shown in Figure 4.10. The output voltage is

$$V_o = g_m V_{sg}(r_o \| R_D) \tag{4.19}$$

The control voltage $V_{sg}$, given in terms of the input signal voltage, is

$$V_{sg} = -V_i \tag{4.20}$$

and the small-signal voltage gain is

$$A_v = \frac{V_o}{V_i} = -g_m(r_o \| R_D) \tag{4.21}$$

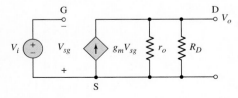

**Figure 4.10** Small-signal equivalent circuit of common-source amplifier with PMOS transistor model

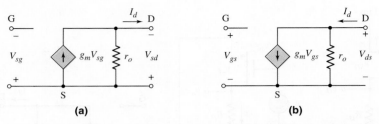

**Figure 4.11** Small signal equivalent circuit of a p-channel MOSFET showing (a) the conventional voltage polarities and current directions and (b) the case when the voltage polarities and current directions are reversed.

This expression for the small-signal voltage gain of the p-channel MOSFET amplifier is exactly the same as that for the n-channel MOSFET amplifier. The negative sign indicates that a 180-degree phase reversal exists between the ouput and input signals, for both the PMOS and the NMOS circuit.

We may note that if the polarity of the small-signal gate-to-source voltage is reversed, then the small-signal drain current direction is reversed. This change of polarity is shown in Figure 4.11. Figure 4.11(a) shows the conventional voltage polarity and current directions in a PMOS transistor. If the control voltage polarity is reversed as shown in Figure 4.11(b), then the dependent current direction is also reversed. The equivalent circuit shown in Figure 4.11(b) is then the same as that of the NMOS transistor. However, the author prefers to use the small-signal equivalent circuit in Figure 4.9 to be consistent with the voltage polarities and current directions of the PMOS transistor.

### 4.1.3 Modeling the Body Effect

As mentioned in Section 3.1.9, Chapter 3, the body effect occurs in a MOSFET in which the substrate, or body, is not directly connected to the source. For an NMOS device, the body is connected to the most negative potential in the circuit and will be at signal ground. Figure 4.12(a) shows the four-terminal MOSFET with dc voltages and Figure 4.12(b) shows the device with ac voltages. Keep in mind that $v_{SB}$ must be greater than or equal to zero. The simplified current-voltage relation is

$$i_D = K_n(v_{GS} - V_{TN})^2 \tag{4.22}$$

and the threshold voltage is given by

$$V_{TN} = V_{TNO} + \gamma\left[\sqrt{2\phi_f + v_{SB}} - \sqrt{2\phi_f}\right] \tag{4.23}$$

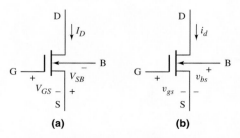

**Figure 4.12** The four-terminal NMOS device with (a) dc voltages and (b) ac voltages

If an ac component exists in the source-to-body voltage, $v_{SB}$, there will be an ac component induced in the threshold voltage, which causes an ac component in the drain current. Thus, a back-gate transconductance can be defined as

$$g_{mb} = \frac{\partial i_D}{\partial v_{BS}}\bigg|_{Q\text{-}pt} = \frac{-\partial i_D}{\partial v_{SB}}\bigg|_{Q\text{-}pt} = -\left(\frac{\partial i_D}{\partial V_{TN}}\right) \cdot \left(\frac{\partial V_{TN}}{\partial v_{SB}}\right)\bigg|_{Q\text{-}pt} \tag{4.24}$$

Using Equation (4.22), we find

$$\frac{\partial i_D}{\partial V_{TN}} = -2K_n(v_{GS} - V_{TN}) = -g_m \tag{4.25(a)}$$

and using Equation (4.23), we find

$$\frac{\partial V_{TN}}{\partial v_{SB}} = \frac{\gamma}{2\sqrt{2\phi_f + v_{SB}}} \equiv \eta \tag{4.25(b)}$$

The back-gate transconductance is then

$$g_{mb} = -(-g_m) \cdot (\eta) = g_m \eta \tag{4.26}$$

Including the body effect, the small-signal equivalent circuit of the MOSFET is shown in Figure 4.13. We note the direction of the current and the polarity of the small-signal source-to-body voltage. If $v_{bs} > 0$, then $v_{SB}$ decreases, $V_{TN}$ decreases, and $i_D$ increases. The current direction and voltage polarity are thus consistent.

For $\phi_f = 0.35$ V and $\gamma = 0.35$ V$^{1/2}$, the value of $\eta$ from Equation (4.25(b)) is $\eta \cong 0.23$. Therefore, $\eta$ will be in the range $0 \le \eta \le 0.23$. The value of $v_{bs}$ will depend on the particular circuit.

In general, we will neglect $g_{mb}$ in our hand analyses and designs, but will investigate the body effect in PSpice analyses.

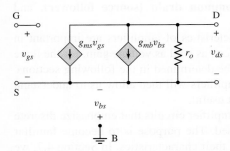

**Figure 4.13** Small-signal equivalent circuit of NMOS device including body effect

## Test Your Understanding

**TYU 4.1** The parameters of an n-channel MOSFET are: $V_{TN} = 0.6$ V, $k'_n = 100\ \mu$A/V$^2$, and $\lambda = 0.015$ V$^{-1}$. The transistor is biased in the saturation region with $I_{DQ} = 1.2$ mA. (a) Design the width-to-length ratio such that the transconductance is $g_m = 2.5$ mA/V. (b) Determine the small-signal output resistance $r_o$. (Ans. (a) 26.0, (b) 55.6 k$\Omega$).

**TYU 4.2** For the circuit shown in Figure 4.1, $V_{DD} = 3.3$ V and $R_D = 8$ k$\Omega$. The transistor parameters are $V_{TN} = 0.4$ V, $K_n = 0.5$ mA/V$^2$, and $\lambda = 0.02$ V$^{-1}$. (a) Determine $V_{GSQ}$ and $V_{DSQ}$ for $I_{DQ} = 0.15$ mA. (b) Calculate $g_m, r_o$, and the small-signal voltage gain. (Ans. (a) $V_{GSQ} = 0.948$ V, $V_{DSQ} = 2.1$ V; (b) $g_m = 0.548$ mA/V, $r_o = 333$ k$\Omega$, $A_v = -4.28$).

**TYU 4.3** For the circuit in Figure 4.1, the circuit and transistor parameters are given in Exercise TYU 4.2. If $v_i = 25 \sin \omega t$ (mV), determine $i_D$ and $v_{DS}$. (Ans. $i_D = 0.15 + 0.0137 \sin \omega t$ (mA), $v_{DS} = 2.1 - 0.11 \sin \omega t$ (V)).

**TYU 4.4** The parameters for the circuit in Figure 4.8 are $V_{DD} = 5$ V and $R_D = 5$ k$\Omega$. The transistor parameters are $V_{TP} = -0.4$ V, $K_p = 0.4$ mA/V$^2$, and $\lambda = 0$. (a) Determine $V_{SGQ}$ and $I_{DQ}$ such that $V_{SDQ} = 3$ V. (b) Calculate $g_m$ and the small-signal voltage gain. (Ans. (a) $I_{DQ} = 0.4$ mA, $V_{SGQ} = 1.4$ V; (b) $g_m = 0.8$ mA/V, $A_v = -4$).

**TYU 4.5** A transistor has the same parameters as those given in Exercise Ex4.1. In addition, the body effect coefficient is $\gamma = 0.40$ V$^{1/2}$ and $\phi_f = 0.35$ V. Determine the value of $\eta$ and the back-gate transconductance $g_{mb}$ for (a) $v_{SB} = 1$ V and (b) $v_{SB} = 3$ V. (Ans. (a) $\eta = 0.153$, (b) $\eta = 0.104$).

## 4.2    BASIC TRANSISTOR AMPLIFIER CONFIGURATIONS

**Objective:** • Discuss the three basic transistor amplifier configurations.

As we have seen, the MOSFET is a three-terminal device. Three basic single-transistor amplifier configurations can be formed, depending on which of the three transistor terminals is used as signal ground. These three basic configurations are appropriately called **common source, common drain (source follower),** and **common gate.**

The input and output resistance characteristics of amplifiers are important in determining loading effects. These parameters, as well as voltage gain, for the three basic MOSFET circuit configurations will be determined in the following sections. The characteristics of the three types of amplifiers will then allow us to understand under what condition each amplifier is most useful.

Initially, we will consider MOSFET amplifier circuits that emphasize discrete designs, in that resistor biasing will be used. The purpose is to become familiar with basic MOSFET amplifier designs and their characteristics. In Section 4.7, we will begin to consider integrated circuit MOSFET designs that involve all-transistor circuits and current source biasing. These initial designs provide an introduction to more advanced MOS amplifier designs that will be considered in Part 2 of the text.

## 4.3    THE COMMON-SOURCE AMPLIFIER

**Objective:** • Analyze the common-source amplifier and become familiar with the general characteristics of this circuit.

In this section, we consider the first of the three basic circuits—the common-source amplifier. We will analyze several basic common-source circuits, and will determine small-signal voltage gain and input and output impedances.

### 4.3.1 A Basic Common-Source Configuration

Figure 4.14 shows the basic common-source circuit with voltage-divider biasing. We see that the source is at ground potential—hence the name common source. The signal from the signal source is coupled into the gate of the transistor through the coupling capacitor $C_C$, which provides dc isolation between the amplifier and the signal source. The dc transistor biasing is established by $R_1$ and $R_2$, and is not disturbed when the signal source is capacitively coupled to the amplifier.

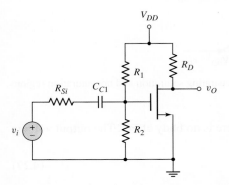

**Figure 4.14** Common-source circuit with voltage divider biasing and coupling capacitor

If the signal source is a sinusoidal voltage at frequency $f$, then the magnitude of the capacitor impedance is $|Z_C| = [1/(2\pi f C_C)]$. For example, assume that $C_C = 10\,\mu\text{F}$ and $f = 2$ kHz. The magnitude of the capacitor impedance is then

$$|Z_C| = \frac{1}{2\pi f C_C} = \frac{1}{2\pi(2 \times 10^3)(10 \times 10^{-6})} \cong 8\,\Omega$$

The magnitude of this impedance is generally much less than the Thevenin resistance at the capacitor terminals. We can therefore assume that the capacitor is essentially a short circuit to signals with frequencies greater than 2 kHz. We will also neglect, in this chapter, any capacitance effects within the transistor.

For the circuit shown in Figure 4.14, assume that the transistor is biased in the saturation region by resistors $R_1$ and $R_2$, and that the signal frequency is sufficiently large for the coupling capacitor to act essentially as a short circuit. The signal source is represented by a Thevenin equivalent circuit, in which the signal voltage source $v_i$ is in series with an equivalent source resistance $R_{Si}$. As we will see, $R_{Si}$ should be much less than the amplifier input resistance, $R_i = R_1 \| R_2$, in order to minimize loading effects.

Figure 4.15 shows the resulting small-signal equivalent circuit. The small-signal variables, such as the input signal voltage $V_i$, are given in phasor form.

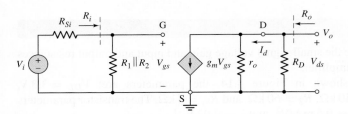

**Figure 4.15** Small-signal equivalent circuit, assuming coupling capacitor acts as a short circuit

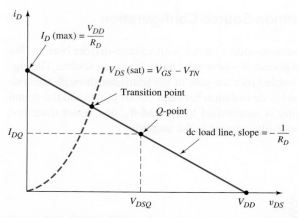

**Figure 4.16** DC load line and transition point separating saturation and nonsaturation regions

Since the source is at ground potential, there is no body effect. The output voltage is

$$V_o = -g_m V_{gs}(r_o \| R_D) \tag{4.27}$$

The input gate-to-source voltage is

$$V_{gs} = \left( \frac{R_i}{R_i + R_{Si}} \right) \cdot V_i \tag{4.28}$$

so the small-signal voltage gain is

$$A_v = \frac{V_o}{V_i} = -g_m(r_o \| R_D) \cdot \left( \frac{R_i}{R_i + R_{Si}} \right) \tag{4.29}$$

We can also relate the ac drain current to the ac drain-to-source voltage, as $V_{ds} = -I_d(R_D)$.

Figure 4.16 shows the dc load line, the transition point (that separates the saturation bias region and nonsaturation bias region), and the Q-point, which is in the saturation region. In order to provide the maximum symmetrical output voltage swing and keep the transistor biased in the saturation region, the Q-point must be near the middle of the saturation region. At the same time, the input signal must be small enough for the amplifier to remain linear.

The input and output resistances of the amplifier can be determined from Figure 4.15. The input resistance to the amplifier is $R_i = R_1 \| R_2$. Since the low-frequency input resistance looking into the gate of the MOSFET is essentially infinite, the input resistance is only a function of the bias resistors. The output resistance looking back into the output terminals is found by setting the independent input source $V_i$ equal to zero, which means that $V_{gs} = 0$. The output resistance is therefore $R_o = R_D \| r_o$.

## EXAMPLE 4.3

**Objective:** Determine the small-signal voltage gain and input and output resistances of a common-source amplifier.

For the circuit shown in Figure 4.14, the parameters are: $V_{DD} = 3.3$ V, $R_D = 10$ kΩ, $R_1 = 140$ kΩ, $R_2 = 60$ kΩ, and $R_{Si} = 4$ kΩ. The transistor parameters are: $V_{TN} = 0.4$ V, $K_n = 0.5$ mA/V², and $\lambda = 0.02$ V⁻¹.

**Solution  (dc calculations):** The dc or quiescent gate-to-source voltage is

$$V_{GSQ} = \left(\frac{R_2}{R_1 + R_2}\right)(V_{DD}) = \left(\frac{60}{140 + 60}\right)(3.3) = 0.99 \text{ V}$$

The quiescent drain current is

$$I_{DQ} = K_n(V_{GSQ} - V_{TN})^2 = (0.5)(0.99 - 0.4)^2 = 0.174 \text{ mA}$$

and the quiescent drain-to-source voltage is

$$V_{DSQ} = V_{DD} - I_{DQ}R_D = 3.3 - (0.174)(10) = 1.56 \text{ V}$$

Since $V_{DSQ} > V_{GSQ} - V_{TN}$, the transistor is biased in the saturation region.

**Small-signal Voltage Gain:**  The small-signal transconductance $g_m$ is then

$$g_m = 2\sqrt{K_n I_{DQ}} = 2\sqrt{(0.5)(0.174)} = 0.590 \text{ mA/V}$$

and the small-signal output resistance is

$$r_o = \frac{1}{\lambda I_Q} = \frac{1}{(0.02)(0.174)} = 287 \text{ k}\Omega$$

The input resistance to the amplifier is

$$R_i = R_1 \| R_2 = 140 \| 60 = 42 \text{ k}\Omega$$

From Figure 4.15 and Equation (4.29), the small-signal voltage gain is

$$A_v = -g_m(r_o \| R_D)\left(\frac{R_i}{R_i + R_{Si}}\right) = -(0.59)(287 \| 10)\left(\frac{42}{42 + 4}\right)$$

or

$$A_v = -5.21$$

**Input and Output Resistances:**  As already calculated, the amplifier input resistance is

$$R_i = R_1 \| R_2 = 140 \| 60 = 42 \text{ k}\Omega$$

and the amplifier output resistance is

$$R_o = R_D \| r_o = 10 \| 287 = 9.66 \text{ k}\Omega$$

**Comment:**  The resulting $Q$-point is not in the center of the saturation region. Therefore, this circuit does not achieve the maximum symmetrical output voltage swing in this case.

**Discussion:**  The small-signal input gate-to-source voltage is

$$V_{gs} = \left(\frac{R_i}{R_i + R_{Si}}\right) \cdot V_i = \left(\frac{42}{42 + 4}\right) \cdot V_i = (0.913) \cdot V_i$$

Since $R_{Si}$ is not zero, the amplifier input signal $V_{gs}$ is approximately 91 percent of the signal voltage. This is called a loading effect. Even though the input resistance to the gate of the transistor is essentially infinite, the bias resistors greatly influence the amplifier input resistance and loading effect. This loading effect can be eliminated or minimized when current source biasing is considered.

## EXERCISE PROBLEM

**Ex 4.3:** The parameters of the circuit shown in Figure 4.14 are $V_{DD} = 5$ V, $R_1 = 520$ kΩ, $R_2 = 320$ kΩ, $R_D = 10$ kΩ, and $R_{Si} = 0$. Assume transistor parameters of $V_{TN} = 0.8$ V, $K_n = 0.20$ mA/V$^2$, and $\lambda = 0$. (a) Determine the small-signal transistor parameters $g_m$ and $r_o$. (b) Find the small-signal voltage gain. (c) Calculate the input and output resistances $R_i$ and $R_o$ (see Figure 4.15). (Ans. (a) $g_m = 0.442$ mA/V, $r_o = \infty$; (b) $A_v = -4.42$; (c) $R_i = 198$ kΩ, $R_o = R_D = 10$ kΩ)

## DESIGN EXAMPLE 4.4

**Objective:** Design the bias of a MOSFET circuit such that the $Q$-point is in the middle of the saturation region. Determine the resulting small-signal voltage gain.

**Specifications:** The circuit to be designed has the configuration shown in Figure 4.17. Let $R_1 \| R_2 = 100$ kΩ. Design the circuit such that the $Q$-point is $I_{DQ} = 2$ mA and the $Q$-point is in the middle of the saturation region.

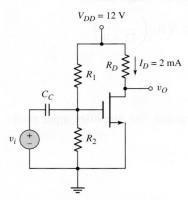

**Figure 4.17** Common-source NMOS transistor circuit

**Choices:** A transistor with nominal parameters $V_{TN} = 1$ V, $k'_n = 80\ \mu$A/V$^2$, $W/L = 25$, and $\lambda = 0.015$ V$^{-1}$ is available.

**Solution (dc design):** The load line and the desired $Q$-point are given in Figure 4.18. If the $Q$-point is to be in the middle of the saturation region, the current at the transition point must be 4 mA.

The conductivity parameter is

$$K_n = \frac{k'_n}{2} \cdot \frac{W}{L} = \left(\frac{0.080}{2}\right)(25) = 1 \text{ mA/V}^2$$

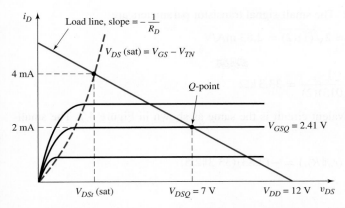

**Figure 4.18** DC load line and transition point for NMOS circuit shown in Figure 4.17

We can now calculate $V_{DS}(\text{sat})$ at the transition point. The subscript $t$ indicates transition point values. To determine $V_{GSt}$, we use

$$I_{Dt} = 4 = K_n(V_{GSt} - V_{TN})^2 = 1(V_{GSt} - 1)^2$$

which yields

$$V_{GSt} = 3 \text{ V}$$

Therefore

$$V_{DSt} = V_{GSt} - V_{TN} = 3 - 1 = 2 \text{ V}$$

If the $Q$-point is in the middle of the saturation region, then $V_{DSQ} = 7$ V, which would yield a 10 V peak-to-peak symmetrical output voltage. From Figure 4.17, we can write

$$V_{DSQ} = V_{DD} - I_{DQ}R_D$$

or

$$R_D = \frac{V_{DD} - V_{DSQ}}{I_{DQ}} = \frac{12 - 7}{2} = 2.5 \text{ k}\Omega$$

We can determine the required quiescent gate-to-source voltage from the current equation, as follows:

$$I_{DQ} = 2 = K_n(V_{GSQ} - V_{TN})^2 = (1)(V_{GSQ} - 1)^2$$

or

$$V_{GSQ} = 2.41 \text{ V}$$

Then

$$V_{GSQ} = 2.41 = \left(\frac{R_2}{R_1 + R_2}\right)(V_{DD}) = \left(\frac{1}{R_1}\right)\left(\frac{R_1 R_2}{R_1 + R_2}\right)(V_{DD})$$

$$= \frac{R_i}{R_1} \cdot V_{DD} = \frac{(100)(12)}{R_1}$$

which yields

$$R_1 = 498 \text{ k}\Omega \quad \text{and} \quad R_2 = 125 \text{ k}\Omega$$

**Solution (ac analysis):** The small-signal transistor parameters are

$$g_m = 2\sqrt{K_n I_{DQ}} = 2\sqrt{(1)(2)} = 2.83 \text{ mA/V}$$

and

$$r_o = \frac{1}{\lambda I_{DQ}} = \frac{1}{(0.015)(2)} = 33.3 \text{ k}\Omega$$

The small-signal equivalent circuit is the same as shown in Figure 4.7. The small-signal voltage gain is

$$A_v = \frac{V_o}{V_i} = -g_m \left(r_o \| R_D\right) = -(2.83)(33.3 \| 2.5)$$

or

$$A_v = -6.58$$

**Comment:** Establishing the $Q$-point in the middle of the saturation region allows the maximum symmetrical swing in the output voltage, while keeping the transistor biased in the saturation region.

### EXERCISE PROBLEM

**Ex 4.4:** Consider the circuit shown in Figure 4.14. Assume transistor parameters of $V_{TN} = 0.8$ V, $K_n = 0.20$ mA/V$^2$, and $\lambda = 0$. Let $V_{DD} = 5$ V, $R_i = R_1 \| R_2 = 200$ k$\Omega$, and $R_{Si} = 0$. Design the circuit such that $I_{DQ} = 0.5$ mA and the $Q$-point is in the center of the saturation region. Find the small-signal voltage gain. (Ans. $R_D = 2.76$ k$\Omega$, $R_1 = 420$ k$\Omega$, $R_2 = 382$ k$\Omega$, $A_v = -1.75$)

### 4.3.2   Common-Source Amplifier with Source Resistor

A source resistor $R_S$ tends to stabilize the $Q$-point against variations in transistor parameters (Figure 4.19). If, for example, the value of the conduction parameter varies from one transistor to another, the $Q$-point will not vary as much if a source

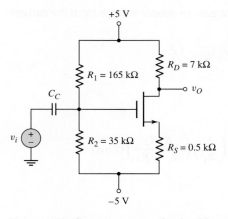

**Figure 4.19** Common-source circuit with source resistor and positive and negative supply voltages

resistor is included in the circuit. However, as shown in the following example, a source resistor also reduces the signal gain.

The circuit in Figure 4.19 is an example of a situation in which the body effect should be taken into account. The substrate (not shown) would normally be connected to the $-5$ V supply, so that the body and substrate terminals are not at the same potential. However, in the following example, we will neglect this effect.

The circuit shown in Figure 4.20(a) is a PMOS version of the common-source amplifier with a source resistor $R_S$ included.

## EXAMPLE 4.5

**Objective:** Determine the small-signal voltage gain of a PMOS transistor circuit.

Consider the circuit shown in Figure 4.20(a). The transistor parameters are $K_p = 0.80$ mA/V$^2$, $V_{TP} = -0.5$ V, and $\lambda = 0$. The quiescent drain current is found to be $I_{DQ} = 0.297$ mA.

The small-signal equivalent circuit is shown in Figure 4.20(b). To sketch the small-signal equivalent circuit, start with the three terminals of the transistor, draw in the transistor equivalent circuit between these three terminals, and then sketch in the other circuit elements around the transistor.

**Solution:** The small-signal output voltage is

$$V_o = +g_m V_{sg} R_D$$

Writing a KVL equation from the input around the gate–source loop, we find

$$V_i = -V_{sg} - g_m V_{sg} R_S$$

or

$$V_{sg} = \frac{-V_i}{1 + g_m R_S}$$

Substituting this expression for $V_{sg}$ into the output voltage equation, we find the small-signal voltage gain as

$$A_v = \frac{V_o}{V_i} = \frac{-g_m R_D}{1 + g_m R_S}$$

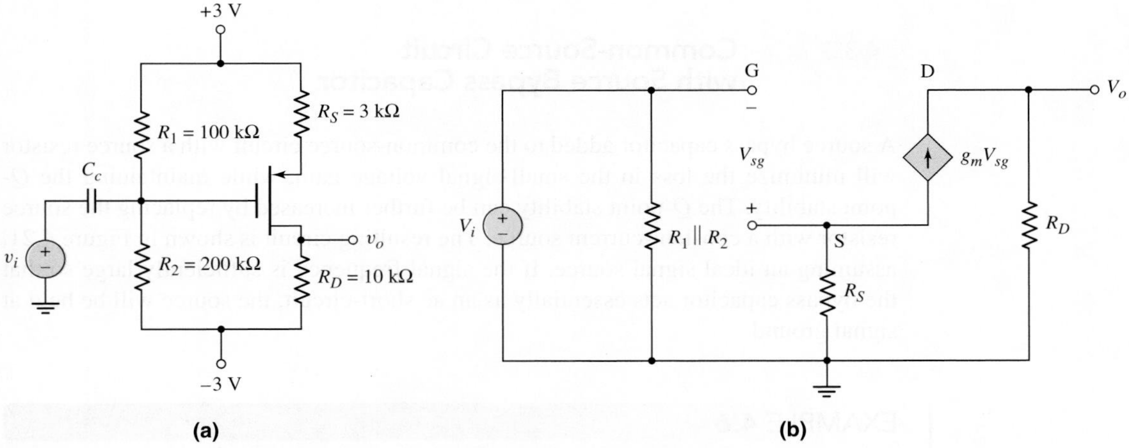

**Figure 4.20** (a) PMOS circuit for Example 4.5, and (b) small-signal equivalent circuit

The small-signal transconductance is

$$g_m = 2\sqrt{K_p I_{DQ}} = 2\sqrt{(0.80)(0.297)} = 0.975 \text{ mA/V}$$

We then find the small-signal voltage gain as

$$A_v = \frac{-(0.975)(10)}{1 + (0.975)(3)} \quad \text{or} \quad A_v = -2.48$$

**Comment:** The analysis of a PMOS transistor circuit is essentially the same as that of an NMOS transistor circuit. The voltage gain of a MOS transistor circuit that contains a source resistor is degraded compared to a circuit without a source resistor. However, the $Q$-point tends to be stabilized.

**Discussion:** We mentioned that including a source resistor tends to stabilize the circuit characteristics against any changes in transistor parameters. If, for example, the conduction parameter $K_p$ varies by ±10 percent, we find the following results.

| $K_p$(mA/V²) | $g_m$(mA/V) | $A_v$(V/V) |
|---|---|---|
| 0.72 | 0.9121 | −2.441 |
| 0.80 | 0.9749 | −2.484 |
| 0.88 | 1.035 | −2.521 |

With a ±10 percent variation in $K_p$, there is less than a ±1.8 percent variation in the voltage gain.

### EXERCISE PROBLEM

**Ex 4.5:** For the circuit shown in Figure 4.19, the transistor parameters are $V_{TN} = 0.8$ V, $K_n = 1$ mA/V², and $\lambda = 0$. (a) From the dc analysis, find $I_{DQ}$ and $V_{DSQ}$. (b) Determine the small-signal voltage gain. (Ans. (a) $I_{DQ} = 0.494$ mA, $V_{DSQ} = 6.30$ V; (b) $A_v = -5.78$).

### 4.3.3 Common-Source Circuit with Source Bypass Capacitor

A source bypass capacitor added to the common-source circuit with a source resistor will minimize the loss in the small-signal voltage gain, while maintaining the $Q$-point stability. The $Q$-point stability can be further increased by replacing the source resistor with a constant-current source. The resulting circuit is shown in Figure 4.21, assuming an ideal signal source. If the signal frequency is sufficiently large so that the bypass capacitor acts essentially as an ac short-circuit, the source will be held at signal ground.

### EXAMPLE 4.6

**Objective:** Determine the small-signal voltage gain of a circuit biased with a constant-current source and incorporating a source bypass capacitor.

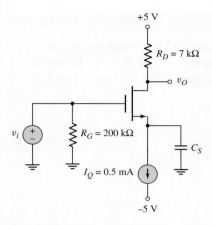

**Figure 4.21** NMOS common-source circuit with source bypass capacitor

For the circuit shown in Figure 4.21, the transistor parameters are: $V_{TN} = 0.8$ V, $K_n = 1$ mA/V$^2$, and $\lambda = 0$.

**Solution:** Since the dc gate current is zero, the dc voltage at the source terminal is $V_S = -V_{GSQ}$, and the gate-to-source voltage is determined from

$$I_{DQ} = I_Q = K_n(V_{GSQ} - V_{TN})^2$$

or

$$0.5 = (1)(V_{GSQ} - 0.8)^2$$

which yields

$$V_{GSQ} = -V_S = 1.51 \text{ V}$$

The quiescent drain-to-source voltage is

$$V_{DSQ} = V_{DD} - I_{DQ}R_D - V_S = 5 - (0.5)(7) - (-1.51) = 3.01 \text{ V}$$

The transistor is therefore biased in the saturation region.

The small-signal equivalent circuit is shown in Figure 4.22. The output voltage is

$$V_o = -g_m V_{gs} R_D$$

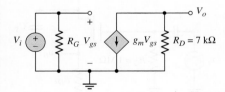

**Figure 4.22** Small-signal equivalent circuit, assuming the source bypass capacitor acts as a short circuit

Since $V_{gs} = V_i$, the small-signal voltage gain is

$$A_v = \frac{V_o}{V_i} = -g_m R_D = -(1.414)(7) = -9.9$$

**Comment:** Comparing the small-signal voltage gain of 9.9 in this example to the 2.48 calculated in Example 4.5, we see that the magnitude of the gain increases when a source bypass capacitor is included.

**Ex 4.6:** The common-source amplifier in Figure 4.23 has transistor parameters $k_p' = 40$ $\mu$A/V$^2$, $W/L = 40$, $V_{TP} = -0.4$ V, and $\lambda = 0.02$ V$^{-1}$. (a) Determine $I_{DQ}$ and $V_{SDQ}$. (b) Find the small-signal voltage gain. (Ans. (a) $I_{DQ} = 1.16$ mA, $V_{SDQ} = 2.29$ V; (b) $A_v = -3.68$)

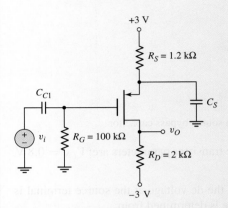

**Figure 4.23** Figure for Exercise Ex 4.6

## Test Your Understanding

**TYU 4.6** Consider the common-source amplifier in Figure 4.24 with transistor parameters $V_{TN} = 1.8$ V, $K_n = 0.15$ mA/V$^2$, and $\lambda = 0$. (a) Calculate $I_{DQ}$ and $V_{DSQ}$. (b) Determine the small-signal voltage gain. (c) Discuss the purpose of $R_G$ and its effect on the small-signal operation of the amplifier. (Ans. (a) $I_{DQ} = 1.05$ mA, $V_{DSQ} = 4.45$ V; (b) $A_v = -2.65$)

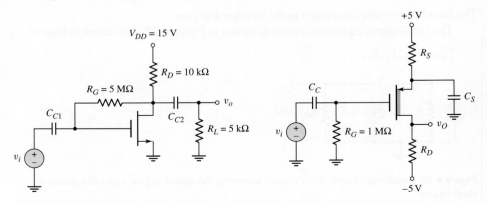

**Figure 4.24** Figure for Exercise TYU 4.6        **Figure 4.25** Figure for Exercise TYU 4.7

**TYU 4.7** The parameters of the transistor shown in Figure 4.25 are: $V_{TP} = +0.8$ V, $K_p = 0.5$ mA/V$^2$, and $\lambda = 0.02$ V$^{-1}$. (a) Determine $R_S$ and $R_D$ such that $I_{DQ} = 0.8$ mA and $V_{SDQ} = 3$ V. (b) Find the small-signal voltage gain. (Ans. (a) $R_S = 5.67$ k$\Omega$, $R_D = 3.08$ k$\Omega$; (b) $A_v = -3.71$)

## 4.4    THE COMMON-DRAIN (SOURCE-FOLLOWER) AMPLIFIER

**Objective:** • Analyze the common-drain (source-follower) amplifier and become familiar with the general characteristics of this circuit.

The second type of MOSFET amplifier to be considered is the **common-drain circuit.** An example of this circuit configuration is shown in Figure 4.26. As seen in the figure, the output signal is taken off the source with respect to ground and the drain is connected directly to $V_{DD}$. Since $V_{DD}$ becomes signal ground in the ac equivalent circuit, we have the name common drain. The more common name is **source follower.** The reason for this name will become apparent as we proceed through the analysis.

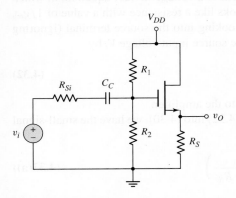

**Figure 4.26** NMOS source-follower or common-drain amplifier

### 4.4.1    Small-Signal Voltage Gain

The dc analysis of the circuit is exactly the same as we have already seen, so we will concentrate on the small-signal analysis. The small-signal equivalent circuit, assuming the coupling capacitor acts as a short circuit, is shown in Figure 4.27(a). The drain is at signal ground, and the small-signal resistance $r_o$ of the transistor is in

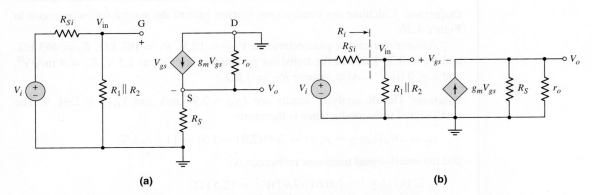

(a)                                                                              (b)

**Figure 4.27** (a) Small-signal equivalent circuit of NMOS source follower and (b) small-signal equivalent circuit of NMOS source follower with all signal grounds at a common point

parallel with the dependent current source. Figure 4.27(b) is the same equivalent circuit, but with all signal grounds at a common point. We are again neglecting the body effect. The output voltage is

$$V_o = (g_m V_{gs})(R_S \| r_o) \tag{4.30}$$

Writing a KVL equation from input to output results in the following:

$$V_{in} = V_{gs} + V_o = V_{gs} + g_m V_{gs}(R_S \| r_o) \tag{4.31(a)}$$

Therefore, the gate-to-source voltage is

$$V_{gs} = \frac{V_{in}}{1 + g_m(R_S \| r_o)} = \left[ \frac{\frac{1}{g_m}}{\frac{1}{g_m} + (R_S \| r_o)} \right] \cdot V_{in} \tag{4.31(b)}$$

Equation (4.31(b)) is written in the form of a voltage-divider equation, in which the gate-to-source of the NMOS device looks like a resistance with a value of $1/g_m$. More accurately, the effective resistance looking into the source terminal (ignoring $r_o$) is $1/g_m$. The voltage $V_{in}$ is related to the source input voltage $V_i$ by

$$V_{in} = \left( \frac{R_i}{R_i + R_{Si}} \right) \cdot V_i \tag{4.32}$$

where $R_i = R_1 \| R_2$ is the input resistance to the amplifier.

Substituting Equations (4.31(b)) and (4.32) into (4.30), we have the small-signal voltage gain:

$$A_v = \frac{V_o}{V_i} = \frac{g_m(R_S \| r_o)}{1 + g_m(R_S \| r_o)} \cdot \left( \frac{R_i}{R_i + R_{Si}} \right) \tag{4.33(a)}$$

or

$$A_v = \frac{R_S \| r_o}{\frac{1}{g_m} + R_S \| r_o} \cdot \left( \frac{R_i}{R_i + R_{Si}} \right) \tag{4.33(b)}$$

which again is written in the form of a voltage-divider equation. An inspection of Equation 4.33(b) shows that the magnitude of the voltage gain is always less than unity.

### EXAMPLE 4.7

**Objective:** Calculate the small-signal voltage gain of the source-follower circuit in Figure 4.26.

Assume the circuit parameters are $V_{DD} = 12$ V, $R_1 = 162$ kΩ, $R_2 = 463$ kΩ, and $R_S = 0.75$ kΩ, and the transistor parameters are $V_{TN} = 1.5$ V, $K_n = 4$ mA/V$^2$, and $\lambda = 0.01$ V$^{-1}$. Also assume $R_{Si} = 4$ kΩ.

**Solution:** The dc analysis results are $I_{DQ} = 7.97$ mA and $V_{GSQ} = 2.91$ V. The small-signal transconductance is therefore

$$g_m = 2K_n(V_{GSQ} - V_{TN}) = 2(4)(2.91 - 1.5) = 11.3 \text{ mA/V}$$

and the small-signal transistor resistance is

$$r_o \cong [\lambda I_{DQ}]^{-1} = [(0.01)(7.97)]^{-1} = 12.5 \text{ k}\Omega$$

The amplifier input resistance is

$$R_i = R_1 \| R_2 = 162 \| 463 = 120 \text{ k}\Omega$$

The small-signal voltage gain then becomes

$$A_v = \frac{g_m(R_S \| r_o)}{1 + g_m(R_S \| r_o)} \cdot \frac{R_i}{R_i + R_{Si}}$$

$$= \frac{(11.3)(0.75 \| 12.5)}{1 + (11.3)(0.75 \| 12.5)} \cdot \frac{120}{120 + 4} = +0.860$$

**Comment:** The magnitude of the small-signal voltage gain is less than 1. An examination of Equation (4.33(b)) shows that this is always true. Also, the voltage gain is positive, which means that the output signal voltage is in phase with the input signal voltage. Since the output signal is essentially equal to the input signal, the circuit is called a source follower.

## EXERCISE PROBLEM

**Ex 4.7:** The source-follower circuit in Figure 4.26 has transistor parameters $V_{TN} = +0.8$ V, $K_n = 1$ mA/V$^2$, and $\lambda = 0.015$ V$^{-1}$. Let $V_{DD} = 10$ V, $R_{Si} = 200$ $\Omega$, and $R_1 + R_2 = 400$ k$\Omega$. Design the circuit such that $I_{DQ} = 1.5$ mA and $V_{DSQ} = 5$ V. Determine the small-signal voltage gain. (Ans. $R_S = 3.33$ k$\Omega$, $R_1 = 119$ k$\Omega$, $R_2 = 281$, k$\Omega$, $A_v = 0.884$)

Although the voltage gain is slightly less than 1, the source follower is an extremely useful circuit because the output resistance is less than that of a common-source circuit, as we will show in the next section. A small output resistance is desirable when the circuit is to act as an ideal voltage source and drive a load circuit without suffering any loading effects.

## DESIGN EXAMPLE 4.8

**Objective:** Design a source-follower amplifier with a p-channel enhancement-mode MOSFET to meet a set of specifications.

**Specifications:** The circuit to be designed has the configuration shown in Figure 4.28 with circuit parameters $V_{DD} = 20$ V and $R_{Si} = 4$ k$\Omega$. The $Q$-point values are to be in the center of the load line with $I_{DQ} = 2.5$ mA. The input resistance is to

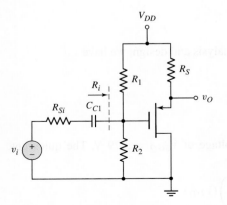

**Figure 4.28** PMOS source follower

be $R_i = 200$ kΩ. The transistor $W/L$ ratio is to be designed such that the small signal voltage gain is $A_v = 0.90$.

**Choices:** A transistor with nominal parameters $V_{TP} = -2$ V, $k'_p = 40\ \mu A/V^2$, and $\lambda = 0$ is available.

**Solution (dc analysis):** From a KVL equation around the source-to-drain loop, we have

$$V_{DD} = V_{SDQ} + I_{DQ}R_S$$

or

$$20 = 10 + (2.5)R_S$$

which yields the required source resistor to be $R_S = 4$ kΩ.

**Solution (ac design):** The small-signal voltage gain of this circuit is the same as that of a source follower with an NMOS device. From Equation (4.33(a)), we have

$$A_v = \frac{V_o}{V_i} = \frac{g_m R_S}{1 + g_m R_S} \cdot \frac{R_i}{R_i + R_{Si}}$$

which yields

$$0.90 = \frac{g_m(4)}{1 + g_m(4)} \cdot \frac{200}{200 + 4}$$

We find that the required transconductance must be $g_m = 2.80$ mA/V. The transconductance can be written as

$$g_m = 2\sqrt{K_p I_{DQ}}$$

We have

$$2.80 \times 10^{-3} = 2\sqrt{K_p(2.5 \times 10^{-3})}$$

which yields

$$K_p = 0.784 \times 10^{-3} A/V^2$$

The conduction parameter, as a function of width-to-length ratio, is

$$K_p = 0.784 \times 10^{-3} = \frac{k'_p}{2} \cdot \frac{W}{L} = \left(\frac{40 \times 10^{-6}}{2}\right) \cdot \left(\frac{W}{L}\right)$$

which means that the required width-to-length ratio must be

$$\frac{W}{L} = 39.2$$

**Solution (dc design):** Completing the dc analysis and design, we have

$$I_{DQ} = K_p(V_{GSQ} + V_{TP})^2$$

or

$$2.5 = 0.784(V_{SGQ} - 2)^2$$

which yields a quiescent source-to-gate voltage of $V_{SGQ} = 3.79$ V. The quiescent source-to-gate voltage can also be written as

$$V_{SGQ} = (V_{DD} - I_{DQ}R_S) - \left(\frac{R_2}{R_1 + R_2}\right)(V_{DD})$$

Since

$$\left(\frac{R_2}{R_1 + R_2}\right) = \left(\frac{1}{R_1}\right)\left(\frac{R_1 R_2}{R_1 + R_2}\right) = \left(\frac{1}{R_1}\right) \cdot R_i$$

we have

$$3.79 = [20 - (2.5)(4)] - \left(\frac{1}{R_1}\right)(200)(20)$$

The bias resistor $R_1$ is then found to be

$$R_1 = 644 \, \text{k}\Omega$$

Since $R_i = R_1 \| R_2 = 200 \, \text{k}\Omega$, we find

$$R_2 = 290 \, \text{k}\Omega$$

**Comment:** In order to achieve the desired specifications, a relatively large transconductance is required, which means that a relatively large transistor is needed. A large value of input resistance $R_i$ has minimized the effect of loading due to the output resistance, $R_{Si}$, of the signal source.

---

EXERCISE PROBLEM

**Ex 4.8:** The circuit and transistor parameters for the source-follower amplifier shown in Figure 4.29 are $R_S = 2 \, \text{k}\Omega$, $V_{TP} = -1.2 \, \text{V}$, $k_p' = 40 \, \mu\text{A/V}^2$, and $\lambda = 0$. (a) Design the transistor width-to-length ratio such that $I_{DQ} = 1.5 \, \text{mA}$. (b) Find the small-signal voltage gain. (c) Using the results of part (a), determine the value of $R_L$ that will result in a 10 percent reduction in voltage gain. (Ans. (a) $W/L = 117$, (b) $A_v = 0.882$, (c) $R_L = 2.12 \, \text{k}\Omega$)

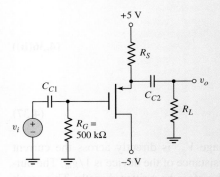

**Figure 4.29** Figure for Exercise Ex 4.8

## 4.4.2 Input and Output Impedance

The small-signal input resistance $R_i$ as defined in Figure 4.27(b), for example, is the Thevenin equivalent resistance of the bias resistors. Even though the input resistance to the gate of the MOSFET is essentially infinite, the input bias resistances do provide a loading effect. This same effect was seen in the common-source circuits.

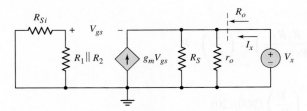

**Figure 4.30** Equivalent circuit of NMOS source follower, for determining output resistance

To calculate the small-signal output resistance, we set all independent small-signal sources equal to zero, apply a test voltage to the output terminals, and measure a test current. Figure 4.30 shows the circuit we will use to determine the output resistance of the source follower shown in Figure 4.26. We set $V_i = 0$ and apply a test voltage $V_x$. Since there are no capacitances in the circuit, the output impedance is simply an output resistance, which is defined as

$$R_o = \frac{V_x}{I_x} \tag{4.34}$$

Writing a KCL equation at the output source terminal produces

$$I_x + g_m V_{gs} = \frac{V_x}{R_S} + \frac{V_x}{r_o} \tag{4.35}$$

Since there is no current in the input portion of the circuit, we see that $V_{gs} = -V_x$. Therefore, Equation (4.35) becomes

$$I_x = V_x \left( g_m + \frac{1}{R_S} + \frac{1}{r_o} \right) \tag{4.36(a)}$$

or

$$\frac{I_x}{V_x} = \frac{1}{R_o} = g_m + \frac{1}{R_S} + \frac{1}{r_o} \tag{4.36(b)}$$

The output resistance is then

$$R_o = \frac{1}{g_m} \| R_S \| r_o \tag{4.37}$$

From Figure 4.30, we see that the voltage $V_{gs}$ is directly across the current source $g_m V_{gs}$. This means that the effective resistance of the device is $1/g_m$. The output resistance given by Equation (4.37) can therefore be written directly. This result also means that the resistance looking into the source terminal (ignoring $r_o$) is $1/g_m$, as previously noted.

## EXAMPLE 4.9

**Objective:** Calculate the output resistance of a source-follower circuit.

Consider the circuit shown in Figure 4.26 with circuit and transistor parameters given in Example 4.7.

**Solution:** The results of Example 4.7 are: $R_S = 0.75$ k$\Omega$, $r_o = 12.5$ k$\Omega$, and $g_m = 11.3$ mA/V. Using Figure 4.30 and Equation (4.37), we find

$$R_o = \frac{1}{g_m} \| R_S \| r_o = \frac{1}{11.3} \| 0.75 \| 12.5$$

or

$$R_o = 0.0787 \text{ k}\Omega = 78.7 \text{ }\Omega$$

**Comment:** The output resistance of a source-follower circuit is dominated by the transconductance parameter. Also, because the output resistance is very low, the source follower tends to act like an ideal voltage source, which means that the output can drive another circuit without significant loading effects.

**Ex 4.9:** Consider the circuit shown in Figure 4.28 with circuit parameters $V_{DD} = 5$ V, $R_S = 5$ k$\Omega$, $R_1 = 70.7$ k$\Omega$, $R_2 = 9.3$ k$\Omega$, and $R_{Si} = 500$ $\Omega$. The transistor parameters are: $V_{TP} = -0.8$ V, $K_p = 0.4$ mA/V$^2$, and $\lambda = 0$. Calculate the small-signal voltage gain $A_v = v_o/v_i$ and the output resistance $R_o$ seen looking back into the circuit. (Ans. $A_v = 0.817$, $R_o = 0.915$ k$\Omega$)

## Test Your Understanding

**TYU 4.8** For an NMOS source-follower circuit, the parameters are $g_m = 4$ mA/V and $r_o = 50$ k$\Omega$. (a) Find the no load ($R_S = \infty$) small-signal voltage gain and the output resistance. (b) Determine the small-signal voltage gain when a 4 k$\Omega$ load is connected to the output. (Ans. (a) $A_v = 0.995$, $R_o \cong 0.25$ k$\Omega$; (b) $A_v = 0.937$)

**TYU 4.9** The transistor in the source-follower circuit shown in Figure 4.31 is biased with a constant current source. The transistor parameters are: $V_{TN} = 2$ V, $k'_n = 40$ $\mu$A/V$^2$, and $\lambda = 0.01$ V$^{-1}$. The load resistor is $R_L = 4$ k$\Omega$. (a) Design the transistor width-to-length ratio such that $g_m = 2$ mA/V when $I = 0.8$ mA. What is the corresponding value for $V_{GS}$? (b) Determine the small-signal voltage gain and the output resistance $R_o$. (Ans. (a) $W/L = 62.5$, $V_{GS} = 2.8$ V; (b) $A_v = 0.886$, $R_o \cong 0.5$ k$\Omega$)

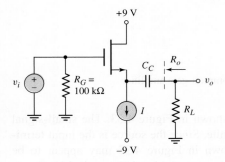

**Figure 4.31** Figure for Exercise TYU 4.9

## 4.5   THE COMMON-GATE CONFIGURATION

**Objective:** • Analyze the common-gate amplifier and become familiar with the general characteristics of this circuit.

The third amplifier configuration is the **common-gate circuit.** To determine the small-signal voltage and current gains, and the input and output impedances, we will use the same small-signal equivalent circuit for the transistor that was used previously. The dc analysis of the common-gate circuit is the same as that of previous MOSFET circuits.

### 4.5.1   Small-Signal Voltage and Current Gains

In the common-gate configuration, the input signal is applied to the source terminal and the gate is at signal ground. The common-gate configuration shown in Figure 4.32 is biased with a constant-current source $I_Q$. The gate resistor $R_G$ prevents the buildup of static charge on the gate terminal, and the capacitor $C_G$ ensures that the gate is at signal ground. The coupling capacitor $C_{C1}$ couples the signal to the source, and coupling capacitor $C_{C2}$ couples the output voltage to load resistance $R_L$.

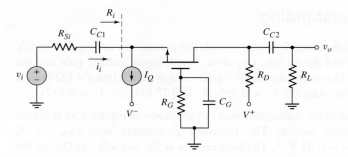

**Figure 4.32** Common-gate circuit

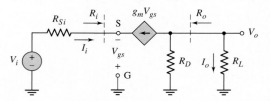

**Figure 4.33** Small-signal equivalent circuit of common-gate amplifier

The small-signal equivalent circuit is shown in Figure 4.33. The small-signal transistor resistance $r_o$ is assumed to be infinite. Since the source is the input terminal, the small-signal equivalent circuit shown in Figure 4.33 may appear to be different from those considered previously. However, to sketch the equivalent circuit, we can use the same technique as used previously. Sketch in the three terminals of the transistor with the source at the input for this case. Then draw in the transistor equivalent circuit between the three terminals and then sketch in the remaining circuit elements around the transistor.

The output voltage is

$$V_o = -(g_m V_{gs})(R_D \| R_L) \tag{4.38}$$

Writing the KVL equation around the input, we find

$$V_i = I_i R_{Si} - V_{gs} \tag{4.39}$$

where $I_i = -g_m V_{gs}$. The gate-to-source voltage can then be written as

$$V_{gs} = \frac{-V_i}{1 + g_m R_{Si}} \tag{4.40}$$

The small-signal voltage gain is found to be

$$A_v = \frac{V_o}{V_i} = \frac{g_m (R_D \| R_L)}{1 + g_m R_{Si}} \tag{4.41}$$

Also, since the voltage gain is positive, the output and input signals are in phase.

In many cases, the signal input to a common-gate circuit is a current. Figure 4.34 shows the small-signal equivalent common-gate circuit with a Norton equivalent circuit as the signal source. We can calculate a current gain. The output current $I_o$ can be written

$$I_o = \left( \frac{R_D}{R_D + R_L} \right) (-g_m V_{gs}) \tag{4.42}$$

At the input we have

$$I_i + g_m V_{gs} + \frac{V_{gs}}{R_{Si}} = 0 \tag{4.43}$$

or

$$V_{gs} = -I_i \left( \frac{R_{Si}}{1 + g_m R_{Si}} \right) \tag{4.44}$$

The small-signal current gain is then

$$A_i = \frac{I_o}{I_i} = \left( \frac{R_D}{R_D + R_L} \right) \cdot \left( \frac{g_m R_{Si}}{1 + g_m R_{Si}} \right) \tag{4.45}$$

We may note that if $R_D \gg R_L$ and $g_m R_{Si} \gg 1$, then the current gain is essentially unity.

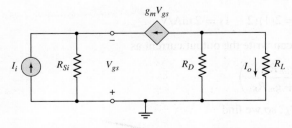

**Figure 4.34** Small-signal equivalent circuit of common-gate amplifier with a Norton equivalent signal source

### 4.5.2    Input and Output Impedance

In contrast to the common-source and source-follower amplifiers, the common-gate circuit has a low input resistance because of the transistor. However, if the input signal is a current, a low input resistance is an advantage. The input resistance is defined, using Figure 4.33, as

$$R_i = \frac{-V_{gs}}{I_i} \tag{4.46}$$

Since $I_i = -g_m V_{gs}$, the input resistance is

$$R_i = \frac{1}{g_m} \tag{4.47}$$

This result has been obtained previously.

We can find the output resistance by setting the input signal voltage equal to zero. From Figure 4.33, we see that $V_{gs} = -g_m V_{gs} R_{Si}$, which means that $V_{gs} = 0$. Consequently, $g_m V_{gs} = 0$. The output resistance, looking back from the load resistance, is therefore

$$R_o = R_D \tag{4.48}$$

---

## EXAMPLE 4.10

**Objective:** For the common-gate circuit, determine the output voltage for a given input current.

For the circuits shown in Figures 4.32 and 4.34, the circuit parameters are: $I_Q = 1$ mA, $V^+ = 5$ V, $V^- = -5$ V, $R_G = 100$ kΩ, $R_D = 4$ kΩ, and $R_L = 10$ kΩ. The transistor parameters are: $V_{TN} = 1$ V, $K_n = 1$ mA/V$^2$, and $\lambda = 0$. Assume the input current in Figure 4.34 is $100 \sin \omega t$ $\mu$A and assume $R_{Si} = 50$ kΩ.

**Solution:** The quiescent gate-to-source voltage is determined from

$$I_Q = I_{DQ} = K_n (V_{GSQ} - V_{TN})^2$$

or

$$1 = 1(V_{GSQ} - 1)^2$$

which yields

$$V_{GSQ} = 2 \text{ V}$$

The small-signal transconductance is

$$g_m = 2K_n(V_{GSQ} - V_{TN}) = 2(1)(2-1) = 2 \text{ mA/V}$$

From Equation (4.45), we can write the output current as

$$I_o = I_i \left( \frac{R_D}{R_D + R_L} \right) \cdot \left( \frac{g_m R_{Si}}{1 + g_m R_{Si}} \right)$$

The output voltage is $V_o = I_o R_L$, so we find

$$V_o = I_i \left( \frac{R_L R_D}{R_D + R_L} \right) \cdot \left( \frac{g_m R_{Si}}{1 + g_m R_{Si}} \right)$$

$$= \left[ \frac{(10)(4)}{4+10} \right] \cdot \left[ \frac{(2)(50)}{1+(2)(50)} \right] \cdot (0.1) \sin \omega t$$

or

$$V_o = 0.283 \sin \omega t \text{ V}$$

**Comment:** The MOSFET common-gate amplifier is useful if the input signal is a current.

---

### EXERCISE PROBLEM

**Ex 4.10:** Consider the circuit shown in Figure 4.35 with circuit parameters $V^+ = 5$ V, $V^- = -5$ V, $R_S = 4$ k$\Omega$, $R_D = 2$ k$\Omega$, $R_L = 4$ k$\Omega$ and $R_G = 50$ k$\Omega$. The transistor parameters are: $K_p = 1$ mA/V$^2$, $V_{TP} = -0.8$ V, and $\lambda = 0$. Draw the small-signal equivalent circuit, determine the small-signal voltage gain $A_v = V_o/V_i$, and find the input resistance $R_i$. (Ans. $A_v = 2.41$, $R_i = 0.485$ k$\Omega$)

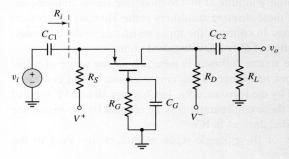

**Figure 4.35**  Figure for Exercise Ex 4.10

## Test Your Understanding

**TYU 4.10**  For the circuit shown in Figure 4.32, the circuit parameters are: $V^+ = 5$ V, $V^- = -5$ V, $R_G = 100$ k$\Omega$, $R_L = 4$ k$\Omega$, and $I_Q = 0.5$ mA. The transistor parameters are $V_{TN} = 1$ V and $\lambda = 0$. The circuit is driven by a signal current source $I_i$. Redesign $R_D$ and $g_m$ such that the transfer function $V_o/I_i$ is 2.4 k$\Omega$ and the input resistance is $R_i = 350$ $\Omega$. Determine $V_{GSQ}$ and show that the transistor is biased in the saturation region. (Ans. $g_m = 2.86$ mA/V, $R_D = 6$ k$\Omega$, $V_{GSQ} = 1.35$ V)

---

## 4.6    THE THREE BASIC AMPLIFIER CONFIGURATIONS: SUMMARY AND COMPARISON

**Objective:** • Compare the general characteristics of the three basic amplifier configurations.

| Table 4.2 | Characteristics of the three MOSFET amplifier configurations | | | |
|---|---|---|---|---|
| Configuration | Voltage gain | Current gain | Input resistance | Output resistance |
| Common source | $A_v > 1$ | — | $R_{TH}$ | Moderate to high |
| Source follower | $A_v \cong 1$ | — | $R_{TH}$ | Low |
| Common gate | $A_v > 1$ | $A_i \cong 1$ | Low | Moderate to high |

Table 4.2 is a summary of the small-signal characteristics of the three amplifier configurations.

The common-source amplifier voltage gain magnitude is generally greater than 1. The voltage gain of the source follower is slightly less than 1, and that of the common-gate circuit is generally greater than 1.

The input resistance looking directly into the gate of the common-source and source-follower circuits is essentially infinite at low to moderate signal frequencies. However, the input resistance of these discrete amplifiers is the Thevenin equivalent resistance $R_{TH}$ of the bias resistors. In contrast, the input resistance to the common-gate circuit is generally in the range of only a few hundred ohms.

The output resistance of the source follower is generally in the range of a few hundred ohms or less. The output resistance of the common-source and common-gate configurations is dominated by the resistance $R_D$. In Chapters 10 and 11, we will see that the output resistance of these configurations is dominated by the resistance $r_o$ when transistors are used as load devices in ICs.

The specific characteristics of these single-stage amplifiers are used in the design of multistage amplifiers.

## 4.7    SINGLE-STAGE INTEGRATED CIRCUIT MOSFET AMPLIFIERS

**Objective: • Analyze all-MOS transistor circuits that become the foundation of integrated circuits.**

In the last chapter, we considered three all-MOSFET inverters and plotted the voltage transfer characteristics. All three inverters use an n-channel enhancement-mode driver transistor. The three types of load devices are an n-channel enhancement-mode device, an n-channel depletion-mode device, and a p-channel enhancement-mode device. The MOS transistor used as a load device is referred to as an **active load.** We mentioned that these three circuits can be used as amplifiers.

In this section, we revisit these three circuits and consider their amplifier characteristics. We will emphasize the small-signal equivalent circuits. This section serves as an introduction to more advanced MOS integrated circuit amplifier designs considered in Part 2 of the text.

### 4.7.1    Load Line Revisited

In dealing with all-transistor circuits, it will be instructive to consider the equivalent load lines that we have considered previously in circuits with resistive loads. Before we deal with the nonlinear load lines or load curves, it may be worthwhile to revisit the load line concept of a single transistor with a resistive load.

Figure 4.36 shows a single MOSFET with a resistive load. The current–voltage characteristic of the resistive load device is given by Ohm's law, or $V_R = I_D R_D$. This curve is plotted in the top portion of Figure 4.37. The load line is given by the KVL equation around the drain-source loop, or $V_{DS} = V_{DD} - I_D R_D$, and is superimposed on the transistor characteristics in the lower portion of Figure 4.37. We may note that the last term in the load line equation, $I_D R_D$, is the voltage across the load device.

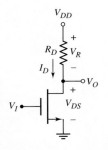

**Figure 4.36**  Single MOSFET circuit with resistive load

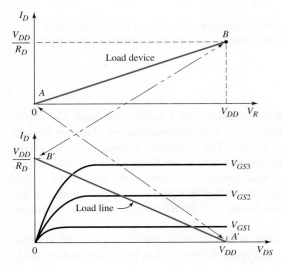

**Figure 4.37**  The $I$–$V$ curve for the resistor load device (top) and the load line superimposed on the transistor characteristics (bottom)

We may compare two points on the load device characteristic to the load line. When $I_D = 0$, $V_R = 0$ on the load characteristic curve denoted by point $A$. On the load line, the $I_D = 0$ point corresponds to $V_{DS} = V_{DD}$, denoted by the point $A'$. The maximum current on the load characteristic curve occurs when $V_R = V_{DD}$ and is denoted by point $B$. On the load line, the maximum current point corresponds to $V_{DS} = 0$, denoted by point $B'$. The load line can be created by taking the mirror image of the load characteristic curve and superimposing this curve on the plot of transistor characteristics. We will see this same effect in the following sections.

### 4.7.2    NMOS Amplifiers with Enhancement Load

The characteristics of an n-channel enhancement load device were presented in the last chapter. Figure 4.38(a) shows an NMOS enhancement load transistor, and Figure 4.38(b) shows the current–voltage characteristics. The threshold voltage is $V_{TNL}$.

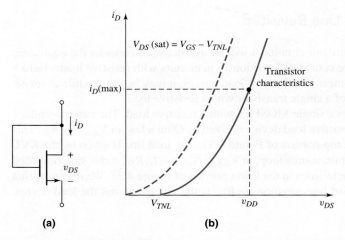

**(a)**                                    **(b)**

**Figure 4.38** (a) NMOS enhancement-mode transistor with gate and drain connected in a load device configuration and (b) current–voltage characteristics of NMOS enhancement load transistor

Figure 4.39(a) shows an NMOS amplifier with enhancement load. The driver transistor is $M_D$ and the load transistor is $M_L$. The characteristics of transistor $M_D$ and the load curve are shown in Figure 4.39(b). The load curve is essentially the mirror image of the $i$–$v$ characteristic of the load device, as we discussed in the last section. Since the $i$–$v$ characteristics of the load device are nonlinear, the load curve is also nonlinear. The load curve intersects the voltage axis at $V_{DD} - V_{TNL}$, which is the point where the current in the enhancement load device goes to zero. The transition point is also shown on the curve.

The voltage transfer characteristic is also useful in visualizing the operation of the amplifier. This curve is shown in Figure 4.39(c). When the enhancement-mode driver first begins to conduct, it is biased in the saturation region. For use as an amplifier, the circuit $Q$-point should be in this region, as shown in both Figures 4.39(b) and (c).

We can now apply the small-signal equivalent circuits to find the voltage gain. In the discussion of the source follower, we found that the equivalent resistance looking into the source terminal (with $R_S = \infty$) was $R_o = (1/g_m)\|r_o$. The small-signal equivalent circuit of the inverter is given in Figure 4.40, where the subscripts $D$ and $L$ refer to the driver and load transistors, respectively. We are again neglecting the body effect of the load transistor.

The small-signal voltage gain is then

$$A_v = \frac{V_o}{V_i} = -g_{mD}\left(r_{oD}\left\|\frac{1}{g_{mL}}\right\|r_{oL}\right) \tag{4.49}$$

Since, generally, $1/g_{mL} \ll r_{oL}$ and $1/g_{mD} \ll r_{oD}$, the voltage gain, to a good approximation is given by

$$A_v = \frac{-g_{mD}}{g_{mL}} = -\sqrt{\frac{K_{nD}}{K_{nL}}} = -\sqrt{\frac{(W/L)_D}{(W/L)_L}} \tag{4.50}$$

The voltage gain, then, is related to the size of the two transistors.

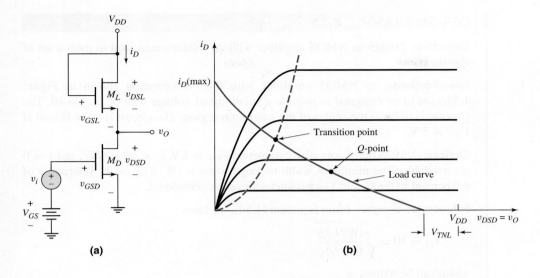

**(a)**     **(b)**

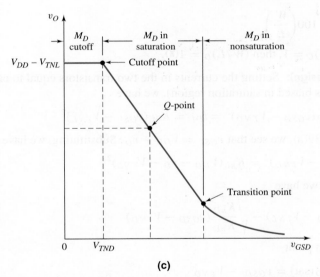

**(c)**

**Figure 4.39** (a) NMOS amplifier with enhancement load device; (b) driver transistor characteristics and enhancement load curve with transition point; and (c) voltage transfer characteristics of NMOS amplifier with enhancement load device

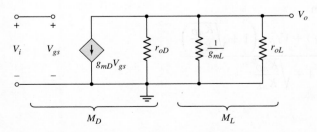

**Figure 4.40** Small-signal equivalent circuit of NMOS inverter with enhancement load device

## DESIGN EXAMPLE 4.11

**Objective:** Design an NMOS amplifier with an enhancement load to meet a set of specifications.

**Specifications:** An NMOS amplifier with the configuration shown in Figure 4.39(a) is to be designed to provide a small-signal voltage gain of $|A_v| = 10$. The Q-point is to be in the center of the saturation region. The circuit is to be biased at $V_{DD} = 5$ V.

**Choices:** NMOS transistors with parameters $V_{TN} = 1$ V, $k_n' = 60\ \mu\text{A/V}^2$, and $\lambda = 0$ are available. The minimum width-to-length ratio is $(W/L)_{min} = 1$. Tolerances of $\pm 5$ percent in the $k_n'$ and $V_{TN}$ parameters must be considered.

**Solution (ac design):** From Equation (4.50), we have

$$|A_v| = 10 = \sqrt{\frac{(W/L)_D}{(W/L)_L}}$$

which can be written as

$$\left(\frac{W}{L}\right)_D = 100 \left(\frac{W}{L}\right)_L$$

If we set $(W/L)_L = 1$, then $(W/L)_D = 100$.

**Solution (dc design):** Setting the currents in the two transistors equal to each other (both transistors biased in saturation region), we have

$$i_{DD} = K_{nD}(v_{GSD} - V_{TND})^2 = i_{DL} = K_{nL}(v_{GSL} - V_{TNL})^2$$

From Figure 4.39(a), we see that $v_{GSL} = V_{DD} - v_O$. Substituting, we have

$$K_{nD}(v_{GSD} - V_{TND})^2 = K_{nL}(V_{DD} - v_O - V_{TNL})^2$$

Solving for $v_O$, we have

$$v_O = (V_{DD} - V_{TNL}) - \sqrt{\frac{K_{nD}}{K_{nL}}}(v_{GSD} - V_{TND})$$

At the transition point,

$$v_{Ot} = v_{DSD}(\text{sat}) = v_{GSDt} - V_{TND}$$

where $v_{GSDt}$ is the gate-to-source voltage of the driver at the transition point. Then

$$v_{GSDt} - V_{TND} = (V_{DD} - V_{TNL}) - \sqrt{\frac{K_{nD}}{K_{nL}}}(v_{GSDt} - V_{TND})$$

Solving for $v_{GSDt}$, we obtain

$$v_{GSDt} = \frac{(V_{DD} - V_{TNL}) + V_{TND}\left(1 + \sqrt{\dfrac{K_{nD}}{K_{nL}}}\right)}{1 + \sqrt{\dfrac{K_{nD}}{K_{nL}}}}$$

Noting that

$$\sqrt{\frac{K_{nD}}{K_{nL}}} = \sqrt{\frac{(W/L)_D}{(W/L)_L}} = 10$$

we find

$$v_{GSDt} = \frac{(5-1)+(1)(1+10)}{1+10} = 1.36 \text{ V}$$

and

$$v_{Ot} = v_{DSDt} = v_{GSDt} - V_{TND} = 1.36 - 1 = 0.36 \text{ V}$$

Considering the transfer characteristics shown in Figure 4.41, we see that the center of the saturation region is halfway between the cutoff point ($v_{GSD} = V_{TND} = 1$ V) and the transition point ($v_{GSdt} = 1.36$ V), or

$$V_{GSQ} = \frac{1.36 - 1.0}{2} + 1.0 = 1.18 \text{ V}$$

Also

$$V_{DSDQ} = \frac{4 - 0.36}{2} + 0.36 = 2.18 \text{ V}$$

**Trade-offs:** Considering the tolerances in the $k_n'$ parameter, we find the range in the small-signal voltage gain to be

$$|A_v|_{max} = \sqrt{\frac{k_{nD}'}{k_{nL}'} \cdot \frac{(W/L)_D}{(W/L)_L}} = \sqrt{\frac{1.05}{0.95} \cdot (100)} = 10.5$$

and

$$|A_v|_{min} = \sqrt{\frac{k_{nD}'}{k_{nL}'} \cdot \frac{(W/L)_D}{(W/L)_L}} = \sqrt{\frac{0.95}{1.05} \cdot (100)} = 9.51$$

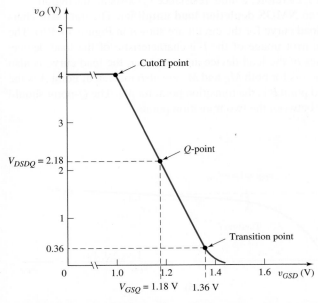

**Figure 4.41** Voltage transfer characteristics and $Q$-point of NMOS amplifier with enhancement load, for Example 4.11

The tolerances in the $k'_n$ and $V_{TN}$ parameters will also affect the $Q$-point. This analysis is left as an end-of-chapter problem.

**Comment:** These results show that a very large difference is required in the sizes of the two transistors to produce a gain of 10. In fact, a gain of 10 is about the largest practical gain that can be produced by an enhancement load device. A larger small-signal gain can be obtained by using a depletion-mode MOSFET as a load device, as shown in the next section.

**Design Pointer:** The body effect of the load transistor was neglected in this analysis. The body effect will actually lower the small-signal voltage gain from that determined in the example.

---

### EXERCISE PROBLEM

**Ex 4.11:** The bias voltage for the enhancement-load amplifier shown in Figure 4.39(a) is $V_{DD} = 3.3$ V. The transistor parameters are $V_{TND} = V_{TNL} = 0.4$ V, $k'_n = 100\,\mu\text{A/V}^2$, $(W/L)_L = 1.2$, and $\lambda = 0$. (a) Design the circuit such that the small-signal voltage gain is $|A_v| = 8$. (b) Determine $V_{GSDQ}$ such that the $Q$-point is in the center of the saturation region. (Ans. (a) $(W/L)_D = 76.8$, (b) $V_{GSDQ} = 0.561$ V).

---

### 4.7.3    NMOS Amplifier with Depletion Load

Figure 4.42(a) shows the NMOS depletion-mode transistor connected as a load device and Figure 4.42(b) shows the current–voltage characteristics. The transition point is also indicated. The threshold voltage $V_{TNL}$ of this device is negative, which means that the $v_{DS}$ value at the transition point is positive. Also, the slope of the curve in the saturation region is not zero; therefore, a finite resistance $r_o$ exists in this region.

Figure 4.43(a) shows an **NMOS depletion load amplifier.** The transistor characteristics of $M_D$ and the load curve for the circuit are shown in Figure 4.43(b). The load curve, again, is the mirror image of the $i$–$v$ characteristic of the load device. Since the $i$–$v$ characteristics of the load device are nonlinear, the load curve is also nonlinear. The transition points for both $M_D$ and $M_L$ are also indicated. Point $A$ is the transition point for $M_D$, and point $B$ is the transition point for $M_L$. The $Q$-point should be approximately midway between the two transition points.

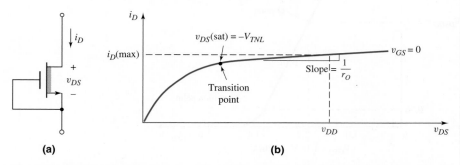

**(a)**                                          **(b)**

**Figure 4.42** (a) NMOS depletion-mode transistor with gate and source connected in a load device configuration and (b) current–voltage characteristic of NMOS depletion load transistor

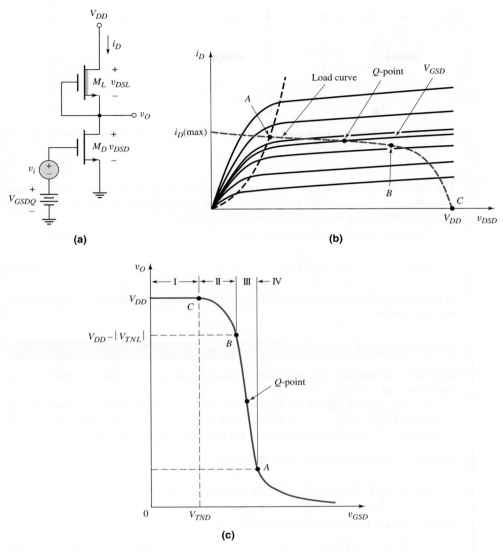

**Figure 4.43** (a) NMOS amplifier with depletion load device; (b) driver transistor characteristics and depletion load curve, with transition points between saturation and nonsaturation regions; (c) voltage transfer characteristics

The dc voltage $V_{GSDQ}$ biases transistor $M_D$ in the saturation region at the $Q$-point. The signal voltage $v_i$ superimposes a time-varying gate-to-source voltage on the dc value, and the bias point moves along the load curve about the $Q$-point. Again, both $M_D$ and $M_L$ must be biased in their saturation regions at all times.

The voltage transfer characteristic of this circuit is shown in Figure 4.43(c). Region III corresponds to the condition in which both transistors are biased in the saturation region. The desired $Q$-point is indicated.

We can again apply the small-signal equivalent circuit to find the small-signal voltage gain. Since the gate-to-source voltage of the depletion-load device is held at zero, the equivalent resistance looking into the source terminal is $R_o = r_o$.

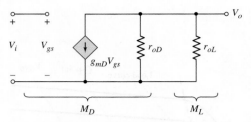

**Figure 4.44** Small-signal equivalent circuit of NMOS inverter with depletion load device

The small-signal equivalent circuit of the inverter is given in Figure 4.44, where the subscripts $D$ and $L$ refer to the driver and load transistors, respectively. We are again neglecting the body effect of the load device.

The small-signal voltage gain is then

$$A_v = \frac{V_o}{V_i} = -g_{mD}(r_{oD} \| r_{oL}) \tag{4.51}$$

In this circuit, the voltage gain is directly proportional to the output resistances of the two transistors.

## EXAMPLE 4.12

**Objective:** Determine the small-signal voltage gain of the NMOS amplifier with depletion load.

For the circuit shown in Figure 4.43(a), assume transistor parameters of $V_{TND} = +0.8$ V, $V_{TNL} = -1.5$ V, $K_{nD} = 1$ mA/V$^2$, $K_{nL} = 0.2$ mA/V$^2$, and $\lambda_D = \lambda_L = 0.01$ V$^{-1}$. Assume the transistors are biased at $I_{DQ} = 0.2$ mA.

**Solution:** The transconductance of the driver is

$$g_{mD} = 2\sqrt{K_{nD}I_{DQ}} = 2\sqrt{(1)(0.2)} = 0.894 \text{ mA/V}$$

Since $\lambda_D = \lambda_L$, the output resistances are

$$r_{oD} = r_{oL} = \frac{1}{\lambda I_{DQ}} = \frac{1}{(0.01)(0.2)} = 500 \text{ k}\Omega$$

The small-signal voltage gain is then

$$A_v = -g_{mD}(r_{oD} \| r_{oL}) = -(0.894)(500\|500) = -224$$

**Comment:** The voltage gain of the NMOS amplifier with depletion load is, in general, significantly larger than that with the enhancement load device. The body effect will lower the ideal gain factor.

**Discussion:** One aspect of this circuit design that we have not emphasized is the dc biasing. We mentioned that both transistors need to be biased in their saturation regions. From Figure 4.43(a), this dc biasing is accomplished with the dc source $V_{GSDQ}$. However, because of the steep slope of the transfer characteristics (Figure 4.43(c)), applying the "correct" voltage becomes difficult. As we will see in the next section, dc biasing is generally accomplished with current source biasing.

**Ex 4.12:** Assume the depletion-load amplifier in Figure 4.43(a) is biased at $I_{DQ} = 0.1$ mA. The transistor parameters are $K_{nD} = 250\,\mu\text{A/V}^2$, $K_{nL} = 25\,\mu\text{A/V}^2$, $V_{TND} = 0.4$ V, $V_{TNL} = -0.8$ V, and $\lambda_1 = \lambda_2 = 0.02$ V$^{-1}$. Determine the small-signal voltage gain. (Ans. $A_v = -79.1$)

### 4.7.4   NMOS Amplifier with Active Loads

#### CMOS Common-Source Amplifier

An amplifier using an n-channel enhancement-mode driver and a p-channel enhancement mode active load is shown in Figure 4.45(a) in a common-source configuration. The p-channel active load transistor $M_2$ is biased from $M_3$ and $I_{\text{Bias}}$. This configuration is similar to the MOSFET current source shown in Figure 3.49 in Chapter 3. With both n- and p-channel transistors in the same circuit, this circuit is now referred to as a CMOS amplifier. The CMOS configuration is used almost exclusively rather than the NMOS enhancement load or depletion load devices.

The $i$–$v$ characteristic curve for $M_2$ is shown in Figure 4.45(b). The source-to-gate voltage is a constant and is established by $M_3$. The driver transistor characteristics and the load curve are shown in Figure 4.45(c). The transition points of both $M_1$

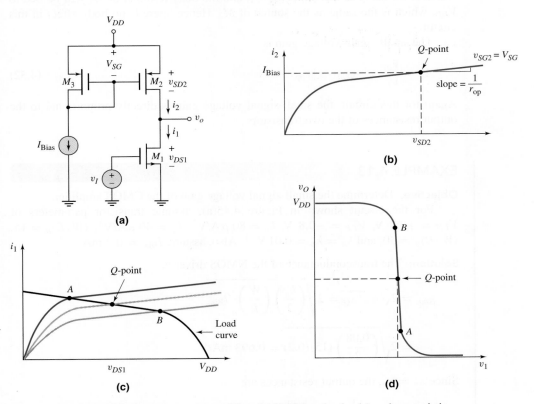

**Figure 4.45** (a) CMOS common-source amplifier; (b) PMOS active load $i$–$v$ characteristic, (c) driver transistor characteristics with load curve, (d) voltage transfer characteristics

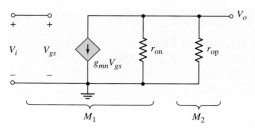

**Figure 4.46**  Small-signal equivalent circuit of the CMOS common-source amplifier

and $M_2$ are shown. Point $A$ is the transition point for $M_1$ and point $B$ is the transition point for $M_2$. The $Q$-point, to establish an amplifier, should be approximately halfway between points $A$ and $B$, so that both transistors are biased in their saturation regions. The voltage transfer characteristics are shown in Figure 4.45(d). Shown on the curve are the same transition points $A$ and $B$ and the desired $Q$-point.

We again apply the small-signal equivalent circuits to find the small-signal voltage gain. With $v_{SG2}$ held constant, the equivalent resistance looking into the drain of $M_2$ is just $R_o = r_{op}$. The small-signal equivalent circuit of the inverter is then as given in Figure 4.46. The subscripts $n$ and $p$ refer to the n-channel and p-channel transistors, respectively. We may note that the body terminal of $M_1$ will be tied to ground, which is the same as the source of $M_1$, and the body terminal of $M_2$ will be tied to $V_{DD}$, which is the same as the source of $M_2$. Hence, there is no body effect in this circuit.

The small-signal voltage gain is

$$A_v = \frac{V_o}{V_i} = -g_{mn}(r_{on}\|r_{op}) \tag{4.52}$$

Again for this circuit, the small-signal voltage gain is directly proportional to the output resistances of the two transistors.

## EXAMPLE 4.13

**Objective:**  Determine the small-signal voltage gain of the CMOS amplifier.

For the circuit shown in Figure 4.45(a), assume transistor parameters of $V_{TN} = +0.8$ V, $V_{TP} = -0.8$ V, $k'_n = 80 \ \mu\text{A/V}^2$, $k'_p = 40 \ \mu\text{A/V}^2$, $(W/L)_n = 15$, $(W/L)_p = 30$, and $\lambda_n = \lambda_p = 0.01$ V$^{-1}$. Also, assume $I_{\text{Bias}} = 0.2$ mA.

**Solution:**  The transconductance of the NMOS driver is

$$g_{mn} = 2\sqrt{K_n I_{DQ}} = 2\sqrt{\left(\frac{k'_n}{2}\right)\left(\frac{W}{L}\right)_n I_{\text{Bias}}}$$

$$= 2\sqrt{\left(\frac{0.08}{2}\right)(15)(0.2)} = 0.693 \ \text{mA/V}$$

Since $\lambda_n = \lambda_p$, the output resistances are

$$r_{on} = r_{op} = \frac{1}{\lambda I_{DQ}} = \frac{1}{(0.01)(0.2)} = 500 \ \text{k}\Omega$$

The small-signal voltage gain is then

$$A_v = -g_m(r_{on}\|r_{op}) = -(0.693)(500\|500) = -173$$

**Comment:** The voltage gain of the CMOS amplifier is on the same order of magnitude as the NMOS amplifier with depletion load. However, the CMOS amplifier does not suffer from the body effect.

**Discussion:** In the circuit configuration shown in Figure 4.45(a), we must again apply a dc voltage to the gate of $M_1$ to achieve the "proper" $Q$-point. We will show in later chapters using more sophisticated circuits how the $Q$-point is more easily established with current-source biasing. However, this circuit demonstrates the basic principles of the CMOS common-source amplifier.

**Ex 4.13:** For the circuit shown in Figure 4.45(a), assume transistor parameters of $V_{TN} = +0.5$ V, $V_{TP} = -0.5$ V, $k'_n = 80$ $\mu$A/V$^2$, $k'_p = 40$ $\mu$A/V$^2$, and $\lambda_n = \lambda_p = 0.015$ V$^{-1}$. Assume $I_{Bias} = 0.1$ mA. Assume $M_2$ and $M_3$ are matched. Find the width-to-length ratio of $M_1$ such that the small-signal voltage gain is $A_v = -250$. (Ans. $(W/L)_1 = 35.2$)

### CMOS Source-Follower Amplifier

The same basic CMOS circuit configuration can be used to form a CMOS source-follower amplifier. Figure 4.47(a) shows a source-follower circuit. We see that for this source-follower circuit, the active load, which is $M_2$, is an n-channel rather than a p-channel device. The input signal is applied to the gate of $M_1$ and the output is at the source of $M_1$.

The small-signal equivalent circuit of this source-follower is shown in Figure 4.47(b). This circuit, with two signal grounds, is redrawn as shown in Figure 4.47(c) to combine the signal grounds.

**EXAMPLE 4.14**

**Objective:** Determine the small-signal voltage gain and output resistance of the source-follower amplifier shown in Figure 4.47(a).

Assume the reference bias current is $I_{Bias} = 0.20$ mA and the bias voltage is $V_{DD} = 3.3$ V. Assume that all transistors are matched (identical) with parameters $V_{TN} = 0.4$ V, $K_n = 0.20$ mA/V$^2$, and $\lambda = 0.01$ V$^{-1}$.

We may note that since $M_3$ and $M_2$ are matched transistors and have the same gate-to-source voltages, the drain current in $M_1$ is $I_{D1} = I_{Bias} = 0.2$ mA.

**Solution (voltage gain):** From Figure 4.47(c), we find the small-signal output voltage to be

$$V_o = g_{m1}V_{gs}(r_{o1}\|r_{o2})$$

A KVL equation around the outside loop produces

$$V_i = V_{gs} + V_o = V_{gs} + g_{m1}V_{gs}(r_{o1}\|r_{o2})$$

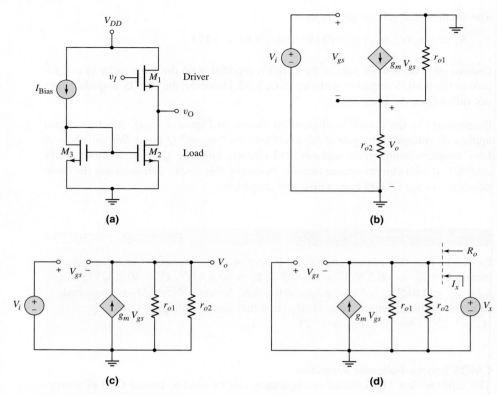

Figure 4.47 (a) All NMOS source-follower circuit, (b) small-signal equivalent circuit, (c) reconfiguration of small-signal equivalent circuit, and (d) small-signal equivalent circuit for determining the output resistance

or

$$V_{gs} = \frac{V_i}{1 + g_{m1}(r_{o1}\|r_{o2})}$$

Substituting this equation for $V_{gs}$ into the output voltage expression, we obtain the small-signal voltage gain as

$$A_v = \frac{V_o}{V_i} = \frac{g_{m1}(r_{o1}\|r_{o2})}{1 + g_{m1}(r_{o1}\|r_{o2})}$$

The small-signal equivalent circuit parameters are determined to be

$$g_{m1} = 2\sqrt{K_n I_{D1}} = 2\sqrt{(0.20)(0.20)} = 0.40 \text{ mA/V}$$

and

$$r_{o1} = r_{o2} = \frac{1}{\lambda I_D} = \frac{1}{(0.01)(0.20)} = 500 \text{ k}\Omega$$

The small-signal voltage gain is then

$$A_v = \frac{(0.40)(500\|500)}{1 + (0.40)(500\|500)}$$

or

$$A_v = 0.990$$

**Solution (output resistance):** The output resistance can be determined from the equivalent circuit shown in Figure 4.47(d). The independent source $V_i$ is set equal to zero and a test voltage $V_x$ is applied to the output.

Summing currents at the output node, we find

$$I_x + g_{m1}V_{gs} = \frac{V_x}{r_{o2}} + \frac{V_x}{r_{o1}}$$

From the circuit, we see that $V_{gs} = -V_x$. We then have

$$I_x = V_x\left(g_{m1} + \frac{1}{r_{o2}} + \frac{1}{r_{o1}}\right)$$

The output resistance is then given as

$$R_o = \frac{V_x}{I_x} = \frac{1}{g_{m1}}\|r_{o2}\|r_{o1}$$

We find

$$R_o = \frac{1}{0.40}\|500\|500$$

or

$$R_o = 2.48\,\text{k}\Omega$$

**Comment:** A voltage gain of $A_v = 0.99$ is typical of a source-follower circuit. An output resistance of $R_o = 2.48\text{ k}\Omega$ is relatively small for a MOSFET circuit and is also a characteristic of a source-follower circuit.

---

EXERCISE PROBLEM

**Ex 4.14:** The transconductance $g_m$ of the transistor in the circuit of Figure 4.47 is to be changed by changing the bias current such that the output resistance of the circuit is $R_o = 2$ k$\Omega$. Assume all other parameters are as given in Example 4.14. (a) What are the required values of $g_m$ and $I_{Bias}$? (b) Using the results of part (a), what is the small-signal voltage gain? (Ans. (a) $I_D = 0.3125$ mA; (b) $A_v = 0.988$)

---

COMPUTER ANALYSIS EXERCISE

**PS 4.1:** Using a PSpice analysis, investigate the small-signal voltage gain and output resistance of the source-follower circuit shown in Figure 4.47 taking into account the body effect.

**CMOS Common-Gate Amplifier**

Figure 4.48(a) shows a common-gate circuit. We see that in this common-gate circuit, the active load is the PMOS device $M_2$. The input signal is applied to the source of $M_1$ and the output is at the drain of $M_1$.

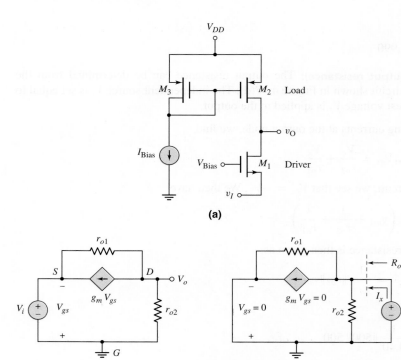

**Figure 4.48** (a) CMOS common-gate amplifier, (b) small-signal equivalent circuit, and (c) small-signal equivalent circuit for determining the output resistance

The small-signal equivalent circuit of the common-gate circuit is shown in Figure 4.48(b).

## EXAMPLE 4.15

**Objective:** Determine the small-signal voltage gain and output resistance of the common-gate circuit shown in Figure 4.48(a).

Assume the reference bias current is $I_{Bias} = 0.20$ mA and the bias voltage is $V_{DD} = 3.3$ V. Assume that the transistor parameters are $V_{TN} = +0.4$ V, $V_{TP} = -0.4$ V, $K_n = 0.20$ mA/V$^2$, $K_p = 0.20$ mA/V$^2$, and $\lambda_n = \lambda_p = 0.01$ V$^{-1}$.

We may note that, since $M_2$ and $M_3$ are matched transistors and have the same source-to-gate voltage, the bias current in $M_1$ is $I_{D1} = I_{Bias} = 0.20$ mA.

**Solution (voltage gain):** From Figure 4.48(b), we can sum currents at the output node and obtain

$$\frac{V_o}{r_{o2}} + g_{m1}V_{gs} + \frac{V_o - (-V_{gs})}{r_{o1}} = 0$$

or

$$V_o \left( \frac{1}{r_{o2}} + \frac{1}{r_{o1}} \right) + V_{gs} \left( g_{m1} + \frac{1}{r_{o1}} \right) = 0$$

From the circuit, we see that $V_{gs} = -V_i$. We then find the small-signal voltage gain to be

$$A_v = \frac{\left( g_{m1} + \dfrac{1}{r_{o1}} \right)}{\left( \dfrac{1}{r_{o2}} + \dfrac{1}{r_{o1}} \right)}$$

We find the small-signal equivalent circuit parameters to be

$$g_{m1} = 2\sqrt{K_n I_{D1}} = 2\sqrt{(0.20)(0.20)} = 0.40 \text{ mA/V}$$

and

$$r_{o1} = r_{o2} = \frac{1}{\lambda I_{D1}} = \frac{1}{(0.01)(0.20)} = 500 \text{ k}\Omega$$

We then find

$$A_v = \frac{\left( 0.40 + \dfrac{1}{500} \right)}{\left( \dfrac{1}{500} + \dfrac{1}{500} \right)}$$

or

$$A_v = 101$$

**Solution (output resistance):** The output resistance can be found from Figure 4.48(c). Summing currents at the output node, we find

$$I_x = \frac{V_x}{r_{o2}} + g_{m1} V_{gs} + \frac{V_x - (-V_{gs})}{r_{o1}}$$

However, $V_{gs} = 0$ so that $g_{m1} V_{gs} = 0$. We then find

$$R_o = \frac{V_x}{I_x} = r_{o1} \| r_{o2} = 500 \| 500$$

or

$$R_o = 250 \text{ k}\Omega$$

**Comment:** A voltage gain of $A_v = +101$ is typical of a common-gate amplifier. The output signal is in phase with respect to the input signal and the gain is relatively large. Also, a large output resistance of $R_o = 250 \text{ k}\Omega$ is typical of a common-gate amplifier in that the circuit acts like a current source.

## EXERCISE PROBLEM

**Ex 4.15:** The transconductance $g_m$ of the transistor in the circuit of Figure 4.48 is to be changed by changing the bias current such that the small-signal voltage gain is $A_v = 120$. Assume all other parameters are as given in Example 4.15. (a) What are the required values of $g_m$ and $I_{Bias}$? (b) Using the results of part (a), what is the output resistance? (Ans. (a) $I_D = 0.14 \text{ mA}$, $g_m = 0.335 \text{ mA/V}$; (b) $R_o = 357 \text{ k}\Omega$)

**PS 4.2:** Using a PSpice analysis, investigate the small-signal voltage gain and output resistance of the common-gate amplifier shown in Figure 4.48 taking into account the body effect.

## Test Your Understanding

**TYU 4.11** For the enhancement load amplifier shown in Figure 4.39(a), the parameters are: $V_{TND} = V_{TNL} = 0.8$ V, $k'_n = 40$ $\mu$A/V$^2$, $(W/L)_D = 80$, $(W/L)_L = 1$, and $V_{DD} = 5$ V. Determine the small-signal voltage gain. Determine $V_{GS}$ such that the $Q$-point is in the middle of the saturation region. (Ans. $A_v = -8.94$, $V_{GS} = 1.01$ V)

## 4.8    MULTISTAGE AMPLIFIERS

**Objective:** • Analyze multitransistor or multistage amplifiers and understand the advantages of these circuits over single-transistor amplifiers.

In most applications, a single-transistor amplifier will not be able to meet the combined specifications of a given amplification factor, input resistance, and output resistance. For example, the required voltage gain may exceed that which can be obtained in a single-transistor circuit. We will consider, here, the ac analysis of the two multitransistor circuits investigated in Chapter 3.

### 4.8.1    Multistage Amplifier: Cascade Circuit

The circuit shown in Figure 4.49 is a cascade of a common-source amplifier followed by a source-follower amplifier. As shown previously, the common-source amplifier

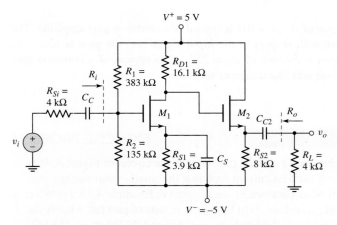

**Figure 4.49** Common-source amplifier in cascade with source follower

provides a small-signal voltage gain and the source follower has a low output impedance and provides the required output current. The resistor values are those determined in Section 3.5.1 of the previous chapter.

The midband small-signal voltage gain of the multistage amplifier is determined by assuming that all external coupling capacitors act as short circuits and inserting the small-signal equivalent circuits for the transistors.

## EXAMPLE 4.16

**Objective:** Determine the small-signal voltage gain of a multistage cascade circuit.

Consider the circuit shown in Figure 4.49. The transistor parameters are $K_{n1} = 0.5$ mA/V$^2$, $K_{n2} = 0.2$ mA/V$^2$, $V_{TN1} = V_{TN2} = 1.2$ V, and $\lambda_1 = \lambda_2 = 0$. The quiescent drain currents are $I_{D1} = 0.2$ mA and $I_{D2} = 0.5$ mA.

**Solution:** The small-signal equivalent circuit is shown in Figure 4.50. The small-signal transconductance parameters are

$$g_{m1} = 2\sqrt{K_{n1}I_{D1}} = 2\sqrt{(0.5)(0.2)} = 0.632 \text{ mA/V}$$

and

$$g_{m2} = 2\sqrt{K_{n2}I_{D2}} = 2\sqrt{(0.2)(0.5)} = 0.632 \text{ mA/V}$$

The output voltage is

$$V_o = g_{m2}V_{gs2}(R_{S2}\|R_L)$$

Also,

$$V_{gs2} + V_o = -g_{m1}V_{gs1}R_{D1}$$

where

$$V_{gs1} = \left(\frac{R_i}{R_i + R_{Si}}\right) \cdot V_i$$

Then

$$V_{gs2} = -g_{m1}R_{D1}\left(\frac{R_i}{R_i + R_{Si}}\right) \cdot V_i - V_o$$

Therefore

$$V_o = g_{m2}\left[-g_{m1}R_{D1}\left(\frac{R_i}{R_i + R_{Si}}\right) \cdot V_i - V_o\right](R_{S2}\|R_L)$$

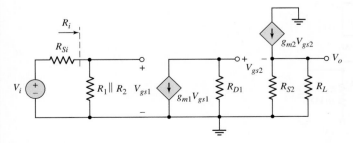

**Figure 4.50** Small-signal equivalent circuit of NMOS cascade circuit

The small-signal voltage gain is then

$$A_v = \frac{V_o}{V_i} = \frac{-g_{m1}g_{m2}R_{D1}(R_{S2}\|R_L)}{1 + g_{m2}(R_{S2}\|R_L)} \cdot \left(\frac{R_i}{R_i + R_{Si}}\right)$$

or

$$A_v = \frac{-(0.632)(0.632)(16.1)(8\|4)}{1 + (0.632)(8\|4)} \cdot \left(\frac{100}{100 + 4}\right) = -6.14$$

**Comment:** Since the small-signal voltage gain of the source follower is slightly less than 1, the overall gain is due essentially to the common-source input stage. Also, as shown previously, the output resistance of the source follower is small, which is desirable in many applications.

### EXERCISE PROBLEM

**Ex 4.16:** For the cascade circuit shown in Figure 4.49, the transistor and circuit parameters are given in Example 4.16. Calculate the small-signal output resistance $R_o$. (The small-signal equivalent circuit is shown in Figure 4.50.) (Ans. $R_o = 1.32\ k\Omega$)

### 4.8.2    Multistage Amplifier: Cascode Circuit

Figure 4.51 shows a cascode circuit with n-channel MOSFETs. Transistor $M_1$ is connected in a common-source configuration and $M_2$ is connected in a common-gate configuration. The advantage of this type of circuit is a higher frequency response, which will be discussed in Chapter 7. The resistor values are those determined in Section 3.5.2 of the previous chapter.

We will consider additional multistage and multitransistor circuits in Chapters 11 and 13.

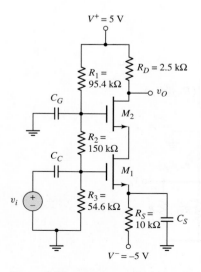

**Figure 4.51** NMOS cascode circuit

## EXAMPLE 4.17

**Objective:** Determine the small-signal voltage gain of a cascode circuit.

Consider the cascode circuit shown in Figure 4.51. The transistor parameters are $K_{n1} = K_{n2} = 0.8$ mA/V$^2$, $V_{TN1} = V_{TN2} = 1.2$ V, and $\lambda_1 = \lambda_2 = 0$. The quiescent drain current is $I_D = 0.4$ mA in each transistor. The input signal to the circuit is assumed to be an ideal voltage source.

**Solution:** Since the transistors are identical and since the current in the two transistors is the same, the small-signal transconductance parameters are

$$g_{m1} = g_{m2} = 2\sqrt{K_n I_D} = 2\sqrt{(0.8)(0.4)} = 1.13 \text{ mA/V}$$

The small-signal equivalent circuit is shown in Figure 4.52. Transistor $M_1$ supplies the source current of $M_2$ with the signal current $(g_{m1}V_i)$. Transistor $M_2$ acts as a current follower and passes this current on to its drain terminal. The output voltage is therefore

$$V_o = -g_{m1}V_{gs1}R_D$$

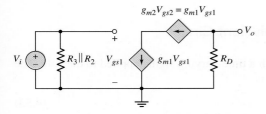

$$g_{m2}V_{gs2} = g_{m1}V_{gs1}$$

**Figure 4.52** Small-signal equivalent circuit of NMOS cascode circuit

Since $V_{gs1} = V_i$, the small-signal voltage gain is

$$A_v = \frac{V_o}{V_i} = -g_{m1}R_D$$

or

$$A_v = -(1.13)(2.5) = -2.83$$

**Comment:** The small-signal voltage gain is essentially the same as that of a single common-source amplifier stage. The addition of a common-gate transistor will increase the frequency bandwidth, as we will see in a later chapter.

## EXERCISE PROBLEM

**Ex 4.17:** The transistor parameters of the NMOS cascode circuit in Figure 4.51 are $V_{TN1} = V_{TN2} = 0.8$ V, $K_{n1} = K_{n2} = 3$ mA/V$^2$, and $\lambda_1 = \lambda_2 = 0$. (a) Determine $I_{DQ}$, $V_{DSQ1}$, and $V_{DSQ2}$. (b) Determine the small-signal voltage gain. (Ans. (a) $I_{DQ} = 0.471$ mA, $V_{DSQ1} = 2.5$ V, $V_{DSQ2} = 1.61$ V; (b) $A_v = -5.94$)

## *Test Your Understanding

**TYU 4.12** The transistor parameters of the circuit in Figure 4.49 are $V_{TN1} = V_{TN2} = 0.6$ V, $K_{n1} = 1.5$ mA/V$^2$, $K_{n2} = 2$ mA/V$^2$, and $\lambda_1 = \lambda_2 = 0$. (a) Find $I_{DQ1}$, $I_{DQ2}$, $V_{DSQ1}$, and $V_{DSQ2}$. (b) Determine the small-signal voltage gain. (c) Find the output resistance $R_o$. (Ans. (a) $I_{DQ1} = 0.3845$ mA, $I_{DQ2} = 0.349$ mA, $V_{DSQ1} = 2.31$ V, $V_{DSQ2} = 7.21$ V; (b) $A_v = -20.3$; (c) $R_o = 402$ $\Omega$)

---

## 4.9    BASIC JFET AMPLIFIERS

**Objective:** • Develop the small-signal model of JFET devices and analyze basic JFET amplifiers.

Like MOSFETs, JFETs can be used to amplify small time-varying signals. Initially, we will develop the small-signal model and equivalent circuit of the JFET. We will then use the model in the analysis of JFET amplifiers.

### 4.9.1    Small-Signal Equivalent Circuit

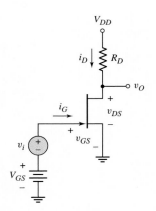

**Figure 4.53** JFET common-source circuit with time-varying signal source in series with gate dc source

Figure 4.53 shows a JFET circuit with a time-varying signal applied to the gate. The instantaneous gate-to-source voltage is

$$v_{GS} = V_{GS} + v_i = V_{GS} + v_{gs} \qquad (4.53)$$

where $v_{gs}$ is the small-signal gate-to-source voltage. Assuming the transistor is biased in the saturation region, the instantaneous drain current is

$$i_D = I_{DSS}\left(1 - \frac{v_{GS}}{V_P}\right)^2 \qquad (4.54)$$

where $I_{DSS}$ is the saturation current and $V_P$ is the pinchoff voltage. Substituting Equation (4.53) into (4.54), we obtain

$$i_D = I_{DSS}\left[\left(1 - \frac{V_{GS}}{V_P}\right) - \left(\frac{v_{gs}}{V_P}\right)\right]^2 \qquad (4.55)$$

If we expand the squared term, we have

$$i_D = I_{DSS}\left(1 - \frac{V_{GS}}{V_P}\right)^2 - 2I_{DSS}\left(1 - \frac{V_{GS}}{V_P}\right)\left(\frac{v_{gs}}{V_P}\right) + I_{DSS}\left(\frac{v_{gs}}{V_P}\right)^2 \qquad (4.56)$$

The first term in Equation (4.56) is the dc or quiescent drain current $I_{DQ}$, the second term is the time-varying drain current component, which is linearly related to the signal voltage $v_{gs}$, and the third term is proportional to the square of the signal voltage. As in the case of the MOSFET, the third term produces a nonlinear distortion in the output current. To minimize this distortion, we will usually impose the following condition:

$$\left|\frac{v_{gs}}{V_P}\right| \ll 2\left(1 - \frac{V_{GS}}{V_P}\right) \qquad (4.57)$$

Equation (4.57) represents the small-signal condition that must be satisfied for JFET amplifiers to be linear.

Neglecting the term $v_{gs}^2$ in Equation (4.56), we can write

$$i_D = I_{DQ} + i_d \qquad (4.58)$$

where the time-varying signal current is

$$i_d = +\frac{2I_{DSS}}{(-V_P)}\left(1 - \frac{V_{GS}}{V_P}\right)v_{gs} \qquad (4.59)$$

The constant relating the small-signal drain current and small-signal gate-to-source voltage is the transconductance $g_m$. We can write

$$i_d = g_m v_{gs} \qquad (4.60)$$

where

$$g_m = +\frac{2I_{DSS}}{(-V_P)}\left(1 - \frac{V_{GS}}{V_P}\right) \qquad (4.61)$$

Since $V_P$ is negative for n-channel JFETs, the transconductance is positive. A relationship that applies to both n-channel and p-channel JFETs is

$$g_m = \frac{2I_{DSS}}{|V_P|}\left(1 - \frac{V_{GS}}{V_P}\right) \qquad (4.62)$$

We can also obtain the transconductance from

$$g_m = \left.\frac{\partial i_D}{\partial v_{GS}}\right|_{v_{GS}=V_{GSQ}} \qquad (4.63)$$

Since the transconductance is directly proportional to the saturation current $I_{DSS}$, the transconductance is also a function of the width-to-length ratio of the transistor.

Since we are looking into a reverse-biased pn junction, we assume that the input gate current $i_g$ is zero, which means that the small-signal input resistance is infinite. Equation (4.54) can be expanded to take into account the finite output resistance of a JFET biased in the saturation region. The equation becomes

$$i_D = I_{DSS}\left(1 - \frac{v_{GS}}{V_P}\right)^2 (1 + \lambda v_{DS}) \qquad (4.64)$$

The small-signal output resistance is

$$r_o = \left.\left(\frac{\partial i_D}{\partial v_{DS}}\right)^{-1}\right|_{v_{GS}=\text{const.}} \qquad (4.65)$$

Using Equation (4.64), we obtain

$$r_o = \left[\lambda I_{DSS}\left(1 - \frac{V_{GS}}{V_P}\right)^2\right]^{-1} \qquad (4.66(a))$$

or

$$r_o \cong [\lambda I_{DQ}]^{-1} = \frac{1}{\lambda I_{DQ}} \qquad (4.66(b))$$

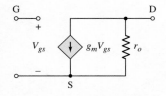

**Figure 4.54** Small-signal equivalent circuit of n-channel JFET

The small-signal equivalent circuit of the n-channel JFET, shown in Figure 4.54, is exactly the same as that of the n-channel MOSFET. The small-signal equivalent circuit of the p-channel JFET is also the same as that of the p-channel MOSFET. However, the polarity of the controlling gate-to-source voltage and the direction of the dependent current source are reversed from those of the n-channel device.

### 4.9.2 Small-Signal Analysis

Since the small-signal equivalent circuit of the JFET is the same as that of the MOSFET, the small-signal analyses of the two types of circuits are identical. For illustration purposes, we will analyze two JFET circuits.

### EXAMPLE 4.18

**Objective:** Determine the small-signal voltage gain of a JFET amplifier.

Consider the circuit shown in Figure 4.55 with transistor parameters $I_{DSS} = 12$ mA, $V_P = -4$ V, and $\lambda = 0.008$ V$^{-1}$. Determine the small-signal voltage gain $A_v = v_o/v_i$.

**Solution:** The dc quiescent gate-to-source voltage is determined from

$$V_{GSQ} = \left(\frac{R_2}{R_1 + R_2}\right)V_{DD} - I_{DQ}R_S$$

where

$$I_{DQ} = I_{DSS}\left(1 - \frac{V_{GSQ}}{V_P}\right)^2$$

Combining these two equations produces

$$V_{GSQ} = \left(\frac{180}{180 + 420}\right)(20) - (12)(2.7)\left(1 - \frac{V_{GSQ}}{(-4)}\right)^2$$

which reduces to

$$2.025V_{GSQ}^2 + 17.25V_{GSQ} + 26.4 = 0$$

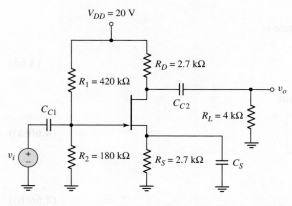

**Figure 4.55** Common-source JFET circuit with source resistor and source bypass capacitor

The appropriate solution is

$$V_{GSQ} = -2.0\,\text{V}$$

The quiescent drain current is

$$I_{DQ} = I_{DSS}\left(1 - \frac{V_{GSQ}}{V_P}\right)^2 = (12)\left(1 - \frac{(-2.0)}{(-4)}\right)^2 = 3.00\,\text{mA}$$

The small-signal parameters are then

$$g_m = \frac{2I_{DSS}}{(-V_P)}\left(1 - \frac{V_{GS}}{V_P}\right) = \frac{2(12)}{(4)}\left(1 - \frac{(-2.0)}{(-4)}\right) = 3.00\,\text{mA/V}$$

and

$$r_o = \frac{1}{\lambda I_{DQ}} = \frac{1}{(0.008)(3.00)} = 41.7\,\text{k}\Omega$$

The small-signal equivalent circuit is shown in Figure 4.56.
Since $V_{gs} = V_i$, the small-signal voltage gain is

$$A_v = \frac{V_o}{V_i} = -g_m(r_o \| R_D \| R_L)$$

or

$$A_v = -(3.0)(41.7 \| 2.7 \| 4) = -4.66$$

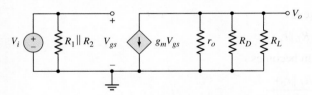

**Figure 4.56** Small-signal equivalent circuit of common-source JFET, assuming bypass capacitor acts as a short circuit

**Comment:** The voltage gain of JFET amplifiers is the same order of magnitude as that of MOSFET amplifiers.

## EXERCISE PROBLEM

**Ex 4.18:** For the JFET amplifier shown in Figure 4.55, the transistor parameters are: $I_{DSS} = 4$ mA, $V_P = -3$ V, and $\lambda = 0.005$ V$^{-1}$. Let $R_L = 4$ k$\Omega$, $R_S = 2.7$ k$\Omega$, and $R_1 + R_2 = 500$ k$\Omega$. Redesign the circuit such that $I_{DQ} = 1.2$ mA and $V_{DSQ} = 12$ V. Calculate the small-signal voltage gain. (Ans. $R_D = 3.97$ k$\Omega$, $R_1 = 453$ k$\Omega$, $R_2 = 47$ k$\Omega$, $A_v = -2.87$)

## DESIGN EXAMPLE **4.19**

**Objective:** Design a JFET source-follower circuit with a specified small-signal voltage gain.

For the source-follower circuit shown in Figure 4.57, the transistor parameters are: $I_{DSS} = 12$ mA, $V_P = -4$ V, and $\lambda = 0.01$ V$^{-1}$. Determine $R_S$ and $I_{DQ}$ such that the small-signal voltage gain is at least $A_v = v_o/v_i = 0.90$.

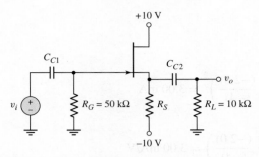

Figure 4.57  JFET source-follower circuit

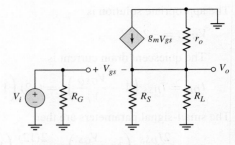

Figure 4.58  Small-signal equivalent circuit of JFET source-follower circuit

**Solution:** The small-signal equivalent circuit is shown in Figure 4.58. The output voltage is

$$V_o = g_m V_{gs}(R_S \| R_L \| r_o)$$

Also

$$V_i = V_{gs} + V_o$$

or

$$V_{gs} = V_i - V_o$$

Therefore, the output voltage is

$$V_o = g_m(V_i - V_o)(R_S \| R_L \| r_o)$$

The small-signal voltage gain becomes

$$A_v = \frac{V_o}{V_i} = \frac{g_m(R_S \| R_L \| r_o)}{1 + g_m(R_S \| R_L \| r_o)}$$

As a first approximation, assume $r_o$ is sufficiently large for the effect of $r_o$ to be neglected.

The transconductance is

$$g_m = \frac{2I_{DSS}}{(-V_P)}\left(1 - \frac{V_{GS}}{V_P}\right) = \frac{2(12)}{4}\left(1 - \frac{V_{GS}}{(-4)}\right)$$

If we pick a nominal transconductance value of $g_m = 2$ mA/V, then $V_{GS} = -2.67$ V and the quiescent drain current is

$$I_{DQ} = I_{DSS}\left(1 - \frac{V_{GS}}{V_P}\right)^2 = (12)\left(1 - \frac{(-2.67)}{(-4)}\right)^2 = 1.335 \text{ mA}$$

The value of $R_S$ is then determined from

$$R_S = \frac{-V_{GS} - (-10)}{I_{DQ}} = \frac{2.67 + 10}{1.335} = 9.49 \text{ k}\Omega$$

Also, the value of $r_o$ is

$$r_o = \frac{1}{\lambda I_{DQ}} = \frac{1}{(0.01)(1.335)} = 74.9 \text{ k}\Omega$$

The small-signal voltage gain, including the effect of $r_o$, is

$$A_v = \frac{g_m(R_S\|R_L\|r_o)}{1 + g_m(R_S\|R_L\|r_o)} = \frac{(2)(9.49\|10\|74.9)}{1 + (2)(9.49\|10\|74.9)} = 0.902$$

**Comment:** This particular design meets the design criteria, but the solution is not unique.

**Ex 4.19:** Reconsider the source-follower circuit shown in Figure 4.57 with transistor parameters $I_{DSS} = 8$ mA, $V_P = -3.5$ V, and $\lambda = 0.01$ V$^{-1}$. (a) Design the circuit such that $I_{DQ} = 2$ mA. (b) Calculate the small-signal voltage gain if $R_L$ approaches infinity. (c) Determine the value of $R_L$ at which the small-signal gain is reduced by 20 percent from its value for (b). (Ans. (a) $R_S = 5.88$ k$\Omega$, (b) $A_v = 0.923$, $R_L = 1.61$ k$\Omega$)

In Example 4.19, we chose a value of transconductance and continued through the design. A more detailed examination shows that both $g_m$ and $R_S$ depend upon the drain current $I_{DQ}$ in such a way that the product $g_m R_S$ is approximately a constant. This means the small-signal voltage gain is insensitive to the initial value of the transconductance.

## Test Your Understanding

**TYU 4.13** Reconsider the JFET amplifier shown in Figure 4.55 with transistor parameters given in Example 4.18. Determine the small-signal voltage gain if a 20 k$\Omega$ resistor is in series with the signal source $v_i$. (Ans. $A_v = -3.98$)

**\*TYU 4.14** For the circuit shown in Figure 4.59, the transistor parameters are: $I_{DSS} = 6$ mA, $|V_P| = 2$ V, and $\lambda = 0$. (a) Calculate the quiescent drain current and drain-to-source voltage of each transistor. (b) Determine the overall small-signal voltage gain $A_v = v_o/v_i$. (Ans. (a) $I_{DQ1} = 1$ mA, $V_{SDQ1} = 12$ V, $I_{DQ2} = 1.27$ mA, $V_{DSQ2} = 14.9$ V; (b) $A_v = -2.05$)

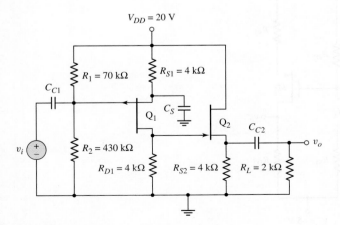

**Figure 4.59** Figure for Exercise TYU 4.14

### 4.10    DESIGN APPLICATION: A TWO-STAGE AMPLIFIER

**Objective:** • Design a two-stage MOSFET circuit to amplify the output of a sensor.

**Specifications:** Assume the resistance $R_2$ in the voltage divider circuit in Figure 4.60 varies linearly as a function of temperature, pressure, or some other variable. The output of the amplifier is to be zero volts when $\delta = 0$.

**Design Approach:** The amplifier configuration to be designed is shown in Figure 4.60. A resistor $R_1$ will be chosen such that the voltage divider between $R_1$ and $R_2$ will produce a dc voltage $v_I$ that is negative. A negative gate voltage to $M_1$ then means that the resistance $R_{S1}$ does not need to be so large.

**Choices:** Assume NMOS and PMOS transistors are available with parameters $V_{TN} = 1$ V, $V_{TP} = -1$ V, $K_n = K_p = 2$ mA/V$^2$, and $\lambda_n = \lambda_p \cong 0$.

**Solution (voltage divider analysis):** The voltage $v_I$ can be written as

$$v_I = \left[ \frac{R(1+\delta)}{R(1+\delta)+3R} \right](10) - 5 = \frac{(1+\delta)(10)}{4+\delta} - 5$$

or

$$v_I = \frac{(1+\delta)(10) - 5(4+\delta)}{4+\delta} = \frac{-10+5\delta}{4+\delta}$$

Assuming that $\delta \ll 4$, we then have

$$v_I = -2.5 + 1.25\delta$$

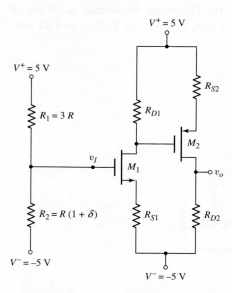

Figure 4.60    Two-stage MOSFET amplifier for design application

**Solution (DC Design):** We will choose $I_{D1} = 0.5$ mA and $I_{D2} = 1$ mA. The gate-to-source voltages are determined to be:

$$0.5 = 2(V_{GS1} - 1)^2 \Rightarrow V_{GS1} = 1.5 \text{ V}$$

and

$$1 = 2(V_{SG2} - 1)^2 \Rightarrow V_{SG2} = 1.707 \text{ V}$$

We find $V_{S1} = V_I - V_{GS1} = -2.5 - 1.5 = -4$ V. The resistor $R_{S1}$ is then

$$R_{S1} = \frac{V_{S1} - V^-}{I_{D1}} = \frac{-4 - (-5)}{0.5} = 2 \text{ k}\Omega$$

Letting $V_{D1} = 1.5$ V, we find the resistor $R_{D1}$ to be

$$R_{D1} = \frac{V^+ - V_{D1}}{I_{D1}} = \frac{5 - 1.5}{0.5} = 7 \text{ k}\Omega$$

We have $V_{S2} = V_{D1} + V_{SG2} = 1.5 + 1.707 = 3.207$ V. Then

$$R_{S2} = \frac{V^+ - V_{S2}}{I_{D2}} = \frac{5 - 3.207}{1} = 1.79 \text{ k}\Omega$$

For $V_O = 0$, we find

$$R_{D2} = \frac{V_O - V^-}{I_{D2}} = \frac{0 - (-5)}{1} = 5 \text{ k}\Omega$$

**Solution (ac Analysis):** The small-signal equivalent circuit is shown in Figure 4.61. We find $V_2 = -g_{m1}V_{gs1}R_{D1}$ and $V_{gs1} = V_i/(1 + g_{m1}R_{S1})$. We also find $V_o = g_{m2}V_{sg2}R_{D2}$ and $V_{sg2} = -V_2/(1 + g_{m2}R_{S2})$. Combining terms, we find

$$V_o = \frac{g_{m1}g_{m2}R_{D1}R_{D2}}{(1 + g_{m1}R_{S1})(1 + g_{m2}R_{S2})} V_i$$

The ac input signal is $V_i = 1.25\,\delta$, so we have

$$V_o = \frac{(1.25)g_{m1}g_{m2}R_{D1}R_{D2}}{(1 + g_{m1}R_{S1})(1 + g_{m2}R_{S2})}\,\delta$$

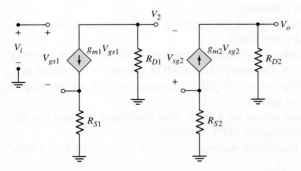

**Figure 4.61** Small-signal equivalent circuit of two-stage MOSFET amplifier for design application

We find that

$$g_{m1} = 2\sqrt{K_n I_{D1}} = 2\sqrt{(2)(0.5)} = 2 \text{ mA/V}$$

and

$$g_{m2} = 2\sqrt{K_p I_{D2}} = 2\sqrt{(2)(1)} = 2.828 \text{ mA/V}$$

We then find

$$V_o = \frac{(1.25)(2)(2.828)(7)(5)}{[1 + (2)(2)][1 + (2.828)(1.79)]}\delta$$

or

$$V_o = 8.16\delta$$

**Comment:** Since the low-frequency input impedance to the gate of the NMOS is essentially infinite, there is no loading effect on the voltage divider circuit.

**Design Pointer:** As mentioned previously, by choosing the value of $R_1$ to be larger than $R_2$, the dc voltage to the gate of $M_1$ is negative. A negative gate voltage implies that the required value of $R_{S1}$ is reduced and can still establish the required current. Since the drain voltage at $M_1$ is positive, then by using a PMOS transistor in the second stage, the source resistor value of $R_{S2}$ is also reduced. Smaller source resistances generate larger voltage gains.

 ## 4.11   SUMMARY

- The application of MOSFET transistors in linear amplifiers was emphasized in this chapter. The basic process by which a transistor circuit can amplify a small time-varying input signal was discussed.
- A small-signal equivalent circuit for the MOSFET transistor, which is used in the analysis and design of linear amplifiers, was developed.
- The three basic amplifier configurations were considered: the common-source, source-follower, and common-gate. These three circuits form the basic building blocks for more complex integrated circuits.
- The common-source circuit amplifies a time-varying voltage.
- The small-signal voltage gain of a source-follower circuit is approximately unity, but has a low output resistance.
- The common-gate circuit amplifies a time-varying voltage, and has a low input resistance and a large output resistance.
- Analysis of n-channel circuits with enhancement-load or depletion-load devices was performed. In addition, the analysis of CMOS circuits was carried out. These circuits are examples of all MOSFET circuits, which are developed throughout the remainder of the text.
- The small-signal equivalent circuit of a JFET was developed and used in the analysis of several configurations of JFET amplifiers.
- As an application, MOS transistors were incorporated in the design of a two-stage amplifier.

CHECKPOINT

After studying this chapter, the reader should have the ability to:

✓ Explain graphically the amplification process in a simple MOSFET amplifier circuit.
✓ Describe the small-signal equivalent circuit of the MOSFET and to determine the values of the small-signal parameters.
✓ Apply the small-signal equivalent circuit to various MOSFET amplifier circuits to obtain the time-varying circuit characteristics.
✓ Characterize the small-signal voltage gain and output resistance of the common-source, source-follower, and common-gate amplifiers.
✓ Describe the operation of an NMOS amplifier with either an enhancement load, a depletion load, or a PMOS load.
✓ Apply the MOSFET small-signal equivalent circuit in the analysis of multistage amplifier circuits.
✓ Describe the operation and analyze basic JFET amplifier circuits.

## REVIEW QUESTIONS

1. Discuss, using the concept of a load line, how a simple common-source circuit can amplify a time-varying signal.
2. How does the transistor width-to-length ratio affect the small-signal voltage gain of a common-source amplifier?
3. Discuss the physical meaning of the small-signal circuit parameter $r_o$.
4. How does the body effect change the small-signal equivalent circuit of the MOSFET?
5. Sketch a simple common-source amplifier circuit and discuss the general ac circuit characteristics (voltage gain and output resistance).
6. Discuss the general conditions under which a common-source amplifier would be used.
7. Why, in general, is the magnitude of the voltage gain of a common-source amplifier relatively small?
8. What are the changes in dc and ac characteristics of a common-source amplifier when a source resistor and a source bypass capacitor are incorporated in the design?
9. Sketch a simple source-follower amplifier circuit and discuss the general ac circuit characteristics (voltage gain and output resistance).
10. Sketch a simple common-gate amplifier circuit and discuss the general ac circuit characteristics (voltage gain and output resistance).
11. Discuss the general conditions under which a source-follower or a common-gate amplifier would be used.
12. Compare the ac circuit characteristics of the common-source, source-follower, and common-gate circuits.
13. State the advantage of using transistors in place of resistors in MOSFET integrated circuits.
14. State at least two reasons why a multistage amplifier circuit would be required in a design compared to using a single-stage circuit.

**PROBLEMS**

### Section 4.1 The MOSFET Amplifier

4.1    An NMOS transistor has parameters $V_{TN} = 0.4$ V, $k_n' = 100\,\mu\text{A/V}^2$, and $\lambda = 0.02$ V$^{-1}$. (a) (i) Determine the width-to-length ratio $W/L$ such that $g_m = 0.5$ mA/V at $I_{DQ} = 0.5$ mA when biased in the saturation region. (ii) Calculate the required value of $V_{GSQ}$. (b) Repeat part (a) for $I_{DQ} = 0.15$ mA.

4.2    A PMOS transistor has parameters $V_{TP} = -0.6$ V, $k_p' = 40\,\mu\text{A/V}^2$, and $\lambda = 0.015$ V$^{-1}$. (a) (i) Determine the width-to-length ratio $(W/L)$ such that $g_m = 1.2$ mA/V at $I_{DQ} = 0.15$ mA. (ii) What is the required value of $V_{SGQ}$? (b) Repeat part (a) for $I_{DQ} = 0.50$ mA.

4.3    An NMOS transistor is biased in the saturation region at a constant $V_{GS}$. The drain current is $I_D = 3$ mA at $V_{DS} = 5$ V and $I_D = 3.4$ mA at $V_{DS} = 10$ V. Determine $\lambda$ and $r_o$.

4.4    The minimum value of small-signal resistance of a PMOS transistor is to be $r_o = 100$ k$\Omega$. If $\lambda = 0.012$ V$^{-1}$, calculate the maximum allowed value of $I_D$.

4.5    An n-channel MOSFET is biased in the saturation region at a constant $V_{GS}$. (a) The drain current is $I_D = 0.250$ mA at $V_{DS} = 1.5$ V and $I_D = 0.258$ mA at $V_{DS} = 3.3$ V. Determine the value of $\lambda$ and $r_o$. (b) Using the results of part (a), determine $I_D$ at $V_{DS} = 5$ V.

4.6    The value of $\lambda$ for a MOSFET is 0.02 V$^{-1}$. (a) What is the value of $r_o$ at (i) $I_D = 50\,\mu\text{A}$ and at (ii) $I_D = 500\,\mu\text{A}$? (b) If $V_{DS}$ increases by 1 V, what is the percentage increase in $I_D$ for the conditions given in part (a)?

4.7    A MOSFET with $\lambda = 0.01$ V$^{-1}$ is biased in the saturation region at $I_D = 0.5$ mA. If $V_{GS}$ and $V_{DS}$ remain constant, what are the new values of $I_D$ and $r_o$ if the channel length $L$ is doubled?

4.8    The parameters of the circuit in Figure 4.1 are $V_{DD} = 3.3$ V and $R_D = 5$ k$\Omega$. The transistor parameters are $k_n' = 100\,\mu\text{A/V}^2$, $W/L = 40$, $V_{TN} = 0.4$ V, and $\lambda = 0.025$ V$^{-1}$. (a) Find $I_{DQ}$ and $V_{GSQ}$ such that $V_{DSQ} = 1.5$ V. (b) Determine the small-signal voltage gain.

4.9    The circuit shown in Figure 4.1 has parameters $V_{DD} = 2.5$ V and $R_D = 10$ k$\Omega$. The transistor is biased at $I_{DQ} = 0.12$ mA. The transistor parameters are $V_{TN} = 0.3$ V, $k_n' = 100\,\mu\text{A/V}^2$, and $\lambda = 0$. (a) Design the $W/L$ ratio of the transistor such that the small-signal voltage gain is $A_v = -3.8$. (b) Repeat part (a) for $A_v = -5.0$.

4.10   For the circuit shown in Figure 4.1, the transistor parameters are $V_{TN} = 0.6$ V, $k_n' = 80\,\mu\text{A/V}^2$, and $\lambda = 0.015$ V$^{-1}$. Let $V_{DD} = 5$ V. (a) Design the transistor width-to-length ratio $W/L$ and the resistance $R_D$ such that $I_{DQ} = 0.5$ mA, $V_{GSQ} = 1.2$ V, and $V_{DSQ} = 3$ V. (b) Determine $g_m$ and $r_o$. (c) Determine the small-signal voltage gain $A_v = v_o/v_i$.

*4.11  In our analyses, we assumed the small-signal condition given by Equation (4.4). Now consider Equation (4.3(b)) and let $v_{gs} = V_{gs}\sin\omega t$. Show that the ratio of the signal at frequency $2\omega$ to the signal at frequency $\omega$ is given by $V_{gs}/[4(V_{GS} - V_{TN})]$. This ratio, expressed in a percentage, is called the **second-harmonic distortion.** [Hint: Use the trigonometric identity $\sin^2\theta = \frac{1}{2} - \frac{1}{2}\cos 2\theta$.]

4.12   Using the results of Problem 4.11, find the peak amplitude $V_{gs}$ that produces a second-harmonic distortion of 1 percent if $V_{GS} = 3$ V and $V_{TN} = 1$ V.

## Section 4.3 The Common-Source Amplifier

4.13   Consider the circuit in Figure 4.14 in the text. The circuit parameters are $V_{DD} = 3.3$ V, $R_D = 8$ k$\Omega$, $R_1 = 240$ k$\Omega$, $R_2 = 60$ k$\Omega$, and $R_{Si} = 2$ k$\Omega$. The transistor parameters are $V_{TN} = 0.4$ V, $k'_n = 100\,\mu$A/V$^2$, $W/L = 80$, and $\lambda = 0.02$ V$^{-1}$. (a) Determine the quiescent values $I_{DQ}$ and $V_{DSQ}$. (b) Find the small-signal parameters $g_m$ and $r_o$. (c) Determine the small-signal voltage gain.

4.14   A common-source amplifier, such as shown in Figure 4.14 in the text, has parameters $r_o = 100$ k$\Omega$ and $R_D = 5$ k$\Omega$. Determine the transconductance of the transistor if the small-signal voltage gain is $A_v = -10$. Assume $R_{Si} = 0$.

4.15   For the NMOS common-source amplifier in Figure P4.15, the transistor parameters are: $V_{TN} = 0.8$ V, $K_n = 1$ mA/V$^2$, and $\lambda = 0$. The circuit parameters are $V_{DD} = 5$ V, $R_S = 1$ k$\Omega$, $R_D = 4$ k$\Omega$, $R_1 = 225$ k$\Omega$, and $R_2 = 175$ k$\Omega$. (a) Calculate the quiescent values $I_{DQ}$ and $V_{DSQ}$. (b) Determine the small-signal voltage gain for $R_L = \infty$. (c) Determine the value of $R_L$ that will reduce the small-signal voltage gain to 75 percent of the value found in part (b).

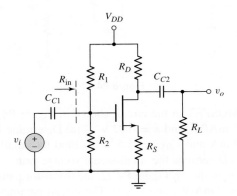

**Figure P4.15**

4.16   The parameters of the circuit shown in Figure P4.15 are $V_{DD} = 12$ V, $R_S = 0.5$ k$\Omega$, $R_{in} = 250$ k$\Omega$, and $R_L = 10$ k$\Omega$. The transistor parameters are $V_{TN} = 1.2$ V, $K_n = 1.5$ mA/V$^2$, and $\lambda = 0$. (a) Design the circuit such that $I_{DQ} = 2$ mA and $V_{DSQ} = 5$ V. (b) Determine the small-signal voltage gain.

4.17   Repeat Problem 4.15 if the source resistor is bypassed by a source capacitor $C_S$.

4.18   The ac equivalent circuit of a common-source amplifier is shown in Figure P4.18. The small-signal parameters of the transistor are $g_m = 2$ mA/V and $r_o = \infty$. (a) The voltage gain is found to be $A_v = V_o/V_i = -15$ with $R_S = 0$. What is the value of $R_D$? (b) A source resistor $R_S$ is inserted. Assuming the transistor parameters do not change, what is the value of $R_S$ if the voltage gain is reduced to $A_v = -5$.

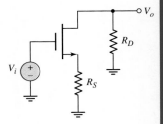

**Figure P4.18**

4.19   Consider the ac equivalent circuit shown in Figure P4.18. Assume $r_o = \infty$ for the transistor. The small-signal voltage gain is $A_v = -8$ for the case when $R_S = 1$ k$\Omega$. (a) When $R_S$ is shorted ($R_S = 0$), the magnitude of the voltage gain doubles. Assuming the small-signal transistor parameters do not change, what are the values of $g_m$ and $R_D$? (b) A new value of $R_S$ is inserted into the circuit and the voltage gain becomes $A_v = -10$. Using the results of part (a), determine the value of $R_S$.

4.20 The transistor in the common-source amplifier in Figure P4.20 has parameters $V_{TN} = 0.8$ V, $k'_n = 100\,\mu\text{A/V}^2$, $W/L = 50$, and $\lambda = 0.02\text{ V}^{-1}$. The circuit parameters are $V^+ = 5$ V, $V^- = -5$ V, $I_Q = 0.5$ mA, and $R_D = 6\text{ k}\Omega$. (a) Determine $V_{GSQ}$ and $V_{DSQ}$. (b) Find the small-signal voltage gain for $R_L = \infty$. (c) Repeat part (b) for $R_L = 20\text{ k}\Omega$. (d) Repeat part (b) for $R_L = 6\text{ k}\Omega$.

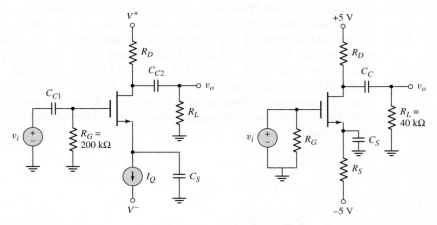

Figure P4.20               Figure P4.21

4.21 The parameters of the MOSFET in the circuit shown in Figure P4.21 are $V_{TN} = 0.8$ V, $K_n = 0.85\text{ mA/V}^2$, and $\lambda = 0.02\text{ V}^{-1}$. (a) Determine $R_S$ and $R_D$ such that $I_{DQ} = 0.1$ mA and $V_{DSQ} = 5.5$ V. (b) Find the small-signal transistor parameters. (c) Determine the small-signal voltage gain.

4.22 For the common-source amplifier in Figure P4.22, the transistor parameters are $V_{TN} = -0.8$ V, $K_n = 2\text{ mA/V}^2$, and $\lambda = 0$. The circuit parameters are $V_{DD} = 3.3$ V and $R_L = 10\text{ k}\Omega$. (a) Design the circuit such that $I_{DQ} = 0.5$ mA and $V_{DSQ} = 2$ V. (b) Determine the small-signal voltage gain.

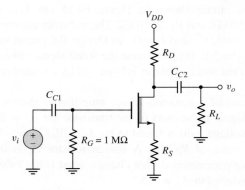

Figure P4.22

*4.23 The transistor in the common-source circuit in Figure P4.22 has the same parameters as given in Problem 4.22. The circuit parameters are $V_{DD} = 5$ V and $R_D = R_L = 2$ k$\Omega$. (a) Find $R_S$ for $V_{DSQ} = 2.5$ V. (b) Determine the small-signal voltage gain.

*4.24   Consider the PMOS common-source circuit in Figure P4.24 with transistor parameters $V_{TP} = -2$ V and $\lambda = 0$, and circuit parameters $R_D = R_L = 10$ kΩ. (a) Determine the values of $K_p$ and $R_S$ such that $V_{SDQ} = 6$ V. (b) Determine the resulting value of $I_{DQ}$ and the small-signal voltage gain. (c) Can the values of $K_p$ and $R_S$ from part (a) be changed to achieve a larger voltage gain, while still meeting the requirements of part (a)?

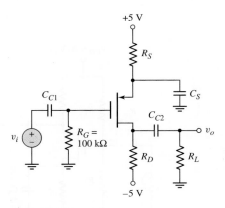

**Figure P4.24**

D4.25   For the common-source circuit in Figure P4.24, the bias voltages are changed to $V^+ = 3$ V and $V^- = -3$ V. The PMOS transistor parameters are: $V_{TP} = -0.5$ V, $K_p = 0.8$ mA/V$^2$, and $\lambda = 0$. The load resistor is $R_L = 2$ kΩ. (a) Design the circuit such that $I_{DQ} = 0.25$ mA and $V_{SDQ} = 1.5$ V. (b) Determine the small-signal voltage gain $A_v = v_o/v_i$.

*D4.26   Design the common-source circuit in Figure P4.26 using an n-channel MOSFET with $\lambda = 0$. The quiescent values are to be $I_{DQ} = 6$ mA, $V_{GSQ} = 2.8$ V, and $V_{DSQ} = 10$ V. The transconductance is $g_m = 2.2$ mA/V. Let $R_L = 1$ kΩ, $A_v = -1$, and $R_{in} = 100$ kΩ. Find $R_1$, $R_2$, $R_S$, $R_D$, $K_n$, and $V_{TN}$.

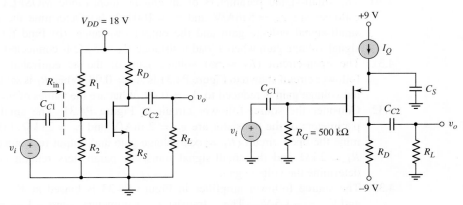

**Figure P4.26**            **Figure P4.27**

4.27   For the common-source amplifier shown in Figure P4.27, the transistor parameters are $V_{TP} = -1.2$ V, $K_p = 2$ mA/V$^2$, and $\lambda = 0.03$ V$^{-1}$. The drain resistor is $R_D = 4\,\text{k}\Omega$. (a) Determine $I_Q$ such that $V_{SDQ} = 5$ V. (b) Find the small-signal voltage gain for $R_L = \infty$. (c) Repeat part (b) for $R_L = 8\,\text{k}\Omega$.

D4.28   For the circuit shown in Figure P4.28, the transistor parameters are: $V_{TP} = 0.8$ V, $K_p' = 0.25\,\text{mA/V}^2$, and $\lambda = 0$. (a) Design the circuit such that $I_{DQ} = 0.5$ mA and $V_{SDQ} = 3$ V. (b) Determine the small-signal voltage gain $A_v = v_o/v_i$.

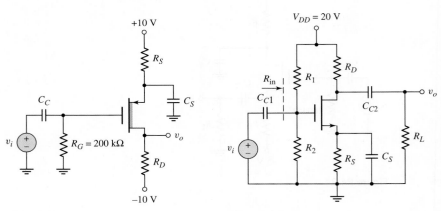

**Figure P4.28**                    **Figure P4.29**

*D4.29   Design a common-source amplifier, such as that in Figure P4.29, to achieve a small-signal voltage gain of at least $A_v = v_o/v_i = -10$ for $R_L = 20\,\text{k}\Omega$ and $R_{in} = 200\,\text{k}\Omega$. Assume the Q-point is chosen at $I_{DQ} = 1$ mA and $V_{DSQ} = 10$ V. Let $V_{TN} = 2$ V, and $\lambda = 0$.

### Section 4.4 The Source-Follower Amplifier

4.30   The small-signal parameters of an enhancement-mode MOSFET source follower are $g_m = 5\,\text{mA/V}$ and $r_o = 100\,\text{k}\Omega$. (a) Determine the no-load small-signal voltage gain and the output resistance. (b) Find the small-signal voltage gain when a load resistance $R_S = 5\,\text{k}\Omega$ is connected.

4.31   The open-circuit ($R_L = \infty$) voltage gain of the ac equivalent source-follower circuit shown in Figure P4.31 is $A_v = 0.98$. When $R_L$ is set to 1 k$\Omega$, the voltage gain is reduced to $A_v = 0.49$. What are the values of $g_m$ and $r_o$?

4.32   Consider the source-follower circuit in Figure P4.31. The small-signal parameters of the transistor are $g_m = 2\,\text{mA/V}$ and $r_o = 25\,\text{k}\Omega$. (a) Determine the open-circuit ($R_L = \infty$) voltage gain and output resistance. (b) If $R_L = 2\,\text{k}\Omega$ and the small-signal transistor parameters remain constant, determine the voltage gain.

4.33   The source follower amplifier in Figure P4.33 is biased at $V^+ = 1.5$ V and $V^- = -1.5$ V. The transistor parameters are $V_{TN} = 0.4$ V, $k_n' = 100\,\mu\text{A/V}^2$, $W/L = 80$, and $\lambda = 0.02\,\text{V}^{-1}$. (a) The dc value of $v_O$ is to be zero volts. What is the current $I_{DQ}$ and the required value of $V_{GSQ}$? (b) Determine the small-signal voltage gain. (c) Find the output resistance $R_o$.

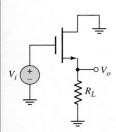

**Figure P4.31**

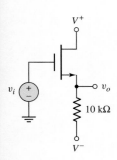

**Figure P4.33**

4.34   Consider the circuit in Figure P4.34. The transistor parameters are $V_{TN} = 0.6$ V, $k_n' = 100\,\mu\text{A/V}^2$, and $\lambda = 0$. The circuit is to be designed such that $V_{DSQ} = 1.25$ V and such that the small-signal voltage gain is $A_v = 0.85$. (a) Find $I_{DQ}$. (b) Determine the width-to-length ratio of the transistor. (c) What is the required dc value of the input voltage?

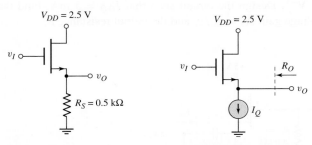

**Figure P4.34**          **Figure P4.35**

4.35   The quiescent power dissipation in the circuit in Figure P4.35 is to be limited to 2.5 mW. The parameters of the transistor are $V_{TN} = 0.6$ V, $k_n' = 100\,\mu\text{A/V}^2$, and $\lambda = 0.02$ V$^{-1}$. (a) Determine $I_Q$. (b) Determine $W/L$ such that the output resistance is $R_o = 0.5$ k$\Omega$. (c) Using the results of parts (a) and (b), determine the small-signal voltage gain. (d) Determine the output resistance if the transistor width-to-length ratio is $W/L = 100$.

4.36   The parameters of the circuit in Figure P4.36 are $R_S = 4$ k$\Omega$, $R_1 = 850$ k$\Omega$, $R_2 = 350$ k$\Omega$, and $R_L = 4$ k$\Omega$. The transistor parameters are $V_{TP} = -1.2$ V, $k_p' = 40\,\mu\text{A/V}^2$, $W/L = 80$, and $\lambda = 0.05$ V$^{-1}$. (a) Determine $I_{DQ}$ and $V_{SDQ}$. (b) Find the small-signal voltage gain $A_v = v_o/v_i$. (c) Determine the small-signal circuit transconductance gain $A_g = i_o/v_i$. (d) Find the small-signal output resistance $R_o$.

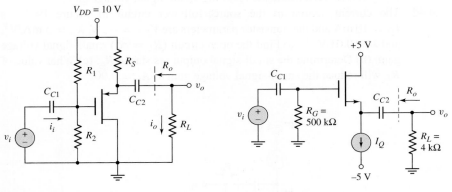

**Figure P4.36**          **Figure P4.37**

4.37   Consider the source follower circuit in Figure P4.37 with transistor parameters $V_{TN} = 0.8$ V, $k_n' = 100\,\mu\text{A/V}^2$, $W/L = 20$, and $\lambda = 0.02$ V$^{-1}$. (a) Let $I_Q = 5$ mA. (i) Determine the small-signal voltage gain. (ii) Find the output resistance $R_o$. (b) Repeat part (a) for $I_Q = 2$ mA.

4.38    For the source-follower circuit shown in Figure P4.37, the transistor parameters are: $V_{TN} = 1$ V, $k'_n = 60$ $\mu$A/V$^2$, and $\lambda = 0$. The small-signal voltage gain is to be $A_v = v_o/v_i = 0.95$. (a) Determine the required width-to-length ratio $(W/L)$ for $I_Q = 4$ mA. (b) Determine the required $I_Q$ if $(W/L) = 60$.

*D4.39    In the source-follower circuit in Figure P4.39 with a depletion NMOS transistor, the device parameters are: $V_{TN} = -2$ V, $K_n = 5$ mA/V$^2$, and $\lambda = 0.01$ V$^{-1}$. Design the circuit such that $I_{DQ} = 5$ mA. Find the small-signal voltage gain $A_v = v_o/v_i$ and the output resistance $R_o$.

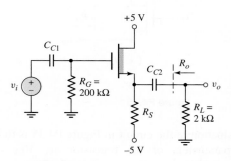

**Figure P4.39**

4.40    For the circuit in Figure P4.39, $R_S = 1$ k$\Omega$ and the quiescent drain current is $I_{DQ} = 5$ mA. The transistor parameters are $V_{TN} = -2$ V, $k'_n = 100$ $\mu$A/V$^2$, and $\lambda = 0.01$ V$^{-1}$. (a) Determine the transistor width-to-length ratio. (b) Using the results of part (a), find the small-signal voltage gain for $R_L = \infty$. (c) Find the small-signal output resistance $R_o$. (d) Using the results of part (a), find $A_v$ for $R_L = 2$ k$\Omega$.

D4.41    For the source-follower circuit in Figure P4.39, the transistor parameters are: $V_{TN} = -2$ V, $K_n = 4$ mA/V$^2$, and $\lambda = 0$. Design the circuit such that $R_o \leq 200$ $\Omega$. Determine the resulting small-signal voltage gain.

4.42    The current source in the source-follower circuit in Figure P4.42 is $I_Q = 10$ mA and the transistor parameters are $V_{TP} = -2$ V, $K_p = 5$ mA/V$^2$, and $\lambda = 0.01$ V$^{-1}$. (a) Find the open circuit $(R_L = \infty)$ small-signal voltage gain. (b) Determine the small-signal output resistance $R_o$. (c) What value of $R_L$ will reduce the small-signal voltage gain to $A_v = 0.90$?

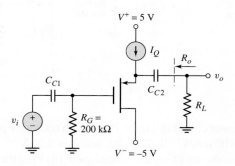

**Figure P4.42**

4.43 Consider the source-follower circuit shown in Figure P4.43. The most negative output signal voltage occurs when the transistor just cuts off. Show that this output voltage $v_o(\min)$ is given by

$$v_o(\min) = \frac{-I_{DQ}R_S}{1 + \frac{R_S}{R_L}}$$

Show that the corresponding input voltage is given by

$$v_i(\min) = -\frac{I_{DQ}}{g_m}(1 + g_m(R_S \| R_L))$$

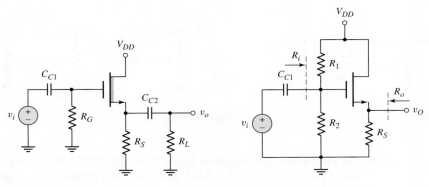

**Figure P4.43**      **Figure P4.44**

4.44 The transistor in the circuit in Figure P4.44 has parameters $V_{TN} = 0.4$ V, $K_n = 0.5$ mA/V$^2$, and $\lambda = 0$. The circuit parameters are $V_{DD} = 3$ V and $R_i = 300\,\text{k}\Omega$. (a) Design the circuit such that $I_{DQ} = 0.25$ mA and $V_{DSQ} = 1.5$ V. (b) Determine the small-signal voltage gain and the output resistance $R_o$.

## Section 4.5 The Common-Gate Configuration

4.45 Figure P4.45 is the ac equivalent circuit of a common-gate amplifier. The transistor parameters are $V_{TN} = 0.4$ V, $k_n' = 100\,\mu\text{A/V}^2$, and $\lambda = 0$. The quiescent drain current is $I_{DQ} = 0.25$ mA. Determine the transistor $W/L$ ratio and the value of $R_D$ such that the small-signal voltage gain is $A_v = V_o/V_i = 20$ and the input resistance is $R_i = 500\,\Omega$.

4.46 The transistor in the common-gate circuit in Figure P4.46 has the same parameters that are given in Problem 4.45. The output resistance $R_o$ is to

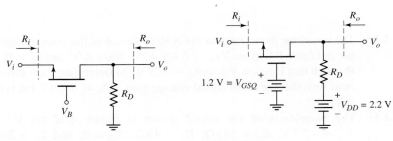

**Figure P4.45**      **Figure P4.46**

be 500 $\Omega$ and the drain-to-source quiescent voltage is to be $V_{DSQ} = V_{DS}(\text{sat}) + 0.3$ V. (a) What is the value of $R_D$? (b) What is the quiescent drain current $I_{DQ}$? (c) Find the input resistance $R_i$. (d) Determine the small-signal voltage gain $A_v = V_o/V_i$.

4.47 The small-signal parameters of the NMOS transistor in the ac equivalent common-gate circuit shown in Figure P4.47 are $V_{TN} = 0.4$ V, $k'_n = 100\ \mu\text{A/V}^2$, $W/L = 80$, and $\lambda = 0$. The quiescent drain current is $I_{DQ} = 0.5$ mA. Determine the small-signal voltage gain and the input resistance.

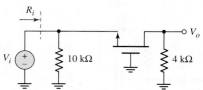

**Figure P4.47**

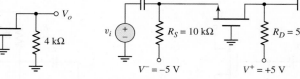

**Figure P4.48**

4.48 For the common-gate circuit in Figure P4.48, the NMOS transistor parameters are: $V_{TN} = 1$ V, $K_n = 3$ mA/V$^2$, and $\lambda = 0$. (a) Determine $I_{DQ}$ and $V_{DSQ}$. (b) Calculate $g_m$ and $r_o$. (c) Find the small-signal voltage gain $A_v = v_o/v_i$.

4.49 Consider the PMOS common-gate circuit in Figure P4.49. The transistor parameters are: $V_{TP} = -1$ V, $K_p = 0.5$ mA/V$^2$, and $\lambda = 0$. (a) Determine $R_S$ and $R_D$ such that $I_{DQ} = 0.75$ mA and $V_{SDQ} = 6$ V. (b) Determine the input impedance $R_i$ and the output impedance $R_o$. (c) Determine the load current $i_o$ and the output voltage $v_o$, if $i_i = 5 \sin \omega t\ \mu$A.

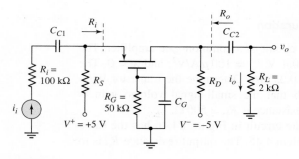

**Figure P4.49**

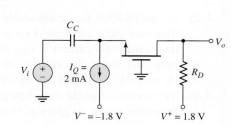

**Figure P4.50**

4.50 The transistor parameters of the NMOS device in the common-gate amplifier in Figure P4.50 are $V_{TN} = 0.4$ V, $k'_n = 100\ \mu\text{A/V}^2$, and $\lambda = 0$. (a) Find $R_D$ such that $V_{DSQ} = V_{DS}$ (sat) $+ 0.25$ V. (b) Determine the transistor $W/L$ ratio such that the small-signal voltage gain is $A_v = 6$. (c) What is the value of $V_{GSQ}$?

4.51 The parameters of the circuit shown in Figure 4.32 are $V^+ = 3.3$ V, $V^- = -3.3$ V, $R_G = 50$ k$\Omega$, $R_L = 4$ k$\Omega$, $R_{Si} = 0$, and $I_Q = 2$ mA. The transistor parameters are $V_{TN} = 0.6$ V, $K_n = 4$ mA/V$^2$, and $\lambda = 0$. (a) Find

$R_D$ such that $V_{DSQ} = 3.5$ V. (b) Determine the small-signal parameters $g_m$ and $R_i$. (c) Find the small-signal voltage gain $A_v$.

4.52 For the common-gate amplifier in Figure 4.35 in the text, the PMOS transistor parameters are $V_{TP} = -0.8$ V, $K_p = 2.5$ mA/V$^2$, and $\lambda = 0$. The circuit parameters are $V^+ = 3.3$ V, $V^- = -3.3$, $R_G = 100$ kΩ, and $R_L = 4$ kΩ. (a) Determine $R_S$ and $R_D$ such that $I_{DQ} = 1.2$ mA and $V_{SDQ} = 3$ V. (b) Determine the small-signal voltage gain $A_v = v_o/v_i$.

## Section 4.7 Amplifiers with MOSFET Load Devices

4.53 Consider the NMOS amplifier with saturated load in Figure 4.39(a). The transistor parameters are $V_{TND} = V_{TNL} = 0.6$ V, $k_n' = 100\,\mu$A/V$^2$, $\lambda = 0$, and $(W/L)_L = 1$. Let $V_{DD} = 3.3$ V. (a) Design the circuit such that the small-signal voltage gain is $|A_v| = 5$ and the $Q$-point is in the center of the saturation region. (b) Determine $I_{DQ}$ and $V_{DSDQ}$.

4.54 For the NMOS amplifier with depletion load in Figure 4.43(a), the transistor parameters are $V_{TND} = 0.6$ V, $V_{TNL} = -0.8$ V, $K_{nD} = 1.2$ mA/V$^2$, $K_{nL} = 0.2$ mA/V$^2$, and $\lambda_D = \lambda_L = 0.02$ V$^{-1}$. Let $V_{DD} = 5$ V. (a) Determine the transistor voltages at the transition points A and B. (b) Find $V_{GSDQ}$ and $V_{DSDQ}$ such that the $Q$-point is in the middle of the saturation region. (c) Determine $I_{DQ}$. (d) Find the small-signal voltage gain.

4.55 Consider a saturated load device in which the gate and drain of an enhancement-mode MOSFET are connected together. The transistor drain current becomes zero when $V_{DS} = 0.6$ V. (a) At $V_{DS} = 1.5$ V, the drain current is 0.5 mA. Determine the small-signal resistance at this operating point. (b) What is the drain current and small-signal resistance at $V_{DS} = 3$ V?

4.56 The parameters of the transistors in the circuit in Figure P4.56 are $V_{TND} = -1$ V, $K_{nD} = 0.5$ mA/V$^2$ for transistor $M_D$, and $V_{TNL} = +1$V, $K_{nL} = 30\,\mu$A/V$^2$ for transistor $M_L$. Assume $\lambda = 0$ for both transistors. (a) Calculate the quiescent drain current $I_{DQ}$ and the dc value of the output voltage. (b) Determine the small-signal voltage gain $A_v = v_o/v_i$ about the $Q$-point.

4.57 A source-follower circuit with a saturated load is shown in Figure P4.57. The transistor parameters are $V_{TND} = 1$ V, $K_{nD} = 1$ mA/V$^2$ for $M_D$, and $V_{TNL} = 1$ V, $K_{nL} = 0.1$ mA/V$^2$ for $M_L$. Assume $\lambda = 0$ for both transistors. Let $V_{DD} = 9$ V. (a) Determine $V_{GG}$ such that the quiescent value of $v_{DSL}$ is 4 V. (b) Show that the small-signal open-circuit ($R_L = \infty$) voltage gain about this $Q$-point is given by $A_v = 1/[1 + \sqrt{K_{nL}/K_{nD}}]$. (c) Calculate the small-signal voltage gain for $R_L = 4$ kΩ.

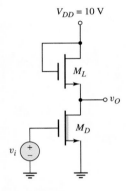

Figure P4.56

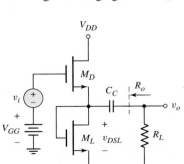

Figure P4.57

4.58    For the source-follower circuit with a saturated load as shown in Figure P4.57, assume the same transistor parameters as given in Problem 4.57. (a) Determine the small-signal voltage gain if $R_L = 10\ \text{k}\Omega$. (b) Determine the small-signal output resistance $R_o$.

4.59    The transistor parameters for the common-source circuit in Figure P4.59 are $V_{TND} = 0.4\ \text{V}$, $V_{TPL} = -0.4\ \text{V}$, $(W/L)_L = 50$, $\lambda_D = 0.02\ \text{V}^{-1}$, $\lambda_L = 0.04\ \text{V}^{-1}$, $k'_n = 100\ \mu\text{A/V}^2$, and $k'_p = 40\ \mu\text{A/V}^2$. At the $Q$-point, $I_{DQ} = 0.5\ \text{mA}$. (a) Determine $(W/L)_D$ such that the small-signal voltage gain is $A_v = V_o/V_i = -40$. (b) What is the required value of $V_B$? (c) What is the value of $V_{GSDQ}$?

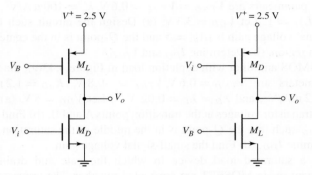

Figure P4.59                                Figure P4.60

4.60    Consider the circuit in Figure P4.60. The transistor parameters are $V_{TPD} = -0.6\ \text{V}$, $V_{TNL} = 0.4\ \text{V}$, $k'_n = 100\ \mu\text{A/V}^2$, $k'_p = 40\ \mu\text{A/V}^2$, $\lambda_L = 0.02\ \text{V}^{-1}$, $\lambda_D = 0.04\ \text{V}^{-1}$, and $(W/L)_L = 10$. (a) At the $Q$-point, the quiescent drain current is $I_{DQ} = 0.25\ \text{mA}$. (i) Determine $(W/L)_D$ such that the small-signal voltage gain is $A_v = V_o/V_i = -25$. (ii) What is the required value of $V_B$? (iii) What is the value of $V_{SGDQ}$? (b) Repeat part (a) for $I_{DQ} = 0.1\ \text{mA}$.

4.61    The ac equivalent circuit of a CMOS common-source amplifier is shown in Figure P4.61. The transistor parameters for $M_1$ are $V_{TN} = 0.5\ \text{V}$, $k'_n = 85\ \mu\text{A/V}^2$, $(W/L)_1 = 50$, and $\lambda = 0.05\ \text{V}^{-1}$, and for $M_2$ and $M_3$ are

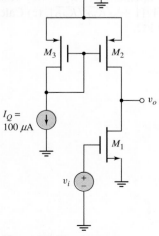

Figure P4.61

$V_{TP} = -0.5\,\text{V}$,   $k'_p = 40\,\mu\text{A/V}^2$,   $(W/L)_{2,3} = 50$,   and   $\lambda = 0.075\,\text{V}^{-1}$.
Determine the small-signal voltage gain.

4.62   Consider the ac equivalent circuit of a CMOS common-source amplifier shown in Figure P4.62. The parameters of the NMOS and PMOS transistors are the same as given in Problem 4.61. Determine the small-signal voltage gain.

4.63   The parameters of the transistors in the circuit in Figure P4.63 are $V_{TND} = V_{TNL} = 0.4\,\text{V}$,  $K_{nD} = 2\,\text{mA/V}^2$,  $K_{nL} = 0.5\,\text{mA/V}^2$,  and  $\lambda_D = \lambda_L = 0$.
(a) Plot $V_o$ versus $V_I$ over the range $0.8 \le V_I \le 2.5\,\text{V}$. (b) Plot $I_D$ versus $V_I$ over the same voltage range as part (a). (c) At $I_{DQ} = 0.20\,\text{mA}$, find the small-signal voltage gain $A_v = V_o/V_i = dV_O/dV_I$.

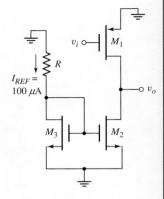

**Figure P4.62**

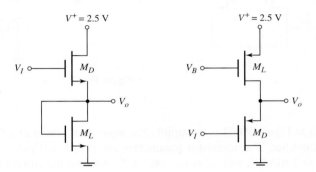

**Figure P4.63**          **Figure P4.64**

4.64   Consider the source-follower circuit in Figure P4.64. The transistor parameters are $V_{TP} = -0.4\,\text{V}$, $k'_p = 40\,\mu\text{A/V}^2$, $(W/L)_L = 5$, $(W/L)_D = 50$, and $\lambda_D = \lambda_L = 0.025\,\text{V}^{-1}$. Assume $V_B = 1\,\text{V}$. (a) What is the maximum value of $V_o$ such that $M_L$ remains biased in the saturation region? (b) For $M_L$ biased in the saturation region, determine $I_D$. (c) Using the results of parts (a) and (b), find $V_{SGD}$. (d) Determine the small-signal voltage gain when the dc value of $V_I = 0.2\,\text{V}$.

4.65   Figure P4.65 shows a common-gate amplifier. The transistor parameters are $V_{TN} = 0.6\,\text{V}$, $V_{TP} = -0.6\,\text{V}$, $K_n = 2\,\text{mA/V}^2$, $K_p = 0.5\,\text{mA/V}^2$, and $\lambda_n = \lambda_p = 0$. (a) Find the values of $V_{SGLQ}$, $V_{GSDQ}$, and $V_{DSDQ}$. (b) Derive the expression for the small-signal voltage gain in terms of $K_n$ and $K_p$. (c) Calculate the value of the small-signal voltage gain $A_v = V_o/V_i$.

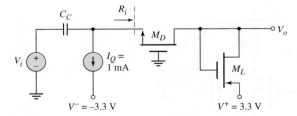

**Figure P4.65**

4.66   The ac equivalent circuit of a CMOS common-gate circuit is shown in Figure P4.66. The parameters of the NMOS and PMOS transistors are the same as given in Problem 4.61. Determine the (a) small-signal parameters of the transistors, (b) small-signal voltage gain $A_v = v_o/v_i$, (c) input resistance $R_i$, and (d) output resistance $R_o$.

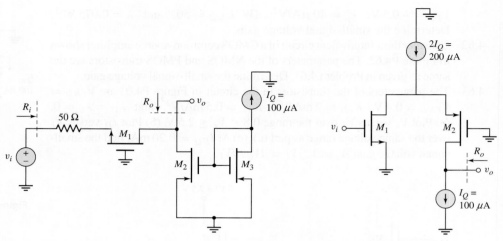

**Figure P4.66**

**Figure P4.67**

4.67 The circuit in Figure P4.67 is a simplified ac equivalent circuit of a folded-cascode amplifier. The transistor parameters are $|V_{TN}| = |V_{TP}| = 0.5$ V, $K_n = K_p = 2$ mA/V$^2$, and $\lambda_n = \lambda_p = 0.1$ V$^{-1}$. Assume the current source $2I_Q = 200$ $\mu$A is ideal and the resistance looking into the current source $I_Q = 100$ $\mu$A is 50 k$\Omega$. Determine the (a) small-signal parameters of each transistor, (b) small-signal voltage gain, and (c) output resistance $R_o$.

### Section 4.8 Multistage Amplifiers

4.68 The transistor parameters in the circuit in Figure P4.68 are $V_{TN1} = 0.6$ V, $V_{TP2} = -0.6$ V, $K_{n1} = 0.2$ mA/V$^2$, $K_{p2} = 1.0$ mA/V$^2$, and $\lambda_1 = \lambda_2 = 0$. The circuit parameters are $V_{DD} = 5$ V and $R_{in} = 400$ k$\Omega$. (a) Design the circuit such that $I_{DQ1} = 0.2$ mA, $I_{DQ2} = 0.5$ mA, $V_{DSQ1} = 2$ V, and $V_{SDQ2} = 3$ V. The voltage across $R_{S1}$ is to be 0.6 V. (b) Determine the small-signal voltage gain $A_v = v_o/v_i$.

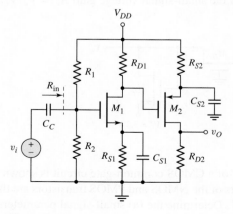

**Figure P4.68**

4.69    The transistor parameters in the circuit in Figure P4.68 are the same as those given in Problem 4.68. The circuit parameters are $V_{DD} = 3.3$ V, $R_{S1} = 1$ k$\Omega$, and $R_{in} = 250$ k$\Omega$. (a) Design the circuit such that $I_{DQ1} = 0.1$ mA, $I_{DQ2} = 0.25$ mA, $V_{DSQ1} = 1.2$ V, and $V_{SDQ2} = 1.8$ V. (b) Determine the small-signal voltage gain $A_v = v_o/v_i$.

4.70    Consider the circuit shown in Figure P4.70. The transistor parameters are $V_{TP1} = -0.4$ V, $V_{TN2} = 0.4$ V, $(W/L)_1 = 20$, $(W/L)_2 = 80$, $k'_p = 40$ $\mu$A/V$^2$, $k'_n = 100$ $\mu$A/V$^2$, and $\lambda_1 = \lambda_2 = 0$. Let $R_{in} = 200$ k$\Omega$. (a) Design the circuit such that $I_{DQ1} = 0.1$ mA, $I_{DQ2} = 0.3$ mA, $V_{SDQ1} = 1.0$ V, and $V_{DSQ2} = 2.0$ V. The voltage across $R_{S1}$ is to be 0.6 V. (b) Determine the small-signal voltage gain $A_v = v_o/v_i$. (c) Find the small-signal output resistance $R_o$.

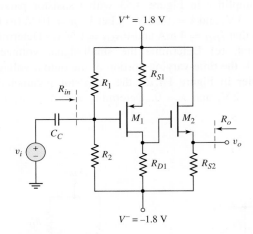

**Figure P4.70**

4.71    For the circuit in Figure P4.71, the transistor parameters are: $K_{n1} = K_{n2} = 4$ mA/V$^2$, $V_{TN1} = V_{TN2} = 2$ V, and $\lambda_1 = \lambda_2 = 0$. (a) Determine $I_{DQ1}$, $I_{DQ2}$, $V_{DSQ1}$, and $V_{DSQ2}$. (b) Determine $g_{m1}$ and $g_{m2}$. (c) Determine the overall small-signal voltage gain $A_v = v_o/v_i$.

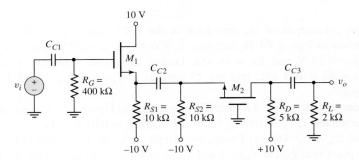

**Figure P4.71**

D4.72   For the cascode circuit in Figure 4.51 in the text, the transistor parameters are: $V_{TN1} = V_{TN2} = 1$ V, $K_{n1} = K_{n2} = 2$ mA/V$^2$, and $\lambda_1 = \lambda_2 = 0$. (a) Let $R_S = 1.2$ k$\Omega$ and $R_1 + R_2 + R_3 = 500$ k$\Omega$. Design the circuit such that $I_{DQ} = 3$ mA and $V_{DSQ1} = V_{DSQ2} = 2.5$ V. (b) Determine the small-signal voltage gain $A_v = v_o/v_i$.

D4.73   The supply voltages to the cascode circuit in Figure 4.51 in the text are changed to $V^+ = 10$ V and $V^- = -10$ V. The transistor parameters are: $K_{n1} = K_{n2} = 4$ mA/V$^2$, $V_{TN1} = V_{TN2} = 1.5$ V, and $\lambda_1 = \lambda_2 = 0$. (a) Let $R_S = 2$ k$\Omega$, and assume the current in the bias resistors is 0.1 mA. Design the circuit such that $I_{DQ} = 5$ mA and $V_{DSQ1} = V_{DSQ2} = 3.5$ V. (b) Determine the resulting small-signal voltage gain.

## Section 4.9 Basic JFET Amplifiers

4.74   Consider the JFET amplifier in Figure 4.53 with transistor parameters $I_{DSS} = 6$ mA, $V_P = -3$ V, and $\lambda = 0.01$ V$^{-1}$. Let $V_{DD} = 10$ V. (a) Determine $R_D$ and $V_{GS}$ such that $I_{DQ} = 4$ mA and $V_{DSQ} = 6$ V. (b) Determine $g_m$ and $r_o$ at the Q-point. (c) Determine the small-signal voltage gain $A_v = v_o/v_i$ where $v_o$ is the time-varying portion of the output voltage $v_O$.

4.75   For the JFET amplifier in Figure P4.75, the transistor parameters are: $I_{DSS} = 2$ mA, $V_P = -2$ V, and $\lambda = 0$. Determine $g_m$, $A_v = v_o/v_i$, and $A_i = i_o/i_i$.

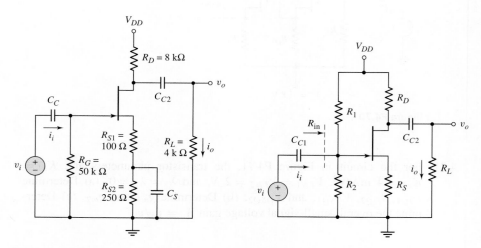

**Figure P4.75**                    **Figure P4.76**

D4.76   The parameters of the transistor in the JFET common-source amplifier shown in Figure P4.76 are: $I_{DSS} = 8$ mA, $V_P = -4.2$ V, and $\lambda = 0$. Let $V_{DD} = 20$ V and $R_L = 16$ k$\Omega$. Design the circuit such that $V_S = 2$ V, $R_1 + R_2 = 100$ k$\Omega$, and the Q-point is at $I_{DQ} = I_{DSS}/2$ and $V_{DSQ} = V_{DD}/2$.

*D4.77   Consider the source-follower JFET amplifier in Figure P4.77 with transistor parameters $I_{DSS} = 10$ mA, $V_P = -5$ V, and $\lambda = 0.01$ V$^{-1}$. Let $V_{DD} = 12$ V and $R_L = 0.5$ k$\Omega$. (a) Design the circuit such that $R_{in} = 100$ k$\Omega$, and the Q-point is at $I_{DQ} = I_{DSS}/2$ and $V_{DSQ} = V_{DD}/2$. (b) Determine the resulting small-signal voltage gain $A_v = v_o/v_i$ and the output resistance $R_o$.

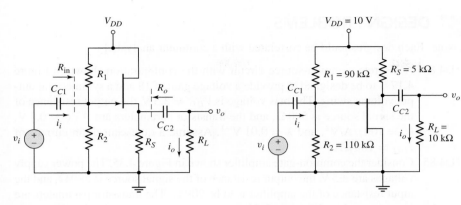

Figure P4.77                                    Figure P4.78

4.78 For the p-channel JFET source-follower circuit in Figure P4.78, the transistor parameters are: $I_{DSS} = 2$ mA, $V_P = +1.75$ V, and $\lambda = 0$. (a) Determine $I_{DQ}$ and $V_{SDQ}$. (b) Determine the small-signal gains $A_v = v_o/v_i$ and $A_i = i_o/i_i$. (c) Determine the maximum symmetrical swing in the output voltage.

D4.79 The p-channel JFET common-source amplifier in Figure P4.79 has transistor parameters $I_{DSS} = 8$ mA, $V_P = 4$ V, and $\lambda = 0$. Design the circuit such that $I_{DQ} = 4$ mA, $V_{SDQ} = 7.5$ V, $A_v = v_o/v_i = -3$, and $R_1 + R_2 = 400$ kΩ.

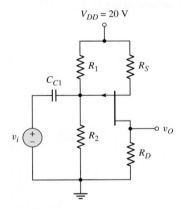

Figure P4.79

## COMPUTER SIMULATION PROBLEMS

4.80 Consider the common-source circuit described in Example 4.5. (a) Using a computer simulation, verify the results obtained in Example 4.5. (b) Determine the change in the results when the body effect is taken into account.

4.81 Using a computer simulation, verify the results of Example 4.7 for the source-follower amplifier.

4.82 Using a computer simulation, verify the results of Example 4.10 for the common-gate amplifier.

4.83 Using a computer simulation, verify the results of Example 4.17 for the cascode amplifier.

**DESIGN PROBLEMS**

[Note: Each design should be correlated with a computer analysis.]

*D4.84 A discrete common-source circuit with the configuration shown in Figure 4.17 is to be designed to provide a voltage gain of 18 and a symmetrical output voltage swing. The bias voltage is $V_{DD} = 3.3$ V, the output resistance of the signal source is 500 $\Omega$, and the transistor parameters are: $V_{TN} = 0.4$ V, $k'_n = 100\,\mu A/V^2$, and $\lambda = 0.01$ V$^{-1}$. Assume a quiescent drain current of $I_{DQ} = 100\,\mu A$.

*D4.85 Consider the common-gate amplifier shown in Figure 4.35. The power supply voltages are $\pm 5$ V, the output resistance of the signal source is 500 $\Omega$, and the input resistance of the amplifier is to be 200 $\Omega$. The transistor parameters are $k'_p = 40\,\mu A/V^2$, $V_{TP} = -0.6$ V, and $\lambda = 0$. The output load resistance is $R_L = 10$ k$\Omega$. Design the circuit such that the output voltage has a peak-to-peak symmetrical swing of at least 4 V.

*D4.86 A source-follower amplifier with the configuration shown in Figure 4.31 is to be designed. The power supplies are to be $\pm 12$ V. The transistor parameters are $V_{TN} = 1.2$ V, $k'_n = 100\,\mu A/V^2$, and $\lambda = 0$. The load resistance is $R_L = 200$ $\Omega$. Design the circuit, as well as a constant-current source, to deliver 250 mW of signal power to the load.

*D4.87 Consider the multitransistor circuit in Figure 4.49. Assume transistor parameters of $V_{TN} = 0.6$ V, $k'_n = 100\,\mu A/V^2$, and $\lambda = 0$. Design the transistors such that the small-signal voltage gain of the first stage is $A_{v1} = -10$ and the small-signal voltage gain of the second stage is $A_{v2} = 0.9$.

# The Bipolar Junction Transistor

In Chapter 2, we saw that the rectifying current–voltage characteristics of the diode are useful in electronic switching and waveshaping circuits. However, diodes are not capable of amplifying currents or voltages. As was shown in Chapter 4, the electronic device that is capable of current and voltage amplification, or gain, in conjunction with other circuit elements, is the transistor. The development of the transistor by Bardeen, Brattain, and Schockley at Bell Telephone Laboratories in the late 1940s started the first electronics revolution of the 1950s and 1960s. This invention led to the development of the first integrated circuit in 1958 and to the transistor operational amplifier (op-amp), which is one of the most widely used electronic circuits.

The bipolar transistor, which is introduced in this chapter, is one of the two major types of transistors. The second type of transistor, the field-effect transistor (FET), was introduced in Chapter 3. These two device types are the basis of modern microelectronics. Each device type is equally important and each has particular advantages for specific applications.

## PREVIEW

In this chapter, we will:

- Discuss the physical structure and operation of the bipolar junction transistor.
- Understand and become familiar with the dc analysis and design techniques of bipolar transistor circuits.
- Examine three basic applications of bipolar transistor circuits.
- Investigate various dc biasing schemes of bipolar transistor circuits, including integrated circuit biasing.
- Consider the dc biasing of multistage or multi-transistor circuits.
- As an application, incorporate the bipolar transistor in a circuit design that enhances the simple diode electronic thermometer discussed in Chapter 1.

## 5.1    BASIC BIPOLAR JUNCTION TRANSISTOR

**Objective:** • Understand the physical structure, operation, and characteristics of the bipolar junction transistors (BJT), including the npn and pnp devices.

The **bipolar junction transistor (BJT)** has three separately doped regions and contains two pn junctions. A single pn junction has two modes of operation—forward bias and reverse bias. The bipolar transistor, with two pn junctions, therefore has four possible modes of operation, depending on the bias condition of each pn junction, which is one reason for the versatility of the device. With three separately doped regions, the bipolar transistor is a three-terminal device. The basic transistor principle is that *the voltage between two terminals controls the current through the third terminal.*

Our discussion of the bipolar transistor starts with a description of the basic transistor structure and a qualitative description of its operation. To describe its operation, we use the pn junction concepts presented in Chapter 1. However, the two pn junctions are sufficiently close together to be called interacting pn junctions. The operation of the transistor is therefore totally different from that of two back-to-back diodes.

Current in the transistor is due to the flow of both electrons and holes, hence the name **bipolar.** Our discussion covers the relationship between the three terminal currents. In addition, we present the circuit symbols and conventions used in bipolar circuits, the bipolar transistor current–voltage characteristics, and finally, some non-ideal current–voltage characteristics.

### 5.1.1    Transistor Structures

Figure 5.1 shows simplified block diagrams of the basic structure of the two types of bipolar transistor: npn and pnp. The **npn bipolar transistor** contains a thin p-region between two n-regions. In contrast, the **pnp bipolar transistor** contains a thin n-region sandwiched between two p-regions. The three regions and their terminal connections are called the **emitter, base,** and **collector.**[1] The operation of the device depends on the two pn junctions being in close proximity, so the width of the base must be very narrow, normally in the range of tenths of a micrometer ($10^{-6}$ m).

The actual structure of the bipolar transistor is considerably more complicated than the block diagrams of Figure 5.1. For example, Figure 5.2 is the cross section of

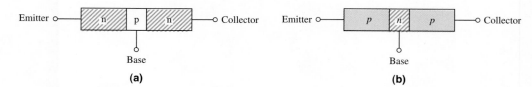

**Figure 5.1**  Simple geometry of bipolar transistors: (a) npn and (b) pnp

---

[1]The reason for the names **emitter** and **collector** for the terminals will become obvious as we go through the operation of the transistor. The term **base** refers to the structure of the original transistor.

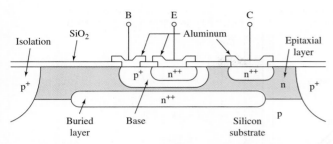

**Figure 5.2** Cross section of a conventional integrated circuit npn bipolar transistor

a classic npn bipolar transistor fabricated in an integrated circuit. One important point is that the device is not symmetrical electrically. This asymmetry occurs because the geometries of the emitter and collector regions are not the same, and the impurity doping concentrations in the three regions are substantially different. For example, the impurity doping concentrations in the emitter, base, and collector may be on the order of $10^{19}$, $10^{17}$, and $10^{15}$ cm$^{-3}$, respectively. Therefore, even though both ends are either p-type or n-type on a given transistor, switching the two ends makes the device act in drastically different ways.

Although the block diagrams in Figure 5.1 are highly simplified, they are still useful for presenting the basic transistor characteristics.

### 5.1.2 npn Transistor: Forward-Active Mode Operation

Since the transistor has two pn junctions, four possible bias combinations may be applied to the device, depending on whether a forward or reverse bias is applied to each junction. For example, if the transistor is used as an amplifying device, the **base–emitter (B–E) junction** is forward biased and the **base–collector (B–C) junction** is reverse biased, in a configuration called the **forward-active operating mode,** or simply the **active region.** The reason for this bias combination will be illustrated as we look at the operation of such transistors and the characteristics of circuits that use them.

**Transistor Currents**
Figure 5.3 shows an idealized npn bipolar transistor biased in the forward-active mode. Since the B–E junction is forward biased, electrons from the emitter are injected across

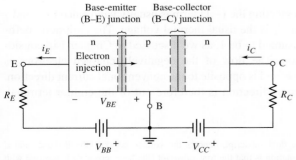

**Figure 5.3** An npn bipolar transistor biased in the forward-active mode; base–emitter junction forward biased and base–collector junction reverse biased

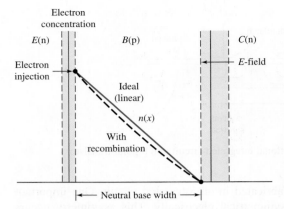

**Figure 5.4** Minority carrier electron concentration across the base region of an npn bipolar transistor biased in the forward-active mode. Minority carrier concentration is a linear function versus distance for an ideal transistor (no carrier recombination), and is a nonlinear function versus distance for a real device (with carrier recombination).

the B–E junction into the base, creating an excess minority carrier concentration in the base. Since the B–C junction is reverse biased, the electron concentration at the edge of that junction is approximately zero.

The base region is very narrow so that, in the ideal case, the injected electrons will not recombine with any of the majority carrier holes in the base. In this case, the electron distribution versus distance through the base is a straight line as shown in Figure 5.4. Because of the large gradient in this concentration, electrons that are injected, or *emitted,* from the emitter region diffuse across the base, are swept across the base–collector space-charge region by the electric field, and are *collected* in the collector region creating the collector current. However, if some carrier recombination does occur in the base, the electron concentration will deviate from the ideal linear curve, as shown in the figure. To minimize recombination effects, the width of the neutral base region must be small compared to the minority carrier diffusion length.

**Emitter Current:** Since the B–E junction is forward biased, we expect the current through this junction to be an exponential function of B–E voltage, just as we saw that the current through a pn junction diode was an exponential function of the forward-biased diode voltage. We can then write the current at the emitter terminal as

$$i_E = I_{EO}[e^{v_{BE}/V_T} - 1] \cong I_{EO}e^{v_{BE}/V_T} \tag{5.1}$$

where the approximation of neglecting the $(-1)$ term is usually valid since $v_{BE} \gg V_T$ in most cases.[2] The parameter $V_T$ is the usual thermal voltage. The emission coefficient $n$ that multiplies $V_T$ is assumed to be 1, as we discussed in Chapter 1 in considering the ideal diode equation. The flow of the negatively charged electrons is through the emitter into the base and is opposite to the conventional current direction. The conventional emitter current direction is therefore out of the emitter terminal.

---

[2]The voltage notation $v_{BE}$, with the dual subscript, denotes the voltage between the $B$ (base) and $E$ (emitter) terminals. Implicit in the notation is that the first subscript (the base terminal) is positive with respect to the second subscript (the emitter terminal).

We will assume that the ideality factor $n$ in this diode equation is unity (see Chapter 1).

The multiplying constant, $I_{EO}$, contains electrical parameters of the junction, but in addition is directly proportional to the active B–E cross-sectional area. Therefore, if two transistors are identical except that one has twice the area of the other, then the emitter currents will differ by a factor of two for the same applied B–E voltage. Typical values of $I_{EO}$ are in the range of $10^{-12}$ to $10^{-16}$ A, but may, for special transistors, vary outside of this range.

**Collector Current:** Since the doping concentration in the emitter is much larger than that in the base region, the vast majority of emitter current is due to the injection of electrons into the base. The number of these injected electrons reaching the collector is the major component of collector current.

The number of electrons reaching the collector per unit time is proportional to the number of electrons injected into the base, which in turn is a function of the B–E voltage. To a first approximation, the collector current is proportional to $e^{v_{BE}/V_T}$ and is independent of the reverse-biased B–C voltage. The device therefore looks like a **constant-current source.** The collector current is controlled by the B–E voltage; in other words, the current at one terminal (the collector) is controlled by the voltage across the other two terminals. *This control is the basic transistor action.*

We can write the collector current as

$$i_C = I_S e^{v_{BE}/V_T} \tag{5.2}$$

The collector current is slightly smaller than the emitter current, as we will show. The emitter and collector currents are related by $i_C = \alpha i_E$. We can also relate the coefficients by $I_S = \alpha I_{EO}$. The parameter $\alpha$ is called the **common-base current gain** whose value is always slightly less than unity. The reason for this name will become clearer as we proceed through the chapter.

**Base Current:**  Since the B–E junction is forward biased, holes from the base are injected across the B–E junction into the emitter. However, because these holes do not contribute to the collector current, they are not part of the transistor action. Instead, the flow of holes forms one component of the base current. This component is also an exponential function of the B–E voltage, because of the forward-biased B–E junction. We can write

$$i_{B1} \propto e^{v_{BE}/V_T} \tag{5.3(a)}$$

A few electrons recombine with majority carrier holes in the base. The holes that are lost must be replaced through the base terminal. The flow of such holes is a second component of the base current. This "recombination current" is directly proportional to the number of electrons being injected from the emitter, which in turn is an exponential function of the B–E voltage. We can write

$$i_{B2} \propto e^{v_{BE}/V_T} \tag{5.3(b)}$$

The total base current is the sum of the two components from Equations (5.3(a)) and (5.3(b)):

$$i_B \propto e^{v_{BE}/V_T} \tag{5.4}$$

Figure 5.5 shows the flow of electrons and holes in an npn bipolar transistor, as well as the terminal currents.[3] (Reminder: the conventional current direction is the

---

[3]A more thorough study of the physics of the bipolar transistor shows that there are other current components, in addition to the ones mentioned. However, these additional currents do not change the basic properties of the transistor and can be neglected for our purposes.

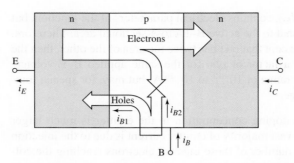

**Figure 5.5** Electron and hole currents in an npn bipolar transistor biased in the forward-active mode. Emitter, base, and collector currents are proportional to $e^{v_{BE}/V_T}$.

same as the flow of positively charged holes and opposite to the flow of negatively charged electrons.)

If the concentration of electrons in the n-type emitter is much larger than the concentration of holes in the p-type base, then the number of electrons injected into the base will be much larger than the number of holes injected into the emitter. This means that the $i_{B1}$ component of the base current will be much smaller than the collector current. In addition, if the base width is small, then the number of electrons that recombine in the base will be small, and the $i_{B2}$ component of the base current will also be much smaller than the collector current.

### Common-Emitter Current Gain

In the transistor, the rate of flow of electrons and the resulting collector current are an exponential function of the B–E voltage, as is the resulting base current. This means that the collector current and the base current are linearly related. Therefore, we can write

$$\frac{i_C}{i_B} = \beta \tag{5.5}$$

or

$$i_B = I_{BO}e^{v_{BE}/V_T} = \frac{i_C}{\beta} = \frac{I_S}{\beta}e^{v_{BE}/V_T} \tag{5.6}$$

The parameter $\beta$ is the **common-emitter current gain**[4] and is a key parameter of the bipolar transistor. In this idealized situation, $\beta$ is considered to be a constant for any given transistor. The value of $\beta$ is usually in the range of $50 < \beta < 300$, but it can be smaller or larger for special devices.

The value of $\beta$ is highly dependent upon transistor fabrication techniques and process tolerances. Therefore, the value of $\beta$ varies between transistor types and also between transistors of a given type, such as the discrete 2N2222. In any example or problem, we generally assume that $\beta$ is a constant. However, it is important to realize that $\beta$ can and does vary.

Figure 5.6 shows an npn bipolar transistor in a circuit. Because the emitter is the common connection, this circuit is referred to as a **common-emitter configuration.** When the transistor is biased in the forward-active mode, the B–E junction is forward

---

[4]Since we are considering the case of a transistor biased in the forward-active mode, the common–base current gain and common-emitter current gain parameters are often denoted as $\alpha_F$ and $\beta_F$, respectively. For ease of notation, we will simply define these parameters as $\alpha$ and $\beta$.

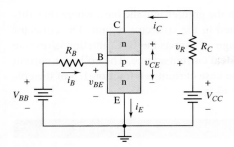

**Figure 5.6** An npn transistor circuit in the common-emitter configuration. Shown are the current directions and voltage polarities for the transistor biased in the forward-active mode.

biased and the B–C junction is reverse biased. Using the piecewise linear model of a pn junction, we assume that the B–E voltage is equal to $V_{BE}(\text{on})$, the junction turn-on voltage. Since $V_{CC} = v_{CE} + i_C R_C$, the power supply voltage must be sufficiently large to keep the B–C junction reverse biased. The base current is established by $V_{BB}$ and $R_B$, and the resulting collector current is $i_C = \beta i_B$.

If we set $V_{BB} = 0$, the B–E junction will have zero applied volts; therefore, $i_B = 0$, which implies that $i_C = 0$. This condition is called **cutoff.**

### Current Relationships

If we treat the bipolar transistor as a single node, then, by Kirchhoff's current law, we have

$$i_E = i_C + i_B \qquad (5.7)$$

If the transistor is biased in the forward-active mode, then

$$i_C = \beta i_B \qquad (5.8)$$

Substituting Equation (5.8) into (5.7), we obtain the following relationship between the emitter and base currents:

$$i_E = (1 + \beta)i_B \qquad (5.9)$$

Solving for $i_B$ in Equation (5.8) and substituting into Equation (5.9), we obtain a relationship between the collector and emitter currents, as follows:

$$i_C = \left(\frac{\beta}{1+\beta}\right)i_E \qquad (5.10)$$

We can write $i_C = \alpha i_E$ so

$$\alpha = \frac{\beta}{1+\beta} \qquad (5.11)$$

The parameter $\alpha$ is called the common-base current gain and is always slightly less than 1. We may note that if $\beta = 100$, then $\alpha = 0.99$, so $\alpha$ is indeed close to 1. From Equation (5.11), we can state the common-emitter current gain in terms of the common-base current gain:

$$\beta = \frac{\alpha}{1-\alpha} \qquad (5.12)$$

### Summary of Transistor Operation

We have presented a first-order model of the operation of the npn bipolar transistor biased in the forward-active region. The forward-biased B–E voltage, $v_{BE}$, causes an exponentially related flow of electrons from the emitter into the base where they diffuse across the base region and are collected in the collector region. The collector current, $i_C$, is independent of the B–C voltage as long as the B–C junction is reverse biased. The collector, then, behaves as an ideal current source. The collector current is a fraction $\alpha$ of the emitter current, and the base current is a fraction $1/\beta$ of the collector current. If $\beta \gg 1$, then $\alpha \cong 1$ and $i_C \cong i_E$.

## EXAMPLE 5.1

**Objective:** Calculate the collector and emitter currents, given the base current and current gain.

Assume a common-emitter current gain of $\beta = 150$ and a base current of $i_B = 15\ \mu\text{A}$. Also assume that the transistor is biased in the forward-active mode.

**Solution:** The relation between collector and base currents gives

$$i_C = \beta i_B = (150)(15\ \mu\text{A}) \Rightarrow 2.25\ \text{mA}$$

and the relation between emitter and base currents yields

$$i_E = (1 + \beta)i_B = (151)(15\ \mu\text{A}) \Rightarrow 2.27\ \text{mA}$$

From Equation (5.11), the common-base current gain is

$$\alpha = \frac{\beta}{1+\beta} = \frac{150}{151} = 0.9934$$

**Comment:** For reasonable values of $\beta$, the collector and emitter currents are nearly equal, and the common-base current gain is nearly 1.

## EXERCISE PROBLEM

**Ex 5.1:** An npn transistor is biased in the forward-active mode. The base current is $I_B = 8.50\ \mu\text{A}$ and the emitter current is $I_E = 1.20\ \text{mA}$. Determine $\beta$, $\alpha$, and $I_C$. (Ans. $\beta = 140.2$, $\alpha = 0.9929$, $I_C = 1.1915\ \text{mA}$)

### 5.1.3 pnp Transistor: Forward-Active Mode Operation

We have discussed the basic operation of the npn bipolar transistor. The complementary device is the pnp transistor. Figure 5.7 shows the flow of holes and electrons in a pnp device biased in the forward-active mode. Since the B–E junction is forward biased, the p-type emitter is positive with respect to the n-type base, holes flow from the emitter into the base, the holes diffuse across the base, and they are swept into the collector. The collector current is a result of this flow of holes.

Again, since the B–E junction is forward biased, the emitter current is an exponential function of the B–E voltage. Noting the direction of emitter current and the polarity of the foward-biased B–E voltage, we can write

$$i_E = I_{EO}e^{v_{EB}/V_T} \tag{5.13}$$

where $v_{EB}$ is the voltage between the emitter and base, and now implies that the emitter is positive with respect to the base. We are again assuming the $-1$ term in the ideal diode equation is negligible.

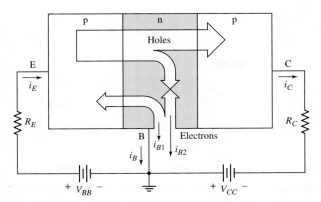

**Figure 5.7** Electron and hole currents in a pnp bipolar transistor biased in the forward-active mode. Emitter, base, and collector currents are proportional to $e^{v_{EB}/V_T}$.

The collector current is an exponential function of the E–B voltage, and the direction is out of the collector terminal, which is opposite to that in the npn device. We can now write

$$i_C = \alpha i_E = I_S e^{v_{EB}/V_T} \tag{5.14}$$

where $\alpha$ is again the common-base current gain.

The base current in a pnp device is the sum of two components. The first component, $i_{B1}$, comes from electrons flowing from the base into the emitter as a result of the forward-biased E–B junction. We can then write $i_{B1} \propto \exp(v_{EB}/V_T)$. The second component, $i_{B2}$, comes from the flow of electrons supplied through the base terminal to replace those lost by recombination with the minority carrier holes injected into the base from the emitter. This component is proportional to the number of holes injected into the base, so $i_{B2} \propto \exp(v_{EB}/V_T)$. Therefore the total base current is $i_B = i_{B1} + i_{B2} \propto \exp(v_{EB}/V_T)$. The direction of the base current is out of the base terminal. Since the total base current in the pnp device is an exponential function of the E–B voltage, we can write

$$i_B = I_{BO} e^{v_{EB}/V_T} = \frac{i_C}{\beta} = \frac{I_S}{\beta} e^{v_{EB}/V_T} \tag{5.15}$$

The parameter $\beta$ is also the common-emitter current gain of the pnp bipolar transistor.

The relationships between the terminal currents of the pnp transistor are exactly the same as those of the npn transistor and are summarized in Table 5.1 in the next section. Also the relationships between $\beta$ and $\alpha$ are the same as given in Equations (5.11) and (5.12).

### 5.1.4    Circuit Symbols and Conventions

The block diagram and conventional circuit symbol of an npn bipolar transistor are shown in Figures 5.8(a) and 5.8(b). The arrowhead in the circuit symbol is always placed on the emitter terminal, and it indicates the direction of the emitter current. For the npn device, this direction is out of the emitter. The simplified block diagram and conventional circuit symbol of a pnp bipolar transistor are shown in Figures 5.9(a) and 5.9(b). Here, the arrowhead on the emitter terminal indicates that the direction of the emitter current is into the emitter.

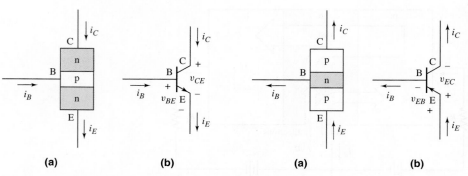

**Figure 5.8** npn bipolar transistor: (a) simple block diagram and (b) circuit symbol. Arrow is on the emitter terminal and indicates the direction of emitter current (out of emitter terminal for the npn device).

**Figure 5.9** pnp bipolar transistor: (a) simple block diagram and (b) circuit symbol. Arrow is on the emitter terminal and indicates the direction of emitter current (into emitter terminal for the pnp device).

| Table 5.1 | Summary of the bipolar current–voltage relationships in the active region |
|---|---|
| **npn** | **pnp** |
| $i_C = I_S e^{v_{BE}/V_T}$ | $i_C = I_S e^{v_{EB}/V_T}$ |
| $i_E = \frac{i_C}{\alpha} = \frac{I_S}{\alpha} e^{v_{BE}/V_T}$ | $i_E = \frac{i_C}{\alpha} = \frac{I_S}{\alpha} e^{v_{EB}/V_T}$ |
| $i_B = \frac{i_C}{\beta} = \frac{I_S}{\beta} e^{v_{BE}/V_T}$ | $i_B = \frac{i_C}{\beta} = \frac{I_S}{\beta} e^{v_{EB}/V_T}$ |
| **For both transistors** | |
| $i_E = i_C + i_B$ | $i_C = \beta i_B$ |
| $i_E = (1 + \beta) i_B$ | $i_C = \alpha i_E = \left(\frac{\beta}{1+\beta}\right) i_E$ |
| $\alpha = \frac{\beta}{1+\beta}$ | $\beta = \frac{\alpha}{1-\alpha}$ |

Referring to the circuit symbols given for the npn (Figure 5.8(b)) and pnp (Figure 5.9(b)) transistors showing current directions and voltage polarities, we can summarize the current–voltage relationships as given in Table 5.1.

Figure 5.10(a) shows a common-emitter circuit with an npn transistor. The figure includes the transistor currents, and the base-emitter (B–E) and collector–emitter

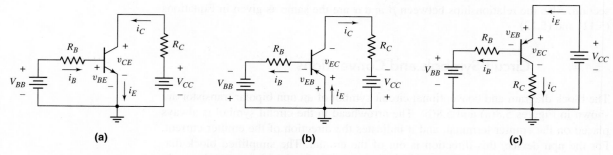

**Figure 5.10** Common-emitter circuits: (a) with an npn transistor, (b) with a pnp transistor, and (c) with a pnp transistor biased with a positive voltage source

(C–E) voltages. Figure 5.10(b) shows a common-emitter circuit with a pnp bipolar transistor. Note the different current directions and voltage polarities in the two circuits. A more usual circuit configuration using the pnp transistor is shown in Figure 5.10(c). This circuit allows positive voltage supplies to be used.

## Test Your Understanding

**TYU 5.1** (a) The common-emitter current gains of two transistors are $\beta = 60$ and $\beta = 150$. Determine the corresponding common-base current gains. (b) The common-base current gains of two transistors are $\alpha = 0.9820$ and $\alpha = 0.9925$. Determine the corresponding common-emitter current gains. (Ans. (a) $\alpha = 0.9836$, $\alpha = 0.9934$; (b) $\beta = 54.6$, $\beta = 132.3$)

**TYU 5.2** An npn transistor is biased in the forward-active mode. The base current is $I_B = 5.0\,\mu\text{A}$ and the collector current is $I_C = 0.62\,\text{mA}$. Determine $I_E$, $\beta$, and $\alpha$. (Ans. $I_E = 0.625\,\text{mA}$, $\beta = 124$, and $\alpha = 0.992$)

**TYU 5.3** The emitter current in a pnp transistor biased in the forward-active mode is $I_E = 1.20\,\text{mA}$. The common-base current gain of the transistor is $\alpha = 0.9915$. Determine $\beta$, $I_B$, and $I_C$. (Ans. $\beta = 117$, $I_B = 10.2\,\mu\text{A}$, $I_C = 1.19\,\text{mA}$)

---

### 5.1.5    Current–Voltage Characteristics

Figures 5.11(a) and 5.11(b) are **common-base circuit configurations** for an npn and a pnp bipolar transistor, respectively. The current sources provide the emitter current. Previously, we stated that the collector current $i_C$ was nearly independent of the C–B voltage as long as the B–C junction was reverse biased. When the B–C junction becomes forward biased, the transistor is no longer in the forward-active mode, and the collector and emitter currents are no longer related by $i_C = \alpha i_E$.

Figure 5.12 shows the typical common-base current–voltage characteristics. When the collector–base junction is reverse biased, then for constant values of emitter

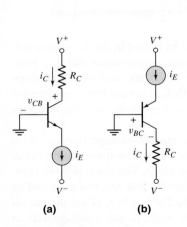

**Figure 5.11** Common-base circuit configuration with constant current source biasing: (a) an npn transistor and (b) a pnp transistor

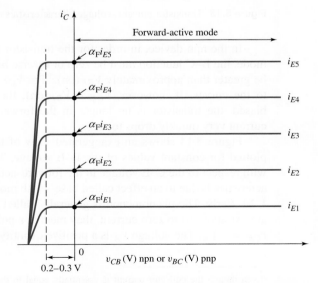

**Figure 5.12** Transistor current–voltage characteristics of the common-base circuit

current, the collector current is nearly equal to $i_E$. These characteristics show that the common-base device is nearly an ideal constant-current source.

The C–B voltage can be varied by changing the $V^+$ voltage (Figure 5.11(a)) or the $V^-$ voltage (Figure 5.11(b)). When the collector–base junction becomes forward biased in the range of 0.2 and 0.3 V, the collector current $i_C$ is still essentially equal to the emitter current $i_E$. In this case, the transistor is still basically biased in the forward-active mode. However, as the forward-bias C–B voltage increases, the linear relationship between the collector and emitter currents is no longer valid, and the collector current very quickly drops to zero.

The common-emitter circuit configuration provides a slightly different set of current–voltage characteristics, as shown in Figure 5.13. For these curves, the collector current is plotted against the collector–emitter voltage, for various constant values of the base current. These curves are generated from the common-emitter circuits shown in Figure 5.10. In this circuit, the $V_{BB}$ source forward biases the B–E junction and controls the base current $i_B$. The C–E voltage can be varied by changing $V_{CC}$.

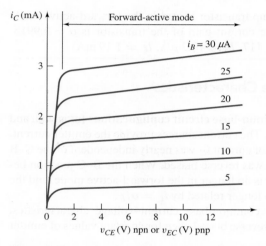

**Figure 5.13** Transistor current–voltage characteristics of the common-emitter circuit

In the npn device, in order for the transistor to be biased in the forward-active mode, the B–C junction must be zero or reverse biased, which means that $V_{CE}$ must be greater than approximately $V_{BE}$(on).[5] For $V_{CE} > V_{BE}$(on), there is a finite slope to the curves. If, however, $V_{CE} < V_{BE}$(on), the B–C junction becomes forward biased, the transistor is no longer in the forward-active mode, and the collector current very quickly drops to zero.

Figure 5.14 shows an exaggerated view of the current–voltage characteristics plotted for constant values of the B–E voltage. The curves are theoretically linear with respect to the C–E voltage in the forward-active mode. The slope in these characteristics is due to an effect called base-width modulation that was first analyzed by J. M. Early. The phenomenon is generally called the *Early effect*. When the curves are extrapolated to zero current, they meet at a point on the negative voltage axis, at $v_{CE} = -V_A$. The voltage $V_A$ is a positive quantity called the **Early voltage**. Typical

---

[5]Even though the collector current is essentially equal to the emitter current when the B–C junction becomes slightly forward biased, as was shown in Figure 5.12, the transistor is said to be biased in the forward-active mode when the B–C junction is zero or reverse biased.

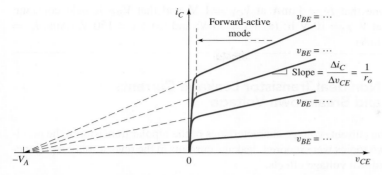

**Figure 5.14** Current-voltage characteristics for the common-emitter circuit, showing the Early voltage and the finite output resistance, $r_o$, of the transistor

values of $V_A$ are in the range $50 < V_A < 300$ V. For a pnp transistor, this same effect is true except the voltage axis is $v_{EC}$.

For a given value of $v_{BE}$ in an npn transistor, if $v_{CE}$ increases, the reverse-bias voltage on the collector–base junction increases, which means that the width of the B–C space-charge region also increases. This in turn reduces the neutral base width $W$ (see Figure 5.4). A decrease in the base width causes the gradient in the minority carrier concentration to increase, which increases the diffusion current through the base. The collector current then increases as the C–E voltage increases.

The linear dependence of $i_C$ versus $v_{CE}$ in the forward-active mode can be described by

$$i_C = I_S(e^{v_{BE}/V_T}) \cdot \left(1 + \frac{v_{CE}}{V_A}\right) \tag{5.16}$$

where $I_S$ is assumed to be constant.

In Figure 5.14, the nonzero slope of the curves indicates that the **output resistance** $r_o$ looking into the collector is finite. This output resistance is determined from

$$\frac{1}{r_o} = \left.\frac{\partial i_C}{\partial v_{CE}}\right|_{v_{BE}=\text{const.}} \tag{5.17}$$

Using Equation (5.16), we can show that

$$r_o \cong \frac{V_A}{I_C} \tag{5.18}$$

where $I_C$ is the quiescent collector current when $v_{BE}$ is a constant and $v_{CE}$ is small compared to $V_A$.

In most cases, the dependence of $i_C$ on $v_{CE}$ is not critical in the dc analysis or design of transistor circuits. However, the finite output resistance $r_o$ may significantly affect the amplifier characteristics of such circuits. This effect is examined more closely in Chapter 6 of this text.

## Test Your Understanding

**TYU 5.4** The output resistance of a bipolar transistor is $r_o = 225$ k$\Omega$ at $I_C = 0.8$ mA. (a) Determine the Early voltage. (b) Using the results of part (a), find $r_o$ at (i) $I_C = 0.08$ mA and (ii) $I_C = 8$ mA. (Ans. (a) $V_A = 180$ V; (b) (i) $r_o = 2.25$ M$\Omega$, (ii) $r_o = 22.5$ k$\Omega$)

**TYU 5.5** Assume that $I_C = 1$ mA at $V_{CE} = 1$ V, and that $V_{BE}$ is held constant. Determine $I_C$ at $V_{CE} = 10$ V if: (a) $V_A = 75$ V; and (b) $V_A = 150$ V. (Ans. $I_C = 1.12$ mA, 1.06 mA)

### 5.1.6     Nonideal Transistor Leakage Currents and Breakdown Voltage

In discussing the current–voltage characteristics of the bipolar transistor in the previous sections, two topics were ignored: leakage currents in the reverse-biased pn junctions and breakdown voltage effects.

**Leakage Currents**
In the common-base circuits in Figure 5.11, if we set the current source $i_E = 0$, transistors will be cut off, but the B–C junctions will still be reverse biased. A reverse-bias leakage current exists in these junctions, and this current corresponds to the reverse-bias saturation current in a diode, as described in Chapter 1. The direction of these reverse-bias leakage currents is the same as that of the collector currents. The term $I_{CBO}$ is the collector leakage current in the common-base configuration, and is the collector-base leakage current when the emitter is an open circuit. This leakage current is shown in Figure 5.15(a).

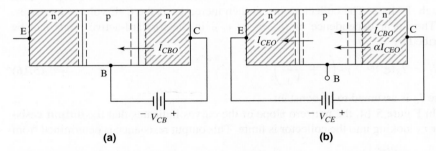

**(a)**                                    **(b)**

**Figure 5.15** Block diagram of an npn transistor in an (a) open-emitter configuration showing the junction leakage current $I_{CBO}$ and (b) open-base configuration showing the leakage current $I_{CEO}$

Another leakage current can exist between the emitter and collector with the base terminal an open circuit. Figure 5.15(b) is a block diagram of an npn transistor in which the base is an open circuit ($i_B = 0$). The current component $I_{CBO}$ is the normal leakage current in the reverse-biased B–C pn junction. This current component causes the base potential to increase, which forward biases the B–E junction and induces the B–E current $I_{CEO}$. The current component $\alpha I_{CEO}$ is the normal collector current resulting from the emitter current $I_{CEO}$. We can write

$$I_{CEO} = \alpha I_{CEO} + I_{CBO} \tag{5.19(a)}$$

or

$$I_{CEO} = \frac{I_{CBO}}{1 - \alpha} \cong \beta I_{CBO} \tag{5.19(b)}$$

This relationship indicates that the open-base configuration produces different characteristics than the open-emitter configuration.

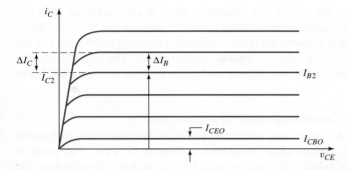

**Figure 5.16** Transistor current–voltage characteristics for the common-emitter circuit including leakage currents. The dc beta and ac beta for the transistor can be determined from this set of characteristics. The Early voltage for this set of characteristics is assumed to be $V_A = \infty$.

When the transistor is biased in the forward-active mode, the various leakage currents still exist. Common-emitter current–voltage characteristics are shown in Figure 5.16, in which the leakage current has been included. A dc beta or dc common-emitter current gain can be defined, for example, as

$$\beta_{\text{dc}} = \frac{I_{C2}}{I_{B2}} \qquad (5.20)$$

where the collector current $I_{C2}$ includes the leakage current as shown in the figure. An ac $\beta$ is defined as

$$\beta_{\text{ac}} = \frac{\Delta I_C}{\Delta I_{B|V_{CE}=\text{const.}}} \qquad (5.21)$$

This definition of beta excludes the leakage current as shown in the figure.

If the leakage currents are negligible, the two values of beta are equal. We will assume in the remainder of this text that the leakage currents can be neglected and beta can simply be denoted as $\beta$ as previously defined.

### Breakdown Voltage: Common-Base Characteristics

The common-base current–voltage characteristics shown in Figure 5.12 are ideal in that breakdown is not shown. Figure 5.17 shows the same $i_C$ versus $v_{CB}$ characteristics with the breakdown voltage.

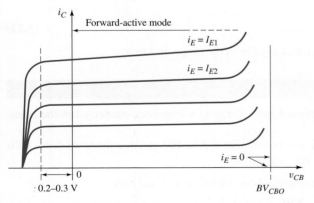

**Figure 5.17** The $i_C$ versus $v_{CB}$ common-base characteristics, showing the collector–base junction breakdown

Consider the curve for $i_E = 0$ (the emitter terminal is effectively an open circuit). The collector–base junction breakdown voltage is indicated as $BV_{CBO}$. This is a simplified figure in that it shows breakdown occurring abruptly at $BV_{CBO}$. For the curves in which $i_E > 0$, breakdown actually begins earlier. The carriers flowing across the junction initiate the breakdown avalanche process at somewhat lower voltages.

### Breakdown Voltage: Common-Emitter Characteristics

Figure 5.18 shows the $i_C$ versus $v_{CE}$ characteristics of an npn transistor, for various constant base currents, and an ideal breakdown voltage of $BV_{CEO}$. The value of $BV_{CEO}$ is less than the value of $BV_{CBO}$ because $BV_{CEO}$ includes the effects of the transistor action, while $BV_{CBO}$ does not. This same effect was observed in the $I_{CEO}$ leakage current.

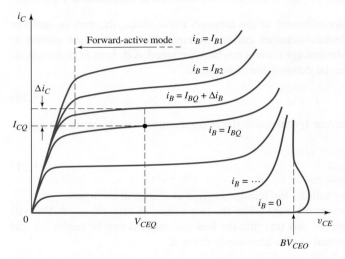

**Figure 5.18** Common-emitter characteristics showing breakdown effects

The breakdown voltage characteristics for the two configurations are also different. The breakdown voltage for the open-base case is given by

$$BV_{CEO} = \frac{BV_{CBO}}{\sqrt[n]{\beta}}$$

(5.22)

where $n$ is an empirical constant usually in the range of 3 to 6.

---

### EXAMPLE 5.2

**Objective:** Calculate the breakdown voltage of a transistor connected in the open-base configuration.

Assume that the transistor current gain is $\beta = 100$ and that the breakdown voltage of the B–C junction is $BV_{CBO} = 120$ V.

**Solution:** If we assume an empirical constant of $n = 3$, we have

$$BV_{CEO} = \frac{BV_{CBO}}{\sqrt[n]{\beta}} = \frac{120}{\sqrt[3]{100}} = 25.9 \text{ V}$$

**Comment:** The breakdown voltage of the open-base configuration is substantially less than that of the C–B junction. This represents a worst-case condition, which must be considered in any circuit design.

**Design Pointer:** The designer must be aware of the breakdown voltage of the specific transistors used in a circuit, since this will be a limiting factor in the size of the dc bias voltages that can be used.

---

### EXERCISE PROBLEM

**Ex 5.2:** The open-emitter breakdown voltage is $BV_{CBO} = 200$ V, the current gain is $\beta = 120$, and the empirical constant is $n = 3$. Determine $BV_{CEO}$. (Ans. 40.5 V)

---

Breakdown may also occur in the B–E junction if a reverse-bias voltage is applied to that junction. The junction breakdown voltage decreases as the doping concentrations increase. Since the emitter doping concentration is usually substantially larger than the doping concentration in the collector, the B–E junction breakdown voltage is normally much smaller than that of the B–C junction. Typical B–E junction breakdown voltage values are in the range of 6 to 8 V.

## Test Your Understanding

**TYU 5.6** A particular transistor circuit requires a minimum open-base breakdown voltage of $BV_{CEO} = 30$ V. If $\beta = 100$ and $n = 3$, determine the minimum required value of $BV_{CBO}$. (Ans. 139 V)

---

## 5.2    DC ANALYSIS OF TRANSISTOR CIRCUITS

**Objective:** • Understand and become familiar with the dc analysis and design techniques of bipolar transistor circuits.

We've considered the basic transistor characteristics and properties. We can now start analyzing and designing the dc biasing of bipolar transistor circuits. A primary purpose of the rest of the chapter is to become familiar and comfortable with the bipolar transistor and transistor circuits. The dc biasing of transistors, the focus of this chapter, is an important part of designing bipolar amplifiers, the focus of the next chapter.

The piecewise linear model of a pn junction can be used for the dc analysis of bipolar transistor circuits. We will first analyze the common-emitter circuit and introduce the load line for that circuit. We will then look at the dc analysis of other bipolar transistor circuit configurations. Since a transistor in a linear amplifier must be biased in the forward-active mode, we emphasize, in this section, the analysis and design of circuits in which the transistor is biased in this mode.

**Common-Emitter Circuit**

One of the basic transistor circuit configurations is called the **common-emitter circuit.** Figure 5.19(a) shows one example of a common-emitter circuit. The emitter terminal is obviously at ground potential. This circuit configuration will appear in many amplifiers that will be considered in Chapter 6.

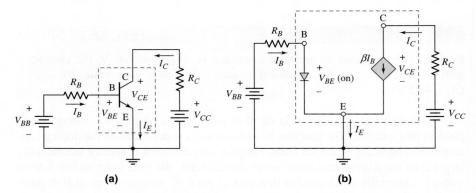

**Figure 5.19** (a) Common-emitter circuit with npn transistor and (b) dc equivalent circuit. Transistor equivalent circuit is shown within the dotted lines with piecewise linear transistor parameters.

Figure 5.19(a) shows a common-emitter circuit with an npn transistor, and Figure 5.19(b) shows the dc equivalent circuit. We will assume that the B–E junction is forward biased, so the voltage drop across that junction is the cut-in or turn-on voltage $V_{BE}$(on). When the transistor is biased in the forward-active mode, the collector current is represented as a dependent current source that is a function of the base current. We are neglecting the reverse-biased junction leakage current and the Early effect in this case. In the following circuits, we will be considering dc currents and voltages, so the dc notation for these parameters will be used.

The base current is

$$I_B = \frac{V_{BB} - V_{BE}(\text{on})}{R_B}$$

(5.23)

Implicit in Equation (5.23) is that $V_{BB} > V_{BE}$(on), which means that $I_B > 0$. When $V_{BB} < V_{BE}$(on), the transistor is cut off and $I_B = 0$.

In the collector–emitter portion of the circuit, we can write

$$I_C = \beta I_B$$

(5.24)

and

$$V_{CC} = I_C R_C + V_{CE}$$

(5.25(a))

or

$$V_{CE} = V_{CC} - I_C R_C$$

(5.25(b))

In Equation (5.25(b)), we are also implicitly assuming that $V_{CE} > V_{BE}$(on), which means that the B–C junction is reverse biased and the transistor is biased in the forward-active mode.

Considering Figure 5.19(b), we can see that the power dissipated in the transistor is given by

$$P_T = I_B V_{BE}(\text{on}) + I_C V_{CE} \qquad (5.26(a))$$

In most cases, $I_C \gg I_B$ and $V_{CE} > V_{BE}(\text{on})$ so that a good first approximation of the power dissipated is given as

$$P_T \cong I_C V_{CE} \qquad (5.26(b))$$

The principal condition where this approximation is not valid is for a transistor biased in the saturation mode (discussed later).

## EXAMPLE 5.3

**Objective:** Calculate the base, collector, and emitter currents and the C–E voltage for a common-emitter circuit. Calculate the transistor power dissipation.

For the circuit shown in Figure 5.19(a), the parameters are: $V_{BB} = 4$ V, $R_B = 220$ k$\Omega$, $R_C = 2$ k$\Omega$, $V_{CC} = 10$ V, $V_{BE}(\text{on}) = 0.7$ V, and $\beta = 200$. Figure 5.20(a) shows the circuit without explicitly showing the voltage sources.

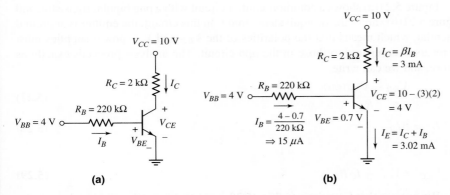

**Figure 5.20** Circuit for Example 5.3: (a) circuit and (b) circuit showing current and voltage values

**Solution:** Referring to Figure 5.20(b), the base current is found as

$$I_B = \frac{V_{BB} - V_{BE}(\text{on})}{R_B} = \frac{4 - 0.7}{220} \Rightarrow 15 \ \mu\text{A}$$

The collector current is

$$I_C = \beta I_B = (200)(15 \ \mu\text{A}) \Rightarrow 3 \ \text{mA}$$

and the emitter current is

$$I_E = (1 + \beta) \cdot I_B = (201)(15\mu\text{A}) \Rightarrow 3.02 \ \text{mA}$$

From Equation (5.25(b)), the collector-emitter voltage is

$$V_{CE} = V_{CC} - I_C R_C = 10 - (3)(2) = 4 \ \text{V}$$

The power dissipated in the transistor is found to be

$$P_T = I_B V_{BE}(\text{on}) + I_C V_{CE} = (0.015)(0.7) + (3)(4) \cong I_C V_{CE}$$

or

$$P_T \cong 12 \text{ mW}$$

**Comment:** Since $V_{BB} > V_{BE}(\text{on})$ and $V_{CE} > V_{BE}(\text{on})$, the transistor is indeed biased in the forward-active mode. As a note, in an actual circuit, the voltage across a B–E junction may not be exactly 0.7 V, as we have assumed using the piecewise linear approximation. This may lead to slight inaccuracies between the calculated currents and voltages and the measured values. Also note that, if we take the difference between $I_E$ and $I_C$, which is the base current, we obtain $I_B = 20 \ \mu\text{A}$ rather than 15 $\mu$A. The difference is the result of roundoff error in the emitter current.

## EXERCISE PROBLEM

**Ex 5.3:** The circuit elements in Figure 5.20(a) are changed to $V_{CC} = 3.3$ V, $V_{BB} = 2$ V, $R_C = 3.2 \text{ k}\Omega$, and $R_B = 430 \text{ k}\Omega$. The transistor parameters are $\beta = 150$ and $V_{BE} = 0.7$ V. Calculate $I_B$, $I_C$, $V_{CE}$, and the power dissipated in the transistor. (Ans. $I_B = 3.02 \ \mu\text{A}$, $I_C = 0.453$ mA, $V_{CE} = 1.85$ V, $P = 0.838$ mW)

Figure 5.21(a) shows a common-emitter circuit with a pnp bipolar transistor, and Figure 5.21(b) shows the dc equivalent circuit. In this circuit, the emitter is at ground potential, which means that the polarities of the $V_{BB}$ and $V_{CC}$ power supplies must be reversed compared to those in the npn circuit. The analysis proceeds exactly as before, and we can write

$$I_B = \frac{V_{BB} - V_{EB}(\text{on})}{R_B} \tag{5.27}$$

$$I_C = \beta I_B \tag{5.28}$$

and

$$V_{EC} = V_{CC} - I_C R_C \tag{5.29}$$

We can see that Equations (5.27), (5.28), and (5.29) for the pnp bipolar transistor in the common-emitter configuration are exactly the same as Equations (5.23), (5.24), and (5.25(b)) for the npn bipolar transistor in a similar circuit, if we properly define the current directions and voltage polarities.

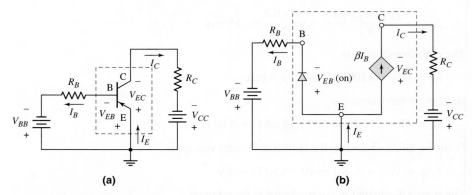

**(a)**                                    **(b)**

**Figure 5.21** (a) Common-emitter circuit with pnp transistor and (b) dc equivalent circuit. Transistor equivalent circuit is shown within the dotted lines with piecewise linear transistor parameters.

In many cases, the pnp bipolar transistor will be reconfigured in a circuit so that positive voltage sources, rather than negative ones, can be used. We see this in the following example.

## EXAMPLE 5.4

**Objective:** Analyze the common-emitter circuit with a pnp transistor.

For the circuit shown in Figure 5.22(a), the parameters are: $V_{BB} = 1.5$ V, $R_B = 580$ kΩ, $V^+ = 5$ V, $V_{EB}(\text{on}) = 0.6$ V, and $\beta = 100$. Find $I_B$, $I_C$, $I_E$, and $R_C$ such that $V_{EC} = \left(\frac{1}{2}\right)V^+$.

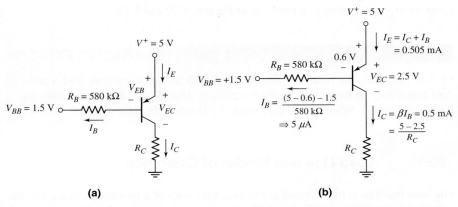

**Figure 5.22** Circuit for Example 5.4; (a) circuit and (b) circuit showing current and voltage values

**Solution:** Writing a Kirchhoff voltage law equation around the E–B loop, we find the base current to be

$$I_B = \frac{V^+ - V_{EB}(\text{on}) - V_{BB}}{R_B} = \frac{5 - 0.6 - 1.5}{580} \Rightarrow 5\mu A$$

The collector current is

$$I_C = \beta I_B = (100)(5\mu A) \Rightarrow 0.5\,\text{mA}$$

and the emitter current is

$$I_E = (1 + \beta)I_B = (101)(5\mu A) \Rightarrow 0.505\,\text{mA}$$

For a C–E voltage of $V_{EC} = \frac{1}{2}V^+ = 2.5$ V, $R_C$ is

$$R_C = \frac{V^+ - V_{EC}}{I_C} = \frac{5 - 2.5}{0.5} = 5\,\text{k}\Omega$$

**Comment:** In this case, the difference between $V^+$ and $V_{BB}$ is greater than the transistor turn-on voltage, or $(V^+ - V_{BB}) > V_{EB}(\text{on})$. Also, because $V_{EC} > V_{EB}(\text{on})$, the pnp bipolar transistor is biased in the forward-active mode.

**Discussion:** In this example, we used an emitter-base turn-on voltage of $V_{EB}(\text{on}) = 0.6$ V, whereas previously we used a value of 0.7 V. We must keep in mind that the turn-on voltage is an approximation and the actual base–emitter voltage will depend on the type of transistor used and the current level. In most situations, choosing a value of 0.6 V or 0.7 V will make only minor differences. However, most people tend to use the value of 0.7 V.

**Ex 5.4:** The circuit elements in Figure 5.22(a) are $V^+ = 3.3$ V, $V_{BB} = 1.2$ V, $R_B = 400$ k$\Omega$, and $R_C = 5.25$ k$\Omega$. The transistor parameters are $\beta = 80$ and $V_{EB}(\text{on}) = 0.7$ V. Determine $I_B$, $I_C$, and $V_{EC}$. (Ans. $I_B = 3.5\ \mu$A, $I_C = 0.28$ mA, $V_{EC} = 1.83$ V)

The dc equivalent circuits, such as those given in Figures 5.19(b) and 5.21(b), are useful initially in analyzing transistor circuits. From this point on, however, we will not explicitly draw the equivalent circuit. We will simply analyze the circuit using the transistor circuit symbols, as in Figures 5.20 and 5.22.

**PS 5.1:** (a) Verify the results of Example 5.3 with a PSpice analysis. Use a standard transistor. (b) Repeat the analysis for $R_B = 180$ k$\Omega$. (c) Repeat the analysis for $R_B = 260$ k$\Omega$. What can be said about $R_B$ limiting the base current?

### 5.2.2   Load Line and Modes of Operation

The load line can help us visualize the characteristics of a transistor circuit. For the common-emitter circuit in Figure 5.20(a), we can use a graphical technique for both the B–E and C–E portions of the circuit. Figure 5.23(a) shows the piecewise linear characteristics for the B–E junction and the input load line. The input load line is

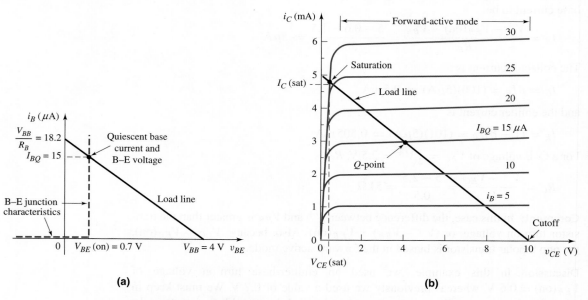

**(a)**                                                                 **(b)**

**Figure 5.23** (a) Base–emitter junction piecewise linear $i$–$v$ characteristics and the input load line, and (b) common-emitter transistor characteristics and the collector–emitter load line showing the $Q$-point for the circuit shown in Example 5.3 (Figure 5.20)

obtained from Kirchhoff's voltage law equation around the B–E loop, written as follows:

$$I_B = \frac{V_{BB}}{R_B} - \frac{V_{BE}}{R_B} \qquad\qquad \textbf{(5.30)}$$

Both the load line and the quiescent base current change as either or both $V_{BB}$ and $R_B$ change. The load line in Figure 5.23(a) is essentially the same as the load line characteristics for diode circuits, as shown in Chapter 1.

For the C–E portion of the circuit in Figure 5.20(a), the load line is found by writing Kirchhoff's voltage law equation around the C–E loop. We obtain

$$V_{CE} = V_{CC} - I_C R_C \qquad\qquad \textbf{(5.31(a))}$$

which can be written in the form

$$I_C = \frac{V_{CC}}{R_C} - \frac{V_{CE}}{R_C} = 5 - \frac{V_{CE}}{2}\,(\text{mA}) \qquad\qquad \textbf{(5.31(b))}$$

Equation (5.31(b)) is the load line equation, showing a linear relationship between the collector current and collector–emitter voltage. Since we are considering the dc analysis of the transistor circuit, this relationship represents the dc load line. The ac load line is presented in the next chapter.

Figure 5.23(b) shows the transistor characteristics for the transistor in Example 5.3, with the load line superimposed on the transistor characteristics. The two end points of the load line are found by setting $I_C = 0$, yielding $V_{CE} = V_{CC} = 10$ V, and by setting $V_{CE} = 0$, yielding $I_C = V_{CC}/R_C = 5$ mA.

The quiescent point, or $Q$-point, of the transistor is given by the dc collector current and the collector–emitter voltage. The $Q$-point is the intersection of the load line and the $I_C$ versus $V_{CE}$ curve corresponding to the appropriate base current. The $Q$-point also represents the simultaneous solution to two expressions. The load line is useful in visualizing the bias point of the transistor. In the figure, the $Q$-point shown is for the transistor in Example 5.3.

As previously stated, if the power supply voltage in the base circuit is smaller than the turn-on voltage, then $V_{BB} < V_{BE}(\text{on})$ and $I_B = I_C = 0$, and the transistor is in the cutoff mode. In this mode, all transistor currents are zero, neglecting leakage currents, and for the circuit shown in Figure 5.20(a), $V_{CE} = V_{CC} = 10$ V.

As $V_{BB}$ increases ($V_{BB} > V_{BE}(\text{on})$), the base current $I_B$ increases and the $Q$-point moves up the load line. As $I_B$ continues to increase, a point is reached where the collector current $I_C$ can no longer increase. At this point, the transistor is biased in the **saturation mode;** that is, the transistor is said to be in saturation. The B–C junction becomes forward biased, and the relationship between the collector and base currents is no longer linear. The transistor C–E voltage in saturation, $V_{CE}(\text{sat})$, is less than the B–E cut-in voltage. The forward-biased B–C voltage is always less than the forward-biased B–E voltage, so the C–E voltage in saturation is a small positive value. Typically, $V_{CE}(\text{sat})$ is in the range of 0.1 to 0.3 V.

## EXAMPLE 5.5

**Objective:** Calculate the currents and voltages in a circuit when the transistor is driven into saturation.

For the circuit shown in Figure 5.24, the transistor parameters are: $\beta = 100$, and $V_{BE}(\text{on}) = 0.7$ V. If the transistor is biased in saturation, assume $V_{CE}(\text{sat}) = 0.2$ V.

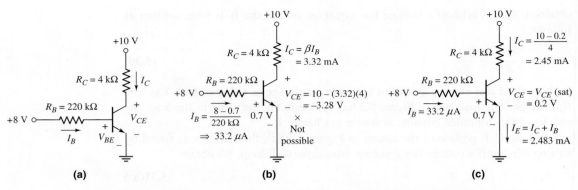

**Figure 5.24** Circuit for Example 5.5: (a) circuit; (b) circuit showing current and voltage values, assuming the transistor is biased in the forward-active mode (an incorrect assumption); and (c) circuit showing current and voltage values, assuming the transistor is biased in the saturation mode (correct assumption)

**Solution:** Since +8 V is applied to the input side of $R_B$, the base–emitter junction is certainly forward biased, so the transistor is turned on. The base current is

$$I_B = \frac{V_{BB} - V_{BE}(\text{on})}{R_B} = \frac{8 - 0.7}{220} \Rightarrow 33.2\ \mu\text{A}$$

If we first assume that the transistor is biased in the active region, then the collector current is

$$I_C = \beta I_B = (100)(33.2\ \mu\text{A}) \Rightarrow 3.32\ \text{mA}$$

The collector–emitter voltage is then

$$V_{CE} = V_{CC} - I_C R_C = 10 - (3.32)(4) = -3.28\ \text{V}$$

However, the collector–emitter voltage of the npn transistor in the common-emitter configuration shown in Figure 5.24(a) cannot be negative. Therefore, our initial assumption of the transistor being biased in the forward-active mode is incorrect. Instead, the transistor must be biased in saturation.

As given in the "objective" statement, set $V_{CE}(\text{sat}) = 0.2$ V. The collector current is

$$I_C = I_C(\text{sat}) = \frac{V_{CC} - V_{CE}(\text{sat})}{R_C} = \frac{10 - 0.2}{4} = 2.45\ \text{mA}$$

Assuming that the B–E voltage is still equal to $V_{BE}(\text{on}) = 0.7$ V, the base current is $I_B = 33.2\ \mu\text{A}$, as previously determined. If we take the ratio of collector current to base current, then

$$\frac{I_C}{I_B} = \frac{2.45}{0.0332} = 74 < \beta$$

The emitter current is

$$I_E = I_C + I_B = 2.45 + 0.033 = 2.48\ \text{mA}$$

The power dissipated in the transistor is found to be

$$P_T = I_B V_{BE}(\text{on}) + I_C V_{CE} = (0.0332)(0.7) + (2.45)(0.2)$$

or

$$P_T = 0.513\ \text{mW}$$

**Comment:** When a transistor is driven into saturation, we use $V_{CE}(\text{sat})$ as another piecewise linear parameter. In addition, when a transistor is biased in the saturation mode, we have $I_C < \beta I_B$. This condition is very often used to prove that a transistor is indeed biased in the saturation mode.

---

**EXERCISE PROBLEM**

**Ex 5.5:** Consider the pnp circuit in Figure 5.22(a). Assume transistor parameters of $V_{EB}(\text{on}) = 0.7$ V, $V_{EC}$ (sat) $= 0.2$ V, and $\beta = 110$. Assume circuit parameters of $V^+ = 3.3$ V, $R_C = 5$ k$\Omega$, and $R_B = 150$ k$\Omega$. Calculate $I_B$, $I_C$, and $V_{EC}$ for (a) $V_{BB} = 2$ V and (b) $V_{BB} = 1$ V. (Ans. (a) $I_B = 4\,\mu$A, $I_C = 0.44$ mA, $V_{EC} = 1.1$ V; (b) $I_B = 10.7\,\mu$A, $I_C = 0.62$ mA, $V_{EC} = 0.2$ V)

---

### Problem-Solving Technique: Bipolar DC Analysis

Analyzing the dc response of a bipolar transistor circuit requires knowing the mode of operation of the transistor. In some cases, the mode of operation may not be obvious, which means that we have to guess the state of the transistor, then analyze the circuit to determine if we have a solution consistent with our initial guess. To do this, we can:

1. Assume that the transistor is biased in the forward-active mode in which case $V_{BE} = V_{BE}(\text{on})$, $I_B > 0$, and $I_C = \beta I_B$.
2. Analyze the "linear" circuit with this assumption.
3. Evaluate the resulting state of the transistor. If the initial assumed parameter values and $V_{CE} > V_{CE}(\text{sat})$ are true, then the initial assumption is correct. However, if the calculation shows $I_B < 0$, then the transistor is probably cut off, and if the calculation shows $V_{CE} < 0$, the transistor is likely biased in saturation.
4. If the initial assumption is proven incorrect, then a new assumption must be made and the new "linear" circuit must be analyzed. Step 3 must then be repeated.

Because it is not always clear whether a transistor is biased in the forward-active or saturation mode, we may initially have to make an educated guess as to the state of the transistor and then verify our initial assumption. This is similar to the process we used for the analysis of multidiode circuits. For instance, in Example 5.5, we assumed a forward-active mode, performed the analysis, and showed that $V_{CE} < 0$. However, a negative $V_{CE}$ for an npn transistor in the common-emitter configuration is not possible. Therefore, our initial assumption was disproved, and the transistor was biased in the saturation mode. Using the results of Example 5.5, we also see that when a transistor is in saturation, the ratio of $I_C$ to $I_B$ is always less than $\beta$, or

$$I_C / I_B < \beta$$

This condition is true for both the npn and the pnp transistor biased in the saturation mode. When a bipolar transistor is biased in saturation, we may define

$$\frac{I_C}{I_B} \equiv \beta_{\text{Forced}} \tag{5.32}$$

where $\beta_{\text{Forced}}$ is called the "forced beta." We then have that $\beta_{\text{Forced}} < \beta$.

Another mode of operation for a bipolar transistor is the **inverse-active mode.** In this mode, the B–E junction is reverse biased and the B–C junction is forward biased. In effect, the transistor is operating "upside down"; that is, the emitter is acting as the collector and the collector is operating as the emitter. We will postpone discussions on this operating mode until we discuss digital electronic circuits later in this text.

To summarize, the four modes of operation for an npn transistor are shown in Figure 5.25. The four possible combinations of B–E and B–C voltages determine the modes of operation. If $v_{BE} > 0$ (forward-biased junction) and $v_{BC} < 0$ (reverse-biased junction), the transistor is biased in the forward-active mode. If both junctions are zero or reverse biased, the transistor is in cutoff. If both junctions are forward biased, the transistor is in saturation. If the B–E junction is reverse biased and the B–C junction is forward biased, the transistor is in the inverse-active mode.

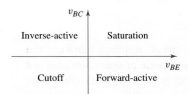

**Figure 5.25**  Bias conditions for the four modes of operation of an npn transistor

The piecewise linear parameter model of the transistor that we have used in the dc analysis of transistor circuits is adequate for many applications. Another transistor model is known as the **Ebers–Moll model.** This model can be used to describe the transistor in each of its possible operating modes and is used in the SPICE computer simulation program. However, we will not consider the Ebers–Moll model here.

## Test Your Understanding

In the following exercise problems, assume $V_{BE}(\text{on}) = 0.7$ V and $V_{CE}(\text{sat}) = 0.2$ V.

**TYU 5.7** For the circuit shown in Figure 5.26, assume $\beta = 50$. Determine $V_O$, $I_B$, and $I_C$ for: (a) $V_I = 0.2$ V, and (b) $V_I = 3.6$ V. Then, calculate the power dissipated in the transistor for the two conditions. (Ans. (a) $I_B = I_C = 0$, $V_O = 5$ V, $P = 0$; (b) $I_B = 4.53$ mA, $I_C = 10.9$ mA, $P = 5.35$ mW)

**TYU 5.8** For the circuit shown in Figure 5.26, let $\beta = 50$, and determine $V_I$ such that $V_{BC} = 0$. Calculate the power dissipated in the transistor. (Ans. $V_I = 0.825$ V, $P = 6.98$ mW)

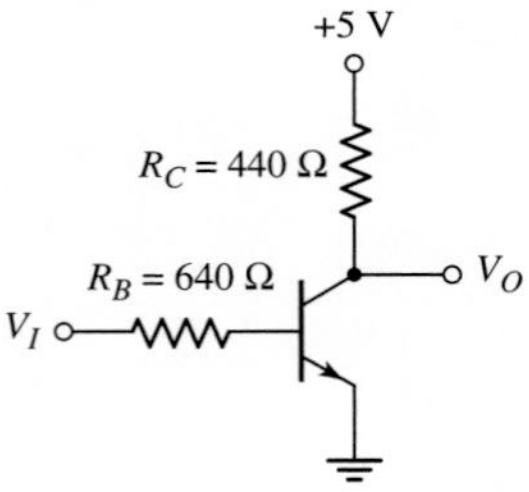

**Figure 5.26**  Figure for Exercise TYU 5.7 and TYU 5.8

+5 V

$R_C = 440\ \Omega$

$R_B = 640\ \Omega$

$V_I$ ○———

○ $V_O$

### 5.2.3    Voltage Transfer Characteristics

A plot of the voltage transfer characteristics (output voltage versus input voltage) can also be used to visualize the operation of a circuit or the state of a transistor. The following example considers both an npn and a pnp transistor circuit.

## EXAMPLE 5.6

**Objective:** Develop the voltage transfer curves for the circuits shown in Figures 5.27(a) and 5.27(b).

Assume npn transistor parameters of $V_{BE}(\text{on}) = 0.7$ V, $\beta = 120$, $V_{CE}(\text{sat}) = 0.2$ V, and $V_A = \infty$, and pnp transistor parameters of $V_{EB}(\text{on}) = 0.7$ V, $\beta = 80$, $V_{EC}(\text{sat}) = 0.2$ V, and $V_A = \infty$.

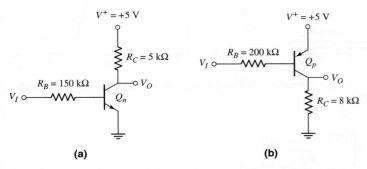

**Figure 5.27** Circuits for Example 5.6; (a) npn circuit and (b) pnp circuit

**Solution (npn Transistor Circuit):** For $V_I \leq 0.7$ V, the transistor $Q_n$ is cut off, so that $I_B = I_C = 0$. The output voltage is then $V_O = V^+ = 5$ V.

For $V_I > 0.7$ V, the transistor $Q_n$ turns on and is initially biased in the forward-active mode. We have

$$I_B = \frac{V_I - 0.7}{R_B}$$

and

$$I_C = \beta I_B = \frac{\beta(V_I - 0.7)}{R_B}$$

Then

$$V_O = 5 - I_C R_C = 5 - \frac{\beta(V_I - 0.7)R_C}{R_B}$$

This equation is valid for $0.2 \leq V_O \leq 5$ V. When $V_O = 0.2$ V, the transistor $Q_n$ goes into saturation. When $V_O = 0.2$ V, the input voltage is found from

$$0.2 = 5 - \frac{(120)(V_I - 0.7)(5)}{150}$$

which yields $V_I = 1.9$ V. For $V_I \geq 1.9$ V, the transistor $Q_n$ remains biased in the saturation region.

The voltage transfer curve is shown in Figure 5.28(a).

**Solution (pnp transistor circuit):** For $4.3 \leq V_I \leq 5$ V, the transistor $Q_p$ is cut off, so that $I_B = I_C = 0$. The output voltage is then $V_O = 0$.

For $V_I < 4.3$ V, the transistor $Q_p$ turns on and is biased in the forward-active mode. We have

$$I_B = \frac{(5 - 0.7) - V_I}{R_B}$$

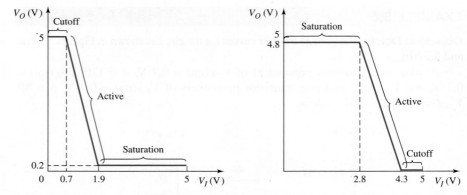

Figure 5.28  Voltage transfer characteristics for (a) npn circuit in Figure 5.27(a) and (b) pnp circuit in Figure 5.27(b)

and

$$I_C = \beta I_B = \beta \left[ \frac{(5 - 0.7) - V_I}{R_B} \right]$$

The output voltage is then

$$V_O = I_C R_C = \beta R_C \left[ \frac{(5 - 0.7) - V_I}{R_B} \right]$$

This equation is valid for $0 \le V_O \le 4.8$ V. When $V_O = 4.8$ V, the transistor $Q_p$ goes into saturation.

When $V_O = 4.8$ V, the input voltage is found from

$$4.8 = (80)(8) \left[ \frac{(5 - 0.7) - V_I}{200} \right]$$

which yields $V_I = 2.8$ V. For $V_I \le 2.8$ V, the transistor $Q_p$ remains biased in the saturation mode.

The voltage transfer curve is shown in Figure 5.28(b).

**Computer Simulation:**  Figure 5.29 shows the voltage transfer characteristics from a PSpice simulation using a standard 2N3904 transistor. One result that may be

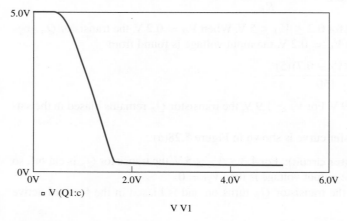

Figure 5.29  Voltage transfer characteristic for the circuit in Figure 5.27(a) generated by a PSpice simulation

observed from the computer simulation is that the output voltage in the forward-active mode is not exactly a linear function of input voltage as the hand analysis suggested. In addition, the base-emitter voltage when $v_I = 1.3$ V is $v_{BE} = 0.649$ V in the computer analysis results rather than the assumed value of 0.7 V in the hand analysis. However, the hand analysis gives a good first approximation.

**Comment:** As shown in this example, the voltage transfer characteristics are determined by finding the range of input voltage values that biases the transistor in cutoff, the forward-active mode, or the saturation mode.

### EXERCISE PROBLEM

**Ex 5.6:** The circuit elements in Figure 5.27(a) are changed to $R_B = 200$ k$\Omega$, $R_C = 4$ k$\Omega$, and $V^+ = 9$ V. The transistor parameters are $\beta = 100$, $V_{BE}(\text{on}) = 0.7$ V, and $V_{CE}(\text{sat}) = 0.2$ V. Plot the voltage transfer characteristics for $0 \leq V_I \leq 9$ V. (Ans. For $0 \leq V_I \leq 0.7$ V, $Q_n$ is cut off, $V_O = 9$ V; For $V_I \geq 5.1$ V, $Q_n$ is in saturation, $V_O = 0.2$ V)

### COMPUTER ANALYSIS EXERCISE

**PS 5.2:** Using a PSpice simulation, plot the voltage transfer characteristics of the circuit shown in Figure 5.27(b). Use a standard transistor. What is the value of $v_{EB}$ when the transistor is biased in the forward-active region?

### 5.2.4    Commonly Used Bipolar Circuits: dc Analysis

There are a number of other bipolar transistor circuit configurations, in addition to the common-emitter circuits shown in Figures 5.20 and 5.22, that are commonly used. Several examples of such circuits are presented in this section. BJT circuits tend to be very similar in terms of dc analysis procedures, so that the same basic analysis approach will work regardless of the appearance of the circuit. We continue our dc analysis and design of bipolar circuits to increase our proficiency and to become more comfortable with these types of circuits.

### EXAMPLE 5.7

**Objective:** Calculate the characteristics of a circuit containing an emitter resistor.
  For the circuit shown in Figure 5.30(a), let $V_{BE}(\text{on}) = 0.7$ V and $\beta = 75$. Note that the circuit has both positive and negative power supply voltages.

**Solution (Q-point values):** Writing Kirchhoff's voltage law equation around the B–E loop, we have

$$V_{BB} = I_B R_B + V_{BE}(\text{on}) + I_E R_E + V^- \tag{5.33}$$

Assuming the transistor is biased in the forward-active mode, we can write $I_E = (1 + \beta) I_B$. We can then solve Equation (5.33) for the base current:

$$I_B = \frac{V_{BB} - V_{BE}(\text{on}) - V^-}{R_B + (1 + \beta)R_E} = \frac{1 - 0.7 - (-1.8)}{560 + (76)(3)} \Rightarrow 2.665 \ \mu\text{A}$$

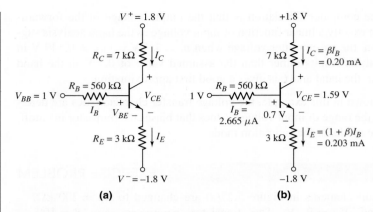

**Figure 5.30** Circuit for Example 5.7: (a) circuit and (b) circuit showing current and voltage values

The collector and emitter currents are

$$I_C = \beta I_B = (75)(2.665\,\mu A) \Rightarrow 0.20\,\text{mA}$$

and

$$I_E = (1 + \beta)\,I_B = (76)(2.665\,\mu A) \Rightarrow 0.203\,\text{mA}$$

From Figure 5.30(b), the collector–emitter voltage is

$$V_{CE} = V^+ - I_C R_C - I_E R_E - V^- = 1.8 - (0.20)(7) - (0.203)(3) - (-1.8)$$

or

$$V_{CE} = 1.59\,\text{V}$$

**Solution (load line):** We again use Kirchhoff's voltage law around the C–E loop. From the relationship between the collector and emitter currents, we find

$$V_{CE} = (V^+ - V^-) - I_C\left[R_C + \left(\frac{1+\beta}{\beta}\right)R_E\right]$$

$$= [1.8 - (-1.8)] - I_C\left[7 + \left(\frac{76}{75}\right)(3)\right]$$

or

$$V_{CE} = 3.6 - I_C(10.04)$$

The load line and the calculated $Q$-point are shown in Figure 5.31. A few transistor characteristics of $I_C$ versus $V_{CE}$ are superimposed on the figure.

**Comment:** Since the C–E voltage is 1.59 V, $V_{CE} > V_{BE}(\text{on})$ and the transistor is biased in the forward-active mode, as initially assumed. We will see, later in the chapter, the value of including an emitter resistor in a circuit.

EXERCISE PROBLEM

**Ex 5.7:** The parameters of the circuit shown in Figure 5.30(a) are changed to $V^+ = 3.3\,\text{V}$, $V^- = -3.3\,\text{V}$, $V_{BB} = 0$, $R_B = 640\,\text{k}\Omega$, $R_E = 2.4\,\text{k}\Omega$, and $R_C = 10\,\text{k}\Omega$. The transistor parameters are $\beta = 80$ and $V_{BE}(\text{on}) = 0.7\,\text{V}$. Calculate all transistor currents and $V_{CE}$. (Ans. $I_B = 3.116\,\mu A$, $I_C = 0.249\,\text{mA}$, $I_E = 0.252\,\text{mA}$, $V_{CE} = 3.51\,\text{V}$)

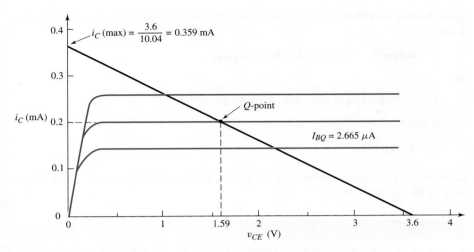

Figure 5.31  Load line and $Q$-point for the circuit shown in Figure 5.30 for Example 5.7

## DESIGN EXAMPLE 5.8

**Objective:** Design the common-base circuit shown in Figure 5.32 such that $I_{EQ} = 0.50$ mA and $V_{ECQ} = 4.0$ V.

Assume transistor parameters of $\beta = 120$ and $V_{EB}(\text{on}) = 0.7$ V.

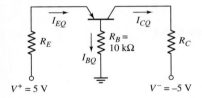

Figure 5.32  Common-base circuit for Example 5.8

**Solution:** Writing Kirchhoff's voltage law equation around the base–emitter loop (assuming the transistor is biased in the forward-active mode), we have

$$V^+ = I_{EQ}R_E + V_{EB}(\text{on}) + \left(\frac{I_{EQ}}{1+\beta}\right)R_B$$

or

$$5 = (0.5)R_E + 0.7 + \left(\frac{0.5}{121}\right)(10)$$

which yields

$$R_E = 8.52 \text{ k}\Omega$$

We can find

$$I_{CQ} = \left(\frac{\beta}{1+\beta}\right)I_{EQ} = \left(\frac{120}{121}\right)(0.5) = 0.496 \text{ mA}$$

Now, writing Kirchhoff's voltage law equation around the emitter–collector loop, we have

$$V^+ = I_{EQ}R_E + V_{ECQ} + I_{CQ}R_C + V^-$$

or

$$5 = (0.5)(8.52) + 4 + (0.496)R_C + (-5)$$

which yields

$$R_C = 3.51 \text{ k}\Omega$$

**Comment:** The circuit analysis of the common-base circuit proceeds in the same way as all previous circuits.

EXERCISE PROBLEM

**Ex 5.8:** Design the common-base circuit shown in Figure 5.33 such that $I_{EQ} = 0.125$ mA and $V_{ECQ} = 2.2$ V. The transistor parameters are $\beta = 110$ and $V_{EB}(\text{on}) = 0.7$ V. (Ans. $R_E = 18.4$ k$\Omega$, $R_C = 12.1$ k$\Omega$)

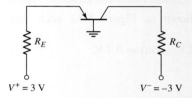

**Figure 5.33** Common-base circuit for Exercise Problem Ex 5.8

## Test Your Understanding

**TYU 5.9** The bias voltages in the circuit shown in Figure 5.34 are $V^+ = 3.3$ V and $V^- = -3.3$ V. The measured value of the collector voltage is $V_C = 2.27$ V. Determine $I_B, I_C, I_E, \beta,$ and $\alpha$. (Ans. $I_B = 2.50\ \mu$A, $I_C = 0.2575$ mA, $I_E = 0.26$ mA, $\beta = 103, \alpha = 0.99038$)

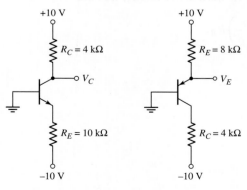

**Figure 5.34** Figure for Exercise TYU 5.9

**Figure 5.35** Figure for Exercise TYU 5.10

**TYU 5.10** The bias voltages in the circuit shown in Figure 5.35 are $V^+ = 5$ V and $V^- = -5$ V. Assume that $\beta = 85$. Determine $I_B$, $I_C$, $I_E$, and $V_{EC}$. (Ans. $I_B = 6.25\ \mu\text{A}$, $I_C = 0.531$ mA, $I_E = 0.5375$ mA, $V_{EC} = 3.575$ V)

## DESIGN EXAMPLE 5.9

**Objective:** Design a pnp bipolar transistor circuit to meet a set of specifications.

**Specifications:** The circuit configuration to be designed is shown in Figure 5.36(a). The quiescent emitter-collector voltage is to be $V_{ECQ} = 2.5$ V.

**Choices:** Discrete resistors with tolerances of $\pm 10$ percent are to be used, an emitter resistor with a nominal value of $R_E = 2$ k$\Omega$ is to be used, and a transistor with $\beta = 60$ and $V_{EB}(\text{on}) = 0.7$ V is available.

**Solution (ideal Q-point value):** Writing the Kirchhoff's voltage law equation around the C–E loop, we obtain

$$V^+ = I_{EQ}R_E + V_{ECQ}$$

or

$$5 = I_{EQ}(2) + 2.5$$

which yields $I_{EQ} = 1.25$ mA. The collector current is

$$I_{CQ} = \left(\frac{\beta}{1+\beta}\right) \cdot I_{EQ} = \left(\frac{60}{61}\right)(1.25) = 1.23\,\text{mA}$$

The base current is

$$I_{BQ} = \frac{I_{EQ}}{1+\beta} = \frac{1.25}{61} = 0.0205\,\text{mA}$$

Writing the Kirchhoff's voltage law equation around the E–B loop, we find

$$V^+ = I_{EQ}R_E + V_{EB}(\text{on}) + I_{BQ}R_B + V_{BB}$$

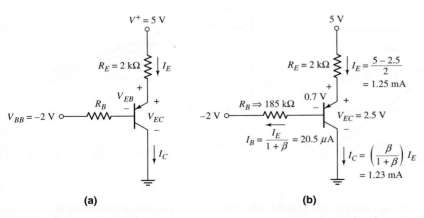

**(a)**                                    **(b)**

**Figure 5.36** Circuit for Design Example 5.9: (a) circuit and (b) circuit showing current and voltage values

or

$$5 = (1.25)(2) + 0.7 + (0.0205)R_B + (-2)$$

which yields $R_B = 185 \, \text{k}\Omega$.

**Solution (ideal load line):** The load line equation is

$$V_{EC} = V^+ - I_E R_E = V^+ - I_C \left( \frac{1+\beta}{\beta} \right) R_E$$

or

$$V_{EC} = 5 - I_C \left( \frac{61}{60} \right)(2) = 5 - I_C(2.03)$$

The load line, using the nominal value of $R_E$, and the calculated $Q$-point are shown in Figure 5.37(a).

**Trade-offs:** As shown in Appendix C, a standard resistor value of $185 \, \text{k}\Omega$ is not available. We will pick a value of $180 \, \text{k}\Omega$. We will consider $R_B$ and $R_E$ resistor tolerances of $\pm 10$ percent.

The quiescent collector current is given by

$$I_{CQ} = \beta \left[ \frac{V^+ - V_{EB}(\text{on}) - V_{BB}}{R_B + (1+\beta)R_E} \right] = (60) \left[ \frac{6.3}{R_B + (61)R_E} \right]$$

and the load line is given by

$$V_{EC} = V^+ - I_C \left( \frac{1+\beta}{\beta} \right) R_E = 5 - \left( \frac{61}{60} \right) I_C R_E$$

The extreme values of $R_E$ are:

$$2 \, \text{k}\Omega - 10\% = 1.8 \, \text{k}\Omega \qquad 2 \, \text{k}\Omega + 10\% = 2.2 \, \text{k}\Omega.$$

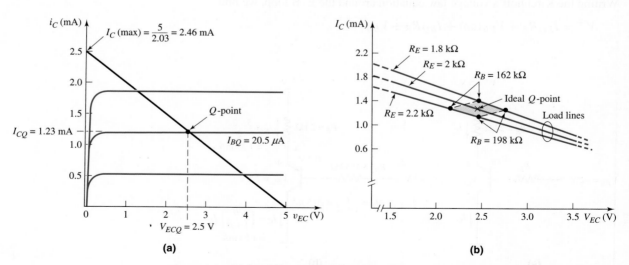

**(a)**                                                                                    **(b)**

**Figure 5.37** (a) Load line and $Q$-point value for the ideal designed circuit shown in Figure 5.36 used in Example 5.9; (b) load lines and $Q$-point values for the extreme tolerance values of resistors

The extreme values of $R_B$ are:

$$180 \text{ k}\Omega - 10\% = 162 \text{ k}\Omega \qquad 180 \text{ k}\Omega + 10\% = 198 \text{ k}\Omega.$$

The Q-point values for the extreme values of $R_B$ and $R_E$ are given in the following table.

| $R_B$ | $R_E$ | |
|---|---|---|
| | **1.8 k$\Omega$** | **2.2 k$\Omega$** |
| 162 k$\Omega$ | $I_{CQ} = 1.39$ mA | $I_{CQ} = 1.28$ mA |
| | $V_{ECQ} = 2.46$ V | $V_{ECQ} = 2.14$ V |
| 198 k$\Omega$ | $I_{CQ} = 1.23$ mA | $I_{CQ} = 1.14$ mA |
| | $V_{ECQ} = 2.75$ V | $V_{ECQ} = 2.45$ V |

Figure 5.37(b) shows the Q-points for the various possible extreme values of emitter and base resistances. The shaded area shows the region in which the Q-point will occur over the range of resistor values.

**Comment:** This example shows that an ideal Q-point can be determined based on a set of specifications, but, because of resistor tolerance, the actual Q-point will vary over a range of values. Other examples will consider the tolerances involved in transistor parameters.

### EXERCISE PROBLEM

**Ex 5.9:** The circuit elements in Figure 5.36(a) are $V^+ = 5$ V, $V_{BB} = -2$ V, $R_E = 2$ k$\Omega$, and $R_B = 180$ k$\Omega$. Assume $V_{EB}(\text{on}) = 0.7$ V. Plot the Q-point on the load line for (a) $\beta = 40$, (b) $\beta = 60$, (c) $\beta = 100$, and (d) $\beta = 150$. (Ans. (a) $I_{CQ} = 0.962$ mA, (b) $I_{CQ} = 1.25$ mA, (c) $I_{CQ} = 1.65$ mA, (d) $I_{CQ} = 1.96$ mA)

### EXAMPLE 5.10

**Objective:** Calculate the characteristics of an npn bipolar transistor circuit with a load resistance. The load resistance can represent a second transistor stage connected to the output of a transistor circuit.

For the circuit shown in Figure 5.38(a), the transistor parameters are: $V_{BE}(\text{on}) = 0.7$ V, and $\beta = 100$.

**Solution (Q-Point Values):** Kirchhoff's voltage law equation around the B–E loop yields

$$I_B R_B + V_{BE}(\text{on}) + I_E R_E + V^- = 0$$

Again assuming $I_E = (1 + \beta)I_B$, we find

$$I_B = \frac{-(V^- + V_{BE}(\text{on}))}{R_B + (1 + \beta)R_E} = \frac{-(-5 + 0.7)}{10 + (101)(5)} \Rightarrow 8.35 \ \mu\text{A}$$

The collector and emitter currents are

$$I_C = \beta I_B = (100)(8.35 \ \mu\text{A}) \Rightarrow 0.835 \text{ mA}$$

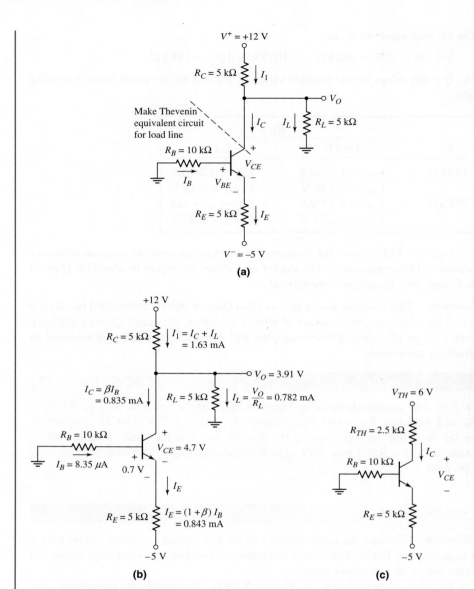

**Figure 5.38** Circuit for Example 5.10: (a) circuit; (b) circuit showing current and voltage values; and (c) Thevenin equivalent circuit

and

$$I_E = (1 + \beta)I_B = (101)(8.35\,\mu\text{A}) \Rightarrow 0.843\,\text{mA}$$

At the collector node, we can write

$$I_C = I_1 - I_L = \frac{V^+ - V_O}{R_C} - \frac{V_O}{R_L}$$

or

$$0.835 = \frac{12 - V_O}{5} - \frac{V_O}{5}$$

Solving for $V_O$, we find $V_O = 3.91$ V. The currents are then $I_1 = 1.62$ mA and $I_L = 0.782$ mA. Referring to Figure 5.38(b), the collector–emitter voltage is

$$V_{CE} = V_O - I_E R_E - (-5) = 3.91 - (0.843)(5) - (-5) = 4.70 \text{ V}$$

**Solution (Load Line):**  The load line equation for this circuit is not as straightforward as for previous circuits. The easiest approach to finding the load line is to make a "Thevenin equivalent circuit" of $R_L$, $R_C$, and $V^+$, as indicated in Figure 5.38(b). (Thevenin equivalent circuits are also covered later in this chapter, in Section 5.4.) The Thevenin equivalent resistance is

$$R_{TH} = R_L \| R_C = 5 \| 5 = 2.5 \text{ k}\Omega$$

and the Thevenin equivalent voltage is

$$V_{TH} = \left( \frac{R_L}{R_L + R_C} \right) \cdot V^+ = \left( \frac{5}{5+5} \right) \cdot (12) = 6 \text{ V}$$

The equivalent circuit is shown in Figure 5.38(c). The Kirchhoff voltage law equation around the C–E loop is

$$V_{CE} = (6 - (-5)) - I_C R_{TH} - I_E R_E = 11 - I_C(2.5) - I_C \left( \frac{101}{100} \right) \cdot (5)$$

or

$$V_{CE} = 11 - I_C(7.55)$$

The load line and the calculated $Q$-point values are shown in Figure 5.39.

**Comment:**  Remember that the collector current, determined from $I_C = \beta I_B$, is the current into the collector terminal of the transistor; it is not necessarily the current in the collector resistor $R_C$.

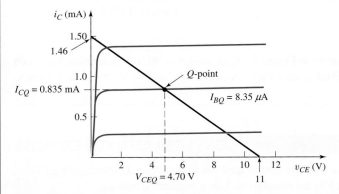

**Figure 5.39**  Load line and $Q$-point for the circuit shown in Figure 5.38(a) for Example 5.10

**Ex 5.10:** For the transistor shown in the circuit of Figure 5.40, the common-base current gain is $\alpha = 0.9920$. Determine $R_E$ such that the emitter current is limited to $I_E = 1.0$ mA. Also determine $I_B$, $I_C$, and $V_{BC}$. (Ans. $R_E = 3.3$ kΩ, $I_C = 0.992$ mA, $I_B = 8.0$ $\mu$A, $V_{BC} = 4.01$ V)

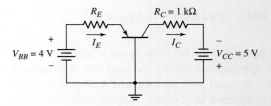

**Figure 5.40**  Figure for Exercise Ex 5.10

## Test Your Understanding

**TYU 5.11** For the circuit shown in Figure 5.41, determine $I_E$, $I_B$, $I_C$, and $V_{CE}$, if $\beta = 75$. (Ans. $I_B = 15.1$ $\mu$A, $I_C = 1.13$ mA, $I_E = 1.15$ mA, $V_{CE} = 6.03$ V)

**TYU 5.12** Assume $\beta = 120$ for the transistor in Figure 5.42. Determine $R_E$ such that $V_{CE} = 2.2$ V. (Ans. $R_E = 154$ Ω)

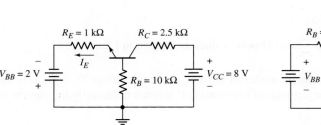

**Figure 5.41**  Figure for Exercise TYU 5.11

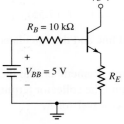

**Figure 5.42**  Figure for Exercise TYU 5.12

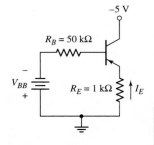

**Figure 5.43**  Figure for Exercise TYU 5.13

**TYU 5.13** For the transistor in Figure 5.43, assume $\beta = 90$. (a) Determine $V_{BB}$ such that $I_E = 1.2$ mA. (b) Find $I_C$ and $V_{EC}$. (Ans. (a) $V_{BB} = 2.56$ V; (b) $I_C = 1.19$ mA, $V_{EC} = 3.8$ V)

**PS 5.3:** Verify the common-base circuit analysis in Test Your Understanding Exercise TYU 5.11 with a PSpice simulation. Use a standard transistor.

## 5.3    BASIC TRANSISTOR APPLICATIONS

**Objective:** • Examine three applications of bipolar transistor circuits: a switch circuit, digital logic circuit, and an amplifier circuit.

Transistors can be used to: switch currents, voltages, and power; perform digital logic functions; and amplify time-varying signals. In this section, we consider the switching properties of the bipolar transistor, analyze a simple transistor digital logic circuit, and then show how the bipolar transistor is used to amplify time-varying signals.

### 5.3.1    Switch

Figure 5.44 shows a bipolar circuit called an **inverter,** in which the transistor in the circuit is switched between cutoff and saturation. The load, for example, could be a motor, a light-emitting diode or some other electrical device. If $v_I < V_{BE}(\text{on})$, then $i_B = i_C = 0$ and the transistor is cut off. Since $i_C = 0$, the voltage drop across the load is zero, so the output voltage is $v_O = V_{CC}$. Also, since the currents in the transistor are zero, the power dissipation in the transistor is zero. If the load were a motor, the motor would be off with zero current. Likewise, if the load were a light-emitting diode, the light output would be zero with zero current.

If we let $v_I = V_{CC}$ and if the ratio of $R_B$ to $R_C$, where $R_C$ is the effective resistance of the load, is less than $\beta$, then the transistor is usually driven into saturation, which means that

$$i_B \cong \frac{v_I - V_{BE}(\text{on})}{R_B} \tag{5.34}$$

$$i_C = I_C(\text{sat}) = \frac{V_{CC} - V_{CE}(\text{sat})}{R_C} \tag{5.35}$$

and

$$v_O = V_{CE}(\text{sat}) \tag{5.36}$$

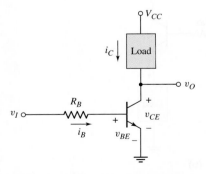

**Figure 5.44**  An npn bipolar inverter circuit used as a switch

In this case, a collector current is induced that would turn on the motor or the LED, depending on the type of load.

Equation (5.34) assumes that the B–E voltage can be approximated by the turn-on voltage. This approximation will be modified slightly when we discuss bipolar digital logic circuits in Chapter 17.

## EXAMPLE 5.11

**Objective:** Calculate the appropriate resistance values and transistor power dissipation for the two inverter switching configurations shown in Figure 5.45.

**Specifications (Figure 5.45(a)):** The transistor in the inverter circuit in Figure 5.45(a) is used to turn the light-emitting diode (LED) on and off. The required LED current is $I_{C1} = 12 \, \text{mA}$ to produce the specified output light. Assume transistor parameters of $\beta = 80$, $V_{BE}(\text{on}) = 0.7 \, \text{V}$, and $V_{CE}(\text{sat}) = 0.2 \, \text{V}$, and assume the diode cut-in voltage is $V_\gamma = 1.5 \, \text{V}$. [Note: LEDs are fabricated with compound semiconductor materials and have a larger cut-in voltage compared to silicon diodes.]

**Specifications (Figure 5.45(b)):** The inverter circuit in Figure 5.45(b) uses a pnp transistor. In this case, one side of the load (for example a motor) can be connected to ground potential. The required load current is $I_{C2} = 5 \, \text{A}$. Assume transistor parameters of $\beta = 40$, $V_{EB}(\text{on}) = 0.7 \, \text{V}$, and $V_{EC}(\text{sat}) = 0.2 \, \text{V}$.

**Solution (Figure 5.45(a)):** For $v_{I1} = 0$, transistor $Q_1$ is cut off so that $I_{B1} = I_{C2} = 0$ and the LED is also off.

For $v_{I1} = 5 \, \text{V}$, we require $I_{C1} = 12 \, \text{mA}$ and want the transistor to be driven into saturation. Then

$$R_1 = \frac{V^+ - (V_\gamma + V_{CE}(\text{sat}))}{I_{C1}} = \frac{5 - (1.5 + 0.2)}{12} \Rightarrow R_1 = 275 \, \Omega$$

We may let $I_{C1}/I_{B1} = 40$. Then $I_{B1} = 12/40 = 0.3 \, \text{mA}$. Now

$$R_{B1} = \frac{v_{I1} - V_{BE}(\text{on})}{I_{B1}} = \frac{5 - 0.7}{0.3} = 14.3 \, \text{k}\Omega$$

The power dissipation in $Q_1$ is

$$P_1 = I_{B1}V_{BE}(\text{on}) + I_{C1}V_{CE}(\text{sat}) = (0.3)(0.7) + (12)(0.2) = 2.61 \, \text{mW}$$

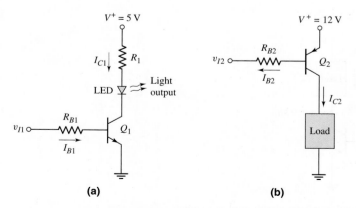

**(a)**    **(b)**

**Figure 5.45** Figures for Example 5.11

**Solution (Figure 5.45(b)):** For $v_{I2} = 12$ V, transistor $Q_2$ is cut off so that $I_{B2} = I_{C2} = 0$ and the voltage across the load is zero.

For $v_{I2} = 0$, transistor $Q_2$ is to be driven into saturation so that $V_{EC2} = V_{EC}(\text{sat}) = 0.2$ V. The voltage across the load is 11.8 V, the current is 5 A, which means the effective load resistance is 2.36 $\Omega$

If we let $I_{C2}/I_{B2} = 20$, then $I_{B2} = 5/20 = 0.25$ A. Now

$$R_{B2} = \frac{V^+ - V_{EB}(\text{on}) - v_{I2}}{I_{B2}} = \frac{12 - 0.7 - 0}{0.25} = 45.2 \ \Omega$$

The power dissipation in transistor $Q_2$ is

$$P_2 = I_{B2}V_{EB}(\text{on}) + I_{C2}V_{EC}(\text{sat}) = (0.25)(0.7) + (5)(0.2) = 1.175 \text{ W}$$

**Comment:** As with most electronic circuit designs, there are some assumptions that need to be made. The assumption to let $I_C/I_B = (1/2)\beta$ in each case ensures that each transistor will be driven into saturation even if variations in circuit parameters occur. At the same time, base currents are limited to reasonable values.

We may note that for the circuit in Figure 5.45(a), a base current of only 0.3 mA induces a load current of 12 mA. For the circuit in Figure 5.45(b), a base current of only 0.25 A induces a load current of 5 A. The advantage of transistor switches is that large load currents can be switched with relatively small base currents.

---

### EXERCISE PROBLEM

**Ex 5.11:** (a) Redesign the LED circuit in Figure 5.45(a) such that $I_{C1} = 15$ mA and $I_{C1}/I_{B1} = 50$ for $v_I = 5$ V. Use the same $Q_1$ transistor parameters given in Example 5.11. (b) Redesign the circuit in Figure 5.45(b) such that $I_{C2} = 2$ A and $I_{C2}/I_{B2} = 25$ for $v_I = 0$. Use the same $Q_2$ transistor parameters as given in Example 5.11. (Ans. (a) $R_1 = 220 \ \Omega$, $R_{B1} = 14.3$ k$\Omega$; (b) $R_{B2} = 141 \ \Omega$)

---

When a transistor is biased in saturation, the relationship between the collector and base currents is no longer linear. Consequently, this mode of operation cannot be used for linear amplifiers. On the other hand, switching a transistor between cutoff and saturation produces the greatest change in output voltage, which is especially useful in digital logic circuits, as we will see in the next section.

### 5.3.2    Digital Logic

Consider the simple transistor inverter circuit shown in Figure 5.46(a). If the input $V_I$ is approximately zero volts, the transistor is cut off and the output $V_O$ is high and equal to $V_{CC}$. If, on the other hand, the input is high and equal to $V_{CC}$, the transistor can be driven into saturation, in which case the output is low and equal to $V_{CE}(\text{sat})$.

Now consider the case when a second transistor is connected in parallel, as shown in Figure 5.46(b). When the two inputs are zero, both transistors $Q_1$ and $Q_2$ are in cutoff, and $V_O = 5$ V. When $V_1 = 5$ V and $V_2 = 0$, transistor $Q_1$ can be driven into saturation, and $Q_2$ remains in cutoff. With $Q_1$ in saturation, the output voltage is $V_O = V_{CE}(\text{sat}) \cong 0.2$ V. If we reverse the input voltages so that $V_1 = 0$ and $V_2 = 5$ V, then $Q_1$ is in cutoff, $Q_2$ can be driven into saturation, and $V_O = V_{CE}(\text{sat}) \cong 0.2$ V. If both inputs are high, meaning $V_1 = V_2 = 5$ V, then both transistors can be driven into saturation, and $V_O = V_{CE}(\text{sat}) \cong 0.2$ V.

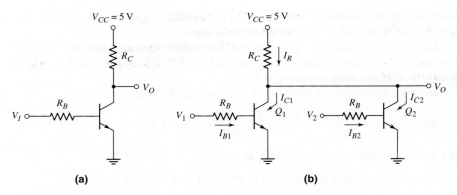

**Figure 5.46** A bipolar (a) inverter circuit and (b) NOR logic gate

Table 5.2 shows these various conditions for the circuit in Figure 5.46(b). In a **positive logic system,** meaning that the larger voltage is a logic 1 and the lower voltage is a logic 0, this circuit performs the **NOR logic function.** The circuit of Figure 5.46(b) is then a two-input bipolar NOR logic circuit.

| Table 5.2 | The bipolar NOR logic circuit response | |
|---|---|---|
| $V_1$ (V) | $V_2$ (V) | $V_O$ (V) |
| 0 | 0 | 5 |
| 5 | 0 | 0.2 |
| 0 | 5 | 0.2 |
| 5 | 5 | 0.2 |

## EXAMPLE 5.12

**Objective:** Determine the currents and voltages in the circuit shown in Figure 5.46(b).

Assume the transistor parameters are: $\beta = 50$, $V_{BE}(\text{on}) = 0.7$ V, and $V_{CE}(\text{sat}) = 0.2$ V. Let $R_C = 1$ k$\Omega$ and $R_B = 20$ k$\Omega$. Determine the currents and output voltage for various input conditions.

**Solution:** The following table indicates the equations and results for this example.

| Condition | $V_O$ | $I_R$ | $Q_1$ | $Q_2$ |
|---|---|---|---|---|
| $V_1 = 0$, $V_2 = 0$ | 5 V | 0 | $I_{B1} = I_{C1} = 0$ | $I_{B2} = I_{C2} = 0$ |
| $V_1 = 5$ V, $V_2 = 0$ | 0.2 V | $\dfrac{5 - 0.2}{1} = 4.8$ mA | $I_{B1} = \dfrac{5 - 0.7}{20}$ $= 0.215$ mA $I_{C1} = I_R = 4.8$ mA | $I_{B2} = I_{C2} = 0$ |
| $V_1 = 0$, $V_2 = 5$ V | 0.2 V | 4.8 mA | $I_{B1} = I_{C1} = 0$ | $I_{B2} = 0.215$ mA $I_{C2} = I_R = 4.8$ mA |
| $V_1 = 5$ V, $V_2 = 5$ V | 0.2 V | 4.8 mA | $I_{B1} = 0.215$ mA $I_{C1} = \dfrac{I_R}{2} = 2.4$ mA | $I_{B2} = 0.215$ mA $I_{C2} = \dfrac{I_R}{2} = 2.4$ mA |

**Comment:** In this example, we see that whenever a transistor is conducting, the ratio of collector current to base current is always less than $\beta$. This shows that the transistor is in saturation, which occurs when either $V_1$ or $V_2$ is 5 V.

---

**EXERCISE PROBLEM**

**Ex 5.12:** The transistor parameters in the circuit in Figure 5.46(b) are: $\beta = 40$, $V_{BE}(\text{on}) = 0.7$ V, and $V_{CE}(\text{sat}) = 0.2$ V. Let $R_C = 600 \ \Omega$ and $R_B = 950 \ \Omega$. Determine the currents and output voltage for: (a) $V_1 = V_2 = 0$; (b) $V_1 = 5$ V, $V_2 = 0$; and (c) $V_1 = V_2 = 5$ V. (Ans. (a) The currents are zero, $V_O = 5$ V; (b) $I_{B2} = I_{C2} = 0$, $I_{B1} = 4.53$ mA, $I_{C1} = I_R = 8$ mA, $V_O = 0.2$ V; (c) $I_{B1} = I_{B2} = 4.53$ mA, $I_{C1} = I_{C2} = 4$ mA $= I_R/2$, $V_O = 0.2$ V)

---

This example and the accompanying discussion illustrate that bipolar transistor circuits can be configured to perform logic functions. In Chapter 17, we will see that this circuit can experience loading effects when load circuits or other digital logic circuits are connected to the output. Therefore, logic circuits must be designed to minimize or eliminate such loading effects.

### 5.3.3    Amplifier

The bipolar inverter circuit can also be used to amplify a time-varying signal. Figure 5.47(a) shows an inverter circuit including a time-varying voltage source $\Delta v_I$ in the base circuit. The voltage transfer characteristics are shown in Figure 5.47(b). The dc voltage source $V_{BB}$ is used to bias the transistor in the forward-active region. The Q-point is shown on the transfer characteristics.

The voltage source $\Delta v_I$ introduces a time-varying signal on the input. A change in the input voltage then produces a change in the output voltage. These time-varying input and output signals are shown in Figure 5.47(b). If the magnitude of the slope of the transfer characteristics is greater than unity, then the time-varying output signal will be larger than the time-varying input signal—thus an amplifier.

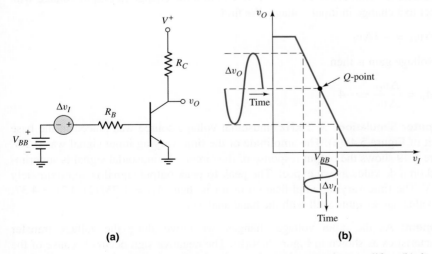

**(a)**                                **(b)**

**Figure 5.47** (a) A bipolar inverter circuit to be used as a time-varying amplifier; (b) the voltage transfer characteristics

## EXAMPLE 5.13

**Objective:** Determine the amplification factor for the circuit given in Figure 5.48(a). The transistor parameters are $\beta = 120$, $V_{BE}(\text{on}) = 0.7$ V, and $V_A = \infty$.

**DC Solution:** The voltage transfer characteristics were developed in Example 5.6 for this same circuit. The voltage transfer curve is repeated for convenience in Figure 5.48(b).

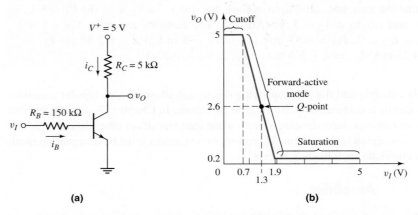

**Figure 5.48** (a) A bipolar inverter used as an amplifier; (b) the inverter voltage transfer characteristics

For $0.7 \leq v_I \leq 1.9$ V, the transistor is biased in the forward-active mode and the output voltage is given by

$$v_O = 7.8 - 4v_I$$

Now bias the transistor in the center of the active region with an input voltage of $v_I = V_{BB} = 1.3$ V. The dc output voltage is $v_O = 2.6$ V. The $Q$-point is shown on the transfer characteristics.

**AC Solution:** From $v_O = 7.8 - 4v_I$, we can find the change in output voltage with respect to a change in input voltage. We find

$$\Delta v_O = -4\Delta v_I$$

The voltage gain is then

$$A_v = \frac{\Delta v_O}{\Delta v_I} = -4$$

**Computer Simulation:** A 2 kHz sinusoidal voltage source was placed in the base circuit of Figure 5.48(a). The amplitude of the time-varying input signal was 0.2 V. Figure 5.49 shows the output response of the circuit. A sinusoidal signal is superimposed on a dc value as we expect. The peak-to-peak output signal is approximately 1.75 V. The time-varying amplification factor is then $|A_v| = 1.75/(2)(0.2) = 4.37$. This value agrees quite well with the hand analysis.

**Comment:** As the input voltage changes, we move along the voltage transfer characteristics as shown in Figure 5.50(b). The negative sign occurs because of the inverting property of the circuit.

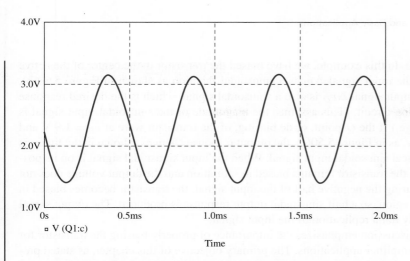

**Figure 5.49** Output signal from the circuit shown in Figure 5.48 for input signals of $V_{BB} = 1.3$ V and $\Delta v_I = 0.2 \sin \omega t$ (V)

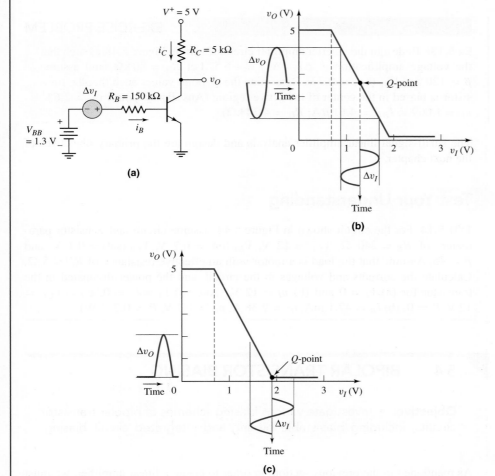

**Figure 5.50** (a) The inverter circuit with both a dc and an ac input signal; (b) the dc voltage transfer characteristics, $Q$-point, and sinusoidal input and output signals; (c) the transfer characteristics showing improper dc biasing

**Discussion:** In this example, we have biased the transistor in the center of the active region. If the input signal $\Delta v_I$ is a sinusoidal function as shown in Figure 5.50(b), then the output signal $\Delta v_O$ is also a sinusoidal signal, which is the desired response for an analog circuit. (This assumes the magnitude of the sinusoidal input signal is not too large.) If the $Q$-point, or dc biasing, of the transistor were at $v_I = 1.9$ V and $v_O = 0.2$ V, as in Figure 5.50(c), the output response changes. Shown in the figure is a symmetrical sinusoidal input signal. When the input sinusoidal signal is on its positive cycle, the transistor remains biased in saturation and the output voltage does not change. During the negative half of the input signal, the transistor becomes biased in the active region, so a half sinusoidal output response is produced. The output signal is obviously not a replication of the input signal.

This discussion emphasizes the importance of properly biasing the transistor for analog or amplifier applications. The primary objective of this chapter, as stated previously, is to help readers become familiar with transistor circuits, but it is also to enable them to design the dc biasing of transistor circuits that are to be used in analog applications.

### EXERCISE PROBLEM

**Ex 5.13:** Redesign the inverter amplifier circuit shown in Figure 5.48(a) such that the voltage amplification is $\Delta v_O / \Delta v_I = -6.5$. Let $R_B = 80\,\text{k}\Omega$, and assume $\beta = 120$ and $V_{BE}(\text{on}) = 0.7$ V. Determine the $Q$-point values such that the transistor is biased in the center of the active region. (Ans. For $Q$-point: $v_O = 2.6$ V, $v_I = 1.069$ V, $I_{BQ} = 4.61\,\mu\text{A}$; $R_C = 4.34\,\text{k}\Omega$)

The small-signal linear amplifier analysis and design are the primary objectives of the next chapter.

## Test Your Understanding

**TYU 5.14** For the circuit shown in Figure 5.44, assume circuit and transistor parameters of $R_B = 240\ \Omega$, $V_{CC} = 12$ V, $V_{BE}(\text{on}) = 0.7$ V, $V_{CE}(\text{sat}) = 0.1$ V, and $\beta = 75$. Assume that the load is a motor with an effective resistance of $R_C = 5\ \Omega$. Calculate the currents and voltages in the circuit, and the power dissipated in the transistor for (a) $v_I = 0$ and (b) $v_I = 12$ V. (Ans. (a) $i_B = i_C = 0$, $v_O = V_{CC} = 12$ V, $P = 0$; (b) $i_B = 47.1$ mA, $i_C = 2.38$ A, $v_O = 0.1$ V, $P = 0.271$ W)

 **5.4    BIPOLAR TRANSISTOR BIASING**

**Objective:** • Investigate various biasing schemes of bipolar transistor circuits, including bias-stable biasing and integrated circuit biasing.

As mentioned in the previous section, in order to create a linear amplifier, we must keep the transistor in the forward-active mode, establish a $Q$-point near the center of the load line, and couple the time-varying input signal to the base. The circuit in Figure 5.47(a) may be impractical for two reasons: (1) the signal source is not connected to ground, and (2) there may be situations where we do not want a dc base current flowing through the signal source. In this section, we will examine several alternative biasing schemes. These basic biasing circuits illustrate some desirable and some undesirable biasing characteristics. More sophisticated biasing circuits that use additional transistors and that are used in integrated circuits are discussed in Chapter 10.

### 5.4.1  Single Base Resistor Biasing

The circuit shown in Figure 5.51(a) is one of the simplest transistor circuits. There is a single dc power supply, and the quiescent base current is established through the resistor $R_B$. The **coupling capacitor** $C_C$ acts as an open circuit to dc, isolating the signal source from the dc base current. If the frequency of the input signal is large enough and $C_C$ is large enough, the signal can be coupled through $C_C$ to the base with little attenuation. Typical values of $C_C$ are generally in the range of 1 to 10 $\mu$F, although the actual value depends upon the frequency range of interest (see Chapter 7). Figure 5.51(b) is the dc equivalent circuit; the $Q$-point values are indicated by the additional subscript $Q$.

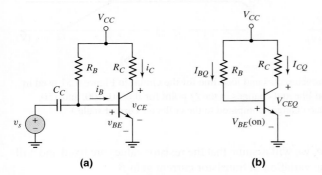

(a)                                (b)

**Figure 5.51**  (a) Common-emitter circuit with a single bias resistor in the base and (b) dc equivalent circuit

---

### DESIGN EXAMPLE 5.14

**Objective:** Design a circuit with a single-base resistor to meet a set of specifications.

**Specifications:** The circuit configuration to be designed is shown in Figure 5.51(b). The circuit is to be biased with $V_{CC} = +12$ V. The transistor quiescent values are to be $I_{CQ} = 1$ mA and $V_{CEQ} = 6$ V.

**Choices:** The transistor used in the design has nominal values of $\beta = 100$ and $V_{BE}(\text{on}) = 0.7$ V, but the current gain for this type of transistor is assumed to be in the range $50 \leq \beta \leq 150$ because of fairly wide fabrication tolerances. We will assume, in this example, that the designed resistor values are available.

**Solution:** The collector resistor is found from

$$R_C = \frac{V_{CC} - V_{CEQ}}{I_{CQ}} = \frac{12 - 6}{1} = 6\,\text{k}\Omega$$

The base current is

$$I_{BQ} = \frac{I_{CQ}}{\beta} = \frac{1 \text{ mA}}{100} \Rightarrow 10 \,\mu\text{A}$$

and the base resistor is determined to be

$$R_B = \frac{V_{CC} - V_{BE}(\text{on})}{I_{BQ}} = \frac{12 - 0.7}{10 \,\mu\text{A}} = 1.13 \text{ M}\Omega$$

The transistor characteristics, load line, and Q-point for this set of conditions are shown in Figure 5.52(a).

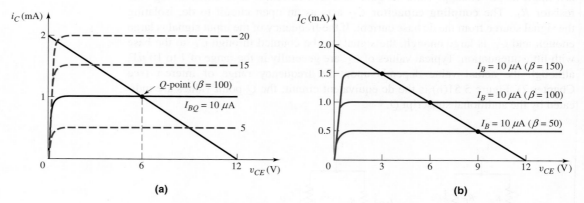

**Figure 5.52** (a) Transistor characteristics and load line for the circuit in Figure 5.51 used in Design Example 5.14; (b) load line and changes in the Q-point for $\beta = 50, 100,$ and $150$. (Note that the base current scale changes compared to the collector current scale.)

**Trade-offs:** In this example, we will assume that the resistor values are fixed and will investigate the effects of the variation in transistor current gain $\beta$.

The base current is given by

$$I_{BQ} = \frac{V_{CC} - V_{BE}(\text{on})}{R_B} = \frac{12 - 0.7}{1.13 \text{ M}\Omega} = 10 \,\mu\text{A} \text{ (unchanged)}$$

The base current for this circuit configuration is independent of the transistor current gain.

The collector current is

$$I_{CQ} = \beta I_{BQ}$$

and the load line is found from

$$V_{CE} = V_{CC} - I_C R_C = 12 - I_C(6)$$

The load line is fixed. However, the Q-point will change. The transistor Q-point values for three values of $\beta$ are given as:

| $\beta$ | 50 | 100 | 150 |
|---|---|---|---|
| Q-point values | $I_{CQ} = 0.50$ mA $V_{CEQ} = 9$ V | $I_{CQ} = 1$ mA $V_{CEQ} = 6$ V | $I_{CQ} = 1.5$ mA $V_{CEQ} = 3$ V |

The various $Q$-points are plotted on the load line shown in Figure 5.52(b). In this figure, the collector current scale and load line are fixed. The base current scale changes as $\beta$ changes.

**Comment:** In this circuit configuration with a single base resistor, the $Q$-point is not stabilized against variations in $\beta$; as $\beta$ changes, the $Q$-point varies significantly. In our discussion of the amplifier in Example 5.13 (see Figure 5.50), we noted the importance of the placement of the $Q$-point. In the following two examples, we will analyze and design bias-stable circuits.

Although a value of 1.13 M$\Omega$ for $R_B$ will establish the required base current, this resistance is too large to be used in integrated circuits. The following two examples will also demonstrate a circuit design to alleviate this problem.

---

### EXERCISE PROBLEM

**Ex 5.14:** Consider the circuit shown in Figure 5.51(b). Assume $V_{CC} = 2.8$ V, $\beta = 150$, and $V_{BE}(\text{on}) = 0.7$ V. Design the circuit such that $I_{CQ} = 0.12$ mA and $V_{CEQ} = 1.4$ V. (Ans. $R_C = 11.7$ k$\Omega$, $R_B = 2.625$ M$\Omega$)

---

## Test Your Understanding

[Note: In the following exercises, assume the B–E cut-in voltage is 0.7 V. Also assume the C–E saturation voltage is 0.2 V.]

**TYU 5.15** Consider the circuit shown in Figure 5.53. (a) If $\beta = 120$, determine $R_B$ such that $V_{CEQ} = 2.5$ V. (b) If the current gain varies over the range $80 \leq \beta \leq 160$, determine the variation in $V_{CEQ}$. (Ans. (a) $R_B = 413$ k$\Omega$, (b) $1.67 \leq V_{CEQ} \leq 3.33$)

**TYU 5.16** For the circuit shown in Figure 5.53, let $R_B = 800$ k$\Omega$. If the range of $\beta$ is between 75 and 150, determine a new value of $R_C$ such that the $Q$-point will always be in the range $1 \leq V_{CEQ} \leq 4$ V. What will be the actual range of $V_{CEQ}$ for the new value of $R_C$? (Ans. For $V_{CEQ} = 2.5$ V, $R_C = 4.14$ k$\Omega$; (b) $1.66 \leq V_{CEQ} \leq 3.33$ V)

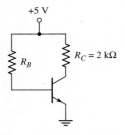

**Figure 5.53** Figure for Exercises TYU 5.15 and TYU 5.16

### 5.4.2    Voltage Divider Biasing and Bias Stability

The circuit in Figure 5.54(a) is a classic example of discrete transistor biasing. ($IC$ biasing is different and will be discussed in Chapter 10.) The single bias resistor $R_B$ in the previous circuit is replaced by a pair of resistors $R_1$ and $R_2$, and an emitter resistor $R_E$ is added. The ac signal is still coupled to the base of the transistor through the coupling capacitor $C_C$.

The circuit is most easily analyzed by forming a **Thevenin equivalent circuit** for the base circuit. The coupling capacitor acts as an open circuit to dc. The equivalent Thevenin voltage is

$$V_{TH} = [R_2/(R_1 + R_2)]V_{CC}$$

and the equivalent Thevenin resistance is

$$R_{TH} = R_1 \| R_2$$

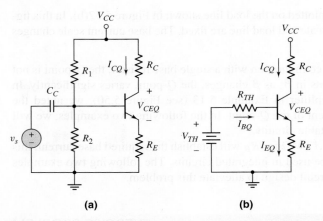

**Figure 5.54** (a) A common-emitter circuit with an emitter resistor and voltage divider bias circuit in the base; (b) the dc circuit with a Thevenin equivalent base circuit

where the symbol ‖ indicates the parallel combination of resistors. Figure 5.54(b) shows the equivalent dc circuit. As we can see, this circuit is similar to those we have previously considered.

Applying Kirchhoff's law around the B–E loop, we obtain

$$V_{TH} = I_{BQ}R_{TH} + V_{BE}(\text{on}) + I_{EQ}R_E \tag{5.37}$$

If the transistor is biased in the forward-active mode, then

$$I_{EQ} = (1 + \beta)I_{BQ}$$

and the base current, from Equation (5.37), is

$$I_{BQ} = \frac{V_{TH} - V_{BE}(\text{on})}{R_{TH} + (1 + \beta)R_E} \tag{5.38}$$

The collector current is then

$$I_{CQ} = \beta I_{BQ} = \frac{\beta(V_{TH} - V_{BE}(\text{on}))}{R_{TH} + (1 + \beta)R_E} \tag{5.39}$$

## EXAMPLE 5.15

**Objective:** Analyze a circuit using a voltage divider bias circuit, and determine the change in the $Q$-point with a variation in $\beta$ when the circuit contains an emitter resistor.

For the circuit given in Figure 5.54(a), let $R_1 = 56$ k$\Omega$, $R_2 = 12.2$ k$\Omega$, $R_C = 2$ k$\Omega$, $R_E = 0.4$ k$\Omega$, $V_{CC} = 10$ V, $V_{BE}(\text{on}) = 0.7$ V, and $\beta = 100$.

**Solution:** Using the Thevenin equivalent circuit in Figure 5.54(b), we have

$$R_{TH} = R_1\|R_2 = 56\|12.2 = 10.0 \text{ k}\Omega$$

and

$$V_{TH} = \left(\frac{R_2}{R_1 + R_2}\right) \cdot V_{CC} = \left(\frac{12.2}{56 + 12.2}\right)(10) = 1.79 \text{ V}$$

Writing the Kirchhoff voltage law equation around the B–E loop, we obtain

$$I_{BQ} = \frac{V_{TH} - V_{BE}(\text{on})}{R_{TH} + (1 + \beta)R_E} = \frac{1.79 - 0.7}{10 + (101)(0.4)} \Rightarrow 21.6\,\mu\text{A}$$

The collector current is

$$I_{CQ} = \beta I_{BQ} = (100)(21.6\,\mu\text{A}) \Rightarrow 2.16\,\text{mA}$$

and the emitter current is

$$I_{EQ} = (1 + \beta)I_{BQ} = (101)(21.6\,\mu\text{A}) \Rightarrow 2.18\,\text{mA}$$

The quiescent C–E voltage is then

$$V_{CEQ} = V_{CC} - I_{CQ}R_C - I_{EQ}R_E = 10 - (2.16)(2) - (2.18)(0.4) = 4.81\,\text{V}$$

These results show that the transistor is biased in the active region.

If the current gain of the transistor were to decrease to $\beta = 50$ or increase to $\beta = 150$, we obtain the following results:

| $\beta$ | 50 | 100 | 150 |
|---|---|---|---|
| Q-point values | $I_{BQ} = 35.9\,\mu\text{A}$ $I_{CQ} = 1.80\,\text{mA}$ $V_{CEQ} = 5.67\,\text{V}$ | $I_{BQ} = 21.6\,\mu\text{A}$ $I_{CQ} = 2.16\,\text{mA}$ $V_{CEQ} = 4.81\,\text{V}$ | $I_{BQ} = 15.5\,\mu\text{A}$ $I_{CQ} = 2.32\,\text{mA}$ $V_{CEQ} = 4.40\,\text{V}$ |

The load line and Q-points are plotted in Figure 5.55. The variation in Q-points for this circuit configuration is to be compared with the variation in Q-point values shown previously in Figure 5.52(b).

For a 3 : 1 ratio in $\beta$, the collector current and collector–emitter voltage change by only a 1.29 : 1 ratio.

**Comment:** The voltage divider circuit of $R_1$ and $R_2$ can bias the transistor in its active region using resistor values in the low kilohm range. In contrast, single resistor biasing requires a resistor in the megohm range. In addition, the change in $I_{CQ}$ and $V_{CEQ}$ with a change in $\beta$ has been substantially reduced compared to the change shown in Figure 5.52(b). Including an emitter resistor $R_E$ has tended to **stabilize** the Q-point. This means that including the emitter resistor helps to stabilize the Q-point with respect to variations in $\beta$. Including the resistor $R_E$ introduces negative feedback, as we will see in Chapter 12. Negative feedback tends to stabilize circuits.

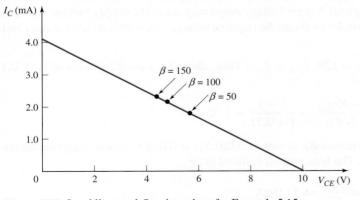

**Figure 5.55**  Load lines and Q-point values for Example 5.15

**Ex 5.15:** For the circuit shown in Figure 5.54(a), let $V_{CC} = 3.3$ V, $R_E = 500\,\Omega$, $R_C = 4\,\text{k}\Omega$, $R_1 = 85\,\text{k}\Omega$, $R_2 = 35\,\text{k}\Omega$, and $\beta = 150$. (a) Determine $R_{TH}$ and $V_{TH}$. (b) Find $I_{BQ}$, $I_{CQ}$, and $V_{CEQ}$. (c) Repeat part (b) for $\beta = 75$. (Ans. (a) $R_{TH} = 24.8\,\text{k}\Omega$, $V_{TH} = 0.9625$ V; (b) $I_{BQ} = 2.62\,\mu\text{A}$, $I_{CQ} = 0.393$ mA, $V_{CEQ} = 1.53$ V; (c) $I_{BQ} = 4.18\,\mu\text{A}$, $I_{CQ} = 0.314$ mA, $V_{CEQ} = 1.89$ V)

Considering Equation (5.39), the design requirement for bias stability is $R_{TH} \ll (1 + \beta)R_E$. Consequently, the collector current is approximately

$$I_{CQ} \cong \frac{\beta(V_{TH} - V_{BE}(\text{on}))}{(1 + \beta)R_E} \tag{5.40}$$

Normally, $\beta \gg 1$; therefore, $\beta/(1 + \beta) \cong 1$, and

$$I_{CQ} \cong \frac{(V_{TH} - V_{BE}(\text{on}))}{R_E} \tag{5.41}$$

Now the quiescent collector current is essentially a function of only the dc voltages and the emitter resistance, and the $Q$-point is stabilized against $\beta$ variations. However, if $R_{TH}$ is too small, then $R_1$ and $R_2$ are small, and excessive power is dissipated in these resistors. The general rule is that a circuit is considered **bias stable** when

$$R_{TH} \cong 0.1(1 + \beta)R_E \tag{5.42}$$

## DESIGN EXAMPLE 5.16

**Objective:** Design a bias-stable circuit to meet a set of specifications.

**Specifications:** The circuit configuration to be designed is shown in Figure 5.54(a). Let $V_{CC} = 5$ V and $R_C = 1$ k$\Omega$. Choose $R_E$ and determine the bias resistors $R_1$ and $R_2$ such that the circuit is considered bias stable and that $V_{CEQ} = 3$ V.

**Choices:** Assume the transistor has nominal values of $\beta = 120$ and $V_{BE}(\text{on}) = 0.7$ V. We will choose standard resistor values and will assume that the transistor current gain varies over the range $60 \le \beta \le 180$.

**Design Pointer:** Typically, the voltage across $R_E$ should be on the same order of magnitude as $V_{BE}(\text{on})$. Larger voltage drops may mean the supply voltage $V_{CC}$ has to be increased in order to obtain the required voltage across the collector-emitter and across $R_C$.

**Solution:** With $\beta = 120$, $I_{CQ} \approx I_{EQ}$. Then, choosing a standard value of 0.51 k$\Omega$ for $R_E$, we find

$$I_{CQ} \cong \frac{V_{CC} - V_{CEQ}}{R_C + R_E} = \frac{5 - 3}{1 + 0.51} = 1.32\,\text{mA}$$

The voltage drop across $R_E$ is now $(1.32)(0.51) = 0.673$ V, which is approximately the desired value. The base current is found to be

$$I_{BQ} = \frac{I_{CQ}}{\beta} = \frac{1.32}{120} \Rightarrow 11.0\,\mu\text{A}$$

Using the Thevenin equivalent circuit in Figure 5.54(b), we find

$$I_{BQ} = \frac{V_{TH} - V_{BE}(\text{on})}{R_{TH} + (1 + \beta)R_E}$$

For a bias-stable circuit, $R_{TH} = 0.1(1 + \beta)R_E$, or

$$R_{TH} = (0.1)(121)(0.51) = 6.17\,\text{k}\Omega$$

Then,

$$I_{BQ} = 11.0\,\mu\text{A} \Rightarrow \frac{V_{TH} - 0.7}{6.17 + (121)(0.51)}$$

which yields

$$V_{TH} = 0.747 + 0.70 = 1.447\,\text{V}$$

Now

$$V_{TH} = \left(\frac{R_2}{R_1 + R_2}\right)V_{CC} = \left(\frac{R_2}{R_1 + R_2}\right)(5) = 1.447\,\text{V}$$

or

$$\left(\frac{R_2}{R_1 + R_2}\right) = \frac{1.447}{5} = 0.2894$$

Also,

$$R_{TH} = \frac{R_1 R_2}{R_1 + R_2} = 6.17\,\text{k}\Omega = R_1\left(\frac{R_2}{R_1 + R_2}\right) = R_1(0.2894)$$

which yields

$$R_1 = 21.3\,\text{k}\Omega \quad \text{and} \quad R_2 = 8.69\,\text{k}\Omega$$

From Appendix C, we can choose standard resistor values of $R_1 = 20$ kΩ and $R_2 = 8.2$ kΩ.

**Trade-offs:** We will neglect, in this example, the tolerance effects of the resistors (end-of-chapter problems such as Problems 5.18 and 5.40 do include tolerance effects). We will consider the effect on the transistor $Q$-point values of the common-emitter current gain variation.

Using the standard resistor values, we have

$$R_{TH} = R_1 \| R_2 = 20\|8.2 = 5.82\,\text{k}\Omega$$

and

$$V_{TH} = \left(\frac{R_2}{R_1 + R_2}\right)(V_{CC}) = \left(\frac{8.2}{20 + 8.2}\right)(5) = 1.454\,\text{V}$$

The base current is given by

$$I_{BQ} = \left[\frac{V_{TH} - V_{BE}(\text{on})}{R_{TH} + (1 + \beta)R_E}\right]$$

while the collector current is $I_{CQ} = \beta I_{BQ}$, and the collector–emitter voltage is given by

$$V_{CEQ} = V_{CC} - I_{CQ}\left[R_C + \left(\frac{1+\beta}{\beta}\right)R_E\right]$$

The $Q$-point values for three values of $\beta$ are shown in the following table.

| $\beta$ | 60 | 120 | 180 |
|---|---|---|---|
| $Q$-point values | $I_{BQ} = 20.4\ \mu\mathrm{A}$<br>$I_{CQ} = 1.23\ \mathrm{mA}$<br>$V_{CEQ} = 3.13\ \mathrm{V}$ | $I_{BQ} = 11.2\ \mu\mathrm{A}$<br>$I_{CQ} = 1.34\ \mathrm{mA}$<br>$V_{CEQ} = 2.97\ \mathrm{V}$ | $I_{BQ} = 7.68\ \mu\mathrm{A}$<br>$I_{CQ} = 1.38\ \mathrm{mA}$<br>$V_{CEQ} = 2.91\ \mathrm{V}$ |

**Comment:** The $Q$-point in this example is now considered to be stabilized against variations in $\beta$, and the voltage divider resistors $R_1$ and $R_2$ have reasonable values in the kilohm range. We see that the collector current changes by only $-8.2$ percent when $\beta$ changes by a factor of 2 (from 120 to 60), and changes by only $+3.0$ percent when $\beta$ changes by $+50$ percent (from 120 to 180). Compare these changes to those of the single-base resistor design in Example 5.14.

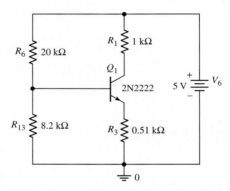

**Figure 5.56** PSpice circuit schematic for Design Example 5.16

**Computer Simulation:** Figure 5.56 shows the PSpice circuit schematic diagram with the standard resistor values and with a standard 2N2222 transistor from the PSpice library for the circuit designed in this example. A dc analysis was performed and the resulting transistor $Q$-point values are shown. The collector–emitter voltage is $V_{CE} = 2.80$ V, which is close to the design value of 3 V. One reason for the difference is that the standard-valued resistors are not exactly equal to the design values. Another reason for the slight difference is that the effective $\beta$ of the 2N2222 is 157 compared to the assumed value of 120.

```
**** BIPOLAR JUNCTION TRANSISTORS
NAME        Q_Q1
MODEL       Q2N2222
IB          9.25E-06
IC          1.45E-03
VBE         6.55E-01
VBC         -2.15E+00
VCE         2.80E+00
BETADC      1.57E+02
```

**Ex 5.16:** In the circuit shown in Figure 5.54(a), let $V_{CC} = 5$ V, $R_E = 0.2$ k$\Omega$, $R_C = 1$ k$\Omega$, $\beta = 150$, and $V_{BE}(\text{on}) = 0.7$ V. Design a bias-stable circuit such that the $Q$-point is in the center of the load line. (Ans. $R_1 = 13$ k$\Omega$, $R_2 = 3.93$ k$\Omega$)

Another advantage of including an emitter resistor is that it stabilizes the $Q$-point with respect to temperature. To explain, we noted in Figure 1.20 that the current in a pn junction increases with increasing temperature, for a constant junction voltage. We then expect the transistor current to increase as the temperature increases. If the current in a junction increases, the junction temperature increases (because of $I^2R$ heating), which in turn causes the current to increase, thereby further increasing the junction temperature. This phenomenon can lead to thermal runaway and to device destruction. However, from Figure 5.54(b), we see that as the current increases, the voltage drop across $R_E$ increases. The Thevenin equivalent voltage and resistance are assumed to be essentially independent of temperature, and the temperature-induced change in the voltage drop across $R_{TH}$ will be small. The net result is that the increased voltage drop across $R_E$ reduces the B–E junction voltage, which then tends to stabilize the transistor current against increases in temperature.

## Test Your Understanding

**TYU 5.17** The parameters of the circuit shown in Figure 5.54(a) are $V_{CC} = 5$ V, $R_E = 1$ k$\Omega$, $R_C = 4$ k$\Omega$, $R_1 = 440$ k$\Omega$, and $R_2 = 230$ k$\Omega$. The transistor parameters are $\beta = 150$ and $V_{BE}(\text{on}) = 0.7$ V. (a) Find $V_{TH}$ and $R_{TH}$. (b) Determine $I_{CQ}$ and $V_{CEQ}$. (c) Repeat parts (a) and (b) for $\beta = 90$. (Ans. (a) $V_{TH} = 1.716$ V, $R_{TH} = 151$ k$\Omega$;   (b) $I_{CQ} = 0.505$ mA,   $V_{CEQ} = 2.47$ V;   (c)   $I_{CQ} = 0.378$ mA, $V_{CEQ} = 3.11$ V)

**TYU 5.18** Consider the circuit in Figure 5.54(a). The circuit parameters are $V_{CC} = 5$ V and $R_E = 1$ k$\Omega$. The transistor parameters are $\beta = 150$ and $V_{BE}(\text{on}) = 0.7$ V. (a) Design a bias-stable circuit such that $I_{CQ} = 0.40$ mA and $V_{CEQ} = 2.7$ V. (b) Using the results of part (a), determine $I_{CQ}$ and $V_{CEQ}$ for $\beta = 90$. (Ans. (a) $R_1 = 66$ k$\Omega$, $R_2 = 19.6$ k$\Omega$, $R_C = 4.74$ k$\Omega$; (b) $I_{CQ} = 0.376$ mA, $V_{CEQ} = 2.84$ V)

### 5.4.3     Positive and Negative Voltage Biasing

There are applications in which biasing a transistor with both positive and negative dc voltages is desirable. We will see this especially in Chapter 11 when we are discussing the differential amplifier. Biasing with dual supplies allows us, in some applications, to eliminate the coupling capacitor and allows dc input voltages as input signals. The following example demonstrates this biasing scheme.

## EXAMPLE 5.17

**Objective:** Design a bias-stable pnp transistor circuit to meet a set of specifications.

**Specifications:** The circuit configuration to be designed is shown in Figure 5.57(a). The transistor $Q$-point values are to be: $V_{ECQ} = 7$ V, $I_{CQ} \cong 0.5$ mA, and $V_{RE} \cong 1$ V.

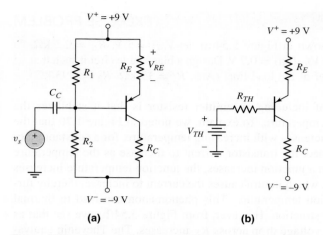

**Figure 5.57** (a) Circuit for Example 5.17 and (b) Thevenin equivalent circuit

**Choices:** Assume transistor parameters of $\beta = 80$ and $V_{EB}(\text{on}) = 0.7$ V. Standard resistor values are to be used in the final design.

**Solution:** The Thevenin equivalent circuit is shown in Figure 5.57(b). The Thevenin equivalent resistance is $R_{TH} = R_1 \| R_2$ and the Thevenin equivalent voltage, measured with respect to ground, is given by

$$V_{TH} = \left( \frac{R_2}{R_1 + R_R} \right)(V^+ - V^-) + V^-$$

$$= \frac{1}{R_1} \left( \frac{R_1 R_2}{R_1 + R_2} \right)(V^+ - V^-) + V^-$$

For $V_{RE} \cong 1$ V and $I_{CQ} \cong 0.5$ mA, then we can set

$$R_E = \frac{1}{0.5} = 2\,\text{k}\Omega$$

For a bias stable circuit, we want

$$R_{TH} = \frac{R_1 R_2}{R_1 + R_2} = (0.1)(1 + \beta)R_E$$

$$= (0.1)(81)(2) = 16.2\,\text{k}\Omega$$

Then the Thevenin equivalent voltage can be written as

$$V_{TH} = \frac{1}{R_1}(16.2)[9 - (-9)] + (-9) = \frac{1}{R_1}(291.6) - 9$$

The KVL equation around the E–B loop is given by

$$V^+ = I_{EQ}R_E + V_{EB}(\text{on}) + I_{BQ}R_{TH} + V_{TH}$$

The transistor is to be biased in the forward-active mode so that $I_{EQ} = (1 + \beta)I_{BQ}$. We then have

$$V^+ = (1 + \beta)I_{BQ}R_E + V_{EB}(\text{on}) + I_{BQ}R_{TH} + V_{TH}$$

For $I_{CQ} = 0.5$ mA, then $I_{BQ} = 0.00625$ mA so we can write

$$9 = (81)(0.00625)(2) + 0.7 + (0.00625)(16.2) + \frac{1}{R_1}(291.6) - 9$$

We find $R_1 = 18.0$ k$\Omega$. Then, from $R_{TH} = R_1 \| R_2 = 16.2$ k$\Omega$, we find $R_2 = 162$ k$\Omega$.

For $I_{CQ} = 0.5$ mA, then $I_{EQ} = 0.506$ mA. The KVL equation around the E–C loop yields

$$V^+ = I_{EQ}R_E + V_{ECQ} + I_{CQ}R_C + V^-$$

or

$$9 = (0.506)(2) + 7 + (0.50)R_C + (-9)$$

which yields

$$R_C \cong 20\,\text{k}\Omega$$

**Trade-offs:** All resistor values are standard values except for $R_2 = 162$ k$\Omega$. A standard discrete value of 160 k$\Omega$ is available. However, because of the bias-stable design, the $Q$-point will not change significantly. The change in $Q$-point values with a change in transistor current gain $\beta$ is considered in end-of-chapter problems such as Problems 5.31 and 5.34.

**Comment:** In many cases, specifications such as a collector current level or an emitter–collector voltage value are not absolute, but are given as approximate values. For this reason, the emitter resistor, for example, is determined to be 2 k$\Omega$, which is a standard discrete resistor value. The final bias resistor values are also chosen to be standard values. However, these small changes compared to the calculated resistor values will not change the $Q$-point values significantly.

## EXERCISE PROBLEM

**Ex 5.17:** Consider the circuit shown in Figure 5.58. The transistor parameters are $\beta = 150$ and $V_{BE}(\text{on}) = 0.7$ V. The circuit parameters are $R_E = 2$ k$\Omega$ and $R_C = 10$ k$\Omega$. Design a bias-stable circuit such that the quiescent output voltage is zero. What are the values of $I_{CQ}$ and $V_{CEQ}$? (Ans. $I_{CQ} = 0.5$ mA, $V_{CEQ} = 3.99$ V, $R_1 = 167$ k$\Omega$, $R_2 = 36.9$ k$\Omega$)

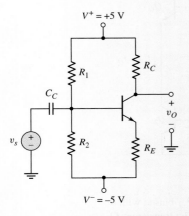

**Figure 5.58**  Figure for Exercise Ex 5.17

**Integrated Circuit Biasing**

The resistor biasing of transistor circuits considered up to this point is primarily applied to discrete circuits. For integrated circuits, we would like to eliminate as many resistors as possible since, in general, they require a larger surface area than transistors.

A bipolar transistor can be biased by using a constant-current source $I_Q$, as shown in Figure 5.59. The advantages of this circuit are that the emitter current is independent of $\beta$ and $R_B$, and the collector current and C–E voltage are essentially independent of transistor current gain, for reasonable values of $\beta$. The value of $R_B$ can be increased, thus increasing the input resistance at the base, without jeopardizing the bias stability.

The constant current source can be implemented by using transistors as shown in Figure 5.60. The transistor $Q_1$ is a diode-connected transistor, but still operates in the forward-active mode. The transistor $Q_2$ must also operate in the forward-active mode ($V_{CE} \geq V_{BE}(\text{on})$).

Current $I_1$ is called the reference current and is found by writing Kirchhoff's voltage law equation around the $R_1$–$Q_1$ loop. We have

$$0 = I_1 R_1 + V_{BE}(\text{on}) + V^-  \tag{5.43(a)}$$

which yields

$$I_1 = \frac{-(V^- + V_{BE}(\text{on}))}{R_1}  \tag{5.43(b)}$$

Since $V_{BE1} = V_{BE2}$, the circuit mirrors the reference current in the left branch into the right branch. The circuit of $R_1$, $Q_1$, and $Q_2$ is then referred to as a current mirror.

Summing the currents at the collector of $Q_1$ gives

$$I_1 = I_{C1} + I_{B1} + I_{B2}  \tag{5.44}$$

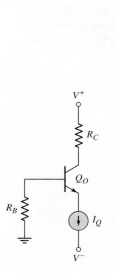

**Figure 5.59** Bipolar transistor biased with a constant-current source

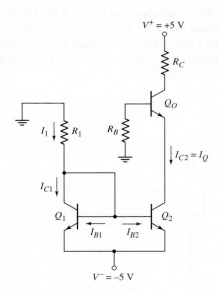

**Figure 5.60** Transistor $Q_O$ biased with a constant current source. The transistors $Q_1$ and $Q_2$ form a current mirror.

The B–E voltages of $Q_1$ and $Q_2$ are equal. If $Q_1$ and $Q_2$ are identical transistors and are held at the same temperature, then $I_{B1} = I_{B2}$ and $I_{C1} = I_{C2}$. Equation (5.44) can then be written as

$$I_1 = I_{C1} + 2I_{B2} = I_{C2} + \frac{2I_{C2}}{\beta} = I_{C2}\left(1 + \frac{2}{\beta}\right) \tag{5.45}$$

Solving for $I_{C2}$, we find

$$I_{C2} = I_Q = \frac{I_1}{\left(1 + \dfrac{2}{\beta}\right)} \tag{5.46}$$

This current biases the transistor $Q_O$ in the active region.

The circuit with $Q_1$, $Q_2$, and $R_1$ is referred to as a two-transistor current source.

## EXAMPLE 5.18

**Objective:** Determine the currents in a two-transistor current source.

For the circuit in Figure 5.60, the circuit and transistor parameters are: $R_1 = 10\ \text{k}\Omega$, $\beta = 50$, and $V_{BE}(\text{on}) = 0.7\ \text{V}$.

**Solution:** The reference current is

$$I_1 = \frac{-(V^- + V_{BE}\,(\text{on}))}{R_1} = \frac{-((-5) + 0.7)}{10} = 0.43\ \text{mA}$$

From Equation (5.46), the bias current $I_Q$ is

$$I_{C2} = I_Q = \frac{I_1}{\left(1 + \dfrac{2}{\beta}\right)} = \frac{0.43}{\left(1 + \dfrac{2}{50}\right)} = 0.413\ \text{mA}$$

The base currents are then

$$I_{B1} = I_{B2} = \frac{I_{C2}}{\beta} = \frac{0.413}{50} \Rightarrow 8.27\ \mu\text{A}$$

**Comment:** For relatively large values of current gain $\beta$, the bias current $I_Q$ is essentially the same as the reference current $I_1$.

### EXERCISE PROBLEM

**Ex 5.18:** In the circuit shown in Figure 5.60, the parameters are $V^+ = 3.3\ \text{V}$, $V^- = -3.3\ \text{V}$, and $R_B = 0$. The transistor parameters are $\beta = 60$ and $V_{BE}(\text{on}) = 0.7\ \text{V}$. Design the circuit such that $I_{CQ}(Q_O) = 0.12\ \text{mA}$ and $V_{CEQ}(Q_O) = 1.6\ \text{V}$. What are the values of $I_Q$ and $I_1$? (Ans. $I_Q = 0.122\ \text{mA}$, $I_1 = 0.126\ \text{mA}$, $R_1 = 20.6\ \text{k}\Omega$, $R_C = 20\ \text{k}\Omega$)

As mentioned, constant-current biasing is used almost exclusively in integrated circuits. As we will see in Part 2 of the text, circuits in integrated circuits use a minimum number of resistors, and transistors are often used to replace these resistors. Transistors take up much less area than resistors on an IC chip, so it's advantageous to minimize the number of resistors.

## Test Your Understanding

**TYU 5.19** The parameters of the circuit shown in Figure 5.57(a) are $V^+ = 5$ V, $V^- = -5$ V, $R_E = 0.5$ k$\Omega$, and $R_C = 4.5$ k$\Omega$. The transistor parameters are $\beta = 120$ and $V_{EB}$(on) $= 0.7$ V. Design a bias-stable circuit such that the $Q$-point is in the center of the load line. What are the values of $I_{CQ}$ and $V_{ECQ}$? (Ans. $I_{CQ} = 1$ mA, $V_{ECQ} = 5$ V, $R_1 = 6.92$ k$\Omega$, $R_2 = 48.1$ k$\Omega$)

**TYU 5.20** For Figure 5.59, the circuit parameters are $I_Q = 0.25$ mA, $V^+ = 2.5$ V, $V^- = -2.5$ V, $R_B = 75$ k$\Omega$, and $R_C = 4$ k$\Omega$. The transistor parameters are $I_S = 3 \times 10^{-14}$ A and $\beta = 120$. (a) Determine the dc voltage at the base of the transistor and also $V_{CEQ}$. (b) Repeat part (a) for $\beta = 60$. (Ans. (a) $V_B = -0.155$ V, $V_{CEQ} = 2.26$V; (b) $V_B = -0.307$ V, $V_{CEQ} = 2.42$ V)

## 5.5 MULTISTAGE CIRCUITS

**Objective:** • Consider the dc biasing of multistage or multitransistor circuits.

Most transistor circuits contain more than one transistor. We can analyze and design these multistage circuits in much the same way as we studied single-transistor circuits. As an example, Figure 5.61 shows an npn transistor, $Q_1$, and a pnp bipolar transistor, $Q_2$, in the same circuit.

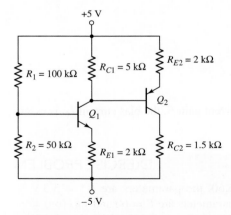

**Figure 5.61** A multistage transistor circuit

### EXAMPLE 5.19

**Objective:** Calculate the dc voltages at each node and the dc currents through the elements in a multistage circuit.

For the circuit in Figure 5.61, assume the B–E turn-on voltage is 0.7 V and $\beta = 100$ for each transistor.

**Solution:** The Thevenin equivalent circuit of the base circuit of $Q_1$ is shown in Figure 5.62. The various currents and nodal voltages are defined as shown. The Thevenin resistance and voltage are

$$R_{TH} = R_1 \| R_2 = 100 \| 50 = 33.3 \text{ k}\Omega$$

and

$$V_{TH} = \left( \frac{R_2}{R_1 + R_2} \right)(10) - 5 = \left( \frac{50}{150} \right)(10) - 5 = -1.67 \text{ V}$$

Kirchhoff's voltage law equation around the B–E loop of $Q_1$ is

$$V_{TH} = I_{B1}R_{TH} + V_{BE} \text{ (on)} + I_{E1}R_{E1} - 5$$

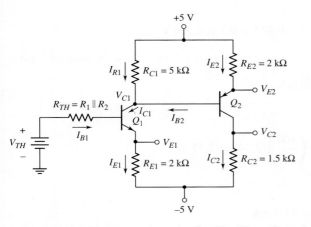

**Figure 5.62** Multistage transistor circuit with a Thevenin equivalent circuit in the base of $Q_1$

Noting that $I_{E1} = (1 + \beta)I_{B1}$, we have

$$I_{B1} = \frac{-1.67 + 5 - 0.7}{33.3 + (101)(2)} \Rightarrow 11.2 \, \mu\text{A}$$

Therefore,

$$I_{C1} = 1.12 \text{ mA}$$

and

$$I_{E1} = 1.13 \text{ mA}$$

Summing the currents at the collector of $Q_1$, we obtain

$$I_{R1} + I_{B2} = I_{C1}$$

which can be written as

$$\frac{5 - V_{C1}}{R_{C1}} + I_{B2} = I_{C1} \tag{5.47}$$

The base current $I_{B2}$ can be written in terms of the emitter current $I_{E2}$, as follows:

$$I_{B2} = \frac{I_{E2}}{1 + \beta} = \frac{5 - V_{E2}}{(1 + \beta)R_{E2}} = \frac{5 - (V_{C1} + 0.7)}{(1 + \beta)R_{E2}} \tag{5.48}$$

Substituting Equation (5.48) into (5.47), we obtain

$$\frac{5 - V_{C1}}{R_{C1}} + \frac{5 - (V_{C1} + 0.7)}{(1 + \beta)R_{E2}} = I_{C1} = 1.12 \, \text{mA}$$

which can be solved for $V_{C1}$ to yield

$$V_{C1} = -0.482 \, \text{V}$$

Then,

$$I_{R1} = \frac{5 - (-0.482)}{5} = 1.10 \, \text{mA}$$

To find $V_{E2}$, we have

$$V_{E2} = V_{C1} + V_{EB}(\text{on}) = -0.482 + 0.7 = 0.218 \, \text{V}$$

The emitter current $I_{E2}$ is

$$I_{E2} = \frac{5 - 0.218}{2} = 2.39 \, \text{mA}$$

Then,

$$I_{C2} = \left(\frac{\beta}{1 + \beta}\right) I_{E2} = \left(\frac{100}{101}\right)(2.39) = 2.37 \, \text{mA}$$

and

$$I_{B2} = \frac{I_{E2}}{1 + \beta} = \frac{2.39}{101} \Rightarrow 23.7 \, \mu\text{A}$$

The remaining nodal voltages are

$$V_{E1} = I_{E1}R_{E1} - 5 = (1.13)(2) - 5 \Rightarrow V_{E1} = -2.74 \, \text{V}$$

and

$$V_{C2} = I_{C2}R_{C2} - 5 = (2.37)(1.5) - 5 = -1.45 \, \text{V}$$

We then find that

$$V_{CE1} = V_{C1} - V_{E1} = -0.482 - (-2.74) = 2.26 \, \text{V}$$

and that

$$V_{EC2} = V_{E2} - V_{C2} = 0.218 - (-1.45) = 1.67 \, \text{V}$$

**Comment:** These results show that both $Q_1$ and $Q_2$ are biased in the forward-active mode, as originally assumed. However, when we consider the ac operation of this circuit as an amplifier in the next chapter, we will see that a better design would increase the value of $V_{EC2}$.

### EXERCISE PROBLEM

**Ex 5.19:** In the circuit shown in Figure 5.61, determine new values of $R_{C1}$ and $R_{C2}$ such that $V_{CEQ1} = 3.25$ V and $V_{ECQ2} = 2.5$ V. (Ans. $R_{C1} = 4.08$ kΩ, $R_{C2} = 1.97$ kΩ)

## EXAMPLE 5.20

**Objective:** Design the circuit shown in Figure 5.63, called a cascode circuit, to meet the following specifications: $V_{CE1} = V_{CE2} = 2.5$ V, $V_{RE} = 0.7$ V, $I_{C1} \cong I_{C2} \cong$ 1 mA, and $I_{R1} \cong I_{R2} \cong I_{R3} \cong 0.10$ mA.

**Solution:** The initial design will neglect base currents. We can then define $I_{Bias} = I_{R1} = I_{R2} = I_{R3} = 0.10$ mA. Then

$$R_1 + R_2 + R_3 = \frac{V^+}{I_{Bias}} = \frac{9}{0.10} = 90 \, k\Omega$$

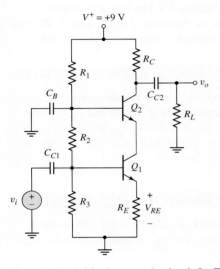

Figure 5.63 A bipolar cascode circuit for Example 5.20

The voltage at the base of $Q_1$ is

$$V_{B1} = V_{RE} + V_{BE}(on) = 0.7 + 0.7 = 1.4 \, V$$

Then

$$R_3 = \frac{V_{B1}}{I_{Bias}} = \frac{1.4}{0.10} = 14 \, k\Omega$$

The voltage at the base of $Q_2$ is

$$V_{B2} = V_{RE} + V_{CE1} + V_{BE}(on) = 0.7 + 2.5 + 0.7 = 3.9 \, V$$

Then

$$R_2 = \frac{V_{B2} - V_{B1}}{I_{Bias}} = \frac{3.9 - 1.4}{0.10} = 25 \, k\Omega$$

We then obtain

$$R_1 = 90 - 25 - 14 = 51 \, k\Omega$$

The emitter resistor $R_E$ can be found as

$$R_E = \frac{V_{RE}}{I_{C1}} = \frac{0.7}{1} = 0.7 \, k\Omega$$

The voltage at the collector of $Q_2$ is

$$V_{C2} = V_{RE} + V_{CE1} + V_{CE2} = 0.7 + 2.5 + 2.5 = 5.7 \, \text{V}$$

Then

$$R_C = \frac{V^+ - V_{C2}}{I_{C2}} = \frac{9 - 5.7}{1} = 3.3 \, \text{k}\Omega$$

**Comment:** By neglecting base currents, the design of this circuit is straightforward. A computer analysis using PSpice, for example, will verify the design or show that small changes need to be made to meet the design specifications.

We will see the cascode circuit again in Section 6.9.3 of the next chapter.

One advantage of the cascode circuit will be determined in Chapter 7. The cascode circuit has a larger bandwidth than just a simple common-emitter amplifier.

### EXERCISE PROBLEM

**Ex 5.20:** For the circuit shown in Figure 5.63, the circuit parameters are $V^+ = 12 \, \text{V}$ and $R_E = 2 \, \text{k}\Omega$, and the transistor parameters are $\beta = 120$ and $V_{BE}(\text{on}) = 0.7 \, \text{V}$. Redesign the circuit such that $I_{C1} \cong I_{C2} \cong 0.5 \, \text{mA}$, $I_{R1} \cong I_{R2} \cong I_{R3} \cong 0.05 \, \text{mA}$, and $V_{CE1} \cong V_{CE2} \cong 4 \, \text{V}$. (Ans. $R_1 = 126 \, \text{k}\Omega$, $R_2 = 80 \, \text{k}\Omega$, $R_3 = 34 \, \text{k}\Omega$, and $R_C = 6 \, \text{k}\Omega$)

### COMPUTER ANALYSIS EXERCISE

**PS 5.4:** (a) Verify the cascode circuit design in Example 5.20 using a PSpice simulation. Use standard transistors. (b) Repeat part (a) using standard resistor values.

##  5.6    DESIGN APPLICATION: DIODE THERMOMETER WITH A BIPOLAR TRANSISTOR

**Objective:** • Incorporate a bipolar transistor in a design application that enhances the simple diode thermometer design discussed in Chapter 1.

**Specifications:** The electronic thermometer is to operate over a temperature range of 0 to 100 °F.

**Design Approach:** The output-diode voltage developed in the diode thermometer in Figure 1.48 is to be applied to the base–emitter junction of an npn bipolar transistor to enhance the voltage over the temperature range. The bipolar transistor will be held at a constant temperature.

**Choices:** Assume an npn bipolar transistor with $I_S = 10^{-12} \, \text{A}$ is available.

**Solution:** From the design in Chapter 1, the diode voltage is given by

$$V_D = 1.12 - 0.522 \left( \frac{T}{300} \right)$$

where $T$ is in kelvins.

Consider the circuit shown in Figure 5.64. We assume that the diode is in a variable temperature environment while the rest of the circuit is held at room temperature. Neglecting the bipolar transistor base current, we have

$$V_D = V_{BE} + I_C R_E \tag{5.49}$$

We can write

$$I_C = I_S e^{V_{BE}/V_T} \tag{5.50}$$

so that Equation (5.49) becomes

$$\frac{V_D - V_{BE}}{R_E} = I_S e^{V_{BE}/V_T} \tag{5.51}$$

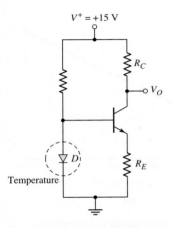

**Figure 5.64** Design application circuit to measure output voltage of diode versus temperature

and

$$V_O = 15 - I_C R_C \tag{5.52}$$

From Chapter 1, we have the following:

| $T$ (°F) | $V_D$ (V) |
|---|---|
| 0 | 0.6760 |
| 40 | 0.6372 |
| 80 | 0.5976 |
| 100 | 0.5790 |

If we assume that $I_S = 10^{-12}$ A for the transistor, then from Equations (5.50), (5.51), and (5.52), we find

| $T$ (°F) | $V_{BE}$ (V) | $I_C$ (mA) | $V_O$ (V) |
|---|---|---|---|
| 0 | 0.5151 | 0.402 | 4.95 |
| 40 | 0.5092 | 0.320 | 7.00 |
| 80 | 0.5017 | 0.240 | 9.00 |
| 100 | 0.4974 | 0.204 | 9.90 |

**Comment:** Figure 5.65(a) shows the diode voltage versus temperature and Figure 5.65(b) now shows the output voltage versus temperature from the bipolar transistor circuit. We can see that the transistor circuit provides a voltage gain. This voltage gain is the desired characteristic of the transistor circuit.

**Discussion:** We can see from the equations that the collector current is not a linear function of the base–emitter voltage or diode voltage. This effect implies that the transistor output voltage is also not exactly a linear function of temperature. The line drawn in Figure 5.65(b) is a good linear approximation. We will obtain a better circuit design using operational amplifiers in Chapter 9.

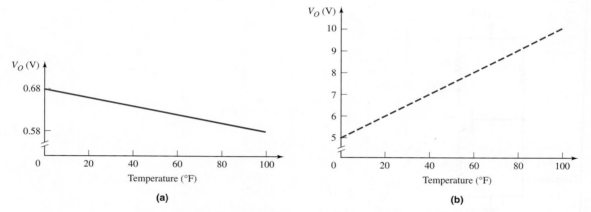

**Figure 5.65** (a) Diode voltage versus temperature and (b) circuit output voltage versus temperature

 **5.7 SUMMARY**

- In this chapter, we considered the structure, characteristics, and properties of the bipolar transistor. Both npn and pnp complementary bipolar transistors can be formed. The defining transistor action is that the voltage across two terminals (base and emitter) controls the current in the third terminal (collector).
- The four modes of operation are: forward-active, cutoff, saturation, and inverse-active. In the forward-active mode, the B–E junction is forward biased, the B–C junction is reverse biased, and the collector and base currents are related by the common-emitter current gain $\beta$. When the transistor is cut off, all currents are zero. In the saturation mode, the collector current is no longer a function of base current.
- The dc analysis and design techniques of bipolar transistor circuits were emphasized in this chapter. We continued to use the piecewise linear model of the pn junction in these analyses and designs. Techniques to design a transistor circuit with a stable $Q$-point were developed.
- An introduction to dc biasing of integrated circuits using constant current circuits was presented.

- Basic applications of the transistor include switching currents and voltages, performing digital logic functions, and amplifying time-varying signals. The amplifying characteristics of bipolar transistor circuits are considered in detail in the next chapter.
- An introduction to dc biasing in mutistage circuits was given.
- As an application, the bipolar transistor was incorporated in a circuit design that enhances the simple diode thermometer discussed in Chapter 1.

 CHECKPOINT

After studying this chapter, the reader should have the ability to:

✓ Understand and describe the structure and general current–voltage characteristics for both the npn and pnp bipolar transistors.
✓ Apply the piecewise linear model to the dc analysis and design of various bipolar transistor circuits, including the understanding of the load line.
✓ Define the four modes of operation of a bipolar transistor.
✓ Qualitatively understand how a transistor circuit can be used to switch currents and voltages, to perform digital logic functions, and to amplify time-varying signals.
✓ Design the dc biasing of a transistor circuit to achieve specified dc currents and voltages, and to stabilize the $Q$-point against transistor parameter variations.
✓ Apply the dc analysis and design techniques to multistage transistor circuits.

REVIEW QUESTIONS

1. Describe the basic structure and operation of npn and pnp bipolar transistors.
2. What are the bias voltages that need to be applied to an npn bipolar transistor such that the transistor is biased in the forward-active mode?
3. Define the conditions for cutoff, forward-active mode, and saturation mode for a pnp bipolar transistor.
4. Define common-base current gain and common-emitter current gain.
5. Discuss the difference between the ac and dc common-emitter current gains.
6. State the relationships between collector, emitter, and base currents in a bipolar transistor biased in the forward-active mode.
7. Define Early voltage and collector output resistance.
8. Describe a simple common-emitter circuit with an npn bipolar transistor and discuss the relation between collector–emitter voltage and input base current.
9. Describe the parameters that define a load line. Define $Q$-point.
10. What are the steps used to analyze the dc response of a bipolar transistor circuit?
11. Describe how an npn transistor can be used to switch an LED diode on and off.
12. Describe a bipolar transistor NOR logic circuit.
13. Describe how a transistor can be used to amplify a time-varying voltage.
14. Discuss the advantages of using resistor voltage divider biasing compared to a single base resistor.
15. How can the $Q$-point be stabilized against variations in transistor parameters?
16. What is the principal difference between biasing techniques used in discrete transistor circuits and integrated circuits?

##  PROBLEMS

[Note: In the following problems, unless otherwise stated, assume $V_{BE}(\text{on}) = 0.7$ V and $V_{CE}(\text{sat}) = 0.2$ V for npn transistors, and assume $V_{EB}(\text{on}) = 0.7$ V and $V_{EC}(\text{sat}) = 0.2$ V for pnp transistors.]

### Section 5.1 Basic Bipolar Junction Transistor

5.1  (a) In a bipolar transistor biased in the forward-active region, the base current is $i_B = 2.8\,\mu$A and the emitter current is $i_E = 325\,\mu$A. Determine $\beta$, $\alpha$, and $i_C$. (b) Repeat part (a) if $i_B = 20\,\mu$A and $i_E = 1.80$ mA.

5.2  (a) A bipolar transistor is biased in the forward-active mode. The collector current is $i_C = 726\,\mu$A and the emitter current is $i_E = 732\,\mu$A. Determine $\beta, \alpha$, and $i_B$. (b) Repeat part (a) if $i_C = 2.902$ mA and $i_E = 2.961$ mA.

5.3  (a) The range of $\beta$ for a particular type of transistor is $110 \leq \beta \leq 180$. Determine the corresponding range of $\alpha$. (b) If the base current is $50\,\mu$A, determine the range of collector current.

5.4  (a) A bipolar transistor is biased in the forward-active mode. The measured parameters are $i_E = 1.25$ mA and $\beta = 150$. Determine $i_B$, $i_C$, and $\alpha$. (b) Repeat part (a) for $i_E = 4.52$ mA and $\beta = 80$.

5.5  (a) For the following values of common-base current gain $\alpha$, determine the corresponding common-emitter current gain $\beta$:

| $\alpha$ | 0.90 | 0.950 | 0.980 | 0.990 | 0.995 | 0.9990 |
|---|---|---|---|---|---|---|
| $\beta$ | | | | | | |

(b) For the following values of common-emitter current gain $\beta$, determine the corresponding common-base current gain $\alpha$:

| $\beta$ | 20 | 50 | 100 | 150 | 220 | 400 |
|---|---|---|---|---|---|---|
| $\alpha$ | | | | | | |

5.6  An npn transistor with $\beta = 80$ is connected in a common-base configuration as shown in Figure P5.6. (a) The emitter is driven by a constant-current source with $I_E = 1.2$ mA. Determine $I_B$, $I_C$, $\alpha$, and $V_C$. (b) Repeat part (a) for $I_E = 0.80$ mA. (c) Repeat parts (a) and (b) for $\beta = 120$.

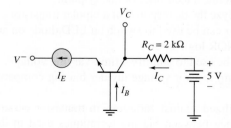

Figure P5.6

5.7   The emitter current in the circuit in Figure P5.6 is $I_E = 0.80$ mA. The transistor parameters are $\alpha = 0.9910$ and $I_{EO} = 5 \times 10^{-14}$ A. Determine $I_B$, $I_C$, $V_{BE}$, and $V_C$.

5.8   A pnp transistor with $\beta = 60$ is connected in a common-base configuration as shown in Figure P5.8. (a) The emitter is driven by a constant-current source with $I_E = 0.75$ mA. Determine $I_B$, $I_C$, $\alpha$, and $V_C$. (b) Repeat part (a) if $I_E = 1.5$ mA. (c) Is the transistor biased in the forward-active mode for both parts (a) and (b)? Why or why not?

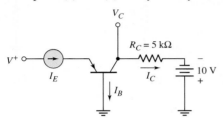

**Figure P5.8**

5.9   (a) The pnp transistor shown in Figure P5.8 has a common-base current gain $\alpha = 0.9860$. Determine the emitter current such that $V_C = -1.2$ V. What is the base current? (b) Using the results of part (a) and assuming $I_{EO} = 2 \times 10^{-15}$ A, determine $V_{EB}$.

5.10  An npn transistor has a reverse-saturation current of $I_S = 5 \times 10^{-15}$ A and a current gain of $\beta = 125$. The transistor is biased at $v_{BE} = 0.615$ V. Determine $i_B$, $i_C$, and $i_E$.

5.11  Two pnp transistors, fabricated with the same technology, have different junction areas. Both transistors are biased with an emitter-base voltage of $v_{EB} = 0.650$ V and have emitter currents of 0.50 and 12.2 mA. Find $I_{EO}$ for each device. What are the relative junction areas?

5.12  The collector currents in two transistors, $A$ and $B$, are both $i_C = 275$ $\mu$A. For transistor $A$, $I_{SA} = 8 \times 10^{-16}$ A. The base–emitter area of transistor $B$ is 4 times that of transistor $A$. Determine $I_{SB}$ and the base–emitter voltage of each transistor.

5.13  A BJT has an Early voltage of 80 V. The collector current is $I_C = 0.60$ mA at a collector–emitter voltage of $V_{CE} = 2$ V. (a) Determine the collector current at $V_{CE} = 5$ V. (b) What is the output resistance?

5.14  The open-emitter breakdown voltage of a B–C junction is $BV_{CBO} = 60$ V. If $\beta = 100$ and the empirical constant is $n = 3$, determine the C–E breakdown voltage in the open-base configuration.

5.15  In a particular circuit application, the minimum required breakdown voltages are $BV_{CBO} = 220$ V and $BV_{CEO} = 56$ V. If $n = 3$, determine the maximum allowed value of $\beta$.

5.16  A particular transistor circuit design requires a minimum open-base breakdown voltage of $BV_{CEO} = 50$ V. If $\beta = 50$ and $n = 3$, determine the minimum required value of $BV_{CBO}$.

## Section 5.2  DC Analysis of Transistor Circuits

5.17  For all the transistors in Figure P5.17, $\beta = 75$. The results of some measurements are indicated on the figures. Find the values of the other labeled currents, voltages, and/or resistor values.

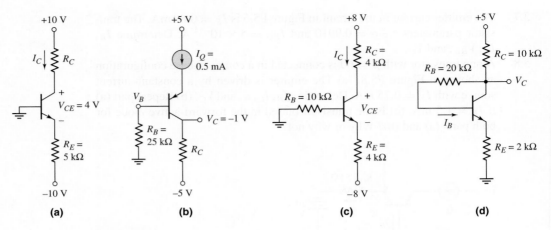

**Figure P5.17**

5.18 The emitter resistor values in the circuits show in Figures P5.17(a) and (c) may vary by ±5 percent from the given value. Determine the range of calculated parameters.

5.19 Consider the two circuits in Figure P5.19. The parameters of each transistor are $I_S = 5 \times 10^{-16}$ A and $\beta = 90$. Determine $V_{BB}$ in each circuit such that $V_{CE} = 1.10$ V.

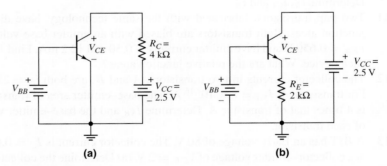

**Figure P5.19**

5.20 The current gain for each transistor in the circuits shown in Figure P5.20 is $\beta = 120$. For each circuit, determine $I_C$ and $V_{CE}$.

5.21 Consider the circuits in Figure P5.21. For each transistor, $\beta = 120$. Determine $I_C$ and $V_{EC}$ for each circuit.

5.22 (a) The circuit and transistor parameters for the circuit shown in Figure 5.20(a) are $V_{CC} = 3$ V, $V_{BB} = 1.3$ V, and $\beta = 100$. Redesign the circuit such that $I_{BQ} = 5\ \mu$A and $V_{CEQ} = 1.5$ V. (b) Using the results of part (a), determine the variation in $V_{CEQ}$ if $\beta$ is in the range $75 \leq \beta \leq 125$.

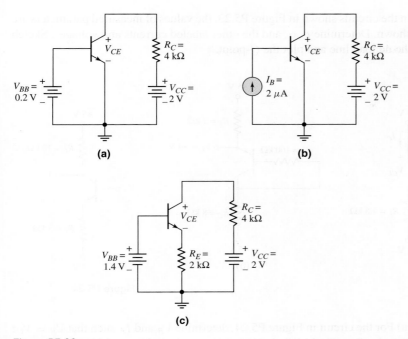

**Figure P5.20**

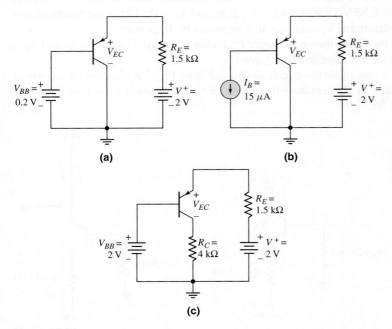

**Figure P5.21**

5.23    In the circuits shown in Figure P5.23, the values of measured parameters are shown. Determine $\beta$, $\alpha$, and the other labeled currents and voltages. Sketch the dc load line and plot the $Q$-point.

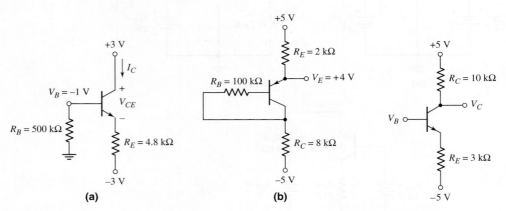

**Figure P5.23**                                                                                    **Figure P5.24**

5.24    (a) For the circuit in Figure P5.24, determine $V_B$ and $I_E$ such that $V_B = V_C$. Assume $\beta = 90$. (b) What value of $V_B$ results in $V_{CE} = 2$ V?

5.25    (a) The bias voltages in the circuit shown in Figure P5.25 are changed to $V^+ = 3.3$ V and $V^- = -3.3$ V. The measured value of emitter voltage is $V_E = 0.85$ V. Determine $I_E$, $I_C$, $\beta$, $\alpha$, and $V_{EC}$. (b) Using the results of part (a), determine $V_E$ and $V_{EC}$ if $\beta$ increases by 10 percent.

5.26    The transistor shown in Figure P5.26 has $\beta = 120$. Determine $I_C$ and $V_{EC}$. Plot the load line and the $Q$-point.

5.27    The transistor in the circuit shown in Figure P5.27 is biased with a constant current in the emitter. If $I_Q = 1$ mA, determine $V_C$ and $V_E$. Assume $\beta = 50$.

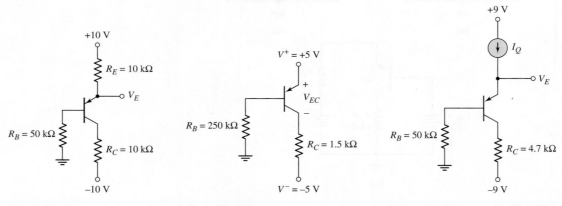

**Figure P5.25**                          **Figure P5.26**                          **Figure P5.27**

5.28    In the circuit in Figure P5.27, the constant current is $I = 0.5$ mA. If $\beta = 50$, determine the power dissipated in the transistor. Does the constant current source supply or dissipate power? What is the value?

5.29 For the circuit shown in Figure P5.29, if $\beta = 200$ for each transistor, determine: (a) $I_{E1}$, (b) $I_{E2}$, (c) $V_{C1}$, and (d) $V_{C2}$.

5.30 The circuit shown in Figure P5.30 is to be designed such that $I_{CQ} = 0.8$ mA and $V_{CEQ} = 2$ V for the case when (a) $R_E = 0$ and (b) $R_E = 1$ k$\Omega$. Assume $\beta = 80$. (c) The transistor in Figure P5.30 is replaced with one with a value of $\beta = 120$. Using the results of parts (a) and (b), determine the $Q$-point values $I_{CQ}$ and $V_{CEQ}$. Which design shows the smallest change in $Q$-point values?

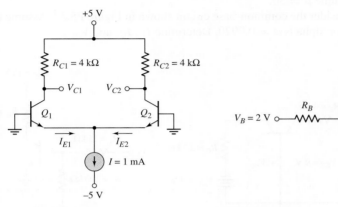

Figure P5.29

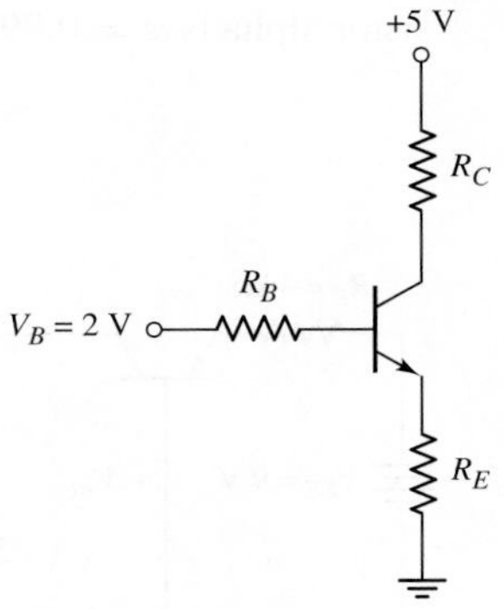

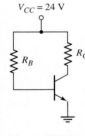

Figure P5.30

D5.31 (a) The bias voltage in the circuit in Figure P5.31 is changed to $V_{CC} = 9$ V. The transistor current gain is $\beta = 80$. Design the circuit such that $I_{CQ} = 0.25$ mA and $V_{CEQ} = 4.5$ V. (b) If the transistor is replaced by a new one with $\beta = 120$, find the new values of $I_{CQ}$ and $V_{CEQ}$. (c) Sketch the load line and $Q$-point for both parts (a) and (b).

Figure P5.31

5.32 The current gain of the transistor in the circuit shown in Figure P5.32 is $\beta = 150$. Determine $I_C$, $I_E$, and $V_C$ for (a) $V_B = 0.2$ V, (b) $V_B = 0.9$ V, (c) $V_B = 1.5$ V, and (d) $V_B = 2.2$ V.

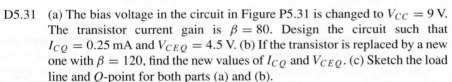

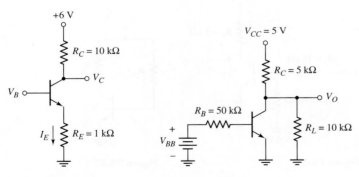

Figure P5.32

Figure P5.33

5.33 (a) The current gain of the transistor in Figure P5.33 is $\beta = 75$. Determine $V_O$ for: (i) $V_{BB} = 0$, (ii) $V_{BB} = 1$ V, and (iii) $V_{BB} = 2$ V. (b) Verify the results of part (a) with a computer simulation.

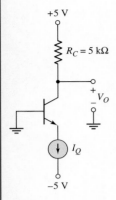

**Figure P5.34**

5.34 (a) The transistor shown in Figure P5.34 has $\beta = 100$. Determine $V_O$ for (i) $I_Q = 0.1$ mA, (ii) $I_Q = 0.5$ mA, and (iii) $I_Q = 2$ mA. (b) Determine the percent change in $V_O$ for the conditions in part (a) if the current gain increases to $\beta = 150$.

5.35 Assume $\beta = 120$ for the transistor in the circuit shown in Figure P5.34. Determine $I_Q$ such that (a) $V_O = 4$ V, (b) $V_O = 2$ V, and (c) $V_O = 0$.

5.36 For the circuit shown in Figure P5.27, calculate and plot the power dissipated in the transistor for $I_Q = 0$, 0.5, 1.0, 1.5, 2.0, 2.5, and 3.0 mA. Assume $\beta = 50$.

5.37 Consider the common-base circuit shown in Figure P5.37. Assume the transistor alpha is $\alpha = 0.9920$. Determine $I_E$, $I_C$, and $V_{BC}$.

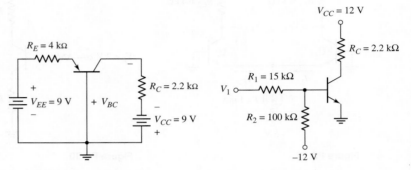

**Figure P5.37**  **Figure P5.38**

5.38 (a) For the transistor in Figure P5.38, $\beta = 80$. Determine $V_1$ such that $V_{CEQ} = 6$ V. (b) Determine the range in $V_1$ that produces $3 \leq V_{CEQ} \leq 9$ V.

5.39 Let $\beta = 25$ for the transistor in the circuit shown in Figure P5.39. Determine the range of $V_1$ such that $1.0 \leq V_{CE} \leq 4.5$ V. Sketch the load line and show the range of the $Q$-point values.

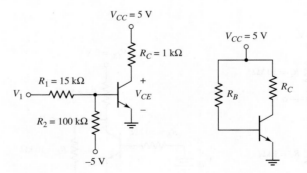

**Figure P5.39**  **Figure P5.40**

D5.40 (a) The circuit shown in Figure P5.40 is to be designed such that $I_{CQ} = 0.5$ mA and $V_{CEQ} = 2.5$ V. Assume $\beta = 120$. Sketch the load line and plot the $Q$-point. (b) Pick standard values of resistors that are close to the designed values. Assume that the standard resistor values vary by $\pm 10$

percent. Plot the load lines and $Q$-point values for the maximum and minimum values of $R_B$ and $R_C$ values (four $Q$-point values).

5.41 The circuit shown in Figure P5.41 is sometimes used as a thermometer. Assume the transistors $Q_1$ and $Q_2$ in the circuit are identical. Writing the emitter currents in the form $I_E = I_{EO} \exp(V_{BE}/V_T)$, derive the expression for the output voltage $V_O$ as a function of temperature $T$.

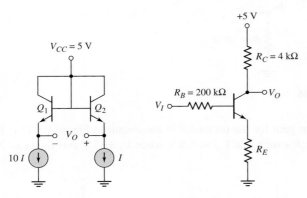

Figure P5.41

Figure P5.42

5.42 The transistor in Figure P5.42 has $\beta = 120$. (a) Determine $V_I$ that produces $V_O = 4\,\text{V}$ for (i) $R_E = 0$ and (ii) $R_E = 1\,\text{k}\Omega$. (b) Repeat part (a) for $V_O = 2.5\,\text{V}$. (c) Determine $V_O$ for $V_I = 3.5\,\text{V}$ and for $R_E = 1\,\text{k}\Omega$.

5.43 The common-emitter current gain of the transistor in Figure P5.43 is $\beta = 80$. Plot the voltage transfer characteristics over the range $0 \le V_I \le 5\,\text{V}$.

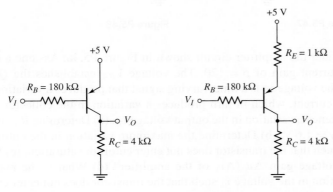

Figure P5.43

Figure P5.44

5.44 For the circuit shown in Figure P5.44, plot the voltage transfer characteristics over the range $0 \le V_I \le 5\,\text{V}$. Assume $\beta = 100$.

## Section 5.3 Basic Transistor Applications

5.45 The transistor in the circuit shown in Figure P5.45 has a current gain of $\beta = 40$. Determine $R_B$ such that $V_O = 0.2\,\text{V}$ and $I_C/I_B = 20$ when $V_I = 5\,\text{V}$.

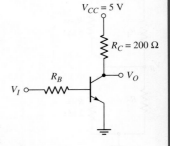

Figure P5.45

5.46   Consider the circuit in Figure P5.46. For the transistor, $\beta = 50$. Find $I_B$, $I_C$, $I_E$, and $V_O$ for (a) $V_I = 0$, (b) $V_I = 2.5$ V, and (c) $V_I = 5$ V.

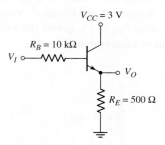

**Figure P5.46**

5.47   The current gain for the transistor in the circuit in Figure P5.47 is $\beta = 60$. Determine $R_B$ such that $V_O = 8.8$ V when $V_I = 5$ V and $I_C/I_B = 25$.

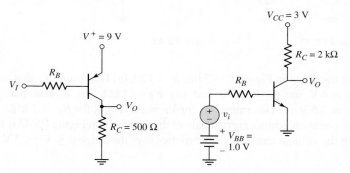

**Figure P5.47**                    **Figure P5.48**

5.48   Consider the amplifier circuit shown in Figure P5.48. Assume a transistor current gain of $\beta = 120$. The voltage $V_{BB}$ establishes the $Q$-point, and the voltage $v_i$ is a time-varying signal that produces a variation in the base current, which in turn produces a variation in the collector current and hence a variation in the output voltage $v_o$. (a) Determine $R_B$ such that $V_{CEQ} = 1.6$ V. (b) Determine the maximum variation in the output voltage such that the transistor does not enter cutoff or saturation. (c) What is the voltage gain $\Delta v_o / \Delta v_i$ of the amplifier? (d) What is the maximum variation in the voltage $v_i$ such that the transistor does not enter cutoff or saturation?

## Section 5.4   Bipolar Transistor Biasing

D5.49   For the transistor in the circuit shown in Figure P5.49, assume $\beta = 120$. Design the circuit such that $I_{CQ} = 0.15$ mA and $R_{TH} = 200$ k$\Omega$. What is the value of $V_{CEQ}$?

5.50   Reconsider Figure P5.49. The transistor current gain is $\beta = 150$. The circuit parameters are changed to $R_{TH} = 120$ k$\Omega$ and $R_E = 1$ k$\Omega$. Determine the values of $R_C$, $R_1$, and $R_2$ such that $V_{CEQ} = 1.5$ V and $I_{CQ} = 0.20$ mA.

5.51   The current gain of the transistor shown in the circuit of Figure P5.51 is $\beta = 100$. Determine $V_B$ and $I_{EQ}$.

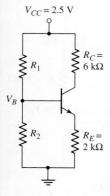

**Figure P5.49**

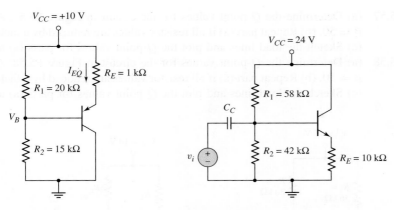

Figure P5.51                    Figure P5.52

5.52    For the circuit shown in Figure P5.52, let $\beta = 125$. (a) Find $I_{CQ}$ and $V_{CEQ}$.
        Sketch the load line and plot the $Q$-point. (b) If the resistors $R_1$ and $R_2$ vary
        by $\pm 5$ percent, determine the range in $I_{CQ}$ and $V_{CEQ}$. Plot the various
        $Q$-points on the load line.
5.53    Consider the circuit shown in Figure P5.53. (a) Determine $I_{BQ}$, $I_{CQ}$, and
        $V_{CEQ}$ for $\beta = 80$. (b) What is the percent change in $I_{CQ}$ and $V_{CEQ}$ if $\beta$ is
        changed to $\beta = 120$?
5.54    (a) Redesign the circuit shown in Figure P5.49 using $V_{CC} = 9$ V such
        that the voltage drop across $R_C$ is $(\frac{1}{3})V_{CC}$ and the voltage drop across $R_E$ is
        $(\frac{1}{3})V_{CC}$. Assume $\beta = 100$. The quiescent collector current is to be $I_{CQ} =$
        0.4 mA, and the current through $R_1$ and $R_2$ should be approximately
        $0.2I_{CQ}$. (b) Replace each resistor in part (a) with the closest standard value
        (Appendix C). What is the value of $I_{CQ}$ and what are the voltage drops
        across $R_C$ and $R_E$?
5.55    For the circuit shown in Figure P5.55, let $\beta = 100$. (a) Find $R_{TH}$ and $V_{TH}$
        for the base circuit. (b) Determine $I_{CQ}$ and $V_{CEQ}$. (c) Draw the load line and
        plot the $Q$-point. (d) If the resistors $R_C$ and $R_E$ vary by $\pm 5$ percent, deter-
        mine the range in $I_{CQ}$ and $V_{CEQ}$. Draw the load lines corresponding to the
        maximum and minimum resistor values and plot the $Q$-points.
5.56    Consider the circuit shown in Figure P5.56. (a) Determine $R_{TH}$, $V_{TH}$, $I_{BQ}$,
        $I_{CQ}$, and $V_{ECQ}$ for $\beta = 90$. (b) Determine the percent change in $I_{CQ}$ and
        $V_{ECQ}$ if $\beta$ is changed to $\beta = 150$.

Figure P5.53

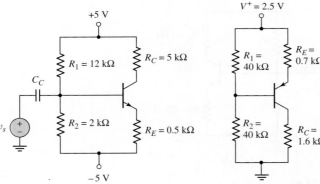

Figure P5.55                    Figure P5.56

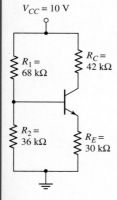

$V_{CC} = 10$ V

$R_1 = 68$ kΩ     $R_C = 42$ kΩ

$R_2 = 36$ kΩ     $R_E = 30$ kΩ

**Figure P5.57**

5.57   (a) Determine the $Q$-point values for the circuit in Figure P5.57. Assume $\beta = 50$. (b) Repeat part (a) if all resistor values are reduced by a factor of 3. (c) Sketch the load lines and plot the $Q$-point values for parts (a) and (b).

5.58   (a) Determine the $Q$-point values for the circuit in Figure P5.58. Assume $\beta = 50$. (b) Repeat part (a) if all resistor values are reduced by a factor of 3. (c) Sketch the load lines and plot the $Q$-point values for parts (a) and (b).

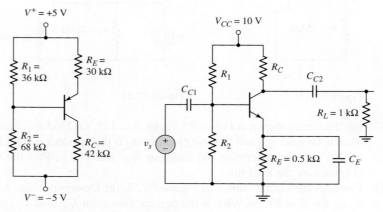

$V^+ = +5$ V

$R_1 = 36$ kΩ     $R_E = 30$ kΩ

$R_2 = 68$ kΩ     $R_C = 42$ kΩ

$V^- = -5$ V

**Figure P5.58**

$V_{CC} = 10$ V

$R_1$     $R_C$

$C_{C1}$     $C_{C2}$

$R_L = 1$ kΩ

$v_s$     $R_2$

$R_E = 0.5$ kΩ     $C_E$

**Figure P5.59**

D5.59   (a) For the circuit shown in Figure P5.59, design a bias-stable circuit such that $I_{CQ} = 0.8$ mA and $V_{CEQ} = 5$ V. Let $\beta = 100$. (b) Using the results of part (a), determine the percentage change in $I_{CQ}$ if $\beta$ is in the range $75 \le \beta \le 150$. (c) Repeat parts (a) and (b) if $R_E = 1$ kΩ.

D5.60   Design a bias-stable circuit in the form of Figure P5.59 with $\beta = 120$ such that $I_{CQ} = 0.8$ mA, $V_{CEQ} = 5$ V, and the voltage across $R_E$ is approximately 0.7 V.

D5.61   Using the circuit in Figure P5.61, design a bias-stable amplifier such that the $Q$-point is in the center of the load line. Let $\beta = 125$. Determine $I_{CQ}$, $V_{CEQ}$, $R_1$, and $R_2$.

D5.62   For the circuit shown in Figure P5.61, the bias voltages are changed to $V^+ = 3$ V and $V^- = -3$ V. (a) Design a bias-stable circuit for $\beta = 120$

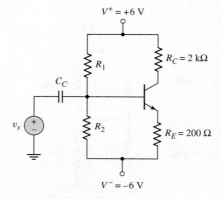

$V^+ = +6$ V

$R_1$     $R_C = 2$ kΩ

$C_C$

$v_s$     $R_2$     $R_E = 200$ Ω

$V^- = -6$ V

**Figure P5.61**

such that $V_{CEQ} = 2.8$ V. Determine $I_{CQ}$, $R_1$, and $R_2$. (b) If the resistors $R_1$ and $R_2$ vary by $\pm 5$ percent, determine the range in $I_{CQ}$ and $V_{CEQ}$. Plot the various $Q$-points on the load line.

5.63 (a) A bias-stable circuit with the configuration shown in Figure P5.61 is to be designed such that $I_{CQ} = (3 \pm 0.1)$ mA and $V_{CEQ} \cong 5$ V using a transistor with $75 \leq \beta \leq 150$. (b) Sketch the load line and plot the range of $Q$-point values for part (a).

D5.64 (a) For the circuit shown in Figure P5.64, assume that the transistor current gain is $\beta = 90$ and that the circuit parameter is $R_{TH} = 2.4$ k$\Omega$. Design the circuit such that $V_{ECQ} = 1.5$ V. Find $I_{BQ}$, $I_{CQ}$, $R_1$, and $R_2$. (b) Determine the values of $I_{BQ}$, $I_{CQ}$, and $V_{ECQ}$ if the current gain is changed to $\beta = 130$.

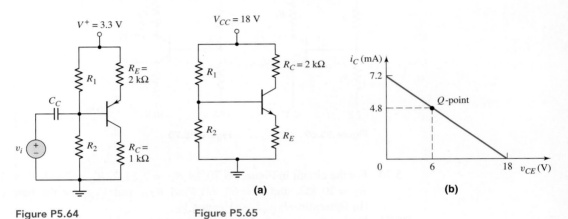

Figure P5.64          Figure P5.65

5.65 The dc load line and $Q$-point of the circuit in Figure P5.65(a) are shown in Figure P5.65(b). For the transistor, $\beta = 120$. Find $R_E$, $R_1$, and $R_2$ such that the circuit is bias stable.

D5.66 The range of $\beta$ for the transistor in the circuit in Figure P5.66 is $80 \leq \beta \leq 120$. Design a bias-stable circuit such that the nominal $Q$-point values are $I_{CQ} = 0.2$ mA and $V_{CEQ} = 1.6$ V. The value of $I_{CQ}$ must fall in the range $0.19 \leq I_{CQ} \leq 0.21$ mA. Determine $R_E$, $R_1$, and $R_2$.

D5.67 The nominal $Q$-point of the circuit in Figure P5.67 is $I_{CQ} = 1$ mA and $V_{CEQ} = 5$ V, for $\beta = 60$. The current gain of the transistor is in the range

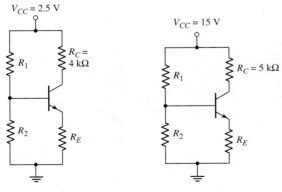

Figure P5.66          Figure P5.67

$45 \leq \beta \leq 75$. Design a bias-stable circuit such that $I_{CQ}$ does not vary by more than 5 percent from its nominal value.

D5.68  (a) For the circuit in Figure P5.67, the value of $V_{CC}$ is changed to 3 V. Let $R_C = 5R_E$ and $\beta = 120$. Redesign a bias-stable circuit such that $I_{CQ} = 100 \ \mu A$ and $V_{CEQ} = 1.4$ V. (b) Using the results of part (a), determine the dc power dissipation in the circuit.

D5.69  For the circuit in Figure P5.69, let $\beta = 100$ and $R_E = 3 \ k\Omega$. Design a bias-stable circuit such that $V_E = 0$.

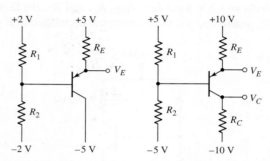

Figure P5.69          Figure P5.70

5.70   For the circuit in Figure P5.70, let $R_C = 2.2 \ k\Omega$, $R_E = 2 \ k\Omega$, $R_1 = 10 \ k\Omega$, $R_2 = 20 \ k\Omega$, and $\beta = 60$. (a) Find $R_{TH}$ and $V_{TH}$ for the base circuit. (b) Determine $I_{BQ}$, $I_{CQ}$, $V_E$, and $V_C$.

D5.71  Design the circuit in Figure P5.70 to be bias stable and to provide nominal $Q$-point values of $I_{CQ} = 0.5$ mA and $V_{ECQ} = 8$ V. Let $\beta = 60$. The maximum current in $R_1$ and $R_2$ is to be limited to $40 \ \mu A$.

D5.72  Consider the circuit shown in Figure P5.72. (a) The nominal transistor current gain is $\beta = 80$. Design a bias-stable circuit such that $I_{CQ} = 0.15$ mA and $V_{ECQ} = 2.7$ V. (b) Using the results of part (a), determine the percent change in $I_{CQ}$ and $V_{ECQ}$ if the transistor current gain is in the range $60 \leq \beta \leq 100$.

5.73   For the circuit in Figure P5.73, let $\beta = 100$. (a) Find $V_{TH}$ and $R_{TH}$ for the base circuit. (b) Determine $I_{CQ}$ and $V_{CEQ}$.

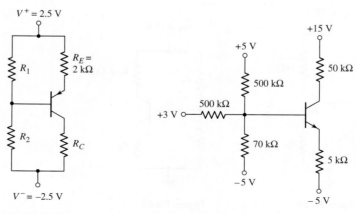

Figure P5.72          Figure P5.73

**D5.74** Design a bias-stable four-resistor bias network for an npn transistor such that $I_{CQ} = 0.8$ mA, $V_{CEQ} = 4$ V, and $V_E = 1.5$ V. The circuit and transistor parameters are $V_{CC} = 10$ V and $\beta = 120$, respectively.

**D5.75** (a) Design a four-resistor bias network with the configuration shown in Figure P5.61 to yield Q-point values of $I_{CQ} = 50$ $\mu$A and $V_{CEQ} = 5$ V. The bias voltages are $V^+ = +5$ V and $V^- = -5$ V. Assume a transistor with $\beta = 80$ is available. The voltage across the emitter resistor should be approximately 1 V. (b) The transistor in part (a) is replaced by one with $\beta = 120$. Determine the resulting Q-point.

**D5.76** (a) Design a four-resistor bias network with the configuration shown in Figure P5.61 to yield Q-point values of $I_{CQ} = 0.50$ mA and $V_{CEQ} = 2.5$ V. The bias voltages are $V^+ = 3$ V and $V^- = -3$ V. The transistor current gain is $\beta = 120$. The voltage across the emitter resistor should be approximately 0.7 V. (b) Replace the designed resistors in part (a) with standard resistors with values closest to the designed values. Determine the resulting Q-point.

**D5.77** (a) A four-resistor bias network is to be designed with the configuration shown in Figure P5.77. The Q-point values are to be $I_{CQ} = 100$ $\mu$A and $V_{ECQ} = 3$ V. The bias voltages are $V^+ = 3$ V and $V^- = -3$ V. A transistor with $\beta = 110$ is available. The voltage across the emitter resistor should be approximately 0.7 V. (b) The transistor in part (a) is replaced with one with $\beta = 150$. What is the resulting Q-point?

**D5.78** (a) Design a four-resistor bias network with the configuration shown in Figure P5.77 such that the Q-point values are $I_{CQ} = 1.2$ mA and $V_{ECQ} = 6$ V. The bias voltages are $V^+ = 9$ V and $V^- = -9$ V. A transistor with $\beta = 75$ is available. The voltage across the emitter resistor should be approximately 1.5 V. (b) Replace the designed resistors in part (a) with standard resistors with values closest to the designed values. Determine the resulting Q-point.

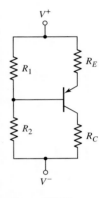

**Figure P5.77**

## Section 5.5 Multistage Circuits

**5.79** For each transistor in the circuit in Figure P5.79, $\beta = 120$ and the B–E turn-on voltage is 0.7 V. Determine the quiescent base, collector, and emitter currents in $Q_1$ and $Q_2$. Also determine $V_{CEQ1}$ and $V_{CEQ2}$.

**5.80** The parameters for each transistor in the circuit in Figure P5.80 are $\beta = 80$ and $V_{BE}(\text{on}) = 0.7$ V. Determine the quiescent values of base, collector, and emitter currents in $Q_1$ and $Q_2$.

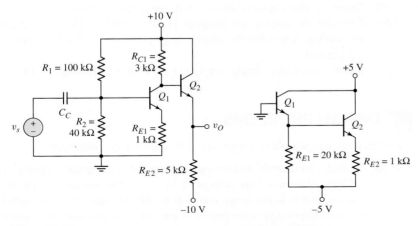

**Figure P5.79**                          **Figure P5.80**

D5.81 The bias voltage in the circuit shown in Figure 5.63 is changed to $V^+ = 5$ V. Design the circuit to meet the following specifications: $V_{CE1} = V_{CE2} = 1.2$ V, $V_{RE} = 0.5$ V, $I_{C1} \cong I_{C2} \cong 0.2$ mA, and $I_{R1} \cong I_{R2} \cong I_{R3} \cong 20\,\mu$A.

5.82 Consider the circuit shown in Figure P5.82. The current gain for the npn transistor is $\beta_n = 120$ and for the pnp transistor is $\beta_p = 80$. Determine $I_{B1}$, $I_{C1}$, $I_{B2}$, $I_{C2}$, $V_{CE1}$, and $V_{EC2}$.

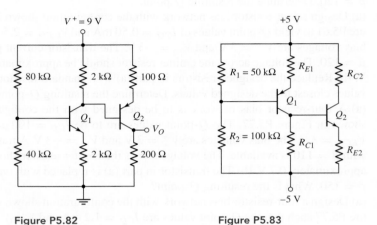

Figure P5.82                                Figure P5.83

5.83 (a) For the transistors in the circuit shown in Figure P5.83, the parameters are: $\beta = 100$ and $V_{BE}(\text{on}) = V_{EB}(\text{on}) = 0.7$ V. Determine $R_{C1}$, $R_{E1}$, $R_{C2}$, and $R_{E2}$ such that $I_{C1} = I_{C2} = 0.8$ mA, $V_{ECQ1} = 3.5$ V, and $V_{CEQ2} = 4.0$ V. (b) Correlate the results of part (a) with a computer simulation.

## COMPUTER SIMULATION PROBLEMS

5.84 Using a computer simulation, plot $V_{CE}$ versus $V_1$ over the range $0 \le V_I \le 8$ V for the circuit in Figure 5.24(a). At what voltage does the transistor turn on and at what voltage does the transistor go into saturation?

5.85 Using a computer simulation, verify the results of Example 5.7.

5.86 Consider the circuit and parameters in Example 5.15. Using a computer simulation, determine the change in $Q$-point values if all resistors vary by $\pm 5$ percent.

5.87 Using a computer simulation, verify the results of Example 5.19.

## DESIGN PROBLEMS

[Note: Each design should be correlated with a computer simulation.]

*D5.88 Consider a common-emitter circuit with the configuration shown in Figure 5.54(a). Assume a bias voltage of $V_{CC} = 3.3$ V and assume the transistor current gain is in the range $100 \le \beta \le 160$. Design the circuit such that the nominal $Q$-point is in the center of the load line and that the $Q$-point values do not vary by more than $\pm 3$ percent. Determine appropriate values for $R_1$ and $R_2$.

*D5.89 The emitter-follower circuit shown in Figure P5.89 is biased at $V^+ = 2.5\,\text{V}$ and $V^- = -2.5\,\text{V}$. Design a bias-stable circuit such that the nominal $Q$-point values are $I_{CQ} \cong 5\,\text{mA}$ and $V_{CEQ} \cong 2.5\,\text{V}$. The transistor current gain values are in the range $100 \le \beta \le 160$. Select standard 5 percent tolerance resistance values in the final design. What is the range in $Q$-point values?

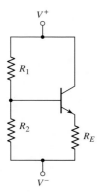

**Figure P5.89**

*D5.90 The bias voltages for the circuit in Figure 5.57(a) are $V^+ = 3.3\,\text{V}$ and $V^- = -3.3\,\text{V}$. The transistor current gain is $\beta = 100$. Design a bias-stable circuit such that $I_{CQ} \cong 120\,\mu\text{A}$, $V_{RE} \cong 0.7\,\text{V}$, and $V_{ECQ} \cong 3\,\text{V}$. Use standard resistor values in the final design.

*D5.91 The multitransistor circuit in Figure 5.61 is to be redesigned. The bias voltages are to be $\pm 3.3\,\text{V}$ and the nominal transistor current gains are $\beta = 120$. Design a bias-stable circuit such that $I_{CQ1} = 100\,\mu\text{A}$, $I_{CQ2} = 200\,\mu\text{A}$, and $V_{CEQ1} \cong V_{ECQ2} \cong 3\,\text{V}$.

# Basic BJT Amplifiers

In the previous chapter, we described the structure and operation of the bipolar junction transistor, and analyzed and designed the dc response of circuits containing these devices. In this chapter, we emphasize the use of the bipolar transistor in linear amplifier applications. Linear amplifiers imply that, for the most part, we are dealing with analog signals. The magnitude of an analog signal may have any value, within limits, and may vary continuously with respect to time. A linear amplifier then means that the output signal is equal to the input signal multiplied by a constant, where the magnitude of the constant of proportionality is, in general, greater than unity.

## PREVIEW

In this chapter, we will:

- Investigate the process by which a transistor circuit can amplify a small, time-varying input signal, and develop the small-signal models of the transistor that are used in the analysis of linear amplifiers.
- Discuss the three basic transistor amplifier configurations.
- Analyze the common-emitter amplifier and become familiar with the general characteristics of this circuit.
- Understand the concept of the ac load line and determine the maximum symmetrical swing of the output signal.
- Analyze the emitter-follower amplifier and become familiar with the general characteristics of this circuit.
- Analyze the common-base amplifier and become familiar with the general characteristics of this circuit.
- Compare the general characteristics of the three basic amplifier configurations.
- Analyze multitransistor or multistage amplifiers and understand the advantages of these circuits over single-transistor circuits.
- Understand the concept of signal power gain in an amplifier circuit.
- As an application, incorporate bipolar transistors in a design of a multistage amplifier circuit configuration to provide a specified output signal power.

## 6.1   ANALOG SIGNALS AND LINEAR AMPLIFIERS

**Objective:** • Understand the concept of an analog signal and the principle of a linear amplifier.

In this chapter, we will be considering **signals, analog circuits,** and **amplifiers.** A signal contains some type of information. For example, sound waves produced by a speaking human contain the information the person is conveying to another person. A sound wave is an analog signal. The magnitude of an **analog signal** can take on any value, within limits, and may vary continuously with time. Electronic circuits that process analog signals are called analog circuits. One example of an analog circuit is a linear amplifier. A **linear amplifier** magnifies an input signal and produces an output signal whose magnitude is larger and directly proportional to the input signal.

Time-varying signals from a particular source very often need to be amplified before the signal is capable of being "useful." For example, Figure 6.1 shows a signal source that may be the output of a microphone. The output of the microphone will need to be amplified in order to drive the speakers at the output. The amplifier is the circuit that performs this function. A dc voltage source is also an input to the amplifier. The amplifier contains transistors that must be biased so that the transistors can act as amplifying devices.

In this chapter, we analyze and design linear amplifiers that use bipolar transistors as the amplifying device. The term **small-signal** means that we can linearize the ac equivalent circuit. We will define what is meant by *small signal* in the case of BJT circuits. The term *linear amplifier* means that we can use superposition so that the dc analysis and ac analysis of the circuits can be performed separately and the total response is the sum of the two individual responses.

The mechanism by which BJT circuits amplify small time-varying signals was introduced in the last chapter. In this section, we will expand that discussion, using the graphical technique, dc load line, and ac load line. In the process, we will develop the various small-signal parameters of linear circuits and the corresponding equivalent circuits.

Figure 6.1 suggests that there are two types of analyses of the amplifier that we must consider. The first is a dc analysis because of the applied dc voltage source, and the second is a time-varying or ac analysis because of the time-varying signal source.

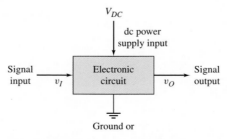

**Figure 6.1** Schematic of an electronic circuit with two input signals: the dc power supply input, and the signal input

A linear amplifier means that the superposition principle applies. The principle of superposition states: *The response of a linear circuit excited by multiple independent input signals is the sum of the responses of the circuit to each of the input signals alone.*

For the linear amplifier, then, the dc analysis can be performed with the ac source set to zero. This analysis, called a *large signal analysis,* establishes the $Q$-point of the transistors in the amplifier. This analysis and design was the primary objective of the previous chapter. The ac analysis, called a *small-signal analysis,* can be performed with the dc source set to zero. The total response of the amplifier circuit is the sum of the two individual responses.

##  6.2   THE BIPOLAR LINEAR AMPLIFIER

**Objective:** • Investigate the process by which a single-transistor circuit can amplify a small, time-varying input signal and develop the small-signal models of the transistor that are used in the analysis of linear amplifiers.

The transistor is the heart of an amplifier. In this chapter, we will consider bipolar transistor amplifiers. Bipolar transistors have traditionally been used in linear amplifier circuits because of their relatively high gain.

We begin our discussion by considering the same bipolar circuit that was discussed in the last chapter. Figure 6.2(a) shows the circuit where the input signal $v_I$ contains both a dc and an ac signal. Figure 6.2(b) shows the same circuit where $V_{BB}$ is a dc voltage to bias the transistor at a particular $Q$-point and $v_s$ is the ac signal that is to be amplified. Figure 6.2(c) shows the voltage transfer characteristics that were

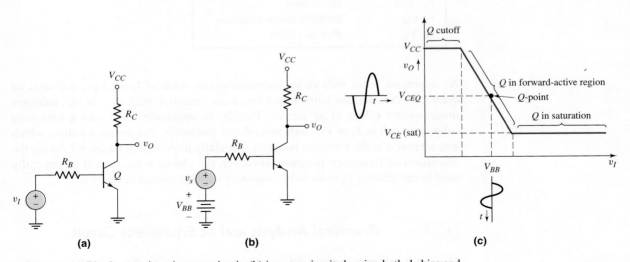

**Figure 6.2** (a) Bipolar transistor inverter circuit, (b) inverter circuit showing both dc bias and ac signal sources in the base circuit, and (c) transistor inverter voltage transfer characteristics showing desired $Q$-point

developed in Chapter 5. To use the circuit as an amplifier, the transistor needs to be biased with a dc voltage at a quiescent point (Q-point), as shown in the figure, such that the transistor is biased in the forward-active region. This dc analysis or design of the circuit was the focus of our attention in Chapter 5. If a time-varying (e.g., sinusoidal) signal is superimposed on the dc input voltage, $V_{BB}$, the output voltage will change along the transfer curve producing a time-varying output voltage. If the time-varying output voltage is directly proportional to and larger than the time-varying input voltage, then the circuit is a linear amplifier. From this figure, we see that if the transistor is not biased in the active region (biased either in cutoff or saturation), the output voltage does not change with a change in the input voltage. Thus, we no longer have an amplifier.

In this chapter, we are interested in the ac analysis and design of bipolar transistor amplifiers, which means that we must determine the relationships between the time-varying output and input signals. We will initially consider a graphical technique that can provide an intuitive insight into the basic operation of the circuit. We will then develop a small-signal equivalent circuit that will be used in the mathematical analysis of the ac signals. In general, we will be considering a steady-state, sinusoidal analysis of circuits. We will assume that any time-varying signal can be written as a sum of sinusoidal signals of different frequencies and amplitudes (Fourier series), so that a sinusoidal analysis is appropriate.

We will be dealing with time-varying as well as dc currents and voltages in this chapter. Table 6.1 gives a summary of notation that will be used. This notation was discussed in the Prologue, but is repeated here for convenience. A lowercase letter with an uppercase subscript, such as $i_B$ or $v_{BE}$, indicates *total instantaneous values*.

| Table 6.1 | Summary of notation |
|---|---|
| **Variable** | **Meaning** |
| $i_B$, $v_{BE}$ | Total instantaneous values |
| $I_B$, $V_{BE}$ | DC values |
| $i_b$, $v_{be}$ | Instantaneous ac values |
| $I_b$, $V_{be}$ | Phasor values |

An uppercase letter with an uppercase subscript, such as $I_B$ or $V_{BE}$, indicates *dc quantities*. A lowercase letter with a lowercase subscript, such as $i_b$ or $v_{be}$, indicates instantaneous values of *ac signals*. Finally, an uppercase letter with a lowercase subscript, such as $I_b$ or $V_{be}$, indicates *phasor quantities*. The phasor notation, which was reviewed in the Prologue becomes especially important in Chapter 7 during the discussion of frequency response. However, the phasor notation will be generally used in this chapter in order to be consistent with the overall ac analysis.

### 6.2.1    Graphical Analysis and ac Equivalent Circuit

Figure 6.3 shows the same basic bipolar inverter circuit that has been discussed, but now includes a sinusoidal signal source in series with the dc source as was shown in Figure 6.2(b).

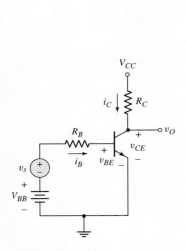

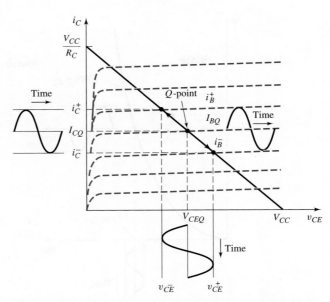

**Figure 6.3**  A common-emitter circuit with a time-varying signal source in series with the base dc source

**Figure 6.4**  Common-emitter transistor characteristics, dc load line, and sinusoidal variation in base current, collector current, and collector–emitter voltage

Figure 6.4 shows the transistor characteristics, the dc load line, and the $Q$-point. The sinusoidal signal source, $v_s$, will produce a time-varying or ac base current superimposed on the quiescent base current as shown in the figure. The time-varying base current will induce an ac collector current superimposed on the quiescent collector current. The ac collector current then produces a time-varying voltage across $R_C$, which induces an ac collector–emitter voltage as shown in the figure. The ac collector–emitter voltage, or output voltage, in general, will be larger than the sinusoidal input signal, so that the circuit has produced signal amplification—that is, the circuit is an amplifier.

We need to develop a mathematical method or model for determining the relationships between the sinusoidal variations in currents and voltages in the circuit. As already mentioned, a linear amplifier implies that superposition applies so that the dc and ac analyses can be performed separately. To obtain a linear amplifier, the time-varying or ac currents and voltages must be small enough to ensure a linear relation between the ac signals. To meet this objective, the time-varying signals are assumed to be *small signals,* which means that the amplitudes of the ac signals are small enough to yield linear relations. The concept of "small enough," or small signal, will be discussed further as we develop the small-signal equivalent circuits.

A time-varying signal source, $v_s$, in the base of the circuit in Figure 6.3 generates a time-varying component of base current, which implies there is also a time-varying component of base–emitter voltage. Figure 6.5 shows the exponential relationship between base-current and base–emitter voltage. If the magnitudes of the time-varying signals that are superimposed on the dc quiescent point are small, then we can develop a linear relationship between the ac base–emitter voltage and ac base current. This relationship corresponds to the slope of the curve at the $Q$-point.

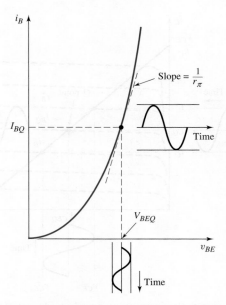

**Figure 6.5** Base current versus base–emitter voltage characteristic with superimposed sinusoidal signals. Slope at the $Q$-point is inversely proportional to $r_\pi$, a small-signal parameter.

### Small Signal

Using Figure 6.5, we can now determine one quantitative definition of small signal. From the discussion in Chapter 5, in particular, Equation (5.6), the relation between base–emitter voltage and base current can be written as

$$i_B = \frac{I_S}{\beta} \cdot \exp\left(\frac{v_{BE}}{V_T}\right) \tag{6.1}$$

If $v_{BE}$ is composed of a dc term with a sinusoidal component superimposed, i.e., $v_{BE} = V_{BEQ} + v_{be}$, then

$$i_B = \frac{I_S}{\beta} \cdot \exp\left(\frac{V_{BEQ} + v_{be}}{V_T}\right) = \frac{I_S}{\beta} \cdot \exp\left(\frac{V_{BEQ}}{V_T}\right) \cdot \exp\left(\frac{v_{be}}{V_T}\right) \tag{6.2}$$

where $V_{BEQ}$ is normally referred to as the base–emitter turn-on voltage, $V_{BE}$(on). The term $[I_S/\beta] \cdot \exp(V_{BEQ}/V_T)$ is the quiescent base current, so we can write

$$i_B = I_{BQ} \cdot \exp\left(\frac{v_{be}}{V_T}\right) \tag{6.3}$$

The base current, given in this form, is not linear and cannot be written as an ac current superimposed on a dc quiescent value. However, if $v_{be} \ll V_T$, then we can expand the exponential term in a Taylor series, keeping only the **linear term.** This approximation is what is meant by **small signal.** We then have

$$i_B \cong I_{BQ}\left(1 + \frac{v_{be}}{V_T}\right) = I_{BQ} + \frac{I_{BQ}}{V_T} \cdot v_{be} = I_{BQ} + i_b \tag{6.4(a)}$$

where $i_b$ is the time-varying (sinusoidal) base current given by

$$i_b = \left( \frac{I_{BQ}}{V_T} \right) v_{be} \qquad \text{(6.4(b))}$$

The sinusoidal base current, $i_b$, is linearly related to the sinusoidal base–emitter voltage, $v_{be}$. In this case, the term small-signal refers to the condition in which $v_{be}$ is sufficiently small for the linear relationships between $i_b$ and $v_{be}$ given by Equation (6.4(b)) to be valid. As a general rule, if $v_{be}$ is less than 10 mV, then the exponential relation given by Equation (6.3) and its linear expansion in Equation (6.4(a)) agree within approximately 10 percent. Ensuring that $v_{be} < 10$ mV is another useful rule of thumb in the design of linear bipolar transistor amplifiers.

If the $v_{be}$ signal is assumed to be sinusoidal, but if its magnitude becomes too large, then the output signal will no longer be a pure sinusoidal voltage but will become distorted and contain harmonics (see box "Harmonic Distortion").

## Harmonic Distortion

If an input sinusoidal signal becomes too large, the output signal may no longer be a pure sinusoidal signal because of nonlinear effects. A nonsinusoidal output signal may be expanded into a Fourier series and written in the form

$$v_O(t) = \underset{\text{dc}}{V_O} + \underset{\substack{\text{desired} \\ \text{linear output}}}{V_1 \sin(\omega t + \phi_1)} + \underset{\substack{\text{2nd harmonic} \\ \text{distortion}}}{V_2 \sin(2\omega t + \phi_2)} + \underset{\substack{\text{3rd harmonic} \\ \text{distortion}}}{V_3 \sin(3\omega t + \phi_3)} + \cdots$$

$$\text{(6.5)}$$

The signal at the frequency $\omega$ is the desired linear output signal for a sinusoidal input signal at the same frequency.

The time-varying input base-emitter voltage is contained in the exponential term given in Equation (6.3). Expanding the exponential function into a Taylor series, we find

$$e^x = 1 + x + \frac{x^2}{2} + \frac{x^3}{6} + \cdots \qquad \text{(6.6)}$$

where, from Equation (6.3), we have $x = v_{be}/V_T$. If we assume the input signal is a sinusoidal function, then we can write

$$x = \frac{v_{be}}{V_T} = \frac{V_\pi}{V_T} \sin \omega t \qquad \text{(6.7)}$$

The exponential function can then be written as

$$e^x = 1 + \frac{V_\pi}{V_T} \sin \omega t + \frac{1}{2} \cdot \left( \frac{V_\pi}{V_T} \right)^2 \sin^2 \omega t + \frac{1}{6} \cdot \left( \frac{V_\pi}{V_T} \right)^3 \sin^3 \omega t + \cdots \quad \text{(6.8)}$$

From trigonometric identities, we can write

$$\sin^2 \omega t = \frac{1}{2}[1 - \cos(2\omega t)] = \frac{1}{2}[1 - \sin(2\omega t + 90°)] \qquad \text{(6.9a)}$$

and

$$\sin^3 \omega t = \frac{1}{4}[3 \sin \omega t - \sin(3\omega t)] \qquad \text{(6.9b)}$$

Substituting Equations (6.9a) and (6.9b) into Equation (6.8), we obtain

$$e^x = \left[1 + \frac{1}{4}\left(\frac{V_\pi}{V_T}\right)^2\right] + \frac{V_\pi}{V_T}\left[1 + \frac{1}{8}\left(\frac{V_\pi}{V_T}\right)^2\right]\sin\omega t$$

$$-\frac{1}{4}\left(\frac{V_\pi}{V_T}\right)^2\sin(2\omega t + 90°) - \frac{1}{24}\left(\frac{V_\pi}{V_T}\right)^3\sin(3\omega t) + \cdots \qquad \textbf{(6.10)}$$

Comparing Equation (6.10) to Equation (6.8), we find the coefficients as

$$V_O = \left[1 + \frac{1}{4}\left(\frac{V_\pi}{V_T}\right)^2\right] \qquad V_1 = \frac{V_\pi}{V_T}\left[1 + \frac{1}{8}\left(\frac{V_\pi}{V_T}\right)^2\right]$$

$$V_2 = -\frac{1}{4}\left(\frac{V_\pi}{V_T}\right)^2 \qquad V_3 = -\frac{1}{24}\left(\frac{V_\pi}{V_T}\right)^3 \qquad \textbf{(6.11)}$$

We see that as $(V_\pi/V_T)$ increases, the second and third harmonic terms become non-zero. In addition, the dc and first harmonic coefficients also become nonlinear. A figure of merit is called the percent total harmonic distortion (THD) and is defined as

$$\text{THD}(\%) = \frac{\sqrt{\sum_2^\infty V_n^2}}{V_1} \times 100\% \qquad \textbf{(6.12)}$$

Considering only the second and third harmonic terms, the THD is plotted in Figure 6.6. We see that, for $V_\pi \leq 10$ mV, the THD is less than 10 percent. This total harmonic distortion value may seem excessive, but as we will see later in Chapter 12, distortion can be reduced when feedback circuits are used.

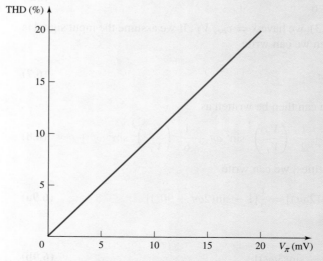

**Figure 6.6** Total harmonic distortion of the function $e^{v_{BE}/V_T}$, where $v_{BE} = V_\pi \sin\omega t$, as a function of $V_\pi$

## The AC Equivalent Circuit

From the concept of small signal, all the time-varying signals shown in Figure 6.4 will be linearly related and are superimposed on dc values. We can write (refer to notation given in Table 6.1)

$$i_B = I_{BQ} + i_b \tag{6.13(a)}$$

$$i_C = I_{CQ} + i_c \tag{6.13(b)}$$

$$v_{CE} = V_{CEQ} + v_{ce} \tag{6.13(c)}$$

and

$$v_{BE} = V_{BEQ} + v_{be} \tag{6.13(d)}$$

**The Base–Emitter Loop:** If the signal source, $v_s$, is zero, then the base-emitter loop equation is

$$V_{BB} = I_{BQ} R_B + V_{BEQ} \tag{6.14}$$

Taking into account the time-varying signals, we find the base–emitter loop equation is

$$V_{BB} + v_s = i_B R_B + v_{BE} \tag{6.15(a)}$$

or

$$V_{BB} + v_s = (I_{BQ} + i_b) R_B + (V_{BEQ} + v_{be}) \tag{6.15(b)}$$

Rearranging terms, we find

$$V_{BB} - I_{BQ} R_B - V_{BEQ} = i_b R_B + v_{be} - v_s \tag{6.15(c)}$$

From Equation (6.14), we see that the left side of Equation (6.15(c)) is zero. Equation (6.15(c)) can then be written as

$$v_s = i_b R_B + v_{be} \tag{6.16}$$

which is the base-emitter loop equation with all dc term effectively set equal to zero.

**The Collector–Emitter Loop:** Again, if the signal source, $v_s$, is zero, then the collector-emitter loop equation is

$$V_{CC} = I_{CQ} R_C + V_{CEQ} \tag{6.17}$$

Taking into account the time-varying signals, the collector-emitter loop equation becomes

$$V_{CC} = i_C R_C + v_{CE} = (I_{CQ} + i_c) R_C + (V_{CEQ} + v_{ce}) \tag{6.18(a)}$$

Rearranging terms, we find

$$V_{CC} - I_{CQ} R_C - V_{CEQ} = i_c R_C + v_{ce} \tag{6.18(b)}$$

From Equation (6.17), we see that the left side of Equation (6.18(b)) is zero. Equation (6.18(b)) can be written as

$$i_c R_C + v_{ce} = 0 \tag{6.19}$$

which is the collector–emitter loop equation with all dc terms set equal to zero.

Equations (6.16) and (6.19) relate the ac parameters in the circuit. These equations can be obtained directly by setting all dc currents and voltages equal to zero, so the dc voltage sources become short circuits and any dc current sources would

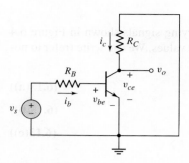

**Figure 6.7** The ac equivalent circuit of the common-emitter circuit shown in Figure 6.3. The dc voltage sources have been set equal to zero.

become open circuits. *These results are a direct consequence of applying superposition to a linear circuit.* The resulting BJT circuit, shown in Figure 6.7, is called the *ac equivalent circuit,* and all currents and voltages shown are time-varying signals. We should stress that this circuit is an equivalent circuit. We are implicitly assuming that the transistor is still biased in the forward-active region with the appropriate dc voltages and currents.

Another way of looking at the ac equivalent circuit is as follows. In the circuit in Figure 6.3, the base and collector currents are composed of ac signals superimposed on dc values. These currents flow through the $V_{BB}$ and $V_{CC}$ voltage sources, respectively. Since the voltages across these sources are assumed to remain constant, the sinusoidal currents do not produce any sinusoidal voltages across these elements. Then, since the sinusoidal voltages are zero, the equivalent ac impedances are zero, or short circuits. In other words, the dc voltage sources are ac short circuits in an equivalent ac circuit. We say that the node connecting $R_C$ and $V_{CC}$ is at signal ground.

### 6.2.2   Small-Signal Hybrid-$\pi$ Equivalent Circuit of the Bipolar Transistor

We developed the ac equivalent circuit shown in Figure 6.7. We now need to develop a **small-signal equivalent circuit** for the transistor. One such circuit is the **hybrid-$\pi$** model, which is closely related to the physics of the transistor. This effect will become more apparent in Chapter 7 when a more detailed hybrid-$\pi$ model is developed to take into account the frequency response of the transistor.

We can treat the bipolar transistor as a two-port network as shown in Figure 6.8. The input port is between the base and emitter, and the output port is between the collector and emitter.

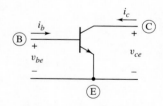

**Figure 6.8** The BJT as a small-signal, two-port network

#### Input Base–Emitter Port

One element of the hybrid-$\pi$ model has already been described. Figure 6.5 showed the base current versus base–emitter voltage characteristic, with small time-varying signals superimposed at the $Q$-point. Since the sinusoidal signals are small, we can treat the slope at the $Q$-point as a constant, which has units of conductance. The inverse of this conductance is the small-signal resistance defined as $r_\pi$. We can then relate the small-signal input base current to the small-signal input voltage by

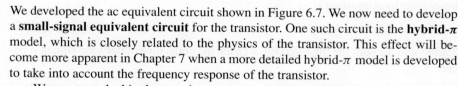

$$v_{be} = i_b r_\pi \tag{6.20}$$

where $1/r_\pi$ is equal to the slope of the $i_B$–$v_{BE}$ curve, as shown in Figure 6.5. From Equation (6.2), we then find $r_\pi$ from

$$\frac{1}{r_\pi} = \left.\frac{\partial i_B}{\partial v_{BE}}\right|_{Q\text{-}pt} = \left.\frac{\partial}{\partial v_{BE}}\left[\frac{I_S}{\beta}\cdot\exp\left(\frac{v_{BE}}{V_T}\right)\right]\right|_{Q\text{-}pt} \qquad \textbf{(6.21(a))}$$

or

$$\frac{1}{r_\pi} = \frac{1}{V_T}\cdot\left.\left[\frac{I_S}{\beta}\cdot\exp\left(\frac{v_{BE}}{V_T}\right)\right]\right|_{Q\text{-}pt} = \frac{I_{BQ}}{V_T} \qquad \textbf{(6.21(b))}$$

Then

$$\frac{v_{be}}{i_b} = r_\pi = \frac{V_T}{I_{BQ}} = \frac{\beta V_T}{I_{CQ}} \qquad \textbf{(6.22)}$$

The resistance $r_\pi$ is called the **diffusion resistance** or base–emitter input resistance. We see that $r_\pi$ is a function of the $Q$-point parameters. Note that this is the same expression obtained in Equation (6.4(b)).

## Output Collector–Emitter Port

We can consider the output terminal characteristics of the bipolar transistor. If we initially consider the case in which the output collector current is independent of the collector–emitter voltage, then the collector current is a function only of the base–emitter voltage, as discussed in Chapter 5. We can then write

$$\Delta i_C = \left.\frac{\partial i_C}{\partial v_{BE}}\right|_{Q\text{-}pt}\cdot\Delta v_{BE} \qquad \textbf{(6.23(a))}$$

or

$$i_c = \left.\frac{\partial i_C}{\partial v_{BE}}\right|_{Q\text{-}pt}\cdot v_{be} \qquad \textbf{(6.23(b))}$$

From Chapter 5, in particular Equation (5.2), we had written

$$i_C = I_S\exp\left(\frac{v_{BE}}{V_T}\right) \qquad \textbf{(6.24)}$$

Then

$$\left.\frac{\partial i_C}{\partial v_{BE}}\right|_{Q\text{-}pt} = \left.\frac{1}{V_T}\cdot I_S\exp\left(\frac{v_{BE}}{V_T}\right)\right|_{Q\text{-}pt} = \frac{I_{CQ}}{V_T} \qquad \textbf{(6.25)}$$

The term $I_S\exp(v_{BE}/V_T)$ evaluated at the $Q$-point is just the quiescent collector current. The term $I_{CQ}/V_T$ is a conductance. Since this conductance relates a current in the collector to a voltage in the B–E circuit, the parameter is called a **transconductance** and is written

$$g_m = \frac{I_{CQ}}{V_T} \qquad \textbf{(6.26)}$$

We can then write the small-signal collector current as

$$i_c = g_m v_{be} \qquad \textbf{(6.27)}$$

The small-signal transconductance is also a function of the $Q$-point parameters and is directly proportional to the dc bias current. The variation of transconductance with quiescent collector current will prove to be useful in amplifier design.

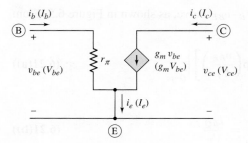

**Figure 6.9** A simplified small-signal hybrid-$\pi$ equivalent circuit for the npn transistor. The ac signal currents and voltages are shown. The phasor signals are shown in parentheses.

### Hybrid-$\pi$ Equivalent Circuit

Using these new parameters, we can develop a simplified small-signal hybrid-$\pi$ equivalent circuit for the npn bipolar transistor, as shown in Figure 6.9. The phasor components are given in parentheses. This circuit can be inserted into the ac equivalent circuit previously shown in Figure 6.7.

### Alternative Form of Equivalent Circuit

We can develop a slightly different form for the output of the equivalent circuit. We can relate the small-signal collector current to the small-signal base current as

$$\Delta i_C = \left.\frac{\partial i_C}{\partial i_B}\right|_{Q\text{-}pt} \cdot \Delta i_B \tag{6.28(a)}$$

or

$$i_c = \left.\frac{\partial i_C}{\partial i_B}\right|_{Q\text{-}pt} \cdot i_b \tag{6.28(b)}$$

where

$$\left.\frac{\partial i_C}{\partial i_B}\right|_{Q\text{-}pt} \equiv \beta \tag{6.28(c)}$$

and is called an incremental or ac common-emitter current gain. We can then write

$$i_c = \beta i_b \tag{6.29}$$

The small-signal equivalent circuit of the bipolar transistor in Figure 6.10 uses this parameter. The parameters in this figure are also given as phasors. This circuit can also be inserted in the ac equivalent circuit given in Figure 6.7. Either equivalent circuit, Figure 6.9 or 6.10, may be used. We will use both circuits in the examples that follow in this chapter.

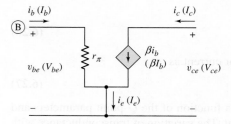

**Figure 6.10** BJT small-signal equivalent circuit using the common-emitter current gain. The ac signal currents and voltages are shown. The phasor signals are shown in parentheses.

## Common-Emitter Current Gain

The common-emitter current gain defined in Equation (6.28(c)) is actually defined as an ac beta and does not include dc leakage currents. We discussed the common-emitter current gain in Chapter 5. We defined a dc beta as the ratio of a dc collector current to the corresponding dc base current. In this case leakage currents are included. However, we will assume in this text that leakage currents are negligible so that the two definitions of beta are equivalent.

The small-signal hybrid-$\pi$ parameters $r_\pi$ and $g_m$ were defined in Equations (6.22) and (6.26). If we multiply $r_\pi$ and $g_m$, we find

$$r_\pi g_m = \left(\frac{\beta V_T}{I_{CQ}}\right) \cdot \left(\frac{I_{CQ}}{V_T}\right) = \beta \tag{6.30}$$

In general, we will assume that the common-emitter current gain $\beta$ is a constant for a given transistor. However, we must keep in mind that $\beta$ may vary from one device to another and that $\beta$ does vary with collector current. This variation with $I_C$ will be specified on data sheets for specific discrete transistors.

### 6.2.3  Small-Signal Voltage Gain

Continuing our discussion of equivalent circuits, we may now insert the bipolar, equivalent circuit in Figure 6.9, for example, into the ac equivalent circuit in Figure 6.7. The result is shown in Figure 6.11. Note that we are using the phasor notation. When incorporating the small-signal hybrid-$\pi$ model of the transistor (Figure 6.9) into the ac equivalent circuit (Figure 6.7), it is generally helpful to start with the three terminals of the transistor as shown in Figure 6.11. Then sketch the hybrid-$\pi$ equivalent circuit between these three terminals. Finally, connect the remaining circuit elements, such as $R_B$ and $R_C$, to the transistor terminals. As the circuits become more complex, this technique will minimize errors in developing the small-signal equivalent circuit.

The **small-signal voltage gain,** $A_v = V_o/V_s$, of the circuit is defined as the ratio of output signal voltage to input signal voltage. We may note a new variable in Figure 6.11. The conventional phasor notation for the small-signal base-emitter voltage is $V_\pi$, called the control voltage. The dependent current source is then given by $g_m V_\pi$. The dependent current $g_m V_\pi$ flows through $R_C$, producing a negative collector–emitter voltage, or

$$V_o = V_{ce} = -(g_m V_\pi) R_C \tag{6.31}$$

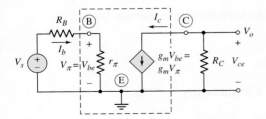

**Figure 6.11**  The small-signal equivalent circuit of the common-emitter circuit shown in Figure 6.3. The small-signal hybrid-$\pi$ model of the npn bipolar transistor is shown within the dotted lines.

and, from the input portion of the circuit, we find

$$V_\pi = \left( \frac{r_\pi}{r_\pi + R_B} \right) \cdot V_s \qquad (6.32)$$

The small-signal voltage gain is then

$$A_v = \frac{V_o}{V_s} = -(g_m R_C) \cdot \left( \frac{r_\pi}{r_\pi + R_B} \right) \qquad (6.33)$$

## EXAMPLE 6.1

**Objective:** Calculate the small-signal voltage gain of the bipolar transistor circuit shown in Figure 6.3.

Assume the transistor and circuit parameters are: $\beta = 100$, $V_{CC} = 12$ V, $V_{BE} = 0.7$ V, $R_C = 6$ k$\Omega$, $R_B = 50$ k$\Omega$, and $V_{BB} = 1.2$ V.

**DC Solution:** We first do the dc analysis to find the $Q$-point values. We obtain $I_{CQ} = 1$ mA and $V_{CEQ} = 6$ V. The transistor is biased in the forward-active mode.

**AC Solution:** The small-signal hybrid-$\pi$ parameters are

$$r_\pi = \frac{\beta V_T}{I_{CQ}} = \frac{(100)(0.026)}{1} = 2.6 \text{ k}\Omega$$

and

$$g_m = \frac{I_{CQ}}{V_T} = \frac{1}{0.026} = 38.5 \text{ mA/V}$$

The small-signal voltage gain is determined using the small-signal equivalent circuit shown in Figure 6.11. From Equation (6.33), we find

$$A_v = \frac{V_o}{V_s} = -(g_m R_C) \cdot \left( \frac{r_\pi}{r_\pi + R_B} \right)$$

or

$$= -(38.5)(6) \left( \frac{2.6}{2.6 + 50} \right) = -11.4$$

**Comment:** We see that the magnitude of the sinusoidal output voltage is 11.4 times the magnitude of the sinusoidal input voltage. We will see that other circuit configurations result in even larger small-signal voltage gains.

**Discussion:** We may consider a specific sinusoidal input voltage. Let

$$v_s = 0.25 \sin \omega t \text{ V}$$

The sinusoidal base current is given by

$$i_b = \frac{v_s}{R_B + r_\pi} = \frac{0.25 \sin \omega t}{50 + 2.6} \rightarrow 4.75 \sin \omega t \ \mu\text{A}$$

The sinusoidal collector current is

$$i_c = \beta i_b = (100)(4.75 \sin \omega t) \rightarrow 0.475 \sin \omega t \text{ mA}$$

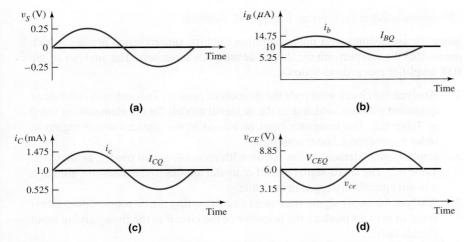

Figure 6.12  The dc and ac signals in the common-emitter circuit: (a) input voltage signal, (b) input base current, (c) output collector current, and (d) output collector-emitter voltage. The ac output voltage is 180° out of phase with respect to the input voltage signal.

and the sinusoidal collector-emitter voltage is

$$v_{ce} = -i_c R_C = -(0.475)(6) \sin \omega t = -2.85 \sin \omega t \text{ V}$$

Figure 6.12 shows the various currents and voltages in the circuit. These include the sinusoidal signals superimposed on the dc values. Figure 6.12(a) shows the sinusoidal input voltage, and Figure 6.12(b) shows the sinusoidal base current superimposed on the quiescent value. The sinusoidal collector current superimposed on the dc quiescent value is shown in Figure 6.12(c). Note that, as the base current increases, the collector current increases.

Figure 6.12(d) shows the sinusoidal component of the C–E voltage superimposed on the quiescent value. As the collector current increases, the voltage drop across $R_C$ increases so that the C–E voltage decreases. Consequently, the sinusoidal component of the output voltage is 180 degrees out of phase with respect to the input signal voltage. The minus sign in the voltage gain expression represents this 180-degree **phase shift.** In summary, the signal was both amplified and inverted by this amplifier.

**Analysis Method:**  To summarize, the analysis of a BJT amplifier proceeds as shown in the box "Problem Solving Method: Bipolar AC Analysis."

## EXERCISE PROBLEM

**Ex 6.1:** The circuit parameters for the circuit in Figure 6.3 are $V_{CC} = 3.3$ V, $V_{BB} = 0.850$ V, $R_B = 180$ k$\Omega$, and $R_C = 15$ k$\Omega$. The transistor parameters are $\beta = 120$ and $V_{BE}(\text{on}) = 0.7$ V. (a) Determine the $Q$-point values $I_{CQ}$ and $V_{CEQ}$. (b) Find the small-signal hybrid-$\pi$ parameters $g_m$ and $r_\pi$. (c) Calculate the small-signal voltage gain. (Ans. (a) $I_{CQ} = 0.1$ mA, $V_{CEQ} = 1.8$ V; (b) $g_m = 3.846$ mA/V, $r_\pi = 31.2$ k$\Omega$; (c) $A_v = -8.52$).

### Problem-Solving Technique: Bipolar AC Analysis

Since we are dealing with linear amplifier circuits, superposition applies, which means that we can perform the dc and ac analyses separately. The analysis of the BJT amplifier proceeds as follows:

1. Analyze the circuit with only the dc sources present. This solution is the dc or quiescent solution, which uses the dc signal models for the elements, as listed in Table 6.2. The transistor must be biased in the forward-active region in order to produce a linear amplifier.
2. Replace each element in the circuit with its small-signal model, as shown in Table 6.2. The small-signal hybrid-$\pi$ model applies to the transistor although it is not specifically listed in the table.
3. Analyze the small-signal equivalent circuit, setting the dc source components equal to zero, to produce the response of the circuit to the time-varying input signals only.

| **Table 6.2** | Transformation of elements in dc and small-signal analysis | | |
|---|---|---|---|
| **Element** | **I–V relationship** | **DC model** | **AC model** |
| Resistor | $I_R = \dfrac{V}{R}$ | $R$ | $R$ |
| Capacitor | $I_C = sCV$ | Open | $C$ |
| Inductor | $I_L = \dfrac{V}{sL}$ | Short | $L$ |
| Diode | $I_D = I_S(e^{v_D/V_T} - 1)$ | $+V_\gamma - r_f$ | $r_d = V_T/I_D$ |
| Independent voltage source | $V_S = $ constant | $+V_S-$ | Short |
| Independent current source | $I_S = $ constant | $I_S$ | Open |

Table suggested by Richard Hester of Iowa State University.

In Table 6.2, the dc model of the resistor is a resistor, the capacitor model is an open circuit, and the inductor model is a short circuit. The forward-biased diode model includes the cut-in voltage $V_\gamma$ and the forward resistance $r_f$.

The small-signal models of $R$, $L$, and $C$ remain the same. However, if the signal frequency is sufficiently high, the impedance of a capacitor can be approximated by a short circuit. The small-signal, low-frequency model of the diode becomes the diode diffusion resistance $r_d$. Also, the independent dc voltage source becomes a short circuit, and the independent dc current source becomes an open circuit.

6.2.4     ## Hybrid-$\pi$ Equivalent Circuit, Including the Early Effect

So far in the small-signal equivalent circuit, we have assumed that the collector current is independent of the collector–emitter voltage. We discussed the Early effect in the last chapter in which the collector current does vary with collector–emitter voltage. Equation (5.16) in the previous chapter gives the relation

$$i_C = I_S \left[ \exp\left( \frac{v_{BE}}{V_T} \right) \right] \cdot \left( 1 + \frac{v_{CE}}{V_A} \right) \tag{6.34}$$

where $V_A$ is the Early voltage and is a positive quantity. The equivalent circuits in Figures 6.9 and 6.10 can be expanded to take into account the Early voltage.

The output resistance $r_o$ is defined as

$$r_o = \left. \frac{\partial v_{CE}}{\partial i_C} \right|_{Q\text{-}pt} \tag{6.35}$$

Using Equations (6.34) and (6.35), we can write

$$\frac{1}{r_o} = \left. \frac{\partial i_C}{\partial v_{CE}} \right|_{Q\text{-}pt} = \left. \frac{\partial}{\partial v_{CE}} \left\{ I_S \left[ \exp\left( \frac{v_{BE}}{V_T} \right)\left( 1 + \frac{v_{CE}}{V_A} \right) \right] \right\} \right|_{Q\text{-}pt} \tag{6.36(a)}$$

or

$$\frac{1}{r_o} = I_S \left[ \exp\left( \frac{v_{BE}}{V_T} \right) \right] \cdot \left. \frac{1}{V_A} \right|_{Q\text{-}pt} \cong \frac{I_{CQ}}{V_A} \tag{6.36(b)}$$

Then

$$r_o = \frac{V_A}{I_{CQ}} \tag{6.37}$$

and is called the **small-signal transistor output resistance.**

This resistance can be thought of as an equivalent Norton resistance, which means that $r_o$ is in parallel with the dependent current sources. Figure 6.13(a) and (b) show the modified bipolar equivalent circuits including the output resistance $r_o$.

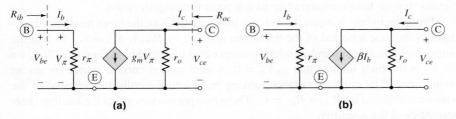

**Figure 6.13** Expanded small-signal model of the BJT, including output resistance due to the Early effect, for the case when the circuit contains the (a) transconductance and (b) current gain parameters

## EXAMPLE 6.2

**Objective:** Determine the small-signal voltage gain, including the effect of the transistor output resistance $r_o$.

Reconsider the circuit shown in Figure 6.3, with the parameters given in Example 6.1. In addition, assume the Early voltage is $V_A = 50$ V.

**Solution:** The small-signal output resistance $r_o$ is determined to be

$$r_o = \frac{V_A}{I_{CQ}} = \frac{50}{1 \text{ mA}} = 50 \text{ k}\Omega$$

Applying the small-signal equivalent circuit in Figure 6.13 to the ac equivalent circuit in Figure 6.7, we see that the output resistance $r_o$ is in parallel with $R_C$. The small-signal voltage gain is therefore

$$A_v = \frac{V_o}{V_s} = -g_m(R_C \| r_o)\left(\frac{r_\pi}{r_\pi + R_B}\right) = -(38.5)(6\|50)\left(\frac{2.6}{2.6 + 50}\right) = -10.2$$

**Comment:** Comparing this result to that of Example 6.1, we see that $r_o$ reduces the magnitude of the small-signal voltage gain. In many cases, the magnitude of $r_o$ is much larger than that of $R_C$, which means that the effect of $r_o$ is negligible.

### EXERCISE PROBLEM

**Ex 6.2:** For the circuit in Figure 6.3, assume transistor parameters of $\beta = 150$, $V_{BE}(\text{on}) = 0.7$ V, and $V_A = 150$ V. The circuit parameters are $V_{CC} = 5$ V, $V_{BB} = 1.025$ V, $R_B = 100 \text{ k}\Omega$, and $R_C = 6 \text{ k}\Omega$. (a) Determine the small-signal hybrid-$\pi$ parameters $g_m$, $r_\pi$, and $r_o$. (b) Find the small-signal voltage gain $A_v = V_o/V_s$. (Ans. (a) $g_m = 18.75$ mA/V, $r_\pi = 8 \text{ k}\Omega$, $r_o = 308 \text{ k}\Omega$; (b) $A_v = -8.17$)

The hybrid-$\pi$ model derives its name, in part, from the hybrid nature of the parameter units. The four parameters of the equivalent circuits shown in Figures 6.13(a) and 6.13(b) are: input resistance $r_\pi$ (ohms), current gain $\beta$ (dimensionless), output resistance $r_o$ (ohms), and transconductance $g_m$ (mhos).

### Input and Output Resistance

Two other parameters that affect the performance of an amplifier are the small-signal input and output resistances. The determination of these parameters for the simple circuits that we have considered up to this point is straightforward.

From the hybrid-$\pi$ equivalent circuit in Figure 6.13(a), the input resistance looking into the base terminal of the transistor, denoted by $R_{ib}$, is $R_{ib} = r_\pi$. To find the output resistance, set all independent sources equal to zero. So, in Figure 6.13(a), we set $V_\pi = 0$ which implies that $g_m V_\pi = 0$. A zero-valued current source means an open circuit. The output resistance looking back into the collector terminal of the transistor, denoted by $R_{oc}$, is $R_{oc} = r_o$. These two parameters affect the loading characteristics of the amplifier.

### Equivalent Circuit for a pnp Transistor

Up to this point, we have considered only circuits with npn bipolar transistors. However, the same basic analysis and equivalent circuit also applies to the pnp transistor. Figure 6.14(a) shows a circuit containing a pnp transistor. Here again, we see the change of current directions and voltage polarities compared to the circuit containing the npn transistor. Figure 6.14(b) is the ac equivalent circuit, with the dc voltage sources replaced by an ac short circuit, and all current and voltages shown are only the sinusoidal components.

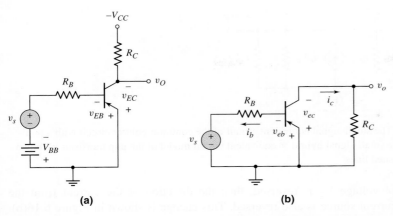

**(a)**                                                **(b)**

**Figure 6.14** (a) A common-emitter circuit with a pnp transistor and (b) the corresponding ac equivalent circuit

The transistor in Figure 6.14(b) can now be replaced by either of the hybrid-$\pi$ equivalent circuits shown in Figure 6.15. The hybrid-$\pi$ equivalent circuit of the pnp transistor is the same as that of the npn device, except that again all current directions and voltage polarities are reversed. The hybrid-$\pi$ parameters are determined by using exactly the same equations as for the npn device; that is, Equation (6.22) for $r_\pi$, Equation (6.26) for $g_m$, and Equation (6.37) for $r_o$.

We can note that, in the small-signal equivalent circuits in Figure 6.15, if we define currents of opposite direction and voltages of opposite polarity, the equivalent circuit model is exactly the same as that of the npn bipolar transistor. Figure 6.16(a) is a repeat of Figure 6.15(a) showing the conventional voltage polarities and current directions in the hybrid-$\pi$ equivalent circuit for a pnp transistor. Keep in mind that these voltages and currents are small-signal parameters. If the polarity of the

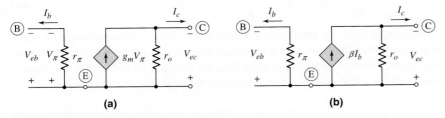

**(a)**                                                **(b)**

**Figure 6.15** The small-signal hybrid-$\pi$ equivalent circuit for the pnp transistor with the (a) transconductance and (b) current gain parameters. The ac voltage polarities and current directions are consistent with the dc parameters.

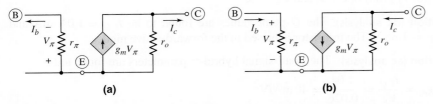

**(a)**                                                **(b)**

**Figure 6.16** Small-signal hybrid-$\pi$ models of the pnp transistor: (a) original circuit shown in Figure 6.15 and (b) equivalent circuit with voltage polarities and current directions reversed

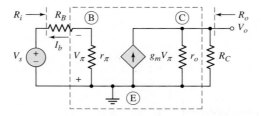

**Figure 6.17** The small-signal equivalent circuit of the common-emitter circuit with a pnp transistor. The small-signal hybrid-$\pi$ equivalent circuit model of the pnp transistor is shown within the dashed lines.

input control voltage $V_\pi$ is reversed, then the direction of the current from the dependent current source is also reversed. This change is shown in Figure 6.16(b). We may note that this small-signal equivalent circuit is the same as the hybrid-$\pi$ equivalent circuit for the npn transistor.

However, the author prefers to use the models shown in Figure 6.15 because the current directions and voltage polarities are consistent with the pnp device.

Combining the hybrid-$\pi$ model of the pnp transistor (Figure 6.15(a)) with the ac equivalent circuit (Figure 6.14(b)), we obtain the small-signal equivalent circuit shown in Figure 6.17. The output voltage is given by

$$V_o = (g_m V_\pi)(r_o \| R_C) \tag{6.38}$$

The control voltage $V_\pi$ can be expressed in terms of the input signal voltage $V_s$ using a voltage divider equation. Taking into account the polarity, we find

$$V_\pi = -\frac{V_s r_\pi}{R_B + r_\pi} \tag{6.39}$$

Combining Equations (6.38) and (6.39), we obtain the small-signal voltage gain:

$$A_v = \frac{V_o}{V_s} = \frac{-g_m r_\pi}{R_B + r_\pi}(r_o \| R_C) = \frac{-\beta}{R_B + r_\pi}(r_o \| R_C) \tag{6.40}$$

The expression for the small-signal voltage gain of the circuit containing a pnp transistor is exactly the same as that for the npn transistor circuit. Taking into account the reversed current directions and voltage polarities, the voltage gain still contains a negative sign indicating a 180-degree phase shift between the input and output signals.

## EXAMPLE 6.3

**Objective:** Analyze a pnp amplifier circuit.

Consider the circuit shown in Figure 6.18. Assume transistor parameters of $\beta = 80$, $V_{EB}(\text{on}) = 0.7$ V, and $V_A = \infty$.

**Solution (dc analysis):** The $Q$-point values are found to be $I_{CQ} = 1.04$ mA and $V_{ECQ} = 1.88$ V. The transistor is biased in the forward-active mode.

**Solution (ac analysis):** The small-signal hybrid-$\pi$ parameters are found to be

$$g_m = \frac{I_{CQ}}{V_T} = \frac{1.04}{0.026} = 40 \text{ mA/V}$$

$$r_\pi = \frac{\beta V_T}{I_{CQ}} = \frac{(80)(0.026)}{1.04} = 2 \text{ k}\Omega$$

**Figure 6.18** pnp common-emitter circuit for Example 6.3

and

$$r_o = \frac{V_A}{I_{CQ}} = \frac{\infty}{1.04} = \infty$$

The small-signal equivalent circuit is the same as shown in Figure 6.17. With $r_o = \infty$, the small-signal output voltage is

$$V_o = (g_m V_\pi) R_C$$

and we have

$$V_\pi = -\left(\frac{r_\pi}{r_\pi + R_B}\right) \cdot V_s$$

Noting that $\beta = g_m r_\pi$, we find the small-signal voltage gain to be

$$A_v = \frac{V_o}{V_s} = \frac{-\beta R_C}{r_\pi + R_B} = \frac{-(80)(3)}{2 + 50}$$

or

$$A_v = -4.62$$

The small-signal input resistance seen by the signal source (see Figure 6.17) is

$$R_i = R_B + r_\pi = 50 + 2 = 52 \text{ k}\Omega$$

The small-signal output resistance looking back into the output terminal is

$$R_o = R_C \| r_o = 3 \| \infty = 3 \text{ k}\Omega$$

**Comment:** We again note the $-180°$ phase shift between the output and input signals. We may also note that the base resistance $R_B$ in the denominator substantially reduces the magnitude of the small-signal voltage gain. We can also note that placing the pnp transistor in this configuration allows us to use positive power supplies.

## EXERCISE PROBLEM

**Ex 6.3:** For the circuit in Figure 6.14(a), let $\beta = 90$, $V_A = 120$ V, $V_{CC} = 5$ V, $V_{EB}(\text{on}) = 0.7$ V, $R_C = 2.5$ k$\Omega$, $R_B = 50$ k$\Omega$, and $V_{BB} = 1.145$ V. (a) Determine the small-signal hybrid-$\pi$ parameters $r_\pi$, $g_m$, and $r_o$. (b) Find the small-signal voltage gain $A_v = V_o/V_s$. (Ans. (a) $g_m = 30.8$ mA/V, $r_\pi = 2.92$ k$\Omega$, $r_o = 150$ k$\Omega$ (b) $A_v = -4.18$)

## Test Your Understanding

**TYU 6.1** Using the circuit and transistor parameters given in Exercise Ex 6.1, find $i_B$, $v_{BE}$, and $v_{CE}$ for $v_s = 0.065 \sin \omega t$ V. (Ans. $i_B = 0.833 + 0.308 \sin \omega t$ $\mu$A, $v_{BE} = 0.7 + 0.00960 \sin \omega t$ V, $v_{CE} = 1.8 - 0.554 \sin \omega t$ V)

**TYU 6.2** Consider the circuit in Figure 6.18. The circuit parameters are $V^+ = 3.3$ V, $V_{BB} = 2.455$ V, $R_B = 80$ k$\Omega$, and $R_C = 7$ k$\Omega$. The transistor parameters are $\beta = 110$, $V_{EB}(\text{on}) = 0.7$ V, and $V_A = 80$ V. (a) Determine $I_{CQ}$ and $V_{ECQ}$. (b) Find $g_m$, $r_\pi$, and $r_o$. (c) Determine the small-signal voltage gain $A_v = v_o/v_s$. (d) Find the small-signal input and output resistances $R_i$ and $R_o$, respectively. (Ans. (a) $I_{CQ} = 0.2$ mA, $V_{ECQ} = 1.9$ V; (b) $g_m = 7.692$ mA/V, $r_\pi = 14.3$ k$\Omega$, $r_o = 400$ k$\Omega$; (c) $A_v = -8.02$; (d) $R_i = 94.3$ k$\Omega$, $R_o = 6.88$ k$\Omega$)

## *6.2.5    Expanded Hybrid-$\pi$ Equivalent Circuit

Figure 6.19 shows an expanded hybrid-$\pi$ equivalent circuit, which includes two additional resistances, $r_b$ and $r_\mu$.

The parameter $r_b$ is the **series resistance** of the semiconductor material between the external base terminal B and an idealized internal base region B'. Typically, $r_b$ is a few tens of ohms and is usually much smaller than $r_\pi$; therefore, $r_b$ is normally negligible (a short circuit) at low frequencies. However, at high frequencies, $r_b$ may not be negligible, since the input impedance becomes capacitive, as we will see in Chapter 7.

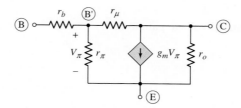

**Figure 6.19** Expanded hybrid-$\pi$ equivalent circuit

The parameter $r_\mu$ is the **reverse-biased diffusion resistance** of the base–collector junction. This resistance is typically on the order of megohms and can normally be neglected (an open circuit). However, the resistance does provide some feedback between the output and input, meaning that the base current is a slight function of the collector–emitter voltage.

In this text, when we use the hybrid-$\pi$ equivalent circuit model, we will neglect both $r_b$ and $r_\mu$, unless they are specifically included.

## *6.2.6    Other Small-Signal Parameters and Equivalent Circuits

Other small-signal parameters can be developed to model the bipolar transistor or other transistors described in the following chapters.

One common equivalent circuit model for bipolar transistor uses the **h-parameters,** which relate the small-signal terminal currents and voltages of a two-port network. These parameters are normally given in bipolar transistor data sheets, and are convenient to determine experimentally at low frequency.

Figure 6.20(a) shows the small-signal terminal current and voltage phasors for a common-emitter transistor. If we assume the transistor is biased at a $Q$-point in the forward-active region, the linear relationships between the small-signal terminal currents and voltages can be written as

$$V_{be} = h_{ie}I_b + h_{re}V_{ce} \tag{6.41(a)}$$

$$I_c = h_{fe}I_b + h_{oe}V_{ce} \tag{6.41(b)}$$

These are the defining equations of the common–emitter $h$-parameters, where the subscripts are: $i$ for input, $r$ for reverse, $f$ for forward, $o$ for output, and $e$ for common emitter.

---

*Sections can be skipped without loss of continuity.

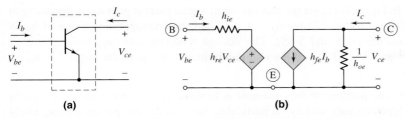

**Figure 6.20** (a) Common-emitter npn transistor and (b) the $h$-parameter model of the common-emitter bipolar transistor

These equations can be used to generate the small-signal $h$-parameter equivalent circuit, as shown in Figure 6.20(b). Equation (6.41(a)) represents a Kirchhoff voltage law equation at the input, and the resistance $h_{ie}$ is in series with a dependent voltage source equal to $h_{re}V_{ce}$. Equation (6.41(b)) represents a Kirchhoff current law equation at the output, and the conductance $h_{oe}$ is in parallel with a dependent current source equal to $h_{fe}I_b$.

Since both the hybrid-$\pi$ and $h$-parameters can be used to model the characteristics of the same transistor, these parameters are not independent. We can relate the hybrid-$\pi$ and $h$-parameters using the equivalent circuit shown in Figure 6.19.

We can show the *small-signal input resistance* $h_{ie}$ is

$$h_{ie} = r_b + r_\pi \| r_\mu \cong r_\pi \tag{6.42}$$

The parameter $h_{fe}$ is the *small-signal current gain* and is found to be

$$h_{fe} = g_m r_\pi = \beta \tag{6.43}$$

The *small-signal output admittance* $h_{oe}$ is given by

$$h_{oe} \cong \frac{1}{r_o} \tag{6.44}$$

The fourth $h$-parameter, $h_{re}$, is called the *voltage feedback ratio* and can be written as

$$h_{re} = \frac{r_\pi}{r_\pi + r_\mu} \approx 0 \tag{6.45}$$

The $h$-parameters for a pnp transistor are defined in the same way as those for an npn device. Also, the small-signal equivalent circuit for a pnp transistor using $h$-parameters is identical to that of an npn device, except that the current directions and voltage polarities are reversed.

## EXAMPLE 6.4

**Objective:** Determine the $h$-parameters of a specific transistor.

The 2N2222A transistor is a commonly used discrete npn transistor. Data for this transistor are shown in Figure 6.21. Assume the transistor is biased at $I_C = 1$ mA and let $T = 300$ K.

**Solution:** In Figure 6.21, we see that the small-signal current gain $h_{fe}$ is generally in the range $100 < h_{fe} < 170$ for $I_C = 1$ mA, and the corresponding value of $h_{ie}$ is generally between 2.5 and 5 k$\Omega$. The voltage feedback ratio $h_{re}$ varies between $1.5 \times 10^{-4}$ and $5 \times 10^{-4}$, and the output admittance $h_{oe}$ is in the range $8 < h_{oe} < 18$ $\mu$mhos.

**Comment:** The purpose of this example is to show that the parameters of a given transistor type can vary widely. In particular, the current gain parameter can easily vary by a factor of two. These variations are due to tolerances in the initial semiconductor properties and in the production process variables.

**Design Pointer:** This example clearly shows that there can be a wide variation in transistor parameters. Normally, a circuit is designed using nominal parameter values, but the allowable variations must be taken into account. In Chapter 5, we noted

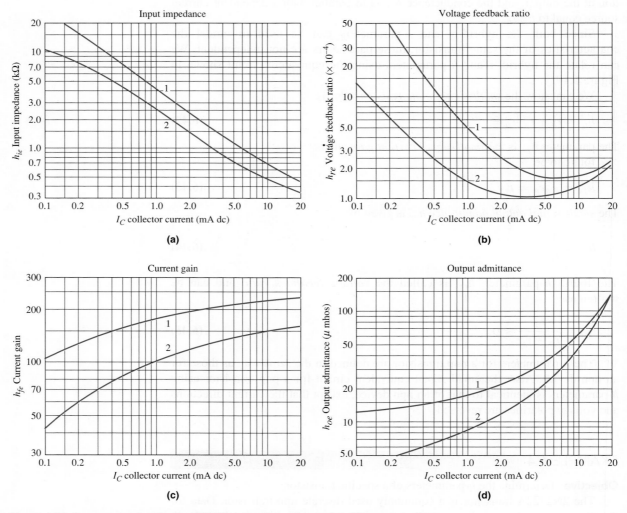

(a)

(b)

(c)

(d)

**Figure 6.21** $h$-parameter data for the 2N2222A transistor. Curves 1 and 2 represent data from high-gain and low-gain transistors, respectively.

how a variation in $\beta$ affects the $Q$-point. In this chapter, we will see how the variations in small-signal parameters affect the small-signal voltage gain and other characteristics of a linear amplifier.

### EXERCISE PROBLEM

**Ex 6.4:** Repeat Example 6.4 if the quiescent collector current is (a) $I_{CQ} = 0.2$ mA and (b) $I_{CQ} = 5$ mA. [Ans. (a) $7.8 < h_{ie} < 15$ k$\Omega$, $6.2 \times 10^{-4} < h_{re} < 50 \times 10^{-4}$, $60 < h_{fe} < 125$, $5 < h_{oe} < 13$ $\mu$mhos; (b) $0.7 < h_{ie} < 1.1$ k$\Omega$, $1.05 \times 10^{-4} < h_{re} < 1.6 \times 10^{-4}$, $140 < h_{fe} < 210$, $22 < h_{oe} < 35$ $\mu$mhos)

In the previous discussion, we indicated that the $h$-parameters $h_{ie}$ and $1/h_{oe}$ are essentially equivalent to the hybrid-$\pi$ parameters $r_\pi$ and $r_o$, respectively, and that $h_{fe}$ is essentially equal to $\beta$. The transistor circuit response is independent of the transistor model used. This reinforces the concept of a relationship between hybrid-$\pi$ parameters and $h$-parameters. In fact, this is true for any set of small-signal parameters; that is, any given set of small-signal parameters is related to any other set of parameters.

### Data Sheet

In the previous example, we showed some data for the 2N2222 discrete transistor. Figure 6.22 shows additional data from the data sheet for this transistor. Data sheets contain a lot of information, but we can begin to discuss some of the data at this time.

The first set of parameters pertains to the transistor in cutoff. The first two parameters listed are $V_{(BR)CEO}$ and $V_{(BR)CBO}$, which are the collector–emitter breakdown voltage with the base terminal open and the collector–base breakdown voltage with the emitter open. These parameters were discussed in Section 5.1.6 in the last chapter. In that section, we argued that $V_{(BR)CBO}$ was larger than $V_{(BR)CEO}$, which is supported by the data shown. These two voltages are measured at a specific current in the breakdown region. The third parameter, $V_{(BR)EBO}$, is the emitter–base breakdown voltage, which is substantially less than the collector–base or collector–emitter breakdown voltages.

The current $I_{CBO}$ is the reverse-biased collector–base junction current with the emitter open ($I_E = 0$). This parameter was also discussed in Section 5.1.6. In the data sheet, this current is measured at two values of collector–base voltage and at two temperatures. The reverse-biased current increases with increasing temperature, as we would expect. The current $I_{EBO}$ is the reverse-biased emitter–base junction current with the collector open ($I_C = 0$). This current is also measured at a specific reverse-bias voltage. The other two current parameters, $I_{CEX}$ and $I_{BL}$, are the collector current and base current measured at given specific cutoff voltages.

The next set of parameters applies to the transistor when it is turned on. As was shown in Example 6.4, the data sheets give the $h$-parameters of the transistor. The first parameter, $h_{FE}$, is the dc common-emitter current gain and is measured over a wide range of collector current. We discussed, in Section 5.4.2, stabilizing the $Q$-point against variations in current gain. The data presented in the data sheet show that the current gain for a given transistor can vary significantly, so that stabilizing the $Q$-point is indeed an important issue.

We have used $V_{CE}(sat)$ as one of the piecewise linear parameters when a transistor is driven into saturation and have always assumed a particular value in our

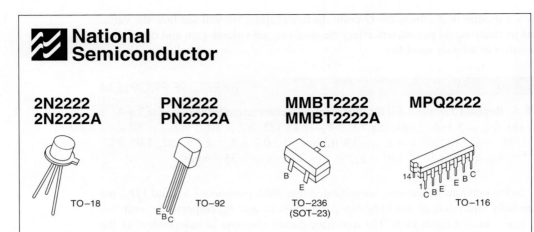

**National Semiconductor**

| 2N2222 | PN2222 | MMBT2222 | MPQ2222 |
|---|---|---|---|
| 2N2222A | PN2222A | MMBT2222A | |

TO–18        TO–92        TO–236 (SOT–23)        TO–116

## NPN General Purpose Amplifier

## Electrical Characteristics  $T_A = 25\ °C$ unless otherwise noted

| Symbol | Parameter | | Min | Max | Units |
|---|---|---|---|---|---|
| **OFF CHARACTERISTICS** | | | | | |
| $V_{(BR)CEO}$ | Collector-Emitter Breakdown Voltage (Note 1) $(I_C = 10\ mA,\ I_B = 0)$ | 2222 2222A | 30 40 | | V |
| $V_{(BR)CBO}$ | Collector-Base Breakdown Voltage $(I_C = 10\ \mu A,\ I_E = 0)$ | 2222 2222A | 60 75 | | V |
| $V_{(BR)EBO}$ | Emitter Base Breakdown Voltage $(I_E = 10\ \mu A,\ I_C = 0)$ | 2222 2222A | 5.0 6.0 | | V |
| $I_{CEX}$ | Collector Cutoff Current $(V_{CE} = 60\ V,\ V_{EB}(\text{off}) = 3.0\ V)$ | 2222A | | 10 | nA |
| $I_{CBO}$ | Collector Cutoff Current $(V_{CB} = 50\ V,\ I_E = 0)$ $(V_{CB} = 60\ V,\ I_E = 0)$ $(V_{CB} = 50\ V,\ I_E = 0,\ T_A = 150\ °C)$ $(V_{CB} = 60\ V,\ I_E = 0,\ T_A = 150\ °C)$ | 2222 2222A 2222 2222A | | 0.01 0.01 10 10 | $\mu A$ |
| $I_{EBO}$ | Emitter Cutoff Current $(V_{EB} = 3.0\ V,\ I_C = 0)$ | 2222A | | 10 | nA |
| $I_{BL}$ | Base Cutoff Current $(V_{CE} = 60\ V,\ V_{EB}(\text{off}) = 3.0)$ | 2222A | | 20 | nA |
| **ON CHARACTERISTICS** | | | | | |
| $h_{FE}$ | DC Current Gain $(I_C = 0.1\ mA,\ V_{CE} = 10\ V)$ $(I_C = 1.0\ mA,\ V_{CE} = 10\ V)$ $(I_C = 10\ mA,\ V_{CE} = 10\ V)$ $(I_C = 10\ mA,\ V_{CE} = 10\ V,\ T_A = -55\ °C)$ $(I_C = 150\ mA,\ V_{CE} = 10\ V)$ (Note 1) $(I_C = 150\ mA,\ V_{CE} = 1.0\ V)$ (Note 1) $(I_C = 500\ mA,\ V_{CE} = 10\ V)$ (Note 1) | 2222 2222A | 35 50 75 35 100 50 30 40 | 300 | |

Note 1: Pulse Test: Pulse Width ≤ 300 $\mu$s, Duty Cycle ≤ 2.0%.

**Figure 6.22**  Basic data sheet for the 2N2222 bipolar transistor

## NPN General Purpose Amplifier (Continued)

### Electrical Characteristics $T_A = 25\,°C$ unless otherwise noted (Continued)

| Symbol | Parameter | | Min | Max | Units |
|---|---|---|---|---|---|
| **ON CHARACTERISTICS** (Continued) | | | | | |
| $V_{CE}$ (sat) | Collector-Emitter Saturation Voltage (Note 1) | | | | |
| | ($I_C = 150$ mA, $I_B = 15$ mA) | 2222 | | 0.4 | |
| | | 2222A | | 0.3 | |
| | ($I_C = 500$ mA, $I_B = 50$ mA) | 2222 | | 1.6 | |
| | | 2222A | | 1.0 | V |
| $V_{BE}$ (sat) | Base-Emitter Saturation Voltage (Note 1) | | | | |
| | ($I_C = 150$ mA, $I_B = 15$ mA) | 2222 | 0.6 | 1.3 | |
| | | 2222A | 0.6 | 1.2 | |
| | ($I_C = 500$ mA, $I_B = 50$ mA) | 2222 | | 2.6 | |
| | | 2222A | | 2.0 | V |
| **SMALL-SIGNAL CHARACTERISTICS** | | | | | |
| $f_T$ | Current Gain—Bandwidth Product (Note 3) | | | | |
| | ($I_C = 20$ mA, $V_{CE} = 20$ V, $f = 100$ MHz) | 2222 | 250 | | |
| | | 2222A | 300 | | MHz |
| $C_{obo}$ | Output Capacitance (Note 3) | | | | |
| | ($V_{CB} = 10$ V, $I_E = 0$, $f = 100$ kHz) | | | 8.0 | pF |
| $C_{ibo}$ | Input Capacitance (Note 3) | | | | |
| | ($V_{EB} = 0.5$ V, $I_C = 0$, $f = 100$ kHz) | 2222 | | 30 | |
| | | 2222A | | 25 | pF |
| $rb'C_C$ | Collector Base Time Constant | | | | |
| | ($I_E = 20$ mA, $V_{CB} = 20$ V, $f = 31.8$ MHz) | 2222A | | 150 | ps |
| $NF$ | Noise Figure | | | | |
| | ($I_C = 100\ \mu$A, $V_{CE} = 10$ V, $R_S = 1.0$ kΩ, $f = 1.0$ kHz) | 2222A | | 4.0 | dB |
| $Re(h_{ie})$ | Real Part of Common-Emitter High Frequency Input Impedance | | | | |
| | ($I_C = 20$ mA, $V_{CE} = 20$ V, $f = 300$ MHz) | | | 60 | Ω |
| **SWITCHING CHARACTERISTICS** | | | | | |
| $t_D$ | Delay Time | ($V_{CC} = 30$ V, $V_{BE}$(off) $= 0.5$ V, $I_C = 150$ mA, $I_{B1} = 15$ mA) | except MPQ2222 | | 10 | ns |
| $t_R$ | Rise Time | | | 25 | ns |
| $t_S$ | Storage Time | ($V_{CC} = 30$ V, $I_C = 150$ mA, $I_{B1} = I_{B2} = 15$ mA) | except MPQ2222 | | 225 | ns |
| $t_F$ | Fall Time | | | 60 | ns |

Note 1: Pulse Test: Pulse Width < 300 $\mu$s, Duty Cycle ≤ 2.0%.
Note 2: For characteristics curves, see Process 19.
Note 3: $f_T$ is defined as the frequency at which $h_{fe}$ extrapolates to unity.

**Figure 6.22** (*continued*)

analysis or design. This parameter, listed in the data sheet, is not a constant but varies with collector current. If the collector current becomes relatively large, then the collector–emitter saturation voltage also becomes relatively large. The larger $V_{CE}$(sat) value would need to be taken into account in large-current situations. The base–emitter voltage for a transistor driven into saturation, $V_{BE}$(sat), is also given.

C

$i_c$

$g_m v_{be}$
$= \alpha i_e$

B

$i_b$

$+$
$v_{be}$    $r_e = \dfrac{V_T}{I_E}$
$-$

$i_e$

E

**Figure 6.23** The T-model of an npn bipolar transistor

Up to this point in the text, we have not been concerned with this parameter; however, the data sheet shows that the base–emitter voltage can increase significantly when a transistor is driven into saturation at high current levels.

The other parameters listed in the data sheet become more applicable later in the text when the frequency response of transistors is discussed. The intent of this short discussion is to show that we can begin to read through data sheets even though there are a lot of data presented.

**The T-model:** The hybrid-pi model can be used to analyze the time-varying characteristics of all transistor circuits. We have briefly discussed the h-parameter model of the transistor. The h-parameters of this model are often given in data sheets for discrete transistors. Another small-signal model of the transistor, the T-model, is shown in Figure 6.23. This model might be convenient to use in specific applications. However, to avoid introducing too much confusion, we will concentrate on using the hybrid-$\pi$ model in this text and leave the T-model to more advanced electronics courses.

## 6.3    BASIC TRANSISTOR AMPLIFIER CONFIGURATIONS

**Objective:** • Discuss the three basic transistor amplifier configurations and discuss the four equivalent two-port networks.

As we have seen, the bipolar transistor is a three-terminal device. Three basic single-transistor amplifier configurations can be formed, depending on which of the three transistor terminals is used as signal ground. These three basic configurations are appropriately called **common emitter, common collector (emitter follower),** and **common base.** Which configuration or amplifier is used in a particular application depends to some extent on whether the input signal is a voltage or current and whether the desired output signal is a voltage or current. The characteristics of the three types of amplifiers will be determined to show the conditions under which each amplifier is most useful.

The input signal source can be modeled as either a Thevenin or Norton equivalent circuit. Figure 6.24(a) shows the Thevenin equivalent source that would represent a voltage signal, such as the output of a microphone. The voltage source $v_s$

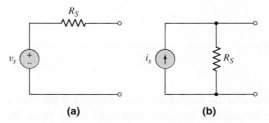

$R_S$

$v_s$

$i_s$    $R_S$

(a)                                    (b)

**Figure 6.24.** Input signal source modeled as (a) Thevenin equivalent circuit and (b) Norton equivalent circuit

represents the voltage generated by the microphone. The resistance $R_S$ is called the output resistance of the source and takes into account the change in output voltage as the source supplies current. Figure 6.24(b) shows the Norton equivalent source that would represent a current signal, such as the output of a photodiode. The current source $i_s$ represents the current generated by the photodiode and the resistance $R_S$ is the output resistance of this signal source.

Each of the three basic transistor amplifiers can be modeled as a two-port network in one of four configurations as shown in Table 6.3. We will determine the gain parameters, such as $A_{vo}$, $A_{io}$, $G_{mo}$, and $R_{mo}$, for each of the three transistor amplifiers. These parameters are important since they determine the amplification of the amplifier. However, we will see that the input and output resistances, $R_i$ and $R_o$, are also important in the design of these amplifiers. Although one configuration shown in Table 6.3 may be preferable for a given application, any one of the four can be used to model a given amplifier. Since each configuration must produce the same terminal characteristics for a given amplifier, the various gain parameters are not independent, but are related to each other.

| Table 6.3 | Four equivalent two-port networks | |
|---|---|---|
| **Type** | **Equivalent circuit** | **Gain property** |
| Voltage amplifier | | Output voltage proportional to input voltage |
| Current amplifier | | Output current proportional to input current |
| Transconductance amplifier | | Output current proportional to input voltage |
| Transresistance amplifier | | Output voltage proportional to input current |

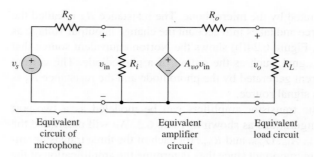

**Figure 6.25** Equivalent preamplifier circuit

If we wish to design a voltage amplifier (preamp) so that the output voltage of a microphone, for example, is amplified, the total equivalent circuit may be that shown in Figure 6.25. The input voltage to the amplifier is given by

$$v_{\text{in}} = \frac{R_i}{R_i + R_S} \cdot v_s \tag{6.46}$$

In general, we would like the input voltage $v_{\text{in}}$ to the amplifier to be as nearly equal to the source voltage $v_s$ as possible. This means, from Equation (6.46), that we need to design the amplifier such that the input resistance $R_i$ is much larger than the signal source output resistance $R_S$. (The output resistance of an ideal voltage source is zero, but is not zero for most practical voltage sources.) To provide a particular voltage gain, the amplifier must have a gain parameter $A_{vo}$ of a certain value. The output voltage supplied to the load (where the load may be a second power amplifier stage) is given by

$$v_o = \frac{R_L}{R_L + R_o} \cdot A_{vo} v_{\text{in}} \tag{6.47}$$

Normally, we would like the output voltage to the load to be equal to the Thevenin equivalent voltage generated by the amplifier. This means that we need $R_o \ll R_L$ for the voltage amplifier. So again, for a voltage amplifier, the output resistance should be very small. The input and output resistances are significant in the design of an amplifier.

For a current amplifier, we would like to have $R_i \ll R_S$ and $R_o \gg R_L$. We will see as we proceed through the chapter that each of the three basic transistor amplifier configurations exhibits characteristics that are desirable for particular applications.

We should note that, in this chapter, we will be primarily using the two-port equivalent circuits shown in Table 6.3 to model single-transistor amplifiers. However, these equivalent circuits are also used to model multitransistor circuits. This will become apparent as we get into Part 2 of the text.

## 6.4 COMMON-EMITTER AMPLIFIERS

**Objective:** • Analyze the common-emitter amplifier and become familiar with the general characteristics of this circuit.

In this section, we consider the first of the three basic amplifiers—the **common-emitter** circuit. We will apply the equivalent circuit of the bipolar transistor that was previously developed. In general, we will use the hybrid-$\pi$ model throughout the text.

### 6.4.1    Basic Common-Emitter Amplifier Circuit

Figure 6.26 shows the basic common-emitter circuit with voltage-divider biasing. We see that the emitter is at ground potential—hence the name common emitter. The signal from the signal source is coupled into the base of the transistor through the coupling capacitor $C_C$, which provides dc isolation between the amplifier and the signal source. The dc transistor biasing is established by $R_1$ and $R_2$, and is not disturbed when the signal source is capacitively coupled to the amplifier.

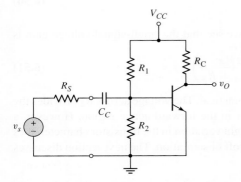

**Figure 6.26** A common-emitter circuit with a voltage-divider biasing circuit and a coupling capacitor

If the signal source is a sinusoidal voltage at frequency $f$, then the magnitude of the capacitor impedance is $|Z_c| = [1/(2\pi f C_C)]$. For example, assume that $C_C = 10\ \mu$F and $f = 2$ kHz. The magnitude of the capacitor impedance is then

$$|Z_c| = \frac{1}{2\pi f C_C} = \frac{1}{2\pi(2 \times 10^3)(10 \times 10^{-6})} \cong 8\ \Omega \qquad \textbf{(6.48)}$$

The magnitude of this impedance is in general much less than the Thevenin resistance at the capacitor terminals, which in this case is $R_1 \| R_2 \| r_\pi$. We can therefore assume that the capacitor is essentially a short circuit to signals with frequencies greater than 2 kHz. We are also neglecting any capacitance effects within the transistor. Using these results, our analyses in this chapter assume that the signal frequency is sufficiently high that any coupling capacitance acts as a perfect short circuit, and is also sufficiently low that the transistor capacitances can be neglected. Such frequencies are in the midfrequency range, or simply at the midband of the amplifier.

The small-signal equivalent circuit in which the coupling capacitor is assumed to be a short circuit is shown in Figure 6.27. The small-signal variables, such as the

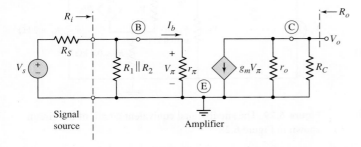

**Figure 6.27** The small-signal equivalent circuit, assuming the coupling capacitor is a short circuit

input signal voltage and input base current, are given in phasor form. The control voltage $V_\pi$ is also given as a phasor.

The output voltage is

$$V_o = -g_m V_\pi (r_o \| R_C) \qquad (6.49)$$

and the control voltage $V_\pi$ is found to be

$$V_\pi = \frac{R_1 \| R_2 \| r_\pi}{R_1 \| R_2 \| r_\pi + R_S} \cdot V_s \qquad (6.50)$$

Combining Equations (6.49) and (6.50), we see that the small-signal voltage gain is

$$A_v = \frac{V_o}{V_s} = -g_m (r_o \| R_C) \left( \frac{R_1 \| R_2 \| r_\pi}{R_1 \| R_2 \| r_\pi + R_S} \right) \qquad (6.51)$$

The circuit in Figure 6.26 is not very practical. The voltage across $R_2$ provides the base–emitter voltage to bias the transistor in the forward-active region. However, a slight variation in the resistor value or a slight variation in the transistor characteristics may cause the transistor to be biased in cutoff or saturation. The next section discusses an improved circuit configuration.

### 6.4.2 Circuit with Emitter Resistor

In the last chapter, we found that the $Q$-point was stabilized against variations in $\beta$ if an emitter resistor were included in the circuit, as shown in Figure 6.28. We will find a similar property for the ac signals, in that the voltage gain of a circuit with $R_E$ will be less dependent on the transistor current gain $\beta$. Even though the emitter of this circuit is not at ground potential, this circuit is still referred to as a common-emitter circuit.

Assuming that $C_C$ acts as a short circuit, Figure 6.29 shows the small-signal hybrid-$\pi$ equivalent circuit. As we have mentioned previously, to develop the small-signal equivalent circuit, start with the three terminals of the transistor. Sketch the hybrid-$\pi$ equivalent circuit between the three terminals and then sketch

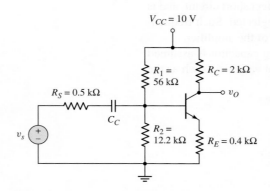

**Figure 6.28** An npn common-emitter circuit with an emitter resistor, a voltage-divider biasing circuit, and a coupling capacitor

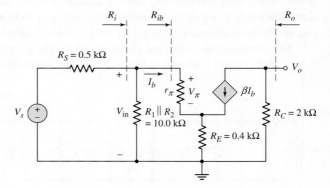

**Figure 6.29** The small-signal equivalent circuit of the circuit shown in Figure 6.28

in the remaining circuit elements around these three terminals. In this case, we are using the equivalent circuit with the current gain parameter $\beta$, and we are assuming that the Early voltage is infinite so the transistor output resistance $r_o$ can be neglected (an open circuit). The ac output voltage is

$$V_o = -(\beta I_b) R_C \tag{6.52}$$

To find the small-signal voltage gain, it is worthwhile finding the input resistance first. The resistance $R_{ib}$ is the input resistance looking into the base of the transistor. We can write the following loop equation

$$V_{in} = I_b r_\pi + (I_b + \beta I_b) R_E \tag{6.53}$$

The input resistance $R_{ib}$ is then defined as, and found to be,

$$R_{ib} = \frac{V_{in}}{I_b} = r_\pi + (1 + \beta) R_E \tag{6.54}$$

In the common-emitter configuration that includes an emitter resistance, the small-signal input resistance looking into the base of the transistor is $r_\pi$ plus the emitter resistance multiplied by the factor $(1 + \beta)$. This effect is called the **resistance reflection rule.** We will use this result throughout the text without further derivation.

The input resistance to the amplifier is now

$$R_i = R_1 \| R_2 \| R_{ib} \tag{6.55}$$

We can again relate $V_{in}$ to $V_s$ through a voltage-divider equation as

$$V_{in} = \left( \frac{R_i}{R_i + R_S} \right) \cdot V_s \tag{6.56}$$

Combining Equations (6.52), (6.54), and (6.56), we find the small-signal voltage gain is

$$A_v = \frac{V_o}{V_s} = \frac{-(\beta I_b) R_C}{V_s} = -\beta R_C \left( \frac{V_{in}}{R_{ib}} \right) \cdot \left( \frac{1}{V_s} \right) \tag{6.57}$$

or

$$A_v = \frac{-\beta R_C}{r_\pi + (1 + \beta) R_E} \left( \frac{R_i}{R_i + R_S} \right) \tag{6.58}$$

From this equation, we see that if $R_i \gg R_S$ and if $(1 + \beta) R_E \gg r_\pi$, then the small-signal voltage gain is approximately

$$A_v \cong \frac{-\beta R_C}{(1 + \beta) R_E} \cong \frac{-R_C}{R_E} \tag{6.59}$$

Equations (6.58) and (6.59) show that the voltage gain for this circuit is less dependent on the current gain $\beta$ than in the previous circuit (Equation (6.51), which means that there is a smaller change in voltage gain when the transistor current gain changes. The circuit designer now has more control in the design of the voltage gain, but this advantage is at the expense of a smaller gain.

In Chapter 5, we discussed the variation in the $Q$-point with variations or tolerances in resistor values. Since the voltage gain is a function of resistor values, it is also a function of the tolerances in those values. This must be considered in a circuit design.

## EXAMPLE 6.5

**Objective:** Determine the small-signal voltage gain and input resistance of a common-emitter circuit with an emitter resistor.

For the circuit in Figure 6.28, the transistor parameters are: $\beta = 100$, $V_{BE}(\text{on}) = 0.7$ V, and $V_A = \infty$.

**DC Solution:** From a dc analysis of the circuit, we can determine that $I_{CQ} = 2.16$ mA and $V_{CEQ} = 4.81$ V, which shows that the transistor is biased in the forward-active mode.

**AC Solution:** The small-signal hybrid-$\pi$ parameters are determined to be

$$r_\pi = \frac{V_T \beta}{I_{CQ}} = \frac{(0.026)(100)}{(2.16)} = 1.20\,\text{k}\Omega$$

$$g_m = \frac{I_{CQ}}{V_T} = \frac{2.16}{0.026} = 83.1\,\text{mA/V}$$

and

$$r_o = \frac{V_A}{I_{CQ}} = \infty$$

The input resistance to the base can be determined as

$$R_{ib} = r_\pi + (1 + \beta)R_E = 1.20 + (101)(0.4) = 41.6\,\text{k}\Omega$$

and the input resistance to the amplifier is now found to be

$$R_i = R_1 \| R_2 \| R_{ib} = 10 \| 41.6 = 8.06\,\text{k}\Omega$$

Using the exact expression for the voltage gain, we find

$$A_v = \frac{-(100)(2)}{1.20 + (101)(0.4)} \left( \frac{8.06}{8.06 + 0.5} \right) = -4.53$$

If we use the approximation given by Equation (6.59), we obtain

$$A_v = \frac{-R_C}{R_E} = \frac{-2}{0.4} = -5.0$$

**Comment:** The magnitude of the small-signal voltage gain is substantially reduced when an emitter resistor is included because of the $(1 + \beta)R_E$ term in the denominator. Also, Equation (6.59) gives a good first approximation for the gain, which means that it can be used in the initial design of a common-emitter circuit with an emitter resistor.

**Discussion:** The amplifier gain is nearly independent of changes in the current gain parameter $\beta$. This fact is shown in the following calculations:

| $\beta$ | $A_v$ |
|---|---|
| 50 | $-4.41$ |
| 100 | $-4.53$ |
| 150 | $-4.57$ |

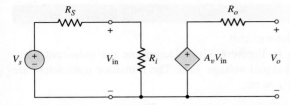

**Figure 6.30**  Two-port equivalent circuit for the amplifier in Example 6.5

The two-port equivalent circuit along with the input signal source for the common-emitter amplifier analyzed in this example is shown in Figure 6.30. We can determine the effect of the source resistance $R_S$ in conjunction with the amplifier input resistance $R_i$. Using a voltage-divider equation, we find the input voltage to the amplifier is

$$V_{in} = \left( \frac{R_i}{R_i + R_S} \right) \cdot V_s = (0.942) \cdot V_s$$

The actual input voltage to the amplifier $V_{in}$ is reduced compared to the input signal. This is called a **loading effect.** In this case, the input voltage is approximately 94 percent of the signal voltage.

## EXERCISE PROBLEM

**Ex 6.5:** For the circuit in Figure 6.31, let $R_E = 0.6\,\text{k}\Omega$, $R_C = 5.6\,\text{k}\Omega$, $\beta = 120$, $V_{BE}(\text{on}) = 0.7\,\text{V}$, $R_1 = 250\,\text{k}\Omega$, and $R_2 = 75\,\text{k}\Omega$. (a) For $V_A = \infty$, determine the small-signal voltage gain $A_v$. (b) Determine the input resistance looking into the base of the transistor. (Ans. (a) $A_v = -8.27$, (b) $R_{ib} = 80.1\,\text{k}\Omega$)

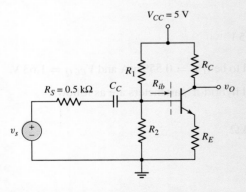

**Figure 6.31**  Figure for Exercise Ex6.5

## COMPUTER ANALYSIS EXERCISE

**PS 6.1:** (a) Verify the results of Example 6.5 with a PSpice analysis. Use a standard 2N2222 transistor, for example. (b) Repeat part (a) for $R_E = 0.3\,\text{k}\Omega$.

## EXAMPLE 6.6

**Objective:** Analyze a pnp transistor circuit.

Consider the circuit shown in Figure 6.32(a). Determine the quiescent parameter values and then the small-signal voltage gain. The transistor parameters are $V_{EB}(\text{on}) = 0.7$ V, $\beta = 80$, and $V_A = \infty$.

**Solution (dc analysis):** The dc equivalent circuit with the Thevenin equivalent circuit of the base biasing is shown in Figure 6.32(b). We find

$$R_{TH} = R_1 \| R_2 = 40 \| 60 = 24 \text{ k}\Omega$$

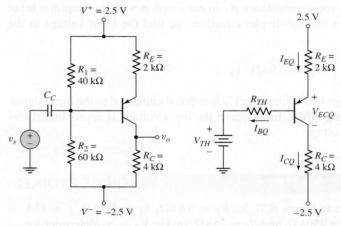

**Figure 6.32** (a) pnp transistor circuit for Example 6.6 and (b) Thevenin equivalent circuit for Example 6.6

and

$$V_{TH} = \left( \frac{R_2}{R_1 + R_2} \right)(5) - 2.5 = 0.5 \text{ V}$$

The transistor quiescent values are found to be $I_{CQ} = 0.559$ mA and $V_{ECQ} = 1.63$ V.

**Solution (ac analysis):** The small-signal hybrid-$\pi$ parameters are as follows:

$$r_\pi = \frac{\beta V_T}{I_{CQ}} = \frac{(80)(0.026)}{0.559} = 3.72 \text{ k}\Omega$$

$$g_m = \frac{I_{CQ}}{V_T} = \frac{0.559}{0.026} = 21.5 \text{ mA/V}$$

and

$$r_o = \frac{V_A}{I_Q} = \infty$$

The small-signal equivalent circuit is shown in Figure 6.33. As noted before, we start with the three terminals of the transistor, sketch the hybrid-$\pi$ equivalent circuit between these three terminals, and then put in the other circuit elements around the transistor.

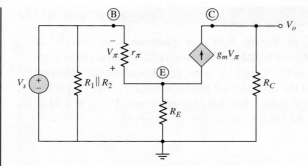

**Figure 6.33** Small-signal equivalent circuit for circuit shown in Figure 6.32(a) used in Example 6.6

The output voltage is

$$V_o = g_m V_\pi R_C$$

Writing a KVL equation from the input around the B–E loop, we find

$$V_s = -V_\pi - \left( \frac{V_\pi}{r_\pi} + g_m V_\pi \right) R_E$$

The term in the parenthesis is the total current through the $R_E$ resistor. Solving for $V_\pi$ and recalling that $g_m r_\pi = \beta$, we obtain

$$V_\pi = \frac{-V_s}{1 + \left( \dfrac{1+\beta}{r_\pi} \right) R_E}$$

Substituting into the expression for the output voltage, we find the small-signal voltage gain as

$$A_v = \frac{V_o}{V_s} = \frac{-\beta R_C}{r_\pi + (1+\beta) R_E}$$

Then

$$A_v = \frac{-(80)(4)}{3.72 + (81)(2)} = -1.93$$

The negative sign indicates that the output voltage is 180 degrees out of phase with respect to the input voltage. This same result was found in common-emitter circuits using npn transistors.

Using the approximation given by Equation (6.59), we have

$$A_v \cong -\frac{R_C}{R_E} = -\frac{4}{2} = -2$$

This approximation is very close to the actual value of gain calculated.

**Comment:** In the previous chapter, we found that including an emitter resistor provided stability in the $Q$-point. However, we may note that in the small-signal analysis, the $R_E$ resistor reduces the small-signal voltage gain substantially. There are almost always trade-offs to be made in electronic design.

**Ex 6.6:** The circuit shown in Figure 6.34 has parameters $R_E = 0.3\,k\Omega$, $R_C = 4\,k\Omega$, $R_1 = 14.4\,k\Omega$, $R_2 = 110\,k\Omega$ and $R_L = 10\,k\Omega$. The transistor parameters are $\beta = 100$, $V_{EB}(\text{on}) = 0.7\,V$, and $V_A = \infty$. (a) Determine the quiescent values $I_{CQ}$ and $V_{ECQ}$. (b) Find the small-signal parameters $g_m, r_\pi$, and $r_o$. (c) Determine the small-signal voltage gain. (Ans. (a) $I_{CQ} = 1.6\,\text{mA}$, $V_{ECQ} = 5.11\,V$; (b) $g_m = 61.54\,\text{mA/V}$, $r_\pi = 1.625\,k\Omega$, $r_o = \infty$; (c) $A_v = -8.95$).

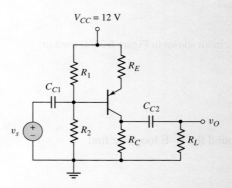

**Figure 6.34**  Figure for Exercise Ex 6.6

## Test Your Understanding

**TYU 6.3** The parameters of the circuit shown in Figure 6.28 are $V_{CC} = 5\,V$, $R_C = 4\,k\Omega$, $R_E = 0.25\,k\Omega$, $R_S = 0.25\,k\Omega$, $R_1 = 100\,k\Omega$, and $R_2 = 25\,k\Omega$. The transistor parameters are $\beta = 120$, $V_{BE}(\text{on}) = 0.7\,V$, and $V_A = \infty$. Determine the small-signal voltage gain. (Ans. $A_v = -13.6$)

**TYU 6.4** For the circuit shown in Figure 6.31, let $\beta = 100$, $V_{BE}(\text{on}) = 0.7\,V$, and $V_A = \infty$. Design a bias-stable circuit such that $I_{CQ} = 0.5$ mA, $V_{CEQ} = 2.5\,V$, and $A_v = -8$. (Ans. To a good approximation: $R_C = 4.54\,k\Omega$, $R_E = 0.454\,k\Omega$, $R_1 = 24.1\,k\Omega$, and $R_2 = 5.67\,k\Omega$)

**TYU 6.5** Design the circuit in Figure 6.35 such that it is bias stable and the small-signal voltage gain is $A_v = -8$. Let $I_{CQ} = 0.6$ mA, $V_{ECQ} = 3.75\,V$, $\beta = 100$,

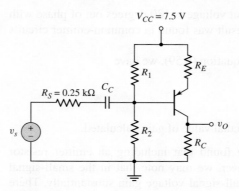

**Figure 6.35**  Figure for Exercise TYU 6.5

$V_{EB}(\text{on}) = 0.7$ V, and $V_A = \infty$. (Ans. To a good approximation: $R_C = 5.62\,\text{k}\Omega$, $R_E = 0.625\,\text{k}\Omega$, $R_1 = 7.41\,\text{k}\Omega$, and $R_2 = 42.5\,\text{k}\Omega$)

**TYU 6.6** For the circuit in Figure 6.28, the small-signal voltage gain is given approximately by $-R_C/R_E$. For the case of $R_C = 2\,\text{k}\Omega$, $R_E = 0.4\,\text{k}\Omega$, and $R_S = 0$, what must be the value of $\beta$ such that the approximate value is within 5 percent of the actual value? (Ans. $\beta = 76$)

## COMPUTER ANALYSIS EXERCISE

**PS 6.2:** Verify the results of Example 6.6 with a PSpice analysis. Use a standard transistor.

### 6.4.3   Circuit with Emitter Bypass Capacitor

There may be times when the emitter resistor must be large for the purposes of dc design, but degrades the small-signal voltage gain too severely. We can use an emitter bypass capacitor to effectively short out a portion or all of the emitter resistance as seen by the ac signals. Consider the circuit shown in Figure 6.36 biased with both positive and negative voltages. Both emitter resistors $R_{E1}$ and $R_{E2}$ are factors in the dc design of the circuit, but only $R_{E1}$ is part of the ac equivalent circuit, since $C_E$ provides a short circuit to ground for the ac signals. To summarize, the ac gain stability is due only to $R_{E1}$ and most of the dc stability is due to $R_{E2}$.

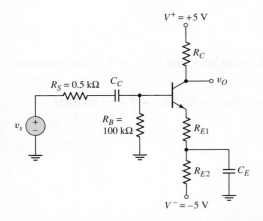

**Figure 6.36** A bipolar circuit with an emitter resistor and an emitter bypass capacitor

## DESIGN EXAMPLE 6.7

**Objective:** Design a bipolar amplifier to meet a set of specifications.

**Specifications:** The circuit configuration to be designed is shown in Figure 6.36 and is to amplify a 12 mV sinusoidal signal from a microphone to a 0.4 V sinusoidal output signal. We will assume that the output resistance of the microphone is $0.5\,k\Omega$ as shown.

**Choices:** The transistor used in the design has nominal values of $\beta = 100$ and $V_{BE}(\text{on}) = 0.7$ V, but the current gain for this type of transistor is assumed to be in the range $75 \leq \beta \leq 125$ because of tolerance effects. We will assume that $V_A = \infty$. Standard resistor values are to be used in the final design, but we will assume, in this example, that the actual resistor values are available (no tolerance effects).

**Solution (Initial Design Approach):** The magnitude of the voltage gain of the amplifier needs to be

$$|A_v| = \frac{0.4\,\text{V}}{12\,\text{mV}} = 33.3$$

From Equation (6.59), the approximate voltage gain of the amplifier is

$$|A_v| \cong \frac{R_C}{R_{E1}}$$

Noting from the last example that this value of gain produces an optimistically high value, we can set $R_C/R_{E1} = 40$ or $R_C = 40\,R_{E1}$.

The dc base-emitter loop equation is

$$5 = I_B R_B + V_{BE}(\text{on}) + I_E(R_{E1} + R_{E2})$$

Assuming $\beta = 100$ and $V_{BE}(\text{on}) = 0.7$ V, we can design the circuit to produce a quiescent emitter current of, for example, 0.20 mA. We then have

$$5 = \frac{(0.20)}{(101)}(100) + 0.70 + (0.20)(R_{E1} + R_{E2})$$

which yields

$$R_{E1} + R_{E2} = 20.5\,k\Omega$$

Assuming $I_E \cong I_C$ and designing the circuit such that $V_{CEQ} = 4$ V, the collector–emitter loop equation produces

$$5 + 5 = I_C R_C + V_{CEQ} + I_E(R_{E1} + R_{E2}) = (0.2)R_C + 4 + (0.2)(20.5)$$

or

$$R_C = 9.5\,k\Omega$$

Then

$$R_{E1} = \frac{R_C}{40} = \frac{9.5}{40} = 0.238\,k\Omega$$

and $R_{E2} = 20.3\,k\Omega$.

**Trade-offs:** From Appendix C, we will pick standard resistor values of $R_{E1} = 240\,\Omega$, $R_{E2} = 20\,k\Omega$, and $R_C = 10\,k\Omega$. We will assume that these resistor values are available and will investigate the effects of the variation in transistor current gain $\beta$.

The various parameters of the circuit for three values of $\beta$ are shown in the following table. The output voltage $V_o$ is the result of a 12 mV input signal.

| $\beta$ | $I_{CQ}$ (mA) | $r_\pi$ (k$\Omega$) | $|A_v|$ | $V_o$ (V) |
|---------|---------------|---------------------|---------|-----------|
| 75      | 0.197         | 9.90                | 26.1    | 0.313     |
| 100     | 0.201         | 12.9                | 26.4    | 0.317     |
| 125     | 0.203         | 16.0                | 26.6    | 0.319     |

One important point to note is that, the output voltage is less than the design objective of 0.4 V for a 12 mV input signal. This effect will be discussed further in the next section involving the computer simulation.

A second point to note is that the quiescent collector current, small-signal voltage gain, and output voltage are relatively insensitive to the current gain $\beta$. This stability is a direct result of including the emitter resistor $R_{E1}$.

**Computer Simulation:** Since we used approximation techniques in our design, we can use PSpice to give us a more accurate valuation of the circuit for the standard resistor values that were chosen. Figure 6.37 shows the PSpice circuit schematic diagram.

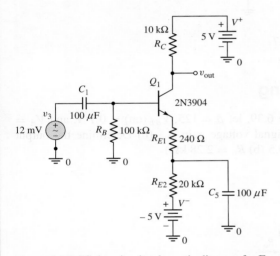

**Figure 6.37** PSpice circuit schematic diagram for Example 6.7

Using the standard resistor values and the 2N3904 transistor, the output signal voltage produced by a 12 mV input signal is 323 mV. A frequency of 2 kHz and capacitor values of 100 $\mu$F were used in the simulation. The magnitude of the output signal is slightly less than the desired value of 400 mV. The principal reason for the difference is that the $r_\pi$ parameter of the transistor was neglected in the design. For a collector current of approximately $I_C = 0.2$ mA, $r_\pi$ can be significant.

In order to increase the small-signal voltage gain, a smaller value of $R_{E1}$ is necessary. For $R_{E1} = 160 \, \Omega$, the output signal voltage is 410 mV, which is very close to the desired value.

**Design Pointer:** Approximation techniques are extremely useful in an initial electronic circuit design. A computer simulation, such as PSpice, can then be used to verify the design. Slight changes in the design can then be made to meet the required specifications.

**Ex 6.7:** The circuit in Figure 6.38 has parameters $V^+ = 5$ V, $V^- = -5$ V, $R_E = 4$ k$\Omega$, $R_C = 4$ k$\Omega$, $R_B = 100$ k$\Omega$, and $R_S = 0.5$ k$\Omega$. The transistor parameters are $\beta = 120$, $V_{BE}(\text{on}) = 0.7$ V, and $V_A = 80$ V. (a) Determine the input resistance seen by the signal source. (b) Find the small-signal voltage gain. (Ans. (a) $R_i = 3.91$ k$\Omega$, (b) $A_v = -114$)

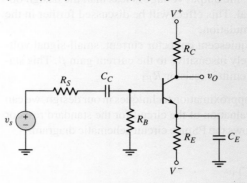

**Figure 6.38** Figure for Exercise Ex 6.7

## Test Your Understanding

**TYU 6.7** For the circuit in Figure 6.39, let $\beta = 125$, $V_{BE}(\text{on}) = 0.7$ V, and $V_A = 200$ V. (a) Determine the small-signal voltage gain $A_v$. (b) Determine the output resistance $R_o$. (Ans. (a) $A_v = -50.5$ (b) $R_o = 2.28$ k$\Omega$)

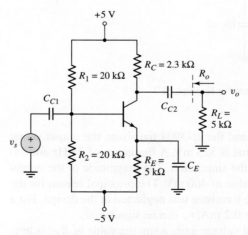

**Figure 6.39** Figure for Exercise TYU 6.7

**PS 6.3:** (a) Using a PSpice simulation, determine the voltage gain of the circuit shown in Figure 6.39. (b) Repeat Part (a) if $R_L = 50$ k$\Omega$. What can be said about loading effects?

**Advanced Common-Emitter Amplifier Concepts**

Our previous analysis of common-emitter circuits assumed constant load or collector resistances. The common-emitter circuit shown in Figure 6.40(a) is biased with a constant-current source and contains a nonlinear, rather than a constant, collector resistor. Assume the current–voltage characteristics of the nonlinear resistor are described by the curve in Figure 6.40(b). The curve in Figure 6.40(b) can be generated using the pnp transistor as shown in Figure 6.40(c). The transistor is biased at a constant $V_{EB}$ voltage. This transistor is now the load device and, since transistors are active devices, this load is referred to as an **active load.** We will encounter active loads in much more detail in Part 2 of the text.

Neglecting the base current in Figure 6.40(a), we can assume the quiescent current and voltage values of the load device are $I_Q = I_{CQ}$ and $V_{RQ}$ as shown in Figure 6.40(b). At the $Q$-point of the load device, assume the incremental resistance $\Delta v_R / \Delta i_C$ is $r_c$.

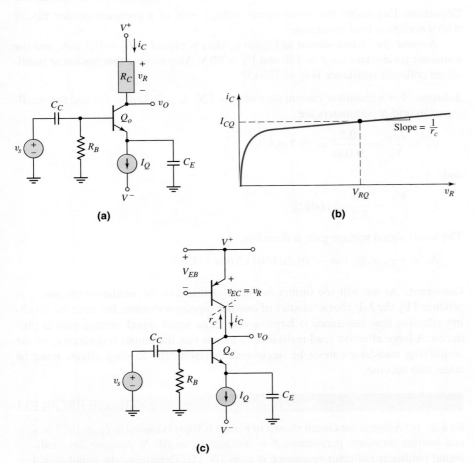

**Figure 6.40** (a) A common-emitter circuit with current source biasing and a nonlinear load resistor, (b) current–voltage characteristics of the nonlinear load resistor, and (c) pnp transistor that can be used to generate the nonlinear load characteristics

The small-signal equivalent circuit of the common-emitter amplifier circuit in Figure 6.40(a) is shown in Figure 6.41. The collector resistor $R_C$ is replaced by the small-signal equivalent resistance $r_c$ that exists at the Q-point. The small-signal voltage gain is then, assuming an ideal voltage signal source,

$$A_v = \frac{V_o}{V_s} = -g_m(r_o \| r_c) \tag{6.60}$$

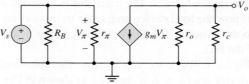

**Figure 6.41** Small-signal equivalent circuit of the circuit in Figure 6.40(a)

## EXAMPLE 6.8

**Objective:** Determine the small-signal voltage gain of a common-emitter circuit with a nonlinear load resistance.

Assume the circuit shown in Figure 6.40(a) is biased at $I_Q = 0.5$ mA, and the transistor parameters are $\beta = 120$ and $V_A = 80$ V. Also assume that nonlinear small-signal collector resistance is $r_c = 120\,\text{k}\Omega$.

**Solution:** For a transistor current gain of $\beta = 120$, $I_{CQ} \cong I_{EQ} = I_Q$, and the small-signal hybrid-$\pi$ parameters are

$$g_m = \frac{I_{CQ}}{V_T} = \frac{0.5}{0.026} = 19.2 \text{ mA/V}$$

and

$$r_o = \frac{V_A}{I_{CQ}} = \frac{80}{0.5} = 160\,\text{k}\Omega$$

The small-signal voltage gain is therefore

$$A_v = -g_m(r_o \| r_c) = -(19.2)(160 \| 120) = -1317$$

**Comment:** As we will see further in Part 2 of this text, the nonlinear resistor $r_c$ is produced by the I–V characteristics of another bipolar transistor. Because the resulting effective load resistance is large, a very large small-signal voltage gain is produced. A large effective load resistance $r_c$ means that the output resistance $r_o$ of the amplifying transistor cannot be neglected; therefore, the loading effects must be taken into account.

## EXERCISE PROBLEM

**Ex 6.8:** (a) Assume the circuit shown in Figure 6.40(a) is biased at $I_Q = 0.25$ mA and assume transistor parameters $\beta = 100$ and $V_A = 100$ V. Assume the small-signal nonlinear collector resistance is $r_c = 100\,\text{k}\Omega$. Determine the small-signal voltage gain. (b) Repeat part (a) assuming that a small-signal load resistance of $r_L = 100\,\text{k}\Omega$ is connected between the output terminal and ground. (Ans. (a) $A_v = -769$; (b) $A_v = -427$)

## 6.5     AC LOAD LINE ANALYSIS

**Objective:** • Understand the concept of the ac load line and calculate the maximum symmetrical swing of the output signal.

A dc load line gives us a way of visualizing the relationship between the $Q$-point and the transistor characteristics. When capacitors are included in a transistor circuit, a new effective load line, called an **ac load line,** may exist. The ac load line helps in visualizing the relationship between the small-signal response and the transistor characteristics. The ac operating region is on the ac load line.

### 6.5.1     AC Load Line

The circuit in Figure 6.36 has emitter resistors and an emitter bypass capacitor. The dc load line is found by writing a Kirchhoff voltage law (KVL) equation around the collector–emitter loop, as follows:

$$V^+ = I_C R_C + V_{CE} + I_E(R_{E1} + R_{E2}) + V^- \tag{6.61}$$

Noting that $I_E = [(1 + \beta)/\beta]I_C$, Equation (6.61) can be written as

$$V_{CE} = (V^+ - V^-) - I_C\left[R_C + \left(\frac{1 + \beta}{\beta}\right)(R_{E1} + R_{E2})\right] \tag{6.62}$$

which is the equation of the dc load line. For the parameters and standard resistor values found in Example 6.7, the dc load line and the $Q$-point are plotted in Figure 6.42. If $\beta \gg 1$, then we can approximate $(1 + \beta)/\beta \cong 1$.

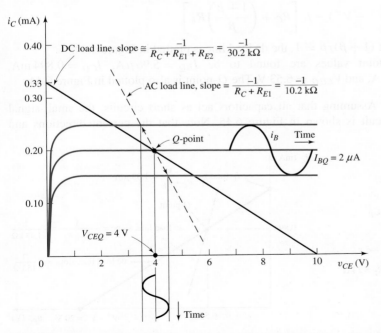

**Figure 6.42** The dc and ac load lines for the circuit in Figure 6.36, and the signal responses to input signal

From the small-signal analysis in Example 6.7, the KVL equation around the collector–emitter loop is

$$i_c R_C + v_{ce} + i_e R_{E1} = 0 \tag{6.63(a)}$$

or, assuming $i_c \cong i_e$, then

$$v_{ce} = -i_c(R_C + R_{E1}) \tag{6.63(b)}$$

This equation is the ac load line. The slope is given by

$$\text{Slope} = \frac{-1}{R_C + R_{E1}}$$

The ac load line is shown in Figure 6.42. When $v_{ce} = i_c = 0$, we are at the $Q$-point. When ac signals are present, we deviate about the $Q$-point on the ac load line.

The slope of the ac load line differs from that of the dc load line because the emitter resistor $R_{E2}$ is not included in the small-signal equivalent circuit. The small-signal C–E voltage and collector current response are functions of the resistor $R_C$ and $R_{E1}$ only.

### EXAMPLE 6.9

**Objective:** Determine the dc and ac load lines for the circuit shown in Figure 6.43. Assume the transistor parameters are: $V_{EB}(\text{on}) = 0.7$ V, $\beta = 150$, and $V_A = \infty$.

**DC Solution:** The dc load line is found by writing a KVL equation around the C–E loop, as follows:

$$V^+ = I_E R_E + V_{EC} + I_C R_C + V^-$$

The dc load line equation is then

$$V_{EC} = (V^+ - V^-) - I_C \left[ R_C + \left( \frac{1+\beta}{\beta} \right) R_E \right]$$

Assuming that $(1 + \beta)/\beta \cong 1$, the dc load line is plotted in Figure 6.44.

The $Q$-point values are found to be $I_{BQ} = 5.96\,\mu\text{A}$, $I_{CQ} = 0.894\,\text{mA}$, $I_{EQ} = 0.90\,\text{mA}$, and $V_{ECQ} = 6.53$ V. The $Q$-point is also plotted in Figure 6.44.

**AC Solution:** Assuming that all capacitors act as short circuits, the small-signal equivalent circuit is shown in Figure 6.45. Note that the current directions and

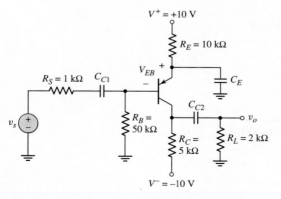

**Figure 6.43** Circuit for Example 6.9

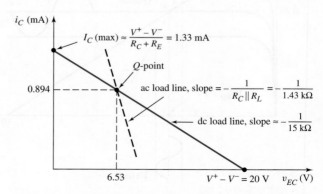

**Figure 6.44** Plots of dc and ac load lines for Example 6.9

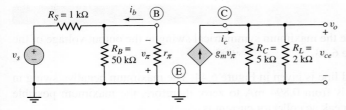

**Figure 6.45** The small-signal equivalent circuit for Example 6.9

voltage polarities in the hybrid-$\pi$ equivalent circuit of the pnp transistor are reversed compared to those of the npn device. The small-signal hybrid-$\pi$ parameters are

$$r_\pi = \frac{V_T \beta}{I_{CQ}} = \frac{(0.026)(150)}{0.894} = 4.36\,\text{k}\Omega$$

$$g_m = \frac{I_{CQ}}{V_T} = \frac{0.894}{0.026} = 34.4\,\text{mA/V}$$

and

$$r_o = \frac{V_A}{I_{CQ}} = \frac{\infty}{I_{CQ}} = \infty$$

The small-signal output voltage, or C–E voltage, is

$$v_o = v_{ce} = +(g_m v_\pi)(R_C \| R_L)$$

where

$$g_m v_\pi = i_c$$

The ac load line, written in terms of the E–C voltage, is defined by

$$v_{ec} = -i_c (R_C \| R_L)$$

The ac load line is also plotted in Figure 6.44.

**Comment:** In the small-signal equivalent circuit, the large 10 k$\Omega$ emitter resistor is effectively shorted by the bypass capacitor $C_E$, the load resistor $R_L$ is in parallel with $R_C$ as a result of the coupling capacitor $C_{C2}$, so that the slope of the ac load line is substantially different than that of the dc load line.

EXERCISE PROBLEM

**Ex 6.9:** For the circuit in Figure 6.39, let $\beta = 125$, $V_{BE}(\text{on}) = 0.7$ V, and $V_A = 200$ V. Plot the dc and ac load lines on the same graph. (Ans. $I_{CQ} = 0.840$ mA, dc load line, $V_{CE} = 10 - I_C(7.3)$; ac load line, $V_{ce} = -I_c(1.58)$)

| 6.5.2 | **Maximum Symmetrical Swing** |
|---|---|

When symmetrical sinusoidal signals are applied to the input of an amplifier, symmetrical sinusoidal signals are generated at the output, as long as the amplifier operation remains linear. We can use the ac load line to determine the **maximum output symmetrical swing**. If the output exceeds this limit, a portion of the output signal will be clipped and signal distortion will occur.

## EXAMPLE 6.10

**Objective:** Determine the maximum symmetrical swing in the output voltage of the circuit given in Figure 6.43.

**Solution:** The ac load line is given in Figure 6.44. The maximum negative swing in the collector current is from 0.894 mA to zero; therefore, the maximum possible symmetrical peak-to-peak ac collector current is

$$\Delta i_c = 2(0.894) = 1.79 \, \text{mA}$$

The maximum symmetrical peak-to-peak output voltage is given by

$$|\Delta v_{ec}| = |\Delta i_c|(R_C \| R_L) = (1.79)(5 \| 2) = 2.56 \, \text{V}$$

Therefore, the maximum instantaneous collector current is

$$i_C = I_{CQ} + \tfrac{1}{2}|\Delta i_c| = 0.894 + 0.894 = 1.79 \, \text{mA}$$

**Comment:** Considering the $Q$-point and the maximum swing in the C–E voltage, the transistor remains biased in the forward-active region. Note that the maximum instantaneous collector current, 1.79 mA, is larger than the maximum dc collector current, 1.33 mA, as determined from the dc load line. This apparent anomaly is due to the different resistance in the C–E circuit for the ac signal and the dc signal.

### EXERCISE PROBLEM

**Ex 6.10:** Reconsider the circuit in Figure 6.38. Let $\beta = 120$, $V_{BE}(\text{on}) = 0.7$ V, and $V_A = \infty$. The circuit parameters are given in Exercise Ex 6.7. (a) Plot the dc and ac load lines on the same graph. (b) Determine the Q-point values. (c) Determine the maximum symmetrical swing in the output voltage for $i_C > 0$ and $0.5 \leq v_{CE} \leq 9$ V. (Ans. (b) $I_{CQ} = 0.884$ mA, $V_{CEQ} = 2.9$ V; (c) $\Delta v_{ce} = 4.8$ V, peak-to-peak)

**Note:** In considering Figure 6.42, it appears that the ac output signal is smaller for the ac load line compared to the dc load line. This is true for a given sinusoidal input base current. However, the required input signal voltage $v_s$ is substantially smaller for the ac load line to generate the given ac base current. This means the voltage gain for the ac load line is larger than that for the dc load line.

### Problem-Solving Technique: Maximum Symmetrical Swing

Again, since we are dealing with linear amplifier circuits, superposition applies so that we can add the dc and ac analysis results. To design a BJT amplifier for maximum symmetrical swing, we perform the following steps.

1.  Write the dc load line equation that relates the quiescent values $I_{CQ}$ and $V_{CEQ}$.
2.  Write the ac load line equation that relates the ac values $i_c$ and $v_{ce}$ : $v_{ce} = -i_c R_{eq}$ where $R_{eq}$ is the effective ac resistance in the collector–emitter circuit.
3.  In general, we can write $i_c = I_{CQ} - I_C(\text{min})$, where $I_C(\text{min})$ is zero or some other specified minimum collector current.

4.  In general, we can write $v_{ce} = V_{CEQ} - V_{CE}(\text{min})$, where $V_{CE}(\text{min})$ is some specified minimum collector-emitter voltage.
5.  The above four equations can be combined to yield the optimum $I_{CQ}$ and $V_{CEQ}$ values to obtain the maximum symmetrical swing in the output signal.

## DESIGN EXAMPLE 6.11

**Objective:** Design a circuit to achieve a maximum symmetrical swing in the output voltage.

**Specifications:** The circuit configuration to be designed is shown in Figure 6.46a. The circuit is to be designed to be bias stable. The minimum collector current is to be $I_C(\text{min}) = 0.1$ mA and the minimum collector-emitter voltage is to be $V_{CE}(\text{min}) = 1$ V.

**Choices:** Assume nominal resistance values of $R_E = 2\,\text{k}\Omega$ and $R_C = 7\,\text{k}\Omega$. Let $R_{TH} = R_1 \| R_2 = (0.1)(1 + \beta)R_E = 24.2\,\text{k}\Omega$. Assume transistor parameters of $\beta = 120$, $V_{BE}(\text{on}) = 0.7$ V, and $V_A = \infty$.

**Solution (Q-Point):** The dc equivalent circuit is shown in Figure 6.46(b) and the midband small-signal equivalent circuit is shown in Figure 6.46(c).

The dc load line, from Figure 6.46(b), is (assuming $I_C \cong I_E$)

$$V_{CE} = 10 - I_C(R_C + R_E) = 10 - I_C(9)$$

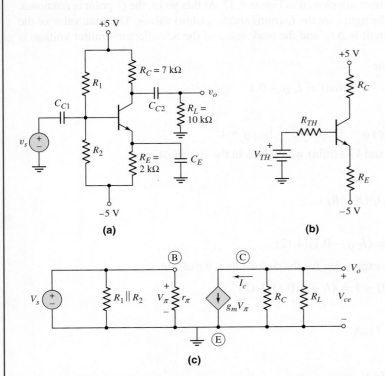

**(a)**

**(b)**

**(c)**

**Figure 6.46** (a) Circuit for Example 6.11, (b) Thevenin equivalent circuit, and (c) small-signal equivalent circuit

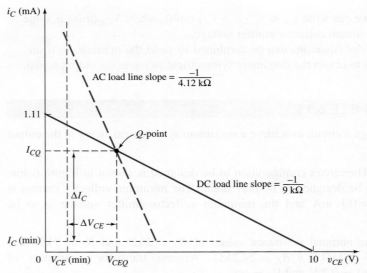

**Figure 6.47** The ac and dc load lines to find the maximum symmetrical swing for the circuit shown in Figure 6.46(a) used in Example 6.11

The ac load line, from Figure 6.46(c), is

$$V_{ce} = -I_c(R_C \| R_L) = -I_c(4.12)$$

These two load lines are plotted in Figure 6.47. At this point, the Q-point is unknown. Also shown in the figure are the $I_C$(min) and $V_{CE}$(min) values. The peak value of the ac collector current is $\Delta I_C$ and the peak value of the ac collector–emitter voltage is $\Delta V_{CE}$.

We can write

$$\Delta I_C = I_{CQ} - I_C(\text{min}) = I_{CQ} - 0.1$$

and

$$\Delta V_{CE} = V_{CEQ} - V_{CE}(\text{min}) = V_{CEQ} - 1$$

where $I_C$(min) and $V_{CE}$(min) were given in the specifications.

Now

$$\Delta V_{CE} = \Delta I_C(R_C \| R_L)$$

or

$$V_{CEQ} - 1 = (I_{CQ} - 0.1)(4.12)$$

Substituting the expression for the dc load line, we obtain

$$10 - I_{CQ}(9) - 1 = (I_{CQ} - 0.1)(4.12)$$

which yields

$$I_{CQ} = 0.717 \, \text{mA}$$

and then

$$V_{CEQ} = 3.54 \, \text{V}$$

**Solution (Bias Resistors):** We can now determine $R_1$ and $R_2$ to produce the desired $Q$-point.

From the dc equivalent circuit, we have

$$V_{TH} = \left(\frac{R_2}{R_1 + R_2}\right)[5 - (-5)] - 5$$

$$= \frac{1}{R_1}(R_{TH})(10) - 5 = \frac{1}{R_1}(24.2)(10) - 5$$

Then, from a KVL equation around the B–E loop, we obtain

$$V_{TH} = \left(\frac{I_{CQ}}{\beta}\right) R_{TH} + V_{BE}(\text{on}) + \left(\frac{1+\beta}{\beta}\right) I_{CQ} R_E - 5$$

or

$$\frac{1}{R_1}(24.2)(10) - 5 = \left(\frac{0.717}{120}\right)(24.2) + 0.7 + \left(\frac{121}{120}\right)(0.717)(2) - 5$$

which yields

$$R_1 = 106 \,\text{k}\Omega$$

We then find

$$R_2 = 31.4 \,\text{k}\Omega$$

**Symmetrical Swing Results:** We then find the peak ac collector current to be

$$\Delta I_C = I_{CQ} - I_C(\text{min}) = 0.717 - 0.1 = 0.617 \,\text{mA}$$

or a peak-to-peak ac collector current to be 1.234 mA. The peak ac collector-emitter voltage is

$$\Delta V_{CE} = V_{CEQ} - V_{CE}(\text{min}) = 3.54 - 1 = 2.54 \,\text{V}$$

or the peak-to-peak ac collector–emitter voltage is 5.08 V.

**Comment:** We have found the $Q$-point to yield the maximum undistorted ac output signal. However, tolerances in resistor values or transistor parameters may change the $Q$-point such that this maximum ac output signal may not be possible without inducing distortion. The effect of tolerances is most easily determined from a computer analysis.

## EXERCISE PROBLEM

**Ex 6.11:** For the circuit shown in Figure 6.48, let $\beta = 120$, $V_{EB}(\text{on}) = 0.7$ V, and $r_o = \infty$. (a) Design a bias-stable circuit such that $I_{CQ} = 1.6$ mA. Determine $V_{ECQ}$. (b) Determine the value of $R_L$ that will produce the maximum symmetrical swing in the output voltage and collector current for $i_C \geq 0.1$ mA and $0.5 \leq v_{EC} \leq 11.5$ V. (Ans. (a) $R_1 = 15.24 \,\text{k}\Omega$, $R_2 = 58.7 \,\text{k}\Omega$, $V_{ECQ} = 3.99$ V (b) $R_L = 5.56 \,\text{k}\Omega$)

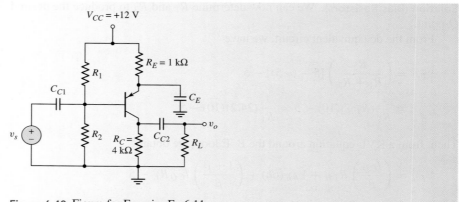

**Figure 6.48** Figure for Exercise Ex 6.11

## Test Your Understanding

**TYU 6.8** For the circuit in Figure 6.31, use the parameters given in Exercise Ex 6.5. If the total instantaneous current must always be greater than 0.1 mA and the total instantaneous C–E voltage must be in the range $0.5 \leq v_{CE} \leq 5$ V, determine the maximum symmetrical swing in the output voltage. (Ans. 3.82 V peak-to-peak)

**TYU 6.9** Consider the circuit in Figure 6.38. Assume transistor and circuit parameters as given in Exercise Ex 6.7, except $R_B$ is a variable and $V_A = \infty$. Assume $i_C \geq 0.1$ mA and $v_{CE} \geq 0.7$ V. (a) Determine the $Q$-point values to yield the maximum symmetrical swing. (b) What is the maximum swing in the collector current and the output voltage? (Ans. (a) $I_{CQ} = 0.808$ mA, $V_{CEQ} = 3.53$ V; (b) peak-to-peak values: $\Delta I_C = 1.42$ mA, $\Delta V_{CE} = 5.67$ V)

## 6.6 COMMON-COLLECTOR (EMITTER-FOLLOWER) AMPLIFIER

**Objective:** • Analyze the emitter-follower amplifier and become familiar with the general characteristics of this circuit.

The second type of transistor amplifier to be considered is the **common-collector circuit.** An example of this circuit configuration is shown in Figure 6.49. As seen in the figure, the output signal is taken off of the emitter with respect to ground and the collector is connected directly to $V_{CC}$. Since $V_{CC}$ is at signal ground in the ac equivalent circuit, we have the name common-collector. The more common name for this circuit is **emitter follower.** The reason for this name will become apparent as we proceed through the analysis.

### 6.6.1 Small-Signal Voltage Gain

The dc analysis of the circuit is exactly the same as we have already seen, so we will concentrate on the small-signal analysis. The hybrid-$\pi$ model of the bipolar

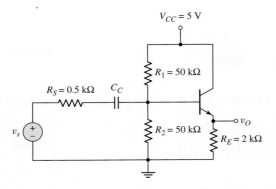

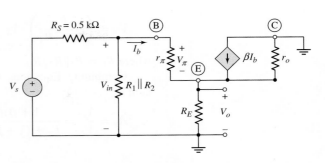

**Figure 6.49** Emitter-follower circuit. Output signal is at the emitter terminal with respect to ground.

**Figure 6.50** Small-signal equivalent circuit of the emitter-follower

transistor can also be used in the small-signal analysis of this circuit. Assuming the coupling capacitor $C_C$ acts as a short circuit, Figure 6.50 shows the small-signal equivalent circuit of the circuit shown in Figure 6.49. The collector terminal is at signal ground and the transistor output resistance $r_o$ is in parallel with the dependent current source.

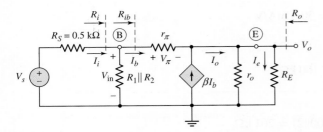

**Figure 6.51** Small-signal equivalent circuit of the emitter-follower with all signal grounds connected together

Figure 6.51 shows the equivalent circuit rearranged so that all signal grounds are at the same point.

We see that

$$I_o = (1 + \beta)I_b \tag{6.64}$$

so the output voltage can be written as

$$V_o = I_b(1 + \beta)(r_o \| R_E) \tag{6.65}$$

Writing a KVL equation around the base-emitter loop, we obtain

$$V_{in} = I_b[r_\pi + (1 + \beta)(r_o \| R_E)] \tag{6.66(a)}$$

or

$$R_{ib} = \frac{V_{in}}{I_b} = r_\pi + (1 + \beta)(r_o \| R_E) \tag{6.66(b)}$$

We can also write

$$V_{\text{in}} = \left( \frac{R_i}{R_i + R_S} \right) \cdot V_s \tag{6.67}$$

where $R_i = R_1 \| R_2 \| R_{ib}$.

Combining Equations (6.65), (6.66(b)), and (6.67), the small-signal voltage gain is

$$A_v = \frac{V_o}{V_s} = \frac{(1 + \beta)(r_o \| R_E)}{r_\pi + (1 + \beta)(r_o \| R_E)} \cdot \left( \frac{R_i}{R_i + R_S} \right) \tag{6.68}$$

---

### EXAMPLE 6.12

**Objective:** Calculate the small-signal voltage gain of an emitter-follower circuit.

For the circuit shown in Figure 6.49, assume the transistor parameters are: $\beta = 100$, $V_{BE}(\text{on}) = 0.7$ V, and $V_A = 80$ V.

**Solution:** The dc analysis shows that $I_{CQ} = 0.793$ mA and $V_{CEQ} = 3.4$ V. The small-signal hybrid-$\pi$ parameters are determined to be

$$r_\pi = \frac{V_T \beta}{I_{CQ}} = \frac{(0.026)(100)}{0.793} = 3.28 \text{ k}\Omega$$

$$g_m = \frac{I_{CQ}}{V_T} = \frac{0.793}{0.026} = 30.5 \text{ mA/V}$$

and

$$r_o = \frac{V_A}{I_{CQ}} = \frac{80}{0.793} \cong 100 \text{ k}\Omega$$

We may note that

$$R_{ib} = 3.28 + (101)(100 \| 2) = 201 \text{ k}\Omega$$

and

$$R_i = 50 \| 50 \| 201 = 22.2 \text{ k}\Omega$$

The small-signal voltage gain is then

$$A_v = \frac{(101)(100 \| 2)}{3.28 + (101)(100 \| 2)} \cdot \left( \frac{22.2}{22.2 + 0.5} \right)$$

or

$$A_v = +0.962$$

**Comment:** The magnitude of the voltage gain is slightly less than 1. An examination of Equation (6.68) shows that this is always true. Also, the voltage gain is positive, which means that the output signal voltage at the emitter is in phase with the input signal voltage. The reason for the terminology emitter-follower is now clear. The ac output voltage at the emitter is essentially equal to the ac input voltage.

At first glance, a transistor amplifier with a voltage gain essentially of 1 may not seem to be of much value. However, the input and output resistance characteristics make this circuit extremely useful in many applications, as we will show in the next section.

**Ex 6.12:** For the circuit shown in Figure 6.49, let $V_{CC} = 12$ V, $R_E = 30\,\Omega$, $R_1 = 1.3\,\text{k}\Omega$, $R_2 = 4.2\,\text{k}\Omega$, and $R_S = 0$. The transistor parameters are $\beta = 80$, $V_{BE}(\text{on}) = 0.7$ V, and $V_A = 75$ V. (a) Determine the quiescent values $I_{EQ}$ and $V_{CEQ}$. (b) Find the small-signal voltage gain $A_v = V_o/V_s$. (c) Determine the input resistance looking into the base of the transistor. (Ans. (a) $I_{EQ} = 0.2$ A, $V_{CEQ} = 6$ V; (b) $A_v = 0.9954$; (c) $R_{ib} = 2.27\,\text{k}\Omega$)

**PS 6.4:** Perform a PSpice simulation on the circuit in Figure 6.49. (a) Determine the small-signal voltage gain and (b) find the effective resistance seen by the signal source $v_s$.

### 6.6.2    Input and Output Impedance

**Input Resistance**

The input impedance, or small-signal input resistance for low-frequency signals, of the emitter-follower is determined in the same manner as for the common-emitter circuit. For the circuit in Figure 6.49, the input resistance looking into the base is denoted $R_{ib}$ and is indicated in the small-signal equivalent circuit shown in Figure 6.51.

The input resistance $R_{ib}$ was given by Equation (6.66(b)) as

$$R_{ib} = r_\pi + (1 + \beta)(r_o \| R_E)$$

Since the emitter current is $(1 + \beta)$ times the base current, the effective impedance in the emitter is multiplied by $(1 + \beta)$. We saw this same effect when an emitter resistor was included in a common-emitter circuit. This multiplication by $(1 + \beta)$ is again called the **resistance reflection rule.** The input resistance at the base is $r_\pi$ plus the effective resistance in the emitter multiplied by the $(1 + \beta)$ factor. This resistance reflection rule will be used extensively throughout the remainder of the text.

**Output Resistance**

Initially, to find the output resistance of the emitter-follower circuit shown in Figure 6.49, we will assume that the input signal source is ideal and that $R_S = 0$. The circuit shown in Figure 6.52 can be used to determine the output resistance looking back into the output terminals. The circuit is derived from the small-signal equivalent

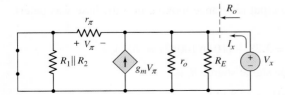

**Figure 6.52** Small-signal equivalent circuit of the emitter-follower used to determine the output resistance. The source resistance $R_S$ is assumed to be zero (an ideal signal source).

circuit shown in Figure 6.51 by setting the independent voltage source $V_s$ equal to zero, which means that $V_s$ acts as a short circuit. A test voltage $V_x$ is applied to the output terminal and the resulting test current is $I_x$. The output resistance, $R_o$, is given by

$$R_o = \frac{V_x}{I_x} \tag{6.69}$$

In this case, the control voltage $V_\pi$ is not zero, but is a function of the applied test voltage. From Figure 6.52, we see that $V_\pi = -V_x$. Summing currents at the output node, we have

$$I_x + g_m V_\pi = \frac{V_x}{R_E} + \frac{V_x}{r_o} + \frac{V_x}{r_\pi} \tag{6.70}$$

Since $V_\pi = -V_x$, Equation (6.70) can be written as

$$\frac{I_x}{V_x} = \frac{1}{R_o} = g_m + \frac{1}{R_E} + \frac{1}{r_o} + \frac{1}{r_\pi} \tag{6.71}$$

or the output resistance is given by

$$R_o = \frac{1}{g_m} \| R_E \| r_o \| r_\pi \tag{6.72}$$

The output resistance may also be written in a slightly different form. Equation (6.71) can be written in the form

$$\frac{1}{R_o} = \left( g_m + \frac{1}{r_\pi} \right) + \frac{1}{R_E} + \frac{1}{r_o} = \left( \frac{1+\beta}{r_\pi} \right) + \frac{1}{R_E} + \frac{1}{r_o} \tag{6.73}$$

or the output resistance can be written in the form

$$R_o = \frac{r_\pi}{1+\beta} \| R_E \| r_o \tag{6.74}$$

Equation (6.74) says that the output resistance looking back into the output terminals is the effective resistance in the emitter, $R_E \| r_o$, in parallel with the resistance looking back into the emitter. In turn, the resistance looking into the emitter is the total resistance in the base circuit divided by $(1+\beta)$. This is an important result and is called the **inverse resistance reflection rule** and is the inverse of the reflection rule looking to the base.

## EXAMPLE 6.13

**Objective:** Calculate the input and output resistance of the emitter-follower circuit shown in Figure 6.49. Assume $R_S = 0$.

The small-signal parameters, as determined in Example 6.12, are $r_\pi = 3.28\,k\Omega$, $\beta = 100$, and $r_o = 100\,k\Omega$.

**Solution (Input Resistance):** The input resistance looking into the base was determined in Example 6.12 as

$$R_{ib} = r_\pi + (1+\beta)(r_o \| R_E) = 3.28 + (101)(100\|2) = 201\ k\Omega$$

and the input resistance seen by the signal source $R_i$ is

$$R_i = R_1 \| R_2 \| R_{ib} = 50\|50\|201 = 22.2\ k\Omega$$

**Comment:** The input resistance of the emitter-follower looking into the base is substantially larger than that of the simple common-emitter circuit because of the

$(1 + \beta)$ factor. This is one advantage of the emitter-follower circuit. However, in this case, the input resistance seen by the signal source is dominated by the bias resistors $R_1$ and $R_2$. To take advantage of the large input resistance of the emitter-follower circuit, the bias resistors must be designed to be much larger.

**Solution (Output Resistance):** The output resistance is found from Equation (6.74) as

$$R_o = \left(\frac{r_\pi}{1 + \beta}\right) \| R_E \| r_o = \left(\frac{3.28}{101}\right) \| 2 \| 100$$

or

$$R_o = 0.0325 \| 2 \| 100 = 0.0320 \text{ k}\Omega \Rightarrow 32.0 \; \Omega$$

The output resistance is dominated by the first term that has $(1 + \beta)$ in the denominator.

**Comment:** The emitter-follower circuit is sometimes referred to as an **impedance transformer,** since the input impedance is large and the output impedance is small. The very low output resistance makes the *emitter-follower act almost like an ideal voltage source,* so the output is not loaded down when used to drive another load. Because of this, the emitter-follower is often used as the output stage of a multistage amplifier.

**EXERCISE PROBLEM**

**Ex 6.13:** Consider the circuit and transistor parameters described in Exercise Ex 6.12 for the circuit shown in Figure 6.49. For the case of $R_S = 0$, determine the output resistance looking into the output terminals. (Ans. $R_o = 0.129 \; \Omega$)

We can determine the output resistance of the emitter-follower circuit taking into account a nonzero source resistance. The circuit in Figure 6.53 is derived from the small-signal equivalent circuit shown in Figure 6.51 and can be used to find $R_o$. The independent source $V_s$ is set equal to zero and a test voltage $V_x$ is applied to the output terminals. Again, the control voltage $V_\pi$ is not zero, but is a function of the test voltage. Summing currents at the output node, we have

$$I_x + g_m V_\pi = \frac{V_x}{R_E} + \frac{V_x}{r_o} + \frac{V_x}{r_\pi + R_1 \| R_2 \| R_S} \tag{6.75}$$

The control voltage can be written in terms of the test voltage by a voltage divider equation as

$$V_\pi = -\left(\frac{r_\pi}{r_\pi + R_1 \| R_2 \| R_S}\right) \cdot V_x \tag{6.76}$$

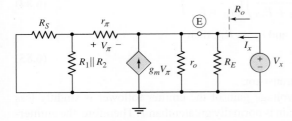

**Figure 6.53** Small-signal equivalent circuit of the emitter-follower used to determine the output resistance including the effect of the source resistance $R_S$

Equation (6.75) can then be written as

$$I_x = \left( \frac{g_m r_\pi}{r_\pi + R_1 \| R_2 \| R_S} \right) \cdot V_x + \frac{V_x}{R_E} + \frac{V_x}{r_o} + \frac{V_x}{r_\pi + R_1 \| R_2 \| R_S} \tag{6.77}$$

Noting that $g_m r_\pi = \beta$, we find

$$\frac{I_x}{V_x} = \frac{1}{R_o} = \frac{1 + \beta}{r_\pi + R_1 \| R_2 \| R_S} + \frac{1}{R_E} + \frac{1}{r_o} \tag{6.78}$$

or

$$R_o = \left( \frac{r_\pi + R_1 \| R_2 \| R_S}{1 + \beta} \right) \| R_E \| r_o \tag{6.79}$$

In this case, the source resistance and bias resistances contribute to the output resistance.

### 6.6.3 Small-Signal Current Gain

We can determine the small-signal current gain of an emitter-follower by using the input resistance and the concept of current dividers. For the small-signal emitter-follower equivalent circuit shown in Figure 6.51, the small signal current gain is defined as

$$A_i = \frac{I_e}{I_i} \tag{6.80}$$

where $I_e$ and $I_i$ are the output and input current phasors.

Using a current divider equation, we can write the base current in terms of the input current, as follows:

$$I_b = \left( \frac{R_1 \| R_2}{R_1 \| R_2 + R_{ib}} \right) I_i \tag{6.81}$$

Since $g_m V_\pi = \beta I_b$, then,

$$I_o = (1 + \beta) I_b = (1 + \beta) \left( \frac{R_1 \| R_2}{R_1 \| R_2 + R_{ib}} \right) I_i \tag{6.82}$$

Writing the load current in terms of $I_o$ produces

$$I_e = \left( \frac{r_o}{r_o + R_E} \right) I_o \tag{6.83}$$

Combining Equations (6.82) and (6.83), we obtain the small-signal current gain, as follows:

$$A_i = \frac{I_e}{I_i} = (1 + \beta) \left( \frac{R_1 \| R_2}{R_1 \| R_2 + R_{ib}} \right) \left( \frac{r_o}{r_o + R_E} \right) \tag{6.84}$$

If we assume that $R_1 \| R_2 \gg R_{ib}$ and $r_o \gg R_E$, then

$$A_i \cong (1 + \beta) \tag{6.85}$$

which is the current gain of the transistor.

Although the small-signal voltage gain of the emitter follower is slightly less than 1, the small-signal current gain is normally greater than 1. Therefore, the emitter-follower circuit produces a small-signal power gain.

Although we did not explicitly calculate a current gain in the common-emitter circuit previously, the analysis is the same as that for the emitter-follower and in general the current gain is also greater than unity.

## DESIGN EXAMPLE 6.14

**Objective:** To design an emitter-follower amplifier to meet an output resistance specification.

**Specifications:** Consider the output signal of the amplifier designed in Example 6.7. We now want to design an emitter-follower circuit with the configuration shown in Figure 6.54 such that the output signal from this circuit does not vary by more than 5 percent when a load in the range $R_L = 4\,\text{k}\Omega$ to $R_L = 20\,\text{k}\Omega$ is connected to the output.

**Choices:** We will assume that a transistor with nominal parameter values of $\beta = 100$, $V_{BE}(\text{on}) = 0.7$ V, and $V_A = 80$ V is available.

**Discussion:** The output resistance of the common-emitter circuit designed in Example 6.7 is $R_o = R_C = 10$ k$\Omega$. Connecting a load resistance between 4 k$\Omega$ and 20 k$\Omega$ will load down this circuit, so that the output voltage will change substantially. For this reason, an emitter-follower circuit with a low output resistance must be designed to minimize the loading effect. The Thevenin equivalent circuit is shown in Figure 6.55. The output voltage can be written as

$$v_o = \left( \frac{R_L}{R_L + R_o} \right) \cdot v_{TH}$$

where $v_{TH}$ is the ideal voltage generated by the amplifier. In order to have $v_o$ change by less than 5 percent as a load resistance $R_L$ is added, we must have $R_o$ less than or equal to approximately 5 percent of the minimum value of $R_L$. In this case, then, we need $R_o$ to be approximately 200 $\Omega$.

**Initial Design Approach:** Consider the emitter-follower circuit shown in Figure 6.54. Note that the source resistance is $R_S = 10$ k$\Omega$, corresponding to the output resistance of the circuit designed in Example 6.7.

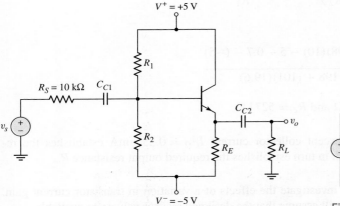

Figure 6.54 Figure for Example 6.14

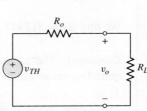

Figure 6.55 Thevenin equivalent of the output of an amplifier

The output resistance, given by Equation (6.79), is

$$R_o = \left(\frac{r_\pi + R_1 \| R_2 \| R_S}{1 + \beta}\right) \| R_E \| r_o$$

The first term, with $(1 + \beta)$ in the denominator, dominates, and if $R_1 \| R_2 \| R_S \cong R_S$, then we have

$$R_o \cong \frac{r_\pi + R_S}{1 + \beta}$$

For $R_o = 200\ \Omega$, we find

$$0.2 = \frac{r_\pi + 10}{101}$$

or $r_\pi = 10.2\ \mathrm{k}\Omega$. Since $r_\pi = (\beta V_T)/I_{CQ}$, the quiescent collector current must be

$$I_{CQ} = \frac{\beta V_T}{r_\pi} = \frac{(100)(0.026)}{10.2} = 0.255\ \mathrm{mA}$$

Assuming $I_{CQ} \cong I_{EQ}$ and letting $V_{CEQ} = 5\ \mathrm{V}$, we find

$$R_E = \frac{V^+ - V_{CEQ} - V^-}{I_{EQ}} = \frac{5 - 5 - (-5)}{0.255} = 19.6\ \mathrm{k}\Omega$$

The term $(1 + \beta)R_E$ is

$$(1 + \beta)R_E = (101)(19.6) \Rightarrow 1.98\ \mathrm{M}\Omega$$

With this large resistance, we can design a bias-stable circuit as defined in Chapter 5 and still have large values for bias resistances. Let

$$R_{TH} = (0.1)(1 + \beta)R_E = (0.1)(101)(19.6) = 198\ \mathrm{k}\Omega$$

The base current is

$$I_B = \frac{V_{TH} - V_{BE}(\mathrm{on}) - V^-}{R_{TH} + (1 + \beta)R_E}$$

where

$$V_{TH} = \left(\frac{R_2}{R_1 + R_2}\right)(10) - 5 = \frac{1}{R_1}(R_{TH})(10) - 5$$

We can then write

$$\frac{0.255}{100} = \frac{\dfrac{1}{R_1}(198)(10) - 5 - 0.7 - (-5)}{198 + (101)(19.6)}$$

We find $R_1 = 317\ \mathrm{k}\Omega$ and $R_2 = 527\ \mathrm{k}\Omega$.

**Comment:** The quiescent collector current $I_{CQ} = 0.255$ mA establishes the required $r_\pi$ value which in turn establishes the required output resistance $R_o$.

**Trade-offs:** We will investigate the effects of a variation in transistor current gain. In this example, we will assume that the designed resistor values are available.

The Thevenin equivalent resistance is $R_{TH} = R_1 \| R_2 = 198\,\text{k}\Omega$ and the Thevenin equivalent voltage is $V_{TH} = 1.244$ V. The base current is found by the KVL equation around the B–E loop. We find

$$I_{BQ} = \frac{1.244 - 0.7 - (-5)}{198 + (1 + \beta)(19.6)}$$

The collector current is $I_{CQ} = \beta I_{BQ}$ and we find $r_\pi = (\beta V_T)/I_{CQ}$. Finally, the output resistance is approximately

$$R_o \cong \frac{r_\pi + R_{TH} \| R_S}{1 + \beta} = \frac{r_\pi + 198 \| 10}{1 + \beta}$$

The values of these parameters for several values of $\beta$ are shown in the following table.

| $\beta$ | $I_{CQ}$ (mA) | $r_\pi$ (k$\Omega$) | $R_o$ ($\Omega$) |
|---|---|---|---|
| 50 | 0.232 | 5.62 | 297 |
| 75 | 0.246 | 7.91 | 229 |
| 100 | 0.255 | 10.2 | 195 |
| 125 | 0.260 | 12.5 | 175 |

From these results, we see that the specified maximum output resistance of $R_o \cong 200\,\Omega$ is met only if the current gain of the transistor is at least $\beta = 100$. In this design, then, we must specify that the minimum current gain of a transistor is 100.

**Computer Simulation:** We again used approximation techniques in our design. For this reason, it is useful to verify our design with a PSpice analysis, since the computer simulation will take into account more details than our hand design.

Figure 6.56 shows the PSpice circuit schematic diagram. A 1 mV sinusoidal signal source is capacitively coupled to the output of the emitter follower. The input signal source has been set equal to zero. The current from the output signal source was found to be 5.667 $\mu$A. The output resistance of the emitter follower is then $R_o = 176\,\Omega$, which means that we have met our desired specification that the output resistance should be less than 200 $\Omega$.

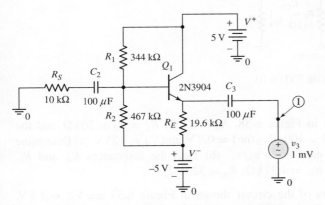

**Figure 6.56** PSpice circuit schematic for Example 6.14

**Discussion:** From the computer simulation, the quiescent collector current is $I_{CQ} = 0.239$ mA compared to the designed value of $I_{CQ} = 0.255$ mA. The principal reason for the difference in value is the difference in base-emitter voltage and current gain between the hand analysis and computer simulation.

The output resistance specification is met in the computer simulation. In the PSpice analysis, the ac beta is 135 and the output resistance is $R_o = 176\,\Omega$. This value correlates very well with the hand analysis in which $R_o = 184\,\Omega$ for $\beta = 125$.

**Ex 6.14:** For the circuit in Figure 6.54, the transistor parameters are: $\beta = 100$, $V_{BE}(\text{on}) = 0.7$ V, and $V_A = 125$ V. Assume $R_S = 0$ and $R_L = 1$ k$\Omega$. (a) Design a bias-stable circuit such that $I_{CQ} = 1.25$ mA and $V_{CEQ} = 4$ V. (b) What is the small-signal current gain $A_i = i_o/i_i$? (c) What is the output resistance looking back into the output terminals? (Ans. (a) $R_E = 4.76$ k$\Omega$, $R_1 = 65.8$ k$\Omega$, $R_2 = 178.8$ k$\Omega$; (b) $A_i = 29.9$, (c) $R_o = 20.5\,\Omega$)

## Test Your Understanding

**TYU 6.10** Assume the circuit in Figure 6.57 uses a 2N2222 transistor. Assume a nominal dc current gain of $\beta = 130$. Using the average $h$-parameter values (assume $h_{re} = 0$) given in the data sheets, determine $A_v = v_o/v_s$, $A_i = i_o/i_s$, $R_{ib}$, and $R_o$ for $R_S = R_L = 10$ k$\Omega$. (Ans. $A_v = 0.891$, $A_i = 8.59$, $R_{ib} = 641$ k$\Omega$, $R_o = 96\,\Omega$)

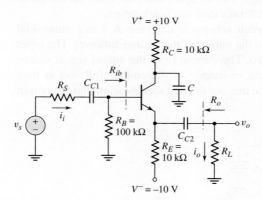

**Figure 6.57** Figure for Exercise TYU 6.10

**TYU 6.11** For the circuit in Figure 6.58, $R_E = 2$ k$\Omega$, $R_1 = R_2 = 50$ k$\Omega$ and the transistor parameters are $\beta = 100$, $V_{EB}(\text{on}) = 0.7$ V, and $V_A = 125$ V. (a) Determine the small-signal voltage gain $A_v = v_o/v_s$. (b) Find the resistances $R_{ib}$ and $R_o$. (Ans. (a) $A_v = 0.925$, (b) $R_{ib} = 4.37$ k$\Omega$, $R_o = 32.0\,\Omega$)

**TYU 6.12** The parameters of the circuit shown in Figure 6.57 are $V^+ = 3.3$ V, $V^- = -3.3$ V, $R_E = 15$ k$\Omega$, $R_L = 2$ k$\Omega$, $R_S = 2$ k$\Omega$, and $R_C = 0$. The transistor parameters are $\beta = 120$ and $V_A = \infty$. (a) Determine the quiescent values $I_{EQ}$ and

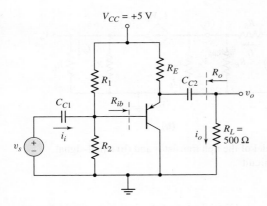

**Figure 6.58** Figure for Exercises TYU 6.11.

$V_{CEQ}$. (b) Find the small-signal voltage gain and small-signal current gain. (c) Calculate the small-signal input resistance $R_{ib}$ and the small-signal output resistance $R_o$. (Ans. (a) $I_{EQ} = 0.163\,\text{mA}$, $V_{CEQ} = 4.14\,\text{V}$; (b) $A_v = 0.892$, $A_i = 32.1$; (c) $R_{ib} = 232.7\,\text{k}\Omega$, $R_o = 172\,\Omega$)

## 6.7   COMMON-BASE AMPLIFIER

**Objective:** • Analyze the common-base amplifier and become familiar with the general characteristics of this circuit.

A third amplifier circuit configuration is the **common-base circuit.** To determine the small-signal voltage and current gains, and the input and output impedances, we will use the same hybrid-$\pi$ equivalent circuit for the transistor that was used previously. The dc analysis of the common-base circuit is essentially the same as for the common-emitter circuit.

### 6.7.1   Small-Signal Voltage and Current Gains

Figure 6.59 shows the basic common-base circuit, in which the base is at signal ground and the input signal is applied to the emitter. Assume a load is connected to the output through a coupling capacitor $C_{C2}$.

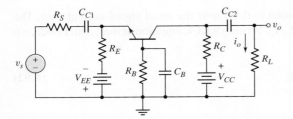

**Figure 6.59** Basic common-base circuit. The input signal is applied to the emitter terminal and the output signal is measured at the collector terminal.

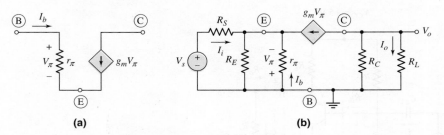

**Figure 6.60** (a) Simplified hybrid-$\pi$ model of the npn transistor and (b) small-signal equivalent circuit of the common-base circuit

Figure 6.60(a) again shows the hybrid-$\pi$ model of the npn transistor, with the output resistance $r_o$ assumed to be infinite. Figure 6.60(b) shows the small-signal equivalent circuit of the common-base circuit, including the hybrid-$\pi$ model of the transistor. As a result of the common-base configuration, the hybrid-$\pi$ model in the small-signal equivalent circuit may look a little strange.

The small signal output voltage is given by

$$V_o = -(g_m V_\pi)(R_C \| R_L) \tag{6.86}$$

Writing a KCL equation at the emitter node, we obtain

$$g_m V_\pi + \frac{V_\pi}{r_\pi} + \frac{V_\pi}{R_E} + \frac{V_s - (-V_\pi)}{R_S} = 0 \tag{6.87}$$

Since $\beta = g_m r_\pi$, Equation (6.87) can be written

$$V_\pi \left( \frac{1+\beta}{r_\pi} + \frac{1}{R_E} + \frac{1}{R_S} \right) = -\frac{V_s}{R_S} \tag{6.88}$$

Then,

$$V_\pi = -\frac{V_s}{R_S} \left[ \left( \frac{r_\pi}{1+\beta} \right) \| R_E \| R_S \right] \tag{6.89}$$

Substituting Equation (6.89) into (6.86), we find the small-signal voltage gain, as follows:

$$A_v = \frac{V_o}{V_s} = +g_m \left( \frac{R_C \| R_L}{R_S} \right) \left[ \left( \frac{r_\pi}{1+\beta} \right) \| R_E \| R_S \right] \tag{6.90}$$

We can show that as $R_S$ approaches zero, the small-signal voltage gain becomes

$$A_v = g_m (R_C \| R_L) \tag{6.91}$$

Figure 6.60(b) can also be used to determine the small-signal current gain. The current gain is defined as $A_i = I_o / I_i$. Writing a KCL equation at the emitter node, we have

$$I_i + \frac{V_\pi}{r_\pi} + g_m V_\pi + \frac{V_\pi}{R_E} = 0 \tag{6.92}$$

Solving for $V_\pi$, we obtain

$$V_\pi = -I_i \left[ \left( \frac{r_\pi}{1+\beta} \right) \| R_E \right] \tag{6.93}$$

The load current is given by

$$I_o = -(g_m V_\pi)\left(\frac{R_C}{R_C + R_L}\right) \qquad \textbf{(6.94)}$$

Combining Equations (6.93) and (6.94), we obtain an expression for the small-signal current gain, as follows:

$$A_i = \frac{I_o}{I_i} = g_m\left(\frac{R_C}{R_C + R_L}\right)\left[\left(\frac{r_\pi}{1+\beta}\right)\middle\| R_E\right] \qquad \textbf{(6.95)}$$

If we take the limit as $R_E$ approaches infinity and $R_L$ approaches zero, then the current gain becomes the short-circuit current gain given by

$$A_{io} = \frac{g_m r_\pi}{1+\beta} = \frac{\beta}{1+\beta} = \alpha \qquad \textbf{(6.96)}$$

where $\alpha$ is the common-base current gain of the transistor.

Equations (6.90) and (6.96) indicate that, for the common-base circuit, the small-signal voltage gain is usually greater than 1 and the small-signal current gain is slightly less than 1. However, we still have a small-signal power gain. The applications of a common-base circuit take advantage of the input and output resistance characteristics.

### 6.7.2    Input and Output Impedance

Figure 6.61 shows the small-signal equivalent circuit of the common-base configuration looking into the emitter. In this circuit, for convenience only, we have reversed the polarity of the control voltage, which reverses the direction of the dependent current source.

The input resistance looking into the emitter is defined as

$$R_{ie} = \frac{V_\pi}{I_i} \qquad \textbf{(6.97)}$$

If we write a KCL equation at the input, we obtain

$$I_i = I_b + g_m V_\pi = \frac{V_\pi}{r_\pi} + g_m V_\pi = V_\pi\left(\frac{1+\beta}{r_\pi}\right) \qquad \textbf{(6.98)}$$

Therefore,

$$R_{ie} = \frac{V_\pi}{I_i} = \frac{r_\pi}{1+\beta} \equiv r_e \qquad \textbf{(6.99)}$$

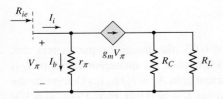

**Figure 6.61** Common-base equivalent circuit for input resistance calculations

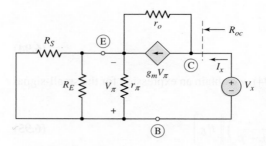

**Figure 6.62** Common-base equivalent circuit for output resistance calculations

The resistance looking into the emitter, with the base grounded, is usually defined as $r_e$ and is quite small, as already shown in the analysis of the emitter-follower circuit. When the input signal is a current source, a small input resistance is desirable.

Figure 6.62 shows the circuit used to calculate the output resistance looking back into the collector terminal. The small-signal resistance $r_o$ is included. The independent source $v_s$ has been set equal to zero. We may define an equivalent resistance $R_{eq} = R_S \| R_E \| r_\pi$.

Writing a KCL equation at the output node, we obtain

$$I_x = g_m V_\pi + \frac{V_x - (-V_\pi)}{r_o} \qquad \textbf{(6.100(a))}$$

A KCL equation at the emitter node yields

$$\frac{V_\pi}{R_{eq}} + g_m V_\pi + \frac{V_x - (-V_\pi)}{r_o} = 0 \qquad \textbf{(6.100(b))}$$

Combining Equations (6.100(a)) and (6.100(b)), we find that output resistance as

$$\frac{V_x}{I_x} = R_{oc} = r_o \left( 1 + g_m R_{eq} \right) + R_{eq} \qquad \textbf{(6.101)}$$

If the input resistance $R_S = 0$, then $R_{eq} = 0$ and the output resistance is just given by $R_{oc} = r_o$. Including a collector resistor and a load resistor, the output resistance looking back into the output terminal is $R_o = R_{oc} \| R_C \| R_L$.

Since the output resistance looking back into the collector terminal is very large, the common-base circuit looks almost like an ideal current source. The circuit is also referred to as a **current buffer.**

### Discussion

The common-base circuit is very useful when the input signal is a current. We will see this type of application when we discuss the cascode circuit in Section 6.9.

## Test Your Understanding

**TYU 6.13** For the circuit shown in Figure 6.63, the transistor parameters are: $\beta = 100$, $V_{EB}(\text{on}) = 0.7$ V, and $r_o = \infty$. (a) Calculate the quiescent values of $I_{CQ}$ and $V_{ECQ}$. (b) Determine the small-signal current gain $A_i = i_o / i_i$. (c) Determine the small-signal voltage gain $A_v = v_o / v_s$. (Ans. (a) $I_{CQ} = 0.921$ mA, $V_{ECQ} = 6.1$ V (b) $A_i = 0.987$ (c) $A_v = 177$)

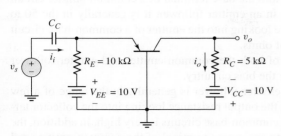

**Figure 6.63** Figure for Exercise TYU 6.13

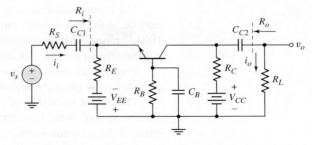

**Figure 6.64** Figure for Exercises TYU 6.14 and TYU 6.15

**TYU 6.14** The transistor parameters for the circuit shown in Figure 6.64 are $\beta = 120$, $V_{BE}(\text{on}) = 0.7\,\text{V}$, and $V_A = \infty$. The circuit parameters are $V_{CC} = V_{EE} = 3.3\,\text{V}$, $R_S = 500\,\Omega$, $R_L = 6\,\text{k}\Omega$, $R_B = 100\,\text{k}\Omega$, $R_E = 12\,\text{k}\Omega$, and $R_C = 12\,\text{k}\Omega$. (a) Determine the small-signal transistor parameters $g_m$, $r_\pi$, and $r_o$. (b) Find the small-signal current gain $A_i = i_o/i_i$ and the small-signal voltage gain $A_v = v_o/v_s$. (c) Determine the input resistance $R_i$ and the output resistance $R_o$. (Ans. (a) $g_m = 7.73\,\text{mA/V}$, $r_\pi = 15.5\,\text{k}\Omega$, $r_o = \infty$; (b) $A_i = 0.654$, $A_v = 6.26$; (c) $R_i = 127\,\Omega$, $R_o = 12\,\text{k}\Omega$)

**TYU 6.15** For the circuit shown in Figure 6.64, let $R_S = 0$, $C_B = 0$, $R_C = R_L = 2\,\text{k}\Omega$, $V_{CC} = V_{EE} = 5\,\text{V}$, $\beta = 100$, $V_{BE}(\text{on}) = 0.7\,\text{V}$, and $V_A = \infty$. Design $R_E$ and $R_B$ for a dc quiescent collector current of 1 mA and a small-signal voltage gain of 20. (Ans. $R_B = 2.4\,\text{k}\Omega$, $R_E = 4.23\,\text{k}\Omega$)

## 6.8  THE THREE BASIC AMPLIFIERS: SUMMARY AND COMPARISON

**Objective:** • Compare the general characteristics of the three basic amplifier configurations.

The basic small-signal characteristics of the three single-stage amplifier configurations are summarized in Table 6.4.

For the common-emitter circuit, the voltage and current gains are generally greater than 1. For the emitter-follower, the voltage gain is slightly less than 1, while the current gain is greater than 1. For the common-base circuit, the voltage gain is greater than 1, while the current gain is less than 1.

**Table 6.4**  Characteristics of the three BJT amplifier configurations

| Configuration | Voltage gain | Current gain | Input resistance | Output resistance |
|---|---|---|---|---|
| Common emitter | $A_v > 1$ | $A_i > 1$ | Moderate | Moderate to high |
| Emitter follower | $A_v \cong 1$ | $A_i > 1$ | High | Low |
| Common base | $A_v > 1$ | $A_i \cong 1$ | Low | Moderate to high |

The input resistance looking into the base terminal of a common-emitter circuit may be in the low kilohm range; in an emitter follower, it is generally in the 50 to 100 kΩ range. The input resistance looking into the emitter of a common-base circuit is generally on the order of tens of ohms.

The overall input resistance of both the common-emitter and emitter-follower circuits can be greatly affected by the bias circuitry.

The output resistance of the emitter follower is generally in the range of a few ohms to tens of ohms. In contrast, the output resistance looking into the collector terminal of the common-emitter and common-base circuits is very high. In addition, the output resistance looking back into the output terminal of the common-emitter and common-base circuits is a strong function of the collector resistance. For these circuits, the output resistance can easily drop to a few kilohms.

The characteristics of these single-stage amplifiers will be used in the design of multistage amplifiers.

## 6.9    MULTISTAGE AMPLIFIERS

**Objective:** • Analyze multitransistor or multistage amplifiers and understand the advantages of these circuits over single-transistor amplifiers.

In most applications, a single transistor amplifier will not be able to meet the combined specifications of a given amplification factor, input resistance, and output resistance. For example, the required voltage gain may exceed that which can be obtained in a single transistor circuit. We also saw an illustration of this effect in Example 6.14, in which a low output resistance was required in a particular design.

Transistor amplifier circuits can be connected in series, or **cascaded,** as shown in Figure 6.65. This may be done either to increase the overall small-signal voltage gain or to provide an overall voltage gain greater than 1, with a very low output resistance. The overall voltage or current gain, in general, is not simply the product of the individual amplification factors. For example, the gain of stage 1 is a function of the input resistance of stage 2. In other words, loading effects may have to be taken into account.

There are many possible multistage configurations; we will examine a few here, in order to understand the type of analysis required.

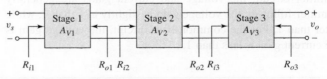

Figure 6.65  A generalized three-stage amplifier

### 6.9.1    Multistage Analysis: Cascade Configuration

In Figure 6.66, the circuit is a cascade configuration of two common-emitter circuits. The dc analysis of this circuit, done in Example 5.19 of Chapter 5, showed that both transistors are biased in the forward-active mode. Figure 6.67 shows the small-signal

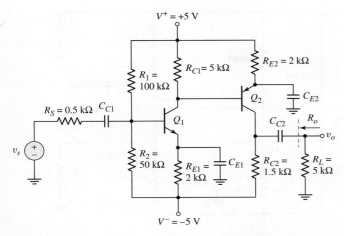

**Figure 6.66** A two-stage common-emitter amplifier in a cascade configuration with npn and pnp transistors

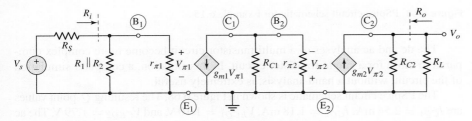

**Figure 6.67** Small-signal equivalent circuit of the cascade circuit shown in Figure 6.66

equivalent circuit, assuming all capacitors act as short circuits and each transistor output resistance $r_o$ is infinite.

We may start the analysis at the output and work back to the input, or start at the input and work toward the output.

The small-signal voltage gain is

$$A_v = \frac{V_o}{V_s} = g_{m1}g_{m2}(R_{C1}\|r_{\pi 2})(R_{C2}\|R_L)\left(\frac{R_i}{R_i + R_S}\right) \tag{6.102}$$

The input resistance of the amplifier is

$$R_i = R_1\|R_2\|r_{\pi 1}$$

which is identical to that of a single-stage common-emitter amplifier. Similarly, the output resistance looking back into the output terminals is $R_o = R_{C2}$. To determine the output resistance, the independent source $V_s$ is set equal to zero, which means that $V_{\pi 1} = 0$. Then $g_{m1}V_{\pi 1} = 0$, which gives $V_{\pi 2} = 0$ and $g_{m2}V_{\pi 2} = 0$. The output resistance is therefore $R_{C2}$. Again, this is the same as the output resistance of a single-stage common-emitter amplifier.

## COMPUTER EXAMPLE 6.15

**Objective:** Determine the small-signal voltage gain of the multitransistor circuit shown in Figure 6.66 using a PSpice analysis.

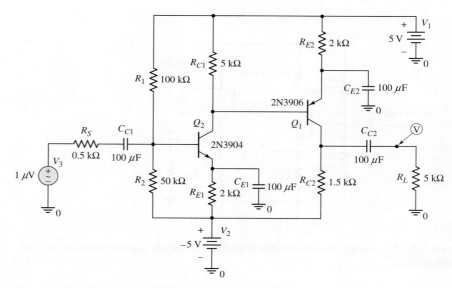

**Figure 6.68** PSpice circuit schematic for Example 6.15

The dc and ac analyses of a multitransistor circuit become more complex compared to those for a single-transistor circuit. In this situation, a computer simulation of the circuit, without a hand analysis, is extremely useful.

The PSpice circuit schematic is shown in Figure 6.68. The resulting $Q$-point values are $I_{CQ1} = 2.54$ mA, $I_{CQ2} = 1.18$ mA, $V_{ECQ1} = 1.10$ V, and $V_{CEQ2} = 1.79$ V. The ac common-emitter current gains are $\beta_1 = 173$ and $\beta_2 = 157$.

A 1 $\mu$V sinusoidal signal is applied. The sinusoidal voltage at the collector of $Q_2$ is 51 $\mu$V and the sinusoidal voltage at the output is 4.79 mV. The overall voltage gain is then 4790. We will show, in later chapters, that even larger voltage gains can be obtained by replacing the discrete collector resistors with active loads.

**Comment:** We can see from the $Q$-point values that the collector–emitter voltage of each transistor is quite small. This implies that the maximum symmetrical swing in the output voltage is limited to a fairly small value. These $Q$-point values can be increased by a slight redesign of the circuit.

**Discussion:** The transistors used in this PSpice analysis of the circuit were standard bipolar transistors from the PSpice library. We must keep in mind that, for the computer simulation to be valid, the models of the devices used in the simulation must match those of the actual devices used in the circuit. If the actual transistor characteristics were substantially different from those used in the computer simulation, then the results of the computer analysis would not be accurate.

### EXERCISE PROBLEM

**Ex 6.15:** For each transistor in the circuit in Figure 6.69, the parameters are: $\beta = 125$, $V_{BE}$(on) $= 0.7$ V, and $r_o = \infty$. (a) Determine the $Q$-points of each transistor. (b) Find the overall small-signal voltage gain $A_v = V_o/V_s$. (c) Determine the input resistance $R_i$ and the output resistance $R_o$. (Ans. (a) $I_{CQ1} = 0.364$ mA, $V_{CEQ1} = 7.92$ V, $I_{CQ2} = 4.82$ mA, $V_{CEQ2} = 2.71$ V (b) $A_v = -17.7$ (c) $R_i = 4.76$ k$\Omega$, $R_o = 43.7$ $\Omega$)

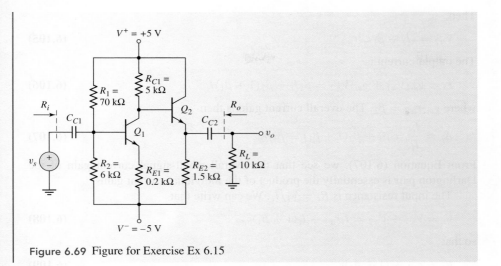

**Figure 6.69** Figure for Exercise Ex 6.15

---

| 6.9.2 | **Multistage Circuit: Darlington Pair Configuration** |

In some applications, it would be desirable to have a bipolar transistor with a much larger current gain than can normally be obtained. Figure 6.70(a) shows a multitransistor configuration, called a **Darlington pair** or a **Darlington configuration,** that provides increased current gain.

The small-signal equivalent in which the input signal is assumed to be a current source, is shown in Figure 6.70(b). We will use the input current source to determine the current gain of the circuit. To determine the small-signal current gain $A_i = I_o/I_i$, we see that

$$V_{\pi 1} = I_i r_{\pi 1} \tag{6.103}$$

Therefore,

$$g_{m1}V_{\pi 1} = g_{m1}r_{\pi 1}I_i = \beta_1 I_i \tag{6.104}$$

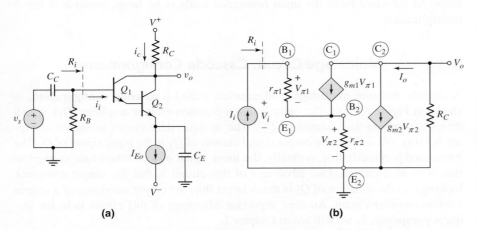

**(a)**    **(b)**

**Figure 6.70** (a) A Darlington pair configuration; (b) small-signal equivalent circuit

Then,

$$V_{\pi 2} = (I_i + \beta_1 I_i) r_{\pi 2} \tag{6.105}$$

The output current is

$$I_o = g_{m1} V_{\pi 1} + g_{m2} V_{\pi 2} = \beta_1 I_i + \beta_2 (1 + \beta_1) I_i \tag{6.106}$$

where $g_{m2} r_{\pi 2} = \beta_2$. The overall current gain is then

$$A_i = \frac{I_o}{I_i} = \beta_1 + \beta_2 (1 + \beta_1) \cong \beta_1 \beta_2 \tag{6.107}$$

From Equation (6.107), we see that the overall small-signal current gain of the Darlington pair is essentially the product of the individual current gains.

The input resistance is $R_i = V_i / I_i$. We can write that

$$V_i = V_{\pi 1} + V_{\pi 2} = I_i r_{\pi 1} + I_i (1 + \beta_1) r_{\pi 2} \tag{6.108}$$

so that

$$R_i = r_{\pi 1} + (1 + \beta_1) r_{\pi 2} \tag{6.109}$$

The base of transistor $Q_2$ is connected to the emitter of $Q_1$, which means that the input resistance to $Q_2$ is multiplied by the factor $(1 + \beta_1)$, as we saw in circuits with emitter resistors. We can write

$$r_{\pi 1} = \frac{\beta_1 V_T}{I_{CQ1}} \tag{6.110}$$

and

$$I_{CQ1} \cong \frac{I_{CQ2}}{\beta_2} \tag{6.111}$$

Therefore,

$$r_{\pi 1} = \beta_1 \left( \frac{\beta_2 V_T}{I_{CQ2}} \right) = \beta_1 r_{\pi 2} \tag{6.112}$$

From Equation (6.109), the input resistance is then approximately

$$R_i \cong 2\beta_1 r_{\pi 2} \tag{6.113}$$

We see from these equations that the overall gain of the Darlington pair is large. At the same time, the input resistance tends to be large, because of the $\beta$ multiplication.

### 6.9.3 Multistage Circuit: Cascode Configuration

A slightly different multistage configuration, called a **cascode configuration,** is shown in Figure 6.71(a). The input is into a common-emitter amplifier ($Q_1$), which drives a common-base amplifier ($Q_2$). The ac equivalent circuit is shown in Figure 6.71(b). We see that the output signal current of $Q_1$ is the input signal of $Q_2$. We mentioned previously that, normally, the input signal of a common-base configuration is to be a current. One advantage of this circuit is that the output resistance looking into the collector of $Q_2$ is much larger than the output resistance of a simple common-emitter circuit. Another important advantage of this circuit is in the frequency response, as we will see in Chapter 7.

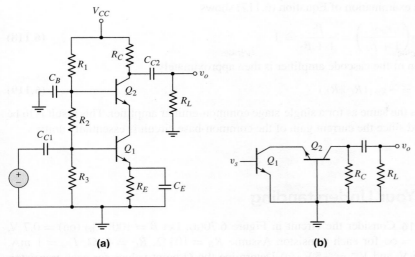

**Figure 6.71** (a) Cascode amplifier and (b) the ac equivalent circuit

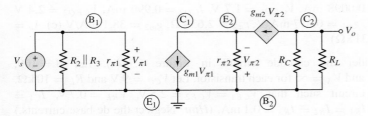

**Figure 6.72** Small-signal equivalent circuit of the cascode configuration

The small-signal equivalent circuit is shown in Figure 6.72 for the case when the capacitors act as short circuits. We see that $V_{\pi 1} = V_s$ since we are assuming an ideal signal voltage source. Writing a KCL equation at $E_2$, we have

$$g_{m1}V_{\pi 1} = \frac{V_{\pi 2}}{r_{\pi 2}} + g_{m2}V_{\pi 2} \qquad \text{(6.114)}$$

Solving for the control voltage $V_{\pi 2}$ (noting that $V_{\pi 1} = V_s$), we find

$$V_{\pi 2} = \left(\frac{r_{\pi 2}}{1+\beta_2}\right)(g_{m1}V_s) \qquad \text{(6.115)}$$

where $\beta_2 = g_{m2}r_{\pi 2}$. The output voltage is

$$V_o = -(g_{m2}V_{\pi 2})(R_C \| R_L) \qquad \text{(6.116(a))}$$

or

$$V_o = -g_{m1}g_{m2}\left(\frac{r_{\pi 2}}{1+\beta_2}\right)(R_C \| R_L)V_s \qquad \text{(6.116(b))}$$

Therefore, the small-signal voltage gain is

$$A_v = \frac{V_o}{V_s} = -g_{m1}g_{m2}\left(\frac{r_{\pi 2}}{1+\beta_2}\right)(R_C \| R_L) \qquad \text{(6.117)}$$

An examination of Equation (6.117) shows

$$g_{m2}\left(\frac{r_{\pi 2}}{1 + \beta_2}\right) = \frac{\beta_2}{1 + \beta_2} \cong 1 \tag{6.118}$$

The gain of the cascode amplifier is then approximately

$$A_v \cong -g_{m1}(R_C \| R_L) \tag{6.119}$$

which is the same as for a single-stage common-emitter amplifier. This result is to be expected since the current gain of the common-base circuit is essentially unity.

## Test Your Understanding

**TYU 6.16** Consider the circuit in Figure 6.70(a). Let $\beta = 100$, $V_{BE}(\text{on}) = 0.7$ V, and $V_A = \infty$ for each transistor. Assume $R_B = 10\,\text{k}\Omega$, $R_C = 4\,\text{k}\Omega$, $I_{Eo} = 1$ mA, $V^+ = 5$ V, and $V^- = -5$ V. (a) Determine the $Q$-point values for each transistor. (b) Calculate the small-signal hybrid-$\pi$ parameters for each transistor. (c) Find the overall small-signal voltage gain $A_v = V_o/V_s$. (d) Find the input resistance $R_i$. (Ans. (a) $I_{CQ1} = 0.0098$ mA, $V_{CEQ1} = 1.7$ V, $I_{CQ2} = 0.990$ mA, $V_{CEQ2} = 2.4$ V (b) $r_{\pi 1} = 265\,\text{k}\Omega$, $g_{m1} = 0.377$ mA/V, $r_{\pi 2} = 2.63\,\text{k}\Omega$, $g_{m2} = 38.1$ mA/V (c) $A_v = -77.0$ (d) $R_i = 531\,\text{k}\Omega$)

**TYU 6.17** Consider the cascode circuit in Figure 6.71(a). Let $\beta = 100$, $V_{BE}(\text{on}) = 0.7$ V, and $V_A = \infty$ for each transistor. Let $V_{CC} = 9$ V and $R_L = 10\,\text{k}\Omega$. (a) Design the circuit such that $V_{CE1} = V_{CE2} = 2.5$ V, $V_{RE} = 0.7$ V, $I_{C1} \cong I_{C2} \cong 1$ mA, and $I_{R1} \cong I_{R2} \cong I_{R3} \cong 0.1$ mA. (*Hint:* Neglect the dc base currents.) (b) Determine the small-signal hybrid-$\pi$ parameters for each transistor. (c) Determine the small-signal voltage gain $A_v = V_o/V_s$. (Ans. (a) $R_1 = 51\,\text{k}\Omega$, $R_2 = 25\,\text{k}\Omega$, $R_3 = 14\,\text{k}\Omega$, $R_E = 0.7\,\text{k}\Omega$, $R_C = 3.3\,\text{k}\Omega$; (b) $g_m = 38.46$ mA/V, $r_\pi = 2.6\,\text{k}\Omega$; (c) $A_v = -94.5$)

 ## 6.10   POWER CONSIDERATIONS

**Objective:** • Analyze the ac and dc power dissipation in a transistor amplifier and understand the concept of signal power gain.

As mentioned previously, an amplifier produces a **small-signal power gain.** Since energy must be conserved, the question naturally arises as to the source of this "extra" signal power. We will see that the "extra" signal power delivered to a load is a result of a redistribution of power between the load and the transistor.

Consider the simple common-emitter circuit shown in Figure 6.73 in which an ideal signal voltage source is connected at the input. The dc power supplied by the $V_{CC}$ voltage source $P_{CC}$, the dc power dissipated or supplied to the collector resistor $P_{RC}$, and the dc power dissipated in the transistor $P_Q$ are given, respectively, as

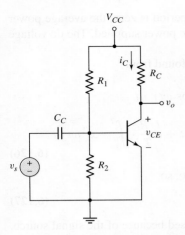

**Figure 6.73** Simple common-emitter amplifier for power calculations

$$P_{CC} = I_{CQ}V_{CC} + P_{\text{Bias}} \qquad\qquad\qquad\qquad (6.120(a))$$

$$P_{RC} = I_{CQ}^2 R_C \qquad\qquad\qquad\qquad (6.120(b))$$

and

$$P_Q = I_{CQ}V_{CEQ} + I_{BQ}V_{BEQ} \cong I_{CQ}V_{CEQ} \qquad\qquad\qquad\qquad (6.120(c))$$

The term $P_{\text{Bias}}$ is the power dissipated in the bias resistors $R_1$ and $R_2$. Normally in a transistor $I_{CQ} \gg I_{BQ}$, so the power dissipated is primarily a function of the collector current and collector–emitter voltage.

If the signal voltage is given by

$$v_s = V_p \cos \omega t \qquad\qquad\qquad\qquad (6.121)$$

then the total base current is given by

$$i_B = I_{BQ} + \frac{V_p}{r_\pi} \cos \omega t = I_{BQ} + I_b \cos \omega t \qquad\qquad\qquad\qquad (6.122)$$

and the total collector current is

$$i_C = I_{CQ} + \beta I_b \cos \omega t = I_{CQ} + I_c \cos \omega t \qquad\qquad\qquad\qquad (6.123)$$

The total instantaneous collector-emitter voltage is

$$v_{CE} = V_{CC} - i_C R_C = V_{CC} - (I_{CQ} + I_c \cos \omega t)R_C = V_{CEQ} - I_c R_C \cos \omega t \qquad (6.124)$$

The average power, including ac signals, supplied by the voltage source $V_{CC}$ is given by

$$\begin{aligned}
\bar{p}_{cc} &= \frac{1}{T} \int_0^T V_{CC} \cdot i_C \, dt + P_{\text{Bias}} \\
&= \frac{1}{T} \int_0^T V_{CC} \cdot [I_{CQ} + I_c \cos \omega t] \, dt + P_{\text{Bias}} \\
&= V_{CC} I_{CQ} + \frac{V_{CC} I_c}{T} \int_0^T \cos \omega t \, dt + P_{\text{Bias}}
\end{aligned} \qquad (6.125)$$

Since the integral of the cosine function over one period is zero, the average power supplied by the voltage source is the same as the dc power supplied. The dc voltage source does not supply additional power.

The average power delivered to the load $R_C$ is found from

$$\bar{p}_{RC} = \frac{1}{T} \int_0^T i_C^2 R_C \, dt = \frac{R_C}{T} \int_0^T [I_{CQ} + I_c \cos \omega t]^2 \, dt$$

$$= \frac{I_{CQ}^2 R_C}{T} \int_0^T dt + \frac{2 I_{CQ} I_c}{T} \int_0^T \cos \omega t \, dt + \frac{I_c^2 R_C}{T} \int_0^T \cos^2 \omega t \, dt \tag{6.126}$$

The middle term of this last expression is again zero, so

$$\bar{p}_{RC} = I_{CQ}^2 R_C + \tfrac{1}{2} I_c^2 R_C \tag{6.127}$$

The average power delivered to the load has increased because of the signal source. This is expected in an amplifier.

Now, the average power dissipated in the transistor is

$$\bar{p}_Q = \frac{1}{T} \int_0^T i_C \cdot v_{CE} \, dt$$

$$= \frac{1}{T} \int_0^T [I_{CQ} + I_c \cos \omega t] \cdot [V_{CEQ} - I_c R_C \cos \omega t] \, dt \tag{6.128}$$

which produces

$$\bar{p}_Q = I_{CQ} V_{CEQ} - \frac{I_c^2 R_C}{T} \int_0^T \cos^2 \omega t \, dt \tag{6.129(a)}$$

or

$$\bar{p}_Q = I_{CQ} V_{CEQ} - \tfrac{1}{2} I_c^2 R_C \tag{6.129(b)}$$

From Equation (6.129(b)), we can deduce that the average power dissipated in the transistor decreases when an ac signal is applied. The $V_{CC}$ source still supplies all of the power, but the input signal changes the relative distribution of power between the transistor and the load.

## Test Your Understanding

**TYU 6.18** In the circuit in Figure 6.74 the transistor parameters are: $\beta = 80$, $V_{BE}(\text{on}) = 0.7$ V, and $V_A = \infty$. Determine the average power dissipated in $R_C$, $R_L$, and $Q$ for: (a) $v_s = 0$, and (b) $v_s = 18 \cos \omega t$ mV. (Ans. (a) $\bar{p}_{RC} = 8$ mW, $\bar{p}_{RL} = 0$, $\bar{p}_Q = 14$ mW (b) $\bar{p}_Q = 13.0$ mW, $\bar{p}_{RL} = 0.479$ mW, $\bar{p}_{RC} = 8.48$ mW)

**TYU 6.19** For the circuit in Figure 6.75, the transistor parameters are: $\beta = 100$, $V_{BE}(\text{on}) = 0.7$ V, and $V_A = \infty$. (a) Determine $R_C$ such that the Q-point is in the center of the load line. (b) Determine the average power dissipated in $R_C$ and $Q$ for $v_s = 0$. (c) Considering the maximum symmetrical swing in the output voltage, determine the ratio of maximum signal power delivered to $R_C$ to the total power dissipated in $R_C$ and the transistor. (Ans. (a) $R_C = 2.52 \, \text{k}\Omega$ (b) $\bar{p}_{RC} = \bar{p}_Q = 2.48$ mW (c) 0.25)

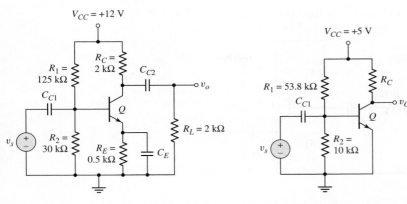

**Figure 6.74** Figure for Exercise TYU 6.18    **Figure 6.75** Figure for Exercise TYU 6.19

 ## 6.11    DESIGN APPLICATION: AUDIO AMPLIFIER

**Objective:** • Design a bipolar transistor audio amplifier to meet a set of specifications.

**Specifications:** An audio amplifier is to deliver an average power of 0.1 W to an 8 Ω speaker from a microphone that produces a 10 mV peak sinusoidal signal and has a source resistance of 10 kΩ.

**Design Approach:** A direct, perhaps brute force, approach will be taken in this design. The generalized multistage amplifier configuration that will be designed is shown in Figure 6.76. An input buffer stage, which will be an emitter-follower circuit, is to be used to reduce the loading effect of the 10 kΩ source resistance. The output stage will also be an emitter-follower circuit to provide the necessary output current and output signal power. The gain stage will actually be composed of a 2-stage common-emitter amplifier that will provide the necessary voltage gain. We will assume that the entire amplifier system is biased with a 12 volt power supply.

**Solution (Input Buffer Stage):** The input buffer stage, an emitter-follower amplifier, is shown in Figure 6.77. We will assume that the transistor has a current gain of $\beta_1 = 100$. We will design the circuit so that the quiescent collector current is $I_{CQ1} = 1$ mA, the quiescent collector-emitter voltage is $V_{CEQ1} = 6$ V, and $R_1 \| R_2 = 100$ kΩ.

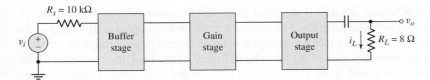

**Figure 6.76** Generalized multistage amplifier for design application

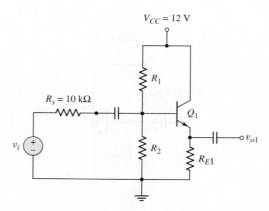

**Figure 6.77** Input signal source and input buffer stage (emitter-follower) for design application

We find

$$R_{E1} \cong \frac{V_{CC} - V_{CEQ1}}{I_{CQ1}} = \frac{12 - 6}{1} = 6 \text{ k}\Omega$$

We obtain

$$r_{\pi 1} = \frac{\beta_1 V_T}{I_{CQ1}} = \frac{(100)(0.026)}{1} = 2.6 \text{ k}\Omega$$

We also have, neglecting the loading effect of the next stage,

$$R_{i1} = R_1 \| R_2 \| [r_{\pi 1} + (1 + \beta_1) R_{E1}]$$
$$= 100 \| [2.6 + (101)(6)] = 85.9 \text{ k}\Omega$$

The small-signal voltage gain, from Equation (6.68) and assuming that $r_o = \infty$, is (again neglecting the loading effect from the next stage)

$$A_{v1} = \frac{v_{o1}}{v_i} = \frac{(1 + \beta_1) R_{E1}}{r_{\pi 1} + (1 + \beta_1) R_{E1}} \cdot \left( \frac{R_{i1}}{R_{i1} + R_S} \right)$$
$$= \frac{(101)(6)}{2.6 + (101)(6)} \cdot \left( \frac{85.9}{85.9 + 10} \right)$$

or

$$A_{v1} = 0.892$$

For a 10 mV peak input signal voltage, the peak voltage at the output of the buffer stage is now $v_{o1} = 8.92$ mV.

We find the bias resistors to be $R_1 = 155 \text{ k}\Omega$ and $R_2 = 282 \text{ k}\Omega$.

**Solution (Output Stage):** The output stage, another emitter-follower amplifier circuit, is shown in Figure 6.78. The 8 Ω speaker is capacitively coupled to the output of the amplifier. The coupling capacitor ensures that no dc current flows through the speaker.

For an average power of 0.1 W to be delivered to the load, the rms value of the load current is found from $P_L = i_L^2(\text{rms}) \cdot R_L$ or $0.1 = i_L^2(\text{rms}) \cdot 8$ which yields $i_L(\text{rms}) = 0.112$ A. For a sinusoidal signal, the peak output current is then

$$i_L(\text{peak}) = 0.158 \text{ A}$$

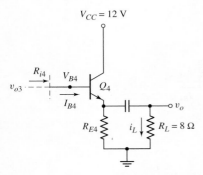

**Figure 6.78**  Output stage (emitter-follower) for design application

and the peak output voltage is

$$v_o(\text{peak}) = (0.158)(8) = 1.26 \text{ V}$$

We will assume that the output power transistor has a current gain of $\beta_4 = 50$. We will set the quiescent transistor parameters at

$$I_{EQ4} = 0.3 \text{ A} \qquad \text{and} \qquad V_{CEQ4} = 6 \text{ V}$$

Then

$$R_{E4} = \frac{V_{CC} - V_{CEQ4}}{I_{EQ4}} = \frac{12 - 6}{0.3} = 20 \text{ }\Omega$$

We find

$$I_{CQ4} = \left(\frac{\beta_4}{1 + \beta_4}\right) \cdot I_{EQ4} = \left(\frac{50}{51}\right)(0.3) = 0.294 \text{ A}$$

Then

$$r_{\pi4} = \frac{\beta_4 V_T}{I_{CQ4}} = \frac{(50)(0.026)}{0.294} = 4.42 \text{ }\Omega$$

The small-signal voltage gain of the output stage is

$$A_{v4} = \frac{v_o}{v_{o3}} = \frac{(1 + \beta_4)(R_{E4} \| R_L)}{r_{\pi4} + (1 + \beta_4)(R_{E4} \| R_L)}$$

$$= \frac{(51)(20\|8)}{4.42 + (51)(20\|8)} = 0.985$$

which is very close to unity, as we would expect. For a required peak output voltage of $v_o = 1.26$ V, we then need a peak voltage at the output of the gain stage to be $v_{o3} = 1.28$ V.

**Solution (Gain Stage):** The gain stage, which will actually be a two-stage common-emitter amplifier, is shown in Figure 6.79. We will assume that the buffer stage is capacitively coupled to the input of the amplifier, the two stages of the amplifier are capacitively coupled, and the output of this amplifier is directly coupled to the output stage.

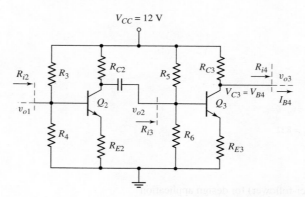

**Figure 6.79** Gain stage (two-stage common-emitter amplifier) for design application

We include emitter resistors to help stabilize the voltage gain of the amplifier. Assume that each transistor has a current gain of $\beta = 100$.

The overall gain (magnitude) of this amplifier must be

$$\left| \frac{v_{o3}}{v_{o1}} \right| = \frac{1.28}{0.00892} = 144$$

We will design the amplifier so that the individual gains (magnitudes) are

$$|A_{v3}| = \left| \frac{v_{o3}}{v_{o2}} \right| = 5 \qquad \text{and} \qquad |A_{v2}| = \left| \frac{v_{o2}}{v_{o1}} \right| = 28.8$$

The dc voltage at the collector of $Q_3$ (with $V_{BE4}(\text{on}) = 0.7$ V) is $V_{C3} = V_{B4} = 6 + 0.7 = 6.7$ V. The quiescent base current to the output transistor is $I_{B4} = 0.294/50$ or $I_{B4} = 5.88$ mA. If we set the quiescent collector current in $Q_3$ to be $I_{CQ3} = 15$ mA, then $I_{RC3} = 15 + 5.88 = 20.88$ mA. Then

$$R_{C3} = \frac{V_{CC} - V_{C3}}{I_{RC3}} = \frac{12 - 6.7}{20.88} \Rightarrow 254 \ \Omega$$

Also

$$r_{\pi 3} = \frac{\beta_3 V_T}{I_{CQ3}} = \frac{(100)(0.026)}{15} \Rightarrow 173 \ \Omega$$

We also find

$$R_{i4} = r_{\pi 4} + (1 + \beta_4)(R_{E4} \| R_L)$$
$$= 4.42 + (51)(20 \| 8) = 296 \ \Omega$$

The small-signal voltage gain, for a common-emitter amplifier with an emitter resistor, can be written as

$$|A_{v3}| = |\frac{v_{o3}}{v_{o2}}| = \frac{\beta_3 (R_{C3} \| R_{i4})}{r_{\pi 3} + (1 + \beta_3) R_{E3}}$$

Setting $|A_{v3}| = 5$, we have

$$5 = \frac{(100)(254 \| 296)}{173 + (101) R_{E3}}$$

which yields $R_{E3} = 25.4 \ \Omega$.

If we set $R_5 \| R_6 = 50 \ \text{k}\Omega$, we find $R_5 = 69.9 \ \text{k}\Omega$ and $R_6 = 176 \ \text{k}\Omega$.

Finally, if we set $V_{C2} = 6 \ \text{V}$ and $I_{CQ2} = 5 \ \text{mA}$, then

$$R_{C2} = \frac{V_{CC} - V_{C2}}{I_{CQ2}} = \frac{12 - 6}{5} = 1.2 \ \text{k}\Omega$$

Also

$$r_{\pi2} = \frac{\beta_2 V_T}{I_{CQ2}} = \frac{(100)(0.026)}{5} = 0.52 \ \text{k}\Omega$$

and

$$R_{i3} = R_5 \| R_6 \| [r_{\pi3} + (1 + \beta_3)R_{E3}]$$
$$= 50 \| [0.173 + (101)(0.0254)] = 2.60 \ \text{k}\Omega$$

The expression for the voltage gain can be written as

$$|A_{v2}| = \left| \frac{v_{o2}}{v_{o1}} \right| = \frac{\beta_2(R_{C2} \| R_{i3})}{r_{\pi2} + (1 + \beta_2)R_{E2}}$$

Setting $|A_{v2}| = 28.8$, we find

$$28.8 = \frac{(100)(1.2 \| 2.6)}{0.52 + (101)R_{E2}}$$

which yields $R_{E2} = 23.1 \ \Omega$.

If we set $R_3 \| R_4 = 50 \ \text{k}\Omega$, we find $R_3 = 181 \ \text{k}\Omega$ and $R_4 = 69.1 \ \text{k}\Omega$.

**Comment:** We may note that, as with any design, there is no unique solution. In addition, to actually build this circuit with discrete components, we would need to use standard values for resistors, which means the quiescent current and voltage values will change, and the overall voltage gain will probably change from the designed value. Also, the current gains of the actual transistors used will probably not be exactly equal to the assumed values. Therefore some slight modifications will likely need to be made in the final design.

**Discussion:** We implicitly assumed that we were designing an audio amplifier, but we have not discussed the frequency response. For example, the coupling capacitors in the design must be large enough to pass audio signal frequencies. The frequency response of amplifiers will be discussed in detail in Chapter 7.

We will also see in later chapters, in particular Chapter 8, that a more efficient output stage can be designed. The efficiency of the output stage in this design is relatively small; that is, the average signal power delivered to the load is small compared to the average power dissipated in the output stage. However, this design is a first approximation in the design process.

 **6.12    SUMMARY**

- This chapter emphasized the application of bipolar transistors in linear amplifier circuits. The basic process by which a transistor circuit can amplify a small time-varying signal was discussed.

- The hybrid-$\pi$ small-signal equivalent circuit of the bipolar transistor was developed. This equivalent circuit is used in the analysis and design of transistor linear amplifiers.
- Three basic circuit configurations were considered: the common-emitter, emitter-follower, and common-base. These three configurations form the basic building blocks for more complex integrated circuits.
- The common-emitter circuit amplifies both time-varying voltages and currents.
- The emitter-follower circuit amplifies time-varying currents, and has a large input resistance and low output resistance.
- The common-base circuit amplifies time-varying voltages, and has a low input resistance and large output resistance.
- Three multitransistor circuits were considered: a cascade configuration of two common-emitter circuits, a Darlington pair, and a cascode configuration formed by common-emitter and common-base circuits. Each configuration provides specialized characteristics such as overall larger voltage gain or an overall larger current gain.
- The concept of signal power gain in amplifier circuits was discussed. There is a redistribution of power within the amplifier circuit.
- As an application, bipolar transistors were incorporated in the design of a multi-stage amplifier circuit configuration to provide a specified output signal power.

##  CHECKPOINT

After studying this chapter, the reader should have the ability to:

✓ Explain graphically the amplification process in a simple bipolar amplifier circuit.
✓ Describe the small-signal hybrid-$\pi$ equivalent circuit of the bipolar transistor and to determine the values of the small-signal hybrid-$\pi$ parameters.
✓ Apply the small-signal hybrid-$\pi$ equivalent circuit to various bipolar amplifier circuits to obtain the time-varying circuit characteristics.
✓ Characterize the small-signal voltage and current gains and the input and output resistances of the common-emitter, emitter-follower, and common-base amplifiers.
✓ Determine the maximum symmetrical swing in the output signal of an amplifier.
✓ Apply the bipolar small-signal equivalent circuit in the analysis of multistage amplifier circuits.

##  REVIEW QUESTIONS

1. Discuss, using the concept of a load line, how a simple common-emitter circuit can amplify a time-varying signal.
2. Why can the analysis of a transistor circuit be split into a dc analysis, with all ac sources set equal to zero, and an ac analysis, with all dc sources set equal to zero?
3. What does the term small-signal imply?
4. Sketch the hybrid-$\pi$ equivalent circuit of an npn and a pnp bipolar transistor.
5. State the relationships of the small-signal hybrid-$\pi$ parameters $g_m$, $r_\pi$, and $r_o$ to the transistor dc quiescent values.
6. What are the physical meanings of the hybrid-$\pi$ parameters $r_\pi$ and $r_o$?
7. Sketch a simple common-emitter amplifier circuit and discuss the general ac circuit characteristics (voltage gain, current gain, input and output resistances).

8. What are the changes in the dc and ac characteristics of a common-emitter amplifier when an emitter resistor and an emitter bypass capacitor are incorporated in the design?

9. Discuss the concepts of a dc load line and an ac load line.

10. Sketch a simple emitter-follower amplifier circuit and discuss the general ac circuit characteristics (voltage gain, current gain, input and output resistances).

11. Sketch a simple common-base amplifier circuit and discuss the general ac circuit characteristics (voltage gain, current gain, input and output resistances).

12. Compare the ac circuit characteristics of the common-emitter, emitter-follower, and common-base circuits.

13. Discuss the general conditions under which a common-emitter amplifier, an emitter-follower amplifier, and a common-base amplifier would be used in an electronic circuit design.

14. State at least two reasons why a multistage amplifier circuit would be required in a design rather than a single-stage circuit.

 ## PROBLEMS

[Note: In the following problems, assume that the B–E turn-on voltage is 0.7 V for both npn and pnp transistors and that $V_A = \infty$ unless otherwise stated. Also assume that all capacitors act as short circuits to the signal.]

### Section 6.2 The Bipolar Linear Amplifier

6.1 (a) Determine the small-signal parameters $g_m$, $r_\pi$, and $r_o$ of a transistor with parameters $\beta = 180$ and $V_A = 150$ V for bias currents of (i) $I_{CQ} = 0.5$ mA and (ii) $I_{CQ} = 2$ mA. (b) Repeat part (a) for $\beta = 80$ and $V_A = 100$ V when biased at (i) $I_{CQ} = 0.25$ mA and (ii) $I_{CQ} = 80\,\mu$A.

6.2 (a) The transistor parameters are $\beta = 125$ and $V_A = 200$ V. A value of $g_m = 95$ mA/V is desired. Determine the required collector current and then find $r_\pi$ and $r_o$. (b) A second transistor has small-signal parameters of $g_m = 120$ mA/V and $r_\pi = 1.2$ kΩ. What is the quiescent collector current and the transistor current gain?

6.3 A transistor has a current gain in the range $90 \leq \beta \leq 180$ and the quiescent collector current is in the range $0.8 \leq I_{CQ} \leq 1.2$ mA. What is the possible range in the small-signal parameters $g_m$ and $r_\pi$?

6.4 The transistor in Figure 6.3 has parameters $\beta = 120$ and $V_A = \infty$. The circuit parameters are $V_{CC} = 3.3$ V, $R_C = 15$ kΩ, and $I_{CQ} = 0.12$ mA. A small signal $v_{be} = 5\sin\omega t$ mV is applied. (a) Determine $i_C$ and $v_{CE}$. (b) What is the small-signal voltage gain $A_v = v_{ce}/v_{be}$?

6.5 For the circuit in Figure 6.3, the transistor parameters are $\beta = 120$, $V_{BE}(\text{on}) = 0.7$ V, and $V_A = 80$ V. The circuit parameters are $V_{CC} = 3.3$ V, $V_{BB} = 1.10$ V, $R_C = 4$ kΩ, and $R_B = 110$ kΩ. (a) Determine the hybrid-$\pi$ parameters. (b) Find the small-signal voltage gain $A_v = v_o/v_s$. (c) If the time-varying output signal is given by $v_o = 0.5\sin(100t)$ V, what is $v_s(t)$?

6.6 For the circuit in Figure 6.3, $\beta = 120$, $V_{CC} = 5$ V, $V_A = 100$ V, and $R_B = 25$ kΩ. (a) Determine $V_{BB}$ and $R_C$ such that $r_\pi = 5.4$ kΩ and the Q-point is in the center of the load line. (b) Find the resulting small-signal voltage gain $A_v = v_o/v_s$.

6.7 The parameters of each transistor in the circuits shown in Figure P6.7 are $\beta = 120$ and $I_{CQ} = 0.5$ mA. Determine the input resistance $R_i$ for each circuit.

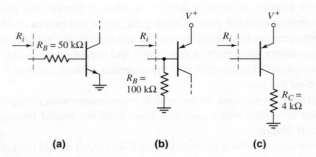

(a)                (b)                (c)

Figure P6.7

6.8 The parameters of each transistor in the circuits shown in Figure P6.8 are $\beta = 130$, $V_A = 80$ V, and $I_{CQ} = 0.2$ mA. Determine the output resistance $R_o$ for each circuit.

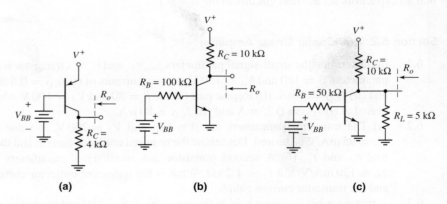

(a)                (b)                (c)

Figure P6.8

6.9 The circuit in Figure 6.3 is biased at $V_{CC} = 10$ V and has a collector resistor of $R_C = 4$ kΩ. The voltage $V_{BB}$ is adjusted such that $V_C = 4$ V. The transistor has $\beta = 100$. The signal voltage between the base and emitter is $v_{be} = 5 \sin \omega t \text{(mV)}$. Determine the total instantaneous values of $i_B(t)$, $i_C(t)$, and $v_C(t)$, and determine the small-signal voltage gain $A_v = v_c(t)/v_{be}(t)$.

6.10 For the circuit in Figure 6.14, $\beta = 100$, $V_A = \infty$, $V_{CC} = 10$ V, and $R_B = 50$ kΩ. (a) Determine $V_{BB}$ and $R_C$ such that $I_{CQ} = 0.5$ mA and the Q-point is in the center of the load line. (b) Find the small-signal parameters $g_m$, $r_\pi$, and $r_o$. (c) Determine the small-signal voltage gain $A_v = v_o/v_s$.

6.11 The ac equivalent circuit shown in Figure 6.7 has $R_C = 2$ kΩ. The transistor parameters are $g_m = 50$ mA/V and $\beta = 100$. The time-varying output voltage is given by $v_o = 1.2 \sin \omega t$ (V). Determine $v_{be}(t)$ and $i_b(t)$.

## Section 6.4 Common-Emitter Amplifier

6.12 The parameters of the transistor in the circuit in Figure P6.12 are $\beta = 150$ and $V_A = \infty$. (a) Determine $R_1$ and $R_2$ to obtain a bias-stable circuit with the Q-point in the center of the load line. (b) Determine the small-signal voltage gain $A_v = v_o/v_s$.

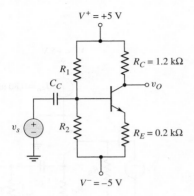

**Figure P6.12**

6.13 Assume that $\beta = 100$, $V_A = \infty$, $R_1 = 33\,k\Omega$, and $R_2 = 50\,k\Omega$ for the circuit in Figure P6.13. (a) Plot the Q-point on the dc load line. (b) Determine the small-signal voltage gain. (c) Determine the range in voltage gain if $R_1$ and $R_2$ vary by $\pm 5$ percent.

D6.14 The transistor parameters for the circuit in Figure P6.13 are $\beta = 100$ and $V_A = \infty$. (a) Design the circuit such that it is bias stable and that the Q-point is in the center of the load line. (b) Determine the small-signal voltage gain of the designed circuit.

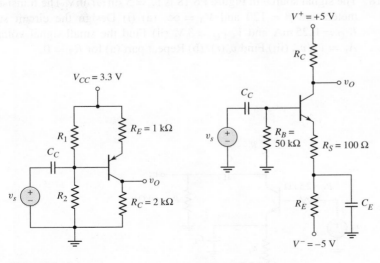

**Figure P6.13**                **Figure P6.15**

D6.15 For the circuit in Figure P6.15, the transistor parameters are $\beta = 100$ and $V_A = \infty$. Design the circuit such that $I_{CQ} = 0.25$ mA and $V_{CEQ} = 3$ V.

Find the small-signal voltage gain $A_v = v_o/v_s$. Find the input resistance seen by the signal source $v_s$.

D6.16    Assume the transistor in the circuit in Figure P6.16 has parameters $\beta = 120$ and $V_A = 100$ V. (a) Design a bias-stable circuit such that $V_{CEQ} = 5.20$ V. (b) Determine the small-signal transresistance function $R_m = v_o/i_s$. (c) Using the results of part (a), determine the variation in $R_m$ if $100 \leq \beta \leq 150$.

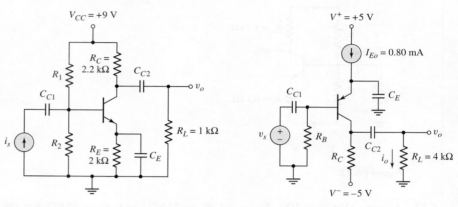

Figure P6.16                          Figure P6.17

D6.17    (a) For transistor parameters $\beta = 80$ and $V_A = 100$ V, (i) design the circuit in Figure P6.17 such that the dc voltages at the base and collector terminals are 0.20 V and $-3$ V, respectively, and (ii) determine the small-signal transconductance function $G_f = i_o/v_s$. (b) Repeat part (a) for $\beta = 120$ and $V_A = 80$ V.

6.18    The signal source in Figure P6.18 is $v_s = 5 \sin \omega t$ mV. The transistor parameters are $\beta = 120$ and $V_A = \infty$. (a) (i) Design the circuit such that $I_{CQ} = 0.25$ mA and $V_{CEQ} = 3$ V. (ii) Find the small-signal voltage gain $A_v = v_o/v_s$. (iii) Find $v_o(t)$. (b) Repeat part (a) for $R_S = 0$.

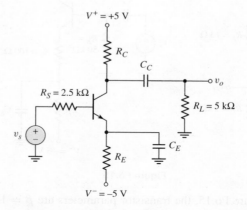

Figure P6.18

6.19  Consider the circuit shown in Figure P6.19 where the signal-source is $v_s = 4 \sin \omega t$ mV. (a) For transistor parameters of $\beta = 80$ and $V_A = \infty$, (i) find the small-signal voltage gain $A_v = v_o/v_s$ and the transconductance function $G_f = i_o/v_s$, and (ii) calculate $v_o(t)$ and $i_o(t)$. (b) Repeat part (a) for $\beta = 120$.

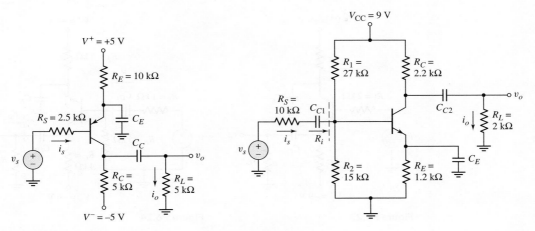

Figure P6.19                                  Figure P6.20

6.20  Consider the circuit shown in Figure P6.20. The transistor parameters are $\beta = 100$ and $V_A = 100$ V. Determine $R_i$, $A_v = v_o/v_s$, and $A_i = i_o/i_s$.

6.21  The parameters of the transistor in the circuit in Figure P6.21 are $\beta = 100$ and $V_A = 100$ V. (a) Find the dc voltages at the base and emitter terminals. (b) Find $R_C$ such that $V_{CEQ} = 3.5$ V. (c) Assuming $C_C$ and $C_E$ act as short circuits, determine the small-signal voltage gain $A_v = v_o/v_s$. (d) Repeat part (c) if a 500 $\Omega$ source resistor is in series with the $v_s$ signal source.

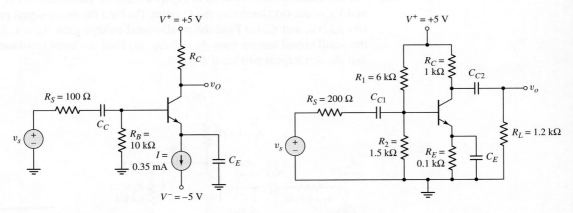

Figure P6.21                                  Figure P6.22

6.22  For the circuit in Figure P6.22, the transistor parameters are $\beta = 180$ and $r_o = \infty$. (a) Determine the Q-point values. (b) Find the small-signal hybrid-$\pi$ parameters. (c) Find the small-signal voltage gain $A_v = v_o/v_s$.

6.23    For the circuit in Figure P6.23, the transistor parameters are $\beta = 80$ and $V_A = 80$ V. (a) Determine $R_E$ such that $I_{EQ} = 0.75$ mA. (b) Determine $R_C$ such that $V_{ECQ} = 7$ V. (c) For $R_L = 10$ k$\Omega$, determine the small-signal voltage gain $A_v = v_o/v_s$. (d) Determine the impedance seen by the signal source $v_s$.

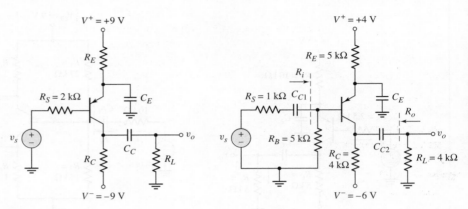

Figure P6.23                        Figure P6.24

6.24    The transistor in the circuit in Figure P6.24 has parameters $V_{EB}$(on) $= 0.7$ V, $V_A = 50$ V, and a current gain in the range $80 \leq \beta \leq 120$. Determine (a) the range in the small-signal voltage gain $A_v = v_o/v_s$, (b) the range in the input resistance $R_i$, and (c) the range in the output resistance $R_o$.

D6.25    Design a one-transistor common-emitter preamplifier that can amplify a 10 mV (rms) microphone signal and produce a 0.5 V (rms) output signal. The source resistance of the microphone is 1 k$\Omega$. Use standard resistor values in the design and specify the value of $\beta$ required.

6.26    For the transistor in the circuit in Figure P6.26, the parameters are $\beta = 100$ and $V_A = \infty$. (a) Determine the Q-point. (b) Find the small-signal parameters $g_m$, $r_\pi$, and $r_o$. (c) Find the small-signal voltage gain $A_v = v_o/v_s$ and the small-signal current gain $A_i = i_o/i_s$. (d) Find the input resistances $R_{ib}$ and $R_{is}$. (e) Repeat part (c) if $R_S = 0$.

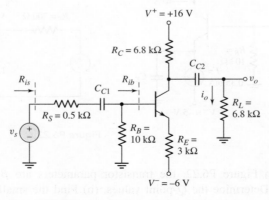

Figure P6.26

6.27 If the collector of a transistor is connected to the base terminal, the transistor continues to operate in the forward-active mode, since the B–C junction is not reverse biased. Determine the small-signal resistance, $r_e = v_{ce}/i_e$, of this two-terminal device in terms of $g_m$, $r_\pi$, and $r_o$.

D6.28 (a) Design an amplifier with the configuration similar to that shown in Figure 6.31. The circuit is to be biased with $V_{CC} = 3.3$ V and the source resistance is $R_S = 100\,\Omega$. The minimum small-signal voltage gain is to be $|A_v| = 10$. The available npn transistors have parameters of $\beta = 120$ and $V_A = \infty$. (b) Using the results of part (a), what is the resulting input resistance seen by the signal source and what is the resulting output resistance?

D6.29 An ideal signal voltage source is given by $v_s = 5\sin(5000t)$ (mV). The peak current that can be supplied by this source is $0.2\,\mu$A. The desired output voltage across a 10 k$\Omega$ load resistor is $v_o = 100\sin(5000t)$ (mV). Design a one-transistor common-emitter amplifier to meet this specification. Use standard resistor values and specify the required value of $\beta$.

D6.30 Design a one-transistor common-emitter amplifier with a small-signal voltage gain of approximately $A_v = -10$. The circuit is to be biased from a single power supply of $V_{CC} = 5$ V that can supply a maximum current of 0.8 mA. The input resistance is to be greater than 20 k$\Omega$ and the output resistance is to be 5 k$\Omega$. The available transistor is a pnp device with $\beta = 90$ and $V_A = \infty$.

D6.31 Design a common-emitter circuit whose output is capacitively coupled to a load resistor $R_L = 10\,\text{k}\Omega$. The minimum small-signal voltage gain is to be $|A_v| = 50$. The circuit is to be biased at $\pm5$ V and each voltage source can supply a maximum of 0.5 mA. The parameters of the available transistors are $\beta = 120$ and $V_A = \infty$.

## Section 6.5 AC Load Line Analysis

6.32 Consider the circuit shown in Figure P6.13. Assume $R_1 = 33\,\text{k}\Omega$ and $R_2 = 50\,\text{k}\Omega$. The transistor parameters are $\beta = 100$ and $V_A = \infty$. Determine the maximum undistorted swing in the output voltage if the total instantaneous E–C voltage is to remain in the range $0.5 \le v_{EC} \le 3$ V.

6.33 For the circuit in Figure P6.15, let $\beta = 100$, $V_A = \infty$, $R_E = 12.9\,\text{k}\Omega$, and $R_C = 6\,\text{k}\Omega$. Determine the maximum undistorted swing in the output voltage if the total instantaneous C–E voltage is to remain in the range $1 \le v_{CE} \le 9$ V and if the total instantaneous collector current is to remain greater or equal to 50 $\mu$A.

6.34 Consider the circuit in Figure P6.19. The transistor parameters are $\beta = 80$ and $V_A = \infty$. (a) Determine the maximum undistorted swing in the output voltage if the total instantaneous C–E voltage is to remain in the range $0.7 \le v_{CE} \le 9$ V and the instantaneous collector is to be $i_C \ge 0$. (b) Using the results of part (a), determine the range in collector current.

6.35 The parameters of the circuit shown in Figure P6.17 are $R_B = 20\,\text{k}\Omega$ and $R_C = 2.5\,\text{k}\Omega$. The transistor parameters are $\beta = 80$ and $V_A = \infty$. Determine the maximum undistorted swing in the output current $i_o$ if the total instantaneous collector current is to be $i_C \ge 0.08$ mA and the total instantaneous E–C voltage is to be in the range $1 \le v_{EC} \le 9$ V.

6.36 Consider the circuit in Figure P6.26 with transistor parameters described in Problem 6.26. Determine the maximum undistorted swing in the output current $i_C$ if the total instantaneous collector current is $i_C \geq 0.1$ mA and the total instantaneous C–E voltage is in the range $1 \leq v_{CE} \leq 21$ V.

6.37   For the circuit in Figure P6.20, the transistor parameters are $\beta = 100$ and $V_A = 100$ V. The values of $R_C$, $R_E$, and $R_L$ are as shown in the figure. Design a bias-stable circuit to achieve the maximum undistorted swing in the output voltage if the total instantaneous C–E voltage is to remain in the range $1 \leq v_{CE} \leq 8$ V and the minimum collector current is to be $i_C(\text{min}) = 0.1$ mA.

6.38   In the circuit in Figure P6.22 with transistor parameters $\beta = 180$ and $V_A = \infty$, redesign the bias resistors $R_1$ and $R_2$ to achieve maximum symmetrical swing in the output voltage and to maintain a bias-stable circuit. The total instantaneous C–E voltage is to remain in the range $0.5 \leq v_{CE} \leq 4.5$ V and the total instantaneous collector current is to be $i_C \geq 0.25$ mA.

6.39   For the circuit in Figure P6.24, the transistor parameters are $\beta = 100$ and $V_A = \infty$. (a) Determine the maximum undistorted swing in the output voltage if the total instantaneous E–C voltage is to remain in the range $1 \leq v_{EC} \leq 9$ V. (b) Using the results of part (a), determine the range of collector current.

### Section 6.6   Common-Collector (Emitter-Follower) Amplifier

6.40   Figure P6.40 shows the ac equivalent circuit of an emitter follower. (a) The transistor parameters are $\beta = 120$ and $V_A = \infty$. For $R_E = 500\,\Omega$, determine $I_{CQ}$ such that the small-signal voltage gain is $A_v = 0.92$. (b) Using the results of part (a), determine the voltage gain if $V_A = 20$ V. (c) Determine the small-signal output resistance $R_o$ for both parts (a) and (b).

6.41   Consider the ac equivalent circuit in Figure P6.40. The transistor parameters are $\beta = 80$ and $V_A = \infty$. (a) Design the circuit (find $I_{CQ}$ and $R_E$) such that $R_{ib} = 50\,\text{k}\Omega$ and $A_v = 0.95$. (b) Using the results of part (a), find $R_o$.

6.42   For the ac equivalent circuit in Figure P6.42, $R_S = 1\,\text{k}\Omega$ and the transistor parameters are $\beta = 80$ and $V_A = 50$ V. (a) For $I_{CQ} = 2$ mA, find $A_v$, $R_i$, and $R_o$. (b) Repeat part (a) for $I_{CQ} = 0.2$ mA.

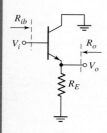

**Figure P6.40**

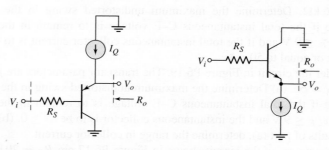

**Figure P6.42**          **Figure P6.43**

6.43   The circuit and transistor parameters for the ac equivalent circuit in Figure P6.43 are $R_S = 0.5\,\text{k}\Omega$, $\beta = 120$, and $V_A = \infty$. (a) Determine the required value of $I_Q$ to produce a small-signal output resistance of $R_o = 15\,\Omega$. (b) Using the results of part (a), find the small-signal voltage gain if $V_A = 50$ V.

6.44   The transistor parameters for the circuit in Figure P6.44 are $\beta = 180$ and $V_A = \infty$. (a) Find $I_{CQ}$ and $V_{CEQ}$. (b) Plot the dc and ac load lines. (c) Calculate the small-signal voltage gain. (d) Determine the input and output resistances $R_{ib}$ and $R_o$.

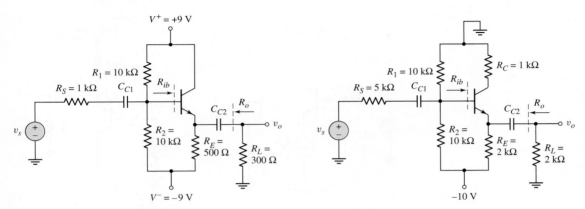

**Figure P6.44**                                    **Figure P6.45**

6.45   Consider the circuit in Figure P6.45. The transistor parameters are $\beta = 120$ and $V_A = \infty$. Repeat parts (a)–(d) of Problem 6.44.

6.46   For the circuit shown in Figure P6.46, let $V_{CC} = 3.3$ V, $R_L = 4$ k$\Omega$, $R_1 = 585$ k$\Omega$, $R_2 = 135$ k$\Omega$, and $R_E = 12$ k$\Omega$. The transistor parameters are $\beta = 90$ and $V_A = 60$ V. (a) Determine the quiescent values $I_{CQ}$ and $V_{ECQ}$. (b) Plot the dc and ac load lines. (c) Determine $A_v = v_o/v_s$ and $A_i = i_o/i_s$. (d) Find $R_{ib}$ and $R_o$.

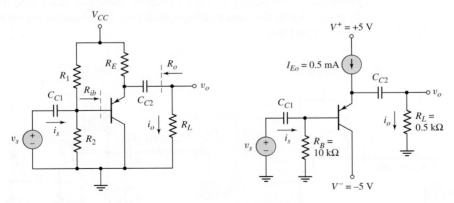

**Figure P6.46**                                    **Figure P6.47**

6.47   For the transistor in Figure P6.47, $\beta = 80$ and $V_A = 150$ V. (a) Determine the dc voltages at the base and emitter terminals. (b) Calculate the small-signal parameters $g_m$, $r_\pi$, and $r_o$. (c) Determine the small-signal voltage gain and current gain. (d) Repeat part (c) if a 2 k$\Omega$ source resistor is in series with the $v_s$ signal source.

6.48 Consider the emitter-follower amplifier shown in Figure P6.48. The transistor parameters are $\beta = 100$ and $V_A = 100$ V. (a) Find the output resistance $R_o$. (b) Determine the small-signal voltage gain for (i) $R_L = 500$ $\Omega$ and (ii) $R_L = 5$ k$\Omega$.

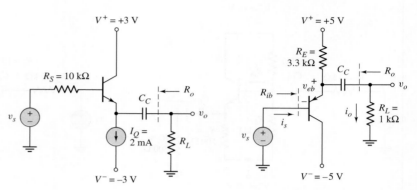

Figure P6.48                    Figure P6.49

6.49 The transistor parameters for the circuit in Figure P6.49 are $\beta = 110$, $V_A = 50$ V, and $V_{EB}(\text{on}) = 0.7$ V. (a) Determine the quiescent values $I_{CQ}$ and $V_{ECQ}$. (b) Find $A_v$, $R_{ib}$, and $R_o$. (c) The signal source is $v_s(t) = 2.8 \sin \omega t$ (V). Determine $i_s(t)$, $i_o(t)$, $v_o(t)$, and $v_{eb}(t)$.

D6.50 For the transistor in Figure P6.50, the parameters are $\beta = 100$ and $V_A = \infty$. (a) Design the circuit such that $I_{EQ} = 1$ mA and the Q-point is in the center of the dc load line. (b) If the peak-to-peak sinusoidal output voltage is 4 V, determine the peak-to-peak sinusoidal signals at the base of the transistor and the peak-to-peak value of $v_s$. (c) If a load resistor $R_L = 1$ k$\Omega$ is connected to the output through a coupling capacitor, determine the peak-to-peak value in the output voltage, assuming $v_s$ is equal to the value determined in part (b).

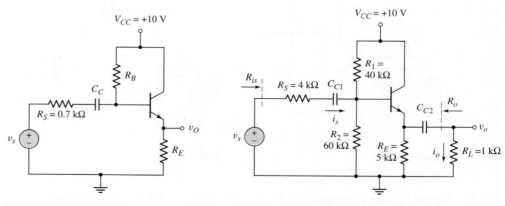

Figure P6.50                    Figure P6.51

6.51 In the circuit shown in Figure P6.51, determine the range in small-signal voltage gain $A_v = v_o/v_s$ and current gain $A_i = i_o/i_s$ if $\beta$ is in the range $75 \leq \beta \leq 150$.

6.52    The transistor current gain $\beta$ in the circuit shown in Figure P6.52 is in the range $50 \leq \beta \leq 200$. (a) Determine the range in the dc values of $I_E$ and $V_E$. (b) Determine the range in the values of input resistance $R_i$ and voltage gain $A_v = v_o/v_s$.

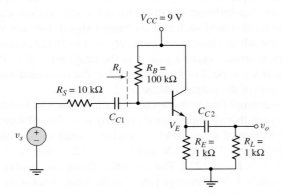

**Figure P6.52**

6.53    Consider the circuit shown in Figure P6.47. The transistor current gain is in the range $100 \leq \beta \leq 180$ and the Early voltage is $V_A = 150$ V. Determine the range in small-signal voltage gain if the load resistance varies from $R_L = 0.5$ k$\Omega$ to $R_L = 500$ k$\Omega$.

6.54    For the circuit in Figure P6.54, the parameters are $V_{CC} = 5$ V and $R_E = 500\ \Omega$. The transistor parameters are $\beta = 120$ and $V_A = \infty$. (a) Design the circuit to obtain a small-signal current gain of $A_i = i_o/i_s = 10$ for $R_L = 500\ \Omega$. Find $R_1$, $R_2$, and also the small-signal output resistance $R_o$. (b) Using the results of part (a), determine the current gain for $R_L = 2$ k$\Omega$.

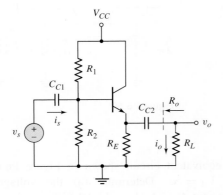

**Figure P6.54**

D6.55    Design an emitter-follower circuit with the configuration shown in Figure 6.49 such that the input resistance $R_i$, as defined in Figure 6.51, is 120 k$\Omega$. Assume transistor parameters of $\beta = 120$ and $V_A = \infty$. Let $V_{CC} = 5$ V and $R_E = 2$ k$\Omega$. Find new values of $R_1$ and $R_2$. The $Q$-point should be approximately in the center of the load line.

D6.56   (a) For the emitter-follower circuit in Figure P6.54, assume $V_{CC} = 24$ V, $\beta = 75$, and $A_i = i_o/i_s = 8$. Design the circuit to drive an 8 $\Omega$ load. (b) Determine the maximum undistorted swing in the output voltage. (c) Determine the output resistance $R_o$.

*D6.57   The output of an amplifier can be represented by $v_s = 4 \sin \omega t$ (V) and $R_S = 4$ k$\Omega$. An emitter-follower circuit, with the configuration shown in Figure 6.54, is to be designed such that the output signal does not vary by more than 5 percent when a load in the range $R_L = 4$ to 10 k$\Omega$ is connected to the output. The transistor current gain is in the range $90 \leq \beta \leq 130$ and the Early voltage is $V_A = \infty$. For your design, find the minimum and maximum possible value of the output voltage.

*D6.58   An emitter-follower amplifier, with the configuration shown in Figure 6.54, is to be designed such that an audio signal given by $v_s = 5 \sin(3000t)$ V but with a source resistance of $R_S = 10$ k$\Omega$ can drive a small speaker. Assume the supply voltages are $V^+ = +12$ V and $V^- = -12$ V. The load, representing the speaker, is $R_L = 12$ $\Omega$. The amplifier should be capable of delivering approximately 1 W of average power to the load. What is the signal power gain of your amplifier?

### Section 6.7   Common-Base Amplifier

6.59   Figure P6.59 is an ac equivalent circuit of a common-base amplifier. The transistor parameters are $\beta = 120$, $V_A = \infty$, and $I_{CQ} = 1$ mA. Determine (a) the voltage gain $A_v = V_o/V_i$, (b) the current gain $A_i = I_o/I_i$, (c) the input resistance $R_i$, and (d) the output resistance $R_o$.

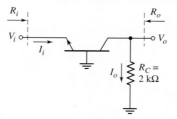

Figure P6.59

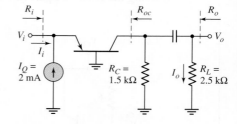

Figure P6.60

6.60   The transistor in the ac equivalent circuit shown in Figure P6.60 has parameters $\beta = 80$ and $V_A = \infty$. Determine (a) the voltage gain $A_v = V_o/V_i$, (b) the current gain $A_i = I_o/I_i$, and (c) the input resistance $R_i$. (d) If $V_A = 80$ V, find (i) the output resistance $R_{oc}$ and (ii) the output resistance $R_o$.

6.61   Consider the ac equivalent common-base circuit shown in Figure P6.61. The transistor has parameters $\beta = 110$ and $V_A = \infty$. Determine (a) the voltage gain $A_v = V_o/V_i$, (b) the current gain $A_i = I_o/I_i$, (c) the input resistance $R_i$, and (d) the output resistance $R_o$.

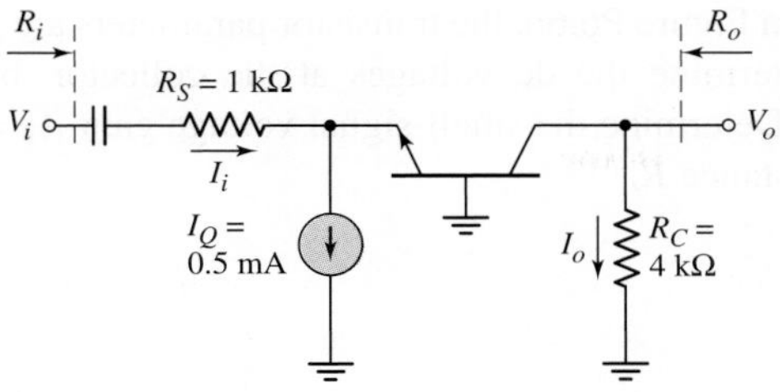

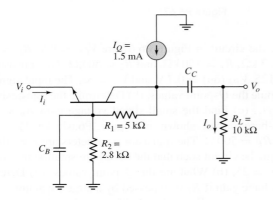

**Figure P6.61**

6.62  Figure P6.62 shows an ac equivalent circuit of a common-base amplifier. The parameters of the transistor are $\beta = 120$, $V_{BE}(\text{on}) = 0.7$ V, and $V_A = \infty$. (a) Determine the quiescent values $I_{CQ}$ and $V_{CEQ}$. (b) Find the small-signal voltage gain $A_v = V_o/V_i$. (c) Find the small-signal current gain $A_i = I_o/I_i$.

**Figure P6.62**

6.63  The transistor in the circuit shown in Figure P6.63 has $\beta = 100$ and $V_A = \infty$. (a) Determine the quiescent values $I_{CQ}$ and $V_{ECQ}$. (b) Determine the small-signal voltage gain $A_v = v_o/v_s$.

6.64  Repeat Problem 6.63 with a 100 $\Omega$ resistor in series with the $v_s$ signal source.

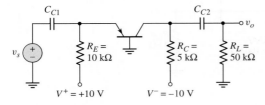

**Figure P6.63**

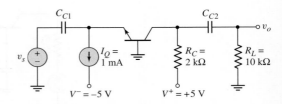

**Figure P6.65**

6.65  Consider the common-base circuit in Figure P6.65. The transistor has parameters $\beta = 120$ and $V_A = \infty$. (a) Determine the quiescent $V_{CEQ}$. (b) Determine the small-signal voltage gain $A_v = v_o/v_s$.

6.66 For the circuit shown in Figure P6.66, the transistor parameters are $\beta = 100$ and $V_A = \infty$. (a) Determine the dc voltages at the collector, base, and emitter terminals. (b) Determine the small-signal voltage gain $A_v = v_o/v_s$. (c) Find the input resistance $R_i$.

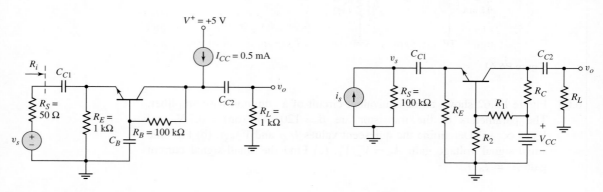

**Figure P6.66**          **Figure P6.67**

6.67 The parameters of the circuit in Figure P6.67 are $V_{CC} = 9\,\text{V}$, $R_L = 4\,\text{k}\Omega$, $R_C = 6\,\text{k}\Omega$, $R_E = 3\,\text{k}\Omega$, $R_1 = 150\,\text{k}\Omega$, and $R_2 = 50\,\text{k}\Omega$. The transistor parameters are $\beta = 125$, $V_{BE}(\text{on}) = 0.7\,\text{V}$, and $V_A = \infty$. The input signal is a current. (a) Determine the $Q$-point values. (b) Determine the transresistance function $R_m = v_o/i_s$. (c) Find the small-signal voltage gain $A_v = v_o/v_s$.

6.68 For the common-base circuit shown in Figure P6.67, let $V_{CC} = 5\,\text{V}$, $R_L = 12\,\text{k}\Omega$, and $R_E = 500\,\Omega$. The transistor parameters are $\beta = 100$ and $V_A = \infty$. (a) Design the circuit such that the minimum small-signal voltage gain is $A_v = v_o/v_s = 25$. (b) What are the $Q$-point values? (c) Determine the small-signal voltage gain if $R_2$ is bypassed by a large capacitor.

6.69 Consider the circuit shown in Figure P6.69. The transistor has parameters $\beta = 60$ and $V_A = \infty$. (a) Determine the quiescent values of $I_{CQ}$ and $V_{CEQ}$. (b) Determine the small-signal voltage gain $A_v = v_o/v_s$.

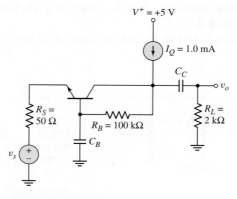

**Figure P6.69**

*D6.70 A photodiode in an optical transmission system, such as shown in Figure 1.40, can be modeled as a Norton equivalent circuit with $i_s$ in parallel with $R_S$ as shown in Figure P6.67. Assume that the current source is given by $i_s = 2.5 \sin \omega t \, \mu A$ and $R_S = 50 \, k\Omega$. Design the common-base circuit of Figure P6.67 such that the output voltage is $v_o = 5 \sin \omega t \, mV$. Assume transistor parameters of $\beta = 120$ and $V_A = \infty$. Let $V_{CC} = 5 \, V$.

6.71  In the common-base circuit shown in Figure P6.71, the transistor is a 2N2907A, with a nominal dc current gain of $\beta = 80$. (a) Determine $I_{CQ}$ and $V_{ECQ}$. (b) Using the $h$-parameters (assuming $h_{re} = 0$), determine the range in small-signal voltage gain $A_v = v_o/v_s$. (c) Determine the range in input and output resistances $R_i$ and $R_o$.

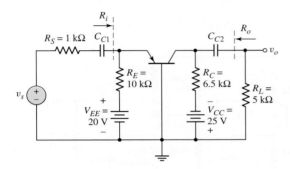

**Figure P6.71**

*D6.72  In the circuit of Figure P6.71, let $V_{EE} = V_{CC} = 5 \, V$, $\beta = 100$, $V_A = \infty$, $R_L = 1 \, k\Omega$, and $R_S = 0$. (a) Design the circuit such that the small-signal voltage gain is $A_v = v_o/v_s = 25$ and $V_{ECQ} = 3 \, V$. (b) What are the values of the small-signal parameters $g_m$, $r_\pi$, and $r_o$?

## Section 6.9 Multistage Amplifiers

6.73  Consider the ac equivalent circuit in Figure P6.73. The transistor parameters are $\beta_1 = 120$, $\beta_2 = 80$, $V_{A1} = V_{A2} = \infty$, and $I_{CQ1} = I_{CQ2} = 1 \, mA$. (a) Find the small-signal voltage gain $A_{v1} = V_{o1}/V_i$. (b) Determine the small-signal voltage gain $A_{v2} = V_{o2}/V_{o1}$. (c) Find the overall small-signal voltage gain $A_v = V_{o2}/V_i$.

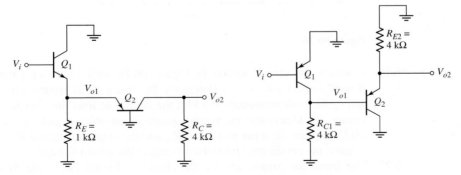

**Figure P6.73**                              **Figure P6.74**

6.74  The transistor parameters in the ac equivalent circuit shown in Figure P6.74 are $\beta_1 = \beta_2 = 100$, $V_{A1} = V_{A2} = \infty$, $I_{CQ1} = 0.5 \, mA$, and $I_{CQ2} = 2 \, mA$.

(a) Find the small-signal voltage gain $A_{v1} = V_{o1}/V_i$. (b) Determine the small-signal voltage gain $A_{v2} = V_{o2}/V_{o1}$. (c) Determine the overall small-signal voltage gain $A_v = V_{o2}/V_i$.

*6.75 The parameters for each transistor in the circuit shown in Figure P6.75 are $\beta = 100$ and $V_A = \infty$. (a) Determine the small-signal parameters $g_m$, $r_\pi$, and $r_o$ for both transistors. (b) Determine the small-signal voltage gain $A_{v1} = v_{o1}/v_s$, assuming $v_{o1}$ is connected to an open circuit, and determine the gain $A_{v2} = v_o/v_{o1}$. (c) Determine the overall small-signal voltage gain $A_v = v_o/v_s$. Compare the overall gain with the product $A_{v1} \cdot A_{v2}$, using the values calculated in part (b).

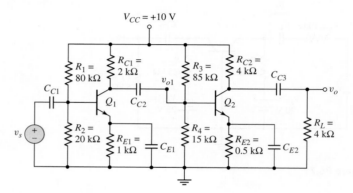

**Figure P6.75**

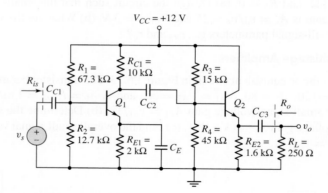

**Figure P6.76**

*6.76 Consider the circuit shown in Figure P6.76 with transistor parameters $\beta = 120$ and $V_A = \infty$. (a) Determine the small-signal parameters $g_m$, $r_\pi$, and $r_o$ for both transistors. (b) Plot the dc and ac load lines for both transistors. (c) Determine the overall small-signal voltage gain $A_v = v_o/v_s$. (d) Determine the input resistance $R_{is}$ and the output resistance $R_o$. (e) Determine the maximum undistorted swing in the output voltage.

6.77 The transistor parameters for the circuit in Figure P6.77 are $\beta_1 = 120$, $\beta_2 = 80$, $V_{BE1}(\text{on}) = V_{BE2}(\text{on}) = 0.7$ V, and $V_{A1} = V_{A2} = \infty$. (a) Determine the quiescent collector current in each transistor. (b) Find the small-signal voltage gain $A_v = v_o/v_s$. (c) Determine the input and output resistances $R_{ib}$ and $R_o$.

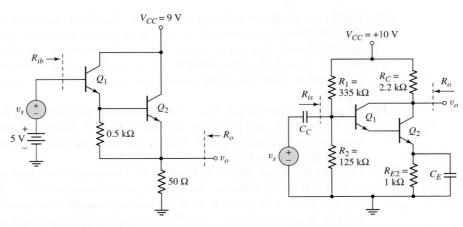

**Figure P6.77**                           **Figure P6.78**

*6.78   For each transistor in Figure P6.78, the parameters are $\beta = 100$ and $V_A = \infty$. (a) Determine the Q-point values for both $Q_1$ and $Q_2$. (b) Determine the overall small-signal voltage gain $A_v = v_o/v_s$. (c) Determine the input and output resistances $R_{is}$ and $R_o$.

6.79   An ac equivalent circuit of a Darlington pair configuration is shown in Figure P6.79. The transistor parameters are $\beta_1 = 120$, $\beta_2 = 80$, $V_{A1} = 80$ V, and $V_{A2} = 50$ V. Determine the output resistance $R_o$ for (a) $I_{C2} = I_{Bias} = 1$ mA; (b) $I_{C2} = 1$ mA, $I_{Bias} = 0.2$ mA; and (c) $I_{C2} = 2$ mA, $I_{Bias} = 0$.

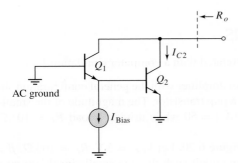

**Figure P6.79**

## Section 6.10  Power Considerations

6.80   Consider the circuit in Figure 6.31. The circuit and transistor parameters are given in Exercise Ex 6.5. (a) Determine the average power dissipated in the transistor, $R_C$, and $R_E$ for $v_s = 0$. (b) Repeat part (a) for $v_s = 100 \sin \omega t$ (mV).

6.81   Consider the circuit shown in Figure 6.38. The transistor parameters are given in Exercise Ex 6.7. (a) Calculate the average power dissipated in the transistor, $R_C$, and $R_E$ for $v_s = 0$. (b) Determine the maximum undistorted signal power that can be delivered to $R_C$ for the condition that $i_C \geq 0$ and $0.5 \leq v_{CE} \leq 9$ V.

6.82    For the circuit shown in Figure 6.43, use the circuit and transistor parameters described in Example 6.9. (a) Calculate the average power dissipated in the transistor, $R_E$, and $R_C$, for $v_s = 0$. (b) Determine the maximum signal power that can be delivered to $R_L$. What are the signal powers dissipated in $R_E$ and $R_C$, and what is the average power dissipated in the transistor in this case?

6.83    For the circuit shown in Figure 6.57, the transistor parameters are $\beta = 100$ and $V_A = 100$ V, and the source resistor is $R_S = 0$. Determine the maximum undistorted signal power that can be delivered to $R_L$ if: (a) $R_L = 1$ k$\Omega$, and (b) $R_L = 10$ k$\Omega$.

6.84    Consider the circuit shown in Figure 6.64 with parameters given in Exercise TYU 6.14. (a) Calculate the average power dissipated in the transistor and $R_C$, for $v_s = 0$. (b) Determine the maximum undistorted signal power that can be delivered to $R_L$, and the resulting average power dissipated in the transistor and $R_C$.

 **COMPUTER SIMULATION PROBLEMS**

6.85    (a) Using a computer simulation, verify the results of Exercise Ex 6.5. (b) Repeat part (a) for Early voltages of (i) $V_A = 100$ V and (ii) $V_A = 50$ V.

6.86    (a) Using a computer simulation, verify the results of Exercise TYU 6.7. (b) Repeat part (a) for an Early voltage of $V_A = 50$ V.

6.87    Using a computer simulation, verify the results of Example 6.10.

6.88    Using a computer simulation, verify the results of Exercise Ex 6.15 for the multitransistor amplifier.

**DESIGN PROBLEMS**

[Note: Each design should be correlated with a computer simulation.]

*D6.89    Design a common-emitter amplifier with the general configuration shown in Figure 6.39 except with a pnp transistor. The magnitude of the small-signal voltage gain should be $|A_v| = 50$ while driving a load $R_L = 10$ k$\Omega$. Bias the circuit at $\pm 3.3$ V.

*D6.90    Consider the circuit in Figure 6.20. Let $V_{CC} = 5$ V, $R_L = 10$ k$\Omega$, $\beta = 120$, and $V_A = \infty$. Design the circuit such that the small-signal current gain is $A_i = 20$ and such that the maximum undistorted swing in the output voltage is achieved.

*D6.91    A microphone puts out a peak voltage of 2 mV and has an output resistance of 5 k$\Omega$. Design an amplifier system to drive a 24 $\Omega$ speaker, producing 0.5 W of signal power. Assume a current gain of $\beta = 50$ for all available transistors. Specify the current and power ratings of the transistors.

*D6.92    Redesign the two-stage amplifier in Figure 6.66 such that the voltage gain of each stage is $A_{v1} = A_{v2} = -50$. Assume transistor current gains of $\beta_{npn} = 150$ and $\beta_{pnp} = 110$. The total power dissipated in the circuit should be limited to 25 mW.

# Frequency Response

Thus far in our linear amplifier analyses, we have assumed that coupling capacitors and bypass capacitors act as short circuits to the signal voltages and open circuits to dc voltages. However, capacitors do not change instantaneously from a short circuit to an open circuit as the frequency approaches zero. We have also assumed that transistors are ideal in that output signals respond instantaneously to input signals. However, there are internal capacitances in both the bipolar transistor and field-effect transistor that affect the frequency response. The major goal of this chapter is to determine the frequency response of amplifier circuits due to circuit capacitors and transistor capacitances.

## PREVIEW

In this chapter, we will:

- Discuss the general frequency response characteristics of amplifiers.
- Derive the system transfer functions of two simple $R-C$ circuits, develop the Bode plots for the magnitude and phase of the transfer functions, and become familiar with sketching the Bode diagrams.
- Analyze the frequency response of transistor circuits with capacitors.
- Determine the frequency response of the bipolar transistor, and determine the Miller effect and Miller capacitance.
- Determine the frequency response of the MOS transistor, and determine the Miller effect and Miller capacitance.
- Determine the high-frequency response of basic transistor circuit configurations including the cascode circuit.
- As an application, design a two-stage BJT amplifier with coupling capacitors such that the 3 dB frequencies associated with each stage are equal.

## 7.1 AMPLIFIER FREQUENCY RESPONSE

**Objective: •** Discuss the general frequency response characteristics of amplifiers.

All amplifier gain factors are functions of signal frequency. These gain factors include voltage, current, transconductance, and transresistance. Up to this point, we have assumed that the signal frequency is high enough that coupling and bypass capacitors can be treated as short circuits and, at the same time, we have assumed that the signal frequency is low enough that parasitic, load, and transistor capacitances can be treated as open circuits. In this chapter, we consider the amplifier response over the entire frequency range.

In general, an amplifier gain factor versus frequency will resemble that shown in Figure 7.1.[1] Both the gain factor and frequency are plotted on logarithmic scales (the gain factor in terms of decibels). Three frequency ranges, low, midband, and high, are indicated. In the **low-frequency range,** $f < f_L$, the gain decreases as the frequency decreases because of coupling and bypass capacitor effects. In the **high-frequency range,** $f > f_H$, stray capacitance and transistor capacitance effects cause the gain to decrease as the frequency increases. The **midband range** is the region where coupling and bypass capacitors act as short circuits, and stray and transistor capacitances act as open circuits. In this region, the gain is almost a constant. As we will show, the gain at $f = f_L$ and at $f = f_H$ is 3 dB less than the maximum midband gain. The bandwidth of the amplifier (in Hz) is defined as $f_{BW} = f_H - f_L$.

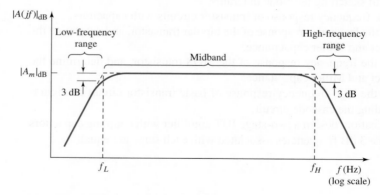

**Figure 7.1** Amplifier gain versus frequency

For an audio amplifier, for example, signal frequencies in the range of 20 Hz < $f$ < 20 kHz need to be amplified equally so as to reproduce the sound as accurately as possible. Therefore, in the design of a good audio amplifier, the frequency $f_L$ must be designed to be less than 20 Hz and $f_H$ must be designed to be greater than 20 kHz.

---

[1]In many references, the gain is plotted as a function of the radian frequency $\omega$. All curves in this chapter, for consistency, will be plotted as a function of cyclical frequency $f$ (Hz). We note that $\omega = 2\pi f$. The amplifier gain is also plotted in terms of decibels (dB), where $|A|_{dB} = 20 \log_{10} |A|$.

### 7.1.1          Equivalent Circuits

Each capacitor in a circuit is important at only one end of the frequency spectrum. For this reason, we can develop specific equivalent circuits that apply to the low-frequency range, to midband, and to the high-frequency range.

#### Midband Range
The equivalent circuits used for calculations in the midband range are the same as those considered up to this point in the text. As already mentioned, the coupling and bypass capacitors in this region are treated as short circuits. The stray and transistor capacitances are treated as open circuits. In this frequency range, there are no capacitances in the equivalent circuit. These circuits are referred to as midband equivalent circuits.

#### Low-Frequency Range
In this frequency range, we use a low-frequency equivalent circuit. In this region, coupling and bypass capacitors must be included in the equivalent circuit and in the amplification factor equations. The stray and transistor capacitances are treated as open circuits. The mathematical expressions obtained for the amplification factor in this frequency range must approach the midband results as $f$ approaches the midband frequency range, since in this limit the capacitors approach short-circuit conditions.

#### High-Frequency Range
In the high-frequency range, we use a high-frequency equivalent circuit. In this region, coupling and bypass capacitors are treated as short circuits. The transistor and any parasitic or load capacitances must be taken into account in this equivalent circuit. The mathematical expressions obtained for the amplification factor in this frequency range must approach the midband results as $f$ approaches the midband frequency range, since in this limit the capacitors approach open-circuit conditions.

### 7.1.2          Frequency Response Analysis

Using the three equivalent circuits just considered rather than a complete circuit is an approximation technique that produces useful hand-analysis results while avoiding complex transfer functions. This technique is valid if there is a large separation between $f_L$ and $f_H$, that is $f_H \gg f_L$. This condition is satisfied in many electronic circuits that we will consider.

Computer simulations, such as PSpice, can take into account all capacitances and can produce frequency response curves that are more accurate than the hand-analysis results. However, the computer results do not provide any physical insight into a particular result and hence do not provide any suggestions as to design changes that can be made to improve a particular frequency response. A hand analysis can provide insight into the "whys and wherefores" of a particular response. This basic understanding can then lead to a better circuit design.

In the next section, we introduce two simple circuits to begin our frequency analysis study. We initially derive the mathematical expressions relating output voltage to input voltage (transfer function) as a function of signal frequency. From these functions, we can develop the response curves. The two frequency response curves give the magnitude of the transfer function versus frequency and the phase of the transfer function versus frequency. The phase response relates the phase of the output signal to the phase of the input signal.

We will then develop a technique by which we can easily sketch the frequency response curves without resorting to a full analysis of the transfer function. This simplified approach will lead to a general understanding of the frequency response of electronic circuits. We will then rely on a computer simulation to provide more detailed calculations when required.

## 7.2     SYSTEM TRANSFER FUNCTIONS

**Objective:** • Derive the system transfer functions of two circuits, develop the Bode diagrams of the magnitude and phase of the transfer functions, and become familiar with sketching the Bode diagrams.

The frequency response of a circuit is usually determined by using the **complex frequency $s$.** Each capacitor is represented by its complex impedance, $1/sC$, and each inductor is represented by its complex impedance, $sL$. The circuit equations are then formulated in the usual way. Using the complex frequency, the mathematical expressions obtained for voltage gain, current gain, input impedance, or output impedance are ratios of polynomials in $s$.

We will be concerned in many cases with system transfer functions. These will be in the form of ratios of, for example, output voltage to input voltage (voltage transfer function) or output current to input voltage (transconductance function). The four general transfer functions are listed in Table 7.1.

Once a transfer function is found, we can find the result of a steady-state sinusoidal excitation by setting $s = j\omega = j2\pi f$. The ratio of polynomials in $s$ then reduces to a complex number for each frequency $f$. The complex number can be reduced to a magnitude and a phase.

| Table 7.1 | Transfer functions of the complex frequency $s$ | |
|---|---|
| **Name of function** | **Expression** |
| Voltage transfer function | $T(s) = V_o(s)/V_i(s)$ |
| Current transfer function | $I_o(s)/I_i(s)$ |
| Transresistance function | $V_o(s)/I_i(s)$ |
| Transconductance function | $I_o(s)/V_i(s)$ |

### 7.2.1     s-Domain Analysis

In general, a transfer function in the $s$-domain can be expressed in the form

$$T(s) = K \frac{(s - z_1)(s - z_2) \cdots (s - z_m)}{(s - p_1)(s - p_2) \cdots (s - p_n)} \tag{7.1}$$

where $K$ is a constant, $z_1, z_2, \ldots, z_m$ are the transfer function "zeros," and $p_1, p_2, \ldots, p_n$ are the transfer function "poles." When the complex frequency is equal to a zero, $s = z_i$, the transfer function is zero; when the complex frequency is equal to a pole, $s = p_i$, the transfer function diverges and becomes infinite. The transfer function can be evaluated for physical frequencies by replacing $s$ with $j\omega$. In general, the resulting transfer function $T(j\omega)$ is a complex function, that is, its magnitude and phase are both functions of frequency. These topics are usually discussed in a basic circuit analysis course.

To introduce the frequency response analysis of transistor circuits, we will examine the circuits shown in Figures 7.2 and 7.3. The voltage transfer function for the circuit in Figure 7.2 can be expressed in a voltage divider format, as follows:

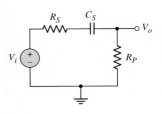

**Figure 7.2** Series coupling capacitor circuit

$$\frac{V_o(s)}{V_i(s)} = \frac{R_P}{R_S + R_P + \dfrac{1}{sC_S}} \tag{7.2}$$

The elements $R_S$ and $C_S$ are in series between the input and output signals, and the element $R_P$ is in parallel with the output signal. Equation (7.2) can be written in the form

$$\frac{V_o(s)}{V_i(s)} = \frac{sR_PC_S}{1 + s(R_S + R_P)C_S} \tag{7.3}$$

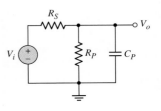

**Figure 7.3** Parallel load capacitor circuit

which can be rearranged and written as

$$\frac{V_o(s)}{V_i(s)} = \left(\frac{R_P}{R_S + R_P}\right)\left[\frac{s(R_S + R_P)C_S}{1 + s(R_S + R_P)C_S}\right] = K\left(\frac{s\tau_s}{1 + s\tau_s}\right) \tag{7.4}$$

where $\tau_S$ is a time constant and is given by $\tau_S = (R_S + R_P)C_S$.

Writing a Kirchhoff current law (KCL) equation at the output node, we can determine the voltage transfer function for the circuit shown in Figure 7.3, as follows:

$$\frac{V_o - V_i}{R_S} + \frac{V_o}{R_P} + \frac{V_o}{(1/sC_P)} = 0 \tag{7.5}$$

In this case, the element $R_S$ is in series between the input and output signals, and the elements $R_P$ and $C_P$ are in parallel with the output signal. Rearranging the terms in Equation (7.5) produces

$$\frac{V_o(s)}{V_i(s)} = \left(\frac{R_P}{R_S + R_P}\right)\left[\frac{1}{1 + s\left(\dfrac{R_SR_P}{R_S + R_P}\right)C_P}\right] \tag{7.6}$$

or

$$\frac{V_o(s)}{V_i(s)} = \left(\frac{R_P}{R_S + R_P}\right)\left[\frac{1}{1 + s(R_S\|R_P)C_P}\right] = K\left(\frac{1}{1 + s\tau_P}\right) \tag{7.7}$$

where $\tau_P$ is also a time constant and is given by $\tau_P = (R_S\|R_P)C_P$.

### 7.2.2    First-Order Functions

In our hand analysis of transistor circuits in this chapter, we will, in general, limit ourselves to the consideration of only one capacitance at a time. We will therefore be dealing with **first-order transfer functions** that, in most cases, will have the general form of either Equation (7.4) or (7.7). This simplified analysis will allow us to present the frequency responses of specific capacitances and of the transistors themselves. We will then compare our hand analysis results with more rigorous solutions, using a computer simulation.

### 7.2.3    Bode Plots

A simplified technique for obtaining approximate plots of the magnitude and phase of a transfer function, given the poles and zeros or the equivalent time constants, was developed by H. Bode, and the resulting diagrams are called **Bode plots.**

**Qualitative Discussion:** Initially, we will consider the magnitude of the voltage transfer function versus frequency. Before we delve into the mathematics, we can qualitatively determine the general characteristics of this plot. The capacitor $C_S$ in Figure 7.2 is in series between the input and output terminals. This capacitor then behaves as a coupling capacitor.

In the limit of zero frequency (the input signal is a constant dc voltage), the impedance of the capacitor is infinite (an open circuit). In this case, then, the input signal does not get coupled to the output terminal so the output voltage is zero. In this case, the magnitude of the voltage transfer function is zero.

In the limit of a very high frequency, the impedance of the capacitor becomes very small (approaching a short circuit). In this situation, the magnitude of the output voltage reaches a constant value given by a voltage divider, or $V_o = [R_P/(R_P + R_S)] \cdot V_i$.

We therefore expect the magnitude of the transfer function to start at zero for zero frequency, increase for increasing frequency, and reach a constant value at a relatively high frequency.

### Bode Plot for Figure 7.2

**Mathematical Derivation:** For the transfer function in Equation (7.4), corresponding to the circuit in Figure 7.2, if we replace $s$ by $j\omega$ and define a time constant $\tau_s$ as $\tau_s = (R_S + R_P)C_S$, we obtain

$$T(j\omega) = \frac{V_o(j\omega)}{V_i(j\omega)} = \left(\frac{R_P}{R_S + R_P}\right)\left[\frac{j\omega\tau_S}{1 + j\omega\tau_S}\right] \qquad (7.8)$$

The magnitude of Equation (7.8) is

$$|T(j\omega)| = \left(\frac{R_P}{R_S + R_P}\right)\left[\frac{\omega\tau_S}{\sqrt{1 + \omega^2\tau_S^2}}\right] \qquad (7.9(a))$$

or

$$|T(jf)| = \left(\frac{R_P}{R_S + R_P}\right)\left[\frac{2\pi f\tau_S}{\sqrt{1 + (2\pi f\tau_S)^2}}\right] \qquad (7.9(b))$$

We can develop the Bode plot of the gain magnitude versus frequency. We may note that $|T(jf)|_{dB} = 20\log_{10}|T(jf)|$. From Equation (7.9(b)), we can write

$$|T(jf)|_{dB} = 20\log_{10}\left[\left(\frac{R_P}{R_S + R_P}\right) \cdot \frac{2\pi f\tau_S}{\sqrt{1 + (2\pi f\tau_S)^2}}\right] \qquad (7.10(a))$$

or

$$|T(jf)|_{dB} = 20\log_{10}\left(\frac{R_P}{R_S + R_P}\right) + 20\log_{10}(2\pi f \tau_S)$$
$$-20\log_{10}\sqrt{1 + (2\pi f \tau_S)^2}$$

**(7.10(b))**

We can plot each term of Equation (7.10(b)) and then combine the three plots to form the final Bode plot of the gain magnitude.

Figure 7.4(a) is the plot of the first term of equation (7.10(b)), which is just a constant independent of frequency. We may note that, since $[R_P/(R_S + R_P)]$ is less than unity, the dB value is less than zero.

Figure 7.4(b) is the plot of the second term of Equation (7.10(b)). When $f = 1/2\pi\tau_S$, we have $20\log_{10}(1) = 0$. The slopes in Bode plot magnitudes are described in units of either dB/octave or dB/decade. An **octave** means that frequency is increased by a factor of two, and a **decade** implies that the frequency is increased by a factor of 10. The value of the function $20\log_{10}(2\pi f \tau_S)$ increases by a factor of $6.02 \cong 6$ dB for every factor of 2 increase in frequency, and the value of the function increases by a factor of 20 dB for every factor of 10 increase in frequency. Hence, we can consider a slope of 6 dB/octave or 20 dB/decade.

Figure 7.4(c) is the plot of the third term in Equation (7.10(b)). For $f \ll 1/2\pi\tau_S$, the value of the function is essentially 0 dB and when $f = 1/2\pi\tau_S$, the value is $-3$ dB. For $f \gg 1/2\pi\tau_S$, the function becomes $-20\log_{10}(2\pi f \tau_S)$, so the slope becomes $-6$ dB/octave or $-20$ dB/decade. A straight-line projection of this slope passes through 0 dB at $f = 1/2\pi\tau_S$. We can then approximate the Bode plot for this term by two straight line asymptotes intersecting at 0 dB and $f = 1/2\pi\tau_S$. This particular frequency is known as a **break-point frequency, corner frequency, or −3 dB frequency.**

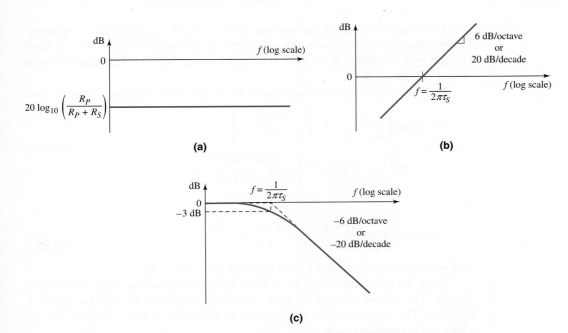

**Figure 7.4** Plots of (a) the first term, (b) the second term, and (c) the third term of Equation (7.10(b))

The complete Bode plot of Equation (7.10(b)) is shown in Figure 7.5. For $f \gg 1/2\pi \tau_S$, the second and third terms of Equation (7.10(b)) cancel, and for $f \ll 1/2\pi \tau_S$, the large negative dB value from Figure 7.4(b) dominates.

The transfer function given by Equation (7.9) is for the circuit shown in Figure 7.2. The series capacitor $C_S$ is a coupling capacitor between the input and output signals. At a high enough frequency, capacitor $C_S$ acts as a short circuit, and the output voltage, from a voltage divider, is

$$V_o = [R_P/(R_S + R_P)]V_i$$

For very low frequencies, the impedance of $C_S$ approaches that of an open circuit, and the output voltage approaches zero. This circuit is called a **high-pass network** since the high-frequency signals are passed through to the output. We can now understand the form of the Bode plot shown in Figure 7.5.

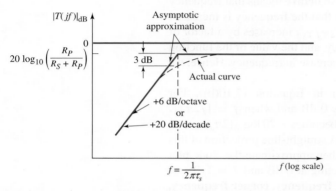

**Figure 7.5** Bode plot of the voltage transfer function magnitude for the circuit in Figure 7.2

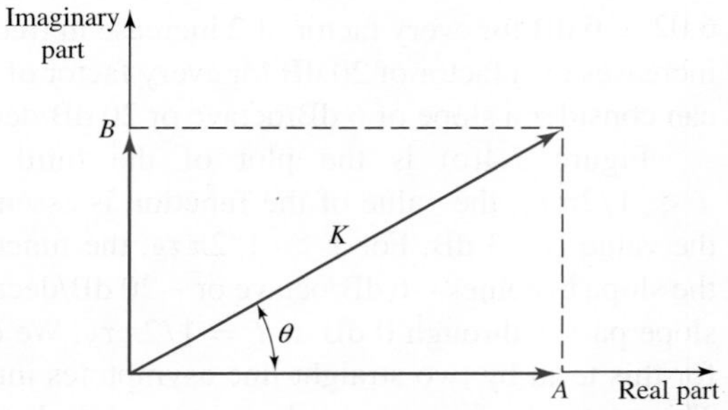

**Figure 7.6** Relation between rectangular and polar forms of a complex number

The Bode plot of the phase function can be easily developed by recalling the relation between the rectangular and polar form of a complex number. We can write $A + jB = Ke^{j\theta}$, where $K = \sqrt{A^2 + B^2}$ and $\theta = \tan^{-1}(B/A)$. This relationship is shown in Figure 7.6.

For the function given in Equation (7.8), we can write the function in the form

$$T(jf) = \left(\frac{R_P}{R_S + R_P}\right) \cdot \left[\frac{j2\pi f \tau_S}{1 + j2\pi f \tau_S}\right]$$

$$= \left[\left|\frac{R_P}{R_S + R_P}\right| e^{j\theta_1}\right] \frac{[|j2\pi f \tau_S|e^{j\theta_2}]}{[|1 + j2\pi f \tau_S|e^{j\theta_3}]}$$

**(7.11(a))**

or

$$T(jf) = [K_1 e^{j\theta_1}]\frac{[K_2 e^{j\theta_2}]}{[K_3 e^{j\theta_3}]} = \frac{K_1 K_2}{K_3} e^{j(\theta_1 + \theta_2 - \theta_3)}$$

**(7.11(b))**

The net phase of the function $T(jf)$ is then $\theta = \theta_1 + \theta_2 - \theta_3$.

Since the first term, $[R_P/(R_S + R_P)]$, is a positive real quantity, the phase is $\theta_1 = 0$. The second term, $(j2\pi f \tau_S)$, is purely imaginary so that the phase is $\theta_2 = 90°$. The third term is complex so that its phase is $\theta_3 = \tan^{-1}(2\pi f \tau_S)$. The net phase of the function is now

$$\theta = 90 - \tan^{-1}(2\pi f \tau_S)$$

**(7.12)**

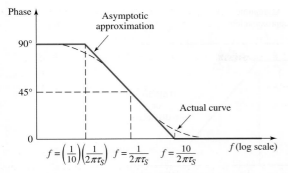

**Figure 7.7** Bode plot of the voltage transfer function phase for the circuit in Figure 7.2

For the limiting case of $f \to 0$, we have $\tan^{-1}(0) = 0$, and for $f \to \infty$, we have $\tan^{-1}(\infty) = 90°$. At the corner frequency of $f = 1/(2\pi\tau_S)$, the phase is $\tan^{-1}(1) = 45°$. The Bode plot of the phase of the function given in Equation (7.11(a)) is given in Figure 7.7. The actual plot as well as an asymptotic approximation is shown. The phase is especially important in feedback circuits since this can influence stability. We will see this effect in Chapter 12.

### Bode Plot for Figure 7.3

**Qualitative Discussion:** Again, we will initially consider the magnitude of the voltage function versus frequency. The capacitor $C_P$ in Figure 7.3 is in parallel with the output and then behaves as a load capacitor on the output of a circuit, or may represent the input capacitance of a follow-on amplifier stage.

In the limit of zero frequency (the input signal is a constant dc voltage), the impedance of the capacitor is infinite (an open circuit). In this case the output signal is a constant value given by a voltage divider, or $V_o = [R_P/(R_P + R_S)] \cdot V_i$.

In the limit of a very high frequency, the impedance of the capacitor becomes very small (approaching a short circuit). In this situation, the output voltage will be zero, or the magnitude of the voltage transfer function will be zero.

We therefore expect the magnitude of the transfer function to start at a constant value for zero and low frequencies, and then decrease toward zero at high frequencies.

**Mathematical Derivation:** The transfer function given by Equation (7.7) is for the circuit that was shown in Figure 7.3. If we replace $s$ by $s = j\omega = j2\pi f$ and define a time constant $\tau_P$ as $\tau_P = (R_S \| R_P)C_P$, then the transfer function is

$$T(jf) = \left(\frac{R_P}{R_S + R_P}\right)\left[\frac{1}{1 + j2\pi f \tau_P}\right] \tag{7.13}$$

The magnitude of Equation (7.13) is

$$|T(jf)| = \left(\frac{R_P}{R_S + R_P}\right) \cdot \left[\frac{1}{\sqrt{1 + (2\pi f \tau_P)^2}}\right] \tag{7.14}$$

A Bode plot of this magnitude expression is shown in Figure 7.8. The low-frequency asymptote is a horizontal line, and the high-frequency asymptote is a

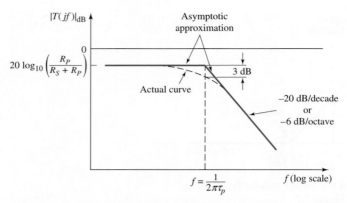

**Figure 7.8** Bode plot of the voltage transfer function magnitude for the circuit in Figure 7.3

straight line with a slope of $-20$ dB/decade, or $-6$ dB/octave. The two asymptotes meet at the frequency $f = 1/2\pi\tau_P$, which is the corner, or 3 dB, frequency for this circuit. Again, the actual magnitude of the transfer function at the corner frequency differs from the maximum asymptotic value by 3 dB.

Again, the magnitude of the transfer function given by Equation (7.14) is for the circuit shown in Figure 7.3. The parallel capacitor $C_P$ is a load, or parasitic, capacitance. At low frequencies, $C_P$ acts as an open circuit, and the output voltage, from a voltage divider, is

$$V_o = [R_P/(R_S + R_P)]V_i$$

As the frequency increases, the magnitude of the impedance of $C_P$ decreases and approaches that of a short circuit, and the output voltage approaches zero. This circuit is called a **low-pass network,** since the low-frequency signals are passed through to the output.

The phase of the transfer function given by Equation (7.13) is

$$\text{Phase} = -\angle \tan^{-1}(2\pi f \tau_P) \tag{7.15}$$

The Bode plot of the phase is shown in Figure 7.9. The phase is $-45$ degrees at the corner frequency and 0 degrees at the low-frequency asymptote, where $C_P$ is effectively out of the circuit.

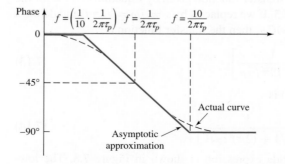

**Figure 7.9** Bode plot of the voltage transfer function phase for the circuit in Figure 7.3

## EXAMPLE **7.1**

**Objective:** Determine the corner frequencies and maximum-magnitude asymptotes of the Bode plots for a specified circuit.

For the circuits in Figures 7.2 and 7.3, the parameters are: $R_S = 1 \text{ k}\Omega$, $R_P = 10 \text{ k}\Omega$, $C_S = 1 \text{ }\mu\text{F}$, and $C_P = 3 \text{ pF}$.

**Solution:** (Figure 7.2) The time constant is

$$\tau_S = (R_S + R_P)C_S = (10^3 + 10 \times 10^3)(10^{-6}) = 1.1 \times 10^{-2} \text{ s} \Rightarrow 11 \text{ ms}$$

The corner frequency of the Bode plot shown in Figure 7.5 is then

$$f = \frac{1}{2\pi\tau_S} = \frac{1}{(2\pi)(11 \times 10^{-3})} = 14.5 \text{ Hz}$$

The maximum magnitude is

$$\frac{R_P}{R_S + R_P} = \frac{10}{1 + 10} = 0.909$$

or

$$20 \log_{10}\left(\frac{R_P}{R_S + R_P}\right) = -0.828 \text{ dB}$$

**Solution:** (Figure 7.3) The time constant is

$$\tau_P = (R_S \| R_P)C_P = (10^3 \| (10 \times 10^3))(3 \times 10^{-12}) \Rightarrow 2.73 \text{ ns}$$

The corner frequency of the Bode plot in Figure 7.8 is then

$$f = \frac{1}{2\pi\tau_P} = \frac{1}{(2\pi)(2.73 \times 10^{-9})} \Rightarrow 58.3 \text{ MHz}$$

The maximum magnitude is the same as just calculated: 0.909 or −0.828 dB.

**Comment:** Since the two capacitance values are substantially different, the two time constants differ by orders of magnitude, which means that the two corner frequencies also differ by orders of magnitude. Later in this text, we will take advantage of these differences in our analysis of transistor circuits.

## EXERCISE PROBLEM

**Ex 7.1:** (a) For the circuit shown in Figure 7.2, the parameters are $R_S = 2 \text{ k}\Omega$ and $R_P = 8 \text{ k}\Omega$. (i) If the corner frequency is $f_L = 50 \text{ Hz}$, determine the value of $C_S$. (ii) Find the magnitude of the transfer function at $f = 20 \text{ Hz}$, 50 Hz, and 100 Hz. (Ans. (i) $C_S = 0.318 \text{ }\mu\text{F}$; (ii) 0.297, 0.566, and 0.716)

(b) Consider the circuit shown in Figure 7.3 with parameters $R_S = 4.7 \text{ k}\Omega$, $R_P = 25 \text{ k}\Omega$, and $C_P = 120 \text{ pF}$. (i) Determine the corner frequency $f_H$. (ii) Determine the magnitude of the transfer function at $f = 0.2f_H$, $f = f_H$, and $f = 8f_H$. (Ans. (i) $f_H = 335 \text{ kHz}$; (ii) 0.825, 0.595, 0.104)

# Short-Circuit and Open-Circuit Time Constants

The two circuits shown in Figures 7.2 and 7.3 each contain only one capacitor. The circuit in Figure 7.10 is the same basic configuration but contains both capacitors. Capacitor $C_S$ is the coupling capacitor and is in series with the input and output; capacitor $C_P$ is the load capacitor and is in parallel with the output and ground.

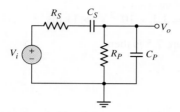

**Figure 7.10** Circuit with both a series coupling and a parallel load capacitor

We can determine the voltage transfer function of this circuit by writing a KCL equation at the output node. The result is

$$\frac{V_o(s)}{V_i(s)} = \left(\frac{R_P}{R_S + R_P}\right) \times \frac{1}{\left[1 + \left(\frac{R_P}{R_S + R_P}\right)\left(\frac{C_P}{C_S}\right) + \frac{1}{s\tau_S} + s\tau_P\right]} \tag{7.16}$$

where $\tau_S$ and $\tau_P$ are the same time constants as previously defined.

Although Equation (7.16) is the exact transfer function, it is awkward to deal with in this form.

We have seen in the previous analysis, however, that $C_S$ affects the low-frequency response and $C_P$ affects the high-frequency response. Further, if $C_P \ll C_S$ and if $R_S$ and $R_P$ are of the same order of magnitude, then the corner frequencies of the Bode plots created by $C_S$ and $C_P$ will differ by orders of magnitude. (We actually encounter this situation in real circuits.) Consequently, when a circuit contains both coupling and load capacitors, and when the values of the capacitors differ by orders of magnitude, then we can determine the effect of each capacitor individually.

At low frequencies, we can treat the load capacitor $C_P$ as an open circuit. To find the equivalent resistance seen by a capacitor, set all independent sources equal to zero. Therefore, the effective resistance seen by $C_S$ is the series combination of $R_S$ and $R_P$. The time constant associated with $C_S$ is

$$\tau_S = (R_S + R_P)C_S \tag{7.17}$$

Since $C_P$ was made an open circuit, $\tau_S$ is called an **open-circuit time constant.** The subscript $S$ is associated with the coupling capacitor, or the capacitor in series with the input and output signals.

At high frequencies, we can treat the coupling capacitor $C_S$ as a short circuit. The effective resistance seen by $C_P$ is the parallel combination of $R_S$ and $R_P$, and the associated time constant is

$$\tau_P = (R_S \| R_P)C_P \tag{7.18}$$

which is called the **short-circuit time constant.** The subscript $P$ is associated with the load capacitor, or the capacitor in parallel with the output and ground.

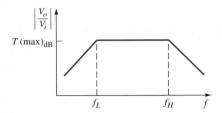

**Figure 7.11** Bode plot of the voltage transfer function magnitude for the circuit in Figure 7.10

We can now define the corner frequencies of the Bode plot. The **lower corner,** or 3 dB frequency, which is at the low end of the frequency scale, is a function of the open-circuit time constant and is defined as

$$f_L = \frac{1}{2\pi \tau_S} \qquad \text{(7.19(a))}$$

The **upper corner,** or 3 dB, frequency, which is at the high end of the frequency scale, is a function of the short-circuit time constant and is defined as

$$f_H = \frac{1}{2\pi \tau_P} \qquad \text{(7.19(b))}$$

The resulting Bode plot of the magnitude of the voltage transfer function for the circuit in Figure 7.9 is shown in Figure 7.11.

This Bode plot is for a passive circuit; the Bode plots for transistor amplifiers are similar. The amplifier gain is constant over a wide frequency range, called the **midband.** In this frequency range, all capacitance effects are negligible and can be neglected in the gain calculations. At the high end of the frequency spectrum, the gain drops as a result of the load capacitance and, as we will see later, the transistor effects. At the low end of the frequency spectrum, the gain decreases because coupling capacitors and bypass capacitors do not act as perfect short circuits.

The midband range, or **bandwidth,** is defined by the corner frequencies $f_L$ and $f_H$, as follows:

$$f_{BW} = f_H - f_L \qquad \text{(7.20(a))}$$

Since $f_H \gg f_L$, as we have seen in our examples, the bandwidth is essentially given by

$$f_{BW} \cong f_H \qquad \text{(7.20(b))}$$

## EXAMPLE 7.2

**Objective:** Determine the corner frequencies and bandwidth of a passive circuit containing two capacitors.

Consider the circuit shown in Figure 7.10 with parameters $R_S = 1$ k$\Omega$, $R_P = 10$ k$\Omega$, $C_S = 1$ $\mu$F, and $C_P = 3$ pF.

**Solution:** Since $C_P$ is less than $C_S$ by approximately six orders of magnitude, we can treat the effect of each capacitor separately. The open-circuit time constant is

$$\tau_S = (R_S + R_P)C_S = (10^3 + 10 \times 10^3)(10^{-6}) = 1.1 \times 10^{-2} \text{ s}$$

and the short-circuit time constant is

$$\tau_P = (R_S \| R_P)C_P = [10^3 \| (10 \times 10^3)](3 \times 10^{-12}) = 2.73 \times 10^{-9} \text{ s}$$

The corner frequencies are then

$$f_L = \frac{1}{2\pi\,\tau_S} = \frac{1}{2\pi\,(1.1 \times 10^{-2})} = 14.5 \text{ Hz}$$

and

$$f_H = \frac{1}{2\pi\,\tau_P} = \frac{1}{2\pi\,(2.73 \times 10^{-9})} \Rightarrow 58.3 \text{ MHz}$$

Finally, the bandwidth is

$$f_{\text{BW}} = f_H - f_L = 58.3 \text{ MHz} - 14.5 \text{ Hz} \cong 58.3 \text{ MHz}$$

**Comment:** The corner frequencies in this example are exactly the same as those determined in Example 7.1. This occurred because the two corner frequencies are far apart. The maximum magnitude of the voltage function is again

$$\frac{R_P}{R_S + R_P} = \frac{10}{1 + 10} = 0.909 \Rightarrow -0.828 \text{ dB}$$

The Bode plot of the magnitude of the voltage transfer function is shown in Figure 7.12.

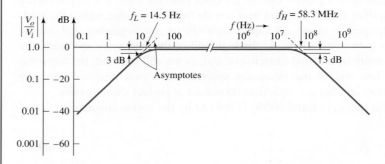

**Figure 7.12** Bode plot of the magnitude of the voltage transfer function for the circuit in Figure 7.10

---

EXERCISE PROBLEM

**Ex 7.2:** The circuit shown in Figure 7.10 has parameters of $R_P = 7.5 \text{ k}\Omega$ and $C_P = 80 \text{ pF}$. The midband gain is –2 dB and the lower corner frequency is $f_L = 200 \text{ Hz}$ (a) Determine $R_S$, $C_S$, and the upper corner frequency $f_H$. (b) Determine the open-circuit and short-circuit time constants. (Ans. (a) $R_S = 1.94 \text{ k}\Omega$, $C_S = 0.0843\,\mu\text{F}$, $f_H = 1.29 \text{ MHz}$; (b) $\tau_S = 0.796 \text{ ms}$, $\tau_P = 0.123\,\mu\text{s}$)

We will continue, in the following sections of the chapter, to use the concept of open-circuit and short-circuit time constants to determine the corner frequencies of the Bode plots of transistor circuits. An implicit assumption in this technique is that coupling and load capacitance values differ by many orders of magnitude.

# Test Your Understanding

**TYU 7.1** For the equivalent circuit shown in Figure 7.13, the parameters are: $R_S = 1 \text{ k}\Omega$, $r_\pi = 2 \text{ k}\Omega$, $R_L = 4 \text{ k}\Omega$, $g_m = 50 \text{ mA/V}$, and $C_C = 1 \mu\text{F}$. (a) Determine the expression for the circuit time constant. (b) Calculate the 3 dB frequency and maximum gain asymptote. (c) Sketch the Bode plot of the transfer function magnitude. (Ans. (a) $\tau = (r_\pi + R_S)C_C$, (b) $f_{3\text{dB}} = 53.1 \text{ Hz}$, $|T(j\omega)|_{\max} = 133$)

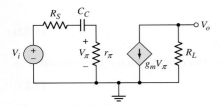

**Figure 7.13** Figure for Exercise TYU 7.1

**TYU 7.2** The equivalent circuit in Figure 7.14 has circuit parameters $R_S = 100 \Omega$, $r_\pi = 2.4 \text{ k}\Omega$, $g_m = 50 \text{ mA/V}$, $R_L = 10 \text{ k}\Omega$, and $C_L = 2 \text{ pF}$. (a) Determine the expression for and the value of the circuit time constant. (b) Calculate the 3 dB frequency and the maximum voltage gain. (c) Sketch the Bode plot of the transfer function magnitude. (Ans. (a) $\tau = R_L C_L = 0.02 \mu\text{s}$; (b) $f_{3\text{dB}} = 7.96 \text{ MHz}$, $|A_v| = 480$)

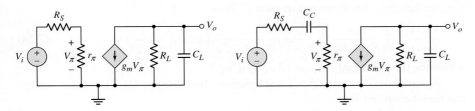

**Figure 7.14** Figure for Exercise TYU 7.2    **Figure 7.15** Figure for Exercise TYU 7.3

**TYU 7.3** The parameters in the circuit shown in Figure 7.15 are $R_S = 100 \Omega$, $r_\pi = 2.4 \text{ k}\Omega$, $g_m = 50 \text{ mA/V}$, $R_L = 10 \text{ k}\Omega$, $C_C = 5 \mu\text{F}$, and $C_L = 4 \text{ pF}$. (a) Find the open-circuit and short-circuit time constants. (b) Calculate the midband voltage gain. (c) Determine the lower and upper 3 dB frequencies. (Ans. (a) $\tau_S = 12.5 \text{ ms}$, $\tau_P = 0.04 \mu\text{s}$; (b) $A_v = -480$; (c) $f_L = 12.7 \text{ Hz}$, $f_H = 3.98 \text{ MHz}$)

## 7.2.5    Time Response

Up to this point, we have been considering the steady-state sinusoidal frequency response of circuits. In some cases, however, we may need to amplify nonsinusoidal signals, such as square waves. This situation might occur if digital signals are to be amplified. In these cases, we need to consider the time response of the output signals. In addition, such signals as pulses or square waves may be used in testing the frequency response of circuits.

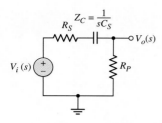

**Figure 7.16** Repeat of Figure 7.2 (coupling capacitor circuit), but showing complex $s$ parameters

To gain some understanding, consider the circuit shown in Figure 7.16, which is a repeat of Figure 7.2. As mentioned previously, the capacitor represents a coupling capacitor. The transfer function was given in Equation (7.4) as

$$\frac{V_o(s)}{V_i(s)} = \left(\frac{R_P}{R_S + R_P}\right)\left[\frac{s(R_S + R_P)C_S}{1 + s(R_S + R_P)C_S}\right] = K_2\left(\frac{s\tau_S}{1 + s\tau_S}\right) \tag{7.21}$$

where the time constant is $\tau_S = (R_S + R_P)C_S$.

If the input voltage is a step function, then $V_i(s) = 1/s$. The output voltage can then be written as

$$V_o(s) = K_2\left(\frac{\tau_S}{1 + s\tau_S}\right) = K_2\left(\frac{1}{s + 1/\tau_S}\right) \tag{7.22}$$

Taking the inverse Laplace transform, we find the output voltage time response as

$$v_o(t) = K_2 e^{-t/\tau_S} \tag{7.23}$$

If we are trying to amplify an input voltage pulse using a coupling capacitor, the voltage applied to the amplifier (load) will begin to droop. In this case, we would need to ensure that the time constant $\tau_S$ is large compared to the input pulse width $T$. The output voltage versus time for a square wave input signal is shown in Figure 7.17. A large time constant implies a large coupling capacitor.

If the cutoff frequency of the transfer function is $f_{3\text{-dB}} = 1/2\pi\tau_S = 5$ kHz, then the time constant is $\tau_S = 3.18$ $\mu$s. For a pulse width of $T = 0.1$ $\mu$s, the output voltage will droop by only 0.314 percent at the end of the pulse.

Consider, now, the circuit shown in Figure 7.18, which is a repeat of Figure 7.3. In this case, the capacitor $C_P$ may represent the input capacitance of an amplifier. The transfer function was given in Equation (7.7) as

$$\frac{V_o(s)}{V_i(s)} = \left(\frac{R_P}{R_S + R_P}\right)\left[\frac{1}{1 + s(R_P\|R_S)C_P}\right] = K_1\left(\frac{1}{1 + s\tau_P}\right) \tag{7.24}$$

where the time constant is $\tau_P = (R_P\|R_S)C_P$.

Again, if the input signal is a step function, then $V_i(s) = 1/s$. The output voltage can then be written as

$$V_o(s) = \frac{K_1}{s}\left(\frac{1}{1 + s\tau_P}\right) = \frac{K_1}{s}\left(\frac{1/\tau_P}{s + 1/\tau_P}\right) \tag{7.25}$$

Taking the inverse Laplace transform, we find the output voltage time response as

$$v_o(t) = K_1(1 - e^{-t/\tau_P}) \tag{7.26}$$

If we are trying to amplify an input voltage pulse, we need to ensure that the time constant $\tau_P$ is short compared to the pulse width $T$, so that the signal $v_O(t)$ reaches a

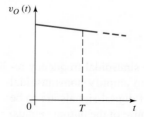

**Figure 7.17** Output response of circuit in Figure 7.16 for a square-wave input signal and for a large time constant

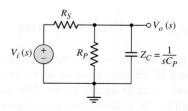

**Figure 7.18** Repeat of Figure 7.3 (load capacitor circuit), but showing complex $s$ parameters

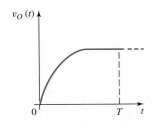

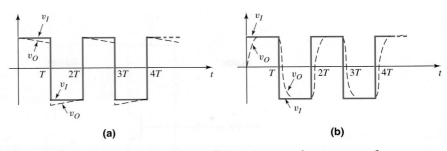

**Figure 7.19** Output response of circuit in Figure 7.18 for a square-wave input signal and for a short time constant

**Figure 7.20** Steady-state output response for a square-wave input response for (a) circuit in Figure 7.16 (coupling capacitor) and a large time constant, and (b) circuit in Figure 7.18 (load capacitor) and a short time constant

steady-state value. The output voltage is shown in Figure 7.19 for a square wave input signal. A short time constant implies a very small capacitor $C_P$ as an input capacitance to an amplifier.

In this case, if the cutoff frequency of the transfer function is $f_{3\text{-dB}} = 1/2\pi\tau_P = 10$ MHz, then the time constant is $\tau_P = 15.9$ ns.

Figure 7.20 summarizes the steady-state output responses for square wave input signals of the two circuits we've just been considering. Figure 7.20(a) shows the steady-state output response of the circuit in Figure 7.16 (coupling capacitor) for a long time constant, and Figure 7.20(b) shows the steady-state output response of the circuit in Figure 7.18 (load capacitor) for a short time constant.

## 7.3 FREQUENCY RESPONSE: TRANSISTOR AMPLIFIERS WITH CIRCUIT CAPACITORS

**Objective:** • Analyze the frequency response of transistor circuits with capacitors.

In this section, we will analyze the basic single-stage amplifier that includes circuit capacitors. Three types of capacitors will be considered: coupling capacitor, load capacitor, and bypass capacitor. In our hand analysis, we will consider each type of capacitor individually and determine its frequency response. In the last part of this section, we will consider the effect of multiple capacitors using a PSpice analysis.

The frequency response of multistage circuits will be considered in Chapter 12 when the stability of amplifiers is considered.

### 7.3.1 Coupling Capacitor Effects

**Input Coupling Capacitor: Common-Emitter Circuit**
Figure 7.21(a) shows a bipolar common-emitter circuit with a coupling capacitor. Figure 7.21(b) shows the corresponding small-signal equivalent circuit, with the transistor small-signal output resistance $r_o$ assumed to be infinite. This assumption is valid since $r_o \gg R_C$ and $r_o \gg R_E$ in most cases. Initially, we will use a current–voltage analysis to determine the frequency response of the circuit. Then, we will use the equivalent time constant technique.

From the analysis in the previous section, we note that this circuit is a high-pass network. At high frequencies, the capacitor $C_C$ acts as a short circuit, and the input

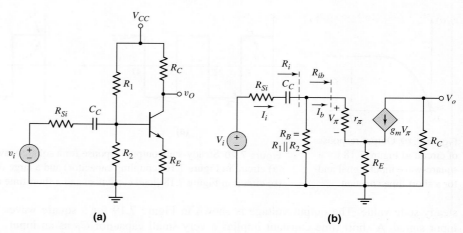

**Figure 7.21** (a) Common-emitter circuit with coupling capacitor and (b) small-signal equivalent circuit

signal is coupled through the transistor to the output. At low frequencies, the impedance of $C_C$ becomes large and the output approaches zero.

**Current–Voltage Analysis:** The input current can be written as

$$I_i = \frac{V_i}{R_{Si} + \dfrac{1}{sC_C} + R_i} \tag{7.27}$$

where the input resistance $R_i$ is given by

$$R_i = R_B \| [r_\pi + (1+\beta)R_E] = R_B \| R_{ib} \tag{7.28}$$

In writing Equation (7.28), we used the resistance reflection rule given in Chapter 6. To determine the input resistance to the base of the transistor, we multiplied the emitter resistance by the factor $(1+\beta)$.

Using a current divider, we determine the base current to be

$$I_b = \left( \frac{R_B}{R_B + R_{ib}} \right) I_i \tag{7.29}$$

and then

$$V_\pi = I_b r_\pi \tag{7.30}$$

The output voltage is given by

$$V_o = -g_m V_\pi R_C \tag{7.31}$$

Combining Equations (7.27) through (7.31) produces

$$V_o = -g_m R_C (I_b r_\pi) = -g_m r_\pi R_C \left( \frac{R_B}{R_B + R_{ib}} \right) I_i$$

$$= -g_m r_\pi R_C \left( \frac{R_B}{R_B + R_{ib}} \right) \left( \frac{V_i}{R_{Si} + \dfrac{1}{sC_C} + R_i} \right) \tag{7.32}$$

Therefore, the small-signal voltage gain is

$$A_v(s) = \frac{V_o(s)}{V_i(s)} = -g_m r_\pi R_C \left( \frac{R_B}{R_B + R_{ib}} \right) \left( \frac{sC_C}{1 + s(R_{Si} + R_i)C_C} \right) \qquad (7.33)$$

which can be written in the form

$$A_v(s) = \frac{V_o(s)}{V_i(s)} = \frac{-g_m r_\pi R_C}{(R_{Si} + R_i)} \left( \frac{R_B}{R_B + R_{ib}} \right) \left( \frac{s\tau_S}{1 + s\tau_S} \right) \qquad (7.34)$$

where the time constant is

$$\tau_S = (R_{Si} + R_i)C_C \qquad (7.35)$$

The form of the voltage transfer function as given in Equation (7.34) is the same as that of Equation (7.4), for the coupling capacitor circuit in Figure 7.2. The Bode plot is therefore similar to that shown in Figure 7.5. The corner frequency is

$$f_L = \frac{1}{2\pi \tau_S} = \frac{1}{2\pi (R_{Si} + R_i)C_C} \qquad (7.36)$$

and the maximum magnitude, in decibels, is

$$|A_v(\max)|_{dB} = 20 \log_{10} \left( \frac{g_m r_\pi R_C}{R_{Si} + R_i} \right) \left( \frac{R_B}{R_B + R_{ib}} \right) \qquad (7.37)$$

## EXAMPLE 7.3

**Objective:** Calculate the corner frequency and maximum gain of a bipolar common-emitter circuit with a coupling capacitor.

For the circuit shown in Figure 7.21, the parameters are: $R_1 = 51.2\ k\Omega$, $R_2 = 9.6\ k\Omega$, $R_C = 2\ k\Omega$, $R_E = 0.4\ k\Omega$, $R_{Si} = 0.1\ k\Omega$, $C_C = 1\ \mu F$, and $V_{CC} = 10\ V$. The transistor parameters are: $V_{BE}(\text{on}) = 0.7\ V$, $\beta = 100$, and $V_A = \infty$.

**Solution:** From a dc analysis, the quiescent collector current is $I_{CQ} = 1.81\ mA$. The small-signal parameters are $g_m = 69.6\ mA/V$ and $r_\pi = 1.44\ k\Omega$.

The input resistance is

$$R_i = R_1 \| R_2 \| [r_\pi + (1 + \beta)R_E]$$
$$= 51.2 \| 9.6 \| [1.44 + (101)(0.4)] = 6.77\ k\Omega$$

and the time constant is therefore

$$\tau_S = (R_{Si} + R_i)C_C = (0.1 \times 10^3 + 6.77 \times 10^3)(1 \times 10^{-6}) \Rightarrow 6.87\ ms$$

The corner frequency is

$$f_L = \frac{1}{2\pi \tau_S} = \frac{1}{2\pi (6.87 \times 10^{-3})} = 23.2\ Hz$$

Finally, the maximum voltage gain magnitude is

$$|A_v|_{max} = \frac{g_m r_\pi R_C}{(R_{Si} + R_i)} \left( \frac{R_B}{R_B + R_{ib}} \right)$$

where

$$R_{ib} = r_\pi + (1 + \beta)R_E = 1.44 + (101)(0.4) = 41.8\ k\Omega$$

Therefore,

$$|A_v|_{max} = \frac{(69.6)(1.44)(2)}{(0.1 + 6.775)} \left( \frac{8.084}{8.084 + 41.84} \right) = 4.72$$

**Comment:** The coupling capacitor produces a high-pass network. In this circuit, if the signal frequency is approximately two octaves above the corner frequency, the coupling capacitor acts as a short circuit.

EXERCISE PROBLEM

**Ex 7.3:** For the circuit shown in Figure 7.21(a), the parameters are: $V_{CC} = 3$ V, $R_{Si} = 0$, $R_1 = 110\,k\Omega$, $R_2 = 42\,k\Omega$, $R_E = 0.5\,k\Omega$, $R_C = 7\,k\Omega$, and $C_C = 0.47\,\mu F$. The transistor parameters are $\beta = 150$, $V_{BE}(on) = 0.7$ V, and $V_A = \infty$. (a) Determine the expression for and the value of the time constant $\tau_S$. (b) Determine the corner frequency and midband voltage gain. (Ans. (a) $\tau_S = R_i C_C = 10.87$ ms; (b) $f_L = 14.6$ Hz, $A_v = -10.84$)

**Time Constant Technique:** In general, we do not need to derive the complete circuit transfer function including capacitance effects in order to complete the Bode plot and determine the frequency response. By looking at a circuit with, initially, only one capacitor, we can determine if the amplifier is a low-pass or high-pass circuit. We can then specify the Bode plot if we know the time constant and the maximum midband gain. The time constant determines the corner frequency. The midband gain is found in the usual way when capacitances are eliminated from the circuit.

This time constant technique yields good results when all poles are real, as will be the case in this chapter. In addition, this technique does not determine the corner frequencies due to system zeros. The major benefit of using the time constant approach is that it gives information about which circuit elements affect the −3 dB frequency of the circuit. A coupling capacitor produces a high-pass network, so the form of the Bode plot will be the same as that shown in Figure 7.5. Also, the maximum gain is determined when the coupling capacitor acts as a short circuit, as was assumed in Chapters 4 and 6.

The time constant for the circuit is a function of the equivalent resistance seen by the capacitor. The small-signal equivalent circuit is shown in Figure 7.21(b). If we set the independent voltage source equal to zero, the equivalent resistance seen by the coupling capacitor $C_C$ is $(R_{Si} + R_i)$. The time constant is then

$$\tau_S = (R_{Si} + R_i)C_C \tag{7.38}$$

This is the same as Equation (7.35), which was determined by using a current–voltage analysis.

**Output Coupling Capacitor: Common-Source Circuit**

Figure 7.22(a) shows a common-source MOSFET amplifier. We assume that the resistance of the signal generator is much less than $R_G$ and can therefore be neglected. In this case, the output signal is connected to the load through a coupling capacitor.

The small-signal equivalent circuit, assuming $r_o$ is infinite, is shown in Figure 7.22(b). The maximum output voltage, assuming $C_C$ is a short circuit, is

$$|V_o|_{max} = g_m V_{gs}(R_D \| R_L) \tag{7.39}$$

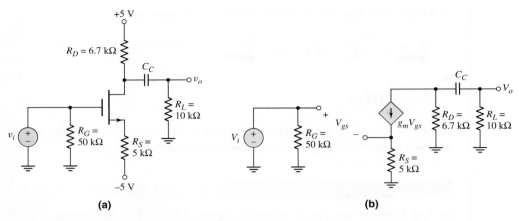

**Figure 7.22** (a) Common-source circuit with output coupling capacitor and (b) small-signal equivalent circuit

and the input voltage can be written as

$$V_i = V_{gs} + g_m R_S V_{gs} \tag{7.40}$$

Therefore, the maximum small-signal gain is

$$|A_v|_{max} = \frac{g_m(R_D \| R_L)}{1 + g_m R_S} \tag{7.41}$$

Even though the coupling capacitor is in the output portion of the circuit, the Bode plot will still be that of a high-pass network, as shown in Figure 7.5. Using the time constant technique to determine the corner frequency will substantially simplify the circuit analysis, since we do not specifically need to determine the transfer function for the frequency response.

The time constant is a function of the effective resistance seen by capacitor $C_C$, which is determined by setting all independent sources equal to zero. Since $V_i = 0$, then $V_{gs} = 0$ and $g_m V_{gs} = 0$, and the effective resistance seen by $C_C$ is $(R_D + R_L)$. The time constant is then

$$\tau_S = (R_D + R_L)C_C \tag{7.42}$$

and the corner frequency is $f_L = 1/2\pi\tau_S$.

## DESIGN EXAMPLE 7.4

**Objective:** The circuit in Figure 7.22(a) is to be used as a simple audio amplifier. Design the circuit such that the lower corner frequency is $f_L = 20$ Hz.

**Solution:** The corner frequency can be written in terms of the time constant, as follows:

$$f_L = \frac{1}{2\pi\tau_S}$$

The time constant is then

$$\tau_S = \frac{1}{2\pi f} = \frac{1}{2\pi(20)} \Rightarrow 7.96 \text{ ms}$$

Therefore, from Equation (7.42) the coupling capacitance is

$$C_C = \frac{\tau_S}{R_D + R_L} = \frac{7.96 \times 10^{-3}}{6.7 \times 10^3 + 10 \times 10^3} = 4.77 \times 10^{-7} \text{ F}$$

or

$$C_C = 0.477 \ \mu\text{F}$$

**Comment:** Using the time constant technique to find the corner frequency is substantially easier than using the circuit analysis approach.

### EXERCISE PROBLEM

**Ex 7.4:** Consider the circuit shown in Figure 7.22(a). The bias voltages are changed to $V^+ = 3 \text{ V}$ and $V^- = -3 \text{ V}$. Other circuit parameters are $R_L = 20 \text{ k}\Omega$ and $R_G = 100 \text{ k}\Omega$. The transistor parameters are $V_{TN} = 0.4 \text{ V}$, $K_n = 100 \ \mu\text{A/V}^2$, and $\lambda = 0$. (a) Design the circuit such that $I_{DQ} = 250 \ \mu\text{A}$ and $V_{DSQ} = 1.7 \text{ V}$. (b) If $C_C = 0.7 \ \mu\text{F}$, determine the corner frequency. (Ans. (a) $R_S = 4.08 \text{ k}\Omega$, $R_D = 13.1 \text{ k}\Omega$; (b) $f_L = 6.87 \text{ Hz}$)

**Output Coupling Capacitor: Emitter-Follower Circuit:** An emitter follower with a coupling capacitor in the output portion of the circuit is shown in Figure 7.23(a). We assume that coupling capacitor $C_{C1}$, which is part of the original emitter follower, is very large, and that it acts as a short circuit to the input signal.

The small-signal equivalent circuit, including the small-signal transistor resistance $r_o$, is shown in Figure 7.23(b). The equivalent resistance seen by the coupling capacitor $C_{C2}$ is $[R_o + R_L]$, and the time constant is

$$\tau_S = [R_o + R_L]C_{C2} \tag{7.43}$$

where $R_o$ is the output resistance as defined in Figure 7.23(b). As shown in Chapter 6, the output resistance is

$$R_o = R_E \| r_o \| \left\{ \frac{[r_\pi + (R_S \| R_B)]}{1 + \beta} \right\} \tag{7.44}$$

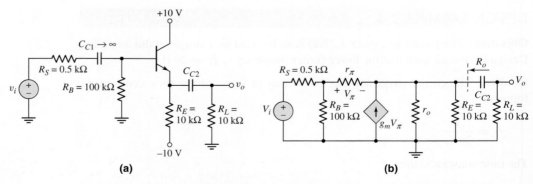

**Figure 7.23** (a) Emitter-follower circuit with output coupling capacitor and (b) small-signal equivalent circuit

If we combine Equations (7.44) and (7.43), the time constant expression becomes fairly complicated. However, the current–voltage analysis of this circuit including $C_{C2}$ is even more cumbersome. The time constant technique again simplifies the analysis substantially.

## EXAMPLE 7.5

**Objective:** Determine the 3 dB frequency of an emitter-follower amplifier circuit with an output coupling capacitor.

Consider the circuit shown in Figure 7.23(a) with transistor parameters $\beta = 100$, $V_{BE}(\text{on}) = 0.7$ V, and $V_A = 120$ V. The output coupling capacitance is $C_{C2} = 1\ \mu\text{F}$.

**Solution:** A dc analysis shows that $I_{CQ} = 0.838$ mA. Therefore, the small-signal parameters are: $r_\pi = 3.10\ \text{k}\Omega$, $g_m = 32.2$ mA/V, and $r_o = 143\ \text{k}\Omega$.

From Equation (7.44), the output resistance $R_o$ of the emitter follower is

$$R_o = R_E \| r_o \| \left\{ \frac{[r_\pi + (R_S \| R_B)]}{1 + \beta} \right\}$$

$$= 10 \| 143 \| \left\{ \frac{[3.10 + (0.5 \| 100)]}{101} \right\} = 10 \| 143 \| 0.0356\ \text{k}\Omega \cong 35.5\ \Omega$$

From Equation (7.43), the time constant is

$$\tau_S = [R_o + R_L]C_{C2} = [35.5 + 10^4](10^{-6}) \cong 1 \times 10^{-2}\ \text{s}$$

The 3 dB frequency is then

$$f_L = \frac{1}{2\pi \tau_S} = \frac{1}{2\pi (10^{-2})} = 15.9\ \text{Hz}$$

**Comment:** Determining the 3 dB or corner frequency is very direct with the time constant technique.

**Computer Verification:** Based on PSpice, analysis, Figure 7.24 is a Bode plot of the voltage gain magnitude of the emitter-follower circuit shown in Figure 7.23(a). The corner frequency is essentially identical to that obtained by the time constant technique. Also, the asymptotic value of the small-signal voltage gain is $A_v = 0.988$, as expected for an emitter-follower circuit.

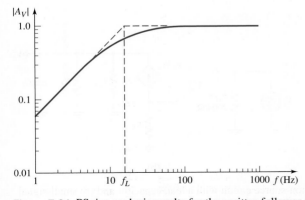

**Figure 7.24** PSpice analysis results for the emitter-follower circuit in Figure 7.23(a)

EXERCISE PROBLEM

**Ex 7.5:** For the emitter-follower circuit shown in Figure 7.23(a), determine the required value of $C_{C2}$ to yield a corner frequency of 10 Hz. (Ans. $C_{C2} = 1.59 \ \mu F$)

Problem-Solving Technique: Bode Plot of Gain Magnitude

1. For a particular capacitor in a circuit, determine whether the capacitor is producing a low-pass or high-pass circuit. From this, sketch the general shape of the Bode plot.
2. The corner frequency is found from $f = 1/2\pi\tau$ where the time constant is $\tau = R_{eq}C$. The equivalent resistance $R_{eq}$ is the equivalent resistance seen by the capacitor.
3. The maximum gain magnitude is the midband gain. Coupling and bypass capacitors act as short circuits and load capacitors act as open circuits.

### 7.3.2   Load Capacitor Effects

An amplifier output may be connected to a load or to the input or another amplifier. The model of the load circuit input impedance is generally a capacitance in parallel with a resistance. In addition, there is a parasitic capacitance between ground and the line that connects the amplifier output to the load circuit.

Figure 7.25(a) shows a MOSFET common-source amplifier with a load resistance $R_L$ and a load capacitance $C_L$ connected to the output, and Figure 7.25(b) shows the small-signal equivalent circuit. The transistor small-signal output resistance $r_o$ is assumed to be infinite. This circuit configuration is essentially the same as that in Figure 7.3, which is a low-pass network. At high frequencies, the impedance of $C_L$ decreases and acts as a shunt between the output and ground, and the output voltage tends toward zero. The Bode plot is similar to that shown in Figure 7.8, with an upper corner frequency and a maximum gain asymptote.

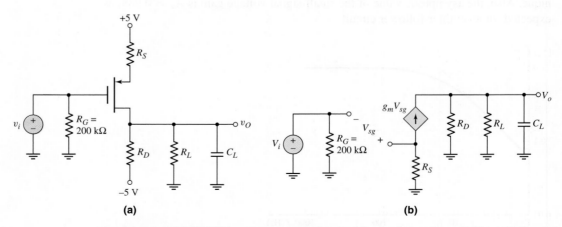

**(a)**                                                        **(b)**

**Figure 7.25** (a) MOSFET common-source circuit with a load capacitor and (b) small-signal equivalent circuit

The equivalent resistance seen by the load capacitor $C_L$ is $R_D \| R_L$. Since we set $V_i = 0$, then $g_m V_{sg} = 0$, which means that the dependent current source does not affect the equivalent resistance.

The time constant for this circuit is

$$\tau_P = (R_D \| R_L) C_L \tag{7.45}$$

The maximum gain asymptote, which is found by assuming $C_L$ is an open circuit, is

$$|A_v|_{max} = \frac{g_m (R_D \| R_L)}{1 + g_m R_S} \tag{7.46}$$

### 7.3.3  Coupling and Load Capacitors

A circuit with both a coupling capacitor and a load capacitor is shown in Figure 7.26(a). Since the values of the coupling capacitance and load capacitance differ by orders of magnitude, the corner frequencies are far apart and can be treated separately as discussed previously. The small-signal equivalent circuit is shown in Figure 7.26(b), assuming the transistor small-signal resistance $r_o$ is infinite.

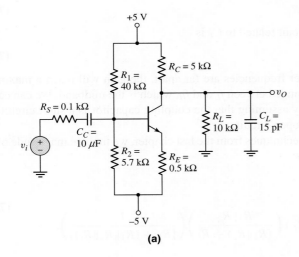

**(a)**

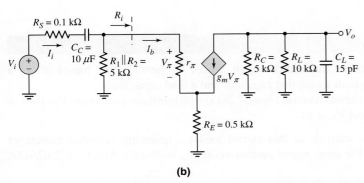

**(b)**

**Figure 7.26** (a) Circuit with both a coupling and a load capacitor and (b) small-signal equivalent circuit

The Bode plot of the voltage gain magnitude is similar to that shown in Figure 7.11. The lower corner frequency $f_L$ is given by

$$f_L = \frac{1}{2\pi\tau_S} \tag{7.47}$$

where $\tau_S$ is the time constant associated with the coupling capacitor $C_C$, and the upper corner frequency $f_H$ is given by

$$f_H = \frac{1}{2\pi\tau_P} \tag{7.48}$$

where $\tau_P$ is the time constant associated with the load capacitor $C_L$. It should be emphasized that Equations (7.47) and (7.48) are valid only as long as the two corner frequencies are far apart.

Using the small-signal equivalent circuit in Figure 7.26(b), we set the signal source equal to zero to find the equivalent resistance associated with the coupling capacitor. The related time constant is

$$\tau_S = [R_S + (R_1\|R_2\|R_i)C_C \tag{7.49}$$

where

$$R_i = r_\pi + (1 + \beta)R_E \tag{7.50}$$

Similarly, the time constant related to $C_L$ is

$$\tau_P = (R_C\|R_L)C_L \tag{7.51}$$

Since the two corner frequencies are far apart, the gain will reach a maximum value in the frequency range between $f_L$ and $f_H$, which is the midband. We can calculate the midband gain by assuming that the coupling capacitor is a short circuit and the load capacitor is an open circuit.

Using the analysis techniques from the last chapter, we find the magnitude of the midband gain as follows:

$$|A_v| = \left|\frac{V_o}{V_i}\right|$$

$$= g_m r_\pi (R_C\|R_L)\left(\frac{R_1\|R_2}{(R_1\|R_2) + R_i}\right)\left(\frac{1}{[R_S + (R_1\|R_2\|R_i)]}\right) \tag{7.52}$$

---

## EXAMPLE 7.6

**Objective:** Determine the midband gain, corner frequencies, and bandwidth of a circuit containing both a coupling capacitor and a load capacitor.

Consider the circuit shown in Figure 7.26(a) with transistor parameters $V_{BE}(\text{on}) = 0.7\,\text{V}$, $\beta = 100$, and $V_A = \infty$.

**Solution:** The dc analysis of this circuit yields a quiescent collector current of $I_{CQ} = 0.99\,\text{mA}$. The small-signal parameters are $g_m = 38.08\,\text{mA/V}$, $r_\pi = 2.626\,\text{k}\Omega$, and $r_o = \infty$.

The input resistance $R_i$ is therefore

$$R_i = r_\pi + (1 + \beta)R_E = 2.63 + (101)(0.5) = 53.1\,\text{k}\Omega$$

From Equation (7.52), the midband gain is

$$|A_v|_{\max} = \left|\frac{V_o}{V_i}\right|_{\max} = g_m r_\pi (R_C \| R_L)\left(\frac{R_1 \| R_2}{(R_1 \| R_2) + R_i}\right)\left(\frac{1}{[R_S + (R_1 \| R_2 \| R_i)]}\right)$$

$$= (38.08)(2.626)(5\|10)\left(\frac{40\|5.7}{(40\|5.7) + 53.1}\right)\left(\frac{1}{[0.1 + (40\|5.7\|53.1)]}\right)$$

or

$$|A_v|_{\max} = 6.14$$

The time constant $\tau_S$ is

$$\tau_S = (R_S + R_1\|R_2\|R_i)C_C$$
$$= (0.1 \times 10^3 + (5.7 \times 10^3)\|(40 \times 10^3)\|(53.1 \times 10^3))(10 \times 10^{-6})$$
$$= 4.67 \times 10^{-2}\ \text{s} \Rightarrow 46.6\ \text{ms}$$

and the time constant $\tau_P$ is

$$\tau_P = (R_C \| R_L)C_L = ((5 \times 10^3)\|(10 \times 10^3))(15 \times 10^{-12}) = 5 \times 10^{-8}\ \text{s}$$

or

$$\tau_P = 50\ \text{ns}$$

The lower corner frequency is

$$f_L = \frac{1}{2\pi \tau_S} = \frac{1}{2\pi(46.6 \times 10^{-3})} = 3.42\ \text{Hz}$$

and the upper corner frequency is

$$f_H = \frac{1}{2\pi \tau_P} = \frac{1}{2\pi(50 \times 10^{-9})} \Rightarrow 3.18\ \text{MHz}$$

Finally, the bandwidth is

$$f_{BW} = f_H - f_L = 3.18\ \text{MHz} - 3.4\ \text{Hz} \cong 3.18\ \text{MHz}$$

**Comment:** The two corner frequencies differ by approximately six orders of magnitude; therefore, considering one capacitor at a time is a valid approach.

## EXERCISE PROBLEM

**Ex 7.6:** Consider the circuit in Figure 7.26(a). The load resistance value is changed to $R_L = 5\ \text{k}\Omega$, and the capacitor values are changed to $C_L = 5$ pF and $C_C = 5\ \mu\text{F}$. Other circuit and transistor parameters are the same as given in Example 7.6. (a) Determine the new values of the collector current and small-signal hybrid-pi parameters. (b) Determine the value of midband voltage gain. (c) Find the corner frequencies of the circuit. (Ans. (a) $I_{CQ} = 0.986$ mA, (b) $A_v = -4.60$, (c) $f_L = 6.82$ Hz, $f_H = 12.7$ MHz)

A figure of merit for an amplifier is the **gain–bandwidth product.** Assuming the corner frequencies are far apart, the bandwidth is

$$f_{BW} = f_H - f_L \cong f_H \tag{7.53}$$

and the maximum gain is $|A_v|_{max}$. The gain–bandwidth product is therefore

$$GB = |A_v|_{max} \cdot f_H \tag{7.54}$$

Later we will show that, for a given load capacitance, this product is essentially a constant. We will also describe how trade-offs must be made between gain and bandwidth in amplifier design.

### 7.3.4    Bypass Capacitor Effects

In bipolar and FET discrete amplifiers, emitter and source bypass capacitors are often included so that emitter and source resistors can be used to stabilize the $Q$-point without sacrificing the small-signal gain. The bypass capacitors are assumed to act as short circuits at the signal frequency. However, to guide us in choosing a bypass capacitor, we must determine the circuit response in the frequency range where these capacitors are neither open nor short circuits.

Figure 7.27(a) shows a common-emitter circuit with an emitter bypass capacitor. The small-signal equivalent circuit is shown in Figure 7.27(b). We can find the small-signal voltage gain as a function of frequency. Using the impedance reflection rule, the small-signal input current is

$$I_b = \frac{V_i}{R_S + r_\pi + (1 + \beta)\left(R_E \left\| \dfrac{1}{sC_E}\right.\right)} \tag{7.55}$$

The total impedance in the emitter is multiplied by the factor $(1 + \beta)$. The control voltage is

$$V_\pi = I_b r_\pi \tag{7.56}$$

and the output voltage is

$$V_o = -g_m V_\pi R_C \tag{7.57}$$

Combining equations produces the small-signal voltage gain, as follows:

$$A_v(s) = \frac{V_o(s)}{V_i(s)} = \frac{-g_m r_\pi R_C}{R_S + r_\pi + (1 + \beta)\left(R_E \left\| \dfrac{1}{sC_E}\right.\right)} \tag{7.58}$$

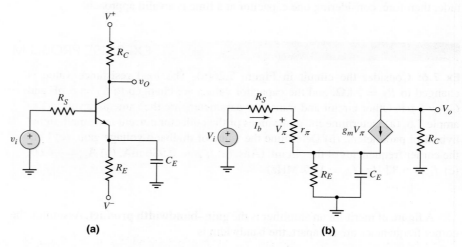

**(a)**                                    **(b)**

**Figure 7.27** (a) Circuit with emitter bypass capacitor and (b) small-signal equivalent circuit

Expanding the parallel combination of $R_E$ and $1/sC_E$ and rearranging terms, we find

$$A_v = \frac{-g_m r_\pi R_C}{[R_S + r_\pi + (1+\beta)R_E]} \times \frac{(1 + sR_E C_E)}{\left\{1 + \dfrac{sR_E(R_S + r_\pi)C_E}{[R_S + r_\pi + (1+\beta)R_E]}\right\}} \qquad (7.59)$$

Equation (7.59) can be written in terms of time constants as

$$A_v = \frac{-g_m r_\pi R_C}{[R_S + r_\pi + (1+\beta)R_E]} \left\{\frac{1 + s\tau_A}{1 + s\tau_B}\right\} \qquad (7.60)$$

The form of this transfer function is somewhat different from what we have previously encountered in that we have both a zero and a pole.

The Bode plot of the voltage gain magnitude has two limiting horizontal asymptotes. If we set $s = j\omega$, we can then consider the limit as $\omega \to 0$ and the limit as $\omega \to \infty$. For $\omega \to 0$, $C_E$ acts as an open circuit; for $\omega \to \infty$, $C_E$ acts as a short circuit. From Equation (7.59), we have

$$|A_v|_{\omega \to 0} = \frac{g_m r_\pi R_C}{[R_S + r_\pi + (1+\beta)R_E]} \qquad (7.61(a))$$

and

$$|A_v|_{\omega \to \infty} = \frac{g_m r_\pi R_C}{R_S + r_\pi} \qquad (7.61(b))$$

From these results, we see that for $\omega \to 0$, $R_E$ is included in the gain expression, and for $\omega \to \infty$, $R_E$ is not part of the gain expression, since it has been effectively shorted out by $C_E$.

If we assume that the time constants $\tau_A$ and $\tau_B$ in Equation (7.60) differ substantially in magnitude, then the corner frequency due to $\tau_B$ is

$$f_B = \frac{1}{2\pi \tau_B} \qquad (7.62(a))$$

and the corner frequency due to $\tau_A$ is

$$f_A = \frac{1}{2\pi \tau_A} \qquad (7.62(b))$$

The resulting Bode plot of the voltage gain magnitude is shown in Figure 7.28.

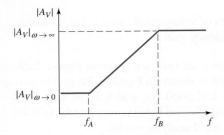

**Figure 7.28** Bode plot of the voltage gain magnitude for the circuit with an emitter bypass capacitor

## EXAMPLE 7.7

**Objective:** Determine the corner frequencies and limiting horizontal asymptotes of a common-emitter circuit with an emitter bypass capacitor.

Consider the circuit in Figure 7.27(a) with parameters $R_E = 4$ k$\Omega$, $R_C = 2$ k$\Omega$, $R_S = 0.5$ k$\Omega$, $C_E = 1$ $\mu$F, $V^+ = 5$ V, and $V^- = -5$ V. The transistor parameters are: $\beta = 100$, $V_{BE}$(on) $= 0.7$ V, and $r_o = \infty$.

**Solution:** From the dc analysis, we find the quiescent collector current as $I_{CQ} = 1.06$ mA. The small-signal parameters are $g_m = 40.77$ mA/V, $r_\pi = 2.45$ k$\Omega$, and $r_o = \infty$.

The time constant $\tau_A$ is

$$\tau_A = R_E C_E = (4 \times 10^3)(1 \times 10^{-6}) = 4 \times 10^{-3} \text{ s}$$

and the time constant $\tau_B$ is

$$\tau_B = \frac{R_E(R_S + r_\pi)C_E}{[R_S + r_\pi + (1 + \beta)R_E]}$$

$$= \frac{(4 \times 10^3)(0.5 \times 10^3 + 2.45 \times 10^3)(1 \times 10^{-6})}{[0.5 \times 10^3 + 2.45 \times 10^3 + (101)(4 \times 10^3)]}$$

or

$$\tau_B = 2.90 \times 10^{-5} \text{ s}$$

The corner frequencies are then

$$f_A = \frac{1}{2\pi \tau_A} = \frac{1}{2\pi(4 \times 10^{-3})} = 39.8 \text{ Hz}$$

and

$$f_B = \frac{1}{2\pi \tau_B} = \frac{1}{2\pi(2.9 \times 10^{-5})} \Rightarrow 5.49 \text{ kHz}$$

The limiting low-frequency horizontal asymptote, given by Equation (7.61(a)) is

$$|A_v|_{\omega \to 0} = \frac{g_m r_\pi R_C}{[R_S + r_\pi + (1 + \beta)R_E]} = \frac{(40.8)(2.45)(2)}{[0.5 + 2.45 + (101)(4)]} = 0.491$$

The limiting high-frequency horizontal asymptote, given by Equation (7.61(b)) is

$$|A_v|_{\omega \to \infty} = \frac{g_m r_\pi R_C}{R_S + r_\pi} = \frac{(40.77)(2.45)(2)}{0.5 + 2.45} = 67.7$$

**Comment:** Comparing the two limiting values of voltage gain, we see that including a bypass capacitor produces a large high-frequency gain.

**Computer Verification:** The results of a PSpice analysis are given in Figure 7.29. The magnitude of the small-signal voltage gain is shown in Figure 7.29(a). The two corner frequencies are approximately 39 Hz and 5600 Hz, which agree very well with the results from the time constant analysis. The two limiting magnitudes of 0.49 and 68 also correlate extremely well with the hand analysis results.

Figure 7.29(b) is a plot of the phase response versus frequency. At very low and very high frequencies, where the capacitor acts as either an open circuit or short

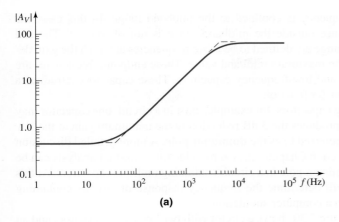

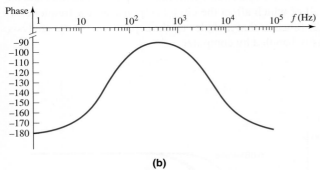

**Figure 7.29** PSpice analysis results for the circuit with an emitter bypass capacitor: (a) voltage gain magnitude response and (b) phase response

circuit, the phase is $-180$ degrees, as expected for a common-emitter circuit. Between the two corner frequencies, the phase changes substantially, approaching $-90$ degrees.

## EXERCISE PROBLEM

**Ex 7.7:** The circuit shown in Figure 7.27(a) has parameters $V^+ = 10$ V, $V^- = -10$ V, $R_S = 0.5$ k$\Omega$, $R_E = 4$ k$\Omega$, and $R_C = 2$ k$\Omega$. The transistor parameters are: $V_{BE}(\text{on}) = 0.7$ V, $V_A = \infty$, and $\beta = 100$. (a) Determine the value of $C_E$ such that the low-frequency 3 dB point is $f_B = 200$ Hz. (b) Using the results from part (a), determine $f_A$. (Ans. (a) $C_E = 49.5$ $\mu$F, (b) $f_A = 0.80$ Hz)

The analysis of an FET amplifier with a source bypass capacitor is essentially the same as for the bipolar circuit. The general form of the voltage gain expression is the same as Equation (7.60), and the Bode plot of the gain is essentially the same as that shown in Figure 7.28.

### 7.3.5    Combined Effects: Coupling and Bypass Capacitors

When a circuit contains multiple capacitors, the frequency response analysis becomes more complex. In many amplifier applications, the circuit is to amplify an input signal whose frequency is confined to the midband range. In this case, the actual frequency response outside the midband range is not of interest. The end points of the midband range are defined to be those frequencies at which the gain decreases by 3 dB from the maximum midband value. These endpoint frequencies are a function of the high- and low-frequency capacitors. These capacitors introduce a pole to the amplifier transfer function.

If multiple coupling capacitors, for example, exist in a circuit, one capacitor may introduce the pole that produces the 3 dB reduction in the maximum gain at the low frequency. This pole is referred to as the **dominant pole.** A more detailed discussion of dominant poles is given in Chapter 12. A zero-value time constant analysis can be used to estimate the dominant pole in a circuit containing multiple capacitors. At this point in the text, we will determine the frequency response of circuits containing multiple capacitors with a computer simulation.

As an example, Figure 7.30 shows a circuit with two coupling capacitors and an emitter bypass capacitor, all of which affect the circuit response at low frequencies. We could develop a transfer function that includes all the capacitors, but the analysis of such circuits is most easily handled by computer.

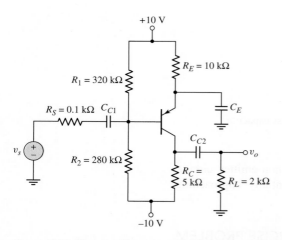

**Figure 7.30** Circuit with two coupling capacitors and an emitter bypass capacitor

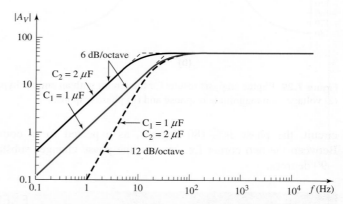

**Figure 7.31** PSpice results for each coupling capacitor, and the combined effect for the circuit in Figure 7.30 ($C_E \rightarrow \infty$)

Figure 7.31 is the Bode plot of the voltage gain magnitude for the example circuit, taking into account the effects of the two coupling capacitors. In this case, the bypass capacitor is assumed to be a short circuit. The plots consider $C_1$ and $C_2$ individually, as well as together. As expected, with two capacitors both acting at the same time, the slope is 40 dB/decade or 12 dB/octave. Since the poles are not far apart, in the actual circuit, we cannot consider the effect of each capacitor individually.

Figure 7.32 is the Bode plot of the voltage gain magnitude, taking into account the emitter bypass capacitor and the two coupling capacitors. The plot shows the effect of the bypass capacitor, the effect of the two coupling capacitors, and the net effect of the three capacitors together. When all three capacitors are taken into account, the slope is continually changing; there is no definitive corner frequency. However, at approximately $f = 150$ Hz, the curve is 3 dB below the maximum asymptotic

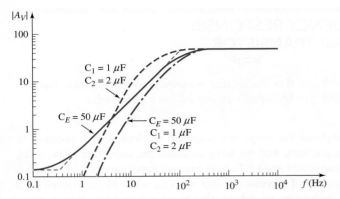

**Figure 7.32** PSpice results for the two coupling capacitors, the bypass capacitor, and the combined effects

value, and this frequency is defined as the **lower corner frequency,** or **lower cutoff frequency.**

## Test Your Understanding

**TYU 7.4** Consider the common-base circuit shown in Figure 7.33. Can the two coupling capacitors be treated separately? (a) From a computer analysis, determine the cutoff frequency. Assume the parameter values are $\beta = 100$ and $I_S = 2 \times 10^{-15}$ A. (b) Determine the midband small-signal voltage gain. (Ans. (a) $f_{3\,db} = 1.2$ kHz, (b) $A_v = 118$)

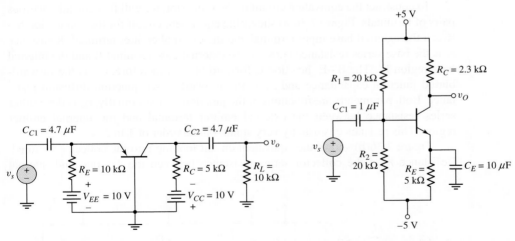

**Figure 7.33** Figure for Exercise TYU 7.4

**Figure 7.34** Figure for Exercise TYU 7.5

**TYU 7.5** The common-emitter circuit shown in Figure 7.34 contains both a coupling capacitor and an emitter bypass capacitor. (a) From a computer analysis, determine the 3 dB frequency. Assume the parameter values are $\beta = 100$ and $I_S = 2 \times 10^{-15}$ A. (b) Determine the midband small-signal voltage gain. (Ans. (a) $f_{3\,dB} \approx 575$ Hz, (b) $|A_v|_{max} = 74.4$)

📐 7.4 **FREQUENCY RESPONSE: BIPOLAR TRANSISTOR**

**Objective:** • Determine the frequency response of the bipolar transistor, and determine the Miller effect and Miller capacitance.

Thus far, we have considered the frequency response of circuits as a function of external resistors and capacitors, and we have assumed the transistor to be ideal. However, both bipolar transistors and FETs have internal capacitances that influence the high-frequency response of circuits. In this section, we will first develop an expanded small-signal hybrid-$\pi$ model of the bipolar transistor, taking these capacitances into account. We will then use this model to analyze the frequency characteristics of the bipolar transistor.

### 7.4.1    Expanded Hybrid-$\pi$ Equivalent Circuit

When a bipolar transistor is used in a linear amplifier circuit, the transistor is biased in the forward-active region, and small sinusoidal voltages and currents are superimposed on the dc voltages and currents. Figure 7.35(a) shows an npn bipolar transistor in a common-emitter configuration, along with the small-signal voltages and currents. Figure 7.35(b) is a cross section of the npn transistor in a classic integrated circuit configuration. The C, B, and E terminals are the external connections to the transistor, and the C′, B′, and E′ points are the idealized internal collector, base, and emitter regions.

To construct the equivalent circuit of the transistor, we will first consider various pairs of terminals. Figure 7.36(a) shows the equivalent circuit for the connection between the external base input terminal and the external emitter terminal. Resistance $r_b$ is the base series resistance between the external base terminal B and the internal base region B′. The B′–E′ junction is forward biased; therefore, $C_\pi$ is the forward-biased junction capacitance and $r_\pi$ is the forward-biased junction diffusion resistance. Both parameters are functions of the junction current. Finally, $r_{ex}$ is the emitter series resistance between the external emitter terminal and the internal emitter region. This resistance is usually very small, on the order of 1 to 2 $\Omega$.

Figure 7.36(b) shows the equivalent circuit looking into the collector terminal. Resistance $r_c$ is the collector series resistance between the external and internal

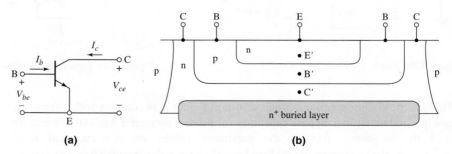

**Figure 7.35** (a) Common-emitter npn bipolar transistor with small-signal currents and voltages and (b) cross section of an npn bipolar transistor, for the hybrid-$\pi$ model

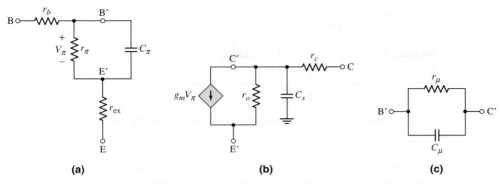

**(a)**                    **(b)**                    **(c)**

**Figure 7.36** Components of the hybrid-$\pi$ equivalent circuit: (a) base to emitter, (b) collector to emitter, and (c) base to collector

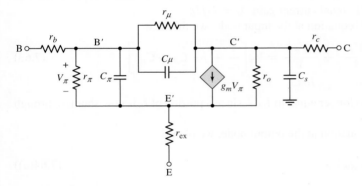

**Figure 7.37** Hybrid-$\pi$ equivalent circuit

collector connections, and capacitance $C_s$ is the junction capacitance of the reverse-biased collector–substrate junction. The dependent current source, $g_m V_\pi$, is the transistor collector current controlled by the internal base–emitter voltage. Resistance $r_o$ is the inverse of the output conductance $g_o$ and is due primarily to the Early effect.

Finally, Figure 7.36(c) shows the equivalent circuit of the reverse-biased B′–C′ junction. Capacitance $C_\mu$ is the reverse-biased junction capacitance, and $r_\mu$ is the reverse-biased diffusion resistance. Normally, $r_\mu$ is on the order of megohms and can be neglected. The value of $C_\mu$ is usually much smaller than $C_\pi$; however, because of a phenomenon known as the Miller effect, $C_\mu$ usually cannot be neglected. (We will consider the Miller effect later in this chapter.)

The complete hybrid-$\pi$ equivalent circuit for the bipolar transistor is shown in Figure 7.37. The capacitances lead to frequency effects in the transistor. One result is that the gain is a function of the input signal frequency. Because of the large number of elements, a computer simulation of this complete model is easier than a hand analysis. However, we can make some simplifications in order to evaluate some fundamental frequency effects of bipolar transistors.

## 7.4.2  Short-Circuit Current Gain

We can begin to understand the frequency effects of the bipolar transistor by first determining the **short-circuit current gain,** after simplifying the hybrid-$\pi$ model.

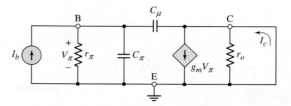

**Figure 7.38** Simplified hybrid-$\pi$ equivalent circuit for determining the short-circuit current gain

Figure 7.38 shows a simplified equivalent circuit for the transistor, in which we neglect the parasitic resistances $r_b$, $r_c$, and $r_{ex}$, the B–C diffusion resistance $r_\mu$, and the substrate capacitance $C_s$. Also, the collector is connected to signal ground. Keep in mind that the transistor must still be biased in the forward-active region. We will determine the small-signal current gain $A_i = I_c/I_b$.

Writing a KCL equation at the input node, we find that

$$I_b = \frac{V_\pi}{r_\pi} + \frac{V_\pi}{\dfrac{1}{j\omega C_\pi}} + \frac{V_\pi}{\dfrac{1}{j\omega C_\mu}} = V_\pi\left[\frac{1}{r_\pi} + j\omega(C_\pi + C_\mu)\right] \tag{7.63}$$

We see that $V_\pi$ is no longer equal to $I_b r_\pi$, since a portion of $I_b$ is now shunted through $C_\pi$ and $C_\mu$.

From a KCL equation at the output node, we obtain

$$\frac{V_\pi}{\dfrac{1}{j\omega C_\mu}} + I_c = g_m V_\pi \tag{7.64(a)}$$

or

$$I_c = V_\pi(g_m - j\omega C_\mu) \tag{7.64(b)}$$

The input voltage $V_\pi$ can then be written as

$$V_\pi = \frac{I_c}{(g_m - j\omega C_\mu)} \tag{7.64(c)}$$

Substituting this expression for $V_\pi$ into Equation (7.63) yields

$$I_b = I_c \cdot \frac{\left[\dfrac{1}{r_\pi} + j\omega(C_\pi + C_\mu)\right]}{(g_m - j\omega C_\mu)} \tag{7.65}$$

The small-signal current gain usually designated as $h_{fe}$, becomes

$$A_i = \frac{I_c}{I_b} = h_{fe} = \frac{(g_m - j\omega C_\mu)}{\left[\dfrac{1}{r_\pi} + j\omega(C_\pi + C_\mu)\right]} \tag{7.66}$$

If we assume typical circuit parameter values of $C_\mu = 0.05$ pF, $g_m = 50$ mA/V, and a maximum frequency of $f = 500$ MHz, then we see that $\omega C_\mu \ll g_m$. Therefore, to a good approximation, the small-signal current gain is

$$h_{fe} \cong \frac{g_m}{\left[\dfrac{1}{r_\pi} + j\omega(C_\pi + C_\mu)\right]} = \frac{g_m r_\pi}{1 + j\omega r_\pi(C_\pi + C_\mu)} \tag{7.67}$$

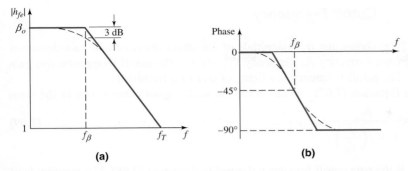

**Figure 7.39** Bode plots for the short-circuit current gain: (a) magnitude and (b) phase

Since $g_m r_\pi = \beta$, then the low frequency current gain is just $\beta$, as we previously assumed. Equation (7.67) shows that, in a bipolar transistor, the magnitude and phase of the current gain are both functions of the frequency.

Figure 7.39(a) is a Bode plot of the short-circuit current gain magnitude. The corner frequency, which is also the **beta cutoff frequency** $f_\beta$ in this case, is given by

$$f_\beta = \frac{1}{2\pi r_\pi (C_\pi + C_\mu)} \qquad (7.68)$$

Figure 7.39(b) shows the phase of the current gain. As the frequency increases, the small-signal collector current is no longer in phase with the small-signal base current. At high frequencies, the collector current lags the input current by 90 degrees.

## EXAMPLE 7.8

**Objective:** Determine the 3 dB frequency of the short-circuit current gain of a bipolar transistor.

Consider a bipolar transistor with parameters $r_\pi = 2.6\,\text{k}\Omega$, $C_\pi = 0.5\,\text{pF}$, and $C_\mu = 0.025\,\text{pF}$.

**Solution:** From Equation (7.68), we find

$$f_\beta = \frac{1}{2\pi r_\pi (C_\pi + C_\mu)} = \frac{1}{2\pi (2.6 \times 10^3)(0.5 + 0.025)(10^{-12})}$$

or

$$f_\beta = 117\,\text{MHz}$$

**Comment:** High-frequency transistors must have small capacitances; therefore, small devices must be used.

## EXERCISE PROBLEM

**Ex 7.8:** A bipolar transistor has parameters $\beta_o = 120$, $C_\mu = 0.02\,\text{pF}$, and $f_\beta = 90\,\text{MHz}$ and is biased at $I_{CQ} = 0.2\,\text{mA}$. (a) Determine $C_\pi$. (b) What is the magnitude of the short-circuit current gain at (i) $f = 50\,\text{MHz}$, (ii) $f = 125\,\text{MHz}$, and (iii) $f = 500\,\text{MHz}$. (Ans. (a) $C_\pi = 0.093\,\text{pF}$; (b) (i) 105, (ii) 70.1, (iii) 21.3)

### 7.4.3    Cutoff Frequency

Figure 7.39(a) shows that the magnitude of the small-signal current gain decreases with increasing frequency. At frequency $f_T$, which is the **cutoff frequency,** this gain goes to 1. The cutoff frequency is a figure of merit for transistors.

From Equation (7.67), we can write the small-signal current gain in the form

$$h_{fe} = \frac{\beta_o}{1 + j\left(\dfrac{f}{f_\beta}\right)} \tag{7.69}$$

where $f_\beta$ is the beta cutoff frequency defined by Equation (7.68). The magnitude of $h_{fe}$ is

$$|h_{fe}| = \frac{\beta_o}{\sqrt{1 + \left(\dfrac{f}{f_\beta}\right)^2}} \tag{7.70}$$

At the cutoff frequency $f_T$, $|h_{fe}| = 1$, and Equation (7.70) becomes

$$|h_{fe}| = 1 = \frac{\beta_o}{\sqrt{1 + \left(\dfrac{f_T}{f_\beta}\right)^2}} \tag{7.71}$$

Normally, $\beta_o \gg 1$, which implies that $f_T \gg f_\beta$. Then Equation (7.71) can be written as

$$1 \cong \frac{\beta_o}{\sqrt{\left(\dfrac{f_T}{f_\beta}\right)^2}} = \frac{\beta_o f_\beta}{f_T} \tag{7.72(a)}$$

or

$$f_T = \beta_o f_\beta \tag{7.72(b)}$$

Frequency $f_\beta$ is also called the bandwidth of the transistor. Therefore, from Equation (7.72(b)), the cutoff frequency $f_T$ is the gain–bandwidth product of the transistor, or more commonly the **unity-gain bandwidth.** From Equation (7.68), the unity-gain bandwidth is

$$f_T = \beta_o \left[ \frac{1}{2\pi r_\pi (C_\pi + C_\mu)} \right] = \frac{g_m}{2\pi (C_\pi + C_\mu)} \tag{7.73}$$

Since the capacitances are a function of the size of the transistor, we see again that high frequency transistors imply small device sizes.

The cutoff frequency $f_T$ is also a function of the dc collector current $I_C$, and the general characteristic of $f_T$ versus $I_C$ is shown in Figure 7.40. The transconductance $g_m$ is directly proportional to $I_C$, but only a portion of $C_\pi$ is related to $I_C$. The cutoff frequency is therefore lower at low collector current levels. However, the cutoff frequency also decreases at high current levels, in the same way that $\beta$ decreases at large currents.

The cutoff frequency or unity-gain bandwidth of a transistor is usually specified on the device data sheets. Since the low-frequency current gain is also given, the beta cutoff frequency, or bandwidth, of the transistor can be determined from

$$f_\beta = \frac{f_T}{\beta_o} \tag{7.74}$$

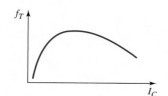

**Figure 7.40** Cutoff frequency versus collector current

The cutoff frequency of the general-purpose 2N2222A discrete bipolar transistor is $f_T = 300$ MHz. For the MSC3130 discrete bipolar transistor, which has a special surface mount package, the cutoff frequency is $f_T = 1.4$ GHz. This tells us that very small transistors fabricated in integrated circuits can have cutoff frequencies in the low GHz range.

## EXAMPLE **7.9**

**Objective:** Calculate the bandwidth $f_\beta$ and capacitance $C_\pi$ of a bipolar transistor.

Consider a bipolar transistor that has parameters $f_T = 20$ GHz at $I_C = 1$ mA, $\beta_o = 120$, and $C_\mu = 0.08$ pF.

**Solution:** From Equation (7.74), we find the bandwidth as

$$f_\beta = \frac{f_T}{\beta_o} = \frac{20 \times 10^9}{120} \rightarrow f_\beta = 167 \text{ MHz}$$

The transconductance is

$$g_m = \frac{I_C}{V_T} = \frac{1}{0.026} = 38.46 \text{ mA/V}$$

The $C_\pi$ capacitance is determined from Equation (7.73). We have

$$f_T = \frac{g_m}{2\pi (C_\pi + C_\mu)}$$

or

$$20 \times 10^9 = \frac{38.5 \times 10^{-3}}{2\pi (C_\pi + 0.08 \times 10^{-12})}$$

which yields $C_\pi = 0.226$ pF.

**Comment:** Although the value of $C_\pi$ may be much larger than that of $C_\mu$, $C_\mu$ cannot be neglected in circuit applications as we will see in the next section.

## EXERCISE PROBLEM

**Ex 7.9:** A BJT is biased at $I_C = 0.15$ mA, and has parameters $\beta_o = 150$, $C_\pi = 0.8$ pF, and $C_\mu = 0.012$ pF. Determine $f_\beta$ and $f_T$. (Ans. $f_\beta = 7.54$ MHz, $f_T = 1.13$ GHz)

The hybrid-$\pi$ equivalent circuit for the bipolar transistor uses discrete or lumped elements. However, when cutoff frequencies are on the order of $f_T \cong 10$ GHz and the transistor is operated at microwave frequencies, other parasitic elements and distributed parameters must be included in the transistor model. For simplicity, we will assume in this text that the hybrid-$\pi$ model is sufficient to model the transistor characteristics up through the beta cutoff frequency.

### 7.4.4    Miller Effect and Miller Capacitance

As previously mentioned, the $C_\mu$ capacitance cannot in reality be ignored. The **Miller effect,** or feedback effect, is a multiplication effect of $C_\mu$ in circuit applications.

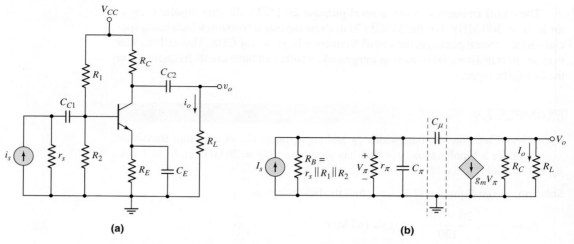

**Figure 7.41** (a) Common-emitter circuit with current source input; (b) small-signal equivalent circuit with simplified hybrid-$\pi$ model

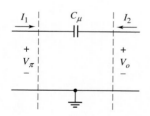

**Figure 7.42** Two-terminal network of capacitor $C_\mu$

Figure 7.41(a) is a common-emitter circuit with a signal current source at the input. We will determine the small-signal current gain $A_i = i_o/i_s$ of the circuit. Figure 7.41(b) is the small-signal equivalent circuit, assuming the frequency is sufficiently high for the coupling and bypass capacitors to act as short circuits. The transistor model is the simplified hybrid-$\pi$ circuit shown in Figure 7.38 (assuming $r_o = \infty$). Capacitor $C_\mu$ is a feedback element that connects the output back to the input. The output voltage and current will therefore influence the input characteristics.

The presence of $C_\mu$ complicates the analysis. Previously, we could write KCL equations at the input and output nodes and derive an expression for the current gain. Here, however, we will approach the problem differently. We will treat capacitor $C_\mu$ as a two-port network and will develop an equivalent circuit, with elements between the input base and ground and between the output collector and ground. This procedure may appear more complicated, but it will demonstrate the effect of $C_\mu$ more clearly.

Consider the circuit segment between the two dotted lines in Figure 7.41(b). We can treat this section as a two-port network, as shown in Figure 7.42. The input voltage is $V_\pi$ and the output voltage is $V_o$. Also, the input and output currents, $I_1$ and $I_2$, are defined as shown in the figure.

Writing KVL equations at the input and output terminals, we now have

$$V_\pi = I_1\left(\frac{1}{j\omega C_\mu}\right) + V_o \tag{7.75(a)}$$

and

$$V_o = I_2\left(\frac{1}{j\omega C_\mu}\right) + V_\pi \tag{7.75(b)}$$

Using Equations (7.75(a)) and (7.75(b)), we can form a two-port equivalent circuit, as shown in Figure 7.43(a). We then convert the Thevenin equivalent circuit on the output to a Norton equivalent circuit, as shown in Figure 7.43(b).

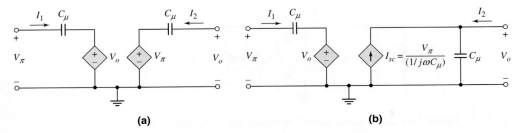

**Figure 7.43** (a) Two-port equivalent circuit of capacitor $C_\mu$ with equivalent output circuits: (a) Thevenin equivalent and (b) Norton equivalent

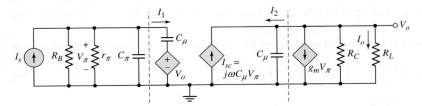

**Figure 7.44** Small-signal equivalent circuit, including the two-port equivalent model of capictor $C_\mu$

The equivalent circuit in Figure 7.43(b) replaces the circuit segment between the dotted lines in Figure 7.41(b), and the modified circuit is shown in Figure 7.44. To evaluate this circuit, we will make some simplifying approximations.

Typical values of $g_m$ and $C_\mu$ are $g_m = 50\text{ mA/V}$ and $C_\mu = 0.05\text{ pF}$. From these values, we can calculate the frequency at which the magnitudes of the two dependent current sources are equal. If

$$\omega C_\mu V_\pi = g_m V_\pi \qquad\qquad \textbf{(7.76(a))}$$

then

$$f = \frac{g_m}{2\pi C_\mu} = \frac{50 \times 10^{-3}}{2\pi(0.05 \times 10^{-12})} = 1.59 \times 10^{11}\text{ Hz} \Rightarrow 159\text{ GHz} \qquad \textbf{(7.76(b))}$$

Since the frequency of operation of bipolar transistors is far less than 159 GHz, the current source $I_{sc} = j\omega C_\mu V_\pi$ is negligible compared to the $g_m V_\pi$ source.

We can now calculate the frequency at which the magnitude of the impedance of $C_\mu$ is equal to $R_C \| R_L$. If

$$\frac{1}{\omega C_\mu} = R_C \| R_L \qquad\qquad \textbf{(7.77(a))}$$

then

$$f = \frac{1}{2\pi C_\mu (R_C \| R_L)} \qquad\qquad \textbf{(7.77(b))}$$

If we assume $R_C = R_L = 4\text{ k}\Omega$, which are typical values for discrete bipolar circuits, then

$$f = \frac{1}{2\pi(0.05 \times 10^{-12})[(4 \times 10^3)\|(4 \times 10^3)]} = 1.59 \times 10^9\text{ Hz} \qquad \textbf{(7.78)}$$

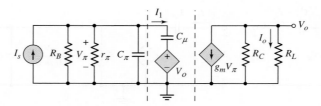

**Figure 7.45** Small-signal equivalent circuit, including approximations

If the frequency of operation of the bipolar transistor is very much smaller than 1.59 GHz, then the impedance of $C_\mu$ will be much greater than $R_C \| R_L$ and $C_\mu$ can be considered an open circuit. Using these approximations, the circuit in Figure 7.44 reduces to that shown in Figure 7.45.

The $I_1$ versus $V_\pi$ characteristic of the circuit segment between the dotted lines is

$$I_1 = \frac{V_\pi - V_o}{\dfrac{1}{j\omega C_\mu}} = j\omega C_\mu (V_\pi - V_o) \tag{7.79}$$

The output voltage is

$$V_o = -g_m V_\pi (R_C \| R_L) \tag{7.80}$$

Substituting Equation (7.80) into (7.79), we obtain

$$I_1 = j\omega C_\mu [1 + g_m(R_C \| R_L)] V_\pi \tag{7.81}$$

In Figure 7.45, the circuit segment between the dotted lines can be replaced by an equivalent capacitance given by

$$C_M = C_\mu [1 + g_m(R_C \| R_L)] \tag{7.82}$$

as shown in Figure 7.46. Capacitance $C_M$ is called the **Miller capacitance,** and the multiplication effect of $C_\mu$ is the Miller effect.

For the equivalent circuit in Figure 7.46, the input capacitance is now $C_\pi + C_M$, rather than just $C_\pi$ if $C_\mu$ had been ignored.

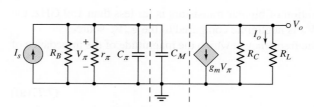

**Figure 7.46** Small-signal equivalent circuit, including the equivalent Miller capacitance

## EXAMPLE 7.10

**Objective:** Determine the 3 dB frequency of the current gain for the circuit shown in Figure 7.46, both with and without the effect of $C_M$.

The circuit parameters are: $R_C = R_L = 4\ \text{k}\Omega$, $r_\pi = 2.6\ \text{k}\Omega$, $R_B = 200\ \text{k}\Omega$, $C_\pi = 0.8\ \text{pF}$, $C_\mu = 0.05\ \text{pF}$, and $g_m = 38.5\ \text{mA/V}$.

**Solution:** The output current can be written as

$$I_o = -(g_m V_\pi) \left( \frac{R_C}{R_C + R_L} \right)$$

Also, the input voltage is

$$V_\pi = I_s \left[ R_B \| r_\pi \left\| \frac{1}{j\omega C_\pi} \right\| \frac{1}{j\omega C_M} \right]$$

$$= I_s \left[ \frac{R_B \| r_\pi}{1 + j\omega(R_B \| r_\pi)(C_\pi + C_M)} \right]$$

Therefore, the current gain is

$$A_i = \frac{I_o}{I_s} = -g_m \left( \frac{R_C}{R_C + R_L} \right) \left[ \frac{R_B \| r_\pi}{1 + j\omega(R_B \| r_\pi)(C_\pi + C_M)} \right]$$

The 3 dB frequency is

$$f_{3\,dB} = \frac{1}{2\pi(R_B \| r_\pi)(C_\pi + C_M)}$$

Neglecting the effect of $C_\mu(C_M = 0)$, we find that

$$f_\beta = \frac{1}{2\pi[(200 \times 10^3)\|(2.6 \times 10^3)](0.8 \times 10^{-12})} \Rightarrow 77.5\,\text{MHz}$$

The Miller capacitance is

$$C_M = C_\mu[1 + g_m(R_C \| R_L)] = (0.05)[1 + (38.5)(4\|4)] = 3.9\,\text{pF}$$

Taking into account the Miller capacitance, the 3 dB frequency is

$$f_{3-dB} = \frac{1}{2\pi(R_B \| r_\pi)(C_\pi + C_M)}$$

$$= \frac{1}{2\pi[(200 \times 10^3)\|(2.6 \times 10^3)][0.8 + 3.9](10^{-12})} \Rightarrow 13.2\,\text{MHz}$$

**Comment:** The Miller effect, or multiplication factor of $C_\mu$, is 78, giving a Miller capacitance of $C_M = 3.9$ pF. The Miller capacitance, in this case, is approximately a factor of five larger than $C_\pi$. This means that the actual transistor bandwidth is approximately six times less than the bandwidth expected if $C_\mu$ is neglected.

The Miller capacitance, from Equation (7.82), can be written in the form

$$C_M = C_\mu(1 + |A_v|) \tag{7.83}$$

where $A_v$ is the internal base-to-collector voltage gain. The physical origin of the Miller effect is in the voltage gain factor appearing across the feedback element $C_\mu$. A small input voltage $V_\pi$ produces a large output voltage $V_o = -|A_v| \cdot V_\pi$ of the opposite polarity at the output of $C_\mu$. Thus the voltage across $C_\mu$ is $(1 + |A_v|)V_\pi$, which induces a large current through $C_\mu$. For this reason, the effect of $C_\mu$ on the input portion of the circuit is significant.

We can now see one of the trade-offs that can be made in an amplifier design. The tradeoff is between amplifier gain and bandwidth. If the gain is reduced, then the Miller capacitance will be reduced and the bandwidth will be increased. We will consider this tradeoff again when we consider the cascode amplifier later in the chapter.

**Discussion:** In Equation (7.80), we assumed that $|j\omega C_\mu| \ll g_m$, which is valid even for frequencies in the 100 MHz range. If $j\omega C_\mu$ is not negligible, we can write

$$g_m V_\pi + V_o \left( \frac{1}{R_C \| R_L} + j\omega C_\mu \right) = 0 \qquad\qquad (7.84)$$

Equation (7.84) implies that a capacitance $C_\mu$ should be in parallel with $R_C$ and $R_L$ in the output portion of the equivalent circuit in Figure 7.44. For $R_C = R_L = 4\ \text{k}\Omega$ and $C_\mu = 0.05$ pF, we indicated that this capacitance is negligible for $f \ll 1.5$ GHz. However, in special circuits involving, for example, active loads, the equivalent $R_C$ and $R_L$ resistances may be on the order of $100\ \text{k}\Omega$. This means that the $C_\mu$ capacitance in the output part of the circuit is not negligible for frequencies even in the low-megahertz range. We will consider a few special cases in which $C_\mu$ in the output circuit is not negligible.

---

### EXERCISE PROBLEM

**Ex 7.10:** For the circuit in Figure 7.41(a), the parameters are $R_1 = 200\ \text{k}\Omega$, $R_2 = 220\ \text{k}\Omega$, $R_C = 2.2\ \text{k}\Omega$, $R_L = 4.7\ \text{k}\Omega$, $R_E = 1\ \text{k}\Omega$, $r_s = 100\ \text{k}\Omega$, and $V_{CC} = 5$ V. The transistor parameters are $\beta_o = 100$, $V_{BE}(\text{on}) = 0.7$ V, $V_A = \infty$, and $C_\pi = 1$ pF. Consider the simplified hybrid-$\pi$ model of the transistor. (a) Determine the midband current gain $A_i = I_o/I_i$. (b) Find the Miller capacitance $C_M$ for (i) $C_\mu = 0$ and (ii) $C_\mu = 0.08$ pF. (c) Determine the upper 3 dB frequency for (i) $C_\mu = 0$ and (ii) $C_\mu = 0.08$ pF. (Ans. (a)$A_i = -30.24$; (b) (i) $C_M = 0$, (ii) $C_M = 4.38$ pF; (c) (i) $f_{3dB} = 60.2$ MHz, (ii) $f_{3dB} = 11.2$ MHz)

---

### 7.4.5    Physical Origin of the Miller Effect

Figure 7.47(a) shows the hybrid-$\pi$ equivalent circuit of the bipolar transistor with a load resistor $R_C$ connected at the output. Figure 7.47(b) shows the equivalent circuit with the Miller capacitance. As a first approximation, the output voltage is $v_o = -g_m v_\pi R_C$. Assuming sinusoidal signals, Figure 7.48 shows the input signal $v_\pi$ and the output signal $v_o$. As we have noted previously, the output signal is 180 degrees out of phase with respect to the input signal. In addition, because of the gain, the magnitude of the output voltage is larger than the input voltage. The difference between $v_\pi$ and $v_o$ is the voltage across the $C_\mu$ capacitor as seen in Figure 7.47(a).

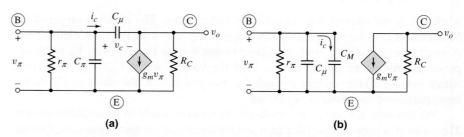

**(a)**          **(b)**

**Figure 7.47** (a) Hybrid-$\pi$ equivalent circuit of a bipolar transistor with a load resistor $R_C$ connected to the output. (b) Equivalent circuit with the Miller capacitance.

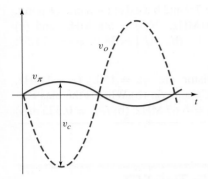

**Figure 7.48** Input signal voltage $v_\pi$ and output signal voltage $v_o$ for the circuits in Figure 7.47

We may write the sinusoidal signals as $v_\pi = V_\pi e^{j\omega t}$, $v_o = V_o e^{j\omega t}$, and $v_c = V_c e^{j\omega t}$. The current $i_c$ through the capacitor $C_\mu$ can be written as

$$i_c = C_\mu \frac{dv_c}{dt} \tag{7.85(a)}$$

Using phasor notation, we find

$$I_c = j\omega C_\mu V_c \tag{7.85(b)}$$

This current influences the input impedance looking into the base terminal of the transistor.

For the two circuits shown in Figures 7.47(a) and 7.47(b) to be equivalent, the current $i_c$ in the two circuits must be the same. From Figure 7.47(b), we can write

$$i_C = C_M \frac{dv_\pi}{dt} \tag{7.86(a)}$$

or using phasors, we have

$$I_c = j\omega C_M V_\pi \tag{7.86(b)}$$

For the two capacitor currents in Equations (7.85(b)) and (7.86(b)) to be equal, we must have

$$C_\mu V_c = C_M V_\pi \tag{7.87}$$

From the signals shown in Figure 7.48, we see that $V_\pi < V_c$ so that we must have $C_M > C_\mu$. Because of the 180 degree phase shift and voltage gain, the voltage across $C_\mu$ is quite large leading to a relatively significant value of capacitor current $i_c$. In order to have the current in the Miller capacitor $C_M$ be the same with a smaller voltage across $C_M$, the value of $C_M$ must be relatively large. This, then, is the physical origin of the Miller multiplication effect.

## Test Your Understanding

**TYU 7.6** A bipolar transistor is biased at $I_{CQ} = 120\,\mu\text{A}$ and its parameters are $\beta_o = 120$, $C_\mu = 0.08$ pF, and $f_\beta = 15$ MHz. Determine the capacitance $C_\pi$. (Ans. $C_\pi = 0.328$ pF)

**TYU 7.7** For the transistor described in Example 7.9 and biased at the same $Q$-point, determine $|h_{fe}|$ and the phase at (a) $f = 150$ MHz, (b) $f = 500$ MHz, and (c) $f = 4$ GHz. (Ans. (a) $|h_{fe}| = 89.3$, $\phi = -41.9°$; (b) $|h_{fe}| = 38.0$, $\phi = -71.5°$; (c) $|h_{fe}| = 5.0$, $\phi = -87.6°$)

**TYU 7.8** The parameters of a bipolar transistor are: $\beta_o = 150$, $f_T = 1$ GHz, $r_\pi = 12$ k$\Omega$, and $C_\mu = 0.15$ pF. (a) Determine $C_\pi$ and $f_\beta$. (b) What is the bias current in the transistor? (Ans. (a) $C_\pi = 1.84$ pF, $f_\beta = 6.67$ MHz; (b) $I_{CQ} = 0.325$ mA)

## 7.5    FREQUENCY RESPONSE: THE FET

**Objective:** • Determine the frequency response of the MOS transistor, and determine the Miller effect and Miller capacitance.

We have considered the expanded hybrid-$\pi$ equivalent circuit of the bipolar transistor that models the high-frequency response of the transistor. We will now develop the high-frequency equivalent circuit of the FET that takes into account various capacitances in the device. We will develop the model for a MOSFET, but it also applies to JFETs and MESFETs.

### 7.5.1    High-Frequency Equivalent Circuit

We will construct the small-signal equivalent circuit of a MOSFET from the basic MOSFET geometry, as described in Chapter 3. Figure 7.49 shows a model based on the inherent capacitances and resistances in an n-channel MOSFET, as well as the elements representing the basic device equations. We make one simplifying assumption in the equivalent circuit: The source and substrate are both tied to ground.

Two capacitances connected to the gate are inherent in the transistor. These capacitances, $C_{gs}$ and $C_{gd}$, represent the interaction between the gate and the channel

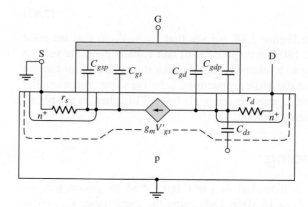

**Figure 7.49** Inherent resistances and capacitances in the n-channel MOSFET structure

inversion charge near the source and drain terminals, respectively. If the device is biased in the nonsaturation region and $v_{DS}$ is small, the channel inversion charge is approximately uniform, which means that

$$C_{gs} \cong C_{gd} \cong \left(\tfrac{1}{2}\right) WLC_{ox}$$

where $C_{ox}(\text{F/cm}^2) = \epsilon_{ox}/t_{ox}$. The parameter $\epsilon_{ox}$ is the oxide permittivity, which for silicon MOSFETs is $\epsilon_{ox} = 3.9\epsilon_o$, where $\epsilon_o = 8.85 \times 10^{-14}$ F/cm is the permittivity of free space. The parameter $t_{ox}$ is the oxide thickness in cm.

However, when the transistor is biased in the saturation region, the channel is pinched off at the drain and the inversion charge is no longer uniform. The value of $C_{gd}$ essentially goes to zero, and $C_{gs}$ approximately equals $(2/3)WLC_{ox}$. As an example, if a device has an oxide thickness of 100 Å, a channel length of $L = 0.18\,\mu$m, and a channel width of $W = 20\,\mu$m, the value of $C_{gs}$ is $C_{gs} \cong 8.3$ fF. The value of $C_{gs}$ changes as the device size changes, but typical values are in the tens of femtofarads range.

The remaining two gate capacitances, $C_{gsp}$ and $C_{gdp}$, are parasitic or **overlap capacitances,** so called because, in actual devices, the gate oxide overlaps the source and drain contacts, because of tolerances or other fabrication factors. As we will see, the drain overlap capacitance $C_{gdp}$ lowers the bandwidth of the FET. The parameter $C_{ds}$ is the drain-to-substrate pn junction capacitance, and $r_s$ and $r_d$ are the series resistances of the source and drain terminals. The internal gate-to-source voltage controls the small-signal channel current through the transconductance.

The small-signal equivalent circuit for the n-channel common-source MOSFET is shown in Figure 7.50. Voltage $V'_{gs}$ is the internal gate-to-source voltage that controls the channel current. We will assume that the gate-to-source and gate-to-drain capacitances, $C_{gs}$ and $C_{gd}$, contain the parasitic overlap capacitances. One parameter, $r_o$, shown in Figure 7.50 is not shown in Figure 7.49. This resistance is associated with the slope of $I_D$ versus $V_{DS}$. In the ideal MOSFET biased in the saturation region, $I_D$ is independent of $V_{DS}$, which means that $r_o$ is infinite. However, $r_o$ is finite in short-channel-length devices, because of channel-length modulation, and is therefore included in the equivalent circuit.

Source resistance $r_s$ can have a significant effect on the transistor characteristics. To illustrate, Figure 7.51 shows a simplified low-frequency equivalent circuit including $r_s$ but not $r_o$. For this circuit, the drain current is

$$I_d = g_m V'_{gs} \qquad\qquad (7.88)$$

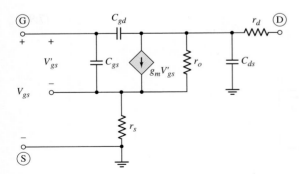

Figure 7.50  Equivalent circuit of the n-channel common-source MOSFET

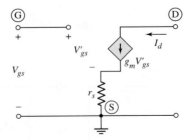

Figure 7.51  Simplified low-frequency equivalent circuit of the n-channel common-source MOSFET including source resistance $r_s$ but not resistance $r_o$

and the relationship between $V_{gs}$ and $V'_{gs}$ is

$$V_{gs} = V'_{gs} + (g_m V'_{gs})r_s = (1 + g_m r_s)V'_{gs} \qquad \textbf{(7.89)}$$

From Equation (7.88), the drain current can now be written as

$$I_d = \left(\frac{g_m}{1 + g_m r_s}\right)V_{gs} = g'_m V_{gs} \qquad \textbf{(7.90)}$$

Equation (7.90) shows that the source resistance reduces the effective transconductance, or the transistor gain.

The equivalent circuit of a p-channel MOSFET is exactly the same as that of an n-channel device, except that all voltage polarities and current directions are reversed. The capacitances and resistances are the same for both models.

### 7.5.2  Unity-Gain Bandwidth

As for the bipolar transistor, the unity-gain frequency or bandwidth is a figure of merit for the FETs. If we neglect $r_s$, $r_d$, $r_o$, and $C_{ds}$, and connect the drain to signal ground, the resulting equivalent small-signal circuit is shown in Figure 7.52. Since the input gate impedance is no longer infinite at high frequency, we can define the short-circuit current gain. From that we can define and calculate the unity-gain bandwidth.

Writing a KCL equation at the input node, we find that

$$I_i = \frac{V_{gs}}{\dfrac{1}{j\omega C_{gs}}} + \frac{V_{gs}}{\dfrac{1}{j\omega C_{gd}}} = V_{gs}[j\omega(C_{gs} + C_{gd})] \qquad \textbf{(7.91)}$$

From a KCL equation at the output node, we obtain

$$\frac{V_{gs}}{\dfrac{1}{j\omega C_{gd}}} + I_d = g_m V_{gs} \qquad \textbf{(7.92(a))}$$

or

$$I_d = V_{gs}(g_m - j\omega C_{gd}) \qquad \textbf{(7.92(b))}$$

Solving Equation (7.92(b)) for $V_{gs}$ produces

$$V_{gs} = \frac{I_d}{(g_m - j\omega C_{gd})} \qquad \textbf{(7.93)}$$

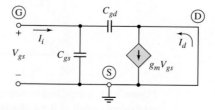

**Figure 7.52** Equivalent high-frequency small-signal circuit of a MOSFET, for calculating short-circuit current gain

Substituting Equation (7.93) into (7.91) yields

$$I_i = I_d \cdot \frac{[j\omega(C_{gs} + C_{gd})]}{(g_m - j\omega C_{gd})} \tag{7.94}$$

Therefore, the small-signal current gain is

$$A_i = \frac{I_d}{I_i} = \frac{g_m - j\omega C_{gd}}{j\omega(C_{gs} + C_{gd})} \tag{7.95}$$

If we assume typical values of $C_{gd} = 10$ fF and $g_m = 1$ mA/V, and a maximum frequency of $f = 1$ GHz, we find that $\omega C_{gd} \ll g_m$. The small-signal current gain, to a good approximation, is then

$$A_i = \frac{I_d}{I_i} \cong \frac{g_m}{j\omega(C_{gs} + C_{gd})} \tag{7.96}$$

The unity-gain frequency $f_T$ is defined as the frequency at which the magnitude of the short-circuit current gain goes to 1. From Equation (7.96) we find that

$$f_T = \frac{g_m}{2\pi(C_{gs} + C_{gd})} \tag{7.97}$$

The unity-gain frequency or bandwidth is a parameter of the transistor and is independent of the circuit.

## EXAMPLE 7.11

**Objective:** Determine the unity-gain bandwidth of an FET.

Consider an n-channel MOSFET with parameters $K_n = 1.5$ mA/V$^2$, $V_{TN} = 0.4$ V, $\lambda = 0$, $C_{gd} = 10$ fF, and $C_{gs} = 50$ fF. Assume the transistor is biased at $V_{GS} = 0.8$ V.

**Solution:** The transconductance is

$$g_m = 2K_n(V_{GS} - V_{TN}) = 2(1.5)(0.8 - 0.4) = 1.2 \text{ mA/V}$$

From Equation (7.97), the unity-gain bandwidth, or frequency, is

$$f_T = \frac{g_m}{2\pi(C_{gs} + C_{gd})} = \frac{1.2 \times 10^{-3}}{2\pi(50 + 10) \times 10^{-15}} = 3.18 \times 10^9 \text{ Hz}$$

or

$$f_T = 3.18 \text{ GHz}$$

**Comment:** As with bipolar transistors, high-frequency FETs require small capacitances and a small device size.

## EXERCISE PROBLEM

**Ex 7.11:** The parameters of an n-channel MOSFET are $K_n = 1.2$ mA/V$^2$, $V_{TN} = 0.5$ V, $\lambda = 0$, $C_{gd} = 8$ fF, and $C_{gs} = 60$ fF. The unity-gain frequency is found to be $f_T = 3$ GHz. Determine the transconductance and the bias current of the MOSFET. (Ans. $g_m = 1.282$ mA/V, $I_{DQ} = 0.342$ mA)

Typically, values of $C_{gs}$ for MOSFETs are in the range of 10 to 50 fF and values of $C_{gd}$ are typically from 0.1 to 0.5 fF.

As previously stated, the equivalent circuit is the same for MOSFETs, JFETS, and MESFETs. For JFETs, and MESFETS, capacitances $C_{gs}$ and $C_{gd}$ are depletion capacitances rather than oxide capacitances. Typically, for JFETs, $C_{gs}$ and $C_{gd}$ are larger than for MOSFETs, while the values for MESFETs are smaller. Also, for MESFETs fabricated in gallium arsenide, the unity-gain bandwidths may be in the range of tens of GHz. For this reason, gallium arsenide MESFETs are often used in microwave amplifiers.

### 7.5.3   Miller Effect and Miller Capacitance

As for the bipolar transistor, the Miller effect and Miller capacitance are factors in the high-frequency characteristics of FET circuits. Figure 7.53 is a simplified high-frequency transistor model, with a load resistor $R_L$ connected to the output. We will determine the current gain in order to demonstrate the impact of the Miller effect.

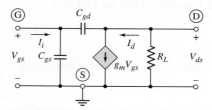

**Figure 7.53** Equivalent high-frequency small-signal circuit of a MOSFET with a load resistance $R_L$

Writing a Kirchhoff current law (KCL) equation at the input gate node, we have

$$I_i = j\omega C_{gs} V_{gs} + j\omega C_{gd}(V_{gs} - V_{ds}) \tag{7.98}$$

where $I_i$ is the input current. Likewise, summing currents at the output drain node, we have

$$\frac{V_{ds}}{R_L} + g_m V_{gs} + j\omega C_{gd}(V_{ds} - V_{gs}) = 0 \tag{7.99}$$

We can combine Equations (7.98) and (7.99) to eliminate voltage $V_{ds}$. The input current is then

$$I_i = j\omega\left\{ C_{gs} + C_{gd}\left[ \frac{1 + g_m R_L}{1 + j\omega R_L C_{gd}} \right] \right\} V_{gs} \tag{7.100}$$

Normally, $(\omega R_L C_{gd})$ is much less than 1; therefore, we can neglect $(j\omega R_L C_{gd})$ and Equation (7.100) becomes

$$I_i = j\omega[C_{gs} + C_{gd}(1 + g_m R_L)]V_{gs} \tag{7.101}$$

Figure 7.54 shows the equivalent circuit described by Equation (7.101). The parameter $C_M$ is the Miller capacitance and is given by

$$C_M = C_{gd}(1 + g_m R_L) \tag{7.102}$$

Equation (7.102) clearly shows the effect of the parasitic drain overlap capacitance. When the transistor is biased in the saturation region, as in an amplifier circuit, the major contribution to the total gate-to-drain capacitance $C_{gd}$ is the overlap

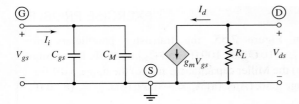

**Figure 7.54** MOSFET high-frequency circuit, including the equivalent Miller capacitance

capacitance. This overlap capacitance is multiplied because of the Miller effect and may become a significant factor in the bandwidth of an amplifier. Minimizing the overlap capacitance is one of the challenges of fabrication technology.

The cutoff frequency $f_T$ of a MOSFET is defined as the frequency at which the short circuit current gain magnitude is 1, or the magnitude of the input current $I_i$ is equal to the ideal current $I_d$. From Figure 7.54, we see that

$$I_i = j\omega(C_{gs} + C_M)V_{gs} \tag{7.103}$$

and the ideal short-circuit output current is

$$I_d = g_m V_{gs} \tag{7.104}$$

The magnitude of the current gain is therefore

$$|A_i| = \left| \frac{I_d}{I_i} \right| = \frac{g_m}{2\pi f(C_{gs} + C_M)} \tag{7.105}$$

Setting Equation (7.105) equal to 1, we find the cutoff frequency

$$f_T = \frac{g_m}{2\pi(C_{gs} + C_M)} = \frac{g_m}{2\pi C_G} \tag{7.106}$$

where $C_G$ is the equivalent input gate capacitance.

## EXAMPLE 7.12

**Objective:** Determine the Miller capacitance and cutoff frequency of an FET circuit.

The n-channel MOSFET described in Example 7.11 is biased at the same current, and a 10 kΩ load is connected to the output.

**Solution:** From Example 7.11, the transconductance is $g_m = 1.2 \, \text{mA/V}$. The Miller capacitance is therefore

$$C_M = C_{gd}(1 + g_m R_L) = (10)[1 + (1.2)(10)] = 130 \, \text{fF}$$

From Equation (7.106), the cutoff frequency is

$$f_T = \frac{g_m}{2\pi(C_{gs} + C_M)} = \frac{1.2 \times 10^{-3}}{2\pi(50 + 130) \times 10^{-15}} = 1.06 \times 10^9 \, \text{Hz}$$

or

$$f_T = 1.06 \, \text{MHz}$$

**Comment:** The Miller effect and equivalent Miller capacitance reduce the cutoff frequency of an FET circuit, just as they do in a bipolar circuit.

**Ex 7.12:** For the circuit in Figure 7.55, the transistor parameters are $K_n = 0.8 \text{ mA/V}^2$, $V_{TN} = 2 \text{ V}$, $\lambda = 0$, $C_{gs} = 100 \text{ fF}$, and $C_{gd} = 20 \text{ fF}$. Determine (a) the midband voltage gain, (b) the Miller capacitance, and (c) the upper 3 dB frequency of the small-signal voltage gain. (Ans. (a) $A_v = -6.69$, (b) $C_M = 167.6 \text{ fF}$, (c) $f_{3dB} = 1.32 \text{ GHz}$)

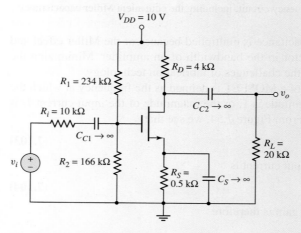

**Figure 7.55** Figure for Exercise Ex 7.12

## Test Your Understanding

**TYU 7.9** An n-channel MOSFET has parameters $K_n = 0.4 \text{ mA/V}^2$, $V_{TN} = 1 \text{ V}$, and $\lambda = 0$. (a) Determine the maximum source resistance such that the transconductance is reduced by no more than 20 percent from its ideal value when $V_{GS} = 3 \text{ V}$. (b) Using the value of $r_s$ calculated in part (a), determine how much $g_m$ is reduced from its ideal value when $V_{GS} = 5 \text{ V}$. (Ans. (a) $r_s = 156 \ \Omega$, (b) 33.3%)

**TYU 7.10** An n-channel MOSFET has a unity-gain bandwidth of $f_T = 1.2 \text{ GHz}$. Assume overlap capacitances of $C_{gsp} = C_{gdp} = 3 \text{ fF}$, and assume $k_n' = 100 \ \mu\text{A/V}^2$, $W/L = 15$, and $V_{TN} = 0.4 \text{ V}$. If the transistor is biased at $I_{DQ} = 100 \ \mu\text{A}$, determine $C_{gs}$. (Assume $C_{gd}$ is equal to zero.) (Ans. $C_{gs} = 66.6 \text{ fF}$)

**TYU 7.11** For a MOSFET, assume that $g_m = 1.2 \text{ mA/V}$. The basic gate capacitances are $C_{gs} = 60 \text{ fF}$, $C_{gd} = 0$, and the overlap capacitances are $C_{gsp} = C_{gdp}$. Determine the minimum overlap capacitance for a unity-gain bandwidth of 2.5 GHz. (Ans. $C_{gsp} = C_{gdp} = 8.2 \text{ fF}$)

## 7.6   HIGH-FREQUENCY RESPONSE OF TRANSISTOR CIRCUITS

**Objective:** • Determine the high-frequency response of basic transistor circuit configurations including the cascode circuit.

In the last sections, we developed the high-frequency equivalent circuits for the bipolar and field-effect transistors. We also discussed the Miller effect, which occurs
when transistors are operating in a circuit configuration. In this section, we will expand our analysis of the high-frequency characteristics of transistor circuits.

Initially, we will look at the high-frequency response of the common-emitter and
common-source configurations. We will then examine common-base and common-
gate circuits, and a cascode circuit that is a combination of the common-emitter and
common-base circuits. Finally, we will analyze the high-frequency characteristics of
emitter-follower and source-follower circuits. In the following examples, we will use
the same basic bipolar transistor circuit so that a good comparison can be made between the three circuit configurations.

## 7.6.1 Common-Emitter and Common-Source Circuits

The transistor capacitances and the load capacitance in the common-emitter amplifier shown in Figure 7.56 affect the high-frequency response of the circuit. Initially,
we will use a hand analysis to determine the effects of the transistor on the high-
frequency response. In this analysis, we will assume that $C_C$ and $C_E$ are short circuits, and $C_L$ is an open circuit. A computer analysis will then be used to determine
the effect of both the transistor and load capacitances.

The high-frequency small-signal equivalent circuit of the common-emitter
circuit is shown in Figure 7.57(a) in which $C_L$ is assumed to be an open circuit. We
replace the capacitor $C_\mu$ with the equivalent Miller capacitance $C_M$ as shown in
Figure 7.57(b). From our previous analysis of the Miller capacitance, we can write

$$C_M = C_\mu(1 + g_m R_L')$$

(7.107)

where the output resistance $R_L'$ is $r_o \| R_C \| R_L$.

The upper 3 dB frequency can be determined by using the time constant technique. We can write

$$f_H = \frac{1}{2\pi \tau_P}$$

(7.108)

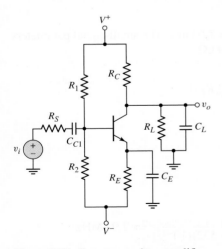

**Figure 7.56** Common-emitter amplifier

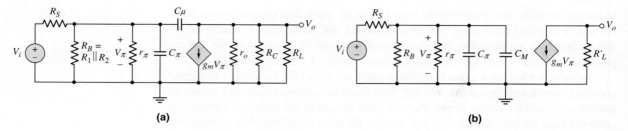

**Figure 7.57** (a) High-frequency equivalent circuit of common-emitter amplifier; (b) high-frequency equivalent circuit of common-emitter amplifier, including the Miller capacitance

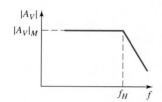

**Figure 7.58** Bode plot of the high-frequency voltage gain magnitude for the common-emitter amplifier

where $\tau_P = R_{eq}C_{eq}$. In this case, the equivalent capacitance is $C_{eq} = C_\pi + C_M$, and the equivalent resistance is the effective resistance seen by the capacitance, $R_{eq} = r_\pi \| R_B \| R_S$. The upper corner frequency is therefore

$$f_H = \frac{1}{2\pi [r_\pi \| R_B \| R_S](C_\pi + C_M)} \tag{7.109}$$

We determine the midband voltage gain magnitude by assuming $C_\pi$ and $C_M$ are open circuits. We find that

$$|A_v|_M = \left|\frac{V_o}{V_i}\right|_M = g_m R'_L \left[\frac{R_B \| r_\pi}{R_B \| r_\pi + R_S}\right] \tag{7.110}$$

The Bode plot of the high-frequency voltage gain magnitude is shown in Figure 7.58.

### EXAMPLE 7.13

**Objective:** Determine the upper corner frequency and midband gain of a common-emitter circuit.

For the circuit in Figure 7.56, the parameters are: $V^+ = 5$ V, $V^- = -5$ V, $R_S = 0.1$ kΩ, $R_1 = 40$ kΩ, $R_2 = 5.72$ kΩ, $R_E = 0.5$ kΩ, $R_C = 5$ kΩ, and $R_L = 10$ kΩ. The transistor parameters are: $\beta = 150$, $V_{BE}(\text{on}) = 0.7$ V, $V_A = \infty$, $C_\pi = 35$ pF, and $C_\mu = 4$ pF.

**Solution:** From a dc analysis, we find $I_{CQ} = 1.03$ mA. The small-signal parameters are therefore $g_m = 39.6$ mA/V and $r_\pi = 3.79$ kΩ.

The Miller capacitance is then

$$C_M = C_\mu(1 + g_m R'_L) = C_\mu[1 + g_m(R_C \| R_L)]$$

or

$$C_M = (4)[1 + (39.6)(5\|10)] = 532 \text{ pF}$$

and the upper 3 dB frequency is therefore

$$f_H = \frac{1}{2\pi [r_\pi \| R_B \| R_S](C_\pi + C_M)}$$

$$= \frac{1}{2\pi [3.79\|40\|5.72\|0.1](10^3)(35 + 532)(10^{-12})} \Rightarrow 2.94 \text{ MHz}$$

Finally, the midband gain is

$$|A_v|_M = g_m R'_L \left[ \frac{R_B \| r_\pi}{R_B \| r_\pi + R_S} \right]$$

$$= (39.6)(5\|10) \left[ \frac{40\|5.72\|3.79}{40\|5.72\|3.79 + 0.1} \right] = 126$$

**Comments:** This example demonstrates the importance of the Miller effect. The feedback capacitance $C_\mu$ is multiplied by a factor of 133 (from 4 pF to 532 pF), and the resulting Miller capacitance $C_M$ is approximately 15 times larger than $C_\pi$. The actual corner frequency is therefore approximately 15 times smaller than it would be if $C_\mu$ were neglected.

**PSpice Verification:** Figure 7.59 shows the results of a PSpice analysis of this common-emitter circuit. The computer values are: $C_\pi = 35.5$ pF and $C_\mu = 3.89$ pF. The curve marked "$C_\pi$ only" is the circuit frequency response if $C_\mu$ is neglected; the curve marked "$C_\pi$ and $C_\mu$ only" is the response due to $C_\pi$, $C_\mu$, and the Miller effect. These curves illustrate that the bandwidth of this circuit is drastically reduced by the Miller effect.

The corner frequency is approximately 2.5 MHz and the midband gain is 125, which agree very well with the hand analysis results.

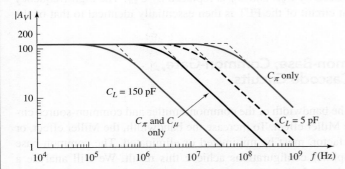

**Figure 7.59** PSpice analysis results for common-emitter amplifier

The curves marked "$C_L = 5$ pF" and "$C_L = 150$ pF" show the circuit response if the transistor is ideal, with zero $C_\pi$ and $C_\mu$ capacitances and a load capacitance connected to the output. These results show that, for $C_L = 5$ pF, the circuit response is dominated by the $C_\pi$ and $C_\mu$ capacitances of the transistor. However, if a large load capacitance, such as $C_L = 150$ pF, is connected to the output, the circuit response is dominated by the $C_L$ capacitance.

## EXERCISE PROBLEM

**\*Ex 7.13:** The transistor in the circuit in Figure 7.60 has parameters $\beta = 125$, $V_{BE}(\text{on}) = 0.7$ V, $V_A = 200$ V, $C_\pi = 24$ pF, and $C_\mu = 3$ pF. (a) Calculate the Miller capacitance. (b) Determine the upper 3 dB frequency. (c) Determine the small-signal midband voltage gain. (Ans. (a) $C_M = 155$ pF, (b) $f_H = 1.21$ MHz, (c) $|A_v| = 37.3$)

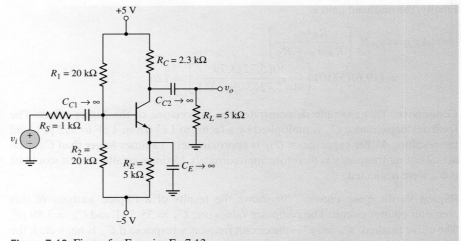

**Figure 7.60** Figure for Exercise Ex 7.13

The high-frequency response of the common-source circuit is similar to that of the common-emitter circuit, and the discussion and conclusions are the same. Capacitance $C_\pi$ is replaced by $C_{gs}$, and $C_\mu$ is replaced by $C_{gd}$. The high-frequency small-signal equivalent circuit of the FET is then essentially identical to that of the bipolar transistor.

### 7.6.2 Common-Base, Common-Gate, and Cascode Circuits

As we have just seen, the bandwidth of the common-emitter and common-source circuits is reduced by the Miller effect. To increase the bandwidth, the Miller effect, or the $C_\mu$ multiplication factor, must be minimized or eliminated. The common-base and common-gate amplifier configurations achieve this result. We will analyze a common-base circuit; the analysis is the same for the common-gate circuit.

### Common-Base Circuit

Figure 7.61 shows a common-base circuit. The circuit configuration is the same as the common-emitter circuit considered previously, except a bypass capacitor is added to the base and the input is capacitively coupled to the emitter.

Figure 7.62(a) shows the high-frequency equivalent circuit, with the coupling and bypass capacitors replaced by short circuits. Resistors $R_1$ and $R_2$ are then effectively short circuited. Also, resistance $r_o$ is assumed to be infinite. Capacitance $C_\mu$, which led to the multiplication effect, is no longer between the input and output terminals. One side of capacitor $C_\mu$ is tied to signal ground.

Writing a KCL equation at the emitter, we find that

$$I_e + g_m V_\pi + \frac{V_\pi}{(1/sC_\pi)} + \frac{V_\pi}{r_\pi} = 0 \tag{7.111}$$

Since $V_\pi = -V_e$, Equation (7.111) becomes

$$\frac{I_e}{V_e} = \frac{1}{Z_i} = \frac{1}{r_\pi} + g_m + sC_\pi \tag{7.112}$$

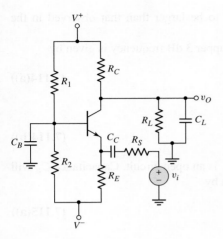

**Figure 7.61** Common-base amplifier

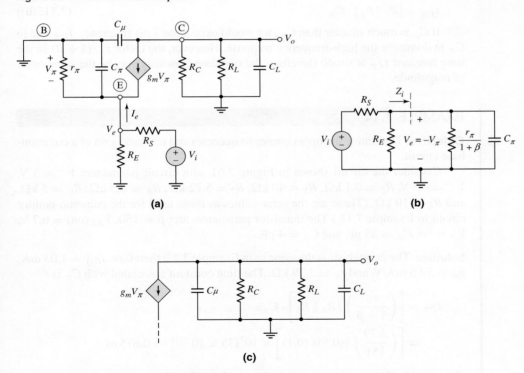

**Figure 7.62** (a) High-frequency common-base equivalent circuit, (b) equivalent input circuit, and (c) equivalent output circuit

where $Z_i$ is the impedance looking into the emitter. Rearranging terms, we have

$$\frac{1}{Z_i} = \frac{1 + r_\pi g_m}{r_\pi} + sC_\pi = \frac{1 + \beta}{r_\pi} + sC_\pi \qquad (7.113)$$

The equivalent input portion of the circuit is shown in Figure 7.62(b).

Figure 7.62(c) shows the equivalent output portion of the circuit. Again, one side of $C_\mu$ is tied to ground, which eliminates the feedback or Miller multiplication effect.

We then expect the upper 3 dB frequency to be larger than that observed in the common-emitter configuration.

For the input portion of the circuit, the upper 3 dB frequency is given by

$$f_{H\pi} = \frac{1}{2\pi \tau_{p\pi}} \tag{7.114(a)}$$

where the time constant is

$$\tau_{P\pi} = \left[\left(\frac{r_\pi}{1+\beta}\right) \| R_E \| R_S\right] \cdot C_\pi \tag{7.114(b)}$$

In the hand analysis, we assume that $C_L$ is an open circuit. Capacitance $C_\mu$ will also produce an upper 3 dB frequency, given by

$$f_{H\mu} = \frac{1}{2\pi \tau_{P\mu}} \tag{7.115(a)}$$

where the time constant is

$$\tau_{P\mu} = [R_C \| R_L] \cdot C_\mu \tag{7.115(b)}$$

If $C_\mu$ is much smaller than $C_\pi$, we would expect the 3 dB frequency $f_{H\pi}$ due to $C_\pi$ to dominate the high-frequency response. However, the factor $r_\pi/(1+\beta)$ in the time constant $\tau_{P\pi}$ is small; therefore, the two time constants may be the same order of magnitude.

## EXAMPLE 7.14

**Objective:** Determine the upper corner frequencies and midband gain of a common-base circuit.

Consider the circuit shown in Figure 7.61 with circuit parameters $V^+ = 5$ V, $V^- = -5$ V, $R_S = 0.1$ k$\Omega$, $R_1 = 40$ k$\Omega$, $R_2 = 5.72$ k$\Omega$, $R_E = 0.5$ k$\Omega$, $R_C = 5$ k$\Omega$, and $R_L = 10$ k$\Omega$. (These are the same values as those used for the common-emitter circuit in Example 7.13.) The transistor parameters are: $\beta = 150$, $V_{BE}(\text{on}) = 0.7$ V, $V_A = \infty$, $C_\pi = 35$ pF, and $C_\mu = 4$ pF.

**Solution:** The dc analysis is the same as in Example 7.13; therefore, $I_{CQ} = 1.03$ mA, $g_m = 39.6$ mA/V, and $r_\pi = 3.79$ k$\Omega$. The time constant associated with $C_\pi$ is

$$\begin{aligned}
\tau_{P\pi} &= \left[\left(\frac{r_\pi}{1+\beta}\right) \| R_E \| R_S\right] \cdot C_\pi \\
&= \left[\left(\frac{3.79}{151}\right) \| (0.5) \| (0.1)\right] \times 10^3 (35 \times 10^{-12}) \Rightarrow 0.675 \text{ ns}
\end{aligned}$$

The upper 3 dB frequency corresponding to $C_\pi$ is therefore

$$f_{H\pi} = \frac{1}{2\pi \tau_{P\pi}} = \frac{1}{2\pi(0.675 \times 10^{-9})} \Rightarrow 236 \text{ MHz}$$

The time constant associated with $C_\mu$ in the output portion of the circuit is

$$\tau_{P\mu} = [R_C \| R_L] \cdot C_\mu = [5\|10] \times 10^3 (4 \times 10^{-12}) \Rightarrow 13.33 \text{ ns}$$

The upper 3 dB frequency corresponding to $C_\mu$ is therefore

$$f_{H\mu} = \frac{1}{2\pi \tau_{P\mu}} = \frac{1}{2\pi(13.3 \times 10^{-9})} \Rightarrow 11.9 \text{ MHz}$$

So in this case, $f_{H\mu}$ is the dominant pole frequency.

The magnitude of the midband voltage gain is

$$|A_v|_M = g_m(R_C \| R_L) \left[ \frac{R_E \left\| \left( \dfrac{r_\pi}{1+\beta} \right)}{R_E \left\| \left( \dfrac{r_\pi}{1+\beta} \right) + R_S} \right]$$

$$= (39.6)(5\|10) \left[ \frac{0.5 \left\| \left( \dfrac{3.79}{151} \right)}{0.5 \left\| \left( \dfrac{3.79}{151} \right) + 0.1} \right] = 25.5$$

**Comment:** The results of this example show that the bandwidth of the common-base circuit is limited by the capacitance $C_\mu$ in the output portion of the circuit. The bandwidth of this particular circuit is 12 MHz, which is approximately a factor of four greater than the bandwidth of the common-emitter circuit in Example 7.14.

**Computer Verification:** Figure 7.63 shows the results of a PSpice analysis of the common-base circuit. The computer values are $C_\pi = 35.5$ pF and $C_\mu = 3.89$ pF, which are the same as those in Example 7.13. The curve marked "$C_\pi$ only" is the circuit frequency response if $C_\mu$ is neglected. The curve marked "$C_\pi$ and $C_\mu$ only" includes the effect of both $C_\pi$ and $C_\mu$. As the hand analysis predicted, $C_\mu$ dominates the circuit high-frequency response.

The corner frequency is approximately 13.5 MHz and the midband gain is 25.5, both of which agree very well with the hand analysis results.

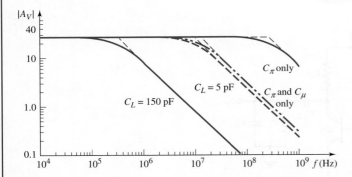

**Figure 7.63** PSpice analysis results for common-base circuit

The curves marked "$C_L = 5$ pF" and "$C_L = 150$ pF" are the circuit response if the transistor is ideal and only a load capacitance is included. These results again show that if a load capacitance of $C_L = 150$ pF were connected to the output, the circuit response would be dominated by this capacitance. However, if a 5 pF load capacitor were connected to the output, the circuit response would be a function of both the $C_L$ and $C_\mu$ capacitances, since the two response characteristics are almost identical.

**\*Ex 7.14:** Consider the common-base circuit in Figure 7.64. The transistor parameters are $\beta = 100$, $V_{BE}(\text{on}) = 0.7$ V, $V_A = \infty$, $C_\pi = 24$ pF, and $C_\mu = 3$ pF. (a) Determine the upper 3 dB frequencies corresponding to the input and output portions of the equivalent circuit. (b) Calculate the small-signal midband voltage gain. (Ans. (a) $f_{H\pi} = 223$ MHz, $f_{H\mu} = 58.3$ MHz, (b) $A_v = 0.869$)

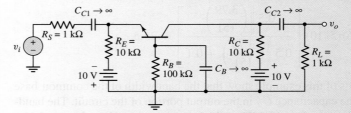

**Figure 7.64** Figure for Exercise Ex 7.14

### Cascode Circuit

The cascode circuit, as shown in Figure 7.65, combines the advantages of the common-emitter and common-base circuits. The input signal is applied to the common-emitter circuit ($Q_1$), and the output signal from the common emitter is fed into the common-base circuit ($Q_2$). The input impedance to the common-emitter circuit ($Q_1$) is relatively large, and the load resistance seen by $Q_1$ is the input impedance to the emitter of $Q_2$ and is fairly small. The low output resistance seen by $Q_1$ reduces the Miller multiplication factor on $C_{\mu 1}$ and therefore extends the bandwidth of the circuit.

Figure 7.66(a) shows the high-frequency small-signal equivalent circuit. The coupling and bypass capacitors are equivalent to short circuits, and resistance $r_o$ for $Q_2$ is assumed to be infinite.

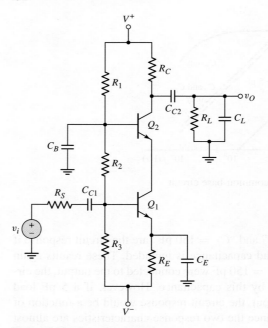

**Figure 7.65** Cascode circuit

The input impedance to the emitter of $Q_2$ is $Z_{ie2}$. From Equation (7.113) in our previous analysis, we have

$$Z_{ie2} = \left(\frac{r_{\pi2}}{1+\beta}\right) \middle\| \left(\frac{1}{sC_{\pi2}}\right) \tag{7.116}$$

The input portion of the small-signal equivalent circuit can be transformed to that shown in Figure 7.66(b). The input impedance $Z_{ie2}$ is again shown.

The input portion of the circuit shown in Figure 7.66(b) can be transformed to that given in Figure 7.66(c), which shows the Miller capacitance. The Miller capacitance $C_{M1}$ is included in the input, and capacitance $C_{\mu1}$ is included in the output portion of the $Q_1$ model. The possibility of including $C_\mu$ in the output circuit was discussed previously in Section 7.4.4.

In the center of this equivalent circuit, $r_{o1}$ is in parallel with $r_{\mu2}/(1+\beta)$. Since $r_{o1}$ is usually large, it can be approximated as an open circuit. The Miller capacitance is then

$$C_{M1} = C_{\mu1}\left[1 + g_{m1}\left(\frac{r_{\pi2}}{1+\beta}\right)\right] \tag{7.117}$$

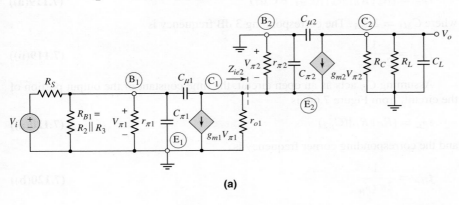

**(a)**

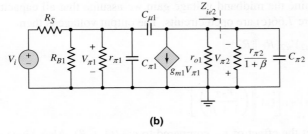

**(b)**

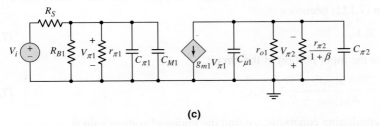

**(c)**

**Figure 7.66** (a) High-frequency equivalent circuit of cascode configuration, (b) rearranged high-frequency equivalent circuit, and (c) variation of the high-frequency circuit, including the Miller capacitance

Transistors $Q_1$ and $Q_2$ are biased with essentially the same current; therefore,

$$r_{\pi 1} \cong r_{\pi 2} \quad \text{and} \quad g_{m1} \cong g_{m2}$$

Then

$$g_{m1}r_{\pi 2} = \beta$$

which yields

$$C_{M1} \cong 2C_{\mu 1} \tag{7.118}$$

Equation (7.118) shows that this cascode circuit greatly reduces the Miller multiplication factor.

The time constant related to $C_{\pi 2}$ involves resistance $r_{\pi 2}/(1 + \beta)$. Since this resistance is small, the time constant is small, and the corner frequency related to $C_{\pi 2}$ is very large. We can therefore neglect the effects of $C_{\mu 1}$ and $C_{\pi 2}$ in the center portion of the circuit.

The time constant for the input portion of the circuit is

$$\tau_{P\pi} = [R_S \| R_{B1} \| r_{\pi 1}](C_{\pi 1} + C_{M1}) \tag{7.119(a)}$$

where $C_{M1} = 2C_{\mu 1}$. The corresponding 3 dB frequency is

$$f_{H\pi} = \frac{1}{2\pi \tau_{P\pi}} \tag{7.119(b)}$$

Assuming $C_L$ acts as an open circuit, the time constant of the output portion of the circuit, from Figure 7.66, is

$$\tau_{P\mu} = [R_C \| R_L](C_{\mu 2}) \tag{7.120(a)}$$

and the corresponding corner frequency is

$$f_{H\mu} = \frac{1}{2\pi \tau_{P\mu}} \tag{7.120(b)}$$

To determine the midband voltage gain we assume that all capacitances in the circuit in Figure 7.66(c) are open circuits. The output voltage is then

$$V_o = -g_{m2}V_{\pi 2}(R_C \| R_L) \tag{7.121}$$

and

$$V_{\pi 2} = g_{m1}V_{\pi 1}\left[r_{o1} \left\| \left(\frac{r_{\pi 2}}{1 + \beta}\right)\right.\right] \tag{7.122}$$

We can neglect the effect of $r_{o1}$ compared to $r_{\pi 2}/(1 + \beta)$. Also, since $g_{m1}r_{\pi 2} = \beta$, Equation (7.122) becomes

$$V_{\pi 2} \cong V_{\pi 1} \tag{7.123}$$

and, from the input portion of the circuit,

$$V_{\pi 1} = \frac{R_{B1} \| r_{\pi 1}}{R_{B1} \| r_{\pi 1} + R_S} \times V_i \tag{7.124}$$

Finally, combining equations, we find the midband voltage gain is

$$A_{vM} = \frac{V_o}{V_i} = -g_{m2}(R_C \| R_L)\left[\frac{R_{B1} \| r_{\pi 1}}{R_{B1} \| r_{\pi 1} + R_S}\right] \tag{7.125}$$

If we compare Equation (7.125) to Equation (7.110) for the common-emitter circuit, we see that the expression for the midband gain of the cascode circuit is identical to that of the common-emitter circuit. The cascode circuit achieves a relatively large voltage gain, while extending the bandwidth.

## EXAMPLE 7.15

**Objective:** Determine the 3 dB frequencies and midband gain of a cascode circuit.

For the circuit in Figure 7.65, the parameters are: $V^+ = 10$ V, $V^- = -10$ V, $R_S = 0.1$ k$\Omega$, $R_1 = 42.5$ k$\Omega$, $R_2 = 20.5$ k$\Omega$, $R_3 = 28.3$ k$\Omega$, $R_E = 5.4$ k$\Omega$, $R_C = 5$ k$\Omega$, $R_L = 10$ k$\Omega$, and $C_L = 0$. The transistor parameters are: $\beta = 150$, $V_{BE}(\text{on}) = 0.7$ V, $V_A = \infty$, $C_\pi = 35$ pF, and $C_\mu = 4$ pF.

**Solution:** Since $\beta$ is large for each transistor, the quiescent collector current is essentially the same in each transistor and is $I_{CQ} = 1.02$ mA. The small-signal parameters are: $r_{\pi 1} = r_{\pi 2} \equiv r_\pi = 3.82$ k$\Omega$ and $g_{m1} = g_{m2} \equiv g_m = 39.2$ mA/V.

From Equation (7.119(a)), the time constant related to the input portion of the circuit is

$$\tau_{P\pi} = [R_S \| R_{B1} \| r_{\pi 1}](C_{\pi 1} + C_{M1})$$

Since $R_{B1} = R_2 \| R_3$ and $C_{M1} = 2C_{\mu 1}$, then

$$\tau_{P\pi} = [(0.1) \| 20.5 \| 28.3 \| 3.82] \times 10^3 [35 + 2(4)] \times 10^{-12} \Rightarrow 4.16 \text{ ns}$$

The corresponding 3 dB frequency is

$$f_{H\pi} = \frac{1}{2\pi \tau_{P\pi}} = \frac{1}{2\pi(4.16 \times 10^{-9})} \Rightarrow 38.3 \text{ MHz}$$

From Equation (7.120(a)), the time constant of the output portion of the circuit is

$$\tau_{P\mu} = [R_C \| R_L] C_{\mu 2} = [5 \| 10] \times 10^3 (4 \times 10^{-12}) \Rightarrow 13.3 \text{ ns}$$

and the corresponding 3 dB frequency is

$$f_{H\mu} = \frac{1}{2\pi \tau_{P\mu}} = \frac{1}{2\pi(13.3 \times 10^{-9})} \Rightarrow 12 \text{ MHz}$$

From Equation (7.125), the midband voltage gain is

$$|A_v|_M = g_{m2}(R_C \| R_L) \left[ \frac{R_{B1} \| r_{\pi 1}}{R_{B1} \| r_{\pi 1} + R_S} \right]$$

$$= (39.2)(5 \| 10) \left[ \frac{(20.5 \| 28.3 \| 3.82)}{(20.5 \| 28.3 \| 3.82) + (0.1)} \right] = 126$$

**Comment:** As was the case for the common-base circuit, the 3 dB frequency for the cascode circuit is determined by capacitance $C_\mu$ in the output stage. The bandwidth of the cascode circuit is 12 Mz, compared to approximately 3 MHz for the common-emitter circuit. The midband voltage gains for the two circuits are essentially the same.

**Computer Verification:** Figure 7.67 shows the results of a PSpice analysis of the cascode circuit. From the hand analysis, the two corner frequencies are 12 Mz and 38.3 MHz. Since these frequencies are fairly close, we expect the actual response to show the effects of both capacitances. This hypothesis is verified and demonstrated

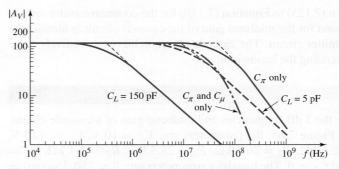

**Figure 7.67** PSpice analysis results for cascode circuit

in the computer analysis results. The curves marked "$C_\pi$ only" and "$C_\pi$ and $C_\mu$ only" are fairly close together, and their slopes are steeper than −6 dB/octave, which shows that more than one capacitor is involved in the response. At a frequency of 12 MHz, the response curve is 3 dB below the maximum asymptotic gain, and the midband gain is 120. These values closely agree with the hand analysis results.

The curves marked "$C_L = 5$ pF" and "$C_L = 150$ pF" show the circuit response if the transistor is ideal and only a load capacitance is included.

### EXERCISE PROBLEM

**\*Ex 7.15:** The cascode circuit in Figure 7.65 has parameters $V^+ = 12$ V, $V^- = 0$, $R_1 = 58.8$ kΩ, $R_2 = 33.3$ kΩ, $R_3 = 7.92$ kΩ, $R_C = 7.5$ kΩ, $R_S = 1$ kΩ, $R_E = 0.5$ kΩ, and $R_L = 2$ kΩ. The transistor parameters are: $\beta = 100$, $V_{BE}$(on) $= 0.7$ V, $V_A = \infty$, $C_\pi = 24$ pF, and $C_\mu = 3$ pF. Let $C_L$ be an open circuit. (a) Determine the 3 dB frequencies corresponding to the input and output portions of the equivalent circuit. (b) Calculate the small-signal midband voltage gain. (c) Correlate the results from parts (a) and (b) with a computer analysis. (Ans. (a) $f_{H\pi} = 7.15$ MHz, $f_{H\mu} = 33.6$ MHz, (b) $|A_v| = 22.5$)

### 7.6.3 Emitter- and Source-Follower Circuits

In this section, we analyze the high-frequency response of the emitter follower. We will analyze the same basic circuit configuration that we have considered previously. The results and discussions also apply to the source follower.

Figure 7.68 shows an emitter-follower circuit with the output signal at the emitter capacitively coupled to a load. Figure 7.69(a) shows the high-frequency small-signal equivalent circuit, with the coupling capacitors acting effectively as short circuits.

We will rearrange the circuit so that we can gain a better insight into the circuit behavior. We see that $C_\mu$ is tied to ground potential and also that $r_o$ is in parallel with $R_E$ and $R_L$. We may define

$$R'_L = R_E \| R_L \| r_o$$

In this analysis we neglect the effect of $C_L$. Figure 7.69(b) shows a rearrangement of the circuit.

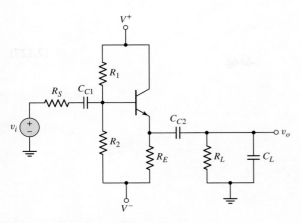

**Figure 7.68** Emitter-follower circuit

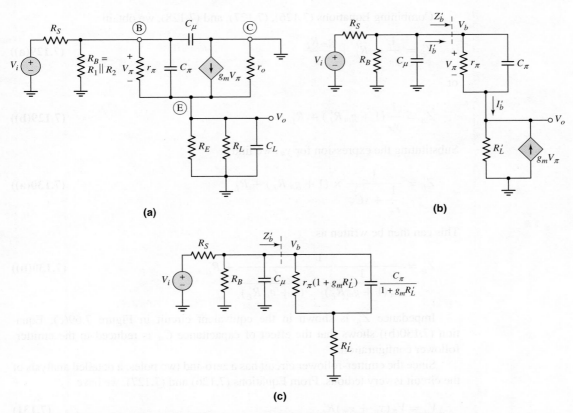

**Figure 7.69** (a) High-frequency equivalent circuit of emitter follower, (b) rearranged high-frequency equivalent circuit, and (c) high-frequency equivalent circuit with effective input base impedance

We can find the impedance $Z_b'$ looking into the base without capacitance $C_\mu$. The current $I_b'$ entering the parallel combination of $r_\pi$ and $C_\pi$ is the same as that coming out of the combination. The output voltage is then

$$V_o = (I_b' + g_m V_\pi) R_L' \tag{7.126}$$

Voltage $V_\pi$ is given by

$$V_\pi = \frac{I_b'}{y_\pi} \tag{7.127}$$

where

$$y_\pi = (1/r_\pi) + sC_\pi$$

Voltage $V_b$ is

$$V_b = V_\pi + V_o$$

Therefore,

$$Z_b' = \frac{V_b}{I_b'} = \frac{V_\pi + V_o}{I_b'} \tag{7.128}$$

Combining Equations (7.126), (7.127), and (7.128), we obtain

$$Z_b' = \frac{1}{y_\pi} + R_L' + \frac{g_m R_L'}{y_\pi} \tag{7.129(a)}$$

or

$$Z_b' = \frac{1}{y_\pi}(1 + g_m R_L') + R_L' \tag{7.129(b)}$$

Substituting the expression for $y_\pi$, we find

$$Z_b' = \frac{1}{\dfrac{1}{r_\pi} + sC_\pi} \times (1 + g_m R_L') + R_L' \tag{7.130(a)}$$

This can then be written as

$$Z_b' = \frac{1}{\dfrac{1}{r_\pi(1 + g_m R_L')} + \dfrac{sC_\pi}{(1 + g_m R_L')}} + R_L' \tag{7.130(b)}$$

Impedance $Z_b'$ is shown in the equivalent circuit in Figure 7.69(c). Equation (7.130(b)) shows that the effect of capacitance $C_\pi$ is reduced in the emitter-follower configuration.

Since the emitter-follower circuit has a zero and two poles, a detailed analysis of the circuit is very tedious. From Equations (7.126) and (7.127), we have

$$V_o = V_\pi(y_\pi + g_m)R_L' \tag{7.131}$$

which yields a zero when $y_\pi + g_m = 0$. Using the definition of $y_\pi$, the zero occurs at

$$f_o = \frac{1}{2\pi C_\pi \left(\dfrac{r_\pi}{1 + \beta}\right)} \tag{7.132}$$

Since $r_\pi/(1 + \beta)$ is small, frequency $f_o$ is usually very high.

If we make a simplifying assumption, we can determine an approximate value of one pole. In many applications, the impedance of $r_\pi(1 + g_m R'_L)$ in parallel with $C_\pi/(1 + g_m R'_L)$ is large compared to $R'_L$. If we neglect $R'_L$, then the time constant is

$$\tau_P = [R_S \| R_B \| (1 + g_m R'_L) r_\pi]\left(C_\mu + \frac{C_\pi}{1 + g_m R'_L}\right) \tag{7.133(a)}$$

and the 3 dB frequency (or pole) is

$$f_H = \frac{1}{2\pi \tau_P} \tag{7.133(b)}$$

## EXAMPLE 7.16

**Objective:** Determine the frequency of a zero and a pole in the high-frequency response of an emitter follower.

Consider the emitter-follower circuit in Figure 7.68 with parameters $V^+ = 5$ V, $V^- = -5$ V, $R_S = 0.1$ kΩ, $R_1 = 40$ kΩ, $R_2 = 5.72$ kΩ, $R_E = 0.5$ kΩ, and $R_L = 10$ kΩ. The transistor parameters are: $\beta = 150$, $V_{BE}(\text{on}) = 0.7$ V, $V_A = \infty$, $C_\pi = 35$ pF, and $C_\mu = 4$ pF.

**Solution:** As in previous examples, the dc analysis yields $I_{CQ} = 1.02$ mA. Therefore, $g_m = 39.2$ mA/V and $r_\pi = 3.82$ kΩ.

From Equation (7.132), the zero occurs at

$$f_o = \frac{1}{2\pi C_\pi\left(\dfrac{r_\pi}{1+\beta}\right)} = \frac{1}{2\pi(35 \times 10^{-12})\left(\dfrac{3.82 \times 10^3}{151}\right)} \Rightarrow 180\,\text{MHz}$$

To determine the time constant for the high-frequency pole calculation, we know that

$$1 + g_m R'_L = 1 + g_m(R_E \| R_L) = 1 + (39.2)(0.5\|10) = 19.7$$

and

$$R_B = R_1 \| R_2 = 40\|5.72 = 5\,\text{k}\Omega$$

The time constant is therefore

$$\tau_P = [R_S \| R_B \| (1 + g_m R'_L) r_\pi]\left(C_\mu + \frac{C_\pi}{1 + g_m R'_L}\right)$$

$$= [(0.1)\|5\|(19.7)(3.82)] \times 10^3\left(4 + \frac{35}{19.7}\right) \times 10^{-12} \Rightarrow 0.566\,\text{ns}$$

The 3 dB frequency (or pole) is then

$$f_H = \frac{1}{2\pi \tau_P} = \frac{1}{2\pi(0.566 \times 10^{-9})} \Rightarrow 281\,\text{MHz}$$

**Comment:** The frequencies for the zero and the pole are very high and are not far apart. This makes the calculations suspect. However, since the frequencies are high, the emitter follower is a wide-bandwidth circuit.

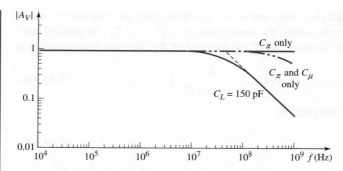

**Figure 7.70** PSpice analysis results for emitter follower

**Computer Verification:** Figure 7.70 shows the results of a PSpice analysis of the emitter follower. From the hand analysis, the 3 dB frequency is on the order of 281 MHz. However, the computer results show the 3 dB frequency to be approximately 400 MHz. We must keep in mind that at these high frequencies, distributed parameter effects may need to be considered in the transistor to more accurately predict the frequency response.

Also shown in the figure is the frequency response due to a 150 pF load capacitance. Comparing this result to the common-emitter circuit, for example, we see that the bandwidth of the emitter-follower circuit is approximately two orders of magnitude larger.

### 7.6.4 High-Frequency Amplifier Design

Our analysis shows that the frequency response, or the high-frequency cutoff point of an amplifier, depends on the transistor used, the circuit parameters, and the amplifier configuration.

We also saw that a computer simulation is easier than a hand analysis, particularly for the emitter-follower circuit. However, the parameters of the actual transistor used in the circuit must be used in the simulation if it is to predict the circuit frequency response accurately. Also, at high frequencies, additional parasitic capacitances, such as the collector–substrate capacitance, may need to be included. This was not done in our examples. Finally, in high-frequency amplifiers, the parasitic capacitances of the interconnect lines between the devices in an IC may also be a factor in the overall circuit response.

## Test Your Understanding

**\*TYU 7.12** For the circuit in Figure 7.71, the transistor parameters are: $K_n = 1$ mA/V$^2$, $V_{TN} = 0.8$ V, $\lambda = 0$, $C_{gs} = 2$ pF, and $C_{gd} = 0.2$ pF. Determine: (a) the Miller capacitance, (b) the upper 3 dB frequency, and (c) the midband voltage gain. (d) Correlate the results from parts (b) and (c) with a computer analysis. (Ans. (a) $C_M = 1.62$ pF, (b) $f_H = 3.38$ MHz, (c) $|A_v| = 4.60$)

**\*TYU 7.13** For the circuit in Figure 7.72, the transistor parameters are: $V_{TN} = 1$ V, $K_n = 1$ mA/V$^2$, $\lambda = 0$, $C_{gd} = 0.4$ pF, and $C_{gs} = 5$ pF. Perform a computer simulation to determine the upper 3 dB frequency and the midband small-signal voltage gain. (Ans. $f_H = 64.5$ MHz, $|A_v| = 0.127$)

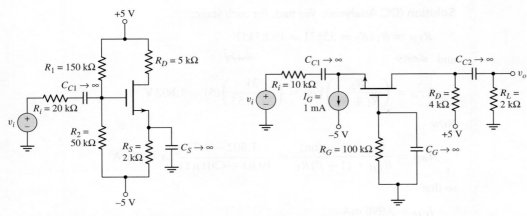

Figure 7.71 Figure for Exercise TYU 7.12        Figure 7.72 Figure for Exercise TYU 7.13

## 7.7 DESIGN APPLICATION: A TWO-STAGE AMPLIFIER WITH COUPLING CAPACITORS

**Objective:** • Design a two-stage BJT amplifier with coupling capacitors such that the 3 dB frequencies associated with each stage are equal.

**Specifications:** The first two stages of a multistage BJT amplifier are to be capacitively coupled and the 3 dB frequency of each stage is to be 20 Hz.

**Design Approach:** The circuit configuration to be designed is shown in Figure 7.73. This circuit represents the first two stages of a discrete multistage amplifier.

**Choices:** Assume the BJTs have parameters $V_{BE}(\text{on}) = 0.7$ V, $\beta = 200$, and $V_A = \infty$.

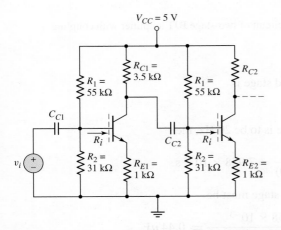

Figure 7.73 Two-stage BJT amplifier with coupling capacitors for design application

**Solution (DC Analysis):** We find, for each stage,

$$R_{TH} = R_1 \| R_2 = 55 \| 31 = 19.83 \, \text{k}\Omega$$

and

$$V_{TH} = \left( \frac{R_2}{R_1 + R_2} \right) V_{CC} = \left( \frac{31}{31 + 55} \right)(5) = 1.802 \, \text{V}$$

Now

$$I_{BQ} = \frac{V_{TH} - V_{BE}(\text{on})}{R_{TH} + (1 + \beta)R_E} = \frac{1.802 - 0.7}{19.83 + (201)(1)} \Rightarrow 4.99 \, \mu\text{A}$$

so that

$$I_{CQ} = 0.998 \, \text{mA}$$

**Solution (AC Analysis):** The small-signal diffusion resistance is

$$r_\pi = \frac{\beta V_T}{I_{CQ}} = \frac{(200)(0.026)}{0.988} = 5.21 \, \text{k}\Omega$$

The input resistance looking into each base terminal is

$$R_i = r_\pi + (1 + \beta)R_E = 5.21 + (201)(1) = 206.2 \, \text{k}\Omega$$

**Solution (AC Design):** The small-signal equivalent circuit is shown in Figure 7.74. The time constant of the first stage is

$$\tau_A = (R_1 \| R_2 \| R_i) C_{C1}$$

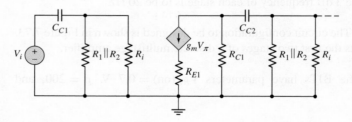

**Figure 7.74** Small-signal equivalent circuit of two-stage BJT amplifier with coupling capacitors for design application

and the time constant of the second stage is

$$\tau_B = (R_{C1} + R_1 \| R_2 \| R_i) C_{C2}$$

If the 3 dB frequency of each stage is to be 20 Hz, then

$$\tau_A = \tau_B = \frac{1}{2\pi f_{3\text{-dB}}} = \frac{1}{2\pi (20)} = 7.958 \times 10^{-3} \, \text{s}$$

The coupling capacitor of the first stage must be

$$C_{C1} = \frac{\tau_A}{R_1 \| R_2 \| R_i} = \frac{7.958 \times 10^{-3}}{(55 \| 31 \| 206.2) \times 10^3} \Rightarrow 0.44 \, \mu\text{F}$$

and the coupling capacitor of the second stage must be

$$C_{C2} = \frac{\tau_B}{R_{C1} + R_1 \| R_2 \| R_i} = \frac{7.958 \times 10^{-3}}{(2.5 + 55 \| 31 \| 206.2) \times 10^3} \Rightarrow 0.386 \, \mu F$$

**Comment:** This circuit design using two coupling capacitors is a brute-force approach to a two-stage amplifier design and would not be used in an IC design.

Since the 3 dB frequency for each capacitor is 20 Hz, this circuit is referred to as a two-pole high-pass filter.

 ## 7.8 SUMMARY

- In this chapter, the frequency response of transistor circuits was discussed. The effects due to circuit capacitors, such as coupling, bypass, and load capacitors, were determined. In addition, expanded equivalent circuits of BJTs and MOSFETs were analyzed to determine the frequency response of the transistors.
- A time-constant technique was developed so that Bode plots can be constructed without the need of deriving complex transfer functions. The high and low corner frequencies, or 3 dB frequencies, can be determined directly from the time constants.
- Coupling and bypass capacitors affect the low-frequency characteristics of a circuit, while load capacitors affect the high-frequency characteristics of a circuit.
- The capacitances included in the small-signal equivalent circuits of both the bipolar and MOS transistors result in reduced transistor gain at high frequencies. The cutoff frequency is a figure of merit for the transistor and is defined as the frequency at which the magnitude of the current gain is unity.
- The Miller effect is a multiplication of the base–collector or gate–drain capacitance due to feedback between the output and input of the transistor. The bandwidth of the amplifier is reduced by this affect.
- The common-emitter (common-source) amplifier, in general, shows the greatest reduction in bandwidth due to the Miller effect. The common-base (common-gate) amplifier has a larger bandwidth because of a smaller multiplication factor. The cascode configuration, a combination of a common emitter and common base, combines the advantages of high gain and wide bandwith.
- As an application, a two-stage BJT amplifier was designed to meet specified 3 dB frequencies.

 ## CHECKPOINT

After studying this chapter, the reader should have the ability to:

✓ Construct the Bode plots of the gain magnitude and phase from a transfer function written in terms of the complex frequency $s$.

✓ Construct the Bode plots of the gain magnitude and phase of electronic amplifier circuits, taking into account circuit capacitors, using the time constant technique.

✓ Determine the short-circuit current gain versus frequency of a BJT and determine the Miller capacitance of a BJT circuit using the expanded hybrid-$\pi$ equivalent circuit.

✓ Determine the unity-gain bandwidth of an FET and determine the Miller capacitance of an FET circuit using the expanded small-signal equivalent circuit.

✓ Describe the relative frequency responses of the three basic amplifier configurations and the cascode amplifier.

## REVIEW QUESTIONS

1. Describe the general frequency response of an amplifier and define the low-frequency, midband, and high-frequency ranges.
2. Describe the general characteristics of the equivalent circuits that apply to the low-frequency, midband, and high-frequency ranges.
3. Describe what is meant by a system transfer function in the s-domain.
4. What is the criterion that defines a corner, or 3 dB, frequency?
5. Describe what is meant by the phase of the transfer function.
6. Describe the time constant technique for determining the corner frequencies.
7. Describe the general frequency response of a coupling capacitor, a bypass capacitor, and a load capacitor.
8. Sketch the expanded hybrid-$\pi$ model of the BJT.
9. Describe the short-circuit current gain versus frequency response of a BJT and define the cutoff frequency.
10. Describe the Miller effect and the Miller capacitance.
11. What effect does the Miller capacitance have on the amplifier bandwidth?
12. Sketch the expanded small-signal equivalent circuit of a MOSFET.
13. Define the cutoff frequency for a MOSFET.
14. What is the major contribution to the Miller capacitance in a MOSFET?
15. Why is there not a Miller effect in a common-base circuit?
16. Describe the configuration of a cascode amplifier.
17. Why is the bandwidth of a cascode amplifier larger, in general, than that of a simple common-emitter amplifier?
18. Why is the bandwidth of the emitter-follower amplifier the largest of the three basic BJT amplifiers?

## PROBLEMS

### Section 7.2 System Transfer Functions

7.1   (a) Determine the voltage transfer function $T(s) = V_o(s)/V_i(s)$ for the circuit shown in Figure P7.1. (b) Sketch the Bode magnitude plot and determine the corner frequency. (c) Determine the time response of the circuit to an input step function of magnitude $V_{Io}$.

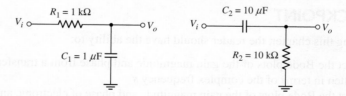

Figure P7.1                                    Figure P7.2

7.2   Repeat Problem 7.1 for the circuit in Figure P7.2.
7.3   Consider the circuit in Figure P7.3. (a) Derive the expression for the voltage transfer function $T(s) = V_o(s)/V_i(s)$. (b) What is the time constant associated with this circuit? (c) Find the corner frequency. (d) Sketch the Bode magnitude plot of the voltage transfer function.
7.4   Consider the circuit in Figure P7.4 with a signal current source. The circuit parameters are $R_i = 30 \text{ k}\Omega$, $R_P = 10 \text{ k}\Omega$, $C_S = 10 \text{ } \mu\text{F}$, and $C_P = 50 \text{ pF}$.

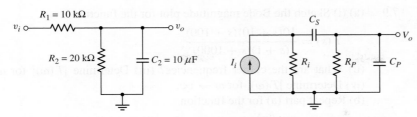

Figure P7.3                  Figure P7.4

(a) Determine the open-circuit time constant associated with $C_S$ and the short-circuit time constant associated with $C_P$. (b) Determine the corner frequencies and the magnitude of the transfer function $T(s) = V_o(s)/I_i(s)$ at midband. (c) Sketch the Bode magnitude plot.

7.5     Consider the circuit shown in Figure P7.5. (a) What is the value of the voltage transfer function $V_o/V_i$ at very low frequencies? (b) Determine the voltage transfer function at very high frequencies. (c) Derive the expression for the voltage transfer function $T(s) = V_o(s)/V_i(s)$. Put the expression in the form $T(s) = K(1 + s\tau_A)/(1 + s\tau_B)$. What are the values of $K$, $\tau_A$, and $\tau_B$?

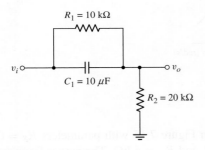

Figure P7.5

*7.6     (a) Derive the voltage transfer function $T(s) = V_o(s)/V_i(s)$ for the circuit shown in Figure 7.10, taking both capacitors into account. (b) Let $R_S = R_P = 10\ \text{k}\Omega$, $C_S = 1\ \mu\text{F}$, and $C_P = 10\ \text{pF}$. Calculate the actual magnitude of the transfer function at $f_L = 1/[(2\pi)(R_S + R_P)C_S]$ and at $f_H = 1/[(2\pi)(R_S \| R_P)C_P]$. How do these magnitudes compare to the maximum magnitude of $R_P/(R_S + R_P)$? (c) Repeat part (b) for $R_S = R_P = 10\ \text{k}\Omega$ and $C_S = C_P = 0.1\ \mu\text{F}$.

7.7     A voltage transfer function is given by $T(f) = 1/(1 + jf/f_T)^3$. (a) Show that the actual response at $f = f_T$ is approximately 9 dB below the maximum value. What is the phase angle at this frequency? (b) What is the slope of the magnitude plot for $f \gg f_T$? What is the phase angle in this frequency range?

7.8     Sketch the Bode magnitude plots for the following functions:

(a) $T_1(s) = \dfrac{s}{s + 100}$

(b) $T_2(s) = \dfrac{5}{s/2000 + 1}$

(c) $T_3(s) = \dfrac{200(s + 10)}{(s + 1000)}$

7.9    (a) (i) Sketch the Bode magnitude plot for the function

$$T(s) = \frac{10(s + 10)(s + 100)}{(s + 1)(s + 1000)}$$

(ii) What are the corner frequencies? (iii) Determine $|T(\omega)|$ for $\omega \to 0$.
(iv) Determine $|T(\omega)|$ for $\omega \to \infty$.
(b) Repeat part (a) for the function

$$T(s) = \frac{8s^2}{(0.2s + 1)^2}$$

7.10    (a) Determine the transfer function corresponding to the Bode plot of the magnitude shown in Figure P7.10. (b) What is the actual gain at (i) $\omega = 50$ rad/s, (ii) $\omega = 150$ rad/s, and (iii) $\omega = 100$ krad/s.

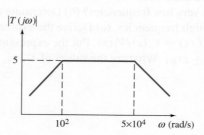

**Figure P7.10**

7.11    Consider the circuit shown in Figure 7.15 with parameters $R_S = 0.5$ k$\Omega$, $r_\pi = 5.2$ k$\Omega$, $g_m = 29$ mA/V, and $R_L = 6$ k$\Omega$. The corner frequencies are $f_L = 30$ Hz and $f_H = 480$ kHz. (a) Calculate the midband voltage gain. (b) What are the open-circuit and short-circuit time constants? (c) Determine $C_C$ and $C_L$.

*7.12    For the circuit shown in Figure P7.12, the parameters are $R_1 = 10$ k$\Omega$, $R_2 = 10$ k$\Omega$, $R_3 = 40$ k$\Omega$, and $C = 10$ $\mu$F. (a) What is the value of the voltage transfer function $V_o/V_i$ at very low frequencies? (b) Determine the value of the voltage transfer function at very high frequencies. (c) Derive the expression for the voltage transfer function $T(s) = V_o(s)/V_i(s)$. Put the expression in the form $T(s) = K(1 + s\tau_A)/(1 + s\tau_B)$. What are the values of $K$, $\tau_A$, and $\tau_B$?

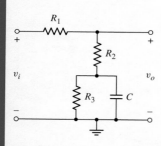

**Figure P7.12**

7.13    The circuit shown in Figure 7.10 has parameters $R_S = 1$ k$\Omega$, $R_P = 10$ k$\Omega$, and $C_S = C_P = 0.01$ $\mu$F. Using PSpice, plot the magnitude and phase of the voltage transfer function. Determine the maximum value of the voltage transfer function. Determine the frequencies at which the magnitude is $1/\sqrt{2}$ of the peak value.

## Section 7.3 Frequency Response: Transistor Circuits

7.14    The transistor shown in Figure P7.14 has parameters $V_{TN} = 0.4$ V, $K_n = 0.4$ mA/V$^2$, and $\lambda = 0$. The transistor is biased at $I_{DQ} = 0.8$ mA. (a) What is the maximum voltage gain? (b) What is the bandwidth?

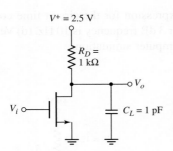

Figure P7.14

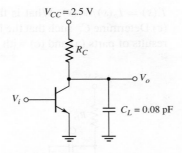

Figure P7.15

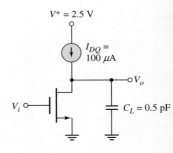

Figure P7.16

7.15 Consider the circuit shown in Figure P7.15. The transistor has parameters $\beta = 120$ and $V_A = \infty$. The circuit bandwidth is 800 MHz and the quiescent collector–emitter voltage is $V_{CEQ} = 1.25$ V. (a) Determine $R_C$, (b) find $I_{CQ}$, and (c) determine the maximum gain.

7.16 The transistor in the circuit shown in Figure P7.16 has parameters $V_{TN} = 0.4$V, $K_n = 50\ \mu A/V^2$, and $\lambda = 0.01$ V$^{-1}$. (a) Derive the expression for the voltage transfer function $T(s) = V_o(s)/V_i(s)$. (b) Determine the maximum voltage gain. (c) What is the bandwidth?

7.17 For the common-emitter circuit in Figure P7.17, the transistor parameters are: $\beta = 100$, $V_{BE}(on) = 0.7$ V, and $V_A = \infty$. (a) Calculate the lower corner frequency. (b) Determine the midband voltage gain. (c) Sketch the Bode plot of the voltage gain magnitude.

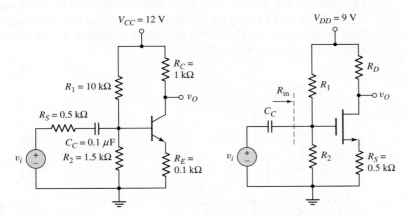

Figure P7.17

Figure P7.18

D7.18 (a) Design the circuit shown in Figure P7.18 such that $I_{DQ} = 0.8$ mA, $V_{DSQ} = 3.2$ V, $R_{in} = 160$ kΩ, and $f_L = 16$ Hz. The transistor parameters are $K_n = 0.5$ mA/V$^2$, $V_{TN} = 1.2$ V, and $\lambda = 0$. (b) What is the midband voltage gain? (c) Determine the magnitude of the voltage gain at (i) $f = 5$ Hz, (ii) $f = 14$ Hz, and (iii) $f = 25$ Hz. (d) Sketch the Bode plot of the voltage gain magnitude and phase.

D7.19 The transistor in the circuit in Figure P7.19 has parameters $K_n = 0.5$ mA/V$^2$, $V_{TN} = 1$ V, and $\lambda = 0$. (a) Design the circuit such that $I_{DQ} = 1$ mA and $V_{DSQ} = 3$ V. (b) Derive the expression for the transfer function

$T(s) = I_o(s)/V_i(s)$. What is the expression for the circuit time constant? (c) Determine $C_C$ such that the lower 3 dB frequency is 10 Hz. (d) Verify the results of parts (a) and (c) with a computer simulation.

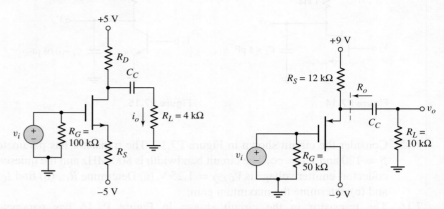

Figure P7.19                          Figure P7.20

*D7.20  The transistor in the circuit in Figure P7.20 has parameters $K_p = 0.5$ mA/V², $V_{TP} = -2$ V, and $\lambda = 0$. (a) Determine $R_o$. (b) What is the expression for the circuit time constant? (c) Determine $C_C$ such that the lower 3 dB frequency is 20 Hz.

7.21   For the circuit in Figure P7.21, the transistor parameters are $\beta = 120$, $V_{BE}(\text{on}) = 0.7$ V, and $V_A = 50$ V. (a) Design a bias-stable circuit such that $I_{EQ} = 1.5$ mA. (b) Using the results of part (a), find the small-signal midband voltage gain. (c) Determine the output resistance $R_o$. (d) What is the lower 3 dB corner frequency?

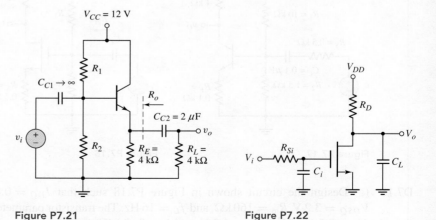

Figure P7.21                          Figure P7.22

7.22   (a) For the circuit shown in Figure P7.22, write the voltage transfer function $T(s) = V_o(s)/V_i(s)$. Assume $\lambda > 0$ for the transistor. (b) What is the expression for the time constant associated with the input portion of the circuit? (c) What is the expression for the time constant associated with the output portion of the circuit?

7.23   Consider the circuit shown in Figure P7.23. (a) Write the transfer function $T(s) = V_o(s)/V_i(s)$. Assume $\lambda = 0$ for the transistor. (b) Determine the expression for the time constant associated with the input portion of the circuit. (c) Determine the expression for the time constant associated with the output portion of the circuit.

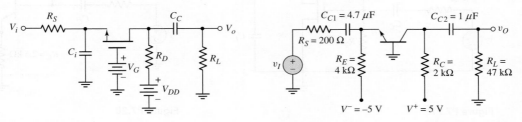

**Figure P7.23**                                   **Figure P7.24**

7.24   The parameters of the transistor in the circuit in Figure P7.24 are $V_{BE}(\text{on}) = 0.7$ V, $\beta = 100$, and $V_A = \infty$. (a) Determine the quiescent and small-signal parameters of the transistor. (b) Find the time constants associated with $C_{C1}$ and $C_{C2}$. (c) Is there a dominant $-3$ dB frequency? Estimate the $-3$ dB frequency.

7.25   A capacitor is placed in parallel with $R_L$ in the circuit in Figure P7.24. The capacitance is $C_L = 10$ pF. The transistor parameters are the same as given in Problem 7.24. (a) Determine the upper $-3$ dB frequency. (b) Find the high frequency value at which the small-signal voltage gain magnitude is one-tenth the midband value.

7.26   The parameters of the transistor in the circuit in Figure P7.26 are $K_p = 1$ mA/V$^2$, $V_{TP} = -1.5$ V, and $\lambda = 0$. (a) Determine the quiescent and small-signal parameters of the transistor. (b) Find the time constants associated with $C_{C1}$ and $C_{C2}$. (c) Is there a dominant pole frequency? Estimate the $-3$ dB frequency.

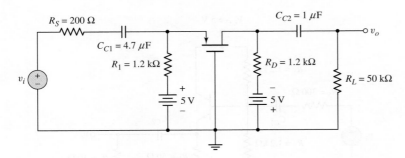

**Figure P7.26**

*D7.27  A MOSFET amplifier with the configuration in Figure P7.27 is to be designed for use in a telephone circuit. The magnitude of the voltage gain should be 10 in the midband range, and the midband frequency range should extend from 200 Hz to 3 kHz. [Note: A telephone's frequency range does not correspond to a high-fidelity system's.] All resistor, capacitor, and MOSFET parameters should be specified.

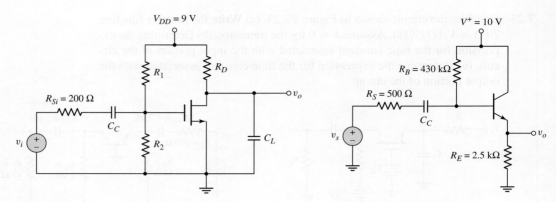

**Figure P7.27**                                            **Figure P7.28**

7.28   The circuit in Figure P7.28 is a simple output stage of an audio amplifier. The transistor parameters are $\beta = 200$, $V_{BE}(\text{on}) = 0.7$ V, and $V_A = \infty$. Determine $C_C$ such that the lower $-3$ dB frequency is 15 Hz.

7.29   Reconsider the circuit in Figure P7.28. The transistor parameters are $\beta = 120$, $V_{BE}(\text{on}) = 0.7$ V, and $V_A = \infty$. The circuit parameters are $V^+ = 3.3$ V and $R_S = 100\,\Omega$. (a) Find $R_B$ and $R_E$ such that $I_{EQ} = 0.25$ mA and $V_{CEQ} = 1.8$ V. (b) Using the results of part (a), find the value of $C_C$ such that $f_L = 20$ Hz. (c) Determine the midband voltage gain.

D7.30  The parameters of the transistor in the circuit in Figure P7.30 are $\beta = 100$, $V_{BE}(\text{on}) = 0.7$ V, and $V_A = \infty$. The time constant associated with $C_{C1}$ is a factor of 100 larger than the time constant associated with $C_{C2}$. (a) Determine $C_{C2}$ such that the $-3$ dB frequency associated with this capacitor is 25 Hz. (b) Determine $C_{C1}$.

D7.31  Consider the circuit shown in Figure P7.30. The time constant associated with $C_{C2}$ is a factor of 100 larger than the time constant associated with $C_{C1}$. (a) Determine $C_{C1}$ such that the $-3$ dB frequency associated with this capacitor is 20 Hz. (b) Find $C_{C2}$.

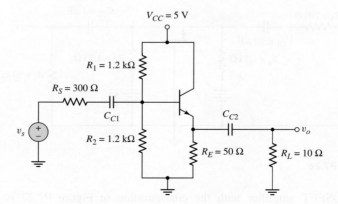

**Figure P7.30**

7.32   Consider the circuit shown in Figure P7.32. The transistor parameters are $\beta = 120$, $V_{BE}(\text{on}) = 0.7$ V, and $V_A = \infty$. (a) Find $R_C$ such that $V_{CEQ} = 2.2$ V. (b) Determine the midband gain. (c) Derive the expression for the corner frequencies associated with $C_C$ and $C_E$. (d) Determine $C_C$

and $C_E$ such that the corner frequency associated with $C_E$ is $f_E = 10\,\text{Hz}$ and the corner frequency associated with $C_C$ is $f_C = 50\,\text{Hz}$.

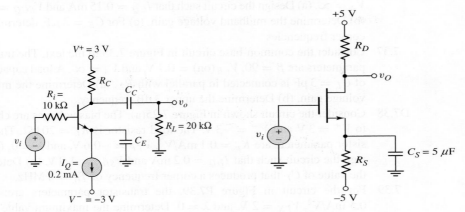

**Figure P7.32**                    **Figure P7.33**

*D7.33 For the transistor in the circuit in Figure P7.33, the parameters are: $K_n = 0.5\,\text{mA/V}^2$, $V_{TN} = 0.8\,\text{V}$, and $\lambda = 0$. (a) Design the circuit such that $I_{DQ} = 0.5\,\text{mA}$ and $V_{DSQ} = 4\,\text{V}$. (b) Determine the 3 dB frequencies. (c) If the $R_S$ resistor is replaced by a constant-current source producing the same $I_{DQ}$ quiescent current, determine the 3 dB corner frequencies.

7.34 Figure P7.34 shows the ac equivalent circuit of two identical common-source circuits in cascade. The transistor parameters are $K_{n1} = K_{n2} = 0.8\,\text{mA/V}^2$, $\lambda_1 = \lambda_2 = 0.02\,\text{V}^{-1}$, and $I_{DQ1} = I_{DQ2} = 0.5\,\text{mA}$. The circuit parameters are $R_D = 5\,\text{k}\Omega$ and $C_L = 12\,\text{pF}$. (a) Derive the expressions for the voltage transfer functions (i) $T_1(s) = V_{o1}(s)/V_i(s)$, (ii) $T_2(s) = V_o(s)/V_{o1}(s)$, and (iii) $T(s) = V_o(s)/V_i(s)$. (b) Determine the −3 dB frequencies for (i) $T_1(s)$, (ii) $T_2(s)$, and (iii) $T(s)$. (c) Sketch the Bode plot for the magnitude of the transfer function $T(s)$.

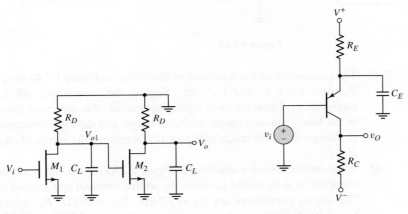

**Figure P7.34**                    **Figure P7.35**

*7.35 The common-emitter circuit in Figure P7.35 has an emitter bypass capacitor. (a) Derive the expression for the small-signal voltage gain $A_v(s) = V_o(s)/V_i(s)$. Write the expression in a form similar to that of Equation (7.60). (b) What are the expressions for the time constants $\tau_A$ and $\tau_B$?

**D7.36** Consider the circuit in Figure P7.35. The bias voltages are $V^+ = 3\,\text{V}$ and $V^- = -3\,\text{V}$. The transistor parameters are $\beta = 90$, $V_{EB}(\text{on}) = 0.7$ V and $V_A = \infty$. (a) Design the circuit such that $I_{CQ} = 0.15\,\text{mA}$ and $V_{ECQ} = 2.2\,\text{V}$. (b) Determine the midband voltage gain. (c) For $C_E = 3\,\mu\text{F}$, determine the corner frequencies.

**7.37** Consider the common-base circuit in Figure 7.33 in the text. The transistor parameters are $\beta = 90$, $V_{EB}(\text{on}) = 0.7$ V, and $V_A = \infty$. A load capacitance of $C_L = 3\,\text{pF}$ is connected in parallel with $R_L$. (a) Determine the midband voltage gain. (b) Determine the upper 3 dB frequency.

**D7.38** Consider the circuit shown in Figure 7.25(a). The bias voltages are changed to $V^+ = 3\,\text{V}$ and $V^- = -3\,\text{V}$. The load resistor is $R_L = 20\,\text{k}\Omega$. The transistor parameters are $K_p = 0.1\,\text{mA/V}^2$, $V_{TP} = -0.6\,\text{V}$, and $\lambda = 0$. (a) Design the circuit such that $I_{DQ} = 0.2\,\text{mA}$ and $V_{SDQ} = 1.9\,\text{V}$. (b) Determine the value of $C_L$ that produces a corner frequency of $f_H = 4\,\text{MHz}$.

**7.39** For the circuit in Figure P7.39, the transistor parameters are: $K_n = 0.5\,\text{mA/V}^2$, $V_{TN} = 2\,\text{V}$, and $\lambda = 0$. Determine the maximum value of $C_L$ such that the bandwidth is at least $BW = 5\,\text{MHz}$. State any approximations or assumptions that you make. What is the magnitude of the small-signal midband voltage gain? Verify the results with a computer simulation.

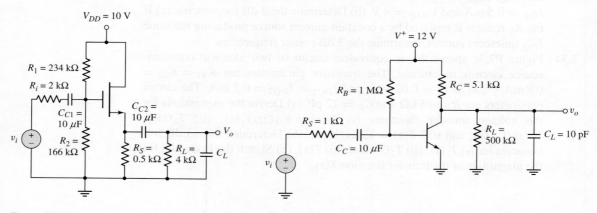

Figure P7.39

Figure P7.40

**7.40** The parameters of the transistor in the circuit in Figure P7.40 are $\beta = 100$, $V_{BE}(\text{on}) = 0.7$ V, and $V_A = \infty$. Neglect the capacitance effects of the transistor. (a) Draw the three equivalent circuits that represent the amplifier in the low-frequency range, midband range, and the high frequency range. (b) Sketch the Bode magnitude plot. (c) Determine the values of $|A_m|_{\text{dB}}$, $f_L$, and $f_H$.

**7.41** In the common-source amplifier in Figure 7.25(a) in the text, a source bypass capacitor is to be added between the source terminal and ground potential. The circuit parameters are $R_S = 3.2\,\text{k}\Omega$, $R_D = 10\,\text{k}\Omega$, $R_L = 20\,\text{k}\Omega$, and $C_L = 10\,\text{pF}$. The transistor parameters are $V_{TP} = -2\,\text{V}$, $K_P = 0.25\,\text{mA/V}^2$, and $\lambda = 0$. (a) Derive the small-signal voltage gain expression, as a function of $s$, that describes the circuit behavior in the high-frequency range. (b) What is the expression for the time constant associated with the upper 3 dB frequency? (c) Determine the time constant, upper 3 dB frequency, and small-signal midband voltage gain.

*7.42    Consider the common-base circuit in Figure P7.42. Choose appropriate transistor parameters. (a) Using a computer analysis, generate the Bode plot of the voltage gain magnitude from a very low frequency to the midband frequency range. At what frequency is the voltage gain magnitude 3 dB below the maximum value? What is the slope of the curve at very low frequencies? (b) Using the PSpice analysis, determine the voltage gain magnitude, input resistance $R_i$, and output resistance $R_o$ at midband.

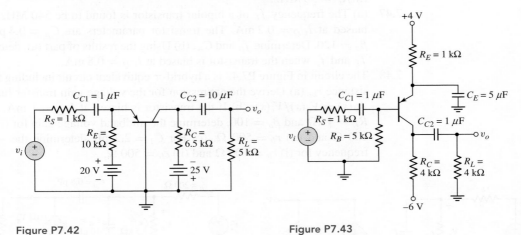

**Figure P7.42**                          **Figure P7.43**

*7.43    For the common-emitter circuit in Figure P7.43, choose appropriate transistor parameters and perform a computer analysis. Generate the Bode plot of the voltage gain magnitude from a very low frequency to the midband frequency range. At what frequency is the voltage gain magnitude 3 dB below the maximum value? Does one capacitor dominate this 3 dB frequency? If so, which one?

*7.44    For the multitransistor amplifier in Figure P7.44, choose appropriate transistor parameters. The lower 3 dB frequency is to be less than or equal to 20 Hz. Assume that all three coupling capacitors are equal. Let $C_B \rightarrow \infty$. Using a computer analysis, determine the maximum values of the coupling capacitors. Determine the slope of the Bode plot of the voltage gain magnitude at very low frequencies.

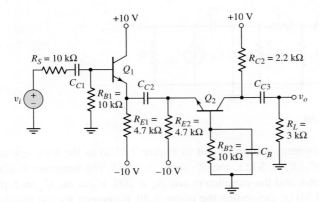

**Figure P7.44**

## Section 7.4 Frequency Response: Bipolar Transistor

7.45   A bipolar transistor has $f_T = 4\,\text{GHz}$, $\beta_o = 120$, and $C_\mu = 0.08\,\text{pF}$ when operated at $I_{CQ} = 0.25\,\text{mA}$. Determine $g_m$, $f_\beta$, and $C_\pi$.

7.46   A high-frequency bipolar transistor is biased at $I_{CO} = 0.4\,\text{mA}$ and has parameters $C_\mu = 0.075\,\text{pF}$, $f_T = 2\,\text{GHz}$, and $\beta_o = 120$. (a) Determine $C_\pi$ and $f_\beta$. (b) Determine $|h_{fe}|$ at (i) $f = 10\,\text{MHz}$, (ii) $f = 20\,\text{MHz}$, and (iii) $f = 50\,\text{MHz}$.

7.47   (a) The frequency $f_T$ of a bipolar transistor is found to be 540 MHz when biased at $I_{CQ} = 0.2\,\text{mA}$. The transistor parameters are $C_\mu = 0.4\,\text{pF}$ and $\beta_o = 120$. Determine $f_\beta$ and $C_\pi$. (b) Using the results of part (a), determine $f_T$ and $f_\beta$ when the transistor is biased at $I_{CQ} = 0.8\,\text{mA}$.

7.48   The circuit in Figure P7.48 is a hybrid-$\pi$ equivalent circuit including the resistance $r_b$. (a) Derive the expression for the voltage gain transfer function $A_v(s) = V_o(s)/V_i(s)$. (b) If the transistor is biased at $I_{CQ} = 1\,\text{mA}$, and if $R_L = 4\,\text{k}\Omega$ and $\beta_o = 100$, determine the midband voltage gain for (i) $r_b = 100\,\Omega$ and (ii) $r_b = 500\,\Omega$. (c) For $C_1 = 2.2\,\text{pF}$, determine the $-3\,\text{dB}$ frequency for (i) $r_b = 100\,\Omega$ and (ii) $r_b = 500\,\Omega$.

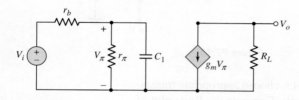

**Figure P7.48**

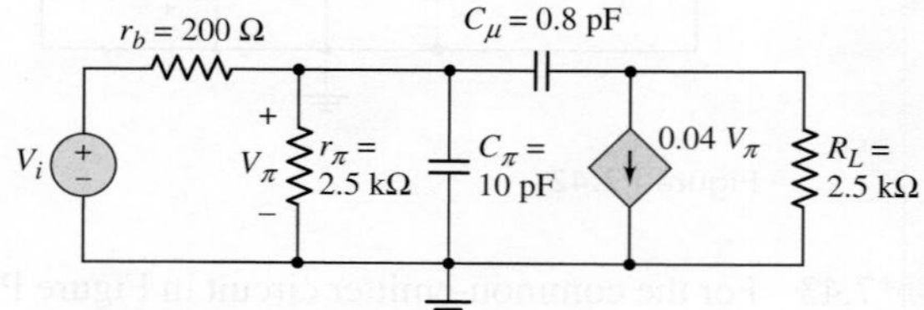

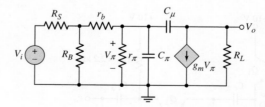

**Figure P7.49**

7.49   Consider the circuit in Figure P7.49. Calculate the impedance seen by the signal source $V_i$ at (a) $f = 1\,\text{kHz}$, (b) $f = 10\,\text{kHz}$, (c) $f = 100\,\text{kHz}$, and (d) $f = 1\,\text{MHz}$.

*7.50   A common-emitter equivalent circuit is shown in Figure P7.50. (a) What is the expression for the Miller capacitance? (b) Derive the expression for the voltage gain $A_v(s) = V_o(s)/V_i(s)$ in terms of the Miller capacitance and other circuit parameters. (c) What is the expression for the upper 3 dB frequency?

**Figure P7.50**

7.51   For the common-emitter circuit in Figure 7.41(a) in the text, assume that $r_s = \infty$, $R_1 \| R_2 = 5\,\text{k}\Omega$, and $R_C = R_L = 1\,\text{k}\Omega$. The transistor is biased at $I_{CQ} = 5\,\text{mA}$ and the parameters are: $\beta_o = 200$, $V_A = \infty$, $C_\mu = 5\,\text{pF}$, and $f_T = 250\,\text{MHz}$. Determine the upper 3 dB frequency for the small-signal current gain.

*7.52  For the common-emitter circuit in Figure P7.52, assume the emitter bypass capacitor $C_E$ is very large, and the transistor parameters are: $\beta_o = 100$, $V_{BE}(\text{on}) = 0.7$ V, $V_A = \infty$, $C_\mu = 2$ pF, and $f_T = 400$ MHz. Determine the lower and upper 3 dB frequencies for the small-signal voltage gain. Use the simplified hybrid-$\pi$ model for the transistor.

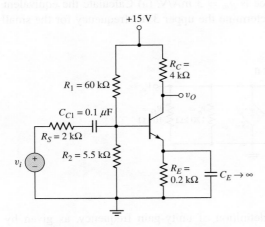

**Figure P7.52**

7.53  Consider the circuit in Figure P7.52. The resistor $R_S$ is changed to 500 $\Omega$ and all other resistor values are increased by a factor of 10. The transistor parameters are the same as listed in Problem 7.52. Determine the lower and upper $-3$ dB frequencies for the voltage gain magnitude and find the mid-band gain.

7.54  The parameters of the circuit shown in Figure P7.52 are changed to $V^+ = 5$ V, $R_S = 0$, $R_1 = 33$ k$\Omega$, $R_2 = 22$ k$\Omega$, $R_C = 5$ k$\Omega$, and $R_E = 4$ k$\Omega$. The transistor parameters are $\beta_o = 150$, $C_\mu = 0.45$ pF, and $f_T = 800$ MHz. (a) Determine $I_{CQ}$ and $V_{CEQ}$. (b) Determine $C_\pi$, $f_\beta$, and the Miller capacitance $C_M$. (c) Find the upper 3 dB frequency.

## Section 7.5  Frequency Response: The FET

7.55  The parameters of an n-channel MOSFET are $k'_n = 80$ $\mu$A/V$^2$, $W = 4$ $\mu$m, $L = 0.8$ $\mu$m, $C_{gs} = 50$ fF, and $C_{gd} = 10$ fF. The transistor is biased at $I_{DQ} = 0.6$ mA. Determine $f_T$.

7.56  Find $f_T$ for a MOSFET biased at $I_{DQ} = 120$ $\mu$A and $V_{GS} - V_{TN} = 0.20$ V. The transistor parameters are $C_{gs} = 40$ fF and $C_{gd} = 10$ fF.

7.57  Fill in the missing parameter values in the following table for a MOSFET. Let $K_n = 1.5$ mA/V$^2$.

| $I_D$ ($\mu$A) | $f_T$ (GHz) | $C_{gs}$(fF) | $C_{gd}$ (fF) |
|---|---|---|---|
| 50 | | 60 | 10 |
| 300 | | 60 | 10 |
| | 3 | 60 | 10 |
| 250 | 2.5 | | 8 |

7.58   (a) An n-channel MOSFET has an electron mobility of 450 cm$^2$/V–s and a channel length of 1.2 $\mu$m. Let $V_{GS} - V_{TN} = 0.5$ V. Determine the cutoff frequency $f_T$. (b) Repeat part (a) if the channel length is reduced to 0.18 $\mu$m.

7.59   A common-source equivalent circuit is shown in Figure P7.59. The transistor transconductance is $g_m = 3$ mA/V. (a) Calculate the equivalent Miller capacitance. (b) Determine the upper 3 dB frequency for the small-signal voltage gain.

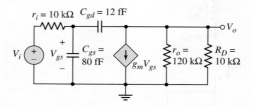

Figure P7.59

7.60   Starting with the definition of unity-gain frequency, as given by Equation (7.97), neglect the overlap capacitance, assume $C_{gd} \cong 0$ and $C_{gs} \cong \left(\frac{2}{3}\right) WLC_{ox}$, and show that

$$f_T = \frac{3}{2\pi L} \cdot \sqrt{\frac{\mu_n I_D}{2C_{ox} W L}}$$

Since $I_D$ is proportional to $W$, this relationship indicates that to increase $f_T$, the channel length $L$ must be small.

7.61   The parameters of an ideal n-channel MOSFET are $W/L = 8$, $\mu_n = 400$ cm$^2$/V–s, $C_{ox} = 6.9 \times 10^{-7}$ F/cm$^2$, and $V_{TN} = 0.4$ V. (a) Determine the maximum source resistance such that the transconductance $g_m$ is reduced by no more that 20 percent from its ideal value when $V_{GS} = 3$ V. (b) Using the results of part (a), find how much $g_m$ is reduced from its ideal value when $V_{GS} = 1$ V.

*7.62   Figure P7.62 shows the high-frequency equivalent circuit of an FET, including a source resistance $r_s$. (a) Derive an expression for the low-frequency current gain $A_i = I_o/I_i$. (b) Assuming $R_i$ is very large, derive an expression for the current gain transfer function $A_i(s) = I_o(s)/I_i(s)$. (c) How does the magnitude of the current gain behave as $r_s$ increases?

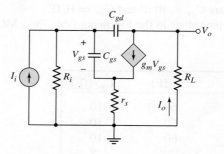

Figure P7.62

7.63 For the FET circuit in Figure P7.63, the transistor parameters are: $K_n = 1 \text{ mA/V}^2$, $V_{TN} = 2 \text{ V}$, $\lambda = 0$, $C_{gs} = 50 \text{ fF}$, and $C_{gd} = 8 \text{ fF}$. (a) Draw the simplified high-frequency equivalent circuit. (b) Calculate the equivalent Miller capacitance. (c) Determine the upper 3 dB frequency for the small-signal voltage gain and find the midband voltage gain.

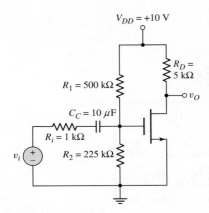

**Figure P7.63**

## Section 7.6 High-Frequency Response of Transistor Circuits

7.64 The midband voltage gain of a common-source MOSFET amplifier is $A_v = -15 \text{ V/V}$. The capacitances of the transistor are $C_{gs} = 0.2 \text{ pF}$ and $C_{gd} = 0.04 \text{ pF}$. (a) Determine the input Miller capacitance. (b) What equivalent input resistance (bias resistance and signal source resistance) would result in an upper corner frequency of 5 MHz?

7.65 In the circuit in Figure P7.65, the transistor parameters are: $\beta = 120$, $V_{BE}(\text{on}) = 0.7 \text{ V}$, $V_A = 100 \text{ V}$, $C_\mu = 1 \text{ pF}$, and $f_T = 600 \text{ MHz}$. (a) Determine $C_\pi$ and the equivalent Miller capacitance $C_M$. State any approximations or assumptions that you make. (b) Find the upper 3 dB frequency and the midband voltage gain.

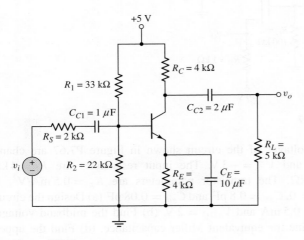

**Figure P7.65**

*7.66 In the circuit in Figure P7.66, the transistor parameters are: $\beta = 120$, $V_{BE}(\text{on}) = 0.7$ V, $V_A = \infty$, $C_\mu = 3$ pF, and $f_T = 250$ MHz. Assume the emitter bypass capacitor $C_E$ and the coupling capacitor $C_{C2}$ are very large. (a) Determine the lower and upper 3 dB frequencies. Use the simplified hybrid-$\pi$ model for the transistor. (b) Sketch the Bode plot of the voltage gain magnitude.

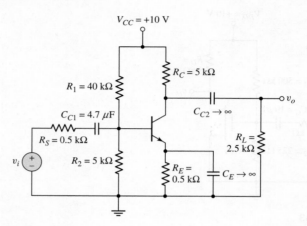

**Figure P7.66**

7.67 The parameters of the transistor in the common-source circuit in Figure P7.67 are: $K_p = 2$ mA/V$^2$, $V_{TP} = -2$ V, $\lambda = 0.01$ V$^{-1}$, $C_{gs} = 10$ pF, and $C_{gd} = 1$ pF. (a) Determine the equivalent Miller capacitance $C_M$. (b) Find the upper 3 dB frequency and midband voltage gain.

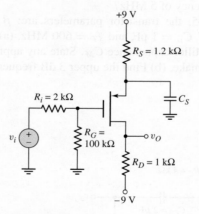

**Figure P7.67**

7.68 The bias voltages of the circuit shown in Figure P7.67 are changed to $V^+ = 3$ V and $V^- = -3$ V. The input resistances are $R_i = 4$ kΩ and $R_G = 200$ kΩ. The transistor parameters are $K_p = 0.5$ mA/V$^2$, $V_{TP} = -0.5$ V, $\lambda = 0, C_{gs} = 0.8$ pF, and $C_{gd} = 0.08$ pF. (a) Design the circuit such that $I_{DQ} = 0.5$ mA and $V_{SDQ} = 2$ V. (b) Find the midband voltage gain. (c) Determine the equivalent Miller capacitance. (d) Find the upper 3 dB frequency.

7.69   For the PMOS common-source circuit shown in Figure P7.69, the transistor
       parameters are: $V_{TP} = -2$ V, $K_p = 1$ mA/V$^2$, $\lambda = 0$, $C_{gs} = 15$ pF, and
       $C_{gd} = 3$ pF. (a) Determine the upper 3 dB frequency. (b) What is the equiv-
       alent Miller capacitance? State any assumptions or approximations that you
       make. (c) Find the midband voltage gain.

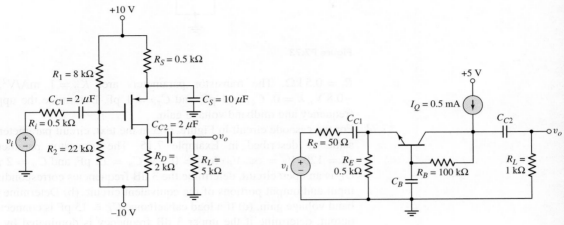

Figure P7.69                                          Figure P7.70

*7.70  In the common-base circuit shown in Figure P7.70, the transistor parame-
       ters are: $\beta = 100$, $V_{BE}(\text{on}) = 0.7$ V, $V_A = \infty$, $C_\pi = 10$ pF, and $C_\mu = 1$ pF.
       (a) Determine the upper 3 dB frequencies corresponding to the input and
       output portions of the equivalent circuit. (b) Calculate the small-signal mid-
       band voltage gain. (c) If a load capacitor $C_L = 15$ pF is connected between
       the output and ground, determine if the upper 3 dB frequency will be dom-
       inated by the $C_L$ load capacitor or by the transistor characteristics.
*7.71  Repeat Problem 7.70 for the common-base circuit in Figure P7.71. Assume
       $V_{EB}(\text{on}) = 0.7$ for the pnp transistor. The remaining transistor parameters
       are the same as given in Problem 7.70.

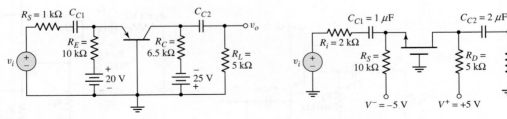

Figure P7.71                                          Figure P7.72

*7.72  In the common-gate circuit in Figure P7.72, the transistor parameters are:
       $V_{TN} = 1$ V, $K_n = 3$ mA/V$^2$, $\lambda = 0$, $C_{gs} = 15$ pF, and $C_{gd} = 4$ pF. Deter-
       mine the upper 3 dB frequency and midband voltage gain.
7.73   Consider the common-gate circuit in Figure P7.73 with parameters $V^+ =$
       5 V, $V^- = -5$ V, $R_S = 4$ k$\Omega$, $R_D = 2$ k$\Omega$, $R_L = 4$ k$\Omega$, $R_G = 50$ k$\Omega$, and

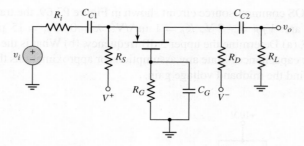

**Figure P7.73**

$R_i = 0.5$ k$\Omega$. The transistor parameters are: $K_p = 1$ mA/V$^2$, $V_{TP} = -0.8$ V, $\lambda = 0$, $C_{gs} = 4$ pF, and $C_{gd} = 1$ pF. Determine the upper 3 dB frequency and midband voltage gain.

*7.74 For the cascode circuit in Figure 7.65 in the text, circuit parameters are the same as described in Example 7.15. The transistor parameters are: $\beta_o = 120$, $V_A = \infty$, $V_{BE}$(on) $= 0.7$ V, $C_\pi = 12$ pF, and $C_\mu = 2$ pF. (a) If $C_L$ is an open circuit, determine the 3 dB frequencies corresponding to the input and output portions of the equivalent circuit. (b) Determine the midband voltage gain. (c) If a load capacitance $C_L = 15$ pF is connected to the output, determine if the upper 3 dB frequency is dominated by the load capacitance or by the transistor characteristics.

## COMPUTER SIMULATION PROBLEMS

7.75 An emitter-follower amplifier is shown in Figure P7.75. Using a computer simulation, determine the upper 3 dB frequency and the midband voltage gain for: (a) $R_L = 0.2$ k$\Omega$, (b) $R_L = 2$ k$\Omega$, and (c) $R_L = 20$ k$\Omega$. Use a standard transistor. Explain any differences between the results of the three parts.

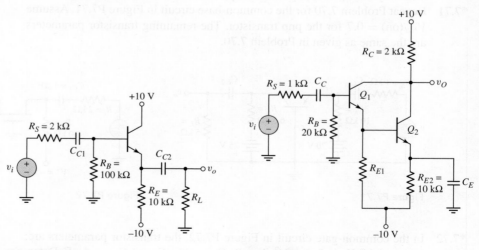

**Figure P7.75**

**Figure P7.76**

7.76 The transistor circuit in Figure P7.76 is a Darlington pair configuration. Using a computer simulation, determine the upper 3 dB frequency and the midband voltage gain for (a) $R_{E1} = 10\,\text{k}\Omega$, (b) $R_{E1} = 40\,\text{k}\Omega$, and (c) $R_{E1} = \infty$. Use standard transistors. Explain any differences between the results of the three parts.

7.77  Consider the common-source amplifier in Figure P7.77(a) and the cascode amplifier in Figure P7.77(b). Using standard transistors, determine the upper 3 dB frequency and the midband voltage gain for each circuit using a computer simulation. Compare the 3 dB frequencies and midband voltage gains.

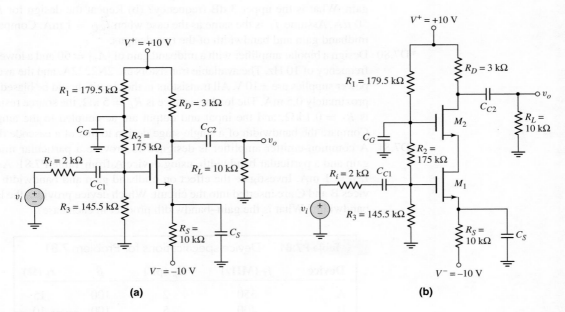

(a)                                                     (b)

**Figure P7.77**

7.78  Consider identical transistors in the circuit in Figure P7.78. Assume the two coupling capacitors are both equal to $C_C = 4.7\,\mu\text{F}$. Using a computer simulation, determine the lower and upper 3 dB frequencies as well as the midband gain. What value of load capacitance will change the bandwidth by a factor of two?

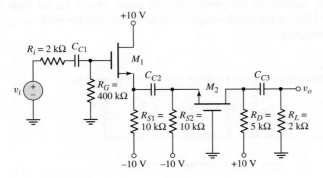

**Figure P7.78**

**DESIGN PROBLEMS**

[Note: Each design should be verified with a computer analysis.]

*D7.79 (a) Design a common-emitter amplifier using a 2N2222A transistor biased at $I_{CQ} = 1$ mA and $V_{CEQ} = 10$ V. The available power supplies are $\pm 15$ V, the load resistance is $R_L = 20$ k$\Omega$, the source resistance is $R_S = 0.5$ k$\Omega$, the input and output are ac coupled to the amplifier, and the lower 3 dB frequency is to be less than 10 Hz. Design the circuit to maximize the midband gain. What is the upper 3 dB frequency? (b) Repeat the design for $I_{CQ} = 50$ $\mu$A. Assume $f_T$ is the same as the case when $I_{CQ} = 1$ mA. Compare the midband gain and bandwidth of the two designs.

*D7.80 Design a bipolar amplifier with a midband gain of $|A_v| = 50$ and a lower 3 dB frequency of 10 Hz. The available transistors are 2N2222A, and the available power supplies are $\pm 10$ V. All transistors in the circuit should be biased at approximately 0.5 mA. The load resistance is $R_L = 5$ k$\Omega$, the source resistance is $R_S = 0.1$ k$\Omega$, and the input and output are ac coupled to the amplifier. Compare the bandwidth of a single-stage design to that of a cascode design.

*D7.81 A common-emitter amplifier is designed to provide a particular midband gain and a particular bandwidth, using device A from Table P7.81. Assume $I_{CQ} = 1$ mA. Investigate the effect on midband gain and bandwidth if devices B and C are inserted into the circuit. Which device provides the largest bandwidth? What is the gain–bandwidth product in each case?

| **Table P7.81** | Device specifications for Problem 7.81 | | | |
|---|---|---|---|---|
| **Device** | $f_T$ **(MHz)** | $C_\mu$ **(pF)** | $\beta$ | $r_b$ **($\Omega$)** |
| A | 350 | 2 | 100 | 15 |
| B | 400 | 5 | 100 | 10 |
| C | 500 | 2 | 50 | 5 |

*D7.82 A simplified high-frequency equivalent circuit of a common-emitter amplifier is shown in Figure P7.82. The input signal is coupled into the amplifier through $C_{C1}$, the output signal is coupled to the load through $C_{C2}$, and the amplifier provides a midband gain of $|A_m|$ and an upper 3 dB frequency of $f_H$. Compare this single-stage amplifier design to one in which three amplifier stages are used between the signal and load. In the three-stage amplifier, assume all parameters are the same, except $g_m$ for each stage is one-third that of the single-stage amplifier. Compare the midband gains and the bandwidths.

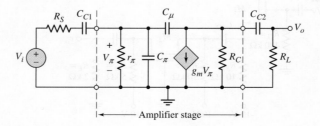

**Figure P7.82**

# Output Stages and Power Amplifiers

In previous chapters, we dealt mainly with small-signal voltage gains, current gains, and impedance characteristics of linear amplifiers. In this chapter, we analyze and design circuits that must deliver a specified power to a load. We will, therefore, be concerned with power dissipation in transistors, especially in the output stage, since the output stage must deliver the signal power. Linearity in the output signal is still a priority, however. A figure of merit for the output stage linearity characteristic is the total harmonic distortion that is present.

Various classes of power amplifiers are defined. The ideal and actual power efficiencies of these classes of amplifiers are determined.

## PREVIEW

In this chapter, we will:

- Describe the concept of a power amplifier.
- Describe the characteristics of BJT and MOSFET power transistors, and analyze the temperature and heat flow characteristics of devices using heat sinks.
- Define the various classes of power amplifiers and determine the maximum power efficiency of each class of amplifier.
- Analyze several circuit configurations of class-A power amplifiers.
- Discuss the characteristics of an ideal class-B output stage.
- Analyze and design various circuit configurations of class-AB output stages.
- As an application, design an output stage using MOSFETs as the output devices.

## 8.1    POWER AMPLIFIERS

**Objective:** • Describe the concept of a power amplifier.

A multistage amplifier may be required to deliver a large amount of power to a passive load. This power may be in the form of a large current delivered to a relatively small load resistance such as an audio speaker, or may be in the form of a large voltage delivered to a relatively large load resistance such as in a switching power supply. The output stage of the power amplifier must be designed to meet the power requirements. In this chapter, we are interested only in power amplifiers using BJTs or MOSFETS, and will not consider other types of power electronics that, for example, use thyristors.

Two important functions of the output stage are to provide a low output resistance so that it can deliver the signal power to the load without loss of gain and to maintain linearity in the output signal. A low output resistance implies the use of emitter-follower or source-follower circuit configurations. A measure of the linearity of the output signal is the **total harmonic distortion (THD).** This figure of merit is the rms value of the harmonic components of the output signal, excluding the fundamental, expressed as a percentage of the fundamental.

A particular concern in the design of the output stage is to deliver the required signal power to the load efficiently. This specification implies that the power dissipated in the transistors of the output stage should be as small as possible. The output transistors must be capable of delivering the required current to the load, and must be capable of sustaining the required output voltage.

We will initially discuss power transistors and will then consider several output stages of power amplifiers.

## 8.2    POWER TRANSISTORS

**Objective:** • Describe the characteristics of BJT and MOSFET power transistors, and analyze the temperature and heat flow characteristics of devices using heat sinks.

In our previous discussions, we have ignored any physical transistor limitations in terms of maximum current, voltage, and power. We implicitly assumed that the transistors were capable of handling the current and voltage, and could handle the power dissipated within the transistor without suffering any damage.

However, since we are now discussing power amplifiers, we must be concerned with transistor limitations. The limitations involve: maximum rated current (on the order of amperes), maximum rated voltage (on the order of 100 V), and maximum rated power (on the order of watts or tens of watts).[1] We will consider these effects in the BJT and then in the MOSFET. The maximum power limitation is related to the

---

[1] We must note that, in general, the maximum rated current and maximum rated voltage cannot occur at the same time.

maximum allowed temperature of the transistor, which in turn is a function of the rate at which heat is removed. We will therefore briefly consider heat sinks and heat flow.

### 8.2.1    Power BJTs

Power transistors are large-area devices. Because of differences in geometry and doping concentrations, their properties tend to vary from those of the small-signal devices. Table 8.1 compares the parameters of a general-purpose small-signal BJT to those of two power BJTs. The current gain is generally smaller in the power transistors, typically in the range of 20 to 100, and may be a strong function of collector current and temperature. Figure 8.1 shows typical current gain versus collector current characteristics for the 2N3055 power BJT at various temperatures. At high current levels, the current gain tends to drop off significantly, and parasitic resistances in the base and collector regions may become significant, affecting the transistor terminal characteristics.

The **maximum rated collector current** $I_{C,\text{rated}}$ may be related to: the maximum current that the wires connecting the semiconductor to the external terminals can handle; the collector current at which the current gain falls below a minimum specified value; or the current that leads to the maximum power dissipation when the transistor is in saturation.

| | | | **Power** |
|---|---|---|---|
| | **Small-Signal BJT** | **Power BJT** | **BJT** |
| **Parameter** | **(2N2222A)** | **(2N3055)** | **(2N6078)** |
| $V_{CE}$ (max) (V) | 40 | 60 | 250 |
| $I_C$ (max) (A) | 0.8 | 15 | 7 |
| $P_D$ (max) (W) (at $T = 25\,°C$) | 1.2 | 115 | 45 |
| $\beta$ | 35–100 | 5–20 | 12–70 |
| $f_T$ (MHz) | 300 | 0.8 | 1 |

Table 8.1    Comparison of the characteristics and maximum ratings of a small-signal and power BJT

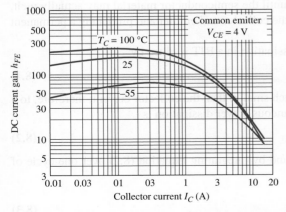

**Figure 8.1** Typical dc beta characteristics ($h_{FE}$ versus $I_C$) for 2N3055

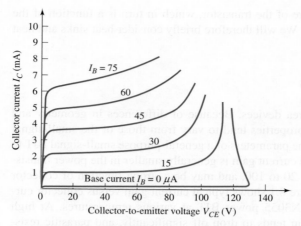

**Figure 8.2** Typical collector current versus collector–emitter voltage characteristics of a bipolar transistor, showing breakdown effects

The maximum voltage limitation in a BJT is generally associated with avalanche breakdown in the reverse-biased base–collector junction. In the common-emitter configuration, the breakdown voltage mechanism also involves the transistor gain, as well as the breakdown phenomenon on the pn junction. Typical $I_C$ versus $V_{CE}$ characteristics are shown in Figure 8.2. The breakdown voltage when the base terminal is open circuited ($I_B = 0$) is $V_{CEO}$. From the data in Figure 8.2, this value is approximately 130 V.

When the transistor is biased in the active region, the collector current begins to increase significantly before breakdown voltage $V_{CEO}$ is reached, and all the curves tend to merge to the same collector–emitter voltage once breakdown has occurred. The voltage at which these curves merge is denoted $V_{CE(\text{sus})}$ and is the minimum voltage necessary to sustain the transistor in breakdown. From the data in Figure 8.2, the value of $V_{CE(\text{sus})}$ is approximately 115 V.

Another breakdown effect is called **second breakdown,** which occurs in a BJT operating at high voltage and a fairly high current. Slight nonuniformities in current density produce local regions of increased heating that decreases the resistance of the semiconductor material, which in turn increases the current in those regions. This effect results in positive feedback, and the current continues to increase, producing a further increase in temperature, until the semiconductor material may actually melt, creating a short circuit between the collector and emitter and producing a permanent failure.

The instantaneous power dissipation in a BJT is given by

$$p_Q = v_{CE} i_C + v_{BE} i_B \tag{8.1}$$

The base current is generally much smaller than the collector current; therefore, to a good approximation, the instantaneous power dissipation is

$$p_Q \cong v_{CE} i_C \tag{8.2}$$

The average power, which is found by integrating Equation (8.2) over one cycle of the signal, is

$$\bar{P}_Q = \frac{1}{T} \int_0^T v_{CE} i_C \, dt \tag{8.3}$$

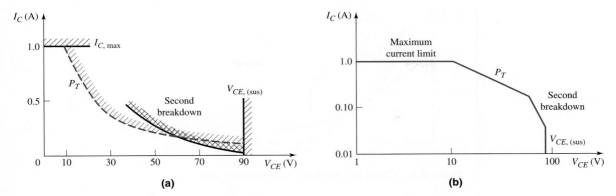

**Figure 8.3**  The safe operating area of a bipolar transistor plotted on: (a) linear scales and (b) logarithmic scales

The average power dissipated in a BJT must be kept below a specified maximum value, to ensure that the temperature of the device remains below a maximum value. If we assume that the collector current and collector–emitter voltage are dc quantities, then at the **maximum rated power** $P_T$ for the transistor, we can write

$$P_T = V_{CE} I_C \tag{8.4}$$

The maximum current, voltage, and power limitations can be illustrated on the $I_C$ versus $V_{CE}$ characteristics, as shown in Figure 8.3. The average power limitation $P_T$ is a hyperbola described by Equation (8.4). The region where the transistor can be operated safely is known as the **safe operating area (SOA)** and is bounded by $I_{C,\max}$, $V_{CE(\text{sus})}$, $P_T$, and the transistor's second breakdown characteristics curve. Figure 8.3(a) shows the safe operating area, using linear current and voltage scales; Figure 8.3(b) shows the same characteristics using logarithmic scales.

The $i_C$–$v_{CE}$ operating point may move momentarily outside the safe operating area without damaging the transistor, but this depends on how far the $Q$-point moves outside the area and for how long. For our purposes, we will assume that the device must remain within the safe operating area at all times.

## EXAMPLE **8.1**

**Objective:** Determine the required current, voltage, and power ratings of a power BJT.

Consider the common-emitter circuit in Figure 8.4. The parameters are $R_L = 8\ \Omega$ and $V_{CC} = 24$ V.

**Solution:** For $V_{CE} \cong 0$, the maximum collector current is

$$I_C(\max) = \frac{V_{CC}}{R_L} = \frac{24}{8} = 3\ \text{A}$$

For $I_C = 0$, the maximum collector–emitter voltage is

$$V_{CE}(\max) = V_{CC} = 24\ \text{V}$$

The load line is given by

$$V_{CE} = V_{CC} - I_C R_L$$

and must remain within the safe operating area, as shown in Figure 8.5.

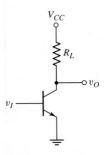

**Figure 8.4**  Figure for Example 8.1

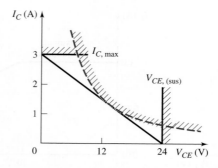

**Figure 8.5** DC load line within the safe operating area

The transistor power dissipation is therefore

$$P_T = V_{CE}I_C = (V_{CC} - I_C R_L)I_C = V_{CC}I_C - I_C^2 R_L$$

The current at which the maximum power occurs is found by setting the derivative of this equation equal to zero as follows:

$$\frac{dP_T}{dI_C} = 0 = V_{CC} - 2I_C R_L$$

which yields

$$I_C = \frac{V_{CC}}{2R_L} = \frac{24}{2(8)} = 1.5 \text{ A}$$

The C–E voltage at the maximum power point is

$$V_{CE} = V_{CC} - I_C R_L = 24 - (1.5)(8) = 12 \text{ V}$$

The maximum power dissipation in the transistor occurs at the center of the load line. The maximum transistor power dissipation is therefore

$$P_T = V_{CE}I_C = 12(1.5) = 18 \text{ W}$$

**Comment:** To find a transistor for a given application, safety factors are normally used. For this example, a transistor with a current rating greater than 3 A, a voltage rating greater than 24 V, and a power rating greater than 18 W would be required.

### EXERCISE PROBLEM

**Ex 8.1:** Consider the common-emitter circuit shown in Figure 8.4. Assume the transistor has limiting factors of: $I_{C,\max} = 5$ A, $V_{CE(\text{sus})} = 30$ V, and $P_T = 25$ W. Neglecting second breakdown effects, determine the minimum $R_L$ such that the Q-point of the transistor stays within the safe operating area for: (a) $V_{CC} = 24$ V and (b) $V_{CC} = 12$ V. In each case, determine the maximum collector current and maximum transistor power dissipation. (Ans. (a) $R_L = 5.76 \Omega$, $I_{C,\max} = 4.17$ A, $P_{Q,\max} = 25$ W; (b) $R_L = 2.4 \Omega$, $I_{C,\max} = 5$ A, $P_{Q,\max} = 15$ W)

Power transistors, which are designed to handle large currents, require large emitter areas to maintain reasonable current densities. These transistors are usually designed with narrow emitter widths to minimize the parasitic base resistance,

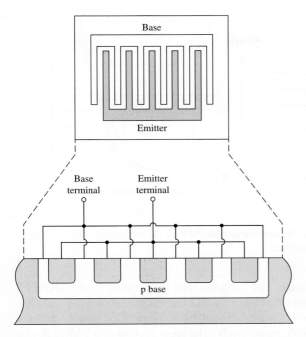

**Figure 8.6** An interdigitated bipolar transistor structure showing the top view and cross-sectional view

and may be fabricated as an **interdigitated structure,** as shown in Figure 8.6. Also, emitter ballast resistors, which are small resistors in each emitter leg, are usually incorporated in the design. These resistors help maintain equal currents in each B–E junction.

### 8.2.2 Power MOSFETs

Table 8.2 lists the basic parameters of two n-channel power MOSFETs. The drain currents are in the ampere range and the breakdown voltages are in the hundreds of volts range. These transistors must also operate within a safe operating area as discussed for the BJTs.

| Table 8.2 | Characteristics of two power MOSFETs | |
|---|---|---|
| **Parameter** | **2N6757** | **2N6792** |
| $V_{DS}$(max) (V) | 150 | 400 |
| $I_D$(max) (at $T = 25\,°C$) | 8 | 2 |
| $P_D$ (W) | 75 | 20 |

Power MOSFETs differ from bipolar power transistors both in operating principles and performance. The superior performance characteristics of power MOSFETs are: faster switching times, no second breakdown, and stable gain and response time over a wide temperature range. Figure 8.7(a) shows the transconductance

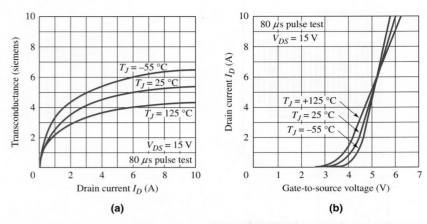

**Figure 8.7** Typical characteristics for high-power MOSFETs: (a) transconductance versus drain current; (b) transfer characteristics

of the 2N6757 versus temperature. The variation with temperature of the MOSFET transconductance is less than the variation in the BJT current gain shown in Figure 8.1.

Power MOSFETs are often manufactured by a vertical or double-diffused process, called VMOS or DMOS, respectively. The cross section of a VMOS device is shown in Figure 8.8(a) and the cross section of the DMOS device is shown in Figure 8.8(b). The DMOS process can be used to produce a large number of closely packed hexagonal cells on a single silicon chip, as shown in Figure 8.8(c). Also, such MOSFETs can be paralleled to form large-area devices, without the need of an equivalent emitter ballast resistance to equalize the current density. A single power MOSFET chip may contain as many as 25,000 paralleled cells.

Since the path between the drain and the source is essentially resistive, the **on resistance** $r_{ds}(\text{on})$ is an important parameter in the power capability of a MOSFET. Figure 8.9 shows a typical $r_{ds}(\text{on})$ characteristic as a function of drain current. Values in the tens of milliohm range have been obtained.

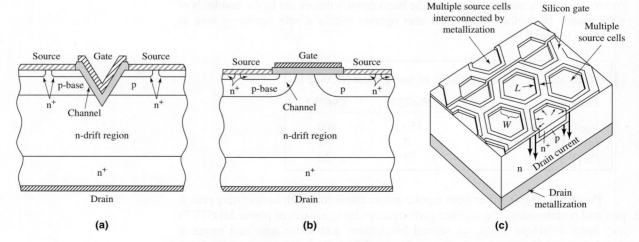

**Figure 8.8** (a) Cross section of a VMOS device; (b) cross section of DMOS device; (c) HEXFET structure

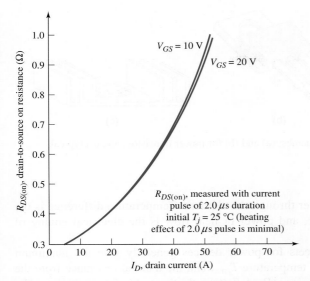

**Figure 8.9** Typical drain-to-source resistance versus drain current characteristics of a MOSFET

### 8.2.3 Comparison of Power MOSFETs and BJTs

Since a MOSFET is a high input impedance, voltage-controlled device, the drive circuitry is simpler. The gate of a 10 A power MOSFET may be driven by the output of a standard logic circuit. In contrast, if the current gain of a 10 A BJT is $\beta = 10$, then a base current of 1 A is required for a collector current of 10 A. However, this required input current is much larger than the output drive capability of most logic circuits, which means that the drive circuitry for power BJTs is more complicated.

The MOSFET is a majority carrier device. Majority carrier mobility decreases with increasing temperature, which makes the semiconductor more resistive. This means that MOSFETs are more immune to the thermal runaway effects and second breakdown phenomena experienced in bipolars. Figure 8.7(b) shows typical $I_D$ versus $V_{GS}$ characteristics at several temperatures, clearly demonstrating that at high current levels, the current actually decreases with increasing temperature, for a given gate-to-source voltage.

### 8.2.4 Heat Sinks

The power dissipated in a transistor increases its internal temperature above the ambient temperature. If the device or junction temperature $T_j$ becomes too high, the transistor may suffer permanent damage. Special precautions must be taken in packaging power transistors and in providing heat sinks so that heat can be conducted from the transistor. Figures 8.10(a) and (b) show two packaging schemes, and Figure 8.10(c) shows a typical heat sink.

To design a heat sink for a power transistor, we must first consider the concept of **thermal resistance** $\theta$, which has units of °C/W. The temperature difference, $T_2 - T_1$, across an element with a thermal resistance $\theta$ is

$$T_2 - T_1 = P\theta \qquad \textbf{(8.5)}$$

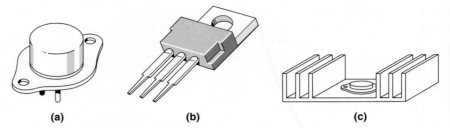

**(a)**                    **(b)**                    **(c)**

**Figure 8.10** Two packaging schemes: (a) and (b) for power transistors and (c) typical heat sink

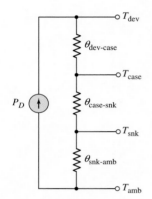

**Figure 8.11** Electrical equivalent circuit for heat flow from the device to the ambient

where $P$ is the thermal power through the element. Temperature difference is the electrical analog of voltage, and power or heat flow is the electrical analog of current.

Manufacturers' data sheets for power devices generally give the maximum operating junction or device temperature $T_{j,\text{max}}$ and the thermal resistance from the junction to the case $\theta_{jc} = \theta_{\text{dev}-\text{case}}$.[2] By definition, the thermal resistance between the case and heat sink is $\theta_{\text{case}-\text{snk}}$, and between the heat sink and ambient is $\theta_{\text{snk}-\text{amb}}$.

The temperature difference between the device and the ambient can now be written as follows, when a heat sink is used:

$$T_{\text{dev}} - T_{\text{amb}} = P_D(\theta_{\text{dev}-\text{case}} + \theta_{\text{case}-\text{snk}} + \theta_{\text{snk}-\text{amb}}) \qquad \textbf{(8.6)}$$

where $P_D$ is the power dissipated in the device. Equation (8.6) may also be modeled by its equivalent electrical elements, as shown in Figure 8.11. The temperature difference across the elements, such as the case and heat sink, is the dissipated power $P_D$ multiplied by the applicable thermal resistance, which is $\theta_{\text{case}-\text{snk}}$ for this example.

If a heat sink is not used, the temperature difference between the device and ambient is written as

$$T_{\text{dev}} - T_{\text{amb}} = P_D(\theta_{\text{dev}-\text{case}} + \theta_{\text{case}-\text{amb}}) \qquad \textbf{(8.7)}$$

where $\theta_{\text{case}-\text{amb}}$ is the thermal resistance between the case and ambient.

## EXAMPLE 8.2

**Objective:** Determine the maximum power dissipation in a transistor and determine the temperature of the transistor case and heat sink.

Consider a power MOSFET for which the thermal resistance parameters are:

$$\theta_{\text{dev}-\text{case}} = 1.75\,°\text{C/W} \qquad \theta_{\text{case}-\text{snk}} = 1\,°\text{C/W}$$

$$\theta_{\text{snk}-\text{amb}} = 5\,°\text{C/W} \qquad \theta_{\text{case}-\text{amb}} = 50\,°\text{C/W}$$

The ambient temperature is $T_{\text{amb}} = 30\,°\text{C}$, and the maximum junction or device temperature is $T_{j,\text{max}} = T_{\text{dev}} = 150\,°\text{C}$.

---

[2]In this short discussion, we use a more descriptive subscript notation to help clarify the discussion.

**Solution (Maximum Power):** When no heat sink is used, the maximum device power dissipation is found from Equation (8.7) as

$$P_{D,\text{max}} = \frac{T_{j,\text{max}} - T_{\text{amb}}}{\theta_{\text{dev−case}} + \theta_{\text{case−amb}}} = \frac{150 - 30}{1.75 + 50} = 2.32 \, \text{W}$$

When a heat sink is used, the maximum device power dissipation is found from Equation (8.6) as

$$P_{D,\text{max}} = \frac{T_{j,\text{max}} - T_{\text{amb}}}{\theta_{\text{dev−case}} + \theta_{\text{case−snk}} + \theta_{\text{snk−amb}}}$$

$$= \frac{150 - 30}{1.75 + 1 + 5} = 15.5 \, \text{W}$$

**Solution (Temperature):** The device temperature is $T = 150 \, °\text{C}$ and the ambient temperature is $T_{\text{amb}} = 30 \, °\text{C}$. The heat flow is $P_D = 15.5$ W. The heat sink temperature (see Figure 8.11) is found from

$$T_{\text{snk}} - T_{\text{amb}} = P_D \cdot \theta_{\text{snk−amb}}$$

or

$$T_{\text{snk}} = 30 + (15.5)(5) \Rightarrow T_{\text{snk}} = 107.5 \, °\text{C}$$

The case temperature is found from

$$T_{\text{case}} - T_{\text{amb}} = P_D \cdot (\theta_{\text{case−snk}} + \theta_{\text{snk−case}})$$

or

$$T_{\text{case}} = 30 + (15.5)(1 + 5) \Rightarrow T = 123 \, °\text{C}$$

**Comment:** These results illustrate that the use of a heat sink allows more power to be dissipated in the device, while keeping the device temperature at or below its maximum limit.

## EXERCISE PROBLEM

**Ex 8.2:** A power BJT is operating with an average collector current of $I_C = 2$ A and an average collector–emitter voltage of $V_{CE} = 8$ V. The device parameters are $\theta_{\text{dev−case}} = 3 \, °\text{C/W}$, $\theta_{\text{case−snk}} = 1 \, °\text{C/W}$, and $\theta_{\text{snk−amb}} = 4 \, °\text{C/W}$. The ambient temperature is $25 \, °\text{C}$. Determine the temperatures of the (a) device, (b) case, and (c) heat sink. (Ans. (a) $153 \, °\text{C}$, (b) $105 \, °\text{C}$, (c) $89 \, °\text{C}$)

The maximum safe power dissipation in a device is a function of: (1) the temperature difference between the junction and case, and (2) the thermal resistance between the device and the case $\theta_{\text{dev−case}}$, or

$$P_{D,\text{max}} = \frac{T_{j,\text{max}} - T_{\text{case}}}{\theta_{\text{dev−case}}} \tag{8.8}$$

A plot of $P_{D,\text{max}}$ versus $T_{\text{case}}$, called the **power derating curve** of the transistor, is shown in Figure 8.12. The temperature at which the power derating curve crosses the horizontal axis corresponds to $T_{j,\text{max}}$. At this temperature, no additional temperature rise in the device can be tolerated; therefore, the allowed power dissipation must be zero, which implies a zero input signal.

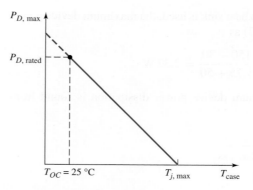

**Figure 8.12** A power derating curve

The rated power of a device is generally defined as the power at which the device reaches its maximum temperature, while the case temperature remains at room or ambient temperature, that is, $T_{case} = 25\ °C$. Maintaining the case at ambient temperature implies that the thermal resistance between the case and ambient is zero, or that an infinite heat sink is used. However, an infinite heat sink is not possible. With nonzero values of $\theta_{case-snk}$ and $\theta_{snk-amb}$, the case temperature rises above the ambient, and the maximum rated power of the device cannot be achieved. This effect can be seen by examining the equivalent circuit model in Figure 8.11. If the device temperature is at its maximum allowed value of $T_{dev} = T_{j,max}$, then as $T_{case}$ increases, the temperature difference across $\theta_{dev-case}$ decreases, which means that the power through the element must decrease.

## EXAMPLE 8.3

**Objective:** Determine the maximum safe power dissipation in a transistor.

Consider a BJT with a rated power of 20 W and a maximum junction temperature of $T_{j,max} = 175\ °C$. The transistor is mounted on a heat sink with parameters $\theta_{case-snk} = 1\ °C/W$ and $\theta_{snk-amb} = 5\ °C/W$.

**Solution:** From Equation (8.8), the device-to-case thermal resistance is

$$\theta_{dev-case} = \frac{T_{j,max} - T_{OC}}{P_{D,rated}} = \frac{175 - 25}{20} = 7.5\ °C/W$$

From Equation (8.6), the maximum power dissipation is

$$P_{D,max} = \frac{T_{j,max} - T_{amb}}{\theta_{dev-case} + \theta_{case-snk} + \theta_{snk-amb}}$$

$$= \frac{175 - 25}{7.5 + 1 + 5} = 11.1\ W$$

**Comment:** The actual maximum safe power dissipation in a device may be less than the rated value. This occurs when the case temperature cannot be held at the ambient temperature, because of the nonzero thermal resistance factors between the case and ambient.

EXERCISE PROBLEM

Ex 8.3: The rated power of a power BJT is $P_{D,\text{rated}} = 50$ W, the maximum allowed junction temperature is $T_{j,\text{max}} = 200$ °C, and the ambient temperature is $T_{\text{amb}} = 25$ °C. The thermal resistance between the heat sink and air is $\theta_{\text{snk}-\text{amb}} = 2$ °C/W, and that between the case and heat sink is $\theta_{\text{case}-\text{snk}} = 0.5$ °C/W. Find the maximum safe power dissipation and the temperature of the case. (Ans. $P_{D,\text{max}} = 29.2$ W, $T_{\text{case}} = 98$ °C)

## Test Your Understanding

TYU 8.1  Consider the common-source circuit shown in Figure 8.13. The parameters are $R_D = 20$ Ω and $V_{DD} = 24$ V. Determine the required current, voltage, and power ratings of the MOSFET. (Ans. $I_D(\text{max}) = 1.2$ A, $V_{DS}(\text{max}) = 24$ V, $P_D(\text{max}) = 7.2$ W)

TYU 8.2  The emitter-follower circuit shown in Figure 8.14 is biased with $V_{CC} = 12$ V. The transistor current gain is $\beta = 80$, and the transistor limiting factors are $I_{C,\text{max}} = 250$ mA and $V_{CE(\text{sus})} = 30$ V. The transistor $Q$-point is to remain within the safe operating area at all times. (a) Determine the minimum value of $R_E$. (b) Determine the minimum required transistor power rating. (Ans. (a) $R_E = 96$ Ω, (b) $P_Q = 1.5$ W)

TYU 8.3  (a) Assume the power flow through a material with a thermal resistance parameter of $\theta = 1.8$ °C/W is $P = 6$ W. Determine the resulting temperature difference across the material. (b) The thermal resistance of a material is $\theta = 2.5$ °C/W. If the temperature difference across the material is $\Delta T = 100$ °C, find the power flow through the material. (Ans. (a) $\Delta T = 10.8$ °C, (b) $P = 40$ W)

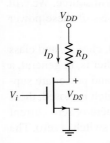

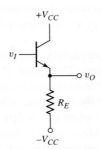

Figure 8.13  Figure for Exercise TYU 8.1 and Example 8.4

Figure 8.14  Figure for Exercise TYU 8.2

# 8.3    CLASSES OF AMPLIFIERS

**Objective:** • Define various classes of power amplifiers, and investigate the characteristics, including power efficiency, of a few of these amplifiers.

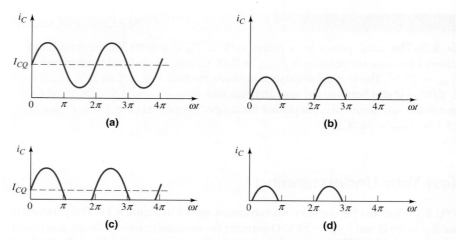

**Figure 8.15** Collector current versus time characteristics: (a) class-A amplifier, (b) class-B amplifier, (c) class-AB amplifier, and (d) class-C amplifier

Some power amplifiers are classified according to the percent of time the output transistors are conducting, or "turned on." Four of the principal classifications are: class A, class B, class AB, and class C. These classifications are illustrated in Figure 8.15 for a sinusoidal input signal. In **class-A operation,** an output transistor is biased at a quiescent current $I_Q$ and conducts for the entire cycle of the input signal. For **class-B operation,** an output transistor conducts for only one-half of each sine wave input cycle. In **class-AB operation,** an output transistor is biased at a small quiescent current $I_Q$ and conducts for slightly more than half a cycle. In contrast, in **class-C operation** an output transistor conducts for less than half a cycle. These four types of power amplifiers use the output transistors as a current source. We will analyze the biasing, load lines, and power efficiency of each class of these power amplifiers.

Another classification of power amplifiers, including class D, class E, and class F, uses the output transistors as switches. The output of the amplifier is, in general, a high-$Q$ resonant $RLC$ circuit. When the switch is closed, current and power are supplied to the output resonant circuit. In the ideal case, when the switch is closed, there is zero voltage across the switch, and when the switch is open, there is zero current through the switch. In both cases, the ideal power dissipated in the switch is zero. The power efficiency of these amplifiers can then approach 100 percent.

The intent of this chapter is to provide the basic characteristics of a few of these power amplifiers. As usual, there are other types of power amplifiers and power electronics that are beyond the scope of this text, including high-frequency radio-frequency (RF) circuit design.

### 8.3.1    Class-A Operation

The small-signal amplifiers considered in Chapters 4 and 6 were all biased for class-A operation. A basic common-emitter configuration is shown in Figure 8.16(a). The bias circuitry has been omitted, for convenience. Also, in this **standard class-A amplifier** configuration, no inductors or transformers are used.

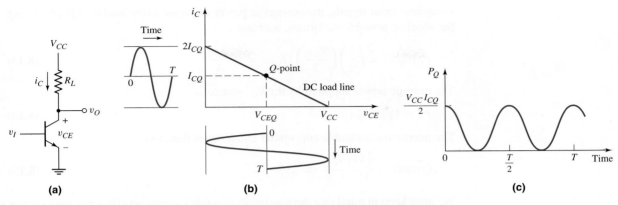

**Figure 8.16** (a) Common-emitter amplifier, (b) dc load line, and (c) instantaneous power dissipation versus time in the transistor

The dc load line is shown in Figure 8.16(b). The $Q$-point is assumed to be in the center of the load line, so that $V_{CEQ} = V_{CC}/2$. If a sinusoidal input signal is applied, sinusoidal variations are induced in the collector current and collector–emitter voltage. The absolute possible variations are shown in the figure, although values of $v_{CE} = 0$ and $i_C = 2I_{CQ}$ cannot actually be attained.

The instantaneous power dissipation in the transistor, neglecting the base current, is

$$p_Q = v_{CE}i_C \tag{8.9}$$

For a sinusoidal input signal, the collector current and collector-emitter voltage can be written

$$i_C = I_{CQ} + I_p \sin \omega t \tag{8.10(a)}$$

and

$$v_{CE} = \frac{V_{CC}}{2} - V_p \sin \omega t \tag{8.10(b)}$$

If we consider the absolute possible variations, then $I_p = I_{CQ}$ and $V_p = V_{CC}/2$. Therefore, the instantaneous power dissipation in the transistor, from Equation (8.9), is

$$p_Q = \frac{V_{CC}I_{CQ}}{2}(1 - \sin^2 \omega t) \tag{8.11}$$

Figure 8.16(c) is a plot of the instantaneous transistor power dissipation. Since the maximum power dissipation corresponds to the quiescent value (see Figure 8.5), the transistor must be capable of handling a continuous power dissipation of $V_{CC}I_{CQ}/2$ when the input signal is zero.

The **power conversion efficiency** is defined as

$$\eta = \frac{\text{signal load power}(\bar{P}_L)}{\text{supply power}(\bar{P}_S)} \tag{8.12}$$

where $\bar{P}_L$ is the average ac power delivered to the load and $\bar{P}_S$ is the average power supplied by the $V_{CC}$ power source(s). For the standard class-A amplifier and sinusoidal input signals, the average ac power delivered to the load is $\left(\frac{1}{2}\right)V_p I_p$. Using the absolute possible variations, we have

$$\bar{P}_L(\text{max}) = \left(\frac{1}{2}\right)\left(\frac{V_{CC}}{2}\right)(I_{CQ}) = \frac{V_{CC} I_{CQ}}{4} \tag{8.13}$$

The average power supplied by the $V_{CC}$ source is

$$\bar{P}_S = V_{CC} I_{CQ} \tag{8.14}$$

The maximum attainable conversion efficiency is therefore

$$\eta(\text{max}) = \frac{\frac{1}{4}V_{CC} I_{CQ}}{V_{CC} I_{CQ}} \Rightarrow 25\% \tag{8.15}$$

We must keep in mind that the maximum possible conversion efficiency may change when a load is connected to the output of the amplifier. This efficiency is relatively low; therefore, standard class-A amplifiers are normally not used when signal powers greater than approximately 1 W are required.

We must also emphasize that in practice, a maximum signal voltage of $V_{CC}/2$ and a maximum signal current of $I_{CQ}$ are not possible. The output signal voltage must be limited to smaller values in order to avoid transistor saturation and cutoff, and the resulting nonlinear distortion. The calculation for the maximum possible efficiency also neglects power dissipation in the bias circuitry. Consequently, the realistic maximum conversion efficiency in a standard class-A amplifier is on the order of 20 percent or less.

**Design Pointer:** In circuit analysis, the maximum power transfer theorem stated that the load impedance should be matched to the amplifier output impedance, which provides a 50 percent power conversion efficiency. However, in the design of power amplifiers, this theorem is not practical. For example, if 50 kW is to be delivered to an antenna, then the circuit would also dissipate 50 kW if the power conversion efficiency were only 50 percent. In general, this amount of power being dissipated in the amplifier would be unacceptable. Power conversion efficiencies as close to 100 percent as possible are desirable in very high power amplifiers.

## EXAMPLE 8.4

**Objective:** Calculate the actual efficiency of a class-A output stage.

Consider the common-source circuit in Figure 8.13. The circuit parameters are $V_{DD} = 10$ V and $R_D = 5$ k$\Omega$, and the transistor parameters are: $K_n = 1$ mA/V$^2$, $V_{TN} = 1$ V, and $\lambda = 0$. Assume the output voltage swing is limited to the range between the transition point and $v_{DS} = 9$ V, to minimize nonlinear distortion.

**Solution:** The load line is given by

$$V_{DS} = V_{DD} - I_D R_D$$

At the transition point, we have

$$V_{DS}(\text{sat}) = V_{GS} - V_{TN}$$

and

$$I_D = K_n(V_{GS} - V_{TN})^2$$

Combining these expressions, the transition point is determined from

$$V_{DS}(\text{sat}) = V_{DD} - K_n R_D V_{DS}^2(\text{sat})$$

or

$$(1)(5)V_{DS}^2(\text{sat}) + V_{DS}(\text{sat}) - 10 = 0$$

which yields

$$V_{DS}(\text{sat}) = 1.32 \text{ V}$$

To obtain the maximum symmetrical swing under the conditions specified, we want the Q-point midway between $V_{DS} = 1.32$ V and $V_{DS} = 9$ V, or

$$V_{DSQ} = 5.16 \text{ V}$$

The maximum ac component of voltage across the load resistor is then

$$v_r = 3.84 \sin \omega t$$

and the average power delivered to the load is

$$\bar{P}_L = \frac{1}{2} \cdot \frac{(3.84)^2}{5} = 1.47 \text{ mW}$$

The quiescent drain current is found to be

$$I_{DQ} = \frac{10 - 5.16}{5} = 0.968 \text{ mA}$$

The average power supplied by the $V_{DD}$ source is

$$\bar{P}_S = V_{DD}I_{DQ} = (10)(0.968) = 9.68 \text{ mW}$$

and the power conversion efficiency, from Equation (8.12), is

$$\eta = \frac{\bar{P}_L}{\bar{P}_S} = \frac{1.47}{9.68} \Rightarrow 15.2\%$$

**Comment:** By limiting the swing in the drain–source voltage, to avoid nonsaturation and cutoff and the resulting nonlinear distortion, we reduce the output stage power conversion efficiency considerably, compared to the theoretical maximum possible value of 25 percent for the standard class-A amplifier.

---

### EXERCISE PROBLEM

**\*Ex 8.4:** For the common-source circuit shown in Figure 8.17, the Q-point is $V_{DSQ} = 4$ V. (a) Find $I_{DQ}$. (b) The minimum value of the instantaneous drain current must be no less than $\left(\frac{1}{10}\right)I_{DQ}$, and the minimum value of the instantaneous drain–source voltage must be no less than $v_{DS} = 1.5$ V. Determine the

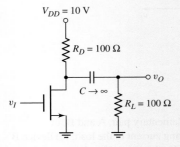

**Figure 8.17**  Figure for Exercise Ex 8.4

maximum peak-to-peak amplitude of a symmetrical sinusoidal output voltage. (c) For the conditions of part (b), calculate the power conversion efficiency, where the signal power is the power delivered to $R_L$. (Ans. (a) $I_{DQ} = 60$ mA (b) $V_{p-p} = 5.0$ V (c) $\bar{P}_L = 31.25$ mW, $\eta = 5.2\%$)

Class-A operation also applies to the emitter-follower, common-base, source-follower, and common-gate configurations. As previously stated, the circuits considered in Figures 8.13 and 8.16(a) are standard class-A amplifiers in that no inductors or transformers are used. Later in this chapter, we will analyze inductively-coupled and transformer-coupled power amplifiers that also operate in the class-A mode. We will show that, for these circuits, the maximum conversion efficiency is 50 percent.

### 8.3.2    Class-B Operation

**Idealized Class-B Operation**

Figure 8.18(a) shows an idealized class-B output stage that consists of a complementary pair of electronic devices. When $v_I = 0$, both devices are off, the bias currents are zero, and $v_O = 0$. For $v_I > 0$, device A turns on and supplies current to the load as shown in Figure 8.18(b). For $v_I < 0$, device B turns on and sinks current

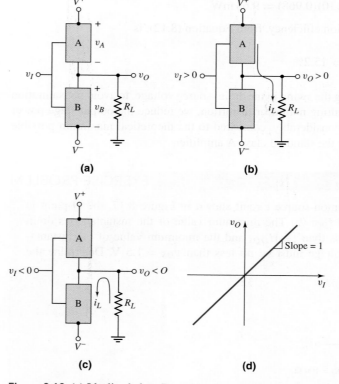

**Figure 8.18** (a) Idealized class-B output stage with complementary pair, A and B, of electronic devices; (b) device A turns on for $v_I > 0$, supplying current to the load; (c) device B turns on for $v_I < 0$, sinking current from the load; (d) ideal voltage transfer characteristics

from the load as shown in Figure 8.18(c). Figure 8.18(d) shows the voltage transfer characteristics. The ideal voltage gain is unity.

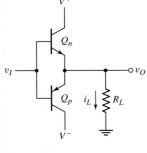

**Figure 8.19** Basic complementary push–pull output stage

### Approximate Class-B Circuit

Figure 8.19 shows an output stage that consists of a complementary pair of bipolar transistors. When the input voltage is $v_I = 0$, both transistors are cut off and the output voltage is $v_O = 0$. If we assume a B–E cut-in voltage of 0.6 V, then the output voltage $v_O$ remains zero as long as the input voltage is in the range $-0.6 \leq v_I \leq +0.6$ V.

If $v_I$ becomes positive and is greater than 0.6 V, then $Q_n$ turns on and operates as an emitter follower. The load current $i_L$ is positive and is supplied through $Q_n$, and the B–E junction of $Q_p$ is reverse biased. If $v_I$ becomes negative by more than 0.6 V, then $Q_p$ turns on and operates as an emitter follower. Transistor $Q_p$ is a sink for the load current, which means that $i_L$ is negative.

This circuit is called a **complementary push–pull** output stage. Transistor $Q_n$ conducts during the positive half of the input cycle, and $Q_p$ conducts during the negative half-cycle. The transistors do not both conduct at the same time.

Figure 8.20 shows the voltage transfer characteristics for this circuit. When either transistor is conducting, the voltage gain, which is the slope of the curve, is essentially unity as a result of the emitter follower. Figure 8.21 shows the output voltage for a sinusoidal input signal. When the output voltage is positive, the npn transistor is conducting, and when the output voltage is negative, the pnp transistor is conducting. We can see from this figure that each transistor actually conducts for slightly less than half the time. Thus the bipolar push–pull circuit shown in Figure 8.19 is not exactly a class-B circuit.

We will see that an output stage using NMOS and PMOS transistors will produce the same general voltage transfer characteristics.

### Crossover Distortion

From Figure 8.20, we see that there is a range of input voltage around zero volts where both transistors are cut off and $v_O$ is zero. This portion of the curve is called the *dead band*. The output voltage for a sinusoidal input voltage is shown in Figure 8.21. The output voltage is not a perfect sinusoidal signal, which means that *crossover distortion* is produced by the dead band region.

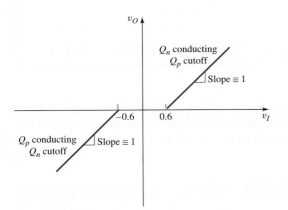

**Figure 8.20** Voltage transfer characteristics of basic complementary push–pull output stage

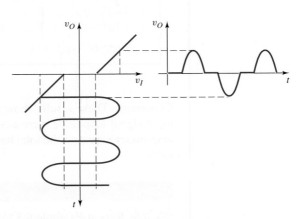

**Figure 8.21** Crossover distortion of basic complementary push–pull output stage

Crossover distortion can be virtually eliminated by biasing both $Q_n$ and $Q_p$ with a small quiescent collector current when $v_I$ is zero. This technique is discussed in the next section. The crossover distortion effect can also be minimized with an op-amp used in a feedback configuration. Op-amps are discussed in Chapter 9 and feedback is discussed in Chapter 12, so this technique is not discussed here.

## EXAMPLE 8.5

**Objective:** Determine the total harmonic distortion (THD) of the class B complementary push–pull output stage in Figure 8.19.

A PSpice analysis was performed, which yielded the harmonic content of the output signal.

**Solution:** A 1 kHz sinusoidal signal with an amplitude of 2 V was applied to the input of the circuit shown in Figure 8.19. The circuit was biased at $\pm 10$ V. The transistors used in the circuit were 2N3904 npn and 2N3906 pnp devices. A 1 k$\Omega$ load was connected to the output.

The harmonic content for the first nine harmonics is shown in Table 8.3. We see that the output is rich in odd harmonics with the 3 kHz third harmonic being 18 percent as large as the 1 kHz principal output signal. The total harmonic distortion is 19.7 percent, which is large.

| Table 8.3 | Harmonic content for Example 8.5 | | |
|---|---|---|---|
| **Frequency (Hz)** | **Fourier component** | **Normalized component** | **Phase (degrees)** |
| 1.000E+03 | 1.151E+00 | 1.000E+00 | −1.626E−01 |
| 2.000E+03 | 6.313E−03 | 5.485E−03 | −9.322E+01 |
| 3.000E+03 | 2.103E−01 | 1.827E−01 | −1.793E+02 |
| 4.000E+03 | 4.984E−03 | 4.331E−03 | −9.728E+01 |
| 5.000E+03 | 8.064E−02 | 7.006E−02 | −1.792E+02 |
| 6.000E+03 | 3.456E−03 | 3.003E−03 | −9.702E+01 |
| 7.000E+03 | 2.835E−02 | 2.464E−02 | 1.770E+02 |
| 8.000E+03 | 2.019E−03 | 1.754E−03 | −8.029E+01 |
| 9.000E+03 | 6.679E−03 | 5.803E−03 | 1.472E+02 |
| TOTAL HARMONIC DISTORTION = 1.974899E+01 PERCENT | | | |

**Comment:** These results show the obvious effects of the dead band region. If the input signal amplitude increases, the total harmonic distortion decreases, but if the amplitude decreases, the total harmonic distortion will increase above the 19 percent value.

## EXERCISE PROBLEM

**\*Ex 8.5:** Repeat Example 8.5 for the case when an NMOS transistor replaces the npn transistor and a PMOS transistor replaces the pnp transistor in Figure 8.19.

## Idealized Power Efficiency

If we consider an idealized version of the circuit in Figure 8.19 in which the base–emitter turn-on voltages are zero, then each transistor would conduct for exactly one-half cycle of the sinusoidal input signal. This circuit would be an ideal class-B output stage, and the output voltage and load current would be replicas of the input signal. The collector–emitter voltages would also show the same sinusoidal variation.

Figure 8.22 illustrates the applicable dc load line. The $Q$-point is at zero collector current, or at cutoff for both transistors. The quiescent power dissipation in each transistor is then zero.

The output voltage for this idealized class-B output stage can be written

$$v_O = V_p \sin \omega t \tag{8.16}$$

where the maximum possible value of $V_p$ is $V_{CC}$.

The instantaneous power dissipation in $Q_n$ is

$$p_{Qn} = v_{CEn} i_{Cn} \tag{8.17}$$

and the collector current is

$$i_{Cn} = \frac{V_p}{R_L} \sin \omega t \tag{8.18(a)}$$

for $0 \le \omega t \le \pi$, and

$$i_{Cn} = 0 \tag{8.18(b)}$$

for $\pi \le \omega t \le 2\pi$, where $V_p$ is the peak output voltage.

From Figure 8.22, we see that the collector–emitter voltage can be written as

$$v_{CEn} = V_{CC} - V_p \sin \omega t \tag{8.19}$$

Therefore, the total instantaneous power dissipation in $Q_n$ is

$$p_{Qn} = (V_{CC} - V_p \sin \omega t) \left( \frac{V_p}{R_L} \sin \omega t \right) \tag{8.20}$$

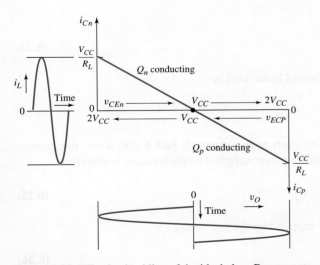

**Figure 8.22**  Effective load line of the ideal class-B output stage

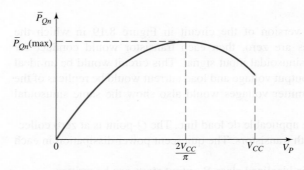

**Figure 8.23** Average power dissipation in each transistor versus peak output voltage for class-B output stage

for $0 \leq \omega t \leq \pi$, and

$$p_{Qn} = 0$$

for $\pi \leq \omega t \leq 2\pi$. The average power dissipation is therefore

$$\bar{P}_{Qn} = \frac{V_{CC} V_p}{\pi R_L} - \frac{V_p^2}{4 R_L} \tag{8.21}$$

The average power dissipation in transistor $Q_p$ is exactly the same as that for $Q_n$, because of symmetry.

A plot of the average power dissipation in each transistor, as a function of $V_p$, is shown in Figure 8.23. The power dissipation first increases with increasing output voltage, reaches a maximum, and finally decreases with increasing $V_p$. We determine the maximum average power dissipation by setting the derivative of $\bar{P}_{Qn}$ with respect to $V_p$ equal to zero, producing

$$\bar{P}_{Qn}(\text{max}) = \frac{V_{CC}^2}{\pi^2 R_L} \tag{8.22}$$

which occurs when

$$V_p \big|_{\bar{P}_{Qn}(\text{max})} = \frac{2 V_{CC}}{\pi} \tag{8.23}$$

The average power delivered to the load is

$$\bar{P}_L = \frac{1}{2} \cdot \frac{V_p^2}{R_L} \tag{8.24}$$

Since the current supplied by each power supply is half a sine wave, the average current is $V_p/(\pi R_L)$. The average power supplied by each source is therefore

$$\bar{P}_{S+} = \bar{P}_{S-} = V_{CC} \left( \frac{V_p}{\pi R_L} \right) \tag{8.25}$$

and the total average power supplied by the two sources is

$$\bar{P}_S = 2 V_{CC} \left( \frac{V_p}{\pi R_L} \right) \tag{8.26}$$

From Equation (8.12), the conversion efficiency is

$$\eta = \frac{\frac{1}{2} \cdot \frac{V_p^2}{R_L}}{2V_{CC}\left(\dfrac{V_p}{\pi R_L}\right)} = \frac{\pi}{4} \cdot \frac{V_p}{V_{CC}} \tag{8.27}$$

The maximum possible efficiency, which occurs when $V_p = V_{CC}$, is

$$\eta(\text{max}) = \frac{\pi}{4} \Rightarrow 78.5\% \tag{8.28}$$

This maximum efficiency value is substantially larger than that of the standard class-A amplifier.

From Equation (8.24), we find the maximum possible average power that can be delivered to the load, as follows:

$$\bar{P}_L(\text{max}) = \frac{1}{2} \cdot \frac{V_{CC}^2}{R_L} \tag{8.29}$$

The actual conversion efficiency obtained in practice is less than the maximum value because of other circuit losses, and because the peak output voltage must remain less than $V_{CC}$ to avoid transistor saturation. As the output voltage amplitude increases, output signal distortion also increases. To limit this distortion to an acceptable level, the peak output voltage is usually limited to several volts below $V_{CC}$. From Figure 8.23 and Equation (8.23), we see that the maximum transistor power dissipation occurs when $V_p = 2V_{CC}/\pi$. At this peak output voltage, the conversion efficiency of the class-B amplifier is, from Equation (8.27),

$$\eta = \frac{\pi}{4V_{CC}} \cdot V_p = \left(\frac{\pi}{4V_{CC}}\right) \cdot \left(\frac{2V_{CC}}{\pi}\right) = \frac{1}{2} \Rightarrow 50\% \tag{8.30}$$

### 8.3.3    Class-AB Operation

Crossover distortion can be virtually eliminated by applying a small quiescent bias on each output transistor, for a zero input signal. This is called a class-AB output stage and is shown schematically in the circuit in Figure 8.24. If $Q_n$ and $Q_p$ are matched, then for $v_I = 0$, $V_{BB}/2$ is applied to the B–E junction of $Q_n$, $V_{BB}/2$ is applied to the E–B junction of $Q_p$, and $v_O = 0$. The quiescent collector currents in each transistor are given by

$$i_{Cn} = i_{Cp} = I_S e^{V_{BB}/2V_T} \tag{8.31}$$

As $v_I$ increases, the voltage at the base of $Q_n$ increases and $v_O$ increases. Transistor $Q_n$ operates as an emitter follower, supplying the load current to $R_L$. The output voltage is given by

$$v_O = v_I + \frac{V_{BB}}{2} - v_{BEn} \tag{8.32}$$

and the collector current of $Q_n$ (neglecting base currents) is

$$i_{Cn} = i_L + i_{Cp} \tag{8.33}$$

Since $i_{Cn}$ must increase to supply the load current, $v_{BEn}$ increases. Assuming $V_{BB}$ remains constant, as $v_{BEn}$ increases, $v_{EBp}$ decreases resulting in a decrease in $i_{Cp}$.

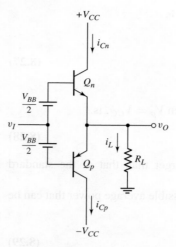

**Figure 8.24**  Bipolar class-AB output stage

As $v_I$ goes negative, the voltage at the base of $Q_p$ decreases and $v_O$ decreases. Transistor $Q_p$ operates as an emitter follower, sinking current from the load. As $i_{Cp}$ increases, $v_{EBp}$ increases, causing a decrease in $v_{BEn}$ and $i_{Cn}$.

Figure 8.25(a) shows the voltage transfer characteristics for this class-AB output stage. If $v_{BEn}$ and $v_{EBp}$ do not change significantly, then the voltage gain, or the slope of the transfer curve, is essentially unity. A sinusoidal input signal voltage and the resulting collector currents and load current are shown in Figures 8.25(b), (c), and (d). Each transistor conducts for more than one-half cycle, which is the definition of class-AB operation.

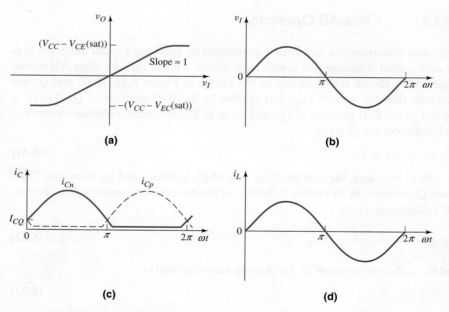

**Figure 8.25**  Characteristics of a class-AB output stage: (a) voltage transfer curve, (b) sinusoidal input signal, (c) collector currents, and (d) output current

There is a relationship between $i_{Cn}$ and $i_{Cp}$. We know that

$$v_{BEn} + v_{EBp} = V_{BB} \tag{8.34(a)}$$

which can be written

$$V_T \ln\left(\frac{i_{Cn}}{I_S}\right) + V_T \ln\left(\frac{i_{Cp}}{I_S}\right) = 2V_T \ln\left(\frac{I_{CQ}}{I_S}\right) \tag{8.34(b)}$$

Combining terms in Equation (8.34(b)), we find

$$i_{Cn}i_{Cp} = I_{CQ}^2 \tag{8.35}$$

The product of $i_{Cn}$ and $i_{Cp}$ is a constant; therefore, if $i_{Cn}$ increases, $i_{Cp}$ decreases, but does not go to zero.

Since, for a zero input signal, quiescent collector currents exist in the output transistors, the average power supplied by each source and the average power dissipated in each transistor are larger than for a class-B configuration. This means that the power conversion efficiency for a class-AB output stage is less than that for an idealized class-B circuit. In addition, the required power handling capability of the transistors in a class-AB circuit must be slightly larger than in a class-B circuit. However, since the quiescent collector currents $I_{CQ}$ are usually small compared to the peak current, this increase in power dissipation is not great. The advantage of eliminating crossover distortion in the class-AB output stage greatly outweighs the slight disadvantage of reduced conversion efficiency and increased power dissipation.

## EXAMPLE 8.6

**Objective:** Determine the total harmonic distortion (THD) of the class AB complementary push–pull output stage shown in Figure 8.24.

A PSpice analysis was performed, which yielded the harmonic content of the output signal.

**Solution:** A 1 kHz sinusoidal signal with an amplitude of 2 V was applied to the input of the circuit. The bias voltages $V_{BB}/2$ were varied. The circuit was biased at $\pm 10$ V and a 1 k$\Omega$ load was connected to the output. Shown in Table 8.4 are the $V_{BB}/2$ bias voltages applied, the quiescent transistor currents, and the total harmonic distortion (THD).

| Table 8.4 | Quiescent collector currents and total harmonic distortion of class-AB circuit | |
|---|---|---|
| $V_{BB}/2$ (V) | $I_{CQ}$ (mA) | THD (%) |
| 0.60 | 0.048 | 1.22 |
| 0.65 | 0.33 | 0.244 |
| 0.70 | 2.20 | 0.0068 |
| 0.75 | 13.3 | 0.0028 |

**Discussion:** With a peak input voltage of 2 V and a 1 k$\Omega$ load, the peak load current is on the order of 2 mA. From the results shown in Table 8.4, the THD decreases as the ratio of quiescent transistor current to peak load current increases. In other words, for a given input voltage, the smaller the variation in collector current when the signal is applied compared to the quiescent collector current, the smaller the distortion. However, there is a trade-off. As the quiescent transistor current increases, the power efficiency is reduced. The circuit should be designed such that the transistor quiescent current is the smallest value while meeting the maximum total harmonic distortion specification.

**Comment:** We see that the class-AB output stage results in a much smaller THD value than the class-B circuit, but as with most circuits, there are no uniquely specified bias voltages.

A class-AB output stage using enhancement-mode MOSFETs is shown in Figure 8.26. If $M_n$ and $M_p$ are matched, and if $v_I = 0$, then $V_{BB}/2$ is applied across the gate–source terminals of $M_n$ and the source–gate terminals of $M_p$. The quiescent drain currents established in each transistor are given by

$$i_{Dn} = i_{Dp} = I_{DQ} = K \left( \frac{V_{BB}}{2} - |V_T| \right)^2 \tag{8.36}$$

As $v_I$ increases, the voltage at the gate of $M_n$ increases and $v_O$ increases. Transistor $M_n$ operates as a source follower, supplying the load current to $R_L$. Since $i_{Dn}$ must increase to supply the load current, $v_{GSn}$ must also increase. Assuming $V_{BB}$ remains constant, an increase in $v_{GSn}$ implies a decrease in $v_{SGp}$ and a resulting decrease in $i_{Dp}$. As $v_I$ goes negative, the voltage at the base of $M_p$ decreases and $v_O$ decreases. Transistor $M_p$ then operates as a source follower, sinking current from the load.

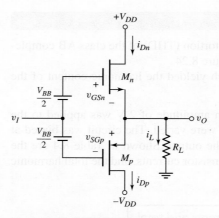

**Figure 8.26** MOSFET class-AB output stage

## EXAMPLE 8.7

**Objective:** Determine the required biasing in a MOSFET class-AB output stage.

The circuit is shown in Figure 8.26. The parameters are $V_{DD} = 10$ V and $R_L = 20$ Ω. The transistors are matched, and the parameters are $K = 0.20$ A/V$^2$ and $|V_T| = 1$ V. The quiescent drain current is to be 20 percent of the load current when $v_O = 5$ V.

**Solution:** For $v_O = 5$ V,

$$i_L = 5/20 = 0.25 \, \text{A}$$

Then, for $I_Q = 0.05$ A when $v_O = 0$, we have

$$I_{DQ} = 0.05 = K\left(\frac{V_{BB}}{2} - |V_T|\right)^2 = (0.20)\left(\frac{V_{BB}}{2} - 1\right)^2$$

which yields

$$V_{BB}/2 = 1.50 \text{ V}$$

The input voltage for $v_O$ positive is

$$v_I = v_O + v_{GSn} - \frac{V_{BB}}{2}$$

For $v_O = 5$ V and $i_{Dn} \cong i_L = 0.25$ A, we have

$$v_{GSn} = \sqrt{\frac{i_{Dn}}{K}} + |V_T| = \sqrt{\frac{0.25}{0.20}} + 1 = 2.12 \text{ V}$$

The source-to-gate voltage of $M_p$ is

$$v_{SGp} = V_{BB} - V_{GSn} = 3 - 2.12 = 0.88 \text{ V}$$

which means that $M_p$ is cut off and $i_{Dn} = i_L$. Finally, the input voltage is

$$v_I = 5 + 2.12 - 1.5 = 5.62 \text{ V}$$

**Comment:** Since $v_I > v_O$, the voltage gain of this output stage is less than unity, as expected.

---

### EXERCISE PROBLEM

**Ex 8.7:** Consider the MOSFET class-AB output stage shown in Figure 8.26. The circuit parameters are $V_{DD} = 15$ V and $R_L = 25\ \Omega$. The transistors are matched with parameters $K = 0.25$ A/V$^2$ and $|V_T| = 1.2$ V. The quiescent drain currents are to be 20 percent of the load current when $v_O = 8$ V. (a) Determine $V_{BB}$ and (b) find the small-signal voltage gain $A_v = dv_O/dv_I$ at (i) $v_O = 0$ and (ii) $v_O = 8$ V. (Ans. (a) $V_{BB} = 3.412$ V; (b) (i) $A_v = 0.927$, (ii) $A_v = 0.934$)

---

Voltage $V_{BB}$ can be established in a MOSFET class-AB circuit by using additional enhancement-mode MOSFETs and a constant current $I_{Bias}$. This will be considered in a problem at the end of the chapter.

### 8.3.4 Class-C Operation

The transistor circuit ac load line, including an extension beyond cutoff, is shown in Figure 8.27. For class-C operation, the transistor has a reverse-biased B–E voltage at the Q-point. This effect is illustrated in Figure 8.27. Note that the collector current is not negative, but is zero at the quiescent point. The transistor conducts only when the input signal becomes sufficiently positive during its positive half-cycle. The transistor therefore conducts for less than a half-cycle, which defines class-C operation.

Class-C amplifiers are capable of providing large amounts of power, with conversion efficiencies larger than 78.5 percent. These amplifiers are normally used for radio-frequency (RF) circuits, with tuned *RLC* loads that are commonly used in radio

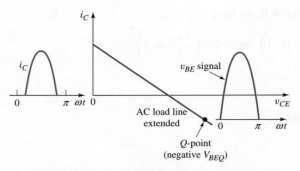

Figure 8.27 Effective ac load line of a class-C amplifier

and television transmitters. The *RLC* circuits convert drive current pulses into sinusoidal signals. Since this is a specialized area, we will not analyze these circuits here.

## Test Your Understanding

**TYU 8.4** For the common-emitter output stage shown in Figure 8.16(a), let $V_{CC} = 12$ V and $R_L = 1$ k$\Omega$. Assume the transistor $Q$-point is in the center of the load line. (a) Determine the quiescent power dissipated in the transistor. (b) Assume the sinusoidal output voltage is limited to a 9 V peak-to-peak value. Determine (i) the average signal power delivered to the load, (ii) the power conversion efficiency, and (iii) the average power dissipated in the transistor. (Ans. (a) $P_Q = 36$ mW; (b) (i) $\overline{P_L} = 10.1$ mW, (ii) $\eta = 14.1\%$, (iii) $\overline{P_Q} = 25.9$ mW)

**TYU 8.5** Design an idealized class-B output stage, as shown in Figure 8.18, to deliver an average of 25 W to an 8 $\Omega$ speaker. The peak output voltage must be no larger than 80 percent of supply voltages $V_{CC}$. Determine: (a) the required value of $V_{CC}$, (b) the peak current in each transistor, (c) the average power dissipated in each transistor, and (d) the power conversion efficiency. (Ans. (a) $V_{CC} = 25$ V (b) $I_p = 2.5$ A (c) $\bar{P}_Q = 7.4$ W (d) $\eta = 62.8\%$)

**TYU 8.6** For the idealized class-B output stage shown in Figure 8.18, the parameters are $V_{CC} = 5$ V and $R_L = 100$ $\Omega$. The measured output signal is $v_o = 4 \sin \omega t$ (V). Determine: (a) the average signal load power, (b) the peak current in each transistor, (c) the average power dissipated in each transistor, and (d) the power conversion efficiency. (Ans. (a) $\bar{P}_L = 80$ mW (b) $I_p = 40$ mA (c) $\bar{P}_Q = 23.7$ mW (d) $\eta = 62.8\%$)

##  8.4 CLASS-A POWER AMPLIFIERS

**Objective:** • Analyze several circuit configurations of class-A power amplifiers.

The standard class-A amplifier was analyzed previously, and the maximum possible power conversion efficiency was found to be 25 percent. This conversion efficiency can be increased with the use of inductors and transformers.

### 8.4.1  Inductively Coupled Amplifier

Delivering a large power to a load generally requires both a large voltage and a high current. In a common-emitter circuit, this requirement can be met by replacing the collector resistor with an inductor, as shown in Figure 8.28(a). The inductor is a short circuit to a dc current, but acts as an open circuit to an ac signal operating at a sufficiently high frequency. The entire ac current is therefore coupled to the load. We assume that $\omega L \gg R_L$ at the lowest signal frequency.

The dc and ac load lines are shown in Figure 8.28(b). We assume that the resistance of the inductor is negligible, and that the emitter resistor value is small. The quiescent collector–emitter voltage is then approximately $V_{CEQ} \cong V_{CC}$. The ac collector current is

$$i_c = \frac{-v_{ce}}{R_L} \tag{8.37}$$

To obtain the maximum symmetrical output-signal swing, which will in turn produce the maximum power, we want

$$I_{CQ} \cong \frac{V_{CC}}{R_L} \tag{8.38}$$

For this condition, the ac load line intersects the $v_{CE}$ axis at $2V_{CC}$.

The use of an inductor or storage device results in an output ac voltage swing that is larger than $V_{CC}$. The polarity of the induced voltage across the inductor may be such that the voltage adds to $V_{CC}$, producing an output voltage that is larger than $V_{CC}$.

The absolute maximum amplitude of the signal current in the load is $I_{CQ}$; therefore, the maximum possible average signal power delivered to the load is

$$\bar{P}_L(\text{max}) = \frac{1}{2}I_{CQ}^2 R_L = \frac{1}{2} \cdot \frac{V_{CC}^2}{R_L} \tag{8.39}$$

If we neglect the power dissipation in the bias resistors $R_1$ and $R_2$, the average power supplied by the $V_{CC}$ source is

$$\bar{P}_S = V_{CC}I_{CQ} = \frac{V_{CC}^2}{R_L} \tag{8.40}$$

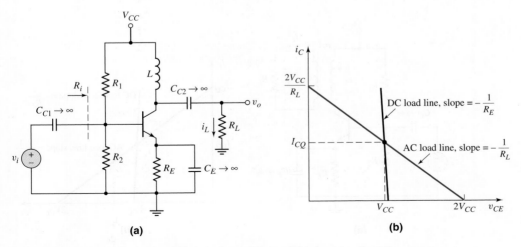

**(a)**            **(b)**

**Figure 8.28** (a) Inductively coupled class-A amplifier and (b) dc and ac load lines

The maximum possible power conversion efficiency is then

$$\eta(\text{max}) = \frac{\bar{P}_L(\text{max})}{\bar{P}_S} = \frac{\frac{1}{2} \cdot \frac{V_{CC}^2}{R_L}}{\frac{V_{CC}^2}{R_L}} = \frac{1}{2} \Rightarrow 50\% \tag{8.41}$$

This demonstrates that, in a standard class-A amplifier, replacing the collector resistor with an inductor doubles the maximum possible power conversion efficiency.

### 8.4.2   Transformer-Coupled Common-Emitter Amplifier

The design of an inductively coupled amplifier to achieve high power conversion efficiency may be difficult, depending on the relationship between the supply voltage $V_{CC}$ and the load resistance $R_L$. The effective load resistance can be optimized by using a transformer with the proper turns ratio.

Figure 8.29(a) shows a common-emitter amplifier with a transformer-coupled load in the collector circuit.

The dc and ac load lines are shown in Figure 8.29(b). If we neglect any resistance in the transformer and assume that $R_E$ is small, the quiescent collector–emitter voltage is

$$V_{CEQ} \cong V_{CC}$$

Assuming an ideal transformer, the currents and voltages in Figure 8.29(a) are related by $i_L = ai_C$ and $v_2 = v_1/a$ where $a$ is the ratio of primary to secondary turns, or simply the turns ratio. Dividing voltages by currents, we find

$$\frac{v_2}{i_L} = \frac{v_1/a}{ai_C} = \frac{v_1}{i_C} \cdot \frac{1}{a^2} \tag{8.42}$$

The load resistance is $R_L = v_2/i_L$. We can define a transformed load resistance as

$$R_L' = \frac{v_1}{i_C} = a^2 \cdot \frac{v_2}{i_L} = a^2 R_L \tag{8.43}$$

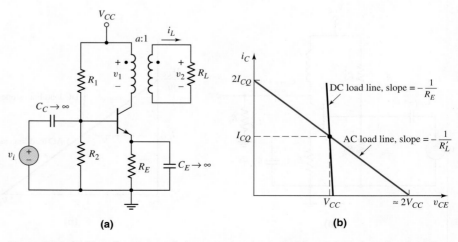

**(a)**                    **(b)**

**Figure 8.29** (a) Transformer-coupled common-emitter amplifier and (b) dc and ac load lines

The turns ratio is designed to produce the maximum symmetrical swing in the output current and voltage; therefore,

$$R'_L = \frac{2V_{CC}}{2I_{CQ}} = \frac{V_{CC}}{I_{CQ}} = a^2 R_L \tag{8.44}$$

The maximum average power delivered to the load is equal to the maximum average power delivered to the primary of the ideal transformer, as follows:

$$\bar{P}_L(\max) = \tfrac{1}{2} V_{CC} I_{CQ} \tag{8.45}$$

where $V_{CC}$ and $I_{CQ}$ are the maximum possible amplitudes of the sinusoidal signals. If we neglect the power dissipation in the bias resistors $R_1$ and $R_2$, the average power supplied by the $V_{CC}$ source is

$$\bar{P}_S = V_{CC} I_{CQ}$$

and the maximum possible power conversion efficiency is again

$$\eta(\max) = 50\%$$

## 8.4.3 Transformer-Coupled Emitter-Follower Amplifier

Since the emitter follower has a low output impedance, it is often used as the output stage of an amplifier. A transformer-coupled emitter follower is shown in Figure 8.30(a). The dc and ac load lines are shown in Figure 8.30(b). As before, the resistance of the transformer is assumed to be negligible.

The transformed load resistance is again $R'_L = a^2 R_L$. By correctly designing the turns ratio, we can achieve the maximum symmetrical swing in the output voltage and current.

The average power delivered to the load is

$$\bar{P}_L = \frac{1}{2} \cdot \frac{V_p^2}{R_L} \tag{8.46}$$

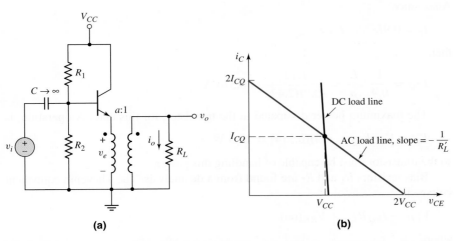

**(a)** **(b)**

**Figure 8.30** (a) Transformer-coupled emitter-follower amplifier and (b) dc and ac load lines

where $V_p$ is the peak amplitude of the sinusoidal output voltage. The maximum peak amplitude of the emitter voltage is $V_{CC}$, so that the maximum peak amplitude of the output signal is

$$V_p(\text{max}) = V_{CC}/a$$

The maximum average output signal power is therefore

$$\bar{P}_L(\text{max}) = \frac{1}{2} \cdot \frac{[V_p(\text{max})]^2}{R_L} = \frac{V_{CC}^2}{2a^2 R_L} \tag{8.47}$$

The maximum power conversion efficiency for this circuit is also 50 percent.

## DESIGN EXAMPLE 8.8

**Objective:** Design a transformer-coupled emitter-follower amplifier to deliver a specified signal power.

Consider the circuit shown in Figure 8.30(a), with parameters $V_{CC} = 24$ V and $R_L = 8\ \Omega$. The average power delivered to the load is to be 5 W, the peak amplitude of the signal emitter current is to be no more than $0.9I_{CQ}$, and that of the signal emitter voltage is to be no more than $0.9V_{CC}$. Let $\beta = 100$.

**Solution:** The average power delivered to the load is given by Equation (8.46). The peak output voltage must therefore be

$$V_p = \sqrt{2R_L \bar{P}_L} = \sqrt{2(8)(5)} = 8.94 \text{ V}$$

and the peak output current is

$$I_p = \frac{V_p}{R_L} = \frac{8.94}{8} = 1.12 \text{ A}$$

Since

$$V_e = 0.9 V_{CC} = aV_p$$

then

$$a = \frac{0.9 V_{CC}}{V_p} = \frac{(0.9)(24)}{8.94} = 2.42$$

Also, since

$$I_e = 0.9I_{CQ} = I_p/a$$

then

$$I_{CQ} = \frac{1}{0.9} \cdot \frac{I_p}{a} = \frac{1.12}{(0.9)(2.42)} = 0.514 \text{ A}$$

The maximum power dissipated in the transistor, for this class-A operation, is

$$P_Q = V_{CC}I_{CQ} = (24)(0.514) = 12.3 \text{ W}$$

so the transistor must be capable of handling this power.

Bias resistors $R_1$ and $R_2$ are found from a dc analysis. The Thevenin equivalent voltage is

$$V_{TH} = I_{BQ}R_{TH} + V_{BE}(\text{on})$$

where

$$R_{TH} = R_1 \| R_2 \quad \text{and} \quad V_{TH} = [R_2/(R_1 + R_2)] \cdot V_{CC}$$

We also have

$$I_{BQ} = \frac{I_{CQ}}{\beta} = \frac{0.514}{100} \Rightarrow 5.14 \text{ mA}$$

Since $V_{TH} < V_{CC}$ and $I_{BQ} \cong 5$ mA, then $R_{TH}$ cannot be unduly large. However, if $R_{TH}$ is small, then the power dissipation in $R_1$ and $R_2$ becomes unacceptably high. We choose $R_{TH} = 2.5$ k$\Omega$, so that

$$V_{TH} = \frac{1}{R_1}(R_{TH})V_{CC} = \frac{1}{R_1}(2.5)(24) = (5.14)(2.5) + 0.7$$

Therefore, $R_1 = 4.43$ k$\Omega$ and $R_2 = 5.74$ k$\Omega$.

**Comment:** The average power delivered by $V_{CC}$ (neglecting bias resistor effects) is $\bar{P}_S = V_{CC}I_{CQ} = 12.3$ W, which means that the power conversion efficiency is $\eta = 5/12.3 \Rightarrow 40.7\%$. The efficiency will always be less than the 50% maximum value, if transistor saturation and distortion are to be minimized.

---

### EXERCISE PROBLEM

**\*Ex 8.8:** A transformer-coupled emitter-follower amplifier is shown in Figure 8.30(a). The parameters are: $V_{CC} = 18$ V, $V_{BE}(\text{on}) = 0.7$ V, $\beta = 100$, $a = 10$, and $R_L = 8$ $\Omega$. (a) Design $R_1$ and $R_2$ to deliver the maximum power to the load. The input resistance seen by the $v_i$ source is to be 1.5 k$\Omega$. (b) If the peak amplitude of the emitter voltage $v_E$ is limited to $0.9V_{CC}$, and the peak amplitude of the emitter current $i_E$ is limited to $0.9I_{CQ}$, determine the maximum amplitude of the output signal voltage, and the average power delivered to the load. (Ans. (a) $R_1 = 26.4$ k$\Omega$, $R_2 = 1.62$ k$\Omega$ (b) $V_p = 1.62$ V, $I_p = 203$ mA, $\bar{P}_L = 0.164$ W)

---

## Test Your Understanding

**\*TYU 8.7** For the inductively coupled amplifier shown in Figure 8.28(a), the parameters are: $V_{CC} = 12$ V, $V_{BE}(\text{on}) = 0.7$ V, $R_E = 0.1$ k$\Omega$, $R_L = 1.5$ k$\Omega$, and $\beta = 75$. (a) Design $R_1$ and $R_2$ for maximum symmetrical swing in the output current and voltage. (Let $R_{TH} = (1 + \beta)R_E$.) (b) If the peak output voltage amplitude is limited to $0.9V_{CC}$, and the peak output current amplitude is limited to $0.9I_{CQ}$, determine the average power delivered to the load, the average power dissipated in the transistor, and the power conversion efficiency. (Ans. (a) $R_1 = 39.1$ k$\Omega$, $R_2 = 9.43$ k$\Omega$ (b) $\bar{P}_L = 38.9$ mW, $\bar{P}_Q = 57.1$ mW, $\eta = 40.5\%$)

---

## 8.5    CLASS-AB PUSH–PULL COMPLEMENTARY OUTPUT STAGES

**Objective:** • Analyze several circuit configurations of class-AB power amplifiers.

A class-AB output stage eliminates the crossover distortion that occurs in a class-B circuit. In this section, we will analyze several circuits that provide a small quiescent bias to the output transistors. Such circuits are used as the output stage of power amplifiers, as well as the output stage of operational amplifiers, and will be discussed in Chapter 13.

### 8.5.1    Class-AB Output Stage with Diode Biasing

In a class-AB circuit, the $V_{BB}$ voltage that provides the quiescent bias for the output transistors can be established by voltage drops across diodes, as shown in Figure 8.31. A constant current $I_{Bias}$ is used to establish the required voltage across the pair of diodes, or the diode-connected transistors, $D_1$ and $D_2$. Since $D_1$ and $D_2$ are not necessarily matched with $Q_n$ and $Q_p$, the quiescent transistor currents may not be equal to $I_{Bias}$.

As the input voltage increases, the output voltage increases, causing an increase in $i_{Cn}$. This in turn produces an increase in the base current $i_{Bn}$. Since the increase in base current is supplied by $I_{Bias}$, the current through $D_1$ and $D_2$, and hence the voltage $V_{BB}$, decreases slightly. Since voltage $V_{BB}$ does not remain constant in this circuit, the relationship between $i_{Cn}$ and $i_{Cp}$, as given by Equation (8.35), is not precisely valid for this situation. The analysis in the previous section must therefore be modified slightly, but the basic operation of this class-AB circuit is the same.

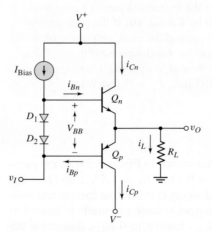

**Figure 8.31**  Class-AB output stage with quiescent bias established by diodes

### DESIGN EXAMPLE 8.9

**Objective:**  Design the class-AB output stage in Figure 8.31 to meet specific design criteria.

Assume $I_{SD} = 3 \times 10^{-14}$ A for $D_1$ and $D_2$, $I_{SQ} = 10^{-13}$ A for $Q_n$ and $Q_p$, and $\beta_n = \beta_p = 75$. Let $R_L = 8\ \Omega$. The average power delivered to the load is to be 5 W. The peak output voltage is to be no more than 80 percent of $V_{CC}$, and the minimum value of diode current $I_D$ is to be no less than 5 mA.

**Solution:**  The average power delivered to the load, from Equation (8.24), is

$$\bar{P}_L = \frac{1}{2} \cdot \frac{V_p^2}{R_L}$$

Therefore,

$$V_p = \sqrt{2R_L \bar{P}_L} = \sqrt{2(8)(5)} = 8.94 \text{ V}$$

The supply voltages must then be

$$V_{CC} = \frac{V_p}{0.8} = \frac{8.94}{0.8} = 11.2 \text{ V}$$

At this peak output voltage, the emitter current of $Q_n$ is approximately equal to the load current, or

$$i_{En} \cong i_L(\text{max}) = \frac{V_p(\text{max})}{R_L} = \frac{8.94}{8} = 1.12 \text{ A}$$

and the base current is

$$i_{Bn} = \frac{i_{En}}{1 + \beta_n} = \frac{1.12}{76} \Rightarrow 14.7 \text{ mA}$$

For a minimum $I_D = 5$ mA, we can choose $I_{\text{Bias}} = 20$ mA. For a zero input signal, neglecting base currents, we find that

$$V_{BB} = 2V_T \ln\left(\frac{I_D}{I_{SD}}\right) = 2(0.026) \ln\left(\frac{20 \times 10^{-3}}{3 \times 10^{-14}}\right) = 1.416 \text{ V}$$

The quiescent collector currents are then

$$I_{CQ} = I_{SQ} e^{(V_{BB}/2)V_T} = 10^{-13} e^{1.416/2(0.026)} \Rightarrow 67.0 \text{ mA}$$

For $v_O = 8.94$ V and $i_L = 1.12$ A, the base current is $i_{Bn} = 14.7$ mA, and

$$I_D = I_{\text{Bias}} - i_{Bn} = 5.3 \text{ mA}$$

The new value of $V_{BB}$ is then

$$V'_{BB} = 2V_T \ln\left(\frac{I_D}{I_{SD}}\right) = 2(0.026) \ln\left(\frac{5.3 \times 10^{-3}}{3 \times 10^{-14}}\right) = 1.347 \text{ V}$$

The B–E voltage of $Q_n$ is

$$v_{BEn} = V_T \ln\left(\frac{i_{Cn}}{I_{SQ}}\right) = (0.026) \ln\left(\frac{1.12}{10^{-13}}\right) = 0.781 \text{ V}$$

The emitter–base voltage of $Q_p$ is then

$$v_{EBp} = V'_{BB} - v_{BEn} = 1.347 - 0.781 = 0.566 \text{ V}$$

and

$$i_{Cp} = I_{SQ} e^{v_{EBp}/V_T} = (10^{-13}) e^{0.566/0.026} \Rightarrow 0.285 \text{ mA}$$

**Comment:** When the output goes positive, the current in $Q_p$ decreases significantly, as expected, but it does not go to zero. There is a factor of approximately $10^3$ difference in the currents between $Q_n$ and $Q_p$.

**Design Pointer:** If the output signal currents are large, the base currents in the output transistors may become significant compared to the bias current through the diodes $D_1$ and $D_2$. The change in the diode bias current should be minimized in order to keep the small-signal voltage gain of the output stage close to unity.

**Ex 8.9:** The BJT class-AB output stage shown in Figure 8.31 is biased at $V^+ = 5$ V and $V^- = -5$ V. The load resistance is $R_L = 1$ k$\Omega$. The device parameters are $I_{SD} = 1.2 \times 10^{-14}$ A for the diodes and $I_{SQ} = 2 \times 10^{-14}$ A for the transistors. (a) Neglecting base currents, determine the value of $I_{Bias}$ that induces quiescent collector currents of $I_{CQ} = 1$ mA. (b) Assuming $\beta_n = \beta_p = 100$, determine $i_{Cn}, i_{Cp}, v_{BEn}, v_{EBp}$, and $I_D$ for $v_O = 1.2$ V. (c) Repeat part (b) for $v_O = 3$ V. (Ans. (a) $I_{Bias} = 0.6$ mA; (b) $i_{Cn} = 1.73$ mA, $i_{Cp} = 0.547$ mA, $v_{BEn} = 0.6547$ V, $v_{EBp} = 0.6248$ V, $I_D = 0.5827$ mA; (c) $i_{Cn} = 3.24$ mA, $i_{Cp} = 0.276$ mA, $v_{BEn} = 0.671$ V, $v_{EBp} = 0.607$ V, $I_D = 0.5676$ mA)

### 8.5.2    Class-AB Biasing Using the $V_{BE}$ Multiplier

An alternative biasing scheme, which provides more flexibility in the design of the output stage, is shown in Figure 8.32. The bias circuit that provides voltage $V_{BB}$ consists of transistor $Q_1$ and resistors $R_1$ and $R_2$, biased by a constant-current source $I_{Bias}$.

If we neglect the base current in $Q_1$, then

$$I_R = \frac{V_{BE1}}{R_2} \tag{8.48}$$

and voltage $V_{BB}$ is

$$V_{BB} = I_R(R_1 + R_2) = V_{BE1}\left(1 + \frac{R_1}{R_2}\right) \tag{8.49}$$

Since voltage $V_{BB}$ is a multiplication of the junction voltage $V_{BE1}$, the circuit is called a $V_{BE}$ **multiplier.** The multiplication factor can be designed to yield the required value of $V_{BB}$.

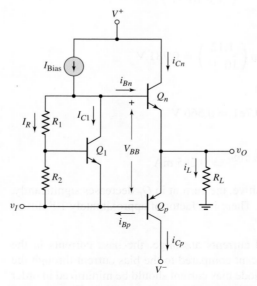

**Figure 8.32** Class-AB output stage with $V_{BE}$ multiplier bias circuit

A fraction of the constant current $I_{\text{Bias}}$ flows through $Q_1$, so that

$$V_{BE1} = V_T \ln\left(\frac{I_{C1}}{I_{S1}}\right) \tag{8.50}$$

Also, the quiescent bias currents $i_{Cn}$ and $i_{Cp}$ are normally small; therefore, we can neglect $i_{Bn}$ and $i_{Bp}$. Current $I_{\text{Bias}}$ divides between $I_R$ and $I_{C1}$, satisfying both Equations (8.48) and (8.50).

As $v_I$ increases, $v_O$ becomes positive, and $i_{Cn}$ and $i_{Bn}$ increase, which reduces the collector current in $Q_1$. However, the logarithmic dependence of $I_{C1}$, shown in Equation (8.50), means that $V_{BE1}$ and, in turn $V_{BB}$ remain essentially constant as the output voltage changes.

## DESIGN EXAMPLE 8.10

**Objective:** Design a Class-AB output stage using the $V_{BE}$ multiplier circuit to meet a specified total harmonic distortion.

Assume the circuit in Figure 8.32, biased at $V^+ = 15$ V and $V^- = -15$ V, is the output stage of an audio amplifier that is to drive another power amplifier whose input resistance is 1 k$\Omega$. The maximum peak sinusoidal output voltage is to be 10 V and the total harmonic distortion is to be less than 0.1 percent.

**Solution:** Standard 2N3904 and 2N3906 transistors are to be used. From the results of Example 8.6, the THD is a function of the output transistor quiescent currents. For the basic circuit in Figure 8.24, the THD is found to be 0.097 percent for $V_{BB} = 1.346$ V, quiescent collector currents of 0.88 mA, and a peak sinusoidal output voltage of 10 V.

Figure 8.33 is the PSpice circuit schematic. For a peak output voltage of 10 V, the peak load current is 10 mA. Assuming $\beta \cong 100$, the peak base current is 0.1 mA.

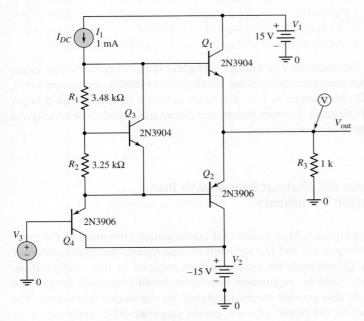

**Figure 8.33** PSpice circuit schematic for Example 8.10

A bias current of 1 mA is chosen to bias the $V_{BE}$ multiplier. The peak 0.1 mA base current, then, will not greatly disturb the current through the multiplier circuit.

We may select $I_R = 0.2$ mA (current through $R_1$ and $R_2$) and $I_{C3} = 0.8$ mA. We then have

$$R_1 + R_2 = \frac{V_{BB}}{I_R} = \frac{1.346}{0.2} = 6.73 \text{ k}\Omega$$

For the 2N3904, we find that $V_{BE} \cong 0.65$ V for a quiescent collector current of approximately 0.8 mA. Therefore

$$R_2 = \frac{V_{BE3}}{I_R} = \frac{0.65}{0.2} = 3.25 \text{ k}\Omega$$

so that $R_1 = 3.48$ k$\Omega$.

From the PSpice results, we find that the voltage at the base of $Q_1$ to be 0.6895 V and the voltage at the base of $Q_2$ to be $-0.6961$ V, which means that $V_{BB} = 1.3856$ V. This voltage is slightly greater than the design value of $V_{BB} = 1.346$ V. Listed below are the quiescent transistor parameters. The quiescent collector currents of the output transistors are 1.88 mA, approximately twice the design value of 0.88 mA. The total harmonic distortion is 0.0356 percent, which is well within the design specification.

| NAME | Q_Q1 | Q_Q2 | Q_Q3 | Q_Q4 |
|---|---|---|---|---|
| MODEL | Q2N3904 | Q2N3906 | Q2N3904 | Q2N3906 |
| IB | 1.12E-05 | -5.96E-06 | 6.01E-06 | -3.20E-06 |
| IC | 1.88E-03 | -1.88E-03 | 7.80E-04 | -9.92E-04 |
| VBE | 6.78E-01 | -7.08E-01 | 6.59E-01 | -6.92E-01 |
| VBC | -1.43E+01 | 1.43E+01 | -7.27E-01 | 1.36E+01 |
| VCE | 1.50E+01 | -1.50E+01 | 1.39E+00 | -1.43E+01 |
| BETADC | 1.67E+02 | 3.15E+02 | 1.30E+02 | 3.10E+02 |
| GM | 7.11E-02 | 7.15E-02 | 2.98E-02 | 3.80E-02 |
| RPI | 2.66E+03 | 4.34E+03 | 5.01E+03 | 8.09E+03 |

**Comment:** Since the resulting $V_{BB}$ voltage is slightly larger than the design value, the quiescent output transistor currents are approximately double the design value. Although the THD specification is met, the larger collector currents mean a larger quiescent power dissipation. For this reason, the circuit may need to be redesigned slightly to lower the quiescent currents.

### 8.5.3 Class-AB Output Stage with Input Buffer Transistors

The output stage in Figure 8.34 is a class-AB configuration composed of the complementary transistor pair $Q_3$ and $Q_4$. Resistors $R_1$ and $R_2$ and the emitter-follower transistors $Q_1$ and $Q_2$ establish the quiescent bias required in this configuration. Resistors $R_3$ and $R_4$, used in conjunction with short-circuit protection devices not shown in the figure, also provide thermal stability for the output transistors. The input signal $v_I$ may be the output of a low-power amplifier. Also, since this is an emitter follower, the output voltage is approximately equal to the input voltage.

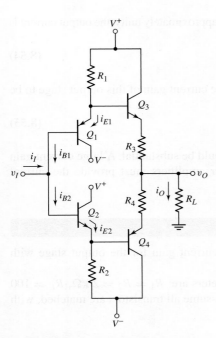

**Figure 8.34** Class-AB output stage with input buffer transistors

When the input voltage $v_I$ increases from zero, the base voltage of $Q_3$ increases, and the output voltage $v_O$ increases. The load current $i_O$ is positive, and the emitter current in $Q_3$ increases to supply the load current, which causes an increase in the base current into $Q_3$. Since the base voltage of $Q_3$ increases, the voltage drop across $R_1$ decreases, resulting in a smaller current in $R_1$. This means that $i_{E1}$ and $i_{B1}$ also decrease. As $v_I$ increases, the voltage across $R_2$ increases, and $i_{E2}$ and $i_{B2}$ increase. A net input current $i_I$ is then produced, to account for the reduction in $i_{B1}$ and the increase in $i_{B2}$.

The net input current is

$$i_I = i_{B2} - i_{B1} \tag{8.51}$$

Neglecting the voltage drops across $R_3$ and $R_4$ and the base currents in $Q_3$ and $Q_4$, we have

$$i_{B2} = \frac{(v_I - V_{BE}) - V^-}{(1 + \beta_n)R_2} \tag{8.52(a)}$$

and

$$i_{B1} = \frac{V^+ - (v_I + V_{EB})}{(1 + \beta_p)R_1} \tag{8.52(b)}$$

where $\beta_n$ and $\beta_p$ are the current gains of the npn and pnp transistors, respectively. If $V^+ = -V^-$, $V_{BE} = V_{EB}$, $R_1 = R_2 \equiv R$, and $\beta_n = \beta_p \equiv \beta$, then combining Equations (8.52(a)), (8.52(b)), and (8.51) produces

$$i_I = \frac{(v_I - V_{BE} - V^-)}{(1 + \beta)R} - \frac{(V^+ - v_I - V_{EB})}{(1 + \beta)R} = \frac{2v_I}{(1 + \beta)R} \tag{8.53}$$

Since the voltage gain of this output stage is approximately unity, the output current is

$$i_O = \frac{v_O}{R_L} \cong \frac{v_I}{R_L} \qquad\qquad (8.54)$$

Using Equations (8.53) and (8.54), we find the current gain of this output stage to be

$$A_i = \frac{i_O}{i_I} = \frac{(1+\beta)R}{2R_L} \qquad\qquad (8.55)$$

With $\beta$ in the numerator, this current gain should be substantial. A large current gain is desirable, since the output stage of power amplifiers must provide the current necessary to meet the power requirements.

## EXAMPLE 8.11

**Objective:** Determine the currents and the current gain for the output stage with input buffer transistors.

For the circuit in Figure 8.34, the parameters are: $R_1 = R_2 = 2$ k$\Omega$, $R_L = 100$ $\Omega$, $R_3 = R_4 = 0$, and $V^+ = -V^- = 15$ V. Assume all transistors are matched, with $\beta = 60$ and $V_{BE}(\text{npn}) = V_{EB}(\text{pnp}) = 0.6$ V.

**Solution:** For $v_I = 0$,

$$i_{R1} = i_{R2} \cong i_{E1} = i_{E2} = \frac{15 - 0.6}{2} = 7.2 \,\text{mA}$$

Assuming all transistors are matched, the bias currents in $Q_3$ and $Q_4$ are also approximately 7.2 mA, since the base–emitter voltages of $Q_1$ and $Q_3$ are equal and those of $Q_2$ and $Q_4$ are equal.

**Solution:** For $v_I = 10$ V, the output current is approximately

$$i_O = \frac{v_O}{R_L} \cong \frac{10}{0.1} = 100 \,\text{mA}$$

The emitter current in $Q_3$ is essentially equal to the load current, which means that the base current in $Q_3$ is approximately

$$i_{B3} = 100/61 = 1.64 \,\text{mA}$$

The current in $R_1$ is

$$i_{R1} = \frac{15 - (10 + 0.6)}{2} = 2.2 \,\text{mA}$$

which means that

$$i_{E1} = i_{R1} - i_{B3} = 0.56 \,\text{mA}$$

and

$$i_{B1} = i_{E1}/(1+\beta) = 0.56/61 \Rightarrow 9.18 \,\mu\text{A}$$

Since $Q_4$ tends to turn off when $v_O$ increases, we have

$$i_{E2} \cong i_{R2} = \frac{10 - 0.6 - (-15)}{2} = 12.2 \,\text{mA}$$

and

$$i_{B2} = i_{E2}/(1 + \beta) = 12.2/61 \Rightarrow 200\,\mu A$$

The input current is then

$$i_I = i_{B2} - i_{B1} = 200 - 9.18 \cong 191\,\mu A$$

The current gain is then

$$A_i = \frac{i_O}{i_I} = \frac{100}{0.191} = 524$$

From Equation (8.55), the predicted current gain is

$$A_i = \frac{i_O}{i_I} = \frac{(1 + \beta)R}{2R_L} = \frac{(61)(2)}{2(0.1)} = 610$$

**Comment:** Since the current gain determined from Equation (8.55) neglects base currents in $Q_3$ and $Q_4$, the actual current gain is less than the predicted value, as expected. The input current of 191 $\mu A$ can easily be supplied by a low-power amplifier.

---

**EXERCISE PROBLEM**

**Ex 8.11:** Consider the class-AB output stage in Figure 8.34. The parameters are: $V^+ = -V^- = 12\,\text{V}$, $R_1 = R_2 = 250\;\Omega$, $R_L = 8\;\Omega$, and $R_3 = R_4 = 0$. Assume all transistors are matched, with $\beta = 40$ and $V_{BE}(\text{npn}) = V_{EB}(\text{pnp}) = 0.7$ V. (a) For $v_I = 0$, determine $i_{E1}$, $i_{E2}$, $i_{B1}$, and $i_{B2}$. (b) For $v_I = 5$ V, find $i_O$, $i_{E1}$, $i_{E2}$, $i_{B1}$, $i_{B2}$, and $i_I$. (c) Using the results of part (b), determine the current gain of the output stage. Compare this value to that found using Equation (8.55). (Ans. (a) $i_{E1} = i_{E2} = 44.1$ mA, $i_{B1} = i_{B2} = 1.08$ mA (b) $i_O = 0.625$ A, $i_{E1} = 10.0$ mA, $i_{B1} = 0.244$ mA, $i_{E2} = 65.2$ mA, $i_{B2} = 1.59$ mA, $i_I = 1.35$ mA (c) $A_i = 463$, from Equation (8.55) $A_i = 641$)

---

**8.5.4**  **Class-AB Output Stage Utilizing the Darlington Configuration**

The complementary push–pull output stage uses npn and pnp bipolar transistors. Usually in IC design, the pnp transistors are fabricated as lateral devices with low $\beta$ values that are typically in the range of 5 to 10, and the npn transistors are fabricated as vertical devices with $\beta$ values on the order of 200. This means that the npn and pnp transistors are not well matched, as we have assumed in our analyses.

Consider the two-transistor configuration shown in Figure 8.35(a). Assume the transistor current gains are $\beta_n$ and $\beta_p$ for the npn and pnp transistors, respectively. We can write

$$i_{Cp} = i_{Bn} = \beta_p i_{Bp} \tag{8.56}$$

and

$$i_2 = (1 + \beta_n)i_{Bn} = (1 + \beta_n)\beta_p i_{Bp} \cong \beta_n \beta_p i_{Bp} \tag{8.57}$$

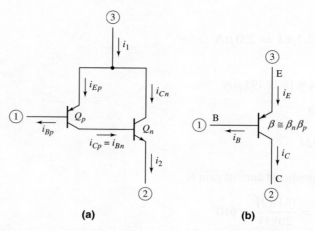

**(a)**     **(b)**

**Figure 8.35** (a) A two-transistor configuration of an equivalent pnp transistor; (b) the equivalent pnp transistor

Terminal 1 acts as the base of the composite three-terminal device, terminal 2 acts as the collector, and terminal 3 is the emitter. The current gain of the device is then approximately $\beta_n\beta_p$. The equivalent circuit is shown in Figure 8.35(b). We can use the two-transistor configuration in Figure 8.35(a) as a single equivalent pnp transistor with a current gain on the same order of magnitude as that of an npn device.

In Figure 8.36, the output stage uses Darlington pairs to provide the necessary current gain. Transistors $Q_1$ and $Q_2$ constitute the npn Darlington emitter-follower that sources current to the load. Transistors $Q_3$, $Q_4$, and $Q_5$ constitute a composite pnp Darlington emitter follower that sinks current from the load. The three diodes $D_1$, $D_2$, and $D_3$ establish the quiescent bias for the output transistors.

The effective current gain of the three-transistor configuration $Q_3$–$Q_4$–$Q_5$ is essentially the product of the three individual gains. With the low current gain of the pnp device $Q_3$, the overall current gain of the $Q_3$–$Q_4$–$Q_5$ configuration is similar to that of the $Q_1$–$Q_2$ pair.

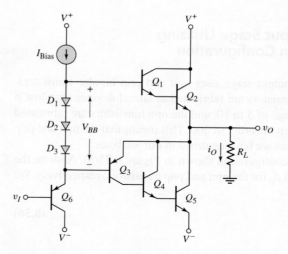

**Figure 8.36** Class-AB output stage with Darlington pairs

## Test Your Understanding

**TYU 8.8**  Consider the class-AB output stage shown in Figure 8.37. The transistor parameters are $\beta_n = 100$ and $I_{Sn} = 5 \times 10^{-16}$ A for the npn device, and $\beta_p = 100$ and $I_{Sp} = 8 \times 10^{-16}$ A for the pnp device. (a) What value of $V_{BB}$ will establish quiescent collector currents of $I_{CQ} = 1$ mA with $v_O = 0$. (b) What are the values of $v_{BEn}$ and $v_{EBp}$? (c) What must be the value of $v_I$ such that $v_O = 0$. (Ans. (a) $V_{BB} = 1.4606$ V; (b) $v_{BEn} = 0.7364$ V, $v_{EBp} = 0.7242$ V; (c) $v_I = 6.1$ mV)

**TYU 8.9**  From Figure 8.36, show that the overall current gain of the three-transistor configuration composed of $Q_3$, $Q_4$, and $Q_5$ is approximately $\beta = \beta_3\beta_4\beta_5$.

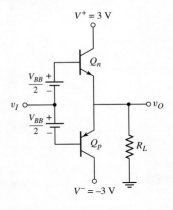

**Figure 8.37** Figure for exercise TYU 8.8

## 8.6    DESIGN APPLICATION: AN OUTPUT STAGE USING MOSFETs

**Objective:** • Design an output stage using power MOSFETs as the output devices.

**Specifications:** The output stage configuration to be designed is shown in Figure 8.38. The current $I_{Bias}$ is 5 mA and the zero output quiescent current in $M_n$ and $M_p$ is to be 0.5 mA.

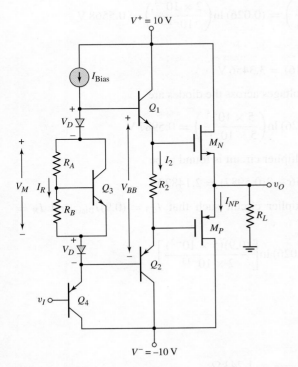

**Figure 8.38** Output stage for design application

**Design Pointer:** The output devices are to be MOSFETs because of their superior power characteristics. The low output resistance of the emitter follower transistors $Q_1$ and $Q_2$ tends to increase switching speed of the output transistors. The voltage drop across resistor $R_2$ provides the bias to $M_n$ and $M_p$ so that crossover distortion is minimized.

**Choices:** MOSFETs with parameters $V_{TN} = 0.8$ V, $V_{TP} = -0.8$ V, $K_n = K_p = 5$ mA/V$^2$, and $\lambda = 0$ are available. BJTs with parameters $I_{S1} = I_{S2} = 10^{-12}$ A, $I_{S3} = I_{S4} = 2 \times 10^{-13}$ A, and $\beta = 150$ are available. Also diodes with parameters $I_{SD} = 5 \times 10^{-13}$ A are available.

**Solution:** For $I_{NP} = 0.5$ mA, the gate-to-source voltages are found from

$$I_{NP} = K_n(V_{GSn} - V_{TN})^2$$

or

$$0.5 = 5(V_{GSn} - 0.8)^2$$

Since the two output transistors are matched, we have

$$V_{GSn} = V_{SGp} = 1.116 \text{ V}$$

If we design for $I_2 = 2$ mA, then the value of resistor $R_2$ is

$$R_2 = \frac{2(1.116)}{2} = 1.116 \text{ k}\Omega$$

Considering the BJTs, we find

$$V_{BE1} = V_{EB2} = V_T \ln\left(\frac{I_2}{I_{S1}}\right) = (0.026) \ln\left(\frac{2 \times 10^{-3}}{10^{-12}}\right) = 0.5568 \text{ V}$$

We then have

$$V_{BB} = 2(0.5568) + 2(1.116) = 3.3456 \text{ V}$$

Neglecting base currents, the voltages across the diodes are

$$V_D = V_T \ln\left(\frac{I_D}{I_{SD}}\right) = (0.026) \ln\left(\frac{5 \times 10^{-3}}{5 \times 10^{-13}}\right) = 0.5987 \text{ V}$$

The voltage across the $V_{BE}$ multiplier circuit is found to be

$$V_M = V_{BB} - 2V_D = 3.3456 - 2(0.5987) = 2.1482 \text{ V}$$

We will design the $V_{BE}$ multiplier circuit such that $I_{C3} = (0.9)I_{\text{Bias}}$ and $I_R = (0.1)I_{\text{Bias}}$. Then

$$V_{BE3} = V_T \ln\left(\frac{I_{C3}}{I_{S3}}\right) = (0.026) \ln\left[\frac{(0.9)(5 \times 10^{-3})}{2 \times 10^{-13}}\right]$$

or

$$V_{BE3} = 0.6198 \text{ V}$$

We also have

$$R_B = \frac{V_{BE3}}{I_R} = \frac{0.6198}{(0.1)(5 \times 10^{-3})} = 1.24 \text{ k}\Omega$$

From Equation (8.48), we have

$$V_M = V_{BE3}\left(1 + \frac{R_A}{R_B}\right)$$

or

$$2.1482 = (0.6198)\left(1 + \frac{R_A}{R_B}\right)$$

which yields $R_A/R_B = 2.466$, so that $R_A = 2.466 \, R_B = 3.06 \text{ k}\Omega$.

We see that

$$V_{EB4} = V_T \ln\left(\frac{I_{Bias}}{I_{C4}}\right) = (0.026) \ln\left(\frac{5 \times 10^{-3}}{2 \times 10^{-13}}\right) = 0.6225 \text{ V}$$

Then, for $v_O = 0$, the input voltage $v_I$ must be

$$v_I = -V_{SGP} - V_{EB2} - V_{EB4} = -1.116 - 0.5568 - 0.6225$$

or

$$v_I = -2.295 \text{ V}$$

**Comment:** The required input voltage $v_I$ to yield $v_O = 0$ would be designed from the previous stage of the amplifier. In addition, the circuit required to establish the $I_{Bias}$ current will be considered in Chapter 10. We may notice that, except for $I_{Bias}$, all the design parameters are independent of the bias voltages $V^+$ and $V^-$.

## 8.7   SUMMARY

- In this chapter, we analyzed and designed amplifiers and output stages capable of delivering a substantial amount of power to a load.
- The current, voltage, and power ratings of BJTs and MOSFETs were considered, and the safe operating area for the transistors was defined in terms of these limiting parameters. The maximum power rating of a transistor is related to the maximum allowed device temperature at which the device can operate without being damaged.
- Several classes of power amplifiers were defined.
- In a class-A amplifier, the output transistor conducts 100 percent of the time. The theoretical maximum power efficiency for a standard class-A amplifier is 25 percent. This efficiency can be theoretically increased to 50 percent by incorporating inductors and transformers in the circuit.
- Class-B output stages are composed of complementary pairs of transistors operating in a push-pull manner. In an ideal class-B operation, each output transistor conducts 50 percent of the time. For an idealized class-B output stage, the theoretical maximum power conversion efficiency is 78.5 percent. However, practical class-B output stages tend to suffer from crossover distortion effects when the output is in the vicinity of zero volts.
- The class-AB output stage is similar to the class-B circuit, except that each output transistor is provided with a small quiescent bias and conducts slightly more than 50 percent of the time. The power conversion efficiency of this circuit is less than that of a class-B circuit, but is substantially larger than that of the class-A circuit. In addition, the crossover distortion is greatly reduced.
- As an application, a class-AB output stage using MOSFETs was designed.

 CHECKPOINT

After studying this chapter, the reader should have the ability to:

✓ Describe what factors are related to the maximum transistor current and maximum transistor voltage.
✓ Define the safe operating area of a transistor and define the power derating curve.
✓ Define the power conversion efficiency of an output stage.
✓ Describe the operation of a class-A output stage.
✓ Describe the operation of an ideal class-B output stage and discuss the concept of crossover distortion.
✓ Describe and design a class-AB output stage and discuss why crossover distortion is essentially eliminated.

 REVIEW QUESTIONS

1. Discuss the limiting factors for the maximum rated current and maximum rated voltage in a BJT and MOSFET.
2. Describe the safe operating area for a transistor.
3. Why is an interdigitated structure typically used in a high-power BJT design?
4. Discuss the role of thermal resistance between various junctions in a high-power transistor structure.
5. Define and describe the power derating curve for a transistor.
6. Define power conversion efficiency for an output stage.
7. Describe the operation of a class-A output stage.
8. Describe the operation of an ideal class-B output stage.
9. Discuss crossover distortion.
10. What is meant by harmonic distortion?
11. Describe the operation of a class-AB output stage and why a class-AB output stage is important.
12. Describe the operation of a transformer-coupled class-A common-emitter amplifier.
13. Sketch a class-AB complementary BJT push–pull output stage using a $V_{BE}$ multiplier circuit.
14. Sketch a class-AB complementary MOSFET push-pull output stage using all MOSFETs.
15. What are the advantages of a Darlington pair configuration?
16. Sketch a two-transistor configuration using npn and pnp BJTs that are equivalent to a single pnp BJT.

 PROBLEMS

### Section 8.2 Power Transistors

8.1 The maximum current, voltage, and power ratings of a power MOSFET are 4 A, 40 V, and 30 W, respectively. (a) Sketch and label the safe operating area for this transistor, using linear current and voltage scales. (b) For the common-source circuit shown in Figure P8.1, determine $R_D$ and sketch the load line that produces a maximum power in the transistor for (i) $V_{DD} = 24$ V and (ii) $V_{DD} = 40$ V. (c) Using the results of part (b), determine the maximum possible drain current for (i) $V_{DD} = 24$ V and (ii) $V_{DD} = 40$ V.

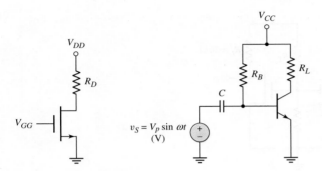

Figure P8.1            Figure P8.2

8.2    The common-emitter circuit in Figure P8.2 is biased at $V_{CC} = 24$ V. The maximum transistor power is rated at $P_{Q,\max} = 25$ W. The other parameters of the transistor are $\beta = 60$ and $V_{BE}(\text{on}) = 0.7$ V. (a) Determine $R_L$ and $R_B$ such that the transistor is biased at the maximum power point. (b) For $V_p = 12$ mV, determine the average power dissipated in the transistor.

8.3    For the transistor in the common-emitter circuit in Figure P8.2, the parameters are: $\beta = 80$, $P_{D,\max} = 10$ W, $V_{CE(\text{sus})} = 30$ V, and $I_{C,\max} = 1.2$ A. (a) Design the values of $R_L$ and $R_B$ for $V_{CC} = 30$ V. What is maximum power dissipated in the transistor? (b) Using the value of $R_L$ in part (a), find $I_{C,\max}$ and $V_{CC}$ if $P_{D,\max} = 5$ W. (c) Calculate the maximum undistorted ac power that can be delivered to $R_L$ in parts (a) and (b) for the assumption that $i_C \geq 0$ and $0 \leq v_{CE} \leq V_{CC}$.

8.4    Sketch the safe operating region for a MOSFET. Label three arbitrary points on the maximum hyperbola. Assume each of the labeled points is a $Q$-point and draw a tangent load line through each point. Discuss the advantages or disadvantages of each point relative to the maximum possible signal amplitude.

8.5    A power MOSFET is connected in a common-source configuration as shown in Figure P8.1. The parameters are: $I_{D,\max} = 4$ A, $V_{DS,\max} = 50$ V, $P_{D,\max} = 35$ W, $V_{TN} = 4$ V, and $K_n = 0.25$ A/V$^2$. The circuit parameters are $V_{DD} = 40$ V and $R_L = 10$ Ω. (a) Sketch and label the safe operating area for this transistor, using linear current and voltage scales. Also sketch the load line on the same graph. (b) Calculate the power dissipated in the transistor for $V_{GG} = 5$, 6, 7, 8, and 9 V. (c) Is there a possibility of damaging the transistor? Explain.

D8.6   Consider the common-source circuit shown in Figure P8.6. The transistor parameters are $V_{TN} = 4$ V and $K_n = 0.2$ A/V$^2$. (a) Design the bias circuit such that the $Q$-point is in the center of the load line. (b) What is the power dissipated in the transistor at the $Q$-point? (c) Determine the minimum rated $I_{D,\max}$, $V_{DS,\max}$, and $P_{D,\max}$ values. (d) If $v_i = 0.5 \sin \omega t$ V, calculate the ac power delivered to $R_L$, and determine the average power dissipated in the transistor.

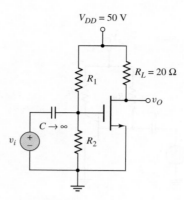

**Figure P8.6**

8.7   A particular transistor is rated for a maximum power dissipation of 60 W if the case temperature is at 25 °C. Above 25 °C, the allowed power dissipation is reduced by 0.5W/°C. (a) Sketch the power derating curve. (b) What is the maximum allowed junction temperature? (c) What is the value of $\theta_{\text{dev}-\text{case}}$?

8.8   A MOSFET has a rated power of 50 W and a maximum specified junction temperature of 150 °C. The ambient is $T_{\text{amb}} = 25$ °C. Find the relationship between the actual operating power and $\theta_{\text{case}-\text{amb}}$.

8.9   For a power MOSFET, $\theta_{\text{dev}-\text{case}} = 1.5$ °C/W, $\theta_{\text{snk}-\text{amb}} = 2.8$ °C/W, and $\theta_{\text{case}-\text{snk}} = 0.6$ °C/W. The ambient temperature is 25 °C. (a) If the maximum junction temperature is limited to $T_{j,\text{max}} = 120$ °C, determine the maximum allowed power dissipation. (b) Using the results of part (a), determine the temperature of the case and heat sink.

8.10   A power BJT must dissipate 30 W of power. The maximum allowed junction temperature is $T_{j,\text{max}} = 150$ °C, the ambient temperature is 25 °C, and the device-to-case thermal resistance is $\theta_{\text{dev}-\text{case}} = 2.8$ °C/W. (a) Find the maximum permissible thermal resistance between the case and ambient. (b) Using the results of part (a), determine the junction temperature if the power dissipated in the transistor is 20 W.

8.11   The quiescent collector current in a BJT is $I_{CQ} = 3$ A. The maximum allowed junction temperature is $T_{j,\text{max}} = 150$ °C and the ambient temperature is $T_{\text{amb}} = 25$ °C. Other parameters are $\theta_{\text{snk}-\text{amb}} = 3.8$ °C/W, $\theta_{\text{case}-\text{snk}} = 1.5$ °C/W, and $\theta_{\text{dev}-\text{case}} = 4$ °C/W. (a) Determine the power that can be safely dissipated in the transistor. (b) Using the results of part (a), determine the maximum collector-emitter voltage that may be applied.

## Section 8.3 Classes of Amplifiers

8.12   For the class-A amplifier shown in Figure 8.16(a), show that the maximum theoretical conversion efficiency for a symmetrical square-wave input signal is 50 percent.

8.13   Consider the emitter-follower amplifier shown in Figure P8.13. (a) Assuming $\beta \gg 1$, show that the small-signal voltage gain can be written in the form

$$A_v = \frac{I_C R_L}{I_C R_L + V_T} = \frac{R_L}{R_L + \dfrac{1}{g_m}}$$

(b) If $R_L = 8\ \Omega$, determine the minimum collector current that produces a small-signal voltage gain of (i) $A_v = 0.9$, (ii) $A_v = 0.95$, and (iii) $A_v = 0.9970$.

8.14   Consider the emitter-follower amplifier shown in Figure P8.13. An average power of 0.5 W is to be delivered to a load of $R_L = 8\ \Omega$. (a) What are the peak values of ac output voltage and ac load current? (b) The minimum collector current occurs when $V_O$ reaches the maximum negative value. If the minimum collector current is to be 10 percent of $I_O$, determine $I_O$. (Use the results of part (a).)

8.15   Consider the emitter-follower amplifier in Figure P8.13. Since the base-emitter voltage is a function of collector current, the voltage gain changes as the collector current changes. This effect results in distortion of the output signal. Assume $R_L = 8\ \Omega$ and $I_O = 0.25$ A. Defining the voltage gain as (Problem 8.13)

$$A_v = \frac{R_L}{R_L + \dfrac{1}{g_m}}$$

find the voltage gain for (a) $V_O = +1.6$ V, (b) $V_O = 0$, and (c) $V_O = -1.6$ V.

8.16   Consider the class-A emitter-follower circuit shown in Figure P8.16. Assume all transistors are matched with $V_{BE}(\text{on}) = 0.7$ V, $V_{CE}(\text{sat}) = 0.2$ V, and $V_A = \infty$. Neglect base currents. Determine the maximum and minimum values of output voltage and the corresponding input voltages for the circuit to operate in the linear region.

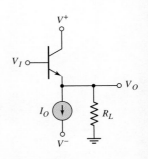

**Figure P8.13**

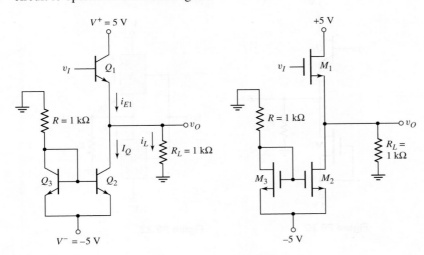

**Figure P8.16**        **Figure P8.17**

8.17   Consider the class-A source-follower circuit shown in Figure P8.17. The transistors are matched with parameters $V_{TN} = 0.5$ V, $K_n = 12$ mA/V$^2$, and $\lambda = 0$. Determine the maximum and minimum values of output voltage and the corresponding input voltages for the circuit to operate in the linear region.

8.18    A class-A emitter follower biased with a constant current source is shown in Figure P8.16. Assume circuit parameters of $V^+ = 12$ V, $V^- = -12$ V, and $R_L = 20\,\Omega$. The transistor parameters are $\beta = 40$ and $V_{BE}(\text{on}) = 0.7$ V. The minimum current in $Q_1$ is to be $i_{E1} = 50$ mA and the minimum collector-emitter voltage is to be $v_{CE}(\text{min}) = 0.7$ V. (a) Determine the value of $R$ that will produce the maximum possible output voltage swing. What is the value of $I_Q$? What are the maximum and minimum values of $i_{E1}$? (b) Using the results of part (a), calculate the conversion efficiency.

8.19    The circuit parameters for the class-A emitter follower shown in Figure P8.16 are $V^+ = 24$ V, $V^- = -24$ V, and $R_L = 200\,\Omega$. The transistor parameters are $\beta = 50$, $V_{BE}(\text{on}) = 0.7$ V, and $V_{CE}(\text{sat}) = 0.2$ V. The output voltage is to vary between $+20$ V and $-20$ V. The minimum current in $Q_1$ is to be $i_{E1} = 20$ mA. (a) Find the minimum required $I_Q$ and the minimum value of $R$. (b) For $v_O = 0$, find the power dissipated in the transistor $Q_1$ and the power dissipated in the current source ($Q_2$, $Q_3$, and $R$). (c) Determine the conversion efficiency for a symmetrical sine-wave output voltage with a peak value of 20 V.

8.20    Consider the BiCMOS follower circuit shown in Figure P8.20. The BJT transistor parameters are $V_{BE}(\text{on}) = 0.7$ V, $V_{CE}(\text{sat}) = 0.2$ V, $V_A = \infty$, and the MOSFET parameters are $V_{TN} = -1.8$ V, $K_n = 12$ mA/V$^2$, $\lambda = 0$. Determine the maximum and minimum values of output voltage and the corresponding input voltages for the circuit to operate in the linear region for (a) $R_L = \infty$ and (b) $R_L = 500\,\Omega$. (c) What is the smallest value of $R_L$ possible if a 2 V peak sine wave is produced at the output? What is the corresponding conversion efficiency?

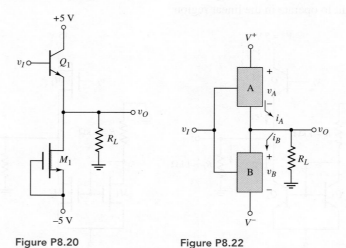

Figure P8.20                    Figure P8.22

8.21    For the idealized class-B output stage in Figure 8.18 in the text, show that the maximum theoretical conversion efficiency for a symmetrical square-wave input signal is 100 percent.

8.22    Consider an idealized class-B output stage shown in Figure P8.22. (The effective turn-on voltages of devices A and B are zero, and the effective "saturation" voltages of $v_A$ and $v_B$ are zero.) Assume $V^+ = 5$ V and $V^- = -5$ V. Assume a symmetrical sine wave is produced at the output. (a) What is the peak output voltage at maximum power conversion efficiency? (b) What is the peak output voltage when each device dissipates the maximum power? (c) If the maximum allowed power dissipation in each device is 2 W and the output voltage is at its maximum value, what is the smaller permitted value of output load resistance?

8.23   Consider an idealized class-B output stage shown in Figure P8.22. (See Problem 8.22 for definitions of "ideal.") The output stage is to deliver 50 W of average power to a 24 Ω load for a symmetrical input sine wave. Assume the supply voltages are $\pm n$ volts, where $n$ is an integer. (a) The power supply voltages are to be at least 3 V greater than the maximum output voltage. What must be the power supply voltages? (b) What is the peak current in each device? (c) What is the power conversion efficiency?

8.24   Consider the class-B output stage with complementary MOSFETs shown in Figure P8.24. The transistor parameters are $V_{TN} = V_{TP} = 0$ and $K_n = K_p = 0.4$ mA/V². Let $R_L = 5$ kΩ. (a) Find the maximum output voltage such that $M_n$ remains biased in the saturation region. What are the corresponding values of $i_L$ and $v_I$ for this condition? (b) Determine the conversion efficiency for a symmetrical sine-wave output signal with the peak value found in part (a).

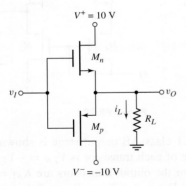

**Figure P8.24**

8.25   For the class-B output stage shown in Figure P8.24, the bias voltages are $V^+ = 12$ V and $V^- = -12$ V. The load resistance is $R_L = 50$ Ω, and the transistor parameters are $V_{TN} = V_{TP} = 0$ and $K_n = K_p = 4$ mA/V². (a) Plot $v_O$ versus $v_I$ for $-10 \leq v_I \leq +10$ V. (b) What is the voltage gain, $A_v = dv_O/dv_I$, at (i) $v_I = 0$, (ii) $v_I = 1$ V, and (iii) $v_I = 10$ V?

8.26   A simplified class-AB output stage with BJTs is shown in Figure 8.24. The circuit parameters are $V_{CC} = 5$ V and $R_L = 1$ kΩ. For each transistor, $I_S = 2 \times 10^{-15}$ A. (a) Determine the value of $V_{BB}$ that produces $i_{Cn} = i_{Cp} = 1$ mA when $v_I = 0$. What is the power dissipated in each transistor? (b) For $v_O = -3.5$ V, determine $i_L$, $i_{Cn}$, $i_{Cp}$, and $v_I$. What is the power dissipated in $Q_n$, $Q_p$, and $R_L$?

8.27   A simplified class-AB output stage with enhancement-mode MOSFETs is shown in Figure 8.26. The circuit parameters are $V_{DD} = 12$ V and

$R_L = 1 \text{ k}\Omega$. The transistor parameters are $V_{TN} = -V_{TP} = 1.5$ V and $K_n = K_p = 4 \text{ mA/V}^2$. (a) (i) Determine the value of $V_{BB}$ such that $i_{Dn} = i_{Dp} = 1\text{mA}$ when $v_I = 0$. (ii) What is the power dissipated in each transistor? (b) (i) Determine the maximum output voltage such that $M_n$ remains biased in the saturation region. (ii) What are the values of $i_{Dn}, i_{Dp}, i_L$, and $v_I$ for this case? (iii) Calculate the power dissipated in $M_n$, $M_p$, and $R_L$ for this case.

8.28 Consider the class-AB output stage in Figure P8.28. The diodes and transistors are matched, with parameters $I_S = 6 \times 10^{-12}$ A, and $\beta = 40$. (a) Determine $R_1$ such that the minimum current in the diodes is 25 mA when $v_O = 24$ V. Find $i_N$ and $i_P$ for this condition. (b) Using the results of part (a), determine the diode and transistor currents when $v_O = 0$.

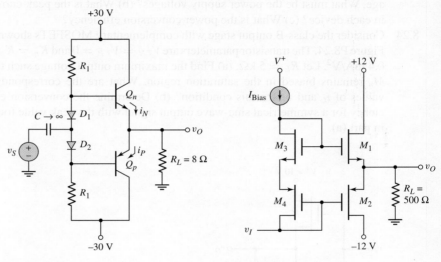

Figure P8.28          Figure P8.29

*8.29 An enhancement-mode MOSFET class-AB output stage is shown in Figure P8.29. The threshold voltage of each transistor is $V_{TN} = -V_{TP} = 1$V and the conduction parameters of the output transistors are $K_{n1} = K_{p2} = 5 \text{ mA/V}^2$. Let $I_{\text{Bias}} = 200 \ \mu\text{A}$. (a) Determine $K_{n3} = K_{p4}$ such that the quiescent drain currents in $M_1$ and $M_2$ are 5 mA. (b) Using the results of part (a), find the small-signal voltage gain $A_v = dv_O/dv_I$ evaluated at: (i) $v_O = 0$, and (ii) $v_O = 5$ V.

D8.30 Consider the MOSFET class-AB output stage in Figure 8.26. The parameters are: $V_{DD} = 10$ V and $R_L = 100 \ \Omega$. For transistors $M_n$ and $M_p$, $V_{TN} = -V_{TP} = 1$ V. The peak amplitude of the output voltage is limited to 5 V. Design the circuit such that the small-signal voltage gain is $A_v = dv_O/dv_I = 0.95$ when $v_O = 0$.

8.31 The parameters of the amplifier shown in Figure 8.28(a) are $V_{CC} = 12$ V, $R_E = 20 \ \Omega$, $R_1 = 14 \text{ k}\Omega$, and $R_2 = 10 \text{ k}\Omega$. The transistor parameters are $\beta = 90$ and $V_{BE}(\text{on}) = 0.7$V. (a) Determine the quiescent value $I_{CQ}$. (b) Find the value of $R_L$ such that the maximum power will be delivered to the load. (c) What is the maximum power that can be delivered to the load if the output voltage is to remain in the range $1 \leq v_O \leq 23$ V? (d) Using the results of part (c) and neglecting currents in the bias resistors, what is the conversion efficiency?

**D8.32**   For the inductively coupled amplifier in Figure 8.28(a), the parameters are: $V_{CC} = 15$ V, $R_E = 0.1$ kΩ, and $R_L = 1$ kΩ. The transistor parameters are $\beta = 100$ and $V_{BE} = 0.7$ V. Design $R_1$ and $R_2$ to deliver the maximum power to the load. What is the maximum power that can be delivered to the load?

**8.33**   Consider the transformer-coupled common-emitter circuit shown in Figure P8.33 with parameters $V_{CC} = 12$ V, $R_E = 20$ Ω, $R_L = 8$ Ω, $R_1 = 2.3$ kΩ, and $R_2 = 1.75$ kΩ. The transistor parameters are $\beta = 40$ and $V_{BE}(\text{on}) = 0.7$ V. (a) Determine the quiescent value $I_{CQ}$. (b) Determine the turns ratio $a$ such that the maximum power is delivered to the load. (c) Determine the maximum power that can be delivered to the load if the voltage $v_1$ is to remain in the range $2 \leq v_1 \leq 20$ V. (d) Using the results of part (c) and neglecting currents in the bias resistors, find the conversion efficiency.

**8.34**   The parameters for the transformer-coupled common-emitter circuit in Figure P8.33 are $V_{CC} = 36$ V and $n_1 : n_2 = 4 : 1$. The signal power delivered to the load is 2 W. Determine: (a) the rms voltage across the load; (b) the rms voltage across the transformer primary; and (c) the primary and secondary currents. (d) If $I_{CQ} = 150$ mA, what is the conversion efficiency?

**8.35**   A BJT emitter follower is coupled to a load with an ideal transformer, as shown in Figure P8.35. The bias circuit is not shown. The transistor current gain is $\beta = 49$, and the transistor is biased such that $I_{CQ} = 100$ mA. (a) Derive the expressions for the voltage transfer functions $v_e/v_i$ and $v_o/v_i$. (b) Find $n_1 : n_2$ for maximum ac power transfer to $R_L$. (c) Determine the small-signal output resistance looking back into the emitter.

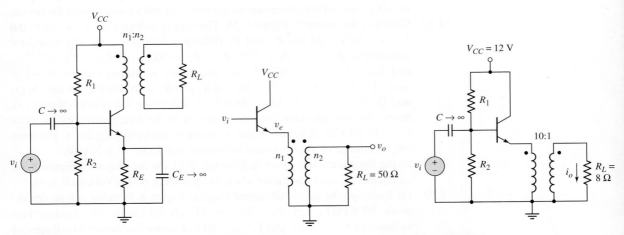

Figure P8.33              Figure P8.35              Figure P8.36

**D8.36**   Consider the transformer-coupled emitter follower in Figure P8.36. Assume an ideal transformer. The transistor parameters are $\beta = 100$ and $V_{BE} = 0.7$ V. (a) Design the circuit to provide a current gain at $A_i = i_o/i_i = 80$. (b) If the magnitude of the signal emitter current is limited to $0.9\,I_{CQ}$ to prevent distortion, determine the power delivered to the load, and the conversion efficiency.

**D8.37**   A class-A transformer-coupled emitter follower must deliver 2 W to an 8 Ω speaker. Let $V_{CC} = 18$ V, $\beta = 100$, and $V_{BE} = 0.7$ V. (a) Determine the required transformer ratio $n_1 : n_2$. (b) Determine the minimum transistor power rating.

D8.38 Repeat Problem 8.36 if the primary side of the transformer has a resistance of 100 $\Omega$.

## Section 8.5 Class-AB Push–Pull Complementary Output Stages

8.39 Consider the circuit in Figure 8.31. The circuit parameters are $I_{\text{Bias}} = 1\text{mA}$, $R_L = 100\,\Omega$, $V^+ = 10$ V, and $V^- = -10$ V. The diode and transistor parameters are $I_{SD} = 5 \times 10^{-16}$ A and $I_{SQ} = 7 \times 10^{-15}$ A, respectively. Neglecting base currents, find (a) $V_{BB}$ and (b) the transistor quiescent collector currents (for $v_O = 0$).

D8.40 The circuit in Figure 8.31 is to be designed such that the quiescent collector currents are 4 mA ($v_O = 0$). Assume $I_{SQ} = 2 \times 10^{-15}$ A and $I_{SD} = 4 \times 10^{-16}$ A. Neglecting base currents, (a) determine the required value of $I_{\text{Bias}}$, (b) the resulting value of $V_{BB}$, and (c) the required value of $v_I$.

8.41 The value of $I_{\text{Bias}}$ in the circuit shown in Figure 8.31 is 0.5 mA. Assume diode and transistor parameters of $I_{SD1} = 10^{-16}$ A, $I_{SD2} = 4 \times 10^{-16}$ A, $I_{SQn} = 8 \times 10^{-16}$ A, and $I_{SQp} = 1.6 \times 10^{-15}$ A. For $v_O = 0$ and neglecting base currents, determine (a) $V_{BB}$, (b) $v_{BEn}$ and $v_{EBp}$, (c) the quiescent collector currents, and (d) the required value of $v_I$.

8.42 The transistors in the output stage in Figure 8.34 are all matched. Their parameters are $\beta = 60$ and $I_S = 5 \times 10^{-13}$ A. Resistors $R_1$ and $R_2$ are replaced by 3 mA ideal current sources, and $R_3 = R_4 = 0$. Let $V^+ = 10$ V and $V^- = -10$ V. (a) Determine the quiescent collector currents in the four transistors for $v_I = v_O = 0$. (b) For a load resistance of $R_L = 200\,\Omega$ and a peak output voltage of 6 V, determine the current gain and voltage gain of the circuit.

*8.43 Consider the circuit in Figure 8.34. The supply voltages are $V^+ = 10$ V and $V^- = -10$ V, and the $R_3$ and $R_4$ resistor values are zero. The transistor parameters are: $\beta_1 = \beta_2 = 120$, $\beta_3 = \beta_4 = 50$, $I_{S1} = I_{S2} = 2 \times 10^{-13}$ A, and $I_{S3} = I_{S4} = 2 \times 10^{-12}$ A. (a) The range in output current is $-1 \leq i_O \leq +1$ A. Determine the values of $R_1$ and $R_2$ such that the currents in $Q_1$ and $Q_2$ do not vary by more than 2 : 1. (b) Using the results of part (a), determine the quiescent collector currents in the four transistors for $v_I = v_O = 0$. (c) Calculate the output resistance, excluding $R_L$, for a quiescent output voltage of zero. Assume the source resistance of $v_I$ is zero.

8.44 Using the parameters given in Example 8.11 for the circuit in Figure 8.34, calculate the input resistance when the quiescent output voltage is zero.

8.45 (a) Redesign the class-AB output stage in Figure 8.34 using enhancement-mode MOSFETs. Let $R_3 = R_4 = 0$. Sketch the circuit. (b) Assume bias voltages of $V^+ = 10$ V and $V^- = -10$ V. Assume the threshold voltages of the n-channel devices are $V_{TN} = 1$ V and the threshold voltages of the p-channel devices are $V_{TP} = -1$ V. Also assume the conduction parameters are $K_{p1} = K_{n2} = 2$ mA/V$^2$ and $K_{n3} = K_{p4} = 5$ mA/V$^2$. Determine $R_1$ and $R_2$ such that the quiescent drain currents in the output transistors are 5 mA (for $v_I = v_O = 0$). (c) Using the results of part (b), find the currents in $M_1$ and $M_2$. (d) If $R_L = 150\,\Omega$, determine the current in each transistor, the input voltage $v_I$, and the power delivered to the load if $v_O = 3.5$ V.

8.46 Consider the class-AB MOSFET output stage shown in Figure P8.46. The circuit parameters are $I_{\text{Bias}} = 0.2$ mA and $R_L = 1$ k$\Omega$. The transistor parameters are $k'_n = 100\,\mu\text{A/V}^2$, $k'_p = 40\,\mu\text{A/V}^2$, $V_{TN} = 0.8$ V, and $V_{TP} = -0.8$ V. For the quiescent condition, assume $v_{GS3} = v_{SG4}$ and $v_{GS1} = v_{SG2}$. When $v_I = -1.5$ V, $v_O = 0$ and $i_{D1} = i_{D2} = 0.5$ mA. Determine the width-to-length ratio of each transistor.

8.47   Figure P8.47 shows a composite pnp Darlington emitter follower that sinks current from a load. Parameter $I_Q$ is the equivalent bias current and $Z$ is the equivalent impedance in the base of $Q_1$. Assume the transistor parameters are: $\beta(\text{pnp}) = 10$, $\beta(\text{npn}) = 50$, $V_{AP} = 50$ V, and $V_{AN} = 100$ V, where $V_{AP}$ and $V_{AN}$ are the Early voltages of the pnp and npn devices, respectively. Calculate the output resistance $R_o$.

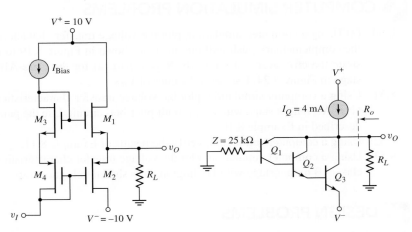

**Figure P8.46**                     **Figure P8.47**

*8.48   Consider the class-AB output stage in Figure P8.48. The parameters are: $V^+ = 12$ V, $V^- = -12$ V, $R_L = 100\ \Omega$, and $I_{\text{Bias}} = 5$ mA. The transistor and diode parameters are $I_S = 10^{-13}$ A. The transistor current gains are $\beta_n = 100$ and $\beta_p = 20$ for the npn and pnp devices, respectively. (a) For $v_O = 0$, determine $V_{BB}$, and the quiescent collector current and base–emitter voltage for each transistor. (b) Repeat part (a) for $v_O = 10$ V. What is the power delivered to the load and what is the power dissipated in each transistor?

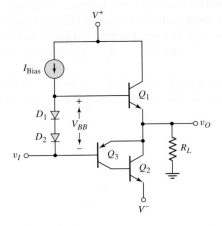

**Figure P8.48**

*8.49    For the class-AB output stage in Figure 8.36, the parameters are: $V^+ = 24$ V, $V^- = -24$ V, $R_L = 20$ Ω, and $I_{\text{Bias}} = 10$ mA. The diode and transistor parameters are $I_S = 2 \times 10^{-12}$ A. The transistor current gains are $\beta_n = 20$ and $\beta_p = 5$ for the npn and pnp devices, respectively. (a) For $v_O = 0$, determine $V_{BB}$, and the quiescent collector current and base–emitter voltage for each transistor. (b) An average power of 10 watts is to be delivered to the load. Determine the quiescent collector current in each transistor and the instantaneous power dissipated in $Q_2$, $Q_5$, and $R_L$ when the output voltage is at its peak negative amplitude.

 ## COMPUTER SIMULATION PROBLEMS

8.50    (a) Using a computer simulation, plot the voltage transfer characteristics of the complementary push-pull output stage shown in Figure 8.19 to demonstrate the crossover distortion. (b) Repeat part (a) for the class-AB output stage in Figure 8.24. Use several values of $V_{BB}$.

8.51    Using a computer simulation, plot the voltage transfer characteristics of the class-AB output stage with a $V_{BE}$ multiplier bias circuit. Use the parameters described in Example 8.10.

8.52    Using a computer simulation, verify the results of Example 8.11.

8.53    Using a computer simulation, plot the voltage transfer characteristics of the class-AB output stage with Darlington pairs shown in Figure 8.36.

## DESIGN PROBLEMS

[Note: Each design should be correlated with a computer analysis.]

*D8.54    Design an audio amplifier to deliver an average of 10 W to an 8 Ω speaker. The bandwidth is to cover the range from 20 Hz to 18 kHz. Specify minimum current gains, and current, voltage, and power ratings of all transistors.

*D8.55    Design a class-A transformer-coupled emitter-follower amplifier to deliver 10 W to an 8 Ω speaker. The ambient temperature is 25 °C and the maximum junction temperature is $T_{j,\text{max}} = 150$ °C. Assume the thermal resistance values are: $\theta_{\text{dev–case}} = 3.2$ °C/W, $\theta_{\text{case–snk}} = 0.8$ °C/W, and $\theta_{\text{snk–amb}} = 4$ °C/W. Specify the power supply voltage, transformer turns ratio, bias resistor values, and transistor current, voltage, and power ratings.

*D8.56    Design the class-AB output stage with the $V_{BE}$ multiplier in Figure 8.32 to deliver an average of 1 W to an 8 Ω load. The peak output voltage must be no more than 80 percent of $V^+$. Let $V^- = -V^+$. Specify the circuit and transistor parameters.

*D8.57    Design the circuit shown in Figure P8.46 to deliver 2 W to a 20 ohm load. The maximum output voltage should be a symmetrical 8 V sine wave.

# PROLOGUE TO ELECTRONIC DESIGN

## PREVIEW

In Part 1, we dealt, for the most part, with discrete electronic circuits; that is, circuits containing discrete resistors, capacitors, and transistors. The analysis of these fundamental circuits provided a basic understanding of circuit operation and characteristics. Some design discussions were also included to introduce the concept of electronic circuit design. As part of the design discussion, various tradeoffs were considered.

In Part 2, we will develop, analyze, and design more complex analog electronic circuits. We will combine and expand the basic circuits considered in Part 1, to form these more complex circuits. Although, for the most part, we will continue to analyze and design discrete circuits, these circuits are usually fabricated as integrated circuits. In this short prologue, we discuss some fundamental aspects of the electronic design process.

## DESIGN APPROACH

The design process can be viewed from two directions, as indicated in Figure PR2.1. The top-down design process begins with a proposed overall system concept. The whole system is divided into subsystems, of which one may include the electronics associated with the project. The electronics is then divided into its own set of subsystems.

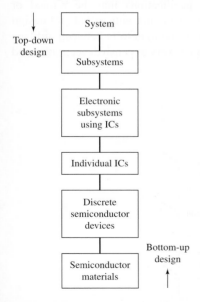

**Figure PR2.1** Top-down and bottom-up approaches to electronic design

The top-down approach usually relies on existing technologies and devices, which means that the electronic subsystems are usually designed with existing ICs. New or customized ICs may be designed and fabricated for a specialized application, although this may increase the cost of the system.

The design engineer must be able to evaluate existing design strategies and technologies to determine if they are able to meet the design or performance objectives. Insight into the operation and characteristics of basic circuits is essential for circuit design, and for being able to make appropriate choices in a top-down design process.

The bottom-up design process usually begins in a research laboratory with the development of new and unique semiconductor materials. Silicon-based devices and circuits still dominate electronics technology, but compound semiconductors are gaining importance in specialized applications. These compound semiconductor materials are being used in the development and design of new discrete devices, such as high-performance JFETs and improved optoelectronic devices. These new devices may be incorporated into integrated circuits, which may eventually lead to the development of new systems based on the characteristics and properties of the new devices.

## SYSTEM DESIGN

Consider a top-down approach in which the design of an electronic circuit or system begins with a proposed design for a large system, such as a new airplane. Designing and building such a system may involve hundreds or even thousands of engineers from the initial concept to the final working system. The concept begins with a set of specifications or performance objectives. The large total airplane system can be broken into subsystems, such as those shown in Figure PR2.2.

The specifications for the electronics subsystem are usually dictated by the overall system specifications, which may include such things as size, weight, and power consumption. Design is an iterative process, and trade-offs are an integral part of the process, all the way from the overall system to each individual circuit. As work progresses, the overall system or subsystem specifications may be refined or modified. During the design, issues may arise that were not anticipated and design trade-offs may be required. For example, there may be trade-offs between airplane performance and cost. High performance may require very expensive electronics and

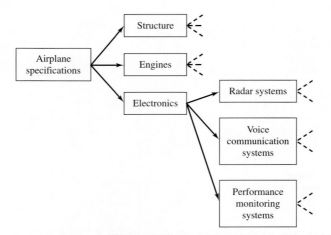

**Figure PR2.2** System and subsystem block diagrams for an airplane

higher than expected costs in the development of high-performance engines. Once the system or subsystem requirements are finalized, design engineers then evaluate various approaches for meeting the design specifications. There is seldom a unique solution for a design, and engineering creativity is an integral part of this phase.

Once suitable approaches are selected for an overall subsystem, such as the electronics subsystem, it may then be broken down into smaller subsystems. For example, the initial electronics breakdown may include radar systems, voice communications, and aircraft performance monitoring systems, such as shown in Figure PR2.2. The specifications for each subsystem are developed from the overall set of specifications.

# ELECTRONIC DESIGN

A flowchart of the general electronic design process is shown in Figure PR2.3. This chart can apply to an entire system or to an individual circuit. A set of specifications is developed for each electronic system, and then each system is divided into many simpler circuits. For example, one relatively simple electronic system may be a high temperature warning indicator. If the temperature of an engine or a particular engine part becomes greater than some predetermined value, a warning light would go on in the cockpit.

Initial design approaches are considered and a circuit configuration is proposed, based on the experience and creativity of the circuit design engineer. This is where experience in the analysis of many different types of electronic circuits becomes important. Knowledge of particular characteristics, such as input impedance, output impedance, gain characteristics, and bandwidth, for many types of circuits, is used to choose a particular circuit configuration.

Figure PR2.4 shows a block diagram for a particular circuit configuration that can serve as a starting point for the design of the temperature indicator. The block showing the amplifier may be further divided to show a proposed configuration for the circuit. Component values can then be chosen.

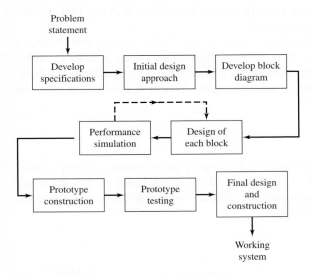

**Figure PR2.3** Flowchart of the design process

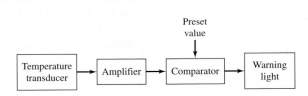

**Figure PR2.4** Block diagram of a temperature warning light circuit

The choice of a temperature transducer is based in part on the required temperature range. The design of the amplifier begins with the anticipated output signal of the temperature transducer, which in turn implies a required level of input impedance and signal gain for the amplifier. The necessary stability of the amplifier gain may determine whether a simple amplifier design may be used, or a more complex amplifier using feedback may be necessary. In addition, the location of the amplifier on the aircraft may determine the temperature range over which the amplifier must function. The comparator compares the amplifier output, which corresponds to a particular temperature, to a preset value. If the amplifier output is greater than the preset value, the output of the comparator must then be able to activate the warning light. The voltage and current levels required to activate the warning light are determined and are incorporated into the design of the comparator.

In proposing the initial circuit configuration and component values, the design engineer may use an intuitive approach based to a large extent on experience. However, once this initial design is completed, the design must be verified with a mathematical analysis or computer simulation. The initial design process may include calculations using simple models for the transistors and circuits. Normally, a more sophisticated analysis is required to take into account such things as temperature variations, tolerances in component values, and other parameter variations based on a particular application.

The circuit performance estimation or simulation is usually a very important phase of the design process. To validate the final design, it is necessary to simulate, as precisely as possible, the performance of the discrete devices and ICs used in the design. Simulation models are required for each circuit component in order to predict the operation and characteristics of the designed circuits. On the basis of these simulation models, trade-offs between technologies and devices may be evaluated to obtain the optimum performance. With improved simulation models, the breadboard development stage may be eliminated and the design process may proceed directly to the construction of a prototype circuit. Since the prototype circuit may involve the fabrication of specialized or customized integrated circuits, this phase of the design process may be expensive; therefore, costly mistakes in the design must be avoided. A good circuit simulation may identify potential problems that can be corrected before the prototype circuit is fabricated.

The prototype circuit is then tested and evaluated. At this point, a minor redesign may involve only selecting slightly different component values. A more extensive redesign may require selecting an entirely different circuit configuration in order to meet the system requirements. Finally, the entire system is constructed from the operating subsystems.

## CONCLUSION

Design involves creativity, and it can be challenging and rewarding. Design is based on experience. The design process in Part 2 of the text is based on the experience gained in Part 1. Our design experience should continue to grow as we proceed through the remainder of this book.

# Analog Electronics

Part 1 dealt with basic electronic devices and fundamental circuit configurations. Part 2 now deals with more complex analog circuits, including more sophisticated amplifiers.

Chapter 9 introduces the ideal op-amp and related circuits. The op-amp is one of the most common analog integrated circuits and can be used in a plethora of electronic applications. IC biasing techniques, which primarily use constant-current sources, are described in Chapter 10. One of the most widely used amplifier configurations is the differential amplifier, which is analyzed in Chapter 11. Chapter 12 covers the fundamentals of feedback. Feedback is used extensively in analog circuits to set or control gain values more precisely, and to alter, in a favorable way, input and output impedance values.

More complex analog integrated circuits, including circuits that form operational amplifiers, are discussed in Chapter 13. These circuits are composed of the fundamental configurations, such as the differential amplifier, constant-current biasing, active load, and output stage, that have been previously analyzed. Then Chapter 14 considers nonideal effects in operational amplifier circuits, and discusses the effects of these nonideal characteristics on op-amp circuit performance. Additional integrated circuit applications and designs are considered in Chapter 15. Such applications include active filters, oscillators, and integrated circuit power amplifiers.

# Ideal Operational Amplifiers and Op-Amp Circuits[1]

An operational amplifier (op-amp) is an integrated circuit that amplifies the difference between two input voltages and produces a single output. The op-amp is prevalent in analog electronics, and can be thought of as another electronic device, in much the same way as the bipolar or field-effect transistor.

The term operational amplifier comes from the original applications of the device in the early 1960s. Op-amps, in conjunction with resistors and capacitors, were used in analog computers to perform mathematical operations to solve differential and integral equations. The applications of op-amps have expanded significantly since those early days.

The main reason for postponing the discussion of op-amp circuits until now is that we can use a relatively simple transistor circuit to develop the ideal characteristics of the op-amps, instead of simply stating the ideal parameters as postulates. Once the ideal properties have been developed, the reader can then be more comfortable applying these ideal characteristics in the design of op-amp circuits. Just as we developed equivalent circuits of transistors that include dependent sources representing gain factors, we will develop a basic op-amp equivalent circuit with a dependent source that represents the device gain that can be used to determine some of the nonideal properties of op-amp circuits.

For the most part, this chapter deals with ideal op-amps. Nonideal op-amp effects are considered in Chapter 14.

## PREVIEW

In this chapter, we will:

- Discuss and develop the parameters and characteristics of the ideal operational amplifier, and determine the analysis method of ideal op-amp circuits.
- Analyze and understand the characteristics of the inverting operational amplifier.
- Analyze and understand the characteristics of the summing operational amplifier.
- Analyze and understand the characteristics of the noninverting operational amplifier, including the voltage follower or buffer.
- Analyze several ideal op-amp circuits including the difference amplifier and the instrumentation amplifier.
- Discuss the operational transconductance amplifier.
- Design several ideal op-amp circuits with given design specifications.
- As an application, design an electronic thermometer in conjunction with an instrumentation amplifier that will provide the necessary amplification.

---

[1]This chapter, through Section 9.5.5, can easily be studied as a first chapter in electronics for those who wish to cover op-amp circuits first.

## 9.1    THE OPERATIONAL AMPLIFIER

**Objective:** • Discuss and develop the parameters and characteristics of the ideal operational amplifier, and determine the analysis method of ideal op-amp circuits.

The integrated circuit operational amplifier evolved soon after development of the first bipolar integrated circuit. The $\mu$A-709 was introduced by Fairchild Semiconductor in 1965 and was one of the first widely used general-purpose op-amps. The now classic $\mu$A-741, also by Fairchild, was introduced in the late 1960s. Since then, a vast array of op-amps with improved characteristics, using both bipolar and MOS technologies, have been designed. Most op-amps are very inexpensive (less than a dollar) and are available from a wide range of suppliers.

From a signal point of view, the op-amp has two input terminals and one output terminal, as shown in the small-signal circuit symbol in Figure 9.1(a). The op-amp also requires dc power, as do all transistor circuits, so that the transistors are biased in the active region. Also, most op-amps are biased with both a positive and a negative voltage supply, as indicated in Figure 9.1(b). As before, the positive voltage is indicated by $V^+$ and the negative voltage by $V^-$.

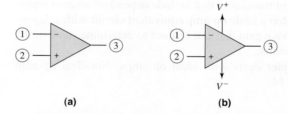

**(a)**                    **(b)**

**Figure 9.1** (a) Small-signal circuit symbol of the op-amp; (b) op-amp with positive and negative supply voltages

There are normally 20 to 30 transistors that make up an op-amp circuit. The typical IC op-amp has parameters that approach the ideal characteristics. For this reason, then, we can treat the op-amp as a "simple" electronic device, which means that it is quite easy to design a wide range of circuits using the IC op-amp.

In this chapter, we develop the ideal set of op-amp parameters and then consider the analysis and design of a wide variety of op-amp circuits, which will aid in our understanding of the design process of electronic circuits. We generally assume, in this chapter, that the op-amp is ideal. In the following chapters, we consider the differential amplifier, current-source biasing, and feedback, which leads to the development of the actual operational amplifier circuit in Chapter 13. Once the actual op-amp circuit is studied, then the source of nonideal characteristics can be understood. The effect of nonideal op-amp parameters is then considered in Chapter 14. Additional op-amp applications are given in Chapter 15.

### 9.1.1    Ideal Parameters

The ideal op-amp senses the difference between two input signals and amplifies this difference to produce an output signal. The terminal voltage is the voltage at a terminal

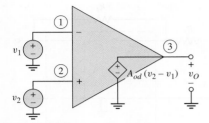

**Figure 9.2** Ideal op-amp equivalent circuit

measured with respect to ground. The ideal op-amp equivalent circuit is shown in Figure 9.2.

Ideally, the input resistance $R_i$ between terminals 1 and 2 is infinite, which means that the input current at each terminal is zero. The output terminal of the ideal op-amp acts as the output of an ideal voltage source, meaning that the small-signal output resistance $R_o$ is zero.

The parameter $A_{od}$ shown in the equivalent circuit is the open-loop **differential voltage gain** of the op-amp. The output is out of phase with respect to $v_1$ and in phase with respect to $v_2$. Terminal (1) then is the **inverting input terminal,** designated by the "−" notation, and terminal (2) is the **noninverting input terminal,** designated by the "+" notation. In the ideal op-amp, the open-loop gain $A_{od}$ is very large and approaches infinity.

Since the ideal op-amp responds only to the difference between the two input signals $v_1$ and $v_2$, the ideal op-amp maintains a zero output signal for $v_1 = v_2$. When $v_1 = v_2 \neq 0$, there is what is called a **common-mode input signal.** For the ideal op-amp, the common-mode output signal is zero. This characteristic is referred to as **common-mode rejection.**

Because the device is biased with both positive and negative power supplies, most op-amps are direct-coupled devices (i.e., no coupling capacitors are used on the input). Therefore, the input voltages $v_1$ and $v_2$ shown in Figure 9.2 can be dc voltages, which will produce a dc output voltage $v_O$.

Another characteristic of the op-amp that must be considered in any design is the bandwidth or frequency response. In the ideal op-amp, this parameter is neglected. The frequency response of practical op-amps and other nonideal characteristics are discussed in Chapters 13 and 14. These nonideal parameters are considered after the actual operational amplifier circuits are analyzed in Chapter 13.

The ideal op-amp is being considered in this chapter in order to gain an appreciation of the properties and characteristics of op-amp circuits.

## 9.1.2  Development of the Ideal Parameters

To develop the ideal op-amp parameters, we start with the basic equivalent circuit shown in Figure 9.2.[2] We may note that this equivalent circuit is very similar to the MOSFET small-signal equivalent circuit. Figure 9.3(a) shows an n-channel enhancement-mode MOSFET, and Figure 9.3(b) shows the simplified low-frequency small-signal equivalent circuit. In our analysis, the transistor small-signal output resistance $r_o$ is assumed to be infinite.

---

[2]For those readers studying this chapter as the first topic in electronics, concentrate on the analysis using the equivalent circuit and ignore any reference to the MOSFET.

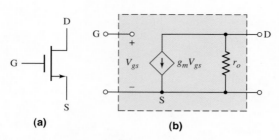

**(a)** **(b)**

**Figure 9.3** (a) n-channel enhancement-mode MOSFET and (b) small-signal equivalent circuit

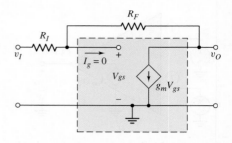

**Figure 9.4** Simplified small-signal equivalent circuit of a MOSFET with input and feedback resistors

Figure 9.4 shows the equivalent circuit including two external circuit resistors $R_I$ and $R_F$. The voltage at the noninverting terminal is set equal to zero, so that the noninverting terminal is at ground potential. An input voltage $v_I$ is applied. Resistor $R_F$ is a feedback resistor that connects the output back to the input of the amplifier. This circuit is therefore called a feedback circuit. In this example, we use a single device (transistor) as the basic amplifier of the op-amp circuit.

Writing a KCL equation at the gate, or inverting, terminal, we obtain

$$\frac{v_I - V_{gs}}{R_I} = \frac{V_{gs} - v_O}{R_F} \tag{9.1(a)}$$

which can be arranged as

$$\frac{v_I}{R_I} + \frac{v_O}{R_F} = V_{gs}\left(\frac{1}{R_I} + \frac{1}{R_F}\right) \tag{9.1(b)}$$

Since the input impedance to the transistor is infinite, the current into the device is zero. A KCL equation at the output node yields

$$\frac{V_{gs} - v_O}{R_F} = g_m V_{gs} \tag{9.2(a)}$$

which can be solved for $V_{gs}$, as follows:

$$V_{gs} = \frac{v_O}{R_F} \cdot \frac{1}{\left(\dfrac{1}{R_F} - g_m\right)} \tag{9.2(b)}$$

Substituting Equation (9.2(b)) into (9.1(b)) results in the overall voltage gain of the circuit

$$\frac{v_O}{v_I} = -\frac{R_F}{R_I} \cdot \frac{\left(1 - \dfrac{1}{g_m R_F}\right)}{\left(1 + \dfrac{1}{g_m R_F}\right)} \tag{9.3}$$

If we let the gain $g_m$ of the basic amplifier (i.e., the transistor) go to infinity, then the overall voltage gain becomes

$$\frac{v_O}{v_I} = -\frac{R_F}{R_I} \tag{9.4}$$

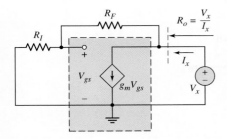

**Figure 9.5** Equivalent circuit determining output resistance

Equation (9.4) shows that the overall voltage gain is the ratio of two external circuit resistors, which is one result of using an ideal op-amp. The negative sign indicates a 180 degree phase shift between the input and the output, which means that the input to the transistor corresponds to the inverting terminal of an op-amp. The voltage gain given by Equations (9.3) and (9.4) is called a **closed-loop voltage gain,** since feedback is incorporated into the circuit. Conversely, the voltage gain $A_{od}$ is an **open-loop gain.**

Voltage $V_{gs}$ at the input of the basic amplifier (transistor) is given by Equation (9.2(b)). Again, if we let the gain $g_m$ go to infinity, then $V_{gs} \cong 0$; that is, the voltage at the input terminal to the basic amplifier is almost at ground potential. This terminal is said to be at **virtual ground,** which is another characteristic that we will observe in ideal op-amp circuits. The concept of virtual ground will be discussed in more detail in later sections.

The output resistance of this circuit can be determined from the equivalent circuit shown in Figure 9.5. The input signal source is set at zero. A KCL equation at the output node, written in phasor notation, is

$$I_x = g_m V_{gs} + \frac{V_x}{R_I + R_F} \tag{9.5}$$

Voltage $V_{gs}$ can be written in terms of the test voltage $V_x$, as

$$V_{gs} = V_x \left( \frac{R_I}{R_I + R_F} \right) \tag{9.6}$$

Substituting Equation (9.6) into (9.5), we find that

$$\frac{I_x}{V_x} = \frac{1}{R_o} = \frac{1 + g_m R_I}{R_I + R_F} \tag{9.7(a)}$$

or

$$R_o = \frac{R_I + R_F}{1 + g_m R_I} \tag{9.7(b)}$$

If the gain $g_m$ goes to infinity, then $R_o \rightarrow 0$. The output resistance of the circuit with negative feedback included goes to zero. This is also a property of an ideal op-amp circuit.

A simplified MOSFET model with a large gain has thus provided the properties of an ideal op-amp.

### 9.1.3   Analysis Method

Usually, an op-amp is not used in the open-loop configuration shown in Figure 9.2(a). Instead, feedback is added to close the loop between the output and the input. In this

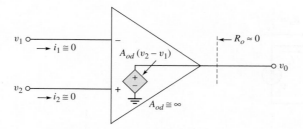

**Figure 9.6** Parameters of the ideal op-amp

chapter, we will limit our discussion to **negative feedback,** in which the connection from the output goes to the inverting terminal, or terminal (1). As we will see later, this configuration produces stable circuits; positive feedback, in which the output is connected to the noninverting terminal, can be used to produce oscillators.

The ideal op-amp characteristics resulting from our negative feedback analysis are shown in Figure 9.6 and summarized below.

1. The internal differential gain $A_{od}$ is considered to be infinite.
2. The differential input voltage $(v_2 - v_1)$ is assumed to be zero. If $A_{od}$ is very large and if the output voltage $v_O$ is finite, then the two input voltages must be nearly equal.
3. The effective input resistance to the op-amp is assumed to be infinite, so the two input currents, $i_1$ and $i_2$, are essentially zero.
4. The output resistance $R_o$ is assumed to be zero in the ideal case, so the output voltage is connected directly to the dependent voltage source, and the output voltage is independent of any load connected to the output.

We use these ideal characteristics in the analysis and design of op-amp circuits.

### 9.1.4    Practical Specifications

In the previous discussion, we have considered the properties of an ideal op-amp. Practical op-amps are not ideal, although their characteristics approach those of an ideal op-amp. Figure 9.7(a) is a more accurate equivalent circuit of an op-amp. Also included is a load resistance connected to the output terminal. This load resistance may actually represent another op-amp circuit connected to the output terminal.

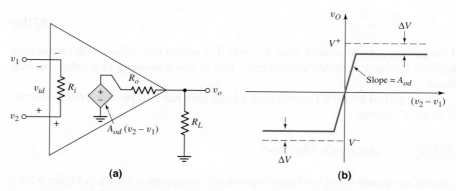

**(a)**                                                     **(b)**

**Figure 9.7** (a) Equivalent circuit of the op-amp and (b) simplified voltage transfer characteristic

## Output Voltage Swing

Since the op-amp is composed of transistors biased in the active region by the dc input voltages $V^+$ and $V^-$, the output voltage is limited. When $v_O$ approaches $V^+$, it will saturate, or be limited to a value nearly equal to $V^+$, since it cannot go above the positive bias voltage. Similarly, when the output voltage approaches $V^-$, it will saturate at a value nearly equal to $V^-$. The output voltage is limited to $V^- + \Delta V < v_O < V^+ - \Delta V$, as shown in Figure 9.7(b). Figure 9.7(b) is a simplified voltage transfer characteristic for the op-amp, showing the saturation effect. In older op-amp designs, such as the 741, the value of $\Delta V$ is between 1 and 2 V. We will see this property in Chapter 13. However, in newer CMOS op-amp designs, the value of $\Delta V$ may be as low as 10 mV.

## Output Currents

As we can see from Figure 9.7(a), if the output voltage $v_O$ becomes either positive or negative, a current is induced in the load resistance. If the output voltage is positive, the load current is supplied by the output of the op-amp. If the output voltage is negative, then the output of the op-amp sinks the load current. A limitation of practical op-amps is the maximum current that an op-amp can supply or sink. A typical value of the maximum current is on the order of $\pm 20$ mA for a general-purpose op-amp.

### 9.1.5    PSpice Modeling

Three general purpose op-amps are included in the PSpice library. The PSpice circuit simulation uses a macromodel, which is a simplified version of the op-amp, to model the op-amp characteristics. For example, the $\mu$A-741 op-amp has parameters $R_i = 2$ M$\Omega$, $R_o = 75$ $\Omega$, $A_{od} = 2 \times 10^5$, and a unity-gain bandwidth of $f_{BW} = 1$ MHz. This device is also capable of producing output voltages of $\pm 14$ V with dc power supply voltages of $\pm 15$ V. We will see in several examples as to whether these nonideal parameters affect actual circuit properties.

## 9.2    INVERTING AMPLIFIER

**Objective:** • Analyze and understand the characteristics of the inverting operational amplifier.

One of the most widely used op-amp circuits is the **inverting amplifier.** Figure 9.8 shows the closed-loop configuration of this circuit. We must keep in mind that the op-amp is biased with dc voltages, although those connections are seldom explicitly shown.

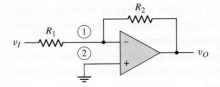

**Figure 9.8** Inverting op-amp circuit

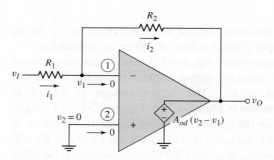

**Figure 9.9** Inverting op-amp equivalent circuit

### Basic Amplifier

We analyze the circuit in Figure 9.8 by considering the ideal equivalent circuit shown in Figure 9.9. The **closed-loop voltage gain,** or simply the voltage gain, is defined as

$$A_v = \frac{v_O}{v_I} \tag{9.8}$$

We stated that if the open-loop gain $A_{od}$ is very large, then the two inputs $v_1$ and $v_2$ must be nearly equal. Since $v_2$ is at ground potential, voltage $v_1$ must also be approximately zero volts. We must point out, however, that having $v_1$ be essentially at ground potential does not imply that terminal (1) is grounded. Rather, terminal (1) is said to be at **virtual ground;** that is, it is essentially zero volts, but it does not provide a current path to ground. The virtual ground concept will be used in the analysis of ideal op-amp circuits. To repeat this important concept, with terminal 1 being at virtual ground means that terminal 1 is essentially at zero volts, but is not connected to ground potential.

From Figure 9.9, we can write

$$i_1 = \frac{v_I - v_1}{R_1} = \frac{v_I}{R_1} \tag{9.9}$$

Since the current into the op-amp is assumed to be zero, current $i_1$ must flow through resistor $R_2$ to the output terminal, which means that $i_1 = i_2$.

The output voltage is given by

$$v_O = v_1 - i_2 R_2 = 0 - \left(\frac{v_I}{R_1}\right) R_2 \tag{9.10}$$

Therefore, the closed-loop voltage gain is

$$A_v = \frac{v_O}{v_I} = -\frac{R_2}{R_1} \tag{9.11}$$

For the ideal op-amp, the closed-loop voltage gain is a function of the ratio of two resistors; it is not a function of the transistor parameters within the op-amp circuit. Again, the minus sign implies a phase reversal. If the input voltage $v_I$ is positive, then, because $v_1$ is essentially at ground potential, the output voltage $v_O$ must be negative, or below ground potential. Also note that if the output terminal is open-circuited, current $i_2$ must flow back into the op-amp. However, since the output impedance for the ideal case is zero, the output voltage is not a function of this current that flows back into the op-amp and is not dependent on the load.

We can also determine the input resistance seen by the voltage source $v_I$. Because of the virtual ground, we have, from Equation (9.9)

$$i_1 = v_I/R_1$$

The **input resistance** is then defined as

$$R_i = \frac{v_I}{i_1} = R_1 \tag{9.12}$$

This shows that the input resistance seen by the source is a function of $R_1$ only, and is a result of the "virtual ground" concept. Figure 9.10 summarizes our analysis of the inverting amplifier circuit.

Since there are no coupling capacitors in the op-amp circuit, the input and output voltages, as well as the currents in the resistors, can be dc signals. The inverting op-amp can then amplify dc voltages.

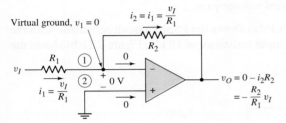

**Figure 9.10**  Currents and voltages in the inverting op-amp

## DESIGN EXAMPLE 9.1

**Objective:**  Design an inverting amplifier with a specified voltage gain.

**Specifications:**  The circuit configuration to be designed is shown in Figure 9.10. Design the circuit such that the voltage gain is $A_v = -5$. Assume the op-amp is driven by an ideal sinusoidal source, $v_s = 0.1 \sin \omega t$ (V), that can supply a maximum current of 5 $\mu$A. Assume that frequency $\omega$ is low so that any frequency effects can be neglected.

**Design Pointer:**  If the sinusoidal input signal source has a nonzero output resistance, the op-amp must be redesigned to provide the specified voltage gain.

**Initial Solution:**  The input current is given by

$$i_1 = \frac{v_I}{R_1} = \frac{v_s}{R_1}$$

If $i_1(\text{max}) = 5\ \mu$A, then we can write

$$R_1 = \frac{v_s(\text{max})}{i_1(\text{max})} = \frac{0.1}{5 \times 10^{-6}} \Rightarrow 20\,\text{k}\Omega$$

The closed-loop gain is given by

$$A_v = \frac{-R_2}{R_1} = -5$$

We then have

$$R_2 = 5R_1 = 5(20) = 100 \text{ k}\Omega$$

**Trade-offs:** If the signal source has a finite output resistance and the desired output voltage is $v_o = -0.5 \sin \omega t$, the circuit must be redesigned. Assume the output resistance of the source is $R_S = 1 \text{ k}\Omega$.

**Redesign Solution:** The output resistance of the signal source is now part of the input resistance to the op-amp. We now write

$$R_1 + R_S = \frac{v_s(\text{max})}{i_1(\text{max})} = \frac{0.1}{5 \times 10^{-6}} \Rightarrow 20 \text{ k}\Omega$$

Since $R_S = 1 \text{ k}\Omega$, we then have $R_1 = 19 \text{ k}\Omega$. The feedback resistor is then $R_2 = 5(R_1 + R_S) = 5(19 + 1) = 100 \text{ k}\Omega$.

**Comment:** The output resistance of the signal source must be included in the design of the op-amp to provide a specified voltage gain.

**Computer Verification:** Figure 9.11(a) shows the PSpice circuit schematic with the source resistance of 1 k$\Omega$ and an input resistance of 19 k$\Omega$. Figure 9.11(b) shows the

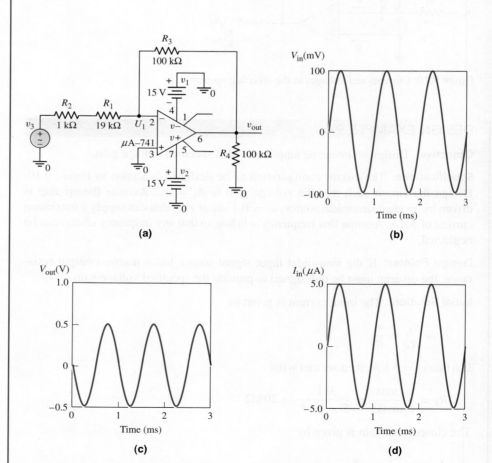

**Figure 9.11** (a) PSpice circuit schematic, (b) input signal, (c) output signal, and (d) input current signal for Example 9.1

100 mV sinusoidal input signal. Figure 9.11(c) is the output signal which shows that a gain of 5 (magnitude) has been achieved and also shows that the output signal is 180 degrees out of phase with respect to the input signal. Finally, the input current is shown in Figure 9.11(d) with a maximum value of 5 $\mu$A. The actual circuit characteristics are not influenced to any great extent by the nonideal parameters of the $\mu$A-741 op-amp used in the circuit simulation.

---

### EXERCISE PROBLEM

**Ex 9.1:** Design an ideal inverting op-amp circuit such that the voltage gain is $A_v = -25$. The maximum current in any resistor is to be limited to $10\,\mu$A with the input voltage in the range $-25 \le v_I \le +25$ mV. (a) What are the values of $R_1$ and $R_2$? (b) What is the range of output voltage $v_O$? (Ans. (a) $R_1 = 2.5$ k$\Omega$, $R_2 = 62.5$ k$\Omega$; (b) $-0.625 \le v_O \le 0.625$ V)

---

### Problem-Solving Technique: Ideal Op-Amp Circuits

1. If the noninverting terminal of the op-amp is at ground potential, then the inverting terminal is at virtual ground. Sum currents at this node, assuming zero current enters the op-amp itself.
2. If the noninverting terminal of the op-amp is not at ground potential, then the inverting terminal voltage is equal to that at the noninverting terminal. Sum currents at the inverting terminal node, assuming zero current enters the op-amp itself.
3. For the ideal op-amp circuit, the output voltage is determined from either step 1 or step 2 above and is independent of any load connected to the output terminal.

### 9.2.2    Amplifier with a T-Network

Assume that an inverting amplifier is to be designed having a closed-loop voltage gain of $A_v = -100$ and an input resistance of $R_i = R_1 = 50$ k$\Omega$. The feedback resistor $R_2$ would then have to be 5 M$\Omega$. However this resistance value is too large for most practical circuits.

Consider the op-amp circuit shown in Figure 9.12 with a T-network in the feedback loop. The analysis of this circuit is similar to that of the inverting op-amp circuit of Figure 9.10. At the input, we have

$$i_1 = \frac{v_I}{R_1} = i_2 \tag{9.13}$$

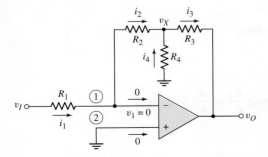

**Figure 9.12**  Inverting op-amp with T-network

We can also write that

$$v_X = 0 - i_2 R_2 = -v_I \left( \frac{R_2}{R_1} \right) \tag{9.14}$$

If we sum the currents at the node $v_X$, we have

$$i_2 + i_4 = i_3$$

which can be written

$$-\frac{v_X}{R_2} - \frac{v_X}{R_4} = \frac{v_X - v_O}{R_3} \tag{9.15}$$

or

$$v_X \left( \frac{1}{R_2} + \frac{1}{R_4} + \frac{1}{R_3} \right) = \frac{v_O}{R_3} \tag{9.16}$$

Substituting the expression for $v_X$ from Equation (9.14), we obtain

$$-v_I \left( \frac{R_2}{R_1} \right) \left( \frac{1}{R_2} + \frac{1}{R_4} + \frac{1}{R_3} \right) = \frac{v_O}{R_3} \tag{9.17}$$

The closed-loop voltage gain is therefore

$$A_v = \frac{v_O}{v_I} = -\frac{R_2}{R_1} \left( 1 + \frac{R_3}{R_4} + \frac{R_3}{R_2} \right) \tag{9.18}$$

The advantage of using a T-network is demonstrated in the following example.

## DESIGN EXAMPLE 9.2

**Objective:** An op-amp with a T-network is to be designed as a microphone preamplifier.

**Specifications:** The circuit configuration to be designed is shown in Figure 9.12. The maximum microphone output voltage is 12 mV (rms) and the microphone has an output resistance of 1 kΩ. The op-amp circuit is to be designed such that the maximum output voltage is 1.2 V (rms). The input amplifier resistance should be fairly large, but all resistance values should be less that 500 kΩ.

**Choices:** The final design should use standard resistor values. In addition, standard resistors with tolerances of ±2 percent are to be considered.

**Solution:** We need a voltage gain of

$$|A_v| = \frac{1.2}{0.012} = 100$$

Equation (9.18) can be written in the form

$$A_v = -\frac{R_2}{R_1} \left( 1 + \frac{R_3}{R_4} \right) - \frac{R_3}{R_1}$$

If, for example, we arbitrarily choose $\dfrac{R_2}{R_1} = \dfrac{R_3}{R_1} = 8$, then

$$-100 = -8 \left( 1 + \frac{R_3}{R_4} \right) - 8$$

which yields

$$\frac{R_3}{R_4} = 10.5$$

The effective $R_1$ must include the $R_S$ resistance of the microphone. If we set $R_1 = 49 \text{ k}\Omega$ so that $R_{1,\text{eff}} = 50 \text{ k}\Omega$, then

$$R_2 = R_3 = 400 \text{ k}\Omega$$

and

$$R_4 = 38.1 \text{ k}\Omega$$

**Design Pointer:** If we need to use standard resistance values in our design, then, using Appendix C, we can choose $R_1 = 51 \text{ k}\Omega$ so that $R_{1,\text{eff}} = 52 \text{ k}\Omega$, and we can choose $R_2 = R_3 = 390 \text{ k}\Omega$. Then, using Equation (9.18), we have

$$A_v = -100 = \frac{-R_2}{R_{1,\text{eff}}}\left(1 + \frac{R_3}{R_4}\right) - \frac{R_3}{R_{1,\text{eff}}} = \frac{-390}{52}\left(1 + \frac{390}{R_4}\right) - \frac{390}{52}$$

which yields $R_4 = 34.4 \text{ k}\Omega$. We may use a standard resistor of $R_4 = 33 \text{ k}\Omega$. This resistance value then produces a voltage gain of $A_v = -103.6$.

**Trade-offs:** If we consider $\pm 2$ percent tolerances in the standard resistor values, the voltage gain can be written as

$$A_v = \frac{-R_2(1 \pm 0.02)}{1 \text{ k}\Omega + R_1(1 \pm 0.02)}\left[1 + \frac{R_3(1 \pm 0.02)}{R_4(1 \pm 0.02)}\right] - \frac{R_3(1 \pm 0.02)}{1 \text{ k}\Omega + R_1(1 \pm 0.02)}$$

or

$$A_v = \frac{-390(1 \pm 0.02)}{1 + 51(1 \pm 0.02)}\left[1 + \frac{390(1 \pm 0.02)}{33(1 \pm 0.02)}\right] - \frac{390(1 \pm 0.02)}{1 + 51(1 \pm 0.02)}$$

Analyzing this equation, we find the maximum magnitude as $|A_v|_{\max} = 111.6$ or $+7.72$ percent, and the minimum magnitude as $|A_v|_{\min} = 96.3$ or $-7.05$ percent.

**Comment:** As required, all resistor values are less than $500 \text{ k}\Omega$. Also the resistance ratios in the voltage gain expression are approximately equal. As with most design problems, there is no unique solution. We must keep in mind that, because of resistor value tolerances, the actual gain of the amplifier will have a range of values.

---

### EXERCISE PROBLEM

**Ex 9.2:** Design an ideal inverting op-amp circuit with a T-network that has a closed-loop voltage gain of $A_v = -75$ and an input resistance of $R = 20 \text{ k}\Omega$. The maximum resistor value is to be limited to $200 \text{ k}\Omega$. (Ans. Let $R_1 = 20 \text{ k}\Omega$ and $R_2 = R_3 = 160 \text{ k}\Omega$. Then $R_4 = 21.7 \text{ k}\Omega$.)

The amplifier with a T-network allows us to obtain a large gain using reasonably sized resistors.

### 9.2.3    Effect of Finite Gain

A **finite open-loop gain** $A_{od}$, also called the finite differential-mode gain, affects the closed-loop gain of an inverting amplifier. We will consider nonideal effects in

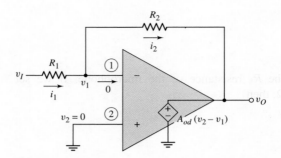

**Figure 9.13** Equivalent circuit of the inverting op-amp with a finite differential-mode gain

op-amps in a later chapter; here, we will determine the magnitude of $A_{od}$ required to approach the ideal case.

Consider the inverting op-amps shown in Figure 9.13. As before, we assume an infinite input resistance at terminals (1) and (2), which means the input currents to the op-amp are zero.

The current through $R_1$ can be written as

$$i_1 = \frac{v_I - v_1}{R_1} \tag{9.19}$$

and the current through $R_2$ is

$$i_2 = \frac{v_1 - v_O}{R_2} \tag{9.20}$$

The output voltage is now given by

$$v_O = -A_{od}v_1$$

so that the terminal (1) voltage can be written as

$$v_1 = -\frac{v_O}{A_{od}} \tag{9.21}$$

Combining Equations (9.21), (9.19), and (9.20), and setting $i_1 = i_2$, we obtain

$$i_1 = \frac{v_I + \dfrac{v_O}{A_{od}}}{R_1} = i_2 = \frac{-\dfrac{v_O}{A_{od}} - v_O}{R_2} \tag{9.22}$$

Solving for the closed-loop voltage gain, we find that

$$A_v = \frac{v_O}{v_I} = -\frac{R_2}{R_1}\left[\frac{1}{1 + \dfrac{1}{A_{od}}\left(1 + \dfrac{R_2}{R_1}\right)}\right] \tag{9.23}$$

Equation (9.23) shows that if $A_{od} \to \infty$, the ideal closed-loop voltage gain reduces to that given by Equation (9.11).

## EXAMPLE 9.3

**Objective:** Determine the deviation from the ideal due to a finite differential gain.

Consider an inverting op-amp with $R_1 = 10\,\text{k}\Omega$ and $R_2 = 100\,\text{k}\Omega$. Determine the closed-loop gain for: $A_{od} = 10^2, 10^3, 10^4, 10^5$, and $10^6$. Calculate the percent deviation from the ideal gain.

**Solution:** The ideal closed-loop gain is

$$A_v = -\frac{R_2}{R_1} = -\frac{100}{10} = -10$$

If $A_{od} = 10^2$, we have, from Equation (9.23),

$$A_v = -\frac{100}{10} \cdot \frac{1}{\left[1 + \frac{1}{10^2}\left(1 + \frac{100}{10}\right)\right]} = \frac{-10}{(1 + 0.11)} = -9.01$$

which is a 9.9 percent deviation from the ideal. For the other differential gain values we have the following results:

| $A_{od}$ | $A_v$ | Deviation (%) |
|---|---|---|
| $10^2$ | $-9.01$ | 9.9 |
| $10^3$ | $-9.89$ | 1.1 |
| $10^4$ | $-9.989$ | 0.11 |
| $10^5$ | $-9.999$ | 0.01 |
| $10^6$ | $-9.9999$ | 0.001 |

**Comment:** For this case, the open-loop gain must be on the order of at least $10^3$ in order to be within 1 percent of the ideal gain. If the ideal closed-loop gain changes, a new value of open-loop gain must be determined in order to meet the specified requirements. As we will see in Chapter 14, at low frequencies, most op-amp circuits have gains on the order of $10^5$, so achieving the required accuracy is not difficult.

### EXERCISE PROBLEM

**Ex 9.3:** (a) An inverting op-amp circuit is to be designed using an op-amp with a finite differential voltage gain of $A_{od} = 10^4$. The closed-loop voltage gain is to be $A_v = -15.0$ and the input resistance is to be $R = 25\,\text{k}\Omega$. What is the required value of $R_2$? (b) Using the results of part (a), what is the closed-loop voltage gain if (i) $A_{od} = 10^5$ and (ii) $A_{od} = 10^3$? (Ans. (a) $R_1 = 25\,\text{k}\Omega$, $R_2 = 375.6\,\text{k}\Omega$; (b) (i) $A_v = -15.0216$, (ii) $A_v = -14.787$)

## Test Your Understanding

**TYU 9.1** (a) Design an ideal inverting op-amp circuit such that $A_v = -12$. Let $R_2 = 240\,\text{k}\Omega$. (b) Using the results of part (a), find $i_1$ when (i) $v_I = -0.15\,\text{V}$ and (ii) $v_I = +0.25\,\text{V}$. (Ans. (a) $R_1 = 20\,\text{k}\Omega$; (b) (i) $i_1 = -7.5\,\mu\text{A}$, (ii) $i_1 = 12.5\,\mu\text{A}$)

**TYU 9.2** Consider Example 9.1. Suppose the source resistance is not a constant, but varies within the range $0.7\,\text{k}\Omega \leq R_S \leq 1.3\,\text{k}\Omega$. Using the results of Example 9.1, what is the range in (a) the voltage gain $A_v$ and (b) the input current $i_1$. (c) Is the specified maximum input current still maintained? (Ans. (a) $4.926 \leq A_v \leq 5.076$, (b) $4.926 \leq i_1 \leq 5.076\,\mu\text{A}$)

**TYU 9.3** Consider an inverting op-amp circuit as shown in Figure 9.13 with $R_1 = 20\,\text{k}\Omega$ and $R_2 = 200\,\text{k}\Omega$. The op-amp is ideal except the open-loop gain is $A_{od} = 10^4$. Determine (a) $v_1$ and $v_O$ when $v_I = 50\,\text{mV}$, (b) $v_1$ and $v_I$ when $v_O = 5\,\text{V}$, and (c) $v_I$ and $v_O$ when $v_1 = 0.20\,\text{mV}$. (Ans. (a) $v_O = -0.49945\,\text{V}$, $v_1 = +49.945\,\mu\text{V}$; (b) $v_I = -0.50055\,\text{V}$, $v_1 = -0.5\,\text{mV}$; (c) $v_O = -2.0\,\text{V}$, $v_I = 0.20022\,\text{V}$)

# 9.3    SUMMING AMPLIFIER

**Objective: •** Analyze and understand the characteristics of the summing operational amplifier.

To analyze the op-amp circuit shown in Figure 9.14(a), we will use the superposition theorem and the concept of virtual ground. Using the superposition theorem, we will determine the output voltage due to each input acting alone. We will then algebraically sum these terms to determine the total output.

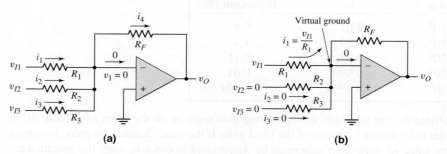

**Figure 9.14** (a) Summing op-amp amplifier circuit and (b) currents and voltages in the summing amplifier

If we set $v_{I2} = v_{I3} = 0$, the current $i_1$ is

$$i_1 = \frac{v_{I1}}{R_1} \tag{9.24}$$

Since $v_{I2} = v_{I3} = 0$ and the inverting terminal is at virtual ground, the currents $i_2$ and $i_3$ must both be zero. Current $i_1$ does not flow through either $R_2$ or $R_3$, but the entire current must flow through the feedback resistor $R_F$, as indicated in Figure 9.14(b). The output voltage due to $v_{I1}$ acting alone is

$$v_O(v_{I1}) = -i_1 R_F = -\left(\frac{R_F}{R_1}\right) v_{I1} \tag{9.25}$$

Similarly, the output voltages due to $v_{I2}$ and $v_{I3}$ acting individually are

$$v_O(v_{I2}) = -i_2 R_F = -\left(\frac{R_F}{R_2}\right) v_{I2} \tag{9.26}$$

and

$$v_O(v_{I3}) = -i_3 R_F = -\left(\frac{R_F}{R_3}\right) v_{I3} \tag{9.27}$$

The total output voltage is the algebraic sum of the individual output voltages, or

$$v_O = v_O(v_{I1}) + v_O(v_{I2}) + v_O(v_{I3}) \tag{9.28}$$

which becomes

$$v_O = -\left(\frac{R_F}{R_1} v_{I1} + \frac{R_F}{R_2} v_{I2} + \frac{R_F}{R_3} v_{I3}\right) \tag{9.29}$$

The output voltage is the sum of the three input voltages, with different weighting factors. This circuit is therefore called the **inverting summing amplifier.** The number of input terminals and input resistors can be changed to add more or fewer voltages.

A special case occurs when the three input resistances are equal. When $R_1 = R_2 = R_3 \equiv R$, then

$$v_O = -\frac{R_F}{R_1}(v_{I1} + v_{I2} + v_{I3}) \tag{9.30}$$

This means that the output voltage is the sum of the input voltages, with a single amplification factor.

**Discussion:** Up to this point, we have seen that op-amps can be used to multiply a signal by a constant and sum a number of signals with prescribed weights. These are mathematical operations. Later in the chapter, we will see that op-amps can also be used to integrate and differentiate. These circuits are the building blocks needed to perform analog computations—hence the original name of operational amplifier. Op-amps, however, are versatile and can do much more than just perform mathematical operations, as we will continue to observe through the remainder of the chapter.

## DESIGN EXAMPLE 9.4

**Objective:** Design a summing amplifier to produce a specified output signal.

**Specifications:** The output signal generated from an ideal amplifier circuit is $v_{O1} = 1.2 - 0.5 \sin \omega t$ (V). Design a summing amplifier to be connected to the amplifier circuit such that the output signal is $v_O = 2 \sin \omega t$ (V).

**Choices:** Standard precision resistors with tolerances of $\pm 1$ percent are to be used in the final design. Assume an ideal op-amp is available.

**Solution:** In this case, we need only two inputs to the summing amplifier, as shown in Figure 9.14. One input to the summing amplifier is the output of the ideal amplifier circuit and the second input should be a dc voltage to cancel the $+1.2$ V signal from the amplifier circuit. If the voltage gains of each input to the summing amplifier are equal, then an input of $-1.2$ V at the second input will cancel the $+1.2$ V from the amplifier circuit.

For a $-0.5$ V sinusoidal input signal and a desired 2 V sinusoidal output signal, the summing amplifier gain must be

$$A_v = \frac{-R_F}{R_1} = \frac{2}{-0.5} = -4$$

If we choose the input resistances to be $R_1 = R_2 = 30 \, \text{k}\Omega$, then the feedback resistance must be $R_F = 120 \, \text{k}\Omega$.

**Trade-offs:** From Appendix C, we can choose precision resistor values of $R_F = 124 \, \text{k}\Omega$ and $R_1 = R_2 = 30.9 \, \text{k}\Omega$. The ratio of the ideal resistors is 4.013. Considering the $\pm 1$ percent tolerance values, the output of the summing amplifier is

$$v_O = \frac{-R_F(1 \pm 0.01)}{R_1(1 \pm 0.01)} \cdot (1.2 - 0.5 \sin \omega t) - \frac{R_F(1 \pm 0.01)}{R_2(1 \pm 0.01)} \cdot (-1.2)$$

The dc output voltage is in the range $-0.1926 \leq v_O\,(\text{dc}) \leq 0.1926$ V and the peak ac output voltage is in the range $1.967 \leq v_O\,(\text{ac}) \leq 2.047$ V.

**Comment:** In this example, we have used a summing amplifier to amplify a time-varying signal and eliminate a dc voltage (ideally).

**Ex 9.4:** (a) Design an inverting summing amplifier that will produce an output voltage of $v_O = -3(v_{I1} + 2v_{I2} + 0.3v_{I3} + 4v_{I4})$. The maximum resistance is to be limited to 400 kΩ. (b) Using the results of part (a), determine $v_O$ for (i) $v_{I1} = 0.1$ V, $v_{I2} = -0.2$ V, $v_{I3} = -1$ V, $v_{I4} = 0.05$ V; and for (ii) $v_{I1} = -0.2$ V, $v_{I2} = 0.3$ V, $v_{I3} = 1.5$ V, $v_{I4} = -0.1$ V. (Ans. (a) Let $R_3 = 400$ kΩ, $R_F = 360$ kΩ, $R_1 = 120$ kΩ, $R_2 = 60$ kΩ, $R_4 = 30$ kΩ; (b) (i) $v_O = +1.2$ V, (ii) $v_O = -1.35$ V)

## Test Your Understanding

**TYU 9.4** Consider an ideal summing amplifier as shown in Figure 9.14(a) with $R_1 = 20$ kΩ, $R_2 = 40$ kΩ, $R_3 = 50$ kΩ, and $R_F = 200$ kΩ. Determine the output voltage $v_O$ for (a) $v_{I1} = -0.25$ mV, $v_{I2} = +0.30$ mV, $v_{I3} = -0.50$ mV; and (b) $v_{I1} = +10$ mV, $v_{I2} = -40$ mV, $v_{I3} = +25$ mV. (Ans. (a) $v_O = 3$ mV, (b) $v_O = 0$)

**TYU 9.5** Design the summing amplifier in Figure 9.14 to produce the average (magnitude) of three input voltages, i.e., $v_O = (v_{I1} + v_{I2} + v_{I3})/3$. The amplifier is to be designed such that each input signal sees the maximum possible input resistance under the condition that the maximum allowed resistance in the circuit is 1 M. (Ans. $R_1 = R_2 = R_3 = 1$ MΩ, $R_F = 333$ kΩ)

## 9.4    NONINVERTING AMPLIFIER

**Objective:** • Analyze and understand the characteristics of the non-inverting operational amplifier, including the voltage follower or buffer.

In our previous discussions, the feedback element was connected between the output and the inverting terminal. However, a signal can be applied to the noninverting terminal while still maintaining negative feedback.

### 9.4.1    Basic Amplifier

Figure 9.15 shows the basic **noninverting amplifier.** The input signal $v_I$ is applied directly to the noninverting terminal, while one side of resistor $R_1$ is connected to the inverting terminal and the other side is at ground.

Previously, when $v_2$ was at ground potential, we argued that $v_1$ was also essentially at ground potential, and we stated that terminal (1) was at virtual ground. The same principle applies to the circuit in Figure 9.15, with slightly different terminology. The negative feedback connection forces the terminal voltages $v_1$ and $v_2$ to be essentially equal. Such a condition is referred to as a **virtual short.** This condition exists since a change in $v_2$ will cause the output voltage $v_O$ to change in such a way that $v_1$ is forced to track $v_2$. The virtual short means that the voltage difference

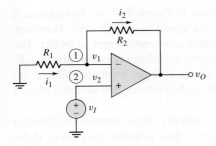

**Figure 9.15** Noninverting op-amp circuit

between $v_1$ and $v_2$ is, for all practical purposes, zero. However, unlike a true short circuit, there is no current flow directly from one terminal to the other. We use the virtual short concept, i.e. $v_1 = v_2$, as an ideal op-amp characteristic and use this property in our circuit analysis.

The analysis of the noninverting amplifier is essentially the same as for the inverting amplifier. We assume that no current enters the input terminals. Since $v_1 = v_2$, then $v_1 = v_I$, and current $i_1$ is given by

$$i_1 = -\frac{v_1}{R_1} = -\frac{v_I}{R_1} \tag{9.31}$$

Current $i_2$ is given by

$$i_2 = \frac{v_1 - v_O}{R_2} = \frac{v_I - v_O}{R_2} \tag{9.32}$$

As before, $i_1 = i_2$, so that

$$-\frac{v_I}{R_1} = \frac{v_I - v_O}{R_2} \tag{9.33}$$

Solving for the closed-loop voltage gain, we find

$$A_v = \frac{v_O}{v_I} = 1 + \frac{R_2}{R_1} \tag{9.34}$$

From this equation, we see that the output is in phase with the input, as expected. Also note that the gain is always greater than unity.

The input signal $v_I$ is connected directly to the noninverting terminal; therefore, since the input current is essentially zero, the input impedance seen by the source is very large, ideally infinite. The ideal equivalent circuit of the noninverting op-amp is shown in Figure 9.16.

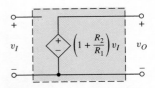

**Figure 9.16** Equivalent circuit of ideal noninverting op-amp

### 9.4.2    Voltage Follower

An interesting property of the noninverting op-amp occurs when $R_1 = \infty$, an open circuit. The closed-loop gain then becomes

$$A_v = \frac{v_O}{v_I} = 1 \tag{9.35}$$

Since the output voltage follows the input, this op-amp circuit is called a **voltage follower**. The closed-loop gain is independent of resistor $R_2$ (except when $R_2 = \infty$), so we can set $R_2 = 0$ to create a short circuit.

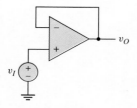

**Figure 9.17** Voltage-follower op-amp

The voltage-follower op-amp circuit is shown in Figure 9.17. At first glance, it might seem that this circuit, with unity voltage gain, would be of little value. However, other terms used for the voltage follower are **impedance transformer** or **buffer.** The input impedance is essentially infinite, and the output impedance is essentially zero. If, for example, the output impedance of a signal source is large, a voltage follower inserted between the source and a load will prevent loading effects, that is, it will act as a buffer between the source and the load.

Consider the case of a voltage source with a 100 kΩ output impedance driving a 1 kΩ load impedance, as shown in Figure 9.18(a). This situation may occur if the source is a transducer. (We will see an example of this later in the chapter when we consider a temperature-sensitive resistor, or thermistor, in a bridge circuit.) The ratio of output voltage to input voltage is

$$\frac{v_O}{v_I} = \frac{R_L}{R_L + R_S} = \frac{1}{1 + 100} \cong 0.01$$

This equation indicates that, for this case, there is a severe loading effect, or **attenuation,** in the signal voltage.

Figure 9.18(b) shows a voltage follower inserted between the source and the load. Since the input impedance to the noninverting terminal is usually much greater than 100 kΩ, then $v_O \cong v_I$ and the loading effect is eliminated.

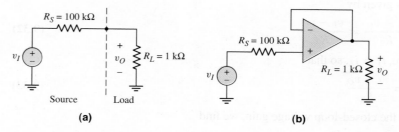

**Figure 9.18**  (a) Source with a 100 kΩ output resistance driving a 1 kΩ load and (b) source with a 100 kΩ output resistance, voltage follower, and 1 kΩ load

## Test Your Understanding

**TYU 9.6** (a) Design a noninverting amplifier such that the closed-loop gain is $A_v = 10$. The maximum resistance is to be 180 kΩ and the output voltage is to be in the range $-9 \leq v_O \leq +9$ V. (b) Repeat part (a) for a closed-loop gain of $A_v = 5$. The maximum current in any resistor is to be limited to 100 μA when the output voltage is in the range $-5 \leq v_O \leq +5$ V. (Ans. (a) $R_2 = 180\,\text{k}\Omega$, $R_1 = 20\,\text{k}\Omega$; (b) $R_2 = 40\,\text{k}\Omega$, $R_1 = 10\,\text{k}\Omega$)

**TYU 9.7**  The noninverting op-amp in Figure 9.15 has a finite differential gain of $A_{od}$. Show that the closed-loop gain is

$$A_v = \frac{v_O}{v_I} = \frac{\left(1 + \dfrac{R_2}{R_1}\right)}{\left[1 + \dfrac{1}{A_{od}}\left(1 + \dfrac{R_2}{R_1}\right)\right]}$$

**TYU 9.8** Use superposition to determine the output voltage $v_O$ in the ideal op-amp circuit in Figure 9.19. (Ans. $v_O = 10v_{I1} + 5v_{I2}$)

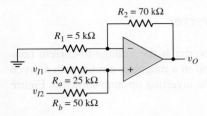

**Figure 9.19** Figure for Exercise TYU 9.8

##  9.5    OP-AMP APPLICATIONS

> **Objective:** • Analyze several ideal op-amp circuits including the difference amplifier and the instrumentation amplifier.

The summing amplifier is one example of special functional capabilities that can be provided by the op-amp. In this section, we will look at other examples of op-amp versatility.

### 9.5.1    Current-to-Voltage Converter

In some situations, the output of a device or circuit is a current. An example is the output of a photodiode or photodetector. We may need to convert this output current to an output voltage.

Consider the circuit in Figure 9.20. The input resistance $R_i$ at the virtual ground node is

$$R_i = \frac{v_1}{i_1} \cong 0 \qquad (9.36)$$

In most cases, we can assume that $R_S \gg R_i$; therefore, current $i_1$ is essentially equal to the signal current $i_S$. Then,

$$i_2 = i_1 = i_S \qquad (9.37)$$

and

$$v_O = -i_2 R_F = -i_S R_F \qquad (9.38)$$

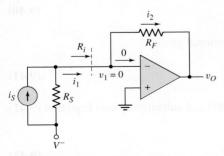

**Figure 9.20** Current-to-voltage converter

The output voltage is directly proportional to the signal current, and the feedback resistance $R_F$ is the magnitude of the ratio of the output voltage to the signal current.

<h3>9.5.2    Voltage-to-Current Converter</h3>

The complement of the current-to-voltage converter is the voltage-to-current converter. For example, we may want to drive a coil in a magnetic circuit with a given current, using a voltage source. We could use the inverting op-amp shown in Figure 9.21. For this circuit,

$$i_2 = i_1 = \frac{v_I}{R_1} \tag{9.39}$$

which means that current $i_2$ is directly proportional to input voltage $v_I$ and is independent of the load impedance or resistance $R_2$. However, one side of the load device might need to be at ground potential, so the circuit in Figure 9.21 would not be practical for such applications.

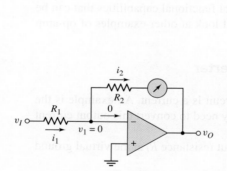

**Figure 9.21** Simple voltage-to-current converter

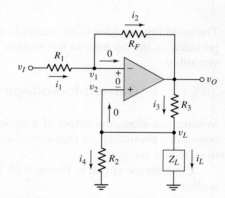

**Figure 9.22** Voltage-to-current converter

Consider the circuit in Figure 9.22. In this case, one terminal of the load device, which has an impedance of $Z_L$, is at ground potential. The inverting terminal (1) is not at virtual ground. From the virtual short concept, $v_1 = v_2$. We also note that $v_1 = v_2 = v_L = i_L Z_L$. Equating the currents $i_1$ and $i_2$, we have

$$\frac{v_I - i_L Z_L}{R_1} = \frac{i_L Z_L - v_O}{R_F} \tag{9.40}$$

Summing the currents at the noninverting terminal gives

$$\frac{v_O - i_L Z_L}{R_3} = i_L + \frac{i_L Z_L}{R_2} \tag{9.41}$$

Solving for $(v_O - i_L Z_L)$ from Equation (9.40) and substituting into Equation (9.41) produces

$$\frac{R_F}{R_1} \cdot \frac{(i_L Z_L - v_I)}{R_3} = i_L + \frac{i_L Z_L}{R_2} \tag{9.42}$$

Combining terms in $i_L$, we obtain

$$i_L \left( \frac{R_F Z_L}{R_1 R_3} - 1 - \frac{Z_L}{R_2} \right) = v_I \left( \frac{R_F}{R_1 R_3} \right) \tag{9.43}$$

In order to make $i_L$ independent of $Z_L$, we can design the circuit such that the coefficient of $Z_L$ is zero, or

$$\frac{R_F}{R_1 R_3} = \frac{1}{R_2} \tag{9.44}$$

Equation (9.43) then reduces to

$$i_L = -v_I \left( \frac{R_F}{R_1 R_3} \right) = \frac{-v_I}{R_2} \tag{9.45}$$

which means that the load current is proportional to the input voltage and is independent of the load impedance $Z_L$, as long as the output voltage remains between allowed limits.

We may note that the input resistance seen by the source $v_I$ is finite, and is actually a function of the load impedance $Z_L$. For a constant $i_L$, a change in $Z_L$ produces a change in $v_L = v_2 = v_1$, which causes a change in $i_1$. A voltage follower may be inserted between the voltage source $v_I$ and the resistor $R_1$ to eliminate any loading effects due to a variable input resistance.

## EXAMPLE 9.5

**Objective:** Determine a load current in a voltage-to-current converter.

Consider the circuit in Figure 9.22. Let $Z_L = 100\ \Omega$, $R_1 = 10\ k\Omega$, $R_2 = 1\ k\Omega$, $R_3 = 1\ k\Omega$, and $R_F = 10\ k\Omega$. If $v_I = -5$ V, determine the load current $i_L$ and the output voltage $v_O$.

**Solution:** We note first that the condition expressed by Equation (9.44) is satisfied; that is,

$$\frac{1}{R_2} = \frac{R_F}{R_1 R_3} = \frac{10}{(10)(1)} \rightarrow \frac{1}{1}$$

The load current is

$$i_L = \frac{-v_I}{R_2} = \frac{-(-5)}{1\ k\Omega} = 5\ \text{mA}$$

and the voltage across the load is

$$v_L = i_L Z_L = (5 \times 10^{-3})(100) = 0.5\ \text{V}$$

Currents $i_4$ and $i_3$ are

$$i_4 = \frac{v_L}{R_2} = \frac{0.5}{1} = 0.5\ \text{mA}$$

and

$$i_3 = i_4 + i_L = 0.5 + 5 = 5.5\ \text{mA}$$

The output voltage is then

$$v_O = i_3 R_3 + v_L = (5.5 \times 10^{-3})(10^3) + 0.5 = 6 \text{ V}$$

We could also calculate $i_1$ and $i_2$ as

$$i_1 = i_2 = -0.55 \text{ mA}$$

**Comment:** In this example, we implicitly assume that the op-amp is not in saturation, which means that the applied dc bias voltage must be greater than 6 V. In addition, since currents $i_2$ (which is negative) and $i_3$ must be supplied by the op-amp, we are assuming that the op-amp is capable of supplying 6.05 mA.

**Computer Verification:** The PSpice circuit schematic of the voltage-to-current converter is shown in Figure 9.23(a). The input voltage was varied between 0 and $-10$ V. Figure 9.23(b) shows the current through the 100 $\Omega$ load and Figure 9.23(c)

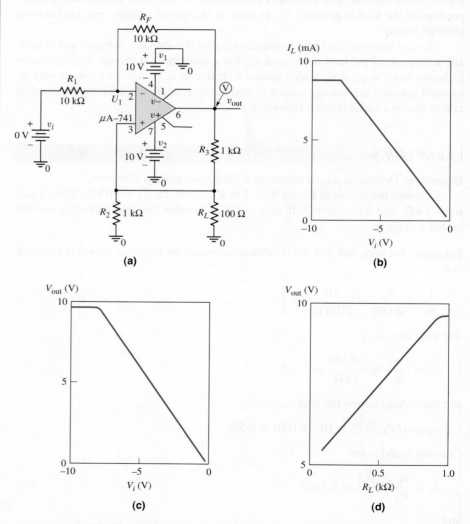

**Figure 9.23** (a) PSpice circuit schematic; (b) load current and (c) op-amp output voltage versus input voltage; (d) op-amp output voltage versus load resistance for $v_1 = -5$ V

shows the op-amp output voltage as a function of the input voltage. At approximately $v_I = -7.5$ V, the op-amp saturates, so the load current and output voltage no longer increase with input voltage. This result demonstrates that the ideal voltage-to-current conversion is valid only if the op-amp is operating in its linear region. Figure 9.23(d) shows the output voltage as a function of load resistance for an input voltage of $v_I = -5$ V. At a load resistance greater than approximately 900 $\Omega$, the op-amp saturates. The range over which the op-amp remains linear could be increased by increasing the bias to $\pm 15$ V, for example.

### EXERCISE PROBLEM

**Ex 9.5:** Consider the voltage-to-current converter shown in Figure 9.22. The load impedance is $Z_L = 200$ $\Omega$ and the input voltage is $v_I = -3$ V. Determine $i_L$ and $v_O$ if $R_1 = 10$ k$\Omega$, $R_2 = 1.5$ k$\Omega$, $R_3 = 3$ k$\Omega$, and $R_F = 20$ k$\Omega$. (Ans. $i_L = 2$ mA, $v_O = 7.2$ V)

### 9.5.3 Difference Amplifier

An ideal difference amplifier amplifies only the difference between two signals; it rejects any common signals to the two input terminals. For example, a microphone system amplifies an audio signal applied to one terminal of a difference amplifier, and rejects any 60 Hz noise signal or "hum" existing on both terminals. The basic op-amp also amplifies the difference between two input signals. However, we would like to make a difference amplifier, in which the output is a function of the ratio of resistors, as we had for the inverting and noninverting amplifiers.

Consider the circuit shown in Figure 9.24(a), with inputs $v_{I1}$ and $v_{I2}$. To analyze the circuit, we will use superposition and the virtual short concept. Figure 9.24(b)

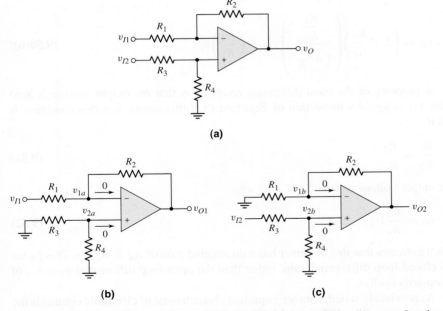

**Figure 9.24** (a) Op-amp difference amplifier, (b) difference amplifier with $v_{I2} = 0$ and (c) difference amplifier with $v_{I1} = 0$

shows the circuit with input $v_{I2} = 0$. There are no currents in $R_3$ and $R_4$; therefore, $v_{2a} = 0$. The resulting circuit is the inverting amplifier previously considered, for which

$$v_{O1} = -\frac{R_2}{R_1} v_{I1} \tag{9.46}$$

Figure 9.24(c) shows the circuit with $v_{I1} = 0$. Since the current into the op-amp is zero, $R_3$ and $R_4$ form a voltage divider. Therefore,

$$v_{2b} = \frac{R_4}{R_3 + R_4} v_{I2} \tag{9.47}$$

From the virtual short concept, $v_{1b} = v_{2b}$ and the circuit becomes a noninverting amplifier, for which

$$v_{O2} = \left(1 + \frac{R_2}{R_1}\right) v_{1b} = \left(1 + \frac{R_2}{R_1}\right) v_{2b} \tag{9.48}$$

Substituting Equation (9.47) into (9.48), we obtain

$$v_{O2} = \left(1 + \frac{R_2}{R_1}\right)\left(\frac{R_4}{R_3 + R_4}\right) v_{I2} \tag{9.49(a)}$$

which can be rearranged as follows:

$$v_{O2} = (1 + R_2/R_1)\left(\frac{R_4/R_3}{1 + R_4/R_3}\right) v_{I2} \tag{9.49(b)}$$

Since the net output voltage is the sum of the individual terms, we have

$$v_O = v_{O1} + v_{O2} \tag{9.50(a)}$$

or

$$v_O = \left(1 + \frac{R_2}{R_1}\right)\left(\frac{\dfrac{R_4}{R_3}}{1 + \dfrac{R_4}{R_3}}\right) v_{I2} - \left(\frac{R_2}{R_1}\right) v_{I1} \tag{9.50(b)}$$

A property of the ideal difference amplifier is that the output voltage is zero when $v_{I1} = v_{I2}$. An inspection of Equation (9.50(b)) shows that this condition is met if

$$\frac{R_4}{R_3} = \frac{R_2}{R_1} \tag{9.51}$$

The output voltage is then

$$v_O = \frac{R_2}{R_1}(v_{I2} - v_{I1}) \tag{9.52}$$

which indicates that this amplifier has a differential gain of $A_d = R_2/R_1$. This factor is a closed-loop differential gain, rather than the open-loop differential gain $A_{od}$ of the op-amp itself.

As previously stated, another important characteristic of electronic circuits is the input resistance. The **differential input resistance** of the differential amplifier can be determined by using the circuit shown in Figure 9.25. In the figure, we have

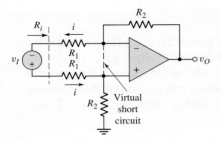

**Figure 9.25** Circuit for measuring differential input resistance of op-amp difference amplifier

imposed the condition given in Equation (9.51) and have set $R_1 = R_3$ and $R_2 = R_4$. The input resistance is then defined as

$$R_i = \frac{v_I}{i} \tag{9.53}$$

Taking into account the virtual short concept, we can write a loop equation, as follows:

$$v_I = iR_1 + iR_1 = i(2R_1) \tag{9.54}$$

Therefore, the input resistance is

$$R_i = 2R_1 \tag{9.55}$$

## DESIGN EXAMPLE 9.6

**Objective:** Design a difference amplifier with a specified gain.

**Specifications:** Design the difference amplifier with the configuration shown in Figure 9.24 such that the differential gain is 30. Standard valued resistors are to be used and the maximum resistor value is to be 500 kΩ.

**Choices:** An ideal op-amp is available.

**Solution:** The differential gain is given by

$$\frac{R_2}{R_1} = \frac{R_4}{R_3} = 30$$

From Appendix C, we can use standard resistors of

$$R_2 = R_4 = 390\,\text{k}\Omega \qquad \text{and} \qquad R_1 = R_3 = 13\,\text{k}\Omega$$

These resistor values are obviously less than 500 kΩ and will give an input resistance of $R_i = 2R_1 = 2(13) = 26\,\text{k}\Omega$.

**Trade-offs:** Resistor tolerances must be considered as we have in other designs. This effect is considered in end-of-chapter Problem 9.62. Resistor tolerances also affect the common-mode rejection ratio, as analyzed in the following example.

**Comment:** This example illustrated one disadvantage of this differential amplifier design. It cannot achieve both high gain and high input impedance without using extremely large resistor values.

EXERCISE PROBLEM

**Ex 9.6:** Consider the difference amplifier in Figure 9.24(a). (a) Design the circuit with $R_2 = R_4$, $R_1 = R_3$, and such that the differential voltage gain is $A_d = 50$. For input voltages in the range of $-50$ mV to $+50$ mV, the maximum current in $R_2$ is to be limited to $50\,\mu A$. (b) Using the results of part (a), what is the maximum current in $R_3$? (Ans. (a) Set $R_2 = R_4 = 100\,k\Omega$, then $R_1 = R_2 = 2\,k\Omega$; (b) $0.49\,\mu A$)

In the ideal difference amplifier, the output $v_O$ is zero when $v_{I1} = v_{I2}$. However, an inspection of Equation (9.50(b)) shows that this condition is not satisfied if $R_4/R_3 \neq R_2/R_1$. When $v_{I1} = v_{I2}$, the input is called a **common-mode input signal.** The common-mode input voltage is defined as

$$v_{cm} = (v_{I1} + v_{I2})/2 \qquad\qquad\qquad (9.56)$$

The common-mode gain is then defined as

$$A_{cm} = \frac{v_O}{v_{cm}} \qquad\qquad\qquad (9.57)$$

Ideally, when a common-mode signal is applied, $v_O = 0$ and $A_{cm} = 0$.

A nonzero common-mode gain may be generated in actual op-amp circuits. This is discussed in Chapter 14.

A figure of merit for a difference amplifier is the **common-mode rejection ratio (CMRR),** which is defined as the magnitude of the ratio of differential gain to common-mode gain, or

$$\mathrm{CMRR} = \left| \frac{A_d}{A_{cm}} \right| \qquad\qquad\qquad (9.58)$$

Usually, the CMRR is expressed in decibels, as follows:

$$\mathrm{CMRR(dB)} = 20 \log_{10} \left| \frac{A_d}{A_{cm}} \right| \qquad\qquad\qquad (9.59)$$

Ideally, the common-mode rejection ratio is infinite. In an actual differential amplifier, we would like the common-mode rejection ratio to be as large as possible.

EXAMPLE **9.7**

**Objective:** Calculate the common-mode rejection ratio of a difference amplifier.

Consider the difference amplifier shown in Figure 9.24(a). Let $R_2/R_1 = 10$ and $R_4/R_3 = 11$. Determine CMRR(dB).

**Solution:** From Equation (9.50(b)), we have

$$v_O = (1 + 10) \left( \frac{11}{1 + 11} \right) v_{I2} - (10)v_{I1}$$

or

$$v_O = 10.0833 v_{I2} - 10 v_{I1} \qquad\qquad\qquad (9.60)$$

The differential-mode input voltage is defined as

$$v_d = v_{I2} - v_{I1}$$

and the common-mode input voltage is defined as

$$v_{cm} = (v_{I1} + v_{I2})/2$$

Combining these two equations produces

$$v_{I1} = v_{cm} - \frac{v_d}{2} \tag{9.61(a)}$$

and

$$v_{I2} = v_{cm} + \frac{v_d}{2} \tag{9.61(b)}$$

If we substitute Equations (9.61(a)) and (9.61(b)) in Equation (9.60), we obtain

$$v_O = (10.0833)\left(v_{cm} + \frac{v_d}{2}\right) - (10)\left(v_{cm} - \frac{v_d}{2}\right)$$

or

$$v_O = 10.042 v_d + 0.0833 v_{cm} \tag{9.62}$$

The output voltage is also

$$v_O = A_d v_d + A_{cm} v_{cm} \tag{9.63}$$

If we compare Equations (9.62) and (9.63), we see that

$$A_d = 10.042 \quad \text{and} \quad A_{cm} = 0.0833$$

Therefore, from Equation (9.59), the common-mode rejection ratio, is

$$\text{CMRR(dB)} = 20 \log_{10}\left(\frac{10.042}{0.0833}\right) = 41.6 \, \text{dB}$$

**Comment:** For good differential amplifiers, typical CMRR values are in the range of 80–100 dB. This example shows how close the ratios $R_2/R_1$ and $R_4/R_3$ must be in order to achieve a CMRR value in that range.

**Computer Verification:** A PSpice analysis was performed on the differential amplifier in this example with a $\mu$A-741 op-amp. For input voltages of $v_{I1} = -50$ mV and $v_{I2} = +50$ mV, the output voltage is $v_O = 1.0043$ V, which gives a differential voltage gain of 10.043. For input voltages of $v_{I1} = v_{I2} = 5$ V, the output voltage is $v_O = 0.4153$ V, which gives a common-mode voltage gain of $A_{cm} = 0.4153/5 = 0.0831$. The common-mode rejection ratio is then CMRR $= 10.043/0.0831 = 120.9 \Rightarrow 41.6$ dB, which agrees with the hand analysis. This result demonstrates that at this point, the nonideal characteristics of the $\mu$A-741 op-amp do not affect these results.

---

EXERCISE PROBLEM

*Ex 9.7:* In the difference amplifier shown in Figure 9.24(a), $R_1 = R_3 = 10$ k$\Omega$, $R_2 = 20$ k$\Omega$, and $R_4 = 21$ k$\Omega$. Determine $v_O$ when: (a) $v_{I1} = +1$ V, $v_{I2} = -1$ V; and (b) $v_{I1} = v_{I2} = +1$ V. (c) Determine the common-mode gain. (d) Determine the CMRR(dB). (Ans. (a) $v_O = -4.032$ V, (b) $v_O = 0.0323$ V, (c) $A_{cm} = 0.0323$, (d) CMRR(db) = 35.9 dB)

### 9.5.4 Instrumentation Amplifier

We saw in the last section that it is difficult to obtain a high input impedance and a high gain in a difference amplifier with reasonable resistor values. One solution is to insert a voltage follower between each source and the corresponding input. However, a disadvantage of this design is that the gain of the amplifier cannot easily be changed. We would need to change two resistance values and still maintain equal ratios between $R_2/R_1$ and $R_4/R_3$. Optimally, we would like to be able to change the gain by changing only a single resistance value. The circuit in Figure 9.26, called an instrumentation amplifier, allows this flexibility. Note that two noninverting amplifiers, $A_1$ and $A_2$, are used as the input stage, and a difference amplifier, $A_3$ is the second, or amplifying, stage.

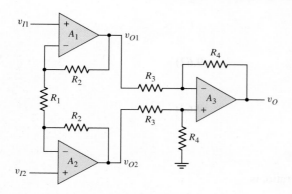

**Figure 9.26** Instrumentation amplifier

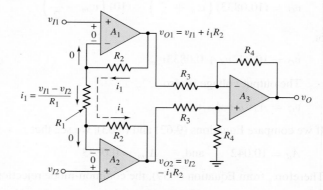

**Figure 9.27** Voltages and currents in instrumentation amplifier

We begin the analysis using the virtual short concept. The voltages at the inverting terminals of the voltage followers are equal to the input voltages. The currents and voltages in the amplifier are shown in Figure 9.27. The current in resistor $R_1$ is then

$$i_1 = \frac{v_{I1} - v_{I2}}{R_1} \tag{9.64}$$

The current in resistors $R_2$ is also $i_1$, as shown in the figure, and the output voltages of op-amps $A_1$ and $A_2$ are, respectively,

$$v_{O1} = v_{I1} + i_1 R_2 = \left(1 + \frac{R_2}{R_1}\right) v_{I1} - \frac{R_2}{R_1} v_{I2} \tag{9.65(a)}$$

and

$$v_{O2} = v_{I2} - i_1 R_2 = \left(1 + \frac{R_2}{R_1}\right) v_{I2} - \frac{R_2}{R_1} v_{I1} \tag{9.65(b)}$$

From previous results, the output of the difference amplifier is given as

$$v_O = \frac{R_4}{R_3}(v_{O2} - v_{O1}) \tag{9.66}$$

Substituting Equations (9.65(a)) and (9.65(b)) into Equation (9.66), we find the output voltage, as follows:

$$v_O = \frac{R_4}{R_3}\left(1 + \frac{2R_2}{R_1}\right)(v_{I2} - v_{I1}) \tag{9.67}$$

Since the input signal voltages are applied directly to the noninverting terminals of $A_1$ and $A_2$, the input impedance is very large, ideally infinite, which is one desirable characteristic of the instrumentation amplifier. Also, the differential gain is a function of resistor $R_1$, which can easily be varied by using a potentiometer, thus providing a variable amplifier gain with the adjustment of only one resistance.

## EXAMPLE 9.8

**Objective:** Determine the range required for resistor $R_1$, to realize a differential gain adjustable from 5 to 500.

The instrumentation amplifier circuit is shown in Figure 9.26. Assume that $R_4 = 2R_3$, so that the difference amplifier gain is 2.

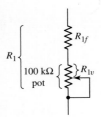

**Figure 9.28** Equivalent resistance $R_1$ in instrumentation amplifier

**Solution:** Assume that resistance $R_1$ is a combination of a fixed resistance $R_{1f}$ and a variable resistance $R_{1v}$, as shown in Figure 9.28. The fixed resistance ensures that the gain is limited to a maximum value, even if the variable resistance is set equal to zero. Assume the variable resistance is a 100 kΩ potentiometer.

From Equation (9.67), the maximum differential gain is

$$500 = 2\left(1 + \frac{2R_2}{R_{1f}}\right)$$

and the minimum differential gain is

$$5 = 2\left(1 + \frac{2R_2}{R_{1f} + 100}\right)$$

From the maximum gain expression, we find that

$$2R_2 = 249R_{1f}$$

Substituting this $R_2$ value into the minimum gain expression, we have

$$1.5 = \frac{2R_2}{R_{1f} + 100} = \frac{249R_{1f}}{R_{1f} + 100}$$

The resulting value of $R_{1f}$ is $R_{1f} = 0.606$ kΩ, which yields $R_2 = 75.5$ kΩ.

**Comment:** We can select standard resistance values that are close to the values calculated, and the range of the gain will be approximately in the desired range.

**Design Pointer:** An amplifier with a wide range of gain and designed with a potentiometer would normally not be used with standard integrated circuits in electronic systems. However, such a circuit might be very useful in specialized test equipment.

## EXERCISE PROBLEM

**Ex 9.8:** For the instrumentation amplifier in Figure 9.26, the parameters are $R_4 = 90$ kΩ, $R_3 = 30$ kΩ, and $R_2 = 50$ kΩ. Resistance $R_1$ is a series combination of a fixed 2 kΩ resistor and a 100 kΩ potentiometer. (a) Determine the range of the differential voltage gain. (b) Determine the maximum current in $R_1$ for input voltages in the range $-25$ mV to $+25$ mV. (Ans. (a) $5.94 \leq A_d \leq 153$, (b) $25 \mu A$)

### 9.5.5    Integrator and Differentiator

In the op-amp circuits previously considered, the elements exterior to the op-amp have been resistors. Other elements can be used, with differing results. Figure 9.29 shows a generalized inverting amplifier for which the voltage transfer function has the same general form as before, that is,

$$\frac{v_O}{v_I} = -\frac{Z_2}{Z_1} \tag{9.68}$$

where $Z_1$ and $Z_2$ are generalized impedances. Two special circuits can be developed from this generalized inverting amplifier.

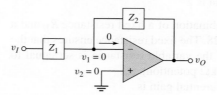

**Figure 9.29** Generalized inverting amplifier

In the first, $Z_1$ corresponds to a resistor and $Z_2$ to a capacitor. The impedances are then $Z_1 = R_1$ and $Z_2 = 1/sC_2$, where $s$ again is the complex frequency. The output voltage is

$$v_O = -\frac{Z_2}{Z_1} v_I = \frac{-1}{s R_1 C_2} v_I \tag{9.69}$$

Equation (9.69) represents integration in the time domain. If $V_C$ is the voltage across the capacitor at $t = 0$, the output voltage is

$$v_O = V_C - \frac{1}{R_1 C_2} \int_0^t v_I(t') \, dt' \tag{9.70}$$

where $t'$ is the variable of integration. Figure 9.30 summarizes these results.

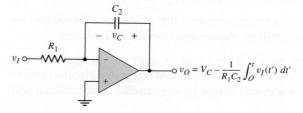

**Figure 9.30** Op-amp integrator

Equation (9.70) is the output response of the integrator circuit, shown in Figure 9.30, for any input voltage $v_I$. Note that if $v_I(t)$ is a finite step function, output $v_O$ will be a linear function of time. The output $v_O$ will be a ramp function and will eventually saturate at a voltage near either the positive or negative supply voltage. We will use the integrator in filter circuits, which are covered in Chapter 15.

We will show in Chapter 14 that nonzero bias currents into the op-amp greatly influence the characteristics of this circuit. A dc current through the capacitor will cause the output voltage to linearly change with time until the positive or negative

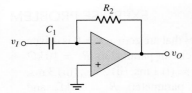

**Figure 9.31** Op-amp differentiator

supply voltage is reached. In many applications, a transistor switch needs to be added in parallel with the capacitor to periodically set the capacitor voltage to zero.

The second generalized inverting op-amp uses a capacitor for $Z_1$ and a resistor for $Z_2$, as shown in Figure 9.31. The impedances are $Z_1 = 1/sC_1$ and $Z_2 = R_2$, and the voltage transfer function is

$$\frac{v_O}{v_I} = -\frac{Z_2}{Z_1} = -sR_2C_1 \qquad\qquad \textbf{(9.71(a))}$$

The output voltage is

$$v_O = -sR_2C_1v_I \qquad\qquad \textbf{(9.71(b))}$$

Equation (9.71(b)) represents differentiation in the time domain, as follows:

$$v_O(t) = -R_2C_1\frac{dv_I(t)}{dt} \qquad\qquad \textbf{(9.72)}$$

The circuit in Figure 9.31 is therefore a differentiator.

Differentiator circuits are more susceptible to noise than are the integrator circuits. Input noise fluctuations of small amplitudes may have large derivatives. When differentiated, these noise fluctuations may generate large noise signals at the output, creating a poor output signal to noise ratio. This problem may be alleviated by placing a resistor in series with the input capacitor. This modified circuit then differentiates low-frequency signals but has a constant high-frequency gain.

## EXAMPLE 9.9

**Objective:** Determine the time constant required in an integrator.

Consider the integrator shown in Figure 9.30. Assume that voltage $V_C$ across the capacitor is zero at $t = 0$. A step input voltage of $v_I = -1$ V is applied at $t = 0$. Determine the time constant required such that the output reaches $+10$ V at $t = 1$ ms.

**Solution:** From Equation (9.70), we have

$$v_o = \frac{-1}{R_1C_2}\int_0^t (-1)\,dt' = \frac{1}{R_1C_2}t'\Big|_0^t = \frac{t}{R_1C_2}$$

At $t = 1$ ms, we want $v_O = 10$ V. Therefore,

$$10 = \frac{10^{-3}}{R_1C_2}$$

which means the time constant is $R_1C_2 = 0.1$ ms.

**Comment:** As an example, for a time constant of 0.1 ms, we could have $R_1 = 10\ \text{k}\Omega$ and $C_2 = 0.01\ \mu\text{F}$, which are reasonable values of resistance and capacitance.

**Ex 9.9:** An integrator with input and output voltages that are zero at $t = 0$ is driven by the input signal shown in Figure 9.32. (a) For circuit parameters $R_1 = 10\,\text{k}\Omega$ and $C_2 = 0.1\,\mu\text{F}$, determine the output voltage at $t =$ (i) 1 ms, (ii) 2 ms, (iii) 3 ms, and (iv) 4 ms. (b) Repeat part (a) for circuit parameters $R_1 = 10\,\text{k}\Omega$ and $C_2 = 1\,\mu\text{F}$. (Ans. (a) (i) $-1$ V, (ii) 0, (iii) $-1$ V, (iv) 0; (b) (i) $-0.1$ V, (ii) 0, (iii) $-0.1$ V, (iv) 0)

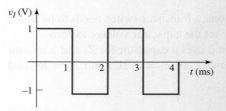

**Figure 9.32** Figure for Exercise Ex 9.9

| 9.5.6 | **Nonlinear Circuit Applications** |
|---|---|

Up to this point in the chapter, we have used linear passive elements in conjunction with the op-amp. Many useful circuits can be fabricated if nonlinear elements, such as diodes or transistors, are used in the op-amp circuits. We will consider three simple examples to illustrate the types of nonlinear characteristics that can be generated and to illustrate the general analysis technique.

**Precision Half-Wave Rectifier**

An op-amp and diode are combined as shown in Figure 9.33 to form a precision half-wave rectifier. For $v_I > 0$, the circuit behaves as a voltage follower. The output voltage is $v_O = v_I$, the load current $i_L$ is positive, and a positive diode current is induced such that $i_D = i_L$. The feedback loop is closed through the forward-biased diode. The output voltage of the op-amp, $v_{O1}$, adjusts itself to exactly absorb the forward voltage drop of the diode.

For $v_I < 0$, the output voltage tends to go negative, which tends to produce negative load and diode currents. However, a negative diode current cannot exist, so the diode cuts off, the feedback loop is broken, and $v_O = 0$.

The voltage transfer characteristics are shown in Figure 9.34. The rectification is precise in that, even at small positive input voltages, $v_O = v_I$ and we do not observe a diode cut-in voltage.

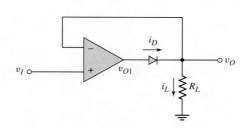

**Figure 9.33** Precision half-wave rectifier circuit

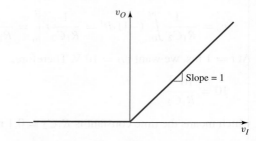

**Figure 9.34** Voltage transfer characteristics of precision half-wave rectifier

A potential problem in this circuit exists for $v_I < 0$. The feedback loop is broken so that the op-amp output voltage $v_{O1}$ will saturate near the negative supply voltage. When $v_I$ switches positive, it will take time for the internal circuit to recover, so the response time of the output voltage may be relatively slow. In addition, for $v_I < 0$ and $v_O = 0$, there is now a voltage difference applied across the input terminals of the op-amp. Most op-amps provide input voltage protection so the op-amp will not be damaged in this case. However, if the op-amp does not have input protection, the op-amp may be damaged if the input voltage is larger than 5 or 6 V.

## Log Amplifier

Consider the circuit in Figure 9.35. The diode is to be forward biased, so the input signal voltage is limited to positive values. The diode current is

$$i_D = I_S(e^{v_D/V_T} - 1) \tag{9.73(a)}$$

If the diode is sufficiently forward biased, the $(-1)$ term is negligible, and

$$i_D \cong I_S e^{v_D/V_T} \tag{9.73(b)}$$

The input current can be written

$$i_1 = \frac{v_I}{R_1} \tag{9.74}$$

and the output voltage, since $v_1$ is at virtual ground, is given by

$$v_O = -v_D \tag{9.75}$$

Noting that $i_1 = i_D$, we can write

$$i_1 = \frac{v_I}{R_1} = i_D = I_S e^{-v_O/V_T} \tag{9.76}$$

If we take the natural log of both sides of this equation, we obtain

$$\ln\left(\frac{v_I}{I_S R_1}\right) = -\frac{v_O}{V_T} \tag{9.77(a)}$$

or

$$v_O = -V_T \ln\left(\frac{v_I}{I_S R_1}\right) \tag{9.77(b)}$$

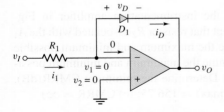

**Figure 9.35** Simple log amplifier

Equation (9.77(b)) indicates that, for this circuit, the output voltage is proportional to the log of the input voltage. One disadvantage of this circuit is that the reverse-saturation current $I_S$ is a strong function of temperature, and it varies substantially from one diode to another. A more sophisticated circuit uses bipolar transistors to eliminate the $I_S$ parameter in the log term. This circuit will not be considered here.

### Antilog or Exponential Amplifier

The complement, or inverse function, of the log amplifier is the antilog, or exponential, amplifier. A simple example using a diode is shown in Figure 9.36. Since $v_1$ is at virtual ground, we can write for $v_I > 0$

$$i_D \cong I_S e^{v_I/V_T} \tag{9.78}$$

and

$$v_O = -i_2 R = -i_D R \tag{9.79(a)}$$

or

$$v_O = -I_S R \cdot e^{v_I/V_T} \tag{9.79(b)}$$

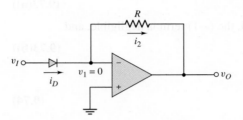

**Figure 9.36**  A simple antilog, or exponential, amplifier

The output voltage is an exponential function of the input voltage. Again, there are more sophisticated circuits that perform this function, but they will not be considered here.

## Test Your Understanding

**TYU 9.9** A current source has an output impedance of $R_S = 100\,\text{k}\Omega$. Design a current-to-voltage converter with an output voltage of $v_O = -10\,\text{V}$ when the signal current is $i_S = 100\,\mu\text{A}$. (Ans. Figure 9.20 with $R_F = 100\,\text{k}\Omega$)

**TYU 9.10** Design the voltage-to-current converter shown in Figure 9.22 such that the load current in a $500\,\Omega$ load can be varied between 0 and 1 mA with an input voltage between 0 and $-5\,\text{V}$. Assume the op-amp is biased at $\pm 10\,\text{V}$. (Ans. $R_2 = 5\,\text{k}\Omega$; for example, let $R_3 = 7\,\text{k}\Omega$, $R_1 = 10\,\text{k}\Omega$, $R_F = 14\,\text{k}\Omega$)

**TYU 9.11** All parameters associated with the instrumentation amplifier in Figure 9.26 are as given in Exercise Ex 9.8, except that resistor $R_2$ associated with the $A_1$ op-amp is $R_2 = 50\,\text{k}\Omega \pm 5\%$. (a) Determine the maximum and minimum possible values of the common-mode gain. (b) Determine the maximum and minimum possible values of the differential-mode gain. (c) Determine the minimum CMRR(dB). (Ans. (a) $A_{cm} = 0$; (b) $A_d(\text{min}) = 5.87$, $A_d(\text{max}) = 156.75$; (c) CMRR $= \infty$)

**TYU 9.12** Design the instrumentation amplifier in Figure 9.26 such that the variable differential voltage gain is in the range of 5 to 500. The range of the input voltages is between $-2$ mV and $+2$ mV, and the maximum current in $R_1$ is to be limited to $2\,\mu\text{A}$. Set the gain of the difference amplifier to 2.5. (Ans. $R_1(\text{fixed}) = 2\,\text{k}\Omega$, $R_2 = 199\,\text{k}\Omega$, $R_1(\text{var}) = 396\,\text{k}\Omega$)

**TYU 9.13** An integrator is driven by the series of pulses shown in Figure 9.37. At the end of the tenth pulse, the output voltage is to be $v_O = -5\,\text{V}$. Assume $V_C = 0$ at $t = 0$.

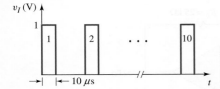

**Figure 9.37**  Figure for Exercise TYU 9.13

Determine the time constant and values of $R_1$ and $C_2$ that will meet these specifications. (Ans. $\tau = 20\,\mu s$; for example, let $C_2 = 0.01\,\mu F$, $R_1 = 2\,k\Omega$)

---

## 9.6   OPERATIONAL TRANSCONDUCTANCE AMPLIFIERS

**Objective:** • Discuss the operational transconductance amplifier.

The operational amplifiers considered up to this point have been voltage amplifiers. The input signal is a voltage and the output signal is a voltage.

Another type of op-amp is an operational transconductance amplifier (OTA). This op-amp is a voltage-input, current-output amplifier. Its circuit symbol is shown in Figure 9.38(a) and the equivalent circuit model is given in Figure 9.38(b). For the ideal OTA, both the input and output impedance is infinite. (The output impedance of an ideal current source is infinite.) The output current for the ideal circuit can be written as

$$i_O = g_m v_d \tag{9.80}$$

where $g_m$ is called the *unloaded transconductance*, with units of amperes per volt. The transconductance can be varied by changing the control current in the op-amp circuit. The OTA can then be used to electronically program functions.

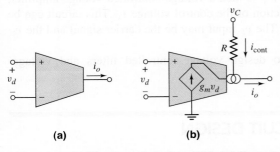

**(a)**          **(b)**

**Figure 9.38**  (a) Circuit symbol of the OTA. (b) Equivalent circuit model of the OTA.

We will see examples of actual OTA circuits in Chapter 13.

One example of an OTA application is shown in Figure 9.39. This circuit is a simple voltage-controlled amplifier. The output op-amp is configured as a current-to-voltage converter. We see that

$$v_O = -i_O R_F = -i_O(25\,k\Omega) \tag{9.81}$$

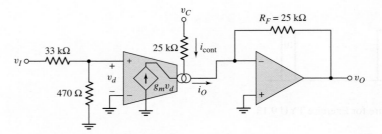

**Figure 9.39** Example of a voltage-controlled voltage amplifier using an OTA

and

$$i_O = g_m v_d \tag{9.82}$$

where

$$v_d = \frac{470}{470 + 33,000} \cdot v_I = 0.014 v_I \tag{9.83}$$

From the OTA circuit, we have

$$g_m = \frac{i_{\text{cont}}}{2V_T} \tag{9.84}$$

where $V_T = 0.026$ V at room temperature. The control current is given by

$$i_{\text{cont}} = \frac{v_C}{25 \, \text{k}\Omega} \tag{9.85}$$

where $v_C$ may be in the range $2 \le v_C \le 10$ V. The transconductance of the transconductance operational amplifier is controlled by the control voltage $v_C$.

Combining equations, we can write the voltage gain as

$$A_v = \frac{v_O}{v_I} = 0.269 v_C \tag{9.86}$$

The amplifier shown in Figure 9.39 is then a voltage-controlled voltage amplifier. The amplification factor is a function of the control voltage $v_C$. This circuit can be used as an amplitude modulator. The $v_I$ input may be the carrier signal and the $v_C$ input may be the audio signal.

OTAs can also be used to design voltage-controlled filters and voltage-controlled oscillators.

# 9.7   OP-AMP CIRCUIT DESIGN

**Objective:** • Design several ideal op-amp circuits with given design specifications.

Up to this point, we have mainly been concerned with analyzing ideal op-amp circuits and designing a few basic op-amp circuits. In this section, we will design three specific op-amp circuits. We will assume that these circuits will be fabricated as integrated circuits so that we are not limited to standard resistor values.

### 9.7.1      Summing Op-Amp Circuit Design

In an inverting summing op-amp, each input is connected to the inverting terminal through a resistor. The summing op-amp can be designed such that the output is

$$v_O = -a_1 v_{I1} - a_2 v_{I2} + a_3 v_{I3} + a_4 v_{I4} \tag{9.87}$$

where the coefficients $a_i$ are all positive. In one design, we could apply voltages $v_{I3}$ and $v_{I4}$ to inverter amplifiers and use the summing op-amp considered previously. This design would require three such op-amps. Alternatively, we could use the results of Exercise TYU 9.8 to design a summing circuit that uses only one op-amp and is more versatile.

Consider the circuit shown in Figure 9.40. Resistor $R_C$ provides more versatility in the design. When we consider nonideal effects, such as bias currents, in op-circuits, in Chapter 14, we will impose a design constraint on the relationship between the resistors connected to the inverting and noninverting terminals. In this section, we will continue to use the ideal op-amp.

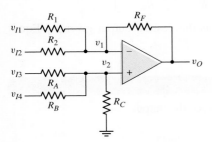

**Figure 9.40** Generalized op-amp summing amplifier

To determine the output voltage of our circuit, we use superposition. The inputs $v_{I1}$ and $v_{I2}$ produce the usual outputs, as follows:

$$v_O(v_{I1}) = -\frac{R_F}{R_1} v_{I1} \tag{9.88(a)}$$

and

$$v_O(v_{I2}) = -\frac{R_F}{R_2} v_{I2} \tag{9.88(b)}$$

We then determine the output due to $v_{I3}$, with all other inputs set equal to zero. We can write

$$v_2(v_{I3}) = \frac{R_B \| R_C}{R_A + R_B \| R_c} v_{I3} = v_1(v_{I3}) \tag{9.89}$$

Since $v_{I1} = v_{I2} = 0$, the voltage $v_2(v_{I3})$ is the input to a noninverting op-amp with $R_1$ and $R_2$ in parallel.

Then,

$$v_O(v_{I3}) = \left(1 + \frac{R_F}{R_1 \| R_2}\right) v_1(v_{I3}) = \left(1 + \frac{R_F}{R_1 \| R_2}\right)\left(\frac{R_B \| R_C}{R_A + R_B \| R_C}\right) v_{I3} \tag{9.90}$$

which can be rearranged as follows:

$$v_O(v_{I3}) = \left(1 + \frac{R_F}{R_N}\right)\left(\frac{R_P}{R_A}\right)v_{I3} \tag{9.91}$$

Here, we define

$$R_N = R_1 \| R_2 \tag{9.92(a)}$$

and

$$R_P = R_A \| R_B \| R_C \tag{9.92(b)}$$

The output voltage due to $v_{I4}$ is similarly determined and is

$$v_O(v_{I4}) = \left(1 + \frac{R_F}{R_N}\right)\left(\frac{R_P}{R_B}\right)v_{I4} \tag{9.93}$$

The total output voltage is then the sum of the individual terms, or

$$v_O = -\frac{R_F}{R_1}v_{I1} - \frac{R_F}{R_2}v_{I2} + \left(1 + \frac{R_F}{R_N}\right)\left[\frac{R_P}{R_A}v_{I3} + \frac{R_P}{R_B}v_{I4}\right] \tag{9.94}$$

This form of the output voltage is the same as the desired output given by Equation (9.87).

## DESIGN EXAMPLE 9.10

**Objective:** Design a summing op-amp to produce the output

$$v_O = -10v_{I1} - 4v_{I2} + 5v_{I3} + 2v_{I4}$$

The smallest resistor value allowable is $20\,\text{k}\Omega$. Consider the circuit in Figure 9.40.

**Solution:** First we determine the values of resistors $R_1$, $R_2$, and $R_F$, and then we can determine the noninverting terms. We know that

$$\frac{R_F}{R_1} = 10 \quad \text{and} \quad \frac{R_F}{R_2} = 4$$

Resistor $R_1$ will be the smallest value, so we can set $R_1 = 20\,\text{k}\Omega$. Then,

$$R_F = 200\,\text{k}\Omega \quad \text{and} \quad R_2 = 50\,\text{k}\Omega$$

The multiplying factor in the noninverting terms becomes

$$\left(1 + \frac{R_F}{R_1 \| R_2}\right) = \left(1 + \frac{200}{20 \| 50}\right) = 15$$

We then need

$$(15)\left(\frac{R_P}{R_A}\right) = 5 \quad \text{and} \quad (15)\left(\frac{R_P}{R_B}\right) = 2$$

If we take the ratio of these two expressions, we have

$$\frac{R_B}{R_A} = \frac{5}{2}$$

If we choose $R_A = 80\,\text{k}\Omega$, then $R_B = 200\,\text{k}\Omega$, $R_P = 26.67\,\text{k}\Omega$, and $R_C$ becomes $R_C = 50\,\text{k}\Omega$.

**Comment:** We could change the number of inputs to either the inverting or noninverting terminal, depending on the desired output versus input voltage response.

**Ex 9.10:** Consider the summing op-amp in Figure 9.40. Let $R_F = 80\,\text{k}\Omega$, $R_1 = 40\,\text{k}\Omega$, $R_2 = 20\,\text{k}\Omega$, $R_A = R_B = 50\,\text{k}\Omega$, and $R_C = 100\,\text{k}\Omega$. (a) Determine the output voltage in terms of the input voltages. (b) Determine $v_O$ for (i) $v_{I1} = 0.1$ V, $v_{I2} = 0.15$ V, $v_{I3} = 0.2$ V, $v_{I4} = 0.3$ V; and for (ii) $v_{I1} = -0.2$ V, $v_{I2} = 0.25$ V, $v_{I3} = -0.1$ V, $v_{I4} = 0.2$ V. (Ans. (a) $v_O = -2v_{I1} - 4v_{I2} + 2.8v_{I3} + 2.8v_{I4}$; (b) (i) $v_O = 0.6$ V, (ii) $v_O = -0.32$ V)

### 9.7.2    Reference Voltage Source Design

In Chapter 2, we discussed the use of Zener diodes to provide a constant or reference voltage source. A limitation, however, was that the reference voltage could never be greater than the Zener voltage. Now, we can combine a Zener diode with an op-amp to provide more flexibility in the design of reference voltage sources.

Consider the circuit shown in Figure 9.41. Voltage source $V_S$ and resistor $R_S$ bias the Zener diode in the breakdown region. The op-amp is then used as a noninverting amplifier. The output voltage is

$$V_O = \left(1 + \frac{R_2}{R_1}\right)V_Z \tag{9.95}$$

The output current to the load circuit is supplied by the op-amp. A change in the load current will not produce a change in the Zener diode current; consequently, voltage regulation is much improved compared to the simple Zener diode voltage source previously considered.

Since the incremental Zener resistance is not zero, the Zener diode voltage is a slight function of the diode current. The circuit shown in Figure 9.42 is less affected by variations in $V_S$, since $V_S$ is used only to start up the circuit. The Zener diode begins to conduct when

$$\frac{R_4}{R_3 + R_4}V_S > V_Z + V_D \cong V_Z + 0.7 \tag{9.96}$$

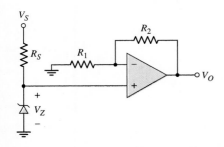

**Figure 9.41** Simple op-amp voltage reference circuit

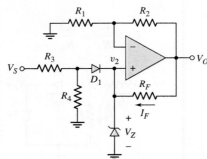

**Figure 9.42** Op-amp voltage reference circuit

At this specific voltage, we have

$$V_O = \left(1 + \frac{R_2}{R_1}\right)V_Z \tag{9.97}$$

and

$$I_F = \frac{V_O - V_Z}{R_F} = \frac{R_2 V_Z}{R_1 R_F} \tag{9.98}$$

If $V_S$ decreases and diode $D_1$ becomes reverse biased, the Zener diode continues to conduct; the Zener diode current is then constant. However, if diode $D_1$ is conducting, the circuit can be designed such that variations in Zener diode current will be small.

## DESIGN EXAMPLE 9.11

**Objective:** Design a voltage reference source with an output of 10.0 V. Use a Zener diode with a breakdown voltage of 5.6 V. Assume the voltage regulation will be within specifications if the Zener diode is biased between 1–1.2 mA.

**Solution:** Consider the circuit shown in Figure 9.42. For this example, we need

$$\frac{V_O}{V_Z} = \left(1 + \frac{R_2}{R_1}\right) = \frac{10.0}{5.6}$$

Therefore,

$$\frac{R_2}{R_1} = 0.786$$

We know that

$$I_F = \frac{V_O - V_Z}{R_F}$$

If we set $I_F$ equal to the minimum bias current, we have

$$1\,\text{mA} = \frac{10 - 5.6}{R_F}$$

which means that $R_F = 4.4\,\text{k}\Omega$. If we choose $R_2 = 30\,\text{k}\Omega$, then $R_1 = 38.17\,\text{k}\Omega$.

Resistors $R_3$ and $R_4$ can be determined from Figure 9.43. The maximum Zener current supplied by $V_S$, $R_3$, and $R_4$ should be no more than 0.2 mA. We set the current through $D_1$ equal to 0.2 mA, for $V_S = 10\,\text{V}$. We then have

$$V_2' = V_Z + 0.7 = 5.6 + 0.7 = 6.3\,\text{V}$$

Also,

$$I_4 = \frac{V_2'}{R_4} = \frac{6.3}{R_4}$$

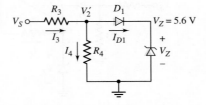

**Figure 9.43** Input circuit of the op-amp voltage reference circuit

and

$$I_3 = \frac{V_S - V_2'}{R_3} = \frac{10 - 6.3}{R_3} = \frac{3.7}{R_3}$$

If we set $I_4 = 0.2\,\text{mA}$, then

$$I_3 = 0.4\,\text{mA} \qquad R_3 = 9.25\,\text{k}\Omega \qquad R_4 = 31.5\,\text{k}\Omega$$

**Comment:** Voltage $V_S$ is used as a start-up source. Once the Zener diode is biased in breakdown, the output will be maintained at 10.0 V, even if $V_S$ is reduced to zero.

EXERCISE PROBLEM

**Ex 9.11:** Consider the op-amp voltage reference circuit in Figure 9.42 with parameters given in Example 9.11. Initially set $V_S = 10\,\text{V}$ and then plot, using PSpice, $v_O$ and $I_F$ versus $V_S$ as $V_S$ decreases from 10 to 0 V. Bias the op-amp at $\pm 15\,\text{V}$.

### 9.7.3   Difference Amplifier and Bridge Circuit Design

A transducer is a device that transforms one form of energy into another form. One type of transducer uses nonelectrical inputs to produce electrical outputs. For example, a microphone converts acoustical energy into electrical energy. A pressure transducer is a device in which, for example, a resistance is a function of pressure, so that pressure can be converted to an electrical signal. Often, the output characteristics of these transducers are measured with a bridge circuit.

Figure 9.44 shows a bridge circuit. Resistance $R_3$ represents the transducer, and parameter $\delta$ is the deviation of $R_3$ from $R_2$ due to the input response of the transducer. The output voltage $v_{O1}$ is a measure of $\delta$. If $v_{O1}$ is an open-circuit voltage, then

$$v_{O1} = \left[ \frac{R_2(1 + \delta)}{R_2(1 + \delta) + R_1} - \frac{R_2}{R_1 + R_2} \right] V^+ \tag{9.99}$$

which reduces to

$$v_{O1} = \delta \left( \frac{R_1 \| R_2}{R_1 + R_2} \right) V^+ \tag{9.100}$$

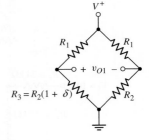

**Figure 9.44** Bridge circuit

Since neither side of voltage $v_{O1}$ is at ground potential, we must connect $v_{O1}$ to an instrumentation amplifier. In addition, $v_{O1}$ is directly proportional to supply voltage $V^+$; therefore, this bias should be a well-defined voltage reference.

DESIGN EXAMPLE **9.12**

**Objective:** Design an amplifier system that will produce an output voltage of $\pm 5\,\text{V}$ when the resistance $R_3$ deviates by $\pm 1\%$ from the value of $R_2$. This would occur, for example, in a system where $R_3$ is a thermistor whose resistance is given by

$$R_3 = 200 \left[ 1 + \frac{(0.040)(T - 300)}{300} \right] \text{k}\Omega$$

where $T$ is the absolute temperature. For $R_3$ to vary by $\pm 1\%$ means the temperature is in the range $225 \leq T \leq 375\,\text{K}$.

Consider biasing the bridge circuit at $V^+ = 7.5\,\text{V}$ using a 5.6 V Zener diode. Assume $\pm 10\,\text{V}$ is available for biasing the op-amp and reference voltage source, and that $R_1 = R_2 = 200\,\text{k}\Omega$.

**Solution:** With $R_1 = R_2$, from Equation (9.100), we have

$$v_{O1} = \left(\frac{\delta}{4}\right) V^+$$

For $V^+ = 7.5\,\text{V}$ and $\delta = 0.01$, the maximum output of the bridge circuit is $v_{O1} = 0.01875\,\text{V}$. If the output of the amplifier system is to be $+5\,\text{V}$, the gain of the instrumentation amplifier must be $5/0.01875 = 266.7$. Consider the instrumentation amplifier shown in Figure 9.26. The output voltage is given by Equation (9.67), which can be written

$$\frac{v_O}{v_{O1}} = \frac{R'_4}{R'_3}\left(1 + \frac{2R'_2}{R'_1}\right) = 266.7$$

We would like the ratios $R'_4/R'_3$ and $R'_2/R'_1$ to be the same order of magnitude. If we let $R'_3 = 15.0\,\text{k}\Omega$ and $R'_4 = 187.0\,\text{k}\Omega$, then $R'_4/R'_3 = 12.467$ and $R'_2/R'_1 = 10.195$. If we set $R'_2 = 200.0\,\text{k}\Omega$, then $R'_1 = 19.62\,\text{k}\Omega$.

Resistance $R'_1$ can be a combination of a fixed resistance in series with a potentiometer, to permit adjustment of the gain.

**Comment:** The complete design of this instrumentation amplifier is shown in Figure 9.45. Correlation of the reference voltage source design is left as an exercise.

**Design Pointer:** The design of fairly sophisticated op-amp circuits is quite straightforward when the ideal op-amp parameters are used.

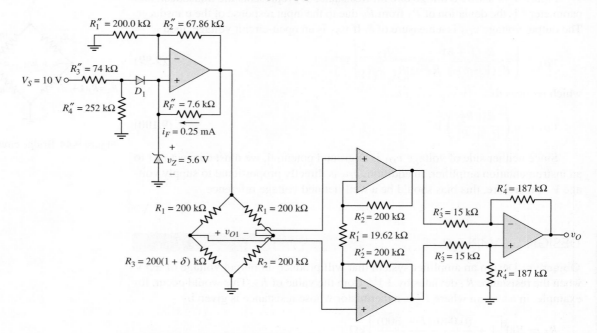

**Figure 9.45** Complete amplifier system

## Test Your Understanding

**TYU 9.14** Consider the bridge circuit in Figure 9.46. The resistance is $R = 20\,\text{k}\Omega$ and the variable resistance $\Delta R$ ranges between $-100\,\Omega$ and $+100\,\Omega$. The circuit is biased at $V^+ = 5\,\text{V}$. (a) Find $v_{O1}$ as a function of $\Delta R$. (b) Design an amplifier system such that the output is $-3\,\text{V}$ when $\Delta R = -100\,\Omega$. (Ans. (a) $|v_{O1}| = 2.5 \times 10^{-4}(\Delta R)$. (b) For an instrumentation amplifier, let $R_4/R_3 = 10$ and $R_2/R_1 = 5.5$)

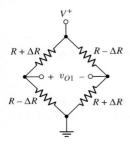

**Figure 9.46**  Figure for Exercise TYU 9.14

**Figure 9.47**  Figure for Exercise TYU 9.15

**TYU 9.15** The resistance $R$ in the bridge circuit in Figure 9.47 is $50\,\text{k}\Omega$. The circuit is biased at $V^+ = 3\,\text{V}$. (a) Find $v_{O1}$ as a function of $\delta$. (b) Design an amplifier system such that the output varies between $+3\,\text{V}$ and $-3\,\text{V}$ as the parameter $\delta$ varies between $+0.025$ and $-0.025$. (Ans. (a) $v_{O1} \cong 0.75\delta$. (b) For an instrumentation amplifier, let $R_4/R_3 = 10$ and $R_2/R_1 = 7.5$)

## 9.8   DESIGN APPLICATION: ELECTRONIC THERMOMETER WITH AN INSTRUMENTATION AMPLIFIER

**Objective:** • Design an electronic thermometer with an instrumentation amplifier to provide the necessary amplification.

**Specifications:** The temperature range to be measured is 0 to 100 °F. The output voltage is to be in the range of 0 to 5 V with 0 V corresponding to 0 °F and 5 V corresponding to 100 °F.

**Design Approach:** In Chapter 1, we began a design of an electronic thermometer using the temperature characteristics of a pn junction diode. Here, we expand on that design.

Figure 9.48(a) shows a circuit with two diodes, each biased with a constant current source. Figure 9.48(b) shows the same circuit, but with the constant current sources implemented with transistor circuits. The current source circuits were briefly

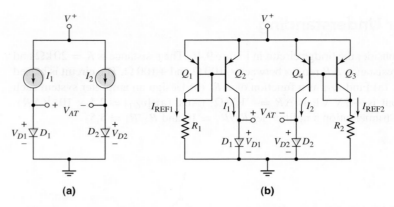

**Figure 9.48** (a) Two diodes biased with constant current sources. (b) The same circuit with the constant current sources implemented with transistor circuits.

described and analyzed in Chapter 5. The two diodes, $D_1$ and $D_2$, are assumed to be matched or identical devices. We also assume that all transistors are matched. Neglecting base currents, we have $I_1 = I_{\text{REF1}}$ and $I_2 = I_{\text{REF2}}$.

**Choices:** Ideal matched silicon diodes and bipolar transistors are available. In addition, ideal op-amps are available.

**Solution (Diodes):** From Chapter 1, we can write the voltage drops across each diode as

$$V_{D1} = V_T \ln\left(\frac{I_1}{I_S}\right) \tag{9.101(a)}$$

and

$$V_{D2} = V_T \ln\left(\frac{I_2}{I_S}\right) \tag{9.101(b)}$$

We may note that, since the two diodes are matched, the reverse-saturation current, $I_S$, is the same in the two expressions.

The output voltage is defined as the difference between the voltages across the two diodes, or

$$V_{AT} = V_{D1} - V_{D2} = V_T\left[\ln\left(\frac{I_1}{I_S}\right) - \ln\left(\frac{I_2}{I_S}\right)\right] \tag{9.102(a)}$$

or

$$V_{AT} = V_T \ln\left(\frac{I_1}{I_2}\right) = \frac{kT}{e}\ln\left(\frac{I_{\text{REF1}}}{I_{\text{REF2}}}\right) \tag{9.102(b)}$$

The output voltage, $V_{AT}$, is now directly proportional to absolute temperature $T$, hence the subscript $AT$.

If we let $I_{\text{REF1}}/I_{\text{REF2}} = 5$, then Equation (9.102(b)) can be written as

$$V_{AT} = (0.0259)\left(\frac{T}{300}\right)\ln(5) = (1.3895 \times 10^{-4})T \tag{9.103}$$

Letting $I_{REF1}/I_{REF2} > 0$ provides a small amount of gain. Converting absolute temperature to degrees Celsius and then to degrees Fahrenheit, we find

$$T = T_C + 273.15 \qquad \textbf{(9.104)}$$

and

$$T_F = 32 + \frac{9}{5}T_C \Rightarrow T_C = (T_F - 32)\left(\frac{5}{9}\right) \qquad \textbf{(9.105)}$$

where $T_C$ and $T_F$ are temperatures in degrees Celsius and degrees Fahrenheit, respectively.

Combining Equations (9.104) and (9.105), we obtain

$$T = (T_F - 32)\left(\frac{5}{9}\right) + 273.15 = \frac{5}{9}T_F + 255.37 \qquad \textbf{(9.106)}$$

The output voltage from Equation (9.103) can now be written as

$$V_{AT} = (1.3895 \times 10^{-4})\left(\frac{5}{9}T_F + 255.37\right)$$
$$= (7.719 \times 10^{-5})T_F + 3.5484 \times 10^{-2} \qquad \textbf{(9.107)}$$

**Solution (Instrumentation Amplifier):** Since neither terminal of the output voltage is at ground potential, we can apply this voltage to an instrumentation amplifier to obtain a voltage gain. The output of the instrumentation amplifier will be applied to a summing amplifier in addition to an offset voltage. The objective of the design is to obtain an output voltage of zero volts at $T_F = 0$ and an output voltage of 5 V at $T_F = 100\,°F$.

If the gain of the instrumentation amplifier is $A = -129.55$, then the output of the instrumentation amplifier is as follows:

| $T_F$ | $V_{AT}$ | $V_{O1}$ |
|-------|----------|----------|
| 0 | 0.035484 | −4.5970 |
| 100 | 0.043203 | −5.5970 |

**Solution (Output Stage):** The offset voltage can be generated by using the noninverting op-amp circuit with a Zener diode, as shown in Figure 9.49. If we use a Zener diode with a breakdown voltage of 3.60 V and if we set $R_3/R_4 = 0.277$, then the output voltage is $V_{O2} = +4.597$ V. Applying the output voltage of the instrumentation amplifier, $V_{O1}$, and the offset voltage, $V_{O2}$, to a summing amplifier with a gain of $-5$ as shown in Figure 9.49, we achieve the desired specifications. That is $V_O = 0$ at $T_F = 0$ and $V_O = 5$ V at $T_F = 100\,°F$.

**Comment:** The primary advantage of this system is that the output voltage is a linear function of temperature.

In Chapter 16, we can apply the analog output voltage $V_O$ to an A/D converter and use a seven-segment display so that the output signal is actually displayed in terms of degrees Fahrenheit.

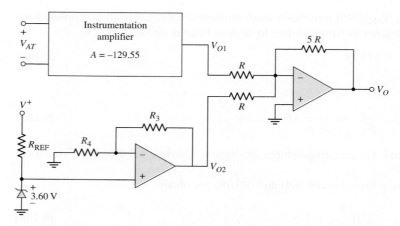

**Figure 9.49** The output voltage $V_{AT}$ applied to an instrumentation amplifier, an offset voltage generated by a Zener diode and a noninverting amplifier, and the final output voltage obtained from a summing amplifier

 **9.9 SUMMARY**

- In this chapter, we considered the ideal operational amplifier (op-amp) and various op-amp applications. The op-amp is a three-terminal device (three signal terminals) that ideally amplifies only the difference between two input signals. The op-amp, then, is a high-gain differential amplifier.
- The ideal op-amp model has infinite input impedance (zero input bias currents), infinite open-loop differential voltage gain (zero voltage between the two input terminals), and zero output impedance.
- Two basic op-amp circuits are the inverting amplifier and the noninverting amplifier. For an ideal op-amp, the voltage gain of these circuits is just a function of the ratio of resistors.
- Other amplifier configurations considered were the summing amplifier, voltage follower, current-to-voltage converter, and voltage-to-current converter.
- A versatile circuit is the instrumentation amplifier. The input resistance is essentially infinite and the amplifier gain can be varied by changing a single resistor value.
- If a capacitor is included as a feedback element, the output voltage is the integral of the input voltage. If a capacitor is included as an input element, the output voltage is the derivative of the input voltage. Nonlinear feedback elements, such as diodes or transistors, produce nonlinear transfer functions such as a logarithmic function.
- As an application, an electronic thermometer in conjunction with an instrumentation amplifier was designed to yield a given amplification.

 **CHECKPOINT**

After studying this chapter, the reader should have the ability to:

✓ Describe the characteristics of an ideal op-amp.
✓ Analyze various op-amp circuits using the ideal op-amp model.
✓ Analyze various op-amp circuits, taking into account the finite gain of the op-amp.

✓ Understand and describe the characteristics and operation of various op-amp circuits, such as the summing amplifier, difference amplifier, and instrumentation amplifier.
✓ Design various op-amp circuits to perform specific functions using the ideal op-amp model.
✓ Analyze and design op-amp circuits using nonlinear feedback elements.

## REVIEW QUESTIONS

1. Describe the ideal op-amp model and describe the implications of this ideal model in terms of input currents and voltages.
2. Describe the op-amp model including the effect of a finite op-amp voltage gain.
3. Describe the operation and characteristics of the ideal inverting amplifier.
4. What is the concept of virtual ground?
5. What is the significance of a zero output resistance?
6. When a finite op-amp gain is taken into account, is the magnitude of the resulting amplifier voltage gain less than or greater than the ideal value?
7. Describe the operation and characteristics of the ideal summing amplifier.
8. Describe the operation and characteristics of the ideal noninverting amplifier.
9. Describe the voltage follower. What are the advantages of using this circuit.
10. What is the input resistance of an ideal current-to-voltage converter?
11. Describe the operation and characteristics of a difference amplifier.
12. Describe the operation and characteristics of an instrumentation amplifier.
13. Describe the operation and characteristics of an op-amp circuit using a capacitor as a feedback element.
14. Describe the operation and characteristics of an op-amp circuit using a diode as a feedback element.

## PROBLEMS

### Section 9.1 The Operational Amplifier

9.1 Assume an op-amp is ideal, except for having a finite open-loop differential gain. Measurements were made with the op-amp in the open-loop mode. Determine the open-loop gain and complete the following table, which shows the results of those measurements.

| $v_1$ | $v_2$ | $v_O$ |
|---|---|---|
| −1 mV | +1 mV | 1 V |
| +1 mV | | 1 V |
| | 1 V | 5 V |
| −1 V | −1 V | |
| −0.5 V | | −3 V |

9.2 The op-amp in the circuit shown in Figure P9.2 is ideal except it has a finite open-loop gain. (a) If $A_{od} = 10^4$ and $v_O = -2\,\text{V}$, determine $v_I$. (b) If $v_I = 2\,\text{V}$ and $v_O = 1\,\text{V}$, determine $A_{od}$.

9.3 An op-amp is in an open-loop configuration as shown in Figure 9.2. (a) If $v_1 = 2.0010\,\text{V}$, $v_2 = 2.000\,\text{V}$, and $A_{od} = 5 \times 10^3$, determine $v_O$.

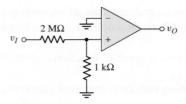

Figure P9.2

(b) If $v_2 = 3.0025$ V, $v_O = -3.00$ V, and $A_{od} = 2 \times 10^4$, what is $v_1$? (c) If $v_1 = -0.01$ mV, $v_2 = +0.01$ mV, and $v_O = 1.80$ V, determine $A_{od}$.

9.4    Consider the equivalent circuit of the op-amp shown in Figure 9.7(a). Assume terminal $v_1$ is grounded and the input to terminal $v_2$ is from a transducer that can be represented by a 0.8 mV voltage source in series with a 25 k$\Omega$ resistance. What is the minimum input resistance $R_i$ such that the minimum differential input voltage is $v_{id} = 0.790$ mV?

### Section 9.2 Inverting Amplifier

9.5    Consider the ideal inverting op-amp circuit shown in Figure 9.8. Determine the voltage gain $A_v = v_O/v_I$ for (a) $R_2 = 200$ k$\Omega$, $R_1 = 20$ k$\Omega$; (b) $R_2 = 120$ k$\Omega$, $R_1 = 40$ k$\Omega$; and (c) $R_2 = 40$ k$\Omega$, $R_1 = 40$ k$\Omega$.

9.6    Assume the op-amps in Figure P9.6 are ideal. Find the voltage gain $A_v = v_O/v_I$ and the input resistance $R_i$ of each circuit.

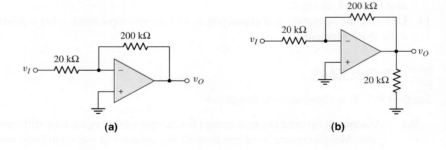

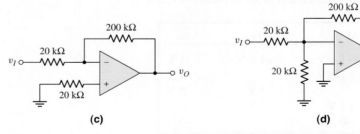

Figure P9.6

9.7    Consider an ideal inverting op-amp with $R_2 = 100$ k$\Omega$ and $R_1 = 10$ k$\Omega$. (a) Determine the ideal voltage gain and input resistance $R_i$. (b) Repeat part (a) for a second 100 k$\Omega$ resistor connected in parallel with $R_2$. (c) Repeat part (a) for a second 10 k$\Omega$ resistance connected in series with $R_1$.

D9.8   (a) Design an inverting op-amp circuit with a closed-loop voltage gain of $A_v = v_O/v_I = -12$. The current in each resistor is to be no larger than $20\,\mu\text{A}$ when the output voltage is $-4.0\,\text{V}$. (b) Using the results of part (a), determine $v_I$ and the current in each resistor when $v_O = +1.5\,\text{V}$.

9.9   Consider an ideal op-amp used in an inverting configuration as shown in Figure 9.8. Determine the closed-loop voltage gain for the following resistor values.
(a) $R_1 = 20\,\text{k}\Omega$, $R_2 = 200\,\text{k}\Omega$
(b) $R_1 = 20\,\text{k}\Omega$, $R_2 = 20\,\text{k}\Omega$
(c) $R_1 = 20\,\text{k}\Omega$, $R_2 = 4\,\text{k}\Omega$
(d) $R_1 = 50\,\text{k}\Omega$, $R_2 = 500\,\text{k}\Omega$
(e) $R_1 = 50\,\text{k}\Omega$, $R_2 = 100\,\text{k}\Omega$
(f) $R_1 = 50\,\text{k}\Omega$, $R_2 = 50\,\text{k}\Omega$

9.10   Consider the inverting amplifier shown in Figure 9.8. Assume the op-amp is ideal. Determine the resistor values $R_1$ and $R_2$ to produce a closed-loop voltage gain of (a) $-3.0$, (b) $-8.0$, (c) $-20$, and (d) $-0.50$. In each case the largest resistor is to be limited to $200\,\text{k}\Omega$.

D9.11   (a) Design an inverting op-amp circuit with a closed-loop voltage gain of $A_v = -6.5$. When in the input voltage is $v_I = -0.25\,\text{V}$, the magnitude of the currents is to be $50\,\mu\text{A}$. Determine $R_1$ and $R_2$. (b) Using the results of part (a), find $v_I$, $i_1$, and $i_2$ when $v_O = -4.0\,\text{V}$.

D9.12   (a) Design an inverting op-amp circuit such that the closed-loop voltage gain is $A_v = -20$ and the smallest resistor value is $25\,\text{k}\Omega$. (b) Repeat part (a) for the case when the largest resistor value is $1\,\text{M}\Omega$. (c) Determine $i_1$ in both parts (a) and (b) when the input voltage is $v_I = -0.20\,\text{V}$.

9.13   (a) In an inverting op-amp circuit, the nominal resistance values are $R_2 = 300\,\text{k}\Omega$ and $R_1 = 15\,\text{k}\Omega$. The tolerance of each resistor is $\pm5\%$, which means that each resistance can deviate from its nominal value by $\pm5\%$. What is the maximum deviation in the voltage gain from its nominal value? (b) Repeat part (a) if the resistor tolerance is reduced to $\pm1\%$.

9.14   (a) The input to the circuit shown in Figure P9.14 is $v_I = -0.20\,\text{V}$. (i) What is $v_O$? (ii) Determine $i_2$, $i_O$, and $i_L$. (b) Repeat part (a) for $v_I = +0.05\,\text{V}$. (c) Repeat part (a) for $v_I = 8\sin\omega t$ mV.

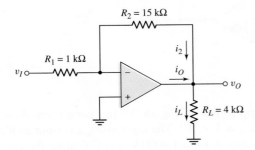

**Figure P9.14**

D9.15   Design an inverting amplifier to provide a nominal closed-loop voltage gain of $A_v = -30$. The maximum input voltage signal is 25 mV with a source resistance in the range $1\,\text{k}\Omega \le R_S \le 2\,\text{k}\Omega$. The variable source resistance should introduce no more than a 5 percent difference in the gain factor. What is the range in output voltage?

9.16    The parameters of the two inverting op-amp circuits connected in cascade in Figure P9.16 are $R_1 = 10\,\text{k}\Omega$, $R_2 = 80\,\text{k}\Omega$, $R_3 = 20\,\text{k}\Omega$, and $R_4 = 100\,\text{k}\Omega$. For $v_I = -0.15\,\text{V}$, determine $v_{O1}$, $v_O$, $i_1$, $i_2$, $i_3$, and $i_4$. Also determine the current into or out of the output terminal of each op-amp.

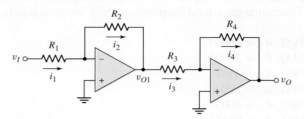

Figure P9.16

D9.17   Design the cascade inverting op-amp circuit in Figure P9.16 such that the overall closed-loop voltage gain is $A_v = v_O/v_I = 100$ and such that the maximum current in any resistor is limited to $50\,\mu\text{A}$ when $v_I = 50\,\text{mV}$, but under the condition that the minimum resistance is $10\,\text{k}\Omega$.

D9.18   Design an amplifier system with three inverting op-amps circuits in cascade such that the overall closed-loop voltage gain is $A_v = v_O/v_I = -300$. The maximum resistance is limited to $200\,\text{k}\Omega$ and the minimum resistance is limited to $20\,\text{k}\Omega$. In addition, the maximum current in any resistor is to be limited to $60\,\mu\text{A}$ when $v_O = 6\,\text{V}$.

9.19    Consider the circuit shown in Figure P9.19. (a) Determine the ideal output voltage $v_O$ if $v_I = -0.40\,\text{V}$. (b) Determine the actual output voltage if the open-loop gain of the op-amp is $A_{od} = 5 \times 10^3$. (c) Determine the required value of $A_{od}$ in order that the actual voltage gain be within 0.2 percent of the ideal value.

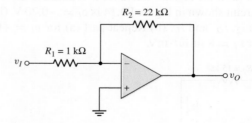

Figure P9.19

9.20    The inverting op-amp shown in Figure 9.9 has parameters $R_1 = 25\,\text{k}\Omega$, $R_2 = 100\,\text{k}\Omega$, and $A_{od} = 5 \times 10^3$. The input voltage is from an ideal voltage source whose value is $v_I = 1.0000\,\text{V}$. (a) Calculate the closed-loop voltage gain. (b) Determine the actual output voltage. (c) What is the percentage difference between the actual output voltage and the ideal output voltage. (d) What is the voltage at the inverting terminal of the op-amp?

9.21    (a) An op-amp with an open-loop gain of $A_{od} = 7 \times 10^3$ is to be used in an inverting op-amp circuit. Let $R_2 = 100\,\text{k}\Omega$ and $R_1 = 10\,\text{k}\Omega$. If the output voltage is $v_O = 7\,\text{V}$, determine the input voltage and the voltage at the inverting terminal of the op-amp. (b) If the output voltage is $v_O = -5\,\text{V}$ and the voltage at the inverting terminal of the op-amp is 0.2 mV, what is the input voltage and the value of $A_{od}$?

9.22 (a) For the ideal inverting op-amp circuit with T-network, shown in Figure 9.12, the circuit parameters are $R_1 = 10\,\text{k}\Omega$, $R_2 = R_3 = 50\,\text{k}\Omega$, and $R_4 = 5\,\text{k}\Omega$. Determine the closed-loop voltage gain. (b) Determine a new value of $R_4$ to produce a voltage gain of (i) $A_v = -100$ and (ii) $A_v = -150$.

D9.23 Consider the ideal inverting op-amp circuit with T-network in Figure 9.12. (a) Design the circuit such that the input resistance is $500\,\text{k}\Omega$ and the gain is $A_v = -80$. Do not use resistor values greater than $500\,\text{k}\Omega$. (b) For the design in part (a), determine the current in each resistor if $v_I = -0.05\,\text{V}$.

9.24 An ideal inverting op-amp circuit is to be designed with a closed-loop voltage gain of $A_v = -1000$. The largest resistor value to be used is $500\,\text{k}\Omega$. (a) If the simple two-resistor design shown in Figure 9.8 is used, what is the input resistance? (b) If the T-network design shown in Figure 9.12 with $R_3 = 500\,\text{k}\Omega$ and $R_2 = R_4 = 250\,\text{k}\Omega$ is used, what is the input resistance?

9.25 For the op-amp circuit shown in Figure P9.25, determine the gain $A_v = v_O/v_I$. Compare this result to the gain of the circuit shown in Figure 9.12, assuming all resistor values are equal.

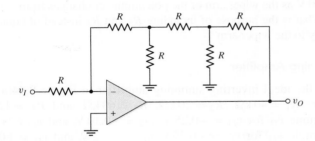

**Figure P9.25**

9.26 The inverting op-amp circuit in Figure 9.9 has parameters $R_1 = 20\,\text{k}\Omega$, $R_2 = 200\,\text{k}\Omega$, and $A_{od} = 5 \times 10^4$. The output voltage is $v_O = -4.80\,\text{V}$. (a) Determine the closed-loop voltage gain. (b) Find the input voltage. (c) Determine the voltage at the inverting terminal of the op-amp. (d) Using $v_I$ from part (b), find the percent error in output voltage compared to the ideal value.

9.27 (a) Consider the op-amp circuit in Figure P9.27. The open-loop gain of the op-amp is $A_{od} = 2.5 \times 10^3$. (i) Determine $v_O$ when $v_I = -0.80\,\text{V}$. (ii) What is the percent error in output voltage compared to the ideal value? (b) Repeat part (a) for $A_{od} = 200$.

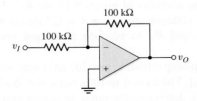

**Figure P9.27**

*9.28    The circuit in Figure P9.28 is similar to the inverting amplifier except the resistor $R_3$ has been added. (a) Derive the expression for $v_O$ in terms of $v_I$ and the resistors. (b) Derive the expression for $i_3$ in terms of $v_I$ and the resistors.

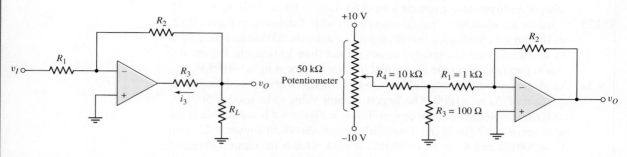

Figure P9.28                                    Figure P9.29

*D9.29   Design the amplifier in Figure P9.29 such that the output voltage varies between $\pm 10$ V as the wiper arm of the potentiometer changes from $-10$ V to $+10$ V. What is the purpose of including $R_3$ and $R_4$ instead of connecting $R_1$ directly to the wiper arm?

### Section 9.3 Summing Amplifier

9.30    Consider the ideal inverting summing amplifier in Figure 9.14(a) with parameters $R_1 = 40\,\text{k}\Omega$, $R_2 = 20\,\text{k}\Omega$, $R_3 = 60\,\text{k}\Omega$, and $R_F = 120\,\text{k}\Omega$. (a) Determine $v_O$ for $v_{I1} = -0.25$ V, $v_{I2} = +0.10$ V, and $v_{I3} = +1.5$ V. (b) Determine $v_{I1}$ for $v_{I2} = +0.25$ V, $v_{I3} = -1.2$ V, and $v_O = +0.50$ V.

D9.31   (a) Design an ideal inverting summing amplifier to produce an output voltage of $v_O = -2.5(1.2v_{I1} + 2.5v_{I2} + 0.25v_{I3})$. Design the circuit to produce the largest possible input resistance, assuming the largest resistance in the circuit is limited to 400 k$\Omega$. (b) Using the results of part (a), determine the current in the feedback resistor for $v_{I1} = -1.0$ V, $v_{I2} = +0.25$ V, and $v_{I3} = +2$ V.

D9.32   Design an ideal inverting summing amplifier to produce an output voltage of $v_O = -2\,(v_{I1} + 3v_{I2})$. The input voltages are limited to the ranges of $-1 \le v_{I1} \le +1$ V and $-0.5 \le v_{I2} \le +0.2$ V. The current in any resistor is to be limited to a maximum of 80 $\mu$A.

9.33    Consider the summing amplifier in Figure 9.14 with $R_F = 10\,\text{k}\Omega$, $R_1 = 1\,\text{k}\Omega$, $R_2 = 5\,\text{k}\Omega$, and $R_3 = 10\,\text{k}\Omega$. If $v_{I1}$ is a 1 kHz sine wave with an rms value of 50 mV, if $v_{I2}$ is a 100 Hz square wave with an amplitude of $\pm 1$ V, and if $v_{I3} = 0$, sketch the output voltage $v_O$.

9.34    The parameters for the summing amplifier in Figure 9.14 are $R_F = 100\,\text{k}\Omega$ and $R_3 = \infty$. The two input voltages are $v_{I1} = 4 + 125 \sin \omega t$ mV and $v_{I2} = -6$ mV. Determine $R_1$ and $R_2$ to produce an output voltage of $v_O = -0.5 \sin \omega t$ V.

D9.35   (a) Design an ideal summing op-amp circuit to provide an output voltage of $v_O = -2\,[(v_{I1}/4) + 2v_{I2} + v_{I3}]$. The largest resistor value is to be 250 k$\Omega$. (b) Using the results of part (a), determine the range in output voltage and the maximum current in $R_F$ if the input voltages are in the ranges $-2 \le v_{I1} \le +2$ V , $0 \le v_{I2} \le 0.5$ V, and $-1 \le v_{I3} \le 0$ V.

**D9.36** An ideal three-input inverting summing amplifier is to be designed. The input voltages are $v_{I1} = 2 + 2 \sin \omega t$ V, $v_{I2} = 0.5 \sin \omega t$ V, and $v_{I3} = -4$ V. The desired output voltage is $v_O = -6 \sin \omega t$ V. The maximum current in any resistor is to be limited to $120 \, \mu A$.

**9.37** A summing amplifier can be used as a digital-to-analog converter (DAC). An example of a 4-bit DAC is shown in Figure P9.37. When switch $S_3$ is connected to the $-5$ V supply, the most significant bit is $a_3 = 1$; when $S_3$ is connected to ground, the most significant bit is $a_3 = 0$. The same condition applies to the other switches $S_2$, $S_1$, and $S_o$, corresponding to bits $a_2$, $a_1$, and $a_o$, where $a_o$ is the least significant bit. (a) Show that the output voltage is given by

$$v_O = \frac{R_F}{10} \left[ \frac{a_3}{2} + \frac{a_2}{4} + \frac{a_1}{8} + \frac{a_o}{16} \right] (5)$$

where $R_F$ is in k$\Omega$. (b) Find the value of $R_F$ such that $v_O = 2.5$ V when the digital input is $a_3 a_2 a_1 a_o = 1000$. (c) Using the results of part (b), find $v_o$ for: (i) $a_3 a_2 a_1 a_o = 0001$, and (ii) $a_3 a_2 a_1 a_o = 1111$.

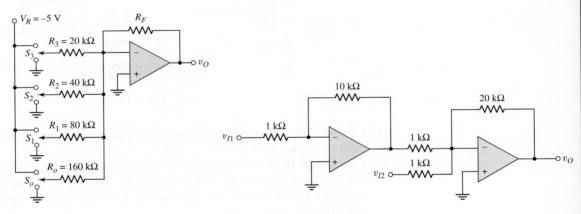

**Figure P9.37**                    **Figure P9.38**

**9.38** Consider the circuit in Figure P9.38. (a) Derive the expression for the output voltage $v_O$ in terms of $v_{I1}$ and $v_{I2}$. (b) Determine $v_O$ for $v_{I1} = +5$ mV and $v_{I2} = -25 - 50 \sin \omega t$ mV. (c) Determine the peak currents in the 10 k$\Omega$ and 20 k$\Omega$ resistors.

**\*9.39** Consider the summing amplifier in Figure 9.14 (a). Assume the op-amp has a finite open-loop differential gain $A_{od}$. Using the principle of superposition, show that the output voltage is given by

$$v_O = \frac{-1}{1 + \dfrac{(1 + R_F / R_P)}{A_{od}}} \left[ \frac{R_F}{R_1} v_{I1} + \frac{R_F}{R_2} v_{I2} + \frac{R_F}{R_3} v_{I3} \right]$$

where $R_P = R_1 \| R_2 \| R_3$. Demonstrate how the expression will change if more or fewer inputs are included.

## Section 9.4 Noninverting Amplifier

9.40   Consider the ideal noninverting op-amp circuit in Figure 9.15. Determine the closed-loop gain for the following circuit parameters: (a) $R_1 = 15\,\text{k}\Omega$, $R_2 = 150\,\text{k}\Omega$; (b) $R_1 = 50\,\text{k}\Omega$, $R_2 = 150\,\text{k}\Omega$; (c) $R_1 = 50\,\text{k}\Omega$, $R_2 = 20\,\text{k}\Omega$; and (d) $R_1 = 20\,\text{k}\Omega$, $R_2 = 20\,\text{k}\Omega$.

D9.41  (a) Design an ideal noninverting op-amp circuit with the configuration shown in Figure 9.15 to have a closed-loop gain of $A_v = 15$. When $v_O = -7.5$ V, the current in any resistor is to be limited to a maximum value of $120\,\mu\text{A}$. (b) Using the results of part (a), determine the output voltage $v_O$ and the currents in the resistors for $v_I = 0.25$ V.

9.42   Consider the noninverting amplifier in Figure 9.15. Assume the op-amp is ideal. Determine the resistor values $R_1$ and $R_2$ to produce a closed-loop gain of (a) 3, (b) 9, (c) 30, and (d) 1.0. The maximum resistor value is to be limited to $290\,\text{k}\Omega$.

9.43   For the circuit in Figure P9.43, the input voltage is $v_I = 5$ V. (a) If $v_O = 2.5$ V, determine the finite open-loop differential gain of the op-amp. (b) If the open-loop differential gain of the op-amp is 5000, determine $v_O$.

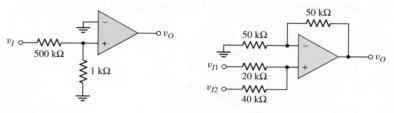

Figure P9.43                  Figure P9.44

9.44   Determine $v_O$ as a function of $v_{I1}$ and $v_{I2}$ for the ideal noninverting op-amp circuit in Figure P9.44.

9.45   Consider the ideal noninverting op-amp circuit in Figure P9.45. (a) Derive the expression for $v_O$ as a function of $v_{I1}$ and $v_{I2}$. (b) Find $v_O$ for $v_{I1} = 0.2$ V and $v_{I2} = 0.3$ V. (c) Find $v_O$ for $v_{I1} = +0.25$ V and $v_{I2} = -0.40$ V.

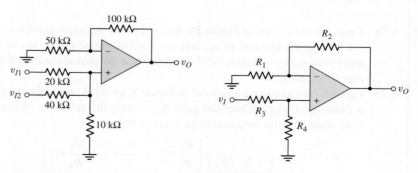

Figure P9.45                  Figure P9.46

9.46   (a) Derive the expression for the closed-loop voltage gain $A_v = v_O/v_I$ for the circuit shown in Figure P9.46. Assume an ideal op-amp. (b) Let

$R_4 = 50\,\text{k}\Omega$ and $R_3 = 25\,\text{k}\Omega$. Determine $R_1$ and $R_2$ such that $A_v = 6$, assuming the maximum resistor value is limited to $200\,\text{k}\Omega$.

9.47   The circuit shown in Figure P9.47 can be used as a variable noninverting amplifier. The circuit uses a $50\,\text{k}\Omega$ potentiometer in conjunction with an ideal op-amp. (a) Derive the expression for the closed-loop voltage gain $v_O/v_I$ in terms of the potentiometer setting $x$. (b) What is the range of closed-loop voltage gain? (c) Is there a potential problem with this circuit? If so, what is the problem?

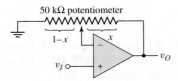

**Figure P9.47**

9.48   (a) Determine the closed-loop voltage gain $A_v = v_O/v_I$ for the ideal op-amp circuit in Figure P9.48. (b) Determine $v_O$ for $v_I = 0.25\,\text{V}$. (c) Let $R = 30\,\text{k}\Omega$. For $v_I = -0.15\,\text{V}$, determine the current in the resistor $R$ in the T-network.

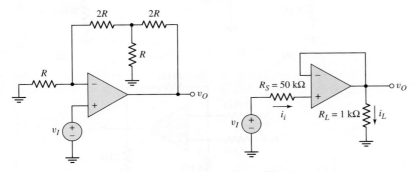

**Figure P9.48**

**Figure P9.49**

9.49   For the amplifier in Figure P9.49, determine (a) the ideal closed-loop voltage gain, (b) the actual closed-loop voltage gain if the open-loop gain is $A_{od} = 150,000$, and (c) the open-loop gain such that the actual closed-loop gain is within 1 percent of the ideal.

9.50   Consider the voltage-follower circuit in Figure 9.17. Determine the closed-loop voltage gain if the op-amp open-loop voltage gain $A_{od}$ is (a) 20, (b) 200, (c) $2 \times 10^3$, and (d) $2 \times 10^4$.

9.51   (a) Consider the ideal op-amp circuit shown in Figure P9.51. Determine the voltage gains $A_{v1} = v_{O1}/v_I$ and $A_{v2} = v_{O2}/v_I$. What is the relationship between $v_{O1}$ and $v_{O2}$? (b) For $R_2 = 60\,\text{k}\Omega$, $R_1 = 20\,\text{k}\Omega$, and $R = 50\,\text{k}\Omega$, determine $v_{O1}$ and $v_{O2}$ for $v_I = -0.50\,\text{V}$. (c) Determine $(v_{O1} - v_{O2})$ for $v_I = +0.8\,\text{V}$.

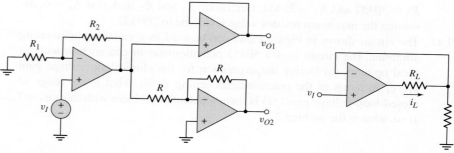

**Figure P9.51**

**Figure P9.52**

9.52   (a) Assume the op-amp in the circuit in Figure P9.52 is ideal. Determine $i_L$ as a function of $v_I$. (b) Let $R_1 = 9\,k\Omega$ and $R_L = 1\,k\Omega$. If the op-amp saturates at $\pm 10\,V$, determine the maximum value of $v_I$ and $i_L$ before the op-amp saturates.

9.53   Consider the three circuits shown in Figure P9.53. Determine each output voltage for (i) $v_I = 3\,V$ and (ii) $v_I = -5\,V$.

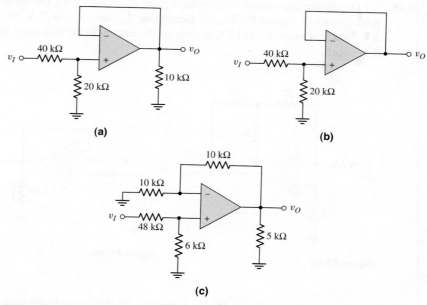

**Figure P9.53**

## Section 9.5 Op-Amp Applications

*9.54   A current-to-voltage converter is shown in Figure P9.54. The current source has a finite output resistance $R_S$, and the op-amp has a finite open-loop differential gain $A_{od}$. (a) Show that the input resistance is given by

$$R_{in} = \frac{R_F}{1 + A_{od}}$$

(b) If $R_F = 10\,k\Omega$ and $A_{od} = 1000$, determine the range of $R_S$ such that the output voltage deviates from its ideal value by less than 1 percent.

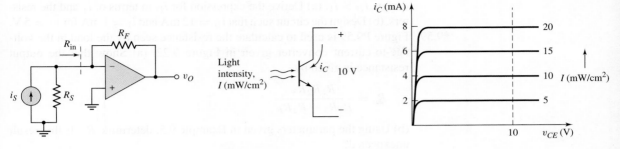

Figure P9.54                     Figure P9.55

*D9.55  Figure P9.55 shows a phototransistor that converts light intensity into an output current. The transistor must be biased as shown. The transistor output versus input characteristics are shown. Design a current-to-voltage converter to produce an output voltage between 0 and 8 V for an input light intensity between 0 and 20 mW/cm$^2$. Power supplies of $+10$ V and $-10$ V are available.

D9.56  The circuit in Figure P9.56 is an analog voltmeter in which the meter reading is directly proportional to the input voltage $v_I$. Design the circuit such that a 1 mA full-scale reading corresponds to $v_I = 10$ V. Resistance $R_2$ corresponds to the meter resistance, and $R_1$ corresponds to the source resistance. How do these resistances influence the design?

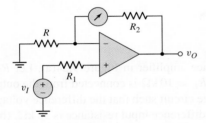

Figure P9.56

D9.57  Consider the voltage-to-current converter in Figure 9.22 using an ideal op-amp. (a) Design the circuit such that the current in a 200 $\Omega$ load can be varied between 0 and 5 mA with an input voltage between 0 and $-5$ V. Assume the op-amp is biased at $\pm15$ V. (b) Using the results of part (a), determine voltage $v_O$; currents $i_2$, $i_3$, $i_4$; and the output current of the op-amp for $v_I = -5$ V.

D9.58  The circuit in Figure P9.58 is used to drive an LED with a voltage source. The circuit can also be thought of as a current amplifier in that, with the proper

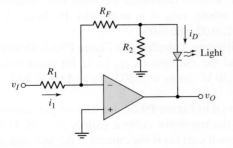

Figure P9.58

design, $i_D > i_1$. (a) Derive the expression for $i_D$ in terms of $i_1$ and the resistors. (b) Design the circuit such that $i_D = 12\,\text{mA}$ and $i_1 = 1\,\text{mA}$ for $v_I = 5\,\text{V}$.

*9.59    Figure P9.59 is used to calculate the resistance seen by the load in the voltage-to-current converter given in Figure 9.22. (a) Show that the output resistance is given by

$$R_o = \frac{R_1 R_2 R_3}{R_1 R_3 - R_2 R_F}$$

(b) Using the parameters given in Example 9.5, determine $R_o$. Is this result unexpected?

(c) Consider the design specification given by Equation (9.44). What is the expected value of $R_o$?

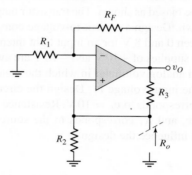

**Figure P9.59**

D9.60    Consider the op-amp difference amplifier in Figure 9.24(a). Let $R_1 = R_3$ and $R_2 = R_4$. A load resistor $R_L = 10\,\text{k}\Omega$ is connected from the output terminal to ground. (a) Design the circuit such that the difference voltage gain is $A_d = 15$ and the minimum difference input resistance is $30\,\text{k}\Omega$. (b) If the load current is $i_L = 0.25\,\text{mA}$, what is the differential input voltage $(v_{I2} - v_{I1})$? (c) If $v_{I1} = 1.5\,\text{V}$ and $v_{I2} = 1.2\,\text{V}$, determine $i_L$. (d) If $i_L = 0.5\,\text{mA}$ when $v_{I2} = 2.0\,\text{V}$, determine $v_{I1}$.

D9.61    Consider the differential amplifier shown in Figure 9.24(a). Let $R_1 = R_3$ and $R_2 = R_4$. Design the amplifier such that the differential voltage gain is (a) 40, (b) 25, (c) 5, and (d) 0.5. In each case the differential input resistance should be as large as possible but under the condition that the largest resistor value is limited to $250\,\text{k}\Omega$.

*9.62    Consider the differential amplifier shown in Figure 9.24(a). Assume that each resistor is $50(1 \pm x)\,\text{k}\Omega$. (a) Determine the worst case common-mode gain $A_{CM} = v_O/v_{CM}$, where $v_{CM} = v_1 = v_2$. (b) Evaluate $A_{CM}$ and CMRR(dB) for $x = 0.01, 0.02,$ and $0.05$.

9.63    Let $R = 10\,\text{k}\Omega$ in the differential amplifier in Figure P9.63. Determine the voltages $v_X$, $v_Y$, $v_O$ and the currents $i_1$, $i_2$, $i_3$, $i_4$ for input voltages of (a) $v_1 = 1.80\,\text{V}$, $v_2 = 1.40\,\text{V}$; (b) $v_1 = 3.20\,\text{V}$, $v_2 = 3.60\,\text{V}$; and (c) $v_1 = -1.20\,\text{V}$, $v_2 = -1.35\,\text{V}$.

9.64    Consider the circuit shown in Figure P9.64. (a) The output current of the op-amp is 1.2 mA and the transistor current gain is $\beta = 75$. Determine the resistance $R$. (b) Repeat part (a) if the current is 0.2 mA and the transistor current gain is $\beta = 100$. (c) Using the results of part (a), determine

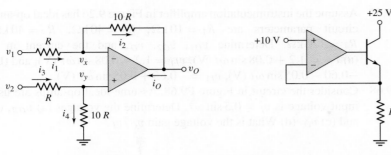

Figure P9.63                              Figure P9.64

the op-amp output current if the input voltage is 6 V. (d) Using the results of part (b), determine the op-amp output current if the input voltage is 4 V.

*9.65  The circuit in Figure P9.65 is a representation of the common-mode and differential-input signals to a difference amplifier. The output voltage can be written as

$$v_O = A_d v_d + A_{cm} v_{cm}$$

where $A_d$ is the differential-mode gain and $A_{cm}$ is the common-mode gain. (a) Setting $v_d = 0$, show that the common-mode gain is given by

$$A_{cm} = \frac{\left(\dfrac{R_4}{R_3} - \dfrac{R_2}{R_1}\right)}{(1 + R_4/R_3)}$$

(b) Determine $A_{cm}$ if $R_1 = 10.4\,\text{k}\Omega$, $R_2 = 62.4\,\text{k}\Omega$, $R_3 = 9.6\,\text{k}\Omega$, and $R_4 = 86.4\,\text{k}\Omega$. (c) Determine the maximum value of $|A_{cm}|$ if $R_1 = 20\,\text{k}\Omega \pm 1\%$, $R_2 = 80\,\text{k}\Omega \pm 1\%$, $R_3 = 20\,\text{k}\Omega \pm 1\%$, and $R_4 = 80\,\text{k}\Omega \pm 1\%$.

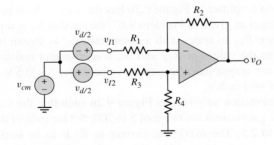

Figure P9.65

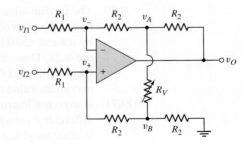

Figure P9.66

*9.66  Consider the adjustable gain difference amplifier in Figure P9.66. Variable resistor $R_V$ is used to vary the gain. Show that the output voltage $v_O$, as a function of $v_{I1}$ and $v_{I2}$, is given by

$$v_O = \frac{2R_2}{R_1}\left(1 + \frac{R_2}{R_V}\right)(v_{I2} - v_{I1})$$

9.67   Assume the instrumentation amplifier in Figure 9.26 has ideal op-amps. The circuit parameters are $R_1 = 10\,\text{k}\Omega$, $R_2 = 40\,\text{k}\Omega$, $R_3 = 40\,\text{k}\Omega$, and $R_4 = 120\,\text{k}\Omega$. Determine $v_{O1}$, $v_{O2}$, $v_O$, and the current in $R_1$ for (a) $v_{I2} = 1.2 + 0.08\sin\omega t$ (V), $v_{I1} = 1.2 - 0.08\sin\omega t$ (V); and (b) $v_{I2} = -0.60 - 0.05\sin\omega t$ (V), $v_{I1} = -0.65 + 0.05\sin\omega t$ (V).

9.68   Consider the circuit in Figure P9.68. Assume ideal op-amps are used. The input voltage is $v_I = 0.5\sin\omega t$. Determine the voltages (a) $v_{OB}$, (b) $v_{OC}$, and (c) $v_O$. (d) What is the voltage gain $v_O/v_I$?

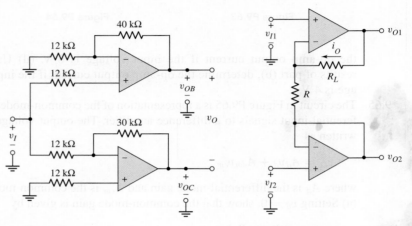

Figure P9.68                                    Figure P9.69

9.69   Consider the circuit in Figure P9.69. Assume ideal op-amps are used. (a) Derive the expression for the current $i_O$ as a function of input voltages $v_{I1}$ and $v_{I2}$. (b) Design the circuit such that $i_O = 5\,\text{mA}$ for $v_{I1} = 0.25$ V and $v_{I2} = -0.25$ V. (c) Using the results of part (b), determine $v_{O1}$ and $v_{O2}$ if $R_L = 1\,\text{k}\Omega$. (d) Determine $i_O$, $v_{O1}$, and $v_{O2}$ for $R = 500\,\Omega$, $R_L = 3\,\text{k}\Omega$, $v_{I1} = 1.25$ V, and $v_{I2} = 1.75$ V.

9.70   The instrumentation amplifier in Figure 9.26 has the same circuit parameters and input voltages as given in Problem 9.67, except that $R_1$ is replaced by a fixed resistance $R_{1f}$ in series with a potentiometer, as shown in Figure 9.28. Determine the values of $R_{1f}$ and the potentiometer resistance if the magnitude of the output has a minimum value of $|v_O| = 0.5$ V and a maximum value of $|v_O| = 8$ V.

D9.71  Design the instrumentation amplifier in Figure 9.26 such that the variable differential voltage gain covers the range of 5 to 200. Set the gain of the difference amplifier to 2.5. The maximum current in $R_1$ is to be limited to $50\,\mu\text{A}$ for an output voltage of 10 V. What value of potentiometer is required?

9.72   All parameters associated with the instrumentation amplifier in Figure 9.26 are the same as given in Exercise Ex 9.8, except that resistor $R_3$, which is connected to the inverting terminal of A3, is $R_3 = 30\,\text{k}\Omega \pm 5\%$. Determine the maximum common-mode gain.

9.73   The parameters in the integrator circuit shown in Figure 9.30 are $R_1 = 20\,\text{k}\Omega$ and $C_2 = 0.02\,\mu\text{F}$. The input signal is $v_I = 0.25\cos\omega t$ (V). (a) Determine the frequency at which the input and output signals have equal amplitudes. At this frequency, what is the phase of the output signal with respect to the input? (b) At what frequency will the output signal amplitude be (i) $|v_O| = 1.5$ V and (ii) $|v_O| = 0.15$ V?

9.74   Consider the ideal op-amp integrator. Assume the capacitor is initially uncharged. (a) The output voltage is $v_O = -5$ V at $t = 1.2$ s after a $+0.25$ V pulse is applied to the input. What is the $RC$ time constant? (b) Use the results of part (a). At $t = 1.2$ s, the input changes to $-0.10$ V. (i) At what time does $v_O = 0$? (ii) At what time does $v_O = +5$ V.

9.75   The circuit in Figure P9.75 is a first-order low-pass active filter. (a) Show that the voltage transfer function is given by

$$A_v = \frac{-R_2}{R_1} \cdot \frac{1}{1 + j\omega R_2 C_2}$$

(b) What is the voltage gain at dc ($\omega = 0$)? (c) At what frequency is the magnitude of the voltage gain a factor of $\sqrt{2}$ less that the dc value? (This is the $-3$ dB frequency.)

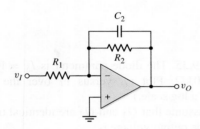

**Figure P9.75**

D9.76   (a) Using the results of Problem 9.75, design the low-pass active filter in Figure P9.75 such that the input resistance is $20\,\text{k}\Omega$, the low-frequency gain is $-15$, and the $-3$ dB frequency is 5 kHz. (b) Repeat part (a) such that the input resistance is $15\,\text{k}\Omega$, the low-frequency gain is $-25$, and the $-3$ dB frequency is 15 kHz.

9.77   The circuit shown in Figure P9.77 is a first-order high-pass active filter. (a) Show that the voltage transfer function is given by

$$A_v = \frac{-R_2}{R_1} \cdot \frac{j\omega R_1 C_1}{1 + j\omega R_1 C_1}$$

(b) What is the voltage gain as the frequency becomes large?
(c) At what frequency is the magnitude of the gain a factor of $\sqrt{2}$ less than the high-frequency limiting value?

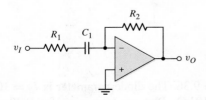

**Figure P9.77**

D9.78    (a) Using the results of Problem 9.77, design the high-pass active filter in Figure P9.77 such that the high-frequency voltage gain is $-15$ and the $-3$ dB frequency is 20 kHz. The maximum resistance value is to be limited to $350 \, \mathrm{k}\Omega$. (b) Repeat part (a) such that the high-frequency gain is $-25$ and the $-3$ dB frequency is 35 kHz. The minimum resistance value is to be limited to $20 \, \mathrm{k}\Omega$.

9.79    Consider the voltage reference circuit shown in Figure P9.79. Determine $v_O$, $i_2$, and $i_Z$.

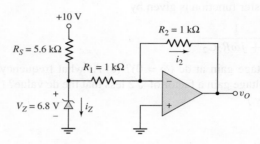

**Figure P9.79**

9.80    Consider the circuit in Figure 9.35. The diode parameter is $I_S = 10^{-14} \mathrm{A}$ and the resistance is $R_1 = 10 \, \mathrm{k}\Omega$. Plot $v_O$ versus $v_I$ over the range $20 \, \mathrm{mV} \le v_I \le 2 \, \mathrm{V}$. (Plot $v_I$ on a log scale.)

*9.81    In the circuit in Figure P9.81, assume that $Q_1$ and $Q_2$ are identical transistors. If $T = 300 \, \mathrm{K}$, show that the output voltage is

$$v_O = 1.0 \log_{10} \left( \frac{v_2 R_1}{v_1 R_2} \right)$$

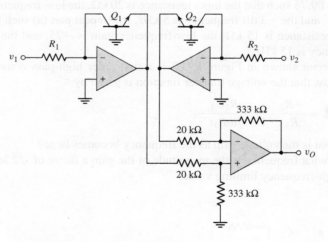

**Figure P9.81**

9.82    Consider the circuit in Figure 9.36. The diode parameter is $I_S = 10^{-14} \mathrm{A}$ and the resistance is $R_1 = 10 \, \mathrm{k}\Omega$. Plot $v_O$ versus $v_I$ for $0.30 \le v_I \le 0.60 \, \mathrm{V}$. (Plot $v_O$ on a log scale.)

## Section 9.7 Op-Amp Circuit Design

*D9.83  Design an op-amp summer to produce the output voltage $v_O = 2v_{I1} - 10v_{I2} + 3v_{I3} - v_{I4}$. Assume the largest resistor value is $500\,k\Omega$, and the input impedance seen by each source is the largest value possible.

*D9.84  Design an op-amp summer to produce an output voltage of $v_O = 3v_{I1} + 1.5v_{I2} + 2v_{I3} - 4v_{I4} - 6v_{I5}$. The largest resistor value is to be $250\,k\Omega$.

*D9.85  Design a voltage reference source as shown in Figure 9.42 to have an output voltage of 12.0 V. A Zener diode with a breakdown voltage of 5.6 V is available. Assume the voltage regulation will be within specifications if the Zener diode current is within the range of $1.2 \le I_Z \le 1.35$ mA. The start-up voltage $V_S$ is to be 10 V.

*D9.86  Consider the voltage reference circuit in Figure P9.86. Using a Zener diode with a breakdown voltage of 5.6 V, design the circuit to produce an output voltage of 12.0 V. Assume the input voltage is 15 V and the Zener diode current is $I_Z = 2$ mA.

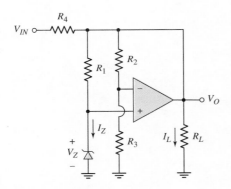

**Figure P9.86**

*D9.87  Consider the bridge circuit in Figure P9.87. The resistor $R_T$ is a thermistor with values of $20\,k\Omega$ at $T = 300\,K$ and $21\,k\Omega$ at $T = 350\,K$. Assume that the thermistor resistance is linear with temperature, and that the bridge is biased at $V^+ = 10$ V. Design an amplifier system with an output of 0 V at $T = 300\,K$ and 5 V at $T = 350\,K$.

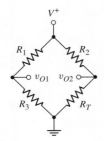

**Figure P9.87**

*D9.88  Consider the bridge circuit in Figure 9.46. The resistance $R$ is $20\,k\Omega$ and the bias is $V^+ = 9$ V. (a) Determine $v_{O1}$ as a function of $\Delta R$. (b) Design an amplifier system such that the output varies from $-5$ V to $+5$ V as $\Delta R$ varies from $+200\,\Omega$ to $-200\,\Omega$.

##  COMPUTER SIMULATION PROBLEMS

9.89    Using a computer simulation, verify the design in Example 9.4.

9.90    Using a computer simulation, verify the design in Example 9.8.

9.91    Using a computer simulation, verify the design in Problem 9.76(b). Plot $v_O$ versus frequency over the range $2 \leq f \leq 50\,\text{kHz}$.

9.92    Using a computer simulation, verify the design in Problem 9.78(a). Plot $v_O$ versus frequency over the range $2 \leq f \leq 100\,\text{kHz}$.

##  DESIGN PROBLEMS

See Design Problems 9.83 to 9.88.

[Note: Each design should be correlated with a computer analysis.]

# Integrated Circuit Biasing and Active Loads

The biasing techniques in Chapters 3 through 6 for FET and BJT amplifiers for the most part used voltage-divider resistor networks. While this technique can be used for discrete circuits, it is not suitable for integrated circuits. Resistors require relatively large areas on an IC compared to transistors; therefore, a resistor-intensive circuit would necessitate a large chip area. Also, the resistor biasing technique uses coupling and bypass capacitors extensively. On an IC, it is almost impossible to fabricate capacitors in the microfarad range, as would be required for the coupling capacitors.

Biasing transistors and transistor circuits in ICs is considerably different from that in discrete transistor designs. Essentially, biasing integrated circuit amplifiers involves the use of constant-current sources. In this chapter, we will analyze and design both bipolar and FET circuits that form these constant-current sources. We will begin to see for the first time in this chapter the use of matched or identical transistor characteristics as a specific design parameter. Transistors can easily be fabricated in ICs with matched or identical parameters. A principal goal of this chapter is to help the reader understand how matched transistor characteristics are used in design and to be able to design BJT and MOSFET current source circuits.

Transistors are also used as load devices in amplifier circuits. These transistors, called active loads, replace the discrete drain and collector resistors in FET and BJT circuits. Using an active load eliminates resistors from the IC and achieves a higher small-signal voltage gain. The active load is essentially an "upside down" constant-current source, so an initial discussion of active loads is entirely appropriate in this chapter.

## PREVIEW

In this chapter, we will:

- Analyze and understand the characteristics of various bipolar circuits used to provide a constant output current.
- Analyze and understand the characteristics of various MOSFET (and a few JFET) circuits used to provide a constant output current.
- Analyze the dc characteristics of amplifier circuits using transistors as load devices (active loads).
- Analyze the small-signal characteristics of amplifier circuits with active loads.
- As an application, design an MOS current source circuit to provide a specified bias current and output resistance.

### 10.1   BIPOLAR TRANSISTOR CURRENT SOURCES

**Objective:** • Analyze and understand the characteristics of various bipolar circuits used to provide a constant output current.

As we saw in previous chapters, when the bipolar transistor is used as a linear amplifying device, it must be biased in the forward-active mode. The bias may be a current source that establishes the quiescent collector current as shown in Figure 10.1. We now need to consider the types of circuits that can be designed to establish the bias current $I_O$. We will discuss a simple two-transistor current-source circuit and then two improved versions of the constant-current source. We will then analyze another current-source circuit, known as the Widlar current source. Finally, we will discuss a multitransistor current source.

#### 10.1.1   Two-Transistor Current Source

The **two-transistor current source,** also called a **current mirror,** is the basic building block in the design of integrated circuit current sources. Figure 10.2(a) shows the basic current-source circuit, which consists of two *matched* or *identical* transistors, $Q_1$ and $Q_2$, operating at the same temperature, with their base terminals and emitter terminals connected together. The B–E voltage is therefore the same in the two transistors. Transistor $Q_1$ is connected as a diode; consequently, when the supply voltages are applied, the B–E junction of $Q_1$ is forward biased and a reference current $I_{REF}$ is established. Although there is a specific relationship between $I_{REF}$ and $V_{BE1}$, we can think of $V_{BE1}$ as being the result of $I_{REF}$. Once $V_{BE1}$ is established, it is applied to the B–E junction of $Q_2$. The applied $V_{BE2}$ turns $Q_2$ on and generates the load current $I_O$, which is used to bias a transistor or transistor circuit.

The reference current in the two-transistor current source can be established by connecting a resistor to the positive voltage source, as shown in Figure 10.2(b). The reference current is then

$$I_{REF} = \frac{V^+ - V_{BE} - V^-}{R_1} \tag{10.1}$$

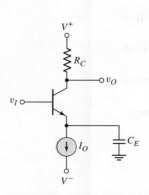

**Figure 10.1** Bipolar circuit with ideal current-source biasing

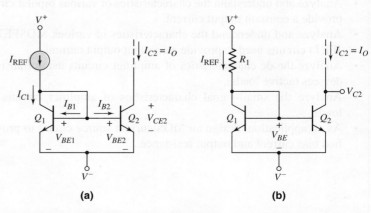

**Figure 10.2** (a) Basic two-transistor current source; (b) two-transistor current source with reference resistor $R_1$

where $V_{BE}$ is the B–E voltage corresponding to the collector current, which is essentially equal to $I_{\text{REF}}$.

Connecting the base and collector terminals of a bipolar transistor effectively produces a two-terminal device with *I–V* characteristics that are identical to the $i_C$ versus $v_{BE}$ characteristic of the BJT. For $v_{CB} = 0$, the transistor is still biased in the forward-active mode, and the base, collector, and emitter currents are related through the current gain $\beta$. In constant-current source circuits, $\beta$ is a dc term that is the ratio of the dc collector current to the dc base current. However, as discussed in Chapter 5, we assume the dc leakage currents are negligible; therefore, the dc beta and ac beta are essentially the same. We do not distinguish between the two values.

## Current Relationships

Figure 10.2(a) shows the currents in the two-transistor current source. Since $V_{BE}$ is the same in both devices, and the transistors are identical, then $I_{B1} = I_{B2}$ and $I_{C1} = I_{C2}$. Transistor $Q_2$ is assumed to be biased in the forward-active region. If we sum the currents at the collector node of $Q_1$, we have

$$I_{\text{REF}} = I_{C1} + I_{B1} + I_{B2} = I_{C1} + 2I_{B2} \tag{10.2}$$

Replacing $I_{C1}$ by $I_{C2}$ and noting that $I_{B2} = I_{C2}/\beta$, Equation (10.2) becomes

$$I_{\text{REF}} = I_{C2} + 2\frac{I_{C2}}{\beta} = I_{C2}\left(1 + \frac{2}{\beta}\right) \tag{10.3}$$

The output current is then

$$I_{C2} = I_O = \frac{I_{\text{REF}}}{1 + \dfrac{2}{\beta}} \tag{10.4}$$

Equation (10.4) gives the ideal output current of the two-transistor current source, taking into account the finite current gain of the transistors. Implicit in Equation (10.4) is that $Q_2$ is biased in the forward-active region (the base–collector junction is zero or reverse biased, meaning $V_{CE2} > V_{BE2}$[1]) and the Early voltage is infinite, or $V_A = \infty$. We will consider the effects of a finite Early voltage later in this chapter.

## DESIGN EXAMPLE 10.1

**Objective:** Design a two-transistor current source to meet a set of specifications.

**Specifications:** The circuit to be designed has the configuration shown in Figure 10.2(b). Assume that matched transistors are available with parameters $V_{BE}(\text{on}) = 0.6$ V, $\beta = 100$, and $V_A = \infty$. The designed output $I_O$ is to be 200 $\mu$A. The bias voltages are to be $V^+ = 5$ V and $V^- = 0$.

**Choices:** The circuit will be fabricated as an integrated circuit so that a standard resistor value is not required and matched transistors can be fabricated.

---

[1] In actual circuits, the collector–emitter voltage may decrease to values as low as 0.2 or 0.3 V, and the circuit will still behave as a constant-current source.

**Solution:** The reference current can be written as

$$I_{REF} = I_O\left(1 + \frac{2}{\beta}\right) = (200)\left(1 + \frac{2}{100}\right) = 204\,\mu A$$

From Equation (10.1), the resistor $R_1$ is found to be

$$R_1 = \frac{V^+ - V_{BE}\,(\text{on})}{I_{REF}} = \frac{5 - 0.6}{0.204} = 21.6\,\text{k}\Omega$$

**Trade-offs:** The design assumes that matched transistors exist. The effect of mismatched transistors will be discussed later in this section.

**Comment:** In this example, we assumed a B–E voltage of 0.6 V. This approximation is satisfactory for most cases. The B–E voltage is involved in the reference current or resistor calculation. If a value of $V_{BE}(\text{on}) = 0.7$ V is assumed, the value of $I_{REF}$ or $R_1$ will change, typically, by only 1 to 2 percent.

**Design Pointer:** We see in this example that, for $\beta = 100$, the reference and load currents are within 2 percent of each other in this two-transistor current source. In most circuit applications, we can use the approximation that $I_O \cong I_{REF}$.

---

### EXERCISE PROBLEM

**Ex 10.1:** The circuit parameters for the two-transistor current source shown in Figure 10.2(b) are $V^+ = 3$ V, $V^- = -3$ V, and $R_1 = 47$ k$\Omega$. The transistor parameters are $\beta = 120$, $V_{BE}(\text{on}) = 0.7$ V, and $V_A = \infty$. Determine $I_{REF}$, $I_O$, and $I_{B1}$. (Ans. $I_{REF} = 0.1128$ mA, $I_O = 0.1109$ mA, $I_{B1} = 0.9243\,\mu A$)

---

### Output Resistance

In our previous analysis, we assumed the Early voltage was infinite, so that $r_O = \infty$. In actual transistors, the Early voltage is finite, which means that the collector current is a function of the collector–emitter voltage. The stability of a load current generated in a constant-current source is a function of the output resistance looking back into the output transistor.

Figure 10.3 shows the dc equivalent circuit of a simple transistor circuit biased with a two-transistor current source. The voltage $V_I$ applied to the base of $Q_o$ is a dc voltage. If the value of $V_I$ changes, the collector–emitter voltage $V_{CE2}$ changes since the B–E voltage of $Q_o$ is essentially a constant. A variation in $V_{CE2}$ in turn changes the output current $I_O$, because of the Early effect. Figure 10.4 shows that $I_O$ versus $V_{CE2}$ characteristic at a constant B–E voltage.

The ratio of load current to reference current, taking the Early effect into account, is

$$\frac{I_O}{I_{REF}} = \frac{1}{\left(1 + \dfrac{2}{\beta}\right)} \times \frac{\left(1 + \dfrac{V_{CE2}}{V_A}\right)}{\left(1 + \dfrac{V_{CE1}}{V_A}\right)} \tag{10.5}$$

where $V_A$ is the Early voltage and the factor $(1 + 2/\beta)$ accounts for the finite gain. From the circuit configuration, we see that $V_{CE1} = V_{BE}$, which is essentially a constant, and

$$V_{CE2} = V_I - V_{BEo} - V^- \tag{10.6}$$

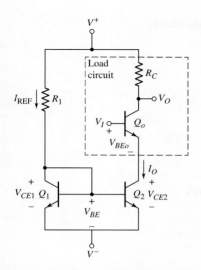

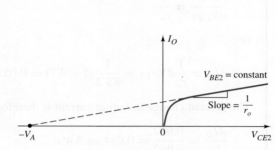

**Figure 10.3** The dc equivalent circuit of simple amplifier biased with two-transistor current source

**Figure 10.4** Output current versus collector–emitter voltage, showing the Early voltage

The differential change in $I_O$ with respect to a change in $V_{CE2}$, is, from Equation (10.5),

$$\frac{dI_O}{dV_{CE2}} = \frac{I_{\text{REF}}}{\left(1 + \frac{2}{\beta}\right)} \times \frac{1}{V_A} \times \frac{1}{\left(1 + \frac{V_{BE}}{V_A}\right)} \tag{10.7}$$

If we assume $V_{BE} \ll V_A$, then Equation (10.7) becomes

$$\frac{dI_O}{dV_{CE2}} \cong \frac{I_O}{V_A} = \frac{1}{r_o} \tag{10.8}$$

where $r_o$ is the small-signal output resistance looking into the collector of $Q_2$.

## EXAMPLE 10.2

**Objective:** Determine the change in load current produced by a change in collector–emitter voltage in a two-transistor current source.

Consider the circuit shown in Figure 10.3. The circuit parameters are: $V^+ = 5$ V, $V^- = -5$ V, and $R_1 = 9.3$ kΩ. Assume the transistor parameters are: $\beta = 50$, $V_{BE}(\text{on}) = 0.7$ V, and $V_A = 80$ V. Determine the change in $I_O$ as $V_{CE2}$ changes from 0.7 V to 5 V.

**Solution:** The reference current is

$$I_{\text{REF}} = \frac{V^+ - V_{BE}(\text{on}) - V^-}{R_1} = \frac{5 - 0.7 - (-5)}{9.3} = 1.0 \text{ mA}$$

For $V_{CE2} = 0.7$ V, transistors $Q_1$ and $Q_2$ are identically biased. From Equation (10.5), we then have

$$I_O = \frac{I_{\text{REF}}}{1 + \frac{2}{\beta}} = \frac{1.0}{1 + \frac{2}{50}} = 0.962 \text{ mA}$$

From Equation (10.8), the small-signal output resistance is

$$r_o = \frac{V_A}{I_O} = \frac{80}{0.962} = 83.2\,\text{k}\Omega$$

The change in load current is determined from

$$\frac{dI_O}{dV_{CE2}} = \frac{1}{r_o}$$

or

$$dI_O = \frac{1}{r_o}dV_{CE2} = \frac{1}{83.2}(5 - 0.7) = 0.052\,\text{mA}$$

The percent change in output current is therefore

$$\frac{dI_O}{I_O} = \frac{0.052}{0.962} = 0.054 \Rightarrow 5.4\%$$

**Comment:** Although in many circuits a 5 percent change in bias current is insignificant, there are cases, such as digital-to-analog converters, in which the bias current must be held to very tight tolerances. The stability of the load current can be significantly affected by a change in collector–emitter voltage. The stability is a function of the output impedance of the current source.

### EXERCISE PROBLEM

**Ex 10.2:** Consider the circuit shown in Figure 10.3. The circuit parameters are: $V^+ = 5$ V, $V^- = -5$ V, and $R_1 = 12$ k$\Omega$. The transistor parameters are $\beta = 75$ and $V_{BE}(\text{on}) = 0.7$ V. The percentage change in load current $\Delta I_O/I_O$ must be no more than 2 percent for a change in $V_{CE2}$ from 1 V to 5 V. Determine the minimum required value of Early voltage. (Ans. $V_A \cong 200$ V)

### Integrated Circuit Fabrication

We have assumed in the previous analysis that the two transistors in the current source circuit are matched or identical. When fabricated as an integrated circuit, the two transistors will be directly adjacent to each other. The material properties will therefore be essentially identical, and any ion implant dose and thermal anneal characteristics will be essentially identical. So, the two adjacent transistors can be very well matched. There may be some variation in transistor characteristics from one circuit to another but, again, the characteristics of the adjacent transistors are closely matched. In practice, the characteristics of $Q_1$ and $Q_2$ may be mismatched by 1 or 2 percent.

### Mismatched Transistors

If $\beta \gg 1$, we can neglect base currents. The current–voltage relationship for the circuit in Figure 10.2(b) is then

$$I_{\text{REF}} \cong I_{C1} = I_{S1}e^{V_{BE}/V_T} \tag{10.9(a)}$$

and

$$I_O = I_{C2} = I_{S2}e^{V_{BE}/V_T} \tag{10.9(b)}$$

Here, we are neglecting the Early effect. The parameters $I_{S1}$ and $I_{S2}$ contain both the electrical and geometric parameters of $Q_1$ and $Q_2$. If $Q_1$ and $Q_2$ are not identical, then $I_{S1} \neq I_{S2}$.

Combining Equations (10.9(a)) and (10.9(b)), we obtain the relationship between the bias and reference currents, neglecting base currents, as follows:

$$I_O = I_{REF}\left(\frac{I_{S2}}{I_{S1}}\right) \tag{10.10}$$

Any deviation in bias current from the ideal, as a function of mismatch between $Q_1$ and $Q_2$, is directly related to the ratio of the reverse-saturation currents $I_{S1}$ and $I_{S2}$. The parameter $I_S$ is a strong function of temperature. The temperatures of $Q_1$ and $Q_2$ must be the same in order for the circuit to operate properly. Therefore, $Q_1$ and $Q_2$ must be close to one another on the semiconductor chip. If $Q_1$ and $Q_2$ are not maintained at the same temperature, then the relationship between $I_O$ and $I_{REF}$ is a function of temperature, which is undesirable.

Also, the parameters $I_{S1}$ and $I_{S2}$ are functions of the cross-sectional area of the B–E junctions. Therefore, we can use Equation (10.10) to our advantage. By using different sizes of transistors, we can design the circuit such that $I_O \neq I_{REF}$. This is discussed further later in this chapter.

Integrated circuit resistors are a function of the resistivity of the semiconductor material as well as the geometry of the device. Since the geometry of each IC resistor can be individually designed, resistor values are not limited to standard values. So, IC resistors of any value (within reason) can be fabricated.

## 10.1.2    Improved Current-Source Circuits

In many IC designs, critical current-source characteristics are the changes in bias current with variations in $\beta$ and with changes in the output transistor collector voltage. In this section, we will look at two constant-current circuits that have improved load current stability against changes in $\beta$ and changes in output collector voltage.

### Basic Three-Transistor Current Source
A basic three-transistor current source is shown in Figure 10.5. We again assume that all transistors are identical; therefore, since the B–E voltage is the same for $Q_1$ and

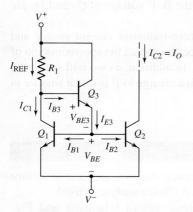

**Figure 10.5** Basic three-transistor current source

$Q_2$, $I_{B1} = I_{B2}$ and $I_{C1} = I_{C2}$. Transistor $Q_3$ supplies the base currents to $Q_1$ and $Q_2$, so these base currents should be less dependent on the reference current. Also, since the current in $Q_3$ is substantially smaller than that in either $Q_1$ or $Q_2$, we expect the current gain of $Q_3$ to be less than those of $Q_1$ and $Q_2$. We define the current gains of $Q_1$ and $Q_2$ as $\beta_1 = \beta_2 \equiv \beta$, and the current gain of $Q_3$ as $\beta_3$. Summing the currents at the collector node of $Q_1$, we obtain

$$I_{\text{REF}} = I_{C1} + I_{B3} \tag{10.11}$$

Since

$$I_{B1} = I_{B2} = 2I_{B2} = I_{E3} \tag{10.12}$$

and

$$I_{E3} = (1 + \beta_3)I_{B3} \tag{10.13}$$

then combining Equations (10.11), (10.12), and (10.13) produces

$$I_{\text{REF}} = I_{C1} + \frac{I_{E3}}{(1 + \beta_3)} = I_{C1} + \frac{2I_{B2}}{(1 + \beta_3)} \tag{10.14}$$

Replacing $I_{C1}$ by $I_{C2}$ and noting that $I_{B2} = I_{C2}/\beta$, we can rewrite Equation (10.14) as

$$I_{\text{REF}} = I_{C2} + \frac{2I_{C2}}{\beta(1 + \beta_3)} = I_{C2}\left[1 + \frac{2}{\beta(1 + \beta_3)}\right] \tag{10.15}$$

The output or bias current is then

$$I_{C2} = I_O = \frac{I_{\text{REF}}}{\left[1 + \dfrac{2}{\beta(1 + \beta_3)}\right]} \tag{10.16}$$

The reference current is given by

$$I_{\text{REF}} = \frac{V^+ - V_{BE3} - V_{BE} - V^-}{R_1} \cong \frac{V^+ - 2V_{BE} - V^-}{R_1} \tag{10.17}$$

As a first approximation, we usually assume that the B–E voltage of $Q_3$ and $Q_1$ are equal, as indicated in Equation (10.17).

A comparison of Equation (10.16) for the three-transistor current source and Equation (10.4) for the two-transistor current source shows that the approximation of $I_O \cong I_{\text{REF}}$ is better for the three-transistor circuit. In addition, as we will see in the following example, the change in load current with a change in $\beta$ is much smaller in the three-transistor current source.

---

### EXAMPLE 10.3

**Objective:** Compare the variation in bias current between the two- and three-transistor current-source circuits as a result of variations in $\beta$. A PSpice analysis is used.

Figure 10.6(a) shows the two-transistor PSpice circuit schematic and Figure 10.6(b) shows the three-transistor PSpice circuit schematic used in this analysis.

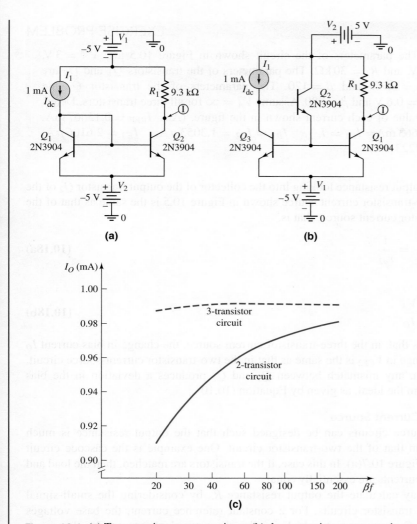

**Figure 10.6** (a) Two-transistor current mirror; (b) three-transistor current mirror; (c) variation in bias currents with a change in $\beta$

**Solution:** In both circuits, the current gain $\beta$ of all transistors was assumed to be equal, but the actual value was varied between 20 and 200. Since the change in $\beta$ is very large, we cannot use derivatives to determine the changes in bias currents. Standard 2N3904 transistors were used, which means that the Early voltage is 74 V, and not infinite as in the ideal circuit. The Early voltage will influence the actual value of bias current, but has very little effect in terms of the change in bias current with a change in current gain.

Figure 10.6(c) shows the bias current versus current gain for both the two-transistor and three-transistor current-source circuits.

**Comment:** There is a significant decrease in the variation in bias current for the three-transistor circuit compared to that of the two-transistor circuit. For values of $\beta$ greater than approximately 50, there is no perceptible change in bias current for the three-transistor current mirror.

**Ex 10.3:** The parameters of the circuit shown in Figure 10.5 are: $V^+ = 3\,\text{V}$, $V^- = -3\,\text{V}$, and $R_1 = 30\,\text{k}\Omega$. The parameters of the transistors $Q_1$ and $Q_2$ are $V_{BE1,2}(\text{on}) = 0.7\,\text{V}$ and $\beta = 120$. The parameters of the transistor $Q_3$ are $V_{BE3}(\text{on}) = 0.6\,\text{V}$ and $\beta_3 = 80$. Assume $V_A = \infty$ for all three transistors. Determine the value of each current shown in the figure. (Ans. $I_{\text{REF}} = 0.15667\,\text{mA}$, $I_O = 0.15663\,\text{mA} = I_{C1} = I_{C2}$, $\quad I_{B1} = I_{B2} = 1.3053\,\mu\text{A}$, $\quad I_{E3} = 2.6106\,\mu\text{A}$, $I_{B3} = 0.03223\,\mu\text{A}$)

The output resistance looking into the collector of the output transistor $Q_2$ of the basic three-transistor current source shown in Figure 10.5 is the same as that of the two-transistor current source; that is,

$$\frac{dI_O}{dV_{CE2}} = \frac{1}{r_{o2}} \tag{10.18a}$$

where

$$r_{o2} = \frac{V_A}{I_O} \tag{10.18b}$$

This means that, in the three-transistor current source, the change in bias current $I_O$ with a change in $V_{CE2}$ is the same as that in the two-transistor current-source circuit. In addition, any mismatch between $Q_1$ and $Q_2$ produces a deviation in the bias current from the ideal, as given by Equation (10.10).

### Cascode Current Source

Current-source circuits can be designed such that the output resistance is much greater than that of the two-transistor circuit. One example is the cascode circuit shown in Figure 10.7(a). In this case, if the transistors are matched, then the load and reference currents are essentially equal.

We may calculate the output resistance $R_o$ by considering the small-signal equivalent transistor circuits. For a constant reference current, the base voltages of $Q_2$ and $Q_4$ are constant, which implies these terminals are at signal ground.

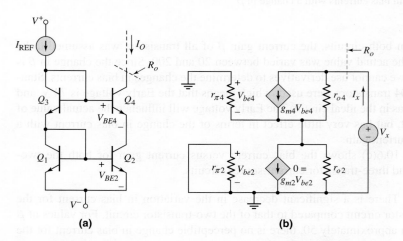

**Figure 10.7** (a) Bipolar cascode current mirror; (b) small-signal equivalent circuit

The equivalent circuit is then shown in Figure 10.7(b). Since $g_{m2}V_{be2} = 0$, then $V_{be4} = -I_x(r_{o2}\|r_{\pi4})$. Summing currents at the output node yields

$$I_x = g_{m4}V_{be4} + \left(\frac{V_x - I_x(r_{o2}\|r_{\pi4})}{r_{o4}}\right)$$

$$= -g_{m4}I_x(r_{o2}\|r_{\pi4}) + \left(\frac{V_x - I_x(r_{o2}\|r_{\pi4})}{r_{o4}}\right) \tag{10.19}$$

Combining terms and assuming $r_{\pi4} \ll r_{o2}$, we find

$$R_o = \frac{V_x}{I_x} = r_{o4}(1 + \beta) + r_{\pi4} \cong \beta r_{o4} \tag{10.20}$$

The output resistance has increased by a factor of $\beta$ compared to the two-transistor current source, which increases the stability of the current source with changes in output voltage.

## Wilson Current Source

Another configuration of a three-transistor current source, called a **Wilson current source,** is shown in Figure 10.8. This circuit also has a large output resistance. Our analysis again assumes identical transistors, with $I_{B1} = I_{B2}$ and $I_{C1} = I_{C2}$. The current levels in all three transistors are nearly the same; therefore, we can assume that the current gains of the three transistors are equal. Nodal equations at the collector of $Q_1$ and the emitter of $Q_3$ yield

$$I_{REF} = I_{C1} + I_{B3} \tag{10.21}$$

and

$$I_{E3} = I_{C2} + 2I_{B2} = I_{C2}\left(1 + \frac{2}{\beta}\right) \tag{10.22}$$

Using the relationships between the base, collector, and emitter currents in $Q_3$, we can write the collector current $I_{C2}$, from Equation (10.22), as follows:

$$I_{C2} = \frac{I_{E3}}{\left(1 + \frac{2}{\beta}\right)} = \frac{1}{\left(1 + \frac{2}{\beta}\right)} \times \left(\frac{1+\beta}{\beta}\right)I_{C3} = \left(\frac{1+\beta}{2+\beta}\right)I_{C3} \tag{10.23}$$

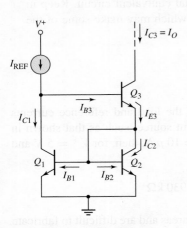

**Figure 10.8** Wilson current source

If we replace $I_{C1}$ by $I_{C2}$ in Equation (10.21), the reference current becomes

$$I_{REF} = I_{C2} + I_{B3} = \left(\frac{1+\beta}{2+\beta}\right) I_{C3} + \frac{I_{C3}}{\beta} \tag{10.24}$$

Rearranging terms, we can solve for the output current,

$$I_{C3} = I_O = I_{REF} \times \frac{1}{1 + \dfrac{2}{\beta(2+\beta)}} \tag{10.25}$$

This current relationship is essentially the same as that of the previous three-transistor current source.

The difference between the two three-transistor current-source circuits is the output resistance. In the Wilson current source, the output resistance looking into the collector of $Q_3$ is $R_o \cong \beta r_{o3}/2$, which is approximately a factor $\beta/2$ larger than that of either the two-transistor source or the basic three-transistor source. This means that, in the Wilson current source, the change in bias current $I_O$ with a change in output collector voltage is much smaller.

### Output Voltage Swing

If we consider the equivalent circuit in Figure 10.3, we see that the maximum possible swing in the output voltage is a function of the minimum possible collector–emitter voltage of $Q_2$. For the two-transistor current source in this figure, the minimum value of $V_{CE2} = V_{CE}(\text{sat})$, which may be on the order of 0.1 to 0.3 V.

For the cascode and Wilson current sources, the minimum output voltage is $V_{BE} + V_{CE}(\text{sat})$ above the negative power supply voltage, which may be on the order of 0.7 to 0.9 V. For circuits biased at ±5 V, for example, this increased minimum voltage may not be a serious problem. However, as the voltages decrease in low-power circuits, this minimum voltage effect may become more serious.

---

### Problem-Solving Technique: BJT Current Source Circuits

1. Sum currents at the various nodes in the circuit to find the relation between the reference current and the bias current.
2. To find the output resistance of the current source circuit, place a test voltage at the output node and analyze the small-signal equivalent circuit. Keep in mind that the reference current is a constant, which may make some of the base voltages constant or at ac ground.

---

### 10.1.3   Widlar Current Source

In the current-source circuits considered thus far, the load and reference currents have been nearly equal. For a two-transistor current source, such as that shown in Figure 10.2(a), if we require a load current of $I_O = 10 \ \mu\text{A}$, then, for $V^+ = 5$ V and $V^- = -5$ V, the required resistance value is

$$R_1 = \frac{V^+ - V_{BE} - V^-}{I_{REF}} \cong \frac{5 - 0.7 - (-5)}{10 \times 10^{-6}} = 930 \text{ k}\Omega$$

In ICs, resistors on the order of 1 MΩ require large areas and are difficult to fabricate accurately. We therefore need to limit IC resistor values to the low kilohm range.

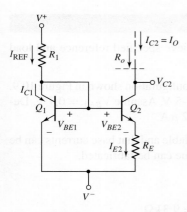

**Figure 10.9**  Widlar current source

The transistor circuit in Figure 10.9, called a **Widlar current source,** meets this objective. A voltage difference is produced across resistor $R_E$, so that the B–E voltage of $Q_2$ is less than the B–E voltage of $Q_1$. A smaller B–E voltage produces a smaller collector current, which in turn means that the load current $I_O$ is less than the reference current $I_{\text{REF}}$.

### Current Relationship
If $\beta \gg 1$ for $Q_1$ and $Q_2$, and if the two transistors are identical, then

$$I_{\text{REF}} \cong I_{C1} = I_S e^{V_{BE1}/V_T} \tag{10.26(a)}$$

and

$$I_O = I_{C2} = I_S e^{V_{BE2}/V_T} \tag{10.26(b)}$$

Solving for the B–E voltages, we have

$$V_{BE1} = V_T \ln\left(\frac{I_{\text{REF}}}{I_S}\right) \tag{10.27(a)}$$

and

$$V_{BE2} = V_T \ln\left(\frac{I_O}{I_S}\right) \tag{10.27(b)}$$

Combining Equations (10.27(a)) and (10.27(b)) yields

$$V_{BE1} - V_{BE2} = V_T \ln\left(\frac{I_{\text{REF}}}{I_O}\right) \tag{10.28}$$

From the circuit, we see that

$$V_{BE1} - V_{BE2} = I_{E2} R_E \cong I_O R_E \tag{10.29}$$

When we combine Equations (10.28) and (10.29), we obtain:

$$I_O R_E = V_T \ln\left(\frac{I_{\text{REF}}}{I_O}\right) \tag{10.30}$$

This equation gives the relationship between the reference and bias currents.

## DESIGN EXAMPLE 10.4

**Objective:** Design a Widlar current source to achieve specified reference and load currents.

**Specifications:** The circuit to be designed has the configuration shown in Figure 10.9. Assume bias voltages of $V^+ = +5$ V and $V^- = -5$ V. Assume $V_{BE1} = 0.7$ V. Design the circuit such that $I_{REF} = 1$ mA and $I_O = 12\ \mu$A.

**Choices:** Assume that matched transistors are available and that base currents can be neglected. Also assume that IC resistors of any value can be fabricated.

**Solution:** Resistance $R_1$ is

$$R_1 = \frac{V^+ - V_{BE1} - V^-}{I_{REF}} = \frac{5 - 0.7 - (-5)}{1} = 9.3\ \text{k}\Omega$$

Resistance $R_E$ is, from Equation (10.30),

$$R_E = \frac{V_T}{I_O} \ln\left(\frac{I_{REF}}{I_O}\right) = \frac{0.026}{0.012} \ln\left(\frac{1}{0.012}\right) = 9.58\ \text{k}\Omega$$

From Equation (10.29), we can determine the difference between the two B–E voltages, as follows:

$$V_{BE1} - V_{BE2} = I_O R_E = (12 \times 10^{-6})(9.58 \times 10^3) = 0.115\ \text{V}$$

**Trade-offs:** A slight variation in $V_{BE1}$ and slight tolerance variations in resistor values will change the current values slightly. These effects are evaluated in end-of-chapter problems.

**Comment:** A difference of 115 mV in the B–E voltages of $Q_1$ and $Q_2$ produces approximately two orders of magnitude difference between the reference and load currents. Therefore, we can produce a very low bias current using resistors in the low kilohm range. These resistors can easily be fabricated in an IC. Including the resistor $R_E$ gives the designer additional versatility in adjusting the load to reference current ratio.

### EXERCISE PROBLEM

**Ex 10.4:** Consider the Widlar current source in Figure 10.9. The bias voltages are $V^+ = 3$ V and $V^- = -3$ V. Design the circuit such that $I_O = 20\ \mu$A and $I_{REF} = 100\ \mu$A. Assume $V_{BE1} = 0.6$ V and $V_A = \infty$, and neglect base currents. Determine $R_1$, $R_E$, and $V_{BE2}$. (Ans. $R_1 = 54\ \text{k}\Omega$, $R_E = 2.09\ \text{k}\Omega$, $V_{BE2} = 0.558$ V)

In our analysis of constant-current source circuits, we have assumed a piecewise linear approximation for the B–E voltage, $V_{BE}(\text{on})$. However, in the Widlar current source and other current-source circuits, the piecewise linear approximation is not adequate, since the B–E voltages are not all equal. With the exponential relationship between collector current and base–emitter voltage, as shown in Equations (10.26(a)) and (10.26(b)), a small change in B–E voltage produces a large change in collector current. To take this variation into account, either the reverse-biased saturation current $I_S$ or the B–E voltage at a particular collector current must be known.

Also in our analysis, we have assumed that the temperatures of all transistors are equal. Maintaining equal temperatures is important for proper circuit operation.

## EXAMPLE 10.5

**Objective:** To determine the currents in a Widlar current source circuit.

Assume the Widlar source is biased at $V^+ = +5$ V and $V^- = -5$ V, and assume resistor values $R_1 = 7$ kΩ and $R_E = 4$ kΩ. Also assume $V_{BE1} = 0.7$ V.

**Solution:** The reference current is found to be

$$I_{REF} = \frac{V^+ - V_{BE1} - V^-}{R_1} = \frac{5 - 0.7 - (-5)}{7} = 1.33 \, \text{mA}$$

The load current is found from the relation

$$I_O R_E = V_T \ln\left(\frac{I_{REF}}{I_O}\right)$$

or

$$I_O(4) = 0.026 \ln\left(\frac{1.33}{I_O}\right)$$

A transcendental equation cannot be solved directly. A computer solution or a trial and error solution yields

$$I_O \cong 25.7 \, \mu\text{A}$$

**Comment:** In this case, the difference between the two base–emitter voltages is $I_O R_E \cong 103$ mV. Again, a relatively small difference in the two base–emitter voltages can produce a relatively large difference between the reference and load currents.

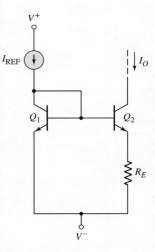

**Figure 10.10** Figure or Exercise Ex 10.5

## EXERCISE PROBLEM

**Ex 10.5:** Consider the circuit in Figure 10.10. Assume the reference current is $I_{REF} = 120 \, \mu\text{A}$ and assume the transistor parameters are $I_{S1} = I_{S2} = 2 \times 10^{-16}$ A. Neglect base currents. (a) Find $V_{BE1}$. (b) If $I_O = 50 \, \mu\text{A}$, determine $V_{BE2}$ and $R_E$. (c) Find $I_O$ if $R_E = 700 \, \Omega$. What is $V_{BE2}$? (Ans. (a) $V_{BE1} = 0.7051$ V; (b) $R_E = 455 \, \Omega$, $V_{BE2} = 0.6824$ V; (c) $I_O = 40.4 \, \mu\text{A}$, $V_{BE2} = 0.6768$ V)

### Output Resistance

The change in load current with a change in voltage $V_{C2}$ of the Widlar current source in Figure 10.9 can be expressed as

$$\frac{dI_O}{dV_{C2}} = \frac{1}{R_o} \tag{10.31}$$

where $R_o$ is the output resistance looking into the collector of $Q_2$. This output resistance can be determined by using the small-signal equivalent circuit in Figure 10.11(a). (Again, we use the phasor notation in small-signal analyses.) The base, collector, and emitter terminals of each transistor are indicated on the figure.

First, we calculate the resistance $R_{o1}$ looking into the base of $Q_1$. Writing a KCL equation at the base of $Q_1$, we obtain

$$I_{x1} = \frac{V_{x1}}{r_{\pi 1}} + g_{m1} V_{\pi 1} + \frac{V_{x1}}{r_{o1} \| R_1} \tag{10.32}$$

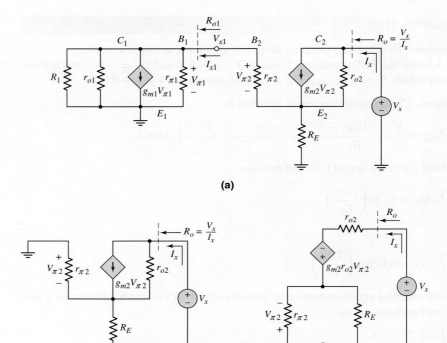

**Figure 10.11** (a) Small-signal equivalent circuit for determining output resistance of Widlar current source, (b) simplified equivalent circuit for determining output resistance, and (c) equivalent circuit after a Norton transformation

Noting that $V_{\pi 1} = V_{x1}$, we have

$$\frac{1}{R_{o1}} = \frac{I_{x1}}{V_{x1}} = \frac{1}{r_{\pi 1}} + g_{m1} + \frac{1}{r_{o1} \| R_1} \tag{10.33(a)}$$

or

$$R_{o1} = r_{\pi 1} \left\| \frac{1}{g_{m1}} \right\| r_{o1} \| R_1 \tag{10.33(b)}$$

Next, we calculate the approximate value for $R_{o1}$. If $I_{REF} = 1$ mA, then for $\beta = 100$, $r_{\pi 1} = 2.6$ kΩ and $g_{m1} = 38.5$ mA/V. Assume that $R_1 = 9.3$ kΩ and $r_{o1} = \infty$. For these conditions, $R_{o1} \cong 0.026$ kΩ $= 26$ Ω. For a load current of $I_O = 12$ μA, we find $r_{\pi 2} = 217$ kΩ. Resistance $R_{o1}$ is in series with $r_{\pi 2}$, and since $R_{o1} \ll r_{\pi 2}$, we can neglect the effect of $R_{o1}$, which means that the base of $Q_2$ is essentially at signal ground.

Now we determine the output resistance at the collector of $Q_2$, using the simplified equivalent circuit in Figure 10.11(b). The Norton equivalent of the current source $g_{m2}V_{\pi 2}$ and resistance $r_{o2}$ can be transformed into a Thevenin equivalent circuit, as shown in Figure 10.11(c). Resistances $r_{\pi 2}$ and $R_E$ are in parallel; therefore, we define $R_E' = R_E \| r_{\pi 2}$. Since the current through the parallel combination of $R_E$ and $r_{\pi 2}$ is $I_x$, we have

$$V_{\pi 2} = -I_x R_E' \tag{10.34}$$

Writing a KVL equation, we obtain

$$V_x = I_x r_{o2} - g_{m2} r_{o2} V_{\pi 2} + I_x R_E' \tag{10.35}$$

Substituting Equation (10.34) into (10.35) yields

$$\frac{V_x}{I_x} = R_o = r_{o2} \left[ 1 + R_E' \left( g_{m2} + \frac{1}{r_{o2}} \right) \right] \tag{10.36}$$

Normally, $(1/r_{o2}) \ll g_{m2}$; therefore,

$$R_o \cong r_{o2}(1 + g_{m2} R_E') \tag{10.37}$$

The output resistance of the Widlar current source is a factor $(1 + g_{m2} R_E')$ larger than that of the simple two-transistor current source.

## EXAMPLE 10.6

**Objective:** Determine the change in load current with a change in collector voltage in a Widlar current source.

Consider the circuit in Figure 10.9. The parameters are: $V^+ = 5$ V, $V^- = -5$ V, $R_1 = 9.3$ k$\Omega$, and $R_E = 9.58$ k$\Omega$. Let $V_A = 80$ V and $\beta = 100$. Determine the change in $I_O$ as $V_{C2}$ changes by 4 V.

**Solution:** From Example 10.4, we have $I_O = 12$ $\mu$A. The small-signal collector resistance is

$$r_{o2} = \frac{V_A}{I_O} = \frac{80}{0.012} \Rightarrow 6.67 \, \text{M}\Omega$$

We can determine that

$$g_{m2} = \frac{I_O}{V_T} = \frac{0.012}{0.026} = 0.462 \, \text{mA/V}$$

and

$$r_{\pi 2} = \frac{\beta V_T}{I_O} = \frac{(100)(0.026)}{0.012} = 217 \, \text{k}\Omega$$

The output resistance of the circuit is

$$R_o = r_{o2}[1 + g_{m2}(R_E \| r_{\pi 2})] = (6.67) \cdot [1 + (0.462)(9.58 \| 217)] = 34.9 \, \text{M}\Omega$$

From Equation (10.31), the change in load current is

$$dI_O = \frac{1}{R_o} dV_{C2} = \frac{1}{34.9 \times 10^6} \times 4 \Rightarrow 0.115 \, \mu\text{A}$$

The percentage change in output current is then

$$\frac{dI_O}{I_O} = \frac{0.115}{12} = 0.0096 \Rightarrow 0.96\%$$

**Comment:** The stability of the load current, as a function of a change in output voltage, is improved in the Widlar current source, compared to the simple two-transistor current source.

**Ex 10.6:** A Widlar current source is shown in Figure 10.9. The parameters are: $V^+ = 5$ V, $V^- = 0$, $I_{REF} = 0.70$ mA, and $I_O = 25$ $\mu$A at $V_{C2} = 1$ V. The transistor parameters are: $\beta = 150$, $V_{BE1}(\text{on}) = 0.7$ V, and $V_A = 100$ V. Determine the change in $I_O$ when $V_{C2}$ changes from 1 V to 4 V. (Ans. $dI_O = 0.176$ $\mu$A)

### 10.1.4    Multitransistor Current Mirrors

In the previous current sources, we established a reference current and one load current. In the two-transistor current source in Figure 10.2(a), the B–E junction of the diode-connected transistor $Q_1$ is forward biased when the bias voltages $V^+$ and $V^-$ are applied. Once $V_{BE}$ is established, the voltage is applied to the B–E junction of $Q_2$, which turns $Q_2$ on and produces the load current $I_O$.

The B–E voltage of $Q_1$ can also be applied to additional transistors, to generate multiple load currents. Consider the circuit in Figure 10.12. Transistor $Q_R$, which is the reference transistor, is connected as a diode. The resulting B–E voltage of $Q_R$, established by $I_{REF}$, is applied to $N$ output transistors, creating $N$ load currents. The relationship between each load current and the reference current, assuming all transistors are matched and $V_A = \infty$, is

$$I_{O1} = I_{O2} = \cdots = I_{ON} = \frac{I_{REF}}{1 + \frac{(1 + N)}{\beta}} \qquad (10.38)$$

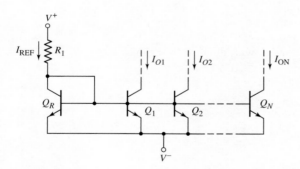

**Figure 10.12** Multitransistor current mirror

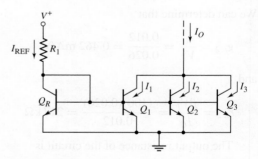

**Figure 10.13** Multioutput transistor current source

The collectors of multiple output transistors can be connected together, changing the load current versus reference current relationship. As an example, the circuit in Figure 10.13 has three output transistors with common collectors and a load current $I_O$. We assume that transistors $Q_R$, $Q_1$, $Q_2$, and $Q_3$ are all matched. If the current gain $\beta$ is very large, the base currents can be neglected, $I_1 = I_2 = I_3 = I_{REF}$, and the load current is $I_O = 3I_{REF}$. [Note: This process is not recommended for discrete devices, since a mismatch between devices will generally cause one device to carry more current than the other devices.]

Connecting transistors in parallel increases the effective B–E area of the device. In actual IC fabrication, the B–E area would be doubled or tripled to provide a load current twice or three times the value of $I_{REF}$.

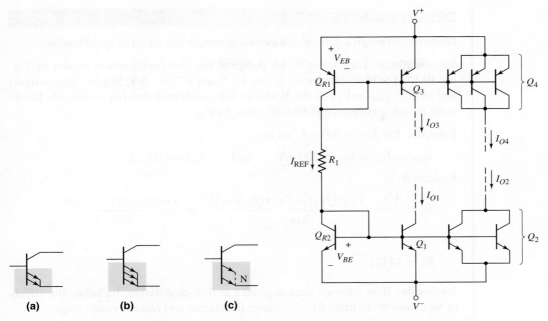

**Figure 10.14** Equivalent circuit symbols (a) two transistors in parallel, (b) three transistors in parallel, and (c) $N$ transistors in parallel

**Figure 10.15** Generalized current mirror

Rather than drawing each set of parallel output transistors, we can use the circuit symbols in Figure 10.14. Figure 10.14(a) is the equivalent symbol for two transistors connected in parallel, Figure 10.14(b) is for three transistors in parallel, and Figure 10.14(c) is for $N$ transistors in parallel. Although the transistors appear to be multiemitter devices, we are simply indicating devices with different B–E junction areas.

A generalized current mirror is shown in Figure 10.15. We can use pnp transistors to establish the load currents, as shown in the figure. Transistors $Q_{R1}$ and $Q_{R2}$ are connected as diodes. The reference current is established in the branch of the circuit that has the diode-connected transistors, resistor $R_1$, and bias voltages, and is given by

$$I_{REF} = \frac{V^+ - V_{EB}(Q_{R1}) - V_{BE}(Q_{R2}) - V^-}{R_1} \quad (10.39)$$

If $\beta$ for each transistor is very large, the base current effects can be neglected. Then the load current $I_{O1}$ generated by output transistor $Q_1$ is equal to $I_{REF}$. Likewise, $Q_3$ generates a load current $I_{O3}$ equal to $I_{REF}$. Implicitly, all transistors are identical, all load transistors are biased in their forward-active region, and all transistor Early voltages are infinite. Transistor $Q_2$ is effectively two transistors in parallel; then, since all transistors are identical, $I_{O2} = 2I_{REF}$. Similarly, $Q_4$ is effectively three transistors connected in parallel, which means that the load current is $I_{O4} = 3I_{REF}$.

In the above discussion, we neglected the effect of base currents. However, a finite $\beta$ causes the collector currents in each load transistor to be smaller than $I_{REF}$ since the reference current supplies all base currents. This effect becomes more severe as more load transistors are added.

### DESIGN EXAMPLE 10.7

**Objective:** Design a generalized current mirror to meet a set of specifications.

**Specifications:** The circuit to be designed has the configuration shown in Figure 10.15. The bias voltages are $V^+ = +5$ V and $V^- = -5$ V. Neglect base currents and assume $V_{BE} = V_{EB} = 0.6$ V. Design the circuit such that $I_{O2} = 400$ $\mu$A. Determine the other currents and find the value for $R_1$.

**Solution:** For $I_{O2} = 400$ $\mu$A, we have

$$I_{REF} = I_{O1} = I_{O3} = 200 \ \mu A \qquad \text{and} \qquad I_{O4} = 600 \ \mu A$$

Resistor $R_1$ is

$$R_1 = \frac{V^+ - V_{EB}(Q_{R1}) - V_{BE}(Q_{R2}) - V^-}{I_{REF}} = \frac{5 - 0.6 - 0.6 - (-5)}{0.2}$$

or

$$R_1 = 44 \ \text{k}\Omega$$

**Trade-offs:** Base currents were neglected in this ideal design. Including the effects of base currents (a finite $\beta$) will change the current and resistor values slightly.

**Comment:** If the load and reference currents are to be within a factor of approximately four of each other, it is more efficient, from an IC point of view, to adjust the B–E areas of the transistors to achieve the specified currents rather than use the Widlar current source with its additional resistors.

**Design Pointer:** This example demonstrates that a single reference current can be used to induce multiple load currents, which can be used to bias various stages of a complex circuit. We will see specific examples of this technique in Chapter 13 when we consider actual operational amplifier circuits.

### EXERCISE PROBLEM

**\*Ex 10.7:** Figure 10.12 shows the $N$-output current mirror. Assuming all transistors are matched, with a finite gain and $V_A = \infty$, derive Equation (10.38). If each load current must be within 10 percent of $I_{REF}$, and if $\beta = 50$, determine the maximum number of load transistors that can be connected. (Ans. $N = 4$)

## Test Your Understanding

**TYU 10.1** The circuit parameters for the current source shown in Figure 10.2(b) are $V^+ = 2.5$ V and $V^- = -2.5$, and the transistor parameters are $V_{BE}(\text{on}) = 0.7$ V, $\beta = 120$, and $V_A = \infty$. Design the circuit such that $I_O = 0.20$ mA. What is the value of $I_{REF}$? (Ans. $R_1 = 21.15$ k$\Omega$, $I_{REF} = 0.2033$ mA)

**TYU 10.2** Consider the circuit in Figure 10.2(a). The current source is $I_{REF} = 150 \ \mu$A. The transistor parameters are $I_{S1} = 8 \times 10^{-15}$ A, $I_{S2} = 5 \times 10^{-15}$ A, and $\beta = 150$. Determine $V_{BE1}$ and $I_O$. (Ans. $V_{BE1} = 0.6150$ V, $I_O = 93.75 \ \mu$A)

**TYU 10.3** For the Wilson current source in Figure 10.8, the transistor parameters are: $V_{BE}(\text{on}) = 0.7$ V, $\beta = 50$, and $V_A = \infty$. For $I_{REF} = 0.50$ mA, determine all currents shown in the figure. (Ans. $I_O = 0.4996$ mA, $I_{B3} = 9.99\ \mu\text{A}$, $I_{E3} = 0.5096$ mA, $I_{C2} = 0.490$ mA $= I_{C1}$, $I_{B1} = I_{B2} = 9.80\ \mu\text{A}$)

**TYU 10.4** The circuit and transistor parameters for the circuits in Figures 10.2(b) and 10.9 are $V^+ = 3$ V, $V^- = -3$ V, $I_{\text{REF}} = 1$ mA, $\beta = 200$, and $V_A = 50$ V. For the circuit in Figure 10.9, let $R_E = 2$ k$\Omega$. For each circuit, determine (a) $I_O$, (b) $R_o$, and (c) $dI_O/I_O$ (in percent) for $\Delta V_{C2} = 3$ V. (Ans. (a) $I_O = 1$ mA, $I_O = 41.4\ \mu\text{A}$; (b) $R_o = 50$ k$\Omega$, $R_o = 5.0$ M$\Omega$; (c) 6%, 1.45%)

# 10.2   FET CURRENT SOURCES

**Objective:** • Analyze and understand the characteristics of various MOSFET (and a few JFET) circuits used to provide a constant output current.

Field-effect transistor integrated circuits are biased with current sources in much the same way as bipolar circuits. We will examine the relationship between the reference and load currents, and will determine the output impedance of the basic two-transistor MOSFET current source. We will then analyze multi-MOSFET current-source circuits to determine reference and load current relationships and output impedance. Finally, we will discuss JFET constant-current source circuits.

## 10.2.1   Basic Two-Transistor MOSFET Current Source

**Current Relationship**

Figure 10.16 shows a basic two-transistor NMOS current source. The drain and source terminals of the enhancement-mode transistor $M_1$ are connected, which means that $M_1$ is always biased in the saturation region. Assuming $\lambda = 0$, we can write the reference current as

$$I_{\text{REF}} = K_{n1}(V_{GS} - V_{TN1})^2 \qquad (10.40)$$

Solving for $V_{GS}$ yields

$$V_{GS} = V_{TN1} + \sqrt{\frac{I_{\text{REF}}}{K_{n1}}} \qquad (10.41)$$

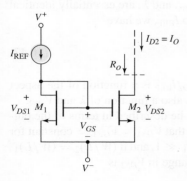

**Figure 10.16**  Basic two-transistor MOSFET current source

For the drain current to be independent of the drain-to-source voltage (for $\lambda = 0$), transistor $M_2$ should always be biased in the saturation region. The load current is then

$$I_O = K_{n2}(V_{GS} - V_{TN2})^2 \tag{10.42}$$

Substituting Equation (10.41) into (10.42), we have

$$I_O = K_{n2} \left[ \sqrt{\frac{I_{\text{REF}}}{K_{n1}}} + V_{TN1} - V_{TN2} \right]^2 \tag{10.43}$$

If $M_1$ and $M_2$ are identical transistors, then $V_{TN1} = V_{TN2}$ and $K_{n1} = K_{n2}$, and Equation (10.43) becomes

$$I_O = I_{\text{REF}} \tag{10.44}$$

Since there are no gate currents in MOSFETs, the induced load current is identical to the reference current, provided the two transistors are matched. The relationship between the load current and the reference current changes if the width-to-length ratios, or **aspect ratios,** of the two transistors change.

If the transistors are matched except for the aspect ratios, we find

$$I_O = \frac{(W/L)_2}{(W/L)_1} \cdot I_{\text{REF}} \tag{10.45}$$

The ratio between the load and reference currents is directly proportional to the aspect ratios and gives designers versatility in their circuit designs.

### Output Resistance

The stability of the load current as a function of the drain-to-source voltage is an important consideration in many applications. The drain current versus drain-to-source voltage is similar to the bipolar characteristic shown in Figure 10.4. Taking into account the finite output resistance of the transistors, we can write the load and reference currents as follows:

$$I_O = K_{n2}(V_{GS} - V_{TN2})^2(1 + \lambda_2 V_{DS2}) \tag{10.46(a)}$$

and

$$I_{\text{REF}} = K_{n1}(V_{GS} - V_{TN1})^2(1 + \lambda_1 V_{DS1}) \tag{10.46(b)}$$

Since transistors in the current mirror are processed on the same integrated circuit, all physical parameters, such as $V_{TN}$, $\mu_n$, $C_{\text{ox}}$, and $\lambda$, are essentially identical for both devices. Therefore, taking the ratio of $I_O$ to $I_{\text{REF}}$, we have

$$\frac{I_O}{I_{\text{REF}}} = \frac{(W/L)_2}{(W/L)_1} \cdot \frac{(1 + \lambda V_{DS2})}{(1 + \lambda V_{DS1})} \tag{10.47}$$

Equation (10.47) again shows that the ratio $I_O/I_{\text{REF}}$ is a function of the aspect ratios, which is controlled by the designer, and it is also a function of $\lambda$ and $V_{DS2}$.

As before, the stability of the load current can be described in terms of the output resistance. Note from the circuit in Figure 10.16 that $V_{DS1} = V_{GS1} = $ constant for a given reference current. Normally, $\lambda V_{DS1} = \lambda V_{GS1} \ll 1$, and if $(W/L)_2 = (W/L)_1$, then the change in bias current with respect to a change in $V_{DS2}$ is

$$\frac{1}{R_o} \equiv \frac{dI_O}{dV_{DS2}} = \frac{1}{r_o} \tag{10.48(a)}$$

where

$$r_o = \frac{1}{\lambda I_O} \tag{10.48(b)}$$

where $r_o$ is the output resistance of the transistor. As we found with bipolar current-source circuits, MOSFET current sources require a large output resistance for excellent stability.

### Reference Current

The reference current in bipolar current-source circuits is generally established by the bias voltages and a resistor. Since MOSFETs can be configured to act like a resistor, the reference current in MOSFET current mirrors is usually established by using additional transistors.

Consider the current mirror shown in Figure 10.17. Transistors $M_1$ and $M_3$ are in series; assuming $\lambda = 0$, we can write,

$$K_{n1}(V_{GS1} - V_{TN1})^2 = K_{n3}(V_{GS3} - V_{TN3})^2 \tag{10.49}$$

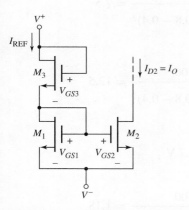

**Figure 10.17** MOSFET current source

If we again assume that $V_{TN}$, $\mu_n$, and $C_{ox}$ are identical in all transistors, then Equation (10.49) can be rewritten

$$V_{GS1} = \sqrt{\frac{(W/L)_3}{(W/L)_1}} \cdot V_{GS3} + \left(1 - \sqrt{\frac{(W/L)_3}{(W/L)_1}}\right) \cdot V_{TN} \tag{10.50}$$

where $V_{TN}$ is the threshold voltage of both transistors.

From the circuit, we see that

$$V_{GS1} + V_{GS3} = V^+ - V^- \tag{10.51}$$

Therefore,

$$V_{GS1} = \frac{\sqrt{\dfrac{(W/L)_3}{(W/L)_1}}}{1 + \sqrt{\dfrac{(W/L)_3}{(W/L)_1}}} \cdot (V^+ - V^-) + \frac{\left(1 - \sqrt{\dfrac{(W/L)_3}{(W/L)_1}}\right)}{\left(1 + \sqrt{\dfrac{(W/L)_3}{(W/L)_1}}\right)} \cdot V_{TN} = V_{GS2} \tag{10.52}$$

Finally, the load current, for $\lambda = 0$, is given by

$$I_O = \frac{k_n'}{2} \cdot \left(\frac{W}{L}\right)_2 (V_{GS2} - V_{TN})^2 \tag{10.53}$$

Since the designer has control over the width-to-length ratios of the transistors, there is considerable flexibility in the design of MOSFET current sources.

## DESIGN EXAMPLE 10.8

**Objective:** Design a MOSFET current source circuit to meet a set of specifications.

**Specifications:** The circuit to be designed has the configuration shown in Figure 10.17. The bias voltages are $V^+ = 2.5$ V and $V^- = 0$. Transistors are available with parameters $k_n' = 100\ \mu\text{A/V}^2$, $V_{TN} = 0.4$ V, and $\lambda = 0$. Design the circuit such that $I_{\text{REF}} = 100\ \mu\text{A}$, $I_O = 60\ \mu\text{A}$, and $V_{DS2}(\text{sat}) = 0.4$ V.

**Solution:** We have $V_{DS2}(\text{sat}) = 0.4 = V_{GS2} - 0.4$, so that $V_{GS2} = V_{GS1} = 0.8$ V. Then for transistor $M_2$,

$$\left(\frac{W}{L}\right)_2 = \frac{I_0}{\left(\frac{k_n'}{2}\right)(V_{GS2} - V_{TN})^2} = \frac{60}{\left(\frac{100}{2}\right)(0.8 - 0.4)^2} = 7.5$$

For transistor $M_1$,

$$\left(\frac{W}{L}\right)_1 = \frac{I_{\text{REF}}}{\left(\frac{k_n'}{2}\right)(V_{GS1} - V_{TN})^2} = \frac{100}{\left(\frac{100}{2}\right)(0.8 - 0.4)^2} = 12.5$$

The value of $V_{GS3}$ is found as

$$V_{GS3} = \left(V^+ - V^-\right) - V_{GS1} = 2.5 - 0.8 = 1.7 \text{ V}$$

Then for transistor $M_3$ we find

$$\left(\frac{W}{L}\right)_3 = \frac{I_{\text{REF}}}{\left(\frac{k_n'}{2}\right)(V_{GS3} - V_{TN})^2} = \frac{100}{\left(\frac{100}{2}\right)(1.7 - 0.4)^2} = 1.18$$

**Trade-offs:** As with other designs, slight variations in transistor parameters ($k_n'$, $W/L$, and $V_{TN}$) will change the current values slightly. See Test Your Understanding exercise TYU 10.5.

**Comment:** In this design, the output transistor remains biased in the saturation region for

$$V_{DS} > V_{DS}(\text{sat}) = V_{GS} - V_{TN} = 0.8 - 0.4 = 0.4 \text{ V}$$

**Design Pointer:** As with most design problems, there is not a unique solution. The general design criterion was that $M_2$ was biased in the saturation region over a wide range of $V_{DS2}$ values. Letting $V_{GS2} = 0.8$ V was somewhat arbitrary. If $V_{GS2}$ were smaller, the width-to-length ratios of $M_1$ and $M_2$ would need to be larger. Larger values of $V_{GS2}$ would result in smaller width-to-length ratios.

The value of $V_{GS3}$ is the difference between the bias voltage and $V_{GS1}$. If $V_{GS3}$ becomes too large, the ratio $(W/L)_3$ will become unreasonably small (much less than 1). Two or more transistors in series can be used in place of $M_3$ to divide the voltage in order to provide reasonable $W/L$ ratios (see end-of-chapter problems).

**Ex 10.8:** For the circuit shown in Figure 10.17, the bias voltages are $V^+ = 1.8$ V and $V^- = -1.8$ V, and the transistor parameters are $V_{TN} = 0.4$ V, $k_n' = 100\ \mu\text{A/V}^2$, and $\lambda = 0$. Design the circuit such that $I_{REF} = 0.5$ mA and $I_O = 0.1$ mA, and that $M_2$ remains biased in the saturation region for $V_{DS2} \geq 0.4$ V. (Ans. $(W/L)_1 = 62.5$, $(W/L)_2 = 12.5$, $(W/L)_3 = 1.74$)

### Problem-Solving Technique: MOSFET Current-Source Circuit

1. Analyze the reference side of the circuit to determine gate-to-source voltages. Using these gate-to-source voltages, determine the bias current in terms of the reference current.
2. To find the output resistance of the current source circuit, place a test voltage at the output node and analyze the small-signal equivalent circuit. Keep in mind that the reference current is a constant, which may make some of the gate voltages constant or at ac ground.

## 10.2.2  Multi-MOSFET Current-Source Circuits

### Cascode Current Mirror

In MOSFET current-source circuits, the output resistance is a measure of the stability with respect to changes in the output voltage. This output resistance can be increased by modifying the circuit, as shown in Figure 10.18, which is a **cascode current mirror.** The reference current is established by including another MOSFET in the reference branch of the circuit as was done in the basic two-transistor current mirror. Assuming all transistors are identical, then $I_O = I_{REF}$.

To determine the output resistance at the drain of $M_4$, we use the small-signal equivalent circuit. Since $I_{REF}$ is a constant, the gate voltages to $M_1$ and $M_3$, and hence to $M_2$ and $M_4$, are constant. This is equivalent to an ac short circuit. The ac equivalent circuit for calculating the output resistance is shown in Figure 10.19(a).

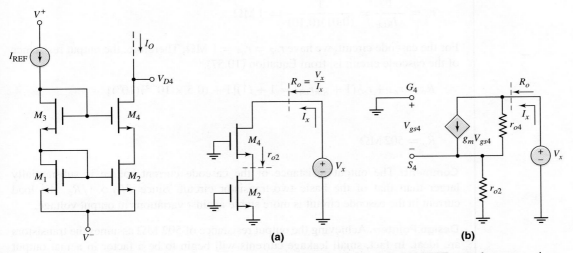

**Figure 10.18** MOSFET cascode current mirror

**Figure 10.19** Equivalent circuits of the MOSFET cascode current mirror for determining output resistance

The small-signal equivalent circuit is given in Figure 10.19(b). The small-signal resistance looking into the drain of $M_2$ is $r_{O2}$.

Writing a KCL equation, in phasor form, at the output node, we have

$$I_x = g_m V_{gs4} + \frac{V_x - (-V_{gs4})}{r_{o4}} \tag{10.54}$$

Also,

$$V_{gs4} = -I_x r_{o2} \tag{10.55}$$

Substituting Equation (10.55) into (10.54), we obtain

$$I_x + \frac{r_{o2}}{r_{o4}} I_x + g_m r_{o2} I_x = \frac{V_x}{r_{o4}} \tag{10.56}$$

The output resistance is then

$$R_o = \frac{V_x}{I_x} = r_{o4} + r_{o2}(1 + g_m r_{o4}) \tag{10.57}$$

Normally, $g_m r_{o4} \gg 1$, which implies that the output resistance of this cascode configuration is much larger than that of the basic two-transistor current source.

### EXAMPLE 10.9

**Objective:** Compare the output resistance of the cascode MOSFET current source to that of the two-transistor current source.

Consider the two-transistor current source in Figure 10.17 and the cascode current source in Figure 10.18. Assume $I_{REF} = I_O = 100\ \mu A$ in both circuits, $\lambda = 0.01\ V^{-1}$ for all transistors, and $g_m = 0.5$ mA/V.

**Solution:** The output resistance of the two-transistor current source is, from Equation (10.48(b)),

$$r_o = \frac{1}{\lambda I_{REF}} = \frac{1}{(0.01)(0.10)} \Rightarrow 1\ M\Omega$$

For the cascode circuit, we have $r_{o2} = r_{o4} = 1\ M\Omega$. Therefore, the output resistance of the cascode circuit is, from Equation (10.57),

$$R_o = r_{o4} + r_{o2}(1 + g_m r_{o4}) = 1 + (1)[1 + (0.5 \times 10^{-3})(10^6)]$$

or

$$R_o = 502\ M\Omega$$

**Comment:** The output resistance of the cascode current source is substantially larger than that of the basic two-transistor circuit. Since $dI_O \propto 1/R_o$, the load current in the cascode circuit is more stable against variations in output voltage.

**Design Pointer:** Achieving the output resistance of 502 M$\Omega$ assumes the transistors are ideal. In fact, small leakage currents will begin to be a factor in actual output resistance values, so a value of 502 M$\Omega$ may not be achieved in reality.

**Ex 10.9:** In the MOSFET cascode current source shown in Figure 10.18, all transistors are identical, with parameters: $V_{TN} = 1$ V, $K_n = 80$ $\mu$A/V$^2$, and $\lambda = 0.02$ V$^{-1}$. Let $I_{REF} = 20$ $\mu$A. The circuit is biased at $V^+ = 5$ V and $V^- = -5$ V. Determine: (a) $V_{GS}$ of each transistor, (b) the lowest possible voltage value $V_{D4}$, and (c) the output resistance $R_o$. (Ans. (a) $V_{GS} = 1.5$ V (b) $V_{D4}(\text{min}) = -3.0$ V (c) $R_o = 505$ M$\Omega$)

## Wilson Current Mirror

Two additional multi-MOSFET current sources are shown in Figures 10.20(a) and 10.20(b). The circuit in Figure 10.20(a) is the **Wilson current source.** Note that the $V_{DS}$ values of $M_1$ and $M_2$ are not equal. Since $\lambda$ is not zero, the ratio $I_O/I_{REF}$ is slightly different from the aspect ratios. This problem is solved in the **modified Wilson current source,** shown in Figure 10.20(b), which includes transistor $M_4$. For a constant reference current, the drain-to-source voltages of $M_1$, $M_2$, and $M_4$ are held constant. The primary advantage of these circuits is the increase in output resistance, which further stabilizes the load current.

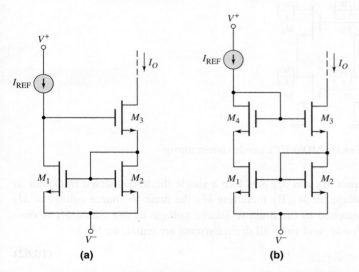

**(a)**                    **(b)**

**Figure 10.20** (a) MOSFET Wilson current source and (b) modified MOSFET Wilson current source

## Wide-Swing Current Mirror

If we consider the cascode current mirror in Figure 10.18, we can determine the minimum value of $V_{D4}$, which will influence the maximum symmetrical swing of the voltage in the load circuit being biased. The gate voltage of $M_4$ is

$$V_{G4} = V^- + V_{GS1} + V_{GS3} \tag{10.58}$$

The minimum $V_{D4}$ is then

$$V_{D4}(\text{min}) = V_{G4} - V_{GS4} + V_{DS4}(\text{sat}) \tag{10.59}$$

Assuming matched transistors, $V_{GS1} = V_{GS2} = V_{GS4} \equiv V_{GS}$. We then find

$$V_{D4}(\text{min}) = V^- + (V_{GS} + V_{DS4}(\text{sat})) \tag{10.60}$$

In considering the simple two-transistor current mirror, the minimum output voltage is

$$V_O(\text{min}) = V^- + V_{DS}(\text{sat}) \tag{10.61}$$

If, for example, $V_{GS} = 0.75$ V and $V_{TN} = 0.50$ V, then from Equation (10.60), $V_{D4}(\text{min}) = 1.0$ V above $V^-$, and from Equation (10.61), $V_O(\text{min})$ is only 0.25 V above $V^-$. For bias voltages in the range of $\pm 3.5$ V, this additional required voltage across the output of the cascode current mirror can have a significant effect on the output of the load circuit.

One current mirror circuit that does not limit the output voltage swing as severely as the cascode circuit, but retains the high output resistance, is shown in Figure 10.21. Width-to-length ratios of the transistors are shown. Otherwise, the transistors are assumed to be identical.

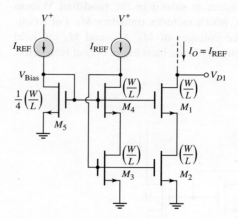

**Figure 10.21** A wide-swing MOSFET cascode current mirror

The transistor pair $M_3$ and $M_4$ acts like a single diode-connected transistor in creating the gate voltage for $M_3$. By including $M_4$, the drain-to-source voltage of $M_3$ is reduced and is matched to the drain-to-source voltage of $M_2$. Since $M_5$ is one-fourth the size of $M_1–M_4$ and since all drain currents are equal, we have

$$(V_{GS5} - V_{TN}) = 2(V_{GSi} - V_{TN}) \tag{10.62}$$

where $V_{GSi}$ corresponds to the gate-to-source voltage of $M_1 - M_4$.

The voltage at the gate of $M_1$ is

$$V_{G1} = V_{GS5} = (V_{GS5} - V_{TN}) + V_{TN} \tag{10.63}$$

The minimum output voltage at the drain of $M_1$ is

$$V_{D1}(\text{min}) = V_{G1} - V_{GS1} + V_{DS1}(\text{sat})$$
$$= [(V_{GS5} - V_{TN}) + V_{TN}] - V_{GS1} + (V_{GS1} - V_{TN}) \tag{10.64}$$

or

$$V_{D1}(\text{min}) = V_{GS5} - V_{TN} = 2(V_{GSi} - V_{TN}) = 2V_{DSi}(\text{sat}) \tag{10.65}$$

If we have $V_{GSi} = 0.75$ V and $V_{TN} = 0.5$ V, then $V_{D1}(\text{min}) = 0.50$ V, which is one-half the value for the cascode circuit. At the same time, the high output resistance is maintained.

**Discussion:** In the ideal circuit design in Figure 10.21, the transistors $M_3$ and $M_4$ are biased exactly at the transition point between the saturation and non-saturation regions. The analysis has neglected the body effect, so threshold voltages will not be exactly equal. In an actual circuit design, therefore, the size of $M_5$ will be made slightly smaller to ensure transistors are biased in the saturation region. This design change then means that the minimum output voltage increases by perhaps 0.1 to 0.15 V.

### 10.2.3 Bias-Independent Current Source

In all of the current mirror circuits considered up to this point (both BJT and MOSFET), the reference current is a function of the applied supply voltages. This implies that the load current is also a function of the supply voltages. In most cases, the supply voltage dependence is undesirable. Circuit designs exist in which the load currents are essentially independent of the bias. One such MOSFET circuit is shown in Figure 10.22. The width-to-length ratios are given.

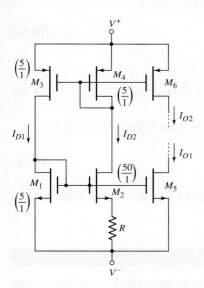

**Figure 10.22** Bias-independent MOSFET current mirror

Since the PMOS devices are matched, the currents $I_{D1}$ and $I_{D2}$ must be equal. Equating the currents in $M_1$ and $M_2$, we find

$$I_{D1} = \frac{k_n'}{2}\left(\frac{W}{L}\right)_1 (V_{GS1} - V_{TN})^2 = I_{D2} = \frac{k_n'}{2}\left(\frac{W}{L}\right)_2 (V_{GS2} - V_{TN})^2 \quad (10.66)$$

Also

$$V_{GS2} = V_{GS1} - I_{D2}R \quad (10.67)$$

Substituting Equation (10.67) into Equation (10.66) and solving for $R$, we obtain

$$R = \frac{1}{\sqrt{K_{n1}I_{D1}}}\left(1 - \sqrt{\frac{(W/L)_1}{(W/L)_2}}\right) \quad (10.68)$$

This value of resistance $R$ will establish the drain currents $I_{D1} = I_{D2}$. These currents establish the gate-to-source voltage across $M_1$ and source-to-gate voltage across $M_3$.

These voltages, in turn, can be applied to $M_5$ and $M_6$ to establish load currents $I_{O1}$ and $I_{O2}$.

The currents $I_{D1}$ and $I_{D2}$ are independent of the supply voltages $V^+$ and $V^-$ as long as $M_2$ and $M_3$ are biased in the saturation region. As the difference, $V^+ - V^-$, increases, the values of $V_{DS2}$ and $V_{SD3}$ increase but the currents remain essentially constant.

Similar bipolar bias-independent current mirror designs exist, but will not be covered here.

### 10.2.4    JFET Current Sources

Current sources are also fundamental elements in JFET integrated circuits. The simplest method of forming a current source is to connect the gate and source terminals of a depletion-mode JFET, as shown in Figure 10.23 for an n-channel device. The device will remain biased in the saturation region as long as

$$v_{DS} \geq v_{DS}(\text{sat}) = v_{GS} - V_P = |V_P| \tag{10.69}$$

In the saturation region, the current is

$$i_D = I_{DSS}\left(1 - \frac{v_{GS}}{V_P}\right)^2 (1 + \lambda v_{DS}) = I_{DSS}(1 + \lambda v_{DS}) \tag{10.70}$$

The output resistance looking into the drain is, from Equation (10.70),

$$\frac{1}{r_o} = \frac{di_D}{dv_{DS}} = \lambda I_{DSS} \tag{10.71}$$

This expression for the output resistance of a JFET current source is the same as that of the MOSFET current source.

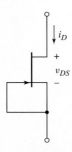

**Figure 10.23** Depletion-mode JFET connected as a current source

### EXAMPLE 10.10

**Objective:** Determine the currents and voltages in a simple JFET circuit biased with a constant-current source.

Consider the circuit shown in Figure 10.24. The transistor parameters are: $I_{DSS1} = 2$ mA, $I_{DSS2} = 1$ mA, $V_{P1} = V_{P2} = -1.5$ V, and $\lambda_1 = \lambda_2 = 0.05$ V$^{-1}$. Determine the minimum values of $V_S$ and $V_I$ such that $Q_2$ is biased in the saturation region. What is the value of $I_O$?

**Solution:** In order for $Q_2$ to remain biased in the saturation region, we must have $v_{DS} \geq |V_P| = 1.5$ V, from Equation (10.69). The minimum value of $V_S$ is then

$$V_S(\text{min}) - V^- = v_{DS}(\text{min}) = 1.5 \text{ V}$$

or

$$V_S(\text{min}) = 1.5 + V^- = 1.5 + (-5) = -3.5 \text{ V}$$

From Equation (10.70), the output current is

$$i_D = I_O = I_{DSS2}(1 + \lambda v_{DS}) = (1)[1 + (0.05)(1.5)] = 1.08 \text{ mA}$$

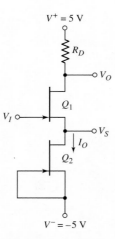

**Figure 10.24** The dc equivalent circuit of simple JFET amplifier biased with JFET current source

As a first approximation in calculating the minimum value of $V_I$, we neglect the effect of $\lambda$ in transistor $Q_1$. Then, assuming $Q_1$ is biased in the saturation region, we have

$$i_D = I_{DSS1}\left(1 - \frac{v_{GS1}}{V_{P1}}\right)^2$$

or

$$1.08 = 2\left(1 - \frac{v_{GS1}}{(-1.5)}\right)^2$$

which yields

$$v_{GS1} = -0.40 \text{ V}$$

We see that

$$v_{GS1} = -0.40 \text{ V} = V_I - V_S = V_I - (-3.5)$$

or

$$V_I = -3.90 \text{ V}$$

**Comment:** Since $Q_1$ is an n-channel device, the voltage at the gate is negative with respect to the source.

---

### EXERCISE PROBLEM

**\*Ex 10.10:** Consider the JFET circuit in Figure 10.24. The transistor parameters are: $I_{DSS2} = 0.5$ mA, $I_{DSS1} = 0.8$ mA, $V_{P1} = V_{P2} = -2$ V, and $\lambda_1 = \lambda_2 = 0.15$ V$^{-1}$. Determine the minimum values of $V_S$ and $V_I$ such that $Q_2$ is biased in the saturation region. What is the value of $I_O$? What is the output impedance looking into the drain of $Q_2$? (Ans. $V_S(\text{min}) = -3$ V, $I_O = 0.65$ mA, $V_I(\text{min}) = -3.2$ V, $r_o = 1.09$ k$\Omega$)

---

The output resistance of a JFET current source can be increased by using a cascode configuration. A simple JFET cascode current source with two n-channel depletion-mode devices is shown in Figure 10.25. The current–voltage relationship, assuming $Q_1$ and $Q_2$ are identical, is given by

$$i_D = I_{DSS}(1 + \lambda v_{DS1}) = I_{DSS}\left(1 - \frac{v_{GS2}}{V_P}\right)^2(1 + \lambda v_{DS2}) \qquad \textbf{(10.72)}$$

From the circuit, we see that $v_{GS2} = -v_{DS1}$. We define

$$V_{DS} = v_{DS1} + v_{DS2} \qquad \textbf{(10.73(a))}$$

so that

$$v_{DS2} = V_{DS} - v_{DS1} \qquad \textbf{(10.73(b))}$$

From Equation (10.72), we obtain

$$(1 + \lambda v_{DS1}) = \left(1 + \frac{v_{DS1}}{V_P}\right)^2[1 + \lambda(V_{DS} - v_{DS1})] \qquad \textbf{(10.74)}$$

For a given application, the value of $V_{DS}$ will usually be known, and the value of $v_{DS1}$ can then be determined. The load current $i_D$ can then be calculated by using Equation (10.72).

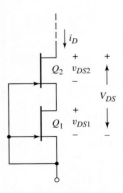

**Figure 10.25** JFET cascode current source

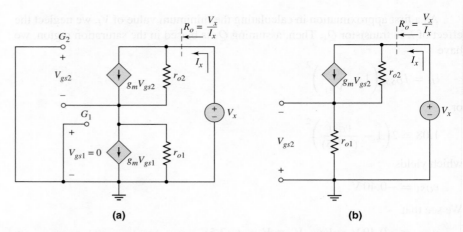

**(a)**                                **(b)**

**Figure 10.26** (a) Equivalent circuit, using phasor notation, of the JFET cascode current source for determining output resistance and (b) final configuration

We can determine the output resistance by using the small-signal equivalent circuit of the composite two-transistor configuration, as shown in Figure 10.26(a), which includes the phasor variables. Since the gate and source of $Q_1$ are connected together, the small-signal voltage $V_{gs1}$ is zero, which means that the dependent current source $g_m V_{gs1}$ is zero. This corresponds to an open circuit. Figure 10.26(b) shows the final configuration.

The analysis is the same as for the MOSFET cascode circuit in Figure 10.19. Writing a KCL equation at the output node, we have

$$I_x = g_m V_{gs2} + \frac{V_x - (-V_{gs2})}{r_{o2}} \tag{10.75}$$

Noting that

$$V_{gs2} = -I_x r_{o1} \tag{10.76}$$

Equation (10.75) becomes

$$I_x = -(g_m r_{o1})I_x + \frac{V_x}{r_{o2}} - \left(\frac{r_{o1}}{r_{o2}}\right)I_x \tag{10.77}$$

The output resistance is then

$$R_o = \frac{V_x}{I_x} = r_{o2} + r_{o1} + g_m r_{o1} r_{o2} = r_{o2} + r_{o1}(1 + g_m r_{o2}) \tag{10.78}$$

From Equation (10.78), we see that the output resistance relationship for the JFET cascode current source has the same form as that of the MOSFET cascode current source.

## Test Your Understanding

**TYU 10.5** Consider Design Example 10.8. Assume transistor parameters of $k'_{n1} = 100\,\mu\text{A/V}^2$, $k'_{n2} = 105\,\mu\text{A/V}^2$, $k'_{n3} = 95\,\mu\text{A/V}^2$, $V_{TN1} = 0.38$ V, $V_{TN2} = 0.40$ V, $V_{TN3} = 0.42$ V, and $\lambda_1 = \lambda_2 = \lambda_3 = 0$. (a) Using the designed values of $W/L$ for each transistor, determine the values of $I_{\text{REF}}$ and $I_O$. (b) What is the percent change in $I_{\text{REF}}$ and $I_O$ from Example 10.8? (Ans. (a) $I_{\text{REF}} = 95.93\,\mu\text{A}$, $I_O = 54.43\,\mu\text{A}$; (b) −4.06%, −9.28%)

**TYU 10.6** The bias voltages of the MOSFET current source in Figure 10.17 are $V^+ = 3$ V and $V^- = -3$ V. The transistor parameters are $V_{TN} = 0.5$ V, $k'_n = 80\,\mu\text{A/V}^2$, and $\lambda = 0.02$ V$^{-1}$. The transistor width-to-length ratios are $(W/L)_3 = 3$, $(W/L)_1 = 12$, and $(W/L)_2 = 6$. Determine: (a) $I_{REF}$, (b) $I_O$ at $V_{DS2} = 2$ V, and (c) $I_O$ at $V_{DS2} = 4$ V. (Ans. (a) $I_{REF} = 1.33$ mA, (b) $I_O = 0.6936$ mA, (c) $I_O = 0.7203$ mA)

**TYU 10.7** Consider the circuit shown in Figure 10.27. The bias voltages are changed to $V^+ = 3$ V and $V^- = -3$ V. The transistor parameters are $V_{TN} = 0.7$ V, $K_{n1} = 0.35$ mA/V$^2$, $K_{n2} = 0.30$ mA/V$^2$, $K_{n3} = 0.10$ mA/V$^2$, and $\lambda = 0$. Determine $I_{REF}$ and $I_O$. (Note: All transistors labeled $M_2$ are identical.) (Ans. $I_{REF} = 0.8986$ mA, $I_O = 2.31$ mA)

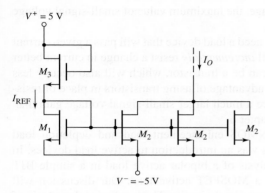

$V^+ = 5$ V

$I_O$

$M_3$

$I_{REF}$

$M_1$

$M_2$   $M_2$   $M_2$

$V^- = -5$ V

**Figure 10.27** Figure for Exercise TYU 10.7

**TYU 10.8** All transistors in the MOSFET modified Wilson current source in Figure 10.20(b) are identical. The parameters are: $V_{TN} = 1$ V, $K_n = 0.2$ mA/V$^2$, and $\lambda = 0$. If $I_{REF} = 250\,\mu\text{A}$, determine $I_O$ and $V_{GS}$ for each transistor. (Ans. $I_O = I_{REF} = 250\,\mu\text{A}$, $V_{GS} = 2.12$ V)

 ## 10.3 CIRCUITS WITH ACTIVE LOADS

**Objective:** • Analyze the dc characteristics of amplifier circuits using transistors as load devices (active loads).

In bipolar amplifiers, such as that shown in Figure 10.28, the small-signal voltage gain is directly proportional to the collector resistor $R_C$. To increase the gain, we need to increase the value of $R_C$, but there is a practical limitation. We can show that the voltage gain (assuming $C_C$ acts as a short circuit to the signal frequency) of this circuit is given by

$$A_v = -g_m R_C$$

where

$$g_m = \frac{I_{CQ}}{V_T}$$

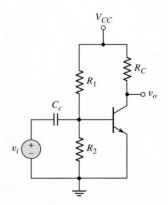

**Figure 10.28** Bipolar common-emitter circuit

Assuming the $Q$-point is in the center of the load line, then

$$I_{CQ} = \frac{V_{CC}}{2R_C}$$

or

$$R_C = \frac{V_{CC}}{2I_{CQ}}$$

Substituting into the voltage gain expression, we have

$$|A_v| = \frac{V_{CC}}{2V_T}$$

So for reasonable values of bias voltage, the maximum value of small-signal voltage gain is essentially fixed.

To get around this limitation, we need a load device that will pass a given current at a given bias voltage, but which will *incrementally* resist a change in current better than the fixed $R_C$. This load device can be a transistor, which will also occupy less area in an integrated circuit, another advantage of using transistors in place of resistors. In addition, active loads produce a much larger small-signal voltage gain than discrete resistors, as discussed in Chapter 6.

In Chapter 4, we introduced NMOS enhancement load and depletion load devices in MOSFET amplifiers. This was an introduction to active load devices. In this section, we consider the dc analysis of a bipolar active load in a simple BJT circuit and then the dc analysis of a MOSFET active load. Our discussion will include the voltage gains of these active load circuits. The small-signal analysis of active load circuits is covered in the next section.

The discussion of active loads here can be considered an introduction. The use of active loads with differential amplifiers is considered in detail in the next chapter.

### 10.3.1 DC Analysis: BJT Active Load Circuit

Consider the circuit shown in Figure 10.29. The elements $R_1$, $Q_1$, and $Q_2$ form the active load circuit, and $Q_2$ is referred to as the **active load device** for driver transistor $Q_0$. The combination of $R_1$, $Q_1$, and $Q_2$ forms the pnp version of the two-transistor

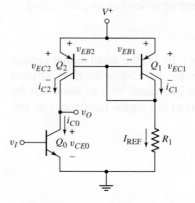

**Figure 10.29** Simple BJT amplifier with active load, showing currents and voltages

current mirror. For the dc analysis of this circuit, we will use the dc symbols for the currents and voltages. The objective of this analysis is to obtain the voltage transfer function $V_O$ versus $V_I$.

The B–E voltage of $Q_0$ is the dc input voltage $V_I$; therefore, the collector current in $Q_0$ is

$$I_{C0} = I_{S0}[e^{V_I/V_T}]\left(1 + \frac{V_{CE0}}{V_{AN}}\right) \tag{10.79}$$

where $I_{S0}$ is the reverse-saturation current, $V_T$ is the thermal voltage, and $V_{AN}$ is the Early voltage of the npn transistor. Similarly, the collector current in $Q_2$ is

$$I_{C2} = I_{S2}[e^{V_{EB2}/V_T}]\left(1 + \frac{V_{EC2}}{V_{AP}}\right) \tag{10.80}$$

where $V_{AP}$ is the Early voltage of the pnp transistors.

If we neglect base currents, then

$$I_{REF} = I_{C1} = I_{S1}[e^{V_{EB1}/V_T}]\left(1 + \frac{V_{EC1}}{V_{AP}}\right) \tag{10.81}$$

Assuming $Q_1$ and $Q_2$ are identical, then $I_{S1} = I_{S2}$ and the Early voltages of the pnp transistors are equal. Also note that $V_{EC1} = V_{EB1} = V_{EB2}$. We can also assume that $V_{CE} \ll V_{AN}$ and $V_{EC} \ll V_{AP}$. Combining equations, we find the output voltage is given as

$$V_O = \frac{V_{AN}V_{AP}}{V_{AN} + V_{AP}}\left[1 - \frac{I_{S0}e^{V_I/V_T}}{I_{REF}}\right] + \frac{V_{AN}}{V_{AN} + V_{AP}}(V^+ - V_{EB2}) \tag{10.82}$$

Equation (10.82) is valid as long as $Q_0$ and $Q_2$ remain biased in the forward-active region, which means that the output voltage must remain in the range

$$V_{CE0}(\text{sat}) < V_O < (V^+ - V_{EC2}(\text{sat})) \tag{10.83}$$

A sketch of $V_O$ versus $V_I$ is shown in Figure 10.30. If the circuit is to be used as a small-signal amplifier, a $Q$-point must be established, as indicated in the figure, for maximum symmetrical swing. Because of the exponential input voltage function, as given in Equation (10.82), the input voltage range over which both $Q_0$ and $Q_2$ remain in their active regions is very small. A sinusoidal variation in the input voltage produces a sinusoidal variation in the output voltage as shown in the figure.

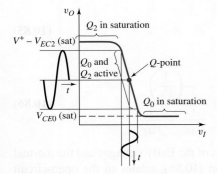

**Figure 10.30** Voltage transfer characteristics of bipolar circuit with active load

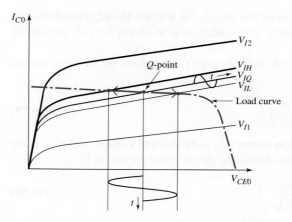

**Figure 10.31** Driver transistor characteristics and load curve for BJT circuit with active load

In addition to the voltage transfer function, we can also consider the load curve. Figure 10.31 shows the transistor characteristics of the driver transistor $Q_0$ for several values of B–E or $V_I$ voltages. Superimposed on these curves is the load curve, which essentially is the $I_C$ versus $V_{EC}$ characteristic of the active load $Q_2$ at a constant $V_{EB}$ voltage.

The $Q$-point shown corresponds to a quiescent input voltage $V_{IQ}$. From the curve, we see that as the input changes between $V_{IH}$ and $V_{IL}$, the $Q$-point moves up and down the load curve producing a change in output voltage. Also, as $V_I$ increases to $V_{I2}$, the driver transistor $Q_0$ is driven into saturation; as $V_I$ decreases to $V_{I1}$, the load transistor $Q_2$ is driven into saturation.

10.3.2   **Voltage Gain: BJT Active Load Circuit**

The small-signal voltage gain of a circuit is the slope of the voltage transfer function curve at the $Q$-point. For the bipolar circuit with an active load, the voltage gain can be found by taking the derivative of Equation (10.82) with respect to $V_I$, as follows:

$$A_v = \frac{dV_O}{dV_I} = -\left(\frac{V_{AN}\,V_{AP}}{V_{AN} + V_{AP}}\right)\left(\frac{I_{S0}}{I_{REF}}\right)\left(\frac{1}{V_T}\right)e^{V_I/V_T} \tag{10.84}$$

As a good approximation, we can write that

$$I_{REF} \cong I_{S0}\,e^{V_I/V_T} \tag{10.85}$$

Equation (10.84) then becomes

$$A_v = \frac{dV_O}{dV_I} = -\left(\frac{V_{AN}\,V_{AP}}{V_{AN} + V_{AP}}\right)\left(\frac{1}{V_T}\right) = \frac{-\left(\dfrac{1}{V_T}\right)}{\dfrac{1}{V_{AN}} + \dfrac{1}{V_{AP}}} \tag{10.86}$$

The small-signal voltage gain is a function of the Early voltages and the thermal voltage. The voltage gain, given by Equation (10.86), relates to the open-circuit condition. When a load is connected to the output, the voltage gain is degraded, as we will see in the next section.

## EXAMPLE **10.11**

**Objective:** Calculate the open-circuit voltage gain of a simple BJT amplifier with an active load.

Consider the circuit shown in Figure 10.29. The transistor parameters are $V_{AN} = 120$ V and $V_{AP} = 80$ V. Let $V_T = 0.026$ V.

**Solution:** From Equation (10.86), the small-signal, open-circuit voltage gain is

$$A_v = \frac{-\left(\dfrac{1}{V_T}\right)}{\dfrac{1}{V_{AN}} + \dfrac{1}{V_{AP}}} = \frac{-\left(\dfrac{1}{0.026}\right)}{\dfrac{1}{120} + \dfrac{1}{80}} = \frac{-38.46}{0.00833 + 0.0125} = -1846$$

**Comment:** For a circuit with an active load, the magnitude of the small-signal, open-circuit voltage gain is substantially larger than the resulting gain when a discrete resistor load is used.

**Computer Verification:** The voltage transfer characteristics of the active load circuit in Figure 10.29 were determined for a standard 2N3904 transistor as the npn device and standard 2N3906 transistors as the pnp devices. The circuit was biased at 5 V and the resistor was set at $R = 1$k$\Omega$. The transfer curve is shown in Figure 10.32.

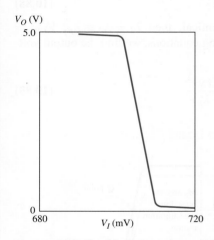

**Figure 10.32** Graphical output from a PSpice analysis, showing voltage transfer characteristics of bipolar active load circuit

The input transition region, during which both $Q_0$ and $Q_2$ remain biased in the forward-active mode, is indeed very narrow. The slope of the curve, which is the voltage gain, is found to be $-572$. The reason for the smaller value compared to the hand calculation is that the Early voltages of these standard transistors are smaller than assumed in the previous calculation. The Early voltage of the npn device is 74 V and that of the pnp devices is only 18.7 V.

**Design Pointer:** From the transfer characteristics in Figure 10.32, we can see that, for this circuit, it would be very difficult to apply the required input voltage to bias both $Q_0$ and $Q_2$ in the active region. This particular circuit, therefore, is not practical as an amplifier. However, the circuit does demonstrate the basic properties of an active load. In Chapters 11 and 13, we will see how an active load is applied to actual circuits.

EXERCISE PROBLEM

**\*Ex 10.11:** A simple BJT amplifier with active load is shown in Figure 10.29. The transistor parameters are: $I_{S0} = I_{S1} = I_{S2} = 10^{-12}$ A and $V_{AN} = V_{AP} = 100$ V. Let $V^+ = 5$ V. (a) Determine the value of $V_{EB2}$ such that $I_{REF} = 0.5$ mA. (b) Find the value of $R_1$. (c) What value of $V_I$ will produce $V_{CE0} = V_{EC2}$? (d) Determine the open-circuit, small-signal voltage gain. (Ans. (a) $V_{EB2} = 0.521$ V (b) $R_1 = 8.96$ kΩ (c) $V_I = 0.521$ V (d) $A_V = -1923$)

### 10.3.3    DC Analysis: MOSFET Active Load Circuit

Consider the circuit in Figure 10.33. Transistors $M_1$ and $M_2$ form a PMOS active load circuit, and $M_2$ is the active load device. We will consider the voltage transfer function of $V_O$ versus $V_I$ for this circuit.

The reference current may be written in the form

$$I_{REF} = K_{p1}(V_{SG} + V_{TP1})^2(1 + \lambda_1 V_{SD1}) \qquad (10.87)$$

The drain current $I_2$ is

$$I_2 = K_{p2}(V_{SG} + V_{TP2})^2(1 + \lambda_2 V_{SD2}) \qquad (10.88)$$

If we assume that $M_1$ and $M_2$ are identical, then $\lambda_1 = \lambda_2 \equiv \lambda_p$, $V_{TP1} = V_{TP2} \equiv V_{TP}$, and $K_{p1} = K_{p2} \equiv K_p$. Combining equations, we find the output voltage as

$$V_O = \frac{[1 + \lambda_p(V^+ - V_{SG})]}{\lambda_n + \lambda_p} - \frac{K_n(V_I - V_{TN})^2}{I_{REF}(\lambda_n + \lambda_p)} \qquad (10.89)$$

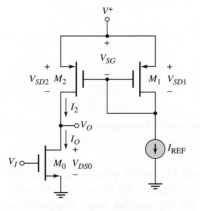

**Figure 10.33**  Simple MOSFET amplifier with active load, showing currents and voltages

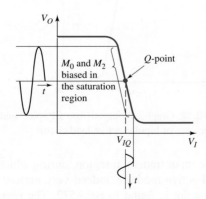

**Figure 10.34**  Voltage transfer characteristic of MOSFET circuit with active load

Equation (10.89) describes that $V_O$ versus $V_I$ characteristic of the circuit, provided that both $M_0$ and $M_2$ remain biased in their saturation regions. Figure 10.34 shows a sketch of the voltage transfer characteristics. If the circuit is to be used as a small-signal amplifier, then a $Q$-point must be established, as indicated on the figure, for maximum symmetrical swing. As before, the input transition region in which both $M_0$ and $M_2$ are

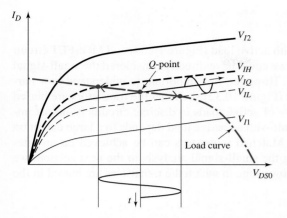

**Figure 10.35** Driver transistor characteristics and load curve for MOSFET circuit with active load

biased in the saturation region is quite narrow. A sinusoidal variation in the input voltage produces a sinusoidal variation in the output voltage as shown in the figure.

We can also consider the load curve for this device. Figure 10.35 shows the transistor characteristics of the driver transistor $M_0$ for several values of gate-to-source or $V_I$ voltages. Superimposed on these curves is the load curve, which essentially is the $I_D$ versus $V_{SD}$ characteristic of the active load $M_2$ at a constant $V_{SG}$ voltage.

The $Q$-point shown corresponds to a quiescent input voltage $V_{IQ}$. From the curve, we see that as the input changes between $V_{IH}$ and $V_{IL}$, the $Q$-point moves up and down the load curve producing a change in output voltage. Also, as $V_I$ increases to $V_{I2}$, the driver transistor $M_0$ is driven into the nonsaturation region; as $V_I$ decreases to $V_{I1}$, the load transistor $M_2$ is driven into the non-saturation region.

### 10.3.4    Voltage Gain: MOSFET Active Load Circuit

The small-signal voltage gain of a MOSFET circuit with an active load is also the slope of the voltage transfer function curve at the $Q$-point. Taking the derivative of Equation (10.89) with respect to $V_I$, we obtain

$$A_v = \frac{dV_O}{dV_I} = \frac{-2K_n(V_I - V_{TN})}{I_{\text{REF}}(\lambda_n + \lambda_p)} \tag{10.90}$$

The transconductance of the driver transistor is $g_m = 2K_n(V_I - V_{TN})$. Since $M_1$ and $M_2$ are assumed to be identical, then $I_O = I_{\text{REF}}$, and the small-signal transistor resistances are $r_{on} = 1/\lambda_n I_{\text{REF}}$ and $r_{op} = 1/\lambda_p I_{\text{REF}}$. From Equation (10.90), the small-signal, open-circuit voltage gain can now be written

$$A_v = \frac{-g_m}{\left(\dfrac{1}{r_{on}} + \dfrac{1}{r_{op}}\right)} = -g_m(r_{on}\|r_{op}) \tag{10.91}$$

In general, the transconductance $g_m$ of a MOSFET is less than that of a BJT; therefore, the voltage gain of a MOSFET amplifier with an active load is less than that of a BJT amplifier with an active load. However, the active load still produces a significant increase in the voltage gain.

### 10.3.5 Discussion

In considering the BJT circuit with active load (Figure 10.29) and MOSFET circuit with active load (Figure 10.33), we could have directly considered the small-signal analysis without the dc analysis. However, it is important to understand how narrow the input transition width is (Figure 10.32) such that the transistors are biased correctly. For this reason, the use of active loads in discrete circuits is almost impossible. The biasing of the circuit with an active load depends to a large extent on the use of matched transistors. Matched transistors can be achieved on an integrated circuit. So in considering the small-signal analysis in the next section, we must keep in mind the very narrow range in which the transistors are biased in the active region.

## Test Your Understanding

**TYU 10.9** Repeat Exercise Problem 10.11 if the transistor parameters are $I_{SO} = I_{S1} = I_{S2} = 5 \times 10^{-14}$ A and if $I_{REF} = 0.1$ mA. (Ans. (a) $V_{EB2} = 0.557$ V (b) $R_1 = 44.4$ kΩ (c) $V_I = 0.557$ V (d) $A_v = -1923$)

**TYU 10.10** Consider the simple MOSFET amplifier with active load in Figure 10.33. The transistor parameters are $V_{TN} = 0.7$ V, $V_{TP} = -0.7$ V, $K_n = K_p = 0.12$ mA/V$^2$, and $\lambda_n = \lambda_p = 0.02$ V$^{-1}$. Let $V^+ = 5$ V and $I_{REF} = 0.15$ mA. (a) Determine $V_{SG}$. (b) Find the value of $V_I$ that produces $V_{DSO} = V_{SD2}$. (c) Determine the open-circuit small-signal voltage gain. (Ans. (a) $V_{SG} = 1.818$ V, (b) $V_I = 1.798$ V, (c) $A_v = -43.9$)

**TYU 10.11** Repeat Exercise TYU 10.10 if the transistor parameters are $K_n = K_p = 50 \, \mu$A/V$^2$, and if $I_{REF} = 80 \, \mu$A. Other transistor parameters are as given in TYU 10.10. (Ans. (a) $V_{SG} = 1.965$ V, (b) $V_I = 1.940$ V, (c) $A_v = -38.74$)

## 10.4 SMALL-SIGNAL ANALYSIS: ACTIVE LOAD CIRCUITS

**Objective:** • Analyze the small-signal characteristics of amplifier circuits with active loads.

The small-signal voltage gain of a circuit with an active load can be determined from the small-signal equivalent circuit. This is probably the easiest and most direct method of obtaining the gain of such circuits. Again, the dc analysis of these circuits, as shown in the previous section, clearly demonstrates the narrow range of input voltages over which the transistors will remain biased in the active region. The load curves in Figure 10.31 for the BJT circuit and in Figure 10.35 for the MOSFET circuit also help in visualizing the operation of these circuits. Even though a small-signal analysis is extremely useful for determining the voltage gain, we must not lose sight of the physical operation of these circuits, which is described through the dc analysis. If the BJTs are not biased in the active region or the MOSFETs are not biased in the saturation region, the small-signal analysis is not valid.

## Small-Signal Analysis: BJT Active Load Circuit

To find the small-signal voltage gain of the BJT circuit with an active load, we must determine the resistance looking into the collector of the active load device. Figure 10.36 is the small-signal equivalent circuit of the entire active load circuit in Figure 10.29, which uses pnp transistors. The base, collector, and emitter terminals of the two transistors are indicated on the figure.

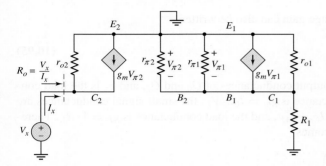

**Figure 10.36** Small-signal equivalent circuit of BJT active load circuit

In the $Q_1$ portion of the equivalent circuit, there are no independent ac sources to excite any currents or voltages. Therefore, $V_{\pi 1} = V_{\pi 2} = 0$, which means that the dependent source $g_m V_{\pi 2}$ is zero and is equivalent to an open circuit. The resistance looking into the collector of $Q_2$ is just

$$R_o = r_{o2} \qquad\qquad (10.92)$$

We will use this equivalent resistance to calculate the small-signal voltage gain of the amplifier.

Figure 10.37(a) shows a simple amplifier with an active load and the output voltage capacitively coupled to passive load $R_L$. The small-signal equivalent circuit, shown in Figure 10.37(b), includes the load resistance $R_L$, the resistance $r_{o2}$ of the active load, and the output resistance $r_o$ of the amplifying transistor $Q_0$.

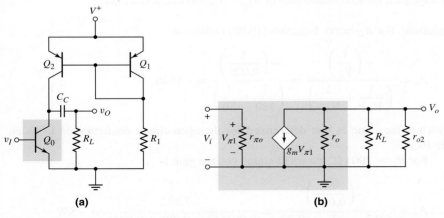

**(a)**          **(b)**

**Figure 10.37** (a) Simple BJT amplifier with active load and load resistance and (b) small-signal equivalent circuit

The output voltage is

$$V_o = -(g_m V_{\pi 1})(r_o \| R_L \| r_{o2}) \tag{10.93}$$

Since $V_{\pi 1} = V_i$, where $V_i$ is the ac input voltage, the small-signal voltage gain is

$$A_v = \frac{V_o}{V_i} = -g_m(r_o \| R_L \| r_{o2}) = \frac{-g_m}{\left(\dfrac{1}{r_o} + \dfrac{1}{R_L} + \dfrac{1}{r_{o2}}\right)} \tag{10.94}$$

The small-signal voltage gain can also be written

$$A_v = \frac{-g_m}{g_o + g_L + g_{o2}} \tag{10.95}$$

where $g_o$ and $g_{o2}$ are the output conductances of $Q_0$ and $Q_2$, and $g_L$ is the load conductance. The transconductance is $g_m = I_{Co}/V_T$, the small-signal conductances are $g_o = I_{Co}/V_{AN}$ and $g_{o2} = I_{Co}/V_{AP}$, and the load conductance is $g_L = 1/R_L$. Therefore, Equation (10.95) becomes

$$A_v = \frac{-\left(\dfrac{I_{Co}}{V_T}\right)}{\left(\dfrac{I_{Co}}{V_{AN}} + \dfrac{1}{R_L} + \dfrac{I_{Co}}{V_{AP}}\right)} \tag{10.96}$$

If the passive load is an open circuit ($R_L \to \infty$), the small-signal voltage gain is identical to that determined from the dc analysis as given by Equation (10.86). If the load resistance $R_L$ is not an open circuit, then the magnitude of the small-signal voltage gain is reduced.

## EXAMPLE 10.12

**Objective:** Calculate the small-signal voltage gain of an amplifier with an active load and a load resistance $R_L$.

For the circuit in Figure 10.37(a), the transistor parameters are $V_{AN} = 120$ V and $V_{AP} = 80$ V. Let $V_T = 0.026$ V and $I_{Co} = 0.2$ mA. Determine the small-signal voltage gain for load resistances of $R_L = \infty$, 200 k$\Omega$, and 20 k$\Omega$.

**Solution:** For $R_L = \infty$, Equation (10.96) reduces to

$$A_v = \frac{-\left(\dfrac{1}{V_T}\right)}{\left(\dfrac{1}{V_{AN}} + \dfrac{1}{V_{AP}}\right)} = \frac{-\left(\dfrac{1}{0.026}\right)}{\left(\dfrac{1}{120} + \dfrac{1}{80}\right)} = -1846$$

which is the same as that determined for the open-circuit configuration in Example 10.11.

For $R_L = 200$ k$\Omega$, the small-signal voltage gain is

$$A_v = \frac{-\left(\dfrac{0.2}{0.026}\right)}{\left(\dfrac{0.2}{120} + \dfrac{1}{200} + \dfrac{0.2}{80}\right)} = \frac{-7.692}{0.001667 + 0.005 + 0.0025} = -839$$

and for $R_L = 20\,\text{k}\Omega$, the voltage gain is

$$A_v = \frac{-\left(\dfrac{0.2}{0.026}\right)}{\left(\dfrac{0.2}{120} + \dfrac{1}{20} + \dfrac{0.2}{80}\right)} = \frac{-7.692}{0.001667 + 0.05 + 0.0025} = -142$$

**Comment:** The small-signal voltage gain is a strong function of the load resistance $R_L$. As the value of $R_L$ decreases, the loading effect becomes more severe.

**Design Pointer:** If an amplifier with an active load is to drive another amplifier stage, the loading effect must be taken into account when the small-signal voltage gain is determined. Also, the input resistance of the next stage must be large in order to minimize the loading effect.

---

EXERCISE PROBLEM

**Ex 10.12:** For the circuit shown in Figure 10.37(a), the transistor parameters are $V_{AN} = 100\,\text{V}$ and $V_{AP} = 60\,\text{V}$. Let $I_{Co} = 0.25\,\text{mA}$. (a) Determine the open-circuit small-signal voltage gain. (b) Find the value of $R_L$ such that the voltage gain is 60 percent of the open-circuit value. (Ans. (a) $A_v = -1442$, (b) $R_L = 225\,\text{k}\Omega$)

---

The small-signal voltage gain of an active-load circuit can be increased by increasing the effective resistance of the active load. Figure 10.38 shows the same type of BJT amplifier in which the active load is an "upside down" modified Widlar current source. The small-signal voltage gain can be written in the same form as Equation (10.94), or

$$A_v = \frac{-g_m}{\left(\dfrac{1}{r_o} + \dfrac{1}{R_L} + \dfrac{1}{R_{o2}}\right)} \tag{10.97}$$

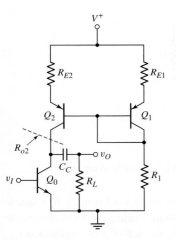

**Figure 10.38** BJT amplifier with a modified Widlar current source as an active load

where $R_{o2}$ is now the effective resistance looking into the collector of $Q_2$. We analyzed the output resistance of a Widlar current source in the Section 10.1. In this case, however, we are including a resistor in the emitter of both transistors $Q_1$ and $Q_2$. We can show that

$$R_{o2} = r_{o2}\left[1 + g_{m2} R_E''\right] \tag{10.98a}$$

where

$$R_E'' = R_E \| [r_{\pi 2} + R_1 \| (R_{o1} + R_E)] \tag{10.98b}$$

and where

$$R_{o1} = r_{\pi 1} \left\| \frac{1}{g_{m1}} \right\| r_{o1} \tag{10.98c}$$

We may note that, if $Q_1$ and $Q_2$ are matched, then $g_{m1} = g_{m2}$ and $r_{\pi 1} = r_{\pi 2}$ since the resistors in the emitters of $Q_1$ and $Q_2$ are the same value.

### Problem-Solving Technique: Active Loads

1. Ensure that the active load devices are biased in the forward-active mode.
2. The small-signal analysis of the circuit with an active load then simply involves considering the output resistance looking into the output of the active load device as well as the equivalent circuit of the amplifying transistor.

### 10.4.2    Small-Signal Analysis: MOSFET Active Load Circuit

The small-signal voltage gain of a MOSFET amplifier with an active load can also be determined from the small-signal equivalent circuit. Figure 10.39 is the small-signal equivalent circuit of the entire MOSFET active load in Figure 10.33. The signal voltages $V_{sg1}$ and $V_{sg2}$ are zero, since there is no ac excitation in this part of the circuit. This means that $g_m V_{sg2} = 0$ and

$$R_o = r_{o2} \tag{10.99}$$

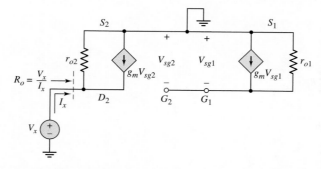

**Figure 10.39** Small-signal equivalent circuit of the MOSFET active load circuit

A simple MOSFET amplifier with an active load, and a load resistor $R_L$ capacitively coupled to the output, is shown in Figure 10.40(a). Figure 10.40(b) shows the small-signal equivalent circuit, in which the load $R_L$, the active load resistance $r_{o2}$, and the output resistance $r_o$ of transistor $M_0$ are included.

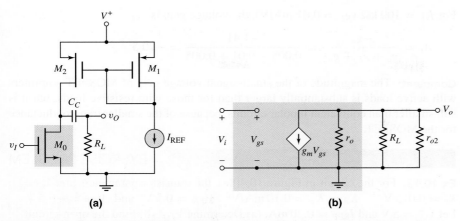

**Figure 10.40** (a) Simple MOSFET amplifier with active load and load resistance and (b) small-signal equivalent circuit

The output voltage is

$$V_o = -g_m \, V_{gs}(r_o \| R_L \| r_{o2}) \tag{10.100}$$

and since $V_{gs} = V_i$, where $V_i$ is the ac voltage, the small-signal voltage gain is

$$A_v = \frac{V_o}{V_i} = -g_m \, (r_o \| R_L \| r_{o2}) = \frac{-g_m}{g_o + g_L + g_{o2}} \tag{10.101}$$

The parameters $g_o$ and $g_{o2}$ are the output conductances of $M_0$ and $M_2$, and $g_L$ is the load conductance. This expression for the small-signal voltage gain of a MOSFET amplifier with active load is the same as that of the BJT amplifier.

A load resistance $R_L$ tends to degrade the gain and to cause a loading effect, as it did in the bipolar circuit with an active load. However, in MOSFET amplifiers, the output may be connected to the gate of another MOSFET amplifier in which the effective $R_L$ is very large.

## EXAMPLE 10.13

**Objective:** Calculate the small-signal voltage gain of an NMOS amplifier with an active load.

For the amplifier shown in Figure 10.40(a) the transistor parameters are: $\lambda_n = \lambda_p = 0.01 \text{ V}^{-1}$, $V_{TN} = 1$ V, and $K_n = 1 \text{ mA/V}^2$. Assume $M_1$ and $M_2$ are matched and $I_{REF} = 0.5$ mA. Calculate the small-signal voltage gain for load resistances of $R_L = \infty$ and $100 \text{ k}\Omega$.

**Solution:** Since $M_1$ and $M_2$ are matched, then $I_O = I_{REF}$, and the transconductance is

$$g_m = 2\sqrt{K_n I_{REF}} = 2\sqrt{(1)(0.5)} = 1.41 \text{ mA/V}$$

The small-signal transistor conductances are

$$g_o = g_{o2} = \lambda I_{REF} = (0.01)(0.5) = 0.005 \text{ mA/V}$$

For $R_L = \infty$, Equation (10.101) reduces to

$$A_v = \frac{-g_m}{g_o + g_{o2}} = \frac{-1.41}{0.005 + 0.005} = -141$$

For $R_L = 100\,\text{k}\Omega$ $(g_L = 0.01\,\text{mA/V})$, the voltage gain is

$$A_v = \frac{-g_m}{g_o + g_L + g_{o2}} = \frac{-1.41}{0.005 + 0.01 + 0.005} = -70.5$$

**Comment:** The magnitude of the small-signal voltage gain of MOSFET amplifiers with active loads is substantially larger than for those with resistive loads, but it is still smaller than equivalent bipolar circuits, because of the smaller transconductance for the MOSFET.

### EXERCISE PROBLEM

**Ex 10.13:** For the circuit in Figure 10.40(a), the transistor parameters are: $\lambda_n = \lambda_p = 0.015\,\text{V}^{-1}$, $K_n = K_p = 0.10\,\text{mA/V}^2$, $V_{TN} = 0.5\,\text{V}$, and $V_{TP} = -0.5\,\text{V}$. Let $V^+ = 5\,\text{V}$ and $I_{\text{REF}} = 0.20\,\text{mA}$. (a) Determine $V_{IQ}$. (b) Find the open-circuit small-signal voltage gain. (c) Find the value of $R_L$ that results in a voltage gain of one-half the open-circuit value. (Ans. (a) $V_{IQ} = 1.914\,\text{V}$, (b) $A_v = -47.1$, (c) $R_L = 166\,\text{k}\Omega$)

### 10.4.3 Small-Signal Analysis: Advanced MOSFET Active Load

The active loads considered in the BJT (Figure 10.37) and MOSFET (Figure 10.40(a)) circuits correspond to the simple two-transistor current mirrors. We may use a more advanced current mirror with a high output resistance as an active load to increase the amplifier gain. Figure 10.41(a) shows a MOSFET cascode amplifying stage with a cascode active load. The small-signal equivalent circuit is shown in Figure 10.41(b),

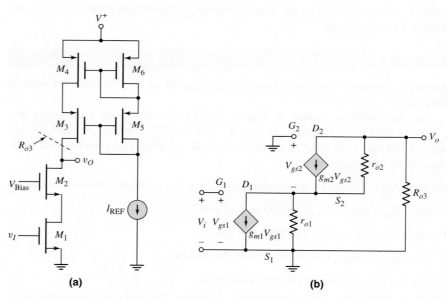

**(a)**                                                    **(b)**

**Figure 10.41** (a) MOSFET cascode amplifying stage with cascode active load; (b) small-signal equivalent circuit

where $R_{o3}$ is the effective resistance looking into the drain of $M_3$. From our discussion of the cascode current mirror, we found $R_{o3} = r_{o3} + r_{o4}(1 + g_m r_{o3})$ (Equation (10.57)).

We can assume all transistors are matched so that the currents in all transistors are equal. Summing currents at $D_1$, we have

$$g_m V_{gs1} + \frac{(-V_{gs2})}{r_{o1}} = g_m V_{gs2} + \frac{V_o - (-V_{gs2})}{r_{o2}} \qquad \textbf{(10.102)}$$

Summing currents at the output node, we find

$$\frac{V_o}{R_{o3}} + \frac{V_o - (-V_{gs2})}{r_{o2}} + g_m V_{gs2} = 0 \qquad \textbf{(10.103)}$$

Eliminating $V_{gs2}$ from the two equations, noting that $V_{gs1} = V_i$, and assuming $g_m \gg 1/r_o$, we find the small-signal voltage gain is

$$A_v = \frac{V_o}{V_i} = \frac{-g_m^2}{\dfrac{g_m}{R_{o3}} + \dfrac{1}{r_{o1} r_{o2}}} \qquad \textbf{(10.104)}$$

The resistance $R_{o3}$ is approximately $R_{o3} \cong g_m r_{o3} r_{o4}$, so the gain can be written as

$$A_v = \frac{-g_m^2}{\dfrac{1}{r_{o3} r_{o4}} + \dfrac{1}{r_{o1} r_{o2}}} \qquad \textbf{(10.105)}$$

For the same transistor parameters given in Example 10.13, the small-signal voltage gain of this circuit would be 39,762! However, a word of warning is in order. As we mentioned previously, output resistances in the hundreds of megohm range are ideal and will, in reality, be limited by leakage currents. For this reason, a voltage gain of 39,000 in a one-stage amplifier will probably not be achieved. However, the voltage gain of this amplifier should be substantially larger than the amplifier using a simple active load.

## Test Your Understanding

**TYU 10.12** In the circuit shown in Figure 10.37(a), the transistor parameters are $V_{AN} = 120$ V and $V_{AP} = 80$ V. Let $I_{C_o} = 0.5$ mA and $R_L = 50$ k$\Omega$. (a) Determine the small-signal parameters $g_m$, $r_o$, and $r_{o2}$. (b) Find the small-signal voltage gain. (Ans. (a) $g_m = 19.2$ mA/V, $r_o = 240$ k$\Omega$, $r_{o2} = 160$ k$\Omega$ (b) $A_v = -631$)

**TYU 10.13** Repeat Example 10.12 for the case where a resistor $R_E = 1$ k$\Omega$ is included in the emitters of $Q_1$ and $Q_2$ as shown in Figure 10.38. (Ans. For $R_L = \infty$, $A_v = -4404$; for $R_L = 200$ k$\Omega$, $A_v = -2800$; for $R_L = 20$ k$\Omega$, $A_v = -655$)

**TYU 10.14** In the circuit in Figure 10.40(a), the transistor parameters are: $K_p = 0.1$ mA/V$^2$, $K_n = 0.2$ mA/V$^2$, $V_{TN} = 1$ V, $V_{TP} = -1$ V, $\lambda_n = 0.01$ V$^{-1}$, and $\lambda_p = 0.02$ V$^{-1}$. Let $V^+ = 10$ V, $I_{REF} = 0.25$ mA, and $R_L = 100$ k$\Omega$. (a) Determine the small-signal parameters $g_m$ (for $M_0$), $r_{on}$, and $r_{op}$. (b) Find the small-signal voltage gain. (Ans. (a) $g_m = 0.448$ mA/V, $r_{on} = 400$ k$\Omega$, $r_{op} = 200$ k$\Omega$ (b) $A_v = -25.6$)

## 10.5    DESIGN APPLICATION: AN NMOS CURRENT SOURCE

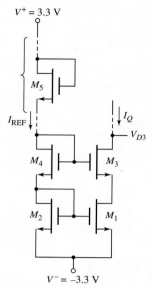

$V^+ = 3.3$ V

**Figure 10.42** MOSFET cascode current source circuit for design application

**Objective:** • Design an NMOS current source circuit to provide a specified bias current and output resistance.

**Specifications:** Design an NMOS current source to provide a bias current of $I_Q = 100$ $\mu$A and an output resistance greater than 20 M$\Omega$. The reference current is to be $I_{REF} = 150$ $\mu$A. The circuit is to be biased at $\pm 3.3$ V and the voltage at the drain of the current source transistor is to be no smaller than $-2.2$ V.

**Design Approach:** A simple two-transistor current source would yield an output resistance of

$$R_o = r_o = \frac{1}{\lambda I_Q} = \frac{1}{(0.01)(0.1)} \Rightarrow 1 \text{ M}\Omega$$

Therefore, to obtain a larger output resistance, a cascode current source is required. The basic circuit is shown in Figure 10.42. The transistor $M_5$ may actually need to be two or more transistors in series.

**Choices:** NMOS transistors are available with the following parameters: $V_{TN} = 0.5$ V, $k'_n = 80$ $\mu$A/V$^2$, and $\lambda = 0.01$ V$^{-1}$. The minimum width-to-length ratio of any transistor is to be unity.

**Solution:** The minimum voltage $V_{D3}$ is to be $-2.2$ V. This voltage is given by

$$V_{D3} = V_{GS1} + V_{DS3}(\text{sat}) + V^- = V_{GS1} + V_{GS3} - V_{TN} + V^-$$

Assuming that $M_1$ and $M_3$ are matched, we find

$$V_{D3} = -2.2 = 2V_{GS1} - 0.5 + (-3.3)$$

or

$$V_{GS1} = V_{GS3} = 0.8 \text{ V}$$

Now

$$I_Q = \frac{k'_n}{2} \frac{W}{L} (V_{GS1} - V_{TN})^2$$

or

$$100 = \frac{80}{2} \left(\frac{W}{L}\right)_1 (0.8 - 0.5)^2$$

which yields

$$\left(\frac{W}{L}\right)_1 = 27.8$$

If we set

$$\left(\frac{W}{L}\right)_1 = \left(\frac{W}{L}\right)_3 = 28$$

then we find that $V_{GS1} = V_{GS3} = 0.799$ V and $V_{D3}(\text{min}) = -2.202$ V.

Assuming that $M_2$ and $M_4$ are matched, we have

$$\frac{I_{REF}}{I_Q} = \frac{(W/L)_2}{(W/L)_1}$$

or

$$\frac{150}{100} = \frac{(W/L)_2}{28}$$

which yields

$$\left(\frac{W}{L}\right)_2 = \left(\frac{W}{L}\right)_4 = 42$$

Now the equivalent $V_{GS5}$ is given by

$$V_{GS5} = V^+ - 2V_{GS2} - V^- = 3.3 - 2(0.799) - (-3.3)$$

or

$$V_{GS5} = 5.0 \text{ V}$$

The width-to-length ratio is found from

$$I_{REF} = 150 = \frac{80}{2}\left(\frac{W}{L}\right)_5 (5.0 - 0.5)^2$$

which yields $(W/L)_5 = 0.185$. A width-to-length ratio less than unity is unacceptable. Putting two equivalent $M_5$ transistors in series yields a gate-to-source voltage of $V_{GS5} = 5.0/2$ V. Then

$$I_{REF} = 150 = \frac{80}{2}\left(\frac{W}{L}\right)_5 \left(\frac{5.0}{2} - 0.5\right)^2$$

which yields $(W/L)_5 = 0.938$. This value is still less than unity. Putting three equivalent $M_5$ transistors in series yields a gate-to-source voltage of $V_{GS5} = 5.0/3$ V. Then

$$I_{REF} = 150 = \frac{80}{2}\left(\frac{W}{L}\right)_5 \left(\frac{5.0}{3} - 0.5\right)^2$$

which yields $(W/L)_5 = 2.76$. This is an acceptable solution.

The output resistance is given by

$$R_o = r_{o3} + r_{o1}(1 + g_{m3} r_{o3})$$

We find

$$r_{o1} = r_{o3} = \frac{1}{\lambda I_Q} = \frac{1}{(0.01)(0.1)} \Rightarrow 1 \text{ M}\Omega$$

and

$$g_{m3} = 2\sqrt{\frac{k_n'}{2}\left(\frac{W}{L}\right)_3 I_Q} = 2\sqrt{\left(\frac{80}{2}\right)(28)(100)} = 669 \ \mu\text{A/V}$$

We then find

$$R_o = 1 + 1[1 + (669)(1)] = 671 \text{ M}\Omega$$

This value certainly meets the design criteria.

**Comment:** The very large output resistance of 671 MΩ assumes that we have ideal MOS transistors. In fact there are leakage currents that, in reality, will lower the output resistance. However, the cascode current source does provide a very high output resistance that is useful in differential amplifiers as we will see in Chapter 11.

 ## 10.6 SUMMARY

- This chapter addressed the biasing of bipolar and FET circuits with constant-current sources. The current source biasing technique eliminates the need for resistor-intensive biasing used up to this point.
- The basic bipolar current source is the simple two-transistor circuit with a resistor to establish the reference current. The basic FET current source is also a simple two-transistor circuit but includes additional transistors in the reference portion of the circuit. The relation between the bias current and reference current was determined.
- One parameter of interest in the current source circuit is the output resistance, which determines the stability of the bias current. More sophisticated current-source circuits, such as the Widlar and Wilson circuits in the BJT configuration and the Wilson and cascode circuits in the FET configuration, have larger output resistance parameters and increased bias-current stability.
- Multitransistor output stages, in both bipolar and FET constant-current circuits, are used to bias multiple amplifier stages with a single reference current. These circuits, called current mirrors, reduce the number of elements required to bias amplifier stages throughout an IC.
- Both bipolar and MOSFET active load circuits were analyzed. Active loads are essentially "upside down" current-source circuits that replace the discrete collector and drain resistors. The active loads produce a much larger small-signal voltage gain compared to discrete resistor circuits.
- As an application, a MOSFET current source circuit was designed to provide a specified bias current and output resistance.

 ## CHECKPOINT

After studying this chapter, the reader should have the ability to:

- ✓ Analyze and design a simple two-transistor BJT current-source circuit to produce a given bias current.
- ✓ Analyze and design more sophisticated BJT current-source circuits, such as the three-transistor circuit, cascode circuit, Wilson circuit, and Widlar circuit.
- ✓ Design a BJT current-source circuit to yield a specified output resistance.
- ✓ Analyze and design a basic two-transistor MOSFET current-source circuit with additional MOSFET devices in the reference portion of the circuit to yield a given bias current.
- ✓ Analyze and design more sophisticated MOSFET current-source circuits, such as the cascode circuit, Wilson circuit, and wide-swing cascode circuit.
- ✓ Design a MOSFET current-source circuit to yield a specified output resistance.
- ✓ Describe the operation and characteristics of a BJT and MOSFET active load circuit.
- ✓ Discuss the reason for the increased small-signal voltage gain when an active load is used.

# REVIEW QUESTIONS

1. Sketch the basic BJT two-transistor current source and explain the operation.
2. Explain the significance of the output resistance of the current-source circuit.
3. Discuss the effect of mismatched transistors on the characteristics of the BJT two-transistor current source.
4. Sketch the BJT three-transistor current source and discuss the advantages of this circuit.
5. What is the primary advantage of a BJT cascode current source?
6. Sketch a Widlar current source and explain the operation.
7. Can a piecewise linear model of the transistor be used in the analysis of the Widlar current source? Why or why not?
8. Discuss the operation and significance of a multiple-output transistor current mirror.
9. Sketch the basic MOSFET two-transistor current source and explain the operation.
10. Discuss the effect of mismatched transistors on the characteristics of the MOSFET two-transistor current source.
11. Discuss how the reference portion of a MOSFET current source can be designed with MOSFETs only.
12. Sketch a MOSFET cascode current source circuit and discuss the advantages of this design.
13. Discuss the operation of an active load.
14. What is the primary advantage of using an active load?
15. Sketch the voltage transfer characteristics of a simple amplifier with an active load. Where should the $Q$-point be placed?
16. What is the impedance seen looking into a simple active load?
17. What is the advantage of using a cascode active load?

# PROBLEMS

## Section 10.1 Bipolar Transistor Current Sources

10.1    Figure P10.1 shows another form of a bipolar current source. (a) Neglecting base currents, derive the expression for $I_C$ in terms of the circuit, transistor, and diode parameters. (b) If the transistor B–E and diode voltages are equal, show that, for $R_1 = R_2$, the expression for $I_C$ reduces to

$$I_C = \frac{(-V^-)}{2R_3}$$

(c) For $V^- = -10$ V and $V_{BE}(\text{on}) = V_\gamma = 0.7$ V, design the circuit such that $I_C = I_1 = I_2 = 2$ mA.

10.2    The matched transistors $Q_1$ and $Q_2$ in Figure 10.2(a) have parameters $I_S = 10^{-16}$ A. (a) For $\beta = \infty$, determine $I_O$ and $V_{BE1}$ for (i) $I_{REF} = 50\,\mu$A, (ii) $I_{REF} = 150\,\mu$A, and (iii) $I_{REF} = 1.5$ mA. (b) Repeat part (a) for $\beta = 50$.

10.3    Consider the circuit in Figure 10.2(a). Let $I_{REF} = 200\,\mu$A. Assume transistor parameters of $\beta = 80$, $I_{S1} = 5 \times 10^{-15}$ A, and $I_{S2} = 2 \times 10^{-15}$ A. Find $V_{BE1}$, $V_{BE2}$, and $I_O$.

10.4    Reconsider the circuit in Figure 10.2(a). Let $I_{REF} = 150\,\mu$A. Assume transistor parameters of $\beta = 120$, $I_{S1} = 10^{-16}$ A, and $I_{S2} = 3 \times 10^{-16}$ A. Find $V_{BE1}$, $V_{BE2}$, and $I_O$.

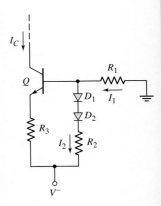

**Figure P10.1**

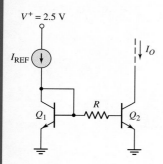

$V^+ = 2.5$ V

$I_{REF}$

$I_O$

$Q_1$ $R$ $Q_2$

**Figure P10.5**

**10.5** Consider the circuit shown in Figure P10.5. Assume $I_{REF} = 200\,\mu$A and $R = 2\,$k$\Omega$. The transistor parameters are $\beta = 40$, $I_{S1} = I_{S2} = 5 \times 10^{-15}$ A. Find $V_{BE1}$, $V_{BE2}$, and $I_O$.

**10.6** The transistor and circuit parameters for the circuit in Figure 10.2(b) are: $V_{BE}(\text{on}) = 0.7$ V, $\beta = 60$, $V_A = \infty$, $V^+ = +3$ V, $V^- = -3$ V, and $I_{REF} = 0.250$ mA. Determine the value of $R_1$ and determine $I_{C1}$, $I_{B1}$, $I_{B2}$, and $I_{C2}$.

**10.7** The bias voltages in the circuit shown in Figure 10.2(b) are $V^+ = +5$ V, $V^- = -5$ V and the resistor value is $R_1 = 18.3$ k$\Omega$. Assume transistor parameters of $V_{BE}$ (on) $= 0.7$ V, $\beta = 80$, and $V_A = \infty$. Determine $I_{REF}$, $I_{C1}$, $I_{B1}$, $I_{B2}$, and $I_{C2}$.

**10.8** Consider the current source in Figure 10.2(b). The circuit is biased at $V^+ = 2.5$ V and $V^- = -2.5$ V. The transistor parameters are $\beta \cong \infty$, $V_A = \infty$, and $I_{S2} = 10^{-15}$ A. The circuit is to be designed such that $I_O = 0.25$ mA and the power dissipated in the circuit is no greater than 1.8 mW. (a) Determine the maximum value of $I_{REF}$, (b) the required value of $I_{S1}$, and (c) the required value of $R_1$.

**10.9** For the basic two-transistor current source in Figure 10.2(b), the transistor parameters are: $\beta = 120$, $V_{BE}(\text{on}) = 0.7$ V, and $V_A = 100$ V. The bias voltages are $V^+ = 5$ V and $V^- = -5$ V. (a) Design the circuit such that $I_O = 0.5$ mA when $V_{CE2} = 0.7$ V. (b) What is the percent change in $I_O$ as $V_{CE2}$ varies between 0.7 V and 7 V?

**10.10** The transistors in the basic current mirror in Figure 10.2(b) have a finite $\beta$ and an infinite Early voltage. The B–E area of $Q_2$ is $n$ times that of $Q_1$. Derive the expression for $I_O$ in terms of $I_{REF}$, $\beta$, and $n$.

**D10.11** Figure P10.11 shows a basic two-transistor pnp current source. The transistor parameters are $V_{EB}(\text{on}) = 0.7$ V, $\beta = 40$, and $V_A = \infty$. Design the circuit such that $I_O = 0.20$ mA and determine the value of $I_{REF}$.

**D10.12** In the circuit in Figure P10.11, the transistor parameters are $\beta = 80$ and $V_{EB}(\text{on}) = 0.7$ V. (a) Design the circuit such that $I_O = 120\,\mu$A for $V_{EC2} = 0.7$ V. (b) If $V_A = 80$ V, determine the change in $I_O$ for (i) $V_{EC2} = 2$ V and (ii) $V_{EC2} = 4$ V.

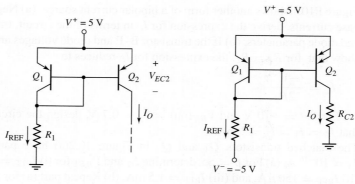

$V^+ = 5$ V

$Q_1$ $Q_2$ $V_{EC2}$ $+$ $-$

$I_{REF}$ $R_1$

$I_O$

**Figure P10.11**

$V^+ = 5$ V

$Q_1$ $Q_2$

$I_{REF}$ $R_1$

$I_O$ $R_{C2}$

$V^- = -5$ V

**Figure P10.13**

**D10.13** Consider the pnp current source in Figure P10.13, with transistor parameters $\beta = \infty$, $V_A = \infty$, and $V_{EB}(\text{on}) = 0.7$ V. (a) Design the circuit such that $I_{REF} = 1$ mA. (b) What is the value of $I_O$? (c) What is the maximum value of $R_{C2}$ such that $Q_2$ remains biased in the forward-active mode?

10.14 Consider the circuit shown in Figure P10.14. The transistor $Q_2$ is equivalent to two identical transistors in parallel, each of which is matched to $Q_1$. Assume the transistor parameters are $V_{BE}(\text{on}) = 0.7$ V, $\beta = 60$, and $V_A = \infty$, and assume the bias voltage is $V^+ = 2.5$ V. Design the circuit such that $I_O = 0.50$ mA and determine the value of $I_{REF}$.

D10.15 Design a basic two-transistor current source circuit configuration such that $I_O = 0.40$ mA and $I_{REF} = 0.20$ mA. The circuit is to be biased at $V^+ = 2.5$ V and $V^- = -2.5$ V. Neglect base currents and assume that $V_{BE}(\text{on}) = 0.7$ V and $V_A = \infty$.

10.16 The values of $\beta$ for the transistors in Figure P10.16 are very large. (a) If $Q_1$ is diode-connected with $I_1 = 0.5$ mA, determine the collector currents in the other two transistors. (b) Repeat part (a) if $Q_2$ is diode-connected with $I_2 = 0.5$ mA. (c) Repeat part (a) if $Q_3$ is diode-connected with $I_3 = 0.5$ mA.

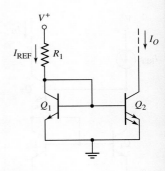

Figure P10.14

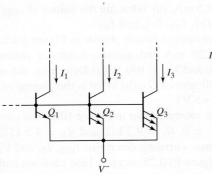

Figure P10.16

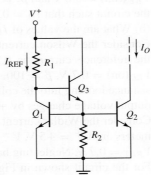

Figure P10.17

10.17 Consider the circuit in Figure P10.17. The transistor parameters are: $\beta = 80$, $V_{BE}(\text{on}) = 0.7$ V, and $V_A = \infty$. (a) Derive the expression for $I_O$ in terms of $I_{REF}$, $\beta$, and $R_2$. (b) For $R_2 = 10$ k$\Omega$ and $V^+ = 10$ V, design the circuit such that $I_O = 0.70$ mA. What is the value of $I_{REF}$?

10.18 All transistors in the $N$ output current mirror in Figure P10.18 are matched, with a finite $\beta$ and $V_A = \infty$. (a) Derive the expression for each load current in terms of $I_{REF}$ and $\beta$. (b) If the circuit parameters are $V^+ = 5$ V and $V^- = -5$ V, and the transistor parameter is $\beta = 50$, determine $R_1$ such that each load current is 0.5 mA for $N = 5$. Assume that $V_{EB}(Q_R) = V_{BE}(Q_S) = 0.7$ V.

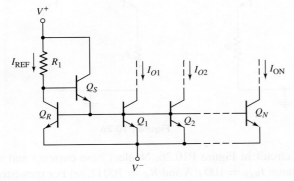

Figure P10.18

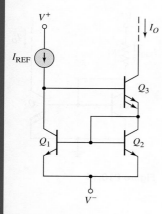

**Figure P10.21**

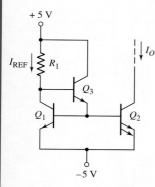

**Figure P10.22**

**D10.19** Design a pnp version of the basic three-transistor current source circuit, using a resistor to establish $I_{REF}$. The bias current is to be $I_O = 0.15$ mA, and the circuit is to be biased at $V^+ = 3$ V and $V^- = -3$ V. The transistor parameters are $\beta = 40$, $V_{EB}(on) = 0.7$ V, and $V_A = \infty$.

**D10.20** Design a pnp version of the Wilson current source, using a resistor to establish $I_{REF}$. The circuit parameters are $V^+ = 9$ V and $V^- = -9$ V, and the transistor parameters are: $V_{EB}(on) = 0.7$ V, $\beta = 25$, and $V_A = \infty$. If the load current is 0.8 mA, what is $I_{REF}$?

**\*10.21** Consider the Wilson current source in Figure P10.21. The transistors have a finite $\beta$ and an infinite Early voltage. Derive the expression for $I_O$ in terms of $I_{REF}$ and $\beta$.

**10.22** Consider the circuit in Figure P10.22. The transistor parameters for $Q_1$ and $Q_2$ are $V_{BE1,2}(on) = 0.7$ V and $\beta_{1,2} = 90$. The parameters for $Q_3$ are $V_{BE3}(on) = 0.6$ V and $\beta_3 = 60$. Assume $V_A = \infty$ for all transistors. Design the circuit such that $I_O = 0.5$ mA. (a) What are the values of $I_{REF}$ and $R_1$? (b) What are the values of $I_{B1}$, $I_{B2}$, $I_{E3}$, and $I_{B3}$?

**10.23** Consider the Wilson current-source circuit shown in Figure 10.8. Assume the reference current is 0.25 mA and assume transistor parameters of $V_{BE}(on) = 0.7$ V, $\beta = 100$, and $V_A = 100$ V. (a) Determine the output resistance looking into the collector of $Q_3$. (b) What is the change in $I_O$ as the output voltage changes by $+5$ V?

**10.24** Consider the Widlar current source shown in Figure 10.9. The circuit parameters are $V^+ = +5$ V, $V^- = 0$, $R_1 = 9.3$ k$\Omega$, and $R_E = 1.5$ k$\Omega$. Assume $V_{BE1} = 0.7$ V. Neglecting base currents, determine $I_{REF}$, $I_O$, and $V_{BE2}$.

**10.25** For the circuit shown in Figure P10.25, neglect base currents and assume $V_A = \infty$. Let $I_{REF} = 200 \,\mu$A and $R_E = 500 \,\Omega$. (a) Assume the transistor parameters are $I_{S1} = I_{S2} = 5 \times 10^{-15}$ A. Find $V_{BE1}$, $V_{BE2}$, and $I_O$. (b) Repeat part (a) if the transistor parameters are $I_{S1} = 5 \times 10^{-15}$ A and $I_{S2} = 7 \times 10^{-15}$ A.

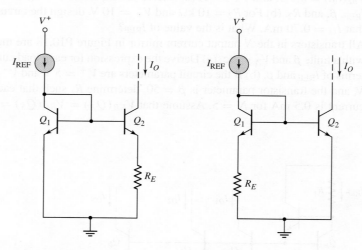

**Figure P10.25**                    **Figure P10.26**

**10.26** Consider the circuit in Figure P10.26. Neglect base currents and assume $V_A = \infty$. Assume $I_{REF} = 100 \,\mu$A and $R_E = 700 \,\Omega$. (a) For transistor parameters of $I_{S1} = I_{S2} = 5 \times 10^{-15}$ A, find $V_{BE1}$, $V_{BE2}$, and $I_O$. (b) Repeat part (a) if the transistor parameters are $I_{S1} = 5 \times 10^{-15}$ A and $I_{S2} = 2 \times 10^{-15}$ A.

10.27 (a) For the Widlar current source shown in Figure 10.9, find $I_{\text{REF}}$, $I_O$, and $V_{BE2}$ if $R_1 = 50 \,\text{k}\Omega$, $R_E = 3 \,\text{k}\Omega$, $V^+ = 5$ V, and $V^- = -5$ V. The transistor parameters are $\beta = 120$ and $V_{BE1}(\text{on}) = 0.7$ V. (b) Determine $R_o$ for $V_A = 80$ V.

*10.28 Consider the Widlar current source in Problem 10.27. For $\beta = 80$ and $V_A = 80$ V, determine the change in $I_O$ corresponding to a 5 V change in the output voltage.

D10.29 (a) Design the Widlar current source such that $I_{\text{REF}} = 0.50$ mA and $I_O = 50 \,\mu$A. Assume that $V^+ = +5$ V, $V^- = -5$ V, $V_{BE1} = 0.7$ V, and neglect base currents. (b) If $\beta = 75$ and $V_A = 100$ V, determine the output resistance looking into the collector of $Q_2$. (c) What is the percent change in $I_O$ if the voltage at the collector of $Q_2$ changes by +5 V?

D10.30 Design a Widlar current source to provide a bias current of $I_O = 50 \,\mu$A. The circuit is to be biased at $V^+ = 3$ V and $V^- = -3$ V. Assume $V_{BE1}(\text{on}) = 0.7$ V and $V_A = \infty$. The maximum resistor value is to be limited to $10 \,\text{k}\Omega$.

D10.31 Design the Widlar current source shown in Figure 10.9 such that $I_{\text{REF}} = 2$ mA and $I_O = 50 \,\mu$A. Let $V^+ = 15$ V and $V^- = 0$. The transistors are matched, and $V_{BE} = 0.7$ V at 1 mA.

10.32 The circuit parameters of the Widlar current source in Figure 10.9 are $V^+ = 3$ V, $V^- = -3$ V, and $R_1 = 20 \,\text{k}\Omega$. Assume $V_{BE1}(\text{on}) = 0.7$ V and $V_A = \infty$. (a) Determine $I_{\text{REF}}$ and (b) $R_E$ such that $I_O = 100 \,\mu$A.

10.33 Consider the Widlar current source in Figure 10.9. The circuit parameters are: $V^+ = 10$ V, $V^- = -10$ V, $R_1 = 40 \,\text{k}\Omega$, and $R_E = 12 \,\text{k}\Omega$. Neglect base currents and assume $V_{BE1} = 0.7$ V at 1 mA. Determine $I_{\text{REF}}$, $I_O$, $V_{BE1}$, and $V_{BE2}$.

10.34 Consider the circuit in Figure P10.34. The transistors are matched. Assume that base currents are negligible and that $V_A = \infty$. Using the current–voltage relationships given by Equations (10.26(a)) and (10.26(b)), show that

$$I_O R_{E2} - I_{\text{REF}} R_{E1} = V_T \ln\left(\frac{I_{\text{REF}}}{I_O}\right)$$

If $R_{E1} = R_{E2} \neq 0$ and $V_A \neq \infty$, explain the advantage of this circuit over the basic two-transistor current source in Figure 10.2(b).

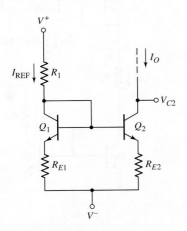

**Figure P10.34**

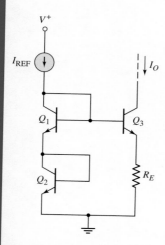

Figure P10.36

10.35   The modified Widlar current-source circuit shown in Figure P10.34 is biased at $V^+ = 3$ V and $V^- = -3$ V. (a) For $I_{S1} = I_{S2} = 10^{-15}$ A and $R_{E1} = 500\,\Omega$, design the circuit such that $I_{REF} = 0.5$ mA and $I_O = 0.2$ mA. Neglect base currents. What are the values of $V_{BE1}$ and $V_{BE2}$? (b) Repeat part (a) for $I_{S1} = 10^{-15}$ A and $I_{S2} = 2 \times 10^{-15}$ A.

*10.36   Consider the circuit in Figure P10.36. Neglect base currents and assume $V_A = \infty$. (a) Derive the expression for $I_O$ in terms of $I_{REF}$ and $R_E$. (b) Determine the value of $R_E$ such that $I_O = I_{REF} = 100\,\mu$A. Assume $V_{BE} = 0.7$ V at a collector current of 1 mA.

10.37   Consider the Widlar current-source circuit with multiple output transistors shown in Figure P10.37. Assume $V_{BE1} = 0.7$ V. (a) For circuit parameters $R_1 = 10$ k$\Omega$, $R_{E2} = 1$ k$\Omega$, and $R_{E2} = 2$ k$\Omega$, find $I_{REF}$, $I_{O2}$, and $I_{O3}$. (b) Determine new values of $R_{E2}$ and $R_{E3}$ such that $I_{O2} = 20\,\mu$A and $I_{O3} = 80\,\mu$A.

10.38   Assume that all transistors in the circuit in Figure P10.38 are matched and that $\beta = \infty$ (neglect base currents). (a) Derive an expression for $I_O$ in terms of bias voltages and resistor values. (b) Show that if $R_1 = R_2$ and $I_O = I_{REF}$, then $I_O = (V^+ - V^-)/2R_E$, which means that the currents are independent of $V_{BE}$. (c) For $V^+ = +5$ V and $V^- = -5$ V, design the circuit such that $I_O = 0.5$ mA.

10.39   In the circuit in Figure P10.39, the transistor parameters are: $\beta = \infty$, $V_A = \infty$, and $V_{BE} = V_{EB} = 0.7$ V. Let $R_{C1} = 2$ k$\Omega$, $R_{C2} = 3$ k$\Omega$, $R_{C3} = 1$ k$\Omega$, and $R_1 = 12$ k$\Omega$. (a) Determine $I_{O1}$, $I_{O2}$, and $I_{O3}$. (b) Calculate $V_{CE1}$, $V_{EC2}$, and $V_{EC3}$.

10.40   Consider the circuit in Figure P10.39, with transistor parameters $\beta = \infty$, $V_A = \infty$, and $V_{BE}(\text{on}) = V_{EB}(\text{on}) = 0.7$ V. Let $R_1 = 24$ k$\Omega$. (a) Find $I_{REF}$, $I_{O1}$, $I_{O2}$, and $I_{O3}$. (b) Determine the maximum values of $R_{C1}$, $R_{C2}$, and $R_{C3}$ such that $Q_1$, $Q_2$, and $Q_3$ remain biased in the forward-active region. Assume $V_{CE}(\text{min}) = V_{EC}(\text{min}) = 0.7$ V.

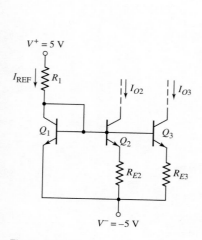

Figure P10.37

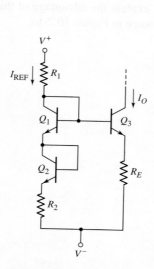

Figure P10.38

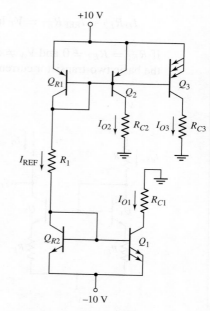

Figure P10.39

**10.41**  Consider the circuit shown in Figure P10.41. Assume $V_{BE} = V_{EB} = 0.7$ V for all transistors except $Q_5$ and let $\beta = \infty$. Determine all collector currents, and find $V_{CE3}, V_{CE5}$, and $V_{EC7}$.

**10.42**  For the circuit shown in Figure P10.42, assume transistor parameters $V_{BE} = V_{EB} = 0.7$ V for all transistors except $Q_3$ and $Q_6$, and let $\beta = \infty$. Find the collector current in each transistor.

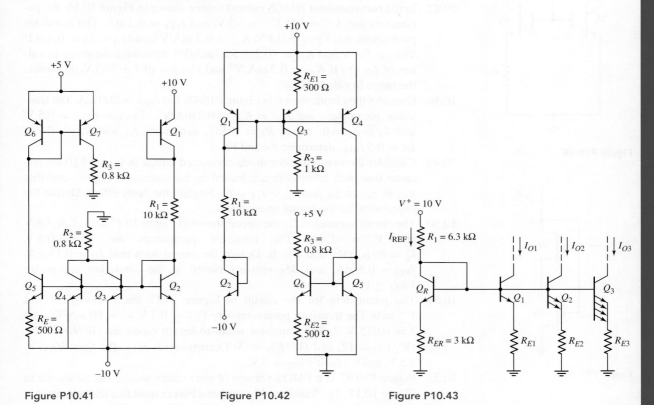

| Figure P10.41 | Figure P10.42 | Figure P10.43 |
| --- | --- | --- |

**\*D10.43**  Consider the circuit in Figure P10.43. The transistor parameters are: $\beta = \infty$, $V_A = \infty$, and $V_{BE} = 0.7$ V. Design the circuit such that the B–E voltages of $Q_1$, $Q_2$, and $Q_3$ are identical to that of $Q_R$. What are the values of $I_{O1}$, $I_{O2}$, and $I_{O3}$?

## Section 10.2 FET Current Sources

**10.44**  Consider the MOSFET current-source circuit in Figure P10.44 with $V^+ = +2.5$ V and $R = 15$ k$\Omega$. The transistor parameters are $V_{TN} = 0.5$ V, $k'_n = 80$ $\mu$A/V$^2$, $W/L = 6$, and $\lambda = 0$. Determine $I_{REF}$, $I_O$, and $V_{DS2}(\text{sat})$.

**\*D10.45**  The MOSFET current-source circuit in Figure P10.44 is biased at $V^+ = 2.0$ V. The transistor parameters are $V_{TN} = 0.5$ V, $k'_n = 80$ $\mu$A/V$^2$, and $\lambda = 0.015$ V$^{-1}$. (a) Design the circuit such that $I_{REF} = 50$ $\mu$A and the nominal bias current is $I_O = 100$ $\mu$A. (b) Find the output resistance $R_o$. (c) Determine the percentage change in $I_O$ for a change in drain-to-source voltage of $\Delta V_{DS2} = 1$ V.

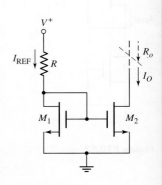

Figure P10.44

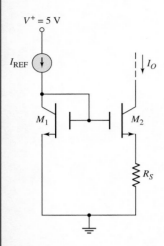

Figure P10.48

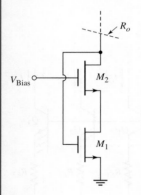

Figure P10.49

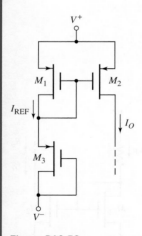

Figure P10.52

10.46   Consider the basic two-transistor NMOS current source in Figure 10.16. The circuit parameters are $V^+ = +5$ V, $V^- = -5$ V, and $I_{REF} = 250$ $\mu$A. The transistor parameters are $V_{TN} = 1$ V, $k'_n = 80$ $\mu$A/V$^2$, and $\lambda = 0.02$ V$^{-1}$. (a) For $(W/L)_1 = (W/L)_2 = 3$, find $I_O$ for (i) $V_{DS2} = 3$ V, (ii) $V_{DS2} = 4.5$ V, and (iii) $V_{DS2} = 6$ V. (b) Repeat part (a) for $(W/L)_1 = 3$ and $(W/L)_2 = 4.5$.

10.47   In the two-transistor NMOS current source shown in Figure 10.16, the parameters are: $V^+ = 3$ V, $V^- = -3$ V, and $I_{REF} = 0.2$ mA. The transistor parameters are: $V_{TN1} = 0.4$ V, $K_{n1} = 0.2$ mA/V$^2$, and $\lambda_1 = \lambda_2 = 0$. (a) If $V_{TN2} = 0.4$ V and $K_{n2} = (0.2 \pm 5\%)$ mA/V$^2$, determine the range in values of $I_O$. (b) If $K_{n2} = 0.2$ mA/V$^2$ and $V_{TN2} = (0.4 \pm 5\%)$ V, determine the range in values of $I_O$.

10.48   Consider the circuit shown in Figure P10.48. Let $I_{REF} = 200$ $\mu$A. The transistor parameters are $K_{n1} = K_{n2} = 0.2$ mA/V$^2$, $V_{TN1} = V_{TN2} = 0.5$ V, and $\lambda_1 = \lambda_2 = 0$. (a) If $R_S = 10$ k$\Omega$, determine $I_O$ and $V_{GS2}$. (b) If $I_O = 0.5 I_{REF}$, determine $R_S$ and $V_{GS2}$.

10.49   Consider the two-transistor diode-connected circuit in Figure P10.49. Assume that both transistors are biased in the saturation region, and that $g_{m1} = g_{m2} \equiv g_m$ and $r_{o1} = r_{o2} \equiv r_o$. Neglect the body effect. Derive the expression for the output resistance $R_o$.

10.50   The circuit parameters for the circuit shown in Figure 10.17 are $V^+ = 1.8$ V and $V^- = -1.8$ V. The transistor parameters are $V_{TN} = 0.5$ V, $k'_n = 80$ $\mu$A/V$^2$, and $\lambda = 0$. Design the circuit such that $I_O = 0.15$ mA, $I_{REF} = 0.5$ mA, and $M_2$ remains biased in the saturation region for $V_{DS2} \geq 1$ V.

10.51   The parameters for the circuit in Figure 10.17 are $V^+ = +5$ V and $V^- = 0$. The transistor parameters are $V_{TN} = 0.7$ V, $k'_n = 60$ $\mu$A/V$^2$, and $\lambda = 0.015$ V$^{-1}$. The transistor width-to-length ratios are $(W/L)_1 = 20$, $(W/L)_2 = 12$, and $(W/L)_3 = 3$. Determine (a) $I_{REF}$, (b) $I_O$ at $V_{DS2} = 1.5$ V, and (c) $I_O$ at $V_{DS2} = 3$ V.

10.52   Figure P10.52 is a PMOS version of the current-source circuit shown in Figure 10.17. The transistor $M_2$ sources a bias current to a load circuit. Assume the circuit is biased at $V^+ = +5$ V and $V^- = -5$ V, and assume the transistor parameters are $V_{TP} = -0.5$ V, $k'_p = 50$ $\mu$A/V$^2$, $(W/L)_1 = (W/L)_2 = 15$, $(W/L)_3 = 3$, and $\lambda = 0$. Determine $I_{REF}$, $I_O$, and $V_{SD2}$(sat).

D10.53   The circuit shown in Figure P10.52 is biased at $V^+ = +2$ V and $V^- = -2$ V. Assume the transistor parameters are $V_{TP} = -0.35$ V, $k'_p = 50$ $\mu$A/V$^2$, and $\lambda = 0$. Design the circuit such that $I_{REF} = 200$ $\mu$A, $I_O = 100$ $\mu$A, and $V_{SD2}$(sat) = 1.2 V.

10.54   The transistor circuit shown in Figure P10.54 is biased at $V^+ = +5$ V and $V^- = -5$ V. The transistor parameters are $V_{TP} = -1.2$ V, $k'_p = 80$ $\mu$A/V$^2$, $\lambda = 0$, $(W/L)_1 = (W/L)_2 = 25$, and $(W/L)_3 = (W/L)_4 = 4$. Determine $I_{REF}$, $I_O$, and $V_{SD2}$(sat).

D10.55   Assume the circuit shown in Figure P10.54 is biased at $V^+ = 3$ V and $V^- = -3$ V. The transistor parameters are $V_{TP} = -0.5$ V, $k'_p = 60$ $\mu$A/V$^2$, and $\lambda = 0$. Design the circuit such that $I_{REF} = 250$ $\mu$A, $I_O = 80$ $\mu$A, and $V_{SD2}$(sat) = 1.0 V. Assume $M_3$ and $M_4$ are matched.

10.56   The circuit in Figure P10.56 is a PMOS version of a two-transistor MOS current mirror. Assume transistor parameters of $V_{TP} = -0.4$ V, $k'_p = 60$ $\mu$A/V$^2$, and $\lambda = 0$. The transistor width-to-length ratios are

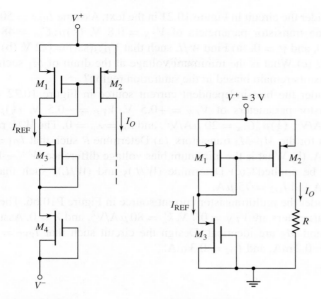

Figure P10.54          Figure P10.56

$(W/L)_1 = 25$, $(W/L)_2 = 15$, and $(W/L)_3 = 5$. (a) Determine $I_O$, $I_{REF}$, $V_{SG1}$, and $V_{SG3}$. (b) What is the largest value of $R$ such that $M_2$ remains biased in the saturation region?

D10.57  The transistors in Figure P10.56 have the same parameters as in Problem 10.56 except for the $W/L$ ratios. Design the circuit such that $I_O = 80\ \mu$A, $I_{REF} = 220\ \mu$A, and $V_{SD2}$(sat) $= 0.35$ V.

10.58  Consider the NMOS cascode current source in Figure 10.18. The circuit parameters are $V^+ = 5$ V, $V^- = -5$ V, and $I_{REF} = 100\ \mu$A. All transistors are matched with parameters $V_{TN} = 0.5$ V, $K_n = 100\ \mu$A/V$^2$, and $\lambda = 0.02$ V$^{-1}$. (a) Determine $I_O$ for $V_{D4} = -2$ V. (b) Determine the percent change in $I_O$ as $V_{D4}$ changes from $-2$ to $+2$ V.

*10.59  Consider the NMOS current source in Figure P10.59. Let $I_{REF} = 0.2$ mA, $K_n = 0.2$ mA/V$^2$, $V_{TN} = 1$ V, and $\lambda = 0.02$ V$^{-1}$. (All transistors are matched.) Determine the output resistance looking into the drain of $M_6$.

10.60  The transistors in the circuit shown in Figure P10.60 have parameters $V_{TN} = 0.4$ V, $V_{TP} = -0.4$ V, $k'_n = 100\ \mu$A/V$^2$, $k'_p = 60\ \mu$A/V$^2$, and $\lambda_n = \lambda_p = 0$. The transistor width-to-length ratios are $(W/L)_1 = (W/L)_2 = 20$, $(W/L)_3 = 5$, and $(W/L)_4 = 10$. Determine $I_O$, $I_{REF}$, and $V_{DS2}$(sat). What are the values of $V_{GS1}$, $V_{GS3}$, and $V_{SG4}$?

D10.61  The transistors in the circuit shown in Figure P10.60 have the same parameters as in Problem 10.60 except for the $(W/L)$ ratios. Design the circuit such that $I_O = 50\ \mu$A, $I_{REF} = 500\ \mu$A, $V_{DS2}$(sat) $= 0.5$ V, and $V_{GS3} = V_{SG4}$.

*10.62  A Wilson current mirror is shown in Figure 10.20(a). The parameters are: $V^+ = 5$ V, $V^- = -5$ V, and $I_{REF} = 80\ \mu$A. The transistor parameters are: $V_{TN} = 1$ V, $K_n = 80\ \mu$A/V$^2$, and $\lambda = 0.02$ V$^{-1}$. Determine $I_O$ at: (a) $V_{D3} = -1$ V, and (b) $V_{D3} = +3$ V.

*10.63  Repeat Problem 10.62 for the modified Wilson current mirror in Figure 10.20(b).

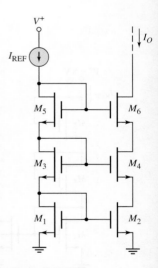

Figure P10.59

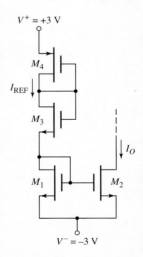

Figure P10.60

*10.64   Consider the circuit in Figure 10.21 in the text. Assume $I_{REF} = 50\ \mu A$ and assume transistor parameters of $V_{TN} = 0.8$ V, $(\frac{1}{2})\mu_n C_{ox} = 48\ \mu A/V^2$, $\lambda = 0$, and $\gamma = 0$. (a) Find $W/L$ such that $V_{DS3}(sat) = 0.2$ V. (b) What is $V_{GS5}$? (c) What is the minimum voltage at the drain of $M_1$ such that all transistors remain biased in the saturation region?

10.65   Consider the bias-independent current source in Figure 10.22. Assume transistor parameters of $V_{TN} = +0.5$ V, $V_{TP} = -0.5$ V, $(\frac{1}{2})\mu_n C_{ox} = 50\ \mu A/V^2$, $(\frac{1}{2})\mu_p C_{ox} = 20\ \mu A/V^2$, and $\lambda_n = \lambda_p = 0$. The $W/L$ ratios are given for the $M_1$–$M_4$ transistors. (a) Determine $R$ such that $I_{D1} = I_{D2} = 50\ \mu A$. (b) What is the minimum bias voltage difference $(V^+ - V^-)$ that must be applied? (c) Determine $(W/L)_5$ and $(W/L)_6$ such that $I_{O1} = 25\ \mu A$ and $I_{O2} = 75\ \mu A$.

D10.66   Consider the multitransistor current source in Figure P10.66. The transistor parameters are $V_{TN} = 0.7$ V, $k_n' = 80\ \mu A/V^2$, and $\lambda = 0$. Assume $M_3$, $M_4$, and $M_5$ are identical. Design the circuit such that $I_{REF} = 0.1$ mA, $I_{O1} = 0.2$ mA, and $I_{O2} = 0.3$ mA.

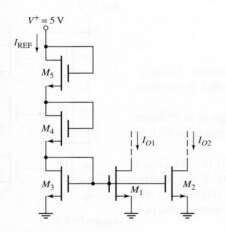

**Figure P10.66**

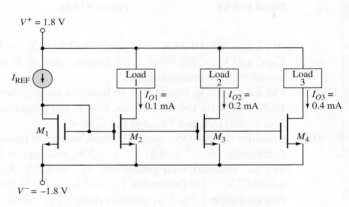

**Figure P10.67**

D10.67   Consider the circuit shown in Figure P10.67. The transistor parameters are $V_{TN} = 0.4$ V, $k_n' = 100\ \mu A/V^2$, and $\lambda = 0$. Design the $(W/L)$ ratios of the transistors such that the total power dissipated in the circuit is 5 mW and $V_{DS2}(sat) = 0.4$ V.

10.68   The parameters of the transistors in the circuit in Figure P10.68 are $V_{TN} = 0.8$ V, $V_{TP} = -0.8$ V, $k_n' = 100\ \mu A/V^2$, $k_p' = 60\ \mu A/V^2$, and $\lambda_n = \lambda_p = 0$. The transistor $(W/L)$ ratios are given in the figure. For $R = 100\ k\Omega$, determine $I_{REF}$, $I_1$, $I_2$, $I_3$, and $I_4$.

10.69   Repeat Problem 10.68 if the bias voltages are changed to $V^+ = 5$ V and $V^- = -5$ V.

10.70   Consider the circuit shown in Figure P10.70. The NMOS transistor parameters are $V_{TN} = 0.4$ V, $k_n' = 100\ \mu A/V^2$, $\lambda_n = 0$ and the PMOS transistor parameters are $V_{TP} = -0.6$ V, $k_p' = 40\ \mu A/V^2$, $\lambda_p = 0$. The width-to-length ratios are $(W/L)_1 = 15$, $(W/L)_2 = (W/L)_3 = 9$, and $(W/L)_4 = 20$. Assume $I_{REF} = 200\ \mu A$. Determine $I_{D2}$, $I_O$, and $V_{SD4}(sat)$.

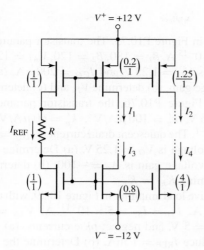

**Figure P10.68**

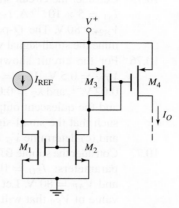

**Figure P10.70**

10.71 For the circuit shown in Figure P10.70, $I_{REF} = 100\,\mu A$. The transistor parameters are $V_{TN} = 0.4$ V, $V_{TP} = -0.4$ V, $k'_n = 100\,\mu A/V^2$, $k'_p = 60\,\mu A/V^2$, and $\lambda_n = \lambda_p = 0$. The transistor width-to-length ratios are $(W/L)_1 = 4$, $(W/L)_2 = 2.5$, $(W/L)_3 = 6$, and $(W/L)_4 = 4$. Determine $I_{D2}$, $I_O$, and all gate-to-source voltages.

D10.72 The parameters of the NMOS transistors in the circuit in Figure P10.72 are $V_{TN} = 0.4$ V, $k'_n = 100\,\mu A/V^2$, $\lambda_n = 0$ and the parameters of the PMOS transistors in the circuit are $V_{TP} = -0.6$ V, $k'_p = 40\,\mu A/V^2$, $\lambda_p = 0$. Design the circuit such that $I_{REF} = 50\,\mu A$, $I_{O1} = 120\,\mu A$, $I_{D3} = 25\,\mu A$, $I_{O2} = 150\,\mu A$, $V_{SD2}(sat) = 0.35$ V, and $V_{DS5}(sat) = 0.35$ V.

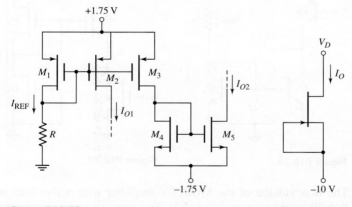

**Figure P10.72**                    **Figure P10.73**

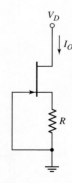

**Figure P10.74**

10.73 For the JFET in Figure P10.73, the parameters are: $I_{DSS} = 2$ mA, $V_P = -2$ V, and $\lambda = 0.05$ V$^{-1}$. Determine $I_O$ for: (a) $V_D = -5$ V, (b) $V_D = 0$ V, and (c) $V_D = +5$ V.

D10.74 A JFET circuit is biased with the current source in Figure P10.74. The transistor parameters are: $I_{DSS} = 4$ mA, $V_P = -4$ V, and $\lambda = 0$. Design the circuit such that $I_O = 2$ mA. What is the minimum value of $V_D$ such that the transistor is biased in the saturation region?

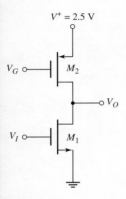

**Figure P10.75**

**Figure P10.76**

### Section 10.3 Active Load Circuits

10.75 Consider the circuit shown in Figure P10.75. The transistor parameters are $I_{S1} = 5 \times 10^{-16}$ A, $I_{S2} = 10^{-15}$ A, $\beta_1 = 180$, $\beta_2 = 120$, $V_{A1} = 120$ V, and $V_{A2} = 80$ V. The $Q$-point is $V_O = 1.25$ V and $I_{CQ} = 200\,\mu$A. (a) Determine the small-signal voltage gain, (b) determine $V_I$, and (c) determine $V_B$.

10.76 For the circuit shown in Figure P10.76, the transistor parameters are $V_{TN} = 0.5$ V, $V_{TP} = -0.5$ V, $k'_n = 100\,\mu$A/V$^2$, $k'_p = 60\,\mu$A/V$^2$, $\lambda_1 = 0.02$ V$^{-1}$, and $\lambda_2 = 0.03$ V$^{-1}$. The quiescent drain current is $I_{DQ} = 200\,\mu$A and the quiescent output voltage is $V_O = 1.25$ V. (a) Determine $(W/L)_1$ such that the small-signal voltage gain is $A_v = -100$, (b) determine $V_I$, and (c) determine $V_G$ assuming $K_{n1} = K_{p2}$.

10.77 Consider the simple BJT active load amplifier in Figure 10.29, with transistor parameters: $I_{SO} = 10^{-12}$ A, $I_{S1} = I_{S2} = 5 \times 10^{-13}$ A, $V_{AN} = 120$ V, and $V_{AP} = 80$ V. Let $V^+ = 5$ V, and neglect base currents. (a) Find the value of $V_{EB}$ that will produce $I_{REF} = 1$ mA. (b) Determine the value of $R_1$. (c) What value of $V_I$ will produce $V_{CEO} = V_{EC2}$? (d) Determine the open-circuit small-signal voltage gain.

10.78 The amplifier shown in Figure P10.78 uses a pnp driver and an npn active load circuit. The transistor parameters are: $I_{S0} = 5 \times 10^{-13}$ A, $I_{S1} = I_{S2} = 10^{-12}$ A, $V_{AN} = 120$ V, and $V_{AP} = 80$ V. Let $V^+ = 5$ V, and neglect base currents. (a) Find the value of $V_{BE}$ that will produce $I_{REF} = 0.5$ mA. (b) Determine the value of $R_1$. (c) What value of $V_I$ will produce $V_{ECO} = V_{CE2}$? (d) Determine the open-circuit small-signal voltage gain.

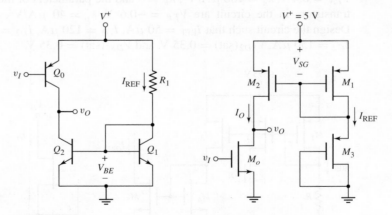

**Figure P10.78**   **Figure P10.79**

D10.79 The bias voltage of the MOSFET amplifier with active load in Figure P10.79 is changed to $V^+ = 3$ V. The transistor parameters are $V_{TN} = 0.5$ V, $V_{TP} = -0.5$ V, $k'_n = 100\,\mu$A/V$^2$, $k'_p = 60\,\mu$A/V$^2$, and $\lambda_n = \lambda_p = 0.02$ V$^{-1}$. The quiescent values are $V_O = 1.5$ V and $V_I = 1.2$ V. (a) Design the circuit $W/L$ ratios, such that $I_{REF} = I_O = 100\,\mu$A. Assume $M_1$ and $M_2$ are matched. (b) Determine the small-signal voltage gain.

10.80 The simple MOSFET amplifier with active load shown in Figure 10.33 is biased at $V^+ = 3$ V. The reference current is $I_{REF} = 80\,\mu$A. The transistor parameters are $V_{TN} = 0.5$ V, $V_{TP} = -0.5$ V, $K_n = K_p = 0.1$ mA, and $\lambda_n = \lambda_p = 0.02$ V$^{-1}$. (a) Find $V_{SG}$. (b) What value of $V_I$ will produce $V_{DSO} = V_{SD2}$? (c) Determine the small-signal voltage gain.

## Section 10.4  Small-Signal Analysis: Active Load Circuits

10.81 Consider the circuit shown in Figure 10.37(a). Let $V^+ = 3$ V and $R_1 = 47$ k$\Omega$. The transistors $Q_1$ and $Q_2$ are matched with $V_{EB}(\text{on}) = 0.6$ V. Neglect base currents, and assume $V_{AP} = 90$ V and $V_{AN} = 120$ V. Determine the small-signal voltage gain for (a) $R_L = \infty$, (b) $R_L = 300$ k$\Omega$, and (c) $R_L = 150$ k$\Omega$.

10.82 Again consider the circuit shown in Figure 10.37(a). Let $V^+ = 5$ V and $R_1 = 35$ k$\Omega$. Let $V_{EB1}(\text{on}) = 0.6$ V. Neglect dc base currents. The base-emitter area of $Q_2$ is twice that of $Q_1$. The Early voltages are $V_{AN} = 120$ V and $V_{AP} = 80$ V. Determine the small-signal voltage gain for (a) $R_L = \infty$ and (b) $R_L = 250$ k$\Omega$.

10.83 A BJT amplifier with active load is shown in Figure P10.83. The circuit contains emitter resistors $R_E$ and a load resistor $R_L$. (a) Derive the expression for the output resistance looking into the collector of $Q_2$. (b) Using the small-signal equivalent circuit, derive the equation for the small-signal voltage gain. Express the relationship in a form similar to Equation (10.94).

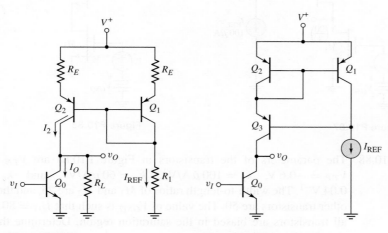

Figure P10.83                     Figure P10.84

10.84 In the circuit in Figure P10.84, the active load circuit is replaced by a Wilson current source. Assume that $\beta = 80$ for all transistors, and that $V_{AN} = 120$ V, $V_{AP} = 80$ V, and $I_{REF} = 0.2$ mA. Determine the open-circuit small-signal voltage gain.

10.85 For the circuit in Figure 10.40(a), the transistor parameters are $k'_n = 80$ $\mu$A/V$^2$, $k'_p = 40$ $\mu$A/V$^2$, $V_{TN} = 0.8$ V, $V_{TP} = -0.6$ V, $\lambda_n = 0.015$ V$^{-1}$, and $\lambda_p = 0.02$ V$^{-1}$. Also, assume $(W/L)_o = 20$ and $(W/L)_1 = (W/L)_2 = 35$. The circuit parameters are $V^+ = 5$ V and $I_{REF} = 200$ $\mu$A. (a) Determine the $g_m$ and $r_o$ parameters of each transistor. (b) Determine the open-circuit small-signal voltage gain. (c) Determine the value of $R_L$ that results in a voltage gain of one-half the open-circuit value.

10.86 Consider the circuit in Figure 10.40(a). The transistor and circuit parameters are the same as given in Problem 10.85 except for the width-to-length ratios of the transistors. Determine the $W/L$ ratios such that the open-circuit small-signal voltage gain is $A_v = -100$. Also let the dc voltage values be $V_{GSo} = V_{SG2}$.

10.87   The parameters of the transistors in Figure P10.87 are $V_{TN} = 0.6$ V, $V_{TP} = -0.6$ V, $k'_n = 100\,\mu\text{A/V}^2$, $k'_p = 60\,\mu\text{A/V}^2$, and $\lambda_n = \lambda_p = 0.02\,\text{V}^{-1}$. The width-to-length ratios are shown in the figure. The value of $V_{GSO}$ is such that $I_{D1} = 100\,\mu\text{A}$, and $M_1$ and $M_2$ are biased in the saturation region. Determine the small-signal voltage gain $A_v = v_o/v_i$.

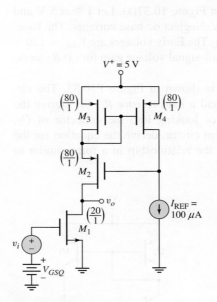

Figure P10.87

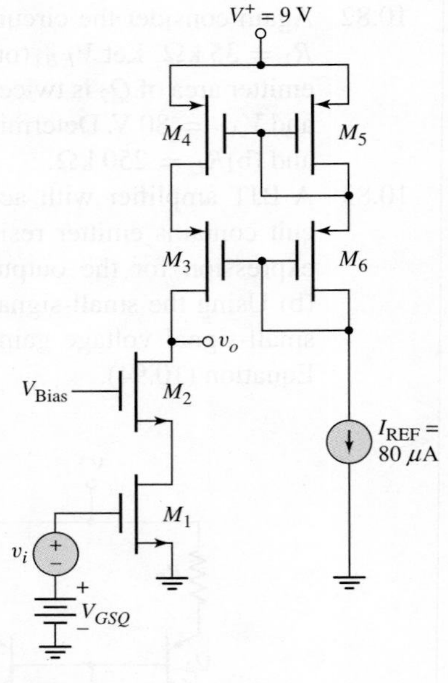

Figure P10.88

10.88   The parameters of the transistors in Figure P10.88 are $V_{TN} = 0.6$ V, $V_{TP} = -0.6$ V, $k'_n = 100\,\mu\text{A/V}^2$, $k'_p = 60\,\mu\text{A/V}^2$, and $\lambda_n = \lambda_p = 0.04\,\text{V}^{-1}$. The width-to-length ratios of $M_1$ and $M_2$ are 25, and those of all other transistors are 50. The value of $V_{GSQ}$ is such that $I_{D1} = 80\,\mu\text{A}$, and all transistors are biased in the saturation region. Determine the small-signal voltage gain $A_v = v_o/v_i$.

10.89   A BJT cascode amplifier with a cascode active load is shown in Figure P10.89. Assume transistor parameters of $\beta = 120$ and $V_A = 80$ V. The $V_{BB}$ voltage is such that all transistors are biased in the active region. Determine the small-signal voltage gain $A_v = v_o/v_i$.

D10.90   Design a bipolar cascode amplifier with a cascode active load similar to that in Figure P10.89 except the amplifying transistors are to be pnp and the load transistors are to be npn. Bias the circuit at $V^+ = 10$ V and incorporate a reference current of $I_{REF} = 200\,\mu\text{A}$. If all transistors are matched with $\beta = 100$ and $V_A = 60$ V, determine the small-signal voltage gain.

D10.91   Design a MOSFET cascode amplifier with a cascode active load similar to that shown in Figure P10.88 except that the amplifying transistors are to be PMOS and the load transistors are to be NMOS. Assume transistor parameters similar to those in Problem 10.88. Determine the small-signal voltage gain.

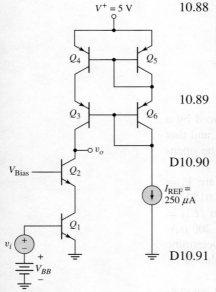

Figure P10.89

#  COMPUTER SIMULATION PROBLEMS

10.92  Consider the Widlar current source in Figure 10.9, with parameters given in Example 10.5. Choose appropriate transistor parameters. Connect a 50 kΩ resistor between $V^+$ and the collector of $Q_2$ as a load. Using a computer simulation, determine $I_{REF}$, $I_O$, $V_{BE1}$, and $V_{BE2}$.

10.93  Using a computer simulation, verify the results of Example 10.9.

10.94  Using a computer simulation, verify the results of Example 10.12. In each case, plot $v_O$ versus $v_I$ over the range $0 \leq v_I \leq 1.0$ V.

10.95  Using a computer simulation, verify the results of Problem 10.87.

# DESIGN PROBLEMS

[Note: Each design should be verified with a computer analysis.]

*D10.96  Design a generalized Widlar current source (Figure P10.34) to provide a bias current of $I_O = 100\,\mu$A and an output resistance of $R_o = 10$ MΩ. The circuit is to be biased at $V^+ = 3$ V and $V^- = -3$ V. The transistor parameters are $I_S = 10^{-15}$ A and $V_A = 120$ V.

*D10.97  The current source to be designed has the general configuration shown in Figure 10.17. The bias voltages are $V^+ = 2.5$ V and $V^- = -2.5$ V. The bias current is to be $I_O = 80\,\mu$A and $V_{DS2}(\text{sat}) = 0.5$ V. The total power dissipated in the circuit is to be limited to 1 mW. Use appropriate transistor parameters.

*D10.98  Design a PMOS version of the current source circuit shown in Figure 10.27. The circuit is to be biased at $V^+ = 2.5$ V and $V^- = 2.5$ V. The currents are to be $I_O = 0.6$ mA and $I_{REF} = 0.25$ mA. Use appropriate transistor parameters.

*D10.99  Consider Exercise TYU 10.10. Redesign the circuit such that the small-signal voltage gain is $A_v = -120$.

# Differential and Multistage Amplifiers

In this chapter, we introduce a special multitransistor circuit configuration called the differential amplifier, or diff-amp. We have encountered a diff-amp previously in our discussion of op-amp circuits. However, the diff-amp, in the context of this chapter, is at the basic transistor level.

The diff-amp is a fundamental building block of analog circuits. It is the input stage of virtually every op-amp, and is the basis of a high-speed digital logic circuit family, called emitter-coupled logic, which will be addressed in Chapter 17.

The design of diff-amps for integrated circuits, in general, incorporates current-source biasing and active loads, which were analyzed in the last chapter. At the end of this chapter, the reader should be able to design both BJT and MOSFET diff-amps to meet particular specifications.

Basic BiCMOS analog circuits are also considered. BiCMOS circuits combine bipolar and MOS transistors on the same semiconductor chip. The advantages of the MOSFET's high input impedance and the bipolar high gain can be utilized in the same circuit.

Up to this point, we have concentrated primarily on the analysis and design of single-stage amplifiers. However, these circuits have limited gain, input resistance, and output resistance characteristics. Multistage or cascaded-stage amplifiers can be designed to produce high gain and specified input and output resistance properties. In this chapter, we begin to consider these multistage amplifiers.

## PREVIEW

In this chapter, we will:

- Describe the characteristics and terminology of the ideal differential amplifier.
- Analyze and determine the characteristics of the basic bipolar differential amplifier.
- Analyze and determine the characteristics of the basic MOSFET differential amplifier.
- Determine the characteristics of BJT and MOSFET differential amplifiers with active loads.
- Describe the characteristics of and analyze various BiCMOS circuits.
- Analyze an example of a gain stage and output stage of a multistage amplifier.
- Analyze a simplified multistage bipolar amplifier.
- Analyze the frequency response of the differential amplifier.
- As an application, design a CMOS diff-amp with an output gain stage to meet a set of specifications.

## 11.1 THE DIFFERENTIAL AMPLIFIER

**Objective:** • Describe the characteristics and terminology of the ideal differential amplifier.

In Chapters 4 and 6, we discussed the reasons linear amplifiers are necessary in analog electronic systems. In these chapters, we analyzed and designed several configurations of MOSFET and bipolar amplifiers. In these circuits, there was one input terminal and one output terminal.

In this chapter, we introduce another basic transistor circuit configuration called the differential amplifier. This amplifier, also called a diff-amp, is the input stage to virtually all op-amps and is probably the most widely used amplifier building block in analog integrated circuits. Figure 11.1 is a block diagram of the diff-amp. There are two input terminals and one output terminal. Ideally, the output signal is proportional to only the difference between the two input signals.

The ideal output voltage can be written as

$$v_o = A_{\text{vol}}(v_1 - v_2) \tag{11.1}$$

where $A_{\text{vol}}$ is called the open-loop voltage gain. In the ideal case, if $v_1 = v_2$, the output voltage is zero. We only obtain a nonzero output voltage if $v_1$ and $v_2$ are not equal.

We define the **differential-mode input voltage** as

$$v_d = v_1 - v_2 \tag{11.2}$$

and the **common-mode input voltage** as

$$v_{cm} = \frac{v_1 + v_2}{2} \tag{11.3}$$

These equations show that if $v_1 = v_2$, the differential-mode input signal is zero and the common-mode input signal is $v_{cm} = v_1 = v_2$.

If, for example, $v_1 = +10 \, \mu\text{V}$ and $v_2 = -10 \, \mu\text{V}$, then the differential-mode voltage is $v_d = 20 \, \mu\text{V}$ and the common-mode voltage is $v_{cm} = 0$. However, if $v_1 = 110 \, \mu\text{V}$ and $v_2 = 90 \, \mu\text{V}$, then the differential-mode input signal is still $v_d = 20 \, \mu\text{V}$, but the common-mode input signal is $v_{cm} = 100 \, \mu\text{V}$. If each pair of input voltages were applied to the ideal difference amplifier, the output voltage in each case would be exactly the same. However, amplifiers are not ideal, and the common-mode input signal does affect the output. One goal of the design of differential amplifiers is to minimize the effect of the common-mode input signal.

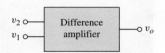

$v_2 \circ$ — Difference amplifier — $\circ \, v_o$
$v_1 \circ$

**Figure 11.1** Difference amplifier block diagram

## 11.2 BASIC BJT DIFFERENTIAL PAIR

**Objective:** • Describe the characteristics of and analyze the basic bipolar differential amplifier.

In this section, we consider the basic bipolar **difference amplifier** or **diff-amp.** We introduce the terminology, qualitatively describe the operation of the circuit, and analyze the dc and small-signal characteristics of the diff-amp.

## BJT Diff-Amp Operation—Qualitative Description

Figure 11.2 shows the basic BJT differential-pair configuration. Two identical transistors, $Q_1$ and $Q_2$, whose emitters are connected together, are biased by a constant-current source $I_Q$, which is connected to a negative supply voltage $V^-$. The collectors of $Q_1$ and $Q_2$ are connected through resistors $R_C$ to a positive supply voltage $V^+$. By design, transistors $Q_1$ and $Q_2$ are to remain biased in the forward-active region. We assume that the two collector resistors $R_C$ are equal, and that $v_{B1}$ and $v_{B2}$ are ideal sources, meaning that the output resistances of these sources are negligibly small.

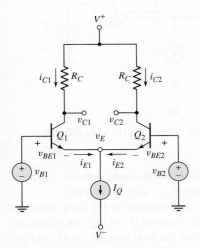

**Figure 11.2** Basic BJT differential-pair configuration

Since both positive and negative bias voltages are used in the circuit, the need for coupling capacitors and voltage divider biasing resistors at the inputs of $Q_1$ and $Q_2$ has been eliminated. If the input signal voltages $v_{B1}$ and $v_{B2}$ in the circuit shown in Figure 11.2 are both zero, $Q_1$ and $Q_2$ are still biased in the active region by the current source $I_Q$. The common-emitter voltage $v_E$ would be on the order of $-0.7$ V. This circuit, then, is referred to as a dc-coupled differential amplifier, so differences in dc input voltages can be amplified. Although the diff-amp contains two transistors, it is considered a single-stage amplifier. The analysis will show that it has characteristics similar to those of the common-emitter amplifier.

First, we consider the circuit in which the two base terminals are connected together and a common-mode voltage $v_{cm}$ is applied as shown in Figure 11.3(a). The transistors are biased "on" by the constant-current source, and the voltage at the common emitters is $v_E = v_{cm} - V_{BE}(\text{on})$. Since $Q_1$ and $Q_2$ are matched or identical, current $I_Q$ splits evenly between the two transistors, and

$$i_{E1} = i_{E2} = \frac{I_Q}{2} \qquad \textbf{(11.4)}$$

If base currents are negligible, then $i_{C1} \cong i_{E1}$ and $i_{C2} \cong i_{E2}$, and

$$v_{C1} = V^+ - \frac{I_Q}{2}R_C = v_{C2} \qquad \textbf{(11.5)}$$

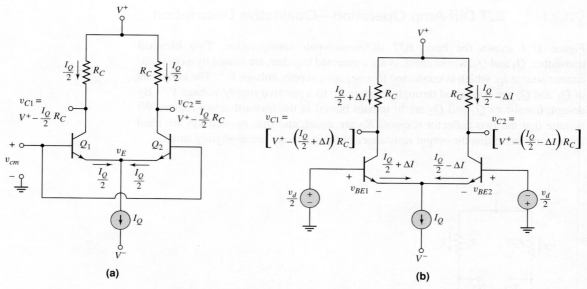

**Figure 11.3** Basic diff-amp with applied common-mode voltage and (b) basic diff-amp with applied differential-mode voltage

We see from Equation (11.5) that, for an applied common-mode voltage, $I_Q$ splits evenly between $Q_1$ and $Q_2$ and the difference between $v_{C1}$ and $v_{C2}$ is zero.

Now, if $v_{B1}$ increases by a few millivolts and $v_{B2}$ decreases by the same amount, or $v_{B1} = v_d/2$ and $v_{B2} = -v_d/2$, the voltages at the bases of $Q_1$ and $Q_2$ are no longer equal. Since the emitters are common, this means that the B–E voltages on $Q_1$ and $Q_2$ are no longer equal. Since $v_{B1}$ increases and $v_{B2}$ decreases, then $v_{BE1} > v_{BE2}$, which means that $i_{C1}$ increases by $\Delta I$ above its quiescent value and $i_{C2}$ decreases by $\Delta I$ below its quiescent value. This is shown in Figure 11.3(b). A potential difference now exits between the two collector terminals. We can write

$$v_{C2} - v_{C1} = \left[V^+ - \left(\frac{I_{CQ}}{2} - \Delta I\right)R_C\right]$$
$$- \left[V^+ - \left(\frac{I_{CQ}}{2} + \Delta I\right)R_C\right] = 2\Delta I R_C \tag{11.6}$$

A voltage difference is created between $v_{C2}$ and $v_{C1}$ when a differential-mode input voltage is applied.

## EXAMPLE 11.1

**Objective:** Determine the quiescent collector current and collector-emitter voltage in a difference amplifier.

Consider the diff-amp in Figure 11.2, with circuit parameters: $V^+ = 10$ V, $V^- = -10$ V, $I_Q = 1$ mA, and $R_C = 10$ k$\Omega$. The transistor parameters are: $\beta = \infty$ (neglect base currents), $V_A = \infty$, and $V_{BE}(\text{on}) = 0.7$ V. Determine $i_{C1}$ and $v_{CE1}$ for common-mode voltages $v_{B1} = v_{B2} = v_{CM} = 0, -5$ V, and $+5$ V.

**Solution:** We know that

$$i_{C1} = i_{C2} = \frac{I_Q}{2} = 0.5\,\text{mA}$$

therefore,

$$v_{C1} = v_{C2} = V^+ - i_{C1}R_C = 10 - (0.5)(10) = 5 \text{ V}$$

From $v_{CM} = 0$, $v_E = -0.7 \text{ V}$ and

$$v_{CE1} = v_{C1} - v_E = 5 - (-0.7) = 5.7 \text{ V}$$

For $v_{CM} = -5 \text{ V}$, $v_E = -5.7 \text{ V}$ and

$$v_{CE1} = v_{C1} - v_E = 5 - (-5.7) = 10.7 \text{ V}$$

For $v_{CM} = +5 \text{ V}$, $v_E = 4.3 \text{ V}$ and

$$v_{CE1} = v_{C1} - v_E = 5 - 4.3 = 0.7 \text{ V}$$

**Comment:** As the common-mode input voltage varies, the ideal constant current $I_Q$ still splits evenly between $Q_1$ and $Q_2$, but the collector–emitter voltage varies, which means that the $Q$-point changes. The variation in $Q$-point as a function of common-mode input voltage is shown in Figure 11.4(a). In this example, if $v_{CM}$ were to increase about $+5 \text{ V}$, then $Q_1$ and $Q_2$ would be driven into saturation. This demonstrates that there is a limited range of applied common-mode voltage over which $Q_1$ and $Q_2$ will remain biased in the forward-active mode.

Figure 11.4(b) shows the $Q$-point when $v_{CM} = 0$ and also shows the variation in $i_{C1}$ and $v_{CE1}$ when an 18 mV sinusoidal differential voltage is applied.

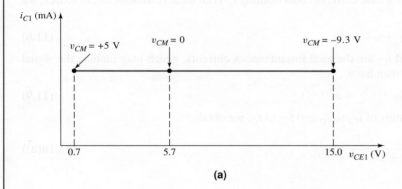

**(a)**

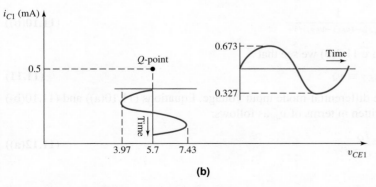

**(b)**

**Figure 11.4** (a) Variation of $Q$-point for transistor $Q_1$ in the BJT diff-amp as the common-mode input voltage varies from $+5$ to $-9.3 \text{ V}$; (b) change in collector current and collector–emitter voltage versus time for transistor $Q_1$ in the BJT diff-amp when a sinusoidal 18 mV differential voltage is applied

**Ex 11.1:** The circuit parameters for the differential amplifier shown in Figure 11.2 are $V^+ = 5$ V, $V^- = -5$ V, $I_Q = 0.3$ mA, and $R_C = 20$ kΩ. The transistor parameters are $\beta = 180$, $V_{BE}(\text{on}) = 0.7$ V, and $V_A = \infty$. Determine the voltages $v_E$, $v_{C1}$, $v_{C2}$, $v_{CE1}$, and $v_{CE2}$ for (a) $v_1 = v_2 = 0$, (b) $v_1 = v_2 = -1$ V, and (c) $v_1 = v_2 = +1$ V. (Ans. (a) $v_E = -0.7$ V, $v_{C1} = v_{C2} = 2$ V, $v_{CE1} = v_{CE2} = 2.7$ V; (b) $v_E = -1.7$ V, $v_{C1} = v_{C2} = 2$ V, $v_{CE1} = v_{CE2} = 3.7$ V; (c) $v_E = +0.3$ V, $v_{C1} = v_{C2} = 2$ V, $v_{CE1} = v_{CE2} = 1.7$ V)

### 11.2.2 DC Transfer Characteristics

We can perform a general analysis of the differential-pair configuration by using the exponential relationship between collector current and B–E voltage. To begin, we know that

$$i_{C1} = I_S e^{v_{BE1}/V_T} \tag{11.7(a)}$$

and

$$i_{C2} = I_S e^{v_{BE2}/V_T} \tag{11.7(b)}$$

We assume $Q_1$ and $Q_2$ are matched and are operating at the same temperature, so the coefficient $I_S$ is the same in each expression.

Neglecting base currents and assuming $I_Q$ is an ideal constant-current source, we have

$$I_Q = i_{C1} + i_{C2} \tag{11.8}$$

where $i_{C1}$ and $i_{C2}$ are the total instantaneous currents, which may include the signal currents. We then have

$$I_Q = I_S[e^{v_{BE1}/V_T} + e^{v_{BE2}/V_T}] \tag{11.9}$$

Taking the ratios of $i_{C1}$ to $I_Q$ and $i_{C2}$ to $I_Q$, we obtain

$$\frac{i_{C1}}{I_Q} = \frac{1}{1 + e^{(v_{BE2} - v_{BE1})/V_T}} \tag{11.10(a)}$$

and

$$\frac{i_{C2}}{I_Q} = \frac{1}{1 + e^{-(v_{BE2} - v_{BE1})/V_T}} \tag{11.10(b)}$$

From Figure 11.3(b) we see that

$$v_{BE1} - v_{BE2} \equiv v_d \tag{11.11}$$

where $v_d$ is the differential-mode input voltage. Equations (11.10(a)) and (11.10(b)) can then be written in terms of $v_d$, as follows:

$$i_{C1} = \frac{I_Q}{1 + e^{-v_d/V_T}} \tag{11.12(a)}$$

and

$$i_{C2} = \frac{I_Q}{1 + e^{+v_d/V_T}} \tag{11.12(b)}$$

Equations (11.12(a)) and (11.12(b)) describe the basic current–voltage characteristics of the differential amplifier. If the differential-mode input voltage is zero, then the current $I_Q$ splits evenly between $i_{C1}$ and $I_{C2}$, as we discussed. However, when a differential-mode signal $v_d$ is applied, a difference occurs between $i_{C1}$ and $i_{C2}$ which in turn causes a change in the collector terminal voltage. This is the fundamental operation of the diff-amp. If a common-mode signal $v_{CM} = v_{B1} = v_{B2}$ is applied, the bias current $I_Q$ still splits evenly between the two transistors.

Figure 11.5 is the normalized plot of the **dc transfer characteristics** for the differential amplifier. We can make two basic observations. First, the gain of the differential amplifier is proportional to the slopes of the transfer curves about the point $v_d = 0$. In order to maintain a linear amplifier, the excursion of $v_d$ about zero must be kept small.

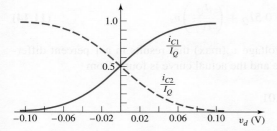

**Figure 11.5** Normalized dc transfer characteristics for BJT differential amplifier

Second, as the magnitude of $v_d$ becomes sufficiently large, essentially all of current $I_Q$ goes to one transistor, and the second transistor effectively turns off. This particular characteristic is used in the emitter-coupled logic (ECL) family of digital logic circuits, which is discussed in Chapter 17.

## EXAMPLE 11.2

**Objective:** Determine the maximum differential-mode input signal that can be applied and still maintain linearity in the differential amplifier.

Figure 11.6 shows an expanded view of the normalized $i_{C1}$ versus $v_d$ characteristic. A linear approximation that corresponds to the slope at $v_d = 0$ is superimposed on the curve. Determine $v_d(\text{max})$ such that the difference between the linear approximation and the actual curve is 1 percent.

**Solution:** The actual expression for $i_{C1}$ versus $v_d$ is, from Equation (11.12(a)),

$$i_{C1}(\text{actual}) = \frac{I_Q}{1 + e^{-v_d/V_T}}$$

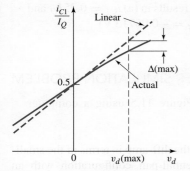

**Figure 11.6** Expanded view, normalized $i_{C1}$ versus $v_d$ transfer characteristic

The slope at $v_d = 0$ is found to be

$$g_f = \frac{di_{C1}}{dv_d}\bigg|_{v_d=0} = I_Q(-1)[1 + e^{-v_d/V_T}]^{-2}\left(\frac{-1}{V_T}\right)[e^{-v_d/V_T}]\bigg|_{v_d=0}$$

or

$$g_f = \frac{I_Q}{4V_T} \tag{11.13}$$

where $g_f$ is the **forward transconductance.** The linear approximation for $i_{C1}$ versus $v_d$ can be written

$$i_{C1}(\text{linear}) = 0.5I_Q + g_f v_d = 0.5I_Q + \left(\frac{I_Q}{4V_T}\right)v_d \tag{11.14}$$

The differential-mode input voltage $v_d(\text{max})$ that results in a 1 percent difference between the ideal linear curve and the actual curve is found from

$$\frac{i_{C1}(\text{linear}) - i_{C1}(\text{actual})}{i_{C1}(\text{linear})} = 0.01$$

or

$$\frac{\left[0.5I_Q + \left(\dfrac{I_Q}{4V_T}\right)v_d(\text{max})\right] - \dfrac{I_Q}{1 + e^{-v_d(\text{max})/V_T}}}{\left[0.5I_Q + \left(\dfrac{I_Q}{4V_T}\right)v_d(\text{max})\right]} = 0.01$$

If we rearrange terms, this expression becomes

$$0.99\left[0.5 + \left(\frac{1}{4V_T}\right)v_d(\text{max})\right] = \frac{1}{1 + e^{-v_d(\text{max})/V_T}}$$

Assuming $V_T = 26$ mV, and using trial and error, we find that

$$v_d(\text{max}) \cong 18\,\text{mV}$$

**Comment:** The differential-mode input voltage must be held to within $\pm 18$ mV in order for the output signal of this diff-amp to be within 1 percent of a linear response.

## EXERCISE PROBLEM

**Ex 11.2:** Consider the dc transfer characteristics shown in Figure 11.5. Determine the value of the differential-mode input voltage that results in (a) $i_{C1} = 0.25I_Q$ and (b) $i_{C2} = 0.9I_Q$. (Ans. (a) $v_d = -0.02856$ V, (b) $v_d = -0.05713$ V)

## COMPUTER SIMULATION PROBLEM

**PS 11.1** Plot the dc transfer characteristics in Figure 11.5 using a computer simulation.

We can now begin to consider the operation of the diff-amp in terms of the small-signal parameters. Figure 11.7 shows the differential-pair configuration with an

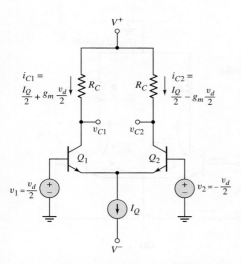

**Figure 11.7** BJT differential amplifier with differential-mode input signal

applied differential-mode input signal. Note that the polarity of the input voltage at $Q_1$ is opposite to that at $Q_2$. The forward-transconductance $g_f$ can be written in terms of the individual transistor transconductances $g_m$. From Equation (11.13), we have

$$g_f = \frac{I_Q}{4V_T} = \frac{1}{2}\frac{I_Q/2}{V_T} = \frac{1}{2}g_m \tag{11.15}$$

where $(I_Q/2)$ is the quiescent collector current in $Q_1$ and $Q_2$. The magnitude of the small-signal collector current in each transistor is then $(g_m v_d)/2$.

Figure 11.7 also shows the linear approximations for the collector currents in terms of the transistor transconductances $g_m$. The slope of $i_{C1}$ versus $v_d$ is the same magnitude as that of $i_{C2}$ versus $v_d$, but it has the opposite sign. This is the reason for the negative sign in the expression for $i_{C2}$ versus $v_d$.

We can define the output signal voltage as

$$v_o = v_{C2} - v_{C1} \tag{11.16}$$

When the output is defined as the difference between the two collector voltages, we have a **two-sided output.** From Figure 11.7, we can write the output voltage as

$$v_o = [V^+ - i_{C2}R_C] - [V^+ - i_{C1}R_C] = (i_{C1} - i_{C2})R_C \tag{11.17(a)}$$

or

$$v_o = \left[\left(\frac{I_Q}{2} + \frac{g_m v_d}{2}\right) - \left(\frac{I_Q}{2} - \frac{g_m v_d}{2}\right)\right]R_C = g_m R_C v_d \tag{11.17(b)}$$

Figure 11.8 shows the ac equivalent circuit of the diff-amp configuration, as well as the signal voltages and currents as functions of the transistor transconductances $g_m$. Since we are assuming an ideal current source, the output resistance looking into the current source is infinite (represented by the dashed line). Using the equivalent circuit in Figure 11.8(a), we find the signal output voltage to be

$$v_o = v_{c2} - v_{c1} = \left(\frac{g_m v_d}{2}\right)R_C - \left(\frac{-g_m v_d}{2}\right)R_C = g_m R_C v_d \tag{11.18}$$

which is the same as Equation (11.17(b)).

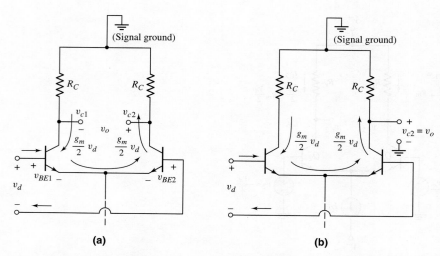

**Figure 11.8** (a) Equivalent ac circuit, diff-amp with differential-mode input signal and two-sided output voltage and (b) ac equivalent circuit with one-sided output

The ratio of the output signal voltage to the differential-mode input signal is called the **differential-mode gain,** $A_d$, which is

$$A_d = \frac{v_o}{v_d} = g_m R_C = \frac{I_Q R_C}{2V_T} \tag{11.19}$$

If the output voltage is the difference between the two collector terminal voltages, then neither side of the output voltage is at ground potential. In many cases, the output voltage is taken at one collector terminal with respect to ground. The resulting voltage output is called a **one-sided output.** If we define the output to be $v_{c2}$, then from Figure 11.8(b), the signal output voltage is

$$v_o = \left(\frac{g_m v_d}{2}\right) R_C \tag{11.20}$$

The differential gain for the one-sided output is then given by

$$A_d = \frac{v_o}{v_d} = \frac{g_m R_C}{2} = \frac{I_Q R_C}{4V_T} \tag{11.21}$$

The differential gain for the one-sided output is one-half that of the two-sided output. However, as we will see in our discussion on active loads, only a one-sided output is available.

We have assumed that the transistors $Q_1$ and $Q_2$, and the two collector resistors $R_C$, are matched. The effects of mismatched elements are discussed in the next section.

### 11.2.3    Small-Signal Equivalent Circuit Analysis

The dc transfer characteristics derived in the last section provide insight into the operation of the differential amplifier. Assuming we are operating in the linear range, we can also derive the gain and other characteristics of the diff-amp, using the small-signal equivalent circuit.

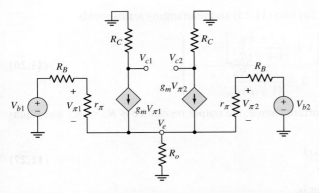

**Figure 11.9** Small-signal equivalent circuit, bipolar differential amplifier

Figure 11.9 shows the small-signal equivalent circuit of the bipolar differential-pair configuration. We assume that the Early voltage is infinite for the two emitter-pair transistors, and that the constant-current source is not ideal but can be represented by a finite output impedance $R_o$. Resistances $R_B$ are also included. These represent the output resistance of the signal voltage sources. All voltages are represented by their phasor components. Since the two transistors are biased at the same quiescent current, we have

$$r_{\pi 1} = r_{\pi 2} \equiv r_\pi \qquad \text{and} \qquad g_{m1} = g_{m2} \equiv g_m$$

Writing a KCL equation at node $V_e$, using phasor notation, we have

$$\frac{V_{\pi 1}}{r_\pi} + g_m V_{\pi 1} + g_m V_{\pi 2} + \frac{V_{\pi 2}}{r_\pi} = \frac{V_e}{R_o} \qquad \text{(11.22(a))}$$

or

$$V_{\pi 1}\left(\frac{1+\beta}{r_\pi}\right) + V_{\pi 2}\left(\frac{1+\beta}{r_\pi}\right) = \frac{V_e}{R_o} \qquad \text{(11.22(b))}$$

where $g_m r_\pi = \beta$. From the circuit, we see that

$$\frac{V_{\pi 1}}{r_\pi} = \frac{V_{b1} - V_e}{r_\pi + R_B} \qquad \text{and} \qquad \frac{V_{\pi 2}}{r_\pi} = \frac{V_{b2} - V_e}{r_\pi + R_B}$$

Solving for $V_{\pi 1}$ and $V_{\pi 2}$ and substituting into Equation (11.22(b)), we find

$$(V_{b1} + V_{b2} - 2V_e)\left(\frac{1+\beta}{r_\pi + R_B}\right) = \frac{V_e}{R_o} \qquad \text{(11.23)}$$

Solving for $V_e$, we obtain

$$V_e = \frac{V_{b1} + V_{b2}}{2 + \dfrac{r_\pi + R_B}{(1+\beta)R_o}} \qquad \text{(11.24)}$$

### One-Sided Output

If we consider a one-sided output at the collector of $Q_2$, then

$$V_o = V_{c2} = -(g_m V_{\pi 2})R_C = -\frac{\beta R_C (V_{b2} - V_e)}{r_\pi + R_B} \qquad \text{(11.25)}$$

Substituting Equation (11.24) into (11.25) and rearranging terms yields

$$V_o = \frac{-\beta R_C}{r_\pi + R_B} \left\{ \frac{V_{b2}\left[1 + \dfrac{r_\pi + R_B}{(1+\beta)R_o}\right] - V_{b1}}{2 + \dfrac{r_\pi + R_B}{(1+\beta)R_o}} \right\} \tag{11.26}$$

In an ideal constant-current source, the output resistance is $R_o = \infty$, and Equation (11.26) reduces to

$$V_o = -\frac{\beta R_C (V_{b2} - V_{b1})}{2(r_\pi + R_B)} \tag{11.27}$$

The differential-mode input is

$$V_d = V_{b1} - V_{b2}$$

and the differential-mode gain is

$$A_d = \frac{V_o}{V_d} = \frac{\beta R_C}{2(r_\pi + R_B)} \tag{11.28}$$

which for $R_B = 0$ is identical to Equation (11.21), which was developed from the voltage transfer characteristics.

Equation (11.26) includes a finite output resistance for the current source. We can see that when a common-mode signal $V_{cm} = V_{b1} = V_{b2}$ is applied, the output voltage is no longer zero.

Differential- and common-mode voltages are defined in Equations (11.2) and (11.3). Using phasor notation, we can solve these equations for $V_{b1}$ and $V_{b2}$ in terms of $V_d$ and $V_{cm}$. We obtain

$$V_{b1} = V_{cm} + \frac{V_d}{2} \tag{11.29(a)}$$

and

$$V_{b2} = V_{cm} - \frac{V_d}{2} \tag{11.29(b)}$$

Since we are dealing with a linear amplifier, superposition applies. Equations (11.29(a)) and (11.29(b)) then simply state that the two input signals can be written as the sum of a differential-mode input signal component and a common-mode input signal component.

Substituting Equations (11.29(a)) and (11.29(b)) into Equation (11.26) and rearranging terms results in the following:

$$V_o = \frac{\beta R_C}{2(r_\pi + R_B)} \cdot V_d - \frac{\beta R_C}{r_\pi + R_B + 2(1+\beta)R_o} \cdot V_{cm} \tag{11.30}$$

We can write the output voltage in the general form

$$V_o = A_d V_d + A_{cm} V_{cm} \tag{11.31}$$

where $A_d$ is the differential-mode gain and $A_{cm}$ is the common-mode gain. Comparing Equations (11.30) and (11.31), we see that the differential-mode gain is

$$A_d = \frac{\beta R_C}{2(r_\pi + R_B)} \tag{11.32(a)}$$

and the common-mode gain is

$$A_{cm} = \frac{-\beta R_C}{r_\pi + R_B + 2(1+\beta)R_o} \qquad \textbf{(11.32(b))}$$

We again observe that the common-mode gain goes to zero for an ideal current source in which $R_o = \infty$. For a nonideal current source, $R_o$ is finite and the common-mode gain is not zero for this case of a one-sided output. A nonzero common-mode gain implies that the diff-amp is not ideal.

### 11.2.4    Common-Mode Rejection Ratio

The ability of a differential amplifier to reject a common-mode signal is described in terms of the common-mode rejection ratio (CMRR). The CMRR is a figure of merit for the diff-amp and is defined as

$$\text{CMRR} = \left| \frac{A_d}{A_{cm}} \right| \qquad \textbf{(11.33)}$$

For an ideal diff-amp, $A_{cm} = 0$ and CMRR $= \infty$. Usually, the CMRR is expressed in decibels, as follows:

$$\text{CMMR}_{dB} = 20 \log_{10} \left| \frac{A_d}{A_{cm}} \right| \qquad \textbf{(11.34)}$$

For the diff-amp in Figure 11.2, the one-sided differential- and common-mode gains are given by Equations (11.32(a)) and (11.32(b)). Using these equations, we can express the CMRR as

$$\text{CMRR} = \left| \frac{A_d}{A_{cm}} \right| = \frac{1}{2} \left[ 1 + \frac{2(1+\beta)R_o}{r_\pi + R_B} \right] \qquad \textbf{(11.35)}$$

The common-mode gain decreases as $R_o$ increases. Therefore, we see that the CMRR increases as $R_o$ increases.

### EXAMPLE 11.3

**Objective:** Determine the differential- and common-mode gains and the common-mode rejection ratio of a diff-amp.

Consider the circuit in Figure 11.2, with parameters: $V^+ = 10$ V, $V^- = -10$ V, $I_Q = 0.8$ mA, and $R_C = 12$ k$\Omega$. The transistor parameters are $\beta = 100$ and $V_A = \infty$. Assume the output resistance looking into the constant-current source is $R_o = 25$ k$\Omega$. Assume the source resistors $R_B$ are zero. Use a one-sided output at $v_{C2}$.

**Solution:** From Equation (11.32(a)), the differential-mode gain can be written as

$$A_d = \frac{g_m R_C}{2} = \frac{I_{CQ} R_C}{2V_T} = \frac{I_Q R_C}{4V_T} = \frac{(0.8)(12)}{4(0.026)} = 92.3$$

From Equation (11.32(b)), the common-mode gain can be written as

$$A_{cm} = \frac{-\left( \dfrac{I_Q R_C}{2V_T} \right)}{1 + \dfrac{(1+\beta)I_Q R_o}{V_T \beta}} = \frac{-\left[ \dfrac{(0.8)(12)}{(2)(0.026)} \right]}{1 + \dfrac{(101)(0.8)(25)}{(0.026)(100)}} = -0.237$$

The common-mode rejection ratio is

$$\text{CMRR} = \left| \frac{A_d}{A_{cm}} \right| = \left| \frac{92.3}{-0.237} \right| = 389$$

In many cases, the value of CMRR is expressed in decibels, or

$$\text{CMRR}|_{dB} = 20 \log_{10} \text{CMRR}$$

which, for this example, becomes

$$\text{CMRR}|_{dB} = 20 \log_{10}(389) = 51.8 \, \text{dB}$$

**Comment:** The common-mode gain is less than the differential-mode gain, but is not zero as determined for the ideal diff-amp with an ideal current source. In general, a common-mode rejection ratio of $\text{CMRR}|_{dB} > 80 \, \text{dB}$ is a design goal for a diff-amp. The aim, then, is to design a better diff-amp than considered in this example.

### EXERCISE PROBLEM

**Ex 11.3:** Consider the diff-amp described in Example 11.3. Assume the same circuit and transistor parameters except for the current source output resistance $R_o$. Determine the required value of $R_o$ to produce a $\text{CMRR}_{dB}$ of (a) 75 dB and (b) 95 dB. (Ans. (a) $R_o = 362 \, \text{k}\Omega$, (b) $R_o = 3.62 \, \text{M}\Omega$)

### DESIGN EXAMPLE 11.4

**Objective:** Design a differential amplifier to meet the specifications of an experimental system.

**Specifications:** Figure 11.10 shows a Hall-effect experiment to measure semiconductor material parameters. A Hall voltage $V_H$, which is perpendicular to both a current $I_X$ and a magnetic field $B_Z$, is to be measured by using a diff-amp. The range of $V_H$ is $-8 \le V_H \le +8 \, \text{mV}$ and the desired range of the diff-amp output signal is to be $-0.8 \le V_O \le +0.8 \, \text{V}$. The probes that make contact to the semiconductor have an effective resistance of 500 $\Omega$, and each probe has an induced 60 Hz signal with a magnitude of 100 mV. The diff-amp output 60 Hz signal is to be no larger than 10 mV. Typically, $V_X = 5 \, \text{V}$, so that the quiescent or common-mode voltage of the Hall probes is 2.5 V.

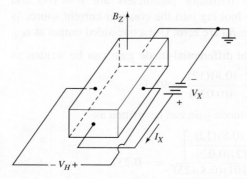

**Figure 11.10** Experimental arrangement for measuring Hall voltage

**Choices:** The bipolar diff-amp with the configuration in Figure 11.7 is to be designed with bias voltages of $\pm 10$ V. The diff-amp transistors are matched with $\beta = 100$ and matched integrated collector resistors of any value can be fabricated. Assume the transistors in the current source are matched with very large $\beta$ values, $V_{BE}(\text{on}) = 0.7$ V, and $V_A = 80$ V. A bias current of $I_Q = 0.5$ mA is to be used.

**Solution (Differential-Mode Gain):** The differential-mode voltage gain requirement is

$$A_d = \frac{V_o}{V_d} = \frac{0.8}{0.008} = 100$$

The small-signal parameters are then

$$r_\pi = \frac{\beta V_T}{I_{CQ}} = \frac{(100)(0.026)}{0.25} = 10.4\,\text{k}\Omega$$

and

$$g_m = \frac{I_{CQ}}{V_T} = \frac{0.25}{0.026} = 9.62 \text{ mA/V}$$

The differential gain is

$$A_d = \frac{\beta R_C}{2(r_\pi + R_B)}$$

or

$$100 = \frac{(100)R_C}{2(10.4 + 0.5)}$$

which means that $R_C = 21.8$ k$\Omega$. We may note that the voltage drop across $R_C$ under quiescent conditions is 5.45 V. With a 2.5 V common-mode input voltage, the quiescent collector-emitter voltages of $Q_1$ and $Q_2$ are approximately 3.65 V. The two input transistors will then remain in the active region.

**Solution (Common-Mode Gain):** The common-mode voltage gain requirement is

$$A_{cm} = \frac{V_o}{V_{cm}} = \frac{10\,\text{mV}}{100\text{ mV}} = 0.10$$

The common-mode gain is given by

$$|A_{cm}| = \frac{\beta R_C}{r_\pi + R_B + 2(1 + \beta)R_o}$$

or

$$0.10 = \frac{(100)(21.8)}{10.4 + 0.5 + 2(101)R_o}$$

which means that $R_o = 108$ k$\Omega$. If we consider a simple two-transistor current source as discussed in the last chapter, the output resistance is $R_o = r_o = V_A/I_Q$, where $V_A$ is the Early voltage. With $I_Q = 0.5$ mA, then $V_A = 54$ V is the Early voltage requirement. This specification is not difficult to achieve for most bipolar transistors.

**Trade-offs:** If the common-mode gain requirement had been more stringent, a different current source circuit might be required to provide a larger output resistance. The effects of mismatched devices and elements are considered in the next section.

**Computer Simulation Verification:** Figure 11.11 shows the circuit used in the computer simulation for this example. The bias current $I_Q$ supplied by the $Q_3$ current source transistor is 0.568 mA. A 2.5 V common-mode input voltage is applied, a 500 Ω source (probe) resistance is included, and an 8 mV differential-mode input signal is applied. The differential output signal voltage measured at the collector of $Q_2$ is 0.84 V, which is just slightly larger than the designed value. The current gains of the standard 2N3904 transistors used in the computer simulation are larger than the values of 100 used in the hand analysis and design. A common-mode signal voltage of 100 mV replaced the differential-mode signals. The common-mode output signal is 7.11 mV, which is within the design specification.

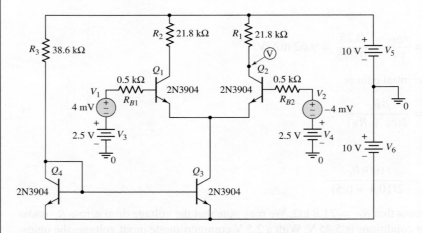

**Figure 11.11** Circuit used in the computer simulation of Design Example 11.4

## EXERCISE PROBLEM

**Ex 11.4:** Consider the diff-amp configuration shown in Figure 11.7. Assume $Q_1$ and $Q_2$ are matched, let $V_A = \infty$, and neglect base currents. Let $I_Q = 200\,\mu A$. (a) Design the circuit such that the differential-mode gain at $v_{C1}$ is $-150$, the differential-mode gain at $v_{C2}$ is $+100$, and the common-mode voltage is in the range $-1.5 \le v_{cm} \le 1.5$ V. (b) Using the results of part (a), what are the minimum bias voltages $V^+ = -V^-$ such that the input transistors always remain biased in the forward-active region. (Ans. (a) $R_{C1} = 78.0\,k\Omega$, $R_{C2} = 52.0\,k\Omega$; (b) $V^+ = -V^- = 9.3$ V)

## Test Your Understanding

**TYU 11.1** Find the differential- and common-mode components of the input signal applied to a diff-amp for input voltages of (a) $v_1 = 2.100$ V and $v_2 = 2.120$ V; and (b) $v_1 = 0.25 - 0.002 \sin \omega t$ V and $v_2 = 0.50 + 0.002 \sin \omega t$ V. (Ans. (a) $v_d = -0.02$ V, $v_{cm} = 2.110$ V; (b) $v_d = -0.25 - 0.004 \sin \omega t$ V, $v_{cm} = 0.375$ V)

**TYU 11.2** Consider the diff-amp in Figure 11.2, with parameters $V^+ = 5$ V, $V^- = -5$ V, and $I_Q = 0.4$ mA. (a) Redesign the circuit such that the common-mode input voltage is in the range $-3 \leq v_{cm} \leq 3$ V, while $Q_1$ and $Q_2$ remain biased in the forward-active region. (b) Using the results of part (a), find the differential-mode voltage gain $A_d = (v_{c2} - v_{c1})/v_d$. (Ans. (a) $R_C = 10$ kΩ, (b) $A_d = 76.9$)

**TYU 11.3** Assume the differential-mode gain of a diff-amp is $A_d = 80$ and the common-mode gain is $A_{cm} = -0.20$. Determine the output voltage for input signals of: (a) $v_1 = 0.995 \sin \omega t$ V and $v_2 = 1.005 \sin \omega t$ V; and (b) $v_1 = 2 - 0.005 \sin \omega t$ V and $v_2 = 2 + 0.005 \sin \omega t$ V. (Ans. (a) $v_o = -1.0 \sin \omega t$ V, (b) $v_o = -0.4 - 0.8 \sin \omega t$ V)

## 11.2.5 Two-Sided Output

If we consider the two-sided output of an ideal op-amp and define the output voltage as $V_o = V_{c2} - V_{c1}$, we can show that the differential-mode voltage gain is given by

$$A_d = \frac{\beta R_C}{r_\pi + R_B} \tag{11.36(a)}$$

and the common-mode voltage gain is given by

$$A_{cm} = 0 \tag{11.36(b)}$$

The result of $A_{cm} = 0$ for the two-sided output is a consequence of using matched devices and elements in the diff-amp circuit. We will reconsider a two-sided output and discuss the effects of mismatched elements.

### Effect of $R_C$ Mismatch—Two-Sided Output

We assume that $R_{C1}$ and $R_{C2}$ are the resistors in the collectors of $Q_1$ and $Q_2$. If the two resistors are not matched, we assume that we can write $R_{C1} = R_C + \Delta R_C$ and $R_{C2} = R_C - \Delta R_C$. For simplicity, let $R_B = 0$.

From Figure 11.9, the output voltage for a two-sided output is given by

$$V_o = V_{c2} - V_{c1} = (-g_m V_{\pi 2} R_{C2}) - (-g_m V_{\pi 1} R_{C1}) \tag{11.37}$$

We also see from the figure (with $R_B = 0$) that $V_{\pi 1} = V_{b1} - V_e$ and $V_{\pi 2} = V_{b2} - V_e$. Using the expressions for $V_e$ (Equation (11.24)), $V_{b1}$ (Equation (11.29(a))), and $V_{b2}$ (Equation (11.29(b))), we find the differential voltage gain as

$$A_d = g_m R_C \tag{11.38}$$

and the common-mode gain as

$$A_{cm} = g_m(2\Delta R_C) \cdot \frac{1}{\left[1 + \dfrac{2(1+\beta)R_o}{r_\pi}\right]} \tag{11.39}$$

In general, $2(1 + \beta)R_o/r_\pi \gg 1$, so that

$$A_{cm} \cong g_m(2\Delta R_C) \cdot \frac{r_\pi}{2(1+\beta)R_o} \tag{11.40(a)}$$

Noting that $g_m r_\pi = \beta$ and $\beta/(1 + \beta) \cong 1$, we have the common-mode gain as

$$A_{cm} \cong \frac{\Delta R_C}{R_o} \tag{11.40(b)}$$

The common-mode rejection ratio is then

$$\text{CMRR} = \left| \frac{A_d}{A_{cm}} \right| = \frac{g_m R_o}{(\Delta R_C / R_C)} \tag{11.41}$$

### Effect of $g_m$ Mismatch—Two-Sided Output

We can consider the effect of transistor mismatch by considering the effect of a mismatch in the transconductance $g_m$. We assume $g_{m1}$ and $g_{m2}$ are the transconductance parameters of the two transistors in the diff-amp. We will assume that we can write $g_{m1} = g_m + \Delta g_m$ and $g_{m2} = g_m - \Delta g_m$. Again, for simplicity, let $R_B = 0$.

Again, from Figure 11.9, the output voltage for a two-sided output is

$$V_o = V_{c2} - V_{c1} = (-g_{m2} V_{\pi 2} R_C) - (-g_{m1} V_{\pi 1} R_C) \tag{11.42}$$

Applying a differential input voltage, we find $V_{\pi 1} = V_d / 2$ and $V_{\pi 2} = -V_d / 2$. The differential voltage gain is then

$$A_d = \frac{V_o}{V_d} = g_m R_C \tag{11.43}$$

Applying a common-mode input voltage, we have $V_{\pi 1} = V_{\pi 2} = V_{cm} - V_e$. The output voltage is again given by

$$V_o = V_{c2} - V_{c1} = (-g_{m2} V_{\pi 2} R_C) - (-g_{m1} V_{\pi 1} R_C) \tag{11.44(a)}$$

or

$$V_o = (V_{cm} - V_e) R_C (g_{m1} - g_{m2}) \tag{11.44(b)}$$

Summing currents at the $V_e$ node in Figure 11.9, we have

$$\frac{V_{\pi 1}}{r_{\pi 1}} + g_{m1} V_{\pi 1} + g_{m2} V_{\pi 2} + \frac{V_{\pi 2}}{r_{\pi 2}} = \frac{V_e}{R_o} \tag{11.45}$$

In general, we have $g_m \gg 1/r_\pi$. Then Equation (11.45) becomes

$$(V_{cm} - V_e)(g_{m1} + g_{m2}) = \frac{V_e}{R_o} \tag{11.46(a)}$$

or

$$V_e = \frac{V_{cm}(g_{m1} + g_{m2})}{\dfrac{1}{R_o} + g_{m1} + g_{m2}} \tag{11.46(b)}$$

The output voltage is then

$$V_o = \left[ V_{cm} - \frac{V_{cm}(g_{m1} + g_{m2})}{(1/R_o) + g_{m1} + g_{m2}} \right] \cdot R_C (g_{m1} - g_{m2}) \tag{11.47}$$

Noting that $g_{m1} + g_{m2} = 2g_m$ and $g_{m1} - g_{m2} = 2(\Delta g_m)$, the common-mode gain is

$$A_{cm} = \frac{R_C (2\Delta g_m)}{1 + 2R_o g_m} \tag{11.48}$$

The common-mode rejection ratio now becomes

$$\text{CMRR} = \left| \frac{A_d}{A_{cm}} \right| = \frac{1 + 2R_o g_m}{2(\Delta g_m / g_m)} \tag{11.49}$$

## Differential- and Common-Mode Gains—Further Observations

For greater insight into the mechanism that causes differential- and common-mode gains, we reconsider the diff-amp as pure differential- and common-mode signals are applied.

Figure 11.12(a) shows the ac equivalent circuit of the diff-amp with two sinusoidal input signals. The two input voltages are 180 degrees out of phase, so a pure differential-mode signal is being applied to the diff-amp. We see that $v_{b1} + v_{b2} = 0$. From Equation (11.24), we find $v_e = 0$, so the common emitters of $Q_1$ and $Q_2$ remain at signal ground. In essence, the circuit behaves like a balanced seesaw. As the base voltage of $Q_1$ goes into its positive-half cycle, the base voltage of $Q_2$ is in its negative half-cycle. Then, as the base voltage of $Q_1$ goes into its negative half-cycle, the base voltage of $Q_2$ is in its positive half-cycle. The signal current directions shown in the figure are valid for $v_{b1}$ in its positive half-cycle.

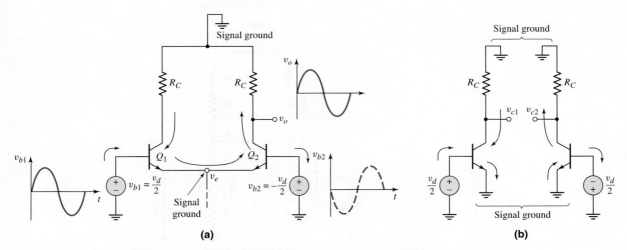

**Figure 11.12** (a) Equivalent ac circuit, diff-amp with applied sinusoidal differential-mode input signal, and resulting signal current directions and (b) differential-mode half-circuits

Since $v_e$ is always at ground potential, we can treat each half of the diff-amp as a common-emitter circuit. Figure 11.12(b) shows the differential half-circuits, clearly depicting the common-emitter configuration. The differential-mode characteristics of the diff-amp can be determined by analyzing the half-circuit. In evaluating the small-signal hybrid-$\pi$ parameters, we must keep in mind that the half-circuit is biased at $I_Q/2$.

Figure 11.13(a) shows the ac equivalent circuit of the diff-amp with a pure common-mode sinusoidal input signal. In this case, the two input voltages are in phase. The current source is represented as an ideal source $I_Q$ in parallel with its output resistance $R_o$. Current $i_q$ is the time-varying component of the source current. As the two input signals increase, voltage $v_e$ increases and current $i_q$ increases. Since this current splits evenly between $Q_1$ and $Q_2$, each collector current also increases. The output voltage $v_o$ then decreases below its quiescent value.

As the two input voltages go through the negative half-cycle, all signal currents shown in the figure reverse direction, and $v_o$ increases above its quiescent value.

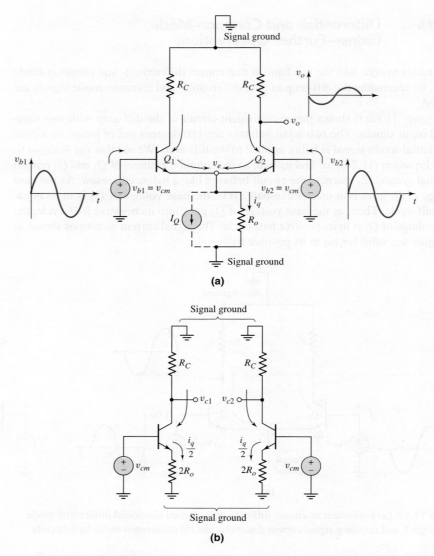

**Figure 11.13** (a) Equivalent ac circuit of diff-amp with common-mode input signal, and resulting signal current directions and (b) common-mode half-circuits

Consequently, a common-mode sinusoidal input signal produces a sinusoidal output voltage, which means that the diff-amp has a nonzero common-mode voltage gain. If the value of $R_o$ increases, the magnitude of $i_q$ decreases for a given common-mode input signal, producing a smaller output voltage and hence a smaller common-mode gain.

With an applied common-mode voltage, the circuit shown in Figure 11.13(a) is perfectly symmetrical. The circuit can therefore be split into the identical common-mode half-circuits shown in Figure 11.13(b). The common-mode characteristics of the diff-amp can then be determined by analyzing the half-circuit, which is a common-emitter configuration with an emitter resistor. Each half-circuit is biased at $I_Q/2$.

The following examples further illustrate the effect of a nonzero common-mode gain on circuit performance.

## EXAMPLE 11.5

**Objective:** Determine the output of a diff-amp when both differential- and common-mode signals are applied.

Consider the circuit shown in Figure 11.2. Use the transistor and circuit parameters described in Example 11.3. Assume that four sets of inputs are applied, as described in the following table, which also includes the differential- and common-mode voltages.

| | Input signal ($\mu$V) | Differential- and common-mode input signals ($\mu$V) |
|---|---|---|
| Case 1 | $v_1 = 10 \sin \omega t$ | $v_d = 20 \sin \omega t$ |
| | $v_2 = -10 \sin \omega t$ | $v_{cm} = 0$ |
| Case 2 | $v_1 = 20 \sin \omega t$ | $v_d = 40 \sin \omega t$ |
| | $v_2 = -20 \sin \omega t$ | $v_{cm} = 0$ |
| Case 3 | $v_1 = 210 \sin \omega t$ | $v_d = 20 \sin \omega t$ |
| | $v_2 = 190 \sin \omega t$ | $v_{cm} = 200 \sin \omega t$ |
| Case 4 | $v_1 = 220 \sin \omega t$ | $v_d = 40 \sin \omega t$ |
| | $v_2 = 180 \sin \omega t$ | $v_{cm} = 200 \sin \omega t$ |
| Case 5 | $v_1 = 200 \sin \omega t$ | $v_d = 0$ |
| | $v_2 = 200 \sin \omega t$ | $v_{cm} = 200 \sin \omega t$ |

**Solution:** The output voltage is given by Equation (11.31), as follows:

$$v_o = A_d v_d + A_{cm} v_{cm}$$

From Example 11.3, the differential- and common-mode gains are $A_d = 92.3$ and $A_{cm} = -0.237$. The output voltages for the four sets of inputs are:

| | Output signal (mV) |
|---|---|
| Case 1 | $v_o = 1.846 \sin \omega t$ |
| Case 2 | $v_o = 3.692 \sin \omega t$ |
| Case 3 | $v_o = 1.799 \sin \omega t$ |
| Case 4 | $v_o = 3.645 \sin \omega t$ |
| Case 5 | $v_o = -0.0474 \sin \omega t$ |

**Comment:** In cases 1 and 2, the common-mode input is zero, and the output is directly proportional to the differential input signal. Comparing cases 1 and 3 and cases 2 and 4, we see that the output voltages are not equal, even though the differential input signals are the same. This shows that the common-mode signal affects the output. Also, even though the differential signal is doubled, in cases 4 and 3, the ratio of the output signals is not 2.0. If a common-mode signal is present, the output is not exactly linear with respect to the differential input signal.

For Case 5, the differential-input voltage is zero, but the output voltage is not zero, since a common-mode input voltage exists and $|A_{cm}| \neq 0$.

EXERCISE PROBLEM

**Ex 11.5:** Assume a diff-amp has a differential-mode gain of $A_d = 150$ and a common-mode rejection ratio of $\text{CMMR}_{\text{dB}} = 50\,\text{dB}$. Assume $A_{cm}$ is positive. Determine the output voltage if the input voltages are (a) $v_1 = -10\,\mu\text{V}$, $v_2 = +10\,\mu\text{V}$ and (b) $v_1 = 190\,\mu\text{V}$, $v_2 = 210\,\mu\text{V}$. (Ans. (a) $v_o = -3.0\,\text{mV}$, (b) $v_o = -2.905\,\text{mV}$)

As mentioned previously, the common-mode gain is a function of the output resistance of the current source. If the required common-mode gain needs to be reduced, then the current source output resistance must be increased, which may require the design of a more sophisticated current source.

Problem-Solving Technique: Diff-Amps with Resistive Loads

1. To determine the differential-mode voltage gain, apply a pure differential-mode input voltage and use the differential-mode half-circuit in the analysis.
2. To determine the common-mode voltage gain, apply a pure common-mode input voltage and use the common-mode half-circuit in the analysis.

### 11.2.7    Differential- and Common-Mode Input Impedances

The input impedance, or resistance, of an amplifier is as important a property as the voltage gain. The input resistance determines the loading effect of the circuit on the signal source. We will look at two input resistances for the difference amplifier: the **differential-mode input resistance,** which is the resistance seen by a differential-mode signal source; and the **common-mode input resistance,** which is the resistance seen by a common-mode input signal source.

**Differential-Mode Input Resistance**

The differential-mode input resistance is the effective resistance between the two input base terminals when a differential-mode signal is applied. A diff-amp with a pure differential input signal is shown in Figure 11.14. The applicable differential-mode half-circuits were shown in Figure 11.12(b). For this circuit, we have

$$\frac{v_d/2}{i_b} = r_\pi \tag{11.50}$$

The differential-mode input resistance is therefore

$$R_{id} = \frac{v_d}{i_b} = 2r_\pi \tag{11.51}$$

Another common diff-amp configuration uses emitter resistors, as shown in Figure 11.15. With a pure applied differential-mode voltage, similar differential-mode half-circuits are applicable to this configuration. We can then use the resistance reflection rule to find the differential-mode input resistance. We have

$$\frac{v_d/2}{i_b} = r_\pi + (1 + \beta)R_E \tag{11.52}$$

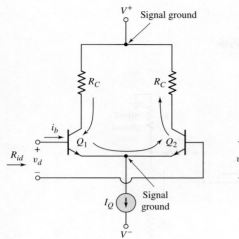

**Figure 11.14** BJT differential amplifier with differential-mode input signal, showing differential input resistance

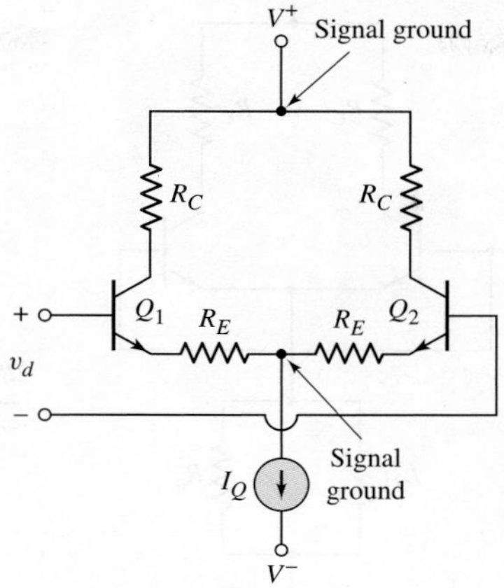

**Figure 11.15** BJT differential amplifier with emitter resistors

Therefore,

$$R_{id} = \frac{v_d}{i_b} = 2[r_\pi + (1 + \beta)R_E] \tag{11.53}$$

Equation (11.53) implies that the differential-mode input resistance increases significantly when emitter resistors are included. We will see that the differential-mode gain decreases when emitter resistors are included in the same way that the voltage gain of a common-emitter amplifier decreases when an emitter resistor is included in the design. However, a larger differential-mode voltage (greater than 18 mV) may be applied to the diff-amp in Figure 11.15 and the amplifier remains linear.

### Common-Mode Input Resistance

Figure 11.16(a) shows a diff-amp with an applied common-mode voltage. The small-signal output resistance $R_o$ of the constant-current source is also shown. The equivalent common-mode half-circuits were given in Figure 11.13(b). Since the half-circuits are in parallel, we can write

$$2R_{icm} = r_\pi + (1 + \beta)(2R_o) \cong (1 + \beta)(2R_o) \tag{11.54}$$

Equation (11.54) is a first approximation for determining the common-mode input resistance.

Normally, $R_o$ is large, and $R_{icm}$ is typically in the megohm range. Therefore, the transistor output resistance $r_o$ and the base–collector resistance $r_\mu$ may need to be included in the calculation. Figure 11.16(b) shows the more complete equivalent half-circuit model. For this model, we have

$$2R_{icm} = r_\mu \,\|\, [(1 + \beta)(2R_o)] \,\|\, [(1 + \beta)r_o] \tag{11.55(a)}$$

Therefore,

$$R_{icm} = \left(\frac{r_\mu}{2}\right) \,\|\, [(1 + \beta)(R_o)] \,\|\, \left[(1 + \beta)\left(\frac{r_o}{2}\right)\right] \tag{11.55(b)}$$

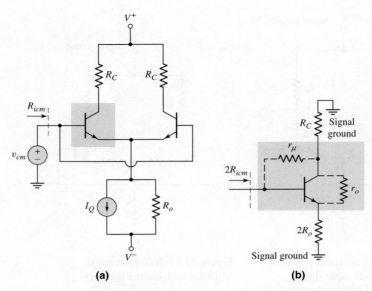

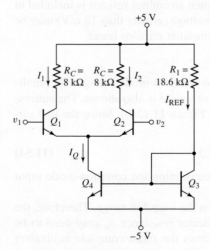

**Figure 11.16** (a) BJT differential amplifier with common-mode input signal, including finite current source resistance and (b) equivalent common-mode half-circuit

---

## EXAMPLE 11.6

**Objective:** Determine the differential- and common-mode input resistances of a differential amplifier.

Consider the circuit in Figure 11.17, with transistor parameters $\beta = 100$, $V_{BE}(\text{on}) = 0.7$ V, and $V_A = 100$ V. Determine $R_{id}$ and $R_{icm}$.

**Figure 11.17** BJT differential amplifier for Example 11.6

**Solution:** From the circuit, we find

$$I_{\text{REF}} = 0.5\,\text{mA} \cong I_Q$$

and

$$I_1 = I_2 \cong I_Q/2 = 0.25\,\text{mA}$$

The small-signal parameters for $Q_1$ and $Q_2$ are then

$$r_\pi = \frac{\beta V_T}{I_{CQ}} = \frac{(100)(0.026)}{0.25} = 10.4\,\text{k}\Omega$$

and

$$r_o = \frac{V_A}{I_{CQ}} = \frac{100}{0.25} = 400\,\text{k}\Omega$$

and the output resistance of $Q_4$ is

$$R_o = \frac{V_A}{I_Q} = \frac{100}{0.5} = 200\,\text{k}\Omega$$

From Equation (11.51), the differential-mode input resistance is

$$R_{id} = 2r_\pi = 2(10.4) = 20.8\,\text{k}\Omega$$

From Equation (11.55(b)), neglecting the effect of $r_\mu$, the common-mode input resistance is

$$R_{icm} = (1+\beta)\left[(R_o)\left\|\left(\frac{r_o}{2}\right)\right.\right] = (101)\left\{200\left\|\left(\frac{400}{2}\right)\right.\right\}\text{k}\Omega \rightarrow 10.1\,\text{M}\Omega$$

**Comment:** If a differential-mode input voltage with a peak value of 15 mV is applied, the source must be capable of supplying a current of $15 \times 10^{-3}/20.8 \times 10^{+3} = 0.72\,\mu\text{A}$ without any severe loading effect. However, the input current from a 15 mV common-mode signal would only be approximately 1.5 nA.

---

### EXERCISE PROBLEM

**Ex 11.6:** Consider the diff-amp shown in Figure 11.15. Assume the current source has a value of $I_Q = 0.5$ mA, the transistor current gains are $\beta = 100$, and the emitter resistors are $R_E = 500\,\Omega$. Find the differential input resistance. (Ans. $R_{id} = 122\,\text{k}\Omega$)

---

### Differential-Mode Voltage Gain with Emitter Degeneration

We may determine the differential-mode voltage gain of the circuit shown in Figure 11.15. Figure 11.18 shows the differential-mode half circuits. For a one-sided output and for matched elements, we have

$$V_o = V_{c2} = -g_m V_{\pi 2} R_C \qquad (11.56)$$

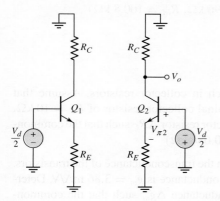

**Figure 11.18** Differential half-circuits with emitter degeneration

Writing a KVL equation around the B–E loop, we have

$$\frac{V_d}{2} + V_{\pi 2} + g_m V_{\pi 2} R_E = 0 \tag{11.57}$$

which yields

$$V_{\pi 2} = \frac{-(V_d/2)}{1 + g_m R_E} \tag{11.58}$$

Substituting Equation (11.58) into (11.56), we find the differential-mode voltage gain as

$$A_d = \frac{V_o}{V_d} = \frac{g_m R_C}{2(1 + g_m R_E)} \tag{11.59}$$

## EXAMPLE 11.7

**Objective:** Determine the one-sided differential-mode voltage gain of the circuit shown in Figure 11.15.

Assume $I_Q = 0.5\,\text{mA}$, $\beta = 100$, and $R_C = 10\,\text{k}\Omega$. Find the differential-mode voltage gain for (a) $R_E = 0$ and (b) $R_E = 500\,\Omega$.

**Solution:** The small-signal transconductance is found to be $g_m = 9.62\,\text{mA/V}$. We find the differential-mode voltage gain to be (a) for $R_E = 0$:

$$A_d = \frac{g_m R_C}{2} = \frac{(9.62)(10)}{2} = 48.1$$

and (b) for $R_E = 500\,\Omega$:

$$A_d = \frac{g_m R_C}{2(1 + g_m R_E)} = \frac{(9.62)(10)}{2[1 + (9.62)(0.5)]} = 8.28$$

**Comment:** As with any design problem, there are trade-offs. Including an emitter resistor $R_E$ decreases the voltage gain but increases the input differential-mode resistance.

## EXERCISE PROBLEM

**Ex 11.7:** Consider the diff-amp described in Example 11.7. Assume the same parameters except the value of $R_E$. Determine the value of $R_E$ that results in a differential-mode voltage gain of $A_d = 10$. What is the corresponding value of differential-input resistance? (Ans. $R_E = 0.396\,\text{k}\Omega$, $R_{id} = 100.8\,\text{k}\Omega$)

## Test Your Understanding

**TYU 11.4** Consider the effect of a mismatch in collector resistors. Assume that $g_m = 3.86\,\text{mA/V}^2$, $R_o = 100\,\text{k}\Omega$, and a nominal collector resistor of $R_C = 10\,\text{k}\Omega$. Determine the minimum mismatch in the collector resistor $\Delta R_C$ such that the common-mode rejection ratio is 75 dB. (Ans. $\Delta R_C = 0.686\,\text{k}\Omega$)

**TYU 11.5** Consider the effect of a mismatch in the transconductance of the transistors. Assume $R_o = 100\,\text{k}\Omega$ and the nominal transconductance is $g_m = 3.86\,\text{mA/V}$. Determine the minimum mismatch in the transconductance $\Delta g_m$ such that the common-mode rejection ratio is 90 dB. (Ans. $\Delta g_m = 0.0472\,\text{mA/V}$ or $\Delta g_m/g_m \rightarrow 1.22\%$)

**TYU 11.6** The parameters of the diff-amp shown in Figure 11.2 are $V^+ = 5$ V, $V^- = -5$ V, $I_Q = 0.4$ mA, and $R_C = 10$ k$\Omega$. The output resistance of the constant-current source is $R_o = 100$ k$\Omega$. The transistor parameters are $\beta = 150$, $V_{BE}(\text{on}) = 0.7$ V, and $V_A = \infty$. (a) Determine the dc input base currents. (b) Determine the differential signal input currents if a differential-mode input voltage $v_d = 10 \sin \omega t$ mV is applied. (c) If a common-mode input voltage $v_{cm} = 3 \sin \omega t$ V is applied, determine the common-mode signal input base currents. (Ans. (a) $I_{B1} = I_{B2} = 1.32$ $\mu$A, (b) $I_b = 0.256 \sin \omega t$ $\mu$A, (c) $I_b = 0.0993 \sin \omega t$ $\mu$A)

---

## 11.3 BASIC FET DIFFERENTIAL PAIR

**Objective:** • Describe the characteristics of and analyze the basic FET differential amplifier.

In this section, we will evaluate the basic FET differential amplifier, concentrating on the MOSFET diff-amp. As we did for the bipolar diff-amp, we will develop the dc transfer characteristics, and determine the differential- and common-mode gains.

Differential amplifiers using JFETs are also available. Since the analysis is almost identical to that for the MOSFET diff-amp, we will only briefly consider the JFET differential pair. A few of the problems at the end of this chapter are based on these circuits.

### 11.3.1 DC Transfer Characteristics

Figure 11.19 shows the basic MOSFET differential pair, with matched transistors $M_1$ and $M_2$ biased with a constant current $I_Q$. We assume that $M_1$ and $M_2$ are always biased in the saturation region.

Like the basic bipolar configuration, the basic MOSFET diff-amp uses both positive and negative bias voltages, thereby eliminating the need for coupling capacitors and voltage divider biasing resistors at the gate terminals. Even with $v_{G1} = v_{G2} = 0$, the transistors $M_1$ and $M_2$ can be biased in the saturation region by the current source $I_Q$. This circuit, then, is also a **dc-coupled** diff-amp.

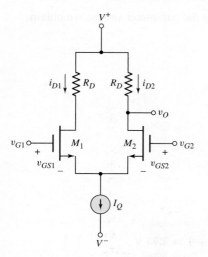

**Figure 11.19** Basic MOSFET differential pair configuration

## EXAMPLE 11.8

**Objective:** Calculate the dc characteristics of a MOSFET diff-amp.

Consider the differential amplifier shown in Figure 11.20. The transistor parameters are: $K_{n1} = K_{n2} = 0.1 \text{ mA/V}^2$, $K_{n3} = K_{n4} = 0.3 \text{ mA/V}^2$, and for all transistors, $\lambda = 0$ and $V_{TN} = 1$ V. Determine the maximum range of common-mode input voltage.

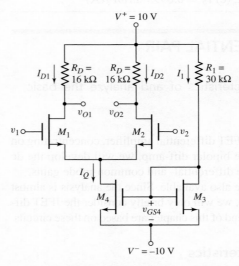

**Figure 11.20** MOSFET differential amplifier for Example 11.8

**Solution:** The reference current can be determined from

$$I_1 = \frac{20 - V_{GS4}}{R_1}$$

and from

$$I_1 = K_{n3}(V_{GS4} - V_{TN})^2$$

Combining these two equations and substituting the parameter values, we obtain

$$9V_{GS4}^2 - 17V_{GS4} - 11 = 0$$

which yields

$$V_{GS4} = 2.40 \text{ V} \quad \text{and} \quad I_1 = 0.587 \text{ mA}$$

Since $M_3$ and $M_4$ are identical, we also find

$$I_Q = 0.587 \text{ mA}$$

The quiescent drain currents in $M_1$ and $M_2$ are

$$I_{D1} = I_{D2} = I_Q/2 \cong 0.293 \text{ mA}$$

The gate-to-source voltages are then

$$V_{GS1} = V_{GS2} = \sqrt{\frac{I_{D1}}{K_{n1}}} + V_{TN} = \sqrt{\frac{0.293}{0.1}} + 1 = 2.71 \text{ V}$$

The quiescent values of $v_{O1}$ and $v_{O2}$ are

$$v_{O1} = v_{O2} = 10 - I_{D1}R_D = 10 - (0.293)(16) = 5.31 \text{ V}$$

The maximum common-mode input voltage is the value when $M_1$ and $M_2$ reach the transition point, or

$$V_{DS1} = V_{DS2} = V_{DS1}(\text{sat}) = V_{GS1} - V_{TN} = 2.71 - 1 = 1.71 \text{ V}$$

Therefore,

$$v_{CM}(\text{max}) = v_{O1} - V_{DS1}(\text{sat}) + V_{GS1} = 5.31 - 1.71 + 2.71$$

or

$$v_{CM}(\text{max}) = 6.31 \text{ V}$$

The minimum common-mode input voltage is the value when $M_4$ reaches the transition point, or

$$V_{DS4} = V_{DS4}(\text{sat}) = V_{GS4} - V_{TN} = 2.4 - 1 = 1.4 \text{ V}$$

Therefore,

$$v_{CM}(\text{min}) = V_{GS1} + V_{DS4}(\text{sat}) - 10 = 2.71 + 1.4 - 10$$

or

$$v_{CM}(\text{min}) = -5.89 \text{ V}$$

**Comment:** For this circuit the maximum range for the common-mode input voltage is $-5.89 \leq v_{CM} \leq 6.31$ V.

### EXERCISE PROBLEM

**\*Ex 11.8:** For the differential amplifier in Figure 11.20, the parameters are: $V^+ = 5$ V, $V^- = -5$ V, $R_1 = 80$ k$\Omega$, and $R_D = 40$ k$\Omega$. The transistor parameters are $\lambda = 0$ and $V_{TN} = 0.8$ V for all transistors, and $K_{n3} = K_{n4} = 100 \ \mu\text{A/V}^2$ and $K_{n1} = K_{n2} = 50 \ \mu\text{A/V}^2$. Determine the range of the common-mode input voltage. (Ans. $-2.18 \leq v_{cm} \leq 3.76$ V)

The dc transfer characteristics of the MOSFET differential pair can be determined from the circuit in Figure 11.19. Neglecting the output resistances of $M_1$ and $M_2$, and assuming the two transistors are matched, we can write

$$i_{D1} = K_n(v_{GS1} - V_{TN})^2 \tag{11.60(a)}$$

and

$$i_{D2} = K_n(v_{GS2} - V_{TN})^2 \tag{11.60(b)}$$

Taking the square roots of Equations (11.60(a)) and (11.60(b)), and subtracting the two equations, we obtain

$$\sqrt{i_{D1}} - \sqrt{i_{D2}} = \sqrt{K_n}(v_{GS1} - v_{GS2}) = \sqrt{K_n} \cdot v_d \tag{11.61}$$

where $v_d = v_{G1} - v_{G2} = v_{GS1} - v_{GS2}$ is the differential-mode input voltage. If $v_d > 0$, then $v_{G1} > v_{G2}$ and $v_{GS1} > v_{GS2}$, which implies that $i_{D1} > i_{D2}$. Since

$$i_{D1} + i_{D2} = I_Q \tag{11.62}$$

then Equation (11.61) becomes

$$\left(\sqrt{i_{D1}} - \sqrt{I_Q - i_{D1}}\right)^2 = \left(\sqrt{K_n} \cdot v_d\right)^2 = K_n v_d^2 \tag{11.63}$$

when both sides of the equation are squared. After the terms are rearranged, Equation (11.63) becomes

$$\sqrt{i_{D1}(I_Q - i_{D1})} = \frac{1}{2}\left(I_Q - K_n v_d^2\right) \tag{11.64}$$

If we square both sides of this equation, we develop the quadratic equation

$$i_{D1}^2 - I_Q i_{D1} + \frac{1}{4}\left(I_Q - K_n v_d^2\right)^2 = 0 \tag{11.65}$$

Applying the quadratic formula, rearranging terms, and noting that $i_{D1} > I_Q/2$ and $v_d > 0$, we obtain

$$i_{D1} = \frac{I_Q}{2} + \sqrt{\frac{K_n I_Q}{2}} \cdot v_d \sqrt{1 - \left(\frac{K_n}{2I_Q}\right)v_d^2} \tag{11.66}$$

Using Equation (11.62), we find that

$$i_{D2} = \frac{I_Q}{2} - \sqrt{\frac{K_n I_Q}{2}} \cdot v_d \sqrt{1 - \left(\frac{K_n}{2I_Q}\right)v_d^2} \tag{11.67}$$

The normalized drain currents are

$$\frac{i_{D1}}{I_Q} = \frac{1}{2} + \sqrt{\frac{K_n}{2I_Q}} \cdot v_d \sqrt{1 - \left(\frac{K_n}{2I_Q}\right)v_d^2} \tag{11.68}$$

and

$$\frac{i_{D2}}{I_Q} = \frac{1}{2} - \sqrt{\frac{K_n}{2I_Q}} \cdot v_d \sqrt{1 - \left(\frac{K_n}{2I_Q}\right)v_d^2} \tag{11.69}$$

These equations describe the dc transfer characteristics for this circuit. They are plotted in Figure 11.21 as a function of a normalized differential input voltage $v_d/\sqrt{(2I_Q/K_n)}$.

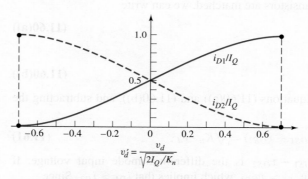

**Figure 11.21** Normalized dc transfer characteristics, MOSFET differential amplifier

We can see from Equations (11.68) and (11.69) that, at a specific differential input voltage, bias current $I_Q$ is switched entirely to one transistor or the other. This occurs when

$$|v_d|_{max} = \sqrt{\frac{I_Q}{K_n}}$$    **(11.70)**

The forward transconductance is defined as the slope of the $i_{D1}$ versus $v_d$ transfer characteristic evaluated at $v_d = 0$, or

$$g_f(max) = \frac{di_{D1}}{dv_d}\bigg|_{v_d=0}$$    **(11.71)**

Using Equation (11.66), we find that

$$g_f(max) = \sqrt{\frac{K_n I_Q}{2}} = \frac{g_m}{2}$$    **(11.72)**

where $g_m$ is the transconductance of each transistor. The slope of the $i_{D2}$ characteristic curve at $v_d = 0$ is the same, except it is negative.

We can perform an analysis similar to that in Example 11.2 to determine the maximum differential-mode input signal that can be applied and still maintain linearity. If we let $I_Q = 1$ mA and $K_n = 1$ mA/V$^2$, then for differential input voltages less than 0.34 V, the difference between the linear approximation and the actual curve is less than 1 percent. The maximum differential input signal for the MOSFET diff-amp is much larger than for the bipolar diff-amp. The principal reason is that the gain of the MOSFET diff-amp, as we will see, is much smaller than the gain of the bipolar diff-amp.

Figure 11.22 is the ac equivalent circuit of the diff-amp configuration, showing only the differential voltage and signal currents as a function of the transistor transconductance $g_m$. We assume that the output resistance looking into the current source is infinite. Using this equivalent circuit, we find the one-sided output voltage at $v_{o2}$, as follows:

$$v_{o2} \equiv v_o = +\left(\frac{g_m v_d}{2}\right) R_D$$    **(11.73)**

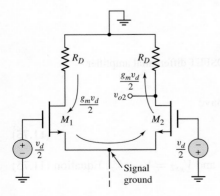

**Figure 11.22** AC equivalent circuit, MOSFET differential amplifier

The differential voltage gain is then

$$A_d = \frac{v_o}{v_d} = \frac{g_m R_D}{2} = \sqrt{\frac{K_n I_Q}{2}} \cdot R_D \qquad \textbf{(11.74)}$$

### 11.3.2    Differential- and Common-Mode Input Impedances

At low frequencies, the input impedance of a MOSFET is essentially infinite, which means that both the differential- and common-mode input resistances of a MOSFET diff-amp are infinite. Also, we know that the differential input resistance of a bipolar pair can be in the low kilohm range. A design trade-off, then, would be to use a MOSFET diff-amp with infinite input resistance, and sacrifice the differential-mode voltage gain.

### 11.3.3    Small-Signal Equivalent Circuit Analysis

We can determine the basic relationships for the differential-mode gain, common-mode gain, and common-mode rejection ratio from an analysis of the small-signal equivalent circuit.

Figure 11.23 shows the small-signal equivalent circuit of the MOSFET differential pair configuration. We assume the transistors are matched, with $\lambda = 0$ for each transistor, and that the constant-current source is represented by a finite output resistance $R_o$. All voltages are represented by their phasor components. The two transistors are biased at the same quiescent current, and $g_{m1} = g_{m2} \equiv g_m$.

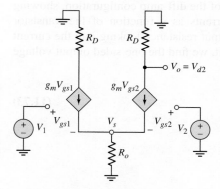

**Figure 11.23** Small-signal equivalent circuit, MOSFET differential amplifier

Writing a KCL equation at node $V_s$, we have

$$g_m V_{gs1} + g_m V_{gs2} = \frac{V_s}{R_o} \qquad \textbf{(11.75)}$$

From the circuit, we see that $V_{gs1} = V_1 - V_s$ and $V_{gs2} = V_2 - V_s$. Equation (11.75) then becomes

$$g_m(V_1 + V_2 - 2V_s) = \frac{V_s}{R_o} \qquad \textbf{(11.76)}$$

Solving for $V_s$ we obtain

$$V_s = \frac{V_1 + V_2}{2 + \frac{1}{g_m R_o}} \tag{11.77}$$

For a one-sided output at the drain of $M_2$, we have

$$V_o = V_{d2} = -(g_m V_{gs2})R_D = -(g_m R_D)(V_2 - V_s) \tag{11.78}$$

Substituting Equation (11.77) into (11.78) and rearranging terms yields

$$V_o = -g_m R_D \left[ \frac{V_2 \left(1 + \frac{1}{g_m R_o}\right) - V_1}{2 + \frac{1}{g_m R_o}} \right] \tag{11.79}$$

Based on the relationships between the input voltages $V_1$ and $V_2$ and the differential- and common-mode voltages, as given by Equation (11.29), Equation (11.79) can be written

$$V_o = \frac{g_m R_D}{2} V_d - \frac{g_m R_D}{1 + 2g_m R_o} V_{cm} \tag{11.80}$$

The output voltage, in general form, is

$$V_o = A_d V_d + A_{cm} V_{cm} \tag{11.81}$$

The transconductance $g_m$ of the MOSFET is

$$g_m = 2\sqrt{K_n I_{DQ}} = \sqrt{2K_n I_Q}$$

Comparing Equations (11.80) and (11.81), we develop the relationships for the differential-mode gain,

$$A_d = \frac{g_m R_D}{2} = \sqrt{2K_n I_Q} \left(\frac{R_D}{2}\right) = \sqrt{\frac{K_n I_Q}{2}} \cdot R_D \tag{11.82(a)}$$

and the common-mode gain

$$A_{cm} = \frac{-g_m R_D}{1 + 2g_m R_o} = \frac{-\sqrt{2K_n I_Q} \cdot R_D}{1 + 2\sqrt{2K_n I_Q} \cdot R_o} \tag{11.82(b)}$$

We again see that for an ideal current source, the common-mode gain is zero since $R_o = \infty$.

From Equations (11.82(a)) and (11.82(b)), the common-mode rejection ratio, CMRR $= |A_d/A_{cm}|$, is found to be

$$\text{CMRR} = \frac{1}{2}\left[1 + 2\sqrt{2K_n I_Q} \cdot R_o\right] \tag{11.83}$$

This demonstrates that the CMRR for the MOSFET diff-amp is also a strong function of the output resistance of the constant-current source.

## EXAMPLE 11.9

**Objective:** Determine the differential-mode voltage gain, common-mode voltage gain, and CMRR for a MOSFET diff-amp.

Consider a MOSFET diff-amp with the configuration in Figure 11.20. Assume the same transistor parameters as given in Example 11.8 except assume $\lambda = 0.01$ V$^{-1}$ for $M_4$.

**Solution:** From Example 11.8, we found the bias current to be $I_Q = 0.587$ mA. The output resistance of the current source is then

$$R_o = \frac{1}{\lambda I_Q} = \frac{1}{(0.01)(0.587)} = 170 \text{ k}\Omega$$

The differential-mode voltage gain is

$$A_d = \sqrt{\frac{K_n I_Q}{2}} \cdot R_D = \sqrt{\frac{(1)(0.587)}{2}} \cdot (16) = 8.67$$

and the common-mode voltage gain is

$$A_{cm} = -\frac{\sqrt{2K_n I_Q} \cdot R_D}{1 + 2\sqrt{2K_n I_Q} \cdot R_o} = -\frac{\sqrt{2(1)(0.587)} \cdot (16)}{1 + 2\sqrt{2(1)(0.587)} \cdot (170)} = -0.0469$$

The common-mode rejection ratio is then

$$\text{CMRR}_{dB} = 20 \log_{10}\left(\frac{8.67}{0.0469}\right) = 45.3 \text{ dB}$$

**Comment:** As mentioned earlier, the differential-mode voltage gain of the MOSFET diff-amp is considerably less than that of the bipolar diff-amp, since the value of the MOSFET transconductance is, in general, much smaller than that of the BJT.

EXERCISE PROBLEM

**Ex 11.9:** The parameters of the circuit shown in Figure 11.19 are $V^+ = 3$ V, $V^- = -3$ V, $I_Q = 0.2$ mA, and $R_D = 15$ k$\Omega$. Assume $M_1$ and $M_2$ are matched with parameters $V_{TN} = 0.4$ V, $k'_n = 100 \mu\text{A/V}^2$, and $\lambda = 0$. (a) Design the width-to-length ratios of the transistors such that the one-sided differential voltage gain is $A_d = 15$. (b) Using the results of part (a), what is the value of $g_f(\text{max})$? (Ans. (a) $W/L = 200$, (b) $g_f(\text{max}) = 1.0$ mA/V)

The value of the common-mode rejection ratio can be increased by increasing the output resistance of the current source. An increase in the output resistance can be accomplished by using a more sophisticated current source circuit. Figure 11.24

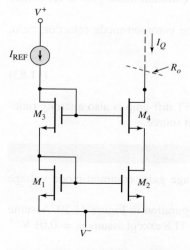

**Figure 11.24** MOSFET cascode current source

shows a MOSFET cascode current mirror that was discussed in the last chapter. The output resistance, as given by Equation (10.57), is $R_o = r_{o4} + r_{o2}(1 + g_m r_{o4})$. For the parameters of Example 11.9, $r_{o2} = r_{o4} = 170 \, \text{k}\Omega$ and $g_m = 2\sqrt{K_n I_Q} = 1.53 \, \text{mA/V}$. Then

$$R_o = 170 + 170[1 + (1.53)(170)] \Rightarrow 44.6 \, \text{M}\Omega$$

Again, using the parameters of Example 11.9, the common-mode voltage gain of the diff-amp with a cascode current mirror would be

$$A_{cm} = -\frac{\sqrt{2K_n I_Q} \cdot R_D}{1 + 2\sqrt{2K_n I_Q} \cdot R_o} = -\frac{\sqrt{2(1)(0.587)} \cdot (16)}{1 + 2\sqrt{2(1)(0.587)} \cdot (44600)} = -0.000179$$

so that the CMRR would be

$$\text{CMRR}_{dB} = 20 \log_{10}\left(\frac{8.67}{0.000179}\right) = 93.7 \, \text{dB}$$

We increased the common-mode rejection ratio dramatically by using the cascode current mirror instead of the single two-transistor current source. Note, however, that the differential-mode voltage gain is unchanged.

To gain an appreciation of the difference in CMRR between 45.3 dB and 93.7 dB, we can reconsider the linear scale. For a $\text{CMRR}_{dB} = 45.3 \, \text{dB}$, the differential gain is a factor of 185 times larger than the common-mode gain, while for a $\text{CMRR}_{dB} = 93.7 \, \text{dB}$, the differential gain is a factor of 48,436 times larger than the common-mode gain.

## 11.3.4 Two-Sided Output

If we consider the two-sided output of an ideal MOSFET op-amp and define the output voltage as $V_o = V_{d2} - V_{d1}$, we can show that the differential-mode voltage gain is given by

$$A_d = g_m R_D \tag{11.84(a)}$$

and the common-mode voltage gain is given by

$$A_{cm} = 0 \tag{11.84(b)}$$

The result of $A_{cm} = 0$ for the two-sided output is a consequence of using matched devices and elements in the diff-amp circuit. We will reconsider a two-sided output and discuss the effects of mismatched elements in the next section.

### Effect of $R_D$ Mismatch—Two-Sided Output

We assume that $R_{D1}$ and $R_{D2}$ are the resistors in the drains of $M_1$ and $M_2$. If the two resistors are not matched, we assume that we can write $R_{D1} = R_D + \Delta R_D$ and $R_{D2} = R_D - \Delta R_D$. Using the small-signal equivalent circuit in Figure 11.23, we can find

$$A_d = g_m R_D \tag{11.85(a)}$$

and

$$A_{cm} \cong \frac{\Delta R_D}{R_o} \tag{11.85(b)}$$

The common-mode rejection ratio is then

$$\text{CMRR} = \left| \frac{A_d}{A_{cm}} \right| = \frac{g_m R_D}{(\Delta R_D / R_o)} \tag{11.86}$$

This result is essentially the same as the BJT diff-amp.

### Effect of $g_m$ Mismatch—Two-Sided Output

We can consider the effect of transistor mismatch by considering the effect of a mismatch in the transconductance $g_m$. We assume $g_{m1}$ and $g_{m2}$ are the transconductance parameters of the two transistors in the diff-amp. We will assume that we can write $g_{m1} = g_m + \Delta g_m$ and $g_{m2} = g_m - \Delta g_m$. Again, using the small-signal equivalent circuit shown in Figure 11.23, we find the differential-mode voltage gain is

$$A_d = g_m R_D \tag{11.87(a)}$$

and the common-mode gain is

$$A_{cm} = \frac{R_D(2\Delta g_m)}{1 + 2R_o g_m} \tag{11.87(b)}$$

The common-mode rejection ratio now becomes

$$\text{CMRR} = \left| \frac{A_d}{A_{cm}} \right| = \frac{1 + 2R_o g_m}{2(\Delta g_m / g_m)} \tag{11.88}$$

The CMRR of mismatched elements in the MOSFET diff-amp is identical with the results of mismatched elements in the BJT diff-amp.

### 11.3.5   JFET Differential Amplifier

Figure 11.25 shows a basic JFET differential pair biased with a constant-current source. If a pure differential-mode input signal is applied such that $v_{G1} = +v_d/2$ and $v_{G2} = -v_d/2$, then drain currents $I_{D1}$ and $I_{D2}$ increase and decrease, respectively, in exactly the same way as in the MOSFET diff-amp.

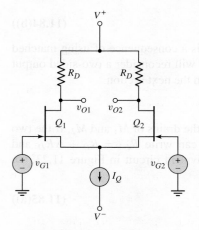

**Figure 11.25**  Basic JFET differential pair configuration

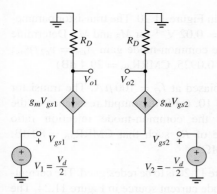

**Figure 11.26** Small-signal equivalent circuit, JFET differential amplifier

We can determine the differential-mode voltage gain by analyzing the small-signal equivalent circuit. Figure 11.26 shows the equivalent circuit, with the output resistance of the constant-current source and the small-signal resistances of $Q_1$ and $Q_2$ assumed to be infinite. The small-signal equivalent circuit of the JFET diff-amp is identical to that of the MOSFET diff-amp in Figure 11.23 for the case when the current-source output resistance is infinite. A KCL equation at the common-source node, in phasor notation, is

$$g_m V_{gs1} + g_m V_{gs2} = 0 \tag{11.89(a)}$$

or

$$V_{gs1} = -V_{gs2} \tag{11.89(b)}$$

The differential-mode input voltage is

$$V_d \equiv V_1 - V_2 = V_{gs1} - V_{gs2} = -2V_{gs2} \tag{11.90}$$

A one-sided output at $V_{o2}$ is given by

$$V_{o2} = -g_m V_{gs2} R_D = -g_m \left( \frac{-V_d}{2} \right) R_D \tag{11.91}$$

and the differential-mode voltage gain is

$$A_d = \frac{V_{o2}}{V_d} = +\frac{g_m R_D}{2} \tag{11.92}$$

The expression for the differential-mode voltage gain for the JFET diff-amp (Equation (11.92)) is exactly the same as that of the MOSFET diff-amp (Equation 11.82(a)). If the constant-current source output resistance is finite, then the JFET diff-amp will also have a nonzero common-mode voltage gain.

## Test Your Understanding

**TYU 11.7** The circuit parameters of the diff-amp shown in Figure 11.19 are $V^+ = 3$ V, $V^- = -3$ V, $I_Q = 0.40$ mA, and $R_D = 7.5$ kΩ. The transistor parameters are $V_{TN} = 0.5$ V, $k'_n = 100 \, \mu\text{A/V}^2$, and $\lambda = 0$. (a) Design the transistor $W/L$ ratio such that the differential voltage gain is $A_d = 12$. (b) What is the maximum positive common-mode voltage that can be applied such that the transistors remain biased in the saturation region. (Ans. (a) $W/L = 256$, (b) $v_{cm} = 2$ V)

**TYU 11.8** Consider the differential amplifier in Figure 11.20. The transistor parameters are given in Example 11.8, except that $\lambda = 0.02 \text{ V}^{-1}$ for $M_3$ and $M_4$. Determine the differential voltage gain $A_d = v_{o2}/v_d$, the common-mode gain $A_{cm} = v_{o2}/v_{cm}$, and the CMRR$_{dB}$. (Ans. $A_d = 2.74$, $A_{cm} = -0.0925$, CMRR$_{dB} = 29.4$ dB)

**TYU 11.9** The diff-amp in Figure 11.19 is biased at $I_Q = 100 \,\mu\text{A}$. The transistor parameters are $k'_n = 100 \,\mu\text{A/V}^2$ and $W/L = 10$. (a) If the output resistance of the current source is $R_o = 1 \text{ M}\Omega$, determine the common-mode rejection ratio CMRR$_{dB}$. (b) Determine the required value of $R_o$ such that CMRR$_{dB} = 80$ dB. (Ans. (a) CMRR$_{dB} = 50$ dB, (b) $R_o = 31.6 \text{ M}\Omega$)

**\*TYU 11.10** The differential amplifier in Figure 11.20 is to be redesigned. The current-source biasing is to be replaced with the cascode current source in Figure 11.24. The reference current is $I_{\text{REF}} = 100 \,\mu\text{A}$ and $\lambda$ for transistors in the current source circuit is $0.01 \text{ V}^{-1}$. The parameters of the differential pair $M_1$ and $M_2$ are the same as described in Example 11.8. The range of the common-mode input voltage is to be $-4 \le v_{cm} \le +4$ V. Redesign the diff-amp to achieve the highest possible differential-mode voltage gain. Determine the values of $A_d$, $A_{cm}$, and CMRR$_{dB}$. (Ans. $A_d = 9.90$, $A_{cm} = 0.0003465$, CMRR$_{dB} = 89.1$ dB)

---

## 11.4    DIFFERENTIAL AMPLIFIER WITH ACTIVE LOAD

**Objective: •** Analyze the characteristics of BJT and FET differential amplifiers with active loads.

In Chapter 10, we considered an active load in conjunction with a simple transistor amplifier. Active loads can also be used in diff-amp circuits to increase the differential-mode gain.

Active loads are essentially transistor current sources used in place of resistive loads. The transistors in the active load circuit are biased at a $Q$-point in the forward-active mode as shown in Figure 11.27. A change in collector current is induced by the differential-pair, which, in turn, produces a change in the emitter–collector voltage as shown in the figure. The relation between the change in current and change in voltage is proportional to the small-signal output resistance $r_o$ of the transistor. The value of $r_o$ is, in general, much larger than that of a discrete resistive load, so the small-signal voltage gain will be larger with the active load.

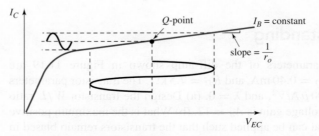

**Figure 11.27** Current–voltage characteristic of active load device

## 11.4.1 BJT Diff-Amp with Active Load

Figure 11.28 shows a differential amplifier with an active load. Transistors $Q_1$ and $Q_2$ are the differential pair biased with a constant current $I_Q$, and transistors $Q_3$ and $Q_4$ form the load circuit. From the collectors of $Q_2$ and $Q_4$, we obtain a one-sided output.

If we assume all transistors are matched, then a pure applied common-mode voltage means that $v_{B1} = v_{B2} = v_{CM}$, and current $I_Q$ splits evenly between $Q_1$ and $Q_2$. Neglecting base currents, $I_4 = I_3$ through the current-source circuit and $I_1 = I_2 = I_3 = I_4 = I_Q/2$ with no load connected at the output.

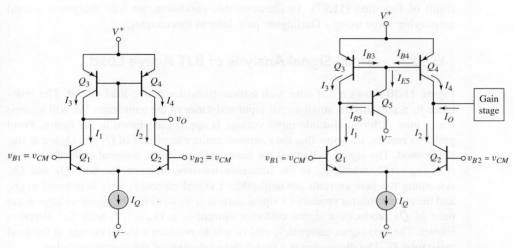

**Figure 11.28** BJT differential amplifier with active load

**Figure 11.29** BJT differential amplifier with three-transistor active load and second gain stage

In actual diff-amp circuits, base currents are not zero. In addition, a second amplifier stage is connected at the diff-amp output. Figure 11.29 shows a diff-amp with an active load circuit, corresponding to a three-transistor current source, as well as a second amplifying stage. In general, the common–emitter current gain $\beta$ is a function of collector current, as was shown in Figure 6.21(c). However, for simplicity, we assume all transistor current gains are equal, even though the current level in $Q_5$ is much smaller than in the other transistors. Current $I_O$ is the dc bias current from the gain stage. Assuming all transistors are matched and $v_{B1} = v_{B2} = v_{CM}$, current $I_Q$ splits evenly and $I_1 = I_2$. To ensure that $Q_2$ and $Q_4$ are biased in the forward-active mode, the dc currents must be balanced, or $I_3 = I_4$. We see that

$$I_{E5} = I_{B3} + I_{B4} = \frac{I_3}{\beta} + \frac{I_4}{\beta} \tag{11.93}$$

Then

$$I_{B5} = \frac{I_{E5}}{1 + \beta} = \frac{I_3 + I_4}{\beta(1 + \beta)} \tag{11.94}$$

If the base currents and $I_O$ are small, then

$$I_3 + I_4 \cong I_Q \tag{11.95}$$

Therefore,

$$I_{B5} \cong \frac{I_Q}{\beta(1+\beta)} \tag{11.96}$$

For the circuit to be balanced, that is, for $I_1 = I_2$ and $I_3 = I_4$, we must have

$$I_O = I_{B5} = \frac{I_Q}{\beta(1+\beta)} \tag{11.97}$$

Equation (11.97) implies that the second amplifying stage must be designed and biased such that the direction of the dc bias current is as shown and is equal to the result of Equation (11.97). To illustrate this condition, we will analyze a second amplifying stage using a Darlington pair, later in this chapter.

<table><tr><td>11.4.2</td><td>**Small-Signal Analysis of BJT Active Load**</td></tr></table>

Figure 11.30 shows a diff-amp with a three-transistor active load circuit. The resistance $R_L$ represents the small-signal input resistance of the gain stage. We will assume that a pure differential-mode input voltage is applied as shown in the figure. From previous results, we know that the common-emitter terminals of $Q_1$ and $Q_2$ are at signal ground. The signal voltage at the base of $Q_1$ produces a signal collector current $i_1 = (g_m v_d)/2$, where $g_m$ is the transistor transconductance for both $Q_1$ and $Q_2$. Assuming the base currents are negligible, a signal current $i_3 = i_1$ is induced in $Q_3$, and the current mirror produces a signal current $i_4$ equal to $i_3$. The signal voltage at the base of $Q_2$ produces a signal collector current $i_2 = (g_m v_d)/2$, with the direction shown. The two signal currents, $i_2$ and $i_4$, add to produce a signal current in the load resistance $R_L$. The discussion is a first-order evaluation of the circuit operation.

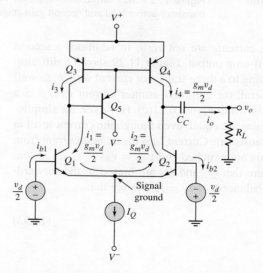

**Figure 11.30** BJT differential amplifier with three-transistor active load, showing the signal currents

From the above discussion, we know the induced currents in $Q_2$ and $Q_4$. To more accurately determine the output voltage, we need to consider the equivalent small-signal collector–emitter output circuit of the two transistors. Figure 11.31(a)

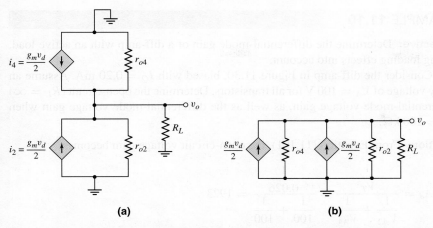

**Figure 11.31** (a) Small-signal equivalent circuit BJT differential amplifier with active load and (b) rearrangement of small-signal equivalent circuit

shows the small-signal equivalent circuit at the collector nodes of $Q_2$ and $Q_4$. The circuit can be rearranged to combine the signal grounds at a common point, as in Figure 11.31(b). From this figure, we determine that

$$v_o = 2\left(\frac{g_m v_d}{2}\right)(r_{o2}\|r_{o4}\|R_L) \tag{11.98}$$

and the small-signal differential-mode voltage gain is

$$A_d = \frac{v_o}{v_d} = g_m(r_{o2}\|r_{o4}\|R_L) \tag{11.99}$$

Equation (11.99) can be rewritten in the form

$$A_d = \frac{g_m}{\dfrac{1}{r_{o2}} + \dfrac{1}{r_{o4}} + \dfrac{1}{R_L}} = \frac{g_m}{g_{o2} + g_{o4} + G_L} \tag{11.100}$$

We recall that $g_m = I_Q/2V_T, r_{o2} = V_{A2}/I_2$, and $r_{o4} = V_{A4}/I_4$. The parameters $g_{o2}$, $g_{o4}$, and $G_L$ are the corresponding conductances. Assuming $I_2 = I_4 = I_Q/2$, we can write Equation (11.100) in the form

$$A_d = \frac{\dfrac{I_Q}{2V_T}}{\dfrac{I_Q}{2V_{A2}} + \dfrac{I_Q}{2V_{A4}} + \dfrac{1}{R_L}} \tag{11.101}$$

This expression of the differential-mode voltage gain of the diff-amp with an active load is very similar to that obtained in the last chapter for a simple amplifier with an active load.

The output resistance looking back into the common collector node is $R_o = r_{o2}\|r_{o4}$. To minimize loading effects, we need $R_L > R_o$. However, since $R_o$ is generally large for active loads, we may not be able to satisfy this condition. We can determine the severity of the loading effect by comparing $R_L$ and $R_o$.

## EXAMPLE 11.10

**Objective:** Determine the differential-mode gain of a diff-amp with an active load, taking loading effects into account.

Consider the diff-amp in Figure 11.30, biased with $I_Q = 0.20$ mA. Assume an Early voltage of $V_A = 100$ V for all transistors. Determine the open-circuit ($R_L = \infty$) differential-mode voltage gain, as well as the differential-mode voltage gain when $R_L = 100$ kΩ.

**Solution:** From Equation (11.101), the open-circuit voltage gain becomes

$$A_d = \frac{\dfrac{1}{V_T}}{\dfrac{1}{V_{A2}} + \dfrac{1}{V_{A4}}} = \frac{\dfrac{1}{0.026}}{\dfrac{1}{100} + \dfrac{1}{100}} = 1923$$

When $R_L = 100$ kΩ, the voltage gain is

$$A_d = \frac{\dfrac{0.20 \times 10^{-3}}{2(0.026)}}{\dfrac{0.20 \times 10^{-3}}{2(100)} + \dfrac{0.20 \times 10^{-3}}{2(100)} + \dfrac{1}{100 \times 10^3}}$$

which can be written

$$A_d = \frac{\dfrac{0.20}{2(0.026)}}{\dfrac{0.20}{2(100)} + \dfrac{0.20}{2(100)} + \dfrac{1}{100}} = \frac{3.85}{0.001 + 0.001 + 0.01} = 321$$

An inspection of this last equation shows that the external load factor, $1/R_L$, dominates the denominator term and thus has a tremendous influence on the gain.

**Comment:** The open-circuit differential-mode voltage gain, for a diff-amp with an active load, is large. However, a finite load resistance $R_L$ causes severe loading effects, as shown in this example. A 100 kΩ load caused almost an order of magnitude decrease in the gain.

## EXERCISE PROBLEM

**Ex 11.10:** The diff-amp circuit in Figure 11.30 is biased at $I_Q = 0.4$ mA. The transistor parameters are $\beta = 120$, $V_{A1} = V_{A2} = 150$ V, and $V_{A3} = V_{A4} = 90$ V. (a) Determine the open-circuit ($R_L = \infty$) differential-mode voltage gain. (b) Find the differential-mode voltage gain when $R_L = 250$ kΩ. (c) Determine the differential-mode input resistance. (d) Find the output resistance looking back from the load $R_L$. (Ans. (a) 2163, (b) 1018, (c) 31.2 kΩ, (d) 281 kΩ)

### 11.4.3    MOSFET Differential Amplifier with Active Load

We can use an active load in conjunction with a MOSFET differential pair, as we did for the bipolar differential amplifier. Figure 11.32 shows a MOSFET diff-amp with an active load. Transistors $M_1$ and $M_2$ are n-channel devices and form the differential

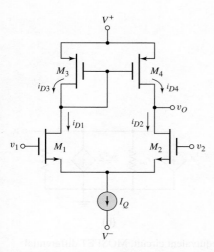

**Figure 11.32** MOSFET differential amplifier with active load

pair biased with $I_Q$. The load circuit consists of transistors $M_3$ and $M_4$, both p-channel devices, connected in a current mirror configuration. A one-sided output is taken from the common drains of $M_2$ and $M_4$. When a common-mode voltage of $v_1 = v_2 = v_{cm}$ is applied, the current $I_Q$ splits evenly between $M_1$ and $M_2$, and $i_{D1} = i_{D2} = I_Q/2$. There are no gate currents; therefore, $i_{D3} = i_{D1}$ and $i_{D4} = i_{D2}$.

If a small differential-mode input voltage $v_d = v_1 - v_2$ is applied, then from Equation (11.66) and (11.67), we can write

$$i_{D1} = \frac{I_Q}{2} + i_d \qquad\qquad \textbf{(11.102(a))}$$

and

$$i_{D2} = \frac{I_Q}{2} - i_d \qquad\qquad \textbf{(11.102(b))}$$

where $i_d$ is the signal current. For small values of $v_d$, we have $i_d = (g_m v_d)/2$. Since $M_1$ and $M_3$ are in series, we see that

$$i_{D3} = i_{D1} = \frac{I_Q}{2} + i_d \qquad\qquad \textbf{(11.103)}$$

Finally, the current mirror consisting of $M_3$ and $M_4$ produces

$$i_{D4} = i_{D3} = \frac{I_Q}{2} + i_d \qquad\qquad \textbf{(11.104)}$$

Figure 11.33 is the ac equivalent circuit of the diff-amp with active load, showing the signal currents. The negative sign for $i_{D2}$ in Equation (11.102(b)) shows up as a change in current direction in $M_2$, as indicated in the figure.

Figure 11.34(a) shows the small-signal equivalent circuit at the drain node of $M_2$ and $M_4$. If the output is connected to the gate of another MOSFET, which is equivalent to an infinite impedance at low frequency, the output terminal is effectively an open circuit. The circuit can be rearranged by combining the signal grounds at a common point, as shown in Figure 11.34(b). Then,

$$v_o = 2\left(\frac{g_m v_d}{2}\right)(r_{o2} \| r_{o4}) \qquad\qquad \textbf{(11.105)}$$

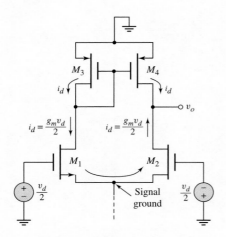

**Figure 11.33** The ac equivalent circuit, MOSFET differential amplifier with active load

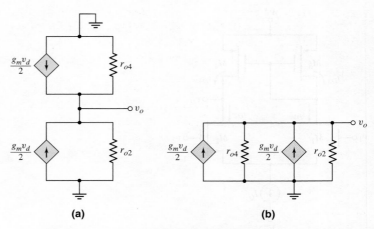

**Figure 11.34** (a) Small-signal equivalent circuit, MOSFET differential amplifier with active load and (b) rearranged small-signal equivalent circuit

and the small-signal differential-mode voltage gain is

$$A_d = \frac{v_o}{v_d} = g_m(r_{o2} \| r_{o4}) \tag{11.106}$$

Equation (11.106) can be rewritten in the form

$$A_d = \frac{g_m}{\dfrac{1}{r_{o2}} + \dfrac{1}{r_{o4}}} = \frac{g_m}{g_{o2} + g_{o4}} \tag{11.107}$$

If we recall that $g_m = 2\sqrt{K_n I_D} = \sqrt{2K_n I_Q}$, $g_{o2} = \lambda_2 I_{DQ2} = (\lambda_2 I_Q)/2$, and $g_{o4} = \lambda_4 I_{DQ4} = (\lambda_4 I_Q)/2$, then Equation (11.107) becomes

$$A_d = \frac{2\sqrt{2K_n I_Q}}{I_Q(\lambda_2 + \lambda_4)} = 2\sqrt{\frac{2K_n}{I_Q}} \cdot \frac{1}{\lambda_2 + \lambda_4} \tag{11.108}$$

## DESIGN EXAMPLE 11.11

**Objective:** Design a MOSFET diff-amp with the configuration in Figure 11.32 to meet the specifications of the experimental system in Example 11.4.

**Design Approach:** We need not only to try to obtain the necessary differential-mode gain and minimize the common-mode gain in our design, but we must also be cognizant of the swing in the output voltage. In the circuit in Figure 11.32, if the corresponding PMOS and NMOS transistors are matched, then the quiescent value of $V_{SD4}$ is equal to $V_{SG4} = V_{SG3}$. As the signal output voltage increases, the source-to-drain voltage of $M_4$ decreases. The minimum value of this voltage such that $M_4$ remains biased in the saturation region is $V_{SD4}(\text{min}) = V_{SD4}(\text{sat}) = V_{SG} + V_{TP}$. This means that the maximum swing in the output voltage is equal to the magnitude of the threshold voltage of $M_4$. In this example, the maximum swing in the output voltage is 0.8 V, so that the magnitude of the threshold voltages of the PMOS devices must be greater than 0.8 V. Assume that NMOS devices are available with the following parameters: $V_{TN} = 0.5$ V, $k_n' = 80$ $\mu$A/V$^2$, and $\lambda_n = 0.02$ V$^{-1}$. Assume that

PMOS devices are available with the following parameters: $V_{TP} = -1.0$ V, $k'_p = 40$ $\mu$A/V$^2$, and $\lambda_p = 0.02$ V$^{-1}$. Choose supply voltages of $\pm 5$ V and choose a bias current of approximately $I_Q = 200$ $\mu$A.

Figure 11.35 is the diff-amp and current-source network used for the design in this example.

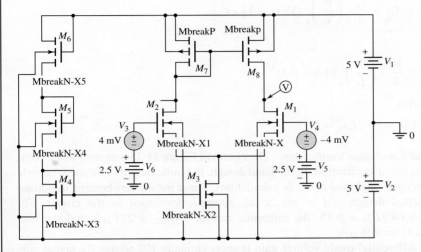

**Figure 11.35** CMOS differential amplifier and current source network for Example 11.11

**Design, Differential Amplifier: Differential-Mode Gain:** From Equation (11.108), the differential-mode gain is

$$A_d = 2\sqrt{2\left(\frac{k'_n}{2}\right)\left(\frac{W}{L}\right)_n \frac{1}{I_Q}} \cdot \frac{1}{\lambda_n + \lambda_p}$$

or

$$100 = 2\sqrt{2\left(\frac{80}{2}\right)\left(\frac{W}{L}\right)\frac{1}{200}} \cdot \frac{1}{0.02 + 0.02}$$

which yields a width-to-length ratio of $(W/L)_n = 10$ for the NMOS differential pair. Since the width-to-length ratios of the other transistors do not directly affect the gain of the diff-amp, we may arbitrarily choose width-to-length ratios of 10 for all other transistors except $M_5$ and $M_6$. The $W/L$ ratio of 10 means that the other devices are reasonably small and do not lead to a large circuit area.

**Design, Current-Source Network:** For the transistor $M_3$ in the current source, we have

$$I_Q = \frac{k'}{2} \cdot \frac{W}{L} \cdot (V_{GS3} - V_{TN})^2$$

or

$$200 = \frac{80}{2}(10)(V_{GS3} - 0.5)^2$$

which means that the required gate-to-source voltage of $M_3$ is $V_{GS3} = 1.21$ V. We may choose $M_4$ and $M_3$ to be identical so that the current in the reference portion of the circuit is also 200 $\mu$A. Assuming that $M_5$ and $M_6$ are identical, then each transistor must have a gate-to-source voltage of

$$V_{GS5} = V_{GS6} = (10 - 1.21)/2 \cong 4.4 \text{ V}$$

The width-to-length of these transistors is now found from

$$I_{REF} = I_Q = \frac{k'_n}{2} \cdot \left(\frac{W}{L}\right)_5 (V_{GS5} - V_{TN})^2$$

or

$$200 = \frac{80}{2} \cdot \left(\frac{W}{L}\right)_5 (4.4 - 0.5)^2$$

which yields

$$(W/L)_5 = (W/L)_6 = 0.33$$

**Computer Simulation Verification:** The circuit in Figure 11.35 was used in the computer simulation verification. In the hand design, the finite output resistance (lambda parameter) was neglected in the dc calculations. These parameters became important in the actual design and in the actual currents developed in the circuit. For $(W/L)_5 = (W/L)_6 = 0.75$, the reference current is $I_{REF} = 231$ $\mu$A and the bias current is $I_Q = 208$ $\mu$A.

The differential-mode voltage gain is approximately 102 so that the signal output voltage is 0.82 V for a differential-mode input signal voltage of 8 mV. The common-mode output signal is approximately 0.86 mV, which is well within the specified 10 mV maximum value.

**Design Pointer:** The body effect has been neglected in this design. In actual integrated circuits, the differential pair transistors may actually be fabricated within their own p-type substrate region (for NMOS devices). This p-type substrate region is then directly connected to the source terminals so that the body effect in the NMOS differential pair devices can be neglected.

---

### EXERCISE PROBLEM

**Ex 11.11:** Determine $I_{REF}$, $I_Q$, and $A_d$ of the diff-amp designed in Example 11.11 for the case when the bias voltages are changed to $V^+ = +3$ V and $V^- = -3$ V. (Ans. $I_{REF} = I_Q = 56.37$ $\mu$A, $A_d = 188$)

---

### 11.4.4    MOSFET Diff-Amp with Cascode Active Load

The differential-mode voltage gain is proportional to the output resistance looking into the active load transistor. The voltage gain can be increased, therefore, if the output resistance can be increased. An increase in output resistance can be achieved by using, for example, a cascode active load. This configuration is shown in Figure 11.36.

The output resistance $R_o$ was considered in the last section in the discussion of the cascode current source. As applied to Figure 11.36, the output resistance is given by

$$R_o = r_{o4} + r_{o6}(1 + g_m r_{o4}) \cong g_m r_{o4} r_{o6} \tag{11.109}$$

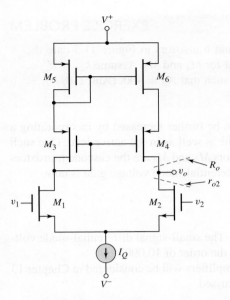

**Figure 11.36** MOSFET diff-amp with cascode active load

The small-signal differential-mode voltage gain is then

$$A_d = \frac{v_o}{v_d} = g_m(r_{o2} \| R_o) \qquad\qquad\qquad \textbf{(11.110)}$$

## EXAMPLE 11.12

**Objective:** Calculate the differential-mode voltage gain of a MOSFET diff-amp with a cascode active load.

Consider the diff-amp shown in Figure 11.36. Assume the circuit and transistor parameters are the same as in Example 11.11.

**Solution:** The transistor transconductance is

$$g_m = 2\sqrt{K_n I_{DQ}} = 2\sqrt{\left(\frac{0.08}{2}\right)(10)(0.1)} = 0.40 \text{ mA/V}$$

The output resistance of the individual transistors is

$$r_o = \frac{1}{\lambda I_{DQ}} = \frac{1}{(0.02)(0.1)} = 500 \text{ k}\Omega$$

The output resistance of the cascode active load is then

$$R_o = r_{o4} + r_{o6}(1 + g_m r_{o4}) = 0.5 + 0.5[1 + (0.40)(500)] = 101 \text{ M}\Omega$$

The differential-mode voltage gain is then found to be

$$A_d = g_m(r_{o2} \| R_o) = (0.40)(500 \| 101000) = 200$$

**Comment:** Since $R_o \gg r_{o2}$, the voltage gain is now essentially equal to $A_d = g_m r_{o2}$ which is twice as large as the gain calculated in Example 11.11.

**Ex 11.12:** The parameters of the circuit and transistors in Figure 11.36 are the same as described in Example 11.12 except for $M_1$ and $M_2$. Assume $k'_{n1} = k'_{n2} = 80\ \mu\text{A/V}^2$. Determine $(W/L)_1 = (W/L)_2$ such that $A_d = 400$. (Ans. $(W/L)_1 = (W/L)_2 = 40$)

The differential-mode voltage gain can be further increased by incorporating a cascode configuration in the differential pair as well as in the active load. One such example is shown in Figure 11.37. Transistors $M_3$ and $M_4$ are the cascode transistors for the differential pair $M_1$ and $M_2$. The differential-mode voltage gain is now

$$A_d = \frac{v_o}{v_d} = g_m(R_{o4} \| R_{o6})$$

where $R_{o4} \cong g_m r_{o2} r_{o4}$ and $R_{o6} \cong g_m r_{o6} r_{o8}$. The small-signal differential-mode voltage gain of this type of amplifier can be on the order of 10,000.

Other types of MOSFET differential amplifiers will be considered in Chapter 13 when operational amplifier circuits are discussed.

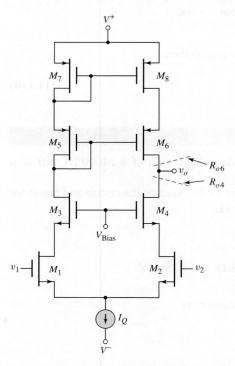

**Figure 11.37** A MOSFET cascode diff-amp with a cascode active load

## Test Your Understanding

**TYU 11.11** Consider the diff-amp in Figure 11.29, with parameters: $V^+ = 10$ V, $V^- = -10$ V, and $I_Q = 0.5$ mA. The transistor parameters are: $\beta = 180$, $V_{BE}(\text{on}) = 0.7$ V, and $V_A = 100$ V. (a) Find $I_O$ such that the circuit is balanced. (b) For the balanced condition, what are the values of $V_{EC4}$ and $V_{CE2}$, for $v_1 = v_2 = 0$? (Ans. (a) $I_O = 15.3$ nA (b) $V_{EC4} = 1.4$ V, $V_{CE2} = 9.3$ V)

**TYU 11.12** The circuit parameters of the diff-amp in Figure 11.28 are $V^+ = 5$ V, $V^- = -5$ V, and $I_Q = 0.1$ mA. The npn transistor parameters are $\beta_{npn} = 180$, $V_{AN} = 120$ V, and $V_{BE}(on) = 0.7$ V; and the pnp transistor parameters are $\beta_{pnp} = 120$, $V_{AP} = 80$ V, and $V_{EB}(on) = 0.7$ V. Determine the differential-mode voltage gain. (Ans. $A_d = 1846$)

**TYU 11.13** Redesign the circuit in Figure 11.30 using a Widlar current source and bias voltages of $\pm 5$ V. The bias current $I_Q$ is to be no less than $100\ \mu$A and the total power dissipated in the circuit (including the current-source circuit) is to be no more than 10 mW. The diff-amp transistor parameters are the same as in Exercise Ex11.10. The circuit is to provide a minimum loading effect when a second stage with an input resistance of $R = 90$ k$\Omega$ is connected to the diff-amp. Determine the differential-mode voltage gain for this circuit. (Ans. $R_1 = 10.3$ k$\Omega$, $R_E = 0.571$ k$\Omega$, $A_d = 158$)

**TYU 11.14** Consider the diff-amp in Figure 11.28, using the parameters described in Exercise TYU11.12. (a) For a differential-mode input signal, determine the output resistance $R_o$ at the output terminal. (b) Determine the load resistance $R_L$ that would reduce the differential-mode voltage gain to one-half the open-circuit value. (Ans. (a) $R_o = 0.96$ M$\Omega$, (b) $R_L = 0.96$ M$\Omega$)

**TYU 11.15** The circuit parameters of the diff-amp in Figure 11.32 are $V^+ = 5$ V, $V^- = -5$ V, and $I_Q = 0.2$ mA. The NMOS transistor parameters are $K_n = 180\ \mu$A/V$^2$, $V_{TN} = 0.5$ V, and $\lambda_n = 0.015$ V$^{-1}$ and the PMOS transistor parameters are $K_p = 120\ \mu$A/V$^2$, $V_{TP} = -0.5$ V, and $\lambda_p = 0.025$ V$^{-1}$. Determine the differential-mode voltage gain $A_d = v_o/v_d$. (Ans. $A_d = 67.1$)

---

 ## 11.5 BiCMOS CIRCUITS

> **Objective:** • Describe the characteristics of and analyze various BiCMOS circuits.

Thus far, we have considered two basic amplifier design technologies: the bipolar technology, which uses npn and pnp bipolar junction transistors; and the MOS technology, which uses NMOS and PMOS field-effect transistors. We showed that bipolar transistors have a larger transconductance than MOSFETs biased at the same current levels, and that, in general, bipolar amplifiers have larger voltage gains. We also showed that MOSFET circuits have an essentially infinite input impedance at low frequencies, which implies a zero input bias current.

These advantages of the two technologies can be exploited by combining bipolar and MOS transistors in the same integrated circuit. The technology is called **BiCMOS**. BiCMOS technology is especially useful in digital circuit design, but also has applications in analog circuits. In this section, we will examine basic BiCMOS analog circuit configurations.

### 11.5.1 Basic Amplifier Stages

A bipolar multitransistor circuit previously studied is the Darlington pair configuration. Figure 11.38(a) shows a modified Darlington pair configuration, in which the bias current $I_{BIAS}$, or some equivalent element, is used to control the quiescent

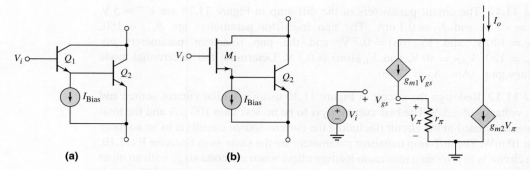

**(a)**    **(b)**

**Figure 11.38** (a) Bipolar Darlington pair configuration and (b) BiCMOS Darlington pair configuration

**Figure 11.39** Small-signal equivalent circuit, BiCMOS Darlington pair configuration

current in $Q_1$. This Darlington pair circuit is used to boost the effective current gain of bipolar transistors. There is no comparable configuration in FET circuits.

A potentially useful BiCMOS circuit is shown in Figure 11.38(b). Transistor $Q_1$ in the Darlington pair is replaced with a MOSFET. The advantages of this configuration are an infinite input resistance, and a large transconductance due to the bipolar transistor $Q_2$.

To analyze the circuit, we consider the small-signal equivalent circuit shown in Figure 11.39. We assume that $r_o = \infty$ in both transistors.

The output signal current is

$$I_o = g_{m1}V_{gs} + g_{m2}V_\pi \tag{11.111}$$

We see that

$$V_i = V_{gs} + V_\pi \tag{11.112}$$

and

$$V_\pi = g_{m1}V_{gs}r_\pi \tag{11.113}$$

Combining Equations (11.112) and (11.113) produces

$$V_{gs} = \frac{V_i}{1 + g_{m1}r_\pi} \tag{11.114}$$

From Equation (11.111), the output current can now be written

$$I_o = g_{m1}V_{gs} + g_{m2}(g_{m1}r_\pi)V_{gs} = (g_{m1} + g_{m2}g_{m1}r_\pi)V_{gs} \tag{11.115}$$

Substituting Equation (11.114) into (11.115), we obtain

$$I_o = \frac{g_{m1}(1 + g_{m2}r_\pi)}{(1 + g_{m1}r_\pi)} \cdot V_i = g_m^c \cdot V_i \tag{11.116}$$

where $g_m^c$ is the **composite transconductance.** Since $g_{m2}$ of the bipolar transistor is usually at least an order of magnitude greater than $g_{m1}$ of the MOSFET, the composite transconductance is approximately an order of magnitude larger than that of the MOSFET alone. We now have the advantages of a large transconductance and an infinite input resistance.

A bipolar cascode circuit is shown in Figure 11.40(a); a corresponding BiCMOS configuration is shown in Figure 11.40(b). The output resistance of the cascode circuit is very high, as we saw in Chapter 10. Also, the cascode amplifier has a wider

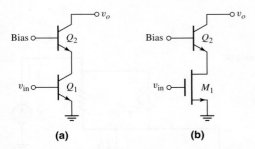

**Figure 11.40** (a) Bipolar cascode configuration and (b) BiCMOS cascode configuration

frequency bandwidth than the common-emitter circuit, since the input resistance looking into the emitter of $Q_2$ is very low, thereby minimizing the Miller multiplication effect. This effect was observed in Chapter 7.

Again, the advantage of the BiCMOS circuit is the infinite input resistance of $M_1$. The equivalent resistance looking into the emitter of a bipolar transistor is much less than the resistance looking into the source of a MOSFET; therefore, the frequency response of a BiCMOS cascode circuit is superior to that of an all-MOSFET cascode circuit.

### 11.5.2 Current Sources

In our previous discussions of constant-current sources, we mentioned that cascode current sources increase the output resistance, as well as the stability of the bias current. Figure 11.41 shows a bipolar cascode configuration in which the output resistance is $R_o \cong \beta r_{o4}$. The bias current in this circuit is much more stable against variations in output voltage than the basic two-transistor current source.

A BiCMOS double cascode constant-current source is shown in Figure 11.42. The small-signal equivalent circuit for determining output resistance is shown in

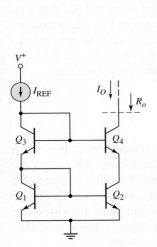

**Figure 11.41** Bipolar cascode constant-current source

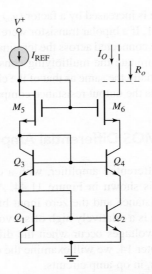

**Figure 11.42** BiCMOS double cascode constant-current source

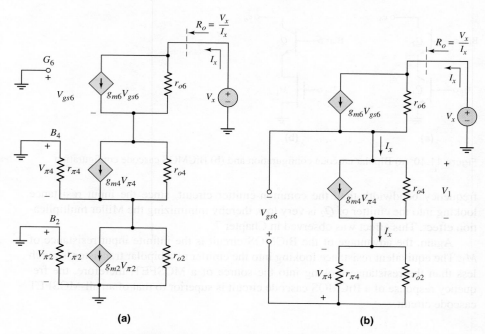

**Figure 11.43** (a) Equivalent circuit for determining output impedance of BiCMOS double cascode current source and (b) rearranged equivalent circuit

Figure 11.43(a). The gate voltage to $M_6$ and the base voltages to $Q_2$ and $Q_4$ are constants, equivalent to signal ground. Also, since $V_{\pi 2} = 0$, then $g_{m6}V_{\pi 2} = 0$, and the equivalent circuit can be rearranged as shown in Figure 11.43(b).

The output resistance of this circuit is extremely large. A detailed analysis shows that the output resistance is given approximately by

$$R_o \cong (g_{m6}r_{o6})(\beta r_{o4}) \tag{11.117}$$

The output resistance is increased by a factor $(g_m r_{o6})$ compared to the bipolar cascode circuit in Figure 11.41. If a bipolar transistor were to be used in place of $M_6$, then a resistance $r_{\pi 6}$ would be connected across the terminals indicated by $V_{gs6}$. This resistance would effectively eliminate the multiplying constant $(g_{m6}r_{o6})$, and the output resistance would be essentially the same as that of the circuit in Figure 11.41. The BiCMOS circuit, then, increases the output resistance compared to an all-bipolar circuit.

### 11.5.3 BiCMOS Differential Amplifier

A basic BiCMOS differential amplifier, with a constant-current source bias and a bipolar active load, is shown in Figure 11.44. Again, the primary advantages are the infinite input resistance and the zero input bias current. One disadvantage of a MOSFET input stage is a relatively high offset voltage compared to that of a bipolar input circuit. Offset voltages occur when the differential-pair input transistors are mismatched. In Chapter 14, we will examine the effect of offset voltages, as well as nonzero bias currents, in op-amp circuits.

We will consider additional BiCMOS op-amp circuits in Chapter 13, when we discuss the analysis and design of full op-amp circuits.

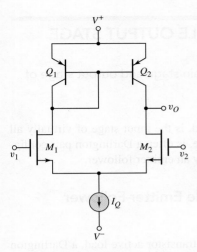

**Figure 11.44**  Basic BiCMOS differential amplifier

# Test Your Understanding

**TYU 11.16** Consider the BiCMOS Darlington pair in Figure 11.45. The NMOS transistor parameters are $K_n = 50\,\mu\text{A/V}^2$, $V_{TN} = 0.5$ V, and $\lambda = 0$. The BJT parameters are $\beta = 150$, $V_{BE}(\text{on}) = 0.7$ V, and $V_A = \infty$. Determine the small-signal parameters for each transistor, as well as the composite transconductance. (Ans. $g_{m1} = 71.4\,\mu\text{A/V}$, $g_{m2} = 2.865$ mA/V, $r_{\pi 2} = 52.3$ k$\Omega$, $g_m^c = 2.275$ mA/V)

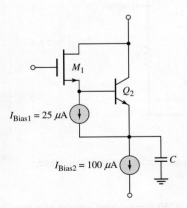

**Figure 11.45**  Figure for Exercise TYU 11.16

**TYU 11.17** The reference current in each of the constant-current source circuits shown in Figures 11.41 and 11.42 is $I_{\text{REF}} = 0.5$ mA. All bipolar transistor parameters are $\beta = 150$ and $V_A = 80$ V, and all MOSFET parameters are: $K_n = 500\,\mu\text{A/V}^2$, $V_{TN} = 1$ V, and $\lambda = 0.0125$ V$^{-1}$. Neglecting bipolar base currents, determine the output resistance $R_o$ of each constant-current source. (Ans. For Figure 11.41, $R_o \cong 24$ M$\Omega$; for Figure 11.42, $R_o = 3840$ M$\Omega$)

## 11.6    GAIN STAGE AND SIMPLE OUTPUT STAGE

**Objective:** • Analyze an example of a gain stage and output stage of a multistage amplifier.

A diff-amp, including those previously discussed, is the input stage of virtually all op-amps. The second op-amp stage, or gain stage, is often a Darlington pair configuration, and the third, or output, stage is normally an emitter follower.

### 11.6.1    Darlington Pair and Simple Emitter-Follower Output

Figure 11.46 shows a BJT diff-amp with a three-transistor active load, a Darlington pair connected to the diff-amp output, and a simple emitter-follower output stage.

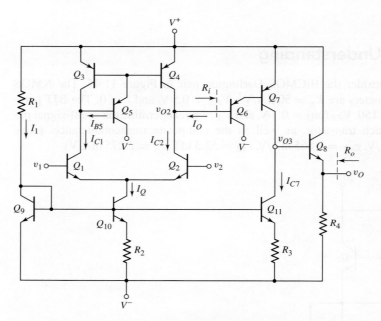

**Figure 11.46** BJT diff-amp with three-transistor active load, Darlington pair gain stage, and simple emitter-follower output stage

The differential-pair transistors are biased with a Widlar current source at a bias current $I_Q$. We noted previously that, for the diff-amp dc currents to be balanced, we must have

$$I_O = I_{B5} = \frac{I_Q}{\beta(1+\beta)} \tag{11.118}$$

From the figure, we see that

$$I_O = \frac{I_{E6}}{(1+\beta)} = \frac{I_{C7}}{\beta(1+\beta)} \tag{11.119}$$

In order for $I_O = I_{B5}$, we must require that $I_{C7} = I_Q$. This means that the emitter resistors of $Q_{10}$ and $Q_{11}$ should have the same value. Transistor $Q_{11}$ also acts as an active load for the Darlington pair gain stage.

Transistor $Q_8$ and resistor $R_4$ form the simple emitter-follower output stage. The emitter-follower amplifier minimizes loading effects because its output resistance is small.

Ideally, when the diff-amp input is a pure common-mode signal, the output $v_o$ is zero. The combination of $Q_7$ and $Q_{11}$ allows the dc level to shift. By slightly changing the bias current $I_{C7}$, we can vary voltages $V_{EC7}$ and $V_{CE11}$ such that $v_o = 0$. The small variation of $I_{C7}$ required to achieve the necessary dc level shift will not significantly change the balance between $I_O$ and $I_{B5}$. As we will see in later chapters, other forms of level shifters could also be used.

## 11.6.2   Input Impedance, Voltage Gain, and Output Impedance

The input resistance of the Darlington pair determines the loading effect on the basic diff-amp. In addition, the gain of the Darlington pair affects the overall gain of the op-amp circuit, and the output resistance of the emitter follower determines any loading effects on the output signal.

Figure 11.47(a) is the ac equivalent circuit of the Darlington pair, where $R_{L7}$ is the effective resistance connected between the collector of $Q_7$ and signal ground. Figure 11.47(b) shows the simple hybrid-$\pi$ model of the Darlington pair. We see that the equivalent circuits for $Q_6$ and $Q_7$ have been effectively turned upside down compared to the transistors in Figure 11.47(a).

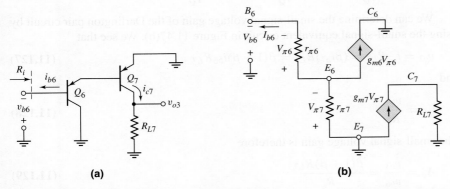

**(a)**   **(b)**

**Figure 11.47**  (a) The ac equivalent circuit, Darlington pair, and (b) small-signal equivalent circuit, Darlington pair

Writing a KVL equation around the B–E loop of $Q_6$ and $Q_7$, we have

$$V_{b6} = V_{\pi 6} + V_{\pi 7} \tag{11.120}$$

We can also write that

$$V_{\pi 6} = I_{b6} r_{\pi 6} \tag{11.121}$$

and the KCL equation is

$$\frac{V_{\pi 7}}{r_{\pi 7}} = \frac{V_{\pi 6}}{r_{\pi 6}} + g_{m6} V_{\pi 6} \tag{11.122(a)}$$

or

$$V_{\pi 7} = r_{\pi 7} \left[ \frac{(1+\beta)}{r_{\pi 6}} \right] V_{\pi 6} = r_{\pi 7}(1+\beta) I_{b6} \tag{11.122(b)}$$

where $r_{\pi 6} g_{m6} = \beta$. Substituting Equations (11.122(b)) and (11.121) into Equation (11.120), we obtain

$$V_{b6} = I_{b6} r_{\pi 6} + r_{\pi 7}(1+\beta) I_{b6} \tag{11.123}$$

The input resistance is therefore

$$R_i = \frac{V_{b6}}{I_{b6}} = r_{\pi 6} + r_{\pi 7}(1+\beta) \tag{11.124}$$

Assuming $I_{C7} = I_Q$, the hybrid-$\pi$ parameters are

$$r_{\pi 7} = \frac{\beta V_T}{I_{C7}} = \frac{\beta V_T}{I_Q} \tag{11.125(a)}$$

and

$$r_{\pi 6} = \frac{\beta V_T}{I_{C6}} = \frac{(1+\beta)\beta V_T}{I_Q} \tag{11.125(b)}$$

Combining Equations (11.125(a)), (11.125(b)), and Equation (11.124) yields an expression for the input resistance, as follows:

$$R_i = \frac{(1+\beta)\beta V_T}{I_Q} + \frac{(1+\beta)\beta V_T}{I_Q} = \frac{2(1+\beta)\beta V_T}{I_Q} \tag{11.126}$$

We can determine the small-signal voltage gain of the Darlington pair circuit by using the small-signal equivalent circuit in Figure 11.47(b). We see that

$$v_{o3} = i_{c7} R_{L7} = (\beta i_{b7}) R_{L7} = \beta(1+\beta) i_{b6} R_{L7} \tag{11.127}$$

and

$$i_{b6} = \frac{v_{b6}}{R_i} \tag{11.128}$$

The small-signal voltage gain is therefore

$$A_v = \frac{v_{o3}}{v_{b6}} = \frac{\beta(1+\beta) R_{L7}}{R_i} \tag{11.129}$$

Substituting Equation (11.126) into (11.129), we find that

$$A_v = \frac{\beta(1+\beta) R_{L7}}{\dfrac{2(1+\beta)\beta V_T}{I_Q}} = \left( \frac{I_Q}{2V_T} \right) R_{L7} \tag{11.130}$$

In Figure 11.46, we see that resistance $R_{L7}$ is the parallel combination of the resistance looking into the collector of $Q_{11}$ and the resistance looking into the base of $Q_8$. From Chapter 10, the resistance looking into the collector of $Q_{11}$ is

$$R_{c11} = r_{o11}(1 + g_{m11} R_E') \tag{11.131}$$

where $R'_E = r_{\pi 11} \| R_3$. The resistance looking into the base of $Q_8$ is

$$R_{b8} = r_{\pi 8} + (1 + \beta)R_4. \qquad \textbf{(11.132)}$$

Equations (11.131) and (11.132) indicate that resistances $R_{c11}$ and $R_{b8}$ are large, which means that the effective resistance $R_{L7}$ is also large.

## EXAMPLE 11.13

**Objective:** Calculate the input resistance and the small-signal voltage gain of a Darlington pair.

Consider the circuit shown in Figure 11.46, with parameters $I_{C7} = I_Q = 0.2$ mA, $I_{C8} = 1$ mA, $R_4 = 10$ k$\Omega$, and $R_3 = 0.2$ k$\Omega$. Assume $\beta = 100$ for all transistors, and the Early voltage for $Q_{11}$ is 100 V.

**Solution:** The input resistance, given by Equation (11.126), is

$$R_i = \frac{2(1+\beta)\beta V_T}{I_Q} = \frac{2(101)(100)(0.026)}{0.2} \Rightarrow 2.63 \text{ M}\Omega$$

The small-signal voltage gain is a function of $R_{L7}$, which in turn is a function of $R_{c11}$ and $R_{b8}$. We can find that

$$r_{\pi 11} = \beta V_T / I_Q = (100)(0.026)/0.2 = 13 \text{ k}\Omega$$

such that

$$R'_E = 13\|0.2 = 0.197 \text{ k}\Omega$$

Also

$$g_{m11} = I_Q/V_T = 0.2/0.026 = 7.69 \text{ mA/V}$$

and

$$r_{o11} = V_A/I_Q = 100/0.2 = 500 \text{ k}\Omega$$

Therefore,

$$R_{c11} = r_{o11}(1 + g_{m11}R'_E) = 500\,[1 + (7.69)(0.197)] \Rightarrow 1.26 \text{ M}\Omega$$

We can determine that

$$r_{\pi 8} = \beta V_T / I_{C8} = (100)(0.026)/1 = 2.6 \text{ k}\Omega$$

Then

$$R_{b8} = r_{\pi 8} + (1 + \beta)R_4 = 2.6 + (101)(10) \Rightarrow 1.01 \text{ M}\Omega$$

Consequently, resistance $R_{L7}$ is

$$R_{L7} = R_{c11}\|R_{b8} = 1.26\|1.01 = 0.561 \text{ M}\Omega$$

Finally, from Equation (11.130), the small-signal voltage gain is

$$A_v = \left(\frac{I_Q}{2V_T}\right)R_{L7} = \left[\frac{0.2}{2(0.026)}\right](561) = 2158$$

**Comment:** The input resistance of the Darlington pair is in the megohm range, which should minimize severe loading effects on the diff-amp. In addition, the small-signal gain is large because of the active load ($Q_{11}$) and the large input resistance of the emitter-follower output stage.

EXERCISE PROBLEM

**Ex 11.13:** Consider the Darlington pair $Q_6$ and $Q_7$ in Figure 11.46. Determine the current gain of the Darlington pair, $I_{c7}/I_{b6}$. Use the parameters described in Example 11.13. (Ans. $(101)(100) = 1.01 \times 10^4$)

We can use the results of Chapter 6 to determine the output resistance of the emitter follower. The output resistance is

$$R_o = R_4 \left\| \left(\frac{r_{\pi 8} + Z}{(1 + \beta)}\right) \right. \tag{11.133}$$

where $Z$ is the equivalent impedance, or resistance, in the base of $Q_8$. In this case, $Z = R_{c11} \| R_{c7}$, where $R_{c7}$ is the resistance looking into the collector of $Q_7$. Because of the factor $(1 + \beta)$ in the denominator, the output resistance of the emitter follower is normally small, as previously determined.

## EXAMPLE 11.14

**Objective:** Calculate the output resistance of the circuit in Figure 11.46.
    Consider the same circuit and transistor parameters described in Example 11.13. Assume the Early voltage of $Q_7$ is 100 V.

**Solution:** From Example 11.13, we have that $R_{c11} = 1.26$ M$\Omega$ and $r_{\pi 8} = 2.6$ k$\Omega$. We can then determine that

$$R_{c7} = \frac{V_A}{I_Q} = \frac{100}{0.2} = 500 \text{ k}\Omega$$

Then,

$$Z = R_{c11} \| R_{c7} = 1260 \| 500 = 358 \text{ k}\Omega$$

Therefore,

$$R_o = R_4 \left\| \left[\frac{r_{\pi 8} + Z}{(1 + \beta)}\right] = 10 \right\| \left(\frac{2.6 + 358}{101}\right) = 2.63 \text{ k}\Omega$$

**Comment:** The output resistance is obviously less than $R_4$ and is substantially less than the equivalent resistance $Z$ in the base of $Q_8$. In a later chapter, we will examine a Darlington pair emitter-follower output stage in which the output resistance is on the order of 100 $\Omega$.

EXERCISE PROBLEM

**Ex 11.14:** The circuit shown in Figure 11.48 is an ac equivalent circuit of a Darlington pair output stage. Assume the transistor current gains are $\beta_A = 90$ and $\beta_B = 180$. Assuming transistor $Q_B$ is biased at $I_{CQB} = 0.5$ mA, determine the output resistance $R_o$. (Ans. $R_o = 120 \, \Omega$)

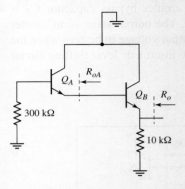

**Figure 11.48** Figure for Exercise Ex11.14

A BJT diff-amp with an active load can produce a small-signal differential-mode voltage gain on the order of $10^3$, and the Darlington pair can also provide a voltage gain on the order of $10^3$. Since the emitter follower has a gain of essentially unity, the overall voltage gain of the op-amp circuit is on the order of $10^6$. This value is typical for the low-frequency, open-loop gain of op-amp circuits.

## Test Your Understanding

**TYU 11.18**  Consider the Darlington pair and emitter-follower portions of the circuit in Figure 11.46. The parameters are: $I_{C7} = I_Q = 0.5$ mA, $I_{C8} = 2$ mA, $R_4 = 5$ k$\Omega$, and $R_3 = 0.1$ k$\Omega$. For all transistors, the current gain is $\beta = 120$, and for $Q_{11}$ and $Q_7$, the Early voltage is $V_A = 120$ V. Calculate the input resistance and small-signal voltage gain of the Darlington pair, and the output resistance of the emitter follower. (Ans. $R_i = 1.51$ M$\Omega$, $A_v = 3115$, $R_o = 1.14$ k$\Omega$)

**TYU 11.19**  In the circuit in Figure 11.46, the Darlington pair and emitter-follower transistor parameters are the same as in Exercise TYU 11.18. Determine the effective resistance $R_{L7}$ (see Figure 11.47(a)) such that the small-signal voltage gain is $10^3$. (Ans. $R_{L7} = 104$ k$\Omega$)

## 11.7  SIMPLIFIED BJT OPERATIONAL AMPLIFIER CIRCUIT

**Objective:** • Analyze a simplified multistage bipolar amplifier.

An operational amplifier (op-amp) is a multistage circuit composed of a differential amplifier input stage, a gain stage, and an output stage. In this section, we will consider a simplified BJT op-amp circuit.

Although active load devices increase the gain of an amplifier, in this discussion, we will consider resistive loads, in order to simplify the analysis and design. For the bipolar circuit, all component values are given; we will analyze both the dc and ac circuit characteristics.

Figure 11.49. depicts a simple bipolar operational amplifier. The differential amplifier stage is biased with a Widlar current source, and a one-sided output is connected to the Darlington pair gain stage. An emitter bypass capacitor $C_E$ is included to increase the small-signal voltage gain. The output stage is an emitter follower. In general, we want the dc value of the output voltage to be zero when the input voltage is zero. To accomplish this, we need to insert a dc level shifting circuit between the voltage $v_{O3}$ and the output voltage $v_O$.

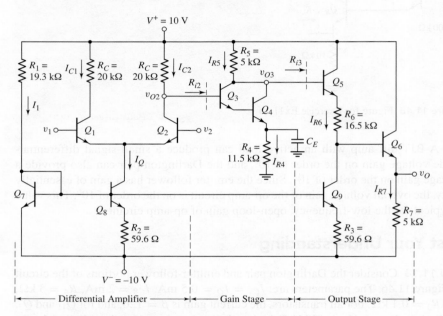

**Figure 11.49**  Bipolar operational amplifier circuit

## EXAMPLE 11.15

**Objective:** Analyze the dc characteristics of the bipolar op-amp circuit.

Consider the circuit in Figure 11.49. Neglect base currents and, as a simplification, assume $V_{BE}(\text{on}) = 0.7$ V for all transistors except $Q_8$ and $Q_9$ in the Widlar circuit.

**Solution:** The reference current $I_1$ is

$$I_1 = \frac{10 - 0.7 - (-10)}{19.3} = 1 \text{ mA}$$

The bias current $I_Q$ is determined from

$$I_Q R_2 = V_T \ln\left(\frac{I_1}{I_Q}\right)$$

and is

$$I_Q = 0.4 \text{ mA}$$

The collector currents are then

$$I_{C1} = I_{C2} = 0.2 \, \text{mA}$$

The dc voltage at the collector of $Q_2$ is

$$V_{O2} = 10 - I_{C2}R_C = 10 - (0.2)(20) = 6 \, \text{V}$$

With these circuit parameters, the common-mode input voltage is limited to the range $-8.6 \le v_{CM} \le 6$ V, which will keep all transistors biased in the forward-active mode.
The current $I_{R4}$ is determined to be

$$I_{R4} = \frac{V_{O2} - 2V_{BE}(\text{on})}{R_4} = \frac{6 - 1.4}{11.5} = 0.4 \, \text{mA}$$

Since base currents are assumed negligible, the current $I_{R5}$ is $I_{R5} \cong I_{R4}$.
The dc voltage at the collectors of $Q_3$ and $Q_4$ is then

$$V_{O3} = 10 - I_{R5}R_5 = 10 - (0.4)(5) = 8 \, \text{V}$$

This shows us that the dc voltage $V_{O3}$ is midway between the 10 V supply voltage and the dc input voltage $V_{O2} = 6$ V to $Q_3$. This allows a maximum symmetrical swing in the time-varying voltage at $v_{o3}$.

Transistor $Q_5$ and resistor $R_6$ form the dc voltage level shifting function. Since $R_3 = R_2$, we have

$$I_{R6} = I_Q = 0.4 \, \text{mA}$$

The dc voltage at the base of $Q_6$ is found to be

$$V_{B6} = V_{O3} - V_{BE}(\text{on}) - I_{R6}R_6 = 8 - 0.7 - (0.4)(16.5) = 0.7 \, \text{V}$$

This relationship produces a zero dc output voltage when a zero differential-mode voltage is applied at the input.

Finally, current $I_{R7}$ is

$$I_{R7} = \frac{v_o - (-10)}{R_7} = \frac{10}{5} = 2 \, \text{mA}$$

**Comment:** The dc analysis of this simplified op-amp circuit proceeds in much the same way as in previous examples. We observe that all transistors are biased in the forward-active mode.

### EXERCISE PROBLEM

**Ex 11.15:** Consider the simple bipolar op-amp circuit in Figure 11.49. The transistor parameters are: $\beta = 100$, $V_{BE}(\text{on}) = 0.7$ V (except for $Q_8$ and $Q_9$), and $V_A = \infty$. Redesign the circuit such that $I_{C1} = I_{C2} = 0.1$ mA, $I_{R7} = 5$ mA, $I_1 = I_{R4} = I_{R6} = 0.6$ mA, $V_{CE1} = V_{CE2} = 4$ V, $V_{CE4} = 3$ V, and $v_O = 0$. (Ans. $R_1 = 32.2 \, \text{k}\Omega$, $R_2 = 143 \, \Omega$, $R_3 = 0$, $R_C = 67 \, \text{k}\Omega$, $R_4 = 3.17 \, \text{k}\Omega$, $R_5 = 8.5 \, \text{k}\Omega$, $R_6 = 5.83 \, \text{k}\Omega$, and $R_7 = 2 \, \text{k}\Omega$)

## EXAMPLE 11.16

**Objective:** Determine the small-signal differential-mode voltage gain of the bipolar op-amp circuit.

Consider the circuit in Figure 11.49, with transistor parameters $\beta = 100$ and $V_A = \infty$.

**Solution:** The overall differential-mode voltage gain can be written

$$A_d = A_{d1} \cdot A_2 \cdot A_3 = \left( \frac{v_{o2}}{v_1 - v_2} \right) \cdot \left( \frac{v_{o3}}{v_{o2}} \right) \cdot \left( \frac{v_o}{v_{o3}} \right)$$

The overall small-signal voltage gain is the product of the individual stage gains *only* if the load resistance of the following stage is taken into account.

We will rely on previous results to determine the individual voltage gains. The input resistances to the Darlington pair $R_{i2}$ and to the output stage $R_{i3}$ are indicated in Figure 11.49. The one-sided differential-mode voltage gain of the diff-amp is given by

$$A_{d1} = \frac{V_{o2}}{v_d} = \frac{g_m}{2} (R_C \| R_{i2})$$

where $R_{i2}$ is the input resistance of the Darlington pair, as follows:

$$R_{i2} = r_{\pi3} + (1 + \beta) r_{\pi4}$$

where

$$r_{\pi4} = \beta V_T / I_{R4} = (100)(0.026)/0.4 = 6.5 \text{ k}\Omega$$

and

$$r_{\pi3} \cong \beta^2 V_T / I_{R4} = (100)^2 (0.026)/0.4 = 650 \text{ k}\Omega$$

Therefore,

$$R_{i2} = 650 + (101)(6.5) = 1307 \text{ k}\Omega$$

The transistor transconductance is

$$g_m = \frac{I_Q}{2V_T} = \frac{0.4}{2(0.026)} = 7.70 \text{ mA/V}$$

The gain of the differential amplifier stage is therefore

$$A_{d1} = \frac{g_m}{2} (R_C \| R_{i2}) = \left( \frac{7.70}{2} \right) [20 \| 1307] = 75.8$$

Since the load resistance $R_{i2} \gg R_C$, there is no significant loading effect of the second stage on the diff-amp stage.

From previous results, we know the voltage gain of the Darlington pair is given by

$$A_2 = \left( \frac{I_{R4}}{2V_T} \right) (R_5 \| R_{i3})$$

where

$$R_{i3} = r_{\pi5} + (1 + \beta)[R_6 + r_{\pi6} + (1 + \beta)R_7]$$

We find that

$$r_{\pi5} = \beta V_T / I_{R6} = (100)(0.026)/0.4 = 6.5 \text{ k}\Omega$$

and

$$r_{\pi6} = \beta V_T / I_{R7} = (100)(0.026)/2 = 1.3 \text{ k}\Omega$$

Therefore

$$R_{i3} = 6.5 + (101)[16.5 + 1.3 + (101)(5)] \Rightarrow 52.8 \text{ M}\Omega$$

Since $R_{i3} \gg R_5$, the output stage does not load down the gain stage, and the small-signal voltage gain is approximately

$$A_2 \cong \left( \frac{I_{R4}}{2V_T} \right) R_5 = \left[ \frac{0.4}{2(0.026)} \right] (5) = 38.5$$

The combination of $Q_5$ and $Q_6$ forms an emitter follower, and the gain of the output stage is

$$A_3 = v_o/v_{o3} \cong 1$$

The overall small-signal voltage gain is therefore

$$A_d = A_{d1} \cdot A_2 \cdot A_3 = (75.8)(38.5)(1) = 2918$$

**Comment:** From our previous discussion, we know that the overall gain can be increased substantially by using active loads. Yet, the analysis of this simplified circuit provides some insight into the design of multistage circuits, as well as the overall small-signal voltage gain of op-amp circuits.

**Computer Correlation:** A PSpice analysis was performed on the bipolar op-amp circuit in Figure 11.49. The dc output voltage from this analysis was $V_O = -0.333$ V, rather than the desired value of zero. This occurred because the B–E voltages were not exactly 0.7 V, as assumed in the hand analysis. A zero output voltage can be obtained by slightly adjusting $R_6$. The differential voltage gain was $A_d = 2932$, which agrees very well with the hand analysis.

### EXERCISE PROBLEM

**Ex 11.16:** Consider the simple bipolar op-amp circuit in Figure 11.49 with circuit and transistor parameters given in Exercise Problem Ex11.15. Determine the input resistances $R_{i2}$ and $R_{i3}$, and the differential-mode voltage gain $A_d = v_o/v_d$. (Ans. $R_{i2} = 870 \text{ k}\Omega$, $R_{i3} = 21.0 \text{ M}\Omega$, $A_d = 11{,}729$)

### Problem-Solving Technique: Multistage Circuits

1. Perform the dc analysis of the circuit to determine the small-signal parameters of the transistors. In most cases BJT base currents can be neglected. This assumption will normally provide sufficient accuracy for a hand analysis.
2. Perform the ac analysis on each stage of the circuit, *taking into account the loading effect of the following stage.* (In many cases, previous results of small-signal analyses can be used directly.)
3. The overall small-signal voltage gain or current gain is the product of the gains of the individual stages *as long as the loading effect of each stage is taken into account.*

## 11.8    DIFF-AMP FREQUENCY RESPONSE

**Objective:** • Analyze the frequency response of the differential amplifier.

In Chapter 7, we considered the frequency responses of the three basic amplifier configurations. In this section, we will analyze the frequency response of the differential amplifier. Since the diff-amp is a linear circuit, we can determine the frequency response due to: (a) a pure differential-mode input signal, (b) a pure common-mode input signal, and (c) the total or net result, using superposition.

### 11.8.1    Due to Differential-Mode Input Signal

Consider the basic bipolar diff-amp shown in Figure 11.50(a). The input is a pure differential-mode input signal. We know from Equation (11.24) that the small-signal voltage $v_e$ is at signal ground when a differential-mode input signal is applied. To determine the frequency response, we evaluate the equivalent common-emitter half-circuit in Figure 11.50(b).

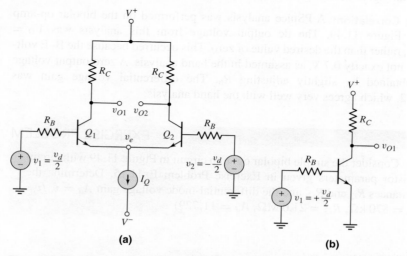

**(a)**                                          **(b)**

**Figure 11.50**  (a) BJT differential amplifier with differential-mode input signal and (b) equivalent common-emitter half-circuit of differential amplifier

Since the diff-amp is a direct-coupled amplifier, the midband voltage gain extends to zero frequency. This one-sided midband gain is

$$A_{v1} = \frac{V_{o1}}{V_d/2} = -g_m R_C \left( \frac{r_\pi}{r_\pi + R_B} \right) \qquad (11.134(a))$$

or

$$A_{v1} = \frac{-\beta R_C}{r_\pi + R_B} \qquad (11.134(b))$$

From the high-frequency common-emitter characteristics determined in Chapter 7 we know that the upper 3 dB frequency is

$$f_H = \frac{1}{2\pi [r_\pi \| R_B](C_\pi + C_M)} \qquad (11.135)$$

where $C_M$ is the equivalent Miller capacitance given by

$$C_M = C_\mu (1 + g_m R_C) \qquad (11.136)$$

Equation (11.136) implies that, if the value of $R_C$ is fairly large, the Miller capacitance will significantly affect the bandwidth of the differential amplifier.

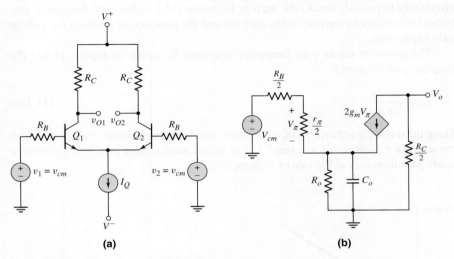

## 11.8.2    Due to Common-Mode Input Signal

Figure 11.51(a) shows the basic diff-amp with a pure common-mode input signal. The circuit is symmetrical, which means that resistors $R_B$, resistors $R_C$, and the transistors are effectively in parallel. Figure 11.51(b) is the small-signal equivalent circuit, with the constant-current source replaced by its output resistance $R_o$ and capacitance $C_o$.

**Figure 11.51** (a) BJT differential amplifier with common-mode input signal and (b) small-signal equivalent circuit, common-mode configuration

We will justify neglecting the transistor parameters $C_\pi$ and $C_\mu$. The output voltage is

$$V_o = -(2g_m V_\pi)\left(\frac{R_C}{2}\right) \tag{11.137}$$

A KVL equation around the B–E loop produces

$$V_{cm} = \left(\frac{V_\pi}{r_\pi/2}\right)\left(\frac{R_B}{2}\right) + V_\pi + \left(\frac{V_\pi}{r_\pi/2} + 2g_m V_\pi\right)\left[R_o \,\middle\|\, \left(\frac{1}{sC_o}\right)\right] \tag{11.138(a)}$$

or

$$V_{cm} = V_\pi\left\{\frac{R_B}{r_\pi} + 1 + 2\left(\frac{1+\beta}{r_\pi}\right)\left(\frac{R_o}{1+sR_oC_o}\right)\right\} \tag{11.138(b)}$$

Solving for $V_\pi$ and substituting the result into Equation (11.137) yields the common-mode gain, which is

$$A_{cm} = \frac{V_o}{V_{cm}} = \frac{-g_m R_C}{\dfrac{R_B}{r_\pi} + 1 + \dfrac{2(1+\beta)}{r_\pi}\left(\dfrac{R_o}{1+sR_oC_o}\right)} \tag{11.139(a)}$$

or

$$A_{cm} = \frac{-g_m R_C(1 + sR_oC_o)}{\left(1 + \dfrac{R_B}{r_\pi}\right)(1 + sR_oC_o) + \dfrac{2(1 + \beta)R_o}{r_\pi}}$$  **(11.139(b))**

Equation (11.139(b)) shows that there is a zero in the common-mode gain. To explain, capacitor $C_o$ is in parallel with $R_o$, and it acts as a bypass capacitor. At very low frequency, $C_o$ is effectively an open circuit and the common-mode signal "sees" $R_o$. As the frequency increases, the impedance of the capacitor decreases and $R_o$ is effectively bypassed; hence, the zero in Equation (11.139(b)). The frequency analysis of an emitter bypass capacitor also showed the presence of a zero in the voltage gain expression.

The common-mode gain frequency response is shown in Figure 11.52. The frequency of the zero is

$$f_z = \frac{1}{2\pi R_o C_o}$$  **(11.140)**

Since the output resistance $R_o$ of a constant-current source is normally large, a small capacitance $C_o$ can result in a small $f_z$. For frequencies greater than $f_z$, the common-mode gain increases at the rate of 6 dB/octave.

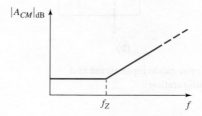

**Figure 11.52** Frequency response of common-mode gain

Equation (11.139(b)) also shows that there is a pole associated with the common-mode gain. Rearranging the terms in that equation, we see that the frequency of the pole is

$$f_p = \frac{1}{2\pi R_{eq} C_o}$$  **(11.141)**

where

$$R_{eq} = \frac{R_o\left(1 + \dfrac{R_B}{r_\pi}\right)}{1 + \dfrac{R_B}{r_\pi} + \dfrac{2(1 + \beta)R_o}{r_\pi}}$$  **(11.142)**

The denominator of Equation (11.142) is very large, because of the term $(1 + \beta)R_o$. This implies that $R_{eq}$ is small, which means that the frequency $f_p$ of the pole is very large.

The differential-mode gain is shown in Figure 11.53. The frequency response of the common-mode rejection ratio is found by combining Figures 11.52 and 11.53, and is shown in Figure 11.54.

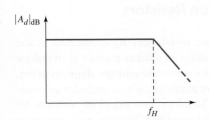

**Figure 11.53** Frequency response of differential-mode gain

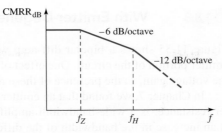

**Figure 11.54** Frequency response of common-mode rejection ratio

## EXAMPLE 11.17

**Objective:** Determine the zero and pole frequencies in the common-mode gain.

Consider a diff-amp biased with a constant-current source. The output resistance is $R_o = 10 \text{ M}\Omega$ and the output capacitance is $C_o = 1$ pF. Assume the circuit and transistor parameters are $R_B = 0.5 \text{ k}\Omega$, $r_\pi = 10 \text{ k}\Omega$, and $\beta = 100$.

**Solution:** In the common-mode gain, the frequency of the zero is

$$f_z = \frac{1}{2\pi R_o C_o} = \frac{1}{2\pi (10 \times 10^6)(1 \times 10^{-12})} \Rightarrow 15.9 \text{ kHz}$$

Also in the common-mode gain, the frequency of the pole is

$$f_P = 1/(2\pi R_{eq} C_o)$$

where

$$R_{eq} = \frac{R_o \left(1 + \dfrac{R_B}{r_\pi}\right)}{1 + \dfrac{R_B}{r_\pi} + \dfrac{2(1+\beta)R_o}{r_\pi}} = \frac{(10 \times 10^6)\left(1 + \dfrac{0.5}{10}\right)}{1 + \dfrac{0.5}{10} + \dfrac{2(101)(10 \times 10^6)}{10 \times 10^3}}$$

or

$$R_{eq} = 51.98 \ \Omega$$

The frequency of the pole is therefore

$$f_P = \frac{1}{2\pi (51.98)(1 \times 10^{-12})} \Rightarrow 3.06 \text{ GHz}$$

**Comment:** The frequency of the zero in the common-mode gain is fairly low, while the frequency of the pole is extremely large. The relatively low frequency of the zero justifies neglecting the effect of $C_\pi$ and $C_\mu$. The CMRR frequency response is shown in Figure 11.54, where $f_z$ is the zero frequency of the common-mode gain and $f_H$ is the upper 3 dB frequency of the differential-mode gain.

## EXERCISE PROBLEM

**Ex 11.17:** Repeat Example 11.17 for the case when the output capacitance of the constant current source is $C_o = 0.2$ pF. (Ans. $f_z = 79.6$ kHz, $f_p = 15.3$ GHz)

### 11.8.3 With Emitter-Degeneration Resistors

Figure 11.55 shows a bipolar diff-amp with two resistances $R_E$ connected in the emitter portion of the circuit. One effect of including an emitter resistor is to reduce the voltage gain, so the presence of these resistors is termed **emitter degeneration.**

In Chapter 7, we found that an emitter-follower circuit, which includes an emitter resistance, is a wide-bandwidth amplifier. Therefore, one effect of resistors $R_E$ is an increase in the bandwidth of the differential amplifier. We rely on a computer simulation to evaluate emitter degeneration effects.

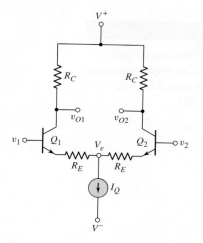

**Figure 11.55** BJT differential amplifier with emitter-degeneration resistors

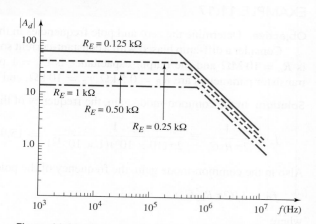

**Figure 11.56** PSpice results for frequency response of diff-amp with emitter-degeneration

Figure 11.56 shows the frequency response of a one-sided differential-mode gain, obtained from a PSpice analysis for four $R_E$ resistance values. The diff-amp is biased at $I_Q = 0.5$ mA and the $R_C$ resistors are $R_C = 30$ k$\Omega$. The transistor capacitances are $C_\pi = 34.6$ pF and $C_\mu = 4.3$ pF. As the emitter degeneration increases, the differential-mode voltage gain decreases, but the bandwidth increases, as previously indicated. The figure-of-merit for amplifiers, the gain-bandwidth product, is approximately a constant for the results shown in Figure 11.56.

### 11.8.4 With Active Load

Figure 11.57 shows a bipolar diff-amp with an active load and a single input at $v_1$. The base and collector junctions of $Q_3$ are connected together, and a one-sided output is taken at $v_{O2}$.

With the connection of $Q_3$, the equivalent load resistance in the collector of $Q_1$ is on the order of $r_\pi/(1 + \beta)$. This small resistance minimizes the Miller multiplication factor in $Q_1$. Also, with the base of $Q_2$ at ground potential, one side of $C_{\mu2}$ is grounded, and the Miller multiplication in $Q_2$ is zero. Therefore, we expect the bandwidth of the diff-amp with an active load to be relatively wide. At high frequencies, however, the effective impedance in the collector of $Q_1$ also includes the input

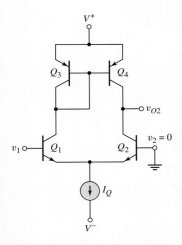

**Figure 11.57**  BJT diff-amp with active load and single-sided input

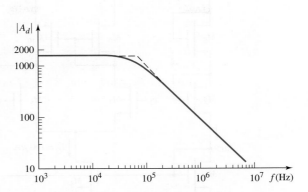

**Figure 11.58**  PSpice results for frequency response of diff-amp with active load and single-sided input

capacitances of $Q_3$ and $Q_4$. These additional capacitances also affect the frequency response of the diff-amp, potentially narrowing the bandwidth.

Again, we rely on a computer analysis to determine the frequency characteristics of the diff-amp with an active load. Figure 11.58 shows the results of the computer simulation. The diff-amp is biased at $I_Q = 0.5$ mA, and the Early voltage of each transistor is assumed to be 80 V. The transistor capacitances are $C_\pi = 34.6$ pF for each transistor, $C_\mu = 3.8$ pF in $Q_1$ and $Q_2$, and $C_\mu = 7$ pF and 5.5 pF in $Q_3$ and $Q_4$, respectively.

The low-frequency voltage gain is 1560 and the upper 3 dB frequency is 64 kHz. The large gain is as expected for an active load amplifier, but the 3 dB frequency is lower than expected. However, the gain–bandwidth product for the active load diff-amp is approximately four times that of the diff-amp shown in Figure 11.55. The increased gain–bandwidth product implies a reduced Miller multiplication factor in the active load diff-amp, as predicted.

## 11.9    DESIGN APPLICATION: A CMOS DIFF-AMP

**Objective:** • Design a CMOS diff-amp with an output gain stage to meet a set of specifications.

**Specifications:**  Design a CMOS diff-amp with an output stage. The magnitude of voltage gain of each stage is to be at least 600. Bias currents are to be $I_Q = I_{REF} = 100$ $\mu$A, and biasing of the circuit is to be $V^+ = 2.5$ V and $V^- = -2.5$ V.

**Design Approach:**  The circuit to be designed has the configuration shown in Figure 11.59. The diff-amp has NMOS amplifying transistors and a PMOS active load. The diff-amp is biased with a cascode current source to provide a large output resistance. The gain stage is a PMOS transistor in a common source configuration that also has an active load.

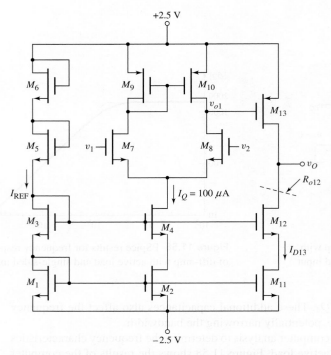

**Figure 11.59** A CMOS diff-amp with an output stage for the design application

We will assume that several sets of transistors are matched. In particular, we will assume that $M_1$ to $M_4$, $M_{11}$, and $M_{12}$ are matched; $M_7$ and $M_8$ are matched; $M_5$ and $M_6$ are matched; and $M_9$ and $M_{10}$ are matched.

**Choices:** Assume NMOS and PMOS transistors are available with parameters $V_{TN} = 0.5$ V, $V_{TP} = -0.5$ V, $k_n' = 80$ $\mu$A/V$^2$, $k_p' = 40$ $\mu$A/V$^2$, and $\lambda_n = \lambda_p = 0.01$ V$^{-1}$.

**Solution (Differential Pair):** The differential gain of the diff-amp is given by

$$A_d = g_m(r_{o8} \| r_{o10})$$

We find

$$r_{o8} = r_{o10} = \frac{1}{\lambda I_D} = \frac{1}{(0.01)(0.05)} = 2000 \text{ k}\Omega$$

Then, for $A_d = 600$, we have

$$600 = g_m(2000 \| 2000)$$

which yields $g_m = 0.6$ mA/V. Then

$$g_m = 2\sqrt{\frac{k_n'}{2}\frac{W}{L}I_D} = 0.6 = 2\sqrt{\left(\frac{0.08}{2}\right)\left(\frac{W}{L}\right)_7 (0.05)}$$

which yields

$$\left(\frac{W}{L}\right)_7 = \left(\frac{W}{L}\right)_8 = 45$$

We will also, somewhat arbitrarily, make the width-to-length ratios of all other transistors, except $M_5$, $M_6$, and $M_{13}$, the same value of 45.

**Solution (Current Source):** We need to consider the two transistors $M_5$ and $M_6$. The gate-to-source voltages of $M_1$ and $M_3$ are found from

$$I_{REF} = 100 = \frac{80}{2}(45)(V_{GS1} - 0.5)^2$$

which yields $V_{GS1} = V_{GS3} = 0.736$ V. Since $M_5$ and $M_6$ are matched, we find

$$V_{GS5} = V_{GS6} = \frac{2.5 - (-2.5) - 2(0.736)}{2} = 1.76 \text{ V}$$

The width-to-length ratios of $M_5$ and $M_6$ are found from

$$I_{REF} = 100 = \frac{80}{2}\left(\frac{W}{L}\right)_{5,6}(1.76 - 0.5)^2$$

which yields

$$\left(\frac{W}{L}\right)_5 = \left(\frac{W}{L}\right)_6 = 1.57$$

**Solution (Second Stage):** The source-to-gate voltage applied to the common-source transistor $M_{13}$ is equal to the source-to-drain voltage on $M_{10}$ which is the same as the source-to-gate voltage of $M_9$ since the diff-amp is balanced. We find

$$I_{D9} = 50 = \frac{40}{2}(45)(V_{SG9} - 0.5)^2$$

or

$$V_{SG9} = V_{SG13} = 0.736 \text{ V}$$

The drain current in $M_{13}$ is $I_{D13} = I_Q = 100 \ \mu$A because of the matched transistors in the current source circuit. We now find

$$\frac{I_{D13}}{I_{D9}} = \frac{(W/L)_{13}}{(W/L)_9} \Rightarrow \frac{100}{50} = \frac{(W/L)_{13}}{45}$$

which yields $(W/L)_{13} = 90$. The gain of the second stage is given by

$$A_2 = -g_{m13}(r_{o13} \| R_{o12})$$

We find

$$g_{m13} = 2\sqrt{\frac{k_p'}{2}\left(\frac{W}{L}\right)_{13} I_{D13}} = 2\sqrt{\left(\frac{0.04}{2}\right)(90)(0.1)} = 0.849 \text{ mA/V}$$

and

$$r_{o13} = \frac{1}{\lambda I_{D13}} = \frac{1}{(0.01)(0.1)} = 1000 \text{ k}\Omega$$

The output resistance $R_{o12}$ is given by

$$R_{o12} = r_{o12} + r_{o11}(1 + g_{m12}r_{o12})$$

We find

$$r_{o11} = r_{o12} = \frac{1}{\lambda I_{D12}} = \frac{1}{(0.01)(0.1)} = 1000\,\text{k}\Omega$$

and

$$g_{m12} = 2\sqrt{\frac{k_n'}{2}\left(\frac{W}{L}\right)_{12} I_{D12}} = 2\sqrt{\left(\frac{0.08}{2}\right)(45)(0.1)} = 0.849\,\text{mA/V}$$

Then

$$R_{o12} = 1000 + 1000[1 + (0.849)(1000)] = 851{,}000\,\text{k}\Omega$$

The second stage voltage gain is then

$$A_2 = -0.849(1000\|851{,}000) = -849$$

**Solution (Overall Voltage Gain):** Since there is no loading of the second stage on the diff-amp circuit, the overall voltage gain is

$$A_v = A_d A_2 = (600)(-849) = -5.094 \times 10^5$$

**Comment:** We may note that the amplifier we have just designed is an all MOSFET circuit. The circuit contains no resistors. An all-transistor circuit is one of the advantages of MOS transistors.

We may also note that a large voltage gain can be obtained from a circuit using active loads.

## 11.10   SUMMARY

- The ideal transistor differential amplifier amplifies only the difference between two input signals.
- The differential-mode input voltage is defined as the difference between the two input signals and the common-mode input voltage is defined as the average of the two input signals.
- When a differential input voltage is applied, one transistor of the differential pair turns on more than the second transistor of the differential pair so that the currents become unbalanced, producing a signal output voltage.
- A common-mode output signal is generated when the output resistance of the current source is finite rather than ideally infinite.
- The common-mode rejection ratio, CMRR, is defined in terms of decibels as $\text{CMRR}_{\text{dB}} = 20\log_{10}|A_d/A_{cm}|$, where $A_d$ and $A_{cm}$ are the differential-mode voltage gain and common-mode voltage gain, respectively.
- Differential amplifiers are usually designed with active loads to increase the differential-mode voltage gain.
- BiCMOS circuits may be designed to incorporate the best parameters and characteristics of BJTs and MOSFETs in the same circuit.
- A BJT Darlington pair is typically used as a second stage in a BJT diff-amp. The input impedance is large, which tends to minimize loading effects on the diff-amp, and the effective current gain of the pair is the product of the individual gains.
- As an application, a CMOS diff-amp with an output gain stage was designed to meet a set of specifications.

 CHECKPOINT

After studying this chapter, the reader should have the ability to:

✓ Describe the mechanism by which a differential-mode signal and common-mode signal are produced in a BJT diff-amp.
✓ Describe the dc transfer characteristics of a BJT diff-amp.
✓ Define common-mode rejection ratio.
✓ Describe the mechanism by which a differential-mode signal and common-mode signal are produced in a MOSFET diff-amp.
✓ Describe the dc transfer characteristics of a MOSFET diff-amp.
✓ Design a MOSFET diff-amp with an active load to yield a specified differential-mode voltage gain.
✓ Analyze a simplified BJT operational amplifier circuit.
✓ Design a simplified MOSFET operational amplifier circuit.

## REVIEW QUESTIONS

1. Define differential-mode and common-mode input voltages.
2. Sketch the dc transfer characteristics of a BJT differential amplifier.
3. From the dc transfer characteristics, qualitatively define the linear region of operation for a differential amplifier.
4. What is meant by matched transistors and why are matched transistors important in the design of diff-amps?
5. Explain how a differential-mode output signal is generated.
6. Explain how a common-mode output signal is generated.
7. Define the common-mode rejection ratio, CMRR. What is the ideal value?
8. What design criteria will yield a large value of CMRR in an emitter-coupled pair?
9. Sketch the differential-mode and common-mode half-circuit models for an emitter-coupled diff-amp.
10. Define differential-mode and common-mode input resistances.
11. Sketch the dc transfer characteristics of a MOSFET differential amplifier.
12. Sketch and describe the advantages of a MOSFET cascode current source used with a MOSFET differential amplifier.
13. Sketch a simple MOSFET differential amplifier with an active load.
14. Explain the advantages of an active load.
15. Describe the loading effects of connecting a second stage to the output of a BJT diff-amp.
16. Explain the frequency response of the differential-mode voltage gain.
17. Sketch a BJT Darlington pair circuit and explain the advantages.
18. Describe the three stages of a simple BJT operational amplifier.

##  PROBLEMS

**Section 11.2 Basic BJT Differential Pair**

11.1 (a) A differential-amplifier has a differential-mode gain of $A_d = 250$ and a common-mode rejection ratio of $CMRR_{dB} = \infty$. A differential-mode input signal of $v_d = 1.5 \sin \omega t$ mV is applied along with a common-mode input signal of $v_{cm} = 3 \sin \omega t$ V. Assuming the common-mode gain is positive, determine the output voltage. (b) Repeat part (a) if the common-mode rejection ratio is $CMRR_{dB} = 80$ dB. (c) Repeat part (a) if the common-mode rejection ratio is $CMRR_{dB} = 50$ dB.

11.2 Consider the circuit shown in Figure P11.2. Assume $g_m = 1.0$ mA/V. Assume the input signal voltages are $v_1 = 0.7 + 0.1 \sin \omega t$ V and $v_2 = 0.7 - 0.1 \sin \omega t$ V. (a) Determine the signal voltages (i) $v_{o1}$, (ii) $v_{o2}$, and (iii) $v_{o1} - v_{o2}$. (b) Using the results of part (a), determine the small-signal voltage gains (i) $A_{d1} = \Delta v_{o1}/\Delta(v_1 - v_2)$, (ii) $A_{d2} = \Delta v_{o2}/\Delta(v_1 - v_2)$, and (iii) $A_{d3} = \Delta(v_{o1} - v_{o2})/\Delta(v_1 - v_2)$.

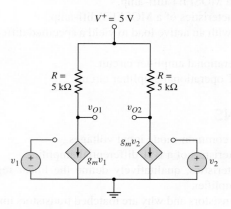

**Figure P11.2**

11.3 Consider the differential amplifier shown in Figure P11.3 with transistor parameters $\beta = 150$, $V_{BE}(on) = 0.7$ V, and $V_A = \infty$. (a) Design the circuit such that the $Q$-point values are $I_{C1} = I_{C2} = 100$ $\mu$A and $v_{O1} = v_{O2} = 1.2$ V for $v_1 = v_2 = 0$. (b) Draw the dc load line and plot the $Q$-point for transistor $Q_2$. (c) What are the maximum and minimum values of the common-mode input voltage?

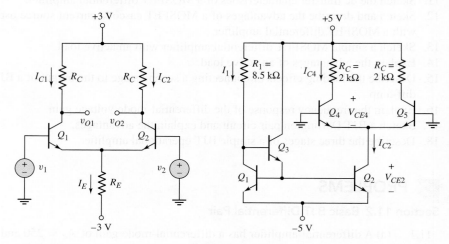

**Figure P11.3**                    **Figure P11.4**

11.4 The differential amplifier in Figure P11.4 is biased with a three-transistor current source. The transistor parameters are: $\beta = 100$, $V_{BE}(on) = 0.7$ V, and

$V_A = \infty$. (a) Determine $I_1$, $I_{C2}$, $I_{C4}$, $V_{CE2}$, and $V_{CE4}$. (b) Determine a new value of $R_1$ such that $V_{CE4} = 2.5$ V. What are the values of $I_{C4}$, $I_{C2}$, $I_1$, and $R_1$?

*D11.5  For the transistors in the circuit in Figure P11.5, the parameters are $\beta = 100$ and $V_{BE}(\text{on}) = 0.7$ V. The Early voltage is $V_A = \infty$ for $Q_1$ and $Q_2$, and is $V_A = 50$ V for $Q_3$ and $Q_4$. (a) Design resistor values such that $I_3 = 400 \ \mu A$ and $V_{CE1} = V_{CE2} = 10$ V. (b) Find $A_d$, $A_{cm}$, and $\text{CMRR}_{\text{dB}}$ for a one-sided output at $v_{O2}$. (c) Determine the differential- and common-mode input resistances.

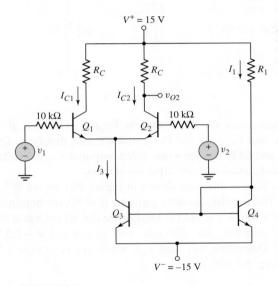

**Figure P11.5**

11.6  The diff-amp in Figure 11.3 of the text has parameters $V^+ = +5$ V, $V^- = -5$ V, $R_C = 8 \ k\Omega$, and $I_Q = 0.5$ mA. The transistor parameters are $\beta = 120$, $V_{BE}(\text{on}) = 0.7$ V, and $V_A = \infty$. (a) Using Figure 11.3(a), determine the maximum common-mode input voltage $v_{cm}$ that can be applied such that the transistors $Q_1$ and $Q_2$ remain biased in the active region. (b) Using Figure 11.3(b), determine the change in $v_{C2}$ from its dc value if $v_d = 18$ mV. (c) Repeat part (b) if $v_d = 10$ mV.

D11.7  The diff-amp configuration shown in Figure P11.7 is biased at $\pm 3$ V. The maximum power dissipation in the entire circuit is to be no more than 1.2 mW when $v_1 = v_2 = 0$. The available transistors have parameters: $\beta = 120$, $V_{BE}(\text{on}) = 0.7$ V, and $V_A = \infty$. Design the circuit to produce the maximum possible differential-mode voltage gain, but such that the common-mode input voltage can be within the range $-1 \leq v_{CM} \leq 1$ V and the transistors are still biased in the forward-active region. What is the value of $A_d$? What are the current and resistor values?

11.8  Consider the circuit in Figure P11.8, with transistor parameters: $\beta = 100$, $V_{BE}(\text{on}) = 0.7$ V, and $V_A = \infty$. (a) For $v_1 = v_2 = 0$, find $I_{C1}$, $I_{C2}$, $I_E$, $V_{CE1}$, and $V_{CE2}$. (b) Determine the maximum and minimum values of the common-mode input voltage. (c) Calculate $A_d$ for a one-sided output at the collector of $Q_2$.

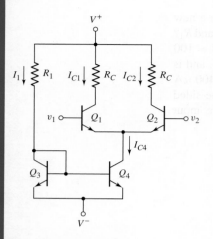

Figure P11.7

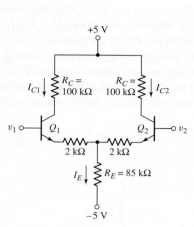

Figure P11.8

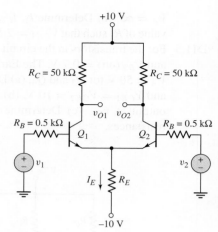

Figure P11.9

*11.9    The transistor parameters for the circuit in Figure P11.9 are: $\beta = 100$, $V_{BE}(\text{on}) = 0.7$ V, and $V_A = \infty$. (a) Determine $R_E$ such that $I_E = 150\ \mu\text{A}$. (b) Find $A_d$, $A_{cm}$, and $\text{CMRR}_{\text{dB}}$ for a one-sided output at $v_{O2}$. (c) Determine the differential- and common-mode input resistances.

11.10    The bias voltages for the diff-amp shown in Figure P11.10 are $V^+ = 3$ V and $V^- = -3$ V. The transistor current gains are $\beta = 80$, the nominal value of $V_{BE}(\text{on})$ is 0.6 V, and $V_A = \infty$. (a) Design the circuit such that the quiescent collector currents are 50 $\mu\text{A}$ and $v_{C1} = v_{C2} = -1.5$ V for $v_1 = v_2 = 0$. (b) Determine $v_{C1}$ and $v_{C2}$ when (i) $v_1 = v_2 = 1$ V and (ii) $v_1 = 0.994$ V, $v_2 = 1.006$ V.

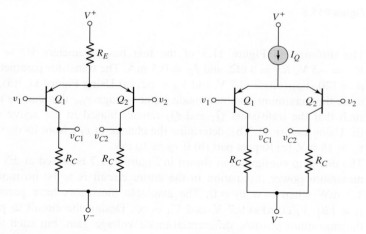

Figure P11.10                                    Figure P11.11

11.11    Consider the circuit shown in Figure P11.11. The circuit and transistor parameters are $V^+ = +3$ V, $V^- = -3$ V, $R_C = 360$ k$\Omega$, $I_Q = 12\ \mu\text{A}$, $\beta = 60$, $V_{EB}(\text{on}) = 0.6$ V, and $V_A = \infty$. The output resistance of the current source is $R_o = 4$ M$\Omega$. (a) Determine the $Q$-points of the transistors for $v_1 = v_2 = 0$. (b) Determine the differential- and common-mode voltage gains for (i) $v_O = v_{C1} - v_{C2}$ and (ii) $v_O = v_{C2}$.

11.12 The circuit and transistor parameters for the circuit shown in Figure P11.11 are $V^+ = 5$ V, $V^- = -5$ V, $I_Q = 0.2$ mA, $\beta = 80$, $V_{EB}(\text{on}) = 0.6$ V, and $V_A = \infty$. (a) Design the circuit such that the minimum common-mode voltage is $v_{cm} = -2.5$ V. (b) Using the results of part (a), what is the magnitude of the differential-mode gain, $A_d = |(v_{C1} - v_{C2})/(v_1 - v_2)|$? (c) Determine $v_{C1}$ and $v_{C2}$ for $v_1 = 0.507$ V and $v_2 = 0.493$ V. (d) What is the minimum output resistance of the current source such that $\text{CMRR}_{\text{dB}} = 60$ dB for a one-sided output.

11.13 Consider the differential amplifier shown in Figure P11.13 with mismatched collector resistors. The circuit and transistor parameters are $V^+ = 5$ V, $V^- = -5$ V, $\beta = 120$, $V_{BE}(\text{on}) = 0.7$ V, and $V_A = \infty$. (a) For $\Delta R = 0$, design the circuit such that $I_{CQ1} = I_{CQ2} = 120 \,\mu\text{A}$ and $v_{C1} = v_{C2} = 3$ V for $v_1 = v_2 = 0$. (b) Using the results of part (a), determine $|A_d|$ for a two-sided output. (c) For $\Delta R = 500\,\Omega$, determine $A_d$, $A_{cm}$, and $\text{CMRR}_{\text{dB}}$ for $v_o = \Delta(v_{C1} - v_{C2})$.

11.14 Consider the differential amplifier shown in Figure P11.14 with mismatched transistors. The mismatched transistors result in mismatched transconductances as shown. The circuit and transistor parameters are $V^+ = +10$ V, $V^- = -10$ V, $R_C = 50$ k$\Omega$, $R_E = 75$ k$\Omega$, $\beta = 120$, $V_{BE}(\text{on}) = 0.7$ V, and $V_A = \infty$. Determine $A_d$, $A_{cm}$, and $\text{CMRR}|_{\text{dB}}$ for $\Delta g_m / g_m = 0.01$ and for $v_O = v_{C1} - v_{C2}$. Assume $v_1 = v_2 = 0$ in the quiescent condition.

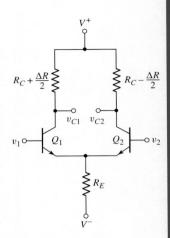

**Figure P11.13**

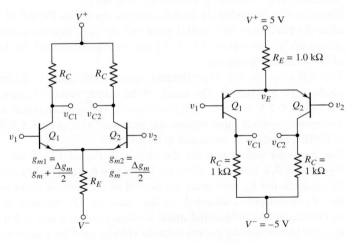

**Figure P11.14**          **Figure P11.15**

11.15 Consider the circuit in Figure P11.15. The transistor parameters are $\beta = 120$, $V_{EB}(\text{on}) = 0.7$ V, and $V_A = \infty$. Determine $v_E$, $v_{C1}$, and $v_{C2}$ for (a) $v_1 = v_2 = 0$; (b) $v_1 = 0.5$ V, $v_2 = 0$; and (c) $v_1 = 0$, $v_2 = 0.015$ V.

11.16 (a) Design the circuit shown in Figure P11.16 such that $v_O = v_{C1} - v_{C2} = 1$ V when $v_1 = -5$ mV and $v_2 = +5$ mV. The transistor parameters are $\beta = 180$, $V_{BE}(\text{on}) = 0.7$ V, and $V_A = \infty$. (b) Using the results of part (a), determine the maximum common-mode input voltage.

11.17 Consider the differential amplifier in Figure P11.17 with parameters $V^+ = 5$ V, $V^- = -5$ V, and $I_O = 0.8$ mA. Neglect base currents and assume $V_A = \infty$ for all transistors. The emitter currents can be written as

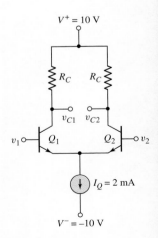

**Figure P11.16**

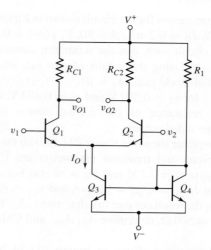

**Figure P11.17**

$I_{E1} = I_{S1}e^{V_{BE1}/V_T}$ and $I_{E2} = I_{S2}e^{V_{BE2}/V_T}$. (a) If $v_1 = v_2 = 0$ and $I_{S1} = I_{S2} = 3 \times 10^{-15}$ A, determine $(v_{O1} - v_{O2})$ for (i) $R_{C1} = R_{C2} = 7.5\,\text{k}\Omega$ and (ii) $R_{C1} = 7.4\,\text{k}\Omega$, $R_{C2} = 7.6\,\text{k}\Omega$. (b) Repeat part (a) for $I_{S1} = 2.9 \times 10^{-15}$ A and $I_{S2} = 3.1 \times 10^{-15}$ A.

11.18 For the diff-amp in Figure 11.2, determine the value of $v_d = v_1 - v_2$ that produces (a) $i_{C1} = 0.20 I_Q$ and (b) $i_{C2} = 0.90 I_Q$.

11.19 Consider the expanded dc transfer curves shown in Figure 11.6. Determine the maximum differential input voltage such that the actual curve is within (a) 0.5 percent of the ideal linear extrapolation and (b) 1.5 percent of the ideal extrapolation.

*D11.20 The diff-amp for the experimental system described in Example 11.4 needs to be redesigned. The range of the output voltage has increased to $-2 \le V_O \le 2$ V while the differential-mode voltage gain is still $A_d = 100$. The common-mode input voltage has increased to $v_{CM} = 3.5$ V. The value of CMRR needs to be increased to 80 dB.

*11.21 The transistor parameters for the circuit in Figure P11.9 are: $\beta = 120$, $V_{BE}(\text{on}) = 0.7$ V, and $V_A = \infty$. (a) Determine $R_E$ such that $I_E = 0.25$ mA. (b) Assume the $R_B$ resistance connected to the base of $Q_2$ is zero while the $R_B$ resistance connected to the base of $Q_1$ remains at $0.5\,\text{k}\Omega$. (i) Determine the differential-mode voltage gain for a one-sided output at $v_{O2}$. (ii) Determine the common-mode voltage gain for a one-sided output at $v_{O2}$.

11.22 The circuit parameters of the diff-amp shown in Figure 11.2 are $V^+ = 3$ V, $V^- = -3$ V, and $I_Q = 0.25$ mA. Base currents are negligible and $V_A = \infty$ for each transistor. (a) Design the circuit such that a differential-mode output voltage of $v_o = v_{C1} - v_{C2} = 1.2$ V is produced when a differential-mode input voltage of $v_d = v_1 - v_2 = 16$ mV is applied. (b) What is the maximum possible common-mode input voltage that can be applied such that the input transistors remain biased in the forward-active mode? (c) For a one-sided output, what is the value of $\text{CMRR}_{\text{dB}}$ if the output resistance of the current source is $R_o = 4\,\text{M}\Omega$?

*11.23 Consider the circuit in Figure P11.23. Assume the Early voltage of $Q_1$ and $Q_2$ is $V_A = \infty$, and assume the current source $I_Q$ is ideal. Derive the

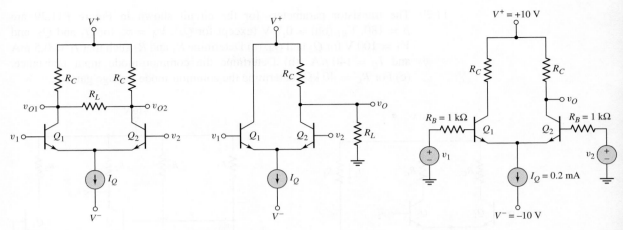

Figure P11.23  Figure P11.24  Figure P11.26

expressions for the one-sided differential-mode gain $A_{v1} = v_{o1}/v_d$ and $A_{v2} = v_{o2}/v_d$, and for the two-sided differential-mode gain $A_d = (v_{o2} - v_{o1})/v_d$.

11.24  The Early voltage of transistors $Q_1$ and $Q_2$ in the circuit in Figure P11.24 is $V_A = \infty$. Assuming an ideal current source $I_Q$, derive the expression for the differential-mode gain $A_d = v_o/v_d$.

*11.25  Consider the small-signal equivalent circuit of the differential-pair configuration shown in Figure 11.9. Derive the expressions for the differential- and common-mode voltage gains if the output is a two-sided output defined as $V_o = V_{c2} - V_{c1}$.

*D11.26  Consider a BJT diff-amp with the configuration in Figure P11.26. The signal sources have nonzero source resistances as shown. The transistor parameters are: $\beta = 150$, $V_{BE}(\text{on}) = 0.7$ V, and $V_A = \infty$. The range of the common-mode input voltage is to be $-3 \leq v_{CM} \leq 3$ V and the CMRR is to be 75 dB. (a) Design the diff-amp to produce the maximum possible differential-mode voltage gain. (b) Design the current source to produce the desired bias current and CMRR.

11.27  The bridge circuit in Figure P11.27 is a temperature transducer in which the resistor $R_A$ is a thermistor (a resistor whose resistance varies with temperature). The value of $\delta$ varies over the range of $-0.01 \leq \delta \leq 0.01$ as temperature varies over a particular range. Assume the value of $R = 40\,\text{k}\Omega$. The bridge circuit is to be connected to the diff-amp in Figure 11.2. The diff-amp circuit parameters are $V^+ = 5$ V, $V^- = -5$ V, $I_Q = 0.2$ mA, and $R_C = 15\,\text{k}\Omega$. The transistor parameters are $\beta = 120$, $V_{BE}(\text{on}) = 0.7$ V, and $V_A = \infty$. Terminal A of the bridge circuit is connected to the base of $Q_1$ and terminal B is connected to the base of $Q_2$. Determine the range of output voltage $v_{O2}$ as $\delta$ changes. [Hint: Make a Thevenin equivalent circuit at terminals A and B of the bridge circuit.]

11.28  A diff-amp is biased with a constant-current source $I_Q = 0.25$ mA that has an output resistance of $R_o = 8\,\text{M}\Omega$. The bipolar transistor parameters are $\beta = 120$ and $V_A = \infty$. Determine (a) the differential-mode input resistance and (b) the common-mode input resistance.

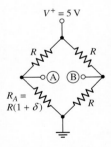

Figure P11.27

**11.29** The transistor parameters for the circuit shown in Figure P11.29 are $\beta = 180$, $V_{BE}(\text{on}) = 0.7$ V (except for $Q_4$), $V_A = \infty$ for $Q_1$ and $Q_2$, and $V_A = 100$ V for $Q_3$ and $Q_4$. (a) Determine $R_1$ and $R_2$ such that $I_1 = 0.5$ mA and $I_Q = 140\ \mu$A. (b) Determine the common-mode input resistance. (c) For $R_C = 40$ k$\Omega$, determine the common-mode voltage gain.

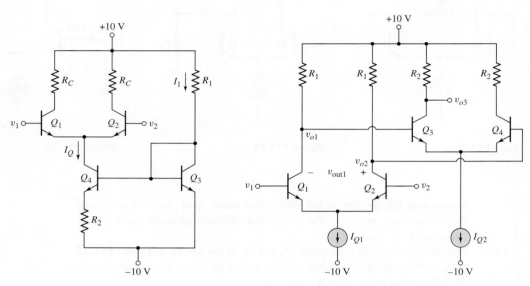

Figure P11.29                  Figure P11.30

**D11.30** Figure P11.30 shows a two-stage cascade diff-amp with resistive loads. Power supply voltages of $\pm 10$ V are available. Assume transistor parameters of: $\beta = 100$, $V_{BE}(\text{on}) = 0.7$ V, and $V_A = \infty$. Design the circuit such that the two-sided differential-mode voltage gain is $A_{d1} = (v_{o2} - v_{o1})/(v_1 - v_2) = 20$ for the first stage, and that the one-sided differential-mode voltage gain is $A_{d2} = v_{o3}/(v_{o2} - v_{o1}) = 30$ for the second stage. The circuit is to be designed such that the maximum differential-mode voltage swing is obtained in each stage.

### Section 11.3   Basic FET Differential Pair

**11.31** For the differential amplifier in Figure P11.31 the parameters are $R_1 = 50$ k$\Omega$ and $R_D = 24$ k$\Omega$. The transistor parameters are: $K_n = 0.25$ mA/V$^2$, $\lambda = 0$, and $V_{TN} = 2$ V. (a) Determine $I_1$, $I_Q$, $I_{D1}$, $V_{DS1}$, and $V_{DS4}$ when $v_1 = v_2 = 0$. (b) Draw the dc load line and plot the $Q$-point for transistor $M_2$. (c) What are the maximum and minimum values of the common-mode input voltage?

**11.32** The bias voltages in the diff-amp shown in Figure P11.31 are changed to $V^+ = 3$ V and $V^- = -3$ V. The transistor parameters are $K_{n1} = K_{n2} = 100\ \mu$A/V$^2$, $K_{n3} = K_{n4} = 200\ \mu$A/V$^2$, $\lambda_1 = \lambda_2 = 0$, $\lambda_3 = \lambda_4 = 0.01$ V$^{-1}$ and $V_{TN} = 0.3$ V (all transistors). (a) Design the circuit such that $V_{DS1} = V_{DS2} = 4$ V and $I_{D1} = I_{D2} = 60\ \mu$A when $v_1 = v_2 = -1.15$ V. (i) What are the values of $I_Q$ and $I_1$? (ii) What are the values of $R_D$ and $R_1$? (iii) What are the values of $V_{GS1}$ and $V_{GS4}$? (b) Calculate the change in $I_Q$ if $v_1 = v_2 = +1.15$ V.

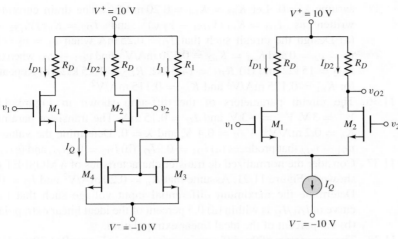

**Figure P11.31**

**Figure P11.33**

11.33 The transistor parameters for the differential amplifier shown in Figure P11.33 are $V_{TN} = 0.5$ V, $k_n' = 80 \ \mu\text{A/V}^2$, $W/L = 4$, and $\lambda = 0$. (a) Find $R_D$ and $I_Q$ such that $I_{D1} = I_{D2} = 80 \ \mu\text{A}$ and $v_{O2} = 2$ V when $v_1 = v_2 = 0$. (b) Draw the dc load line, and plot the $Q$-point for $M_2$. (c) What is the maximum common-mode input voltage?

11.34 The diff-amp in Figure P11.34 has parameters $V^+ = 3$ V, $V^- = -3$ V, and $I_Q = 0.18$ mA. The transistor parameters are $V_{TN} = 0.35$ V, $k_n' = 100 \ \mu\text{A/V}^2$, $W/L = 4$, and $\lambda = 0$. (a) Using Figure P11.34(a), determine $R_D$ such that the maximum value of the common-mode input voltage is $v_{cm} \text{(max)} = 2.25$ V. The input transistors $M_1$ and $M_2$ must remain biased in the saturation region. (b) Using Figure P11.34(b), determine the value of $v_{D2}$ for (i) $v_d = 0$, (ii) $v_d = +120$ mV, and (iii) $v_d = -50$ mV.

11.35 The bias voltages of the diff-amp shown in Figure P11.35 are $V^+ = 5$ V and $V^- = -5$ V. The threshold voltage of each transistor is $V_{TN} = 0.4$ V and

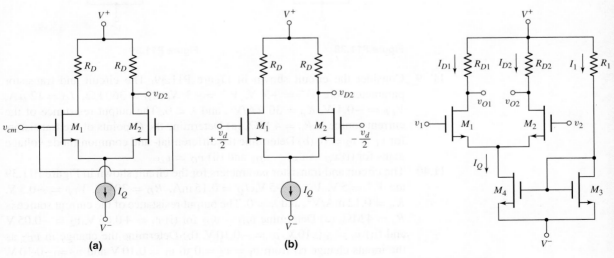

**(a)**

**(b)**

**Figure P11.34**

**Figure P11.35**

assume $\lambda = 0$. Let $K_{n3} = K_{n4} = 0.20\,\text{mA/V}^2$. The drain currents can be written as $I_{D1} = K_{n1}(V_{GS1} - V_{TN})^2$ and $I_{D2} = K_{n2}(V_{GS2} - V_{TN})^2$. (a) Design the circuit such that $I_Q = 0.25\,\text{mA}$ when $v_1 = v_2 = 0$. (b) If $v_1 = v_2 = 0$ and $K_{n1} = K_{n2} = 0.120\,\text{mA/V}^2$, find $v_{O1} - v_{O2}$ when (i) $R_{D1} = R_{D2} = 15\,\text{k}\Omega$ and (ii) $R_{D1} = 14.5\,\text{k}\Omega$, $R_{D2} = 15.5\,\text{k}\Omega$. (c) Repeat part (b) for $K_{n1} = 0.125\,\text{mA/V}^2$ and $K_{n2} = 0.115\,\text{mA/V}^2$.

11.36  The circuit parameters of the diff-amp shown in Figure 11.19 are $V^+ = 3\,\text{V}$, $V^- = -3\,\text{V}$, and $I_Q = 0.15\,\text{mA}$. The transistor parameters are $K_n = 0.2\,\text{mA/V}^2$, $V_{TN} = 0.4\,\text{V}$, and $\lambda = 0$. Determine the value of $v_d = v_{G1} - v_{G2}$ that produces (a) $i_{D1} = 0.2I_Q$, (b) $i_{D2} = 0.8I_Q$, and (c) $i_{D1} = I_Q$.

11.37  Consider the normalized dc transfer characteristics of a MOSFET diff-amp shown in Figure 11.21. Assume that $K_n = 0.20\,\text{mA/V}^2$ and $I_Q = 0.10\,\text{mA}$. Determine the maximum differential input voltage such that the actual curve of $i_{D1}/I_Q$ is within (a) 0.5 percent of the ideal linear extrapolation and (b) 1.5 percent of the ideal linear extrapolation.

11.38  The parameters of the diff-amp circuit shown in Figure P11.38 are $V^+ = 9\,\text{V}$, $V^- = -9\,\text{V}$, $R_D = 510\,\text{k}\Omega$, and $R_S = 390\,\text{k}\Omega$. The transistor parameters are $V_{TP} = -0.8\,\text{V}$, $K_p = 50\,\mu\text{A/V}^2$, and $\lambda = 0$. Determine $v_{D1}$ and $v_{D2}$ for (a) $v_1 = v_2 = 1\,\text{V}$ and (b) $v_1 = 1.050\,\text{V}$, $v_2 = 0.950\,\text{V}$.

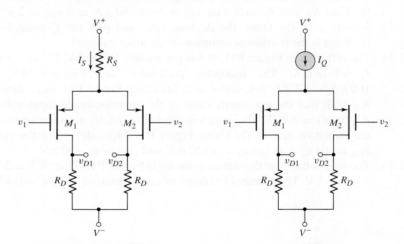

Figure P11.38                    Figure P11.39

11.39  Consider the circuit shown in Figure P11.39. The circuit and transistor parameters are $V^+ = +3\,\text{V}$, $V^- = -3\,\text{V}$, $R_D = 360\,\text{k}\Omega$, $I_Q = 12\,\mu\text{A}$, $V_{TP} = -0.4\,\text{V}$, $K_p = 30\,\mu\text{A/V}^2$, and $\lambda = 0$. The output resistance of the current source is $R_o = 4\,\text{M}\Omega$. (a) Determine the $Q$-points of the transistors for $v_1 = v_2 = 0$. (b) Determine the differential- and common-mode voltage gains for (i) $v_O = v_{D1} - v_{D2}$ and (ii) $v_O = v_{D2}$.

11.40  The circuit and transistor parameters for the circuit shown in Figure P11.39 are $V^+ = 5\,\text{V}$, $V^- = -5\,\text{V}$, $I_Q = 0.15\,\text{mA}$, $R_D = 30\,\text{k}\Omega$, $V_{TP} = -0.5\,\text{V}$, $K_p = 0.12\,\text{mA/V}^2$, and $\lambda = 0$. The output resistance of the current source is $R_o = 4\,\text{M}\Omega$. (a) Determine $v_{D1} - v_{D2}$ for (i) $v_1 = +0.05\,\text{V}$, $v_2 = -0.05\,\text{V}$ and (ii) $v_1 = +0.10\,\text{V}$, $v_2 = -0.10\,\text{V}$. (b) Determine the change in $v_{D2}$ as the inputs change (i) from $v_1 = v_2 = 0$ to $v_1 = 0.10\,\text{V}$ and $v_2 = -0.10\,\text{V}$, and (ii) from $v_1 = v_2 = 0$ to $v_1 = 1.10\,\text{V}$ and $v_2 = 0.90\,\text{V}$.

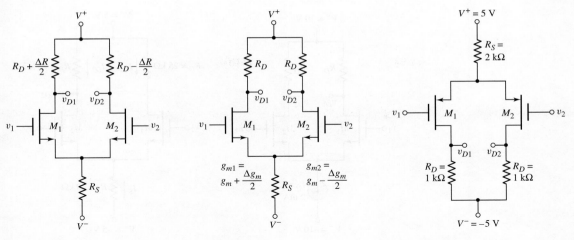

Figure P11.41          Figure P11.42          Figure P11.43

11.41 Consider the differential amplifier shown in Figure P11.41 with mismatched drain resistors. The circuit and transistor parameters are $V^+ = +10$ V, $V^- = -10$ V, $R_D = 50\,\text{k}\Omega$, $R_S = 75\,\text{k}\Omega$, $V_{TN} = 1$ V, $K_n = 0.15\,\text{mA/V}^2$, and $\lambda = 0$. Determine $A_d$, $A_{cm}$, and CMRR|$_{\text{dB}}$ for $\Delta R = 500\,\Omega$ and for $v_O = v_{D1} - v_{D2}$. Assume $v_1 = v_2 = 0$ in the quiescent condition.

11.42 Consider the differential amplifier shown in Figure P11.42 with mismatched transistors. The mismatched transistors result in mismatched transconductances as shown. The circuit and transistor parameters are $V^+ = +10$ V, $V^- = -10$ V, $R_D = 50\,\text{k}\Omega$, $R_S = 75\,\text{k}\Omega$, $V_{TN} = 1$ V, $K_n = 0.15\,\text{mA/V}^2$, and $\lambda = 0$. Determine $A_d$, $A_{cm}$, and CMRR|$_{\text{dB}}$ for $\Delta g_m/g_m = 0.01$ and for $v_O = v_{D1} - v_{D2}$. Assume $v_1 = v_2 = 0$ in the quiescent condition.

11.43 Consider the circuit in Figure P11.43. The transistor parameters are $K_p = 1.2\,\text{mA/V}^2$, $V_{TP} = -0.6$ V, and $\lambda = 0$. Determine $v_S$, $v_{D1}$, and $v_{D2}$ for (a) $v_1 = v_2 = 0$; (b) $v_1 = v_2 = 1$ V; (c) $v_1 = -0.1$ V, $v_2 = 0.1$ V; and (d) $v_1 = 0.9$ V, $v_2 = 1.1$ V.

D11.44 (a) Design the circuit shown in Figure P11.44 such that $v_O = v_{D1} - v_{D2} = 1$ V when $v_1 = -50$ mV and $v_2 = +50$ mV. The transistor parameters are $V_{TN} = 0.8$ V, $K_n = 0.4\,\text{mA/V}^2$, and $\lambda = 0$. (b) Using the results of part (a), determine the maximum common-mode input voltage.

*D11.45 The Hall effect experimental arrangement was described in Example 11.4. The required diff-amp is to be designed in the circuit configuration in Figure P11.35. The transistor parameters are $V_{TN} = 0.8$ V, $k'_n = 80\,\mu\text{A/V}^2$, $\lambda_1 = \lambda_2 = 0$, and $\lambda_3 = \lambda_4 = 0.01\,\text{V}^{-1}$. If the CMRR requirement cannot be met, a more sophisticated current source may have to be designed.

*11.46 Consider the diff-amp in Figure P11.46. The transistor parameters are: $K_{n1} = K_{n2} = 50\,\mu\text{A/V}^2$, $\lambda_1 = \lambda_2 = 0.02\,\text{V}^{-1}$, and $V_{TN1} = V_{TN2} = 1$ V. (a) Determine $I_S$, $I_{D1}$, $I_{D2}$, and $v_{O2}$ for $v_1 = v_2 = 0$. (b) Using the small-signal equivalent circuit, determine the differential-mode voltage gain $A_d = v_{o2}/v_d$, the common-mode voltage gain $A_{cm} = v_{o2}/v_{cm}$, and the CMRR$_{\text{dB}}$.

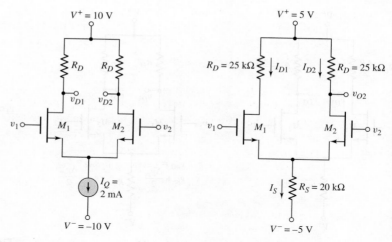

**Figure P11.44**                           **Figure P11.46**

**11.47**  Consider the circuit shown in Figure P11.47. Assume that $\lambda = 0$ for $M_1$
and $M_2$. Also assume an ideal current source $I_Q$. Derive the expression for
the one-sided differential mode gains $A_{d1} = v_{o1}/v_d$ and $A_{d2} = v_{o2}/v_d$,
and the two-sided differential-mode gain $A_d = (v_{o2} - v_{o1})/v_d$.

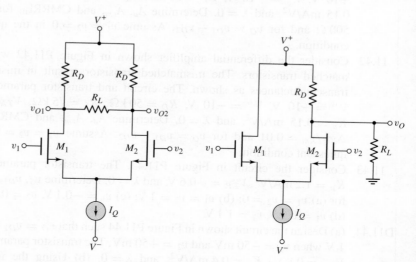

**Figure P11.47**                           **Figure P11.48**

**11.48**  Consider the diff-amp shown in Figure P11.48. Assume $\lambda_1 = \lambda_2 = 0$ and
assume the current source has an output resistance of $R_o$. (a) Derive the
expression for the differential-mode voltage gain $A_d = \Delta v_O/v_d$ where
$v_d = v_1 - v_2$. (b) Derive the expression for the common-mode voltage
gain $A_{cm} = \Delta v_O/v_{cm}$ where $v_{cm} = (v_1 + v_2)/2$.

**11.49**  The bias voltages of the diff-amp circuit shown in Figure 11.19 are
$V^+ = 5$ V and $V^- = -5$ V, and the bias current is $I_Q = 0.2$ mA. The transistor parameters are $V_{TN} = 0.4$ V, $K_n = 0.15$ mA/V$^2$, and $\lambda = 0$. (a) Design the circuit such that a differential-mode output voltage of $\Delta v_O = 0.5$ V is

produced when a differential-mode input voltage of $v_d = v_1 - v_2 = 100\,\text{mV}$ is applied. (b) Using the results of part (a), determine the maximum possible common-mode input voltage that can be applied such that the transistors remain biased in the saturation region.

11.50 Consider the small-signal equivalent circuit in Figure 11.23. Assume the output is a two-sided output defined as $V_o = V_{d2} - V_{d1}$, where $V_{d2}$ and $V_{d1}$ are the signal voltages at the drains of $M_2$ and $M_1$, respectively. Derive expressions for the differential- and common-mode voltage gains.

11.51 Consider the MOSFET diff-amp with the configuration in Figure P11.33. The circuit parameters are $V^+ = 3\,\text{V}$, $V^- = -3\,\text{V}$, and $I_Q = 0.2\,\text{mA}$. The transistor parameters are $V_{TN} = 0.4\,\text{V}$, $k'_n = 100\,\mu\text{A/V}^2$, $W/L = 10$, and $\lambda = 0$. The range of the common-mode input voltage is to be $-1.5 \leq v_{cm} \leq +1.5\,\text{V}$, and the common-mode rejection ratio is to be $\text{CMRR}_{\text{dB}} = 50\,\text{dB}$. (a) Design the diff-amp to produce the maximum possible differential-mode voltage gain. (b) Design an all MOSFET current source to produce the desired bias current and CMRR. The minimum $W/L$ ratio of any transistor is to be 0.8, and assume $\lambda = 0.02\,\text{V}^{-1}$ for all transistors in the current source circuit.

11.52 Consider the bridge circuit and diff-amp described in Problem 11.27. The BJT diff-amp is to be replaced with a MOSFET diff-amp as shown in Figure 11.19. The transistor parameters are $V_{TN} = 0.4\,\text{V}$, $K_n = 1\,\text{mA/V}^2$, and $\lambda = 0$. The bias voltages of the MOSFET diff-amp are $V^+ = 5\,\text{V}$ and $V^- = -5\,\text{V}$, and the reference current is $I_Q = 0.2\,\text{mA}$. Let $R_D = 20\,\text{k}\Omega$. Terminal A of the bridge circuit is to be connected to the gate of $M_1$ and terminal B is to be connected to the gate of $M_2$. (a) Determine the range of output voltage $v_O$ as $\delta$ changes. (b) Explain the advantages and disadvantages of this circuit configuration compared to that in Problem 11.27.

*D11.53 Figure P11.53 shows a two-stage cascade diff-amp with resistive loads. Power supply voltages of $\pm 10\,\text{V}$ are available. Assume transistor parameters of $V_{TN} = 1\,\text{V}$, $k'_n = 60\,\mu\text{A/V}^2$, and $\lambda = 0$. Design the circuit such

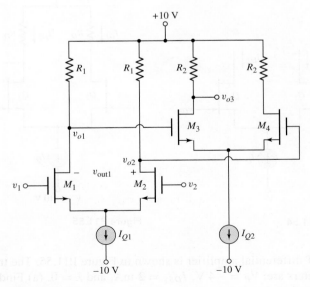

**Figure P11.53**

that the two-sided differential-mode voltage gain is $A_{d1} = (v_{o2} - v_{o1})/(v_1 - v_2) = 20$ for the first stage, and that the one-sided differential-mode voltage gain is $A_{d2} = v_{o3}/(v_{o2} - v_{o1}) = 30$ for the second stage. The circuit is to be designed such that the maximum differential-mode voltage swing is obtained in each stage.

*11.54   Figure P11.54 shows a matched JFET differential pair biased with a current source $I_Q$. (a) Starting with

$$i_D = I_{DSS}\left(1 - \frac{v_{GS}}{V_P}\right)^2$$

show that

$$\frac{i_{D1}}{I_Q} = \frac{1}{2} + \left(\frac{1}{-2V_P}\right) v_d \sqrt{2\left(\frac{I_{DSS}}{I_Q}\right) - \left(\frac{I_{DSS}}{I_Q}\right)^2 \left(\frac{v_d}{V_P}\right)^2}$$

and

$$\frac{i_{D2}}{I_Q} = \frac{1}{2} - \left(\frac{1}{-2V_P}\right) v_d \sqrt{2\left(\frac{I_{DSS}}{I_Q}\right) - \left(\frac{I_{DSS}}{I_Q}\right)^2 \left(\frac{v_d}{V_P}\right)^2}$$

(b) Show that the $I_Q$ bias current is switched entirely to one transistor or the other when

$$|v_d| = |V_P|\sqrt{\frac{I_Q}{I_{DSS}}}$$

(c) Show that the maximum forward transconductance is given by

$$g_f(\text{max}) = \left.\frac{di_{D1}}{dv_d}\right|_{v_d=0} = \left(\frac{1}{-V_P}\right)\sqrt{\frac{I_Q \cdot I_{DSS}}{2}}$$

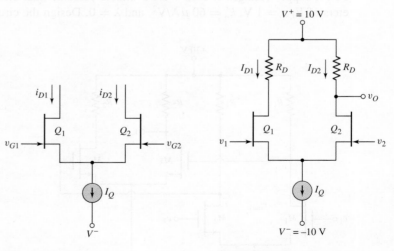

Figure P11.54                    Figure P11.55

11.55   A JFET differential amplifier is shown in Figure P11.55. The transistor parameters are: $V_P = -4$ V, $I_{DSS} = 2$ mA, and $\lambda = 0$. (a) Find $R_D$ and $I_Q$ such that $I_{D1} = I_{D2} = 0.5$ mA and $v_{o2} = 7$ V when $v_1 = v_2 = 0$.

(b) Calculate the maximum forward transconductance. (c) Determine the one-sided differential-mode voltage gain $A_d = v_o/v_d$.

*11.56   Consider the JFET diff-amp shown in Figure P11.56. The transistor parameters are: $I_{DSS} = 0.8$ mA, $\lambda = 0$, and $V_P = -2$ V. (a) Determine $I_S$, $I_{D1}$, $I_{D2}$, and $v_{o2}$ for $v_1 = v_2 = 0$. (b) Using the small-signal equivalent circuit, determine the differential-mode voltage gain $A_d = v_{o2}/v_d$, the common-mode voltage gain $A_{cm} = v_o/v_{cm}$, and the CMRR$_{dB}$.

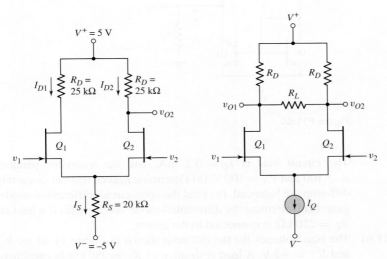

Figure P11.56          Figure P11.57

*11.57   Consider the circuit in Figure P11.57. Assume that $\lambda = 0$ for the transistors, and assume an ideal current source $I_Q$. Derive the expressions for the one-sided differential-mode gains $A_{d1} = v_{o1}/v_d$ and $A_{d2} = v_{o2}/v_d$, and for the two-sided differential-mode gain $A_d = (v_{o2} - v_{o1})/v_d$.

## Section 11.4 Differential Amplifier with Active Load

11.58   The circuit parameters for the diff-amp shown in Figure 11.30 are $V^+ = 3.3$ V, $V^- = -3.3$ V, and $I_Q = 0.4$ mA. The transistor parameters are $\beta = 120$, $V_{A1} = V_{A2} = 120$ V, $V_{A3} = V_{A4} = 80$ V, and $V_{A5} = \infty$. (a) Determine the open-circuit differential-mode voltage gain. (b) What is the output resistance of the diff-amp? (c) Find the value of load resistance $R_L$ that reduces the differential-mode gain to 75 percent of the open-circuit value.

D11.59   Design a differential amplifier with the configuration shown in Figure 11.28 incorporating a basic two-transistor current source to establish $I_Q$. The bias voltages are to be $V^+ = +5$ V and $V^- = -5$ V, the bias current is to be $I_Q = 250$ $\mu$A, and the available transistors have parameters $\beta = 180$, $V_{BE}(\text{on}) = V_{EB}(\text{on}) = 0.7$ V, $V_{AN} = 150$ V, and $V_{AP} = 100$ V. (a) Show the complete circuit. (b) What is the open-circuit differential-mode voltage gain. (c) Determine the differential-mode input resistance and the output resistance. (d) Determine the common-mode input voltage range.

11.60   The differential amplifier shown in Figure P11.60 has a pair of pnp bipolars as input devices and a pair of npn bipolars connected as an active load.

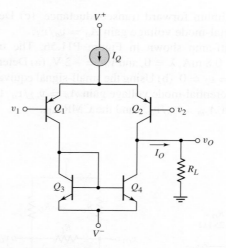

**Figure P11.60**

The circuit bias is $I_Q = 0.2$ mA, and the transistor parameters are $\beta = 100$ and $V_A = 100$ V. (a) Determine $I_0$ such that the dc currents in the diff-amp are balanced. (b) Find the open-circuit differential-mode voltage gain. (c) Determine the differential-mode voltage gain if a load resistance $R_L = 250$ k$\Omega$ is connected to the output.

11.61 The bias voltages for the diff-amp shown in Figure 11.30 are $V^+ = 5$ V and $V^- = -5$ V. A load resistance of $R_L = 250$ k$\Omega$ is capacitively coupled to the output. The transistor parameters are $\beta = 120$, $V_{A1} = V_{A2} = 90$ V, and $V_{A3} = V_{A4} = 60$ V. (a) Determine the bias current $I_Q$ that will produce a differential-mode voltage gain of $A_d = 1000$. (b) If $V_{EB}(\text{on}) = 0.6$ V, what is the maximum common-mode voltage that can be applied such that all transistors are biased in the forward-active mode?

11.62 Consider the diff-amp shown in Figure P11.62. The circuit parameters are $V^+ = 3$ V, $V^- = -3$ V, and $I_Q = 0.4$ mA. The npn transistor parameters are $\beta_{\text{npn}} = 180$, $V_{BE}(\text{on}) = 0.7$ V, and $V_{AN} = 120$ V, and the pnp transistor parameters are $\beta_{\text{pnp}} = 120$, $V_{EB}(\text{on}) = 0.7$ V, and $V_{AP} = 80$ V. (a) Sketch the small-signal equivalent circuit for the diff-amp assuming an ideal differential-mode input signal. (b) Determine the one-sided differential-mode gain $A_{d1} = \Delta v_{O1}/v_d$. (c) Determine the one-sided differential-mode gain $A_{d2} = \Delta v_{O2}/v_d$. (d) Find the two-sided differential-mode gain $A_{d3} = \Delta (v_{O2} - v_{O1})/v_d$.

11.63 Consider the MOSFET diff-amp shown in Figure P11.63. The bias voltages are $V^+ = 3$ V and $V^- = -3$ V. The current source is $I_Q = 200$ $\mu$A and has an output resistance of $R_o = 2$ M$\Omega$. The transistor parameters are $V_{TN} = 0.4$ V, $V_{TP} = -0.4$ V, $K_n = K_p = 0.5$ mA/V$^2$, $\lambda_2 = 0.02$ V, V$^{-1}$, $\lambda_4 = 0.03$ V$^{-1}$ and $\lambda_1 = \lambda_3 = 0$. (a) Determine the voltage gain $A = v_o/v_d$ for $v_1 = v_d$ and $v_2 = 0$. (b) Determine the voltage gain $A = v_o/v_d$ for $v_1 = 0$ and $v_2 = -v_d$. (c) Determine the voltage gain $A = v_o/v_d$ for $v_1 = v_d/2$ and $v_2 = -v_d/2$.

11.64 The differential amplifier in Figure P11.64 has a pair of PMOS transistors as input devices and a pair of NMOS transistors connected as an active load. The circuit is biased with $I_Q = 0.2$ mA, and the transistor parameters are: $K_n = K_p = 0.1$ mA/V$^2$, $\lambda_n = 0.01$ V$^{-1}$, $\lambda_p = 0.015$ V$^{-1}$,

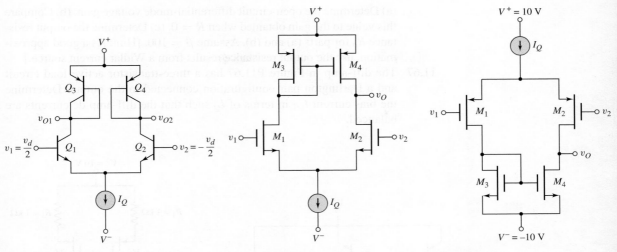

Figure P11.62                    Figure P11.63                    Figure P11.64

$V_{TN} = 1$ V, and $V_{TP} = -1$ V. (a) Determine the quiescent drain-to-source voltage in each transistor. (b) Find the open-circuit differential-mode voltage gain. (c) What is the output resistance?

11.65  The circuit parameters for the diff-amp shown in Figure 11.32 are $V^+ = 1.8$ V, $V^- = -1.8$ V, and $I_Q = 120\,\mu\text{A}$. The NMOS transistor parameters are $V_{TN} = 0.3$ V, $k_n' = 100\,\mu\text{A/V}^2$, $(W/L)_n = 8$, and $\lambda_n = 0.025$ V$^{-1}$. The parameters of the PMOS transistors are $V_{TP} = -0.3$ V, $k_p' = 40\,\mu\text{A/V}^2$, $(W/L)_p = 10$, and $\lambda_p = 0.04$ V$^{-1}$. (a) Determine the small-signal differential-mode voltage gain $A_d = v_o/v_d$. (b) What is the maximum common-mode voltage gain that can be applied such that all transistors are still biased in the saturation region?

*11.66  Consider the diff-amp with active load in Figure P11.66. The Early voltages are $V_{AN} = 120$ V for $Q_1$ and $Q_2$ and $V_{AP} = 80$ V for $Q_3$ and $Q_4$.

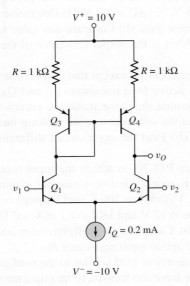

Figure P11.66

(a) Determine the open-circuit differential-mode voltage gain. (b) Compare this value to the gain obtained when $R = 0$. (c) Determine the output resistance $R_o$ for parts (a) and (b). Assume $\beta = 100$. [Hint: As a good approximation, use the output resistance results from a Widlar current source.]

11.67 The diff-amp in Figure P11.67 has a three-transistor active load circuit and a Darlington pair configuration connected to the output. Determine the bias current $I_{Q1}$ in terms of $I_Q$ such that the diff-amp dc currents are balanced.

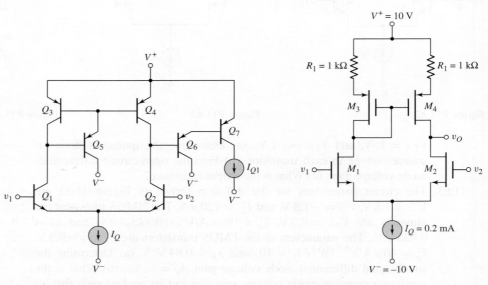

**Figure P11.67**                    **Figure P11.68**

*11.68 Consider the diff-amp in Figure P11.68. The PMOS parameters are: $K_p = 80\ \mu\text{A/V}^2$, $\lambda_p = 0.02\ \text{V}^{-1}$, $V_{TP} = -2\ \text{V}$. The NMOS parameters are: $K_n = 80\ \mu\text{A/V}^2$, $\lambda_n = 0.015\ \text{V}^{-1}$, $V_{TN} = +2\ \text{V}$. (a) Determine the open-circuit differential-mode voltage gain. (b) Compare this value to the gain obtained when $R_1 = 0$. (c) What is the output resistance of the diff-amp for parts (a) and (b)?

*11.69 Reconsider the circuit in Figure P11.60 except that 1 k$\Omega$ resistors are inserted at the emitters of the active load transistors $Q_3$ and $Q_4$ as in the circuit in Figure P11.66. Assume the same transistor parameters as in Problem 11.60. (a) Determine the output resistance looking into the output of the diff-amp circuit. (b) Find the open-circuit differential-mode voltage gain.

*11.70 Consider the circuit in Figure P11.70, in which the input transistors to the diff-amp are Darlington pairs. Assume transistor parameters of $\beta(\text{npn}) = 120$, $\beta(\text{pnp}) = 80$, $V_A(\text{npn}) = 100\ \text{V}$, and $V_A(\text{pnp}) = 80\ \text{V}$. Let the power supply voltages be $\pm 10\ \text{V}$ and let $I_Q = 1\ \text{mA}$. (a) Determine the output resistance $R_o$. (b) Calculate the differential-mode voltage gain. (c) Find the differential-mode input resistance $R_{id}$.

*D11.71 Design a BJT diff-amp with an active load similar to the configuration in Figure P11.70 except that the input devices are to be pnp transistors and

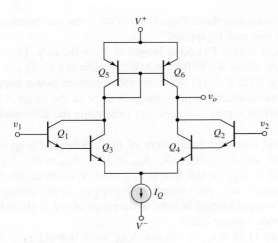

**Figure P11.70**

the active load will have npn transistors. Using the same parameters as in Problem 11.70, determine the small-signal differential-mode voltage gain.

D11.72  Reconsider the diff-amp specifications listed in Problem 11.45. Design an all-CMOS diff-amp with the configuration in Figure 11.32 to meet the specifications. The NMOS transistor parameters are $V_{TN} = 0.4$ V, $k'_n = 100 \, \mu A/V^2$, and $\lambda_n = 0.025 \, V^{-1}$. The parameters of the PMOS transistors are $V_{TP} = -0.4$ V, $k'_p = 40 \, \mu A/V^2$, and $\lambda_p = 0.04 \, V^{-1}$.

D11.73  An all-CMOS diff-amp, including the current source circuit, with the configuration in Figure 11.32 is to be designed to have a differential-mode gain of $A_d = 240$. The bias voltages are $V^+ = 3$ V and $V^- = -3$ V. The total power dissipation in the circuit is to be limited to 0.8 mW. Assume the NMOS transistor parameters are $V_{TN} = 0.4$ V, $k'_n = 100 \, \mu A/V^2$, and $\lambda_n = 0.02 \, V^{-1}$. Assume PMOS transistor parameters of $V_{TP} = -0.4$ V, $k'_p = 40 \, \mu A/V^2$, and $\lambda_p = 0.03 \, V^{-1}$.

D11.74  The differential amplifier with the configuration shown in Figure 11.36 is to be designed to achieve a differential-mode voltage gain of $A_d = 400$. The circuit parameters are to be $V^+ = +5$ V, $V^- = -5$ V, and $I_Q = 200 \, \mu A$. The available transistors have parameters for the PMOS of $V_{TP} = -0.5$ V, $k'_p = 40 \, \mu A/V^2$, and $\lambda_p = 0.02 \, V^{-1}$, and for the NMOS of $V_{TN} = +0.5$ V, $k'_n = 80 \, \mu A/V^2$, and $\lambda_n = 0.015 \, V^{-1}$.

*11.75  Consider the fully cascoded diff-amp in Figure 11.37. Assume $I_Q = 80 \, \mu A$ and transistor parameters of: $V_{TN} = 0.8$ V, $k'_n = 60 \, \mu A/V^2$, $\lambda_n = 0.015 \, V^{-1}$, $V_{TP} = -0.8$ V, $k'_p = 25 \, \mu A/V^2$, and $\lambda_p = 0.02 \, V^{-1}$. The transistor width-to-length ratios are $W/L = 60/4$ for transistors $M_1$–$M_4$, $W/L = 40/4$ for transistors $M_5$–$M_6$, and $W/L = 4/4$ for transistors $M_7$–$M_8$. (a) Determine the output resistance of the diff-amp. (b) Calculate the differential-mode voltage gain of the diff-amp. (c) Find the common-mode voltage gain of the diff-amp using a computer simulation.

11.76  Consider the diff-amp that was shown in Figure P11.63. The circuit and transistor parameters are $V^+ = 2.8$ V, $V^- = -2.8$ V, $I_Q = 120 \, \mu A$, $K_n = K_p = 0.2 \, mA/V^2$, $V_{TN} = +0.3$ V, $V_{TP} = -0.3$ V, and $\lambda_n = \lambda_p = 0.025 \, V^{-1}$. (a) Determine the differential-mode voltage gain. (b) What is the output resistance of the diff-amp? (c) What is the maximum common-mode voltage that may be applied?

11.77  The diff-amp in Figure P11.63 is biased at $I_Q = 0.5$ mA. The transistor parameters are $K_n = K_p = 0.25$ mA/V$^2$, $V_{TN} = 0.4$ V, $V_{TP} = -0.4$ V, and $\lambda_n = \lambda_p = 0.02$ V$^{-1}$. (a) What are the minimum power supply voltages if the common-mode input voltage is to be in the range $\pm 3$ V? Assume symmetrical supply voltages. (b) Determine the differential-mode voltage gain.

11.78  The circuit and transistor parameters of the bipolar diff-amp shown in Figure P11.78 are $I_Q = 200 \ \mu$A, $\beta_{npn} = 125$, $\beta_{pnp} = 80$, $V_{BE}$(on) = $V_{EB}$(on) = 0.7 V, $V_{AN} = 100$ V, and $V_{AP} = 60$ V. (a) What are the minimum power supply voltages (assume symmetrical supply voltages) if the common-mode input voltage is to be in the range of $\pm 2$ V? (b) What is the differential-mode voltage gain?

11.79  Repeat Problem 11.78 if $I_Q = 120 \ \mu$A, $V_{AN} = 75$ V, and $V_{AP} = 40$ V. All other parameters remain the same.

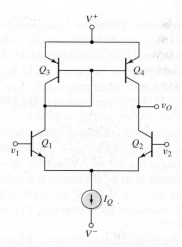

**Figure P11.78**

## Section 11.5   BiCMOS Circuits

11.80  (a) The Darlington pair circuit in Figure 11.45 has new bias current levels of $I_{BIAS1} = 0.25$ mA and $I_{BIAS2} = 0.50$ mA. The transistor parameters are $K_n = 0.2$ mA/V$^2$, $V_{TN} = 0.4$ V, and $\lambda = 0$ for $M_1$; and $\beta = 150$, $V_{BE}$(on) = 0.7 V, and $V_A = \infty$ for $Q_2$. Determine the small-signal parameters for each transistor and find the composite transconductance. (b) Repeat part (a) for bias currents of $I_{BIAS1} = 0.05$ mA and $I_{BIAS2} = 0.50$ mA.

11.81  Consider the BiCMOS diff-amp in Figure 11.44, biased at $I_Q = 0.4$ mA. The transistor parameters for $M_1$ and $M_2$ are: $K_n = 0.2$ mA/V$^2$, $V_{TN} = 1$ V, and $\lambda = 0.01$ V$^{-1}$. The parameters for $Q_1$ and $Q_2$ are: $\beta = 120$, $V_{EB}$(on) = 0.7 V, and $V_A = 80$ V. (a) Determine the differential-mode voltage gain. (b) If the output resistance of the current source is $R_o = 500$ k$\Omega$, determine the common-mode voltage gain using a computer simulation analysis.

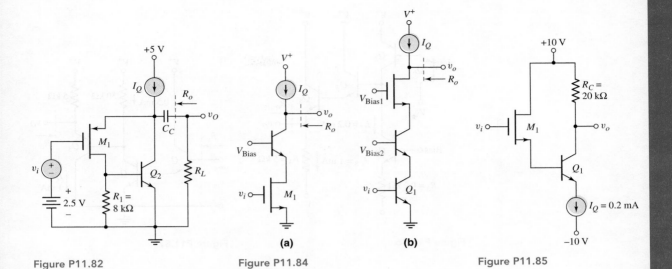

Figure P11.82         Figure P11.84         Figure P11.85

11.82   The BiCMOS circuit shown in Figure P11.82 is equivalent to a pnp bipolar transistor with an infinite input impedance. The bias current is $I_Q = 0.5$ mA. The MOS transistor parameters are $V_{TP} = -0.5$ V, $K_p = 0.7$ mA/V$^2$, and $\lambda = 0$, and the BJT parameters are $\beta = 180$, $V_{BE}$(on) $= 0.7$ V, and $V_A = \infty$. (a) Sketch the small-signal equivalent circuit. (b) Calculate the small-signal parameters for each transistor. (c) Determine the small-signal voltage gain $A_v = v_o/v_i$ for (i) $R_L = 10$ k$\Omega$ and (ii) $R_L = 100$ k$\Omega$.

11.83   The bias current in the BiCMOS circuit shown in Figure P11.82 is $I_Q = 0.8$ mA. The transistor parameters are the same as described in Problem 11.82. (a) Sketch the small-signal equivalent circuit and calculate the small-signal parameters for each transistor. (b) Determine the output resistance as defined in the figure.

*11.84   The bias current $I_Q$ is 25 $\mu$A in each circuit in Figure P11.84. The BJT parameters are $\beta = 100$ and $V_A = 50$ V, and the MOSFET parameters are $V_{TN} = 0.8$ V, $K_n = 0.25$ mA/V$^2$, and $\lambda = 0.02$ V$^{-1}$. Assume the two amplifying transistors $M_1$ and $Q_1$ are biased in the saturation region and forward-active region, respectively. Determine the small-signal voltage gain $A_v = v_o/v_i$ and the output resistance $R_o$ for each circuit.

11.85   For the circuit shown in Figure P11.85, determine the small-signal voltage gain, $A_v = v_o/v_i$. Assume transistor parameters of $V_{TN} = 1$ V, $K_n = 0.2$ mA/V$^2$, and $\lambda = 0$ for $M_1$ and $\beta = 80$ and $V_A = \infty$ for $Q_1$.

## Section 11.6 Gain Stage and Simple Output Stage

11.86   The output stage in the circuit shown in Figure P11.86 is a Darlington pair emitter-follower configuration. Assume $\beta = 120$ for all npn transistors and $\beta = 90$ for all pnp transistors. Let $V_{A7} = 60$ V for $Q_7$, $V_{A11} = 120$ V for $Q_{11}$, and $V_A = \infty$ for all other transistors. Determine the output resistance $R_o$.

*11.87   For the circuit in Figure P11.87, the transistor parameters are $\beta = 100$ and $V_A = \infty$. The bias currents in the transistors are indicated on the figure.

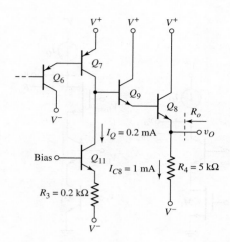

Figure P11.86

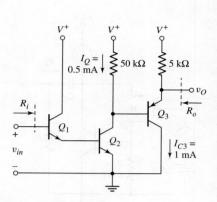

Figure P11.87

Determine the input resistance $R_i$, the output resistance $R_o$, and the small-signal voltage gain $A_v = v_o/v_{in}$.

11.88    Consider the circuit in Figure P11.88. The bias currents $I_1$ and $I_2$ are such that a zero dc output voltage is established. The transistor parameters are: $K_p = 0.2$ mA/V$^2$, $K_n = 0.5$ mA/V$^2$, $V_{TP} = -0.8$ V, $V_{TN} = +0.8$ V, and $\lambda_n = \lambda_p = 0.01$ V$^{-1}$. Determine the small-signal voltage gain $A_v = v_o/v_{in}$ and the output resistance $R_o$.

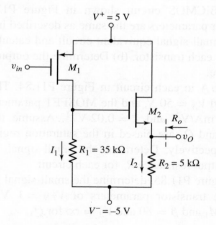

Figure P11.88

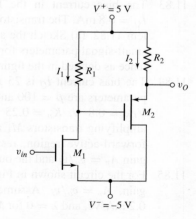

Figure P11.89

11.89    The bias currents in the circuit shown in Figure P11.89 are $I_1 = 0.25$ mA and $I_2 = 1.0$ mA. The transistor parameters are $K_n = 0.5$ mA/V$^2$, $K_p = 1.0$ mA/V$^2$, $V_{TN} = 0.8$ V, $V_{TP} = -0.8$ V, and $\lambda_n = \lambda_p = 0.02$ V$^{-1}$. (a) Determine the resistor values $R_1$ and $R_2$ such that the dc value of the output voltage is zero. (b) Sketch the small-signal equivalent circuit and find the small-signal transistor parameters. (c) Determine the small-signal voltage gain $A_v = v_o/v_{in}$. (d) Determine the output resistance $R_o$.

## Section 11.7 Simplified Op-Amp Circuits

*11.90    Consider the multistage bipolar circuit in Figure P11.90, in which dc base currents are negligible. Assume the transistor parameters are $\beta = 120$, $V_{BE}(\text{on}) = 0.7$ V, and $V_A = \infty$. The output resistance of the constant current source is $R_o = 200\,\text{k}\Omega$. (a) For $v_1 = v_2 = -1.5$ V, design the circuit such that $v_{O2} = v_O = 0$, $I_{CQ3} = 0.25\,\text{mA}$, and $I_{CQ4} = 2\,\text{mA}$. (b) Assuming $C_E$ acts as a short circuit, determine the differential-mode voltage gains $A_{d1} = v_{o2}/v_d$ and $A_d = v_o/v_d$. (c) Determine the common-mode gains $A_{cm1} = v_{o2}/v_d$ and $A_{cm} = v_o/v_d$, and the overall $\text{CMRR}_{\text{dB}}$.

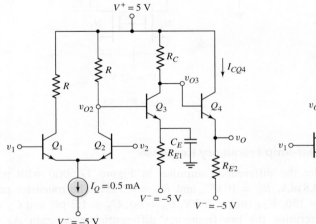

**Figure P11.90**

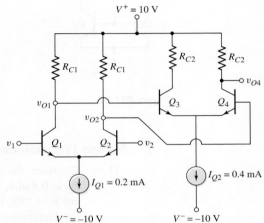

**Figure P11.91**

*D11.91    The circuit in Figure P11.91 has two bipolar differential amplifiers in cascade, biased with ideal current sources $I_{Q1}$ and $I_{Q2}$. Assume the transistor parameters are $\beta = 180$ and $V_A = \infty$. (a) Design the circuit such that $v_{o1} = v_{o2} = 2$ V and $v_{O4} = 6$ V when $v_1 = v_2 = 0$. (b) Determine the differential-mode voltage gains $A_{d1} = (v_{o1} - v_{o2})/v_d$ and $A_d = v_{o4}/v_d$.

*11.92    The transistor parameters for the circuit in Figure P11.92 are: $\beta = 200$, $V_{BE}(\text{on}) = 0.7$ V, and $V_A = 80$ V. (a) Determine the differential-mode voltage gain $A_d = v_{o3}/v_d$ and the common-mode voltage gain $A_{cm} = v_{o3}/v_{cm}$. (b) Determine the output voltage $v_{o3}$ if $v_1 = 2.015 \sin \omega t$ V and $v_2 = 1.985 \sin \omega t$ V. Compare this output to the ideal output that would be obtained if $A_{cm} = 0$. (c) Find the differential-mode and common-mode input resistances.

*11.93    For the transistors in the circuit in Figure P11.93, the parameters are: $K_n = 0.2\,\text{mA/V}^2$, $V_{TN} = 2$ V, and $\lambda = 0.02\,\text{V}^{-1}$. (a) Determine the differential-mode voltage gain $A_d = v_{o3}/v_d$ and the common-mode voltage gain $A_{cm} = v_{o3}/v_{cm}$. (b) Determine the output voltage $v_{o3}$ if $v_1 = 2.15 \sin \omega t$ V and $v_2 = 1.85 \sin \omega t$ V. Compare this output to the ideal output that would be obtained if $A_{cm} = 0$.

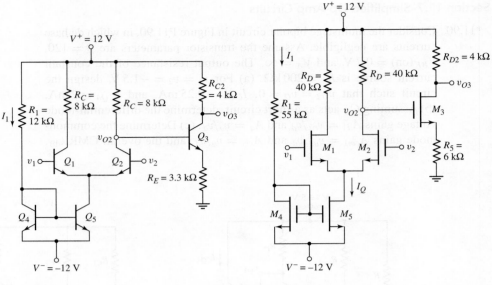

Figure P11.92

Figure P11.93

### Section 11.8 Diff-Amp Frequency Response

11.94 Consider the differential amplifier in Figure 11.50(a) with parameters $I_Q = 0.8\,\text{mA}$, $R_C = 10\,\text{k}\Omega$, and $R_B = 0.5\,\text{k}\Omega$. The transistor parameters are $\beta = 150$, $V_{BE}$ (on) $= 0.7\,\text{V}$, $V_A = \infty$, $C_\pi = 1.2\,\text{pF}$, and $C_\mu = 0.2\,\text{pF}$. (a) Determine the low-frequency differential-mode gain $A_d = v_{o2}/v_d$. (b) Find the equivalent Miller capacitance of each transistor. (c) Determine the upper 3 dB frequency.

11.95 The differential amplifier in Figure 11.51(a) has the same circuit and transistor parameters as described in Problem 11.94. The equivalent impedance of the current source is $R_o = 10\,\text{M}\Omega$ and $C_o = 0.4\,\text{pF}$. (a) Determine the frequency of the zero in the common-mode gain. (b) Find the frequency of the pole in the common-mode gain.

11.96 A BJT diff-amp is biased with a current source $I_Q = 2\,\text{mA}$, and the circuit parameters are $R_C = 10\,\text{k}\Omega$ and $R_B = 1\,\text{k}\Omega$. The transistor parameters are: $\beta = 120$, $f_T = 800\,\text{MHz}$, and $C_\mu = 1\,\text{pF}$. (a) Determine the upper 3 dB frequency of the differential-mode gain. (b) If the current source impedance parameters are $R_o = 10\,\text{M}\Omega$ and $C_o = 1\,\text{pF}$, find the frequency of the zero in the common-mode gain.

11.97 Consider the diff-amp in Figure 11.55. The circuit and transistor parameters are the same as in Problem 11.6. For a one-sided output at $v_{o2}$, determine the differential-mode gain for: (a) $R_E = 100\,\Omega$, and (b) $R_E = 250\,\Omega$.

### 💿 COMPUTER SIMULATION PROBLEMS

11.98 Using a computer simulation, verify the results of Example 11.12.

11.99 Using a computer simulation, verify the results of Example 11.13 for the simple op-amp circuit.

11.100  Consider the circuit in Figure P11.100. Use standard transistors. Using a computer simulation, determine the small-signal differential-mode voltage gain and common-mode voltage gain for (a) $R_L = 10\,M\Omega$ and (b) $R_L = 200\,k\Omega$.

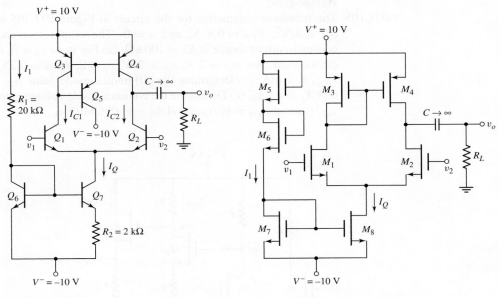

**Figure P11.100**                                    **Figure P11.101**

11.101  Consider the circuit in Figure P11.101. Use standard transistors. Using a computer simulation, determine the small-signal differential-mode voltage gain and common-mode voltage gain for (a) $R_L = 10\,M\Omega$ and (b) $R_L = 200\,k\Omega$.

# DESIGN PROBLEMS

[Note: Each design is to be correlated with a computer simulation analysis.]

\*D11.102  Design a basic BJT diff-amp with an active load and constant current-source biasing. The bias voltages are to be $\pm 3$ V and the maximum power dissipation is to be limited to 2 mW. The open-circuit differential-mode voltage gain should be $|A_d| = 1800$ and the common-mode rejection ratio should be $CMRR_{dB} = 75$ dB. Specify bias currents, minimum Early voltages, and the minimum output impedance of the current source. Design the current source to achieve the required output impedance.

\*D11.103  Design a basic MOSFET diff-amp with an active load and constant current-source biasing. The bias voltages are to be $\pm 3$ V and the maximum power dissipation is to be limited to 2 mW. The open-circuit differential-mode voltage gain should be $|A_d| = 180$ and the common-mode rejection ratio should be $CMRR_{dB} = 80$ dB. Use appropriate transistor parameters. Specify bias currents, minimum $\lambda$ values, and the minimum output impedance of the current source. Design the current source to achieve the required output impedance.

*D11.104 Consider the bipolar op-amp configuration in Figure 11.49. The bias voltages are $\pm 10$ V, as shown, the current $I_{R7}$ is to be $I_{R7} = 3$ mA, and the maximum dc power dissipation in the circuit is to be 120 mW. The output voltage is to be $v_o = 0$ for $v_1 = v_2 = 0$. Design the circuit, using reasonable resistance and current values. What is the overall differential-mode voltage gain?

*D11.105 The transistor parameters for the circuit in Figure P11.105 are: $K_n = 0.2$ mA/V$^2$, $V_{TN} = 0.8$ V, and $\lambda = 0$. The output resistance of the constant-current source is $R_o = 100$ k$\Omega$. (a) For $v_1 = v_2 = 0$, design the circuit such that: $v_{o2} = 2$ V, $v_{o3} = 3$ V, $v_o = 0$, $I_{DQ3} = 0.25$ mA, and $I_{DQ4} = 2$ mA. (b) Determine the differential-mode gains $A_{d1} = v_{o2}/v_d$ and $A_d = v_o/v_d$. (c) Determine the common-mode voltage gains $A_{cm1} = v_{o2}/v_{cm}$ and $A_{cm} = v_o/v_{cm}$, and the overall CMRR$_{dB}$.

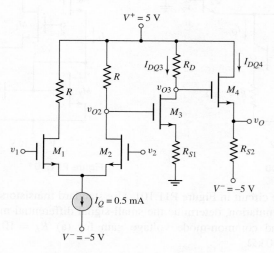

Figure P11.105

# Feedback and Stability

Previously, we found that the small-signal voltage gain and other characteristics of BJT and MOSFET circuit amplifiers are functions of, for example, the bipolar current gain and the MOSFET conduction parameter. In general, these transistor parameters vary with temperature and have a range of values for a given type of transistor, because of processing and material property tolerances. These parameter variations mean that the $Q$-point, voltage gain, and other circuit properties can vary from one circuit to another, and can be functions of temperature.

Transistor circuit characteristics can be made essentially independent of the individual transistor parameters by using feedback. The feedback process takes a portion of the output signal and returns it to the input to become part of the input excitation. We previously encountered feedback in our study of ideal op-amps and op-amp circuits. For example, resistors are connected between the output and input terminals of an ideal op-amp to form a feedback network. The voltage gain of these ideal circuits is a function only of the ratio of resistors and not of any individual transistor parameters. In this chapter, we formally study feedback and feedback circuits.

## PREVIEW

In this chapter, we will:

- Introduce feedback concepts and discuss, in general terms, advantages and disadvantages of using feedback in electronic circuits.
- Derive the transfer function of the ideal feedback system and determine a few characteristics of the feedback system.
- Analyze the four ideal feedback circuit configurations and determine circuit characteristics including input and output resistances.
- Analyze op-amp and discrete transistor circuit examples of voltage feedback amplifiers, current feedback amplifiers, transconductance feedback amplifiers, and transresistance feedback amplifiers.
- Derive the loop-gain of ideal and practical feedback circuits.
- Determine the stability criteria of feedback circuits.
- Consider frequency compensation techniques, methods by which unstable feedback circuits can be stabilized.
- As an application, redesign a BJT feedback circuit using MOSFETs.

# 12.1 INTRODUCTION TO FEEDBACK

**Objective:** • Introduce feedback concepts and discuss, in general terms, a few advantages and disadvantages of using feedback in electronic circuits.

Feedback is used in virtually all amplifier systems. Harold Black, an electronics engineer with the Western Electric Company, invented the feedback amplifier in 1928 while searching for methods to stabilize the gain of amplifiers for use in telephone repeaters. In a feedback system, a signal that is proportional to the output is fed back to the input and combined with the input signal to produce a desired system response. As we will see, external feedback is used deliberately to achieve particular system benefits. However, feedback may be unintentional and an undesired system response may be produced.

We have already seen examples of feedback in previous chapters, although the term feedback may not have been used. For example, in Chapters 3 and 5 we introduced resistors at the emitter of BJT common-emitter circuits and at the source of MOSFET common-source circuits to stabilize the $Q$-point against variations in transistor parameters. This technique introduces *negative feedback* in the circuit. An increase in collector or drain current produces an increase in the voltage across these resistors which produces a decrease in the base-emitter or gate-source voltage. The decrease in these device voltages tends to reduce or oppose the change in collector or drain current. Opposition to change is suggested by use of the term negative feedback.

Feedback can be either negative or positive. In **negative feedback,** a portion of the output signal is subtracted from the input signal; in **positive feedback,** a portion of the output signal is added to the input signal. Negative feedback, for example, tends to maintain a constant value of amplifier voltage gain against variations in transistor parameters, supply voltages, and temperature. Positive feedback is used in the design of oscillators and in a number of other applications. In this chapter, we will concentrate on negative feedback.

## 12.1.1 Advantages and Disadvantages of Negative Feedback

Before we actually get into the analysis and design of feedback circuits, we will list some of the advantages and disadvantages of negative feedback. Although these characteristics and properties of negative feedback are not obvious at this point, they are listed here so that the reader can anticipate these results during the derivations and analysis.

### Advantages

1. *Gain sensitivity*. Variations in the circuit transfer function (gain) as a result of changes in transistor parameters are reduced by feedback. This reduction in sensitivity is one of the most attractive features of negative feedback.
2. *Bandwidth extension*. The bandwidth of a circuit that incorporates negative feedback is larger than that of the basic amplifier.

3.  *Noise sensitivity.* Negative feedback may increase the signal-to-noise ratio if noise is generated within the feedback loop.
4.  *Reduction of nonlinear distortion.* Since transistors have nonlinear characteristics, distortion may appear in the output signals, especially at large signal levels. Negative feedback reduces this distortion.
5.  *Control of impedance levels.* The input and output impedances can be increased or decreased with the proper type of negative feedback circuit.

### Disadvantages

1.  *Circuit gain.* The overall amplifier gain, with negative feedback, is reduced compared to the basic amplifier used in the circuit.
2.  *Stability.* There is a possibility that the feedback circuit may become unstable (oscillate) at high frequencies.

These advantages and disadvantages will be further discussed as we develop the feedback theory.

In the course of our discussion, we will analyze several feedback circuits, in both discrete and op-amp circuit configurations. First, however, we will consider the ideal feedback theory and derive the general characteristics of feedback amplifiers. In this section, we discuss the ideal signal gain, gain sensitivity, bandwidth extension, noise sensitivity, and reduction of nonlinear distortion of a generalized feedback amplifier.

### 12.1.2    Use of Computer Simulation

Conventional methods of analysis that have been used in the previous chapters apply directly to feedback circuits. That is, the same dc analysis techniques and the same small-signal transistor equivalent circuits apply directly to feedback circuits in this chapter. However, in the analysis of feedback circuits, several simultaneous equations can be obtained, the time involved may be quite long and the probability of introducing errors may become almost certain.

Therefore, computer simulation of feedback circuits may prove to be very useful and is used fairly often throughout this chapter. As always, a word of warning is in order concerning computer simulation. Computer simulation does not replace basic understanding. It is important for the reader to understand the concepts and characteristics of the basic types of feedback circuits. Computer simulation is used only as a tool for obtaining specific results.

## 12.2    BASIC FEEDBACK CONCEPTS

**Objective: •** Analyze and obtain the transfer function of the ideal feedback system, and determine a few characteristics (advantages) of the feedback system.

Figure 12.1 shows the basic configuration of a feedback amplifier. In the diagram, the various signals $S$ can be either currents or voltages. The circuit contains a basic amplifier with an open-loop gain $A$ and a feedback circuit that samples the output

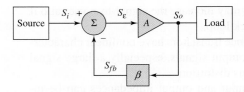

**Figure 12.1** Basic configuration of a feedback amplifier

signal and produces a feedback signal $S_{fb}$. The feedback signal is subtracted from the input source signal, which produces an error signal $S_\varepsilon$. The error signal is the input to the basic amplifier and is the signal that is amplified to produce the output signal. The subtraction property produces the negative feedback.

Implicit in the diagram in Figure 12.1 is the assumption that the input signal is transmitted through the amplifier only, none through the feedback network, and that the output signal is transmitted back through the feedback network only, none through the amplifier. Also, there are no loading effects in the ideal feedback system. The feedback network does not load down the output of the basic amplifier, and the basic amplifier and feedback network do not produce a loading effect on the input signal source. In actual feedback circuits, these assumptions and conditions are not entirely accurate. We will see later how nonideal conditions change the characteristics of actual feedback circuits with respect to those of the ideal feedback network.

## 12.2.1   Ideal Closed-Loop Signal Gain

From Figure 12.1, the output signal is

$$S_o = AS_\varepsilon \tag{12.1}$$

where $A$ is the **amplification factor,** and the feedback signal is

$$S_{fb} = \beta S_o \tag{12.2}$$

where $\beta$ in this case is the **feedback transfer function.**[1] At the summing node, we have

$$S_\varepsilon = S_i - S_{fb} \tag{12.3}$$

where $S_i$ is the input signal. Equation (12.1) then becomes

$$S_o = A(S_i - \beta S_o) = AS_i - \beta AS_o \tag{12.4}$$

Equation (12.4) can be rearranged to yield the **closed-loop transfer function,** or gain, which is

$$A_f = \frac{S_o}{S_i} = \frac{A}{(1 + \beta A)} \tag{12.5}$$

As mentioned, signals $S_i$, $S_o$, $S_{fb}$, and $S_\varepsilon$ can be either currents or voltages; however, they do not need to be all voltages or all currents in a given feedback amplifier.

---

[1] In this chapter, $\beta$ is the feedback transfer function, rather than the transistor current gain. The parameter $h_{FE}$ will be used as the transistor current gain. Normally, $h_{FE}$ indicates the dc current gain and $h_{fe}$ indicates the ac current gain. However, as usual, we neglect any difference between the two parameters and assume $h_{FE} = h_{fe}$.

In other words, there may be a combination of current and voltage signals in the same circuit.

Equation (12.5) can be written

$$A_f = \frac{A}{(1 + \beta A)} = \frac{A}{1 + T} \tag{12.6}$$

where $T = \beta A$ is the **loop gain.** For negative feedback, we assume $T$ to be a positive real factor. We will see later that the loop gain can become a complex function of frequency, but for the moment, we will assume that $T$ is positive for negative feedback. We will also see that in some cases the gain will be negative (180 degree phase difference between input and output signals) which means that the feedback transfer function $\beta$ will also be a negative quantity for a negative feedback circuit.

Combining Equations (12.1) and (12.2), we obtain the loop gain relationship

$$T = A\beta = \frac{S_{fb}}{S_\varepsilon} \tag{12.7}$$

Normally, the error signal is small, so the expected loop gain is large. If the loop gain is large so that $\beta A \gg 1$, then, from Equation (12.6), we have

$$A_f \cong \frac{A}{\beta A} = \frac{1}{\beta} \tag{12.8}$$

and the gain or transfer function of the feedback amplifier essentially becomes a function of the feedback network only.

The feedback circuit is usually composed of passive elements, which means that the feedback amplifier gain is almost completely independent of the basic amplifier properties, including individual transistor parameters. Since the feedback amplifier gain is a function of the feedback elements only, the closed-loop gain can be designed to be a given value. This property was demonstrated in Chapter 9, where we showed that the closed-loop gain of ideal op-amp circuits is a function of the feedback elements only. The individual transistor parameters may vary widely, and may depend on temperature and frequency, but the feedback amplifier gain is constant. The net results of negative feedback is stability in the amplifier characteristics.

In general, the magnitude and phase of the loop gain are functions of frequency, and they become important when we discuss the stability of feedback circuits.

## EXAMPLE 12.1

**Objective:** Calculate the feedback transfer function $\beta$, given $A$ and $A_f$.

**Case A.** Assume that the open-loop gain of a system is $A = 10^5$ and the closed-loop gain is $A_f = 50$.

**Solution:** From Equation (12.5), the closed-loop gain is

$$A_f = \frac{A}{(1 + \beta A)} \quad \text{or} \quad 50 = \frac{10^5}{1 + \beta(10^5)}$$

which yields $\beta = 0.01999$ or $1/\beta = 50.025$.

**Case B.** Now assume that the open-loop gain is $A = -10^5$ and the close-loop gain is $A_f = -50$.

**Solution:** Again, from Equation (12.5), the closed-loop gain is

$$A_f = \frac{A}{(1 + \beta A)} \quad \text{or} \quad -50 = \frac{-10^5}{1 + \beta(-10^5)}$$

which yields $\beta = -0.01999$ or $1/\beta = -50.025$.

**Comment:** From these typical parameter values, we see that $A_f \cong 1/\beta$, as Equation (12.8) predicts. We also see that if the open-loop gain $A$ is negative, then the closed-loop gain $A_f$ and feedback transfer function $\beta$ will also be negative for a negative feedback network.

---

**EXERCISE PROBLEM**

**Ex 12.1:** (a) The open-loop gain of an amplifier is $A = 5 \times 10^4$ and the closed-loop gain is $A_f = 50$. (i) What is the feedback transfer function? (ii) What is the ratio of $A_f$ to $1/\beta$? (b) Repeat part (a) for $A = 100$ and $A_f = 20$. (Ans. (a) (i) 0.01998, (ii) 0.9990; (b) (i) 0.04, (ii) 0.80)

---

Assuming a large loop gain, the output signal, from Equation (12.5), becomes

$$S_o = \left( \frac{A}{1 + \beta A} \right) S_i \cong \frac{1}{\beta} \cdot S_i \tag{12.9}$$

Substituting Equation (12.9) into (12.3), we obtain the error signal,

$$S_\varepsilon = S_i - \beta S_o \cong S_i - \beta \left( \frac{S_i}{\beta} \right) = 0 \tag{12.10}$$

With a large loop gain, the error signal decreases to almost zero. We will see this result again as we consider specific feedback circuits throughout the chapter.

### 12.2.2    Gain Sensitivity

As previously stated, if the loop gain $T = \beta A$ is very large, the overall gain of the feedback amplifier is essentially a function of the feedback network only. We can quantify this characteristic.

If the feedback transfer function $\beta$ is a constant, then taking the derivative of $A_f$ with respect to $A$, from Equation (12.5), produces

$$\frac{dA_f}{dA} = \frac{1}{(1 + \beta A)} - \frac{A}{(1 + \beta A)^2} \cdot \beta = \frac{1}{(1 + \beta A)^2} \tag{12.11(a)}$$

or

$$dA_f = \frac{dA}{(1 + \beta A)^2} \tag{12.11(b)}$$

Dividing both sides of Equation (12.11(b)) by the closed-loop gain yields

$$\frac{dA_f}{A_f} = \frac{\dfrac{dA}{(1 + \beta A)^2}}{\dfrac{A}{1 + \beta A}} = \frac{1}{(1 + \beta A)} \cdot \frac{dA}{A} = \left( \frac{A_f}{A} \right) \frac{dA}{A} \tag{12.12}$$

Equation (12.12) shows that the percent change in the closed-loop gain $A_f$ is less than the corresponding percent change in the open-loop gain $A$ by the factor $(1 + \beta A)$. The change in open-loop gain may result from variations in individual transistor parameters in the basic amplifier.

## EXAMPLE **12.2**

**Objective:** Calculate the percent change in the closed-loop gain $A_f$, given a change in the open-loop gain $A$.

Using the same parameter values as in Example 12.1, we have $A = 10^5$, $A_f = 50$, and $\beta = 0.01999$. Assume that the change in the open-loop gain is $dA = 10^4$ (a 10 percent change).

**Solution:** From Equation (12.12), we have

$$dA_f = \frac{A_f}{(1 + \beta A)} \cdot \frac{dA}{A} = \frac{50}{[1 + (0.01999)(10^5)]} \cdot \frac{10^4}{10^5} = 2.5 \times 10^{-3}$$

The percent change is then

$$\frac{dA_f}{A_f} = \frac{2.5 \times 10^{-3}}{50} = 5 \times 10^{-5} \Rightarrow 0.005\%$$

compared to the 10 percent change assumed in the open-loop gain.

**Comment:** From this example, we see that the resulting percent change in the closed-loop gain is substantially less than the percent change in the open-loop gain. This is one of the principal advantages of negative feedback.

## EXERCISE PROBLEM

**Ex 12.2:** (a) Consider a general feedback system with parameters $A = 5 \times 10^5$ and $A_f = 50$. If the magnitude of $A$ decreases by 15 percent, what is the new value of $A_f$ and what is the corresponding percent change in $A_f$? (b) Repeat part (a) if $A = 100$ and $A_f = 20$. (Ans. (a) $A_f = 49.99912$, $-1.76 \times 10^{-3}$ %; (b) $A_f = 19.318$, $-3.41\%$)

From Equation (12.12), the change in $A_f$ is reduced by the factor $(1 + \beta A)$ compared to the change in $A$. The term $(1 + \beta A)$ is called the **desensitivity factor.**

### 12.2.3    Bandwidth Extension

The amplifier bandwidth is a function of feedback. Assume the frequency response of the basic amplifier can be characterized by a single pole. We can then write

$$A(s) = \frac{A_o}{1 + \dfrac{s}{\omega_H}} \tag{12.13}$$

where $A_o$ is the low-frequency or midband gain, and $\omega_H$ is the upper 3 dB or corner frequency.

The closed-loop gain of the feedback amplifier can be expressed as

$$A_f(s) = \frac{A(s)}{(1 + \beta A(s))} \tag{12.14}$$

where we assume that the feedback transfer function $\beta$ is independent of frequency. Substituting Equation (12.13) into Equation (12.14), we can write the closed-loop gain in the form

$$A_f(s) = \frac{A_o}{(1 + \beta A_o)} \cdot \frac{1}{1 + \dfrac{s}{\omega_H(1 + \beta A_o)}} \tag{12.15}$$

From Equation (12.15), we see that the low-frequency closed-loop gain is smaller than the open-loop gain by a factor of $(1 + \beta A_o)$, but the closed-loop 3 dB frequency is larger than the open-loop value by a factor of $(1 + \beta A_o)$.

If we multiply the low-frequency open-loop gain $A_o$ by the bandwidth (3 dB frequency) $\omega_H$, we obtain $A_o \omega_H$, which is the gain–bandwidth product. The product of the low-frequency closed-loop gain and the closed-loop band-width is

$$\frac{A_o}{(1 + \beta A_o)}[\omega_H(1 + \beta A_o)] = A_o \omega_H \tag{12.16}$$

Equation (12.16) states that the gain-bandwidth product of a feedback amplifier is a constant. That is, for a given circuit, we can increase the gain at the expense of a reduced bandwidth, or we can increase the bandwidth at the expense of a reduced gain. This property is illustrated in Figure 12.2.

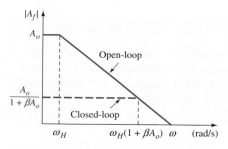

**Figure 12.2** Open-loop and closed-loop gain versus frequency, illustrating bandwidth extension

## EXAMPLE 12.3

**Objective:** Determine the bandwidth of a feedback amplifier.

Consider a feedback amplifier with an open-loop low-frequency gain of $A_o = 10^4$, an open-loop bandwidth of $\omega_H = (2\pi)(100)$ rad/s, and a closed-loop low-frequency gain of $A_f(0) = 50$.

**Solution:** From Equation (12.15), the low-frequency closed-loop gain is

$$A_f(0) = \frac{A_o}{(1 + \beta A_o)} \qquad \text{or} \qquad 50 = \frac{10^4}{(1 + \beta A_o)}$$

which yields

$$(1 + \beta A_o) = \frac{10^4}{50} = 200$$

From Equation (12.15), the closed-loop bandwidth is

$$\omega_{fH} = \omega_H(1 + \beta A_o) = (2\pi)(100)(200) = (2\pi)(20 \times 10^3)$$

**Comment:** The bandwidth increases from 100 Hz to 20 kHz as the gain decreases from $10^4$ to 50.

EXERCISE PROBLEM

**Ex 12.3:** (a) A feedback amplifier has an open-loop low-frequency gain of $A_O = 5 \times 10^4$, an open-loop bandwidth of $\omega_H = (2\pi)(5)$ rad/s, and a closed-loop low-frequency gain of $A_f(0) = 80$. Determine (i) $\beta$ and (ii) the closed-loop bandwidth. (b) Using the results of part (a), if $\beta$ is reduced by 50 percent, determine the percent change in (i) $A_f(0)$ and (ii) $\omega_{fH}$? (Ans. (a) (i) $\beta = 0.01248$, (ii) $\omega_{fH} = (2\pi)(3.125 \times 10^3)$ rad/s; (b) (i) $+100\%$, (ii) $-50\%$)

### 12.2.4   Noise Sensitivity

In any electronic system, unwanted random and extraneous signals may be present in addition to the desired signal. These random signals are called **noise.** Electronic noise can be generated within an amplifier, or may enter the amplifier along with the input signal. Negative feedback may reduce the noise level in amplifiers; more accurately, it may increase the **signal-to-noise ratio.** More precisely, feedback can help reduce the effect of noise generated in an amplifier, but it cannot reduce the effect when the noise is part of the input signal.

The input signal-to-noise ratio is defined as

$$(\text{SNR})_i = \frac{S_i}{N_i} = \frac{v_i}{v_n} \qquad (12.17)$$

where $S_i = v_i$ is the input source signal and $N_i = v_n$ is the input noise signal. The output signal-to-noise ratio is

$$(\text{SNR})_o = \frac{S_o}{N_o} = \frac{A_{Ti}S_i}{A_{Tn}N_i} \qquad (12.18)$$

where the desired output signal is $S_o = A_{Ti}S_i$ and the output noise signal is $N_o = A_{Tn}N_i$. The parameter $A_{Ti}$ is the amplification factor that multiplies the source signal, and the parameter $A_{Tn}$ is the amplification factor that multiplies the noise signal. A large signal-to-noise ratio allows the signal to be detected without any loss of information. This is a desirable characteristic.

The following example compares the signal and noise amplification factors, which may or may not be equal.

## EXAMPLE 12.4

**Objective:** Determine the effect of feedback on the source signal and noise signal levels.

Consider the four possible amplifier configurations shown in Figure 12.3. The amplifiers are designed to provide the same output signal voltage. Determine the effect of the noise signal $v_n$.

**Solution (Figure 12.3(a)):** Two open-loop amplifiers are in a cascade configuration, and the noise signal is generated between the two amplifiers. The output voltage is

$$v_{oa} = A_1 A_2 v_i + A_2 v_n = 100 v_i + 10 v_n$$

Therefore, the output signal-to-noise ratio is

$$\frac{S_o}{N_o} = \frac{100 v_i}{10 v_n} = 10 \frac{S_i}{N_i}$$

**Solution (Figure 12.3(b)):** Two open-loop amplifiers are in a cascade configuration, and the noise is part of the input signal. The output voltage is

$$v_{ob} = A_1 A_2 v_i + A_1 A_2 v_n = 100 v_i + 100 v_n$$

Therefore, the output signal-to-noise ratio is

$$\frac{S_o}{N_o} = \frac{100 v_i}{100 v_n} = \frac{S_i}{N_i}$$

**Solution (Figure 12.3(c)):** Two amplifiers are in a feedback configuration, and the noise signal is generated between the two amplifiers. The output voltage is

$$v_{oc} = A_1 A_2 v_\varepsilon + A_2 v_n$$

and the feedback signal is

$$v_{fb} = \beta v_{oc}$$

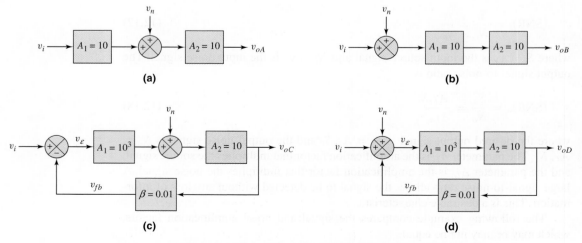

**Figure 12.3** Four amplifier configurations with different input noise sources

Then,

$$v_\varepsilon = v_i - v_{fb} = v_i - \beta v_{oc}$$

therefore,

$$v_{oc} = A_1 A_2 (v_i - \beta v_{oc}) + A_2 v_n$$

or

$$v_{oc} = \frac{A_1 A_2}{(1 + \beta A_1 A_2)} \cdot v_i + \frac{A_2}{(1 + \beta A_1 A_2)} \cdot v_n \cong 100 v_i + 0.1 v_n$$

The output signal-to-noise ratio is

$$\frac{S_o}{N_o} = \frac{100 v_i}{0.1 v_n} = 1000 \frac{S_i}{N_i}$$

**Solution (Figure 12.3(d)):** A basic feedback configuration, and the noise is part of the input signal. The output voltage is

$$v_{od} = \frac{A_1 A_2}{(1 + \beta A_1 A_2)} (v_i + v_n) \cong 100 v_i + 100 v_n$$

Therefore, the output signal-to-noise ratio is

$$\frac{S_o}{N_o} = \frac{100 v_i}{100 v_n} = \frac{S_i}{N_i}$$

**Comment:** Comparing the four configurations, we see that Figure 12.3(c) produces the largest output signal-to-noise ratio. This configuration may occur when amplifier $A_2$ is an audio power-amplifier stage, in which large currents can produce excessive noise, and when amplifier $A_1$ corresponds to a low-noise preamplifier, which provides most of the voltage gain.

EXERCISE PROBLEM

**Ex 12.4:** (a) Consider the circuit shown in Figure 12.3(a). Assume $A_1 = 100$ and $A_2 = 10$. Determine the output signal-to-noise ratio in terms of the input signal-to-noise ratio. (b) Consider the circuit shown in Figure 12.3(c). Assume $A_1 = 10^4$, $A_2 = 10$, and $\beta = 0.001$. Determine the output signal-to-noise ratio in terms of the input signal-to-noise ratio. (Ans. (a) $S_o/N_o = 100(S_i/N_i)$, (b) $S_o/N_o = 10^4(S_i/N_i)$)

We must emphasize that the increased signal-to-noise ratio due to feedback occurs only in specific situations. As indicated in Figure 12.3(d), when noise is effectively part of the amplifier input signal, the feedback mechanism does not improve the ratio.

## 12.2.5  Reduction of Nonlinear Distortion

Distortion in an output signal is caused by a change in the basic amplifier gain or a change in the slope of the basic amplifier transfer function. The change in gain is a function of the nonlinear properties of bipolar and MOS transistors used in the basic amplifier.

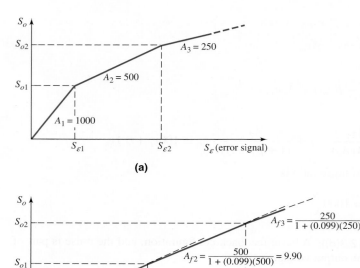

**Figure 12.4** (a) Basic amplifier (open-loop) transfer characteristics; (b) closed-loop transfer characteristics

Assume the basic amplifier, or open-loop, transfer function is as shown in Figure 12.4(a), which shows changes in gain as the input signal amplitude changes. The gain values are shown on the figure. When this amplifier is incorporated in a feedback circuit with a feedback transfer function of $\beta = 0.099$, the resulting closed-loop transfer characteristics are shown in Figure 12.4(b). This transfer function also has changes in gain but, whereas the open-loop gain changes by a factor of 2, the closed-loop gain changes by only 1 percent and 2 percent, respectively. A smaller change in gain means less distortion in the output signal of the negative feedback amplifier.

## Test Your Understanding

**TYU 12.1** (a) The closed-loop gain of a feedback amplifier is $A_f = 50$ and the feedback transfer function is $\beta = 0.019$. Determine the open-loop gain $A$. (b) If the open-loop gain is $A = 5 \times 10^5$ and $\beta = 0.019$, find the closed-loop gain $A_f$. (Ans. (a) $A = 10^3$, (b) $A_f = 52.63$)

**TYU 12.2** The gain factors in a feedback system are $A = 5 \times 10^5$ and $A_f = 100$. Parameter $A_f$ must not change more than $\pm 0.001$ percent because of a change in $A$. What is the maximum allowable variation in $A$? (Ans. $\pm 5\%$)

**TYU 12.3** In a feedback system, the basic amplifier open-loop low-frequency gain is $A_o = 5 \times 10^5$ and the open-loop 3 dB frequency is 6 Hz. (a) If the required closed-loop bandwidth is $f = 200\,\text{kHz}$, determine the maximum closed-loop low-frequency gain $A_f(0)$. (b) If the required closed-loop bandwidth is $f = 100\,\text{kHz}$, what is the maximum closed-loop low-frequency gain $A_f(0)$? (Ans. (a) $A_f(0) = 15$, (b) $A_f(0) = 30$)

##  12.3    IDEAL FEEDBACK TOPOLOGIES

**Objective:** • Analyze the four ideal feedback circuit configurations and determine circuit characteristics including input and output resistances.

There are four basic feedback topologies, based on the parameter to be amplified (voltage or current) and the output parameter (voltage or current). The four feedback circuit categories can be described by the types of connections at the input and output of circuit. The four types of connections are shown in Figure 12.5. The four connections are referred to as: series–shunt (voltage amplifier), shunt–series (current amplifier), series–series (transconductance amplifier), and shunt–shunt (transresistance amplifier). The first term refers to the connection at the amplifier input, and the second term refers to the connection at the output. Also, the type of connection determines which parameter (voltage or current) is sampled at the output and which parameter is amplified. The connections also determine the feedback amplifier characteristics—in particular, the input and output resistances. The resistance parameters become an important circuit property, when, for example, we consider voltage amplifiers versus current amplifiers.

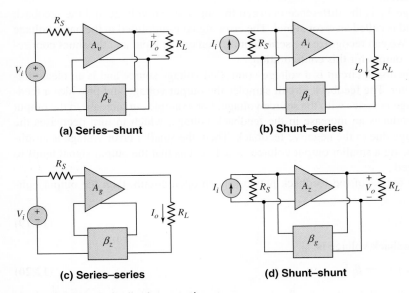

**Figure 12.5** Basic feedback connections

In this section, we will determine the ideal transfer functions and the ideal input and output resistances of each of the four feedback topologies. In later sections, we will compare actual versus ideal feedback circuit characteristics.

As a note, the ideal topologies are small-signal equivalent circuits; therefore, phasor notation is used throughout this analysis.

### 12.3.1    Series–Shunt Configuration

The configuration of an ideal **series–shunt** feedback amplifier is shown in Figure 12.6. The circuit consists of a basic voltage amplifier with an input resistance $R_i$

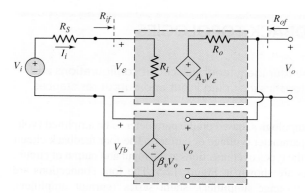

**Figure 12.6** Ideal series–shunt feedback topology

and an open-loop voltage gain $A_v$. The feedback circuit samples the output voltage and produces a feedback voltage $V_{fb}$, which is in series with the input signal voltage $V_i$. In this ideal configuration, the input resistance to the feedback circuit is infinite; therefore, there is no loading effect on the output of the basic amplifier due to the feedback circuit.

Voltage $V_\varepsilon$ is the difference between the input signal voltage and the feedback voltage and is called an error signal. The error signal is amplified in the basic voltage amplifier. We can recognize the series connection on the input and the shunt connection of the output for this configuration.

The feedback circuit is a voltage-controlled voltage source and is an ideal voltage amplifier. The feedback circuit samples the output voltage and provides a feedback voltage in series with the source voltage. For example, an increase in the output voltage produces an increase in the feedback voltage, which in turn decreases the error voltage due to the negative feedback. Then, the smaller error voltage is amplified producing a smaller output voltage, which means that the output signal tends to be stabilized.

If the output of the feedback network is an open circuit, then the output voltage is

$$V_o = A_v V_\varepsilon \tag{12.19}$$

and the feedback voltage is

$$V_{fb} = \beta V_o = \beta_v V_o \tag{12.20}$$

Parameter $\beta_v$ is the voltage feedback transfer function, which is the ratio of the feedback voltage to the output voltage. The notation is similar to the voltage gain $A_v$, which is also the ratio of two voltages.

The error voltage, assuming the source resistance $R_S$ is negligible, is

$$V_\varepsilon = V_i - V_{fb} \tag{12.21}$$

Combining Equations (12.19), (12.20), and (12.21), we find the closed-loop voltage transfer function is

$$A_{vf} = \frac{V_o}{V_i} = \frac{A_v}{(1 + \beta_v A_v)} \tag{12.22}$$

Equation (12.22) is the **closed-loop voltage gain** of the feedback amplifier, and it has the same form as the ideal feedback transfer function given by Equation (12.5).

The input resistance including feedback is denoted by $R_{if}$. Starting with Equation (12.21), using Equations (12.19) and (12.20), we find that

$$V_i = V_\varepsilon + V_{fb} = V_\varepsilon + \beta_v V_o = V_\varepsilon + \beta_v(A_v V_\varepsilon) \qquad \textbf{(12.23(a))}$$

or

$$V_\varepsilon = \frac{V_i}{(1 + \beta_v A_v)} \qquad \textbf{(12.23(b))}$$

The input current is

$$I_i = \frac{V_\varepsilon}{R_i} = \frac{V_i}{R_i(1 + \beta_v A_v)} \qquad \textbf{(12.24)}$$

and the input resistance with feedback is then

$$R_{if} = \frac{V_i}{I_i} = R_i(1 + \beta_v A_v) \qquad \textbf{(12.25)}$$

Equation (12.25) shows that a series input connection results in an increased input resistance compared to that of the basic voltage amplifier. A large input resistance is a desirable property of a voltage amplifier. This eliminates loading effects on the input signal source.

The output resistance of the feedback circuit can be determined from the equivalent circuit in Figure 12.7. The input signal voltage source is set equal to zero (a short circuit), and a test voltage is applied to the output terminals.

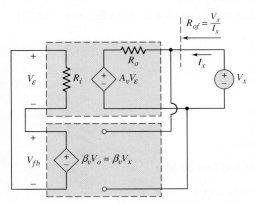

**Figure 12.7** Ideal series–shunt feedback configuration for determining output resistance

From the circuit, we see that

$$V_\varepsilon + V_{fb} = V_\varepsilon + \beta_v V_x = 0 \qquad \textbf{(12.26(a))}$$

or

$$V_\varepsilon = -\beta_v V_x \qquad \textbf{(12.26(b))}$$

The output current is

$$I_x = \frac{V_x - A_v V_\varepsilon}{R_o} = \frac{V_x - A_v(-\beta_v V_x)}{R_o} = \frac{V_x(1 + \beta_v A_v)}{R_o} \qquad \textbf{(12.27)}$$

and the output resistance, including feedback, is

$$R_{of} = \frac{V_x}{I_x} = \frac{R_o}{(1 + \beta_v A_v)} \qquad \textbf{(12.28)}$$

Equation (12.28) shows that a shunt output connection results in a decreased output resistance compared to that of the basic voltage amplifier. A small output resistance is a desirable property of a voltage amplifier. This eliminates loading effects on the output signal when an output load is connected.

The equivalent circuit of this feedback voltage amplifier is shown in Figure 12.8.

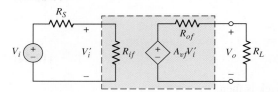

**Figure 12.8**  Equivalent circuit of the series–shunt feedback circuit or voltage amplifier

## EXAMPLE 12.5

**Objective:** Determine the input resistance of a series input connection and the output resistance of a shunt output connection for an ideal feedback voltage amplifier.

Consider a series–shunt feedback amplifier in which the open-loop gain is $A_v = 10^5$ and the closed-loop gain is $A_{vf} = 50$. Assume the input and output resistances of the basic amplifier are $R_i = 10 \text{ k}\Omega$ and $R_o = 20 \text{ k}\Omega$, respectively.

**Solution:** The ideal closed-loop voltage transfer function is, from Equation (12.22),

$$A_{vf} = \frac{A_v}{(1 + \beta_v A_v)}$$

or

$$(1 + \beta_v A_v) = \frac{A_v}{A_{vf}} = \frac{10^5}{50} = 2 \times 10^3$$

From Equation (12.25), the input resistance is

$$R_{if} = R_i(1 + \beta_v A_v) = (10)(2 \times 10^3) \text{ k}\Omega \Rightarrow 20 \text{ M}\Omega$$

and, from Equation (12.28), the output resistance is

$$R_{of} = \frac{R_o}{(1 + \beta_v A_v)} = \frac{20}{2 \times 10^3} \text{ k}\Omega \Rightarrow 10 \ \Omega$$

**Comment:** With a series input connection, the input resistance increases drastically, and with a shunt output connection, the output resistance decreases substantially, with negative feedback. These are the desired characteristics of a voltage amplifier.

### EXERCISE PROBLEM

**Ex 12.5:** An ideal series–shunt feedback amplifier is shown in Figure 12.6. Assume $R_S$ is negligibly small. (a) If $V_i = 100$ mV, $V_{fb} = 99$ mV, and $V_o = 5$ V, determine $A_v$, $\beta_v$, and $A_{vf}$, including units. (b) Using the results of part (a), determine $R_{if}$ and $R_{of}$, for $R_i = 5 \text{ k}\Omega$ and $R_o = 4 \text{ k}\Omega$. (Ans. (a) $A_v = 5000$ V/V, $\beta_v = 0.0198$ V/V, $A_{vf} = 50$ V/V (b) $R_{if} = 500 \text{ k}\Omega$, $R_{of} = 40 \ \Omega$)

### 12.3.2    Shunt–Series Configuration

The configuration of an ideal **shunt–series** feedback amplifier is shown in Figure 12.9. The circuit consists of a basic current amplifier with an input resistance $R_i$ and an open-loop current gain $A_i$. The feedback circuit samples the output current and produces a feedback current $I_{fb}$, which is in shunt with an input signal current $I_i$. In this ideal configuration, the feedback circuit does not load down the basic amplifier output; therefore, the load current $I_o$ is not affected.

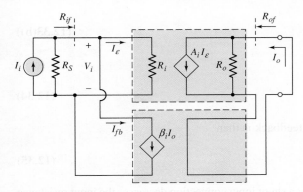

**Figure 12.9** Ideal shunt–series feedback topology

Current $I_\varepsilon$ is the difference between the input signal current and the feedback current and is the error signal. The error signal is amplified in the basic current amplifier. We can recognize the shunt connection on the input and the series connection on the output for this configuration.

This circuit is a current-controlled current source and is an ideal current amplifier. The feedback circuit samples the output current and provides a feedback signal in shunt with the signal current. An increase in output current produces an increase in feedback current, which in turn decreases the error current. The smaller error current is then amplified, producing a smaller output current and stabilizing the output signal.

The input source shown is a Norton equivalent circuit; it could be converted to a Thevenin equivalent circuit.

If the output is essentially a short circuit, then the output current is

$$I_o = A_i I_\varepsilon \tag{12.29}$$

and the feedback current is

$$I_{fb} = \beta I_o = \beta_i I_o \tag{12.30}$$

The parameter $\beta_i$ is the feedback current transfer function. The input signal current, assuming $R_S$ is large, is

$$I_i = I_\varepsilon + I_{fb} \tag{12.31}$$

Combining Equations (12.29), (12.30), and (12.31) yields the closed-loop current transfer function

$$A_{if} = \frac{I_o}{I_i} = \frac{A_i}{(1 + \beta_i A_i)} \tag{12.32}$$

Equation (12.32) is the **closed-loop current gain** of the feedback amplifier.

The form of the equation for the current transfer function of the current amplifier (shunt–series connection) is the same as that for the voltage transfer function of the voltage amplifier (series–shunt connection). We will show that this will be the same for the two feedback connections yet to be discussed.

The input resistance of the shunt–series configuration is $R_{if}$. Starting with Equation (12.31), using Equations (12.29) and (12.30), we find that

$$I_i = I_\varepsilon + I_{fb} = I_\varepsilon + \beta_i I_o = I_\varepsilon + \beta_i (A_i I_\varepsilon) \qquad \textbf{(12.33(a))}$$

or

$$I_\varepsilon = \frac{I_i}{(1 + \beta_i A_i)} \qquad \textbf{(12.33(b))}$$

The input voltage is

$$V_i = I_\varepsilon R_i = \frac{I_i R_i}{(1 + \beta_i A_i)} \qquad \textbf{(12.34)}$$

The input resistance with feedback is then

$$R_{if} = \frac{V_i}{I_i} = \frac{R_i}{(1 + \beta_i A_i)} \qquad \textbf{(12.35)}$$

Equation (12.35) shows that a shunt input connection decreases the input resistance compared to that of the basic amplifier. A small input resistance is a desirable property of a current amplifier, to avoid loading effects on the input signal current source.

The output resistance of the feedback circuit can be determined from the equivalent circuit in Figure 12.10. The input signal current is set equal to zero (an open circuit) and a test current is applied to the output terminals. Since the input signal current source is assumed to be ideal we have $R_S = \infty$.

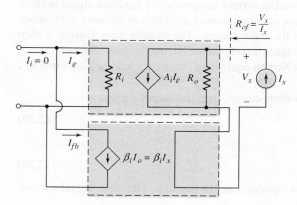

Figure 12.10 Ideal shunt–series feedback configuration for determining output resistance

From the circuit, we see that

$$I_\varepsilon + I_{fb} = I_\varepsilon + \beta_i I_x = 0 \qquad \textbf{(12.36(a))}$$

or

$$I_\varepsilon = -\beta_i I_x \qquad \textbf{(12.36(b))}$$

The output voltage can be written as

$$V_x = (I_x - A_i I_\varepsilon)R_o = [I_x - A_i(-\beta_i I_x)]R_o$$
$$= I_x(1 + \beta_i A_i)R_o \qquad \textbf{(12.37)}$$

Therefore,

$$R_{of} = \frac{V_x}{I_x} = (1 + \beta_i A_i)R_o \qquad \textbf{(12.38)}$$

Equation (12.38) shows that a series output connection increases the output resistance compared to that of the basic amplifier. A large output resistance is a desirable property of a current amplifier, to avoid loading effects on the output signal due to a load connected to the amplifier output.

The equivalent circuit of this feedback current amplifier is shown in Figure 12.11.

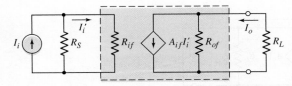

**Figure 12.11** Equivalent circuit of shunt–series feedback circuit, or current amplifier

## EXAMPLE 12.6

**Objective:** Determine the input resistance of a shunt input connection and the output resistance of a series output connection, for a feedback current amplifier.

Consider a shunt–series feedback amplifier in which the open-loop gain is $A_i = 10^5$ and the closed-loop gain is $A_{if} = 50$. Assume the input and output resistances of the basic amplifier are $R_i = 10 \text{ k}\Omega$ and $R_o = 20 \text{ k}\Omega$, respectively.

**Solution:** The ideal closed-loop current transfer function, from Equation (12.32), is

$$A_{if} = \frac{A_i}{(1 + \beta_i A_i)}$$

or

$$(1 + \beta_i A_i) = \frac{A_i}{A_{if}} = \frac{10^5}{50} = 2 \times 10^3$$

From Equation (12.35), the input resistance is

$$R_{if} = \frac{R_i}{(1 + \beta_i A_i)} = \frac{10}{2 \times 10^3} \text{ k}\Omega \Rightarrow 5\ \Omega$$

and from Equation (12.38), the output resistance is

$$R_{of} = (1 + \beta_i A_i)R_o = (2 \times 10^3)(20) \text{ k}\Omega \Rightarrow 40 \text{ M}\Omega$$

**Comment:** With a shunt input connection, the input resistance decreases drastically, and with a series output connection, the output resistance increases substantially, assuming negative feedback. These are the desired characteristics of a current amplifier.

**Ex 12.6:** Consider the ideal shunt–series feedback amplifier in Figure 12.9. Assume that the source resistance is $R_S = \infty$. (a) If $I_i = 100\ \mu A$, $I_{fb} = 99\ \mu A$, and $I_o = 5$ mA, determine $A_i$, $\beta_i$, and $A_{if}$, including units. (b) Using the results of part (a), determine $R_{if}$ and $R_{of}$, for $R_i = 5\ k\Omega$ and $R_o = 4\ k\Omega$. (Ans. (a) $A_i = 5000$ A/A, $\beta_i = 0.0198$ A/A, $A_{if} = 50$ A/A (b) $R_{if} = 50\ \Omega$, $R_{of} = 400\ k\Omega$)

### 12.3.3   Series–Series Configuration

The configuration of an ideal **series–series** feedback amplifier is shown in Figure 12.12. The feedback samples a portion of the output current and converts it to a voltage. This feedback circuit can therefore be thought of as a voltage-to-current amplifier.

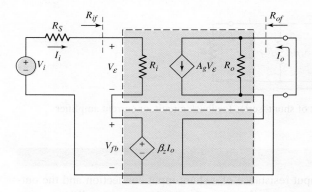

**Figure 12.12** Ideal series–series feedback topology

The circuit consists of a basic amplifier that converts the error voltage to an output current with a gain factor $A_g$ and that has an input resistance $R_i$. The feedback circuit samples the output current and produces a feedback voltage $V_{fb}$, which is in series with the input signal voltage $V_i$.

Assuming the output is essentially a short circuit, the output current is

$$I_o = A_g V_\varepsilon$$

and the feedback voltage is

$$V_{fb} = \beta_z I_o$$

where $\beta_z$ is called a resistance feedback transfer function, with units of resistance. The input signal voltage, neglecting the effect of $R_S$, is

$$V_i = V_\varepsilon + V_{fb}$$

Combining these equations, as we have in previous analyses, yields the closed-loop current-to-voltage transfer function,

$$A_{gf} = \frac{I_o}{V_i} = \frac{A_g}{(1 + \beta_z A_g)} \tag{12.39}$$

The units of the transfer function given by Equation (12.39) are amperes/volt, or conductance. We may note that the term $\beta_z A_g$ is dimensionless. This particular feedback circuit is therefore called a **transconductance amplifier.**

The input and output resistances are a function of the specific types of input and output connections, respectively. The input resistance for the series connection is given by Equation (12.25), which shows that with this configuration, the input resistance increases compared to that of the basic amplifier. The output resistance for the series connection is given by Equation (12.38), which shows that with this configuration, the output resistance increases compared to that of the basic amplifier. The equivalent circuit for the series–series feedback amplifier is shown in Figure 12.13.

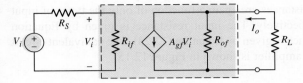

**Figure 12.13**  Equivalent circuit of series–series feedback circuit, or transconductance amplifier

### 12.3.4  Shunt–Shunt Configuration

The configuration of the ideal **shunt–shunt** feedback amplifier is shown in Figure 12.14. The feedback samples a portion of the output voltage and converts it to a current. This feedback circuit can therefore be thought of as a current-to-voltage amplifier.

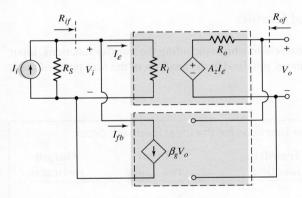

**Figure 12.14**  Ideal shunt–shunt feedback topology

The circuit consists of a basic amplifier that converts the error current to an output voltage with a gain factor $A_z$ and that has an input resistance $R_i$. The feedback circuit samples the output voltage and produces a feedback current $I_{fb}$, which is in shunt with the input signal current $I_i$.

Assuming the output is essentially an open circuit, the output voltage is

$$V_o = A_z I_\varepsilon$$

and the feedback current is

$$I_{fb} = \beta_g V_o$$

where $\beta_g$ is the conductance feedback transfer function, with units of conductance. The input signal current, assuming $R_S$ is very large, is

$$I_i = I_\varepsilon + I_{fb}$$

Combining these equations yields the closed-loop voltage-to-current transfer function,

$$A_{zf} = \frac{V_o}{I_i} = \frac{A_z}{(1 + \beta_g A_z)} \qquad (12.40)$$

The units of the transfer function given by Equation (12.40) are volts/ampere, or resistance. We may note that the term $\beta_g A_z$ is dimensionless. This particular feedback circuit is therefore referred to as a **transresistance amplifier.**

The input and output resistances are again a function of only the types of input and output connections, respectively. The input resistance is given by Equation (12.35) and the output resistance is given by Equation (12.28). The equivalent circuit for the shunt–shunt feedback amplifier is shown in Figure 12.15.

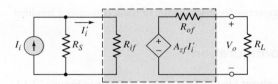

**Figure 12.15** Equivalent circuit of shunt–shunt feedback circuit or, transresistance amplifier

### 12.3.5  Summary of Results

Table 12.1 summarizes the ideal relationships, including the transfer functions, input resistances, and output resistances, obtained in the analysis of the four types of feedback amplifiers.

| Table 12.1 | Summary results of feedback amplifier functions for the ideal feedback circuit | | | | |
|---|---|---|---|---|---|
| Feedback amplifier | Source signal | Output signal | Transfer function | Input resistance | Output resistance |
| Series–shunt (voltage amplifier) | Voltage | Voltage | $A_{vf} = \dfrac{V_o}{V_i} = \dfrac{A_v}{(1 + \beta_v A_v)}$ | $R_i(1 + \beta_v A_v)$ | $\dfrac{R_o}{(1 + \beta_v A_v)}$ |
| Shunt–series (current amplifier) | Current | Current | $A_{if} = \dfrac{I_o}{I_i} = \dfrac{A_i}{(1 + \beta_i A_i)}$ | $\dfrac{R_i}{(1 + \beta_i A_i)}$ | $R_o(1 + \beta_i A_i)$ |
| Series–series (transconductance amplifier) | Voltage | Current | $A_{gf} = \dfrac{I_o}{V_i} = \dfrac{A_g}{(1 + \beta_z A_g)}$ | $R_i(1 + \beta_z A_g)$ | $R_o(1 + \beta_z A_g)$ |
| Shunt–shunt (transresistance amplifier) | Current | Voltage | $A_{zf} = \dfrac{V_o}{I_i} = \dfrac{A_z}{(1 + \beta_g A_z)}$ | $\dfrac{R_i}{(1 + \beta_g A_z)}$ | $\dfrac{R_o}{(1 + \beta_g A_z)}$ |

Having analyzed the characteristics of the four ideal feedback topologies, we will next derive the transfer functions and resistance characteristics of op-amp and discrete transistor representations of each type of feedback configuration. We will compare actual results with the ideal results, discussing any deviations from the ideal.

## Test Your Understanding

**TYU 12.4** An ideal series–series feedback amplifier is shown in Figure 12.12. Assume $R_S$ is negligibly small. If $V_i = 100$ mV, $V_{fb} = 99$ mV, and $I_o = 5$ mA, determine $A_g$, $\beta_z$, and $A_{gf}$, including units. (Ans. $A_g = 5$ A/V, $\beta_z = 19.8$ V/A, $A_{gf} = 50$ mA/V)

**TYU 12.5** Consider the ideal shunt–shunt feedback amplifier in Figure 12.14. Assume that the source resistance is $R_S = \infty$. If $I_i = 100$ $\mu$A, $I_{fb} = 99$ $\mu$A, and $V_o = 5$ V, determine $A_z$, $\beta_g$, and $A_{zf}$, including units. (Ans. $A_z = 5 \times 10^6$ V/A, $\beta_g = 1.98 \times 10^{-5}$ A/V, $A_{zf} = 50$ V/mA)

## 12.4    VOLTAGE (SERIES–SHUNT) AMPLIFIERS

**Objective:** • Analyze op-amp and discrete transistor circuit examples of series–shunt (voltage) feedback amplifiers.

In this section, we will analyze an op-amp and a discrete circuit representation of the series–shunt feedback configuration. Since the series–shunt circuit is a voltage amplifier, we will derive the transfer function relating the output signal voltage to the input signal voltage. For the ideal configuration, this function is shown in Equation (12.22) and is

$$A_{vf} = \frac{A_v}{(1 + \beta_v A_v)}$$

where $A_v$ is the basic amplifier voltage gain and $\beta_v$ is the voltage feedback transfer function. We found that, in this feedback configuration, the input resistance increases and the output resistance decreases compared to the basic amplifier values.

### 12.4.1    Op-Amp Circuit Representation

Figure 12.16 shows a noninverting op-amp circuit, which is an example of the series–shunt configuration. The input signal is the input voltage $V_i$, the feed-back voltage is $V_{fb}$, and the error signal is the voltage $V_\varepsilon$. Since the shunt output samples the output voltage, the feedback voltage is a function of the output voltage.

In the ideal feedback circuit, the amplification factor $A_v$ is very large; from Equation (12.22), the transfer function is then

$$A_{vf} = \frac{V_o}{V_i} \cong \frac{1}{\beta_v} \qquad \textbf{(12.41)}$$

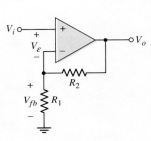

**Figure 12.16** Example of an op-amp series–shunt feedback circuit

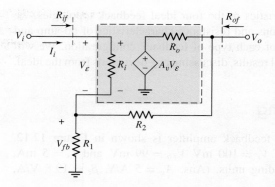

**Figure 12.17** Equivalent circuit, op-amp series–shunt feedback configuration

For the ideal noninverting op-amp amplifier, we found in Chapter 9 that

$$A_{vf} = \frac{V_o}{V_i} = \left(1 + \frac{R_2}{R_1}\right) \tag{12.42}$$

Therefore, the feedback transfer function $\beta_v$ is

$$\beta_v = \frac{1}{\left(1 + \frac{R_2}{R_1}\right)} \tag{12.43}$$

We can take a finite amplifier gain into account by considering the equivalent circuit in Figure 12.17. The parameter $A_v$ is the open-loop voltage gain of the basic amplifier. We can write, for $R_o \approx 0$,

$$V_o = A_v V_\varepsilon \tag{12.44}$$

and

$$V_\varepsilon = V_i - V_{fb} \tag{12.45}$$

therefore,

$$V_o = A_v \left(V_i - V_{fb}\right) \tag{12.46}$$

Assuming the input resistance $R_i$ is very large, the feedback voltage is given by

$$V_{fb} \cong \left(\frac{R_1}{R_1 + R_2}\right) V_o \tag{12.47}$$

Substituting Equation (12.47) into (12.46) and rearranging terms, we obtain

$$A_{vf} = \frac{V_o}{V_i} = \frac{A_v}{1 + \dfrac{A_v}{\left(1 + \dfrac{R_2}{R_1}\right)}} \tag{12.48}$$

The voltage feedback transfer function $\beta_v$ is given by Equation (12.43), and the closed-loop voltage transfer function can be written

$$A_{vf} = \frac{A_v}{(1 + \beta_v A_v)} \tag{12.49}$$

The voltage transfer function for the noninverting op-amp circuit has the same form as that for the ideal series–shunt configuration, assuming the input resistance $R_i$ is very large.

We may note in this case that the voltage gain $A_v$ of the basic amplifier is positive and that the feedback transfer function $\beta_v$ is also positive, so that the loop gain $T = \beta_v A_v$ is positive for negative feedback.

We can now derive the expression for the input resistance $R_{if}$. We see from the figure that $V_\varepsilon = I_i R_i$, $V_o = A_v V_\varepsilon$, and $V_i = V_\varepsilon + V_{fb}$. The approximate feedback voltage is given by Equation (12.47). Therefore, the input voltage is

$$V_i = V_\varepsilon + \left( \frac{R_1}{R_1 + R_2} \right) V_o = V_\varepsilon + \frac{A_v V_\varepsilon}{\left( 1 + \dfrac{R_2}{R_1} \right)}$$

$$= V_\varepsilon \left[ 1 + \frac{A_v}{(1 + R_2/R_1)} \right] \qquad (12.50)$$

The input resistance is then

$$R_{if} = \frac{V_i}{I_i} = \frac{V_i}{(V_\varepsilon/R_i)}$$

$$= R_i \left[ 1 + \frac{A_v}{(1 + (R_2/R_1))} \right] = R_i (1 + \beta_v A_v) \qquad (12.51)$$

The expression for the input resistance for the op-amp circuit has the same form as that for the ideal series input connection, as given in Equation (12.25). In the ideal case in which the gain is $A_v = \infty$, the input resistance of the noninverting op-amp is also infinite. However, if the gain is finite, the input resistance will also be finite.

## EXAMPLE 12.7

**Objective:** Determine the expected input resistance of the noninverting op-amp circuit.

Consider the noninverting op-amp in Figure 12.16, with parameters $R_i = 50 \text{ k}\Omega$, $R_1 = 10 \text{ k}\Omega$, $R_2 = 90 \text{ k}\Omega$, and $A_v = 10^4$.

**Solution:** The feedback transfer function $\beta_v$ is

$$\beta_v = \frac{1}{\left( 1 + \dfrac{R_2}{R_1} \right)} = \frac{1}{\left( 1 + \dfrac{90}{10} \right)} = 0.10$$

The input resistance is therefore

$$R_{if} = R_i(1 + \beta_v A_v) = (50)[1 + (0.10)(10^4)]$$

or

$$R_{if} \cong 50 \times 10^3 \text{ k}\Omega = 50 \text{ M}\Omega$$

**Comment:** Even with a moderate differential input resistance $R_i$ to the op-amp, the closed-loop input resistance $R_{if}$ is very large, because of the series input feedback connection.

**Ex 12.7:** Consider the noninverting op-amp circuit shown in Figure 12.16, with parameters $R_1 = 15\,\text{k}\Omega$, $R_2 = 60\,\text{k}\Omega$, and $A_v = 5 \times 10^4$. Assume $R_i = \infty$. Let the input signal voltage be $V_i = 0.10\,\text{V}$. (a) What is the ideal voltage gain and the ideal output voltage? (b) (i) Determine the actual closed-loop gain and the actual output voltage. (ii) What is the error voltage $V_\varepsilon$? (c) If the open-loop gain increases by a factor of 10, what are the values of (i) the closed-loop gain and (ii) the error voltage? (Ans. (a) $A_f = 5.00$, $V_o = 0.500\,\text{V}$; (b) (i) $A_f = 4.9995$, $V_o = 0.49995\,\text{V}$, (ii) $V_\varepsilon = 9.999\,\mu\text{V}$; (c) (i) $A_f = 4.99995$, (ii) $V_\varepsilon = 0.99999\,\mu\text{V}$)

The analysis results for the noninverting op-amp circuit are consistent with the ideal series–shunt feedback characteristics.

## 12.4.2    Discrete Circuit Representation

Figures 12.18(a) and (b) show the basic emitter-follower and source-follower circuits, which we examined in previous chapters. These are examples of discrete-circuit series–shunt feedback topologies. The input signal is the voltage $v_i$, the error signal is the base-emitter voltage in the emitter follower and the gate-source voltage in the source follower, and the feedback voltage is equal to the output voltage, which means that the feedback transfer function is $\beta_v = 1$.

The small-signal equivalent circuit of the emitter follower is shown in Figure 12.18(c). Since we have already analyzed the emitter-follower circuit, we will simply state the results here. The small-signal voltage gain is

$$A_{vf} = \frac{V_o}{V_i} = \frac{\left(\dfrac{1}{r_\pi} + g_m\right) R_E}{1 + \left(\dfrac{1}{r_\pi} + g_m\right) R_E} = \frac{\dfrac{R_E}{r_e}}{1 + \dfrac{R_E}{r_e}} \tag{12.52}$$

where

$$r_e = \frac{r_\pi}{(1 + g_m r_\pi)}$$

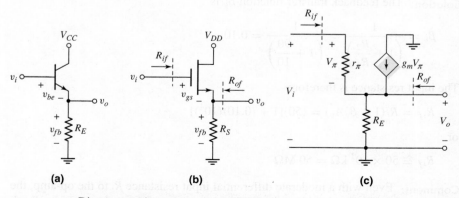

**Figure 12.18** Discrete transistor series–shunt feedback circuits: (a) emitter-follower, (b) source-follower, and (c) small-signal equivalent circuit of emitter follower

The voltage gain of the emitter follower can be written as a voltage divider equation. Since the feedback transfer function is unity, the form of the voltage gain expression is the same as that for the ideal series–shunt configuration, as given in Equation (12.22). The open-loop voltage gain corresponds to

$$A_v = \left( \frac{1}{r_\pi} + g_m \right) R_E = \frac{R_E}{r_e} \tag{12.53}$$

The closed-loop input resistance is[2]

$$R_{if} = r_\pi + (1 + h_{FE})R_E = r_\pi \left[ 1 + \left( \frac{1}{r_\pi} + g_m \right) R_E \right] \tag{12.54}$$

The form of the input resistance is also the same as that of the ideal expression, given by Equation (12.25). The input resistance increases with a series input connection.

The output resistance of the emitter-follower circuit is given by

$$R_{of} = R_E \left\| \frac{r_\pi}{1 + h_{FE}} = R_E \right\| r_e \tag{12.55}$$

which can be written in the form

$$R_{of} = \frac{R_E}{1 + \left( \dfrac{1}{r_\pi} + g_m \right) R_E} \tag{12.56}$$

The output resistance decreases with a shunt output connection. For the emitter-follower circuit, the form of the output resistance is also the same as that of the ideal expression, given by Equation (12.28).

Even though the magnitude of the emitter-follower voltage gain is slightly less than unity, this circuit is a classic example of a series–shunt feedback configuration, which represents a voltage amplifier.

## DESIGN EXAMPLE 12.8

**Objective:** Design a feedback amplifier to amplify the output signal of a microphone to meet a set of specifications.

**Specifications:** The output signal from the microphone is 10 mV and the output signal from the feedback amplifier is to be 0.5 V in order to drive a power amplifier that in turn will drive the speakers. The nominal output resistance of the microphone is $R_S = 5$ k$\Omega$ and the nominal input resistance of the power amplifier is $R_L = 75$ $\Omega$.

**Choices:** An op-amp with parameters $R_i = 10$ k$\Omega$, $R_o = 100$ $\Omega$, and a low-frequency gain of $A_v = 10^4$ is available. [Note: In this simple design, neglect frequency response.]

**Solution (Design Approach):** Since the source resistance is fairly large, an amplifier with a large input resistance is required to minimize loading at the input. Also, since the load resistance is low, an amplifier with a low output resistance is required to minimize loading at the output. To satisfy these requirements, a series–shunt feedback configuration, or voltage amplifier, should be used.

---

[2]Reminder: In this chapter, the parameter $h_{FE}$ is used as the transistor current gain to avoid confusion with $\beta$, which is used as the feedback transfer function. Again, we assume that the dc and ac current gains are equal; therefore, $h_{FE} = h_{fe} = g_m r_\pi$.

The closed-loop voltage gain must be $A_{vf} = 0.5/0.01 = 50$. For the ideal case, $A_{vf} = 1/\beta_v$, so the feedback transfer function is $\beta_v = 1/50 = 0.02$. The loop gain is then

$$T = \beta_v A_v = (0.02)(10^4) = 200$$

Referring to Table 12.1, we expect the input resistance to be

$$R_{if} \cong (10)(200) \text{ k}\Omega \rightarrow 2 \text{ M}\Omega$$

and the output resistance to be

$$R_{of} \cong (100/200) \ \Omega = 0.5 \ \Omega$$

These input and output resistance values will minimize any loading effects at the amplifier input and output terminals.

If we use the noninverting amplifier configuration in Figure 12.16, then we have

$$\frac{1}{\beta_v} = 1 + \frac{R_2}{R_1} = 50$$

and

$$\frac{R_2}{R_1} = 49$$

The feedback network loads the output of the amplifier; consequently, we need $R_1 + R_2$ to be much larger than $R_o$. However, the output resistance of the feedback network is in series with the input terminals, so extremely large values of $R_1$ and $R_2$ will reduce the actual signal applied to the op-amp because of voltage divider action. Initially, then, we choose $R_1 = 1 \text{ k}\Omega$ and $R_2 = 49 \text{ k}\Omega$.

**Computer Simulation Verification:** The circuit in Figure 12.19 was used in a PSpice analysis of the voltage amplifier. A standard 741 op-amp was used in the circuit. For a 10 mV input signal, the output signal was 499.6 mV, for a gain of 49.96. This result is within 0.08 percent of the ideal designed value. The input resistance $R_{if}$ was found to be approximately 580 M$\Omega$ and the output resistance $R_{of}$ was determined to be approximately 0.042 $\Omega$. The differences between the measured input and output

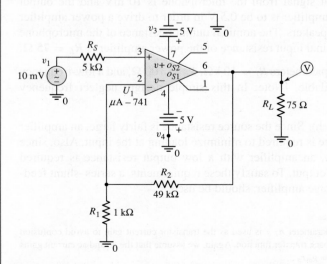

**Figure 12.19** Circuit used in the computer simulation analysis in Example 12.8

resistances compared to the predicted values are due to the differences between the actual $\mu$A-741 op-amp parameters and the assumed parameters. However, the measured input resistance is larger than predicted and the measured output resistance is smaller than predicted, which is desired and more in line with an ideal op-amp circuit.

**Comment:** An almost ideal feedback voltage amplifier can be realized if an op-amp is used in the circuit.

---

**EXERCISE PROBLEM**

**\*Ex 12.8:** Design a feedback voltage amplifier to provide a voltage gain of 15. The nominal voltage source resistance is $R_S = 2\,\text{k}\Omega$, and the nominal load is $R_L = 100\,\Omega$. An op-amp with parameters $R_i = 5\,\text{k}\Omega$, $R_o = 50\,\Omega$, and a low-frequency open-loop gain of $A_v = 5 \times 10^3$ is available. Correlate the design with a computer simulation analysis to determine the voltage gain, input resistance, and output resistance.

## Test Your Understanding

**TYU 12.6** Assume the transistor in the emitter-follower circuit in Figure 12.18(a) is biased such that $I_{CQ} = 1.2\,\text{mA}$. Let $R_E = 1.5\,\text{k}\Omega$. (a) If the transistor current gain is $h_{FE} = 120$, determine $A_{vf}$, $R_{if}$, and $R_{of}$. (b) Determine the percent change in $A_{vf}$, $R_{if}$, and $R_{of}$ if the transistor current gain increases to $h_{FE} = 180$. Assume the quiescent collector current remains unchanged. (Ans. (a) $A_{vf} = 0.985877$, $R_{if} = 184.1\,\text{k}\Omega$, $R_{of} = 21.18\,\Omega$; (b) $A_{vf}$: $-0.00386\%$, $R_{if}$: $+49.6\%$, $R_{of}$: $+0.283\%$)

**TYU 12.7** (a) Assume the transistor in the source-follower circuit shown in Figure 12.18(b) is biased at $I_{DQ} = 250\,\mu\text{A}$. Let $R_S = 3\,\text{k}\Omega$. If the transistor parameters are $K_n = 0.5\,\text{mA/V}^2$, $V_{TN} = 0.8\,\text{V}$, and $\lambda = 0$, determine $A_{vf}$ and $R_{of}$. (b) Determine the percent change in $A_{vf}$ and $R_{of}$ if the quiescent drain current is increased to $I_{DQ} = 1\,\text{mA}$. (Ans. (a) $A_{vf} = 0.6796$, $R_{of} = 961\,\Omega$; (b) $A_{vf}$: $+19.1\%$, $R_{of}$: $-40.5\%$)

---

## 12.5  CURRENT (SHUNT–SERIES) AMPLIFIERS

**Objective:** • Analyze op-amp and discrete transistor circuit examples of shunt–series (current) feedback amplifiers.

In this section, we will analyze an op-amp and a discrete circuit representation of the shunt–series feedback amplifier. The shunt–series circuit is a current amplifier; therefore, we must derive the output current to input current transfer function. For the ideal configuration, this function is given in Equation (12.32):

$$A_{if} = \frac{A_i}{(1 + \beta_i A_i)}$$

where $A_i$ is the basic amplifier current gain and $\beta_i$ is the current feedback transfer function. For this amplifier, the input resistance decreases and the output resistance increases compared to the basic amplifier values.

### 12.5.1 Op-Amp Circuit Representation

Figure 12.20 shows an op-amp current amplifier, which is a shunt–series configuration. The input signal is the current $I_i'$ from the Norton equivalent source of $I_i$ and $R_S$. The feedback current is $I_{fb}$, the error signal is the current $I_\varepsilon$, and the output signal is the current $I_o$. With the shunt input connection, the input resistance $R_{if}$ is small, as previously stated. Resistance $R_S$ is the output resistance of the current source and is normally large. If $R_S \gg R_{if}$, then $I_i' \cong I_i$.

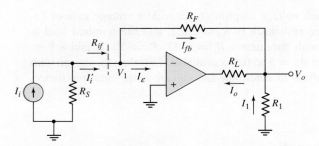

**Figure 12.20** Example of an op-amp shunt–series feedback circuit

If we assume initially that $I_\varepsilon$ is negligible, then, from Figure 12.20, we have

$$I_i \cong I_i' = I_{fb}$$

The output voltage $V_o$, assuming $V_1$ is at virtual ground, is

$$V_o = -I_{fb}R_F = -I_i R_F$$

and current $I_1$ is

$$I_1 = -V_o/R_1$$

The output current can be expressed

$$I_o = I_{fb} + I_1 = I_i + \left(-\frac{1}{R_1}\right)(-I_i R_F) = I_i\left(1 + \frac{R_F}{R_1}\right) \qquad (12.57)$$

Therefore, the ideal current gain is

$$\frac{I_o}{I_i} = 1 + \frac{R_F}{R_1} \qquad (12.58)$$

In the ideal feedback circuit, the amplification factor $A_i$ is very large; consequently, the current transfer function, from Equation (12.32), becomes

$$A_{if} = \frac{I_o}{I_i} \cong \frac{1}{\beta_i} \qquad (12.59)$$

Comparing Equation (12.59) with (12.58), we see that the current feedback transfer function for the ideal op-amp current amplifier is

$$\beta_i = \frac{1}{\left(1 + \dfrac{R_F}{R_1}\right)} \qquad (12.60)$$

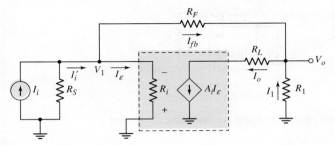

**Figure 12.21** Equivalent circuit, op-amp shunt–series feedback configuration

We can take the finite amplifier gain into account by considering the equivalent circuit in Figure 12.21. The parameter $A_i$ is the open-loop current gain. We have

$$I_o = A_i I_\varepsilon \tag{12.61}$$

and

$$I_\varepsilon = I_i' - I_{fb} \cong I_i - I_{fb} \tag{12.62}$$

therefore,

$$I_o = A_i(I_i - I_{fb}) \tag{12.63}$$

If we again assume that $V_1$ is at virtual ground, voltage $V_o$ is given by

$$V_o = -I_{fb}R_F \tag{12.64}$$

We can then write

$$I_1 = -\frac{V_o}{R_1} = -\left(\frac{1}{R_1}\right)(-I_{fb}R_F) = I_{fb}\left(\frac{R_F}{R_1}\right) \tag{12.65}$$

The output current is also expressed as

$$I_o = I_{fb} + I_1 = I_{fb} + I_{fb}\left(\frac{R_F}{R_1}\right) \tag{12.66}$$

Solving for $I_{fb}$ from Equation (12.66), substituting that into Equation (12.63), and rearranging terms yields the closed-loop current gain

$$A_{if} = \frac{I_o}{I_i} = \frac{A_i}{1 + \dfrac{A_i}{\left(1 + \dfrac{R_F}{R_1}\right)}} \tag{12.67}$$

Since the current feedback transfer function is $\beta_i = 1/[1 + (R_F/R_1)]$, the closed-loop current gain expression for the op-amp current amplifier has the same form as that for the ideal shunt–series configuration.

### 12.5.2 Simple Discrete Circuit Representation

Figure 12.22(a) shows the ac equivalent circuit of a common-base circuit, which is an example of a simple discrete shunt–series configuration. Figure 12.22(b) is the same circuit rearranged to demonstrate more clearly the input, feedback, and error components of

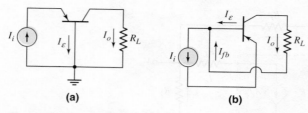

**(a)**                              **(b)**

**Figure 12.22** (a) Equivalent circuit for simple common-base circuit and (b) reconfigured circuit

the currents. The output current is equal to the feedback current, which means that the feedback transfer function is $\beta_i = 1$. The basic amplifier gain is

$$I_o/I_\varepsilon = A_i = h_{FE}$$

which is simply the common-emitter current gain of the transistor.

From Figure 12.22(b), we see that the closed-loop current transfer function or gain is

$$A_{if} = \frac{I_o}{I_i} = \frac{h_{FE}}{1 + h_{FE}} = \frac{A_i}{1 + A_i} \tag{12.68}$$

Since the current feedback transfer function $\beta_i$ is unity, Equation (12.68) has the same form as that for the ideal shunt–series transfer function.

Figure 12.23(a) is a more realistic common-base circuit. Resistor $R_E$ and the supply voltages $V^+$ and $V^-$ bias the transistor in the forward-active mode. The ac equivalent circuit is in Figure 12.23(b). We can show that the current gain is

$$A_{if} = \frac{I_o}{I_i} = \frac{h_{FE}}{\left(1 + \dfrac{r_\pi}{R_E}\right) + h_{FE}} = \frac{A_i}{\left(1 + \dfrac{r_\pi}{R_E}\right) + A_i} \tag{12.69}$$

Equation (12.69) does not have the same form as the ideal shunt–series feedback transfer function. This is common in many discrete transistor feedback circuits. The reason is that resistor $R_E$ introduces loading effects that are not present in the ideal configuration. Typically, then, the transfer functions of actual discrete circuits are not the same as for the ideal case.

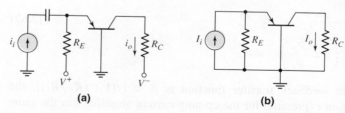

**(a)**                              **(b)**

**Figure 12.23** (a) Common-base circuit, including biasing and (b) ac equivalent circuit

### 12.5.3    Discrete Circuit Representation

Figure 12.24(a) shows a two-stage discrete transistor circuit example of a shunt–series feedback configuration. While the large number of capacitors makes this circuit somewhat impractical, it can be used to illustrate the basic concepts of feedback.

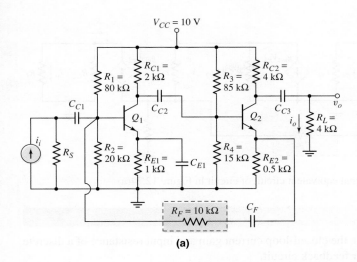

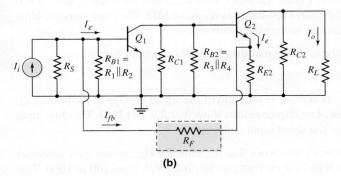

**Figure 12.24** (a) Example of a discrete transistor shunt–series feedback circuit and (b) ac equivalent circuit

Figure 12.24(b) shows the ac equivalent circuit, in which all capacitors act as short circuits. With the shunt input connection, the input signal current is essentially $I_i$ (assuming $R_S$ is large), the feedback current is $I_{fb}$, and the error signal is $I_\varepsilon$. The signal emitter current $I_e$ is directly proportional to the load current $I_o$, and the feedback current is directly proportional to $I_e$, demonstrating that this series output connection samples the output current $I_o$.

[Note: It may be argued that, even though $I_e$ is related to the output current $I_o$, the output current is not part of the feedback circuit. In particular, the output resistance $r_{o2}$ of $Q_2$ is not within the feedback network. For this reason, the output connection may be thought of as a shunt connection with the output signal being a voltage at the emitter of $Q_2$. However, we are assuming the output signal is a current so we will treat this circuit as a shunt-series amplifier.]

The small-signal equivalent circuit is shown in Figure 12.25. We assume that the small-signal output resistance $r_o$ of each transistor is infinite. We could derive the expression for the closed-loop current gain by writing and solving a set of simultaneous nodal equations. However, as with most discrete transistor feedback circuits, the transfer function cannot be arranged exactly in the ideal form without several approximations. For this circuit, then, we rely on a computer analysis to provide the required results.

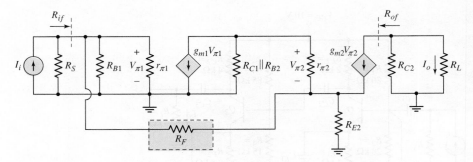

**Figure 12.25** Small-signal equivalent circuit of circuit in Figure 12.24(a)

## EXAMPLE 12.9

**Objective:** Determine the closed-loop current gain and input resistance of a discrete shunt–series transistor feedback circuit.

Consider the circuit in Figure 12.24(a), with transistor parameters $h_{FE} = 100$ and $V_A = \infty$. Assume the source resistance is $R_S = 10$ M$\Omega$. The capacitors are large enough to act as short circuits to the signal currents.

**Solution:** A PSpice analysis shows that the closed-loop current gain is

$$A_{if} = I_o/I_i = 9.58$$

The input resistance $R_{if}$ is defined as the ratio of the signal voltage at the base of $Q_1$ to the input signal current. The PSpice results show that $R_{if} = 134 \ \Omega$. This low input resistance is expected for the shunt input connection.

**Comment:** The PSpice analysis shows that the closed-loop current gain increases from 9.58 to 10.2 as the transistor current gain $h_{FE}$ increases from 100 to 1000. This result again demonstrates a principal characteristic of feedback circuits, which is that the transfer function is relatively insensitive to changes in the individual transistor parameters.

## EXERCISE PROBLEM

**\*Ex 12.9:** Consider the common-base circuit in Figure 12.23(a), with transistor parameters $h_{FE} = 80$, $V_{EB}(\text{on}) = 0.7$ V, and $V_A = \infty$. Assume the transistor is biased at $I_{CQ} = 0.5$ mA. Redesign the circuit such that the closed-loop current gain is greater than 0.95. (Ans. $R_E(\text{min}) = 1.30$ k$\Omega$, and $V^+(\text{min}) = 1.36$ V)

From the small-signal equivalent circuit in Figure 12.25, we find that the output resistance $R_{of}$ looking into the collector of $Q_2$ is very large. If $r_o$ of $Q_2$ is assumed to be infinite, then $R_{of}$ is also infinite. We expect a large output impedance for the series output connection of this feedback circuit.

## DESIGN EXAMPLE 12.10

**Objective:** Design a feedback amplifier to provide a given current gain.

**Specifications:** Assume that a signal current source has a nominal output resistance of $R_S = 10$ k$\Omega$ and that the amplifier will drive a nominal load of $R_L = 50 \ \Omega$. A current gain of 10 is required.

**Choices:** An op-amp with the same characteristics described in Example 12.8 is available.

**Solution (Design Approach):** An amplifier with a low input resistance and a large output resistance is required, to minimize loading effects at the input and output. For these reasons, a shunt–series feedback configuration, or current amplifier, will be used.

The closed-loop gain is

$$A_{if} = 10 \cong 1/\beta_i$$

and the feedback transfer function is $\beta_i = 0.1$.

The dependent open-loop voltage source of the op-amp, as shown in Figure 12.17, can be transformed to an equivalent dependent open-loop current source, as shown in Figure 12.9. We find that

$$A_i = A_v R_i / R_o$$

Using the parameters specified for the op-amp, we find $A_i = 10^6$. The loop gain for the shunt–series configuration is

$$A_i \beta_i = (10^6)(0.1) = 10^5$$

Referring to Table 12.1, we expect the input resistance to be

$$R_{if} = 10/10^5 \text{ k}\Omega \rightarrow 0.1 \ \Omega$$

and the output resistance to be

$$R_{of} = (100)(10^5) \ \Omega \rightarrow 10 \text{ M}\Omega$$

These resistance values will minimize any loading effects at the amplifier input and output.

For the shunt–series configuration in Figure 12.20, we have

$$\frac{1}{\beta_i} = 1 + \frac{R_F}{R_1} = 10$$

or

$$R_F / R_1 = 9$$

For our purposes, $R_1$ must be fairly small, to avoid a loading effect at the output. However, $R_1$ must not be too small, to avoid large currents in the amplifier. Therefore, we choose $R_1 = 1 \text{ k}\Omega$ and $R_F = 9 \text{ k}\Omega$.

**Computer Simulation Verification:** Figure 12.26 shows the circuit used in the computer simulation. A standard $\mu$A-741 op-amp was used in the circuit. The current gain was found to be exactly 10.0. The input resistance $R_{if}$ looking into the op-amp with feedback was found to be 0.056 $\Omega$, which compares favorably to the predicted value of 0.1 $\Omega$. The output resistance seen by the load resistor was found to be approximately 200 M$\Omega$. This value is on the order of 20 times larger than the predicted value, but is closer to the ideal value. The differences between predicted and measured values are due to the differences in assumed op-amp parameters and the $\mu$A-741 op-amp parameters.

**Comment:** This design also produces an almost ideal feedback current amplifier, if reasonable values of feedback resistors are used.

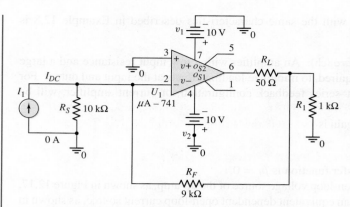

**Figure 12.26** Circuit used in the computer simulation analysis in Example 12.10

**Ex 12.10:** Design a feedback current amplifier to provide a current gain of 15. The nominal current source resistance is $R_S = 500 \ \Omega$, and the nominal load is $R_L = 200 \ \Omega$. An op-amp with parameters $R_i = 5 \ \text{k}\Omega$, $R_o = 50 \ \Omega$, and a low-frequency open-loop voltage gain of $A_v = 5 \times 10^3$ is available. Correlate the design with a PSpice analysis to determine the current gain, input resistance, and output resistance.

## Test Your Understanding

**TYU 12.8** Consider the shunt–series feedback circuit in Figure 12.24(a). Using a computer simulation analysis, investigate the magnitude of the current gain $A_{if}$ as the emitter resistor $R_{E2}$ is varied between 0.4 k$\Omega$ and 1.6 k$\Omega$. What is the relationship between $R_F$, $R_{E2}$, and $A_{if}$?

**TYU 12.9** Consider the shunt–series feedback circuit in Figure 12.24(a). Using a computer simulation analysis, investigate the magnitude of the input resistance $R_{if}$ as the feedback resistor $R_F$ is varied between 5 k$\Omega$ and 50 k$\Omega$. What is the influence of $R_F$ on the input resistance $R_{if}$?

---

 **12.6   TRANSCONDUCTANCE (SERIES–SERIES) AMPLIFIERS**

**Objective:** • Analyze op-amp and discrete transistor circuit examples of series–series (transconductance) feedback amplifiers.

In this section, we will analyze an op-amp and a discrete circuit representation of the series–series feedback amplifier. The series–series circuit is a transconductance amplifier; therefore, we must derive the output current to input voltage transfer function. For the ideal configuration, this function is, from Equation (12.39),

$$A_{gf} = \frac{A_g}{(1 + \beta_z A_g)}$$

where $A_g$ is the basic amplifier transconductance gain and $\beta_z$ is the resistance feedback transfer function. We found that with this feedback configuration, both the input and output resistances increase compared to the basic amplifier values.

### 12.6.1    Op-Amp Circuit Representation

The op-amp circuit in Figure 12.27 is an example of the series–series feedback configuration. The input signal is the input voltage $V_i$, the feedback voltage is $V_{fb}$, and the error signal is the voltage $V_\varepsilon$. The series output connection samples the output current, which means that the feedback voltage is a function of the output current.

In the ideal feedback circuit, the amplification factor $A_g$ is very large; therefore, from Equation (12.39), the transfer function is

$$A_{gf} = \frac{I_o}{V_i} \cong \frac{1}{\beta_z} \tag{12.70}$$

Assuming an ideal op-amp circuit and neglecting the transistor base current, we have

$$V_i = V_{fb} = I_o R_E$$

and

$$A_{gf} = \frac{I_o}{V_i} = \frac{1}{R_E} \tag{12.71}$$

Comparing Equations (12.70) and (12.71), we see that the ideal feedback transfer function is

$$\beta_z = R_E \tag{12.72}$$

We can take a finite amplifier gain into account by considering the equivalent circuit in Figure 12.28. The parameter $A_g$ is the open-loop transconductance gain of

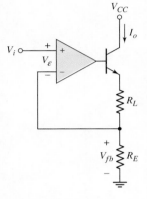

**Figure 12.27** Example of an op-amp series–series feedback circuit

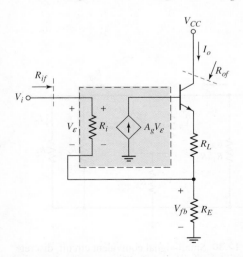

**Figure 12.28** Equivalent circuit, op-amp series–series feedback configuration

the amplifier. Assuming the collector and emitter currents are nearly equal and $R_i$ is very large, we can write that

$$I_o = \frac{V_{fb}}{R_E} = h_{FE} I_b = h_{FE} A_g V_\varepsilon \tag{12.73}$$

Also,

$$V_\varepsilon = V_i - V_{fb} = V_i - I_o R_E \tag{12.74}$$

Substituting Equation (12.74) into Equation (12.73) yields

$$I_o = h_{FE} A_g (V_i - I_o R_E) \tag{12.75}$$

which can be rearranged to yield the closed-loop transfer function,

$$A_{gf} = \frac{I_o}{V_i} = \frac{(h_{FE} A_g)}{1 + (h_{FE} A_g) R_E} \tag{12.76}$$

which has the same form as that of the ideal theory. In this example, we see that in this feedback network, the transistor current gain is part of the basic amplifier gain.

### 12.6.2    Discrete Circuit Representation

Figure 12.29 shows a single bipolar transistor circuit that is an example of a series–series feedback configuration. This circuit is similar to those evaluated in Chapters 5 and 6. The input signal is the input voltage $v_i$, the feedback voltage is $v_{fb}$, and the error signal is the base–emitter voltage. The series output connection samples the output current; therefore, the feedback voltage is a function of the output current.

The small-signal equivalent circuit is shown in Figure 12.30. The Early voltage of the transistor is assumed to be infinite. The output current can be written

$$I_o = -(g_m V_\pi)\left(\frac{R_C}{R_C + R_L}\right) \tag{12.77}$$

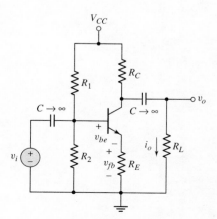

Figure 12.29  Example of a discrete transistor series–series feedback circuit

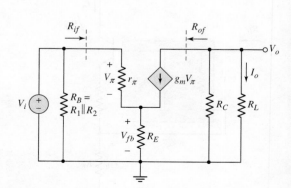

Figure 12.30  Small-signal equivalent circuit, discrete transistor series–series feedback configuration

and the feedback voltage is

$$V_{fb} = \left( \frac{V_\pi}{r_\pi} + g_m V_\pi \right) R_E \qquad (12.78)$$

A KVL equation around the B–E loop yields

$$V_i = V_\pi + V_{fb} = V_\pi \left[ 1 + \left( \frac{1}{r_\pi} + g_m \right) R_E \right] \qquad (12.79)$$

Solving Equation (12.79) for $V_\pi$, substituting that into Equation (12.77), and rearranging terms produces the expression for the transconductance transfer function,

$$A_{gf} = \frac{I_o}{V_i} = \frac{-g_m \left( \dfrac{R_C}{R_C + R_L} \right)}{1 + \left( \dfrac{1}{r_\pi} + g_m \right) R_E} \qquad (12.80)$$

Again, the closed-loop transfer function of the discrete transistor feedback circuit cannot be put in exactly the same form as that of the ideal series–series feedback network. Resistor $R_C$ introduces loading on the output, and $r_\pi$ introduces loading on the input. If both $R_C$ and $r_\pi$ become large, then Equation (12.80) changes to the ideal form, where the feedback transfer function is $\beta_z = -R_E$ and the basic amplifier transconductance is $A_g = -g_m$.

## EXAMPLE 12.11

**Objective:** Determine the transconductance gain of a transistor feedback circuit.

Consider the circuit in Figure 12.29, with transistor parameters $h_{FE} = 100$, $V_{BE}(\text{on}) = 0.7$ V, and $V_A = \infty$. The circuit parameters are: $V_{CC} = 10$ V, $R_1 = 55$ k$\Omega$, $R_2 = 12$ k$\Omega$, $R_E = 1$ k$\Omega$, $R_C = 4$ k$\Omega$, and $R_L = 4$ k$\Omega$.

**Solution:** From a dc analysis of the circuit, the quiescent values are $I_{CQ} = 0.983$ mA and $V_{CEQ} = 5.08$ V. The transistor small-signal parameters are found to be $r_\pi = 2.64$ k$\Omega$ and $g_m = 37.8$ mA/V.

From Equation (12.80), the transconductance transfer function is

$$A_{gf} = \frac{-(37.8)\left( \dfrac{4}{4+4} \right)}{1 + \left( \dfrac{1}{2.64} + 37.8 \right)(1)} = -0.482 \text{ mA/V}$$

As a first approximation, we have

$$A_{gf} = \frac{1}{\beta_z} = \frac{1}{-R_E} = \frac{1}{-1 \text{ k}\Omega} = -1 \text{ mA/V}$$

The term $R_C/(R_C + R_L)$ introduces the largest discrepancy between the actual and ideal transconductance values.

This circuit is often used as a voltage amplifier. The output voltage is directly proportional to the output current. Therefore,

$$A_{vf} = \frac{v_o}{v_i} = \frac{i_o R_L}{v_i} = A_{gf} R_L$$

which yields

$$A_{vf} = (-0.482)(4) = -1.93$$

**Comment:** The circuit in Figure 12.29 is an example of a series–series feedback topology, even though in many cases we treat this circuit as a voltage amplifier. When an emitter resistor is included, the small-signal voltage gain decreases, because of the feedback effect of $R_E$. However, the transconductance and voltage gain become insensitive to the transistor parameters, also a result of the feedback effect of $R_E$. A 100 percent increase in the transistor current gain $h_{FE}$ produces a 0.5 percent change in the closed-loop voltage gain.

---

EXERCISE PROBLEM

**Ex 12.11:** For the circuit in Figure 12.31, the transistor parameters are $K_n = 2\,\text{mA/V}^2$, $V_{TN} = 2\,\text{V}$, and $\lambda = 0$. (a) Determine (i) $I_{DQ}$ and (ii) the transconductance transfer function $A_{gf} = i_o/v_i$. (b) If the conductance parameter decreases by 10 percent to $K_n = 1.8\,\text{mA/V}^2$, determine (i) the new value of $I_{DQ}$ and (ii) the percent change in $A_{gf}$. (Ans. (a) (i) $I_{DQ} = 2.31\,\text{mA}$, (ii) $A_{gf} = -0.7904\,\text{mA/V}$; (b) (i) $I_{DQ} = 2.22\,\text{mA}$, (ii) $-2.68\%$)

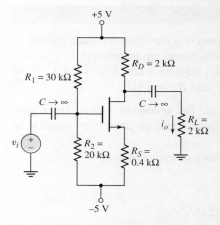

**Figure 12.31** Figure for Exercise Ex 12.11

The input resistance $R_{if}$ of the series input feedback connection includes $R_E$ multiplied by $(1 + h_{FE})$, where $h_{FE}$ is the transistor current gain. The input resistance increases significantly because of the series connection.

The output resistance of a series output feedback connection is usually very large. However, resistance $R_C$ reduces the output resistance and introduces a loading effect. The reduced output resistance demonstrates that discrete transistor feedback circuits do not conform exactly to ideal feedback circuits. Nevertheless, overall circuit characteristics improve when feedback is used.

## DESIGN EXAMPLE 12.12

**Objective:** Design a driver amplifier to supply current to an LED.

**Specifications:** The available voltage source is variable from 0 to 5 V and has an output resistance of 200 $\Omega$. The required diode current is 10 mA when the maximum input voltage is applied. The required closed-loop transconductance gain is then $A_{gf} = I_o/V_i = (10 \times 10^{-3})/5 \to 2$ mS.

**Choices:** An op-amp with the characteristics described in Example 12.8 and a BJT with $h_{FE} = 100$ are available.

**Solution (Design Approach):** To minimize loading effects on the input, an amplifier with a large input resistance is required; to minimize loading effects on the output, a large output resistance is required. For these reasons, a series–series feedback configuration, or transconductance amplifier, is selected.

The closed-loop gain is

$$A_{gf} = 2 \times 10^{-3} \cong 1/\beta_z$$

and the resistance feedback transfer function is

$$\beta_z = 500 \ \Omega$$

The dependent open-loop voltage source of the op-amp, as shown in Figure 12.17, can be transformed to an equivalent dependent op-loop transconductance source for the transconductance amplifier, as shown in Figure 12.12. We find that

$$A_g = A_v/R_o$$

The parameters specified for the op-amp yield

$$A_g = 100 \ \text{A/V}$$

The loop gain for the series–series configuration is

$$A_g\beta_z = (100)(500) = 5 \times 10^4$$

Referring to Table 12.1, the expected input resistance is

$$R_{if} = (10)(5 \times 10^4) \ \text{k}\Omega \to 500 \ \text{M}\Omega$$

and the expected output resistance is

$$R_{of} = (100)(5 \times 10^4) \ \Omega \to 5 \ \text{M}\Omega$$

These input and output resistances should minimize any loading effects at the amplifier input and output.

For this example, we may use the amplifier configuration shown in Figure 12.27, in which the load resistor $R_L$ is replaced by an LED. In the ideal case,

$$\beta_z = R_E = 500 \ \Omega$$

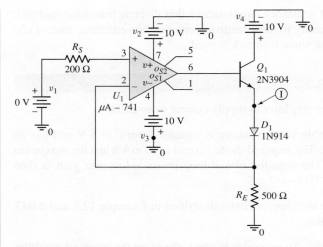

**Figure 12.32** Circuit used in the computer simulation analysis for Example 12.12

**Computer Simulation Verification:** Figure 12.32 shows the circuit used in the computer simulation. Again, a standard $\mu$A-741 op-amp was used in the circuit and a standard diode was used in place of an LED. When the input voltage reached 5 V, the current through the diode was 10.0 mA, which was the design value. The input resistance $R_{if}$ was found to be approximately 2400 M$\Omega$ and the output resistance $R_{of}$ was found to be approximately 60 M$\Omega$. Both of these values are larger than predicted because of the differences in the assumed op-amp parameters and those of the $\mu$A-741 op-amp.

**Comment:** Again, an almost ideal feedback circuit can be designed by using an op-amp.

### EXERCISE PROBLEM

**\*Ex 12.12:** Design a transconductance feedback amplifier with a gain of $A_{gf} = 10$ mS. The source resistance is $R_S = 500$ $\Omega$, and the load is an LED. State any necessary assumptions. Use an op-amp with the characteristics described in Example 12.8. From a computer simulation analysis, determine the closed-loop transconductance, input resistance, and output resistance of your design.

## Test Your Understanding

**TYU 12.10** Consider the op-amp circuit in Figure 12.27 with parameters $R_E = 1$ k$\Omega$ and $A_g = 10^2$ mA/V. Assume the transistor current gain is $h_{FE} = 180$. The input voltage is $V_i = 1.5$ V. (a) (i) Determine the transfer function $A_{gf} = I_o/V_i$ and the output current $I_o$. (ii) Determine the value of the error voltage $V_\varepsilon$. (b) If the transistor current gain decreases by 20 percent to $h_{FE} = 144$, (i) determine the new values of $A_{gf}$ and $I_o$, and the percent change in these values and (ii) determine the new value of $V_\varepsilon$. (Ans. (a) (i) $A_{gf} = 0.9999444$ mA/V, $I_o = 1.4999166$ mA, (ii) $V_\varepsilon = 83.4$ $\mu$V; (b) (i) $A_{gf} = 0.9999306$ mA/V, $I_o = 1.4998959$ mA, $-0.00138\%$ change, (ii) $V_\varepsilon = 104$ $\mu$V)

# 12.7 TRANSRESISTANCE (SHUNT–SHUNT) AMPLIFIERS

**Objective:** • Analyze op-amp and discrete transistor circuit examples of shunt–shunt (transresistance) feedback amplifiers.

In this section, we will analyze an op-amp and a discrete circuit representation of the shunt–shunt feedback amplifier. The shunt-shunt circuit is a transresistance amplifier; therefore, we must derive the output voltage to input current transfer function. For the ideal configuration, this function is given by Equation (12.40) as

$$A_{zf} = \frac{A_z}{(1 + \beta_g A_z)}$$

where $A_z$ is the basic amplifier transresistance gain, and $\beta_g$ is the feedback transfer function. With this feedback connection, both the input and output resistance decrease compared to the basic amplifier values.

## 12.7.1 Op-Amp Circuit Representation

Figure 12.33(a) shows the basic inverting op-amp circuit that we analyzed in Chapter 9. We treated this as a voltage amplifier whose voltage gain is $A_v = -V_o/V_i$. However, this circuit is actually an example of a shunt–shunt configuration. The defining input signal is the input current $I_i$.

Figure 12.33(b) shows the same circuit without the input resistance. From this configuration, we see the input shunt connection. The input current splits between the feedback current $I_{fb}$ and the error current $I_\varepsilon$. The shunt output connection samples the output voltage; therefore, the feedback current is a function of the output voltage.

In the ideal feedback circuit, the amplification factor $A_z$ is very large, and the transresistance transfer function is, from Equation (12.40),

$$A_{zf} = \frac{V_o}{I_i} \cong \frac{1}{\beta_g} \tag{12.81}$$

For the ideal inverting op-amp circuit, $V_1$ is at virtual ground, and

$$V_o = -I_{fb}R_2$$

Also for the ideal op-amp, $I_{fb} = I_i$, and the ideal transresistance transfer function is

$$A_{zf} = \frac{V_o}{I_i} = -R_2 \tag{12.82}$$

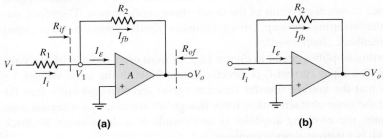

**(a)**        **(b)**

**Figure 12.33** (a) The basic inverting op-amp circuit and (b) the circuit showing the shunt input connection

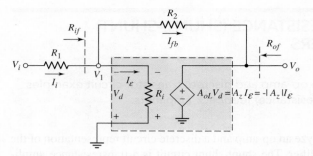

**Figure 12.34** Equivalent circuit, op-amp shunt–shunt feedback configuration

Comparing Equation (12.82) to Equation (12.81), we see that the feedback transfer function for the ideal inverting op-amp circuit is

$$\beta_g = -\frac{1}{R_2} \tag{12.83}$$

We can take a finite amplifier gain into account by considering the equivalent circuit in Figure 12.34. The parameter $A_z$ is the open-loop transresistance gain factor, and is negative since the error current $I_\varepsilon$ is considered to be positive entering the inverting terminal. We can write $V_o = A_z I_\varepsilon = -|A_z| I_\varepsilon$, $I_\varepsilon = I_i - I_{fb}$, and $V_o = A_z (I_i - I_{fb}) = -|A_z| (I_i - I_{fb})$. If we assume that voltage $V_1$ is at virtual ground, then $I_{fb} = -V_o/R_2$. Combining equations, we see that the closed-loop transresistance transfer function is

$$A_{zf} = \frac{V_o}{I_i} = \frac{-|A_z|}{1 + \dfrac{|A_z|}{R_2}} \tag{12.84}$$

From Equation (12.83), the feedback transfer function is $\beta_g = -1/R_2$, and Equation (12.84) becomes

$$A_{zf} = \frac{V_o}{I_i} = \frac{-|A_z|}{1 + (-|A_z|)\beta_g} = \frac{A_z}{1 + A_z \beta_g} \tag{12.85}$$

This feedback circuit is one example in which the open-loop gain of the basic amplifier, $A_z = V_o/I_\varepsilon$, is negative. The feedback transfer function, $\beta_g = -1/R_2$, is also negative, but the loop gain $T = A_z \beta_g$ is positive for this negative feedback circuit.

The transresistance transfer function for the inverting op-amp circuit has the same form as that for the ideal shunt–shunt configuration. In addition, since $V_1$ is at virtual ground, the input resistance including feedback, $R_{if}$, is essentially zero, and we have shown that the output resistance with feedback, $R_{of}$, is very small. These small resistance values are a result of the shunt–shunt configuration. Therefore, our analysis of the inverting op-amp circuit produces results consistent with ideal shunt–shunt feedback characteristics.

The inverting amplifier circuit in Figure 12.33 is most often thought of as a voltage amplifier. The input current $I_i$ is directly proportional to the input voltage $V_i$, which means that the voltage transfer function (gain) and transresistance transfer function have the same characteristics. Even though we are usually concerned with the voltage gain, the inverting amplifier is an example of a shunt–shunt feedback topology which is a transresistance amplifier.

### 12.7.2    Discrete Circuit Representation

Figure 12.35 shows a single bipolar transistor circuit, which is an example of a shunt–shunt feedback configuration. The input signal current is $i_i$, the feedback current is $i_{fb}$, and the error signal current is $i_\varepsilon$ and is the signal base current. The shunt output samples the output voltage; therefore, the feedback current is a function of $v_o$.

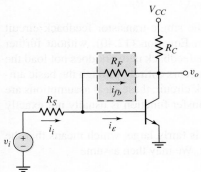

**Figure 12.35** Example of a discrete transistor shunt–shunt feedback circuit

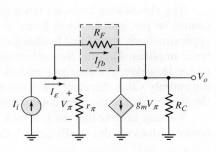

**Figure 12.36** Small-signal equivalent circuit, discrete transistor shunt–shunt feedback configuration

The small-signal equivalent circuit is shown in Figure 12.36. The input signal is assumed to be an ideal signal current source. Also the Early voltage of the transistor is assumed to be infinite.

Writing a KCL equation at the output node, we find

$$\frac{V_o}{R_C} + g_m V_\pi + \frac{V_o - V_\pi}{R_F} = 0 \tag{12.86}$$

A KCL equation at the input node yields

$$I_i = \frac{V_\pi}{r_\pi} + \frac{V_\pi - V_o}{R_F} \tag{12.87}$$

Solving Equation (12.87) for $V_\pi$ and substituting that result into Equation (12.86), we obtain

$$V_o\left(\frac{1}{R_C} + \frac{1}{R_F}\right)\left(\frac{1}{r_\pi} + \frac{1}{R_F}\right) + \left(g_m - \frac{1}{R_F}\right)\left(I_i + \frac{V_o}{R_F}\right) = 0 \tag{12.88}$$

The transresistance transfer function is then

$$A_{zf} = \frac{V_o}{I_i} = \frac{-\left(g_m - \dfrac{1}{R_F}\right)}{\left(\dfrac{1}{R_C} + \dfrac{1}{R_F}\right)\left(\dfrac{1}{r_\pi} + \dfrac{1}{R_F}\right) + \dfrac{1}{R_F}\left(g_m - \dfrac{1}{R_F}\right)} \tag{12.89}$$

The open-loop transresistance gain factor $A_z$ is found by setting $R_F = \infty$. We find

$$A_z = \frac{-g_m}{\left(\dfrac{1}{R_C}\right)\left(\dfrac{1}{r_\pi}\right)} = -g_m r_\pi R_C = -h_{FE} R_C \tag{12.90}$$

where $h_{FE}$ is the common-emitter transistor current gain. Multiplying both numerator and denominator of Equation (12.89) by $(r_\pi R_C)$, we obtain the closed-loop transresistance gain,

$$A_{zf} = \frac{V_o}{I_i} = \frac{+\left(A_z + \dfrac{r_\pi R_C}{R_F}\right)}{\left(1 + \dfrac{R_C}{R_F}\right)\left(1 + \dfrac{r_\pi}{R_F}\right) - \dfrac{1}{R_F}\left(A_z + \dfrac{r_\pi R_C}{R_F}\right)} \tag{12.91}$$

The closed-loop transresistance gain for the single-transistor feedback circuit cannot be put into the ideal form, as given in Equation (12.40), without further approximations. In an ideal feedback circuit, the feedback network does not load the basic amplifier. Also, the forward transmission occurs entirely through the basic amplifier. However, in a discrete transistor feedback circuit, these ideal assumptions are not entirely valid; therefore, the form of the transfer function is usually not exactly the same as that of the ideal configuration.

We may assume that the feedback resistor is fairly large, which means that the feedback does not drastically perturb the circuit. We may then assume

$$h_{FE} = g_m r_\pi \gg (r_\pi / R_F)$$

If we also assume that $R_C \ll R_F$ and $r_\pi \ll R_F$, then Equation (12.91) reduces to

$$A_{zf} = \frac{V_o}{I_i} \cong \frac{A_z}{1 + (A_z)\left(\dfrac{-1}{R_F}\right)} \tag{12.92}$$

Consequently, the feedback transfer function is approximately

$$\beta_g \cong \frac{-1}{R_F} \tag{12.93}$$

Equation (12.93) demonstrates that the approximate value of the feedback transfer function depends only on a resistance value.

Although the actual closed-loop transfer function does not fit the ideal form, the magnitude of that function depends less on the individual transistor parameters than does the open-loop gain. This characteristic is one of the general properties of feedback circuits.

Also, since the input current is proportional to the input voltage, we can use this circuit as a voltage amplifier.

### EXAMPLE 12.13

**Objective:** Determine the transresistance and voltage gain of a single-transistor shunt–shunt feedback circuit.

Consider the circuit in Figure 12.37(a). The transistor parameters are: $h_{FE} = 100$, $V_{BE}(\text{on}) = 0.7$ V, and $V_A = \infty$. Since the input signal current is directly proportional to the input voltage, the voltage gain of this shunt–shunt configuration has the same general properties as the transresistance transfer function.

As with many circuits considered in this chapter, several capacitors are included. In the circuit in Figure 12.37(a), $R_1$ and $C_{C2}$ may be removed. Resistor $R_F$ can be used for biasing, and the circuit can be redesigned to provide the same feedback properties.

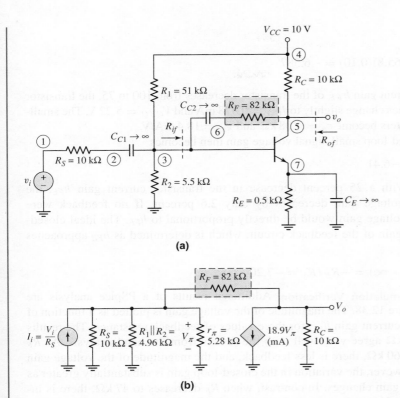

(a)

(b)

**Figure 12.37** (a) Circuit for Example 12.13 and (b) small-signal equivalent circuit

**Solution:** By including $C_{C2}$ in the circuit, the feedback is a function of the ac signal only, which means that the transistor quiescent values are not affected by feedback. The quiescent parameters are found to be $I_{CQ} = 0.492$ mA and $V_{CEQ} = 5.08$ V, and the small-signal parameters are $r_\pi = 5.28$ kΩ and $g_m = 18.92$ mA/V.

In the small-signal equivalent circuit, which is shown in Figure 12.37(b), the Thevenin equivalent input source is converted to a Norton equivalent circuit. Writing a KCL equation at the output, we obtain

$$\frac{V_o}{10} + (18.9)V_\pi + \frac{V_o - V_\pi}{82} = 0$$

A KCL equation at the input yields

$$I_i = \frac{V_\pi}{10} + \frac{V_\pi}{4.96} + \frac{V_\pi}{5.28} + \frac{V_\pi - V_o}{82}$$

Combining these two equations and eliminating $V_\pi$, we find the small-signal transresistance gain, which is

$$A_{zf} = \frac{V_o}{I_i} = -65.87 \text{ k}\Omega$$

Since this unit of gain is not as familiar as voltage gain, we determine the voltage gain from

$$I_i = V_i/R_S = V_i/10$$

Therefore,

$$\frac{V_o}{V_i} = -(65.8)(0.10) = -6.587$$

If the current gain $h_{FE}$ of the transistor decreases from 100 to 75, the transistor quiescent values change slightly to $I_{CQ} = 0.478$ mA and $V_{CEQ} = 5.22$ V. The small-signal parameters become $r_\pi = 4.08$ k$\Omega$ and $g_m = 18.4$ mA/V.

The closed-loop small-signal voltage gain then becomes

$$V_o/V_i = -6.41$$

**Comment:** With a 25 percent decrease in the transistor current gain $h_{FE}$, the closed-loop voltage gain decreases by only 2.6 percent. If no feedback were present, the voltage gain would be directly proportional to $h_{FE}$. The ideal closed-loop voltage gain of the feedback circuit, which is determined as $h_{FE}$ approaches infinity, is

$$A_v(h_{FE} \to \infty) = -R_F/R_S = -7.20$$

**Computer Simulation Verification:** Additional results of a PSpice analysis are shown in Figure 12.38. The magnitude of the voltage gain is plotted as a function of the transistor current gain $h_{FE}$, for three values of feedback resistance. The results for $R_F = 82$ k$\Omega$ agree very well with the results from the hand analysis. As $R_F$ increases to 160 k$\Omega$, there is less feedback, and the magnitude of the voltage gain increases. However, the variation in the closed-loop gain is substantially greater as the transistor gain changes. In contrast, when $R_F$ decreases to 47 k$\Omega$, there is increased feedback, and the magnitude of the voltage gain decreases. However, there is very little variation in closed-loop gain as the transistor gain changes. In all cases, as the gain of the transistor increases, there is less change in closed-loop gain. This result demonstrates the need for a large gain in the basic amplifier in the feedback network.

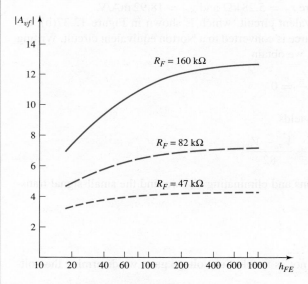

**Figure 12.38** Voltage gain magnitude versus transistor current gain, for three values of feedback resistance, from a PSpice analysis of the circuit in Figure 12.37(a)

Expressions for the input and output resistances of the ideal shunt–shunt configuration are given in Equations (12.35) and (12.28), respectively. As with the loop gain function, the input and output resistance expressions for the single-transistor feedback circuit cannot be put in exactly the same form as that for the ideal configuration. However, the same general characteristics are obtained; that is, both input and output resistances decrease, predicted by the ideal case.

## EXERCISE PROBLEM

**Ex 12.13:** Consider the circuit in Figure 12.39, with transistor parameters $V_{TN} = 0.8$ V, $K_n = 1.5\,\text{mA/V}^2$, and $\lambda = 0$. (a) (i) Find the open-loop gain for $R_F = \infty$. (ii) Find the closed-loop gain for $R_F = 47\,\text{k}\Omega$. (b) Repeat part (a) if the conductance parameter decreases by 15 percent to $K_n = 1.275\,\text{mA/V}^2$. What is the percent change in the magnitude of each gain factor? (Ans. (a) (i) $A_v = -3.528$, (ii) $A_{vf} = -1.204$; (b) (i) $A_v = -3.0$, $-15\%$ change; (ii) $A_{vf} = -1.107$, $-8.06\%$ change)

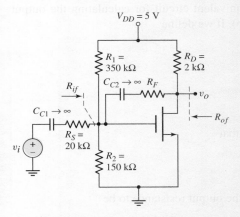

**Figure 12.39**  Circuit for Exercises Ex12.13 and Ex12.14

## EXAMPLE 12.14

**Objective:** Determine the input and output resistances of a single-transistor shunt–shunt feedback circuit.

Consider the circuit in Figure 12.37(a), with transistor parameters: $h_{FE} = 100$, $V_{BE}(\text{on}) = 0.7$ V, and $V_A = \infty$.

**Solution: Input Resistance:** The small-signal equivalent circuit for calculating the input resistance $R_{if}$ is shown in Figure 12.40(a). The small-signal transistor parameters were determined in Example 12.13.

Writing a KCL equation at the input, we have

$$I_x = \frac{V_\pi}{r_\pi} + \frac{V_\pi - V_o}{R_F} = \frac{V_\pi}{5.28} + \frac{V_\pi - V_o}{82}$$

From a KCL equation at the output node, we have

$$\frac{V_o}{R_C} + g_m V_\pi + \frac{V_o - V_\pi}{R_F} = \frac{V_o}{10} + (18.9)V_\pi + \frac{V_o - V_\pi}{82} = 0$$

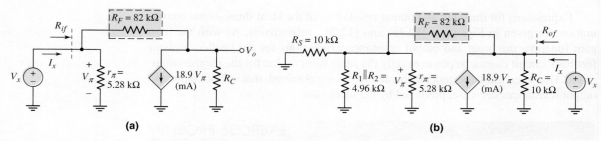

**Figure 12.40** Small-signal equivalent circuits of the circuit in Figure 12.37(a) for calculating (a) input resistance and (b) output resistance

Combining these two equations, eliminating $V_o$, and noting that $V_\pi = V_x$, we find that

$$R_{if} = \frac{V_x}{I_x} = 0.443 \text{ k}\Omega$$

**Output Resistance:** The small-signal equivalent circuit for calculating the output resistance $R_{of}$ is shown in Figure 12.40(b). If we define

$$R_{eq} = r_\pi \| R_1 \| R_2 \| R_S$$

then a KCL equation at node $V_x$ yields

$$I_x = \frac{V_x}{R_C} + g_m V_\pi + \frac{V_x}{R_F + R_{eq}}$$

From a voltage divider equation, we find that

$$V_\pi = \left( \frac{R_{eq}}{R_{eq} + R_F} \right) V_x$$

Combining these two equations, we find the output resistance to be

$$R_{of} = \frac{V_x}{I_x} = 1.75 \text{ k}\Omega$$

**Comment:** The input resistance with no feedback would be $r_\pi = 5.28 \text{ k}\Omega$. The shunt input feedback connection has lowered the input resistance to $R_{if} = 0.443 \text{ k}\Omega$. Similarly, the output resistance with no feedback would be $R_C = 10 \text{ k}\Omega$. The shunt output feedback connection has lowered the output resistance to $R_{of} = 1.75 \text{ k}\Omega$. The decrease in both the input and output resistances agrees with the ideal feedback theory.

**Ex 12.14:** Consider the feedback circuit in Figure 12.39, with transistor parameters $V_{TN} = 0.8$ V and $K_n = 1.5 \text{ mA/V}^2$. Let $R_F = 47 \text{ k}\Omega$. (a) Determine the output resistance $R_{of}$ for $\lambda = 0$. (b) Repeat part (a) for $\lambda = 0.04 \text{ V}^{-1}$. (Ans. (a) $R_{of} = 0.9358 \text{ k}\Omega$, (b) $R_{of} = 0.9107 \text{ k}\Omega$)

The magnitude of the transfer function, input resistance, and output resistance of the discrete transistor feedback circuit all tend to approach the ideal values if additional transistor stages are included to increase the basic amplifier gain. As an example, a multistage shunt–shunt connection is shown in Figure 12.41. Once again, several

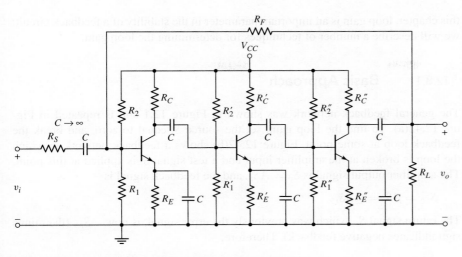

**Figure 12.41** Example of multistage shunt–shunt feedback circuit

capacitors are included, which simplifies the dc analysis. However, the capacitors may adversely affect the circuit frequency response.

Since negative feedback is desired, there must be an odd number of negative gain stages. As the number of stages increases, the open-loop gain increases, and the circuit characteristics approach those of the ideal shunt–shunt configuration. The analysis of this circuit is left as a computer simulation problem at the end of the chapter.

## Test Your Understanding

**TYU 12.11** Consider the BJT feedback circuit in Figure 12.37(a). The transistor parameters are $h_{FE} = 180$, $V_{BE}(\text{on}) = 0.7$ V, and $V_A = \infty$. (a) Determine the voltage gain $A_v = v_o/v_i$ for (i) $R_F = \infty$ and (ii) $R_F = 60\,k\Omega$. (b) Repeat part (a) if $h_{FE}$ decreases to $h_{FE} = 120$. (c) By what percent do the magnitudes of the voltage gains change from part (a) to part (b)? (Ans. (a) (i) $A_v = -48.19$, (ii) $A_{vf} = -5.212$; (b) (i) $A_v = -41.72$, (ii) $A_{vf} = -5.111$; (c) (i) $-13.4\%$, (ii) $-1.93\%$)

**TYU 12.12** The transistor parameters for the circuit shown in Figure 12.37(a) are $h_{FE} = 180$, $V_{BE}(\text{on}) = 0.7$ V, and $V_A = \infty$. (a) Determine the output resistance $R_{of}$ for (i) $R_F = \infty$ and (ii) $R_F = 60\,k\Omega$. (b) Repeat part (a) if $h_{FE}$ decreases to $h_{FE} = 120$. (Ans. (a) (i) $R_o = 10\,k\Omega$, (ii) $R_{of} = 1.126\,k\Omega$; (b) (i) $R_o = 10\,k\Omega$, (ii) $R_{of} = 1.27\,k\Omega$)

## 12.8    LOOP GAIN

**Objective:** • Derive the loop gain of ideal and practical feedback circuits.

In previous sections, the loop gain $T$ was easily determined for circuits involving ideal op-amps. For discrete transistor circuits, however, the loop gain usually cannot be obtained directly from the closed-loop transfer function. As we will see later in this chapter, loop gain is an important parameter in the stability of a feedback circuit; we will describe a number of techniques for determining the loop gain.

### 12.8.1 Basic Approach

The general feedback network was shown in Figure 12.1 and is repeated in Figure 12.42(a). To find the loop gain, set the source $S_i$ equal to zero, and break the feedback loop at some point. Figure 12.42(b) shows a feedback network in which the loop is broken at the amplifier input and a test signal $S_t$ is applied at this point. The amplifier output signal is $S_o = AS_t$, and the feedback signal is

$$S_{fb} = \beta S_o = A\beta S_t$$

The return signal $S_r$, which was previously the error signal, is now $-S_{fb}$ (the minus sign indicates negative feedback). Therefore,

$$\frac{S_r}{S_t} = -A\beta \tag{12.94}$$

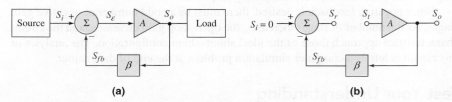

**(a)**                         **(b)**

**Figure 12.42** (a) Ideal configuration of a feedback amplifier; (b) basic feedback network with loop broken at amplifier input

The ratio of the return signal $S_r$ to the test signal $S_t$ is the negative of the loop gain factor.

    As the feedback loop is broken, the conditions that existed prior to the loop being broken must remain unchanged. These conditions include: maintaining the same transistor biasing and maintaining the same impedance at the return point. An equivalent impedance must therefore be inserted at the point where the loop is broken. This is shown in Figure 12.43. Figure 12.43(a) shows the amplifier input

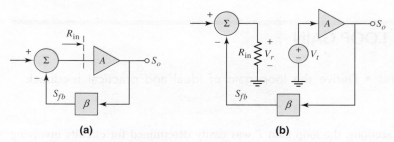

**(a)**                         **(b)**

**Figure 12.43** (a) Basic feedback network, showing amplifier input resistance and (b) feedback network after the loop is broken, showing test voltage and load resistance

impedance $R_{in}$ prior to the loop being broken. Figure 12.43(b) shows the configuration after the loop is broken. A test voltage $V_t$ is applied, and a load impedance $R_{in}$ is inserted at the output of the broken loop. The return voltage is then measured at this output terminal. The loop gain is found to be

$$T = A\beta = -\frac{V_r}{V_t} \qquad (12.95)$$

Also, a test current $I_t$ may be applied and a return current signal $I_r$ measured, to find the loop gain as

$$T = -\frac{I_r}{I_t} \qquad (12.96)$$

As an example, consider the circuit shown in Figure 12.44(a). The circuit is similar to the one considered in Examples 12.13 and 12.14. The feedback loop is broken at the input to the transistor, at the point marked $X$. The small-signal equivalent circuit is shown in Figure 12.44(b). A test voltage is applied to the base of the transistor and the equivalent load resistance $r_\pi$ is connected at the return point. The input signal current is set equal to zero.

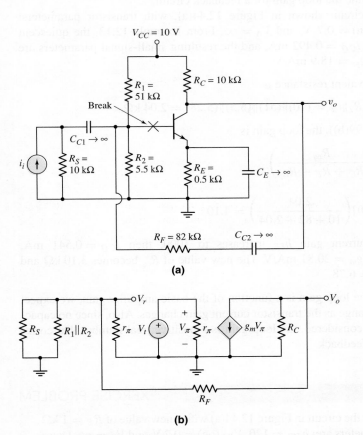

(a)

(b)

**Figure 12.44** (a) Feedback circuit prior to breaking the loop and (b) small-signal equivalent circuit after breaking the loop

Since $V_\pi = V_t$, if we define $R_{eq} = R_S \| R_1 \| R_2 \| r_\pi$, then the output voltage can be written

$$V_o = -g_m V_t [R_C \| (R_F + R_{eq})] \tag{12.97}$$

From a voltage divider, the return voltage $V_r$ expression is

$$V_r = \left( \frac{R_{eq}}{R_F + R_{eq}} \right) V_o \tag{12.98}$$

Substituting Equation (12.97) into Equation (12.98) yields the loop gain

$$T = -\frac{V_r}{V_t} = +g_m \left( \frac{R_{eq}}{R_F + R_{eq}} \right) [R_C \| (R_F + R_{eq})] \tag{12.99(a)}$$

which can be written as

$$T = (g_m R_c) \left( \frac{R_{eq}}{R_C + R_F + R_{eq}} \right) \tag{12.99(b)}$$

## EXAMPLE 12.15

**Objective:** Determine the loop gain for a feedback circuit.

Consider the circuit shown in Figure 12.44(a), with transistor parameters: $h_{FE} = 100$, $V_{BE}(\text{on}) = 0.7$ V, and $V_A = \infty$. From Example 12.13, the quiescent collector current is $I_{CQ} = 0.492$ mA, and the resulting small-signal parameters are $r_\pi = 5.28$ k$\Omega$ and $g_m = 18.9$ mA/V.

**Solution:** The equivalent resistance is

$$R_{eq} = R_S \| R_1 \| R_2 \| r_\pi = (10) \| (51) \| (5.5) \| (5.28) = 2.04 \text{ k}\Omega$$

From Equation (12.99(b)), the loop gain is

$$T = (g_m R_C) \left( \frac{R_{eq}}{R_C + R_F + R_{eq}} \right)$$

$$= [(18.9)(10)] \left( \frac{2.04}{10 + 82 + 2.04} \right) = 4.10$$

If the transistor current gain $h_{FE}$ increases to 1000, then $I_{CQ} = 0.541$ mA, $r_\pi = 48.1$ k$\Omega$, and $g_m = 20.81$ mA/V. The new value of $R_{eq}$ becomes 3.10 k$\Omega$ and the loop gain is $T = 6.78$.

**Comment:** Since the loop gain is a function of the basic amplifier gain, we expect this parameter to change as the transistor current gain changes. Also, since no capacitance effects were considered, the loop gain is a positive, real number that corresponds to negative feedback.

## EXERCISE PROBLEM

**Ex 12.15:** Consider the circuit in Figure 12.44(a) with a new value of $R_E = 1$ k$\Omega$. The transistor parameters are: $h_{FE} = 120$, $V_{BE}(\text{on}) = 0.7$ V, and $V_A = \infty$. Determine the loop gain $T$. (Ans. $T = 2.75$)

### 12.8.2  Computer Analysis

The loop gain can also be determined from a computer analysis of the feedback circuit. In Example 12.16, we demonstrate a direct approach to determining the loop gain. First, we consider the circuit analyzed in the last example, to correlate the results of a computer analysis to those of a hand analysis. Then, we determine the loop gain of a feedback circuit when taking capacitance effects into account.

### EXAMPLE 12.16

**Objective:** Determine the loop gain factor for a feedback circuit, using a computer simulation analysis.

Consider the circuit in Figure 12.44(a).

**Solution:** We determine the loop gain factor by using the circuit in Figure 12.45, in which the loop is effectively broken at the base of the transistor. The circuit conditions, however, must remain unchanged from those prior to breaking the loop. This includes maintaining the same bias currents in the transistor and terminating the broken loop with the proper impedance.

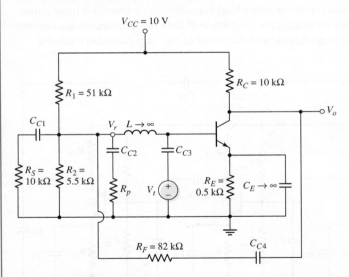

**Figure 12.45** Feedback circuit with the loop effectively broken, for determining the loop gain from a computer analysis

A large inductance is inserted in the transistor base connection, to act as a short circuit for dc signals, so that the proper dc bias can be maintained on the transistor, and to act as an open circuit for ac signals, so that the loop appears to be broken for the ac signal. A test voltage $V_t$ is applied to the base of the transistor through a coupling capacitor, and a load resistance $R_p$ is connected through a coupling capacitor at the return point. These coupling capacitors act as short circuits to the ac signals, but as open circuits to dc signals, so that the dc bias is not disturbed by these elements.

From the computer simulation, the loop gain for a transistor current gain of $h_{FE} = 100$ is

$$T = -V_r/V_t = 5.04$$

For a current gain of 1000, the loop gain is $T = 9.37$. These values differ slightly from the hand analysis results in Example 12.15. The slight difference arises because the quiescent collector currents determined in the hand analysis and the computer analysis are not quite the same, leading to different values of $g_m$ and $r_\pi$.

**Comment:** The analysis of this circuit is straightforward. In the next example, we demonstrate another advantage of a computer analysis.

**Ex 12.16:** Consider the feedback circuit described in Exercise Problem Ex 12.15. Determine the loop gain from a PSpice analysis.

When capacitances are part of the feedback circuit, the phase of the loop gain becomes a factor in determining whether the feedback is negative or positive. Figure 12.46 shows a three-stage amplifier with feedback. Each stage is the same as the circuit given in Figure 12.44(a). For an odd number of stages at low frequency, the loop gain is a positive, real quantity, and negative feedback is applied. The coupling and emitter bypass capacitors are assumed to be very large, and capacitors $C_1$, $C_2$, and $C_3$ between the stages can represent either load capacitances or transistor input capacitances. As the frequency increases, the magnitude of the loop gain decreases, because of decreasing capacitor impedances, and the phase of the loop gain also changes.

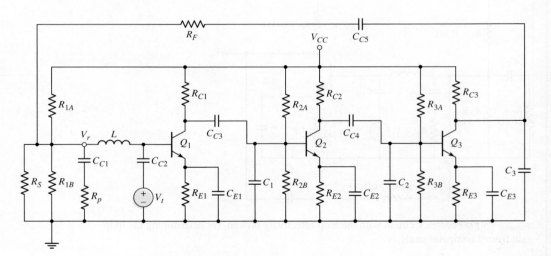

**Figure 12.46** The ac equivalent circuit of three-stage feedback amplifier, including load capacitors

## EXAMPLE 12.17

**Objective:** Determine the magnitude and phase of the loop gain of a multistage feedback circuit.

Consider the circuit in Figure 12.46, with parameters: $R_S = 10\ \text{M}\Omega$, $R_A = 51\ \text{k}\Omega$, $R_B = 5.5\ \text{k}\Omega$, $R_F = 82\ \text{k}\Omega$, $R_C = 10\ \text{k}\Omega$, and $C = 100\ \text{pF}$. The transistor current gains are assumed to be $h_{FE} = 15$, which keeps the overall gain fairly small.

**Solution:** The loop is broken at the base of $Q_1$, and the ratio of the return signal to the test signal is measured by the same technique shown in Figure 12.45.

The magnitude of $V_r/V_t$ versus frequency is shown in Figure 12.47(a). The magnitude of loop gain drops off with frequency, as expected, and is equal to unity at approximately 5.5 MHz.

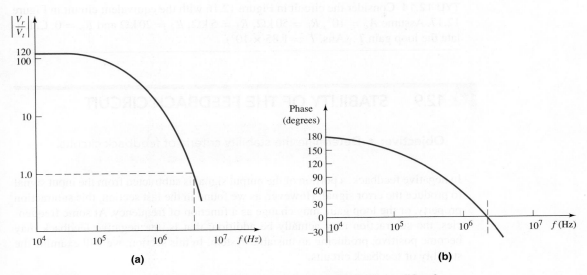

**Figure 12.47** (a) Bode plot of loop gain magnitude for three-stage feedback amplifier, from Example 12.17; (b) phase of the return signal for the three-stage amplifier

The phase of the return signal is shown in Figure 12.47(b). Since the loop gain is given by $T = -V_r/V_t$, then the phase of the loop gain is $\angle T = -180° + \angle V_r - \angle V_t$ where the $-180°$ corresponds to the minus sign. Since the phase of the input signal was set to zero, then the phase of the loop gain is $\angle T = -180° + \angle V_r$. At low frequencies, where the phase of the return signal is approximately $+180°$, the phase of the loop gain is essentially zero, corresponding to negative feedback. At approximately $f = 2.5$ MHz, the phase of the return signal is zero so that the phase of the loop gain is $-180°$, which corresponds to positive feedback.

**Comment:** For this circuit, the loop gain magnitude is greater than unity at the frequency at which the phase of $T$ is $-180$ degrees. As discussed in the next section, this condition means that the circuit is unstable and will oscillate.

**EXERCISE PROBLEM**

**Ex 12.17:** Consider the feedback circuit in Figure 12.16, with the equivalent circuit given in Figure 12.17. Break the feedback loop at an appropriate point, and derive the expression for the loop gain. (Ans. $T = A_v/[1 + R_2/(R_1 \| R_i)]$)

A hand analysis of the three-stage amplifier just considered would be tedious, especially taking the frequency response into account. In this case, a computer analysis is more suitable.

## Test Your Understanding

**TYU 12.13** Consider the circuit in Figure 12.44(a) with parameters described in Example 12.15. Determine the percentage change in the loop gain $T$ as $h_{FE}$ increases from $h_{FE} = 100$ to $h_{FE} = 150$. (Ans. +17.3% change)

**TYU 12.14** Consider the circuit in Figure 12.16 with the equivalent circuit in Figure 12.17. Assume $A_v = 10^4$, $R_i = 50\,\text{k}\Omega$, $R_1 = 5\,\text{k}\Omega$, $R_2 = 20\,\text{k}\Omega$ and $R_o = 0$. Calculate the loop gain $T$. (Ans. $T = 1.85 \times 10^3$)

## 12.9    STABILITY OF THE FEEDBACK CIRCUIT

**Objective:** • Determine the stability criteria of feedback circuits.

In negative feedback, a portion of the output signal is subtracted from the input signal to produce the error signal. However, as we found in the last section, this subtraction property, or the loop gain, may change as a function of frequency. At some frequencies, the subtraction may actually be addition; that is, the negative feedback may become positive, producing an unstable system. In this section, we will examine the stability of feedback circuits.

### 12.9.1    The Stability Problem

The basic feedback configuration is shown in Figure 12.1, and the ideal closed-loop transfer function is given by Equation (12.5), which is repeated here:

$$A_f = \frac{S_o}{S_i} = \frac{A}{(1 + \beta A)} \tag{12.5}$$

The open-loop gain is a function of the individual transistor parameters and capacitances, and is therefore a function of frequency. The closed-loop gain can then be written as

$$A_f(s) = \frac{A(s)}{(1 + \beta A(s))} = \frac{A(s)}{1 + T(s)} \tag{12.100}$$

where $T(s)$ is the loop gain. For physical frequencies, $s = j\omega$, and the loop gain is $T(j\omega)$, which is a complex function. The loop gain can be represented by its magnitude and phase, as follows:

$$T(j\omega) = |T(j\omega)|\angle\phi \tag{12.101}$$

The closed-loop gain can be written

$$A_f(j\omega) = \frac{A(j\omega)}{1 + T(j\omega)} \tag{12.102}$$

The stability of the feedback circuit is a function of the loop gain $T(j\omega)$. If the loop gain magnitude is unity when the phase is 180 degrees, then $T(j\omega) = -1$ and the closed-loop gain goes to infinity. This implies that an output will exist for a zero input, which means that the circuit will oscillate. If we are trying to build a linear amplifier, an oscillator is considered an unstable circuit. We will show that if $|T(j\omega)| < 1$ when the phase is 180 degrees, the system is stable, whereas if $|T(j\omega)| \geq 1$ when the phase is 180 degrees, the system is unstable. To study the stability of feedback circuits, we must therefore analyze the frequency response of the loop gain factor.

12.9.2     ## Bode Plots: One-, Two-, and Three-Pole Amplifiers

Figure 12.48(a) shows a simple single-stage common-emitter current amplifier. The high-frequency small-signal equivalent circuit is shown in Figure 12.48(b). The capacitance $C_1$ includes the forward-biased base-emitter junction capacitance as well as the effective Miller capacitance. The Miller capacitance and Miller effect were discussed in Chapter 7. The equivalent circuit shown in Figure 12.48(b) is identical to that developed in Figure 7.46. The output current in Figure 12.48(b) is given by

$$I_o = \left(\frac{R_C}{R_C + R_L}\right) g_m V_\pi \tag{12.103}$$

and the voltage $V_\pi$ is

$$V_\pi = I_i \left[ R_\pi \left\| \left(\frac{1}{sC_1}\right) \right. \right] \tag{12.104}$$

where $R_\pi = r_\pi \| R_B = r_\pi \| R_1 \| R_2$. Equation (12.104) can be expanded to

$$V_\pi = I_i \left[\frac{R_\pi}{1 + sR_\pi C_1}\right] \tag{12.105}$$

Substituting Equation (12.105) into (12.103), we get an expression for the small-signal current gain,

$$A_i = g_m R_\pi \left(\frac{R_C}{R_C + R_L}\right)\left[\frac{1}{1 + sR_\pi C_1}\right] \tag{12.106}$$

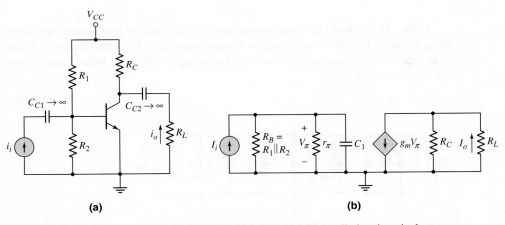

**(a)**                         **(b)**

**Figure 12.48**  (a) Single-stage common-emitter amplifier and (b) small-signal equivalent circuit, including input capacitance

When we set $s = j\omega = j(2\pi f)$, Equation (12.106) can be written as

$$A_i = \frac{A_{io}}{1 + j\left(\dfrac{f}{f_1}\right)}$$ (12.107)

where $A_{io}$ is the low-frequency or midband gain and $f_1$ is the upper 3 dB frequency. The gain is a complex function that can be written

$$A_i = \frac{A_{io}}{\sqrt{1 + \left(\dfrac{f}{f_1}\right)^2}} \angle -\tan^{-1}\left(\frac{f}{f_1}\right)$$ (12.108)

Figure 12.49(a) is a Bode plot of the current gain magnitude, and Figure 12.49(b) is a Bode plot of the current gain phase. Note that, from the definition of the directions of input and output currents, the output current is in phase with the input current at low frequencies. At high frequencies, the output current becomes 90 degrees out of phase with respect to the input current. This single-stage circuit is an example of a one-pole amplifier. As we have previously shown, similar expressions can be obtained for voltage gain, the transresistance transfer function, and the transconductance transfer function.

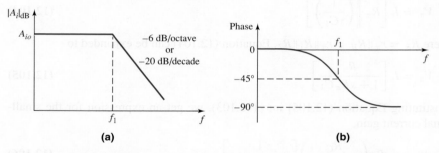

**Figure 12.49** Bode plots of current gain for single-stage common-emitter amplifier: (a) magnitude and (b) phase

Figure 12.50 shows the small-signal equivalent circuit of a two-stage amplifier, using the same hybrid-$\pi$ configuration for the transistors. The capacitance $C_2$ is the input capacitance of the second transistor, including the effective Miller capacitance. The output current is

$$I_o = -g_{m2}V_{\pi 2}$$ (12.109)

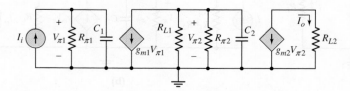

**Figure 12.50** Small-signal equivalent circuit, two-stage amplifier including input capacitances

and $V_{\pi 2}$ is

$$V_{\pi 2} = -g_{m1}V_{\pi 1}\left[R_{L1}\left\|R_{\pi 2}\right\|\left(\frac{1}{sC_2}\right)\right] \tag{12.110}$$

The voltage $V_{\pi 1}$ is

$$V_{\pi 1} = I_i\left[R_{\pi 1}\left\|\left(\frac{1}{sC_1}\right)\right.\right] \tag{12.111}$$

Combining Equations (12.109), (12.110), and (12.111) yields an expression for the small-signal current gain, as follows:

$$A_i = \frac{I_o}{I_i} = (g_{m1}g_{m2})(R_{\pi 1})(R_{L1}\|R_{\pi 2})\left[\frac{1}{1+sR_{\pi 1}C_1}\right]\left[\frac{1}{1+s(R_{L1}\|R_{\pi 2})C_2}\right] \tag{12.112}$$

Setting $s = j\omega = j(2\pi f)$, we can write Equation (12.112)

$$A_i = \frac{A_{io}}{\left(1+j\dfrac{f}{f_1}\right)\left(1+j\dfrac{f}{f_2}\right)} \tag{12.113}$$

where $f_1 = 1/2\pi R_{\pi 1}C_1$ and $f_2 = 1/2\pi(R_{L1}\|R_{\pi 2})C_2$. Frequency $f_1$ is the upper 3 dB frequency of the first stage, and $f_2$ is the upper 3 dB frequency of the second stage. This two-stage circuit is an example of a two-pole amplifier.

Equation (12.113) can be written

$$A_i = \frac{A_{io}}{\sqrt{1+\left(\dfrac{f}{f_1}\right)^2}\sqrt{1+\left(\dfrac{f}{f_2}\right)^2}} \angle -\left[\tan^{-1}\left(\frac{f}{f_1}\right)+\tan^{-1}\left(\frac{f}{f_2}\right)\right] \tag{12.114}$$

Figure 12.51(a) is a Bode plot of the current gain magnitude, assuming $f_1 \ll f_2$. This assumption implies that the two poles are far apart. The Bode plot of the current gain phase is shown in Figure 12.51(b). Again the phase of the output current is in phase with the input current at low frequency. This phase relation is a direct result of the way the directions of current were defined. At high frequencies, the output current becomes 180 degrees out of phase with respect to the input current.

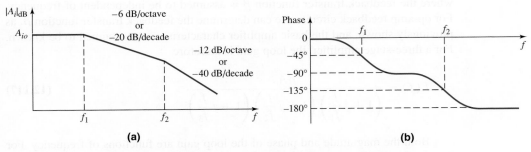

**Figure 12.51** Bode plots of current gain for two-stage amplifier: (a) magnitude and (b) phase

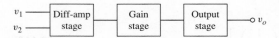

**Figure 12.52** Three-stage amplifier

An op-amp is a three-stage amplifier, as shown in Figure 12.52. Since each stage has an equivalent input resistance and capacitance, this circuit is an example of a three-pole amplifier. The overall gain can be expressed as

$$A = \frac{A_o}{\left(1 + j\dfrac{f}{f_1}\right)\left(1 + j\dfrac{f}{f_2}\right)\left(1 + j\dfrac{f}{f_3}\right)} \tag{12.115}$$

where $A_o$ is the low-frequency gain factor. Assuming the poles are far apart (let $f_1 \ll f_2 \ll f_3$), the Bode plots of the gain magnitude and phase are shown in Figure 12.53. At very high frequencies, the phase difference between the output and input signals is −270 degrees.

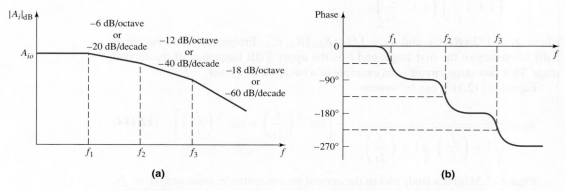

**Figure 12.53** Bode plots of three-stage amplifier gain: (a) magnitude and (b) phase

If we assume an ideal feedback amplifier, the loop gain is

$$T(j\omega) = \beta A(j\omega) \tag{12.116}$$

where the feedback transfer function $\beta$ is assumed to be independent of frequency. For op-amp feedback circuits, we can determine the feedback transfer function $\beta$, as previously shown, and the basic amplifier characteristics are assumed to be known. For a three-stage amplifier, the loop gain is therefore

$$T(f) = \frac{\beta A_o}{\left(1 + j\dfrac{f}{f_1}\right)\left(1 + j\dfrac{f}{f_2}\right)\left(1 + j\dfrac{f}{f_3}\right)} \tag{12.117}$$

Both the magnitude and phase of the loop gain are functions of frequency. For the three-stage amplifier, the phase will be −180 degrees at some particular frequency, which means that the amplifier may become unstable.

### 12.9.3    Nyquist Stability Criterion

In the last section, we saw that a feedback system can become unstable. Several methods can be used to determine whether a system is stable or unstable. The method we will consider is called the **Nyquist stability criterion.** This method not only determines if a system is stable, it also indicates the degree of system stability.

To apply this method, we must plot a **Nyquist diagram,** which is a polar plot of the loop gain factor $T(j\omega)$. The loop gain, which is a complex function, can be written in terms of its magnitude and phase, $T(j\omega) = |T(j\omega)|\angle\phi$, as shown in Equation (12.101). The Nyquist diagram is a plot of the real and imaginary components of $T(j\omega)$ as the frequency $\omega$ varies from minus infinity to plus infinity. Although negative frequencies have no physical meaning, they are not mathematically excluded in the loop gain function. The polar plot for negative frequencies, as we will see, is the complex conjugate of the polar plot for positive frequencies.

The loop gain for a two-pole amplifier is, from Equation (12.113),

$$T(j\omega) = \frac{\beta A_{io}}{\left(1 + j\dfrac{\omega}{\omega_1}\right)\left(1 + j\dfrac{\omega}{\omega_2}\right)} \tag{12.118}$$

where $\omega_1$ and $\omega_2$ are the upper 3 dB radian frequencies of the first and second stages, respectively. We can also write Equation (12.118) in the form

$$T(j\omega) = \frac{\beta A_{io}}{\sqrt{1 + \left(\dfrac{\omega}{\omega_1}\right)^2}\sqrt{1 + \left(\dfrac{\omega}{\omega_2}\right)^2}} \angle - \left[\tan^{-1}\left(\frac{\omega}{\omega_1}\right) + \tan^{-1}\left(\frac{\omega}{\omega_2}\right)\right] \tag{12.119}$$

The Nyquist plot of Equation (12.119) is shown in Figure 12.54. At $\omega = 0$, the magnitude of $T(j\omega)$ is $\beta A_{io}$ and the phase is zero. As $\omega$ increases, the magnitude decreases and the phase is negative. From Equation (12.119), we see that for negative values of $\omega$, the magnitude also decreases, but the phase becomes positive. This means that the loop gain function for negative frequencies is the complex conjugate of the loop gain function for positive frequencies, and the real axis is the axis of symmetry. As $\omega$ approaches $+\infty$, the magnitude approaches zero and the phase approaches $-180$ degrees.

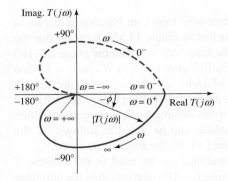

**Figure 12.54** Nyquist plot, loop gain for two-stage amplifier

The loop gain for a three-pole amplifier is, from Equation (12.117),

$$T(j\omega) = \frac{\beta A_o}{\left(1 + j\dfrac{\omega}{\omega_1}\right)\left(1 + j\dfrac{\omega}{\omega_2}\right)\left(1 + j\dfrac{\omega}{\omega_3}\right)} \tag{12.120}$$

This loop gain function can also be written in the form

$$T(j\omega) = \frac{\beta A_o}{\sqrt{1 + \left(\dfrac{\omega}{\omega_1}\right)^2}\sqrt{1 + \left(\dfrac{\omega}{\omega_2}\right)^2}\sqrt{1 + \left(\dfrac{\omega}{\omega_3}\right)^2}}\angle\phi \tag{12.121(a)}$$

where $\phi$ is the phase, given by

$$\phi = -\left[\tan^{-1}\left(\frac{\omega}{\omega_1}\right) + \tan^{-1}\left(\frac{\omega}{\omega_2}\right) + \tan^{-1}\left(\frac{\omega}{\omega_3}\right)\right] \tag{12.121(b)}$$

Figure 12.55(a) shows one possible Nyquist plot. For $\omega = 0$, the magnitude is $\beta A_o$ and the phase is zero. As $\omega$ increases in the positive direction, the magnitude decreases and the phase becomes negative. As the Bode plot in Figure 12.53 shows, the phase goes through $-90$ degrees, then through $-180$ degrees, and finally approaches $-270$ degrees as the magnitude approaches zero. This same effect is shown in the Nyquist diagram. The plot approaches the origin and is tangent to the imaginary axis as $\omega \to \infty$. Again, the plot for negative frequencies is the mirror image of the positive frequency plot about the real axis.

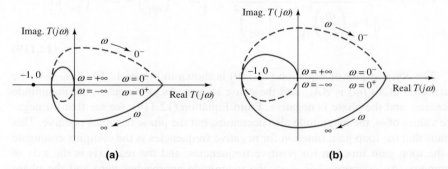

**Figure 12.55** Nyquist plot, loop gain for three-stage amplifier, for: (a) stable system and (b) unstable system

Another possible Nyquist plot for the three-pole loop gain function is shown in Figure 12.55(b). The basic plot is the same as that in Figure 12.55(a), except that the position of the point $(-1, 0)$ is different. At the frequency at which the phase is $-180$ degrees, the curve crosses the negative real axis. In Figure 12.55(a), $|T(j\omega)| < 1$ when the phase is $-180$ degrees, whereas in Figure 12.55(b), $|T(j\omega)| > 1$ when the phase is $-180$ degrees. The Nyquist diagram encircles the point $(-1, 0)$ in Figure 12.55(b), and this has particular significance for stability. For this treatment of a three-pole amplifier, the Nyquist criterion for stability of the amplifier can be stated as follows: "If the Nyquist plot encircles or goes through the point $(-1, 0)$, the amplifier is unstable."

Using the criterion, a simpler test for stability can be used in most cases. If $|T(j\omega)| \geq 1$ at the frequency at which the phase is $-180$ degrees, then the amplifier is unstable. This simpler test allows us to use the Bode plots considered previously, instead of explicitly constructing the Nyquist diagram.

## EXAMPLE 12.18

**Objective:** Determine the stability of an amplifier, given the loop gain function.

Consider a three-pole feedback amplifier with a loop gain given by

$$T(f) = \frac{\beta(100)}{\left(1 + j\dfrac{f}{10^5}\right)^3}$$

In this case, the three poles all occur at the same frequency. Determine the stability of the amplifier for $\beta = 0.20$ and $\beta = 0.02$.

**Solution:** The loop gain can be written in terms of its magnitude and phase,

$$T(f) = \frac{\beta(100)}{\left[\sqrt{1 + \left(\dfrac{f}{10^5}\right)^2}\right]^3} \angle -3\ \tan^{-1}\left(\frac{f}{10^5}\right)$$

The frequency $f_{180}$ at which the phase becomes $-180$ degrees is

$$-3\ \tan^{-1}\left(\frac{f_{180}}{10^5}\right) = -180°$$

which yields

$$f_{180} = 1.73 \times 10^5\ \text{Hz}$$

The magnitude of the loop gain at this frequency for, $\beta = 0.20$, is then

$$|T(f_{180})| = \frac{(0.20)(100)}{8} = 2.5$$

For $\beta = 0.02$, the magnitude is

$$|T(f_{180})| = \frac{(0.020)(100)}{8} = 0.25$$

**Comment:** The loop gain magnitude at the frequency at which the phase is $-180$ degrees is 2.5 when $\beta = 0.20$ and 0.25 when $\beta = 0.02$. The system is therefore unstable for $\beta = 0.20$ and stable for $\beta = 0.02$.

## EXERCISE PROBLEM

**Ex 12.18:** The loop gain function of a feedback amplifier is given by

$$T(f) = \frac{\beta(3000)}{\left(1 + j\dfrac{f}{10^3}\right)\left(1 + j\dfrac{f}{10^5}\right)^2}$$

Determine the value of $\beta$ at which the amplifier becomes unstable. (Ans. $\beta = 0.0667$)

We can also consider the stability of the feedback system in terms of Bode plots. The Bode plot of the loop gain magnitude from the previous example is shown in Figure 12.56(a), for $\beta = 0.20$ and $\beta = 0.02$. The low-frequency loop gain magnitude is dependent on $\beta$, but the 3 dB frequency is the same in both cases. Since the three poles all occur at the same frequency, the magnitude of $T(f)$ decreases at the rate of $-18$ dB/octave at the higher frequencies. The frequencies at which $|T(f)| = 1$ are indicated on the figure.

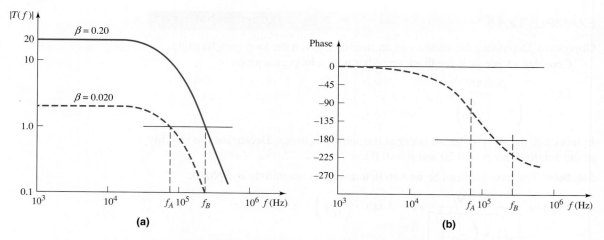

**Figure 12.56** Bode plots of loop gain of function described in Example 12.18, for two values of feedback transfer function: (a) magnitude and (b) phase

The phase of the loop gain function is shown in Figure 12.56(b). The two frequencies at which $|T(f)| = 1$, for the two values of $\beta$, are also indicated. We see that $|\phi| > 180°$ at $|T(f)| = 1$, when $\beta = 0.20$. This is equivalent to $|T(f)| > 1$ when $\phi = -180°$, which makes the system unstable. However, $|\phi| < 180°$ at $|T(f)| = 1$, when $\beta = 0.02$, so the feedback circuit is stable for this feedback transfer factor.

### 12.9.4    Phase and Gain Margins

From the discussion in the previous section, we can determine whether a feedback amplifier is stable or unstable by examining the loop gain as a function of frequency. This can be done from a Nyquist diagram or from the Bode plots. We can also use this technique to determine the degree of stability of a feedback amplifier.

At the frequency at which the loop gain magnitude is unity, if the magnitude of the phase is less than 180 degrees, the system is stable. This is illustrated in Figure 12.57. The difference (magnitude) between the phase angle at this frequency and 180 degrees is called the **phase margin.** The loop gain can change due, for example, to temperature

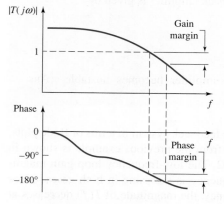

**Figure 12.57** Bode plots of loop gain magnitude and phase, indicating phase margin and gain margin

variations, and the phase margin indicates how much the loop gain can increase and still maintain stability. A typical desired phase margin is in the range of 45 to 60 degrees.

A second term that describes the degree of stability is the **gain margin,** which is also illustrated in Figure 12.57. This function is defined to be $|T(j\omega)|$ in decibels at the frequency where the phase is $-180$ degrees. This value is usually expressed in dB and also gives an indication of how much the loop gain can increase and still maintain stability.

## EXAMPLE 12.19

**Objective:** Determine the required feedback transfer function $\beta$ to yield a specific phase margin, and determine the resulting closed-loop low-frequency gain.

Consider a three-pole feedback amplifier with a loop gain function given by

$$T(f) = \frac{\beta(1000)}{\left(1 + j\dfrac{f}{10^3}\right)\left(1 + j\dfrac{f}{5 \times 10^4}\right)\left(1 + j\dfrac{f}{10^6}\right)}$$

Determine the value of $\beta$ that yields a phase margin of 45 degrees.

**Solution:** A phase margin of 45 degrees implies that the phase of the loop gain is $-135$ degrees at the frequency at which the magnitude of the loop gain is unity. The phase of the loop gain is

$$\phi = -\left[\tan^{-1}\left(\frac{f}{10^3}\right) + \tan^{-1}\left(\frac{f}{5 \times 10^4}\right) + \tan^{-1}\left(\frac{f}{10^6}\right)\right]$$

Since the three poles are far apart, the frequency at which the phase is $-135$ degrees is approximately equal to the frequency of the second pole, as shown in Figure 12.53. In this example, $f_{135} \cong 5 \times 10^4$ Hz, so we have that

$$\phi = -\left[\tan^{-1}\left(\frac{5 \times 10^4}{10^3}\right) + \tan^{-1}\left(\frac{5 \times 10^4}{5 \times 10^4}\right) + \tan^{-1}\left(\frac{5 \times 10^4}{10^6}\right)\right]$$

or

$$\phi = -[88.9° + 45° + 2.86°] \cong -135°$$

Since we want the loop gain magnitude to be unity at this frequency, we have

$$|T(f)| = 1 = \frac{\beta(1000)}{\sqrt{1 + \left(\dfrac{5 \times 10^4}{10^3}\right)^2}\sqrt{1 + \left(\dfrac{5 \times 10^4}{5 \times 10^4}\right)^2}\sqrt{1 + \left(\dfrac{5 \times 10^4}{10^6}\right)^2}}$$

or

$$1 \cong \frac{\beta(1000)}{(50)(1.41)(1)}$$

which yields $\beta = 0.0707$.

The closed-loop low-frequency gain for this case is

$$A_{fo} = \frac{A_o}{1 + \beta A_o} = \frac{1000}{1 + (0.0707)(1000)} = 13.9$$

**Comment:** For this value of $\beta$, if the frequency is greater than $5 \times 10^4$ Hz, the loop gain magnitude is less than unity. If the frequency is less than $5 \times 10^4$ Hz, the phase of the loop gain is $|\phi| < 135°$ (phase margin of 45 degrees). These conditions imply that the system is stable.

**Ex 12.19:** Consider the loop gain function

$$T(f) = \frac{\beta(3000)}{\left(1 + j\dfrac{f}{10^3}\right)\left(1 + j\dfrac{f}{10^5}\right)^2}$$

For $\beta = 0.008$, determine the low-frequency closed-loop gain and the phase margin. (Ans. $A_f(0) = 120$, phase margin $= 66.8°$)

## Test Your Understanding

**TYU 12.15** Consider the loop gain function given in Exercise Ex 12.19. Determine the value of $\beta$ that produces a phase margin of $45°$. (Ans. $\beta = 0.0167$)

**TYU 12.16** A two-pole feedback amplifier has an open-loop gain given by Equation (12.113), with parameters: $A_{io} = 10^5$ A/A, $f_1 = 10^4$ Hz, and $f_2 = 10^5$ Hz. The basic amplifier is connected to a feedback circuit, for which the feedback transfer ratio is $\beta$. Determine the value of $\beta$ that results in a phase margin of 60 degrees. (Ans. $\beta = 9.73 \times 10^{-5}$ A/A)

**TYU 12.17** For the loop gain function given in Example 12.18, determine the value of $\beta$ that produces a phase margin of 60 degrees. (Ans. $\beta = 0.0222$)

## 12.10   FREQUENCY COMPENSATION

**Objective:** • Consider frequency compensation techniques, methods by which unstable feedback circuits can be stabilized.

In the previous section, we presented a method for determining whether a feedback system is stable or unstable. In this section, we will discuss a method for modifying the loop gain of a feedback amplifier, to make the system stable. The general technique of making a feedback system stable is called **frequency compensation.**

### 12.10.1   Basic Theory

One basic method of frequency compensation involves introducing a new pole in the loop gain function, at a sufficiently low frequency that $|T(f)| = 1$ occurs when $|\phi| < 180°$. As an example, consider the Bode plots of a three-pole loop gain magnitude and phase given in Figure 12.58 and shown by the solid lines. In this case, when the magnitude of the loop gain is unity, the phase is nearly $-270$ degrees and the system is unstable.

If we introduce a new pole $f_{PD}$ at a very low frequency, and if we assume that the original three poles do not change, the new Bode plots of the magnitude and phase will be as shown by the dotted lines in Figure 12.58. In this situation, the

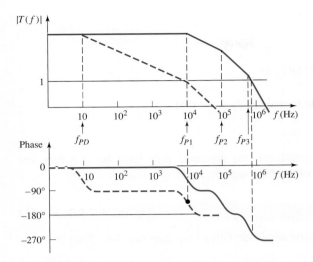

**Figure 12.58** Bode plots of loop gain magnitude and phase for three-stage amplifier, before frequency compensation (solid curves), and after frequency compensation (dotted curves)

magnitude of the loop gain becomes unity when the phase is $|\phi| < 180°$, and the system is stable. Since the pole is introduced at a low frequency and since it dominates the frequency response, it is called a **dominant pole.** This fourth pole can be introduced by adding a fourth stage with an extremely large input capacitance. Though not practical, this method demonstrates the basic idea of stabilizing a circuit.

## EXAMPLE 12.20

**Objective:** Determine the dominant pole required to stabilize a feedback system.
 Consider a three-pole feedback amplifier with a loop gain given by

$$T(f) = \frac{1000}{\left(1 + j\dfrac{f}{10^4}\right)\left(1 + j\dfrac{f}{10^6}\right)\left(1 + j\dfrac{f}{10^8}\right)}$$

Insert a dominant pole, assuming the original poles do not change, such that the phase margin is at least 45 degrees.

**Solution:** By inserting a dominant pole, we change the loop gain function to

$$T_{PD}(f) = \frac{1000}{\left(1 + j\dfrac{f}{f_{PD}}\right)\left(1 + j\dfrac{f}{10^4}\right)\left(1 + j\dfrac{f}{10^6}\right)\left(1 + j\dfrac{f}{10^8}\right)}$$

We assume that $f_{PD} \ll 10^4$ Hz. A phase of $-135$ degrees, giving a phase margin of 45 degrees, occurs approximately at $f_{135} = 10^4$ Hz.
 Since we want the loop gain magnitude to be unity at this frequency, we have

$$|T_{PD}(f_{135})| = 1 = \frac{1000}{\sqrt{1 + \left(\dfrac{10^4}{f_{PD}}\right)^2}\sqrt{1 + \left(\dfrac{10^4}{10^4}\right)^2}\sqrt{1 + \left(\dfrac{10^4}{10^6}\right)^2}\sqrt{1 + \left(\dfrac{10^4}{10^8}\right)^2}}$$

or

$$1 = \frac{1000}{\sqrt{1 + \left(\frac{10^4}{f_{PD}}\right)^2}(1.414)(1)(1)}$$

Solving for the dominant pole frequency $f_{PD}$, we find

$$f_{PD} = 14.14 \text{ Hz}$$

**Comment:** With high-gain amplifiers, the dominant pole must be at a very low frequency to ensure stability of the feedback circuit.

EXERCISE PROBLEM

**Ex 12.20:** Consider a three-pole amplifier with a loop gain function given by

$$T(f) = \frac{250}{\left(1 + j\frac{f}{10^3}\right)\left(1 + j\frac{f}{10^5}\right)^2}$$

(a) Show that the system is unstable. (b) Stabilize the circuit by inserting a new dominant pole. Assume the original poles are not altered. At what frequency must the new pole be placed to achieve a phase margin of 60°. (Ans. (a) For $\phi = -180°$, $|T| = 1.25 > 1$; (b) $f_{PD} = 2.67$ Hz)

Problem-Solving Technique: Frequency Compensation

1. To stabilize a circuit, insert a dominant pole or move an existing pole to a dominant pole position (see next section). Assume that the dominant pole frequency is small. Determine the frequency of the resulting loop gain function to achieve the required phase margin.
2. Set the magnitude of the loop gain function equal to unity at the frequency determined in step 1 to find the required dominant pole frequency.
3. To actually achieve the required dominant pole frequency in the circuit, a number of techniques are available (for example, see Miller compensation).

One disadvantage of this frequency compensation method is that the loop gain magnitude, and in turn the open-loop gain magnitude, is drastically reduced over a very wide frequency range. This affects the closed-loop response of the feedback amplifier. However, the advantage of maintaining a stable amplifier greatly outweighs the disadvantage of a reduced gain, demonstrating another trade-off in design criteria.

### 12.10.2 Closed-Loop Frequency Response

Inserting a dominant pole to obtain the open-loop characteristics (dotted lines, Figure 12.58) is not as extreme or devastating to the circuit as it might first appear. Amplifiers are normally used in a closed-loop configuration, for which we briefly considered the bandwidth extension, in Section 12.2.3.

For the region in which the frequency response is characterized by the dominant pole, the open-loop amplifier gain is

$$A(f) = \frac{A_o}{1 + j\dfrac{f}{f_{PD}}} \tag{12.122}$$

where $A_o$ is the low-frequency gain and $f_{PD}$ is the dominant-pole frequency. The feedback amplifier closed-loop gain can be expressed as

$$A_f(f) = \frac{A(f)}{(1 + \beta A(f))} \tag{12.123}$$

where $\beta$ is the feedback transfer ratio, which is assumed to be independent of frequency. Substituting Equation (12.122) into (12.123), we can write the closed-loop gain as

$$A_f(f) = \frac{A_o}{(1 + \beta A_o)} \times \frac{1}{1 + j\dfrac{f}{f_{PD}(1 + \beta A_o)}} \tag{12.124}$$

The term $A_o/(1 + \beta A_o)$ is the closed-loop low-frequency gain, and $f_{PD}(1 + \beta A_o) = f_C$ is the 3 dB frequency of the closed-loop system.

Figure 12.59 shows the Bode plot of the gain magnitude for the open-loop parameters $A_o = 10^6$ and $f_{PD} = 10$ Hz, at several feedback transfer ratios. As the closed-loop gain decreases, the bandwidth increases. As previously determined, the gain–bandwidth product is essentially a constant.

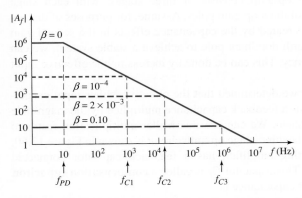

**Figure 12.59** Bode plot, gain magnitude for open-loop and three closed-loop conditions

## EXAMPLE 12.21

**Objective:** Determine the shift in the 3 dB frequency when an amplifier is operated in a closed-loop system.

Consider an amplifier with a low-frequency open-loop gain of $A_o = 10^6$ and an open-loop 3 dB frequency of $f_{PD} = 10$ Hz. The feedback transfer ratio is $\beta = 0.01$.

**Solution:** The low-frequency closed-loop gain is

$$A_f(0) = \frac{A_o}{(1 + \beta A_o)} = \frac{10^6}{1 + (0.01)(10^6)} \cong 100$$

From Equation (12.124), the closed-loop 3dB frequency is

$$f_C = f_{PD}(1 + \beta A_o) = (10)[1 + (0.01)(10^6)]$$

or

$$f_C \cong 10^5 \text{ Hz} = 100 \text{ kHz}$$

**Comment:** Even though the open-loop 3 dB frequency is only 10 Hz, the closed-loop bandwidth is extended to 100 kHz. This effect is due to the fact that the gain–bandwidth product is a constant.

**Ex 12.21:** An amplifier has an open loop response given by

$$A(f) = \frac{10^5}{\left(1 + j\dfrac{f}{10}\right)}$$

The amplifier is connected in a closed-loop configuration with $\beta = 0.025$. Determine the closed-loop low-frequency gain and closed-loop bandwidth. (Ans. $A_f(0) \cong 40$, $f_C \cong 25$ kHz)

### 12.10.3 Miller Compensation

As previously discussed, an op-amp consists of three stages, with each stage normally responsible for one of the loop gain poles. Assume, for purposes of discussion, that the first pole $f_{P1}$ is created by the capacitance effects in the second gain stage. Instead of adding a fourth dominant pole to achieve a stable system, we can move pole $f_{P1}$ to a low frequency. This can be done by increasing the effective input capacitance to the gain stage.

Previously in Chapter 7, we determined that the effective Miller input capacitance to a transistor amplifier is a feedback capacitance multiplied by the magnitude of the gain of the amplifier stage. We can use this Miller multiplication factor to stabilize a feedback system. The three-stage op-amp circuit is shown in Figure 12.60. The second stage, an inverting amplifier, has a feedback capacitor connected between the output and input. This capacitor $C_F$ is called a **compensation capacitor.**

The effective input Miller capacitance is

$$C_M = C_F(1 + A) \tag{12.125}$$

Since the gain of the second stage is large, the equivalent Miller capacitance will normally be very large. The pole introduced by the second stage is approximately

$$f_{P1} = \frac{1}{2\pi R_2 C_M} \tag{12.126}$$

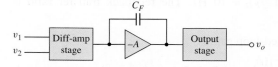

**Figure 12.60** Three-stage amplifier, including Miller compensation capacitor

where $R_2$ is the effective resistance between the amplifier input node and ground. Resistance $R_2$, then, is the parallel combination of the input resistance to the amplifier and the output resistance of the diff-amp stage.

## EXAMPLE 12.22

**Objective:** Determine the pole of the gain stage that includes a feedback capacitor.

Consider a gain stage with an amplification $A = 10^3$, a feedback capacitor $C_F = 30$ pF, and a resistance $R_2 = 5 \times 10^5 \ \Omega$.

**Solution:** The effective input Miller capacitance is

$$C_M = C_F(1 + A) \cong (30)(1000) \ \text{pF} = 3 \times 10^{-8} \ \text{F}$$

The dominant-pole frequency is therefore

$$f_{P1} = \frac{1}{2\pi R_2 C_M} = \frac{1}{2\pi(5 \times 10^5)(3 \times 10^{-8})} = 10.6 \ \text{Hz}$$

**Comment:** The pole of the second stage can be moved to a significantly lower frequency by using the Miller effect.

EXERCISE PROBLEM

**Ex 12.22:** The loop gain function of an amplifier is described in Exercise Ex 12.20. To stabilize the circuit, the first pole at $f_{P1} = 10^3$ Hz is to be moved by introducing a compensation capacitor. Assume the other two poles remain fixed. Determine the frequency to which the first pole must be moved to achieve a phase margin of 45°. (Ans. $f_{PD} = 194$ Hz)

The effect of moving pole $f_{P1}$, using the Miller compensation technique, is shown in Figure 12.61. We assume at this point that the other two poles $f_{P2}$ and $f_{P3}$

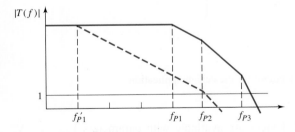

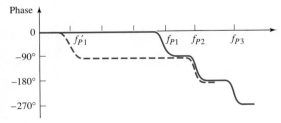

**Figure 12.61** Bode plots of loop gain for three-stage amplifier, before (solid curves) and after (dotted curves) incorporating Miller compensation capacitor: (a) magnitude and (b) phase

are not affected. Moving the pole $f_{P1}$ to $f'_{P1}$ means that the frequency at which $|T(f)| = 1$ is lower, and that the phase is $|\phi| < 180°$, which means that the amplifier is stabilized.

A detailed analysis of the system using Miller compensation shows that pole $f_{P2}$ does not remain constant; it increases. This phenomenon is called **pole-splitting.** The increase in $f_{P2}$ is actually beneficial, because it increases the phase margin, or the frequency at which a particular phase margin is achieved.

## 12.11 DESIGN APPLICATION: A MOSFET FEEDBACK CIRCUIT

**Objective:** • Redesign a BJT feedback circuit using MOSFETs.

**Specifications:** The circuit in Figure P12.36 is to be redesigned using MOSFETs. The new circuit configuration is shown in Figure 12.62. The output voltage is to be zero for $v_i = 0$.

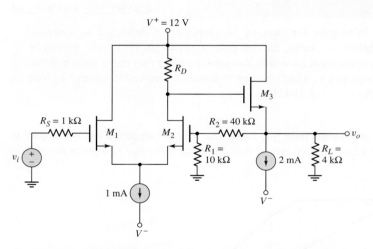

**Figure 12.62** A MOSFET feedback circuit for the design application

**Choices:** Assume that NMOS devices are available with parameters $V_{TN} = 1$ V, $K_n = 1$ mA/V$^2$, and $\lambda = 0$.

**Solution (DC Design):** For $v_O = 0$, the current in $M_3$ is $I_{D3} = 2$ mA. Then

$$I_D = K_n(V_{GS3} - V_{TN})^2$$

or

$$2 = (1)(V_{GS3} - 1)^2$$

which yields

$$V_{GS3} = 2.414 \text{ V}$$

The voltage at the gate of $M_3$ is then to be $V_{G3} = 2.414$ V. The current in $M_2$ is 0.5 mA, so the resistance $R_D$ is

$$R_D = \frac{12 - 2.414}{0.5} = 19.2 \text{ k}\Omega$$

**Solution (AC Analysis):** We can find the small-signal parameters as

$$g_{m1} = g_{m2} \equiv g_m = 2\sqrt{K_n I_{D1}} = 2\sqrt{(1)(0.5)} = 1.414 \text{ mA/V}$$

and

$$g_{m3} = 2\sqrt{K_n I_{D3}} = 2\sqrt{(1)(2)} = 2.828 \text{ mA/V}$$

The small-signal equivalent circuit is shown in Figure 12.63. Summing currents at the $V_1$ node, we have

$$g_m V_{gs1} + g_m V_{gs2} = 0 \Rightarrow V_{gs2} = -V_{gs1}$$

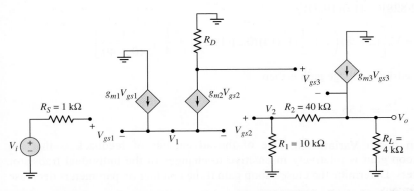

**Figure 12.63** Small-signal equivalent circuit of the MOSFET feedback circuit for the design application

Writing a KVL equation from the input, we find

$$V_i = V_{gs1} - V_{gs2} + V_2 = -2V_{gs2} + V_2$$

or

$$V_{gs2} = \frac{V_2 - V_i}{2}$$

We see that

$$V_{gs3} = -g_m V_{gs2} R_D - V_o = -\frac{1}{2}(V_2 - V_i) R_D - V_o$$

Also

$$V_2 = \left(\frac{R_1}{R_1 + R_2}\right) V_o = \left(\frac{10}{10 + 40}\right) V_o = 0.2 V_o$$

so that

$$V_{gs3} = -\frac{1}{2} g_m [(0.2) V_o - V_i] R_D - V_o$$

Summing currents at the output node, we obtain

$$g_{m3}V_{gs3} = \frac{V_o}{R_L} + \frac{V_o}{R_1 + R_2}$$

or

$$g_{m3}\left\{-\frac{1}{2}g_m\,[(0.2)V_o - V_i]\,R_D - V_o\right\} = \frac{V_o}{R_L} + \frac{V_o}{R_1 + R_2}$$

Combining terms, we obtain

$$\frac{1}{2}g_{m3}g_mR_DV_i = V_o\left[g_{m3}\left(1 + \frac{1}{2}g_m(0.2)R_D\right) + \frac{1}{R_L} + \frac{1}{R_1 + R_2}\right] \qquad \textbf{(12.127)}$$

Substituting parameters, we find

$$\frac{1}{2}(2.828)(1.414)(19.2)V_i$$
$$= V_o\left[(2.828)\left(1 + \frac{1}{2}(1.414)(0.2)(19.2)\right) + \frac{1}{4} + \frac{1}{10 + 40}\right]$$

The closed-loop voltage gain is then

$$A_v = \frac{V_o}{V_i} = 3.56$$

**Solution (Gain Variations):** One of the advantages of feedback is that the closed-loop gain is relatively insensitive to changes in the individual transistor parameters. Determine the closed-loop gain if the conduction parameters decrease by 10 percent.

The new values of the small-signal parameters are

$$g_{m1} = g_{m2} \equiv g_m = 2\sqrt{K_nI_{D1}} = 2\sqrt{(0.9)(0.5)} = 1.342 \text{ mA/V}$$

and

$$g_{m3} = 2\sqrt{K_nI_{D3}} = 2\sqrt{(0.9)(2)} = 2.683 \text{ mA/V}$$

Substituting these values into Equation (12.127), we obtain

$$\frac{1}{2}(2.683)(1.342)(19.2)V_i$$
$$= V_o\left[(2.683)\left(1 + \frac{1}{2}(1.342)(0.2)19.2\right) + \frac{1}{4} + \frac{1}{10 + 40}\right]$$

The closed-loop gain is then

$$A_v = \frac{V_o}{V_i} = 3.50$$

**Comment:** With a decrease of 10 percent in the transistor conduction parameters, the closed-loop gain has decreased by less than 2 percent. Even though we are considering a relatively simple feedback circuit with only three transistors, the advantage of feedback is observed.

**12.12    SUMMARY**

- In a feedback circuit, a portion of the output signal is fed back to the input and combined with the input signal. In negative feedback, a portion of the output signal is subtracted from the input signal. In positive feedback, a portion of the output signal is added to the input signal.
- An important advantage of negative feedback is that the closed-loop amplifier gain is essentially independent of individual transistor parameters and is a function only of the feedback elements.
- Negative feedback increases bandwidth, may increase the signal-to-noise ratio, reduces nonlinear distortion, and controls input and output impedance values at the expense of reduced gain magnitude.
- There are four basic feedback topologies. A series input connection is used when the input signal is a voltage, and a shunt input connection is used when the input signal is a current. A series output connection is used when the output signal is a current, and a shunt output connection is used when the output signal is a voltage.
- The loop gain factor of a feedback amplifier is defined as $T = A\beta$, which is dimensionless and where $A$ is the gain of the basic amplifier and $\beta$ is the feedback factor. The loop gain is a function of frequency and is complex when the input capacitance of each transistor stage is taken into account.
- A three-stage negative feedback amplifier is guaranteed to be stable when, at the frequency for which the phase of the loop gain is $-180$ degrees, the magnitude of loop gain is less than unity.
- A common technique of frequency compensation utilizes the Miller multiplication effect by incorporating a feedback capacitor across, usually, the second stage of the amplifier.
- As an application, a MOSFET feedback circuit was designed.

 **CHECKPOINT**

After studying this chapter, the reader should have the ability to:

✓ Describe the ideal feedback circuit configuration.
✓ Describe some of the advantages and disadvantages of negative feedback.
✓ Discuss the general characteristics of the four basic feedback configurations in terms of input and output signals and input and output resistances.
✓ Design a feedback circuit given the input signal and desired output signal.
✓ Determine the loop gain of a feedback circuit.
✓ Determine whether or not a three-stage feedback amplifier is stable.
✓ Stabilize a three-stage amplifier using frequency compensation techniques.
✓ Analyze op-amp and discrete transistor circuits that are examples of the four basic feedback configurations.

 **REVIEW QUESTIONS**

1. What are the two general types of feedback and what are the advantages and disadvantages of each type?
2. Write the ideal form of the general feedback transfer function.
3. Define the loop gain factor.
4. What is the difference between open-loop gain and closed-loop gain?

5. Describe what is meant by the terms (a) gain sensitivity and (b) bandwidth extension.
6. Describe the series and shunt input connections of a feedback amplifier.
7. Describe the series and shunt output connections of a feedback amplifier.
8. Describe the effect of a series or shunt input connection on the value of input resistance.
9. Describe the effect of a series or shunt output connection on the value of output resistance.
10. Consider a noninverting op-amp circuit. Describe the type of input and output feedback connections.
11. Consider an inverting op-amp circuit. Describe the type of input and output feedback connections.
12. What is the Nyquist stability criterion for a feedback amplifier?
13. Using Bode plots, describe the conditions of stability and instability in a feedback amplifier.
14. Define phase margin.
15. What is meant by frequency compensation?
16. What is a dominant pole?
17. What is a common technique of frequency compensation in a feedback amplifier?

##  PROBLEMS

### Section 12.2 Basic Feedback Concepts

12.1  (a) A negative-feedback amplifier has a closed-loop gain of $A_f = 100$ and an open-loop gain of $A = 5 \times 10^4$. Determine the feedback transfer function $\beta$. (b) If $\beta = 0.012$ and $A_f = 80$, determine the open-loop gain $A$.

12.2  (a) The closed-loop gain of a negative-feedback amplifier is $A_f = -80$ and the open-loop gain is $A = -10^5$. Find the feedback transfer function $\beta$. (b) If $\beta = -0.015$ and $A = -5 \times 10^4$, determine the closed-loop gain $A_f$.

12.3  The ideal feedback transfer function is given by Equation (12.5). (a) Assume the feedback transfer function is $\beta = 0.15$. Determine the loop gain $T$ and the closed-loop gain $A_f$ for (i) $A = \infty$, (ii) $A = 80$ dB, and (c) $A = 10^2$. (b) Repeat part (a) for $\beta = 0.25$.

12.4  (a) The closed-loop gain of a feedback amplifier using an ideal feedback amplifier $(A \rightarrow \infty)$ is $A_f = 125$. What is the value of $\beta$? (b) If the basic amplifier has a finite open-loop gain, what must be the value of $A$ such that the closed-loop gain is within 0.25 percent of the ideal value. Use the results of part (a).

12.5  Consider the feedback system shown in Figure 12.1. The closed-loop gain is $A_f = -80$ and the open-loop gain is $A = -2 \times 10^4$. (a) Determine the feedback transfer function $\beta$. (b) The closed-loop gain is not to change by more than 0.01 percent when the open-loop gain changes. What is the maximum allowed change in the open-loop gain?

12.6  The open-loop gain of an amplifier is $A = 5 \times 10^4$. If the open-loop gain decreases by 10 percent, the closed-loop gain must not change by more than 0.1 percent. Determine the required value of the feedback transfer function $\beta$ and the closed-loop gain $A_f$.

12.7  Two feedback configurations are shown in Figures P12.7(a) and P12.7(b). The closed-loop gain in each case is $A_{vf} = v_o/v_i = 50$. (a) Determine $\beta_1$

and $\beta_2$ for the two circuits. (b) The gain $A_2$ decreases by 10 percent in both circuits. Using the results of part (a), determine the percent change in the closed-loop gain for each circuit. (c) What conclusion can be made as to the "better" feedback configuration?

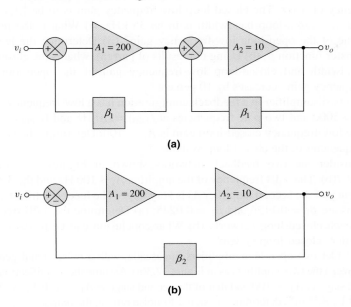

**(a)**

**(b)**

**Figure P12.7**

12.8    Three voltage amplifiers are in cascade as shown in Figure P12.8 with various amplification factors. The 180 degree phase shift for negative feedback actually occurs in the basic amplifier itself. (a) Determine the value of $\beta$ such that the closed-loop voltage gain is $A_{vf} = V_o/V_s = -120$. (b) Using the results of part (a), determine the percent change in $A_{vf}$ if each individual amplifier gain decreases by 10 percent.

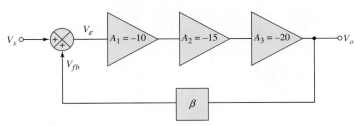

**Figure P12.8**

12.9    (a) The open-loop low-frequency voltage gain of an amplifier is $A_v = 5 \times 10^4$ and the open-loop 3 dB frequency is $f_H = 10$ Hz. If the closed-loop low-frequency gain is $A_{vf} = 25$, what is the closed-loop bandwidth? (b) The minimum closed-loop bandwidth of a feedback amplifier is to be 20 kHz. The open-loop low-frequency voltage gain is $A_v = 10^5$ and the open-loop 3 dB frequency is $f_H = 8$ Hz. What is the maximum closed-loop voltage gain?

12.10   (a) Determine the closed-loop bandwidth of a noninverting amplifier with a closed-loop low-frequency gain of 50. The op-amp has the characteristics described in Problem 12.9(a). (b) If the open-loop gain decreases to $A_v = 10^4$ but the closed-loop low-frequency gain remains fixed at 50, what is the resulting closed-loop bandwidth?

12.11   (a) An inverting amplifier uses an op-amp with an open-loop 3 dB frequency of 5 Hz. The closed-loop low-frequency gain is to be $|A_{vf}| = 75$ and the closed-loop bandwidth is to be 35 kHz. (i) What is the required value of the open-loop low-frequency gain? (ii) Determine the feedback transfer function $\beta$. (b) Using the results of part (a), what is the closed-loop bandwidth and closed-loop low-frequency gain if the open-loop low-frequency gain decreases by 10 percent?

12.12   The basic amplifier in a feedback configuration has a low-frequency gain of $A = 5000$ and two pole frequencies at $f_{3\text{-dB1}} = 10$ Hz and $f_{3\text{-dB2}} = 2$ kHz. The low-frequency closed-loop gain is $A_f = 100$. Determine the two 3 dB frequencies of the closed-loop system.

12.13   Consider the two feedback networks shown in Figures P12.7(a) and P12.7(b). The 3 dB frequency of the amplifier $A_1$ is 100 Hz and the 3 dB frequency of the second amplifier $A_2$ is very large. The feedback transfer functions are $\beta_1 = 0.1126$ and $\beta_2 = 0.0245$. (a) Determine the 3 dB frequency of each closed-loop network. (b) What conclusion can be made as to the "better" closed-loop system?

12.14   Consider two open-loop amplifiers in cascade, with a noise signal generated between the two amplifiers as in Figure 12.3(a). Assume the amplification of the first stage is $A_2 = 100$ and that of the second stage is $A_1 = 1$. If $V_{\text{in}} = 10$ mV and $V_n = 1$ mV, determine the signal-to-noise ratio at the output.

12.15   Two feedback configurations are shown in Figures P12.15(a) and (b). At low input voltages, the two gains are $A_1 = A_2 = 90$ and at higher input voltages, the gains change to $A_1 = A_2 = 60$. Determine the change in closed-loop gain, $A_f = V_o/V_i$, for the two feedback circuits. (See Figure 12.4.) Which feedback configuration will result in less distortion in the output signal?

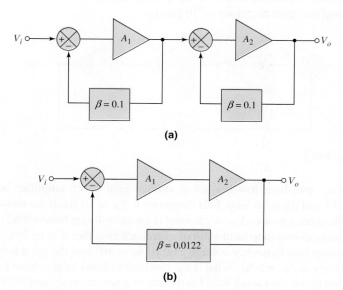

(a)

(b)

Figure P12.15

## Section 12.3 Ideal Feedback Topologies

12.16    Consider the ideal series-shunt circuit shown in Figure 12.6. Let $A_v = 5 \times 10^3$ V/V, $\beta = 0.0080$ V/V, $R_i = 10$ k$\Omega$, and $R_o = 1$ k$\Omega$. Determine the ideal values of $A_{vf} = V_o/V_i$, $R_{if}$, and $R_{of}$.

12.17    The parameters of the ideal series–shunt circuit shown in Figure 12.6 are $V_i = 25$ mV, $V_o = 2.5$ V, and $\beta = 0.0096$ V/V. Determine the values and units of $A_v$, $A_{vf}$, $V_{fb}$, and $V_\varepsilon$.

12.18    For the noninverting op-amp circuit in Figure P12.18, the parameters are: $A = 10^5$, $A_{vf} = 20$, $R_i = 100$ k$\Omega$, and $R_o = 100$ $\Omega$. Determine the ideal closed-loop input and output resistances, $R_{if}$ and $R_{of}$, respectively.

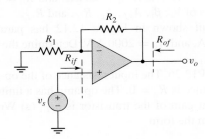

**Figure P12.18**

12.19    Consider the noninverting op-amp circuit in Figure P12.18. The input resistance of the op-amp is $R_i = \infty$ and the output resistance is $R_o = 0$, but the op-amp has a finite gain $A$. (a) Write the closed-loop transfer function in the form

$$A_{vf} = \frac{v_o}{v_s} = \frac{A}{(1 + \beta A)}$$

(b) What is the expression for $\beta$? (c) If $A = 10^5$ and $A_{vf} = 20$, what is the required $\beta$ and $R_2/R_1$? (d) If $A$ decreases by 10 percent, what is the percent change in $A_{vf}$?

12.20    The circuit parameters of the ideal shunt–series amplifier shown in Figure 12.9 are $I_i = 20\,\mu$A, $I_{fb} = 19\,\mu$A, $R_i = 500$ $\Omega$, $R_o = 20$ k$\Omega$, and $\beta_i = 0.0095$ A/A. Determine the values and units of $I_\varepsilon$, $I_o$, $A_i$, $A_{if}$, $R_{if}$, and $R_{of}$.

12.21    Consider the ideal shunt–series amplifier shown in Figure 12.9. The parameters are $I_i = 25\,\mu$A, $I_\varepsilon = 0.8\,\mu$A, and $A_{if} = 125$. Determine the values and units of $I_{fb}$, $I_o$, $\beta_i$, and $A_i$.

12.22    Consider the op-amp circuit in Figure P12.22. The op-amp has a finite gain, so that $i_o = Ai_\varepsilon$, and a zero output impedance. (a) Write the closed-loop transfer function in the form

$$A_{if} = \frac{i_o}{i_s} = \frac{A_i}{(1 + \beta_i A_i)}$$

(b) What is the expression for $\beta_i$? (c) If $A_i = 10^5$ and $A_{if} = 25$, what is the required $\beta_i$ and $R_F/R_3$? (d) If $A_i$ decreases by 15 percent, what is the percent change in $A_{if}$?

12.23    An op-amp circuit is shown in Figure P12.22. Its parameters are as described in Problem 12.22, except that $R_i = 2$ k$\Omega$ and $R_o = 20$ k$\Omega$. Determine the closed-loop input and output resistances, $R_{if}$ and $R_{of}$, respectively.

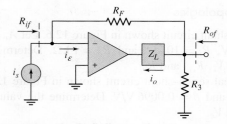

**Figure P12.22**

12.24 The parameters of the ideal series–series amplifier in Figure 12.12 are $V_i = 0.2\,\text{mV}$, $I_o = 5\,\text{mA}$, $V_{fb} = 0.195\,\text{mV}$, $R_i = 20\,\text{k}\Omega$, and $R_o = 10\,\text{k}\Omega$. Determine the values and units of $V_\varepsilon$, $\beta_z$, $A_{gf}$, $A_g$, $R_{if}$, and $R_{of}$.

12.25 The ideal series–series circuit shown in Figure 12.12 has parameters $V_i = 150\,\mu\text{V}$, $\beta_z = 0.0245\,\text{V/A}$, and $A_g = 2000\,\text{A/V}$. Determine the values and units of $V_{fb}$, $V_\varepsilon$, $I_o$, and $A_{gf}$.

12.26 Consider the circuit in Figure P12.26. The input resistance of the op-amp is $R_i = \infty$ and the output resistance is $R_o = 0$. The op-amp has a finite gain, so that $i_o' = A_g v_\varepsilon$. The current gain of the transistor is $h_{FE}$. (a) Write the closed-loop transfer function in the form

$$A_{gf} = \frac{i_o}{v_s} = \frac{A_g}{(1 + \beta_z A_g)}$$

where $A_g$ is the open-loop gain of the system. (b) What is the expression for $\beta_z$? (c) If $A_g = 5 \times 10^5\,\text{mS}$ and $A_{gf} = 10\,\text{mS}$, what is the required $\beta_z$ and $R_E$? (d) If $A_g$ increases by 10 percent, what is the corresponding percent change in $A_{gf}$?

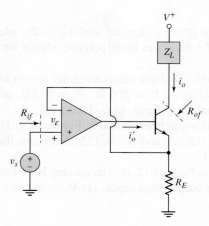

**Figure P12.26**

12.27 The circuit shown in Figure P12.26 has the same parameters as described in Problem 12.26, except that $R_i = 20\,\text{k}\Omega$ and $R_o = 50\,\text{k}\Omega$. Determine the closed-loop input and output resistances, $R_{if}$ and $R_{of}$, respectively.

12.28 The circuit parameters of the ideal shunt–shunt amplifier shown in Figure 12.14 are $A_{zf} = 0.20\,\text{V}/\mu\text{A}$, $\beta_g = 4.25/\mu\text{A/V}$, and $R_i = R_o = 500\,\Omega$. Determine the values and units of $A_z$, $R_{if}$, and $R_{of}$.

12.29 Voltage and current values in the ideal shunt–shunt circuit shown in Figure 12.14 are $I_i = 40\ \mu A$, $I_{fb} = 38\ \mu A$, and $V_o = 8\ V$. Determine the values and units of $A_z$, $A_{zf}$, and $\beta_g$.

12.30 Consider the current-to-voltage converter circuit shown in Figure P12.30. The input resistance $R_{if}$ is assumed to be small, the output resistance is $R_o = 0$, and the op-amp gain $A_z$ is large. (a) Write the closed-loop transfer function in the form

$$A_{zf} = \frac{v_o}{i_s} = \frac{A_z}{(1 + \beta_g A_z)}$$

(b) What is the expression for $\beta_g$? (c) If $A_z = 5 \times 10^6\ \Omega$ and $A_{zf} = 5 \times 10^4\ \Omega$, what is the required $\beta_g$ and $R_F$? (d) If $A_z$ decreases by 10 percent, what is the percent change in $A_{zf}$?

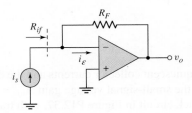

**Figure P12.30**

12.31 For the current-to-voltage converter circuit in Figure P12.30, the parameters are as described in Problem 12.30. If $R_i = 10\ k\Omega$, determine the closed-loop input resistance $R_{if}$.

D12.32 Determine the type of feedback configuration that should be used in a design to achieve the following objectives: (a) low input resistance and low output resistance, (b) high input resistance and high output resistance, (c) low input resistance and high output resistance, and (d) high input resistance and low output resistance.

12.33 Consider a series of amplifiers and feedback circuits connected in the ideal feedback configurations. In each case the input resistance to the basic amplifier is $R_i = 10\ k\Omega$, the output resistance of the basic amplifier is $R_o = 1\ k\Omega$, and the loop gain is $T = 10^4$. (a) Determine the maximum possible input resistance and minimum possible input resistance to the feedback circuit. (b) Determine the maximum possible output resistance and minimum possible output resistance to the feedback circuit.

D12.34 A compound transconductance amplifier is to be designed by connecting two basic feedback amplifiers in cascade. What two amplifiers should be connected in cascade to form the compound circuit? Is there more than one possible design?

## Section 12.4 Voltage (Series–Shunt) Amplifiers

*12.35 The parameters of the op-amp in the circuit shown in Figure P12.35 are $A_v = 10^5$, $R_i = 30\ k\Omega$, and $R_o = 500\ \Omega$. The transistor parameters are $h_{FE} = 140$ and $V_A = \infty$. Assume that $v_O = 0$ at the quiescent point. Determine (a) $A_{vf}$, (b) $R_{if}$, and (c) $R_{of}$.

12.36 The circuit in Figure P12.36 is an example of a series–shunt feedback circuit. Assume the transistor parameters are: $h_{FE} = 100$, $V_{BE}(\text{on}) = 0.7\ V$,

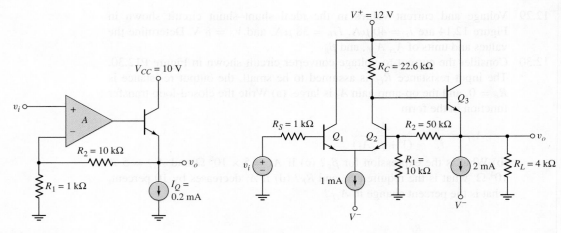

**Figure P12.35**

**Figure P12.36**

and $V_A = \infty$. (a) Determine the quiescent collector currents and the dc voltage at the output. (b) Determine the small-signal voltage gain $A_{vf} = v_o/v_i$.

12.37 Consider the series–shunt feedback circuit in Figure P12.37, with transistor parameters: $h_{FE} = 120$, $V_{BE}(\text{on}) = 0.7$ V, and $V_A = \infty$. (a) Determine the small-signal parameters for $Q_1$, $Q_2$, and $Q_3$. Using nodal analysis, determine: (b) the small-signal voltage gain $A_{vf} = v_o/v_i$, (c) the input resistance $R_{if}$, and (d) the output resistance $R_{of}$.

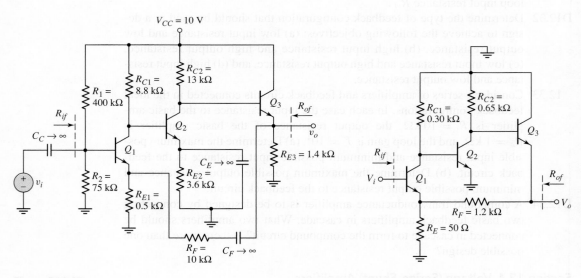

**Figure P12.37**

**Figure P12.38**

*12.38 The circuit shown in Figure P12.38 is an ac equivalent circuit of a feedback amplifier. The transistor parameters are $h_{FE} = 100$ and $V_A = \infty$. The quiescent collector currents are $I_{C1} = 14.3$ mA, $I_{C2} = 4.62$ mA, and $I_{C3} =$

4.47 mA. (a) Determine the closed-loop voltage gain $A_{vf} = V_o/V_i$. Compare this value to the approximate ideal value of $A_{vf} \cong (R_F + R_E)/R_E$. (b) Determine the values of $R_{if}$ and $R_{of}$.

12.39   Consider the MOSFET feedback amplifier shown in Figure P12.39. The transistor parameters are $V_{TN} = 0.5$ V, $K_n = 0.5$ mA/V$^2$, and $\lambda = 0$. Determine the small-signal voltage gain $A_v = v_o/v_i$.

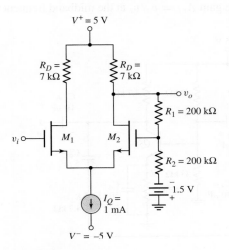

**Figure P12.39**

*12.40   The parameters of the BiCMOS circuit in Figure P12.40 are $V^+ = 5$ V, $V_{GG} = 2.5$ V, $R_{D1} = 5$ k$\Omega$, $R_{E2} = 1.6$ k$\Omega$, and $R_L = 1.2$ k$\Omega$. The transistor parameters are $K_n = 1.5$ mA/V$^2$, $V_{TN} = 0.5$ V, $\lambda = 0$ for $M_1$ and $h_{FE} = 120$, $V_{EB}(\text{on}) = 0.7$ V, $V_A = \infty$ for $Q_2$. (a) Determine the quiescent values $I_{DQ1}$ and $I_{CQ2}$. (b) Find the small-signal voltage gain $A_v = v_o/v_i$. (c) Determine the small-signal output resistance $R_{of}$.

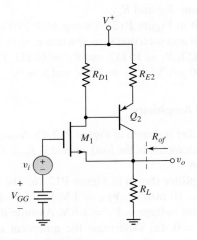

**Figure P12.40**

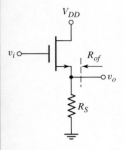

$V_{DD}$

$v_i$

$R_{of}$

$v_o$

$R_S$

**Figure P12.41**

12.41 The parameters of the basic source-follower circuit in Figure P12.41 are $R_S = 1.5\,\text{k}\Omega$, $V_{TN} = 1.2\,\text{V}$, and $\lambda = 0$. Assume the transistor is biased at $I_{DQ} = 1.2\,\text{mA}$. (a) If the transistor conduction parameter is $K_n = 1.5\,\text{mA/V}^2$, determine (i) $A_{vf} = v_o/v_i$ and (ii) $R_{of}$. (b) If the conduction parameter increases by 50 percent to $K_n = 2.25\,\text{mA/V}^2$, determine the percent change in (i) $A_{vf} = v_o/v_i$ and (ii) $R_{of}$.

12.42 The transistor parameters for the circuit in Figure P12.42 are: $h_{FE} = 50$, $V_{BE}(\text{on}) = 0.7\,\text{V}$, and $V_A = \infty$. Using nodal analysis, determine the closed-loop small-signal voltage gain $A_{vf} = v_o/v_s$ at the midband frequency.

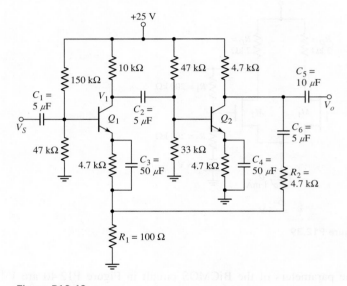

**Figure P12.42**

*D12.43 Design a discrete transistor feedback voltage amplifier to provide a voltage gain of 50. Assume the available transistors have parameters: $h_{FE} = 120$ and $V_A = \infty$. The signal voltage source has a source resistance of $R_S = 2\,\text{k}\Omega$ and the load is $R_L = 3\,\text{k}\Omega$. Verify the design with a computer simulation. Determine $R_{if}$ and $R_{of}$.

*D12.44 Redesign the feedback circuit in Figure P12.36 using MOSFETs to provide a voltage gain of $A_{vf} = 8$ and such that $v_o = 0$ when $v_i = 0$. Assume circuit parameters of $R_S = 1\,\text{k}\Omega$, $R_1 = 15\,\text{k}\Omega$, and $R_L = 10\,\text{k}\Omega$. The transistor parameters are $k'_n = 100\,\mu\text{A/V}^2$, $V_{TN} = 1.5\,\text{V}$, and $\lambda = 0$.

## Section 12.5 Current (Shunt–Series) Amplifiers

12.45 An op-amp current gain amplifier is shown in Figure P12.45. Assuming an ideal op-amp, design the circuit such that the load current is $I_o = 5\,\text{mA}$ for an input current of $I_s = 60\,\mu\text{A}$.

*12.46 Consider the current gain amplifier shown in Figure P12.46. The transistor parameters are $K_n = K_p = 10\,\text{mA/V}^2$, $V_{TN} = 1\,\text{V}$, $V_{TP} = -1\,\text{V}$, and $\lambda_n = \lambda_p = 0$. The LED turn-on voltage is $V_\gamma = 1.6\,\text{V}$. Assume the LED small-signal resistance is $r_f = 0$. (a) Determine the quiescent currents $I_{DQ1}$ and $I_{DQ2}$. (b) Show that the small-signal current gain is given by

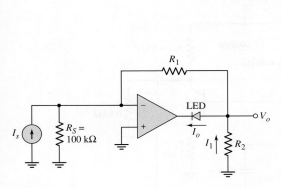

Figure P12.45

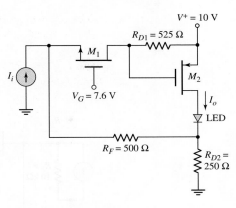

Figure P12.46

$$A_i = \frac{I_o}{I_i} = \frac{-g_{m2}R_{D1}}{1 + \dfrac{1}{g_{m1}(R_F + R_{D2})} + \dfrac{g_{m2}R_{D1}R_{D2}}{R_F + R_{D2}}}$$

(c) Calculate the value of $A_i$.

*12.47 A MOSFET current gain amplifier is shown in Figure P12.47. The transistor parameters are $K_n = K_p = 10$ mA/V$^2$, $V_{TN} = 1$ V, $V_{TP} = -1$ V, and $\lambda_n = \lambda_p = 0$. The LED turn-on voltage is $V_\gamma = 1.6$ V. Assume the LED small-signal resistance is $r_f = 0$. (a) Determine the quiescent currents $I_{DQ1}$ and $I_{DQ2}$. (b) Determine the small-signal current gain $A_i = I_o/I_i$.

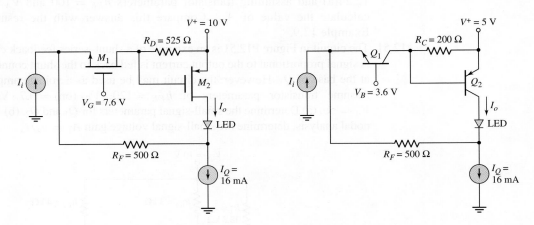

Figure P12.47

Figure P12.48

*12.48 A BJT current gain amplifier is shown in Figure P12.48. The transistor parameters are $\beta_1 = \beta_2 = 180$, $V_{A1} = V_{A2} = \infty$, and $I_{S1} = I_{S2} = 10^{-15}$ A. The LED turn-on voltage is $V_\gamma = 1.6$ V. Assume the LED small-signal resistance is $r_f = 0$. (a) Determine the quiescent currents $I_{CQ1}$ and $I_{CQ2}$. (b) Derive the expression for the small-signal current gain $A_i = I_o/I_i$. (c) Calculate the value of $A_i = I_o/I_i$.

12.49 The circuit in Figure P12.49 has transistor parameters: $h_{FE} = 100$, $V_{BE}(\text{on}) = 0.7$ V, and $V_A = \infty$. (a) From the quiescent values, determine

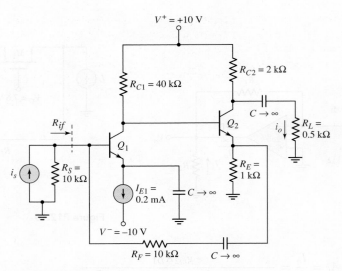

**Figure P12.49**

the small-signal parameters for $Q_1$ and $Q_2$. (b) Using nodal analysis, determine the small-signal closed-loop current gain $A_{if} = i_o/i_s$. (c) Using nodal analysis, find the input resistance $R_{if}$.

*12.50 (a) Using the small-signal equivalent circuit in Figure 12.25 for the circuit in Figure 12.24(a), derive the expression for the small-signal current gain $A_{if} = I_o/I_s$. (b) Using the circuit parameters given in Figure 12.24(a) and assuming transistor parameters $h_{FE} = 100$ and $V_A = \infty$, calculate the value of $A_{if}$. Compare this answer with the results of Example 12.9.

*12.51 The circuit in Figure P12.51 is an example of a shunt–series feedback circuit. A signal proportional to the output current is fed back to the shunt connection at the base of $Q_1$. However, the circuit may be used as a voltage amplifier. Assume transistor parameters of $h_{FE} = 120$, $V_{BE}(\text{on}) = 0.7$ V, and $V_A = \infty$. (a) Determine the small-signal parameters for $Q_1$ and $Q_2$. (b) Using nodal analysis, determine the small-signal voltage gain $A_v = v_o/v_s$.

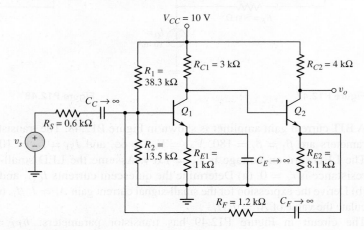

**Figure P12.51**

12.52 Consider the circuit in Figure P12.51 with transistor parameters, $h_{FE} = 120$, $V_{BE}(\text{on}) = 0.7$ V, and $V_A = \infty$. Using nodal analysis, determine the input resistance $R_{if}$.

12.53 For the transistors in the circuit in Figure P12.53, the parameters are: $h_{FE} = 50$, $V_{BE}(\text{on}) = 0.7$ V, and $V_A = \infty$. Using nodal analysis, determine the closed-loop current gain $A_{if} = i_o/i_s$.

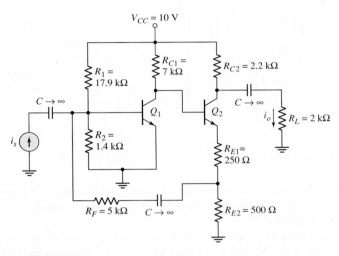

**Figure P12.53**

*D12.54 Design a discrete transistor feedback current amplifier to provide a current gain of 30. Assume the available transistors have parameters $h_{FE} = 120$ and $V_A = \infty$. The signal current source has a source resistance of $R_S = 25$ k$\Omega$ and the load is $R_L = 500$ $\Omega$. Verify the design with a computer simulation. Determine $R_{if}$ and $R_{of}$.

## Section 12.6 Transconductance (Series–Series) Amplifiers

12.55 Consider the transconductance amplifier shown in Figure P12.55. Assume the op-amp is ideal. (a) Derive the expression for the transconductance function $A_{gf} = I_o/V_i$. (b) If the circuit is designed such that $R_F/R_1 = R_3/R_2$, show that $I_o/V_i = -1/R_2$. (c) Design the circuit such that $I_o/V_i = -0.5$ mA/V.

12.56 Consider the transconductance feedback amplifier shown in Figure P12.56 with $R_D = 1.6$ k$\Omega$ and $R_L = 248$ $\Omega$. The transistor parameters are $V_{TN} = 0.5$ V, $V_{TP} = -0.5$ V, $K_n = 2$ mA/V$^2$, $K_p = 10$ mA/V$^2$, and $\lambda_n = \lambda_p = 0$. The LED turn-on voltage is $V_\gamma = 1.6$ V. Assume the LED small-signal resistance is $r_f = 0$. The current source is ideal. (a) Determine the quiescent values of $V_{D1}$, $I_{DQ3}$ and $V_{G2}$. (b) Derive the small-signal transconductance function $A_{gf} = I_o/V_i$. (c) Calculate the value of $A_{gf} = I_o/V_i$.

12.57 The circuit in Figure P12.57 is the ac equivalent circuit of a series–series feedback amplifier. Assume that the bias circuit, which is not shown, results in quiescent collector currents of $I_{C1} = 0.5$ mA, $I_{C2} = 1$ mA, and $I_{C3} = 2$ mA. Assume transistor parameters of $h_{FE} = 120$ and $r_o = \infty$. Determine the transconductance transfer function $A_{gf} = I_o/V_s$.

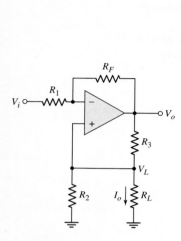

Figure P12.55

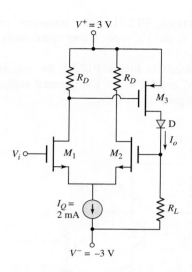

Figure P12.56

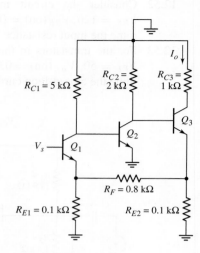

Figure P12.57

D12.58  Using a computer simulation analysis, redesign the circuit in Figure P12.57 by changing the value of $R_F$ to achieve a transconductance gain of $A_{gf} = I_o/V_s = 120$ mA/V.

12.59  In the circuit in Figure P12.59, the transistor parameters are: $h_{FE} = 100$, $V_{BE}(\text{on}) = 0.7$ V, and $V_A = \infty$. Determine the transconductance transfer function $A_{gf} = i_o/v_s$.

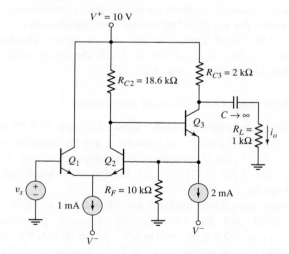

Figure P12.59

D12.60  Design a feedback amplifier to supply a current to an LED. Use the configuration shown in Figure 12.27 where $R_L$ is replaced by the LED. (a) Assuming an ideal op-amp is available, design the circuit such that the diode current is $I_O = 3 \times 10^{-3} V_i$ where $V_i$ is in the range 0 to 5 V. (b) If $V_i = 5$ V, what is the error in diode current if the op-amp has a finite gain of $A_g = 10^3$ mA/V and $h_{FE} = 80$?

## Section 12.7 Transresistance (Shunt–Shunt) Amplifiers

12.61  Consider the common-emitter circuit in Figure P12.61 driven by an ideal signal current source. The transistor parameters are $h_{FE} = 80$, $V_{EB}(\text{on}) = 0.7\,\text{V}$, and $V_A = 100\,\text{V}$. (a) Determine the quiescent values $I_{CQ}$ and $V_{ECQ}$. (b) Find the transresistance transfer function $A_{zf} = v_o/i_s$. (c) Determine the input resistance $R_{if}$. (d) Find the output resistance $R_{of}$.

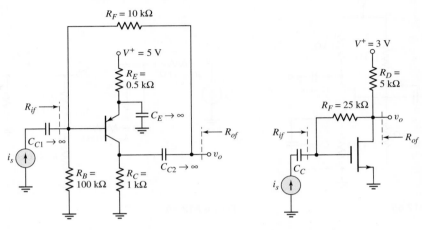

Figure P12.61                                              Figure P12.62

12.62  The transistor parameters for the circuit shown in Figure P12.62 are $V_{TN} = 0.4\,\text{V}$, $K_n = 0.5\,\text{mA/V}^2$, and $\lambda = 0$. (a) Find (i) the quiescent drain current $I_{DQ}$ and (ii) the small-signal transistor parameters. (b) Determine the transresistance transfer function $A_{zf} = v_o/i_s$. (c) Find the output resistance $R_{of}$.

12.63  Consider Problem 12.62. (a) What is the magnitude of the ideal transresistance transfer function $\left| A_{zf} \right|$ as the transistor transconductance parameter $g_m \to \infty$. (b) Determine the value of $g_m$ so that $\left| A_{zf} \right|$ is within 95 percent of the ideal magnitude.

12.64  For the circuit in Figure P12.64, the transistor parameters are: $h_{FE} = 150$, $V_{BE}(\text{on}) = 0.7\,\text{V}$, and $V_A = \infty$. Determine the value of $R_F$ that will result in a closed-loop voltage gain of $A_v = V_o/V_s = -5.0$.

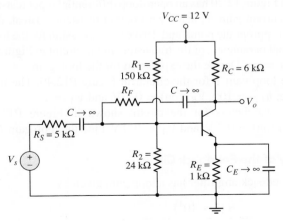

Figure P12.64

12.65 Consider the three-stage cascade feedback circuit in Figure 12.41. Each stage corresponds to the circuit in Figure P12.65, with transistor parameters: $h_{FE} = 180$, $V_{BE}(\text{on}) = 0.7$ V, and $V_A = \infty$. The source resistor is $R_S = 10$ k$\Omega$, and the load resistor is $R_L = 4$ k$\Omega$. Determine the value of $R_F$ such that the closed-loop gain is $A_v = v_o/v_i = -80$.

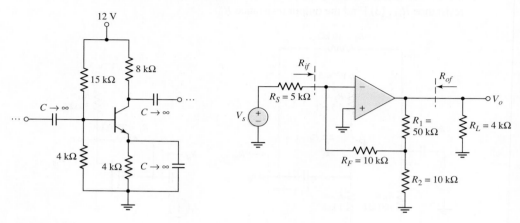

**Figure P12.65**                    **Figure P12.66**

12.66 The op-amp in the circuit in Figure P12.66 has an open-loop differential voltage gain of $A_d = 10^4$. Neglect the current into the op-amp, and assume the output resistance looking back into the op-amp is zero. Determine: (a) the closed-loop voltage gain $A_v = V_o/V_s$, (b) the input resistance $R_{if}$, and (c) the output resistance $R_{of}$.

D12.67 Design a feedback transresistance amplifier using an op-amp with parameters $R_i = 10$ k$\Omega$, $R_o = 100$ $\Omega$, and a low-frequency open-loop gain of $A_v = 10^4$ to produce a gain of 5 k$\Omega$. The source resistance is $R_S = 500$ $\Omega$ and the load resistance is $R_L = 2$ k$\Omega$. Determine the actual gain, input resistance, and output resistance using a computer simulation.

## Section 12.8 Loop Gain

12.68 The op-amp in Figure 12.20 has an open-loop differential input resistance $R_i$, an open-loop current gain $A_i$, and a zero output resistance. Break the feedback loop at an appropriate point, and derive the expression for the loop gain.

12.69 The small-signal parameters of the transistors in the circuit in Figure P12.37 are $h_{FE}$ and $V_A = \infty$. Derive the expression for the loop gain.

12.70 Determine the loop gain $T$ for the circuit in Figure P12.49. The transistor parameters are: $h_{FE} = 100$, $V_{BE}(\text{on}) = 0.7$ V, and $V_A = \infty$.

12.71 The transistor parameters for the circuit shown in Figure P12.64 are: $h_{FE} = 50$, $V_{BE}(\text{on}) = 0.7$ V, and $V_A = 100$ V. Find the loop gain $T$.

## Section 12.9 Stability of the Feedback Circuit

12.72 A three-pole feedback amplifier has a loop gain given by

$$T(f) = \frac{\beta \left(5 \times 10^4\right)}{\left(1 + j\dfrac{f}{10^3}\right)\left(1 + j\dfrac{f}{5 \times 10^4}\right)^2}$$

(a) Determine the frequency $f_{180}$ at which the phase is $-180$ degrees. (b) At the frequency $f_{180}$, determine the value of $\beta$ such that $|T(f_{180})| = 0.25$.

12.73   The open-loop voltage gain of an amplifier is given by

$$A_v = \frac{10^4}{\left(1 + j\dfrac{f}{5 \times 10^3}\right)^2 \left(1 + j\dfrac{f}{5 \times 10^5}\right)}$$

(a) Assuming the feedback transfer function is not a function of frequency, determine the frequency at which the phase of the loop gain is 180 degrees. (b) At what value of $\beta$ will the feedback amplifier become unstable? (c) Using the value of $\beta$ found in part (b), what is the low-frequency closed-loop gain? (d) Is the closed-loop feedback system stable for smaller or larger values of $\beta$?

12.74   A loop gain function is given by

$$T(f) = \frac{\beta(10^3)}{\left(1 + j\dfrac{f}{10^4}\right)\left(1 + j\dfrac{f}{5 \times 10^4}\right)\left(1 + j\dfrac{f}{10^5}\right)}$$

Sketch the Nyquist plot for: (a) $\beta = 0.005$, and (b) $\beta = 0.05$. (c) Is the system stable or unstable in each case?

12.75   A three-pole feedback amplifier has a loop gain function given by

$$T(f) = \frac{\beta(5 \times 10^3)}{\left(1 + j\dfrac{f}{10^3}\right)^2 \left(1 + j\dfrac{f}{5 \times 10^4}\right)}$$

(a) Sketch the Nyquist diagram for $\beta = 0.20$. (b) Determine the value of $\beta$ that produces a phase margin of 80 degrees.

12.76   A three-pole feedback amplifier has a loop gain given by

$$T(f) = \frac{\beta(10^4)}{\left(1 + j\dfrac{f}{10^3}\right)\left(1 + j\dfrac{f}{10^4}\right)\left(1 + j\dfrac{f}{10^5}\right)}$$

Sketch Bode plots of the loop gain magnitude and phase for: (a) $\beta = 0.005$, and (b) $\beta = 0.05$. (c) Is the system stable or unstable in each case? If the system is stable, what is the phase margin?

12.77   A feedback system has an amplifier with a low-frequency open-loop gain of $5 \times 10^4$ and has poles at $10^3$ Hz, $10^5$ Hz, and $10^7$ Hz. (a) Determine the frequency $f_{180}$ at which the phase of the loop gain is 180 degrees. (b) Determine the feedback transfer function $\beta$ for which the phase margin of the system is $45°$. (c) Using the results of part (b), determine the low-frequency closed-loop gain.

12.78   The open-loop voltage gain of an amplifier is given by

$$A_v = \frac{10^5}{\left(1 + j\dfrac{f}{10^3}\right)\left(1 + j\dfrac{f}{10^5}\right)}$$

(a) If the low-frequency, closed-loop gain is 100, is this amplifier stable? (b) If so, determine the phase margin.

12.79    The loop gain function of a feedback system is described by

$$T(f) = \frac{\beta\,(10^3)}{\left(1 + j\dfrac{f}{10^4}\right)\left(1 + j\dfrac{f}{10^5}\right)\left(1 + j\dfrac{f}{10^6}\right)}$$

(a) Determine the frequency $f_{180}$ at which the phase of $T(f)$ is $-180$ degrees. (b) For $\beta = 0.019$, (i) find $|T(f_{180})|$ and (ii) find the phase at which $|T| = 1$. (c) Using the results of part (b), determine the low-frequency closed-loop gain $A_f(0)$.

12.80    Consider a feedback amplifier for which the open-loop gain is given by

$$A(f) = \frac{2 \times 10^3}{\left(1 + j\dfrac{f}{5 \times 10^3}\right)\left(1 + j\dfrac{f}{10^5}\right)^2}$$

(a) Determine the frequency $f_{180}$ at which the phase of $A(f)$ is $-180$ degrees. (b) For $\beta = 0.0045$, determine the magnitude of the loop gain $T(f)$ at the frequency $f = f_{180}$ and determine the phase of $A(f)$ when $|T(f)| = 1$. Determine the closed-loop, low-frequency gain. Is the system stable or unstable? (c) Repeat part (b) for $\beta = 0.15$.

12.81    Consider a four-pole feedback system with a loop gain given by

$$T(f) = \frac{\beta(10^3)}{\left(1 + j\dfrac{f}{10^3}\right)\left(1 + j\dfrac{f}{10^4}\right)\left(1 + j\dfrac{f}{10^5}\right)\left(1 + j\dfrac{f}{10^6}\right)}$$

Determine the value of $\beta$ that produces a phase margin of 45 degrees.

## Section 12.10 Frequency Compensation

12.82    A feedback amplifier has a low-frequency open-loop gain of 4000 and three poles at $f_{P1} = 400\,\text{kHz}$, $f_{P2} = 4\,\text{MHz}$, and $f_{P3} = 40\,\text{MHz}$. A dominant pole is to be inserted such that the phase margin is 60 degrees. Assuming the original poles remain fixed, determine the dominant pole frequency.

12.83    The loop gain of a three-pole amplifier is given by

$$T(f) = \frac{10^3}{\left(1 + j\dfrac{f}{10^4}\right)^2\left(1 + j\dfrac{f}{10^6}\right)}$$

(a) Show that this function will lead to an unstable feedback system. (b) Insert a dominant pole such that the phase margin is 45 degrees. Assume the original poles remain fixed. What is the dominant pole frequency.

12.84    A loop gain function is given by

$$T(f) = \frac{500}{\left(1 + j\dfrac{f}{10^4}\right)\left(1 + j\dfrac{f}{5 \times 10^4}\right)\left(1 + j\dfrac{f}{10^5}\right)}$$

(a) Determine the frequency $f_{180}$ (to a good approximation) at which the phase of $T(f)$ is $-180$ degrees. (b) What is the magnitude of $T(f)$ at the frequency $f = f_{180}$ found in part (a)? (c) Insert a dominant pole such that the phase margin is approximately 60 degrees. Assume the original poles are fixed. What is the dominant pole frequency?

12.85 An open-loop amplifier can be described by

$$A_v = \frac{10^4}{\left(1 + j\dfrac{f}{10^5}\right)}$$

A dominant pole is to be inserted such that a closed-loop amplifier with a low-frequency gain of 50 has a phase margin of 45 degrees. (a) Determine $\beta$ and the required dominant pole frequency. (b) The feedback transfer function is increased such that the closed-loop, low-frequency gain of the amplifier in part (a) is 20. Determine the phase margin of this new amplifier.

12.86 The open-loop amplifier of a feedback system has its first two poles at $f_{P1} = 1$ MHz and $f_{P2} = 10$ MHz, and has a low-frequency open-loop gain of $|A_o| = 100$ dB. (a) A dominant pole is to be added such that the closed-loop amplifier with a low-frequency gain of 20 has a phase margin of 45 degrees. What is the dominant pole frequency? (b) If the feedback transfer function from part (a) is increased such that the closed-loop low-frequency gain is 5, determine the phase margin of the amplifier.

12.87 A feedback amplifier with a compensation capacitor has a low-frequency loop gain of $T(0) = 100$ dB and poles at $f'_{P1} = 10$ Hz, $f_{P2} = 5$ MHz, and $f_{P3} = 10$ MHz. (a) Find the frequency at which $|T(f)| = 1$, and determine the phase margin. (b) If the frequency $f'_{P1}$ is due to a compensation capacitor $C_F = 20$ pF, determine the new dominant pole frequency $f'_{P1}$ and phase margin if the compensation capacitor is increased to $C_F = 75$ pF.

12.88 The equivalent circuit at the interface between the first and second stages of an op-amp is shown in Figure P12.88. The parameters are $R_{o1} = 2$ MΩ, $R_{i2} = 750$ kΩ, and $C_i = 1.2$ pF. (a) Determine the pole frequency for this part of the circuit. (b) Determine the additional Miller capacitance $C_M$ that would need to be added so that the pole frequency is moved to $f_{PD} = 6$ Hz. (c) If the gain of the second-stage amplifier is 1000, what is the required value of a feedback capacitor around the second-stage amplifier to produce this Miller capacitance.

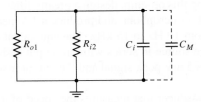

Figure P12.88

12.89 The amplifier described in Problem 12.82 is to be stabilized by moving the first pole by using Miller compensation. Assuming that $f_{P2}$ and $f_{P3}$ remain fixed, determine the frequency to which $f_{P1}$ must be moved such that the phase margin is 60 degrees.

12.90 The loop gain of an amplifier is given by

$$T(f) = \frac{\beta\left(5 \times 10^5\right)}{\left(1 + j\dfrac{f}{10^4}\right)\left(1 + j\dfrac{f}{5 \times 10^5}\right)\left(1 + j\dfrac{f}{10^7}\right)}$$

The pole at $f_{P1} = 10^4$ is to be moved such that the feedback amplifier with a closed-loop low-frequency gain of 40 has a phase margin of 60 degrees. (a) Find the value of $\beta$. (b) Determine the new pole frequency.

##  COMPUTER SIMULATION PROBLEMS

12.91   Consider the circuit shown in Figure 12.24(a). Replace the input signal source with an ideal signal voltage source. Using a computer simulation, investigate the small-signal voltage gain, input resistance $R_{if}$, and output resistance $R_{of}$ as a function of the feedback resistance $R_F$.

12.92   For the circuit shown in Figure 12.39, investigate the small-signal voltage gain, input resistance, and output resistance as a function of the transistor width-to-length ratio, using a computer simulation.

12.93   Consider the circuit shown in Figure 12.37(a). Using a computer simulation, plot the loop gain as a function of frequency.

12.94   In the circuit shown in Figure P12.42, use a computer simulation to plot the small-signal voltage gain versus frequency. Determine the low-frequency and high-frequency cutoff values.

## DESIGN PROBLEMS

[Note: Each design should be correlated with a computer simulation analysis.]

*D12.95   The circuit shown in Figure P12.46 is to have a minimum loop gain of $T = 200$. Design the width-to-length ratios of the transistors to meet this requirement. Use appropriate transistor parameters.

*D12.96   Op-amps with low-frequency open-loop gains of $5 \times 10^4$ and dominant-pole frequencies of 8 Hz are available. Design a cascade of noninverting amplifiers such that the overall voltage gain is 500 and the bandwidth is 15 kHz

*D12.97   An op-amp has a low-frequency open-loop gain of $5 \times 10^4$ and a dominant-pole frequency of 10 Hz. Using this op-amp, design a preamplifier system that can amplify the output of a microphone and produce a 1 V peak signal over a frequency range from 10 Hz to 15 kHz. The equivalent circuit of the microphone is a voltage source in series with an output resistance. The voltage source produces a 5 mV peak signal and the output resistance is 10 kΩ.

*D12.98   The equivalent circuit of a transducer that measures the speed of a motor is a current source in parallel with an output resistance. The current source produces an output of 1 $\mu$A per revolution per second of the motor and the output resistance is 50 kΩ. Design a discrete transistor circuit that produces a full-scale output of 5 V for a maximum motor speed of 60 revolutions per second. The nominal transistor current gain is $h_{FE} = 100$ with tolerances of $\pm 20$ percent. The accuracy of the output signal is to remain within $\pm 1$ percent.

# Operational Amplifier Circuits

Thus far, we have considered basic circuit configurations, such as the common emitter, emitter follower, and diff-amp, among others. We have discussed the basic concepts in design and analysis, including biasing techniques, frequency response, and feedback effects. In this chapter, we combine basic circuit configurations to form larger analog circuits that are fabricated as integrated circuits. Operational amplifiers are used extensively in electronic systems, so we concentrate on several configurations of operational amplifier circuits.

We introduced the ideal op-amp in Chapter 9. Now, we analyze and design the circuitry of the op-amp, to determine how the various circuit configurations can be combined to form a nearly ideal op-amp.

## PREVIEW

In this chapter, we will:

- Discuss the general design philosophy of an operational amplifier circuit.
- Describe and analyze the dc and ac characteristics of the classic 741 bipolar operational amplifier circuit.
- Describe and analyze the dc and ac characteristics of CMOS operational amplifier circuits.
- Describe and analyze the dc and ac characteristics of BiCMOS operational amplifier circuits.
- Describe the characteristics of two hybrid JFET operational amplifier circuits.
- As an application, design a two-stage CMOS op-amp to match a given output stage.

## 13.1    GENERAL OP-AMP CIRCUIT DESIGN

**Objective:** • Discuss the general design philosophy of an operational amplifier circuit.

An operational amplifier, in general, is a three-stage circuit, as shown in Figure 13.1, and is fabricated as an integrated circuit. The first stage is a differential amplifier, the second stage provides additional voltage gain, and the third stage provides current gain and low output impedance. A feedback capacitor is often included in the second stage to provide frequency compensation as discussed in the last chapter. In some cases, in particular with MOSFET op-amp circuits, only the first two stages are used.

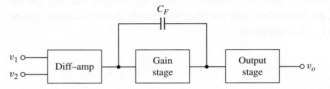

**Figure 13.1** General block diagram of an operational amplifier

We have on numerous occasions made reference to the op-amp. In Chapter 9, we analyzed and designed op-amp circuits using the ideal op-amp model. In Chapter 10, we introduced current-source biasing and introduced the active load. The differential amplifier, using current source biasing and active loads was considered in Chapter 11. We also introduced the bipolar Darlington pair in Chapter 11, which is often used as a second gain stage. Previously, in Chapter 8, we considered the class-AB output stage that is often used in operational amplifier circuits. These individual building blocks will now be combined to form the operational amplifier.

In Chapter 9, as mentioned, we analyzed and designed ideal op-amp circuits. Practical operational amplifiers, as we will see in this chapter, exhibit characteristics that deviate from the ideal characteristics. Once we have analyzed these practical op-amp circuits and determined some of their nonideal properties, we will then consider, in the next chapter, the effect of these nonideal characteristics on the op-amp circuits.

### 13.1.1    General Design Philosophy

All stages of the operational amplifier circuit are direct coupled. There are no coupling capacitors and there are also no bypass capacitors. These types of capacitors would require extremely large areas on the IC chip and hence are impractical. In addition, resistors whose values are over approximately 50 k$\Omega$ are avoided in ICs, since they also require large areas and introduce parasitic effects. Op-amp circuits are designed with transistors having matching characteristics.

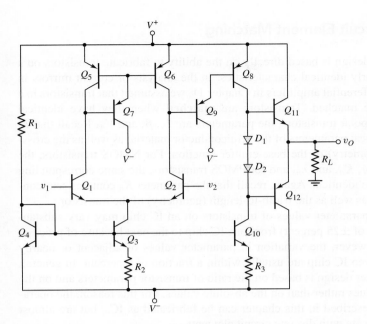

**Figure 13.2** A simple bipolar operational amplifier

We may begin to design a simple bipolar operational amplifier by using the knowledge gained in the previous chapters. Figure 13.2 shows the general configuration of the circuit. The first stage will be a differential pair, $Q_1$ and $Q_2$, biased with a Widlar current source, $Q_3$, $Q_4$, and $R_2$, and using a three-transistor active load. Assuming matched transistors, we expect the dc voltage at the collector of $Q_6$ to be two base–emitter voltage drops below the positive bias voltage. Therefore, the Darlington pair, $Q_8$ and $Q_9$, that forms the second stage should be properly biased. The bias current for $Q_8$ is supplied by the Widlar current source, $Q_4$, $Q_{10}$, and $R_3$. The output stage is the complementary push–pull, emitter-follower configuration of $Q_{11}$ and $Q_{12}$. The crossover distortion is eliminated by including the diodes $D_1$ and $D_2$. The emitter-follower configuration provides low output resistance so that the op-amp can drive a load with minimal loading effect. By changing the value of $R_3$ slightly, the current through $Q_{10}$ and $Q_8$ can be changed, which will change the collector–emitter voltages across these transistors. This part of the circuit then acts as a dc voltage shifter such that the output voltage, $v_O$, can be set equal to zero for zero input voltages.

From results that we have derived previously, we expect the differential-mode voltage gain of the first stage to be in the range of $10^2$–$10^3$, depending on the specific transistor parameters and the voltage gain of the second stage to also be the range of $10^2$–$10^3$. The voltage gain of the output stage, an emitter follower, is essentially unity. The overall voltage gain of the op-amp circuit is then expected to be in the range of $10^4$–$10^6$. From our study in Chapter 9, this magnitude of voltage gain is required for the circuit to act essentially as an ideal op-amp.

The same op-amp configuration can be designed with MOS transistors. In general, as we have seen, BJT circuits have higher voltage gains, whereas MOSFET circuits have higher input resistances. So, whether a bipolar or MOSFET design is used depends to a large extent on the specific application of the op-amp.

### 13.1.2    Circuit Element Matching

Integrated circuit design is based directly on the ability to fabricate transistors on a chip that have nearly identical characteristics. In the analysis of current mirrors in Chapter 10 and differential amplifiers in Chapter 11, we assumed that transistors in a given circuit were matched. Transistors are **matched** when they have identical parameters. For bipolar transistors, the parameters are $I_S$, $\beta$, and $V_A$. Recall that $I_S$ includes the electrical parameters of the semiconductor material as well as the cross-sectional area (geometry) of the base-emitter junction. For NMOS transistors, the parameters are $V_{TN}$, $K_n$, and $\lambda_n$, and for PMOS transistors, the same corresponding parameters must be identical. Again, recall that the parameter $K_n$ contains semiconductor parameters as well as the width-to-length (geometry) of the transistor.

The absolute parameter values of transistors on an IC chip may vary substantially (on the order of ±25 percent) from one IC chip to the next because of processing variations. However, the variation in parameter values of adjacent or nearby transistors on a given IC chip are usually within a fraction of a percent. In general, much of an amplifier design is based on the ratio of transistor parameters and on the ratio of resistor values rather than on the absolute values. For this reason, the operational amplifiers described in this chapter can be fabricated as ICs, but are almost impossible to fabricate with discrete circuit elements.

## Test Your Understanding

**TYU 13.1** Using a computer simulation, determine the dc voltages and currents in the bipolar op-amp circuit in Figure 13.2. Use reasonable resistor values. Adjust the value of $R_3$ such that the output voltage is nearly zero for zero input voltages.

**TYU 13.2** Consider the basic diff-amp with active load and current biasing in Figure 13.2. Using a computer simulation, investigate the change in the voltage at the collector of $Q_2$ as $Q_1$ and $Q_2$, and also $Q_5$ and $Q_6$, become slightly mismatched.

##   13.2    A BIPOLAR OPERATIONAL AMPLIFIER CIRCUIT

**Objective:** • Describe and analyze the dc and ac characteristics of the classic 741 bipolar operational amplifier circuit.

The **741 op-amp** has been produced since 1966 by many semiconductor device manufacturers. Since then, there have been many advances in op-amp design, but the 741 is still a widely used general-purpose op-amp. Even though the 741 is a fairly old design, it still provides a useful case study to describe the general circuit configuration and to perform a detailed dc and small-signal analysis. From the ac analysis, we determine the voltage gain and the frequency response of this circuit.

### 13.2.1    Circuit Description

Figure 13.3 shows the equivalent circuit of the 741 op-amp. For easier analysis, we break the overall circuit down into its basic circuits and consider each one individually.

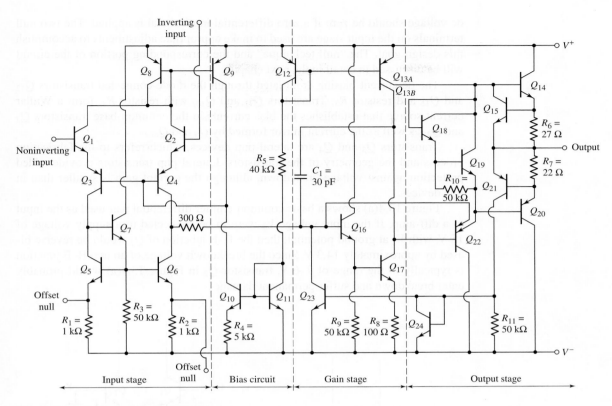

**Figure 13.3** Equivalent circuit, 741 op-amp

As with most op-amps, this circuit consists of three stages: the input differential amplifier, the gain stage, and the output stage. Figure 13.3 also shows a separate bias circuit, which establishes the bias currents throughout the op-amp. Like most op-amps, the 741 is biased with both positive and negative supply voltages. This eliminates the need for input coupling capacitors, which in turn means that the circuit is also a dc amplifier. The dc output voltage is zero when the applied differential input signal is zero. Typical supply voltages are $V^+ = 15$ V and $V^- = -15$ V, although input voltages as low as $\pm 5$ V can be used.

### Input Diff-Amp
The input diff-amp stage is more complex than those previously covered. The input stage consists of transistors $Q_1$ through $Q_7$, with biasing established by transistors $Q_8$ through $Q_{12}$. The two input transistors $Q_1$ and $Q_2$ act as emitter followers, which results in a high differential input resistance. The differential output currents from $Q_1$ and $Q_2$ are the inputs to the common-base amplifier formed by $Q_3$ and $Q_4$, which provides a relatively large voltage gain.

Transistors $Q_5$, $Q_6$, and $Q_7$, with associated resistors $R_1$, $R_2$, and $R_3$, form the active load for the diff-amp. A single-sided output at the common collectors of $Q_4$ and $Q_6$ is the input signal to the following gain stage.

The dc output voltage at the collector of $Q_6$ is at a lower potential than the inputs at the bases of $Q_1$ and $Q_2$. As the signal passes through the op-amp, the dc voltage level shifts several times. By design, when the signal reaches the output terminal, the dc voltage should be zero if a zero differential input signal is applied. The two null terminals on the input stage are used to make appropriate adjustments to accomplish this design goal. The "null technique" and the corresponding portion of the circuit will be discussed in detail in the next chapter.

The dc current biasing is initiated through the diode-connected transistors $Q_{12}$ and $Q_{11}$ and resistor $R_5$. Transistors $Q_{11}$ and $Q_{10}$, with resistor $R_4$, form a Widlar current source that establishes the bias currents in the common-base transistors $Q_3$ and $Q_4$, as well as the current mirror formed by $Q_9$ and $Q_8$.

Transistors $Q_3$ and $Q_4$ are lateral pnp devices, which refers to the fabrication process and the geometry of the transistors. Lateral pnp transistors provide added protection against voltage breakdown, although the current gain is smaller than in npn devices.

Figure 13.4(a) shows a basic common-emitter differential pair used as the input to a diff-amp. If the input voltage $V_1$ were to be connected to a supply voltage of 15 V, with $V_2$ at ground potential, then the B–E junction of $Q_2$ would be reverse biased by approximately 14.3 V. Since the breakdown voltage of an npn B–E junction is typically in the range of 3–6 V, transistor $Q_2$ in Figure 13.4(a) would probably enter breakdown and suffer permanent damage.

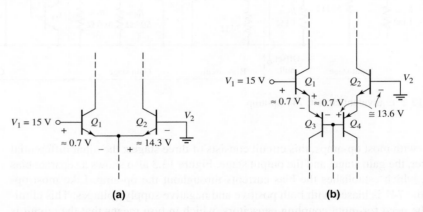

**(a)**                                **(b)**

**Figure 13.4** (a) Basic common-emitter differential pair, with a large differential voltage and (b) the 741 input stage, with a large differential voltage

By comparison, Figure 13.4(b) shows the input stage of the 741 op-amp with the same input voltages. The B–E junctions of $Q_1$ and $Q_3$ are forward biased, which means that the series combination of B–E junctions of $Q_2$ and $Q_4$ is reverse biased by approximately 13.6 V. The breakdown voltage of a lateral pnp B–E junction is typically on the order of 50 V, which means that for this input voltage polarity, the B–E junction of $Q_4$ provides the necessary breakdown protection for the input diff-amp stage.

### Gain Stage

The second, or gain, stage consists of transistors $Q_{16}$ and $Q_{17}$. Transistor $Q_{16}$ operates as an emitter follower; therefore, the input resistance of the gain stage is large. As previously discussed, a large input resistance to the gain stage minimizes loading effects on the diff-amp stage.

Transistor $Q_{13}$ is effectively two transistors connected in parallel, with common base and emitter terminals. The area of $Q_{13A}$ is effectively one-fourth the area of $Q_{12}$, and the area of $Q_{13B}$ is effectively three-fourths that of $Q_{12}$. Transistor $Q_{13B}$ provides the bias current for $Q_{17}$ and also acts as an active load to produce a high-voltage gain. Transistor $Q_{17}$ operates in a common-emitter configuration; therefore, the voltage at the collector of $Q_{17}$ is the input signal to the output stage. The signal undergoes another dc level shift as it goes through this gain stage.

The 741 is internally compensated by the feedback capacitor $C_1$ connected between the output and input terminals of the gain stage. This Miller compensation technique assures that the 741 op-amp forms stable feedback circuits.

### Output Stage

The output stage of an op-amp should provide a low output resistance, as well as a current gain, if it is to drive relatively large load currents. The output stage is therefore a class-AB circuit consisting of the complementary emitter-follower pair $Q_{14}$ and $Q_{20}$.

The output of the gain stage is connected to the base of $Q_{22}$, which operates as an emitter follower and provides a very high input resistance; the gain stage therefore suffers no significant loading effects due to the output stage. Transistor $Q_{13A}$ provides a bias current for $Q_{22}$, as well as for $Q_{18}$ and $Q_{19}$, which are used to establish a quiescent bias current in the output transistors $Q_{14}$ and $Q_{20}$. Transistors $Q_{15}$ and $Q_{21}$ are referred to as short-circuit protection devices. These transistors are normally off; they conduct only if the output is inadvertently connected to ground, resulting in a very large output current. We will consider the characteristics of the output stage in Section 13.2.2.

An abbreviated data sheet for the 741 is shown in Table 13.1. During our discussions in this chapter, we will compare our analysis results to the values in the table. A more complete data sheet for the 741 op-amp is given in Appendix B.

**Table 13.1**  Data for 741 at $T = 300$ K and supply voltage of $\pm 15$ V

| Parameter | Minimum | Typical | Maximum | Units |
|---|---|---|---|---|
| Input bias current | | 80 | 500 | nA |
| Differential-mode input resistance | 0.3 | 2.0 | | MΩ |
| Input capacitance | | 1.4 | | pF |
| Output short-circuit current | | 25 | | mA |
| Open-loop gain ($R_L \geq 2$ kΩ) | 50,000 | 200,000 | | V/V |
| Output resistance | | 75 | | Ω |
| Unity-gain frequency | | 1 | | MHz |

## 13.2.2  DC Analysis

In this section, we will analyze the dc characteristics of the 741 op-amp to determine the dc bias currents. We assume that both the noninverting and inverting input terminals are at ground potential, and that the dc supply voltages are $V^+ = 15$ V and $V^- = -15$ V. As an approximation, we assume $V_{BE} = 0.6$ V for npn transistors and $V_{EB} = 0.6$ V for pnp transistors. In most dc calculations, we neglect dc base currents, although we include base current effects in a few specific cases.

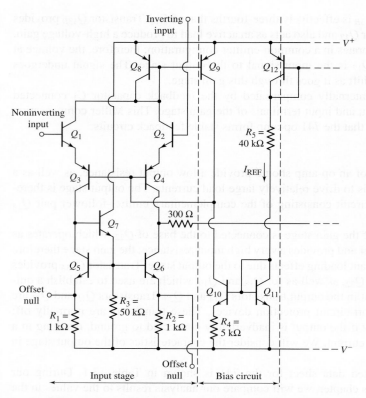

**Figure 13.5**  Bias circuit and input stage portion of 741 op-amp circuit

## Bias Circuit and Input Stage

Figure 13.5 shows the bias circuit and input stage portion of the 741 circuit. The reference current, which is established in the bias circuit branch composed of $Q_{12}$, $Q_{11}$, and $R_5$, is

$$I_{\text{REF}} = \frac{V^+ - V_{EB12} - V_{BE11} - V^-}{R_5} \tag{13.1}$$

Transistors $Q_{11}$ and $Q_{10}$ and resistor $R_4$ form a Widlar current source. Therefore, $I_{C10}$ is determined from the relationship

$$I_{C10}R_4 = V_T \ln\left(\frac{I_{\text{REF}}}{I_{C10}}\right) \tag{13.2}$$

where $V_T$ is the thermal voltage and $Q_{10}$ and $Q_{11}$ are assumed to be matched transistors.

Neglecting base currents, $I_{C8} = I_{C9} = I_{C10}$. The quiescent collector currents in $Q_1$ through $Q_4$ are then

$$I_{C1} = I_{C2} = I_{C3} = I_{C4} = \frac{I_{C10}}{2} \tag{13.3}$$

Assuming the dc currents in the input stage are exactly balanced, the dc voltage at the collector of $Q_6$, which is the input to the second stage, is the same as the dc voltage at the collector of $Q_5$. We can write

$$V_{C6} = V_{BE7} + V_{BE6} + I_{C6}R_2 + V^- \tag{13.4}$$

As previously discussed, the dc level shifts through the op-amp.

## EXAMPLE 13.1

**Objective:** Calculate the dc currents in the bias circuit and input stage of the 741 op-amp.

The bias circuit and input stage are shown in Figure 13.5.

**Solution:** From Equation (13.1), the reference current is

$$I_{REF} = \frac{V^+ - V_{EB12} - V_{BE11} - V^-}{R_5} = \frac{15 - 0.6 - 0.6 - (-15)}{40} = 0.72 \text{ mA}$$

Current $I_{C10}$ is found from Equation (13.2), as follows:

$$I_{C10}(5) = (0.026) \ln\left(\frac{0.72}{I_{C10}}\right)$$

By trial and error, we find that $I_{C10} = 19\ \mu A$. The bias currents in the input stage are then

$$I_{C1} = I_{C2} = I_{C3} = I_{C4} = 9.5\ \mu A$$

From Equation (13.4), the voltage at the collector of $Q_6$ is

$$V_{C6} = V_{BE7} + V_{BE6} + I_{C6}R_2 + V^- = 0.6 + 0.6 + (0.0095)(1) + (-15)$$

or

$$V_{C6} \cong -13.8 \text{ V}$$

**Comment:** The bias currents in the input stage are quite small; the input base currents at the noninverting and inverting terminals are generally in the nanoampere range. Small bias currents mean that the differential input resistance is large.

---

EXERCISE PROBLEM

**Ex 13.1:** Consider the input stage and bias circuit of the 741 op-amp shown in Figure 13.5. The resistor $R_5$ is changed to $R_5 = 25\ \text{k}\Omega$, and the bias voltages are $V^+ = 5$ V and $V^- = -5$ V. Let $V_{BE}(\text{on}) = V_{EB}(\text{on}) = 0.6$ V and neglect base currents. Determine the currents $I_{REF}$, $I_{C10}$, $I_{C1}$, and $I_{C2}$. (Ans. $I_{REF} = 0.352\ \text{mA}$, $I_{C10} = 16\ \mu A$, $I_{C1} = I_{C2} = 8\ \mu A$)

The transistor current gain of the lateral pnp transistors $Q_3$, $Q_4$, $Q_8$, and $Q_9$ may be relatively small, which means that the base currents in these transistors may not be negligible. To determine the effect of the base currents, consider the expanded input stage shown in Figure 13.6. The base currents in the npn transistors are still assumed to be negligible. Current $I_{C10}$ establishes the base currents in $Q_3$ and $Q_4$, which then establish the emitter currents designated as $I$. At the $Q_8$ collector, we have

$$2I = I_{C8} + \frac{2I_{C9}}{\beta_p} = I_{C9}\left(1 + \frac{2}{\beta_p}\right) \tag{13.5}$$

Since $Q_8$ and $Q_9$ are matched, $I_{C8} = I_{C9}$. Then,

$$I_{C10} = \frac{2I}{1 + \beta_p} + I_{C9} = \frac{2I}{1 + \beta_p} + \frac{2I}{\left(1 + \dfrac{2}{\beta_p}\right)} = 2I\left[\frac{\beta_p^2 + 2\beta_p + 2}{\beta_p^2 + 3\beta_p + 2}\right] \tag{13.6}$$

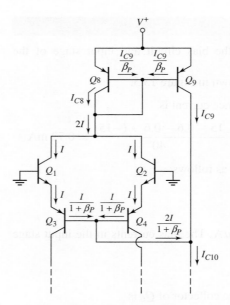

Figure 13.6  Expanded input stage, 741 op-amp, showing base currents

Even if the pnp transistor base currents are not negligible, the bias currents in $Q_1$ and $Q_2$ are, from Equation (13.6), very nearly

$$I = \frac{I_{C10}}{2} \tag{13.7}$$

This bias current is essentially the same as originally assumed in Equation (13.3).

### Gain Stage

Figure 13.7 shows the reference portion of the bias circuit and the gain stage. The reference current is given by Equation (13.1). Transistors $Q_{12}$ and $Q_{13}$ form a current mirror, and $Q_{13B}$ has a scale factor 0.75 times that of $Q_{12}$. Neglecting base currents, current $I_{C13B}$ is then

$$I_{C13B} = 0.75 I_{\text{REF}} \tag{13.8}$$

The emitter current in $Q_{16}$ is the sum of the base current in $Q_{17}$ and the current in $R_9$, as follows:

$$I_{C16} \cong I_{E16} = I_{B17} + \frac{I_{E17} R_8 + V_{BE17}}{R_9} \tag{13.9}$$

### EXAMPLE 13.2

Objective: Calculate the bias currents in the gain stage of the 741 op-amp in Figure 13.7. Assume bias voltages of $\pm 15$ V.

Solution: In Example 13.1, we determined the reference current to be $I_{\text{REF}} = 0.72$ mA. From Equation (13.8), the collector current in $Q_{17}$ is

$$I_{C17} = I_{C13B} = 0.75 I_{\text{REF}} = (0.75)(0.72) = 0.54 \text{ mA}$$

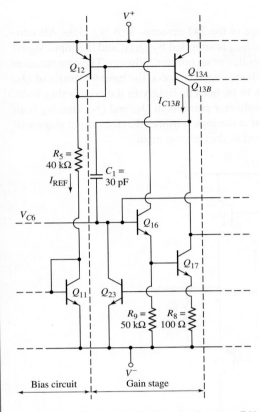

**Figure 13.7**  Reference circuit and gain stage, 741 op-amp

Assuming $\beta = 200$ for the npn transistor, the collector current in $Q_{16}$ is, from Equation (13.9),

$$I_{C16} \cong I_{B17} + \frac{I_{E17}R_8 + V_{BE17}}{R_9} = \frac{0.54}{200} + \frac{(0.54)(0.1) + 0.6}{50}$$

or

$$I_{C16} = 15.8 \ \mu A$$

**Comment:** The small bias current in $Q_{16}$, in conjunction with the resistor $R_9$, ensures that the input resistance to the gain stage is large, which minimizes loading effects on the diff-amp stage. The small bias current in $Q_{16}$ also means that the base current in $Q_{16}$ is negligible, as assumed in the dc analysis of the input stage.

EXERCISE PROBLEM

**Ex 13.2:**  Repeat Example 13.2 for bias voltages of $\pm 5$ V. (Ans. $I_{REF} = 0.22$ mA, $I_{C17} = 0.165$ mA, $I_{C16} = 13.2 \ \mu A$)

## Output Stage

Figure 13.8 shows the basic output stage of the 741 op-amp. This is a class-AB configuration, discussed in Chapter 8. The $I_{\text{Bias}}$ is supplied by $Q_{13A}$, and the input signal is applied to the base of $Q_{22}$, which operates as an emitter follower. The combination of $Q_{18}$ and $Q_{19}$ establishes two B–E voltage drops between the base terminals of $Q_{14}$ and $Q_{20}$, causing the output transistors to be biased slightly in the conducting state. This $V_{BB}$ voltage produces quiescent collector currents in $Q_{14}$ and $Q_{20}$. Biasing both $Q_{14}$ and $Q_{20}$ "on" with no signal present at the input ensures that the output stage will respond linearly when a signal is applied to the op-amp input.

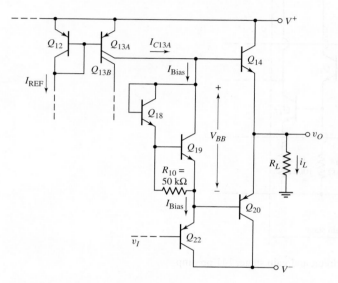

**Figure 13.8** Basic output stage, 741 op-amp, showing currents and voltages

The collector of $Q_{13A}$ has a scale factor of 0.25 times that of $Q_{12}$. Neglecting base currents, current $I_{C13A}$ is

$$I_{C13A} = 0.25 I_{\text{REF}} = I_{\text{Bias}} \tag{13.10}$$

where $I_{\text{REF}}$ is given by Equation (13.1). Neglecting base currents, the collector current in $Q_{22}$ is also equal to $I_{\text{Bias}}$. The collector current in $Q_{18}$ is

$$I_{C18} \cong \frac{V_{BE19}}{R_{10}} \tag{13.11}$$

Therefore,

$$I_{C19} = I_{\text{Bias}} - I_{C18} \tag{13.12}$$

## EXAMPLE 13.3

**Objective:** Calculate the bias currents in the output stage of the 741 op-amp.

Consider the output stage shown in Figure 13.8. Assume the reverse saturation currents of $Q_{18}$ and $Q_{19}$ are $I_S = 10^{-14}$ A, and the reverse saturation currents of $Q_{14}$ and $Q_{20}$ are $I_S = 3 \times 10^{-14}$ A. Neglect base currents.

**Solution:** The reference current, from Example 13.1, is $I_{REF} = 0.72$ mA. Current $I_{C13A}$ is then

$$I_{C13A} = (0.25)I_{REF} = (0.25)(0.72) = 0.18 \text{ mA} \cong I_{Bias}$$

If we assume $V_{BE19} = 0.6$ V, then the current in $R_{10}$ is

$$I_{R10} = \frac{V_{BE19}}{R_{10}} = \frac{0.6}{50} = 0.012 \text{ mA}$$

The current in $Q_{19}$ is

$$I_{C19} \cong I_{E19} = I_{C13A} - I_{R10} = 0.18 - 0.012 = 0.168 \text{ mA}$$

For this value of collector current, the B–E voltage of $Q_{19}$ is

$$V_{BE19} = V_T \ln\left(\frac{I_{C19}}{I_S}\right) = (0.026) \ln\left(\frac{0.168 \times 10^{-3}}{10^{-14}}\right) = 0.612 \text{ V}$$

which is close to the assumed value of 0.6 V. Assuming $\beta_n = 200$ for the npn devices, the base current in $Q_{19}$ is

$$I_{B19} = \frac{I_{C19}}{\beta_n} = \frac{168 \ \mu A}{200} = 0.84 \ \mu A$$

The current in $Q_{18}$ is now

$$I_{C18} \cong I_{E18} = I_{R10} + I_{B19} = 12 + 0.84 = 12.84 \ \mu A$$

The B–E voltage of $Q_{18}$ is therefore

$$V_{BE18} = V_T \ln\left(\frac{I_{C18}}{I_S}\right) = (0.026) \ln\left(\frac{12.84 \times 10^{-6}}{10^{-14}}\right) = 0.545 \text{ V}$$

The voltage difference $V_{BB}$ is thus

$$V_{BB} = V_{BE18} + V_{BE19} = 0.545 + 0.612 = 1.157 \text{ V}$$

Since the output transistors $Q_{14}$ and $Q_{20}$ are identical, one-half of $V_{BB}$ is across each B–E junction. The quiescent currents in $Q_{14}$ and $Q_{20}$ are

$$I_{C14} = I_{C20} = I_S e^{(V_{BB}/2)/V_T} = 3 \times 10^{-14} e^{(1.157/2)/0.026}$$

or

$$I_{C14} = I_{C20} = 138 \ \mu A$$

**Comment:** Using the piecewise linear approximation of 0.6 V for the B–E junction voltage does not allow us to determine the quiescent currents in $Q_{14}$ and $Q_{20}$. For a more accurate analysis, the exponential relationship must be used, since the base–emitter areas of the output transistors are larger than those of the other transistors, and because the output transistors are biased at a low quiescent current.

## EXERCISE PROBLEM

**Ex 13.3:** Calculate the bias currents $I_{C13A}$, $I_{R10}$, $I_{C19}$, $I_{C18}$, and $I_{C14}$ in the output stage of the 741 op-amp for $I_{REF} = 0.50$ mA, $V^+ = 5$ V, and $V^- = -5$ V. All other parameters are the same as described in Example 13.3. (Ans. $I_{C13A} = 0.125$ mA, $I_{R10} = 0.012$ mA, $I_{C19} = 0.113$ mA, $I_{C18} = 12.565 \ \mu A$, $I_{C14} = 0.113$ mA)

As the input signal $v_I$ increases, the base voltage of $Q_{14}$ increases since the $V_{BB}$ voltage remains almost constant. The output voltage increases at approximately the same rate as the input signal. As $v_I$ decreases, the base voltage of $Q_{20}$ decreases, and the output voltage also decreases, again at approximately the same rate as the input signal. The small-signal voltage gain of the output stage is essentially unity.

### Short-Circuit Protection Circuitry

The output stage includes a number of transistors that are off during the normal operation of the amplifier. If the output terminal is at a positive voltage because of an applied input signal, and if the terminal is inadvertently shorted to ground potential, a large current will be induced in output transistor $Q_{14}$. A large current can produce sufficient heating to cause transistor burnout.

The complete output stage of the 741, including the **short-circuit protection devices,** is shown in Figure 13.9. Resistor $R_6$ and transistor $Q_{15}$ limit the current in $Q_{14}$ in the event of a short circuit. If the current in $Q_{14}$ reaches 20 mA, the voltage drop across $R_6$ is 540 mV, which is sufficient to bias $Q_{15}$ in the conducting stage. As $Q_{15}$ turns on, excess base current into $Q_{14}$ is shunted through the collector of $Q_{15}$. The base current into $Q_{14}$ is then limited to a maximum value, which limits the collector current.

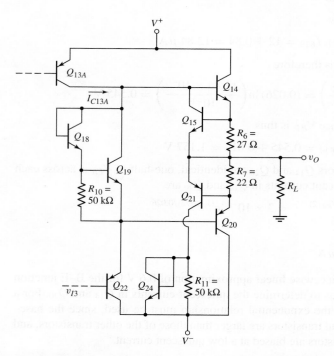

**Figure 13.9** Output stage, 741 op-amp with short-circuit protection devices

The maximum current in $Q_{20}$ is limited by components $R_7$, $Q_{21}$, and $Q_{24}$, in much the same way as just discussed. A large output current will result in a voltage drop across $R_7$, which will be sufficient to bias $Q_{21}$ in its conducting state. Transistors $Q_{21}$ and $Q_{24}$ will shunt excessive output current away from $Q_{20}$, to protect this output transistor.

### 13.2.3    Small-Signal Analysis

We can analyze the small-signal voltage gain of the 741 op-amp by dividing it into its basic circuits and using results previously obtained.

#### Input Stage

Figure 13.10 shows the ac equivalent circuit of the input stage with a differential voltage $v_d$ applied between the input terminals. The constant-current biasing at the base of $Q_3$ and $Q_4$ means that the effective impedance connected to the base terminal of $Q_3$ and $Q_4$ is ideally infinite, or an open circuit. Resistance $R_{act1}$ is the effective resistance of the active load and $R_{i2}$ is the input resistance of the gain stage.

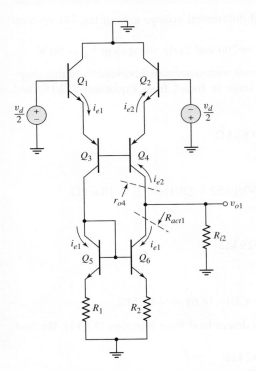

**Figure 13.10** Simplified ac equivalent circuit of input stage of 741 op-amp

From the results in Chapter 11, the small-signal differential voltage gain can be written as

$$A_d = \frac{v_{o1}}{v_d} = -g_m(r_{o4} \| R_{act1} \| R_{i2}) = -\left(\frac{I_{CQ}}{V_T}\right)(r_{o4} \| R_{act1} \| R_{i2}) \qquad \textbf{(13.13)}$$

where $I_{CQ}$ is the quiescent collector current in each of the transistors $Q_1$ through $Q_4$, and $r_{o4}$ is the small-signal output resistance looking into the collector of $Q_4$. Using $r_{o4}$ as the resistance looking into the collector of $Q_4$ neglects the effective resistance in the emitter of $Q_4$. This effective resistance is simply the resistance looking into the emitter of $Q_2$, which is normally very small. The minus sign in the voltage gain expression results from the applied signal voltage polarity and resulting current directions.

The effective resistance of the active load is given by

$$R_{act1} = r_{o6} \left[ 1 + g_{m6}(R_2 \| r_{\pi 6}) \right] \tag{13.14}$$

as determined in Chapter 10 for the output resistance of a Widlar current source. From Figure 13.7, the input resistance of the gain stage is

$$R_{i2} = r_{\pi 16} + (1 + \beta_n) R'_E \tag{13.15}$$

where $R'_E$ is the effective resistance in the emitter of $Q_{16}$, as given by

$$R'_E = R_9 \| [r_{\pi 17} + (1 + \beta_n) R_8] \tag{13.16}$$

## EXAMPLE 13.4

**Objective:** Determine the small-signal differential voltage gain of the 741 op-amp input stage.

Assume npn transistor gains of $\beta_n = 200$ and Early voltages of $V_A = 50$ V.

**Solution:** The quiescent collector currents were determined previously in this chapter. The input resistance to the gain stage is found from Equations (13.15) and (13.16), as follows:

$$r_{\pi 17} = \frac{\beta_n V_T}{I_{C17}} = \frac{(200)(0.026)}{0.54} = 9.63 \text{ k}\Omega$$

Therefore,

$$R'_E = R_9 \| [r_{\pi 17} + (1 + \beta_n) R_8] = 50 \| [9.63 + (201)(0.1)] = 18.6 \text{ k}\Omega$$

Also,

$$r_{\pi 16} = \frac{\beta_n V_T}{I_{C16}} = \frac{(200)(0.026)}{0.0158} = 329 \text{ k}\Omega$$

Consequently,

$$R_{i2} = r_{\pi 16} + (1 + \beta_n) R'_E = 329 + (201)(18.6) \Rightarrow 4.07 \text{ M}\Omega$$

The resistance of the active load is determined from Equation (13.14). We find

$$r_{\pi 6} = \frac{\beta_n V_T}{I_{C6}} = \frac{(200)(0.026)}{0.0095} = 547 \text{ k}\Omega$$

$$g_{m6} = \frac{I_{C6}}{V_T} = \frac{0.0095}{0.026} = 0.365 \text{ mA/V}$$

and

$$r_{o6} = \frac{V_A}{I_{C6}} = \frac{50}{0.0095} \Rightarrow 5.26 \text{ M}\Omega$$

Then,

$$R_{act1} = r_{o6} \left[ 1 + g_{m6}(R_2 \| r_{\pi 6}) \right] = 5.26 \left[ 1 + (0.365)(1 \| 547) \right] = 7.18 \text{ M}\Omega$$

Resistance $r_{o4}$ is

$$r_{o4} = \frac{V_A}{I_{C4}} = \frac{(50)}{(0.0095)} \Rightarrow 5.26 \text{ M}\Omega$$

Finally, from Equation (13.13), the small-signal differential voltage gain is

$$A_d = -\left(\frac{I_{CQ}}{V_T}\right)(r_{o4}\|R_{act1}\|R_{i2}) = -\left(\frac{9.5}{0.026}\right)(5.26\|7.18\|4.07)$$

or

$$A_d = -636$$

**Comment:** The relatively large gain results from the use of an active load and the fact that the gain stage does not drastically load the input stage.

**Ex 13.4:** Repeat Example 13.4 assuming Early voltages of $V_A = 100$ V. (Ans. $A_d = -889$)

**Gain Stage**

Figure 13.11 shows the ac equivalent circuit of the gain stage. Resistance $R_{act2}$ is the effective resistance of the active load and $R_{i3}$ is the input resistance of the output stage.

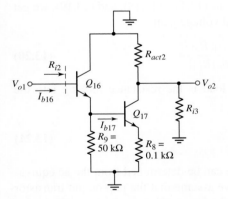

**Figure 13.11** The ac equivalent circuit, gain stage of 741 op-amp

We develop the small-signal voltage gain using Figure 13.11 directly. The input base current to $Q_{16}$ is

$$i_{b16} = \frac{v_{o1}}{R_{i2}} \tag{13.17}$$

where $R_{i2}$ is the input resistance to the gain stage. The base current into $Q_{17}$ is

$$i_{b17} = \frac{R_9}{R_9 + [r_{\pi 17} + (1+\beta_n)R_8]} \times i_{e16} \tag{13.18}$$

where $i_{e16}$ is the emitter current from $Q_{16}$. The output voltage is

$$v_{o2} = -i_{c17}(R_{act2}\|R_{i3}\|R_{o17}) \tag{13.19}$$

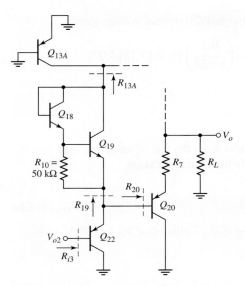

**Figure 13.12** The ac equivalent circuit, 741 op-amp output stage, for calculating input resistance

where $i_{c17}$ is the ac collector current in $Q_{17}$ and $R_{o17}$ is the output impedance looking into the collector of $Q_{17}$. Combining Equations (13.17), (13.18), and (13.19), we get the following expression for the small-signal voltage gain:

$$A_{v2} = \frac{v_{o2}}{v_{o1}} = \frac{-\beta_n(1+\beta_n)R_9(R_{act2} \| R_{i3} \| R_{o17})}{R_{i2}\{R_9 + [r_{\pi17} + (1+\beta_n)R_8]\}} \qquad (13.20)$$

The effective resistance of the active load is the resistance looking into the collector of $Q_{13B}$, or

$$R_{act2} = r_{o13B} = \frac{V_A}{I_{C13B}} \qquad (13.21)$$

The input resistance of the output stage can be determined from the ac equivalent circuit in Figure 13.12. In this figure, we assume that the pnp output transistor $Q_{20}$ is active and the npn output transistor $Q_{14}$ is cut off. A load resistor $R_L$ is also included. Transistor $Q_{22}$ operates as an emitter follower, which means that the input resistance is

$$R_{i3} = r_{\pi22} + (1+\beta_p)[R_{19} \| R_{20}] \qquad (13.22)$$

Resistance $R_{19}$ is the series combination of the resistance looking into the emitters of $Q_{19}$ and $Q_{18}$, and the resistance looking into the collector of $Q_{13A}$. The effective resistance of the combination of $Q_{18}$ and $Q_{19}$ is small compared to $R_{13A}$; therefore,

$$R_{19} \cong R_{13A} = r_{o13A} = \frac{V_A}{I_{C13A}} \qquad (13.23)$$

The output transistor $Q_{20}$ is also an emitter follower; therefore,

$$R_{20} = r_{\pi20} + (1+\beta_p)R_L \qquad (13.24)$$

where the load resistance $R_L$ is assumed to be much larger than $R_7$.

## EXAMPLE 13.5

**Objective:** Determine the small-signal voltage gain of the second stage of the 741 op-amp.

Assume the current gains of the pnp transistors are $\beta_p = 50$ and the gains of the npn transistors are $\beta_n = 200$. Also assume the Early voltage is 50 V for all transistors and the load resistance connected to the output is $R_L = 2 \text{ k}\Omega$. The dc quiescent currents were determined previously.

**Solution:** First, we calculate the various resistances. To begin,

$$r_{\pi 20} = \frac{\beta_p V_T}{I_{C20}} = \frac{(50)(0.026)}{0.138} = 9.42 \text{ k}\Omega$$

which means that

$$R_{20} = r_{\pi 20} + (1 + \beta_p)R_L = 9.42 + (51)(2) \cong 111 \text{ k}\Omega$$

Also,

$$R_{19} = r_{o13A} = \frac{V_A}{I_{C13A}} = \frac{50}{0.18} = 278 \text{ k}\Omega$$

and

$$r_{\pi 22} = \frac{\beta_p V_T}{I_{C13A}} = \frac{(50)(0.026)}{0.18} = 7.22 \text{ k}\Omega$$

The input resistance to the output stage is therefore

$$R_{i3} = r_{\pi 22} + (1 + \beta_p)[R_{19} \| R_{20}] = 7.22 + (51)[278 \| 111] \Rightarrow 4.05 \text{ M}\Omega$$

The effective resistance of the active load is

$$R_{act2} = \frac{V_A}{I_{C13B}} = \frac{50}{0.54} = 92.6 \text{ k}\Omega$$

and the output resistance $R_{o17}$ is

$$R_{o17} \cong \frac{V_A}{I_{C17}} = \frac{50}{0.54} = 92.6 \text{ k}\Omega$$

This calculation neglects the very small value of $R_8$ in the emitter.

From Equation (13.20), the small-signal voltage gain is as follows (all resistances are given in kilohms):

$$A_{v2} = \frac{-\beta_n(1 + \beta_n)R_9(R_{act2} \| R_{i3} \| R_{o17})}{R_{i2}\{R_9 + [r_{\pi 17} + (1 + \beta_n)R_8]\}}$$

$$= \frac{-(200)(201)(50)(92.6 \| 4050 \| 92.6)}{4070\{50 + [9.63 + (201)(0.1)]\}}$$

or

$$A_{v2} = -285$$

**Comment:** The voltage gain of the second stage is fairly large, again because an active load is used and because there is no severe loading effect from the output stage.

**Ex 13.5:** Repeat Example 13.5 assuming Early voltages of $V_A = 100$ V. (Ans. $A_2 = -562$)

## Overall Gain

In calculating the voltage gain of each stage, we took the loading effect of the following stage into account. Therefore, the overall voltage gain is the product of the individual gain factors, or

$$A_v = A_d A_{v2} A_{v3} \tag{13.25}$$

where $A_{v3}$ is the voltage gain of the output stage. If we assume that $A_{v3} \approx 1$, as previously discussed, then the overall gain of the 741 op-amp is

$$A_v = A_d A_{v2} A_{v3} = (-636)(-285)(1) = 181{,}260 \tag{13.26}$$

Typical voltage gain values for the 741 op-amp are in the range of 200,000. The value determined in our calculations illustrates the magnitude of voltage gains that can be obtained in op-amp circuits.

## Output Resistance

The output resistance can be determined by using the ac equivalent circuit in Figure 13.13. In this case, we assume the output transistor $Q_{20}$ is conducting and $Q_{14}$ is cut off. The same basic result is obtained if $Q_{14}$ is conducting and $Q_{20}$ is cut off.

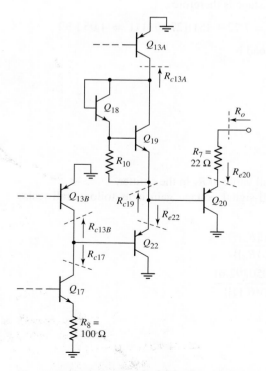

**Figure 13.13** The ac equivalent circuit, 741 op-amp output stage, for calculating output resistance

We again rely on results obtained previously for output resistances of basic amplifier stages.

The output resistance is

$$R_o = R_7 + R_{e20} \tag{13.27}$$

where

$$R_{e20} = \frac{r_{\pi 20} + R_{e22}\|R_{c19}}{(1 + \beta_p)} \tag{13.28}$$

Previously we argued that the series resistance due to $Q_{18}$ and $Q_{19}$ is small compared to $R_{c13A}$, so that $R_{c19} \cong R_{c13A}$. We also have

$$R_{e22} = \frac{r_{\pi 22} + R_{c17}\|R_{c13B}}{(1 + \beta_p)} \tag{13.29}$$

where

$$R_{c13B} = r_{o13B}$$

and

$$R_{c17} = r_{o17}[1 + g_{m17}(R_8 \| r_{\pi 17})]$$

The output resistance of the op-amp is then found by combining all the resistance terms.

## EXAMPLE 13.6

**Objective:**  Calculate the output resistance of the 741 op-amp.

Consider the output stage configuration in Figure 13.13. Assume the output current is $I_{c20} = 2$ mA and all other bias currents are as previously determined.

**Solution:**  Using $\beta_n = 200$, $\beta_p = 50$, and $V_A = 50$ V, we find the following:

$$r_{\pi 17} = 9.63 \text{ k}\Omega \qquad r_{\pi 22} = 7.22 \text{ k}\Omega \qquad r_{\pi 20} = 0.65 \text{ k}\Omega$$
$$g_{m17} = 20.8 \text{ mA/V} \qquad r_{o17} = 92.6 \text{ k}\Omega \qquad r_{o13B} = 92.6 \text{ k}\Omega$$

Then,

$$R_{c17} = r_{o17}[1 + g_{m17}(R_8 \| r_{\pi 17})] = 92.6[1 + (20.8)(0.1\|9.63)] = 283 \text{ k}\Omega$$

and

$$R_{e22} = \frac{r_{\pi 22} + R_{c17}\|R_{c13B}}{(1 + \beta_p)} = \frac{7.22 + 283\|92.6}{51} = 1.51 \text{ k}\Omega$$

Also,

$$R_{c19} \cong R_{c13A} = r_{o13A} = \frac{V_A}{I_{C13A}} = \frac{50}{0.18} = 278 \text{ k}\Omega$$

Therefore

$$R_{e20} = \frac{r_{\pi 20} + R_{e22}\|R_{c19}}{(1 + \beta_p)} = \frac{0.65 + 1.51\|278}{51} = 0.0422 \text{ k}\Omega \Rightarrow 42.2 \ \Omega$$

Consequently, the output resistance is

$$R_o = R_7 + R_{e20} = 22 + 42.2 = 64.2 \ \Omega$$

**Comment:** We showed previously that the output resistance of an emitter-follower circuit is low. For comparison, typical output resistance values for the 741 op-amp are 75 $\Omega$. This correlates well with our analysis.

EXERCISE PROBLEM

**Ex 13.6:** Repeat Example 13.6 assuming Early voltages of $V_A = 100$ V. (Ans. $R_o = 90.9 \, \Omega$)

### 13.2.4    Frequency Response

The 741 op-amp is internally compensated by the Miller compensation technique to introduce a dominant low-frequency pole. From Miller's theorem, the effective input capacitance of the second gain stage is

$$C_i = C_1(1 + |A_{v2}|) \tag{13.30}$$

The dominant low-frequency pole is

$$f_{PD} = \frac{1}{2\pi R_{eq} C_i} \tag{13.31}$$

where $R_{eq}$ is the equivalent resistance between the second-stage input node and ground, and is

$$R_{eq} = R_{o1} \| R_{i2} \tag{13.32}$$

Here $R_{i2}$ is the input resistance of the gain stage and $R_{o1}$ is the output resistance of the diff-amp stage. From Figure 13.10, we see that

$$R_{o1} = R_{act1} \| r_{o4} \tag{13.33}$$

### EXAMPLE 13.7

**Objective:** Determine the dominant-pole frequency of the 741 op-amp.
Use appropriate results from previous calculations.

**Solution:** Previously, we determined that $|A_{v2}| = 285$, which means that the effective input capacitance is

$$C_i = C_1(1 + |A_{v2}|) = (30)(1 + 285) = 8580 \text{ pF}$$

The gain stage input resistance was found to be $R_{i2} = 4.07$ M$\Omega$. We find

$$R_{o1} = R_{act1} \| r_{o4} = 7.18 \| 5.26 = 3.04 \text{ M}\Omega$$

The equivalent resistance is then

$$R_{eq} = R_{o1} \| R_{i2} = 3.04 \| 4.07 = 1.74 \text{ M}\Omega$$

Finally, the dominant-pole frequency is

$$f_{PD} = \frac{1}{2\pi R_{eq} C_i} = \frac{1}{2\pi (1.74 \times 10^6)(8580 \times 10^{-12})} = 10.7 \text{ Hz}$$

**Comment:** The very large equivalent input capacitance $C_i$ justifies neglecting any other capacitance effects at the gain stage input.

**Ex 13.7:** Repeat Example 13.7 assuming Early voltages of $V_A = 100$ V. See Exercise Problems Ex 13.4, Ex 13.5, and Ex 13.6. (Ans. 3.88 Hz)

If all other poles of the op-amp circuit are at very high frequencies, then the unity-gain bandwidth is

$$f_T = A_o f_{PD} \tag{13.34}$$

Using our results, we find that

$$f_T = (181,260)(10.7) \cong 1.9 \text{ MHz} \tag{13.35}$$

A typical unity-gain bandwidth value for the 741 op-amp is 1 MHz. With all the approximations and assumptions, such as the value of reverse saturation current and Early voltage, used in the calculations, a factor of two between the actual and predicted cutoff frequency is not significant.

If the frequencies of the other poles of the 741 op-amp are greater than 1.9 MHz, the phase margin is 90 degrees. This phase margin ensures that any closed-loop amplifier circuit using the 741 op-amp will be stable for any feedback transfer function.

## Problem-Solving Technique: Operational Amplifier Circuits

1. *DC analysis.* The bias portion of the op-amp circuit must be identified. A reference current must be determined and then the bias currents in the individual building blocks of the overall circuit can be determined.
2. *AC analysis.* The small-signal properties of the building blocks of the overall circuit can be analyzed individually, provided that the loading effects of follow-on stages are taken into account.

## Test Your Understanding

**TYU 13.3** Using the results of Example 13.1 and assuming $\beta_n = 200$, determine the input base currents to $Q_1$ and $Q_2$. (Ans. $I_{B1} = I_{B2} = 47.5$ nA)

**TYU 13.4** The 741 op-amp in Figure 13.3 is biased at $V^+ = 15$ V and $V^- = -15$ V. Assume $V_{BE}(\text{npn}) = V_{EB}(\text{pnp}) = 0.6$ V. Determine the input common-mode voltage range, neglecting voltage drops across $R_1$ and $R_2$. (Ans. $-12.6 < v_{in}(\text{cm}) \leq 14.4$ V)

**TYU 13.5** (a) If the 741 op-amp in Figure 13.3 is biased at $V^+ = 15$ and $V^- = -15$ V, estimate the maximum and minimum output voltages such that the op-amp remains biased in its linear region. (b) Repeat part (a) if $V^+ = 5$ V and $V^- = -5$ V. (Ans. (a) $-13.2 \leq v_O \leq 13.8$ V (b) $-3.2 \leq v_O \leq 3.8$ V)

**TYU 13.6** Consider the input stage and bias circuit in Figure 13.5 with supply voltages $V^+ = 5$ V and $V^- = -5$ V. If $I_S = 5 \times 10^{-15}$ A for each transistor, determine $I_{REF}$, $V_{BE11}$, $V_{BE10}$, $V_{BE6}$, and $I_{C10}$. (Ans. $I_{REF} = 0.218$ mA, $V_{BE11} = 0.637$ V, $I_{C10} = 14.2 \mu$A, $V_{BE10} = 0.566$ V, $V_{BE6} = 0.548$ V)

**TYU 13.7** The power supply voltages for the 741 op-amp in Figure 13.3 are $V^+ = 10$ V and $V^- = -10$ V. Neglect base currents and assume $V_{BE}(\text{npn}) = V_{EB}(\text{pnp}) = 0.6$ V. Calculate the bias currents $I_{REF}$, $I_{C10}$, $I_{C6}$, $I_{C13B}$, and $I_{C13A}$.

(Ans. $I_{REF} = 0.47$ mA, $I_{C10} = 17.2$ $\mu$A, $I_{C6} = 8.6$ $\mu$A, $I_{C13B} = 0.353$ mA, $I_{C13A} = 0.118$ mA)

**\*TYU 13.8** In the 741 op-amp output stage in Figure 13.3, the combination of $Q_{18}$, $Q_{19}$, and $R_{10}$ is replaced by two series diodes with $I_S = 10^{-14}$ A. The transistor parameters are: $\beta_n = 200$, $\beta_p = 50$, and $V_A = 50$ V. Assume the same dc bias currents calculated previously. Calculate the output resistance, assuming $Q_{14}$ is conducting, producing a load current of 5 mA. (Ans. 41 $\Omega$)

---

##  13.3 CMOS OPERATIONAL AMPLIFIER CIRCUITS

**Objective:** • Describe and analyze the dc and ac characteristics of CMOS operational amplifier circuits.

The 741 bipolar op-amp is a general-purpose op-amp capable of sourcing and sinking reasonably large load currents. The output stage is an emitter follower capable of supplying the necessary load current, with a low output resistance to minimize loading effects.

In contrast, most CMOS op-amps are designed for specific on-chip applications and are only required to drive capacitive loads of a few picofarads. Most CMOS op-amps therefore do not need a low-resistance output stage, and, if the op-amp inputs are not connected directly to the IC external terminals, they also do not need electrostatic input protection devices.

In this section, we consider four designs of a CMOS op-amp. Initially we consider a simple CMOS design to begin to understand the basic concepts of a CMOS op-amp. We then analyze a three-stage CMOS op-amp with a complementary push-pull output stage. The third CMOS op-amp is a more sophisticated design, called a folded cascode op-amp. Finally, we consider a current-mirror CMOS op-amp. In each case, we will do a dc analysis/design and a small-signal analysis/design.

### 13.3.1 MC14573 CMOS Operational Amplifier Circuit

**Circuit Description**
An example of an all-CMOS op-amp is the MC14573, for which a simplified circuit diagram is shown in Figure 13.14. The p-channel transistors $M_1$ and $M_2$ form the input differential pair, and the n-channel transistors $M_3$ and $M_4$ form the active load. The diff-amp input stage is biased by the current mirror $M_5$ and $M_6$, in which the reference current is determined by an external resistor $R_{set}$.

The second stage, which is also the output stage, consists of the common-source-connected transistor $M_7$. Transistor $M_8$ provides the bias current for $M_7$ and acts as the active load. An internal compensation capacitor $C_1$ is included to provide stability.

**DC Analysis**
Assuming transistors $M_5$ and $M_6$ are matched, the reference and input-stage bias currents are given by

$$I_{set} = I_Q = \frac{V^+ - V^- - V_{SG5}}{R_{set}} \tag{13.36}$$

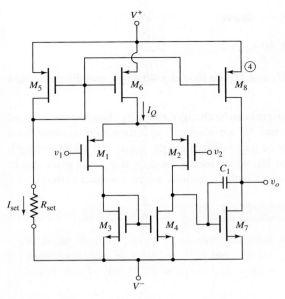

**Figure 13.14** MC14573 CMOS op-amp equivalent circuit

The reference current and source-to-gate voltage are also related by

$$I_{set} = K_{p5}(V_{SG5} + V_{TP})^2 \tag{13.37}$$

where $V_{TP}$ and $K_{p5}$ are the threshold voltage and conduction parameter of the p-channel transistor $M_5$.

## EXAMPLE 13.8

**Objective:** Determine the dc bias currents in the MC14573 op-amp.

Assume transistor parameters of $|V_T| = 0.5$ V (all transistors), $k'_n = 100\,\mu\text{A/V}^2$, $k'_p = 40\,\mu\text{A/V}^2$, and circuit parameters of $V^+ = 5$ V, $V^- = -5$ V, and $R_{set} = 225$ kΩ. Assume transistor width-to-length ratios of 6.25 for $M_3$ and $M_4$, and 12.5 for all other transistors.

**Solution:** For transistors $M_5$ and $M_6$, the conduction parameters are:

$$K_p = \left(\frac{k'_p}{2}\right)\left(\frac{W}{L}\right)_5 = \left(\frac{0.04}{2}\right)(12.5) = 0.25 \text{ mA/V}^2$$

Combining Equations (13.36) and (13.37) yields the source-to-gate voltage of $M_5$:

$$K_p \left(V_{SG5} + V_{TP}\right)^2 = \frac{V^+ - V_{SG5} - V^-}{R_{set}}$$

or

$$0.25 \left(V_{SG5} - 0.5\right)^2 = \frac{5 - V_{SG5} - (-5)}{225}$$

which yields

$$V_{SG5} = 0.9022 \text{ V}$$

From Equation (13.36), we have

$$I_{\text{set}} = I_Q = \frac{10 - 0.9022}{225} \Rightarrow 40.4\,\mu A$$

The quiescent drain currents in $M_7$ and $M_8$ are then also $40.4\,\mu A$, and the currents in $M_1$ through $M_4$ are $20.2\,\mu A$.

**Comment:** The quiescent bias currents can be changed easily by changing the external resistor $R_{\text{set}}$. Transistors $M_5$, $M_6$, and $M_8$ are identical, so the currents in these three devices are equal since the source-to-gate voltages are the same. The width-to-length ratio of $M_7$ is twice that of $M_3$ and $M_4$, which means the current in $M_7$ is twice that in $M_3$ and $M_4$. However, this is consistent with the current-source transistor currents.

## EXERCISE PROBLEM

**Ex 13.8:** Repeat Example 13.8 assuming transistor width-to-length ratios of $(W/L)_{3,4} = 10$ for transistors $M_3$ and $M_4$, and $W/L = 20$ for all other transistors. Let $R_{\text{set}} = 150\,\text{k}\Omega$. (Ans. $I_{\text{REF}} = I_Q = I_{D7} = I_{D8} = 60.74\,\mu A$, $I_{D1} - I_{D4} = 30.37\,\mu A$)

### Small-Signal Analysis

The small-signal differential voltage gain of the input stage can be written as

$$A_d = g_m\,(r_{o2}\|r_{o4}) = 2\sqrt{K_{p1}I_{DQ1}}\,(r_{o2}\|r_{o4}) = \sqrt{2K_{p1}I_Q}\,(r_{o2}\|r_{o4}) \tag{13.38}$$

where $r_{o2}$ and $r_{o4}$ are the output resistances of $M_2$ and $M_4$, respectively. The input impedance to the second stage is essentially infinite; therefore, there is no loading effect due to the second stage. If we assume that the parameter $\lambda$ is the same for all transistors, then

$$r_{o2} = r_{o4} = \frac{1}{\lambda I_D} \tag{13.39}$$

where $I_D$, which is the quiescent drain current in $M_2$ and $M_4$, is $I_D = I_Q/2$.

The magnitude of the gain of the second stage is

$$A_{v2} = g_{m7}\,(r_{o7}\|r_{o8}) \tag{13.40}$$

where

$$g_{m7} = 2\sqrt{K_{n7}I_{D7}}$$

and

$$r_{o7} = r_{o8} = 1/\lambda I_{D7}$$

Equation (13.40) implies that there is no loading effect due to an external load connected at the output.

## EXAMPLE 13.9

**Objective:** Determine the small-signal voltage gains of the input and second stages, and the overall voltage gain, of the MC14573 op-amp.

Assume the same transistor and circuit parameters as in Example 13.8. Let $\lambda = 0.02\,\text{V}^{-1}$ for all transistors.

**Solution:** The conduction parameters of $M_1$ and $M_2$ are

$$K_{p1} = K_{p2} = \left(\frac{k_p'}{2}\right)\left(\frac{W}{L}\right)_1 = \left(\frac{0.04}{2}\right)(12.5) = 0.25\,\text{mA/V}^2$$

and the output resistances are

$$r_{o2} = r_{o4} = \frac{1}{\lambda I_{D2}} = \frac{1}{(0.02)(0.0202)} \Rightarrow 2.475\,\text{M}\Omega$$

From Equation (13.38), the differential voltage gain of the input stage is then

$$A_d = \sqrt{2K_{p1}I_Q}\,(r_{o2}\|r_{o4}) = \sqrt{2\,(0.25)\,(0.0404)}(2475\|2475)$$

or

$$A_d = 176$$

The transconductance of $M_7$ is

$$g_{m7} = 2\sqrt{\left(\frac{k_n'}{2}\right)\left(\frac{W}{L}\right)_7 I_{D7}} = 2\sqrt{\left(\frac{0.1}{2}\right)(12.5)(0.0404)}$$

$$= 0.3178\,\text{mA/V}$$

and the output resistances of $M_7$ and $M_8$ are

$$r_{o7} = r_{o8} = \frac{1}{\lambda I_{D7}} = \frac{1}{(0.02)(0.0404)} \Rightarrow 1.238\,\text{M}\Omega$$

From Equation (13.40), the magnitude of the voltage gain of the second stage is then

$$A_{v2} = g_{m7}\,(r_{o7}\|r_{o8}) = (0.3178)(1238\|1238) = 197$$

Finally, the overall voltage gain magnitude of the op-amp is

$$A_v = A_d A_{v2} = (176)(197) = 34{,}672$$

**Comment:** The calculated overall voltage gain is 91 dB, which correlates very well with typical values of 90 dB, as listed in the data sheet for the MC14573 op-amp. The open-loop gain of a CMOS op-amp is generally less than that of a bipolar op-amp, but the use of active loads provides acceptable results.

EXERCISE PROBLEM

**Ex 13.9:** Repeat Example 13.9 using circuit and transistor parameters given in Exercise Ex 13.8. (Ans. $A_d = 181.4$, $A_{v2} = 202.9$, $A_v = 91.3$ dB)

### 13.3.2   Three-Stage CMOS Operational Amplifier

Figure 13.15 shows a three-stage CMOS op-amp circuit. The differential input stage consists of the differential pair $M_1$ and $M_2$ with active load transistors $M_3$ and $M_4$. The input stage is biased with the constant-current source $M_{10}$ and $M_{11}$. As shown in Chapter 10, the reference current can be established with additional NMOS transistors.

The output of the input stage is connected to the common-source amplifier consisting of $M_5$. The transistor $M_9$ establishes the bias current $I_{Q2}$ and also acts as the active load for the common-source amplifier.

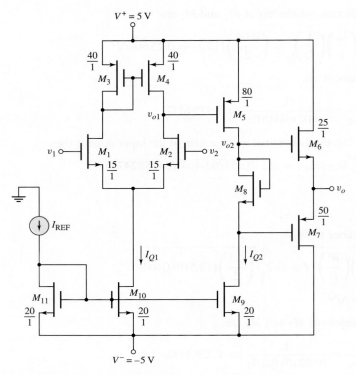

**Figure 13.15** A three-stage CMOS operational amplifier

Transistors $M_6$ and $M_7$ form the complementary push–pull output stage. Transistor $M_8$ acts as a resistor and provides a potential difference between the gates of the output transistors to minimize crossover distortion in the output signal.

Example width-to-length ratios of the transistors in the CMOS op-amp circuit are shown in the figure. These ratios will be used in the following example.

## EXAMPLE 13.10

**Objective:** Determine the dc and ac characteristics of a three-stage CMOS op-amp.

Consider the three-stage CMOS op-amp shown in Figure 13.15. The NMOS transistor parameters are $V_{TN} = 0.7$ V, $k'_n = 80\ \mu\text{A/V}^2$, $\lambda_n = 0.01$ V$^{-1}$, and the PMOS transistor parameters are $V_{TP} = -0.7$ V, $k'_p = 40\ \mu\text{A/V}^2$, $\lambda_p = 0.015$ V$^{-1}$. Assume the reference current is $I_{\text{REF}} = 160\ \mu\text{A}$.

**Solution (DC Analysis):** Since $M_9$, $M_{10}$, and $M_{11}$ are matched transistors, then $I_{Q1} = I_{Q2} = I_{\text{REF}} = 160\ \mu\text{A}$.

Transistors $M_3$ and $M_4$ are matched so that in the quiescent condition, $V_{SG3} = V_{SD3} = V_{SD4}$. Since $V_{SG5} = V_{SD4}$ and since the current in $M_5$ is twice as large as that in $M_4$, the width-to-length ratio of $M_5$ must be twice as large as that of $M_3$ and $M_4$.

If we provide dc biases of $V_{GS6} = V_{SG7} = 0.85$ V to the output transistors, then the dc quiescent current in the output transistors will be

$$I_{D6} = I_{D7} = \frac{k'_n}{2}\left(\frac{W}{L}\right)_6 (V_{GS6} - V_{TN})^2 = \left(\frac{80}{2}\right)(25)(0.85 - 0.7)^2$$

or

$$I_{D6} = I_{D7} = 22.5\ \mu\text{A}$$

The potential difference across $M_8$ must then be $V_{DS8} = 2(0.85) = 1.7$ V. We then have

$$I_{D8} = I_{Q2} = 160 = \left(\frac{80}{2}\right)\left(\frac{W}{L}\right)_8 (1.7 - 0.7)^2$$

which yields a required width-to-length ratio of $(W/L)_8 = 4$.

**Solution (AC Analysis):** Since there is no loading effect between stages of the CMOS op-amp, we can write the overall differential voltage gain as

$$A_v = A_{d1} A_2 A_3$$

where the gains $A_{d1}$, $A_2$, and $A_3$ are the voltage gains of each individual stage. Since the output stage is a source-follower circuit, we can write that $A_3 \cong 1$.

Defining the differential input voltage as $v_d = v_1 - v_2$, the differential voltage gain of the input stage (using results from Chapter 11) is

$$A_{d1} = \frac{v_{o1}}{v_d} = g_{m1}(r_{o2} \| r_{o4})$$

We find

$$g_{m1} = 2\sqrt{\left(\frac{k_n'}{2}\right)\left(\frac{W}{L}\right)_1 \left(\frac{I_{Q1}}{2}\right)} = 2\sqrt{\left(\frac{0.08}{2}\right)(15)\left(\frac{0.16}{2}\right)}$$

or

$$g_{m1} = 0.438 \text{ mA/V}$$

Also

$$r_{o2} = \frac{1}{\lambda_n(I_{Q1}/2)} = \frac{1}{(0.01)(0.08)} = 1250 \text{ k}\Omega$$

and

$$r_{o4} = \frac{1}{\lambda_p(I_{Q1}/2)} = \frac{1}{(0.015)(0.08)} = 833.3 \text{ k}\Omega$$

We then find

$$A_{d1} = (0.438)(1250 \| 833.3) = 219$$

The resistance of $M_8$ is relatively small, so the voltage gain of the second common-source stage is given by

$$A_2 = -g_{m5}(r_{o5} \| r_{o9})$$

We find

$$g_{m5} = 2\sqrt{\left(\frac{k_p'}{2}\right)\left(\frac{W}{L}\right)_5 I_{Q2}} = 2\sqrt{\left(\frac{0.04}{2}\right)(80)(0.16)}$$

or

$$g_{m5} = 1.012 \text{ mA/V}$$

Also

$$r_{o5} = \frac{1}{\lambda_p I_{Q2}} = \frac{1}{(0.015)(0.16)} = 416.7 \text{ k}\Omega$$

and

$$r_{o9} = \frac{1}{\lambda_n I_{Q2}} = \frac{1}{(0.01)(0.16)} = 625 \text{ k}\Omega$$

The voltage gain of the second stage is then

$$A_2 = -(1.012)(416.7\|625) = -253$$

The overall differential voltage gain of this three-stage CMOS op-amp is then

$$A_v = A_{d1}A_2 = (219)(-253) = -55{,}407$$

**Comment:** A reasonable differential voltage gain is obtained in this three-stage CMOS amplifier.

### EXERCISE PROBLEM

**Ex 13.10:** (a) Calculate the differential voltage gains of the first and second stages, and the overall voltage gain of the three-stage CMOS op-amp in Figure 13.15 if $(W/L)_{1,2} = 22.5$ and $I_{\text{REF}} = 200 \, \mu\text{A}$. All other parameters are the same as given in Example 13.10. (b) Recalculate $(W/L)_8$ if the quiescent current in the output transistors is to be $40 \, \mu\text{A}$. (Ans. (a) $A_d = 240$, $A_2 = -226.2$, $A = -54{,}288$; (b) $(W/L)_8 = 4.13$)

### 13.3.3    Folded Cascode CMOS Operational Amplifier Circuit

As we have mentioned previously, the voltage gain of an amplifier can be increased by using a cascode configuration. In its simplest form, the conventional cascode configuration consists of two transistors in series, as shown in Figure 13.16(a). The

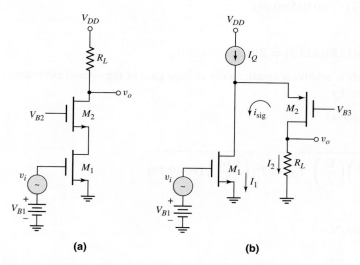

**Figure 13.16** (a) Classical cascode stage. (b) folded-cascode stage

transistor $M_1$ is the common-source amplifying device whose current is determined by the input voltage. This current is the input signal to $M_2$, which is connected in a common-gate configuration. The output is taken off the drain of the cascode transistor. The circuit in Figure 13.16(b) has a slightly different configuration. The dc current $I_1$ in $M_1$ is determined by the input voltage. The dc current in $M_2$ is the difference between the bias current $I_Q$ and $I_1$.

The ac current in the conventional cascode circuit of Figure 13.16(a) is through both transistors and the dc power supply. The ac current in the cascode circuit in Figure 13.16(b) is through both transistors and ground as indicated in the figure. The ac current in $M_2$ of this circuit is equal in magnitude but in the opposite direction to $M_1$. Thus the current is said to be folded back and the circuit in Figure 13.16(b) is called a folded cascode circuit.

The folded cascode configuration can be applied to the diff-amp as shown in Figure 13.17. The transistors $M_1$ and $M_2$ are the differential pair, as usual, and transistors $M_5$ and $M_6$ are the cascode transistors. Transistors $M_7$–$M_{10}$ form a modified Wilson current mirror acting as an active load. This configuration was discussed in Chapter 10.

Assuming that transistors $M_3$, $M_4$, and $M_{11}$–$M_{13}$ are all matched, then the dc currents in $M_1$ and $M_2$ are $I_{REF}/2$ and those in $M_3$ and $M_4$ are $I_{REF}$. This means that the dc currents in the cascode transistors $M_5$ and $M_6$ are $I_{REF}/2$.

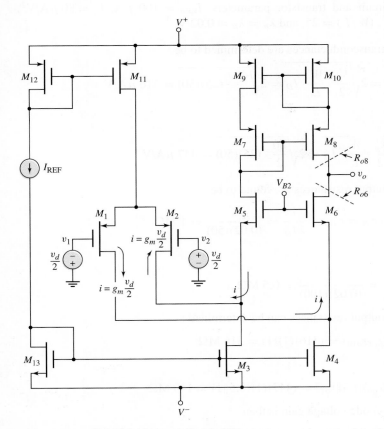

**Figure 13.17**  CMOS folded cascode amplifier

If a differential-mode input voltage is applied, then ac currents are induced in the differential pair as shown in the figure. The ac current in $M_1$ flows through $M_6$ to the output. The ac current in $M_2$ flows through $M_5$ and is induced in $M_8$ by the current-mirror action of the active load. From previous work on diff-amps, the differential-mode voltage gain is

$$A_d = g_{m1}(R_{o6} \| R_{o8}) \tag{13.41}$$

where

$$R_{o8} = g_{m8}(r_{o8}r_{o10}) \tag{13.42(a)}$$

and

$$R_{o6} = g_{m6}(r_{o6})(r_{o4} \| r_{o1}) \tag{13.42(b)}$$

We may note that we are neglecting the body effect. Normally the substrates of all NMOS devices are tied to $V^-$ and the substrates of all PMOS devices are tied to $V^+$.

## EXAMPLE 13.11

**Objective:** Determine the differential-mode voltage gain of the folded cascode diff-amp in Figure 13.17.

Assume circuit and transistor parameters: $I_{\text{REF}} = 100 \ \mu\text{A}$, $k'_n = 80 \ \mu\text{A/V}^2$, $k'_p = 40 \ \mu\text{A/V}^2$, $(W/L) = 25$, and $\lambda_n = \lambda_p = 0.02 \ \text{V}^{-1}$.

**Solution:** The transconductances are determined to be

$$g_{m1} = g_{m8} = 2\sqrt{\frac{k'_p}{2} \cdot \frac{W}{L} \cdot I_D} = 2\sqrt{\frac{40}{2} \cdot (25)(50)} = 316 \ \mu\text{A/V}$$

and

$$g_{m6} = 2\sqrt{\frac{k'_n}{2} \cdot \frac{W}{L} \cdot I_D} = 2\sqrt{\frac{80}{2} \cdot (25)(50)} = 447 \ \mu\text{A/V}$$

The transistor output resistances are found to be

$$r_{o1} = r_{o6} = r_{o8} = r_{o10} = \frac{1}{\lambda I_D} = \frac{1}{(0.02)(50)} = 1 \ \text{M}\Omega$$

and

$$r_{o4} = \frac{1}{\lambda I_{D4}} = \frac{1}{(0.02)(100)} = 0.5 \ \text{M}\Omega$$

The composite output resistances can be determined as

$$R_{o8} = g_{m8}(r_{o8}r_{o10}) = (316)(1)(1) = 316 \ \text{M}\Omega$$

and

$$R_{o6} = g_{m6}(r_{o6})(r_{o4} \| r_{o1}) = (447)(1)(0.5 \| 1) = 149 \ \text{M}\Omega$$

The differential-mode voltage gain is then

$$A_d = g_{m1}(R_{o6} \| R_{o8}) = (316)(149 \| 316) \cong 32{,}000$$

**Comment:** This example shows that very high differential-mode voltage gains can be achieved in a folded cascode CMOS circuit. In actual circuits, the output resistances may be limited by leakage currents so the very ideal values may not be realizable. However, substantially higher differential-mode voltage gains can be achieved in the folded cascode configuration than in the simpler diff-amp circuits.

### 13.3.4  CMOS Current-Mirror Operational Amplifier Circuit

Another CMOS op-amp circuit is shown in Figure 13.18. The differential pair is formed by $M_1$ and $M_2$. The induced ac currents from these transistors drive transistors $M_3$ and $M_4$, which are the inputs of two current mirrors with a current multiplication factor $B$. The current output of $M_5$ is then induced in $M_8$ by the current-mirror action of $M_7$ and $M_8$. The output signal currents then have a multiplication factor $B$. The differential-mode voltage gain is then given by

$$A_d = \frac{v_o}{v_d} = B g_{m1}(r_{o6} \| r_{o8}) \tag{13.43}$$

The factor of $B$ in the gain expression of Equation (13.43) may be slightly misleading. Recall that the individual transistor output resistance is inversely proportional to the drain current. If the current in the output transistors increases by the factor $B$, then

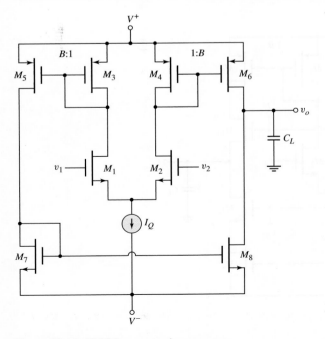

**Figure 13.18**  CMOS current-mirror op-amp

$R_o = r_{o6} \| r_{o8}$ decreases by the factor $B$ so the differential-mode voltage gain remains unchanged.

The advantage of the current-mirror op-amp is an increase in the gain–bandwidth product. The dominant-pole frequency will be determined by the parameters at the output node. The dominant-pole frequency is given by

$$f_{pd} = \frac{1}{2\pi R_o (C_L + C_p)} \tag{13.44}$$

where $R_o$ is the output resistance, $C_L$ is the load capacitance, and $C_p$ is the sum of all other capacitances at the output node. If $R_o$ decreases by the factor $B$, then the dominant-pole frequency increases by the same factor $B$. The gain–bandwidth product is

$$\text{GBW} = A_d \cdot f_{pd} \tag{13.45}$$

Since $A_d$ is now independent of $B$ and $f_{pd}$ increases by $B$, then the gain–bandwidth product increases by $B$.

Further analysis of this circuit shows that the phase margin decreases with increasing $B$. As a practical limit, the maximum value of $B$ is limited to approximately 3.

### 13.3.5 CMOS Cascode Current-Mirror Op-Amp Circuit

As we have already seen, the differential-mode gain can be increased by adding cascode transistors in the output portion of the circuit. Figure 13.19 shows the same current-mirror configuration considered previously but with cascode transistors added to the output. Transistors $M_9$–$M_{12}$ are the cascode transistors. The differential-mode voltage gain is given by

$$A_d = \frac{v_o}{v_d} = B g_{m1} (R_{o10} \| R_{o12}) \tag{13.46}$$

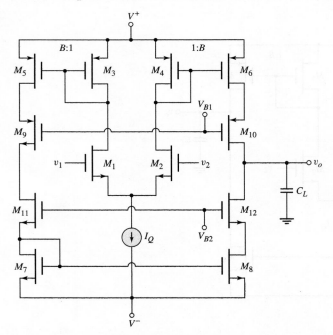

Figure 13.19 CMOS cascode current-mirror op-amp

where

$$R_{o10} = g_{m10}(r_{o10}r_{o6}) \qquad (13.47)$$

and

$$R_{o12} = g_{m12}(r_{o12}r_{o8}) \qquad (13.48)$$

The advantage of this circuit is the increased gain at low frequency. The gain–bandwidth product of this circuit is not changed from that of the simple current-mirror op-amp considered previously.

## Test Your Understanding

**\*TYU 13.9** Using the parameters given in Example 13.8, determine the input common-mode voltage range for the MC14573 op-amp. (Ans. $-4.75 \leq v_{cm} \leq 3.81$ V)

**TYU 13.10** Using the parameters given in Example 13.8, determine the maximum and minimum output voltage in the MC14573 circuit such that the op-amp remains biased in its linear region. (Ans. $-4.75 \leq v_o \leq 4.6$ V)

**\*TYU 13.11** Consider the MC14573 op-amp in Figure 13.14. Assume the same circuit and transistor parameters as given in Examples 13.8 and 13.9, except change $R_{set}$ to 100 kΩ. (a) Calculate all dc bias currents. (b) Determine the overall voltage gain of the op-amp. (Ans. (a) $I_{set} = I_Q = I_{D7} = I_{D8} = 89.03 \, \mu A$, $I_{D1} - I_{D4} = 44.52 \, \mu A$; (b) $A_v = 15,701$)

**TYU 13.12** Consider the CMOS current-gain op-amp in Figure 13.18. Assume the bias current is $I_Q = 200 \, \mu A$ and assume transistor parameters $k'_n = 100 \, \mu A/V^2$, $k'_p = 40 \, \mu A/V^2$, and $\lambda_n = \lambda_p = 0.02 \, V^{-1}$. Assume the basic $W/L$ ratio of the transistors is 40 and let $B = 3$. (a) Determine the small-signal voltage gain. (b) If the effective capacitance at the output node is $C_L + C_p = 2 \, pF$, determine the dominant pole frequency and the gain-bandwidth product. (Ans. (a) $A_d = 223.6$; (b) $f_{PD} = 955 \, kHz$, GBW $= 213.5 \, MHz$)

**TYU 13.13** Consider the CMOS cascode current-mirror op-amp in Figure 13.19. Assume the bias current and transistor parameters are the same as in Exercise TYU 13.12. Repeat parts (a) and (b) of Exercise TYU 13.12 for this circuit. (Ans. (a) $A_d = 44,751$; (b) $f_{PD} = 4.77 \, kHz$, GBW $= 213.5 \, MHz$)

---

 **13.4    BiCMOS OPERATIONAL AMPLIFIER CIRCUITS**

**Objective:** • Describe and analyze the dc and ac characteristics of BiCMOS operational amplifier circuits.

As discussed in Chapter 11, BiCMOS circuits combine the advantages of bipolar and MOSFET devices in the same circuit. One advantage of MOSFETs is the very high input impedance. Therefore, when MOSFETs form the input differential pair of an op-amp, the input bias currents are extremely small. However, the equivalent noise of the input stage may be greater than for an all-BJT op-amp.

In this section, we will examine two BiCMOS op-amp circuits. The first is a variation of the folded cascode configuration analyzed in the last section and the second is the CA3140 BiCMOS op-amp. Since we previously fully analyzed the folded cascode circuit, we will discuss, here, the advantages of using the BiCMOS technology. Many features of the CA3140 BiCMOS op-amp are similar to those of the 741. Therefore, we will not analyze this op-amp in as great a detail as we did the 741. Instead, we will concentrate on some of its unique features.

### 13.4.1 BiCMOS Folded Cascode Op-Amp

Figure 13.20 shows an example of a BiCMOS folded cascode op-amp. The cascode transistors, $Q_5$ and $Q_6$, are now bipolar devices, replacing n-channel MOSFETs. The small-signal voltage gain expression for this circuit is identical to that of the all-CMOS design. We have mentioned that the dominant-pole frequency is determined by the circuit parameters at the output node because of the very large output resistance. Nondominant-pole frequencies are then a function of the parameters at the other circuit nodes. In particular, one node of interest is at the drain of an input transistor and emitter of a cascode transistor. The nondominant-pole frequency can be written as

$$f_{3-\text{dB}} = \frac{g_{m6}}{2\pi C_{p6}} \qquad\qquad (13.49)$$

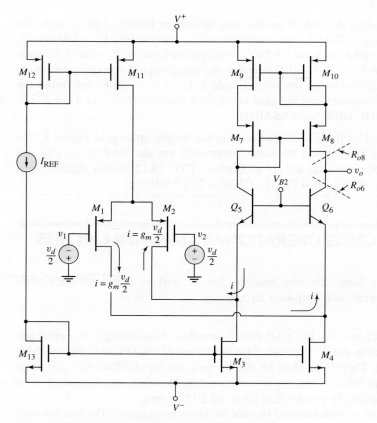

**Figure 13.20** BiCMOS folded cascode amplifier

where $g_{m6}$ is the transconductance of the cascode transistor $Q_6$ and $C_{p6}$ is the effective capacitance at this node. Since the transconductance of a bipolar is usually greater than that of a MOSFET, this 3 dB frequency is larger for the BiCMOS circuit than for the all-CMOS design. This result means that the phase margin of the BiCMOS op-amp circuit is larger than that of the all-CMOS op-amp.

## 13.4.2    CA3140 BiCMOS Circuit Description

Figure 13.21 shows the basic equivalent circuit of the CA3140 op-amp. Like the 741, this op-amp consists of three basic stages: the input differential stage, the gain stage, and

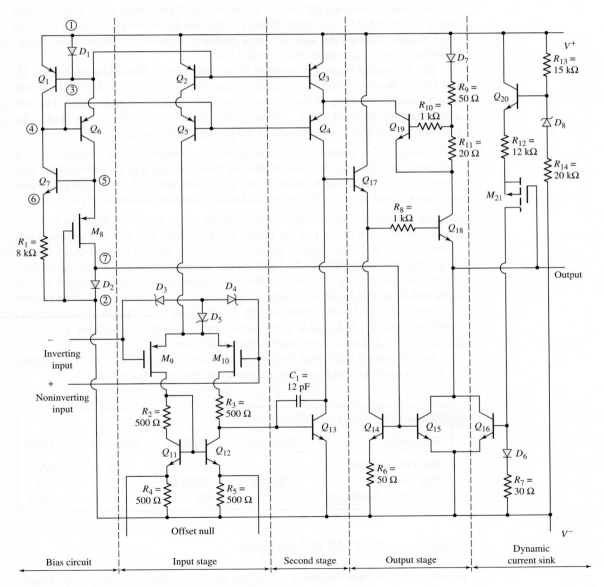

**Figure 13.21**  CA3140 BiCMOS op-amp equivalent circuit

the output stage. Also shown in the figure are: the bias circuit, which establishes the dc bias currents in the op-amp; and a section referred to as a dynamic current sink, which will be explained later. Typical supply voltages are $V^+ = 15$ V and $V^- = -15$ V.

### Input Diff-Amp

The input differential pair consists of p-channel transistors $M_9$ and $M_{10}$, and transistors $Q_{11}$ and $Q_{12}$ form the active load for the diff-amp. A single-sided output at the collector of $Q_{12}$ is the input signal to the following gain stage. Two offset null terminals are also shown, and will be discussed in the next chapter.

MOS transistors are very susceptible to damage from electrostatic charge. For example, electrostatic voltage can be inadvertently induced on the gate of a MOSFET during routine handling. These voltages may be great enough to induce breakdown in the gate oxide, destroying the device. Therefore, input protection against electrostatic damage is provided by the Zener diodes $D_3$, $D_4$, and $D_5$. If the gate voltage becomes large enough, these diodes will provide a discharge path for the electrostatic charge, thus protecting the gate oxide from breakdown.

The dc current biasing is initiated in the bias circuit. The elements labeled $D_1$ and $D_2$ are diode-connected transistors. Transistor $Q_1$ and diode $D_1$ are matched, which forces the currents in the two branches of the bias circuit to be equal. The current is determined from $Q_7$, $R_1$, and $M_8$. The combination of $Q_6$ and $Q_7$ makes the bias current essentially independent of the power supply voltages.

**Gain Stage:** The second stage consists of $Q_{13}$ connected in a common-emitter configuration. The cascode configuration of transistors $Q_3$ and $Q_4$ provides the bias current for $Q_{13}$, in addition to acting as the active load. Since $Q_3$ and $Q_4$ are connected in a cascode configuration, the resistance looking into the collector of $Q_4$ is very high.

**Output Stage:** The basic output stage consists of the npn transistors $Q_{17}$ and $Q_{18}$. During the positive portion of the output voltage cycle, $Q_{18}$ acts as an emitter follower, supplying a load current. During the negative portion of the output voltage cycle, $Q_{16}$ sinks current from the load. As the output voltage decreases, the source-to-gate voltage on the p-channel $M_{21}$ MOSFET increases, producing a larger current in $D_6$ and $R_7$ so that the base voltage on $Q_{16}$ increases. The increase B–E voltage of $Q_{16}$ allows increased load current sinking. Short-circuit protection is provided by the combination of $R_{11}$ and $Q_{19}$. If a sufficiently large voltage is developed across $R_{11}$, $Q_{19}$ turns on and shunts excess base current away from $Q_{17}$.

An abbreviated data sheet for the CA3140 op-amp is in Table 13.2. As before, we will compare the results of our analysis to the values listed in the table.

**Table 13.2**   CA3140 BiCMOS data

| Parameter | Minimum | Typical | Maximum | Units |
|---|---|---|---|---|
| Input bias current | | 10 | 50 | pA |
| Open-loop gain | 20,000 | 100,000 | | V/V |
| Unity-gain frequency | | 4.5 | | MHz |

### 13.4.3    CA3140 DC Analysis

In this section, we will determine the dc bias currents in the CA3140 op-amp. As previously stated, we will concentrate on the features that are unique to the CA3140 compared to the 741.

The basic bias circuit is shown in Figure 13.22. The current mirror consisting of $Q_1$ and $D_1$ ensures that the two branch currents $I_1$ and $I_2$ are equal, since $Q_1$ and $D_1$ are matched. The p-channel MOSFET $M_8$ is to operate in the saturation region, so that we must have

$$V_{SD} > V_{SG} - |V_{TP}| \tag{13.50}$$

From the figure, we see that

$$V_{SG} = V_{SD} + V_D \tag{13.51}$$

or

$$V_{SD} = V_{SG} - V_D \tag{13.52}$$

Combining Equations (13.52) and (13.50) yields

$$V_{SG} - V_D > V_{SG} - |V_{TP}| \tag{13.53}$$

which implies that $|V_{TP}| > V_D$. In other words, for $M_8$ to remain biased in the saturation region, the magnitude of the threshold voltage must be greater than the diode voltage. From the left branch of the bias circuit, we see that the current can be written

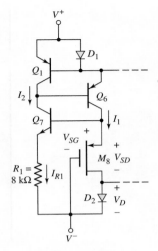

**Figure 13.22** Bias circuit, CA3140 BiCMOS op-amp

$$I_2 \cong I_{R1} = \frac{V_{SG} - V_{BE7}}{R_1} \tag{13.54}$$

and from the right branch, we have

$$I_1 = K_p(V_{SG} - |V_{TP}|)^2 \tag{13.55}$$

Since $I_1 = I_2$, a simultaneous solution of Equations (13.54) and (13.55) determines the currents and voltages in this bias circuit.

## EXAMPLE 13.12

**Objective:** Determine the currents and voltages in the bias circuit of the CA3140 op-amp.

Consider the bias circuit in Figure 13.22, with parameters: $V^+ = 15$ V, $V^- = -15$ V, and $R_1 = 8$ k$\Omega$. Assume transistor parameters of $V_{BE}$ (npn) $= V_{EB}$ (pnp) $= 0.6$ V for the bipolars, and $K_p = 0.2$ mA/V$^2$ and $|V_{TP}| = 1.4$ V for the MOSFET $M_8$.

**Solution:** Set $I_1 = I_2$. Then, from Equations (13.54) and (13.55), we find

$$V_{SG} = 2.49 \text{ V} \quad \text{and} \quad I_1 = I_2 = 0.236 \text{ mA}$$

The voltage at the collector of $Q_6$ is

$$V_{C6} = V_{SG8} + V^- = 2.49 - 15 = -12.5 \text{ V}$$

and the voltage at the collector of $Q_7$ is

$$V_{C7} = V^+ - V_{EB1} - V_{EB6} = 15 - 0.6 - 0.6 = 13.8 \text{ V}$$

Therefore, the collector–base junctions of both $Q_6$ and $Q_7$ are reverse biased by $13.8 - (-12.5) = 26.3$ V, and both $Q_6$ and $Q_7$ are biased in the active region.

**Comment:** The nominal bias current listed in Table 13.2 is 200 $\mu$A, which correlates well with our calculated value of 236 $\mu$A. As long as the B–C junctions of $Q_6$ and $Q_7$ remain reverse biased, the bias currents remain constant. This means that the bias current is independent of $V^+$ and $V^-$ over a wide range of voltages.

The PSpice analysis, using $I_S = 2 \times 10^{-15}$ A for the BJTs shows that the currents in the two branches of the current source are essentially 220 $\mu$A. This compares very favorably with the 236 $\mu$A obtained by the hand analysis.

---

**EXERCISE PROBLEM**

**Ex 13.12:** (a) Assume the bias circuit of the CA3140 op-amp circuit shown in Figure 13.22 has supply voltages of $V^+ = 5$ V and $V^- = -5$ V. Let $R_1 = 10$ k$\Omega$. The transistor parameters are $V_{TP} = -0.8$ V and $K_p = 0.15$ mA/V$^2$ for $M_8$; and $V_{BE7}(\text{on}) = V_{EB6} = 0.6$ V for the bipolars. Find the currents $I_1$ and $I_2$; and the voltages $V_{SG8}$, $V_{C7}$, $V_{C6}$, $V_{CB7}$, and $V_{BC6}$. (b) Using the results of part (a), determine the minimum supply voltages that will still maintain the bipolar transistors biased in the forward active region. Assume $V^+ = -V^-$. (Ans. (a) $I_1 = I_2 = 0.1028$ mA, $V_{SG8} = 1.628$ V, $V_{C7} = 3.8$ V, $V_{C6} = -3.37$ V, $V_{CB7} = V_{BC6} = 7.17$ V; (b) $V^+ = -V^- = 1.414$ V)

---

Transistors $Q_1$ through $Q_6$ and diode $D_1$ in Figure 13.21 are all matched, which means that $I_{C5} = I_{C4} \cong 200$ $\mu$A. The current in $D_2$ establishes the diode voltage that also biases $Q_{14}$ and $Q_{15}$. The nominal value of $I_{C18}$ is 2 mA.

### 13.4.4 CA3140 Small-Signal Analysis

We analyze the small-signal voltage gain of the CA3140 op-amp by dividing the configuration into its basic circuits and using results previously obtained.

**Input Stage**

From the results in Chapter 11, the small-signal differential voltage gain can be written

$$A_d = \sqrt{2 K_p I_{Q5}} (r_{o10} \| R_{act1} \| R_{i2}) \tag{13.56}$$

where $I_{Q5}$ is the bias current supplied by $Q_2$ and $Q_5$. Resistance $r_{o10}$ is the output resistance looking into the drain of $M_{10}$, $R_{act1}$ is the effective resistance of the active load, and $R_{i2}$ is the input resistance of the gain stage.

---

**EXAMPLE 13.13**

**Objective:** Calculate the small-signal differential voltage gain of the CA3140 op-amp input stage. Assume a bias current of $I_Q = 0.2$ mA.

Assume a conduction parameter value of $K_p = 0.6$ mA/V$^2$ for $M_{10}$, an npn bipolar current gain of $\beta_n = 200$, and a bipolar Early voltage of $V_A = 50$ V.

**Solution:** The input resistance to the gain stage is $R_{i2} = r_{\pi13}$; therefore,

$$R_{i2} = r_{\pi13} = \frac{\beta_n V_T}{I_{C13}} = \frac{(200)(0.026)}{0.20} = 26 \text{ k}\Omega$$

Resistances $r_{o10}$ and $R_{act1}$ are normally in the hundreds of kilohms or megohm range, so the small value of $R_{i2}$ dominates the parallel resistance value in the gain expression. We then have

$$A_d \cong \sqrt{2 K_p I_{Q5}} (R_{i2}) = \sqrt{2(0.6)(0.2)} (26) = 12.7$$

**Comment:** The low input resistance of the gain stage severely loads the input stage, which in turn results in a relatively low voltage gain for the input stage.

**Ex 13.13:** Repeat Example 13.13 for the case when $K_p = 1$ mA/V$^2$ for $M_{10}$ and when the Early voltage of a bipolar transistor is $V_A = 120$ V. All other circuit and transistor parameters are the same as given in Example 13.13. (Ans. $A_d = 16.4$)

### Gain Stage

The magnitude of the small-signal voltage gain for the second stage is

$$|A_{v2}| = g_{m13}(r_{o13}\|R_{o4}\|R_{i3}) \tag{13.57}$$

where $R_{i3}$ is the input resistance of the output stage and $R_{o4}$ is the output resistance of the cascode configuration of $Q_3$ and $Q_4$. Transistor $Q_{17}$, which is the input transistor of the output stage, is connected as an emitter follower, which means that $R_{i3}$ is typically in the megohm range. Similarly, the output resistance $R_{o4}$ of the cascode configuration is typically in the megohm range.

The voltage gain of the second stage is then approximately

$$|A_{v2}| \cong g_{m13}r_{o13} \tag{13.58}$$

### EXAMPLE 13.14

**Objective:** Calculate the small-signal voltage gain of the second stage of the CA3140 op-amp.

Assume an Early voltage of $V_A = 50$ V for $Q_{13}$.

**Solution:** The transconductance is

$$g_{m13} = \frac{I_{C13}}{V_T} = \frac{0.20}{0.026} = 7.69 \text{ mA/V}$$

and the output resistance is

$$r_{o13} = \frac{V_A}{I_{C13}} = \frac{50}{0.20} = 250 \text{ k}\Omega$$

The voltage gain is therefore

$$|A_{v2}| = g_{m13}r_{o13} = (7.69)(250) = 1923$$

**Comment:** The second stage of the CA3140 operational amplifier provides the majority of the voltage gain.

**\*Ex 13.14:** Assume the gain stage of the CA3140 op-amp is modified to include an emitter resistor, as shown in Figure 13.23. Let $\lambda = 0.02$ V$^{-1}$ for $M_{10}$. Assume all other transistor parameters are the same as those in Example 13.13. If the transistor bias currents in $M_{10}$ and $Q_{12}$ are 100 $\mu$A and the current in $Q_{13}$ is 200 $\mu$A, determine the new value of the small-signal differential voltage gain of the input stage. (Ans. 69.1)

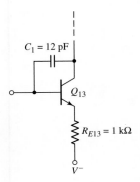

Figure 13.23 Figure for Exercise Ex 13.14

### Overall Gain

Since we have taken the loading effects of each following stage into account, the overall voltage gain is the product of the individual gain factors, or

$$A_v = A_d A_{v2} A_{v3} \qquad (13.59)$$

where $A_{v3}$ is the voltage gain of the output stage. If we assume that $A_{v3} \cong 1$ for the emitter-follower output stage, then the overall gain of the CA3140 op-amp is

$$A_v = A_d A_{v2} A_{v3} = (12.7)(1923)(1) = 24,422 \qquad (13.60)$$

Typical values of the gain of the CA3140 op-amp are in the area of 100,000; thus, our calculations give a somewhat smaller value.

### Frequency Response

The CA3140 op-amp is internally compensated by the Miller compensation technique to introduce a dominant pole, as was done in the 741 op-amp. The feedback capacitor $C_1$ is 12 pF and is connected between the collector and the base of $Q_{13}$, as shown in Figure 13.20. From Miller's theorem, the effective input capacitance of the second stage is

$$C_i = C_1(1 + |A_{v2}|) \qquad (13.61)$$

The low-frequency dominant pole is

$$f_{PD} = \frac{1}{2\pi R_{eq} C_i} \qquad (13.62)$$

where $R_{eq}$ is the equivalent resistance between the second-stage input node and ground. Since this resistance is dominated by the input resistance to $Q_{13}$, we have

$$R_{eq} \cong R_{i2} = r_{\pi 13} \qquad (13.63)$$

## EXAMPLE 13.15

**Objective:** Determine the dominant-pole frequency and unity-gain bandwidth of the CA3140 op-amp.

Again, we will use results from previous calculations.

**Solution:** Previously, we determined that $|A_{v2}| = 1923$; therefore, the effective input capacitance is

$$C_i = C_1(1 + |A_{v2}|) = 12(1 + 1923) = 23,088 \text{ pF}$$

The gain stage input resistance is

$$R_{i2} = r_{\pi 13} = 26 \text{ k}\Omega$$

which means that

$$f_{PD} \cong \frac{1}{2\pi R_{i2} C_i} = \frac{1}{2\pi(26 \times 10^3)(23,088 \times 10^{-12})} = 265 \text{ Hz}$$

Finally, the unity-gain bandwidth is

$$f_T = f_{PD} A_v = (265)(24,422) \Rightarrow 6.47 \text{ MHz}$$

**Comment:** This unity-gain bandwidth value compares favorably with typical values of 4.5 MHz listed in the data sheet.

**Ex 13.15:** If the gain of the input stage of the CA3140 op-amp is increased to $A_d = 16.4$, determine the unity-gain bandwidth. All other parameters are the same as given in Example 13.15. (Ans. $f_T = 8.32$ MHz)

## Test Your Understanding

**TYU 13.14** Consider the BiCMOS folded cascode amplifier in Figure 13.20. Assume the circuit and MOS transistor parameters are the same as in Example 13.11. Assume BJT parameters of $\beta = 120$ and $V_A = 80$ V. (a) Determine the small-signal voltage gain. (b) If the effective capacitance at the output node is 2 pF, determine the dominant-pole frequency and the gain–bandwidth product. (Ans. (a) 76,343, (b) 329 Hz, 25.1 MHz)

**TYU 13.15** Consider the CA3140 op-amp bias circuit in Figure 13.22. Assume that $V_{BE7} = 0.6$ V and $R_1 = 5$ k$\Omega$. If the p-channel MOSFET parameters are $K_p = 0.3$ mA/V$^2$ and $|V_{TP}| = 1.4$ V, determine $I_1, I_2,$ and $V_{SG}$. (Ans. $V_{SG} = 2.54$ V, $I_1 = I_2 = 0.388$ mA)

---

## 13.5    JFET OPERATIONAL AMPLIFIER CIRCUITS

**Objective:** • Describe the characteristics of two hybrid JFET operational amplifier circuits.

The advantage of using MOSFETs as input devices in a BiCMOS op-amp is that extremely small input bias currents can be achieved. However, MOSFET gates connected to outside terminals of an IC must be protected against electrostatic damage. Typically, this is accomplished by using back-biased diodes on the input, as was shown in Figure 13.21. Unfortunately, the input op-amp bias currents are then dominated by the leakage currents in the protection diodes, which means that the small input bias currents cannot be fully realized. JFETs as input devices also offer the advantage of low input currents, and they do not need electrostatic protection devices. Input gate currents in a JFET are usually well below 1 nA, and are often on the order of 10 pA. In addition, JFETs offer greatly reduced noise properties.

In this section, we will examine two op-amp configurations using JFETs as input devices. Since the analysis is essentially identical to that given in the last two sections, we will limit ourselves to a general discussion of the circuit characteristics.

### 13.5.1    Hybrid FET Op-Amp, LH002/42/52 Series

Figure 13.24 is a simplified circuit diagram of an LH002/42/52 series op-amp, which uses a pair of JFETs for the input differential pair. Note that the general layout of the circuit is essentially the same as that of the 741 op-amp.

The input diff-amp stage consists of transistors $J_1, J_2, Q_3,$ and $Q_4$; $J_1$ and $J_2$ are n-channel JFETs operating in a source-follower configuration. The differential output

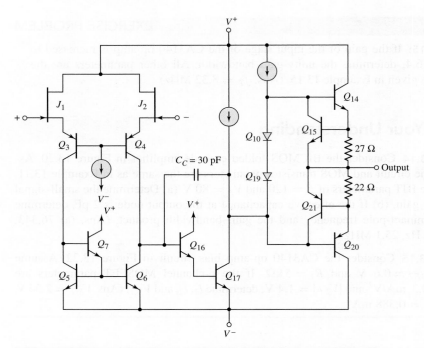

**Figure 13.24** Equivalent circuit, LH0022/42/52 series hybrid JFET op-amp

signal from $J_1$ and $J_2$ is the input to the common-base amplifier formed by $Q_3$ and $Q_4$, which provides a large voltage gain. Transistors $Q_5$, $Q_6$, and $Q_7$ form the active load for the input stage.

The gain stage is composed of $Q_{16}$ and $Q_{17}$ connected in a Darlington pair configuration. This stage also includes a 30 pF compensation capacitor. The output stage consists of the complementary push–pull emitter-follower configuration of $Q_{14}$ and $Q_{20}$. Transistors $Q_{14}$ and $Q_{20}$ are biased slightly "on" by diodes $Q_{10}$ and $Q_{19}$, to minimize crossover distortion. Transistors $Q_{15}$ and $Q_{21}$ and the associated 27 $\Omega$ and 22 $\Omega$ resistors provide the short-circuit protection.

An abbreviated data sheet for an LH0042C op-amp is shown in Table 13.3. Note the very large differential-mode input resistance and the low input bias current.

**Table 13.3** LH0042C data

| Parameter | Minimum | Typical | Maximum | Units |
|---|---|---|---|---|
| Input bias current | | 15 | 50 | pA |
| Differential-mode input resistance | | $10^{12}$ | | $\Omega$ |
| Input capacitance | | 4 | | pF |
| Open-loop gain ($R_L = 1$ k$\Omega$) | 25,000 | 100,000 | | V/V |
| Unity-gain frequency | | 1 | | MHz |

**13.5.2 Hybrid FET Op-Amp, LF155 Series**

Another example of a JFET op-amp is the LF155 BiFET op-amp. A simplified circuit diagram showing the input stage is in Figure 13.25. The input BiFET op-amp stage

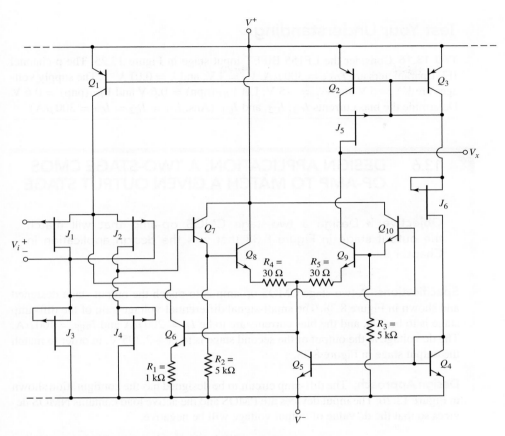

**Figure 13.25** Equivalent circuit, LF155 BiFET op-amp input stages

consists of p-channel JFETs $J_1$ and $J_2$ biased by the bipolar transistor $Q_1$. The active load for the input diff-amp consists of the p-channel JFETs $J_3$ and $J_4$, for which $V_{GS} = 0$.

A two-sided output from the input diff-amp stage is connected to a second diff-amp stage consisting of Darlington pairs $Q_7$ through $Q_{10}$. The second, or gain, stage is biased by bipolar transistor $Q_5$. The cascode configuration of $J_5$ and $Q_2$ form the active load for the gain stage.

The circuit has a common-mode feedback loop in the bias circuit. The base of $Q_6$ is connected to the collector of $Q_5$. If the drain voltages of $J_1$ and $J_2$ increase, the Darlington second stage drives the base voltage of $Q_6$ higher. The current in $Q_6$ then increases, reducing the drain currents in $J_1$ and $J_2$, since $I_{C1}$ is a constant current. Smaller drain currents cause the voltages at the $J_1$ and $J_2$ drains to decrease, which then stabilizes the drain voltages.

JFET $J_6$ is connected as a current source, which establishes a reference current in $Q_3$, $Q_4$, and $J_6$. This reference current then produces the bias currents in the current mirrors $Q_4$–$Q_5$ and $Q_1$–$Q_2$–$Q_3$.

In this BiFET op-amp, we see the advantages of incorporating both JFET and bipolars in the same circuit. The JFET input devices provide a very high input impedance, normally in the range of $10^{12}$ $\Omega$. The current-connected transistor $J_6$ allows the reference bias current to be controlled without the use of a resistor. Incorporating bipolar transistors in the second stage takes advantage of their higher transconductance values compared to JFETs, to produce a high second-stage gain.

## Test Your Understanding

**TYU 13.16** Consider the LF155 BiFET input stage in Figure 13.25. The p-channel JFET parameters are $I_{DSS} = 300\,\mu A$, $V_p = 1$ V, and $\lambda = 0.01$ V$^{-1}$. The supply voltages are $V^+ = 5$ V and $V^- = -5$ V. Let $V_{BE}$(npn) $= 0.6$ V and $V_{EB}$(pnp) $= 0.6$ V. Determine the bias currents $I_{C3}$, $I_{C2}$, and $I_{C1}$. (Ans. $I_{C1} = I_{C2} = I_{C3} = 300\,\mu A$)

---

## 13.6   DESIGN APPLICATION: A TWO-STAGE CMOS OP-AMP TO MATCH A GIVEN OUTPUT STAGE

**Objective:** • Design a two-stage CMOS op-amp that will match the output stage in Figure 8.38 that was the design application in Chapter 8.

**Specifications:** A two-stage CMOS op-amp is to match the output stage designed and shown in Figure 8.38. The small-signal differential-voltage gain of the diff-amp stage is to be 300, and the bias currents are to be $I_Q = 200\,\mu A$ and $I_{REF} = 400\,\mu A$. The dc voltage at the output of the second stage is to be $-2.295$ V, in order to match the output stage in Figure 8.38.

**Design Approach:** The diff-amp circuit to be designed has the configuration shown in Figure 13.26. The input devices are PMOS and the active load contains NMOS devices so that the dc value of output voltage will be negative.

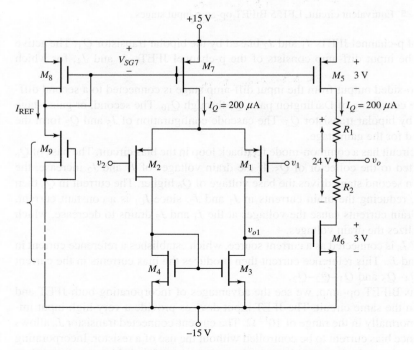

**Figure 13.26** A two-stage CMOS op-amp for the design application

**Choices:** MOS transistors are available with parameters $V_{TN} = 1$ V, $V_{TP} = -1$ V, $k'_n = 80$ $\mu$A/V$^2$, $k'_p = 40$ $\mu$A/V$^2$, and $\lambda_n = \lambda_p = 0.01$ V$^{-1}$.

**Solution (Diff-Amp Design):** From previous results, the differential voltage gain is

$$A_d = g_{m1}(r_{o1} \| r_{o3})$$

We find

$$r_{o1} = r_{o3} = \frac{1}{\lambda I_{DQ}} = \frac{1}{(0.01)(0.1)} = 1000 \text{ k}\Omega$$

We then find

$$300 = g_{m1}(1000 \| 1000)$$

so we must have $g_{m1} = 0.6$ mA/V. We then find the required width-to-length values of the input PMOS devices from

$$g_{m1} = 2\sqrt{\left(\frac{k'_p}{2}\right)\left(\frac{W}{L}\right)_1 I_{DQ1}}$$

or

$$0.60 = 2\sqrt{\left(\frac{0.04}{2}\right)\left(\frac{W}{L}\right)_1 (0.1)}$$

which yields

$$\left(\frac{W}{L}\right)_1 = \left(\frac{W}{L}\right)_2 = 45$$

We may also set

$$\left(\frac{W}{L}\right)_3 = \left(\frac{W}{L}\right)_4 = 45$$

**Solution (Current Source Design):** If we set $(W/L)_7 = 45$, then $V_{SG7}$ is found from

$$I_Q = 200 = \left(\frac{k'_p}{2}\right)\left(\frac{W}{L}\right)_7 (V_{SG7} + V_{TP})^2 = \left(\frac{40}{2}\right)(45)(V_{SG7} - 1)^2$$

We obtain $V_{SG7} = 1.47$ V.

We can write

$$\frac{I_{REF}}{I_Q} = \frac{(W/L)_8}{(W/L)_7}$$

or

$$\frac{0.4}{0.2} = \frac{(W/L)_8}{45}$$

which yields $(W/L)_8 = 90$.

If we assume the minimum width-to-length ratio of a MOSFET is unity, then we can show that six transistors are required in place of $M_9$. The total voltage drop across the six transistors is $30 - 1.47 = 28.53$ V. The voltage drop across each transistor is then $V_{SG9} = 28.53/6$ V. The width-to-length ratios are then found from

$$I_{REF} = 400 = \left(\frac{40}{2}\right)\left(\frac{W}{L}\right)_9 \left(\frac{28.53}{6} - 1\right)^2$$

which yields $(W/L)_9 = 1.42$ for each of the six transistors.

**Solution (Second Stage—DC Design):** The transistor $M_5$ must match $M_7$, so $(W/L)_5 = 45$. Since the current in $M_6$ is twice as large as in $M_3$, then the width-to-length of $M_6$ must be twice that of $M_3$ and $M_4$, or $(W/L)_6 = 90$.

The resistors $R_1$ and $R_2$ are used to produce the required dc output voltage. Since $\lambda_n = \lambda_p$, then $V_{SD5} = V_{DS6}$. If we choose $V_{SD5} = V_{DS6} = 3$ V, then $\Delta V_1 + \Delta V_2 = 24$ V. In order for $v_O = -2.295$ V, then $\Delta V_1 = 14.3$ V and $\Delta V_2 = 9.7$ V. The resistors are then found to be

$$R_1 = \frac{\Delta V_1}{I_Q} = \frac{14.3}{0.2} = 71.5 \text{ k}\Omega$$

and

$$R_2 = \frac{\Delta V_2}{I_Q} = \frac{9.7}{0.2} = 48.5 \text{ k}\Omega$$

**Solution (Second Stage—AC Analysis):** The small-signal equivalent circuit for the second stage is shown in Figure 13.27. Summing currents at the $V_a$ node, we find

$$g_{m6}V_{o1} + \frac{V_a}{r_{o6}} + \frac{V_a}{R_2 + R_1 + r_{o5}} = 0 \qquad (13.64)$$

The output voltage $V_{o2}$ can be written as

$$V_{o2} = \left(\frac{R_1 + r_{o5}}{R_1 + R_2 + r_{o5}}\right)V_a \qquad (13.65)$$

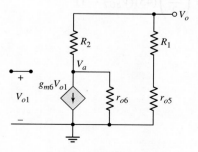

**Figure 13.27** Small-signal equivalent circuit of the second stage of the CMOS op-amp for the design application

Combining Equations (13.64) and (13.65), we obtain

$$g_{m6}V_{o1} + \left(\frac{R_1 + R_2 + r_{o5}}{R_1 + r_{o5}}\right)\left(\frac{1}{r_{o6}} + \frac{1}{R_1 + R_2 + r_{o5}}\right)V_{o2} = 0 \qquad \textbf{(13.66)}$$

The small-signal parameters are found to be

$$g_{m6} = 2\sqrt{\left(\frac{k_n'}{2}\right)\left(\frac{W}{L}\right)_6 I_Q} = 2\sqrt{\left(\frac{0.08}{2}\right)(90)(0.2)} = 1.697\,\text{mA/V}$$

and

$$r_{o5} = r_{o6} = \frac{1}{\lambda I_Q} = \frac{1}{(0.01)(0.2)} = 500\,\text{k}\Omega$$

Then, substituting the parameters into Equation (13.66), we find

$$1.697V_{o1} + \left(\frac{71.5 + 48.5 + 500}{71.5 + 500}\right)\left(\frac{1}{500} + \frac{1}{71.5 + 48.5 + 500}\right)V_{o2} = 0$$

The voltage gain of the second stage is then

$$A_2 = \frac{V_{o2}}{V_{o1}} = -433$$

The overall voltage gain of the circuit is

$$A_v = A_d A_2 = (300)(-433) = -1.3 \times 10^5$$

**Comment:** Achieving the required dc output voltage of $-2.295$ V will be difficult because of device and circuit element tolerances. A circuit similar to the one to be discussed in the design application of Chapter 14 would be required to provide for offset voltage compensation.

## 13.7   SUMMARY

- In this chapter, we combined various basic circuit configurations to form larger operational amplifier circuits. In general, an op-amp circuit consists of a diff-amp input stage, a second gain stage, and an output stage. The design of integrated circuit operational amplifier circuits depends on the use of matched devices.
- The LM741 op-amp is a widely used, general-purpose, bipolar op-amp. This circuit serves as a good case study for a detailed discussion of the circuit design, including a discussion of the input stage design, the Darlington pair gain stage, and a class-AB complementary output stage with the protection circuitry.
- A detailed dc analysis of each stage of the 741 was performed to determine the dc currents and voltages. A detailed small-signal analysis determined the gain of each stage and the overall small-signal voltage gain. The calculated results agree well with the typical values given in data sheets.
- In many cases, all-CMOS operational amplifier circuits require only two stages. These circuits typically drive only low capacitive loads on an IC chip, so the low

output impedance of a third stage is not required. The voltage gain of CMOS amplifiers is generally smaller than that of typical bipolar op-amps, but CMOS op-amps are useful in specialized on-chip applications.

- An all-CMOS folded cascode operational amplifier was found to have a very high differential-mode voltage gain. An all-CMOS current-mirror operational amplifier was found to have an increased gain–bandwidth product.
- The bias current in a BiCMOS op-amp was found to be independent of bias voltage over a wide range of applied bias voltages.
- As an application, a two-stage CMOS op-amp was designed to match a given output stage.

 ## CHECKPOINT

After studying this chapter, the reader should have the ability to:

✓ Understand the general topology and biasing technique of an operational amplifier circuit.

✓ Analyze and understand the operation and characteristics of the LM741 op-amp circuit.

✓ Design a basic bipolar or MOSFET operational amplifier circuit.

✓ Analyze and understand the operation and characteristics of CMOS op-amp circuits, including the folded cascode and the CMOS current-mirror circuits.

✓ Analyze and understand the operation and characteristics of BiCMOS operational amplifier circuits.


## REVIEW QUESTIONS

1. Describe the principal stages of a general-purpose operational amplifier.
2. What is meant by the term matched transistors? What parameters in BJTs and MOSFETs are identical in matched devices?
3. Describe the operation and characteristics of a BJT complementary push–pull output stage. What are the advantages of this circuit?
4. Describe the operation and characteristics of a MOSFET complementary push–pull output stage. What are the advantages of this circuit?
5. Describe the configuration and operation of the input diff-amp stage of the 741 op-amp.
6. What is the purpose of the resistor $R_3$ in the active load of the 741 op-amp?
7. Describe the configuration of the output stage of the 741 op-amp.
8. Describe the operation of the short-circuit protection circuitry in the 741 op-amp.
9. Describe the frequency compensation technique in the 741 op-amp circuit.
10. Sketch and describe the general characteristics of a folded cascode circuit.
11. Sketch and describe the general characteristics of a current–mirror op-amp circuit. Why is the gain not increased? What is the principal advantage of this circuit?
12. Sketch and describe the principal advantage of a BiCMOS folded cascode op-amp circuit.
13. Explain why an output resistance on the order of five hundred megohms may not be achieved in practice.
14. What are the principal factors limiting the unity-gain bandwidth of an op-amp circuit?

# PROBLEMS

## Section 13.1 General Op-Amp Circuit Design

13.1 Consider the simple MOS op-amp circuit shown in Figure P13.1. The bias current is $I_Q = 200\,\mu A$. Transistor parameters are $k_n' = 100\,\mu A/V^2$, $k_p' = 40\,\mu A/V^2$, $V_{TN} = 0.4\,V$, $V_{TP} = -0.4\,V$, and $\lambda_n = \lambda_p = 0$. The width-to-length ratio $(W/L)$ for $M_1$ and $M_2$ is 20 and for $M_3$ is 40. (a) Design the circuit such that $I_{D3} = 200\,\mu A$ and $v_o = 0$ when $v_1 = v_2 = 0$. (b) Find the small-signal voltage gains (i) $A_d = v_{o1}/v_d$ and (ii) $A_2 = v_o/v_{o1}$. (c) Determine the overall small-signal voltage gain $A = v_o/v_d$.

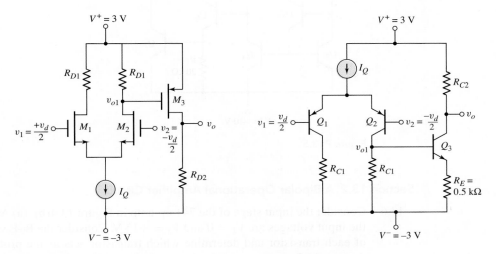

**Figure P13.1**　　　　　　　**Figure P13.2**

13.2 Consider the simple bipolar op-amp circuit shown in Figure P13.2. The bias current is $I_Q = 0.5\,mA$. Transistor parameters are $\beta_n = 180$, $\beta_p = 120$, $V_{BE}(\text{on}) = V_{EB}(\text{on}) = 0.7\,V$, and $V_{An} = V_{Ap} = \infty$. (a) Design the circuit such that $I_{C3} = 0.4\,mA$ and $v_o = 0$ when $v_1 = v_2 = 0$. (b) Find the small-signal voltage gains (i) $A_d = v_{o1}/v_d$ and (ii) $A_2 = v_o/v_{o1}$. (c) Determine the overall small-signal voltage gain $A = v_o/v_d$.

D13.3 Design the circuit in Figure 13.2 such that the maximum power dissipated in the circuit is 15 mW and such that the common-mode input voltage is in the range $-3 \le v_{CM} \le 3\,V$. Using a computer simulation, adjust the value of $R_3$ such that the output voltage is zero for zero input signal voltages.

13.4 Using the results of Problem 13.3, determine, from a computer simulation, the differential-mode voltage gain of the diff-amp and the voltage gain of the second stage of the op-amp circuit in Figure 13.2. Use standard transistor models in the circuit.

*13.5 Consider the BJT op-amp circuit in Figure P13.5. The transistor parameters are: $\beta(\text{npn}) = 120$, $\beta(\text{pnp}) = 80$, $V_A = 80\,V$ (all transistors), and base–emitter turn-on voltage $= 0.6\,V$ (all transistors). (a) Determine the small-signal differential-mode voltage gain. (b) Find the differential-mode input resistance. (c) Determine the unity-gain bandwidth.

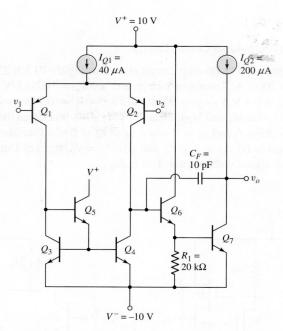

Figure P13.5

## Section 13.2 A Bipolar Operational Amplifier Circuit

13.6    Consider the input stage of the 741 op-amp in Figure 13.4(b). (a) Assume the input voltages are $V_1 = 0$ and $V_2 = +15$ V. Consider the B–E voltage of each transistor and determine which transistor acts as the protection device. (b) Repeat part (a) for $V_1 = -15$ V and $V_2 = 0$.

13.7    For the input stage of the 741 op-amp, assume B–E breakdown voltages of 5 V for the npn devices and 50 V for the pnp devices. Estimate the differential input voltage at which breakdown will occur.

13.8    Consider the bias circuit portion of the 741 op-amp in Figure 13.5. Assume transistor parameters of $I_S = 5 \times 10^{-16}$ A. Neglect base currents. (a) Redesign the circuit such that $I_{\text{REF}} = 0.5$ mA and $I_{C10} = 30\,\mu$A for bias voltages of $\pm 15$ V. What are the values of $V_{BE11}$, $V_{EB12}$, and $V_{BE10}$? (b) Using the resistor values found in part (a) and assuming $V_{BE}(\text{on}) = V_{EB}(\text{on}) = 0.6$ V, determine the values of $I_{\text{REF}}$ and $I_{C10}$. (c) What are the percent differences in the current values between parts (a) and (b).

13.9    Repeat Problem 13.8 for bias voltages of $\pm 5$ V.

13.10   Consider the bias circuit shown in Figure P13.10. Let $V^+ = 3$ V, $V^- = -3$ V, $R_1 = 80\,\text{k}\Omega$, and $R_E = 3.5\,\text{k}\Omega$. Assume transistor parameters of $I_S = 5 \times 10^{-15}$ A for $Q_1$, $Q_2$, $Q_3$; $I_S = 3 \times 10^{-15}$ A for $Q_4$; and $I_S = 10^{-15}$ A for $Q_5$. (a) Find the currents $I_{\text{REF}}$, $I_3$, $I_4$, and $I_5$. (b) Repeat part (a) for $I_S = 8 \times 10^{-15}$ A for $Q_4$ and $I_S = 2 \times 10^{-15}$ A for $Q_5$.

13.11   The minimum recommended supply voltages for the 741 op-amp are $V^+ = 5$ V and $V^- = -5$ V. Using these lower supply voltages, calculate: $I_{\text{REF}}$, $I_{C10}$, $I_{C6}$, $I_{C17}$, and $I_{C13A}$.

13.12   An expanded circuit diagram of the 741 input stage is shown in Figure 13.6. Assume $I_{C10} = 50\,\mu$A. If the current gain of the npn transistors is $\beta_n = 90$

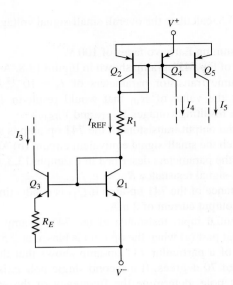

Figure P13.10

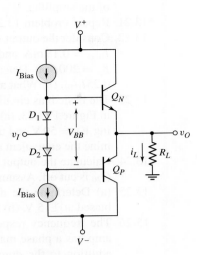

Figure P13.16

and the current gain of the pnp transistors is $\beta_p = 40$, determine $I_{C9}$, $I_{C2}$, $I_{C4}$, $I_{B9}$, and $I_{B4}$. Do not neglect npn transistor base currents.

13.13   Consider the 741 op-amp in Figure 13.3, biased with $V^+ = 15$ V and $V^- = -15$ V. Assume that no load is connected at the output, and let the input voltages be zero. Calculate the total power dissipated in the op-amp circuit. What are the currents supplied by $V^+$ and $V^-$?

13.14   Consider the 741 circuit in Figure 13.3. (a) Determine the maximum range of common-mode input voltage if the bias voltages are $\pm 15$ V. (b) Repeat part (a) if the bias voltages are $\pm 5$ V.

13.15   Consider the output stage of the 741 op-amp shown in Figure 13.8. Assume $v_1 = v_2 = 0$ at the input and assume the bias voltages are $V^+ = 5$ V and $V^- = -5$ V. Let $I_{REF} = 0.5$ mA. All other circuit and transistor parameters are described in Example 13.3. Find $I_{C13A}$, $I_{R10}$, $I_{C19}$, $I_{C18}$, $V_{BE19}$, $V_{BE18}$, and $I_{C14}$.

*13.16   Consider the output stage in Figure P13.16 with parameters $V^+ = 5$ V, $V^- = -5$ V, $R_L = 10$ kΩ, and $I_{Bias} = 80\,\mu$A. Assume the diode parameters are $I_{SD} = 5 \times 10^{-15}$ A, and assume the transistor parameters are $I_{SQ} = 8 \times 10^{-15}$ A and $\beta_n = \beta_p = 120$. (a) For $v_I = 0$, determine $V_{BB}$, $I_{CN}$, and $I_{CP}$. (b) For $v_I = 3$ V, determine $v_O$, $i_L$, $V_{BB}$, $I_{CN}$, and $I_{CP}$.

D13.17   Figure P13.17 shows a circuit often used to provide the $V_{BB}$ voltage in the op-amp output stage. Assume $I_S = 5 \times 10^{-15}$ A for the transistor, $I_{Bias} = 120\,\mu$A, and $I_C = 0.9I_{Bias}$. Neglect the base current. Design the circuit such that $V_{BB} = 1.160$ V.

13.18   Assume bias voltages on the 741 op-amp of $\pm 15$ V. (a) Determine the differential-mode voltage gain of the first stage if $R_1 = R_2 = 0$. (b) Determine the voltage gain of the second stage if $R_8 = 0$.

13.19   Recalculate the voltage gain of the 741 op-amp input stage if $I_{C10} = 40\,\mu$A.

*13.20   Assume the 741 op-amp shown in Figure 13.3 is biased at $\pm 5$ V. Using the circuit parameters given in the figure and transistor parameters given in

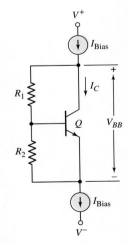

Figure P13.17

Examples 13.1 through 13.5, calculate the overall small-signal voltage gain of the amplifier.

*13.21  Repeat Problem 13.20 assuming Early voltages of 100 V.

13.22  Consider the output stage of the 741 op-amp shown in Figure 13.8. Assume $I_{Bias} = 0.18 \, mA$ and assume transistor parameters of $I_S = 10^{-14} \, A$ and $\beta_n = 200$. (a) Determine the value of $R_{10}$ that would result in $I_{C18} = 0.25 I_{C19}$. (b) What are the resulting voltages $V_{BE18}$ and $V_{BE19}$?

13.23  The basic bias circuit of the output transistors of the 741 op-amp is shown in Figure P13.23. (a) Sketch the small-signal equivalent circuit. (b) Assuming $V_A = 50 \, V$ and using the parameters described in Example 13.3, determine the equivalent small-signal resistance $R_{eq} = v_x/i_x$.

13.24  Calculate the output resistance of the 741 op-amp if $Q_{14}$ is conducting and $Q_{20}$ is cut off. Assume an output current of 2 mA.

13.25  (a) Determine the differential input resistance of the 741 op-amp when biased at $\pm 15$ V. (b) Repeat part (a) when the op-amp is biased at $\pm 5$ V.

13.26  The frequency response of a particular 741 op-amp shows that the op-amp has a phase margin of 70 degrees. If a second single pole exists, in addition to the dominant pole, determine the frequency of the second pole. Use the overall gain and dominant-pole parameters calculated in Section 13.2.

13.27  An op-amp that is internally compensated by Miller compensation has a unity-gain bandwidth of 10 MHz and a low-frequency gain of $10^6$. (a) What is the dominant pole frequency? (b) The feedback capacitor is across the second stage, which has a gain of $-10^3$. The effective resistance at the input of the second stage is $R_{eq} = 1.2 \, M\Omega$. What is the value of the feedback capacitor?

13.28  A three-stage 741 op-amp has a low-frequency open-loop gain of 200,000 and a dominant pole frequency of 10 Hz. The second and third poles are at the same frequency. If the phase margin is 70 degrees, determine the frequency of the second and third poles.

### Section 13.3 CMOS Operational Amplifier Circuits

13.29  Consider the simple CMOS op-amp circuit in Figure P13.29 biased with $I_Q = 200 \, \mu A$. The transistor parameters are $k'_n = 100 \, \mu A/V^2$, $k'_p = 40 \, \mu A/V^2$, $V_{TN} = 0.4 \, V$, $V_{TP} = -0.4 \, V$, and $\lambda_n = \lambda_p = 0$. The transistor width-to-length ratios are $(W/L)_{1,2} = 20$, $(W/L)_3 = 50$, and $(W/L)_4 = 40$. (a) Design the circuit such that $I_{D3} = 150 \, \mu A$, $I_{D4} = 200 \, \mu A$, and $v_o = 0$ for $v_1 = v_2 = 0$. (b) Find the small-signal voltage gains (i) $A_d = v_{o1}/v_d$, (ii) $A_2 = v_{o2}/v_{o1}$, and (iii) $A_3 = v_o/v_{o2}$. (c) Determine the overall small-signal voltage gain $A = v_o/v_d$.

13.30  A simple CMOS op-amp circuit is shown in Figure P13.30 with $I_Q = 100 \, \mu A$. The transistor parameters are the same as given in Problem 13.29 except for the width-to-length ratios. The width-to-length ratios are $(W/L)_{1,2} = 80$, $(W/L)_3 = 25$, and $(W/L)_4 = 100$. (a) The circuit is to be designed such that $I_{DQ3} = 100 \, \mu A$, $I_{DQ4} = 200 \, \mu A$, and $v_o = 0$ for $v_1 = v_2 = 0$. (b) Determine the small-signal voltage gains (i) $A_d = v_{o1}/v_d$, (ii) $A_2 = v_{o2}/v_{o1}$, and (iii) $A_3 = v_o/v_{o2}$. (c) Find the overall small-signal voltage gain $A = v_o/v_d$.

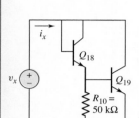

Figure P13.23

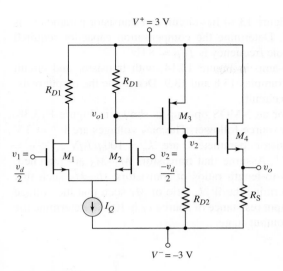

Figure P13.29

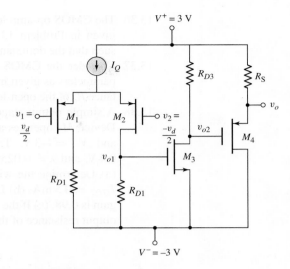

Figure P13.30

D13.31 Consider the MC14573 op-amp in Figure 13.14. The dc bias currents and small-signal voltage gains were determined in Examples 13.8 and 13.9. Redesign the circuit such that the width-to-length ratio of $M_1$ and $M_2$ is increased from 12.5 to 50. All other circuit and transistor parameters remain the same. (a) Determine the original transconductance of $M_1$ and $M_2$, and the new transconductance value. (b) Determine the new values of voltage gain for the input and second stages, and the overall voltage gain.

13.32 Consider the basic diff-amp with active load and current biasing in Figure 13.14. Using the parameters and results of Example 13.8, determine the maximum range of common-mode input voltage under the conditions that the minimum source-to-drain voltage for a PMOS is limited to $v_{SD}(\text{min}) = v_{SD}(\text{sat}) + 0.2$ V and the minimum drain-to-source voltage for an NMOS is limited to $v_{DS}(\text{min}) = v_{DS}(\text{sat}) + 0.2$ V.

13.33 The CMOS op-amp in Figure 13.14 is biased at $V^+ = 5$ V and $V^- = -5$ V. Let $R_{\text{set}} = 50\,\text{k}\Omega$. Assume transistor parameters of $V_{TN} = 0.7$ V, $V_{TP} = -0.7$ V, $k_n' = 100\,\mu\text{A/V}^2$, $k_p' = 40\,\mu\text{A/V}^2$, $\lambda_n = 0.02\,\text{V}^{-1}$, and $\lambda_p = 0.04\,\text{V}^{-1}$. The transistor width-to-length ratios are $(W/L)_{3,4} = 15$, $(W/L)_7 = 30$, and $(W/L) = 50$ for all other transistors. (a) Determine $I_{\text{set}}$, $I_Q$, and $I_{DQ7}$. (b) Find the small-signal voltage gains of the input and second stages, and the overall voltage gain.

13.34 For the CMOS op-amp in Figure 13.14, the dc biasing is designed such that $I_{\text{set}} = I_Q = I_{DQ8} = 200\,\mu\text{A}$. The transistor parameters are $V_{TN} = 0.5$ V, $V_{TP} = -0.5$ V, $k_n' = 100\,\mu\text{A/V}^2$, $k_p' = 40\,\mu\text{A/V}^2$, $\lambda_n = 0.015\,\text{V}^{-1}$, and $\lambda_p = 0.025\,\text{V}^{-1}$. The transistor width-to-length ratios are $(W/L)_{1,2} = 50$, $(W/L)_{3,4} = 15$, $(W/L)_{5,6,8} = 10$, and $(W/L)_7 = 30$. Determine the small-signal voltage gains of the input and second stages, and the overall voltage gain.

13.35 Consider the MC14573 op-amp in Figure 13.14, with circuit and transistor parameters as given in Examples 13.8 and 13.9. If the compensation capacitor is $C_1 = 12$ pF, determine the dominant-pole frequency.

13.36 The CMOS op-amp in Figure 13.14 has circuit and transistor parameters as given in Problem 13.33. Determine the compensation capacitor required such that the dominant-pole frequency is $f_{PD} = 8$ Hz.

13.37 Consider the CMOS op-amp in Figure 13.14, with transistor and circuit parameters as given in Examples 13.8 and 13.9. Determine the output resistance $R_o$ of the open-loop circuit.

13.38 A simple output stage for an NMOS op-amp is shown in Figure P13.38. Device $M_1$ operates as a source follower. The bias voltages are $V^+ = 3$ V and $V^- = -3$ V. Transistor parameters are $k'_n = 100 \, \mu$A/V$^2$, $V_{TN} = 0.4$ V, and $\lambda = 0.025$ V$^{-1}$. Assume that transistors $M_2$–$M_5$ are matched. (a) Determine the width-to-length ratios of transistors $M_2$–$M_5$ such that $I_{DQ2} = 0.5$ mA. (b) Determine the $W/L$ ratio of $M_1$ such that the voltage gain is 0.98. (c) If the output resistance of source $v_I$ is 10 k$\Omega$, determine the output resistance of this output stage.

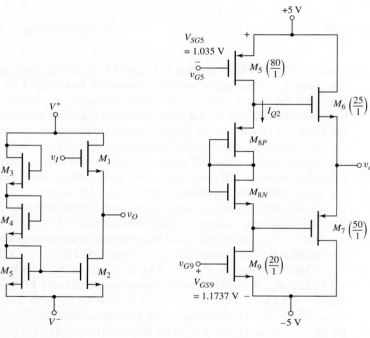

Figure P13.38                     Figure P13.39

13.39 The circuit in Figure P13.39 is another form of an output stage for the CMOS op-amp shown in Figure 13.15. Assume the same transistor parameters as given in Example 13.10. The width-to-length values of some transistors are given and the applied gate-to-source voltages of $M_5$ and $M_9$ are shown. (a) What is the bias current $I_{Q2}$? (b) Determine the $W/L$ ratios of $M_{8P}$ and $M_{8N}$ such that the quiescent currents in $M_6$ and $M_7$ are 25 $\mu$A.

D13.40 Consider the three-stage CMOS op-amp in Figure 13.15. Design an all-NMOS transistor current source circuit to establish $I_{Q1} = 150 \, \mu$A. The NMOS transistor parameters are $k'_n = 100 \, \mu$A/V$^2$ and $V_{TN} = 0.5$ V. Assume the minimum width-to-length ratio of any transistor is 2. Assume $(W/L)_{10} = (W/L)_{11} = 20$ as shown in the figure.

13.41 Assume $I_{REF} = 250\ \mu A$ and $(W/L)_8 = 5$ in the CMOS op-amp shown in Figure 13.15. Determine (a) the quiescent currents in $M_6$ and $M_7$ and (b) the overall small-signal voltage gain. Assume transistor parameters as given in Example 13.10.

*13.42 The CMOS folded cascode circuit in Figure 13.17 is biased at $\pm 5$ V and the reference current is $I_{REF} = 50\ \mu A$. The transistor parameters are $V_{TN} = 0.5$ V, $V_{TP} = -0.5$ V, $K_n = K_p = 0.5$ mA/V$^2$, and $\lambda_n = \lambda_p = 0.015$ V$^{-1}$. (a) Determine the small-signal differential voltage gain. (b) Find the output resistance of the circuit. (c) If the capacitance at the output node is $C_L = 5$ pF, determine the unity-gain bandwidth of the amplifier.

*D13.43 The CMOS folded cascode amplifier in Figure 13.17 is to be redesigned to provide a differential voltage gain of 10,000. The biasing is the same as described in Problem 13.42. The transistor parameters are $V_{TN} = 0.5$ V, $V_{TP} = -0.5$ V, $k'_n = 80\ \mu A/V^2$, $k'_p = 35\ \mu A/V^2$, $\lambda_n = 0.015$ V$^{-1}$, and $\lambda_p = 0.02$ V$^{-1}$. Assume $(W/L)_p = 2.2(W/L)_n$ where appropriate so that the electrical parameters of PMOS and NMOS devices are nearly identical.

*D13.44 The CMOS folded cascode amplifier of Figure 13.17 is to be designed to provide a differential voltage gain of 25,000. The maximum power dissipated in the circuit is to be limited to 3 mW. Assume transistor parameters as described in Problem 13.43, except the relation between NMOS and PMOS width-to-length ratios need not be maintained.

13.45 The bias current in the CMOS current-gain op-amp in Figure 13.18 is $I_Q = 120\ \mu A$. The transistor parameters are $V_{TN} = 0.5$ V, $V_{TP} = -0.5$ V, $k'_n = 100\ \mu A/V^2$, $k'_p = 40\ \mu A/V^2$, $\lambda_n = 0.02$ V$^{-1}$, and $\lambda_p = 0.04$ V$^{-1}$. The transistor width-to-length ratios are 20 except for $M_5$ and $M_6$. Let $B = 3$. (a) Determine the small-signal differential voltage gain. (b) Find the output resistance of the circuit. (c) If the total capacitance at the output terminal is 5 pF, determine the dominant-pole frequency and the unity-gain bandwidth.

D13.46 The CMOS current gain op-amp in Figure 13.18 is to be redesigned to provide a differential voltage gain of 400. The transistor parameters are $V_{TN} = 0.5$ V, $V_{TP} = -0.5$ V, $k'_n = 80\ \mu A/V^2$, $k'_p = 35\ \mu A/V^2$, $\lambda_n = 0.015$ V$^{-1}$, and $\lambda_p = 0.02$ V$^{-1}$. The bias current is to be $I_Q = 80\ \mu A$. Let $B = 2.5$. (a) Design the basic amplifier to provide the specified voltage gain. (b) Design a current source to provide the necessary bias current. (c) Determine the unity-gain bandwidth if the capacitance at the output terminal is 3 pF.

D13.47 Redesign the CMOS cascode current mirror in Figure 13.19 to provide a differential voltage gain of 20,000. The bias current and transistor parameters are the same as in Problem 13.46. (a) Design the basic amplifier to provide the specified voltage gain. (b) Design a current source to provide the necessary bias current. (c) Determine the unity gain bandwidth if the capacitance at the output terminal is 3 pF.

## Section 13.4 BiCMOS Operational Amplifier Circuits

13.48 A simple BiCMOS amplifier is shown in Figure P13.48. The MOS transistor parameters are $k'_p = 40\ \mu A/V^2$, $V_{TP} = -0.4$ V, $\lambda = 0$, and $(W/L)_{1,2} = 50$, and the bipolar transistor parameters are $V_{BE}(\text{on}) = 0.7$ V, $\beta = 120$, and $V_A = \infty$. (a) Design the circuit such that $I_{CQ} = 300\ \mu A$ and $v_o = 0$ for

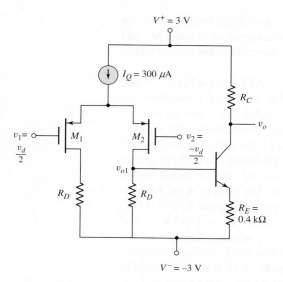

**Figure P13.48**

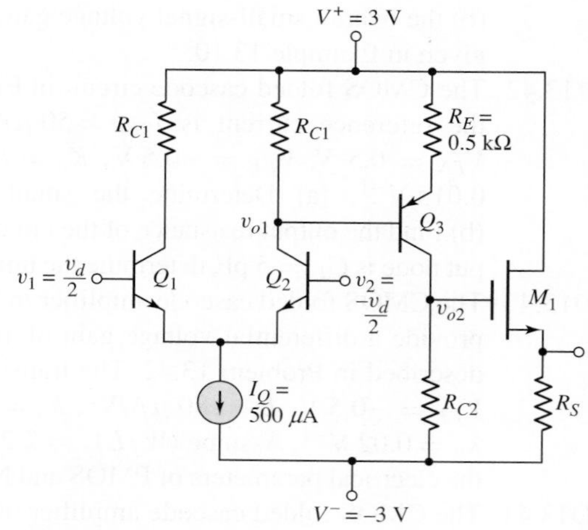

**Figure P13.49**

$v_1 = v_2 = 0$. (b) Determine the small-signal differential voltage gain of the first stage $A_d = v_{o1}/v_d$. (c) Find the small-signal voltage gain of the second stage $A_2 = v_o/v_{o1}$. (d) Determine the overall small-signal voltage gain $A = v_o/v_d$.

13.49   Consider the simple BiCMOS amplifier shown in Figure P13.49. The bipolar transistor parameters are $\beta_n = \beta_p = 120$, $V_{BE}(\text{on}) = V_{EB}(\text{on}) = 0.7$ V, $V_A = \infty$, and the MOS transistor parameters are $V_{TN} = 0.4$ V, $K_n = 3 \text{ mA/V}^2$, $\lambda = 0$. (a) Design the circuit such that $I_{CQ3} = I_{DQ1} = 300\,\mu\text{A}$ and $v_o = 0$ for $v_1 = v_2 = 0$. (b) Determine the small-signal voltage gain of the first stage $A_d = v_{o1}/v_d$. (c) Find the small-signal voltage gain of the second stage $A_2 = v_{o2}/v_{o1}$. (d) Determine the small-signal voltage gain of the third stage $A_3 = v_o/v_{o2}$. (e) Find the overall differential voltage gain $A = v_o/v_d$.

13.50   A BiCMOS amplifier is shown in Figure P13.50. The transistor parameters are $V_{TP} = -0.4$ V, $k'_p = 40\,\mu\text{A/V}^2$, $W/L = 40$, $\lambda = 0.035\text{ V}^{-1}$, $\beta = 120$, and $V_A = 150$ V. The bias current is $I_Q = 250\,\mu\text{A}$. (a) Determine the small-signal parameters of the transistors. (b) Find the small-signal differential voltage gain.

13.51   Design a BiCMOS amplifier that is complementary to the one in Figure P13.50 in that the input devices are NMOS and the load transistors are pnp. Assume transistor parameters of $V_{TN} = 0.4$ V, $k'_n = 100\,\mu\text{A/V}^2$, $W/L = 40$, $\lambda = 0.02\text{ V}^{-1}$, $\beta = 80$, and $V_A = 100$ V. Assume the bias current is $I_Q = 250\,\mu\text{A}$. (a) Determine the small-signal parameters of the transistors. (b) Find the small-signal differential voltage gain.

*13.52   The reference current in the BiCMOS folded cascode amplifier in Figure 13.20 is $I_{REF} = 200\,\mu\text{A}$ and the circuit bias voltages are $\pm10$ V. The MOS transistor parameters are the same as in Problem 13.42. The BJT parameters are $\beta = 120$ and $V_A = 80$ V. (a) Determine the small-signal differential voltage gain. (b) Find the output resistance of the circuit. (c) If

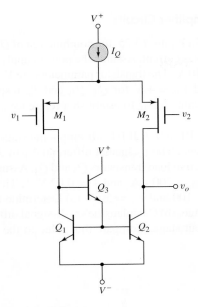

**Figure P13.50**

the capacitance at the output node is 5 pF, determine the unity-gain band-width of the amplifier.

*D13.53   The BiCMOS folded cascode amplifier in Figure 13.20 is to be designed to provide a differential voltage gain of 25,000. The maximum power dissipated in the circuit is to be limited to 10 mW. Assume MOS transistor parameters as described in Problem 13.43. The BJT parameters are $\beta = 120$ and $V_A = 80$ V.

13.54   If the CA3140 op-amp is biased at $V^+ = 15$ V and $V^- = -15$ V, determine the input common-mode voltage range. Assume B–E voltages of 0.6 V for the bipolar transistors and $|V_{TP}| = 1.4$ V for the MOSFETs.

13.55   Consider the bias circuit portion of the CA3140 op-amp in Figure 13.22. If $V_{BE7} = 0.6$ V for $Q_7$ and $V_{TP} = -1.0$ V for $M_8$, determine the required value of the conduction parameter for $M_8$ such that $I_1 = I_2 = 150\,\mu$A.

13.56   In the bias portion of the CA1340 op-amp in Figure 13.22, the bipolar transistor parameters are $V_{BE}(\text{npn}) = 0.6$ V and $V_{EB}(\text{pnp}) = 0.6$ V and the MOSFET parameters are $V_{TP} = -1.2$ V and $K_p = 0.15$ mA/V$^2$. (a) Determine the currents $I_1 = I_2$. (b) If the bias voltages are $V^+ = -V^- \equiv V_S$, determine the minimum value of $V_S$ such that the bias currents are independent of the supply voltage.

13.57   Consider the CA3140 op-amp in Figure 13.21. If the bias currents change such that $I_{C5} = I_{C4} = 300\,\mu$A, determine the voltage gains of the input and second stages, and find the overall voltage gain.

13.58   Assume the gain stage of the CA3140 op-amp is modified to include an emitter resistor, as shown in Figure 13.23. Let $\lambda = 0.02$ V$^{-1}$ for $M_{10}$. If the transistor bias currents in $M_{10}$ and $Q_{12}$ are 150 $\mu$A and the current in $Q_{13}$ is 300 $\mu$A, determine the dominant-pole frequency and unity-gain bandwidth.

### Section 13.5 JFET Operational Amplifier Circuits

13.59   In the LF155 BiFET op-amp in Figure 13.25, the combination of $Q_3, J_6$, and $Q_4$ establishes the reference bias current. Assume the power supply voltages are $V^+ = 10$ V and $V^- = -10$ V. The transistor parameters are $V_{EB}(\text{on}) = 0.6$ V, $V_{BE}(\text{on}) = 0.6$ V, and $V_P = 4$ V for $Q_3$, $Q_4$, and $J_6$, respectively. Determine the required $I_{DSS}$ value for $J_6$ to establish a reference current of $I_{\text{REF}} = 0.8$ mA.

13.60   Consider the circuit in Figure P13.60. A JFET diff-amp input stage drives a bipolar Darlington second stage. The p-channel differential pair $J_1$ and $J_2$ are connected to the bipolar active load transistors $Q_3$ and $Q_4$. Assume JFET parameters of $V_P = 3$ V, $I_{DSS} = 200\ \mu$A, and $\lambda = 0.02$ V$^{-1}$. The bipolar transistor parameters are $\beta = 100$ and $V_A = 50$ V. (a) Determine the input resistance $R_{i2}$ to the second stage. (b) Calculate the small-signal differential-mode voltage gain of the input stage. Compare this value to the 741 and CA3140 input stage results.

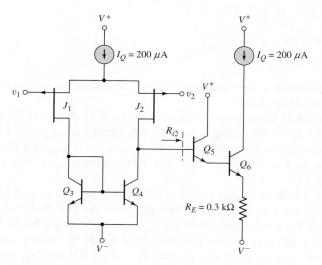

**Figure P13.60**

D13.61   Consider the BiFET differential input stage in Figure P13.61, biased with power supply voltages $V^+$ and $V^-$. Let $V^+ = -V^- \equiv V_S$. (a) Design the bias circuit such that $I_{\text{REF2}} = 100\ \mu$A for supply voltages in the range $3 \le V_S \le 12$ V. Determine $V_{ZK}$, $R_3$, and the JFET parameters. (b) Determine the value of $R_4$ such that $I_{O1} = 500\ \mu$A when $V^+ = 12$ V.

13.62   The BiFET diff-amp input stage in Figure P13.61 is biased at $I_{O1} = 1$ mA. The JFET parameters are $V_P = 4$ V, $I_{DSS} = 1$ mA, and $\lambda = 0.02$ V$^{-1}$. The bipolar transistor parameters are $\beta = 200$ and $V_A = 100$ V. (a) For $R_1 = R_2 = 500\ \Omega$, determine the minimum load resistance $R_L$ such that a differential-mode voltage gain of $A_d = 500$ is obtained in the input stage. (b) If $R_L = 500$ kΩ, determine the range of resistance values $R_1 = R_2$ such that a differential-mode voltage gain of $A_d = 700$ is obtained in this input stage.

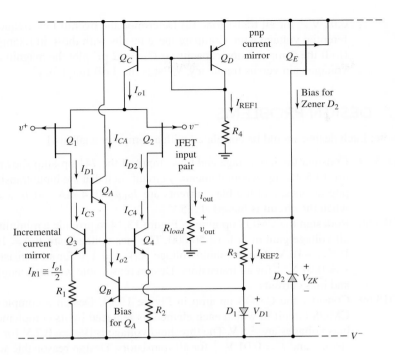

**Figure P13.61**

 **COMPUTER SIMULATION PROBLEMS**

13.63  Consider the input stage and bias circuit of the 741 op-amp in Figure 13.5. Transistor $Q_{10}$ may be replaced by a constant-current source equal to 19 $\mu$A. Assume: the npn devices have parameters $\beta = 200$ and $V_A = 150$ V; the pnp devices have parameters $\beta = 50$ and $V_A = 50$ V; and all transistors have $I_S = 10^{-14}$ A. Place an appropriate ac load at the collector of $Q_6$. (a) Using a computer simulation, determine the differential voltage gain of the input stage. (b) Determine the differential-mode input resistance. (c) Determine the common-mode input resistance.

13.64  The output stage of the 741 op-amp is shown in Figure 13.9. Transistor $Q_{13}$ may be replaced with a constant-current source equal to 0.18 mA. Use standard transistors. (a) Using a computer simulation, plot the voltage transfer function $v_o$ versus $v_{13}$. What is the voltage gain? Has the crossover distortion been eliminated? (b) Apply an input voltage $v_{13}$ that establishes an output voltage of $v_o = 5$ V. Then set $R_L = 0$. Find the output short-circuit current and the transistor currents.

13.65  Consider the BiCMOS input stage of the CA3140 op-amp in Figure 13.21. Transistor $Q_5$ can be replaced with a constant-current source of 200 $\mu$A. Assume: bipolar transistor parameters of $\beta = 200$, $I_{EO} = 10^{-14}$ A, and $V_A = 50$ V; and MOSFET parameters of $K_p = 0.6$ mA/V$^2$, $|V_{TP}| = 1$ V, and $\lambda = 0.01$ V$^{-1}$. Using an appropriate ac load at the collector of $Q_{12}$, determine the differential gain of the input stage. Compare the computer analysis results with those in Example 13.12.

13.66  Consider the CMOS op-amp in Figure 13.14. Assume the circuit and transistor parameters are as given in Example 13.8. In addition, let $\lambda =$

$0.01 \text{ V}^{-1}$ for all transistors. (a) Determine the overall low-frequency differential voltage gain. Compare these results with those in Example 13.9. (b) If the compensation capacitor is $C_1 = 12 \text{ pF}$, plot the magnitude of the voltage gain versus frequency. What is the 3 dB frequency?

## DESIGN PROBLEMS

[Note: Each design should be correlated with a computer analysis.]

*D13.67  Consider the input stage and bias circuit of the 741 op-amp shown in Figure 13.5. Design a complementary circuit such that the input transistors are pnp devices, and the bias currents are $I_{REF} = 0.4 \text{ mA}$ and $I_{C10} = 24\mu\text{A}$ when the circuit is biased at $\pm 5$ V.

*D13.68  Redesign the CMOS op-amp in Figure 13.14 to provide a minimum overall voltage gain of at least 50,000. The bias voltages are $V^+ = 10$ V and $V^- = -10$ V. The threshold voltage is $|V_T| = 1$ V for all transistors, and $\lambda = 0.01 \text{ V}^{-1}$ for all transistors. Design reasonable width-to-length ratios and bias currents.

*D13.69  Consider the CMOS op-amp in Figure 13.14. Design a complementary CMOS circuit in which each element is replaced by its complement. The bias voltages are $\pm 5$ V. The threshold voltage is $|V_T| = 0.7$ V for all transistors, and $\lambda = 0.01 \text{ V}^{-1}$ for all transistors. Design reasonable width-to-length ratios and bias currents to provide a minimum overall voltage gain of at least 20,000.

*D13.70  Consider the bipolar op-amp circuit in Figure P13.70. Design the circuit such that the differential gain is at least 800, and the output voltage is zero when the input voltages are zero. The transistor current gains are 120 for all transistors, and the base–emitter voltages are 0.6 V, where appropriate.

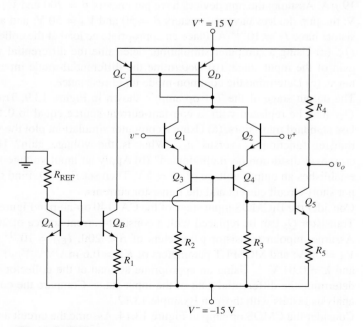

**Figure P13.70**

# Nonideal Effects in Operational Amplifier Circuits

Chapter 9 introduced the ideal operational amplifier and covered a few of its many applications. In the previous chapter, we analyzed actual operational amplifier circuits, including the classic 741 op-amp. From these discussions, we can identify sources of nonideal properties in actual op-amps. In particular, we consider the effects of a finite open-loop gain, reconsider the frequency response, consider the source and effects of offset voltage, and consider the source and effects of input bias currents.

Although nonideal effects could have been introduced in Chapter 9, that discussion would have been less meaningful since the source of any nonideal effect would not have been completely understood at that time. In particular, the reason for a very low dominant-pole frequency in the basic amplifier would have been a mystery. Therefore, the discussion of nonideal effects in op-amp circuits has been postponed until now.

## PREVIEW

In this chapter, we will:

- Define and discuss various practical op-amp parameters.
- Analyze the effect of finite open-loop gain.
- Analyze the open-loop and closed-loop frequency response.
- Define and analyze sources and effects of offset voltage.
- Define and analyze effects of input bias currents.
- Discuss and analyze additional nonideal properties, such as temperature and common-mode rejection effects.
- As an application, design an offset voltage compensation network for a CMOS diff-amp.

## 14.1  PRACTICAL OP-AMP PARAMETERS

> **Objective:** • Define and discuss various practical op-amp parameters.

In ideal op-amps, we assume, for example, that the differential voltage gain is infinite, the input resistance is infinite, and the output resistance is zero. In practical op-amp circuits, these ideal parameter values are not realized. In this section, we define some of the practical op-amp parameters that will be considered in detail throughout the chapter. We will discuss and analyze the effect of these nonideal parameters in op-amp circuits.

### 14.1.1  Practical Op-Amp Parameter Definitions

*Input voltage limits.* Two input voltage limitations must be considered—a dc input voltage limit and a differential signal input voltage. All transistors in the input diff-amp stage must be properly biased, so there is a limit in the range of common-mode input voltage that can be applied and still maintain the proper transistor biasing. The maximum differential input signal voltage that can be applied and still maintain linear circuit operation is limited primarily by the maximum allowed output signal voltage.

*Output voltage limits.* The output voltage of the op-amp can never exceed the limits of the dc supply voltages. In practice, the difference between the bias voltage and the maximum output voltage depends on the design of the output stage. In older designs, this difference was on the order of 1 to 2 volts. In newer designs, this difference can be on the order of millivolts. If $V_{out} = A_v \cdot V_{in}$ (where $A_v$ is the overall voltage gain) is greater than the bias voltage, then the output voltage would saturate and would no longer be a linear function of the input voltage.

*Output current limitation.* The maximum current out of or into the op-amp is determined by the current ratings of the output transistors. Practical op-amp circuits cannot source or sink an infinite amount of current.

*Finite open-loop voltage gain.* The open-loop gain of the ideal op-amp is assumed to be infinite. In practice, the open-loop gain of any op-amp circuit is always finite. This nonideal parameter value will affect circuit performance.

*Input resistance.* The input resistance $R_i$ is the small-signal resistance between the inverting and noninverting terminals when a differential voltage is applied. Ideally, this parameter is infinite, but, especially for BJT circuits, this parameter is finite.

*Output resistance.* The output resistance is the Thevenin equivalent small-signal resistance looking back into the output terminal of the op-amp measured with respect to ground. The ideal output resistance is zero, which means there is no loading effect at the output. In practice, this value is not zero.

*Finite bandwidth.* In the ideal op-amp, the bandwidth is infinite. In practical op-amps, the bandwidth is finite because of capacitances within the op-amp circuit.

*Slew rate.* The slew rate is defined as the maximum rate of change in output voltage per unit of time. The maximum rate at which the output voltage can change is also a function of capacitances within the op-amp circuit.

*Input offset voltage.* In an ideal op-amp, the output voltage is zero for zero differential input signal voltage. However, mismatches between input devices, for example, may create an output voltage with zero input. The input offset voltage is the applied differential input voltage required to induce a zero output voltage.

**Table 14.1**  Nonideal parameter values for three op-amp circuits

| | 741E | | | CA3140 | | | LH0042C | | |
| --- | --- | --- | --- | --- | --- | --- | --- | --- | --- |
| | Typ. | Max. | Unit | Typ. | Max. | Unit | Typ. | Max. | Unit |
| Input offset voltage | 0.8 | 3 | mV | 5 | 15 | mV | 6 | 20 | mV |
| Average input offset voltage drift | | 15 | $\mu V/°C$ | | | | | 10 | $\mu V/°C$ |
| Input offset current | 3.0 | 30 | nA | 0.5 | 30 | pA | 2 | | pA |
| Average input offset current drift | | 0.5 | $nA/°C$ | | | | | | |
| Input bias current | 30 | 80 | nA | 10 | 50 | pA | 2 | 10 | pA |
| Slew rate | 0.7 | | $V/\mu s$ | 9 | | $V/\mu s$ | 3 | | $V/\mu s$ |
| CMRR | 95 | | dB | 90 | | dB | 80 | | dB |

*Input bias currents.* In an ideal op-amp, the input current to the op-amp circuit is assumed to be zero. However, in practical op-amps, especially with BJT input devices, the input bias currents are not zero.

The cause of these nonideal op-amp parameters will be discussed in the following sections, as well as the effect these nonideal parameters have on op-amp circuit performance. A few other nonideal parameters will be considered in the last section of the chapter.

Table 14.1 lists a few of the nonideal parameter values for three of the op-amps considered in the previous chapter. We will refer to this table as we discuss each of the nonideal parameters.

## 14.1.2  Input and Output Voltage Limitations

For linear circuit operation, all BJTs in an op-amp circuit must be biased in the forward-active region and all MOSFETs must be biased in the saturation region. For these reasons, there are limitations to the range of input and output voltages in op-amp circuits.

Figure 14.1(a) shows the simple all-BJT op-amp circuit discussed at the beginning of Chapter 13 and Figure 14.1(b) shows the all-CMOS folded cascode op-amp circuit discussed in the last chapter. We will use these two circuits to discuss the input and output voltage limitations.

### Input Voltage Limitations

Assume that in the BJT circuit of Figure 14.1(a) we apply a common-mode input voltage such that $v_{cm} = v_1 = v_2$. As $v_{cm}$ increases, the base–collector voltages of $Q_1$ and $Q_2$ decrease, since the collector voltages are fixed at two base–emitter voltage drops below $V^+$. If we assume the minimum base–collector voltage is zero so that the transistor is still biased in the active mode, then the maximum value of $v_{cm}$ is $v_{cm}(\max) = V^+ - 2V_{EB}(\text{on})$.

As $v_{cm}$ decreases, the collector–emitter voltage of $Q_3$ decreases. If we again assume the minimum base–collector voltage is zero, or the minimum collector–emitter voltage is $V_{BE}(\text{on})$, then, taking into account the base–emitter voltage of the input transistors, the minimum value of $v_{cm}$ is $v_{cm}(\min) = V^- + 2V_{BE}(\text{on})$. So the maximum range of $v_{cm}$ is within approximately 1.4 V of each bias voltage.

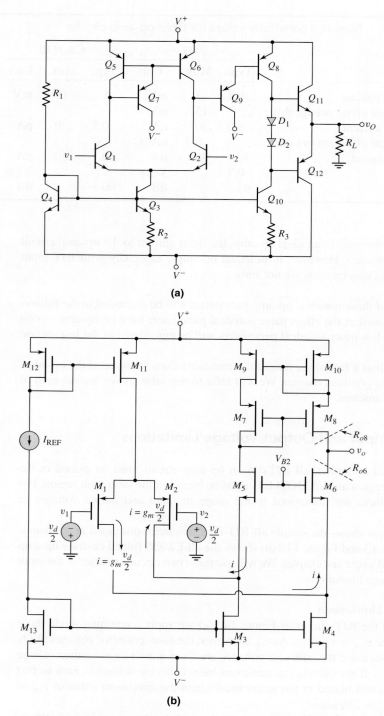

**Figure 14.1** (a) Simple all-bipolar op-amp circuit; (b) all-CMOS folded cascode op-amp circuit

The same range of common-mode input voltage can be found for the all-MOSFET diff-amp in Figure 14.1(b). In this case, all MOSFETs must be biased in the saturation region. We can again define the common-mode input voltage as $v_{cm} = v_1 = v_2$. Now, as $v_{cm}$ increases, $V_{SD}$ of $M_{11}$ decreases. The minimum value of $V_{SD}$ is $V_{SD11}(\text{sat}) = V_{SG11} + V_{TP11}$. The maximum value of $v_{cm}$ is then $v_{cm}(\text{max}) = V^+ - [V_{SG1} + (V_{SG11} + V_{TP11})]$. The gate-to-source voltages can be determined from the transistor parameters and currents.

As $v_{cm}$ decreases, the source-to-drain voltage of the input transistors decreases. Assuming that $M_3$ and $M_4$ are matched to $M_{13}$, then the drain-to-source voltage of these transistors is equal to $V_{GS13}$. The minimum common-mode input voltage is then $v_{cm}(\text{min}) = V^- + [V_{GS11} + (V_{SG1} + V_{TP1}) - V_{SG1}]$. The $V_{SG1}$ terms cancel, so $v_{cm}(\text{min}) = V^- + [V_{GS11} + V_{TP1}]$.

### Output Voltage Limitations

As the output voltage of the BJT circuit in Figure 14.1(a) increases or decreases, the collector–emitter voltages of the output transistors change. Again, assuming the minimum base–collector voltage is zero for a BJT biased in the forward active region, then the maximum output voltage is $v_O(\text{max}) = V^+ - [V_{EB8}(\text{on}) + V_{BE11}(\text{on})]$. The minimum output voltage is similarly found to be $v_O(\text{min}) = V^- + [V_{BE4}(\text{on}) + V_{EB12}(\text{on})]$.

For the all-CMOS circuit in Figure 14.1(b), the maximum output voltage is $v_O(\text{max}) = V^+ - [(V_{SG8} + V_{TP8}) + V_{SG10}]$. The minimum output voltage is $v_O(\text{min}) = V^- + [(V_{GS6} - V_{TN6}) + V_{GS13}]$.

## Test Your Understanding

**TYU 14.1** Using the circuit and transistor parameters of Example 13.11, and assuming threshold voltages of $V_{TN} = 0.5$ V and $V_{TP} = -0.5$ V, determine the maximum range of common-mode input voltage for the all-CMOS folded cascode circuit of Figure 14.1(b). (Ans. $V^- - 0.184 \leq v_{CM} \leq V^+ - 1.13$ V)

**TYU 14.2** Using the same circuit and transistor parameters as in Exercise TYU14.1, calculate the maximum range of output voltage for the all-CMOS folded cascode circuit of Figure 14.1(b). (Ans. $V^- + 0.54$ V $\leq v_O \leq V^+ - 1.13$ V)

## 14.2    FINITE OPEN-LOOP GAIN

**Objective:** • Analyze the effect of finite open-loop gain.

In the ideal op-amp, the open-loop gain is infinite, the input differential resistance is infinite, and the output resistance is zero. None of these conditions exists in actual operational amplifiers. In the last chapter, we determined that the open-loop gain and input differential resistance may be large but finite, and the output resistance may be small but nonzero. In this section, we will determine the effect of a finite open-loop gain and input resistance on both the inverting and noninverting amplifier characteristics. We will then calculate the output resistance.

In this section, we limit our discussion of the finite open-loop gain to low frequency. In the next section, we consider the effect of finite gain as well as the frequency response of the amplifier.

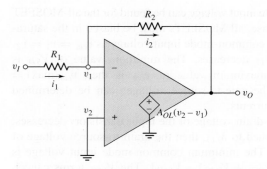

**Figure 14.2** Equivalent circuit, inverting amplifier with finite open-loop gain

### 14.2.1    Inverting Amplifier Closed-Loop Gain

The equivalent circuit of the inverting amplifier with a finite open-loop gain is shown in Figure 14.2. If the open-loop input resistance is assumed to be infinite, then $i_1 = i_2$, or

$$\frac{v_I - v_1}{R_1} = \frac{v_1 - v_O}{R_2} \tag{14.1(a)}$$

or

$$\frac{v_I}{R_1} = v_1 \left( \frac{1}{R_1} + \frac{1}{R_2} \right) - \frac{v_O}{R_2} \tag{14.1(b)}$$

Since $v_2 = 0$, the output voltage is

$$v_O = -A_{OL} v_1 \tag{14.2}$$

where $A_{OL}$ is the low-frequency open-loop gain. Solving for $v_1$ from Equation (14.2) and substituting the result into Equation (14.1(b)), we find

$$\frac{v_I}{R_1} = -\left( \frac{v_O}{A_{OL}} \right) \left( \frac{1}{R_1} + \frac{1}{R_2} \right) - \frac{v_O}{R_2} \tag{14.3}$$

The closed-loop voltage gain is then

$$A_{CL} = \frac{v_O}{v_I} = \frac{-\dfrac{R_2}{R_1}}{1 + \dfrac{1}{A_{OL}} \left( 1 + \dfrac{R_2}{R_1} \right)} \tag{14.4}$$

---

## EXAMPLE 14.1

**Objective:** Determine the minimum open-loop voltage gain to achieve a particular accuracy.

A pressure transducer produces a maximum dc voltage signal of 2 mV and has an output resistance of $R_S = 2\,\text{k}\Omega$. The maximum dc current from the transducer is to be limited to 0.2 μA. An inverting amplifier is to be used in conjunction with the transducer to produce an output voltage of $-0.10\,\text{V}$ for a 2 mV transducer signal. The error in the output voltage cannot be greater than 0.1 percent. Determine the minimum open-loop gain of the amplifier to meet this specification.

**Solution:** We must first determine the resistor values to be used in the inverting amplifier. The source resistor is in series with $R_1$, so let

$$R_1' = R_1 + R_S$$

The minimum input resistance is found from the maximum input current as

$$R_1'(\text{min}) = \frac{v_i}{i_i(\text{max})} = \frac{2 \times 10^{-3}}{0.2 \times 10^{-6}} = 10 \times 10^3\,\Omega = 10\,\text{k}\Omega$$

The resistor $R_1$ then needs to be $8\,\text{k}\Omega$. The closed-loop voltage gain required is

$$A_{CL} = \frac{v_O}{v_i} = \frac{-0.10}{2 \times 10^{-3}} = -50 = \frac{-R_F}{R_1'}$$

The required value of the feedback resistor is then $R_F = 500\,\text{k}\Omega$.

For the voltage gain to be within 0.1 percent, the minimum gain (magnitude) is 49.95. Using Equation (14.4), we can determine the minimum value of the open-loop gain. We have

$$A_{CL} = \frac{\dfrac{-R_2}{R_1'}}{1 + \dfrac{1}{A_{OL}}\left(1 + \dfrac{R_2}{R_1'}\right)} = -49.95 = \frac{-50}{1 + \dfrac{1}{A_{OL}}(51)}$$

which yields $A_{OL}(\text{min}) = 50{,}949$.

**Comment:** If the open-loop gain is greater than the value of $A_{OL}(\text{min}) = 50{,}949$, then the error in the voltage gain will be less than 0.1 percent.

## EXERCISE PROBLEM

**Ex 14.1:** Consider an inverting amplifier in which the op-amp open-loop gain is $A_{OL} = 2 \times 10^5$ and the ideal closed-loop amplifier gain is $A_{CL}(\infty) = -40$. (a) Determine the actual closed-loop gain. (b) Repeat part (a) if the open-loop gain is $A_{OL} = 5 \times 10^4$. (c) What is the percent change between the magnitudes of the actual gains from part (a) to part (b)? (Ans. (a) $-39.9918$, (b) $-39.9672$, (c) $-0.0615\%$)

In the limit as $A_{OL} \to \infty$, the closed-loop gain is equal to the ideal value, designated $A_{CL}(\infty)$, which for the inverting amplifier is

$$A_{CL}(\infty) = -\frac{R_2}{R_1} \tag{14.5}$$

as previously determined. Equation (14.4) is then

$$A_{CL} = \frac{A_{CL}(\infty)}{1 + \dfrac{1 - A_{CL}(\infty)}{A_{OL}}} \tag{14.6}$$

To determine the variation in closed-loop gain with changes in open-loop gain, we take the derivative of $A_{CL}$ with respect to $A_{OL}$. We find

$$\frac{dA_{CL}}{dA_{OL}} = \frac{A_{CL}(\infty)(1 - A_{CL}(\infty))}{[A_{OL} + (1 - A_{CL}(\infty))]^2} \tag{14.7}$$

which can be rearranged in the form

$$\frac{dA_{CL}}{A_{CL}} = \frac{dA_{OL}}{A_{OL}} \frac{\frac{1 - A_{CL}(\infty)}{A_{OL}}}{1 + \left(\frac{1 - A_{CL}(\infty)}{A_{OL}}\right)} \tag{14.8}$$

Normally, $A_{CL}(\infty)| \ll |A_{OL}|$ and Equation (14.8) is approximately

$$\frac{dA_{CL}}{A_{CL}} \cong \frac{dA_{OL}}{A_{OL}} \frac{1 - A_{CL}(\infty)}{A_{OL}} \tag{14.9}$$

Equation (14.9) relates the percent change in the closed-loop gain of the inverting amplifier as the result of a change in open-loop gain. Open-loop gain variations occur when individual transistor parameters change from one circuit to another or with temperature.

From Equation (14.9), we see that changes in closed-loop gain become smaller as the open-loop gain becomes larger.

### 14.2.2    Noninverting Amplifier Closed-Loop Gain

Figure 14.3 shows the equivalent circuit of the noninverting amplifier with a finite open-loop gain. Again, the open-loop input differential resistance is assumed to be infinite. The analysis proceeds in much the same way as in the previous section. We have $i_1 = i_2$, and

$$-\frac{v_1}{R_1} = \frac{v_1 - v_O}{R_2} \tag{14.10(a)}$$

or

$$\frac{v_O}{R_2} = v_1 \left(\frac{1}{R_1} + \frac{1}{R_2}\right) \tag{14.10(b)}$$

The output voltage is

$$v_O = A_{OL}(v_2 - v_1) \tag{14.11}$$

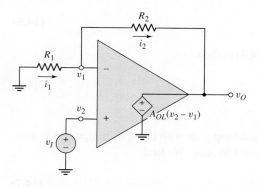

**Figure 14.3** Equivalent circuit, noninverting amplifier with finite open-loop gain

Since $v_2 = v_I$, voltage $v_1$ can be written

$$v_1 = v_I - \frac{v_O}{A_{OL}} \tag{14.12}$$

Combining Equations (14.12) and (14.10(b)) and rearranging terms, we have an expression for the closed-loop voltage gain:

$$A_{CL} = \frac{v_O}{v_I} = \frac{1 + \dfrac{R_2}{R_1}}{1 + \dfrac{1}{A_{OL}}\left(1 + \dfrac{R_2}{R_1}\right)} \tag{14.13}$$

In the limit as $A_{OL} \rightarrow \infty$, the ideal closed-loop gain is

$$A_{CL}(\infty) = 1 + \frac{R_2}{R_1} \tag{14.14}$$

and Equation (14.13) becomes

$$A_{CL} = \frac{A_{CL}(\infty)}{1 + \dfrac{A_{CL}(\infty)}{A_{OL}}} \tag{14.15}$$

Taking the derivative of the closed-loop gain with respect to the open-loop gain and rearranging terms, we obtain

$$\frac{dA_{CL}}{A_{CL}} = \frac{dA_{OL}}{A_{OL}}\left(\frac{A_{CL}}{A_{OL}}\right) \tag{14.16}$$

Equation (14.16) yields the fractional change in the closed-loop gain of the non-inverting amplifier as a result of a change in the open-loop gain. The result for the noninverting amplifier is very similar to that for the inverting amplifier.

### 14.2.3   Inverting Amplifier Closed-Loop Input Resistance

The closed-loop input resistance $R_{if}$ of the inverting amplifier is defined in Figure 14.4(a), and it includes the effect of feedback. The equivalent circuit, including a finite open-loop gain $A_{OL}$, finite open-loop input differential resistance $R_i$, and nonzero output resistance $R_o$, is shown in Figure 14.4(b).

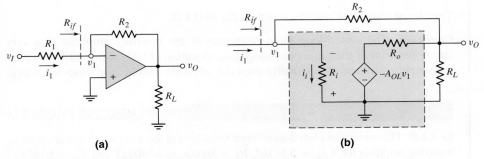

(a)                                                  (b)

**Figure 14.4**  (a) Inverting amplifier and (b) inverting amplifier equivalent circuit, for calculating closed-loop input resistance

A KCL equation at the output node yields

$$\frac{v_O}{R_L} + \frac{v_O - (-A_{OL}v_1)}{R_o} + \frac{v_o - v_1}{R_2} = 0 \tag{14.17}$$

Solving for the output voltage, we have

$$v_O = \frac{-v_1 \left( \dfrac{A_{OL}}{R_o} - \dfrac{1}{R_2} \right)}{\dfrac{1}{R_L} + \dfrac{1}{R_o} + \dfrac{1}{R_2}} \tag{14.18}$$

A KCL equation at the input node yields

$$i_1 = \frac{v_1}{R_i} + \frac{v_1 - v_O}{R_2} \tag{14.19}$$

Combining Equations (14.18) and (14.19) and rearranging terms produces

$$\frac{i_1}{v_1} = \frac{1}{R_{if}} = \frac{1}{R_i} + \frac{1}{R_2} \frac{1 + A_{OL} + \dfrac{R_o}{R_L}}{1 + \dfrac{R_o}{R_L} + \dfrac{R_o}{R_2}} \tag{14.20}$$

Equation (14.20) describes the closed-loop input resistance of the inverting amplifier, with a finite open-loop gain, finite open-loop input resistance, and nonzero output resistance. In the limit as $A_{OL} \to \infty$, we see that $1/R_{if} \to \infty$, or $R_{if} \to 0$, which means that $v_1 \to 0$, or $v_1$ is at virtual ground. This is a characteristic of an ideal inverting op-amp.

## EXAMPLE 14.2

**Objective:** Determine the closed-loop input resistance at the inverting terminal of an inverting amplifier.

Consider an inverting amplifier with a feedback resistor $R_2 = 10\,\text{k}\Omega$, and an op-amp with parameters $A_{OL} = 10^5$ and $R_i = 10\,\text{k}\Omega$. Assume the output resistance $R_o$ of the op-amp is negligible.

**Solution:** If $R_o = 0$, then Equation (14.20) becomes

$$\frac{1}{R_{if}} = \frac{1}{R_i} + \frac{1 + A_{OL}}{R_2} = \frac{1}{10^4} + \frac{1 + 10^5}{10^4} \cong 10^{-4} + 10 \tag{14.21}$$

The closed-loop input resistance is then $R_{if} \cong 0.1\,\Omega$.

**Comment:** The closed-loop input resistance of the inverting amplifier is a very strong function of the finite open-loop gain. Equation (14.21) shows that the open-loop input resistance $R_i$ essentially does not affect the closed-loop input resistance.

## EXERCISE PROBLEM

**Ex 14.2:** Determine the closed-loop input resistance at the inverting terminal of an inverting amplifier for $A_{OL} = 5 \times 10^4$, $R_2 = 80\,\text{k}\Omega$, $R_i = 40\,\text{k}\Omega$, and $R_L = 10\,\text{k}\Omega$ if (a) $R_o = 0$ and (b) $R_o = 1\,\text{k}\Omega$. (Ans. (a) $R_{if} = 1.6\,\Omega$, (b) $R_{if} = 1.78\,\Omega$)

A nonzero closed-loop input resistance $R_{if}$ in conjunction with a finite open-loop input resistance $R_i$ implies that the signal current into the op-amp is not zero, as assumed in the ideal case. From Figure 14.4(b), we see that

$$v_1 = i_1 R_{if} \tag{14.22}$$

Therefore,

$$i_i = \frac{v_1}{R_i} = i_1 \left( \frac{R_{if}}{R_i} \right) \tag{14.23}$$

The fraction of input signal current shunted away from $R_2$ and into the op-amp is $(R_{if}/R_i)$.

### 14.2.4 Noninverting Amplifier Closed-Loop Input Resistance

A noninverting amplifier is shown in Figure 14.5(a). The input resistance seen by the signal source is designated $R_{if}$. The equivalent circuit, including a finite open-loop gain $A_{OL}$, finite open-loop input differential resistance $R_i$ and non-zero output resistance $R_o$, is shown in Figure 14.5(b).

Writing a KCL equation at the output node yields

$$\frac{v_O}{R_L} + \frac{v_O - A_{OL}v_d}{R_o} + \frac{v_O - v_1}{R_2} = 0 \tag{14.24}$$

Solving for the output voltage, we have

$$v_O = \frac{\dfrac{v_1}{R_2} + \dfrac{A_{OL}v_d}{R_o}}{\dfrac{1}{R_L} + \dfrac{1}{R_o} + \dfrac{1}{R_2}} \tag{14.25}$$

A KCL equation at the $v_1$ node yields

$$i_I = \frac{v_1}{R_1} + \frac{v_1 - v_O}{R_2} \tag{14.26}$$

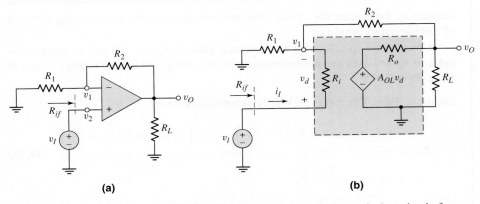

**(a)**             **(b)**

**Figure 14.5** (a) Noninverting amplifier and (b) noninverting amplifier equivalent circuit, for calculating closed-loop input resistance

Combining Equations (14.25) and (14.26) and rearranging terms, we obtain

$$i_I\left(1 + \frac{R_o}{R_L} + \frac{R_o}{R_2}\right) = v_1\left\{\left(\frac{1}{R_1} + \frac{1}{R_2}\right)\left(1 + \frac{R_o}{R_L} + \frac{R_o}{R_2}\right) - \frac{R_o}{R_2^2}\right\} - \frac{A_{OL}v_d}{R_2}$$

(14.27)

From Figure 14.5(b), we see that

$$v_d = i_1 R_i$$

(14.28)

and

$$v_1 = v_I - i_I R_i$$

(14.29)

Substituting Equations (14.28) and (14.29) into (14.27) we obtain an equation in $i_I$ and $v_I$ so that the input resistance $R_{if}$ can be found as

$$R_{if} = v_I / i_I$$

In order to simplify the algebra, we neglect the effect of $R_o$, which is normally small. Setting $R_o = 0$ reduces Equation (14.27) to

$$i_I = v_1\left(\frac{1}{R_1} + \frac{1}{R_2}\right) - \frac{A_{OL}v_d}{R_2}$$

(14.30)

Substituting Equations (14.28) and (14.29) into (14.30), we find that the input resistance can be written in the form

$$R_{if} = \frac{v_I}{i_I} = \frac{R_i(1 + A_{OL}) + R_2\left(1 + \dfrac{R_i}{R_1}\right)}{1 + \dfrac{R_2}{R_1}}$$

(14.31)

Equation (14.31) describes the closed-loop input resistance of the noninverting amplifier with a finite open-loop gain and a finite open-loop input resistance. In the limit as $A_{OL} \to \infty$, or as the open-loop input resistance approaches infinity, we see that $R_{if} \to \infty$, which is a property of the ideal noninverting amplifier.

### EXAMPLE 14.3

**Objective:** Determine the closed-loop input resistance at the noninverting terminal of a noninverting amplifier.

Consider an op-amp with an open-loop gain of $A_{OL} = 10^5$ and an input resistance of $R_i = 10\,\text{k}\Omega$ in a noninverting amplifier configuration with resistor values of $R_1 = R_2 = 10\,\text{k}\Omega$.

**Solution:** From Equation (14.31), the input resistance is

$$R_{if} = \frac{R_i(1 + A_{OL}) + R_2\left(1 + \dfrac{R_i}{R_1}\right)}{1 + \dfrac{R_2}{R_1}} = \frac{10(1 + 10^5) + 10\left(1 + \dfrac{10}{10}\right)}{1 + \dfrac{10}{10}}$$

(14.32)

or

$$R_{if} \cong 5 \times 10^5\,\text{k}\Omega \Rightarrow 500\,\text{M}\Omega$$

**Comment:** As expected, the closed-loop input resistance of the noninverting amplifier is very large. Equation (14.32) shows that the input resistance is dominated by the term $R_i(1 + A_{OL})$. The combination of a large $R_i$ and large $A_{OL}$ produces an extremely large input resistance, as predicted by ideal feedback theory.

---

### EXERCISE PROBLEM

**Ex 14.3:** For a noninverting amplifier, the resistances are $R_2 = 99\,\text{k}\Omega$ and $R_1 = 1\,\text{k}\Omega$. The op-amp properties are: $A_{OL} = 10^4$, $R_i = 40\,\text{k}\Omega$, and $R_o = 0$. Determine the closed-loop input resistance. (Ans. $R_{if} = 4.04\,\text{M}\Omega$)

## 14.2.5     Nonzero Output Resistance

Since the ideal op-amp has a zero output resistance, the output voltage is independent of the load impedance. The op-amp acts as an ideal voltage source and there is no loading effect. An actual op-amp circuit has a nonzero output resistance, which means that the output voltage, and therefore the closed-loop gain, is a function of the load impedance.

Figure 14.6 is the equivalent circuit of both an inverting and noninverting amplifier and is used to find the output resistance. The op-amp has a finite open-loop gain $A_{OL}$, a nonzero output resistance $R_o$, and an infinite input resistance $R_i$. To determine the output resistance, we set the independent input voltages equal to zero. A KCL equation at the output node yields

$$i_o = \frac{v_o - A_{OL}v_d}{R_o} + \frac{v_o}{R_1 + R_2} \qquad (14.33)$$

The differential input voltage is $v_d = -v_1$, where

$$v_1 = \left(\frac{R_1}{R_1 + R_2}\right)v_o \qquad (14.34)$$

Combining Equations (14.34) and (14.33), we have

$$i_o = \frac{v_o}{R_o} - \frac{A_{OL}}{R_o}\left[-\left(\frac{R_1}{R_1 + R_2}\right)v_o\right] + \frac{v_o}{R_1 + R_2} \qquad (14.35(a))$$

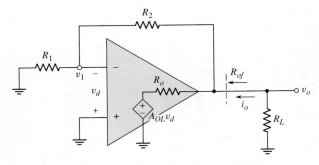

**Figure 14.6** Equivalent circuit for calculating closed-loop output resistance

or

$$\frac{i_o}{v_o} = \frac{1}{R_{of}} = \frac{1}{R_o}\left[1 + \frac{A_{OL}}{(1 + R_2/R_1)}\right] + \frac{1}{R_1 + R_2} \qquad \textbf{(14.35(b))}$$

Since $R_o$ is normally small and $A_{OL}$ is normally large, Equation (14.35b), to a good approximation, is as follows:

$$\frac{1}{R_{of}} \cong \frac{1}{R_o}\left[\frac{A_{OL}}{1 + R_2/R_1}\right] \qquad \textbf{(14.36)}$$

In most op-amp circuits, the open-loop output resistance $R_o$ is on the order of 100 Ω. Since $A_{OL}$ is normally much larger than $(1 + R_2/R_1)$, the closed-loop output resistance can be very small. Output resistance values in the milliohm range are easily attained.

## EXAMPLE 14.4

**Objective:** Determine the output resistance of an op-amp circuit.

**Computer Simulation Solution:** Figure 14.7 shows an inverting amplifier circuit with a standard 741 op-amp. One method of determining the output resistance is to measure the output voltage for two different values of load resistance connected to the output. Then, treating the amplifier as a Thevenin equivalent circuit with a fixed source in series with an output resistance, the output resistance can be determined. A 1 mV signal was applied. For a 10 Ω load, the output voltage is 0.999837 mV, and for a 20 Ω load, the output voltage is 0.9999132 mV. This gives an output resistance of 1.53 mΩ.

**Comment:** As mentioned, the output resistance of a voltage amplifier with negative feedback can be very small. The ideal output resistance is zero, but a practical op-amp circuit can have an output resistance in the milliohm range.

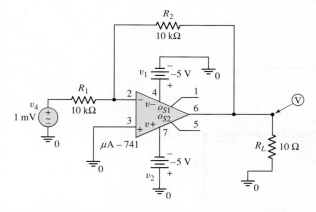

**Figure 14.7** Circuit using 741 op-amp to measure output resistance

EXERCISE PROBLEM

**Ex 14.4:** An op-amp with an open-loop gain of $A_{OL} = 10^5$ is used in a noninverting amplifier configuration with a closed-loop gain of $A_{CL} = 100$. Determine the closed-loop output resistance $R_{of}$ for: (a) $R_o = 100\ \Omega$, and (b) $R_o = 10\,k\Omega$. (Ans. (a) $R_{of} = 0.1\ \Omega$ (b) $R_{of} = 10\ \Omega$)

## Test Your Understanding

**TYU 14.3** The resistors in an inverting amplifier are $R_1 = 25\,k\Omega$ and $R_2 = 250\,k\Omega$. Determine the minimum open-loop op-amp gain if the closed-loop gain must be within (a) 0.1 percent of ideal and (b) 0.05 percent of ideal. (Ans. (a) 10,989, (b) 21,989)

**TYU 14.4** An operational amplifier connected in a noninverting configuration has an open-loop gain of $A_{OL} = 10^5$. The resistors are $R_2 = 495\,k\Omega$ and $R_1 = 5\,k\Omega$. (a) Determine the actual and ideal closed-loop gains. (b) If the open-loop gain decreases by 10 percent, determine the percent change in closed-loop gain and the actual closed-loop gain. (Ans. (a) $A_{CL} = 99.90$, $A_{CL}(\infty) = 100$ (b) 0.01 %, $A_{CL} = 99.89$)

**TYU 14.5** A noninverting amplifier has an op-amp with an open-loop gain of $A_{OL} = 2 \times 10^4$. (a) Determine the maximum ideal closed-loop gain such that the actual closed-loop gain is within 0.1 percent of the ideal closed-loop value. (b) Repeat part (a) for 0.05 percent. (Ans. (a) $A_{CL}(\infty) = 20.02$, (b) $A_{CL}(\infty) = 10.005$)

**TYU 14.6** Consider the equivalent circuit in Figure 14.4(b). If $R_i = 10\,k\Omega$, determine the percentage of input signal current $i_1$ shunted from $R_2$ for: (a) $R_{if} = 0.1\ \Omega$, and (b) $R_{if} = 10\ \Omega$. (Ans. (a) $10^{-3}\%$ (b) 0.1 %)

**TYU 14.7** Find the closed-loop input resistance of a voltage follower with op-amp characteristics $A_{OL} = 5 \times 10^5$, $R_i = 10\,k\Omega$, and $R_o = 0$. (Ans. $R_{if} = 5000\,M\Omega$)

## 14.3 FREQUENCY RESPONSE

**Objective:** • Analyze the open-loop and closed-loop frequency response.

In the previous chapter, we considered the basic op-amp frequency response. Frequency compensation was included as a means of stabilizing the circuit. In this section, we will consider the bandwidth and the transient response of the closed-loop amplifier.

When a step function is applied at the op-amp input, the output voltage cannot change instantaneously with time because of capacitance effects within the op-amp circuit. The maximum rate at which the output changes with time is called the **slew rate**. We will determine the factors that limit the slew rate.

### 14.3.1 Open-Loop and Closed-Loop Frequency Response

The frequency response of the open-loop gain can be written as

$$A_{OL}(f) = \frac{A_O}{1 + j\dfrac{f}{f_{PD}}} \tag{14.37}$$

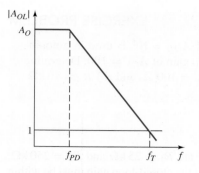

Figure 14.8  Bode plot, open-loop gain magnitude

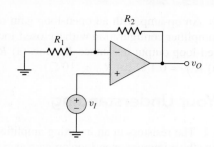

Figure 14.9  Noninverting amplifier

where $A_O$ is the low-frequency open-loop gain and $f_{PD}$ is the dominant-pole frequency. Figure 14.8 shows the Bode plot of the open-loop gain magnitude. The dominant-pole frequency $f_{PD}$ is shown as well as the unity-gain bandwidth $f_T$. We showed previously that the unity-gain bandwidth is

$$f_T = f_{PD} A_O \tag{14.38}$$

and is also called the gain–bandwidth product. Equation (14.38) assumes that additional poles of the open-loop frequency response occur at higher frequencies than $f_T$.

Figure 14.9 shows a noninverting amplifier. In our discussion on feedback theory in Chapter 12, we found that, assuming ideal feedback, the closed-loop gain $A_{CL}$ can be written

$$A_{CL} = \frac{A_{OL}}{(1 + \beta A_{OL})} \tag{14.39}$$

where $\beta$ is the feedback transfer function. For the noninverting amplifier, this feedback transfer function is

$$\beta = \frac{1}{1 + \dfrac{R_2}{R_1}} \tag{14.40}$$

Combining Equations (14.37), (14.40) and (14.39), we find the expression for the closed-loop gain as a function of frequency, as follows:

$$A_{CL}(f) = \frac{A_O}{1 + \dfrac{A_O}{1 + (R_2/R_1)}} \times \frac{1}{1 + j\dfrac{f}{f_{PD}\left[1 + \dfrac{A_O}{1 + (R_2/R_1)}\right]}} \tag{14.41}$$

Normally, $A_O \gg [1 + (R_2/R_1)]$; therefore, the low-frequency closed-loop gain is

$$A_{CLO} = 1 + \frac{R_2}{R_1} \tag{14.42}$$

as previously determined. For $A_O \gg A_{CLO}$, Equation (14.41) is approximately

$$A_{CL}(f) = \frac{A_{CLO}}{1 + j\dfrac{f}{f_{PD}\left(\dfrac{A_O}{A_{CLO}}\right)}} \tag{14.43}$$

The 3 dB frequency, or small-signal bandwidth, is then

$$f_{3\,dB} = f_{PD}\left(\frac{A_O}{A_{CLO}}\right) \tag{14.44}$$

Since in most cases $A_O \gg A_{CLO}$, the bandwidth of the closed-loop system is substantially larger than the open-loop dominant-pole frequency $f_{PD}$. Note also that Equation (14.44) applies to the inverting, as well as the noninverting, amplifier in which $A_{CLO}$ is the magnitude of the closed-loop gain. We have seen this same bandwidth extension for negative feedback several times previously.

## 14.3.2  Gain–Bandwidth Product

We can also determine the unity-gain bandwidth of the closed-loop system. From Equation (14.43), we can write

$$|A_{CL}(f = f_{\text{unity}})| = 1 = \frac{A_{CLO}}{\sqrt{1 + \left[\dfrac{f_{\text{unity}}}{f_{PD}(A_O/A_{CLO})}\right]^2}} \tag{14.45}$$

where $f_{\text{unity}}$ is the unity-gain frequency of the closed-loop system.

If $A_{CLO} \gg 1$, then Equation (14.45) yields

$$\frac{f_{\text{unity}}}{f_{PD}\left(\dfrac{A_O}{A_{CLO}}\right)} \cong A_{CLO} \tag{14.46(a)}$$

which reduces to

$$f_{\text{unity}} = A_{CLO}\, f_{PD}\left(\frac{A_O}{A_{CLO}}\right) = f_{PD} A_O = f_T \tag{14.46(b)}$$

The unity-gain frequency or bandwidth of the closed-loop system is essentially the same as that of the open-loop amplifier.

The open-loop and closed-loop frequency response curves are shown in Figure 14.10. We observed these same results in Chapter 12 in the discussion on ideal feedback theory.

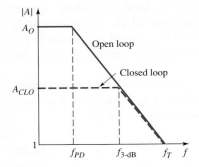

**Figure 14.10** Bode plot, open-loop and closed-loop gain magnitude

### EXAMPLE 14.5

**Objective:** Determine the unity-gain bandwidth and the maximum closed-loop gain for a specified closed-loop bandwidth.

An audio amplifier system is to use an op-amp with an open-loop gain of $A_O = 2 \times 10^5$ and a dominant-pole frequency of 5 Hz. The bandwidth of the audio system is to be 20 kHz. Determine the maximum closed-loop gain for the audio amplifier.

**Solution:** The unity-gain bandwidth is found as

$$f_T = f_{PD} A_O = (5)(2 \times 10^5) = 10^6 \text{ Hz} \Rightarrow 1 \text{ MHz}$$

Since the gain–bandwidth product is a constant, we have

$$f_{3\text{-dB}} \cdot A_{CL} = f_T$$

where $f_{3\text{-dB}}$ is the closed-loop bandwidth and $A_{CL}$ is the closed-loop gain. The maximum closed-loop gain is then

$$A_{CL} = \frac{f_T}{f_{3\text{-dB}}} = \frac{10^6}{20 \times 10^3} = 50$$

**Comment:** If the closed-loop gain is less than or equal to 50, then the required bandwidth of 20 kHz for the audio amplifier will be realized.

### EXERCISE PROBLEM

**Ex 14.5:** An op-amp with open-loop parameters of $A_{OL} = 2 \times 10^5$ and $f_{PD} = 5\,\text{Hz}$ is connected in a noninverting amplifier configuration with a low-frequency closed-loop gain of $A_{CLO} = 30$. An input voltage signal of $v_I = 100 \sin(2\pi f t)\ \mu\text{V}$ is applied. (a) What is the closed-loop bandwidth? (b) What is the low-frequency output voltage? (c) Determine the peak amplitude of the output voltage if the frequency of the signal is (i) $f = 5\,\text{kHz}$, (ii) $f = 50\,\text{kHz}$, and (iii) $f = 200\,\text{kHz}$. (Ans. (a) $f_{3\text{-dB}} = 33.3\,\text{kHz}$; (b) $v_O = 3 \sin(2\pi f t)\,\text{mV}$; (c) (i) $\cong 3\,\text{mV}$, (ii) 1.663 mV, (iii) 0.493 mV)

### 14.3.3    Slew Rate

Implicit in the frequency response analysis for the closed-loop amplifier is the assumption that the sinusoidal input signals are small. If a large sinusoidal signal or step function is applied to an op-amp circuit, the input stage can be overdriven and the small-signal model will no longer apply.

Figure 14.11 shows a simplified op-amp circuit. If a large step voltage (greater than 120 mV) is applied at $v_2$ with $v_1$ held at ground potential, then $Q_2$ is effectively cut off, which means $i_{C2} \cong 0$ and $i_{C1} \cong I_Q$. The entire bias current is switched to $Q_1$. Since $i_{C3} \cong i_{C1}$, then $i_{C3} \cong I_Q$; since $Q_3 - Q_4$ form a current mirror, then we also have $i_{C4} \cong I_Q$.

The base current into $Q_5$ is very small; therefore, the current through the compensation capacitor $C_1$ is $i_O = i_{C4} = I_Q$. Since the voltage gain of the emitter-follower output stage is essentially unity, the capacitor current can be written as

$$i_O = C_1 \frac{d(v_O - v_{O1})}{dt} \tag{14.47}$$

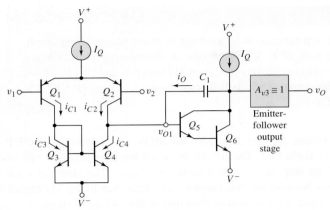

Figure 14.11  Simplified op-amp for calculating slew rate

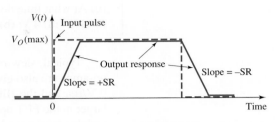

**Figure 14.12** Slew-rate-limited response of voltage follower to rectangular input voltage pulse

The gain of the second stage is large, which means that $v_{O1} \ll v_O$. Equation (14.47) then becomes

$$i_O \cong C_1 \frac{dv_O}{dt} = I_Q \tag{14.48}$$

or

$$\frac{dv_O}{dt} = \frac{I_Q}{C_1} \tag{14.49}$$

The maximum current through the compensation capacitor is limited to the bias current $I_Q$; consequently, the maximum rate at which the output voltage can change is also limited by the bias current $I_Q$.

The maximum rate of change of the output voltage is the slew rate of the op-amp, the units of which are usually given as volts per microsecond. From Equation (14.49), we have

$$\text{Slew rate (SR)} = \left(\frac{dv_O}{dt}\right)_{max} = \frac{I_Q}{C_1} \tag{14.50}$$

Although the rate of change in output voltage can be either positive or negative, the slew rate is *defined* as a positive quantity.

Figure 14.12 shows the slew-rate limited response of an op-amp voltage follower to a rectangular input voltage pulse. Note the trapezoidal shaped output response. The time needed to reach the full-scale response is approximately $V_O(\text{max})/\text{SR}$.

## EXAMPLE 14.6

**Objective:** Calculate the slew rate of the 741 op-amp.

From the previous chapter, the bias current in the 741 op-amp is $I_Q = 19 \, \mu\text{A}$ and the internal frequency compensation capacitor is $C_1 = 30 \, \text{pF}$.

**Solution:** From Equation (14.50), the slew rate is

$$\text{SR} = \frac{I_Q}{C_1} = \frac{19 \times 10^{-6}}{30 \times 10^{-12}} = 0.63 \times 10^6 \, \text{V/s} \Rightarrow 0.63 \, \text{V/}\mu\text{s}$$

**Comment:** The partial data sheet in Table 14.1 for the 741 op-amp lists the typical slew rate as 0.7 V/$\mu$s, which is in close agreement with our calculated value.

**Ex 14.6:** A 0.5 V input step function is applied at $t = 0$ to a noninverting amplifier with a closed-loop gain of 8. The slew rate of the op-amp is 1.25 V/$\mu$s. Determine the output voltage at (a) (i) $t = 2\,\mu$s, (ii) $t = 4\,\mu$s, and (iii) $t = 6\,\mu$s. (b) At what time does the output reach its full scale response? (Ans. (a) (i) 2.5 V, (ii) 4 V, (iii) 4 V; (b) 3.2 $\mu$s)

Typical slew-rate values for the CA3140 BiCMOS and LH0042C BiFET op-amps are also given in Table 14.1. The BiCMOS circuit has a typical slew rate of 9 V/$\mu$s, and the BiFET op-amp has a typical value of 3 V/$\mu$s. The slew rates are larger in the FET op-amps because the bias currents are larger than in the 741 circuit and the gain of the FET input stage is smaller than that of the 741 input stage.

The slew rate is directly related to the unity-gain bandwidth. To explain, the unity-gain bandwidth is directly proportional to the dominant-pole frequency, or $f_T \propto f_{PD}$. In turn, the dominant-pole frequency is inversely proportional to $R_{eq}C_1$, where $R_{eq}$ is the equivalent resistance at the node of the second stage input and $C_1$ is the compensation capacitance. The equivalent resistance $R_{eq}$ is a function of the second stage input resistance and the diff-amp stage output resistance, both of which are inversely proportional to $I_Q$. Then,

$$f_T \propto f_{PD} \propto \frac{1}{R_{eq}C_1} \propto \frac{1}{\left(\dfrac{1}{I_Q}\right)C_1} \propto \frac{I_Q}{C_1} \tag{14.51}$$

where $I_Q/C_1$ is the slew rate. Equation (14.51) shows that the slew rate is directly proportional to the unity-gain bandwidth.

Now consider what happens when a sinosoidal input signal is applied, for example, to the noninverting amplifier shown in Figure 14.9. If $v_I = V_p \sin \omega t$, then

$$v_O(t) = V_P\left(1 + \frac{R_2}{R_1}\right)\sin \omega t = V_{po} \sin \omega t \tag{14.52}$$

where $V_{po}$ is the ideal peak value of the sinusoidal output voltage.

The rate at which the output voltage changes is

$$\frac{dv_O(t)}{dt} = \omega V_{po} \cos \omega t \tag{14.53}$$

Therefore, the maximum rate of change is $\omega V_{po}$. Figure 14.13 shows two sinusoidal waveforms of the same frequency but different peak amplitudes. The maximum rate of

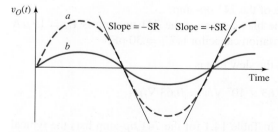

**Figure 14.13** Two sinusoidal waveforms of the same frequency with different peak voltages, showing different maximum slopes

change, or slope, occurs as the curves cross the zero axis. The waveform with the larger peak value has a larger maximum slope. Curve $a$ in Figure 14.13 has a maximum slope corresponding to the slew rate; curve $b$, with a smaller peak value, has a maximum slope less than the slew rate. If the maximum slope, $\omega V_{po}$, is greater than the slew rate SR, then the op-amp is slew-rate-limited and the output signal is distorted.

Thus, the maximum frequency at which the op-amp can operate without being slew-rate-limited is a function of both the frequency and peak amplitude of the signal. We have that

$$\omega_{max} V_{po} = 2\pi f_{max} V_{po} = SR \tag{14.54(a)}$$

or

$$f_{max} = \frac{SR}{2\pi V_{po}} \tag{14.54(b)}$$

As the output voltage peak amplitude increases, the maximum frequency at which slew-rate-limiting occurs decreases. The **full-power bandwidth (FPBW)** is the frequency at which the op-amp output becomes slew-rate-limited. The FPBW is the $f_{max}$ frequency from Equation (l4.54(b)), or

$$FPBW = \frac{SR}{2\pi V_{po}} \tag{14.55}$$

The full-power bandwidth can be considerably less than the small-signal bandwidth.

## EXAMPLE 14.7

**Objective:** Determine the small-signal bandwidth of an amplifier and the full-power bandwidth that will produce an undistorted output voltage.

Consider an amplifier with a unity-gain bandwidth of $f_T = 1$ MHz and a low-frequency closed-loop gain of $A_{CLO} = 10$. Assume the op-amp slew rate is SR $= 1$ V/$\mu$s and the desired peak output voltage is $V_{po} = 10$ V.

**Solution:** The small-signal closed-loop bandwidth is, from Equation (14.44),

$$f_{3\text{-dB}} = \frac{f_T}{A_{CLO}} = \frac{10^6}{10} \Rightarrow 100\,\text{kHz}$$

The full-power bandwidth, based on slew-rate limitation, from Equations (14.54(b)) and (14.55), is

$$f_{max} = FPBW = \frac{SR}{2\pi V_{po}} = \frac{(1\,\text{V}/\mu s)(10^6\,\mu s/s)}{2\pi(10)} \Rightarrow 15.9\,\text{kHz}$$

**Comment:** The full-power bandwidth, or the actual maximum frequency at which the system can be operated and still produce a large, undistorted output signal, is considerably smaller than the bandwidth under small-signal nonslew-rate-limiting conditions.

## EXERCISE PROBLEM

**Ex 14.7:** The slew rate of the 741 op-amp is 0.63 V/$\mu$s. Determine the full-power bandwidth for a peak undistorted output voltage of (a) 0.25 V, (b) 2 V, and (c) 8 V. (Ans. (a) 401 kHz, (b) 50.1 kHz, (c) 12.5 kHz)

## Test Your Understanding

**TYU 14.8** (a) An op-amp is connected in an inverting configuration. The parameters of the op-amp are $A_{OL} = 5 \times 10^4$, $f_{PD} = 15\,\text{Hz}$, and $SR = 0.8\,\text{V/}\mu\text{s}$. The low-frequency closed-loop gain is $|A_{CLO}| = 25$. (i) What is $f_{3\text{-dB}}$ of the closed-loop system? (ii) If $f_{max} = f_{3\text{-dB}}$, determine the maximum undistorted output voltage amplitude. (b) Repeat part (a) if the op-amp parameters are $A_{OL} = 5 \times 10^5$, $f_{PD} = 10\,\text{Hz}$, and $SR = 0.8\,\text{V/}\mu\text{s}$. (Ans. (a) (i) $f_{3\text{-dB}} = 30\,\text{kHz}$, (ii) $V_{po} = 4.24\,\text{V}$; (b) (i) $f_{3\text{-dB}} = 200\,\text{kHz}$, (ii) $V_{po} = 0.637\,\text{V}$)

## 14.4    OFFSET VOLTAGE

**Objective:** • Define and analyze sources and effects of offset voltage.

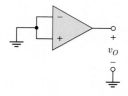

**Figure 14.14** Circuit for measuring output offset voltage

In Chapter 11, we analyzed the basic difference amplifier, which is the input stage of the op-amp. In that analysis, we assumed the input differential-pair transistors to be identical, or matched. If the two input devices are mismatched, the currents in the two branches of the diff-amp are unequal and this affects the diff-amp dc output voltage. In fact, the internal circuitry of the entire op-amp usually contains imbalances and asymmetries, all of which can cause a nonzero output voltage for a zero input differential voltage.

The **output dc offset voltage** is the measured open-loop output voltage when the input voltage is zero. This configuration is shown in Figure 14.14. The **input dc offset voltage** is defined as the input differential voltage that must be applied to the open-loop op-amp to produce a zero output voltage. This configuration is shown in Figure 14.15. The input offset voltage is the parameter most often specified and is usually referred to simply as the offset voltage.

Offset voltage values have a statistical distribution among op-amps of the same type, and the offset voltage polarity may vary from one op-amp to another. The offset voltage specification for an op-amp is the magnitude of the maximum offset voltage for a particular type of op-amp. The offset voltage is a dc value, generally in the range of 1 to 2 mV for bipolar op-amps, although some op-amps may have offset voltages in the range of 5 to 10 mV. Further, the maximum offset voltage specification for a precision op-amp may be as low as 10 $\mu$V.

In this section we will analyze offset voltage effects in the input diff-amp stage and will then consider various techniques used to compensate for offset voltage.

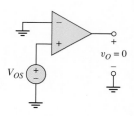

**Figure 14.15** Circuit for measuring input offset voltage

### 14.4.1    Input Stage Offset Voltage Effects

Several possible mismatches in the input diff-amp stage can produce offset voltages. We will analyze offset voltage effects in two bipolar input stages and in a MOSFET input diff-amp circuit.

**Basic Bipolar Diff-Amp Stage**
A basic bipolar diff-amp is shown in Figure 14.16. The differential pair is biased with a constant-current source. If $Q_1$ and $Q_2$ are matched, then for $v_1 = v_2 = 0$, $I_Q$ splits evenly between the two transistors and $i_{C1} = i_{C2}$. If a two-sided output is defined as

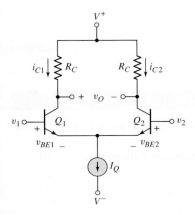

**Figure 14.16** Basic bipolar difference amplifier

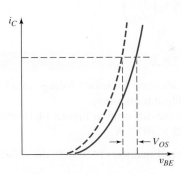

**Figure 14.17** The $i_C$ versus $v_{BE}$ characteristics for two unmatched bipolar transistors

the difference in voltage between the two collector terminals, then $v_O = 0$ when the transistors are matched and the collector resistors are matched, which means that the offset voltage is zero.

The collector currents can be written as

$$i_{C1} = I_{S1}e^{v_{BE1}/V_T} \tag{14.56(a)}$$

and

$$i_{C2} = I_{S2}e^{v_{BE2}/V_T} \tag{14.56(b)}$$

where $I_{S1}$ and $I_{S2}$ are related to the reverse-saturation currents in the B–E junctions and are functions of the electrical and geometric transistor properties. If the two transistors are exactly matched, then $I_{S1} = I_{S2}$; if there is any mismatch in the electrical or geometric parameters, then $I_{S1} \neq I_{S2}$.

The input offset voltage is defined as the input differential voltage required to produce a zero output voltage, or in this case to produce $i_{C1} = i_{C2}$. Figure 14.17 shows the $i_C$ versus $v_{BE}$ characteristics of two unmatched transistors. Slightly different B–E voltages must be applied to produce equal collector currents that will result in a zero output voltage in the diff-amp.

For $i_{C1} = i_{C2}$, we have

$$I_{S1}e^{v_{BE1}/V_T} = I_{S2}e^{v_{BE2}/V_T} \tag{14.57}$$

or

$$e^{(v_{BE1}-v_{BE2})/V_T} = \frac{I_{S2}}{I_{S1}} \tag{14.58}$$

We define the offset voltage as

$$v_{BE1} - v_{BE2} \equiv V_{OS}$$

Since $v_1 - v_2 = v_{BE1} - v_{BE2}$, then the offset voltage $V_{OS}$ is the differential input voltage that must be applied to produce $i_{C1} = i_{C2}$.

Equation (14.58) can then be written as

$$e^{V_{OS}/V_T} = \frac{I_{S2}}{I_{S1}} \tag{14.59(a)}$$

or

$$V_{OS} = V_T \ln\left(\frac{I_{S2}}{I_{S1}}\right)$$                                (14.59(b))

## EXAMPLE 14.8

**Objective:** Calculate the offset voltage in a bipolar diff-amp for a given mismatch between the input transistors.

Consider the diff-amp in Figure 14.16 with transistor parameters $I_{S1} = 10^{-14}$ A and $I_{S2} = 1.05 \times 10^{-14}$ A.

**Solution:** From Equation (14.59(b)), the offset voltage is

$$V_{OS} = V_T \ln\left(\frac{I_{S2}}{I_{S1}}\right) = (0.026) \ln\left(\frac{1.05 \times 10^{-14}}{1 \times 10^{-14}}\right) = 0.00127\, V \Rightarrow 1.27\, mV$$

**Comment:** A 5 percent difference in $I_S$ for $Q_1$ and for $Q_2$ produces an offset voltage of 1.27 mV. Since the offset voltage is defined as a positive quantity, if in the previous example $I_{S1}$ were 5 percent larger than $I_{S2}$, the offset voltage would also be 1.27 mV.

## EXERCISE PROBLEM

**Ex 14.8:** Consider the bipolar diff-amp in Figure 14.16. For $Q_1$, assume $I_{S1} = 2 \times 10^{-15}$ A. If the offset voltage is $V_{OS} = 2$ mV, what is the percent difference in the value of $I_{S2}$ compared to $I_{S1}$? (Ans. 8 percent)

It should be cautioned that the offset voltage in this example is one component of the offset voltage for the entire op-amp. For example, if the two collector resistors are not equal, then the two-sided output voltage $v_O$ will not be zero even if the two transistors are identical. Nevertheless, the calculation provides information on one source of offset voltage, as well as the resulting magnitude of $V_{OS}$.

### Bipolar Active Load Diff-Amp Stage

Figure 14.18 shows a bipolar diff-amp with a simple two-transistor active load. As before, this input stage is biased with a constant-current source. If $Q_1$ and $Q_2$ are matched and if $Q_3$ and $Q_4$ are matched, then $I_Q$ splits evenly between $Q_1$ and $Q_2$ for $v_1 = v_2$, and the E–C voltages of $Q_3$ and $Q_4$ are equal. The one-sided dc output voltage $v_O$ will therefore be one E–B voltage below $V^+$.

If, however, $Q_3$ and $Q_4$ are not exactly matched, then $i_{C1}$ and $i_{C2}$ may not be equal since the active load influences the split in the bias current, even if $Q_1$ and $Q_2$ are matched. This effect is caused by a finite Early voltage. Taking the Early voltages into account, but neglecting base currents, we can write the collector currents as

$$i_{C1} = i_{C3} = I_{S1}(e^{v_{BE1}/V_T})\left(1 + \frac{v_{CE1}}{V_{A1}}\right)$$

$$= I_{S3}(e^{v_{EB3}/V_T})\left(1 + \frac{v_{EC3}}{V_{A3}}\right)$$                                (14.60(a))

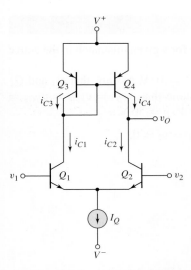

**Figure 14.18**  Basic bipolar diff-amp with active load

and

$$i_{C2} = i_{C4} = I_{S2}(e^{v_{BE2}/V_T})\left(1 + \frac{v_{CE2}}{V_{A2}}\right)$$

$$= I_{S4}(e^{v_{EB4}/V_T})\left(1 + \frac{v_{EC4}}{V_{A4}}\right) \tag{14.60(b)}$$

If we assume that $Q_1$ and $Q_2$ are matched, then $I_{S1} = I_{S2} \equiv I_S$ and $V_{A1} = V_{A2} \equiv V_{AN}$. Assume that $Q_3$ and $Q_4$ are slightly mismatched, so that $I_{S3} \neq I_{S4}$ but still assume that $V_{A3} = V_{A4} \equiv V_{AP}$. For $v_1 = v_2$, we have $v_{BE1} = v_{BE2}$; also, $v_{EB3} = v_{EB4} = v_{EC3} \equiv v_{EB}$. Taking the ratio of Equations (14.60(a)) and (14.60(b)) produces

$$\frac{i_{C1}}{i_{C2}} = \frac{1 + \dfrac{v_{CE1}}{V_{AN}}}{1 + \dfrac{v_{CE2}}{V_{AN}}} = \frac{I_{S3}}{I_{S4}}\frac{1 + \dfrac{v_{EB}}{V_{AP}}}{1 + \dfrac{v_{EC4}}{V_{AP}}} \tag{14.61}$$

Equation (14.61) can be rearranged in the form

$$\frac{1 + \dfrac{v_{CE1}}{V_{AN}}}{1 + \dfrac{v_{EB}}{V_{AP}}} = \frac{I_{S3}}{I_{S4}}\frac{1 + \dfrac{v_{CE2}}{V_{AN}}}{1 + \dfrac{v_{EC4}}{V_{AP}}} \tag{14.62}$$

Since $Q_3$ is connected as a diode, $v_{CE1}$ is a constant for a given bias current and supply voltage, which means that the left side of Equation (14.62) is a constant. If $I_{S3} = I_{S4}$, then $v_{CE2} = v_{CE1}$ and $v_{EC4} = v_{EB} = v_{EC3}$. However, if $I_{S3} \neq I_{S4}$, then the collector–emitter voltages on $Q_2$ and $Q_4$ must change. If, for example, $I_{S3} > I_{S4}$, then $v_{EC4}$ must increase and $v_{CE2}$ must decrease in order to keep Equation (14.62) balanced. If, on the other hand, $I_{S4} > I_{S3}$, then $v_{CE2}$ must increase and $v_{EC4}$ must decrease. A decrease in $v_{EC4}$ means that $Q_4$ may be driven into saturation by the mismatch.

### EXAMPLE 14.9

**Objective:** Calculate the change in output voltage for a given mismatch in the active load transistors.

Consider the diff-amp in Figure 14.18 with $V^+ = 10$ V. Assume that $Q_1$ and $Q_2$ are matched with $v_{BE1} = v_{BE2} = 0.6$ V, and assume that $v_{EB3} = v_{EB4} = v_{EC3} = 0.6$ V. Let $I_{S3} = 1.05 I_{S4}$. Also assume that $V_{AN} = V_{AP} = 50$ V.

**Solution:** Since $v_{EB3} = 0.6$ V $= v_{BE1}$, then for $v_1 = v_2 = 0$,

$$v_{CE1} = V^+ = 10 \text{ V}$$

The left side of Equation (14.62) is therefore

$$\frac{1 + \dfrac{v_{CE1}}{V_{AN}}}{1 + \dfrac{v_{EB}}{V_{AP}}} = \frac{1 + \dfrac{10}{50}}{1 + \dfrac{0.6}{50}} = 1.186$$

We have that

$$v_{EC4} + v_{CE2} = V^+ + v_{BE2} = 10.6 \text{ V}$$

or

$$v_{CE2} = 10.6 - v_{EC4}$$

Equation (14.62) then becomes

$$1.186 = 1.05 \frac{1 + \dfrac{10.6 - v_{EC4}}{50}}{1 + \dfrac{v_{EC4}}{50}}$$

which yields

$$v_{EC4} = 1.94 \text{ V}$$

**Comment:** A 5 percent difference between the properties of $Q_3$ and $Q_4$ produces a change from 0.6 to 1.94 V in the E–C voltage of $Q_4$.

**Computer Simulation Verification:** A PSpice analysis of the offset voltage effects in the active load diff-amp was performed. The two input terminals are at ground potential.

Using $I_S = 5 \times 10^{-15}$ A for all transistors, the PSpice analysis shows that $v_{EB3} = 0.654$ V rather than the assumed value of 0.6 V. Also, $v_{EC4}$ is 1.19 V rather than equal to $v_{EB3}$. This occurs because the circuit is slightly unbalanced; that is, $i_{C1}$ includes the base currents of $Q_3$ and $Q_4$, and $i_{C4}$ does not. When $Q_3$ and $Q_4$ are not matched and $I_{S3} = 1.05 I_{S4} = 5.25 \times 10^{-15}$ A, then $v_{EC4}$ increases to 2.51 V, compared to 1.94 V from the hand analysis. If, however, $I_{S3} = 0.95 I_{S4} = 4.75 \times 10^{-15}$ A, then $Q_4$ goes into saturation.

### EXERCISE PROBLEM

***Ex 14.9:** Consider the active load bipolar diff-amp stage in Figure 14.18. Assume the circuit and transistor parameters are as given in Example 14.9. Using Equations (14.60(a)) and (14.60(b)), determine the offset voltage $V_{OS} = |v_{BE2} - v_{BE1}|$ such that $v_{EC3} = v_{EC4}$ and $v_{CE1} = v_{CE2}$. (Ans. 1.27 mV)

An offset voltage that will slightly change $i_{C1}$ and $i_{C2}$ will allow the E–C voltage of $Q_4$ to be adjusted back to its original value.

As shown in actual op-amp circuits, resistors are usually included in the emitters of the active load transistors. By producing a slight imbalance in the two resistor values, we can change the ratio of $i_{C1}$ to $i_{C2}$, causing a change in the output voltage. This is discussed in the next section when offset voltage null adjustment is discussed.

### MOSFET Diff-Amp Stage

Figure 14.19 shows a basic MOSFET diff-amp in which the differential pair is biased with a constant-current source. If $M_1$ and $M_2$ are matched, then for $v_1 = v_2 = 0$, $I_Q$ splits evenly between the two transistors and $i_{D1} = i_{D2}$. Since a two-sided output is the voltage difference between the two drain terminals, then for this symmetrical situation, $v_O = 0$ and the offset voltage is zero.

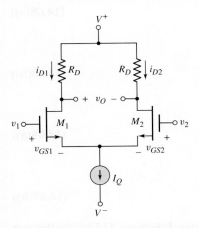

**Figure 14.19** Basic MOSFET diff-amp

The drain currents can be written as

$$i_{D1} = K_{n1}(v_{GS1} - V_{TN1})^2 \tag{14.63(a)}$$

and

$$i_{D2} = K_{n2}(v_{GS2} - V_{TN2})^2 \tag{14.63(b)}$$

As previously stated, the conduction parameters $K_{n1}$ and $K_{n2}$ are functions of the electrical and geometric properties of the two transistors, and the threshold voltages $V_{TN1}$ and $V_{TN2}$ are also functions of the transistor electrical properties. If there is a mismatch in electrical or geometric parameters, then we may have $K_{n1} \neq K_{n2}$ and $V_{TN1} \neq V_{TN2}$.

As with the bipolar diff-amp, the input offset voltage is defined as the input differential voltage that must be applied to produce a zero output voltage, or

$$V_{OS} = v_{GS1} - v_{GS2} \tag{14.64}$$

When the offset voltage is applied, $i_{D1} = i_{D2} = I_Q/2$; when the two drain resistors are equal, then $v_O = 0$. Solving Equations (14.63(a)) and (14.63(b)) for $v_{GS1}$ and $v_{GS2}$ and substituting the results into Equation (14.64), we find

$$V_{OS} = \sqrt{\frac{i_{D1}}{K_{n1}}} + V_{TN1} - \left( \sqrt{\frac{i_{D2}}{K_{n2}}} + V_{TN2} \right) \tag{14.65}$$

The various difference and average quantities are defined as follows:

$$\Delta K_n = K_{n1} - K_{n2} \tag{14.66(a)}$$

$$K_n = \frac{K_{n1} + K_{n2}}{2} \tag{14.66(b)}$$

$$\Delta V_{TN} = V_{TN1} - V_{TN2} \tag{14.67(a)}$$

and

$$V_{TN} = \frac{V_{TN1} + V_{TN2}}{2} \tag{14.67(b)}$$

Combining Equations (14.66(a)) and (14.66(b)), we have

$$K_{n1} = K_n + \frac{\Delta K_n}{2} \tag{14.68(a)}$$

and

$$K_{n2} = K_n - \frac{\Delta K_n}{2} \tag{14.68(b)}$$

Similarly,

$$V_{TN1} = V_{TN} + \frac{\Delta V_{TN}}{2} \tag{14.69(a)}$$

and

$$V_{TN2} = V_{TN} - \frac{\Delta V_{TN}}{2} \tag{14.69(b)}$$

Noting that $i_{D1} = i_{D2} = I_Q/2$ and substituting Equations (14.68(a)) through (14.69(b)) into Equation (14.65), we obtain

$$V_{OS} = \sqrt{\frac{I_Q}{2}} \left[ \frac{1}{\sqrt{K_n + (\Delta K_n/2)}} - \frac{1}{\sqrt{K_n - (\Delta K_n/2)}} \right] + \Delta V_{TN} \tag{14.70}$$

If we assume that $\Delta K_n \ll K_n$ then Equation (14.70) reduces to

$$V_{OS} = -\frac{1}{2} \sqrt{\frac{I_Q}{2K_n}} \cdot \left( \frac{\Delta K_n}{K_n} \right) + \Delta V_{TN} \tag{14.71}$$

Equation (14.71) is the offset voltage in a MOSFET diff-amp as a function of the differences in conduction parameters and threshold voltages.

## EXAMPLE 14.10

**Objective:** Calculate the offset voltage in a MOSFET diff-amp stage for a given mismatch between input transistors.

Consider the diff-amp in Figure 14.19 with transistor parameters $K_{n1} = 105\,\mu\text{A/V}^2$, $K_{n2} = 100\,\mu\text{A/V}^2$, and $V_{TN1} = V_{TN2}$. Assume $I_Q = 200\,\mu\text{A}$.

**Solution:** From Equation (14.66(a)), the difference in conduction parameters is

$$\Delta K_n = K_{n1} - K_{n2} = 105 - 100 = 5\,\mu\text{A/V}^2$$

From Equation (14.66(b)), the average of the conduction parameters is

$$K_n = \frac{K_{n1} + K_{n2}}{2} = \frac{105 + 100}{2} = 102.5 \,\mu\text{A/V}^2$$

The magnitude of the offset voltage is, from Equation (14.71),

$$|V_{OS}| = \frac{1}{2}\sqrt{\frac{I_Q}{2K_n}} \cdot \left(\frac{\Delta K_n}{K_n}\right) = \frac{1}{2}\sqrt{\frac{200}{2(102.5)}} \left(\frac{5}{102.5}\right) = 0.0241 \text{ V} \Rightarrow 24.1 \text{ mV}$$

**Comment:** A 5 percent difference in conduction parameter values between the input MOS transistors produces an offset voltage of 24.1 mV.

---

EXERCISE PROBLEM

**Ex 14.10:** Assume the MOSFET diff-amp shown in Figure 14.19 is biased with a current $I_Q = 150 \,\mu\text{A}$. Let $V_{TN1} = V_{TN2}$. Assume the nominal conduction parameter value is $K_n = 50 \,\mu\text{A/V}^2$. Determine the maximum variation $\Delta K_n$ such that the offset voltage is limited to $V_{OS} = 20$ mV. (Ans. $\Delta K_n = 1.63 \,\mu\text{A/V}^2$)

Comparing the results of Examples 14.8 and 14.10 shows that typically the offset voltage for a MOSFET diff-amp is substantially larger than that of a bipolar diff-amp. The difference can be explained by comparing Equation (14.71) for the MOSFET diff-amp and Equation (14.59(b)) for the bipolar diff-amp. The offset voltage for the MOSFET diff-amp is directly proportional to the percent change in conduction parameter values, whereas the offset voltage for the bipolar diff-amp is proportional to the logarithm of the percent change in the $I_S$ current parameters. In addition, the offset voltage for the MOSFET pair is proportional to

$$\sqrt{I_Q/K_n} = V_{GS} - V_{TN}$$

which is typically in the range of 0.3–2 V. In contrast, the offset voltage for the bipolar pair is proportional to

$$V_T \cong 26 \,\text{mV}$$

which is substantially smaller than $(V_{GS} - V_{TN})$. Thus, a MOSFET diff-amp inherently displays a higher input offset voltage than a bipolar pair for the same level of mismatch.

Partial data sheets showing some of the nonideal characteristics for the op-amps considered in the last chapter are in Table 14.1. The 741 op-amp, an all-bipolar circuit, has a maximum input offset voltage of 3 mV. The CA3140, which has a MOSFET input differential pair, has a maximum input offset voltage of 15 mV; and the LH0042C, which has a JFET input differential pair, has a maximum input offset voltage of 20 mV. This supports our conclusion that op-amps with FET input transistors have substantially larger input offset voltages than the all-bipolar circuit discussed.

## 14.4.2 Offset Voltage Compensation

In many applications, especially those for which the input signal is large compared to the offset voltage $V_{OS}$, the effect of the offset voltage is negligible. However, there are situations in which it is necessary to compensate for, or "null out," the offset voltage. Two such methods are: (a) an externally connected **offset compensation network,** and (2) an operational amplifier with **offset-null terminals.**

### External Offset Compensation Network

Figure 14.20 shows a simple network for offset voltage compensation in an inverting amplifier. The resistive voltage divider of $R_4$ and $R_5$, in conjunction with potentiometer $R_3$, is used to make voltage adjustments of either polarity at the noninverting terminal to cancel the effects of $V_{OS}$. If $R_5 \ll R_4$, then the compensating voltage applied to the noninverting terminal can be in the millivolt range, which is typical of offset voltage values.

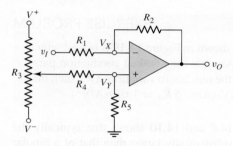

**Figure 14.20**  Offset voltage compensation circuit for inverting amplifier

## EXAMPLE 14.11

**Objective:** Determine the range of voltage produced by an offset voltage compensation network.

Consider the compensation network in Figure 14.20 with $R_5 = 100\,\Omega$, $R_4 = 100\,k\Omega$, and a 100 kΩ potentiometer $R_3$. Let $V^+ = 15$ V and $V^- = -15$ V. Determine the voltage range at $V_Y$.

**Solution:** Assume the potentiometer wiper arm is connected to the $V^+$ supply voltage. The voltage $V_Y$ is then

$$V_Y = \left( \frac{R_5}{R_5 + R_4} \right) V^+ = \left( \frac{0.1}{0.1 + 100} \right)(15) \Rightarrow 15\,\text{mV}$$

**Comment:** For this particular circuit, the compensation voltage range is $-15\,\text{mV}$ to $+15$ mV. A larger resistance $R_5$ will increase the offset voltage compensation range, and a smaller resistance $R_5$ will increase the sensitivity of offset voltage compensation.

## EXERCISE PROBLEM

**Ex 14.11:** Consider the compensation network in Figure 14.20. Assume $V^+ = 10$ V, $V^- = -10$ V, $R_3 = 100\,k\Omega$, and $R_4 = 100\,k\Omega$. Design $R_5$ such that the circuit can compensate for an offset voltage of $V_{OS} = 5$ mV. (Ans. 50 Ω)

Figure 14.21 shows a compensation network that can be used with a noninverting op-amp circuit. The same $R_4$–$R_5$ voltage divider is used with the potentiometer $R_3$. Typically, $R_5$ is on the order of $100\,\Omega$ and $R_4$ on the order of $100\,k\Omega$. If $V^+ = 15$ V

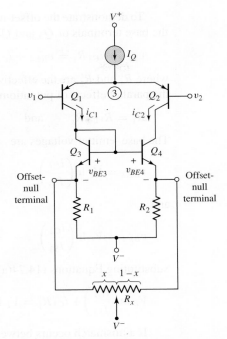

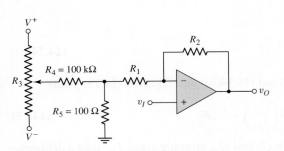

**Figure 14.21** Offset voltage compensation circuit for noninverting amplifier

**Figure 14.22** Basic bipolar input diff-amp stage, including a pair of offset-null terminals connected to a potentiometer

and $V^- = -15$ V, then the compensation voltage is again in the range of $-15$ mV to $+15$ mV.

The voltage gain of the noninverting amplifier becomes a function of the compensation network. Since $R_5 \ll R_4$, then the gain of the amplifier, to a good approximation, is

$$A_v = \frac{v_O}{v_I} = \left(1 + \frac{R_2}{R_1 + R_5}\right) \tag{14.72}$$

Since $R_5$ is small, Equation (14.72) shows that the gain is not a strong function of the compensation network; however, it may still need to be taken into account.

### Offset-Null Terminals

Many op-amps, including the 741 bipolar and the CA3140 BiCMOS circuits studied in Chapter 13, include a pair of external offset-null terminals, which are used to compensate for the offset voltage. Figure 14.22 shows a basic bipolar input diff-amp stage, including a pair of offset-null terminals. An external potentiometer $R_x$ is connected between these terminals, and the wiper arm is connected to supply voltage $V^-$.

If the wiper arm of $R_x$ is centered, then $R_1$ and $R_2$ will each have a resistance $R_x/2$ connected in parallel. When the wiper arm is moved off center, then $R_1$ and $R_2$ will each have a different resistance connected in parallel, and an asymmetry will be introduced into the circuit. This asymmetry in turn introduces an offset voltage, which cancels the input offset voltage effects. In practice, to adjust for offset voltage effects, the op-amp is connected in a feedback configuration with the input differential voltage set equal to zero. The wiper arm of potentiometer $R_x$ is then adjusted until the output voltage becomes zero.

To demonstrate the offset-null technique, we first write a KVL equation between the base terminals of $Q_3$ and $Q_4$ and voltage $V^-$ in Figure 14.22, as follows:

$$v_{BE3} + i_{C1}R_1' = v_{BE4} + i_{C2}R_2' \tag{14.73}$$

where $R_1'$ and $R_2'$ are the effective resistances in the emitters of $Q_3$ and $Q_4$, including the parallel effects of potentiometer $R_x$. We have that

$$R_1' = R_1 \| x R_x \qquad \text{and} \qquad R_2' = R_2 \| (1-x) R_x$$

The base–emitter voltages are

$$v_{BE3} = V_T \ln\left(\frac{i_{C1}}{I_{S3}}\right) \tag{14.74(a)}$$

and

$$v_{BE4} = V_T \ln\left(\frac{i_{C2}}{I_{S4}}\right) \tag{14.74(b)}$$

Substituting Equations (14.74(a)) and (14.74(b)) into Equation (14.73) yields

$$V_T \ln\left(\frac{i_{C1}}{I_{S3}}\right) + i_{C1}R_1' = V_T \ln\left(\frac{i_{C2}}{I_{S4}}\right) + i_{C2}R_2' \tag{14.75}$$

If a mismatch occurs between $Q_3$ and $Q_4$, meaning $I_{S3} \neq I_{S4}$, then a deliberate mismatch between $R_1'$ and $R_2'$ can be introduced to compensate for the transistor mismatch and the adjustment can make $i_{C1} = i_{C2}$. Similarly, a deliberate mismatch between $R_1'$ and $R_2'$ can be used to compensate for a mismatch between $Q_1$ and $Q_2$.

## EXAMPLE 14.12

**Objective:** Determine the required difference between $R_1'$ and $R_2'$, and the value of $x$ in the potentiometer to compensate for a mismatch between active load transistors $Q_3$ and $Q_4$ in the diff-amp in Figure 14.22.

Assume that $I_Q = 200\,\mu\text{A}$, which means that we want $i_{C1} = i_{C2} = 100\,\mu\text{A}$. Let $I_{S3} = 10^{-14}\,\text{A}$ and $I_{S4} = 1.05 \times 10^{-14}\,\text{A}$. Also assume $R_1 = R_2 = 1\,\text{k}\Omega$ and $R_x = 100\,\text{k}\Omega$.

**Solution:** The difference between $R_2'$ and $R_1'$ is determined from Equation (14.75), as follows:

$$V_T \ln\left(\frac{i_{C1}}{I_{S3}}\right) + i_{C1}R_1' = V_T \ln\left(\frac{i_{C2}}{I_{S4}}\right) + i_{C2}R_2'$$

or

$$(0.026)\ln\left(\frac{100 \times 10^{-6}}{10^{-14}}\right) + (0.10)R_1' = (0.026)\ln\left(\frac{100 \times 10^{-6}}{1.05 \times 10^{-14}}\right) + (0.10)R_2'$$

which yields

$$R_2' - R_1' = 0.0127\,\text{k}\Omega \Rightarrow 12.7\,\Omega$$

We can also write the difference between $R_2'$ and $R_1'$ as

$$\frac{R_2(1-x)R_x}{R_2 + (1-x)R_x} - \frac{R_1 x R_x}{R_1 + x R_x} = 0.0127\,\text{k}\Omega$$

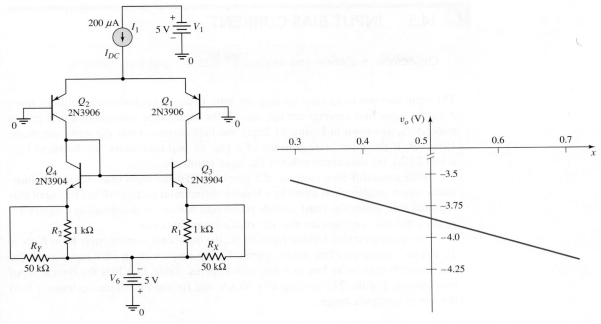

**Figure 14.23** Circuit used in the computer simulation analysis for Example 14.12

**Figure 14.24** Output voltage versus potentiometer setting

Substituting the values for $R_1$, $R_2$, and $R_x$, we find that

$$x = 0.349$$

**Comment:** On the basis of this analysis, the value of $R_1'$ is $1\|34.9 = 0.9721\,\mathrm{k\Omega}$, and the value of $R_2'$ is $1\|(100 - 34.9) = 0.9849\,\mathrm{k\Omega}$.

**Computer Simulation Verification:** Figure 14.23 is the circuit used in PSpice simulation. The values of $R_X$ and $R_Y$ were varied to simulate a change in the variable $x$ in the potentiometer in the circuit in Figure 14.22. The output voltage $v_O$ is taken off the common collectors of $Q_1$ and $Q_3$. This voltage would correspond to the input voltage of a second stage.

A change in the values of $R_X$ and $R_Y$ causes a slight change in the currents in the two sides of the circuit. A change in current causes a change in the collector–emitter voltages of $Q_1$ and $Q_3$, or a change in the output voltage. Figure 14.24 shows the output voltage as a function of $x$, or as a function of the position of the potentiometer. The results show that a change of approximately 0.7 V is possible for this range in potentiometer setting. This change in voltage would represent a large change in input voltage for the second stage, which in turn would cause a large change in the dc value of the output voltage. The dc output voltage could therefore be set to zero by adjusting the potentiometer setting.

## EXERCISE PROBLEM

**\*Ex 14.12:** Consider the diff-amp in Figure 14.22 with a pair of offset-null terminals. Let $R_1 = R_2 = 1\,\mathrm{k\Omega}$. Let $R_x$ be a 100 k$\Omega$ potentiometer. Assume $I_Q = 100\,\mu\mathrm{A}$ and $I_{S3} = 10^{-14}$ A. If the wiper arm on the potentiometer is adjusted such that 25 k$\Omega$ is in parallel with $R_1$ and 75 k$\Omega$ is in parallel with $R_2$, determine the value of $I_{S4}$ for $i_{C1} = i_{C2}$. (Ans. $1.05 \times 10^{-14}$ A)

## ⬛ 14.5    INPUT BIAS CURRENT

**Objective:** • Define and analyze effects of input bias currents.

The input currents to an ideal op-amp are zero. In actual operational amplifiers, however, the input bias currents are not zero. If the input stage consists of a pair of npn transistors, as shown in Figure 14.25(a), the bias currents enter the input terminals. However, if the input state consists of a pair of pnp transistors, as shown in Figure l4.25(b), the bias currents leave the input terminals.

If the input diff-amp consists of a pair of JFETs, the input bias currents are normally much smaller than those in a bipolar differential pair. A MOSFET input differential pair, generally, must include protection devices as discussed in Chapter 13, so the input bias currents are also not zero even in this case.

For op-amps with a bipolar input stage, the input bias currents may be as high as 10 $\mu$A and as low as a few nanoamperes. For op-amps with an FET input stage, the bias currents may be as low as a few picoamperes. Table 14.1 lists the typical input bias current. For the 741 op-amp it is 30 nA, and for the FET input op-amps it is in the low picoampere range.

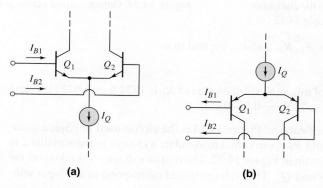

**(a)**                    **(b)**

**Figure 14.25** (a) Pair of npn transistors, showing input bias currents, and (b) pair of pnp transistors, showing input bias currents

### 14.5.1    Bias Current Effects

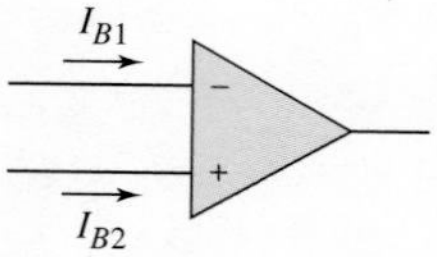

Figure 14.26 schematically shows an op-amp with input bias currents. If the input stage is symmetrical, with all corresponding elements matched, then $I_{B1} = I_{B2}$. However, if the input transistors are not exactly identical, then $I_{B1} \neq I_{B2}$. The **input bias current** is then defined as the average of the two input currents, or

**Figure 14.26** Op-amp with input bias currents

$$I_B = \frac{I_{B1} + I_{B2}}{2} \tag{14.76}$$

The difference between the two input currents is called the **input offset current** $I_{OS}$ and is given by

$$I_{OS} = |I_{B1} - I_{B2}| \tag{14.77}$$

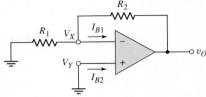

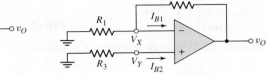

**Figure 14.27** Op-amp with grounded noninverting terminal

**Figure 14.28** Op-amp circuit with resistor connected to noninverting terminal, for input bias current compensation

The algebraic sign of the offset current is usually not important, just as the offset voltage polarity is not critical. The typical input offset current is on the order of 10 percent of the input bias current, although data sheets may list larger values. The typical and maximum input offset currents for the three op-amps analyzed in the last chapter are given in Table 14.1.

Figure 14.27 shows an op-amp and associated resistors for a zero input voltage. Even if $I_{B2} \neq 0$, the noninverting terminal is still at zero volts, or $V_Y = 0$. From the virtual ground concept, we have $V_X = 0$, which means that the current in $R_1$ must be zero. Bias current $I_{B1}$ is therefore supplied by the output of the op-amp and flows through $R_2$, producing an output voltage. If, for example, $I_{B1} = 5\ \mu A$ and $R_2 = 100\ k\Omega$, then $v_O = 0.5\ V$, which is unacceptable in most applications. Smaller input bias currents and a smaller feedback resistor will reduce the bias current effects.

## 14.5.2    Bias Current Compensation

The effect of bias currents in op-amp circuits can be minimized with a simple compensation technique. Consider the circuit in Figure 14.28. We determine $v_O$ as a function of $I_{B1}$ and $I_{B2}$ using superposition. For $I_{B2} = 0$, then $V_Y = V_X = 0$, and the output voltage due to $I_{B1}$ is

$$v_O(I_{B1}) = I_{B1}R_2 \tag{14.78(a)}$$

For $I_{B1} = 0$, we find

$$V_Y = -I_{B2}R_3 = V_X$$

Since

$$v_O = (1 + R_2/R_1)V_X$$

the output voltage due to $I_{B2}$ is

$$v_O(I_{B2}) = -I_{B2}R_3\left(1 + \frac{R_2}{R_1}\right) \tag{14.78(b)}$$

The net output voltage due to both $I_{B1}$ and $I_{B2}$ is the sum of Equations (14.78(a)) and (14.78(b)), or

$$v_O = I_{B1}R_2 - I_{B2}R_3\left(1 + \frac{R_2}{R_1}\right) \tag{14.79}$$

If $I_{B1} = I_{B2} \equiv I_B$ and if the combination of the three resistances can be adjusted to produce $v_O = 0$, then Equation (14.79) becomes

$$0 = I_B\left[R_2 - R_3\left(1 + \frac{R_2}{R_1}\right)\right] \tag{14.80}$$

which means that

$$R_2 = R_3\left(1 + \frac{R_2}{R_1}\right) \tag{14.81}$$

Equation (14.81) can be rearranged as follows:

$$R_3 = \frac{R_1 R_2}{R_1 + R_2} = R_1 \| R_2 \tag{14.82}$$

Equation (14.82) shows that $R_3$ should be made equal to the parallel combination of $R_1$ and $R_2$, to eliminate the effect of equal input bias currents.

If $R_3 = R_1 \| R_2$ and if the bias currents are not equal, then from Equation (14.79), we have

$$v_O = R_2(I_{B1} - I_{B2}) = R_2 I_{OS} \tag{14.83}$$

Since the input offset current is normally a fraction of the input bias current, Equation (14.83) shows that the bias current effect can be reduced by making $R_3 = R_1 \| R_2$.

## EXAMPLE 14.13

**Objective:** Determine the bias current effect in an op-amp circuit, with and without bias current compensation.

Consider the op-amp circuits in Figures 14.27 and 14.28. Let $R_1 = 10$ k$\Omega$ and $R_2 = 100$ k$\Omega$. Assume $I_{B1} = 1.1~\mu$A and $I_{B2} = 1.0~\mu$A.

**Solution:** For the op-amp circuit in Figure 14.27, the output voltage due to the bias currents is

$$v_O = I_{B1}R_2 = (1.1 \times 10^{-6})(100 \times 10^3) = 0.11 \text{ V}$$

For the circuit in Figure 14.28, we design $R_3$ such that

$$R_3 = R_1 \| R_2 = 10 \| 100 = 9.09 \text{ k}\Omega$$

Then, from Equation (14.83), we find

$$v_O = R_2(I_{B1} - I_{B2}) = (100 \times 10^3)(1.1 - 1.0) \times 10^{-6} = 0.010 \text{ V}$$

**Comment:** Even if the input offset current is not zero, the effect of the input bias currents can be reduced substantially by incorporating resistor $R_3$.

## EXERCISE PROBLEM

**Ex 14.13:** For the op-amp circuit shown in Figure 14.28, the parameters are $R_1 = 20$ k$\Omega$ and $R_2 = 120$ k$\Omega$. (a) Let $I_{B1} = I_{B2} = 0.8~\mu$A. (i) Can $R_3$ be adjusted such that $v_O = 0$? (ii) If so, what is the value of $R_3$? (b) Repeat part (a) if $I_{B1} = 0.75~\mu$A and $I_{B2} = 0.85~\mu$A. (Ans. (a) $R_3 = 17.14$ k$\Omega$, (b) $R_3 = 15.13$ k$\Omega$)

Usually the effect of bias currents in op-amp circuits is significant only for circuits with large resistor values. For these situations, an op-amp with an FET input stage may be necessary.

# Test Your Understanding

**TYU 14.9** Consider the inverting summing amplifier in Figure 14.29. Assume input bias currents of $I_{B1} = I_{B2} = 1.1\ \mu A$. (a) For $v_{i1} = v_{i2} = 0$ and $R_4 = 0$, determine $v_O$ due to the bias currents. (b) Find the value of $R_4$ that compensates for the effects of the bias currents. (Ans. (a) $v_O = 0.22$ V (b) $R_4 = 28.6$ k$\Omega$)

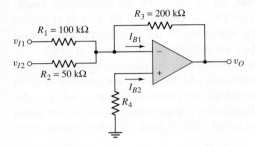

**Figure 14.29** Figure for Exercise TYU14.9

## 14.6   ADDITIONAL NONIDEAL EFFECTS

**Objective:** • Discuss and analyze additional nonideal effects.

Two additional nonideal effects in op-amps are: temperature effects and common-mode rejection ratio. We will look at each of these in this section.

### 14.6.1   Temperature Effects

Individual transistor parameters are functions of temperature. For bipolar transistors, the collector current is

$$i_C = I_S e^{v_{BE}/V_T} \qquad (14.84)$$

where both $I_S$ and $V_T$ are functions of temperature. We expect the open-loop gain to vary with temperature, but as we saw in Section 14.2, the fractional change in the closed-loop gain is orders of magnitude less than the fractional change in the open-loop gain. This then makes the closed-loop gain very insensitive to temperature variations.

**Offset Voltage Temperature Coefficient**
Since the electrical properties of transistors are functions of temperature, the input offset voltage is also a function of temperature. The rate of change of offset voltage with temperature is defined as the **temperature coefficient of offset voltage**, or **input offset voltage drift**, and is given by

$$\mathrm{TC}v_{OS} = \frac{dV_{OS}}{dT} \qquad (14.85)$$

For a bipolar diff-amp input stage, the offset voltage, from Equation (14.59(b)), is

$$V_{OS} = V_T \ln(I_{S2}/I_{S1})$$

The temperature variations of the $I_S$ parameters cancel; therefore, the offset voltage is directly proportional to the thermal voltage $V_T$, which in turn is directly proportional to temperature. From Equation (14.59(b)), the temperature coefficient is then

$$\text{TC}v_{OS} = \frac{V_{OS}}{T} \tag{14.86}$$

where $T$ is the absolute temperature. Thus, for $V_{OS} = 1$ mV, the temperature coefficient is $\text{TC}v_{OS} = 1$ mV/300 K $\Rightarrow 3.3$ $\mu$V/°C. A change of $10\,°$C will therefore result in an offset voltage change of approximately $33$ $\mu$V. The temperature coefficients of offset voltage listed in Table 14.1 are in the range of 10 to 15 $\mu$V/°C.

Consequently, the offset voltage compensation techniques discussed previously are completely effective at only one temperature. As the device temperature drifts in either direction from the temperature at which the compensation network was designed, the offset voltage effect is not completely compensated. However, the offset voltage drift is substantially less than the initial offset voltage, so offset voltage compensation is still desirable.

### Input Offset Current Temperature Coefficient

The input bias currents are functions of temperature. For example, the input bias current of a bipolar input stage has the same functional dependence as the collector current, as given by Equation (14.84). If the input devices are not matched, then an input offset current $I_{OS}$ exists, which is also a function of temperature. The input offset current temperature coefficient is $dI_{OS}/dT$. For the 741 op-amp, the maximum value given in Table 14.1 is 0.5 nA/°C. If the input offset current becomes a problem in a particular design, then a JFET of MOSFET input stage op-amp may be required.

### 14.6.2 Common-Mode Rejection Ratio

We considered the common-mode gain ($A_{cm}$) and common-mode rejection ratio (CMRR) of the difference amplifier in Chapter 11. Since a diff-amp is the op-amp input stage, any common-mode signal produced at the input stage will propagate through the op-amp to the output. Therefore, the CMRR of the op-amp is essentially the same as the CMRR of the input diff-amp.

Figure 14.30(a) shows the open-loop op-amp with a pure differential-mode input signal. The differential-mode gain $A_d$ is the same as the open-loop gain $A_{OL}$. Figure 14.30(b) shows the open-loop op-amp with a pure common-mode input signal.

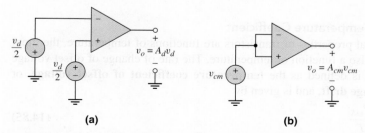

(a)                                    (b)

**Figure 14.30** Open-loop op-amp (a) with pure differential-mode input signal and (b) with pure common-mode input signal

The common-mode rejection ratio, in dB, is

$$\text{CMRR}_{\text{dB}} = 20\log_{10}\left|\frac{A_d}{A_{cm}}\right| \qquad (14.87)$$

Typical values of $\text{CMRR}_{\text{dB}}$ range from 80 to 100 dB. Table 14.1 lists typical $\text{CMRR}_{\text{dB}}$ values for three op-amps.

## 14.7    DESIGN APPLICATION: AN OFFSET VOLTAGE COMPENSATION NETWORK

**Objective:** • Design an offset voltage compensation network for a CMOS diff amp.

**Specifications:** An offset voltage compensation network is to be designed at the active load of a CMOS diff-amp.

**Design Approach:** An offset voltage compensation network with the configuration shown in Figure 14.31 is to be designed. Assume both a 5 percent and $2\frac{1}{2}$ percent difference in conduction parameters between $M_1$ and $M_2$. This mismatch will demonstrate how the network can compensate for an offset voltage.

**Choices:** For both $M_1$ and $M_2$, assume parameters $V_{TN} = 0.5$ V, $W/L = 20$, and $\lambda_n = 0.02$ V$^{-1}$. For $M_1$, assume $k'_{n1} = 80$ $\mu$A/V$^2$ and for $M_2$, assume (i) $k'_{n2} = 76$ $\mu$A/V$^2$ and then (ii) $k'_{n2} = 78$ $\mu$A/V$^2$. A 50 k$\Omega$ center-tapped potentiometer is available.

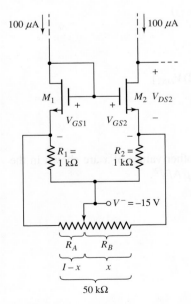

**Figure 14.31** An offset voltage compensation network for the design application

**Solution (for $M_1$):** We can write that

$$I_{D1} = \left(\frac{k'_{n1}}{2}\right)\left(\frac{W}{L}\right)(V_{GS1} - V_{TN})^2(1 + \lambda_n V_{DS1})$$

From the connection, we see that $V_{GS1} = V_{DS1}$. Then

$$100 = \left(\frac{80}{2}\right)(20)(V_{GS1} - 0.5)^2[1 + (0.02)V_{GS1}]$$

We find that $V_{GS1} = V_{DS1} = 0.8506$ V

**Solution (for $M_2$):** Determine the variation in $V_{DS2}$ as a function of $x$ for the potentiometer setting. For $x = 0.3$ and $k'_{n2} = 76$ $\mu$A/V$^2$, we have

$$R'_1 = 1\|35 = 0.97222 \text{ k}\Omega$$

and

$$R'_2 = 1\|15 = 0.93750 \text{ k}\Omega$$

Then

$$V_{GS2} = V_{GS1} + I_{D1}R'_1 - I_{D2}R'_2$$

or

$$V_{GS2} = 0.8506 + (0.1)(0.97222) - (0.1)(0.93750)$$

so

$$V_{GS2} = 0.85407 \text{ V}$$

Now

$$I_{D2} = \left(\frac{k'_{n2}}{2}\right)\left(\frac{W}{L}\right)(V_{GS2} - V_{TN})^2(1 + \lambda_n V_{DS2})$$

or

$$100 = \left(\frac{76}{2}\right)(20)(0.85407 - 0.5)^2[1 + (0.02)V_{DS2}]$$

which yields

$$V_{DS2} = 2.478 \text{ V}$$

Going through the same analysis, the results for other values of $x$ are shown in the following table as well as the results for $k'_{n2} = 78$ $\mu$A/V$^2$.

| $k'_{n2} = 76$ $\mu$A/V$^2$ | | $k'_{n2} = 78$ $\mu$A/V$^2$ | |
|---|---|---|---|
| $x$ | $V_{DS2}$ (V) | $x$ | $V_{DS2}$ (V) |
| 0.3 | 2.478 | 0.4 | 1.696 |
| 0.2 | 1.547 | 0.3 | 1.132 |
| 0.16 | 0.9370 | 0.26 | 0.8330 |
| 0.14 | 0.5311 | 0.22 | 0.4568 |

The results are plotted in Figure 14.32.

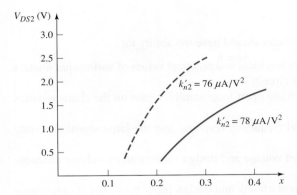

**Figure 14.32** Change in drain-to-source voltage as the compensation network is varied

**Comment:** By varying the voltage $V_{DS2}$, the applied voltage to the second stage changes which in turn will change the dc voltage to the output stage. These results mean that eventually the output voltage can be adjusted to zero for zero input.

We can note that if $k_{n2}' > k_{n1}'$, then the value $x$ of the potentiometer setting would be $x > 0.5$.

## 14.8   SUMMARY

- In this chapter, nonideal effects in op-amp circuits, such as finite open-loop gain, offset voltage, and bias currents, were considered.
- A finite open-loop amplifier gain results in the magnitudes of the inverting amplifier and noninverting amplifier gain being smaller than the ideal values.
- A finite open-loop amplifier gain plus finite input amplifier resistance and nonzero output resistance results in nonideal op-amp input and output resistance values.
- The practical op-amp circuit has a finite bandwidth. With negative feedback, the gain–bandwidth product is essentially constant, so an op-amp circuit with negative feedback has a reduced gain magnitude but an increased small-signal bandwidth.
- Slew rate is defined as the maximum rate at which the op-amp output signal can change per unit time. In general, the slew rate is limited by the internal frequency compensation capacitor. The slew rate is also a function of the bias current in the input diff-amp stage.
- An input offset voltage means that the output voltage is not zero when the input signal voltages are zero. One source of an offset voltage is a mismatch in the differential pair transistor parameters and/or mismatches in active load transistor parameters. Typically, an offset voltage of a few millivolts may occur in a bipolar circuit, whereas an offset voltage of tens of millivolts may occur in a MOSFET circuit.
- Input bias currents to an op-amp may range from a few picoamperes for FET input transistors to as high as a few microamperes for some bipolar input transistors.
- Techniques for offset voltage compensation and input bias current compensation were analyzed.
- As an application, an offset voltage compensation network was designed for a CMOS diff-amp.

**CHECKPOINT**

After studying this chapter the reader should have the ability to:

✓ Understand differences between ideal and practical values of various parameters of the operational amplifier circuit.

✓ Understand the effect of a finite open-loop amplifier gain on the characteristics of the op-amp.

✓ Understand the small-signal frequency response and the large-signal slew-rate response of op-amps.

✓ Understand sources of offset voltage and design various offset voltage compensation circuits for op-amps.

✓ Understand input bias current effects and design input bias current compensation circuits for an op-amp.

 **REVIEW QUESTIONS**

1. List and describe five practical op-amp parameters and discuss the effect they have on op-amp circuit characteristics.
2. What is a typical value of open-loop, low-frequency gain of an op-amp circuit? How does this compare to the ideal value?
3. How does a finite open-loop gain affect the closed-loop gains of the inverting and noninverting amplifiers?
4. How does a finite open-loop gain affect the (a) input resistance of an op-amp circuit and (b) the output resistance of an op-amp circuit? Consider the inverting and noninverting amplifiers.
5. Describe the open-loop amplifier frequency response and define the unity-gain bandwidth.
6. What is a typical corner frequency value, or dominant-pole frequency, in the open-loop frequency response?
7. Describe the gain–bandwidth product property of a closed-loop amplifier response.
8. Define slew rate and define full-power bandwidth.
9. What is the primary source of slew-rate limitation in an op-amp circuit?
10. What is one cause of an offset voltage in the input stage of a BJT op-amp?
11. What is one cause of an offset voltage in the input stage of a CMOS op-amp?
12. Describe an offset voltage compensation technique.
13. What is the source of input bias current in the 741 op-amp?
14. What can be the effect of an input bias current?
15. Describe the effect of input bias currents on an integrator.
16. Describe an input bias current compensation technique.
17. Define and explain common-mode rejection ratio.

 **PROBLEMS**

**Section 14.1 Practical Op-Amp Parameters**

14.1 An op-amp is connected in an inverting amplifier configuration with a voltage gain of −80 and is biased at ±5 V. If the output saturates at ±4.5 V, what is the maximum rms value of an input sine wave that can be applied without causing distortion in the output signal?

14.2    Consider the op-amp described in Problem 14.1. In addition, the maximum output current of the op-amp is $\pm 15$ mA. The resistors used in the configuration are $R_2 = 160$ k$\Omega$ and $R_1 = 2$ k$\Omega$. A load resistor $R_L$ is also connected from the output terminal to ground. (a) If $R_L = 1$ k$\Omega$ and the output voltage is $v_O = 4.5$ V, what is the output current of the op-amp and what is the value of the input voltage? (b) Determine the minimum value of $R_L$ that can be used for $v_O = -4.5$ V.

## Section 14.2 Finite Open-Loop Gain

14.3    Data in the following table were taken for several op-amps operating in the open-loop configuration. Determine the unknown variables in the table.

| Case | $A_{OL}$ | $v_1$ | $v_2$ | $v_O$ |
|---|---|---|---|---|
| 1 | $10^4$ | $-0.1$ mV | $+0.1$ mV | |
| 2 | $2 \times 10^3$ | $+10.0$ mV | | 5 V |
| 3 | | 5.50 mV | 5.00 mV | $-10$ V |
| 4 | $5 \times 10^5$ | | 0 | $-4$ V |
| 5 | | $-2.010$ V | $-2.0050$ V | 5 V |

14.4    (a) An inverting amplifier with resistors $R_1 = 5.6$ k$\Omega$ and $R_2 = 120$ k$\Omega$ is fabricated using an op-amp with an open-loop gain of $10^5$. What is the percent difference between the actual gain and the ideal gain? (b) Repeat part (a) if $R_1$ is changed to $R_1 = 8.2$ k$\Omega$.

14.5    (a) Consider a noninverting amplifier with $R_1 = 6.8$ k$\Omega$, $R_2 = 47$ k$\Omega$, and an op-amp with $A_{OL} = 2 \times 10^4$. (i) What is the closed-loop gain? (ii) Determine the percent difference between the actual gain and the ideal gain. (b) Repeat part (a) if $A_{OL} = 10^3$.

14.6    (a) An op-amp is ideal except it has a finite open-loop gain of $A_{OL} = 2 \times 10^3$. The op-amp is connected in an inverting configuration. Determine $R_2/R_1$ such that the closed-loop voltage gain is $A_{CL} = -15.0$. (b) Using the results of part (a), what is the closed-loop gain if the open-loop gain is $A_{OL} = 5 \times 10^4$?

14.7    A noninverting amplifier is to be fabricated with a specification of an ideal closed-loop gain of 90. What is the minimum open-loop gain of the op-amp such that the closed-loop gain is within 0.01 percent of the ideal value?

14.8    The output of a voltage follower is to be within 0.02 percent of the ideal value. What is the minimum op-amp open-loop gain that is required?

14.9    An inverting amplifier is fabricated using 0.1 percent precision resistors. The nominal resistor values are $R_2 = 210$ k$\Omega$ and $R_1 = 21.0$ k$\Omega$. (a) If the op-amp is ideal, what is the range in the magnitude of voltage gain as a result of the variation in resistor value? (b) Repeat part (a) if the open-loop gain of the op-amp is $A_{OL} = 10^4$.

14.10   For the op-amp used in the inverting amplifier configuration in Figure P14.10, the open-loop parameters are $A_{OL} = 10^3$ and $R_o = 0$. Determine the closed-loop gain $A_{CL} = v_O/v_I$ and input resistance $R_{if}$ for an open-loop input differential-mode resistance of: (a) $R_i = 1$ k$\Omega$, (b) $R_i = 10$ k$\Omega$, and (c) $R_i = 100$ k$\Omega$.

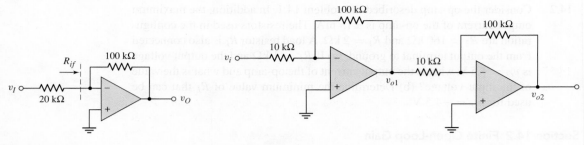

**Figure P14.10**                    **Figure P14.12**

14.11 A pressure transducer, as described in Example 14.1, is to be used in conjunction with a noninverting op-amp circuit. The ideal output voltage is to be +0.10 V for a transducer voltage of 2 mV. Determine the minimum open-loop gain required so that the actual output voltage is within 0.1 percent of the ideal.

14.12 Consider the two inverting amplifiers in cascade in Figure P14.12. The op-amp parameters are $A_{OL} = 5 \times 10^3$, $R_i = 10$ k$\Omega$, and $R_o = 1$ k$\Omega$. Determine the actual closed-loop gains $A_{vf1} = v_{o1}/v_i$ and $A_{vf} = v_{o2}/v_i$. What is the percent error from the ideal values?

14.13 The noninverting amplifier in Figure P14.13 has an op-amp with open-loop properties: $A_{OL} = 10^3$, $R_i = 20$ k$\Omega$, and $R_o = 0.5$ k$\Omega$. (a) Determine the closed-loop values of $A_{CL} = v_O/v_I$, $R_{if}$, and $R_{of}$. (b) If $A_{OL}$ decreases by 10 percent, determine the percentage change in $A_{CL}$.

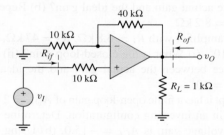

**Figure P14.13**

14.14 For the op-amp in the voltage follower circuit in Figure P14.14, the open-loop parameters are $A_{OL} = 5 \times 10^3$, $R_i = 10$ k$\Omega$, and $R_o = 1$ k$\Omega$. (a) Sketch the small-signal equivalent circuit. (b) Determine the (i) closed-loop voltage gain $v_O/v_I$ and (ii) output resistance $R_{of}$.

14.15 The summing amplifier in Figure P14.15 has an op-amp with open-loop parameters: $A_{OL} = 2 \times 10^3$, $R_i = \infty$, and $R_o = 0$. Determine the actual output voltage as a function of $v_{I1}$ and $v_{I2}$. What is the percent error from the ideal value?

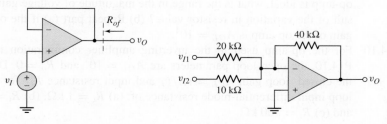

**Figure P14.14**                    **Figure P14.15**

14.16  For the op-amp in the differential amplifier in Figure P14.16, the open-loop parameters are: $A_{OL} = 10^3$, $R_i = \infty$, and $R_o = 0$. Determine the actual differential voltage gain $A_d = v_O/(v_{I2} - v_{I1})$. What is the percentage error from the ideal value?

14.17  Because of a manufacturing error, the open-loop gain of each op-amp in the circuit in Figure P14.17 is only $A_{OL} = 100$. The open-loop input and output resistances are $R_i = 10\,k\Omega$ and $R_o = 1\,k\Omega$, respectively. Determine the closed-loop parameters: (a) $R_{if}$, (b) $R_{of}$, and (c) $A_{CL} = v_{O2}/v_I$. (d) What is the ratio of the actual closed-loop gain to the ideal value?

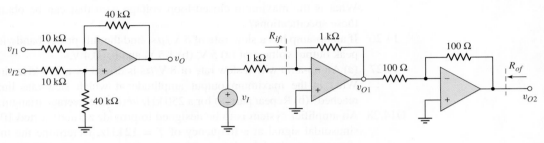

**Figure P14.16**　　　　　　　**Figure P14.17**

## Section 14.3 Frequency Response

14.18  An inverting amplifier has a closed-loop voltage gain of $-25$. The op-amp used has a low-frequency, open-loop gain of $2 \times 10^4$ and has a unity-gain bandwidth of $10^6$ Hz. (a) What is the 3 dB frequency $f_{3\text{-dB}}$ of the op-amp and the 3 dB frequency $f_{3\text{-dB}}$ of the closed-loop amplifier? (b) Using the results of part (a), what is the magnitude of the voltage gain for the open-loop and closed-loop amplifiers at $f = 0.25\,f_{3\text{-dB}}$ and at $f = 5f_{3\text{-dB}}$?

14.19  The open-loop low-frequency gain of an op-amp is $A_o = 100$ dB. At a frequency of $f = 10^4$ Hz, the magnitude of the open-loop gain is 38 dB. Determine the dominant-pole frequency and the unity gain bandwidth.

14.20  A noninverting amplifier uses 5 percent precision resistors with nominal values of $R_2 = 150\,k\Omega$ and $R_1 = 15\,k\Omega$. The op-amp has a low-frequency gain of $A_o = 3 \times 10^4$ and has a unity-gain bandwidth of $f_T = 1.2$ MHz. (a) What is the nominal low-frequency closed-loop gain and bandwidth? (b) Determine the range in low-frequency closed-loop gain and bandwidth.

14.21  The low-frequency open-loop gain of an op-amp is $2 \times 10^5$ and the second pole occurs at a frequency of 5 MHz. An amplifier using this op-amp has a low-frequency closed-loop gain of 100 and a phase margin of 80 degrees. Determine the dominant-pole frequency.

14.22  Two inverting amplifiers are connected in cascade to provide an overall voltage gain of 500. The gain of the first amplifier is $-10$ and the gain of the second amplifier is $-50$. The unity-gain bandwidth of each op-amp is 1 MHz. (a) What is the bandwidth of the overall amplifier system? (b) Redesign the system to achieve the maximum bandwidth. What is the maximum bandwidth?

14.23  Three inverting amplifiers, each with $R_2 = 150\,k\Omega$ and $R_1 = 15\,k\Omega$, are connected in cascade. Each op-amp has a low-frequency gain of $A_o = 5 \times 10^4$ and a unity-gain bandwidth of $f_T = 1.5$ MHz. (a) Determine the low-frequency closed-loop gain and the $-3$ dB frequency of each stage.

(b) Determine the low-frequency closed-loop gain and the −3 dB frequency of the overall system.

14.24 An inverting amplifier circuit has a voltage gain of −25. The op-amp used in the circuit has a low-frequency voltage gain of $5 \times 10^4$ and a unity-gain bandwidth of 1 MHz. Determine the dominant pole frequency of the op-amp and the small-signal bandwidth, $f_{3\text{-dB}}$, of the inverting amplifier. What is the magnitude of the closed-loop voltage gain at $0.5 f_{3\text{-dB}}$ and at $2 f_{3\text{-dB}}$?

14.25 An audio amplifier system, using a noninverting op-amp circuit, needs to have a small-signal bandwidth of 20 kHz. The open-loop low-frequency voltage gain of the op-amp is $10^5$ and the unity-gain bandwidth is 1 MHz. What is the maximum closed-loop voltage gain that can be obtained for these specifications?

14.26 If an op-amp has a slew-rate of 5 V/$\mu$s, find the full-power bandwidth for a peak output voltage of (a) 5 V, (b) 1.5 V, and (c) 0.4 V.

14.27 (a) An op-amp with a slew rate of 8 V/$\mu$s is driven by a 250 kHz sine wave. What is the maximum output amplitude at which slew-rate limiting is reached? (b) Repeat part (a) for a 250 kHz zero time-average triangular wave.

D14.28 An amplifier system is to be designed to provide an undistorted 10 V peak sinusoidal signal at a frequency of $f = 12\,\text{kHz}$. Determine the minimum slew rate required for the amplifier.

14.29 (a) The op-amp to be used in the audio amplifier system in Problem 14.25 has a slew rate of 0.63 V/$\mu$s. Determine the peak value of undistorted output voltage that can be achieved. (b) Repeat part (a) if the slew rate is 3 V/$\mu$s.

14.30 The op-amp in the noninverting amplifier configuration in Figure P14.30 has a slew rate of 1 V/$\mu$s. Sketch the output voltage versus time for each of the three inputs shown. The op-amp is biased at ±10 V.

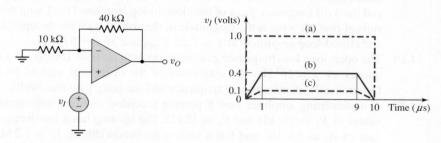

Figure P14.30

14.31 For each op-amp in the circuit shown in Figure P14.31, the bias is ±15 V and the slew rate is 3 V/$\mu$s. Sketch the output voltages $v_{O1}$ and $v_{O2}$ versus time for each input shown.

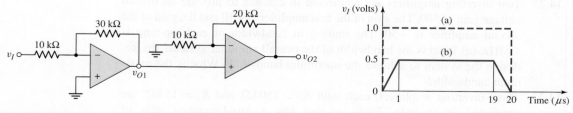

Figure P14.31

## Section 14.4  Offset Voltage

14.32   For the transistors in the diff-amp in Figure 14.16 in the text, the current parameters $I_{S1}$ and $I_{S2}$ can be written as $5 \times 10^{-14}(1 + x)$ A, where $x$ represents the deviation from the ideal due to variations in electrical and geometric characteristics. (The value of $x$ is positive for one transistor and negative for the other transistor.) Determine the maximum value of $x$ such that the maximum offset voltage is limited to $V_{OS} = 2.5$ mV.

14.33   The bipolar active load diff-amp in Figure 14.18 is biased at $V^+ = 5$ V and $V^- = -5$ V. The transistor parameters are $V_{AN} = 120$ V, $V_{AP} = 80$ V, $v_{BE}$ (npn) $= v_{EB}$ (pnp) $= 0.6$ V, $I_{S1} = I_{S2}$, and $I_{S3} = 5 \times 10^{-15}$ A. Let $v_1 = v_2$. Determine the value of $I_{S4}$ for which $Q_4$ has a C–E voltage of (a) $v_{EC4} = 0.6$ V, (b) $v_{EC4} = 1.2$ V, and (c) $v_{EC4} = 2.5$ V.

14.34   For the transistors in the diff-amp in Figure 14.19, the conduction parameters can be written as $150(1 + x)$ $\mu$A/V$^2$, where $x$ represents the deviation from the ideal due to variations in electrical and geometric characteristics. (The value of $x$ is positive for one transistor and negative for the other transistor.) Assume $I_Q = 200$ $\mu$A and $V_{TN1} = V_{TN2} = 0.4$ V. Determine the maximum value of $x$ such that the maximum offset voltage is limited to $V_{OS} = 15$ mV.

14.35   (a) An inverting op-amp circuit has a gain of $-30$. The op-amp used in the circuit has an offset voltage of $\pm 2$ mV. If the input signal voltage to the amplifier is 10 mV, determine the possible range in the output voltage. (b) Repeat part (a) if the input signal voltage is 100 mV.

14.36   Repeat Problem 14.35 for an input signal voltage of $v_I = 25 \sin \omega t$ (mV).

14.37   Consider the integrator circuit in Figure P14.37. The circuit parameters are $R = 10$ k$\Omega$ and $C = 10$ $\mu$F. The op-amp offset voltage is $\pm 5$ mV. For $v_i = 0$, determine the output voltage versus time. For the worst-case offset voltage, determine the time that it would take for the output voltage to reach $\pm 5$ V.

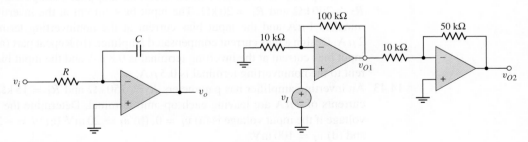

**Figure P14.37**                    **Figure P14.38**

14.38   In the circuit in Figure P14.38, the offset voltage of each op-amp is $\pm 3$ mV. (a) Determine the possible range in output voltages $v_{O1}$ and $v_{O2}$ for $v_I = 0$. (b) Repeat part (a) for $v_I = 10$ mV. (c) Repeat part (a) for $v_I = 100$ mV. (d) Design offset voltage compensation circuit(s) to adjust both $v_{O1}$ and $v_{O2}$ to zero when $v_I = 0$.

14.39 In the circuit shown in Figure P14.39, the op-amp is ideal. For $v_I = 0.5$ V, determine $v_O$ when the wiper arm of the potentiometer is at the $V^+$ node, in the center, and at the $V^-$ node.

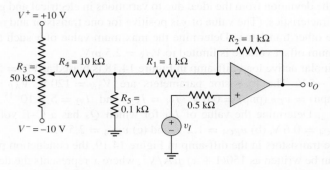

**Figure P14.39**

14.40 Consider the bipolar diff-amp with an active load and a pair of offset-null terminals as shown in Figure 14.22 in the text. Let $R_1 = R_2 = 500$ Ω and let $R_x$ be a 50 kΩ potentiometer. (a) If the wiper arm of the potentiometer is exactly in the center, determine the effective resistances $R_1'$ and $R_2'$. (b) Assume $I_Q = 250$ μA meaning that $i_{C1} = i_{C2} = 125$ μA. Let $I_{S3} = 2 \times 10^{-14}$ A and $I_{S4} = 2.2 \times 10^{-14}$ A. Determine the required values of $x$ and $(1 - x)$ of the potentiometer to compensate for the transistor mismatches.

14.41 The bipolar diff-amp in Figure 14.22 in the text is biased at $I_Q = 500$ μA. Assume all transistors are matched, with $I_S = 10^{-14}$ A. Let $R_1 = R_2 = 500$ Ω, and assume $R_x$ is a 50 kΩ potentiometer. If the wiper arm of the potentiometer is off center such that $x = 15$ kΩ and $(1 - x) = 35$ kΩ, determine the ratio of $i_{C1}/i_{C2}$. What is the corresponding offset voltage?

## Section 14.5 Input Bias Current

14.42 (a) An op-amp is connected in an inverting amplifier configuration with $R_2 = 200$ kΩ and $R_1 = 20$ kΩ. The input bias current at the inverting terminal is 1 μA and the input bias current at the noninverting terminal is 2 μA. Design a bias current compensated amplifier. (b) Repeat part (a) if the input bias current at the inverting terminal is 0.8 μA and the input bias current to the noninverting terminal is 0.5 μA.

14.43 An inverting amplifier has parameters $R_2 = 150$ kΩ and $R_1 = 15$ kΩ. Bias currents of 2 μA are leaving each op-amp terminal. Determine the output voltage if the input voltage is (a) $v_I = 0$, (b) $v_I = 20$ mV, (c) $v_I = -20$ mV, and (d) $v_I = 100$ mV.

14.44 An op-amp is connected in a noninverting amplifier configuration with a voltage gain of $+41$. The feedback resistor is 250 kΩ. The op-amp has input bias currents of $I_{B1} = I_{B2} = 0.6$ μA. Determine the output voltage $v_O$ for input voltages of (a) $v_I = 0$, (b) $v_I = 8$ mV, (c) $v_I = -3.5$ mV, and (c) $v_I = 5 \sin \omega t$ (mV).

D14.45 An op-amp used in a voltage follower configuration is ideal except that the input bias currents are $I_{B1} = I_{B2} = 1$ μA. The source driving the voltage follower has an output resistance of 10 kΩ. (a) Find the output voltage due to the bias current effects when $v_I = 0$. (b) Can the circuit be designed to compensate for the input bias currents? If so, how?

14.46 In the differential amplifier in Figure P14.16, the op-amp is ideal except that the average input bias current is $I_B = 10\ \mu A$ and the input offset current is $I_{OS} = 3\ \mu A$. If $v_{i1} = v_{i2} = 0$, determine the worst-case output voltage $v_O$ due to the input bias current effects.

D14.47 The op-amp bias currents for the circuit in Figure P14.38 are equal at $I_{B1} = I_{B2} = 1\ \mu A$. (a) Find the worst-case output voltages $v_{O1}$ and $v_{O2}$ for $v_I = 0$. (b) Design input bias current compensation circuit(s) to adjust both $v_{O1}$ and $v_{O2}$ to zero when $v_I = 0$.

14.48 (a) For the integrator circuit in Figure P14.48, let the input bias currents be $I_{B1} = I_{B2} = 0.1\ \mu A$. Assume that switch $S$ opens at $t = 0$. Derive an expression for the output voltage versus time for $v_I = 0$. (b) Plot $v_O$ versus time for $0 \le t \le 10$ s. (c) Repeat part (b) for $I_{B1} = I_{B2} = 100$ pA.

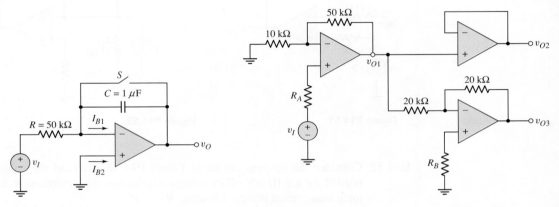

**Figure P14.48**                    **Figure P14.49**

14.49 For the circuit in Figure P14.49, the op-amps are ideal except that the op-amps have bias currents of $I_B = 3\ \mu A$ entering each op-amp terminal. (a) For $v_I = 0$ and $R_A = R_B = 0$, determine the values of $v_{O1}$, $v_{O2}$, and $v_{O3}$. (b) Determine the values of $R_A$ and $R_B$ for input bias current compensation. (c) If the average input bias current is $I_B = 3\ \mu A$ and the input offset current is $I_{OS} = 0.3\ \mu A$, determine the worst-case output values of $v_{O1}$, $v_{O2}$, and $v_{O3}$ using the results of part (b).

14.50 For each circuit in Figure P14.50, the input bias current is $I_B = 0.8\ \mu A$ the input offset current is $I_{OS} = 0.2\ \mu A$. (a) Determine the output voltage due to the average bias current $I_B$. (b) Determine the worst-case output voltage, including the effect of the input offset current.

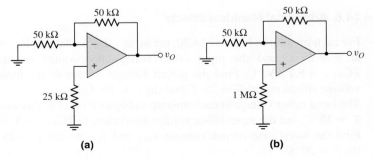

(a)                    (b)

**Figure P14.50**

### Sections 14.4 and 14.5 Offset Voltage and Input Bias Current: Total Effects

14.51 For the op-amp in Figure P14.51, the input offset voltage is $V_{OS} = 3 \, \text{mV}$, the average input bias current is $I_B = 0.4 \, \mu\text{A}$, and the offset bias current is $I_{OS} = 0.06 \, \mu\text{A}$. (a) Determine the possible range in output voltage for $v_I = 0$ and $R = 0$. (b) Repeat part (a) for $v_I = 0$ and $R = 9.09 \, \text{k}\Omega$. (c) Repeat part (a) for $v_I = 0.2 \, \text{V}$ and $R = 9.09 \, \text{k}\Omega$.

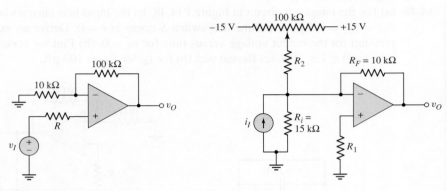

**Figure P14.51**                    **Figure P14.52**

D14.52 Consider the op-amp circuit in Figure P14.52. (a) Find the value of $R_2$ needed for a $\pm 10$ mV offset voltage adjustment. (b) Determine $R_1$ to minimize bias current effects. (Assume $R_2 \gg R_i$.)

D14.53 For each op-amp in the circuit in Figure P14.38, the offset voltage is $V_{OS} = 10$ mV and the input bias currents are $I_{B1} = I_{B2} = 2 \, \mu\text{A}$. (a) Find the worst-case output voltages $v_{O1}$ and $v_{O2}$ for $v_I = 0$. (b) Design compensation circuits to adjust both $v_{O1}$ and $v_{O2}$ to zero when $v_I = 0$.

14.54 The op-amps in the circuit in Figure P14.49 have an offset voltage $V_{OS} = 2$ mV, an average input bias current of $I_B = 0.2 \, \mu\text{A}$, and an offset current of $I_{OS} = 0.02 \, \mu\text{A}$. (a) For $v_I = 0$ and $R_A = R_B = 0$, determine the possible range in output voltages $v_{O1}$, $v_{O2}$, and $v_{O3}$. (b) Repeat part (a) for $R_A = 8.33 \, \text{k}\Omega$ and $R_B = 10 \, \text{k}\Omega$.

14.55 Each op-amp in Figure P14.50 has an offset voltage of $V_{OS} = 2$ mV, an average input bias current of $I_B = 500$ nA, and an input offset current of $I_{OS} = 100$ nA. Determine the worst-case output voltage for each circuit.

### Section 14.6 Additional Nonideal Effects

14.56 For each op-amp in Figure P14.50, the input offset voltage is $V_{OS} = 2$ mV at $T = 25 \, ^\circ\text{C}$ and the input offset voltage temperature coefficient is $\text{TC}v_{OS} = 6.7 \, \mu\text{V}/^\circ\text{C}$. Find the output voltage $v_O$ due to the input offset voltage effects at: (a) $T = 25 \, ^\circ\text{C}$ and (b) $T = 50 \, ^\circ\text{C}$.

14.57 The input offset voltage in each op-amp in Figure P14.57 is $V_{OS} = 1$ mV at $T = 25 \, ^\circ\text{C}$ and the input offset voltage coefficient is $\text{TC}v_{OS} = 3.3 \, \mu\text{V}/^\circ\text{C}$. Find the worst-case output voltages $v_{O1}$ and $v_{O2}$ at: (a) $T = 25 \, ^\circ\text{C}$ and (b) $T = 50 \, ^\circ\text{C}$.

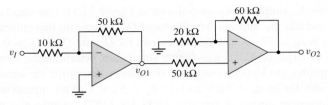

**Figure P14.57**

14.58    For each op-amp in Figure P14.50, the input bias current is $I_B = 500$ nA at $T = 25\,°C$, the input offset current is $I_{OS} = 200$ nA at $T = 25\,°C$, the input bias current temperature coefficient is 8 nA/°C, and the input offset current temperature coefficient is 2 nA/°C. (a) Find the output voltage due to the average input bias currents at $T = 25\,°C$. (b) Find the worst-case output voltage due to the input bias current and input offset current at $T = 25\,°C$. (c) Repeat parts (a) and (b) for $T = 50\,°C$.

14.59    For each op-amp in Figure P14.57, the input bias current is $I_B = 2\ \mu A$ at $T = 25\,°C$, the input offset current is $I_{OS} = 0.2\ \mu A$ at $T = 25\,°C$, the input bias current temperature coefficient is 20 nA/°C, and the input offset current temperature coefficient is 5 nA/°C. (a) Find the worst-case output voltages $v_{O1}$ and $v_{O2}$ due to the average input bias currents at $T = 25\,°C$. (b) Find the worst-case output voltages $v_{O1}$ and $v_{O2}$ due to the input bias currents and input offset current at $T = 25\,°C$. (c) Repeat parts (a) and (b) for $T = 50\,°C$.

14.60    The op-amp in the difference amplifier configuration in Figure P14.60 is ideal. (a) If the tolerance of each resistor is $\pm1.5\%$, determine the minimum value of $\text{CMRR}_{dB}$. (b) Repeat part (a) if the tolerance of each resistor is $\pm3\%$.

14.61    If the tolerance of each resistor in the difference amplifier in Figure P14.60 is $\pm x\%$, what is the maximum value of $x$ if the minimum $\text{CMRR}_{dB}$ is (a) 50 dB and (b) 75 dB.

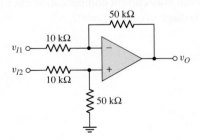

**Figure P14.60**

## COMPUTER SIMULATION PROBLEMS

14.62    Consider an inverting amplifier such as shown in Figure 14.2. Bias a standard op-amp at $\pm5$ V, and let $R_2 = 100\ \text{k}\Omega$ and $R_1 = 10\ \text{k}\Omega$. Using a computer simulation, plot $v_O$ versus $v_I$ over the range $-0.7 \leq v_I \leq 0.7$ V. What is the output saturation voltage?

14.63   Consider the simplified op-amp shown in Figure 14.11. Use standard transistors and take the output at the collector of $Q_6$. Assume the bias current for $Q_1$ and $Q_2$ is $I_Q = 19\,\mu A$ and the bias current for $Q_5$ and $Q_6$ is $I_Q = 0.15$ mA. Let $C_1 = 30$ pF. (a) Using a computer simulation, determine the slew rate of the amplifier. (b) Using a computer simulation, determine the small-signal bandwidth for (i) $v_d = 1\,\mu V$ and (ii) $v_d = 5\,\mu V$. Use an appropriate load.

14.64   The equivalent circuit of the all-CMOS MC14573 op-amp was given in Figure 13.14. Using a computer simulation, determine the slew rate of the op-amp assuming $C_1 = 12$ pF. Use standard transistors.

14.65   A basic bipolar input diff-amp stage is shown in Figure 14.22. Use standard transistors and other appropriate circuit parameters. Let $v_1 = v_2 = 0$. (a) Plot $i_{C1}$ and $i_{C2}$ as a function of the wiper arm position $x$. (b) Plot the collector voltage of $Q_4$ as a function of wiper arm position $x$.

 ## DESIGN PROBLEMS

[Note: Each design should be verified with a computer analysis.]

*D14.66 An amplifier system, using op-amps, is to be designed to provide a low-frequency voltage gain of 50 and a bandwidth of 20 kHz. The only available op-amps have a low-frequency open-loop voltage gain of $3 \times 10^4$ and a bandwidth of 10 Hz. Design an appropriate system.

*D14.67 Consider the simplified op-amp in Figure 14.11. Neglect the emitter-follower output stage. Assume bias voltages of $V^+ = 3$ V and $V^- = -3V$. Let the bias current for $Q_5$ and $Q_6$ be $I_Q = 0.1$ mA. The total power dissipated in the circuit is to be limited to 0.65 mW. Design the circuit such that the slew rate is 2 V/$\mu$s. Determine $I_Q$ for $Q_1$ and $Q_2$, and find the appropriate value for $C_1$.

*D14.68 Consider the op-amp circuit shown in Figure P14.12. Each op-amp has an offset voltage of $V_{OS} = 2$ mV. Design an offset voltage compensation circuit. Assume bias voltages are limited to ±5 V.

*D14.69 Consider the op-amp circuit shown in Figure P14.12. Each op-amp has an average input bias current of $I_B = 1\,\mu A$ and the offset bias current is $I_{OS} = 0.1\,\mu A$. Design an optimum bias-current compensation circuit. What is the possible range of output voltage $v_{O2}$ for $v_I = 0$?

# Applications and Design of Integrated Circuits

In Chapter 9, we introduced the ideal operational amplifier and analyzed and designed basic op-amp circuits. In this chapter, we consider additional applications and designs of op-amp and comparator circuits that may be fabricated as integrated circuits. A comparator is essentially an op-amp operated in an open-loop configuration with either a high or low saturated output signal.

Circuits to be considered include active filters, oscillators, Schmitt trigger circuits, integrated circuit power amplifiers, and voltage regulators.

A general goal of this chapter is to increase our skill at designing electronic circuits to meet particular specifications and to perform particular functions.

## PREVIEW

In this chapter, we will:

- Analyze and design active filters that transmit desired frequency components of an input signal and attenuate undesired frequency components.
- Analyze and design oscillators that provide sinusoidal signals at specified frequencies.
- Analyze and design various Schmitt trigger circuits.
- Analyze and design multivibrator circuits that provide signals with particular waveforms.
- Analyze and design IC power amplifiers that usually consist of high-gain small-signal amplifiers in cascade with an output stage.
- Analyze and design voltage regulators that establish a relatively constant dc voltage generated from an ac signal source.
- As an application, design an active bandpass filter to meet a set of specifications.

# 15.1  ACTIVE FILTERS

**Objective:** • Analyze and design active filters that transmit desired frequency components of an input signal and attenuate undesired frequency components.

An important application of an op-amp is the **active filter.** The word filter refers to the process of removing undesired portions of the frequency spectrum. The word *active* implies the use of one or more active devices, usually an operational amplifier, in the filter circuit. As an example of the application of op-amps in the area of active filters, we will discuss the Butterworth filter. There are many types or classifications of filters. However, the objective here is to concentrate mainly on a single type (Butterworth) in order to demonstrate the use of op-amps in filter design. Additional types of filters are discussed in other references.

Two advantages of active filters over passive filters are:

1.  The maximum gain or the maximum value of the transfer function may be greater than unity.
2.  The loading effect is minimal, which means that the output response of the filter is essentially independent of the load driven by the filter.

## 15.1.1  Active Network Design

From our discussions of frequency response in Chapter 7, we know that $RC$ networks form filters. Figure 15.1(a) is a simple example of a coupling-capacitor circuit. The voltage transfer function for this circuit is

$$T(s) = \frac{V_o(s)}{V_i(s)} = \frac{R}{R + \dfrac{1}{sC}} = \frac{sRC}{1 + sRC} \tag{15.1}$$

The Bode plot of the voltage gain magnitude $|T(j\omega)|$ is shown in Figure 15.1(b). The circuit is called a **high-pass filter.**

Figure 15.2(a) is another example of a simple RC network. Here, the voltage transfer function is

$$T(s) = \frac{V_o(s)}{V_i(s)} = \frac{\dfrac{1}{sC}}{\dfrac{1}{sC} + R} = \frac{1}{1 + sRC} \tag{15.2}$$

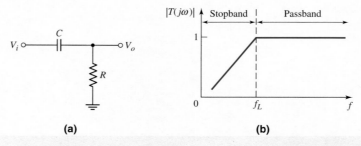

(a)                              (b)

**Figure 15.1** (a) Simple high-pass filter and (b) Bode plot of transfer function magnitude

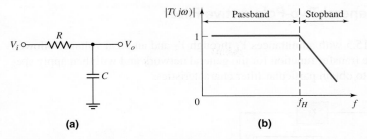

**(a)**                                   **(b)**

**Figure 15.2** (a) Simple low-pass filter and (b) Bode plot of transfer function magnitude

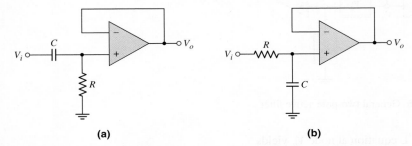

**(a)**                                   **(b)**

**Figure 15.3** (a) High-pass filter with voltage follower and (b) low-pass filter with voltage follower

The Bode plot of the voltage gain magnitude $|T(j\omega)|$ for this circuit is shown in Figure 15.2(b). This circuit is called a **low-pass filter.**

Although these circuits both perform a basic filtering function, they may suffer from loading effects, substantially reducing the maximum gain from the unity value shown in Figures 15.1(b) and 15.2(b). Also, the cutoff frequencies $f_L$ and $f_H$ may change when a load is connected to the output. The loading effect can essentially be eliminated by using a voltage follower as shown in Figure 15.3. In addition, a noninverting amplifier configuration can be incorporated to increase the gain, as well as eliminate the loading effects.

These two filter circuits are called one-pole filters; the slope of the voltage gain magnitude curve outside the passband is 6 dB/octave or 20 dB/decade. This characteristic is called the rolloff. The rolloff becomes sharper or steeper with higher-order filters and is usually one of the specifications given for active filters.

Two other categories of filters are **bandpass** and **band-reject.** The desired ideal frequency characteristics are shown in Figure 15.4.

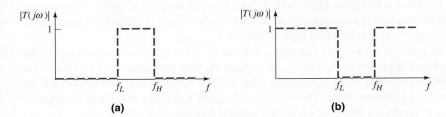

**(a)**                                   **(b)**

**Figure 15.4** Ideal frequency characteristics: (a) bandpass filter and (b) band-reject filter

15.1.2    **General Two-Pole Active Filter**

Consider Figure 15.5 with admittances $Y_1$ through $Y_4$ and an ideal voltage follower. We will derive the transfer function for the general network and will then apply specific admittances to obtain particular filter characteristics.

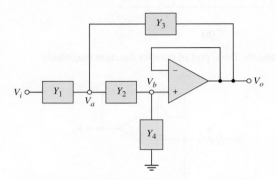

**Figure 15.5** General two-pole active filter

A KCL equation at node $V_a$ yields

$$(V_i - V_a)Y_1 = (V_a - V_b)Y_2 + (V_a - V_o)Y_3 \tag{15.3}$$

A KCL equation at node $V_b$ produces

$$(V_a - V_b)Y_2 = V_b Y_4 \tag{15.4}$$

From the voltage follower characteristics, we have $V_b = V_o$. Therefore, Equation (15.4) becomes

$$V_a = V_b \left( \frac{Y_2 + Y_4}{Y_2} \right) = V_o \left( \frac{Y_2 + Y_4}{Y_2} \right) \tag{15.5}$$

Substituting Equation (15.5) into (15.3) and again noting that $V_b = V_o$, we have

$$V_i Y_1 + V_o(Y_2 + Y_3) = V_a(Y_1 + Y_2 + Y_3)$$

$$= V_o \left( \frac{Y_2 + Y_4}{Y_2} \right)(Y_1 + Y_2 + Y_3) \tag{15.6}$$

Multiplying Equation (15.6) by $Y_2$ and rearranging terms, we get the following expression for the transfer function:

$$T(s) = \frac{V_o(s)}{V_i(s)} = \frac{Y_1 Y_2}{Y_1 Y_2 + Y_4(Y_1 + Y_2 + Y_3)} \tag{15.7}$$

To obtain a low-pass filter, both $Y_1$ and $Y_2$ must be conductances, allowing the signal to pass into the voltage follower at low frequencies. If element $Y_4$ is a capacitor, then the output rolls off at high frequencies.

To produce a two-pole function, element $Y_3$ must also be a capacitor. On the other hand, if elements $Y_1$ and $Y_2$ are capacitors, then the signal will be blocked at low frequencies but will be passed into the voltage follower at high frequencies, resulting in a high-pass filter. Therefore, admittances $Y_3$ and $Y_4$ must both be conductances to produce a two-pole high-pass transfer function.

### 15.1.3  Two-Pole Low-Pass Butterworth Filter

To form a low-pass filter, we set $Y_1 = G_1 = 1/R_1$, $Y_2 = G_2 = 1/R_2$, $Y_3 = sC_3$, and $Y_4 = sC_4$, as shown in Figure 15.6. The transfer function, from Equation (15.7), becomes

$$T(s) = \frac{V_o(s)}{V_i(s)} = \frac{G_1 G_2}{G_1 G_2 + sC_4(G_1 + G_2 + sC_3)} \tag{15.8}$$

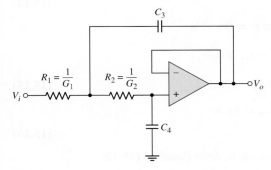

**Figure 15.6** General two-pole low-pass filter

At zero frequency, $s = j\omega = 0$ and the transfer function is

$$T(s = 0) = \frac{G_1 G_2}{G_1 G_2} = 1 \tag{15.9}$$

In the high-frequency limit, $s = j\omega \to \infty$ and the transfer function approaches zero. This circuit therefore acts as a low-pass filter.

A **Butterworth filter** is a **maximally flat magnitude filter.** The transfer function is designed such that the magnitude of the transfer function is as flat as possible within the passband of the filter. This objective is achieved by taking the derivatives of the transfer function with respect to frequency and setting as many as possible equal to zero at the center of the passband, which is at zero frequency for the low-pass filter.

Let $G_1 = G_2 \equiv G = 1/R$. The transfer function is then

$$T(s) = \frac{\dfrac{1}{R^2}}{\dfrac{1}{R^2} + sC_4\left(\dfrac{2}{R} + sC_3\right)} = \frac{1}{1 + sRC_4(2 + sRC_3)} \tag{15.10}$$

We define time constants at $\tau_3 = RC_3$ and $\tau_4 = RC_4$. If we then set $s = j\omega$, we obtain

$$T(j\omega) = \frac{1}{1 + j\omega\tau_4(2 + j\omega\tau_3)} = \frac{1}{(1 - \omega^2\tau_3\tau_4) + j(2\omega\tau_4)} \tag{15.11}$$

The magnitude of the transfer function is therefore

$$|T(j\omega)| = [(1 - \omega^2\tau_3\tau_4)^2 + (2\omega\tau_4)^2]^{-1/2} \tag{15.12}$$

For a maximally flat filter (that is, a filter with a minimum rate of change), which defines a Butterworth filter, we set

$$\frac{d|T|}{d\omega}\bigg|_{\omega=0} = 0 \tag{15.13}$$

Taking the derivative, we find

$$\frac{d|T|}{d\omega} = -\frac{1}{2}[(1 - \omega^2\tau_3\tau_4)^2 + (2\omega\tau_4)^2]^{-3/2}\left[-4\omega\tau_3\tau_4(1 - \omega^2\tau_3\tau_4) + 8\omega\tau_4^2\right]$$

$$\tag{15.14}$$

Setting the derivative equal to zero at $\omega = 0$ yields

$$\frac{d|T|}{d\omega}\bigg|_{\omega=0} = \left[-4\omega\tau_3\tau_4(1 - \omega^2\tau_3\tau_4) + 8\omega\tau_4^2\right]$$

$$= 4\omega\tau_4[-\tau_3(1 - \omega^2\tau_3\tau_4) + 2\tau_4] \tag{15.15}$$

Equation (15.15) is satisfied when $2\tau_4 = \tau_3$, or

$$C_3 = 2C_4 \tag{15.16}$$

For this condition, the transfer magnitude is, from Equation (15.12),

$$|T| = \frac{1}{[1 + 4(\omega\tau_4)^4]^{1/2}} \tag{15.17}$$

The 3 dB, or cutoff, frequency occurs when $|T| = 1/\sqrt{2}$, or when $4(\omega_{3dB}\tau_4)^4 = 1$. We then find that

$$\omega_{3\,dB} = 2\pi f_{3\,dB} = \frac{1}{\tau_4\sqrt{2}} = \frac{1}{\sqrt{2}RC_4} \tag{15.18}$$

In general, we can write the cutoff frequency in the form

$$\omega_{3\,dB} = \frac{1}{RC} \tag{15.19}$$

Finally, comparing Equations (15.19), (15.18), and (15.16) yields

$$C_4 = 0.707C \tag{15.20(a)}$$

and

$$C_3 = 1.414C \tag{15.20(b)}$$

The two-pole low-pass Butterworth filter is shown in Figure 15.7(a). The Bode plot of the transfer function magnitude is shown in Figure 15.7(b). From Equation (15.17), the magnitude of the voltage transfer function for the two-pole low-pass Butterworth filter can be written as

$$|T| = \frac{1}{\sqrt{1 + \left(\dfrac{f}{f_{3\,dB}}\right)^4}} \tag{15.21}$$

Equation (15.15) shows that the derivative of the voltage transfer function magnitude at $\omega = 0$ is zero even without setting $2\tau_4 = \tau_3$. However, the added condition of $2\tau_4 = \tau_3$ produces the maximally flat transfer characteristics of the Butterworth filter.

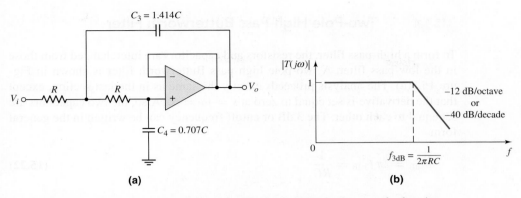

**Figure 15.7**  (a) Two-pole low-pass Butterworth filter and (b) Bode plot, transfer function magnitude

## DESIGN EXAMPLE **15.1**

**Objective:**  Design a two-pole low-pass Butterworth filter for an audio amplifier application.

**Specifications:**  The circuit with the configuration shown in Figure 15.7(a) is to be designed such that the bandwidth is 20 kHz.

**Choices:**  An ideal op-amp is available and standard-valued resistors and capacitors must be used.

**Solution:**  From Equation (15.19), we have

$$f_{3\,dB} = \frac{1}{2\pi RC}$$

or

$$RC = \frac{1}{2\pi f_{3\,dB}} = \frac{1}{2\pi (20 \times 10^3)} = 7.96 \times 10^{-6}$$

If we let $R = 100 \text{ k}\Omega$, then $C = 79.6$ pF, which means that $C_3 = 1.414C = 113$ pF and $C_4 = 0.707C = 56.3$ pF.

**Trade-offs:**  Standard-valued 100 k$\Omega$ resistors can be used. Standard-valued $C_3 = 120$ pF and $C_4 = 56$ pF capacitors can be used. For these elements, a bandwidth of 20.1 kHz is obtained.

**Comment:**  These resistance and capacitance values are generally too large to be fabricated conveniently on an IC. Instead, discrete resistors and capacitors, in conjunction with the IC op-amp, would need to be used.

## EXERCISE PROBLEM

**Ex 15.1:**  Design a two-pole low-pass Butterworth filter with a bandwidth of 25 kHz. The largest capacitor value to be used is 50 pF. (Ans. Set $C_3 = 50$ pF, then $C_4 = 25$ pF, $R = 180 \text{ k}\Omega$)

### 15.1.4   Two-Pole High-Pass Butterworth Filter

To form a high-pass filter, the resistors and capacitors are interchanged from those in the low-pass filter. A two-pole high-pass Butterworth filter is shown in Figure 15.8(a). The analysis proceeds exactly the same as in the last section, except that the derivative is set equal to zero at $s = j\omega = \infty$. Also, the two capacitors are set equal to each other. The 3 dB or cutoff frequency can be written in the general form

$$\omega_{3\,\mathrm{dB}} = 2\pi f_{3\,\mathrm{dB}} = \frac{1}{RC} \tag{15.22}$$

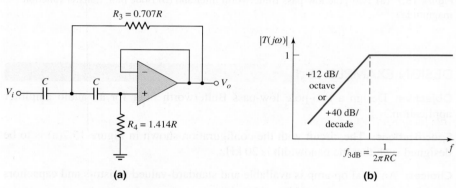

**Figure 15.8**  (a) Two-pole high-pass Butterworth filter and (b) Bode plot, transfer function magnitude

We find that $R_3 = 0.707\,R$ and $R_4 = 1.414\,R$. The magnitude of the voltage transfer function for the two-pole high-pass Butterworth is

$$|T| = \frac{1}{\sqrt{1 + \left(\dfrac{f_{3\,\mathrm{dB}}}{f}\right)^4}} \tag{15.23}$$

The Bode plot of the transfer function magnitude for the two-pole high-pass Butterworth filter is shown in Figure 15.8(b).

### 15.1.5   Higher-Order Butterworth Filters

The filter order is the number of poles and is usually dictated by the application requirements. An $N$-pole active low-pass filter has a high-frequency rolloff rate of $N \times 6$ dB/octave. Similarly, the response of an $N$-pole high-pass filter increases at a rate of $N \times 6$ dB/octave, up to the cutoff frequency. In each case, the 3 dB frequency is defined as

$$f_{3\,\mathrm{dB}} = \frac{1}{2\pi RC} \tag{15.24}$$

The magnitude of the voltage transfer function for a Butterworth $N$th-order low-pass filter is

$$|T| = \frac{1}{\sqrt{1 + \left(\dfrac{f}{f_{3\,\mathrm{dB}}}\right)^{2N}}} \qquad\qquad \textbf{(15.25)}$$

For a Butterworth $N$th-order high-pass filter, the voltage transfer function magnitude is

$$|T| = \frac{1}{\sqrt{1 + \left(\dfrac{f_{3\,\mathrm{dB}}}{f}\right)^{2N}}} \qquad\qquad \textbf{(15.26)}$$

Figure 15.9(a) shows a three-pole low-pass Butterworth filter. The three resistors are equal, and the relationship between the capacitors is found by taking the first and second derivatives of the voltage gain magnitude with respect to frequency and setting those derivatives equal to zero at $s = j\omega = 0$. Figure 15.9(b) shows a three-pole high-pass Butterworth filter. In this case, the three capacitors are equal and the relationship between the resistors is also found through the derivatives.

Higher-order filters can be created by adding additional $RC$ networks. However, the loading effect on each additional $RC$ circuit becomes more severe. The usefulness of active filters is realized when two or more op-amp filter circuits are cascaded to

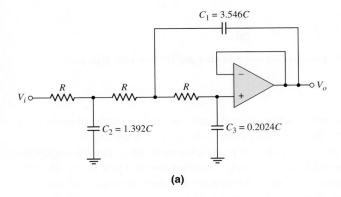

**(a)**

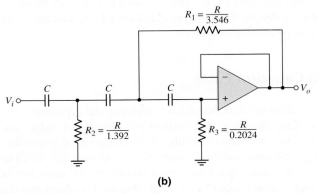

**(b)**

**Figure 15.9** (a) Three-pole low-pass Butterworth filter and (b) three-pole high-pass Butterworth filter

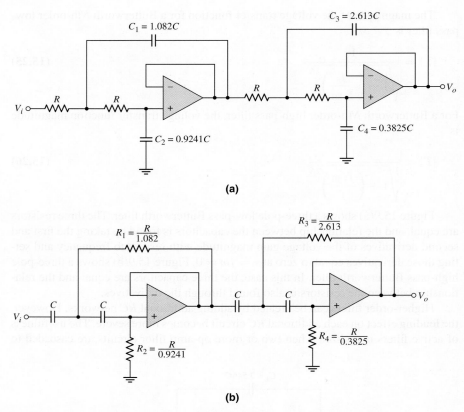

**Figure 15.10**  (a) Four-pole low-pass Butterworth filter and (b) four-pole high-pass Butterworth filter

produce one large higher-order active filter. Because of the low output impedance of the op-amp, there is virtually no loading effect between cascaded stages.

Figure 15.10(a) shows a four-pole low-pass Butterworth filter. The maximally flat response of this filter is *not* obtained by simply cascading two two-pole filters. The relationship between the capacitors is found through the first three derivatives of the transfer function. The four-pole high-pass Butterworth filter is shown in Figure 15.10(b).

Higher-order filters can be designed but are not considered here. Bandpass and band-reject filters use similar circuit configurations.

### 15.1.6    Switched-Capacitor Filter

The results of Example 15.1 demonstrated that discrete resistors and capacitors may be needed in active filters, since the required resistance and capacitance values are too large to be conveniently fabricated on a monolithic IC chip. Large-value resistors ($R > 10$ k$\Omega$) require a large chip area, and the absolute-value tolerance is difficult to maintain. In addition, the maximum capacitance for a monolithic IC capacitor is approximately 100 pF, which is also limited by the large chip area required and the absolute-value tolerance. In these cases, accurate $RC$ time constants may be difficult to maintain.

Conventional active filters usually combine an IC op-amp and discrete resistors and capacitors. However, even with discrete resistors and capacitors, standard components may not be available for the design of a specific cutoff frequency. Design accuracy for a specific cutoff frequency may therefore have to be sacrificed.

**Switched-capacitor filters** have the advantage of an all-IC circuit. The filter uses small capacitance values and realizes large effective resistance values by using a combination of capacitors and MOS switching transistors.

### The Basic Principle of the Switched Capacitor

Figure 15.11 shows a simple circuit in which voltages $V_1$ and $V_2$ are applied at the terminals of a resistance $R$. The current in the resistor is

$$I = \frac{V_1 - V_2}{R} \tag{15.27(a)}$$

The resistance is therefore

$$R = \frac{V_1 - V_2}{I} \tag{15.27(b)}$$

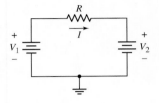

**Figure 15.11**  Voltages applied to resistor terminals, and the current

Since the current is the rate of charge flow, Equation (15.27(b)) states that the resistance is a voltage difference divided by the rate of charge flow. We use this basic definition in switched-capacitor circuits.

The circuit in Figure 15.12(a) consists of two MOSFETs and a capacitor. A two-phase clock provides complementary but nonoverlapping $\phi_1$ and $\phi_2$ gate pulses, as shown in Figure 15.12(b). When a clock pulse is high, the corresponding transistor turns on; when the gate pulse is low, the transistor is off.

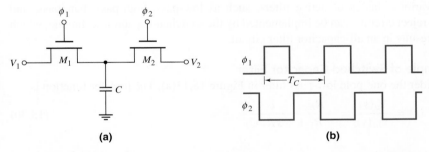

(a)  (b)

**Figure 15.12**  (a) Capacitor with two switching MOSFETs and (b) two-phase clock pulses

When $\phi_1$ goes high, $M_1$ turns on and capacitor $C$ charges up to $V_1$. When $\phi_2$ goes high, $M_2$ turns on and capacitor $C$ discharges to $V_2$ (assuming $V_1 > V_2$). The amount of charge transferred during this process is $Q = C(V_1 - V_2)$ and the transfer occurs during one clock period $T_C$. The equivalent current is then

$$I_{eq} = \frac{Q}{T_C} = \frac{C(V_1 - V_2)}{T_C} = f_C C(V_1 - V_2) = \frac{V_1 - V_2}{R_{eq}} \tag{15.28}$$

where $f_C$ is the clock frequency and $R_{eq}$ is the equivalent resistance given by

$$R_{eq} = \frac{1}{f_C C} \tag{15.29}$$

Using this technique, we can simulate an equivalent resistance by alternately charging and discharging a capacitor between two voltage levels. A large equivalent resistance can be simulated by using a small capacitance and an appropriate clock frequency. The circuit in Figure 15.12(a) is therefore called a switched-capacitor circuit.

## EXAMPLE 15.2

**Objective:** Determine the clock frequency required to simulate a specific resistance.

Consider the switched-capacitor circuit in Figure 15.12(a). Assume a capacitance of $C = 20$ pF. Determine the clock frequency required to simulate a 1 M$\Omega$ resistance.

**Solution:** From Equation (15.29), we find that

$$f_C = \frac{1}{C R_{eq}} = \frac{1}{(20 \times 10^{-12})(10^6)} \Rightarrow 50\,\text{kHz}$$

**Comment:** A very large resistance can be readily simulated by a small capacitance and a reasonable clock frequency.

### EXERCISE PROBLEM

**Ex 15.2:** Consider the switched-capacitor circuit in Figure 15.12(a). (a) If the clock frequency is $f_C = 100$ kHz and $C = 1.2$ pF, what is the value of the simulated resistance? (b) A 50 M$\Omega$ resistor is to be simulated using a clock frequency of $f_C = 50$ kHz. What is the required value of capacitor? (Ans. (a) $R_{eq} = 8.33$ M$\Omega$, (b) $C = 0.4$ pF)

Various classes of active filters, such as low-pass, high-pass, bandpass, and band-reject circuits, can be implemented by the switched-capacitor technique, which then results in an all-capacitor filter circuit.

**Example of Switched-Capacitor Filter**

Consider the one-pole low-pass filter in Figure 15.13(a). The transfer function is

$$T(s) = \frac{V_o(s)}{V_{in}(s)} = -\frac{R_F}{R_1}\frac{1}{1 + sR_FC_F} \tag{15.30}$$

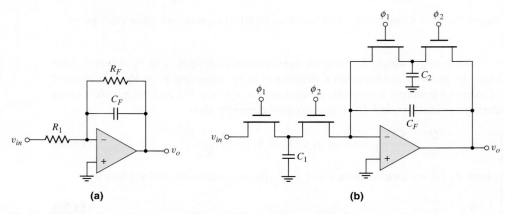

**Figure 15.13** (a) One-pole low-pass filter and (b) equivalent switched-capacitor circuit

and the cutoff frequency is

$$f_{3\,\text{dB}} = \frac{1}{2\pi R_F C_F} \tag{15.31}$$

If a 10 kHz cutoff frequency is required and if $C_F = 10$ pF, then the $R_F$ resistance required is approximately 1.6 M$\Omega$. In addition, if a gain of $-10$ is desired, then resistance $R_1$ must be 160 k$\Omega$.

The equivalent switched-capacitor filter is shown in Figure 15.13(b). The transfer function is still given by Equation (15.30), where $R_{Feq} = 1/(f_C C_2)$ and $R_{1eq} = 1/(f_C C_1)$. The transfer function is then

$$T(j\omega) = -\frac{(1/f_C C_2)}{(1/f_C C_1)} \cdot \frac{1}{1 + j\dfrac{(2\pi f)C_F}{f_C C_2}} = -\frac{C_1}{C_2} \cdot \frac{1}{1 + j\dfrac{f}{f_{3\,\text{dB}}}} \tag{15.32}$$

The low-frequency gain is $-C_1/C_2$, which is just the ratio of two capacitances, and the 3 dB frequency is

$$f_{3\,\text{dB}} = (f_C C_2)/(2\pi C_F)$$

which is also proportional to the ratio of two other capacitances. For MOS IC capacitance values of approximately 10 pF, the ratio tolerance is on the order of 0.1 percent. This means that switched-capacitor filter characteristics can be precisely controlled.

## DESIGN EXAMPLE 15.3

**Objective:** Design a one-pole low-pass switched capacitor filter to meet a set of specifications.

**Specifications:** The circuit with the configuration shown in Figure 15.13(b) is to be designed such that the low-frequency gain is $-1$ and the cutoff frequency is 1 kHz.

**Choices:** An ideal op-amp is available and standard-valued capacitors are to be used.

**Solution:** From Equation (15.32), the low-frequency gain is $-(C_1/C_2)$, and the capacitance ratio must be $(C_1/C_2) = 1$. From Equation (15.32), the cutoff frequency is

$$f_{3\,\text{dB}} = \frac{f_C C_2}{2\pi C_F}$$

If we set the clock frequency to $f_C = 10$ kHz, then

$$\frac{C_2}{C_F} = \frac{2\pi f_{3\,\text{dB}}}{f_C} = \frac{2\pi(10^3)}{10 \times 10^3} = 0.628$$

**Trade-offs:** We can use standard-valued capacitors $C_1 = C_2 = 75$ pF. We would need $C_F = C_2/0.628 = 75/0.628 = 119.4$ pF. A standard-valued capacitor $C_F = 120$ pF can be used.

**Comment:** Since the low-frequency gain and cutoff frequency are both functions of capacitor ratios, the absolute capacitor values can be designed for compatibility with IC fabrication.

**Ex 15.3:** For the switched-capacitor circuit in Figure 15.13(b), the parameters are: $C_1 = 30$ pF, $C_2 = 5$ pF, and $C_F = 12$ pF. The clock frequency is 100 kHz. Determine the low-frequency gain and the cutoff frequency. (Ans. $-C_1/C_2 = -6$, $f_{3\,dB} = 6.63$ kHz)

This discussion of switched-capacitor filters is a short introduction to the topic and is intended only to show another application of operational amplifiers. Switched-capacitor filters are "sampled-data systems"; that is, the analog input signal is not transmitted through the circuit as a continuous signal but passes through the system as a series of pulses. The equivalent resistance given by Equation (15.29) is valid only for clock frequencies much greater than the analog input signal frequency. Switched-capacitor systems can be analyzed and designed by $z$-transform techniques.

## Test Your Understanding

**TYU 15.1** (a) Design a three-pole high-pass Butterworth active filter with a cutoff frequency of 200 Hz and a unity gain at high frequency. (b) Using the results of part (a), determine the magnitude of the voltage transfer function at (i) $f = 100$ Hz and (ii) $f = 300$ Hz. (Ans. (a) Let $C = 0.01\,\mu$F, then $R_1 = 22.44$ k$\Omega$, $R_2 = 57.17$ k$\Omega$, $R_3 = 393.2$ k$\Omega$; (b) (i) $|T| = 0.124 \rightarrow -18.1$ dB, (ii) $|T| = 0.959 \rightarrow -0.365$ dB)

**TYU 15.2** (a) Design a four-pole low-pass Butterworth active filter with a 3 dB frequency of 30 kHz. (b) Determine the frequency at which the voltage transfer function magnitude is 99 percent of its maximum value. (Ans. (a) Let $R = 100$ k$\Omega$, then $C_1 = 57.4$ pF, $C_2 = 49.02$ pF, $C_3 = 138.6$ pF, $C_4 = 20.29$ pF; (b) $f = 18.43$ kHz)

**TYU 15.3** One-, two-, three-, and four-pole low-pass Butterworth active filters are all designed with a cutoff frequency of 10 kHz and unity gain at low frequency. Determine the voltage transfer function magnitude, in dB, at 12 kHz for each filter. (Ans. $-3.87$ dB, $-4.88$ dB, $-6.0$ dB, and $-7.24$ dB)

**TYU 15.4** Simulate a 25 M$\Omega$ resistance using the circuit in Figure 15.12(a). What capacitor value and clock frequency are required? (Ans. For example, for $f_C = 50$ kHz, then $C = 0.8$ pF)

##  15.2 OSCILLATORS

**Objective:** • Analyze and design oscillators that provide sinusoidal signals at specified frequencies.

In this section, we will look at the basic principles of sine-wave oscillators. In our study of feedback in Chapter 12, we emphasized the need for negative feedback to provide a stable circuit. Oscillators, however, use positive feedback and, therefore, are actually nonlinear circuits in some cases. The analysis and design of oscillator circuits are divided into two parts. In the first part, the condition and frequency for oscillation are determined; in the second part, means for amplitude control is addressed. We consider only the first step in this section to gain insight into the basic operation of oscillators.

### 15.2.1    Basic Principles for Oscillation

The basic **oscillator** consists of an amplifier and a **frequency-selective network** connected in a feedback loop. Figure 15.14 shows a block diagram of the fundamental feedback circuit, in which we are implicitly assuming that negative feedback is employed. Although actual oscillator circuits do not have an input signal, we initially include one here to help in the analysis. In previous feedback circuits, we assumed the feedback transfer function $\beta$ was independent of frequency. In oscillator circuits, however, $\beta$ is the principal portion of the loop gain that is dependent on frequency.

For the circuit shown, the ideal closed-loop transfer function is given by

$$A_f(s) = \frac{A(s)}{1 + A(s)\beta(s)} \tag{15.33}$$

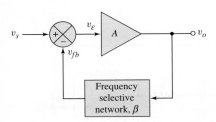

**Figure 15.14**  Block diagram of the fundamental feedback circuit

and the loop gain of the feedback circuit is

$$T(s) = A(s)\beta(s) \tag{15.34}$$

From our discussion of feedback in Chapter 12, we know that the loop gain $T(s)$ is positive for negative feedback, which means that the feedback signal $v_{fb}$ subtracts from the input signal $v_s$. If the loop gain $T(s)$ becomes negative, then the feedback signal phase causes $v_{fb}$ to add to the input signal, increasing the error signal $v_\varepsilon$. If $T(s) = -1$, the closed-loop transfer function goes to infinity, which means that the circuit can have a finite output for a zero input signal.

As $T(s)$ approaches $-1$, an actual circuit becomes nonlinear, which means that the gain does not go to infinity. Assume that $T(s) \approx -1$ so that positive feedback exists over a particular frequency range. If a spontaneous signal (due to noise) is created at $v_s$ in this frequency range, the resulting feedback signal $v_{fb}$ is in phase with $v_s$, and the error signal $v_\varepsilon$ is reinforced and increased. This reinforcement process continues at only those frequencies for which the total phase shift around the feedback loop is zero. Therefore, the condition for oscillation is that, at a specific frequency, we have

$$T(j\omega_o) = A(j\omega_o)\beta(j\omega_o) = -1 \tag{15.35}$$

The condition that $T(j\omega_o) = -1$ is called the **Barkhausen criterion.**

Equation (15.35) shows that two conditions must be satisfied to sustain oscillation:

1.  The total phase shift through the amplifier and feedback network must be $N \times 360°$, where $N = 0, 1, 2, \ldots$.
2.  The magnitude of the loop gain must be unity.

In the feedback circuit block diagram in Figure 15.14, we implicitly assume negative feedback. For an oscillator, the feedback transfer function, or the frequency-selective network, must introduce an additional 180 degree phase shift such that the net phase around the entire loop is zero. For the circuit to oscillate at a single frequency $\omega_o$, the condition for oscillation, from Equation (15.35), should be satisfied at only that one frequency.

### 15.2.2   Phase-Shift Oscillator

An example of an op-amp oscillator is the **phase-shift oscillator.** One configuration of this oscillator circuit is shown in Figure 15.15. The basic amplifier of the circuit is the op-amp $A_3$, which is connected as an inverting amplifier with its output connected to a three-stage $RC$ filter. The voltage followers in the circuit eliminate loading effects between each $RC$ filter stage.

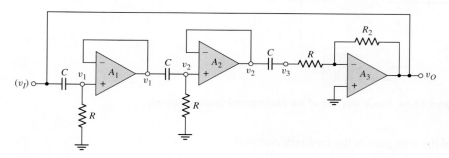

**Figure 15.15**   Phase-shift oscillator circuit with voltage-follower buffer stages

The inverting amplifier introduces a $-180$ degree phase shift, which means that each $RC$ network must provide 60 degrees of phase shift to produce the 180 degrees required of the frequency-sensitive feedback network in order to produce positive feedback. Note that the inverting terminal of op-amp $A_3$ is at virtual ground; therefore, the $RC$ network between op-amps $A_2$ and $A_3$ functions exactly as the other two $RC$ networks. We assume that the frequency effects of the op-amps themselves occur at much higher frequencies than the response due to the $RC$ networks. Also, to aid in the analysis, we assume an input signal ($v_I$) exists at one node as shown in the figure.

The transfer function of the first $RC$ network is

$$v_1 = \left( \frac{sRC}{1 + sRC} \right)(v_I) \tag{15.36}$$

Since the $RC$ networks are assumed to be identical, and since there is no loading effect of one $RC$ stage on another, we have

$$\frac{v_3}{(v_I)} = \left( \frac{sRC}{1 + sRC} \right)^3 = \beta(s) \tag{15.37}$$

where $\beta(s)$ is the feedback transfer function. The amplifier gain $A(s)$ in Equation (15.33) and (15.34) is actually the magnitude of the gain, or

$$A(s) = \left| \frac{v_O}{v_3} \right| = \frac{R_2}{R} \tag{15.38}$$

The loop gain is then

$$T(s) = A(s)\beta(s) = \left(\frac{R_2}{R}\right)\left(\frac{sRC}{1+sRC}\right)^3 \qquad (15.39)$$

From Equation (15.35), the condition for oscillation is that $|T(j\omega_o)| = 1$ and the phase of $T(j\omega_o)$ must be 180 degrees. When these requirements are satisfied, then $v_O$ will equal $(v_I)$ and a separate input signal will not be required.

If we set $s = j\omega$, Equation (15.39) becomes

$$\begin{aligned}
T(j\omega) &= \left(\frac{R_2}{R}\right)\frac{(j\omega RC)^3}{(1+j\omega RC)^3} \\
&= -\left(\frac{R_2}{R}\right)\frac{(j\omega RC)(\omega RC)^2}{[1-3\omega^2 R^2 C^2] + j\omega RC[3-\omega^2 R^2 C^2]}
\end{aligned} \qquad (15.40)$$

To satisfy the condition $T(j\omega_o) = -1$, the imaginary component of Equation (15.40) must equal zero. Since the numerator is purely imaginary, the denominator must become purely imaginary, or

$$[1 - 3\omega_o^2 R^2 C^2] = 0$$

which yields

$$\omega_o = \frac{1}{\sqrt{3}RC} \qquad (15.41)$$

where $\omega_o$ is the oscillation frequency. At this frequency, Equation (15.40) becomes

$$T(j\omega_o) = -\left(\frac{R_2}{R}\right)\frac{(j/\sqrt{3})(1/3)}{0 + (j/\sqrt{3})[3-(1/3)]} = -\left(\frac{R_2}{R}\right)\left(\frac{1}{8}\right) \qquad (15.42)$$

Consequently, the condition $T(j\omega_o) = -1$ is satisfied when

$$\frac{R_2}{R} = 8 \qquad (15.43)$$

Equation (15.43) implies that if the magnitude of the inverting amplifier gain is greater than 8, the circuit will spontaneously begin oscillating and will sustain oscillation.

## EXAMPLE 15.4

**Objective:** Determine the oscillation frequency and required amplifier gain for a phase-shift oscillator.

Consider the phase-shift oscillator in Figure 15.15 with parameters $C = 0.1\ \mu\text{F}$ and $R = 1\ \text{k}\Omega$.

**Solution:** From Equation (15.41), the oscillation frequency is

$$f_o = \frac{1}{2\pi\sqrt{3}RC} = \frac{1}{2\pi\sqrt{3}(10^3)(0.1\times 10^{-6})} = 919\ \text{Hz}$$

The minimum amplifier gain magnitude is 8 from Equation 15.43; therefore, the minimum value of $R_2$ is 8 k$\Omega$.

**Comment:** Higher oscillation frequencies can easily be obtained by using smaller capacitor values.

**Ex 15.4:** Design the phase-shift oscillator shown in Figure 15.15 to oscillate at $f_o = 22.5\,\text{kHz}$. The minimum resistance to be used is $10\,\text{k}\Omega$. (Ans. Set $R = 10\,\text{k}\Omega$, $C = 408\,\text{pF}$, $R_2 = 80\,\text{k}\Omega$)

Using Equation (15.36), we can determine the effect of each $RC$ network in the phase-shift oscillator. At the oscillation frequency $\omega_o$, the transfer function of each $RC$ network stage is

$$\frac{j\omega_o RC}{1 + j\omega_o RC} = \frac{(j/\sqrt{3})}{1 + (j/\sqrt{3})} = \frac{j}{\sqrt{3} + j} \tag{15.44}$$

which can be written in terms of the magnitude and phase, as follows:

$$\frac{1}{\sqrt{3+1}} \times \frac{\angle 90^\circ}{\angle \tan^{-1}(1/\sqrt{3})} = \frac{1}{2} \times [\angle 90^\circ - \angle \tan^{-1}(0.577)] \tag{15.45(a)}$$

or

$$\frac{1}{2} \times (\angle 90^\circ - \angle 30^\circ) = \frac{1}{2} \times \angle 60^\circ \tag{15.45(b)}$$

As required, each $RC$ network introduces a 60 degree phase shift, but they each also introduce an attenuation factor of ($\frac{1}{2}$) for which the amplifier must compensate.

The two voltage followers in the circuit in Figure 15.15 need not be included in a practical phase-shift oscillator. Figure 15.16 shows a phase-shift oscillator without the voltage-follower buffer stages. The three $RC$ network stages and the inverting amplifier are still included. The loading effect of each successive $RC$ network complicates the analysis, but the same principle of operation applies. The analysis shows that the oscillation frequency is

$$\omega_o = \frac{1}{\sqrt{6}RC} \tag{15.46}$$

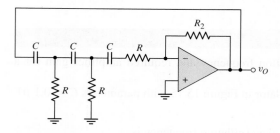

**Figure 15.16** Phase-shift oscillator circuit

and the amplifier resistor ratio must be

$$\frac{R_2}{R} = 29 \tag{15.47}$$

in order to sustain oscillation.

### 15.2.3    Wien-Bridge Oscillator

Another basic oscillator is the **Wien-bridge circuit,** shown in Figure 15.17. The circuit consists of an op-amp connected in a noninverting configuration and two $RC$ networks connected as the frequency-selecting feedback circuit.

Again, we initially assume that an input signal exists at the noninverting terminals of the op-amp. Since the noninverting amplifier introduces zero phase shift, the frequency-selective feedback circuit must also introduce zero phase shift to create the positive feedback condition.

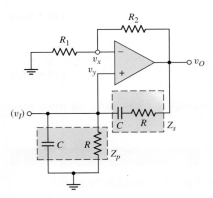

**Figure 15.17** Wien-bridge oscillator

The loop gain is the product of the amplifier gain and the feedback transfer function, or

$$T(s) = \left(1 + \frac{R_2}{R_1}\right)\left(\frac{Z_p}{Z_p + Z_s}\right) \tag{15.48}$$

where $Z_p$ and $Z_s$ are the parallel and series $RC$ network impedances, respectively. These impedances are

$$Z_p = \frac{R}{1 + sRC} \tag{15.49(a)}$$

and

$$Z_s = \frac{1 + sRC}{sC} \tag{15.49(b)}$$

Combining Equations (15.49(a)), (15.49(b)), and (15.48), we get an expression for the loop gain function,

$$T(s) = \left(1 + \frac{R_2}{R_1}\right)\left[\frac{1}{3 + sRC + (1/sRC)}\right] \tag{15.50}$$

Since this circuit has no explicit negative feedback, as was assumed in the general network shown in Figure 15.14, the condition for oscillation is given by

$$T(j\omega_o) = 1 = \left(1 + \frac{R_2}{R_1}\right)\left[\frac{1}{3 + j\omega_o RC + (1/j\omega_o RC)}\right] \tag{15.51}$$

Since $T(j\omega_o)$ must be real, the imaginary component of Equation (15.51) must be zero; therefore,

$$j\omega_o RC + \frac{1}{j\omega_o RC} = 0 \tag{15.52(a)}$$

which gives the frequency of oscillation as

$$\omega_o = \frac{1}{RC} \tag{15.52(b)}$$

The magnitude condition is then

$$1 = \left(1 + \frac{R_2}{R_1}\right)\left(\frac{1}{3}\right) \tag{15.53(a)}$$

or

$$\frac{R_2}{R_1} = 2 \tag{15.53(b)}$$

Equation (15.53(b)) states that to ensure the startup of oscillation, we must have $(R_2/R_1) > 2$.

## DESIGN EXAMPLE 15.5

**Objective:** Design a Wien-bridge circuit to oscillate at a specified frequency.

**Specifications:** Design the Wien-bridge oscillator shown in Figure 15.17 to oscillate at $f_o = 20$ kHz.

**Choices:** An ideal op-amp is available and standard-valued resistors and capacitors are to be used.

**Solution:** The oscillation frequency given by Equation (15.52(b)) yields

$$RC = \frac{1}{2\pi f_o} = \frac{1}{2\pi (20 \times 10^3)} = 7.96 \times 10^{-6}$$

A 10 k$\Omega$ resistor and 796 pF capacitor satisfy this requirement. Since the amplifier resistor ratio must be $R_2/R_1 = 2$, we could, for example, have $R_2 = 20$ k$\Omega$ and $R_1 = 10$ k$\Omega$, which would satisfy the requirement.

**Trade-offs:** Standard-valued resistors $R_1 = 10$ k$\Omega$ and $R_2 = 20$ k$\Omega$. In place of the ideal 796 pF capacitor, a standard-valued capacitor $C = 800$ pF can be used. The oscillation frequency would then be $f_o = 19.9$ kHz. Element tolerance values should also be considered.

**Comment:** As usual in any electronic circuit design, there is no unique solution. Reasonably sized component values should be chosen whenever possible.

**Computer Simulation Verification:** A Computer simulation was performed using the circuit in Figure 15.18(a). Figure 15.18(b) shows the output voltage versus time. Since the ratio of resistances is $R_2/R_1 = 22/10 = 2.2$, the overall gain is greater than unity so the output increases as a function of time. This increase shows the oscillation nature of the circuit. Another characteristic of the circuit is shown in Figure 15.18(c). A 1 mV sinusoidal signal was applied to the input of $R_1$ and the output voltage measured as the frequency was swept from 10 kHz to 30 kHz. The resonant nature of the circuit is observed. The oscillation frequency and the resonant frequency are both at approximately 18.2 kHz, which is below the design value of 20 kHz.

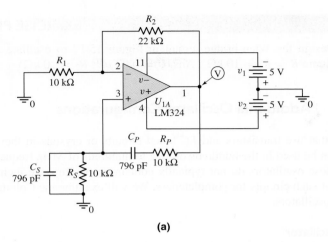

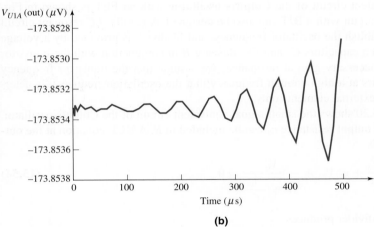

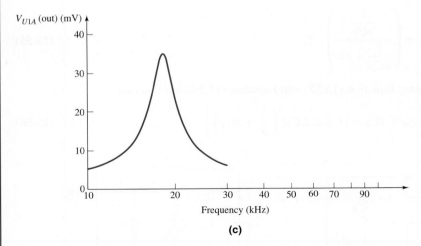

**Figure 15.18** (a) Circuit used in the computer simulation for Example 15.5, (b) output voltage versus time, and (c) output voltage versus input frequency

If the capacitor in the circuit is reduced from 796 pF to 720 pF, the resonant frequency is exactly 20 kHz. This example is one case, then, when the design parameters need to be changed slightly in order to meet the design specifications.

**Ex 15.5:** Design the Wien-bridge circuit in Figure 15.17 to oscillate at $f_o = 800$ Hz. Assume $R = R_1 = 10$ k$\Omega$. (Ans. $C \cong 0.02$ $\mu$F, $R_2 = 20$ k$\Omega$)

### 15.2.4   Additional Oscillator Configurations

Oscillators that use transistors and $LC$ tuned circuits or crystals in their feedback networks can be used in the hundreds of kHz to hundreds of MHz frequency range. Although these oscillators do not typically contain an op-amp, we include a brief discussion of such circuits for completeness. We will examine the Colpitts, Hartley, and crystal oscillators.

#### Colpitts Oscillator

The ac equivalent circuit of the **Colpitts oscillator** with an FET is shown in Figure 15.19. A circuit with a BJT can also be designed. A parallel $LC$ resonant circuit is used to establish the oscillator frequency, and feedback is provided by a voltage divider between capacitors $C_1$ and $C_2$. Resistor $R$ in conjunction with the transistor provides the necessary gain at resonance. We assume that the transistor frequency response occurs at a high enough frequency that the oscillation frequency is determined by the external elements only.

Figure 15.20 shows the small-signal equivalent circuit of the Colpitts oscillator. The transistor output resistance $r_o$ can be included in $R$. A KCL equation at the output node yields

$$\frac{V_o}{\frac{1}{sC_1}} + \frac{V_o}{R} + g_m V_{gs} + \frac{V_o}{sL + \frac{1}{sC_2}} = 0 \tag{15.54}$$

and a voltage divider produces

$$V_{gs} = \left( \frac{\frac{1}{sC_2}}{\frac{1}{sC_2} + sL} \right) \cdot V_o \tag{15.55}$$

Substituting Equation (15.55) into Equation (15.54), we find that

$$V_o \left[ g_m + sC_2 + (1 + s^2 L C_2) \left( \frac{1}{R} + sC_1 \right) \right] = 0 \tag{15.56}$$

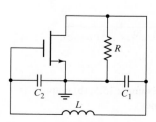

**Figure 15.19** The ac equivalent circuit, MOSFET Colpitts oscillator

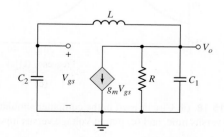

**Figure 15.20** Small-signal equivalent circuit, MOSFET Colpitts oscillator

If we assume that oscillation has started, then $V_o \neq 0$ and can be eliminated from Equation (15.56). We then have

$$s^3 LC_1C_2 + \frac{s^2 LC_2}{R} + s(C_1 + C_2) + \left(g_m + \frac{1}{R}\right) = 0 \tag{15.57}$$

Letting $s = j\omega$, we obtain

$$\left(g_m + \frac{1}{R} - \frac{\omega^2 LC_2}{R}\right) + j\omega[(C_1 + C_2) - \omega^2 LC_1C_2] = 0 \tag{15.58}$$

The condition for oscillation implies that both the real and imaginary components of Equation (15.58) must be zero. From the imaginary component, the oscillation frequency is

$$\omega_o = \frac{1}{\sqrt{L\left(\dfrac{C_1C_2}{C_1 + C_2}\right)}} \tag{15.59}$$

which is the resonant frequency of the $LC$ circuit. From the real part of Equation (15.58), the condition for oscillation is

$$\frac{\omega_o^2 LC_2}{R} = g_m + \frac{1}{R} \tag{15.60}$$

Combining Equations (15.59) and (15.60) yields

$$\frac{C_2}{C_1} = g_m R \tag{15.61}$$

where $g_m R$ is the magnitude of the gain. Equation (15.61) states that to initiate oscillations spontaneously, we must have $g_m R > (C_2/C_1)$.

### Hartley Oscillator

Figure 15.21 shows the ac equivalent circuit of the Hartley oscillator with a BJT. An FET can also be used. Again, a parallel $LC$ resonant circuit establishes the oscillator frequency, and feedback is provided by a voltage divider between inductors $L_1$ and $L_2$.

The analysis of the Hartley oscillator is essentially identical to that of the Colpitts oscillator. The frequency of oscillation, neglecting transistor frequency effects, is

$$\omega_o = \frac{1}{\sqrt{(L_1 + L_2)C}} \tag{15.62}$$

Equation (15.62) also assumes that $r_\pi \gg 1/(\omega C_2)$.

### Crystal Oscillator

A piezoelectric crystal, such as quartz, exhibits electromechanical resonance characteristics in response to a voltage applied across the crystal. The oscillations are very stable over time and temperature, with temperature coefficients on the order of 1 ppm per °C. The oscillation frequency is determined by the crystal dimensions. This means that crystal oscillators are fixed-frequency devices.

The circuit symbol for the piezoelectric crystal is shown in Figure 15.22(a), and the equivalent circuit is shown in Figure 15.22(b). The inductance $L$ can be as high as a few hundred henrys, the capacitance $C_s$ can be on the order of 0.001 pF, and the

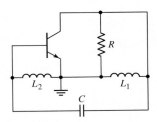

**Figure 15.21** The ac equivalent, BJT Hartley oscillator

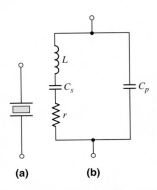

**(a)**          **(b)**

**Figure 15.22** (a) Piezoelectric crystal circuit symbol and (b) piezoelectric crystal equivalent circuit

capacitance $C_p$ can be on the order of a few pF. Also, the $Q$-factor can be on the order of $10^4$, which means that the series resistance $r$ can be neglected.

The impedance of the equivalent circuit in Figure 15.22(b) is

$$Z(s) = \frac{1}{sC_p} \cdot \frac{s^2 + (1/LC_s)}{s^2 + [(C_p + C_s)/(LC_sC_p)]} \tag{15.63}$$

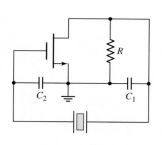

**Figure 15.23** Pierce oscillator in which the inductor in a Colpitts oscillator is replaced by a crystal

Equation (15.63) indicates that the crystal has two resonant frequencies, which are very close together. At the series-resonant frequency $f_s$, the reactance of the series branch is zero; at the parallel-resonant frequency $f_p$, the reactance of the crystal approaches infinity.

Between the resonant frequencies $f_s$ and $f_p$, the crystal reactance is inductive, so the crystal can be substituted for an inductance, such as that in a Colpitts oscillator. Figure 15.23 shows the ac equivalent circuit of a Pierce oscillator, which is similar to the Colpitts oscillator in Figure 15.19 but with the inductor replaced by the crystal. Since the crystal reactance is inductive over a very narrow frequency range, the frequency of oscillation is also confined to this narrow range and is quite constant relative to changes in bias current or temperature. Crystal oscillator frequencies are usually in the range of tens of kHz to tens of MHz.

## Test Your Understanding

**TYU 15.5** Consider the phase-shift oscillator in Figure 15.16. The value of $R$ is 15 kΩ and the frequency of oscillation is $f_o = 20$ kHz. Determine the values of $C$ and $R_2$. (Ans. $C = 217$ pF, $R_2 = 435$ kΩ)

**\*TYU 15.6** For the Colpitts oscillator in Figure 15.19, assume parameters of $L = 1$ μH, $C_1$ and $C_2 = 1$ nF, and $R = 4$ kΩ. Determine the oscillator frequency and the required value of $g_m$. Is this value of $g_m$ reasonable for a MOSFET? Why? (Ans. $f_o = 7.12$ MHz, $g_m = 0.25$ mA/V)

##  15.3    SCHMITT TRIGGER CIRCUITS

**Objective:**  • Analyze and design various Schmitt trigger circuits.

In this section, we will analyze another class of circuits that utilize positive feedback. The basic circuit is commonly called a **Schmitt trigger,** which can be used in the class of waveform generators called multivibrators. The three general types of multivibrators are: bistable, monostable, and astable. In this section, we will examine the bistable multivibrator, which has a comparator with positive feedback and has two stable states. We will discuss the comparator first, and will then describe various applications of the Schmitt trigger.

### 15.3.1    Comparator

The comparator is essentially an op-amp operated in an open-loop configuration, as shown in Figure 15.24(a). As the name implies, a comparator compares two voltages

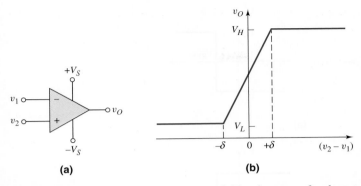

**Figure 15.24** (a) Open-loop comparator and (b) voltage transfer characteristics, open-loop comparator

to determine which is larger. The comparator is usually biased at voltages $+V_S$ and $-V_S$, although other biases are possible.

The voltage transfer characteristics, neglecting any offset voltage effects, are shown in Figure 15.24(b). When $v_2$ is slightly greater than $v_1$, the output is driven to a high saturated state $V_H$; when $v_2$ is slightly less than $v_1$, the output is driven to a low saturated state $V_L$. The saturated output voltages $V_H$ and $V_L$ may be close to the supply voltages $+V_S$ and $-V_S$, respectively, which means that $V_L$ may be negative. The transition region is the region in which the output voltage is in neither of its saturation states. This region occurs when the input differential voltage is in the range $-\delta < (v_2 - v_1) < +\delta$. If, for example, the open-loop gain is $10^5$ and the difference between the two output states is $(V_H - V_L) = 10$ V, then

$$2\delta = 10/10^5 = 10^{-4} \text{ V} = 0.1 \text{ mV}$$

The range of input differential voltage in the transition region is normally very small.

One major difference between a comparator and op-amp is that a comparator need not be frequency compensated. Frequency stability is not a consideration since the comparator is being driven into one of two states. Since a comparator does not contain a frequency compensation capacitor, it is not slew-rate-limited by the compensation capacitor as is the op-amp. Typical response times for the comparator output to change states are in the range of 30 to 200 ns. An expected response time for a 741 op-amp with a slew rate of 0.7 V/$\mu$s would be on the order of 30 $\mu$s, which is a factor of 1000 times greater.

Figure 15.25 shows two comparator configurations along with their voltage transfer characteristics. In both, the input transition region width is assumed to be negligibly small. The reference voltage may be either positive or negative, and the output saturation voltages are assumed to be symmetrical about zero. The crossover voltage is defined as the input voltage at which the output changes states.

Two other comparator configurations, in which the crossover voltage is a function of resistor ratios, are shown in Figure 15.26. Input bias current compensation is also included in this figure. From Figure 15.26(a), we use superposition to obtain

$$v_+ = \left( \frac{R_2}{R_1 + R_2} \right) V_{\text{REF}} + \left( \frac{R_1}{R_1 + R_2} \right) v_I \qquad \textbf{(15.64)}$$

The ideal crossover voltage occurs when $v_+ = 0$, or

$$R_2 V_{\text{REF}} + R_1 v_I = 0 \qquad \textbf{(15.65(a))}$$

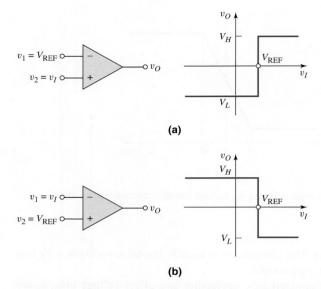

**(a)**

**(b)**

**Figure 15.25** (a) Noninverting comparator circuit and (b) inverting comparator circuit

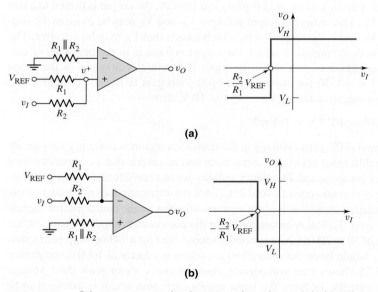

**(a)**

**(b)**

**Figure 15.26** Other comparator circuits: (a) noninverting and (b) inverting

which can be written as

$$v_I = -\frac{R_2}{R_1} V_{\text{REF}} \tag{15.65(b)}$$

The output goes high when $v_+ > 0$. From Equation (15.64), we see that $v_o =$ High when $v_I$ is greater than the crossover voltage. A similar analysis produces the characteristics shown in Figure 15.26(b).

Figure 15.27(a) shows one application of a comparator, to control street lights. The input signal is the output of a photodetector circuit. Voltage $v_I$ is directly

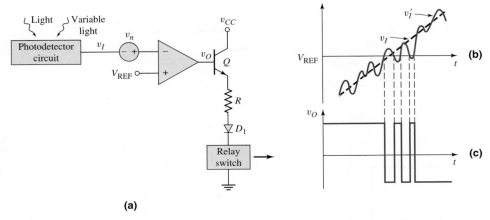

**Figure 15.27** (a) Comparator circuit including input noise source, (b) input signal, and (c) output signal, showing chatter effect

proportional to the amount of light incident on the photodetector. During the night, $v_I < V_{REF}$, and $v_O$ is on the order of $V_S = +15$ V; the transistor turns on. The current in the relay switch then turns the street lights on. During the day, the light incident on the photodetector produces an output signal such that $v_I > V_{REF}$. In this case, $v_O$ is on the order of $-V_S = -15$ V, and the transistor turns off.

Diode $D_1$ is used as a protection device, preventing reverse-bias break-down in the B–E junction. With zero output current, the relay switch is open and the street lights are off. At dusk and dawn, $v_I = V_{REF}$.

The open-loop comparator circuit in Figure 15.27(a) may exhibit unacceptable behavior in response to noise in the system. Figure 15.27(a) shows the comparator circuit, with a variable light source, such as clouds causing the light intensity to fluctuate over a short period of time. A variable light intensity would be equivalent to a noise source $v_n$ in series with the signal source $v_I$. If we assume that $v_I$ is increasing linearly with time (corresponding to dawn), then the total input signal $v_I'$ versus time is shown in Figure 15.27(b). When $v_I' > V_{REF}$, the output switches low; when $v_I' < V_{REF}$, the output switches high, producing a chatter effect in the output signal as shown in Figure 15.27(c). This effect would turn the street lights off and on over a relatively short time period. If the amplitude of the noise signal increases, the chatter effect becomes more severe. This chatter can be eliminated by using a Schmitt trigger.

## 15.3.2    Basic Inverting Schmitt Trigger

The Schmitt trigger or **bistable multivibrator** uses positive feedback with a loop-gain greater than unity to produce a bistable characteristic. Figure 15.28(a) shows one configuration of a Schmitt trigger. Positive feedback occurs because the feedback resistor is connected between the output terminal and noninverting input terminal. Voltage $v_+$, in terms of the output voltage, can be found by using a voltage divider equation to yield

$$v_+ = \left( \frac{R_1}{R_1 + R_2} \right) v_O \qquad\qquad (15.66)$$

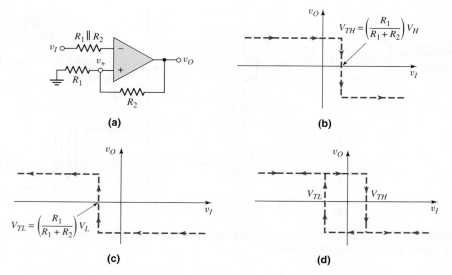

**Figure 15.28** (a) Schmitt trigger circuit, (b) voltage transfer characteristic as input voltage increases, (c) voltage transfer characteristic as input voltage decreases, and (d) net voltage transfer characteristics, showing hysteresis effect

Voltage $v_+$ does not remain constant; rather, it is a function of the output voltage. Input signal $v_I$ is applied to the inverting terminal.

## Voltage Transfer Characteristics

To determine the voltage transfer characteristics, we assume that the output of the comparator is in one state, namely $v_O = V_H$, which is the high state. Then

$$v_+ = \left( \frac{R_1}{R_1 + R_2} \right) V_H \qquad (15.67)$$

As long as the input signal is less than $v_+$, the output remains in its high state. The crossover voltage occurs when $v_I = v_+$ and is defined as $V_{TH}$. We have

$$V_{TH} = \left( \frac{R_1}{R_1 + R_2} \right) V_H \qquad (15.68)$$

When $v_I$ is greater than $V_{TH}$, the voltage at the inverting terminal is greater than that at the noninverting terminal. The differential input voltage $(v_I - V_{TH})$ is amplified by the open-loop gain of the comparator, and the output switches to its low state, or $v_O = V_L$. Voltage $v_+$ then becomes

$$v_+ = \left( \frac{R_1}{R_1 + R_2} \right) V_L \qquad (15.69)$$

Since $V_L < V_H$, the input voltage $v_I$ is still greater than $v_+$, and the output remains in its low state as $v_I$ continues to increase. This voltage transfer characteristic is shown in Figure 15.28(b). Implicit in these transfer characteristics is the assumption that $V_H$ is positive and $V_L$ is negative.

Now consider the transfer characteristic as $v_I$ decreases. As long as $v_I$ is larger than $v_+ = [R_1/(R_1 + R_2)]V_L$, the output remains in its low saturation state. The crossover voltage now occurs when $v_I = v_+$ and is defined as $V_{TL}$. We have

$$V_{TL} = \left( \frac{R_1}{R_1 + R_2} \right) V_L \qquad\qquad (15.70)$$

As $v_I$ drops below this value, the voltage at the noninverting terminal is greater than that at the inverting terminal. The differential voltage at the comparator terminals is amplified by the open-loop gain, and the output switches to its high state, or $v_O = V_H$. As $v_I$ continues to decrease, it remains less than $v_+$; therefore, $v_O$ remains in its high state. This voltage transfer characteristic is shown in Figure 15.28(c).

## Complete Voltage Transfer and Bistable Characteristics

The complete voltage transfer characteristics of the Schmitt trigger in Figure 15.28(a) combine the characteristics in Figures 15.28(b) and 15.28(c). These complete characteristics are shown in Figure 15.28(d). As shown, the crossover voltages depend on whether the input voltage is increasing or decreasing. The complete transfer characteristics therefore show a **hysteresis effect.** The width of the hysteresis is the difference between the two crossover voltages $V_{TH}$ and $V_{TL}$.

The bistable characteristic of the circuit occurs around the point $v_I = 0$, at which the output may be in either its high or low state. The output remains in either state as long as $v_I$ remains in the range $V_{TL} < v_I < V_{TH}$. The output switches states only if the input increases above $V_{TH}$ or decreases below $V_{TL}$.

## EXAMPLE 15.6

**Objective:** Determine the hysteresis width of a particular Schmitt trigger.

Consider the Schmitt trigger in Figure 15.28(a), with parameters $R_1 = 10\ \text{k}\Omega$ and $R_2 = 90\ \text{k}\Omega$. Let $V_H = 10\ \text{V}$ and $V_L = -10\ \text{V}$.

**Solution:** From Equation (15.68), the upper crossover voltage is

$$V_{TH} = \left( \frac{R_1}{R_1 + R_2} \right) V_H = \left( \frac{10}{10 + 90} \right)(10) = 1\ \text{V}$$

and from Equation (15.70), the lower crossover voltage is

$$V_{TL} = \left( \frac{R_1}{R_1 + R_2} \right) V_L = \left( \frac{10}{10 + 90} \right)(-10) = -1\ \text{V}$$

The hysteresis width is therefore $(V_{TH} - V_{TL}) = 2\ \text{V}$.

**Comment:** The hysteresis width can be designed to be larger or smaller for specific applications by adjusting the voltage divider ratio of $R_1$ and $R_2$.

## EXERCISE PROBLEM

**Ex 15.6:** Consider the comparator circuit in Figure 15.28(a). Assume high and low saturated output voltages of $+9\ \text{V}$ and $-9\ \text{V}$, respectively. Design the circuit such that the crossover voltages are $\pm 0.5\ \text{V}$. The minimum resistance is to be $10\ \text{k}\Omega$. (Ans. Set $R_1 = 10\ \text{k}\Omega$, then $R_2 = 170\ \text{k}\Omega$)

The complete voltage transfer characteristics in Figure 15.28(d) show the inverting characteristics of this particular Schmitt trigger. When the input signal becomes sufficiently positive, the output is in its low state; when the input signal is sufficiently negative, the output is in its high state. Since the input signal is applied to the inverting terminal of the comparator, this characteristic is as expected.

### 15.3.3    Additional Schmitt Trigger Configurations

A noninverting Schmitt trigger can be designed by applying the input signal to the network connected to the comparator noninverting terminal. Also, both crossover voltages of a Schmitt trigger circuit can be shifted in either a positive or negative direction by applying a reference voltage. We will study these general circuit configurations, the resulting voltage transfer characteristics, and an application of a Schmitt trigger circuit in this section.

**Noninverting Schmitt Trigger Circuit**
Consider the circuit in Figure 15.29(a). The inverting terminal is held essentially at ground potential, and the input signal is applied to resistor $R_1$, which is connected to the comparator noninverting terminal. Voltage $v_+$ at the noninverting terminal then becomes a function of both the input signal $v_I$ and the output voltage $v_O$. Using superposition, we find that

$$v_+ = \left( \frac{R_2}{R_1 + R_2} \right) v_I + \left( \frac{R_1}{R_1 + R_2} \right) v_O \tag{15.71}$$

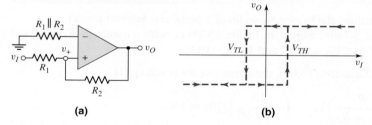

(a)                                              (b)

**Figure 15.29**  (a) Noninverting Schmitt trigger circuit and (b) voltage transfer characteristics

If $v_I$ is negative, and the output is in its low state, then $v_O = V_L$ (assumed to be negative), $v_+$ is negative, and the output remains in its low saturation state. Crossover voltage $v_I = V_{TH}$ occurs when $v_+ = 0$ and $v_O = V_L$, or, from Equation (15.71),

$$0 = R_2 V_{TH} + R_1 V_L \tag{15.72(a)}$$

which can be written

$$V_{TH} = -\left( \frac{R_1}{R_2} \right) V_L \tag{15.72(b)}$$

Since $V_L$ is negative, $V_{TH}$ is positive.
   If we let $v_I = V_{TH} + \delta$, where $\delta$ is a small positive voltage, the input voltage is just greater than the crossover voltage and Equation (15.71) becomes

$$v_+ = \left( \frac{R_2}{R_1 + R_2} \right) (V_{TH} + \delta) + \left( \frac{R_1}{R_1 + R_2} \right) V_L \tag{15.73}$$

Equation (15.73) then becomes

$$v_+ = \left(\frac{R_2}{R_1 + R_2}\right)\left(\frac{-R_1}{R_2}\right)V_L + \left(\frac{R_2}{R_1 + R_2}\right)\delta + \left(\frac{R_1}{R_1 + R_2}\right)V_L \qquad \textbf{(15.74(a))}$$

or

$$v_+ = \left(\frac{R_2}{R_1 + R_2}\right)\delta > 0 \qquad \textbf{(15.74(b))}$$

When $v_+ > 0$, the output switches to its high saturation state.

The lower crossover voltage $v_I = V_{TL}$ occurs when $v_+ = 0$ and $v_O = V_H$. From Equation (15.71), we have

$$0 = R_2\,V_{TL} + R_1\,V_H \qquad \textbf{(15.75(a))}$$

which can be written

$$V_{TL} = -\left(\frac{R_1}{R_2}\right)V_H \qquad \textbf{(15.75(b))}$$

Since $V_H > 0$, then $V_{TL} < 0$.

The complete voltage transfer characteristics are shown in Figure 15.29(b). We again note the hysteresis effect and the bistable characteristic around $v_I = 0$. With $v_I$ sufficiently positive, the output is in its high state; with $v_I$ sufficiently negative, the output is in its low state. The circuit thus exhibits the noninverting transfer characteristic.

### Schmitt Trigger Circuits with Applied Reference Voltages

The switching voltage of a Schmitt trigger is defined as the average value of $V_{TH}$ and $V_{TL}$. For the two circuits in Figure 15.28(a) and 15.29(a), the switching voltages are zero, assuming $V_{TL} = -V_{TH}$. In some applications, the switching voltage must be either positive or negative. Both crossover voltages can be shifted in either a positive or negative direction by applying a reference voltage.

Figure 15.30(a) shows an inverting Schmitt trigger with a reference voltage $V_{REF}$. The complete voltage transfer characteristics are shown in Figure 15.30(b). The switching voltage $V_S$, assuming $V_H$ and $V_L$ are symmetrical about zero, is given by

$$V_S = \left(\frac{R_2}{R_1 + R_2}\right)V_{REF} \qquad \textbf{(15.76)}$$

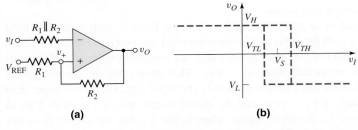

**(a)**                                    **(b)**

**Figure 15.30** (a) Inverting Schmitt trigger circuit with applied reference voltage and (b) voltage transfer characteristics

Note that the switching voltage is not the same as the reference voltage. The upper and lower crossover voltages are

$$V_{TH} = V_S + \left( \frac{R_1}{R_1 + R_2} \right) V_H \qquad (15.77(a))$$

and

$$V_{TL} = V_S + \left( \frac{R_1}{R_1 + R_2} \right) V_L \qquad (15.77(b))$$

A noninverting Schmitt trigger with a reference voltage is shown in Figure 15.31(a), and the complete voltage transfer characteristics are shown in Figure 15.31(b). The switching voltage $V_S$, again assuming $V_H$ and $V_L$ are symmetrical about zero, is given by

$$V_S = \left( 1 + \frac{R_1}{R_2} \right) V_{REF} \qquad (15.78)$$

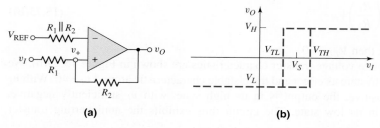

**(a)**                         **(b)**

**Figure 15.31** (a) Noninverting Schmitt trigger circuit with applied reference voltage and (b) voltage transfer characteristics

and the upper and lower crossover voltages are

$$V_{TH} = V_S - \left( \frac{R_1}{R_2} \right) V_L \qquad (15.79(a))$$

and

$$V_{TL} = V_S - \left( \frac{R_1}{R_2} \right) V_H \qquad (15.79(b))$$

If the output saturation voltages are symmetrical such that $V_L = -V_H$, then the crossover voltages are symmetrical about the switching voltage $V_S$.

### Schmitt Trigger Application

Let us reconsider the street light control in Figure 15.27(a), which included a noise source. Figure 15.32(a) shows the same basic circuit, except that a Schmitt trigger is used instead of a simple comparator.

The input signal $v_I$ is again assumed to increase linearly with time. The total input signal $v_I'$ is $v_I$ with the noise signal superimposed, as shown in Figure 15.32(b). At time $t_1$, the input signal becomes greater than the switching voltage $V_S$. The output, however, does not switch, since $v_I' < V_{TH}$. This means that the input signal is less than the upper crossover voltage. At time $t_2$, the input signal becomes larger than the crossover voltage, or $v_I' > V_{TH}$, and the output signal switches from its high to its low state. At time $t_3$, the input signal drops below $V_S$, but the output does not switch states since $v_I' > V_{TL}$. This means that the input signal remains greater than

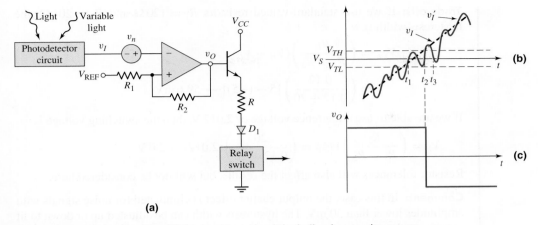

**Figure 15.32** (a) Application of Schmitt trigger circuit including input noise source, (b) input signal, and (c) output signal, showing elimination of chatter effect

the lower crossover voltage. The Schmitt trigger circuit thus eliminates the chatter effect that occurs in the output voltage in Figure 15.27(c). Elimination of the chatter in the output voltage response results directly from the hysteresis effect in the Schmitt trigger characteristics.

## DESIGN EXAMPLE 15.7

**Objective:** Design a Schmitt trigger circuit for the photodetector switch circuit.

**Specifications:** The Schmitt trigger circuit with the configuration shown in Figure 15.32(a) is to be designed such that the switching voltage is $V_S = 2$ V and the hysteresis width is 60 mV. Assume $V_H = 5$ V and $V_L = -5$ V.

**Choices:** An ideal comparator is available and standard-valued resistors are to be used in the final design.

**Solution:** The Schmitt trigger circuit is the inverting type, for which the voltage transfer characteristics are shown in Figure 15.30(b). From Equations (15.77(a)) and (15.77(b)), the hysteresis width is

$$V_{TH} - V_{TL} = \left(\frac{R_1}{R_1 + R_2}\right)(V_H - V_L)$$

so

$$0.060 = \left(\frac{R_1}{R_1 + R_2}\right)[5 - (-5)] = 10\left(\frac{R_1}{R_1 + R_2}\right)$$

which yields $R_2/R_1 = 165.7$. We can find the reference voltage from Equation (15.76), which can be rewritten to obtain

$$V_{REF} = \left(1 + \frac{R_1}{R_2}\right)V_S = \left(1 + \frac{1}{165.7}\right)(2) = 2.012 \text{ V}$$

Resistor values of $R_1 = 100\,\Omega$ and $R_2 = 16.57\text{ k}\Omega$ will satisfy the requirements. The crossover voltages are thus $V_{TH} = 2.03$ V and $V_{TL} = 1.97$ V.

**Trade-offs:** If we use standard-valued resistors $R_1 = 120 \, \Omega$ and $R_2 = 20 \, k\Omega$, the hysteresis width is

$$V_{TH} - V_{TL} = \left( \frac{R_1}{R_1 + R_2} \right)(V_H - V_L)$$

$$= \left( \frac{0.12}{0.12 + 20} \right)[5 - (-5)] \rightarrow 59.6 \, mV$$

If we are able to use a reference voltage of 2.012 V, then the switching voltage is

$$V_S = \left( \frac{R_2}{R_1 + R_2} \right)V_{REF} = \left( \frac{20}{0.12 + 20} \right)(2.012) = 2.0 \, V$$

Resistor tolerances will also affect the results, but will not be considered here.

**Comment:** In this case, the output chatter effect is eliminated for noise signals with amplitudes lower than 30 mV. The hysteresis width can be adjusted up or down to fit specific application requirements in which the noise signal is larger or smaller than that given in this example.

### EXERCISE PROBLEM

**Ex 15.7:** Redesign the street light control circuit shown in Figure 15.32(a) such that the switching voltage is $V_S = 1$ V and the hysteresis width is 100 mV. Assume $V_H = +10$ V and $V_L = -10$ V. Also, find $R$ such that $I = 200 \, \mu A$ when $v_O = V_H$. Assume $V_{BE}(on) = 0.7$ V and $V_\gamma = 0.7$ V, and assume the relay switch resistance is 100 $\Omega$. (Ans. $R_2/R_1 = 199$, $V_{REF} = 1.005$ V, $R = 42.9$ k$\Omega$)

### 15.3.4    Schmitt Triggers with Limiters

In the Schmitt trigger circuits we have thus far considered, the open-loop saturation voltages of the comparator may not be very precise and may also vary from one comparator to another. The output saturation voltages can be controlled and made more precise by adding limiter networks.

A direct approach at limiting the output is shown in Figure 15.33. Two back-to-back Zener diodes are connected between the output and ground. Assuming the two diodes are matched, the output is limited to either the positive or negative value of $(V_\gamma + V_Z)$, where $V_\gamma$ is the forward diode voltage and $V_Z$ is the reverse Zener voltage. Resistor $R$ is chosen to produce a specified current in the diodes.

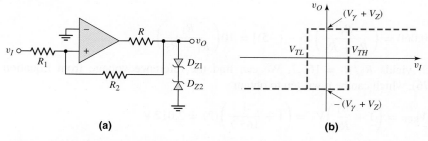

**(a)**                                    **(b)**

**Figure 15.33**  (a) Schmitt trigger with Zener diode limiters and (b) voltage transfer characteristics

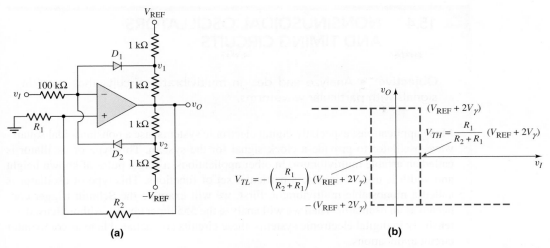

**Figure 15.34** (a) Inverting Schmitt trigger with diode limiters and (b) voltage transfer characteristics

Another Schmitt trigger with a limiter is shown in Figure 15.34(a). If we assume that $v_I = 0$ and $v_O$ is in its high state, then $D_2$ is on and $D_1$ is off. Neglecting currents in the $100\,k\Omega$ resistor, we have $v_2 = +V_\gamma$, where $V_\gamma$ is the forward diode voltage. We can write

$$\frac{v_O - v_2}{1} = \frac{v_2 - (-V_{REF})}{1} \tag{15.80}$$

Solving for $v_O$ yields

$$v_O = V_{REF} + 2V_\gamma \tag{15.81}$$

which means that the output voltage can be controlled and can be designed more accurately. The ideal hysteresis characteristics for this Schmitt trigger are shown in Figure 15.34(b). As $v_I$ increases or decreases, a small current flows in the $100\,k\Omega$ resistor, producing a nonzero slope in the voltage transfer characteristics. The slope is on the order of $1/100$, which is quite small.

## Test Your Understanding

**TYU 15.7** A noninverting Schmitt trigger is shown in Figure 15.29(a) Its saturated output voltages are $\pm 12\,V$. Design the circuit to obtain $\pm 200\,mV$ crossover voltages. The maximum resistance value is to be $200\,k\Omega$. (Ans. Set $R_2 = 200\,k\Omega$, then $R_1 = 3.33\,k\Omega$)

**TYU 15.8** For the Schmitt trigger in Figure 15.30(a), the parameters are: $V_{REF} = 2\,V$, $V_H = 10\,V$, $V_L = -10\,V$, $R_1 = 1\,k\Omega$, and $R_2 = 10\,k\Omega$: (a) Determine $V_S$, $V_{TH}$, and $V_{TL}$. (b) Let $v_I$ be a triangular wave with a zero average voltage, a $10\,V$ peak amplitude, and a $10\,ms$ period. Sketch $v_O$ versus time over two periods. Label the appropriate voltages and times. (Ans. (a) $V_S = 1.82\,V$, $V_{TH} = 2.73\,V$, $V_{TL} = 0.91\,V$)

**TYU 15.9** Consider the Schmitt trigger in Figure 15.31(a). Let $V_H = 9\,V$ and $V_L = -9\,V$. Design the circuit such that $V_S = -2\,V$ and the hysteresis width is $0.5\,V$. The minimum resistance is to be $10\,k\Omega$. (Ans. Set $R_1 = 10\,k\Omega$, then $R_2 = 360\,k\Omega$, $V_{REF} = -1.946\,V$)

## 15.4   NONSINUSOIDAL OSCILLATORS AND TIMING CIRCUITS

**Objective:** • Analyze and design multivibrator circuits that provide signals with particular waveforms.

Many applications, especially digital electronic systems, use a nonsinusoidal square-wave oscillator to provide a clock signal for the system. This type of oscillator is called an astable multivibrator. In other applications, a single pulse of known height and width is used to initiate a particular set of functions. This type of oscillator is called a monostable multivibrator. First, we will examine the Schmitt trigger connected as an oscillator. Then we will analyze the 555 timer circuit. Although used extensively in digital electronic systems, these circuits are included here as comparator circuit applications.

### 15.4.1   Schmitt Trigger Oscillator

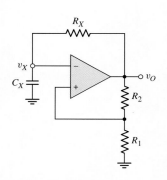

**Figure 15.35** Schmitt trigger oscillator

The Schmitt trigger can be used in an oscillator circuit to generate a square-wave output signal. This is accomplished by adding an $RC$ network to the negative feedback loop of the Schmitt trigger as shown in Figure 15.35. As we will see, this circuit has no stable states. It is therefore called an **astable multivibrator.**

Initially, we set $R_1$ and $R_2$ equal to the same value, or $R_1 = R_2 \equiv R$. We assume that the output switches symmetrically about zero volts, with the high saturated output denoted by $V_H = V_P$ and the low saturated output denoted by $V_L = -V_P$. If $v_O$ is low, or $v_O = -V_P$, then $v_+ = -(\frac{1}{2})V_P$. When $v_X$ drops just slightly below $v_+$, the output switches high so that $v_O = +V_P$ and $v_+ = +(\frac{1}{2})V_P$. The $R_X C_X$ network sees a positive step-increase in voltage, so capacitor $C_X$ begins to charge and voltage $v_X$ starts to increase toward a final value of $V_P$.

The general equation for the voltage across a capacitor in an $RC$ network is

$$v_X = v_{\text{Final}} + (v_{\text{Initial}} - v_{\text{Final}})\, e^{-t/\tau} \tag{15.82}$$

where $v_{\text{Initial}}$ is the initial capacitor voltage at $t = 0$, $v_{\text{Final}}$ is the final capacitor voltage at $t = \infty$, and $\tau$ is the time constant. We can now write

$$v_X = V_P + \left(-\frac{V_P}{2} - V_P\right)e^{-t/\tau_x} \tag{15.83(a)}$$

or

$$v_X = V_P - \frac{3V_P}{2}e^{-t/\tau_x} \tag{15.83(b)}$$

where $\tau_x = R_X C_X$. Voltage $v_X$ increases exponentially with time toward a final voltage $V_P$. However, when $v_X$ becomes just slightly greater than $v_+ = +(\frac{1}{2})V_P$, the output switches to its low state of $v_O = -V_P$ and $v_+ = -(\frac{1}{2})V_P$. The $R_X C_X$ network sees a negative step change in voltage, so capacitor $C_X$ now begins to discharge and voltage $v_X$ starts to decrease toward a final value of $-V_P$. We can now write

$$v_X = -V_P + \left[+\frac{V_P}{2} - (-V_P)\right]e^{-(t-t_1)/\tau_x} \tag{15.84(a)}$$

or

$$v_X = -V_P + \frac{3V_P}{2}e^{-(t-t_1)/\tau_x} \qquad \textbf{(15.84(b))}$$

where $t_1$ is the time at which the output switches to its low state. The capacitor voltage then decreases exponentially with time. When $v_X$ decreases to $v_+ = -(\frac{1}{2})V_P$, the output again switches to its high state. The process continues to repeat itself, which means that this positive-feedback circuit oscillates producing a square-wave output signal. Figure 15.36 shows the output voltage $v_O$ and the capacitor voltage $v_X$ versus time.

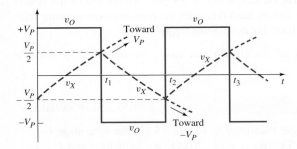

**Figure 15.36** Output voltage and capacitor voltage versus time for Schmitt trigger oscillator

Time $t_1$ can be found from Equation (15.83(b)) by setting $t = t_1$ when $v_X = V_P/2$, or

$$\frac{V_P}{2} = V_P - \frac{3V_P}{2}e^{-t_1/\tau_x} \qquad \textbf{(15.85)}$$

Solving for $t_1$, we find that

$$t_1 = \tau_x \ln 3 = 1.1 R_X C_X \qquad \textbf{(15.86)}$$

From a similar analysis using Equation (15.84(b)), we find that the difference between $t_2$ and $t_1$ is also $1.1R_X C_X$; therefore, the period of oscillation $T$ is

$$T = 2.2 R_X C_X \qquad \textbf{(15.87)}$$

and the frequency of oscillation is

$$f = \frac{1}{T} = \frac{1}{2.2 R_X C_X} \qquad \textbf{(15.88)}$$

As an example of an application of this circuit, a variable frequency oscillator is created by letting $R_X$ be a variable resistor.

The **duty cycle** of the oscillator is defined as the percentage of time that the output voltage $v_O$ is in its high state. For the circuit just considered, the duty cycle is 50 percent, as seen in Figure 15.36. This is a result of the symmetrical output voltages $+V_P$ and $-V_P$. If asymmetrical output voltages are used, then the duty cycle changes from the 50 percent value.

## DESIGN EXAMPLE **15.8**

**Objective:** Design a Schmitt trigger oscillator for a specified frequency.

**Specifications:** Assume that an ideal comparator is available. Use standard-valued resistors and capacitors in the final design.

Consider the oscillator in Figure 15.35. Design the circuit to oscillate at $f_o = 1$ kHz.

**Solution:** Using Equation (15.88), we can write

$$R_X C_X = \frac{1}{2.2 f_o} = \frac{1}{2.2(10^3)} = 4.55 \times 10^{-4}$$

If $C_X = 0.1\,\mu\text{F}$, then $R_X = 4.55\,\text{k}\Omega$.

**Trade-offs:** Using standard-valued elements with values of $C_X = 0.082\,\mu\text{F}$ and $R_X = 5.6\,\text{k}\Omega$ produces an oscillation frequency of 990 Hz, within 1% of the specified value. If element tolerance values are taken into account, a potentiometer may have to be used to produce the 1000 Hz oscillation frequency.

**Comment:** A larger frequency of oscillation can easily be obtained by using a smaller capacitor value.

EXERCISE PROBLEM

**\*Ex 15.8:** For the Schmitt trigger oscillator in Figure 15.35, the saturation output voltages are $+10$ V and $-5$ V. $R_1 = R_2 = 20\,\text{k}\Omega$, $R_X = 50\,\text{k}\Omega$, and $C_X = 0.01\,\mu\text{F}$. Determine the frequency of oscillation and the duty cycle. Sketch $v_O$ and $v_X$ versus time over two periods of the oscillation. (Ans. $f = 866$ Hz, duty cycle $= 39.7\%$)

### 15.4.2    Monostable Multivibrator

A **monostable multivibrator** has one stable state, in which it can remain indefinitely if not disturbed. However, a trigger pulse can force the circuit into a quasi-stable state for a definite time, producing an output pulse with a particular height and width. The circuit then returns to its stable state until another trigger pulse is applied. The monostable multivibrator is also called a **one-shot.**

A monostable multivibrator is created by modifying the Schmitt trigger oscillator as shown in Figure 15.37. A clamping diode $D_1$ is connected in parallel with $C_X$. In the stable state, the output is high and voltage $v_X$ is held low by the conducting diode $D_1$.

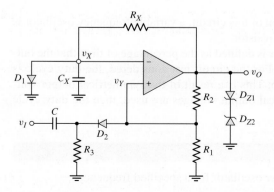

**Figure 15.37** Schmitt trigger monostable multivibrator

The trigger circuit is composed of the capacitor $C$, resistor $R_3$, and diode $D_2$, and is connected to the noninverting terminal of the comparator. The value of $R_3$ is chosen to be much larger than $R_1$, so that voltage $v_Y$ is determined primarily by a voltage divider of $R_1$ and $R_2$. We then have

$$v_Y \cong \left(\frac{R_1}{R_1 + R_2}\right)V_P \equiv \beta V_P \tag{15.89}$$

where $V_P$ is the sum of the forward and breakdown voltages of $D_{Z1}$ and $D_{Z2}$, or $V_P = (V_{\gamma 1} + V_{Z2})$. This voltage is the positive saturated output voltage.

The circuit is triggered by a negative-going step voltage applied to capacitor $C$. This action forward-biases diode $D_2$ and pulls the voltage $v_Y$ below $v_X$. Since the comparator then sees a larger voltage at the inverting terminal, the output switches to its low state of

$$v_O = -V_P = -(V_{\gamma 2} + V_{Z1})$$

Voltage $v_Y$ then becomes

$$v_Y \cong -\left(\frac{R_1}{R_1 + R_2}\right)V_P \equiv -\beta V_P \tag{15.90}$$

causing $D_2$ to become reverse biased, thus isolating the oscillator circuit from the input triggering network. The negative-step change in $v_O$ causes voltage $v_X$ to decrease exponentially with a time constant of $\tau_x = R_X C_X$ toward a final value of $-V_P$. Diode $D_1$ is reverse biased during this time. When $v_X$ drops just below the value of $v_Y$ given by Equation (15.90), the output switches back to its positive saturated value of $+V_P$. The capacitor voltage $v_X$ then starts to increase exponentially toward a final value of $+V_P$. When $v_X$ reaches $V_\gamma$, diode $D_1$ again becomes forward biased, $v_X$ is clamped at $V_\gamma$, and the output remains in its high state.

The waveforms of $v_O$ and $v_X$ versus time are shown in Figure 15.38. After the output has switched back to its high state, the capacitor voltage $v_X$ must return to its quiescent value of $v_X = V_\gamma$. This implies that there is a **recovery time** of $(T' - T)$ during which the circuit should not be retriggered.

For $t > 0$, voltage $v_X$ can be written in the same general form as Equation (15.82), as follows:

$$v_X = -V_P + (V_\gamma - (-V_P))e^{-t/\tau_x} \tag{15.91}$$

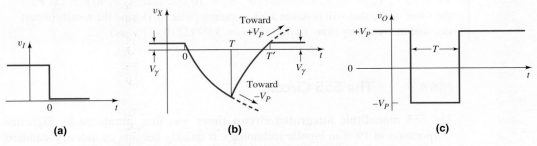

**Figure 15.38** Schmitt trigger monostable multivibrator voltages versus time (a) input trigger pulse, (b) capacitor voltage, and (c) output pulse

where $\tau_x = R_X C_X$. At $t = T$, $v_X = -\beta V_P$ and the output switches high. The pulse width is then

$$T = \tau_x \ln\left[\frac{1 + (V_\gamma / V_P)}{(1 - \beta)}\right] \qquad\qquad (15.92)$$

If we assume $V_\gamma \ll V_P$ and if we let $R_1 = R_2$ such that $\beta = 1/2$, then the pulse width is $T = 0.69\tau_x$. We can show that for $V_\gamma \ll V_P$ and $\beta = 1/2$, the recovery time is $(T' - T) = 0.4\tau_x$. There are alternative circuits with shorter recovery times, but we will not consider them here.

## DESIGN EXAMPLE 15.9

**Objective:** Design a monostable multivibrator to produce a given pulse width.

**Specifications:** The circuit with the configuration shown in Figure 15.37 is to be designed to produce an output pulse that is 1 $\mu$s wide. Assume parameters of $V_P = 10$ V, $V_\gamma = 0.7$ V and $R_1 = R_2 = 20$ k$\Omega$.

**Choices:** Assume an ideal comparator is available. Use standard-valued element values in the final design.

**Solution:** Since $V_\gamma \ll V_P$ and $R_1 = R_2$, then from Equation (15.92), we have

$$T = 0.69\tau_x$$

or

$$\tau_x = R_X C_X = \frac{T}{0.69} = \frac{1}{0.69} = 1.45\,\mu s$$

If $R_X = 10$ k$\Omega$, then $C_X = 145$ pF.

**Trade-offs:** Using standard-valued elements of $R_X = 10$ k$\Omega$ and $C_X = 150$ pF produces a pulse width of $1.035\,\mu$s. Element tolerances must also be taken into account in the final design.

**Comment:** In actual monostable multivibrator ICs, $R_X$ and $C_X$ are external elements to allow for variable times.

## EXERCISE PROBLEM

**\*Ex 15.9:** For the monostable circuit shown in Figure 15.37, the parameters are: $V_P = 12$ V, $V_\gamma = 0.7$ V, $C_X = 0.1\,\mu$F, $R_1 = 10$ k$\Omega$, and $R_2 = 90$ k$\Omega$. (a) Find the value of $R_X$ that will result in a 50 $\mu$s output pulse. (b) Using the results of part (a), find the recovery time. (Ans. (a) $R_X = 3.09$ k$\Omega$ (b) 47.9 $\mu$s)

### 15.4.3    The 555 Circuit

The **555 monolithic integrated circuit timer** was first introduced by Signetics Corporation in 1972 in bipolar technology. It quickly became an industry standard for timing and oscillation functions. Many manufacturers produce a version of a 555 IC, some in CMOS technology. The 555 is a general-purpose IC that can be used for precision timing, pulse generation, sequential timing, time delay generation, pulse width modulation, pulse position modulation, and linear ramp generation. The 555 can operate in both astable and monostable modes, with timing pulses ranging from microseconds to hours. It also has an adjustable duty cycle and can generally source or sink output currents up to 200 mA.

### Basic Operation

The basic block diagram of the 555 IC is shown in Figure 15.39(a). The circuit consists of two comparators, which drive an RS flip-flop, an output buffer, and a transistor that discharges an external timing capacitor. The actual circuit of an LM555 timer is shown in Figure 15.39(b).

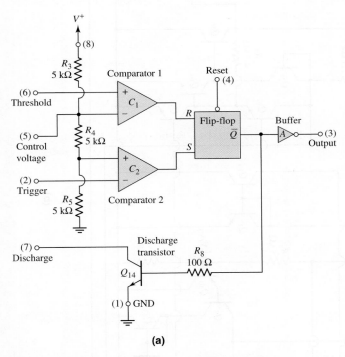

**(a)**

**Figure 15.39** (a) Basic block diagram, 555 IC timer circuit and (b) circuit diagram, LM555 timer circuit

The **RS flip-flop** is a digital circuit that will be considered in detail in a later chapter. Here, we will describe only the basic digital function of the flip-flop, so that the operation of the 555 timer can be explained. When the input $R$ is high and input $S$ is low, output $\bar{Q}$ is high. The complementary state occurs when $R$ is low and $S$ is high, producing a low $\bar{Q}$ output. If both $R$ and $S$ are low, then output $\bar{Q}$ remains in its previous state.

Comparator 1 is called the **threshold comparator,** which compares its input with an internal voltage reference set at $(\frac{2}{3})V^+$ by the voltage divider $R_3$, $R_4$, and $R_5$. When the input level exceeds this reference level, the threshold comparator output goes high, producing a high output at flip-flop terminal $\bar{Q}$. This turns the discharge transistor on and an external timing capacitor (not shown in this figure) starts to discharge.

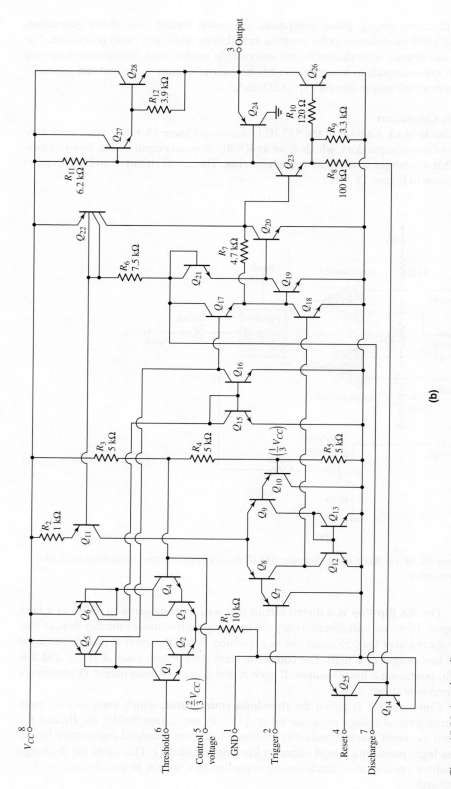

**Figure 15.39** (continued)

The internal control voltage node is connected to an external terminal. This provides external control of the reference level, should the timing period need to be modified. When not in active use, this terminal should be bypassed to ground with a 0.01 $\mu$F capacitor, to improve the circuit's noise immunity.

Comparator 2, called the **trigger comparator,** compares its input trigger voltage to an internal voltage reference set to $(\frac{1}{3})V^+$ by the same voltage divider as before. When the output trigger level is reduced below this reference level, the trigger comparator output goes high, causing the RS flip-flop to reset. Output $\bar{Q}$ goes low and the discharge transistor turns off. This comparator triggers on the leading edge of a negative-going input pulse.

The output stage of the 555 IC is driven by output $\bar{Q}$ of the RS flip-flop. This output is usually a totem-pole push–pull circuit, or a simple buffer, and is generally capable of sourcing or sinking 200 mA.

An external reset input to the RS flip-flop overrides all other inputs and is used to initiate a new timing cycle by turning the discharge transistor on. The reset input must be less than 0.4 V to initiate a reset. When not actively in use, the reset terminal should be connected to $V^+$ to prevent a false reset.

## Monostable Multivibrator

A monostable multivibrator, also called a one-shot, operates by charging a timing capacitor with a current set by an external resistance. When the one-shot is triggered, the charging network cycles only once during the timing interval. The total timing interval includes the recovery time needed for the capacitor to charge up to the threshold level.

The external circuitry and connections for the 555 to be used as a one-shot multivibrator are shown in Figure 15.40. With a high voltage $V^+$ applied to the trigger input, the trigger comparator output is low, the flip-flop output $\bar{Q}$ is high, the discharge transistor is turned on, and the timing capacitor $C$ is discharged to nearly ground potential. The output of the 555 circuit is then low, which is the quiescent state of the one-shot.

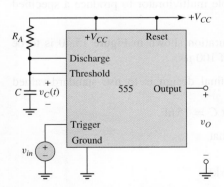

**Figure 15.40** The 555 circuit connected as a monostable multivibrator

When a negative-going pulse is applied to the trigger input, the output of the trigger comparator goes high when the trigger pulse drops below $(\frac{1}{3})V^+$. Output $\bar{Q}$ goes low, which means that the output of the 555 goes high, and the discharge transistor turns off. The output of the 555 remains high even if the trigger pulse returns to its initial high value, because the reset input to the flip-flop is still low. The timing capacitor charges up exponentially toward a final value of $V^+$ through resistor $R$. The capacitor voltage is given by

$$v(t) = V^+(1 - e^{-t/RC})$$  (15.93)

When $v(t) = (\frac{2}{3})V^+$, the threshold comparator output goes high, resetting the flip-flop. Output $\bar{Q}$ then goes high and the output of the 555 goes low. The high output at $\bar{Q}$ also turns on the discharge transistor, allowing the timing capacitor to discharge to near zero volts. The circuit thus returns to its quiescent state.

The width of the output pulse is determined from Equation (15.93). If we set $v(t) = (\frac{2}{3})V^+$ and $t = T$, then

$$\left(\frac{2}{3}\right)V^+ = V^+ (1 - e^{-T/RC})$$  (15.94)

Solving for $T$, we have

$$T = RC \ln(3) = 1.1 \, RC$$  (15.95)

The width of the output pulse is a function of only the external time constant $RC$; it is independent of the supply voltage $V^+$ and any internal circuit parameters. The triggering input pulse must be of a shorter duration than $T$. The output pulse height is a function of $V^+$ as well as of the internal circuitry. For a bipolar 555, the output pulse amplitude is approximately 1.7 V below supply voltage $V^+$.

When the output is high and the timing capacitor is charging, another trigger input pulse will have no effect on the circuit. If desired, the circuit can be reset during this period by applying a low input to the reset terminal. The output will return to zero and will remain in this quiescent state until another trigger pulse is applied.

## DESIGN EXAMPLE 15.10

**Objective:** Design a 555 IC as a monostable multivibrator to produce a specified output pulse width.

**Specifications:** The circuit with the configuration shown in Figure 15.40 is to be designed to produce an output pulse width of 100 $\mu$s.

**Choices:** A 555 circuit is available. The final design is to use standard-valued elements.

Consider the circuit in Figure 15.40. Let $C = 15$ nF.

**Solution:** Using Equation (15.95), we find that

$$R = \frac{T}{1.1C} = \frac{100 \times 10^{-6}}{(1.1)(15 \times 10^{-9})} \Rightarrow 6.06 \, \text{k}\Omega$$

**Trade-offs:** Using standard-valued element values of $C = 12$ nF and $R = 7.5$ k$\Omega$ will produce an output pulse with a width of 99 $\mu$s. Element tolerances also need to be taken into account.

**Comment:** To a very good approximation, the pulse width is a function of only the external resistor and capacitance values. A wide range of pulse widths can be obtained by changing these component values.

**Ex 15.10:** Consider the 555 IC monostable multivibrator. (a) If $R = 20\,k\Omega$ and $C = 0.012\,\mu F$, what is the resulting output pulse width? (b) Design the circuit to produce an output signal with a pulse width of $120\,\mu s$. (Ans. (a) $T = 0.264\,ms$; (b) For example, set $C = 0.01\,\mu F$, then $R = 10.9\,k\Omega$)

## Astable Multivibrator

Figure 15.41 shows a typical external circuit connection for the 555 operating as an astable multivibrator, also called a timer circuit or clock. The threshold input and trigger input terminals are connected together. In the astable mode, the timing capacitor $C$ charges through $R_A = R_B$ until $v(t)$ reaches $(\frac{2}{3})V^+$. The threshold comparator output then goes high, forcing the flip-flop output $\bar{Q}$ to go high. The discharge transistor turns on, and the timing capacitor $C$ discharges through $R_B$ and the discharge transistor. The capacitor voltage decreases until it reaches $(\frac{1}{3})V^+$, at which point the trigger comparator switches states and sends $\bar{Q}$ low. The discharge transistor turns off, and the timing capacitor begins to recharge. When $v(t)$ reaches the threshold level of $(\frac{2}{3})V^+$, the cycle repeats itself.

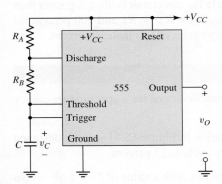

**Figure 15.41** Astable multivibrator 555 circuit

When the timing capacitor is charging, during the time $0 < t < T_C$, the capacitor voltage is

$$v(t) = \frac{1}{3}V^+ + \frac{2}{3}V^+(1 - e^{-t/\tau_A}) \tag{15.96}$$

where $\tau_A = (R_A + R_B)C$. At time $t = T_C$, the capacitor voltage reaches the threshold level, or

$$v(T_C) = \frac{2}{3}V^+ = \frac{1}{3}V^+ + \frac{2}{3}V^+(1 - e^{-T_C/\tau_A}) \tag{15.97}$$

Solving Equation (15.97) for the timing capacitor charging time $T_C$ yields

$$T_C = \tau_A \ln(2) = 0.693(R_A + R_B)C \tag{15.98}$$

When the timing capacitor is discharging, during the time $0 < t' < T_D$, the capacitor voltage is

$$v(t') = \frac{2}{3}V^+ e^{-t'/\tau_B} \tag{15.99}$$

where $\tau_B = R_B C$. At time $t' = T_D$, the capacitor voltage reaches the trigger level and

$$v(T_D) = \frac{1}{3}V^+ = \frac{2}{3}V^+ e^{-T_D/\tau_B} \tag{15.100}$$

Solving Equation (15.100) for the timing capacitor discharge time $T_D$ yields

$$T_D = \tau_B \ln(2) = 0.693 R_B C \tag{15.101}$$

The period $T$ of the astable multivibrator cycle is the sum of the charging period $T_C$ and the discharging period $T_D$. The frequency of oscillation is therefore

$$f = \frac{1}{T} = \frac{1}{T_C + T_D} = \frac{1}{0.693(R_A + 2R_B)C} \tag{15.102}$$

The duty cycle is defined as the percentage of time the output is high during one period of oscillation. During the charging time $T_C$, the output is high; during the discharging time, the output is low. The duty cycle is therefore

$$\text{Duty cycle} = \frac{T_C}{T} \times 100\% = \frac{R_A + R_B}{R_A + 2R_B} \times 100\% \tag{15.103}$$

Equation (15.103) shows that the duty cycle for this circuit is always greater than 50 percent. The duty cycle approaches 50 percent for $R_A \ll R_B$ and 100 percent for $R_B \ll R_A$. Alternative circuits can provide duty cycles of less than 50 percent.

## DESIGN EXAMPLE 15.11

**Objective:** Design the 555 IC as an astable multivibrator for a specified frequency and duty cycle.

**Specifications:** The circuit with the configuration in Figure 15.41 is to be designed such that the frequency is 50 kHz and the duty cycle is 75 percent.

**Choices:** A 555 IC circuit is available. A capacitor with a value of $C = 1$ nF is also available.

**Solution:** The frequency of oscillation, as given by Equation (15.102), is

$$f = \frac{1}{0.693(R_A + 2R_B)C}$$

Therefore,

$$R_A + 2R_B = \frac{1}{(0.693)fC} = \frac{1}{(0.693)(50 \times 10^3)(1 \times 10^{-9})} \Rightarrow 28.9\,\text{k}\Omega \tag{15.104}$$

The duty cycle, given by Equation (15.103), is

$$\text{Duty cycle} = 0.75 = \frac{R_A + R_B}{R_A + 2R_B}$$

which yields

$$R_A = 2R_B \tag{15.105}$$

Combining Equations (15.104) and (15.105), we find that

$$R_A = 14.5\,\text{k}\Omega \qquad \text{and} \qquad R_B = 7.23\,\text{k}\Omega$$

**Trade-offs:** If standard-valued resistors are required, then $R_A = 13\ \text{k}\Omega$ and $R_B = 7.5\ \text{k}\Omega$ would provide a frequency of 51.5 kHz and a duty cycle of 73.2 percent.

**Comment:** A wide range of oscillation frequencies can be obtained by changing the resistance and capacitance values.

---

EXERCISE PROBLEM

**Ex 15.11:** The 555 IC is connected as an astable multivibrator. Let $R_A = 20\ \text{k}\Omega$, $R_B = 80\ \text{k}\Omega$, and $C = 0.01\ \mu\text{F}$. Determine the frequency of oscillation and the duty cycle. (Ans. $f = 802$ Hz, duty cycle $= 55.6\%$)

### Other Applications

When the 555 is connected in the monostable mode, an external signal applied to the control voltage terminal will change the charging time of the timing capacitor and the pulse width. If the one-shot is triggered with a continuous pulse train, the output pulse width will be modulated by the external signal. This circuit is known as a **pulse width modulator (PWM).**

A **pulse position modulator** can also be developed using the astable mode. A modulating signal applied to the control voltage terminal will vary the pulse position, which will be controlled by the modulating signal in a manner similar to the PWM.

Finally, a **linear ramp generator** can be constructed, again using the 555 monostable mode. The normal charging pattern of the timing capacitor is exponential because of the $RC$ circuit. If resistor $R$ is replaced by a constant current source, a linear ramp will be generated.

## Test Your Understanding

**TYU 15.10** (a) The Schmitt trigger oscillator is shown in Figure 15.35. The saturated output voltages are $\pm 5$ V, and $R_1 = R_2 = 15\ \text{k}\Omega$, $R_X = 20\ \text{k}\Omega$, and $C_X = 0.05\ \mu\text{F}$. Determine the frequency of oscillation and the duty cycle. (b) Changing only the value of $R_X$ from part (a), determine the value of $R_X$ such that the frequency of oscillation is $f = 1.2$ kHz. (Ans. (a) $f = 454.5$ Hz, 50% duty cycle; (b) $R_X = 7.576\ \text{k}\Omega$)

**TYU 15.11** Consider the monostable multivibrator in Figure 15.37 with parameters: $V_P = 8$ V, $V_\gamma = 0.7$ V, $C_X = 0.01\ \mu\text{F}$, $R_X = 10\ \text{k}\Omega$, $R_1 = 20\ \text{k}\Omega$, and $R_2 = 40\ \text{k}\Omega$. Determine the output pulse width and recovery time. (Ans. $T = 48.9\ \mu\text{s}, t_2 = 37.8\ \mu\text{s}$)

**TYU 15.12** Design the 555 IC as an astable multivibrator to deliver a 1 kHz signal with a 55 percent duty cycle. (Ans. For example, $C = 0.01\ \mu\text{F}$, $R_A = 14.43\ \text{k}\Omega$, $R_B = 64.9\ \text{k}\Omega$)

---

 ## 15.5 INTEGRATED CIRCUIT POWER AMPLIFIERS

**Objective:** • Analyze and design IC power amplifiers that usually consist of high-gain small-signal amplifiers in cascade with an output stage.

Most IC power amplifiers consist of a high-gain small-signal amplifier cascaded with a class-AB output stage. Some IC power amplifiers are a fixed-gain circuit with negative feedback incorporated on the chip, while others use a current gain output stage and negative feedback external to the chip. We consider three examples of IC power amplifiers in this section.

### 15.5.1   LM380 Power Amplifier

The LM380 is a popular fixed-gain power amplifier capable of an ac power output up to 5 W. Figure 15.42 is a simplified circuit diagram of the amplifier. The input stage is a Darlington pair configuration composed of $Q_1$ through $Q_4$ and an active load formed by $Q_5$ and $Q_6$.

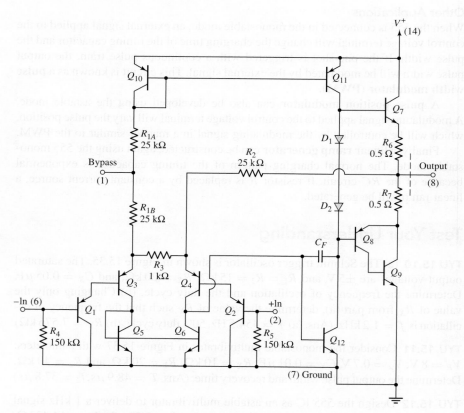

**Figure 15.42** The LM380 power amplifier

The input stage is biased by currents through resistors $R_{1A}$, $R_{1B}$, and $R_2$. Transistor $Q_3$ is biased by a current from power supply $V^+$, through the diode-connected transistor $Q_{10}$ and resistors $R_{1A}$ and $R_{1B}$. Transistor $Q_4$ is biased by a current from the output terminal through $R_2$. For zero input voltages, the currents in $Q_3$ and $Q_4$ are nearly equal. Assuming matched input transistors and neglecting base currents, we find that

$$I_{C3} = \frac{V^+ - 3V_{EB}}{R_{1A} + R_{2A}} \tag{15.106}$$

and

$$I_{C4} = \frac{V_O - 2V_{EB}}{R_2} \qquad (15.107)$$

Since $I_{C3} = I_{C4}$, we can find the quiescent output voltage by combining Equations (15.106) and (15.107), or

$$V_O = 2V_{EB} + \frac{R_2}{R_{1A} + R_{2B}}(V^+ - 3V_{EB}) = \frac{1}{2}V^+ + \frac{1}{2}V_{EB} \qquad (15.108)$$

The quiescent output voltage is approximately half the power supply voltage, which allows for a maximum output voltage swing and for maximum power to be delivered to a load. The feedback from the output to the emitter of $Q_4$, through $R_2$, stabilizes the quiescent output voltage at this value.

The output signal of the diff-amp is the input signal to the base of $Q_{12}$, which is connected in a common-emitter configuration in which $Q_{11}$ acts as an active load. The output signal from the collector of $Q_{12}$ is the input to the class-AB output stage, and capacitor $C_F$ provides frequency compensation.

The class-AB complementary push-pull emitter-follower output stage comprises transistors $Q_7$, $Q_8$, and $Q_9$ and diodes $D_1$ and $D_2$. Transistor $Q_7$, which is the npn half of the push-pull output stage, sources current to the load. Transistors $Q_8$ and $Q_9$ operate as a composite pnp transistor, with the overall current gain equal to the product of the current gains of each transistor. This composite transistor is the pnp half of the push-pull output stage sinking current from the load. Diodes $D_1$ and $D_2$ provide the quiescent bias for class-AB operation.

The closed-loop gain is determined from the ac equivalent circuit in Figure 15.43. A differential-input voltage is applied at the input, with $V_{id}/2$ applied at the noninverting terminal and $-V_{id}/2$ applied at the inverting terminal. An external bypass capacitor is connected at the node between $R_{1A}$ and $R_{1B}$, putting this node at signal ground. The second stage and output stage are represented by amplifier $A$. The input impedance is assumed to be large, which means that the input current is assumed to be negligible.

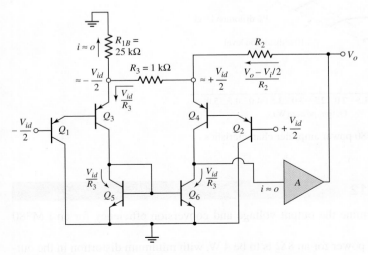

**Figure 15.43** The ac equivalent circuit, LM380 power amplifier

Since the input stage is an emitter-follower configuration, the signal voltage is approximately $+V_{id}/2$ at the emitter of $Q_4$ and is approximately $-V_{id}/2$ at the emitter of $Q_3$. Comparing the resistor values of $R_3$ and $R_{1B}$, we see the signal current in $R_{1B}$ is negligible. The signal current in $Q_3$ is equal to that in $R_3$, and the current-mirror configuration of $Q_5$ and $Q_6$ implies that the current in $Q_6$ is also $V_{id}/R_3$. Summing the currents at the emitter of $Q_4$, we obtain

$$\frac{V_o - V_{id}/2}{R_2} = \frac{V_{id}}{R_3} + \frac{V_{id}}{R_3} \tag{15.109}$$

which yields the closed-loop voltage gain

$$\frac{V_o}{V_{id}} = \frac{1}{2} + \frac{2R_2}{R_3} \cong 50 \tag{15.110}$$

Equation (15.110) shows that the LM380 has a fixed gain of approximately 50.

The LM380 is designed to operate in the range of 12–22 V from a single supply $V^+$. The value of $V^+$ depends on the power requirements. Figure 15.44 shows the relationship between device dissipation, output power, and supply voltage for an 8 Ω load. As the output signal increases, harmonic distortion in the sinusoidal signal increases because the output transistor is approaching the saturation region. The lines marked 3% and 10% are the points at which harmonic distortion reaches 3% and 10%, respectively.

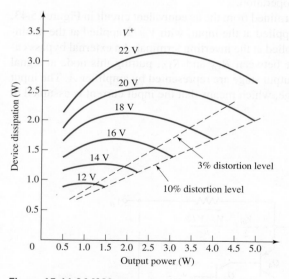

**Figure 15.44** LM380 power amplifier characteristics

---

## EXAMPLE 15.12

**Objective:** Determine the output voltage and conversion efficiency for an LM380 power amplifier.

The required power for an 8 Ω is to be 4 W, with minimum distortion in the output signal.

**Solution:** From the curves in Figure 15.44, for an output of 4 W, minimum distortion occurs when the supply voltage is a maximum, or $V^+ = 22$ V. For 4 W to be delivered to the 8 $\Omega$ load, the peak output signal voltage is determined by

$$\bar{P}_L = 4 = \frac{V_P^2}{2R_L} = \frac{V_P^2}{2(8)}$$

which yields $V_P = 8$ V.

The power dissipated in the device is 3 W, which means that the conversion efficiency is $4/(3+4) \rightarrow 57$ percent.

**Comment:** A reduction in the harmonic distortion means that the conversion efficiency is less than the theoretical value of 78.5 percent for the class-B output stages. However, a conversion efficiency of 57 percent is still substantially larger than would be obtained in any class-A amplifier.

---

EXERCISE PROBLEM

**Ex 15.12:** The supply voltage to an LM380 power amplifier, as shown in Figure 15.42, is 12 V. With a sinusoidal input signal, an average output power of 1 W must be delivered to an 8 $\Omega$ load. (a) Determine the peak output voltage and peak output current. (b) When the output voltage is at its peak value, calculate the instantaneous power being dissipated in $Q_7$. (Ans. (a) $V_P = 4$ V, $I_p = 0.5$ A (b) $P_Q = 4$ W)

---

### 15.5.2    PA12 Power Amplifier

The basic circuit diagram of the PA12 amplifier is shown in Figure 15.45. The input signal to the class-AB output stage is from a small-signal high-gain op-amp. The power supply voltages are in the range of $10 \leq V_S \leq 50$ V, the peak output current is in the range $-15 \leq I_L \leq +15$ A, and the maximum internal power dissipation is 125 W. The output stage is a class-AB configuration using npn and pnp Darlington pair transistors. The bias for the output transistors is established by the $V_{BE}$ multiplier circuit composed of $R_1$, $R_2$, and $Q_4$. Also, external feedback is required.

---

DESIGN EXAMPLE **15.13**

**Objective:** Design the supply voltage required in the PA12 power amplifier to meet a specific conversion efficiency.

**Specifications:** The circuit with the configuration in Figure 15.45 has a load resistance of 10 $\Omega$. The required average power delivered to the load is 20 W. Determine the power supply voltage such that the conversion efficiency is 50 percent.

**Choices:** The circuit shown in Figure 15.45 is available.

**Solution:** For an average of 20 W delivered to the load, the peak output voltage is

$$V_p = \sqrt{2R_L \bar{P}_L} = \sqrt{2(10)(20)} = 20 \text{ V}$$

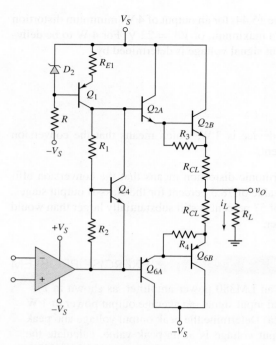

**Figure 15.45** PA12 power amplifier

and the peak load current is

$$I_p = \frac{V_p}{R_L} = \frac{20}{10} = 2 \text{ A}$$

Assuming an ideal class-B condition, for a 50 percent conversion efficiency, the average power supplied by each $V_S$ source must be 20 W. If we neglect power dissipation in the bias circuit, the average power supplied by each source is

$$P_S = V_S \left( \frac{V_p}{\pi R_L} \right)$$

and the required supply voltage is then

$$V_S = \frac{\pi R_L P_S}{V_p} = \frac{\pi (10)(20)}{20} = 31.4 \text{ V}$$

**Trade-offs:** The required power supply must also be able to deliver the required current. For a power of 20 W delivered to the 10 Ω load, the load current (rms value) by itself is 1.41 A.

**Comment:** The actual conversion efficiency for class-AB operation is less than 50 percent. This reduced conversion efficiency ensures that harmonic distortion in the output signal is not severe.

**Computer Simulation Verification:** A computer simulation analysis of the circuit in Figure 15.45 was performed. The supply voltages were set at ±31.4 V and the input sinusoidal signal was adjusted so that the peak sinusoidal output voltage was 19.7 V across a 10 $\Omega$ load resistor. For these settings, the bias supply currents were 1.971 A. The average power delivered by the supply voltage sources is 39.4 W, so that the conversion efficiency is 49.25 percent, which is just slightly below the design value of 50 percent.

---

### EXERCISE PROBLEM

**Ex 15.13:** Consider the power amplifier in Figure 15.45. (a) Assume a load resistance of $R_L = 20\,\Omega$ that must dissipate $P_L = 5\,\text{W}$ is connected to the output. Determine (i) the values $V_p$ and $I_p$ for the load and (ii) the required power supply voltage $V_S$. (b) Repeat part (a) for $R_L = 8\,\Omega$ and $P_L = 10\,\text{W}$. (Ans. (a) (i) $V_p = 14.14\,\text{V}$, $I_p = 0.707\,\text{A}$; (ii) $V_S = 22.2\,\text{V}$; (b) (i) $V_p = 12.65\,\text{V}$, $I_p = 1.58\,\text{A}$; (ii) $V_S = 19.9\,\text{V}$)

---

### 15.5.3    Bridge Power Amplifier

Figure 15.46 shows a bridge power amplifier that uses two op-amps. Amplifier $A_1$ is connected in a noninverting configuration; $A_2$ is connected in an inverting configuration. The magnitudes of the two gains are equal to each other. The load, such as an audio speaker, is connected between the two output terminals and is floating. A sinusoidal input signal produces output voltages $v_{o1}$ and $v_{o2}$, which are equal in magnitude but 180 degrees out of phase. The voltage across the load is therefore twice as large as it would be if produced from a single op-amp.

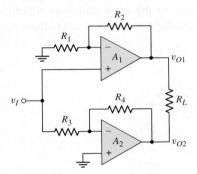

**Figure 15.46** Bridge power amplifier

## Test Your Understanding

**TYU 15.13** (a) Consider the bridge amplifier in Figure 15.46 with parameters $R_1 = R_3 = 20\,\text{k}\Omega$, $R_2 = 40\,\text{k}\Omega$, $R_4 = 60\,\text{k}\Omega$, and $R_L = 500\,\Omega$. Assume the op-amps are biased at $\pm 15\,\text{V}$ and the peak output voltage of each op-amp is limited to $\pm 12\,\text{V}$. Determine (i) the voltage gain of each op-amp circuit, (ii) the average power that can be delivered to the load, and (iii) the peak amplitude of the input voltage $v_I$. (b) From the results of part (a), change the values of $R_2$ and $R_4$ such that the same average power is delivered to the load but the magnitude of the required input voltage is cut in half. What are the required amplifier gains? (Ans. (a) (i) $A_{v1} = +3$, $A_{v2} = -3$, (ii) $\bar{P}_L = 0.576\,\text{W}$, (iii) $v_I = 4\,\text{V}$; (b) $R_2 = 100\,\text{k}\Omega$, $R_4 = 120\,\text{k}\Omega$, $A_{v1} = +6$, $A_{v2} = -6$)

## 15.6    VOLTAGE REGULATORS

**Objective:** • Analyze and design voltage regulators that establish a relatively constant dc voltage generated from an ac signal source.

Another class of analog circuits that is used extensively in electronic systems is the voltage regulator. We briefly considered constant-voltage circuits, or voltage regulators, when we studied diode circuits and when we considered ideal op-amp circuits in Chapter 9. In this section, we will discuss examples of IC voltage regulators.

### 15.6.1    Basic Regulator Description

A **voltage regulator** is a circuit or device that provides a constant voltage to a load. The output voltage is controlled by the internal circuitry and is relatively independent of the load current supplied by the regulator.

A basic diagram of a voltage regulator is shown in Figure 15.47. It consists of three basic parts: a reference voltage circuit; an error amplifier, which is part of a feedback circuit; and a current amplifier, which supplies the required load current. The reference voltage circuit produces a voltage that is essentially independent of both supply voltage $V^+$ and temperature. As shown in the basic circuit of Figure 15.47, a fraction of the output voltage is fed back to the error amplifier which, through negative feedback, maintains the feedback voltage at a value equal to the reference voltage.

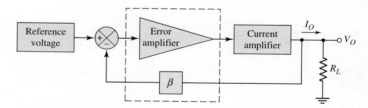

**Figure 15.47** Basic circuit diagram of a voltage regulator

Since the regulator output voltage is derived from the reference voltage, any variation in that reference voltage, as the power supply voltage $V^+$ changes, also affects the output voltage. **Line regulation** is defined as the ratio of the change in output voltage to a given change in the input supply voltage, or

$$\text{Line regulation} = \frac{\Delta V_o}{\Delta V^+} \tag{15.111}$$

Line regulation is one figure of merit of voltage regulators. In many cases, the reference voltage circuit contains one or more Zener diodes. Line regulation is then a function of the Zener diode resistance and the effective resistance of the circuit biasing the diode.

## 15.6.2 Output Resistance and Load Regulation

The ideal voltage regulator is equivalent to an ideal voltage source in that the output voltage is independent of the output current and any output load impedance. In actual voltage regulators, however, the output voltage is a slight function of output current. This dependence is related to the output resistance of the regulator.

The output resistance is defined as the rate of change of output voltage with output current, or

$$R_{of} = -\frac{\Delta V_O}{\Delta I_O} \tag{15.112}$$

The change in $V_O$ and $I_O$ is caused by a change in the load resistance $R_L$. Everything else in the circuit remains constant. The negative sign in Equation (15.112) results from the voltage polarity and current direction, as shown in Figure 15.47. An increase in $I_O$ produces a decrease in $V_O$; therefore, the output resistance $R_{of}$ is positive. The output resistance of a voltage regulator should be small, so that a change in output current $\Delta I_O$ will result in only a small change in output voltage $\Delta V_O$.

The notation $R_{of}$ for the output resistance of the voltage regulator is the same as the term for the output resistance of a feedback circuit. This is appropriate since voltage regulators use feedback.

A second figure-of-merit for voltage regulators is load regulation. **Load regulation** is defined as the change in output voltage between a no-load current condition and a full-load current condition. Load regulation can be expressed as a percentage, or

$$\text{Load regulation} = \frac{V_O(\text{NL}) - V_O(\text{FL})}{V_O(\text{NL})} \times 100\% \tag{15.113}$$

where $V_O(\text{NL})$ is the output voltage for a zero-load current condition and $V_O(\text{FL})$ is the output voltage for a full-load or maximum load current condition.

In some applications, a zero-load current is impractical, and a load current that is approximately 1 percent of the full-load current is used as the no-load condition. In most cases, this condition provides an adequate definition for load regulation.

### EXAMPLE 15.14

**Objective:** Determine the output resistance and load regulation of a voltage regulator.

Assume the output voltage of a regulator is 5.0 V for a load current of 5 mA, and is 4.96 V for a load current of 1.5 A.

**Solution:** If we assume that the output voltage decreases linearly with load current, then the output resistance is

$$R_{of} = -\frac{\Delta V_O}{\Delta I_O} = -\left(\frac{5.0 - 4.96}{0.005 - 1.5}\right) \cong 0.0267 \ \Omega$$

or

$$R_{of} \cong 27 \ \text{m}\Omega$$

The load regulation is then

$$\text{Load regulation} = \frac{V_O(\text{NL}) - V_O(\text{FL})}{V_O(\text{NL})} \times 100\% = \frac{5.0 - 4.96}{5.0} \times 100\% = 0.80\%$$

**Comment:** The output resistance of a voltage regulator is usually not constant at all load currents, but the values are typically in the milliohm range. Also, a load regulation of 0.8% is typical of many voltage regulators.

**Ex 15.14:** The reference voltage for a constant-voltage source is established by the simple combination of $V^+$, $R_1$, and $D_I$, as shown in the regulator circuit in Figure 15.48. If the Zener diode resistance is $R_Z = 10 \ \Omega$ and the zero-current diode voltage is $V_{Zo} = 5.6$ V, determine the line regulation of the voltage regulator. Assume an ideal op-amp. (Ans. 0.454%)

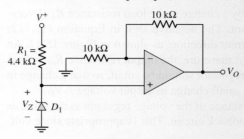

**Figure 15.48** Figure for Exercise Ex15.14

### 15.6.3 Simple Series-Pass Regulator

Figure 15.49 shows a simple voltage regulator that includes an error amplifier (comparator) and series-pass transistors. The series-pass transistors, which are connected in a Darlington emitter-follower configuration, form the current amplifier. A resistive voltage divider allows a portion of the output voltage to be fed back to the error amplifier. The closed-loop feedback system acts to maintain this fraction of the output voltage at a value equal to the reference voltage.

For an ideal system, we can write

$$\left( \frac{R_2}{R_1 + R_2} \right) V_O = V_{\text{REF}} \tag{15.114(a)}$$

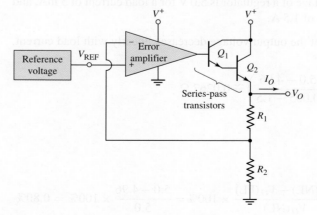

**Figure 15.49** Basic series-pass voltage regulator

or

$$V_O = V_{\text{REF}}\left(1 + \frac{R_1}{R_2}\right) \tag{15.114(b)}$$

Since the output of the feedback circuit is a shunt connection, the output resistance can be written, according to the results from Chapter 12, as

$$R_{of} = \frac{R_o}{1 + T} \tag{15.115}$$

where $R_o$ is the output resistance of the open-loop system and $T$ is the loop gain.

From feedback theory, the closed-loop and open-loop gains are related by

$$A_{CL} = \frac{A_{OL}}{1 + T} \tag{15.116}$$

Combining Equation (15.115) and (15.116), we can write the closed-loop output resistance of the voltage regulator in the form

$$R_{of} = R_o\left(\frac{A_{CL}}{A_{OL}}\right) \tag{15.117}$$

From the circuit in Figure 15.49, the closed-loop gain is

$$A_{CL} = \frac{V_O}{V_{\text{REF}}} \tag{15.118}$$

The open-loop output resistance is the output resistance of the series-pass transistors, which are operating in an emitter-follower configuration. From previous results, we can write

$$R_o = \frac{r_{\pi 2} + R_{o1}}{(1 + \beta_2)} \tag{15.119}$$

where

$$R_{o1} = \frac{r_{\pi 1} + R_{oa}}{(1 + \beta_1)} \tag{15.120}$$

in which $R_{oa}$ is the output resistance of the error amplifier. If the current in $Q_2$ is essentially equal to $I_O$ and if $\beta_1$ and $\beta_2$ are large, then combining Equations (15.119) and (15.120) yields

$$R_o \cong \frac{2V_T}{I_O} + \frac{R_{oa}}{\beta_1 \beta_2} \tag{15.121}$$

Since the product $\beta_1\beta_2$ is large, the second term in Equation (15.121) is generally negligible.

The closed-loop output resistance, given by Equation (15.117), is then

$$R_{of} \cong \left(\frac{2V_T}{I_O}\right)\left(\frac{A_{CL}}{A_{OL}}\right) = \left(\frac{2V_T}{I_O}\right)\left(\frac{V_O}{V_{\text{REF}}}\right)\left(\frac{1}{A_{OL}}\right) \tag{15.122}$$

Equation (15.122) shows that the output resistance of the voltage regulator is not constant, but varies inversely with load current. Also, for very small values of load current, the output resistance may be unacceptably high.

The basic definition of output resistance is given in Equation (15.112). Using this definition and Equation (15.122), and rearranging terms, we obtain

$$\frac{\Delta V_O}{V_O} = -\left(\frac{\Delta I_O}{I_O}\right)\left(\frac{2V_T}{R_{\text{REF}}}\right)\left(\frac{1}{A_{OL}}\right) \tag{15.123}$$

Equation (15.123) relates the fractional change in output voltage to a fractional change in output current. Although valid for only small variations in voltage and current, this equation provides insight into the concept of load regulation.

## EXAMPLE 15.15

**Objective:** Determine the output resistance and the variation in output voltage of a series-pass regulator.

Assume an open-loop gain of $A_{OL} = 1000$, a reference voltage of $V_{REF} = 5$ V, a nominal output voltage of $V_O = 10$ V, and a nominal output current of $I_O = 100$ mA.

**Solution:** From Equation (15.122), the output resistance is

$$R_{of} = \left(\frac{2V_T}{I_O}\right)\left(\frac{V_O}{V_{REF}}\right)\left(\frac{1}{A_{OL}}\right) = \left[\frac{2(0.026)(10)}{(0.10)(5)(1000)}\right] \Rightarrow 1.04 \, \text{m}\Omega$$

From Equation (15.123), the relative change in output voltage is

$$\frac{\Delta V_O}{V_O} = -\left(\frac{\Delta I_O}{I_O}\right)\left(\frac{2V_T}{V_{REF}}\right)\left(\frac{1}{A_{OL}}\right) = -\left(\frac{\Delta I_O}{I_O}\right)\left[\frac{2(0.026)}{(5)(1000)}\right]$$

or

$$\frac{\Delta V_O}{V_O} = -\left(\frac{\Delta I_O}{I_O}\right)(1.04 \times 10^{-5})$$

A 10 percent change in output current results in only a $1.04 \times 10^{-4}$ percent change in output voltage.

**Comment:** An output resistance in the mΩ range is typical of voltage regulators, and a change of only $10^{-4}$ percent in output for a 10 percent change in current is a good load regulation value.

## EXERCISE PROBLEM

**\*Ex 15.15:** Consider the voltage regulator in Figure 15.50. The Zener diode is ideal, with $V_Z = 6.3$ V, and the op-amp has a finite open-loop gain of $A_{OL} = 1000$. The no-load current is $I_O = 1$ mA, and the full-load current is $I_O = 100$ mA. Determine the load regulation. (Ans. 0.786%)

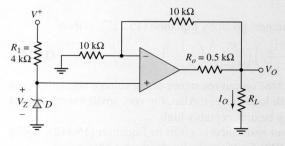

**Figure 15.50** Figure for Exercise Ex15.15

15.6.4 Positive Voltage Regulator

In this section, we will analyze an example of a three-terminal positive voltage regulator fabricated as an IC. The equivalent circuit, shown in Figure 15.51, is part of the LM78LXX series, in which the XX designation indicates the output voltage of the regulator. For example, an LM78L08 is an 8 V regulator.

### Basic Circuit Description

Once the bias current is established, Zener diode $D_2$ provides the basic reference voltage. Transistors $Q_{15}$ and $Q_{16}$ and diode $D_1$ form a start-up circuit that applies the initial bias to the reference voltage circuit. As the voltage across $D_2$ reaches the Zener voltage, transistor $Q_{15}$ turns off, since the B–E voltage goes to zero ($D_1$ and $D_2$ are identical) and, the start-up circuit is then effectively disconnected from the reference voltage circuit.

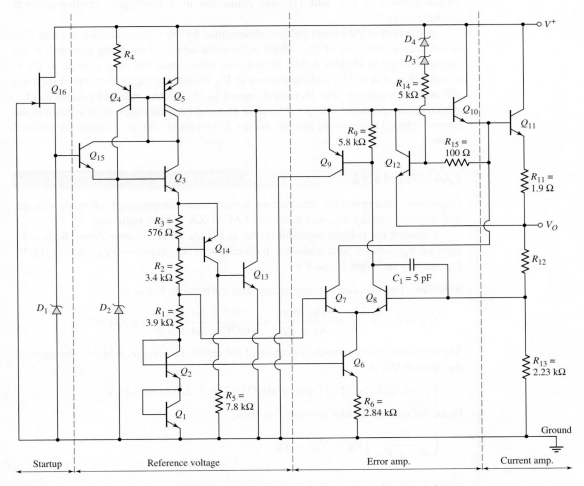

**Figure 15.51** Equivalent circuit, LM78LXX series three-terminal positive voltage regulator

The reference portion of the circuit is composed of Zener diode $D_2$ and transistors $Q_3$, $Q_2$, and $Q_1$, which are used for temperature compensation. The temperature compensation aspects of the circuit are discussed later in this section. Zener diode $D_2$ is biased by the current-source transistor $Q_4$. The temperature-compensated portion of the reference voltage at the node between $R_1$ and $R_2$ is applied to the base of $Q_7$, which is part of the error amplifier.

The bias current in $Q_4$ is established by the current in $Q_5$, which is a multiple-collector, multiple-emitter transistor. Transistor $Q_5$ is biased by the current in $Q_3$, which is controlled by the Zener voltage across $D_2$ and the B–E junction voltages of $Q_3$, $Q_2$, and $Q_1$. Consequently, the bias currents in the reference portion of the circuit become almost independent of the input supply voltage. This in turn means that the reference voltage, and thus the output voltage are essentially independent of the power supply voltage. The overall result is very good line regulation.

The error amplifier is the differential pair $Q_7$ and $Q_8$, biased by $Q_6$ and $R_6$. The error amplifier output is the input to the base of $Q_9$, which is connected as an emitter follower and forms part of the drive for the series-pass transistors. The series-pass output transistors $Q_{10}$ and $Q_{11}$ are connected in a Darlington emitter-follower configuration.

A fraction of the output voltage, determined by the voltage divider $R_{12}$ and $R_{13}$, is fed back to the base of $Q_8$, which is the error-amplifier inverting terminal. If the output voltage is slightly *below* its nominal value, then the base voltage at $Q_8$ is smaller than that at $Q_7$, and the current in $Q_7$ becomes a larger fraction of the total diff-amp bias current. The increased current in $Q_7$ induces a larger current in $Q_{10}$, which in turn produces a larger current in $Q_{11}$ and increases the output voltage to the proper value. The opposite process occurs if the output voltage is *above* its nominal value.

## EXAMPLE 15.16

**Objective:** Determine the bias current, temperature-compensated reference voltage, and required resistor $R_{12}$ in a particular LM78LXX voltage regulator.

Consider the voltage regulator circuit in Figure 15.51. Assume Zener diode voltages of $V_Z = 6.3$ V and transistor parameters of $V_{BE}(\text{npn}) = V_{EB}(\text{pnp}) = 0.6$ V. Design $R_{12}$ such that $V_O = 8$ V.

**Solution:** The bias current, neglecting base currents, is found as

$$I_{C3} = I_{C5} = \frac{V_Z - 3V_{BE}(\text{npn})}{R_3 + R_2 + R_1} = \frac{6.3 - 3(0.6)}{0.576 + 3.4 + 3.9} = 0.571 \text{ mA}$$

The temperature-compensated portion of the reference voltage, which is the input to the base of $Q_7$, is

$$V_{B7} = I_{C3}R_1 + 2V_{BE}(\text{npn}) = (0.571)(3.9) + 2(0.6) = 3.43 \text{ V}$$

From the voltage divider network, we have

$$\left( \frac{R_{13}}{R_{12} + R_{13}} \right) V_O = V_{B8} = V_{B7}$$

or

$$\left( \frac{2.23}{R_{12} + 2.23} \right)(8) = 3.43$$

which yields

$$R_{12} = 2.97 \, \text{k}\Omega$$

**Comment:** The voltage divider of $R_{12}$ and $R_{13}$ is internal to the IC. This means the output voltage of a voltage regulator is fixed.

---

EXERCISE PROBLEM

**Ex 15.16:** Consider the voltage regulator circuit shown in Figure 15.51 with Zener diode voltages of $V_Z = 5.6$ V. Assume transistor parameters of $V_{BE}(\text{npn}) = V_{EB}(\text{pnp}) = 0.6$ V, neglect base currents, and let the resistor in the emitter of $Q_4$ be $R_4 = 100 \, \Omega$. (a) Determine the bias currents $I_{C3}$ and $I_{C4}$, and the temperature-compensated portion of the reference voltage $V_{B7}$. (b) Determine $R_{12}$ such that $V_O = 5$ V. (Ans. (a) $I_{C3} = 0.482$ mA, $I_{C4} = 0.213$ mA, $V_{B7} = 3.08$ V (b) $R_{12} = 1.39 \, \text{k}\Omega$)

## Temperature Compensation

Zener diodes with breakdown voltages greater than approximately 5 V have positive temperature coefficients, and forward-biased pn junctions have negative temperature coefficients. The magnitude of the temperature coefficients in the two devices is nearly the same.

For a given increase in temperature, $V_{Z2}$ increases by $\Delta V$ and each B–E voltage decreases by $\Delta V$, which means that $I_{C3}$ in Figure 15.51 increases by approximately

$$\Delta I_{C3} \cong \frac{4\Delta V}{R_1 + R_2 + R_3} \tag{15.124}$$

The total voltage across the B–E junctions of $Q_1$ and $Q_2$ decreases by approximately $2\Delta V$, and the change in voltage at the base of $Q_7$ is

$$\Delta V_{B7} \cong \Delta I_{C3} R_1 - 2\Delta V = 4\Delta V \left( \frac{R_1}{R_1 + R_2 + R_3} \right) - 2\Delta V \cong 0 \tag{15.125}$$

This indicates that the voltage divider across $R_1$ effectively cancels any temperature variation. The input signal to the error amplifier is thus temperature compensated.

## Protection Devices

Transistors $Q_{13}$ and $Q_{14}$ and resistor $R_3$ in the regulator in Figure 15.51 provide thermal protection. From the results of Example 15.16, the B–E voltage of $Q_{14}$ is approximately 330 mV, which means that both $Q_{14}$ and $Q_{13}$ are effectively cut off. As the temperature increases, the combination of a negative B–E temperature coefficient and an increase in $I_{C3}$ causes $Q_{14}$ to begin conducting, which in turn causes $Q_{13}$ to conduct. The current in $Q_{13}$ shunts current away from the output series-pass transistors and produces thermal shutdown.

Output current limiting is provided by transistor $Q_{12}$ and resistor $R_{11}$, as we saw previously in op-amp output stages. The combination of resistors $R_{14}$ and $R_{15}$ and diodes $D_3$ and $D_4$ produces what is called a **foldback characteristic.** The vast majority of the power dissipated in the regulator is usually due to the output current, or

$$P_D \cong (V^+ - V_O)I_O \tag{15.126}$$

The output current limit, to prevent power dissipation from reaching its maximum value $P_D(\max)$, is given by

$$I_O(\max) = \frac{P_D(\max)}{V^+ - V_O} \qquad (15.127)$$

A current-limiting characteristic of the type described by Equation (15.127) will protect the regulator and allow the maximum output current possible. This type of current limiting is called foldback current limiting.

### Three-Terminal Regulator

The three-terminal voltage regulator is designed with an output voltage set at a predetermined value; external feedback elements and connections are not required. Figure 15.52 shows the basic circuit configuration of a three-terminal regulator. In some applications, capacitors may be inserted across the input and output terminals. The lead inductance between the voltage supply and regulator may cause stability problems. The capacitor across the input terminals is used only if the power supply and regulator are separated by a few centimeters. The load capacitor may improve the response of the regulator to transient changes in load current.

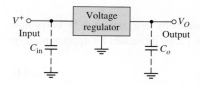

**Figure 15.52** Basic circuit configuration of a three-terminal voltage regulator

## 15.7    DESIGN APPLICATION: AN ACTIVE BANDPASS FILTER

**Objective:** • Design an active bandpass filter to meet a set of specifications.

**Specifications:** The center frequency of the bandpass amplifier is to be $f_o = 2$ kHz, the bandwidth is to be $\Delta f = 10$ Hz, and the maximum voltage gain is to be $|A_v|_{\max} = 40$.

**Design Approach:** The bandpass amplifier configuration to be designed is shown in Figure 15.53.

**Choices:** Ideal op-amps are assumed to be available.

**Solution (Analysis):** Considering the circuit in Figure 15.53, we have

$$\frac{v_{o2}}{v_o} = -\frac{\dfrac{1}{sC}}{R_2} = \frac{-1}{s R_2 C}$$

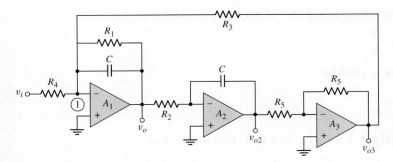

**Figure 15.53** Bandpass filter network for the design application

and

$$\frac{v_{o3}}{v_{o2}} = -1$$

so

$$v_{o3} = \frac{v_o}{s R_2 C} \tag{15.128}$$

Node 1 is at virtual ground. Summing currents at this node, we find

$$\frac{v_i}{R_4} + \frac{v_o}{R_1} + \frac{v_o}{\dfrac{1}{sC}} + \frac{v_{o3}}{R_3} = 0$$

Substituting the expression for $v_{o3}$ from Equation (15.128), we have

$$\frac{v_i}{R_4} = -v_o \left( \frac{1}{R_1} + sC + \frac{1}{s R_2 R_3 C} \right)$$

The overall voltage gain is

$$\frac{v_o}{v_i} = \frac{-\dfrac{1}{R_4}}{\left( \dfrac{1}{R_1} + sC + \dfrac{1}{s R_2 R_3 C} \right)}$$

Setting $s = j\omega$ to obtain the steady-state frequency response, we obtain

$$\frac{v_o}{v_i} = \frac{-\dfrac{1}{R_4}}{\left[ \dfrac{1}{R_1} + j\left( \omega C - \dfrac{1}{\omega R_2 R_3 C} \right) \right]}$$

The center frequency occurs at the point where the imaginary term in the denominator is zero, or

$$\omega_o C = \frac{1}{\omega_o R_2 R_3 C}$$

which can be rewritten as

$$f_o = \frac{1}{2\pi C \sqrt{R_2 R_3}}$$

The maximum voltage gain occurs at the center frequency, so that

$$|A_v|_{\max} = \frac{R_1}{R_4}$$

The bandwidth is given by

$$BW = \frac{1}{2\pi R_1 C}$$

**Solution (Design):** If we let $C = 0.1\,\mu F$, then we can find

$$R_1 = \frac{1}{2\pi(BW)C} = \frac{1}{2\pi(10)(0.1 \times 10^{-6})} = 159\,k\Omega$$

From the maximum gain, we determine

$$|A_v|_{\max} = \frac{R_1}{R_4} \Rightarrow 40 = \frac{159}{R_4}$$

or

$$R_4 = 3.975\,k\Omega$$

If we choose $R_2 = R_3$, then from the center frequency

$$f_o = \frac{1}{2\pi C \sqrt{R_2 R_3}}$$

we find

$$R_2 = R_3 = \frac{1}{2\pi f_o C} = \frac{1}{2\pi(2 \times 10^3)(0.1 \times 10^{-6})}$$

or

$$R_2 = R_3 = 795.8\,\Omega$$

**Solution (Standard Resistor Values):** The closest standard resistor values are $R_2 = 750\,\Omega$, $R_3 = 820\,\Omega$, $R_1 = 160\,k\Omega$, and $R_4 = 3.9\,k\Omega$. A capacitor of $0.1\,\mu F$ is a standard value. Using these circuit elements, we find the center frequency to be

$$f_o = \frac{1}{2\pi C \sqrt{R_2 R_3}} = \frac{1}{2\pi(0.1 \times 10^{-6})\sqrt{(750)(820)}}$$

or

$$f_o = 2.029\,kHz$$

The bandwidth is

$$BW = \frac{1}{2\pi R_1 C} = \frac{1}{2\pi(160 \times 10^3)(0.1 \times 10^{-6})}$$

or

$$BW = 9.947\,Hz$$

The maximum voltage gain at the center frequency is

$$|A_v|_{\max} = \frac{R_1}{R_4} = \frac{160}{3.9} = 41.03$$

**Comment:** Using standard resistor values, the center frequency is within 1.5 percent of the design specification, the bandwidth is within 0.53 percent of the design specification, and the maximum gain is within 2.6 percent of the design specification. The circuit elements, of course, have tolerances that will affect the final circuit performance.

## 15.8 SUMMARY

- This chapter has presented several applications of op-amps and comparators that may be fabricated as integrated circuits.
- An active filter uses an active device, such as an op-amp, so as to minimize the effect of loading on the frequency characteristics of the filter. As an example of an active filter, Butterworth filter design was considered.
- A switched-capacitor filter offers the advantage of an all-IC configuration, since this filter uses small capacitance values in conjunction with MOS switching transistors that simulate large resistance values.
- The basic principles of oscillation are: (1) the net phase through the amplifier and feedback network must be zero and (2) the magnitude of loop gain must be unity. For an oscillator to function, the loop gain of a feedback network must provide sufficient phase shift to produce positive feedback. A phase-shift oscillator, Wien-bridge oscillator, and discrete transistor oscillators were considered.
- A comparator is essentially an op-amp operated in an open-loop configuration. The output signal is either a high or low saturated voltage.
- A Schmitt trigger uses a comparator with positive feedback, which produces a hysteresis in the voltage transfer characteristics. This circuit, with its hysteresis characteristic, can eliminate the chatter effect in an output signal during switching applications in which noise is superimposed on the input signal. Astable and monostable multivibrators were considered.
- Examples of IC power amplifiers and voltage regulators were analyzed.
- As an application, an active bandpass filter to meet a set of specifications was designed.

## CHECKPOINT

After studying this chapter, the reader should have the ability to:

✓ Design a basic active filter.
✓ Design a basic oscillator.
✓ Design a basic Schmitt trigger circuit.
✓ Design a Schmitt trigger square-wave oscillator and use a 555 timer circuit.
✓ Understand the operation and characteristics of examples of integrated circuit power amplifiers.

## REVIEW QUESTIONS

1. Describe the difference between an active filter and a passive filter. What is the primary advantage of an active filter?
2. Sketch the general characteristics of a low-pass filter, a high-pass filter, and a band-pass filter.

3. Consider a low-pass filter. What is the slope of the roll-off with frequency for a (a) one-pole filter, (b) two-pole filter, (c) three-pole filter, and (d) four-pole filter?

4. What characteristic defines a Butterworth filter?

5. Describe how a capacitor in conjunction with two switching transistors can behave as a resistor.

6. Sketch a one-pole low-pass switched-capacitor filter circuit.

7. Explain the two basic principles that must be satisfied in an oscillator circuit.

8. Describe and explain the operation of a phase-shift oscillator.

9. Describe and explain the operation of a Wien-bridge oscillator.

10. Sketch the circuit and characteristics of a basic inverting Schmitt trigger.

11. What is meant by bistable and astable circuits?

12. What is the primary advantage of a Schmitt trigger circuit.

13. Sketch the circuit and explain the operation of a Schmitt trigger oscillator.

14. Describe how an op-amp in conjunction with a class-AB output stage can be used as a power amplifier.

15. Sketch a bridge power amplifier and describe its operation.

16. Sketch the basic circuit block diagram of a voltage regulator and explain the principle of operation.

17. Define load regulation of a voltage regulator.

18. Sketch the basic circuit of a series-pass voltage regulator.

## PROBLEMS

### Section 15.1 Active Filters

D15.1    (a) Design a single-pole high-pass filter with a gain of 8 in the passband and a 3 dB frequency of 30 kHz. The maximum resistance is to be 210 k$\Omega$.
(b) Repeat part (a) for a gain of $-20$ in the passband and a 3 dB frequency of 20 kHz. The minimum input resistance is to be 15 k$\Omega$.

15.2    Consider a Butterworth low-pass filter. Determine the reduction in gain (in dB) at $f = 1.5 f_{3\,dB}$ for a (a) two-pole, (b) three-pole, (c) four-pole, and (d) five-pole filter.

15.3    The specification in a high-pass Butterworth filter design is that the voltage transfer function magnitude at $f = 0.9 f_{3\,dB}$ is 6 dB below the maximum value. Determine the required order of filter.

D15.4    (a) Design a two-pole high-pass Butterworth active filter with a cutoff frequency at $f_{3\,dB} = 25$ kHz and a unity gain magnitude at high frequency.
(b) Determine the magnitude (in dB) of the gain at (i) $f = 22$ kHz, (ii) $f = 25$ kHz, and (iii) $f = 28$ kHz.

D15.5    (a) Design a three-pole low-pass Butterworth active filter with a cutoff frequency at $f_{3\,dB} = 20$ kHz and a unity gain magnitude at low frequency.
(b) Determine the magnitude (in dB) of the gain at (i) $f = 10$ kHz, (ii) $f = 15$ kHz, (iii) $f = 20$ kHz, (iv) $f = 25$ kHz, and (v) $f = 30$ kHz.

15.6    Starting with the general transfer function given by Equation (15.7), derive the relationship between $R_1$ and $R_2$ in the two-pole high-pass Butterworth active filter.

15.7    A low-pass Butterworth filter is to be designed such that the magnitude of the voltage transfer function at $f = 1.2 f_{3\,dB}$ is 14 dB below the maximum gain value. Determine the required order of filter.

15.8   A high-pass Butterworth filter is to be designed with a cutoff frequency of $f_{3\,dB} = 4\,kHz$. The gain magnitude is to be reduced by 12 dB at $f = 3\,kHz$ from the maximum gain value. Determine the required order of filter.

D15.9   A low-pass filter is to be designed to pass frequencies in the 0 to 12 kHz range. The gain of the amplifier is to be $+10$ at the low frequency and change by no more than 10 percent over the frequency range. In addition, the gain of the amplifier for frequencies greater than 14 kHz is to be no greater than 0.1. Determine $f_{3\text{-}dB}$ and the number of poles required in a Butterworth filter.

15.10   Consider a high-pass Butterworth filter. Determine the ratio of the gain magnitude (in dB) of the filter at a frequency $f = 0.8 f_{3\,dB}$ compared to the high-frequency value for a (a) three-pole, (b) five-pole, and (c) seven-pole filter.

15.11   Consider a low-pass Butterworth filter. Determine the ratio of the gain magnitude (in dB) of the filter at a frequency $f = 1.4 f_{3\,dB}$ compared to the low-frequency value for a (a) three-pole, (b) five-pole, and (c) seven-pole filter.

*D15.12  Design a special type of first-order filter (one capacitor) in which the gain magnitude is 25 for frequencies less than approximately 25 kHz and is 1 for frequencies greater than approximately 25 kHz.

*D15.13  An amplitude-modulated radio signal consists of an 80 Hz to 12 kHz audio signal superimposed on a 770 kHz carrier signal. A low-pass filter is to be designed in which the gain in the passband is unity and the carrier signal is attenuated by at least $-100$ dB. What order of filter is required?

D15.14  A band-reject filter may be designed by combining a low-pass filter and a high-pass filter with a summing amplifier. A 60 Hz signal is to be at least $-50$ dB below the maximum gain value of 0 dB with a two-pole low-pass Butterworth filter and a two-pole high-pass Butterworth filter. What is the bandwidth of the reject filter?

15.15   Consider the bandpass filter in Figure P15.15. (a) Show that the voltage transfer function is

$$A_v(s) = \frac{v_O}{v_I} = \frac{-1/R_4}{(1/R_1) + sC + 1/(sCR_2R_3)}$$

(b) For $C = 0.1\,\mu F$, $R_1 = 85\,k\Omega$, $R_2 = R_3 = 300\,\Omega$, $R_4 = 3\,k\Omega$, and $R_5 = 30\,k\Omega$, determine: (i) $|A_v(\text{max})|$; (ii) the frequency $f_o$ at which $|A_v(\text{max})|$ occurs; and (iii) the two 3 dB frequencies.

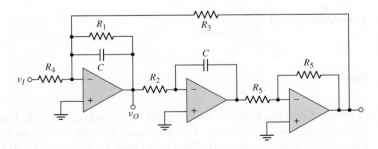

**Figure P15.15**

15.16 Consider the circuit in Figure P15.16. (a) Derive the expressions for the magnitude and phase of the voltage transfer function. (b) Plot the phase versus frequency for $R = 10 \, k\Omega$ and $C = 15.9 \, nF$. [Note: this filter is referred to as an all-pass filter in that the magnitude of the voltage gain is constant, but the phase of the output voltage changes with frequency.]

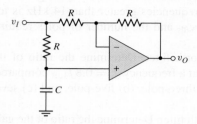

**Figure P15.16**

15.17 For each of the circuits in Figures P15.17, derive the expressions for the voltage transfer function $T(s) = V_o(s)/V_i(s)$ and the cutoff frequency $f_{3 \, dB}$.

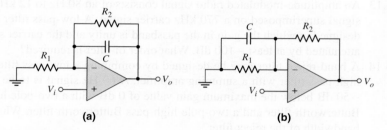

(a)            (b)

**Figure P15.17**

15.18 The circuit in Figure P15.18 is a bandpass filter. (a) Derive the expression for the voltage transfer function $T(s)$. (b) If $R_1 = 10 \, k\Omega$, determine $R_2$, $C_1$, and $C_2$ such that the magnitude of the midband gain is 50 and the cutoff frequencies are 200 Hz and 5 kHz.

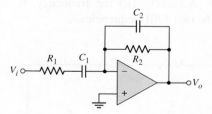

**Figure P15.18**

15.19 A simple bandpass filter can be designed by cascading one-pole high-pass and one-pole low-pass filters. Using op-amp circuits similar to those in Figure 15.3, design a bandpass filter with cutoff frequencies of 150 Hz and 20 kHz and with a midband gain of 30 dB. Resistor values must be no larger than 250 k$\Omega$, but the input resistance must be as large as possible.

15.20 The clock frequency in the switched-capacitor circuit in Figure 15.12(a) is $f_C = 50$ kHz. Find the equivalent resistance that is simulated when (a) $C = 0.5$ pF, (b) $C = 2$ pF, and (c) $C = 10$ pF.

15.21 In the switched-capacitor circuit in Figure 15.12(a), the voltages are $V_1 = 2$ V and $V_2 = 1$ V, the capacitor value is $C = 10$ pF, and the clock frequency is $f_C = 100$ kHz. (a) Determine the charge transferred from $V_1$ to $V_2$ during each clock pulse. (b) What is the average current that source $V_1$ supplies? (c) If the "on" resistance of each MOSFET is 1000 $\Omega$, determine the time required to transfer 99 percent of the charge during each half-clock period.

D15.22 Consider the switched-capacitor filter in Figure 15.13(b). Design the circuit for a low-frequency gain of $-10$ and a cutoff frequency of 10 kHz. The clock frequency must be 10 times the cutoff frequency and the largest capacitance is to be 30 pF. Find the required values of $C_1$, $C_2$, and $C_F$.

15.23 The circuit in Figure P15.23 is a switched-capacitor integrator. Let $C_F = 30$ pF and $C_1 = 5$ pF, and assume the clock frequency is 100 kHz. Also, let $v_I = 1$ V. (a) Determine the integrating $RC$ time constant. (b) Find the change in output voltage during each clock period. (c) If $C_F$ is initially uncharged, how many clock pulses are required for $v_O$ to change by 13 V?

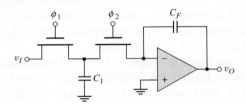

**Figure P15.23**

## Section 15.2 Oscillators

15.24 Consider the phase shift oscillator in Figure 15.15. (a) For $R = 20$ k$\Omega$ and $C = 0.001$ $\mu$F, determine the frequency of oscillation. What is the required value of $R_2$? (b) Design the circuit such that the frequency of oscillation is $f_o = 25$ kHz. Let $R = 20$ k$\Omega$.

15.25 In the phase-shift oscillator in Figure 15.15, the capacitor at the noninverting terminal of op-amp $A_1$ is replaced by a variable capacitor $C_V$. (a) Derive the expression for the frequency of oscillation. (b) If $C = 10$ pF, $R = 10$ k$\Omega$, and $C_V$ is variable between 10 and 50 pF, determine the range of oscillation frequency.

15.26 Consider the phase shift oscillator in Figure 15.16. (a) Determine the frequency of oscillation for $R = 12$ k$\Omega$ and $C = 150$ pF. What is the required value of $R_2$? (b) Design the circuit such that the frequency of oscillation is $f_o = 22$ kHz. Let $C = 0.001$ $\mu$F.

15.27 Analyze the phase-shift oscillator in Figure 15.16. Show that the frequency of oscillation is given by Equation (15.46) and that the condition for oscillation is given by Equation (15.47).

15.28 The circuit in Figure P15.28 is an alternative configuration of a phase-shift oscillator. (a) Assume that $R_1 = R_2 = R_3 = R_{A_1} = R_{A_2} = R_{A_3} \equiv R$ and $C_1 = C_2 = C_3 \equiv C$. Show that the frequency of oscillation is

$\omega_o = \sqrt{3}/RC$. (b) Assume equal magnitudes of gain in each amplifier stage. What is the minimum magnitude of gain required in each stage to sustain oscillation?

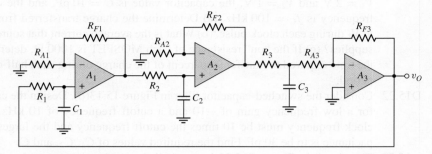

**Figure P15.28**

*15.29  In the circuit in Figure P15.28, let $R_{F1} = R_{F2} = R_{F3} \equiv R_F$, $R_2 = R_3 = R_{A1} = R_{A2} = R_{A3} \equiv R$, $C_1 = C_2 = C_3 \equiv C$, and let $R_1$ be a variable resistance $R_1 = R_V$. (a) Derive the expression for the frequency of oscillation. (b) If $R_V = R$, determine the condition for oscillation. (c) Using the results of part (a), determine the range in the frequency of oscillation for $R = 25 \text{ k}\Omega$, $C = 0.001 \,\mu\text{F}$, and $15 \leq R_V \leq 30 \text{ k}\Omega$.

*15.30  Consider the phase-shift oscillator in Figure P15.30. (a) Derive the expression for the frequency of oscillation. (b) Determine the condition for oscillation. (c) For $R = 20 \text{ k}\Omega$, find $C$ and $R_F$ that will produce sustained oscillations at $f_o = 22 \text{ kHz}$.

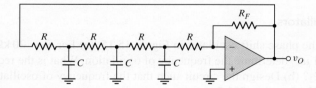

**Figure P15.30**

*15.31  Consider the phase-shift oscillator in Figure P15.30. (a) Assume the first resistor that is connected to $v_o$ is a variable resistor $R_V$. Derive the expression for the frequency of oscillation in terms of $C$, $R$, and $R_V$. (b) Using the results of part (a), determine the range in the frequency of oscillation for $R = 25 \text{ k}\Omega$, $C = 0.001 \,\mu\text{F}$, and $15 \leq R_V \leq 30 \text{ k}\Omega$.

15.32  A Wien-bridge oscillator is shown in Figure P15.32. (a) Derive the expression for the frequency of oscillation. (b) What is the condition for sustained oscillations?

15.33  Consider the oscillator circuit in Figure P15.33. (a) Derive the expression for the loop gain $T(s)$. (b) Determine the expression for the frequency of oscillation. (c) Find the condition for oscillation.

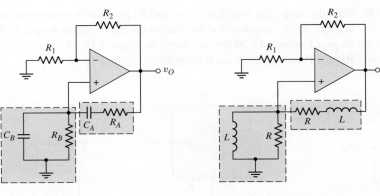

Figure P15.32                                    Figure P15.33

D15.34  Design the Wein-bridge oscillator in Figure 15.17 to oscillate at $f_o = 35\,\text{kHz}$. Choose appropriate component values.

D15.35  The Colpitts oscillator in Figure 15.19 is biased at $I_D = 0.8\,\text{mA}$. The transistor parameters are $V_{TN} = 0.8\,\text{V}$ and $K_n = 0.7\,\text{mA/V}^2$. Let $C_1 = 0.02\,\mu\text{F}$ and $R = 2\,\text{k}\Omega$. Design the circuit to oscillate at $f_o = 350\,\text{kHz}$.

15.36  Figure P15.36 shows a Colpitts oscillator with a BJT. Assume $r_\pi$ and $r_o$ are both very large. Derive the expressions for the frequency of oscillation and the condition of oscillation.

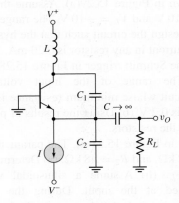

Figure P15.36

15.37  Consider the ac equivalent circuit of the Hartley oscillator in Figure 15.21. (a) Derive the expression for the frequency of oscillation. (b) Determine the condition for sustained oscillations.

*D15.38  For the Hartley oscillator in Figure 15.21, assume $r_\pi \to \infty$ and let $g_m = 30\,\text{mA/V}$. (a) Derive the expression for the frequency of oscillation. (b) Show that the condition for oscillation is given by $g_m R = L_1/L_2$. (c) Design the circuit to oscillate at $f_o = 750\,\text{kHz}$ for $L_1 = L_2 = 50\,\mu\text{H}$.

15.39  Find the loop gain functions $T(s)$ and $T(j\omega)$, the frequency of oscillation, and the $R_2/R_1$ required for oscillation for the circuit in Figure P15.39.

15.40  Repeat Problem 15.39 for the circuit in Figure P15.40.

15.41  Repeat Problem 15.39 for the circuit in Figure P15.41.

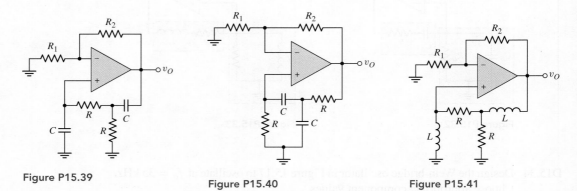

Figure P15.39         Figure P15.40         Figure P15.41

## Section 15.3   Schmitt Trigger Circuits

15.42  For the comparator in the circuit in Figure 15.26(a), the output saturation voltages are $\pm 9$ V. Let $V_{REF} = 5$ V. Let $R_2$ be a fixed resistor in series with a potentiometer. Design the circuit such that the crossover voltage can easily be varied over the range of $-2$ V to $-4$ V. The minimum resistance is to be 20 k$\Omega$.

15.43  Consider the Schmitt trigger shown in Figure 15.29(a). Assume the saturated output voltages are $V_H = +10$ V and $V_L = -10$ V. The range of the input voltage is $-5 \le v_I \le 5$ V. Design the circuit such that the hysteresis width is 0.4 V and the maximum current in any resistor is 0.20 mA.

15.44  The saturated output voltages of the Schmitt trigger in Figure 15.28(a) are $V_H = +9$ V and $V_L = -9$ V. The range of the input voltage is $-9 \le v_I \le 9$ V. (a) Design the circuit whose minimum resistance is 2 k$\Omega$ and such that the hysteresis width is 0.2 V. (b) Using the results of part (a), determine the maximum current in the resistors.

15.45  A Schmitt trigger circuit is shown in Figure 15.28(a). The parameters are $V_H = +10$ V, $V_L = -10$ V, $R_1 = 2$ k$\Omega$, and $R_2 = 48$ k$\Omega$. (a) Determine the crossover voltages $V_{TH}$ and $V_{TL}$. (b) Assume a sinusoidal voltage $v_I = 10 \sin[2\pi (60) t]$ V is applied at the input. During the period $33.3 \le t \le 50$ ms, determine the time periods that the output is high and the time periods that the output is low.

15.46  Consider the Schmitt trigger in Figure P15.46. Assume the saturated output voltages are $\pm V_P$. (a) Derive the expression for the crossover voltages $V_{TH}$ and $V_{TL}$. (b) Let $R_A = 10$ k$\Omega$, $R_B = 20$ k$\Omega$, $R_1 = 5$ k$\Omega$, $R_2 = 20$ k$\Omega$, $V_P = 10$ V, and $V_{REF} = 2$ V. (a) Find $V_{TH}$ and $V_{TL}$. (b) Sketch the voltage transfer characteristics.

15.47  The saturated output voltages are $\pm V_P$ for the Schmitt trigger in Figure P15.47. (a) Derive the expressions for the crossover voltages $V_{TH}$ and $V_{TL}$ (b) If $V_P = 12$ V, $V_{REF} = -10$ V, and $R_3 = 10$ k$\Omega$, find $R_1$ and $R_2$ such that the switching point is $V_S = -5$ V and the hysteresis width is 0.2 V. (c) Sketch the voltage transfer characteristics.

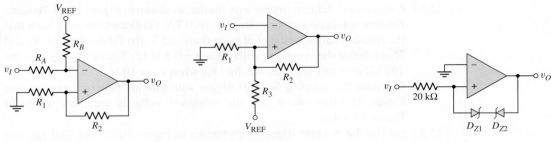

**Figure P15.46**          **Figure P15.47**          **Figure P15.48**

15.48  (a) Plot the voltage transfer characteristics of the comparator circuit in Figure P15.48 assuming the open-loop gain is infinite. Let the reverse Zener voltage be $V_Z = 5.6$ V and the forward diode voltage be $V_\gamma = 0.6$ V. (b) Repeat part (a) for an open-loop gain of $10^3$. (c) Repeat part (a) for 2.5 V applied to the inverting terminal of the comparator.

15.49  Consider the Schmitt trigger in Figure 15.30(a). (a) Derive the expression for the switching point and crossover voltages as given in Equations (15.76) and (15.77). (b) Let $V_H = +12$ V and $V_L = -12$ V. The minimum resistance is to be 4 k$\Omega$. Determine $R_1$, $R_2$, and $V_{REF}$ such that the crossover voltages are $V_{TH} = -1.5$ V and $V_{TL} = -2$ V. (c) What are the currents in the resistors when (i) $v_O = V_H$ and (ii) $v_O = V_L$?

15.50  Consider the Schmitt trigger in Figure 15.31(a). (a) Derive the expressions for the switching point and crossover voltages, as given in Equations (15.78) and (15.79). (b) Let $V_H = 12$ V, $V_L = -12$ V, and $R_2 = 20$ k$\Omega$. Determine $R_1$ and $V_{REF}$ such that $V_{TH} = -1$ V and $V_{TL} = -2$ V.

15.51  Consider the Schmitt trigger in Figure P15.51. The saturated output voltages of the op-amp are $V_H = +10$ V and $V_L = -10$ V. Assume the diode turn-on voltage is 0.7 V. The range of the input voltage is $-2 \leq v_I \leq +2$ V. (a) Determine the crossover voltages. (b) Sketch the voltage transfer characteristics. (c) Determine $I_{D1}$, $I_{D2}$, $I_{R3}$, and $I_{R2}$ for (i) $v_I = 2$ V and (ii) $v_I = -2$ V.

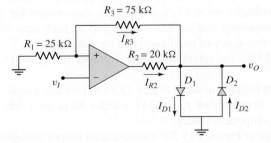

**Figure P15.51**

15.52  The saturated output voltages of the comparator in the circuit shown in Figure 15.33 are $\pm10$ V. Assume forward diode voltages of 0.7 V and reverse Zener voltages of 5.6 V. The range of the input voltage is $-2 \leq v_I \leq +2$ V. (a) Find $R_1$ and $R_2$ such that the hysteresis width is 0.6 V. The minimum resistance value is to be 4 k$\Omega$. (b) Find $R$ such that the maximum current through the diodes is 0.8 mA.

15.53  Consider the Schmitt trigger with limiter, as shown in Figure 15.34. Assume the forward diode turn-on voltage $V_\gamma$ is 0.7 V. (a) Determine $V_{REF}$ such that the bistable output voltages at $v_I = 0$ are $\pm 5$ V. (b) Find values of $R_1$ and $R_2$ such that the crossover voltages are $\pm 0.5$ V. (c) Taking $R_1$, $R_2$, and the 100 k$\Omega$ resistors into account, find $v_O$ when $v_I = 10$ V.

15.54  Consider the inverting Schmitt trigger with limiting network, as shown in Figure 15.34(a). Show that the crossover voltages are those given in Figure 15.34(b).

15.55  (a) For the Schmitt trigger with limiter in Figure P15.55(a), find the two output voltage values at $v_I = 0$ and the two crossover voltages. (b) Derive the expression for the slope of $v_O$ versus $v_I$ for $v_I > V_{TH}$.

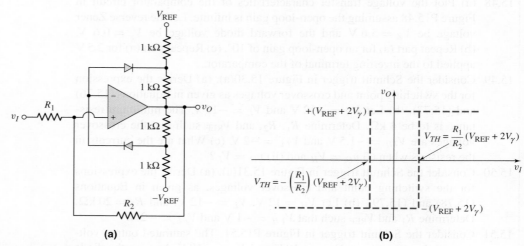

**(a)**

Figure P15.55

**(b)**

## Section 15.4   Nonsinusoidal Oscillators and Timing Circuits

15.56  Consider the Schmitt trigger oscillator in Figure 15.35. The circuit parameters are $R_1 = 10$ k$\Omega$, $R_2 = 20$ k$\Omega$, $R_X = 40$ k$\Omega$, and $C_X = 0.02$ $\mu$F. The saturated output voltages are $\pm 5$ V. (a) Write the expressions for $v_X$ as a function of time assuming that $v_O$ has switched to its high state at $t = 0$. (b) Determine the frequency of oscillation and the duty cycle.

15.57  Repeat Problem 15.56 for saturated output voltages of $V_H = +5$ V and $V_L = -10$ V.

D15.58  Design the Schmitt trigger circuit in Figure 15.35 to produce a square-wave output signal at a frequency of $f_o = 12$ kHz and a 50 percent duty cycle. Choose standard component values.

15.59  Consider the circuit in Figure P15.59. The saturated output voltages of the Schmitt trigger comparator are $\pm 10$ V. Assume that at $t = 0$, output $v_{o1}$ switches from its low state to its high state and $C_Y$ is uncharged. Plot $v_{o1}$ and $v_o$ versus time over two periods of oscillation.

15.60  The saturated output voltages of the comparator in Figure P15.60 are $\pm 10$ V. (a) Find $R_x$ such that the frequency of oscillation is 500 Hz when the potentiometer is connected to point A. (b) Using the results of part (a), determine the oscillator frequency when the potentiometer is connected to point B.

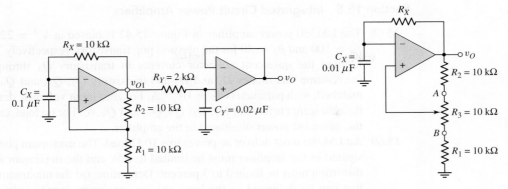

Figure P15.59                                          Figure P15.60

15.61  (a) The monostable multivibrator in Figure 15.37 is to be designed to pro-
       duce a 250 $\mu$s output pulse. Assume saturated output voltages of $\pm 10$ V, and
       let $V_\gamma = 0.7$ V, $R_1 = 20$ k$\Omega$, and $R_2 = 12$ k$\Omega$. (b) Determine the minimum
       input triggering voltage required. (c) What is the recovery time?

15.62  A monostable multivibrator is shown in Figure 15.37. The parameters are
       $R_X = 20$ k$\Omega$, $C_X = 1.2$ $\mu$F, and $R_1 = R_2 = 20$ k$\Omega$. The saturated output
       voltages are $\pm 5$ V. (a) What is the output voltage pulse width? (b) Deter-
       mine the recovery time.

D15.63 Figure 15.40 shows the 555 timer connected in the monostable multivibra-
       tor mode. (a) Design the circuit to provide an output pulse 60 seconds wide.
       (b) Determine the recovery time.

D15.64 Design a 555 monostable multivibrator to provide a 5 $\mu$s pulse. What is the
       recovery time?

15.65  A 555 timer is connected in the astable mode as shown in Figure 15.41.
       Design the circuit such that the frequency of oscillation is $f_o = 80$ kHz and
       the duty cycle is 60 percent. Let $R_A = 25$ k$\Omega$.

15.66  A 555 ICC is connected as shown in Figure P15.66. Determine the range of
       oscillation frequency and the duty cycle.

15.67  Repeat Problem 15.66 for the circuit in Figure P15.67.

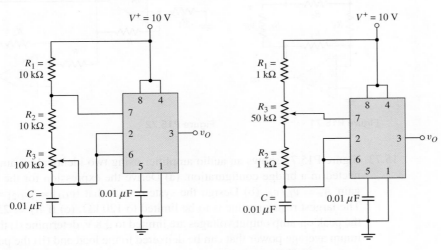

Figure P15.66                          Figure P15.67

## Section 15.5    Integrated Circuit Power Amplifiers

15.68  The LM380 power amplifier in Figure 15.42 is biased at $V^+ = 22$ V. Let $\beta_n = 100$ and $\beta_p = 20$ for the npn and pnp transistors, respectively. (a) Determine the quiescent collector currents in transistors $Q_1$ through $Q_6$. (b) Assume that diodes $D_1$ and $D_2$ and transistors $Q_7$, $Q_8$, and $Q_9$ are all matched, with parameters $I_S = 10^{-13}$ A. For zero input voltages, determine the quiescent currents in $D_1$, $D_2$, $Q_7$, $Q_8$, and $Q_9$. (c) For no load, calculate the quiescent power dissipated in the amplifier.

15.69  An LM380 must deliver ac power to a 10 $\Omega$ load. The maximum power dissipated in the amplifier must be limited to 2 W and the maximum allowed distortion must be limited to 3 percent. Determine: (a) the maximum power that can be delivered to the load, (b) the maximum supply voltage, and (c) the peak amplitude of the sinusoidal output voltage.

15.70  (a) Design the bridge circuit in Figure 15.46 such that the gain magnitude of each op-amp circuit is 12. (b) A load resistance of $R_L = 12 \,\Omega$ that is to dissipate an average power of $\overline{P_L} = 15$ W is connected to the output. Determine the peak output voltage at each op-amp and the peak current that each op-amp must source or sink. (c) If the peak output voltage of each op-amp is limited to $\pm 12$ V and the peak current that each op-amp can source or sink is limited to 0.8 A, determine (i) the maximum average power that can be delivered to the load and (ii) the optimum value of load resistance.

D15.71  Another form of the bridge power amplifier is shown in Figure P15.71. This amplifier has a very high input resistance since the input is to the noninverting terminal of an op-amp. (a) Derive the expression for the voltage gain $A_v = v_L/v_I$. (b) Design the circuit to provide a gain of $A_v = 10$ so that the magnitudes of $v_{o1}$ and $v_{o2}$ are equal. Let $R_1 = 50$ k$\Omega$. (c) If $R_L = 20 \,\Omega$ and if the average power delivered to the load is 10 W, determine the peak amplitude of $v_{o1}$ and $v_{o2}$ and the peak load current.

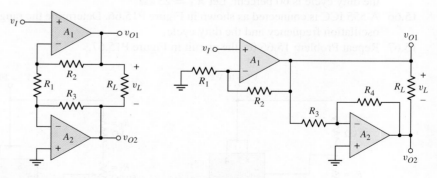

Figure P15.71                 Figure P15.72

15.72  Figure P15.72 shows an audio amplifier using two identical op-amps connected in a bridge configuration. (a) Derive the expression for the voltage gain $A_v = v_L/v_I$. (b) Design the system such that $|v_{O1}| = |v_{O2}| = 12v_I$. The largest resistor value is to be limited to 120 k$\Omega$. (c) If $R_L = 25 \,\Omega$ and the peak op-amp output voltages are limited to $\pm 8$ V, determine (i) the maximum average power that can be delivered to the load and (ii) the peak currents that each op-amp must source or sink.

D15.73  (a) Design the circuit shown in Figure P15.72 such that $v_{O1} = -v_{O2}$ and the voltage gain $A_v = v_L/v_I = 25$. The largest resistor value is to be limited to 100 kΩ. (b) If the peak value of each op-amp output voltage is limited to ±12 V and the peak current that each op-amp can source or sink is limited to 1.2 A, determine (i) the maximum average power that can be delivered to the load and (ii) the optimum load resistance. (c) If $R_L$ is twice the value found in part (b), what is the maximum average power that can be delivered to the load?

## Section 15.6   Voltage Regulators

15.74  Transistors $Q_1$ and $Q_2$ in the voltage regulator circuit in Figure P15.74 have parameters $\beta = 200$, $V_{EB}(\text{on}) = 0.7$ V, and $V_A = 100$ V. The zero-current Zener voltage is $V_{ZO} = 6.3$ V and the Zener resistance is $r_z = 15$ Ω. Assuming an ideal op-amp, calculate the line regulation.

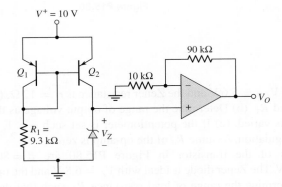

**Figure P15.74**

15.75  (a) The output voltage of a voltage regulator decreases by 8 mV as the load current changes from 0 to 2 A. If the output voltage changes linearly with load current, determine the output resistance of the regulator. (b) If the output resistance of a voltage regulator is $R_{of} = 10$ mΩ and the output current changes by 1.2 A, what is the change in output voltage?

15.76  Consider the three-terminal voltage regulator in Figure 15.51, with parameters as given in Example 15.16. If the maximum load current is $I_O(\text{max}) = 100$ mA, determine the minimum applied power supply voltage $V^+$ that will still maintain all transistors biased in the active region.

D15.77  Consider the three-terminal voltage regulator in Figure 15.51, with Zener diode voltages of $V_Z = 6.3$ V. Assume transistor parameters of $V_{BE}(\text{npn}) = V_{EB}(\text{pnp}) = 0.6$ V, and neglect base currents. (a) Determine resistance $R_4$ such that $I_{Z2} = 0.25$ mA. (b) Determine $R_{12}$ such that $V_O = 12$ V.

15.78  The three-terminal voltage regulator in Figure 15.51 has parameters as described in Example 15.16. Assume $R_4 = 0$, $V_A = 50$ V for $Q_4$, and $r_z = 15$ Ω for $D_2$. Determine the line regulation.

15.79  The voltage regulator shown in Figure P15.79 is a variable voltage, 0 to 5 A power supply. The transistor parameters are $\beta = 80$ and $V_{BE}(\text{on}) = 0.7$ V. The op-amp has a finite open-loop gain of $A_{OL} = 5 \times 10^3$. The zero-current

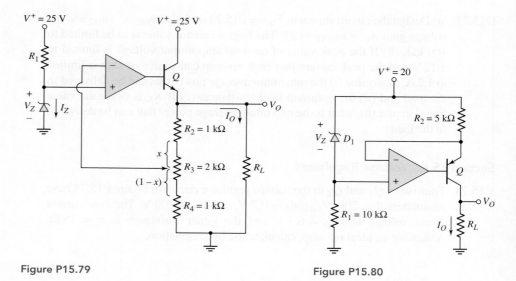

Figure P15.79                                        Figure P15.80

Zener voltage is $V_{ZO} = 5.6$ V and the Zener resistance is $r_z = 12 \, \Omega$. (a) For $I_Z = 12$ mA, find $R_1$. (b) Determine the range of output voltage as the potentiometer $R_3$ is varied. (c) If the potentiometer is set such $x = 1$, determine the load regulation. Assume $R_o$ of the op-amp is zero.

15.80 The parameters of the transistor in Figure P15.80 are $\beta = 80$ and $V_{EB}(\text{on}) = 0.6$ V. The Zener diode is ideal with $V_Z = 6.8$ V and the op-amp is ideal. (a) Determine the range of load resistance $R_L$ such that the load current is a constant. What is the value of the constant load current? (b) If the Zener diode has a resistance $r_z = 20 \, \Omega$ and the power supply is in the range $16 \leq V^+ \leq 20$ V, determine the range in output current for $R_L = 5 \, k\Omega$.

## COMPUTER SIMULATION PROBLEMS

15.81 Consider the three-pole high-pass Butterworth active filter described in Exercise TYU 15.1. Using a computer simulation, plot the magnitude of the voltage transfer function versus frequency and compare these results with those obtained in TYU 15.1.

15.82 A phase shift oscillator is described in Exercise TYU 15.5. Using a computer simulation, plot the output voltage of the oscillator versus time over several cycles. What is the frequency of oscillation?

15.83 Consider the Schmitt trigger oscillator described in Exercise Ex 15.8. Using a computer simulation, plot the voltage $v_X$ versus time over several cycles. What is the frequency of oscillation?

15.84 A bridge power amplifier is described in Exercise TYU 15.13. Using a computer simulation, plot (a) $v_{O1} - v_{O2}$ versus $v_I$ over the range $0 \leq v_I \leq 4$ V and (b) the current in $R_L$ over the same input voltage range.

### DESIGN PROBLEMS

[Note: Each design should be correlated with a computer analysis.]

*D15.85 Design a four-pole high-pass Butterworth active filter such that the low-frequency voltage gain is +20 and the cutoff frequency is 50 Hz.

*D15.86 Consider the Colpitts oscillator in Figure P15.86. The capacitors $C_E$ and $C_B$ are very large bypass and coupling capacitors. Let $V_{CC} = 5$ V. (a) Design the circuit such that the quiescent collector current is $I_{CQ} = 0.5$ mA. (b) Design the circuit such that the frequency of oscillation is $f_o = 650$ kHz.

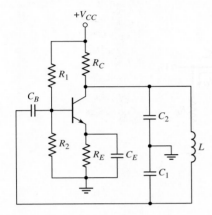

Figure P15.86

*D15.87 Consider the power amplifier in Figure P15.87 with parameters $V^+ = 15$ V, $V^- = -15$ V, and $R_L = 20\ \Omega$. The closed-loop gain must be 10. Design the circuit such that the power delivered to the load is 5 W when $v_I = -1$ V. If the four transistors are matched, determine the minimum $\beta$ required such that the op-amp output current is limited to 2 mA when 5 W is delivered to the load.

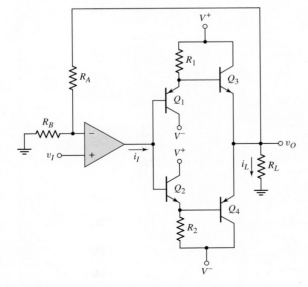

Figure P15.87

*D15.88   Consider the simple series-pass regulator circuit in Figure P15.88. Assume an ideal Zener diode with $V_Z = V_{REF} = 4.7$ V. Let $\beta = 100$ and $V_{BE}(\text{on}) = 0.7$ V for all transistors. (a) Design the circuit such that $V_O = 10$ V and $I_Z = 10$ mA for a nominal supply voltage of $V^+ = 20$ V. (b) Determine the regulator output resistance $R_{of}$.

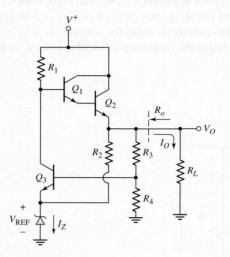

**Figure P15.88**

# Prologue to Digital Electronics

# 111

## PREVIEW

Several basic digital electronics concepts are common to the remaining chapters of this text. These principles, which are usually covered in an introductory course in computer logic design, are reviewed briefly in this prologue.

In a digital system, information is represented solely in discrete or quantized form. Normally, only two discrete states are used, denoted as logic 0 and logic 1. The algebra applicable to the binary system was invented by George Boole (1815–1864) and is known as Boolean algebra. We do not use Boolean algebra directly in this text; however, some familiarity with it is beneficial in the analysis and design of digital integrated circuits. We will be directly concerned with basic Boolean operations and the corresponding logic gates.

Several techniques have been developed to aid in the reduction of Boolean expressions to a minimum set of variables. One common technique is the Karnaugh map. Though not used directly in this text, this technique is helpful in designing digital systems.

## LOGIC FUNCTIONS AND LOGIC GATES

The three basic logic or Boolean operations are: NOT, AND, and OR. These operations can be described using a truth table.

The truth table and logic gate symbol for the NOT function is shown in Figure PR3.1(a). The bar over the output variable indicates the NOT function, or the complement. Since only two states of a variable are permitted, if $A = 0$, then $\bar{A} = 1$. The small circle at the output of the logic gate indicates a logic inversion. As depicted by the figure, this logic gate is also called an inverter.

Figure PR3.1(b) shows the truth table, logic gate symbol, and Boolean expression for the AND function. A logic 1 is produced at the output only when both inputs are a logic 1; otherwise, the output is a logic 0.

The truth table, logic gate symbol, and Boolean expression for the OR operation are shown in Figure PR3.1 (c). In this case, a logic 1 output is produced if either $A = 1$ or $B = 1$, or if both inputs are a logic 1.

Two other commonly used logic functions are the NAND and NOR. The NAND function is the complement of the AND operation, and the NOR function is the complement of the OR operation. The truth tables and logic gate symbols for these functions are shown in Figure PR3.2. Again, the small circle at the output of each logic gate indicates a logic inversion.

Finally, two additional logic functions useful in digital design are the exclusive-OR function and the exclusive-NOR function. Although these logic functions can be derived from a combination of the basic functions, they have their own logic

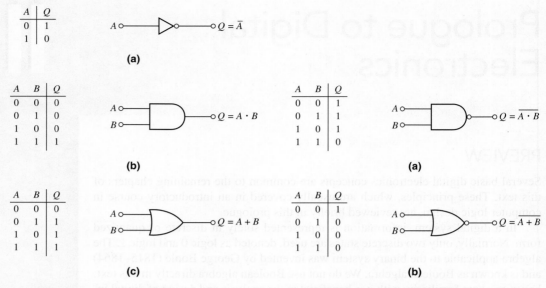

| A | Q |
|---|---|
| 0 | 1 |
| 1 | 0 |

$Q = \bar{A}$

**(a)**

| A | B | Q |
|---|---|---|
| 0 | 0 | 0 |
| 0 | 1 | 0 |
| 1 | 0 | 0 |
| 1 | 1 | 1 |

$Q = A \cdot B$

**(b)**

| A | B | Q |
|---|---|---|
| 0 | 0 | 0 |
| 0 | 1 | 1 |
| 1 | 0 | 1 |
| 1 | 1 | 1 |

$Q = A + B$

**(c)**

**Figure PR3.1** Truth tables, logic gate symbols, and Boolean expressions: (a) NOT function, (b) AND function, and (c) OR function

| A | B | Q |
|---|---|---|
| 0 | 0 | 1 |
| 0 | 1 | 1 |
| 1 | 0 | 1 |
| 1 | 1 | 0 |

$Q = \overline{A \cdot B}$

**(a)**

| A | B | Q |
|---|---|---|
| 0 | 0 | 1 |
| 0 | 1 | 0 |
| 1 | 0 | 0 |
| 1 | 1 | 0 |

$Q = \overline{A + B}$

**(b)**

**Figure PR3.2** Truth tables, logic gate symbols, and Boolean expressions: (a) NAND function and (b) NOR function

gate symbols. The truth tables, logic gate symbols, and Boolean expressions for these operations are shown in Figure PR3.3. In the exclusive-OR operation, the output becomes a logic 1 when either $A = 1$ or $B = 1$, but not when both are a logic 1. The output of the exclusive-NOR is the complement of the exclusive-OR function.

In the following sections of this prologue, we briefly describe the basic logic functions and logic gates with two input variables, although more than two are possible. In practice, the number of input variables is generally limited to a maximum of four because of transistor size and input capacitance effects.

| A | B | Q |
|---|---|---|
| 0 | 0 | 0 |
| 0 | 1 | 1 |
| 1 | 0 | 1 |
| 1 | 1 | 0 |

$Q = A \otimes B$

**(a)**

| A | B | Q |
|---|---|---|
| 0 | 0 | 1 |
| 0 | 1 | 0 |
| 1 | 0 | 0 |
| 1 | 1 | 1 |

$Q = \overline{A \otimes B}$

**(b)**

**Figure PR3.3** Truth tables, logic gate symbols, and Boolean expressions: (a) exclusive-OR function and (b) exclusive-NOR function

# LOGIC LEVELS

The logic 0 and logic 1 states in a digital circuit are represented by two distinct voltage values. In this text, we use positive logic, which means that the more positive voltage represents the logic 1 state and the more negative voltage represents the logic 0 state. The actual voltages may be either positive or negative. Figure PR3.4 shows three possible output voltage combinations that represent positive logic. The condition represented in Figure PR3.4(a) is the most common, although we will see examples of the conditions represented in Figure PR3.4(c). The logic 0 level shown in Figure PR3.4(a) may actually be zero volts in some cases.

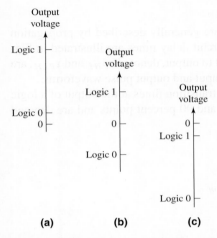

(a)    (b)    (c)

**Figure PR3.4** Three possible output voltage combinations representing positive logic

# NOISE MARGIN

In an ideal digital system, logic 1 would be represented by a well-defined voltage level $V_{OH}$ and logic 0 would be represented by a well-defined voltage level $V_{OL}$. In actual digital systems, however, the voltage values representing the two logic states may change as a result of any number of factors, including variations in temperature, circuit fabrication tolerances, loading effects, and noise.

At the input to a digital circuit, a range of voltages can represent each of the two binary states as illustrated in Figure PR3.5. The amplitude levels that pass through a

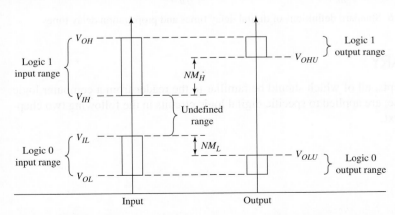

**Figure PR3.5** Voltage ranges representing logic 1 and logic 0, and definition of noise margins

digital system must be regenerated in order that a logic error is not produced. Voltage $V_{IH}$ is the smallest input voltage recognized as a logic 1 and $V_{IL}$ is the largest input voltage recognized as a logic 0. These input levels produce output voltages in the ranges shown in Figure PR3.5. In an inverter circuit, input $V_{IL}$ produces output $V_{OHU}$ and input $V_{IH}$ produces output $V_{OLU}$. The noise margins, then, are defined as shown in the figure. We consider noise margins in more detail in the next two chapters when we analyze specific circuits.

## PROPAGATION DELAY TIMES AND SWITCHING TIMES

The switching characteristics of logic gates are generally described by propagation delay times. Standard definitions of digital circuit delay times are illustrated in Figure PR3.6. Propagation delay times from input to output, denoted $\tau_{PHL}$ and $\tau_{PLH}$, are defined between the 50 percent points of the input and output pulse waveforms.

In addition, high-to-low and low-to-high transition times at the output of a logic gate are defined as the times between the 10 and 90 percent points and are denoted $\tau_{HL}$ and $\tau_{LH}$.

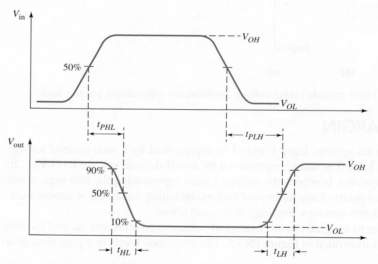

**Figure PR3.6** Standard definitions of digital delay times and propagation delay times

## SUMMARY

These concepts, all of which should be familiar to the reader from a computer logic design course, are applied to specific digital logic circuits in the following two chapters of the text.

# Digital Electronics

Part 2 of the text dealt with analog electronic circuits. Part 3 now deals with digital electronics, another important category of electronics.

Chapter 16 examines field-effect transistor digital circuits. MOSFET digital circuits have revolutionized digital electronics, with CMOS technology producing high-density, low-power digital circuits. Initially, we briefly consider the NMOS inverter and NMOS logic gates. We then analyze the basic CMOS inverter and then develop CMOS logic gates. Finally, we analyze FET shift registers and flip-flops and then discuss some basic A/D and D/A converters.

Bipolar digital circuits are considered in Chapter 17. We initially examine emitter-coupled logic, which is primarily used in specialized high-speed applications. We then briefly consider the basic aspects of transistor-transistor logic (TTL), which was the mainstay of logic design for many years. Low-power Schottky TTL circuits are analyzed in order to obtain a good comparison between FET and bipolar digital technologies.

# MOSFET Digital Circuits

This chapter presents the basic concepts of MOSFET digital integrated circuits, which is the most widely used technology for the fabrication of digital systems. The small transistor size and low power dissipation of CMOS circuits allows for a high level of integration for logic and memory circuits. JFET logic circuits are very specialized and are therefore not considered here.

A discussion of NMOS logic circuits will serve as an introduction to the analysis and design of digital systems. This technology, although old, deals with only one type of transistor (n-channel) and therefore makes the analysis more straightforward than dealing with two types of transistors in the same circuit. This discussion will also serve as a baseline to point out advantages of CMOS technology.

Initially, we consider basic digital logic circuits such as NOR and NAND gates, and then discuss additional logic circuits such as flip flops, shift registers, and adders. Finally, we consider memories, and then A/D and D/A converters.

## PREVIEW

In this chapter, we will:

- Analyze and design NMOS inverters
- Analyze and design NMOS logic gates
- Analyze and design CMOS inverters
- Analyze and design static CMOS logic gates
- Analyze and design clocked CMOS logic gates
- Analyze and understand the characteristics of NMOS and CMOS transmission gates
- Analyze and understand the characteristics of shift registers and various flip-flop designs
- Discuss semiconductor memories
- Analyze and design random-access memory (RAM) cells
- Analyze read-only memories (ROM)
- Discuss the basic concepts in A/D and D/A converters
- As an application, design a static CMOS logic gate to implement a specific logic function.

### 16.1    NMOS INVERTERS

**Objective:** • Analyze and design NMOS inverters.

The inverter is the basic circuit of most MOS logic circuits. The design techniques used in NMOS logic circuits are developed from the dc analysis results for the NMOS inverter. Extending the concepts developed from the inverter to NOR and NAND gates is then direct. Alternative inverter load elements are compared in terms of power consumption, packing density, and transfer characteristics.

### 16.1.1    n-Channel MOSFET Revisited

We studied the structure, operation, and characteristics of MOS transistors in Chapter 3. In this section, we will quickly review the n-channel MOSFET characteristics, emphasizing specific properties important in digital circuit design.

The simplified n-channel MOSFET that we have considered is shown in Figure 16.1(a). A more detailed view of the n-channel MOSFET is shown in Figure 16.1(b). The active transistor region is the surface of the semiconductor and comprises heavily doped $n^+$ source and drain regions and the p-type channel region. The channel length is $L$ and the channel width is $W$. The body, or substrate, is a single-crystal silicon wafer, which is the starting material for circuit fabrication and provides physical support for the integrated circuit.

In an integrated circuit, all n-channel transistors are fabricated in the same p-type substrate material. The substrate is connected to the most negative potential in the circuit, which for digital circuits is normally at ground potential or zero volts. However, the source terminal of many transistors will not be at zero volts, which means that a reverse-biased pn junction will exist between the source and substrate. When the source and body terminals are not at the same potential, the threshold voltage of the transistor becomes a function of the source-to-body voltage. This **body effect** must then be taken into account in determining logic levels in digital circuits.

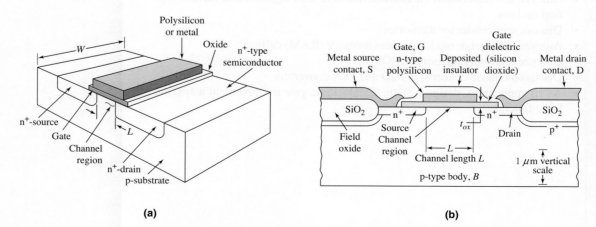

**Figure 16.1** (a) n-channel MOSFET simplified view and (b) n-channel MOSFET detailed cross section

## Current–Voltage Relation

The current–voltage characteristics of the n-channel MOSFET are functions of both the electrical and geometric properties of the device. When the transistor is biased in the nonsaturation region, for $v_{GS} \geq V_{TN}$ and $v_{DS} \leq (v_{GS} - V_{TN})$, we can write

$$i_D = K_n\left[2(v_{GS} - V_{TN})v_{DS} - v_{DS}^2\right] \tag{16.1(a)}$$

In the saturation region, for $v_{GS} \geq V_{TN}$ and $v_{DS} \geq (v_{GS} - V_{TN})$, we have

$$i_D = K_n(v_{GS} - V_{TN})^2 \tag{16.1(b)}$$

The transition point separates the nonsaturation and saturation regions and is the drain-to-source saturation voltage, which is given by

$$v_{DS} = v_{DS}(\text{sat}) = v_{GS} - V_{TN} \tag{16.2}$$

The term $(1 + \lambda v_{DS})$ is sometimes included in Equation (16.1(b)) to account for channel length modulation and the finite output resistance. In most cases, it has little effect on the operating characteristics of MOS digital circuits. In our analysis, the term $\lambda$ is assumed to be zero unless otherwise stated.

The parameter $K_n$ is the NMOS transistor conduction parameter and is given by

$$K_n = \left(\frac{1}{2}\mu_n C_{\text{ox}}\right)\left(\frac{W}{L}\right) = \frac{k_n'}{2}\frac{W}{L} \tag{16.3}$$

The electron mobility $\mu_n$ and oxide capacitance $C_{\text{ox}}$ are assumed to be constant for all devices in a particular IC.

The current–voltage characteristics are directly related to the channel width-to-length ratio, or the size of the transistor. In general, in a given IC, the length $L$ is fixed, but the designer can control the channel width $W$.

Since the MOS transistor is a majority carrier device, the switching speed of MOS digital circuits is limited by the time required to charge and discharge the capacitances between device electrodes and between interconnect lines and ground. Figure 16.2 shows the significant capacitances in a MOSFET. The capacitances $C_{sb}$ and $C_{db}$ are the source-to-body and drain-to-body $n^+p$ junction capacitances. The total input gate capacitance, to a first approximation, is a constant equal to

$$C_g = WLC_{\text{ox}} = WL\left(\frac{\varepsilon_{\text{ox}}}{t_{\text{ox}}}\right) \tag{16.4}$$

where $C_{\text{ox}}$ is the oxide capacitance per unit area, and is a function of the oxide thickness. The parameter $C_{\text{ox}}$ also appears in the expression for the conduction parameter.

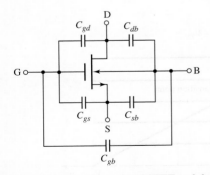

**Figure 16.2** n-channel MOSFET and device capacitances

### Small Geometry Effects

The current–voltage relationships given by Equations (16.1(a)), (16.1(b)), and (16.2) are first-order approximations that apply to "long" channel devices. The tendency in device design is to make the devices as small as possible, which means the channel length is being reduced to values substantially smaller than 1 $\mu$m. The corresponding channel widths are also being reduced. As the channel length is reduced, several effects alter the current–voltage characteristics. First, the threshold voltage becomes a function of the geometry of the device and is dependent on the channel length. This effect must be taken into account in the design of the transistor. Second, carrier velocity saturation reduces the saturation-mode current below the current value predicted by Equation (16.1(b)). The current is no longer a quadratic function of gate-to-source voltage, and tends to become a linear function of voltage. Channel length modulation means that the current tends to be larger than that predicted by the ideal equation. Third, the electron mobility is a function of the gate voltage so that the current tends to be smaller than the predicted value as the gate-to-source voltage increases. All of these effects complicate the analysis considerably.

We can, however, determine the basic operation and behavior of MOSFET logic circuits by using the first-order equations. We will use these first-order equations in our design of logic circuits. To determine the effect of small device size, a computer simulation may be performed in which the appropriate device models are incorporated in the simulation.

### 16.1.2    NMOS Inverter Transfer Characteristics

Since the inverter is the basis for most logic circuits, we will describe the NMOS inverter and will develop the dc transfer characteristics for three types of inverters with different load devices. This discussion will introduce voltage transfer functions and will define the maximum and minimum logic levels.

### NMOS Inverter with Resistor Load

Figure 16.3(a) shows a single NMOS transistor connected to a resistor to form an inverter. The transistor characteristics and load line are shown in Figure 16.3(b), along with the parametric curve separating the saturation and nonsaturation regions. We determine the voltage transfer characteristics of the inverter by examining the various regions in which the transistor can be biased.

When the input voltage is less than or equal to the threshold voltage, or $v_I \le V_{TN}$, the transistor is cut off, $i_D = 0$, and the output voltage is $v_O = V_{DD}$. The maximum output voltage is defined as the logic 1 level. As the input voltage becomes

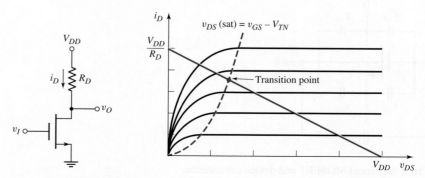

**Figure 16.3**  (a) NMOS inverter with resistor load and (b) transistor characteristics and load line

just greater than $V_{TN}$, the transistor turns on and is biased in the saturation region. The output voltage is then

$$v_O = V_{DD} - i_D R_D \tag{16.5}$$

where the drain current is given by

$$i_D = K_n(v_{GS} - V_{TN})^2 = K_n(v_I - V_{TN})^2 \tag{16.6}$$

Combining Equations (16.5) and (16.6) yields

$$v_O = V_{DD} - K_n R_D(v_I - V_{TN})^2 \tag{16.7}$$

which relates the output and input voltages as long as the transistor is biased in the saturation region.

As the input voltage increases, the $Q$-point of the transistor moves up the load line. At the transition point, we have

$$V_{Ot} = V_{It} - V_{TN} \tag{16.8}$$

where $V_{Ot}$ and $V_{It}$ are the drain-to-source and gate-to-source voltages, respectively, at the transition point. Substituting Equation (16.8) into (16.7), we determine the input voltage at the transition point from

$$K_n R_D(V_{It} - V_{TN})^2 + (V_{It} - V_{TN}) - V_{DD} = 0 \tag{16.9}$$

As the input voltage becomes greater than $V_{It}$, the $Q$-point continues to move up the load line, and the transistor becomes biased in the nonsaturation region. The drain current is then

$$i_D = K_n[2(v_{GS} - V_{TN})v_{DS} - v_{DS}^2] = K_n[2(v_I - v_{TN})v_O - v_O^2] \tag{16.10}$$

Combining Equations (16.5) and (16.10) yields

$$v_O = V_{DD} - K_n R_D[2(v_I - V_{TN})v_O - v_O^2] \tag{16.11}$$

which relates the input and output voltages as long as the transistor is biased in the nonsaturation region.

Figure 16.4 shows the voltage transfer characteristics of this inverter for three resistor values. Also shown is the line, given by Equation (16.8), which separates the

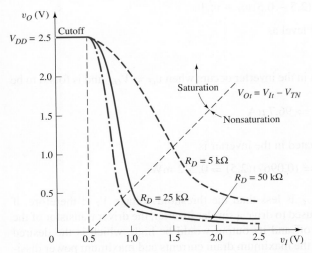

**Figure 16.4** Voltage transfer characteristics of NMOS inverter with resistive load, using parameters from Example 16.1 and for the three resistor values.

saturation and nonsaturation bias regions of the transistor. The figure shows that the minimum output voltage, or the logic 0 level, for a high input decreases with increasing load resistance, and the sharpness of the transition region between a low input and a high input increases with increasing load resistance.

It should be noted that a large resistance is difficult to fabricate in an IC. A large resistor value in the inverter will limit current and power consumption as well as provide a small $V_{OL}$ value. But it would also require a large chip area if fabricated in a standard MOS process. To avoid this problem MOS transistors can be used as load devices, replacing the resistor, as discussed in subsequent paragraphs.

## EXAMPLE 16.1

**Objective:** Determine the transition point, minimum output voltage, maximum drain current, and maximum power dissipation of an NMOS inverter with resistor load.

**Specifications:** Consider the circuit in Figure 16.3(a) with parameters $V_{DD} = 2.5$ V and $R_D = 20$ k$\Omega$. The transistor parameters are $V_{TN} = 0.5$ V and $K_n = 0.3$ mA/V$^2$.

**Solution:** The input voltage at the transition point is found from Equation (16.9). We have

$$(0.3)(25)(V_{It} - 0.5)^2 + (V_{It} - 0.5) - 2.5 = 0$$

which yields

$$V_{It} - 0.5 = 0.515 \text{ V} \quad \text{or} \quad V_{It} = 1.015 \text{ V}$$

The output voltage at the transition point is

$$V_{Ot} = V_{It} - V_{TN} = 1.015 - 0.5 = 0.515 \text{ V}$$

When $v_I$ is high at $v_I = 2.5$ V, the output voltage is found from Equation (16.11). We find

$$v_O = 2.5 - (0.3(25)[2(2.5 - 0.5)v_O - v_O^2]$$

which yields the output low level as

$$v_O = V_{OL} = 82.3 \text{ mV}$$

The maximum drain current in the inverter occurs when $v_O = V_{OL}$ and is found to be

$$i_{D,\max} = \frac{2.5 - 0.0823}{25} \Rightarrow 96.7 \ \mu\text{A}$$

The maximum power dissipated in the inverter is

$$P_{D,\max} = i_{D,\max} \cdot V_{DD} = (0.0967)(2.5) = 0.242 \text{ mW}$$

**Comment:** The level of $V_{OL}$ is less than the threshold voltage $V_{TN}$; therefore, if the output of this inverter is used to drive a similar inverter, the driver transistor of the load inverter would be cut off and its output would be high, which is the desired condition. We will compare the maximum drain currents and maximum power dissipations of the three basic NMOS inverters.

**Ex 16.1:** Consider the NMOS inverter with resistor load in Figure 16.3(a) biased at $V_{DD} = 3$ V. Assume transistor parameters of $k'_n = 100\,\mu\text{A/V}^2$, $W/L = 4$, and $V_{TN} = 0.5$ V. (a) Find the value of $R_D$ such that $v_O = 0.1$ V when $v_I = 3$ V. (b) Using the results of part (a), determine the maximum current and maximum power dissipation in the inverter. (c) Using the results of part (a), determine the transition point for the driver transistor. (Ans. (a) $R_D = 29.6\,\text{k}\Omega$; (b) $i_{D,\text{max}} = 0.098$ mA, $P_{D,\text{max}} = 0.294$ mW; (c) $V_{It} = 1.132$ V, $V_{Ot} = 0.632$ V)

An n-channel enhancement-mode MOSFET with the gate connected to the drain can be used as a load device in an NMOS inverter. This device configuration was analyzed in Chapter 3. We found that, when $v_{GS} = v_{DS} \geq V_{TN}$, the transistor always operates in the saturation region. The drain current is given by

$$i_D = K_n(v_{GS} - V_{TN})^2 = K_n(v_{DS} - V_{TN})^2 \qquad \textbf{(16.12)}$$

We continue to neglect the effect of the output resistance and the $\lambda$ parameter.

Figure 16.5(a) shows an NMOS inverter with the enhancement load device. The driver transistor parameters are denoted by $V_{TND}$ and $K_D$, and the load transistor parameters are denoted by $V_{TNL}$ and $K_L$. The substrate connections are not shown. In the following analysis, we neglect the body effect and we assume all threshold voltages are constant. These assumptions do not seriously affect the basic analysis, nor the inverter characteristics.

The driver transistor characteristics and the load curve are shown in Figure 16.5(b). When the inverter input voltage is less than the driver threshold voltage, the driver is cut off and the drain currents are zero. From Equation (16.12), we have

$$i_{DL} = 0 = K_L(v_{DSL} - V_{TNL})^2 \qquad \textbf{(16.13)}$$

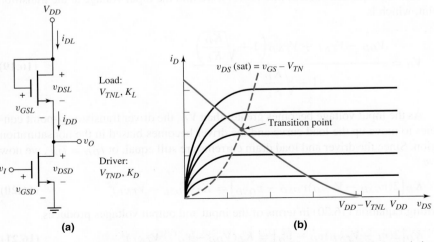

**(a)**          **(b)**

**Figure 16.5** (a) NMOS inverter with saturated load and (b) driver transistor characteristics and load curve

From Figure 16.5(a), we see that $v_{DSL} = V_{DD} - v_O$, which means that

$$v_{DSL} - V_{TNL} = V_{DD} - v_O - V_{TNL} = 0 \tag{16.14(a)}$$

The maximum output voltage is then

$$v_{O,\max} \equiv V_{OH} = V_{DD} - V_{TNL} \tag{16.14(b)}$$

For the enhancement-load NMOS inverter, the maximum output voltage, which is the logic 1 level, does not reach the full $V_{DD}$ value. This cutoff point is shown in the load curve in Figure 16.5(b).

As the input voltage becomes just greater than the driver threshold voltage $V_{TND}$, the driver transistor turns on and is biased in the saturation region. In steady-state, the two drain currents are equal since the output will be connected to the gates of other MOS transistors. We have $i_{DD} = i_{DL}$, which can be written as

$$K_D(v_{GSD} - V_{TND})^2 = K_L(v_{GSL} - V_{TNL})^2 \tag{16.15}$$

Equation (16.15) is expressed in terms of the individual transistor parameters. In terms of the input and output voltages, the expression becomes

$$K_D(v_I - V_{TND})^2 = K_L(V_{DD} - v_O - V_{TNL})^2 \tag{16.16}$$

Solving for the output voltage yields

$$v_O = V_{DD} - V_{TNL} - \sqrt{\frac{K_D}{K_L}}(v_I - V_{TND}) \tag{16.17}$$

As the input voltage increases, the driver $Q$-point moves up the load curve and the output voltage decreases linearly with $v_I$.

At the driver transition point, we have

$$v_{DSD}(\text{sat}) = v_{GSD} - V_{TND}$$

or

$$V_{Ot} = V_{It} - V_{TND} \tag{16.18}$$

Substituting Equation (16.18) into (16.17), we find the input voltage at the transition point, which is

$$V_{It} = \frac{V_{DD} - V_{TNL} + V_{TND}\left(1 + \sqrt{\dfrac{K_D}{K_L}}\right)}{1 + \sqrt{\dfrac{K_D}{K_L}}} \tag{16.19}$$

As the input voltage becomes greater than $V_{It}$, the driver transistor $Q$-point continues to move up the load curve and the driver becomes biased in the nonsaturation region. Since the driver and load drain currents are still equal, or $i_{DD} = i_{DL}$, we now have

$$K_D\left[2(v_{GSD} - V_{TND})v_{DSD} - v_{DSD}^2\right] = K_L(v_{DSL} - V_{TNL})^2 \tag{16.20}$$

Writing Equation (16.20) in terms of the input and output voltages produces

$$K_D\left[2(v_I - V_{TND})v_O - v_O^2\right] = K_L(V_{DD} - v_O - V_{TNL})^2 \tag{16.21}$$

Obviously, the relationship between $v_I$ and $v_O$ in this region is not linear.

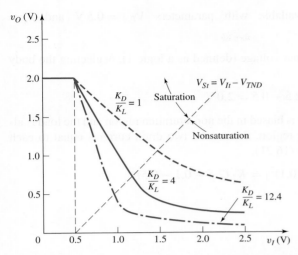

**Figure 16.6** Voltage transfer characteristics of NMOS inverter with saturated load, using parameters from Example 16.2 and for three aspect ratios.

Figure 16.6 shows the voltage transfer characteristics of this inverter for three $K_D$-to-$K_L$ ratios. The ratio $K_D/K_L$ is the aspect ratio and is related to the width-to-length parameters of the driver and load transistors.

The line, given by Equation (16.18), separating the driver saturation and nonsaturation regions is also shown in the figure. We see that the minimum output voltage, or the logic 0 level, for a high input decreases with an increasing $K_D/K_L$ ratio. As the width-to-length ratio of the load transistor decreases, the effective resistance increases, which means that the general behavior of the transfer characteristics is the same as for the resistor load. However, the high output voltage is

$$V_{OH} = V_{DD} - V_{TNL}$$

When the driver is biased in the saturation region, we find the slope of the transfer curve, which is the **inverter gain,** by taking the derivative of Equation (16.17) with respect to $v_I$. We see that

$$dv_O/dv_I = -\sqrt{K_D/K_L}$$

When the aspect ratio is greater than unity, the inverter gain magnitude is greater than unity. A logic circuit family with an inverter transfer curve that exhibits a gain greater than unity for some region is called a **restoring logic family.** Restoring logic is so named because logic signals that are degraded for some reason in one circuit can be restored by the gain of subsequent logic circuits.

## DESIGN EXAMPLE 16.2

**Objective:** Design an NMOS inverter to meet a set of specifications and determine the power dissipation in the inverter.

**Specifications:** The NMOS inverter with saturated load shown in Figure 16.5(a) is to be designed such that $v_O = 0.1$ V when $v_I = 2.0$ V. The circuit is biased at $V_{DD} = 2.5$ V. (Neglect the body effect.)

**Choices:** Transistors are available with parameters $V_{TN} = 0.5$ V and $k'_n = 100 \, \mu A/V^2$.

**Solution:** The maximum output voltage (defined as a logic 1), neglecting the body effect, is

$$V_{OH} = V_{DD} - V_{TNL} = 2.5 - 0.5 = 2.0 \, V$$

For $v_I = 2.0$ V, the driver is biased in the nonsaturation region and the load is always biased in the saturation region. Setting the two drain currents equal to each other, we find, using Equation (16.21),

$$K_D[2(2.0 - 0.5)(0.1) - (0.1)^2] = K_L(2.5 - 0.1 - 0.5)^2$$

which yields

$$\frac{K_D}{K_L} = 12.4$$

If we choose $(W/L)_L = 1$, and since

$$\frac{K_D}{K_L} = \frac{(W/L)_D}{(W/L)_L}$$

then we have

$$\left(\frac{W}{L}\right)_D = 12.4$$

The maximum inverter current occurs for $v_O = V_{OL} = 0.1$ V and is found from

$$i_{D,\max} = \frac{k'_n}{2} \cdot \left(\frac{W}{L}\right)_D [2(v_I - V_{TND})v_O - v_O^2]$$

$$= \left(\frac{0.1}{2}\right)(12.4)[2(2.0 - 0.5)(0.1) - (0.1)^2] = 0.180 \, mA$$

The maximum power dissipated in the inverter is

$$P_{D,\max} = i_{D,\max} \cdot V_{DD} = (0.18)(2.5) = 0.45 \, mW$$

**Comment:** In the NMOS inverter with enhancement-mode load, a relatively large difference in sizes of the driver and load transistors is required to produce a relatively low output voltage $V_{OL}$. The load transistor width-to-length ratio cannot be reduced substantially, so the maximum power dissipation cannot be substantially reduced from the 0.45 mW.

## EXERCISE PROBLEM

**Ex 16.2:** The enhancement-load NMOS inverter in Figure 16.5(a) is biased at $V_{DD} = 3$ V. The transistor parameters are $k'_n = 100 \, \mu A/V^2$, $V_{TND} = V_{TNL} = 0.4$ V, $(W/L)_D = 16$, and $(W/L)_L = 2$. (a) Determine $v_O$ when (i) $v_I = 0.1$ V and (ii) $v_I = 2.6$ V. Neglect the body effect. (b) Determine the maximum current and maximum power dissipation in the inverter. (c) Determine the transition point for the driver transistor. (Ans. (a) (i) $v_O = 2.6$ V, (ii) $v_O = 0.174$ V; (b) $i_{D,\max} = 0.589$ mA, $P_{D,\max} = 1.766$ mW; (c) $V_{It} = 1.08$ V, $V_{Ot} = 0.68$ V)

## NMOS Inverter with Depletion Load

Depletion-mode MOSFETs can also be used as load elements in NMOS inverters. Figure 16.7(a) shows the NMOS inverter with depletion load. The gate and source of the depletion-mode transistor are connected together. The driver transistor is still an enhancement-mode device. As before, the driver transistor parameters are $V_{TND}(V_{TND} > 0)$ and $K_D$, and the load transistor parameters are $V_{TNL}(V_{TNL} < 0)$ and $K_L$. Again, the substrate connections are not shown. The fabrication process for this inverter is slightly more complicated than for the enhancement-load inverter, since the threshold voltages of the two devices are not equal. However, as we will see, the advantages of this inverter make the extra processing steps worthwhile. This inverter has been the basis of many microprocessor and static memory designs.

The current–voltage characteristic curve for the depletion load, neglecting the body effect, is shown in Figure 16.7(b). Since the gate is connected to the source, $v_{GSL} = 0$, and the Q-point of the load is on this particular curve.

The driver transistor characteristics and the ideal load curve are shown in Figure 16.7(c). When the inverter input is less than the driver threshold voltage, the driver is cut off and the drain currents are zero. From Figure 16.7(b), we see that for

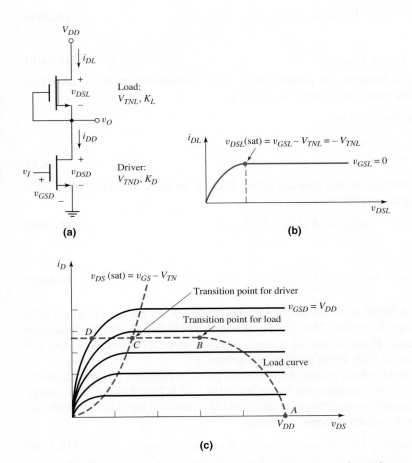

**Figure 16.7** (a) NMOS inverter with depletion load, (b) current–voltage characteristic of depletion load, and (c) driver transistor characteristics and load curve

$i_D = 0$, the drain-to-source voltage of the load transistor must be zero; therefore, $v_O = V_{DD}$ for $v_I \leq V_{TND}$. An advantage of the depletion-load inverter over the enhancement-load inverter is that the high output voltage, or the logic 1 level, is at the full $V_{DD}$ value.

As the input voltage becomes just greater than the driver threshold voltage $V_{TND}$, the driver turns on and is biased in the saturation region; however, the load is biased in the nonsaturation region. The $Q$-point lies between points $A$ and $B$ on the load curve shown in Figure 16.7(c). We again set the two drain currents equal, or $i_{DD} = i_{DL}$, which means that

$$K_D[v_{GSD} - V_{TND}]^2 = K_L\left[2(v_{GSL} - V_{TNL})v_{DSL} - v_{DSL}^2\right] \tag{16.22}$$

Writing Equation (16.22) in terms of the input and output voltages yields

$$K_D[v_I - V_{TND}]^2 = K_L[2(-V_{TNL})(V_{DD} - v_O) - (V_{DD} - v_O)^2] \tag{16.23}$$

This equation relates the input and output voltages as long as the driver is biased in the saturation region and the load is biased in the nonsaturation region.

There are two transition points for the NMOS inverter with a depletion load: one for the load and one for the driver. These are points $B$ and $C$, respectively, in Figure 16.7(c). The transition point for the load is given by

$$v_{DSL} = V_{DD} - V_{Ot} = v_{GSL} - V_{TNL} = -V_{TNL} \tag{16.24(a)}$$

or

$$V_{Ot} = V_{DD} + V_{TNL} \tag{16.24(b)}$$

Since $V_{TNL}$ is negative, the output voltage at the transition point is less than $V_{DD}$. The transition point for the driver is given by

$$v_{DSD} = v_{GSD} - V_{TND}$$

or

$$V_{Ot} = V_{It} - V_{TND} \tag{16.25}$$

When the $Q$-point lies between points $B$ and $C$ on the load curve, both devices are biased in the saturation region, and

$$K_D(v_{GSD} - V_{TND})^2 = K_L(v_{GSL} - V_{TNL})^2 \tag{16.26(a)}$$

or

$$\sqrt{\frac{K_D}{K_L}}(v_I - V_{TND}) = -V_{TNL} \tag{16.26(b)}$$

Equation (16.26(b)) demonstrates that the input voltage is a constant as the $Q$-point passes through this region. This effect is also shown in Figure 16.7(c); the load curve between points $B$ and $C$ lies on a constant $v_{GSD}$ curve. (This characteristic will change when the body effect is taken into account.)

For an input voltage greater than the value given by Equation (16.26(b)), the driver is biased in the nonsaturation region while the load is biased in the saturation region. The $Q$-point is now between points $C$ and $D$ on the load curve in Figure 16.7(c). Equating the two drain currents, we obtain

$$K_D\left[2(v_{GSD} - V_{TND})v_{DSD} - v_{DSD}^2\right] = K_L(v_{GSL} - V_{TNL})^2 \tag{16.27(a)}$$

which becomes

$$\frac{K_D}{K_L}\left[2(v_I - V_{TND})v_O - v_O^2\right] = (-V_{TNL})^2 \tag{16.27(b)}$$

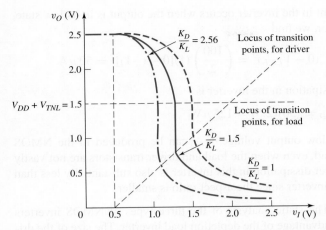

**Figure 16.8** Voltage transfer characteristics of an NMOS inverter with depletion load, using parameters from Example 16.3 and for three aspect ratios

This equation implies that the relationship between the input and output voltages are not linear in this region.

Figure 16.8 shows the voltage transfer characteristics of this inverter for three values of $K_D/K_L$. Also shown are the loci of transition points for the load and driver transistors as given by Equations (16.24(b)) and (16.25), respectively.

## DESIGN EXAMPLE 16.3

**Objective:** Design an NMOS inverter to meet a set of specifications and determine the power dissipation in the inverter.

**Specifications:** The NMOS inverter with depletion load shown in Figure 16.7(a) is to be designed such that $v_O = V_{OL} = 0.10$ V when $v_I = 2.5$ V. The circuit is biased at $V_{DD} = 2.5$ V. (Neglect the body effect.)

**Choices:** Transistors are available with process conduction parameters of $k'_n = 100 \, \mu\text{A/V}^2$. The driver transistor threshold voltage is $V_{TND} = 0.5$ V and the load transistor threshold voltage is $V_{TNL} = -1$ V.

**Solution:** For $v_I = 2.5$ V, the driver transistor is biased in the nonsaturation region and the load transistor is biased in the saturation region. Using Equation (16.27(b), we find

$$K_D[2(2.5 - 0.5)(0.1) - (0.1)^2] = K_L[0 - (-1)]^2$$

which yields

$$\frac{K_D}{K_L} = 2.56$$

If we choose $(W/L)_L = 1$, then

$$\frac{K_D}{K_L} = \frac{(W/L)_D}{(W/L)_L} \Rightarrow 2.56 = \frac{(W/L)_D}{1} \Rightarrow \left(\frac{W}{L}\right)_D = 2.56$$

The maximum current in the inverter occurs when the output is in its low state, so, from the load transistor, we find

$$i_{D,\text{max}} = \frac{k_n'}{2} \cdot \left(\frac{W}{L}\right)_L (0 - V_{TNL})^2 = \left(\frac{100}{2}\right) (1)[0 - (-1)]^2 = 50 \,\mu\text{A}$$

The maximum power dissipation in the inverter is

$$P_{D,\text{max}} = i_{D,\text{max}} \cdot V_{DD} = (50)(2.5) = 125 \,\mu\text{W}$$

**Comment:** A relatively low output voltage $V_{OL}$ can be produced in the NMOS inverter with depletion load, even when the load and driver transistors are not vastly different in size. The power dissipation in this inverter is also substantially less than in the enhancement-load inverter since the aspect ratio is smaller.

**Design Consideration:** The static analysis of the three types of NMOS inverters clearly demonstrates the advantage of the depletion load inverter. The size of the driver transistor is smaller for a given load device size to produce a given low output state. This allows a greater number of inverters to be fabricated in a given chip area. In addition, since the power dissipation is less, more inverters can be fabricated on a chip for a given total power dissipation.

---

### EXERCISE PROBLEM

**Ex 16.3:** The depletion-load NMOS inverter shown in Figure 16.7(a) is biased at $V_{DD} = 3$ V. The transistor parameters are $k_n' = 100 \,\mu\text{A/V}^2$, $V_{TND} = 0.4$ V, $V_{TNL} = -0.8$ V, $(W/L)_D = 6$, and $(W/L)_L = 2$. (a) Determine $v_O$ for $v_I = 3$ V. Neglect the body effect. (b) Determine the maximum current and maximum power dissipation in the inverter. (c) Find the transition points for the driver and load transistors. (Ans. (a) $v_O = 0.0414$ V; (b) $i_{D,\text{max}} = 0.064$ mA, $P_{D,\text{max}} = 0.192$ mW; (c) driver: $V_{It} = 0.862$ V, $V_{Ot} = 0.462$ V; load: $V_{It} = 0.862$ V, $V_{Ot} = 2.2$ V)

---

16.1.3   ## Body Effect

Up to this point, we have neglected the body effect and assumed that all threshold voltages are constant. Figure 16.9 shows enhancement-load and depletion-load

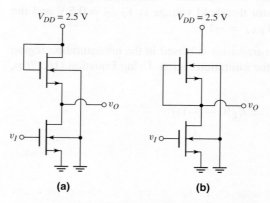

**(a)**                    **(b)**

**Figure 16.9** NMOS inverters, showing substrate connections to ground potential: (a) enhancement-load inverter and (b) depletion-load inverter

NMOS inverters with the substrates of all transistors tied to ground. A nonzero source-to-body voltage will then exist in the load devices. In fact, the source terminal of the depletion load can increase to $V_{DD}$. The threshold voltage equation, including the body effect, must be used in the circuit calculations for the load transistor. This significantly complicates the equations for the voltage transfer calculations, making them very cumbersome for hand analyses.

## EXAMPLE 16.4

**Objective:** Determine the change in the high output voltage of an NMOS inverter with enhancement load, taking the body effect into account.

Consider the NMOS inverter with enhancement load in Figure 16.9(a). The transistor parameters are $V_{TNDO} = V_{TNLO} = 0.5$ V and $K_D/K_L = 16$. Assume the inverter is biased at $V_{DD} = 2.5$ V, assume the body effect coefficient is $\gamma = 0.5$ V$^{1/2}$, and let $\phi_{fp} = 0.365$ V.

**Solution:** When $v_I < V_{TNDO}$, the driver is cut off and the output goes high. From Equation (16.14(b)), the maximum output voltage is

$$v_{O,\max} = V_{OH} = V_{DD} - V_{TNL}$$

where $V_{TNL}$ is given by

$$V_{TNL} = V_{TNLO} + \gamma\left[\sqrt{2\phi_{fp} + V_{SB}} - \sqrt{2\phi_{fp}}\right]$$

From Figure 16.9(a), we see that $V_{SB} = v_O$. Therefore, Equation (16.14(b)) can be written as

$$v_{O,\max} = V_{DD} - \left\{V_{TNLO} + \gamma\left[\sqrt{2\phi_{fp} + v_{O,\max}} - \sqrt{2\phi_{fp}}\right]\right\}$$

Defining $v_{O,\max} \equiv V_{OH}$, we have

$$V_{OH} - 2.427 = -0.5\sqrt{0.73 + V_{OH}}$$

Squaring both sides and rearranging terms yields

$$V_{OH}^2 - 5.1044V_{OH} + 5.7088 = 0$$

Consequently, the maximum output voltage, or the logic 1 level, is

$$V_{OH} = 1.655 \text{ V}$$

**Comment:** Neglecting the body effect, the logic 1 output level is

$$V_{OH} = V_{DD} - V_{TNLO} = 2.5 - 0.5 = 2.0 \text{ V}$$

The body effect, then, can significantly influence the logic high state of the NMOS inverter with enhancement load. These results also impact the inverter noise margins.

The source and body terminals of the depletion load device in the NMOS inverter shown in Figure 16.9(b) are not at the same potential when the output goes high. However, when the driver is cut off, the drain-to-source voltage of the depletion device must be zero in order that $v_{O,\max} = V_{OH} = V_{DD}$.

**Computer Simulation:** A computer analysis of the inverters in Figure 16.9 was performed, neglecting the body effect and taking the body effect into account. The parameters are $V_{DD} = 5$ V, $V_{TNDO} = 0.8$ V for the driver transistors, $V_{TNLO} = 0.8$ V

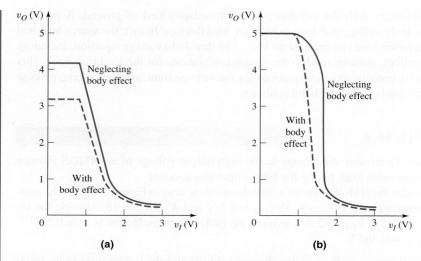

**Figure 16.10** Voltage transfer characteristics of NMOS inverters with and without the body effect (a) enhancement load and (b) depletion load

for the saturated load transistor, and $V_{TNLO} = -2$ V for the depletion load transistor. The body effect coefficient was assumed to be $\gamma = 0.9$ V$^{1/2}$.

The body effect changes the voltage transfer characteristics of both the enhancement load and depletion load inverters. Figure 16.10(a) shows the voltage transfer characteristics for the enhancement load inverter. For $v_I = 0$, the output voltage is 3.15 V when the body effect is taken into account. This compares to 4.2 V when the body effect is neglected.

Figure 16.10(b) shows the voltage transfer characteristics for the depletion load inverter. The output voltage is 5 V in the high state, which is independent of the body effect. However, the characteristics during the transition region are a function of the body effect.

## EXERCISE PROBLEM

**Ex 16.4:** Repeat Example 16.4 for the case when the body effect coefficient is $\gamma = 0.3$ V$^{1/2}$. (Ans. $V_{OH} = 1.781$ V)

## Test Your Understanding

**TYU 16.1** Consider the NMOS inverter with enhancement load, as shown in Figure 16.5(a), biased at $V_{DD} = 1.8$ V. The threshold voltages are $V_{TND} = V_{TNL} = 0.4$ V. Assume $k_n' = 100 \, \mu$A/V$^2$. Design the width-to-length ratios such that the output voltage is 0.12 V and the maximum inverter power dissipation is 0.50 mW when $v_I = 1.4$ V. Neglect the body effect. (Ans. $(W/L)_L = 3.39$, $(W/L)_D = 24.6$)

**TYU 16.2** Consider the depletion load inverter in Figure 16.7(a) biased at $V_{DD} = 1.8$ V. The threshold voltages are $V_{TND} = 0.4$ V and $V_{TNL} = -0.6$ V. Assume $k_n' = 100 \, \mu$A/V$^2$. Design the inverter such that the maximum inverter power dissipation is 0.2 mW and the output voltage is 0.08 V when $v_I = 1.8$ V. Neglect the body effect. (Ans. $(W/L)_L = 6.17$, $(W/L)_D = 10.2$)

**TYU 16.3** (a) Consider the results of Exercise Ex 16.1. Assume 100,000 resistor-load inverters are fabricated on a single chip and the input voltage of each inverter is high. Determine the current that must be supplied to each chip and the maximum power that will be dissipated on each chip. (b) Repeat part (a) for Exercise Ex 16.2 and the enhancement-load inverter. (c) Repeat part (a) for Exercise Ex 16.3 and the depletion-load inverter. (Ans. (a) $I = 9.8\,A$, $P = 29.4\,W$; (b) $I = 58.9\,A$, $P = 176.6\,W$; (c) $I = 6.4\,A$, $P = 19.2\,W$)

## 16.2    NMOS LOGIC CIRCUITS

**Objective:** • Analyze and design NMOS logic gates.

NMOS logic circuits are formed by combining driver transistors in parallel, series, or series–parallel combinations to produce a desired output logic function.

### 16.2.1    NMOS NOR and NAND Gates

The NMOS NOR logic gate contains additional driver transistors connected in parallel. Figure 16.11 shows a two-input NMOS NOR logic gate with a depletion load. If $A = B = $ logic 0, then both $M_{DA}$ and $M_{DB}$ are cut off and $v_O = V_{DD}$. If $A = $ logic 1 and $B = $ logic 0, then $M_{DB}$ is cut off and the NMOS inverter configuration with $M_L$ and $M_{DA}$ is the same as previously considered, and the output voltage goes low. Similarly, if $A = $ logic 0 and $B = $ logic 1, we again have the same inverter configuration.

   If $A = B = $ logic 1, then both $M_{DA}$ and $M_{DB}$ turn on and the two driver transistors are effectively in parallel. The value of the output voltage now changes slightly. Figure 16.12 shows the NOR gate when both input voltages are a logic 1. From our previous analysis, we can assume that the two driver transistors are biased

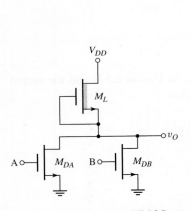

**Figure 16.11** Two-input NMOS NOR logic gate with depletion load

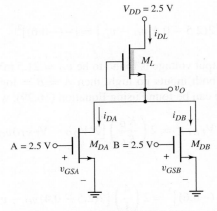

**Figure 16.12** Two-input NMOS NOR logic gate for Example 16.5

in the nonsaturation region and the load device is biased in the saturation region. We then have

$$i_{DL} = i_{DA} + i_{DB}$$

which in general terms can be written

$$K_L[v_{GSL} - V_{TNL}]^2 = K_{DA}\big[2(v_{GSA} - V_{TNA})v_{DSA} - v_{DSA}^2\big]$$
$$+ K_{DB}\big[2(v_{GSB} - V_{TNB})v_{DSB} - v_{DSB}^2\big] \qquad (16.28)$$

If we assume the two driver transistors are identical, then the driver conduction parameters and threshold voltages are also identical, or $K_{DA} = K_{DB} \equiv K_D$ and $V_{TNA} = V_{TNB} \equiv V_{TND}$. Noting that $v_{GSL} = 0$, $v_{GSA} = v_{GSB} = V_{DD}$, and $v_{DSA} = v_{DSB} = v_O$, we can write Equation (16.28) as

$$[-V_{TNL}]^2 = 2\left(\frac{K_D}{K_L}\right)\big[2(V_{DD} - V_{TND})v_O - v_O^2\big] \qquad (16.29)$$

Equation (16.29) shows that when both drivers are conducting, the effective width-to-length ratio of the composite driver transistor doubles. This means that the output voltage becomes slightly smaller when both inputs are high.

---

### EXAMPLE 16.5

**Objective:** Determine the low output voltage of an NMOS NOR circuit.

Consider the NOR logic circuit in Figure 16.12 biased at $V_{DD} = 2.5$ V. Assume transistor parameters of $k_n' = 100\,\mu\text{A/V}^2$, $V_{TND} = 0.4$ V, $V_{TNL} = -0.6$ V, $(W/L)_D = 4$, and $(W/L)_L = 1$. Neglect the body effect.

**Solution:** If, for example, $A = \text{logic } 1 = 2.5$ V and $B = \text{logic } 0$, then $M_{DA}$ is biased in the nonsaturation region and $M_{DB}$ is cut off. The output voltage is determined from Equation (16.27(b)), which is

$$\frac{K_D}{K_L}\big[2(v_I - V_{TND})v_O - v_O^2\big] = (-V_{TNL})^2$$

or

$$\frac{4}{1}\big[2(2.5 - 0.4)v_O - v_O^2\big] = [-(-0.6)]^2$$

The output voltage is found to be $v_O = 21.5$ mV.

If both inputs go high, then $A = B = \text{logic } 1 = V_{DD} = 2.5$ V and the output voltage can be found using Equation (16.29), which is

$$(-V_{TNL})^2 = 2\left(\frac{K_D}{K_L}\right)\big[2(V_{DD} - V_{TND})v_O - v_O^2\big]$$

or

$$[-(-0.6)]^2 = 2\left(\frac{4}{1}\right)\big[2(2.5 - 0.4)v_O - v_O^2\big]$$

The output voltage is found to be $v_O = 10.7$ mV.

**Comment:** An NMOS NOR gate must be designed to achieve a specified $V_{OL}$ output voltage when only one input is high. This will give the largest logic 0 value. When more than one input is high, the output voltage is smaller than the specified $V_{OL}$ value, since the effective width-to-length ratio of the composite driver transistor increases.

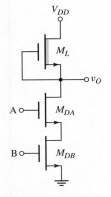

Figure 16.13 Two-input NMOS NAND logic gate with depletion load

### EXERCISE PROBLEM

**Ex 16.5:** Consider the two-input NMOS NOR logic gate shown in Figure 16.11. Let $V_{DD} = 1.8$ V. Assume transistor parameters of $k'_n = 100\,\mu\text{A/V}^2$, $V_{TND} = 0.4$ V, $V_{TNL} = -0.6$ V, $(W/L)_D = 5$, and $(W/L)_L = 1$. Neglect the body effect. (a) Determine $V_{OL}$ when: (i) $A = $ logic 1, $B = $ logic 0, and (ii) $A = B = $ logic 1. (b) Calculate the power dissipation in the circuit for the input condition given in part (a). (Ans. (a) (i) $v_O = 26$ mV, (ii) $v_O = 12.9$ mV; (b) For (i) and (ii), $P = 32.4\,\mu\text{W}$)

The NMOS NAND logic gate contains additional driver transistors connected in series. Figure 16.13 shows a two-input NMOS NAND logic gate with a depletion load. If both $A = B = $ logic 0, or if either $A$ or $B$ is a logic 0, at least one driver is cut off, and the output is high. If both $A = B = $ logic 1, then the composite driver of the NMOS inverter conducts and the output goes low.

Since the gate-to-source voltages of $M_{DA}$ and $M_{DB}$ are not equal, determining the actual voltage $V_{OL}$ of a NAND gate is difficult. The drain-to-source voltages of $M_{DA}$ and $M_{DB}$ must adjust themselves to produce the same current. In addition, if the body effect is also included, the analysis becomes even more difficult. Since the two driver transistors are in series, a good approximation assumes that the width-to-length ratio of the drivers must be twice that of a single driver in an NMOS inverter to achieve a given $V_{OL}$ value.

The composite width-to-length ratios of the driver transistors in the two-input NMOS NOR and NAND gates are shown schematically in Figure 16.14. For the NOR gate, the effective *width* doubles; for the NAND gates, the effective *length* doubles.

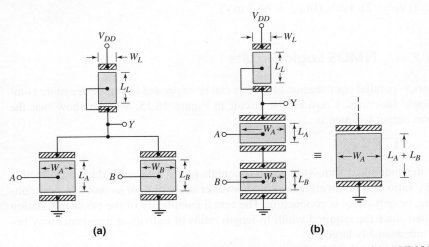

**(a)**                              **(b)**

Figure 16.14 Composite width-to-length ratios of driver transistors in two-input NMOS logic configurations (a) NOR and (b) NAND

## EXAMPLE 16.6

**Objective:** Determine the low output voltage of an NMOS NAND circuit.

Consider the NAND logic circuit shown in Figure 16.13 biased at $V_{DD} = 2.5$ V. Assume transistor parameters of $k'_n = 100\,\mu\text{A/V}^2$, $V_{TND} = 0.4$ V, $V_{TNL} = -0.6$ V, $(W/L)_D = 8$, and $(W/L)_L = 1$. Neglect the body effect.

**Solution:** If either $A$ or $B$ is a logic 0, then $v_O = $ logic $1 = 2.5$ V.

If $A = B = $ logic $1 = 2.5$ V, then both driver transistors are driven in the nonsaturation region and the output goes low. As a good approximation, we will assume the effective length of the driver transistor doubles. Then, using Equation (16.27(b)), we have

$$\frac{\frac{1}{2} \cdot \left(\frac{W}{L}\right)_D}{\left(\frac{W}{L}\right)_L} \left[2(v_I - V_{TND})v_O - v_O^2\right] = (-V_{TNL})^2$$

or

$$\frac{8}{(2)(1)}\left[2(2.5 - 0.4)v_O - v_O^2\right] = [-(-0.6)]^2$$

The output voltage is found to be $v_O = 21.5$ mV.

This output voltage is the same value that would be obtained for a simple inverter with $(W/L)_D = 4$ and $(W/L)_L = 1$.

**Comment:** If an $N$-input NMOS NAND logic gate were to be fabricated then the width-to-length ratio of the drivers would need to be $N$ times that of a single driver in an NMOS inverter to achieve a given value of $V_{OL}$. The increase in the required area of the driver transistors in a NAND logic gate means that logic gates with more than three or four inputs are not attractive.

### EXERCISE PROBLEM

**Ex 16.6:** Repeat Example 16.6 for a three-input NMOS NAND logic gate with depletion load with $(W/L)_L = 1$ and (a) $(W/L)_D = 12$ and (b) $(W/L)_D = 4$. (Ans. (a) $v_O = 21.5$ mV, (b) $v_O = 65.3$ mV)

### 16.2.2    NMOS Logic Circuits

The series–parallel combination of drivers can be expanded to synthesize more complex logic functions. Consider the circuit in Figure 16.15. We can show that the Boolean output function is

$$f = \overline{(A \cdot B + C)}$$

Also, the individual transistor width-to-length ratios shown produce an effective $K_D/K_L$ ratio of 4 for an effective single inverter when only $M_{DA}$ and $M_{DB}$ are conducting, or only $M_{DC}$ is conducting. The actual complexity of the Boolean function is limited since the required width-to-length ratios of individual transistors may become unreasonably large.

Two additional logic functions are the exclusive-OR and exclusive-NOR. Figure 16.16 shows a circuit configuration that produces the exclusive-OR function.

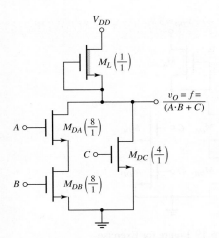

Figure 16.15  NMOS logic circuit example

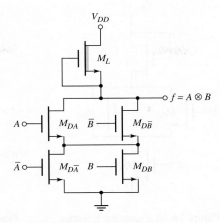

Figure 16.16  NMOS exclusive-OR logic gate

If $A = B =$ logic 1, a path exists from the output to ground through drivers $M_{DA}$ and $M_{DB}$, and the output goes low. Similarly, if $A = B =$ logic 0, which means that $\bar{A} = \bar{B} =$ logic 1, a path exists from the output to ground through the drivers $M_{D\bar{B}}$ and $M_{D\bar{A}}$, and the output again goes low. For all other input logic signal combinations, the output is isolated from ground so the output goes high.

### 16.2.3    Fanout

An NMOS inverter or NMOS logic gate must be capable of driving more than one load, as shown in Figure 16.17. It is assumed that each load is identical to the driver logic circuit. The number of identical-load circuits connected to the output of a driver logic circuit is defined as the **fanout**. For MOS logic circuits, the inputs to the load circuits are the oxide-insulated gates of the MOS transistors; therefore, the static loading caused by multiple driver loads is so small that the dc transfer curve is essentially identical to a no-load condition. The dc characteristics of MOS logic circuits are unaffected by the fanout to other MOS logic inputs. However, the load capacitance due to a large fanout seriously degrades the switching speed and propagation delay times. Consequently, maintaining the propagation delay time below a specified maximum value determines the fanout of MOS digital circuits.

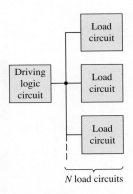

Figure 16.17  Logic circuit driving $N$ load circuits

## Test Your Understanding

**TYU 16.4** (a) Design a three-input NMOS NOR logic gate with depletion load such that $V_{OL}(\text{max}) = 50$ mV and such that the maximum power dissipation is $50\,\mu$W. Let $V_{DD} = 2.5$ V. The transistor parameters are $k_n' = 100\,\mu$A/V$^2$, $V_{TND} = 0.4$ V, and $V_{TNL} = -0.6$ V. (b) Using the results of part (a), determine $V_{OL}$ when all inputs are a logic 1. (Ans. (a) $(W/L)_L = 1.11$, $(W/L)_D = 1.93$; (b) $V_{OL} = 16.5$ mV)

**TYU 16.5** Consider the NMOS logic circuit in Figure 16.18. Assume transistor parameters of $k_n' = 100\,\mu A/V^2$ and $V_{TN} = 0.4$ V. Assume all driver transistors are identical. Neglect the body effect. (a) If $(W/L)_L = 0.5$, determine $(W/L)$ for the drivers such that $V_{OL}(\text{max}) = 80\,\mu$V. Assume logic 1 input voltages are 2.1 V.

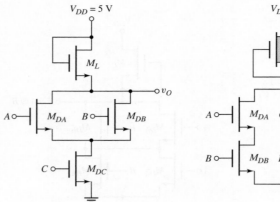

**Figure 16.18** Figure for Exercise TYU 16.5

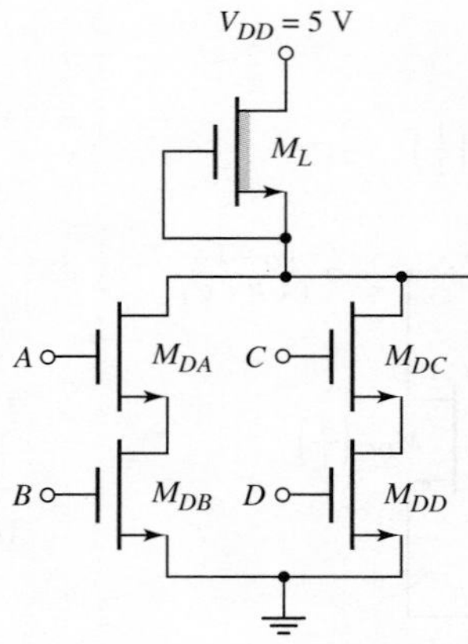

**Figure 16.19** Figure for Exercise TYU 16.6

(b) Determine the maximum power dissipation in the logic circuit. (Ans. (a) $(W/L)_D = 15.4$, (b) $P = 255\,\mu W$)

**TYU 16.6** Repeat Exercise TYU 16.5 for the NMOS logic circuit in Figure 16.19, except assume the threshold voltage of the load device is $V_{TNL} = -0.6$ V. (Ans. (a) $(W/L)_D = 1.09$, (b) $P = 22.5\,\mu W$)

## 16.3    CMOS INVERTER

**Objective:** • Analyze and design CMOS inverters.

Complementary MOS, or CMOS, circuits contain both n-channel and p-channel MOSFETs. As we will see, the power dissipation in CMOS logic circuits is much smaller than in NMOS circuits, which makes CMOS very attractive. We briefly review the characteristics of p-channel transistors, and will then analyze the CMOS inverter, which is the basis of most CMOS logic circuits. We will examine the CMOS NOR and NAND gates and other basic CMOS logic circuits, covering power dissipation, noise margin, fanout, and switching characteristics.

### 16.3.1    p-Channel MOSFET Revisited

Figure 16.20 shows a simplified view of a p-channel MOSFET. The p- and n-regions are reversed from those in an n-channel device. Again, the channel length is $L$ and the channel width is $W$. Usually in any given fabrication process, the channel length is the same for all devices, so the channel width $W$ is the variable in logic circuit design.

Normally, in an integrated circuit, more than one p-channel device will be fabricated in the same n-substrate so the p-channel transistors will exhibit a body effect. The n-substrate is connected to the most positive potential. The source terminal may

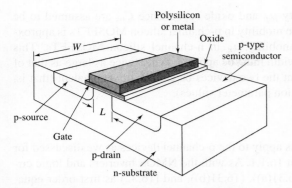

**Figure 16.20** Simplified cross section of p-channel MOSFET

be negative with respect to the substrate; therefore, voltage $V_{BS}$ may exist between the body and the source. The threshold voltage is

$$V_{TP} = V_{TPO} - \frac{\sqrt{2e\varepsilon_s N_d}}{C_{ox}}\left[\sqrt{2\phi_{fn} + V_{BS}} - \sqrt{2\phi_{fn}}\right]$$

$$= V_{TPO} - \gamma\left[\sqrt{2\phi_{fn} + V_{BS}} - \sqrt{2\phi_{fn}}\right] \qquad \textbf{(16.30)}$$

where $V_{TPO}$ is the threshold voltage for zero body-to-source voltage, or $V_{BS} = 0$. The parameter $N_d$ is the n-substrate doping concentration and $\phi_{fn}$ is a potential related to the substrate doping. The parameter $\gamma$ is the body effect coefficient.

### Current–Voltage Relation

The current–voltage characteristics of the p-channel MOSFET are functions of both the electrical and geometric properties of the device. When the transistor is biased in the nonsaturation region, we have $v_{SD} \leq v_{SG} + V_{TP}$. Therefore,

$$i_D = K_p\left[2(v_{SG} + V_{TP})v_{SD} - v_{SD}^2\right] \qquad \textbf{(16.31(a))}$$

In the saturation region, we have $v_{SD} \geq v_{SG} + V_{TP}$, which means that

$$i_D = K_p(v_{SG} + V_{TP})^2 \qquad \textbf{(16.31(b))}$$

The gate potential is negative with respect to the source. For the p-channel transistor to conduct, we must have $v_{GS} < V_{TP}$, where $V_{TP}$ is negative for an enhancement-mode device. We also see that $v_{SG} > |V_{TP}|$ when the p-channel device is conducting.

In most cases, the channel length modulation factor $\lambda$ has very little effect on the operating characteristics of MOS digital circuits. Therefore, the term $\lambda$ is assumed to be zero unless otherwise stated.

The transition point, which separates the nonsaturation and saturation bias regions, is given by

$$v_{SD} = v_{SD}(\text{sat}) = v_{SG} + V_{TP} \qquad \textbf{(16.32)}$$

The parameter $K_p$ is the conduction parameter and is given by

$$K_p = \left(\frac{1}{2}\mu_p C_{ox}\right)\left(\frac{W}{L}\right) = \frac{k_p'}{2}\frac{W}{L} \qquad \textbf{(16.33)}$$

As before, the hole mobility $\mu_p$ and oxide capacitance $C_{\text{ox}}$ are assumed to be constant for all devices. The hole mobility in p-channel silicon MOSFETs is approximately one-half the electron mobility $\mu_n$ in n-channel silicon MOSFETs. This means that a p-channel device width must be approximately twice as large as that of an n-channel device in order that the two devices be electrically equivalent (that is, that they have the same conduction parameter values).

### Small Geometry Effects

The same small geometry effects apply to the p-channel devices as we discussed for the n-channel devices in Section 16.1.1. As with the NMOS inverters and logic circuits, we can use Equations (16.31(a)), (16.31(b)), and (16.32) as first-order equations in the design of CMOS logic circuits. The basic operation and behavior of CMOS logic circuits can be predicted using these first-order equations.

## 16.3.2    DC Analysis of the CMOS Inverter

The **CMOS inverter,** shown in Figure 16.21, is a series combination of a p-channel and an n-channel MOSFET. The gates of the two MOSFETs are connected together to form the input and the two drains are connected together to form the output. Both transistors are enhancement-mode devices. The parameters of the NMOS are denoted by $K_n$ and $V_{TN}$, where $V_{TN} > 0$, and the parameters of the PMOS are denoted by $K_p$ and $V_{TP}$, where $V_{TP} < 0$.

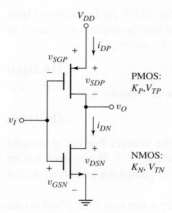

**Figure 16.21**  CMOS inverter

Figure 16.22 shows a simplified cross section of a CMOS inverter. In this process, a separate p-well region is formed within the starting n-substrate. The n-channel device is fabricated in the p-well region and the p-channel device is fabricated in the n-substrate. Although other approaches, such as an n-well in a p-substrate, are also used to fabricate CMOS circuits, the important point is that the processing is more complicated for CMOS circuits than for NMOS circuits.

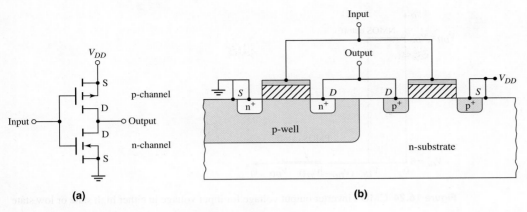

**Figure 16.22** Simplified cross section, CMOS inverter

However, the advantages of CMOS digital logic circuits over NMOS circuits justify their use.

## Voltage Transfer Curve

Figure 16.23 shows the transistor characteristics for both the n- and p-channel devices. We can determine the voltage transfer characteristics of the inverter by evaluating the various transistor bias regions. For $v_I = 0$, the NMOS device is cut off, $i_{DN} = 0$, and $i_{DP} = 0$. The PMOS source-to-gate voltage is $V_{DD}$, which means that the PMOS is biased on the curve marked B in Figure 16.23(b). Since the only point on the curve corresponding to $i_{DP} = 0$ occurs at $v_{SDP} = 0 = V_{DD} - v_O$, the output voltage is $v_O = V_{DD}$. This condition exists as long as the NMOS transistor is cut off, or $v_I \leq V_{TN}$.

For $v_I = V_{DD}$, the PMOS device is cut off, $i_{DP} = 0$, and $i_{DN} = 0$. The NMOS gate-to-source voltage is $V_{DD}$ and the NMOS is biased on the curve marked A in Figure 16.23(a). The only point on the curve corresponding to $i_{DN} = 0$ occurs at $v_{DSN} = v_O = 0$. The output voltage is zero as long as the PMOS transistor is cut off, or $v_{SGP} = V_{DD} - v_I \leq |V_{TP}|$. This means that the input voltage is in the range $V_{DD} - |V_{TP}| \leq v_I \leq V_{DD}$.

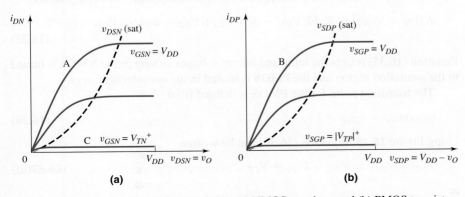

**Figure 16.23** Current–voltage characteristics, (a) NMOS transistor and (b) PMOS transistor

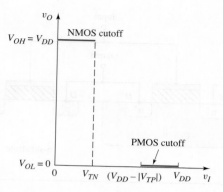

**Figure 16.24** CMOS inverter output voltage for input voltage in either high state or low state

Figure 16.24 shows the voltage transfer characteristics generated thus far for the CMOS inverter. The more positive output voltage corresponds to a logic 1, or $V_{OH} = V_{DD}$, and the more negative output voltage corresponds to a logic 0, or $V_{OL} = 0$. When the output is in the logic 1 state, the NMOS transistor is cut off; when the output is in the logic 0 state, the PMOS transistor is cut off.

Ideally, the current in the CMOS inverter in either steady-state condition is zero, which means that, ideally, the quiescent power dissipation is zero. This result is the attractive feature of CMOS digital circuits. In actuality, CMOS inverter circuits exhibit a small leakage current in both steady-state conditions, due to the reverse-biased pn junctions. However, the power dissipation may be in the nanowatt range rather than in the milliwatt range of NMOS inverters. Without this feature, VLSI would not be possible.

When the input voltage is just greater than $V_{TN}$, or

$$v_I = v_{GSN} = V_{TN}^+$$

the NMOS begins to conduct and the $Q$-point falls on the curve marked C in Figure 16.23(a). The current is small and $v_{DSN} \cong V_{DD}$, which means that the NMOS is biased in the saturation region. The PMOS source-to-drain voltage is small, so the PMOS is biased in the nonsaturation region. Setting $i_{DN} = i_{DP}$, we can write

$$K_n[v_{GSN} - V_{TN}]^2 = K_p\left[2(v_{SGP} + V_{TP})v_{SDP} - v_{SDP}^2\right] \tag{16.34}$$

Relating the gate-to-source and drain-to-source voltages in each transistor to the inverter input and output voltages, respectively, we can rewrite Equation (16.34) as follows:

$$K_n[v_I - V_{TN}]^2 = K_p[2(V_{DD} - v_I + V_{TP})(V_{DD} - v_O) - (V_{DD} - v_O)^2] \tag{16.35}$$

Equation (16.35) relates the input and output voltages as long as the NMOS is biased in the saturation region and the PMOS is biased in the nonsaturation region.

The transition point for the PMOS is defined from

$$v_{SDP}(\text{sat}) = v_{SGP} + V_{TP} \tag{16.36}$$

Using Figure 16.25, Equation (16.36) can be written

$$V_{DD} - V_{OPt} = V_{DD} - V_{IPt} + V_{TP} \tag{16.37(a)}$$

or

$$V_{OPt} = V_{IPt} - V_{TP} \tag{16.37(b)}$$

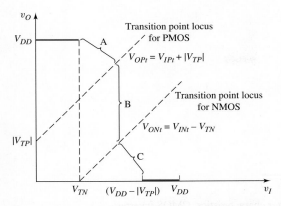

**Figure 16.25** Regions of the CMOS transfer characteristics indicating NMOS and PMOS transistor bias conditions. The NMOS device is biased in the saturation region in areas A and B and in the nonsaturation region in area C. The PMOS device is biased in the saturation region in areas B and C and in the nonsaturation region in area A.

where $V_{OPt}$ and $V_{IPt}$ are the PMOS output and input voltages, respectively, at the transition point.

The transition point for the NMOS is defined from

$$v_{DSN}(\text{sat}) = v_{GSN} - V_{TN} \tag{16.38(a)}$$

or

$$V_{ONt} = V_{INt} - V_{TN} \tag{16.38(b)}$$

where $V_{ONt}$ and $V_{INt}$ are the NMOS output and input voltages, respectively, at the transition point.

On the basis that $V_{TP}$ is negative for an enhancement-mode PMOS, Equations (16.37(b)) and (16.38(b)) are plotted in Figure 16.25. We determine the input voltage at the transition points by setting the two drain currents equal to each other when both transistors are biased in the saturation region. The result is

$$K_n(v_{GSN} - V_{TN})^2 = K_p(v_{SGP} + V_{TP})^2 \tag{16.39}$$

With the gate-to-source voltages related to the input voltage, Equation (16.39) becomes

$$K_n(v_I - V_{TN})^2 = K_p(V_{DD} - v_I + V_{TP})^2 \tag{16.40}$$

For this ideal case, the output voltage does not appear in Equation (16.40), and the input voltage is a constant, as long as the two transistors are biased in the saturation region.

Voltage $v_I$ from Equation (16.40) is the input voltage at the PMOS and NMOS transition points. Solving for $v_I$, we find that

$$v_I = v_{It} = \frac{V_{DD} + V_{TP} + \sqrt{\dfrac{K_n}{K_p}}\, V_{TN}}{1 + \sqrt{\dfrac{K_n}{K_p}}} \tag{16.41}$$

For $v_I > V_{It}$, the NMOS is biased in the nonsaturation region and the PMOS is biased in the saturation region. Again equating the two drain currents, we have

$$K_n\big[2(v_{GSN} - V_{TN})v_{DSN} - v_{DSN}^2\big] = K_p(v_{SGP} + V_{TP})^2 \tag{16.42}$$

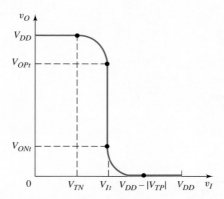

**Figure 16.26** Complete voltage transfer characteristics, CMOS inverter

Also, relating the gate-to-source and drain-to-source voltages to the input and output voltages, respectively, modifies Equation (16.42) as follows:

$$K_n[2(v_I - V_{TN})v_O - v_O^2] = K_p(V_{DD} - v_I + V_{TP})^2 \tag{16.43}$$

Equation (16.43) relates the input and output voltages as long as the NMOS is biased in the nonsaturation region and the PMOS in the saturation region. Figure 16.26 shows the complete voltage transfer curve.

---

## EXAMPLE 16.7

**Objective:** Determine the critical voltages on the voltage transfer curve of a CMOS inverter.

Consider a CMOS inverter biased at $V_{DD} = 5$ V with transistor parameters $K_n = K_p$ and $V_{TN} = -V_{TP} = 0.8$ V. Then consider another CMOS inverter biased at $V_{DD} = 3$ V with transistor parameters $K_n = K_p$ and $V_{TN} = -V_{TP} = 0.6$ V.

**Solution ($V_{DD} = 5$ V):** The input voltage at the transition points is, from Equation (16.41),

$$V_{It} = \frac{5 + (-0.8) + \sqrt{1}(0.8)}{1 + \sqrt{1}} = 2.5 \text{ V}$$

The output voltage at the transition point for the PMOS is, from Equation (16.37(b)),

$$V_{OPt} = V_{It} - V_{TP} = 2.5 - (-0.8) = 3.2 \text{ V}$$

and the output voltage at the transition point or the NMOS is, from Equation (16.38(b)),

$$V_{ONt} = V_{It} - V_{TN} = 2.5 - 0.8 = 1.7 \text{ V}$$

**Solution ($V_{DD} = 3$ V):** The critical voltages are

$$V_{It} = 1.5 \text{ V} \qquad V_{OPt} = 2.1 \text{ V} \qquad V_{ONt} = 0.9 \text{ V}$$

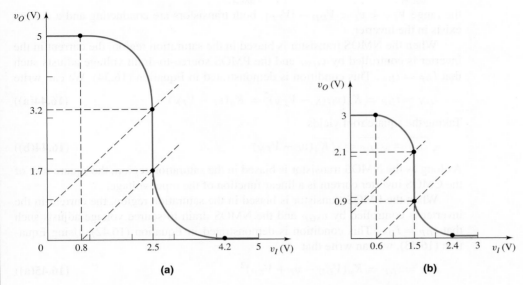

**Figure 16.27** Voltage transfer characteristics of CMOS inverter in Example 16.7 biased at (a) $V_{DD} = 5$ V and (b) $V_{DD} = 3$ V

**Comment:** The two voltage transfer curves are shown in Figure 16.27. These figures depict another advantage of CMOS technology, that is CMOS circuits can be biased over a relatively wide range of voltages.

## EXERCISE PROBLEM

**Ex 16.7:** The CMOS inverter in Figure 16.21 is biased at $V_{DD} = 2.1$ V, and the transistor threshold voltages are $V_{TN} = -V_{TP} = 0.4$ V. Sketch the voltage transfer curve and show the critical voltages as in Figure 16.26 for (a) $K_n/K_p = 1$, (b) $K_n/K_p = 0.5$, and (c) $K_n/K_p = 2$. (Ans. (a) $V_{It} = 1.05$ V, $V_{OPt} = 1.45$ V, $V_{ONt} = 0.65$ V; (b) $V_{It} = 1.16$ V, $V_{OPt} = 1.56$ V, $V_{ONt} = 0.76$ V; (c) $V_{It} = 0.938$ V, $V_{OPt} = 1.338$ V, $V_{ONt} = 0.538$ V)

## Transistor Sizing

We may note that both voltage transfer curves shown in Figure 16.27 are symmetrical about the switching point $V_{DD}/2$. This effect is a direct consequence of the fact that the NMOS and PMOS transistors are matched: that is, $K_n = K_p$ and $V_{TN} = |V_{TP}|$. In general, the process conduction parameters, $k_n'$ and $k_p'$, are not equal. Therefore, in order for the two transistors to be matched, we must adjust the width-to-length ratios. In order for $K_n = K_p$, we have $k_n'(W/L)_n = k_p'(W/L)_p$. In general, $k_p' < k_n'$, so we must have $(W/L)_p > (WL)_n$. The PMOS device must be larger than the NMOS device to make the two devices electrically equivalent.

### CMOS Inverter Currents

When the CMOS inverter input voltage is either a logic 0 or a logic 1, the current in the circuit is zero, since one of the transistors is cut off. When the input voltage is in the range $V_{TN} < v_I < V_{DD} - |V_{TP}|$, both transistors are conducting and a current exists in the inverter.

When the NMOS transistor is biased in the saturation region, the current in the inverter is controlled by $v_{GSN}$ and the PMOS source-to-drain voltage adjusts such that $i_{DP} = i_{DN}$. This condition is demonstrated in Equation (16.34). We can write

$$i_{DN} = i_{DP} = K_n(v_{GSN} - V_{TN})^2 = K_n(v_I - V_{TN})^2 \qquad \textbf{(16.44(a))}$$

Taking the square root yields

$$\sqrt{i_{DN}} = \sqrt{i_{DP}} = \sqrt{K_n}(v_I - V_{TN}) \qquad \textbf{(16.44(b))}$$

As long as the NMOS transistor is biased in the saturation region, the square root of the CMOS inverter current is a linear function of the input voltage.

When the PMOS transistor is biased in the saturation region, the current in the inverter is controlled by $v_{SGP}$ and the NMOS drain-to-source voltage adjusts such that $i_{DP} = i_{DN}$. This condition is demonstrated in Equation (16.42). Using Equation (16.43), we can write that

$$i_{DN} = i_{DP} = K_p(V_{DD} - v_I + V_{TP})^2 \qquad \textbf{(16.45(a))}$$

Taking the square root yields

$$\sqrt{i_{DN}} = \sqrt{i_{DP}} = \sqrt{K_p}(V_{DD} - v_I + V_{TP}) \qquad \textbf{(16.45(b))}$$

As long as the PMOS transistor is biased in the saturation region, the square root of the CMOS inverter current is also a linear function of the input voltage.

Figure 16.28 shows plots of the square root of the inverter current for two values of $V_{DD}$ bias. These curves are quasi-static characteristics in that no current is diverted into a capacitive load. At the inverter switching point, both transistors are biased in the saturation region and both transistors influence the current. At the switching point, the actual current characteristic does not have a sharp discontinuity in the slope. The channel length modulation parameter $\lambda$ also influences the current characteristics at the peak value. However, the curves in Figure 16.28 are excellent approximations.

### 16.3.3    Power Dissipation

In the quiescent or static state, in which the input is either a logic 0 or a logic 1, power dissipation in the CMOS inverter is virtually zero. However, during the switching cycle from one state to another, current flows and power is dissipated. The CMOS

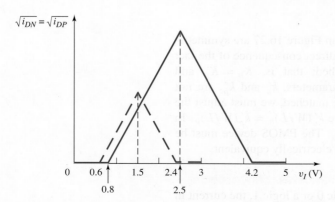

**Figure 16.28** Square root of CMOS inverter current versus input voltage for CMOS inverters described in Example 16.7

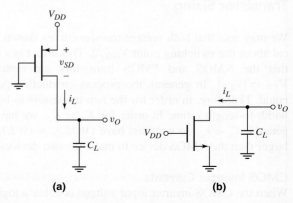

**Figure 16.29** CMOS inverter when the output switches (a) low to high and (b) high to low

inverter and logic circuits are used to drive other MOS devices for which the input impedance is a capacitance. During the switching cycle, then, this load capacitance must be charged and discharged.

In Figure 16.29(a), the output switches from its low to its high state. The input is switched low, the PMOS gate is at zero volts, and the NMOS is cut off. The load capacitance $C_L$ must be charged through the PMOS device. Power dissipation in the PMOS transistor is given by

$$P_P = i_L v_{SD} = i_L(V_{DD} - v_O) \tag{16.46}$$

The current and the output voltage are related by

$$i_L = C_L \frac{dv_O}{dt} \tag{16.47}$$

The energy dissipated in the PMOS device as the output switches from low to high is

$$E_P = \int_0^\infty P_P \, dt = \int_0^\infty C_L(V_{DD} - v_O) \frac{dv_O}{dt} \, dt$$

$$= C_L V_{DD} \int_0^{V_{DD}} dv_O - C_L \int_0^{V_{DD}} v_O \, dv_O \tag{16.48}$$

which yields

$$E_P = C_L V_{DD} v_O \Big|_0^{V_{DD}} - C_L \frac{v_O^2}{2} \Big|_0^{V_{DD}} = \frac{1}{2} C_L V_{DD}^2 \tag{16.49}$$

After the output has switched high, the energy stored in the load capacitance is $(\frac{1}{2})C_L V_{DD}^2$. When the inverter input goes high, the output switches low, as shown in Figure 16.29(b). The PMOS device is cut off, the NMOS transistor conducts, and the load capacitance discharges through the NMOS device. All the energy stored in the load capacitance is dissipated in the NMOS device. As the output switches from high to low, the energy dissipated in the NMOS transistor is

$$E_N = \frac{1}{2} C_L V_{DD}^2 \tag{16.50}$$

The total energy dissipated in the inverter during one switching cycle is therefore

$$E_T = E_P + E_N = \frac{1}{2} C_L V_{DD}^2 + \frac{1}{2} C_L V_{DD}^2 = C_L V_{DD}^2 \tag{16.51}$$

If the inverter is switched at frequency $f$, the power dissipated in the inverter is

$$P = f E_T = f C_L V_{DD}^2 \tag{16.52}$$

Equation (16.52) shows that the power dissipated in a CMOS inverter is directly proportional to the switching frequency and to $V_{DD}^2$. The drive in digital IC design is toward lower supply voltages, such as 3 V or less.

The power dissipation is proportional to $V_{DD}^2$. In some digital circuits, such as digital watches, the CMOS logic circuits are biased at $V_{DD} = 1.5$ V, so the power dissipation is substantially reduced.

## EXAMPLE 16.8

**Objective:** Calculate the power dissipation in a CMOS inverter.

Consider a CMOS inverter with a load capacitance of $C_L = 2$ pF biased at $V_{DD} = 5$ V. The inverter switches at a frequency of $f = 100$ kHz.

**Solution:** From Equation (16.52), power dissipation in the CMOS inverter is

$$P = f C_L V_{DD}^2 = (10^5)(2 \times 10^{-12})(5)^2 \Rightarrow 5 \ \mu W$$

**Comment:** Previously determined values of static power dissipation in NMOS inverters were on the order of 500 $\mu W$; therefore, power dissipation in a CMOS inverter is substantially smaller. In addition, in most digital systems, only a small fraction of the logic gates change state during each clock cycle; consequently, the power dissipation in a CMOS digital system is substantially less than in an NMOS digital system of similar complexity.

EXERCISE PROBLEM

**Ex 16.8:** A CMOS inverter is biased at $V_{DD} = 3$ V. The inverter drives an effective load capacitance of $C_L = 0.5$ pF. Determine the maximum switching frequency such that the power dissipation is limited to $P = 0.10 \ \mu W$. (Ans. $f = 22.2$ kHz)

### 16.3.4    Noise Margin

The word "noise" means transient, unwanted variations in voltages or currents. In digital circuits, if the magnitude of the noise at a logic node is too large, logic errors can be introduced into the system. However, if the noise amplitude is less than a specified value, called the **noise margin,** the noise signal will be attenuated as it passes through a logic gate or circuit, while the logic signals will be transmitted without error.

Noise signals are usually generated outside the digital circuit and transferred to logic nodes or interconnect lines through parasitic capacitances or inductances. The coupling process is usually time dependent, leading to dynamic conditions in the circuit. In digital systems, however, the noise margins are usually defined in terms of static voltages.

#### Noise Margin Definition

For static noise margins, the type of noise usually considered is called series-voltage noise. Figure 16.30 shows two inverters in series in which the output of the second is connected back to the input of the first. Also included are series-voltage noise sources $\delta V_L$ and $\delta V_H$. This type of noise can be developed by inductive coupling. The input voltage levels are indicated by $H$ (high) and $L$ (low). The noise amplitudes $\delta V_L$ and $\delta V_H$ can be different, and the polarities may be such as to increase the low output and reduce the high output. The noise margins are defined as the maximum values of $\delta V_L$ and $\delta V_H$ at which the inverters will remain in the correct state.

The actual definitions of the noise margins $NM_L$ and $NM_H$ are not unique. In addition other types of noise, other than series-voltage source noise, may be present in the system. Dynamic noise sources also complicate the issue. However, in this text, in order to provide some measure of noise margin in a logic circuit, we will use the unity-gain approach to determine the logic threshold levels $V_{IL}$ and $V_{IH}$ and the corresponding noise margins.

Figure 16.31 shows a general voltage transfer function for an inverter. The expected logic 1 and logic 0 output voltages of the inverter are $V_{OH}$ and $V_{OL}$, respec-

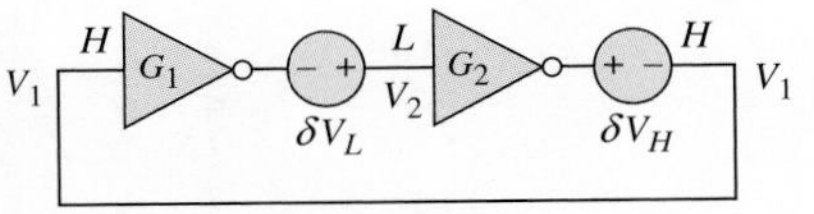

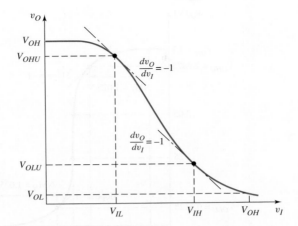

**Figure 16.30** Two-inverter flip-flop, including series-voltage noise sources

**Figure 16.31** Generalized inverter voltage curve and defined voltage limits $V_{IL}$ and $V_{IH}$

tively. The parameters $V_{IH}$ and $V_{IL}$, which determine the noise margins, are defined as the points at which

$$\frac{dv_O}{dv_I} = -1 \qquad \qquad \textbf{(16.53)}$$

For $v_I \leq V_{IL}$, the inverter gain magnitude is less than unity, and the output changes slowly with a change in the input voltage. Similarly, for $v_I \geq V_{IH}$, the output again changes slowly with input voltage since the gain magnitude is less than unity. However, when the input voltage is in the range $V_{IL} < v_I < V_{IH}$, the gain magnitude is greater than one, and the output signal changes rapidly. This region is called the **undefined range.** If the input voltage is inadvertently pushed into this range by a noise signal, the output may change logic states, and a logic error could be introduced into the system. The corresponding output voltages at the unity-gain points are denoted $V_{OHU}$ and $V_{OLU}$, where the last subscript $U$ signifies the unity-gain values.

The noise margins are defined as

$$NM_L = V_{IL} - V_{OLU} \qquad \qquad \textbf{(16.54(a))}$$

and

$$NM_H = V_{OHU} - V_{IH} \qquad \qquad \textbf{(16.54(b))}$$

Figure 16.32 shows the general voltage transfer function of a CMOS inverter. (The numbers in the figure are from Example 16.9 to be considered later.) The parameters $V_{IH}$ and $V_{IL}$ determine the noise margins and are defined as the points at which

$$\frac{dv_O}{dv_I} = -1 \qquad \qquad \textbf{(16.55)}$$

For $v_I \leq V_{IL}$ and $v_I \geq V_{IH}$, the gain is less than unity and the output changes slowly with input voltage. However, when the input voltage is in the range $V_{IL} < v_I < V_{IH}$, the inverter gain is greater than unity, and the output signal changes rapidly with a change in the input voltage. This is the undefined range.

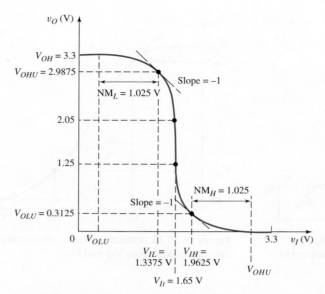

**Figure 16.32** CMOS inverter voltage transfer characteristics with defined noise margins

Point $V_{IL}$ occurs when the NMOS is biased in the saturation region and the PMOS is biased in the nonsaturation region. The relationship between the input and output voltages is given by Equation (16.35). Taking the derivative with respect to $v_I$ yields

$$2K_n[v_I - V_{TN}] = K_p \left[ -2(V_{DD} - v_O) - 2(V_{DD} - v_I + V_{TP})\frac{dv_O}{dv_I} \\ -2(V_{DD} - v_O)\left(-\frac{dv_O}{dv_I}\right) \right] \quad \textbf{(16.56)}$$

Setting the derivative equal $-1$, we have

$$K_n[v_I - V_{TN}] = -K_p[(V_{DD} - v_O) - (V_{DD} - v_I + V_{TP}) + (V_{DD} - v_O)] \quad \textbf{(16.57)}$$

Solving for $v_O$ produces

$$v_O = V_{OHU} = \frac{1}{2}\left\{ \left(1 + \frac{K_n}{K_p}\right)v_I + V_{DD} - \left(\frac{K_n}{K_p}\right)V_{TN} - V_{TP}\right\} \quad \textbf{(16.58)}$$

Combining Equations (16.58) and (16.35), we see that voltage $V_{IL}$ is

$$v_I = V_{IL} = V_{TN} + \frac{(V_{DD} + V_{TP} - V_{TN})}{\left(\frac{K_n}{K_p} - 1\right)}\left[ 2\sqrt{\frac{\frac{K_n}{K_p}}{\frac{K_n}{K_p} + 3}} - 1 \right] \quad \textbf{(16.59)}$$

If $K_n = K_p$, Equation (16.59) becomes indefinite, since a zero would exist in both the numerator and the denominator. However, when $K_n = K_p$, Equation (16.58) becomes

$$v_O = V_{OHU(K_n = K_p)} = \frac{1}{2}\{2v_I + V_{DD} - V_{TN} - V_{TP}\} \quad \textbf{(16.60)}$$

Substituting Equation (16.60) into Equation (16.35) yields a voltage $V_{IL}$ of

$$v_I = V_{IL(K_n=K_p)} = V_{TN} + \tfrac{3}{8}(V_{DD} + V_{TP} - V_{TN}) \qquad (16.61)$$

for $K_n = K_p$.

Point $V_{IH}$ occurs when the NMOS is biased in the nonsaturation region and the PMOS is biased in the saturation region. The relationship between the input and output voltages is given by Equation (16.43). Taking the derivative with respect to $v_I$ yields

$$K_n \left[ 2(v_I - V_{TN})\frac{dv_O}{dv_I} + 2v_O - 2v_O\frac{dv_O}{dv_I} \right] = 2K_p(V_{DD} - v_I + V_{TP})(-1) \qquad (16.62)$$

Setting the derivative equal to $-1$, we find that

$$K_n[-(v_I - V_{TN}) + v_O + v_O] = -K_p[V_{DD} - v_I + V_{TP}] \qquad (16.63)$$

The output voltage $v_O$ is then

$$v_O = V_{OLU} = \frac{v_I\left(1 + \dfrac{K_n}{K_p}\right) - V_{DD} - \left(\dfrac{K_n}{K_p}\right)V_{TN} - V_{TP}}{2\left(\dfrac{K_n}{K_p}\right)} \qquad (16.64)$$

Combining Equations (16.64) and (16.43), yields voltage $V_{IH}$ as

$$v_I = V_{IH} = V_{TN} + \frac{(V_{DD} + V_{TP} - V_{TN})}{\left(\dfrac{K_n}{K_p} - 1\right)} \left[ \frac{2\dfrac{K_n}{K_p}}{\sqrt{3\dfrac{K_n}{K_p} + 1}} - 1 \right] \qquad (16.65)$$

Again, if $K_n = K_p$, Equation (16.65) becomes indefinite, since a zero would exist in both the numerator and the denominator. However, when $K_n = K_p$, Equation (16.64) becomes

$$v_O = V_{OLU(K_n=K_p)} = \tfrac{1}{2}\{2v_I - V_{DD} - V_{TN} - V_{TP}\} \qquad (16.66)$$

Substituting Equation (16.66) into Equation (16.43) yields a voltage $V_{IH}$ of

$$v_I = V_{IH(K_n=K_p)} = V_{TN} + \tfrac{5}{8}(V_{DD} + V_{TP} - V_{TN}) \qquad (16.67)$$

## EXAMPLE 16.9

**Objective:** Determine the noise margins of a CMOS inverter.

Consider a CMOS inverter biased at $V_{DD} = 3.3$ V. Assume the transistors are matched with $K_n = K_p$ and $V_{TN} = -V_{TP} = 0.4$ V.

**Solution:** From Equation (16.41), the input voltage at the transition points, or the inverter switching point, is 1.65 V. Since $K_n = K_p$, $V_{IL}$ is, from Equation (16.61),

$$V_{IL} = V_{TN} + \tfrac{3}{8}(V_{DD} + V_{TP} - V_{TN}) = 0.4 + \tfrac{3}{8}(3.3 - 0.4 - 0.4) = 1.3375 \text{ V}$$

Point $V_{IH}$ is, from Equation (16.67),

$$V_{IH} = V_{TN} + \tfrac{5}{8}(V_{DD} + V_{TP} - V_{TN}) = 0.4 + \tfrac{5}{8}(3.3 - 0.4 - 0.4) = 1.9625 \text{ V}$$

The output voltages at points $V_{IL}$ and $V_{IH}$ are determined from Equations (16.60) and (16.66), respectively. They are

$$V_{OHU} = \tfrac{1}{2}[2V_{IL} + V_{DD} - V_{TN} - V_{TP}]$$
$$= \tfrac{1}{2}[2(1.3375) + 3.3 - 0.4 + 0.4] = 2.9875 \text{ V}$$

and

$$V_{OLU} = \tfrac{1}{2}[2V_{IH} - V_{DD} - V_{TN} - V_{TP}]$$
$$= \tfrac{1}{2}[2(1.9625) - 3.3 - 0.4 + 0.4] = 0.3125 \text{ V}$$

The noise margins are therefore

$$\text{NM}_L = V_{IL} - V_{OLU} = 1.3375 - 0.3125 = 1.025 \text{ V}$$

and

$$\text{NM}_H = V_{OHU} - V_{IH} = 2.9875 - 1.9625 = 1.025 \text{ V}$$

**Comment:** The results of this example are shown in Figure 16.32. Since the two transistors are electrically identical, the voltage transfer curve and the resulting critical voltages are symmetrical. Also, $(V_{OH} - V_{OHU}) = 0.3125$ V, which is less than $|V_{TP}|$, and $(V_{OLU} - V_{OL}) = 0.3125$ V, which is less than $V_{TN}$. As long as the input voltage remains within the limits of the noise margin, no logic error will be transmitted through the digital system.

### EXERCISE PROBLEM

**Ex 16.9:** A CMOS inverter is biased at $V_{DD} = 1.8$ V. The transistor parameters are $V_{TN} = 0.4$ V, $V_{TP} = -0.4$ V, $K_n = 200 \, \mu A/V^2$, and $K_p = 80 \, \mu A/V^2$. (a) Determine the transition points. (b) Find the critical voltages $V_{IL}$ and $V_{IH}$, and the corresponding output voltages. (c) Calculate the noise margins $\text{NM}_L$ and $\text{NM}_H$. (Ans. (a) $V_{It} = 0.7874$ V, $V_{OPt} = 1.187$ V, $V_{ONt} = 0.3874$ V; (b) $V_{IL} = 0.6323$ V, $V_{IH} = 0.8767$ V, $V_{OHU} = 1.7065$ V, $V_{OLU} = 0.1337$ V; (c) $\text{NM}_L = 0.4986$ V, $\text{NM}_H = 0.8298$ V)

## Test Your Understanding

**TYU 16.7** Consider a CMOS inverter biased at $V_{DD} = 5$ V, with transistor threshold voltages of $V_{TN} = +0.8$ V and $V_{TP} = -0.8$ V. Calculate the peak current in the inverter for: (a) $K_n = K_p = 50 \, \mu A/V^2$, and (b) $K_n = K_p = 200 \, \mu A/V^2$. (Ans. (a) $i_D(\text{max}) = 145 \, \mu A$ (b) $i_D(\text{max}) = 578 \, \mu A$)

**TYU 16.8** Repeat Exercise Ex 16.9 for a CMOS inverter biased at $V_{DD} = 5$ V with transistor parameters of $V_{TN} = 0.8$ V, $V_{TP} = -2$ V, and $K_n = K_p = 100 \, \mu A/V^2$. (Ans. (a) $V_{It} = 1.9$ V, $V_{OPt} = 3.9$ V, $V_{ONt} = 1.1$ V; (b) $V_{IL} = 1.625$ V, $V_{IH} = 2.175$ V, $V_{OLU} = 0.275$ V, $V_{OHU} = 4.725$ V; (c) $\text{NM}_L = 1.35$ V, $\text{NM}_H = 2.55$ V)

## 16.4   CMOS LOGIC CIRCUITS

Objective:  • Analyze and design static CMOS logic gates.

Large-scale integrated CMOS circuits are used extensively in digital systems, including watches, calculators, and microprocessors. We will look at the basic CMOS NOR and NAND gates, and will then analyze more complex CMOS logic circuits. Since there is no clock signal applied to these logic circuits, they are referred to as **static CMOS logic** circuits.

### 16.4.1   Basic CMOS NOR and NAND Gates

In the basic or classical CMOS logic circuits, the gates of a PMOS and an NMOS are connected together, and additional PMOS and NMOS transistors are connected in series or parallel to form specific logic circuits. Figure 16.33(a) shows a two-input CMOS NOR gate. The NMOS transistors are in parallel and the PMOS transistors are in series.

If $A = B = $ logic 0, then both $M_{NA}$ and $M_{NB}$ are cut off, and the current in the circuit is zero. The source-to-gate voltage of $M_{PA}$ is $V_{DD}$ but the current is zero; therefore, $v_{SD}$ of $M_{PA}$ is zero. This means that the source-to-gate voltage of $M_{PB}$ is also $V_{DD}$. However, since the current is zero, then $v_{SD}$ of $M_{PB}$ is also zero. The output voltage is therefore $v_O = V_{DD} = $ logic 1.

If the input signals are $A = $ logic $1 = V_{DD}$ and $B = $ logic $0 = 0$ V, then the source-to-gate voltage of $M_{PA}$ is zero, and the current in the circuit is again zero. The gate-to-source voltage of $M_{NA}$ is $V_{DD}$ but the current is zero, so $v_{DS}$ of $M_{NA}$ is zero and $v_O = 0 = $ logic 0. This result also holds for the other two possible input conditions, since at least one PMOS is cut off and at least one NMOS is in a conducting state. The NOR logic function is shown in the truth table of Figure 16.33(b).

In both the CMOS NOR and NAND logic gates, the current in the circuit is essentially zero when the inputs are in any quiescent state. Only very small reverse-bias pn junction currents exist. The quiescent power dissipation is therefore essentially zero. Again, this is the primary advantage of CMOS circuits.

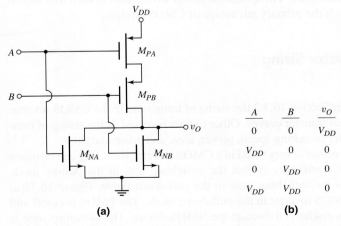

| A | B | $v_O$ |
|---|---|---|
| 0 | 0 | $V_{DD}$ |
| $V_{DD}$ | 0 | 0 |
| 0 | $V_{DD}$ | 0 |
| $V_{DD}$ | $V_{DD}$ | 0 |

**(a)**                                              **(b)**

**Figure 16.33** (a) Two-input CMOS NOR logic circuit and (b) truth table

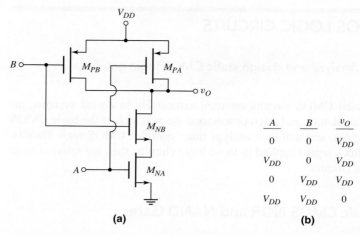

| A | B | $v_O$ |
|---|---|---|
| 0 | 0 | $V_{DD}$ |
| $V_{DD}$ | 0 | $V_{DD}$ |
| 0 | $V_{DD}$ | $V_{DD}$ |
| $V_{DD}$ | $V_{DD}$ | 0 |

**(a)** **(b)**

**Figure 16.34** (a) Two-input CMOS NAND logic circuit and (b) truth table

A two-input CMOS NAND logic gate is shown in Figure 16.34(a). In this case, the NMOS transistors are in series and the PMOS transistors are in parallel. If $A = B = $ logic 0, the two NMOS devices are cut off and the current in the circuit is zero. The source-to-gate voltage of each PMOS device is $V_{DD}$, which means that both PMOS transistors are in a conducting state. However, since the current is zero, $v_{SD}$ for both $M_{PA}$ and $M_{PB}$ is zero and $v_O = V_{DD}$. This result applies if at least one input is a logic 0.

If the input signals are $A = B = $ logic $1 = V_{DD}$, then both PMOS transistors are cut off, and the current in the circuit is zero. With $A = $ logic 1, $M_{NA}$ is in a conducting state; however, since the current is zero, then $v_{DS}$ of $M_{NA}$ is zero. This means that the gate-to-source voltage of $M_{NB}$ is also $V_{DD}$ and $M_{NB}$ is also in a conducting state. However, since the current is zero, then $v_{DS}$ of $M_{NB}$ is zero, and $v_O = $ logic $0 = 0$ V. The NAND logic function is shown in the truth table in Figure 16.34(b).

In both the CMOS NOR and NAND logic gates, the current in the circuit is essentially zero when the inputs are in any quiescent state. Only very small reverse-bias pn junction currents exist. The quiescent power dissipation is therefore essentially zero. Again, this is the primary advantage of CMOS circuits.

### 16.4.2 Transistor Sizing

**CMOS Inverter**

We briefly discussed in Section 16.3.2 the sizing of transistors in the CMOS inverter in terms of symmetrical transfer curves. Other factors involved in the sizing of transistors are, for example, switching speed, power, area, and noise margin.

Since the standby power is very small in a CMOS inverter, the sizing can be based on switching speed. We will specify that the switching time in the pull-up mode should be the same as the switching time in the pull-down mode. Figure 16.35(a) shows the effective CMOS inverter in the pull-down mode. The PMOS is cutoff and the load capacitance is discharged through the NMOS device. The switching time is therefore a function of the current capability of the NMOS transistor. Figure 16.35(b) shows the effective CMOS inverter in the pull-up mode. The NMOS is cutoff and the

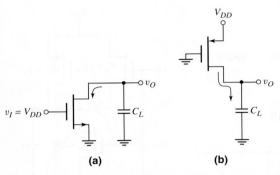

**Figure 16.35** (a) Effective CMOS inverter in pull-down mode and (b) effective CMOS inverter in pull-up mode

load capacitance is charged through the PMOS device. The switching time is a function of the current capability of the PMOS transistor.

Assuming that $V_{TN} = |V_{TP}|$, equal switching times then implies that the conduction parameters of the NMOS and PMOS devices be equal, or

$$\frac{k_n'}{2}\left(\frac{W}{L}\right)_n = \frac{k_p'}{2}\left(\frac{W}{L}\right)_p \tag{16.68}$$

Assuming that $\mu_n \approx 2\mu_p$, we have

$$\frac{(W/L)_p}{(W/L)_n} = \frac{k_n'}{k_p'} = \frac{\mu_n}{\mu_p} \approx 2 \tag{16.69}$$

The width-to-length ratio of the PMOS device must be approximately twice as large as that of the NMOS device to obtain equal switching times.

In any given technology, the channel lengths of the NMOS and PMOS devices are the same. Therefore the channel widths are sized to the desired value. We can write that $W_n = W$ and $W_p = 2W$, where $W_n$ and $W_p$ are the channel widths of the NMOS and PMOS devices, respectively, and $W$ is a standard width.

## CMOS Logic Gates

We can now consider the sizing of transistors in the basic CMOS NAND and NOR logic gates. We will specify, again, equal pull-up and pull-down switching times, and we want the same switching times as the CMOS inverter with a load capacitance $C_L$. We will use the effective 2:1 ratio between PMOS and NMOS sizes from the CMOS inverter.

Consider the two-input CMOS NOR gate shown in Figure 16.33. Assume a load capacitance $C_L$ is connected to the output. In the worst case during a pull-down operation, only one NMOS device will be turned on. To achieve the same switching time as the CMOS inverter, the NMOS channel widths should be $W_n = W$. If both NMOS devices are turned on, the effective channel width will be doubled (see Figure 16.14(a)) and the switching time will be shorter.

During a pull-up operation, both PMOS devices must be turned on. Since the PMOS devices are in series, the effective channel length doubles (see Figure 16. 14(b)). Therefore, to maintain the same effective width-to-length ratio, the channel widths must be doubled. We must therefore have $W_p = 2(2W) = 4W$.

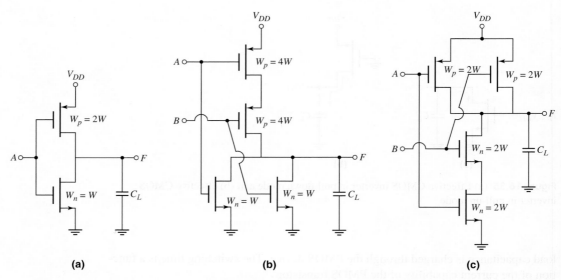

**Figure 16.36** The width-to-length ratios of (a) the CMOS inverter, (b) the CMOS NOR gate, and (c) the CMOS NAND gate

Now consider the two-input NAND logic gate shown in Figure 16.34. Again, assume a load capacitance $C_L$ is connected to the output. In the worst case during a pull-up operation, only one PMOS device will be turned on. This is equivalent to the CMOS inverter, so the channel width should be $W_p = 2W$. If both PMOS devices are turned on, the effective channel width is doubled and the switching time will be shorter.

During the pull-down operation, both NMOS devices must be turned on. Again, since the NMOS devices are in series, the effective channel length doubles. Therefore to maintain the same effective width-to-length ratio, the channel widths must be doubled. We must therefore have $W_n = 2(W) = 2W$.

The results of the transistor sizing for the CMOS inverter, and CMOS NOR and NAND gates are shown in Figure 16.36.

## EXAMPLE 16.10

**Objective:** Determine the transistor width-to-length ratios of a three-input CMOS NAND logic gate.

Symmetrical switching times are desired and the switching times should correspond to the basic CMOS inverter.

**Solution:** There are three p-channel transistors in parallel for the three-input CMOS NAND gate. The worst case is when only one PMOS device is on in the pull-up mode. This corresponds to the basic CMOS inverter, so the effective width should be $W_p = 2W$.

There are three n-channel transistors in series for the three-input CMOS NAND gate. All three transistors must be turned on in the pull-down mode. For three transistors in series, the effective channel length triples. Therefore, to keep the effective NMOS width equal to $W$, we must have $W_n = 3(W) = 3W$.

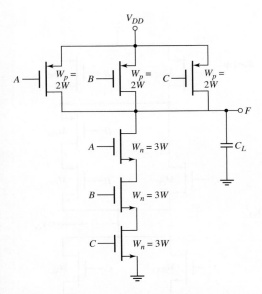

**Figure 16.37**  Width-to-length ratios for a three-input CMOS NAND logic gate

The results are shown in Figure 16.37.

**Comment:**  As the number of inputs to a basic CMOS logic gate increases, the size of the transistors must increase. The increased area of the transistors means that the effective input capacitance increases so that switching times of cascaded logic gates will increase.

EXERCISE PROBLEM

**Ex 16.10:**  Determine the transistor sizes of a 3-input CMOS NOR logic gate. Symmetrical switching times are desired and the switching times should correspond to the basic CMOS inverter. (Ans. $W_p = 6W$, $W_n = W$)

### 16.4.3   Complex CMOS Logic Circuits

Just as with NMOS logic designs, we can form complex logic gates in CMOS, which avoids connecting large numbers of NOR, NAND, and inverter gates to implement the logic function. There are formal methods that can be used to implement the logic circuit. However, we can use the knowledge gained in the analysis and design of the NOR and NAND circuits.

DESIGN EXAMPLE **16.11**

**Objective:**  Design a CMOS logic circuit to implement a particular logic function.
   Implement the logic function $Y = AB + C(D + E)$ in a CMOS design. The signals $A$, $B$, $C$, $D$, and $E$ are available.

**Design Approach:**  The general CMOS design is shown in Figure 16.38, in which the inputs are applied to both the PMOS and NMOS networks. We may start the

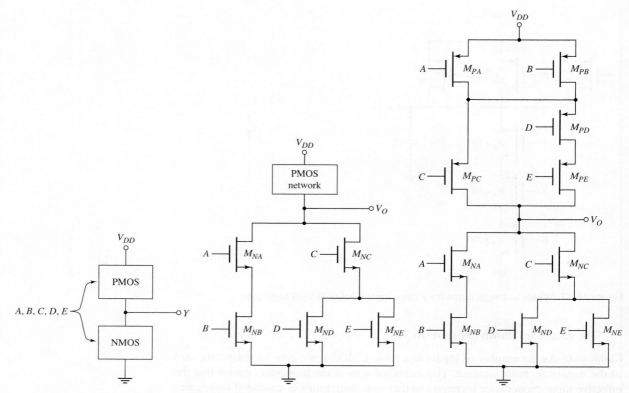

Figure 16.38 General CMOS design

Figure 16.39 NMOS design for Example 16.11

Figure 16.40 Complete CMOS design for Example 16.11

design by considering the NMOS portion of the circuit. To implement a basic OR (NOR) function, the n-channel transistors are in parallel (Figure 16.33) and to implement a basic AND (NAND) function, the n-channel transistors are in series (Figure 16.34). We will consider whether the function or its complement is generated at the end of the design.

**Solution (NMOS Design):** In the overall function, we note the logic OR between the functions $AB$ and $C(D + E)$, so that the NMOS devices used to implement $AB$ will be in parallel with the NMOS devices used to implement $C(D + E)$. There is a logic AND between the inputs $A$ and $B$, so that the NMOS devices with these inputs will be in series. Finally, the NMOS devices with the $D$ and $E$ inputs will be in parallel and this combination will be in series with the NMOS device with the $C$ input. The NMOS implementation of the function is shown in Figure 16.39.

**Solution (PMOS Design):** The arrangement of the PMOS devices is complementary to that of the NMOS devices. PMOS devices that perform the basic OR function are in series and PMOS devices that perform the basic AND function are in parallel. We then see that the PMOS devices used to implement $AB$ will be in series with the devices used to implement $C(D + E)$. The two PMOS devices with the $A$ and $B$ inputs will be in parallel. The two PMOS devices with the $D$ and $E$ inputs will be in series and in turn will be in parallel with the PMOS device with the $C$ input. The completed circuit is shown in Figure 16.40.

**Final Solution:** By considering various inputs, we may note that the output signal of the circuit shown in Figure 16.40 is actually the complement of the desired signal. We may then simply add a CMOS inverter to the output to obtain the desired function.

**Comment:** As mentioned, there are formal ways in which to design circuits. However, in many cases, these circuits can be designed by using the knowledge and intuition gained from previous work. The width-to-length ratios of the various transistors can be determined as we have done in previous examples.

**Ex 16.11:** Design the width-to-length ratios of the transistors in the static CMOS logic circuit of Figure 16.40. Symmetrical switching times are desired and the switching times should correspond to the basic CMOS inverter. (Ans. All NMOS devices, $W_n = 2W$; $W_p(M_{PA}) = W_p(M_{PB}) = W_p(M_{PC}) = 4W$; $W_p(M_{PD}) = W_p(M_{PE}) = 8W$)

Another example of a CMOS logic gate is the exclusive-OR or XOR. The logic function can be written as

$$F_{XOR} = \bar{A}B + A\bar{B} \tag{16.70}$$

We have noticed that the output of the CMOS gates is actually the complement of the input signal. We can therefore write

$$\bar{F}_{XOR} = F_{XNOR} = \bar{A}\bar{B} + AB \tag{16.71}$$

Assuming that input signals $A$, $B$, $\bar{A}$, and $\bar{B}$ are available, Figure 16.41 shows a CMOS static implementation of the logic function.

We may note that $\bar{A}\bar{B}$ as well as $AB$ means two NMOS devices in series and two PMOS devices in parallel. The OR function means the combination of NMOS

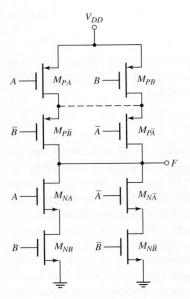

**Figure 16.41** A CMOS static exclusive-OR logic gate

devices is in parallel and the combination of PMOS devices is in series. This design is shown in the figure. In considering the truth table for the exclusive-OR function, we may note that the output of the circuit in Figure 16.41 is indeed the exclusive-OR function. In the design of CMOS logic gates, then, we should actually design the complement of the desired function.

In the PMOS portion of the design, there should be an electrical connection between the drains of $M_{PA}$ and $M_{PB}$. This connection is shown as a dotted line, but is not actually required. The only pull-up conditions are for $A = \bar{B} = 0$ and for $\bar{A} = B = 0$, which are achieved without this connection.

### 16.4.4    Fanout and Propagation Delay Time

**Fanout**

The term *fanout* refers to the number of load gates of similar design connected to the output of a driver gate. The maximum fanout is the maximum number of load gates that may be connected to the output. Since the CMOS logic gate will be driving other CMOS logic gates, the quiescent current required to drive the other CMOS gates is essentially zero. In terms of static characteristics, the maximum fanout is virtually limitless.

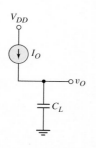

**Figure 16.42** Constant-current source charging a load capacitor

However, each additional load gate increases the load capacitance that must be charged and discharged as the driver gate changes state, and this places a practical limit on the maximum allowable number of load gates. Figure 16.42 shows a constant current charging a load capacitance. The voltage across the capacitance is

$$v_O = \frac{1}{C_L} \int I_O \, dt = \frac{I_O t}{C_L} \qquad (16.72)$$

The load capacitance $C_L$ is proportional to the number $N$ of load gates and to the input gate capacitance of each load. The current $I_O$ is proportional to the conduction parameter of the driver transistor. The switching time is therefore

$$t \propto \frac{N(W \cdot L)_L}{\left(\dfrac{W}{L}\right)_D} \qquad (16.73)$$

where the gate capacitance is directly proportional to the gate area of the load $(W \cdot L)_L$, and the conduction parameter of the driver transistor is proportional to the width-to-length ratio. Equation (16.73) can be rewritten as

$$t \propto N(L_L L_D)\left(\frac{W_L}{W_D}\right) \qquad (16.74)$$

The propagation delay time, which is proportional to the switching time, increases as the fanout increases. The propagation delay time could be reduced by increasing the size of the driver transistor. However, in any given driver logic circuit and load logic circuit, the sizes of the devices are generally fixed. Consequently, the maximum fanout is limited by the maximum acceptable propagation delay time.

Propagation delay times are typically measured with a specified load capacitance. The average propagation delay time of a two-input CMOS NOR gate (such as an SN74HC36) is 25 ns, measured with a load capacitance of $C_L = 50$ pF. Since the input capacitance is $C_I = 10$ pF, a fanout of five would produce a 50 pF load capacitance. A fanout larger than five would increase the load capacitance, and would also increase the propagation delay time above the specified value.

## Propagation Delay Time

Although the propagation delay time of the CMOS inverter can be determined by analytical techniques, it can also be determined by computer simulation. This is especially true when more complex CMOS logic circuits are considered. Using the appropriate transistor models in the simulation, the transient response can be produced. Obtaining an accurate transient response depends on using the correct transistor parameters. Some computer simulation problems in the end-of-chapter problems deal with propagation delay times. However, we will not go into detail here.

## Test Your Understanding

**TYU 16.9** Design a static CMOS logic circuit that implements the logic function $Y = \overline{(ABC + DE)}$. (Ans. NMOS design: $A$, $B$, $C$ inputs to three NMOS devices in series and $D$, $E$ inputs to two NMOS devices in series; then, three NMOS and two NMOS in parallel)

**TYU 16.10** Design the width-to-length ratios of the transistors in the static CMOS exclusive-OR logic gate in Figure 16.41. Symmetrical switching times are desired and the switching times should correspond to the basic CMOS inverter. (Ans. All NMOS, $W_n = 2W$; all PMOS, $W_p = 4W$)

##  16.5    CLOCKED CMOS LOGIC CIRCUITS

**Objective:** • Analyze and design clocked CMOS logic gates.

The CMOS logic circuits considered in the previous section are called static circuits. One characteristic of a static CMOS logic circuit is that the output node always has a low-resistance path to either ground or $V_{DD}$. This implies that the output voltage is well defined and is never left floating.

Static CMOS logic circuits can be redesigned with an added clock signal while at the same time eliminating many of the PMOS devices. In general, the PMOS devices must be larger than NMOS devices. Eliminating as many PMOS devices as possible reduces the required chip area as well as the input capacitance. The low-power dissipation of the CMOS technology, however, is maintained.

Clocked CMOS circuits are dynamic circuits that generally precharge the output node to a particular level when the clock is at a logic 0. Consider the circuit in Figure 16.43. When the clock signal is low, or CLK = logic 0, $M_{N1}$ is cut off and the current in the circuit is zero. Transistor $M_{P1}$ is in a conducting state, but since the current is zero, then $v_{O1}$ charges to $V_{DD}$. A high input to the CMOS inverter means that $v_O = 0$. During this phase of the clock signal, the gate of $M_{P2}$ is precharged.

During the next phase, when the clock signal goes high, or CLK = logic 1, transistor $M_{P1}$ cuts off and $M_{N1}$ is biased in a conducting state. If input $A$ = logic 0, then $M_{NA}$ is cut off and there is no discharge path for voltage $v_{O1}$; therefore, $v_{O1}$ remains charged at $v_{O1} = V_{DD}$. However, if CLK = logic 1 and $A$ = logic 1, then both $M_{N1}$ and $M_{NA}$ are biased in a conducting state, providing a discharge path for voltage $v_{O1}$. As $v_{O1}$ is pulled low, output signal $v_O$ goes high.

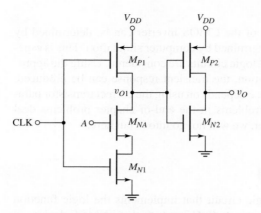

Figure 16.43  Simple clocked CMOS logic circuit

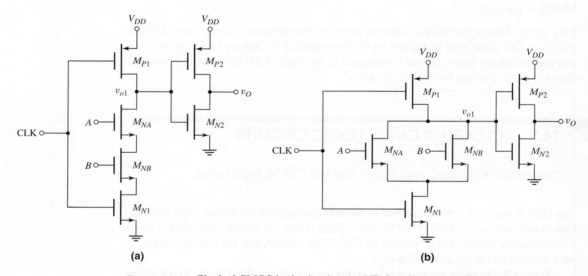

**(a)**                                                        **(b)**

Figure 16.44  Clocked CMOS logic circuit: (a) AND function and (b) OR function

The quiescent power dissipation in this circuit is essentially zero, as it was in the standard CMOS circuits. A small amount of power is required to precharge output $v_{O1}$, if it had been pulled low during the previous half clock cycle.

The single NMOS transistor $M_{NA}$ in Figure 16.43 can be replaced by a more complex NMOS logic circuit. Consider the two circuits in Figure 16.44. When CLK = logic 0, then $M_{N1}$ cuts off and $M_{P1}$ is in its conducting state in both circuits; then, $v_{O1}$ is charged to $v_{O1} = V_{DD}$ and $v_O = 0$. For the circuit in Figure 16.44(a), when CLK = logic 1, voltage $v_{O1}$ is discharged to ground or pulled low only when $A = B = $ logic 1. In this case, $v_O$ goes high. The circuit in Figure 16.44(a) performs the AND function. Similarly, the circuit in Figure 16.44(b) performs the OR function.

The advantage of the precharge technique is that it avoids the use of extensive pull-up networks: Only one PMOS and one NMOS transistor are required. This leads to an almost 50 percent savings in silicon area for larger circuits, and a reduction in capacitance resulting in higher speed. In addition, the static or quiescent power dissipation is essentially zero, so the circuit maintains the characteristics of CMOS circuits.

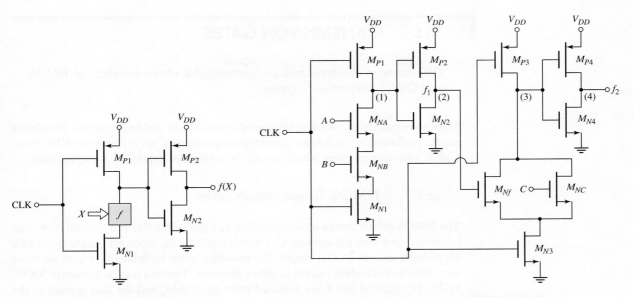

**Figure 16.45** Generalized CMOS clocked logic circuit

**Figure 16.46** Cascaded clocked or domino CMOS logic circuit

The AND and OR logic transistors $M_{NA}$ and $M_{NB}$ in Figures 16.44(a) and 16.44(b) can be replaced by a generalized logic network as indicated in Figure 16.45. The box marked $f$ is an NMOS pull-down network that performs a particular logic function $f(X)$ of $n$ variables, where $X = (x_1, x_2, \ldots, x_n)$. The NMOS circuit is a combination of series–parallel interconnections of $n$ transistors. When the clock signal goes high, the CMOS inverter output is the logic function $f(X)$.

The set of $X$ inputs to the logic circuits $f$ is derived from the outputs of other CMOS inverters and clocked logic circuits. The means that when CLK = logic 0, the outputs of all CMOS inverters are a logic 0 during the precharge cycle. As a result, all $n$ variables $X = (x_1, x_2, \ldots, x_n)$ are a logic 0 during the precharge cycle. During this time, all NMOS transistors are cut off, which guarantees that output $v_{O1}$ can be precharged to $V_{DD}$. There can then be only one possible transition at each node during the evaluation phase. The output of the CMOS buffer may change from a 0 to a 1.

An example of a cascaded domino CMOS circuit is shown in Figure 16.46. During the precharge cycle, in which CLK = logic 0, nodes 1 and 3 are charged high and nodes 2 and 4 are low. Also during this time, the inputs $A$, $B$, and $C$ are all a logic 0. During the evaluation phase, in which CLK = logic 1, if $A = C = $ logic 1 and $B = $ logic 0, then node 1 remains charged high, $f_1 = $ logic 0, and node 3 discharges through $M_{NC}$ causing $f_2$ to go high. However, if, during the evaluation phase, $A = B = $ logic 1 and $C = $ logic 0, then node 1 is pulled low causing $f_1$ to go high, which in turn causes node 3 to go low and forces node 4 high. This chain of actions thus leads to the term **domino circuit.**

## Test Your Understanding

**TYU 16.11** Design a clocked CMOS domino logic circuit, such as shown in Figure 16.45, to generate an output $f(X) = A \cdot B \cdot C + D \cdot E$.

**TYU 16.12** Sketch a clocked CMOS logic circuit that realizes the exclusive OR function.

## 16.6    TRANSMISSION GATES

**Objective:** • Analyze and understand the characteristics of NMOS and CMOS transmission gates.

Transistors can act as switches between driving circuits and load circuits. Transistors used to perform this function are called transmission gates. We will examine NMOS and CMOS transmission gates, which can also be configured to perform logic functions.

### 16.6.1    NMOS Transmission Gate

The NMOS enhancement-mode transistor in Figure 16.47(a) is a transmission gate connected to a load capacitance $C_L$, which could be the input gate capacitance of a MOS logic circuit. In this circuit, the transistor must be bilateral, which means it must be able to conduct current in either direction. This is a natural feature of MOS-FETs. Terminals $a$ and $b$ are assumed to be equivalent, and the bias applied to the transistor determines which terminal acts as the drain and which terminal acts as the source. The substrate must be connected to the most negative potential in the circuit, which is usually ground. Figure 16.47(b) shows a simplified circuit symbol for the **NMOS transmission gate** that is used extensively.

We assume that the NMOS transmission gate is to operate over a voltage range of zero-to-$V_{DD}$. If the gate voltage $\phi$ is zero, then the n-channel transistor is cut off and the output is isolated from the input. The transistor is essentially an open switch.

If $\phi = V_{DD}$, $v_I = V_{DD}$, and $v_O$ is initially zero, then terminal $a$ acts as the drain since its bias is $V_{DD}$, and terminal $b$ acts as the source since its bias is zero. Current enters the drain from the input, charging up the capacitor. The gate-to-source voltage is

$$v_{GS} = \phi - v_O = V_{DD} - v_O \tag{16.75}$$

As the capacitor charges and $v_O$ increases, the gate-to-source voltage decreases. The capacitor stops charging when the current goes to zero. This occurs when the gate-to-source voltage $v_{GS}$ becomes equal to the threshold voltage $V_{TN}$. The maximum output voltage occurs when $v_{GS} = V_{TN}$, therefore, from Equation (16.75), we have

$$v_{GS}(\text{min}) = V_{TN} = V_{DD} - v_O(\text{max}) \tag{16.76(a)}$$

or

$$v_O(\text{max}) = V_{DD} - V_{TN} \tag{16.76(b)}$$

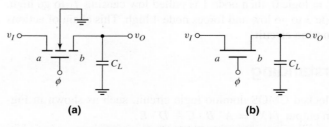

**(a)**                          **(b)**

**Figure 16.47** (a) NMOS transmission gate, showing substrate connection, and (b) simplified diagram

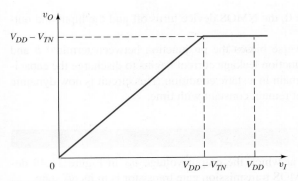

**Figure 16.48** Output voltage versus input voltage characteristics of the NMOS transmission gate

where $V_{TN}$ is the threshold voltage taking into account the body effect.

Equation (16.76(b)) demonstrates one disadvantage of an NMOS transmission gate. A logic 1 level degrades, or attenuates, as it passes through the transmission gate. However, this may not be a serious problem for many applications.

Figure 16.48 shows the quasi-static output voltage versus input voltage of the NMOS transmission gate. As seen in the figure, when $v_I = V_{DD}$, the output voltage is $v_O = V_{DD} - V_{TN}$ as we have discussed. For input voltages in the range $v_I < V_{DD} - V_{TN}$, the figure demonstrates that $v_O = v_I$. In this range of input voltages, the gate-to-source voltage is still greater than the threshold voltage. However, in steady-state, the current must be zero through the capacitor. In this case, the current becomes zero when the drain-to-source voltage is zero, or when $v_O = v_I$.

Now consider the situation in which $\phi = V_{DD}$, $v_I = 0$, and $v_O = V_{DD} - V_{TN}$ initially. Terminal $b$ then acts as the drain and terminal $a$ acts as the source. The gate-to-source voltage is

$$v_{GS} = \phi - v_I = V_{DD} - 0 = V_{DD} \tag{16.77}$$

The value of $v_{GS}$ is a constant, and the capacitor discharges as current enters the NMOS transistor drain. The capacitor stops discharging when the current goes to zero. Since $v_{GS}$ is a constant at $V_{DD}$, the drain current goes to zero when the drain-to-source voltage is zero, which means that the capacitor completely discharges to zero. This implies that a logic 0 is transmitted unattenuated through the NMOS transmission gate.

Using an NMOS transmission gate in a MOS circuit may introduce a dynamic condition. Figure 16.49 shows a cross section of the NMOS transistor in the transmission gate configuration. If $v_I = \phi = V_{DD}$, then the load capacitor charges to

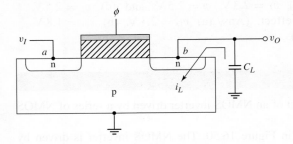

**Figure 16.49** NMOS transmission gate with cross section of NMOS transistor

$v_O = V_{DD} - V_{TN}$. When $\phi = 0$, the NMOS device turns off and the input and output become isolated.

The capacitor voltage reverse biases the pn junction between terminal $b$ and ground. A reverse-biased pn junction leakage current begins to discharge the capacitor, and the circuit does not remain in a static condition. This circuit is now dynamic in that the high output does not remain constant with time.

## EXAMPLE 16.12

**Objective:** Estimate the rate at which the output voltage $v_O$ in Figure 16.49 decreases with time when the NMOS transmission gate transistor is in its off state.

Assume the capacitor is initially charged to $v_O = 2.9$ V. Let $C_L = 0.2$ pF and assume the reverse-biased pn junction leakage current is a constant at $i_L = 100$ pA.

**Solution:** The voltage across the capacitor can be written as

$$v_O = -\frac{1}{C_L}\int i_L \, dt = -\frac{i_L}{C_L}t + K_1$$

where $K_1 = v_O(t=0) = 2.9$ V is the initial condition. Therefore,

$$v_O(t) = 2.9 - \frac{i_L}{C_L}t$$

The rate at which the output voltage decreases is

$$\frac{dv_O}{dt} = -\frac{i_L}{C_L} = -\frac{100 \times 10^{-12}}{0.2 \times 10^{-12}} = -500 \text{ V/s} \Rightarrow -0.5 \text{ V/ms}$$

Therefore, in this example, the capacitor would completely discharge in 5.8 ms.

**Comment:** Even though the NMOS transmission gate may introduce a dynamic condition into a circuit, this gate is still useful in clocked logic circuits in which a clock signal is periodically applied to the NMOS transistor gate. If, for example, the clock frequency is 25 kHz, the clock pulse period is $40\,\mu$s, which means that the output voltage would decay by only approximately 0.7 percent during a clock period.

### EXERCISE PROBLEM

**Ex 16.12:** The threshold voltage of the NMOS transmission gate transistor in Figure 16.47(a) is $V_{TN} = 0.4$ V. Determine the output voltage $v_O$ for: (a) $v_I = \phi = 2.5$ V; (b) $v_I = 1.8$ V, $\phi = 2.5$ V; (c) $v_I = 2.3$ V, $\phi = 2.5$ V; and (d) $v_I = 2.5$ V, $\phi = 1.5$ V. Neglect the body effect. (Ans. (a) $v_O = 2.1$ V, (b) $v_O = 1.8$ V, (c) $v_O = 2.1$ V, (d) $v_O = 1.1$ V)

## EXAMPLE 16.13

**Objective:** Determine the output of an NMOS inverter driven by a series of NMOS transmission gates.

Consider the circuit shown in Figure 16.50. The NMOS inverter is driven by three NMOS transmission gates in series. Assume the threshold voltages of the NMOS transmission gate transistors and the NMOS driver transistor are

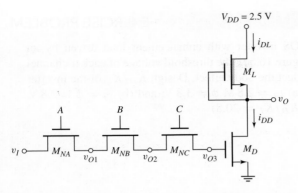

**Figure 16.50** NMOS inverter driven by three NMOS transmission gates
in series

$V_{TN} = 0.4$ V, and the threshold voltage of the load transistor is $V_{TNL} = -0.6$ V. Let
$K_D/K_L = 3$ for the inverter. Determine $v_O$ for $v_I = 0$ and $v_I = 2.5$ V

**Solution:** The three NMOS transmission gates in series act as an AND/NAND function. If $v_I = 0$ and $A = B = C = $ logic $1 = 2.5$ V, the gate capacitance to driver $M_D$
becomes completely discharged, which means that $v_{O1} = v_{O2} = v_{O3} = 0$. Driver
$M_D$ is cut off and $v_O = 2.5$ V.

If $v_I = 2.5$ V and $A = B = C = $ logic $1 = 2.5$ V, the three transmission gates
are biased in their conducting state, and the gate capacitance of $M_D$ becomes
charged. For transistor $M_{NA}$, the current becomes zero when the gate-to-source voltage is equal to the threshold voltage, or, from Equation (16.76(b)),

$$v_{O1} = V_{DD} - V_{TN} = 2.5 - 0.4 = 2.1 \text{ V}$$

Transistors $M_{NB}$ and $M_{NC}$ also cut off when the gate-to-source voltages are equal to
the threshold voltage; therefore

$$v_{O2} = v_{O3} = V_{DD} - V_{TN} = 2.5 - 0.4 = 2.1 \text{ V}$$

This result shows that the drain-to-source voltages of $M_{NB}$ and $M_{NC}$ are also
zero. *A threshold voltage drop is lost in the first transmission gate, but additional
threshold voltage drops are not lost in subsequent NMOS transmission gates in
series.*

For a voltage of $v_{O3} = 2.1$ V applied to the gate of $M_D$, the driver is biased in
the nonsaturation region and the load is biased in the saturation region. From
$i_{DD} = i_{DL}$, we have

$$K_D\left[2(v_{O3} - V_{TN})v_O - v_O^2\right] = K_L[-V_{TNL}]^2$$

The output voltage is found to be $v_O = 35.7$ mV.

If any one of the transmission gate voltages, $A$ or $B$ or $C$, switches to a logic 0, then
$v_{O3}$ will begin to discharge through a reverse-biased pn junction in the transmission
gates, which means that $v_O$ will increase with time.

**Comment:** In this example, the inverter is again in a dynamic condition; that is,
when any transmission gate is cut off, the output voltage changes with time. However, this type of circuit can be used in clocked digital systems.

EXERCISE PROBLEM

**Ex 16.13:** Consider the NMOS inverter with enhancement load driven by an NMOS transmission gate in Figure 16.51. The threshold voltage of each n-channel transistor is $V_{TN} = 0.5$ V. Neglect the body effect. Design $K_D/K_L$ of the inverter such that $v_O = 0.1$ V when: (a) $v_I = 2.8$ V, $\phi = 3.3$ V; and (b) $v_I = \phi = 2.8$ V. (Ans. (a) $K_D/K_L = 16.2$, (b) $K_D/K_L = 20.8$)

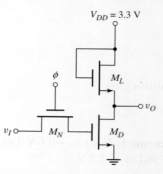

$V_{DD} = 3.3$ V

**Figure 16.51** Figure for Exercise Ex16.13

16.6.2    **NMOS Pass Networks**

As integrated circuit technology advances, one emphasis is on increased circuit density. The maximum number of circuit functions per unit area is determined either by power dissipation density or by the area occupied by transistors and related devices.

One form of NMOS circuit logic that minimizes power dissipation and maximizes device density is called **pass transistor logic.** Pass transistor circuits use minimum-sized transistors, providing high density and high operating speed. The average power dissipation is due only to the switching power consumed by the driver circuits in charging and discharging the pass transistor control gates and driving the pass network inputs.

In this section, we present a few examples of NMOS pass transistor logic circuits. Consider the circuit in Figure 16.52. To determine the output response, we examine the conditions listed in Table 16.1 for the possible states of the input signals $A$ and $B$. We assume that a logic 1 level is $V_{DD}$ volts. In states 1 and 2, transmission gate $M_{N2}$ is biased in its conducting state. For state 1, $\bar{A} = $ logic 1 is transmitted to the output so $f = $ logic $1'$, where the logic $1'$ level is $(V_{DD} - V_{TN})$. The logic 1 level is attenuated by one threshold voltage drop. For state 2, $A = $ logic 0 is transmitted

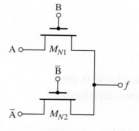

**Figure 16.52** Simple NMOS pass logic network

| Table 16.1 | | Input and output states for the circuit in Figure 16.52 | | | | | |
|---|---|---|---|---|---|---|---|
| State | $A$ | $B$ | $\bar{A}$ | $\bar{B}$ | $M_{N1}$ | $M_{N2}$ | $f$ |
| 1 | 0 | 0 | 1 | 1 | off | on | $1'$ |
| 2 | 1 | 0 | 0 | 1 | off | on | 0 |
| 3 | 0 | 1 | 1 | 0 | on | off | 0 |
| 4 | 1 | 1 | 0 | 0 | on | off | $1'$ |

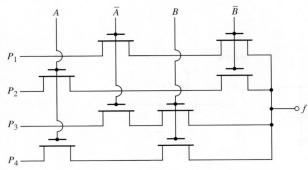

$P_1$
$P_2$
$P_3$
$P_4$

$f$

**Figure 16.53** NMOS pass logic network example

| Table 16.2 | | Input and output states for the circuit in Figure 16.53 | | | |
|---|---|---|---|---|---|
| State | $A$ | $B$ | $\bar{A}$ | $\bar{B}$ | $f$ |
| 1 | 0 | 0 | 1 | 1 | $P_1$ |
| 2 | 1 | 0 | 0 | 1 | $P_2$ |
| 3 | 0 | 1 | 1 | 0 | $P_3$ |
| 4 | 1 | 1 | 0 | 0 | $P_4$ |

unattenuated to the output. In states 3 and 4, transmission gate $M_{N1}$ is biased in its conducting state. The $A$ = logic 0 for state 3 is transmitted unattenuated to the output, and $A$ = logic 1 for state 4 is attenuated during transmission; therefore, $f$ = logic 1'. The output is thus the exclusive-NOR function.

Another example of an NMOS pass transistor logic circuit is shown in Figure 16.53. The output response as a function of the input gate controls $A$ and $B$ is shown in Table 16.2. This circuit is a multiplexer; that is, for a specific set of gate controls, the input signals $P_i$ are individually passed to the output. By using both normal and inverted forms of $A$ and $B$, four inputs can be controlled with just two variables.

A potential problem of NMOS pass transistor logic is that the output may be left floating in a high impedance state and charged high. Consider the circuit shown in Figure 16.54. If, for example, $\bar{B} = C$ = logic 0 and $A$ = logic 1, then $f$ = logic 1', which is the logic 1 level attenuated by $V_{TN}$. When $A$ is switched to logic 0, the output should be low, but there may not be a discharge path to ground, and the output may retain the logic 1' stored at the output capacitance.

The NMOS pass network must be designed to avoid a high impedance output by passing a logic 0 whenever a 0 is required at the output. A logic network that performs the logic function $f = A + \bar{B} \cdot C$, as indicated in Figure 16.54, is shown in Figure 16.55. The complementary function $\bar{f} = \bar{A} \cdot (B + \bar{C})$ attached at the output node drives the output to a logic 0 whenever $f = 0$.

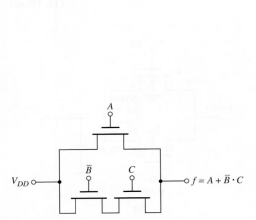

**Figure 16.54** NMOS pass logic network with a potential problem

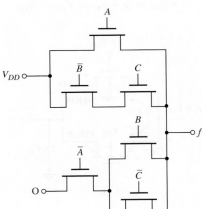

**Figure 16.55** NMOS pass logic network with complementary function in parallel

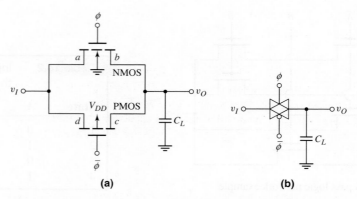

**Figure 16.56** (a) CMOS transmission gate and (b) simplified circuit symbol

### 16.6.3   CMOS Transmission Gate

A **CMOS transmission gate** is shown in Figure 16.56(a). The parallel combination of NMOS and PMOS transistors, with complementary gate signals, allows the input signal to be transmitted to the output without the threshold voltage attenuation. Both transistors must be bilateral; therefore, the NMOS substrate is connected to the most negative potential in the circuit and the PMOS substrate is connected to the most positive potential (usually, ground and $V_{DD}$, respectively). Figure 16.56(b) shows a frequently used simplified circuit symbol for the CMOS transmission gate.

We again assume that the transmission gate is to operate over a voltage range of zero-to-$V_{DD}$. If the control voltages are $\phi = 0$ and $\bar{\phi} = V_{DD}$, then both the NMOS and PMOS transistors are cut off and the output is isolated from the input. In this state, the circuit is essentially an open switch.

If $\phi = V_{DD}$, $\bar{\phi} = 0$, $v_I = V_{DD}$, and $v_O$ is initially zero, then for the NMOS device, terminal $a$ acts as the drain and terminal $b$ acts as the source, whereas for the PMOS device, terminal $c$ acts as the drain and terminal $d$ acts as the source. Current enters the NMOS drain and the PMOS source, as shown in Figure 16.57(a), to charge the load capacitor. The NMOS gate-to-source voltage is

$$v_{GSN} = \phi - v_O = V_{DD} - v_O \qquad\qquad \textbf{(16.78(a))}$$

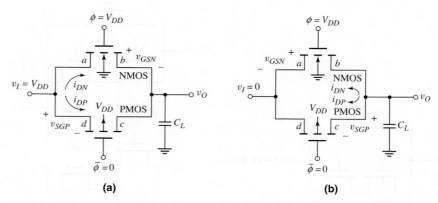

**Figure 16.57** Currents and gate–source voltages in CMOS transmission gate for: (a) input high condition and (b) input low condition

and the PMOS source-to-gate voltage is

$$v_{SGP} = v_I - \bar{\phi} = V_{DD} - 0 = V_{DD} \tag{16.78(b)}$$

As with the NMOS transmission gate, when $v_O = V_{DD} - V_{TN}$, the NMOS transistor cuts off and $i_{DN} = 0$ since $V_{GSN} = V_{TN}$. However, since the source-to-gate voltage of the PMOS device is a constant at $v_{SGP} = V_{DD}$, the PMOS transistor continues to conduct. The drain current $i_{DP}$ goes to zero when the PMOS source-to-drain voltage goes to zero, or $v_{SDP} = 0$. This means that the load capacitor $C_L$ continues to charge through the PMOS device until the output and input voltages are equal, or in this case, $v_O = v_I = V_{DD}$.

Consider what happens if $\phi = V_{DD}, \bar{\phi} = 0, v_I = 0$, and $v_O = V_{DD}$ initially. For the NMOS device, terminal $a$ acts as the source and terminal $b$ acts as the drain, whereas for the PMOS device, terminal $c$ acts as the source and terminal $d$ acts as the drain. Current enters the NMOS drain and the PMOS source, as shown in Figure 16.57(b), to discharge the capacitor. The NMOS gate-to-source voltage is

$$v_{GSN} = \phi - v_I = V_{DD} - 0 = V_{DD} \tag{16.79(a)}$$

and the PMOS source-to-gate voltage is

$$v_{SGP} = v_O - \bar{\phi} = v_O - 0 = v_O \tag{16.79(b)}$$

When $v_{SGP} = v_O = |V_{TP}|$, the PMOS device cuts off and $i_{DP}$ goes to zero. However, since $v_{GSN} = V_{DD}$, the NMOS transistor continues conducting and capacitor $C_L$ completely discharges to zero.

Using a CMOS transmission gate in a MOS circuit may introduce a dynamic condition. Figure 16.58 shows the CMOS transmission gate with simplified cross sections of the NMOS and PMOS transistors. If $\phi = 0$ and $\bar{\phi} = V_{DD}$, then the input and output are isolated. If $v_O = V_{DD}$, then the NMOS substrate-to-terminal $b$ pn junction is reverse biased and capacitance $C_L$ can discharge, as it did in the NMOS transmission gate. If, however, $v_O = 0$, then the PMOS terminal $c$-to-substrate pn junction is reverse biased and capacitance $C_L$ can charge to a positive voltage. This circuit is therefore dynamic in that the output high or low conditions do not remain constant with time.

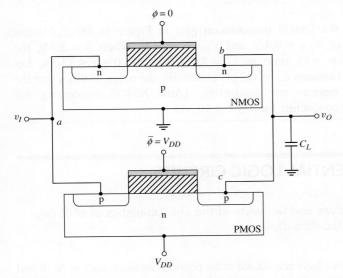

**Figure 16.58** CMOS transmission gate showing cross sections of NMOS and PMOS transistors

16.6.4   **CMOS Pass Networks**

CMOS transmission gates may also be used in pass network logic design. CMOS pass networks use NMOS transistors to pass 0's, PMOS transistors to pass 1's, and CMOS transmission gates to pass a variable to the output. An example is shown in Figure 16.59. One PMOS transistor is used to transmit a logic 1, while transmission gates are used to transmit a variable that may be either a logic 1 or a logic 0. We can show that for any combination of signals, a logic 1 or logic 0 is definitely passed to the output.

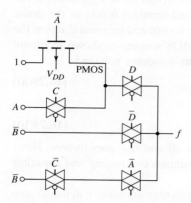

**Figure 16.59** CMOS pass logic network

## Test Your Understanding

**TYU 16.13** Design an NMOS pass network to perform the logic function $f = A(B + C)$.

**TYU 16.14** Consider the CMOS transmission gate in Figure 16.56(a). Assume transistor parameters of $V_{TN} = 0.4$ V and $V_{TP} = -0.4$ V. When $\phi = 2.5$ V, the input voltage $v_I$ varies with time as $v_I = 2.5 - 0.2t$ for $0 \le t \le 12.5$ s. Let $v_O(t = 0) = 2.5$ V and assume $C_L = 0.2$ pF. Determine the range of times that the NMOS and PMOS devices are conducting. (Ans. NMOS conducting for $2 \le t \le 12.5$ s; PMOS conducting for $0 \le t \le 10.5$ s).

## 16.7   SEQUENTIAL LOGIC CIRCUITS

**Objective:** • Analyze and understand the characteristics of shift registers and various flip-flop designs.

In the logic circuits that we have considered in the previous sections, such as NOR and NAND logic gates, the output is determined only by the instantaneous values of the input signals. These circuits are therefore classified as combinational logic circuits.

Another class of circuits is called **sequential logic circuits.** The output depends not only on the inputs, but also on the previous history of its inputs. This feature gives sequential circuits the property of memory. Shift registers and flip-flops are typical examples of such circuits. We will also briefly consider a full-adder circuit. The characteristic of these circuits is that they store information for a short time until the information is transferred to another part of the system.

In this section, we introduce a basic shift register and the basic concept of a flip-flop. These circuits can become very complex and are usually described with logic diagrams. We will also introduce a CMOS full adder circuit in terms of its logic diagram and then provide the transistor implementation of this logic function. Additional information can be found in more advanced texts.

| 16.7.1 | **Dynamic Shift Registers** |
|---|---|

A **shift register** can be formed from transmission gates and inverters. Figure 16.60 shows a combination of NMOS transmission gates and NMOS depletion-load inverters. The clock signals applied to the gates of the NMOS transmission gates must be complementary, nonoverlapping pulses. The effective capacitances at the gates of $M_{D1}$ and $M_{D2}$ are indicated by the dotted connections to $C_{L1}$ and $C_{L2}$.

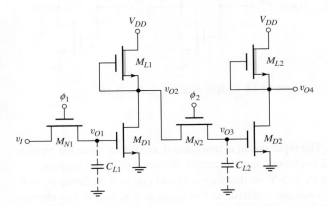

**Figure 16.60** Dynamic shift register with NMOS inverters and transmission gates

If, for example, $C_{L1}$ is initially uncharged when $v_{O1} = 0$ and if $v_I = V_{DD}$ when $\phi_1 = V_{DD}$, then a logic $1' = V_{DD} - V_{TN}$ voltage should exist at $v_{O1}$ at the end of clock pulse $\phi_1$. The capacitance of $C_{L1}$ charges through $M_{N1}$ and the driving circuit of $v_I$. The effective $RC$ time constant must be sufficiently small to achieve this charging effect. As $v_{O1}$ goes high, $v_{O2}$ goes low, but the low is not transmitted through $M_{N2}$ as long as $\phi_2$ remains low.

Figure 16.61 is used to determine the operation of this circuit and the voltages at various times. For simplicity, we assume that $V_{DD} = 5$ V and $V_{TN} = 1$ V for the NMOS drivers and transmission gate transistors.

At $t = t_1$, $v_I = \phi_1 = 5$ V, $v_{O1}$ charges to $v_{DD} - V_{TN} = 4$ V, and $v_{O2}$ goes low. At this time, $M_{N2}$ is still cut off, which means that the values of $v_{O3}$ and $v_{O4}$ depend on the previous history. At $t = t_2$, $\phi_1$ is zero, $M_{N1}$ is cut off, but $v_{O1}$ remains charged. At $t = t_3$, $\phi_2$ is high, and the logic 0 at $v_{O2}$ is transmitted to $v_{O3}$, which forces $v_{O4}$ to 5 V. The input signal $v_I = 5$ V at $t = t_1$ has thus been transmitted to the output; therefore,

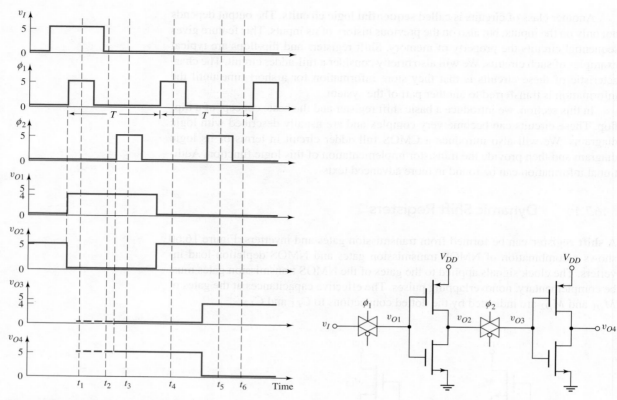

**Figure 16.61** NMOS shift register voltages at various times

**Figure 16.62** CMOS dynamic shift register

$v_{O4} = v_I = 5$ V at $t = t_3$. The input signal is transmitted, or *shifted,* from the input to the output during one clock cycle, making this circuit one stage of a shift register.

At $t = t_4$, $v_I = 0$, and $\phi_1 = 5$ V, so that $v_{O1} = 0$ and $v_{O2} = 5$ V. Since $\phi_2 = 0$, $M_{N2}$ is cut off, and $v_{O2}$ and $v_{O3}$ are isolated. At $t = t_5$, $\phi_2 = 5$ V, so that $v_{O3}$ charges to $V_{DD} - V_{TN} = 4$ V, and $v_{O4}$ goes low (logic 0). At $t = t_6$, both NMOS transmission gates are cut off, and the two inverters remain in their previous states. It is important that $\phi_1$ and $\phi_2$ do not overlap, or the signal would propagate through the whole chain at once and we would no longer have a shift register.

In the dynamic condition of NMOS transmission gates, the high output voltage across the output capacitance does not remain constant with time; it discharges through the transmission gate transistor. This same effect applies to the shift register in Figure 16.60. For example, from Figure 16.61, at $t = t_2$, $v_{O1} = 4$ V, $\phi_1 = 0$, and $M_{N1}$ is cut off. Voltage $v_{O1}$ will start to decay and $v_{O2}$ will begin to increase. To prevent logic errors from being introduced into the system, the clock signal period $T$ must be small compared to the effective $RC$ discharge time constant. The circuit in Figure 16.60 is therefore called a **dynamic shift register.**

A dynamic shift register formed in a CMOS technology is shown in Figure 16.62. Operation of this circuit is very similar to that of the dynamic NMOS shift register, except for the voltage levels. For example, when $v_I = \phi_1 = V_{DD}$, then $v_{O1} = V_{DD}$ and $v_{O2} = 0$. When $\phi_2$ goes high, then $v_{O3}$ goes to zero, $v_{O4} = V_{DD}$, and the input signal is shifted to the output during one clock period.

## 16.7.2   R–S Flip-Flop

**Flip-flops** are bistable circuits usually formed by cross-coupling two NOR gates. Figure 16.63 shows an R–S flip-flop using NMOS NOR logic gates with depletion loads. As shown, $M_1$, $M_2$, and $M_3$ form one NOR gate, and $M_4$, $M_5$, and $M_6$ form the second. The outputs of the two NOR circuits are connected back to the inputs of the opposite NOR gates.

If we assume that $S = $ logic 1 and $R = $ logic 0, then $M_1$ is biased in its conducting state and output $\bar{Q}$ is forced low. The inputs to both $M_4$ and $M_5$ are low, so output $Q$ goes high to a logic $1 = V_{DD}$. Transistor $M_2$ is then also biased in a conducting state. The two outputs $Q$ and $\bar{Q}$ are complementary and, by definition, the flip-flop is in the set state when $Q = $ logic 1 and $\bar{Q} = $ logic 0.

If $S$ returns to logic 0, then $M_1$ turns off, but $M_2$ remains turned on so $\bar{Q}$ remains low and $Q$ remains high. Therefore, when $S$ goes low, nothing in the circuit can force a change and the flip-flop stores this particular logic state.

When $R = $ logic 1 and $S = $ logic 0, then $M_4$ turns on so output $Q$ goes low. With $S = Q = $ logic 0, then both $M_1$ and $M_2$ are cut off and $\bar{Q}$ goes high. Transistor $M_5$ turns on, keeping $Q$ low when $R$ goes low. The flip-flop is now in the reset state.

If both $S$ and $R$ inputs were to go high, then both outputs $Q$ and $\bar{Q}$ would go low. However, this would mean that the outputs would not be complementary. Therefore, a logic 1 at both $S$ and $R$ is considered to be a forbidden or nonallowed condition. If both inputs go high and then return to logic 0, the state of the flip-flop is determined by whichever input goes low last. If both inputs go low simultaneously, then the outputs will flip into one state or the other, as determined by slight imbalances in transistor characteristics.

Figure 16.64 shows an R–S flip-flop using CMOS NOR logic gates. The outputs of the two NOR gates are connected back to the inputs of the opposite NOR gates to form the flip-flop.

If $S = $ logic 1 and $R = $ logic 0, then $M_{N1}$ is turned on, $M_{P1}$ is cut off, and $\bar{Q}$ goes low. With $\bar{Q} = R = $ logic 0, then both $M_{N3}$ and $M_{N4}$ are cut off, both $M_{P3}$ and $M_{P4}$ are biased in a conducting state so that the output $Q$ goes high. With $Q = $ logic 1, $M_{N2}$ is biased on, $M_{P2}$ is biased off, and the flip-flop is in a set condition. When $S$ goes low, $M_{N1}$ turns off, $M_{N2}$ remains conducting, so the state of the flip-flop does not change.

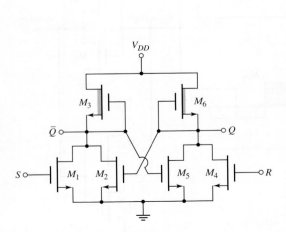

**Figure 16.63** NMOS R–S flip-flop

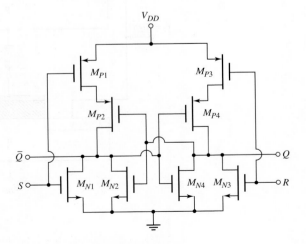

**Figure 16.64** CMOS R–S flip-flop

When $S = $ logic 0 and $R = $ logic 1, then output $Q$ is forced low, output $\bar{Q}$ goes high, and the flip-flop is in a reset condition. Again, a logic 1 at both $S$ and $R$ is considered to be a forbidden or a nonallowed condition, since the resulting outputs are not complementary.

### 16.7.3    D Flip-Flop

A **D-type flip-flop** is used to provide a delay. The logic bit on the D input is transferred to the output at the next clock pulse. This flip-flop is used in counters and shift registers. The basic circuit is similar to the CMOS dynamic shift register in Figure 16.62, except that additional circuitry makes the D flip-flop a static circuit.

Consider the circuit in Figure 16.65. The CMOS inverter composed of $M_{N2}$ and $M_{P2}$ is driven by a CMOS transmission gate composed of $M_{N1}$ and $M_{P1}$. A second CMOS inverter, $M_{N3}$ and $M_{P3}$, is connected in a feedback configuration. If $v_I = $ high, then $v_{O1}$ goes high when the transmission gate is conducting, and output $v_O$, which is the input to the feedback inverter, goes low.

When the CMOS transmission gate turns off, the pn junction in the $M_{N1}$ transmission gate transistor is reverse biased. In this case, however, voltage $v_{O1}$ is not simply across the gate capacitance of inverter $M_{N2}$–$M_{P2}$. Transistor $M_{P3}$ is biased in a conducting state, so the reverse-biased pn junction leakage current $I_L$ is supplied through $M_{P3}$, as indicated in Figure 16.65. Since this leakage current is small, the source-to-drain voltage of $M_{P3}$ will be small, and $v_{O1}$ will remain biased at essentially $V_{DD}$. The circuit will therefore remain in this static condition.

Similarly, when $v_{O1}$ is low and $v_O$ is high, the pn junction in the $M_{P1}$ transmission gate transistor is reverse biased and transistor $M_{N3}$ is biased on. Transistor $M_{N3}$ sinks the pn junction leakage current $I_L'$, and the circuit remains in this static condition until changed by a new input signal through the transmission gate.

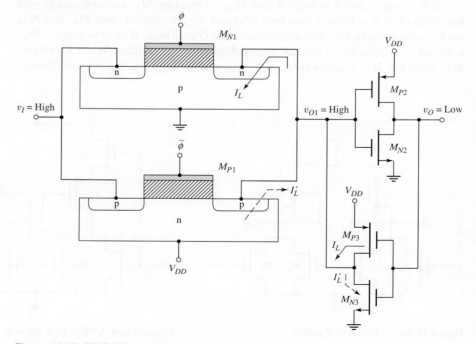

**Figure 16.65** CMOS D-type flip-flop

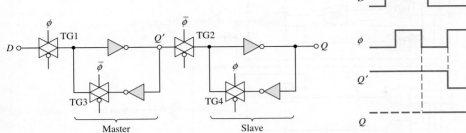

Figure 16.66  CMOS master–slave D flip-flop

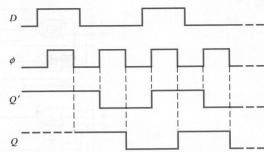

Figure 16.67  D flip-flop signals at various times

The circuit shown in Figure 16.66 is a master–slave configuration of a D flip-flop. When clock pulse $\phi$ is high, transmission gate TG1 is conducting, and data $D$ goes through the first inverter, which means that $Q' = \bar{D}$. Transmission gate TG2 is off, so data stops at $Q'$. When clock pulse $\phi$ goes low, then TG3 turns on, and the master portion of the flip-flip is in a static configuration. Also when $\phi$ goes low, TG2 turns on, the data are transmitted through the slave portion of the flip-flop, and the output is $Q = \bar{Q}' = D$. The data present when $\phi$ is high are transferred to the output of the flip-flop during the negative transition of the clock pulse. The various signals in the D flip-flop are shown in Figure 16.67.

Additional circuitry can be added to the D flip-flop in Figure 16.66 to provide a set and reset capability.

### 16.7.4  CMOS Full-Adder Circuit

One of the most widely used building blocks in arithmetic processing architectures is the one-bit full-adder circuit. We will first consider the logic diagram from the Boolean function and then consider the implementation in a conventional CMOS design.

Assuming that we have two input bits to be added plus a carry signal from a previous stage, the sum-out and carry-out signals are defined by the following two Boolean functions of three input variables $A$, $B$, and $C$.

$$\text{Sum-out} = A \oplus B \oplus C$$
$$= ABC + A\bar{B}\bar{C} + \bar{A}\bar{B}C + \bar{A}B\bar{C} \qquad \textbf{(16.80(a))}$$

$$\text{Carry-out} = AB + AC + BC \qquad \textbf{(16.80(b))}$$

The logic diagrams for these functions are shown in Figure 16.68. As we have seen previously, the implementation at the transistor level can be done with fewer transistors than would be used if all the NOR and NAND gates were actually connected as shown in the logic diagram.

Figure 16.69 is a transistor-level schematic of the one-bit full-adder circuit implemented in a conventional CMOS technology. We can understand the basic design from the logic diagram. For example, we may consider the NMOS portion of the carry-out signal. We see that transistors $M_{NA1}$ and $M_{NB1}$ are in parallel, to perform the basic OR function, and these transistors are in series with transistor $M_{NC1}$, to perform the basic AND function. These three transistors form the NMOS portion of the design of the two gates labeled $G_1$ and $G_2$ in Figure 16.68. We also have transistors $M_{NA2}$ and $M_{NB2}$ in series, to perform the basic AND function of gate $G_3$. This set of

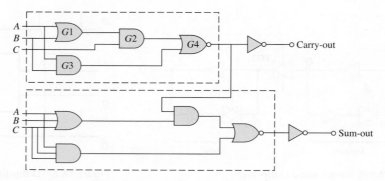

**Figure 16.68** Gate configuration of the one-bit full adder

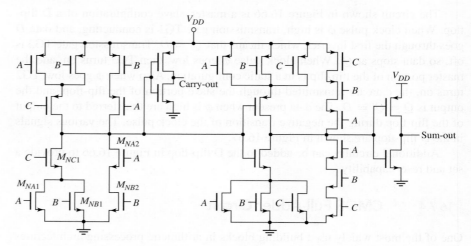

**Figure 16.69** Transistor configuration of the CMOS one-bit full adder

two transistors is in parallel with the previous three transistors, and this configuration performs the basic OR function of gate $G_4$. This output signal goes through an inverter to become the final carry-out signal.

We can go through the same discussion for the design of the NMOS portion of the sum-out signal. The PMOS design is then the complement of the NMOS design. As mentioned, the total number of transistors in the final design is considerably less than would have occurred if the basic OR and AND gates shown in the logic diagram were actually incorporated in the design.

 ## 16.8 MEMORIES: CLASSIFICATIONS AND ARCHITECTURES

**Objective:** • Discuss semiconductor memories.

In the previous sections of this chapter, various logic circuits were considered. Combinations of gates can be used to perform logic functions such as addition, multiplication, and multiplexing. In addition to these combinatorial logic functions, digital computers require some method of storing information. Semiconductor circuits form one type of memory, considered in this chapter, and define a class of digital electronic circuits that are just as important as the logic gates.

A memory cell is a circuit, or in some cases just a single device, that can store a bit of information. A systematic arrangement of memory cells constitutes a memory. The memory must also include peripheral circuits to address and write data into the cells as well as detect data that are stored in the cells.

In this section, we define the various types of semiconductor memories, discuss the memory organization, and briefly consider address decoders. In the next section, we analyze in detail some of the basic memory cells and briefly discuss sense amplifiers.

## 16.8.1 Classifications of Memories

Two basic types of semiconductor memory are considered. The first is the **random access memory** (RAM), a read–write memory, in which each individual cell can be addressed at any particular time. The access time to each cell is virtually the same. Implicit in the definition of the RAM is that both the read and write operations are permissible in each cell with also approximately the same access time. Both static and dynamic RAM cells are considered.

A second class of semiconductor memory is the **read-only memory** (ROM). The set of data in this type of memory is generally considered to be fixed, although in some designs the data can be altered. However, the time required to write new data is considerably longer than the read access time of the memory cell. A ROM may be used, for example, to store the instructions of a system operating program.

A volatile memory is one that loses its data when power is removed from the circuit, while nonvolatile memory retains its data even when power is removed. In general, a random access memory is a volatile memory, while read-only memories are nonvolatile.

### Random Access Memories

Two types of RAM are the static RAM (SRAM) and dynamic RAM (DRAM). A static RAM consists of a basic bistable flip-flop circuit that needs only a dc current or voltage applied to retain its memory. Two stable states exist, defined as logic 1 and logic 0. A dynamic RAM is an MOS memory that stores one bit of information as charge on a capacitor. Since the charge on the capacitor decays with a finite time constant (milliseconds), a periodic refresh is needed to restore the charge so that the dynamic RAM does not lose its memory.

The advantage of the SRAM is that this circuit does not need the additional complexity of a refresh cycle and refresh circuitry, but the disadvantage is that this circuit is fairly large. In general, SRAM requires six transistors. The advantage of a DRAM is that it consists of only one transistor and one capacitor, but the disadvantage is the required refresh circuitry and refresh cycles.

### Read-Only Memories

There are two general types of ROM. The first is programmed either by the manufacturer (mask programmable) or by the user (programmable, or PROM). Once the ROM has been programmed by either method, the data in the memory are fixed and cannot be altered. The second type of ROM may be referred to as an alterable ROM in that the data in the ROM may be reprogrammed if desired. This type of ROM may be called an EPROM (erasable programmable ROM), EEPROM (electrically erasable PROM), or flash memory. As mentioned, the data in these memories can be reprogrammed although the time involved is much longer than the read access time. In some cases, the memory chip may actually have to be removed from the circuit during the reprogramming process.

### 16.8.2 Memory Architecture

The basic memory architecture has the configuration shown in Figure 16.70. The terminal connections may include inputs, outputs, addresses, and read and write controls. The main portion of the memory involves the data storage. A RAM memory will have all of the terminal connections mentioned, whereas a ROM memory will not have the inputs and the write controls.

A typical RAM architecture, shown in Figure 16.71, consists of a matrix of storage bits arranged in an array with $2^M$ columns and $2^N$ rows. The array may be square, in which case $M$ and $N$ are equal. This particular array may be only one of several on a single chip. To read data stored in a particular cell within the array, a row address is inputted and decoded to select one of the row lines. All of the cells along this row are activated. A column address is also inputted and decoded to select one of the columns. The one particular memory cell at the intersection of the row and column addressed is then selected. The logic level stored in the cell is routed down a bit line to a sense amplifier.

Control circuits are used to enable or select a particular memory array on a chip and also to select whether data are to be read from or written into the memory cell. Memory chips or arrays are designed to be paralleled so that the memory capacity can be increased. The additional lines needed to address parallel arrays are called **chip select signals.** If a particular chip or array is not selected, then no memory cell is addressed in that particular array. The chip select signal controls the tristate output of the data-in and data-out buffers. In this way, the data-in and data-out lines to and from several arrays may be connected together without interfering with each other.

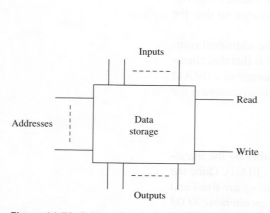

Figure 16.70 Schematic of a basic memory configuration

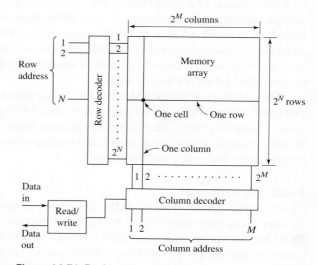

Figure 16.71 Basic random access memory architecture

### 16.8.3 Address Decoders

The row and column decoders in Figure 16.71 are essential elements in all memories. Access time and power consumption of memories may be largely determined by the decoder design. Figure 16.72 shows a simple decoder with a two-bit input. The decoder uses NAND logic circuits, although the same type of decoder may be implemented in NOR gates. The input word goes through input buffers that generate the complement as well as the signal.

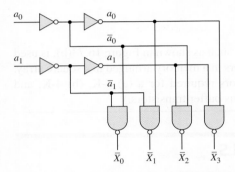

**Figure 16.72** Simplified decoder with two-bit input

Another example of the direct implementation of a decoder is shown in Figure 16.73. Figure 16.73(a) shows a pair of NMOS input buffer-inverters, and Figure 16.73(b) shows a five-input NOR logic address decoder circuit using NMOS enhancement-mode drivers and a depletion load. A pair of input-buffer inverters is required for each input address line. The input signal is then required to drive only an inverter, while the buffer-inverter pair can be designed to drive the remainder of the logic circuits. The output of the NOR decoder goes high only when all inputs are a logic 0. The NOR gate in Figure 16.73(b) would decode the address word 00110 and select the seventh row or column for a read or write operation. (Note: An input of 00000 is used to address the first row or column.)

As the size of the memory increases, the length of the address word must increase. For example, a 64-K (where 1 K = 1024 bits) memory whose cells are

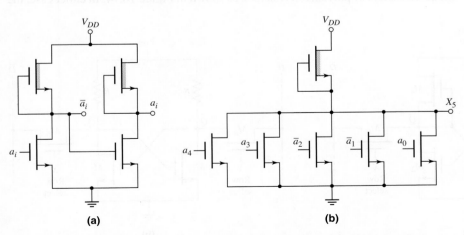

**Figure 16.73** (a) Input buffer-inverter pair; (b) five-input NOR logic address decoder

arranged in a square array would require an 8-bit word for the row address and another 8-bit word for the column address. As the word size increases, the decoder becomes more complex, and the number of transistors and power dissipation may become large. In addition, the total capacitance of MOS decoder transistors and interconnect lines increase so that propagation delay times may become significant. The number of transistors required to design a decoder may be reduced by using a two-stage decoder using both NOR and NAND gates. These circuits may be found in more advanced textbooks on digital circuits.

## Test Your Understanding

**TYU 16.15** A NOR logic address decoder, such as shown in Figure 16.73(b), is used in both the row and column address decoders in a memory arranged in a square array. Calculate the number of decoder transistors required for a (a) 1-K, (b) 4-K, and (c) 16-K memory. (Ans. 384, 896, 2048 plus buffer transistors.)

## 16.9 RAM MEMORY CELLS

**Objective:** • Analyze and design random-access-memory (RAM) cells

In this section, we consider two designs of an NMOS static RAM (SRAM), one design of a CMOS static RAM, and one design of a dynamic RAM (DRAM). We also consider examples of sense amplifiers and read/write circuitry. This section is intended to present the basic concepts used in memory cell design. More advanced designs can again be found in advanced texts on digital circuits.

### 16.9.1 NMOS SRAM Cells

A static RAM cell is designed by cross-coupling the inputs and outputs of two inverters. In the case of an NMOS design, the load devices may be either depletion-mode transistors or polysilicon resistors, as shown in Figure 16.74. In either case, the

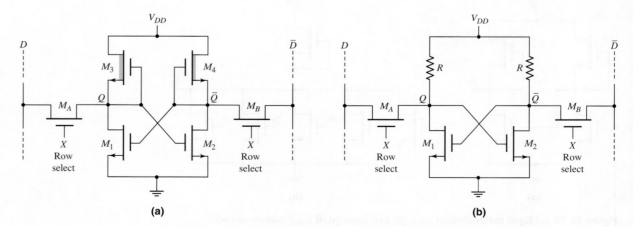

**Figure 16.74** Static NMOS RAM cells with (a) depletion loads and (b) polysilicon resistor loads

inputs and outputs of the two inverters are cross-coupled to form a basic flip-flop. If transistor $M_1$ is turned on, for example, the output $Q$ is low, which means that transistor $M_2$ is cut off. Since $M_2$ is cut off, the output $\bar{Q}$ is high, ensuring that $M_1$ is turned on. Thus, we have a static situation as long as the bias voltage $V_{DD}$ is applied to the circuit.

To access (read or write) the data contained in the memory cell, two NMOS transmission gate transistors, $M_A$ and $M_B$, connect the memory cell to the complementary bit lines. When the word line signal or row select signal is low, both transmission gate transistors are cut off and the memory cell is isolated or in a standby condition. The data stored in the cell remain stored as long as power is applied to the cell. When the row select or word line signal goes high, the memory cell is then connected to the complementary data lines so that the data in the cell can be read or new data can be written into the cell.

One critical parameter in the design of RAM cells is power dissipation. As we will see in the following example, this is one situation in which incorporating a high-valued resistor as a load device improves the design. A lightly doped polysilicon load resistor is formed by ion implantation, which can accurately dope the polysilicon to produce the designed resistance value.

## EXAMPLE 16.14

**Objective:** Determine the currents, voltages, and power dissipation in two NMOS SRAM cells. The first design uses a depletion-load device and the second design uses a resistor-load device.

Assume the following parameters: $V_{DD} = 3$ V and $k'_n = 60\ \mu\text{A/V}^2$; driver transistors: $V_{TND} = 0.5$ V and $(W/L)_D = 2$; load devices: $V_{TNL} = -1.0$ V, $(W/L)_L = 1/2$, and $R = 2$ M$\Omega$.

**Solution (With Depletion Load):** Assume $M_2$ is cut off in the circuit in Figure 16.74(a) so that $\bar{Q} = V_{DD} = 3$ V. $M_1$ is on in the nonsaturation region and $M_3$ is on in the saturation region. The drain current in $M_1$ and $M_3$ is then

$$i_D = \frac{k'_n}{2} \cdot \left(\frac{W}{L}\right)_L (V_{GSL} - V_{TNL})^2 = \frac{60}{2} \cdot \left(\frac{1}{2}\right)(0 - (-1))^2$$

or

$$i_D = 15\ \mu\text{A}$$

The power dissipated in the circuit is then

$$P = i_D \cdot V_{DD} = (15)(3) = 45\ \mu\text{W}$$

The logic 0 value of the $Q$ output is found from

$$i_D = \frac{k'_n}{2} \cdot \left(\frac{W}{L}\right)_D [2(V_{GSD} - V_{TND})V_{DSD} - V_{DSD}^2]$$

or

$$15 = \frac{60}{2} \cdot (2)[2(3 - 0.5)Q - Q^2]$$

which yields

$$Q = 50.5\ \text{mV}$$

**Solution (With Resistor Load):** Again assume $M_2$ is cut off in the circuit in Figure 16.74(b) so that $\bar{Q} = V_{DD} = 3$ V. Again $M_1$ is on in the nonsaturation region. The drain current is found from

$$\frac{V_{DD} - Q}{R} = \frac{k'_n}{2} \cdot \left(\frac{W}{L}\right)_D [2(V_{GSD} - V_{TND})Q - Q^2]$$

or

$$\frac{3 - Q}{2} = \frac{60}{2} \cdot (2)[2(3 - 0.5)Q - Q^2]$$

[Note that dividing by megohms on the left agrees with microamperes on the right.] We find

$$Q \cong 5 \text{ mV}$$

The drain current is then found:

$$i_D = \frac{V_{DD} - Q}{R} = \frac{3 - 0.005}{2} \cong 1.5 \ \mu\text{A}$$

The power dissipated in the circuit is then

$$P = i_D \cdot V_{DD} = (1.5)(3) = 4.5 \ \mu\text{W}$$

**Comment:** We see that the SRAM with the resistive load dissipates 10 times less power than the SRAM with the depletion-load device. Thus, for a given allowed power dissipation per chip, the memory with the resistive load could be 10 times larger than that using the depletion load device.

---

### EXERCISE PROBLEM

**Ex 16.14:** A 16-K NMOS static RAM cell using a resistor load is to be designed. Each cell is to be biased at $V_{DD} = 2.5$ V. Assume transistor parameters as described in Example 16.14. The entire memory is to dissipate no more than 125 mW in standby. Design the value of $R$ in each cell to meet this specification. (Ans. $R = 0.82$ MΩ)

---

Since the value of the load resistance $R$ is, in general, very large, the memory must be designed so that the resistor $R$ is not required to be a pull-up device. We will see this type of design later. The resistors can actually be fabricated on top of the NMOS transistors by a double-polysilicon technology, so that the cell with resistor load devices can be very compact, resulting in a high-density memory.

### 16.9.2    CMOS SRAM Cells

The basic six-transistor CMOS SRAM cell is shown in Figure 16.75. The inputs and outputs of the two CMOS inverters are cross-coupled so that the circuit will be in one of two static conditions. For example, if $\bar{Q}$ is low, then $M_{N1}$ is cut off so that $Q$ is high, which in turn means that $M_{P2}$ is cut off, ensuring that $\bar{Q}$ remains low. The two NMOS transmission gate transistors again connect the basic memory cell to the complementary data lines.

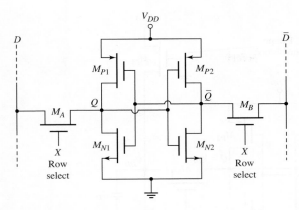

**Figure 16.75** A CMOS static RAM cell

The traditional advantages of CMOS technology include low static power dissipation, superior noise immunity to either bipolar or NMOS, wide operating temperature range, sharp transfer characteristics, and wide voltage supply tolerance.

CMOS is inherently lower power than NMOS, since conducting paths between power and ground do not arise when the circuit is in one logic state or the other. In standard CMOS, the p- and n-channel devices in the memory cell and in the periphery circuits are in series and on at the same time only during switching. Current is, therefore, drawn only during switching. This makes SRAMs and CMOS extremely low power in standby, when there are only surface, junction, and channel leakage currents.

A more complete circuit of the CMOS static RAM is shown in Figure 16.76, which includes PMOS data line pull-up transistors on the complementary bit lines. If all word line signals are zero, then all pass transistors are turned off. The two data lines with the relatively large column capacitances are charged up by the column pull-up transistors, $M_{P3}$ and $M_{P4}$, to the full $V_{DD}$ voltage.

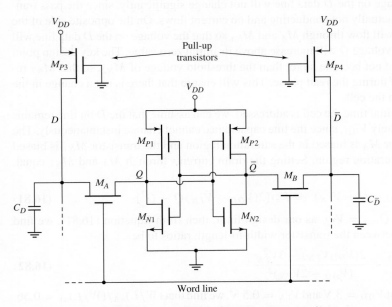

**Figure 16.76** CMOS RAM cell including PMOS pull-up transistors

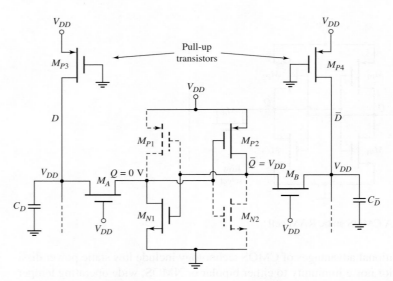

**Figure 16.77** Voltage levels and "on" transistors in CMOS RAM cell at the beginning of the read cycle

To determine the $(W/L)$ ratios of the transistors in a typical CMOS SRAM cell, two basic requirements must be taken into consideration. First, the read operation should not destroy the information stored in the cell, and second, the cell should allow for the modification of the data stored during a write operation. Consider a read operation in which a logic 0 ($Q = 0$ and $\bar{Q} = V_{DD}$) is stored in the cell. The voltage levels in the cell and on the data lines just prior to the read operation are shown in Figure 16.77. Transistors $M_{P1}$ and $M_{N2}$ are turned off while transistors $M_{N1}$ and $M_{P2}$ are biased in the nonsaturation region.

Immediately after the word select signal is applied to the pass transistors $M_A$ and $M_B$, the voltage on the $\bar{D}$ data line will not change significantly, since the pass transistor $M_B$ is actually not conducting and no current flows. On the opposite side of the cell, current will flow through $M_A$ and $M_{N1}$ so that the voltage on the $D$ data line will drop and the voltage $Q$ will increase above its initial zero value. The key design point is that $Q$ must not become larger than the threshold voltage of $M_{N2}$, so that $M_{N2}$ remains cut off during the read phase. This will ensure that there is not a change in the data stored in the cell.

At the initial time the cell is addressed, we can assume that the $D$ bit line remains at approximately $V_{DD}$, since the line capacitance cannot change instantaneously. The pass transistor $M_A$ is biased in the saturation region and the transistor $M_{N1}$ is biased in the nonsaturation region. Setting the drain currents through $M_A$ and $M_{N1}$ equal, we have

$$K_{nA}(V_{DD} - Q - V_{TN})^2 = K_{n1}[2(V_{DD} - V_{TN})Q - Q^2] \tag{16.81}$$

Setting $Q = Q_{\max} = V_{TN}$ as our design limit, then from Equation (16.81), we find the relation between the transistor width-to-length ratios to be

$$\frac{(W/L)_{nA}}{(W/L)_{n1}} < \frac{2(V_{DD}V_{TN}) - 3V_{TN}^2}{(V_{DD} - 2V_{TN})^2} \tag{16.82}$$

Assuming that $V_{DD} = 3$ V and $V_{TN} = 0.5$ V, we find that $(W/L)_{nA}/(W/L)_{n1} < 0.56$. So the width-to-length of the pass transistor should be approximately one-half that of

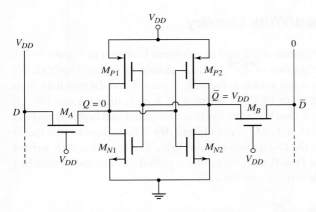

**Figure 16.78** Voltage levels in the CMOS RAM at the beginning of a write cycle

the NMOS device in the memory cell. By symmetry, the same condition applies to the transistors $M_{N2}$ and $M_B$.

We now need to consider the write operation. Assume that a logic 0 is stored and we want to write a logic 1 into the memory cell. Figure 16.78 shows the initial voltage levels in the CMOS SRAM cell when the cell is first addressed at the beginning of the write cycle. Transistors $M_{P1}$ and $M_{N2}$ are initially turned off, and $M_{N1}$ and $M_{P2}$ are biased in the nonsaturation region. The cell voltages are $Q = 0$ and $\bar{Q} = V_{DD}$ just before the pass transistors are turned on. The data line $D$ is held at $V_{DD}$ and the complementary data line $\bar{D}$ is forced to a logic 0 value by the write circuitry. We may assume that $\bar{D} = 0$ V for analysis purposes. The voltage $Q$ will remain below the threshold voltage of $M_{N2}$ because of the condition given by Equation (16.82). Consequently, the voltage at $Q$ is not sufficient to switch the state of the memory cell. To switch the state of the cell, the voltage at $\bar{Q}$ must be reduced below the threshold voltage of $M_{N1}$, so that $M_{N1}$ will turn off. When $\bar{Q} = V_{TN}$, then $M_B$ is biased in the nonsaturation region and $M_{P2}$ is biased in the saturation region. Equating drain currents, we have

$$K_{p2}(V_{DD} + V_{TP})^2 = K_{nB}\left[2(V_{DD} - V_{TN})V_{TN} - V_{TN}^2\right] \tag{16.83(a)}$$

which can be written in the form

$$\frac{K_{p2}}{K_{nB}} < \frac{2(V_{DD}V_{TN}) - 3V_{TN}^2}{(V_{DD} + V_{TP})^2} \tag{16.83(b)}$$

Considering the width-to-length ratios, we find

$$\frac{(W/L)_{p2}}{(W/L)_{nB}} < \frac{k_n'}{k_p'} \cdot \frac{2(V_{DD}V_{TN}) - 3V_{TN}^2}{(V_{DD} + V_{TP})^2} \tag{16.84}$$

Assuming that $V_{DD} = 3$ V, $V_{TN} = 0.5$ V, $V_{TP} = -0.5$ V, and $(k_n'/k_p') = (\mu_n/\mu_p) = 2$, we find that $(W/L)_{p2}/(W/L)_{nB} < 0.72$.

From previous results, if we assume that the width-to-length of the pass transistor is one-half that of the NMOS in the memory cell, and if we assume that the width-to-length of the PMOS in the memory cell is 0.7 that of the pass transistor, then the width-to-length of the PMOS in the cell should be approximately 0.35 that of the NMOS in the memory cell.

### 16.9.3    SRAM Read/Write Circuitry

An example of a read/write circuit at the end of a column is shown in Figure 16.79. We may consider the write portion of the circuit as shown in Figure 16.80(a). We may note that if the column is not selected, then $M_3$ is cut off and the two data lines are held at their precharged value of $V_{DD}$. When $X = Y = 1$, then the one-bit cell shown is addressed. If $\bar{W} = 1$ then the write cycle is deselected and both $M_1$ and $M_2$ are cut off. For $\bar{W} = 0$ and $D = 1$, $M_1$ is cut off and $M_2$ is turned on so that the $\bar{D}$ data line is pulled low while the $D$ data line remains high. The logic 1 is then written into the cell. For $\bar{W} = 0$ and $D = 0$, the $D$ data line is pulled low and the $\bar{D}$ data line is held high so that logic 0 is written into the cell.

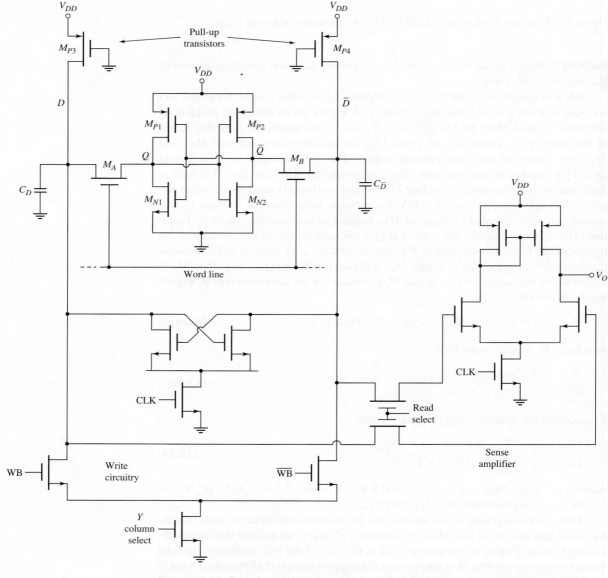

**Figure 16.79** Complete circuit diagram of a CMOS RAM cell with write and read circuitry

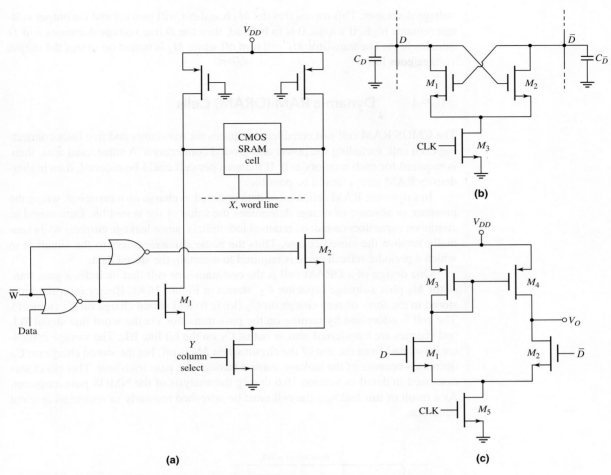

**Figure 16.80** (a) Write circuitry associated with CMOS RAM cell; (b) cross-coupled NMOS sense amplifier; (c) CMOS differential sense amplifier

Figure 16.80(b) shows the NMOS cross-coupled sense amplifier that is in the complete circuit of Figure 16.79. This circuit does not generate an output signal, but rather amplifies the small difference in the data bit lines. Suppose that a logic 1 is to be read from the memory cell. When the cell is addressed, the $D$ bit line is high and the $\bar{D}$ bit line voltage begins to decrease. This means that when the $M_3$ transistor turns on, the $M_2$ transistor turns on harder than $M_1$ so that the $\bar{D}$ bit line voltage is pulled low and the $M_1$ transistor will eventually turn off.

Figure 16.80(c) shows the differential amplifier that senses the output of the memory cell. Note that this sense amplifier is connected to the bit lines through a couple of pass transistors, as seen in Figure 16.79. If the input signal to the pass transistors is also a function of the column select signal, then this configuration enables the use of one main sense amplifier to read the data out of several columns, one at a time. When the clock signal is zero, the $M_3$ transistor in the differential amplifier is cut off and the common source node of $M_1$ and $M_2$ is pulled high, which means the output voltage is pulled high. When a memory cell is selected and the clock goes high, $M_3$ turns on. If a logic 1 level is to be read, then $D$ remains high and the $\bar{D}$ line voltage decreases. This means that the $M_2$ transistor will turn off and the output voltage remains high. If a logic 0 is to be read, then the $D$ line voltage decreases and $\bar{D}$ remains high. The transistor $M_1$ will turn off while $M_2$ is turned on so that the output voltage goes low.

### 16.9.4    Dynamic RAM (DRAM) Cells

The CMOS RAM cell just considered requires six transistors and five lines connecting each cell, including the power and ground connections. A substantial area, then, is required for each memory cell. If the area per cell could be reduced, then higher-density RAM arrays would be possible.

In a dynamic RAM cell, a bit of data is stored as charge on a capacitor, where the presence or absence of charge determines the value of the stored bit. Data stored as charge on capacitors cannot be retained indefinitely, since leakage currents will eventually remove the stored charge. Thus the name *dynamic* refers to the situation in which a periodic refresh cycle is required to maintain the stored data.

One design of a DRAM cell is the one-transistor cell that includes a pass transistor $M_S$ plus a storage capacitor $C_S$, shown in Figure 16.81. Binary information is stored in the form of zero charge on $C_S$ (logic 0) and stored charge on $C_S$ (logic 1). The cell is addressed by turning on the pass transistor via the word line signal WL and charges are transferred into or out of $C_S$ on the bit line BL. The storage capacitor is isolated from the rest of the circuit when $M_S$ is off, but the stored charge on $C_S$ decreases because of the leakage current through the pass transistor. This effect was discussed in detail in Section 16.6 during the analysis of the NMOS pass transistor. As a result of this leakage, the cell must be refreshed regularly to restore its original condition.

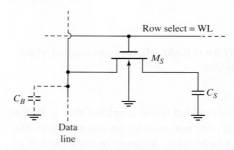

**Figure 16.81** One-transistor dynamic RAM cell

An example of a sense amplifier to detect the charge stored in the memory cell is shown in Figure 16.82. On one side of the amplifier is a memory cell that either stores a full charge or is empty, depending on the binary value of the data. On the other side of the amplifier is a reference cell with a reference or dummy storage capacitor $C_R$ that is one-half the value of the storage capacitor. The charge on $C_R$ will then be one-half the logic 1 charge on $C_S$. A cross-coupled dynamic latch circuit is used to detect the small voltage differences and to restore the signal levels. The capacitors $C_D$ and $C_{DR}$ represent the relatively large parasitic bit line and reference bit line capacitances.

In the standby mode, the bit lines on both sides of the sense amplifier are precharged to the same potential. During the read cycle, both the WL and D–WL

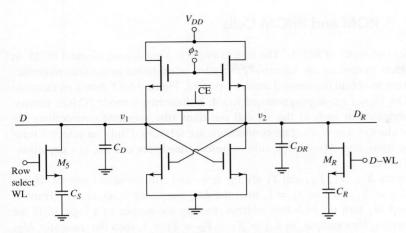

**Figure 16.82** Sense amplifier configuration for dynamic RAM cell

address signals go high allowing the charges in the cells to be redistributed along the bit lines. After the charge equalization and since the charge in the dummy cell is half the full charge, then $v_1 < v_2$ when the memory cell is empty or a logic 0, and $v_1 > v_2$ when the memory cell is full or a logic 1. The sense amplifier detects and amplifies the voltage difference between the bit lines, and will latch at the logic level stored in the basic memory cell.

## Test Your Understanding

**TYU 16.16** A six-transistor CMOS SRAM cell is biased at $V_{DD} = 2.5$ V. The transistor parameters are $V_{TN} = +0.4$ V, $V_{TP} = -0.4$ V, and $(\mu_n/\mu_p) = 2.5$. Determine the relative width-to-length ratios such that Equations (16.81) through (16.84) are satisfied in terms of read/write requirements. (Ans. $[(W/L)_{nA}/(W/L_{n1})] = 0.526, [(W/L)_p/(W/L)_{nB}] = 0.862$)

**TYU 16.17** A one-transistor DRAM cell is composed of a 0.05 pF storage capacitor and an NMOS transistor with a 0.5 V threshold voltage. A logic 1 is written into the cell when both the data line and row-select line are raised to 3 V. Sensing circuitry permits the stored charge to decay to 50 percent of its original value. Refresh occurs every 1.5 ms. Determine the maximum allowed leakage current that can exist. (Ans. $I = 41.7$ pA)

---

 ## 16.10   READ-ONLY MEMORY

> **Objective:** • Analyze read-only memories (ROM)

We consider several examples of read-only memories in this section. The intent is again to provide an introduction to this type of memory. In the case of EPROMs and EEPROMs, the development effort has been directed toward the characteristics of the basic memory cell.

### 16.10.1 ROM and PROM Cells

We consider two types of ROMs. The first example is a mask-programmed ROM, in which contacts to devices are selectively included or excluded in the final manufacturing process to obtain the desired memory pattern. Figure 16.83 shows an example of an NMOS $16 \times 1$ mask-programmed ROM. Enhancement-mode NMOS transistors are fabricated in each of the 16 cell positions (the substrate connections are omitted for clarity). However, gate connections are fabricated only on selected transistors. The transistors $M_1$–$M_4$ are column-select transistors and $M_0$ is a depletion-mode load device.

The inputs $X_O$, $X_1$, $Y_O$, and $Y_1$ are the row- and column-select signals. If, for example, $X_O = \bar{X}_1 = \bar{Y}_O = Y_1 = 1$, then the $M_{12}$ transistor is addressed. Transistors $M_{12}$ and $M_3$ turn on with this address, forcing the output to a logic 0. If the address changes, for example, to $\bar{X}_O = X_1 = \bar{Y}_O = \bar{Y}_1 = 1$, then the transistor $M_{23}$ is addressed. However, this transistor does not have a gate connection and consequently never turns on, so the output is a logic 1.

The mask-programmed memory discussed is only a $16 \times 1$-bit ROM, while a more useful memory would contain many more bits. Memories can be organized in any desired manner, such as a $2048 \times 8$ for a 16-K memory. This ROM is a nonvolatile memory, since the data stored are not lost when power is removed.

The second example of a ROM is a user-programmed ROM. The data pattern is defined by the user after the final manufacture rather than during the manufacture.

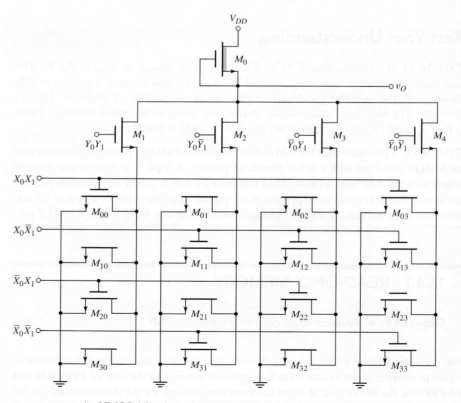

**Figure 16.83** An NMOS $16 \times 1$ mask-programmable ROM

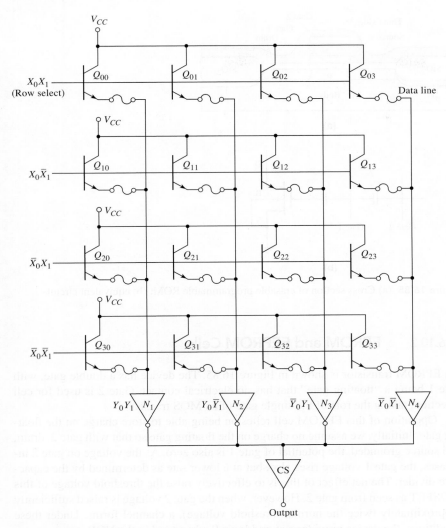

**Figure 16.84** A bipolar fuse-linked user-programmable ROM

One specific type is shown schematically in Figure 16.84. A small fuse is in series with each emitter and can be selectively "blown" or left in place by the user. If, for example, the fuse in $Q_{00}$ is left in place and this transistor is addressed by $X_O = X_1 = Y_O = Y_1 = 1$, then $Q_{00}$ turns on, raising the data line voltage at the emitter of $Q_{00}$. The inverter $N_1$ is enabled, making the output a logic 0. If the fuse is blown in this transistor, then the input to the inverter is a logic 0, so the output is a logic 1.

The polysilicon fuse in the emitter of an npn bipolar transistor has a fairly low resistance, so with the fuse in place and at low currents, there is very little voltage drop across the fuse. When the current through the fuse is increased to the 20 to 30 mA range, the heating of the polysilicon fuse causes the temperature to increase. The silicon oxidizes, forming an insulator that effectively opens the path between the data line and the emitter. The bipolar ROM circuit with the fuses either in place or "blown" form a permanent ROM that is not alterable and is also nonvolatile.

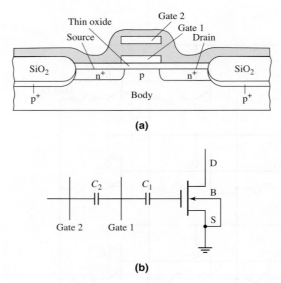

**Figure 16.85** (a) Cross section of erasable programmable ROM; (b) equivalent circuit

**EPROM and EEPROM Cells**

An EPROM transistor is shown in Figure 16.85. The device has a double gate, with gate 1 being a "floating gate" that has no electrical contact. Gate 2 is used for cell selection, taking the role of the single gate of an MOS transistor.

Operation of this EPROM cell relies on being able to store charge on the floating gate. Initially, we assume no charge on the floating gate so that with gate 2, drain, and source grounded, the potential of gate 1 is also zero. As the voltage on gate 2 increases, the gate 1 voltage rises also, but at a lower rate as determined by the capacitive divider. The net effect of this is to effectively raise the threshold voltage of this MOSFET as seen from gate 2. However, when the gate 2 voltage is raised sufficiently (approximately twice the normal threshold voltage), a channel forms. Under these conditions, the device provides a stored logic 0 when used in the NOR array.

To write a logic 1 into this cell, both gate 2 and drain are raised to about 25 V while the source and substrate remain at ground potential. A relatively large drain current flows because of normal device conduction characteristics. In addition, the high field in the drain–substrate depletion region results in avalanche breakdown of the drain–substrate junction, with a considerable additional flow of current. The high field in the drain depletion region accelerates electrons to high velocity such that a small fraction traverse the thin oxide and become trapped on gate 1. When the gate 2 and drain potentials are reduced to zero, the negative charge on gate 1 forces its potential to approximately $-5$ V. If the gate 2 voltage for reading is limited to $+5$ V, then a channel never forms. Thus a logic 1 is stored in the cell.

Gate 1 is completely surrounded by silicon dioxide ($SiO_2$), an excellent insulator, so charge can be stored for many years. Data can be erased, however, by exposing the cells to strong ultraviolet (UV) light. The UV radiation generates electron–hole pairs in the $SiO_2$ making the material slightly conductive. The negative charge on the gate can then leak off, restoring the transistor to its original uncharged condition. These EPROMs must be assembled in packages with transparent covers so

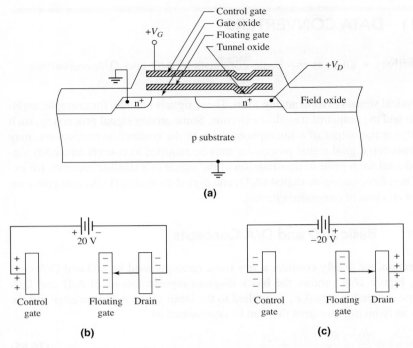

**(a)**

**(b)**                                          **(c)**

**Figure 16.86** (a) Cross section of a floating-gate electrically erasable programmable ROM; (b) charging the floating gate; (c) discharging the floating gate

the silicon chip may be exposed to UV radiation. One disadvantage is that the entire memory must be erased before any reprogramming can be done. In general, reprogramming must also be done on specialized equipment; therefore, the EPROM must be removed from the circuit during this operation.

In the EEPROM, each individual cell can be erased and reprogrammed without disturbing any other cell. The most common form of EEPROM is also a floating gate structure; one example is shown in Figure 16.86(a). The memory transistor is similar to an n-channel MOSFET, but with a physical difference in the gate insulator region. Charge may exist on the floating gate that will alter the threshold voltage of the device. If a net positive charge exists on the floating gate, the n-channel MOSFET is turned on, whereas if zero or negative charge exists on the floating gate, the device is turned off.

The floating gate is capacitively coupled to the control gate with the tunnel oxide thickness less than 200 Å. If 20 V is applied to the control gate while keeping $V_D = 0$, electrons tunnel from the $n^+$-drain region to the floating gate as demonstrated in Figure 16.86(b). This puts the MOSFET in the enhancement mode with a threshold voltage of approximately 10 V, so the device is effectively off. If zero volts is applied to the control gate and 20 V is applied to the drain terminal, then electrons tunnel from the floating gate to the $n^+$-drain terminal as demonstrated in Figure 16.86(c). This leaves a net positive charge on the floating gate that puts the device in the depletion mode with a threshold voltage of approximately $-2$ V, so the device is effectively on. If all voltages are kept to within 5 V during the read cycle, this structure can retain its charge for many years.

# 16.11 DATA CONVERTERS

**Objective:** • Discuss the basic concepts in A/D and D/A converters.

Most physical signals exist in analog form. These signals include, for example, audio or speech and the output of transducer circuits. Some analog signal processing, such as amplifying the output of a microphone prior to the connection to speakers, may occur. However, digital signal processing may be required to convert an analog signal into digital form prior to transmission of the signal to a satellite receiver, for example. Therefore, analog-to-digital (A/D) and digital-to-analog (D/A) converters are an important class of integrated circuits.

## 16.11.1 Basic A/D and D/A Concepts

In this section, we briefly consider a few basic concepts used in A/D and D/A conversions. Figure 16.87 shows the block diagram representations of A/D and D/A converters. An analog signal $v_A$ is applied to the input of the A/D converter and the output is an N-bit digital signal that can be represented as

$$v_D = \frac{b_1}{2^1} + \frac{b_2}{2^2} + \frac{b_3}{2^3} + \cdots + \frac{b_N}{2^N} \tag{16.85}$$

where $b_1$, $b_2$, etc. are the bit coefficients that are either a 1 or 0. The bit $b_1$ is the most significant bit (MSB) and the bit $b_N$ is the least significant bit (LSB). The input to the D/A converter is the N-bit digital signal and the output is an analog signal $v'_A$. Ideally, the output analog signal $v'_A$ is an exact replication of the input analog signal $v_A$.

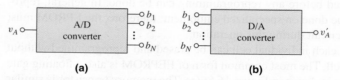

**Figure 16.87** Block diagram representations of (a) A/D converter and (b) D/A converter

The analog signal is to be converted to a digital form as indicated in Equation (16.85). Consider, for example, an analog signal represented by a voltage in the range $0 \leq v_A \leq 5$ V. Assume the digital signal is a 6-bit word. The 6-bit word represents 64 discrete values. The analog signal will then be divided into 64 values, with each bit representing 5 V/64 = 0.078125 V. The analog-to-digital conversion can be visualized in Figure 16.88.

When the analog input voltage is, for example, $v_A = \frac{5}{64}$ V, the digital output is 000001 and when the analog input voltage is $v_A = 2(\frac{5}{64})$ V, the digital output is 000010. However, we see that when the input is in the range $\frac{1}{2}(\frac{5}{64}) < v_A < \frac{3}{2}(\frac{5}{64})$ V, the digital output is constant at $v_D = 000001$. There is an inherent quantization error in the A/D conversion. A larger number of bits in the digital signal reduces the quantization error, but requires a more complex circuit.

The same effect occurs at the output of the D/A converter. Since the digital input signal exists in discrete steps or increments, the output signal will also occur in discrete steps or increments. An example is shown in Figure 16.89. The output signal

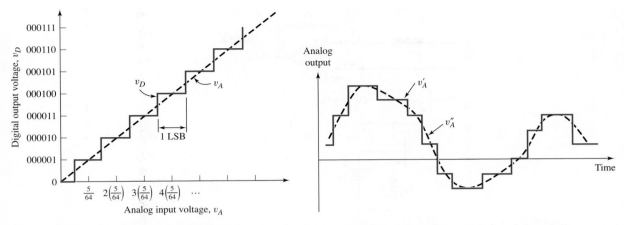

**Figure 16.88** Digital output versus analog input for a 6-bit A/D converter

**Figure 16.89** Discrete analog output $v'_A$ and smoothed output $v''_A$ versus time from a D/A converter

$v'_A$ is in the form of stair steps. Normally, this signal will be fed through a low-pass filter to smooth out the signal to produce the dotted signal $v''_A$ in the figure. The desired result is that the signal $v''_A$ be as close to the original signal $v_A$ as possible.

### 16.11.2   Digital-to-Analog Converters

We will consider a few basic D/A converters to gain an appreciation of the techniques used in these circuits.

#### Weighted-Resistor 4-Bit D/A
A simple circuit for a 4-bit D/A converter was shown in Chapter 9 in Figure P9.37. This circuit is repeated here in Figure 16.90 for convenience. The circuit is a summing amplifier and includes a reference voltage $V_R$, four weighted input resistors, four switches, and an op-amp with a feedback resistor. With $R_F = 10$ k$\Omega$, we find the output voltage to be

$$v_O = \left( \frac{b_1}{2} + \frac{b_2}{4} + \frac{b_3}{8} + \frac{b_4}{16} \right)(5) \text{ V} \tag{16.86}$$

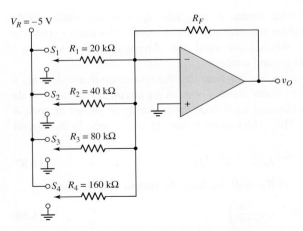

**Figure 16.90** A 4-bit weighted-resistor D/A converter

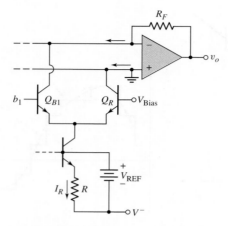

**Figure 16.91** Example of a current switch in the MSB position in a weighted resistor of D/A converter

One factor that determines the accuracy of the circuit is the precision of the weighted input resistors and the feedback resistor. As the number of bits increases, the size of the weighted input resistance increases for the lesser significant bits. The accuracy for large resistance values becomes more difficult to maintain. The size of this D/A converter is in general limited to a 4-bit input.

Another factor that determines the accuracy of the D/A circuit is the precision of the switches. An example of a current switch, showing only the MSB, is shown in Figure 16.91. If the bit $b_1$ is a logic 1 ($> V_{\text{Bias}}$), then $Q_{B1}$ is turned on and $Q_R$ is turned off so that the current $I_R$ is switched through $Q_{B1}$. This current becomes a component of the current through the feedback resistor. If $b_1$ is a logic 0 ($< V_{\text{Bias}}$), then $Q_{B1}$ is turned off and $Q_R$ is turned on so that the current is switched to ground. Because of the virtual ground, we may note that the collector voltages of $Q_{B1}$ and $Q_R$ are essentially identical.

For the circuit to operate properly, the base-emitter voltage of all the transistors must be the same. Since the currents are smaller for the lesser significant bits, the base–emitter areas must be reduced in order to maintain the same current density. The type of circuit shown in Figure 16.91 then requires a wide range of base–emitter areas in the same way that it requires a wide range of weighted resistor values.

### R–2R Ladder Network D/A

A circuit that eliminates the wide spread in weighted input resistor values in the previous circuit is an $R$–$2R$ ladder resistive network. Consider the circuit shown in Figure 16.92. Assuming the switches are ideal, the current in each resistor is a constant because of the virtual ground concept.

Consider node $X$ in the circuit. We can note that the resistance denoted as $R_X$ is $R_X = 2R$. This same resistance occurs at every node in the circuit as indicated on the figure. Therefore, the current entering each node splits evenly as shown at node $X$. We have that $I_{N-1} = \frac{1}{2} I_{N-2}$. This effect again occurs at every node in the circuit. Therefore, we have

$$I_1 = 2I_2 = 4I_3 = \cdots = 2^{N-2} I_{N-1} = 2^{N-1} I_N \qquad (16.87)$$

Setting the feedback resistance to $R_F = R$, we have the output voltage given by

$$v_O = (-V_{\text{REF}}) \left( \frac{b_1}{2} + \frac{b_2}{4} + \cdots + \frac{b_N}{2^N} \right) \qquad (16.88)$$

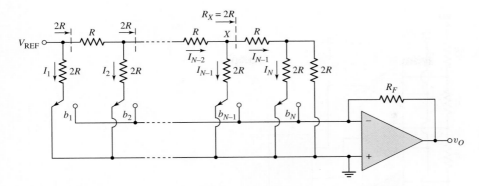

**Figure 16.92** Example of an $R$–$2R$ ladder network in an $N$-bit D/A converter

The circuit in Figure 16.92 requires only two resistance values. These resistance values can then be held to tight tolerances.

There are a variety of D/A converter designs that will not be pursued in this text. This short discussion provides a brief introduction to D/A design.

### 16.11.3  Analog-to-Digital Converters

As in the previous section, we will consider a few basic A/D converters to gain an appreciation of the techniques used in these circuits.

#### Parallel or Flash A/D

The fastest A/D converter, and perhaps simplest in concept, is the parallel A/D or flash converter. Figure 16.93 shows a 3-bit flash A/D converter. The analog input signal $v_A$ is applied to seven comparators at the noninverting terminals. A reference voltage is applied to a resistive ladder network. The outputs of the ladder network are applied to the inverting terminals of the comparators.

The total resistance in the ladder network is $8R$ so $V_{REF}/8R$ represents 1 LSB in terms of current. The smallest output voltage is

$$v_1 = \frac{V_{REF}}{8R}\left(\frac{R}{2}\right) = \frac{V_{REF}}{16} \tag{16.89}$$

which represents $\frac{1}{2}$ LSB. The second output voltage is

$$v_2 = \frac{V_{REF}}{8R}\left(\frac{3R}{2}\right) = 3\left(\frac{V_{REF}}{16}\right) \tag{16.90}$$

which represents $1\frac{1}{2}$ LSB.

If the analog input is $v_A < \frac{1}{2}$ LSB, the output of all comparators will be low. If the analog input is $\frac{1}{2}$ LSB $< v_A < 1\frac{1}{2}$ LSB, the output of the first comparator goes high. As the analog input voltage increases, the outputs of additional comparators go high. The combinational logic network then produces the desired 3-bit output word. We can note that a complete conversion is obtained during one clock period.

A disadvantage of the flash A/D converter is that the number of resistors and comparators increases rapidly as the desired number of output bits increases. We see that $2^N$ resistors and $2^N - 1$ comparators are required. Thus, for a 10 bit word,

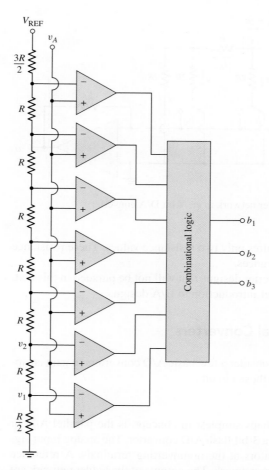

**Figure 16.93** A 3-bit flash or parallel A/D converter

1024 resistors and 1023 comparators are required. However, 10-bit resolution A/D flash converters have been fabricated as ICs.

### Counting A/D

A second type of A/D converter is the counting converter. This system contains a comparator, a counter, and a D/A converter in a feedback configuration as shown in Figure 16.94(a). Additional control circuitry is not shown for simplicity.

Initially the output of the counter is set equal to zero and the output of the D/A converter is set to $v_O = \frac{1}{2}$ LSB. When an analog input voltage $v_A$ is applied, the output of the comparator is high (unless $v_A < \frac{1}{2}$ LSB), which enables the counter. Then for each clock pulse, the output of the counter increases by one, producing an $N$-bit digital output. When the output of the D/A becomes just greater than the analog input voltage, the output of the comparator goes low and the counter is disabled. The $N$-bit digital output then corresponds to the analog input signal.

Figure 16.94(b) shows the timing diagram of a counting converter for a 4-bit digital output. Assume that the analog input signal is in the range $0 \leq v_A \leq 5$ V. A 1 LSB then corresponds to $\frac{5}{16}$ V.

Assume the analog input signal is $v_A = 5.2(\frac{5}{16})$ V. The initial output of the D/A, as mentioned, is a $\frac{1}{2}$ LSB offset voltage. By including the offset voltage, the maximum

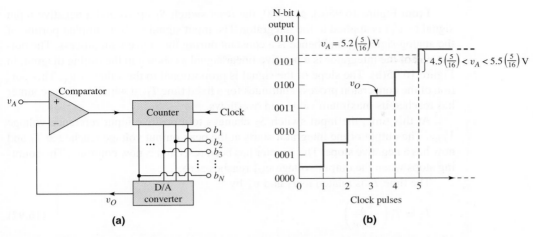

**(a)**                                                                        **(b)**

**Figure 16.94** (a) Block diagram of a counting A/D converter and (b) the timing diagram of a 4-bit A/D counting converter for a specific input voltage

quantization error will then be $\pm\frac{1}{2}$ LSB. We see from Figure 16.94(b) that after the fifth clock pulse, the output of the D/A is

$$v_O = \frac{1}{2}\left(\frac{5}{16}\right) + 5\left(\frac{5}{16}\right) = 5.5\left(\frac{5}{16}\right) \text{V} \tag{16.91}$$

which is larger than $v_A$. The counter then stops counting and the digital output is 0101. We can note that the digital output corresponds to $5(\frac{5}{16})$ V, which is within $\frac{1}{2}$ LSB of the analog input signal.

To complete the conversion process, the clock must go through its complete cycle, which for a 4-bit output is 16 clock periods.

### Dual-Slope A/D

Another type of A/D conversion scheme is the dual-slope A/D converter shown in Figure 16.95(a). This type of converter is found in high-resolution data acquisition systems, for example, since 20-bit conversions can be achieved.

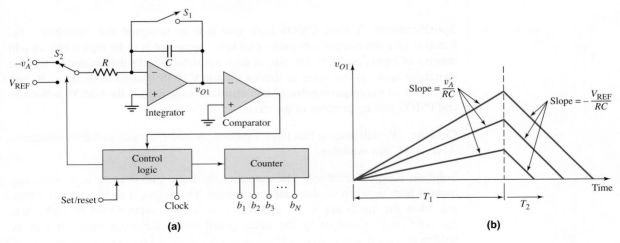

**(a)**                                                                        **(b)**

**Figure 16.95** (a) Block diagram and (b) timing diagram of a dual-slope A/D converter

From Figure 16.95(a), at $t = 0$, the reset switch $S_1$ opens and a negative input signal $(-v'_A)$ is applied to the integrator. The input signal $v'_A$ is a sampled portion of the analog signal $v_A$ and hence is a constant during the conversion process. The output $v_{O1}$ of the integrator is a positive linear signal as shown in the timing diagram in Figure 16.95(b). The slope of the signal is proportional to the value of $v'_A$. This portion of the conversion process continues for a fixed time $T_1$, at which time the counter has reached its maximum value and overflows.

At this time, the input switch $S_2$ changes to a positive input reference voltage $V_{REF}$. The output of the integrator starts at the peak output voltage reached at $T_1$ and now has a negative slope. The counter has been reset and is now counting. The counting stops when the output voltage $v_{O1}$ reaches zero.

The time $T_2$ is related to $T_1$ and $v'_A$ by

$$T_2 = T_1\left(\frac{v'_A}{V_{REF}}\right) \tag{16.92}$$

The counter reading at $T_2$ is given by

$$n = 2^N\left(\frac{v'_A}{V_{REF}}\right) \tag{16.93}$$

The output of the counter is then the digital equivalent of $v'_A$.

The output of the dual-slope A/D converter is independent of the actual values of $R$ and $C$ and hence is very accurate. The disadvantage of this data converter is that it is a fairly slow system. The time $T_1$ requires $2^N$ clock pulses and the maximum possible time $T_2$ would also require $2^N$ clock pulses. For example, a 12-bit A/D converter would require a total of 8192 clock pulses. This corresponds to a conversion time of 8.2 ms for a 1 MHz clock.

## 16.12 DESIGN APPLICATION: A STATIC CMOS LOGIC GATE

**Objective:** • Design a static CMOS logic gate to implement a specific logic function.

**Specifications:** A static CMOS logic gate is to be designed that implements the function of a three-input odd-parity checker. The output is to be high when an odd number of inputs are high. The size of each transistor is to be determined so that the switching speed is the same as that of a basic CMOS inverter with $W_n = W$ and $W_p = 2W$. A minimum number of transistors are to be used in the NMOS pull-down and PMOS pull-up portions of the circuit.

**Choices:** We will assume that input signals $A$, $B$, and $C$ as well as the complements $\bar{A}$, $\bar{B}$, and $\bar{C}$ are available.

**Solution (Logic Function):** The output of the logic gate is to be high when one input is high or when all three inputs are high. The output is to be high, for example, when the inputs are $A = 1$ and $B = C = 0$. The output would be high, then, for $A\bar{B}\bar{C} = 1$. Considering the other possibilities, the logic function can be written as

$$F = A\bar{B}\bar{C} + \bar{A}B\bar{C} + \bar{A}\bar{B}C + ABC \tag{16.94}$$

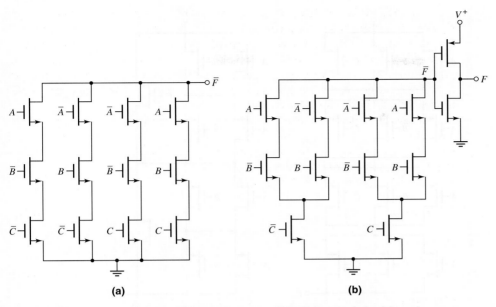

**Figure 16.96** (a) The basic NMOS pull-down portion of the logic gate derived from the logic function; (b) the modified NMOS pull-down portion of the logic gate for the design application

**Solution (NMOS Pull-Down):** Figure 16.96(a) shows the basic NMOS pull-down portion of the logic gate derived from the logic function given in Equation (16.94). However, we may note that the two transistors at the bottom of the first two columns have a common input of $\bar{C}$ and the two transistors at the bottom of the last two columns have a common input of $C$. The two transistors with common inputs can be combined into a single transistor. The final design of the NMOS pull-down portion of the logic gate is shown in Figure 16.96(b).

In order for the NMOS portion of the circuit to be in the pull-down mode, three NMOS devices in series must be turned on. In order for this circuit to be equivalent to the NMOS in the CMOS inverter, each NMOS device must have a width of $W_n = 3W$.

**Solution (PMOS Pull-Up):** Figure 16.97(a) shows the basic PMOS pull-up portion of the logic gate. This circuit is the complement of the NMOS circuit shown in Figure 16.96(a). We may note that two transistors on the right side of the circuit have common inputs $C$ and $\bar{C}$. Each pair of transistors is effectively in series and hence can be replaced by a single transistor. The resulting circuit is shown in Figure 16.97(b). The complete three-input odd-parity checker circuit is then the combination of Figures 16.96(b) and 16.97(b) along with a CMOS inverter on the output.

**Comment:** The basic logic circuit can be derived from the logic function. However, as we have seen, some simplifications can be made in the design. These simplifications can also be obtained from simplifications in the basic logic function also.

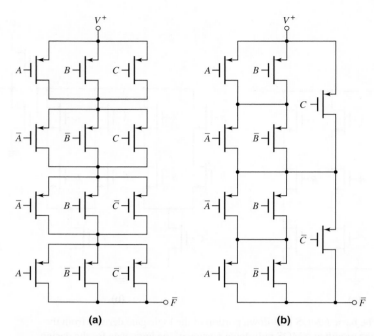

**Figure 16.97** (a) The basic PMOS pull-up portion of the logic gate derived as the complement of the basic NMOS pull-down circuit; (b) the modified PMOS pull-up portion of the logic gate for the design application

## 16.13 SUMMARY

- In this chapter, NMOS and CMOS digital logic circuits were analyzed and designed. These circuits include basic logic gates, shift registers, flip-flops, and memories.
- The NMOS inverter is the basis of NMOS logic circuits. The quasi-static voltage transfer characteristics of NMOS inverters with resistive-load, enhancement-load, and depletion-load devices were generated.
- The basic NMOS NOR and NAND logic gates were analyzed. More sophisticated logic functions can be implemented by combining driver transistors in parallel and series combinations.
- The CMOS inverter is the basis of CMOS logic circuits. The quasi-static voltage transfer characteristics were generated. For the CMOS circuit, the quiescent power dissipation is essentially zero when the input is in either logic state.
- The basic CMOS NOR and NAND logic gates were analyzed. More sophisticated CMOS logic circuits were also analyzed and designed. Transistor width-to-length ratios were designed to provide equal current drive in the NMOS pull-down and PMOS pull-up portions of the circuit.
- CMOS clocked logic circuits can be designed to reduce the number of required PMOS devices. A generalized NMOS logic circuit is inserted between clocked PMOS and NMOS devices. The advantage of low static power dissipation is maintained.
- Sequential logic circuits, such as shift registers, flip-flops, and a full one-bit adder, were analyzed.

- A whole classification of circuits called memories was considered. In the NMOS SRAM, static power is continuously dissipated in the cell, which limits the size of the memory because of the total chip power limitation. The primary advantage of a CMOS SRAM is that there is essentially no static power dissipation. The size of a CMOS memory is limited primarily by chip area requirements.
- Read-only-memory (ROM and PROM) contains fixed data that are implemented by the manufacturer or by the user. In both cases, the data cannot be altered. EPROM and EEPROM cells contain MOSFETs with floating gates that can be either charged or left uncharged by the user depending on whether logic 1 or a logic 0 is to be stored.
- The basic concepts used in A/D and D/A converters were discussed. A few examples of D/A converter circuits and A/D converter systems were analyzed.
- As an application, a static CMOS logic circuit to implement a specific logic function is designed.

## CHECKPOINT

After studying this chapter, the reader should have the ability to:

✓ Analyze the transfer characteristics of NMOS inverters, including the determination of noise margins.
✓ Design an NMOS logic circuit to perform a specific logic function.
✓ Analyze the transfer characteristics of the CMOS inverter, including the determination of switching power and noise margins.
✓ Design a CMOS logic circuit to perform a specific logic function.
✓ Design a clocked CMOS logic circuit to perform a specific logic function.
✓ Design an NMOS or CMOS pass network to perform a specific logic function.
✓ Design an NMOS or CMOS RAM cell and design a simple sense amplifier.
✓ Analyze the $R$–$2R$ ladder network used in a D/A converter circuit.
✓ Describe the characteristics of a 3-bit flash A/D converter and describe the operation of the dual-slope A/D converter.

## REVIEW QUESTIONS

1. Sketch the quasi-static voltage transfer characteristics of an NMOS inverter with depletion load. What effect does changing the transistor $W/L$ ratio have on the transfer characteristics?
2. Sketch an NMOS three-input NOR logic gate. Describe its operation. Discuss the condition under which the maximum logic 0 value is obtained.
3. Discuss how more sophisticated (compared to the basic NOR and NAND) logic functions can be implemented in a single NMOS logic circuit.
4. Sketch the quasi-static voltage transfer characteristics of a CMOS inverter. Discuss the various intervals in terms of transistor bias. What is the effect on the transfer curve of changing the transistor $W/L$ ratios? What is the advantage of the CMOS inverter compared to an NMOS inverter?
5. Discuss the parameters that affect the switching power dissipation in a CMOS inverter.

6. Define the noise margin in a CMOS inverter.
7. Sketch a CMOS three-input NAND logic gate. Describe its operation. Determine the relative transistor $W/L$ ratios to obtain equal pull-up and pull-down switching times.
8. Discuss how more sophisticated (compared to the basic NOR and NAND) logic functions can be implemented in a single CMOS logic circuit.
9. Discuss the basic principles of a clocked CMOS logic circuit.
10. Sketch an NMOS transmission gate and describe its operation. What is the maximum output voltage?
11. Sketch a CMOS transmission gate and describe its operation. Why is the quasi-static output voltage always equal to the quasi-static input voltage?
12. Discuss what is meant by pass transistor logic.
13. If an NMOS or CMOS transmission gate is turned off (an open switch), discuss why the output voltage is, in general, not stable.
14. Sketch an NMOS dynamic shift register and describe its operation.
15. Sketch a CMOS R–S flip flop and describe its operation. Why must the input condition $R = S = 1$ be avoided?
16. Describe the basic architecture of a semiconductor random-access memory.
17. Sketch a CMOS SRAM cell and describe its operation. Discuss any advantages and disadvantages of this design. Describe how the cell is addressed.
18. Sketch a one-transistor DRAM cell and describe its operation. What makes this circuit dynamic?
19. Describe a mask-programmed MOSFET ROM memory.
20. Describe the basic operation of a floating gate MOSFET and how this can be used in an erasable ROM.

 PROBLEMS

[**Note:** In the following problems, unless otherwise stated, assume transistor parameters of $V_{TNO} = 0.5$ V, $V_{TPO} = -0.5$ V, $k'_n = 100\,\mu\text{A/V}^2$, and $k'_p = 40\,\mu\text{A/V}^2$. Neglect the body effect unless otherwise stated and assume $T = 300$ K.]

### Section 16.1 NMOS Inverters

16.1    The load resistor in the NMOS inverter in Figure 16.3(a) is $R_D = 40\,\text{k}\Omega$. The circuit is biased at $V_{DD} = 3.3$ V. (a) Design the transistor width-to-length ratio such that $v_O = 0.1$ V when $v_I = 3.3$ V. (b) Using the results of part (a), determine the transition point for the transistor. (c) Using the results of part (a), find the maximum current and maximum power dissipation in the inverter.

16.2    The inverter circuit in Figure 16.3(a) is biased at $V_{DD} = 3.3$ V. Assume the transistor conduction parameter is $K_n = 50\,\mu\text{A/V}^2$. (a) Let $R_D = 100\,\text{k}\Omega$. (i) Determine the transition point. (ii) Determine $v_O$ for $v_I = 3.3$ V. (b) Repeat part (a) for $R_D = 30\,\text{k}\Omega$. (c) Repeat part (a) for $R_D = 5\,\text{k}\Omega$.

D16.3    (a) Redesign the resistive load inverter in Figure 16.3(a) so that the maximum power dissipation is 0.25 mW with $V_{DD} = 3.3$ V and

$v_O = 0.15$ V when the input is a logic 1. (b) Using the results of part (a), what is the input voltage range when the transistor is biased in the saturation region?

D16.4 (a) Design the saturated load inverter circuit in Figure 16.5(a) such that the power dissipation is 0.30 mW and the output voltage is 0.08 V for $v_I = 1.4$ V. The circuit is biased at $V_{DD} = 1.8$ V and the transistor threshold voltage of each transistor is $V_{TNO} = 0.4$ V. (b) Using the results of part (a), find the range of input voltage such that the driver transistor is biased in the saturation region.

16.5 An NMOS inverter with saturated load is shown in Figure 16.5(a). The bias is $V_{DD} = 3$ V and the transistor threshold voltages are 0.5 V. (a) Find the ratio $K_D/K_L$ such that $v_O = 0.25$ V when $v_I = 3$ V. (b) Repeat part (a) for $v_I = 2.5$ V. (c) If $W/L = 1$ for the load transistor, determine the power dissipation in the inverter for parts (a) and (b).

D16.6 Consider the NMOS inverter with saturated load in Figure 16.5(a). Let $V_{DD} = 3$ V. (a) Design the circuit such that the power dissipation in the circuit is 400 $\mu$W and the output voltage is 0.10 V when the input voltage is a logic 1. (b) Determine the transition point of the driver transistor.

16.7 The NMOS inverter with saturated load in Figure 16.5(a) operates with a supply voltage of $V_{DD}$. The MOSFETs have threshold voltages of $V_{TN} = 0.2\,V_{DD}$. Determine $(W/L)_D/(W/L)_L$ such that $V_O = 0.08\,V_{DD}$. Neglect the body effect.

16.8 The enhancement-load transistor in the NMOS inverter in Figure P16.8 has a separate bias applied to the gate. Assume transistor parameters of $K_n = 1$ mA/V$^2$ for $M_D$, $K_n = 0.4$ mA/V$^2$ for $M_L$, and $V_{TN} = 1$ V for both transistors. Using the appropriate logic 0 and logic 1 input voltages, determine $V_{OH}$ and $V_{OL}$ for: (a) $V_B = 4$ V, (b) $V_B = 5$ V, (c) $V_B = 6$ V, and (d) $V_B = 7$ V.

16.9 For the depletion load inverter shown in Figure 16.7(a), assume parameters of $V_{DD} = 3.3$ V, $V_{TND} = 0.5$ V, $V_{TNL} = -0.8$ V, $K_D = 500\,\mu$A/V$^2$, and $K_L = 100\,\mu$A/V$^2$. (a) Find the transition points of the driver and load transistors. (b) Determine $v_O$ for $v_I = 3.3$ V. (c) Determine the maximum current and maximum power dissipation in the circuit.

16.10 In the depletion-load NMOS inverter circuit in Figure 16.7(a), let $V_{TND} = 0.5$ V and $V_{DD} = 3$ V, $K_L = 50\,\mu$A/V$^2$, and $K_D = 500\,\mu$A/V$^2$. Calculate the value of $V_{TNL}$ such that $v_O = 0.10$ V when $v_I = 3$ V.

D16.11 Consider the NMOS inverter with depletion load in Figure 16.7(a). Let $V_{DD} = 1.8$ V, and assume $V_{TND} = 0.3$ V and $V_{TNL} = -0.6$ V. (a) Design the circuit such that the power dissipation is 80 $\mu$W and the output voltage is $v_O = 0.06$ V when $v_I$ is a logic 1. (b) Using the results of part (a), determine the transition points for the driver and load transistors. (c) If $(W/L)_D$ found in part (a) is doubled, what is the maximum power dissipation in the inverter and what is $v_O$ when $v_I$ is a logic 1?

D16.12 The NMOS inverter with depletion load is shown in Figure 16.7(a). The bias is $V_{DD} = 2.5$ V. The transistor parameters are $V_{TND} = 0.5$ V and $V_{TNL} = -1$ V. The width-to-length ratio of the load device is $W/L = 1$. (a) Design the driver transistor such that $v_O = 0.05$ V when the input is a logic 1. (b) What is the power dissipated in the circuit when $v_I = 2.5$ V?

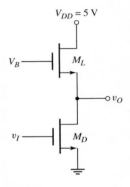

$V_{DD} = 5$ V

$V_B \!-\!|\ M_L$

$\circ\, v_O$

$v_I \!-\!|\ M_D$

Figure P16.8

16.13   Calculate the power dissipated in each inverter circuit in Figure P16.13 for the following input conditions: (a) Inverter a: (i) $v_I = 0.5$ V, (ii) $v_I = 5$ V; (b) Inverter b: (i) $v_I = 0.25$ V, (ii) $v_I = 4.3$ V; (c) Inverter c: (i) $v_I = 0.03$ V, (ii) $v_I = 5$ V.

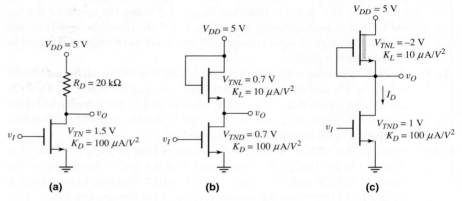

**Figure P16.13**

16.14   For the two inverters in Figure P16.14, assume $(W/L)_L = 1$ for the load devices and $(W/L)_D = 10$ for the driver devices. (a) If $v_I$ is a logic 1, determine the values of $v_{O1}$ and $v_{O2}$. (b) Repeat part (a) if $v_I$ is a logic 0.

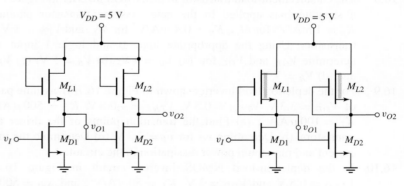

**Figure P16.14**                                        **Figure P16.15**

16.15   Consider the circuit in Figure P16.15. The parameters of the driver transistors are $V_{TND} = 0.8$ V and $(W/L)_D = 4$, and those of the load transistors are $V_{TNL} = -1.2$ V and $(W/L)_L = 1$. (a) If $v_I$ is a logic 1, determine the values of $v_{O1}$ and $v_{O2}$. (b) Repeat part (a) if $v_I$ is a logic 0.

16.16   For the saturated load inverter shown in Figure 16.9(a), assume transistor parameters of $V_{TNDO} = V_{TNLO} = 0.5$ V, $K_D = 200\ \mu A/V^2$, $K_L = 20\ \mu A/V^2$, $\gamma = 0.25\ V^{1/2}$, and $\phi_{fp} = 0.35$ V. Determine $v_O$ for $v_I = 0.12$ V for the case when (a) the body effect is neglected and (b) the body effect is taken into account.

16.17   Consider the NMOS inverter with depletion load in Figure 16.9(b). The transistor parameters are $V_{TNDO} = 0.4$ V, $V_{TNLO} = -0.6$ V, $K_D = 100\ \mu A/V^2$, $K_L = 20\ \mu A/V^2$, $\gamma = 0.25\ V^{1/2}$, and $\phi_{fp} = 0.35$ V. Determine $v_I$ when $v_O = 1.25$ V for the case when (a) the body effect is neglected and (b) the body effect is taken into account.

## Section 16.2 NMOS Logic Circuits

16.18   Consider the circuit with a depletion load device shown in Figure P16.18. (a) For $v_X = 1.8$ V and $v_Y = 0.1$ V, determine $K_D/K_L$ such that $v_O = 0.1$ V. (b) Using the results of part (a), determine $v_O$ when $v_X = v_Y = 1.8$ V. (c) If the width-to-length ratio of the depletion load is $(W/L)_L = 1$, determine the power dissipation in the logic circuit for the input conditions listed in parts (a) and (b).

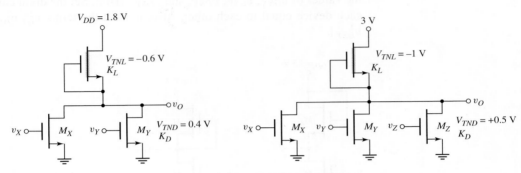

Figure P16.18                         Figure P16.19

D16.19  Consider the three-input NOR logic gate in Figure P16.19. The transistor parameters are $V_{TNL} = -1$ V and $V_{TND} = 0.5$ V. The maximum value of $v_O$ in its low state is to be 0.1 V. (a) Determine $K_D/K_L$. (b) The maximum power dissipation in the NOR logic gate is to be 0.1 mW. Determine the width-to-length ratios of the transistors. (c) Determine $v_O$ when $v_X = v_Y = v_Z = 3$ V.

16.20   Consider a four-input NMOS NOR logic gate with a depletion load similar to the circuit in Figure P16.19. Assume $V_{DD} = 2.5$ V, $V_{TND} = 0.4$ V, and $V_{TNL} = -0.6$ V. The maximum value of $v_O$ in its low state is to be 50 mV. (a) Determine $K_D/K_L$. (b) The maximum power dissipation in this NOR logic gate is to be 50 $\mu$W. Determine the width-to-length ratio of each transistor. (c) Determine $v_O$ when (i) two inputs are a logic 1, (ii) three inputs are a logic 1, and (iii) all inputs are a logic 1.

16.21   The transistor parameters for the circuit in Figure P16.21 are: $V_{TN} = 0.8$ V for all enhancement-mode devices, $V_{TN} = -2$ V for the depletion-mode

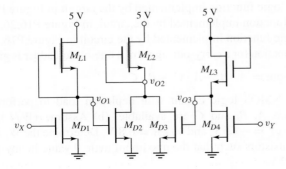

Figure P16.21

devices, and $k'_n = 60 \ \mu A/V^2$ for all devices. The width-to-length ratios of $M_{L2}$ and $M_{L3}$ are 1, and those for $M_{D2}$, $M_{D3}$, and $M_{D4}$ are 8. (a) For $v_X = 5$ V, output $v_{O1}$ is 0.15 V, and the power dissipation in this inverter is to be no more than 250 $\mu$W. Determine $(W/L)_{ML1}$ and $(W/L)_{MD1}$. (b) For $v_X = v_Y = 0$, determine $v_{O2}$.

16.22   Consider the NMOS circuit in Figure P16.22. The transistor parameters are $(W/L)_X = (W/L)_Y = 12$, $(W/L)_L = 1$, and $V_{TN} = 0.4$ V. Neglect the body effect. (a) Determine $v_O$ when $v_X = v_Y = 2.9$ V. (b) What are the values of $v_{GSX}$, $v_{GSY}$, $v_{DSX}$, and $v_{DSY}$? [Hint: Set the drain current in each device equal to each other. Also, neglect the terms $v_O^2$, $v_{DSX}^2$, and $v_{DSY}^2$].

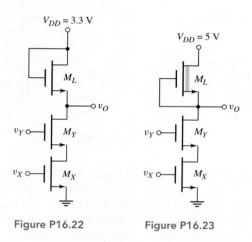

Figure P16.22          Figure P16.23

16.23   In the NMOS circuit in Figure P16.23, the transistor parameters are: $(W/L)_X = (W/L)_Y = 4$, $(W/L)_L = 1$, $V_{TNX} = V_{TNY} = 0.8$ V, and $V_{TNL} = -1.5$ V. (a) Determine $v_O$ when $v_X = v_Y = 5$ V. (b) What are the values of $v_{GSX}$, $v_{GSY}$, $v_{DSX}$, and $v_{DSY}$? Repeat part (a) for $\gamma = 0.5$.

16.24   Consider a four-input NMOS NAND logic gate with a depletion load similar to the circuit in Figure P16.23. The bias voltage is $V_{DD} = 3.3$ V, and the threshold voltages are $V_{TND} = 0.4$ V and $V_{TNL} = -0.6$ V. The logic 0 output voltage is to be 0.10 V. (a) Using approximation methods, determine $K_D/K_L$. (b) The maximum power dissipation in the circuit is to be 100 $\mu$W. Determine $(W/L)_L$ and $(W/L)_D$.

16.25   Determine the logic function implemented by the circuit in Figure P16.25.

16.26   Find the logic function implemented by the circuit in Figure P16.26.

16.27   What is the logic function implemented by the circuit in Figure P16.27.

D16.28   The Boolean function for a carry-out signal of a one-bit full adder is given by

$$\text{Carry-out} = A \cdot B + A \cdot C + B \cdot C$$

(a) Design an NMOS logic circuit with depletion load to perform this function. Signals $A$, $B$, and $C$ are available. (b) Assume $(W/L)_L = 1$, $V_{DD} = 5$ V, $V_{TNL} = -1.5$ V, and $V_{TND} = 0.8$ V. Determine the $W/L$ ratio of the other transistors such that the maximum logic 0 value in any part of the circuit is 0.2 V.

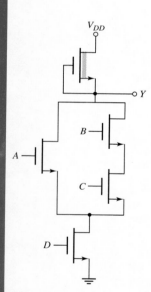

Figure P16.25

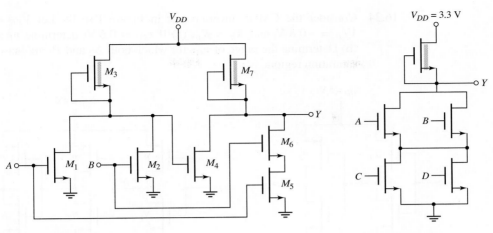

Figure P16.26                                      Figure P16.27

**D16.29**  (a) Design an NMOS depletion-load logic gate that implements the function $\bar{Y} = [A + B \cdot (C + D)]$. (b) Assume $V_{DD} = 2.5\,\text{V}$, $(W/L)_L = 1$, $V_{TND} = 0.4\,\text{V}$, and $V_{TNL} = -0.6\,\text{V}$. Determine $(W/L)_D$ of each transistor such that $V_{OL}\,(\text{max}) = 50\,\text{mV}$.

**D16.30**  Design an NMOS logic circuit with a depletion load that will sound an alarm in an automobile if the ignition is turned off while the headlights are still on and/or the parking brake has not been set. Separate indicator lights are also to be included showing whether the headlights are on or the parking brake needs to be set. State any assumptions that are made.

## Section 16.3  CMOS Inverter

**16.31**  Consider the CMOS inverter in Figure 16.21 biased at $V_{DD} = 2.5\,\text{V}$. The transistor parameters are $V_{TN} = 0.4\,\text{V}$, $V_{TP} = -0.4\,\text{V}$, and $K_n = K_p = 100\,\mu\text{A/V}^2$. (a) Find the transition points for the p-channel and n-channel transistors. (b) Sketch the voltage transfer characteristics, including the appropriate voltage values at the transition points. (c) Determine $v_O$ for $v_I = 1.1\,\text{V}$ and $v_I = 1.4\,\text{V}$.

**16.32**  For the CMOS inverter in Figure 16.21, let $V_{DD} = 3.3\,\text{V}$, $k'_n = 100\,\mu\text{A/V}^2$, $k'_p = 40\,\mu\text{A/V}^2$, $V_{TN} = 0.4\,\text{V}$, and $V_{TP} = -0.4\,\text{V}$. (a) Let $(W/L)_n = 2$ and $(W/L)_p = 5$. (i) Find the transition points for the p-channel and n-channel transistors. (ii) Sketch the voltage transfer characteristics including the appropriate voltage values at the transition points. (iii) Find $v_I$ for $v_O = 0.25\,\text{V}$ and $v_O = 3.05\,\text{V}$. (b) Repeat part (a) for $(W/L)_n = 4$ and $(W/L)_p = 5$.

**16.33**  (a) For the CMOS inverter in Figure 16.21 in the text, let $V_{DD} = 3.3\,\text{V}$, $V_{TN} = +0.4\,\text{V}$, and $V_{TP} = -0.4\,\text{V}$. Assume $(W/L)_n = 4$ and $(W/L)_p = 12$. Determine (i) the input switching voltage, (ii) the input voltage when $v_O = 3.1\,\text{V}$, and (iii) the input voltage when $v_O = 0.2\,\text{V}$. (b) Repeat part (a) for $(W/L)_n = 6$ and $(W/L)_p = 4$.

16.34    Consider the CMOS inverter pair in Figure P16.34. Let $V_{TN} = 0.8$ V, $V_{TP} = -0.8$ V, and $K_n = K_p$. (a) If $v_{O1} = 0.6$ V, determine $v_I$ and $v_{O2}$. (b) Determine the range of $v_{O2}$ for which both $N_2$ and $P_2$ are biased in the saturation region.

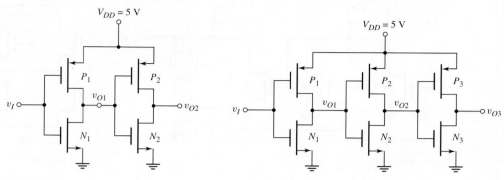

Figure P16.34                                    Figure P16.35

16.35    Consider the series of CMOS inverters in Figure P16.35. The threshold voltages of the n-channel transistors are $V_{TN} = 0.8$ V, and the threshold voltages of the p-channel transistors are $V_{TP} = -0.8$ V. The conduction parameters are all equal. (a) Determine the range of $v_{O1}$ for which both $N_1$ and $P_1$ are biased in the saturation region. (b) If $v_{O2} = 0.6$ V, determine the values of $v_{O3}$, $v_{O1}$, and $v_I$.

16.36    (a) A CMOS inverter is biased at $V_{DD} = 2.5$ V. The transistor parameters are $K_n = K_p = 120\mu$A/V$^2$, $V_{TN} = 0.4$ V, and $V_{TP} = -0.4$ V. Calculate and plot the current in the transistors as a function of the input voltage for $0 \leq v_I \leq 2.5$ V. (b) Repeat part (a) for $V_{DD} = 1.8$ V and $0 \leq v_I \leq 1.8$ V.

16.37    The transistor parameters in the CMOS inverter are $V_{TN} = 0.35$ V, $V_{TP} = -0.35$ V, $k_n' = 80\,\mu$A/V$^2$, and $k_p' = 40\,\mu$A/V$^2$. Let $V_{DD} = 1.8$ V. (a) Determine the peak current in the inverter during a switching cycle for $(W/L)_n = 2$ and $(W/L)_p = 4$. (b) Repeat part (a) for $(W/L)_n = 2$ and $(W/L)_p = 6$. (c) Repeat part (a) for $(W/L)_n = (W/L)_p = 4$.

16.38    A CMOS inverter is biased at $V_{DD} = 3.3$ V. The transistor threshold voltages are $V_{TN} = +0.4$ V and $V_{TP} = -0.4$ V. Determine the peak current in the inverter and the input voltage at which it occurs for (a) $(W/L)_n = 3$, $(W/L)_p = 7.5$; (b) $(W/L)_n = (W/L)_p = 4$; (c) $(W/L)_n = 3$, $(W/L)_p = 12$.

16.39    A load capacitor of 0.2 pF is connected to the output of a CMOS inverter. Determine the power dissipated in the CMOS inverter for a switching frequency of 10 MHz, for inverter parameters described in (a) Problem 16.36 and (b) Problem 16.37.

16.40    (a) A CMOS digital logic circuit contains the equivalent of 4 million CMOS inverters and is biased at $V_{DD} = 1.8$ V. The equivalent load capacitance of each inverter is 0.12 pF and each inverter is switching at 150 MHz. Determine the total average power dissipated in the circuit. (b) If the switching frequency is doubled, but the total power dissipation is to remain the same with the same load capacitance, determine the required bias voltage.

16.41    A particular IC chip can dissipate 3 W and contains 10 million CMOS inverters. Each inverter is being switched at a frequency $f$. (a) Determine the average power that each inverter can dissipate without exceeding the total allowed power. (b) If the switching frequency is $f = 5$ MHz, what is the maximum capacitive load on each inverter if (i) $V_{DD} = 5$ V, (ii) $V_{DD} = 3.3$ V, and (iii) $V_{DD} = 1.5$ V.

16.42   Repeat Problem 16.41 for the case when the chip contains 5 million CMOS inverters being switched at $f = 8$ MHz and the total power dissipated can be 10 W.

16.43   Consider a CMOS inverter. (a) Show that when $v_I \cong V_{DD}$, the resistance of the NMOS device is approximately $1/[k'_n(W/L)_n(V_{DD} - V_{TN})]$, and when $v_I \cong 0$, the resistance of the PMOS device is approximately $1/[k'_p(W/L)_p(V_{DD} + V_{TP})]$. (b) Using the results of part (a), determine the maximum current that the NMOS device can sink such that the output voltage stays below 0.5 V, and determine the maximum current that the PMOS device can source such that the output voltage does not drop more than 0.5 V below $V_{DD}$.

16.44   The CMOS inverter in Figure 16.21 is biased at $V_{DD} = 3.3$ V. Let $K_n = K_p$, $V_{TN} = 0.5$ V, and $V_{TP} = -0.5$ V. (a) Determine the two values of $v_I$ and the corresponding values of $v_O$ for which $(dv_O/dv_I) = -1$ on the voltage transfer characteristics. (b) Find the noise margins.

16.45   Repeat Problem 16.44 if the circuit and transistor parameters are $V_{DD} = 2.5$ V, $V_{TN} = 0.35$ V, $V_{TP} = -0.35$ V, $K_n = 100\,\mu A/V^2$, and $K_p = 50\,\mu A/V^2$.

16.46   (a) Determine the noise margins of a CMOS inverter biased at $V_{DD} = 3.3$ V with $(W/L)_n = 2$ and $(W/L)_p = 5$. Assume $V_{TN} = 0.4$ V and $V_{TP} = -0.4$ V. (b) Repeat part (a) for $(W/L)_n = 4$ and $(W/L)_p = 12$.

## Section 16.4 CMOS Logic Circuits

16.47   Consider the three-input CMOS NAND circuit in Figure P16.47. Assume $k'_n = 2k'_p$ and $V_{TN} = |V_{TP}| = 0.8$ V. (a) If $v_A = v_B = 5$ V, determine $v_C$ such that both $N_3$ and $P_3$ are biased in the saturation region when $(W/L)_p = 2(W/L)_n$. (State any assumptions you make.) (b) If $v_A = v_B = v_C = v_I$, determine the relationship between $(W/L)_p$ and $(W/L)_n$ such that $v_I = 2.5$ V when all transistors are biased in the saturation region. (c) Using the results

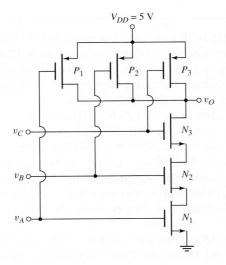

**Figure P16.47**

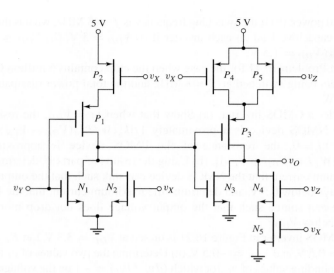

**Figure P16.48**

of part (b) and assuming $v_A = v_B = 5$ V, determine $v_C$ such that both $N_3$ and $P_3$ are biased in the saturation region. (State any assumptions you make.)

16.48 Consider the circuit in Figure P16.48. (a) The inputs $v_X$, $v_Y$, and $v_Z$ listed in the following table are either a logic 0 or a logic 1. These inputs are the outputs from similar-type CMOS logic circuits. The input logic conditions listed are sequential in time. State whether the transistors listed are "on" or "off," and determine the output voltage. (b) What logic function does this circuit implement?

| $v_X$ | $v_Y$ | $v_Z$ | $N_1$ | $N_2$ | $N_3$ | $N_4$ | $N_5$ | $v_O$ |
|---|---|---|---|---|---|---|---|---|
| 1 | 0 | 1 | | | | | | |
| 0 | 0 | 1 | | | | | | |
| 1 | 1 | 0 | | | | | | |
| 1 | 1 | 1 | | | | | | |

16.49 Consider a four-input CMOS NOR logic gate. Determine the $W/L$ ratios of the transistors to provide for symmetrical switching based on the CMOS inverter design with $(W/L)_n = 2$ and $(W/L)_p = 4$. (b) If the load capacitance of the NOR gate doubles, determine the required $W/L$ ratios to provide the same switching speed as the logic gate in part (a).

16.50 Repeat Problem 16.49 for a four-input CMOS NAND logic gate.

16.51 Repeat Problem 16.49 for a three-input CMOS NOR logic gate.

16.52 Repeat Problem 16.49 for a three-input CMOS NAND logic gate.

D16.53 Figure P16.53 shows a classic CMOS logic circuit. (a) What is the logic function performed by the circuit? (b) Design the NMOS network. (c) Determine the transistor $W/L$ ratios to provide symmetrical switching times at twice the switching speed of the basic CMOS inverter with $(W/L)_n = 2$ and $(W/L)_p = 4$.

D16.54 Figure P16.54 is a classic CMOS logic gate. (a) What is the logic function performed by the circuit? (b) Design the PMOS network. (c) Determine the transistor $W/L$ ratios to provide symmetrical switching times at twice

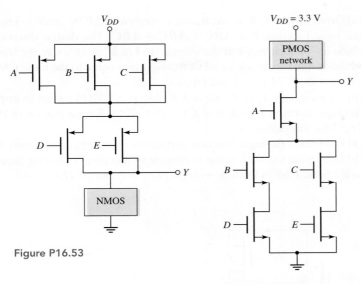

Figure P16.53

Figure P16.54

the switching speed as the basic CMOS inverter with $(W/L)_n = 2$ and $(W/L)_p = 4$.

D16.55 Figure P16.55 is a classic CMOS logic gate. (a) What is the logic function performed by the circuit? (b) Design the NMOS network. (c) Determine the transistor $W/L$ ratios to provide symmetrical switching times equal to the basic CMOS inverter with $(W/L)_n = 2$ and $(W/L)_p = 4$.

D16.56 Consider the classic CMOS logic circuit in Figure P16.56. (a) What is the logic function performed by the circuit? (b) Design the PMOS network. (c) Determine the transistor $W/L$ ratios to provide symmetrical switching times equal to the basic CMOS inverter with $(W/L)_n = 2$ and $(W/L)_p = 4$.

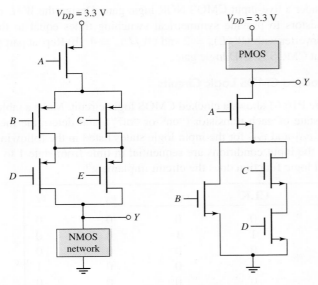

Figure P16.55                    Figure P16.56

D16.57 (a) Given inputs $A$, $B$, $C$, $\bar{A}$, $\bar{B}$, and $\bar{C}$, design a CMOS circuit to implement the logic function $Y = A\bar{B}\bar{C} + \bar{A}\bar{B}C + \bar{A}B\bar{C}$. The design should not include a CMOS inverter at the output. (b) For $k'_n = 2k'_p$, size the transistors in the design to provide equal switching times equal to the basic CMOS inverter with $(W/L)_n = 1$ and $(W/L)_p = 2$.

D16.58 (a) Given inputs $A$, $B$, $C$, $D$, and $E$, design a CMOS circuit to implement the logic function $\bar{Y} = A(B + C) + D + E$. (b) Repeat part (b) of Problem 16.57 for this circuit.

16.59 (a) Determine the logic function performed by the circuit in Figure P16.59. (b) Determine the $W/L$ ratios to provide symmetrical switching times equal to the basic CMOS inverter with $(W/L)_n = 2$ and $(W/L)_p = 4$.

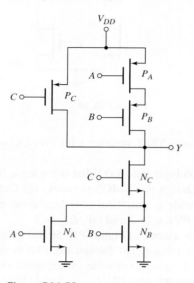

Figure P16.59

D16.60 (a) Consider a five-input CMOS NOR logic gate. Design the $W/L$ ratios of the transistors to provide symmetrical switching times equal to the basic CMOS inverter with $(W/L)_n = 2$ and $(W/L)_p = 4$. (b) Repeat part (a) for a five-input CMOS NAND logic gate.

### Section 16.5 Clocked CMOS Logic Circuits

16.61 (a) Figure P16.61 shows a clocked CMOS logic circuit. Make a table showing the state of each transistor ("on" or "off"), and determine the output voltages $v_{O1}$ and $v_{O2}$ for the input logic states listed in the following table. Assume the input conditions are sequential in time from state 1 to state 6. (b) What logic function does the circuit implement?

| State | CLK | $v_A$ | $v_B$ | $v_C$ |
|-------|-----|-------|-------|-------|
| 1 | 0 | 0 | 0 | 0 |
| 2 | 1 | 1 | 0 | 0 |
| 3 | 0 | 0 | 0 | 0 |
| 4 | 1 | 0 | 0 | 1 |
| 5 | 0 | 0 | 0 | 0 |
| 6 | 1 | 0 | 1 | 1 |

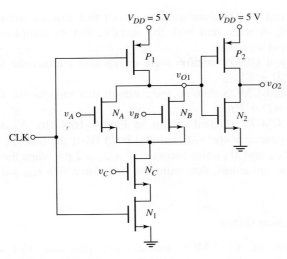

**Figure P16.61**

16.62  (a) For the circuit in Figure P16.62, make a table showing the state of each
transistor ("on" or "off"), and determine the output voltages $v_{O1}$, $v_{O2}$, and
$v_{O3}$ for the input logic states listed in the following table. Assume the input
conditions are sequential in time from state 1 to state 6. (b) What logic function does the circuit implement?

| State | CLK | $v_X$ | $v_Y$ | $v_Z$ |
|-------|-----|-------|-------|-------|
| 1 | 0 | 0 | 0 | 0 |
| 2 | 1 | 1 | 1 | 1 |
| 3 | 0 | 0 | 0 | 0 |
| 4 | 1 | 0 | 1 | 1 |
| 5 | 0 | 0 | 0 | 0 |
| 6 | 1 | 1 | 0 | 1 |

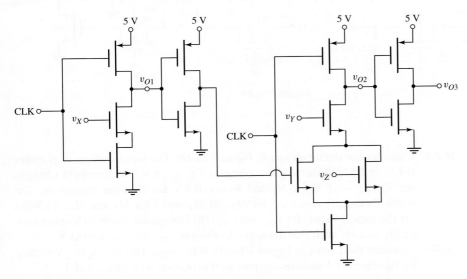

**Figure P16.62**

**D16.63** Sketch a clocked CMOS domino logic circuit that realizes the function $Y = A\bar{B} + \bar{A}B$. Assume that both the variable and its complement are available as input signals.

**D16.64** Sketch a clocked CMOS domino logic circuit that realizes the function $Y = AB + C(D + E)$.

**D16.65** Sketch a clocked CMOS domino logic circuit that realizes the function $Y = A(B + C)(D + E)$.

**16.66** Consider the CMOS clocked circuit in Figure 16.44(b). Assume the effective capacitance at the $v_{O1}$ terminal is 25 fF. If the leakage current through the $M_{NA}$ and $M_{NB}$ transistors is $I_{\text{Leakage}} = 2$ pA when these transistors and $M_{P1}$ are cutoff, determine the time for which $v_{O1}$ will decay by 0.5 V.

## Section 16.6 Transmission Gates

**16.67** The parameters of an NMOS transmission gate are $V_{TN} = 0.4$ V, $K_n = 0.15$ mA/V$^2$, and $C_L = 0.2$ pF. (a) For a gate voltage of $\phi = 3.3$ V, determine the quasi-steady-state output voltage for (i) $v_I = 0$, (ii) $v_I = 3.3$ V, and (iii) $v_I = 2.5$ V. (b) Repeat part (a) for a gate voltage of $\phi = 1.8$ V.

**16.68** The NMOS transistors in the circuit shown in Figure P16.68 have parameters $K_n = 0.2$ mA/V$^2$, $V_{TN} = 0.5$ V, $\lambda = 0$, and $\gamma = 0$. (a) For gate voltages of $\phi = 2.5$ V, determine the quasi-steady-state output voltage for (i) $v_I = 0$, (ii) $v_I = 2.5$ V, and (iii) $v_I = 1.8$ V. (b) Repeat part (a) for gate voltages of $\phi = 2.0$ V.

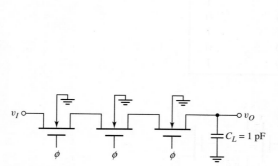

**Figure P16.68**

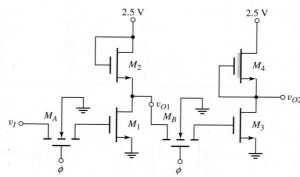

**Figure P16.69**

**16.69** Consider the circuit shown in Figure P16.69. The input voltage $v_I$ is either 0.1 V or 2.5 V. Assume gate voltages of $\phi = 2.5$ V. The threshold voltages are $V_{TN} = -0.6$ V for $M_4$ and $V_{TN} = 0.4$ V for all other transistors. The width-to-length ratios are 1 for $M_2$ and $M_4$, and 5 for $M_A$ and $M_B$. (a) What are the logic 1 values for $v_{O1}$ and $v_{O2}$? (b) Design the width-to-length ratios of $M_1$ and $M_3$ such that the logic 0 values of $v_{O1}$ and $v_{O2}$ are 0.1 V

**16.70** Consider the circuit in Figure P16.70. What logic function is implemented by this circuit? Are there any potential problems with this circuit?

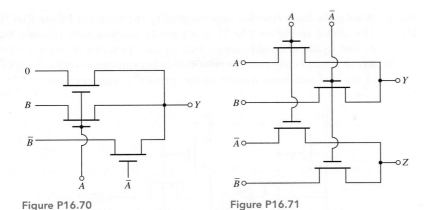

Figure P16.70                         Figure P16.71

16.71   What is the logic function implemented by the circuit in Figure P16.71?

D16.72   (a) Design an NMOS pass transistor logic circuit to perform the function $Y = A + B(C + D)$. Assume that both the variable and its complement are available as input signals. (b) Repeat part (a) for the function $Y = ABC + \bar{A}\bar{B}\bar{C}$.

16.73   Consider the circuit in Figure P16.73. (a) Determine the value of $Y$ for $\phi = 2.5\,\text{V}$ and (i) $A = B = 0$; (ii) $A = 0$, $B = 2.5\,\text{V}$; (iii) $A = 2.5\,\text{V}$, $B = 0$; and (iv) $A = B = 2.5\,\text{V}$. (b) Repeat part (a) for $\phi = 0$. (c) What is the logic function implemented by the circuit?

16.74   What is the logic function implemented by the circuit in Figure P16.74?

16.75   Consider the circuit in Figure P16.75. (a) Determine the value of $Y$ for (i) $A = B = 0$; (ii) $A = 2.5\,\text{V}$, $B = 0$; (iii) $A = 0$, $B = 2.5\,\text{V}$; and (iv) $A = B = 2.5\,\text{V}$. (b) What is the logic function implemented by the circuit?

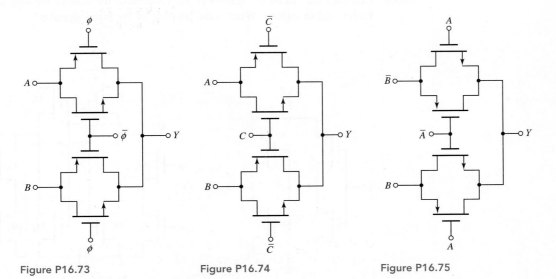

Figure P16.73              Figure P16.74              Figure P16.75

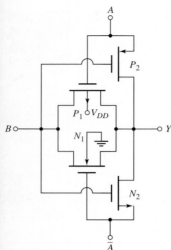

Figure P16.76

16.76 What is the logic function implemented by the circuit in Figure P16.76?

16.77 The circuit in Figure P16.77 is a form of clocked shift register. Signals $\phi_1$ and $\phi_2$ are nonoverlapping clock signals. Describe the operation of the circuit. Discuss any possible relationship between the width-to-length ratios of the load and driver transistors for "proper" circuit operation.

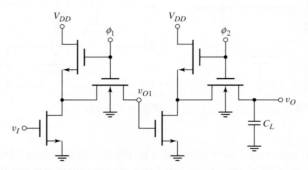

Figure P16.77

## Section 16.7 Sequential Logic Circuits

16.78 Consider the NMOS R-S flip-flop in Figure 16.63 biased at $V_{DD} = 2.5$ V. The threshold voltages are 0.4 V (enhancement-mode devices) and $-0.6$ V (depletion-mode devices). The conduction parameters are $K_3 = K_6 = 40\,\mu\text{A/V}^2$, $K_2 = K_5 = 100\,\mu\text{A/V}^2$, and $K_1 = K_4 = 150\,\mu\text{A/V}^2$. If $Q = $ logic 0 and $\bar{Q} = $ logic 1 initially, determine the voltage at $S$ that will cause the flip-flop to change states.

16.79 Figure P16.79 shows two CMOS inverters in cascade. This circuit can be thought of as an uncoupled CMOS R/S flip flop. The transistor parameters are $K_n = K_p = 0.2\,\text{mA/V}^2$, $V_{TN} = 0.5$ V, $V_{TP} = -0.4$ V, and $\lambda_n = \lambda_p = 0$. Plot $v_{O1}$ and $v_O$ versus $v_I$. In particular, calculate the values of $v_{O1}$ and $v_O$ at $v_I = 1.5$, 1.6, 1.7, and 1.8 V.

16.80 Consider the circuit in Figure P16.80. Determine the state of the outputs for various input signals. What is the purpose of the input signal $\phi$?

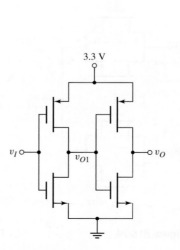

Figure P16.79

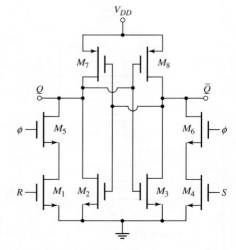

Figure P16.80

**D16.81** The circuit in Figure P16.81 is an example of a D flip-flop. (a) Explain the operation of the circuit. Is this a positive- or negative-edge-triggered flip-flop? (b) Redesign the circuit to make this a static flip-flop.

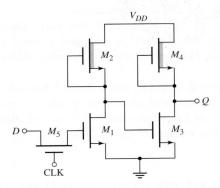

Figure P16.81

**16.82** Show that the circuit in Figure P16.82 is a J–K flip-flop.

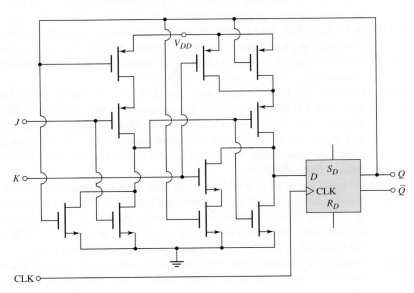

Figure P16.82

**16.83** Reconsider the circuit shown in Figure P16.48. Show that this circuit is a J–K flip-flop with $J = v_X$, $K = v_Y$, and CLK $= v_Z$.

## Section 16.8 Memories: Classifications and Architectures

**16.84** A 256-K memory is organized in a square array and uses the NMOS NOR decoder in Figure 16.73(b) for the row- and column-decoders. (a) How many inputs does each decoder require? (b) What input to the row decoder is required to address row (i) 52, (ii) 129, and (iii) 241? (c) What input to the column decoder is required to address column (i) 24, (ii) 165, and (iii) 203?

16.85    (a) A 1 megabit memory is organized in a square with each memory cell being individually addressed. Determine the number of input address lines required for the row and column decoders. (b) If the 1 megabit memory is organized as 250K words $\times$ 4 bits, determine the minimum number of input address lines required for the row and column decoders.

16.86    A 4096-bit RAM consists of 512 words of 8 bits each. Design the memory array to minimize the number of row and column address decoder transistors required. How many row and column address lines are required?

16.87    Assume that an NMOS address decoder can source 250 $\mu$A when the output goes high. If the effective capacitance of each memory cell is $C_L = 0.8$ pF and the effective capacitance of the address line is $C_{LA} = 5$ pF, determine the rise time of the address line voltage if $V_{IH} = 2.7$ V.

### Section 16.9 RAM Memory Cells

16.88    Consider the NMOS RAM cell with resistor load in Figure 16.74(b). Assume parameters values of $k'_n = 80 \, \mu$A/V$^2$, $V_{TN} = 0.4$ V, $V_{DD} = 2.5$ V, and $R = 1$ M$\Omega$. (a) Design the width-to-length ratio of the driver transistor such that $V_{DS} = 20$ mV for the on transistor. (b) Consider a 16-K memory with the cell described in part (a). Determine the standby cell current and the total memory power dissipation for a standby voltage of $V_{DD} = 1.2$ V.

D16.89    A 16-K NMOS RAM, with the cell design shown in Figure 16.74(b), is to dissipate no more than 200 mW in standby when biased at $V_{DD} = 2.5$ V. Design the width-to-length ratios of the transistors and the resistance value. Assume $V_{TN} = 0.7$ V and $k'_n = 35 \, \mu$A/V$^2$.

16.90    Consider the CMOS RAM cell and data lines in Figure 16.76 biased at $V_{DD} = 2.5$ V. Assume transistor parameters $k'_n = 80 \, \mu$A/V$^2$, $k'_p = 35 \, \mu$A/V$^2$, $V_{TN} = 0.4$ V, $V_{TP} = -0.4$ V, $W/L = 2$ ($M_{N1}$ and $M_{N2}$), $W/L = 4$ ($M_{P1}$ and $M_{P2}$), and $W/L = 1$ (all other transistors). If $Q = 0$ and $\bar{Q} = 1$, determine the steady-state values of $D$ and $\bar{D}$ after the row has been selected. Neglect the body effect.

16.91    Consider the CMOS RAM cell and data lines in Figure 16.76 with circuit and transistor parameters described in Problem 16.90. Assume initially that $Q = 0$ and $\bar{Q} = 1$. Assume the row is selected with $X = 2.5$ V and assume the data lines, through a write cycle, are $\bar{D} = 0$ and $D = 2.5$ V. Determine the values of $Q$ and $\bar{Q}$ just after the row select has been applied.

*16.92    Consider a general sense amplifier configuration shown in Figure 16.82 for a dynamic RAM. Assume that each bit line has a capacitance of 1 pF and is precharged to 4 V. The storage capacitance is 0.05 pF, the reference capacitance is 0.025 pF, and each is charged to 5 V for a logic 1 and to 0 V for a logic 0. The $M_S$ and $M_R$ gate voltages are 5 V when each cell is addressed and the transistor threshold voltages are 0.5 V. Determine the bit line voltages $v_1$ and $v_2$ after the cells are addressed for the case when (a) a logic 1 is stored and (b) a logic 0 is stored.

### Section 16.10 Read-Only Memory

D16.93    Design a 4-word $\times$ 4-bit NMOS mask-programmed ROM to produce outputs of 1011, 1111, 0110, and 1001 when rows 1, 2, 3, and 4, respectively, are addressed.

D16.94  Design an NMOS 16 × 4 mask-programmed ROM that provides the 4-bit product of two 2-bit variables.

D16.95  Design an NMOS mask-programmed ROM that decodes a binary input and produces the output for a seven-segment array. (See Figure 2.45, Chapter 2.) The output is to be high when a particular LED is to be turned on.

## Section 16.11  Data Converters

16.96  An analog signal in the range 0 to 5 V is to be converted to a digital signal with a quantization error of less than one percent. (a) What is the required number of bits? (b) What input voltage value represents 1 LSB? (c) What digital output represents an input voltage of 3.5424 V?

16.97  An analog signal in the range 0 to 3.3 V is to be converted to a digital signal with a quantization error of less than 0.5 percent. (a) What is the required number of bits? (b) What input voltage value represents 1 LSB? (c) What digital output represents an input voltage of 2.5321 V.

16.98  (a) What is the output voltage of the 4-bit weighted-resistor D/A in Figure 16.90 if the input is 0110? Assume $R_F = 10 \text{ k}\Omega$. (b) The input signal changes to 1001. What is the output voltage?

16.99  Consider the 4-bit weighted-resistor D/A converter in Figure 16.90. Let $R_F = 10 \text{ k}\Omega$. (a) What is the maximum allowed tolerance (±percent) in the value of $R_1$ so that the maximum error in the output is limited to $\pm \frac{1}{2}$ LSB? (b) Repeat part (a) for the resistor $R_4$.

16.100  The weighted-resistor D/A converter in Figure 16.90 is to be expanded to an 8-bit device. (a) What are the required resistance values of the additional four input resistors? (b) What is the output voltage if the input is 00000001?

16.101  The $N$-bit D/A converter with an $R$–$2R$ ladder network in Figure 16.92 is to be designed as a 6-bit D/A device. Let $V_{\text{REF}} = -5.0 \text{ V}$ and $R = R_F = 5.0 \text{ k}\Omega$. (a) What are currents $I_1, I_2, I_3, I_4, I_5,$ and $I_6$? (b) The input changes by 1 LSB. What is the change in the output voltage? (c) What is the output voltage if the input is 010011? (d) What is the change in output voltage if the input changes from 101010 to 010101?

16.102  The 3-bit flash A/D converter in Figure 16.93 has a reference voltage of $V_{\text{REF}} = 3.3 \text{ V}$. The 3-bit output is 101. What is the range of $v_A$ that produces this output?

16.103  A 6-bit flash A/D converter, similar to the one in Figure 16.93, is to be fabricated. How many resistors and comparators are required?

16.104  A 10-bit counting A/D converter has an analog input in the range $0 \le v_A \le 5$ V and has a clock frequency of 1 MHz. (a) What is the maximum conversion time? (b) If the output is 0010010010, what is the range of the input signal $v_A$ (assume a quantization error of $\pm \frac{1}{2}$ LSB). (c) How many clock pulses are required to produce an output of 0100100100?

16.105  Consider the 10-bit counting A/D converter described in Problem 16.104. (a) What is the output if the analog input is $v_A = 3.125$ V? (b) Repeat part (a) if $v_A = 1.8613$ V.

## COMPUTER SIMULATION PROBLEMS

16.106  Consider the three types of NMOS inverters shown in Figures 16.3(a), 16.5(a), and 16.7(a). Using a computer simulation, investigate the voltage transfer characteristics and the current versus input voltage characteristics of the three types of inverters as a function of various width-to-length ratios and as a function of the body effect.

16.107 Using a computer simulation, investigate the propagation delay time and switching characteristics of a CMOS inverter by setting up a series of CMOS inverters in cascade. Use standard transistors and assume effective $C_T$ load capacitances of 0.05 pF. Determine the propagation delay time as a function of various transistor width-to-length ratios.

16.108 Consider a three-input CMOS NAND logic circuit similar to the two-input circuit shown in Figure 16.34(a). Using a computer simulation, investigate the voltage transfer characteristics and switching characteristics for various NMOS and PMOS width-to-length ratios. What is the optimum relation between the PMOS and NMOS width-to-length ratios for symmetrical switching speeds?

16.109 Using a computer simulation, investigate the $Q$ and $\bar{Q}$ values in the CMOS RAM cell shown in Figure 16.76 during read and write cycles for various transistor width-to-length ratios. In particular, consider the relations given by Equations (16.82) and (16.84).

## DESIGN PROBLEMS

*D16.110 Design a classic CMOS logic circuit that will implement the logic function $Y = A \cdot (B + C) + D \cdot E$.

*D16.111 Design clocked CMOS logic circuits that will implement the logic functions (a) $Y = [A \cdot B + C \cdot D]$ and (b) $Y = [A \cdot (B + C) + D]$.

*D16.112 Design an NMOS pass logic network that implements the logic functions described in Problem 16.111.

*D16.113 Design a clocked CMOS dynamic shift register in which the output becomes valid on the positive-going edge of a clock signal.

# Bipolar Digital Circuits

In the previous chapter, we presented the basic concepts of MOSFET logic circuits. In this chapter, we discuss the basic principles of bipolar logic circuits. We initially consider emitter-coupled logic (ECL). This technology is based on the differential amplifier and is used in specialized high-speed applications.

Prior to the emergence of MOS digital technology, the bipolar digital family of transistor–transistor logic (TTL) circuits was used extensively. TTL and low-power Schottky TTL logic circuits are analyzed. The basic concepts of BiCMOS logic circuits are presented.

Bipolar digital circuits are now used less frequently because of their relatively large power requirements.

## PREVIEW

In this chapter, we will:

- Analyze the basic emitter-coupled logic circuits.
- Analyze and design modified emitter-coupled logic circuits.
- Analyze transistor–transistor logic circuits.
- Analyze and design Schottky and low-power Schottky transistor–transistor logic circuits.
- Analyze BiCMOS digital logic circuits.
- As an application, design a static ECL gate to implement a specific logic function.

# 17.1    EMITTER-COUPLED LOGIC (ECL)

**Objective:** • Analyze the basic emitter-coupled logic circuits

The emitter-coupled logic (ECL) circuit is based on the differential amplifier circuit, which we studied in Chapter 11 in the context of linear amplifiers. In digital applications, the diff-amp is driven into its nonlinear region. The transistors are either cut off or in the active region. Saturation is avoided in order to minimize switching times and propagation delay times. ECL circuits have the shortest propagation delay times of any bipolar digital technology.

## 17.1.1    Differential Amplifier Circuit Revisited

Consider the basic diff-amp circuit in Figure 17.1. For a linear diff-amp, the difference between the two input voltages is small and both transistors remain biased in the active region at all times. The relationship between collector currents and base–emitter voltages for $Q_1$ and $Q_2$ can be written[1]

$$i_{C1} = I_S e^{v_{BE1}/V_T}$$

**(17.1(a))**

and

$$i_{C2} = I_S e^{v_{BE2}/V_T}$$

**(17.1(b))**

where $Q_1$ and $Q_2$ are assumed to be matched and parameter $I_S$ is the same for both devices. The current–voltage transfer curves are shown in Figure 17.2.

In digital applications, the difference between the two input voltages is large, which means that one transistor remains biased in its active region while the opposite

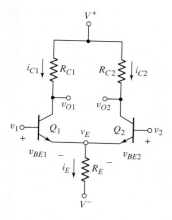

**Figure 17.1** Basic differential amplifier circuit

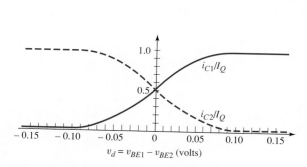

**Figure 17.2** Normalized dc transfer characteristics, BJT differential amplifier

---

[1]In most cases in this chapter, total instantaneous current and voltage parameters are used, even though most analyses of logic circuits involve dc calculations.

transistor is cut off. For example, if $v_{BE1} = v_{BE2} + 0.12$, then the ratio of $i_{C1}$ and $i_{C2}$ is

$$\frac{i_{C1}}{i_{C2}} = \frac{e^{v_{BE1}/V_T}}{e^{v_{BE2}/V_T}} = e^{(v_{BE1} - v_{BE2})/V_T} = e^{0.12/0.026} = 101 \tag{17.2}$$

When the base–emitter voltage of $Q_1$ is 120 mV greater than the base–emitter voltage of $Q_2$, the collector current of $Q_1$ is 100 times that of $Q_2$; for all practical purposes, $Q_1$ is on and $Q_2$ is cut off.

Conversely, if $v_1$ is less than $v_2$ by at least 120 mV, then $Q_1$ is effectively cut off and $Q_2$ is on. The difference amplifier, when operating as a digital circuit, operates as a current switch. When $v_1 > v_2$ by at least 120 mV, it switches an approximately constant current through $R_E$ to $Q_1$; when $v_2 > v_1$ by at least 120 mV, the current goes to $Q_2$.

## EXAMPLE **17.1**

**Objective:** Calculate the currents and voltages in the basic differential amplifier circuit used as a digital circuit.

Consider the circuit in Figure 17.1. Assume that $V^+ = 2.5$ V, $V^- = -2.5$ V, $R_{C1} = R_{C2} \equiv R_C = 5\,k\Omega$, $R_E = 6\,k\Omega$, and $v_2 = 0$. Neglect base currents in the dc analysis.

**Solution:** For $v_1 = v_2 = 0$, both transistors are on. Assume a base–emitter turn-on voltage of 0.7 V; then $v_E = -0.7$ V and

$$i_E = \frac{v_E - V^-}{R_E} = \frac{-0.7 - (-2.5)}{6} = 0.3 \text{ mA}$$

Assuming $Q_1$ and $Q_2$ are matched, we have $i_{C1} = i_{C2} = i_E/2$ so that $i_{C1} = i_{C2} = 0.15$ mA. Then

$$v_{O1} = v_{O2} = V^+ - i_C R_C = 2.5 - (0.15)(5) = 1.75 \text{ V}$$

Both $Q_1$ and $Q_2$ are biased in the active region.

Now let $v_1 = -0.5$ V. Since the base voltage of $Q_1$ is less than the base voltage of $Q_2$ by more than 120 mV, then $Q_1$ is cut off and $Q_2$ is on. In this case, $v_E = v_2 - V_{BE}(\text{on}) = -0.7$ V and $i_E = 0.3$ mA, as before. However, $i_{C1} = 0$ and $i_{C2} = i_E = 0.3$ mA, so that

$$v_{O1} = V^+ = 2.5 \text{ V}$$

and

$$v_{O2} = V^+ - i_{C2} R_C = 2.5 - (0.3)(5) = 1.0 \text{ V}$$

For $v_1 = +0.5$ V, $Q_1$ is on and $Q_2$ is cut off. For this case, $v_E = v_1 - V_{BE}(\text{on}) = 0.5 - 0.7 = -0.2$ V and the current $i_E$ is

$$i_E = i_{C1} = \frac{v_E - V^-}{R_E} = \frac{-0.2 - (-2.5)}{6} = 0.383 \text{ mA}$$

Then

$$v_{O1} = V^+ - i_{C1} R_C = 2.5 - (0.383)(5) = 0.585 \text{ V}$$

and

$$v_{O2} = V^+ = 2.5 \text{ V}$$

**Comment:** For the three conditions given, transistors $Q_1$ and $Q_2$ are biased either in cutoff or in the active region. In terms of digital applications, output $v_{O2}$ is in phase with input $v_1$ and output $v_{O1}$ is 180 degrees out of phase.

When biased on, transistor $Q_1$ conducts slightly more heavily than $Q_2$ when it is conducting. To obtain symmetrical complementary outputs, $R_{C1}$ should therefore be slightly smaller than $R_{C2}$.

---

**Ex 17.1:** Consider the differential amplifier circuit in Figure 17.1 biased at $V^+ = 1.8$ V, $V^- = -1.8$ V, and $v_2 = 0$. Assume $V_{BE}(\text{on}) = 0.7$ V and neglect base currents. (a) Design the circuit such that $i_E = 0.11$ mA and $v_{O1} = v_{O2} = 1.45$ V when $v_1 = 0$. (b) Using the results of part (a), determine $i_E$, $v_{O1}$, and $v_{O2}$ for (i) $v_1 = +0.5$ V and (ii) $v_1 = -0.5$ V. (c) Using the results of parts (a) and (b), calculate the power dissipated in the circuit for (i) $v_1 = +0.5$ V and (ii) $v_1 = -0.5$ V. (Ans. (a) $R_E = 10\,\text{k}\Omega$, $R_C = 6.364\,\text{k}\Omega$; (b) (i) $i_E = 0.16$ mA, $v_{O1} = 0.782$ V, $v_{O2} = 1.8$ V; (ii) $i_E = 0.11$ mA, $v_{O1} = 1.8$ V, $v_{O2} = 1.10$ V; (c) (i) $P = 0.576$ mW, (ii) $P = 0.396$ mW)

---

### 17.1.2     Basic ECL Logic Gate

**Basic ECL Logic Gate**

A basic two-input ECL OR/NOR logic circuit is shown in Figure 17.3. The two input transistors, $Q_1$ and $Q_2$, are connected in parallel. On the basis of the differential amplifier, if both $v_X$ and $v_Y$ are less than the reference voltage $V_R$ (by at least 120 mV), then both $Q_1$ and $Q_2$ are cut off, while the reference transistor $Q_R$ is biased on its active region. In this situation, the output voltage $v_{O1}$ is greater than $v_{O2}$. If either $v_X$ or $v_Y$ becomes greater than $V_R$, then $Q_R$ turns off and $v_{O2}$ becomes larger than $v_{O1}$. The OR logic is at the $v_{O2}$ output and the NOR logic is at the $v_{O1}$ output. An advantage of ECL gates is the availability of complementary outputs, precluding the need for separate inverters to provide the complementary outputs.

One problem with the OR/NOR circuit in Figure 17.3 is that the output voltage levels differ from the required input voltage levels; the output voltages are not

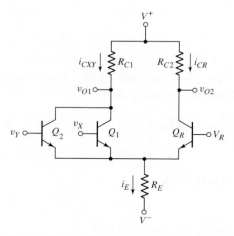

**Figure 17.3** Basic two-input ECL OR/NOR logic circuit

compatible with the input voltages. The mismatch arises because ECL circuit transistors operate between their cutoff and active regions, requiring that the base–collector junctions be reverse biased at all times. We see that a logic 1 voltage of the output is $V_{OH} = V^+$. If this voltage were to be applied to either the $v_X$ or $v_Y$ input, then either $Q_1$ or $Q_2$ would turn on and the collector voltage $v_{O1}$ would decrease below $V^+$; the base–collector voltage would then become forward biased and the transistor would go into saturation. Emitter-follower circuits are added to provide outputs that are compatible with the inputs of similar gates.

### ECL Logic Gate with Emitter Followers

In the ECL circuit in Figure 17.4, emitter followers are added to the OR/NOR outputs, and supply voltage $V^+$ is set equal to zero. The ground and power supply voltages are reversed because analyses show that using the collector–emitter voltage as the output results in less noise sensitivity. If the forward current gain of the transistors is on the order of 100, then the dc base currents may be neglected with little error in the calculations.

If either $v_X$ or $v_Y$ is a logic 1 (defined as greater than $V_R$ by at least 120 mV), then the reference transistor $Q_R$ is cut off, $i_{CR} = 0$, and $v_{O2} = 0$. Output transistor $Q_3$ is biased in the active region, and $v_{OR} = v_{O2} - V_{BE}(\text{on}) = -0.7$ V. If both $v_X$ and $v_Y$ are a logic 0 (defined as less than $V_R$ by at least 120 mV), then both $Q_1$ and $Q_2$ are cut off, $v_{O1} = 0$, and $v_{NOR} = 0 - V_{BE}(\text{on}) = -0.7$ V. The largest possible voltage that can be achieved at either output is $-0.7$ V; therefore, $-0.7$ V is defined as the logic 1 level.

In the following example, we will determine the currents and the logic 0 values in the basic ECL gate.

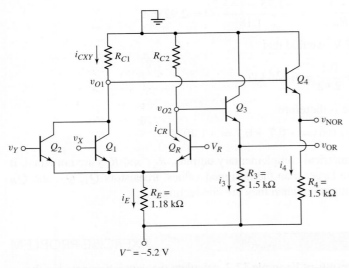

**Figure 17.4** Two-input ECL OR/NOR logic gate with emitter-follower output stages

## EXAMPLE 17.2

**Objective:** Calculate current, resistor, and logic 0 values in the basic ECL logic gate.

Consider the circuit in Figure 17.4. Determine $R_{C1}$ and $R_{C2}$ such that when $Q_1$ and $Q_2$ are conducting, the B–C voltages are zero.

**Solution:** Let $v_X = v_Y = -0.7$ V = logic 1 > $V_R$ such that $Q_1$ and $Q_2$ are on. We find that

$$v_E = v_X - V_{BE}(\text{on}) = -0.7 - 0.7 = -1.4 \text{ V}$$

and the current is

$$i_E = i_{Cxy} = \frac{v_E - V^-}{R_E} = \frac{-1.4 - (-5.2)}{1.18} = 3.22 \text{ mA}$$

In order for the B–C voltages of $Q_1$ and $Q_2$ to be zero, voltage $v_{O1}$ must be $-0.7$ V. Therefore

$$R_{C1} = \frac{-v_{O1}}{I_{Cxy}} = \frac{0.7}{3.22} = 0.217 \text{ k}\Omega$$

The NOR output logic 0 value is then

$$v_{NOR} = v_{O1} - V_{BE}(\text{on}) = -0.70 - 0.7 = -1.40 \text{ V}$$

Input voltages $v_X$ and $v_Y$ are greater than $V_R$ in a logic 1 state and less than $V_R$ in a logic 0 state. If $V_R$ is set at the midpoint between the logic 0 and logic 1 levels, then

$$V_R = \frac{-0.7 - 1.40}{2} = -1.05 \text{ V}$$

When $Q_R$ is on, we have

$$v_E = V_R - V_{BE}(\text{on}) = -1.05 - 0.7 = -1.75 \text{ V}$$

and

$$i_E = i_{CR} = \frac{v_E - V^-}{R_E} = \frac{-1.75 - (-5.2)}{1.18} = 2.92 \text{ mA}$$

For $v_{O2} = -0.7$ V, we find that

$$R_{C2} = \frac{-v_{O2}}{i_{C2}} = \frac{0.7}{2.92} = 0.240 \text{ k}\Omega$$

The OR logic 0 value is therefore

$$v_{OR} = v_{O2} - V_{BE}(\text{on}) = -0.7 - 0.7 = -1.40 \text{ V}$$

**Comment:** For symmetrical complementary outputs, $R_{C1}$ and $R_{C2}$ are not equal. If $R_{C1}$ and $R_{C2}$ become larger than the designed values, transistors $Q_1$, $Q_2$, and $Q_R$ will be driven into saturation when they are conducting.

### EXERCISE PROBLEM

**Ex 17.2:** Using the results of Example 17.2, calculate the power dissipated in the circuit in Figure 17.4; for: (a) $v_x = v_y = $ logic 1, and (b) $v_x = v_y = $ logic 0. (Ans. (a) $P = 45.5$ mW (b) $P = 43.9$ mW)

### The Reference Circuit

Another circuit is required to provide the reference voltage $V_R$. Consider the complete two-input ECL OR/NOR logic circuit shown in Figure 17.5. The reference circuit consists of resistors $R_1$, $R_2$, and $R_5$, diodes $D_1$ and $D_2$, and transistor $Q_5$. The reference portion of the circuit can be specifically designed to provide the desired reference voltage.

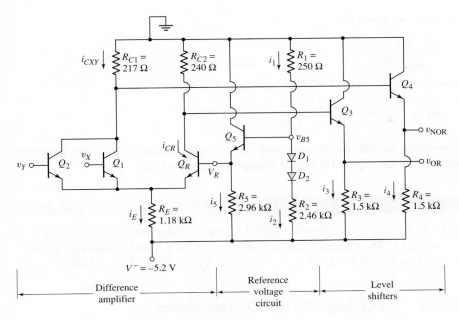

**Figure 17.5**  Basic ECL logic gate with reference circuit

## DESIGN EXAMPLE 17.3

**Objective:** Design the reference portion of the ECL circuit.

Consider the circuit in Figure 17.5. The reference voltage $V_R$ is to be $-1.05$ V.

**Solution:** We know that

$$v_{B5} = V_R + V_{BE}(\text{on}) = -1.05 + 0.7 = -0.35 \text{ V} = -i_1 R_1$$

Since there are two unknowns, we will choose one variable. Let $R_1 = 0.25 \text{ k}\Omega$. Then,

$$i_1 = \frac{0.35}{0.25} = 1.40 \text{ mA}$$

Since this current is on the same order of magnitude as other currents in the circuit, the chosen value of $R_1$ is reasonable. Neglecting base currents, we can now write

$$i_1 = i_2 = \frac{0 - 2V_\gamma - V^-}{R_1 + R_2}$$

where $V_\gamma$ is the diode turn-on voltage and is assumed to be $V_\gamma = 0.7$ V. We then have

$$1.40 = \frac{-1.4 - (-5.2)}{R_1 + R_2}$$

which yields

$$R_1 + R_2 = 2.71 \text{ k}\Omega$$

Since $R_1 = 0.25 \text{ k}\Omega$, resistance $R_2$ is $R_2 = 2.46 \text{ k}\Omega$.

Also, we know that

$$i_5 = \frac{V_R - V^-}{R_5}$$

If we let $i_5 = i_1 = i_2 = 1.40$ mA, then

$$R_5 = \frac{V_R - V^-}{i_5} = \frac{-1.05 - (-5.2)}{1.40} = 2.96 \text{ k}\Omega$$

**Comment:** As with any design, there is no unique solution. The design presented will provide the required reference voltage to the base of $Q_R$.

---

**EXERCISE PROBLEM**

**Ex 17.3:** The reference circuit in Figure 17.5 is to be redesigned with $V^+ = 0$ and $V^- = -3.3$ V. The reference voltage is to be $V_R = -1.0$ V and the currents are to be $i_1 = i_2 = i_5 = 0.5$ mA. (Ans. $R_1 = 0.6$ k$\Omega$, $R_2 = 3.2$ k$\Omega$, $R_5 = 4.6$ k$\Omega$)

---

### 17.1.3 ECL Logic Circuit Characteristics

In this section, we will determine the power dissipation, fanout, and propagation delay times for the ECL logic gate. We will also examine the advantage of using a negative power supply.

**Power Dissipation**

Power dissipation is an important characteristic of a logic circuit. The power dissipated in the basic ECL logic gate in Figure 17.5 is given by

$$P_D = (i_{Cxy} + i_{CR} + i_5 + i_1 + i_3 + i_4)(0 - V^-) \tag{17.3}$$

---

**EXAMPLE 17.4**

**Objective:** Calculate the power dissipated in the ECL logic circuit.

Consider the circuit in Figure 17.5. Let $v_X = v_Y = -0.7$ V = logic 1.

**Solution:** From our previous analysis, we have $i_{Cxy} = 3.22$ mA, $i_{CR} = 0$, $i_5 = 1.40$ mA, and $i_1 = 1.40$ mA, and the output voltages are $v_{OR} = -0.7$ V and $v_{NOR} = -1.40$ V. The currents $i_3$ and $i_4$ are

$$i_3 = \frac{v_{OR} - V^-}{R_3} = \frac{-0.7 - (-5.2)}{1.5} = 3.0 \text{ mA}$$

and

$$i_4 = \frac{v_{NOR} - V^-}{R_4} = \frac{-1.40 - (-5.2)}{1.5} = 2.53 \text{ mA}$$

The power dissipation is then

$$P_D = (3.22 + 0 + 1.40 + 1.40 + 3.0 + 2.53)(5.2) = 60.0 \text{ mW}$$

**Comment:** This power dissipation is significantly larger than that in NMOS and CMOS logic circuits. The advantage of ECL, however, is the short propagation delay times, which can be less than 1 ns.

**Ex 17.4:** Assume the maximum currents in $Q_3$ and $Q_4$ of the ECL circuit in Figure 17.5 are to be 1.0 mA. (a) What are the required values of $R_3$ and $R_4$? (b) Using the results of part (a), calculate the new power dissipated in the circuit for $v_X = v_Y = -0.7$ V. (Ans. (a) $R_3 = R_4 = 4.5$ k$\Omega$, (b) $P = 40.8$ mW)

### Propagation Delay Time

The major advantage of ECL circuits is their small propagation delay time, on the order of 1 ns or less. The two reasons for the short propagation delay times are: (1) the transistors are not driven into saturation, which eliminates any charge storage effects; and (2) the logic swing in the ECL logic gate is small (about 0.7 V), which means that the voltages across the output capacitances do not have to change as much as in other logic circuits. Also, the currents in the ECL circuit are relatively large, which means that these capacitances can charge and discharge quickly. However, the trade-offs for the small propagation delay time are higher power dissipation and smaller noise margins.

ECL circuits are very fast, and they require that special attention be paid to transmission line effects. Improperly designed ECL circuit boards can experience ringing or oscillations. These problems have less to do with the ECL circuits than with the interconnections between the circuits. Care must therefore be taken to terminate the signal lines properly.

### Fanout

Figure 17.6 shows the emitter-follower output stage of the OR output of an ECL circuit used to drive the diff-amp input stage of an ECL load circuit. When $v_{OR}$ is a logic 0, input load transistor $Q_1'$ is cut off, effectively eliminating any load current from the driver output stage. With $v_{OR}$ at a logic 1 level, the input load transistor is

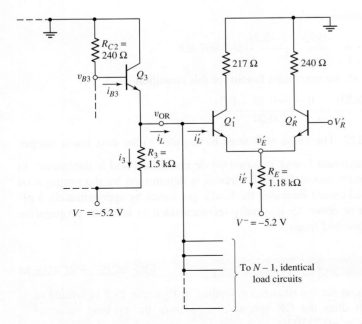

**Figure 17.6** Output stage of ECL logic gate driving $N$ identical ECL input stages

on and an input base current $i'_L$ exists. (Up to this point, we have neglected dc base currents; however, they are not zero.) The load current must be supplied through $Q_3$, whose base current is supplied through $R_{C2}$. As the load current $i_L$ increases with the addition of more load circuits, a voltage drop occurs across $R_{C2}$ and the output voltage decreases. The maximum fanout is determined partially by the maximum amplitude that the output voltage is allowed to drop from its ideal logic 1 value.

## EXAMPLE 17.5

**Objective:** Calculate the maximum fanout of an ECL logic gate, based on dc loading effects.

Consider the circuit in Figure 17.6. Assume the current gain of the transistors is $\beta = 50$, which represents a worst-case scenario. Assume that the logic 1 level at the OR output is allowed to decrease by 50 mV at most from a value of $-0.70$ V to $-0.75$ V.

**Solution:** From the figure, we see that

$$i'_E = \frac{v_{OR} - V_{BE}(\text{on}) - V^-}{R_E} = \frac{-0.75 - 0.7 - (-5.2)}{1.18} = 3.18 \text{ mA}$$

The input base current to the load transistor is

$$i'_B = \frac{i'_E}{(1 + \beta)} = \frac{3.18}{51} \Rightarrow 62.3 \,\mu\text{A} = i'_L$$

The total load current is therefore $i_L = N i'_L$.

The base current $i_{B3}$ required to produce both the load current $i_L$ and current $i_3$ is

$$i_{B3} = \frac{i_3 + i_L}{(1 + \beta)} = \frac{0 - v_{B3}}{R_{C2}} = \frac{0 - (v_{OR} + V_{BE}(\text{on}))}{R_{C2}} \quad \textbf{(17.4)}$$

Also, from the figure we see that

$$i_3 = \frac{v_{OR} - V^-}{R_3} = \frac{-0.75 - (-5.2)}{1.5} = 2.967 \text{ mA}$$

From Equation (17.4), the maximum fanout for this condition is

$$\frac{2.967 + N(0.0623)}{51} = \frac{0 - (-0.75 + 0.7)}{0.24}$$

which yields $N = 122$. The value of $N$ must be rounded to the next lower integer.

**Comment:** This maximum fanout is based on dc conditions and is unrealistic. In practice, the maximum fanout for ECL circuits is determined by the propagation delay time. Each load circuit increases the load capacitance by approximately 3 pF. A maximum fanout of about 15 is usually recommended to keep the propagation delay time within specified limits.

## EXERCISE PROBLEM

**\*Ex 17.5:** If the fanout for the situation described in Example 17.5 is limited to $N = 10$, how much does the OR output change from the no-load value of $-0.70$ V? (Ans. $v_{OR} = -0.7170$ V)

### The Negative Supply Voltage

In classic ECL circuits, it is common practice to ground the positive terminal of the supply voltage, reducing the noise signals at the output terminal. Figure 17.7(a) shows an emitter-follower output stage with the supply voltage $V_{CC}$ in series with a noise source $V_n$. The noise signal may be induced by the effect of switching currents interacting with parasitic inductances and capacitances. The output voltage is measured with respect to ground; therefore, if the positive terminal of $V_{CC}$ is grounded, voltage $V_o$ is taken as the output voltage. If the negative terminal of $V_{CC}$ is at ground, then $V_o'$ is the output voltage.

   To determine the effect of the noise voltage at the output, we assume that $Q_R$ is cut off, and we evaluate the small-signal hybrid-$\pi$ equivalent circuit shown in Figure 17.7(b).

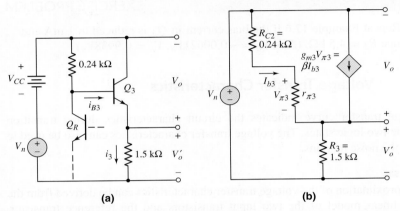

**(a)**                    **(b)**

**Figure 17.7** (a) Equivalent circuit, ECL emitter-follower output stage and noise generator, and the (b) small-signal hybrid-$\pi$ equivalent circuit

---

### EXAMPLE 17.6

**Objective:** Determine the effect of a noise signal on the output of an ECL gate.

   Consider the small-signal equivalent circuit in Figure 17.7(b). Let $\beta = 100$. Find $V_o$ and $V_o'$ as a function of $V_n$.

**Solution:** From a previous analysis, the quiescent collector current in $Q_3$ for $Q_R$ in cutoff is 3 mA. Then,

$$r_{\pi 3} = \frac{\beta V_T}{I_{CQ}} = \frac{(100)(0.026)}{3} = 0.867 \text{ k}\Omega$$

and

$$g_{m3} = \frac{I_{CQ}}{V_T} = \frac{3}{0.026} = 115 \text{ mA/V}$$

   We can also write that

$$V_n = I_{b3}(R_{C2} + r_{\pi 3}) + (1 + \beta)I_{b3}R_3$$

which yields

$$I_{b3} = \frac{V_n}{R_{C2} + r_{\pi 3} + (1 + \beta)R_3} = \frac{V_n}{0.24 + 0.867 + (101)(1.5)} = \frac{V_n}{152.6}$$

The output voltage $V_o$ is

$$V_o = -I_{b3}(R_{C2} + r_{\pi 3}) = -\left(\frac{V_n}{152.6}\right)(0.24 + 0.867) = -0.0073V_n$$

and output voltage $V_o'$ is

$$V_o' = (1 + \beta)I_{b3}R_3 = (101)\left(\frac{V_n}{152.6}\right)(1.5) = 0.99\,V_n$$

**Comment:** The effect of noise on the collector–emitter output voltage $V_o$ is much less than on output voltage $V_o'$. It is advantageous, then, to use $V_o$, which implies that the positive terminal of $V_{CC}$ is grounded. The noise insensitivity gained with a negative power supply may be critical in a logic circuit with a low noise margin.

**Ex 17.6:** Repeat Example 17.6 if the bias current in $Q_3$ is reduced to 1 mA and the resistance $R_3 = 4.5\,\text{k}\Omega$. (Ans. $V_o = -0.00621V_n$, $V_o' = 0.9938V_n$)

### 17.1.4   Voltage Transfer Characteristics

The voltage transfer curve indicates the circuit characteristics during transition between the two logic states. The voltage transfer characteristics can also be used to determine the noise margins.

**DC Analysis**

A good approximation of the voltage transfer characteristics can be derived from the piecewise linear model of the two input transistors and the reference transistor. Consider the ECL gate in Figure 17.5. If inputs $v_X$ and $v_Y$ are a logic 0, or $-1.40$ V, then $Q_1$ and $Q_2$ are cut off and $v_{\text{NOR}} = -0.7$ V. The reference transistor $Q_R$ is on and, as previously seen, $i_E = i_{C2} = 2.92$ mA, $v_{B3} = -0.70$ V, and $v_{\text{OR}} = -1.40$ V. As long as $v_X = v_Y$ remains less than $V_R - 0.12 = -1.17$ V, the output voltages do not change from these values. During the interval when the inputs are within $120\,\text{mV}$ of reference voltage $V_R$, the output voltage levels vary.

When $v_X = v_Y = V_R + 0.12 = -0.93$ V, then $Q_1$ and $Q_2$ are on and $Q_R$ is off. At this point, $i_E = i_{C1} = 3.03$ mA, $v_{B4} = -0.657$ V, and $v_{\text{NOR}} = -1.36$ V. As determined previously, when $v_X = v_Y = -0.7$ V, $v_{\text{NOR}} = -1.40$ V. The voltage transfer curves are shown in Figure 17.8.

**Noise Margin**

For the ECL gate, we define the threshold logic levels $V_{IL}$ and $V_{IH}$ as the points of discontinuity in the voltage transfer curves. These values are $V_{IL} = -1.17$ V and

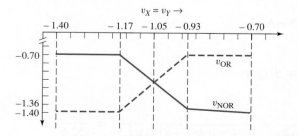

**Figure 17.8** ECL OR/NOR logic gate voltage transfer characteristics

$V_{IH} = -0.93$ V. The high logic level is $V_{OH} = -0.7$ V and the low logic value is $V_{OL} = -1.40$ V.

The noise margins are defined as

$$NM_H = V_{OH} - V_{IH} \qquad \textbf{(17.5(a))}$$

and

$$NM_L = V_{IL} - V_{OL} \qquad \textbf{(17.5(b))}$$

Using the results from Figure 17.8, we find that $NM_H = 0.23$ V and $NM_L = 0.23$ V. The noise margins for the ECL circuit are considerably lower than those calculated for NMOS and CMOS circuits.

## Test Your Understanding

**TYU 17.1** For the ECL logic gate shown in Figure 17.3, the bias voltages are $V^+ = 1.8$ V, $V^- = -1.8\,V$, and $V_R = 0.75$ V. Assume $V_{BE}$(on) $= 0.7$ V and neglect base currents. (a) Determine $R_E$ and $R_{C2}$ such that $i_E = 0.8$ mA and $v_{O2} = 1.1$ V when $v_X = v_Y = $ logic $0 < V_R$. (b) Find $R_{C1}$ such that $v_{O1} = 1.1$ V when $v_X = v_Y = 1.1$ V. What is $i_E$ for this case? (Ans. (a) $R_E = 2.31$ k$\Omega$, $R_{C2} = 0.875$ k$\Omega$; (b) $i_E = 0.951$ mA, $R_{C1} = 0.736$ k$\Omega$)

**TYU 17.2** Redesign the ECL circuit in Figure 17.4 such that the logic 0 values at the $v_{OR}$ and $v_{NOR}$ terminals are $-1.5$ V. The maximum value of $i_E$ is to be 2.5 mA, and the maximum values of $i_3$ and $i_4$ are to be 2.5 mA. The bias voltages are as shown. Determine all resistor values and the value of $V_R$. (Ans. $R_E = 1.52$ k$\Omega$, $R_{C1} = 320$ $\Omega$, $V_R = -1.1$ V, $R_{C2} = 358$ $\Omega$, $R_3 = R_4 = 1.8$ k$\Omega$)

## 17.2 MODIFIED ECL CIRCUIT CONFIGURATIONS

**Objective:** • Analyze and design modified emitter-coupled logic circuits

The large power dissipation in the basic ECL logic gate makes this circuit impractical for large-scale integrated circuits. Certain modifications can simplify the circuit design and decrease the power consumption, making the ECL more compatible with integrated circuits.

### 17.2.1 Low-Power ECL

Figure 17.9(a) shows a basic ECL OR/NOR logic gate with reference voltage $V_R$ and a positive voltage supply. We can make the output voltage states compatible with the input voltages, eliminating the need for the emitter-follower output stages. In some applications, both complementary outputs may not be required. If, for example, only the OR output is required, then we can eliminate resistor $R_{C1}$. Removing this resistor does not reduce the circuit power consumption, but it eliminates one element.

Figure 17.9(b) shows the modified ECL gate. For $v_X = v_Y = $ logic $1 > V_R$, transistors $Q_1$ and $Q_2$ are turned on and $Q_R$ is off. The output voltage is $v_{OR} = V_{CC}$.

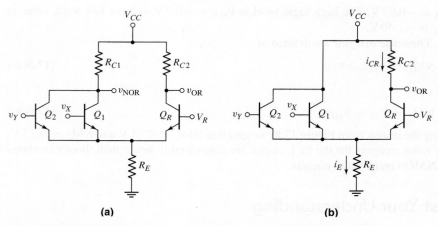

**(a)**                                    **(b)**

**Figure 17.9** (a) Basic ECL OR/NOR logic gate and (b) modified ECL logic gate

For $v_X = v_Y = $ logic $0 < V_R$, then $Q_1$ and $Q_2$ are off and $Q_R$ is on. The currents are

$$i_E = \frac{V_R - V_{BE}(\text{on})}{R_E} \cong i_{CR} \tag{17.6}$$

and the output voltage is

$$v_{OR} = V_{CC} - i_{CR}R_{C2} \tag{17.7}$$

If the resistance values of $R_E$ and $R_{C2}$ vary from one circuit to another because of fabrication tolerances, then current $i_E$ and the logic 0 output voltage will vary from one circuit to another.

To establish a well-defined logic 0 output, we can insert a Schottky diode in parallel with resistor $R_C$, as shown in Figure 17.10. If the two inputs are a logic 0, then $Q_1$ and $Q_2$ are off and $Q_R$ is on. For this condition, we want the Schottky diode to turn on. The output will then be $v_{OR} = V_{CC} - V_\gamma$, where $V_\gamma$ is the turn-on voltage of the Schottky diode. This logic 0 output voltage is a well-defined value. If the diode turns on, then current $i_R$ is limited to $i_R(\text{max}) = V_\gamma/R_C$. Since we must have $i_E > i_R(\text{max})$, the diode current is $i_D = i_E - i_R(\text{max})$.

As usual, we design the reference voltage to be the average of the logic 1 and logic 0 values. The voltage $V_R$ is then $V_R = V_{CC} - V_\gamma/2$. We may assume $V_\gamma = 0.4$ V. When transistor $Q_R$ is off, its collector voltage is $V_{CC}$ and the B–C junction is reverse-

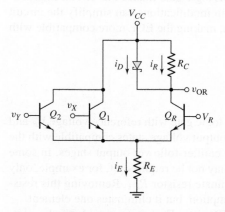

**Figure 17.10** Modified ECL logic gate with Schottky diode

biased by 0.2 V. When $Q_R$ is conducting, its collector voltage is $V_{CC} - V_\gamma$, which means that the B–C junction is now forward-biased by 0.2 V and the transistor is biased slightly in saturation. However, this slight saturation bias does not degrade the switching of $Q_R$, so the fast switching characteristics of the ECL circuit are retained.

## EXAMPLE 17.7

**Objective:** Analyze the modified ECL logic gate.

Consider the circuit in Figure 17.10 with parameters $V_{CC} = 1.7$ V and $R_E = R_C = 8$ k$\Omega$. Assume the diode and transistor piecewise linear parameters are $V_\gamma = 0.4$ V and $V_{BE}(\text{on}) = 0.7$ V.

**Solution:** The output voltage values are

$$v_{OR} = \text{logic } 1 = V_{CC} = 1.7 \text{ V}$$

and

$$v_{OR} = \text{logic } 0 = V_{CC} - V_\gamma = 1.7 - 0.4 = 1.3 \text{ V}$$

For the output voltages to be compatible with the inputs, the reference voltage $V_R$ must be the average of the logic 1 and logic 0 values, or $V_R = 1.5$ V. If $v_X = v_Y = \text{logic } 0 = 1.3$ V, then $Q_R$ is on. Therefore,

$$i_E = \frac{V_R - V_{BE}(\text{on})}{R_E} = \frac{1.5 - 0.7}{8} \Rightarrow 100 \, \mu\text{A}$$

The maximum current in $R_C$ is

$$i_R(\text{max}) = \frac{V_\gamma}{R_C} = \frac{0.4}{8} \Rightarrow 50 \, \mu\text{A}$$

and the current through the diode is

$$i_D = i_E - i_R(\text{max}) = 100 - 50 = 50 \, \mu\text{A}$$

For $v_X = v_Y = \text{logic } 0$, the power dissipation is $P = i_E V_{CC}$, or

$$P = i_E V_{CC} = (100)(1.7) = 170 \, \mu\text{W}$$

For $v_X = v_Y = \text{logic } 1 = 1.7$ V, we have

$$i_E = \frac{v_X - V_{BE}(\text{on})}{R_E} = \frac{1.7 - 0.7}{8} \Rightarrow 125 \, \mu\text{A}$$

Therefore, the power dissipation for this condition is

$$P = i_E V_{CC} = (125)(1.7) = 213 \, \mu\text{W}$$

**Comment:** If the resistance values of $R_E$ and $R_C$ were to change by as much as $\pm 20$ percent as a result of manufacturing tolerances, for example, the currents would still be sufficient to turn the Schottky diode on when $Q_R$ is on. This means that the logic 0 output is well defined. Also, the power dissipation in this ECL gate is considerably less than that in the classic ECL OR/NOR logic circuit. The reduced power is a result of fewer components, lower bias voltage, and smaller currents.

### EXERCISE PROBLEM

**Ex 17.7:** Design the basic ECL logic gate in Figure 17.11 such that the maximum power dissipation is 0.2 mW and the logic swing is 0.4 V. (Ans. $I_Q = 117.6 \, \mu\text{A}$, $R_C = 3.4$ k$\Omega$, $V_R = 1.5$ V)

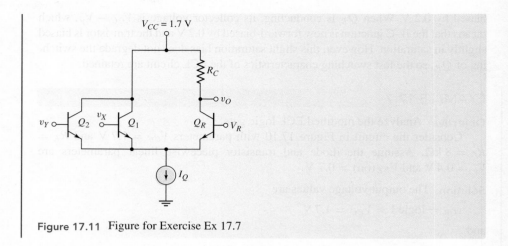

**Figure 17.11** Figure for Exercise Ex 17.7

17.2.2 **Alternative ECL Gates**

In an ECL system, as in all digital systems, a gate is used to drive other logic gates. Connecting load circuits to the basic ECL gate demonstrates changes that can be made to incorporate ECL into integrated circuits more effectively.

Figure 17.12 shows the basic ECL gate with two load circuits. In this configuration, the collectors of $Q_2'$ and $Q_2''$ are at the same potential, as are the bases of

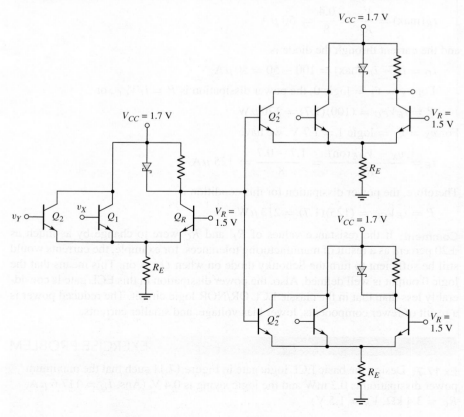

**Figure 17.12** Modified ECL logic gate with two load circuits

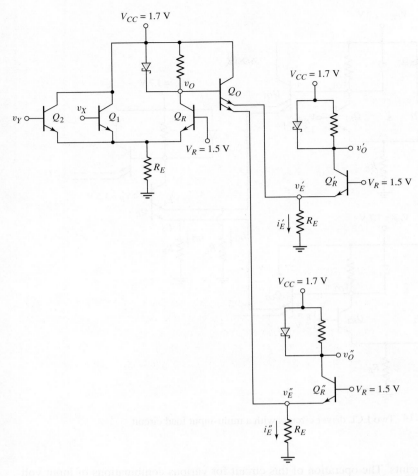

**Figure 17.13** Modified ECL logic gate with multiemitter output transistor and two load circuits

the two transistors. We can therefore replace $Q_2'$ and $Q_2''$ by a single multiemitter transistor.

In Figure 17.13, the multiemitter transistor $Q_O$ is part of the driver circuit. The operation of the circuit is as follows:

- $v_x = v_y = $ logic 1 = 1.7 V: The two input transistors $Q_1$ and $Q_2$ are on, $Q_R$ is off, and $v_O = 1.7$ V. Since the base voltage of $Q_O$ is higher than the base voltages of $Q_R'$ and $Q_R''$, then $Q_O$ is conducting, $Q_R'$ and $Q_R''$ are off, and $v_E' = v_E'' = 1.7 - 0.7 = 1.0$ V. The currents $i_E'$ and $i_E''$ flow through the emitters of $Q_O$. The output voltages are $v_O' = v_O'' = 1.7$ V.
- $v_x = v_y = $ logic 0 = 1.3 V: For this case, the two input transistors $Q_1$ and $Q_2$ are off, $Q_R$ is on, and $v_O = 1.3$ V. The output transistor $Q_O$ is off and both $Q_R'$ and $Q_R''$ are on. The output voltages are then $v_O' = v_O'' = 1.3$ V.

The two load circuits in Figure 17.13 each have only a single input, which limits the circuit functionality. The versatility of the circuit can be further enhanced by making the load transistor $Q_R'$ a multiemitter transistor. This is shown in Figure 17.14. For simplicity, we show only a single input transistor to each of the two

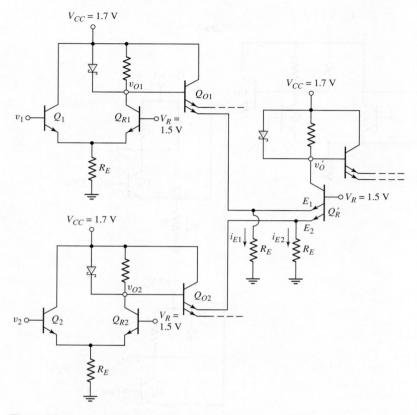

**Figure 17.14** Two ECL driver circuits with a multi-input load circuit

driver circuits. The operation of this circuit for various combinations of input voltages is as follows.

- $v_1 = v_2 = \text{logic } 0 = 1.3 \text{ V}$: The two input transistors $Q_1$ and $Q_2$ are off and the two reference transistors $Q_{R1}$ and $Q_{R2}$ are on. This means that $v_{O1} = v_{O2} = 1.3 \text{ V}$ and both output transistors $Q_{O1}$ and $Q_{O2}$ are off. Both emitters of $Q'_R$ are forward biased, currents $i_{E1}$ and $i_{E2}$ flow through $Q'_R$, and the output voltage is $v'_O = \text{logic } 0 = 1.3 \text{ V}$.
- $v_1 = 1.7 \text{ V}$, $v_2 = 1.3 \text{ V}$: For this case, $Q_1$ is on, $Q_{R1}$ is off, $Q_2$ is off, and $Q_{R2}$ is on. The output voltages are $v_{O1} = 1.7 \text{ V}$ and $v_{O2} = 1.3 \text{ V}$. This means that $Q_{O1}$ is on and $Q_{O2}$ is off. With $Q_{O1}$ on, current $i_{E1}$ flows through $Q_{O1}$ and no current flows in emitter $E_1$. With $Q_{O2}$ off, emitter $E_2$ is forward biased, current $i_{E2}$ flows through $Q'_R$, and the output voltage is $v'_O = \text{logic } 0 = 1.3 \text{ V}$.
- $v_1 = 1.3 \text{ V}$, $v_2 = 1.7 \text{ V}$: This case is the complement of the one just discussed. Here, $Q_{O1}$ is off and $Q_{O2}$ is on. This means that $i_{E1}$ flows through emitter $E_1$ of $Q'_R$, and $i_{E2}$ flows through $Q_{O2}$. The output voltage is $v'_O = \text{logic } 0 = 1.3 \text{ V}$.
- $v_1 = v_2 = 1.7 \text{ V}$: The two input transistors $Q_1$ and $Q_2$ are on, the two reference transistors $Q_{R1}$ and $Q_{R2}$ are off, and $v_{O1} = v_{O2} = 1.7 \text{ V}$. This means that both $Q_{O1}$ and $Q_{O2}$ are on and $Q'_R$ is off. Currents $i_{E1}$ and $i_{E2}$ flow through $Q_{O1}$ and $Q_{O2}$, respectively, and the output voltage is $v'_O = \text{logic } 1 = 1.7 \text{ V}$.

| Table 17.1 | Summary of results for the ECL circuit in Figure 17.14 | |
| --- | --- | --- |
| $v_1$ (V) | $v_2$ (V) | $v'_O$ (V) |
| 1.3 | 1.3 | 1.3 |
| 1.7 | 1.3 | 1.3 |
| 1.3 | 1.7 | 1.3 |
| 1.7 | 1.7 | 1.7 |

These results are summarized in Table 17.1, which shows that this circuit performs the AND logic function. A more complicated or sophisticated logic function can be performed if multiple inputs are used in the driver circuits.

In integrated circuits, resistors $R_E$ are replaced by current sources using transistors. Replacing resistors with transistors in integrated circuits usually results in reduced chip area.

## 17.2.3  Series Gating

Series gating is a bipolar logic circuit technique that allows complex logic functions to be performed with a minimum number of devices and with maximum speed. Series gating is formed by using cascode stages.

Figure 17.15(a) shows the basic emitter-coupled pair, and Figure 17.15(b) shows a cascode stage, also referred to as two-level series gating. Reference voltage $V_{R1}$ is approximately 0.7 V greater than reference voltage $V_{R2}$. The input voltages $v_x$ and $v_y$ must also be shifted approximately 0.7 V with respect to each other.

As an example, we use the multiemitter load circuit from Figure 17.14 as part of a cascode configuration as shown in Figure 17.16. Transistors $Q_{O1}$, $Q_{O2}$, and $Q_{O3}$

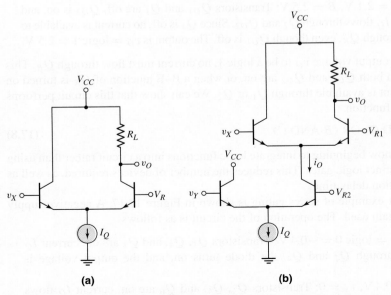

(a)                                        (b)

**Figure 17.15**  (a) Basic emitter-coupled pair and (b) ECL cascode configuration

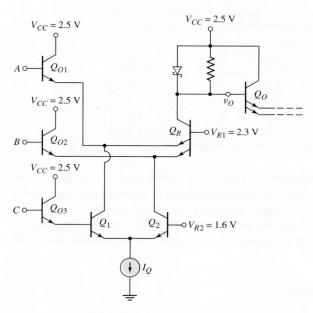

**Figure 17.16** ECL series gating example

represent the output transistors of three ECL driver circuits. We assume a logic 1 level of 2.5 V and a logic 0 level of 2.1 V. The 0.4 V logic swing results from incorporating a Schottky diode in each output stage.

With three input signals, there are eight possible combinations of input states. We will only consider two combinations here:

- $A = B = C = $ logic $0 = 2.1$ V: In this case, transistors $Q_{O1}$ and $Q_{O2}$ are off and transistor $Q_1$ is off. This means that current $I_Q$ flows through $Q_2$ and $Q_R$, and $v_O = $ logic $0 = 2.1$ V.
- $A = C = 2.1$ V, $B = 2.5$ V: Transistors $Q_{O1}$ and $Q_1$ are off, $Q_{O2}$ is on, and current $I_Q$ flows through $Q_2$ and $Q_{O2}$. Since $Q_1$ is off, no current is available to flow through $Q_R$, even though $Q_{O1}$ is off. The output is $v_O = $ logic $1 = 2.5$ V.

For the output voltage $v_O$ to be a logic 1, no current must flow through $Q_R$. This occurs when both $Q_{O1}$ and $Q_{O2}$ are on, or when a B–E junction of $Q_R$ is turned on but no current is available through $Q_1$ or $Q_2$. We can show that this circuit performs the logic of function

$$(A \text{ AND } C) \text{ OR } (B \text{ AND } \bar{C}) \tag{17.8}$$

We are now beginning to integrate logic functions into a circuit rather than using separate, distinct logic gates. This reduces the number of devices required, as well as the propagation delay time.

Another example of series gating is shown in Figure 17.17. A negative supply voltage is again used. The operation of the circuit is as follows.

- $v_x = v_y = $ logic $0 = -0.4$ V: Transistors $Q_1$, $Q_4$, and $Q_7$ are on, current $I_Q$ flows through $Q_7$ and $Q_4$, the diode turns on, and the output voltage is $-0.4$ V.
- $v_x = -0.4$ V, $v_y = 0$: Transistors $Q_1$, $Q_4$, and $Q_6$ are on, current $I_Q$ flows through $Q_6$ and $Q_1$ to ground, and current $I_{Q2}$ flows through $Q_4$ and the

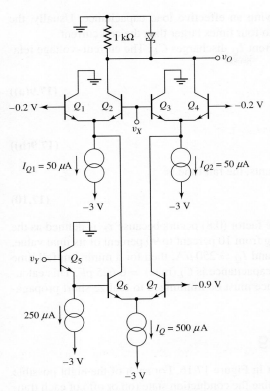

**Figure 17.17** ECL series gating example

| Table 17.2 | | Summary of logic levels for ECL circuit in Figure 17.17 |
| --- | --- | --- |
| $v_x$ | $v_y$ | $v_O$ |
| 0 | 0 | 0 |
| 0 | 1 | 1 |
| 1 | 0 | 1 |
| 1 | 1 | 0 |

resistor. The output voltage is $v_O = -R_C I_{Q2} = -(1)(0.05) = -0.05\,\text{V}$. This voltage is not sufficient to turn the Schottky diode on. Although it is not zero volts, the voltage still represents a logic 1.

- $v_x = 0$, $v_y = -0.4\,\text{V}$: Transistors $Q_2$, $Q_3$, and $Q_7$ are on, current $I_Q$ flows through $Q_7$ and $Q_3$ to ground, and current $I_{Q1}$ flows through $Q_2$ and the resistor. Again, $v_O = -0.05\,\text{V} = \text{logic } 1$.
- $v_x = v_y = \text{logic } 1 \cong 0\,\text{V}$: Transistors $Q_2$, $Q_3$, and $Q_6$ are on, $I_Q$ flows through $Q_6$, $Q_2$, and the Schottky diode, and output voltage is $v_O = -0.4\,\text{V} = \text{logic } 0$.

These results are summarized in Table 17.2, in which the logic levels are given. The results show that the circuit performs the exclusive-OR logic function.

17.2.4    **Propagation Delay Time**

ECL is the fastest bipolar logic technology. Bipolar technology can produce small, very fast transistors with cutoff frequencies in the range of 3 to 15 GHz. Logic gates that use these transistors are so fast that interconnect line delays tend to dominate the propagation delay times. Minimizing these interconnect delays involves minimizing the metal lengths and using sufficient current drive capability.

Speed is derived from low-signal logic swings, nonsaturating logic, and the ability to drive a load capacitance. Figure 17.18 is the emitter-follower output stage

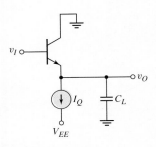

**Figure 17.18** Emitter-follower stage with load capacitance

found in many ECL circuits, showing an effective load capacitance. Usually, the emitter-follower current $I_Q$ is two to four times larger than the cell current.

In the pull-down cycle, the current $I_Q$ discharges $C_L$. The current–voltage relationship of the capacitor is

$$i = C_L \frac{dv_O}{dt} \tag{17.9(a)}$$

or

$$v_O = \frac{1}{C_L} \int i\, dt \tag{17.9(b)}$$

Assuming $C_L$ and $i = I_Q$ are constants, the fall time is

$$\tau_F = (0.8) \frac{C_L V_S}{I_Q} \tag{17.10}$$

where $V_S$ is the logic swing, and the factor (0.8) occurs because $\tau_F$ is defined as the time required for the output to swing from 10 percent to 90 percent of its final value.

As an example, if $V_S = 0.4$ V and $I_Q = 250\,\mu$A, then for a minimum fall time of $\tau_F = 0.8$ ns, the maximum load capacitance is $C_L(\text{max}) = 0.625$ pF. This calculation shows that the load capacitance must be minimized to realize short propagation delay times.

## Test Your Understanding

**TYU 17.3** Consider the ECL circuit in Figure 17.16. For each of the eight possible combinations of input states, determine the conduction state (on or off) of each transistor. Verify that this circuit performs the logic function given by Equation (17.8).

**TYU 17.4** The ECL circuit in Figure 17.19 is an example of three-level series gating. Determine the logic function that the circuit performs. (Ans. $(A \oplus B) \oplus C$)

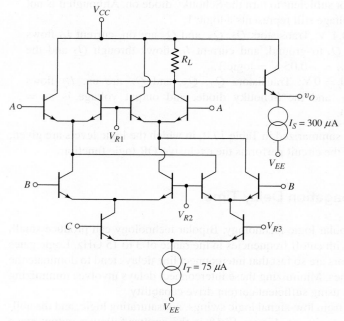

**Figure 17.19** Figure for Exercise TYU 17.4

## 17.3   TRANSISTOR–TRANSISTOR LOGIC

**Objective:** • Analyze transistor–transistor logic circuits

The bipolar inverter is the basic circuit from which most bipolar saturated logic circuits are developed, including diode–transistor logic (DTL) and transistor–transistor logic (TTL). However, the basic bipolar inverter suffers from loading effects. Diode–transistor logic combines diode logic (Chapter 2) and the bipolar inverter to minimize loading effects. Transistor–transistor logic, which evolved directly from DTL, provides reduced propagation delay times, as we will show.

In DTL and TTL circuits, bipolar transistors are driven between cutoff and saturation. Since the transistor is being used essentially as a switch, the current gain is not as important as in amplifier circuits. Typically, for transistors used in these circuits, the current gain is assumed to be in the range of 25 to 50. These transistors need not be fabricated to as tight a tolerance as that of high-gain amplifier transistors.

Table 17.3 lists the piecewise linear parameters used in the analysis of bipolar digital circuits, along with their typical values. Also included is the pn junction diode turn-on voltage $V_\gamma$. Generally, the B–E voltage increases as the transistor is driven into saturation, since the base current increases. When the transistor is biased in the saturation region, the B–E voltage is $V_{BE}(\text{sat})$, where $V_{BE}(\text{sat}) > V_{BE}(\text{on})$.

### 17.3.1   Basic Diode–Transistor Logic Gate

The basic diode–transistor logic (DTL) gate is shown in Figure 17.20. The circuit is designed such that the output transistor operates between cutoff and saturation. This provides the maximum output voltage swing, minimizes loading effects, and produces the maximum noise margins. When $Q_o$ is in saturation, the output voltage is $v_O = V_{CE}(\text{sat}) \cong 0.1$ V and is defined as logic 0 for the DTL circuit. As we will see, the basic DTL logic gate shown in Figure 17.20 performs the NAND logic function.

#### Basic DTL NAND Circuit Operation

If both input signals $v_X$ and $v_Y$ are at logic 0, then the two input diodes $D_X$ and $D_Y$ are forward biased through resistor $R_1$ and voltage source $V_{CC}$. The input diodes

| Table 17.3 | Piecewise linear parameters for a pn junction diode and npn bipolar transistor | |
|---|---|---|
| **Parameter** | | **Value** |
| $V_\gamma$ | | 0.7 V |
| $V_{BE}(\text{on})$ | | 0.7 V |
| $V_{BE}(\text{sat})$ | | 0.8 V |
| $V_{CE}(\text{sat})$ | | 0.1 V |

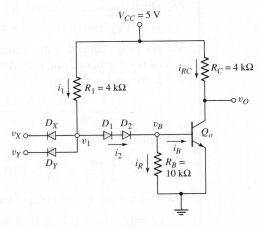

**Figure 17.20**   Basic diode–transistor logic gate

conduct, and voltage $v_1$ is clamped to a value that is one diode drop above the input voltage. If $v_X = v_Y = 0.1$ V and $V_\gamma = 0.7$ V, then $v_1 = 0.8$ V. Diodes $D_1$ and $D_2$ and output transistor $Q_o$ are nonconducting and are off. If $D_1$ and $D_2$ were conducting, then voltage $v_B$ would be $-0.6$ V for $V_\gamma = 0.7$ V. However, no mechanism exists for $v_B$ to become negative and still have a forward-biased diode current. Thus, the current in $D_1$ and $D_2$, the current in $Q_o$, and the voltage $v_B$ are all zero. Since $Q_o$ is cut off, then the output voltage is $v_O = V_{CC}$. This is the largest possible output voltage and is therefore defined as the logic 1 level. This same condition applies as long as at least one input is at logic 0.

When both $v_X$ and $v_Y$ are at logic 1, which is equal to $V_{CC}$, both $D_X$ and $D_Y$ are cut off. Diodes $D_1$ and $D_2$ become forward biased, output transistor $Q_o$ is driven into saturation, and $v_O = V_{CE}(\text{sat})$, which is the smallest possible output voltage and is defined as the logic 0 level.

This circuit is a two-input DTL NAND logic gate. However, the circuit is not limited to two inputs. Additional input diodes may be included to increase the fan-in.

---

### EXAMPLE 17.8

**Objective:** Determine the currents and voltages in the DTL logic circuit.

Consider the DTL circuit in Figure 17.20. Assume the transistor parameters are as given in Table 17.3 and let $\beta = 25$.

**Solution:** Let $v_X = v_Y = $ logic $0 = 0.1$ V. For this case,

$$v_1 = v_X + V_\gamma = 0.1 + 0.7 = 0.8 \text{ V}$$

and

$$i_1 = \frac{V_{CC} - v_1}{R_1} = \frac{5 - 0.8}{4} = 1.05 \text{ mA}$$

Since diodes $D_1$ and $D_2$ and output transistor $Q_o$ are nonconducting, we assume that current $i_1$ divides evenly between the matched diodes $D_X$ and $D_Y$. In this case, the currents $i_2 = i_B = i_C = 0$ and the output voltage is $v_O = 5$ V = logic 1.

If $v_X = 0.1$ V and $v_Y = 5$ V, or $v_X = 5$ V and $v_Y = 0.1$ V, then the output transistor is still cut off and $v_O = 5$ V = logic 1.

If $v_X = v_Y = $ logic $1 = 5$ V, it is impossible for input diodes $D_X$ and $D_Y$ to be forward biased. In this case, diodes $D_1$ and $D_2$ and the output transistor are biased on, which means that, starting at ground potential at the emitter of $Q_o$, $v_1$ is

$$v_1 = V_{BE}(\text{sat}) + 2V_\gamma = 0.8 + 2(0.7) = 2.2 \text{ V}$$

Voltage $v_1$ is clamped at this value and cannot increase. We see that $D_X$ and $D_Y$ are indeed reverse biased and turned off, as assumed.

Currents $i_1$ and $i_2$ are

$$i_1 = i_2 = \frac{V_{CC} - v_1}{R_1} = \frac{5 - 2.2}{4} = 0.70 \text{ mA}$$

and current $i_R$ is

$$i_R = \frac{V_{BE}(\text{sat})}{R_B} = \frac{0.8}{10} = 0.08 \text{ mA}$$

The base current into the output transistor is then

$$i_B = i_2 - i_R = 0.70 - 0.08 = 0.62 \text{ mA}$$

Since the circuit is to be designed such that $Q_o$ is driven into saturation, the collector current is

$$i_C = \frac{V_{CC} - V_{CE}(\text{sat})}{R_C} = \frac{5 - 0.1}{4} = 1.23 \text{ mA}$$

Finally, the ratio of collector to base current is

$$\frac{i_C}{i_B} = \frac{1.23}{0.62} = 1.98 < \beta$$

**Comment:** Since the ratio of the collector current to base current is less than $\beta$, the output transistor is biased in the saturation region. Since the output transistor is biased between cutoff and saturation, the maximum swing between logic 0 and logic 1 is obtained.

EXERCISE PROBLEM

**Ex 17.8:** Consider the basic DTL circuit in Figure 17.20 with circuit and transistor parameters given in Example 17.8. Assume no load is connected to the output. Calculate the power dissipated in the circuit for (a) $v_X = v_Y = 5$ V and (b) $v_X = v_Y = 0$. (Ans. (a) $P = 9.625$ mW, (b) $P = 5.375$ mW)

## Minimum $\beta$

To ensure that the output transistor is in saturation, the common-emitter current gain $\beta$ must be at least as large as the ratio of collector current to base current. For example 17.8, the minimum $\beta$, or $\beta_{\min}$, is 1.98. If the common-emitter current gain were less than 1.98, then $Q_o$ would not be driven into saturation, and the currents and voltages in the circuit would have to be recalculated. A current gain greater than 1.98 ensures that $Q_o$ is driven into saturation for the given circuit parameters and for the no-load condition.

## Pull-Down Resistor

In the basic DTL NAND logic circuit in Figure 17.20, a resistor $R_B$ is connected between the base of the output transistor and ground. This resistor is called a pull-down resistor, and its purpose is to decrease the output transistor switching time as it goes from saturation to cutoff. As previously discussed, excess minority carriers must be removed from the base before a transistor can be switched to cutoff. This base charge removal produces a current out of the transistor base terminal until the transistor is turned off. Without the pull-down resistor, this reverse base current would be limited to the reverse-bias leakage current in diodes $D_1$ and $D_2$, resulting in a relatively long turn-off time. The pull-down resistor provides a path for the reverse base current.

The base charge can be removed more rapidly if the value of $R_B$ is reduced. The larger the reverse base current, the shorter the transistor turn-off time. However, a trade-off must be made in choosing the value of $R_B$. A small $R_B$ provides faster switching, but lowers the base current to the transistor in the on state by diverting some drive current to ground. A lower base current reduces the circuit drive capability, or maximum fanout.

**The Input Transistor of TTL**

Figure 17.21(a) shows a basic DTL circuit with one input diode $D_X$ and one offset diode $D_1$. The structure of these back-to-back diodes is the same as an npn transistor, as indicated in Figure 17.21(b). The base–emitter junction of $Q_1$ corresponds to input diode $D_X$ and the base–collector junction corresponds to offset diode $D_1$.

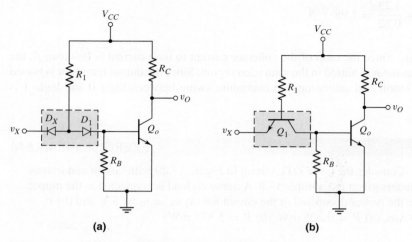

**(a)**                                    **(b)**

**Figure 17.21** (a) Basic DTL gate and (b) basic TTL gate

In isoplanar integrated circuit technology, the emitter of a bipolar transistor is fabricated in the base region. More emitters can then be added in the same base region to form a multiemitter, multi-input device. Figure 17.22(a) shows a simplified cross section of a three-emitter transistor, which is used as the input device in a TTL circuit. Figure 17.22(b) shows the basic TTL circuit with the multiemitter input transistor.

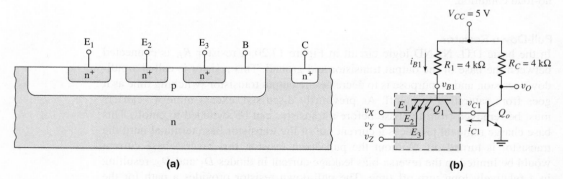

**(a)**                                    **(b)**

**Figure 17.22** (a) Simplified cross section of three-emitter transistor and (b) TTL circuit with three-emitter input transistor

This circuit performs the same NAND operation as its DTL counterpart. The multiemitter transistor reduces the silicon area required, compared to the DTL input diodes, and it increases the switching speed. Transistor $Q_1$ assists in pulling output transistor $Q_o$ out of saturation and into cutoff during a low-to-high transition of the output voltage. Pull-down resistor $R_B$ in Figure 17.21(b) is no longer necessary, since the excess minority carriers in the base of $Q_o$ use transistor $Q_1$ as a path to ground.

The operation of input transistor $Q_1$ is somewhat unconventional. In Figure 17.23(a), if either or both of the two inputs to $Q_1$ are in a low state, the base–emitter junction is forward biased through $R_1$ and $V_{CC}$. The base current enters $Q_1$, and the emitter current exits the specific emitter connected to the low input. Transistor action forces the collector current into $Q_1$, but the only steady-state collector current in this direction is a reverse-bias saturation current out of the base of $Q_o$. The steady-state collector current of $Q_1$ is usually much smaller than the base current, implying that $Q_1$ is biased in saturation.

If at least one input is low such that $Q_1$ is biased in saturation, then from Figure 17.23(a), we see that the base voltage of $Q_1$ is

$$v_{B1} = v_X + V_{BE}(\text{sat}) \tag{17.11}$$

and the base current into $Q_1$ is

$$i_{B1} = \frac{V_{CC} - v_{B1}}{R_1} \tag{17.12}$$

If the forward current gain of $Q_1$ is $\beta_F$, then $Q_1$ will be in saturation as long as $i_{C1} < \beta_F i_{B1}$.

The collector voltage of $Q_1$ is

$$v_{C1} = v_X + V_{CE}(\text{sat}) \tag{17.13}$$

If both $v_X$ and $V_{CE}(\text{sat})$ are approximately 0.1 V, then $v_{C1}$ is small enough for the output transistor to cut off and $v_O = V_{CC} = \text{logic } 1$.

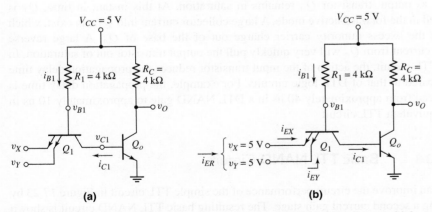

**Figure 17.23** TTL circuit (a) with at least one input low and (b) with all inputs high

If all inputs are high, $v_X = v_Y = 5$ V, as shown in Figure 17.23(b), then the base–emitter junctions of the input transistor are reverse biased. Base voltage $v_{B1}$ increases, which forward-biases the B–C junction of $Q_1$ and drives output transistor $Q_o$ into saturation. Since the B–E junction of $Q_1$ is reverse biased and the B–C junction is forward biased, $Q_1$ is biased in the inverse-active mode. In this bias mode, the roles of the emitter and collector are interchanged.

When input transistor $Q_1$ is biased in the inverse-active mode, base voltage $v_{B1}$ is

$$v_{B1} = V_{BE}(\text{sat})_{Q_o} + V_{BC}(\text{on})_{Q_1} \tag{17.14}$$

where $V_{BC}(\text{on})$ is the B–C junction turn-on voltage. We assume that the B–C junction turn-on voltage is equal to the B–E junction turn-on voltage. The terminal current relationships for $Q_1$ are therefore

$$i_{EX} = i_{EY} = \beta_R i_{B1} \tag{17.15}$$

and

$$i_{C1} = i_{B1} + i_{EX} + i_{EY} = (1 + 2\beta_R)i_{B1} \tag{17.16}$$

where $\beta_R$ is the inverse-active mode current gain of each input emitter of the input transistor.

Since a bipolar transistor is not symmetrical, the inverse and forward current gains are not equal. The inverse current gain is generally quite small, usually less than one. In Figure 17.23(b), the input transistor has a fan-in of two. Transistor $Q_1$ may be considered as two separate transistors with their bases and collectors connected. For simplicity, when all inputs are high, we assume that current $i_{ER}$ splits evenly between the input emitters.

The inverse-active mode current into the emitters of $Q_1$ is not desirable, since this is a load current that must be supplied by a driver logic circuit when its output voltage is in its high state. Because of the transistor action, these currents tend to be larger than the reverse saturation currents of DTL circuit input diodes. The major advantage of TTL over DTL is faster switching of the output transistor from saturation to cutoff.

If all inputs are initially high and then at least one input switches to the logic 0 state, 0.1 V, the B–E junction of $Q_1$ becomes forward biased and base voltage $v_{B1}$ becomes approximately $0.1 + 0.7 = 0.8$ V. Collector voltage $v_{C1}$ is held at 0.8 V as long as output transistor $Q_o$ remains in saturation. At this instant in time, $Q_1$ is biased in the forward-active mode. A large collector current into $Q_1$ can exist, which pulls the excess minority carrier charge out of the base of $Q_o$. A large reverse base current from $Q_o$ will very quickly pull the output transistor out of saturation. In the TTL circuit, the action of the input transistor reduces the propagation delay time compared to that of DTL logic circuits. For example, the propagation delay time is reduced from approximately 40 ns in a DTL NAND gate to approximately 10 ns in an equivalent TTL circuit.

### 17.3.3    Basic TTL NAND Circuit

We can improve the circuit performance of the simple TTL circuit in Figure 17.23 by adding a second current gain stage. The resulting basic TTL NAND circuit is shown in Figure 17.24. In this circuit, both transistors $Q_2$ and $Q_o$ are driven into saturation when $v_X = v_Y = \text{logic } 1$. When at least one input switches from high to low, input transistor $Q_1$ very quickly pulls $Q_2$ out of saturation and pull-down resistor $R_B$ provides a path for the excess charge in $Q_o$, which means that the output transistor can turn off fairly quickly.

#### DC Current–Voltage Analysis
The analysis of the TTL circuit is very similar to that of the DTL circuit, as demonstrated in the following example.

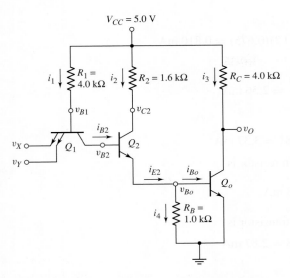

**Figure 17.24** TTL circuit with currents and voltages

## EXAMPLE 17.9

**Objective:** Calculate the currents and voltages for the basic TTL NAND circuit.

Consider the TTL circuit in Figure 17.24. Assume the piecewise linear transistor parameters are as listed in Table 17.3. Assume the forward current gain is $\beta_F \equiv \beta = 25$ and the inverse current gain of each input emitter is $\beta_R = 0.1$.

**Solution:** For $v_X = v_Y = 0.1$ V, $Q_1$ is biased in saturation and

$$v_{B2} = v_X + v_{CE}(\text{sat}) = 0.1 + 0.1 = 0.2 \text{ V}$$

which means that $Q_2$ and $Q_o$ are both cut off. The base voltage $v_{B1}$ is then

$$v_{B1} = v_X + v_{BE}(\text{sat}) = 0.1 + 0.8 = 0.9 \text{ V}$$

and current $i_1$ is

$$i_1 = \frac{V_{CC} - v_{B1}}{R_1} = \frac{5 - 0.9}{4} = 1.03 \text{ mA}$$

This current flows out of the input transistor emitters. Since $Q_2$ and $Q_o$ are cut off, all other currents are zero and the output voltage is $v_O = 5$ V.

If $v_X = v_Y = 5$ V, then the input transistor is biased in the inverse active mode. The base voltage $v_{B1}$ is

$$v_{B1} = V_{BE}(\text{sat})_{Q_o} + V_{BE}(\text{sat})_{Q_2} + V_{BC}(\text{on})_{Q_1}$$

$$= 0.8 + 0.8 + 0.7 = 2.3 \text{ V}$$

and the collector voltage $v_{C2}$ is

$$v_{C2} = V_{BE}(\text{sat})_{Q_o} + V_{CE}(\text{sat})_{Q_2} = 0.8 + 0.1 = 0.9 \text{ V}$$

The currents are

$$i_1 = \frac{V_{CC} - v_{B1}}{R_1} = \frac{5 - 2.3}{4} = 0.675 \text{ mA}$$

and

$$i_{B2} = (1 + 2\beta_R)i_1 = (1 + 0.2)(0.675) = 0.810 \text{ mA}$$

Also,

$$i_2 = \frac{V_{CC} - v_{C2}}{R_2} = \frac{5 - 0.9}{1.6} = 2.56 \text{ mA}$$

which means that

$$i_{E2} = i_2 + i_{B2} = 2.56 + 0.81 = 3.37 \text{ mA}$$

The current in the pull-down resistor is

$$i_4 = \frac{V_{BE}(\text{sat})}{R_B} = \frac{0.8}{1} = 0.8 \text{ mA}$$

and the base drive to the output transistor is

$$i_{Bo} = i_{E2} - i_4 = 3.37 - 0.8 = 2.57 \text{ mA}$$

Current $i_3$ is

$$i_3 = \frac{V_{CC} - V_{CE}(\text{sat})}{R_C} = \frac{5 - 0.1}{4} = 1.23 \text{ mA}$$

**Comment:** As mentioned, the analysis of the basic TTL circuit is essentially the same as that of the DTL circuit. The magnitudes of currents and voltages in the basic TTL circuit are also very similar to the DTL results.

---

### EXERCISE PROBLEM

**Ex 17.9:** The parameters of the TTL NAND circuit in Figure 17.24 are: $R_1 = 12 \text{ k}\Omega$, $R_2 = 4 \text{ k}\Omega$, $R_B = 2 \text{ k}\Omega$, and $R_C = 6 \text{ k}\Omega$. Assume $\beta_F \equiv \beta = 25$ and $\beta_R = 0.1$ (for each input emitter). For a no-load condition, determine the base and collector currents in each transistor for: (a) $v_X = v_Y = 0.1$ V and (b) $v_X = v_Y = 5$ V. (Ans. (a) $i_1 = i_{B1} = 0.342 \text{ mA}$, $i_{C1} \cong 0$, $i_{B2} = i_{C2} = 0$, $i_{Bo} = i_{Co} = 0$; (b) $i_1 = i_{B1} = 0.225 \text{ mA}$, $i_{B2} = |i_{C1}| = 0.27 \text{ mA}$, $i_2 = i_{C2} = 1.025 \text{ mA}$, $i_{Bo} = 0.895 \text{ mA}$, $i_{Co} = 0.8167 \text{ mA}$)

### 17.3.4 TTL Output Stages and Fanout

The propagation delay time can be improved by replacing the output collector resistor with a current source.

When the output changes from low to high, the load capacitance must be charged by a current through the collector pull-up resistor. The total load capacitance is composed of the input capacitances of the load circuits and the capacitances of the interconnect lines. The associated $RC$ time constant for a load capacitance of 15 pF and a collector resistance of 4 k$\Omega$ is 60 ns, which is large compared to the propagation delay time of a commercial TTL circuit.

#### Totem-Pole Output Stage

In Figure 17.25, the combination of $Q_3$, $D_1$, and $Q_o$ forms an output stage called a totem pole. Transistor $Q_2$ forms a phase splitter, because the collector and emitter voltages are 180 degrees out of phase. If $v_X = v_Y = $ logic 1, input transistor $Q_1$ is

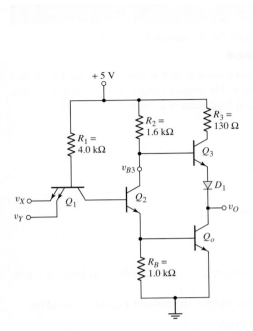

**Figure 17.25** TTL circuit with totem-pole output stage

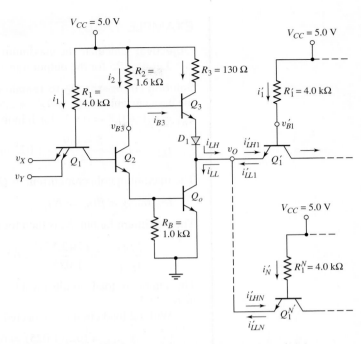

**Figure 17.26** TTL circuit with totem-pole output stage driving $N$ identical TTL stages

biased in the inverse-active mode, and both $Q_2$ and $Q_o$ are driven into saturation. The voltage at the base of $Q_3$ is

$$v_{B3} = V_{C2} = V_{BE}(\text{sat})_{Q_o} + V_{CE}(\text{sat})_{Q_2} \qquad \textbf{(17.17)}$$

which is on the order of 0.9 V, and the output voltage is approximately 0.1 V. The difference between the base voltage of $Q_3$ and the output voltage is not sufficient to turn $Q_3$ and $D_1$ on. The pn junction offset voltage associated with $D_1$ must be included so that $Q_3$ is cut off when the output is low. For this condition, the saturation output transistor discharges the load capacitance and pulls the output low very quickly.

If $v_X = v_Y =$ logic 0, then $Q_2$ and $Q_o$ are cut off, and the base voltage to $Q_3$ goes high. The transistor $Q_3$ and diode $D_1$ turn on so that the output load capacitance can be charged and the output goes high. Since $Q_3$ acts like an emitter follower, the output resistance is small so that the effective $RC$ time constant to charge the load capacitance is now very small.

### Fanout

Logic gates are not operated in isolation, but are used to drive other similar type logic gates to implement a complex logic function. Figure 17.26 shows the TTL NAND gate with a totem-pole output stage connected to $N$ identical TTL NAND gates. The maximum fanout is defined as the maximum number of similar-type logic circuits that can be connected to the logic gate output without affecting proper circuit operation. For example, the output transistor $Q_o$ must remain in saturation when the output goes low to its logic 0 value. For a given value of $\beta$, there is then a maximum allowable load current, and therefore a maximum allowable number of load circuits that can be connected to the output. As another condition, the output transistor is usually rated for a maximum collector current. For an output low condition, the current $i_{LL}$ is the load current that $Q_o$ must sink from the load circuits.

## EXAMPLE 17.10

**Objective:** Calculate the maximum fanout for the output low condition.

Let $\beta = 25$ for the output transistor.

**Solution (Transistor $Q_O$ to remain in saturation):** In Example 17.9, we calculated the base current into $Q_o$ as $i_{Bo} = 2.57$ mA. The output voltage is $v_O = 0.1$ V so that $v'_{B1} = 0.1 + 0.8 = 0.9$ V. Each individual load current is then

$$i'_{LL1} = i'_1 = \frac{5 - 0.9}{4} = 1.025 \text{ mA}$$

The maximum collector current in $Q_o$ is

$$i_{Co}(\text{max}) = \beta i_{Bo} = N i'_{LL1}$$

The maximum fanout, $N$, is then found as

$$N = \frac{\beta i_{Bo}}{i'_{LL1}} = \frac{(25)(2.57)}{1.025} = 62.7$$

The number of load circuits must be an integer, so we round to the next lower integer, or $N = 62$.

With 62 load circuits connected to the output, the collector current would be

$$i_{Co} = N i'_{LL1} = (62)(1.025) = 63.55 \text{ mA}$$

which is a relatively large value. In most cases, the output transistor has a maximum rated collector current that may limit the maximum fanout.

**Solution (Maximum rated output current):** If the maximum rated collector current of the output transistor is $i_{Co}(\text{rated}) = 20$ mA, then the maximum fanout is determined by

$$i_{Co}(\text{rated}) = N i'_{LL1}$$

or

$$N = \frac{i_{Co}(\text{rated})}{i'_{LL1}} = \frac{20}{1.025} = 19.5 \rightarrow 19$$

**Comment:** In the first solution, the resulting fanout of 62 is not realistic since the output transistor current is excessive. In the second solution, a maximum fanout of 19 is more realistic. However, another limitation in terms of proper circuit operation is propagation delay time. For a large number of load circuits connected to the output, the output load capacitance may be quite large which slows down the switching speed to unacceptably large values. The maximum fanout, then, may be limited by the propagation delay time specification.

### EXERCISE PROBLEM

**Ex 17.10:** The TTL circuit shown in Figure 17.25 is redesigned such that $R_1 = 12$ kΩ, $R_2 = 4$ kΩ, $R_3 = 100$ Ω, and $R_B = 2$ kΩ. Assume that $\beta_F \equiv \beta = 25$ and $\beta_R = 0.1$ (for each input emitter). Calculate the fanout for $v_X = v_Y = 3.6$ V under the condition that (a) the output transistor must remain in saturation and (b) the maximum collector current of the output transistor is limited to 12 mA. (Ans. (a) $N = 65$, (b) $N = 35$)

Again, Figure 17.26 shows the TTL circuit with $N$ identical load circuits and the inputs in their low state. The input transistor is biased in saturation, and both $Q_2$ and $Q_o$ are cut off, causing base voltage $v_{B3}$ and the output voltage to go high. The input transistors of the load circuits are biased in the inverse-active mode, and the load currents are supplied through $Q_3$ and $D_1$. In this circuit, the input transistors of the load gates are one-input NAND (inverter) gates, to illustrate the worst-case or maximum load current under the high input condition. Since the load current is supplied through $Q_3$, a base current into $Q_3$ must be supplied from $V_{CC}$ through $R_2$. As the load current increases, the base current through $R_2$ increases, which means that voltage $v_{B3}$ decreases because of the voltage drop across $R_2$. Assuming the B–E voltage of $Q_3$ and the diode voltage across $D_1$ remain essentially constant, the output voltage $v_O$ decreases from its maximum value.

A reasonable fanout of 10 or 15 for the high output condition means that the load current will be small, base current $i_{B3}$ will be very small, and the voltage drop across $R_2$ will be negligible. The output voltage will then be approximately two diode drops below $V_{CC}$. For typical TTL circuits, the logic $1 = V_{OH}$ value is on the order of 3.6 V, rather than the 5 V previously determined.

## Modified Totem-Pole Output Stage

Figure 17.27 shows a modified totem-pole output stage in which transistor $Q_4$ is used in place of a diode. This has several advantages. First, the transistor pair $Q_3$ and $Q_4$ provides greater current gain, which in turn increases the fanout capability of this circuit in its high state. Second, the output impedance in the high state is lower than that of the single transistor, decreasing the switching time. Third, the base–emitter junction of $Q_3$ fulfills the function of diode $D_1$; therefore, the diode is no longer needed to provide a voltage offset. In integrated circuits, the fabrication of transistors is no more complex than the fabrication of diodes.

When the output is switched to its low state, resistor $R_4$ provides a path to ground for the minority carriers that must be pulled out of the base of $Q_3$ to turn the transistor

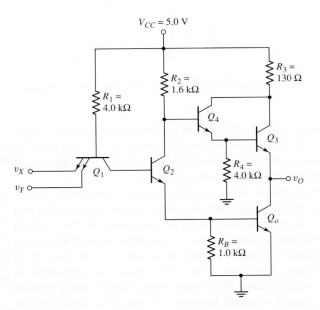

**Figure 17.27** TTL circuit with modified totem-pole output stage

off. Note that when the output is low, with $Q_2$ and $Q_o$ in saturation, the voltage at the base of $Q_4$ is approximately 0.9 V, which is sufficient to bias $Q_4$ in its active region. However, the voltage at the emitter of $Q_4$ is only approximately 0.2 V, which means that the current in $Q_4$ is very small and does not add significantly to the power dissipation.

### 17.3.5   Tristate Output

The output impedances of the totem-pole output TTL logic circuits considered thus far are extremely low when the output voltage is in either the high or low state. In memory circuit applications, situations arise in which the outputs of many TTL circuits must be connected together to form a single output. This creates a serious loading situation, demanding that all other TTL outputs be disabled or put into a high impedance state, as shown symbolically in Figure 17.28. Here, $G_1$ and $G_3$ are disconnected from the output; the output voltage $v_O$ then measures only the output of logic gate $G_2$.

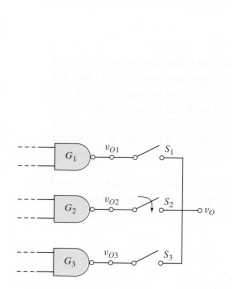

**Figure 17.28** Circuit symbolically showing tristate output

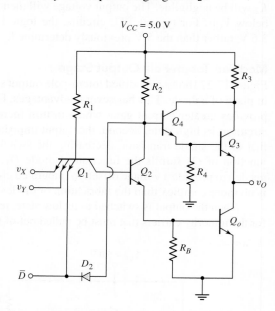

**Figure 17.29** TTL circuit with tristate output stage

The TTL circuit in Figure 17.29 may be used to put the logic output into a high impedance state. When $\bar{D} = 5$ V, the state of input transistor $Q_1$ is controlled by inputs $v_X$ and $v_Y$. Under these circumstances, diode $D_2$ is always reverse biased and the circuit function is the NAND function already considered.

When $\bar{D}$ is driven to a logic 0 state of 0.1 V, the low voltage at the emitter of $Q_1$ ensures that both $Q_2$ and $Q_o$ are cut off, and the low voltage applied to $D_2$ means that $D_2$ is forward biased. The voltage at the base of $Q_4$ is approximately 0.8 V, which means that $Q_3$ is also cut off. In this condition, then, both output transistors $Q_3$ and $Q_o$ are cut off. The impedance looking back into transistors that are cut off is normally in the megohm range. Therefore, when TTL circuits are paralleled to increase the capability of a digital system, the tristate output stage is either enabled or disabled via the $\bar{D}$ select line. The output stage on only one TTL circuit may be enabled at any one time.

## Test Your Understanding

**TYU 17.5** The DTL circuit in Figure 17.20 has new circuit parameters of $R_1 = 15\,k\Omega$, $R_C = 6\,k\Omega$, and $R_B = 15\,k\Omega$. Assume $\beta = 30$ for the transistor. Determine $i_1$, $i_2$, $i_R$, $i_B$, $i_{RC}$, and $v_O$ for: (a) $v_X = v_Y = 0.1$ V; (b) $v_X = 5$ V, $v_Y = 0.1$ V; and (c) $v_X = v_Y = 5$ V. (Ans. (a) $i_1 = 0.28$ mA, $i_2 = i_R = i_B = i_{RC} = 0$, $v_O = 5$ V; (b) Same as part (a); (c) $i_1 = i_2 = 0.1867$ mA, $i_R = 0.0533$ mA, $i_B = 0.1334$ mA, $i_{RC} = 0.8167$ mA, $v_O = 0.1$ V)

**TYU 17.6** For the basic DTL logic circuit in Figure 17.20, the parameters are the same as given in Exercise TYU 17.5. (a) Calculate the maximum fanout for the low output condition such that $Q_o$ remains in saturation. (b) Repeat part (a) for the condition that the maximum rated collector current is $I_{C,\max} = 12$ mA. (Ans. (a) $N = 9$, (b) $N = 9$)

**TYU 17.7** Consider the TTL circuit shown in Figure 17.24 with parameters as given in Exercise Ex 17.9. Calculate the maximum fanout for the low output. For the low output condition, assume that the output transistor must remain in saturation. (Ans. $N = 63$)

**TYU 17.8** For the tristate TTL circuit in Figure 17.29, the parameters are: $R_1 = 6\,k\Omega$, $R_2 = 2\,k\Omega$, $R_3 = 100\,\Omega$, $R_4 = 4\,k\Omega$, and $R_B = 1\,k\Omega$. Assume that $\beta_F \equiv \beta = 20$ and $\beta_R = 0.1$ (for each input emitter). For $\bar{D} = 0.1$ V, calculate the base and collector currents in each transistor. (Ans. $i_{B1} = 0.683$ mA, $|i_{C1}| = i_{B2} = i_{C2} = i_{Bo} = i_{Co} = 0$, $i_{B4} = 1.19\,\mu A$, $i_{C4} = 23.8\,\mu A$, $i_{B3} = i_{C3} = 0$)

## 17.4 SCHOTTKY TRANSISTOR–TRANSISTOR LOGIC

**Objective:** • Analyze and design Schottky and low-power Schottky transistor–transistor logic circuits

The TTL circuits considered thus far drive the output and phase-splitter transistors between cutoff in the high output state and saturation in the low output state. The input transistor is driven between saturation and the inverse-active mode. Since the propagation delay time of a TTL gate is a strong function of the storage time of the saturation transistors, a nonsaturation logic circuit would be an advantage. In the Schottky clamped transistor, the transistor is prevented from being driven into deep saturation and has a storage time of only approximately 50 ps.

### 17.4.1 Schottky Clamped Transistor

The symbol for the Schottky clamped transistor, or simply the Schottky transistor, is shown in Figure 17.30(a); its equivalent configuration is given in Figure 17.30(b). In this transistor, a Schottky diode is connected between the base and collector of an npn bipolar transistor. Two characteristics of the Schottky diode are: a low turn-on voltage and a fast-switching time. When the transistor is in its active region, the base–collector junction is reverse biased, which means that the Schottky diode is reverse biased and effectively out of the circuit. The Schottky transistor then behaves like a normal npn bipolar transistor. As the Schottky transistor goes into saturation, the

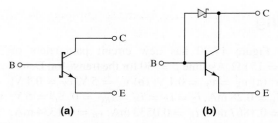

**Figure 17.30** (a) Schottky clamped transistor symbol and (b) Schottky clamped transistor equivalent circuit

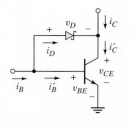

**Figure 17.31** Schottky clamped transistor equivalent circuit, with currents and voltages

base–collector junction becomes forward biased, and the base–collector voltage is effectively clamped at the Schottky diode turn-on voltage, which is normally between 0.3 and 0.4 V. The excess base current is shunted through the diode, and the basic npn transistor is prevented from going deeply into saturation.

Figure 17.31 shows the equivalent circuit of the Schottky transistor with designated currents and voltages. Currents $i_C$ and $i_B$ are the collector and base currents, respectively, of the Schottky transistor, while $i'_C$ and $i'_B$ are the collector and base currents, respectively, of the internal npn transistor.

The three defining equations for the Schottky transistor are

$$i'_C = i_D + i_C \tag{17.18}$$

$$i_B = i'_B + i_D \tag{17.19}$$

and

$$i'_C = \beta i'_B \tag{17.20}$$

Equation (17.20) is appropriate since the internal transistor is clamped at the edge of saturation. If $i_C < \beta i_B$, then the Schottky diode is forward biased, $i_D > 0$, and the Schottky transistor is said to be in saturation. However, the internal transistor is only driven to the edge of saturation in this case.

Combining Equations (17.19) and (17.20), we find that

$$i_D = i_B - i'_B = i_B - \frac{i'_C}{\beta} \tag{17.21}$$

Substituting this equation into Equation (17.18) yields

$$i'_C = i_B - \frac{i'_C}{\beta} + i_C \tag{17.22(a)}$$

or

$$i'_C = \frac{i_B + i_C}{1 + (1/\beta)} \tag{17.22(b)}$$

Equation (17.22(b)) relates the internal transistor collector current to the external Schottky transistor collector and base currents.

### EXAMPLE **17.11**

**Objective:** Determine the currents in a Schottky transistor.

Consider the Schottky transistor in Figure 17.31 with an input base current of $i_B = 1$ mA. Assume that $\beta = 25$. Determine the internal currents in the Schottky transistor for $i_C = 2$ mA, and then for $i_C = 20$ mA.

**Solution:** For $i_C = 2$ mA, the internal collector current is, from Equation (17.22(b)),

$$i'_C = \frac{1+2}{1+(1/25)} = 2.885\,\text{mA}$$

and the internal base current is

$$i'_B = \frac{i'_C}{\beta} = \frac{2.885}{25} = 0.115\,\text{mA}$$

The Schottky diode current is therefore

$$i_D = i_B - i'_B = 1 - 0.115 = 0.885\,\text{mA}$$

Repeating the calculations for $i_C = 20$ mA, we obtain

$$i'_C = 20.2\,\text{mA}$$
$$i'_B = 0.808\,\text{mA}$$
$$i_D = 0.192\,\text{mA}$$

**Comment:** For a relatively small collector current into the Schottky transistor, the majority of the input base current is shunted through the Schottky diode. As the collector current into the Schottky transistor increases, less current is shunted through the Schottky diode and more current flows into the base of the npn transistor.

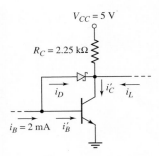

Figure 17.32 Figure for Exercise Ex 17.11

---

EXERCISE PROBLEM

**Ex 17.11:** Consider the Schottky clamped transistor in Figure 17.32. Assume $\beta = 15$, $V_{BE}(\text{on}) = 0.7$ V and $V_\gamma(\text{SD}) = 0.3$ V. (a) For no load, $i_L = 0$, find the currents $i_D$, $i'_B$, and $i'_C$. (b) Repeat part (a) for a load current of $i_L = 10$ mA. (c) Determine the maximum load current $i_L$ that the load transistor can sink and still remain at the edge of saturation. (Ans. (a) $i'_C = 3.791$ mA, $i'_B = 0.253$ mA, $i_D = 1.747$ mA; (b) $i'_C = 13.166$ mA, $i'_B = 0.878$ mA, $i_D = 1.122$ mA; (c) $i_L \cong 28$ mA)

---

Since the internal npn bipolar transistor is not driven deeply into saturation, we assume that the B–E junction voltage remains equal to the turn-on voltage, or $v_{BE} = V_{BE}(\text{on})$. If the Schottky transistor is biased in saturation, then the C–E voltage is

$$v_{CE} = V_{CE}(\text{sat}) = V_{BE}(\text{on}) - V_\gamma(\text{SD}) \tag{17.23}$$

where $V_\gamma(\text{SD})$ is the turn-on voltage of the Schottky diode. Assuming parameter values of $V_{BE}(\text{on}) = 0.7$ V and $V_\gamma(\text{SD}) = 0.3$ V, the collector–emitter saturation voltage of a Schottky transistor is $V_{CE}(\text{sat}) = 0.4$ V. When the Schottky transistor is at the edge of saturation, then $i_D = 0$, $i_C = \beta i_B$, and $v_{CE} = V_{CE}(\text{sat})$.

### 17.4.2 Schottky TTL NAND Circuit

Figure 17.33 shows a Schottky TTL NAND circuit in which all of the transistors except $Q_3$ are Schottky clamped transistors. The connection of $Q_4$ across the base–collector of $Q_3$ prevents this junction from becoming forward biased, ensuring that $Q_3$ never goes into saturation. Another difference between this circuit and the standard TTL circuit is that the pull-down resistor at the base of output transistor $Q_o$ has been replaced by transistor $Q_5$ and two resistors. This arrangement is called a

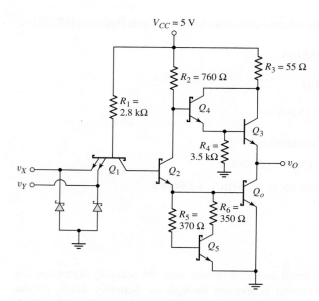

**Figure 17.33**  Schottky TTL NAND logic circuit

**squaring network,** since it squares, or sharpens, the voltage transfer characteristics of the circuit.

Device $Q_2$ is prevented from conducting until the input voltage is large enough to turn on both $Q_2$ and $Q_o$ simultaneously. Recall that the passive pull-down resistor on the TTL circuit provided a pathway for removing stored charge in the base of the output transistor, when the output transistor was turned off from the saturated state. Transistor $Q_5$ now provides an active pull-down network that pulls $Q_o$ out of saturation more quickly.

This is one example of a circuit in which the piecewise linear model of a transistor fails to provide an adequate solution for the circuit analysis. With the piecewise linear model, $Q_5$ would apparently never turn on. However, because of the exponential relationship between collector current and base-emitter voltage, transistor $Q_5$ does turn on and does help pull $Q_o$ out of saturation during switching.

The two Schottky diodes between the input terminals and ground act as clamps to suppress any ringing that might occur from voltage transitions. The input diodes clamp any negative undershoots at approximately −0.3 V.

The dc current–voltage analysis of the Schottky TTL circuit in Figure 17.33 is similar to that for the standard TTL circuit. One minor difference is that when the inputs are high and the input transistor is in the inverse-active mode, the B–C forward bias voltage is 0.3 V, because of the Schottky diode connected between the base and collector junctions.

The major difference between the Schottky circuit and standard TTL circuits is the quantity of excess minority carrier storage in the transistors when they are driven into or near saturation. The internal npn transistor of the Schottky clamped transistor is held at the edge of saturation, and the resulting propagation delay time is on the order of 2 to 5 ns, compared to a nominal 10 to 15 ns for standard TTL circuits.

A slight difference between the Schottky and standard TTL circuits is the value of the output voltage in the logic 0 state. The low output voltage of a standard TTL circuit is in the range of 0.1 to 0.2 V, while the Schottky transistor low output saturation voltage, $V_{OL}$, is approximately 0.4 V. The output voltage in the logic 1 state is essentially the same for both types of logic circuits.

### 17.4.3    Low-Power Schottky TTL Circuits

The Schottky TTL circuit in Figure 17.33 and the standard TTL circuit dissipate approximately the same power, since voltage and resistance values in the two circuits are similar. The advantage of the Schottky TTL circuit is the reduction in propagation delay time by a factor of 3 to 10.

Propagation delay times depend on the type of transistors (Schottky clamped or regular) used in the circuit, and on the current levels in the circuit. The storage time of a regular transistor is a function of the reverse base current that pulls the transistor out of saturation. Also, the transistor turn-on time depends on the current level charging the base–emitter junction capacitance. A desirable trade-off can therefore be made between current levels (power dissipation) and propagation delay times. Smaller current levels lead to lower power dissipation, but at the expense of increased propagation delay times. This trade-off has been successful in commercial applications, where very short propagation delay times are not always necessary, but reduced power requirements are always an advantage.

A **low-power Schottky** TTL NAND circuit is shown in Figure 17.34. With few exceptions, these circuits do not use the multiemitter input transistor of standard TTL circuits. Most low-power Schottky circuits use a DTL type of input circuit, with Schottky diodes performing the AND function. This circuit is faster than the classic multiemitter input transistor circuit, and the input breakdown voltage is also higher.

The dc analysis of the low-power Schottky circuit is identical to that of DTL circuits.

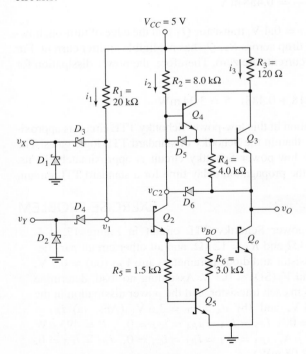

**Figure 17.34** Low-power Schottky TTL NAND logic circuit

## EXAMPLE **17.12**

**Objective:** Calculate the power dissipation in a low-power Schottky TTL circuit.

Consider the circuit shown in Figure 17.34. Assume the Schottky diode turn-on voltage is $V_\gamma(\text{SD}) = 0.3$ V and the transistor parameters are: $V_{BE}(\text{on}) = 0.7$ V, $V_{CE}(\text{sat}) = 0.4$ V, and $\beta = 25$.

**Solution:** For the low input condition, $v_X = v_Y = 0.4$ V and $v_1 = 0.4 + 0.3 = 0.7$ V. Current $i_1$ is

$$i_1 = \frac{V_{CC} - v_1}{R_1} = \frac{5 - 0.7}{20} = 0.215 \, \text{mA}$$

Since $Q_2$ and $Q_o$ are cut off with a no-load condition, all other currents in the circuit are zero. The power dissipation for the low input condition is therefore

$$P_L = i_1(V_{CC} - v_X) = (0.215) \cdot (5 - 0.4) = 0.989 \, \text{mW}$$

For the high input condition, $v_X = v_Y = 3.6$ V, voltage $v_1$ is

$$v_1 = V_{BE}(\text{on})_{Q_o} + V_{BE}(\text{on})_{Q_2} = 0.7 + 0.7 = 1.4 \, \text{V}$$

and voltage $v_{C2}$ is

$$v_{C2} = V_{BE}(\text{on})_{Q_o} + V_{CE}(\text{sat})_{Q_2} = 0.7 + 0.4 = 1.1 \, \text{V}$$

The currents are then

$$i_1 = \frac{V_{CC} - v_1}{R_1} = \frac{5 - 1.4}{20} = 0.18 \, \text{mA}$$

and

$$i_2 = \frac{V_{CC} - v_{C2}}{R_2} = \frac{5 - 1.1}{8} = 0.488 \, \text{mA}$$

When $v_{C2} = 1.1$ V and $v_O = 0.4$ V, transistor $Q_4$ is at the edge of turn-on, however, since there is no voltage drop across $R_4$, $Q_4$ has negligible emitter current. For a no-load condition, all other currents are zero. Therefore, the power dissipation for the high input condition is

$$P_H = (i_1 + i_2)V_{CC} = (0.18 + 0.488) \cdot 5 = 3.34 \, \text{mW}$$

**Comment:** The power dissipation in this low-power Schottky TTL circuit is approximately a factor of five smaller than in the Schottky or standard TTL logic gates. The propagation delay time in the low-power Schottky circuit is approximately 10 ns, which compares closely with the propagation delay time for a standard TTL circuit.

## EXERCISE PROBLEM

**Ex 17.12:** Assume the low-power Schottky TTL circuit in Figure 17.34 is redesigned such that $R_1 = 40 \, \text{k}\Omega$ and $R_2 = 12 \, \text{k}\Omega$, and all other circuit parameters remain the same. The transistor and diode parameters are: $V_{BE}(\text{on}) = 0.7$ V, $V_{CE}(\text{sat}) = 0.4$ V, $\beta = 25$, and $V_\gamma(\text{SD}) = 0.3$ V. Assuming no load, determine the base and collector currents in each transistor, and the power dissipation in the gate, for: (a) $v_X = v_Y = 0.4$ V, and (b) $v_X = v_Y = 3.6$ V. (Ans. (a) $i_{B2} = i_{C2} = i_{Bo} = i_{Co} = i_{B5} = i_{C5} = 0$, $i_{B3} = i_{C3} = i_{B4} = i_{C4} = 0$, $P = 495 \, \mu\text{W}$ (b) $i_{B2} = 90 \, \mu\text{A}$, $i_{C2} = 325 \, \mu\text{A}$, $i_{B4} = i_{C4} = i_{B3} = i_{C3} = 0$, $i_{B5} \cong i_{C5} \cong 0$, $i_{Bo} = 415 \, \mu\text{A}, i_{Co} = 0, P = 2.08 \, \text{mW}$)

Diodes $D_5$ and $D_6$ are called **speedup diodes.** As we showed in the dc analysis, these diodes are reverse biased when the inputs are in either a static logic 0 or a logic 1 mode. When at least one input is in a logic 0 state, the output is high, and $Q_3$ and $Q_4$ tend to turn on, supplying any necessary load current. When both inputs are switched to their logic 1 state, $Q_2$ turns on and $v_{C2}$ decreases, forward biasing $D_5$ and $D_6$. Diode $D_5$ helps to pull charge out of the base of $Q_3$, turning this transistor off more rapidly. Diode $D_6$ helps discharge the load capacitance, which means that output voltage $v_O$ switches low more rapidly.

### 17.4.4 Advanced Schottky TTL Circuits

The advanced low-power Schottky circuit possesses the lowest speed–power product with a propagation delay time short enough to accommodate a large number of digital applications, while still maintaining the low power dissipation of the low-power Schottky family of logic circuits. The major modification lies in the design of the input circuitry. Consider the circuit shown in Figure 17.35. The input circuit contains a pnp transistor $Q_1$, a current amplification transistor $Q_2$, and a Schottky diode $D_2$ from the base of $Q_3$ to the input. Diode $D_2$ provides a low-impedance path to ground when the input makes a high-to-low transition. This enhances the inverter switching time. The current driver transistor $Q_1$ provides a faster transition when the input goes from low to high than if a Schottky diode input stage were used. Transistor $Q_1$ provides the switch element that steers current from $R_1$ either to $Q_2$ or the input source.

When $v_X = 0.4$ V, the E–B junction of $Q_1$ is forward biased, and $Q_1$ is biased in its active region. The base voltage of $Q_2$ is approximately 1.1 V; $Q_2$, $Q_3$, and $Q_5$ are cut off; and the output voltage goes high. Most of the current through $R_1$ goes to ground through $Q_1$, so very little current sinking is required of the driver output transistor. When

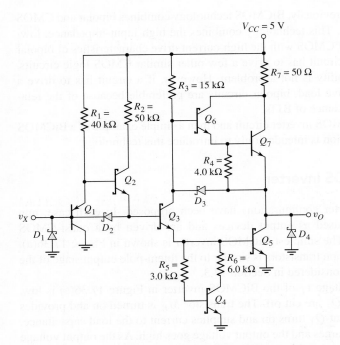

**Figure 17.35** Advanced low-power Schottky (ALS) inverter gate

$v_X = 3.6$ V, transistors $Q_2$, $Q_3$, and $Q_5$ turn on, the voltage at the base of $Q_2$ is clamped at approximately 2.1 V, the E–B junction of $Q_1$ is reverse biased, and $Q_1$ is cut off.

With fast switching circuits, inductances, capacitances, and signal delays may introduce problems requiring the use of transmission line theory. Clamping diodes $D_1$ and $D_4$ at the input and output terminals clamp any negative-going switching transients that result from ringing signals on the interconnect lines.

## Test Your Understanding

**TYU 17.9** In the Schottky TTL NAND circuit in Figure 17.33, assume $\beta_F \equiv \beta = 25$ and $\beta_R = 0$. For a no-load condition, calculate the power dissipation for: (a) $v_X = v_Y = 0.4$ V, and (b) $v_X = v_Y = 3.6$ V. (Ans. $P = 6.41$ mW (b) $P = 31.4$ mW)

**TYU 17.10** Consider the advanced low-power Schottky circuit in Figure 17.35. Determine the currents in $R_1$ and $R_2$ for (a) $v_X = 0.4$ V and (b) $v_X = 3.6$ V. (Ans. (a) $i_{R1} = 97.5\,\mu$A, $i_{R2} = 0$; (b) $i_{R1} = 72.5\,\mu$A, $i_{R2} = 64\,\mu$A)

**TYU 17.11** Let $V_{CC} = 3.5$ V for the advanced low-power Schottky circuit in Figure 17.35. Determine the currents in $R_1$ and $R_2$ for (a) $v_X = 0.4$ V and (b) $v_X = 2.1$ V. (Ans. (a) $i_{R1} = 60\,\mu$A, $i_{R2} = 0$; (b) $i_{R1} = 35\,\mu$A, $i_{R2} = 34\,\mu$A)

# 17.5    BiCMOS DIGITAL CIRCUITS

**Objective: •** Analyze BiCMOS digital logic circuits

As we have discussed previously, BiCMOS technology combines bipolar and CMOS circuits on one IC chip. This technology combines the high-input-impedance, low-power characteristics of CMOS with the high-current drive characteristics of bipolar circuits. If the CMOS circuit has to drive a few other similar CMOS logic circuits, the current drive capability is not a problem. However, if a circuit has to drive a relatively large capacitive load, bipolar circuits are preferable because of the relatively large transconductance of BJTs.

We consider a BiCMOS inverter circuit and then a simple example of a BiCMOS digital circuit. This section is intended only to introduce this technology.

## 17.5.1    BiCMOS Inverter

Several BiCMOS inverter configurations have been proposed. In each case, npn bipolar transistors are used as output devices and are driven by a quasi-CMOS inverter configuration. The simplest BiCMOS inverter is shown in Figure 17.36(a). The output stage of the npn transistors is similar to the totem-pole output stage of the TTL circuits that were considered in Section 17.3.

When the input voltage $v_I$ of the BiCMOS inverter in Figure 17.36(a) is low, the transistors $M_N$ and $Q_2$ are cut off. The transistor $M_P$ is turned on and provides base current to $Q_1$ so that $Q_1$ turns on and supplies current to the load capacitance. The load capacitance charges and the output voltage goes high. As the output voltage goes high, the output current will normally become very small, so that $M_P$ is driven

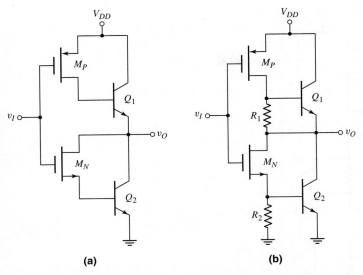

**Figure 17.36** (a) Basic BiCMOS inverter. (b) Improved version of BiCMOS inverter.

into its nonsaturation region and the drain-to-source voltage will become essentially zero. The transistor $Q_1$ will essentially cut off and the output voltage will charge to a maximum value of approximately $v_O(\text{max}) = V_{DD} - V_{BE}(\text{on})$.

When the input voltage $v_I$ goes high, $M_P$ turns off, eliminating any bias current to $Q_1$, so $Q_1$ is also off. The two transistors $M_N$ and $Q_2$ turn on and provide a discharge path for the load capacitance so the output voltage goes low. In steady state, the load current will normally be very small, so $M_N$ will be biased in the nonsaturation region. The drain-to-source voltage will become essentially zero. The transistor $Q_2$ will be essentially off and the output voltage will discharge to a minimum value of approximately $v_O(\text{min}) \cong V_{BE}(\text{on})$.

One serious disadvantage of the inverter in Figure 17.36(a) is that there is no path through which base charge from the npn transistors can be removed when they are turning off. Thus, the turn-off time of the two npn transistors can be relatively long. A solution to this problem is to include pull-down resistors, as shown in the circuit in Figure 17.36(b). Now, when the npn transistors are being turned off, the stored base charge can be removed to ground through $R_1$ or $R_2$. An added advantage of this circuit is, that when $v_I$ goes high and the output goes low, the very small output current through $M_N$ and $R_2$ means the output voltage is pulled to ground potential. Also, as $v_I$ goes low and the output goes high, the very small load current means that the output is pulled up to essentially $V_{DD}$ through the resistor $R_1$. We may note that the two npn output transistors are never on at the same time.

Other circuit designs incorporate other transistors that aid in turning transistors off and increasing switching speed. However, these two examples have demonstrated the basic principle used in BiCMOS inverter circuit designs.

### 17.5.2  BiCMOS Logic Circuit

In BiCMOS logic circuits, the logic function is implemented by the CMOS portion of the circuit and the bipolar transistors again act as a buffered output stage providing the

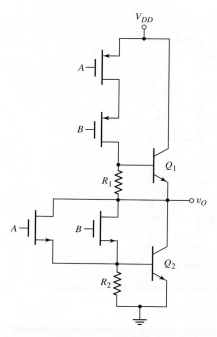

**Figure 17.37** Two-input BiCMOS NOR circuit

necessary current drive. One example of a BiCMOS logic circuit is shown in Figure 17.37. This is a two-input NOR gate. As seen in the figure, the CMOS configuration is the same as the basic CMOS NOR logic gate considered previously. The two npn output transistors and the $R_1$ and $R_2$ resistors have the same configuration and purpose as was seen in the BiCMOS inverter.

Other BiCMOS logic circuits are designed in a manner similar to that shown for the BiCMOS NOR gate.

## 17.6    DESIGN APPLICATION: A STATIC ECL GATE

**Objective:** • Design a static ECL gate to implement a specific logic function.

**Specifications:** A static ECL gate is to be designed to implement the logic function $Y = (A + B)(C + D)$. The circuit is to be designed using constant current sources and the total power dissipation is to be no more than approximately 1 mw.

**Design Approach:** A modified static ECL gate with a Schottky diode similar to the circuit configuration in Figure 17.10 is to be designed.

**Choices:** Inputs $A$, $B$, $C$, and $D$ are assumed to be available. Simple two-transistor current sources will be used.

**Solution (Basic Configuration):** The circuit in Figure 17.10 performs the OR logic function. To implement the AND logic function, we can effectively tie the outputs of two OR logic gates together. Figure 17.38 shows the logic gate configuration. We can show that the output $Y$ is indeed the logic function desired.

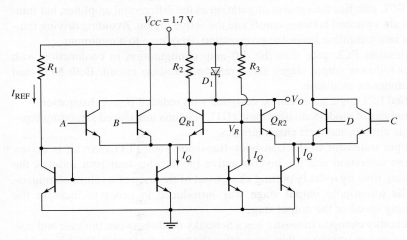

**Figure 17.38** The static ECL gate for the design application

**Solution (DC Circuit Design):** There are four basic currents in the circuit. Assuming that each bias current $I_Q$ is equal to the reference current $I_{REF}$, then from the total power dissipation, we find

$$P_T = 1 = I_T V_{CC} = 4 I_Q (1.7)$$

which yields

$$I_Q \cong 0.15 \, \text{mA}$$

From the reference current leg of the circuit, we have

$$R_1 = \frac{V_{CC} - V_{BE}(\text{on})}{I_{REF}} = \frac{1.7 - 0.7}{0.15}$$

or

$$R_1 = 6.7 \, \text{k}\Omega$$

When either of the reference transistors $Q_{R1}$ or $Q_{R2}$ is turned on, we would like the currents in $R_2$ and $D_1$ to be equal. Assuming the Schottky diode turn-on voltage to be $V_\gamma = 0.4 \, \text{V}$, we then find

$$R_2 = \frac{0.4}{0.075} = 5.3 \, \text{k}\Omega$$

The reference voltage is to be set at $V_R = 1.5 \, \text{V}$ (the average of the logic 0 and logic 1 output voltages). The resistance $R_3$ is then found from

$$R_3 = \frac{V_{CC} - V_R}{I_Q} = \frac{1.7 - 1.5}{0.15} = 1.3 \, \text{k}\Omega$$

**Comment:** The entire circuit will be fabricated as an integrated circuit, so standard-valued resistors are not required in the design.

## 17.7    SUMMARY

- This chapter presented the analysis and design of bipolar digital logic circuits, which were historically the first logic circuit technology used in digital systems.
- Emitter-coupled logic (ECL) is used in specialized high-speed applications. The basic ECL gate has the same configuration as the differential amplifier, but transistors are switched between cutoff and the active region. Avoiding driving transistors into saturation keeps the propagation delay time to a minimum.
- The classical ECL gate uses the diff-amp configuration in conjunction with emitter-follower output stages and a reference voltage circuit. Both NOR and OR outputs are available.
- Modified ECL logic gates can be designed with reduced power dissipation.
- The analysis of diode-transistor logic (DTL) circuits introduced saturating bipolar logic circuits and their characteristics.
- The input transistor of the transistor–transistor logic (TTL) circuit is driven between saturation and the inverse active mode. This transistor reduces the switching time by quickly pulling charge out of the base of a saturated transistor. The totem-pole output stage was introduced in order to increase the switching speed of the output stage.
- The Schottky clamped transistor has a Schottky diode between the base and collector of an npn transistor, thus preventing the transistor from being driven deep into saturation. The propagation delay time of Schottky TTL, then, is shorter than that of regular TTL.
- Low-power Schottky TTL has the same basic configuration as the DTL circuit. Resistor values are increased so as to reduce the currents, which in turn reduce the power dissipated per circuit.
- BiCMOS circuits incorporate the best characteristics of both the CMOS and bipolar technologies. One example is a basic CMOS inverter that drives a bipolar output stage. The high input impedance and low power dissipation of the CMOS design is coupled with the high current drive capability of a bipolar output stage. An example of a BiCMOS NOR logic circuit was considered.
- As an application, a static ECL logic gate to implement a specific logic function was designed.

 ## CHECKPOINT

After studying this chapter, the reader should have the ability to:

✓ Analyze and design a basic ECL OR/NOR logic gate.
✓ Analyze and design modified, lower-power ECL logic gates.
✓ Describe the operation and characteristics of the input transistor of a TTL logic circuit.
✓ Analyze and design a TTL NAND logic gate.
✓ Describe the operation and characteristics of a Schottky transistor, and analyze and design a Schottky TTL logic circuit.
✓ Analyze and design low-power Schottky TTL circuits, and explain tradeoffs between power and switching speed.

 ## REVIEW QUESTIONS

1. Sketch the circuit configuration and discuss the operation of the basic ECL circuit.
2. Why must emitter-follower output stages be added to the diff-amp to make this circuit a practical logic gate?

3. Sketch a modified ECL circuit in which a Schottky diode is incorporated in the collector portion of the circuit. Explain the purpose of the Schottky diode.
4. Explain the concept of series gating for ECL circuits. What are the advantages of this configuration?
5. Sketch a diode–transistor NAND circuit and explain the operation of the circuit. Explain the concept of minimum $\beta$ and the purpose of the pull-down resistor.
6. Explain the operation and purpose of the input transistor in a TTL circuit.
7. Sketch a basic TTL NAND circuit and explain its operation.
8. Sketch a totem-pole output stage and explain its operation and the advantages of incorporating this circuit in the TTL circuit.
9. Explain how maximum fanout can be based on maintaining the output transistor in saturation when the output is low.
10. Explain how maximum fanout can be based on a maximum rated collector current in the output transistor when the output is low.
11. Explain the operation of a Schottky clamped transistor. What are its advantages?
12. What is the primary advantage of a Schottky TTL NAND gate compared to a regular TTL NAND gate.
13. Sketch a low-power Schottky TTL NAND circuit. What are the primary differences between this circuit and the regular DTL circuit considered earlier in the chapter?
14. Sketch a basic BiCMOS inverter and explain its operation. Explain the advantages of this inverter compared to a simple CMOS inverter.

## PROBLEMS

[**Note:** In the following ECL and modified ECL problems, assume $V_{BE}(\text{on}) = V_{EB}(\text{on}) = 0.7\,\text{V}$ and $T = 300\,\text{K}$ unless otherwise stated.

For the TTL problems and Schottky TTL problems, assume transistor and diode parameters listed in Table 17.3. Also assume $V_\gamma = 0.3\,\text{V}$ for a Schottky diode.]

### Section 17.1 Emitter-Coupled Logic (ECL)

17.1   For the differential amplifier circuit in Figure P17.1, neglect the base currents. (a) Determine $R_C$ such that $v_{O1} = v_{O2} = -0.2\,\text{V}$ when $v_1 = -0.7\,\text{V}$. (b) Using the results of part (a), find $v_{O1}$ and $v_{O2}$ when (i) $v_1 = -1.0\,\text{V}$ and (ii) $v_1 = -0.4\,\text{V}$. (c) Using the results of part (a), determine the power dissipated in the circuit for (i) $v_1 = -1.0\,\text{V}$ and (ii) $v_1 = -0.4\,\text{V}$.

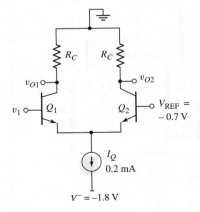

**Figure P17.1**

17.2    Neglect base currents in the circuit in Figure P17.2. (a) Determine $R_E$ and $R_C$ such that $i_E = 80\,\mu A$ and $v_{O1} = v_{O2} = -0.25$ V when $v_1 = -1.0$ V. (b) Using the results of part (a), determine $v_{O1}$ and $v_{O2}$ when (i) $v_1 = -1.3$ V and (ii) $v_1 = -0.7$ V. (c) Using the results of part (a), determine the power dissipation in the circuit for (i) $v_1 = -1.3$ V and (ii) $v_1 = -0.7$ V.

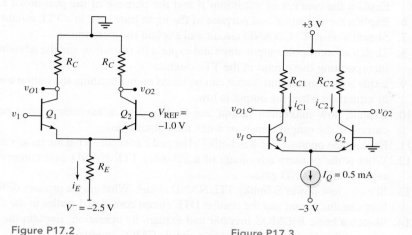

Figure P17.2                                   Figure P17.3

17.3    Neglect base currents in the circuit in Figure P17.3. (a) Determine the value of $R_{C2}$ such that the minimum value of $v_{O2} = 0$. (b) Determine the value of $R_{C1}$ such that $v_{O1} = 1$ V when $v_I = 1$ V. (c) Determine the value of $v_I$ so $i_{C2} = 0.40$ mA and $i_{C1} = 0.10$ mA.

17.4    For the circuit in Figure P17.3, $R_{C1} = R_{C2} = 1$ k$\Omega$. Determine $v_{O1}$ and $v_{O2}$ for (a) $v_I = 0.5$ V and (b) $v_I = -0.5$ V. Neglect base currents.

17.5    Consider the circuit in Figure P17.5. (a) Determine $R_{C2}$ such that $v_2 = -1$ V when $Q_2$ is on and $Q_1$ is off. (b) For $v_{in} = -0.7$, determine $R_{C1}$ such that $v_1 = -1$ V. (c) For $v_{in} = -0.7$ V, find $v_{O1}$ and $v_{O2}$, and for $v_{in} = -1.7$ V, find $v_{O1}$ and $v_{O2}$. (d) Find the power dissipated in the circuit for (i) $v_{in} = -0.7$ V and for (ii) $v_{in} = -1.7$ V.

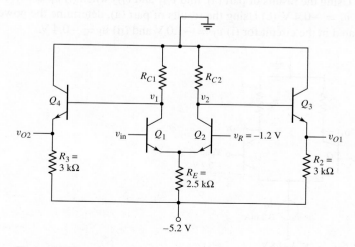

Figure P17.5

17.6   Consider the ECL logic circuit in Figure P17.6. Neglect base currents.
(a) Determine the reference voltage $V_R$. (b) Find the logic 0 and logic 1
voltage values at each output $v_{O1}$ and $v_{O2}$. Assume that inputs $v_X$ and $v_Y$
have the same values as the logic levels at $v_{O1}$ and $v_{O2}$.

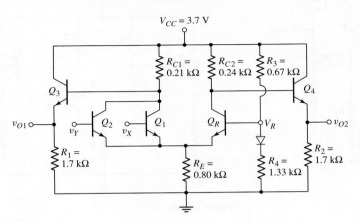

**Figure P17.6**

17.7   Consider the circuit in Figure P17.7. (a) Determine $R_1$ such that
$I_{REF} = 0.20\ \text{mA}$. (b) Determine the values of $R_5$ and $R_6$ such that the maximum currents in $Q_5$ and $Q_6$ are 0.12 mA. (c) Consider $A = B = 0$. What
is $I_Q$? Using the results of part (a), find $R_{C1}$ so that $v_{O1} = -0.7$ V. (d) Consider $A = B = -0.7$ V. What is $I_Q$? Using the results of part (a), find $R_{C1}$
so that $v_{O2} = -0.7$ V.

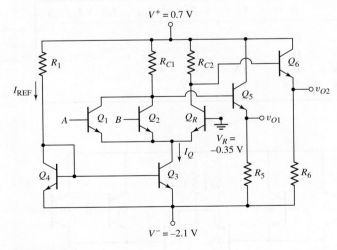

**Figure P17.7**

17.8   Consider the circuit in Figure P17.8. Neglect base currents. Determine all
resistor values such that the following specifications are satisfied: logic
$1 = 0$ V and logic $0 = -1.0$ V; $V_R$ is the average of the logic 1 and logic 0
values; $i_E = 0.4\ \text{mA}$ when $Q_R$ is on; $i_1 = i_2 = 0.4\ \text{mA}$; $i_3 = 0.8\ \text{mA}$ when
$v_{OR} = $ logic 1; and $i_4 = 0.8\ \text{mA}$ when $v_{NOR} = $ logic 1.

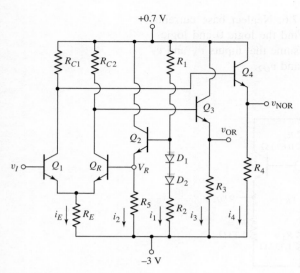

Figure P17.8

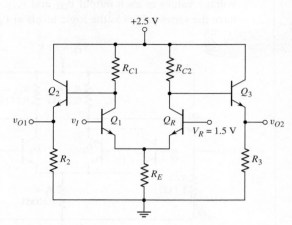

Figure P17.9

**17.9**   In the ECL circuit in Figure P17.9, the outputs have a logic swing of 0.60 V, which is symmetrical about the reference voltage. Neglect base currents. The maximum emitter current for all transistors is 0.8 mA. Assume the input logic voltages $v_I$ are compatible with the output logic voltage. Determine all resistor values.

**17.10**   For the circuit in Figure P17.10, complete the following table. What logic function does the circuit perform?

| A | B | C | D | $I_{E1}$ | $I_{E3}$ | $I_{E5}$ | Y |
|---|---|---|---|---|---|---|---|
| 0 | 0 | 0 | 0 | | | | |
| 5 V | 0 | 0 | 0 | | | | |
| 5 V | 0 | 5 V | 0 | | | | |
| 5 V | 5 V | 5 V | 5 V | | | | |

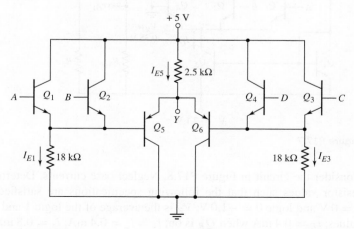

Figure P17.10

17.11   Consider the ECL circuit in Figure P17.11. The input voltages $A$ and $B$ are
compatible with the output voltages $v_{O1}$ and $v_{O2}$. (a) Determine the reference voltage $V_R$. (b) Determine the logic 0 and logic 1 levels at the outputs
$v_{O1}$ and $v_{O2}$. (c) Determine the voltage $V_E$ for $A = B = $ logic 0 and for
$A = B = $ logic 1. (d) Determine the total power dissipated in the circuit
for $A = B = $ logic 0 and for $A = B = $ logic 1.

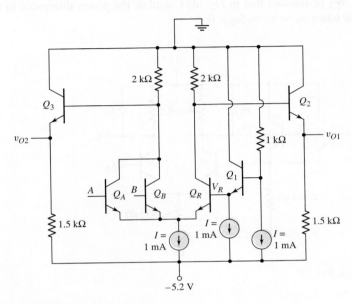

**Figure P17.11**

17.12   A positive-voltage-supply ECL logic gate is shown in Figure P17.12.
Neglect base currents. (a) What logic function is performed by this circuit.
(b) What are the logic 1 and logic 0 values of $v_2$ at the output? (c) When
$v_1 = $ logic 0 for one of the three inputs, determine $i_{E1}, i_{E2}, i_{C3}, i_{C2}$, and $v_2$.
(d) Repeat part (c) when $v_1 = $ logic 1 for all three inputs.

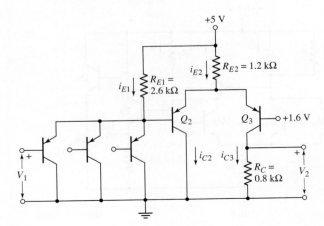

**Figure P17.12**

### Section 17.2  Modified ECL Circuit Configurations

17.13  In the circuit in Figure P17.13, the input voltages $v_X$ and $v_Y$ are compatible with the output voltages $v_{O1}$ and $v_{O2}$. Neglect base currents. (a) Design an appropriate value of $V_R$. State the reason for your selection. (b) Determine the value of $R_{C1}$ such that, when $Q_1$ is on, the current in $R_{C1}$ is one-half that in $D_1$. (c) Determine the value of $R_{C2}$ such that, when $Q_2$ is on, the current in $R_{C2}$ is one-half that in $D_2$. (d) Calculate the power dissipated in the circuit when $v_X = v_Y = $ logic 0.

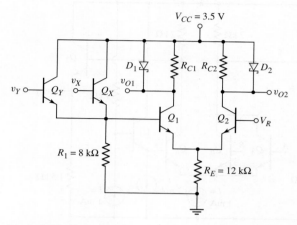

Figure P17.13

17.14  Consider the circuit in Figure P17.14. Neglect base currents. (a) What are the logic 1 and logic 0 values at the output terminals $v_{O1}$ and $v_{O2}$?. (b) For $v_X = v_Y = $ logic 0, determine $R_E$ such that $i_E = 0.25$ mA. (c) Using the results of part (b), determine $R_1$ such that $i_{D1} = 2i_{R1}$ when $Q_R$ is conducting. (d) For $v_X = v_Y = $ logic 1 and $R_1 = R_2$, determine $i_E$, $i_{R2}$, and $i_{D2}$. (e) For $v_X = $ logic 0 and $v_Y = $ logic 1, calculate the power dissipation in the circuit.

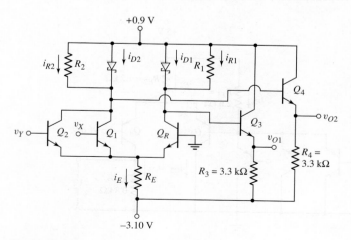

Figure P17.14

17.15   For the circuit in Figure P17.15, assume transistor and diode parameters of $V_{BE}(\text{on}) = 0.7$ V and $V_\gamma = 0.4$ V. Neglect base currents. Find $i_1, i_2, i_3, i_4$, $i_D$, and $v_O$ for: (a) $v_X = v_Y = -0.4$ V, (b) $v_X = 0$, $v_Y = -0.4$ V, (c) $v_X = -0.4$ V, $v_Y = 0$, (d) $v_X = v_Y = 0$.

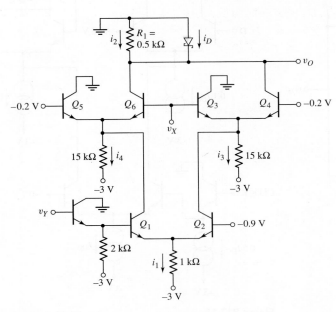

**Figure P17.15**

17.16   Assume the inputs $A, B, C$, and $D$ to the circuit in Figure P17.16 are either 0 or 2.5 V. Let the B–E turn-on voltage be 0.7 V for both the npn and pnp transistors. Assume $\beta = 150$ for the npn devices and $\beta = 90$ for the pnp devices. (a) Determine the voltage at $Y$ for: (i) $A = B = C = D = 0$; (ii) $A = B = 0$, $C = D = 2.5$ V; and (iii) $A = C = 2.5$ V, $B = D = 0$. (b) What logic function does this circuit implement? (c) Determine the power dissipated in the circuit for the conditions given in part (a).

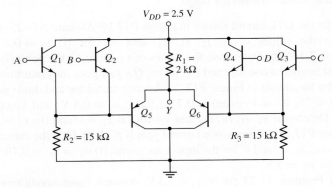

**Figure P17.16**

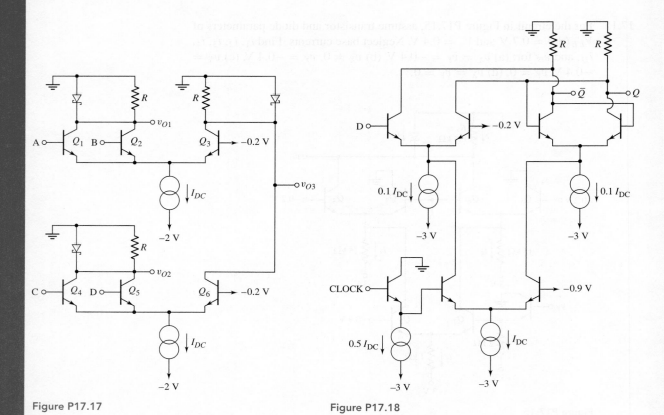

**Figure P17.17**

**Figure P17.18**

17.17 The input and output voltage levels for the circuit in Figure P17.17 are compatible. (a) What are the logic 0 and logic 1 voltage levels? (b) What are the logic functions implemented by this circuit at $v_{O1}$, $v_{O2}$, and $v_{O3}$?

17.18 Consider the circuit in Figure P17.18. (a) Explain the operation of the circuit. Demonstrate that the circuit functions as a clocked D flip-flop. (b) Neglecting base currents, if $i_{DC} = 50\ \mu A$, calculate the maximum power dissipated in the circuit.

### Section 17.3 Transistor–Transistor Logic

17.19 Consider the DTL circuit shown in Figure P17.19. Assume $\beta = 25$. (a) Determine the values of $i_1$, $i_2$, $i_3$, $v_1$, and $v_O$ for (i) $v_I = 0.1\ V$ and (ii) $v_I = 2.5\ V$. (b) Determine the values of $v_I$ and $v_1$ at the point (i) where $Q_O$ just begins to conduct and (ii) where $Q_O$ just goes into saturation.

17.20 Consider the circuit in Figure P17.20. Assume transistor and diode parameters: $\beta = 25$, $V_\gamma = V_{BE}(\text{on}) = 0.7\ V$, $V_{BE}(\text{sat}) = 0.8\ V$, and $V_{CE}(\text{sat}) = 0.1\ V$. Determine $v_1$, $i_1$, $i_B$, $i_C$, and $v_O$ for (a) $v_I = 0$ and (b) $v_I = 3.3\ V$.

17.21 In Figure P17.21, the transistor current gain is $\beta = 20$. Find the currents and voltages $i_1, i_3, i_4$, and $v'$ for the input conditions: (i) $v_X = v_Y = 0.10\ V$, and (ii) $v_X = v_Y = 5\ V$.

17.22 Repeat Problem 17.21 for $V_{CC} = 3.3\ V$. Assume input conditions of (i) $v_X = v_Y = 0.1\ V$ and (ii) $v_X = v_Y = 3.3\ V$.

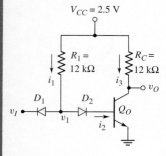

**Figure P17.19**

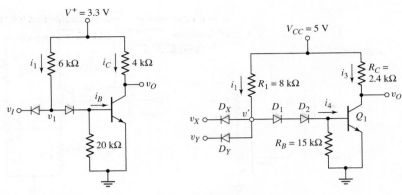

Figure P17.20     Figure P17.21

**17.23** Figure P17.23 shows an improved version of the DTL circuit. One offset diode is replaced by transistor $Q_1$, providing increased current drive to $Q_o$. Assume $\beta = 20$ for both transistors. (a) For $v_X = v_Y = 5$ V, determine the currents and voltages listed in the figure. (b) Calculate the maximum fanout for the low output condition.

**17.24** Repeat Problem 17.23 for $V_{CC} = 3.3$ V. Assume the input condition is $v_X = v_Y = 3.3$ V.

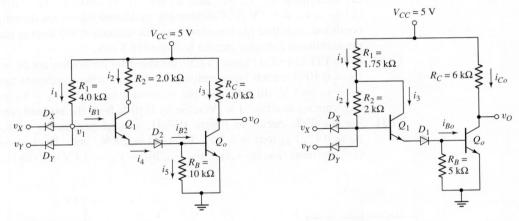

Figure P17.23     Figure P17.25

**17.25** For the modified DTL circuit in Figure P17.25, calculate the indicated currents in the figure for $v_X = v_Y = 5$ V.

**17.26** The transistor $Q_1$ in Figure P17.26 has parameters $\beta = 25$, $\beta_R = 0.5$, and $V_{BE}(\text{on}) = V_{BC}(\text{on}) = 0.7$ V. Find $i_B$, $i_C$, and $i_E$ for (a) $v_I = 0$, (b) $v_I = 0.8$ V, and (c) $v_I = 3.6$ V.

**17.27** The parameters of the transistors in the circuit in Figure P17.27 are $\beta_F \equiv \beta = 25$ and $\beta_R = 0.1$. (a) Determine the values of $i_1$, $i_2$, $i_3$, $v_1$, and $v_O$ for (i) $v_I = 0.1$ V and (ii) $v_I = 2.5$ V. (b) Determine the values of $v_I$ and $v_1$ at the point (i) where $Q_o$ just begins to conduct and (ii) where $Q_o$ just goes into saturation.

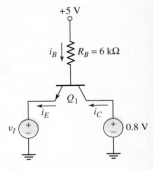

Figure P17.26

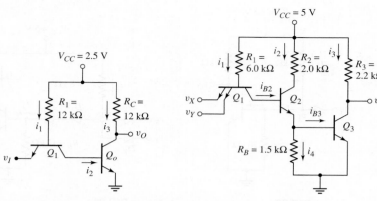

Figure P17.27                    Figure P17.28

17.28   For the transistors in the TTL circuit in Figure P17.28, the parameters
are $\beta_F = 20$ and $\beta_R = 0$. (a) Determine the currents $i_1, i_2, i_3, i_4, i_{B2}$, and $i_{B3}$
for the following input conditions: (i) $v_X = v_Y = 0.1$ V, and (ii) $v_X = v_Y = 5$ V. (b) Show that for $v_X = v_Y = 5$ V, transistors $Q_2$ and $Q_3$ are
biased in saturation.

17.29   The circuit configuration shown in Figure P17.21 is redesigned such that
$V_{CC} = 3.3$ V, $R_1 = 16$ k$\Omega$, $R_C = 6$ k$\Omega$, and $R_B = 20$ k$\Omega$. Let $\beta = 50$.
(a) Determine $i_1, i_3, i_4$, and $v'$ for (i) $v_X = 0.1$ V, $v_Y = 3.3$ V and
(ii) $v_X = v_Y = 3.3$ V. (b) Calculate the maximum fanout for the output low
condition such that $Q_1$ remains biased in saturation. (c) Repeat part (b) if
the maximum collector current is limited to 5 mA.

17.30   In the TTL circuit in Figure P17.30, the transistor parameters are $\beta_F = 20$ and
$\beta_R = 0.10$ (for each input emitter). (a) Calculate the maximum fanout for
$v_X = v_Y = 5$ V. (b) Calculate the maximum fanout for $v_X = v_Y = 0.1$ V.
(Assume $v_O$ is allowed to decrease by 0.10 V from the no-load condition.)

17.31   For the TTL circuit in Figure P17.31, assume parameters of $\beta_F = 50$,
$\beta_R = 0.1$, $V_{BE}(\text{on}) = 0.7$ V, $V_{BE}(\text{sat}) = 0.8$ V, and $V_{CE}(\text{sat}) = 0.1$ V.
(a) Determine $i_{RB}, i_{RCP}, i_{Bo}$, and $V_{out}$ for (i) $V_{in} = 0.1$ V and (ii) $V_{in} = 5$ V.

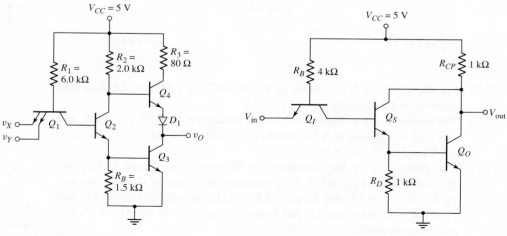

Figure P17.30                              Figure P17.31

(b) For the case when five similar type load circuits are connected to the output, calculate the power dissipated in the circuit shown for (i) $V_{in} = 0.1$ V and (ii) $V_{in} = 5$ V.

17.32 Consider the basic TTL logic gate in Figure P17.32 with a fanout of 5. Assume transistor parameters of $\beta_F = 50$ and $\beta_R = 0.5$ (for each input emitter). Calculate the base and collector currents in each transistor for: (a) $v_X = v_Y = v_Z = 0.1$ V, and (b) $v_X = v_Y = v_Z = 5$ V.

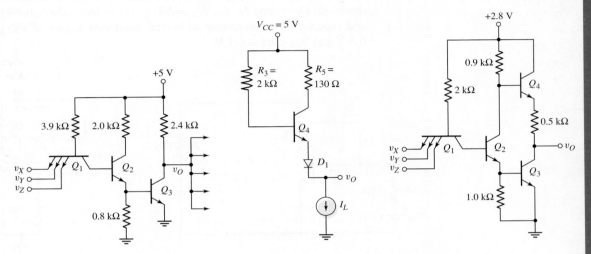

Figure P17.32    Figure P17.33    Figure P17.34

17.33 Consider the portion of the totem-pole output stage shown in Figure P17.33. Let $\beta = 50$. (a) Determine $v_O$ for (i) $I_L = 5\,\mu$A, (ii) $I_L = 5$ mA, and (iii) $I_L = 25$ mA. (b) Determine $I_L$ if the output terminal is accidental shorted to ground.

17.34 For the transistors in the TTL circuit in Figure P17.34, the parameters are $\beta_F = 100$ and $\beta_R = 0.3$ (for each input emitter). (a) For $v_X = v_Y = v_Z = 2.8$ V, determine $i_{B1}$, $i_{B2}$, and $i_{B3}$. (b) For $v_X = v_Y = v_Z = 0.1$ V, determine $i_{B1}$ and $i_{B4}$ for a fanout of 5.

17.35 A low-power TTL logic gate with an active pnp pull-up device is shown in Figure P17.35. The transistor parameters are $\beta_F = 100$ and $\beta_R = 0.2$ (for

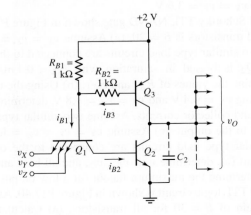

Figure P17.35

each input emitter). Assume a fanout of 5. (a) For $v_X = v_Y = v_Z = 0.1$ V, determine $i_{B1}$, $i_{B2}$, $i_{B3}$, $i_{C2}$, and $i_{C3}$. (b) Repeat part (a) for $v_X = v_Y = v_Z = 2$ V.

## Section 17.4 Schottky Transistor–Transistor Logic

17.36   Consider the Schottky transistor circuit in Figure P17.36. Assume parameter values of $\beta = 50$, $V_{BE}(\text{on}) = 0.7$ V, and $V_\gamma = 0.3$ V for the Schottky diode. (a) Determine $I_B$, $I_D$, $I_C$, and $V_{CE}$. (b) Remove the Schottky diode and repeat part (a) assuming additional parameter values of $V_{BE}(\text{sat}) = 0.8$ V and $V_{CE}(\text{sat}) = 0.1$ V.

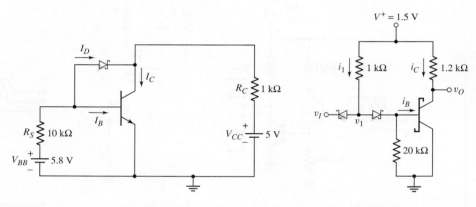

Figure P17.36                                         Figure P17.37

17.37   Let $\beta = 25$ for the transistor in the circuit shown in Figure P17.37. (a) For no load, determine the parameters $i_1$, $i_B$, $i_C$, $v_1$, and $v_O$ when (i) $v_I = 0$ and (ii) $v_I = 1.5$ V. (b) Determine $v_I$, $v_1$, $i_B$, and $i_C$ for the case (i) where the output transistor just begins to conduct and (ii) where the output transistor just goes into saturation. (c) Assume that $N$ similar type load circuits are connected to the output. Determine the maximum number $N$ such that the output transistor remains in saturation.

17.38   Consider the Schottky TTL circuit in Figure 17.33. The transistor parameters are $\beta_F = 30$ and $\beta_R = 0.1$ (for each emitter). (a) Determine all base currents, collector currents, and node voltages for $v_X = v_Y = 0.4$ V. (b) Repeat part (a) for $v_X = v_Y = 3.6$ V.

17.39   Consider the modified Schottky TTL NAND gate shown in Figure P17.39. The current gain of all transistors is $\beta = 20$. (a) Assume $v_X = v_Y = v_Z =$ logic 1 and assume two similar type load circuits are connected to the output. The transistor $Q_2$ is biased in saturation with $i_{B2} = 0.1$ mA and $i_{C2} = 0.2$ mA. Determine the values of $R_{B1}$ and $R_{C1}$. (b) Using the results of part (a), and assuming $v_X = 0.4$ V and $v_Y = v_Z = 1.8$ V, determine $v_{B1}$, $v_{B2}$, $v_O$, and all base and collector currents. Assume two similar type load circuits are connected to the output. (c) Assume $v_X = v_Y = v_Z =$ logic 1 and assume four similar type load circuits are connected to the output. Using the results of part (a), determine $v_{B1}$, $v_{B2}$, $v_O$, and all base and collector currents. (d) Determine the maximum fanout for a low output state.

17.40   A low-power Schottky TTL logic circuit is shown in Figure P17.40. Assume a transistor current gain of $\beta = 30$ for all transistors. (a) Calculate the

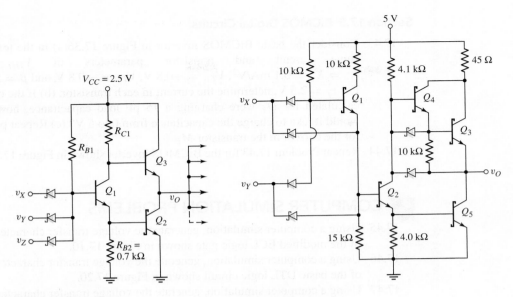

Figure P17.39

Figure P17.40

maximum fanout for $v_X = v_Y = 3.6$ V. (b) Using the results of part (a), determine the power dissipated in the circuit for $v_X = v_Y = 3.6$ V.

17.41 For all transistors in the circuit in Figure 17.35 in the text, the current gain is $\beta = 50$. (a) Calculate the power dissipation in the circuit when the input is at logic 0. (b) Repeat part (a) when the input is at logic 1. (c) Calculate the output short-circuit current. (Assume the input is a logic 0 and the output is inadvertently shorted to ground.)

17.42 Consider the circuit shown in Figure P17.42. Neglect base currents and assume $V_{BE}(\text{on}) = 0.7$ V and $V_\gamma = 0.3$ V. (a) Determine $i_E$ for $v_X = v_Y = 3$ V. (b) Determine $i_E$ and $R_C$ for $v_X = v_Y = 2.4$ V such that $v_O = 2.4$ V. (c) Using the results of parts (a) and (b), determine the power dissipated in the circuit for (i) $v_X = v_Y = 3$ V and (ii) $v_X = v_Y = 2.4$ V.

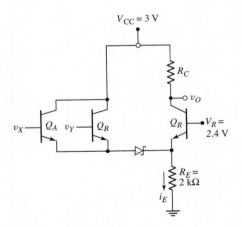

Figure P17.42

### Section 17.5 BiCMOS Digital Circuits

17.43  Consider the basic BiCMOS inverter in Figure 17.36(a) in the text. Assume circuit and transistor parameters of $V_{DD} = 5$ V, $K_n = K_p = 0.1$ mA/V$^2$, $V_{TN} = +0.8$ V, $V_{TP} = -0.8$ V, and $\beta = 50$. (a) For $v_I = 2.5$ V, determine the current in each transistor. (b) If the current calculated for $Q_1$ were charging a 15 pF load capacitance, how long would it take to charge the capacitance from 0 to 5 V? (c) Repeat part (b) for the current in the transistor $M_P$.

17.44  Repeat Problem 17.43 for the BiCMOS inverter shown in Figure 17.36(b).

## COMPUTER SIMULATION PROBLEMS

17.45  Using a computer simulation, generate the voltage transfer characteristics of the modified ECL logic gate shown in Figure 17.10.

17.46  Using a computer simulation, generate the voltage transfer characteristics of the basic DTL logic circuit shown in Figure 17.20.

17.47  Using a computer simulation, generate the voltage transfer characteristics of the advanced low-power Schottky inverter gate shown in Figure 17.35.

17.48  Using a computer simulation, generate the voltage transfer characteristics of the BiCMOS inverter shown in Figure 17.36(b).

## DESIGN PROBLEMS

*D17.49  Design ECL series gating logic circuits, similar to the one shown in Figure 17.16, that will implement the logic functions (a) $Y = [A \cdot (B + C) + D]$ and (b) $Y = [A \cdot B + C \cdot D]$.

*D17.50  Design a clocked D flip-flop, using a modified ECL circuit design, such that the output becomes valid on the negative-going edge of the clock signal.

*D17.51  Design a low-power Schottky TTL exclusive-OR logic circuit.

*D17.52  Design a TTL R–S flip-flop.

# Physical Constants and Conversion Factors

## GENERAL CONSTANTS AND CONVERSION FACTORS

| | | |
|---|---|---|
| Angstrom | Å | $1\ \text{Å} = 10^{-4}\ \mu\text{m} = 10^{-8}\ \text{cm} = 10^{-10}\ \text{m}$ |
| Boltzmann's constant | $k$ | $k = 1.38 \times 10^{-23}\ \text{J/K} = 8.6 \times 10^{-5}\ \text{eV/K}$ |
| Electron–volt | eV | $1\ \text{eV} = 1.6 \times 10^{-19}\ \text{J}$ |
| Electronic charge | $e$ or $q$ | $q = 1.6 \times 10^{-19}\ \text{C}$ |
| Micron | $\mu\text{m}$ | $1\ \mu\text{m} = 10^{-4}\ \text{cm} = 10^{-6}\ \text{m}$ |
| Mil | | $1\ \text{mil} = 0.001\ \text{in.} = 25.4\ \mu\text{m}$ |
| Nanometer | nm | $1\ \text{nm} = 10^{-9}\ \text{m} = 10^{-3}\ \mu\text{m} = 10\ \text{Å}$ |
| Permittivity of free space | $\varepsilon_o$ | $\varepsilon_o = 8.85 \times 10^{-14}\ \text{F/cm}$ |
| Permeability of free space | $\mu_o$ | $\mu_o = 4\pi \times 10^{-9}\ \text{H/cm}$ |
| Planck's constant | $h$ | $h = 6.625 \times 10^{-34}\ \text{J–s}$ |
| Thermal voltage | $V_T$ | $V_T = kT/q \cong 0.026\ \text{V at 300 K}$ |
| Velocity of light in free space | $c$ | $c = 2.998 \times 10^{10}\ \text{cm/s}$ |

## SEMICONDUCTOR CONSTANTS

| | Si | Ge | GaAs | SiO$_2$ |
|---|---|---|---|---|
| Relative dielectric constant | 11.7 | 16.0 | 13.1 | 3.9 |
| Bandgap energy, $E_g(\text{eV})$ | 1.1 | 0.66 | 1.4 | |
| Intrinsic carrier concentration, $n_i$ (cm$^{-3}$ at 300 K) | $1.5 \times 10^{10}$ | $2.4 \times 10^{13}$ | $1.8 \times 10^{6}$ | |

# Selected Manufacturers' Data Sheets

This appendix contains data sheets representative of transistors and op-amps. This appendix is not meant as a substitute for the appropriate data books. In some cases, therefore, only selected information is presented. These data sheets are provided courtesy of National Semiconductor.

## CONTENTS

## National Semiconductor

**2N2222** **PN2222** **MMBT2222** **MPQ2222**
**2N2222A** **PN2222A** **MMBT2222A**

TO–18    TO–92    TO–236 (SOT–23)    TO–116

## NPN General Purpose Amplifier

## Electrical Characteristics $T_A = 25\,°C$ unless otherwise noted

| Symbol | Parameter | | Min | Max | Units |
|---|---|---|---|---|---|
| **OFF CHARACTERISTICS** | | | | | |
| $V_{(BR)CEO}$ | Collector-Emitter Breakdown Voltage (Note 1) ($I_C = 10$ mA, $I_B = 0$) | 2222 2222A | 30 40 | | V |
| $V_{(BR)CBO}$ | Collector-Base Breakdown Voltage ($I_C = 10\ \mu A$, $I_E = 0$) | 2222 2222A | 60 75 | | V |
| $V_{(BR)EBO}$ | Emitter Base Breakdown Voltage ($I_E = 10\ \mu A$, $I_C = 0$) | 2222 2222A | 5.0 6.0 | | V |
| $I_{CEX}$ | Collector Cutoff Current ($V_{CE} = 60$ V, $V_{EB(OFF)} = 3.0$ V) | 2222A | | 10 | nA |
| $I_{CBO}$ | Collector Cutoff Current ($V_{CB} = 50$ V, $I_E = 0$) ($V_{CB} = 60$ V, $I_E = 0$) ($V_{CB} = 50$ V, $I_E = 0$, $T_A = 150\ °C$) ($V_{CB} = 60$ V, $I_E = 0$, $T_A = 150\ °C$) | 2222 2222A 222 2222A | | 0.01 0.01 10 10 | $\mu A$ |
| $I_{EBO}$ | Emitter Cutoff Current ($V_{EB} = 3.0$ V, $I_C = 0$) | 2222A | | 10 | nA |
| $I_{BL}$ | Base Cutoff Current ($V_{CE} = 60$ V, $V_{EB(OFF)} = 3.0$) | 2222A | | 20 | nA |
| **ON CHARACTERISTICS** | | | | | |
| $h_{FE}$ | DC Current Gain ($I_C = 0.1$ mA, $V_{CE} = 10$ V) ($I_C = 1.0$ mA, $V_{CE} = 10$ V) ($I_C = 10$ mA, $V_{CE} = 10$ V) ($I_C = 10$ mA, $V_{CE} = 10$ V, $T_A = -55\ °C$) ($I_C = 150$ mA, $V_{CE} = 10$ V) (Note 1) ($I_C = 150$ mA, $V_{CE} = 1.0$ V) (Note 1) ($I_C = 500$ mA, $V_{CE} = 10$ V) (Note 1) | 2222 2222A | 35 50 75 35 100 50 30 40 | 300 | |

Note 1: Pulse Test: Pulse Width $\leq 300\ \mu s$, Duty Cycle $\leq 2.0\%$.

## NPN General Purpose Amplifier (*Continued*)

### Electrical Characteristics $T_A$ = 25 °C unless otherwise noted (*Continued*)

| Symbol | Parameter | | Min | Max | Units |
|---|---|---|---|---|---|
| **ON CHARACTERISTICS** (*Continued*) | | | | | |
| $V_{CE}$ (sat) | Collector-Emitter Saturation Voltage (Note 1) <br> ($I_C$ = 150 mA, $I_B$ = 15 mA) <br><br> ($I_C$ = 500 mA, $I_B$ = 50 mA) | 2222 <br> 2222A <br> 2222 <br> 2222A | | 0.4 <br> 0.3 <br> 1.6 <br> 1.0 | V |
| $V_{BE}$ (sat) | Base-Emitter Saturation Voltage (Note 1) <br> ($I_C$ = 150 mA, $I_B$ = 15 mA) <br><br> ($I_C$ = 500 mA, $I_B$ = 50 mA) | 2222 <br> 2222A <br> 2222 <br> 2222A | 0.6 <br> 0.6 | 1.3 <br> 1.2 <br> 2.6 <br> 2.0 | V |
| **SMALL-SIGNAL CHARACTERISTICS** | | | | | |
| $f_T$ | Current Gain—Bandwidth Product (Note 3) <br> ($I_C$ = 20 mA, $V_{CE}$ = 20 V, f = 100 MHz) | 2222 <br> 2222A | 250 <br> 300 | | MHz |
| $C_{obo}$ | Output Capacitance (Note 3) <br> ($V_{CB}$ = 10 V, $I_E$ = 0, f = 100 kHz) | | | 8.0 | pF |
| $C_{ibo}$ | Input Capacitance (Note 3) <br> ($V_{EB}$ = 0.5 V, $I_C$ = 0, f = 100 kHz) | 2222 <br> 2222A | | 30 <br> 25 | pF |
| rb'$C_C$ | Collector Base Time Constant <br> ($I_E$ = 20 mA, $V_{CB}$ = 20 V, f = 31.8 MHz) | 2222A | | 150 | ps |
| NF | Noise Figure <br> ($I_C$ = 100 $\mu$A, $V_{CE}$ = 10 V, $R_S$ = 1.0 k$\Omega$, f = 1.0 kHz) | 2222A | | 4.0 | dB |
| Re($h_{ie}$) | Real Part of Common-Emitter <br> High Frequency Input Impedance <br> ($I_C$ = 20 mA, $V_{CE}$ = 20 V, f = 300 MHz) | | | 60 | $\Omega$ |
| **SWITCHING CHARACTERISTICS** | | | | | |
| $t_D$ | Delay Time | ($V_{CC}$ = 30 V, $V_{BE(OFF)}$ = 0.5 V, $I_C$ = 150 mA, $I_{B1}$ = 15 mA) | except MPQ2222 | | 10 | ns |
| $t_R$ | Rise Time | | | 25 | ns |
| $t_S$ | Storage Time | ($V_{CC}$ = 30 V, $I_C$ = 150 mA, $I_{B1}$ = $I_{B2}$ = 15 mA) | except MPQ2222 | | 225 | ns |
| $t_F$ | Fall Time | | | 60 | ns |

Note 1: Pulse Test: Pulse Width <300 $\mu$s, Duty Cycle ≤ 2.0%.
Note 2: For characteristics curves, see Process 19.
Note 3: $f_T$ is defined as the frequency at which $h_{fe}$ extrapolates to unity.

## National Semiconductor

| 2N2907 | PN2907 | MMBT2907 | MPQ2907 |
|--------|--------|----------|---------|
| 2N2907A | PN2907A | MMBT2907A | |

TO–18

TO–92

TO–236
(SOT–23)

TO–116

### PNP General Purpose Amplifier

### Electrical Characteristics $T_A = 25\ °C$ unless otherwise noted

| Symbol | Parameter | | Min | Max | Units |
|--------|-----------|--|-----|-----|-------|
| **OFF CHARACTERISTICS** | | | | | |
| $V_{(BR)CEO}$ | Collector-Emitter Breakdown Voltage (Note 1) $(I_C = 10\ mAdc, I_B = 0)$     2907<br>2907A | | 40<br>60 | | Vdc |
| $V_{(BR)CBO}$ | Collector-Base Breakdown Voltage $(I_C = 10\ \mu Adc, I_E = 0)$ | | 60 | | Vdc |
| $V_{(BR)EBO}$ | Emitter Base Breakdown Voltage $(I_E = 10\ \mu Adc, I_C = 0)$ | | 5.0 | | Vdc |
| $I_{CEX}$ | Collector Cutoff Current $(V_{CE} = 30\ Vdc, V_{BE} = 0.5\ Vdc)$ | | | 50 | nAdc |
| $I_{CBO}$ | Collector Cutoff Current $(V_{CB} = 50\ Vdc, I_E = 0)$    2907<br>2907A<br>$(V_{CB} = 50\ Vdc, I_E = 0, T_A = 150\ °C)$   2907<br>2907A | | | 0.020<br>0.010<br>20<br>10 | $\mu Adc$ |
| $I_B$ | Base Cutoff Current $(V_{CE} = 30\ Vdc, V_{EB} = 0.5\ Vdc)$ | | | 50 | nAdc |

## PNP General Purpose Amplifier (*Continued*)

### Electrical Characteristics $T_A = 25\ °C$ unless otherwise noted (*Continued*)

| Symbol | Parameter | | Min | Max | Units |
|---|---|---|---|---|---|
| **ON CHARACTERISTICS** | | | | | |
| $h_{FE}$ | DC Current Gain<br>($I_C = 0.1$ mAdc, $V_{CE} = 10$ Vdc) | 2907<br>2907A | 35<br>75 | | |
| | ($I_C = 1.0$ mAdc, $V_{CE} = 10$ Vdc) | 2907<br>2907A | 50<br>100 | | |
| | ($I_C = 10$ mAdc, $V_{CE} = 10$ Vdc) | 2907<br>2907A | 75<br>100 | | |
| | ($I_C = 150$ mAdc, $V_{CE} = 10$ Vdc) (Note 1)<br>($I_C = 500$ mAdc, $V_{CE} = 10$ Vdc) (Note 1) | <br>2907<br>2907A | 100<br>30<br>50 | 300 | |
| $V_{CE\ (sat)}$ | Collector-Emitter Saturation Voltage (Note 1)<br>($I_C = 150$ mAdc, $I_B = 15$ mAdc)<br>($I_C = 500$ mAdc, $I_B = 50$ mAdc) | | | 0.4<br>1.6 | Vdc |
| $V_{BE\ (sat)}$ | Base-Emitter Saturation Voltage<br>($I_C = 150$ mAdc, $I_B = 15$ mAdc) (Note 1)<br>($I_C = 150$ mAdc, $I_B = 50$ mAdc) | | | 1.3<br>2.6 | Vdc |
| **SMALL-SIGNAL CHARACTERISTICS** | | | | | |
| $f_T$ | Current Gain—Bandwidth Product<br>($I_C = 50$ mAdc, $V_{CE} = 20$ Vdc, $f = 100$ MHz) | | 200 | | MHz |
| $C_{obo}$ | Output Capacitance<br>($V_{CB} = 10$ Vdc, $I_E = 0$, $f = 100$ kHz) | | | 8.0 | pF |
| $C_{ibo}$ | Input Capacitance<br>($V_{EB} = 2.0$ Vdc, $I_C = 0$, $f = 100$ kHz) | | | 30 | pF |
| **SWITCHING CHARACTERISTICS** | | | | | |
| $t_{on}$ | Turn-On Time | ($V_{CC} = 30$ Vdc, $I_C = 150$ mAdc, $I_{B1} = 15$ mAdc)   Except MPQ2907 | | 45 | ns |
| $t_d$ | Delay Time | | | 10 | ns |
| $t_r$ | Rise Time | | | 40 | ns |
| $t_{off}$ | Turn-Off Time | ($V_{CC} = 6.0$ Vdc, $I_C = 150$ mAdc, $I_{B1} = I_{B2} = 15$ mAdc)   Except MPQ2907 | | 100 | ns |
| $t_s$ | Storage Time | | | 80 | ns |
| $t_f$ | Fall Time | | | 30 | ns |

Note 1: Pulse Test: Pulse Width ≤ 300 $\mu$s, Duty Cycle ≤ 2.0%.

## National Semiconductor

## NDS9410
## Single N-Channel Enhancement Mode Field Effect Transistor

### General Description

These N-channel enhancement mode power field effect transistors are produced using National's proprietary, high cell density, DMOS technology. This very high density process has been especially tailored to minimize on-state resistance, provide superior switching performance, and withstand high energy pulses in the avalanche and commutation modes. These devices are particularly suited to low voltage applications such as laptop computer power management and other battery powered circuits where fast switching, low in-line power loss, and resistance to transients are needed.

### Features

- 7.0 A, 30 V. $R_{DS(ON)} = 0.03\ \Omega$

- Rugged internal source-drain diode can eliminate the need for an external Zener diode transient suppressor

- High density cell design (3.8 million/in$^2$) for extremely low $R_{DS(ON)}$

- High power and current handling capability in a widely used surface mount package

- Critical DC electrical parameters specified at elevated temperature

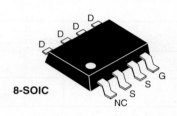

**8-SOIC**

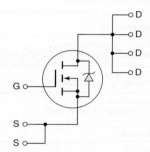

## ABSOLUTE MAXIMUM RATINGS $T_C = 25\ °C$ unless otherwise noted

| Symbol | Parameter | NDS9410 | Units |
|---|---|---|---|
| $V_{DSS}$ | Drain-Source Voltage | 30 | V |
| $V_{DGR}$ | Drain-Gate Voltage ($R_{GS} \leq 1\ M\Omega$) | 30 | V |
| $V_{GSS}$ | Gate-Source Voltage | ±20 | V |
| $I_D$ | Drain Current – Continuous @ $T_A = 25\ °C$ | ±7.0 | A |
| | – Continuous @ $T_A = 70\ °C$ | ±5.8 | A |
| | – Pulsed | ±20 | A |
| $P_D$ | Maximum Power Dissipation @ $T_A = 25\ °C$ | 2.5 (Note 1) | W |
| $T_j T_{STG}$ | Operating and Storage Temperature Range | −55 to 150 | °C |
| **THERMAL CHARACTERISTICS** | | | |
| $R_{\theta JA}(t)$ | Thermal Resistance, Junction-to-Ambient (Pulse = 10 seconds) | 50 (Note 1) | °C/W |
| $R_{\theta JA}$ | Thermal Resistance, Junction-to-Ambient (Steady-State) | 100 (Note 2) | °C/W |

## ELECTRICAL CHARACTERISTICS ($T_C$ = 25 °C unless otherwise noted)

| Symbol | Parameter | Conditions | | Min | Typ | Max | Units |
|--------|-----------|------------|--|-----|-----|-----|-------|
| **OFF CHARACTERISTICS** | | | | | | | |
| $BV_{DSS}$ | Drain-Source Breakdown Voltage | $V_{GS}$ = 0 V, $I_D$ = 250 $\mu$A | | 30 | | | V |
| $I_{DSS}$ | Zero Gate Voltage Drain Current | $V_{DS}$ = 24 V, $V_{GS}$ = 0 V | | | | 2 | $\mu$A |
| | | | $T_C$ = 125 °C | | | 25 | $\mu$A |
| $I_{GSSF}$ | Gate–Body Leakage, Forward | $V_{GS}$ = 20 V, $V_{DS}$ = 0 V | | | | 100 | nA |
| $I_{GSSR}$ | Gate–Body Leakage, Reverse | $V_{GS}$ = –20 V, $V_{DS}$ = 0 V | | | | –100 | nA |
| **ON CHARACTERISTICS (Note 3)** | | | | | | | |
| $V_{GS(h)}$ | Gate Threshold Voltage | $V_{DS}$ = VGS, $I_D$ = 250 $\mu$A | | 1 | 1.4 | 3 | V |
| | | | $T_C$ = 125 °C | 0.7 | 1 | 2.2 | V |
| $R_{DS(ON)}$ | Static Drain-Source On-Resistance | $V_{GS}$ = 10 V, $I_D$ = 7.0 A | | | 0.022 | 0.03 | $\Omega$ |
| | | | $T_C$ = 125 °C | | 0.033 | 0.045 | $\Omega$ |
| | | $V_{GS}$ = 4.5 V, $I_D$ = 3.5 A | | | 0.031 | 0.05 | $\Omega$ |
| | | | $T_C$ = 125 °C | | 0.045 | 0.075 | $\Omega$ |
| $I_{D(ON)}$ | On-State Drain Current | $V_{GS}$ = 10 V, $V_{DS}$ = 5 V | | 20 | | | A |
| | | $V_{GS}$ = 2.7 V, $V_{DS}$ = 2.7 V | | | 7.7 | | A |
| $g_{FS}$ | Forward Transconductance | $V_{DS}$ = 15 V, $I_D$ = 7.0 V | | | 15 | | S |
| **DYNAMIC CHARACTERISTICS** | | | | | | | |
| $C_{ISS}$ | Input Capacitance | $V_{DS}$ = 15 V, $V_{GS}$ = 0 V, f = 1.0 MHz | | | 1250 | | pF |
| $C_{OSS}$ | Output Capacitance | | | | 610 | | pF |
| $C_{RSS}$ | Reverse Transfer Capacitance | | | | 260 | | pF |
| **SWITCHING CHARACTERISTICS (Note 3)** | | | | | | | |
| $t_{D(ON)}$ | Turn-On Delay Time | $V_{DD}$ = 25 V, $I_D$ = 1 A, $V_{GEN}$ = 10 V, $R_{GEN}$ = 6 $\Omega$ | | | 10 | 30 | ns |
| $t_r$ | Turn-On Rise Time | | | | 15 | 60 | ns |
| $t_{D(OFF)}$ | Turn-Off Delay Time | | | | 70 | 150 | ns |
| $t_r$ | Turn-Off Fall Time | | | | 50 | 140 | ns |
| $Q_g$ | Total Gate Charge | $V_{DS}$ = 15 V, $I_D$ = 2.0 A, $V_{GS}$ = 10 V | | | 41 | 50 | nC |
| $Q_{gs}$ | Gate-Source Charge | | | | 2.8 | | nC |
| $Q_{gd}$ | Gate-Drain Charge | | | | 12 | | nC |
| **DRAIN-SOURCE DIODE CHARACTERISTICS AND MAXIMUM RATINGS** | | | | | | | |
| $I_S$ | Maximum Continuous Drain–Source Diode Forward Current | | | | | 2.2 | A |
| $V_{SD}$ | Drain-Source Diode Forward Voltage | $V_{GS}$ = 0 V, $I_S$ = 2.0 A (Note 3) | | | 0.76 | 1.1 | V |
| $t_{rr}$ | Reverse Recovery Time | $V_{GS}$ = 0 V, $I_S$ = 2 A, dI$_S$/dt = 100 A/$\mu$s | | | 100 | | ns |

Notes:
1. Maximum power dissipation and thermal resistance based on an assumption that a 10 second pulse is equivalent to steady-state and using a single-sided maximum-copper mounting board.
2. Junction-to-ambient thermal resistance based on steady-state conditions in still air using mounting board with minimum heat dissipation characteristics.
3. Pulse Test: Pulse Width ≤ 300 $\mu$s, Duty Cycle ≤ 2.0%.

## TYPICAL ELECTRICAL CHARACTERISTICS

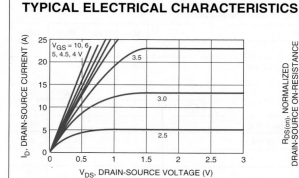

Figure 1. On-Region Characteristics

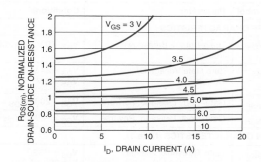

Figure 2. On-Resistance Variation with Drain Current and Gate Voltage

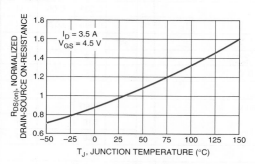

Figure 3. On-Resistance Variation with Temperature

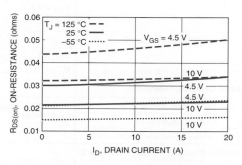

Figure 4. On-Resistance Variation with Drain Current and Temperature

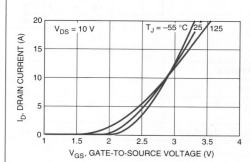

Figure 5. Transfer Characteristics

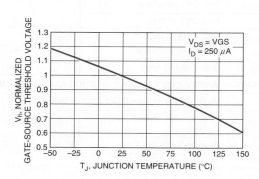

Figure 6. Gate Threshold Variation with Temperature

*National Semiconductor*

# LM741 Operation Amplifier

## General Description

The LM741 series are general purpose operational amplifiers which feature improved performance over industry standards like the LM709. They are direct, plug-in replacements for the 709C, LM201, MC1439 and 748 in most applications. The amplifiers offer many features which make their application nearly foolproof: overload protection on the input and output, no latch-up when the common mode range is exceeded as well as freedom from oscillations.

The LM741C/LM741E are identical to the LM741/LM741A except that the LM741C/LM741E have their performance guaranteed over a 0 °C to +70 °C temperature range, instead of −55 °C to +125 °C.

## Schematic Diagram – (See Figure 13.3 in text)

**Offset Nulling Circuit**

LM741 — OUTPUT

10 kΩ

V −

## Absolute Maximum Ratings

|                              | LM741A            | LM741E            | LM741             | LM741C            |
|------------------------------|-------------------|-------------------|-------------------|-------------------|
| Supply Voltage               | ±22 V             | ±22 V             | ±22 V             | ±18 V             |
| Power Dissipation            | 500 mW            | 500 mW            | 500 mW            | 500 mW            |
| Differential Input Voltage   | ±30 V             | ±30 V             | ±30 V             | ±30 V             |
| Input Voltage (Note 2)       | ±15 V             | ±15 V             | ±15 V             | ±15 V             |
| Output Short Circuit Duration| Continuous        | Continuous        | Continuous        | Continuous        |
| Operating Temperature Range  | −55 °C to +125 °C | 0 °C to +70 °C    | −55 °C to +125 °C | 0 °C to +70 °C    |
| Storage Temperature Range    | −65 °C to +150 °C | −65 °C to +150 °C | −65 °C to +150 °C | −65 °C to +150 °C |
| Junction Temperature         | 150 °C            | 100 °C            | 150 °C            | 100 °C            |

## Electrical Characteristics

| Parameter | Conditions | LM741A/LM741E | | | LM741 | | | LM741C | | | Units |
|-----------|-----------|-----|-----|-----|-----|-----|-----|-----|-----|-----|-------|
| | | Min | Typ | Max | Min | Typ | Max | Min | Typ | Max | |
| Input Offset Voltage | $T_A = 25\ °C$ $R_S \leq 10\ k\Omega$ $R_S \leq 50\ \Omega$ | | 0.8 | 3.0 | | 1.0 | 5.0 | | 2.0 | 6.0 | mV mV |
| | $T_{AMIN} \leq T_A \leq T_{AMAX}$ $R_S \leq 50\ \Omega$ $R_S \leq 10\ k\Omega$ | | | 4.0 | | | 6.0 | | | 7.5 | mV mV |
| Average Input Offset Voltage Drift | | | | 15 | | | | | | | $\mu V/°C$ |
| Input Offset Voltage Adjustment Range | $T_A = 25\ °C$, $V_S = ±20\ V$ | ±10 | | | | ±15 | | | ±15 | | mV |
| Input Offset Current | $T_A = 25\ °C$ | | 3.0 | 30 | | 20 | 200 | | 20 | 200 | nA |
| | $T_{AMIN} \leq T_A \leq T_{AMAX}$ | | | 70 | | 85 | 500 | | | 300 | nA |
| Average Input Offset Current Drift | | | | 0.5 | | | | | | | nA/°C |
| Input Bias Current | $T_A = 25\ °C$ | | 30 | 80 | | 80 | 500 | | 80 | 500 | nA |
| | $T_{AMIN} \leq T_A \leq T_{AMAX}$ | | | 0.210 | | | 1.5 | | | 0.8 | $\mu A$ |
| Input Resistance | $T_A = 25\ °C$, $V_S = ±20\ V$ | 1.0 | 6.0 | | 0.3 | 2.0 | | 0.3 | 2.0 | | $M\Omega$ |
| | $T_{AMIN} \leq T_A \leq T_{AMAX}$, $V_S = ±20\ V$ | 0.5 | | | | | | | | | $M\Omega$ |
| Input Voltage Range | $T_A = 25\ °C$ | | | | | | | ±12 | ±13 | | V |
| | $T_{AMIN} \leq T_A \leq T_{AMAX}$ | | | | ±12 | ±13 | | | | | V |
| Large Signal Voltage Gain | $T_A = 25\ °C$, $R_L \geq 2\ k\Omega$ $V_S = ±20\ V$, $V_O = ±15\ V$ $V_S = ±15\ V$, $V_O = ±10\ V$ | 50 | | | 50 | 200 | | 20 | 200 | | V/mV V/mV |
| | $T_{AMIN} \leq T_A \leq T_{AMAX}$, $R_L \geq 2\ k\Omega$ $V_S = ±20\ V$, $V_O = ±15\ V$ $V_S = ±15\ V$, $V_O = ±10\ V$ $V_S = ±5\ V$, $V_O = ±2\ V$ | 32 10 | | | 25 | | | 15 | | | V/mV V/mV V/mV |

## Electrical Characteristics (Continued)

| Parameter | Conditions | LM741A/LM741E | | | LM741 | | | LM741C | | | Units |
|---|---|---|---|---|---|---|---|---|---|---|---|
| | | Min | Typ | Max | Min | Typ | Max | Min | Typ | Max | |
| Output Voltage Swing | $V_S = \pm20$ V<br>$R_L \geq 10$ k$\Omega$<br>$R_L \geq 2$ k$\Omega$ | $\pm16$<br>$\pm15$ | | | | | | | | | V<br>V |
| | $V_S = \pm15$ V<br>$R_L \geq 10$ k$\Omega$<br>$R_L \geq 2$ k$\Omega$ | | | | $\pm12$<br>$\pm10$ | $\pm14$<br>$\pm13$ | | $\pm12$<br>$\pm10$ | $\pm14$<br>$\pm13$ | | V<br>V |
| Output Short Circuit Current | $T_A = 25$ °C<br>$T_{AMIN} \leq T_A \leq T_{AMAX}$ | 10<br>10 | 25 | 35<br>40 | | 25 | | | 25 | | mA<br>mA |
| Common-Mode Rejection Ratio | $T_{AMIN} \leq T_A \leq T_{AMAX}$<br>$R_S \leq 10$ k$\Omega$, $V_{CM} = \pm12$ V<br>$R_S \leq 50$ $\Omega$, $V_{CM} = \pm12$ V | 80 | 95 | | 70 | 90 | | 70 | 90 | | dB<br>dB |
| Supply Voltage Rejection Ratio | $T_{AMIN} \leq T_A \leq T_{AMAX}$<br>$V_S = \pm20$ V to $V_S = \pm5$ V<br>$R_S \leq 50$ $\Omega$<br>$R_S \leq 10$ $\Omega$ | 86 | 96 | | 77 | 96 | | 77 | 96 | | dB<br>dB |
| Transient Response<br>  Rise Time<br>  Overshoot | $T_A = 25$ °C, Unity Gain | | 0.25<br>6.0 | 0.8<br>20 | | 0.3<br>5 | | | 0.3<br>5 | | $\mu$s<br>% |
| Bandwidth | $T_A = 25$ °C | 0.437 | 1.5 | | | | | | | | MHz |
| Slew Rate | $T_A = 25$ °C, Unity Gain | 0.3 | 0.7 | | | 0.5 | | | 0.5 | | V/$\mu$s |
| Supply Current | $T_A = 25$ °C | | | | | 1.7 | 2.8 | | 1.7 | 2.8 | mA |
| Power Consumption | $T_A = 25$ °C<br>$V_S = \pm20$ V<br>$V_S = \pm15$ V | | 80 | 150 | | 50 | 85 | | 50 | 85 | mW<br>mW |
| LM741A | $V_S = \pm20$ V<br>$T_A = T_{AMIN}$<br>$T_A = T_{AMAX}$ | | | 165<br>135 | | | | | | | mW<br>mW |
| LM741E | $V_S = \pm20$ V<br>$T_A = T_{AMIN}$<br>$T_A = T_{AMAX}$ | | | 150<br>150 | | | | | | | mW<br>mW |
| LM741 | $V_S = \pm15$ V<br>$T_A = T_{AMIN}$<br>$T_A = T_{AMAX}$ | | | | | 60<br>45 | 100<br>75 | | | | mW<br>mW |

Note 2: For supply voltages less than $\pm15$ V, the absolute maximum input voltage is equal to the supply voltage.

# Standard Resistor and Capacitor Values

In this appendix, we list standard component values, which are used for selecting resistor and capacitor values in designing discrete electronic circuits and systems. Low-power carbon and film resistors with 2 percent to 20 percent tolerances have a standard set of values and a standard color-band marking scheme. These tabulated values may vary from one manufacturer to another, so the tables should be considered as typical.

## C.1 CARBON RESISTORS

Standard resistor values are listed in Table C.1. The lightface type indicates 2 percent and 5 percent tolerance values; the boldface type indicates 10% tolerance resistor values.

| Table C.1 | Standard resistance values ($\times 10^n$) | | | |
|---|---|---|---|---|
| **10** | 16 | **27** | 43 | **68** |
| 11 | **18** | 30 | **47** | 75 |
| **12** | 20 | **33** | 51 | **82** |
| 13 | **22** | 36 | **56** | 91 |
| **15** | 24 | **39** | 62 | **100** |

Discrete carbon resistors have a standard color-band marking scheme, which makes it easy to recognize resistor values in a circuit or a parts bin, without having to search for a printed legend. The color bands start at one end of the resistor, as shown in Figure C.1. Two digits and a multiplier digit determine the resistor value. The additional color bands indicate the tolerance and reliability. The digit and multiplier color-code is given in Table C.2.

For example, the first three color bands of a 4.7 kΩ resistor are yellow, violet, and red. The first two digits are 47 and the multiplier is 100. The first three color bands on a 150 kΩ resistor are brown, green, and yellow.

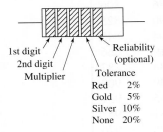

1st digit
2nd digit
Multiplier
Reliability (optional)
Tolerance
Red 2%
Gold 5%
Silver 10%
None 20%

**Figure C.1** Color-band notation of low-power carbon-composition resistors

| Table C.2 | Digit and multiple color code | | |
|---|---|---|---|
| **Digit** | **Color** | **Multiplier** | **Number of zeros** |
| | Silver | 0.01 | −2 |
| | Gold | 0.1 | −1 |
| 0 | Black | 1 | 0 |
| 1 | Brown | 10 | 1 |
| 2 | Red | 100 | 2 |
| 3 | Orange | 1 k | 3 |
| 4 | Yellow | 10 k | 4 |
| 5 | Green | 100 k | 5 |
| 6 | Blue | 1 M | 6 |
| 7 | Violet | 10 M | 7 |
| 8 | Gray | | |
| 9 | White | | |

Ten percent tolerance carbon resistors are available in the following power ratings: $\frac{1}{4}$, $\frac{1}{2}$, 1, and 2 W.

## C.2   PRECISION RESISTORS (ONE PERCENT TOLERANCE)

Metal-film precision resistors can have tolerance levels in the $\frac{1}{2}$ percent to 1 percent range. These resistors use a four-digit code printed on the resistor body, rather than the color-band scheme. The first three digits denote a value, and the last digit is the multiplier for the number of zeros. For example, 2503 denotes a 250 k$\Omega$ resistor, and 2000 denotes a 200 $\Omega$ resistor. If the resistor's value is too small to be described in this way, an $R$ is used to indicate the decimal point; for example, 37R5 is a 37.5 $\Omega$ resistor, and 10R0 is a 10.0 $\Omega$ resistor.

The standard values typically range from 10 $\Omega$ to 301 k$\Omega$. Standard values in each decade are given in Table C.3.

| Table C.3 | Standard precision resistance values | | | | | |
|---|---|---|---|---|---|---|
| 100 | 140 | 196 | 274 | 383 | 536 | 750 |
| 102 | 143 | 200 | 280 | 392 | 549 | 768 |
| 105 | 147 | 205 | 287 | 402 | 562 | 787 |
| 107 | 150 | 210 | 294 | 412 | 576 | 806 |
| 110 | 154 | 215 | 301 | 422 | 590 | 825 |
| 113 | 158 | 221 | 309 | 432 | 604 | 845 |
| 115 | 162 | 226 | 316 | 442 | 619 | 866 |
| 118 | 165 | 232 | 324 | 453 | 634 | 887 |
| 121 | 169 | 237 | 332 | 464 | 649 | 909 |
| 124 | 174 | 243 | 340 | 475 | 665 | 931 |
| 127 | 178 | 249 | 348 | 487 | 681 | 953 |
| 130 | 182 | 255 | 357 | 499 | 698 | 976 |
| 133 | 187 | 261 | 365 | 511 | 715 | |
| 137 | 191 | 267 | 374 | 523 | 732 | |

One percent resistors are often used in applications that require excellent stability and accuracy; a small adjustable trimmer resistor may be connected in series to the 1 percent resistor to set a precise resistance value. It is important to realize that 1 percent resistors are only guaranteed to be within 1 percent of their rated value under a specified set of conditions. Resistance variations due to temperature or humidity changes, and operation at full rated power can exceed the 1 percent tolerance.

## C.3   CAPACITORS

Typical capacitor values for 10 percent tolerance capacitors from one manufacturer are listed in Table C.4. The range of capacitance values for the ceramic-disk capacitor is approximately 10 pF to 1 $\mu$F.

| Table C.4 | Ceramic-disk capacitors | | | |
|---|---|---|---|---|
| 3.3 | 30 | 200 | 600 | 2700 |
| 5 | 39 | 220 | 680 | 3000 |
| 6 | 47 | 240 | 750 | 3300 |
| 6.8 | 50 | 250 | 800 | 3900 |
| 7.5 | 51 | 270 | 820 | 4000 |
| 8 | 56 | 300 | 910 | 4300 |
| 10 | 68 | 330 | 1000 | 4700 |
| 12 | 75 | 350 | 1200 | 5000 |
| 15 | 82 | 360 | 1300 | 5600 |
| 18 | 91 | 390 | 1500 | 6800 |
| 20 | 100 | 400 | 1600 | 7500 |
| 22 | 120 | 470 | 1800 | 8200 |
| 24 | 130 | 500 | 2000 | |
| 25 | 150 | 510 | 2200 | |
| 27 | 180 | 560 | 2500 | |

Tantalum capacitors ($\times 10^{n}$) (to 330 $\mu$F)

| | | |
|---|---|---|
| 0.0047 | 0.010 | 0.022 |
| 0.0056 | 0.012 | 0.027 |
| 0.0068 | 0.015 | 0.033 |
| 0.0082 | 0.018 | 0.039 |

# Reading List

## GENERAL ELECTRONICS TEXTS

Burns, S. G.; and P. R. Bond. *Principles of Electronic Circuits*. 2nd ed. Boston: PWS Publishing Co., 1997.

Hambley, A. R. *Electronics,* 2nd ed. Upper Saddle River, NJ: Prentice-Hall, Inc., 2003.

Hayt, W. H., Jr.; and G. W. Neudeck. *Electronic Circuit Analysis and Design*. 2nd ed. Boston: Houghton Mifflin Co., 1984.

Horenstein, M. N. *Microelectronic Circuits and Devices*. 2nd ed. Englewood Cliffs, NJ: Prentice Hall, Inc., 1995.

Horowitz, P.; and W. Hill. *The Art of Electronics*. 2nd ed. New York: Cambridge University Press, 1989.

Howe, R. T.; and C. G. Sodini. *Microelectronics: An Integrated Approach*. Upper Saddle River, NJ: Prentice-Hall, Inc., 1997.

Jaeger, R. C.; and T. N. Blalock. *Microelectronic Circuit Design,* 3rd ed. New York: McGraw-Hill, 2008.

Malik, N. R. *Electronic Circuits: Analysis, Simulation, and Design*. Englewood Cliffs, NJ: Prentice Hall, Inc., 1995.

Mauro, R. *Engineering Electronics*. Englewood Cliffs, NJ: Prentice Hall, Inc., 1989.

Millman, J.; and A. Graybel. *Microelectronics*. 2nd ed. New York: McGraw-Hill Book Co., 1987.

Mitchell, F. H., Jr.; and F. H. Mitchell, Sr. *Introduction to Electronics Design*. 2nd ed. Englewood Cliffs, NJ: Prentice-Hall, Inc., 1992.

Rashid, M. H. *Microelectronic Circuits: Analysis and Design*. Boston: PWS Publishing Co., 1999.

Razaavi, B. *Fundamentals of Microelectronics*. New York: John Wiley and Sons, Inc., 2008.

Roden, M. S.; and G. L. Carpenter. *Electronic Design: From Concept to Reality*. 3rd ed. Burbank, CA: Discovery Press, 1997.

Schubert, T., Jr; and E. Kim. *Active and Non-Linear Electronics*. New York: John Wiley and Sons, Inc., 1996.

Spencer, R. R.; and M. S. Ghausi. *Introduction to Electronic Circuit Design*. Upper Saddle River, NJ: Prentice-Hall, Inc., 2003.

Sedra, A. S.; and K. C. Smith. *Microelectronic Circuits,* 5th ed. New York: Oxford University Press, 2004.

## LINEAR CIRCUIT THEORY

Alexander, C. K.; and M. N. O. Sadiku. *Fundamentals of Electric Circuits,* 3rd ed. New York: McGraw-Hill, 2007.

Bode, H. W. *Network Analysis and Feedback Amplifier Design*. Princeton, NJ; D. Van Nostrand Co., 1945.

Hayt, W. H., Jr.; J. E. Kemmerly; and S. M. Durbin. *Engineering Circuit Analysis,* 7th ed. New York: McGraw-Hill, 2007.

Irwin, J. D.; and R. M. Nelms. *Basic Engineering Circuit Analysis,* 9th ed. New York: John Wiley and Sons, Inc., 2008.

Johnson, D. E.; J. L. Hillburn; J. R. Johnson; and P. D. Scott. *Basic Electric Circuit Analysis.* 5th ed. Englewood Cliffs, NJ: Prentice Hall, Inc., 1995.

Nilsson, J. W.; and S. A. Riedel. *Electric Circuits,* 8th ed. Upper Saddle River, NJ: Prentice-Hall, Inc., 2007.

Thomas, R. E.; A. J. Rosa; and G. J. Toussaint. *The Analysis and Design of Linear Circuits*, 6th ed. New York: John Wiley and Sons, Inc., 2008.

## SEMICONDUCTOR DEVICES

Neamen, D. A. *An Introduction to Semiconductor Devices.* Boston: McGraw-Hill, 2006.

Neamen, D. A. *Semiconductor Physics and Devices: Basic Principles,* 3rd ed. Boston: McGraw-Hill, 2003.

Streetman, B. G.; and S. Banerjee. *Solid State Electronic Devices,* 6th ed. Upper Saddle River, NJ: Prentice Hall, 2006.

Sze, S. M.; and K. K. Ng. *Physics of Semiconductor Devices*, 3rd ed. New York: John Wiley and Sons, Inc., 2007.

Taur, Y.; and T. H. Ning. *Fundamentals of Modern VLSI Devices.* Cambridge, United Kingdom: Cambridge University Press, 1998.

## ANALOG INTEGRATED CIRCUITS

Allen, P. E.; and D. R. Holberg, *CMOS Analog Circuit Design,* 2nd ed. New York: Oxford University Press, 2002.

Geiger, R. L.; P. E. Allen; and N. R. Strader. *VLSI Design Techniques for Analog and Digital Circuits.* New York: McGraw-Hill Publishing Co., 1990.

Gray, P. R.; P. J. Hurst; S. H. Lewis; and R. G. Meyer. *Analysis and Design of Analog Integrated Circuits,* 5th ed. New York: John Wiley and Sons, Inc., 2009.

Johns, D. A.; and K. Martin. *Analog Integrated Circuit Design.* New York: John Wiley and Sons, Inc., 1997.

Laker, K. R.; and W. M. C. Sansen. *Design of Analog Integrated Circuits and Systems.* New York: McGraw-Hill, Inc., 1994.

Northrop, R. B. *Analog Electronic Circuits.* Reading, MA: Addison-Wesley Publishing Co., 1990.

Razavi, B. *Design of Analog CMOS Integrated Circuits.* Boston: McGraw-Hill, 2001.

Soclof, S. *Design and Applications of Analog Integrated Circuits.* Englewood Cliffs, NJ: Prentice Hall, Inc., 1991.

Solomon, J. E. "The Monolithic Op-Amp: A Tutorial Study," *IEEE Journal of Solid-State Circuits* SC-9, No. 6 (December 1974), pp. 314–32.

Widlar, R. J. "Design Techniques for Monolithic Operational Amplifiers," *IEEE Journal of Solid-State Circuits* SC-4 (August 1969), pp. 184–91.

## OPERATIONAL AMPLIFIER CIRCUITS

Barna, A.; and D. I. Porat. *Operational Amplifiers.* 2nd ed. New York: John Wiley and Sons, Inc., 1989.

Berlin, H. M. *Op-Amp Circuits and Principles.* Carmel, IN: SAMS, A division of Macmillan Computer Publishing, 1991.

Clayton, G.; and B. Newby. *Operational Amplifiers.* London: Butterworth-Heinemann, Ltd., 1992.

Coughlin, R. F.; and F. F. Driscoll. *Operational Amplifiers and Linear Integrated Circuits.* Englewood Cliffs, NJ: Prentice Hall, Inc., 1977.

Fiore, J. M. *Operational Amplifiers and Linear Integrated Circuits: Theory and Application.* New York: West Publishing Co., 1992.

Franco, S. *Design with Operational Amplifiers and Analog Integrated Circuits,* 2nd ed. New York: McGraw-Hill, 1998.

Graeme, J. G.; G. E. Tobey; and L. P. Huelsman. *Operational Amplifiers: Design and Applications.* New York: McGraw-Hill Book Co., 1971.

Helms, H. *Operational Amplifiers 1987 Source Book.* Englewood Cliffs, NJ: Prentice Hall, Inc., 1987.

# DIGITAL CIRCUITS AND DEVICES

Ayers, J. E. *Digital Integrated Circuits: Analysis and Design.* New York: CRC Press, 2004.

Baker, R. J.; H. W. Li, and D. E. Boyce. *CMOS Circuit Design, Layout, and Simulation.* New York: IEEE Press, 1998.

*CMOS/NMOS Integrated Circuits,* RCA Solid State, 1980.

DeMassa, T. A.; and Z. Ciccone. *Digital Integrated Circuits.* New York: John Wiley and Sons, Inc., 1996.

Glasford, G. M. *Digital Electronic Circuits.* Englewood Cliffs, NJ: Prentice Hall, Inc., 1988.

Hauser, J. R., "Noise Margin Criteria for Digital Logic Circuits." *IEEE Transactions on Education* 36, No. 4 (November 1993), pp. 363–68.

Hodges, D. A.; H. G. Jackson, and R. A. Saleh. *Analysis and Design of Digital Integrated Circuits,* 3rd ed. New York; McGraw-Hill, 2004.

Kang, S. M.; and Y. Leblebici. *CMOS Digital Integrated Circuits: Analysis and Design.* 3rd ed. New York; McGraw-Hill, 2003.

Lohstroh, J. "Static and Dynamic Noise Margins of Logic Circuits," *IEEE Journal of Solid-State Circuits* SC-14, No. 3 (June 1979), pp. 591–98.

Mead, C.; and L. Conway. *Introduction to VLSI Systems.* Reading, MA: Addison-Wesley Publishing Co., Inc., 1980.

Mukherjee, A. *Introduction to nMOS and CMOS VLSI Systems Design.* Englewood Cliffs, NJ: Prentice Hall, Inc., 1986.

Prince, B. *Semiconductor Memories: A Handbook of Design, Manufacture and Applications.* 2nd ed. New York: John Wiley and Sons, Inc., 1991.

Rabaey, J. M.; A. Chandrakasan, and B. Nikolic. *Digital Integrated Circuits: A Design Perspective,* 2nd ed. Upper Saddle River, NJ: Prentice-Hall, 2003.

Segura, J.; and C. F. Hawkins. *CMOS Electronics: How It Works, How It Fails.* Piscataway, NJ: IEEE Press, 2004.

Wang, N. *Digital MOS Integrated Circuits.* Englewood Cliffs, NJ: Prentice Hall, Inc., 1989.

Wilson, G. R. "Advances in Bipolar VLSI." *Proceedings of the IEEE* 78, No. 11 (November 1990), pp. 1707–19.

## SPICE AND PSPICE REFERENCES

Banzhap, W. *Computer-Aided Circuit Analysis Using PSpice*. 2nd ed. Englewood Cliffs, NJ: Prentice Hall, Inc., 1992.

Brown, W. L.; and A. Y. J. Szeto. "Verifying Spice Results with Hand Calculations: Handling Common Discrepancies." *IEEE Transactions on Education,* 37, No. 4 (November 1994), pp. 358–68.

Goody, R. W. *MicroSim PSpice for Windows: Volume I: DC, AC, and Devices and Circuits*. 2nd ed. Upper Saddle River, NJ: Prentice-Hall, Inc., 1998.

Goody, R. W. *MicroSim PSpice for Windows: Volume II: Operational Amplifiers and Digital Circuits*. 2nd ed. Upper Saddle River, NJ: Prentice-Hall, 1998.

Herniter, M. E. *Schematic Capture with MicroSim PSpice*. 3rd ed. Upper Saddle River, NJ: Prentice-Hall, 1998.

Meares, L. G.; and C. E. Hymowitz. *Simulating with Spice*. San Pedro, CA: Intusoft, 1988.

MicroSim Staff. *PSpice User's Manual Version 4.03*. Irvine, CA: MicroSim Corporation, 1990.

Natarajan, S. "An Effective Approach to Obtain Model Parameters for BJTs and FETs from Data Books." *IEEE Transactions on Education* 35, No. 2 (May 1992), pp. 164–69.

Rashid, M. H. *SPICE for Circuits and Electronics Using PSpice*. Englewood Cliffs, NJ: Prentice Hall, Inc., 1990.

Roberts, G. W.; and A. S. Sedra. *SPICE for Microelectronic Circuits*. 3rd ed. New York: Saunders College Publishing, 1992.

Thorpe, T. W. *Computerized Circuit Analysis with SPICE*. New York: John Wiley and Sons, Inc., 1992.

Tront, J. G. *PSpice for Basic Microelectronics with CD*. New York: McGraw-Hill, 2008.

Tuinenga, P. W. *SPICE: A Guide to Circuit Simulation and Analysis Using PSpice*. 2nd ed. Englewood Cliffs, NJ: Prentice Hall, Inc., 1992.

## CHAPTER 1

1.1 (a) Silicon
   (i) $n_i = 1.61 \times 10^8$ cm$^{-3}$
   (ii) $n_i = 3.97 \times 10^{11}$ cm$^{-3}$
   (b) GaAs
   (i) $n_i = 6.02 \times 10^3$ cm$^{-3}$
   (ii) $n_i = 1.09 \times 10^8$ cm$^{-3}$

1.3 Silicon
   (a) $n_i = 8.79 \times 10^{-10}$ cm$^{-3}$
   (b) $n_i = 1.5 \times 10^{10}$ cm$^{-3}$
   (c) $n_i = 1.63 \times 10^{14}$ cm$^{-3}$
   Germanium
   (a) $n_i = 35.9$ cm$^{-3}$
   (b) $n_i = 2.40 \times 10^{13}$ cm$^{-3}$
   (c) $n_i = 8.62 \times 10^{15}$ cm$^{-3}$

1.5 (a) p-type, $p_o = 10^{16}$ cm$^{-3}$
       $n_o = 3.24 \times 10^{-4}$ cm$^{-3}$
   (b) p-type, $p_o = 10^{16}$ cm$^{-3}$
       $n_o = 5.76 \times 10^{10}$ cm$^{-3}$

1.7 (a) p-type, $p_o = 5 \times 10^{16}$ cm$^{-3}$
       $n_o = 4.5 \times 10^3$ cm$^{-3}$
   (b) p-type, $p_o = 5 \times 10^{16}$ cm$^{-3}$
       $n_o = 6.48 \times 10^{-5}$ cm$^{-3}$

1.9 (a) $n_o = 5 \times 10^{15}$ cm$^{-3}$
       $p_o = 4.5 \times 10^4$ cm$^{-3}$
   (b) n-type
   (c) $N_d = 5 \times 10^{15}$ cm$^{-3}$

1.11 (a) $I = 0.15$ mA
    (b) E = 2.4 V/cm

1.13 (a) $N_d = 7.69 \times 10^{15}$ cm$^{-3}$
    (b) E = 104 V/cm

1.15 (a) $1.36 \le \sigma \le 1.36 \times 10^4$ ($\Omega$–cm)$^{-1}$
    (b) $0.136 \le J \le 1.36 \times 10^3$ A/cm$^2$

1.17 (a) $J_p = 2.4$ A/cm$^2$
    (b) $J_p = 0.883$ A/cm$^2$
    (c) $J_p = 0.119$ A/cm$^2$

1.19 (a) (i) 0.661 V, (ii) 0.739 V, (iii) 0.937 V
    (b) (i) 1.13 V, (ii) 1.21 V, (iii) 1.41 V

1.21 For $N_a = 10^{15}$ cm$^{-3}$, $V_{bi} = 0.637$ V
    For $N_a = 10^{18}$ cm$^{-3}$, $V_{bi} = 0.817$ V

1.23 (a) $C_j = 0.255$ pF
    (b) $C_j = 0.172$ pF
    (c) $C_j = 0.139$ pF

1.25 (a) $f_o = 6.57$ MHz
    (b) $f_o = 7.95$ MHz
    (c) $f_o = 8.86$ MHz

1.27 (a) (i) 1.03 $\mu$A, (ii) 2.25 mA,
       (iii) 4.93 A, (iv) $-5.37 \times 10^{-12}$ A,
       (v) $-10^{-11}$ A, (vi) $-10^{-11}$ A
    (b) (i) 0.0103 $\mu$A, (ii) 22.5 $\mu$A,
       (iii) 49.3 mA, (iv) $-5.37 \times 10^{-14}$ A,
       (v) $-10^{-13}$ A, (vi) $-10^{-13}$ A

1.29 (a) $I_S = 2.03 \times 10^{-15}$ A
    (b) (i) At $V_D = 0.1$ V,
          $I_D = 9.50 \times 10^{-14}$ A
          At $V_D = 0.7$ V, $I_D = 10^{-3}$ A
       (ii) At $V_D = 0.1$ V,
          $I_D = 1.39 \times 10^{-14}$ A
          At $V_D = 0.7$ V, $I_D = 1.42 \times 10^{-9}$ A

1.31 (a) $\Delta V_D \cong 60$ mV
    (b) $\Delta V_D \cong 120$ mV

1.33 $0.5150 \le V_D \le 0.6347$ V

1.35 (a) 2.31 nA, 5.05 $\mu$A, 11.1 mA,
       $-5.37 \times 10^{-23}$ A, $-10^{-22}$ A, $-10^{-22}$ A
    (b) 115 pA, 0.253 $\mu$A, 0.554 mA,
       $-2.68 \times 10^{-24}$ A, $-5 \times 10^{-24}$ A,
       $-5 \times 10^{-24}$ A

1.37 $\dfrac{I_D(100)}{I_D(-55)} = 2.83 \times 10^3$

1.39 $I_D = 0.4740$ mA, $V_D = 0.5194$ V

1.41 (a) $I_{D1} = I_{D2} = 1$ mA,
       (i) $V_{D1} = V_{D2} = 0.599$ V
       (ii) $V_{D1} = 0.617$ V, $V_{D2} = 0.557$ V
    (b) (i) $V_{D1} = V_{D2} = 0.581$ V,
          $I_{D1} = I_{D2} = 0.5$ mA
       (ii) $V_{D1} = V_{D2} = 0.554$ V,
          $I_{D1} = 0.0909$ mA, $I_{D2} = 0.909$ mA

1.43  (a)  $V_D = 0.7$ V
           $I_D = 26.7$ $\mu$A
      (b)  $V_D = 0.45$ V
           $I_D = 0$

1.45  (a)  $V_O = I_I \cdot (1), I_D = 0;$
           for $0 \leq I_I \leq 0.7$ mA
           $V_O = 0.7$ V, $I_D = (I_I - 0.7)$ mA;
           for $I_I \geq 0.7$ mA
      (b)  $V_O = I_I \cdot (1), I_D = 0;$
           for $0 \leq I_I \leq 1.7$ mA
           $V_O = 1.7$ V, $I_D = (I_I - 1.7)$ mA;
           for $I_I \geq 1.7$ mA
      (c)  $V_O = 0.7$ V, $I_{D1} = I_I$ and $I_{D2} = 0;$
           for $0 \leq I_I \leq 2$ mA

1.47  (a)  (i)   $I = 0.215$ mA, $V_O = 0.7$ V
           (ii)  $I = 0.220$ mA, $V_O = 0.6$ V
      (b)  (i)   $I = 0.2325$ mA, $V_O = -0.35$ V
           (ii)  $I = 0.235$ mA, $V_O = -0.30$ V
      (c)  (i)   $I = 0.372$ mA, $V_O = 0.14$ V
           (ii)  $I = 0.376$ mA, $V_O = 0.12$ V
      (d)  (i)   $I = 0, V_O = -5$ V
           (ii)  $I = 0, V_O = -5$ V

1.49  (a)  $V_D = 0.7$ V, $V = 2.42$ V
      (b)  $P = 0.28$ mW

1.51  (a)  $v_d = 1.30$ mV
      (b)  $v_d = 1.30$ mV

1.53  (b)  For $I = 1$ mA, $v_o/v_s = 0.0909$
           For $I = 0.1$ mA, $v_o/v_s = 0.50$
           For $I = 0.01$ mA, $v_o/v_s = 0.909$

1.55  $I_S = 4.87 \times 10^{-12}$ A

1.57  (a)  $V_O = 5.685$ V
      (b)  $\Delta V_O = 0.0392$ V
      (c)  $V_O = 5.658$ V

1.59  (a)  $V_O = 6.921$ V
      (b)  $\Delta V_O = -0.13$ V

1.61  $V_D = 0, I_D = 0.2$ A
      $V_D = 0.60, I_D = 0.1995$ A
      $V_D = 0.70, I_D = 0.1754$ A
      $V_D = 0.7545, I_D = 0$

# CHAPTER 2

2.1   (a)  $v_O = 0$ for $v_I \leq 0.6$ V
           $$v_O = \left(\frac{1000}{1020}\right)(v_I - 0.6)$$
           for $v_I \geq 0.6$ V
      (b)  (ii)  $v_O(\text{avg}) = 2.89$ V
           (iii) $i_d(\text{max}) = 9.216$ mA
           (iv)  PIV = 10 V

2.3   (a)  $v_O(\text{peak}) = 16.27$ V
      (b)  $i_D(\text{peak}) = 8.14$ mA
      (c)  48.7%
      (d)  $v_O(\text{avg}) = 5.06$ V
      (e)  $i_d(\text{avg}) = 2.53$ mA

2.5   (a)  $R = 4.417$ $\Omega$
      (b)  $i_D$ (avg) = 0.436 A
      (c)  27.6%

2.7   (a)  $\dfrac{N_1}{N_2} = 6.06$
      (b)  $\dfrac{N_1}{N_2} = 1.58$
      (c)  For (a), PIV = 52.1 V
           For (b), PIV = 202.1 V

2.9   (a)  $v_S(\text{rms}) = 8.98$ V
      (b)  $C = 4444$ $\mu$F
      (c)  $i_D(\text{max}) = 4.58$ A

2.11  (a)  $10.3 \leq v_O \leq 12.3$ V
      (b)  $0.490 \leq V_r \leq 0.586$ V
      (c)  $C = 513$ $\mu$F

2.13  (a)  $v_S(\text{rms}) = 11.6$ V
      (b)  $C = 2857$ $\mu$F

2.15  (a)  $\dfrac{N_1}{N_2} = 13.3$
      (b)  $C = 0.0116$ F
      (c)  $i_D(\text{peak}) = 13.3$ A
      (d)  $i_D(\text{avg}) = 0.539$ A
      (e)  PIV = 24.8 V

2.19  (a)  $I_Z = 0.367$ mA, $I_L = 0.975$ mA,
           $P = 1.43$ mW
      (b)  $I_Z = 0.952$ mA, $I_L = 0.39$ mA,
           $P = 3.71$ mW

2.21  (a)  $15 \leq I_Z \leq 259.74$ mA
      (b)  $54.5 \leq R_L \leq 410$ $\Omega$

2.23  (a)  $R_i = 8.08$ $\Omega$
      (b)  $P_Z = 11.9$ W, $P_L = 5$ W

2.25  (a)  $R_i = 257$ $\Omega$
      (b)  $7.450 \leq V_L \leq 7.556$ V; 4.42%
      (c)  $7.50 \leq V_L \leq 7.586$ V; 1.15%

2.27  (a)  $R_i = 200$ $\Omega$
      (b)  $\Delta V_O = 0.35$ V
      (c)  3.5%

2.29  5.0%, $C = 0.0357$ F

2.31  (a)  $v_O = 0$ for $-10 \leq v_I \leq 3$ V
           $v_O = [(0.5) v_I - 1.5]$ V for $v_I \geq 3$ V

(b) $-i_1 = \dfrac{0 - v_I}{10}$ mA for $v_I \le 0$

$i_1 = 0$ for $0 \le v_I \le 3$ V

$i_1 = \dfrac{v_I - 3}{20}$ mA for $v_I \ge 3$ V

2.33 (a) (i) $v_O = v_I$ for $v_I \ge 1.1$ V
$v_O = 1.1$ V for $v_I \le 1.1$ V
(ii) $v_O = v_I$ for $v_I \ge -2.5$ V
$v_O = -2.5$ V for $v_I \le -2.5$ V
(b) (i) $v_O = 2.5$ V for $v_I \ge 2.5$ V
$v_O = v_I$ for $v_I \le 2.5$ V
(ii) $v_O = -1.1$ V for $v_I \ge -1.1$ V
$v_O = v_I$ for $v_I \le -1.1$ V

2.35 (a) (i) $v_O = v_I - 5.7$ V for $v_I \ge 5.7$ V
$v_O = 0$ for $v_I \le 5.7$ V
(ii) $v_O = v_I + 4.3$ V for $v_I \ge -4.3$ V
$v_O = 0$ for $v_I \le -4.3$ V
(b) (i) $v_O = 0$ for $v_I \ge 4.3$ V
$v_O = v_I - 4.3$ V for $v_I \le 4.3$ V
(ii) $v_O = 0$ for $v_I \ge -5.7$ V
$v_O = v_I + 5.7$ V for $v_I \le -5.7$ V

2.37 $v_O(\max) = 6.7$ V, $v_O(\min) = -4.7$ V

2.39 (a) Square wave between $+40$ V and $0$.
(b) Square wave between $+35$ V and $-5$ V.
(c) Square wave between $+5$ V and $-35$ V.

2.41 Circuit similar to Figure 2.31(a)
with $V_B = -10$ V.

2.43 (i) $v_O = (10 \sin \omega t - 5)$ V
(ii) $v_O = (10 \sin \omega t - 15)$ V

2.45 (a) $I = I_{D1} = I_{D2} = 0, V_O = 10$ V
(b) $I = I_{D2} = 0.94$ mA,
$I_{D1} = 0, V_O = 1.07$ V
(c) $I = I_{D2} = 0.44$ mA,
$I_{D1} = 0, V_O = 5.82$ V
(d) $I = 0.964$ mA, $V_O = 0.842$ V,
$I_{D1} = I_{D2} = 0.482$ mA

2.47 (a) $R_1 = 25$ k$\Omega$, $R_2 = 10$ k$\Omega$, $R_3 = 4.4$ k$\Omega$
(b) $V_1 = 4.4$ V, $V_2 = -0.6$ V, $I_{D1} = 0.5$ mA,
$I_{D2} = 0.75$ mA, $I_{D3} = 0.75$ mA
(c) $V_1 = 6.07$ V, $V_2 = -0.6$ V,
$I_{D1} = 1.11$ mA,
$I_{D2} = 0, I_{D3} = 0.65$ mA
(d) $V_1 = 4.4$ V, $V_2 = 1.27$ V,
$I_{D1} = 0.833$ mA,
$I_{D2} = 0.211$ mA, $I_{D3} = 0$

2.49 (a) $I_{D1} = 1.0$ mA, $I_{D2} = 6.0$ mA,
$V_A = 2.30$ V
(b) $I_{D1} = 0, I_{D2} = 5$ mA, $V_A = 8.2$ V

(c) $I_{D1} = 2.15$ mA, $I_{D2} = 7.15$ mA,
$R_2 = 0.60$ k$\Omega$

2.51 $v_O = v_I$ for $-4.65 \le v_I \le 4.65$ V,
$v_O = 4.65$ V for $v_I \ge 4.65$ V,
$v_O = -4.65$ V for $v_I \le -4.65$ V

2.53 (a) $V_O = 0, I_{D1} = 0.86$ mA
(b) $I_{D1} = 0, V_O = -3.57$ V

2.55 (a) $I_{D1} = 0.93$ mA, $V_O = -15$ V
(b) $I_{D1} = 1.86$ mA, $V_O = -15$ V

2.57 (a) $I_D = 0, V_D = 0, V_A = V_B = 3$ V
(b) $I_D = 0.08$ mA, $V_D = 0.7$ V,
$V_A = 1.4$ V, $V_B = 2.1$ V
(c) $I_D = 0, V_D = -0.5$ V, $V_A = 2.5$ V,
$V_B = 2.0$ V
(d) $I_D = 0.23$ mA, $V_D = 0.7$ V,
$V_A = 2.15$ V, $V_B = 2.85$ V

2.59 (a) $I_{D1} = I_{D2} = I_{D3} = 0, v_O = 0.5$ V
(b) $I_{D1} = 0.0667$ mA, $I_{D2} = I_{D3} = 0$,
$v_O = 1.23$ V
(c) $I_{D1} = 0.171$ mA, $I_{D2} = 0.0615$ mA,
$I_{D3} = 0, v_O = 2.069$ V
(d) $I_{D1} = 0.275$ mA, $I_{D2} = 0.20$ mA,
$I_{D3} = 0.05$ mA, $v_O = 2.90$ V

2.61 (a) $V_{O1} = V_{O2} = 0$
(b) $V_{O1} = 4.4$ V, $V_{O2} = 3.8$ V
(c) $V_{O1} = 4.4$ V, $V_{O2} = 3.8$ V
Logic 1 level degrades

2.63 $(V_1 \text{ AND } V_2) \text{ OR } (V_3 \text{ AND } V_4)$

2.65 $V_I = 2.3$ V

2.67 $V_{PS} = 2.6$ V

## CHAPTER 3

3.1 (a) (i) $I_D = 0$
(ii) $I_D = 82.5 \,\mu$A
(iii) $I_D = 0.2325$ mA
(iv) $I_D = 0.3825$ mA
(b) (i) $I_D = 0$
(ii) $I_D = 0.27$ mA
(iii) $I_D = 1.92$ mA
(iv) $I_D = 5.07$ mA

3.3 (a) Enhancement-mode
(b) $V_{TN} = 1.5$ V, $K_n \cong 0.064$ mA/V$^2$
(c) $i_D(\text{sat}) = 0.256$ mA, for $v_{GS} = 3.5$ V
$i_D(\text{sat}) = 0.576$ mA, for $v_{GS} = 4.5$ V

3.5 (a) Saturation
(b) Nonsaturation
(c) Cutoff

3.7 $\dfrac{W}{L} = 5.79$

3.9   (a) $K_n = 1.40 \text{ mA/V}^2$
   (b) $I_D = 3.58 \text{ mA}$
   (c) $V_{DS}(\text{sat}) = 1.6 \text{ V}$

3.11   $W = 3.22 \,\mu\text{m}$

3.13   $V_{TP} = -0.571 \text{ V}, \dfrac{W}{L} = 4.41$

3.15   (a) Enhancement-mode
   (b) $V_{TP} \cong -0.5 \text{ V}, K_p \cong 0.20 \text{ mA/V}^2$
   (c) $i_D(\text{sat}) = 1.8 \text{ mA}, i_D(\text{sat}) = 3.2 \text{ mA}$

3.17   (a) $I_D = 0.21 \text{ mA}$
   (b) $I_D = 0.66 \text{ mA}$
   (c) $I_D = 0.81 \text{ mA}$
   (d) $I_D = 0.844 \text{ mA}$
   (e) $I_D = 0.844 \text{ mA}$

3.19   $K_n = K_p = 41.3 \,\mu\text{A/V}^2$,
   $W_n = 7.09 \,\mu\text{m}, W_p = 12.8 \,\mu\text{m}$

3.21   $\lambda = 0.00926 \text{ V}^{-1}, V_A = 108 \text{ V}$

3.23   (a) $I_D = 0.739 \text{ mA}$
   (b) $I_D = 0.296 \text{ mA}$

3.25   $t_{ox} = 1200 \text{ Å}$

3.27   (a) $V_{DS} = 1.88 \text{ V}, I_D = 2.12 \text{ mA}$
   (b) $V_{DS} = 0.741 \text{ V}, I_D = 1.42 \text{ mA}$

3.29   (a) $V_{SD} = 1.90 \text{ V}, I_D = 1.33 \text{ mA}$
   (b) $V_{SD} = 0.698 \text{ V}, I_D = 1.08 \text{ mA}$

3.31   $V_S = 2.21 \text{ V}, V_{SD} = 5.21 \text{ V}$

3.33   $\dfrac{W}{L} = 20.8, R_1 = 360 \text{ k}\Omega, R_2 = 450 \text{ k}\Omega$

3.35   $V_{GS} = 1.71 \text{ V}, I_D = 2.58 \text{ mA}$,
   $V_{DS} = 5.62 \text{ V}$

3.37   (a) (i) $V_{GS} = 1.516 \text{ V}, V_{DS} = 6.516 \text{ V}$
       (ii) $V_{GS} = 2.61 \text{ V}, V_{DS} = 7.61 \text{ V}$
   (b) (i) $V_{GS} = V_{DS} = 1.516 \text{ V}$
       (ii) $V_{GS} = V_{DS} = 2.61 \text{ V}$

3.39   (a) $R_D = 8 \text{ k}\Omega, R_S = 4.38 \text{ k}\Omega$
   (b) Let $R_D = 8.2 \text{ k}\Omega, R_S = 4.3 \text{ k}\Omega$
       Then $V_{GS} = 2.82 \text{ V}$,
       $I_D = 0.504 \text{ mA}, V_{DS} = 3.70 \text{ V}$
   (c) For $R_S = 4.73 \text{ k}\Omega, I_D = 0.476 \text{ mA}$
       For $R_S = 3.87 \text{ k}\Omega, I_D = 0.548 \text{ mA}$

3.41   $I_{DQ} = 1.25 \text{ mA}, R_2 = 59.4 \text{ k}\Omega$,
   $R_1 = 20.6 \text{ k}\Omega$

3.43   $R_D = 0.8 \text{ k}\Omega, R_1 = 408 \text{ k}\Omega, R_2 = 99.5 \text{ k}\Omega$

3.45   $\left(\dfrac{W}{L}\right)_1 = 2.78$

3.47   (a) $\left(\dfrac{W}{L}\right)_1 = 2.31, \left(\dfrac{W}{L}\right)_2 = 1.59$,
       $\left(\dfrac{W}{L}\right)_3 = 3.69$

   (b) (i) $V_1 = 2.5 \text{ V}, V_2 = 6 \text{ V}$
       (ii) $V_1 = 2.5 \text{ V}, V_2 = 6 \text{ V}$
   (c) $V_1 = 2.469 \text{ V}, V_2 = 5.916 \text{ V}$

3.49   $\left(\dfrac{W}{L}\right)_D = 3.31$

3.51   $R_D = 271 \,\Omega, \dfrac{W}{L} = 231$

3.53   (a) $\dfrac{W}{L} = 0.623$
   (b) $V_O = 0.297 \text{ V}$

3.55   (a) $\left(\dfrac{W}{L}\right)_A = \left(\dfrac{W}{L}\right)_B = 12.5$,
       $\left(\dfrac{W}{L}\right)_C = 7.81$
   (b) $R_D = 11 \text{ k}\Omega$

3.57   $I_D = I_{DSS}$

3.59   $V_P = 3.97 \text{ V}, I_{DSS} = 5.0 \text{ mA}$

3.61   $V_{TN} = 0.221 \text{ V}, K_n = 1.11 \text{ mA/V}^2$

3.63   $V_{GS} = -1.17 \text{ V}, I_D = 5.85 \text{ mA}$,
   $V_{DS} = 7.13 \text{ V}$

3.65   $V_{GS} = 0.838 \text{ V}, V_{SD} = 7.5 \text{ V}$,
   $R_1 = 109 \text{ k}\Omega, R_2 = 1.21 \text{ M}\Omega$

3.67   $V_G = 6 \text{ V}, V_{GS} = -1.30 \text{ V}$,
   $I_D = 3.65 \text{ mA}, V_{DS} = 2.85 \text{ V}$

3.69   $I_D = I_{DSS} = 4 \text{ mA}$,
   $R_D = 1.75 \text{ k}\Omega$

3.71   $R_D = 15 \text{ k}\Omega, R_1 = 100 \text{ k}\Omega$
   $R_2 = 50 \text{ k}\Omega$

## CHAPTER 4

4.1   (a) (i) $\dfrac{W}{L} = 2.5$
       (ii) $V_{GSQ} = 2.4 \text{ V}$
   (b) (i) $\dfrac{W}{L} = 8.33$
       (ii) $V_{GSQ} = 1.0 \text{ V}$

4.3   $\lambda = 0.0308 \text{ V}^{-1}, r_o = 12.5 \text{ k}\Omega$

4.5   (a) $\lambda = 0.01826 \text{ V}^{-1}, r_o = 225 \text{ k}\Omega$
   (b) $I_D = 0.2655 \text{ mA}$

4.7   $I_D = 0.25 \text{ mA}, r_o = 400 \text{ k}\Omega$

4.9   (a) $\dfrac{W}{L} = 6.02$
   (b) $\dfrac{W}{L} = 10.4$

4.13   (a) $I_{DQ} = 0.270 \text{ mA}, V_{DSQ} = 1.14 \text{ V}$
   (b) $g_m = 2.078 \text{ mA/V}, r_o = 185 \text{ k}\Omega$
   (c) $A_v = -15.3$

4.15 (a) $I_{DQ} = 0.608\,\text{mA}, V_{DSQ} = 1.96\,\text{V}$
(b) $A_v = -2.44$
(c) $R_L = 12\,\text{k}\Omega$

4.17 (a) From Problem 4.16, $R_S = 0.5\,\text{k}\Omega$,
$R_D = 3\,\text{k}\Omega, R_1 = 894\,\text{k}\Omega, R_2 = 347\,\text{k}\Omega$
(b) $A_v = -7.99$

4.19 (a) $g_m = 1\,\text{mA/V}, R_D = 16\,\text{k}\Omega$
(b) $R_S = 0.6\,\text{k}\Omega$

4.21 (a) $R_S = 38.6\,\text{k}\Omega, R_D = 6.43\,\text{k}\Omega$
(b) $g_m = 0.583\,\text{mA/V}, r_o = 500\,\text{k}\Omega$
(c) $A_v = -3.19$

4.23 (a) $R_S = 9\,\Omega, I_{DQ} = 1.244\,\text{mA}$
(b) $A_v = -3.10$

4.25 (a) $R_S = 7.76\,\text{k}\Omega, R_D = 10.2\,\text{k}\Omega$
(b) $A_v = -1.50$

4.27 (a) $I_Q = 1.518\,\text{mA}$
(b) $A_v = -11.8$
(c) $A_v = -8.29$

4.29 Let $R_D = 8\,\text{k}\Omega, R_S = 2\,\text{k}\Omega$,
Then, $R_1 = 778\,\text{k}\Omega, R_2 = 269\,\text{k}\Omega$

4.31 $g_m = 0.98\,\text{mA/V}, r_o = 50\,\text{k}\Omega$

4.33 (a) $I_{DQ} = 0.15\,\text{mA}, V_{GSQ} = 0.594\,\text{V}$
(b) $A_v = 0.938$
(c) $R_o = 605\,\Omega$

4.35 (a) $I_Q = 1\,\text{mA}$
(b) $\dfrac{W}{L} = 19.6$
(c) $A_v = 0.990$
(d) $R_o = 223\,\Omega$

4.37 (a) (i) $A_v = 0.927$
(ii) $R_o = 219\,\Omega$
(b) (i) $A_v = 0.907$
(ii) $R_o = 349\,\Omega$

4.39 $R_S = 1.2\,\text{k}\Omega, A_v = 0.878, R_o = 91.9\,\Omega$

4.41 Set $g_m = 5\,\text{mA/V}, R_S = 4.09\,\text{k}\Omega$,
$A_v = 0.870$

4.45 $\dfrac{W}{L} = 80, R_D = 10\,\text{k}\Omega$

4.47 $A_v = 11.3, R_i = 342\,\Omega$

4.49 (a) $R_S = 3.70\,\text{k}\Omega, R_D = 1.63\,\text{k}\Omega$
(b) $R_i = 0.816\,\text{k}\Omega, R_o = 1.63\,\text{k}\Omega$
(c) $i_o = 1.84\sin\omega t\,(\mu\text{A})$,
$v_o = 3.68\sin\omega t\,(\text{mV})$

4.51 (a) $R_D = 0.554\,\text{k}\Omega$
(b) $g_m = 5.657\,\text{mA/V}, R_i = 177\,\Omega$
(c) $A_v = 2.75$

4.53 (a) $\left(\dfrac{W}{L}\right)_D = 25, V_{GSDQ} = 0.825\,\text{V}$

(b) $I_{DQ} = 0.0633\,\text{mA}$,
$V_{DSDQ} = V_O = 1.575\,\text{V}$

4.55 (a) $r = 900\,\Omega$
(b) $I_D = 3.56\,\text{mA}, r = 337\,\Omega$

4.57 (a) $V_{GG} = 5.95\,\text{V}$
(c) $A_v = 0.691$

4.59 (a) $\left(\dfrac{W}{L}\right)_D = 14.4$
(b) $V_B = 1.393\,\text{V}$
(c) $V_{GSDQ} = 1.233\,\text{V}$

4.61 $A_v = -73.7$

4.63 (a) $V_O = \dfrac{2}{3}V_I - \dfrac{1}{3}(0.4)$
(b) $I_D = 0.5\left[\dfrac{2}{3}V_I - 0.533\right]^2$
(c) $A_v = 0.667$

4.65 (a) $V_{GSDQ} = 1.307\,\text{V}, V_{SGLQ} = 2.014\,\text{V}$,
$V_{DSDQ} = 2.593\,\text{V}$
(b) $A_v = \sqrt{\dfrac{K_n}{K_p}}$
(c) $A_v = 2$

4.67 (a) $g_{m1} = g_{m2} = 0.8944\,\text{mA/V}$,
$r_{o1} = r_{o2} = 100\,\text{k}\Omega$
(b) $A_v = -44$
(c) $R_o = 49.7\,\text{k}\Omega$

4.69 (a) $R_1 = 586\,\text{k}\Omega, R_2 = 436\,\text{k}\Omega$,
$R_{D1} = 20\,\text{k}\Omega, R_{S2} = 3.6\,\text{k}\Omega$,
$R_{D2} = 2.4\,\text{k}\Omega$
(b) $A_v = 13.6$

4.71 (a) $I_{DQ1} = 0.757\,\text{mA}, I_{DQ2} = 0.757\,\text{mA}$,
$V_{DSQ1} = 12.4\,\text{V}, V_{DSQ2} = 8.65\,\text{V}$
(b) $g_{m1} = g_{m2} = 3.48\,\text{mA/V}$
(c) $A_v = 2.42$

4.73 (a) $R_1 = 38.8\,\text{k}\Omega, R_2 = 35\,\text{k}\Omega$,
$R_3 = 26.2\,\text{k}\Omega, R_D = 0.6\,\text{k}\Omega$
(b) $A_v = -5.37$

4.75 $g_m = 1.569\,\text{mA/V}, A_v = -3.62, A_i = -45.2$

4.77 (a) $R_S = 1.2\,\text{k}\Omega, R_1 = 265\,\text{k}\Omega$,
$R_2 = 161\,\text{k}\Omega$
(b) $A_v = 0.495, R_o = 0.273\,\text{k}\Omega$

4.79 $R_S = 0.516\,\text{k}\Omega, R_D = 2.61\,\text{k}\Omega$
$R_1 = 65\,\text{k}\Omega, R_2 = 335\,\text{k}\Omega$

## CHAPTER 5

5.1 (a) $\beta = 115, \alpha = 0.9914, i_C = 322\,\mu\text{A}$
(b) $\beta = 89, \alpha = 0.9889, i_C = 1.78\,\text{mA}$

5.3 (a) $0.99099 \le \alpha \le 0.99448$
(b) $5.50 \le I_C \le 9.0\,\text{mA}$

5.5 (a) $\beta = 9,\ 19,\ 49,\ 99,\ 199,\ 999$
(b) $\alpha = 0.9524,\ 0.9804,\ 0.9901,$
$0.9934,\ 0.9955,\ 0.9975$

5.7 $V_{BE} = 0.6109\ \text{V},\ I_C = 0.7928\ \text{mA},$
$I_B = 7.21\ \mu\text{A},\ V_C = 3.41\ \text{V}$

5.9 (a) $I_E = 1.785\ \text{mA},\ I_B = 25\ \mu\text{A}$
(b) $V_{EB} = 0.7154\ \text{V}$

5.11 $I_{EO1} = 6.94 \times 10^{-15}\ \text{A},$
$I_{EO2} = 1.69 \times 10^{-13}\ \text{A, Ratio} = 24.4$

5.13 (a) $I_C = 0.622\ \text{mA}$
(b) $r_o = 137\ \text{k}\Omega$

5.15 $\beta = 60.6$

5.17 (a) $I_C = 1.836\ \text{mA},\ R_C = 3.65\ \text{k}\Omega$
(b) $V_B = 0.164\ \text{V},\ R_C = 8.11\ \text{k}\Omega$
(c) $I_C = 1.744\ \text{mA},\ V_{CE} = 1.96\ \text{V}$
(d) $I_B = 4.61\ \mu\text{A},\ V_C = 1.49\ \text{V}$

5.19 (a) $V_{BB} = 0.7091\ \text{V}$
(b) $V_{BB} = 2.127\ \text{V}$

5.21 (a) $I_C = 0.7273\ \text{mA},\ V_{EC} = 0.9\ \text{V}$
(b) $I_C = 1.185\ \text{mA},\ V_{EC} = 0.2\ \text{V}$
(c) $I_C = 0,\ V_{EC} = 2\ \text{V}$

5.23 (a) $\beta = 134.4,\ \alpha = 0.9926,$
$I_C = 0.269\ \text{mA},\ V_{CE} = 4.7\ \text{V}$
(b) $\beta = 10.63,\ \alpha = 0.9140$

5.25 (a) $I_E = 0.245\ \text{mA},\ I_C = 0.242\ \text{mA},$
$\beta = 80.67,\ \alpha = 0.9878,\ V_{EC} = 1.73\ \text{V}$
(b) $V_E = 0.8371\ \text{V},\ V_{EC} = 1.70\ \text{V}$

5.27 $V_C = -4.39\ \text{V},\ V_E = 1.68\ \text{V}$

5.29 $I_{E1} = I_{E2} = 0.5\ \text{mA},\ V_{C1} = V_{C2} = 3\ \text{V}$

5.31 (a) $R_C = 18\ \text{k}\Omega,\ R_B = 2.656\ \text{M}\Omega$
(b) $I_{CQ} = 0.375\ \text{mA},\ V_{CEQ} = 2.25\ \text{V}$

5.33 (a) (i) $V_O = 3.33\ \text{V}$, (ii) $V_O = 1.83\ \text{V}$,
(iii) $V_O = 0.2\ \text{V}$

5.35 (a) $I_Q = 0.2017\ \text{mA}$
(b) $I_Q = 0.605\ \text{mA}$
(c) $I_Q = 1.008\ \text{mA}$

5.37 $I_E = 2.075\ \text{mA},\ I_C = 2.06\ \text{mA},$
$V_{BC} = 4.47\ \text{V}$

5.39 $1.86 \le V_1 \le 3.96\ \text{V}$

5.41 $V_O = V_T \ln(10)$

5.43 $V_O = 0$ for $4.3 \le V_I \le 5\ \text{V}$,
$V_O = 4.8\ \text{V}$ for $0 \le V_I \le 1.6\ \text{V}$

5.45 $R_B = 3.58\ \text{k}\Omega$

5.47 $R_B = 4.69\ \text{k}\Omega$

5.49 $R_1 = 399\ \text{k}\Omega,\ R_2 = 401\ \text{k}\Omega,$
$V_{CEQ} = 1.30\ \text{V}$

5.51 $I_{EQ} = 4.62\ \text{mA},\ V_B = 4.68\ \text{V}$

5.53 (a) $I_{BQ} = 10.98\ \mu\text{A},\ I_{CQ} = 0.8782\ \text{mA},$
$V_{CEQ} = 3.50\ \text{V}$
(b) For $I_{CQ}$: $+7.21\%$; for $V_{CEQ}$: $-11.2\%$

5.55 (a) $R_{TH} = 1.714\ \text{k}\Omega,\ V_{TH} = -3.571\ \text{V}$
(b) $I_{CQ} = 1.41\ \text{mA},\ V_{CEQ} = 2.32\ \text{V}$
(d) $1.332 \le I_{CQ} \le 1.467\ \text{mA}$
$1.59 \le V_{CEQ} \le 2.97\ \text{V}$

5.57 (a) $I_{CQ} = 0.0888\ \text{mA},\ V_{CEQ} = 3.55\ \text{V}$
(b) $I_{CQ} = 0.266\ \text{mA},\ V_{CEQ} = 3.56\ \text{V}$

5.59 (a) $R_1 = 44.1\ \text{k}\Omega,\ R_2 = 5.70\ \text{k}\Omega,$
$R_C = 5.75\ \text{k}\Omega$
(b) $6.75\%$
(c) $R_1 = 63.6\ \text{k}\Omega,\ R_2 = 12.0\ \text{k}\Omega,\ 6.63\%$

5.61 $I_{CQ} = 2.73\ \text{mA},\ V_{CEQ} = 6\ \text{V},$
$R_1 = 23.2\ \text{k}\Omega,\ R_2 = 2.83\ \text{k}\Omega$

5.63 (a) Let $R_E = 0.333\ \text{k}\Omega,\ R_C = 2\ \text{k}\Omega,$
$R_1 = 22.3\ \text{k}\Omega,\ R_2 = 3.96\ \text{k}\Omega$
(b) $2.90 \le I_{CQ} \le 3.10\ \text{mA}$

5.65 $R_E = 0.496\ \text{k}\Omega,\ R_1 = 32.3\ \text{k}\Omega,$
$R_2 = 7.37\ \text{k}\Omega$

5.67 $R_E = 4.92\ \text{k}\Omega,\ R_1 = 72.5\ \text{k}\Omega,$
$R_2 = 51.2\ \text{k}\Omega$

5.69 $R_1 = 152\ \text{k}\Omega,\ R_2 = 37.8\ \text{k}\Omega$

5.71 Let $R_E = 10\ \text{k}\Omega,\ R_C = 14\ \text{k}\Omega,$
$R_1 = 70.0\ \text{k}\Omega,\ R_2 = 474\ \text{k}\Omega$

5.73 (a) $R_{TH} = 54.7\ \text{k}\Omega,\ V_{TH} = -3.03\ \text{V}$
(b) $I_{CQ} = 0.227\ \text{mA},\ V_{CEQ} = 7.51\ \text{V}$

5.75 (a) $R_E = 19.8\ \text{k}\Omega,\ R_C = 80\ \text{k}\Omega,$
$R_1 = 573\ \text{k}\Omega,\ R_2 = 123\ \text{k}\Omega$
(b) $I_{CQ} = 0.0514\ \text{mA},\ V_{CEQ} = 4.86\ \text{V}$

5.77 (a) $R_E = 7\ \text{k}\Omega,\ R_C = 23\ \text{k}\Omega,$
$R_1 = 103\ \text{k}\Omega,\ R_2 = 316\ \text{k}\Omega$
(b) $I_{CQ} = 0.1027\ \text{mA},\ V_{ECQ} = 2.914\ \text{V}$

5.79 $I_{B1} = 0.0144\ \text{mA},\ I_{C1} = 1.73\ \text{mA},$
$I_{E1} = 1.75\ \text{mA},\ I_{B2} = 0.0232\ \text{mA},$
$I_{C2} = 2.785\ \text{mA},\ I_{E2} = 2.808\ \text{mA},$
$V_{CEQ1} = 2.99\ \text{V},\ V_{CEQ2} = 5.96\ \text{V}$

5.81 $R_1 = 130\ \text{k}\Omega,\ R_2 = 60\ \text{k}\Omega,\ R_3 = 60\ \text{k}\Omega,$
$R_E = 2.5\ \text{k}\Omega,\ R_C = 10.5\ \text{k}\Omega$

5.83 $R_{E1} = 2.93\ \text{k}\Omega,\ R_{E2} = 4.25\ \text{k}\Omega,$
$R_{C1} = 5.21\ \text{k}\Omega,\ R_{C2} = 3.21\ \text{k}\Omega$

## CHAPTER 6

6.1 (a) (i) $g_m = 19.23\ \text{mA/V},\ r_\pi = 9.36\ \text{k}\Omega,$
$r_o = 300\ \text{k}\Omega$
(ii) $g_m = 76.92\ \text{mA/V},\ r_\pi = 2.34\ \text{k}\Omega,$
$r_o = 75\ \text{k}\Omega$
(b) (i) $g_m = 9.615\ \text{mA/V},\ r_\pi = 8.32\ \text{k}\Omega,$
$r_o = 400\ \text{k}\Omega$

(ii) $g_m = 3.077$ mA/V, $r_\pi = 26$ k$\Omega$, $r_o = 1250$ k$\Omega$

6.3   $30.77 \le g_m \le 46.15$ mA/V, $1.95 \le r_\pi \le 5.85$ k$\Omega$

6.5   (a) $g_m = 16.78$ mA/V, $r_\pi = 7.15$ k$\Omega$, $r_o = 183$ k$\Omega$
      (b) $A_v = -4.0$
      (c) $v_s(t) = -0.125 \sin(100t)$ (V)

6.7   (a) $R_i = 56.24$ k$\Omega$
      (b) $R_i = 5.87$ k$\Omega$
      (c) $R_i = 6.24$ k$\Omega$

6.9   $i_B(t) = 15 + 2.89 \sin \omega t$ ($\mu$A)
      $i_C(t) = 1.5 + 0.289 \sin \omega t$ (mA)
      $v_C(t) = 4 - 1.156 \sin \omega t$ (V)
      $A_v = -231$

6.11  $v_{be}(t) = -12 \sin \omega t$ (mV)
      $i_b(t) = -6 \sin \omega t$ ($\mu$A)

6.13  (a) $I_{CQ} = 0.506$ mA, $V_{ECQ} = 1.78$ V
      (b) $A_v = -1.884$
      (c) $1.871 \le |A_v| \le 1.895$

6.15  $R_E = 16.4$ k$\Omega$, $R_C = 11.3$ k$\Omega$, $A_v = -55.1$, $R_i = 14.5$ k$\Omega$

6.17  (a) (i) $R_B = 20.25$ k$\Omega$, $R_C = 2.53$ k$\Omega$
          (ii) $G_f = -11.63$ mA/V
      (b) (i) $R_B = 30.3$ k$\Omega$, $R_C = 2.52$ k$\Omega$
          (ii) $G_f = -11.62$ mA/V

6.19  (a) (i) $A_v = -26.97$, $G_f = -5.39$ mA/V
          (ii) $v_o(t) = -0.108 \sin \omega t$ (V)
               $i_o(t) = -21.6 \sin \omega t$ ($\mu$A)
      (b) (i) $A_v = -30.5$, $G_f = -6.1$ mA/V
          (ii) $v_o(t) = -0.122 \sin \omega t$ (V)
               $i_o(t) = -24.4 \sin \omega t$ ($\mu$A)

6.21  (a) $V_B = -0.0347$ V, $V_E = -0.735$ V
      (b) $R_C = 6.43$ k$\Omega$
      (c) $A_v = -81.7$
      (d) $A_v = -74.9$

6.23  (a) $R_E = 11.0$ k$\Omega$
      (b) $R_C = 3.71$ k$\Omega$
      (c) $A_v = -43.9$
      (d) $R_i = 4.81$ k$\Omega$

6.27  $r_e = r_\pi \left\| \left( \dfrac{1}{g_m} \right) \right\| r_o$

6.29  For Figure 6.28,
      Let $R_E = 0.5$ k$\Omega$, $R_C = 12.7$ k$\Omega$,
      $R_1 = 555$ k$\Omega$, $R_2 = 55$ k$\Omega$

6.31  Let $R_E = 2$ k$\Omega$, $R_C = 15.1$ k$\Omega$,
      $R_1 = 164$ k$\Omega$, $R_2 = 28.4$ k$\Omega$

6.33  3.24 V, peak-to-peak

6.35  $\Delta i_o = 0.546$ mA, peak-to-peak

6.37  $R_1 = 34.9$ k$\Omega$, $R_2 = 18.6$ k$\Omega$

6.39  (a) 2.59 V, peak-to-peak
      (b) $\Delta i_C = 1.294$ mA, peak-to-peak

6.41  (a) $R_E = 0.586$ k$\Omega$, $I_{CQ} = 0.832$ mA
      (b) $R_o = 29.3$ $\Omega$

6.43  (a) $I_Q = 2.37$ mA
      (b) $A_v = 0.9993$

6.45  (a) $I_{CQ} = 2.09$ mA, $V_{CEQ} = 3.69$ V
      (c) $A_v = 0.484$
      (d) $R_{ib} = 122$ k$\Omega$, $R_o = 32.4$ $\Omega$

6.47  (a) $V_B = 0.0617$ V, $V_E = 0.7617$ V
      (b) $g_m = 19$ mA/V, $r_\pi = 4.21$ k$\Omega$, $r_o = 304$ k$\Omega$
      (c) $A_v = 0.906$, $A_i = 14.8$
      (d) $A_v = 0.728$, $A_i = 14.8$

6.49  (a) $I_{CQ} = 1.29$ mA, $V_{ECQ} = 5.7$ V
      (b) $A_v = 0.974$, $R_{ib} = 85.7$ k$\Omega$, $R_o = 19.8$ $\Omega$
      (c) $i_s(t) = 32.7 \sin \omega t$ ($\mu$A)
          $i_o(t) = 2.73 \sin \omega t$ (mA)
          $v_o(t) = 2.73 \sin \omega t$ (V)
          $v_{eb}(t) = -72.5 \sin \omega t$ (mV)

6.51  $0.789 \le A_v \le 0.811$
      $17.0 \le A_i \le 19.6$

6.53  $0.9066 \le A_v \le 0.9999$

6.55  $R_1 = 318$ k$\Omega$, $R_2 = 886$ k$\Omega$

6.57  Let $R_E = 33.3$ k$\Omega$, $R_1 = 594$ k$\Omega$,
      $R_2 = 981$ k$\Omega$, $v_o(\text{min}) = 3.73 \sin \omega t$ (V),
      $v_o(\text{max}) = 3.86 \sin \omega t$ (V)

6.59  (a) $A_v = 76.9$
      (b) $A_i = 0.9917$
      (c) $R_i = 25.8$ $\Omega$
      (d) $R_o = R_C = 2$ k$\Omega$

6.61  (a) $A_v = 3.77$
      (b) $A_i = 0.991$
      (c) $R_i = 1.052$ k$\Omega$
      (d) $R_o = R_C = 4$ k$\Omega$

6.63  (a) $I_{CQ} = 0.921$ mA, $V_{ECQ} = 6.10$ V
      (b) $A_v = 161$

6.65  (a) $V_{CEQ} = 3.72$ V
      (b) $A_v = 63.6$

6.67  (a) $I_{CQ} = 0.4663$ mA, $V_{CEQ} = 4.79$ V
      (b) $R_m = 2.12$ V/mA
      (c) $A_v = 6.73$

6.69  (a) $I_{CQ} = 0.984$ mA, $V_{CEQ} = 2.34$ V
      (b) $A_v = 25.4$

6.71 (a) $I_{CQ} = 1.91$ mA, $V_{ECQ} = 13.3$ V
(b) $2.70 \leq A_v \leq 2.71$
(c) $0.0302 \leq R_i \leq 0.0305$ k$\Omega$
$R_o = R_C = 6.5$ k$\Omega$

6.73 (a) $A_{v1} = 0.4927$
(b) $A_{v2} = 153.8$
(c) $A_v = 75.8$

6.75 (a) $g_{m1} = 42.74$ mA/V, $r_{\pi 1} = 2.34$ k$\Omega$,
$g_{m2} = 48.65$ mA/V, $r_{\pi 2} = 2.06$ k$\Omega$,
$r_{o1} = r_{o2} = \infty$
(b) $A_{v1} = -85.48$, $A_{v2} = -97.3$
(c) $A_v = 3909$, $A'_v = A_{v1} \cdot A_{v2} = 8317$

6.77 (a) $I_{C1} = 2.253$ mA, $I_{C2} = 69.73$
(b) $A_v = 0.990$
(c) $R_{ib} = 467.6$ k$\Omega$, $R_o = 0.512 \, \Omega$

6.79 (a) $R_o = 30.77$ k$\Omega$
(b) $R_o = 31.38$ k$\Omega$
(c) $R_o = 19.17$ k$\Omega$

6.81 (a) $\bar{P}_Q = 2.564$ mW, $\bar{P}_{RC} = 3.123$ mW,
$\bar{P}_{RE} = 3.175$ mW
(b) $\bar{P}_{RC} = 0.721$ mW

6.83 (a) $\bar{P}_{RL} = 0.286$ mW,
(b) $\bar{P}_{RL} = 0.499$ mW

## CHAPTER 7

7.1 (a) $T(s) = \dfrac{1}{1 + sR_1C_1}$
(b) $f_H = 159$ Hz
(c) $v_O(t) = 1 - \exp\left(\dfrac{-t}{R_1C_1}\right)$

7.3 (a) $T(s) = \left(\dfrac{R_2}{R_1 + R_2}\right) \cdot \dfrac{1}{1 + s(R_1 \| R_2)C_1}$
(b) $\tau = 66.7$ ms
(c) $f = 2.39$ Hz

7.5 (a) $\dfrac{V_o}{V_i} = \dfrac{R_2}{R_1 + R_2} = 0.667$
(b) $\dfrac{V_o}{V_i} = 1$
(c) $T(s) = \left(\dfrac{R_2}{R_1 + R_2}\right)$
$\cdot \dfrac{[1 + sR_1C_1]}{[1 + s(R_1 \| R_2)C_1]}$
$K = \dfrac{R_2}{R_1 + R_2}, \tau_A = R_1C_1,$
$\tau_B = (R_1 \| R_2)C_1$

7.7 (a) $|T(f_T)|_{dB} = -9.03$ dB, $\phi = -135°$
(b) Slope $= -18$ dB/octave $=$
$-60$ dB/decade, $\phi = -270°$

7.9 (a) (ii) $\omega_1 = 1$ rad/s, $\omega_2 = 10$ rad/s,
$\omega_3 = 100$ rad/s, $\omega_4 = 1000$ rad/s
(iii) $|T(0)| = 10$
(iv) $|T(\infty)| = 10$
(b) (ii) $\omega = 5$ rad/s
(iii) $|T(0)| = 0$
(iv) $|T(\infty)| = 200$

7.11 (a) $|T|_{midband} = 159$
(b) $\tau_S = 5.31$ ms, $\tau_P = 0.332 \, \mu$s
(c) $C_C = 0.932 \, \mu$F, $C_L = 55.3$ pF

7.15 (a) $R_C = 2.49$ k$\Omega$
(b) $I_{CQ} = 0.503$ mA
(c) $|A_v|_{max} = 48.1$

7.17 (a) $f_L = 959$ Hz
(b) $|A_v| = 6.69$

7.19 (a) $R_S = 2.59$ k$\Omega$, $R_D = 4.41$ k$\Omega$
(b) $T(s) = \dfrac{I_o(s)}{V_i(s)}$
$= \dfrac{-g_m R_D}{(1 + g_m R_S)(R_D + R_L)}$
$\times \dfrac{s(R_D + R_L)C_C}{1 + s(R_D + R_L)C_C}$
$\tau = (R_D + R_L)C_C$
(c) $C_C = 1.89 \, \mu$F

7.21 (a) $R_1 = 79.6$ k$\Omega$, $R_2 = 124$ k$\Omega$
(b) $A_v = 0.991$
(c) $R_o = 17.25 \, \Omega$
(d) $f_L = 19.8$ Hz

7.23 (a) $T(s) = \dfrac{g_m(R_D \| R_L)}{1 + g_m R_S}$
$\cdot \dfrac{1}{\left[1 + s\left(\dfrac{1}{g_m} \| R_S\right)C_i\right]}$
$\cdot \left[\dfrac{s(R_D + R_L)C_C}{1 + s(R_D + R_L)C_C}\right]$
(b) $\tau = \left(\dfrac{1}{g_m} \| R_S\right)C_i$
(c) $\tau = (R_L + R_D)C_C$

7.25 (a) $f_H = 8.30$ MHz
(b) $f = 82.6$ MHz

7.27  Let $I_{DQ} = 0.2$ mA, $V_{DSQ} = 5$ V,
$V_{TN} = 0.5$ V
$R_D = 20$ k$\Omega$, $\dfrac{W}{L} = 6.25$, $R_1 = 192.5$ k$\Omega$,
$R_2 = 32.5$ k$\Omega$, $C_C = 0.0284$ $\mu$F,
$C_L = 2.65$ nF

7.29  (a)  $R_E = 6$ k$\Omega$, $R_B = 532$ k$\Omega$
      (b)  $C_C = 0.0257$ $\mu$F
      (c)  $A_v = 0.983$

7.31  (a)  $C_{C1} = 12$ $\mu$F,    (b)  $C_{C2} = 0.050$ F

7.33  (a)  $R_S = 6.4$ k$\Omega$, $R_D = 5.6$ k$\Omega$
      (b)  $f_A = 4.97$ Hz, $f_B = 36.8$ Hz
      (c)  $f_A \to 0$, $f_B = 31.8$ Hz

7.35  (a)  Same expression as Equation (7.59)
           with $R_S = 0$.
      (b)  $\tau_A = R_E C_E$
      $$\tau_B = \frac{r_\pi R_E C_E}{r_\pi + (1 + \beta) R_E}$$

7.37  (a)  $A_v = 118$
      (b)  $f = 15.9$ MHz

7.39  $C_L = 121$ pF

7.41  (a)  $A_v(s) = -g_m(R_D \| R_L)$
      $$\cdot \left[ \frac{1}{1 + s(R_D \| R_L) C_L} \right]$$
      (b)  $\tau = (R_D \| R_L) C_L$
      (c)  $\tau = 6.67 \times 10^{-8}$ s, $f_H = 2.39$ MHz,
           $A_v = -4.7$

7.45  $g_m = 9.615$ mA/V, $f_\beta = 33.3$ MHz,
      $C_\pi = 0.303$ pF

7.47  (a)  $f_\beta = 4.5$ MHz, $C_\pi = 1.87$ pF
      (b)  $f_T = 2.16$ GHz, $f_\beta = 18.0$ MHz

7.49  (a)  $Z_i = 2700 - j3.33$
      (b)  $Z_i = 2700 - j33.3$
      (c)  $Z_i = 2656 - j327$
      (d)  $Z_i = 1100 - j1200$

7.51  $f = 307$ kHz

7.53  $f_L = 159$ Hz, $f_H = 489$ kHz

7.55  $f_T = 1.84$ GHz

7.57  (a)  $f_T = 1.25$ GHz
      (b)  $f_T = 3.05$ GHz
      (c)  $I_D = 290$ $\mu$A
      (d)  $C_{gs} = 70$ fF

7.59  (a)  $C_M = 344.3$ fF
      (b)  $f_{3\text{-dB}} = 37.5$ MHz

7.61  (a)  $r_s = 43.5$ $\Omega$
      (b)  $\dfrac{g_m'}{g_m} = 94.5\%$

7.63  (b)  $C_M = 96.28$ fF
      (c)  $f_{3\text{-dB}} = 1.095$ GHz, $A_v = -10.96$

7.65  (a)  $C_\pi = 2.20$ pF, $C_M = 27.6$ pF
      (b)  $f_H = 3.61$ MHz, $A_v = -19.7$

7.67  (a)  $C_M = 6.785$ pF
      (b)  $f_H = 4.84$ MHz, $A_v = -5.67$

7.69  (a)  $f_H = -10.4$ MHz
      (b)  $C_M = 18.2$ pF
      (c)  $A_v = -4.66$

7.71  (a)  $f_{P\pi} = 1.20$ GHz, $f_{P\mu} = 56.3$ MHz
      (b)  $A_v = 2.76$
      (c)  $f = 3.75$ MHz, $C_L$ dominates

7.73  $f = 119$ MHz, $A_v = 1.19$

## CHAPTER 8

8.1  (b)  (i)  $R_D = 6$ $\Omega$, $P_{D,\max} = 24$ W
          (ii)  $R_D = 13.3$ $\Omega$, $P_{D,\max} = 30$ W
     (c)  (i)  $I_{D,\max} = 4$ A
          (ii)  $I_{D,\max} = 3$ A

8.3  (a)  $R_L = 25$ $\Omega$, $R_B = 3.91$ k$\Omega$,
          $P_{Q,\max} = 9$ W
     (b)  $I_{C,\max} = 0.8944$ A, $V_{CC} = 22.36$ V
     (c)  For (a): $\bar{P}_L = 4.5$ W
          For (b): $\bar{P}_L = 2.5$ W

8.5  (b)  $V_{GG} = 5$ V, $P = 9.375$ W;
          $V_{GG} = 6$ V, $P = 30$ W;
          $V_{GG} = 7$ V, $P = 39.375$ W;
          $V_{GG} = 8$ V, $P = 10.8$ W;
          $V_{GG} = 9$ V, $P = 7.16$ W

8.7  (b)  $T_{j,\max} = 145$ °C
     (c)  $\theta_{\text{dev}-\text{amb}} = 2$ °C/W

8.9  (a)  $P_T = 19.39$ W
     (b)  $T_{\text{case}} = 90.9$°C, $T_{\text{snk}} = 79.3$°C

8.11  (a)  $P_T = 13.4$ W
      (b)  $V_{CE,\max} = 8.96$ V

8.13  (b)  (i)  $I_C = 29.25$ mA
           (ii)  $I_C = 61.75$ mA
           (iii)  $I_C = 108$ mA

8.15  (a)  $A_v = 0.9928$
      (b)  $A_v = 0.9872$
      (c)  $A_v = 0.939$

8.17  $v_O(\min) = -3.93$ V, $v_I(\min) = -3.43$ V;
      $v_O(\max) = 4.43$ V, $v_I(\max) = 5.76$ V

8.19  (a)  $I_Q = 0.12$ A, $R = 187$ $\Omega$
      (b)  $P_{Q1} = 2.88$ W, $P = 5.88$ W
      (c)  $\eta = 11.4\%$

8.23   (a)   $V^+ = -V^- = 52$ V

      (b)   $I_P = 2.04$ A

      (c)   $\eta = 74\%$

8.25   (b)  (i)   $A_v = 0$

           (ii)   $A_v = 0.472$

           (iii)   $A_v = 0.739$

8.27   (a)  (i)   $V_{BB} = 4$ V

           (ii)   $P = 12$ mW

      (b)  (i)   $v_O(\text{max}) = 10.39$ V

           (ii)   $i_{Dn} = 10.39$ mA, $i_{Dp} = 0$,

                $i_L = 10.39$ mA, $v_I = 11.5$ V

           (iii)   $P_{RL} = 108$ mW, $P_{Mn} = 16.7$ mW,

                $P_{Mp} = 0$

8.29   (a)   $K_{n3} = K_{p4} = 200\ \mu\text{A/V}^2$

      (b)  (i) $A_v = 0$; (ii) $A_v = 0.876$

8.31   (a)   $I_{CQ} = 50.57$ mA

      (b)   $R_L = 237\ \Omega$

      (c)   $\bar{P}_{L,\text{max}} = 255$ mW

      (d)   $\eta = 42\%$

8.33   (a)   $I_{CQ} = 98.91$ mA

      (b)   $a = 3.89$

      (c)   $\bar{P}_L = 333.9$ mW

      (d)   $\eta = 28.1\%$

8.35   (a)   $\dfrac{v_e}{v_i} = \dfrac{(1+\beta)R'_E}{r_\pi + (1+\beta)R'_E}$

            $\dfrac{v_o}{v_i} = \dfrac{1}{(n_1/n_2)} \cdot \dfrac{(1+\beta)\,R'_E}{r_\pi + (1+\beta)R'_E}$

            where $R'_E = \left(\dfrac{n_1}{n_2}\right)^2 R_L$

      (b)   $a^2 = \dfrac{V_{CC}}{5}$

      (c)   $R_o = 0.255\ \Omega$

8.37   (a)   $a = 2.86$

      (b)   $P_Q = 4.95$ W

8.39   (a)   $V_{BB} = 1.473$ V

      (b)   $I_{CQ} = 14$ mA

8.41   (a)   $V_{BB} = 1.4845$ V

      (b)   $v_{BEn} = 0.75124$ V, $v_{EBp} = 0.73322$ V

      (c)   $i_{Cn} = i_{Cp} = 2.828$ mA

      (d)   $v_I = -0.73322$ V

8.43   (a)   For $v_O(\text{max}) = 4$ V,

            $R_1 = R_2 = 32.5\ \Omega$

      (b)   $I_{E1} = I_{E2} = 0.286$ A,

            $I_{E3} = I_{E4} = 2.86$ A

      (c)   $R_o = 5.34$ m$\Omega$

8.45   (b)   $R_1 = R_2 = 4$ k$\Omega$

      (c)   $I_{DQ1} = I_{DQ2} = 2$ mA

8.47   $R_o = 7.46\ \Omega$

8.49   (a)   $V_{BB} = 1.742$ V, $I_{C1} \cong 10$ mA,

            $I_{C2} = 0.20$ A, $I_{C3} \cong 0.5$ mA,

            $V_{BE1} = 0.58065$ V, $V_{BE2} = 0.6585$ V

            $V_{EB3} = 0.50276$ V

      (b)   $i_{C1} = 4.47$ mA, $i_{C2} = 93.9$ mA,

            $i_{C3} = 2.267$ mA, $i_{C4} = 45.3$ mA,

            $i_{C5} = 0.952$ A, $P_{Q2} = 4.13$ W,

            $P_{Q5} = 13.3$ W

      (d)   $I_{Dn3} = 23.33$ mA, $I_{Dp1} = 0.835$ mA,

            $I_{Dn2} = 3.188$ mA, $I_{Dp4} = 0$,

            $\bar{P}_L = 81.7$ mW

## CHAPTER 9

9.1   (a)   $A_{od} = 500$, (b) $v_2 = 3$ mV,

      (c)   $v_1 = 0.990$ V, (d) $v_o = 0$,

      (e)   $v_2 = -0.506$ V

9.3   (a)   $v_O = -5$ V

      (b)   $v_1 = 3.00265$ V

      (c)   $A_{od} = 9 \times 10^4$

9.5   (a)   $A_v = -10$

      (b)   $A_v = -3$

      (c)   $A_v = -1$

9.7   (a)   $A_v = -10$, $R_i = 10$ k$\Omega$

      (b)   $A_v = -5$, $R_i = 10$ k$\Omega$

      (c)   $A_v = -5$, $R_i = 20$ k$\Omega$

9.9   (a)   $-10$, (b) $-1$, (c) $-0.20$,

      (d)   $-10$, (e) $-2$, (f) $-1$

9.11   (a)   $R_1 = 5$ k$\Omega$, $R_2 = 32.5$ k$\Omega$

      (b)   $v_I = 0.6154$ V, $i_1 = i_2 = 0.123$ mA

9.13   (a)   $-9.5\%$ to $+10.5\%$

      (b)   $\pm 2\%$

9.15   $R_1 = 18.5$ k$\Omega$, $R_2 = 599.6$ k$\Omega$

       $0.731 \le |v_O| \le 0.769$ V

9.17   $R_1 = 10$ k$\Omega$, $R_2 = 100$ k$\Omega$,

       $R_3 = 10$ k$\Omega$, $R_4 = 100$ k$\Omega$

9.19   (a)   $v_O = 8.8$ V

      (b)   $v_O = 8.7597$ V

      (c)   $A_{od} = 1.1477 \times 10^4$

9.21   (a)   $v_I = -0.7011$ V, $v_1 = -1$ mV

      (b)   $A_{od} = 2.5 \times 10^4$, $v_I = 0.50022$ V

9.23   (a)   $R_1 = 500$ k$\Omega$, $R_2 = R_3 = 500$ k$\Omega$,

            $R_4 = 6.41$ k$\Omega$

      (b)   $i_1 = i_2 = -0.1\ \mu$A, $i_4 = -7.80\ \mu$A,

            $i_3 = -7.90\ \mu$A

9.25   For Figure P9.25, $A_v = -8$;

       For Figure 9.12, $A_v = -3$

9.27   (a)   (i)   $v_O = 0.79936$ V
              (ii)  0.08%
       (b)   (i)   $v_O = 0.79208$ V
              (ii)  0.99%

9.29   $R_2 = 111$ k$\Omega$

9.31   (a)   $R_1 = 83.3$ k$\Omega$, $R_2 = 40$ k$\Omega$,
              $R_3 = 400$ k$\Omega$, $R_F = 250$ k$\Omega$
       (b)   $|i_F| = 0.75\,\mu$A

9.33   $v_O = [-0.707 \sin(2\pi ft) \pm 2]$ (V)

9.35   (a)   $R_1 = 250$ k$\Omega$, $R_2 = 31.25$ k$\Omega$,
              $R_3 = 62.5$ k$\Omega$, $R_F = 125$ k$\Omega$
       (b)   $-3 \le v_O \le +3$ V, $|i_F|_{\max} = 24\,\mu$A

9.37   (b)   $R_F = 10$ k$\Omega$
       (c)   (i) $v_O = 0.3125$ V, (ii) $v_O = 4.6875$ V

9.41   (a)   $R_1 = 4.17$ k$\Omega$, $R_2 = 58.33$ k$\Omega$
       (b)   $v_O = 3.75$ V, $i_1 = i_2 = 60\,\mu$A

9.43   (a)   $A_{od} = 250.5$
       (b)   $v_O = 49.9$ V

9.45   (a)   $v_O = \dfrac{6}{7} \cdot v_{I1} + \dfrac{3}{7} \cdot v_{I2}$
       (b)   $v_O = 0.3$ V
       (c)   $v_O = 42.86$ mV

9.47   (a)   $A_v = \dfrac{1}{1 - x}$
       (b)   $1 \le A_v \le \infty$
       (c)   $x \ne 1$

9.49   (a)   $A_v = 1$
       (b)   $A_v = 0.999993$
       (c)   $A_{od} = 99$

9.51   (a)   $A_{v1} = \left(1 + \dfrac{R_2}{R_1}\right)$, $A_{v_2} = -\left(1 + \dfrac{R_2}{R_1}\right)$
       (b)   $v_{O1} = -2$ V, $v_{O2} = 2$ V
       (c)   $(v_{O1} - v_{O2}) = 6.4$ V

9.53   (a)   (i)   $v_O = 1$ V
              (ii)  $v_O = 1.67$ V
       (b)   (i)   $v_O = 1$ V
              (ii)  $v_O = 1.67$ V
       (c)   (i)   $v_O = 0.667$ V
              (ii)  $v_O = -1.111$ V

9.55   From Figure P9.54, let $R_F = 1$ k$\Omega$

9.57   (a)   $R_2 = 1$ k$\Omega$, Let $R_3 = 1$ k$\Omega$,
              $R_1 = R_F = 10$ k$\Omega$
       (b)   $i_1 = i_2 = -0.6$ mA, $v_O = 7$ V,
              $i_3 = 6$ mA, $i_4 = 1$ mA, $i_O = 6.6$ mA

9.59   (b)   $R_o = \infty$

9.61   (a)   $R_2 = R_4 = 250$ k$\Omega$,
              $R_1 = R_3 = 6.25$ k$\Omega$

9.61   (b)   $R_2 = R_4 = 250$ k$\Omega$,
              $R_1 = R_3 = 10$ k$\Omega$
       (c)   $R_2 = R_4 = 250$ k$\Omega$,
              $R_1 = R_3 = 50$ k$\Omega$
       (d)   $R_2 = R_4 = 125$ k$\Omega$,
              $R_1 = R_3 = 250$ k$\Omega$

9.63   (a)   $v_O = -4$ V, $v_X = v_Y = 1.273$ V,
              $i_1 = i_2 = 0.0527$ mA,
              $i_3 = i_4 = 0.0127$ mA
       (b)   $v_O = 4$ V, $v_X = v_Y = 3.273$ V,
              $i_1 = i_2 = -0.00727$ mA,
              $i_3 = i_4 = 0.0327$ mA
       (c)   $v_O = -1.5$ V, $v_X = v_Y = -1.227$ V,
              $i_1 = i_2 = 0.00273$ mA,
              $i_3 = i_4 = -0.0123$ mA

9.65   (b)   $A_{cm} = 0.3$
       (c)   $|A_{cm}|_{\max} = 0.0325$

9.67   (a)   $v_{O1} = 1.2 - 0.72 \sin \omega t$ (V)
              $v_{O2} = 1.2 + 0.72 \sin \omega t$ (V)
              $v_O = 4.32 \sin \omega t$ (V)
       (b)   $v_{O1} = -0.85 + 0.45 \sin \omega t$ (V)
              $v_{O2} = -0.40 - 0.45 \sin \omega t$ (V)
              $v_O = 1.35 - 2.7 \sin \omega t$ (V)

9.69   (a)   $i_O = \dfrac{v_{I1} - v_{I2}}{R}$
       (b)   $R = 100\,\Omega$
       (c)   $v_{O1} = 5.25$ V, $v_{O2} = -0.25$ V
       (d)   $i_O = -1$ mA, $v_{O1} = -1.75$ V,
              $v_{O2} = 1.75$ V

9.71   $R_1(\text{fixed}) = 1$ k$\Omega$, $R_2 = 39.5$ k$\Omega$,
       $R_1(\text{var}) = 78$ k$\Omega$

9.73   (a)   $f = 398$ Hz, phase $= 90°$
       (b)   (i) $f = 66.3$ Hz,   (ii) $f = 663$ Hz

9.75   (b)   $A_v(0) = -\dfrac{R_2}{R_1}$
       (c)   $f = \dfrac{1}{2\pi R_2 C_2}$

9.77   (b)   $A_v(\infty) = -\dfrac{R_2}{R_1}$
       (c)   $f = \dfrac{1}{2\pi R_1 C_1}$

9.79   $i_2 = 1.52$ mA, $v_O = -1.52$ V,
       $i_z = 0$

9.83   From Figure 9.40, $R_F = 500$ k$\Omega$,
       $R_1 = 50$ k$\Omega$, $R_2 = 500$ k$\Omega$,
       $R_A = 500$ k$\Omega$, $R_B = 333.3$ k$\Omega$,
       $R_C = 142.8$ k$\Omega$

9.85    $R_F = 5.33\,\text{k}\Omega$, $R_4 = 31.5\,\text{k}\Omega$,
$R_3 = 10.6\,\text{k}\Omega$, $\dfrac{R_2}{R_1} = 1.143$

9.87    For the bridge circuit, $R_T = 20(1 + \delta)\,\text{k}\Omega$,
$R_1 = R_2 = R_3 = 20\,\text{k}\Omega$
For the instrumentation amplifier,
Let $\dfrac{R_4}{R_3} = 4$, $\dfrac{R_2}{R_1} = 4.5$

## CHAPTER 10

10.1    (a)
$$I_C = \frac{1}{R_3}\left[2V_\gamma - (2V_\gamma + V^-)\left(\frac{R_2}{R_1 + R_2}\right) - V_{BE}\right]$$

(c)    $R_3 = 2.5\,\text{k}\Omega$,
$R_1 = R_2 = 2.15\,\text{k}\Omega$

10.3    $V_{BE1} = V_{BE2} = 0.6341\,\text{V}$, $I_O = 78.05\,\mu\text{A}$

10.5    $V_{BE1} = 0.63345\,\text{V}$, $V_{BE2} = 0.62393\,\text{V}$,
$I_O = 132.07\,\mu\text{A}$

10.7    $I_{\text{REF}} = 0.5082\,\text{mA}$,
$I_{C1} = I_{C2} = 0.4958\,\text{mA}$,
$I_{B1} = I_{B2} = 6.198\,\mu\text{A}$

10.9    (a)    $R_1 = 18.3\,\text{k}\Omega$
(b)    6.3%

10.11    $I_{\text{REF}} = 0.210\,\text{mA}$, $R_1 = 20.5\,\text{k}\Omega$

10.13    (a)    $R_1 = 9.3\,\text{k}\Omega$
(b)    $I_O = 2\,\text{mA}$
(c)    $R_{C2} = 2.15\,\text{k}\Omega$

10.15    Similar to Figure P10.14, biased
at $V^+$ and $V^-$. $R_1 = 21.5\,\text{k}\Omega$

10.17    (a)    $I_O = \dfrac{I_{\text{REF}} - \dfrac{V_{BE}}{(1 + \beta)\,R_2}}{\left(1 + \dfrac{2}{\beta(1 + \beta)}\right)}$

(b)    $R_1 = 12.27\,\text{k}\Omega$, $I_{\text{REF}} = 0.7011\,\text{mA}$

10.19    $R_1 = 30.63\,\text{k}\Omega$

10.21    $I_O = \dfrac{I_{\text{REF}}}{\left(1 + \dfrac{2}{\beta(2 + \beta)}\right)}$

10.23    (a)    $R_o = 20\,\text{M}\Omega$
(b)    $\Delta I_O = 0.25\,\mu\text{A}$

10.25    (a)    $V_{BE1} = 0.6347\,\text{V}$, $V_{BE2} = 0.6040\,\text{V}$,
$I_O \cong 61.4\,\mu\text{A}$
(b)    $V_{BE1} = 0.6347\,\text{V}$, $V_{BE2} = 0.5991\,\text{V}$,
$I_O \cong 71.2\,\mu\text{A}$

10.27    (a)    $I_{\text{REF}} = 0.186\,\text{mA}$, $I_O \cong 19.53\,\mu\text{A}$,
$V_{BE2} = 0.6414\,\text{V}$
(b)    $R_o = 13.16\,\text{M}\Omega$

10.29    (a)    $R_1 = 18.6\,\text{k}\Omega$, $R_E = 1.20\,\text{k}\Omega$
(b)    $R_o = 6.477\,\text{M}\Omega$
(c)    1.54%

10.31    $V_{BE1} = 0.718\,\text{V}$ at 2 mA; $R_1 = 7.14\,\text{k}\Omega$,
$R_E = 1.92\,\text{k}\Omega$

10.33    $V_{BE1} = 0.681\,\text{V}$, $I_{\text{REF}} = 0.483\,\text{mA}$,
$I_O \cong 8.7\,\mu\text{A}$, $V_{BE2} = 0.5766\,\text{V}$

10.35    (a)    $R_1 = 10.1\,\text{k}\Omega$, $R_{E2} = 1.37\,\text{k}\Omega$,
$V_{BE1} = 0.70038\,\text{V}$, $V_{BE2} = 0.67656\,\text{V}$
(b)    $R_1 = 10.1\,\text{k}\Omega$, $R_{E2} = 1.46\,\text{k}\Omega$,
$V_{BE2} = 0.65854\,\text{V}$

10.37    (a)    $I_{\text{REF}} = 0.93\,\text{mA}$, $I_{O2} \cong 68\,\mu\text{A}$,
$I_{O3} \cong 40.7\,\mu\text{A}$
(b)    $R_{E2} = 4.99\,\text{k}\Omega$, $R_{E3} = 0.797\,\text{k}\Omega$

10.39    (a)    $I_{O1} = 3.1\,\text{mA}$, $I_{O2} = 1.55\,\text{mA}$,
$I_{O3} = 4.65\,\text{mA}$
(b)    $V_{CE1} = 3.8\,\text{V}$, $V_{EC2} = 5.35\,\text{V}$,
$V_{EC3} = 5.35\,\text{V}$

10.41    $I_{C1} = I_{C2} = 1.86\,\text{mA}$,
$I_{C3} = I_{C4} = 1.86\,\text{mA}$,
$I_{C5} = I_{C6} = I_{C7} = 0.136\,\text{mA}$,
$V_{CE3} = 7.02\,\text{V}$, $V_{CE5} = 14.2\,\text{V}$,
$V_{EC7} = 4.89\,\text{V}$

10.43    $R_{E1} = 3\,\text{k}\Omega$, $R_{E2} = 1.5\,\text{k}\Omega$,
$R_{E3} = 0.75\,\text{k}\Omega$, $I_{O1} = 1\,\text{mA}$,
$I_{O2} = 2\,\text{mA}$, $I_{O3} = 4\,\text{mA}$

10.45    (a)    For $V_{GS} = 0.75\,\text{V}$, $R = 25\,\text{k}\Omega$,
$$\left(\frac{W}{L}\right)_1 = 20, \left(\frac{W}{L}\right)_2 = 40$$
(b)    $R_o = 667\,\text{k}\Omega$
(c)    1.5%

10.47    (a)    $0.19 \le I_O \le 0.21\,\text{mA}$
(b)    $0.1921 \le I_O \le 0.2081\,\text{mA}$

10.49    $R_o = \dfrac{2 + g_m r_o}{g_m(1 + g_m r_o)} \cong \dfrac{1}{g_m}$

10.51    (a)    $I_{\text{REF}} = 606\,\mu\text{A}$
(b)    $I_O = 362.5\,\mu\text{A}$
(c)    $I_O = 370.7\,\mu\text{A}$

10.53    $\left(\dfrac{W}{L}\right)_1 = 5.56$, $\left(\dfrac{W}{L}\right)_2 = 2.78$,
$\left(\dfrac{W}{L}\right)_3 = 1.81$

10.55 $\left(\dfrac{W}{L}\right)_1 = 8.33, \left(\dfrac{W}{L}\right)_2 = 2.67,$

$\left(\dfrac{W}{L}\right)_3 = \left(\dfrac{W}{L}\right)_4 = 2.72$

10.57 $\left(\dfrac{W}{L}\right)_1 = 59.9, \left(\dfrac{W}{L}\right)_2 = 21.8,$

$\left(\dfrac{W}{L}\right)_3 = 2.14$

10.59 $R_o = 2.58 \times 10^9 \ \Omega$

10.61 $\left(\dfrac{W}{L}\right)_1 = 40, \left(\dfrac{W}{L}\right)_2 = 4,$

$\left(\dfrac{W}{L}\right)_3 = 2.16, \left(\dfrac{W}{L}\right)_4 = 3.61$

10.63 (a) $I_O = 80 \ \mu A$
       (b) $I_O = 80.05 \ \mu A$

10.65 (a) $R = 6.12 \ k\Omega$
       (b) $(V^- - V^-)_{min} = 1.66 \ V$

       (c) $\left(\dfrac{W}{L}\right)_5 = 2.5, \left(\dfrac{W}{L}\right)_6 = 7.5$

10.67 $\left(\dfrac{W}{L}\right)_1 = 86.1, \left(\dfrac{W}{L}\right)_2 = 12.5,$

$\left(\dfrac{W}{L}\right)_3 = 25, \left(\dfrac{W}{L}\right)_4 = 50$

10.69 $I_{REF} = 59.09 \ \mu A, I_1 = 11.82 \ \mu A,$
       $I_2 = 73.87 \ \mu A, I_3 = 47.27 \ \mu A,$
       $I_4 = 236.4 \ \mu A$

10.71 $I_{D2} = 62.5 \ \mu A, I_O = 41.67 \ \mu A,$
       $V_{GS1} = V_{GS2} = 1.107 \ V,$
       $V_{SG3} = V_{SG4} = 0.9893 \ V$

10.73 (a) $I_O = 2.5 \ mA$
       (b) $I_O = 3 \ mA$
       (c) $I_O = 3.5 \ mA$

10.75 (a) $A_v = -1846$
       (b) $V_I = 0.6943 \ V$
       (c) $V_B = 1.824 \ V$

10.77 (a) $V_{EB} = 0.5568 \ V$
       (b) $R_1 = 4.44 \ k\Omega$
       (c) $V_I = 0.5389 \ V$
       (d) $A_v = -1846$

10.79 (a) $\left(\dfrac{W}{L}\right)_0 = 3.96,$

$\left(\dfrac{W}{L}\right)_1 = \left(\dfrac{W}{L}\right)_2 = 3.24,$

$\left(\dfrac{W}{L}\right)_3 = 3.24$

       (b) $A_v = -70.35$

10.81 (a) $A_v = -1978$
       (b) $A_v = -454$
       (c) $A_v = -256$

10.83 (a) To a good approximation,
           $R_o = r_{o2}[1 + g_{m2}(r_{\pi 2} \| R_E)]$
       (b) $A_v = -g_{m0}(r_o \| R_L \| R_o)$

10.85 (a) $g_{m0} = 0.80 \ mA/V, r_o = 333 \ k\Omega$
           $g_{m2} = 0.748 \ mA/V, r_{o2} = 250 \ k\Omega$
       (b) $A_v = -114.3$
       (c) $R_L = 143 \ k\Omega$

10.87 $A_v = -316$

10.89 $A_v = -366,165$

## CHAPTER 11

11.1  (a) $v_O = 0.375 \sin \omega t \, (V)$
       (b) $v_O = 0.45 \sin \omega t \, (V)$
       (c) $v_O = 2.75 \sin \omega t \, (V)$

11.3  (a) $R_E = 11.5 \ k\Omega, R_C = 18 \ k\Omega$
       (c) $-2.3 \le v_{cm} \le 0.673 \ V$

11.5  (a) $R_1 = 73.25 \ k\Omega, R_C = 28.5 \ k\Omega$
       (b) $A_d = 62, A_{cm} = -0.113,$
           $CMRR_{dB} = 54.8 \ dB$
       (c) $R_{id} = 46 \ k\Omega, R_{icm} = 12.6 \ M\Omega$

11.7  $R_1 = 53 \ k\Omega, R_C = 40 \ k\Omega, A_d = 38.5$

11.9  (a) $R_E = 62 \ k\Omega$
       (b) $A_d = 71.0, A_{cm} = -0.398,$
           $CMRR_{dB} = 45.0 \ dB$
       (c) $R_{id} = 70.4 \ k\Omega, R_{icm} = 6.28 \ M\Omega$

11.11 (a) $I_{C1} = I_{C2} = 5.90 \ \mu A,$
           $V_{EC1} = V_{EC2} = 1.475 \ V$
       (b) (i) $A_d = 81.7, A_{cm} = 0$
           (ii) $A_d = 40.8, A_{cm} = -0.0442$

11.13 (a) $R_E = 17.9 \ k\Omega, R_C = 16.7 \ k\Omega$
       (b) $|A_d| = 76.9$
       (c) $A_d = 76.9, A_{cm} = 0.0279,$
           $CMRR_{dB} = 68.8 \ dB$

11.15 (a) $v_E = +0.7 \ V, v_{C1} = v_{C2} = -2.87 \ V$
       (b) $v_E = +0.7 \ V, v_{C1} = -5 \ V,$
           $v_{C2} = -0.736 \ V$
       (c) $v_E \cong +0.7 \ V, v_{C1} = -2.255 \ V,$
           $v_{C2} = -3.485 \ V$

11.17 (a) (i) $v_{O1} - v_{O2} = 0,$
           (ii) $v_{O1} - v_{O2} = 0.08 \ V$
       (b) (i) $v_{O1} - v_{O2} = 0.1995 \ V,$
           (ii) $v_{O1} - v_{O2} = 0.2795 \ V$

11.19 (a) $v_d(\text{max}) \cong 14 \ mV$
       (b) $v_d(\text{max}) \cong 21.2 \ mV$

11.21 (a) $R_E = 37.2\,\text{k}\Omega$
    (b) (i) $A_d = 119$, (ii) $A_{cm} = -0.673$

11.23 $A_{v1} = \dfrac{-\dfrac{1}{2}g_m R_L}{\left(2 + \dfrac{R_L}{R_C}\right)}$, $A_{v2} = \dfrac{+\dfrac{1}{2}g_m R_L}{\left(2 + \dfrac{R_L}{R_C}\right)}$

    $A_v = \dfrac{g_m R_L}{\left(2 + \dfrac{R_L}{R_C}\right)}$

11.25 $A_d = g_m R_C$, $A_{cm} = 0$

11.27 $-0.22 \le v_{O2} \le 0.22\,\text{V}$

11.29 (a) $R_1 = 38.6\,\text{k}\Omega$, $R_2 = 236\,\Omega$
    (b) $R_{icm} = 292\,\text{M}\Omega$
    (c) $A_{cm} = -0.0123$

11.31 (a) $I_1 = I_Q = 0.337\,\text{mA}$, $I_{D1} = 0.168\,\text{mA}$,
       $V_{DS1} = 8.79\,\text{V}$, $V_{DS4} = 7.18\,\text{V}$
    (c) $v_{CM}(\text{max}) = 7.97\,\text{V}$,
       $v_{CM}(\text{min}) = -6.02\,\text{V}$

11.33 (a) $I_Q = 160\,\mu\text{A}$, $R_D = 37.5\,\text{k}\Omega$
    (c) $v_{CM} = 2.50\,\text{V}$

11.35 (a) $R_1 = 33.9\,\text{k}\Omega$
    (b) (i) $v_{O1} - v_{O2} = 0$,
      (ii) $v_{O1} - v_{O2} = 0.125\,\text{V}$
    (c) (i) $v_{O1} - v_{O2} = -0.156\,\text{V}$
      (ii) $v_{O1} - v_{O2} = -0.031\,\text{V}$

11.37 (a) $v_d(\text{max}) \cong 0.19\,\text{V}$
    (b) $v_d(\text{max}) \cong 0.285\,\text{V}$

11.39 (a) $I_D = 6\,\mu\text{A}$, $V_{SD} = 1.69\,\text{V}$
    (b) (i) $A_d = 9.66$, $A_{cm} = 0$
      (ii) $A_d = 4.83$, $A_{cm} = -0.0448$

11.41 $A_d = -9.16$, $A_{cm} = -0.003216$,
    $\text{CMRR}_{\text{dB}} = 69.1\,\text{dB}$

11.43 (a) $v_S = 1.459\,\text{V}$, $v_{D1} = v_{D2} = -4.115\,\text{V}$
    (b) $v_S = 2.344\,\text{V}$, $v_{D1} = v_{D2} = -4.336\,\text{V}$
    (c) $v_S \cong 1.459\,\text{V}$, $v_{D1} = -3.909\,\text{V}$,
      $v_{D2} = -4.321\,\text{V}$
    (d) $v_S \cong 2.344\,\text{V}$, $v_{D1} = -4.158\,\text{V}$,
      $v_{D2} = -4.515\,\text{V}$

11.45 Let $I_Q = 0.50\,\text{mA}$; For $V_{D1} = V_{D2} = 3\,\text{V}$,
    then $R_D = 28\,\text{k}\Omega$; $\left(\dfrac{W}{L}\right)_1 = \left(\dfrac{W}{L}\right)_2 = 1274$
    Need $R_o = 200\,\text{k}\Omega$

11.47 $A_{d1} = \dfrac{-\dfrac{1}{2}g_m R_L}{\left(2 + \dfrac{R_L}{R_D}\right)}$, $A_{d2} = \dfrac{+\dfrac{1}{2}g_m R_L}{\left(2 + \dfrac{R_L}{R_D}\right)}$

    $A_d = \dfrac{g_m R_L}{\left(2 + \dfrac{R_L}{R_D}\right)}$

11.49 (a) $R_D = 40.8\,\text{k}\Omega$
    (b) $v_{cm}(\text{max}) = 1.32\,\text{V}$

11.51 (a) $R_D = 19\,\text{k}\Omega$, $R_o = 998\,\text{k}\Omega$
    (b) Use cascode current source similar
      to Figure 10.18 with $v_{DS2}(\text{sat}) = 0.3\,\text{V}$.

11.53 Let $I_{Q1} = I_{Q2} = 0.1\,\text{mA}$,
    then $R_1 = 100\,\text{k}\Omega$, $R_2 = 50\,\text{k}\Omega$
    $\left(\dfrac{W}{L}\right)_1 = \left(\dfrac{W}{L}\right)_2 = 6.67$,
    $\left(\dfrac{W}{L}\right)_3 = \left(\dfrac{W}{L}\right)_4 = 240$

11.55 (a) $R_D = 6\,\text{k}\Omega$, $I_Q = 1\,\text{mA}$
    (b) $g_f(\text{max}) = 0.25\,\text{mA/V}$
    (c) $A_d = 1.5$

11.57 $A_d = \dfrac{g_m R_L}{\left(2 + \dfrac{R_L}{R_D}\right)}$

11.59 (a) $R_1 = 36.8\,\text{k}\Omega$
    (b) $A_d = 2308$
    (c) $R_{id} = 74.9\,\text{k}\Omega$, $R_o = 480\,\text{k}\Omega$
    (d) $-3.6 \le v_{cm} \le 4.3\,\text{V}$

11.61 (a) $I_Q = 0.749\,\text{mA}$
    (b) $v_{cm}(\text{max}) = 3.8\,\text{V}$

11.63 (a) $A_v = 89.72$
    (b) $A_v = 89.32$
    (c) $A_v = 89.52$

11.65 (a) $A_d = 79.44$
    (b) $v_{CM}(\text{max}) = 1.25\,\text{V}$

11.67 $I_{Q1} = I_Q$

11.69 (a) $R_o = 1 \| 4.704 = 0.8247\,\text{M}\Omega$
    (b) $A_d = 3172$

11.73 $I_Q \cong 66\,\mu\text{A}$, $\left(\dfrac{W}{L}\right)_n = 23.76$

11.75 (a) $R_o = 172\,\text{M}\Omega$
    (b) $A_d = 46096$

11.77 (a) $V^+ = -V^- = 4\,\text{V}$
    (b) $A_d = 50$

11.79 (a) $V^+ = -V^- = 3.4\,\text{V}$
    (b) $A_d = 1004$

11.81 (a) $A_d = 88.9$

11.83 (a) $g_{m1} = 0.506\,\text{mA/V}$, $g_{m2} = 27.25\,\text{mA/V}$,
      $r_{\pi2} = 6.605\,\text{k}\Omega$
    (b) $R_o = 19.8\,\Omega$

11.85 $A_v = -48.4$

11.87 $R_i \cong 1.05\,\text{M}\Omega$, $R_o = 0.472\,\text{k}\Omega$,
    $A_v = -438$

11.89  (a)  $R_1 = 27.2 \text{ k}\Omega$, $R_2 = 5 \text{ k}\Omega$
    (b)  $g_{m1} = 0.7071 \text{ mA/V}$, $g_{m2} = 2 \text{ mA/V}$,
      $r_{o1} = 200 \text{ k}\Omega$, $r_{o2} = 50 \text{ k}\Omega$
    (c)  $A_v = -15.25$
    (d)  $R_o = 0.450 \text{ k}\Omega$

11.91  (a)  $R_{C1} = 80 \text{ k}\Omega$, $R_{C2} = 20 \text{ k}\Omega$
    (b)  $A_{d1} = -69.6$, $A_d = -5352$

11.93  (a)  $A_d = -3.73$, $A_{cm} = 0.0718$
    (b)  $v_{O3} = -0.975 \sin \omega t \,(\text{V})$,
      $v_{O3}(\text{ideal}) = -1.12 \sin \omega t \,(\text{V})$

11.95  (a)  $f_z = 39.8 \text{ kHz}$
    (b)  $f_p = 11.7 \text{ GHz}$

11.97  (a)  $A_v = 19.5$
    (b)  $A_v = 11.2$

## CHAPTER 12

12.1  (a)  $\beta = 9.98 \times 10^{-3}$
    (b)  $A = 2000$

12.3  (a)  (i)  $T = \infty$, $A_f = 6.667$
      (ii)  $T = 1.5 \times 10^3$, $A_f = 6.662$
      (iii)  $T = 15$, $A_f = 6.25$
    (b)  (i)  $T = \infty$, $A_f = 4.0$
      (ii)  $T = 2.5 \times 10^3$, $A_f = 3.9984$
      (iii)  $T = 25$, $A_f = 3.846$

12.5  (a)  $\beta = -0.01245$
    (b)  $\dfrac{dA}{A} = 2.5\%$

12.7  (a)  $\beta_1 = 0.096685$, $\beta_2 = 0.0195$
    (b)  For (a), $A_{vf} = 47.33 \,(-5.34\%)$,
      For (b), $A_{vf} = 49.86 \,(-0.28\%)$

12.9  (a)  $f_C = 20 \text{ kHz}$
    (b)  $A_{vf} = 40$

12.11  (a)  (i)  $A_v = 5.25 \times 10^5$
      (ii)  $\beta = 0.01333$
    (b)  (i)  $A_{vf} = 74.99$
      (ii)  $f_C = 31.5 \text{ kHz}$

12.13  (a)  For (a), $f_{3\text{-dB}} = 2.352 \text{ kHz}$,
      For (b), $f_{3\text{-dB}} = 5 \text{ kHz}$
    (b)  Overall feedback $\Rightarrow$ wider bandwidth

12.15  For (a), at low input, gain $= 81$; at high
    input, gain $= 73.4$. For (b), at low input,
    gain $= 81.1$, at high input, gain $= 80.1$

12.17  $V_{fb} = 24 \text{ mV}$, $V_\varepsilon = 1 \text{ mV}$,
    $A_v = 2.5 \times 10^3 \text{ V/V}$, $A_{vf} = 100 \text{ V/V}$

12.19  (a)  $A_{vf} = \dfrac{A}{1 + \left[ A \middle/ \left( 1 + \dfrac{R_2}{R_1} \right) \right]}$

(b)  $\beta = \dfrac{1}{1 + \dfrac{R_2}{R_1}}$

(c)  $\beta = 0.04999$, $\dfrac{R_2}{R_1} = 19.004$

(d)  $-2.222 \times 10^{-3}\%$

12.21  $I_{fb} = 24.2 \,\mu\text{A}$, $I_o = 3.125 \text{ mA}$,
    $\beta_i = 0.007744 \text{ A/A}$, $A_i = 3906 \text{ A/A}$

12.23  $R_{if} = 0.5 \,\Omega$, $R_{of} \cong 2.42 \,\Omega$

12.25  $A_{gf} = 40 \text{ A/V}$, $I_o = 6 \text{ mA}$,
    $V_{fb} = 147 \,\mu\text{V}$, $V_\varepsilon = 3 \,\mu\text{V}$

12.27  $R_{if} = 10^6 \text{ k}\Omega$, $R_{of} = 5.04 \text{ M}\Omega$

12.29  $A_z = 4 \text{ V/}\mu\text{A}$, $A_{zf} = 0.2 \text{ V/}\mu\text{A}$,
    $\beta_g = 4.75 \,\mu\text{A/V}$

12.31  $R_{if} = 99.79 \,\Omega$

12.33  (a)  $1 \,\Omega \le R_i \le 10^5 \text{ k}\Omega$
    (b)  $0.1 \,\Omega \le R_o \le 10^4 \text{ k}\Omega$

12.35  (a)  $A_{vf} = 11.0$
    (b)  $R_{if} = 273 \text{ M}\Omega$
    (c)  $R_{of} = 15 \text{ m}\Omega$

12.37  (a)  $g_{m1} = 32.81 \text{ mA/V}$, $r_{\pi 1} = 3.66 \text{ k}\Omega$
      $g_{m2} = 19.12 \text{ mA/V}$, $r_{\pi 2} = 6.28 \text{ k}\Omega$
      $g_{m3} = 78.08 \text{ mA/V}$, $r_{\pi 3} = 1.54 \text{ k}\Omega$
    (b)  $A_{vf} = 20.7$
    (c)  $R_{if} = 62.4 \text{ k}\Omega$
    (d)  $R_{of} = 1.39 \,\Omega$

12.39  $A_{vf} = 1.273$

12.41  (a)  (i)  $A_{vf} = 0.801$
      (ii)  $R_{of} = 299 \,\Omega$
    (b)  (i)  $+3.78\%$
      (ii)  $-15.4\%$

12.45  For example, $R_2 = 3 \text{ k}\Omega$,
             $R_1 = 247 \text{ k}\Omega$

12.47  (a)  $I_{DQ1} = 3.99 \text{ mA}$, $I_{DQ2} = 12.01 \text{ mA}$
    (b)  $A_i = -0.920$

12.49  (a)  $g_{m1} = 7.62 \text{ mA/V}$, $r_{\pi 1} = 13.1 \text{ k}\Omega$
      $g_{m2} = 52.7 \text{ mA/V}$, $r_{\pi 2} = 1.90 \text{ k}\Omega$
    (b)  $A_{if} = 8.60$
    (c)  $R_{if} = 47.4 \,\Omega$

12.51  (a)  $g_{m1} = 67.31 \text{ mA/V}$, $r_{\pi 1} = 1.78 \text{ k}\Omega$
      $g_{m2} = 19.23 \text{ mA/V}$, $r_{\pi 2} = 6.24 \text{ k}\Omega$
    (b)  $A_v = 7.44$

12.53  $A_{if} = 5.33$

12.55  (a)  $A_{gf} = \dfrac{\dfrac{R_F}{R_1}}{\left( \dfrac{R_L R_F}{R_1} - R_3 - \dfrac{R_L R_3}{R_2} \right)}$

(c) $R_2 = 2\,\text{k}\Omega$, set $R_3 = 2\,\text{k}\Omega$,
   $R_1 = R_F = 10\,\text{k}\Omega$

12.57 $A_{gf} = 98.06\,\text{mA/V}$

12.59 $A_{gf} = -0.0653\,\text{mA/V}$

12.61 (a) $I_{CQ} = 2.448\,\text{mA}$, $V_{ECQ} = 1.31\,\text{V}$
   (b) $A_{zf} = -8.666\,\text{V/mA}$
   (c) $R_{if} = 0.1037\,\text{k}\Omega$
   (d) $R_{of} = 0.1186\,\text{k}\Omega$

12.63 (a) $A_{zf} = -R_F$
   (b) $g_m = 4.6\,\text{mA/V}$

12.65 $R_F = 0.807\,\text{M}\Omega$

12.71 $T = 4.22$

12.73 (a) $f_{180} = 7.1 \times 10^4\,\text{Hz}$
   (b) $\beta = 0.0205$
   (c) $A_{vf}(0) = 48.54$
   (d) Smaller

12.75 (b) $\beta = 4.7 \times 10^{-4}$

12.77 (a) $f_{180} = 10^6\,\text{Hz}$
   (b) $\beta = 2.83 \times 10^{-3}$
   (c) $A_{vf}(0) = 351$

12.79 (a) $f_{180} = 3.33 \times 10^5\,\text{Hz}$
   (b) (i) $|T(f_{180})| = 0.156$
      (ii) $\phi = -142.3°$
   (c) $A_f(0) = 50$

12.81 $\beta = 0.01428$

12.83 (a) $f_{180} = 1.42 \times 10^5\,\text{Hz}$, $|T(f_{180})| = 4.89$
   (b) $f_{PD} = 4.85\,\text{Hz}$

12.85 (a) $\beta = 0.0199$, $f_{PD} = 711\,\text{Hz}$
   (b) Phase margin $= 29.8°$

12.87 (a) $f = 0.976\,\text{MHz}$, phase margin $= 73.4°$
   (b) $f'_{P1} = 2.67\,\text{Hz}$, phase margin $= 85.4°$

12.89 $f_{PD} = 577\,\text{Hz}$

## CHAPTER 13

13.1 (a) $R_{D2} = 15\,\text{k}\Omega$, $R_{D1} = 9\,\text{k}\Omega$
   (b) (i) $A_d = 2.846$,  (ii) $A_2 = -12$
   (c) $A = -34.15$

13.5 (a) $A_v = -1.59 \times 10^6$
   (b) $R_{id} = 208\,\text{k}\Omega$
   (c) GBW $= 12.3\,\text{MHz}$

13.7 $v_d \cong 56.4\,\text{V}$

13.9 (a) $R_5 = 17.13\,\text{k}\Omega$, $R_4 = 2.438\,\text{k}\Omega$,
      $V_{EB12} = V_{BE11} = 0.7184\,\text{V}$,
      $V_{BE10} = 0.6453\,\text{V}$
   (b) $I_{REF} = 0.5137\,\text{mA}$, $I_{C10} \cong 30.22\,\mu\text{A}$
   (c) $I_{REF}$: 2.74%; $I_{C10}$: 0.733%

13.11 $I_{REF} = 0.22\,\text{mA}$, $I_{C10} \cong 14.2\,\mu\text{A}$,
   $I_{C6} = 7.10\,\mu\text{A}$, $I_{C17} = 0.165\,\text{mA}$,
   $I_{C13A} = 0.055\,\text{mA}$

13.13 $P = 48.8\,\text{mW}$, $I = 1.63\,\text{mA}$

13.15 $I_{C13A} = 0.125\,\text{mA}$, $I_{R10} \cong 0.012\,\text{mA}$,
   $I_{C19} = 0.113\,\text{mA}$, $I_{C18} = 12.565\,\mu\text{A}$,
   $V_{BE19} = 0.60185\,\text{V}$, $V_{BE18} = 0.54474\,\text{V}$,
   $I_{C14} = 113\,\mu\text{A}$

13.17 $R_1 = 45.11\,\text{k}\Omega$, $R_2 = 51.56\,\text{k}\Omega$

13.19 $A_d = -882$

13.21 $A_v = 964,650$

13.23 $R_{eq} = 170\,\Omega$

13.25 (a) $R_{id} = 2.095\,\text{M}\Omega$
   (b) $R_{id} = 2.80\,\text{M}\Omega$

13.27 (a) $f_{PD} = 10\,\text{Hz}$
   (b) $C_F = 13.25\,\text{pF}$

13.29 (a) $R_{D1} = 7.87\,\text{k}\Omega$, $R_{D2} = 24.8\,\text{k}\Omega$,
      $R_S = 15\,\text{k}\Omega$
   (b) (i) $A_d = 2.49$
      (ii) $A_2 = -19.21$
      (iii) $A_3 = 0.950$
   (c) $A = -45.4$

13.31 (a) Original: $g_{m1} = g_{m2} = 0.09975\,\text{mA/V}$,
      New: $g_{m1} = g_{m2} = 0.1995\,\text{mA/V}$
   (b) $A_d = 250$, $A_{v2} = 125$, $A_v = 31,250$

13.33 (a) $I_{set} = I_Q = I_{D7} = 0.1776\,\text{mA}$
   (b) $A_d = 111.9$, $A_2 = -96.86$,
      $A = -10,839$

13.35 $f_{PD} = 83.9\,\text{Hz}$

13.37 $R_o = 0.63\,\text{M}\Omega$

13.39 (a) $I_{Q2} = 180\,\mu\text{A}$
   (b) $\left(\dfrac{W}{L}\right)_{8P} = 360$, $\left(\dfrac{W}{L}\right)_{8N} = 180$

13.41 (a) $I_{D6} = I_{D7} = 43.7\,\mu\text{A}$
   (b) $A_v = -35,350$

13.43 $\left(\dfrac{W}{L}\right)_1 = \left(\dfrac{W}{L}\right)_P = 4.06$, $\left(\dfrac{W}{L}\right)_N = 1.85$

13.45 (a) $A_d = 136.1$
   (b) $R_o = 92.6\,\text{k}\Omega$
   (c) $f_{PD} = 343.7\,\text{kHz}$, GBW $= 46.8\,\text{MHz}$

13.47 (a) $\left(\dfrac{W}{L}\right)_5 = \left(\dfrac{W}{L}\right)_6 = 16.4$
   (c) GBW $= 27.2\,\text{MHz}$

13.49 (a) $R_S = 10\,\text{k}\Omega$, $R_{C1} = 3.4\,\text{k}\Omega$,
      $R_{C2} = 12.39\,\text{k}\Omega$
   (b) $A_d = 15.60$

(c) $A_2 = -20.97$

(d) $A_3 = 0.95$

(e) $A = -310.8$

13.51 (a) For NMOS: $g_{mN} = 1.0$ mA/V,

$r_{oN} = 400$ kΩ

For BJT: $r_{o2} = 800$ kΩ

(b) $A_d = 266.7$

13.53 $\left(\dfrac{W}{L}\right)_N = 6.16, \left(\dfrac{W}{L}\right)_P = 13.6$

13.55 $K_p = 0.234$ mA/V$^2$

13.57 $A_d = 10.38, |A_{v2}| = 1917, |A_v| = 19,895$

13.59 $I_{DSS} = 0.8$ mA

13.61 (a) Set $V_P = 3$ V, $V_{ZK} = 3$ V,
$R_3 = 24$ kΩ, Set $I_{DSS} = 0.2$ mA

(b) $R_4 = 22.8$ kΩ

# CHAPTER 14

14.1 $v_i(\text{max})_{\text{rms}} = 39.77$ mV

14.3 (1) $v_O = 2$ V, (2) $v_2 = 12.5$ mV,
(3) $A_{OL} = 2 \times 10^4$, (4) $v_1 = 8$ μV,
(5) $A_{OL} = 1000$

14.5 (a) (i) $A_{CL} = 7.90863$
    (ii) $-0.03956\%$

(b) (i) $A_{CL} = 7.84966$
    (ii) $-0.785\%$

14.7 $A_{OL} = 8.9991 \times 10^5$

14.9 (a) $9.98 \le |A| \le 10.02$
(b) $9.969 \le |A| \le 10.009$

14.11 $A_{OL} = 49,950$

14.13 (a) $A_{CL} = 4.927, R_{if} = 2.687$ MΩ,
$R_{of} = 3.16$ Ω

(b) $-0.05\%$

14.15 $v_O = -1.993v_{I1} - 3.986v_{I2}; -0.35\%$

14.17 (a) $R_{if} = 99.1$ Ω
(b) $R_{of} = 18.4$ Ω
(c) $A_{CL} = 0.650$
(d) $0.650$

14.19 $f_{PD} = 7.94$ Hz, GBW $= 7.94 \times 10^5$ Hz

14.21 $f_{PD} = 448$ Hz

14.23 (a) $A_{CLO} = -9.9978$,
$f_{3\text{-dB}} = 150.033$ kHz

(b) $A_{CLO} = -999.34, f_{3\text{-dB}} = 76.49$ kHz

14.25 $|A_v|_{\text{max}} = 50$

14.27 (a) $V_{PO} = 5.09$ V
(b) $V_{PO} = 8$ V

14.29 (a) $V_{PO} = 5.0$ V
(b) $V_{PO} = 23.87$ V

14.33 (a) $I_{S4} = 5 \times 10^{-15}$ A
(b) $I_{S4} = 4.939 \times 10^{-15}$ A
(c) $I_{S4} = 4.811 \times 10^{-15}$ A

14.35 (a) $-0.360 \le v_O \le -0.240$ V
(b) $-3.06 \le v_O \le -2.94$ V

14.37 $V_o = \dfrac{V_i}{RC} \cdot t, t = 10^3$ s

14.39 At $V^+$ node, $v_O = 0.8645$ V,
At center, $v_O = 0.9545$ V,
At $V^-$ node, $v_O = 1.0445$ V

14.41 $\dfrac{i_{C1}}{i_{C2}} = 1.0155$

14.43 (a) $v_O = -0.30$ V
(b) $v_O = -0.50$ V
(c) $v_O = -0.10$ V
(d) $v_O = -1.30$ V

14.45 (a) $v_O = -0.010$ V

14.47 (a) $v_{O1} = 0.10$ V, $v_{O2} = -0.45$ V

14.49 (a) $v_{O1} = 0.15$ V, $v_{O2} = 0.15$ V,
$v_{O3} = -0.09$ V

(b) $R_A = 8.33$ kΩ, $R_B = 10$ kΩ

(c) $v_{O1} = \pm0.015$ V, $v_{O2} = \pm0.015$ V,
$v_{O3} = \pm0.021$ V

14.51 (a) $4 \le v_O \le 76$ mV
(b) $-39 \le v_O \le 39$ mV
(c) $2.161 \le v_O \le 2.239$ V

14.53 (a) For offset voltage,
$|v_{O1}| = 110$ mV, $|v_{O2}| = 610$ mV
For bias currents,
$v_{O1} = 0.31$ V, $v_{O2} = -1.51$ V

14.55 For circuit (a), $v_O = 9$ mV;
For circuit (b), $v_O = -1.0815$ V

14.57 (a) $v_{O1} = 6$ mV, $v_{O2} = 28$ mV;
(b) $v_{O1} = 6.495$ mV,
$v_{O2} = 30.31$ mV

14.59 (a) $v_{O1} = 0.10$ V, $v_{O2} = 0.12$ V;
(b) $v_{O1} = 0.105$ V,
$v_{O2} = 0.166$ V;

(c) Due to $I_B$, $v_{O1} = 0.125$ V,
$v_{O2} = 0.15$ V;
Due to $I_{OS}$, $v_{O1} = 0.133$ V,
$v_{O2} = 0.224$ V

14.61 (a) $x = 0.4728\%$
(b) $x = 0.0267\%$

# CHAPTER 15

15.1 (a) Noninverting amplifier:
Set $R_1 = 30\,\text{k}\Omega$, $R_2 = 210\,\text{k}\Omega$
At noninverting terminal, let input
$C = 0.001\,\mu\text{F}$, then $R = 5.305\,\text{k}\Omega$

(b) Inverting amplifier:
Set $R_1 = 15\,\text{k}\Omega$, $R_2 = 300\,\text{k}\Omega$;
Capacitor in series with $R_1$,
$C = 530.5\,\text{pF}$

15.3 6-pole filter

15.5 (a) Let $R = 20\,\text{k}\Omega$, then $C = 397.9\,\text{pF}$;
$C_1 = 1411\,\text{pF}$, $C_2 = 553.9\,\text{pF}$,
$C_3 = 80.5\,\text{pF}$

(b) (i) $|T| = -0.0673\,\text{dB}$,
(ii) $|T| = -0.711\,\text{dB}$,
(iii) $|T| = -3.0\,\text{dB}$,
(iv) $|T| = -6.83\,\text{dB}$,
(v) $|T| = -10.9\,\text{dB}$

15.7 9-pole filter

15.9 $N = 35$, $f_{3\text{-dB}} = 12.25\,\text{kHz}$

15.11 (a) $|T| = -9.31\,\text{dB}$
(b) $|T| = -14.8\,\text{dB}$
(c) $|T| = -20.5\,\text{dB}$

15.13 3-pole filter

15.15 (b) (i) $|A_v|_{\text{max}} = 28.3$
(ii) $f_o = 5.305\,\text{kHz}$
(iii) $f_1 = 5.296\,\text{kHz}$, $f_2 = 5.315\,\text{kHz}$

15.17 (a) $T(s) = \left(1 + \dfrac{R_2}{R_1}\right)\left[\dfrac{1 + s(R_1\|R_2)C}{1 + sR_2C}\right]$,

$f_{3\text{-dB1}} = \dfrac{1}{2\pi R_2 C}$,

$f_{3\text{-dB2}} = \dfrac{1}{2\pi (R_1\|R_2)C}$

(b) $T(s) = \left(1 + \dfrac{R_2}{R_1}\right)[1 + s(R_1\|R_2)C]$,

$f_{3\text{-dB}} = \dfrac{1}{2\pi (R_1\|R_2)C}$

15.19 Two noninverting amplifiers:
Let $R_2 = 250\,\text{k}\Omega$, $R_1 = 54.1\,\text{k}\Omega$
For high-pass filter: Set
$R = 250\,\text{k}\Omega$, $C = 31.8\,\text{pF}$
For low-pass filter: Set
$R = 250\,\text{k}\Omega$, $C = 0.00424\,\mu\text{F}$

15.21 (a) $Q = 10\,\text{pC}$
(b) $I_{eq} = 1\,\mu\text{A}$
(c) $t = 4.61 \times 10^{-8}\,\text{s}$

15.23 (a) $\tau = 60\,\mu\text{s}$
(b) $\Delta v_O = 0.167\,\text{V}$
(c) 78 clock pulses

15.25 (a) $f_o = \dfrac{1}{2\pi R\sqrt{C\,(C + 2C_V)}}$
(b) $480 \le f_o \le 919\,\text{kHz}$

15.29 (a) $f_o = \dfrac{1}{2\pi RC} \cdot \sqrt{\dfrac{3R + 2R_V}{R_V}}$

(b) $\dfrac{R_F}{R} = 2$

(c) $13.5 \le f_o \le 16.84\,\text{kHz}$

15.31 (a) $f_o = \dfrac{1}{2\pi RC} \cdot \sqrt{4\left(\dfrac{R}{R\|R_V}\right) + 2}$

(b) $19.45 \le f_o \le 22.66\,\text{kHz}$

15.33 (a) $\dfrac{R_1}{R_1 + R_2} = \dfrac{sRL}{sRL + (R + sL)^2}$

(b) $f_o = \dfrac{R}{2\pi L}$

(c) $\dfrac{R_2}{R_1} = 2$

15.35 $C_2 \cong 0.06\,\mu\text{F}$, $L = 13.8\,\mu\text{H}$

15.37 (a) $f_o = \dfrac{1}{2\pi\,\sqrt{C\,(L_1 + L_2)}}$

(b) $\dfrac{L_1}{L_2} = g_m R$

15.39 $T(s) = \left(1 + \dfrac{R_2}{R_1}\right)\left[\dfrac{sRC}{s^2R^2C^2 + 3sRC + 1}\right]$,
$f_o = \dfrac{1}{2\pi RC}$, $\dfrac{R_2}{R_1} = 2$

15.41 $T(s) = \left(1 + \dfrac{R_2}{R_1}\right)\left[\dfrac{sRL}{s^2L^2 + 3sRL + R^2}\right]$,
$f_o = \dfrac{R}{2\pi L}$, $\dfrac{R_2}{R_1} = 2$

15.43 $R_1 = 1.47\,\text{k}\Omega$, $R_2 = 73.53\,\text{k}\Omega$

15.45 (a) $V_{TH} = 0.4\,\text{V}$, $V_{TL} = -0.4\,\text{V}$
(b) For $33.333 \le t \le 33.439\,\text{ms}$,
$v_O = +10\,\text{V}$,
For $33.439 \le t \le 66.667\,\text{ms}$, $v_O = -10\,\text{V}$

15.47 (a) $V_{TH} = \dfrac{\dfrac{V_{\text{REF}}}{R_3} + \dfrac{V_P}{R_2}}{\dfrac{1}{R_1} + \dfrac{1}{R_2} + \dfrac{1}{R_3}}$,

$V_{TL} = \dfrac{\dfrac{V_{\text{REF}}}{R_3} - \dfrac{V_P}{R_2}}{\dfrac{1}{R_1} + \dfrac{1}{R_2} + \dfrac{1}{R_3}}$

(b) $R_1 = 10.17\,\text{k}\Omega$, $R_2 = 600\,\text{k}\Omega$
(c) $V_{TH} = -4.9\,\text{V}$, $V_{TL} = -5.1\,\text{V}$

15.49 (a) $V_S = \left(\dfrac{R_2}{R_1 + R_2}\right) \cdot V_{\text{REF}}$,

$$V_{TH} = V_S + \left(\dfrac{R_1}{R_1 + R_2}\right) \cdot V_H$$

$$V_{TL} = V_S + \left(\dfrac{R_1}{R_1 + R_2}\right) \cdot V_L$$

(b) $R_1 = 4\,\text{k}\Omega$, $R_2 = 188\,\text{k}\Omega$,
$V_{\text{REF}} = -1.787\,\text{V}$
(c) (i) $|i| = 71.8\,\mu\text{A}$, (ii) $|i| = 53.2\,\mu\text{A}$

15.51 (a) $V_{TH} = 0.175\,\text{V}$, $V_{TL} = -0.175\,\text{V}$
(c) (i) $I_{D1} = 0$, $I_{D2} = 0.458\,\text{mA}$,
$I_{R2} = -0.465\,\text{mA}$, $I_{R3} = 7\,\mu\text{A}$
(ii) $I_{D1} = 0.458\,\text{mA}$, $I_{D2} = 0$,
$I_{R2} = 0.465\,\text{mA}$, $I_{R3} = -7\,\mu\text{A}$

15.53 (a) $V_{\text{REF}} = 3.6\,\text{V}$
(b) Let $R_2 = 90\,\text{k}\Omega$, $R_1 = 10\,\text{k}\Omega$
(c) $v_O = -5.1\,\text{V}$

15.55 (a) $v_O = (V_{\text{REF}} + 2V_\gamma)$,
$v_O = -(V_{\text{REF}} + 2V_\gamma)$,
$$V_{TH} = \dfrac{R_1}{R_2}(V_{\text{REF}} + 2V_\gamma),$$
$$V_{TL} = -\dfrac{R_1}{R_2}(V_{\text{REF}} + 2V_\gamma)$$

15.57 (a) $v_X = 5 - 8.333e^{-t/\tau_X}$ for $0 \le t \le t_1$;
$v_X = -10 + 11.667e^{-(t-t_1)/\tau_X}$ for
$t_1 \le t \le T$
(b) $f = 847\,\text{Hz}$

15.59 For $0 \le t \le t_1$, $t_1 = 1.1\,\text{ms}$,
$v_Y = 10(1 - e^{-t/\tau_Y})$, $\tau_Y = 4 \times 10^{-5}\,\text{s}$

15.61 (a) Set $C_X = 0.01\,\mu\text{F}$, $R_X = 23.84\,\text{k}\Omega$
(b) $v_Y < v_X$
(c) $133\,\mu\text{s}$

15.63 (a) Let $C = 50\,\mu\text{F}$, $R = 1.09\,\text{M}\Omega$
(b) $t \cong 38.7\,\text{ms}$

15.65 $R_B = 50\,\text{k}\Omega$, $C = 144.3\,\text{pF}$

15.67 $1.40 \le f_o \le 2.72\,\text{kHz}$
$50.5\% \le \text{duty cycle} \le 98.1\%$

15.69 (a) $P = 3.7\,\text{W}$
(b) $V^+ \cong 19\,\text{V}$
(c) $V_P = 8.6\,\text{V}$

15.71 (a) $A_v = 1 + \dfrac{R_2}{R_1} + \dfrac{R_3}{R_1}$
(b) $R_1 = 50\,\text{k}\Omega$, $R_2 = 200\,\text{k}\Omega$,
$R_3 = 250\,\text{k}\Omega$
(c) $|v_{O1}| = |v_{O2}| = 10\,\text{V}$, $i_L = 1\,\text{A}$

15.73 (a) Set $R_2 = R_3 = R_4 = 100\,\text{k}\Omega$,
$R_1 = 8.69\,\text{k}\Omega$
(b) (i) $\bar{P}_L = 14.4\,\text{W}$,  (ii) $R_L = 20\,\Omega$
(c) $\bar{P}_L = 7.2\,\text{W}$

15.75 (a) $R_{of} = 4\,\text{m}\Omega$
(b) $\Delta V_O = 12\,\text{mV}$

15.77 (a) $R_4 = 31.5\,\Omega$
(b) $R_{12} = 5.57\,\text{k}\Omega$

15.79 (a) $7.659 \le V_O \le 22.976\,\text{V}$
(b) $4.17 \times 10^{-4}\%$

## CHAPTER 16

16.1 (a) $\left(\dfrac{W}{L}\right) = 2.91$
(b) $V_{It} = 1.172\,\text{V}$, $V_{Ot} = 0.672\,\text{V}$
(c) $i_{D,\max} = 80\,\mu\text{A}$, $P_{D,\max} = 264\,\mu\text{W}$

16.3 (a) $R = 41.6\,\text{k}\Omega$, $\left(\dfrac{W}{L}\right) = 0.193$
(b) $0.5 \le V_{GS} \le 2.38\,\text{V}$

16.5 (a) $\dfrac{K_D}{K_L} = 4.26$

(b) $\dfrac{K_D}{K_L} = 5.4$

(c) $P = 0.759\,\text{mW}$

16.7 $\dfrac{(W/L)_D}{(W/L)_L} = 5.4$

16.9 (a) Driver: $V_{It} = 0.8578\,\text{V}$,
$V_{Ot} = 0.3578\,\text{V}$
Load: $V_{It} = 0.8578\,\text{V}$, $V_{Ot} = 2.5\,\text{V}$
(b) $v_O = 0.0230\,\text{V}$
(c) $i_{D,\max} = 64\,\mu\text{A}$, $P_{D,\max} = 211\,\mu\text{W}$

16.11 (a) $\left(\dfrac{W}{L}\right)_L = 2.47$, $\left(\dfrac{W}{L}\right)_D = 5.04$
(b) Driver: $V_{It} = 0.720\,\text{V}$, $V_{Ot} = 0.420\,\text{V}$
Load: $V_{It} = 0.720\,\text{V}$, $V_{Ot} = 1.20\,\text{V}$
(c) $P_{D,\max} = 80\,\mu\text{W}$, $v_O = 0.0297\,\text{V}$

16.13 (a) (i) $P = 0$,  (ii) $P = 1.16\,\text{mW}$
(b) (i) $P = 0$,  (ii) $P = 0.825\,\text{mW}$
(c) (i) $P = 0$,  (ii) $P = 0.20\,\text{mW}$

16.15 (a) $v_{O1} = 0.0431\,\text{V}$, $v_{O2} = 5\,\text{V}$
(b) $v_{O1} = 5\,\text{V}$, $v_{O2} = 0.0431\,\text{V}$

16.17 (a) $v_I = 0.6683\,\text{V}$
(b) $v_I = 0.6057\,\text{V}$

16.19 (a) $\dfrac{K_D}{K_L} = 2.04$

(b) $\left(\dfrac{W}{L}\right)_L = 0.667, \left(\dfrac{W}{L}\right)_D = 1.36$

(c) $v_O = 0.0329 \text{ V}$

16.21 (a) $\left(\dfrac{W}{L}\right)_{ML1} = 0.417, \left(\dfrac{W}{L}\right)_{MD1} = 1.35$

(b) $v_{O2} = 0.0330 \text{ V}$

16.23 (a) $v_O \cong 0.135 \text{ V}$

(b) $v_{DSX} \cong 0.067 \text{ V}, v_{DSY} \cong 0.068 \text{ V},$
$v_{GSX} = 5 \text{ V}, v_{GSY} \cong 4.933 \text{ V}$

16.25 $\bar{Y} = [A \text{ OR } (B \text{ AND } C)] \text{ AND } D$

16.27 $Y = \overline{(A+B)(C+D)}$

16.29 (a) For NMOS circuit,
$M_{NA}$ in parallel with [$M_{NB}$ in series
with ($M_{NC}$ in parallel with $M_{ND}$)]

(b) $\left(\dfrac{W}{L}\right)_A = 1.735, \left(\dfrac{W}{L}\right)_{B,C,D} = 3.47$

16.31 (a) $V_{It} = 1.25 \text{ V}, V_{OPt} = 1.65 \text{ V},$
$V_{ONt} = 0.85 \text{ V}$

(c) For $v_I = 1.1 \text{ V}, v_O = 2.214 \text{ V}$
For $v_I = 1.4 \text{ V}, v_O = 0.286 \text{ V}$

16.33 (a) (i) $V_{It} = 1.707 \text{ V}$
(ii) $v_I = 1.26 \text{ V}$
(iii) $v_I = 2.157 \text{ V}$

(b) (i) $V_{It} = 1.25 \text{ V}$
(ii) $v_I = 0.855 \text{ V}$
(iii) $v_I = 1.61 \text{ V}$

16.35 (a) $1.7 \le v_{O1} \le 3.3 \text{ V}$

(b) $v_{O3} = 5 \text{ V}, v_{O1} = 2.78 \text{ V},$
$v_I = 2.5 \text{ V}$

16.37 (a) $i_{D,\text{peak}} = 24.2 \,\mu\text{A}$
(b) $i_{D,\text{peak}} = 29.34 \,\mu\text{A}$
(c) $i_{D,\text{peak}} = 33.23 \,\mu\text{A}$

16.39 (a) $P = 12.5 \,\mu\text{W}$
(b) $P = 6.48 \,\mu\text{W}$

16.41 (a) $P = 3 \times 10^{-7} \text{ W}$
(b) (i) $C_L = 0.0024 \text{ pF}$
(ii) $C_L = 0.00551 \text{ pF}$
(iii) $C_L = 0.0267 \text{ pF}$

16.43 (b) Let $\left(\dfrac{W}{L}\right)_N = 2,$

$\left(\dfrac{W}{L}\right)_P = 4, V_{DD} = 5 \text{ V}$

For NMOS: $i_d = 0.45 \text{ mA}$
For PMOS: $i_d = 0.36 \text{ mA}$

16.45 (a) $V_{IL} = 0.8268 \text{ V}, V_{IH} = 1.2713 \text{ V}$
(b) $NM_L = 0.5858 \text{ V}, NM_H = 1.0439 \text{ V}$

16.47 (a) $v_C = 2.5 \text{ V}$

(b) $\left(\dfrac{W}{L}\right)_N = \dfrac{9}{2} \cdot \left(\dfrac{W}{L}\right)_P$

(c) $V_{It} = 1.65 \text{ V}$

16.49 (a) $\left(\dfrac{W}{L}\right)_N = 2, \left(\dfrac{W}{L}\right)_P = 16$

(b) $\left(\dfrac{W}{L}\right)_N = 4, \left(\dfrac{W}{L}\right)_P = 32$

16.51 (a) $\left(\dfrac{W}{L}\right)_N = 2, \left(\dfrac{W}{L}\right)_P = 12$

(b) $\left(\dfrac{W}{L}\right)_N = 4, \left(\dfrac{W}{L}\right)_P = 24$

16.53 (a) $Y = \overline{ABC + DE}$

(b) $M_{NA}, M_{NB}, M_{NC}$ in series and
are in parallel with $M_{ND}, M_{NE}$ in series

(c) $\left(\dfrac{W}{L}\right)_{AN,BN,CN} = 6, \left(\dfrac{W}{L}\right)_{DN,EN} = 4,$

$\left(\dfrac{W}{L}\right)_P = 8$

16.55 (a) $Y = \overline{A + BC + DE}$

(b) $M_{NA}$ in parallel with ($M_{NB}$ in series
with $M_{NC}$) in parallel with ($M_{ND}$ in
series with $M_{NE}$)

(c) NMOS: $\left(\dfrac{W}{L}\right)_A = 2, \left(\dfrac{W}{L}\right)_{B,C,D,E} = 4$

PMOS: $\left(\dfrac{W}{L}\right)_{A,B,C,D,E} = 12$

16.57 (b) All $\left(\dfrac{W}{L}\right)_N = 3$, All $\left(\dfrac{W}{L}\right)_P = 6$

16.59 (a) $\bar{Y} = C(A+B)$

(b) All $\left(\dfrac{W}{L}\right)_N = 4,$

$\left(\dfrac{W}{L}\right)_{CP} = 4, \left(\dfrac{W}{L}\right)_{AP,BP} = 8$

16.61 (a)

| State | $v_{O1}$ | $v_{O2}$ |
|---|---|---|
| 1 | 5 | 0 |
| 2 | 5 | 0 |
| 3 | 5 | 0 |
| 4 | 5 | 0 |
| 5 | 5 | 0 |
| 6 | 0 | 5 |

(b) $v_{O2} = (v_A \text{ OR } v_B) \text{ AND } v_C$

16.63   ($M_{NA}$ in series with $M_{M\bar{B}}$) and in parallel with ($M_{N\bar{A}}$ in series with $M_{NB}$)

16.65   $M_{NA}$ in series with ($M_{NB}$ in parallel with $M_{NC}$) in series with ($M_{ND}$ in parallel with $M_{NE}$)

16.67   (a) (i)  $v_O = 0$,   (ii)  $v_O = 2.9$ V,
              (iii)  $v_O = 2.5$ V
        (b) (i)  $v_O = 0$,   (ii)  $v_O = 1.4$ V,
              (iii)  $v_O = 1.4$ V

16.69   (a)  $v_{O1} = 2.1$ V, $v_{O2} = 2.5$ V
        (b)  $\left(\dfrac{W}{L}\right)_1 = 12.1, \left(\dfrac{W}{L}\right)_3 = 1.09$

16.71   $Y = A + \bar{A}B$
        $Z = \bar{A}\,\bar{B}$

16.73   (a) (i)  $Y = 0$,   (ii)  $Y = 2.5$ V,
              (iii) $Y = 0$,   (iv)  $Y = 2.5$ V
        (b) (i)  $Y = 0$,   (ii)  $Y = 0$,
              (iii) $Y = 2.5$ V,   (iv)  $Y = 2.5$ V
        (c)  Multiplexer

16.75   (a) (i)  $Y = 0$,   (ii)  $Y = 2.5$ V,
              (iii) $Y = 2.5$ V,   (iv)  $Y = 0$
        (b)  Exclusive OR

16.79   For $v_I = 1.5$ V, $v_{O1} = 2.88$ V,
                    $v_O \cong 0$
        $v_I = 1.6$ V, $v_{O1} = 2.693$ V,
                    $v_O = 0.00979$ V
        $v_I = 1.7$ V, Switching point,
        $v_I = 1.8$ V, $v_{O1} = 0.607$ V,
                    $v_O = 3.298$ V

16.81   (a)  Positive edge when $CLK = 1$,
             $Q = \bar{\bar{D}} = D$

16.83   Not actually a J-K flip-flop

16.85   (a)  1 Megabit memory
                    $= 1,048,576$ cells
                    $\Rightarrow 1024 \times 1024$ memory
             Then 10 input row and
             column decoder lines necessary.
        (b)  250 K $\times$ 4 bits $\Rightarrow$
                    $262,144 \times 4$ bits $\Rightarrow$
                    $512 \times 512$ memory
             Then 9 input row and
             column decoder lines
             necessary.

16.87   $t = 62.6$ ns

16.89   $R = 0.512$ M$\Omega$, $\left(\dfrac{W}{L}\right) = 0.797$

16.91   $\bar{Q} = 1.794$ V, $Q = 0.370$ V

16.97   (a)  7 bits
        (b)  0.02578125 V
        (c)  1100010

16.99   (a)  $\Delta R_1 = 5.88\%$
        (b)  $\Delta R_4 = 33.3\%$

16.101  (a)  $I_1 = -0.50$ mA,
             $I_2 = -0.25$ mA,
             $I_3 = -0.125$ mA,
             $I_4 = -0.0625$ mA,
             $I_5 = -0.03125$ mA,
             $I_6 = -0.015625$ mA
        (b)  $\Delta v_O = 0.078125$ V
        (c)  $v_O = 1.484375$ V
        (f)  $\Delta v_O = 1.640625$ V

16.103  64 resistors, 63 comparators

16.105  (a)  1010000000
        (b)  0101111101

## CHAPTER 17

17.1    (a)  $R_C = 2$ k$\Omega$
        (b) (i)  $v_{O1} = 0$, $v_{O2} = -0.4$ V
              (ii)  $v_{O1} = -0.4$ V, $v_{O2} = 0$
        (c)  $P = 0.36$ mW

17.3    (a)  $R_{C2} = 6$ k$\Omega$
        (b)  $R_{C1} = 4$ k$\Omega$
        (c)  $v_I = -0.0360$ V

17.5    (a)  $R_{C2} = 0.758$ k$\Omega$
        (b)  $R_{C1} = 0.658$ k$\Omega$
        (c)  For $v_{in} = -0.7$ V, $v_{O1} = -0.7$ V,
                    $v_{O2} = -1.7$ V
             For $v_{in} = -1.7$ V, $v_{O1} = -1.7$ V,
                    $v_{O2} = -0.7$ V
        (d) (i)  $P = 21.8$ mW,   (ii) $P = 20.7$ mW

17.7    (a)  $R_1 = 10.5$ k$\Omega$
        (b)  $R_5 = R_6 = 17.5$ k$\Omega$
        (c)  $I_Q = 0.20$ mA, $R_{C1} = 3.5$ k$\Omega$
        (d)  $I_Q = 0.20$ mA, $R_{C2} = 3.5$ k$\Omega$

17.9    $R_E = 1.375$ k$\Omega$, $R_{C1} = 0.75$ k$\Omega$,
        $R_{C2} = 1.031$ k$\Omega$, $R_2 = R_3 = 2.25$ k$\Omega$

17.11   (a)  $V_R = -1.7$ V
        (b)  Logic $1 = -0.7$ V, logic $0 = -2.7$ V
        (c)  $V_E = -2.4$ V for logic 0,
             $V_E = -1.4$ V for logic 1
        (d)  $P = 39.9$ mW for logic 0 and logic 1

17.13   (a)  $V_R = 2.6$ V
        (b)  $R_{C1} = 4.57$ k$\Omega$

(c) $R_{C2} = 5.05\ \Omega$

(d) $P = 1.60\ \text{mW}$

17.15 (a) $i_1 = 1.4\ \text{mA}, i_2 = 0.8\ \text{mA},$
$i_3 = 0.14\ \text{mA}, i_4 = 0.14\ \text{mA},$
$i_D = 0.74\ \text{mA}, v_O = -0.4\ \text{V}$

(b) $i_1 = 1.4\ \text{mA}, i_2 = 0.153\ \text{mA},$
$i_3 = 0.153\ \text{mA}, i_4 = 0.153\ \text{mA},$
$i_D = 0, v_O = -0.0765\ \text{V}$

(c) $i_1 = 1.6\ \text{mA}, i_2 = 0.14\ \text{mA},$
$i_3 = 0.14\ \text{mA}, i_4 = 0.14\ \text{mA},$
$i_D = 0, v_O = -0.07\ \text{V}$

(d) $i_1 = 1.6\ \text{mA}, i_2 = 0.8\ \text{mA},$
$i_3 = 0.153\ \text{mA}, i_4 = 0.153\ \text{mA},$
$i_D = 0.953\ \text{mA}, v_O = -0.4\ \text{V}$

17.17 (a) Logic $0 = -0.4\ \text{V}$
Logic $1 = 0$

(b) $v_{O1} = \overline{A + B}, v_{O2} = \overline{C + D},$
$v_{O3} = (A + B) \cdot (C + D)$

17.19 (a) (i) $v_1 = 0.8\ \text{V}, v_O = 2.5\ \text{V},$
$i_1 = 0.1417\ \text{mA}, i_2 = i_3 = 0$

(ii) $v_1 = 1.5\ \text{V}, v_O = 0.1\ \text{V},$
$i_1 = i_2 = 0.0833\ \text{mA}, i_3 = 0.2\ \text{mA}$

(b) (i) $v_1 = 1.4\ \text{V}, v_I = 0.7\ \text{V}$

(ii) $v_1 = 1.5\ \text{V}, v_I = 0.8\ \text{V}$

17.21 (i) $v' = 0.8\ \text{V}, i_1 = 0.525\ \text{mA},$
$i_3 = i_4 = 0$

(ii) $v' = 2.2\ \text{V}, i_1 = 0.35\ \text{mA},$
$i_3 = 2.04\ \text{mA}, i_4 = 0.297\ \text{mA}$

17.23 (a) $v_1 = 2.3\ \text{V}, v_O = 0.1\ \text{V},$
$i_1 = 0.675\ \text{mA}, i_2 = 1.7\ \text{mA},$
$i_3 = 1.225\ \text{mA}, i_4 = 2.375\ \text{mA},$
$i_5 = 0.08\ \text{mA}, i_{B2} = 2.295\ \text{mA}$

(b) $N = 42$

17.25 $i_1 = 1.53\ \text{mA}, i_2 = 0.0589\ \text{mA},$
$i_3 = 1.47\ \text{mA}, i_{Bo} = 1.37\ \text{mA},$
$i_{Co} = 0.817\ \text{mA}$

17.27 (a) (i) $v_1 = 0.9\ \text{V}, v_O = 2.5\ \text{V},$
$i_1 = 0.1333\ \text{mA}, i_2 = i_3 = 0$

(ii) $v_1 = 1.5\ \text{V}, v_O = 0.1\ \text{V},$
$i_1 = 0.0833\ \text{mA}, i_2 = 0.09167\ \text{mA},$
$i_3 = 0.20\ \text{mA}$

(b) (i) $v_1 = 1.4\ \text{V}, v_I = 0.6\ \text{V}$

(ii) $v_1 = 1.5\ \text{V}, v_I = 0.7\ \text{V}$

17.29 (a) (i) $v' = 0.8\ \text{V}, i_1 = 0.156\ \text{mA},$
$i_3 = i_4 = 0$

(ii) $v' = 2.2\ \text{V}, i_1 = 0.06875\ \text{mA},$
$i_3 = 0.5333\ \text{mA}, i_4 = 0.02875\ \text{mA}$

(b) $N = 5$

(c) $N = 5$

17.31 (a) (i) $i_{RB} = 1.025\ \text{mA}, i_{RCP} = i_{Bo} = 0,$
$V_{\text{out}} = 5\ \text{V}$

(ii) $i_{RB} = 0.70\ \text{mA}, i_{RCP} = 4.2\ \text{mA},$
$i_{Bo} = 0.0837\ \text{mA}, V_{\text{out}} = 0.8\ \text{V}$

(b) (i) $P = 5.145\ \text{mW}$

(ii) $P = 25.4\ \text{mW}$

17.33 (a) (i) $v_O = 3.6\ \text{V},$ (ii) $v_O = 3.404\ \text{V},$
(iii) $v_O = 1.11\ \text{V}$

(b) $I_L = 34.05\ \text{mA}$

17.35 (a) $i_{B1} = 1.5\ \text{mA}, i_{B2} = 0,$
$i_{B3} = 0.4\ \text{mA}, i_{C2} = 0,$
$i_{C3} = 0.5\ \text{mA}$

(b) $i_{B1} = 0.5\ \text{mA}, i_{B2} = 0.8\ \text{mA},$
$i_{B3} = 0, i_{C2} = 7.5\ \text{mA},$
$i_{C3} = 0$

17.37 (a) (i) $v_1 = 0.3\ \text{V}, v_O = 1.5\ \text{V},$
$i_1 = 1.2\ \text{mA}, i_B = i_C = 0$

(ii) $v_1 = 1.0\ \text{V}, v_O = 0.4\ \text{V},$
$i_1 = 0.5\ \text{mA}, i_B = 0.465\ \text{mA},$
$i_C = 0.9167\ \text{mA}$

(b) (i) $v_1 = 1.0\ \text{V}, v_I = 0.7\ \text{V},$
$i_B = i_C = 0$

(ii) $v_1 = 1.0\ \text{V}, v_I = 0.7\ \text{V},$
$i_B = 0.03667\ \text{mA}, i_C = 0.9167\ \text{mA}$

(c) $N = 13$

17.39 (a) $R_{B1} = 18\ \text{k}\Omega, R_{C1} = 1.63\ \text{k}\Omega$

(b) $v_{B1} = 0.7\ \text{V}, v_{B2} = 0, v_O \cong 1.8\ \text{V},$
$i_B = i_C = 0$

(c) $v_{B1} = 1.5\ \text{V}, v_{B2} = 0.7\ \text{V}, v_O = 0.4\ \text{V},$
$i_{B1} = 0.0555\ \text{mA}, i_{C1} = 1.043\ \text{mA},$
$i_{B2} = 0.10\ \text{mA}, i_{C2} = 0.40\ \text{mA}$

(d) $N = 20$

17.41 (a) $P = 0.4875\ \text{mW}$

(b) $P = 1.98\ \text{mW}$

(c) $i_{SC} \cong 78\ \text{mA}$

17.43 (a) $i_{DN} = 0.1\ \text{mA}, i_{DP} = 0.289\ \text{mA},$
$i_{C1} = 14.45\ \text{mA}, i_{C2} = 5\ \text{mA}$

(b) $t = 5.19\ \text{ns}$

(c) $t = 260\ \text{ns}$

# Index